2018 2nd IEEE International Conference on Power Electronics, Intelligent Control and Energy Systems (ICPEICES 2018)

Delhi, India
22 – 24 October 2018

Pages 1-616

IEEE Catalog Number: CFP18E68-POD
ISBN: 978-1-5386-6624-1

**Copyright © 2018 by the Institute of Electrical and Electronics Engineers, Inc.
All Rights Reserved**

Copyright and Reprint Permissions: Abstracting is permitted with credit to the source. Libraries are permitted to photocopy beyond the limit of U.S. copyright law for private use of patrons those articles in this volume that carry a code at the bottom of the first page, provided the per-copy fee indicated in the code is paid through Copyright Clearance Center, 222 Rosewood Drive, Danvers, MA 01923.

For other copying, reprint or republication permission, write to IEEE Copyrights Manager, IEEE Service Center, 445 Hoes Lane, Piscataway, NJ 08854. All rights reserved.

*** This is a print representation of what appears in the IEEE Digital Library. Some format issues inherent in the e-media version may also appear in this print version.

IEEE Catalog Number: CFP18E68-POD
ISBN (Print-On-Demand):   978-1-5386-6624-1
ISBN (Online):   978-1-5386-6625-8

Additional Copies of This Publication Are Available From:

Curran Associates, Inc
57 Morehouse Lane
Red Hook, NY 12571 USA
Phone: (845) 758-0400
Fax: (845) 758-2633
E-mail: curran@proceedings.com
Web: www.proceedings.com

TABLE OF CONTENTS

ANALYSIS OF FUZZY LOGIC, ANN AND ANFIS BASED MODELS FOR THE FORECASTING OF WIND POWER 1
Upma Singh ; M. Rizwan

ENERGY AUDIT AND ENERGY CONSERVATION FOR A HOSTEL OF AN ENGINEERING INSTITUTE 8
Srishti Gupta ; Rakhi Kamra ; Manuhar Swaroopa ; Abhishek Sharma

DEVELOPMENT OF A CHARACTERIZATION TOOL FOR DETERMINATION OF SHEET RESISTANCE AT 9 PLACES OF LARGE AREA 156 MM BY 156 MM DIFFUSED SILICON WAFER IN LESS THAN 5 SECONDS AND ITS APPLICATION IN MAKING HIGH EFFICIENCY SILICON SOLAR CELLS 13
M. Shadab Siddiqui ; A. K. Saxena ; Shivangi ; B. K. Pant

AN EMD BASED PROTECTION SCHEME FOR DISTRIBUTION SYSTEM 17
Mahmood Shaik ; Sandeep Kumar Yadav ; Abdul Gafoor Shaik

GREY WOLF OPTIMIZATION ALGORITHM FOR OPTIMAL SITING AND SIZING OF CAPACITORS 23
Vivek Pratap Singh Bhadouria ; Laxmi Srivastava

ORDER REDUCTION OF LINEAR TIME INVARIANT SYSTEM USING PARTICLE SWARM OPTIMIZATION 29
Shubham Kumar Gupta ; Himanshu Gupta ; Tanmay Shrivastava ; Kuldeep Rathore

STATE OF THE ART OF HUMAN FACTORS ANALYSIS APPLIED TO INDUSTRIAL AND COMMERCIAL POWER SYSTEMS 33
Esperanza S. Torres ; David Celeita ; Gustavo Ramos

POWER SYSTEM VULNERABILITY ASSESSMENT USING VOLTAGE COLLAPSE PROXIMITY INDEX 39
A Chaithra ; Sangeeta Modi

TRANSMISSION CONGESTION MANAGEMENT OF IEEE 24-BUS TEST SYSTEM BY OPTIMAL PLACEMENT OF TCSC 44
N Padmini ; Pallavi Choudekar ; Mehtab Fatima

EFFECT OF FACTS DEVICES ON CONGESTION MANAGEMENT USING ACTIVE & REACTIVE POWER RESCHEDULING 50
S. K. Gupta ; N. K. Yadav ; Mukesh Kumar

PITCH ANGLE CONTROL OF VARIABLE SPEED WIND TURBINE BY USING FUZZY LOGIC 56
Akshiv Kumar ; Bhavnesh Kumar

TRANSMISSION LOSS ALLOCATION IN RESTRUCTURED POWER SYSTEM WITH OPTIMIZATION LOSS CRITERION 61
Naimul Hasan ; Ibraheem ; Yudhishthir Pandey

DETERMINATION AND ALLOCATION OF OPTIMAL SIZE OF TCSC AND STATCOM FOR OBSTRUCTION ALLEVIATION IN POWER SYSTEM 65
Jitender Kumar ; Narendra Kumar

POWER FACTOR CORRECTION USING APFC PANEL ON DIFFERENT LOADS 73
Vinay Sehwag ; Vaibhav Dua ; Ajendra Singh ; Jitendra Nath Rai ; Vineet Shekher

SELECTION OF RENEWABLE ENERGY SOURCES FOR OFF-GRID ELECTRIFICATION OF NORTH-EASTERN STATES OF INDIA 78
Kusum Tharani ; Komal Bagai ; Shivam Dabas

INVESTIGATIONS ON EFFECT OF FREQUENCY VARIATION ON THE PERFORMANCE OF COMBINED CYCLE GAS TURBINE 85
J. N. Rai ; Ajendra Singh ; Vineet Shekher ; Pranav Jain ; Nikhil Dwivedi

A TECHNIQUE FOR POWER FACTOR MEASUREMENT OF HOUSEHOLD AND INDUSTRIAL LOAD USING LABVIEW 92
Abhishek Kumar Gupta ; Ravi Saxena ; Rajveer Singh ; Sanjiv Kumar

STUDY OF FORCED OSCILLATIONS IN TWO AREA POWER SYSTEM 96
Vertika Jain ; S. T Nagarajan ; Rachana Garg

AN ANFIS ARTIFICIAL TECHNIQUE BASED MAXIMUM POWER TRACKER FOR STANDALONE PHOTOVOLTAIC POWER GENERATION 102
Neeraj Priyadarshi ; Vigna K. Ramachandaramurthy ; Sanjeevikumar Padmanaban ; Farooque Azam ; Amarjeet Kumar Sharma ; J. P Kesari

LOAD FREQUENCY CONTROL OF TWO AREA INTERCONNECTED POWER SYSTEM USING SSSC WITH PID, FUZZY AND NEURAL NETWORK BASED CONTROLLERS 108

Ashiq Hussain Lone ; Viqar Yousuf ; Surya Prakash ; Mohammad Abid Bazaz

LOAD FREQUENCY CONTROL BASED ON NON-INTEGER TYPE PID CONTROLLER TUNED BY WOA 114

Manoj Kumar Debnath ; Nimai Charan Patel ; Shreeram Choudhury

ENERGY AND EXERGY ANALYSIS FOR HELIOSTAT BASED SOLAR THERMAL POWER PLANT 120

Manoj Kumar ; R. P Saini

EFFICIENT APPROACH FOR DG PLACEMENT AND SIZE IN MEDIUM VOLTAGE DISTRIBUTION SYSTEMS 127

Anand K Pandey ; Sheeraz Kirmani

A COMPARATIVE PERFORMANCE ANALYSIS OF AUTOMATIC GENERATION CONTROL OF MULTI-AREA POWER SYSTEM USING PID, FUZZY AND ANFIS CONTROLLERS 132

Mukesh Kumar Bhaskar ; Nidhi Singh Pal ; Vinod Kumar Yadav

DESIGN OF A HYBRID CONTROLLER BASED ON GA-SMC FOR THE MULTI-TERMINAL VSC-HVDC TRANSMISSION SYSTEM 138

Ramu Srikakulapu ; U. Vinatha

MARKET-BASED LOAD FREQUENCY CONTROL OF INTERCONNECTED POWER SYSTEM 143

Naimul Hasan ; Ibraheem Nasiruddin ; Samiuddin Ahmad

A CONTROL STRATEGY FOR STATCOM IN ALLEVIATION OF SUBSYNCHRONOUS RESONANCE IN POWER SYSTEMS 149

Viqar Yousuf ; Aijaz Ahmad

ANALYTICAL APPROACH FOR MULTIPLE DSTATCOMS ALLOCATION IN RADIAL DISTRIBUTION SYSTEMS FOR ENHANCEMENT OF ENERGY, COST AND EMISSION SAVINGS 154

Imran Ahmad Quadri ; S. Bhowmick ; D. Joshi

APPLICATION OF SINGULAR PERTURBATION METHODOLOGY FOR TRANSIENT STABILITY STUDY OF ELECTRIC POWER SYSTEM 160

Sunitha Anup ; T. S. Bhatti ; Ashu Verma

STUDY OF P&O AND INC PV MPPT TECHNIQUES FOR DIFFERENT ENVIRONMENT CONDITIONS 165

Bipin Singh ; Bharat Verma ; Prabin Kumar Padhy

A UNIFIED APPROACH FOR VOLTAGE ENHANCEMENT IN RADIAL DISTRIBUTION NETWORK WITH DYNAMIC LOADING 170

Swati Arora ; Sandeep Kaur ; Rintu Khanna

LOAD FLOW SOLUTION FOR RADIAL DISTRIBUTION NETWORK 176

Om Pathak ; Prem Prakash

AN EXPONENTIAL SMOOTHING BASED POWER SWING DETECTION TECHNIQUE FOR DISTANCE PROTECTION 182

A. V. Koteswara Rao ; K. M. Soni ; S. K. Sinha ; Ibraheem Nasiruddin

THREE INPUTS AND TWO OUTPUTS BOOST DC-DC CONVERTER FOR DC MICROGRID APPLICATIONS 188

Kumaravel S. ; Arun Sarathi ; Gangavarapu Guru Kumar ; Sivaprasad A.

NOVEL INTELLIGENT CONTROLLERS FOR LOAD FREQUENCY CONTROL OF A MULTI AREA HYBRID POWER SYSTEM 194

Divyank Srivastava ; M. M. Tripathi

A COMPARATIVE STUDY FOR SHORT TERM WIND SPEED FORECASTING USING STATISTICAL AND MACHINE LEARNING APPROACHES 200

Parul Arora ; Himanshu Kumar ; B. K Panigrahi

PERFORMANCE ANALYSIS OF A DISTANCE RELAY FOR ZONE IDENTIFICATION 206

M. Kiruthika ; S Bindu

ARTIFICIAL NEURAL NETWORK BASED MODEL FOR SHORT TERM SOLAR RADIATION FORECASTING CONSIDERING AEROSOL INDEX 212

Astitva Kumar ; Mohammad Rizwan ; Uma Nangia

A NOVEL SCHEME FOR POTENTIAL IDENTIFICATION OF CUSTOMERS FOR DEMAND RESPONSE 218

Pradeep Kumar Mallick ; Arjun Tyagi ; Ashu Verma

PHASOR MEASURING UNITS OF SMART GRID FOR UNINTERRUPTED AND STABLE POWER SUPPLY IN INDIA 223

Punit Mangal

OPTIMAL PLACEMENT OF DISTRIBUTED GENERATION FOR CONGESTION MANAGEMENT: A COMPARATIVE STUDY 229

Md Sarwar ; Anwar Shahzad Siddiqui ; Zainul Abdin Jaffery ; Imran Ahmad Quadri

IMPROVISED BINARY SEQUENCE MPPT METHOD FOR SOLAR PV APPLICATIONS 234

Swapnendu Narayan Ghosh

PLACEMENT OF DISTRIBUTED GENERATION IN RADIAL DISTRIBUTION SYSTEM BY CONSIDERING UNCERTAINTY OF LOAD DEMAND 239

Prem Prakash

APPLICATION OF ALO TO ECONOMIC LOAD DISPATCH WITHOUT NETWORK LOSSES FOR DIFFERENT CONDITIONS 244

A. V. Sudhakra Reddy ; Y. Praveen Kumar Reddy ; B. Bhargava Reddy Reddy ; Y. V. Krishna Reddy

OVERVOLTAGE AND UNDERVOLTAGE PROTECTION OF LOAD USING GSM MODEM SMS ALERT 249

Savita ; Sumit Shrivastava ; Abhishek Arora ; Vikas Varshney

OPTIMAL PLACEMENT AND SIZING OF DISTRIBUTED GENERATION USING FLOWER POLLINATION ALGORITHM FOR POWER LOSS REDUCTION MAXIMIZATION IN DISTRIBUTION NETWORKS 253

G Sabarinath. ; T. Gowri Manohar

ON-LINE MONITORING AND SIMULATION OF TRANSMISSION LINE NETWORK VOLTAGE STABILITY USING FVSI 257

Manish Kumar Meena ; Narendra Kumar

IMPACT OF PRESENCE OF CALENDER FEATURE IN THE PERFORMANCE OF DAY-AHEAD ELECTRIC LOAD POWER FORECASTING WITH NEW FRAMEWORK REALIZED IN ARTIFICIAL NEURAL NETWORK TEHINIQUE USING MATLAB PROGRAMMING 261

M. Vetri Selvi ; Sukumar Mishra

VOLTAGE PROFILE ENHANCEMENT FOR IEEE-14 BUS SYSTEM USING UPFC, TCSC AND SSSC 267

Gurudutt Bagha ; Amit Kumar

PSO TUNED AGC STRATEGY OF MULTI AREA MULTI-SOURCE POWER SYSTEM INCORPORATING SMES 273

Nidhi Gupta ; Narendra Kumar ; Nisha Singh

BIRD SWARM ALGORITHM FOR SOLVING MULTI-OBJECTIVE OPTIMAL POWER FLOW PROBLEM 280

Saket Gupta ; Narendra Kumar ; Laxmi Srivastava

DESIGN AND ANALYSIS OF RENEWABLE ENERGY BASED HYBRID MODEL FOR REMOTE APPLICATIONS 287

P. Anand ; A. H. Quadri ; S. K. Bath ; M. Rizwan ; Narendra Kumar

MAXIMUM POWER POINT TRACKING TECHNIQUES UNDER PARTIAL SHADING CONDITION - A REVIEW 293

Reyaz Hussan ; Adil Sarwar

POWER QUALITY DISTURBANCE PREDICTION USING PNN 299

Richa Sharma ; Laxmi Srivastava

INTELLIGENT APPROACH FOR LOAD BALANCING ON AN 11 KV FEEDER 305

Jagriti Surabhi ; Priyanka Chaudhary ; M. Rizwan

A CONTROL STRATEGY FOR SUBSYNCHRONOUS RESONANCE UNDER VARYING DEGREE OF SERIES COMPENSATION 311

Narendra Kumar ; Shilpa Gupta ; Nisha Singh

DEVELOPMENT OF EMPIRICAL MODELS FOR FORECASTING GLOBAL SOLAR ENERGY 317

Gulnar Perveen ; M. Rizwan ; Nidhi Goel

A COMPARATIVE ANALYSIS OF AGC OF TWO-AREA HYDRO-THERMAL POWER SYSTEM INTERCONNECTED WITH AC-DC PARALLEL LINK IN RESTRUCTURED POWER SYSTEM 325

Nisha Singh ; Narendra Kumar ; Yogendra Arya ; Nidhi Gupta

IMPLEMENTATION OF GRID CONNECTED WIND DRIVEN INDUCTION GENERATOR FROM MATLAB/SIMULINK TO REAL-TIME 331

K. A. Naik ; C. P. Gupta ; E. Fernandez

A MODIFIED PERTURB & OBSERVE ALGORITHM FOR MAXIMUM POWER POINT TRACKING WITH ZERO VOLTAGE SWITCHING BUCK BOOST CONVERTER IN PHOTOVOLTAIC SYSTEM 337

Dubal Amol Jalindar ; Sheela Tiwari

REDUCED DEVICE COUNT ASYMMETRICAL MULTILEVEL INVERTER TOPOLOGY USING DIFFERENT PWM TECHNIQUES 343

Shubham Kumar Gupta ; Praveen Bansal

IMPLEMENTATION AND INFLUENCE OF C-SVPWM AND BC-SVPWM TECHNIQUES USING DSP FOR VOLTAGE ORIENTED CONTROL OF THREE PHASE ACTIVE FRONT END RECTIFIER 349

Jaimin Mehta ; Swathy Pillai ; Atul Gupta ; Rajamanickam R.

A NEW 15-LEVEL ASYMMETRICAL MULTILEVEL INVERTER TOPOLOGY WITH REDUCED NUMBER OF DEVICES FOR DIFFERENT PWM TECHNIQUES 355

Kuldeep Rathore ; Praveen Bansal

GENERATION OF 13-LEVEL OUTPUT VOLTAGE FROM SINGLE-PHASE MULTILEVEL INVERTER CONSISTING OF CASCADED THREE H-BRIDGE UNITS 361

Tapan Kumar Chakraborty ; Ashique Anan ; Sakhawat Hossen Rakib ; Md. Imran Prodhan ; Md. Mostofa Kamal ; Md. Mahabubunnabi

POWER LOSS ANALYSIS IN MULTILEVEL INVERTERS USING MULTI-OBJECTIVE OPTIMIZATION 365

Deepshikha Singla ; Parsh Ram Sharma

PROTECTION OF POWER ELECTRONIC SWITCHES IN A VSI AGAINST OVER-CURRENT AND SHOOT-THROUGH FAULTS 370

V. K. Choubey ; J. Simlai ; T. Ghosh ; A. Mitra

POWER QUALITY IMPROVEMENT OF STANDALONE WIND ENERGY GENERATION SYSTEM FOR NON LINEAR LOAD 374

Seema Agrawal ; Vijay Kumar Gupta ; D. K. Palwalia ; R. K. Somani

HARMONY SEARCH ALGORITHM BASED POWER QUALITY IMPROVEMENT IN MULTILEVEL INVERTER 380

Deepshikha Singla ; Parsh Ram Sharma

POWER QUALITY IMPROVEMENT OF GRID CONNECTED DFIG USING FUZZY CONTROLLER 386

Nitin Goel ; Paras Ram Sharma ; Dheeraj Joshi

A ROBUST ADALINE BASED CONTROL OF SHUNT ACTIVE POWER FILTER WITHOUT LOAD AND FILTER CURRENT MEASUREMENT 392

Shivangni Sharma ; Vimlesh Verma

THD MINIMISATION IN 15- LEVEL HYBRID MULTILEVEL INVERTER USING HARMONIC MINIMIZATION TECHNIQUE 398

Abhishek Azad ; Jagdish Kumar

DESIGN, SIMULATION AND ECONOMICAL ANALYSIS OF SOLAR POWERED IRRIGATION WATER PUMP 403

Archna Aggarwal ; Kriti Srivastava

SIZING AND FINANCIAL OPTIMIZATION OF HYBRID PV/BATTERY/DIESEL SYSTEM 408

Archna Aggarwal ; Divya Nagpal

POWER QUALITY IMPROVEMENT USING FUZZY-PI CONTROLLED D-STATCOM 414

Atma Ram ; Paras Ram Sharma ; Rajesh Kr. Ahuja

NON-LINEAR H-INFINITY CONTROL FOR GRID-INTERACTIVE OFFSHORE WIND FARM AND MARINE CURRENT FARM 420

Satendra Kr Singh Kushwaha ; S. R. Mohanty ; Paulson Samuel

A BRIEF REVIEW REGARDING SENSOR REDUCTION AND FAULTS IN SHUNT ACTIVE POWER FILTER 426

Shivangni Sharma ; Vimlesh Verma

A NOVEL ACTIVE ANTI-ISLANDING SCHEME FOR INVERTER-BASED DISTRIBUTED GENERATION 431

Sheetal Chandak ; Manohar Mishra ; Pravat Kumar Rout

SELECTIVE HARMONIC ELIMINATION FOR CASCADED THREE PHASE MULTILEVEL INVERTER 437

Deepak Singh ; Akhilesh Sharma ; Pukhrambam Devachandra Singh ; S. Gao

AN ANN BASED INTELLIGENT MPPT CONTROL FOR WIND WATER PUMPING SYSTEM 443

Neeraj Priyadarshi ; Vigna K. Ramachandaramurthy ; Sanjeevikumar Padmanaban ; Farooque Azam ; Amarjeet Kumar Sharma ; J. P Kesari

CHARACTERIZATION STUDY OF A GRID CONNECTED DFIG BASED WECS UNDER VARIABLE WIND SPEED AND LOADING CONDITIONS 449

Amit Kumar Roy ; Prasenjit Basak ; Gyan Ranjan Biswal

ECONOMIC SCHEDULING OF HYBRID MICROGRID USING PLANNED MANAGEMENT SCHEME 455

Iram Akhtar ; Sheeraz Kirmani ; Majid Jamil

AN EFFICIENT SOLAR ENERGY HARVESTING SYSTEM FOR WIRELESS SENSOR NODES 461

Himanshu Sharma ; Ahteshamul Haque ; Zainul Abdin Jaffery

HIGH VOLTAGE GAIN SWITCHED CAPACITOR BOOST CONVERTER WITH ANFIS CONTROLLER FOR FUEL CELL ELECTRIC VEHICLE APPLICATIONS 465

K. Jyotheeswara Reddy ; N. Sudhakar

INTERNAL MODEL CONTROLLER DESIGN FOR HVAC SYSTEM 471

Shahzad Hussain ; Sapna Gupta ; Rajeev Gupta

ADAPTIVE PARTICLE SWARM OPTIMIZATION BASED MAXIMUM POWER POINT TRACKING IN GRID CONNECTED PV SYSTEM 477

Majid Jamil ; M. Rizwan ; D. P. Kothari ; Arjun Baliyan

PERFORMANCE EVALUATION OF 3-PHASE 4-WIRE SAPF BASED ON SYNCHRONIZING EPLL WITH FUZZY LOGIC CONTROLLER 483

Seema Agrawal ; Deepika Sharma ; Vijay Kumar Gupta ; R. K. Somani

EVENT-TRIGGERED BASED AUTOMATIC GENERATION CONTROL CONSIDERING PROBABILISTIC ACTUATOR FAILURE 488

Pankaj Dahiya ; Pankaj Mukhija ; Anmol Ratna Saxena

MULTI-INPUT MULTI-LEVEL ISOLATED DC-DC CONVERTER FOR ENHANCED RELIABILITY OF RENEWABLE SOURCES 492

Ravi Vardhan Arya ; Amritesh Kumar ; Aditya Narula

GRID TIED HYBRID PHOTOVOLTAIC-FUEL CELL POWER SYSTEM FOR RESIDENTIAL LOAD 498

Mukul Chankaya ; Aijaz Ahmad

VARIABLE STEP HCS BASED SENSORLESS CONTROL OF WIND DRIVEN DFIG FOR AUTONOMOUS OPERATION 504

Bhim Singh ; Sambasivaiah Puchalapalli ; S. K. Tiwari ; P. K. Goel

AN INTELLIGENT MPPT CONTROLLER BASED AC TO DC BOOST PFC CONVERTER FOR GRID-TIED WIND ENERGY CONVERSION SYSTEM 510

Damodhar Reddy ; Sudha Ramaramy

SYMMETRICAL COMPONENTS ESTIMATION OF UNBALANCED THREE PHASE POWER SYSTEM USING MO- ADALINE STRUCTURE AND HERMITE POLYNOMIAL BASED GAUSS NEWTON ALGORITHM 516

Swastik Acharya ; Umamani Subudhi ; Harish Kumar Sahoo ; Anuj Jena

A NORMALIZED ADAPTIVE FILTER FOR ENHANCED OPTIMAL OPERATION OF GRID INTERFACED PV SYSTEM 522

Vedantham Lakshmi Srinivas ; Shailendra Kumar ; Bhim Singh ; Sukumar Mishra

IFLL BASED CONTROL FOR PV SYSTEM AT ADVERSE GRID CONDITIONS 528

Bhim Singh ; Vandana Jain

INTERNAL MODEL CONTROLLER DESIGN FOR BOOST CONVERTER BY STOCHASTIC OPTIMISATION 534

Kelam Sudheer Kumar ; Bharat Verma ; Prabin Kumar Padhy

OUTPUT VOLTAGE CONTROL OF DC-DC BOOST CONVERTER USING MODEL PREDICTIVE CONTROL APPROACH 539

Kiran Vijay Khunte ; Sreedhar Madichetty ; Sukumar Mishra

DESIGN OF AN EFFICIENT SOLAR PV BATTERY CHARGE CONTROLLER UNDER DYNAMIC MOTOR LOAD CONDITIONS 546

Sourish Ganguly ; Subhrasish Pal ; Imran Khan ; Debabrata Das ; Ahana Ghosh ; Ankur Bhattacharjee

HYBRID VOLTAGE CONTROL FOR STAND ALONE TRANSFORMERLESS INVERTER 552

Mohammed Ali Khan ; Ahteshamul Haque ; Kurukuru Varaha Satya Bharath

PERFORMANCE ANALYSIS OF DIFFERENT MODULATING TECHNIQUES FOR MULTILEVEL INVERTER 558

Allu Bhargav ; Alka Singh ; Ankita Arora

AUTONOMOUS OPERATION OF SINGLE STAGE SOLAR PV-BES BASED MICROGRID 564

Shubhra ; Bhim Singh

WAVELET BASED CONTROL OF SHUNT COMPENSATOR FOR POWER QUALITY ENHANCEMENT 570

Masood Anzar ; Narendra Kumar ; M. Rizwan

ANALYSIS OF DIFFERENT MPPT TECHNIQUES UNDER PARTIAL AND UN-SHADED CONDITION FOR SOLAR PV SYSTEM 576

Anirudh Dube ; Majid Jamil ; M Rizwan

THIRD ORDER SINUSOIDAL INTEGRATOR CONTROL OF PV-HYDRO-BES BASED ISOLATED MICRO-GRID 582

Vineet P. Chandran ; Shadab Murshid ; Bhim Singh

FREQUENCY ESTIMATION OF UNBALANCED THREE PHASE POWER SYSTEM USING LEGENDRE POLYNOMIAL BASED GAUSS NEWTON ALGORITHM 588
Shreeva Pattanaik ; Umamani Subudhi ; Harish Kumar Sahoo ; Saswat Panigrahi

POWER QUALITY IMPROVEMENT USING MULTILAYER GAMMA FILTER BASED CONTROL FOR DSTATCOM UNDER NONIDEAL DISTRIBUTION SYSTEM 593
Pavitra Shukl ; Bhim Singh

A HYSTERESIS CURRENT CONTROLLED GRID CONNECTED FULL BRIDGE INVERTER WITH ZERO CURRENT SWITCHING 599
Shashank Kurm ; Vivek Agarwal

IMPLEMENTATION OF PSO BASED SELECTIVE HARMONIC ELIMINATION TECHNIQUE IN MULTILEVEL INVERTERS 605
Peeyush Kala ; Sudha Arora

MULTI-OBJECTIVE CONTROL ALGORITHM FOR SOLAR PV-BATTERY BASED MICROGRID 611
Yashi Singh ; Bhim Singh ; Sukumar Mishra

HYBRID SWITCHED INDUCTOR / SWITCHED CAPACITOR BASED QUASI-Z-SOURCE DC-DC BOOST CONVERTER 617
Punit Kumar ; Mummadi Veerachary

MODELING AND SIMULATION OF MICROGRID SOLAR PHOTOVOLTAIC SYSTEM WITH ENERGY STORAGE 623
Dinanath Prasad ; Narendra Kumar ; Rakhi Sharma

SINGLE-PHASE GRID INTERFACED WEGS USING FREQUENCY ADAPTIVE NOTCH FILTER FOR POWER QUALITY IMPROVEMENT 630
Tripurari Nath Gupta ; Shadab Murshid ; Bhim Singh

PERFORMANCE ANALYSIS OF DISCONTINUOUS PULSE WIDTH MODULATION SCHEMES ON PUC-5 INVERTER 636
Abdul Azeem ; Mohd Tariq ; Kaif Ahmed Lodi ; C. Bharatiraja

GRID INTERACTIVE MISO CONVERTER BASED PV SYSTEM 642
Anuradha Tomar ; Sukumar Mishra

PERFORMANCE COMPARISON OF DIFFERENT INVERTER STAGES CS-VCO IN 0.18µM CMOS TECHNOLOGY 648
Ashish Mishra ; Indu Prabha Singh

REWEIGHTED SPARSE- LEAST-MEAN MIXED-NORM ADAPTIVE CONTROL FOR SOLAR PV INTEGRATED EV CHARGING STATION 653
Anjeet Verma ; Bhim Singh

ADAPTIVE FREQUENCY ESTIMATION TECHNIQUE FOR GRID CONNECTED PHOTOVOLTAIC SYSTEM 659
Sudip Bhattacharyya ; Shailendra Kumar ; Bhim Singh

ACOUSTIC ECHO CANCELLATION BASED ADAPTIVE CONTROL ALGORITHM FOR GRID INTEGRATED SECS 665
Gaurav Modi ; Shailendra Kumar ; Bhim Singh

DESIGN & IMPLEMENTATION OF SOLAR FED INTENSITY CONTROLLED STREETLIGHT 671
Jaspreet Singh ; Priya Mahajan ; Rachana Garg

MODIFIED GRADIENT SPECTRAL VARIANCE SMOOTHING ADAPTIVE FILTER CONTROL FOR GRID CONNECTED PV SYSTEM 677
Sandeep Kumar Sahoo ; Shailendra Kumar ; Bhim Singh

MULTIFUNCTIONAL ADAPTIVE RZA-NLMF CONTROL TECHNIQUE FOR DOUBLE STAGE SEGS INTERFACED WITH THREE-PHASE DISTRIBUTION FEEDER 683
Amresh Kumar Singh ; Ikhlaq Hussain ; Bhim Singh

IMPLEMENTATION OF MPPT CONTROL IN FUEL CELL FED HIGH STEP UP RATIO DC-DC CONVERTER 689
V. Karthikeyan ; P. Vipin Das ; Frede Blaabjerg

DESIGN AND ANALYSIS OF FIFTH-ORDER BUCK-BOOST CONVERTER 694
M. Veerachary

SOURCE SCHEDULING FOR POWER MATCHING OF CLUSTERED INDUCTION GENERATORS FED VSC SUPPORTED MICRO GRID 700
Peeyush Pant ; Vishal Verma

IMPLEMENTATION OF A MODIFIED DISTRIBUTED NORMALIZED LEAST MEAN SQUARE CONTROL FOR A MULTI-OBJECTIVE SINGLE STAGE SECS 706
Syed Bilal Qaiser Naqvi ; Shailendra Kumar ; Bhim Singh

A DC-DC BOOST CONVERTER FOR SEDIMENT MICROBIAL FUEL CELL ENERGY HARVESTING 712

Jeetendra Prasad ; Ramesh Kumar Tripathi

MODELING OF LED USING PIECEWISE LINEAR APPROXIMATION AND MACLAURIN SERIES EXPANSION 717

Obaidur Rahman ; Priya Mahajan ; Rachana Garg

SOLAR PV-BES BASED MICROGRID SYSTEM WITH SEAMLESS TRANSITION CAPABILITY 722

Vivek Narayanan ; Seema ; Bhim Singh

BATTERY CHARGING OF SMART PHONES USING ORGANIC SOLAR CELLS 729

K. Rahul ; Rakesh K. Meena ; Ritesh K. Gupta ; Parameswar K. Iyer ; M. Arun Tej ; Shabari Nath

IMPROVE COLLISION IN HIGHLY DENSE WIFI ENVIRONMENT 734

Vishal Bhargava ; N. S. Raghava

MICROWAVE CIRCUITS CHARACTERIZATION ON CARBON FIBRE (CFRP) BASED CARRIER PLATES 739

Kamaljeet Singh ; H R Kansara ; V. Venkatesh ; A V Nirmal ; S V Sharma

MORPHOLOGY CONTROL OF MIXED HALIDE PEROVSKITE FOR ITS APPLICATION IN LOW-COST THIN FILM TRANSISTOR 743

Anwesha Choudhury ; Priyanka Dogra ; Biki Teron ; Ritesh Kant Gupta ; Anamika Dey ; Ashish Singh ; Rabindranath Garai ; Parameswar Krishnan Iyer

PV-PIEZO HYBRID GRID CONNECTED SYSTEM 748

Anuradha Tomar ; Ayush Mittal ; Sahil Sharma

DISTRIBUTED CONTROL FOR POWER MANAGEMENT BASED ON FUZZY LOGIC IN DC MICROGRID 754

Rohit R. Deshmukh ; Makarand S. Ballal ; Girish G. Talapur ; H. M. Suryawanshi

SELECTIVE HARMONIC MINIMIZATION SCHEME APPLIED TO CASCADED H-BRIDGE INVERTER FOR SATISFYING CIGRE WG 36-05 AND EN 50160 GRID CODES 760

Sourabh Kundu ; Subrata Banerjee

ADAPTIVE SMC BASED DTC OF POSITION SENSORLESS PMSM DRIVEN SOLAR PV WATER PUMPING SYSTEM 765

Mohd. Kashif ; Shadab Murshid ; Bhim Singh

MULTIPULSE AC-DC CONVERSION FED 3RD HARMONIC INJECTION BASED SPWM CONTROLLED CASCADED MLI DRIVEN VCIMD 771

Piyush Kant ; Bhim Singh

ADAPTIVE NEURAL NETWORK BASED CONTROL OF PV CONNECTED DISTRIBUTION SYSTEM 777

Kanwar Pal ; Shailendra Kumar ; Bhim Singh ; Tara C. Kandpal

IMPROVED LAPLACIAN KERNEL FILTER BASED CONTROL OF MULTIFUNCTIONAL PV SYSTEM WITH ENHANCED POWER QUALITY 783

Sai Pranith ; Shailendra Kumar ; Bhim Singh ; T. S. Bhatti

SINGLE VDBA-BASED VOLTAGE-MODE UNIVERSAL BIQUADRATIC FILTER 788

Kanhaiya Lal Pushkar ; Ghanshyam Singh ; Sushil Kumar

DSOGI BASED GRID SYNCHRONIZATION UNDER ADVERSE GRID CONDITIONS 792

Simar Preet Kaur ; Alka Singh

MODULAR ASSEMBLY SYSTEMS IN INDUSTRY 4.0 MILIEU 798

N. Sumedh ; O. V. L. Narayana ; Madhav Reddy ; N. Sampath ; B. K. Priya ; T. K Ramesh

COMPARATIVE PERFORMANCE ANALYSIS OF RADIAL FLUX AND DUAL AIR-GAP AXIAL FLUX PERMANENT MAGNET BRUSHLESS DC MOTORS FOR ELECTRIC VEHICLE APPLICATION 804

Amit N. Patel ; Bhavik N. Suthar ; Tejas H. Panchal ; Rajesh M. Patel

A COMPLETE FUZZY LOGIC BASED REAL-TIME SIMULATION OF VECTOR CONTROLLED PMSM DRIVE 809

A Mishra ; Garima Dubey ; Dheeraj Joshi ; Pramod Agarwal ; S P Sriavstava

DEVELOPMENT OF THE INTELLIGENT OIL FIELD WITH MANAGEMENT AND CONTROL USING IIOT (INDUSTRIAL INTERNET OF THINGS) 815

Ali S. Allahloh ; Sarfraz Mohammad

SELF EXCITED INDUCTION GENERATOR FOR ISOLATED PICO HYDRO STATION IN REMOTE AREAS 821

Kailash Rana ; Duli Chand Meena

PREDICTING THE POPULARITY OF WEBSITES USING MULTILAYER PERCEPTRON AND EXTREME LEARNING MACHINE 827

Neha Srivastava ; Virendra P. Vishwakarma ; Udayan Ghose

A HYBRID TWENTY FIVE-LEVEL INVERTER FOR AN OPEN-END WINDING INDUCTION MOTOR (OEWIM) DRIVE 831
Anoop Kumar Kanaujia ; Sanjiv Kumar

EFFECTS OF NEUTRAL CONDUCTOR ON INDUCTION MOTOR STEADY-STATE PERFORMANCE UNDER LOSS OF ONE PHASE OF SUPPLY VOLTAGES 837
Pichai Aree

NOVEL AND ROBUST HYSTERESIS CURRENT CONTROL STRATEGIES FOR A BLDC MOTOR: A SIMULATION STUDY AND INVERTER DESIGN 841
Shrivatsan K Chari ; Rishav Dhiman ; Rohit Saxena

MULTI-OBJECTIVE GENETIC AND ADAPTIVE PARTICLE SWARM OPTIMIZATION ALGORITHMS: A PERFORMANCE ANALYSIS WITH BENCHMARK FUNCTIONS 847
Sudarshan K. Valluru ; Madhusudan Singh

FUZZY QUATERNION-BASED PIXEL WISE INFORMATION EXTRACTION FOR FACE RECOGNITION 852
Sudesh Yadav ; Virendra P. Vishwakarma

CONTROLLING OF AVR VOLTAGE AND SPEED OF DC MOTOR USING MODIFIED PI-PD CONTROLLER 858
Vinay Kumar Singh ; Sudeep Sharma ; Prabin Kumar Padhy

DESIGN OF FPI-PD CONTROLLER FOR BRUSHLESS DC MOTOR 864
Roshan Bharti ; Rishika Trivedi ; Prabin Kumar Padhy

CONDITION MONITORING OF PHOTOVOLTAIC SYSTEMS USING MACHINE LEAMING TECHNIQUES 870
Kurukuru Varaha Satya Bharath ; Ahteshamul Haque ; Mohammed Ali Khan

A COMPARATIVE STUDY OF FUZZY SYSTEMS AND NEURAL NETWORKS FOR SYSTEM MODELING AND IDENTIFICATION 876
Karnika Pandey ; Shayari Bhattacharjee ; Shreya Lau ; Meena Tushir

FUZZY C-MEANS BASED MODEL SIMPLIFICATION USING JAYA OPTIMIZATION ALGORITHM 881
Chhabindra Nath Singh ; Akhilesh Kumar Gupta ; Deepak Kumar ; Paulson Samuel

A VOLTAGE CONTROLLED SENSORLESS SPEED CONTROL OF PMBLDC MOTOR DRIVE FOR AN ELECTRIC TWO WHEELER 886
Chitra Bhattacharya ; Shailendra Kumar Sharma ; H. K. Verma

DYNAMIC PERFORMANCE ANALYSIS OF REACTIVE POWER AND IMPROVED ROTOR FLUX BASED MRAS FOR INDUCTION MOTOR DRIVES EMPLOYING PI AND FUZZY CONTROLLER 892
Mogili Ankarao ; M. Vijaya Kumar ; Panditi Dmesh

NEURO-ELO BASED SPEED ESTIMATION OF IMPROVED DESIGNED INDUCTION MOTOR DRIVE FOR SINGLE STAGE PHOTOVOLTAIC FED WATER PUMPING 897
Kuhsro Khan ; Saurabh Shukla ; Bhim Singh

TRANSFORMER HEALTH MONITORING SYSTEM USING INTERNET OF THINGS 903
Divyank Srivastava ; M. M. Tripathi

PARAMETRIC OPTIMIZATION OF PROTON EXCHANGE MEMBRANE FUEL CELL USING SUCCESSIVE TRUST REGION ALGORITHM 909
Shashwati Ray ; Heena Mishra ; T. V. Dixit

DYNAMIC BEHAVIOR CONTROL OF INDUCTION MOTOR WITH STATCOM 915
Majid Dehghani ; Peyman Karimyan ; Mehradad Abedi ; Hadis Karimipour

IMPACT OF PARTICLE SWARM OPTIMIZATION PARAMETERS ON ITS CONVERGENCE 921
N. K. Jain ; Uma Nangia ; Jyoti Jain

AUDIO WATERMARKING USING BEAT DETECTION AND PITCH ESTIMATION 927
Lavi Tanwar ; Radhika Dang ; Satvik Maurya ; Jeebananda Panda

IMPLEMENTATION OF V/F ADJUSTABLE SPEED DRIVE FOR INDUCTION MOTOR USING DSPACE DS1104 938
Chinmay Tigade ; Mini Sreejeth

EVALUATION OF EMISSION AND PERFORMANCE CHARACTERISTICS OF DIFFERENT BIODIESEL BLENDS WITH VARYING FIP ON A SINGLE CYLINDER CI ENGINE 944
Archit Gupta ; Siddhartha Agarwal ; Gianeshwar Aggarwal ; Sumit Roy

DESIGN AND SIMULATION OF SENSORLESS CONTROL ALGORITHMS OF BRUSHLESS DC MOTOR: A REVIEW 948
Ashish Dimri ; R. D. Kulkarni ; S. R. Gurumurthy ; J. Nataraj

BIFURCATION CURVES FOR ELECTRICAL DC MOTORS 953
Sudarshan K Valluru ; Madhusudan Singh ; Aditya Verma ; Anshul Gupta

PERFORMANCE ANALYSIS OF FSO LINK UNDER DIFFERENT CONDITIONS OF FOG IN DELHI, INDIA 958
Aditya Kesarwani ; Anuranjana ; Sanmukh Kaur ; Manpreet Kaur ; Pawan Singh Vohra

FEATURE EXPANDED AND WEIGHT SELECTIVE MODEL TO CLASSIFY THE HEART DISEASE PATIENTS 962
Kapil Juneja ; Chhavi Rana

SYNCHRONIZATION AND ANTI SYNCHRONIZATION OF FRACTIONAL ORDER SYSTEM 967
Lokesh Shankar Singh ; Himesh Handa ; Nitesh Gupta

ENERGY HARVESTING USING D_{33} MODE BY INSOLE EMBEDDED LOW COST PIEZO-SENSORS 972
Sumit Balguvhar ; Sidhartha Singhal ; Suresh Bhalla

GAIN ENHANCEMENT OF CIRCULARLY POLARISED MICROSTRIP PATCH ANTENNA USING METASURFACE 976
Shubham Kumar Mangal ; Sujay Kumar ; Tanuj Agarwal ; Ujjawal Sarkar ; Ankit Sharma ; M Lakshamanan

EFFICIENT FEATURE EXTRACTION USING DWT-DCT FOR ROBUST FACE RECOGNITION UNDER VARYING ILLUMINATIONS 982
Virendra P. Vishwakarma ; Sahil Dalal ; Varsha Sisaudia

ECG CLASSIFICATION USING KERNEL EXTREME LEARNING MACHINE 988
Sahil Dalal ; Virendra P. Vishwakarma ; Varsha Sisaudia

DESIGN AND EXPERIMENTAL ANALYSIS OF INTEGER ORDER PID CONTROLLER FOR CERAMIC IR HEATING SYSTEM 993
Vineet Shekher ; Ajendra Singh ; J. N. Rai

DESIGN AND MATHEMATICAL ANALYSIS OF MICROSTRIP BANDPASS POWER DIVIDER FOR KU BAND COMMUNICATIONS 1000
Karteek Viswanadha ; N. S. Raghava

CONTRAST ENHANCEMENT AND PSEUDO COLORING TECHNIQUES FOR INFRARED THERMAL IMAGES 1005
Mandeep Kaur ; Manminder Singh

LAGUERRE FUNCTION BASED MODEL PREDICTIVE CONTROL FOR VAN-DE-VUSSE REACTOR 1010
Akansha Jain ; Rajashree Taparia

HIGHLY LINEAR CURRENT FOLLOWER TRANSCONDUCTANCE AMPLIFIER (CFTA) DESIGN AND ITS FILTER APPLICATION 1016
Shweta Kumari ; Maneesha Gupta

HIGH GAIN TRANSIMPEDANCE AMPLIFIER USING SELF CASCODE STRUCTURE 1022
Preeti Singh ; Maneesha Gupta

APPLICATION OF LINEAR QUADRATIC METHODS TO STABILIZE CART INVERTED PENDULUM SYSTEMS 1027
Sudarshan K Valluru ; Madhusudan Singh ; Mayank Singh

IMPLEMENTATION OF HIGH PERFORMANCE CLOCK-GATED FLIP-FLOPS 1032
Prakash Kumar ; Kunwar Singh

AN EFFICIENT UNSCENTED KALMAN FILTER FOR JOINT ANGLES ESTIMATION AND CONTROL OF OMNI BUNDLE WITH NOISE 1036
Rohit Rana ; Prerna Gaur ; Vijyant Agarwal ; Harish Parthasarathy

APPLICATION OF WIRELESS TECHNOLOGY TO ENHANCE SAFETY AND PRODUCTIVITY IN STEEL PLANT 1041
Indranil Banerjee ; M Shrujan ; S K Das ; S Singh

OBSERVER BASED CONTROLLER DESIGN FOR INVERTED PENDULUM SYSTEM 1046
Shahida Khatoon ; Devendra Kumar Chaturvedi ; Naimul Hasan ; Md Istiyaque

DESIGN OF FRACTIONAL ORDER BUTTERWORTH FILTER USING GENETIC ALGORITHM 1052
Ashu Soni ; Maneesha Gupta

ANALYSIS OF AQUEOUS SUPERCAPACITOR FOR VARIATION IN SEPARATOR WETNESS. 1056
Ravi Giri ; Saurabh Chaudhari ; Utkarsh Mishra ; A. P. Deshpande ; P. B. Karandikar ; Sahil Sharma

ANALYSING STABILITY OF TIME DELAYED SYNCHRONOUS GENERATOR AND DESIGNING OPTIMAL STABILIZER FRACTIONAL ORDER PID CONTROLLER USING PARTICAL SWARM OPTIMIZATION TECHNIQUE 1062
Mohammad Ali Daftari ; Mohammad Ali Nekoui

INVESTIGATION OF EFFECT OF CHARCOAL PARTICLE SIZE ON EARTH'S RESISTANCE 1067
Meghna R. Yashwante ; P. B. Karandikar ; N. R. Kulkarni ; Sushil B. Dhembare ; Abhijit B. Bhosle

DESIGN AND ANALYSIS OF HIGH PERFORMANCE LINE STARTED PERMANENT MAGNET SYNCHRONOUS MOTOR ... 1073
Arvind Kumar ; Ajay Srivastava

FRACTIONAL ORDER TCHEBICHEF MOMENT AND ITS INVARIANTS ... 1078
Vishal Kumar Pandey ; Jyotsna Singh ; Harish Parthasarathy

MICROCONTROLLER BASED LOAD PRIORITIZATION TECHNIQUE IN RESIDENTIAL SECTOR ... 1083
Mohini Yadav ; Majid Jamil ; M. Rizwan

VISUAL REPRESENTATION OF CHANGE IN VEGETATION AREA OF DEHRADUN, UTTARAKHAND, INDIA USING NORMALIZED DIFFERENCE VEGETATION INDEX (NDVI) ... 1087
Amit Kumar Shakya ; Ayushman Ramola ; Kunal Sawant ; Shalini Tiwari ; Shamshul Aarfin ; Prag Mittal

ELECTRONICALLY TUNABLE CURRENT MODE UNIVERSAL FILTER USING A SINGLE MX CCCII ... 1093
Deepak Agrawal ; Sudhanshu Maheshwari

DESIGN AND IMPLEMENTATION OF CACHE COHERENCE PROTOCOL FOR HIGH-SPEED MULTIPROCESSOR SYSTEM ... 1097
Daman Preet Kaur ; V. Sulochana

MODEL REDUCTION OF CONTINUOUS-TIME INTERVAL SYSTEMS USING EIGEN SPECTRUM ANALYSIS ... 1103
Chhabindra Nath Singh ; Deepak Kumar ; Paulson Samuel ; Ankit Sachan

COMPARATIVE STUDY AND INVESTIGATION OF THE BROADBAND POWERLINE CHANNEL MODEL ... 1109
Zainul Abdin Jaffery ; Ibraheem ; Mukesh Kumar Varma

COGNITIVE RADIO TECHNOLOGY IN 5G WIRELESS COMMUNICATIONS ... 1115
Shruti Bhandari ; Sunil Joshi

DESIGN AND DEVELOPMENT OF DIGITAL SIGNAL CONTROLLER BASED MOTORIZED ZOOM CONTROLLER FOR 16X ZOOM THERMAL IMAGER ... 1121
Himanshu Singh ; Millie Pant ; Sudhir Khare ; Ranabir Mandal ; Kanchan Chandra ; Hirdesh Gangoli

MILLIMETER WAVE RECONFIGURABLE VIVALDI ANTENNA USING POWER DIVIDER FOR 5G APPLICATIONS ... 1126
Akhilesh Verma ; N. S. Raghava

GENERATING ELECTRICITY ON ROADSIDE USING INVELOX ... 1132
Abdullah Abu Sayed ; Md. Zyed Ibn Sadiq ; Quazi Nasrul Rudaba ; Shihab Khondokar ; Abu Hena Md. Shatil

COMPENSATING A THIRD ORDER PROCESS HAVING INVERSE RESPONSE ... 1136
Gaurav Kataria ; Kailash Singh

DESIGN OF PD-PID CONTROLLER WITH DOUBLE DERIVATIVE FILTER FOR FREQUENCY REGULATION ... 1142
Priyambada Satapathy ; Manoj Kumar Debnath ; Pradeep Kumar Mohanty

DESIGN AND SIMULATION OF LOW-POWER CONDITIONAL-DISCHARGING FLIP FLOP ... 1148
Nidhi Gupta ; Krishna Singh

COMPOSITE NONLINEAR FEEDBACK CONTROL FOR INVERTED PENDULUM WITH INPUT SATURATION ... 1154
Bhavna Agarwal ; Manisha Bhandari

INTEGRATING WAVELET COEFFICIENTS AND CNN FOR RECOGNIZING HANDWRITTEN CHARACTERS ... 1160
Madhuri Yadav ; Ravindra Kr. Purwar

REVIEW OF DIFFERENT TRANSFORMS USED IN DIGITAL IMAGE WATERMARKING ... 1165
Lavi Tanwar ; Jeebananda Panda

STATE ESTIMATION OF SINGLE-PHASE RECTIFIER BASED LOAD CIRCUIT USING UNSCENTED KALMAN FILTER ... 1172
Amit Kumar Gautam ; Sudipta Majumdar

A NOVEL METHOD FOR PREDICTING ATTENUATION CAUSED BY CLOUDS FOR HIGHER FREQUENCY BANDS ... 1178
Hitesh Singh ; Boncho Bonev ; Peter Petkov ; Ravinder Kumar

AN EFFICIENT WEIGHTED TRUST METHOD FOR MALICIOUS NODE DETECTION IN CLUSTERED WIRELESS SENSOR NETWORKS ... 1183
Bhavnesh Jaint ; Vishwamitra Singh ; Lalit Kumar Tanwar ; S. Indu ; Neeta Pandey

PERFORMANCE ANALYSIS OF NON-COHERENT MODULATIONS OVER FTR FADING MODEL ... 1188
Veenu Kansal ; Harpreet Kaur ; Simranjit Singh

A NOVEL APPROACH FOR PREDICTING ATTENUATION OF RADIO WAVES CAUSED BY RAIN 1193

Hitesh Singh ; Boncho Bonev ; Peter Petkov ; Ravinder Kumar

REAL TIME ANALYSIS OF MAC BASED AND LEVEL BASED ROUTING PROTOCOL FOR WIRELESS SENSOR NETWORK 1199

Aditi Gaur ; Shweta Jaroli ; Sunil Joshi ; Shruti Bhandari

EVALUATION OF EMISSION CHARACTERISTICS OF GREEN DIESEL IN A SINGLE CYLINDER CI ENGINE 1205

Manu J Nair ; Vaibhav Pahuja ; P. Suvesh ; Sumit Roy

COMPARATIVE RESEARCH FOR MANAGING DELAY IN SIGNAL PROCESSING VIA MULTIPLIERS 1213

Aniket Kumar ; Ekta Gupta ; R. P. Agarwal Shobhit ; R. K. Jain

PERTURBATION BASED NONLINEAR ANALYSIS OF MOSFET CIRCUIT 1219

Rahul Bansal ; Sudipta Majumdar

RESISTORLESS ELECTRONICALLY CONTROLLABLE QUADRATURE SINUSOIDAL OSCILLATOR EMPLOYING VDIBA 1224

Kanhaiya Lal Pushkar ; Komal Rohilla ; Sushil Kumar

DESIGN OF MULTI-LOOP L-PID AND NL-PID CONTROLLERS: AN EXPERIMENTAL VALIDATION 1228

Sudarshan K. Valluru ; Madhusudan Singh ; Arnav Goel ; Manpreet Kaur ; Daksh Dobhal ; Kumar Kartikeya ; Aditya Verma ; Anshul Gupta

RADIO FREQUENCY BASED (RF) CONTROL & OPERATION OF ELECTRICAL/ELECTRONIC APPLIANCES IN HOME/OFFICES 1232

T. Ramachandran ; Sanjiv Kumar ; Ajay Kumar ; Ravi Agarwal

MULTIPLE-INPUT SINGLE-OUTPUT UNIVERSAL BIQUAD FILTER USING SINGLE OUTPUT OTAS 1237

Ajishek Raj ; D. R. Bhaskar ; Pragati Kumar

Author Index

Analysis of Fuzzy Logic, ANN and ANFIS based Models for the Forecasting of Wind Power

Upma Singh
Dept. of Electronics and communication
Maharaja Surajmal Institute of Technology
Delhi,India
upma14@gmail.com

M. Rizwan
Dept. of Electrical Engineering
Delhi Technological University
Delhi,India
rizwan@dce.ac.in

Abstract— Wind Energy Conversion System (WECS) is growing rapidly as one of the most beneficial renewable energy sources available worldwide. The role of WECS is very essential in solving the issues of global warming and depleting fossil fuel. As per the Ministry of New & Renewable Energy (MNRE) statistics (As on 28th February 2018), the contribution of power from renewable energy sources is 107.81GW, which is more than 32.26% of the total installed capacity(334.15GW), out of which 32.85GW is contributed from wind energy. However, the potential for grid connected wind power generation has been estimated about 62053.73 MW considering sites having wind power density greater than 200 W/m2 at 50m hub-height with 1% land availability in potential areas for setting up wind farms. To efficiently utilize the potential of wind energy, prediction of wind power at different locations is utmost important. Keeping in view the aforesaid, intelligent models based on fuzzy logic, ANN, and ANFIS are developed and presented for the prediction of wind power. The proposed models can easily incorporate the uncertainties and nonlinearity associated with climatic variables. Comparative analysis of the above-mentioned models has also been carried out. It is investigated that results obtained from ANFIS model for the prediction of the wind energy are found better and quite accurate. Therefore, ANFIS model may be beneficially utilized in predicting the wind power and better utilization of wind resource.

Keywords—Fuzzy logic(FL), ANN, ANFIS, Wind energy, Renewable energy resources

I. INTRODUCTION

Fast growing wind energy utilization and generation requires extensive research in various fields. The flow of wind is weather dependent and it is discontinuous and intermittent over time-scales. Hence exact prediction of wind power is recognized as a major contribution for large-scale wind power generation [1]. The electricity requirement worldwide is increasing at alarming rate. The power demand is running ahead of supply [2]. Wind energy is one of the economic renewable energy sources characterized by the lowest cost of electricity generation, largely distributed, clean, producing no greenhouse gas-emissions during operation and uses little land. Therefore large number of countries are aware to recognize that wind power can provide a significant opportunity for future power generation. As a result, the installed wind capacity is growing each year. According to wind energy and

green peace organization plan, 12% of all electricity generation should be achieved through wind power by 2022 [3]. It is also now widely recognized that the renewable energy sources like wind, solar, small hydro and bio-mass has huge potential to meet our energy requirements. Wind power generation has grown tremendous rate in the past decade and has been established as an environment friendly and economically competitive means of power generation. The installed capacity of wind power plants in India is about 32.85GW as on 28th Feb. 2018 and there is a target to add more power from wind energy in near future. In view of this it is important to exploit the wind energy power generation. To enhance the wind power penetration, forecasting of wind power plays an important role. Various approaches of wind potential estimation have been reported in the literature. Significant among them are Auto Regressive Moving Average (ARMA), Auto Regressive Integrated Moving Average (ARIMA), ANN based methods, Support Vector Machine (SVM) and adaptive wavelet neural network (AWNN) [4-6]. Artificial intelligence-based wind speed and power estimation has also attracted the focus of recent studies which include artificial neural networks, fuzzy logic, neuro-fuzzy networks and evolutionary optimization methods. M. Carolin Mabel et.al. [7] have developed a model with the help of ANN which utilizes three input variables like wind speed, relative humidity and generation hours and one output variables, i.e. the energy output from wind farms. M. Rizwan et.al. have presented a fuzzy logic based generalized model to forecast wind energy. The proposed model provides significantly less rule base also it has increased the wind speed accuracy as compared with the traditional method. A data driven methodology [8,9] was proposed for the development of virtual models of a wind turbine. The authors have used, two parameters of the wind turbine namely, power output and rotor speed. Shuhui Li et.al. [10] have proposed a model based on neural network to forecast power produced by turbines. A four input ANN is developed, and its performance is shown to be superior to the single parameter traditional model approach. A TSK fuzzy model [11] has been developed to forecast of wind speed and electrical power up to hour ahead. The GA-based learning algorithm is used to train the model. Inputs to the model are wind data that are collected from neighbouring meteorological stations at a radius up to 30km. A new method using fuzzy and ANN was developed [12] to estimate annual energy output for a wind turbine in a site of known wind topology. A novel approach, which exploits two-hidden layer neural network was developed for carefully predicting the wind energy output [13]. Keeping in view of aforesaid literature a FES based model has been developed and analysed to estimate the wind power using wind speed and air density at Chennai and Bangalore stations. This model requires limited data and incorporates most of the uncertainties associated with climatic variables. This paper is structured as follows: Section-II describes the selection of input parameters for the proposed intelligent model. Section-III gives data collection and Section IV describes the fuzzy-logic-based model. Section-V Presents the ANN(Artificial Neural Network) based model. ANFIS (adaptive neuro fuzzy inference system) implementation has been presented in section VI. Results and discussion is presented in Section-VII. This paper is concluded in Section-VIII.

II. SELECTION OF INPUT PARAMETERS

Before developing a model based on expert system, the input parameters must be identified. For the systems of small complexity identification of input and output parameters must be performed by the experts by including all the available inputs. For the systems of higher complexity as it is not possible to consider all the inputs and one may be confined to select only those inputs which have significant contribution to the overall output of the system. Keeping in view of the above, the input meteorological parameters which affects the wind power are as follows: wind speed, relative humidity, temperature, air density and rainfall, dew point temperature, wind direction and pressure etc. The air density incorporates the change in relative humidity and temperature. Therefore, meteorological parameters like wind speed and air density plays an important role. However, the other meteorological parameters such as rainfall, dew point temperature may affects the wind power generation but these parameters are not expected to be significant a priori. Hence these parameters are neglected as input parameters in the present model. However, the impact of wind speed and air density on wind power generation is significant. Hence, wind speed and air density (relative humidity and temperature) have been selected as input meteorological parameters for the prediction of wind power. The air density is given as:

$$\text{Air Density} = D \left(\frac{273.15}{T}\right)\left(\frac{B-0.3783e}{760}\right) \qquad (1)$$

where D is the density of dry air at standard atmospheric temperature (25 °C) and pressure (100 kPa) (D=1.168 kg/m3), B is the barometric pressure in torr, T is the absolute temperature in Kelvin and e is the vapor pressure of moist air in torr.

III. DATA COLLECTION

The wind speed, temperature and relative humidity data has been acquired from IMD Pune [15-16]. The actual and normalized monthly data of wind speed and air density and

computed wind power for New Delhi, Chennai, Bangalore and Vishakhapatnam is given in TABLE-I, II, III and IV respectively.

TABLE I
ACTUAL AND NORMALIZED VALUE OF WIND SPEED, AIR DENSITY AND WIND POWER FOR NEW DELHI

Month	Wind Speed		Air Density		Computed Power (MW)	
	Act.	*Norm.*	*Act.*	*Norm.*	*Act.*	*Norm.*
January	3.590	0.382	1.17	0.901	20.13	0.422
February	4.141	0.520	1.16	0.762	22.91	0.552
March	4.181	0.530	1.13	0.508	22.64	0.538
April	4.160	0.521	1.13	0.250	22.05	0.510
May	4.642	0.645	1.11	0.101	24.23	0.611
June	4.480	0.604	1.12	0.147	23.50	0.577
July	4.10	0.511	1.12	0.225	21.65	0.492
August	3.721	0.415	1.10	0.303	19.81	0.408
September	3.50	0.358	1.10	0.264	18.54	0.351
October	2.70	0.161	1.10	0.321	14.40	0.158
November	2.70	0.174	1.14	0.546	15.01	0.189
December	2.81	0.184	1.10	0.811	15.60	0.214

TABLE II
ACTUAL AND NORMALIZED VALUE OF WIND SPEED, AIR DENSITY AND WIND POWER FOR BANGALORE

Month	Wind Speed		Air Density		Computed Power (MW)	
	Act.	*Norm.*	*Act.*	*Norm.*	*Act.*	*Norm.*
January	2.91	0.212	1.15	0.578	15.87	0.227
February	2.78	0.181	1.14	0.451	15.02	0.187
March	3.00	0.235	1.14	0.500	16.26	0.245
April	2.46	0.100	1.12	0.314	13.11	0.100
May	3.12	0.265	1.13	0.353	16.69	0.265
June	5.38	0.833	1.13	0.481	29.13	0.837
July	5.57	0.882	1.14	0.577	30.47	0.901
August	4.42	0.592	1.14	0.577	24.16	0.608
September	3.33	0.321	1.14	0.577	18.21	0.334
October	2.72	0.167	1.13	0.566	14.91	0.181
November	2.76	0.176	1.14	0.616	15.16	0.193
December	2.88	0.208	1.15	0.664	15.91	0.228

TABLE III
ACTUAL AND NORMALIZED VALUE OF WIND SPEED, AIR DENSITY AND WIND POWER FOR VISHAKHAPATNAM

Month	Wind Speed		Air Density		Computed Power (MW)	
	Act.	*Norm.*	*Act.*	*Norm.*	*Act.*	*Norm.*

Month	Wind Speed Act.	Wind Speed Norm.	Air Density Act.	Air Density Norm.	Computed Power Act.	Computed Power Norm.
January	3.41	0.261	1.13	0.538	18.48	0.271
February	3.24	0.545	1.12	0.491	17.44	0.552
March	3.52	0.632	1.13	0.401	18.66	0.627
April	4.04	0.901	1.10	0.411	21.37	0.894
May	3.91	0.821	1.10	0.362	20.63	0.811
June	4.37	0.650	1.10	0.362	23.10	0.644
July	4.28	0.503	1.11	0.450	22.81	0.507
August	4.08	0.541	1.11	0.401	21.72	0.538
September	3.02	0.284	1.11	0.401	16.04	0.286
October	2.61	0.380	1.12	0.401	13.94	0.381
November	3.56	0.630	1.13	0.501	19.32	0.637
December	4.04	0.552	1.13	0.547	21.87	0.565

TABLE IV
ACTUAL AND NORMALIZED VALUE OF WIND SPEED, AIR DENSITY AND WIND POWER FOR CHENNAI

Month	Wind Speed		Air Density		Computed Power (MW)	
	Act.	Norm.	Act.	Norm.	Act.	Norm.
January	3.41	0.338	1.15	0.481	18.48	0.346
February	3.24	0.297	1.12	0.391	17.44	0.298
March	3.50	0.362	1.11	0.294	18.66	0.355
April	4.04	0.498	1.12	0.206	21.37	0.481
May	3.91	0.465	1.10	0.177	20.63	0.445
June	4.37	0.581	1.10	0.196	23.12	0.561
July	4.28	0.558	1.11	0.284	22.80	0.545
August	4.08	0.508	1.11	0.274	21.71	0.496
September	3.02	0.238	1.11	0.303	16.03	0.234
October	2.61	0.135	1.12	0.391	13.93	0.137
November	3.56	0.378	1.13	0.481	19.32	0.385
December	4.04	0.496	1.13	0.481	21.88	0.503

IV. FUZZY LOGIC BASED MODELLING FOR THE ESTIMATION OF WIND ENERGY

In this section, a Fuzzy Logic based intelligent model has been developed for estimating the wind power potential. The modeling considers the uncertainties present in the atmosphere due to changing weather conditions. The performance of the model is found to be better as compared with the conventional method. Due to non-linear relation between input and output parameters, the percentage error in the prediction of wind power is around 3% with this model.

For Fuzzy-logic based modeling of the wind energy system, the MATLAB Tool box fuzzy-inference-system has been utilized to express the wind energy in terms of wind speed and

air density. The inputs for all the above mentioned stations are fuzzified into five fuzzy subsets such as very low, low, medium, high and very high. The obtained data (input and output) are not of the same order of magnitude, therefore all the data used in the proposed work is expressed in the normalized form (ranges 0.1-0.9) to avoid convergence problem during the learning process.

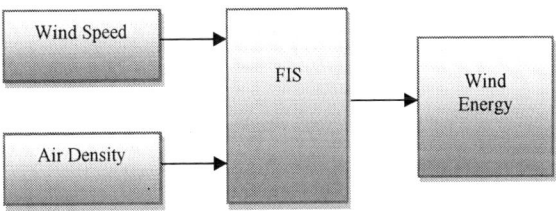

Fig.1. FIS based model for the prediction of wind Energy

Fig.2. shows the membership functions associated with each of the variables. Membership function editor is utilized to express the shape of all membership functions. Fuzzy membership is a curve which defines how each point in the input space is mapped to membership value or degree of membership between 0 and 1. Fuzzy membership functions can be of many forms, but triangular functions with the same base is the easiest possible choice.

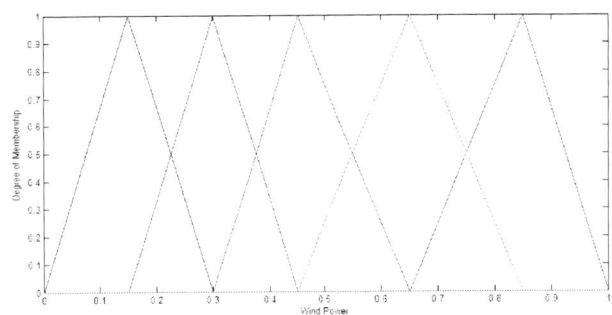

Fig. 2. Fuzzy membership function of wind speed and air density

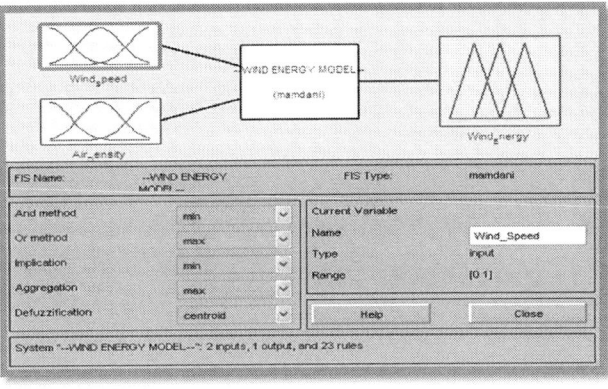

Fig. 3. Proposed fuzzy logic based model

978-1-5386-6624-1/18 $31.00 © 2018 IEEE

Fig.4. shows a MATLAB based fuzzy rule viewer for the estimation of the wind energy. The display of the fuzzy inference diagram consists of a figure window with seven small plots nested in it.

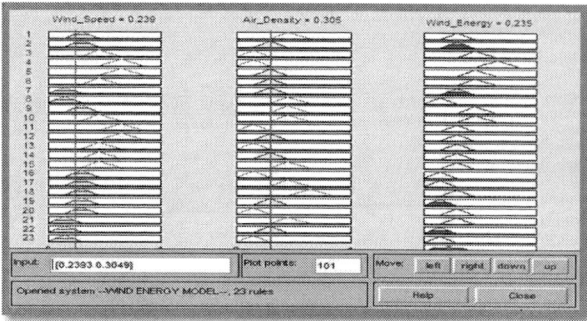

Fig.4. Fuzzy rule viewer for the prediction of wind energy

The two small plots across the top of the figure represent the antecedent and consequent of the first rule. Each rule is a row of plots and each column is a variables. So the first column of plots shows the membership functions referenced and antecedent or if – part of each rule. The second columns of the plots shows the membership functions referenced by the consequent, or then part of each rule. The rule viewer presents a sort of micro view of the fuzzy inference system by showing calculation at a time in detail.

V. ANN BASED MODELLING FOR ESTIMATION OF WIND ENERGY

An artificial neural network-based modeling yields computationally efficient technique for determining an empirical, possibly non-linear relationship between a number of inputs and one or more outputs. The present section is devoted for evolving artificial neural network (ANN) model which can be used for the estimation of wind power at different locations such as New Delhi, Chennai, Bangalore, and Visakhapatnam based on wind speed and air density data.

The data enters the network from the input layer to output layer through hidden layer shown in Fig 5. Neural network have a built incapability to adapt their synaptic weight w.r.t the changing surrounding environment and it can perform tasks that a linear program cannot [17-18]. It gives better performance at large amount of data set. Neural networks learn by example and they cannot be programmed to perform a specific task. They are fault-tolerant, that is, they can handle noisy and incomplete data, are able to deal with non-linear

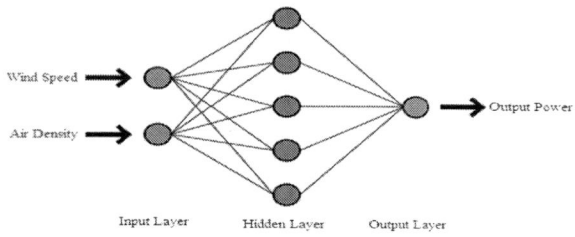

Fig. 5. Proposed artificial neural network architecture

Problems and once trained can assist in prediction and generalization at high speed [19]. It more practical terms, neural network is a non- linear statistical data modeling tool.

VI. ANFIS BASED METHODOLOGY/ANFIS IMPLEMENTATION

From the previous applied techniques we have seen that ANN are simple, but powerful and flexible tools for forecasting, provided that there are enough data for training, an adequate selection of the input-output samples, an appropriate number of hidden units and enough computational resources available. Just like ANN, a fuzzy logic system is a nonlinear mapping of an input vector into a scalar output, but it can handle numerical values and linguistic knowledge. In this section, ANFIS approach is applied for the prediction of wind energy. Jang in 1993 proposed a neuro-fuzzy system known as adaptive neuro-fuzzy inference system (ANFIS). ANFIS is a combination of ANN and Fuzzy and has the advantage of the both. In a neuro-fuzzy system, neural networks extract automatically fuzzy rules from numerical data and, through the learning process, the membership functions are adaptively adjusted. ANFIS is a class of adaptive multi-layer feed-forward networks, applied to nonlinear forecasting where past samples are used to forecast the sample ahead. It incorporates the self-learning ability of NN with the linguistic expression function of fuzzy inference. Therefore, ANFIS model has been developed for the prediction of wind energy based on the same data as used for previous models. The following steps are used for the development of the proposed model. Firstly the input and output data were divided into 2 groups for training and testing purpose, after that a fuzzy model was implemented using the ANFIS editor and training of the data was carried out. At last the test data were used for the validation of the proposed model. The input and output membership function of the model was selected to be gauss2mf and linear. Using this membership function ANFIS model was proposed. A separate data set, not included in the training set, was employed for verifying the ANFIS model generalization capabilities.

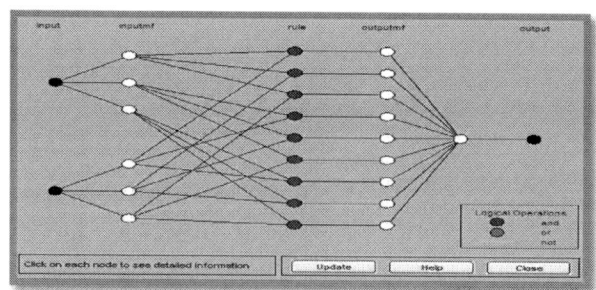

Fig. 6. ANFIS model structure

VII. RESULTS AND DISCUSSIONS

The fuzzy logic, ANN, and ANFIS based models for the forecasting of wind power have been developed and presented. The input parameters are wind speed and air density and output parameter of the proposed model is wind energy. The normalized data of wind speed, air density and wind power is used for the developmet of model. The results obtained using fuzzy logic are presented in Fig.7. However the results obtained using ANN and ANFIS model are presented in Fig. 8 and Fig.9 respectively.

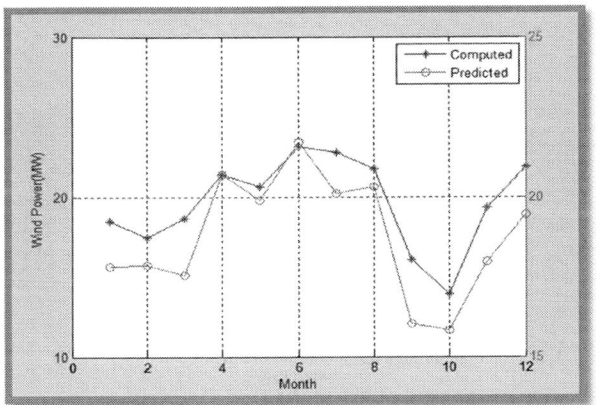

Fig. 7. Predicted value of wind energy in comparison with computed power using fuzzy logic

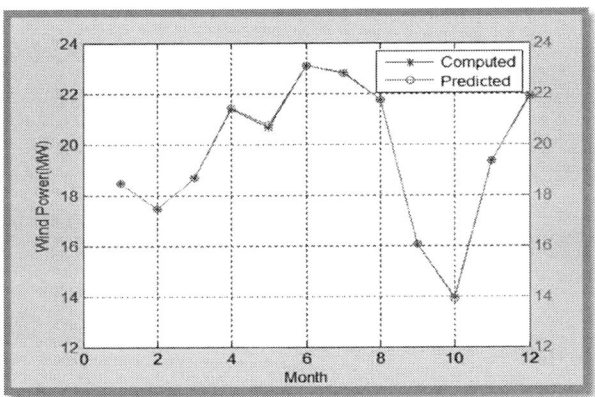

Fig. 8. Predicted value of wind energy in comparison with computed power using artificial neural network

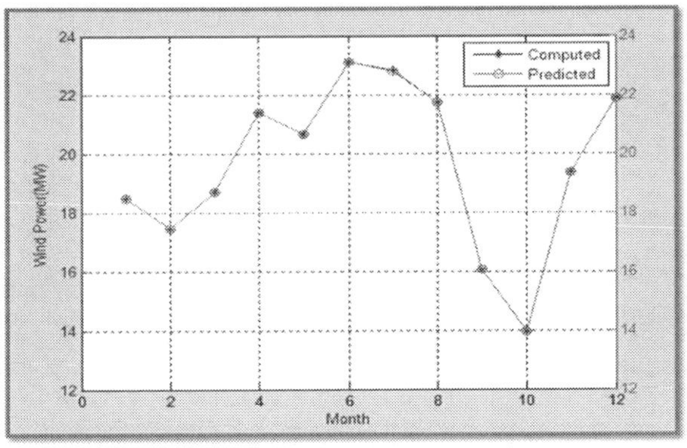

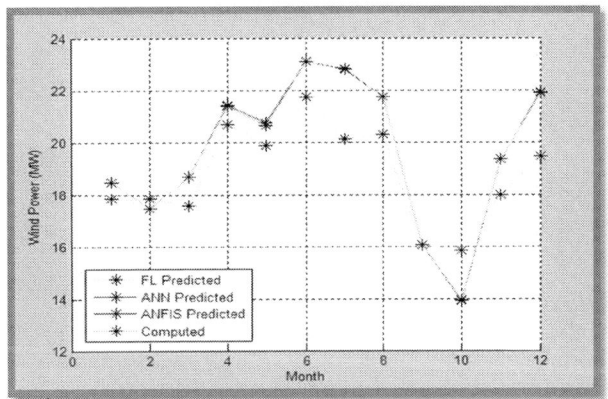

Fig.9. Predicted value of wind energy in comparison with computed power using ANFIS

Fig.10. Performance of Fuzzy Logic, ANN and ANFIS

It is clearly seen from Fig.10 that the most of the predicted values using ANFIS are overlaps or very close to the compared values. It is concluded that the performance of ANFIS model is far better as compared to the ANN and Fuzzy Logic based models on the basis of above mentioned statistical indicators. The predicted wind power in comparison with computed value using fuzzy logic is presented in Table V

TABLE V
MONTHLY MEAN PREDICTED WIND POWER IN COMPARISION WITH COMPUTED VALUE USING FUZZY LOGIC, ANN AND ANFIS

Month	Wind Power (Computed) MW	Wind Power (Predicted) MW		
		FL	ANN	ANFIS
January	18.491	17.825	18.490	18.480
February	17.450	17.847	17.456	17.453
March	18.674	17.565	18.674	18.679
April	21.381	20.692	21.420	21.387
May	20.642	19.891	20.770	20.641
June	23.118	21.737	23.097	23.118
July	22.812	20.106	22.801	22.803
August	21.736	20.301	21.726	21.732
September	16.056	16.044	16.068	16.048
October	13.950	15.827	13.922	13.954
November	19.340	17.977	19.350	19.336
December	21.892	19.454	21.898	21.890

The performance of the models are evaluated on the basis of statistical indicators like absolute relative error (ARE), the mean absolute percentage error (MAPE), the sum squared error (SSE) and the standard deviation of error (SDE).The

978-1-5386-6624-1/18 $31.00 © 2018 IEEE

performance of Fuzzy logic, ANN and ANFIS models for the prediction of wind energy is presented in Table VI in terms of statistical indicators.

$$\text{MAPE} = \frac{100}{N} \sum_{h=1}^{N} \frac{|\hat{p}_h - p_h|}{\bar{p}} \quad (2)$$

$$\bar{p} = \frac{1}{N} \sum_{h=1}^{1} p_h \quad (3)$$

Where $\hat{p}_h$ and p_h are respectively the forecasted an actual wind power at hour h, $\bar{p}$ is the average wind power of the forecasting period and N is the number of forecasted hours.

The SSE criterion is given by:

$$\text{SSE} = \sum_{h=1}^{N} (\hat{p}_h - p_h)^2 \quad (4)$$

The SDE criterion is given by:

$$\text{SDE} = \sqrt{\frac{1}{N} \sum_{h=1}^{N} (e_h - \bar{e})^2} \quad (5)$$

$$e_h = \hat{p}_h - p_h \quad (6)$$

$$\bar{e} = \frac{1}{N} \sum_{h=1}^{N} e_h \quad (7)$$

Where e_h is the forecast error at hour h and $\bar{e}$ is the average error of the forecasting period.

TABLE VI
RESULTS OF FUZZY LOGIC, ANN AND ANFIS MODELS FOR THE PREDICTION OF WIND ENERGY

Techniques	Statistical Indicators			
	ARE	MAPE	SSE	SDE
Fuzzy Logic	3.61	4.225	0.0540	0.0543
ANN	0.05	0.0827	2.232e-006	3.875e-004
ANFIS	0.01	0.0523	7.269e-007	2.279e-004

It is seen from the above table the absolute relative error (ARE), the mean absolute percentage error (MAPE), the sum squared error (SSE) and the standard deviation of error (SDE) using fuzzy logic is 3.61%, 4.22%, 0.054% and 0.054% respectively, whereas the results of artificial neural network (ANN) are 0.05%, 0.083%, 2.232e-006% and 3.875e-004%. Further, adaptive neuro fuzzy (ANFIS) model gave the errors of 0.01%, 0.052%, 7.269e-007% and 2.279e-004%.

VIII. CONCLUSION

An analysis based on Fuzzy logic, artificial neural network (ANN) and Adaptive Neuro-Fuzzy Inference System (ANFIS) models for the estimation of wind energy at different locations has been carried out and presented in this paper. The input parameters are wind speed and air density and output parameter is wind energy. The performance of the models is evaluated on the basis of Statistical indicators such as ARE, MAPE, SSE and SDE. The absolute relative error (ARE), the mean absolute percentage error (MAPE), the sum squared error (SSE) and the Standard deviation of error (SDE) using fuzzy logic is 3.61%, 4.22%, 0.054% and 0.054% respectively, where as the results of ANN are 0.05%, 0.083%, 2.23e-006% and 3.87e-004%. The results of ANFIS model are 0.01%, 0.052%, 7.26e-007 and 2.27e-004% respectively. Therefore, it has been established that ANFIS based model performs better as compared with the fuzzy logic and ANN based model for the estimation of wind energy. Hence, ANFIS based model can be helpful in wind energy prediction.

REFERENCES

[1] Ahmed E. Saleh, Mohamed S. Moustafa, Khaled M. Abo-Al-Ez, Ahmed A. Ahmed A. Abdullah, "A hybrid neuro-fuzzy power prediction system for wind energy generation," Electrical power and energy systems 74, pp. 384-395, 2016.

[2] Tao Hong, Pierre Pinson, Shu Fan, Hamidreza Zareipour, Alberto Troccoli, Rob J. Hyndman, "Probabilistic energy forecasting: Global Energy Forecasting Competition 2014 and beyond," International Journal of Forecasting, Publised by Elsevier B.V., 2016.

[3] Xiaochen Wang, Peng Guo and Xiaobin Huang, "A review of wind power forecasting models," Energy procedia, pp.770-778, sep. 2011, China.

[4] Juan Yan et al., "Hybrid probabilistic wind power forecasting using temporally local Gaussian process," IEEE Transactions on sustainable energy, vol. 7, pp.87-95, Jan 2016.

[5] Yordanos Kassa, J. H. Zhang, D. H. Zheng and Dan Wei, "Short term wind power prediction using ANFIS," IEEE International conference on power and renewable energy, March 2016, Chaina.

[6] Rahu R. Gunjker, Bhupendra Deshmukh and Rakesh K. Jha, "Wind energy scenario and potential in india," International Journal for innovative research in science & technology, vol. 2, Issue 12, May 2016.

[7] M. Carolin Mabel and E. Fernandez, "Estimation of Energy Yield From Wind Farms Using Artificial Neural Networks," IEEE Transactions on Energy Conversion, vol. 24, no. 2, pp.459-464, June 2009.

[8] Amit Kumar Singhal and M.Rizwan, "Generalized Methodology for the Prediction of Wind Powor: A Fuzzy Logic based Approch," International Journal of Engineering Science and Research, vol. 02, July 2011.

[9] A. Kusiak and Wenyan Li, "Virtual Models for Prediction of Wind Turbine Parameters," IEEE Transactions on Energy Conversion, vol. 25, no. 1, pp. 245-252, March 2010.

[10] Shuhui Li, Donald C. Wunsch and Edgar A. O' Hair, "Using Neural Networks to Estimate Wind Turbine Power Generation," IEEE Transactions on Energy Converson, vol. 16, no. 3, pp. 276-282, September 2001.

[11] Ioannis G. Damousis, Minas C. Alexiadis and John B. Theocharis, "A Fuzzy Model for Wind Speed Prediction and Power Generation in Wind Parks Using Spatial Correlation," IEEE Transactions on Energy Conversion, vol. 19, no.2, pp. 352-361, June 2004.

[12] M. Jafarian and A.M. Ranjbar, "Fuzzy Modeling Techniques and Artificial Neural Networks to Estimate Annual Energy Output of a Wind Turbine," Renewable Energy,vol. 35, pp. 2008-2014, 2010.

[13] G. Grassi and P. Vecchio, "Wind Energy Prediction Using a Two-Hidden Layer Neural Network," Commun Nonlinear Science Number Simulation, vol. 15, pp. 2262-2266, 2010.

978-1-5386-6624-1/18 $31.00 © 2018 IEEE

[14] A. Troncoso, S.Salcedo-Sanz. and C.Casanova-Mateo, "Local models-based regression trees for very short-term wind speed prediction," Renewable Energy, vol. 81, pp. 589-598, 2015.

[15] Mani, A., 1980. Handbook of Solar Radiation Data for India, New Delhi: Allied Publishers.

[16] Solar Radiation Handbook, SEC & IMD Pune, 2015.

[17] Stefan Balluff, Jorg Bendfeld and Stefan Krauter, "Short term wind and energy prediction for offshore wind farms using neural networks," 4th IEEE International conference on renewable energy research and applications, pp. 22-25, nov 2015.

[18] Abinet Tesfaye, J. H. Zhang, D.H. Zheng and Dereje Shiferaw, "Short-term wind power forecasting using artificial neural networks for resource scheduling in microgrids," International journal of science and engineering applications, vol. 5, Issue 3, 2016.

[19] Doan Duc Tung and Tuan-Ho Le, " A statistical analysis of short-term wind power forecasting error distribution," International journal of applied engineering research IISN 0973-4562, vol. 12, no.10, pp.2306-2311, 2017.

Energy Audit and Energy Conservation for a Hostel of an Engineering Institute

Srishti Gupta
Department of Electrical and
Electronics Engineering
*Maharaja Surajmal Institute of
Technology(GGSIPU)*
Delhi, India
guptasrishti97@ieee.org

Rakhi Kamra
Department of Electrical and
Electronics Engineering
*Maharaja Surajmal Institute of
Technology(GGSIPU)*
Delhi, India
rakhikamra@msit.in

Manuhar Swaroopa
Department of Electrical and
Electronics Engineering
*Maharaja Surajmal Institute of
Technology(GGSIPU)*
Delhi, India
sswaroopadelhi@gmail.com

Abhishek Sharma
Department of Electrical and
Electronics Engineering
*Maharaja Surajmal Institute of
Technology(GGSIPU)*
Delhi, India
abhisheksharma425@gmail.com

Abstract— The adequate and right kind of energy is necessary for the sustainable development of human society. Due to ever increasing rise in demand we need to take necessary steps to bridge the gap between supply and demand. The paper focuses on the process of energy audit carried out in an academic institution in order to suggest various conservative measures to make energy consumption more economical. Therefore, Energy Audit in hostel and mess of Maharaja Surajmal Institute of Technology, Janakpuri, New Delhi has been conducted. Thus in this paper, efforts have been carried out to augment energy use awareness and encourage energy conservation practices by a thorough analysis of the consumption and wastage of energy through the appliances. The detailed study carried out by us in this paper helped to bring out suggestive measures for cost mitigation , reduction of environmental pollution and monitoring demand supply gap.

Keywords—Energy audit,Energy Conservation Measures, Payback Period, Biomass, Load demand

I. INTRODUCTION

The sustainable development of today's society is dependent on adequate amount of energy. Total installed capacity of power stations in India is at 334,146.31(MW) as on February, 2018. India's energy consumption is set to grow by 4.2% per year by 2035.Owing to such tremendous use in demand, more emphasis is to be laid on renewable sources of energy. When huge figures are seen in the electricity bill, it becomes inevitable to conduct energy audits so as to lessen the burden of energy production to a certain extent. Energy Conservation means to use energy efficiently and hence cutting out waste to zero level.

Unlike other energy audits in which the reduction in consumption and various method to cut down the usage is the main focus areas, this paper deals with the idea of reducing the overall electricity bill by replacing existing devices with those devices which consume less power and are more energy efficient. This paper consists of the complete methodology to conserve energy in best way possible and mainly focuses on implementation of solar water heater and biogas system.

II. LITERATURE REVIEW

Many researchers have done various energy audit in different areas like industry, academic institutions etc. After carefully studying the energy consumption pattern, in an Institution and educational building , researchers proposed some general recommendations like replacing incandescent bulbs with CFLs and use of motion sensors[1-2].Besides this energy audit also holds immense relevance in the residential areas and industrial areas too, as presented by R.Sharma, V. Jadhav,R.Pieterse [3-4][8] . The suggestions made in the paper [6] by Ramkiran et al substantiates the claim in context with financially and technically viable alternatives for reducing energy usage. Moreover, a detailed work on Energy Conservation reiterates the inevitable role of energy in near future and how its conservation holds extreme importance [5][7]. As far as the results and conclusions are concerned, the reference in [9] points out the significant energy savings with fewer changes and adjustments.

III. DATA COLLECTION AND METHODOLOGY

For the collection of data we have taken in account room-wise details of the lighting load, fans, air conditioners, PCs, switches, exhaust load etc. that are present in each floor of the MSIT HOSTEL building, to estimate the present load and the load that can be expanded in the future.

The methodology adopted for this audit was:
i. Grouping of students to perform audit in specific areas for inspections and data collection.
ii. Observation on daily consumptions of various loads and facilities present in MSIT hostel and mess.
iii. Measurements of energy consumptions and other parameters.
iv. Investments and payback period estimations.
v. Recommendations regarding implementation of various kind of energy conservation methods.

978-1-5386-6624-1/18 $31.00 © 2018 IEEE

Some technical data have been gathered regarding the total load present in the MSIT building which can be analysed and studied through the pie charts and table given below in table 1 and figure1.

Table 1: Electrical Distribution in Hostel

Electric Load	Quantity in hostel area	Rating	P.F.
Fan	142	80W	0.5-0.7
Tube Light	185	40W	0.5
Socket	147	90W	
Bulb	118	15W	0.5-0.9
Exhaust Fan	32	65W, 70W	0.6-0.7
AC	1	2kW	0.8

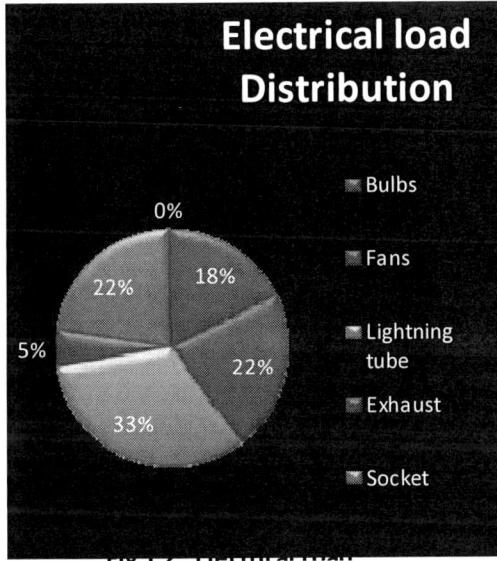

Figure 1:Electrical Load Distribution

IV. MEASUREMENTS PERFORMED

For calculation of energy cost savings and further payback period we first needed to measure the energy consumption in a day and then in a year for different areas.

The following measurements and clear view is explained through table and graphs below:

Measurements were performed in boys hostel and girls hostel and calculations on energy consumptions were done which are tabulated in following table 2 and table 3 respectively.

Table 2:Energy Consumption in Boys Hostel

Electric Load	Wattage (W)	Qty	Daily hours of use	kWh used in a day	kWh used in a year
Tubelight	40	67	10	26.8	7772
Fan	80	67	17	91.12	16857.2
Bulb	15	67	6	6.03	1748.7
Laptop	60	67	3	12.06	3497.4
Mobile Charger	5	120	6	3.6	1044

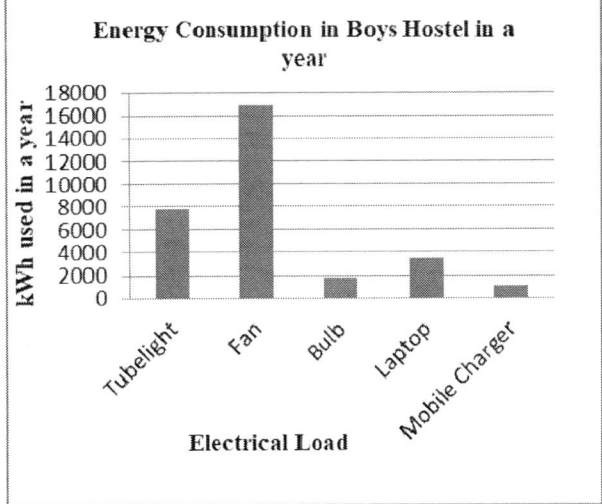

Figure 2:Energy Consumption in boys hostel room

Table 3:Energy consumption in girls hostel

Electrical Load	Wattage (W)	Quantity	Daily hours of use	kWh used in a day	kWh used in a year
Tubelight	40	50	10	20	5800
Fan	80	50	17	68	12580
Bulb	15	50	6	4.5	652.5
Laptop	60	50	3	9	2610
Mobile Charger	5	90	6	2.7	783

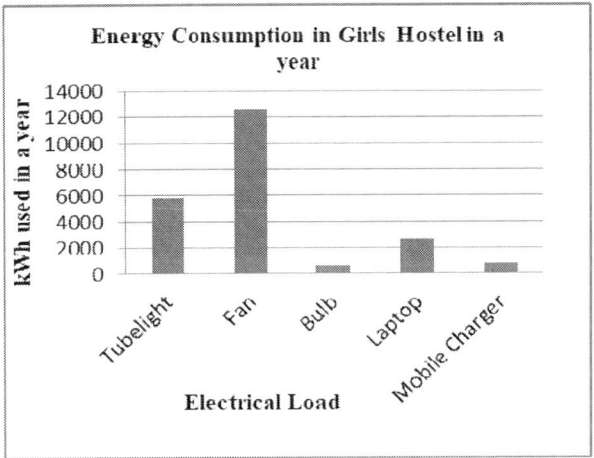

Figure 3:Energy Consumption in Girls hostel rooms

978-1-5386-6624-1/18 $31.00 © 2018 IEEE

Also, further measurements were done in mess to calculate their daily consumptions as shown in table 4.

Table 4:Energy Consumption in Hostel Mess

Electrical Load	Wattage W	Quantity	Daily hours of use	kWh used in a day	kWh used in a year
Tube light	40	28	10.5	11.68	3387.2
Fan	80	16	10.5	13.28	2456.8
Bulb	15	1	24	0.96	278.4
Exhaust	70	2	8	1.12	324.8

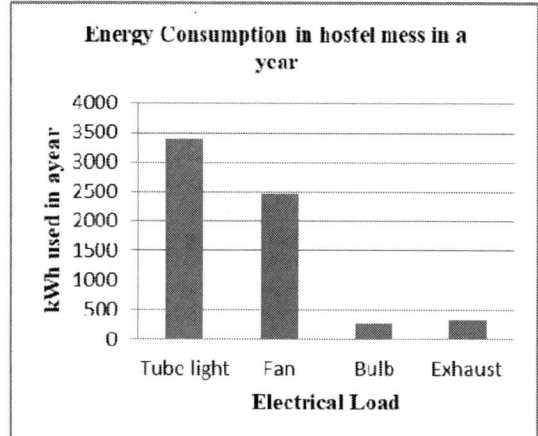

Figure 4: Energy consumption in Mess

V. ENERGY CONSERVATION MEASURES

A. Current Lightning Scheme in Hostel

Total number of tubelights =185
Wattage of Tubelight: 40W which consumes 54W power with 14W conventional chokes
Assumptions of operational hours and days: 10 hours per day for 290 days
Total energy used in a Month:
No.of lights × (Hours ÷Day) × Days in a month × Wattage
=18 × 54 × 10 × 290
=28971 kWh

i. Replacement of 40W tubes with 36W tubes

The 36W tube consume 50W power with conventional choke of 14W

Therefore Total Energy

=185 × 50 × 10 × 290

=26825 kWh

Hence Energy Saved= 2146 kWh

Cost of electricity saved at the rate of ₹11.44

= ₹24550.24

The cost of 36W tube is ₹45 , therefore the total investment of replacing 185 tubes would be ₹8325

Therefore Payback Period would be

= (8325 ÷ 24550.24) × 12

=4 months

ii. Using T5 FTL

Characteristics of Conventional 40W:

Expected life hours: 5000 hours

Energy input per hour: 45 hours

Consumption in 10 hours per day: 0.45 kWh

Annual Consumption in 290 working days: 130.5 units

Annual Consumption at the rate of ₹11.44 per unit
= ₹1493

Characteristics of T5:

Expected life hours: 20000 hours

Energy input per hour: 28 hours

Consumption in 10 hours per day: 0.28 kWh

Annual Consumption 290 working days: 81.2 units

Annual Consumption at the rate of ₹11.44 per unit
= ₹929

Cost of electricity saved = 1493 − 929 = ₹564

Total savings per year considering we replace 185 T8 FTLs with T5 we can save around **₹104340**

Total investment = 185x400 = ₹74000

Therefore Payback Period would be **8.5 months**

iii. Replacement of conventional chokes with electronic chokes

Table 5:Payback Period of electronic chokes

Parameter	Conventional Choke	Electronic Choke
Investment p.a	150x185=27750	220x185=40700
Cost of starters	10x185=1850	-
Net Investment	40700-(27750+1850)=11,100	
Power Rating (W)	(40+14)x185=9990	(40+3)x185=7955
Annual Energy Consumption (kWh)	9990x10x290/1000= 28971	7955x10x290/1000=2 3069.5
Cost of energy@11.4	331428	263915
Net Amount of annual savings	67512	
Simple Payback Period	11,100x12/67512=**2months**	

978-1-5386-6624-1/18 $31.00 © 2018 IEEE

Calculation of payback period when using electronic choke by comparing it with conventional choke in the above table5

B. Energy conservation scheme for Fans

Replacement of Rheostatic regulators with electronic regulators.
Cost of one electronic fan regulator: ₹180
Duration for which fan is used: 16 hours per day
Operational days: 200 days
Rating of Fan: 80W

Annual energy consumed by a fan
$= 80 \times 16 \times 200 = 256$ kWh
If we use electronic fan regulators there is 25% reduction against Conventional Fan Regulators.
Cost of energy saved
$= 0.25 \times 256 \times 11.44 = ₹732.16$

Therefore Payback Period = cost of electric fan regulator ÷ cost of energy saved
$= 180 \div 732.16 = \textbf{0.24 year}$

C. Installation of Solar Water Heater system

Residents: 250

Assuming an average requirement of 20 L of hot water per day
Thus daily amount of hot water used $= 250 \times 20 = 5000$ L

An average flat plate collector area of 2 m^2 gives 125L of hot water per day.

Thus total collector area required $= 5000 \div 125 \times 2 = 20$ m^2

Assuming cost of installation to be around 10,000/m^2 total capital cost comes out to be $= ₹200,000$

Total geyser energy consumption if used in the hostel for approximately 12 hrs on a typical cold day for 2.3 kW geysers installed in the hostel then
$10 \times 2.3 \times 10 = 230$ kWh

Geysers will be typically operational in MSIT for about 130 days from November to February. Thus total energy consumption $= 130 \times 230 = 29,900$ kWh

Total expense with geysers $= 29,900 \times 11.44$ /kWh
$= ₹342,056$
Thus simple payback period $= 200000 \div 342056$
$= \textbf{0.58 years}$

D. Installation of Bio gas plant

Typical waste food density is 30 Kg/m^3

Total volume of waste food per day is about 30 L

Installation cost of Biogas = 1.5Lakhs

Operational and maintenance cost=₹2000

If 1m^3 of biogas produce 4.6kWh energy

Then we need 5.87m^3 of biogas

Therefore we need an urban bio gas unit of approx.200 kg capacity.
Assumptions-
Calorific value of biogas = 6kWh/m^3 =20MJ/m^3

Calorific value of LPG = 26.1kWh/ kg=46.1MJ/kg
Energy output of biogas plant per day:175 MJ
This implies that LPG saved
$= 175 \div 46.1 = 3.79$kg
$= 0.199$ LPG cylinders per day
Therefore, LPG saved
$= 5.99 = 6$ LPG/month

Total working days : 290 days approximately

Total annual savings = 290x0.199x600 = 34626

Payback period =152000/34626 x12 = **4 years approx.**

VI. RECOMMENDATIONS AND CONCLUSIONS

In this paper, efforts have been made to shed light on how we can significantly cut down on energy wastage by making some basic modifications that include :

1. Replacing conventional ballast [chokes] FTL with Electronic Ballasts of high frequency that reduces energy usage by 25%. Also, the capital cost recovery time for replacing it is around 2 months.
2. When it comes to lighting, it was found that replacement of 40W tubes with that of 36W, will also serve the purpose. It shall prove very convenient and economical as payback period is only 4 months.
3. Moving further, we also deduced that using T5 FTL instead of T8 can yet be another measure so as to reduce the energy consumption meanwhile improving the luminous efficiency. It's payback period is found out to be 8.5 months which is more than the above mentioned changes.

Realizing the vital role of renewable energy sources to tackle any energy crisis, in this paper emphasis have been laid on usage of solar water heating system and installation of biogas plant. In the mess, 6 LPG cylinder per month can be saved that will eventually bring down the costs increased.

Energy audit and various efficient tools and technologies that we have introduced in this paper can help bring monumental changes in the energy use pattern. The paper has brought out wastage and conservation of energy. It helps us identifying and addressing the potential of renewable source of energy. It has been shown in table 6 that payback period is relatively less and will compensate the invested amount in a short period of time.

978-1-5386-6624-1/18 $31.00 © 2018 IEEE

Table 6:Summary of different payback period for different energy saving measures

Energy Saving Measure	Payback Period
Replacement of 40W with 36W tubes	4 months
Using T5 FTL instead of T8	8.5 months
Replacement of conventional chokes with electronic chokes	2 months
Replacement of Rheostatic regulators with electronic one	0.24 year
Installation of Solar Water heating System	0.58 year
Installation of Bio Gas Plant	4 year

REFERENCES

[1] A. Rupal, P. Syal, S.Sharma, "Energy conservation opportunities in institutional buildings-a case study in India," IEEE Internation Conference on Power and Renewable Energy (ICPRE), 2017, pp. 664-668

[2] P.S. Magdum, S.R. Lokhande, P.M. Maskar, I.D.Pharne, "A case study:Energy audit at commercial and educational building," IEEE International Conference Signals and Instrumentation Engineering(ICPCSI), 2017, pp. 1360-1364

[3] R. Pieterse, A. Kumar, Gerhard P.Hancke, "Energy consumption audit system for smart building," 40th Annual Conference of the IEEE Electronics Society (IECON), 2014, pp. 3897-3902

[4] R.Sharma and R.K. Jain, "Energy audit of residential buildings to gain energy efficiency credits for LEED certifcation," International Conference on Energy Systems and Applications, 2015, pp.718-722

[5] S.Shenoy, Siddhartha, D.R. Akshay, "Energy Data Analysis," Innovations in Power and Advanced Computing Technologies (i-PACT), 2017, pp. 1-6

[6] Ramkiran et al.,"M Energy audit report," 3rd International Conference on Advances in Electrical, Electronics, Information, Communication and Bio-Informatics(AEEICB), 2017, pp.482-485

[7] S. Zadey, S. Chafle, A.S. Lilhahre, "Analysis of various parameters by energy audit," International Conference on Communication and Signal Processing(ICCSP), 2016, pp.1560-1564

[8] V. Jadhav, R. Jadhav, P. Magar, S. Kharat, S.U. Bagwan, "Energy conservation through energy audit," International Conference on Trends in Electronics and Informatics (ICEI), 2017, pp.481-485

[9] S.N. Chaphekar, R.A. Mohite, A.A.Dharme. "Energy monitoring by energy audit and supply side management," International Conference on Energy Systems and Applications, 2015, pp. 178-183

Development of a Characterization Tool for Determination of Sheet Resistance at 9 Places of Large Area 156 mm by 156 mm Diffused Silicon Wafer in Less than 5 Seconds and Its Application in Making High Efficiency Silicon Solar Cells

M. Shadab Siddiqui
Amorphous Silicon Solar Cell
Plant (ASSCP)
*Bharat Heavy Electricals
Limited (BHEL)*
Gurgaon, India
siddiqui@bhel.in

A. K. Saxena
Amorphous Silicon Solar Cell
Plant (ASSCP)
*Bharat Heavy Electricals
Limited (BHEL)*
Gurgaon, India
aksaxena@bhel.in

Shivangi
Amorphous Silicon Solar Cell
Plant (ASSCP)
*Bharat Heavy Electricals
Limited (BHEL)*
Gurgaon, India
shivangi@bhel.in

B. K. Pant
Amorphous Silicon Solar Cell
Plant (ASSCP)
*Bharat Heavy Electricals
Limited (BHEL)*
Gurgaon, India
bkpant@bhel.in

Abstract— **Sheet resistance measurement using 4 probe method is a very well established technique for measuring the sheet resistance of a thin film on a substrate. This technique is widely used to measure the sheet resistance of diffused silicon wafer in solar photovoltaic industry. However with respect to solar cell industry, the drawback associated with this technique is, it measures the sheet resistance of a very small area. For a large area silicon wafer of size 156 mm by 156 mm, a four probe sheet resistance measurement instrument will take several minutes to take readings at several places on the entire area of the wafer. In the present work a solar cell characterization tool is developed for measuring the sheet resistance at 9 places on a diffused silicon wafer using 9 four probe heads mounted on a single plate. The characterization tool developed in the present work can tell the uniformity of diffusion in a diffused silicon wafer within few seconds on the screen of a computer monitor. Continuous data logging helps in characterizing wafers which are > 50,000/day in any smallest PV industry.**

Keywords— *Four probe technique, Sheet Resistance, Shallow emitter, crystalline silicon solar cell.*

I. INTRODUCTION

There are several steps for converting a raw silicon wafer into a fully functional solar cell. One of the very critical step is the junction formation. Solar cell is a P-N junction device consisting of a single diode on an entire 156 mm by 156 mm silicon wafer. It is very essential to have a uniform junction in order to make a high efficiency cell. The junction is formed by diffusion of either phosphorous (using $POCl_3$ solution in case of p type wafer) or boron (using BBr_3 solution in case of n type wafer) in a tube furnace. The vapours of $POCl_3$ or BBr_3 causes diffusion of phosphorous or boron into the wafer and forms n type or p type regions for the respective wafers. This occurs at the top (and bottom) surfaces of the wafer. The depth to which diffusion of phosphorous or boron happens is estimated as less than a micron thick. This step of diffusion should be very uniform in order to achieve an emitter of uniform sheet resistance. Sheet resistance measurement is most commonly done using the 4 point probe technique [1, 2]. A schematic of how 4 probe technique works is shown in figure 1 (a) and an image of a single four probe head is shown in figure 1 (b). Contact less sheet resistance measurement technique using eddy current is another method to measure sheet resistance [3-5]. However the eddy current based technique is primarily being used currently for sheet resistance measurement of thin films on glass and is still not being widely used by the silicon wafer based solar cell industry.

In order to make a high efficiency cell, the current trend is to make wafers with shallow emitter having sheet resistance in the range of 90 to 100 $\Omega/\square$ [6-7]. In a good and high efficiency solar cell the diffusion process is so tightly controlled that the maximum variation in the sheet resistance is less than 3 %. In the present work a setup is designed that has a gantry of 9 four probe heads arranged in the form of a matrix of 3 by 3. This gantry moves up and down using a stepper motor. With the click of a mouse the probe head gantry comes down and sits on the diffused wafer and using the suitable electronics and relays takes reading from each head and the results are displayed on a computer monitor. In this way the sheet resistance on a large area 156 mm by 156 mm diffused silicon wafer is measured at 9 places with a single touch on the computer screen. The readings are taken symmetrically at 9 places across a silicon wafer and tells about the uniformity and extent of diffusion process. Corrective measures can then be taken in the diffusion process to improve the uniformity of sheet resistance.

Bharat Heavy Electricals Limited (BHEL) is the funding agency.

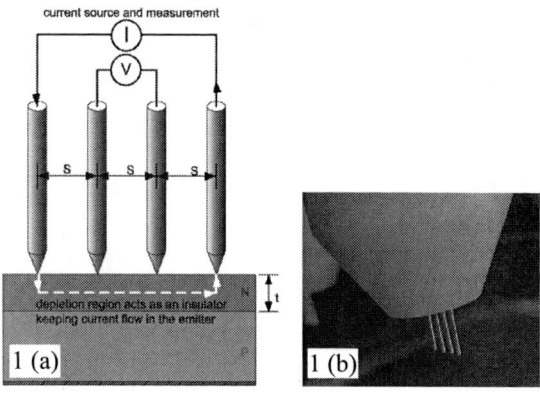

Fig.1 (a). Schematic showing how 4 probe sheet resistance measurement technique works [1]. 1(b) A single four probe head.

II. EXPERIMENTAL

Fig.2. shows a flowchart of the measurement procedure using the Sheet Resistance Measurement Setup (SRMS) developed in the present work.

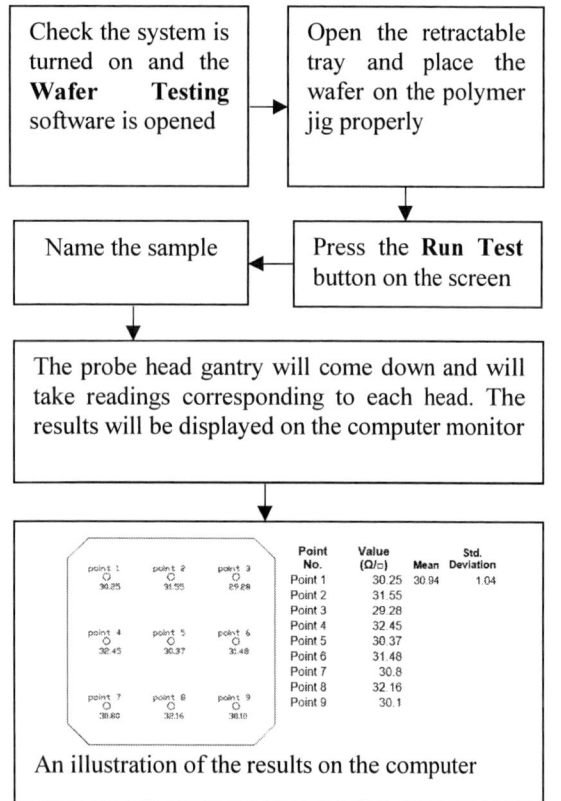

Fig.2. A flowchart of the measurement procedure

The measurement is done after opening the retractable tray having a derlin (polymer) jig for placing the wafer. After the wafer is placed properly in the specified area on the jig and the green light interlock for "wafer placed" comes on, the retractable tray should be inserted inside till the interlock signal for "Fixture tray at home" becomes green. Now touch the "Run test" tab on the screen and the interlock for "Tray locked" will become green and the wafer tray will be locked. Another small window will pop up asking the name for the sample. The operator should type the name of the sample and then click on the check mark. The probe head gantry, shown in Fig.3, driven by a stepper motor will now move downwards and comes in contact with the wafer. The electronic circuitry is designed such that all the 9 heads takes readings one by one till all the readings are completed and the results are displayed on the computer monitor in the graphical as well as tabular form. A flowchart of the measurement process is shown in Fig.4. The mean of all the readings is also displayed on the computer monitor. In the settings window one can set the value for the difference between maximum and minimum values, and based on that, the sample will pass or fail. The equipment can measure the sheet resistance of a diffused wafer of 180 microns as well as of thin films on glass upto 6 mm thickness. Necessary changes in the software settings window need to be made before running the test so that the probe head gantry moves downwards according to the sample thickness.

Fig.3. The SRMS equipment   Fig.4. Probe head gantry and the wafer placement chuck (with wafer placed on it)

The SRMS setup is equipped with movable casters with which the equipment can be taken anywhere at the shop floor of solar cell production unit and can be used there itself.

III. RESULTS AND DISCUSSION

The SRMS setup developed in the present work is a production worthy setup as the system is capable to be used in line. It can measure the sheet resistance at 9 different places in less than 5 seconds. In the solar cell production line of say 85 MW roughly around 70000 wafers of 6 inches by 6 inches are handled every day. The conventional four probe setup if used to map the sheet resistance at several places on a 6 inches by 6 inches wafer will take several minutes and will require writing the values and averaging out, finding standard deviation, and the difference between maximum and minimum values will take a very long time and will not fit in to the requirement of an inline characterization tool. On the other hand SRMS setup developed here takes less than 5 seconds to map the sheet resistance across a full size 6 inches by 6 inches diffused silicon wafer. Also by measuring manually using the conventional four probe setup no two wafers can be mapped on the same spot while in the current setup all the measured wafers are mapped at exactly the same spot and hence can be compared. The tool developed here can successfully be used to enhance the efficiency of solar cell in a production line.

Fig. 4(a) below shows the results of the software screen of computer monitor in graphical form as well as tabular form along with the mean value are shown. Here the mean of the values is coming as 89.15 while the minimum is 84.41 and the maximum value is 94.82. When the operator sets the "acceptable difference from the mean value" as 8 then none of the points falls outside this range and all the points are shown in green colour (it means a pass value). However the difference between the maximum and minimum value is 10.41 which is very high and the control on the diffusion process in not very tight.

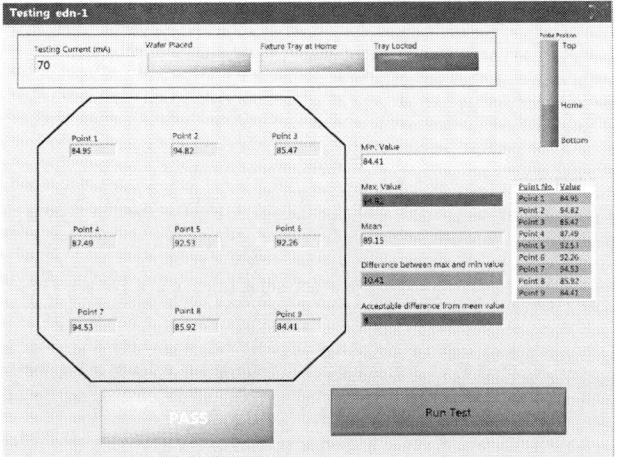

Fig. 4 (a). A screen shot of the testing window showing no value outside the acceptable limits

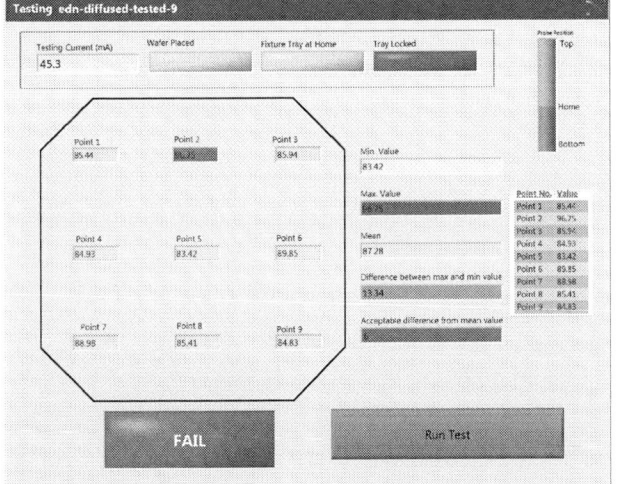

Fig. 4 (b). A screen shot of the testing window showing point number 2 falling outside the acceptable limit

In Fig. 4(b) above, however not all the 9 values are falling within the acceptable difference entered from the mean value as point number 2 is not falling in the acceptable range and hence is shown in red colour (it means a fail value). For the present sample (4 (b)) the acceptable difference from the mean value is kept as 6 which means a tighter control on the diffusion process is required as compared to figure 4 (a). Point number 2 is showing extraordinarily high value of 96.75 and tuning of diffusion process is required so as to control the diffused region in and around point number 2 as per the requirement.

Table 1. Sheet resistance after fine tuning of diffusion process

Point number	Sheet Resistance ($\Omega/\square$)
1	100.4
2	99
3	98.2
4	101.5
5	98.5
6	99.3
7	100.5
8	99.6
9	99.8

After fine tuning of diffusion process, wafers are produced with emitter resistance of ~100 $\Omega/\square$ in the variation of around 3% as shown in table 1 below.

Monocrystalline silicon solar cells made by using these wafers are observed to have efficiencies of ~19.5 %. Monocrystalline silicon wafers which were having variation in sheet resistance of ~ 3% - 10% were also screen printed and cells were made with those cells. However these wafers did not show efficiency more than 18%.

IV. CONCLUSION

At BHEL ASSCP we have developed a characterization tool that can measure the sheet resistance of a large area 156 mm by 156 mm diffused silicon wafer at 9 places on the wafer and displays the results in graphical and tabular form on the computer monitor in less than 5 seconds. The sheet resistance measurement tool is easy to use, and carries out the sheet resistance mapping of the wafer at 9 places, the results could be used to tell the uniformity of diffusion and subsequently to have a tighter control over the diffusion process. By using the sheet resistance measurement setup (SRMS) developed here we were able to fine tune the emitter sheet resistance of large area 156 mm by 156 mm diffused silicon wafer to less than 3% variation and thereby enhanced the efficiency of monocrystalline silicon solar cell to ~19.5 % by using the shallow emitter technique. For monocrystalline diffused silicon wafers which were having variation in sheet resistance of more than 3% were not showing efficiencies of more than 18%. Hence it can be concluded from this study that the uniformity in sheet resistance of diffused silicon wafer is a very critical step to make high efficiency cell and the tool developed in the present work will be very useful for this purpose. The results obtained from each head were repeatable within experimental tolerances.

ACKNOWLEDGMENT

The author would like to thank the BHEL management for giving the opportunity to work on this project. The author would also like to thank Mr. Sudip Bhattacharya, consultant National Institute of solar energy and Dr. S. P. Singh, consultant BHEL ASSCP, for discussions at various stages during the development process.

REFERENCES

[1] http://pveducation.org/pvcdrom/four-point-probe-resistivity-measurementswelcome-to-pvcdrom.

[2] U S Patent number 3609537 A, "Resistance Standard".

[3] https://www.suragus.com/en/products/thin-film-characterization/sheet-resistance/eddycus-tf-map-2525sr/

[4] https://www.semilab.hu/category/products/non-contact-sheet-resistance-measurement

[5] Charles A. Bishop, "Vacuum deposition onto Webs, Films and Foils", Book, 2nd Edition, 2011, Copyright © 2011 Elsevier Inc. All rights reserved, ISBN 978-1-4377-7867-0.

[6] P. PANEK, K. DRABCZYK*, and P. ZIĘBA, "Crystalline silicon solar cells with high resistivity emitter", Opto Electronics Review, Vol. 17, no. 2, 2009.

[7] Vijay Yelundur, Kenta Nakayashiki, Mohamed Hilali, Ajeet Rohatgi, "Implementation of a homogeneous high-sheet-resistance emitter in multicrystalline silicon solar cells", 0-7803-8707-4/05/©2005 IEEE.

An EMD Based Protection Scheme for Distribution System

Mahmood Shaik
Dept. of Electrical Engineering
Indian Institute of Technology
Jodhpur, India
Email: shaik.1@iitj.ac.in

Sandeep Kumar Yadav
Dept. of Electrical Engineering
Indian Institute of Technology
Jodhpur, India
Email: sy@iitj.ac.in

Abdul Gafoor Shaik
Dept. of Electrical Engineering
Indian Institute of Technology
Jodhpur, India
Email: saadgafoor@iitj.ac.in

Abstract—**Fast identification and classification of the faults in distribution network is a crucial task to ensure reliable power supply to the end user. This paper presents a protection scheme which makes use of Empirical Mode Decomposition of three phase current signals of a distribution network. The current signals measured at the substation bus over a moving window of one cycle are decomposed to obtain first level intrinsic mode function IMF_1 and residue R_1. A fault index which is nothing but the absolute mean of R_1 is calculated for each phase and compared with a threshold to detect and categorize the type of fault. The proposed algorithm has been successfully tested on IEEE 13 bus system by varying the type of fault, fault incidence angle and fault location. The selectivity of the proposed scheme has been established by testing the algorithm with non faulty transients such as load switching and capacitor switching.**

Keywords—*EMD, fault, distribution system, fault index, residue, protection.*

I. INTRODUCTION

Protection of the distribution network against faults is extremely important owing to the proximity of the loads. In order to ensure safety and continuous power supply, these unpredictable faults have to be identified as quickly as possible. The faults can be detected either by change in the magnitude of current and voltage after the fault occurrence or by the change of magnitude of phase or frequency. Protective relays are generally used to sense the abnormality in current or voltage signals, and to send a trip signal to the circuit breaker, when the magnitude of sensed quantity exceeds a preset threshold. Fault currents change subjected to a variation in the type of fault, location, inception of the fault and also on the grounding elements of the distribution system. The fault current magnitude is very large in case of solid faults or low impedance faults and very small in case of high impedance faults. Thus detection of faults in distribution network is challenging using conventional over current relays. [1]- [2].

Various features are extracted from the faulty current signals for successful detection of faults. As the current waveform after faults exhibit transient nature containing harmonics, different signal processing techniques are used to detect faults using transients. Kalman filtering theory [1], S-Trasform [3]-[4], Wavelet Transform [5]- [6], Gabor Wigner transform [7] are employed in extracting the frequency information with respect to time. Features based on mathematical morphology (MM) are used for detecting faults in [8].

Empirical Mode Decomposition (EMD) is an effective signal processing technique for identifying signal abnormalities. EMD decomposes a signal into independent and uni-component functions named as Intrinsic Mode Functions (IMF's) and residues. Intrinsic Mode Functions being oscillatory in nature are utilized in the literature for various fault diagnosis applications. IMFs based on kurtosis are utilized in finding faults in internal combustion engines in [9].Power quality events are detected using first three IMFs which contain high frequency content in [10]. Bus bar protection scheme is proposed in [11], utilizing the first intrinsic mode function. Balanced and unbalanced faults in SEIG system are detected using IMF's obtained from negative sequence components of current signals in [12]. IMF's are selected using Mutual Information (MI) for protection of 14 bus microgrid in [13].

In this paper, fault diagnosis technique is proposed using first level residue on IEEE 13 bus distribution system. The novelty of this paper lies in the usage of residue instead of IMF for fault detection. The proposed algorithm is successful in detecting and classifying all types of faults within half cycle. Sequential organization of the paper after introduction is mentioned here. The Empirical mode decomposition technique is explained in section-II. The test system parameters are discussed in section-III. The proposed algorithm is demonstrated in Section-IV. Results and discussion including the threshold criterion is presented in Section-V and the paper is concluded in Section-VI.

II. EMPIRICAL MODE DECOMPOSITION

This is an empirical technique used for analysis of non linear and non stationary data. EMD results in decomposition of a signal into independent oscillating modes known as Intrinsic Mode Functions (IMFs) on the basis of the time varying signal itself. The lower order IMFs hold the signal components corresponding to the high frequency range and higher order IMFs capture the signal components pertaining to low frequency range. The number of IMFs formed using EMD is dependent on the disturbance in the signal [12].

A decomposed signal is designated as an IMF, if the following two properties are satisfied

- The difference in the number of extrema and the number of zero crossing must be zero or atmost one.

- The average value of the envelopes generated by local maxima and local minima is always zero.

The IMFs are extracted from the given signal $S(t)$ using the following steps in EMD [9]

1) Define all the localized maxima (M_i) and minima (m_j) of the signal $S(t)$, where , $i, j = (1,2..)$
2) Interpolate the upper envelope $s_{up}(t)$ and lower envelope $s_{low}(t)$ through cubic spline functions.
3) Compute the average value of the lower and upper envelopes of the signal, $A_{11}(t)$ as

$$A_{11} = \frac{s_{up}(t) + s_{low}(t)}{2} \qquad (1)$$

4) Obtain the first difference between the signal $S(t)$ and the mean $A_{11}(t)$ as

$$d_{11}(t) = S(t) - A_{11}(t) \qquad (2)$$

5) Check if $d_{11}(t)$ is an IMF by using the above proper-ties mentioned and designate it as IMF_1 , otherwise it is considered as the original signal and the steps 1-4 are repeated for p times. If IMF conditions are met after p-iterations, the signal is described as $d_{1p}(t)$, which is obtained as the difference between the signal at $(p-1)^{th}$ iteration and the mean of the envelopes after the pth iteration,

$$d_{1p}(t) = d_{1(p-1)}(t) - A_{1p}(t) \qquad (3)$$

Hence, the first Intrinsic Mode Function is obtained as,

$$IMF_1(t) = d_{1p}(t) \qquad (4)$$

6) Obtain the level-1 residue by subtracting $IMF_1(t)$ from the signal $S(t)$ considered,

$$R_1(t) = S(t) - IMF_1(t) \qquad (5)$$

7) Repeat the the above mentioned steps for extracting further levels of IMFs by treating the residue $R_1(t)$ as the original signal.
8) After k- iterations of steps 1-7 ,
$IMF_1(t), IMF_2(t).. IMF_k(t)$ are extracted along with a residue $R_k(t)$
9) The EMD process ceases if $R_k(t)$ happens to be a monotonic function, where further IMFs cannot be extracted. Hence, the signal $S(t)$ can be reconstructed as

$$S(t) = \sum_{n=1}^{k} IMF_n(t) + R_k(t) \qquad (6)$$

Where IMF$_n$(t) is the intrinsic mode function of level n, and the residue is R$_k$(t).

III. TEST SYSTEM PARAMETERS

An IEEE 13 bus test system is adapted for the proposed protection scheme as reported in [14]. It is a 5 MVA, 60 Hz, radial type distribution feeder with two voltage levels of 0.48 kV and 4.16 kV as shown in Figure 1. The specifications of feeder and loads are provided in Tables I and II respectively. A substation transformer X-SS is utilized to interface the feeder and the utility grid with neutral to ground impedance of (0.5+j0.005). The transformer X-T1 is connected between the buses 633 and 634 for serving low voltage loads. Constant

impedance load of Y-Z type is used at bus 634 and -Z type load is used elsewhere. Transformer specifications are mentioned in Table III. The system contains three phase aerial or underground type feeders with 601 and 606 configurations respectively. The series impedance matrices for simulating feeders are considered from [15]. The capacitance values for 601 configuration is 1.57199 *nF/km* (positive sequence) and 1.3398 *nF/km* (zero sequence). Equal positive and zero sequence capacitance of 15.96979 µF/km is considered for 606 configuration.

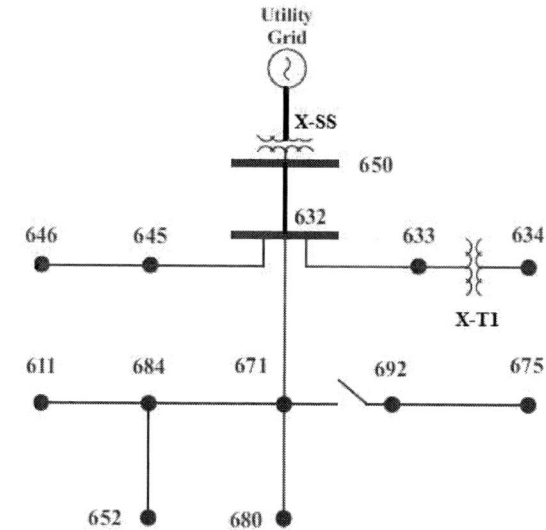

Fig. 1. IEEE 13 Bus Distribution System

TABLE I. FEEDER SPECIFICATIONS

Bus 1	Bus 2	Length (m)	Configuration
632	645	152.4	601
632	633	152.4	601
633	634	0	X-T1
645	646	91.44	601
650	632	609.6	601
684	652	243.84	606
632	671	609.6	601
671	680	304.8	601
671	684	680	601
671	692	0	Switch
684	611	91.44	601
692	675	152.4	606

TABLE II. LOAD SPECIFICATIONS

Bus	Total Load Capacity		Capacitive-kVAr
	kW	kVAr	
632	100	58	-
634	400	290	-
645	170	125	-
646	230	132	-
652	128	86	-
671	1255	718	-
675	843	462	600
692	170	151	-
611	170	80	100

TABLE III. TRANSFORMER SPECIFICATIONS

Transformer	S(KVA)	HV rating(KV)	LV rating(KV)	R%	X%
X-SS	5000	115-D	4.16-N.Y	1	8
X-T1	500	4.16-Gr.Y	0.48-Gr.Y	0.11	0.2

IV. PROPOSED ALGORITHM

The fundamental frequency of the proposed IEEE 13 bus system is 60Hz. All the simulations are carried out for 1sec , at a sampling frequency of 1920 Hz. The faults are simulated at all the buses at 0.5 sec except bus 650 with a ground resistance of 2 Ω. The measurement of the fault current is done at bus 650. The fault currents for an ACG fault at bus 646 are shown in Figure 2. A time period of 0.2 sec (0.4s-0.6s) containing 6 pre-fault cycles and 6 post-fault cycles is shown for demonstration.

The current samples acquired from a moving window of one cycle duration (32 samples) are decomposed to obtain IMF_1 and residue (R_1) using EMD technique. The higher frequency oscillations are associated with IMF_1 while the R_1 would contain low frequency oscillations. The decomposition of phase-A currents in the 16th window after fault occurrence (half-cycle duration after fault) is shown in Figure 3.

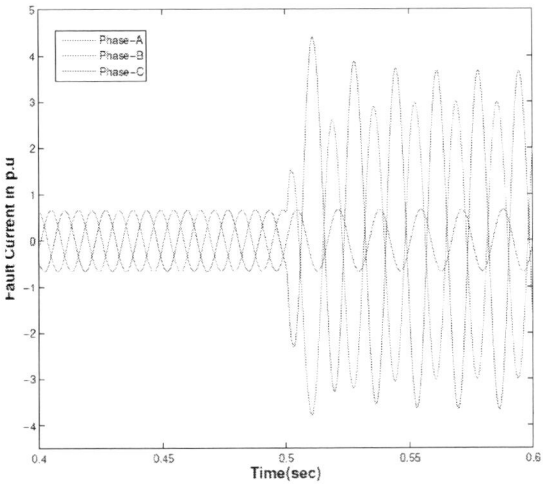

Fig. 2. ACG Fault at Bus 646 measured at bus 650

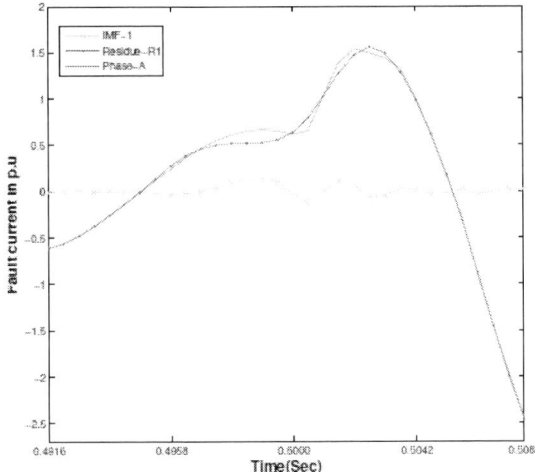

Fig.3. Decomposition of Phase-A currents in 16th window after fault occurrence at bus 646

The absolute mean (AM) of the residue (R_1) is calculated to obtain a fault index value per window for each phase.

Thus a fault index vector is obtained for all the phases (A,B and C) as the moving window surpasses the entire current signal. As the residue for a transient free pre-fault signal is zero, the fault index would also be zero. But whenever a fault occurs, the absolute mean of first level residue would have a finite value above zero which is utilized for detecting a fault. The variation of fault index for phase-A during ACG fault at bus 646 is shown in Figure 4. The maximum value of the fault indices of each phase in half cycle duration is then compared with a threshold to detect the occurrence of a fault. A threshold value of 0.08 is set between the minimum value of the faulty phases and maximum value of healthy phases. The fault type can be classified by the number of phases exceeding the fixed threshold. Thus, if N number of phases exceeds the fixed threshold then it is called an N-phase fault. If 3 phases exceed the threshold then it is designated a three-phase fault. The flow chart of the algorithm proposed is demonstrated in Figure 5.

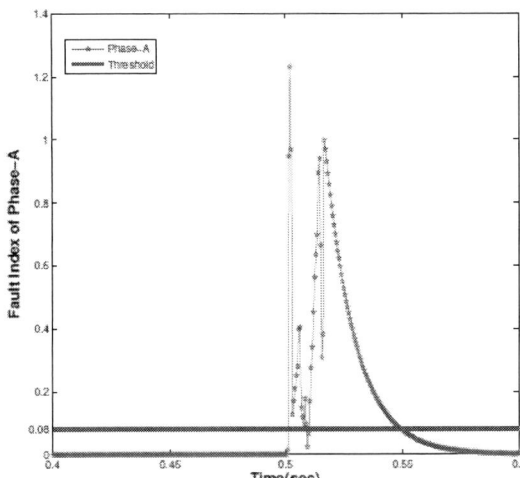

Fig. 4. Fault Index of Phase-A, ACG Fault at 646

V. RESULTS AND DISCUSSION

A. Change of Fault Location

The fault transients vary with the location of the fault. Thus 11 types of faults are simulated at every bus except bus 650 as it is used for measurement of the current signals. The fault index values for three- phase fault, two-phase fault and single-phase fault simulated at all the buses are shown in table IV. It can be noticed that only the faulty phases have higher fault index values whereas the healthy phases have lower fault index values. ABCG, BCG and BG type of faults are only illustrated, the rest of the faults also have similar characteristics.

B. Change of Fault incidence angle

As the fault transients are also dependent on the fault inception, the proposed algorithm is verified for various fault incidence angles (FIA) of 0^0, 30^0, 60^0, 90^0, 120^0, 150^0. It is observed that the proposed algorithm successfully detects the fault for all incidence angles considered. The ABCG fault at a bus 692 is shown in Figure 6, it can be confirmed that the fault index for all the phases A, B and C are clearly above threshold. ABG fault simulated at bus 692 is shown in Figure

TABLE IV.   FAULT INDICES FOR VARIOUS FAULT LOCATIONS

Fault → Bus ↓	Three Phase Fault(ABCG)			Two Phase Fault(BCG)			Single Phase Fault(BG)		
	A	B	C	A	B	C	A	B	C
611	0.684	0.664	0.682	0.002	0.575	0.5525	0.003	0.125	0.003
632	1.146	1.130	1.135	0.003	1.061	0.941	0.005	0.147	0.006
633	1.053	1.039	1.0436	0.003	0.978	0.862	0.006	0.144	0.006
634	0.762	0.751	0.755	0.003	0.722	0.610	0.005	0.141	0.005
645	1.053	1.039	1.043	0.003	0.978	0.862	0.005	0.144	0.005
646	1.005	0.992	0.996	0.003	0.935	0.820	0.005	0.142	0.005
652	0.630	0.576	0.629	0.002	0.536	0.520	0.003	0.122	0.003
671	0.870	0.859	0.863	0.002	0.779	0.707	0.004	0.136	0.004
675	0.796	0.785	0.788	0.002	0.715	0.644	0.003	0.133	0.003
680	0.764	0.756	0.760	0.002	0.687	0.702	0.003	0.131	0.003
684	0.700	0.696	0.694	0.002	0.594	0.589	0.003	0.126	0.003
692	0.869	0.858	0.862	0.002	0.778	0.706	0.004	0.136	0.004

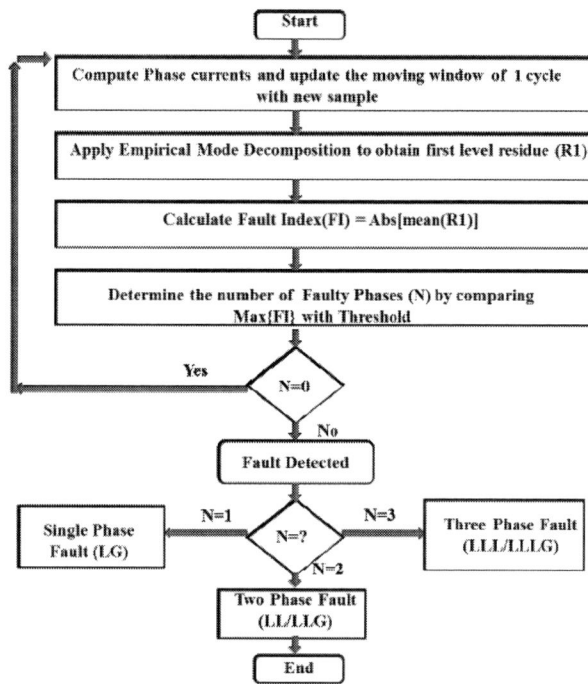

Fig. 5. Flow Chart of the Proposed Algorithm

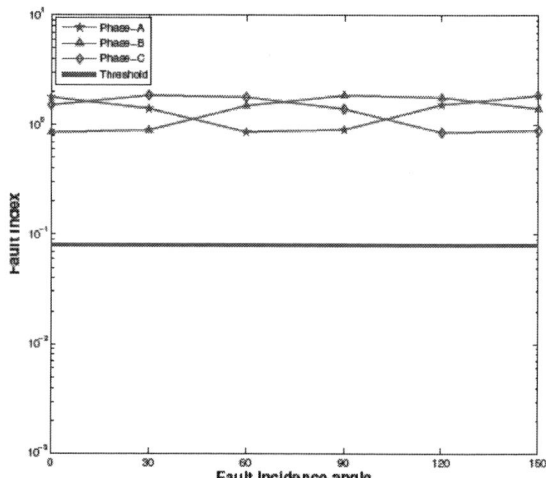

Fig. 6.Fault index variation with FIA, for ABCG fault at bus 692

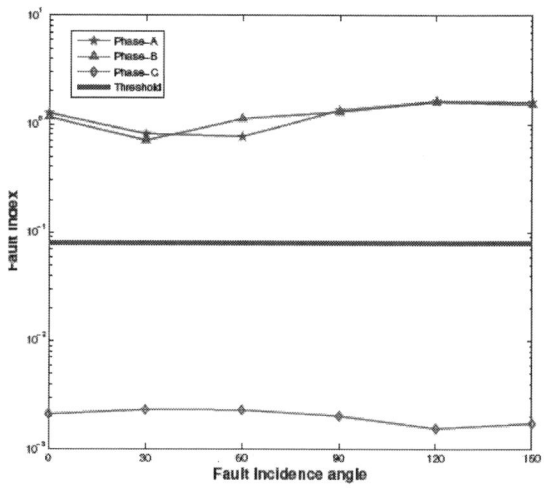

Fig. 7. Fault index variation with FIA, for ABG fault at bus 692

7, it is clearly noted that only the phases A and B are above the threshold whereas the fault index for phase C is below the threshold. Similar observation can be made in Figure 8 for AG fault at Bus 692.

C. Load switching

The proposed algorithm must be tested for classifying non faulty transients also as these can be misinterpreted as faults. Thus, load switching is considered at bus 652 at 0.5 sec. It is compared with the three phase fault at 652 at the same instant. It is observed that the fault index for load switching is below the threshold whereas it is above the threshold for three phase fault as shown in Figure 9 for phase-A. All the other phases follow a similar phenomenon.

D. Capacitor switching

The proposed algorithm is also tested for capacitor switching at bus 675. Capacitor switching is done for improving

the voltage profile but it can also introduce transients. It is compared with the ABCG fault at bus 675 at the same instant. The fault index of phase-A during three phase to ground fault is quite higher than the fault index for capacitor switching as shown in Figure 10.

978-1-5386-6624-1/18 $31.00 © 2018 IEEE

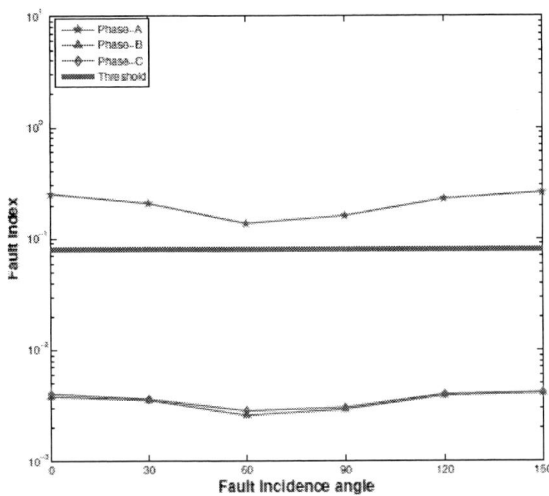

Fig. 8. Fault index variation with FIA, for AG fault at bus 692

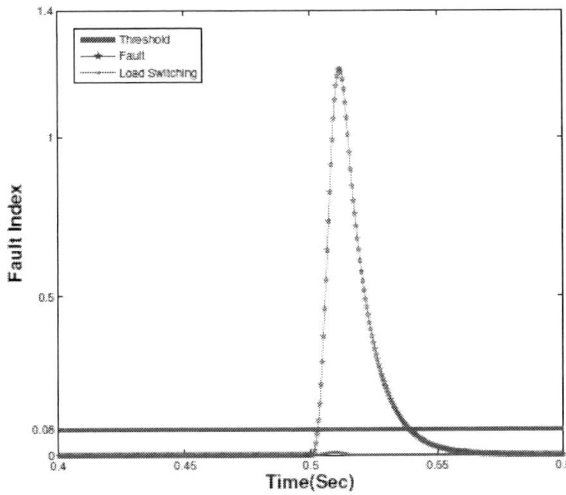

Fig. 9. Comparison of Fault index of Phase-A for ABC Fault and Load Switching at Bus 652

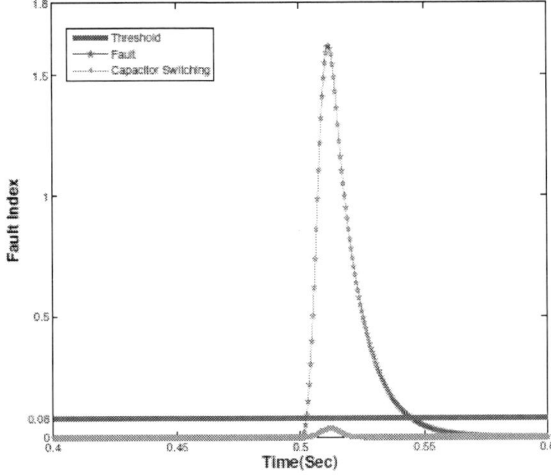

Fig. 10. Comparison of fault index of Phase-A for ABCG Fault and Capacitor Switching at Bus 675

VI. CONCLUSION

An Empirical Mode Decomposition based protection scheme is illustrated for a distribution network in this paper. Three phase currents obtained at substation are decomposed to obtain first level residue whose mean value, termed as fault index is compared with a predefined threshold to detect and classify various faults on an IEEE 13 bus system within half cycle. The case studies which involve changes in fault location, fault incidence angle establishes the robustness of the proposed scheme. Non faulty transients such as load switching and capacitor switching are successfully discriminated from faulty transients demonstrating the selectivity of the proposed algorithm. The unbalanced faults in a grounded distribution network inject transients in healthy phases through neutral. This leads to IMF components in healthy phases also. Hence, the proposed algorithm has preferred to use residue which is free from transients, instead of IMF for the detection of faults to avoid probability of misclassification. Thus the proposed algorithm can be efficiently utilized for fault diagnosis in distribution network.

ACKNOWLEDGEMENT

This research work is supported by Visveswarayya PhD scheme, MietY, India.

REFERENCES

[1] S. Samantaray, P. Dash, and S. Upadhyay, "Adaptive kalman filter and neural network based high impedance fault detection in power distribution networks," International Journal of Electrical Power & Energy Systems, vol. 31, no. 4, pp. 167–172, 2009.

[2] S. Samantaray, B. Panigrahi, and P. Dash, "High impedance fault detection in power distribution networks using time–frequency transform and probabilistic neural network," IET generation, transmission & distribution, vol. 2, no. 2, pp. 261–270, 2008.

[3] S. Kar and S. R. Samantaray, "Time-frequency transform-based differential scheme for microgrid protection," IET Generation, Transmission & Distribution, vol. 8, no. 2, pp. 310–320, 2014.

[4] P. Routray, M. Mishra, and P. Rout, "High impedance fault detection in radial distribution system using s-transform and neural network," in Power, Communication and Information Technology Conference (PCITC), 2015 IEEE. IEEE, 2015, pp. 545–551.

[5] M. Shaik, M. N. Rao, and S. A. Gafoor, "A wavelet based protec-tion scheme for distribution networks with distributed generation," in Emerging Trends in Electrical Engineering and Energy Management (ICETEEEM), 2012 International Conference on. IEEE, 2012, pp. 33–37.

[6] S. A. Gafoor, S. K. Yadav, P. Prashanth, and T. V. Krishna, "Transmis-sion line protection scheme using wavelet based alienation coefficients," in Power and Energy (PECon), 2014 IEEE International Conference on. IEEE, 2014, pp. 32–36.

[7] J.-Y. Cheng, S.-J. Huang, and C.-T. Hsieh, "Application of gabor–wigner transform to inspect high-impedance fault-generated signals," International Journal of Electrical Power & Energy Systems, vol. 73, pp. 192–199, 2015.

[8] S. Gautam and S. M. Brahma, "Detection of high impedance fault in power distribution systems using mathematical morphology," IEEE Transactions on Power Systems, vol. 28, no. 2, pp. 1226–1234, 2013.

[9] S. S. Nidadavolu, S. K. Yadav, and P. K. Kalra, "Condition monitoring of internal combustion engines using empirical mode decomposition and morlet wavelet," in Image and Signal Processing and Analysis, 2009. ISPA 2009. Proceedings of 6th International Symposium on. IEEE, 2009, pp. 65–72.

[10] S. Shukla, S. Mishra, and B. Singh, "Empirical-mode decomposition with hilbert transform for power-quality assessment," IEEE transactions on power delivery, vol. 24, no. 4, pp. 2159–2165, 2009.

[12] Z. Guo, J. Yao, and Z. Tan, "Hilbert–huang transform-based transient busbar protection algorithm," IET Generation, Transmission & Distribution, vol. 9, no. 14, pp. 2032–2039, 2015.

[13] J. Dalei and K. B. Mohanty, "Fault classification in seig system using hilbert-huang transform and least square support vector machine," International Journal of Electrical Power & Energy Systems, vol. 76, pp.11-22, 2016.

[14] A. Gururani, S. R. Mohanty, and J. C. Mohanta, "Microgrid protection using hilbert–huang transform based-differential scheme," IET Generation, Transmission & Distribution, vol. 10, no. 15, pp. 3707–3716, 2016.

[15] O. P. Mahela and A. G. Shaik, "Power quality improvement in distribution network using dstatcom with battery energy storage system," International Journal of Electrical Power & Energy Systems, vol. 83, pp. 229–240, 2016.

[16] M. C. R. Paz, R. G. Ferraz, A. S. Bretas, and R. C. Leborgne, "System unbalance and fault impedance effect on faulted distribution networks," Computers & Mathematics with Applications, vol. 60, no. 4, pp. 1105– 1114, 2010.

2nd IEEE International Conference on Power Electronics, Intelligent Control and Energy Systems (ICPEICES-2018)

Grey Wolf Optimization Algorithm for Optimal Siting and Sizing of Capacitors

Vivek Pratap Singh Bhadouria
Department of Electrical Engineering
Madhav Institute of Technology and Science
Gwalior, India
23368vivek@gmail.com

Laxmi Srivastava
Department of Electrical Engineering
Madhav Institute of Technology and Science
Gwalior, India
srivastaval@hotmail.com

Abstract— **In this paper, the problem of optimal siting and sizing of capacitors is brought up using the Grey Wolf Optimization Algorithm and Loss Sensitivity Factor. This Loss Sensitivity Factor is being used for screening the few most optimal sites (buses) for the placement of capacitors. Then these screened buses are taken as the search space in the optimization algorithm. The objective function is so designed that it reduces the total power loss. The algorithm is separately used on a single as well as three capacitor locations and the results for the same are shown. To check the feasibility of the proposed method, it is applied on standard 69 bus Radial Distribution System. Moreover, the results under different loading conditions (viz. 75% and 50% loading) are also shown to showcase the effectiveness of proposed algorithm.**

Keywords— *Grey wolf Optimization, Loss Sensitivity Factor, Radial Distribution System.*

I. INTRODUCTION

Radial Distribution System (RDS) usually contains diverse loads due to which the load profile from demands will vary time to time and will possibly cause imbalance in power flow or voltage fluctuations and may even lead to collapses. Also the flow of reactive power cause high power losses in the distribution system. It has been figured out that a total of 13% of generated power lost out as ohmic losses [1,2]. In order to diminish or minimize these losses there is a need of such an equipment that could be placed at different locations in the system and can supply required reactive power and thereby accomplish effective voltage control. Incorporating shunt capacitors in the system is one of the possible solutions but their optimal allocation with their proper sizing poses a problem which requires utmost level of search. For the installation of capacitors determination of location and size (kVAr ratings) is required. In this paper Loss Sensitivity Factor (LSF) has been used to screen in some of the potential locations for compensation. LSF is calculated by the sensitivity analysis which is briefly described under the next heading. Grey Wolf Optimization Algorithm (GWO), the algorithm employed in this paper, will choose the optimal capacitor setting to be installed in the respective bus.

In the last few years several works have been done to find out proper locations and sizes of shunt capacitors. Analytical methods [3,4] have been used in the early works of optimal placement of capacitors which do not require any powerful computing resource. The technique of Simulated Annealing (SA) is presented in [5] for optimal placing the capacitors but this technique of SA might have chances of getting trapped in local optimum and therefore might not guarantee the global optimum. A mixed integer linear optimization methodology for the optimal location and sizing of static and switched shunt capacitors in radial distribution system has been presented by Khodr et al. in [6]. Genetic Algorithm (GA) is introduced in [7] for optimal siting and sizing of capacitors but the amount of time it is taking to run is quite more depending on the size of the system under study. Also, the same sub-optimal solutions are getting repeated. A Direct Search Algorithm has been developed in [8] to find out optimal locations and optimal sizes of switched and fixed capacitors. In [9] an evolutionary algorithm called modified cultural algorithm has been implemented to the optimal capacitor allocation problem. In 1989, Baran & Wu [10,11] formulated location, type, size of capacitors, voltage constraints and load variations as a mixed integer programming problem which reduced the problem into a master-slave problem. The former is used to determine the location and the latter is used to determine the type and size of the capacitors. Also the data for 69 bus system is exclusively given there. Few years ago, in 2014, Sultana and Roy [12] developed a very unique optimization algorithm which is called Teaching Learning Based Optimization (TLBO).

II. LOSS SENSITIVITY FACTOR

Loss Sensitivity Factor (LSF) is utilized in this paper as a tool to screen out some best candidate nodes which then becomes the search space for the optimization algorithm.

The voltage at node m + 1 is given by:

$$V_{m+1}=V_m - J_m*(R_m + jX_m) \qquad (1)$$

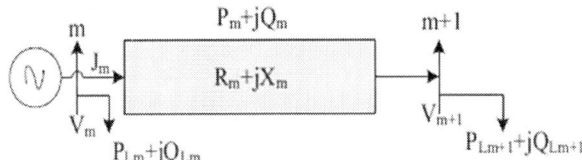

Fig. 1. Sample distribution system

978-1-5386-6624-1/18 $31.00 © 2018 IEEE

The real and reactive power loss in the line section can be determined using the following equations:

$$P_{loss}(m,m+1) = ((P_m^2+Q_m^2) / |V_m|^2) * R_m \qquad (2)$$

$$Q_{loss}(m,m+1) = ((P_m^2+Q_m^2) / |V_m|^2) * X_m \qquad (3)$$

To obtain the mathematical equation for Loss Sensitivity Factor (LSF), the (2) equation is differentiated with respect to reactive power and it is given by

$$\frac{\partial Ploss(m,n)}{\partial Qmn} = \frac{2Qm+1,eff*Rm}{|Vm+1|^2} \qquad (4)$$

This equation gives LSF values which when arranged in descending order gives the optimal location for allocation of capacitors.

Algorithm for sensitivity analysis
- Use the above formula in Eq. (4) to calculate LSF for all buses i.e.

  $LSF = \partial P_{loss} / \partial Q$
- Arrange the values of LSF in descending order and store them.
- Now calculate the normalized voltage magnitudes using this:
  $V_{norm}(i) = V(i)/0.95$
- The buses whose V_{norm} is less than 0.01 are the required sensitive buses and will be used for capacitor allocation.

III. GREY WOLF OPTIMIZER ALGORITHM (GWO)

One of the very recent nature inspired algorithm which is proposed in [13] by Mirjalili and Lewis is Grey Wolf Optimizer algorithm (GWO). This algorithm mimics the behaviour and hunting pattern of grey wolves in day to day life. Grey wolves prefer to live usually in groups of 6 to 12. Their social hierarchy is quite dominant. The top level wolves are known as alpha. They are the leaders of the group and are responsible for making all the decisions within the group.

The second in the list are Beta wolves which are subordinate of the alpha. They can also be thought of as the next alpha in case any alpha gets injured or harmed. Their role is to assist alpha in decision making along with other activities of the group.

The third level is of the subordinate wolves called delta. This consists of sentinels, scouts, hunters, caretakers, etc. Their role can range from safeguarding and guaranteeing the safety of the pack, observing the boundaries for any danger, helping the alphas and betas in hunting to taking care of the weak and wounded wolves.

Omega is the lowest category of wolves. They are the followers of all the other dominant wolves in the pack but their presence is equally important as of the other members.

The ability of grey wolves to memorize prey's position and to encircle them is what is used in developing the Grey Wolf Optimizer algorithm. In mathematically designing the hierarchy of grey wolves the alpha is considered to be the best solution followed by beta, the second and delta as the third. The rest of the candidate solutions are assumed to be omega. Grey wolf hunts in the following three main phases which are Approach, Encircle and Attack.

A. Encircling prey

Grey wolves encircle the prey during their hunt. The equations which nicely model the encircling behaviour of wolves are:

$$\vec{P}(itr + 1) = \vec{P}_P - \vec{J}\vec{K} \qquad (5)$$

$$\vec{M} = |\vec{L}.\vec{P}_P (itr) - \vec{P}(itr)| \qquad (6)$$

where 'itr' is the iteration number, $\vec{P}_P$ is the prey position and $\vec{P}$ is the grey wolf position. The vectors $\vec{L}$ and $\vec{J}$ are calculated as follows:

$$\vec{J} = 2a.\vec{x}_1 - a \qquad (7)$$

$$\vec{L} = 2\vec{x}_2 \qquad (8)$$

where $\vec{x}_1$ and $\vec{x}_2$ are random vectors in the range [0, 1] and the value of a is in a range of [0, 2]. It decreases from 2 to 0 over the course of iterations. The vector $\vec{L}$ is a random value in the range [0, 2]. This vector is used to provide random weights to define attractiveness of prey.

B. Hunting

For mathematically simulating the hunting behaviour of grey wolves, it is assumed that the alpha (α), beta (β) and delta (δ) have better knowledge about the possible locations of prey. So, it is proved useful to save the first three solutions and request the other search agents i.e. the omegas to update their positions according to the position of the best search agents. The wolves' positions are updated as follow:

$$\vec{P}(t + 1) = \frac{\vec{P}_1 + \vec{P}_2 + \vec{P}_3}{3} \qquad (9)$$

where $\vec{P}_1$, $\vec{P}_2$, $\vec{P}_3$ are obtained from Eq. (10) respectively.

$$\vec{P}_1 = |\vec{P}_\alpha - \vec{J}_1 \vec{M}_\alpha|$$

$$\vec{P}_2 = |\vec{P}_\beta - \vec{J}_2 \vec{M}_\beta|$$

$$\vec{P}_3 = |\vec{P}_\delta - \vec{J}_3 \vec{M}_\delta| \qquad (10)$$

where $\vec{P}_\alpha$, $\vec{P}_\beta$, $\vec{P}_\delta$ are the first three best solutions at a given iteration 'itr', $\vec{J}_1, \vec{J}_2, \vec{J}_3$ are determined as in Eq. (7), and the three parts of Eq. (11) gives the values of $\vec{M}_\alpha$, $\vec{M}_\beta$, $\vec{M}_\delta$ respectively.

$$\vec{M}_\alpha = |\vec{L}_1\,\vec{P}_\alpha - \vec{P}|$$

$$\vec{M}_\beta = |\vec{L}_2\,\vec{P}_\beta - \vec{P}|$$

$$\vec{M}_\delta = |\vec{L}_3\,\vec{P}_\delta - \vec{P}| \tag{11}$$

where $\vec{L}_1, \vec{L}_2, \vec{L}_3$ are obtained as in Eq. (8).

The final process is of updating the parameter 'a'. The parameter 'a' that controls the balance between exploration and exploitation is linearly updated to range from 2 to 0 over the course of each iteration as shown in Eq. (12)

$$a = 2 - itr\,\frac{2}{MaxItr} \tag{12}$$

where MaxItr is the total number of iteration. This could be finally concluded that the alphas, betas and deltas find out the position of the prey and other wolves updates their own position randomly according to them.

The pseudo code for the above explained GWO algorithm can be given as:

Initialize the population P_i of grey wolves (i =1,2,...n)

Initialize the values of J, L, a.

Calculate of fitness of each search agents

$\vec{P}_\alpha$ = *the best search agent*

$\vec{P}_\beta$ = *the second best search agent*

$\vec{P}_\delta$ = *third best search agent*

while *(itr < max no of iterations)*

    for *each search agent*

        update the position of current search agent by Eq. (9)

    end for

    update J, L, a

    calculate fitness of all search agents

    update $\vec{P}_\alpha$, $\vec{P}_\beta$ and $\vec{P}_\delta$

    itr = itr+1

end while

Return $\vec{P}_\alpha$

IV. PROBLEM FORMULATION

Proposed methodology used in this paper works with the suitable objective functions and the constraints.

A. Objective Function

The optimal allocation problem generally considers the minimization of power losses in the system. It can be mathematically shown as:

$$\text{Minimize } (F) = \min\,(P_{loss}) \tag{13}$$

where the value of P_{loss} can be obtained from Eq. (2) or can be directly thought of as:

$$P_{loss} = (branchcurrent)^2 * R_m \tag{14}$$

B. Constraints
a. Power Balance

$$P_{swing} = \sum_{m=2}^{n} P_{Lm} + \sum_{m=1}^{nb} P_{Loss}(m,m+1) - \sum_{m=1}^{nb} P_{cap,m} \tag{15}$$

where,

P_{Lm} : real power load at bus m

$P_{Loss}(m,m+1)$: real power loss in line connecting m and m + 1

$P_{cap,m}$: reactive power supplied by capacitor

b. Voltage Constraint

$$V_{m,min} \le |V_m| \le V_{m,max} \tag{16}$$

where,

$V_{m,min}$: min voltage at the mth bus (0.90)

$V_{m,max}$: max voltage at the mth bus (1.05)

V_m : voltage at mth bus

c. Reactive Power Compensation

$$Q_{cm}^{min} \le Q_{cm} \le Q_{cm}^{max}\;,\; m=1,\ldots\ldots,nb \tag{17}$$

where,

Q_{cm}^{min} : minimum reactive power limits of compensated bus m

Q_{cm}^{max} : maximum reactive power limits of compensated bus m

V. RESULTS

The proposed methodology has been tested on IEEE standard 69 bus radial distribution system with a total load of 3.802 MW and 2.694 MVAr [10,11] using MATLAB R2010a software. The system diagram, Fig. 2 below, consists of 69 buses and 68 branches wherein first bus is considered to be the swing bus. The total losses without compensation are 224.9767 kW as shown in Table 2 and Table3. The parameters of GWO algorithm taken are:

Number of search agents = 50

Maximum number of iterations = 60

The load flow method used to calculate the losses is Backward Forward Sweep which is briefly described in [14].

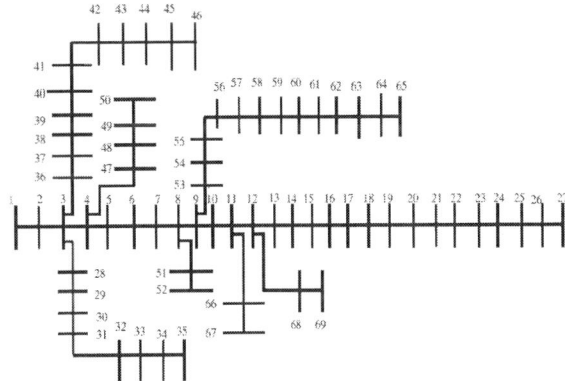

Fig. 2. 69-bus system diagram

978-1-5386-6624-1/18 $31.00 © 2018 IEEE

Fig. 3 shows the graph of LSF values for different buses and also these are clearly reflected in Table 1. The bus numbers are the search space for the algorithm which are 21 in total instead of 69 which clearly shows the reduction in search space. The load flow program is run separately for three different loading conditions i.e. 100%, 75% and 50% and the results are depicted in Table 2. These results are for single capacitor location. In Table 3 the results are compared with other algorithms for three capacitor locations.

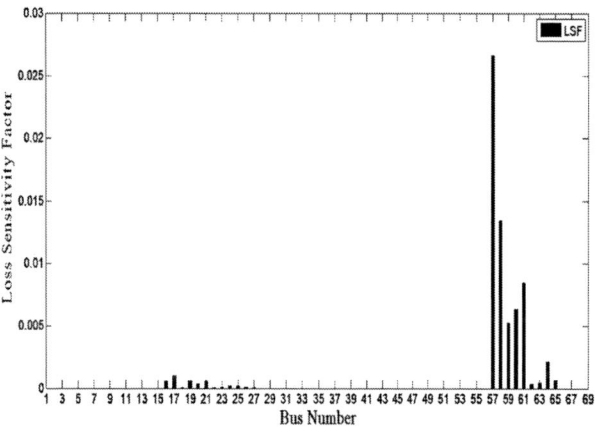

Fig. 3. Loss Sensitivity Factor

TABLE I. OPTIMAL BUSES ACCORDING TO LSF

S. No.	Bus No.	LSF	Normalized Voltage (Vi/0.95)
1	57	0.0266	0.989575
2	58	0.0134	0.977934
3	61	0.0085	0.960355
4	60	0.0064	0.968142
5	59	0.0053	0.973431
6	64	0.0022	0.957642
7	17	0.0010	1.008494
8	65	0.0007	0.957038
9	16	0.0006	1.009421
10	21	0.0006	1.007173
11	19	0.0006	1.007995
12	63	0.0004	0.959642
13	20	0.0004	1.007680
14	62	0.0003	0.960051
15	25	0.0002	1.006747
16	24	0.0002	1.006925
17	23	0.0001	1.007090
18	26	0.0001	1.006674
19	27	0.0000	1.006653
20	18	0.0000	1.008485
21	22	0.0000	1.007166

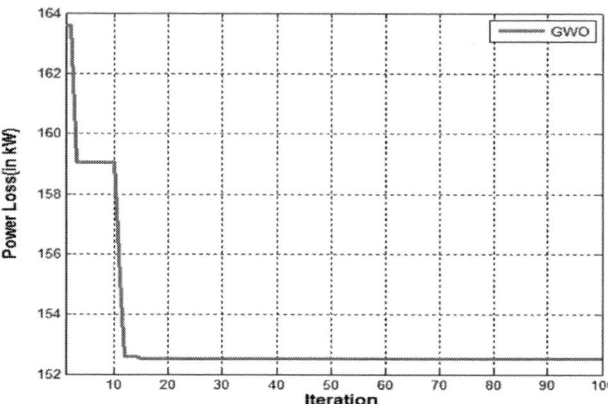

Fig. 4. Convergence Curve for single capacitor location

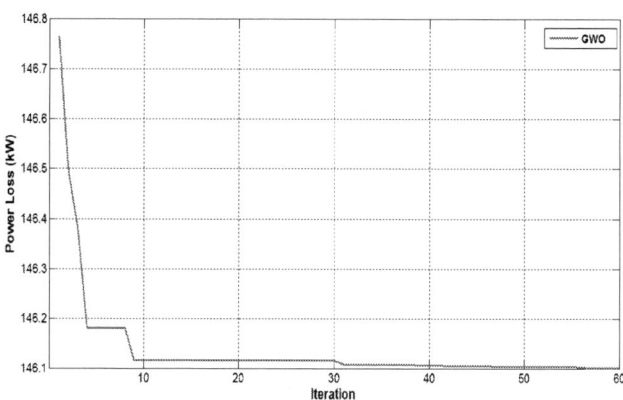

Fig. 5. Convergence curve for three capacitor location

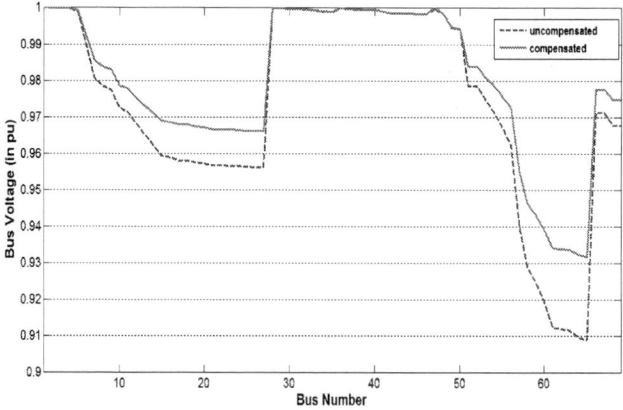

Fig. 6. Effect of single capacitor on voltage profile

TABLE II. RESULT FOR DIFFERENT LOADINGS

Loading	Quantity	Uncompensated	Compensated	Location / Size
100%	Minimum Voltage (pu)	0.9092	0.9308	
	Active power loss (kW)	224.9767	152.06	61 (1335 kVAr)
	Reactive power loss (kVAr)	102.2	70.5	
75%	Minimum Voltage (pu)	0.9335	0.9490	
	Active power loss (kW)	120.9746	82.726	61 (993 kVAr)
	Reactive power loss (kVAr)	55.1	38.5	
50%	Minimum Voltage (pu)	0.9567	0.9665	
	Active power loss (kW)	51.6	35.64	61 (657 kVAr)
	Reactive power loss (kVAr)	23.5	16.6	

TABLE III. RESULT COMPARISION OF PROPOSED ALGORITHM WITH OTHER ALGORITHMS

Quantity	Un-compensated	Compensated					
		Fuzzy-GA [15]	PSO [16]	DSA [8]	Heuristic[17]	TLBO [12]	Proposed GWO
Total Losses (kW)	224.9767	156.62	152.48	147	148.48	146.35	146.1067
Losses reduction (%)	–	30.4	32.2	34.67	34	34.96	35.05
Minimum voltage	0.9092	0.9369	–	–	0.9305	–	0.9318
Optimal location	–	59,61,64	46,47,50	61,15,60	8,58,60	12,61,64	17,61,64
Optimal size (kVAr)	–	100,700, 800	241,365,1015	900,450, 450	600,150,1050	600,1050,150	361,1069, 205
Total kVAr	–	1600	1621	1800	1800	1800	1635

VI. CONCLUSION

This paper presents a recently developed nature inspired optimization algorithm called Grey Wolf Optimizer with the help of which optimal allocation of capacitor banks is done. In this paper single capacitor as well as three capacitor locations have been found out with their proper sizes. A factor known as Loss Sensitivity Factor (LSF) has also been introduced to screen out some candidate buses. Only 69 bus radial distribution system is considered, however it has been divided into three parts i.e. 69 bus system with 100% loading, with 75% loading and with 50% loading. The result section shows that the losses are minimum for proposed algorithm (GWO) signifying its dominance among others. As GWO based approach with LSF method provides better results, it may also be implemented for large size power system networks as well.

ACKNOWLEDGEMENT

The author gratefully acknowledges the guidance and support of Electrical Engineering Department, Madhav Institute of Technology and Science for carrying out this work.

REFERENCES

[1] Rao RS, Narasimham SVL, Ramakingaraju M. Optimal capacitor placement in a radial distribution system using plant growth simulation algorithm. Int J Electr Power Energy Syst 2011;33:1133–9.

[2] Nojavan S, Jalali M, Zare K. Optimal allocation of capacitors in radial/mesh distribution systems using mixed integer nonlinear programming approach. Int J Electr Power Syst Res 2014;107:119–24.

[3] Chang NE. Locating shunt capacitors on primary feeder for voltage control and loss reduction. IEEE Trans Power Appar Syst 1969;88(10):1574–7.

[4] Bae YG. Analytical method of capacitor allocation on distribution primary feeders. IEEE Trans Power Appar Syst 1978;97 (4):1232–8.

[5] Chiang HD, Wang JC, Cockings O, Shin HD. Optimal capacitor placements in distribution systems: Part 1: A new formulation and the overall problem. IEEE Trans Power Delivery 1990;5(2):634–42.

[6] Khodr HM, Olsina FG, Jesus DOD, Yusta JM. Mximum savings approach for location and sizing of capacitors in distribution systems. Elect Power Syst Res 2008;78(7):1192–203.

[7] Sydulu M, Reddy V. Index and GA based optimal location and sizing of distribution system capacitors. IEEE Power Engineering Society General Meeting, 2007; 24–28 June 2007. p. 1–4.

[8] Raju M Ramalinga, Murthy KVS Ramachandra, Ravindra K. Direct search algorithm for capacitive compensation in radial distribution systems. Int J Electr Power Energy Syst 2012;42:24–30.

[9] Haldar V, Chakraborty N. Power loss minimization by optimal capacitor placement in radial distribution system using modified cultural algorithm. Inter Trans Electr Energy Syst 2015;25 (1):54–71.

[10] Baran Mesut E, Wu Felix F. Optimal capacitor placement on radial distribution systems. IEEE Trans Power Delivery 1989;4(1):725–34.

[11] Baran Mesut E, Wu Felix F. Optimal sizing of capacitors placed on a radial distribution system. IEEE Trans Power Delivery 1989;4(1):735–43.

[12] Sultana Sneha, Roy Provas Kumar. Optimal capacitor placement in radial distribution systems using teaching learning based optimization. Elsevier Int. J. Electr. Power Energy Syst. 2014;54:387–98.

[13] S. Mirjalili, S. M. Mirjalili and A. Lewis, "Grey wolf optimizer," Advances in Engineering Software, vol. 69, pp. 46-61, 2014.

[14] B. Ravi Teja1, V.V.S.N. Murty2, Ashwani Kumar, "An Efficient and Simple Load Flow Approach for Radial and Meshed Distribution Networks", International Journal of Grid and Distributed Computing Vol. 9, No. 2 (2016), pp.85-102.

[15] Das D. Optimal placement of capacitors in radial distribution system using a Fuzzy-GA method. Int J Electr Power Energy Syst 2008;30:361–7.

[16] Prakash K, Sydulu M. Particle swarm optimization based capacitor placement on radial distribution systems. In: IEEE power engineering society general meeting 2007; 24th–28th June. p. 1–5.

[17] Hamouda A, Lakehal N, Zehar K. Heuristic method for reactive energy management in distribution feeders. Int J Energy Convers Manage 2010;51:518–23.

Order Reduction of Linear Time Invariant System using Particle Swarm Optimization

Shubham Kumar Gupta[1], Himanshu Gupta[2], Tanmay Shrivastava[3], Kuldeep Rathore[4]

[1,4]Department of Electrical Engineering, *Madhav Institute of Technology and Science*, Gwalior, India
[2]Department of Electronics and Telecommunication, SGSITS, Indore, India
[3]Department of Electrical Engineering, *SGSITS,* Indore, India
[1]shubhamgsti@gmail.com, [2]himanshu.17jan@gmail.com,
[3]tanmaysgsits@gmail.com, [4]kuldeeprathore4545@gmail.com

Abstract— **Computational analysis of Higher Order Control System is a complex, difficult, computationally expensive and tedious task. In recent years the need for order reduction of higher order models to comparatively lower order model for ease of analysis has been greatly increased due to ever increasing complexities of various industrial processes, machines and physical systems. To obtain a reduced order model for higher order system need to satisfy certain conditions so as to preserve the original properties of the system. The reduced model is so constructed that all parameters of the original higher order system are confined within limits for desired accuracy. PSO is a relatively recent population based optimization technique used for order reduction of linear time invariant systems. The mechanism of PSO is inspired from collaborative efforts of biological populations and swarming of bees. PSO method is based on comparing original higher order system and reduced order system and eventually minimizing Integral Square Error (ISE) between their respective transient responses**.

Keywords—Particle Swarm Optimization, Integral Square Error, Model Order Reduction

I. INTRODUCTION

In our day to day life, we come across several complex dynamic systems. The transfer function of these systems is usually of higher order. For analysis of such systems there is a need for simplified models of lower order to provide with ease in computational complexity. Thus, model order reduction finds a monumental significance in engineering applications and theoretical and practical physics, especially in control engineering. Presently many techniques subsist in literature that provides insight on model order reduction of linear time invariant systems [1]-[11]. Integral square error is used for model order reduction [12]-[17]. In some techniques [12], the numerator and denominator for the reduced order model are determined by minimizing the integral square error between transient response of higher order model and reduced order model, whereas in [13]-[17], only numerator coefficients are calculated by means of ISE. Various population based algorithms are also used for optimization purposes like Particle Swarm Optimization (PSO) [18], Big Bang-Big Crunch (BB-BC) [19], Bacterial Foraging Optimization (BFO) [20] and Genetic Algorithm (GA) [21].

Particle Swarm Optimization [18] is an optimization technique developed by Dr. Eberhart and Dr. Kennedy in 1995. This technique is primarily inspired by the social behavior of bird flocking. It is a population based metaheuristic technique, which shares a number of similarities with other population based evolutionary techniques. In PSO, the potential solutions denoted as particles, are assigned arbitrary values in search space following the current optimum solutions. The best optimum solution is taken to improvise results in next iteration.

In this paper, we are discussing the application of Particle Swarm Optimization to propose a method for order reduction of higher order systems to evaluate parameters for numerator and denominator such that the characteristic response of the system remains unchanged.

II. STATEMENT OF PROBLEM

Consider a Higher Order Linear Time Invariant System of order n. Let its transfer function be represented mathematically as G(s):

$$G(s) = \frac{\sum_{i=0}^{n-1} a_i s^i}{\sum_{i=0}^{n} b_i s^i} \qquad (1)$$

Our objective is to determine a reduced order model for above higher order LTI system. Let the reduced order LTI system's transfer function be represented mathematically as R(s):

$$R(s) = \frac{\sum_{i=0}^{r-1} c_i s^i}{\sum_{i=0}^{r} d_i s^i} \qquad (2)$$

The reduced model is determined such that order of R(s) is less than G(s), r<n. This is evaluated in such a way that the transient response of higher order LTI system and reduced order system remains same.

a_i, b_j, d_i, and e_j are scalar constants.

978-1-5386-6624-1/18 $31.00 © 2018 IEEE

III. PARTICLE SWARM OPTIMIZATION

Particle Swarm Optimization is a population based global heuristic optimization method, developed on the concept of swarm intelligence. A swarm is a group of fish and bird flock. Thus, PSO derives its inspiration from the metaphorical behavior of bird and fish flock movement for searching food in a given area [22]. PSO has its many properties similar to other evolutionary computational techniques. PSO as well GA systems are often initialized with a population of random solutions, called particles. Unlike GA, PSO does not suffer from mutation and crossover. The limits in which particles search for solutions is termed as 'search space'.

The PSO algorithm starts by generating a number of particles, called population. The particles are randomly distributed in search space and is assigned random velocity and displacement. After initializing a population the fitness of individual particle is recorder. The fitness value of individual particles is called *pbest,* (personal best). The most optimum value of the population is evaluated from *pbset,* and is called *gbest* (global best). Each particle in PSO flies in search space with an adaptable velocity which is changed according to *pbest* and *gbest* for the population. The *gbest* obtained after completion of process is used to achieve an optimum solution for the problem.

The modified velocity and positions of particles for $(t+1)^{th}$ iteration is determined from velocity and positions evaluated for t^{th} iteration by the following equation [23]:

$$v_{j,g}^{(t+1)} = w * v_{j,g}^t + c_1 * r_1() * (pbest_{j,g} - x_{j,g}^t) + c_2 * r_2() * (gbest_{j,g} - x_{j,g}^t) \quad (3)$$

$$x_{j,g}^{(t+1)} = x_{j,g}^t + v_{j,g}^t \quad (4)$$

with j = 1,2, … , n and g = 1,2, … , m

where,

n = number of particles in a swarm.

m = number of components for vector v_j and x_j

t = number of generations (iteration)

$v_{j,g}^t$ = g^{th} component of velocity of particle j at t^{th} generation.

w = inertia weight factor

c_1, c_2 = cognitive and social acceleration factors

r_1, r_2 = random numbers uniformly distributed in the range (0-1)

$x_{j,g}^t$ = g^{th} component of position of particle j at t^{th} generation

pbest $_j$ = pbest of particle j

gbest = gbest of group

Steps for PSO Algorithm:

1. Initialize population with *p* number of particles and randomly distribute them in search space.
2. Arbitrarily assign them with velocity and positions.

Loop

3. Evaluate the fitness value of each particle, pbest.
4. The particle with the optimum fitness value makes *gbest.*
5. For next iteration modify their velocity and positions as per the equations (3) and (4).
6. Again evaluate the fitness value for each particle, if new fitness is better than fitness for orevious iteration, make it *pbest,* otherwise *pbest* remains unchanged.
7. Comapre the most optimum value in this iteration with *gbest,* if new value provides optimum result compared to *gbest,* promote it to *gbest,* otherwise gbest remains unchanged.
8. Repeat the steps 5-7.
9. Exit the loop when iteration limit is reached.

End of Loop

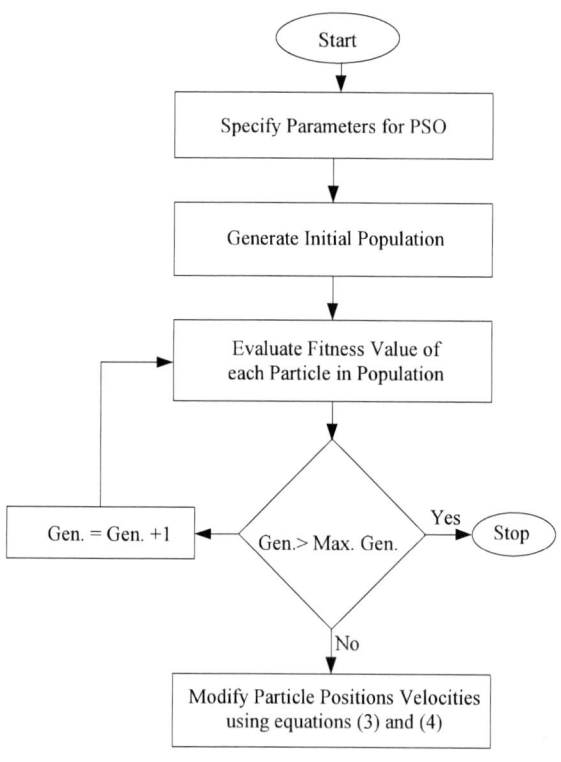

Fig. 1. Flowchart of PSO Algorithm

The Flowchart of Particle Swarm Optimization (PSO) Algorithm is shown in Fig. 1. The objective function used is Integral Square Error (ISE).The Integral Square Error for the difference between the transient responses of higher order model and reduced order model is given by:

$$IES = \int_0^\infty (y(t) - y_r(t))^2 \qquad (5)$$

IV. NUMERICAL EXAMPLES

Example I. Consider the 4th order transfer function

$$G(s) = \frac{s^3 + 7s^2 + 24s + 24}{s^4 + 10s^3 + 35s^2 + 50s + 24}$$

The reduced system obtained is:

$$R(s) = \frac{0.7694s + 1.5694}{s^2 + 2.4614s + 1.5710} \qquad (6)$$

Example II. Consider a 8th order system where:

N(s)=18s^7+514s^6+5982s^5+36380s^4+
122664s^3+ 222088s^2+185760s+40320

D(s)=s^8+36s^7+546s^6+4536s^5+22449s^4+
67284s^3+118124s^2+109584s+40320

The reduced 2nd order system is otained is given by :

$$R(s) = \frac{15.5270s + 5.7399}{s^2 + 6.0661s + 5.6034} \qquad (7)$$

For above examples, the simulation were carried out in MATLAB/Simulink R2013a and the PSO parameters that were used are given in table 1 below:

TABLE I.   PARTICLE SWARM OPTIMIZATION (PSO)PARAMETERS

Parameter	Value
Max. Iterations	500
No. of Particles	30
C1 & C2	1.49
w	Linearly Decreasing from 0.9 to 0.4

The plots of reduced order system and the original system are given in fig 2 and fig 3 for example 1 and example 2 respectively.

V. COMPARISON WITH OTHER METHODS

A comparison of results obtained for example 1 and example 2 with other well-known methods is given in table 2 and table 3 respectively:

TABLE II.   COMPARISON OF REDUCED MODEL FOR EXAMPLE 1

Method	Reduced Model	ISE
Proposed Method	$\dfrac{0.7694s + 1.5694}{s^2 + 2.4614s + 1.5710}$	7.115e-05
Mittal [16]	$\dfrac{0.79980s + 2}{s^2 + 3s + 2}$	2.7209e-04
Parmar [2]	$\dfrac{0.6394s + 4}{s^2 + 5s + 4}$	2.3935e-04
Shamash[3]	$\dfrac{0.8334s + 4}{s^2 + 3s + 2}$	3.3104e-04
Chen [10]	$\dfrac{0.6997(s + 1)}{s^2 + 1.4577s + 0.6996}$	2.8e-03

TABLE III.   COMPARISON OF REDUCED MODEL FOR EXAMPLE 2

Method	Reduced Model	ISE
Proposed Method	$\dfrac{15.5270s + 5.7399}{s^2 + 6.0661s + 5.6034}$	1.1e-03
Panda [23]	$\dfrac{88.0639s + 26.4648}{4.0214s^2 + 28.5882s + 2.6476}$	0.19
Sudhir Y. Kumar[24]	$\dfrac{18s + 4.9954}{s^2 + 6.567s + 5.0135}$	0.0503
Shamash [3]	$\dfrac{15.5270s + 5.7399}{s^2 + 6.0661s + 5.6034}$	0.2792
Parmar [2]	$\dfrac{22.8360s + 8}{s^2 + 9s + 8}$	0.0368

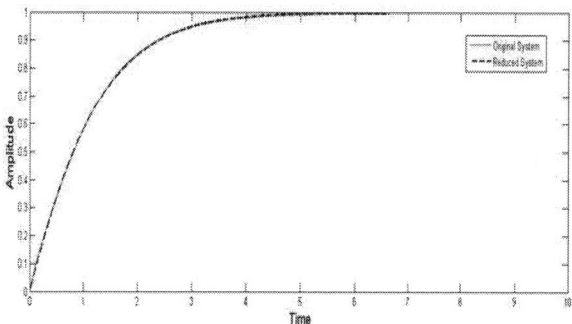

Fig. 2. Step response of original model and reduced model for example 1

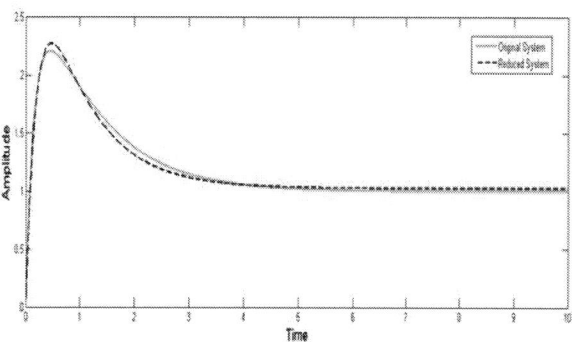

Fig. 3. Step response of original model and reduced model for example 2

VI. CONCLUSION

We have successfully reduced the order of system using Particle Swarm Optimization. The results obtained are compared to those of well-known techniques and we see that order reduction by PSO gives good quality results. Thus, this method can be tried on several other higher order systems for order reduction.

VII. REFERENCES

[1] S. Panda, J. S. Yadav, N. P. Patidar and C. Ardil, "Evolutionary Techniques for Model Order Reduction of Large Scale Linear Systems" International Journal of Applied Science, Engineering and Technology 5:1, pp 22-28, 2009.

[2] G. Parmar, S. Mukharjee, R.Prasad, "System reduction using factor divison algorithm and eigen spectrum analysis", Applied Mathematical Modelling 31,pp 2542-2552, 2007.

[3] Y. Shamash, "Model reduction using the Routh stability criterion and the Pade approximation technique", Int. J. Control, Vol. 21, pp 475-484, 1975.

[4] R. Genesio and M. Milanese, "A note on the derivation and use of reduced order models", IEEE Trans. Automat. Control,Vol. AC-21,No. 1, pp. 118-122, February 1976.

[5] M. J. Bosley and F. P. Lees, "A survey of simple transfer function derivations from high order state variable models", Automatica, Vol. 8, pp. 765-775, !978.

[6] M. Jamshidi, "Large Scale Systems Modelling and Control Series", New York, Amsterdam, Oxford, North Holland, Vol. 9,1983.

[7] R. K. Appiah, "Linear model reduction using Hurwitz polynomial approximation", Int. J. Control, Vol. 28, no. 3, pp 477-488, 1978.

[8] P. O. Gutman, C. F. Mannerfelt and P. Molander, "Contributions to the model reduction problem", IEEE Trans. Auto. Control, Vol. 27, pp 454-455, 1982.

[9] M. F. Hutton and B. Friedland, "Routh approximations for reducing order of linear time- invariant systems", IEEE Trans. Auto. Control, Vol. 20, pp 329-337, 1975.

[10] T.C. Chen, C.Y. Chang and K.W. Han, "Model reduction using the stability equation method and the continued fraction method", Int. J. Control, Vol. 32, No. 1, pp. 81-94, 1980.

[11] S. Mukherjee, Satakshi and R.C.Mittal, "Model order reduction using response-matching technique", Journal of Franklin Inst., Vol. 342 , pp. 503-519, 2005.

[12] G.D. Howitt, and R. Luus, "Model reduction by minimization of integral square error performance indices", Journal of Franklin Inst.,Vol. 327, pp. 343-357, 1990.

[13] S. Mukherjee, and R.N. Mishra, "Order reduction of linear systems using an error minimization technique", Journal of Franklin Inst., Vol. 323,No. 1, pp. 23-32, 1987.

[14] N.N. Puri, and D.P. Lan, "Stable model reduction by impulse response error minimization using Mihailov criterion and Pade's approximation",Trans. ASME, J. Dyn. Syst. Meas. Control, Vol. 110, pp. 389-394, 1988.

[15] P. Vilbe, and L.C. Calvez, "On order reduction of linear systems usingan error minimization technique", Journal of Franklin Inst., Vol. 327,pp. 513-514, 1990.

[16] A.K. Mittal, R. Prasad, and S.P. Sharma, "Reduction of linear dynamic systems using an error minimization technique",Journal of Institution of Engineers IE(I) Journal – EL, Vol. 84, pp. 201-206, March 2004.

[17] C. Hwang, "Mixed method of Routh and ISE criterion approaches for reduced order modelling of continuous time systems", Trans ASME,J.Dyn. Syst. Meas. Control, Vol. 106, pp. 353-356, 1984.

[18] J. Kennedy and R.C. Eberhart, "Particle Swarm Optimization" Proceedings of IEEE Internation Conference on Neural Network, Piscataway, NJ, pp. 1942-1948, 1995.

[19] Y. Labbi and D. Ben Attous, " Big Bang-Big Crunch Optimization Algorithm For Economic Dispatch with Valve-Point Effect", Journal of Theoretical and Applied Information Technology, pp 48-55, 2010.

[20] V. Sharma, S.S. Pattnaik, and T. Garg, "A review of Bacterial Foraging Optimization and its Applications", Intl. Journal of Comp. Applications, pp. 9-12, 2012.

[21] S.K. Mishra, S. Panda, S. Padhy, and C. Ardil, " MIMO System Order Reduction Using Real Coded Genetic Algorithm", Intl. Journal of Electrical, Computer, Energetic, Electronics an

[22] Q. Bai, "Analysis of Particle Swarm Optimization Algorithm", Journal of Computer and Information Science, Vol. 3, No. 1, pp. 180-184, 2010.

[23] S. Panda, S. K. Tomar, R. Prasad, C. Ardil, "Reduction of Linear Time-Invariant Systems Using Routh-Approximation and PSO", International Journal of Electrical, Computer, Energetic, Electronic and Communication Engineering Vol:3, No:9, pp. 1775-1782, 2009.

[24] Sudhir Y Kumar, PK.Ghosh and S Mukherjee," Model Order Reduction using Bio-inspired PSO and BFO Soft -Computing for Comparative Study", Int. J. of Information Systems and Communications Vol. 1, No. 1, pp 43-53 ,June 2011

978-1-5386-6624-1/18 $31.00 © 2018 IEEE

2nd IEEE International Conference on Power Electronics, Intelligent Control and Energy Systems (ICPEICES-2018)

State of the art of Human Factors Analysis Applied to Industrial and Commercial Power Systems

Esperanza S. Torres*, David Celeita** and Gustavo Ramos**
* School of Engineering, *University of Aberdeen Aberdeen*, Scotland
** Department of Electrical and Electronic Engineering, *Universidad de los Andes Bogota* D.C., Colombia
Email: estorres@ieee.org
df.celeita10@uniandes.edu.co
gramos@uniandes.edu.co

Abstract—Electrical power systems are more complex and dynamic. The development of new technologies, equipment, power market transformation, regulations and community demands, are factors that increase uncertainties and risks related to the operation, control, safety and reliability of the power supply networks. There is a historical application of methods for failure analysis and reliability evaluation. However, there is a lack of applied methods focused on power systems reliability including human factors (HF), specifically for industrial and commercial power systems. This paper mainly focuses on a literature review to identify information regarding human reliability analysis (HRA), and human error probability (HEP) applied on reliability analysis of electrical power systems. Also, a sociotechnical system approach of the power systems is introduced.

Keywords—Human reliability assessment, human factors, power systems reliability, human failure, HRA methods history, complex socio-technical systems.

I. INTRODUCTION

Over the past few decades, electric power system reliability has changed the due market, policies, optimization techniques, network configurations and implementation of new technologies, such as the integration of smart grids into power systems. However, either a deficient energy management system or lack of operating practices can lead to power supply failures. Power system outages could last from a few minutes to several hours, or even weeks, depending on the cause of the failure and the structure and configuration of the electrical grid. Furthermore, energy unavailability has serious economic consequences for business and community and will affect hundreds of millions people. There are unexpected causes of power blackouts, such as high winds, snow storms, heavy rains, hurricanes, lightning strikes, substation and transmission lines failures, among others.

Instead, a combination of human factors and power system cascading failures were identified as main causes of some of the major power outages. Such is the case of the Northeast Blackout of 14 August 2003 which affected approximately 50 million people in the United States and Canada, and power was not restored for 4 days in some areas [1]. At the London and the UK's West Midlands blackout incidents on 28 August and 5 September 2003 respectively, human error related to the wrong setting of the protective equipment and protection schemes was also a major factor on both incidents [2]. Moreover, on September 2011 maintenance error and weaknesses in

operations planning caused loss of power which affected at least six million households from Southern California and Arizona to northwestern Mexico [3].

On the above system failures 'human error' is argued as the main cause of failure. Furthermore, the expression 'human error' has been traditionally associated with the attribution of responsibility and blame. Although 'human error' is, usually, a consequence of circumstantial and situational factors that impact on human performance, it is not a cause. According to the Health and Safety Executive (HSE) 'organizations must recognize that they need to consider human factors as an individual element which must be recognized, assessed and managed effectively to control risks' [4]. This approach should be considered to evaluate safety and reliability of power systems. Human Factors Engineering (HFE) covers a range of issues concerning to how people interact with complex technical systems, such as a power system, characterized by a large number of dynamic interactions among its components which result in unpredictable consequences. These interactions can be addressed into one of three main areas: organizational behavior; man-machine interfaces and interactions; and human error and behavior. A brief definition of human failure is given in Section II. Then, state of the art is focused on the evolution of HRA with standards and field applications in Section III. The gap of HRA applications on industrial and commercial power systems is then presented in Section IV, and Section V describes a proposal for further application in a standard case study (IEEE 242-2001) using a complex socio-technical systems approach. Finally, conclusions and future work.

II. HUMAN FAILURE

In order to improve safety and reliability on power systems, is essential to take into account and understand how human failures and behavior can impact the performance of an electric power system.

Human failures are classified as errors and violations. Fig. 2 shows different categories of human failures. A 'human error' is an action or decision which was unintended characterized by an involuntary deviation from a guideline or standard, and which led to an undesirable consequence. A 'violation' is a deliberate deviation from a rule or procedure.

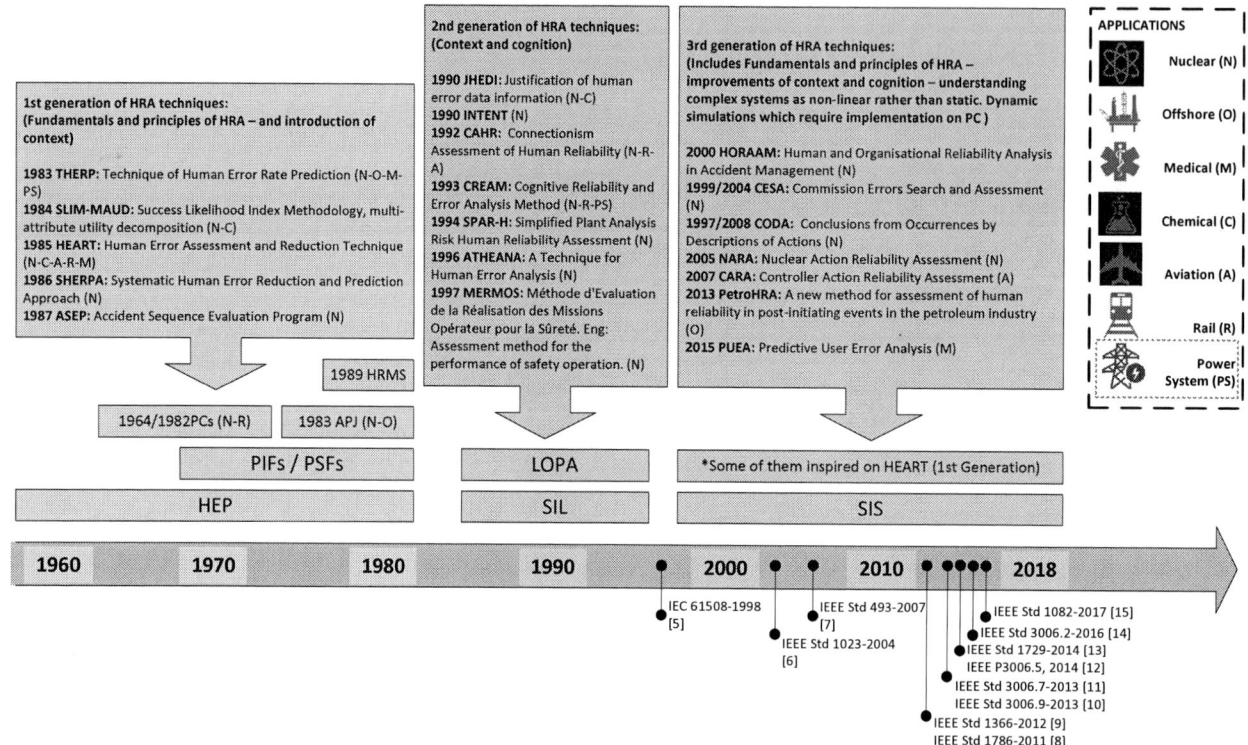

Fig. 1. Evolution of HRA techniques, methods and applications on the industry.

Skill-based errors are actions which occur in ordinary or daily tasks, such as repair, maintenance, calibration or testing work. They are classified as slips and lapses. These errors are made even by experienced and trained people by omitting steps from a task (slips, e.g., operating the wrong switch) or also forgetting to carry out an action (lapses).

Fig. 2. Types of human failure (source [4])

Mistakes are a type of human error where we 'do the wrong thing believing it to be right' [4]. They are classified as rule-based and knowledge-based errors. Violations are any deliberate deviations from rules, procedures, instructions, and regulations [4]. They are classified as routine, situational and exceptional.

III. HISTORY AND BACKGROUND OF HRA TECHNIQUES AND STANDARDS

This section summarizes the historical evolution of HRA tools classified with three traditional generations. Fig. 1 illustrates the timeline of such developments and the introduction of safety standardization and normativity related to reliability assessment including human factors. Some standards are also related to power systems reliability. Each HRA method is identified with bold, approximate year of introduction and existing applications that could be easily read by a single letter as shown in the upper right square.

A. General Timeline of HRA techniques

The role played by humans in complex systems have gained importance into reliability studies since the 1960s. Critical decision making parts of nuclear power plants, aircraft, and chemical plants are some examples of early developments. The effort and evolution of technology have improved complex systems' performance, which results in a direct decreasing of technical failures or accidents through backup strategies or redundant solutions. That could be the reason why the first generation starts near the introduction of the microprocessor because automation-assisted by computers represents the spike of applications in the next two generations [17]. Some of the first generation techniques established the theory and fundamentals approach to human reliability, then improved in the second and third generation.

1) The first generation of HRA techniques (1960s-1980s)

The first tools were developed treating humans as any other part of the system. These early methods established the fundamentals to help risk assessors by quantification and prediction of human error probability. This concept means that the assessment assigns static probabilistic values, for example, to indicate how likely an operator would fail to respond to a critical alarm [21]. These techniques are focused on the skill and the human action. Ironically it is the same reason why first generation methods are not always approved for ignoring a complete context (organizational factors [23]). However, some of them are currently in use for quantitative risk assessments in many industrial applications.

As it can be seen in Fig. 1, first-generation techniques include HEP databases. Then, later between the 1970s and 1980s, the impact of people went further than basic components in the analysis. Since humans are affected by many environmental conditions, and they make choices due to different factors (contextual factors), HEP data was not enough. The inclusion of context began with hybrid methodologies which use Performance influence factors (PIFs) and Performance shaping factors (PSFs). For example, training level, concentration or personnel, pressure conditions of activities, equipment design, and time stress, among others (THERP and HEART classify with these characteristics). In the first three decades since 1960, expert judgment tools enhanced HRA evolution simultaneously, with solutions such as Absolute probability judgment (APJ) and paired comparisons (PCs), both of them widely used in nuclear plants. All these elements, combined allows determining a nominal human error potential (HEP).

2) The second generation of HRA techniques (the 1990s)

State of the art summarized in Fig. 1 shows that the techniques of this generation are currently inspiring novel methods by providing valuable awareness to human reliability challenges [21][23]. To face the gaps of first-generation methods, these techniques attempt to consider context and errors of commission in human error prediction [4]. The context involves a psychology situation behind human failures: causes of occurrence. An additional item was then added to HRA complementing the error prediction and quantification.; Fig. 1 presents the most recognized methods of this generation (JHEDI, SPAR-H, INTENT, CAHR, ATHEANA, CREAM, and MERMOS). However, the majority of these solutions appeared to be no longer applicable without improvements or enhancements nowadays.

3) The third generation of HRA techniques (the 2000s - today)

This generation is developed in parallel with multiple normativity efforts and standardization for safety and reliability assessment. Most of them were based on first-generation traditional methods (THERP and HEART are still the most used methods) including better implementations with context and dynamic conditions of the systems (Layers of protection analysis – LOPA, safety integrity level – SIL, safety instrumented

systems – SIS, etc.). Although the majority of standards have been developed in the last decade (third generation), many industry systems use first generation based methods. It is worth to note that over 90% of the methods have been applied in nuclear, aviation and chemical industries. However, there is a high potential of using HRA in power systems, considering the upcoming challenges and non-linear nature of modern industrial and commercial power systems.

IV. TRADITIONAL HRA METHODOLOGIES IN POWER SYSTEMS

As mentioned in previous sections, based on the literature review and Fig. 2, it could be seen that HRA has been widely used and researched in system design, operation, and optimization for human reliability improvement in many fields. HRA applications on power systems and specifically on industrial and commercial power systems, there are very few studies about HRA. Some of the methods in nuclear plants combine traditional first and second generation techniques (THERP, HEART, CREAM, and ATHEANA) to quantify human reliability by constant transition rates for human failures and using a Markov methodology. Cognitive process and context identification when dealing with system failures require increasing the fidelity of simulated accident scenarios. However, there is a vast lack of appropriate and sufficient performance data, and it is a key factor affecting HRA quality, especially in the estimation of human error probability in power systems.

According to the literature review, HRA on power systems is based on old methods [24], assuming the power system's context as static rather than the dynamic and non-linear system. Then, there is a clear need and challenge to include HRA methods suitable for power system specific situations. The solutions could deal with different operation scenarios, as primary causes of human errors. Table 1 shows the comparison between the two common HRA methods applied in power systems. Section V will describe new methodologies for complex socio-technical systems approach suitable for industrial and commercial power systems.

V. COMPLEX SOCIO-TECHNICAL SYSTEMS APPROACH

Human behavior is too difficult to model. It can be argued that HRA and HEP methods cannot cover the whole uncertainties on human variability neither human task performance nor other human factors (e.g., workload, stress, fatigue). Moreover, these methods assume that the interactions between human roles and power systems are predictable (linear cause and effect). Complex sociotechnical systems (CSTS), such as electric power networks, electricity market, and power plant control room, which include numerous components and interactions within them, are not easy to describe. One of the approaches defines complexity as non-linear interactions within and between organizations, regulations, social and technical factors, systems, humans and their work environment [18][19].

978-1-5386-6624-1/18 $31.00 © 2018 IEEE

*** Formatting Issue - Best Available Paper/Graphic ***

TABLE I.   THERP vs CREAM Analysis for Power Systems based on

Method	Generation	Description	Advantages	Field
Technique for Human Error Rate Prediction THERP	First generation	Process - Kirwan et al (1997) : • Decomposition of tasks into elements • Assignment of nominal HEPs to each element • Determination of effects of PSF on each element • Calculation of effects of dependence between tasks • Modelling in an HRA event tree • Quantification of total task HEP	Decomposition approach due to descriptions of tasks - it has a higher degree of resolution than many other techniques. It is also a logical approach and one that puts a larger degree of emphasis on error recovery than most other techniques.	Nuclear plants, offshore O&G, Medical and power systems.
Cognitive Reliability and Error Analysis Method CREAM	Second generation	The distinction between competence and control is based upon Hollnagel's COCOM (contextual control) model. Common Performance Conditions (CPCs). Genotypes are separated out into three categories. Basic phenotypes (error modes) are divided into four sub-groups: • Action at the wrong time • Action of the wrong type • Action at the wrong object • Action in the wrong place.	Distinction between competence and control. A classification scheme clearly separates genotypes (causes) and phenotypes (manifestations), and furthermore proposes a non-hierarchical organization of categories linked by means of the sub-categories called antecedents and consequents.	Nuclear plants, rail accidents and power systems.

Thus, a CSTS is characterized by a large number of non-linear elements that interact dynamically. There is no direct causality which implies that small changes in the cause may lead to unintended consequences. Then, tightly-coupled elements let rapid escalation of errors and isolating failed components is severe [19].

Considering the type of information exchanged, the shared objectives and the level of cooperation among the components of the system, there are a large number of relationships. Thus, the elements can be categorized into hierarchical levels, a division of tasks, specializations, inputs, and outputs [19]. Given that there is not a direct cause and effect relationship between the elements, uncertainties tend to increase. In fact, as the system changes over time, for the same input there may be a different output. Furthermore, each component of the system can share and receive information from proximal and surroundings element [5]. Also, a CSTS interact with its environment which raises the uncertainty. To manage the uncertain and dynamic environment and perform its functions under both expected and unexpected circumstances, a CSTS is adaptive. Thus, 'the systems can adjust their functioning before, during, or following changes and disturbances' [19]. Moreover, the behavior of a CSTS is adjusted by the continuous feedback, from both recent and past events. Furthermore, according to Levenson [18] is expected that a CSTS have higher accident rates because the interactions among its elements or functions cannot be fully planned, understood or anticipated. On the other hand, a complex socio-technical system does not always involve high risks and catastrophic consequences.

A. Functional Resonance Analysis Method

The Functional Resonance Analysis Method (FRAM) is used for modeling CSTS, which focuses on functions, not on architecture or components. The FRAM is a method to analyze work activities to develop a model which represents how work is done and understand what could go wrong [5].

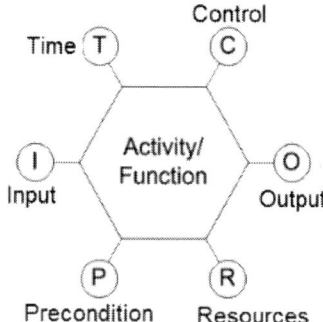

Fig. 3. FRAM functional unit (source [5])

Four principles underlying FRAM [5]:

- The equivalence of failures and successes;

- Everyday performance of CSTS is adjusted to match the conditions ;

- Many outcomes are emergent, not resultant;

- A complement to causality, system relations and dependencies must be described as they develop in a specific situation.

*** **Formatting Issue - Best Available Paper/Graphic** ***

B. Description of case study: IEEE 242-2001

As the first description for complex socio-technical systems approaches applied to power systems, a case study is modeled for further research. It is crucial to make a proper classification of this industrial power system operation scenarios for human reliability analysis. According to the literature review [24], power system operation scenarios are classified into three categories: time-centered scenarios, process-centered scenarios, and emergency-centered scenarios.

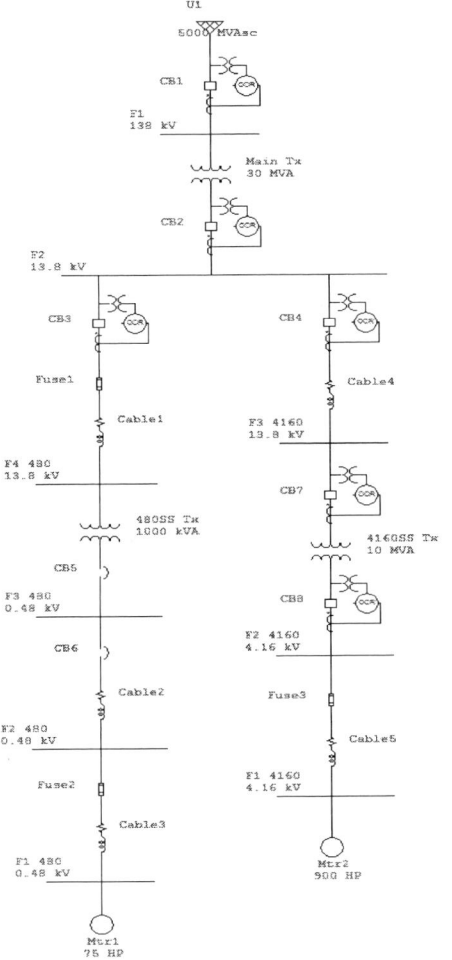

Fig. 4. Case study: IEEE 242-2001 Industrial distribution system (ETAP)

Since this particular and useful case was previously validated and tested in real-time [25][26] for understanding and training with protection coordination in industrial systems, the implementation of a novel complex socio-technical systems approach allows expanding the benefits HRA. The advantage of this application contributes to a better interaction with the standard 242-2001 because the user can now assess different human error scenarios, which means a critical highlight to be included in academia (protection courses), professional training with real distribution systems models and future standardization.

The electrical model was first implemented on ETAP; this system is shown in Fig. 4. The buses' IDs follow an incremental order from the grid to the loads, identifying each branch with its lowest voltage level. The transformers were noted as Main Tx, 4160SS Tx, and 480SS Tx; the relaying labeling system employs the same orderliness from bus naming. Likewise, the naming of the protective devices was performed incrementally, taking the grid connection as the starting point. The parameters for the different impedances, relays, instrumentation equipment and damage curves regarded on ETAPs database were procured to be the closest possible with the standard, without disregarding selectivity criteria. Using the model of IEEE 242-2001 in ETAP and DSSim-PC, the results for different scenarios could be tested using virtual relays.

VI. CONCLUSION AND FURTHER WORK

Human reliability analysis (HRA) and human error probability (HEP) methods are used to predict human performance and quantification of the likelihood of human errors. There are many HRA and HEP assessment methods which involve qualitative and quantitative approaches. However, in the classical view, from transient faults to large blackouts human error is the root or primary cause or the starting event.

THERP and CREAM methodologies for HRA are highly used in a wide range of industries. It is critical to ensure that these methods are useful and reliable considering the current applications and complexity of the systems.

To increase the knowledge and understand of the modern power systems as a CSTS, their interactions with human factors, as well as new methods to assess HRA, the principles of the practical resonance analysis method (FRAM) will be applied to develop a base case to evaluate the reliability of the IEEE-242 model.

ACKNOWLEDGMENT

This work was supported in part by the Administrative Department of Science, Technology, and Innovation of Colombia COLCIENCIAS under the grant 617 for doctoral students and grant 720.

REFERENCES

[1] U.S., Canada Power System Outage Task Force. Final Report on the August 14, 2003 Blackout in the United States and Canada: Causes and Recommendations. Canada, April 2004. Available from: https://www3.epa.gov/region1/npdes/merrimackstation/pdfs/ar/AR-1165.pdf

[2] OFGEM. Report on support investigations into recent blackouts in London and West Midlands. Volume 1, main report. February 2004

[3] FERC/NERC Staff Report on the September 8, 2011 Arizona-Southern California outages. Causes and recommendations. Abril 2012

[4] Health and Safety Executive HSE. HSG48 Reducing error and influencing behaviour. 2009. Available from: http://www.hse.gov.uk

[5] Hollnagel, Erik. FRAM The Functional Resonance Analysis Method. Modelling Complex Socio-technical Systems. London, 2012. CRC Press

[6] IEC 61508:1998(E) IEC 61508 Standard for Functional Safety of Electrical/Electronic/Programmable Electronic Safety-Related Systems

*** Formatting Issue - Best Available Paper/Graphic ***

[7] IEEE Std 1023-2004. IEEE Recommended Practice for the Application of Human Factors Engineering to Systems, Equipment, and Facilities of Nuclear Power Generating Stations and Other Nuclear Facilities.

[8] IEEE Std 493-2007. IEEE Recommended Practice for the Design of Reliable Industrial and Commercial Power Systems

[9] IEEE Std 1786-2011. IEEE Guide for Human Factors Applications of Computerized Operating Procedure Systems (COPS) at Nuclear Power Generating Stations and Other Nuclear Facilities.

[10] IEEE Std 1366-2012. IEEE Guide for Electric Power Distribution Reliability Indices.

[11] IEEE Std 3006.9-2013. IEEE Recommended Practice for Collecting Data for Use in Reliability, Availability, and Maintainability Assessments of Industrial and Commercial Power Systems.

[12] IEEE Std 3006.7-2013. IEEE Recommended Practice for Determining the Reliability of 7x24 Continuous Power Systems in Industrial and Commercial Facilities.

[13] IEEE P3006.5, 2014. IEEE Recommended Practice for the Use of Probability Methods for Conducting a Reliability Analysis of Industrial and Commercial Power Systems

[14] IEEE Std 1729-2014. IEEE Recommended Practice for Electric Power Distribution System Analysis

[15] IEEE Std 3006.2-2016. IEEE Recommended Practice for Evaluating the Reliability of Existing Industrial and Commercial Power Systems.

[16] IEEE Std 1082-2017. IEEE Guide for Incorporating Human Reliability Analysis into Probabilistic Risk Assessments for Nuclear Power Generating Stations and Other Nuclear Facilities

[17] Boring, R.L. (2007) Dynamic human reliability analysis: benefits and challenges of simulating human performance. in European Safety and Reliability Conference (ESREL 2007), INL/CON-07-12773, Idaho National Laboratory.

[18] Leveson, Nancy G. Engineering a Safer World. Systems Thinking Applied to Safety. London, 2012. The MIT Press

[19] Tarcisio Abreu Saurin, Angela WeberRighi. Complex socio-technical systems, characterization and management guidelines. Elsevier. Applied Ergonomics, Volume 50, September 2015, Pages 19-30.

[20] Baber, Chris, Dr; Jenkins, Daniel P, Dr; Walker, Guy H, Dr. Human Factors Methods: a practical guide for engineering and design. Chapter 6. Human error identification and accident analysis methods. Ashgate Publishing Ltd. Second Edition. October, 2013. England.

[21] Health and Safety Executive HSE. Review of human reliability assessment methods. RR679 Research Report. 2009. Available from: http://www.hse.gov.uk

[22] A. D. Swain, H. E. Guttmann. NUREG/CR- 1278. Handbook of human reliability analysis with emphasis on nuclear power plant applications. Final report. Prepared by Sandia National Laboratories Albuquerque, New Mexico. August 1983.Available from: https://www.nrc.gov/docs/ML0712/ML071210299.pdf

[23] EI Guidance on quantified human reliability analysis (QHRA). Energy institute. November 2012. 1st Edition. London, UK.

[24] Y. Bao, J. Guo, J. Tang, Z. Li, S. Pang and C. Guo, "Analysis of power system operation reliability incorporating human errors," *2014 17th International Conference on Electrical Machines and Systems (ICEMS)*, Hangzhou, 2014, pp. 1052-1056.

[25] J.D Pico, D. Celeita and G. Ramos, "Protection Coordination Analysis Under a Real-Time Architecture for Industrial Distribution Systems Based on the Std IEEE 242-2001," in IEEE Transactions on Industry Applications, vol. 52, no. 4, pp. 2826 - 2833, Jul.-Aug. 2016.

[26] D. F. C. Rodriguez, J. D. P. Osorio and G. Ramos, "Virtual Relay Design for Feeder Protection Testing With Online Simulation," in IEEE Transactions on Industry Applications, vol. 54, no. 1, pp. 143-149, Jan.- Feb. 2018.

978-1-5386-6624-1/18 $31.00 © 2018 IEEE

Power System Vulnerability Assessment using Voltage Collapse Proximity Index

Chaithra A
Department of Eletcrical and Electronics
PES University
Bengaluru,India
Email: chaithraanjankumar@gmail.com

Mrs.Sangeeta Modi
Department of Eletcrical and Electronics
PES University
Bengaluru,India
Email: smodi@pes.edu

Abstract— In this paper Voltage Collapse Proximity Index (VCPI) method is discussed. Voltage Collapse Proximity Index method is used to evaluate the voltage stability of various buses connected in a complex power system. This method is useful in predicting the conditions which can lead to voltage instability. In this work the effectiveness of the VCPI is tested through Power system analysis toolbox (PSAT). PSAT (Power System Analysis Toolbox) is a matlab toolbox used for both static and dynamic analysis and control. In this paper, VCPI algorithm is discussed and simulation & analysis of a 6-bus system and an IEEE 14-bus system under disturbances such as load throw-off and line tripping is presented. Using the load flow results which are obtained with the help of PSAT, the VCPI is calculated for all the buses and analyzed. The results thus obtained authenticates that VCPI is a simple and flexible method for determining the stability of the large interconnected power system.

Keywords— Voltage Collapse Proximity Index (VCPI), Power System Stability, Power System Analysis Toolbox (PSAT).

I. INTRODUCTION

THE rapid growth of population in the recent decade has led the electrical power systems to face various problems under heavily loaded conditions and also stress on the system has increased to a greater extent. So with the increased power demand, the major issues like voltage stability and voltage collapse has been seeking more attention. Maintaining voltage stability within in the secure limit has posed a challenge to all the electric utility companies [1, 5].

With the aim of maintaining the stability of the large interconnected complex power system within the secure limits, voltage stability indices are calculated at each bus [2]. Voltage stability indices which are also known as performance indices are considered to be one of the simple ways of determining the proximity of the voltage collapse [3, 6]. These indices will be able to predict a prior or an early susceptible situation of how close a system to voltage instability [4].

One of the simpler and faster indexes that can be used to predict the propinquity of the actual collapse is Voltage Collapse Proximity Index (VCPI) which is derived by using a simple relationship between various factors that are responsible for maintaining stability of the power system [3]. Elements such as reactive power generating devices and tap changing transformers are suitably set at the desired level to avoid the problem of voltage instability and to ensure the system security [7].

The effectiveness of VCPI is verified by a study conducted on IEEE-14 bus system and a 6–bus system using a matlab toolbox called as Power System Analysis Toolbox (PSAT). PSAT is an open source Matlab and GNU/Octave-based software package for design and analysis of electrical power systems. PSAT includes power flow, stability analysis, simulation tools as well as several models including synchronous generators regulators. PSAT is endowed with a simulink based editor for single line diagrams and also has a complete set of user-friendly graphical interfaces. It also has an additional feature such as data conversion capability.

PSAT is available in many versions and is very convenient to use. Since it allows us to use Simulink environment, it is easy to design power networks. PSAT simply reads the data from the Simulink model and writes down the data file. The data stored in it can be changed easily. It is a high level modeling system for mathematical programming problems and the libraries are designed exclusively for voltage stability analysis of power systems. PSAT is useful for research purpose because it allows easy and fast prototyping of new models and algorithm and is also useful for creating bridges between Matlab and other software packages.

II. VOLTAGE COLLAPSE PROXIMITY INDEX

Many performance indices have been formulated to predict the proximity of the voltage collapse. But Voltage Collapse Proximity Index (VCPI) is found to be one of the fastest methods to predict propinquity because in order to calculate this particular index only a modest amount of calculations are required at a particular bus. Moreover, matrix inversions are not required and hence substantial amount of time of CPU time reduces while calculating especially when the system grows bigger in size. VCPI is formulated as follows [3]:

Consider a single line diagram as shown in Fig. 1

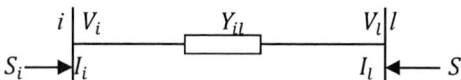

Fig. 1 Single line diagram of two bus system

where S_i is the complex power at bus i, S_l is the complex power at bus l.

For a M bus system, the current injected at i^{th} bus is given by,

$$I_i = V_i \sum_{l=1\,l\neq i}^{M} Y_{il} - \sum_{l=1\,l\neq i}^{M} V_l Y_{il} \qquad \dots (1)$$

Where

- V_i is the voltage at bus i in V
- V_l is the voltage at bus l in V
- Y_{il} is the admittance value between bus i and l in mho.
- l is a real number

Complex power injected into the ith bus is given by,

$$S_i = V_i I_i^* \qquad \dots (2)$$

Substituting (2) into (1) provides,

$$S_i = |V_i|^2 \sum_{l=1\,l\neq i}^{M} Y_{il} - V_i^* \sum_{l=1\,l\neq i}^{M} V_l Y_{il} \qquad \dots (3)$$

Let

$$Y_{ii} = \sum_{l=1\,l\neq i}^{M} Y_{il} \qquad \dots (4)$$

Substituting (4) in (3) provides,

$$S_i = |V_i|^2 Y_{ii} - V_i^* \sum_{l=1\,l\neq i}^{M} V_l' Y_{ii} \qquad \dots (5)$$

Where

$$V_l' = \frac{Y_{il}}{\sum_{n=1\,n\neq i}^{M} Y_{in}} V_l = |V_l'|\delta_l' \qquad \dots (6)$$

Therefore,

$$\frac{S_i^*}{Y_{ii}} = |V_i|^2 - V_i^* \sum_{l=1\,l\neq i}^{M} V_l' \qquad \dots (7)$$

Equation (7) can be written as

$$\frac{S_i^*}{Y_{ii}} = |V_i|^2 - (|V_i|\cos\delta_i - j\,|V_i|\sin\delta_i) *$$

$$\left(\sum_{l=1\,l\neq i}^{M} (|V_l'|\cos\delta_l' + j\,|V_l'|\sin\delta_l') \right)$$

$$\dots (8)$$

where, δ_i is the voltage angle at i^{th} bus.

Re-arranging (8) provides

$$\frac{S_i^*}{Y_{ii}} = |V_i|^2 - \sum_{l=1\,l\neq i}^{M} |V_l'||V_i|\cos(\delta_i - \delta_l')$$

$$+j\left(\sum_{l=1\,l\neq i}^{M} |V_l'||V_i|\sin(\delta_i - \delta_l') \right)$$

$$\dots (9)$$

The RHS of (9) is a complex quantity, therefore it can be written as,

$$\frac{S_i^*}{Y_{ii}} = a - jb \qquad \dots (10)$$

where a and b are any two real numbers.

From (9) and (10),

$$a - jb = |V_i|^2 - \sum_{l=1\,l\neq i}^{M} |V_l'||V_i|\cos(\delta_i - \delta_l')$$

$$+j\left(\sum_{l=1\,l\neq i}^{M} |V_l'||V_i|\sin(\delta_i - \delta_l') \right)$$

$$\dots (11)$$

Comparing real and imaginary parts of (11), gives

$$a = |V_i|^2 - \sum_{l=1\,l\neq i}^{M} |V_l'||V_i|\cos(\delta_i - \delta_l') \quad \dots (12)$$

$$b = \sum_{l=1\,l\neq i}^{M} |V_l'||V_i|\sin(\delta_i - \delta_l') \qquad \dots (13)$$

Let

$$(\delta_i - \delta_l') = \delta \qquad \dots (14)$$

Equation (12) and (13) represents two equations with 2 unknowns V_i and δ. Let,

$$f_1(|V_i|, \delta) = |V_i|^2 - \sum_{l=1\,l\neq i}^{M} |V_l'||V_i|\cos\delta$$

$$\dots (15)$$

$$f_2(|V_i|, \delta) = \sum_{l=1\,l\neq i}^{M} |V_l'||V_i|\sin\delta$$

$$\dots (16)$$

To solve the unknown present in the equations (15) and (16), the partial derivatives are evaluated w. r. t V_i and δ. Therefore matrix J is given as follows

$$= \begin{bmatrix} 2|V_i| - \displaystyle\sum_{l=1\ l\neq i}^{M} |V_l'| \cos \delta & |V_i| \displaystyle\sum_{l=1\ l\neq i}^{M} |V_l'| \sin \delta \\ \displaystyle\sum_{l=1\ l\neq i}^{M} |V_l'| \sin \delta & |V_i| \displaystyle\sum_{l=1\ l\neq i}^{M} |V_l'| \cos \delta \end{bmatrix} \quad \text{.... (17)}$$

The voltage collapse at i^{th} bus means there is no solution to the equation (17) which means that the determinant of the above matrix should be zero. Therefore the matrix becomes singular at the voltage collapse point.

Solving the above matrix and equating to zero gives,

$$\frac{V_i \cos \delta}{\sum_{l=1\ l\neq i}^{M} |V_l'|} = \frac{1}{2} \quad \text{.... (18)}$$

Equation (18) can be written as

$$\frac{\sum_{l=1\ l\neq i}^{M} |V_l'|}{V_i} = \frac{1}{2} + jm \quad \text{.... (19)}$$

where m is a real constant. Equation (19) can be written with the help of complex identities as,

$$\left| 1 - \frac{\sum_{l=1\ l\neq i}^{M} V_l'}{V_i} \right| = 1 \quad \text{.... (20)}$$

Thus VCPI at the ith bus is given by,

$$VCPI_i = \left| 1 - \frac{\sum_{l=1\ l\neq i}^{M} V_l'}{V_i} \right| \quad \text{.... (21)}$$

The VCPI value varies between 0 and 1. The VCPI value of 0.1 or closer to zero is considered to be stable condition and any value that is closer to unity is an indication of possible collapse.

III. SIMULATION AND RESULTS

In order to the effectiveness of the VCPI we conducted a study on IEEE 14 bus system and a 6 bus system for various perturbations such as load throw-off and line tripping.

A. Effect of tripping of the lines on VCPI

The Fig. 2 shows the 6-bus system and is simulated using PSAT. Using the load flow reports, the VCPI is calculated at each bus. Under normal operating conditions, the VCPI values of all the buses are tabulated in Table-I as follows.

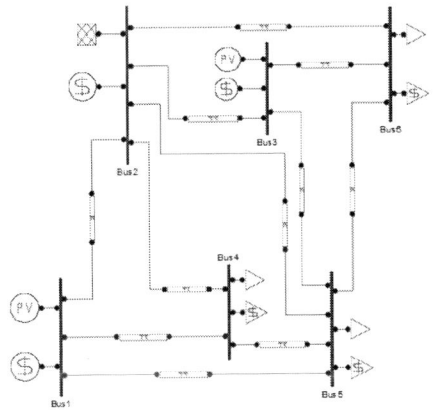

Fig. 2 6-Bus System

TABLE I.   VCPI VALUES OF VARIOUS BUSES IN A 6-BUS SYSTEM

Bus number	VCPI
01	0.041571
02	0.040061
03	0.047615
04	0.047847
05	0.062651
06	0.044617

From the calculation, it was found that bus-5 is weakest of all. In order to study the effect of tripping of a line on VCPI, a line interconnecting bus-5 and bus-4 is eliminated assuming that the line has tripped

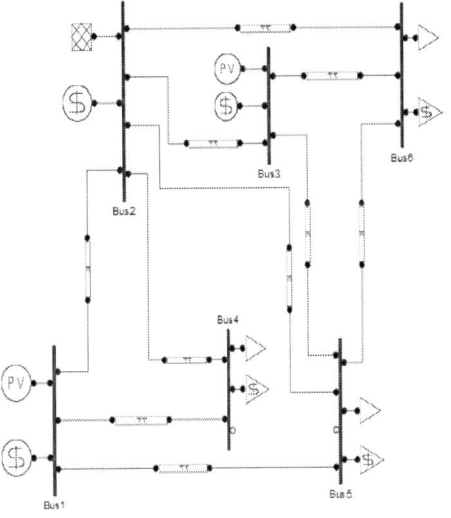

Fig.3: Tripping of a line at Bus 5 in a 6-Bus System

TABLE II.   VCPI VALUES DUE TO TRIPPING OF ONE A LINE IN A 6-BUS SYSTEM

Bus number	VCPI
01	0.04353
02	0.041443
03	0.049481
04	0.64788
05	0.075841
06	0.044853

Due to the tripping of one of the lines from Bus 5, its VCPI value has further increased proving the fact that the Bus 5 is more vulnerable compared to others because the stress on the remaining lines has increased which are present at Bus 5. Now let us consider the case where multiple tripping taking place as shown in the Fig.4.

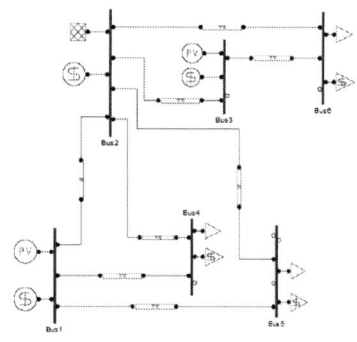

Fig.4 Multiple tripping at Bus 5 in a 6-Bus System

Above figure shows a system where more than one line has tripped at the same time at Bus 5. With reference to that effect, the system is simulated to obtain the voltage magnitudes of all the buses required for the calculation of the index. The results obtained are tabulated in Table-III as follows:

TABLE III.   VCPI VALUES DUE TO MULTIPLE TRIPPING IN A 6-BUS SYSTEM

Bus number	VCPI
01	0.066305
02	0.052281
03	0.036907
04	0.65783
05	0.197993
06	0.054485

When multiple tripping took place at Bus 5, the VCPI value of Bus 5 has increased further indicating that it is prone to collapse because the stress on the remaining lines increased due to the removal of other three lines from the Bus 5 which was present in original system as shown in Fig. 2 So it can be concluded that VCPI is a good agent in determining the dynamic stability of the system.

B.   Verification of Stability of the buses Using VCPI with Incrementing Load.

The Fig.5 shows the IEEE 14-bus system which is simulated with the help of PSAT for various contingencies similar to 6-bus system in order to check the efficacy of the proposed method. Under normal operating conditions, the VCPI values of all the buses are tabulated in Table-IV as follows:

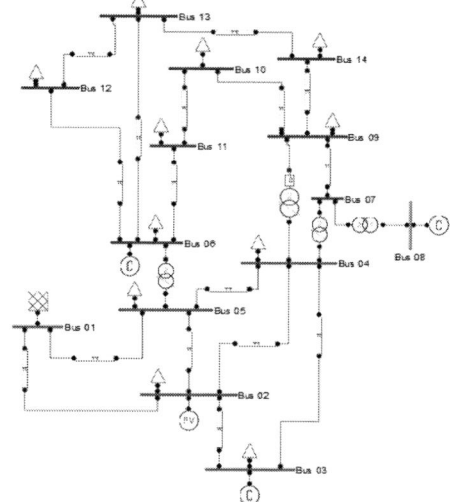

Fig. 5 IEEE 14-Bus System

TABLE IV.   VCPI VALUES OF 14-BUS SYSTEM

Bus number	VCPI
01	0.022735977
02	0.012592008
03	0.010053819
04	0.012769236
05	0.011658508
06	0.029865486
09	0.003967225
10	0.007590403
11	0.004675243
12	0.007057332
13	0.012223261
14	0.026378675

978-1-5386-6624-1/18 $31.00 © 2018 IEEE

From Table-IV, it can be concluded that Bus 6 is weakest of all and Bus 9 is the healthiest compared to other. Since a compensator present at Bus 6, any abnormal condition on Bus 6 will be taken care, so the next weakest bus is Bus 14.

The IEEE 14-Bus system is simulated with p.u addition of loading at the stronger and weaker buses. Initially load of 2p.u is increased at Bus 9 and the system was simulated for load flow analysis with the help of PSAT. It was found that all the bus voltages, real power and reactive power were within the limits and there were no violations.

With further increment say at 2.7p.u addition of load it was found that Bus 9 became more vulnerable and many violations were found in the load flow analysis report. So with this, the stability margin of the Bus 9 is found.

Similarly a load of 1p.u is added at Bus 14 and the simulation is carried out, it was found that there were no violations in voltage and power values. With further 0.2p.u increment in load at Bus 14, the bus became more susceptible to collapse and many violations were seen. As per VCPI calculation it was seen that Bus 14 was more vulnerable compared to Bus 9, with increased loading at both buses.

Bus 14 became vulnerable with a small increment of load compared to Bus 9 which verifies that VCPI is a good agent in determining the steady stability of the system.

IV. CONCLUSION AND FUTURE SCOPE

In this paper, VCPI algorithm is discussed and implemented on the IEEE 14 bus system and a 6 bus system. It was found that VCPI is the fastest and simplest way to predict the early vulnerable condition of the bus and hence preventing the system against voltage collapse. Through simulation using PSAT it was found that VCPI is a good agent in determining the steady state stability and dynamic stability. With the proposed method, the security of the system is ensured because any damage to equipments or line leads to huge economic crisis. The VCPI index can be used for both offline studies and for online implementation. Additional advantage of the VCPI is that it can be applied to modern grids i.e., smart grid.

REFERENCES

[1] Dharmendra Kumar, Nisheet Soni, "Voltage Stability Estimation of Electric Power System Using L-Index" by *IJETMR Journal* Sep 2015.

[2] Fredy A. Sanz, Méxixco Juan M. Ramirez Rosa E.Correa, "Statistical Estimation of Power System Vulnerability" *IEEE* 2013.

[3] V. Balamourougan, T. S. Sidhu, and M. S. Sachdev, "Technique for online prediction of voltage collapse," IEEE Proceedings Generation, Transmission and Distribution, vol. 151, no. 4, pp. 453–460, 2004.

[4] R. Tiwari, K. Niazi, and V. Gupta, "Line collapse proximity index for prediction of voltage collapse in power systems," International Journal of Electrical Power; Energy Systems, vol. 41, no. 1, pp. 105–111, 2012

[5] M.H.Haque, "Online monitoring of maximum permissible loading of a power system within the stability limits", IEEE Proceedings Generation, Transmission and Distribution, vol. 150, No.1, January 2003.

[6] Elfadil.Z. Yahia, Mustafa A. Elsherif , Mahmoud N. Zaggout, "Detection of Proximity to Voltage collapse by using L-index", IJIRSET, Vol 4, Issue 3, March 2015.

[7] Haneesh K M, Arya Vishnu Ram T, "Voltage Stability Analysis Using L-index Under Various Transformer Tap Changer Settings" International Conference on Circuit, Power and Computing Technologies [ICCPCT] 2016.

Transmission Congestion Management of IEEE 24-Bus Test System by Optimal Placement of TCSC

N Padmini
Department of Electrical and Electronics Engineering
Amity University
Noida, Uttar Pradesh
npadhu1996@gmail.com

Pallavi Choudekar
Department of Electrical and Electronics Engineering
Amity University
Noida, Uttar Pradesh
pallaveech@gmail.com

Mehtab Fatima
Department of Electrical and Electronics Engineering
Amity University
Noida, Uttar Pradesh
mehatabfatima@gmail.com

Abstract—In the deregulated power system congestion of transmission network is a key challenge. Any attempt to operate transmission system beyond its line limits leads to congestion. By use of FACTS devices, the power transfer capability of network increases & reactive power compensation is done thereby improving the power system stability and network power quality. This paper presents the optimal placement of Thyristor controlled series compensators (TCSC) to avoid congestion in the transmission network. The optimal placement is figured out by factors like line utilisation factor (LUF), line loading, voltage profile at each bus & real and reactive power losses between buses. An algorithm is presented for finding the optimal location for placing TCSC. The reactance model of TCSC is considered to hike the power capabilities of the line. The potency of the model is tested on IEEE 24-bus system with the standard data using MATLAB-PSAT Simulink.

Keywords— Percentage line loading, LUF, TCSC, voltage stability

I. INTRODUCTION

Industrialization and Urbanisation & hike of life style are the factors to increase the dependency on the electrical energy. Because of this the power sector has seen a rapid growth day by day resulting in few uncertainties [1]. In order to meet the increased energy consumption & trades due to the increase of unplanned power traffic the transmission lines are frequently operated close to or even beyond their respective thermal limits. If the power traffic is not controlled there are chances that some lines may get overloaded and this phenomenon is treated as congestion [2]. In other words, congestion occurs when the transmission network is unable to accommodate all of the desired transactions due to a violation of system operating limits. Utilizing certain physical or financial mechanism in present day competitive market each utility manages the congestion in the system using its own guidelines. Privatization and de-regulation of electrical power markets have shown a vast impact on almost every power systems around the globe [3].. As deregulation is growing rapidly in electrical power sector there is attention on open access. Open access to network provides equal opportunities to use the available transmission system to all buyers and sellers.

Due to increase in electrical energy consumption and because of presence of few uncertainties [4] in the network there are chances that the line loading of the transmission lines increases, voltage profile at buses fall resulting in losses in the system and poor power quality & system stability. So, there is need to manage congestion in the network to achieve improved power quality & system stability.

Basically in a vertically integrated utility market the central agency or single utility directly controls the activities like generation, transmission and distribution. It is a fact that irrespective of relative geographical location of buyer and seller, every buyer wants to buy the power from cheapest price/generator.

In de-regulated parlance Congestion refers to transmission line hitting its limits. There are certain characteristics which limits the network to reliably transfer the electric power, they are

- Thermal limits
- Voltage limits
- Stability limits.

Congestion management is a mechanism to prioritize the transactions and commit to such a schedule which would not overload the network. Despite all these measures following a forced outage of transmission line the congestion can still occur in the transmission network. Only by means of real time congestion management the network operator can handle this situation. Thus the precautionary actions to be taken by network operator as follows

- For keeping transmission network within limits the operator must allow only particular set of transactions

- Even if proper scheduling is done, in real time due to unscheduled flows the transmission corridors may get overloaded. At this point the network operator has to take some remedial actions.

The congestion mainly depends on what type of de-regulation model is employed in that particular area.

With overall market design the implementation of congestion management schemes are influenced by factors such as network topologies, demographic factors & political ideologies. Any congestion management scheme should try to achieve following features

978-1-5386-6624-1/18 $31.00 © 2018 IEEE

a. Economic efficiency

b. Non Discriminative

c. Be transparent

d. Be robust

II. THYRISTOR CONTROLLED SERIES CAPACITOR(TRSC)

First, confirm that you have the correct template for your paper size. This template has been tailored for output on the A4 paper size. If you are using US letter-sized paper, please close this file and download the Microsoft Word, Letter file TCSC comes under one among four types of FACT devices which are used to compensate lines for enhancement of power quality and system stability [5]. It is a series type FACT device uses series compensation technique to govern the line flow in an economic manner by compensating the reactance of the transmission line. The configuration of TCSC consists of parallel combination of controlled reactor along with a capacitor bank. This parallel combination of reactor and capacitor allows smooth control of the basic frequency over a wide range. To stop overvoltage across capacitor a Metal oxide varistor (MOV) is connected across it.

The principle that governs the operation of TCSC is variable series reactance. To control the power flow in the transmission network by installing TCSC in the network is a recent technology to increase system efficiency & to improve stability of the system. And power flow is controlled by varying the series reactance of the device.

The power flow equations between different buses connected by TCSC in between them are given as follows,

$$P = \frac{|V_K|}{X_l}|V_m|\sin(\partial_k - \partial_m) \tag{1}$$

Similarly, the reactive power expression is given by

$$Q = \frac{|V_k|^2}{X_l} - \frac{|V_k|}{X_l}|V_m|Cos(\partial_k - \partial_m) \tag{2}$$

Where

V_k & V_m are respective voltages at buses k & m, $(\delta_k - \delta_m)$ is the angle between the respective voltages, X_l is the line impedance.

A. Operation of TCSC

Thyristor-controlled series capacitor is a series type of FACT device which monitors the power flow in network through series compensation approach. The transmission line reactance is controlled dynamically in order to provide sufficient load compensation. The advantageous functioning of TCSC is seen from two main aspects

- From its ability to operate in different operating modes
- From its ability to control the % of compensation (amount of compensation) of a transmission line

Since the loads in the power sector are constantly changing & it is difficult to predict sometimes the role of TCSC in transmission network is vital.

FSC (Fixed Series Compensation): Because of the effects due to line reactance modification the FSC of line is desirable for power transfer in an efficient manner. FSC is nothing but adding a series capacitance in the line to provide compensation which is shown in Fig.1.The reactive impedance of the line decreases because of the presence of series capacitance in the line, thereby lowering/minimizing the voltage drop across the lines in the network. In this approach the reactance of the line is counteracted by the series capacitance inserted in between buses resulting in overall minimum/lower line impedance thereby minimizing the voltage drops across the line.

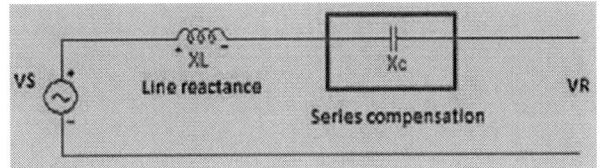

Fig.1 FSC of Transmission line

B. Optimal placement through line loading analysis

This paper deals with congestion management in multiline transmission system by optimal placement of TCSC. The system consists of 24 buses and the power flow is carried out under test data. The statistics of network are as follows

In this paper the optimal location for the placement of TCSC is figured out by the consideration of following factors

- Voltage profile
- Active & reactive power losses in each line
- % line loadings
- LUF

Voltage profile: Voltage profile at each and every bus gives the overview of which bus is having less/lower voltage rating. From the analysis of voltage profile, it is easy to figure out which line needs compensation by placing TCSC in between buses. Thus, voltage profile helps for the optimal placement of TCSC.

Active & reactive power losses: From the power flow, the active power as well as reactive power losses is obtained for each transmission line. From the data of power flow, the process of figuring out which line is having more active/reactive power loss can be done & thereby the optimal placement of TCSC will be done.

Percentage line loadings: It can be defined as the ratio of actual line rating to the maximum line rating. Mathematically it can be expressed as

$$\% \ Line \ Loading = \frac{Actual \ line \ rating}{Maximum \ line \ rating} * 100 \tag{3}$$

Where actual line rating can be calculated as

978-1-5386-6624-1/18 $31.00 © 2018 IEEE

$$Actual\ line\ rating = \sqrt{P_{flow}^2 + Q_{flow}^2} \qquad (4)$$

Whereas the MLR (Maximum Line Rating) of a transmission line is a manufactured/default value assigned to line.

LUF (Line Utilization Factor): In power system concepts the utilization factor is defined as "the ratio of highest/maximum load which could be drawn to the rated capacity of the system. This LUF concept is closely related to the LF (Load Factor)

LUF can also be represented in terms of % line loading concept.

$$LUF = \frac{\%\ Line\ Loading}{100} = \frac{Line\ actual\ rating}{Line\ maximum\ rating} \qquad (5)$$

III. SIMULATION AND RESULTS

The simulation of IEEE 24 bus system is carried out in the MATLAB - 2016a (Matrix-oriented programming) environment using PSAT-2.1.10(Power System Analysis Toolbox).The SLD(single line diagram) of IEEE 24-bus system for simulation analysis is as shown in Fig.2

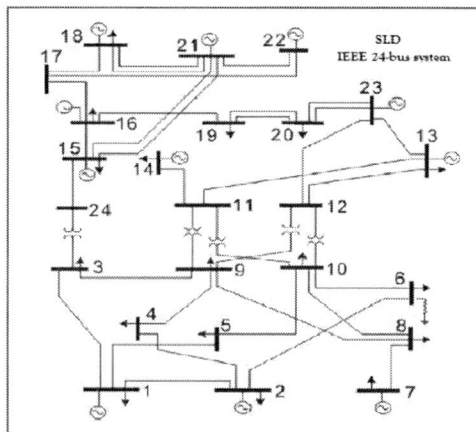

Fig.2 SLD of IEEE 24-bus system

A. Base case

1). The load flow without TCSC: The SLD is composed of two area networks. Load flow is performed on IEEE 24-bus system under base case and from the power flow report the following parameters are calculated,

- Line loadings of each line
- LUF (Line Utilization Factor) of each line

$Base\ power(MVA) = 100\ MVA$

$Base\ KV\ for\ area - 1 = 138\ KV$

$Base\ KV\ for\ area - 2 = 230\ KV$

The line loadings & LUF's are calculated for each transmission line in the network [6]. And the line loading of

line between bus2 & bus6 is coming out to be 117.65%. At the same time the voltage at bus number 6 is marked very low among other bus which is given by 0.721359[p.u]. The active & reactive power losses are high between bus2 & bus6. Thus the line between bus2 & bus6 is the best possible location to install TCSC. Therefore the optimal location for the placement of TCSC is between bus2 & bus6.

Table I shows the parameters cause to choose the optimal location for the placement of TCSC

TABLE I. PARAMETERS RELATED TO BUS 2 & 6

Voltage at bus 2	1.035pu
Voltage at bus 6	0.721359pu
Active power loss between bus 2 & 6	0.200423pu
Reactive power loss between bus 2 & 6	0.732887pu
% Line loading between bus 2 & 6	117.65%
Line Utilization Factor	1.1765

Table II shows the total generation, total load, & total losses of both real & reactive power.

TABLE II. TOTAL GENERATION,LOAD,LOSSES UNDER BASE CASE WITHOUT TCSC

SYSTEM GENERATION	pu
Real power	29.46
Reactive power	9.809
SYSTEM LOAD	**pu**
Real power	28.7082
Reactive power	6.363
SYSTEM LOSSES	**pu**
Real power	0.761055
Reactive power	3.446399

Fig.3 shows the voltage profile at buses under base case before installing TCSC in the network.

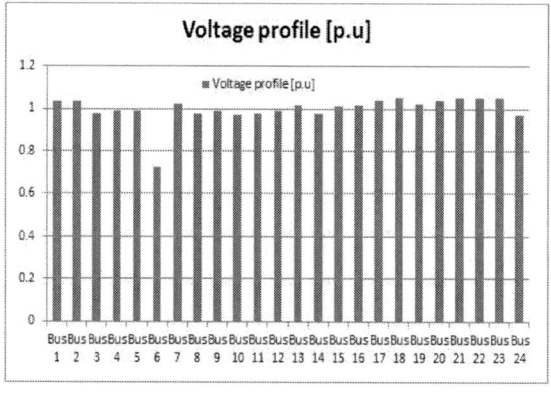

Fig.3 Voltage profile under base case without TCSC

2). Load flow with TCSC: After installing TCSC in between bus 2&6 it is observed that the voltage profile of the system is improved. Reduction in active & reactive power loss is seen. Table III illustrates the improvement in the voltage profile and reduction in the active & reactive power losses between bus 2&6.

TABLE III.   PARAMETERS RELATED TO BUS 2 & 6

Voltage at bus 2	1.035pu
Voltage at bus 6	1.002639pu
Active power loss between bus 2 & 6	0pu
Reactive power loss between bus 2 & 6	0.1703pu

The total generation, total load, total losses of both active & reactive power under base case with TCSC are shown in Table IV.

TABLE IV.   TOTAL GENERATION,LOAD,LOSSES UNDER BASE CASE WITH TCSC

SYSTEM GENERATION	pu
Real power	30.48
Reactive power	10.4499
SYSTEM LOAD	pu
Real power	29.86
Reactive power	7.085
SYSTEM LOSSES	pu
Real power	0.626357
Reactive power	3.364691

The improvement in the voltage profile of the network can be seen by observing Fig.4.

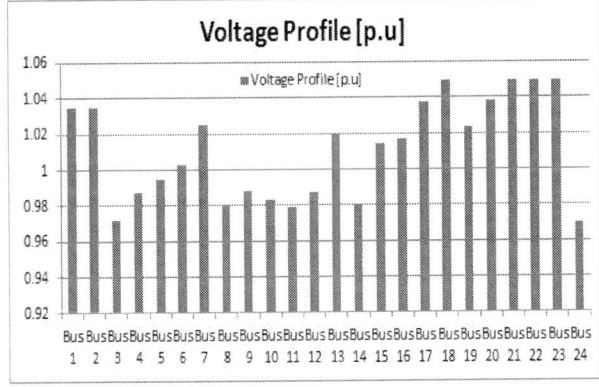

Fig.4 Voltage profile under base case with TCSC

The decrease in % line loadings of lines between different buses can be seen by observing Fig.5.

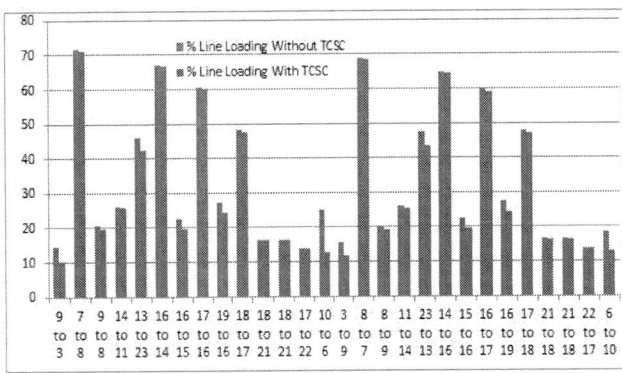

Fig.5 %line loading under base case

B. Contingency Case

Congestion itself means an abnormal condition in the network operation. For congestion management/contingency analysis the heavily loaded line in the base case of system is removed. i.e the line in between the buses 2&6 is removed for congestion analysis[7-11], after taking out the line between bus 2&6, the total number of lines in the IEEE 24-bus system came down to 32.

1). Load flow without TCSC: The power flow is performed by taking IEEE 24-bus system with 32 transmission networks under contingency case. From the power flow report the % line loadings of each line and LUF's of each line are calculated from the formula mentioned above & it is observed that the voltage at bus number 8 is less and its below 1pu value. And also the line between buses 7&8 is loaded heavily. The active power loss between bus 7&8 is coming high. From the calculated %line loading & LUF, observed active power loss & voltage profile at buses, the optimal location for the placement of TCSC is figured out. Thus the TCSC is installed between bus 7&8.

Table V shows parameters of network between bus 7&8 under contingency without TCSC

TABLE V.   PARAMETERS RELATED TO BUS 7 & 8

Voltage at bus 7	1.025pu
Voltage at bus 8	0.9710pu
Active power loss between bus 7&8	0.026pu
Reactive power loss between bus 7&8	0.0847pu

The total active power & reactive power losses of the system are shown in Table VI.

TABLE VI.   TOTAL LOSSES UNDER CONTINGENCY CASE WITHOUT TCSC

TOTAL LOSSES	pu
Real power	0.632812
Reactive power	3.195194

The voltage profile at buses under contingency case without TCSC in the network can be observed from the Fig. 6

978-1-5386-6624-1/18 $31.00 © 2018 IEEE

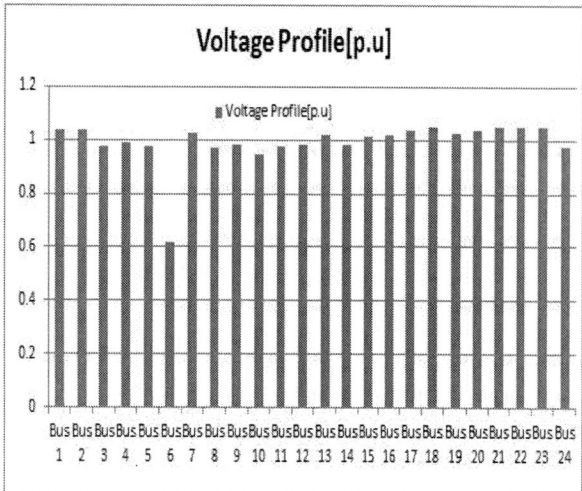

Fig.6 Voltage profile under contingency case without TCSC

2). Load flow with TCSC: From the calculation done on % line loading & LUF the optimal location for placing TCSC is determined and the TCSC is installed between bus 7&8 to improve system performance & stability. After installing TCSC between bus 7&8 it is observed that the voltage at bus 8 has been improved, the power flow through the line between bus 7&8 is increased and the active/reactive power losses of the line as well as system has been reduced.

The voltage at bus 7&8, active as well as reactive power loss profile between bus 7&8 is shown in Table VII

TABLE VII. PARAMETERS RELATED TO BUS 7 & 8

Voltage at bus 7	1.025pu
Voltage at bus 8	1.01896pu
Active power loss between bus 7&8	2.22E-16pu
Reactive power loss between bus 7&8	0.013717pu

The total system losses after installing TCSC can be observed from the Table VIII.

TABLE VIII. TOTAL POWER LOSSES AFTER INSTALLING TCSC

TOTAL LOSSES	pu
Real power	0.610415
Reactive power	3.109922

The improvement in the voltage profile can be seen by observing Fig.7.

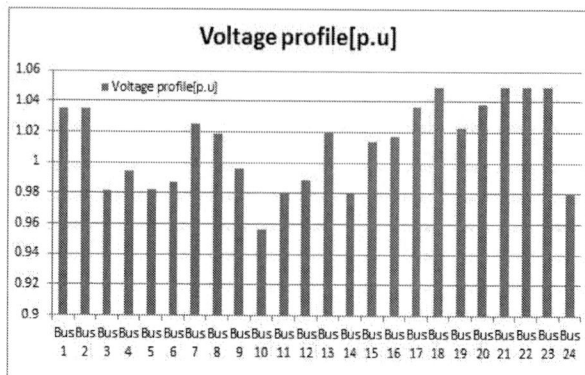

Fig.7 Voltage profile of system under contingency case with TCSC

The reduction in the % line loadings of few lines connected between different buses can be seen by observing Fig.8

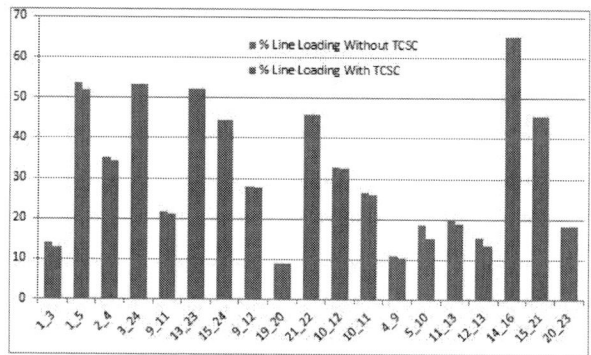

Fig.8 Percentage line loading under base case

IV. CONCLUSION

The main aim of this work is to figure out the optimal placement of TCSC in an IEEE 24-bus system which is performed on MATLAB software under both base & contingency cases. The optimal placement is determined using different factors such %line loading, LUF, voltage profile, losses at different buses. An algorithm is presented to figure out the optimal location for placing TCSC. The reactance model of TCSC is considered to hike the power capabilities of the line. It is observed that the total system losses under considered cases i.e. base case and contingency case are reduced. Under base case condition the percentage reduction in real power losses after installing TCSC is 17.69% & reactive power losses is 2.37%. Under contingency case the reduction in real power loss after installing TCSC is 3.53% & reactive power losses reduced by 2.66%. Apart from reduction in system losses the overall voltage profile is improved with improvement in power quality.

REFERENCES

[1] K.Satyanarayana, B.K.V. Prasad, G.Devanand, N.Siva Prasad, "Optimal Location of TCSC with Minimum Installation Cost using PSO", IJCST Vol. 2, SP 1,pp 156-160, 2011.

[2] Ibrahem Totonchi , Ibrahem Totonchi et. al.,"Sensitivity Analysis for the IEEE 30 Bus System using Load-Flow Studies" , 3rd International Conference on Electric Power and Energy Conversion Systems (EPECS), 2-4 Oct. 2013.

978-1-5386-6624-1/18 $31.00 © 2018 IEEE

[3] Sriparna Roy Ghatak, Debarghya Basu, Parimal Acharjee, "Voltage Profile Improvement and Loss Reduction Using Optimal Allocation of SVC", India Conference (INDICON), 17-20 Dec. 2015.

[4] L. N. Mrunalini Devi1 & A. Surya Prakash Rao, "Optimal Placement Of TCSC For Reactive Power reserve Management With Reactive Power Loss minimization Using Hybrid Psogsa" , International Journal of Power Systems & Microelectronics (TJPRC: IJPSM) Vol. 1, Issue 1,pp 61-72, 2016.

[5] Manasarani Mandala and C. P. Gupta," Transmission Congestion Management with TCSC using Bacterial Foraging-Particle Swarm Optimization", IEEE International Conference on Power Electronics, Drives and Energy Systems (PEDES), 16-19 Dec. 2012

[6] Uma.V , P.Lakshmi ,J.D.Anunciya ,"Congestion Management in Deregulated Power System by Fuzzy Based Optimal Location and Sizing of UPFC" , WSEAS Transactions On Power Systems, Volume 9, pp 258-266,2014.

[7] Sananda Pal, Samarjit Sengupta ,"Congestion Management of a Multi-bus Transmission System using Distributed Smart Wires" ,

International Conference on Control, Instrumentation, Energy and Communication, pp 417-420, 2014.

[8] Kanwardeep Singh1, Vinod K. Yadav, Arvind Dhingra ,"Congestion Management Using Optimal Placement of TCSC in Deregulated Power System", International Journal on Electrical Engineering and Informatics, Volume 4, Number 4, pp 620-632, 2012.

[9] Anwar S. Siddiqui, Manisha Rani ,"Enhancing the Power System Load ability Using TCSC: Improved Gravitational Search Algorithm", International Journal of Innovative Research in Science, Engineering and Technology, Vol. 4, Issue 4, pp 1943-1950, 2015.

[10] Pallavi Choudekar, SK Sinha, Anwar Siddiqui, "Transmission line efficiency improvement and congestion management under critical contingency condition by optimal placement of TCSC", 7th India International Conference on Power Electronics (IICPE), 2016.

[11] H. He, Z. Xu ,"Impacts of Transmission Congestion on Market Power in Electricity Market", Power Systems Conference and Exposition, 10-13 Oct. 2004.

Effect of FACTS Devices on Congestion Management using Active & Reactive Power Rescheduling

S.k. Gupta, N.K.Yadav, Mukesh Kumar

Dept. of Electrical Engineering

DCR University of Sc. & Technology, Murthal (Sonepat)

Haryana, India

e-mail:drskgupta.ee@dcrustm.org

Abstract – **Deregulation has allowed lot of liberty to DISCOS and GENCOS to participate in power market. It has led power transmission congestion problems complex. In this process the scheduled power flows in the transmission line and spontaneous power exchanges have also risen sharply in recent years. Congestion management thus has become an important issue. Transmission congestion is nothing but an operating condition in which there is not enough transmission capability to implement all of the traded transactions. In this study FACTS Devices are used for congestion relieving by the NR Jacobian modification method and their cost comparison is also done so as to determine the most economical device. Congestion Cost is calculated by interfacing MATLAB with GAMS.**

Keywords - Power Rescheduling UPFC IPFC and HVDC

I. INTRODUCTION

In the open-access environment transmission companies are facing the problem of congestion in transmission lines. FACTS devices control and influence the power flow in the network. FACTS devices thus may be good choice for handling congestion of the network.   A.Navabi-Naiki and M.R.Iravani et al. [2] provided mathematical models of unified power flow controller (UPFC) for steady state and transient stability analysis. R.S. Fang et al. [3] developed a congestion management strategy for an open transmission dispatch environment considering both pool and bilateral/multilateral dispatches simultaneously. M. Noroozian and L. Angquist et al. [4] presented optimal power flow analysis considering a unified power flow controller (UPFC) in the network. Authors also presented the method of determining the size of UPFC for power flow applications. Debrata Chattoapadhyaya et al [5] described the optimization problems of the power systems. Suman Bhowmik, Biswarup Das, Senior Member, IEEE, and Narendra Kumar et al [6] described the complexities of computer program codes for implementing Newton–Raphson load flow (NRLF) analysis for enhancing power flow modeling of an interline power flow controller (IPFC). S. K Gupta, R. Bansal [7] described the rescheduling technique for congestion

management. S. Charles Raja, Member IEEE, S. A. Waajitha Banu and Dr. P. Venkatesh et al [8] examined the congestion relief procedure and compared the results of two different objectives i.e. rescheduling cost minimization and real power loss minimization.

This paper is focused on to alleviate congestion in the system and to determine the congestion cost (CC) by rescheduling gencos bids using NR Jacobian modification method. The application of FACTS devices is studied and CC is obtained by rescheduling the bids again. To determine the cost effectiveness of these devices the 75-bus system data is taken from Uttar Pradesh State Electricity Board (UPSEB) and the buses are renumbered. It is observed that IPFC gave the least CC followed by UPFC and then HVDC.

II. PROBLEM FORMULATION

When it is not possible to solve the congestion problem in the transmission network in extreme case only using real and reactive power rescheduling of generators, System Operator (SO) will go for curtailment of loads. The identification of most sensitive generators and loads will reduce the computational burden considerably. In this scheme generators are invited bids for the quantity of power and the cost they would charge for increase as well as decrease their power output. The re-dispatch of transactions for congestion management in a pool model is formulated as a non-linear programming problem; which has been solved using the GAMS/CONOPT solver. The objective function is given as:

$$(\Delta P_{ij} + P_{ij}^o)^2 + (Q_{ij}^o + \Delta Q_{ij})^2 \le (S_{ij}^{max})^2$$

$$\left(P_{ij}^o + \sum_{n=2}^{75} PTCDF_n^k \Delta P_n\right)^2 + \left(Q_{ij}^o + \sum_{n=2}^{75} QTCDF_n^k \Delta Q_n\right)^2 \le \left(S_{ij}^{max}\right)^2 \quad ij \in N_l$$

$$\text{Minimize CC} = \sum_{r=1}^{Ng,up} c_{P_{g,r}}^+ \Delta P_{g,r}^+ + \sum_{s=1}^{Ng,dn} c_{P_{g,s}}^- \Delta P_{g,s}^- + \sum_{t=1}^{Ncl} c_{Pd,t} \Delta P_{D,t} +$$

$$\sum_{v=1}^{Nqg} C_{Qg,v} (\Delta Q_{g,v}) \Delta Q_{g,v}$$

(1)

Subject to

$$\Delta P_n^{min} \le \Delta P_n \le \Delta P_n^{max} \qquad i=1, 2, \ldots\ldots\ldots Nb$$

$$\Delta Q_n^{min} \le \Delta Q_n \le \Delta Q_n^{max}$$

$$\sum_{n=1}^{Nb} \Delta P_n - \sum_{n=1}^{Nb} \frac{\partial P_L}{\partial P_n} \Delta P_n - \sum_{n=1}^{Nb} \frac{\partial P_L}{\partial Q_n} \Delta Q_n = 0$$

$$\sum_{n=1}^{Nb} \Delta Q_n - \sum_{n=1}^{Nb} \frac{\partial Q_L}{\partial P_n} \Delta P_n - \sum_{n=1}^{Nb} \frac{\partial Q_L}{\partial Q_n} \Delta Q_n = 0$$

All the variables in the cost function are described in [16].

III. APPLICATION OF UPFC

The scheme of unified power flow controller (UPFC) is shown in Fig. 1 and its circuit for modeling is sown in Fig. 2. Its modeling and simulation is described in [16].

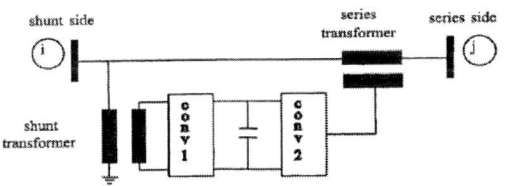

Fig. 1 Basic circuit arrangement

$$\overrightarrow{V_{vR}} = V_{vR}(\cos\theta_{vR} + j\sin\theta_{vR}) \qquad (2)$$

$$\overrightarrow{V_{cR}} = V_{cR}(\cos\theta_{cR} + j\sin\theta_{cR}) \qquad (3)$$

The voltage magnitude and current magnitude constraints should be considered when developing power flow models for the UPFC [11]. The active and reactive powers at bus i can are described in [12]:

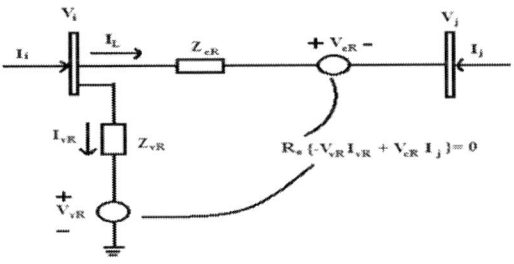

Fig. 2 Circuit for modeling of UPFC

IV. APPLICATION OF HVDC

A scheme of the HVDC-VSC and its equivalent circuit are shown in Fig. 3 and Fig. 4 respectively.

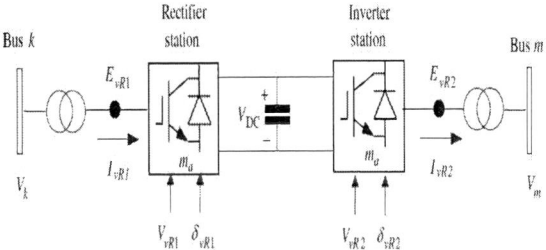

Fig. 3 HVDC -VSC

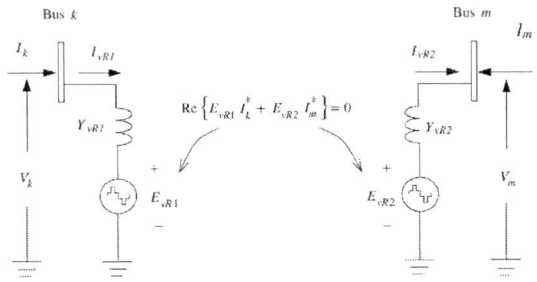

Fig. 4 Equivalent circuit of HVDC -VSC

The power equation for the back-to-back HVDC-VSC (i.e. $R_{DC} = 0$) is given by

$$Re\{-E_{vR1} I_{vR1}^* + E_{vR1}^\rho I_m^*\} = 0 \qquad (4)$$

The power equation when both voltage source converters stations are connected by a DC cable (i.e. $R_{DC} > 0$) is given by

$$Re\{-E_{vR1} I_{vR1}^* + E_{vR1}^\rho I_m^* + P_{DC,loss}\} = 0 \qquad (5)$$

If the power flow equations injected at bus k for the above case are:

$$P_k = V_k^2 G_{vR1} + V_k V_{vR1}[G_{vR1}\cos(\theta_k - \delta_{vR1}) + B_{vR1}\sin(\theta_k - \delta_{vR1})]$$

$$Q_k = V_k^2 B_{vR1} + V_k V_{vR1}[G_{vR1}\sin(\theta_k - \delta_{vR1}) - B_{vR1}\cos(\theta_k - \delta_{vR1})]$$

$$(6 \& 7)$$

The powers flowing into the rectifier are described by the following equations:

978-1-5386-6624-1/18 $31.00 © 2018 IEEE

$$P_{vR1} = V_{vR1}^2 G_{vR1} + V_{vR1}V_k[G_{vR1}\cos(\delta_{vR1} - \theta_k) + B_{vR1}\sin(\delta_{vR1} - \theta_k)]$$

$$Q_{vR1} = -V_{vR1}^2 B_{vR1} + V_{vR1}V_k[G_{vR1}\sin(\delta_{vR1} - \theta_k) - B_{vR1}\cos(\delta_{vR1} - \theta_k)]$$

(8 & 9)

For the case of the full HVDC-VSC, the relevant power equation is:

$$P_{vR1} + P_{vR2} + P_{DC} = 0$$

(10)

$$
\begin{bmatrix} \Delta P_k \\ \Delta Q_k \\ \Delta P_{vR} \\ \Delta Q_{vR} \\ \Delta P_{HVDC} \end{bmatrix} =
\begin{bmatrix}
\frac{\partial P_k}{\partial \theta_k} & \frac{\partial P_k}{\partial V_k}V_k & \frac{\partial P_k}{\partial \delta_{vR1}} & \frac{\partial P_k}{\partial V_{vR1}}V_{vR1} & 0 \\
\frac{\partial Q_k}{\partial \theta_k} & \frac{\partial Q_k}{\partial V_k}V_k & \frac{\partial Q_k}{\partial \delta_{vR1}} & \frac{\partial Q_k}{\partial V_{vR1}}V_{vR1} & 0 \\
\frac{\partial P_{vR1}}{\partial \theta_k} & \frac{\partial P_{vR1}}{\partial V_k}V_k & \frac{\partial P_{vR1}}{\partial \delta_{vR1}} & \frac{\partial P_{vR1}}{\partial V_{vR1}}V_{vR1} & 0 \\
\frac{\partial Q_{vR1}}{\partial \theta_k} & \frac{\partial Q_{vR1}}{\partial V_k}V_k & \frac{\partial Q_{vR1}}{\partial \delta_{vR1}} & \frac{\partial Q_{vR1}}{\partial V_{vR1}}V_{vR1} & 0 \\
\frac{\partial P_{HVDC}}{\partial \theta_k} & \frac{\partial P_{HVDC}}{\partial V_k}V_k & \frac{\partial P_{HVDC}}{\partial \delta_{vR1}} & \frac{\partial P_{HVDC}}{\partial V_{vR1}}V_{vR1} & \frac{\partial P_{HVDC}}{\partial \delta_{vR2}}
\end{bmatrix}
\begin{bmatrix} \Delta\theta_k \\ \frac{\Delta V_k}{V_k} \\ \Delta\delta_{vR1} \\ \frac{\Delta V_{vR1}}{V_{vR1}} \\ \Delta\delta_{vR2} \end{bmatrix},
$$

(11)

Where ΔP_{HVDC} is the active power flow mismatch for the DC link and is given by

$$\Delta P_{HVDC} = \Delta P_{vR1} - \Delta P_{vR2}$$

Notice that, since active power is regulated at the rectifier end, (i.e. $\Delta P_{vr1}^{spec} - \Delta P_{vR2}^{spec}$), the corresponding active power equations of the inverter become redundant (i.e. ΔP_{vR2} and ΔP_m) and are not used in Equation.

$$
[J_{hvdc}] =
\begin{bmatrix}
H_{ii} & N_{ii} & H_{ivR1} & N_{ivR1} & 0 \\
J_{ji} & L_{ii} & J_{ivR1} & L_{ivR1} & 0 \\
H_{vR1i} & N_{vR1i} & H_{vR1vR1} & N_{vR1vR1} & 0 \\
J_{vR1i} & L_{vR1i} & J_{vR1vR1} & L_{vR1vR1} & 0 \\
H_{ii} & N_{ii} & H_{ivR1} & N_{ivR1} & H_{ivR2}
\end{bmatrix}
\begin{Vmatrix} \Delta\theta_i \\ \Delta V_i/V_i \\ \Delta\delta_{vR1} \\ \Delta V_{vR1}/V_{vR2} \\ \Delta\delta_{vR2} \end{Vmatrix}
$$

V. APPLICATION OF IPFC

The Interline Power Flow controller (IPFC) scheme shown in Fig. 5 is used for compensating of transmission line it is described in [9].

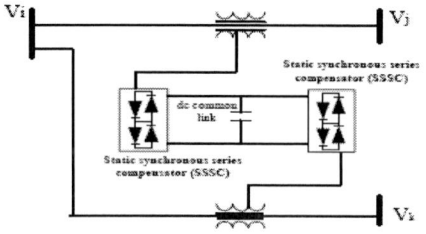

Fig. 5 IPFC schematic diagram

The simplest IPFC is as shown in Fig. 6. Both the systems have been considered identical.

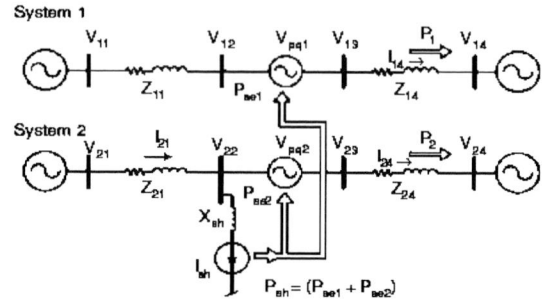

Fig. 6 IPFC DC-to-AC converters

VI. CASE STUDY FOR ACTIVE POWER AND REACTIVE POWER BIDDING

A power system network consisting of 15 generators and 97 lines, 24 transformers as given in [15, 16] is considered. The line limit and power flows in these lines are shown in Table 1.

TABLE I
NO. OF LINES WITH LINE RATING AND POWER FLOWS

No. of Lines	Line Limit (Considered)	Power Flow
6	3.6	3.6584
10	6.5	6.6804
20	17.0	18.3258
71	4.15	4.2831
81	3.91	4.2049
91	3.3	3.3838

Bids for rescheduling of active power and reactive power are taken from [15]. The model for reactive power bidding is given in [17]. The cost function comes out to be as follows:

Min CC= $4000*\Delta Pg,1^+ + 4000*\Delta Pg,12^+ + 5200*\Delta Pg,13^+ + 4000*\Delta Pg,14^+ - 1000*\Delta Pg,1^- - 2000*\Delta Pg,12^- - 2000*\Delta Pg,13^- - 900*\Delta Pg,14^- + 2000*\Delta Pg,20^+ + 1500*\Delta Qg,4^+ + 1000*\Delta Qg,5^+ + 1000*\Delta Qg,7^+ + 1300*\Delta Qg,8^+ + 1500*\Delta Qg,12^+ + 1300*\Delta Qg,15^+ - 650*\Delta Qg,4^- - 700 *\Delta Qg,5^- - 750*\Delta Qg,7^- - 700*\Delta Qg,8^- - 750*\Delta Qg,12^- - 700*\Delta Qg,15^-$

(13)

Another method of rescheduling is given in [18]

VII. RESULTS AFTER ESCHEDULING

The congestion cost comes out to be Rs. 59900/- after change in gencos bidding. NLDC provides the information of total drawl and injection in each of the regions after rescheduling. It may be seen that all of the lines are congestion free as shown in Fig. 7.

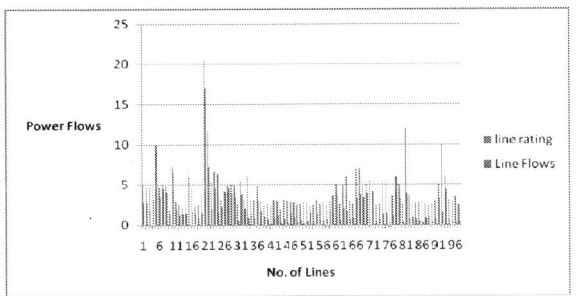

Fig. 7 Power Flow in Lines after Rescheduling

TABLE II
POWER FLOWS AFTER RESCHEDULING

No. of Lines	Line Limit (Considered)	Power Flow (Before Rescheduling)	Power Flow (After Rescheduling)
6	3.6	3.6584	3.5289
10	6.5	6.6804	6.4820
20	17.0	18.3258	16.9949
71	4.15	4.2831	4.0775
81	3.91	4.2049	3.8690
91	3.3	3.3838	3.2291

A. IMPACT OF HVDC ON CONGESTION MANAGEMENT

Line number 26 is obtained suitable for controlling the power flow in a line on the basis of the sensitivity factors. Sensitivity factors are obtained with respect of injected series voltage magnitude and phase angle of HVDC placed in the line. Again the bid structure for active and reactive power is considered as in base case. The congestion cost for the case study (with active and reactive power bidding), with and without HVDC came out to be 54315/-.

TABLE III
POWER FLOWS OF 6 LINES CONSIDERD TO BE CONGESTED AFTER HVDC RESCHEDULING

No. of Lines	Line Limit (Considered)	Power Flow(without HVDC and without Rescheduling)	Power Flow (After HVDC placement)
6	3.6	3.6584	3.0267
10	6.5	6.6804	5.5240
20	17.0	18.3258	14.1007
71	4.15	4.2831	3.2386
81	3.91	4.2049	3.7448

| 91 | 3.3 | 3.3838 | 3.4588 |

As initially six lines were considered to be congested. Fig. 8 below shows the power flows in all the 97 lines.

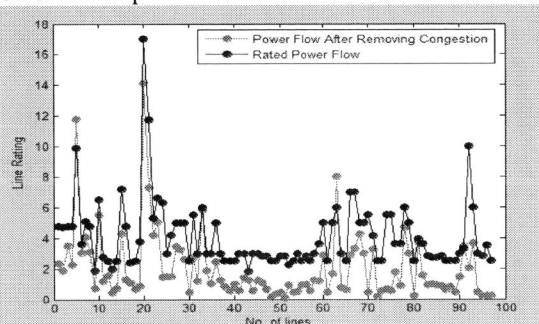

Fig. 8 Power Flow in Lines after HVDC Rescheduling

B. IMPACT OF UPFC ON CONGESTION MANAGEMENT

Considering bus number 26 for controlling the power flow, the congestion cost for the case study (with active and reactive power bidding), with and without UPFC came out to be 48844/-. Fig. 9 below shows the power flows in all the 97 lines in this case.

TABLE IV
POWER FLOWS OF 6 LINES CONSIDERD TO BE CONGESTED IN UPFC

No. of Lines	Line Limit (Considered)	Power Flow	HVDC Power Flow
6	3.6	3.6584	2.4033
10	6.5	6.6804	4.3652
20	17.0	18.3258	10.6477
71	4.15	4.2831	2.4742
81	3.91	4.2049	2.8592
91	3.3	3.3838	2.6545

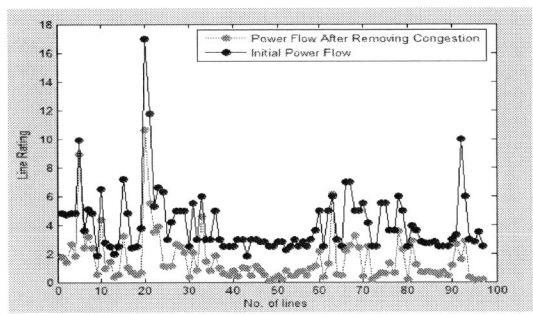

978-1-5386-6624-1/18 $31.00 © 2018 IEEE

Fig. 9 Power Flow in Lines after UPFC Rescheduling

C. IMPACT OF IPFC ON CONGESTION MANAGEMENT

Again by considering line number 26 for controlling the power flow, the congestion cost for the case study (with active and reactive power bidding), with and without IPFC came out to be 47218/-. Fig. 10 below shows the power flows in all the 97 lines.

TABLE V
POWER FLOWS OF SIX LINES CONSIDERED TO BE CONGESTED IN IPFC

No. of Lines	Line Limit (Considered)	Power Flow	Power Flow (After applying IPFC)
6	3.6	3.6584	2.3575
10	6.5	6.6804	4.2916
20	17.0	18.3258	10.7210
71	4.15	4.2831	2.4680
81	3.91	4.2049	2.8420
91	3.3	3.3838	2.6485

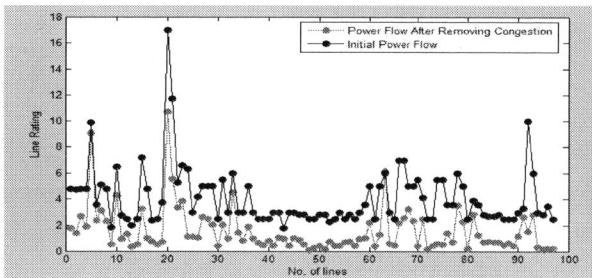

Fig. 10 Power flows after removing congestion in case of IPFC

D. POWER FLOW COMPARISON OF ALL THE DEVICES

Power flow comparison in following cases i.e. power flow in the system when no device is used, power flow with application of UPFC, power flow with application of HVDC, power flow with application of IPFC is done as shown in the Fig. 11.

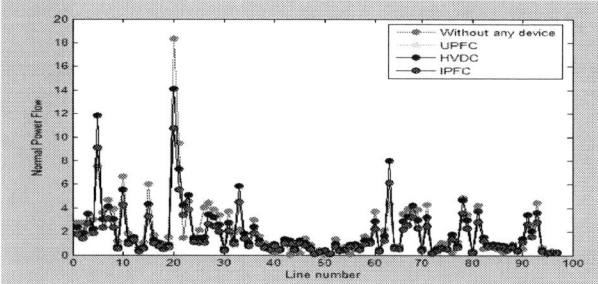

Fig. 11 Power Flow comparison of all the devices

E. CONGESTION COST COMPARISON

Cost comparison of all the devices is shown the figure below. From the Fig. 12 it can be inference that IPFC is the most effective device in reducing the congestion cost.

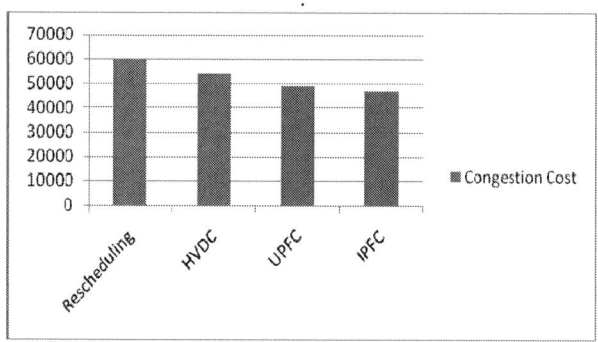

Fig. 12 Cost comparison

VIII. CONCLUSION

This paper presents congestion management using active and reactive power rescheduling and application of FACTS devices. The OPF is performed for optimizing the congestion cost using GAMS software. The studies have been carried out for multiple line congestion case. By rescheduling congestion is relieved. Then FACTS devices UPFC, IPFC and HVDC have been taken into account for the congestion management and rescheduling is carried out to determine the congestion cost in each case. Then cost effective analysis and power flows comparison is done so as to determine which device is most effective for congestion management. IPFC gives comparatively better results than UPFC and HVDC.

REFERENCES

[1] R D Christie B.F. Wollenberg I.Wangensteen, "Transmission management in the deregulated environment", in Proc.IEEE pp.170–194,"2000.

[2] M.R. Iravani and Navabi Naiki, "Steady state dynamic models for UPFC for power system studies", IEEE Transactions on Power Systems, Vol. 11, No. 4, November 1996.

[3] R.S.Fang, A.K.David, "Transmission congestion management in an electricity market", IEEE Transactions on Power System, vol.14. no.3, pp. 877–883, 1999.

[4] M. Noroozian, L Angquist, "Use of UPFC for optimal power flow control", IEEE Transactions on Power Delivery, Vol. 12, No. 4, October 1997.

[5] C.R. Fuerte-Esquivel, E.Acha and H.Ambriz-Perez "A comprehensive Newton- Raphson UPFC model for the quadratic

power flow solution of practical power networks," *IEEE trans. on Power Systems, vol. 15. no.1, Feb 2000.*

[6] Debabrata Chattopadhyay, "*Application of General Algebraic Modeling System to Power System optimization*". *IEEE Transactions on Power Systems, Vol. 14, No. 1, February 1999.*

[7] Suman Bhowmick, Biswarup Das, *Senior Member, IEEE,* and Narendra Kumar, "*An advanced IPFC Model to Reuse Newton Power Flow Codes*" *IEEE Trans on Power Systems , VOL. 24, NO. 2, MAY 2009.*

[8] S. K. Gupta, Richa Bansal, "*ATC in Competitive Electricity Market Using TCSC*", International Journal of Electrical, *Electronic Science and Engineering, Vol. 8, No: 2, 2014.*

[9] S. Charles Raja, Member IEEE, S.A.Waajitha Banu and Dr. P. Venkatesh, "Congestion *Management Using GAMS/CONOPT Solver*", *IEEE- International Conference On Advances In Engineering, Science And Management (ICAESM -2012) March 30, 31, 2012.*

[10] Understanding FACTS. *Concept and Technology of Flexible AC Transmission System.* Gyugyi L. and Hingorani NG. *Piscataway,* NJ: *IEEE Press 2001.*

[11] Gyugyi, L., et al., "*The Unified Power Flow Controller: A New Approach to Power Transmission Control,*" *IEEE Trans. on Power Deliuery, vol. 10, no. 2, April 1995.*

[12] C.R. Fuerte-Esquivel, E.Acha and H.Ambriz-Perez "*A comprehensive Newton- Raphson UPFC model for the quadratic*

power flow solution of practical power networks," *IEEE trans. on Power Systems, vol. 15. no.1, Feb 2000.*

[13] S.N. Singh and I.Erlich, "*Locating Unified power Flow Controller for enhancing power system Loadability,*" *IEEE Trans. on Power Systems, Nov. 2011.*

[14] Modeling and Simulation in Power Networks. *Enrique Acha, Claudio R. Feurte Esquivel, Hugo Embriz- Perez, Cezer Angeles-Camacho, 2004.*

[15] S K Gupta, Rich Bansal, Pratibha and Muksh Saini, "Energy Trading And Congestion Management using Real and Reactive Power Rescheduling and Load Curtailment," Review of energy Technology and Policy Research, 2014, 1(3):28-41.

[16] S K Gupta, Rich Bansal, Pratibha and Muksh Saini, " Power Trading And Congestion management using Real Power Rescheduling using UPFC," International Journal of Electrical Power Systems: Analysis, Security and Deregulation: Volume1, issue1, pp8.

[17] S.K. Gupta, Ratneshawram and Diksha Gupta. "Reactive Power pricing in Deregulated environment." Trend in Electrical Engineering, ISSN: 2249-4774 (online), ISSN: 2321-4260(print), Volume 6.

[18] Saurabh Yadav and S K Gupta, " Optimal Load Dispatch using Flower Pollination Algorithm Technique", i-Manager's Journal on Circuits and Systems, Vol. 5, No. 3, August- October 2017,ISSN Print: 2321-7502, ISSN Online: 2322-035X

Pitch angle Control of Variable Speed Wind Turbine by using Fuzzy Logic

Akshiv Kumar and Bhavnesh Kumar
Division of Instrumentation & Control Engineering
Netaji Subhas Institute of Technology
New Delhi, INDIA
akshivkumar1@yahoo.com , kumar_bhavnesh@yahoo.co.in

Abstract— Wind energy conversion systems have to handle the variations in the wind speed for efficient energy conversion. This paper presents an analysis made on the controller effectiveness system for pitch angle control of a wind turbine which changes performance of the wind energy conversion system. A fuzzy logic controller is proposed in place of classical PI controller for controlling the pitch angle to have maximum available power from the wind energy conversion system and the comparisons are made between PI controller, fractional order PI controller and fuzzy logic controller. The system is tested for different realistic environmental conditions using MATLAB/Simulink Software.

Keywords—fuzzy logic, wind turbine, doubly fed induction generator

I. INTRODUCTION

Due to limited stock of fossil fuels and their degradation because of their combustion, the researchers are compelled to explore the renewable and non-polluting electricity generation systems [1]. Wind energy conversion system (WECS) is one of the most prominent electrical energy generation system since its more eco-friendly. These systems use kinetic energy of the wind and convert it into electrical energy by means of set of turbine and electrical generator [2]. Doubly fed induction generator with appropriate convertor topology is more commonly used in these systems [3]. The output power from WECS depends upon the speed of wind, which varies with time and so is the power. Therefore, to maximize the power output, power coefficient needs to be maximized. To maximize the power coefficient either rotor speed can be controlled or pitch angle of the blades can be controlled.

Wind turbine operates in any of the four regions depending upon the wind speed as shown in Fig. 1. Pitch angle is usually controlled by an appropriate mechanism to get maximum output when the speed of wind is varying in nature. Also, this prevent mechanical power beating the design limits i.e. when wind speed is higher than that of rated then it makes effective regulation of wind power and rotor and also minimize the fatigue loads produced from mechanical parts of wind turbine [4], [5].

In order to have effective control over pitch angle various controllers have been applied in WECS [6-9]. The existing PI control approach is not that much robust to parameter variations and operating condition changes but fractional order PI controller makes it robust to gain changes, it also reduces overshoots. But, fractional order controller is complex in nature and requires high online calculation. In addition, these controllers require exact mathematical model of the WECS for high performance. Fuzzy logic based controller can handle the nonlinearities effectively and does not require mathematical model [10], [11].

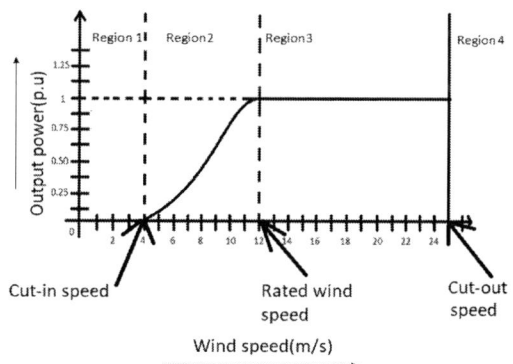

Fig. 1. Power curve of a WECS

This paper describes about comparison between three different control techniques i.e. traditional PI controller, fractional order PI controller and fuzzy logic controller and shows which one of them is giving better results to maximize electrical power output from wind energy conversion system.

II. SYSTEM MODELLING

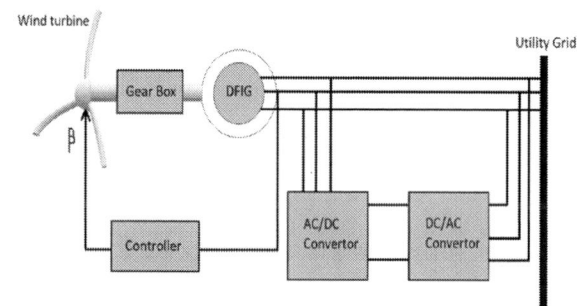

Fig. 2. Basic block diagram of WECS

Wind energy conversion system is available in different type of topologies. The most commonly used type topology is doubly-fed induction generator configuration which is shown in Fig. 2 it is considered in this research. The wind energy conversion system is divided in four major parts:-

a) Aerodynamics

For a given coefficient of power (C_p), electrical power (P_e) and Mechanical power (P_m) are related as

$$P_e = C_p P_m \tag{1}$$

Coefficient (C_p) tells us about the extent to which mechanical power of turbine is converted into electrical power by the generator. The obtained aerodynamic power from the wind turbine can be expressed as:

$$P_w = \frac{1}{2} C_p (\lambda, \beta) R^2 \rho V^3 \pi \tag{2}$$

where: ρ is density of air, v is wind speed and R is radius of blade of wind turbine.

The power coefficient (C_p) is dependent on tip-speed ratio (λ) and blade pitch angle (β). The tip-speed ratio (λ) can be expressed as:

$$\lambda = \frac{w_r R}{v} \tag{3}$$

Therefore, the power coefficient (C_p) can be expressed as:

$$C_p (\lambda, \beta) = 0.22 \left(116 \lambda_i - 0.4\beta - 5\right) e^{\frac{-12.5}{\lambda_i}} \tag{4}$$

Where,

$$\frac{1}{\lambda_i} = \frac{1}{\lambda + 0.08\beta} - \frac{0.035}{\beta^3 + 1} \tag{5}$$

b) Drive train model:

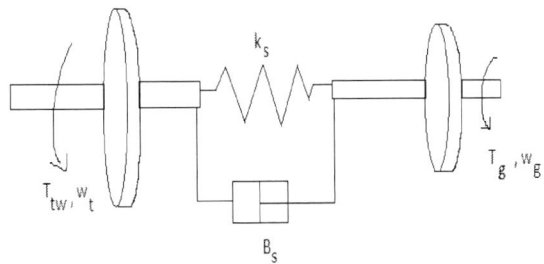

Fig. 3. Drive train model

A flexible shaft representation of two mass is considered for model of drive train as shown in Fig. 3. The following mathematical equations are used to model it:

$$\frac{dw_t}{dt} = -\frac{i}{j_t} T_{tw} + \frac{1}{j_t} T_t \tag{6}$$

$$\frac{dw_g}{dt} = \frac{1}{j_g} T_{tw} - \frac{1}{j_g} T_g \tag{7}$$

$$\frac{dT_{tw}}{dt} = k_s i w_t - k_s w_g - \left(\frac{i^2 B_s}{j_t} + \frac{B_s}{j_g}\right) T_{tw} + \frac{i B_s}{j_t} T_t + \frac{B_s}{j_g} T_g \tag{8}$$

The flexible shaft T_{tw} is experiencing a drive train torsional torque which can be defined as:-

$$T_{tw} = k_s \theta_{tw} + B_s \left(i w_t - w_g\right) \tag{9}$$

Here j_t and j_g are respectively the inertia of turbine and generator; shaft twist angle is expressed as θ_{tw}; gear ratio is i; k_s, B_s are respectively the shaft stiffness and damping coefficient.

c) Generator:-

Doubly-fed induction generator (DFIG) is considered in this research. The modelling of DFIG's in synchronous rotating frame (d-q) of reference with that of rotor (dr-qr) circuit shown in Fig. 4 (a) and 4 (b), could be done by using two phase of stator (ds-qs) variable.

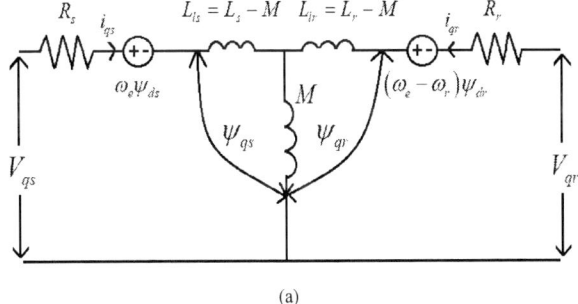

(a)

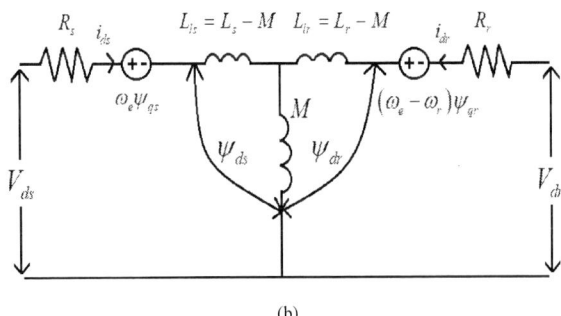

(b)

Fig. 4. DFIG in synchronous equivalent circuit (a) q- axis (b) d- axis

978-1-5386-6624-1/18 $31.00 © 2018 IEEE

The voltage equations of stator side and the rotor side can be expressed as:

$$V_{ds} = R_s i_{ds} - w_s \phi_{qs} \qquad (10)$$

$$V_{qs} = R_s i_{qs} - w_s \phi_{ds} \qquad (11)$$

$$V_{dr} = R_r i_{dr} - s w_s \phi_{qr} + \frac{d\phi_{dr}}{dt} \qquad (12)$$

$$V_{qr} = R_r i_{qr} - s w_s \phi_{dr} + \frac{d\phi_{qr}}{dt} \qquad (13)$$

Where R_r, R_s, l_r and l_s are the respective rotor and stator winding resistances and inductances, main inductance is l_m and the rotor speed is,

$w = P\Omega_{mec}$, $i_{ds}, i_{qs}, i_{dr}, i_{qr}, V_{ds}, V_{qs}, V_{dr}, V_{qr}, \phi_{ds}, \phi_{qs}, \phi_{dr}, \phi_{qr}$ are representing direct and quadrate components of space phasor of respective stator and rotor currents, voltages and flux.

The stator side active and reactive power can be expressed as:

$$P_s = V_{ds} i_{ds} + V_{qs} i_{qs} \qquad (14)$$

$$Q_s = V_{qs} i_{ds} - V_{ds} i_{qs} \qquad (15)$$

The rotor side active and reactive powers are defined as:-

$$P_r = V_{dr} i_{dr} + V_{qr} i_{qr} \qquad (16)$$

$$Q_r = V_{qr} i_{dr} - V_{dr} i_{qr} \qquad (17)$$

The electromagnetic torque developed can be expressed as:

$$T_e = \frac{3P}{4} \left(i_{qs} \phi_{ds} - i_{ds} \phi_{qs} \right) \qquad (18)$$

(d) Controller:

WECS uses wind having complex mass dynamics, random nature of wind and generator is non-linear in nature therefore its mathematical model is of complex nature. In this work, controller is used to control the pitch angle of the blade of the turbine to get maximum power. Effectiveness of WECS system is greatly influenced by the performance of controller.

III. FUZZY LOGIC CONTROLLER

In this research, commonly used configuration of Fuzzy logic controller with two inputs and single output is used. The inputs are the difference in power error and the rate of change of this difference and output is pitch. Membership functions used for them are triangular as shown in Fig. 5, it uses rule base given in table 1 for fuzzifying inputs and to get corresponding defuzzified output for crisp value. This controller is used to replace standard PI controller to control power using pitch.

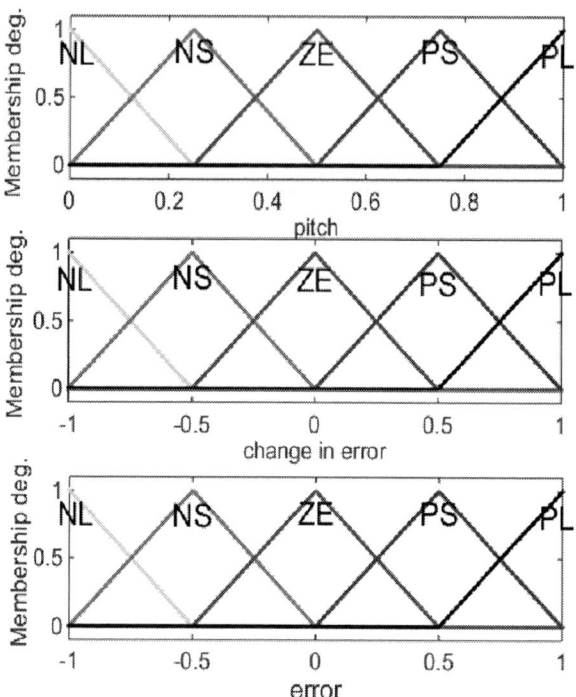

Fig. 5. Input and output membership functions of fuzzy logic controller

Following are Linguistic variables for respective inputs and output used were: NS - Negative Small, NL - Negative Large, ZE - Zero, PS - Positive Small and PL - Positive Large.

Table .1. Rule Base for Fuzzy Logic Controller

		Difference in Power (ΔP)				
		NL	NS	ZE	PS	PL
Rate of change of ΔP	NL	NL	NL	NL	NS	PS
	NS	NL	NL	NS	ZE	PS
	ZE	NL	NS	ZE	PS	PL
	PS	NS	ZE	PS	PL	PL
	PL	NS	PS	PL	PL	PL

IV. SIMULATION RESULTS & DISCUSSION

The turbine optimum speed is in proportion with wind speed. Optimum turbine speed is responsible for production of maximum mechanical energy at a certain speed of wind. To achieve subsyncronous speed of rotor, the wind speed should be less than 10 m/s. If wind speed is high magnitude than 10 m/s then rotor will attain hypersynchronous speed. This model assumes steady state initially and starts to work then.

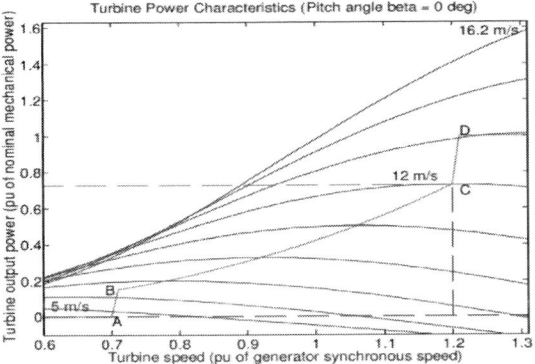

Fig. 6. Wind power characteristics

The mechanical power vs. speed of turbine is shown for a varying wind speed between 5 m/s to 16.2 m/s as shown in Fig. 6. The DFIG gets under control so that it follows curve of red colour on this graph. Turbine speed is optimized between points B and C on this graph. From the developed simulation model, results are obtained with the following controllers:

A. Proportional Integral (PI) controller

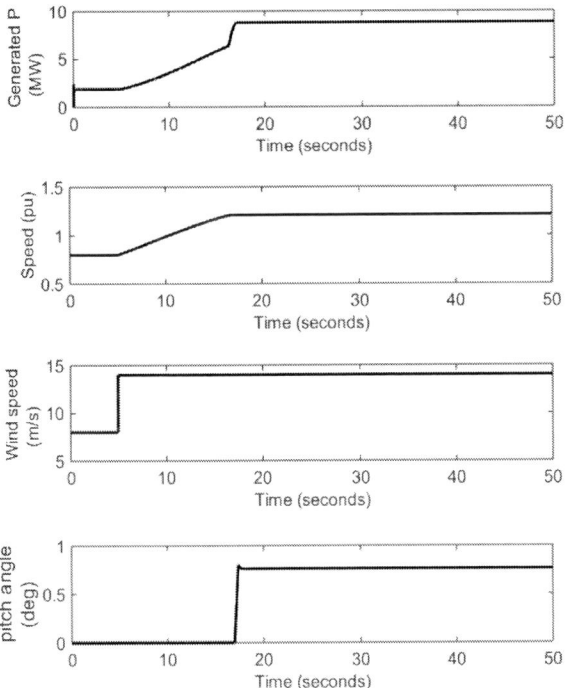

Fig. 7 Response of power, speed, pitch angle with PI controller

Responses of WCES with PI controller for generated power, rotor speed and pitch angle variation for step variation in wind

speed are shown in Fig. 7. Since, there is a step change of wind speed from 8 m/s to 14 m/s at time t = 5 seconds, we can observe that power curve is trying to follow the red curve of Fig. 6. Also variation in pitch angle is observed when rotor speed achieves the final value to protect the turbine from any damage.

B. Fractional order PI controller

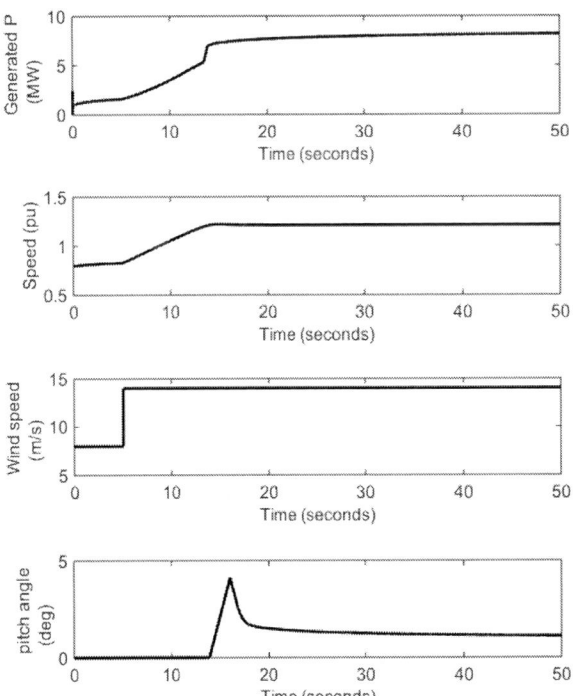

Fig. 8. Response of power, speed, pitch angle with fractional order PI controller

For same step variation in wind speed, responses of WCES with Fractional order PI controller for generated power, rotor speed and pitch angle variation for step variation in wind speed are shown in Fig. 8. From the results, it is observed that it gives better rise time, reduced peak overshoot of power, decreases settling time to increase stability by increasing its pitch peak overshoot and decreases pitch quickly to attain rated power quickly thereby settling time is reduced of pitch.

C. Fuzzy Logic Controller

WCES is tested again for similar step change in wind speed and the responses were recorded. Responses obtained with fuzzy logic controller for the said operating condition are shown in Fig. 9. It is observed that fuzzy logic controller is also making rotor speed linearly increase, giving lesser rise time, reduced peak overshoot of power, decreased settling time and also decreases pitch quickly to attain maximum power quickly thereby settling time is reduced of pitch and also removes nonlinearities apart from the increase in delay time as its drawback faced compared to other controllers used in this work.

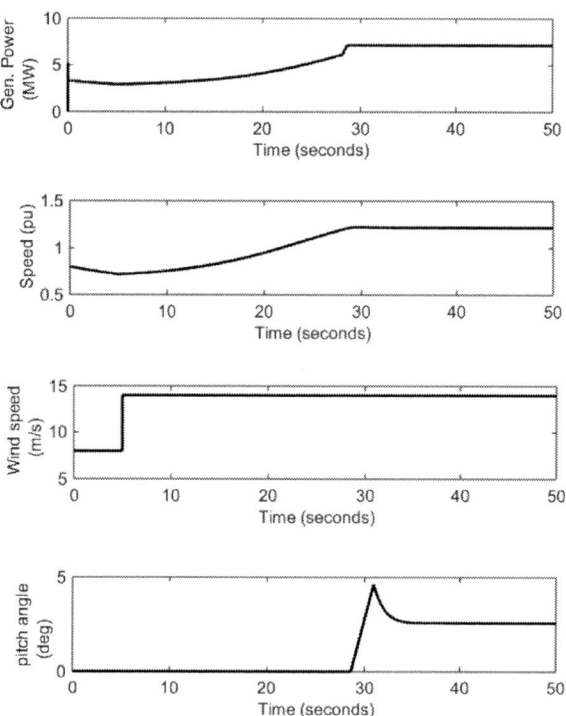

Fig. 9. Response of power, speed, pitch angle with fuzzy logic controller

V. CONCLUSION

In this paper, study on the impact of controller performance in controlling the pitch angle of a wind energy conversion system is presented. Three different controllers that are traditional Proportional- Integral (PI) controller, fractional order PI controller and fuzzy logic controller are used for pitch angle control. Fuzzy logic controller is designed with Mamdani type inference engine utilizing 25 rules for decision-making. Fuzzy logic controller does show better performance in comparison to classical PI Controller and fractional order PI controller on the operating condition considered.

REFERENCE

[1] M. Elnaggar, M. S. Saad, H. A. Abdel Fattah and A. L. Elshafei, "L₁ adaptive fuzzy control of wind energy conversion systems via variable structure adaptation for all wind speed regions," in *IET Renewable Power Generation*, vol. 12, no. 1, pp. 18-27, 1 8 2018.

[2] T. L. Van, T. H. Nguyen and D. C. Lee, "Advanced Pitch Angle Control Based on Fuzzy Logic for Variable-Speed Wind Turbine Systems," in *IEEE Transactions on Energy Conversion*, vol. 30, no. 2, pp. 578-587, June 2015.

[3] M. Ben Smida and A. Sakly, "Different conventional strategies of pitch angle control for variable speed wind turbines," *2014 15th International Conference on Sciences and Techniques of Automatic Control and Computer Engineering (STA)*, Hammamet, 2014, pp. 803-808.

[4] S. Ullah, I. M. Qureshi and M. M. Zohaib, "Design of an Artificial-Intelligence based controller for doubly fed induction generator based wind energy conversion system," *2016 International Conference on Intelligent Systems Engineering (ICISE)*, Islamabad, 2016, pp. 318-325.

[5] M. A. Khan, K. V. S. Bharath, R. Singh, A. Gupta and S. S. Dehury, "Intelligent pitch angle control for wind-doubly fed induction generator system," *2016 7th India International Conference on Power Electronics (IICPE)*, Patiala, 2016, pp. 1-5.

[6] S. Sahoo, B. Subudhi and G. Panda, "Pitch Angle Control for Variable Speed Wind Turbine Using Fuzzy Logic," *2016 International Conference on Information Technology (ICIT)*, Bhubaneswar, 2016, pp. 28-32.

[7] A. Tepljakov, E. Petlenkov, J. Belikov and J. Finajev, "Fractional-order controller design and digital implementation using FOMCON toolbox for MATLAB," *2013 IEEE Conference on Computer Aided Control System Design (CACSD)*, Hyderabad, 2013, pp. 340-345.

[8] N. A. Schinas, N. A. Vovos and G. B. Giannakopoulos, "An Autonomous System Supplied Only by a Pitch-Controlled Variable-Speed Wind Turbine," in *IEEE Transactions on Energy Conversion*, vol. 22, no. 2, pp. 325-331, June 2007.

[9] H. Afghoul, F. Krim, A. Beddar and M. Houabes, "Switched fractional order controller for grid connected wind energy conversion system," *2017 5th International Conference on Electrical Engineering - Boumerdes (ICEE-B)*, Boumerdes, 2017, pp. 1-5.

[10] C. A. M. Amandola and D. P. Gonzaga, "Fuzzy-Logic Control System of a Variable-Speed Variable-Pitch Wind-Turbine and a Double-Fed Induction Generator," *Seventh International Conference on Intelligent Systems Design and Applications (ISDA 2007)*, Rio de Janeiro, 2007, pp. 252-257.

[11] P. Gaur, D. Pathak, B. Kumar and Y. K. Chauhan, "PI and fuzzy logic controller based tip speed ratio control for smoothening of output power fluctuation in a wind energy conversion system," *2016 7th India International Conference on Power Electronics (IICPE)*, Patiala, 2016, pp. 1-6.

Transmission Loss Allocation in Restructured Power system with Optimization Loss Criterion

Naimul Hasan
Department of Electrical Engineering
Qassim University
Buraydah, KSA
naimul_hasan@rediffmail.com

Ibraheem
Department of Electrical Engineering
Qassim University
Buraydah, KSA
ibraheem_2k@yahoo.com

Yudhishthir Pandey
Department of Electrical Engineering
Jamia Millia Islamia
New Delhi, India
yudhishthir.pandey@gmail.com

Abstract— This paper presents an innovative algorithm for transmission loss allocation in restructured power system. Algorithm utilizes the routing algorithm with minimum transmission loss path. This algorithm solves ac power flow then based upon the loss occurring in branches allocates losses to generators and loads. In addition, this algorithm identifies optimal location of DG placement, causing reduction in network losses. Simulation on 6-bus system network proves significant as compared to other existing methods. Results show significant improvement in reducing losses.

Keywords—loss allocation, Routing algorithm, Loss reduction, optimal DG location

I. INTRODUCTION

Restructured power system has given the opportunity to address the issues of market participants in each aspect like fairness and competitive environment and also transmission loss allocation to make power system operation maximally fair. Electrical power systems have been deregulated to make the system more competitive and fair. The competition has led to the evolution of many new factors in generation, transmission and distribution side. Transmission loss is only 3-5% of total generation. But this small percentage becomes a big concern as we see this amount of transmission loss in terms of MW. However, this loss cannot be wiped out. In addition to the concern for the transmission loss, identification of the factors responsible and their contributions is our next concern. Transmission loss allocation algorithm allocates the transmission loss to various participants (generators and loads).

The methods reported in [1-3] on transmission loss allocation are widely accepted and implemented. Power loss-apportioning algorithm depending on proportionality of load flow in transmission line from generator to load is proposed in [1]. In [2] author presented a new procedure of transmission loss allocation in pool operated power network. This new procedure for loss allocation using Z-bus of the network. Research paper [3] compared different algorithms of transmission loss allocation 1) pro rata 2) marginal allocation [5] 3) unsubsidized marginal allocation 4) proportional sharing [1]. [1-3] solved power flow of the power system network then transmission loss allocation methods are applied. No power flow variables (bus voltage, bus angle, active, reactive power flows through lines) change with various transmission line allocation methods. Whereas in [4] transmission cost allocated to each individual generator and load proportional to their power flow and inducted the concept of EBE (equivalent bilateral exchange). The research paper [6] allocated 50-50% losses among generator and load with logic of equal use of the

network, so both should pay for losses. But if the loss has been allocated to generator side, the generator will account pay for losses and it will affect the incremental cost too. Ultimately, the load will have to pay more. Few of the deregulated market like mainland Spain allocates 100% of transmission losses to load side.

Transmission loss has been allocated among various market participants using cooperative game theory, and Shapley value method [6-9], artificial neural network [10-11]. In [6], active and reactive power losses have been allocated in two-step process considering generators as current source in step one and in second step modelling load as current source. Properties of Aumann-shapely ensures equitable allocation and solutions are intitutively considered equitable. However, Aumann-shapely involves combinatorial procedure hence problem size increases exponentially with number of agents. Artificial neural network (ANN) in order to address 1) time changing load variation causing changes in loss allocation 2) game theory is exhaustive for large number of transactions so ANN is better for large number of transactions 3) because of its inherent classification ability, neural network can be trained to generate the allocation, provided a pool of training data is available [11]. But neural network has drawback of black box nature, proness to overfitting, training and detraining is difficult task.

This paper proposed a transmission loss allocation by tracing the power flow path from generator to load (routing) using an algorithm of the minimum loss path. Minimum loss path is by virtue of the nature and supported by two reasons 1) current flow is directly proportional to cross section area of conductor 2) current follows a minimum resistance path. This method will have advantages of 1) locational dependency of transmission loss allocated to load and generator 2) reduced complexity of loss calculation.

Proposed method involves probabilistic approach in routing current flow with the criterion of minimum loss path. This paper will also consider zero loss to loads located at the same bus with generators. In addition to transmission loss allocation, paper suggests the idea of reducing losses of transmission network by reducing large numbers of paths of current or power flow in a network by placing generators of small capacities at various locations rather than installing large bulk power capacity generator at one or two locations for the entire network. This will not only reduce transmission loss, but will also improve power quality (voltage magnitude, frequency, waveform and reliability) of the network along with congestion. Locating generators at various locations can be done only at time of planning an electrical power system network. For existing network, the concept supports employment of distributed generation.

978-1-5386-6624-1/18 $31.00 © 2018 IEEE

II. PHYSICAL FLOW BASED LOSS ALLOCATION

Reference [12] ascertain the loss allocation based on physical power flow through the lines. Each load demand shares the line flows. This knowledge is used as basis for transmission loss allocation. This algorithm formulates allocation with two assumptions, first, currents flowing through a transmission line occupies the cross-section area of transmission line in such a way that the currents causes the least possible transmission losses. Second, if a dc source supplies two currents I_1 & I_2, cross-section area of conductor will be shared by two currents such that

$$\frac{I_1}{I_2} = \frac{a_1}{a_2} \tag{1}$$

If m different loads power flows through a transmission line, losses caused by the jth load can be calculated as follows:

$$P_{Loss_j} = P_{Loss_L} * \frac{S_j . S_{Line}}{\left| S_{Line} \right|^2} \tag{2}$$

Where, S_j is the complex power of jth load, S_{Line} is total complex power through the line and P_{Loss_L} is total active power loss of the line.

III. ROUTING ALGORITHM

In a power system network with n buses and m lines connecting buses, generators and loads. When a generator supplies power to various loads through lines, based on these observations,

Routing decision variable

$$x_{ij}^k \in \{0,1\}, \forall k \in D, \forall (i,j) \in l \tag{3}$$

The variable x_{ij}^k is equal to 1 if and only if the routing path of demand traverses the line between bus i & j. The number of this set of variables is D*L.

IV. PROPOSED METHOD

This paper presents the transmission loss allocation between the participants of the power system network, that is the generator and the load, using the criterion of minimum transmission loss path [15] unless there is congestion in any individual line. This congestion can be managed by putting extra cost. Tracing of electrical power flow and transmission loss allocation would have been become easier if we could have colored the electrical power of each generator with various colors or could have changed one parameter of each generator to identify power of each generator (of course, it should be intrusive to the power system operation) to various loads with transmission line path tracking. Few researchers tried on the same line, but could not address important issues. Out of the many, one major issue was the locational dependence of transmission loss allocation. But this method of routing power flow for minimum transmission loss addresses this issue. In this method we will explore all possible paths from one generator to various loads. For an individual load, we will track all possibilities of the transmission line path. Thereafter, losses evaluated for each possible path. But final path track from generator to load will be one which will have the least loss.

Routing decision variable

$$x_{ij}^k \in \{0,1\}, \forall k \in D, \forall (i,j) \in l$$

Variable x_{ij}^k is equal to 1 if and only if routing path of demand traverses line between bus i & j to supply demand k.

Flow conservation constraints (Kirchhoff's law):

$$P_G = P_D + P_{losses} \tag{4}$$

At a bus,

Sum of incoming flow = sum of outgoing flow

$$\sum P_{in} = \sum P_{out} \tag{5}$$

Transmission line capacity constraints,

$$P_{l,\min} \le P_l \le P_{l,\max} \tag{6}$$

Minimum loss constraint,

For power flow from any generator to load, sum of losses occurring through various transmission line links should be least as compare to any other possible transmission links from the same generator to same load.

$$P_{L(G_i toD_i)} = \min\{ \sum_{m,n \in \{L\}} P_{L,mn} * x_{mn} \} \tag{7}$$

Where, L is the set of branches traversed to supply power contribution from G_i to D_i.

Transmission loss can be calculated by various methods like power flows entering and leaving the transmission line, using equation (6). Either of the methods can be used to calculate transmission loss,

$$P_L = \sum_{m=1}^{n_g} \sum_{n=1}^{n_g} x_{mn}(P_m * B_{mn} * P_n) \tag{8}$$

The resulting complete algorithm presented as

Optimize (7)

Subject to (3), (4), (5) & (6)

V. SIMULATIONS AND RESULTS

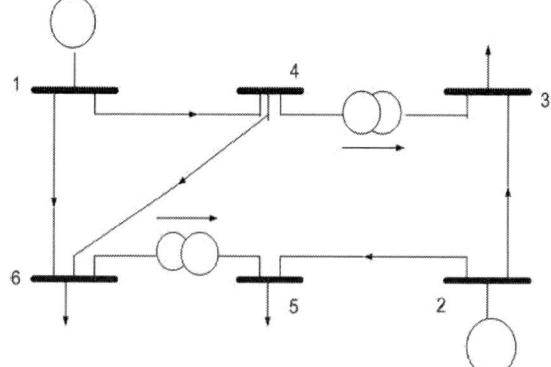

Fig.1. Six-bus system

TABLE-I	BUS DATA FOR FIG. 1 NETWORK			
Bus no	$V(pu)$	$P_G(MW)$	$P_L(MW)$	$Q_L(MW)$
1	1.05	-	0	0
2	1.1	50	0	0
3	-	0	55	13
4	-	0	0	0
5	-	0	30	18
6	-	0	50	10

TABLE-II		LINE DATA FOR FIG. 1 NETWORK		
Line		$R(pu)$	$X(pu)$	$Y_{sh}(pu)$
From	To			
1	4	0.080	0.37	0.014
1	6	0.123	0.518	
2	3	0.723	1.050	
2	5	0.282	0.64	
4	6	0.097	0.407	0.015

TABLE-III		DATA FOR TRANSFORMER OF FIG. 1 NETWORK	
Line		$X(pu)$	Tap
From	To		
4	3	0.133	0.90909
6	5	0.300	0.97650

Results of load flow data shown in table no 4 and 5.

TABLE-IV	BUS VOLTAGES			
Bus no	$V(pu)$	$\delta(degree)$	$P_G - P_L$	$Q_G - Q_L$
1	1.05	0.00	95.45	44.71
2	1.15	-5.14	50.00	29.00
3	1.01	-12.75	-55	-13
4	0.94	-9.87	0	0
5	0.94	-12.48	-30	-18
6	0.92	-12.28	-50	-10

TABLE-V		LINE FLOWS AND POWER LOSSES					
Line		Sending end		Receiving end		Losses	
From	To	P	Q	P	Q	P	Q
1	4	51.01	25.22	-48.66	-14.35	2.35	10.86
1	6	44.44	20.25	-41.78	-9.04	2.66	11.20
2	3	17.41	3.962	-15.67	-1.42	1.74	2.53
2	5	32.58	25.04	-28.98	-16.87	3.60	8.17
4	6	9.33	1.88	-9.23	-1.46	0.10	0.42
4	3	39.32	-58.7	-39.32	65.62	0.00	6.89
6	5	1.01	-6.08	-1.019	6.21	0.00	0.13

TABLE-VI	LOSS ALLOCATIONS
Proposed Method	On physical basis
To generators (Capacity)	To loads (Capacity)
5.111(95.457)	3.853(55.00)
5.346(50.00)	3.638(30.00)
	3.253(50.00)

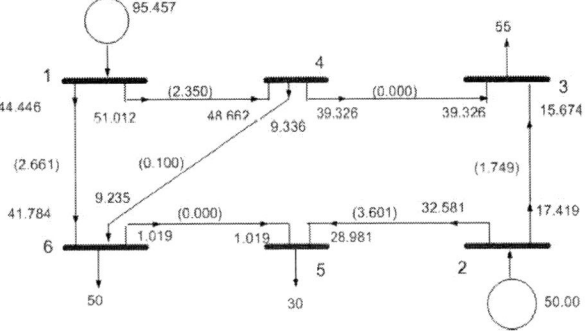

Fig.2. Six- bus system with line power flow and losses

Table-6 shows comparison of losses allocated to generators (by new method) and to loads (by physical flow basis) [7]. Now location of generator at bus no 2 and load at bus no 5 has been interchanged. Now power flow result of bus voltages and line flows are shown in table 7 and 8.

TABLE-VII	BUS VOLTAGES			
Bus no	$V(pu)$	$\delta(degree)$	$P_G - P_L$	$Q_G - Q_L$
1	1.05	0.000	92.801	14.079
2	1.15	-19.1819	-30	37.629
3	1.0745	-14.7665	-55	-13
4	0.9832	-10.9801	0	0
5	1.0881	-7.3706	50	0
6	1.0087	-10.2511	-50	-10

TABLE-VIII		LINE FLOWS AND POWER LOSSES					
Line		Sending end		Receiving end		Losses	
From	To	P	Q	P	Q	P	Q
1	4	55.74	12.01	-53.38	-1.10	2.36	10.91
1	6	37.06	2.83	-35.52	3.65	1.54	6.49
2	3	-2.12	10.08	2.70	-9.24	0.58	0.84
2	5	-27.88	27.55	31.16	-20.12	3.28	7.44
4	6	-4.32	-5.11	4.36	5.30	0.045	0.19
4	3	57.70	-72.34	-57.70	83.05	0.00	10.71
6	5	18.84	-26.89	18.84	29.99	0.00	3.10

TABLE-IX	LOSS ALLOCATIONS
Proposed Method	On physical basis
To generators (Capacity)	To loads (Capacity)
4.525(92.801)	3.856 (30.00)
3.276(50.00)	2.404 (55.00)

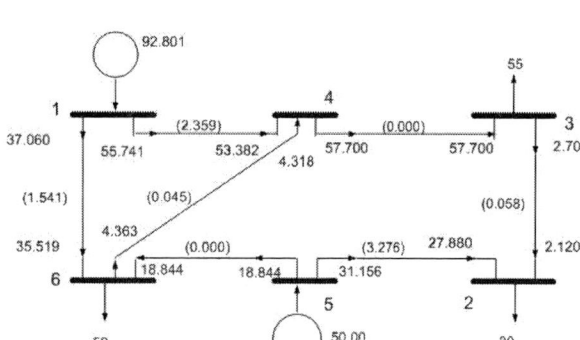

Fig.3. Six- bus system with line power flow and losses

Table 6 and 9 are showing comparative results of transmission loss allocation of proposed with reference [7]. [7] have allocated losses to the loads, However, proposed method is allocating transmission loss to the generators. In table 6, total transmission loss is 10.457 MW. Out of 10.457 MW, 5.111MW is allocated to generator 1 at bus no 1 and 5.346 MW is allocated to generator 2 at bus no 2 as shown in fig. 3.

Table 9 shows transmission loss allocation with a different arrangement of the network as shown in fig.4. In this case, load and generator data have been exchanged at bus no 5 and 2. Now this network arrangement carries generator at bus no 5 and load at bus no 2. Total network losses have been reduced to 7.801 MW. Transmission loss allocated to generator 1 at bus no 1 is 4.527 MW. Transmission loss allocated to generator 2 at bus no 5 is 3.276 MW.

VI. CONCLUSION

Simulation result of 6 bus network shows that power flow from generator to load follows a path that causes minimum transmission loss in that condition. This fact can be verified by changing the positions of generators. By changing location of generator in same power system network losses has been reduced further because the power flow paths have been changed. This shows that losses can be reduced by changing the location of the generator and thereby loads too. Therefore, this paper suggests the installation of small capacity generators rather installation of one generator of large capacity. Since changing the location of the generator changes the transmission loss this shows the locational dependence of loads on the network. The proposed method also shows the congestion management of the power system network.

REFERENCES

[1] J. Bialek, " Tracing the flow of electricity" IEE proc-Gener. Transm, Distrib, vol-143, no-4, pp-313-320, July 1996,
[2] A. J. Conejo, F. D. Galiana, and I. Kockar, "Z-bus loss allocation," *IEEE Trans. Power Syst.,* vol. 16, no. 1, pp. 105–110, Feb. 2001.
[3] A. J. Conejo, J. M. Arroyo, N. Alguacil, and A. L. Guijarro, "Transmission loss allocation: A comparison of different practical algorithms," *IEEE Trans. Power Syst.,* vol. 17, no. 3, pp. 571–576, Aug. 2002.

[4] Francisco D. Galiana, Antonio J. Conejo, Hugo A. Gill, " Transmission network cost allocation Based on Equivalent Bilateral Exchanges", *IEEE Trans. of power system,* vol-18, no-4, pp-1425-1431, November 2003.
[5] F. D. Galiana, A. J. Conejo and I. Kockar, " Incremental transmission loss allocation under pool dispatch," *IEEE Trans. Power syst.,* vol. 17, pp. 26-33, Feb. 2002.
[6] Y. P. Molina, Prada R.B, Saavedra O.R. *et al., "Complex losses allocation to generators and loads based on* circuit theory and Aumann-Shaprouley method," *IEEE Trans. Power Syst., vol. 25, no. 4, pp.* 1928–1936, Nov. 2010.
[7] S. Abdelkader, "Transmission loss allocation in a deregulated electrical energy market," *Elect. Power Syst. Res.,* vol. 76, no. 11, pp. 962–967,Jan. 2006.
[8] H. Shih-Chieh, "Fair transmission loss allocation based on equivalent current injection and Shapley value," in *Proc. IEEE Power Eng. Soc. General Meeting, pp. 6–6,* Feb. 2006
[9] H. Shih-Chieh and W. Hsin-Min, "Allocation of transmission losses based on cooperative game theory and current injection models," in *Proc. IEEE Int. Conf. Industrial Technology (IEEE ICIT '02), vol. 2, pp. 850–853, 2002.*
[9] H. Shih-Chieh and W. Hsin-Min, "Allocation of transmission losses based on cooperative game theory and current injection models," in *Proc. IEEE Int. Conf. Industrial Technology (IEEE ICIT '02), vol. 2, pp. 850–853, 2002.*
[10] X.-Z. Bao and W. Cui, "Allocation of transmission losses in electricity system based on cooperative game theory," in *Proc. 2010 Asia-Pacific Power and Energy Engineering Conf. (APPEEC),Chengdu, China, pp. 1–4, 2010.*
[11] Antonio Gomez, Jeasus Manuel Riquelme Santos, Tomas Gonzalez Garcia and Enrique A. Ruiz Velasco, "Fair Loss Allocation of Transmission Power Losses", *IEEE trans. On Power System,* Vol. 15, no.1,pp.184-188 Feburary 2000.
[12] Naimul Hasan, Ibraheem and Yudhishthir Pandey, "Transmission loss allocation using fairness index in deregulated market", *International IEEE Conference 28-29 November* CIPECH-2014.
[13] Shobhy M. Abdelkader, D. J Morrow and A. J Conejo, "Network usage determination using transformer analogy," *IET Gen. Trans. & Distribution, vol. 8, issue. 1, pp. 81–90,* Feb. 2014.
[14] Changyong Zhang, "A novel Mathematical model for the unique Shortest path routing problem" arXiv:0807.0038v2 [math.OC] 4 March 2010
[15] Sobhy M. Abdelkader, "Transmission Loss Allocation Through Complex Power Flow Tracing" IEEE Transaction on Power Systems, Vol. 22, No. 4, pp. 2240-2248 November 2007

2nd IEEE International Conference on Power Electronics, Intelligent Control and Energy Systems (ICPEICES-2018)

Determination and Allocation of Optimal Size of TCSC and STATCOM for obstruction alleviation in Power System

Jitender Kumar
Department of Electrical Engineering,
Delhi Technological University,
New Delhi, India
jitender3k@gmail.com

Narendra Kumar
Department of Electrical Engineering,
Delhi Technological University,
New Delhi, India
dnk_1963@yahoo.com

Abstract—The traditional technique of power system for measuring, monitoring and identifying the location of the overloaded power line is more complex in nature and time-consuming. The power system basic requirement to limit their parameter within a specified limit and keep close to as Unity either during underloaded or overloaded conditions of Bus. The Flexible AC Transmission System (FACTS) devices are the most effective tool which can be used for the purpose of minimising the impact of variation in load on a power system. Such devices work as the controllers and work effectively in mitigation of varying Voltage profile, enhance steady-state system stability, improve reactive power and power factor, etc to a specific limit. The size of a controller may vary on the basis of their requirement for a specific system. The TCSC and STATCOM are effective controllers of a power system to mitigate the multi-dimension variation of parameters. The proposed tools are implemented and tested on IEEE 14 Bus System individually for its efficacy. The MATLAB software is used for design of IEEE 14 Bus system and implementation of FACTS devices on Simulation Platform.

Keywords— Traditional method, FACTS Devices, TCSC, STATCOM, Congestion Mitigation

I. INTRODUCTION

The traditional power system is more wide and complex in nature. The controller is static in nature and their design is fixed to a specific value. The controller basically uses Static Capacitor or Static Inductor of pre-defined value. As per normal operation of power system are categorised in three distinguish category as Generation, Transmission and Distribution. As per the Indian scenario, there are the different types of organisation which operate their own task and manage their operation optimally. The power generation will be performed by NTPC, NHPC, NPC, etc. The transmission part is handled by PGCIL, Sterlite Power, etc, and Distribution part of electricity is handled by multiple state organisation. So that complexity of power transfer from generation end to distribution end is a complicated task due to an economy of the entire system as per the customer requirement. The cost to consumer per unit is pre-defined and cost incurred during entire process are varied due to different organisation involved. The individual organisation

performance plays a vital role in the economy of the entire power system.

The requirement of electric power for industrial and domestic usage is exponentially increasing as compared to improvement in transmission system capacity. There are bilateral and multilateral contracts between the government and private parties for establishing a proper coordination between generation side and demand side of power supply. The contract is well defined in terms of power transfer from one end to another end with specified power limit. Otherwise, some power lines are either overloaded or under-loaded during normal operation of power system. Such type of operation whenever some lines are overloaded during normal operation is called congestion [2]. It forces the operating mechanism to improve its power transfer capability. Also, to ensure that the system with regards to security, stability and reliability is fruitful for the entire power system which is defined as congestion mitigation management.

In [9] author uses the concept to minimize the transmission line loading by using sensitivity factor and stability indices. The transmission lines during peak hours are under heavy stress of power flow. Such stress can be minimized by using FACTS device like TCSC of a specified value and an optimal location. The sensitivity factor and stability indices are used to allocate the size of FACTS device and for implementing the location. The paper [1] discusses the effect of wind penetration on power generation in association with FACTS devices. The authors apply two types of operation by a reduction in wind and second wind penetration increased and apply FACTS device to compensate voltage and power angle of the system. In paper [3], authors concentrate on voltage profile and control of active & reactive power flow among various buses. They have identified the effect of different FACTS devices on IEEE/CIGRE transmission system model. The different FACTS devices impact on power losses in an interconnected power system.

978-1-5386-6624-1/18 $31.00 © 2018 IEEE

The paper [7] suggests the optimal model of IEEE 14 Bus system on the basis of MATLAB/Simulation software. The authors have proposed [8] a Contingency Severity Index (CSI) and Fast Voltage Stability Index (FVSI) for optimising the size of FACTS devices. Due to that the Cost of Installation of FACTS devices was optimised with reduced overloading on a system.

In [10] optimal location of Shunt FACTS device identified with the help of series FACTS device to get highest possible maximum power transfer and maintain system stability. It has affected the degree of series compensations and decide the optimal allocation of Shunt devices. They proposed the location of FACTS device must be located near to load centre in place of mid-point of a transmission line. The [5] will be focussed on supplementary controller design of STATCOM for oscillation mitigation in power system. The design of controller was based on the requirement of variation applicable to voltage profile and requirement of Reactive power at a specific location of the bus.

The authors have suggested [11] the real power performance index & reactive power loss reduction for identification of the location of FACTS device in a transmission line. The power system will require restructuring for optimal usage of their potential which may lack due to an environmental problem, cost problem and right-of-way during expansion of power network. The FACTS device will help in optimising loaded lines, increase loading capability, minimise system losses, improve system stability, reduce power production cost and fulfil a contractual requirement.

In the literature it has been seen that most of the papers concentrate on the obstruction alleviation during heavy loading of a specific bus of the power system. By use of FACTS devices, will guide in deciding the overloading or under-loading of a congested bus during a loading operation. The TCSC and STATCOM are tested on IEEE 14 Bus System on MATLAB/Simulation platform.

II. FACTS DEVICES MODELLING

A. Thyristor Controlled Series Compensator (TCSC)

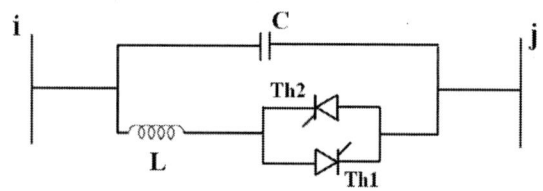

Fig. 1. TCSC

The Thyristor Controlled Series Compensator is known as TCSC [6], which is an advanced version of static Filter

[9] operated with the help of power electronic devices. The parallel combination of capacitor and inductor with bidirectional control performed by power electronic devices. Such type of controller helps in varying the overall impedance of TCSC. The overall reactance of device varied from purely inductive to purely capacitive. The device is dynamic in nature and varied their reactance on the basis of requirement of the power system.

The TCSC is a series compensation device used to compensate the loading condition of the transmission line, improving system stability and enhance the reliability of power system. It never requires any interfacing device like a transformer, so simple in nature and economical as compared to other FACTS devices. The TCSC can mitigate the problem of Voltage compensation [1] and Reactive Power compensation during loading conditions. The variations of net reactance depends upon the firing angle of power electronic devices. The firing angle of TCSC varied from $0°$ to $90°$ for Th$_1$ and $90°$ to $180°$ for Th$_2$ [2]. Due to that, the effect of inductance will vary and accordingly overall impedance of TCSC will be varied.

- If $X_C > X_L$, The TCSC may involve the advantage of both Inductive and Capacitive region.
- If $X_C < X_L$, The TCSC will behave as Capacitive due to shunt connection of Capacitor and low reactance offer by TCSC for power system.
- If $X_C = X_L$, Then a resonance condition achieved which is harmful to power system. It has infinite impedance offer by TCSC and transmission line behave as an open circuit.

The impedance offer by TCSC treated as positive (for inductive) and negative (for capacitive) region. The region of TCSC [4] will varied on the basis of firing angle and can be calculated by

$$X_L = X_1 \frac{\pi}{\pi - 2\alpha - \sin \alpha} \tag{1}$$

The x_{LC} can be correspond to as

$$x_{LC} = \frac{x_C x_L}{x_C - x_L} \tag{2}$$

$$Z_{TCSC} = j\, X_{TCSC} \tag{3}$$

$$X_{TCSC} = -x_C + \frac{x_C + x_{LC}}{\pi}[2(\pi - \alpha_{TCSC}) + \sin\{2(\pi - \alpha_{TCSC})\}]\left[\frac{4x_{LC}^2}{\pi x_L}\right]\cos^2(\pi - \alpha_{TCSC})[\omega \tan\{\omega(\pi - \alpha_{TCSC})\} - \tan(\pi - \alpha_{TCSC})] \tag{4}$$

Where $\omega = \sqrt{\frac{x_C}{x_L}}$ (5)

The value of x_L and x_C are use to designate the fundamental harmonic reactance [4] [6] of TCSC's inductor and capacitor.

So that Net admittance of TCSC will be given as

$$y_{TCSC} = \frac{1}{z_{TCSC}} = g_{TCSC} + jb_{TCSC} \tag{6}$$

Where TCSC's conductance and susceptance are g_{TCSC} and b_{TCSC}, respectively and here g_{TCSC} are Zero.

The flow of Real Power and Reactive Power [3] of power system within two buses are derived as

$$P_i = v_i \sum_{j=1}^{nb} v_j (g_{busij} cos\delta_{ij} + b_{busij} sin\delta_{ij}) \tag{7}$$

$$Q_i = v_i \sum_{j=1}^{nb} v_j (g_{busij} sin\delta_{ij} - b_{busij} cos\delta_{ij}) \tag{8}$$

The linearised equation of newton's law [9] for the compensated transmission lines are specified as in equation. Where reactance of TCSC is taken as the State Variable parameter [4] for matrix

$$\begin{bmatrix} \Delta P_i \\ \Delta P_j \\ \Delta Q_i \\ \Delta Q_j \\ \Delta P_{ij} \end{bmatrix} = \begin{bmatrix} \frac{\partial P_i}{\partial \delta_i} & \frac{\partial P_i}{\partial \delta_j} & \frac{\partial P_i}{\partial v_i} & \frac{\partial P_i}{\partial v_j} & \frac{\partial P_i}{\partial X_C} \\ \frac{\partial P_j}{\partial \delta_i} & \frac{\partial P_j}{\partial \delta_j} & \frac{\partial P_j}{\partial v_i} & \frac{\partial P_j}{\partial v_j} & \frac{\partial P_j}{\partial X_C} \\ \frac{\partial Q_i}{\partial \delta_i} & \frac{\partial Q_i}{\partial \delta_j} & \frac{\partial Q_i}{\partial v_i} & \frac{\partial Q_i}{\partial v_j} & \frac{\partial Q_i}{\partial X_C} \\ \frac{\partial Q_j}{\partial \delta_i} & \frac{\partial Q_j}{\partial \delta_j} & \frac{\partial Q_j}{\partial v_i} & \frac{\partial Q_j}{\partial v_j} & \frac{\partial Q_j}{\partial X_C} \\ \frac{\partial P_{ij}}{\partial \delta_i} & \frac{\partial P_{ij}}{\partial \delta_j} & \frac{\partial P_{ij}}{\partial v_i} & \frac{\partial P_{ij}}{\partial v_j} & \frac{\partial P_{ij}}{\partial X_C} \end{bmatrix} \begin{bmatrix} \Delta \delta_i \\ \Delta \delta_j \\ \Delta V_i \\ \Delta V_j \\ \Delta X_C \end{bmatrix} \tag{9}$$

Where $\Delta P_{ij} = P_{ijreg} - P_{ijcal.}$, is the inequality in flow of real power in the series reactance.

$\Delta X_C = \Delta X_C^i - \Delta X_C^{i-1}$, is the incremental alteration in series reactance.

The state variable of TCSC i.e. X_C, of the series controller, is updated at the ending of every iterative step till the end of an iteration

$$\Delta X_C^i = \Delta X_C^{i-1} + \left(\frac{\Delta X_C}{X_C}\right)^i \Delta X_C^{i-1} \tag{10}$$

The load flow program implemented on Matlab/Simulation as the changes takes place in Xc.

B. Static Synchronous Compensator (STATCOM)

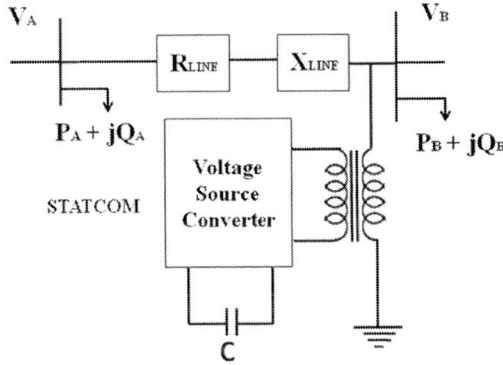

Fig.2. STATCOM

The Static Synchronous Compensator is known as STATCOM, is a power electronic controllable device. The Statcom is a Shunt FACTS Device [5] and sometime used as a voltage source converter with the help of capacitor. So that it may help in controlling of system power factor, voltage profile, steady state stability [5], flow of Power either Real or Reactive Power [7], System Reliability, congestion mitigation of transmission lines, etc. The STATCOM is provide a bidirectional flow of power among a transmission line. The voltage of STATCOM may be represented by $V_{st}\angle\theta_{st}$ and connected with a Bus Voltage of $V_B\angle\theta_B$. Practically STATCOM [8] [10] consists of the busbar, transformer, shunt impedance and system grounding. The devices are signify by:-

$$V_B\angle\theta_B = V_A\angle\theta_A - (R_{LINE} + j\,X_{LINE})I_L\angle\alpha_A - (R + jX)I_{St}\angle(\alpha_{St} + \frac{\pi}{2}) \tag{11}$$

$$E_{St} = V_{St}\angle\delta_{St} = V_{St}(Cos\,\delta_{St} + j\,Sin\,\delta_{St}) \tag{12}$$

$$Y_{St} = 1/(Z_{St}\angle\theta_{St}) = G_{St} + j\,B_{St} \tag{13}$$

So that

$$S_{St} = P_{St} + j\,Q_{St} = I_{St}^* V_{St} = V_{St} Y_{St}^* (V_{St}^* - V_B^*) \tag{14}$$

The following equations are obtained after the realisation of above equations into real & imaginary parts at bus 'a' is as follows:-

$$P_{St} = V_{St}^2 G_{St} + V_{St} V_B [G_{St} Cos(\delta_{St} - \theta_B) + B_{St} Sin(\delta_{St} - \theta_B)] \tag{15}$$

$$Q_{St} = -V_{St}^2 G_{St} + V_{St} V_B [G_{St} Cos(\theta_B - \delta_{St}) + B_{St} Sin(\theta_B - \delta_{St})] \tag{16}$$

$$P_B = V_B^2 G_{St} + V_{St} V_B [G_{St} \cos (\delta_{St} - \theta_B) + B_{St} \sin (\delta_{St} - \theta_B)] \quad (17)$$

$$Q_B = - V_B^2 G_{St} + V_{St} V_B [G_{St} \cos(\theta_B - \delta_{St}) + B_{St} \sin (\theta_B - \delta_{St})] \quad (18)$$

$$\begin{bmatrix} \Delta P_B \\ \Delta Q_B \\ \Delta P_{St} \\ \Delta Q_{St} \end{bmatrix} = \begin{bmatrix} \dfrac{\partial P_B}{\partial \theta_B} & \dfrac{\partial P_B}{\partial V_B} V_B & \dfrac{\partial P_B}{\partial \delta_{St}} & \dfrac{\partial P_B}{\partial V_{St}} V_{St} \\ \dfrac{\partial Q_B}{\partial \theta_B} & \dfrac{\partial Q_B}{\partial V_B} V_B & \dfrac{\partial Q_B}{\partial \delta_{St}} & \dfrac{\partial Q_B}{\partial V_{St}} V_{St} \\ \dfrac{\partial P_{St}}{\partial \theta_B} & \dfrac{\partial P_{St}}{\partial V_B} V_B & \dfrac{\partial P_{St}}{\partial \delta_{St}} & \dfrac{\partial P_{St}}{\partial V_{St}} V_{St} \\ \dfrac{\partial Q_{St}}{\partial \theta_B} & \dfrac{\partial Q_{St}}{\partial V_B} V_B & \dfrac{\partial Q_{St}}{\partial \delta_{St}} & \dfrac{\partial Q_{St}}{\partial V_{St}} V_{St} \end{bmatrix} \begin{bmatrix} \Delta \theta_B \\ \dfrac{\Delta V_B}{V_B} \\ \Delta \delta_{St} \\ \dfrac{\Delta V_{St}}{V_{St}} \end{bmatrix}$$

$$(19)$$

The normal operation parameter of IEEE 14 Bus system [11] obtained, where system operate without any additional device implementation for enhancement of performance of power system [7]. The parameter obtained related to Voltage magnitude, Real Power and Reactive Power on each buses of a system as per-unit basis, are as in Table 1:-

TABLE 1. BASE VALUES OF IEEE 14 BUS SYSTEM

Bus No.	Bus Voltage (V)(pu)	P (W) (pu)	Q (VAR) (pu)
1	1.7012	0.0123	0.0803
2	1.4328	- 0.015	0.0153
3	1.1047	0.0141	0.0057
4	1.0110	- 0.002	- 0.005
5	0.9979	- 0.049	- 0.044
6	1.3807	0.0096	0.0734
7	1.0881	- 0.001	- 0.003
8	1.6916	- 0.002	- 0.032
9	0.9584	0.0172	0.0114
10	0.9464	0.0000	0.0000
11	0.9696	- 0.003	0.006
12	0.9696	0.002	0.0067
13	0.9659	0.0127	0.0165
14	0.9439	- 0.000	0.001

III. SIMULATION RESULT AND DISCUSSION

The implementation of TCSC and STATCOM performed on IEEE 14 Bus System [7] as shown in Figure 3. The table 1 indicate Voltage profile and Reactive Power of power system during normal loading conditions. The parameter on buses indicate that the operational values of parameters are lying below its specific value and need some arrangement to strengthen these parameters. The congestion developed on such buses are mitigating with the help of dynamic controller in place of a fixed controller. So for that purpose, this paper proposed FACTS devices to mitigate such problem of parameter fall below a specific values of power system.

The STATCOM and TCSC are implemented on such specific buses where congestion shows by the system. The implementation of an individual device on such buses and evaluate their impact on the system. The Table 1 shows that the voltage profile of System will be 0.9696 pu, 0.9659 pu & 0.943873 pu on Bus 12, 13 & 14 respectively. So such buses are working on under loaded conditions and needed a device to upgrade the effective parameters of buses during normal as well as loading conditions.

978-1-5386-6624-1/18 $31.00 © 2018 IEEE

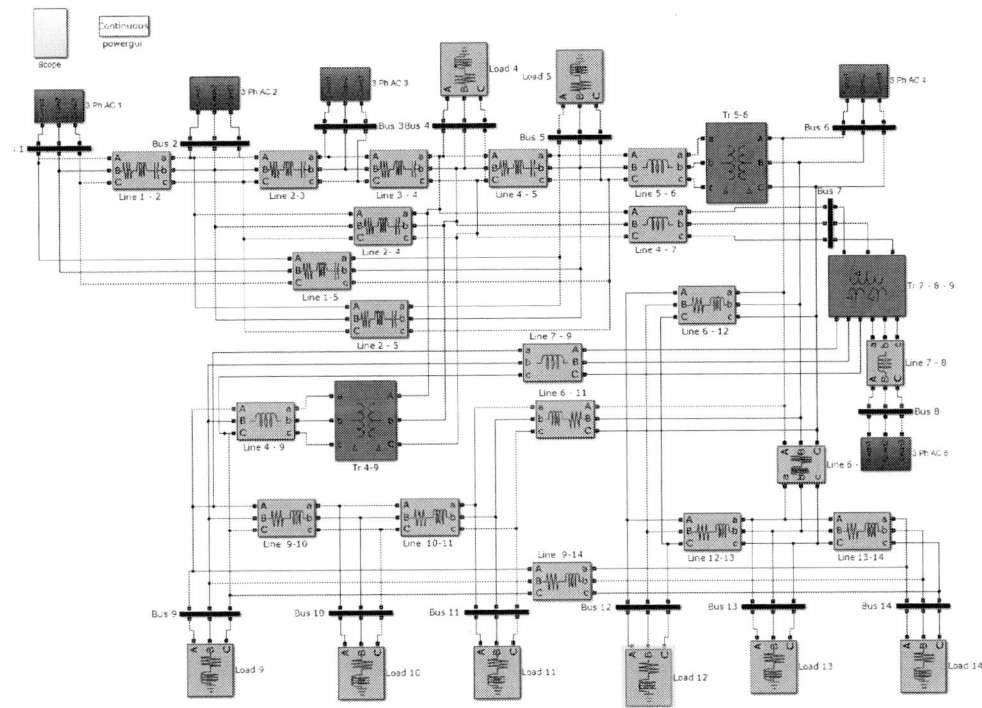

Fig.3. IEEE 14 BUS SYSTEM

The implementation of TCSC at 12 – 13 will improve it to 1.0 pu & 1.0495 pu and the STATCOM improved to 1.15 pu & 1.049 pu. The FACTS device at 13 – 14 will tend to 1.0 pu & 1.28 pu by TCSC and 1.016 pu & 1.237 pu by STATCOM. The effect of FACTS device on Voltage profile is shown in Table 2 & 3 while their graphical view are shown in Figure 4 - 9.

TABLE 2. EFFECT OF TCSC IMPLEMENTATION

Bus No.	Voltage with TCSC		
	Without	12-13	13-14
1	1.70E+00	1.70E+00	1.70E+00
2	1.43E+00	1.43E+00	1.43E+00
3	1.10E+00	1.10E+00	1.10E+00
4	1.01E+00	1.01E+00	1.01E+00
5	9.98E-01	9.98E-01	9.97E-01
6	1.38E+00	1.38E+00	1.38E+00
7	1.09E+00	1.09E+00	1.10E+00
8	1.69E+00	1.69E+00	1.69E+00
9	9.58E-01	9.65E-01	1.02E+00
10	9.46E-01	9.51E-01	9.88E-01
11	9.70E-01	9.71E-01	9.78E-01
12	9.70E-01	1.00E+00	9.80E-01
13	9.66E-01	1.05E+00	1.00E+00
14	9.44E-01	9.85E-01	1.28E+00

TABLE 3. EFFECT OF STATCOM IMPLEMENTATION

Bus No.	Voltage with STATCOM		
	Without	12-13	13-14
1	1.70E+00	1.70E+00	1.70E+00
2	1.43E+00	1.43E+00	1.43E+00
3	1.10E+00	1.10E+00	1.10E+00
4	1.01E+00	1.01E+00	1.01E+00
5	9.98E-01	9.98E-01	9.99E-01
6	1.38E+00	1.38E+00	1.38E+00
7	1.09E+00	1.09E+00	1.10E+00
8	1.69E+00	1.69E+00	1.69E+00
9	9.58E-01	9.62E-01	1.00E+00
10	9.46E-01	9.51E-01	9.81E-01
11	9.70E-01	9.73E-01	9.79E-01
12	9.70E-01	1.15E+00	1.02E+00
13	9.66E-01	1.05E+00	1.07E+00
14	9.44E-01	9.75E-01	1.24E+00

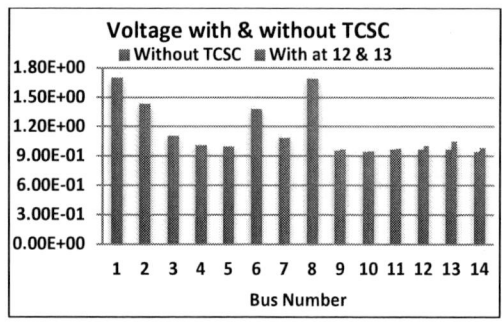

Fig.4. Performance of Voltage with TCSC at 12 & 13

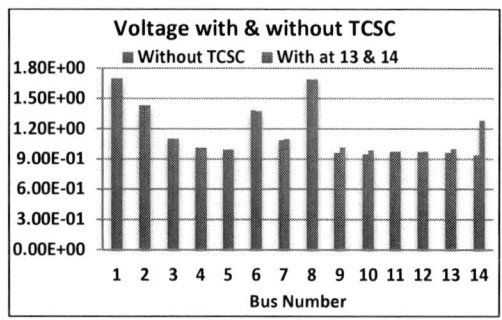

Fig.5. Performance of Voltage with TCSC at 13 & 14

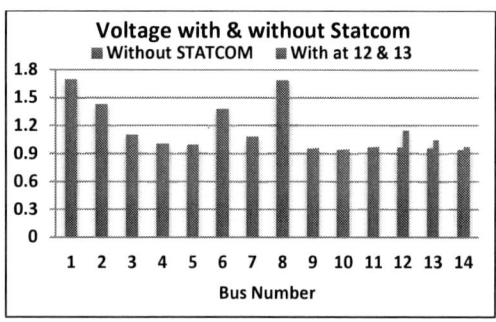

Fig.6. Performance of Voltage with STAT at 12 & 13

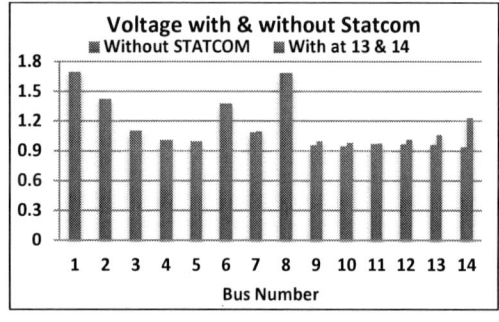

Fig.7. Performance of Voltage with STAT at 13 & 14

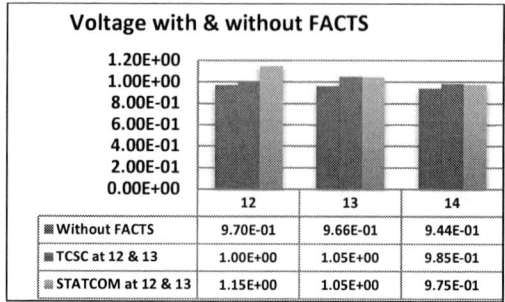

Fig.8. Performance of Voltage with FACTS at 12 & 13

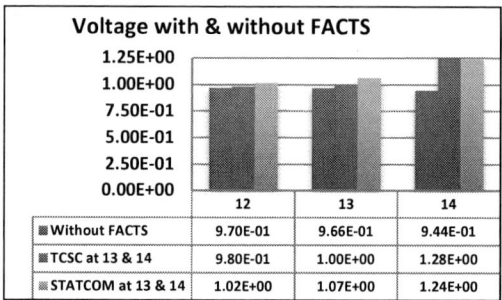

Fig.9. Performance of Voltage with FACTS at 13 & 14

The STATCOM and TCSC are implemented on specific buses where overloading shows by a power system i.e. Table 1. The implementation of an individual device on such buses and evaluate their performance on the IEEE 14 Bus system. The Table 1 shows that the Reactive power profile of System will be 6.71e-03 pu, 1.65e-02 pu & 3.92e-04 pu on Bus 12, 13 & 14 respectively. So such buses are showing that they are working on under loaded conditions and needed a device to improve the performance of buses near to normal conditions. The implementation of TCSC at 12 – 13 will improve it to 7.16e-03 pu & 3.26e-02 pu and STATCOM improved to 1.23e-02 pu & 2.36e-02 pu. The FACTS device at 13 – 14 will tend to 1.78e-02 pu & 5.11e-04 pu by TCSC and 2.39e-02 pu & 5.05e-03 pu by STATCOM. The effect of FACTS devices on Reactive power profile is shown in Table 4 & 5 while their graphical view are shown in Figure. 10 – 15.

TABLE 4. EFFECT OF TCSC IMPLEMENTATION

Bus No.	Reactive Power with TCSC		
	Without	12-13	13-14
1	8.03E-02	8.03E-02	8.05E-02
2	1.53E-02	1.54E-02	1.58E-02
3	5.69E-03	5.75E-03	5.95E-03
4	-5.10E-03	-5.10E-03	-5.12E-03
5	-4.47E-02	-4.47E-02	-4.47E-02
6	7.34E-02	7.45E-02	7.53E-02
7	-2.66E-03	-2.60E-03	-2.49E-03
8	-3.22E-02	-3.24E-02	-3.34E-02
9	1.14E-02	1.16E-02	1.30E-02
10	8.56E-06	8.64E-06	9.34E-06
11	5.70E-03	5.71E-03	5.80E-03
12	6.71E-03	7.16E-03	6.86E-03
13	1.65E-02	3.26E-02	1.78E-02
14	3.92E-04	4.27E-04	5.11E-04

TABLE 5. EFFECT OF STATCOM IMPLEMENTATION

Bus No.	Reactive Power with STATCOM		
	Without	12-13	13-14
1	8.03E-02	8.04E-02	8.05E-02
2	1.53E-02	1.55E-02	1.58E-02
3	5.69E-03	5.77E-03	5.94E-03
4	-5.10E-03	-5.08E-03	-5.03E-03
5	-4.47E-02	-4.45E-02	-4.43E-02
6	7.34E-02	8.37E-02	8.09E-02
7	-2.66E-03	-2.53E-03	-2.06E-03
8	-3.22E-02	-3.21E-02	-3.21E-02
9	1.14E-02	1.17E-02	1.34E-02
10	8.56E-06	8.74E-06	9.66E-06
11	5.70E-03	5.74E-03	5.87E-03
12	6.71E-03	1.23E-02	7.57E-03
13	1.65E-02	2.36E-02	2.39E-02
14	3.92E-04	4.58E-04	5.05E-03

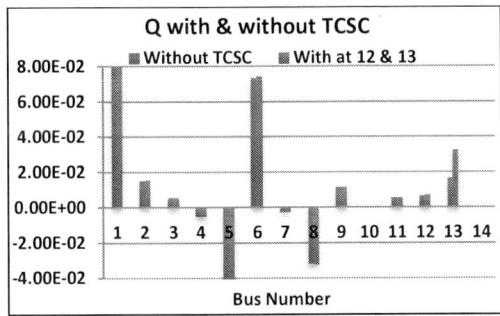

Fig.10. Performance of Q with TCSC at 12 & 13

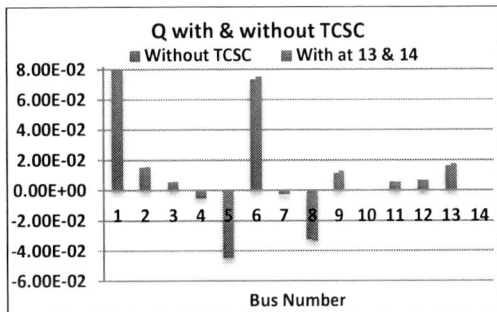

Fig.11. Performance of Q with TCSC at 13 & 14

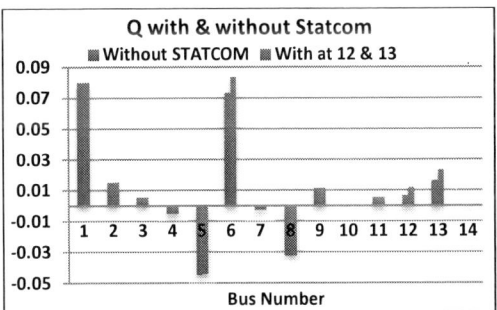

Fig.12. Performance of Q with STAT at 12 & 13

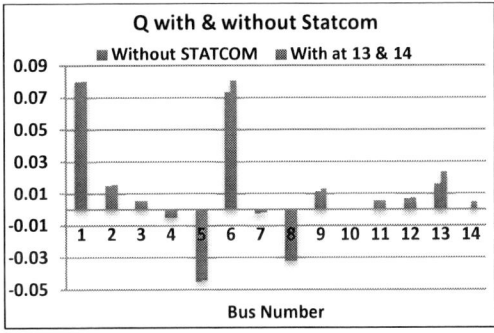

Fig.13. Performance of Q with STAT at 13 & 14

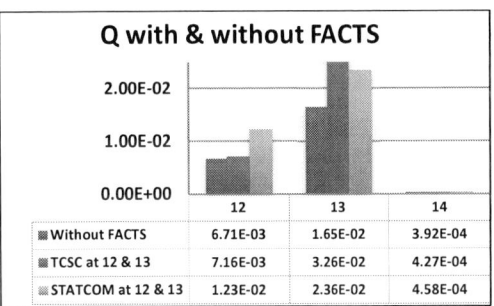

Fig.14. Performance of Q with FACTS at 12 & 13

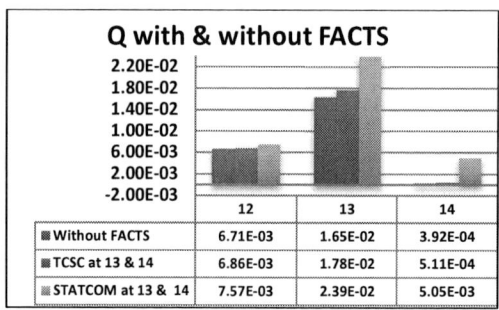

	12	13	14
▪ Without FACTS	6.71E-03	1.65E-02	3.92E-04
▪ TCSC at 13 & 14	6.86E-03	1.78E-02	5.11E-04
▪ STATCOM at 13 & 14	7.57E-03	2.39E-02	5.05E-03

Fig.15. Performance of Q with FACTS at 13 & 14

The Figure 8, 9, 14 and 15 are shows the impact of FACTS devices on the performance of Power System under loading conditions. The Parameter variations are identified on a comparison of Table 1 with Table 2, 3, 4 & 5 with FACTS devices allocation on a IEEE 14 Bus system.

IV. CONCLUSION

In this paper, the power system congestion is discussed with the implementation of FACTS devices. The study is carried out to identified the location and the effect of Series connected FACTS device i.e. TCSC and Shunt connected FACTS device i.e. STATCOM on a power system. The parameter studied are Voltage Profile and flow of Reactive Power on IEEE 14 Bus System. The FACTS device are dynamic in nature and fast to compensate the parameter as per their requirement for the system. They are efficient in delivering or absorbing of either the reactive power of system or system voltage of bus. The main objective of the operation is to maintain all the relevant parameters to a specific value and system become more stable during loading conditions. The result obtained before and after implementation of FACTS Devices will show that entire system are more stable and reliable after FACTS devices implementations. The STATCOM is more effective device as compare with the TCSC implementation on the same type of system.

REFERENCES

[1] Abdelaziz A. Y., Sharkawy M. A. El, Attia M. A., Saadany E. F. El, "Optimal location of series FACTS to improve the performance of power system with wind penetration", 978-1-4799-6415-4/14/$31.00 ©2014 IEEE, pp. 1 – 5.

[2] Gad M, Shinde P, Kulkarni SU (2012) : Optimal location of TCSC by sensitivity methods. International Journal on Computer Engineering Resources Vol. 2, Issue 6, pp : 162–168.

[3] Ghaeth Fandi, Zdenek Muller, Libor Straka, Jan Svec, "FACTS Devices Influence on Power Losses in Transmission Systems", 978-1-4799-3807-0/14/$31.00 ©2014 IEEE, pp. 1 – 5.

[4] Hingorani NG, Gyugyi L (2012) : Understanding FACTS. IEEE Press, New York.

[5] Jain Vipin and Kumar Narendra (2015): 'Designing of supplementary controller for STATCOM for Mitigation of Oscillations in Power Systems', Journal of Engineering, Science & Management Education (Published by: National Institute of Technical Teachers Training & Research, Bhopal & funded by: Ministry of Human Resource Development, Govt. of India), August 2015,Vol.8, Issue 2, pp : 124-133.

[6] Kundur, P.(1994): 'Power System Stability And Control' (McGraw-Hill Press, New York, 1994).

[7] Kumar Jitender, Rai J.N. and Hasan Naimul, "Use of Phasor Measurement Unit (PMU) for Large Scale Power System State Estimation", 2012 IEEE Fifth India International Conference on Power Electronics, IICPE2012, Delhi Technological University, New Delhi, India, ISSN : 2160-3162, PRINT ISBN : 978-1-4673-0931-8, Dec 6 - 8, 2012, 978-1-4673-0934-9/12/$31.00 ©2012 IEEE, pp.1 – 5.

[8] Malathy P. and Shunmugalatha A. : 'Loadability Analysis During Single Contingency With FACTS Devices Using Differential Evolution', IEEE Transactions on Power Systems, 2016, 978-1-5090-0128-6/16/$31.00 ©2016. pp : 1 - 11.

[9] Manganuril Yogasree, Choudekar Pallavi, Abhishek, Asija Divya and Ruchira, "Optimal Location of TCSC using Sensitivity and Stability Indices for Reduction in Losses and Improving the V oltage Profile", 1st IEEE International Conference on Power Electronics. Intelligent Control and Energy Systems (ICPEICES-2016), 978-1-4673-8587-9/16/$31.00 ©2016 IEEE, pp. 1 – 4.

[10] Sharma P.R., Kumar Ashok and Kumar Narendra (2007): 'Optimal Location of Shunt Connected FACTS Devices in a Series compensated Long Transmission Line', Turk. Journal of Electrical Engineering, 2007, Vol.15, No.3, pp : 321-328.

[11] Taher SA, Besharat H (2008) : Transmission congestion management by determining optimal location of FACTS devices in deregulated power systems. American Journal Application Science Vol. 5, Issue 3, pp : 242–247.

Power Factor Correction Using APFC Panel on Different Loads

Vinay Sehwag
Department of Electrical Engineering
Delhi Technological University
Delhi, India
vinaysehwag7@gmail.com

Vaibhav Dua
Department of Electrical Engineering
Delhi Technological University
Delhi, India
vaibhav3025@gmail.com

Ajendra Singh
Department of Electrical Engineering
Delhi Technological University
Delhi, India
ajendrasingh25@gmail.com

Jitendra Nath Rai
Department of Electrical Engineering
Delhi Technological University
Delhi, India
jnrai@dce.ac.in

Vineet Shekhar
Dept. of Electrical and Electronics
Engineering
NIET
Greater Noida, India
vshekher2407@gmail.com

Abstract—A major chunk of losses in electrical power systems is constituted by losses incurred due to poor power factor. A plethora of ways have been implemented to deal with this problem for quite a long time now. The intention behind this paper was to address to this woe which may be used for correction of power factor in various power systems. Employment of Automatic Power Factor Correction panels or more commonly known as APFC panels in electrical distribution networks is gaining prominence owing to the superiority of APFC panels over fixed capacitor banks in maintaining the desired power factor with different loads therefore, improving overall efficiency of the power system. A power factor correction unit enables the system to attain power factor values close to unity for efficient operation. The merits of higher power factor values include increased efficiency, load carrying capacity, better voltage regulation etcetera. The APFC panel was implemented on various loads operating together three phase Induction motor connected in parallel to series combination of a variable resistive and variable inductive load. In accordance to the instantaneous recorded value of power factor, the apposite capacitors are injected into the circuit to improve the power factor values. Improvised instantaneous values of the power factor were observed

Keywords—Automatic Power Factor Correction (APFC), Inductive Load, Resistive Load, Capacitive Load

I. INTRODUCTION

In present world, disparity in electrical power transmission and utilization is a major concern. People are unaware about the amount of energy being wasted by being ignorant of minor factors. Say, to travel short distances, many people turn to private conveyance instead of using public transport being ignorant of the consequences leading to consumption of non-renewable sources of energy like fossil fuels at a disturbing rate plus leading to pollution, global warming and thinning of ozone layer. In an electrical network Power Factor is a clandestine factor, on which depends a great deal of electrical energy losses.

Power factor gives indication of real power presence in a power system [1, 2]. If in a system there is 100% power and the pf is 0.85, then it indicates that net power consists of 85% Active power & 15% Reactive power. Highest possible values of power factor is desired in a system, this leads to

reduction in current demand. Hence, it is intended to maintain power factor close to unity in order to reduce the current demand therefore minimizing the size and the overall conductor costs in the network. Mathematically, power factor is defined as the ratio of the real power (in kW) and apparent load power (in kVA) consumed in a network [4].

$$P.F. = \cos\varphi = \frac{activepower}{reactivepower} \qquad (1)$$

Practical electrical loads are majorly inductive in nature which leads to lagging power factor [1][2]. The problem of lagging power factors can be addressed in a most practical and economical approach by injecting reactive compensation into the circuit by installing power capacitors of appropriate capacity and rating at key points in the electrical network. Power factor improvement holds more relevance in electrical distribution systems. In order to achieve this objective low voltage (LV) capacitors are being substantially used in both conventional fixed capacitor banks and in modern Automatic Power Factor Correction panels. APFC panels are now increasingly becoming more popular than ever due to a number of advantages they offer, namely (1) Capability of maintaining system power factor at the target high value (near unity) [6] and therefore circumvent penalties levied by electricity supply companies due to poor power factor, (2) A void over compensation during low load conditions [6][7] and (3) Enhanced overall efficiency of the power system attributable to reduction in losses due to low power factor [7]. With appropriate design and construction, the required features and rating of APFC panels can be decided, designed and implemented for a variety of loads and power system networks. Power factor correction is the technique of compensating where leading current compensates lagging current through employment of capacitances of suitable rating in the network.

Capacitors present in various power factor correction system extract current that leads voltage and hence produce a leading power factor. A suitable capacitance is introduced in the system to attain highest possible values of power factor i.e. close to unity. Hypothetically speaking, capacitors could provide 100% of the desired reactive power by our network

[5], however, in real world conditions, overcorrecting power factor in extremely close proximity to unity may cause harmonic distortions. If the capacitors connected in a circuit that normally works at a lagging power factor help to reduce the circuit lags quite appreciably. This technique is applied to negate as much of the unwanted magnetizing current as possible in order to reduce losses in the distribution system thus improving operational frequency. The benefits on offer to the commercial electrical consumer include minimized electricity bills by removing charges on reactive power reduced losses making extra, KVA available from the present sources or supply. Hence, improving the system efficiency

A. Power factor improvement through capacitive switching

At No-Load operation the inductive load (3 phase IM, purely inductive load) has a very low pf of about 0.1 Lagging as it consumes a large value of magnetizing current component and relatively a meagre value of real current component to meet the no load losses [3][4] under loaded conditions the inductive load draws a large amount of real current to meet the increased load and losses while the magnetizing component of the current remains constant [3]. As a result, power factor of the system increases. The capacitors in the APFC unit provide required reactive power for the system. Power factor correction capacitors lead to reduction in net current drawn from the distribution system and as a consequence of this they raise the system's overall operational capacity by taking up the power factor values close to unity. [3][4]. By supplying the reactive power requirement at the load end, the industrial and domestic consumer emancipate the utility from having to supply it, hence resulting in reduction in the total amount of apparent power supplied by the utility thus leading to conservation of energy and reduction in energy cost [4].

B. Automatic power factor correction unit

Most of the real world and industrial load is inductive in nature which leads to lagging power factor and hence, there is loss and wastage of energy which further leads to heavy penalties from electricity board combined with large electricity consumption bills. It is a difficult task to maintain ideal unity power factor and to overcome this difficulty we have APFC panels which maintain power factor close to unity [6].

II. HARDWARE DESCRIPTION

A. Moulded Case Circuit Breaker (MCCB)

A moulded circuit breaker is a special kind of electrical protective instrument that can be used for a wide range of voltages and frequencies of both 50Hz and 60Hz. The MCCB is used to control electric energy in distribution network. This circuit Breaker is an electromechanical device which shields a circuit from short circuit and over current. Operational range of MCCB is around 63A for short circuit current and 3000A for overcurrent protection. The chief roles of MCCB are to provide a way to open a circuit both automatically and manually under short circuit or overload condition. Majorly MCCBs have three essential functions, protection against overload, protection against electric faults and lastly switching a circuit ON and OFF.

B. Contactor

It is a special electrical device utilized for isolating or opening and closing or connecting an electric power circuit. They are sometimes called special purpose relays which operate for high values of current. A contactor is controlled by a low power level circuit.

C. Miniature Circuit Breaker

MCB is an electromechanical protective electrical instrument which shields an electrical network from large values of over current, which may a consequence from short circuit, overload or imperfect design. MCBs are superior in operation to conventional fuses since they do not need an alternate once an overload is identified. An MCBs have the ability to be simply rearranged and thus providing improved operational protection and greater handiness without incurring large expenses.

D. Capacitor

A capacitor also known as condenser is a passive electrostatic device where applying a constant potential will gather a fixed amount of charge on both opposing parallel plates, with the magnitude of the charge induced on both ends being equal and opposite to each other. This ratio of the applied voltage to the gathered charge is known as capacitance. Electrical energy in a capacitor is stored in the form of electric field.

E. Power Factor Meter/Relay

It is a five-stage power factor relay which works in five stages as per the system requirement. It is one current transformer operated relay which is connected either in any phase, here we have connected this CT in R-phase of 3-phase supply system.

F. Rotary Switch

It is a special switch which is to be twisted or rotated for its application in an electrical circuit. They are employed in situations where more than two electrical states are desired. It has a rotating axis that has a contact arm or "spoke" which protrudes out from the body. The terminals are arranged like an array around the rotor in a circular fashion, each of which acts like a contact or position for the contact arm to connect various electrical sub circuits in the larger circuit.

G. Current Transformer

A current transformer is a transformer that is used to measure alternate current and shielding associated electrical instruments or devices. An alternate current in its secondary circuit directly proportional to the alternate current flowing in its primary circuit is produced in current transformer.

978-1-5386-6624-1/18 $31.00 © 2018 IEEE

When the current to be measured has exceedingly large value to be measured directly or the system voltage of the circuit has large magnitude, a current transformer can be employed in such conditions.

H. Load Description

A three-phase load which includes three phase resistive and inductive loads in parallel with three phase Induction Motor.

III. DESIGN OF APFC PANEL

APFC unit used here is a five-stage automatic power factor corrections unit, here five condensers two 1 kVAR, one 2 kVAR and two 3 kVAR condensers signifying the five stages constitute the APFC unit. Each condenser is connected through a 32A MCB with 100A MCCB providing power supply to the system. The power factor relay is connected in any one terminal of three phase system and it is energized by single phase supply, S1 and S2 are the terminals of the current transformer connected to S1 and S2

points of power factor relay. There are five points on the power factor relay which are connected to condenser through rotary switch. Power factor correction panel operates in automatic mode when all rotary switches are in position-1 and when the relay does not work due to any reason the rotary switches are turned to position-2, that is in manual mode.

In general, a 440V APFC Panel shall be metal clad, indoor type floor mounted in Non-draw out execution. Fabricated from Sheet steel shall be CRCA of minimum 2.0 mm thickness. Incomer Circuit breaker (MCCB) shall be mounted in a separate compartment and Metering compartment along with APFC Relay etc. shall be separate. The position of various control switches, push buttons, louvers etc. requiring manual operation. The operational Height of Panel shall be at a height not less than 300mm and shall not exceed 1850mm from the finished floor level. Name plate for each incoming and outgoing feeder at front. All equipments of similar rating shall be interchangeable.

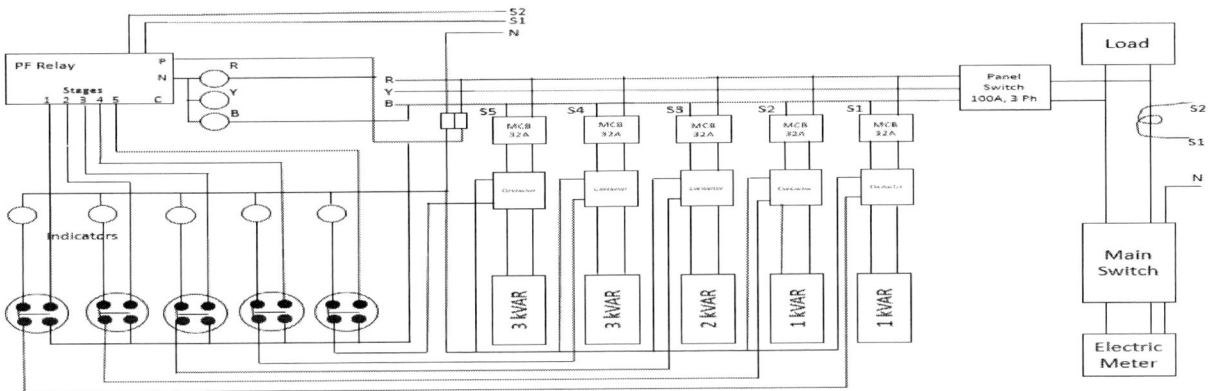

Fig. 1: Circuit Diagram of APFC Panel

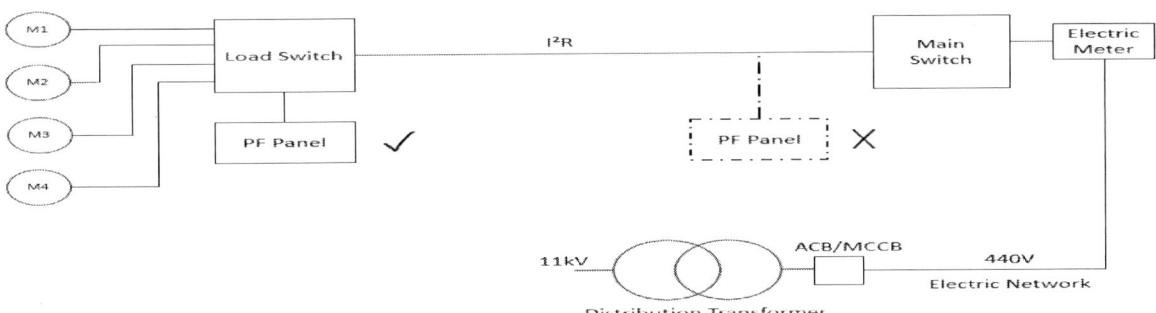

Fig. 2: Location of APFC panel in an electric network

Table 1: Components used in APFC panel

S.No.	Name of Equipments	Qty
1	MCCB	1
2	CONTACTOR LC1D123	2
3	CONTACTOR	3
4	MCB	2
5	MCB	15
6	CAPACITOR	2
7	CAPACITOR	1
8	CAPACITOR	2
9	P.F. METER/RELAY	1
10	ROTARY SWITCH	5
11	CURRENT TRANSFORMER	1
12	LED LIGHT	8
13	PANEL BODY	1

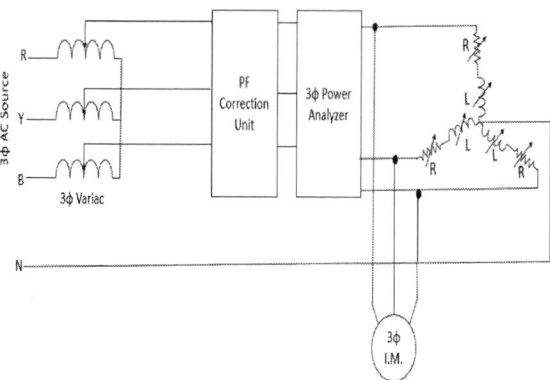

Fig. 4: 3-Phase load with 3-Phase resistance and inductance in parallel with 3-Phase Induction Motor

Fig. 3: Components of APFC Panel

IV. IMPLEMENTATION OF APFC PANEL

Presence of lagging power factor in a system is sensed by power factor relay (the relay of the power factor meter is set in between 0.5 to 0.98) and it switches ON the first stage of the unit. If the first stage is not able to correct the power factor value then second stage comes into play and further stages are engaged till the required value of the power factor is achieved but if the required power factor is maintained then further switching ON of the condenser will stop at that stage. APFC relay automatically switches OFF the stage when the required power factor is maintained. If there is a case of relay becoming faulty due to some reason then we have to manually turn the rotary switches in position-2. In small industries having highly inductive load and lagging power factor, here we have to maintain the pf up to 0.9.

Fig. 5: Hardware Load Depiction

The use of a common motor controller for switching the motor and the capacitor as a unit has become increasingly popular because of the elimination of one switching device and because of the operational convenience [7]. The capacitor is switched in and out of service with its associated motor and therefore tends to eliminate overvoltage usually associated with light load conditions in a plant [7]. This unit switching arrangement is economically attractive when only a few motors are involved or when there is no diversity among motors [7].

Capacitors causes leading power factor as it shifts current ahead of voltage, so to correct lagging power factor, it is a convenient method for which this method is practiced worldwide vastly [5]. Though it has some limitations like the inability to absorb harmonics and doesn't provide step-less correction, it is a popular choice for PFC for its low cost of installation and maintenance [5].

VI. RESULT AND DISCUSSION

The APFC used was for a maximum of 16 kVA and 0.75pf, as we know kW=kVA*pf we can say the maximum real power in consideration was 12 kW, so as to achieve unity power factor we have to add 10 kVAR of reactive power to the circuit.

Table 2 gives the stepped correction of power factor of the system with the application of 5 kW load. Similarly, Table 3 gives the values of the corrected power factor after each stage with the application of 10kW load. A variable resistance connected in series with a variable inductance and this whole combination connected in parallel to a three-phase induction motor constitutes our electrical load.

VII. CONCLUSION

Automatic mode of the APFC swings into action when the power factor on the load side is low due to the inductive effect of the load but when the machine load is increased with the increase in resistive load the power factor shoots up to 0.98 from 0.75. When the power factor of the load is reached to the optimum value the APFC stops working in the automatic mode. The APFC panel drives out large current drawn by the electrical power system and it further leads to decrease in electricity consumption bills. Reduction in electrical energy use results in reduction in fossil fuel consumption by power stations which means protection of ozone layer from depletion and curbing the phenomenon of Green House influence leading to deceleration of global temperature rise. The most common application of these panels are in the small scale industries. The motivation and principle of operation of this paper may further be used in future endeavours involving design of power factor technologies for various loads.

REFERENCES

[1] John J. Grainger, William D. Stevenson, Jr., "Power System Analysis", McGraw-Hill International Editions, ISBN 0-07-113338-0.

[2] P.S. Bhimbra, "Electric Machinery", Khanna Publishers, 7th Edition, 2014, India.

[3] Rishabh Jain, Shashank Sharma, Mini Sreejeth, Madhusuadan Singh, "PLC based Power Factor Correction of 3-phase Induction Motor", 1st IEEE conference on Power Electronics, Intelligent Control and Energy Systems, 2016.

[4] Nagarajan. M, Kandasamy. k. v," Optimal Power Factor Correction for Inductive Load using PIC", International Conference on Modelling, Optimization and Computing, 2012.

[5] Yasin Kabir, Yusuf Mohammad Mohsin, Mohammad Monirujjaman Khan," Automated Power Factor Correction and Energy Monitoring System", IEEE, 2017.

[6] Muhammad Bilal Khan, Muhammad Owais," Automatic Power Factor Correction Unit", IEEE, 2016.

[7] W.C. Bloomquist, W.K. Boice," Application of Capacitors for Power Factor Improvement of Induction Motors", Transactions, Volume 64, May 1945.

2nd IEEE International Conference on Power Electronics, Intelligent Control and Energy Systems (ICPEICES-2018)

Selection of Renewable Energy Sources for Off-grid Electrification of North-Eastern States of India

Kusum Tharani
Electrical and Electronics Engineering Department
Bharati Vidyapeeth's College of Engineering
New Delhi, India
kusum.tharani@rediffmail.com

Komal Bagai
Electrical and Electronics Engineering Department
Bharati Vidyapeeth's College of Engineering
New Delhi, India
komalbagai123@gmail.com

Shivam Dabas
Electrical and Electronics Engineering Department
Bharati Vidyapeeth's College of Engineering
New Delhi, India
dabasshivam41@gmail.com

Abstract— The off grid systems are gaining importance in rural areas situated in remote locations where grid extension is not just difficult but expensive too. The north-eastern states of India have plenty of rivers flowing in and around creating huge opportunities for installing small and micro-hydro plants. Moreover the rural areas have immense biomass potential due to availability of cow dung from cattle and forest waste. With the aim to electrify rural areas in Assam, three different configurations using renewable energy resources-Standalone Micro-hydro plant, Standalone Biogas plant and hybrid Micro-hydro and Biogas plant is simulated using HOMER software. The size and cost optimization results for the three configurations are validated using Hydro-Biogas Optimization (HBO) algorithm developed in MATLAB environment. It was observed that using a standalone micro-hydro system was most economical with lowest cost of energy of 2.56 INR/kWh. But this system was unable to meet the load requirements throughout the year. Hence a slightly costly hybrid configuration of Micro hydro and biogas plant with cost of energy 2.72 INR/kWh is recommended for the study area as using this model, there is no capacity shortage and the supply always meets the load demand.

Keywords— *Micro-Hydro Power Generation, Biogas Generation, Off Grid system, MATLAB*

I. INTRODUCTION

Worldwide, the economy's scenario at present is transforming from fossil energy supplies to inexhaustible i.e. renewable energy sources [1-4]. Originally fossil-based energies evolves toward renewable resources in the prolonged span, pertaining to propel the global energy set-up to a sustainable orientation [5-7]. Out of 8 North-eastern states in India, present electrification rate as revealed by Ministry of Power of India has reached up to 62.45% in Assam, 70.91% in Meghalaya, 75.24% in Arunachal Pradesh, 76.81% in Tripura, 77.66% in Nagaland , 81.23% in Manipur, 84.90% in Sikkim and 94.38% in Mizoram, which is evolving at a good pace. As per the MNRE Annual Report 2016-2017, the Ministry of India has assigned 10 percent of the budgetary reinforced for the deployment of various plants. The National Biogas and Manure Management Programme (NBMMP) is being effectuated in the North-eastern region states through State Government Nodal Departments/ State Nodal Agencies [8, 9]. As many isolated and distant areas have deficient renewable energy resources and it turns out mandatory to use traditional alternatives to complement the accessible energy resources

permissible to satisfy the load claims [10]. Rural electrification is an requisite part for energy supply to distant locations [11], especially in a country like India, which requires cost-effective proposition [11,12,14]. For such remote locations where grid-based power is inaccessible, off-grid Hybrid Energy System (HES) can moderately supply the power [14]. For instance, Tripura, a small state in north-eastern India with a predominantly agrarian economy. For dispensing energy, developments are carried out for effective implementations like a biomass- PV grid-integrated hybrid system designed for a rice husk plant in the absence of any backup[13,15].

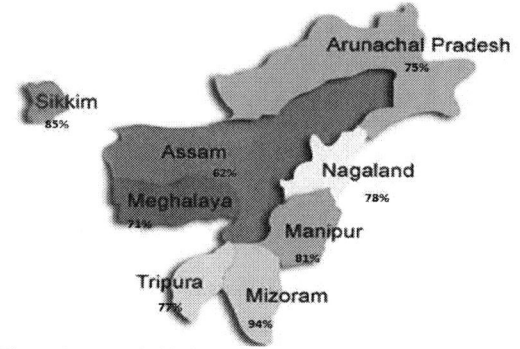

Figure 1: Household electrification status of the north-eastern states

A few combinations of renewable energy sources ascertaining increased system efficiency with greater balance in factual energy supply are to be discovered [16]; by inspecting some unique designs, many limitations of intermittency of renewable sources can be diminished [17]. The North-Eastern States have a quite benign potential to establish compact hydro energy projects. Amongst the N-E States, the highest hydro potential is henceforth engrossed by the state of Arunachal Pradesh followed by Sikkim, Meghalaya, and Mizoram as observed in figure 1. With systematic, well-grounded and economical renewable energy resources, off-grid supply can be used as a substitute to replace diesel based backup for power supply [18,19]. Hydropower generation is a well demonstrated pristine technology and micro hydro schemes are being deployed globally established in standalone power generation [20,21]. For the purpose of generation of electricity, small hydroelectric power plants are taken into account on a large scale from a couple of years. Even from the free flowing

978-1-5386-6624-1/18 $31.00 © 2018 IEEE

water with a velocity of more than 5 m/s, , the electricity can be produced, rather than water from dams or reservoir necessarily [12]. In a country like Sri Lanka, despite having an inordinate degree of grid penetration, various schemes for off grid are still pervasive to enhance the electrification status. Currently, they possess a grid penetration level of 89% and attempts to analyze the prevailing practical issues and challenges [22].

India's approach towards increasing its renewable energy mix has involved financing investigations and evolution [23]. As per MNRE (2015) several support schemes such as the National Biogas and Manure Management Program (NBMMP), off -grid biogas power generation program and over 400 biogas off-grid power plants have been set up with a power generation capacity of about 5.5 MW [24]. At times when proficiency of energy substituents is disputed, it is pertinent to approach to the oldest renewable energy alternatives, biogas which is an expedient alternative, a mixture of gases, generally carbon dioxide and methane [1,24,25]. In 2014-15, about 20,700 lakh cubic meters of biogas is produced in the country which is equivalent to 5% of the total Liquefied Petroleum Gas or autogas (LPG) consumption in the country. More than half of the world's agricultural phytomass comprises of crop residues [26]. The government is also extending the substantial subsidy for setting up of new biogas plants as biomass is one of the most overriding and emerging sources of energy [27-30]. Research and development are constantly ongoing, and improvements in the application are always in mind, but this technology is so well proven to come out as an extremely sustainable option that is being unearthed everyday [31]. Also, hydrogen can be brought into use to perpetuate the temperature range; in winter time biogas is originated in optimum amount for the power generation. The chore of meeting the demands with ample of generation may excite periods of retrenchment or inadequate production thus leading to complications [6, 32].

Figure 1 clearly shows that among the eight north eastern states, Assam has the lowest percentage (62%) of household electrification. With a view to electrify remote rural villages in Assam, an effort is made to test the suitability of renewable energy sources in a particular area. Assam has a solar potential of 13760 MW, wind potential of 112 MW, biomass potential of 212 MW and small hydro potential of 239 MW. Two major rivers Brahmaputra and Barak flow through Assam and more than 40 small rivers pass by the state. Renewable resources like Solar PV and wind being intermittent in nature require a huge battery backup for sustainable generation. The use of batteries with renewable energy sources incurs a huge system Net Present Cost (NPC) and Cost of Energy (COE). Therefore solar PV modules and wind turbines are not considered in the present study. On the other hand standalone small hydro generators and standalone biogas generators can be used without battery backup to suffice the daily load requirements of the small hamlets in the state. Also since the flow of water and biomass production varies according to seasons throughout the year, small hydro generation can be combined with biogas generation to develop a hybrid hydro-biogas system.

In the first part of the paper, a standalone hydro system, a standalone biogas system and a hybrid hydro-biogas are modeled and optimized using HOMER software. The optimization results are validated using an HBO algorithm developed in MATLAB environment considering load requirements and cost constraints.

II. METHODOLOGY

A. Study Area

Out of the 26 districts in Assam, the Golaghat district considered for study has only 44% household electrification. The climatic condition of Golaghat District throughout the summer months and in monsoon season remains tropical with scorching and fuggy environment. As the data depicts, the average rainfall entailed in this region is around 1300 mm. The rainfall is significant in the period ranging from June till the end of July with high degrees of about 38 degree Celsius and around 10 degree Celsius in the months of June and December. The wind flows from north-west to southeast region. There are around 132576 households in the district and only 3286 households are electrified. The Silbheta village located at 26o31 North and 93o57 East in Bokakhat tehsil in Golaghat district is the specific area considered for the present study. Dhansiri with the total catchment area is 1,220 square kilo metres is the major river flowing through the area.

B. Resource Availablity

The designated location in this study has abundant potential of renewable energy sources to satisfy the energy requirement of the particular area. The daily average of solar radiation in this particular area is about 4.57 kWh/m2/day and the mean annual wind speed of 3.11 m/s. The field inspections reveals that there is one site for MHP with discharge of 0.10 m3/s and 15m head and biomass availability of 1.5 tones/day, located in the particular area considered for this study. Evaluation of renewable energy sources in study area is shown in Table 1.

Month	Biomass Availability(tonnes/day)	Discharge(litres/sec)
January	1.2	1220
February	1.2	1340
March	1.4	1450
April	1.5	1560
May	1.5	1600
June	1.5	1660
July	1.8	1720
August	1.9	4140
September	1.8	5970
October	1.5	7960
November	1.5	6600
December	1.5	2200

Table 1. Month-wise availability of Renewable energy resources in the area

C. Load Demand

According to Census 2011, Silbheta village has approximately 70 households with a total geographical area of 113 hectares. The total energy consumed per day for a single household is 2.5 kWh/day. Considering around 100 households after future extension, the average domestic load of the village is 250 kWh/day with a peak load of 35 kW.

The load data shown in figure 2 indicates that during the summer months from May to August, the load requirement increases because of the humid weather in the state. The winter months from October to February are comparatively cool and pleasant and so the load demand decreases.

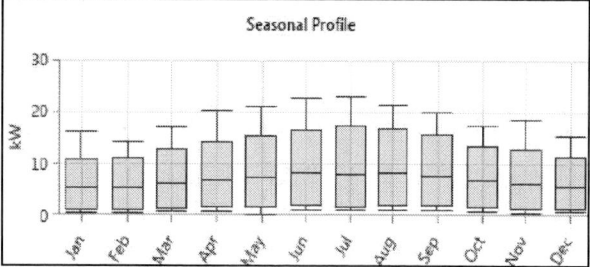

Figure 2: The monthly average load data for the study area.

D. Technology Selection

As per the load demand, resources obtainable and constraints, a standalone micro hydro plant a standalone biogas plant and a hybrid micro hydro-biogas plant are modeled and optimized for the present site location.

III. MATHEMATICAL MODELING

The mathematical modeling of micro-hydro and biogas plants have been discussed in the following sections.

A. Biogas Power Plant

The figure 3 shows the working of a biogas power plant. In a biogas power plant, the cattle/animal dung i.e., the waste material is heated to produce a gas in large amounts. The gas is treated with the gas scrubber and De-humidifier. It is now completely free from any impurity and humidity. Humidity can cause in low calorific value of the fuel, so it is necessary to de-humidify the gas. The treated gas is then collected in the gas collector. The collector sends the gas to the generator which in turn produces the electricity. The waste material i.e., the slurry is used as natural fertilizers.

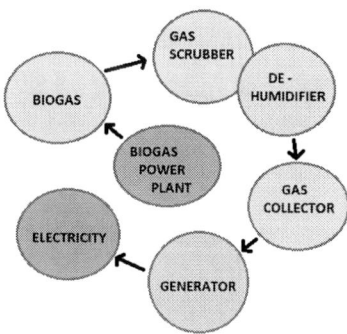

Figure 3: Schematic of a biogas plant

Biogas is generated by the animal/cattle dung or the material that can potentially pollute the environment if not used suitably. The methods of biogas generation uses the unusable dung and convert it into a renewable energy resource. To calculate the biomass potential, forest and agriculture wastes are considered. The biogas energy generated is calculated by the following equations.

Total gas yield (m³) =

$$(0.036)*Total\ dung\ availability\ (kg/day) \quad (1)$$

Energy Yield (kWh/day) =

$$Total\ gas\ yield\ (m^3)*CV*\eta/860 \quad (2)$$

Where, CV is the calorific value of biogas, η is the efficiency of the generator and 860 is the conversion factor of the biomass. CV of biogas is 4700 Kcal/m³. Biogas power plant of 100 cubic metres capacity produces approximately 100-120 units (kWh) `per day working for 10 hours a day. The algorithm uses the energy calculated from equation (2), and plot it against load values to obtain the results.

B. Micro Hydro Power (MHP) Plant

In the MHP plant, the kinetic energy of the flowing water of streams/rivers to the electric energy. The flowing water is passed through the hydraulic turbine; the turbine is connected to the synchronous generator, which produces the electricity as shown in figure 4. The power plant depends upon a river or a stream that acts as an intake system for the diversion of the water. The penstock acts as a canal which carries the diverted water. This arrangement connects to a hydraulic turbine which transforms the kinetic energy of water into the electrical power.

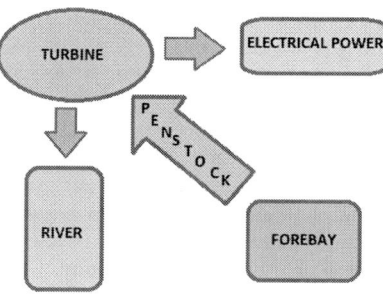

Figure 4: Schematic of a micro hydro power plant

Hydro electricity generation means to use the energy of moving water to generate electricity. This renewable resource is actually pollution free and very promising.
The calculations used in the algorithm to test the load versus supply characteristics were based on the equation (3) and equation (4). The micro hydro power generated is given by:

$$P=\eta*g*Q*H\ (kW)$$

Where, 'η' denotes the efficiency of the rotating turbine, 'g' here stands for the acceleration due to gravity, 'Q' represents the flow rate of the water and 'H' denotes the effective pressure head of water across the turbine. The flow rate of any river is affected monthly. The algorithm uses the flow rate of a stream monthly. The flow rate depends on the velocity of the stream.

$$Q= v*b*h$$

Where, 'Q' is the volumetric discharge, is the mean flow velocity, 'v' is the channel width (breadth) and 'h' is the channel depth. The flow rate per month is taken into account of the algorithm.

978-1-5386-6624-1/18 $31.00 © 2018 IEEE   80

C. Problem Formulation

The system containing Ng number of biogas power plants, 'Nh' number of micro hydro power plants located on the river stream. The total cost of the system comprising of the operating cost and capital cost is given by

$$Can = Can(g) + Can(h) \qquad (5)$$
$$Can(g) = Ca(g)*CRF(g) + Co(g) \qquad (6)$$
$$Can(h) = Ca(h)*CRF(h) + Co(h) \qquad (7)$$

Where, 'Can' represents the total annual cost of the energy produced by the hybrid system. 'Can(g)' and 'Can(h)' are the annual costs of the energy produced by the biogas power plant and small hydro power plant respectively. 'Ca(g)' and 'Ca(h)' are the capital costs of the biogas and small hydro power plants and 'Co(h)' and 'Co(g)' are the operational costs of the energy resources respectively. 'CRF(g)' and CRF(h) are the capital recovery factors of the biogas energy system and micro hydro energy system respectively [33]. The value of 'CRF(g)' is taken as 0.15 and 'CRF(h)' is taken as 0.14. Table 2 shows the different input cost parameters used for the simulation in system models.

Component	Capital Cost (INR/kWh)	Replacement Cost (INR/kWh)	Operation and Maintenance cost
Micro hydro generator	80,000	80,000	2000(INR/kW)
Biogas generator	60,000	60,000	0.8(INR/kW/hr)
Converter	16000	16000	0
Li-Ion Battery	30,000	30,000	0

Table 2. Unit Cost for different components used in the study

The unit costs of the electricity produced by them is given by

$$Cu(h) = C_{an}(h)/E(h) \qquad (8)$$
$$Cu(h) = C_{an}(g)/E(g) \qquad (9)$$

Where, E(h) and E(g) are the costs of net energy generated by small hydro and biogas energy resources over a year. To minimize the cost of the electricity, following constraints are considered [33]:

1. The power that is generated by the hybrid system should always meet the demand at any given time.
$$ag.P_g(t)+ah.P_h(t)>P_d(t) \qquad (10)$$
2. The range of total power generated should lie between the minimum and maximum power that can be generated.
$$P_{min}< ag.P_g(t) +ah.P_h(t) < P_{max} \qquad (11)$$
3. Variables should be bounded within the following ranges:
$$0< ag <N_g$$
$$0< ah <N_h$$
$$Q_{min}< Q <Q_{max}$$

Where, ag and ah are decision variables and Q_{min} and Q_{max} are minimum and maximum flowrates.

IV. PROPOSED SOLUTION

A. Using MATLAB Software

The HBO algorithm is developed in MATLAB environment to find the potential of biogas and hydro power individually using the following steps:

i. Start
ii. Initializing the load values for 12 months
iii. Randomizing the values for all the months between the minimum and maximum bounds
iv. Setting energy parameters
v. For i=1:12
   calculating power supply for all the instances
vi. Plot the load vs supply graph against time for 12 months.

The algorithm is further used to calculate the cost benefit analysis using equations [1]-[11] for the three systems including the stand-alone biogas system, standalone micro hydro power plant and a hybrid system of biogas and micro hydro power plant with the help of 'linprog' function of MATLAB [33].

B. Using HOMER Software

For the present study Homer Pro tool is used to perform simulation and optimization of the three proposed setups. Homer executes a time-series (hourly) simulation that can integrate the effects of uncertainties of various statically changeable quantities for instance size of load, resource availability and the cost of fuel. The feasibility of the system is analyzed by this software. If the rate of generation satisfies the energy needs of the respective load connected across it, the system is declared and accepted as feasible. An estimation study on net present cost (NPC) that is the instantaneous value of combined costs for installation and operation of the system subtracting the present value of total revenue throughout its life span is done. The total NPC is calculated for the system as shown in Equation (12)-

$$NPC = {}_t\Sigma^T (C_{cap,t} + C_{o\&m,t} + C_{replace,t} +C_{fuel,t} - P_{salvage,t}) \qquad (12)$$

Where T denotes the lifetime of the particular project undertaken, is the present capital cost for the year t, is the present running and sustaining cost for year t, is the present fuel cost for the year t, is the present replacement cost for the year t and is the present salvage prize for the year t. Elevated cost of energy LCOE is an estimation of the economic lifetime energy cost as well as the lifetime energy production. Homer software uses Net Present cost (NPC) to assess the LCOE.

V. OPTIMISATION AND RESULT DISCUSSION

The simulation results for size and cost optimization using both HOMER software and HBO algorithm are discussed in this section. From Table 3, it is clear that using a Hybrid Micro hydro-Biogas plant, the COE is minimum of around 2.36 INR/kWh. The use of standalone biogas plant is most expensive with COE accounting to 12.97 INR/kWh.

Configu-ration	Biogas generat-or	Hydro gener-ator	COE (INR /kW h)	Capacity Shortage (kWh/yr)
Standalo-ne Biogas Plant	33kW	-	12.97	77.3
Standalone MHP	-	30kW	2.56	2.63
Hybrid Micro hydro-Biogas Plant	20 kW	15 kW	2.72	0

Table 3. Size and Cost Optimization Results for three
system configurations using HOMER software

Almost similar results for the three configurations are obtained for Cost of Energy using HBO algorithm as shown in figure 5.

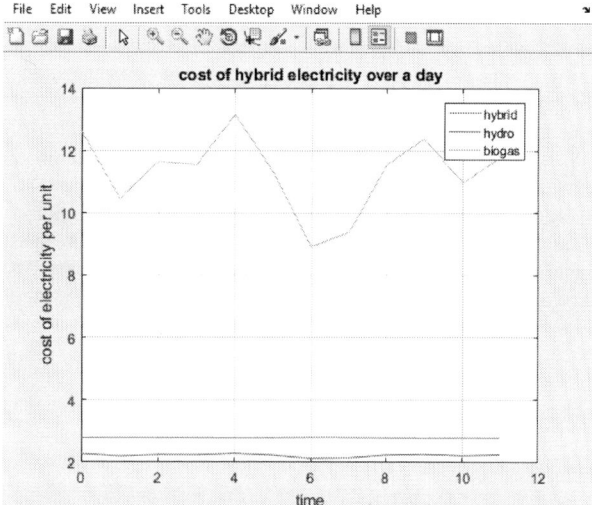

Figure 5. Cost of Energy for three different configurations
using HBO algorithm

Using HBO algorithm, the contribution of MHP is more in comparison to Biogas plant to meet the load demand throughout the year as shown in figure 6.

Figure 7 and figure 8 shows that using Standalone MHP and standalone biogas plant, the supply does not always meet the demand. There are certain periods during the year where the load is unmet and there is a capacity shortage.

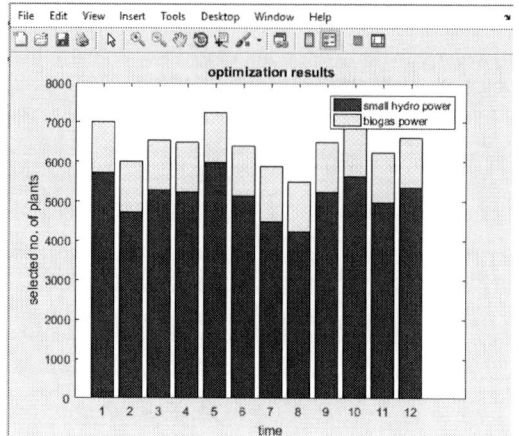

Figure 6. Contribution of Micro-hydro and Biogas plant
using HBO algorithm

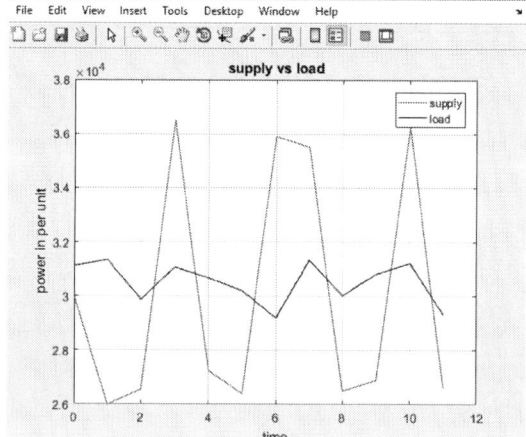

Figure 7. Supply Vs Demand curve for MHP using HBO
algorithm

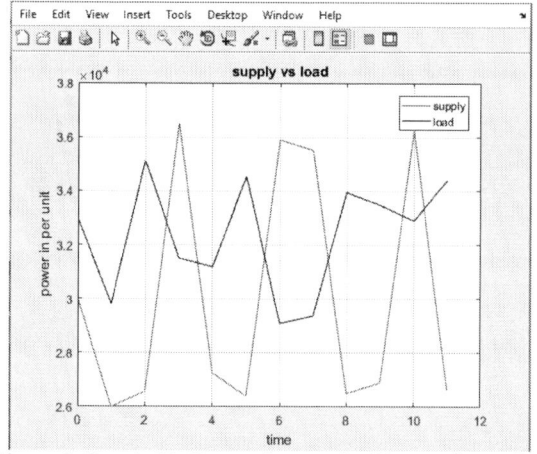

Figure 8. Supply Vs Demand curve for Biogas plant using
HBO algorithm

VI. CONCLUSION

The study in the paper deals with the optimization of three different renewable energy system configurations- Standalone Micro-hydro systems, Standalone Biogas system and a hybrid Micro hydro-Biogas system. The three different configurations are optimized using two different softwares-HOMER and MATLAB. The optimization results for the three different configurations obtained using HOMER software are validated using HBO algorithm developed in MATLAB environment. It is concluded that using standalone Biogas system is not advisable for the present case as using this system, there is a capacity shortage of 77.3 kWh/year and the cost of energy is highest 12.97 INR/kWh. A hybrid Micro hydro-Biogas system is able to serve the load demand continuously for all the 12 months although at a slightly higher cost of energy in comparison to a standalone Micro-hydro plant.

REFERENCES

1. Smruti Ranjan Pradhan, Prajna Pragatika Bhuyan, Sangram Keshari Sahoo, G.R.K.D.Satya Prasad(2013), "Design of Standalone Hybrid Biomass & PV System of an Off-Grid House in a Remote Area", Smruti Ranjan Pradhan et al International Journal of Engineering Research and Applications.
ISSN: 2248-9622, Vol. 3, Issue 6, Nov-Dec 2013, pp.433-437

2. Dalal E. Algapani, Wei Qiao, Francesca di Pumpo, David Bianchi, Simon M. Wandera, Fabrizio Adani, Renjie Dong (2017)"Long-term bio-H2 and bio-CH4 production from food waste in a continuous two-stage system: Energy efficiency and conversion pathways", 0960-8524/2017 Elsevier Ltd.

3. Yu-Jun Zheng, Sheng-Yong Chen, Yao Lin, and Wan-Liang Wang(2013),"Bio-Inspired Optimization of Sustainable Energy Systems: A Review", Hindawi Publishing Corporation, Mathematical Problems in Engineering, Volume 2013, Article ID 354523, 12 pages

4. R.K. Akikur, R.Saidur, H.W.Ping, K.R.Ullah (2013) "Comparative study of stand-alone and hybrid solar energy systems suitable for off-grid rural electrification: A review", 1364-0321/$-see front matter & 2013 Elsevier Ltd.

5. Vita, C. Italiano, D. Previtali, C. Fabiano, A. Palella, F. Freni, G. Bozzano, L. Pino, F. Manenti (2017), "Methanol synthesis from biogas: A thermodynamic 1 analysis", Renewable Energy

6. Pazera, R. Slezak, L. Krzystek, S. Ledakowicz, G. Bochmann, W. Gabauer, S. Helm, S. Reitmeier, L. Marley, F. Gorga, L. Farrant, V. Suchan, J. Kara, Biogas in Europe: Food and beverage (FAB) waste potential for biogas production, Energy Fuels 29 (2015) 4011–4021.

7. S. Specchia, Fuel processing activities at European level: A panoramic overview, Int. J. Hydrogen Energy 39 (2014) 17953–17968.

8. Rajanna Siddaiah, R. P. Saini (2016) "A review on planning, configurations, modeling and optimization Techniques of hybrid renewable energy systems for off grid applications", 1364-0321/& 2016 Elsevier Ltd.

9. Kaushik Rajashekara 92005, "Hybrid Fuel-Cell Strategies for Clean Power Generation", IEEE ON TRANSACTIONS INDUSTRY APPLICATIONS, VOL. 41, NO. 3

10. M.S. Ismail, M. Moghavvemi, T.M.I. Mahlia, K.M. Muttaqi, S. Moghavvemi, "Effective utilization of excess energy in standalone hybrid renewable energy systems for improving comfort ability and reducing cost of energy: A review and analysis", Volume 42, February 2015

11. Nitin Tanwar (2007), "Clean development mechanism and off-grid small-scale hydropower projects: evaluation of additionality"/ Energy Policy 35 (2007) 714–721, 0301-4215/$ - see front matter 2006 Elsevier Ltd.

12. Md Tanbhir. Hoq, Nawshad U. A., Md. N. Islam, IbneaSina, Md. K. Syfullah, Raiyan Rahman (2011) "Micro Hydro Power: Promising Solution for Off-grid Renewable Energy Source", International Journal of Scientific & Engineering Research, Volume 2, Issue 12, December-2011 1 ISSN 2229-5518

13. Subhadeep Bhattacharjee, Anindita Dey (2014) "Techno-economic performance evaluation of grid integrated PV-biomass hybrid power generation for rice mill" 2213-1388/ 2014 Elsevier Ltd.

14. Ankit Bhatt, M.P.Sharma, R.P.Saini (2016) "Feasibility and sensitivity analysis of an off-grid micro hydro–photovoltaic–biomass and biogas–diesel–battery hybrid energy system for a remote area in Uttarakhand state, India", 1364-0321/& 2016 Elsevier Ltd.

15. Jose´ L. Bernal-Agustı´n, Rodolfo Dufo-Lo´ pez (2009) "Simulation and optimization of stand-alone Hybrid renewable energy systems", 1364-0321/$ – see front matter 2009 Elsevier Ltd.

16. M. Fadaee, M.A.M. Radzi (2012) "Multi-objective optimization of a stand-alone hybrid renewable energy system by using evolutionary algorithms: A review", 1364-0321/$ – see front matter © 2012 Elsevier Ltd.

17. Lanre Olatomiwa, Saad Mekhilef, M.S.Ismail, M.Moghavvemi, "Energy management strategies in hybrid renewable energy systems: A review" Volume 62, September 2016

18. Getachew Bekele, Getnet Tadesse (2012) "Feasibility study of small Hydro/PV/Wind hybrid system for off-grid rural electrification in Ethiopia", 0306-2619/$ - see front matter 2011 Elsevier Ltd.

19. J.A. Razak, Y. Ali, M.A. Alghoul, Mohammad Said Zainol, Azami Zaharim, K. Sopian (2010) "Application of Crossflow Turbine in Off-Grid Pico Hydro Renewable Energy System", RECENT ADVANCES in APPLIED MATHEMATICS
ISBN: 978-960-474-150-2, ISSN: 1790-2769

20. Kanzumba Kusakana "Feasibility analysis of river off-grid hydrokinetic systems with pumped hydro storage in rural applications", 0196-8904/ 2015 Elsevier Ltd.

21. Anurag Chauhan, R.P.Saini (2014) "A review on Integrated Renewable Energy System based power Generation for stand-alone applications: Configurations, storage Options, sizing methodologies and control", 13640321/&2014 Elsevier Ltd.

22. Off-Grid Electrification using Micro hydro power schemes- Sri Lankan Experience, A survey and Study on existing off-grid electrification schemes (2012), PUBLIC UTILITIES COMMISION OF SRI LANKA.

23. Debadayita Raha, Pinakeswar Mahanta, Michèle L.Clarke (2014), "The implementation of decentralised biogas plants in Assam, NE India: The impact and effectiveness of the National Biogas and Manure Management Programme", 0301-4215 & 2014 The Authors. Published by Elsevier Ltd.

24. Shivika Mittal, Erik O.Ahlgren, P.R.Shukla (2018), "Barriers to biogas dissemination in India: A review", Volume 112 0301-4215/ © 2017 The Author(s). Published by Elsevier Ltd.

25. Pietro Capaldi, Alfonso Daliento, Renato Rizzo (2014) "An Innovative 10 kW Microco generator Suitable for Off Grid Application and Fed with Syngas or Biogas", 978-1-4799-6557-1/14/$31.00©2014 IEEE

26. Francisco J. Lozano, Rodrigo Lozano (2017) "Assessing the potential sustainability benefits of agricultural residues: Biomass conversion to syngas for energy generation or to chemicals production", Journal of Cleaner Production

27. Karina Ribeiro Salomon, Electo Eduardo Silva Lora(2009) "Estimate of the electric energy generating potential for different sources of biogas in Brazil", Volume 33

28. S. Kumaravel & S. Ashok(2012) "An Optimal Stand-Alone Biomass/Solar-PV/Pico-Hydel Hybrid Energy System for Remote Rural Area Electrification of Isolated Village in Western-Ghats Region of India", International Journal of Green Energy, 9:5, 398-408

29. A. Arya, S. Divekar, R. Rawat, P. Gupta, M.O. Garg, S. Dasgupta, A. Nanoti, R. Singh, P. Xiao, P.A. Webley, Upgrading Biogas at Low Pressure by Vacuum Swing Adsorption, Ind. Eng. Chem. Res. 54 (2015) 404–413.

30. T. Al Seadi, D. Rutz, H. Prassl, M. Köttner, T. Finsterwalder, S. Volk, R. Janssen, Biogas Handbook, University of Southern Denmark Esbjerg, Esbjerg, Denmark, 1972.

31. Cheric Energy, Off Grid and Standby Power Solutions with Solar Energy, Biogas Energy, Wood Gas Energy and LED lightning,© Cheric Energy | v15.08 Maintained By KwaZulu-Natal.co.za

32. Shane McDonagh, Richard O' Shea, David M. Wall, J.P. Deane, Jerry D. Murphy (2018) "Modelling of a power-to-gas system to predict the levelised cost of energy of an advanced renewable gaseous transport fuel", S. McDonagh et al., 0306-2619/© 2018 Published by Elsevier Ltd.

33. Amevi Acakpovi, E.B.Hagan, Mathias Bennet Michael, "Cost Benefit Analysis of Self-Optimized Hybrid Solar-Wind-Hydro Electrical Energy Supply as compared with HOMER Optimization", Volume 114-march 2015, International Journal of Computer Applications.

Investigations on Effect of Frequency Variation on the Performance of Combined Cycle Gas Turbine

J. N. Rai
Department of Electrical Engineering
Delhi Technological University,
Delhi, India
jnraiphd1968@gmail.com

Ajendra Singh
Department of Electrical Engineering
Delhi Technological University,
Delhi, India
ajendrasingh25@gmail.com

Vineet Shekher
Department of Electrical Engineering
NIET,
Greater Noida, UP, India
vshekher2407@gmail.com

Pranav Jain
Department of Electrical Engineering
Delhi Technological University,
Delhi, India
jainpranav97@gmail.com

Nikhil Dwivedi
Department of Electrical Engineering
Delhi Technological University,
Delhi, India
nikhil2360@gmail.com

Abstract – The optimum power output of Combined Cycle Gas Turbine depends upon the system frequency, and its ambient temperature. As a reduction in shaft speed leads to a decrease in airflow, the temperature control of a gas turbine, by lowering the fuel flow, limits the exhaust temperature. The paper shows the dependency of frequency on the power output of the gas turbine model under prevalent assumptions and defines a gas turbine model that facilitates parametric state assumption of the frequency dependency from the ambient temperature dependency. The application of this model includes implementing practical simulations of the system's dynamic performance, including bizarre frequency conditions and variation in temperature.

Keywords-- Gas Turbines, Combined Cycle, Dynamic Modelling, State Estimation

I. INTRODUCTION

Since last few decades the dynamic characteristics of combined cycle gas turbine (CCGT) turned into serious concern. Majority of electricity generation systems worldwide, have an increasing proportion of CCGT plants (Fig. 1). The main reasons for the choice, amongst others are:

- Higher efficiency 50%,
- Greater flexibility,
- Low environmental impact,
- Lower installation time,

- Lower greenhouse gas emissions than conventional thermal generators,
- Greater operating flexibility,
- Reduced staff size.

Considering of dynamic behavior of CCGT units is essential to maintain the authenticity and safety of the system in electrical systems. Nowadays, system operators have limited control over the category and location of new power plants due to the rivalry in the market.

The system operator should keep into consideration the safety and quality standards of the power supply in any electrical system. Contemporary system blackouts in United States, Canada, UK and Italy underline the concern of security and authenticity of the system. If any fault occurs in the system, then system frequency should be kept within established limits is preferred and if there is any encroachment of these limits, the amount of jaunt must be limited and the frequency should reach the nominal value instantaneously.

Large interconnected power system has high inertia, due to this there is only a small change in frequency from the nominal value. This is also true due the fact that any abrupt imbalance between supply and demand is reduced compared to the total capacity of the system. It is likely that the maximum power in a cramped electrical system has higher percentage of total generation while the inertia of the system is appreciably less

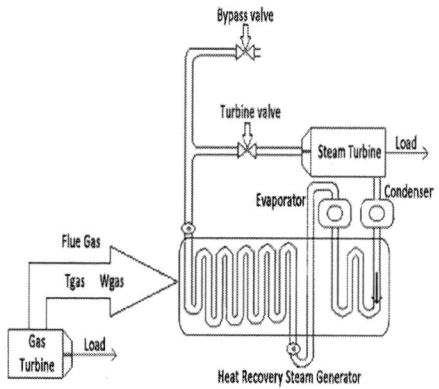

Fig.1: Combined Cycle Power Plant Model

As a result, the effect of an accident, like sudden loss of generation, on the frequency of the system is much more evident. As a result, response characteristics of CCGT generators in relation to frequency change is crucial for small systems which has adverse effect on the system. In larger system, frequency deviation has less impact, with an increase in the percentage of CCGTs, their influence will become more pronounced.

The present day combined cycle plants are operating at capacity factor levels much lower than their owners originally expected. According to *Baxter et al [2004]*, the fleet median capacity factor for large combined cycle power plants in 2002 was 48 %, far below the 80 % expectation. Only 7 % of these units had capacity factors of 80 % and more. In addition to affecting the ability of plant owners to predict revenue and cost streams, uncertainty regarding operating levels affected equipment suppliers by making demand for parts and services more crucial to predict. Suppliers must plan well in advance to assure that sufficient spare parts and trained staff are available to support customers' maintenance requirements. An understanding of operations also makes it possible for the equipment supplier to focus more clearly on improvements that will benefit owners the most.

II. DYNAMIC RESPONSE OF CCGT

Basically, there are two types of turbine in combined cycle gas turbine (CCGT), gas and steam turbine, which are used to run either combined synchronous generator or both the turbines can run their respective generators. Air enters the compressor at high pressure, this compressed air is then passed into the combustion chamber where the fuel and the compressed air are blended together to form an air – gas mixture. This mixture ignites and produces exhaust gases which are responsible for the rotation of gas turbine. (Fig. 2)

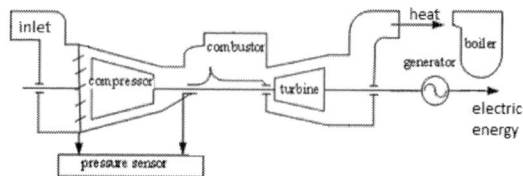

Fig. 2: A contour of Gas Turbine

The gas turbine exhaust gases are passed through HRSG, this steam is crossed through the steam turbine and the energy is produced by generator. If there is any deviation in temperature of exhaust gases it will adversely affect the efficiency of HRSG as well as of steam turbine since there is maximum permissible temperature limit in both OCGT and CCGT power plants. Thenceforth, to accomplish optimal efficiency in CCGTs, the temperature of exhaust gases must be kept at maximum permissible level otherwise which would affect the turbine blade materials badly.

The exhaust temperature is maintained at this optimum level by regulating the air and fuel flows. The variable guide vanes (IGV), installed in the initial gate leading to the compressor, controls the flow of air. As the speed of the gas turbine increases, the IGVs are positioned to guarantee the correct operation of the air compressor (avoiding the loss zones) until the maximum empty speed is accomplished. Thenceforth, IGVs shift from their minimum opening to the no-load position to their maximum opening in line with the intake of the fuel to retain the target temperature of the exhaust gas.

To maintain constant outlet temperature, the change in fuel flow should adjust the airflow. Immediately after the operating point of gas turbine reaches the base load, airflow cannot be increased further. For the frequency of falling system, the speed regulator injects the fuel into the combustion chamber by taking quick action in short interval of time. But if the fuel flow increases expeditiously, it will become difficult for IGV to maintain the accurate air-to-fuel ratio. In general, CCGT are respond units.

CCGT gives an inertial response proportional to both rate of change in frequency and magnitude. This response is of utmost importance which helps in maintaining security of the system by easing the frequency at which frequency falls. Moreover, the compressor slows down as the frequency of system decreases, because it is synchronized with the system. As the compressor speed reduces, this decreases the pressure ratio across the compressor, reducing the flow of air to the combustion chamber. As a result, the pressure ratio along the gas turbine decreases, which results in waning of output power.

If unit operates on incomplete load, the IGV is not completely open, which allows you to open it even more, which compensates for the reduction of airflow in the compressor. The response speed of VAT determines how rapidly this can have an effect. However, the IGVs are in the completely open position when CCGT works with base load and cannot be further adjusted to increase the air flow through the compressor. The reduction in the air flow from the compressor, increments the air-fuel ratio, which determines the intake temperature of the turbine and, consequently, the exhaust temperature. When CCGT works slightly below the base load, if there is an increase in air-fuel ratio, exhaust temperature will rise. As a result, the temperature control system rapidly overwrites the inputs and decreases the fuel flow to restore the appropriate air-fuel ratio. Therefore, reduction in fuel flow results in the reduction of output power.

III. OPERATING REGIMES

According to **Baxter Andrew [2004],** before forecasting power plant operations, it is useful to look at how different operating modes affect plant service and parts need. Maintenance requirements for gas turbines depend on the manner in which the unit is actually operated. Operational variations include the frequency of start-up/shutdown cycles, the number of hours of operation and power level, the number of cycles, the number and duration of over firing events, and others. The number of potential variations is near infinite, making it infeasible to attempt to fully capture the operational detail that might drive maintenance in a predictive model.

The system will identify four different operating modes based on the number of fired hours and the number of starts:

1) **Peaking units** are relatively low usage units with maintenance requirements determined primarily by the number of start cycles.

2) **Cycling units** are high usage units with a significant number of starts, with maintenance requirements determined primarily by the number of start cycles.

3) **Intermediate units** are medium usage units with relatively few starts with maintenance requirements determined primarily by the number of fired hours.

4) **Base Load units** have nearly continuous operation with relatively few starts, with maintenance requirements determined primarily by the number of fired hours.

The precise location of the dividing line between the starts-driven and fired-hours driven groups varies by specific combustion turbine model. Fleet-wide maintenance requirements can be reasonably estimated

based on the number of units in each operating model group and the nominal number of starts and fuel hours for each operating model group.

IV. CCGT MODEL

To evaluate the dynamic response of CCGT to system frequency and consider the consecutive impact on system frequency control we have to develop a CCGT model which can be used to study how system frequency depends on CCGT, which can be further turned to exhibit particular CCGT units.

The following equations describe the gas turbine Thermodynamics:

Assuming the adiabatic compression of air in the compressor, the compressor discharge temperature ($T_d{'}$) is given as:

$$T_d{'} = T_i{'}\left(1 + \frac{x-1}{n_c}\right) \qquad (1)$$

Where, T_i(K) is inlet temperature (ambient temperature), n_c is the compressor efficiency, x is compressor temperature ratio given as

$$x = (P_{ro}W)^{y-1/y} \qquad (2)$$

P_{ro} and y are the design compressor ratio and specific heat ratio respectively and the airflow in per unit of its rated value is denoted by W.
The inlet temperature of Gas turbine T_f' (K) is given by

$$T_f{'} = T_d{'} + (T_{fo} - T_{do})\frac{W_f}{W} \qquad (3)$$

Where, subscript "o" indicates the rated value and W_f is fuel flow in per unit of its rated value.
The exhaust temperature of gas turbine T_e (K) is given below. [11]

$$T_e = T_f{'}[1 - (1 - \frac{1}{x})\eta_t \qquad (4)$$

The turbine efficiency is denoted by n_t.

V. MODEL STRUCTURE

Steady state set point of the unit, the ambient temperature and pressure serve as the inputs to the model. In the earlier models, ambient temperature was overlooked in the model structure but in this model the

the effect of ambient temperature on the rating of the gas turbine is incorporated as a correction factor, which is enforced on the unit set-point input. The effect of ambient pressure on the output of gas turbine was also incorporated in a similar way. Except temperature (^{0}C) and pressure (mbar) all parameters are measured in per unit. On the gas turbine, there are two control loops i.e. speed control and temperature control. Under normal circumstances, the fuel supply is monitored by speed control which takes the form of a simple droop governor. Output from both these controllers (i.e. temperature and speed) is fed into a minimum block and the lesser of these two signals determines the fuel flow into the gas turbine.

The requirement in a gas turbine is of no-load fuel, which is about one fourth of the maximum value, so these controllers regulate the fuel flow between this minimum and maximum point.

The IGV controller regulates the flow of air. First the exhaust temperature is calculated and then compared to the nominal value of the exhaust temperature, if there is a difference then IGV controls the flow of air to return the temperature to the nominal value. However, since the airflow depends on the speed of the compressor, the

expected airflow, due to the IGV position, is modulated using the actual speed, producing the flow. The model is also equipped with an excessive intervention capacity, which allows to increase the temperature limits during a short period of time during a frequency transient. Shalan H. Emam provides some relevant models with their characteristics.

VI. FREQUENCY DEPENDENT MODEL

The above mentioned model does not give appropriate information about the frequency dependency on the gas turbine. We considered the frequency dependency on combustion gas turbine model only if we observe events with anomalous system frequency action. Main motto of the model was to analyze the effects of ambient temperature and axis speed on the calculated flow. When there is any change in shaft speed, it results in change in airflow due to which the frequency alters. This results into change in pressure ratio across compressor and then consequently fuel level. Apposite power output is explicitly affected by these changes. Any change in ambient temperature, it has serious brunt in contrast to changes in rotor speed due to the identical relation between ambient temperature and optimum power output.

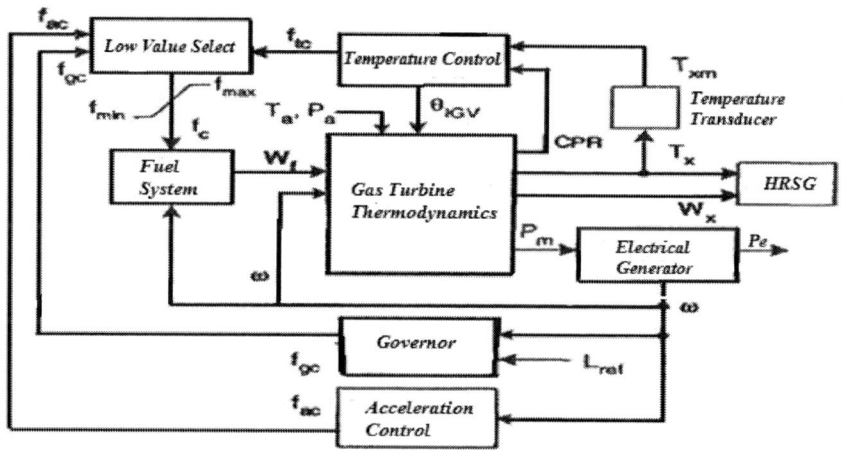

Fig. 3: Frequency dependent Gas Turbine

Fig.3 explains the frequency dependent model. The control scheme is analogous to Rowen's model. In this model, the pressure ratio between the compressor and the exhaust gas flow is calculated along with the exhaust temperature and the mechanical power, while in the Rowen model only the output power and temperature are required. Above model incorporates equations representing the impact of the IGV. The test data of the real machines are used to obtain the various parameters of the model. We have observed that output of gas turbine is dependent on frequency and ambient temperature and that

the temperature control and the governor play vital role while abnormal frequency operations.

A. Structure of FD Model:

As we can see in Fig. 3, the gas turbine comprises of axial compressor, turbine and a combustion system.

Air flow (W_A) and Fuel flow (W_F) are taken as the input variables whereas the Mechanical power output (P_{MG}) and Exhaust heat to the Heat Recovery Steam Generator (HRSG) as determined by the exhaust gas flow (W_X) and exhaust temperature (T_e) are taken as the output variables.

By controlling the air flow and fuel flow, we can get desired output by managing the exhaust temperature to increase the efficiency of heat transfer to HRSG Inlet Guide Vanes (IGV) regulate air flow may and it is also a function of ambient air temperature (Ta) ambient pressure pa and shaft speed (N) [11]

B. FD Model Assumptions:

There are few assumptions described below which we have included in our above model-

- It's valid for shaft speed variation between 90-105%.
- It is not intended for start-up and shutdown simulation.
- W_X is equal to W_A because W_F is lesser than W_A.
- Compressor is running by the constant shaft power.
- W_F solely determines P_{MG} and some significant dynamic effects between their changes are time lags.

VII.   FD MODEL COMPONENTS

After the modelling we can drew thermodynamics equations which shows the dynamic behaviour of gas turbine.

The model can be branched into five blocks:

1)   Air flow block

2)   Compressor Pressure Ratio Block(CPR)

3)   Exhaust Temperature Block

4)   Output Power Block

5)   Maximum output power continuous block

VIII.   RESULTS AND DISCUSSION OF FD MODEL:

This model clarifies the dependency of maximum power output of a gas turbine (PMAX) on shaft speed and ambient temperature. The simulation of the above model is shown in Fig. 4.

As shown in fig. 5, frequency changes are proportionate to changes in shaft speed and leads to changes in airflow. This results into alteration in pressure ratio across the compressor (Fig. 6) and consequently in fuel level. As we can see in fig. 7, the maximum power output is directly affected by these changes. An important relation between the power output and ambient temperature is reported in fig. 8, which shows that atmosphere's ambient temperature is inversely related to the output of gas turbine that is the ambient temperature of atmosphere rises if the output of gas turbine decline.

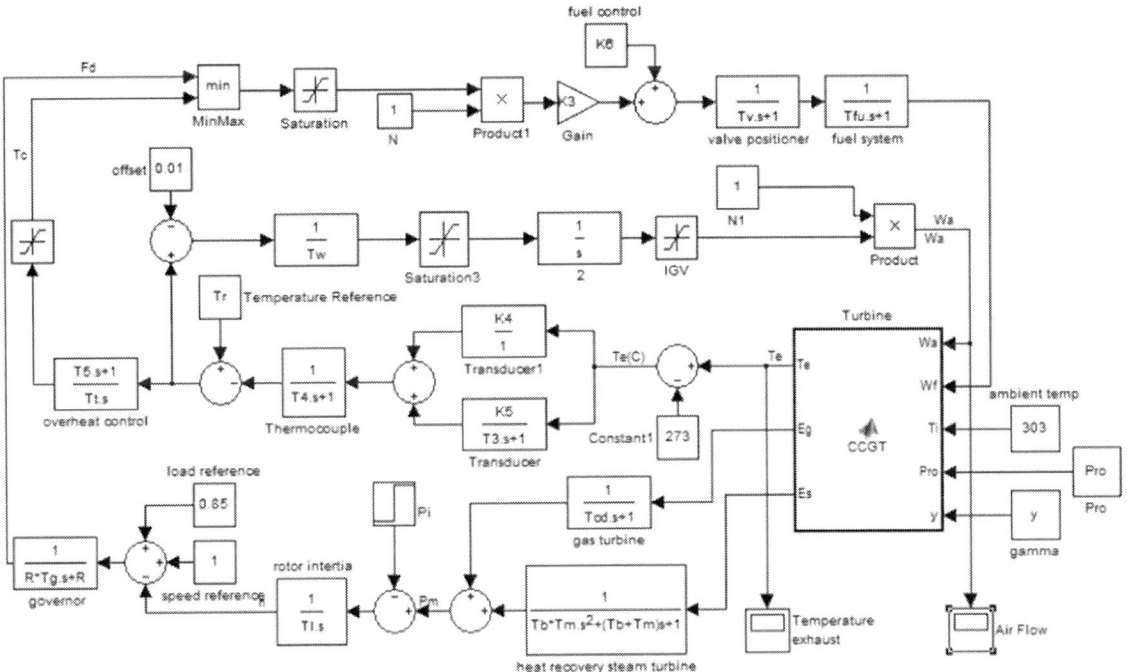

Fig.4: Model Structure Using SIMULINK

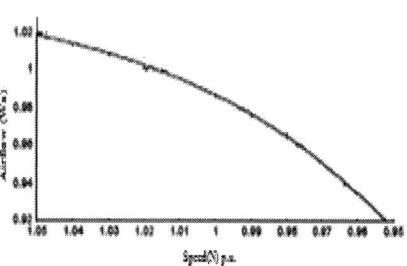

Fig. 5: Response of Speed Changes on Airflow

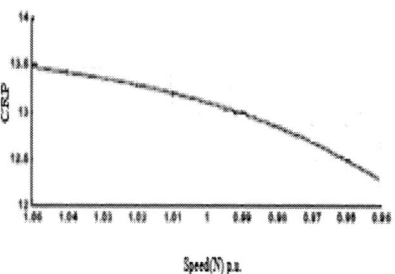

Fig. 6: Response of Speed Changes on Airflow on Compressor Pressure Ratio

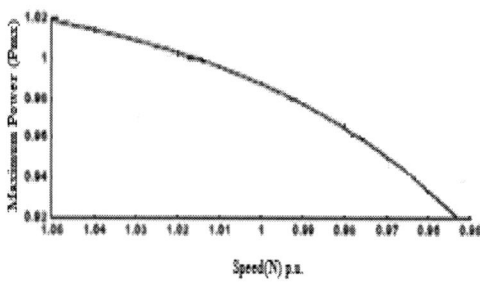

Fig. 7: Response of Speed Changes on Airflow on Maximum Continuous Power

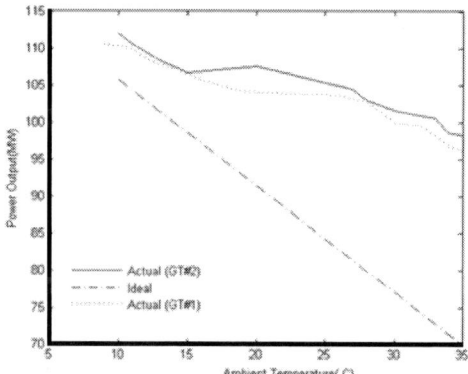

Fig. 8: Response of Speed Changes on Airflow on Maximum Continuous Power

IX. CONCLUSION

From the obtained results we can conclude that the gas turbine's performance is distressed by the ambient circumstances. A parameter entering gas turbine which affects the mass of air flow (W_A), will affect the efficiency of gas turbine. Decrease in ambient temperature makes air dense, which increase the Compressor Discharge Pressure (CDP), so reduction in ambient temperature increases the efficiency of gas turbine. It has great advantage of temperature controlled directly and his quick response delivers information which can be used to set maximum/minimum operating ranges of Automatic Generation Control (AGC) units.

Appendix
GT Model – Frame 6, MS9000 series, 50Hz (speed of rotation 3000 rpm).
Gas Turbine-104 MW, Two Units
Steam Turbine-122 MW, One Unit

REFERENCES

[1] Albanesi C., Bossi M., Magni L., Paderno J., Pretolani F., Kuehl P. and Diehl M., "Optimization of the Start-up Procedure of a Combined Cycle Power Plant",Proceedings of the 45th IEEE Conference on Decision and Control, USA, 2006 , IEEE. pp. 1840 – 1845.

[2] Baxter Andrew, Sanford Mark, Smith Richard and Szczepanski Richard, "Analysis of Combined Cycle Operating Patterns", 8TH INTERNATIONAL CONFERENCE ON PROBABILISTIC METHODS APPLIED TO POWER SYSTEMS, IOWA STATE, AMES, IOWA, 2004 , pp. 850 – 854.

[3] Niu L.X. and Liu X.J., "Multivariable Generalized Predictive Scheme for Gas Turbine Control in Combined Cycle Power Plant",Proceedings of the 45th IEEE Conference on Decision and Control, 2008 IEEE, pp. 791 - 796

[4] Parra M. Sanchez and Verde C., "Analytical Redundancy for a Gas Turbine of a Combined Cycle Power Plant",Proceedings of the 2006 American Control Conference Minneapolis,Minnesots, USA, 2006 IEEE, pp. 4442 – 4447.

[5] Rowen W. I., "Simplified mathematical representations of heavy-duty gas turbines", ASME J. ENG. POWER, 1983, Vol. 105, pp. 865–869.

[6] Shalan H. Emam , Hassan M. A. Moustafa and Bahgat A. B. G., "Parameter Estimation and Dynamic Simulation Of Gas Turbine Model In Combined Cycle Power Plants Based On Actual Operational Data",Journal of American Science , 2011;7(5) [303-310]. (ISSN: 1545-1003), pp. 303 – 310.

[7] Shalan H. Emam, Hassan M. A. Moustafa and Bahgat A. B. G., "Comparative Study On Modelling Of Gas Turbines In Combined Cycle Power Plants",Proceedings of the 14th International Middle East Power Systems Conference (MEPCON'10), CAIRO UNIVERSITY, EGYPT, December 19-21, 2010, Paper ID 317, pp. 970 – 976

[8] Yee Soon Kiat, Milanovic Jovica V., and Hughes F. Michael, "Overview and Comparative Analysis of Gas Turbine Models for System Stability Studies", IEEE Transactions on Power System , Vol. 23, No. 1, Feb 2007 , pp. 108 – 118.

[9] J.N.Rai,Naimul Hasan,B.B. Arora, Garai, R., Gupta, R. K., & Kapoor, R. (2013). "Study the Effect of Temperature Control on the Performance of the Output of Combined Cycle Gas Turbine." International Journal of Theoretical and

Applied Mechanics,ISSN:0973-6085,Vol. 8,No.(1),2013 pp. 15-23.

[10] J. N.Rai, Naimul Hasan, , B.B. Arora, Garai, R., Kapoor, R. & Ibraheem. (2013). "Performance Analysis of CCGT Power Plant using MATLAB/Simulink Based Simulation." International Journal of Advancements in Research & Techno logy,ISSN:2278-7763, Vol.2,Issue 5,May 2013, pp. 285- 290.

[11] Lalor, G., Ritchie, J., Flynn, D., & O'Malley, M. J. (2005). The impact of combined-cycle gas turbine short-term dynamics on frequency control. Power Systems, IEEE Transactions on power system,Vol.20,No.(3)August 2005,pp.1456-1464.

[12] Dilip Kumar Mohanty , Vijay Venkatesh "Performance analysis of a combined cycle gas turbine under varyimg operating conditions" International journal (MEIJ) , vol. 1,No.2, August 2014.

[13] Thamir K. Ibrahim, Marwah N. Mohammed "Thermodynamic evaluation of the performance of combined cycle power plant ", international journal of energy science and engineering, Vol. 1, No. 2, 2015, pp. 60-70.

A Technique for Power Factor Measurement of Household and Industrial Load using LabVIEW

Abhishek Kumar Gupta
Department of Electrical Engineering
Jamia Millia Islamia
New Delhi, India
akeed01@gmail.com

Ravi Saxena
Department of Electrical Engineering
G.B.P.U.A.&T.
Pantnagar, Uttarakhand, India
saxenaravieed@gmail.com

Rajveer Singh
Department of Electrical Engineering
Jamia Millia Islamia
New Delhi, India
rsingh@jmi.ac.in

Sanjiv Kumar
Department of Electrical & Electronics Engineering
Subharti University
Meerut, U.P., India
activesanjiv007@rediffmail.com

Abstract- **Accurate and precise power factor measurement is basic need in various electrical systems. This measurement usually requires special and expensive equipment. Power factor is basically gives the relationship between the input current and voltage waveform to an electrical load. It is defined as the ratio of active power (P) and apparent power (S). Sometimes it is difficult to measure the accurate power factor of low power rating loads. This paper presents an accurate, cost effective and simple power factor measurement system for single-phase ac load using Tektronix-TDS2024B oscilloscope and LabVIEW software. In the proposed technique the current is measured using Hall Effect current sensor probe and voltage is measured directly by connecting probe parallel to load. Using this technique accurate power factor of low power rating single-phase ac loads can be calculated easily.**

Keywords- Power factor, LabVIEW, Tektronix Oscilloscope.

I. Introduction

In electrical circuits, having linear loads the voltages and currents are sine wave and the effects of power factor introduces only when the phase angle between voltage and current increases or decreases in either side. The conventional method of power factor measurement usually requires ammeter, voltmeter and wattmeter. For example, in case of AC current absorbed by non-linear load ac current is to be measured by conventional method like using multimeter, etc the result may differ from actual one. And then the calculated power factor may also differ. In this proposed method a true RMS multimeter is used to measure the true RMS voltage, current and apparent power (VI).

In electrical sectors, power factor plays very important role. The electrical equipments having low power factor leads to draw more power from AC supply. So the electrical equipment designers always have the prime concerns about its power factor. At home or in industries most of the electrical equipment is inductive. The current in the inductive load lag behind the applied voltage. The phase angle between voltage and current is denoted by φ. This angle may decrease or increase according to the connected load. The increase in inductance of the load will lead to increase in the phase angle between supply voltage and current through the load which results low power factor and larger current through it. In this case the power loss will increase. The electrical equipments with low power factor also have larger wires and comparatively bulky equipments. Now a day's the loads found with resistive effect and the current with sinusoidal input cause the phase shift.

Hence, to avoid such conditions it should be the prime concerns of consumers to know about the power factor as well as to know about the measurement of power factor.

$$Power\ Factor = \frac{P}{S} = \cos\varphi \tag{1}$$

Where, P is the Real Power and S is the Apparent Power

II. Mathematical calculation of power factor

The power factor is defined as the ratio of active power and volt-amperes.

$$Power\ Factor = \frac{P}{S} = \frac{P}{VI} \tag{2}$$

For sinusoidal waveforms the power factor is the cosine of the phase angle φ between voltage and current.

$$Power\ Factor = \cos\varphi = \frac{P}{VI} \tag{3}$$

$$P = VI\cos\varphi \tag{4}$$

$$I = \frac{P}{V\cos\varphi} \tag{5}$$

Equation (5) shows that the current is affected by the power factor. The supply voltage is kept constant. Hence, according to load requirement power P, the current I drawn by the load to meet the required power is inversely proportional to load power factor cos φ. Thus, according to the above equations it can be concluded that the current drawn by the load is higher at low power factor than that of at the higher power factor.

III. Power factor measurement issues in conventional method

It is simple to calculate power factor of any electrical load by conventional method using equation 1. This method requires a wattmeter, a voltmeter and an ammeter to measure active power (P), voltage (V) and current (I) respectively. But the major problem of this method is accuracy and rating of load & instrument used. Sometimes it is difficult to measure real power of very low current rating electrical load accurately which results errors in power factor measurement. So a new technique of power factor measurement has been proposed for accurate measurement even under very low current rating load.

978-1-5386-6624-1/18 $31.00 © 2018 IEEE

IV. Experimental setup

Experimental setup is consisting of hardware and software both. An oscilloscope is used to read the voltage and current waveform. And LabVIEW (Laboratory Virtual Instrument Engineering Workbench) is a graphical programming language used to calculate power factor and phase angle of the waveform.

IV.1. Hardware used

In this work an advance oscilloscope named Tektronix TDS2024B 200 MHz 4 Channel Digital Storage Oscilloscope (shown in Figure 1) is used to track the voltage and current waveform. This oscilloscope is easy to use and simple controls make it widely used with reducing response time and increased efficiency. The features of Tektronix TDS2024B like auto set menu, checking of probe, a separate menu for help and colored LCD optimize the difficulties during operations.

In this paper the proposed technique is advance and accurate. The Hall Effect current sensor is used to measure the current accurately.

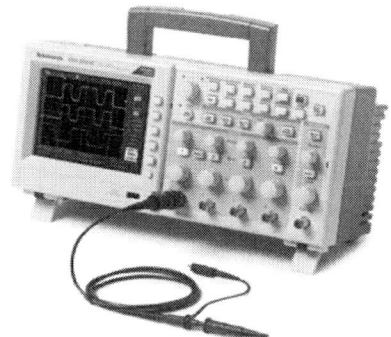

Fig. 1: Tektronix TDS2024B Digital Storage Oscilloscope

There are 4 channels in this oscilloscope out of which only 2 channels are to be used to get the current and voltage waveform. The above oscilloscope capture the waveforms easily, the waveforms can also be saved and can be analyzed for the required purposes. The software provided by this oscilloscope is helpful for remote controlling, advance electrical parameter analysis, on-line analysis of waveforms, etc.

IV.2. Software used

LabVIEW (Laboratory Virtual Instrument Engineering Workbench) is used as software to measure power factor. This software is a graphical programming language. In this software lines of text usually used in programming are replaced by the icons to build applications. In the proposed technique LabVIEW is used to calculate the power factor, phase difference and phase diagram of different single phase ac loads. The LabVIEW software consists of different libraries. The applications of these library functions are to analyze data, to generate waveforms, for function looping, for uninterrupted operations, to read and store the data and for data acquisitions. For multiple tasks different LabVIEW programming would be created in LabVIEW is called a "Virtual Instrument (VI)". The extension of these file is .vi.

and consists of two blocks. One is for user interface is called Front Panel and another one is called Block Diagram.

The current and voltage values given by the oscilloscope are instantaneous values. Figure (2) shows the developed module for measuring the power factor of different ac loads.

V. LabVIEW programming

In this proposed technique, only one hardware (Tektronix TDS2024B Digital Storage Oscilloscope) is used to read both voltage and current waveform so a common Visa Serial Port is selected. As there are four channels in this oscilloscope out of which only two is to be used. Channel 1 is used here for voltage waveform and channel 2 is for current waveform. Phase difference between the voltage and current waveform is calculated using Fundamental Vector block. The current and voltage waveforms are continuously shown on the front panel. There is also power factor meter, voltage vector diagram and current vector diagram which shows the power factor and phase difference of voltage and current at different loads in real time.

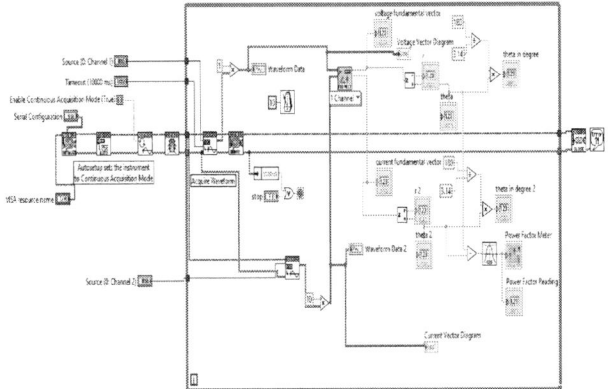

Fig. 2: Developed LabVIEW program

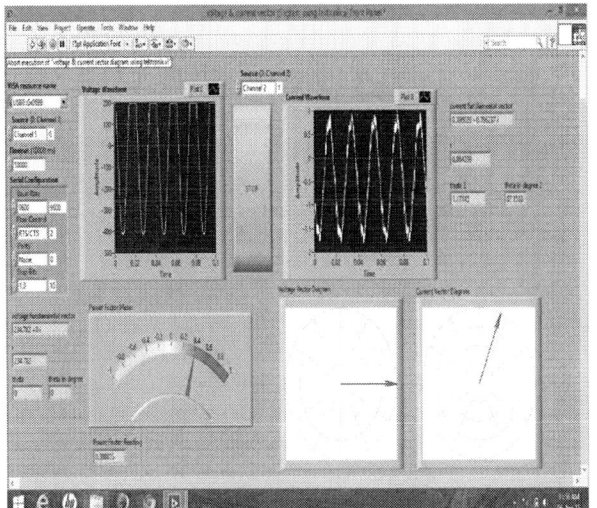

Fig. 3: Front panel of LabVIEW program

VI. Result analysis

The proposed technique of power factor measurement was checked for accuracy. This technique was used to measure the power factor of filament bulb, capacitor and RC load.

978-1-5386-6624-1/18 $31.00 © 2018 IEEE   93

VI.1. Test on filament lamp

The front panel of power factor measurement of filament lamp is shown in figure 4. The waveform of voltage and current can be continuously seen for different loads. The measured power factor reading is shown in figure 5. The reading value is 0.999986 which is nearly equal to 1.

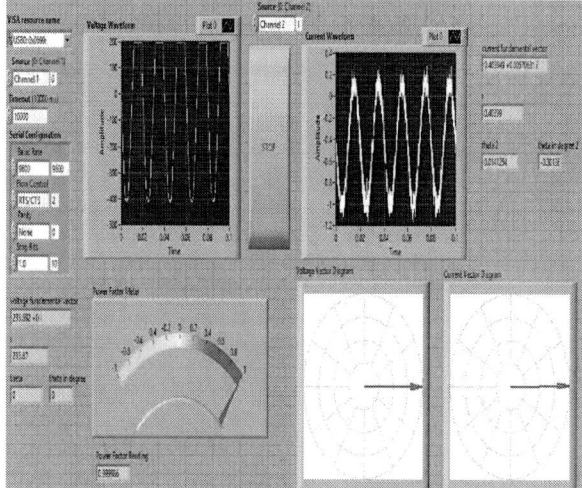

Fig. 4: Power factor measurement of filament lamp

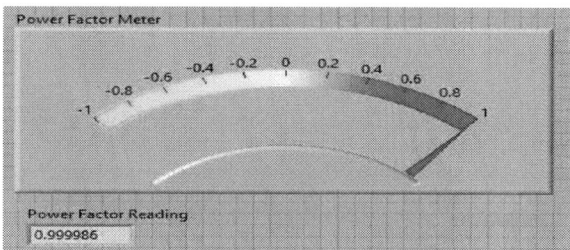

Fig. 5: Power factor reading of filament lamp

The vector diagram is shown in figure 6. The voltage and current waveforms are in same phase can be seen.

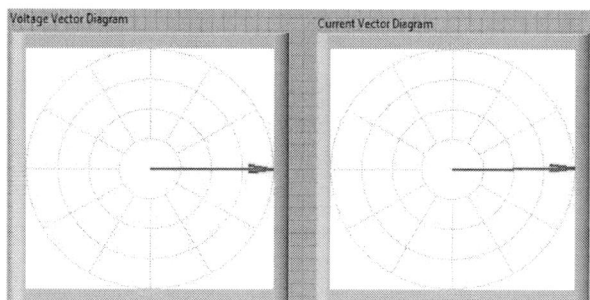

Fig. 6: Vector diagram of voltage and current for filament lamp load

VI.2. Test on capacitor

The front panel of power factor measurement of capacitor is shown in figure 7. The measured power factor reading is shown in figure 8. The reading value is 0.0171747 which is nearly equal to 0.

The vector diagram is shown in figure 9. There is 90^0 phase shift in voltage and current waveform for the capacitor. Current is leading 90^0 to the voltage in this case.

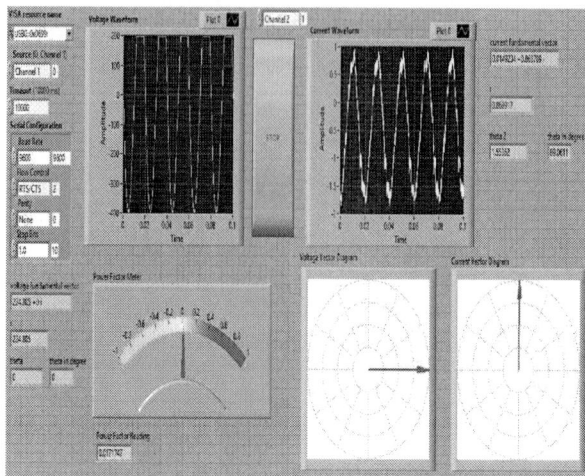

Fig. 7: Power factor measurement of capacitor

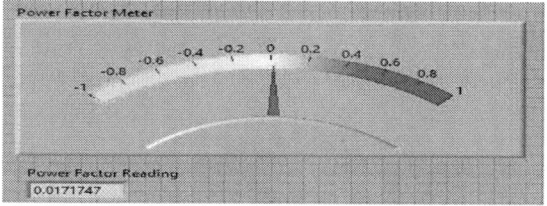

Fig. 8: Power factor reading of capacitor

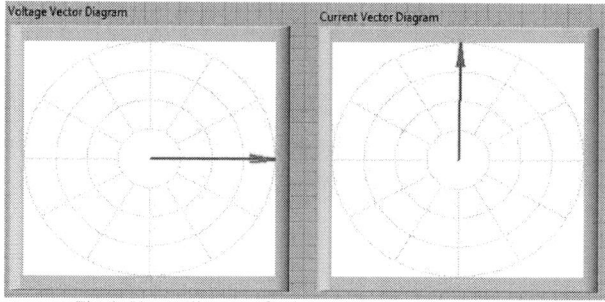

Fig. 9: Vector diagram of voltage and current for capacitor

VI.3. Test on RC load

The front panel of power factor measurement of series RC load is shown in figure 10. The measured power factor reading is shown in figure 11. The reading value is 0.388815.

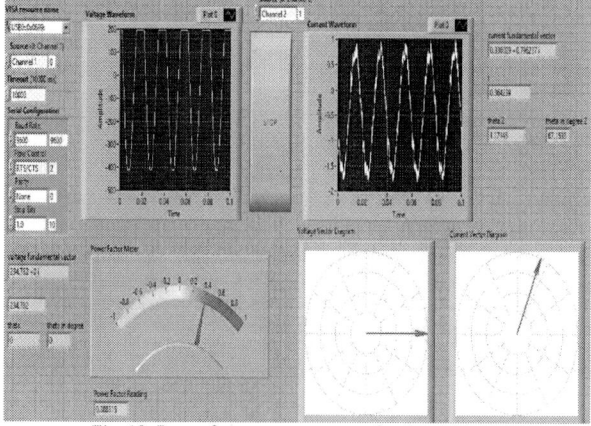

Fig. 10: Power factor measurement of series RC load

978-1-5386-6624-1/18 $31.00 © 2018 IEEE

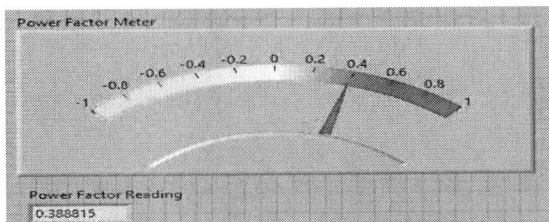

Fig. 11: Power factor reading of series RC load

The vector diagram is shown in figure 12. There is 67^0 phase shift in voltage and current waveform for the RC load. Current is leading 67^0 to the voltage in this case.

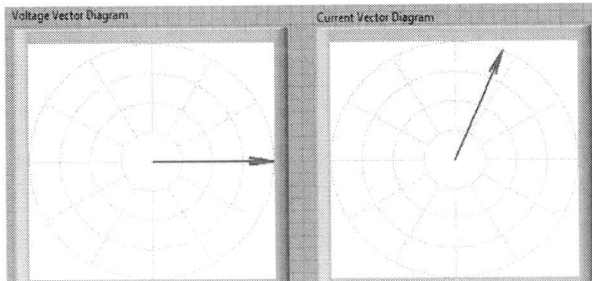

Fig. 12: Vector diagram of voltage and current for series RC load

Table 1. Comparison of Power Factor Measurement Results

Sr. No.	Electrical Load Connected	Actual Power Factor	Power Factor measured using proposed technique
1	Filament Lamp	1.0000	0.999986
2	Capacitor	0.0000	0.0171747
3	Series RC Load	0.4000	0.388815

VII. Conclusion

An efficient and automated power factor measurement circuit is proposed in this paper. From the comparison of these readings it was seen that the proposed measuring system has much accuracy. This system is able to automatically compute and display the phase angle, power factor of the connected load and also draw the vector diagram. In this system Hall Effect sensor for current measurement is providing good sensitivity and accuracy and is able to accurately track the current waveform which makes it possible to measure power factor accurately.

VIII. Further Development

This technique can be used for studying the measurement method of power factor and performing high accuracy power factor measurement. In future, wireless transceivers and receivers may be used for the different electrical measurement system so that the measuring system can measure power factor and other electrical measurement for a remotely placed electrical appliance and may be modified as the requirement.

References

[1] Haddouk, H. Mechergui, A. Ayari, ―Instrumental Platform Controlled by LabVIEW for the Power Measurement and Electric Circuits Characterisation‖, 2011 8th International Multi-Conference on Systems, Signals & Devices, IEEE, 2011.

[2] Jill C. Duplessis, ―Understanding the Limitations of a Traditional Power Factor Measurement and its Analysis‖, Application Note, OMICRON electronics Corp., U.S.A., 2012.

[3] A. K. Gupta, N. S. Chauhan and R. Saxena, "Real time I-V and P-V curve tracer using LabVIEW," 2016 International Conference on Innovation and Challenges in Cyber Security (ICICCS-INBUSH), Noida, 2016, pp. 265-269.

[4] G. Ramm, H. Moser, and A. Braun, "A new scheme for generating and measuring active, reactive, and apparent power at power frequencies with uncertainties of 2:5 _ 10 ," IEEE Trans. Instrum. Meas., vol.48, pp. 422–426, Apr. 1999.

[5] National Instruments: LabVIEW™ 7 Express Measurement Manual, http://ni.com.

[6] Ashfaq Husain, Book 'Electrical Power System' fifth revised edition 2010.

[7] P.S. Filipski, R. Arsenau, Behavior of wattmeters and watthour meters under distorted waveform conditions, IEEE Tutorial Course, 90EH0327-7PWR, 1991.

[8] Jim McDonald, "Adaptive intelligent power systems: Active distribution networks", Journal, Elsevier Science, Energy Policy Vol. 36, pp: 43464351.

[9] J.L. Willems, Reflections on apparent power and power factor in nonsinusoidal and polyphase situations, IEEE Trans. Power Deliv. 19 (April (2)) (2004) 835–840.

[10] N. S. Chauhan, A. K. Gupta and R. Saxena, "Development of Virtual Laboratory for Simulation and Performance Analysis of PV System," International Journal for Research in Emerging Science and Technology, vol. 2, issue. 5, pp. 177-182, May 2015.

2nd IEEE International conference on power Electronics, Intelligent Control and Energy systems (ICPEICES-2018)"

Study of Forced Oscillations in Two Area Power System

Vertika Jain
Post Graduate Student
Electrical Engineering Department
Delhi Technological University,
Delhi dvertikaj28@hotmail.com

S.T Nagarajan
Professor
Electrical Engineering Department
Delhi Technological University,
Delhi stnagarajan@dce.ac.in

Rachana Garg
Professor
Electrical Engineering Department
Delhi Technological University,
Delhi rachnagarg@dce.ac.in

Abstract— **Low frequency oscillations in power systems is a potential threat to the power system as it can lead to blackout of the system. In today's scenario, oscillation analysis has become an area of great interest for the researchers in dealing with stability problems of the power grid. This paper simulates a recent event in western American power system which reported several instances of north-south inter area mode below 3% for several hours. The present work deals with the study of recently detected forced oscillations in power system with the popular Kundur two area test system simulated on Digsilent Power factory and the detection of forced oscillations by measurement based methods namely the Frequency Domain Decomposition (FDD) and Prony Analysis (PA) method.**

Keywords- Power System, Electromechanical modes, Interarea Oscillations, Resonance, Stability.

I. INTRODUCTION

Oscillations in power system is a potential threat to the power system, as oscillations can grow steadily over a period of time leading to blackout of the power system. Power system engineers are more concerned to low frequency oscillations in the system which are below the system nominal frequency which are caused various reasons like sub synchronous resonance and low damped electromechanical modes in the system. The main concern of these low frequency oscillations is the damping of these oscillations by the power system. More recently a new type of low frequency oscillation has been identified to be caused by Forced Oscillation (FO) in the power system. FOs are considered to be injected into the system by mal operation mechanical parts in the system or disoperation of controllers in the system.

For any power system operations and for its reliability it is necessary to have knowledge and estimation of electromechanical oscillation at low frequency. For reliable operation of such mode estimation we have two different methods:

- Using a power system model and linearization of equations about an equilibrium point of operation [1] or by

- Using mode estimation method by measurement techniques [2] [3].

Ring down and ambient data are the two different types of analysis form measurement data. Measurements from power system's response to sudden change or disturbance such as, tripping of line or outage of generator which consequentially results in excitation of oscillatory modes, is treated as ring-down data. Measurements from power system operating in quasi-steady state condition where the measurement system input is from small continuous random oscillations in loads and other system fluctuations which are assumed to be white noise, is treated as ambient data.

Apart from natural electromechanical modes which are excited from load variations, they can be excited from forced oscillations injected by external mechanism such as from cyclic loads or mechanical aspects of generators [4] [5]. They can have adverse effects on generators produced by resonance among forced oscillations. Also, due to widespread nature of power system, forced oscillations are more liable to occur with inter area modes of power system. When a forced oscillation frequency is close to system mode frequency or when inter area mode is poorly damped at a location where inter area mode is associated, oscillations at much higher value as compare to source can appear leading to resonance in system.

Concept of resonance in Physics states that system operating at natural frequencies of system for undamped or poorly damped modes will undergo resonance when subjected to external forced oscillation. The destructive behavior of resonance is strongly identified in power system in the conditions of sub synchronous resonance [5] that originate from opposing interactions from modes of electrical and mechanical torsional behavior. Further, it has been reported in literature that, natural modes of estimation are affected by the presence of forced oscillations by some methods based on measurement estimation [6].

This paper studies the nature of forced oscillations in popular Kundur's two are power system simulated on Digsilent Power factory software in time domain. The different shapes of forced oscillations were presented for different operation scenarios. This work also and analyses the results by measurement based techniques of ambient method, namely the Frequency Domain Decomposition (FDD) method and ring down method namely, Prony Analysis (PA) method.

978-1-5386-6624-1/18 $31.00 © 2018 IEEE

II. RESONANCE IN POWER SYSTEM FROM FORCED OSCILLATIONS

A brief overview on concept of resonance as per theory of physics is explained in this section. Resonance is a process in which the system oscillates with higher amplitude for specific frequencies for the vibrating system or for system which derives external force. When the amplitude of the response of the system is relatively maximum for a frequency then these types of frequencies are known as resonant frequencies. At these resonant frequencies, due to storage of vibrational energy small periodic force has the ability to generate large amplitude oscillations. The phenomenon of resonance occurs between two or more various modes have the capability to store and transfer energy easily. Although, there are various losses which varies from cycle to cycle and is known as damping. When there is no forced oscillation then these resonant frequencies is almost equal to natural frequencies then the damping of the system is small [7].

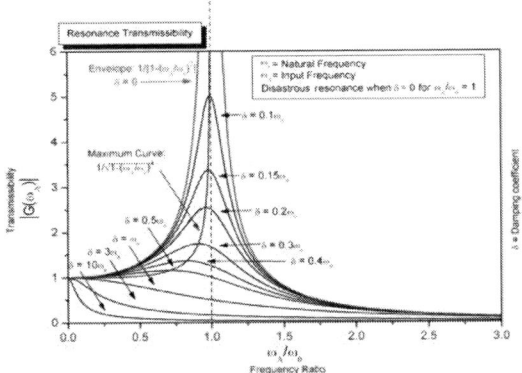

Fig 1: Concept of Resonance

According to the concept of physics in reference to resonance [8], let us consider a mass spring system as shown in Fig 2 for an undamped forced oscillation of $F_o\cos\omega t$. From Newton's law we get:

$$m\ddot{X} + b\dot{X} + KX = F_o\cos\omega t \tag{1}$$

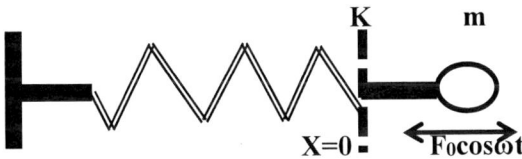

Fig 2: Mass Spring System with undamped Forced oscillation

whereas the system damping γ and system natural frequency ω_o are:

$$\omega_o^2 = \frac{K}{m}, \ \gamma = \frac{b}{m} \tag{2}$$

Damping ratio is given by $\zeta = \dfrac{\gamma}{2\omega_0}$, rewrite equation (1) in complex plane and replacing equation (2) we get,

$$\ddot{Z} + \gamma \dot{Z} + \omega_0^2 Z = \frac{F_0}{m} e^{j\omega t} \tag{3}$$

The solution of equation (3) is given as:

$$Z = A e^{j(\omega t - \delta)} \tag{4}$$

Replacing equation (4) in (3) we solve the equations for A and δ and is given as:

$$\begin{cases} A = \dfrac{\frac{F_0}{m}}{\sqrt{(\omega_0^2 - \omega^2)^2 + (\omega\gamma)^2}} \\[4mm] \tan\delta = \dfrac{\omega\gamma}{(\omega_0^2 - \omega^2)^2} \end{cases} \tag{5}$$

Where δ is the phase difference for the forced oscillation and A is the amplitude of oscillation. Resonance condition arises when amplitude A is at its peak value. From equation (5) for an undamped system i.e. $\gamma = 0$ while forced frequency equal to natural frequency i.e. $\omega = \omega_0$ then resonance occurs due to which amplitude of oscillation becomes infinity. Whereas for damped system i.e. $\gamma = 0$ then amplitude will have some finite value from equation (5) and the amplitude becomes maximum at a frequency ω_{max} which is smaller than ω_0.

There can be three possible conditions:

$$\begin{cases} \omega \to 0, A = \dfrac{F_0}{K} \ \delta \to 0 \\[3mm] \omega \to \omega_0, A = \dfrac{F_0}{K} \ \delta \to \dfrac{\pi}{2} \\[3mm] \omega \to \infty, A = 0 \ \delta \to \pi \end{cases} \tag{6}$$

Fig 1 shows the different damping level of forced oscillation which varies the amplitude A. The lowest damping system $\zeta = 2\%$ has the highest peak and with lowest damping system $\zeta = 27\%$ shows the lowest peak. Therefore the graph clearly shows that for poorly damped system the effect of resonance can be ignored.

III. STUDY OF FORCED OSCILLATIONS TEST SYSTEM

The study system of this work is the popular two area Kundur's test system [9] that provides a standard test case benchmark system. In this system a small mechanical sustained oscillation is induced by governor of generator G4 of Kundur test system which is connected at bus 4 of the system as shown in Fig 3. It is connected to the bus 10 of system through transformer between 4-10. The location of the forced oscillation can be changed to other buses as per requirement of the system or for analysis of the system.

978-1-5386-6624-1/18 $31.00 © 2018 IEEE

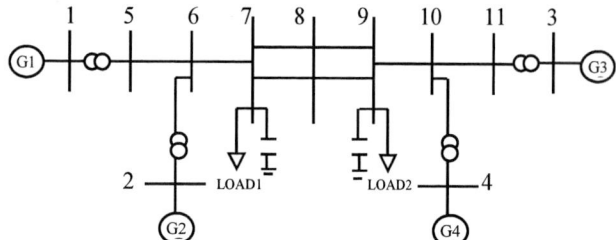

Fig 3: Two Area Kundur Power System

Load flow analysis was performed with Newton Raphson method with constant load on the system. Then modal analysis was performed to find the modal properties of the system. Two operation scenarios were created by varying the damping of the interarea modes of the system.

Scenario 1: Here the interarea modes of the system is at a frequency of 0.2 Hz. The details of the modal analysis are given in table 1.

Table 1: Modal Analysis at 0.2 Hz

Mode No.	Magnitude 1/s	Damped Freq Hz	Angle deg	Damping Ratio
1	0	0	0	0
2	46.90635	7.29357109	102.31599	0.213303
3	46.90635	7.29357109	-102.31599	0.213303
4	37.94158	0	180	1
5	37.94158	0	180	1
6	35.47603	0	180	1
7	34.39755	0	180	1
8	32.9964	0	180	1
9	32.65079	0	180	1
10	26.42154	0	180	1
11	25.93187	0	180	1
12	6.07404	0.96372978	94.50362	0.078522
13	6.07404	0.96372978	-94.50362	0.078522
14	6.10888	0.96413627	97.41143	0.128993
15	6.10888	0.96413627	-97.41143	0.128993
16	5.7795	0	180	1
17	5.29859	0	180	1
18	3.8406	0.07370135	173.07473	0.992704
19	3.8406	0.07370135	-173.07473	0.992704
20	**1.39946**	**0.22270722**	**90.86023**	**0.01501327**
21	**1.39946**	**0.22270722**	**90.86023**	**0.01501327**
22	1.9457	0	180	1
23	0.60331	0	180	1

24	0.26124	0.00191032	177.36658	0.998943
25	0.26124	0.00191032	-177.36658	0.998943
26	0.0498	0.0034	154.59948	0.903331
27	0.0498	0.0034	-154.59948	0.903331
28	0.0003	0	180	1

Scenario 2: Here the interarea modes of the system is at a frequency of 0.3 Hz. The details of the modal analysis are given in table 2, with only relevant mode ie interarea mode.

Table2: Modal Analysis at 0.3 Hz

Mode No.	Magnitude 1/s	Damped Freq Hz	Angle deg	Damping Ratio
20	1.11946	0.34583	90.86022	0.051015
21	1.11946	0.34583	90.86022	0.051015

The interarea mode can be identified with the phasor plot of the mode given below, with pair of generators oscillating opposite to each other in two clusters

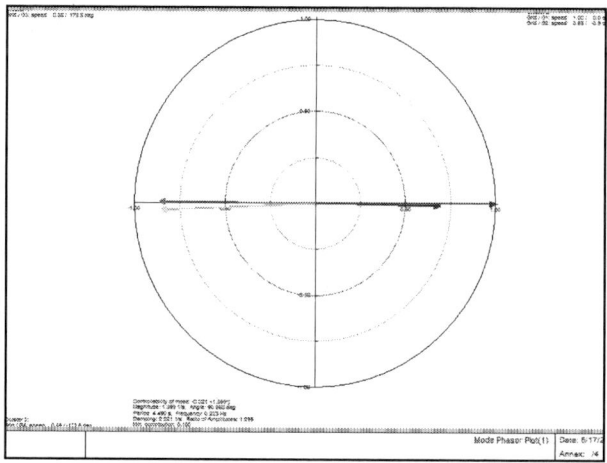

Fig 4: Modal Analysis at 0.2 Hz

IV. DYNAMIC ANALYSIS

Forced oscillations were simulated in governor of generator 4 as explained below.

A. Controller Description:

For the analysis of the dynamic behavior of the system controllers have been implemented in simulation. PowerFactory offers global library for the use of variety of different predefined controllers such as IEEE standard models of power plants i.e. providing a wide range of dynamic regulators like governors, voltage controllers and power system stabilizers.

Power Factory control modes design concept can be summarize as follows. For a control mode designing the fundamental element is 'Model/Block definition' as it constitute specified preconfigured units or controller definition such as generators. Excitation systems, PSS and Hydro Governor are examples of Model/Block Definition described in Fig 6. A composite frame is a general control model which is used to define inputs and outputs of these control models. Hydro power plant schematic diagram of 'composite frame' is shown in Fig 5. Where the forced oscillation is injected.

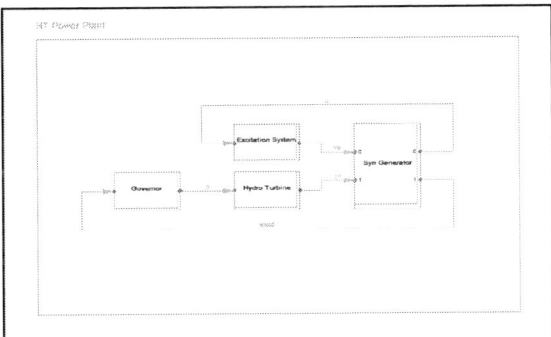

Fig 5: Schematic diagram of a 'Composite Frame' for a hydro power plant

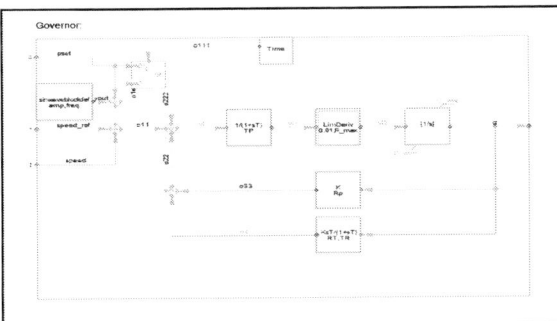

Fig 6: Schematic Diagram Governor

B. Resonance With Inter Area Mode Forced Oscillation Output Graphs

A forced oscillation at bus4 from synchronous generator G4 is injected into the bus 10 with small magnitude of generator governor oscillations at time 5 secs. Now two scenarios were studied in each case we have two cases as case 1 and case 2.

Scenario 1: The interarea mode details is given below for this scenario.

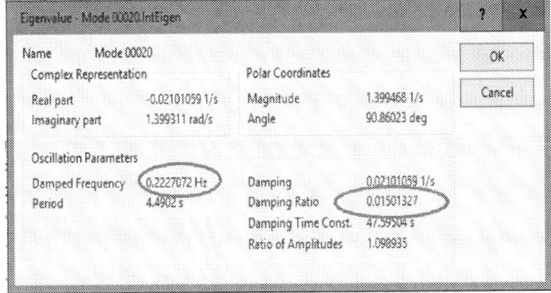

Fig 7: Mode 20 Eigen Value at 0.2Hz interarea mode

Case 1: Resulting oscillations due to forced oscillations in tie line 5 are shown in Fig 8.

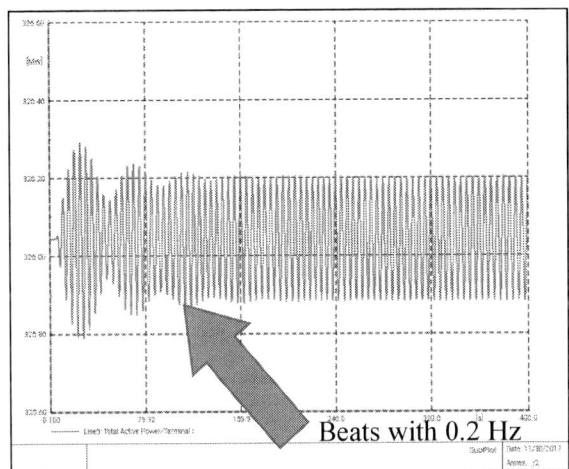

Fig 8: Tie Line Active Power At 0.2Hz FO

Observation: It can be seen form figure above that the beats are visible along with strong oscillations in the tie line power as the forced oscillation frequency of 0.2 Hz is close to the Interarea mode frequency of 0.2227072 Hz. This resembles to the beats in the resonance concept under physics.

Case2: A forced oscillation at G4 is applied with frequency 0.2227072 Hz and the resulting oscillations in tie line 5 are shown in Fig 9.

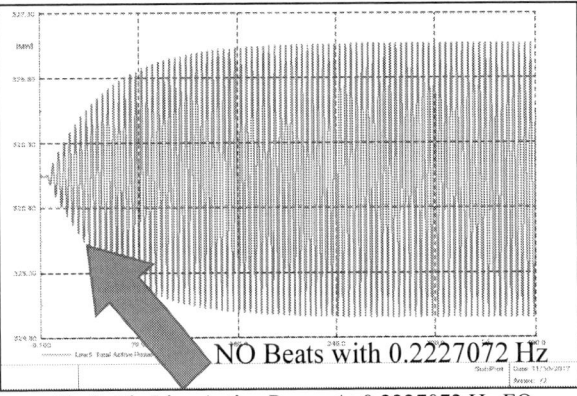

Fig 9: Tie Line Active Power At 0.2227072 Hz FO

Observation: In this case no beats are observed and the envelope is smooth as expected since there is no difference in frequencies of the interarea mode and FO. Further in both cases the amplitude of oscillation is very high which can cause the lines to trip or inflict damage to the system equipment.

Scenario 2: Interarea mode of the two area system as shown in Fig 9 below for this scenario. Also it can be noted from Fig 10 that the Interarea mode is having a frequency of 0.3458351Hz.

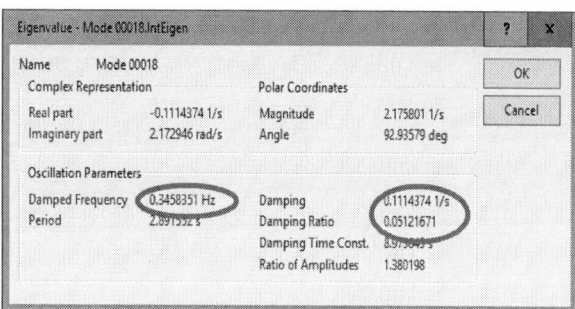

Fig 10: Mode 20 Eigen Value with 0.3Hz interarea mode

Case 1: A forced oscillation at G4 is applied with frequency 0.3 Hz and the resulting oscillations in tie line 5 are shown in Fig 11.

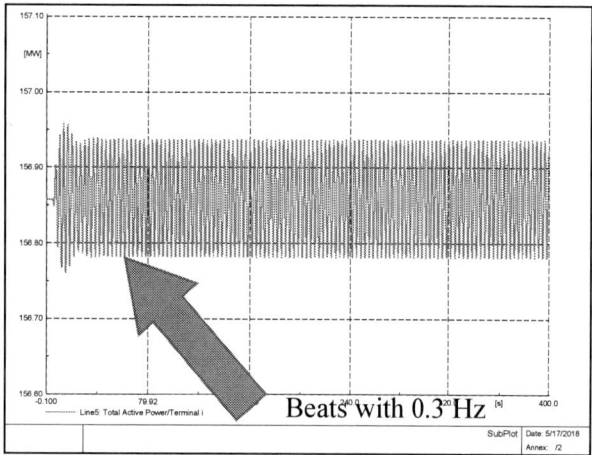

Fig 11: Tie Line Active Power At 0.3Hz FO

Observation: It can be seen form Fig 11 that the beats are visible but subdued due to higher damping along with strong oscillations in the tie line power as the forced oscillation frequency of 0.3 Hz is close to the Interarea mode frequency of 0.3458351 Hz. Compared with scenario1 these beats are subdued due to higher damping of the system.

Case 2: A forced oscillation at G4 is applied with frequency 0. 3458351 Hz and the resulting oscillations in tie line 5 are shown in Fig 12.

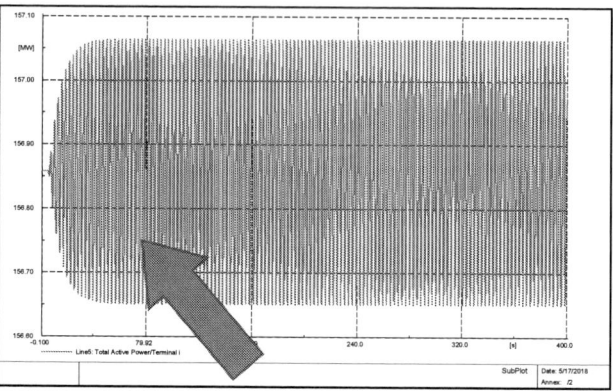

NO Beats with 0.3458351 Hz

Fig 12: Tie Line Active Power At 0.3458351Hz

Observation: There are no beats as in scenario 1. Further, the envelope shape for forced oscillation can be compared. For resonances in scenario 1 and 2 their upper envelopes without overshoot are convex in nature whereas beats are negligible as the forced component dies out rapidly the overshoots are very small in the envelope.

V.   MEASUREMENT BASED STUDIES

Measurement based studies are carried out to identify the method which can detect the presence of forced oscillations in the system. The main target is to identify a method which can detect the zero damping of the modal frequency due to forced oscillations. Two methods are tested as given below.

1)   FDD Analysis

FDD method has been executed by exporting the line oscillation values from Digsilent Power Factory to comma separated value (CSV) file and importing these results in MATLAB. Fig 13 shows the Power spectral density (PSD) at 0.2Hz of frequency. The identified results related to the FDD method is as follows:
Identified frequencies
Mode: 1; Modal Frequency: 0.2 (Hz)
Damping Factor: 0.99
So the system is found estimated to be critically stable at 0.2Hz of frequency of forced oscillations. However, the expected results for forced oscillations has to be undamped or zero damping. Therefore FDD method is found not suitable for detecting forced oscillations.

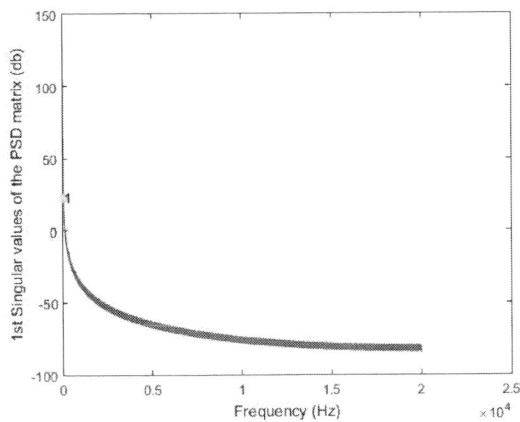

Fig 13: FDD Graph

2) Prony Analysis:

As second method Prony analysis technique has been applied to forced oscillation response data to find an inter-area mode forced oscillation near 0.2Hz case. Power factory has inbuilt Prony analysis engine which was used in this study. By using prony analysis damping ratio was found to be near zero as shown in Table 3, which satisfies the condition for undamped system.

Table 3: Prony Analysis

Line5	
Estimated Prony Frequency in Hz	**Prony Damping Ratio**
-50.000001	13.690756
-41.246751	13.418404
-32.342931	12.575557
-22.941919	10.934012
-0.400688	0.012539
-0.19938	-0.001903
0	0.000001
0.19938	**-0.001903**
0.400688	0.012539
22.941919	10.934012
32.342931	12.575557
41.246751	13.418404

VI. CONCLUSION

In this paper, measurement based oscillation monitoring techniques namely FDD and PA were studied to make use of the PMU data available for small signal stability monitoring with respect to forced oscillations in power system. The objective is to detect the presence of forced oscillations of poorly damped modes that occur in system by making use of Synchrophasor data from PMUs. Modal estimation such as frequency, mode shapes, damping ratio and energy level were estimated for forced oscillations. This work also studies the components, amplitude and envelope of forced oscillation in Two Area Kundur power system.

The nature of the forced oscillations has been studied for Interarea modes, and also how the conditions of resonance occur. Some important conclusions for forced oscillations in power system are:

- Forced oscillations can occur in power system due a malfunctioning of mechanical components in power system as well as power electronic components in the system. If the frequency of the forced oscillations injected in the power system is close to that of the system electromechanical modes, then there is a possibility of resonance in the system.
- The forced one is of the same frequency with zero damping as forced disturbance. Another one is similar as system mode modal properties.
- Forced oscillation shapes are identical as envelope.

It has been found that PA method gives better estimate of forced oscillations than FDD method of estimation. Further research in this area is necessary.

REFERENCES

[1] Pal, Bikash, Chaudhuri, Balarko, "Robust Control in Power Systems", Springer, 2005 [ISBN NO- 9780387259505].

[2] P.Kundur, Power System Stability and Control, New York: McGraw-Hill, Inc., 1994.

[3] J.W.Pierre, D.J.Trudnowskiand, M.K.Donnely, "Initial Results in Electromechanical Mode Identification from Ambient Data", IEEE trans. On Power System, Vol. 12, no.3, pp. 1245-1250, Aug. 1997.

[4] https://en.wikipedia.org/wiki/Resonance#Theory.

[5] Van Ness, "Response of large power systems to cyclic load variations," IEEE trans. On power apparatus and systems, vol.PAS-85, no.7, pp. 723-727, July. 1966.

[6] K.R. Rao and L. Jenkins, "Studies on Power Systems That Are Subjected to Cyclic Loads," IEEE trans. on power syst., vol.3, no.1, pp. 31-37, Feb. 1988.

[7] P.Kundur, "A Survey of Utility Experiences with Power Plant Response During Partial Load Rejections and System Disturbances," IEEE TransPower Apparatus and Systems, vol. PAS-100, pp. 2471–2475, May1981.

[8] J.Follum, and W. Pierre, "Initial Results in the Detection and Estimation of Forced Oscillations in Power Systems," Proc. North American Power Symposium NAPS, Sep 2013, pp. 1-6.

[9] G.Liu, J.Quintero and V.Venkatasubramanian, "Oscillation Monitoring System Based On Wide Area Synchrophasors in Power Systems", Proc. IREP Bulk Power System Dynamics and Control-VII, August 19-24,2007.

[10] S.Arash Nezam Sarmadi, Student Member, IEEE, and Vaithianathan Venkatasbramanian, Fellow IEEE, "Inter-Area Resonance in Power System From Forced Oscillations", January 2016.

[11] W.Lewin, Physics III: Vibration and Waves, Fall 2004, MIT Open Course Ware, http:// ocw.mit.edu (Accessed June 1, 2014). License: Creative Commons BY-NC-SA.

[12] L.Vanfretti, S.Bengtsson, V.Peric and J.Gjerde, "Effects of Forced Oscillations in Power Damping Estimation", In Proc. IEEE International Applied Measurements for Power Systems (AMPS), 2012.

[13] N.Zhou, D.J.Trudnowski, J.W.Pierre and R.T.Guttromoson, "Electromechanical Modes Online Estimation Using Regularized Robust RLS Methods", IEEE trans. On Power System, Vol. 23, No.4, pp. 1670-1680, Nov. 2008

978-1-5386-6624-1/18 $31.00 © 2018 IEEE

2nd IEEE International conference on power Electronics, Intelligent Control and Energy systems (ICPEICES-2018)

An ANFIS Artificial Technique Based Maximum Power Tracker for Standalone Photovoltaic Power Generation

Neeraj Priyadarshi
Department of Electrical Engg.
Birsa Institute of Technology (Trust)
Ranchi, India
neerajrjd@gmail.com

Vigna K. Ramachandaramurthy
Power Quality Research Group,
Dept. of Electrical Power Engg.,
UniversitiTenagaNasional,
Malaysia. vigna@uniten.edu.my

Sanjeevikumar Padmanaban
Dept. of Energy Technology
Aalborg University
Esbjerg, Denmark
san@et.aau.dk

Farooque Azam
Department of Comp.Sc. &Engg.,
Millia Institute of Technology
Purnea, India
farooque53786@gmail.com

Amarjeet Kumar Sharma
Department of Electrical Engg. *Birsa
Institute of Technology (*Trust)
Ranchi, India
maxeramar@gmail.com

J.P Kesari
Department of Mechanical Engg.
Delhi Technological University
Delhi, India
drjpkesari@gmail.com

Abstract— **This paper mainly develops a buck-boost converter based standalone photovoltaic system (PV) for power generation with maximum power point tracking (MPPT).Buck/boost converter is controlled by an adaptive neuro fuzzy inference system (ANFIS) MPPT algorithm which is programmed in a microcontroller. Inverter current controller using dSPACE DS1104 is performed for this purpose. Reliability and validity of standalonephotovoltaicpower generation system is justified using found Simulink and hardware results.**

Keywords—ANFIS, Standalone photovoltaic system, MPPT, DC-DC buck boost converter, dSPACEDS1104.

Solar renewable energy sources are vital technology to produce electric power [1-5].Here the solar based power generation is equipped with solar panels, Buck boost converter, battery, Inverter and lamp load presented in Fig 1. The solar panel receives light from the sun and produces electricity which in turn is stepped up/stepped down according to the panel output. Thebuck/boost operation of the converter is controlled by an adaptive neuro fuzzy inference system (ANFIS) algorithm which isprogrammed in a microcontroller in the buck/boost device. Now converter DC output is given to the battery is charged. And the battery is also connected to the single phase inverter. Finally the lamp load is driven by the output of the single phase inverter.

I. INTRODUCTION

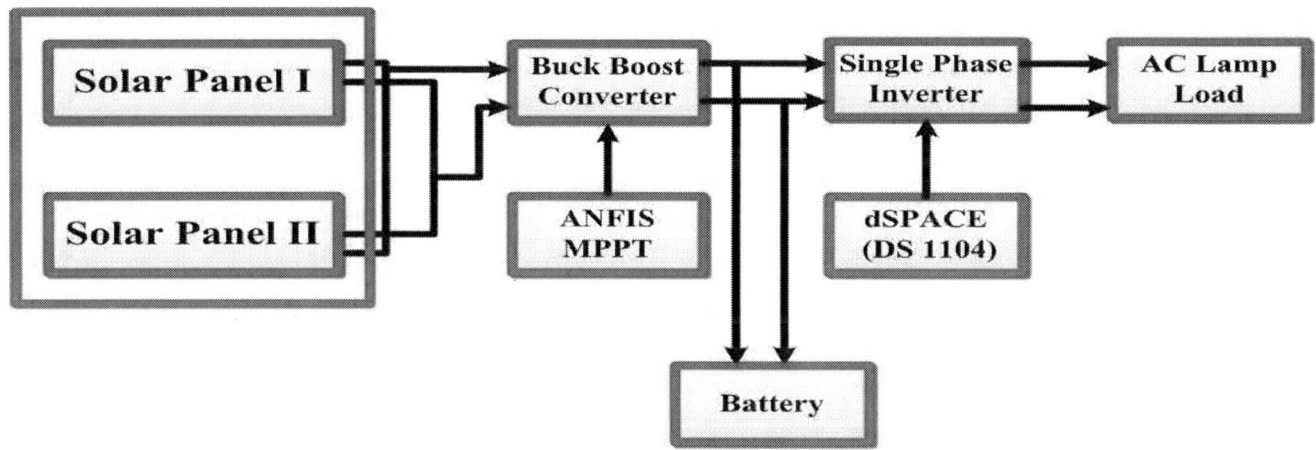

Fig. 1.200W ANFIS controlled PV based power generation

978-1-5386-6624-1/18 $31.00 © 2018 IEEE

II. ANFIS MPPT Algorithm

Plenty MPPT techniques have been reported in the literature of this paper [6-9]. To increase the overall tracking efficiency ANFIS tracked algorithm for MPPT control plays a vital role. It is necessary to implement MPPT control, since the V-I characteristics of PV panel is non-linear. The Adaptive Neuro Fuzzy Inference System (ANFIS) consists of fuzzy parameters trained using ANN which is inspired using neuron and responsible to minimize root mean square error. Back propagation based ANN learning rule is employed for this framework. Fig 2 portrays the hybrid ANFIS flow chart implementation using SIMULINK.

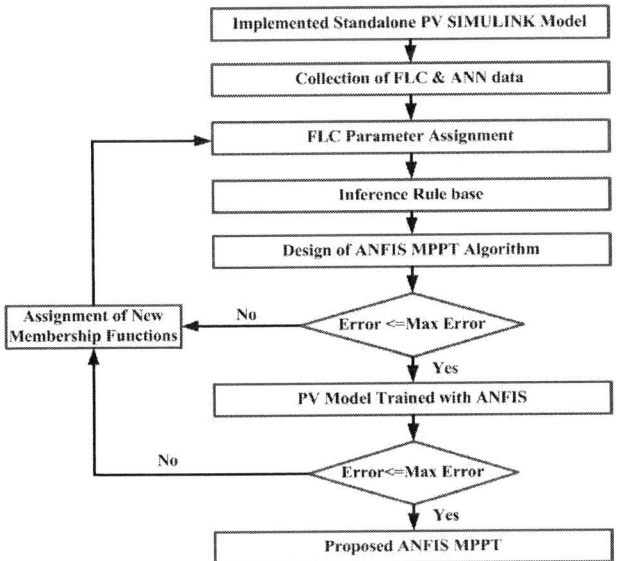

Fig.2.An ANFIS controlled PV

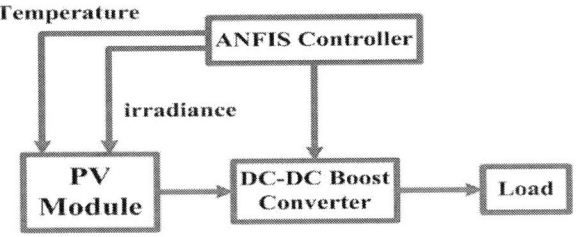

Fig.3.Block diagram of ANFIS based boost converter controlled for standalone PV system

Fig 3 presents the block diagram of ANFIS based boost converter controlled for standalone PV system. Here, temperature and solar radiations are sensed and controlled through ANFIS MPPT which provides proper training with efficient epochs. It produces set of inference rules to generate optimum duty ratio of boost converter by adjusting membership values. Fig 2 depicts the overall flowchart of the employed ANFIS algorithm which is developed using MATLAB/Simulink design.

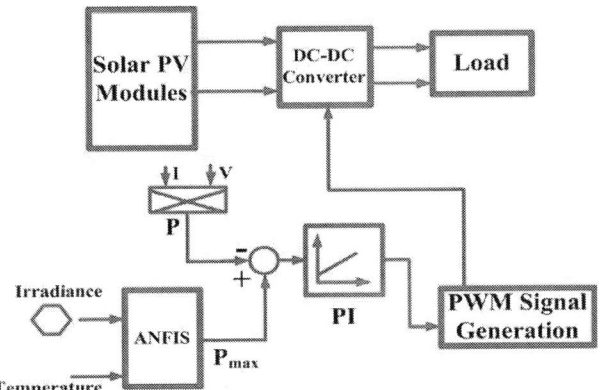

Fig.4.PWM signal generation using ANFIS controller

The training data are collected. With the application of fuzzy data, rule base and neuro model design, the ANFIS controller has been formed and proper trained mapping is embedded using back propagation technique. This process continues till the error is minimized to present value. ANFIS model can be used as a maximum power tracker when a membership value gets optimized. With the variation of duty cycle of boost converter, the optimal power is tracked from PV modules. The PWM signal generation using ANFIS controller is demonstrated with block diagram presented in Fig.4.

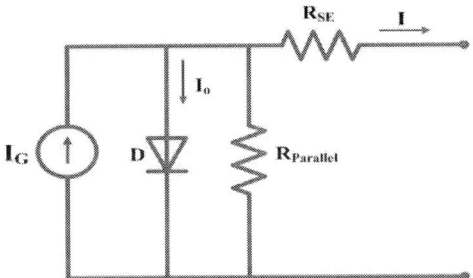

Fig.5. Equivalent PV model

Fig. 5 presents equivalent PV model which is made by combining multiple PV cells (36/72 Cells). The mathematical equation (1) describes the behavior and working of PV equivalent model as:

$$I = I_G - I_{RS}\left[e^{\frac{Q(V_d + IR_{SE})}{N_I K_B T_J}} - 1 \right] - \left(\frac{V_d + IR_{SE}}{R_{Parallel}} \right) \quad (1)$$

Where,

I_G = Photo current

Q = Charge on electron

V_d = Diode voltage

K_B = Boltzmann constant

T_J = Junction temperature

R_{SE} = Resistance in series

$R_{Parallel}$ = Resistance in parallel

N_I = Ideality factor

I_{RS} = Reverse Saturation current

I = PV output current

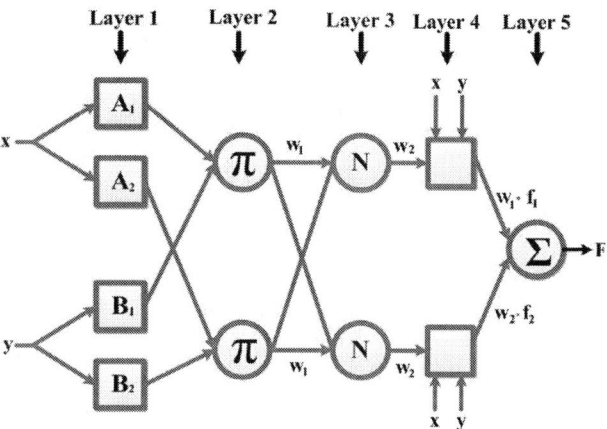

Fig.6. Employed ANFIS architecture

Fig. 6 interprets the employed ANFIS architecture designed by 2 reference rules which comprises total five layers with same functionality. These rules (R_1, R_2) are described in linear connections of two variables as:

$$R_1 = A_1 x + B_1 y + K_1 \qquad (2)$$

$$R_2 = A_2 x + B_2 y + K_2 \qquad (3)$$

Where, A_i, B_i and K_i= training constants parameter

The fuzzy membership values are calculated as, by considering j node of layer I.

$$O_j^1 = \mu_{P_j}(x), \; j = 1, 2 \qquad (4)$$

$$O_j^2 = \mu_{Q_j}(x), j = 1, 2 \qquad (5)$$

Where,

$\mu_{P_j}(x), \mu_{Q_j}(x)$ are membership values.

Firing function $(2_{O_j} = W_j)$ is calculated by multiplying these membership values:

$$2_{O_j} = W_j = \mu_{P_j}(x) * \mu_{Q_j}(y) \qquad (6)$$

Normalized firing strength $(3_{O_j} = \overline{W_j})$ can be expressed mathematical by dividing j^{th} firing strength to total firing strength as :

$$3_{O_j} = \overline{W_j} = \frac{w_j}{w_1 + w_2} \quad , j = 1, 2 \quad (7)$$

4th layer output $(4_{O_j} = \overline{W_j} F_j)$ can be expressed with relation as:

$$4_{O_j} = \overline{W_j} F_j \qquad (8)$$

Final output (5_{O_j}) is calculated mathematically as:

$$5_{O_j} = \sum_j \overline{W_j} F_1 = \frac{\sum_j W_j F_j}{\sum_j W_j} \qquad (9)$$

The overall standalone PV framework has been implemented and metrological data's have been collected. Fuzzy process is implemented and parameters are able to take the decisions based on inference rules. The complete ANFIS model is designed and compared with maximum error produced. Whenever, the trained membership function is obtained, adopted ANFIS is used for PV application. Nevertheless, the measured values are equated with reference value and root mean square error is depreciated. FLC process implementation with proper membership assignment is carried out using Simulink. Membership nature of input and output is demonstrated in Fig 7.Fuzzy parameters are changed to fuzzified values with the help of decision making rules and then defuzzified to accurate duty of power converter.

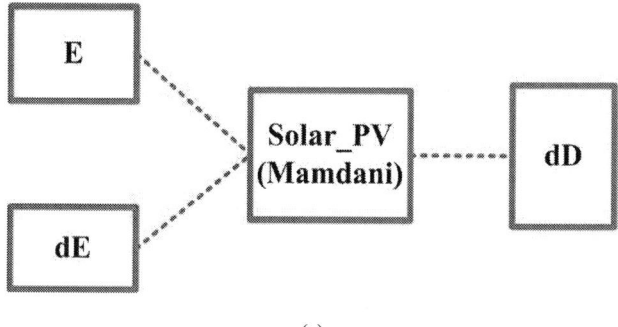

(a)

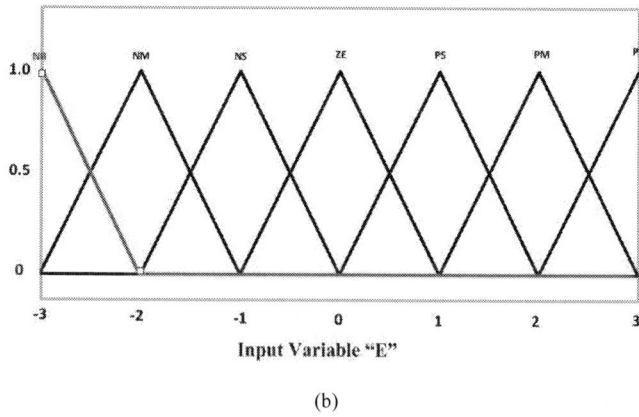

(b)

Fig. 7.(a) Editor Fuzzy inference (b) MF'E

III.   Single Phase Inverter

978-1-5386-6624-1/18 $31.00 © 2018 IEEE   104

Single phase inverter is used to convert 12 V DC input as 230 V AC output. Here double stagesingle phase inverter is used. At first stage a high frequency push pull converter whichconverts 12V to 400V DC is used. At second stage, H bridge inverter converts 400V DC into230V/50Hz AC.The front panel diagram of the single phase inverter is shown below. Here the inverterreceives its input from the battery and the input dc voltage is converted into ac voltage in twosteps as described above. Two LEDs are available in the front panel namely inverter and fault. When the inverter is ON condition the inverter LED will glow, if the inverter is infaulty state then the fault LED will glow.

IV. SIMULINK IMPLEMENTATION

Numerous MPPT techniques [6-8] performances are analyzed with PV standalone system. Variable irradiance and temperature profile have been applied to the PV control with different algorithms of intelligent control. An ANFIS technique provides better computer intelligence with respect to other methods discussed [6-8]. It reveals (Fig. 8) from obtained responses, ANFIS method has zero fluctuation with global point and has fast convergence velocity with good agreement of conversion output and adaptable with any radiation profile.

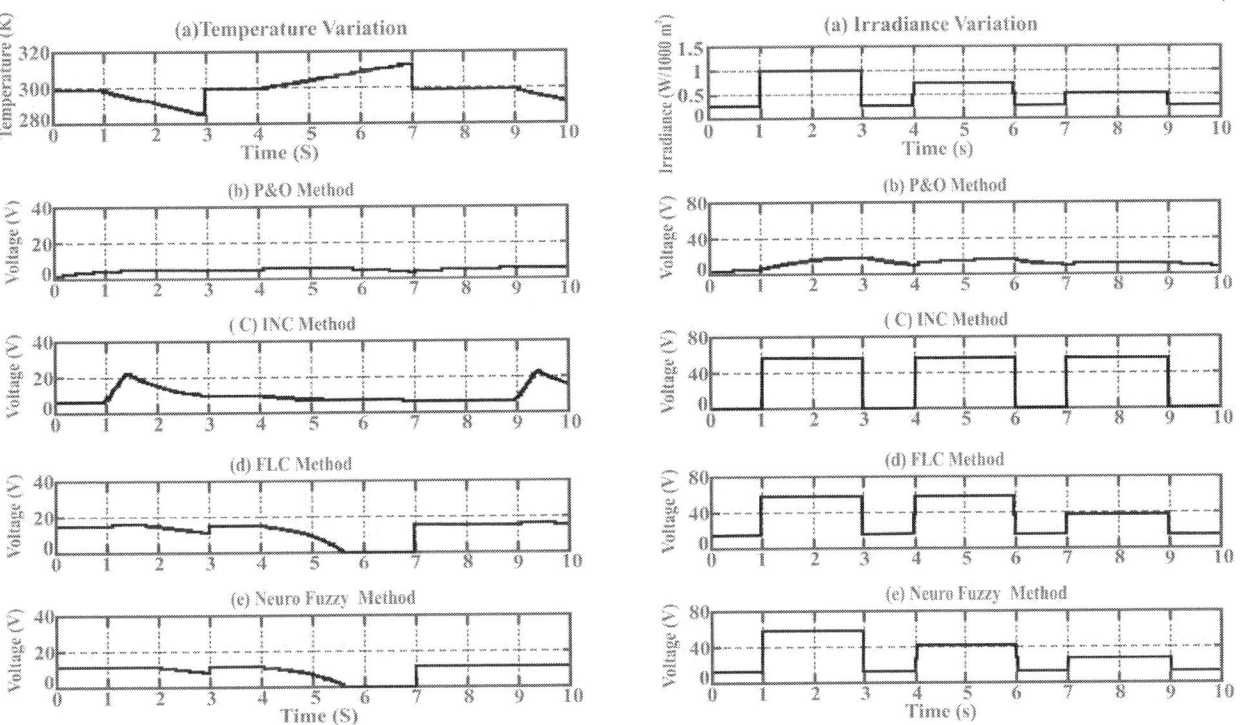

Fig.8. An ANFIS Vs classical and intelligent MPPT

V. EXPERIMENTAL RESULTS

To verify the proposed photovoltaic generation system with MPPT, a 200W laboratory prototype is built .In this section a detailed investigation on the 200 W developed laboratory prototype ANFIS based system is carried out to validate the developed system and its MPPT control under different conditions. Fig 9 shows, the proposed hardware with MPPT.Several responses of PV output, load, inverter output phasor diagram of load voltage and load current, harmonic table for load voltage and current for the developed ANFIS based PV system with resistive load of 40 W is presented in Fig.10.The Inverter output voltage is in phase with the output current since load is resistive type and its shape is quasi square.

Fig.10 interprets hardware results of ANFIS controlled with MPPT for 40 W Resistive load. From experimental results, it is found that the PV output voltage has magnitude 12.3V. The inverter voltage has magnitude 226 V; the black line shows the magnitude of neutral voltage. The load voltage and load current has magnitude 225.8 V and 0.5 A respectively. The phasor diagram of current demonstrated in Fig.10 (e). Load voltage / current have phase difference zero i.e. both are in same phase due to R load. The inverter output current has positive and negative pulses, which is modulated according as sinusoidal reference.

978-1-5386-6624-1/18 $31.00 © 2018 IEEE

Fig. 9. 200 W ANFIS Controlled standalone PV System with MPPT

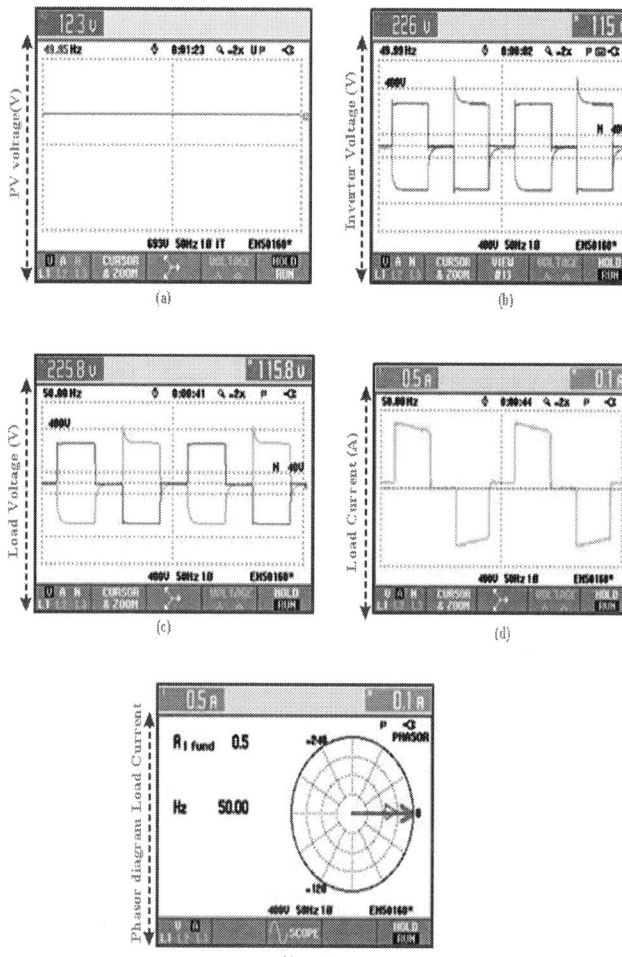

Fig. 10. Experimental waveforms during resistive load of 40 W (a) PV voltage (b) inverter voltage (c) load voltage (d) load current (e) phasor diagram of load current

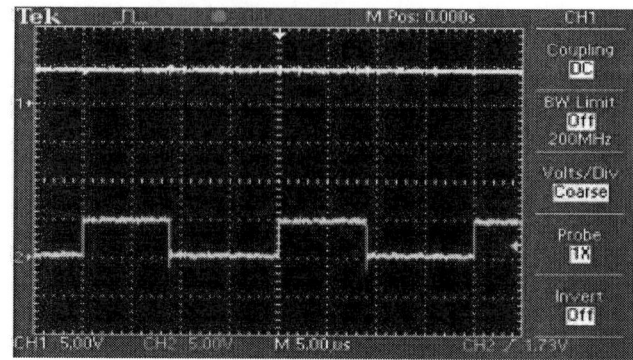

Fig. 11. Generated PWM pulses for buck/boost converter using dSPACE

Fig. 11 shows the generated PWM signal for zeta buck-boost converter using ANFIS based MPPT controller implemented with real time dSPACE DS1104 controller board, which remain constant irrespective of the abrupt weather conditions. The practical instantaneous power tracking efficiency of the proposed MPPT controller is presented in Fig. 12.

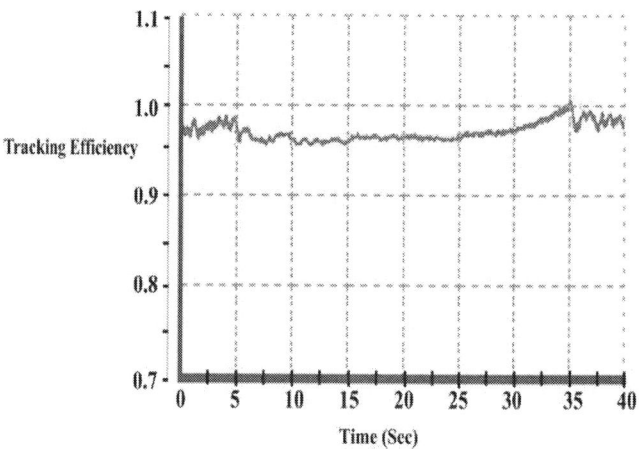

Fig. 12. Instantaneous power tracking efficiency of the proposed ANFIS MPPT controller

VI. CONCLUSION

In this study, the performance analysis of a 200 W ANFIS controlled standalone PV based power generation system is investigated experimentally. The performance of the proposed system is assessed at different load conditions and found to be efficient and have excellent steady state and dynamic performance, which reinforces the validation of system design. Moreover, during varying load condition, it is observed that load voltage is well maintained in spite of load variation. Also the load current is varying as expected. The single phase inverter gives reference current signal which compensates the distortion of load current. It seems evident that the proposed Space vector pulse width modulation current controller

978-1-5386-6624-1/18 $31.00 © 2018 IEEE

regulates the load voltage and frequency efficiently during varying load condition.

REFERENCES

[1] S. Padmaban, G. Grandi, P. W. Wheeler, F. Blaabjerg and J. Loncarski, "A simple MPPT algorithm for novel PV power generation system by high output voltage DC-DC boost converter,"in*IEEE 24th International Symposium on Industrial Electronics (ISIE), Buzios,* pp. 214-220, 2015.

[2] Sachin Jain, Ch. Ramulu, P.Sanjeevikumar, OlorunfemiOjo and AhmetH.Ertas, "Dual MPPT Algorithm for Dual PV Source Fed Open-End Winding Induction Motor Drive for Pumping Application," *Engineering Science and Technology: An International Journal (JESTECH),*vol. 19, no. 4, pp. 1771–1780, 2016.

[3] Zientarski, J.R.R., Martins, M.L.S., Pinheiro, J.R., and Hey, H.L., "Series-Connected Partial-Power Converters Applied to PV Systems: A Design Approach Based on Step-up/down Voltage Regulation Range," *IEEE Trans. on Power Electron.*, vol. 99, pp. 1-1, 2017.

[4] B. singh, C. Jain, and A. Bansal, "An Improved Adjustable Step Adaptive Neuron Based Control Approach for Grid Supportive SPV System," *IEEE Trans. on Industry Applications*, vol. 99, pp. 1-1, 2017.

[5] Sangwongwanich, Y. Yang, F. Blaabjerg and D. Sera, "Delta Power Control Strategy for Multi-String Grid-Connected PV Inverters," *IEEE Trans. on Industry Applications*, vol. 99, pp. 1-7, 2016.

[6] N. Priyadarshi, A. K. Sharma, and S.Priyam, "An Experimental Realization of Grid-Connected PV System with MPPT Using dSPACE DS 1104 Control Board," *Advances in Smart Grid and Renewable Energy, Lecture Notes in Electrical Engineering*, 435, 2018.

[7] N. Priyadarshi, A. K. Sharma, and F.Azam, "A Hybrid Firefly-Asymmetrical Fuzzy Logic Controller based MPPT for PV-Wind-Fuel Grid Integration," *International Journal of Renewable Energy Research*, vol. 7, no.4, 2017.

[8] R.B.A. Koad, A. F.Zobaa and A. El-Shahat, "A Novel MPPT Algorithm Based on Particle Swarm Optimization for Photovoltaic Systems," *IEEE Trans. on Sustainable Energy,*vol. 8, no.4, pp. 468 – 476, 2017.

[9] Manickam, G. R. Raman and G. P. Raman,"A Hybrid Algorithm for Tracking of GMPP Based on P&O and PSO With Reduced Power Oscillation in String Inverters*," IEEE Transactions on Industrial Electron.*, vol. 63, no.10, pp. 6097 – 6106, 2016.

Load Frequency Control of Two Area Interconnected Power System using SSSC with PID, Fuzzy and Neural Network Based Controllers

Ashiq Hussain Lone
Electrical Engineering Department
Thapar Institute of Engineering and Technology
Patiala, India
ashiqlone383@gmail.com

Surya Prakash
Thapar Institute of Engineering and Technology
Patiala, India
surya.prakash@thapar.edu

Viqar Yousuf
Electrical Engineering Department
National Institute of Technology Srinagar
Srinagar, India
viqaryousuf@gmail.com

Mohammad Abid Bazaz
Electrical Engineering Department
National Institute of Technology
Srinagar Srinagar, India
abid@nitsri.net

Abstract—**This paper presents the frequency control of a two area interconnected thermal and reheat thermal systems incorporating static synchronous series compensator (SSSC) to control the tie-line power transfer. Effectiveness of PID to control the system dynamics and minimize area control error (ACE) has been observed. The system dynamics is further improved by using fuzzy logic controller and neural network based controller. It shows the effectiveness of Fuzzy Logic Controller (FLC) over PID controller and neural network controller over FLC. The mathematical model of the system is also explained in this paper.**

Index Terms—**Load Frequency Control (LFC), Area Control Error (ACE), PID Controller, Fuzzy Logic Controller, Adaptive Neuro-inference system, Static Synchronous Series Compensator (SSSC).**

I. INTRODUCTION

The frequency in a power system changes when there is a mismatch between generation and the electrical load. The Load Frequency Control (LFC) provides a way of matching the generation with the load so as the change in frequency and tie-line power is minimized [1] [2]. In the present-day power system hydro and thermal units are the main power producing units. A lot of work in the literature [2] [3] shows that thermal and hydro plants work as base load units. Hydro and thermal power plants differ from each other as the hydro power plant provides non-minimum phase characteristics which are not present in the thermal units. Two interconnected units like thermal and reheat-thermal are considered in this paper.

The interarea oscillations are very common in the interconnected power systems, these oscillations cause change in the frequency. There are different ways to damp out these oscillations and one of them is by using Flexible AC Transmission system (FACTS) devices. Hence LFC in the presence of FACT devices becomes interesting. Different FACTS de-

vices like Static Synchronous Series Compensator (SSSC) [4], Thyristor Control Phase shifter (TCPS) [5], Unified Power Flow Controllers (UPFC) [6] Interline Power Flow (IPFC) and Thyristor Controlled Series Capacitors (TCSCS) [7] are being used to increase the power system reliability and power exchange. Subbaramaiah et al [8] have used SSSC and TCPS in two area power system.

PID controller has always remained the SSSC choice in LFC. PID is preferred because of its reliability, simplicity and low cost and also it is easy to operate. But the major disadvantage of PID is the proportional and derivative kick which causes sharp spikes and sudden overshoots. To overcome these short comings fuzzy logic controller is used, as it gives better results and reduces the overshoot. But there is one disadvantage in using fuzzy controller as it gives desired outputs corresponding to the given input. To overcome these disadvantages and to improve the system performance and make it fast Neural Network Controller is used. This is more advanced adaptive control strategy and it also produces fast response than other controllers (Demiroren et al).

II. SYSTEM INVESTIGATED

An interconnected two area thermal-reheat system is considered.Non-linearities such as generation rate constant GRC is considered for reheat-thermal power plant to get the realistic understanding of AGC problem.The transfer function of different components of power system is given in Figure 1. The parameters of the system considered are taken from ref [9]. Different controllers like PID,fuzzy and neural are considered separately as supplementary controllers. The performance of SSSC with different controllers is evaluated.The parameters considered are given in the appendix. The system performance is evaluated by considering a step load perturbation (SLP)

978-1-5386-6624-1/18 $31.00 © 2018 IEEE

2nd IEEE International Conference on Power Electronics, Intelligent Control and Energy Systems (ICPEICES-2018)

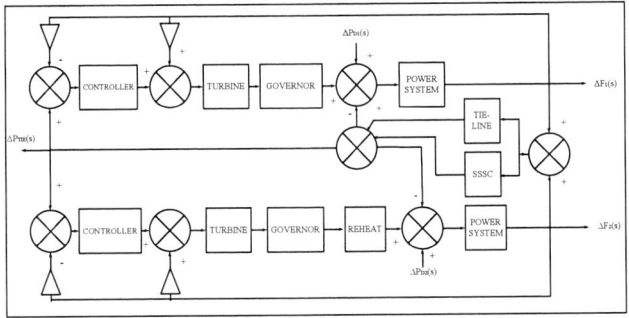

Fig. 1. Block diagram.

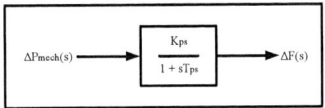

Fig. 3. Block diagram of load.

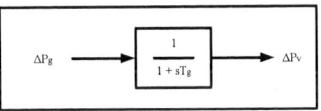

Fig. 4. Block diagram of governor model.

in both areas area1 and area2 for realistic studies. A brief description of generation rate constant (GRC) is given as under:

A. Generation rate constraint

The power generation in a power system can change only at a specified maximum rate which is known as Generation Rate Constraint (GRC). The effect of generation rate constraint is visible on the increment of settling time and overshoots of the system response. Presently a GRC of 3% /min is considered for thermal units [10]. The swing equation of a synchronous machine is given by

$$\frac{2H}{\omega_s}\frac{d^2\Delta\delta}{dt^2} = \Delta P_m - \Delta P_e \qquad (1)$$

When a small change in speed takes place

$$\frac{d\Delta\left(\frac{\omega}{\omega_s}\right)}{dt} = \frac{1}{2H}(\Delta P_m - \Delta P_e) \qquad (2)$$

In per unit we have

$$\frac{d\Delta\omega}{dt} = \frac{1}{2H}(\Delta P_m - \Delta P_e) \qquad (3)$$

Now taking Laplace transform of equation (3) we get

$$\Delta\omega(s) = \frac{1}{2Hs}[\Delta P_m(s) - \Delta P_e(s)] \qquad (4)$$

This can be represented by a block diagram as,

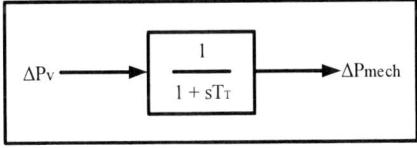

Fig. 2. Turbine Model.

B. LOAD MODELLING

Load on a power system comprises of resistive load which is independent of frequency and inductive load which is sensitive to the frequency. The sensitivity of inductive load depends on speed-load characteristics. This can be shown by a mathematical relation as:

$$\Delta P_e = \Delta P_L + D\Delta\omega \qquad (5)$$

Where ΔP_L represents the resistive load or the load which is independent of frequency, and D, $\Delta\omega$ is the load which depends on frequency. The value of D is written as the ratio of percent change in load to percent change in frequency.

C. PRIME MOVER MODELLING

The prime movers can be hydraulic turbines for waterfalls, steam turbines, gas and nuclear turbines. The modelling of turbine is related to the change in mechanical power output ΔP_m to the change in steam valve position ΔP_V. The modelling of a non-reheat turbine with only one-time constant is given as:

$$G_T(s) = \frac{\Delta P_m(s)}{\Delta P_V(s)} = \frac{1}{1 + T_T(s)} \qquad (6)$$

D. GOVERNOR MODELLING

The transfer function model of governor is given by the following equation:

$$\Delta P_V(s) = \frac{1}{1 + T_g(s)}\Delta P_g(s) \qquad (7)$$

This can be represented by a block diagram as under:

E. TIE-LINE MODELLING

The tie-line connects two power systems together and the power transfer from one area to another area takes place through it. If there is an increase of load in a particular area, power from the other area flows through tie-line to compensate for the change. Mathematically power through the tie-line can be expressed as:

$$P_{12}^0 = \frac{|V_1^0||V_2^0|}{X}\sin(\delta_1^0 - \delta_2^0) \qquad (8)$$

978-1-5386-6624-1/18 $31.00 © 2018 IEEE   109

Where, δ_1^0, δ_2^0 = power angles of the two machines.

When a small disturbance occurs in the angles the tie-line power deviates to

$$\Delta P_{12} = T_{12}(\Delta\delta_1^0 - \Delta\delta_2^0) \tag{9}$$

Where,

$$T_{12} = \frac{|V_1^0||V_2^0|}{X}\cos(\delta_1^0 - \delta_2^0) \tag{10}$$

is the synchronizing torque The relationship between the frequency deviation Δf and angle is given as under

$$\begin{aligned}\Delta f &= \frac{1}{2\pi}\frac{d(\delta_0 + \Delta\delta)}{dt} \\ &= \frac{1}{2\pi}\frac{d(\Delta\delta)}{dt}\end{aligned} \tag{11}$$

$$\Delta\delta = 2\pi\int\Delta f dt \tag{12}$$

$$\Delta P_{12} = 2\pi T_{12}\left(\int\Delta f_1 dt - \int\Delta f_2 dt\right) \tag{13}$$

Taking Laplace transform of above equation we get

$$\Delta P_{12} = \frac{2\pi T_{12}}{s}(\Delta f_1(s) - \Delta f_2(s)) \tag{14}$$

III. CONTROL METHODOLOGY USED

A. SSSC

The static synchronous series compensator is used for stablising the area frequency and tie-line power devations by controlling the tie-line power exchange. The SSSC is connected in series with the tie-line between the areas which are interconnected, hence it can be represented by a series of voltage source which are connected togather. The expression for power when using sssc in the tie-line [10] can be represented in terms of frequency deviation as under:

$$\Delta P_{sssc}(s) = \frac{(1 + T_1 s)(1 + T_3 s)}{(1 + T_2 s)(1 + T_4 s)}\frac{K_{sssc}}{1 + T_{sssc}}\Delta\omega_s(s) \tag{15}$$

Here T_{sssc} and K_{sssc} are the time constants and gain of SSSC and the change in frequency $\Delta\omega_s(s)$ is the input to the SSSC controller.

B. PROPORTIONAL INTEGRAL DERAVATIVE CONTROLLER (PID)

Proportional Integral Derivative (PID) is the most commonly available controller. PID controller improves the dynamic response of the system and eliminates the study state error. A finite zero is added to the open loop transfer function of the plant by the derivative controller and hence it improves the transient response [11]. The system type is increased by one with the help of Integral controller which adds a pole at the origin, that in turn reduces study state error to zero which is caused by the step change. Trail and error method has been used to determine the tuned values of proportional, integral and differential gains The transfer function of PID is given as:

$$G_c(s) = K_P + \frac{K_I}{s} + K_D s \tag{16}$$

C. FUZZY LOGIC CONTROLLER

Fuzzy logic controller has lot of applications in power system. FLC works on the basis of knowledge acquisition process. A fuzzy system has a membership function associated with each fuzzy set and here fuzzy IF-THEN rule is used for controlling the process. The horizontal range of membership functions is obtained by optimisation of error generated by PID controller. In the given system the LFC comprises of sudden load variations in the power system which result in the frequency change and this frequency deviation should be in the permissible limits [11]- [14].

Fuzzification:
It is the way of converting real-valued variable into fuzzy variable.
Rule base:
The rule-base used is IF-THEN rule it consists of a set of rules. The rule base is a combination of a set of fuzzy rules. The information is carried out by the membership functions.
De-Fuzzification:
Defuzzification is the way of converting the fuzzy variable into the real value which is known as crisp-value due to this it is used in the controlling process. The block diagram representation of de-fuzzification is given below. The controlling action of FLC is decided by the fuzzy rule base.

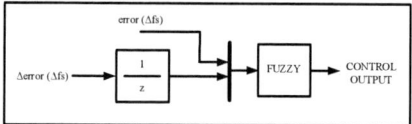

Fig. 5. Block diagram of fuzzy logic controller.

The performance of the controller is based on de-fuzzification, membership function and rule base. The error signal (Δf_s) is generated from the governor, which is given as input to FLC. The fuzzy rule base and the membership-function comprise of five variables these are (NB, NS, ZZ, PS and PB) for the two inputs and two output system as represented in the diagram and the rule-base table.

TABLE I
FUZZY RULE BASE.

ACE/DACE	NB	NS	ZZ	PS	PB
NB	S	S	M	M	B
NS	S	M	M	B	VB
ZZ	M	M	B	VB	VB
PS	M	B	VB	VB	VVB
PB	B	VB	VB	VVB	VVB

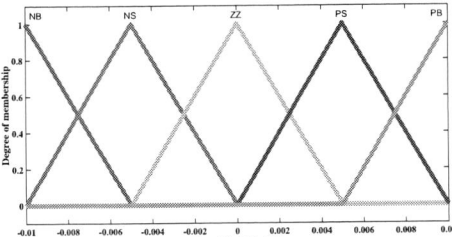

Fig. 7. Membership function of input two DACE.

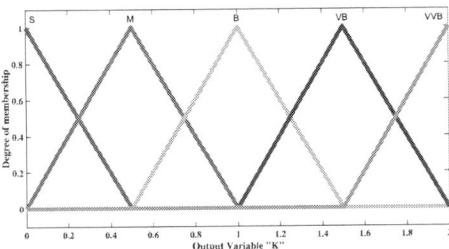

Fig. 8. Membership function of output K.

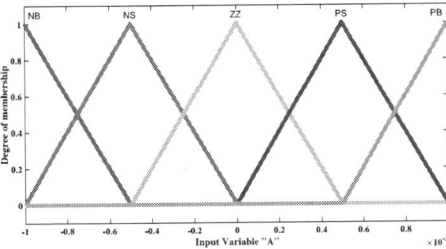

Fig. 6. Membership function of Input one ACE.

D. NEURAL NETWORK CONTROLLER

The information is processed by the neurons in the neural network system nad that is why it is named so. The signal is transmitted from one neuron to another neuron through the connecting links These links are associated with weight and the incoming signal is multiplied with the weights and input signal of the neural network. The output is obtained by giving the activation function to the net input. Neural network has vast applications [15]. The block diagram of NN is shown in fig 10. The network is so adjusted that if the inputs are same, then the response will also be same. The unknowns

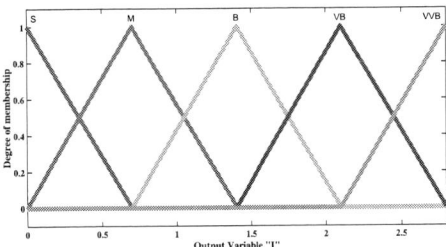

Fig. 9. Membership function of output I.

of the system can also be represented as the inverse of the system which is being controlled for this case we can use NN controller (Hykin 1994) Neuron can have more than one input, this is shown in the fig 10. The inputs $p_1, p_2, p_3 \ldots \ldots p_n$ and the weights $w_{1,1}, w_{1,2} \ldots \ldots w_{1,R}$ of the weight matrix w. The bias b associated with neuron is summed with the weighted inputs to produce the net-input neuron n.

Multi-input neuron

$$n = w_{1,1}p_1 + w_{1,2}p_2 + \ldots \ldots + w_{1,R}p_R + b \quad (17)$$

In matrix form

$$n = w_p + b \quad (18)$$

Matrix w has only one row. This can be written as;

$$a = f(w_p + b) \quad (19)$$

Log-sigmoid function is most commonly used transfer function. This is shown in Fig.12.

This converts input from 0-1.

$$a = \frac{1}{1 + e^{-n}} \quad (20)$$

It is used in multilayer networks which are trained through back propagation algorithm. The control signals for the governor are the output signals of NN controller. The data for the training of NN is obtained by making a reference model NN and applying to the system with step perturbances.

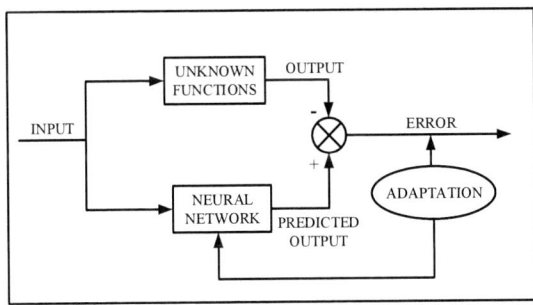

Fig. 10. Neural Network Function Approximator.

978-1-5386-6624-1/18 $31.00 © 2018 IEEE

2nd IEEE International Conference on Power Electronics, Intelligent Control and Energy Systems (ICPEICES-2018)

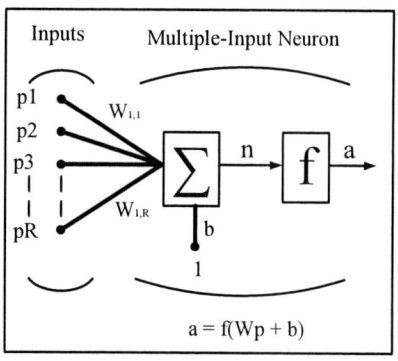

Fig. 11. Neural Stucture.

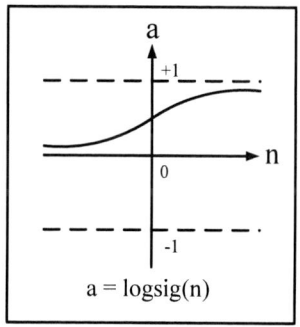

Fig. 12. Log-sigmoid transfer function.

IV. SIMULATION RESULTS

The two area interconnected model is simulated with different controllers and the following results have been obtained from Fig 13 to Fig 20. Various types of controllers used are PID, Fuzzy logic controller and Neural Network based controller. The performance of the controllers have been summed up in Table 2. It has been observed that performance of Neural network based controller is better than Fuzzy Logic controller which in-turn is better than PID controller.

The settling time of system considered is brought to 32s by using PID controller which is further improved and brought down to 20s by replacing PID with FLC and hence improves the system performance. FLC is then replaced by NN controller to Steady the results and it is observed that settling time is further improved and brought down to 5s from 20s. This shows the superiority of NN controller.

TABLE II
SETLING TIME (S).

Controllers	Area 1	Area 2
PID	32	32
Fuzzy	20	20
Neural Network	5	5

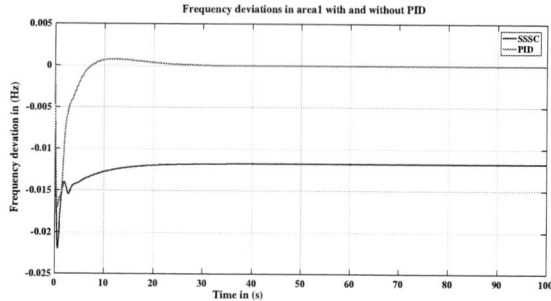

Fig. 13. Comparison of frequency deviations in area1 without and with PID.

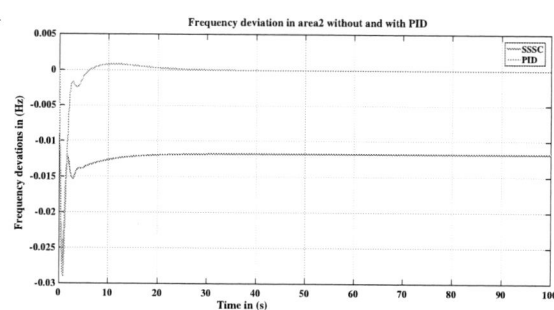

Fig. 14. Comparison of frequency deviations in area1 without and with PID.

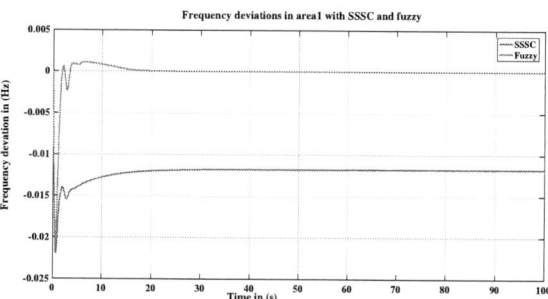

Fig. 15. Comparison of frequency deviations in area1 with and without fuzzy.

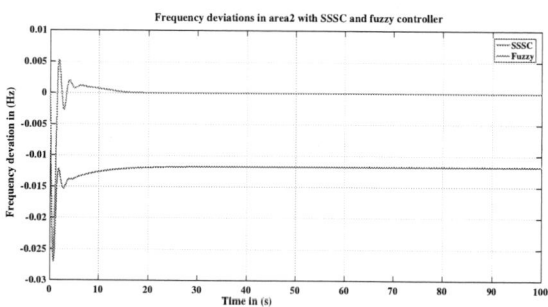

Fig. 16. Comparison of frequency deviations in area2 with and without fuzzy.

978-1-5386-6624-1/18 $31.00 © 2018 IEEE

2nd IEEE International Conference on Power Electronics, Intelligent Control and Energy Systems (ICPEICES-2018)

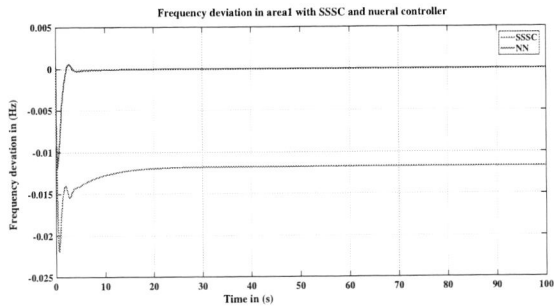

Fig. 17. Comparison of frequency deviations in area1 with and without Neural Network.

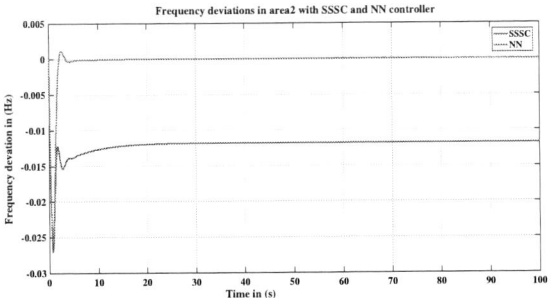

Fig. 18. Comparison of frequency deviations in area2 with and without Neural Network.

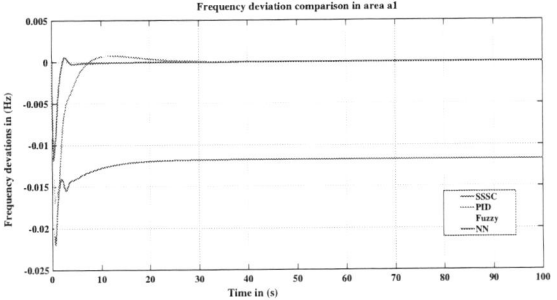

Fig. 19. Comparison of frequency deviation in area 1.

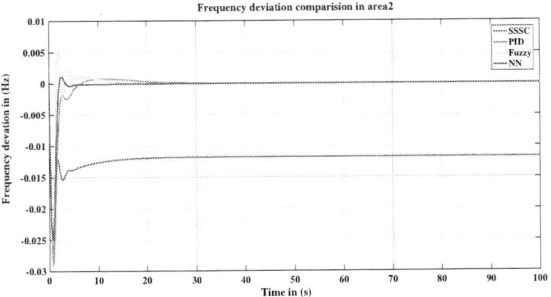

Fig. 20. Comparison of frequency deviation in area 2.

V. CONCLUSION

With the use of FACTS devices like SSSC in the tie-line,not only is the overshoot in frequency and the settling time of the system improved, but it also improves the tie-line power transferred. However the error in the system still present and can be removed by applying different control strategies like PID, Fuzzy and NN Controllers. The steady state error is brought to zero and the system overshoot is minimized by using self tuned PID controller with SSSC. Fuzzy logic controller reduces the settling time and brings the steady state error to zero quickly. The NN controller gives best results among the three controllers used, hence the overshoot is less with NN controller as compared to PID and fuzzy controllers. Further NN controller reduces the settling time and gives better results than PID and fuzzy logic controllers.

REFERENCES

[1] Debdeep Saha and L.C.Saikia "Performance of FACTS and Energy devices in a multi area wind-hydro-thermal system employed with SFS optimized I-PDF Controller" Journal of renewable and sustainable energy 9, 024103 (2017); doi:10.1063/1.4980160.

[2] H.Bevrani and T.Hiyama, Intelligent Automatic Generation Control (CRC Press,2011).

[3] D.P.Kothari and I.J.Nagrath, Modern Power System Analysis, 4th ed.(McGraw Hill, New Delhi,2011)

[4] P.Kundur, Power System Stability and Control .(McGraw Hill, New Delhi,1994)

[5] R.Shankar, R.Bhushan and K.Chatterjee, "Small Signal Stability analysis for two area interconnected power system with load frequency controller in co-ordination with FACTS and energy storage devices," Ain Shamas Eng. J.7.603-612 (2016).

[6] N.G.Hingorani and L.Gyugyi "understsnding FACTS, concept and technology of facts" (IEEE New York 2000).

[7] K.Subbaramaiah, "Improvement of dynamic performance of SSSC and TCPS based hydro-thermal system under deregulated scenario employing PSO based dual mode controller," Eur.J.Sci.Res.57,230-243 (2011).

[8] T.Dash,L.C.Saikia,and N.Sinha "Comparision of performance of several FACTS devices using Cuckoo search algorithm optimized 2DOF controllers in multiarea AGC," Electr.Power Energy Syst,65,316-324(2015).

[9] S.Padhan,R.K.Sahu, and S.Panda, "Automatic generation control with Thyristor controlled series compensator including superconducting magnetic energy storage units," Ain Shams Eng.J.(Elsevier) 5.759-774 (2014).

[10] P.C.Pradhan,R.K.Sahu and S.Panda, "Firefly algorithem optimized fuzzy PID controller for AGC of multi-area multi- source power systems with UPFC and SMES," Int.J.Eng.Sci.Technel,19.238-354(2016).

[11] Aditya "Design of load frequency controllers using genetic algorithm for two area interconnected hydro power system," Electric power components and system, 31:1, 81-94, DOI: 10.1080/1532500390112071.

[12] Y. Arya and N. Kumar, "Fuzzy gain scheduling controllers for AGC of two-area interconnected electrical power systems," Elect. Power Compon. Syst., vol. 44, no. 7, pp. 737–751, Apr. 2016.

[13] Y. Arya and N. Kumar, "Optimal control strategy-based AGC of electrical power systems: A comparative performance analysis," Optim. Contr. Appl. Methods, vol. 38, no. 6, pp. 982–992, Nov.-Dec. 2017.

[14] V. Yousuf, N. Yadav and N. Chopra, "Mitigation of subsynchronous resonance using UPFC with fuzzy logic control for power system stability," 2016 7th India International Conference on Power Electronics (IICPE), Patiala, 2016, pp. 1-6.

[15] Surya Prakash, S.K.Sinha, "Application of artificial intelligence in load frequency control of interconnected power system," International Journal of Engineering, Science and Technology Vol 3 No. 4 2011, pp. 264-275.

978-1-5386-6624-1/18 $31.00 © 2018 IEEE

Load Frequency Control Based on Non-Integer Type PID Controller Tuned by WOA

Manoj Kumar Debnath
Department of Electrical Engineering
Siksha 'O' Anusandhan University
Odisha, India
mkd.odisha@gmail.com

Nimai Charan Patel
Department of Electrical Engineering
Government College of Engineering
Keonjhar, Odisha, India
ncpatel.iter@gmail.com

Shreeram Choudhury
Department of Electrical Engineering
Siksha 'O' Anusandhan University
Odisha, India
sriram.uce@gmail.com

Abstract— **To meet the increasing power demand, various generating sources are introduced in interconnected power system. This research paper takes different generating sources in two different interconnected control areas for LFC (Load frequency Control) analysis. Here a non-integer type PID controller is applied to enhance the control action of conventional type PID controller. For selecting proper parameters for non-integer type PID (NIPID) controller, Whale Optimization Algorithm (WOA) is applied in this analysis. The several generating sources in both areas of the scrutinized dual area interconnected system include gas type generating unit, hydro and thermal type generating unit. WOA algorithm is used to tune the constraints of the system by the application of an abrupt load variation of 1.5% in control area 1. The scrutinized system considers both AC and AC-DC tie-line for LFC analysis in the unified areas. During the investigation the attained outcome is compared with an existing journal paper results so as to establish the governance of NIPID controller over PID controller.**

Keywords— *Non-integer PID controller, Whale Optimization Algorithm, Load Frequency Control, Fractional Calculus, Multi-generation system.*

I. INTRODUCTION

Frequency regulation in modern interconnected power system is a challenging task for power engineers. Modern configuration of power systems consist of multiple areas interconnected together and each area may have multiple generating units. The power exchange between these areas takes place over the tie-lines. Any sudden disturbance or active load change in one area may affect the system constraints mainly frequency and interchange of tie-line power [1], [2]. Thus, the system frequency in such cases must be regulated according to the load change by adjusting the active power generation. The control technique used to serve the purpose is considered as regulation of frequency. Hence, proper design and implementation of AGC facilitates to keep the transient performance indices within the reasonable limits thereby maintaining the interchange of interline power and frequency of the system.

The concept of AGC came into the picture in 1956 [3]. Since then many researchers have paid their attention to solve LFC problems over the last few decades. Several control strategies for AGC have been reported in the literature. Dated back researchers developed a state-variable model for hybrid - interconnected power system and suggested an optimal

feedback control for frequency regulation [4]. Chan and Hsu suggested adjustable designed controllers for AGC of unified system to develop the vigorous responses of the model [5]. Bakken and Grande explained the manual control, ACE based LFC and ramp following control method for AGC of deregulated system [6]. Nanda et al have scrutinized the performance of different classical controllers in AGC [7] and reported some facts on AGC with conventional controller [8]. Design of various controllers using different optimization algorithms have been illustrated in the literatures. Zeynelgil et al have used ANN controller with back propagation algorithm as the learning rule for regulation of frequency of unified power system and demonstrated its superiority over the conventional controllers [9]. Few years back Shayeghi et al have offered a hybrid system for fuzzy-PID controller for AGC over a deregulated power system showing its advantage of less sensitivity to bulky load variations and turbulences in existence of parametric deviations of the system and system nonlinearities [10]. Nanda et al proposed a initial application Bacterial Foraging-Based Optimization Technique for AGC which has faster convergence than GA [11]. B.Mohanty,et al. used DE method to adjust the controller factors for LFC in hybrid source type unified system[12]. For the first time, Sahu et al have invested the application of Teaching Learning Based Optimization (TLBO) algorithm tuned fuzzy-PID controller for load frequency control [13]. Singh et al have implemented JAYA based PID controller for AGC and demonstrated the superior performance of JAYA algorithm over other algorithms such as PSO, TLBO & DE [14]. Till now conventional power sources plays a dominant role in generation of electric energy. Nevertheless the source of renewable energy is greatly encouraged due to environmental concerns and depletion of fossil fuels. Hence, LFC problems in presence of renewable energy sources have drawn the attention of many researchers in recent years. Hasanien has successfully implemented Whale Optimization Algorithm based PID controller to face all the challenges regarding control of frequency including renewable energy sources [15]. A huge integration of renewable energy would require energy storages to deal with the uncertainty of renewable energy sources. For the purpose the use of Electric Vehicles are very convenient. As such many researchers have contributed their research works on use of Electric Vehicles in the field of AGC. Falahati and Taher have used Optimized advaned fuzzy controller for secondary frequency control using EVs in presence of renewable energy sources [16].Y.Arya described fractional order controller [17-18] for frequency regulation analysis in interconnected system.

The literature study reveals that many researchers applied conventional type PID/P/I/PI/PD controllers for LFC analysis. But in our present research analysis LFC is examined with the help of non-integer type PID controller tuned by Whale optimization algorithm.

II.   SYSTEM SCRUTINIZED

To expand the capacity of generation, various types of generating units are employed in unified power system. This research analysis considers hybrid generations like hydro, gas and thermal in each area in a dual area power system. Each unit consists of different non-integer type PID controller having 5 controller factors. The 15 parameters of three NIPID controllers are adjusted with the help of wolf optimization algorithm. The MATLAB Simulink model of the scrutinized system is depicted in Fig.1. The nominal ratings of all the system parameters described in hybrid generation system are considered from reference [12]. The model is examined under two different cases, namely with AC interconnection and AC-DC interconnection. The fitness/objective function used in the WOA is given as follows.

$$ITAE = \int_0^t \left(\left| \Delta f_1 \right| + \left| \Delta f_2 \right| + \left| \Delta P_{tie} \right| \right) t.dt \tag{1}$$

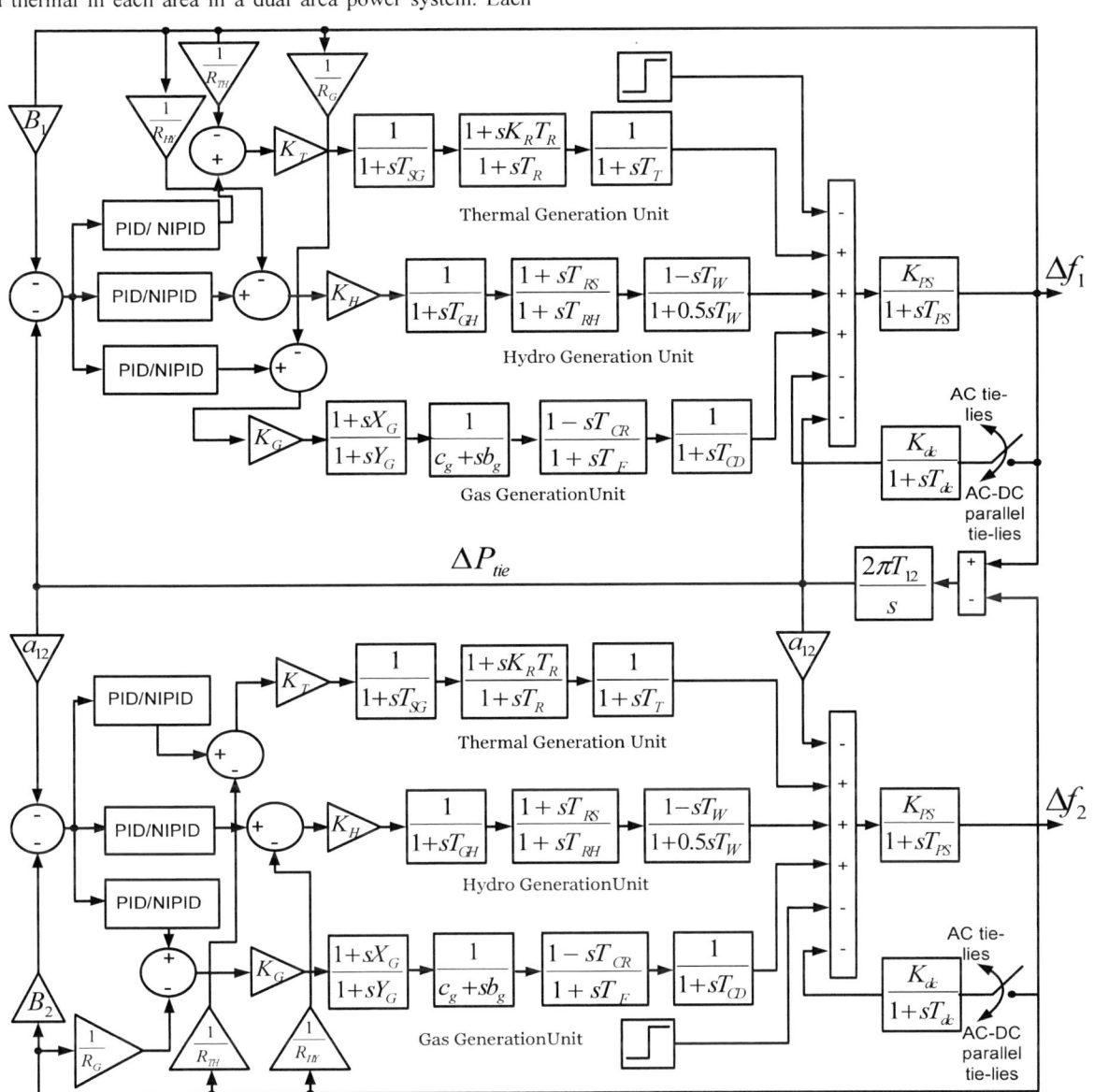

Fig.1. Model of hybrid generation system with AC-DC interconnecting lines.

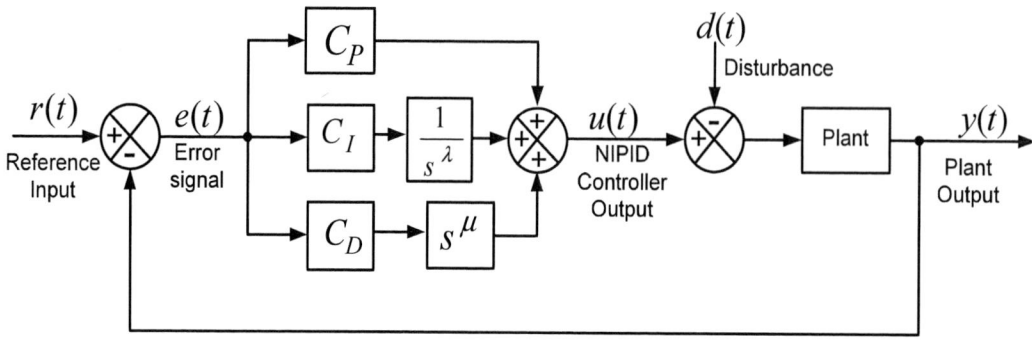

Fig.2. Internal architecture of non-integer type PID controller.

III. MENTIONED APPROACH

A. Construction of NIPID Controller

In this research work novel non-integer type $PI^\lambda D^\mu$ controller helps to face all stabilization problems for a power plant transfer function of any order. The internal design of the implemented NIPID controller is depicted in Fig.2.There is no need of problematical or any time-consuming process or plant constraints to meet all the stabilizing process of this described controller and this is signified in the three planes of C_P, C_I, C_D constraints space. The transfer function of non-integer type PID controller is shown in below

$$G_c(s) = C_P + \frac{C_I}{s^\lambda} + C_D s^\mu$$

Here the proportional gain, integral and derivative gains are denoted by C_P, C_I, C_D respectively. λ and μ are signifying the positive real numbers between 0 & 1. To find out the non-integer type PID controller to be stabilized in all cases, we have to define all the values of C_P, C_I, C_D, λ and μ control parameters and keep the poles of closed loop system on the $j\omega$ axis.

B. Whale Optimization Algorithm

Meta-heuristic optimization algorithms gain the attention due to its easy implementation nature, no need of gradient information and lastly most important can applied widely for complications of several discipline. New Whale optimization meta-heuristic optimization technique based upon the activities of humpback whales. Exploitation and exploration are two phases of search process. Maintaining proper balance is very difficult task between these two search processes during the development of any meta-heuristic algorithm. Generally wheals are the biggest mammals and one of the fancy creatures in the world. Researchers also reveal that there are some cells inside the brain of wheals which is similar to the spindle cell of human being. Due to these fact wheals are known as the smartest animal with emotions. Humpback wheals are one of the biggest wheals and their exceptional hunting method made them unique among all other wheals. Foraging is completed by generating an individual bubbles along a '9'-shaped path called bubble-net feeding method. In this technique firstly wheals dive 12m down and begin to generate bubble in a coiled form neighboring the target/prey and again spin up in the direction of the surface. To execute WOA optimization technique should have an idea about surrounding prey, coiled bubble-net feeding maneuver and hunt of prey which is defined in the below part.

Surrounding prey:
Humpback whales start their hunting by encircling the location of prey. The wanted prey or the solution which is nearer to the best value is considered as the finest candidate solution. The other search agents adjust their locations by moving towards the finest search individual which is followed by the below equation

$$\vec{D} = |\vec{C}.\vec{X}^*(t) - \vec{X}(t)|$$

$$\vec{X}(t+1) = \vec{X}^*(t) - \vec{A}.\vec{D}$$

Here X^*, $\overline{X}$ and t are known as position vector of best solution, position vector of search agents and the present iteration respectively.

The vectors $\vec{A}$ and $\vec{C}$ are known as co-efficient vectors and calculated by using the below equation

$$\vec{A} = 2\vec{a}.\vec{r} - \vec{a}$$

$$\vec{C} = 2.\vec{r}$$

Here random vector $\vec{r}$ is the dimension [0,1] and $\vec{a}$ denotes a vector whose value decreases linearly from 2 to 0 iteration wise.

Bubble-net-attacking method (exploitation phase):
Two methodologies are applied for mathematical modeling of the bubble-net-attacking hunting manner of humpback whales.
(1) Shrinking surrounding method
(2) Spiral way position updation
Shrinking surrounding method: In this approach the value of $\vec{a}$ is decreased which in turn affects the variation range of $\vec{A}$. Earlier the fluctuation of $\vec{A}$ is set in the interval of $[-a,a]$. In order to get the new position of search agent the random values of $\vec{A}$ are set in $\lceil -1,1 \rceil$.

(2) *Spiral way position updation:* In this method we have to calculate the distance amongst the whale location (X,Y) and

978-1-5386-6624-1/18 $31.00 © 2018 IEEE

the target location $((X^*, Y^*))$. To replicate the coil-form behavior of humpback whales a spiral calculation is made amongst the location of whale and target as follows

$$\vec{X}(t+1) = \vec{D}'.e^{bl}.\cos(2\Pi l) + \vec{X}^*(t)$$

In this equation the coefficient b describes the form of logarithmic spiral and the arbitrary number l is in the limit [-1,1]. During hunting humpback whales moves around with a spiral-form path along with decrease circle concurrently. During optimization there is a chance of 50% whether the whales do the reducing encompassing or the twisting/coiled technique to modernize their positions. To solve these problems following equations are formulated.

$$\vec{X}^*(t) - \vec{A}.\vec{D} \quad \text{if } p < 0.5$$

$$\vec{D}'.e^{bl}.\cos(2\Pi l) + \vec{X}^*(t) \text{ if } p \geq 0.5$$

Here arbitrary number p is within limit [0, 1].

Hunt for prey (exploration phase):

During this hunt for prey method the variation of $\vec{A}$ has been taken either more than 1 or smaller than -1. Now update the location of a search individual in the exploitation stage. The below mathematical model is used to perform global search for $|\vec{A}| > 1$.

$$\vec{D} = |\vec{C}.\vec{X}_{rand} - \vec{X}|$$

$$\vec{X}(t+1) = \vec{X}_{rand} - \vec{A}.\vec{D}$$

Here position vector ($\vec{X}_{rand}$) is an arbitrary vector selected from existing population.

IV. SIMULATION AND RESULT

An abrupt load perturbation of 0.015 perunit is subjected in first control area to examine the active behavior of the hybrid generation system. The time dependent function ITAE is considered for optimizing the constraints of NIPID controller in view of whale optimization technique. The structure is simulated by MATLAB toolbox (R2016a). The examination is carried for two types of model, namely (i) Hybrid generation system with AC interconnection and (ii) Hybrid generation system with AC-DC interconnection and

A. Case 1: Hybrid-generation system with AC interconnecting line

In the initial case study the dual area hybrid-generation system is simulated with AC interconnecting lines. Three different controllers are employed for three generating units in each control area. As a whole fifteen controller factors are tuned simultaneously with the help of WOA. The finest controller factors are grouped in Table I. The response (frequencies and inter line power oscillations) of the investigated model are depicted in Fig.3-5. To prove the dominance of the implemented model the time dependent response factors such as minimum undershoots, settling time and peak overshoots are computed and grouped in Table II. In each analysis the outcomes are put to comparison with pre-existing outcomes such as DE tuned conventional type PID

controller. The Figs.3-5 and Table II convey the dominance of NIPID controller over conventional type PID controller.

B. Case 2: Hybrid-generation system with AC-DC interconnecting line

In the further case study a static AC-DC model is implemented as interconnections between the control areas as conveyed by Fig.1. The model is again simulated with distinct NIPID controllers and the finest controller factors are tabulated in Table I. The dynamic responses (frequencies and inter line power oscillations) of the examined model are depicted in Figs.6-8. The time based performance factors for both types of controllers are presented in the Table II. The Figs.6-8 and Table II for a second time verifies the ascendancy of NIPID controller over traditional type PID controller.

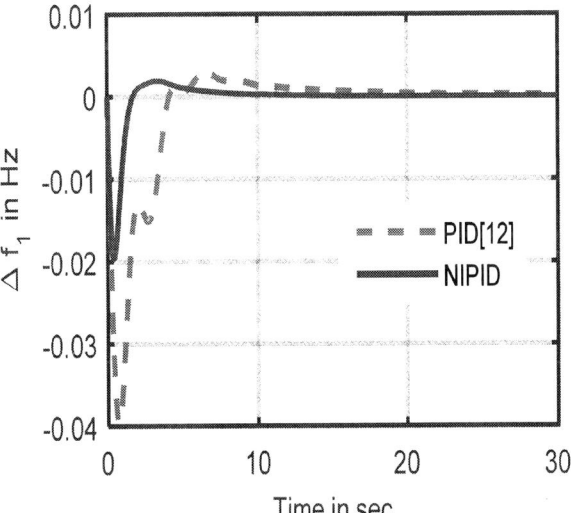

Fig.3. Frequency fluctuations in area 1(AC interconnection).

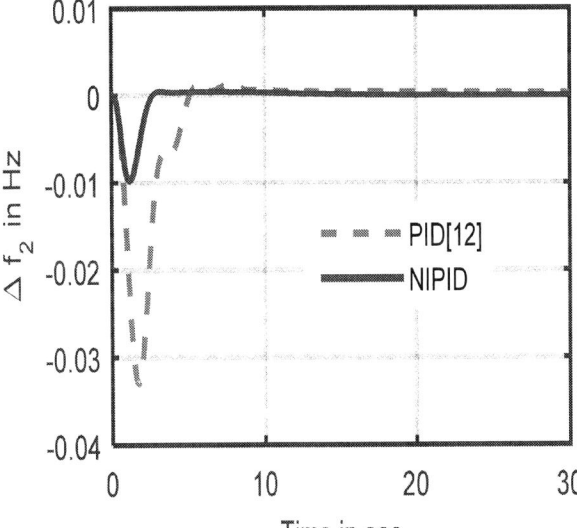

Fig.4. Frequency fluctuations in area 2(AC interconnection).

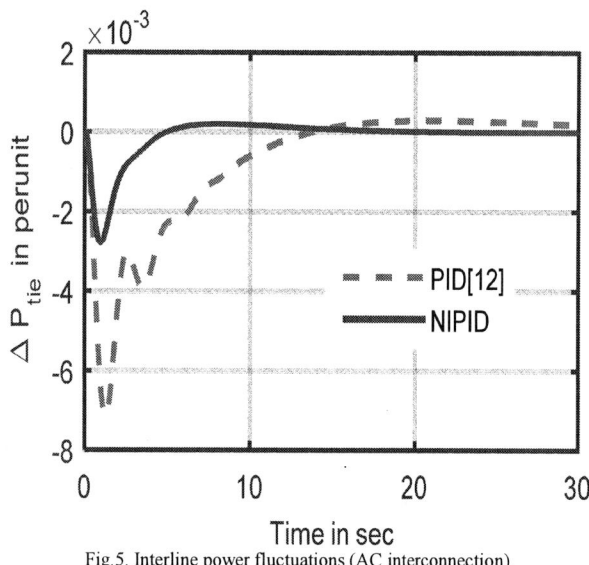

Fig.5. Interline power fluctuations (AC interconnection).

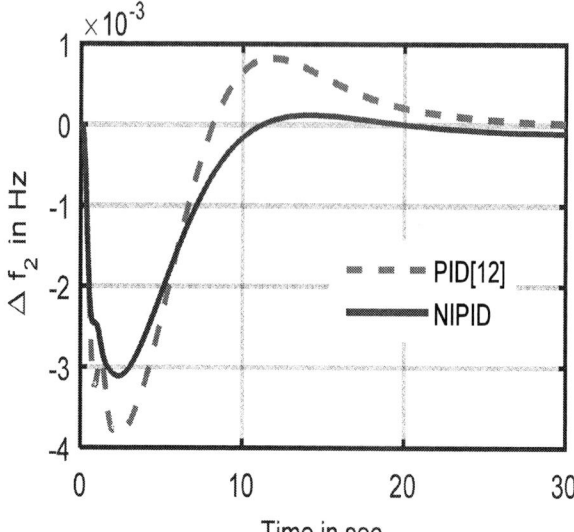

Fig.7. Frequency fluctuations in area 2(AC-DC interconnection).

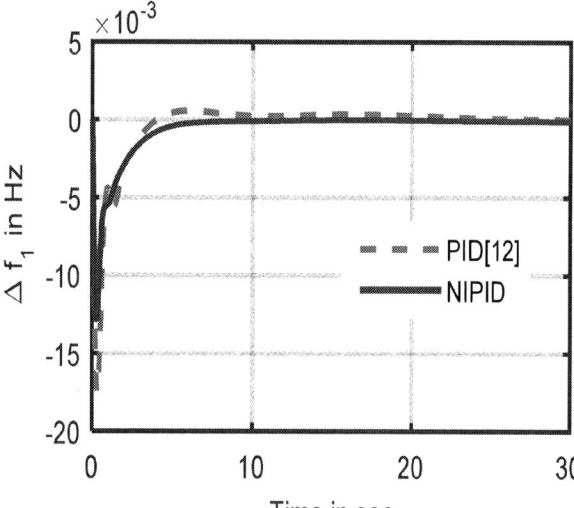

Fig.6. Frequency fluctuations in area 1(AC-DC interconnection).

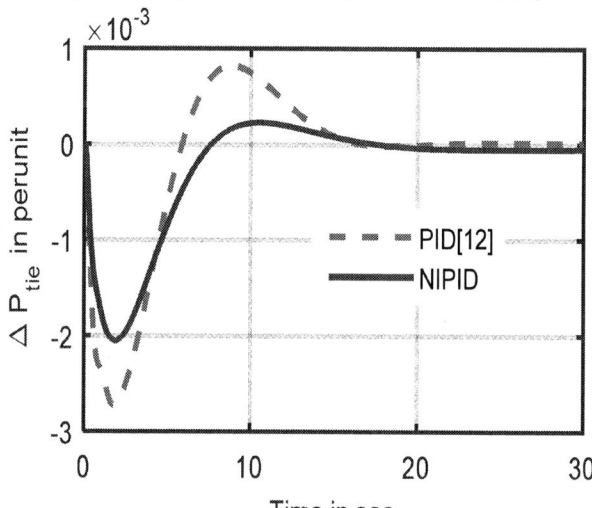

Fig.8. Interline power fluctuations (AC-DC interconnection).

TABLE I. NIPID/PID CONTROLLER FACTORS OBTAINED BY WOA.

Controller Parameters	AC Only						With AC-DC					
	NIPID			PID			NIPID			PID		
	Thermal	Hydro	Gas	Thermal	Hydro	Gas	Thermal	Hydro	Gas	Thermal	Hydro	Gas
λ	0.9900	0.7187	0.7179	-	-	-	0.7923	0.0123	0.4401	-	-	-
μ	0.9900	0.3720	0.9900	-	-	-	0.9731	0.2087	0.9640	-	-	-
C_P	3.4212	4.0000	0.4958	0.779	0.5805	0.5023	2.9454	0.6781	0.0426	1.6929	1.77731	0.9094
C_I	4.0000	0.0100	4.0000	0.2762	0.2291	0.9529	4.0000	0.5316	1.0287	1.9923	0.7091	1.9425
C_D	3.1178	2.9617	4.0000	0.6894	0.7079	0.6569	3.4316	2.8244	2.0462	0.8269	0.4355	0.2513

978-1-5386-6624-1/18 $31.00 © 2018 IEEE

TABLE II. TIME BASED RESPONSE FACTORS LIKE LEAST UNDERSHOOT, SETTLING TIME AND MAXIMUM OVERSHOOTS

Deviances	Undershoot (Ush), Settling time (T_s) & maximum Overshoot(O_{sh})	AC tie-line		AC-DC tie-line	
		NIPID	PID[5]	NIPID	PID[5]
Δf_1	Overshoots(Hz)	0.0019	0.0030	0	0.0005674
	Undershoots(Hz)	-0.0197	-0.0396	-0.0129	-0.0175
	T_s in sec	7.3343	16.7793	4.6565	7.0776
Δf_2	Overshoots(Hz)	0.0005	0.0011	0.000118	0.0008234
	Undershoots(Hz)	-0.0098	-0.0330	-0.0031	-0.0038
	T_s in sec	2.4544	10.3475	8.5208	15.7984
ΔP_{tie}	Overshoots(PU)	0.0002	0.0003	0.0002261	0.0008206
	Undershoots(PU)	-0.0028	-0.0071	-0.0020	-0.0027
	T_s in sec	3.3424	10.4699	5.7923	11.5892

V. CONCLUSION

In this research analysis non-integer type PID controllers were successfully implemented to adjust the frequency and interline power oscillations in multi-generation type unified system. The suitable controller factors were obtained by the application of whale optimization technique taking an abrupt perturbation in one of the control areas. The time based response parameters are computed to analyze the active behavior of the interconnected model. The examined model considered both AC and AC-DC interconnecting line to examine the responses. The outcomes of the proposed model were compared to pre-existing journal results to prove the dominance of non-integer PID controller over PID controller.

REFERENCES

[1] Kundur, Prabha, Neal J. Balu, and Mark G. Lauby. Power system stability and control. Vol. 7. New York: McGraw-hill, 1994.

[2] Elgard OI. Electric energy systems theory. New York: McGraw-Hill; 1982. p. 299–362.

[3] Cohn, N., 1956. Some aspects of tie-line bias control on interconnected power systems. Transactions of the American Institute of Electrical Engineers. Part III: Power Apparatus and Systems, 75(3), pp.1415-1436.

[4] Elgerd, O.I. and Fosha, C.E., 1970. Optimum megawatt-frequency control of multi area electric energy systems. IEEE Transactions on Power Apparatus and Systems, (4), pp.556-563.

[5] Chan, Wah-Chun, and Yuan-Yih Hsu. "Automatic generation control of interconnected power systems using variable-structure controllers." *IEE Proceedings C (Generation, Transmission and Distribution).* Vol. 128. No. 5. IET Digital Library, 1981.

[6] Bakken, Bjorn H., and Ove S. Grande. "Automatic generation control in a deregulated power system." *IEEE Transactions on Power Systems* 13.4 (1998): 1401-1406.

[7] Saikia, Lalit Chandra, J. Nanda, and S. Mishra. "Performance comparison of several classical controllers in AGC for multi-area interconnected thermal system." International Journal of Electrical Power & Energy Systems 33.3 (2011): 394-401.

[8] Nanda, Janardan, Ashish Mangla, and Sanjay Suri. "Some new findings on automatic generation control of an interconnected hydrothermal system with conventional controllers." IEEE Transactions on energy conversion 21.1 (2006): 187-194.

[9] Zeynelgil, H. L., A. Demiroren, and N. S. Sengor. "The application of ANN technique to automatic generation control for multi-area power system." *International journal of electrical power & energy systems* 24.5 (2002): 345-354.

[10] Shayeghi, H., H. A. Shayanfar, and A. Jalili. "Multi-stage fuzzy PID power system automatic generation controller in deregulated environments." Energy Conversion and management 47.18-19 (2006): 2829-2845.

[11] Nanda, Janardan, Sukumar Mishra, and Lalit Chandra Saikia. "Maiden application of bacterial foraging-based optimization technique in multiarea automatic generation control." IEEE Transactions on power systems 24.2 (2009): 602-609.

[12] Mohanty, Banaja, Sidhartha Panda, and P. K. Hota. "Controller parameters tuning of differential evolution algorithm and its application to load frequency control of multi-source power system." International journal of electrical power & energy systems 54 (2014): 77-85.

[13] Sahu, Binod Kumar, et al. "Teaching–learning based optimization algorithm based fuzzy-PID controller for automatic generation control of multi-area power system." Applied Soft Computing 27 (2015): 240-249.

[14] Singh, Sugandh P., et al. "Analytic hierarchy process based automatic generation control of multi-area interconnected power system using Jaya algorithm." Engineering Applications of Artificial Intelligence 60 (2017): 35-44.

[15] Hasanien, Hany M. "Whale optimisation algorithm for automatic generation control of interconnected modern power systems including renewable energy sources." IET Generation, Transmission & Distribution (2017).

[16] Falahati, Saber, Seyed Abbas Taher, and Mohammad Shahidehpour. "Grid Secondary Frequency Control by Optimized Fuzzy Control of Electric Vehicles." IEEE Transactions on Smart Grid (2017).

[17] Arya, Yogendra. "AGC of restructured multi-area multi-source hydrothermal power systems incorporating energy storage units via optimal fractional-order fuzzy PID controller." Neural Computing and Applications (2017): 1-22.

[18] Arya, Yogendra. "AGC performance enrichment of multi-source hydrothermal gas power systems using new optimized FOFPID controller and redox flow batteries." Energy 127 (2017): 704-715.

2nd IEEE International Conference on Power Electronics, Intelligent Control and Energy Systems (ICPEICES-2018)

Energy and Exergy Analysis for Heliostat Based Solar Thermal Power Plant

Manoj Kumar
Department of Mechanical Engineering *Dayalbagh Educational Institute (Deemed University)*
Agra, India
manu.28feb1993@gmail.com

R.P Saini
Alternate Hydro Energy Centre
Indian Institute of Technology, Rookree
Roorkee, India
rajsafah@iitr.ac.in

Abstract— Among all the non-conventional sources of energy solar energy is the most adaptable source because of its nature of all pervasiveness and costless. Solar power tower technology is considered to be the best technology for industrial scale because of its high efficiency and significant scope for improvement. In the literature review it is found that due to the popularity of the solar power tower technology lot of research paper giving enormous contribution in the advancement of this technology have been published.
In the proposed work a mathematical model of the heliostat based solar thermal power plant considering the unsteady variations of solar energy has been developed involving the various laws of thermodynamics and heat transfer. The execution of the model has been done on the Microsoft Excel 2014 and the results have been derived in the graphical form as well as tabular form. An attempt has also been made to validate the results obtained by simulation with the results of performance of DAHAN solar power tower plant situated at Beijing China available in the literature. It is concluded that the energy and exergy analysis for the performance of solar power tower plant may be used to predict the performance of other such plants.

Keywords— Exergy, CSP, Heliostat, CRS, Cavity Receiver

Nomenclature

SYMBOLS	DESCRIPTION	UNITS
DNI	Direct normal irradiance	W/m^2
n	Day of the year	-
δ	Solar declination angle	radians
α_s	Solar altitude angle	radians
ω_s	Hour angle	radians
t_s	Solar time	Hours
A_h	Surface area of heliostat	m^2
$E_{in,hel}$	Power input to the heliostat subsystem	Watt
α_{sun}	Solar elevation angle	radians
A_{sun}	Solar azimuth angle	radians
θ_H	Heliostat azimuth angle	radians
$E_{out,hel}$	Power output from the heliostat	Watt

	subsystem	
$E_{loss,rad}$	Radiation losses from receiver	Watt
α	Absorptivity of receiver surface	-
$E_{loss,reflc}$	Reflection losses	Watt
$E_{loss,conv}$	convection losses	Watt
h_c	Global heat transfer coefficient	$W/m^2 –K$
A_{rec}	Aperture area of the receiver	m^2
$T_{rec,s}$	Receiver surface temperature	^0C
T_a	Ambient temperature	^0C
$\emptyset$	Latitude	Degree
H_{cr}	Height of the receiver	m
η_{cos}	Cosine efficiency	-
$\eta_{s\&b}$	Shading and blocking efficiency	-
η_{int}	Interception efficiency	-
η_{att}	Atmospheric attenuation efficiency	-
η_{ref}	Reflection efficiency	-
η_{opt}	Optical efficiency	-
V	Wind velocity	m/s
η_{rec}	Efficiency of receiver	-
$E_{rec,out}$	Power output from receiver	Watt
$\eta_{heatexchang1}$	Efficiency of heat exchanger 1	-
$E_{outheatexchang1}$	Power output from heat exchanger1	Watt
$E_{loss,net}$	Net loss from receiver subsystem	Watt
$E_{loss,cond,hottank}$	Conduction losses from hot storage tank	Watt
$K_{concrete}$	Thermal conductivity of concrete	$W/m-K$
L	Length of hot storage tank	m
$T_{hottank,s}$	Surface temperature of hot storage tank	^0C
D_2	Outer diameter of hot storage tank	m

978-1-5386-6624-1/18 $31.00 © 2018 IEEE

Symbol	Description	Unit
D_1	Inner diameter of hot storage tank	m
$E_{loss,conv,hottank}$	Convection losses from hot storage tank	Watt
h	Global heat transfer coefficient	W / m²-K
A_{hot}	Surface area of hot storage tank	m²
Nu	Nusselt Number	-
Gr	Grasshoff Number	-
Pr	Prandtl Number	
v	Kinematic viscosity of air	m² / s
μ	Dynamic viscosity of air	Pa.s
c_p	Specific heat of air at constant pressure	kJ / Kg K
k_{air}	Thermal conductivity of air	W / m-K
B	Volume expansivity	K⁻¹
T_f	Film Temperature	K
$E_{loss,rad,hottank}$	Radiation losses from hot storage tank	Watt
E	Surface emissivity	-
$E_{loss,tot,hottank}$	Total loss from hot storage tank	Watt
$\eta_{hot,tank}$	Efficiency of hot storage tank	-
$E_{out,hottank}$	Output power from hot storage tank	Watt
$E_{out,heatexchang2}$	Power output from heat exchanger 2	Watt
$\eta_{heatexchang2}$	Efficiency of heat exchanger 2	-
$E_{out,power}$	Output power from power block	W
η_{power}	Efficiency of power block	-
$\eta_{overall,energy}$	Overall energy efficiency	-
Ψ	Exergy associated with solar irradiation	W
T_0	Temperature of the sun	K
ψ_{rec}	Exergy received by receiver	Watt
ψ_0	Exergy loss from heliostat subsystem	Watt
$\psi_{II,hel}$	Exergy efficiency of heliostat field	-
$\psi_{rec,abs}$	Exergy absorbed	-
$\psi_{rec,loss}$	Exergy losses from receiver	-
IR_1	Irreversibility of receiver	Watt
$\eta_{II,rec}$	Exergy efficiency of receiver	-
m_s	Mass flow rate	Kg / s
C_{pf}	Specific heat of fluid	kJ / kg-K
$\psi_{sgss,abs}$	Exergy absorbed by steam generation subsystem	Watt
IR_2	Exergy loss from SGSS	Watt
$\eta_{II,sgss}$	Exergy efficiency of SGSS	-
$\psi_{power,abs}$	Exergy absorbed by power block	Watt
IR_3	Exergy losses from power block	Watt
$E_{out,power}$	Power output from power block	Watt
$\eta_{II,power}$	Exergy efficiency of power block	-

I. INTRODUCTION AND LITERATURE REVIEW

Energy is fundamental to the quality of our lives. It has become the integral part of human civilisation for it is being utilised in every sphere of life from personal needs to commercial, industrial, social and other spheres of life. The demands of energy are increasing day by day dramatically. India being the fourth largest consumer of the energy in the world is consuming 917.2 kWh per capita [1].

India has enough solar radiations for deriving power from the solar energy. The major reason for this fact is that India lies between tropic of cancer and equator. Factually the solar radiations received by India (more than 5000 trillion kWh per year) are far more that the consumption of the country [2, 3]. The typical solar radiation over the country fluctuates between 4 kWh/day to 7 kWh/day. Fig. 1 displays the solar installation in India.

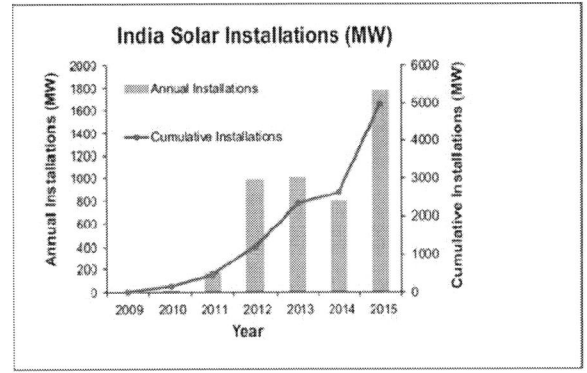

Fig. 1. Solar power installations in India [4]

Basically, solar energy can be harnessed in two ways either using solar photovoltaics systems which produces electricity directly or by using solar thermal systems which utilises the solar energy to produce heat. Solar thermal systems find its applications in low temperature, medium temperature and high temperature domains. In the high temperature application of solar energy four systems are deployed (also known as CSP) to convert the water into steam by utilising the focussed solar radiations.

- Linear fresnel reflector system
- Parabolic trough collector system
- Parabolic dish collector system
- Central receivers or power tower technology system

The solar tower power plant mainly consists of three subsystems namely the heliostat field, the receiver and the power conversion subsystem. The heliostat field is composed of computer-controlled mirrors. These mirrors track the sun in two axes individually and reflect the incoming solar radiation onto the receiver situated at the peak of the tower. The receiver absorbs the solar radiation reflected by the field of the mirrors and utilize this solar radiation into conversion of steam.The temperature of the fluid can range from 250 °C to 1000 °C depending upon the design of the receiver and nature of the fluid. Fig. 2 shows a typical solar power tower plant without storage of thermal energy. It clearly describes the functioning of the plant.

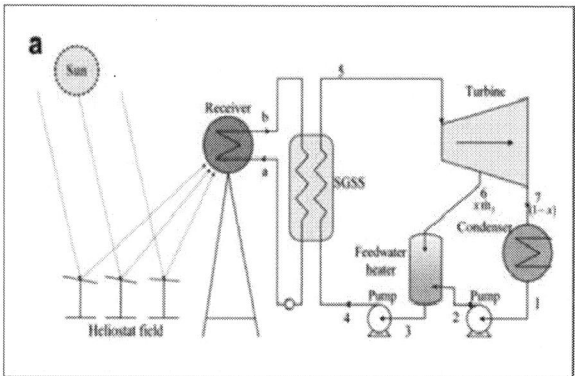

Fig. 2. A representative solar power tower plant without storage of thermal energy [5]

Solar tower technology is emerging out to be the most promising technology throughout the world. Table 1 gives a comparison between the four CSP technologies with respect to different parameters like relative cost, land occupied, thermodynamic efficiency, operating temperature range and concentration ratio etc. The comparison in the Table I along with the following advantages proves that the solar power tower technology is most clean energy for commercial uses.

- Provides large amount of electricity daily
- Constant production of power and supply to grid
- Environment friendly energy
- High concentration ratio
- Significant outlook for improvement
- High temperature operating range

TABLE I. COMPARISON BETWEEN DIFFERENT CSP TECHNOLOGIES [6]

S. No.	Technology Name	Thermodynamic Efficiency	Operating Temperature Range	Solar Concentration Ratio	Outlook for Improvement
1	PTC	Low	20-40	15-45	Limited
2	LFR	Low	50-300	10-40	significant
3	SPT	High	300-365	15-150	Very significant
4	PDC	High	120-1500	100-1000	High potential

A model of the heliostat based solar thermal power plant constrained to ideal situations was presented by Hongfei Zheng et. al [7] to analyse the effect of several parameters on energy and exergy conversion efficiency, with operating temperature of the receiver, concentration ratio, endo reversible heat engine efficiency and so forth. The results disclosed that up to an optimum temperature both thermal and exergy conversion efficiencies was increased by the receiver working temperature initially. In-fact the concentration ratio was also increased by increasing the optimal temperature. And finally, it was shown that the endo reversible engine efficiency will increase the thermal conversion efficiency until it reaches the maximum and optimum value, and then the conversion efficiency will drop dramatically.

Using the modular modelling method a mathematical model of 1MW solar power tower plant for all the working conditions was prepared by Ershu Xu et. al [8]. The steam temperature, power generation, steam flow rate, steam pressure were analysed with respect to disturbance in solar irradiance.

Qiang Yu et. al [9] used the modular modelling method to develop the key parts of "collector and cavity receiver, the two models were coupled together based on the "STAR-90" simulation platform by adopting the area to flux matching method. The transient response curves under different variations and distribution of static heat flux density at equinox noon were simulated. The results demonstrate that it can provide good references for the operation and control system design of the entire solar thermal power tower system.

Wei Han et. al [10] selected the first Chinese solar-power tower demonstration plant DAHAN and presented thermodynamic performance the plant. The high- and low-temperature accumulators are the medium to store sensible heat and latent heat of steam respectively. Some provisions in plant are made in such a way that during the operation the receiver can be connected to a steam turbine or energy storage system. And then the software to simulate the model of the plant was developed and simulated. The results show 8.35% increment in the annual thermal efficiency. The results provide a better understanding for the design and operation of solar-power tower plant.

A software tool HFLD was used by Zhihao Yao et. al [11] to develop the heliostat field layout design and performance of DAHAN, the pioneer 1 MWe CRS (central receiver system) funded by Ministry of Sciences and Technology (MOST), which can be regarded as the milestone in solar thermal power development in China. Development of DAHAN in China comply with the major objective of demonstration the operation of CRS in China. A. Comparison of the simulated results from HFLD with the published heliostat field efficiency data from Spain PS10 has been done and the results are in good approximation.

Incorporating molten salt as the heat transfer fluid the energy and exergy analysis of the solar power tower system was done by Chao Xu et al. [12] theoretically. The results showed that main energy loss occurs in the power cycle subsystem although receiver system has maximum exergy loss followed by the heliostat field system. The energy and the exergy efficiencies of the receiver and overall system increases by increasing DNI and the concentration ratio. Integration of advanced power cycles including reheat Rankine cycles and supercritical Rankine cycle increased overall energy and exergy efficiencies of solar tower system up-to a certain extent.

Designing, Fabrication, and testing of a 5kW cylindrical cavity receiver comprising a tubular absorber was done by Melchior et al. [13] for performing thermo-chemical reaction. Combination of the three heat transfer modes to the chemical kinetics was modelled on a 2D steady-state model. Monte Carlo and finite difference techniques were used for solving it. The prototype has achieved solar-to-chemical energy conversion efficiency of 28.5% at a reactor temperature.

Based on the literature survey it has been found out that the solar power tower technology is gaining popularity day by day and hence lot of researches have been done. But as this technology is in the stage of development so there are lot of scope in the improvement in the different aspect of the technology. Based on the literature survey the following gaps have been identified.

- The energy and exergy analysis of heliostat based solar tower plant is done in many researches but many components like storage tanks, turbines etc. are not considered due to which there are variations in the results. So, there is a need to do the analysis of the whole power plant considering all the major components.
- The thermal analysis of the mathematical model of solar power plant has been done by different ways like energy and exergy analysis and thermodynamic analysis in many researches based on mathematical modelling. However, all the researches have focussed on the steady state analysis of plant i.e. the solar irradiation is constant with time but the solar radiation varies throughout the day. So, there is a need to do the analysis of plant in order to observe the behaviour of the different parameters or variables throughout the day.

In the steady state analysis, it is assumed that the solar insolation is constant for a particular location. But the solar insolation is not constant. While the solar radiation in the extra-terrestrial region is relatively constant, the radiation at the earth's atmosphere varies due to following factors.

- Local variations in the condition of atmosphere such as water vapour, clouds, and pollution.
- Atmospheric phenomena like absorption and scattering effects the solar radiation.

- Latitude angle of the location.
- Day of the year and the time of day.

Solar tower power system consists of mainly four subsystems: heliostat field subsystem, central receiver subsystem, steam generation subsystem and power conversion subsystem. The direct solar radiation coming from the sun is received by the heliostat field which consists of numerous mirrors and is reflected towards the central receiver. As the solar radiation is travelling from sun to the receiver various losses like shading, blocking, cosine losses, attenuation factor, losses due to interception factor etc. occurs as shown in Fig. 3.

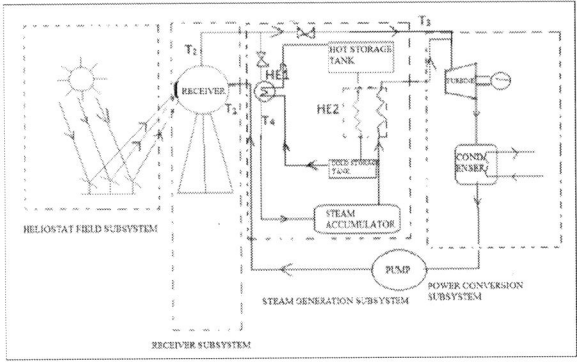

Fig.3. Different components of proposed solar tower power plant

Table II shows the components and functioning of different subsystems of CRS.

TABLE II. COMPONENTS AND FUNCTIONING OF DIFFERENT SUBSYSTEMS OF CENTRAL RECEIVER SYSTEM

S.No.	Subsystem	Component and Function
1.	Heliostat field subsystem	The term heliostat refers to as tracking of the sun. It consists of several computer-controlled mirrors that path the sun separately in two axes and replicate the solar radiation to the receiver situated on the tower.
2.	Receiver subsystem	The main purpose of this subsystem is to receive the reflected solar radiation in most efficient way. It can be of cavity type receiver or external receiver which stores the solar radiation into the working fluid.
3.	Steam generation subsystem	It consists of a heat exchanger. The heat exchanger exchanges the heat of the steam coming from the receiver and transfers it to hot storage tank when not needed. As the solar radiation is not available the hot tank gives up its heat to working fluid. When the heat is extracted from the hot storage tank the fluid is stored in the cold storage tank which is further processed to complete the cycle.

4.	Power conversion subsystem	It is a conventional Rankine cycle. This thermal cycle converts the thermal energy of working fluid into electric power by using all the elements of a typical coal thermal power plant. This subsystem is most critical.

A. Energy Analysis

TABLE III. DEVELOPMENT OF MATHEMATICAL MODEL FOR ENERGY ANALYSIS

S. No.	Particulars	Mathematical Equation
1.	DNI	$$DNI = 1367\left[1 + 0.033\cos\frac{2\pi n}{365}\right]\left[\frac{\sin\alpha_s}{(\sin\alpha_s + 0.33)}\right]$$
2.	Solar altitude angle	$$\alpha_s = \sin^{-1}\big((\sin\delta\,\sin\varnothing) + (\cos\delta\,\cos\omega_s\,\cos\varnothing)\big)$$
3.	Hour angle	$$\omega_s = 15(t_s - 12)$$
4.	Power received by the heliostat subsystem	$$E_{in,hel} = DNI \times A_h$$
5.	Optical efficiency	$$\eta_{opt} = \eta_{cos}\,\eta_{s\,\&\,b}\,\eta_{int}\,\eta_{att}\,\eta_{ref}$$
6.	Solar elevation angle	$$\sin\alpha_{sun} = (\sin\delta\,\sin\phi) + (\cos\delta\,\cos\phi\,\cos\omega_s)$$
7.	Solar azimuth angle	$$\cos A_{sun} = \frac{(\sin\delta\,\cos\phi - \sin\phi\,\cos\delta\,\cos\omega_s)}{\cos\alpha_{sun}}$$
8.	Cosine efficiency	$$\eta_{cos} = \frac{\sqrt{1 + \sin\alpha_{sun}\,\cos\lambda - \big((\cos(\theta_H - A_{sun})\cos\alpha_{sun}\,\sin\lambda)\big)}}{\sqrt{2}}$$
9.	Distance between the selected heliostat and the cavity receiver	$$S_0 = \sqrt{X^2 + (78 - Z)^2}$$
10.	Atmospheric attenuation efficiency	$$\eta_{att} = 0.99321 - 0.0001176\,S_0 + 1.98\,X\,10^{-8}\,S_0^2$$
11.	Power received by receiver	$$E_{out,hel} = E_{in,hel}\,X\,\eta_{opt}$$
12.	Radiation losses from the receiver	$$E_{loss,rad} = \frac{E_{out,hel}\,(3 - 6\alpha + 4\alpha^2 - \alpha^3)}{3}$$
13.	Reflection losses	$$E_{loss,reflec} = 0.05\,X\,DNI\,X\,\eta_{opt}\,X\sum A_H$$
14.	Convection losses	$$E_{loss,conv} = h_C\,A_{ape}\big(T_{rec,s} - T_a\big)$$
15.	Global heat transfer coefficient	$$h_C = 10.45 - V + 10\sqrt{V}$$
16.	Total losses from the receiver subsystem	$$E_{loss,net} = E_{loss,rad} + E_{loss,reflec} + E_{loss,conv}$$
17.	Efficiency of the receiver subsystems	$$\eta_{rec} = 1 - \frac{E_{loss,net}}{E_{out,hel}}$$
18.	Net power outcome the receiver subsystem	$$E_{rec,out} = E_{out,hel}\,\eta_{rec}$$
19.	Energy analysis of heat exchanger	$$E_{outheatexchange1} = E_{heatexchang1}\,E_{rec,out}$$
20.	Conduction losses from the hot storage tank	$$E_{loss,cond,hottank} = \frac{\big(2\pi k_{concrete}\,L(T_{hottank,s} - T_a)\big)}{\ln\left(\frac{R_2}{R_1}\right)}$$
21.	Convection losses from the hot storage tank	$$E_{loss,conv,hottank} = h\,A_{hot}\big(T_{hottank,s} - T_a\big)$$

22.	Nusselt number	$Nu = 0.59\,(Gr\,Pr)^{0.25}$
23.	Grasshop number	$Gr = \dfrac{g\,\beta\,L^3\left(T_{hottank,s} - T_a\right)}{v^2}$
24.	Prandtl number	$Pr = \dfrac{\mu\,c_p}{k_{air}}$
25.	Radiation losses from the tank	$E_{loss,rad,hottank} = \varepsilon\sigma A_{hot}\left(T_{hottank,s}^4 - T_a^4\right)$
26.	Total energy losses from the cold storage tank	$E_{loss,tot,hottank} = E_{loss,cond,hottank} + E_{loss,conv,hottank} + E_{loss,rad,hottank}$
27.	Efficiency of the hot storage tank	$\eta_{hottank} = 1 - \left(\dfrac{E_{loss,tot,hottank}}{E_{heatexchange1,out}}\right)$
28.	Energy coming out of the hot storage tank	$E_{out,hottank} = E_{heatexchange1,out}\,\eta_{hottank}$
29.	Energy analysis of Heat exchanger 2	$E_{outheatexchange2} = \eta_{heatexchange2}\,E_{out,hottank}$
30.	Energy analysis of the power subsystem	$E_{out,power} = \eta_{power}\,E_{heatexchang2}$
31.	Overall energy efficiency	$\eta_{overall,energy} = \eta_{opt}\,\eta_{rec}\,\eta_{heatexchang1}\,\eta_{hot,tank}\,\eta_{heatexchange2}\,\eta_{cold,tank}\,\eta_{power}$

B. Mathematical Model for Exergy Analysis

TABLE IV. DEVELOPMENT OF MATHEMATICAL MODEL FOR EXERGY ANALYSIS

S.No.	Particulars	Mathematical Model
1.	Exergy associated with solar irradiation	$\psi = DNI\left(1 - \dfrac{T_a}{T_0}\right)$
2.	Exergy efficiency of the heliostat subsystem	$\eta_{II,hel} = \dfrac{\psi_{rec}}{\psi}$
3.	Exergy balance for the receiver	$\psi_{rec} = \psi_{rec,abs} + \psi_{rec,loss} + IR_1$
4.	Exergy Loss from receiver	$\psi_{rec,loss} = E_{rec,totloss}\left(1 - \dfrac{T_a}{T_{rec,s}}\right)$
5.	Exergy of absorbing fluid	$\psi_{rec,abs} = m_s c_{pf}\left(T_2 - T_1 - T_a\ln\left(\dfrac{T_2}{T_1}\right)\right)$
6.	Exergy efficiency of the receiver subsystem	$\eta_{II,rec} = \dfrac{\psi_{rec,abs}}{\psi_{rec}}$
7.	Exergy analysis of the steam generation subsystem	$\psi_{rec,abs} = \psi_{sgss,abs} + IR_2$
8.	Exergy efficiency of the steam generation subsystem	$\eta_{II,sgss} = \dfrac{\psi_{power,abs}}{\psi_{sgss,abs}}$

III. RESULT AND DISCUSSION

A. Direct Normal Irradiance Versus Time of the Day

The particular day 22 June 2014 has been chosen in simulating the model in order to simplify the simulation.

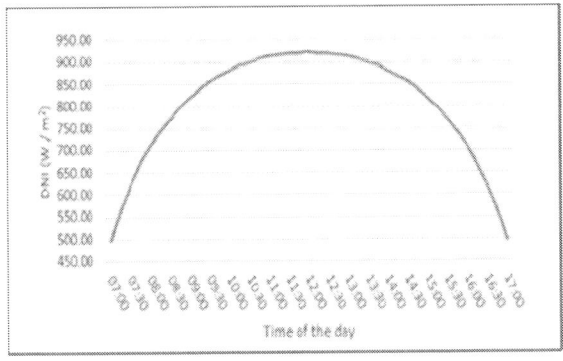

B. Output Power from Heliostat Subsystem Versus Time

Fig. 4. Variation of DNI versus time of the day

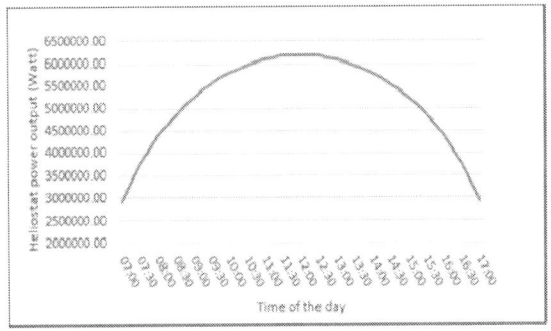

Fig. 5. Variation of output power from heliostat versus time of the day

978-1-5386-6624-1/18 $31.00 © 2018 IEEE

C. Variation of Overall Energy Efficiency Versus Time of the Day

Fig. 6 shows the variation of overall energy efficiency of the power plant considering the effect of the time of the day. Overall energy efficiency increases as the time of the day increases and reaches up-to maximum at noon periods.

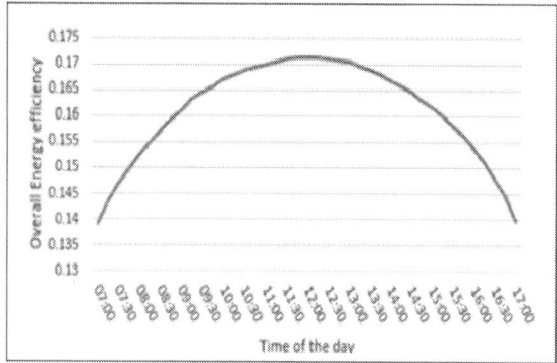

Fig.6. Overall energy efficiency versus time of the day

D. Variation of Exergy Efficiency with Time of the Day

Fig.7 shows the variation of exergy efficiency of the receiver. The variation of other subsystems like heliostat subsystems, power subsystems follow almost same pattern as followed by DNI with time. It can be observed from the figure that exergy efficiency is nearly constant after a particular time period.

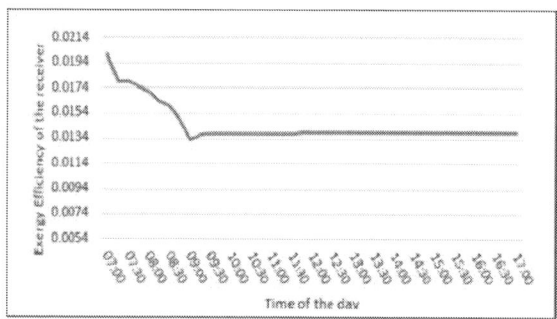

Fig.7. Exergy efficiency of the receiver versus time of the day

IV. CONCLUSIONS

The energy and exergy analysis of a solar tower power plant have been performed. The results obtained are compared with the results data obtained from the literature

- The mathematical model for energy and exergy analysis of a solar power tower plant has been developed which may be used to predict the variation of performance of different solar tower power plants.
- The developed model was simulated in Microsoft excel 2014. The execution of the mathematical model was done by selecting the data of DAHAN power plant situated at Beijing China,

and the results obtained are presented in the tabular as well as graphical form.

- The overall energy efficiency (average) and exergy efficiency (average) of the plant have been obtained as 16.11 % and 23.97 % respectively and the results were validated with the data of a plant and it was found that obtained results are in good agreement with that to the data obtained from the plant.
- The developed mathematical model may be used to predict the behaviour of the different parameters of any solar power tower plant.

REFERENCES

[1] http://www.eia.gov/contries/analysisbriefs/india/india.pdf [accessed on march 2018]

[2] Solar Power in India, Wikipedia, The free encyclopedia [accessed on April 2018]

[3] Ministry of power Available : http://www.powermin.nic.in [accessed on April 2018]

[4] http://mnre.gov.in/file-manager/annual-report/2014-2015/EN/Chapter%204/%20Chapter%20_4.htm.[accessed on april 2018]

[5] Xu C., Wang Z., Li X. and Sun F. ; "Energy and Exergy analysis of solar power tower Plants" Applied Thermal Energy 31, pp3904-3913 (2011)

[6] Zhang H.L., Baeyens J., Degreve J. and Caceres G ;" Concentrated solar power plants:

[7] Zheng H., Yu X., Su Y., Riffat S. and Xiong J. ; "Thermodynamic Analysis of an idealized solar tower thermal power plant" Applied thermal energy 8, pp. 271-278 (2015)

[8] Xu E., Yu Q., Wang Z. and Yang C.; "Modeling and simulation of 1 MW DAHAN solar thermal power tower plant" renewable energy 36, pp 848-857 (2013).

[9] Yu Q., Wang Z., Xu E., Li X. and Guo M, ; "Modeling and dynamic simulation of the Collector and receiver system of 1MWe DAHAN solar thermal power tower plant" Renewable Energy 43, pp.18-29 (2012)

[10] Han W., Hongguang J., Jianfeng S., Rumoul L. and Zhifeng W ; "Design of the first Chinese 1 MW solar-power Tower demonstration plant" International Journal Of Green Energy, September 2009

[11] Yao Z., Wang Z., Lu Z. and Wei X. ; "Modelling and simulation of the pioneer 1MW solar thermal central receiver system in China" Renewable Energy 34, pp.2437–2446(2009)

[12] Xu C., Wang Z., Li X. and Sun F.; "Energy and exergy analysis of solar power tower plants" Applied Thermal Engineering 31, pp. 3904-3913 (2011)

Efficient Approach for DG Placement and Size in Medium Voltage Distribution Systems

Anand K Pandey
Department of Electrical engineering
JamiaMilliaIslamia
New Delhi, India
anand.pandey.42@gmail.com

SheerazKirmani
Department of Electrical engineering
JamiaMilliaIslamia
New Delhi, India
sheerazkirmani@gmail.com

Abstract—**Location and size of Distributed Generation (DG) in distribution lines plays a very important role. Wrongly placed DG may cause distribution losses and poor voltage profile. In this paper a modified analytical method is suggested to find the best location and size of solar photovoltaic (SPV) based DG in medium voltage distribution lines to reduce the power losses and to improve the voltage profile. The proposed method is tested on IEEE 33-bus distribution system. The results showthe validity of the proposed approach for loss reduction and voltage profile improvement and there is an improvement in the result.**

IndexTerms—Analytical approach, SPV based DG unit, power loss minimization and voltage profile improvement.

I. INTRODUCTION

Distributed Generation (DG) based on renewable resources such as solar, wind, ocean, hydro, biomass and geothermal heat are considered as green power because of the negligible impact on greenhouse gas emissions. Such sources have emerged as an alternative energy solution to mitigate the dependence on fossil fuels since the mid-1970s after the first oil crisis [3].However, during this period, due to its high energy production costs along with utility and government disincentives, renewable DG remained relatively dormant until the mid-2000s.There has been a renewed interest in the deployment of renewable DG since the mid-2000s when the issue of global warming came to the forefront of concerns by many parts of the world [11].

The high penetration ofDG resources on the distribution system can has introduced many challenges to distribution network such as high losses, voltage fluctuation and low voltage stability etc. [13]. The high penetration of DGon distribution system can have either positive or negative impact depending on the distribution system operating characteristics and the DG characteristics [6] .High power loss and poor voltage profile is the main concern in developing countries, therefore it becomes necessary to place the DGs of appropriate size at optimal location. Both analytical and intelligent methods have been proposed in literature for optimal placement and sizing of DG in distribution systems. Different analytical methods have been proposed, Willis HL [1] has proposed a "2/3 rule" in distribution systems that can be used for placing DG units at optimal location.

In addition, various methodologies are available in the literature for optimal placement of DGs. These techniques include analytical, intelligent like fuzzy, ANN, GA, PSO etc.,

[7-20].In this paper a repeated load flow method is used to calculate the power losses in the system. The location of the DG is selected on the basis of power loss reduction and voltage profile improvement. SPV (solar photovoltaic) based DGs are selected for testing the method at strategic locations in order to reduce the power loss and to improve voltage profile. In this work, IEEE 33 bus test systemsare considered for simulation purposes and similarly it can be tested on IEEE 69 bus test system also.

This paper is discussed as follows: In section I Introduction, Problem formulation is presented in Section II, in Section III Location and sizing issues, in Section IV description of test system is presented, followed by simulation results in Section V and in Section VI Result and Discussion are presented in Section VII, a conclusion followed by references and appendix.

II. PROBLEM FORMULATION

An efficient load flow solution is used to find the power loss and voltage at each bus. The method proposed for load flow is based on the bus-injection to branch-current (BIBC) and branch-current to bus-voltage (BCBV) matrices which were formulated based on the topological structure of the distribution systems and is widely implemented for the load flow analysis of the distribution systems. The details of both matrices can be found in [4].

In this method we need only one base case load flow to determine the optimum size and location of DG. This method is more practical and easy to use in compare to other methods.In a distribution system at any bus ithe complex powerSi, is given as:

$$S_i = P_i + jQ_i \quad (1)$$

Wherei=1, 2, 3……..N
At any busi, the current injection is given as:

$$I_i = \left(\frac{P_i + jQ_i}{V_i}\right)^* \quad (2)$$

Where V_i is the voltage at bus i, Piis the real power injected at bus i, Q_i is the reactive power injected at bus i. The current injection of bus i can be separated into real and imaginary parts.

$$Real(I_i) = \frac{P_i \cos \theta_i + Q_i \sin \theta_i}{|V_i|}$$

$$Imag(I_i) = \frac{P_i \sin\theta_i - Q_i \cos\theta_i}{|V_i|} \quad (3)$$

We can find the value of B with the help of BIBC matrix. Bus current injections and branch currents also give the value of BIBC. BIBC matrix isconsisting of '0's or '1's:

$$[B]_{nbX1} = [BIBC]_{nbX(n-1)} * [I]_{(n-1)X1} \quad (4)$$

Where number of the branches is represented by nb, [I] is the vector which represents equivalent current injection for each bus except the reference bus.

We can say that the bus voltage can be expressed as a function of branch currents, line parameters, and the substation voltage. Similarly we can proceed for other buses also; therefore, the relationship between branch currents and bus voltages can be expressed as:

$$\Delta V = [Z]_{nbX(n-1)} \cdot [B]_{(n-1)X1} \quad (5)$$

The voltage drop from each bus to the reference bus is obtained with BCBV and BIBC matrices as:

$$[\Delta V]_{(n-1)X1} = [BCBV][BIBC].[I] \quad (6)$$

Where, BCBV matrix is used to find the relations between branch currents and bus voltages. The elements of BCBV matrix consist of the branch impedances.

The detail about the BIBC and BCBV matrix can be found in [30]. With the help of this method, the total power losses can be expressed as a function of the bus current injection:

$$P_{loss} = \sum_{i=1}^{nb} |B_i|^2 . R_i = [R]^T |[BIBC][I]|^2 \quad (7)$$

$$P_{loss} = |R|^T([BIBC][real(I_i)] + j[BIBC][imag(I_i)])^2$$

$$P_{loss} = |R|^T(([BIBC][real(I_i)])^2 + ([BIBC][imag(I_i)])^2)$$

By putting the values of equation (3) in equation (7), we get

$$P_{loss} = |R|^T(([BIBC][\frac{P\cos\theta + Q\sin\theta}{|V|}])^2 + ([BIBC][\frac{P\sin\theta - Q\cos\theta}{|V|}])^2)$$

Power loss at i^{th} branch can be given as:

$$P_{loss,i} = R_i[(\sum_{j=2} BIBC(i,j-1)\frac{P_j \cos\theta_j + Q_j \sin\theta_j}{|V_j|})^2 + (\sum_{j=2} BIBC(i,j-1)\frac{P_j \sin\theta_j + Q_j - \cos\theta_j}{|V_j|})^2]$$

Total power loss can be expressed as:

$$P_{loss} = \sum_{i=1}^{nb} R_i [(\sum_{j=2} BIBC(i,j-1)\frac{P_j \cos\theta_j + Q_j \sin\theta_j}{|V_j|})^2 + (\sum_{j=2} BIBC(i,j-1)\frac{P_j \sin\theta_j + Q_j - \cos\theta_j}{|V_j|})^2]$$
$$(8)$$

To determine the optimal size of DG, the first derivative of equation (8) should be equal to zero i.e.

$$\frac{\partial P_{loss}}{\partial P_i} = 0$$

$$\frac{\partial P_{loss}}{\partial P_i}$$
$$= 2\sum_{i=1}^{nb} R_i \sum_{\substack{j=2 \\ j\neq i}}^{n} dPBIBC_i(i,j-1)[\frac{Cos\theta_i}{|V_i|}real(I_i)$$
$$+ \frac{Sin\theta_i}{|V_i|}imag(I_i)]2\sum_{i=1}^{nb} R_i \sum_{\substack{j=2 \\ j\neq i}}^{n} dPBIBC_i(i,j$$
$$- 1)[P_i\cos^2\theta_i + Q_i\sin\theta_i\cos\theta_i + P_i\sin^2\theta_i$$
$$- \frac{Q_i\sin\theta_i\cos\theta_i}{|V_i|^2} = 0$$

by solving above equation, we get

$$P_i =$$
$$\frac{|V_i|\sum_{j=1}^{nb} R_j \sum_{\substack{k=2 \\ k\neq i}}^{n} dPBIBC_i(j,k-1)[\cos\theta_i real(dI_k) + \sin\theta_i imag(dI_k)]}{\sum_{j=1}^{nb} R_j dPBIBC_i(j,i-1)}$$
$$(9)$$

In matrix form, the optimum size of DG can be written as:

$$P_i$$
$$= \frac{|V_i|[R]^T[dPBIBC_i][\cos\theta_k real(dI_k) + \sin\theta_k imag(dI_k)]}{[R]^T dPBIBC_i[:,k-1]}$$
$$(10)$$

For a five bus systems the value of real(dI₄) and imag(dI₄) can be written as:

$$[real(dI_k)] = [real(I_2) \quad real(I_3) \quad 0 \quad real(I_5)]$$

$$[imag(dI_4)] = [imag(I_2) \quad imag(I_3) \quad 0 \quad imag(I_5)]$$

III. DG LOCATION AND SIZE

The proper location and sizing of DGs in medium voltage distribution systems plays a very important role for power loss reduction and voltage profile improvement. Improper selection of location and size of DG may increase the system losses than the losses without DGs [10]. A simple approach for placement of DGs in the distribution system is that the capacity of DGs should be approximately 2/3 capacity of the incoming generation at approximately 2/3 length of the line [1].

The algorithm for the method proposed in this paper can be written as:

Step1: Initialize all the parameters.

Step2: Find the power loss without DG using equation (7)

Step3: Find the optimum size of DG using equation (10) at each bus except reference bus i.e. bus no. 1.

Step4: Find the bus which has minimum power loss after adding DG at each bus.

Step5: Find the bus voltages are within the range using equation (6)

Step6: If bus voltages are not within the range then go for next value of DG and repeat step (4)

By the application of method proposed in this paper two plots Fig.1 and Fig. 2 have been drawn to find the optimal size and location of DG. Fig. 1 is a plot of size of DG at each bus for minimum loss in case of IEEE 33 bus system. Similarplot can be drawn for IEEE 69 bus system. Fig.2 is a plot between minimum loss occurred and location of DG.Fig. 2 suggests about the location and Fig. 1 gives information about the optimal sizeof DG. Therefore Fig.1 and Fig.2 provides the information about the optimal size and location of DG. Fig.3 is a plot between voltage and location. From both the plot it is clear that if size of DG is more or less than optimal size losses will increase.

IV. THE TEST SYSTEMS

The performance of the modified analytical method has been tested on IEEE33 bus medium voltage distribution system of different size, configurations and having different complexities. The IEEE 33 radial bus systemcomprises of a total load of 3.72 MW and 2.3 MVAR [2].

A modified analytical method to find the distribution system load flow solution [30]is programmed in MATLAB 16 environment. This MATLAB program is used tocompute the optimum sizes of DG at various buses and to calculatethe total losses with DGs at different location to identify the optimumlocation.

V. SIMULATION RESULTS

Load flow analysis using eqn.1 as explained above is performed and two plots have been drawn. First plot i.e. Fig.1 is plotted between DG size vs location while second plot i.e. Fig.2 is plotted between losses and DG size for IEEE 33 bus systems. From fig.2 it is clear that location 6 is having minimum loss i.e. 110.287 KW and from fig.1. At location 6 size of DG is 2.59 MW. Hence we can say that optimum size of DG is 2.59 MW and optimal location of DG is 6. In this paper for study rating of DG varies from 0.40 MW to 4.25 MW for IEEE 33 bus system. Similar study can be performed on IEEE 69 bus test system. Fig.3 is plotted for voltage at each location. It is clear from this plot thatafter placing the DG voltage profile gets improved from 0.945609 pu to 0.972301 pu.

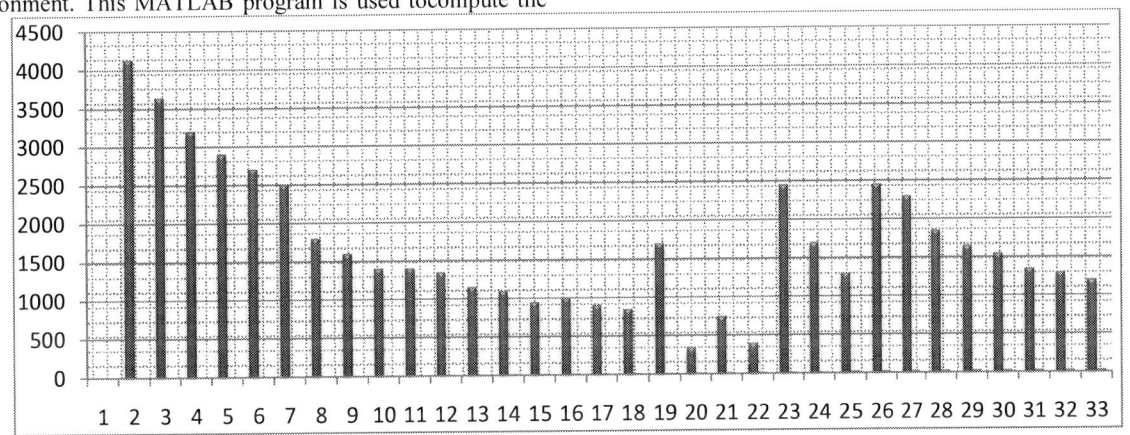

Fig.1. Size (in MW)vs location of DG

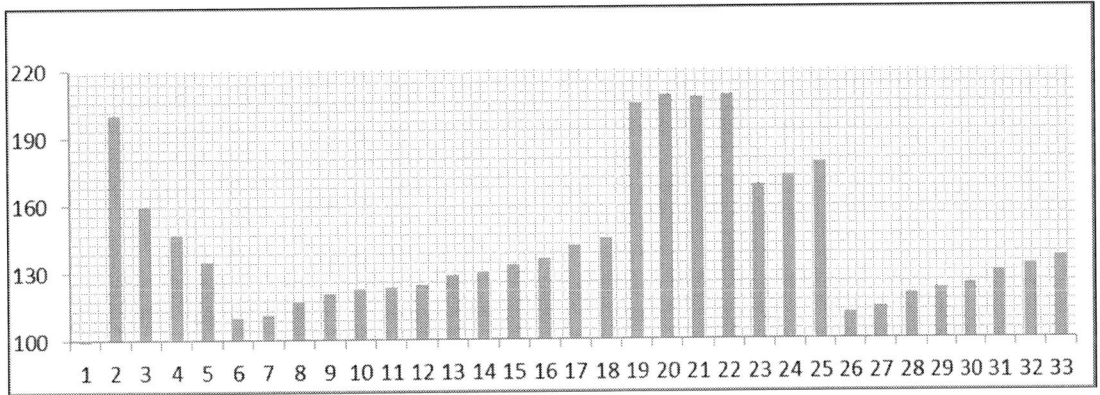

Fig.2. Losses (in KW) vs location of DG

978-1-5386-6624-1/18 $31.00 © 2018 IEEE 129

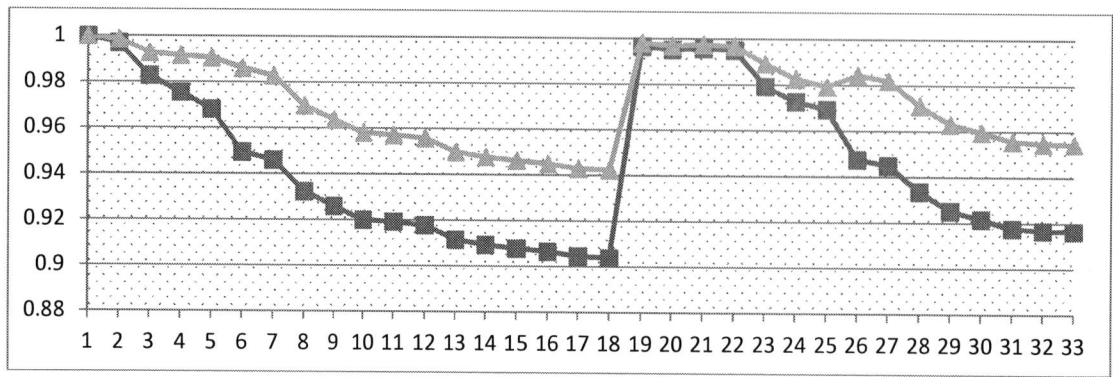

Fig.3.Comparision of voltage profile before and after DG placement

VI. RESULTS AND DISCUSSION

FromTABLE 1we can comment that when the solar photovoltaic based DG is inserted in to IEEE 33 bus medium distribution system then the losses reduced from 210.25 KW to 110.287 KW i.e. 47.545% in saving of losses and average system voltage profile improved from 0.945609 puto 0.972301 pu.It is observed that optimal location is 6 while optimum size is 2590 KW which gives improved results but beyond that no significant change in the reduction in power loss and improvement in average voltage profile.

VII. CONCLUSION

An efficient analytical method to find the best size and location of single DG is proposed in this paper. It is seen that by application of this method losses in a distribution systems can be reduced and voltage profile can be improved. In future this method can be used to find the multiple DG's to find the optimal size and location of DG. This study is important for distribution companies (DISCOM) to find the location and size of solar photovoltaic system to minimize the losses and improve the voltage profile which can be economical.

TABLE 1: COMPARISIONS OF RESULTS

S. No.	Study	Test System	Optimum Location	Optimum DG Size (KW)	Power Loss (KW)		
					Without DG	With DG	% savings
1	Earlier Work[21]	IEEE 33 bus	6	2630	211.20	111.24	47.33
2	Proposed Work	IEEE 33 bus	6	2590	210.25	110.287	47.54

REFERENCES

[1] Willis HL. Analytical methods and rules of thumb for modeling DG distribution interaction. In: Proceeding of IEEE power engineering society summer meeting seattle3; 2000. p. 1643-4.

[2] Kashem MA, Ganapathy V, Jasmon GB, Buhari MI. A novel method for loss minimization in distribution networks. In: Proceedings of international conference on electric utility deregulation and restructuring and power technologies; 2000. p. 251-5.

[3] S. Rahman, "Green power: what is it and where can we find it?," IEEE Power Energy Mag., vol. 1, no. 1, pp. 30-37, Jan./Feb. 2003.

[4] Teng J. A direct approach for distribution system load flow solutions. IEEE Transaction Power Deliv 2003;18:882-7.

[5] Kashem MA, Ledwich G. Multiple distributed generators for distribution feeder voltage support. IEEE Transaction Energy Convers 2005;20(3):676-84.

[6] Naresh A, Pukar M, Mithulananthan N. An analytical approach for DG allocation in primary distribution network. Electr Power Energy Syst 2006;28:669-78.

[7] Venkatesh B, Ranjan R. Fuzzy EP algorithm and dynamic data structure for optimal capacitor allocation in radial distribution systems. Proc InstrumElectrEng Generation Transm Distribution 2006;153(1):80-8.

[8] Carpaneto E, Chicco G, Akilimali JS. Branch current decomposition method for loss allocation in radial distribution systems with distributed generation. IEEE Transaction Power Syst 2006;21(3):1170-9.

[9] Etemadi AH, Fotuhi-Firuzabad M. Distribution system reliability enhancement using optimal capacitor placement. IET GenerTransmDistrib 2008;2(5):621-31.

[10] Hung, N.Mithulananthan DQ, Bansal RC. Analytical expressions for DG allocation in primary distribution networks. IEEE Transaction Energy Convers 2010;25(3):814-20.

[11] C. J. Mozina, "Impact of green power distributed generation," IEEE Indust. Applicat. Mag., vol. 16, no. 4, pp. 55-62, Jul./Aug. 2010.

[12] Hung, N.Mithulananthan DQ, Bansal RC. Analytical expressions for DG allocation in primary distribution networks. IEEE Transaction Energy Convers 2010;25(3):814-20.

[13] F. Katiraei, J. R. Aguero, "Solar PV integration challenges," IEEE Power Energy Mag., vol. 9, no. 3, pp. 62-71, May/Jun. 2011.

[14] Philip P. Barker and Robert W. de Mello, "Determining the Impact of Distributed Generation on Power Systems: Part 1 - Radial Distribution Systems",Power Technologies, Inc.Paudyal S, Canizares CA, Bhattacharya K. Optimal operation of distribution feeders in smart grids. IEEE Transaction Ind Electron 2011;58(10):4495-503.

[15] Sheeraz K, Majid J, Rizwan M. Optimal placement of SPV based DG system for loss reduction in radial distribution network using Heuristiv search strategies. In: Proceedings of IEEE International conference Energy and automation and signal Bhubaneswar; 2011.

[16] Sheeraz K, Majid J. Optimal allocation of SPV based DG system for loss reduction and voltage improvement in radial distribution systems using approximate reasoning. In: Proceedings of IEEE 5th India international conference on power electronics(IICPE); 2012.

[17] Rao RSrinivasa, Ravindra K, Satish K, Narasimham SVL. Power loss minimization in distribution system using network reconfiguration in the presence of distributed generation. IEEE Transaction Power Syst 2013;28(1):317-25.

[18] Rajkumar Viral , D.K. Khatod,"An analytical approach for sizing and siting of DGs in balanced radial distribution networks for loss minimization,"Electrical Power and Energy Systems 67 (2015) 191–201.

[19] Ke-yan Liu, Wanxing Sheng, Yuan Liu, XiaoliMeng , YongmeiLiu,"Optimal sitting and sizing of DGs in distribution system considering time sequence characteristics of loads and DGs," Electrical Power and Energy Systems 69 (2015) 430–440.

[20] Karar Mahmoud, Naoto Yorino, and AbdellaAhmed,"Optimal Distributed Generation Allocation in Distribution Systems for Loss Minimization,"IEEE transactions on power systems, vol. 31, no. 2, march 2016.

[21] NeerajKanwar , Nikhil Gupta , K.R. Niazi , Anil Swarnkar , R.C. Bansal,"Simultaneous allocation of distributed energy resource using improved particle swarm optimization," in press appl energy (2016).

[22] Majid Jamil and Ahmed ShariqueAnees,"Optimal sizing and location of SPV (solar photovoltaic) based MLDG (multiple location distributed generator) in distribution system for loss reduction, voltage profile improvement with economic benefits,"Energy 103 (2016) 231-239.

2nd IEEE International Conference on Power Electronics, Intelligent Control and Energy Systems (ICPEICES-2018)

A Comparative Performance Analysis of Automatic Generation Control of Multi-Area Power System Using PID, Fuzzy and ANFIS Controllers

Mukesh Kumar Bhaskar[1], Nidhi Singh Pal[2], Vinod Kumar Yadav[3]
Department of Electrical Engineering
[1,]Gautam Buddha University Greater Noida, India, [3]Delhi Technological University, Delhi, India

mkb2gbu@gmail.com; ~nidhi@gbu.ac.in; inodiitr07@gmail.com

Abstract—Automatic generation control (AGC) of four-area interlinked power system with generation rate constraints (GRC) is presented in this paper. The four-area interconnected power system having two areas with steam turbines and other areas are nuclear and hydro turbine tied together with power lines is considered in this paper. The AGC of multi-area power system using PID, Fuzzy and ANFIS controllers, and their comparative performance analysis are presented in this paper. ANFIS is a hybrid intelligent controller formed by parallel combination of fuzzy logic and neural network. This controller is able to control the frequency deviation efficiently and at minimum time, so ANFIS controller gives improved result over fuzzy and PID controllers.

Keywords—*Automatic generation control (AGC); Adaptive neuro fuzzy inference system (ANFIS); Fuzzy logic controller (FLC); Generation rate constraints (GRC); Load frequency control (LFC); PID control; Tie-line power.*

I. INTRODUCTION

In electrical power systems both the voltage and frequency are required to be fixed at desired values irrespective of loads changes. The variation of voltage and frequency levels without controls result to maintain both active and reactive power. To drop the impact of load variations and to keep frequency and voltage levels steady a control system is required. In spite of fact that the active and reactive powers combined affect the frequency and voltage levels, thus the control issues of frequency and voltage in power system can be separated. The power system is divided in different control areas depending upon the generation and all these areas are connected through tie-lines. So the load change in one area is divided between different areas and frequency of other areas also gets deviated.

For the most part a power system is made out of a few generating units. To upgrade the adjustment to non-basic disappointment of the whole power system, these generating units are related through tie-lines. This use of tie-line control influences another error in the control to issue, which is the tie-line control exchange botch. When sudden change in unique power stack bounces out at a area, the area will get its energy through tie-lines from various areas. In this paper the comparative performance analysis of automatic generation control of multi-area power system using PID, Fuzzy and ANFIS controllers are presented. Investigated system consists of four area (Thermal-Nuclear-Thermal-Hydro) power system. The tie- line is being used to connect all the areas[1].

II. AUTOMATIC GENERATION CONTROL

AGC is a vital issue in the huge interconnected power system and its control. At whatever point a little load change happens in the framework it causes change in the tie-line power flow and frequency deviation. Numerous explores in the region of AGC of intensity framework has been done in the past [1-5]. In an electric power framework, AGC is a framework for altering the power output of numerous generators at various power plants, because of changes in the load. Load changed in its own particular area or in its neighbor's area can be decide by AGC.

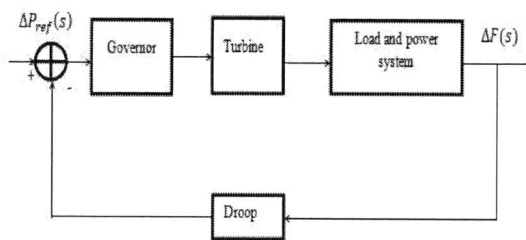

Fig. 1. Block Diagram of Automatic Frequency Control [1]

To guarantee the nature of the power supply, it is important to manage the control of generator loads relying upon frequency with AGC system. In interconnected power system AGC issue isn't just to see that the generation adjusts the request yet in addition to distribute generation between different areas with goal that the aggregate system activity plans are kept up in proper balance. In this way in interconnected system either executed physically or naturally, the limit of AGC is to reallocate the generation changes to pre-picked machines after a basic unpredictable comfort of the load by governor movement. It is vital to get very much wanted frequency consistency over got by speed governor itself. To accomplish this we should vanquish speed changes according to some sensible control technique. Fundamental point of AGC is to limit the transient deviation and give zero consistent state mistakes in short time [6-8]. In this paper, AGC problem of four-area interconnected power system is addressed using PID, fuzzy Mamdani, fuzzy Sugeno, and ANFIS controllers, and a comparative performance analysis of these is presented.

III. FOUR AREA INTERCONNECTED POWER SYSTEM

Investigated system comprises of four-area (Thermal-Nuclear-Thermal-Hydro) power system. The tie-

978-1-5386-6624-1/18 $31.00 © 2018 IEEE   132

line is being used to connect all the areas. For each area, a 1% change in load has been considered. In the power system power generation can change just at a predetermined most extreme rate. The normal estimation of passable rate of generation for hydro plant is substantially higher than reheater type thermal and nuclear unit. A normal estimation of GRC for hydro power plant is 270 %/min for raising generation and 360 %/min for bringing down generation. So limiter being bounded by +0.045 and -0.060 is being utilized inside the automatic generation controller to counteract inordinate control activity. For thermal and nuclear unit GRC is being considered between 3-5%. When we considered GRC to be 3%, two rate limiters bounded by +-0.0005 are used [1][3].

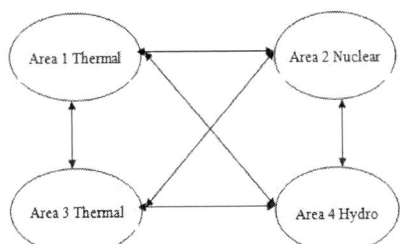

Fig.2. Basic structure of the four-area interlinked power system

IV. AGC Techniques

The different control strategies which are used for automatic generation control are listed as:

A. PID Controller(Manual Tuning Method)

PID controller is a well-developed control technique and being widely used in the industry. The broad utilization of PID controllers is because of their simplicity of design, implementation, and effectiveness for most linear dynamic systems [9]. In mathematically expressed as

$$PID(s) = k_p(1 + \frac{1}{sT_i} + sT_d) \tag{1}$$

where K_p is proportional gain, T_i is integral time and T_d is derivative time. Different value of PID Controller for respective areas in fuzzy-PID are **Area 1,2,3,** Kp=0.23, Ki=0.14, Kd=0.0001 and **Area 4,** Kp=-0.16, Ki=0.08, Kd=0.02

B. Fuzzy Mamdani Controller

This method is regarded widely 'for capturing expert knowledge' and facilitates an intuitively-plausible description of knowledge. This method includes the calculation of a two-dimensional shape by summing, or more accurately integrating across a continuously varying function. The computation can be expensive. For Fuzzy Mamdani Controllers: If $e(k)$ is positive (e) and $\Delta e(k)$ is positive (Δe) then $\Delta u(k)$ is positive (Δu) [10].

C. Fuzzy Sugeno Controller

Takagi Sugeno Kang show replaces the fuzzy consequent, (then part); of Mamdani control with work (condition) of the information factors. The upside of Fuzzy-Sugeno is to diminish the quantity of standards required by the Mamdani show. It is utilized for intricate and high-

dimensional issues build up a methodical way to deal with create fuzzy rules from a given input-output data set. Tagaki-Sugeno controller: If $e(k)$ is positive (e) and $\Delta e(k)$ is positive (Δe) then $\Delta u(k)$ is positive $\Delta u(k) = \alpha e(k) + \beta \Delta e(k)$ [10-11].

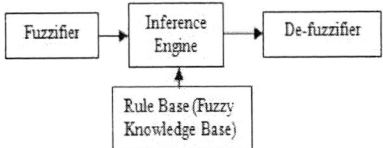

Fig. 3. Basic structure of Fuzzy Controller

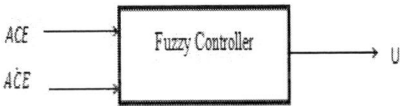

Fig. 3.1.   Block diagram of Fuzzy Logic Controller [12-13]

D. ANFIS Controller

Fuzzy logic and neural frameworks are basic correlative instruments in building shrewd frameworks. While neural frameworks are low-level computational structures that perform well while overseeing rough data, fuzzy method of reasoning oversees thinking on a more raised sum, using etymological information acquired from space authorities. Regardless, fuzzy frameworks don't have the ability to learn and can't change themselves to another condition. Joined neuro-fuzzy frameworks can unite the parallel estimation and learning limits of neural frameworks with the human-like data depiction and illumination limits of fuzzy frameworks. In like manner, neural frameworks end up being clearer, while fuzzy frameworks wind up fit for learning [14-18].

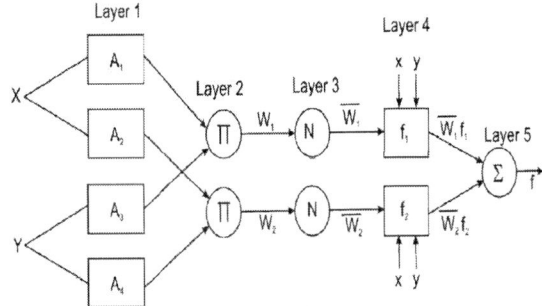

Fig.4. Structure of Adaptive Neuro-Fuzzy Inference System

V. Simulation Results And Analysis

The simulink model of four area automatic generation control is simulated using conventional and artificial intelligent controller. The PID controller has been used as conventional controller. Fuzzy-Mamdani, Fuzzy-Sugeno and ANFIS controllers have been used as artificial intelligent controllers. The result shows the superiority of ANFIS controller over conventional controller. The simulation result has been shown separately for PID, Fuzzy-

978-1-5386-6624-1/18 $31.00 © 2018 IEEE   133

Mamdani, Fuzzy-Sugeno and ANFIS controllers. A comparative analysis has also been performed for better analysis of result.

A. Simulation with PID Controller

PID controller reduces steady state error to zero and also maintain the system stability simultaneously. Here the PID controller has been used as an AGC controller in four-area power system network. For each area, a 1% step change in power demand has been considered. The simulation result of PID controller is shown in Fig. 5.

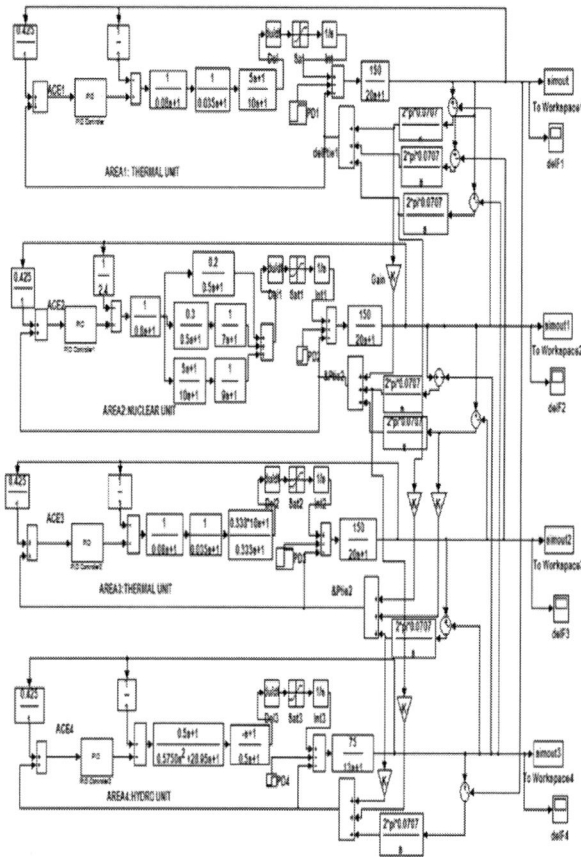

Fig.5. Simulink model of four area PID controller

The different plots for frequency deviation have been drawn after simulating the model shown in figure 5.

Case 1

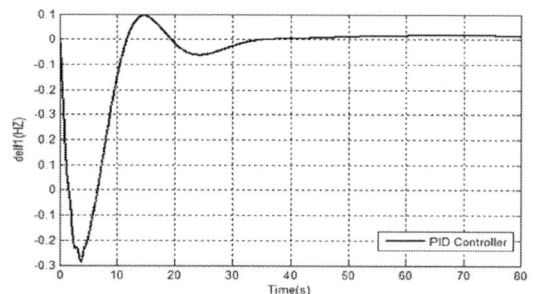

Fig.5.1. Frequency deviation of area 1 w.r.t time for 1% change in load

Fig. 5.1 shows that the system is gaining stability. The system gains steady state condition at 32 seconds with steady state frequency deviation of 0.0215 Hz and the max^{m} deviation in frequency is 0.287 Hz.

Case 2

Fig. 5.2 shows that the system is gaining stability. The system gains steady state condition at 32 seconds with steady state frequency deviation of 0.02 Hz and the max^{m} deviation in frequency is -0.021 Hz.

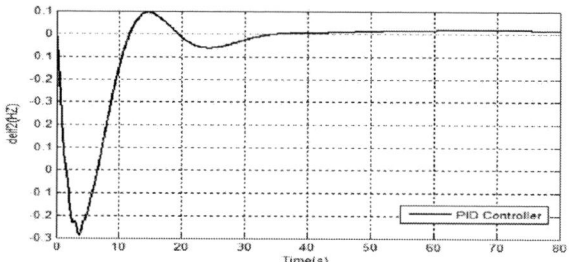

Fig.5.2. Frequency deviation of area 2 w.r.t time for 1% change in load

Case 3

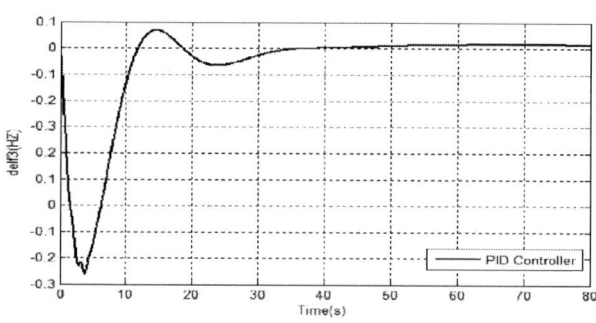

Fig.5.3. Frequency deviation of area 3 w.r.t time for 1% change in load

Fig. 5.3 shows that the system is gaining stability. The system gains steady state condition at 32 seconds with steady state frequency deviation of 0.02 Hz and the max^{m} deviation in frequency is -0.25 Hz.

Case 4

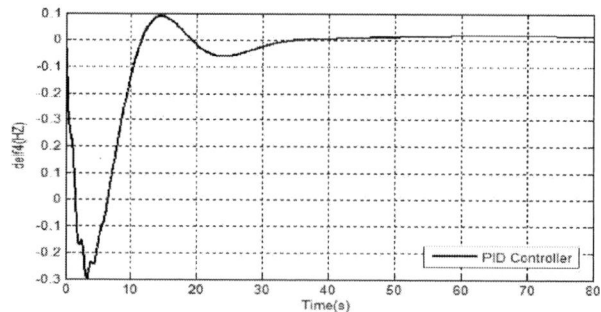

Fig.5.4. Frequency deviation of area 4 w.r.t time for 1% change in load

Fig. 5.4 shows that the system is gaining stability. The system gains steady state condition at 32 seconds with steady state frequency deviation of 0.0215 Hz and the max^{m} deviation in frequency is -0.23 Hz.

B. Simulation with Artificial Intelligent controller

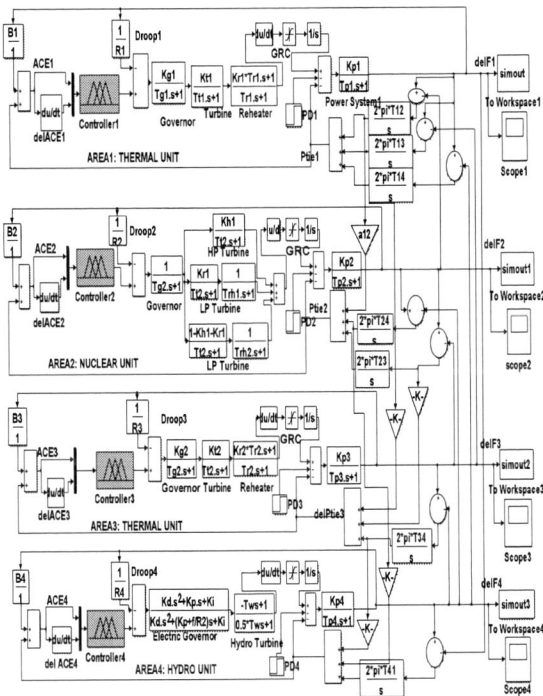

Fig.6 Simulink model of four area Artificial Intelligent controller

1) Simulation with Fuzzy Mamdani

Case 1

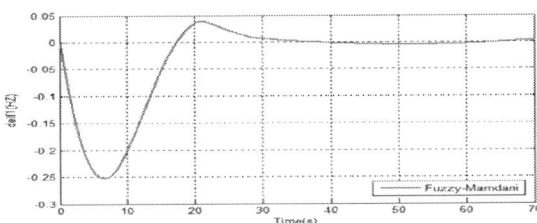

Fig.6.1. Frequency deviation of area 1 w.r.t time for 1% change in load

Fig. 6.1 shows that the system is gaining stability. The system gains steady state condition at 35 seconds with steady state frequency deviation of 0.0015 Hz and the max^m deviation in frequency is -0.25 Hz.

Case 2

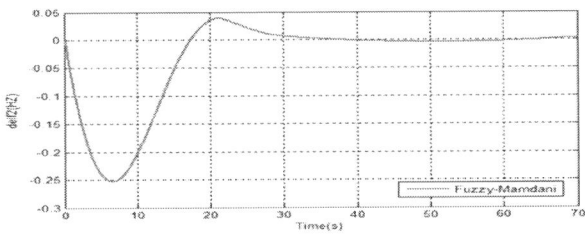

Fig.6.2. Frequency deviation of area 2 w.r.t time for 1% change in load

Fig. 6.2 shows that the system is gaining stability. The system gains steady state condition at 35 seconds with steady state frequency deviation of 0.0015 Hz and the max^m deviation in frequency is -0.25 Hz.

Case 3

Fig. 6.3 shows that the system is gaining stability. The system gains steady state condition at 35 seconds with steady state frequency deviation of 0.0015 Hz and the max^m deviation in frequency is -0.258 Hz.

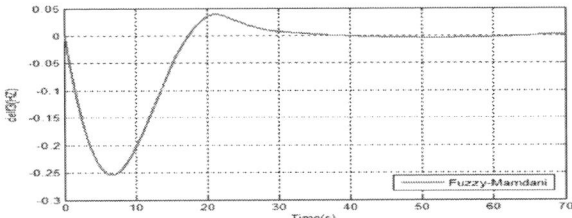

Fig.6.3. Frequency deviation of area 3 w.r.t time for 1% change in load

Case 4

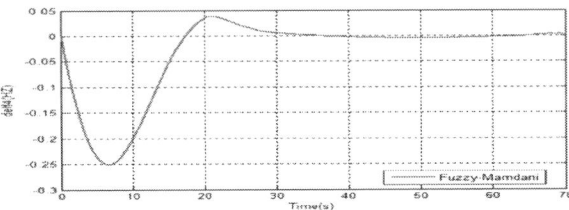

Fig.6.4. Frequency deviation of area 4 w.r.t time for 1% change in load

Fig. 6.4 shows that the system is gaining stability. The system gains steady state condition at 34 seconds with steady state frequency deviation of 0.0015 Hz and the max^m deviation in frequency is -0.25 Hz.

2) Simulation with Fuzzy Sugeno

Case 1

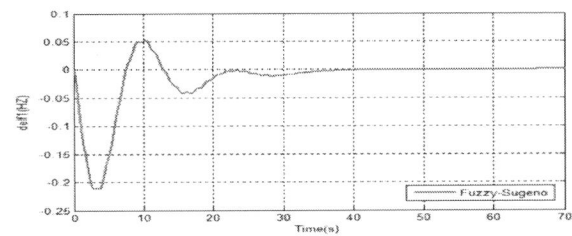

Fig.6.5. Frequency deviation of area 1 w.r.t time for 1% change in load

Fig. 6.5 shows that the system is gaining stability. The system gains steady state condition at 36 seconds and the maximum frequency deviation is -0.22 Hz.

Case 2

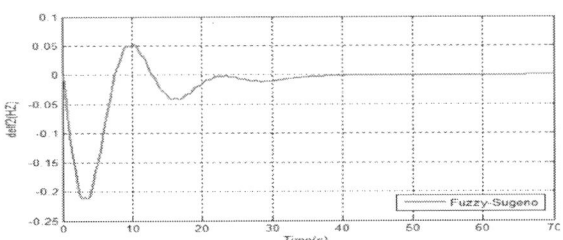

Fig.6.6. Frequency deviation of area 2 w.r.t time for 1% change in load

Fig. 6.6 shows that the system is gaining stability. The system gains steady state condition at 36 seconds and the maximum frequency deviation is -0.22 Hz.

978-1-5386-6624-1/18 $31.00 © 2018 IEEE

Case 3

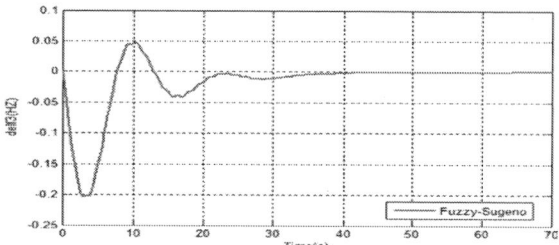

Fig.6.7. Frequency deviation of area 3 w.r.t time for 1% change in load

Fig. 6.7 shows that the system is gaining stability. The system gains steady state condition at 38 seconds and the maximum frequency deviation is -0.20 Hz.

Case 4

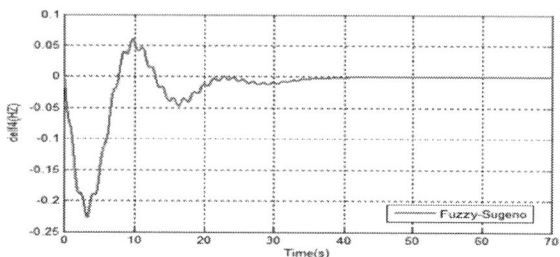

Fig.6.8. Frequency deviation of area 4 w.r.t time for 1% change in load

Fig. 6.8 shows that the system is gaining stability. The system gains steady state condition at 38 seconds and the maximum frequency deviation is -0.225Hz.

3) Simulation with ANFIS Controller

Case 1

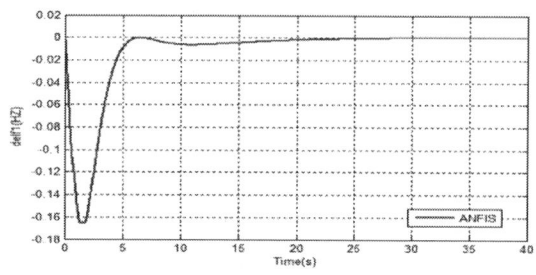

Fig.6.9 Frequency deviation of area 1 w.r.t time for 1% change in load

Fig. 6.9 shows that the system is gaining stability. The system gains steady state condition at 21 seconds and the maximum frequency deviation is -0.168 Hz.

Case 2

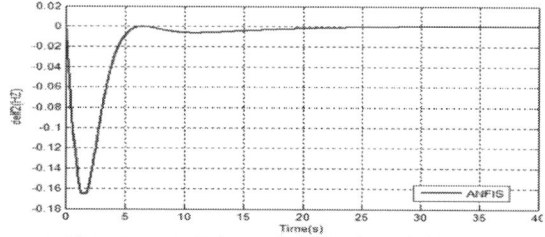

Fig.6.10. Frequency deviation of area 2 w.r.t time for 1% change in load

Fig. 6.10 shows that the system is gaining stability. The system gains steady state condition at 20 seconds and the maximum frequency deviation is -0.168 Hz.

Case 3

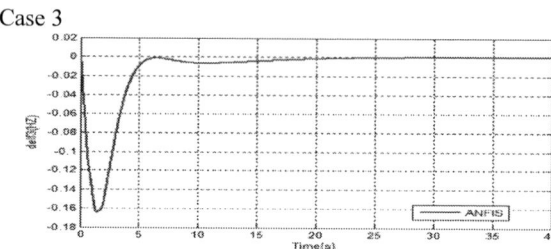

Fig.6.11. Frequency deviation of area 3 w.r.t time for 1% change in load

Fig.6.11 shows that the system is gaining stability. The system gains steady state condition at 20 seconds and the maximum frequency deviation is -0.168Hz.

Case 4

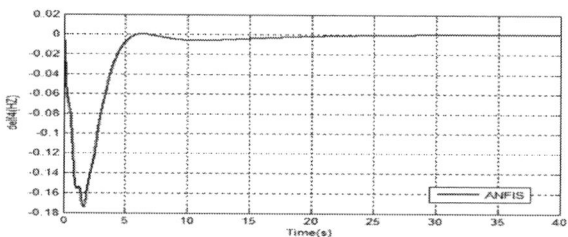

Fig.6.12 Frequency deviation of area 4 w.r.t time for 1% change in load

Fig. 6.12 shows that the system is gaining stability. The system gains steady state condition at 20 seconds and the maximum frequency deviation is -0.172 Hz

VI. COMPARATIVE PERFORMANCE ANALYSIS OF PID, FUZZY MAMDANI, FUZZY SUGENO AND ANFIS CONTROLLERS

Case 1

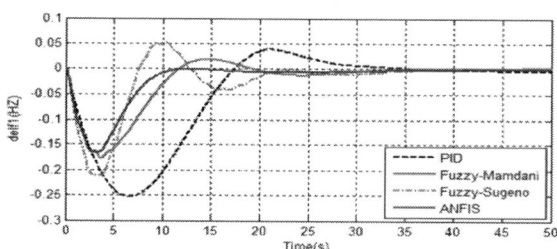

Fig.7. Frequency deviation of area 1 w.r.t time for 1% change in load

Case 2

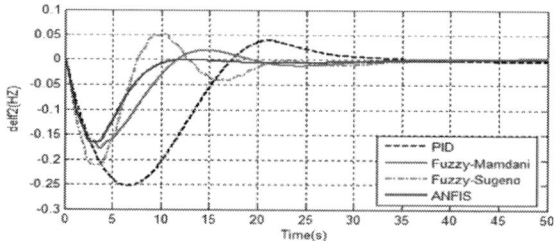

Fig.8. Frequency deviation of area 2 w.r.t time for 1% change in load

Case 3

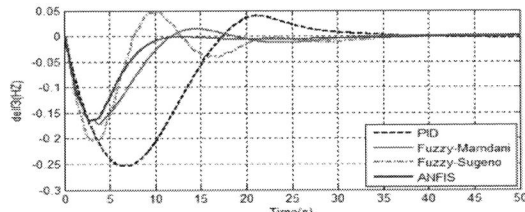

Fig.9. Frequency deviation of area 3 w.r.t time for 1% change in load

Case 4

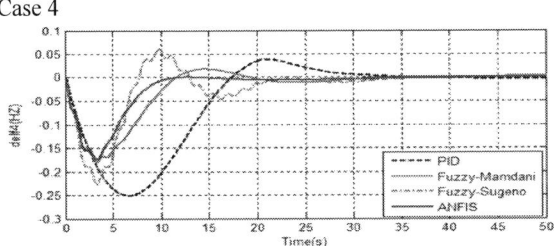

Fig.10. Frequency deviation of area 4 w.r.t time for 1% change in load

The control performance of different controllers can be summarized as-

- For PID controller the system attains steady state at 36 seconds, the maximum deviation in frequency is -0.253 HZ and the steady state error is 0.005.
- For Fuzzy-Mamdani controller the system attains steady state at 33 seconds, the maximum deviation in frequency is -0.213 HZ and the steady state error is -0.8×10^{-3}.
- For Fuzzy-Sugeno controller the system attains steady state at 22 seconds, the maximum deviation in frequency is -0.177 HZ and the steady` state error is -3.1×10^{-3}.
- For ANFIS controller the system attains steady state at 12 seconds, the maximum deviation in frequency is -0.165 HZ and the steady state error is -0.75×10^{-5}.

VII. CONCLUSION AND FUTURE SCOPE

In this paper presented MATLAB Simulink model of four area power systems with PID, Fuzzy and ANFIS controllers is given. Comparative performance analysis of all controllers is done. The ANFIS provides better result comparable to any other controller. This proves the superiority of Hybrid intelligent controller.

The AGC has been successfully implemented for conventional power plant such as thermal, hydro and nuclear generating unit. The next work is to apply the ANFIS in AGC of non-conventional power plants such as tidal power plant, solar power plant and geothermal power plant etc. The multi-area multi-unit system with load scheduling may be considered for further analysis of result.

REFERENCES

[1] J P.Subbaraj and K.Manickavasagm,"Automatic generation control of multi-area power system with generation rate constraints using computational intelligence techniques", International journal of applied power engineering,Vol. 2,No. 1,April 2013, pp. 27-38.

[2] Lal Bahadur Prasad, Barjeev Tyagi and Hari Om Gupta, "Fuzzy-PI based automatic generation control of two-area interconnected nonlinear power system", International. Journal of System Control and Information Processing, Vol. 2, No. 2, 2017, pp. 127-141.

[3] J.Nanda and B.Kaul,"Automatic generation control of an interconnected power system",IEEE proceedings,May 1978, pp. 385-390.

[4] A.Demiroren and E.Yesil, "Automatic generation control with fuzzy logic controllers in power system including SMES unit",International journal of Electrical power and energy systems,2004, pp. 291-305.

[5] EisaBashier and M. Tayeb, "Automation of interconnected powersystem using fuzzy logic generation", IEEE conference on Utility exhibition on power and energy system,2012, pp. 1-5.

[6] D. Das, J. Nanda, M. L. Kothari, and D. P. Kothari, "Automatic generation control of Hydro thermal system with new area control error considering generation rate constraint, "Elect. Mach. Power Syst., Vol. 18, No. 6, Nov/Dec. 1990.

[7] Swasti R. khuntia, Sidhartha panda, "A noval approach for automatic generation control of multi-area power system", IEEE conference on Electrical and Computer Engineering, May 2011, pp.1182-1187.

[8] R.R. Shoults and J.A. Jativa, "Multi area adaptive LFC developed for a comprehensive AGC simulation," *IEEE Trans. Power Systems*, vol. 8, pp. 541-547, 1993.

[9] A.Alwadie, "Stablizing load frequency of a single area power system with uncertain parameters through a genetically tuned PID controller", International journal of engineering and computer science, Dec 2012, pp. 51-57.

[10] A.Demiroren, E.yesil, "Automatic generation control with fuzzy logic controllers in power system including SMES unit",International journal of Electrical power and energy systems, 2004, pp. 291-305.

[11] C. Indulkar and B.Raj, "Application of fuzzy controller to automatic generation control," *Electric Machines and Power Systems*, vol. 23, pp. 209-220, Mar. 1995.

[12] II Kim,Jae-Hyun lee,Eun-oh Bang, "A new approach to adaptive membership function for fuzzy inference system", IEEE conference on knowledge based intelligent information system, Aug1999, pp. 112-116.

[13] Surya Praksah and Sunil Kumar Sinha, "Performance evalution of hybrid intelligent controllers in load frequency of multi-area interconnected power system",World academy of science,engineering and technology, Apr2015, pp. 637-645

[14] P. Li and X. Du, "Multi-area AGC system performance improvement using GA based fuzzy logic control," presented at the Intl. Conf. Elect. Engg., Hong Kong, 2009.

[15] Otman, M.Ahtiwash, Mohz.Z.Abdulmuin and Fatimah siraj, " A neural-fuzzy logic approach for modelling and control of non-linear system"IEEE International symposium on intelligent control, Oct 2002, pp. 270-275.

[16] K.V.SivaReddy, "An adaptive neuro-fuzzy logic controller for a two area load frequency control",International journal of Engineering research and application, Jul 2013, pp. 989-995.

[17] A. Demiroren, H. L. Zeynelgil, and N. S. Sengor, "Application of ANN technique to load frequency control for three area power system," in Proc. 2001 IEEE Porto Power Tech Conf.

2nd IEEE International Conference on Power Electronics, Intelligent Control and Energy Systems (ICPEICES-2018)

Design of A Hybrid Controller Based on GA-SMC for the Multi-Terminal VSC-HVDC Transmission System

Ramu Srikakulapu, *Student Member IEEE*
Department of Electrical and Electronics Engineering
National Institute of Technology Karnataka, Surathkal
Mangalore-575025
Email: ram2314u@gmail.com

Vinatha U, *Senior Member IEEE*
Department of Electrical and Electronics Engineering
National Institute of Technology Karnataka, Surathkal
Mangalore-575025
Email:u vinatha@yahoo.co.in

—

Abstract—**This paper explains a hybrid control approach to model a controller for voltage source converters (VSCs) in offshore wind farm (OSWF) applications, where the OSWFs are integrated to AC grid through the multi-terminal high voltage direct current (MT-HVDC) transmission system. Proper DC link voltage provides the effective power dispatch. So, a constant DC link voltage should be maintained to achieve the effective power dispatch between OSWFs and AC grid. A new control approach is a compound of proportional-integral and sliding mode control to regulate the DC link voltage on the grid side VSC. Also, the proposed control approach can control AC voltage and reactive power on wind farm and grid side VSCs respectively. Time-domain simulations executed in MATLAB/Simulink software are used to verify the proposed and conventional control approaches. Transient stability analysis is carried out for a case study. The three-terminal VSC-HVDC system has taken for study, where two OSWFs are fed to AC grid.**

Index Terms—**Offshore wind farm, Voltage source converter, HVDC, sliding mode control, DC voltage control, AC voltage control.**

I. INTRODUCTION

The wind energy is expected to play a vital role in the global energy sector. In that, offshore wind farms have received the more attention because of higher power generation. The multi-terminal high voltage direct current (MT-HVDC) system is the optimum choice for transmission of enormous rated wind power of OSWFs. The voltage source converter (VSC)-HVDC has more influenced than line commutated converter-HVDC because of independent control of reactive and active power, feasible to connect the weak areas, easy installation at offshore sites, and black start capability [1].

Several studies have been worked on control of MT-HVDC for grid integration of the OSWFs. In [2], authors have used proportional-integral (PI) and DC droop-based DC voltage controls to get proper power dispatch in the 4-terminal HVDC system. The methodology for droop control of MT-HVDC is

discussed and it controls the DC voltage in grid-connected OSWFs [3]. Authors have used optimum voltage control to minimize the loss in MT-HVDC system for OSWFs [4]. In [5], a distributed DC voltage control scheme is discussed for MT-HVDC system. The small signal stability for MT-HVDC system is described and effect of the current, DC voltage, and AC voltage controller are discussed [6]. In [7], authors have worked on stability analysis of grid-integrated OSWFs, where MT-HVDC, VSC-HVDC, and HVAC transmission systems are used for a case study. The sizing of droop control with the output capacitors is clarified for primary control in MT-HVDC grids [8]. The power reduction control in MT-HVDC system is demonstrated to avoid the DC overvoltage [9]. In [10], an adaptive backstepping droop controller is used to control the DC voltage for a 4-terminal HVDC system. The independent local control is proposed to regulate the current and voltage behind the capacitor in a voltage source terminal and the inductor in a current source terminal respectively [11]. A sliding mode control based active power variation for MT-HVDC system is introduced in [12]. In [13], authors have used the DC voltage droop control for both side VSCs in a 4-terminal HVDC system. A communication-less DC voltage control method is proposed to get sufficient power dispatch in MT-HVDC system, where improved master/slave control method is discussed [14]. In [15], authors have presented a control method based on a multi-loop current and voltage control to reach effective power transfer in 3-terminal HVDC system. Where, DC voltage and active power controls are applied at the wind farm side and grid side respectively. The conventional control approaches have restrictions to tune the non-linear system parameters and it reduces the overall system performance [16]. Meanwhile, OSWFs are non-linear systems. So, controller design should be a non-linear, adaptive, and robust. It progresses the transient stability of the system [17].

978-1-5386-6624-1/18 $31.00 © 2018 IEEE

In this paper, VSC based three-terminal HVDC system is used to integrate the two OSWFs and AC grid. The hybrid controllers for VSCs are designed based on sliding mode control (SMC) and PI control. The objective of the proposed SMC-PI based controller is to control the DC link voltage, AC voltage, and reactive power. The proposed control approach is validated by transient stability analysis, where the symmetrical fault is created at the grid end for a case study. The paper is shaped as follows: section II explains the system configuration and sliding mode control. Section III describes the hybrid controller design for MT VSC-HVDC system. The analysis of hybrid controller responses is explicated in Section IV and the final section describes the conclusion.

II. SYSTEM CONFIGURATION

The outline of the studied two OSWFs interconnected to AC grid through a three-terminal VSC-HVDC transmission system is shown in Fig. 1. In the studied system, two 400MW OSWFs are connected to AC grid through wind farm side VSCs (WSVSCs), DC-link, and grid side VSC (GSVSC). The details of MT VSC-HVDC with OSWFs are given in Table I.

TABLE I
MT VSC-HVDC SYSTEM SPECIFICATIONS

System	Specifications	Values
OSWF side	Rated power of OSWF 1&2	400 MW
	Transformer 1&2	33/150 $kV_{ph-ph,rms}$
DC link	Length	100 km
	Capacitor	225 μ F
Grid side	Transformer 3	150/400 $kV_{ph-ph,rms}$
	Grid voltage	400 $kV_{ph-ph,rms}$
AC filter	Inductance	16.7 μ H
	Capacitance	50.368 μ F
	Resistance	0.579 ohms

The dq frame equations of VSC are designed based on Clarke and Park transformation and described as below,

$$C\frac{dV_{DC}}{dt} = \frac{3}{2}(i_q d_q + i_d d_d) \tag{1}$$

Where, C and V_{DC} is the DC-link capacitance and voltage respectively; d_d and d_q are the dq frame duty cycle; and i_{qd} is the dq frame phase current.

$$L\frac{di_{dq}}{dt} \mp \omega L i_{qd} + R i_{dq} = e_{dq} - V_{DC} d_{dq} \tag{2}$$

Where, e_{dq} is the dq frame phase voltages, R is the phase resistance, and L is the phase reactance. The AC filter dq frame equation is given as underneath,

$$C_f \frac{de_{dq}}{dt} = i_{c_{dq}} \pm \omega C_f e_{qd} \tag{3}$$

Where, C_f is the filter capacitance and $i_{c_{dq}}$ is the dq frame filter current. The reference of reactive current $i_q{}^*$ is a function of reference active P and reactive power Q as shown below,

$$i_q{}^* = \frac{Pe_q + Qe_d}{e_d{}^2 + e_q{}^2} \tag{4}$$

Since the grid-connected OSWF is the complex and highly non-linear system the SMC based controller can be a better choice as compared to the conventional controller [18]. The discontinuous control action τ_c is is the function of sliding coefficient α, error signal of x, sliding gain constant K, and edge layer thickness ϕ as shown in (5),

$$\tau_c = K \tanh \frac{\alpha(x_{ref} - x) + \frac{d(x_{ref}-x)}{dt}}{\phi} \tag{5}$$

III. MULTI-TERMINAL VSC-HVDC CONTROLLER DESIGN

Combination of internal (PI) and external (SMC) controllers design the hybrid controller for MT VSC-HVDC. The internal controller intents to regulate the inner current and it tracks the outcome provided by the external controllers. The studied system consists of WSVSC1, WSVSC2, and GSVSC, where the WSVSC controls the AC voltage V_{AC} and GSVSC regulates the V_{DC} and reactive power. The WSVSC contains the AC voltage control (external) and current control (internal). While GSVSC has external controller, it contains the V_{DC} and reactive power controllers. The MT VSC-HVDC controller designs are displayed in Fig. 2 and 3. The AC voltage, DC voltage, current, and reactive power controller are modeled by using (1)-(5).

IV. SIMULATION RESULTS AND STUDY

The 3-terminal VSC-HVDC transmission system with a proposed hybrid controller is executed on the platform of MATLAB/Simulink software. The assumptions for the studied system is OSWF has produced the rated power at the rated voltage. So, the constant voltage source is taken for study instead of the OSWF. The tuned gain values of the PI-based current controller are enumerated by PID tuner tool, and SMC based AC and DC voltage controller are enumerated by using the GA method. Table II has detailed the modulated gain specifications. The fault ride through (FRT) capability is investigated on studied MT VSC-HVDC system, where the symmetrical fault is applied at near to grid at 3s with 0.1s duration. In case of the symmetrical fault, the time taken to regulate the V_{DC} for PI and hybrid controller are 2.2s and 1.2s correspondingly. The responses of source voltage V_S and grid voltage V_{grid} are shown in Fig 4(a) and Fig 5(a) respectively. In Fig 5(a), the fault can be observed at 3s. Fig 4(b) and 5(b) illustrations the V_{DC} response for PI and hybrid controller based studied system, where the V_{DC} is stabilized at one pu. The grid side reactive power Q_{grid} responses of both controllers are exposed in Fig 4(c) and 5(c). The value of Q_{grid} is maintained at a lower level. Therefore, the stable V_{DC} and lower reactive power ensure the stable power transmission between the OSWF and AC grid.

978-1-5386-6624-1/18 $31.00 © 2018 IEEE

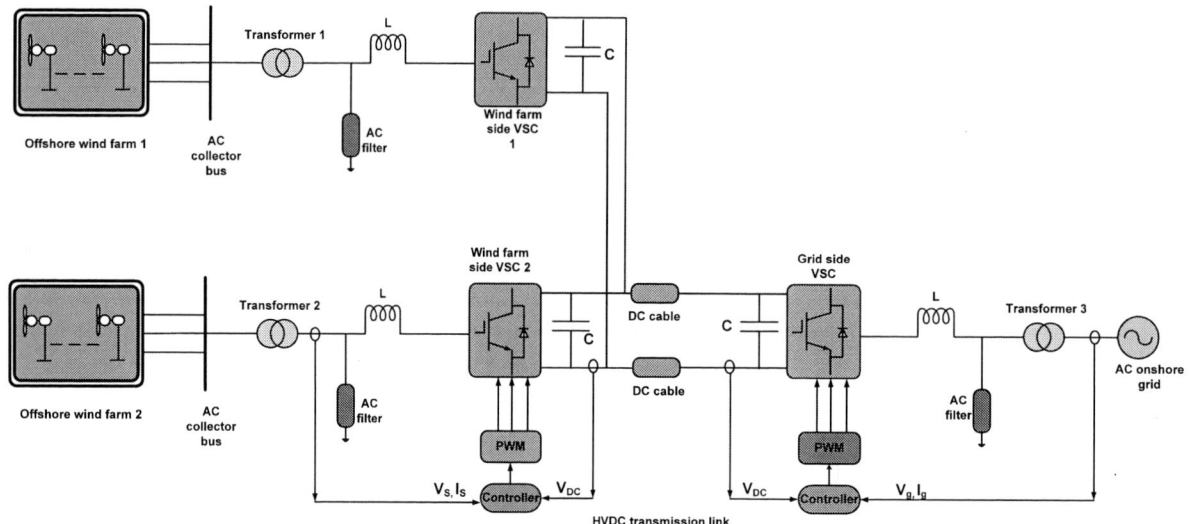

Fig. 1. Single line diagram of MT VSC-HVDC transmission system

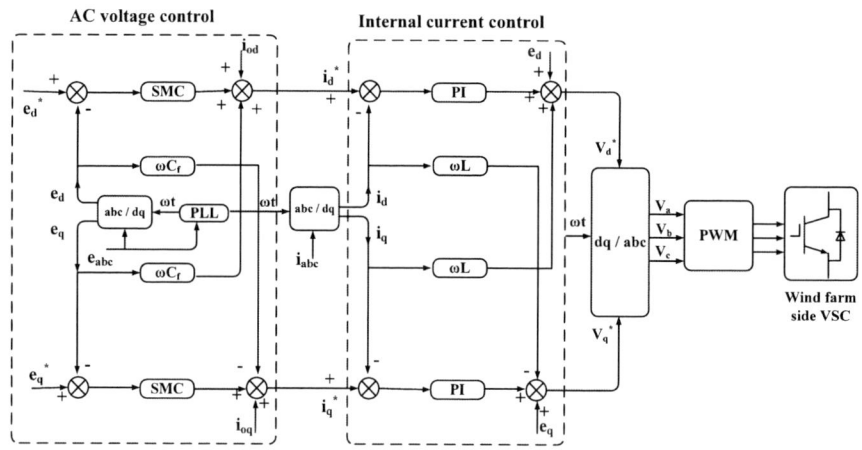

Fig. 2. Hybrid controller design of WSVSC

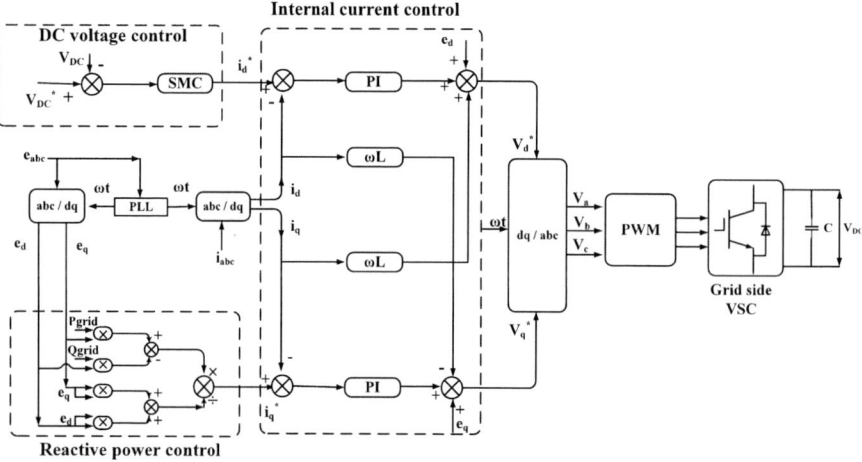

Fig. 3. Hybrid controller design of GSVSC

978-1-5386-6624-1/18 $31.00 © 2018 IEEE

TABLE II
MODULATED GAIN SPECIFICATIONS OF CONTROLLERS

System	Specifications	Values
WSVSC	$K_{P_{AC}}$	2.0
	$K_{I_{AC}}$	18579.0
	$K_{P_{CC}}$	828.6
	$K_{I_{CC}}$	5590900.0
	K_{AC}	10.0
	α_{AC}	3.9
	ϕ_{AC}	1.6
GSVSC	$K_{P_{DC}}$	19.3
	$K_{I_{DC}}$	175500.0
	K_{DC}	10.0
	α_{DC}	50.0
	ϕ_{DC}	1.6

1) $K_{P_{AC}}$ – V_{AC} controller proportional gain
2) $K_{I_{AC}}$ – V_{AC} controller integral gain
3) $K_{P_{CC}}$ – Current controller proportional gain
4) $K_{I_{CC}}$ – Current controller integral gain
5) K_{AC} – V_{AC} controller switching gain
6) α_{AC} – V_{AC} controller sliding co-efficient
7) ϕ_{AC} – V_{AC} controller edge layer thickness
8) $K_{P_{DC}}$ – V_{DC} controller proportional gain
9) $K_{I_{DC}}$ – V_{DC} controller integral gain
10) K_{DC} – V_{DC} controller sliding gain
11) α_{DC} – V_{DC} controller sliding co-efficient
12) ϕ_{DC} – V_{DC} controller edge layer thickness

V. CONCLUSION

This paper has proposed a hybrid controller approach based on a PI-SMC approach for the three-terminal VSC-HVDC system. The hybrid controller is modeled for WSVSC to control the AC voltage and GSVSC to regulate the DC voltage and reactive power. The responses of the conventional controller approach (external and internal are based on PI) and hybrid controller approach (external is SMC-based and internal is PI-based) are observed and compared. The studied system fault ride through capability has been investigated under the symmetrical fault condition, where the hybrid controller responses have taken less time compared to conventional one. The hybrid controller has specified better responses as AC voltage maintained at unity, V_{DC} is regulated to 1 pu, and Q_{grid} is controlled to a minimum value. So, the power dispatch between the OSWFs and onshore AC grid is maximized. From studied system responses, it is concluded that the proposed hybrid control approach provides the transient stability.

REFERENCES

[1] C. Du, "The control of vsc-hvdc and its use for large industrial power systems," 2003.

[2] L. Xu and L. Yao, "Dc voltage control and power dispatch of a multi-terminal hvdc system for integrating large offshore wind farms," *IET renewable power generation*, vol. 5, no. 3, pp. 223–233, 2011.

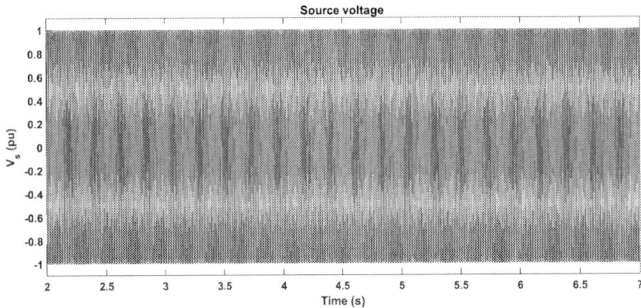

(a) Source Voltage

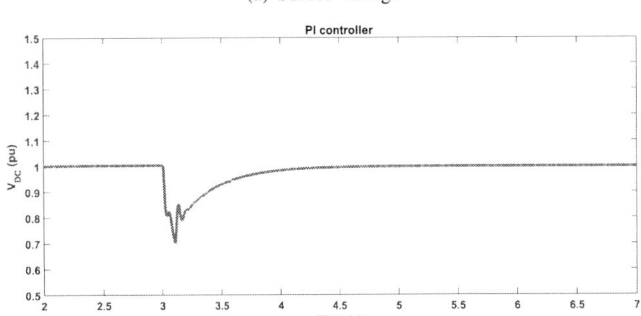

(b) DC-link voltage

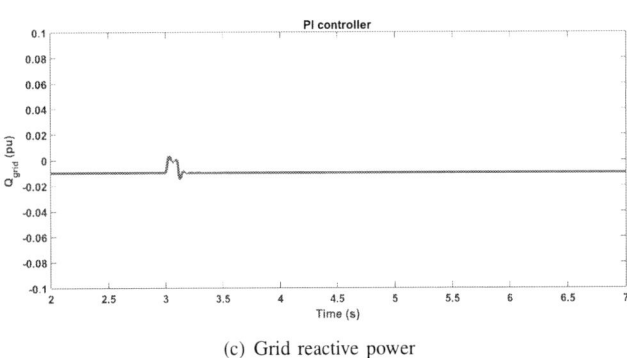

(c) Grid reactive power

Fig. 4. Performance of the MT-HVDC system based on PI controller for symmetrical fault

[3] E. Prieto-Araujo, F. D. Bianchi, A. Junyent-Ferre, and O. Gomis-Bellmunt, "Methodology for droop control dynamic analysis of multiterminal vsc-hvdc grids for offshore wind farms," *IEEE Transactions on power delivery*, vol. 26, no. 4, pp. 2476–2485, 2011.

[4] M. Aragüés-Peñalba, A. Egea-Àlvarez, O. Gomis-Bellmunt, and A. Sumper, "Optimum voltage control for loss minimization in hvdc multi-terminal transmission systems for large offshore wind farms," *Electric Power Systems Research*, vol. 89, pp. 54–63, 2012.

[5] C. Dierckxsens, K. Srivastava, M. Reza, S. Cole, J. Beerten, and R. Belmans, "A distributed dc voltage control method for vsc mtdc systems," *Electric Power Systems Research*, vol. 82, no. 1, pp. 54–58, 2012.

[6] G. O. Kalcon, G. P. Adam, O. Anaya-Lara, S. Lo, and K. Uhlen, "Small-signal stability analysis of multi-terminal vsc-based dc transmission systems," *IEEE Transactions on Power Systems*, vol. 27, no. 4, pp. 1818–1830, 2012.

[7] L. Wang and M. S. N. Thi, "Comparative stability analysis of offshore wind and marine-current farms feeding into a power grid using hvdc links and hvac line," *IEEE Transactions on Power Delivery*, vol. 28, no. 4, pp. 2162–2171, 2013.

[8] C. Gavriluta, I. Candela, C. Citro, A. Luna, and P. Rodriguez, "Design considerations for primary control in multi-terminal vsc-hvdc grids,"

978-1-5386-6624-1/18 $31.00 © 2018 IEEE   141

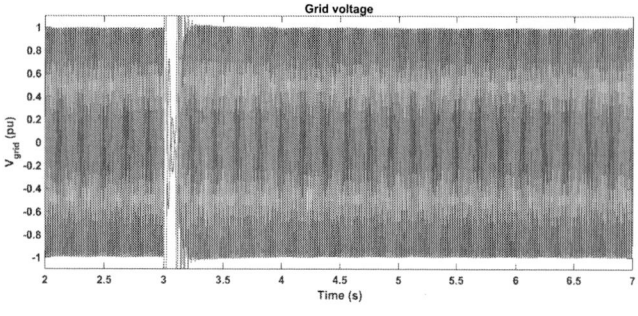

(a) Grid voltage

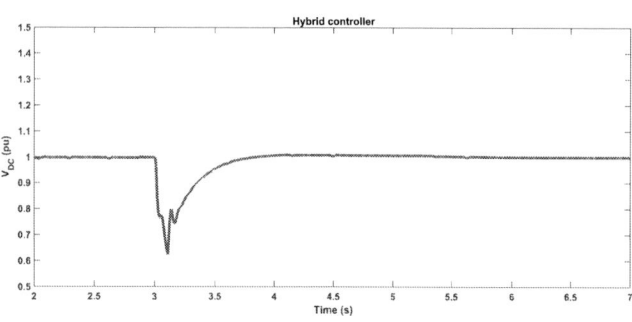

(b) DC-link voltage

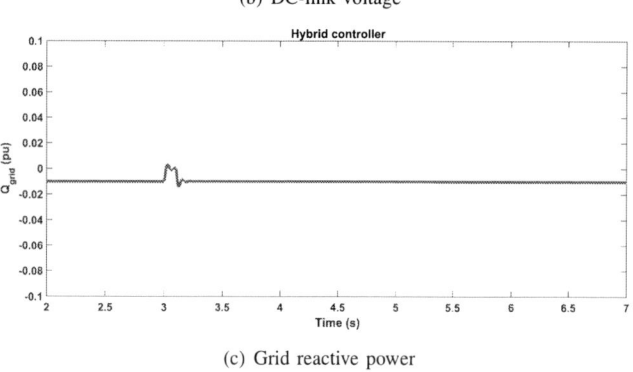

(c) Grid reactive power

Fig. 5. Performance of the MT-HVDC system based on hybrid controller for symmetrical fault

[15] M. Khenar, J. Adabi, E. Pouresmaeil, A. Gholamian, and J. P. Catalão, "A control strategy for a multi-terminal hvdc network integrating wind farms to the ac grid," *International Journal of Electrical Power & Energy Systems*, vol. 89, pp. 146–155, 2017.

[16] H. S. Ramadan, H. Siguerdidjane, M. Petit, and R. Kaczmarek, "Performance enhancement and robustness assessment of vsc–hvdc transmission systems controllers under uncertainties," *International Journal of Electrical Power & Energy Systems*, vol. 35, no. 1, pp. 34–46, 2012.

[17] A. Colbia-Vega, J. de Leon-Morales, L. Fridman, O. Salas-Pena, and M. Mata-Jimenez, "Robust excitation control design using sliding-mode technique for multimachine power systems," *Electric Power Systems Research*, vol. 78, no. 9, pp. 1627–1634, 2008.

[18] S.-C. Tan, Y.-M. Lai, and C.-K. Tse, *Sliding mode control of switching power converters: techniques and implementation.* CRC press, 2011.

Electric Power Systems Research, vol. 122, pp. 33–41, 2015.

[9] O. D. Adeuyi, M. Cheah-Mane, J. Liang, L. Livermore, and Q. Mu, "Preventing dc over-voltage in multi-terminal hvdc transmission," *CSEE Journal of Power and Energy Systems*, vol. 1, no. 1, pp. 86–94, 2015.

[10] X. Zhao and K. Li, "Adaptive backstepping droop controller design for multi-terminal high-voltage direct current systems," *IET Generation, Transmission & Distribution*, vol. 9, no. 10, pp. 975–983, 2015.

[11] J. Sun, "Autonomous local control and stability analysis of multiterminal dc systems," *IEEE Journal of Emerging and Selected Topics in Power Electronics*, vol. 3, no. 4, pp. 1078–1089, 2015.

[12] G. Tang, Z. Xu, H. Dong, and Q. Xu, "Sliding mode robust control based active-power modulation of multi-terminal hvdc transmissions," *IEEE Transactions on Power Systems*, vol. 31, no. 2, pp. 1614–1623, 2016.

[13] A. Raza, X. Dianguo, L. Yuchao, S. Xunwen, B. Williams, and C. Cecati, "Coordinated operation and control of vsc based multiterminal high voltage dc transmission systems," *IEEE Transactions on Sustainable Energy*, vol. 7, no. 1, pp. 364–373, 2016.

[14] R. Sandano, M. Farrell, and M. Basu, "Enhanced master/slave control strategy enabling grid support services and offshore wind power dispatch in a multi-terminal vsc hvdc transmission system," *Renewable Energy*, vol. 113, pp. 1580–1588, 2017.

978-1-5386-6624-1/18 $31.00 © 2018 IEEE

Market-based Load Frequency Control of Interconnected Power System

Naimul Hasan
Department of Electrical Engineering
Qassim University
Buraidah, KSA
nhasan@jmi.ac.in

Ibraheem Nasiruddin
Department of Electrical Engineering
Qassim University
Buraidah, KSA
ibraheem_2k@yahoo.com

Samiuddin Ahmad
Department of Electrical Engineering
Jamia Millia Islamia
New Delhi, India
samiuddinahmad@gmail.com

Abstract—**This paper presents a market-based load frequency control mechanism for the interconnected power system. The scheme is based on the dynamic selection of the optimal bids and activation of the respective secondary reserve to restore the power system frequency and tie-line flow to the nominal value. This paper also discusses the limitation of present ABT/DSM mechanism of the Indian power system. The developed control scheme is investigated through simulation of the two-area interconnected power system. The simulation results indicate that the proposed control scheme is capable to address the limitation of present frequency control approach of Indian power system.**

Keywords—automatic generation control, secondary frequency control, deviation settlement mechanism, availability based tariff, ancillary Services, Indian power system

NOMENCLATURE

AGC	automatic generation control
CERC	central electricity regulatory commission
VIU	vertically integrated utility
ISO	independent system operator
RGMO	restricted governor mode of operation
DSM	deviation settlement mechanism
ABT	availability based tariff
GENCO	generation company
TRANSCO	transmission company
DISCO	distribution company
a	a^{th} Area
b	b^{th} Generator
C_b	bid price of bth Generator
Q_b	total no. of bid quantity
Q_{req}	required generation quantity for AGC
N_{bid}	total no. of bid
K	rank of bid on the basis of price
Q_{sel}	selected bid quantity
MCP	market clearing price
INR	indian rupee
Kr_{ab}	reheat turbine gain
Tr_{ab}	reheat time constant (s)
Tg_{ab}	governor time constant (s)
Tt_{ab}	steam turbine time constant (s)
R_{ab}	speed regulation parameter (Hz/pu MW)
ΔF_a	incremental change in frequency
ΔPm_{ab}	incremental change in power generation
ΔX_{ab}	incremental change in governor valve position
h	h^{th} DISCO
ΔPd_{ah}	load disturbance in area a due to h^{th} DISCO
B_a	frequency bias constant (pu MW/Hz)
Pr_a	rated area capacity (MW)
Kp_a	power system gain constant (Hz/p.u. MW)
Tp_a	power system time constant (s)
CS_a	output of the $area_a$ PID controller
$\Delta Ptie_{1a}$	incremental change in tie-line flow (MW)
ΔPd_{ah}	load disturbance in area a due to h^{th} DISCO

I. INTRODUCTION

The power imbalance caused by disturbance or contingency needs to be rapidly compensated through primary frequency regulation to arrest any possible excursion in frequency. Secondary control is activated to restore the deviated frequency and tie-line flow to the nominal value through set point adjustment. Both primary and the secondary frequency control are faster acting services, with primary frequency response expected activation time being of the order of 1 second to few seconds and secondary frequency response activation time is of the order of 10 seconds to few minutes, i.e. these services are activated immediately. However, tertiary control is activated in the next few time block considering optimization criteria [1].

With envisage of restructuring and deregulation, there have been continuous changes in control and operation philosophy. The services that are required to ensure safety, security and reliability of power system were used to be the inherent part of energy services in Vertically Integrated Utility (VIU). In the present scenario, these services are procured separately. In many systems, frequency related support services are categorized as Ancillary services. In most of the systems, secondary frequency regulation is arranged either by Independent System Operator (ISO), consumer or by both through pool based market and bilateral contracts. While in some systems, primary frequency response from generator has been made mandatory through grid code [2], [3].

Various aspects of market design for procurement and implementation of AGC service in deregulated environment have been reported in [4]-[15]. Christe and Bose [4], presented market-based approach including charged structure, bilateral structure and charged cum bilateral structure for frequency related ancillary service procurement. Kumar, Hoe, and Sheble [5], developed a new market structure to understand price based operation of AGC. Furthermore, they identified and suggested the modification required in conventional AGC software to incorporate load following in price-based market operation. In [6], different market settlement rules and bid selection protocols in ancillary service market are examined. Doorman and

Nygreen [7], recognized interdependency of price and volume of ancillary service and proposed a pool-based market scheme. They estimated volumes of ancillary service and energy simultaneously. They considered self-profit maximization approach while preparing bids for generators and consumers. In [8], a frequency dependent price signal based approach is proposed to remunerate AGC providing generator. Higher price is paid at lower frequency to motivate generators to participate in AGC. ISO New England ancillary service markets along with energy market are cleared simultaneously and their prices are produced in real time. The relationship between the co-optimization of energy and regulation reserve for ISO New England system is discussed in [9], whereas summary of the ancillary service markets of US electricity markets including PJM, ISO New England, Midwest ISO and New York ISO is presented in [10]. Scherer, Zima, and Anderson [11], investigated the potential benefit and feasible approach for common pan-European ancillary services market for frequency control. They concluded that a central ancillary service market can reduce the reserve requirement and cost. A performance-based pricing approach to compensate frequency regulation service provider for faster ramping and accuracy in present energy mix power system is proposed in [12]. Arya, Kumar, and Ibraheem [13], designed an optimal regulators based on optimal control theory for different market contracts. In [14], an Artificial Neural Network (ANN) based optimization method is proposed to consider both capacity and energy bids while selection of Ancillary service providers.

The CERC of India has introduced Restricted Governor Mode of Operation (RGMO) as a primary frequency control mechanism to administer rapid frequency fluctuation. The enactment of RGMO regulation mandates all hydro units with a capacity of 10 MW and thermal units with a capacity of 200 MW and above to respond to any frequency deviation outside the specified range. Whereas Automatic generation control mechanism is still in the implementation phase. However, availability based tariff (ABT) mechanism was introduced to act as a manual load balancing mechanism, which is reintroduced as a deviation settlement mechanism (DSM). It ensures a provision to enumerate generator by a frequency dependent price for the unscheduled generation. Moreover, various emergency action norms are specified in grid code to assure reliability and security of the grid in case of heavy disturbance [15], [16]. Further, CERC has introduced Ancillary services regulation to restore the frequency at the desired level and relieve congestion in the transmission network. Unscheduled available generations are rescheduled to meet the trends of the load. Regulation-up or down instructions are given by the nodal agency to manually ramp up or ramp down the generation [17].

Even though the implementation of RGMO, DSM, and ancillary service regulation has improved the Indian grid frequency condition up to some extent, still, there are following unaddressed issues:

- Generators are not adequately compensated for either over-injection of power above 12% of the schedule limit or for deviation in the generation above 150 MW in an interval. Hence, generators are expected to respond only up to the mandatory limits. This puts a ceiling on the available quantity of frequency reserve and further due to this limitation system response time is also decreased.

- In DSM, Generators are expected to respond when frequency dependent UI cost is higher than their incremental cost. Otherwise, generators responses are not expected. There is no surety of availability of Generators all the time for load generation balancing.

- In DSM, Generators with lower incremental cost than frequency based price is expected to respond, irrespective of the load disturbance area. Since no AGC mechanism exists in the Indian power system, this may lead to undesired tie-line flow.

- Since DSM is a decentralize manual load balancing mechanism and it depends on the choice of individual generators to respond, it is not fast enough and sometimes may act out of merit when grid frequency is changing.

The paper is organized as follows. Following a brief description of single-auction based bid selection mechanism in Section II, Section III presents the power system model under investigation. In Section IV, a market-based AGC mechanism is developed. Section V presents a simulation study to exhibit the effectiveness of the developed market-based AGC mechanism. In Section VI, the results obtained in the present study are discussed, and concluding remarks are presented in Section VII.

II. SINGLE AUCTION POWER POOL MARKET MECHANISM

Since in the event of disturbance, faster activation of AGC services are required to quickly restore the frequency and the tie-line flow to the nominal value, a relatively simple and fast single auction based bid selection mechanism in the charged market structure is considered in the present work.

The power suppliers are invited by ISO to submit their bids. The bids contain the quantity and its associated price. The offered quantity is stacked on the basis of increasing order of their price. Market clearing price is determined by the highest bid price at which stacked quantity intersects with real-time frequency reserve requirement. The basic principle of this bid selection mechanism is illustrated in Fig.1.

III. POWER SYSTEM MODEL

A dynamic model of two area power system interconnected via AC tie-line and operating under a deregulated environment is considered for the investigation.

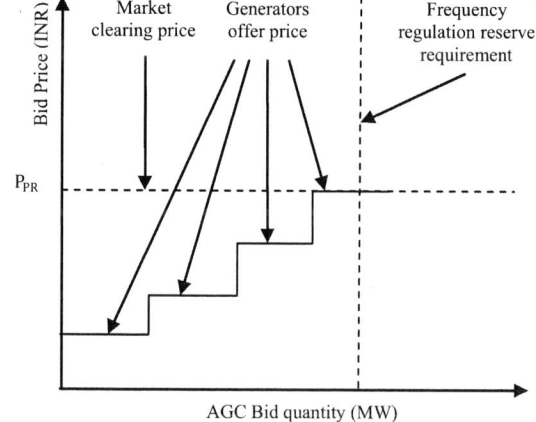

Fig. 1. Secondary regulation reserve market settlement.

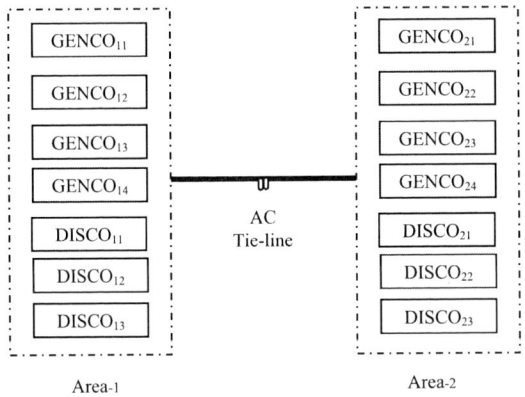

Fig. 2. Schematic of the two-area interconnected power system under deregulated environment.

The schematic diagram of the power system model under investigation is shown in Fig.2. The structure of the power system model consists of four identical reheat thermal power plants as four GENCOs and three distribution system as three DISCOs in each area. The transfer function model of the prime mover, generator, load, governor of the thermal power plant and the tie-line network is reproduced from the published literature in [18]. The secondary frequency reserves bid quantity and corresponding bid price considered for the present study are given in Table I.

IV. MARKET-BASED LOAD FREQUENCY CONTROL MECHANISM

It is considered that the bid window of the next time block for secondary frequency control reserve is remain open till the time block previous to the current time block. The ISO obtains the bids from the GENCOs for the readily available secondary frequency reserve. It ensures that the bids for sufficient reserve are available and updated in the Market-based frequency controller module of the AGC software. In the event of any fault or disturbance, the frequency control module of the AGC software dynamically selects the bids using the algorithm shown in Fig.3. Furthermore, it automatically sends the MW quantity signal to the respective generators for the activation of secondary frequency reserve.

TABLE I. SECONDARY FREQUENCY REGULATION OFFERS FROM GENCOS IN AREA 1 AND AREA 2

Area	Generator	Reserve Quantity (MW)	Price (INR/kWH)
Area-1	G_{11}	7	3.00
	G_{12}	8	3.35
	G_{13}	20	4.50
	G_{14}	12	3.75
Area-2	G_{21}	20	4.50
	G_{22}	10	3.30
	G_{23}	8	3.00
	G_{24}	12	3.65

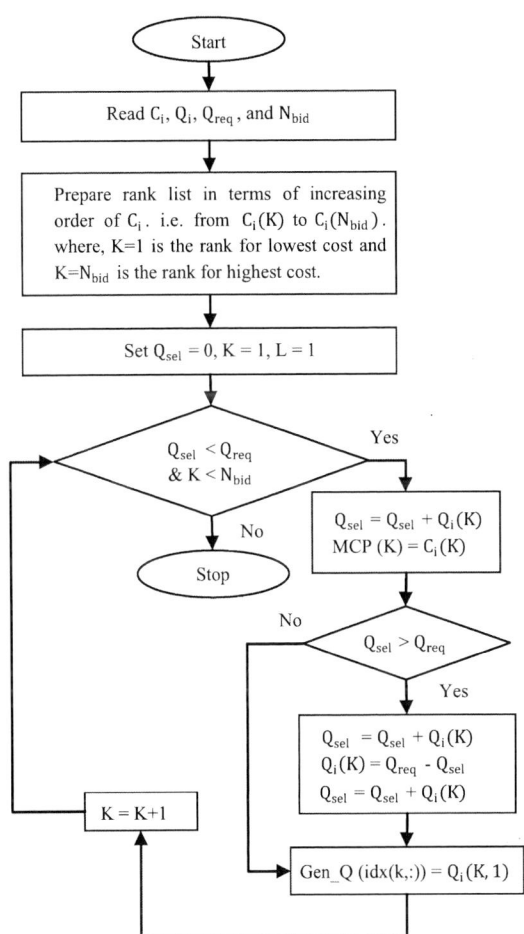

Fig. 3. Flowchart for the procurement of market-based frequency regulation services.

V. SIMULATION STUDY

A transfer function based model of the area-1 power system is shown in Fig.4, while the topology and the power system parameters of the area-2 are same as the area-1. The designed system is simulated in MATLAB/SIMULINK environment for the load perturbation of 20 MW in area-1 and 10 MW in area-2 [19]. A MATLAB program has been developed to implement the proposed market-based frequency control approach. A proportional integral derivative (PID) controller has been employed in both areas for the secondary frequency control. The gains obtained through Ziegler–Nichols gain tuning technique have been presented in Table II [20]. The automatically selected bids for the activation of secondary frequency reserve are given in Table III. The steady state values of the deviation in the generation, frequency, and the tie-line flow are summarized in Table IV. Further, the dynamic responses of the power system model are shown in Fig.5 to Fig.8.

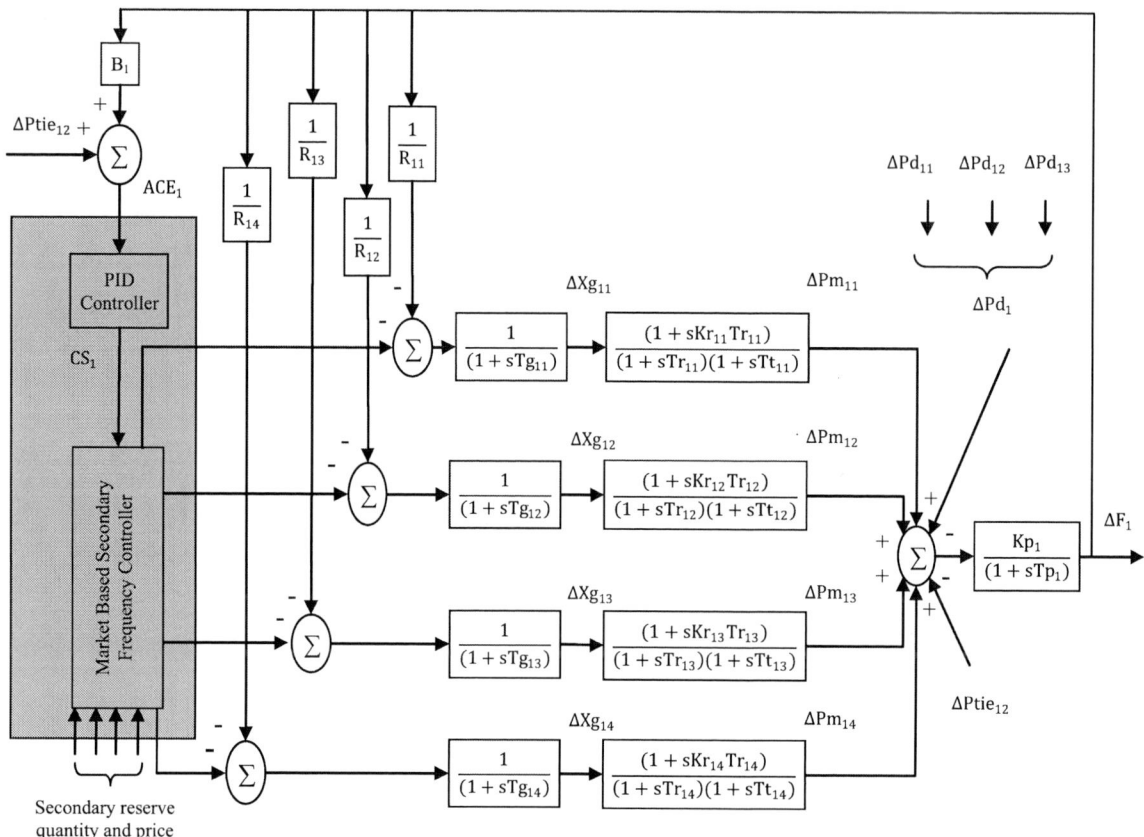

Fig. 4. Area-1 power system transfer function model for market-based AGC.

TABLE II. PID CONTROLLER GAINS

	Proportional Gain	Integral Gain	Derivative Gain
Area-1	1.650	2.3571	0.1473
Area-2	1.650	2.3571	0.1473

VI. RESULTS AND DISCUSSION

The MATLAB tool has been used to obtain the dynamic response of the considered power system model. The dynamic responses of the generators shown in Fig.5 to Fig.8 are investigated in this section.

The Generation and demand of the system is in equilibrium before application of unscheduled load demand of 20 MW in area-1 and 10 MW in area-2. The primary controllers of the all generators responded to the unscheduled demand and generated higher MW than the scheduled generation for the first few seconds to stabilize the system. Later, the frequency control module of the AGC software automatically selected secondary frequency reserve bids on the basis of their offer price and additional generation quantity requirement, and sent MW quantity signals to the respective generators for the activation of secondary frequency control action.

TABLE III. SELECTED SECONDARY FREQUENCY REGULATION OFFERS FROM GENCOs IN AREA 1 AND AREA 2

Area	Generator	Reserve Quantity (MW)	Price (INR/kWH)
Area-1	G_{11}	7	3.00
	G_{12}	8	3.35
	G_{13}	0	---
	G_{14}	5	3.75
Area-2	G_{21}	0	---
	G_{22}	2	3.30
	G_{23}	8	3.00
	G_{24}	0	---

TABLE IV. FINAL DEVIATION IN POWER GENERATION, FREQUENCY, AND TIE-LINE FLOW

Final settling of generation and frequency in Area-1		Final settling of tie-line flow $\Delta Ptie_{12}$ (MW)	Final settling of generation and frequency in Area-2	
ΔPm_{11} (MW)	6.99		ΔPm_{21} (MW)	0.00
ΔPm_{12} (MW)	7.99		ΔPm_{22} (MW)	2.00
ΔPm_{13} (MW)	0.00	0.00	ΔPm_{23} (MW)	7.98
ΔPm_{14} (MW)	4.99		ΔPm_{24} (MW)	0.00
ΔF_1 (Hz)	0.00		ΔF_2 (Hz)	0.00

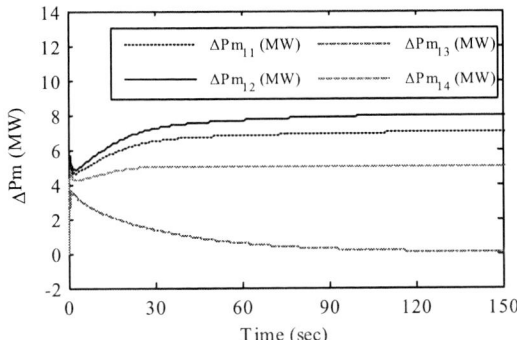

Fig. 5. Response of generators in area-1.

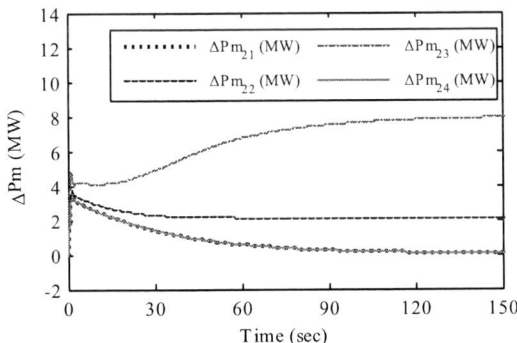

Fig. 6. Response of generators in area-2.

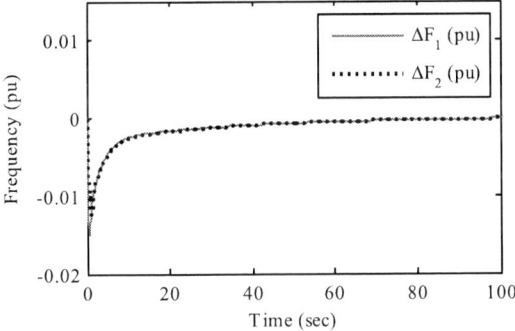

Fig. 7. Frequency deviations in area-1 and area-2.

Generator$_{11}$, Generator$_{12}$, and Generator$_{13}$ offers in area-1, and Generator$_{22}$ and Generator$_{23}$ offers in area-2 are dynamically selected for secondary frequency control. The price and corresponding quantity of the selected offers are mentioned in Table III. The investigation of the plots shown in Fig.5 and Fig.6, and the simulation results given in Table IV reveal that Generator$_{11}$, Generator$_{12}$, and Generator$_{13}$ changed their generation by 6.99 MW, 7.99 MW, and 4.99 MW respectively to meet load perturbation of 20 MW in area-1, while Generator$_{22}$ and Generator$_{23}$ changed their generation by 2.00 MW and 7.98 MW respectively to meet load perturbation of 10 MW in area-2. Further the plots shown in the Fig.7 and Fig.8 show that the frequency deviation and the tie-line deviation are finally settled to zero.

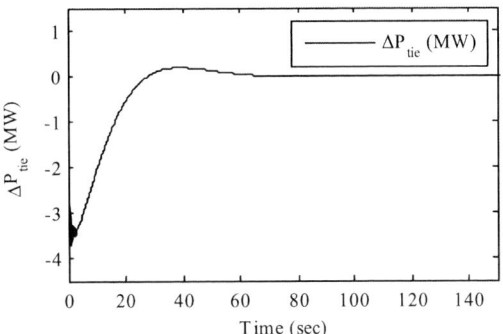

Fig. 8. Tie-line deviation.

VII. CONCLUSION

In this paper, a market based secondary frequency control mechanism is presented. The developed control scheme is validated through simulation of two-area interconnected power system model. The generators with different bid/offer price are considered in the system. The proposed control scheme is found capable to dynamically select the optimal bids and activate the respective secondary reserve to restore the system frequency and tie-line flow to the nominal value. Additionally, it also addresses the limitation of present frequency control approach of the Indian power system.

APPENDIX

The system data of a typical two-area interconnected power system transfer function model having four identical thermal generation units in each area is presented below.

Steam turbine [15]:

time constant of governor, Tg$_{in}$ = 0.08 s
time constant of the steam turbine, Tt$_{in}$ = 0.3 s
time constant of reheat, Tr$_{in}$ = 10 s
gain of reheat turbine, Kr$_{in}$ = 0.3
speed regulation parameter, R$_i$ = 2.4 Hz/pu MW

Power system:

rated capacity of area, P$_{a1}$ = P$_{a2}$ = 2000 MW
nominal system frequency, fr = 50 Hz
inertia constant, H = 5 MWs/MVA

REFERENCES

[1] A. Pappachen, and A. P. Fathima, "Critical research areas on load frequency control issues in a deregulated power system: A state-of-the-art-of-review," *Renewable and Sustainable Energy Reviews*, vol. 72, pp. 163-177, May 2017.

[2] C. Batlle, and I. J. Pérez-Arriaga, *Regulation of the Power Sector*, 1st ed., Springer-Verlag London, 2013, Chapter 7, pp.328-385.

[3] A. M. Pirbazari, "Ancillary services definitions, markets and practices in the world," *Proc. 2010 IEEE/PES Transmission and Distribution Conference and Exposition: Latin America (T&D-LA),* Sao Paulo, Brazil. pp. 32-36, 08-10 November 2010.

[4] R. D. Christie, and A. Bose, "Load frequency control issues in power system operations after deregulation," *IEEE Trans. on Power Systems*, vol. 11, no. 3, pp. 1191-1200, 1996.

[5] J. Kumar, Ng. Hoe and G. Sheble, "AGC simulator for price-based operation Part-II: Case study results", *IEEE Trans.on Power Systems*, vol. 12, No. 2, pp. 533-538, 1997.

[6] S. S. Oren, "Design of ancillary service markets," *Proc. IEEE Int. Conf. on System Sciences*, pp. 1-9, 2001.

978-1-5386-6624-1/18 $31.00 © 2018 IEEE

[7] G. L. Doorman and B. Nygreen, "An integrated model for market pricing of energy and ancillary services," *Electric Power System research*, vol. 61, pp. 169-177, 2002.

[8] J. Zhong and K. Bhattacharya, "Frequency linked pricing as an instrument for frequency regulation in deregulated electricity markets," *Proc. IEEE Power Engineering Society Annual, General Meeting*, Toronto, vol. 2, pp. 566-577, 2003.

[9] K. W. Cheung, M. Xingwang and D. Sun, "Functional design of ancillary service markets under the framework of standard market design for ISO New England," *Proc. Int. Conf. on Power System Technology*, pp. 1-7, 2006.

[10] K. W. Cheung, "Ancillary service market design and implementation in North America: from theory to practice," *In Proceedings of the Third International Conference on Electric Utility Deregulation and Restructuring and Power Technologies, DRPT 2008*, April 2008, pp. 66 – 73.

[11] M. Scherer, M. Zima and G. Andersson,"An integrated pan-European ancillary services market for frequency control," *Energy Policy*, vol. 62, pp. 292-300, 2013.

[12] A. D. Papalexopoulos and P. E. Andrianesis, "Performance-Based Pricing of Frequency Regulation in Electricity Markets," *IEEE Trans. on Power Systems*, vol. 29, no. 1, pp. 441-448, 2014.

[13] Y. Arya, N. Kumar, and Ibraheem, "AGC of a two area multi source power system interconnected via AC/DC parallel link under restructured power system environment," *Optim.contr. Appl.Methods*, vol.37, no4, pp. 590-607, Jul- Aug 2016.

[14] M. Bahmenzadeh and A. A. Foroud, "Reserve market scheduling considering both capacity and energy bids of reserve," *Int. Journal of Electrical Power and Energy Systems*, vol. 81, pp. 1-11, 2016.

[15] N. Hasan, Ibraheem, and S. Ahmad, "ABT based Load Frequency Control of Interconnected Power System," *Electric Power Components and Systems*, Taylor & Francis, vol. 44, Issue 8, April 2016.

[16] "Central Electricity Regulatory Commission (deviation settlement mechanism and related matters) Regulations, 2016, Third Amendment," Central Electricity Regulatory Commission, New Delhi, India, 2016. [Online]. Available: http://www.cercind.gov.in.

[17] "Central Electricity Regulatory Commission (Ancillary Services Operations) Regulations, 2015," Central Electricity Regulatory Commission, New Delhi, India, 2015. [Online]. Available: http://www.cercind.gov.in.

[18] D. Das, *Electrical Power Systems*, New Age International (P) Limited, New Delhi, 2006, Chapter 12, pp.307-338.

[19] MATLAB User Guide. [Online]. Available: http://www.mathworks.in/help/matlab/index.html.

[20] K. J. Astrom, and T. Hagglund, "PID Controllers: Theory, Design, and Tuning," *Instrument Society of American*, 2nd ed., 1995.

2nd IEEE International Conference on Power Electronics, Intelligent Control and Energy Systems (ICPEICES-2018)

A Control Strategy for STATCOM in Alleviation of Subsynchronous Resonance in Power Systems

Viqar Yousuf
Electrical engineering department
National Institute of Technology
Srinagar, India
viqaryousuf@gmail.com

Aijaz Ahmad
Electrical engineering department
National Institute of Technology
Srinagar, India
aijaz54@nitsri.net

Abstract—**Flexible AC transmission systems (FACTS) have been widely utilized as a part of later past for understanding different power system steady state control issues. Moreover, series compensation of transmission lines is being done utilizing capacitor banks with the goal that general power transfer capacity can be enhanced to guarantee stable task. Anyway, this offers ascend to sub-synchronous resonance (SSR) phenomenon under which there might be trade of energy amongst mechanical and electrical system at frequencies underneath the synchronous value, harming the shaft system of generator. This paper explores the impact of FACTS devices like Static Synchronous Compensator (STATCOM) for mitigation of SSR in a single machine infinite bus system. Simulation studies have been done (i) without any FACTS devices and (ii) with FACTS devices. These investigations have demonstrated that SSR in the system can be mitigated, as it were, by utilizing STATCOM with sinusoidal pulse with modulation (SPWM). Further, mitigation of SSR is achieved by outfitting STATCOM with controller for better outcomes.**

Index Terms—**FACTS, Fuzzy logic control, STATCOM, Subsynchronous Resonance (SSR), Torsional Interaction..**

I. INTRODUCTION

This paper discusses a severe problem in power system in present scenario when the power system has to work at the limits because of never ending demand especially in developing countries. In such countries the loss of generation can cause humongous problems. We need to understand the problems that can cause damage to the system or interruption to the supply of energy. It is needed that we should be able to minimize or eradicate the problems in the system before they can cause any damages. We are in the age of information and the industries are very sensitive, thus requiring a power system which is stable, reliable, efficient and adequate. The quality of power has to be increased with passing day [1] [2]. The adaptability in power system has been expanded by the utilization power electronic devices. Subsequently, more dependable and stable power system control is accomplished by the utilization of Flexible AC Transmission System, otherwise called FACTS. These are the new drifting devices which are explanation for the minimized and sound power systems [3]. An IEEE Committee Report (1985) defines SSR as follows: "Sub-Synchronous Resonance is a condition in electric power

system in which the energy in electrical system exchanges with a turbine of a generator at one or more of the frequencies (natural) of the combined system is below the synchronous frequency of the system" [4]. An IEEE second benchmark Simulink model is used to analyze the SSR problem in this paper. STATCOM with SPWM technique and STATCOM with controller is connected to the system in order to mitigate the SSR problem from the system. The execution of the proposed scheme is verified with MATLAB/SIMULINK.

II. STATIC COMPENSATOR (STATCOM)

It is simply a voltage source converter (VSC) incorporated with SVC. Its operation is similar to that of synchronous condenser but it is all the way a more superior device. It uses various thyristor devices such as GTO, IGCT, IGBT, etc. for switching purpose. In STATCOM, we use a dc source or capacitor as shown in Fig.1. Here, any change in magnitude of voltage on the dc side across the capacitor will directly affect

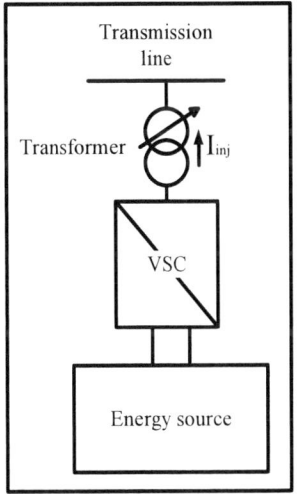

Fig. 1. Schematic diagram of STATCOM

978-1-5386-6624-1/18 $31.00 © 2018 IEEE

the voltage on ac side. The voltage on dc side can be held constant if we make use of a battery. The device can draw both capacitive as well as inductive reactive currents, depending upon the magnitude of the voltage on ac side. Thus, providing the required reactive power compensation to the transmission system and maintaining the appropriate level of voltage in the system [5].

For the purpose of switching, we can use different controllers depending upon our requirement. Since there is an energy source with the STATCOM, the operational range of the device is highly increased and the performance is improved as compared to other shunt devices.

III. OCCURRENCE OF SSR

SSR phenomenon as defined in introduction [4], can be very dangerous problem because of the involvement of the mechanical system. The turbo generators have numerous sorts of rotors relying on the steam turbine they utilize, which might be a high pressure, intermediate pressure and low pressure. These may likewise include an exciter mounted on the shaft, yet all are associated by methods for an adaptable shaft which might be mechanically represented by a spring. In this way the interconnection between the rotor and the shaft is somewhat damped mass-spring system which is similar to capacitors and inductors of an electrical system. This system may have an assortment of torsional modes relying on number of masses (rotors), of frequencies changing from 0-50 hertz, all below synchronous frequency. Mechanical system is viewed as alone while ascertaining these frequencies. A mode in which all rotors equally play a part is named as 'zero mode' and it relates to zero frequency. Outside sources (generators) associated with the transmission system can affect the damping of torsional modes. There is gigantic variety in frequency for zero mode, and we can disregard the frequency changes for residual modes (It is to be remembered that the general variety in frequency is really the variety in zero mode) and extents from 0.2 to 2.0 hertz. For whatever remains of the torsional modes damping is modified by the conduct and the attributes of electrical system. A series compensated power line associated with the generator has the resonant frequency given by [6] [7].

$$f_{rr} = f_{ss} \sqrt{\frac{X_{cc}}{X_{ll}}} \qquad (1)$$

Where,
$X_{ll} = X_n + X_t + X_e$
f_{rr} = resonant frequency,
X_t = leakage reactance of transformer,
X_c and X_e = reactances connected externally,
X_n = reactance of transmission line, and
f_{ss} = system frequency.
Also,

$$f_{tf} = f_{ss} - f_{rr} \qquad (2)$$

f_{tf} = tortional mode frequency

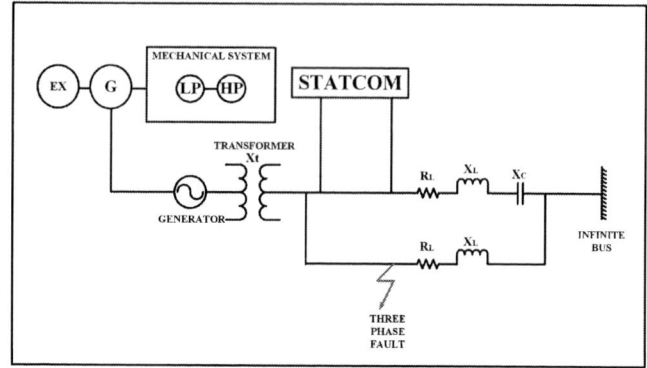

Fig. 2. Single line diagram of the system.

IV. STUDIED SYSTEM

The single line diagram of the model is given in Fig. 2. It is demonstrated that there is a series compensator associated with the transmission line. This compensator in series is in charge of the event of Sub Synchronous Resonance (SSR) in the system. Because of this, the system disturbances are caused and the torsional unsettling influences are observed.

The system examined is an IEEE second benchmark Simulink model for Sub Synchronous Resonance and the same is utilized to study the Sub Synchronous Resonance and in addition methods for damping out the motions caused by it. [8]. The system is modelled by implementing it in MATLAB/Simulink environment.

The model consists of Steam turbine generating unit that is connected to an infinite bus through a step up transformer shown in Fig.2. There are two transmission line that connect the infinite bus and the generator. One of these transmission lines is series compensated and in this study the compensation is of 55%. The mechanical system of the generating unit consists of two mass turbine system and a governor block. Low pressure turbine (LP) and high pressure turbine (HP) are the two masses of turbine system. Three phase fault is applied at transmission line 2 which is responsible for excitation of amplified torque oscillation in the system. STATCOM has been installed in the system as shown in the Fig. 2. for the sole purpose of damping out the oscillations in the system caused by the presence of SSR in the system. It is seen and as shown in the results below, the STATCOM has proved to be very effective in the mitigation of these oscillations. There have been various mitigation measures provided in the study [9] - [19], along with effects of SSR.

V. CONTROL STRATEGY

The STATCOM is designed using simple SPWM technique and a PI based controller. SPWM based technique is shown in the simulation Fig. 3. Along with SPWM technique another technique used for the mitigation of SSR is shown in the Fig.4. This technique utilizes the PI controller. The input to the PI

2nd IEEE International Conference on Power Electronics, Intelligent Control and Energy Systems (ICPEICES-2018)

controller is taken as the error of reference generator speed and measured generator speed as mentioned in equation (3).

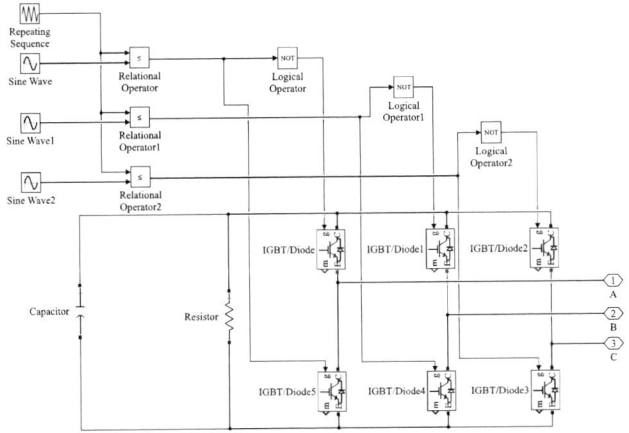

Fig. 3. Simulation diagram of SPWM technique.

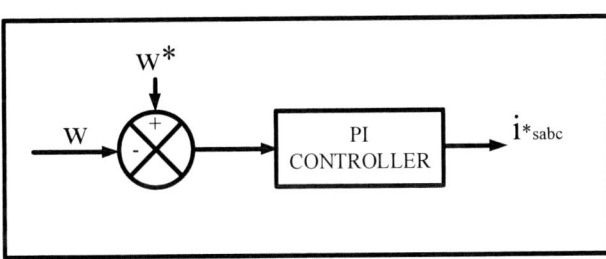

Fig. 4. Control strategy for STATCOM.

$$w_c(n) = w^*(n) - w_{av}(n). \qquad (3)$$

Where,

w_c = the error fed to PI controller,
w^* = the reference value of generator speed, and
w_{av} = the average value of generator speed.
The PI controller gives the output as follow,

$$I_{sabc}(n) = I_{sabc}(n-1) + K_p\{w_c(n) \\ - w_{av}(n-1)\} + K_i w_c(n) \qquad (4)$$

The error calculated between the reference and average value of generator speed is the at the nth sampling instant. Here Kp and Ki are the gain constants of speed controller.

VI. SIMULATION RESULTS AND DISCUSSION

Fast Fourier transform (FFT) analysis of the rotor speed has been done in order to show that there is presence of mode 1 oscillations in the absence of STATCOM device. The Fig.5, shows that there is 24.8 Hz frequency oscillations in speed of generator. Thus,confirming the presence of oscillations in the

system which are at the frequency less than the synchronous frequency.

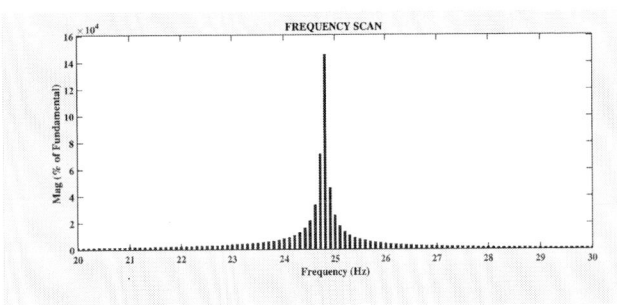

Fig. 5. FFT analysis of rotor speed.

The following results show presence of oscillations in the system when STATCOM is present and when STATCOM is not present. The various parameters taken for analysis are, Generator speed deviation shown in Fig. 7, Generator speed shown in Fig. 6, Generator-LP turbine torque shown in Fig. 9, LP-HP turbine torque shown in Fig. 10. From these all these results it can be seen that the oscillations caused due to occurrence of SSR have been damped out by installing STATCOM in the system.

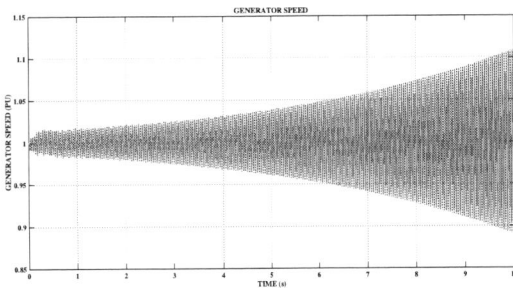

Fig. 6. Generator speed without the use of any mitigation technique.

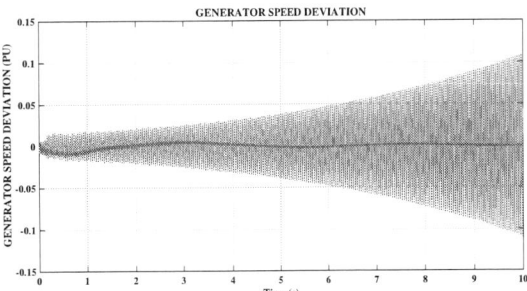

Fig. 7. Generator speed deviation with and without STATCOM.

978-1-5386-6624-1/18 $31.00 © 2018 IEEE   151

2nd IEEE International Conference on Power Electronics, Intelligent Control and Energy Systems (ICPEICES-2018)

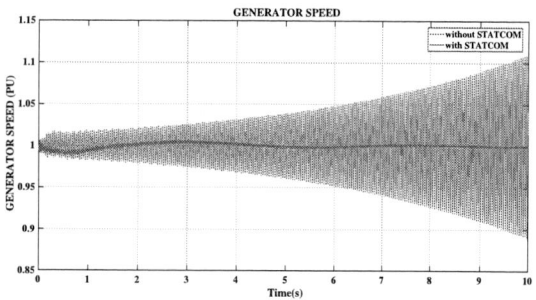

Fig. 8.   Generator speed with and without STATCOM.

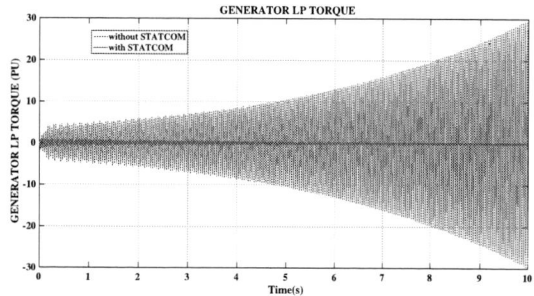

Fig. 9.   Generator LP turbine torque with and without STATCOM.

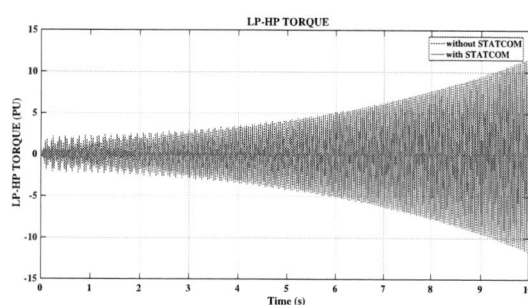

Fig. 10.   LP HP turbine torque with and without STATCOM.

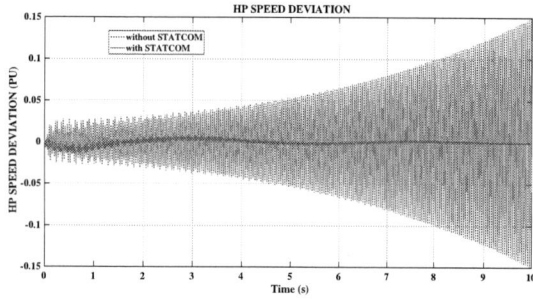

Fig. 11.   Generator-HP turbine speed deviation with and without STATCOM.

The following results show damped oscillations in the system when STATCOM is with and without PI controller. The various parameters taken for analysis are, Genarator speed

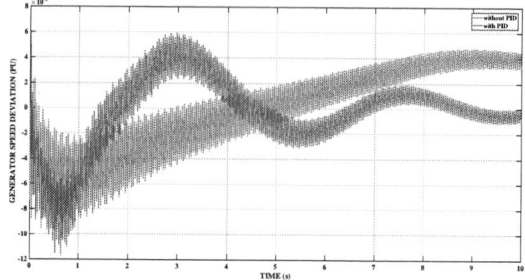

Fig. 12.   Generator speed deviation with PI and without PI.

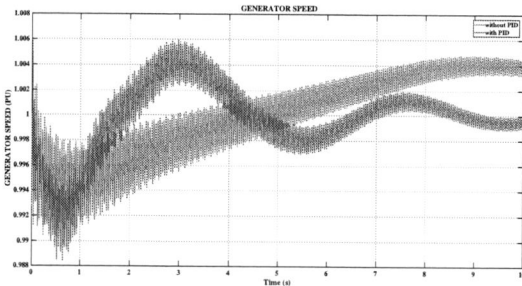

Fig. 13.   Generator speed with PI and without PI..

deviation shown in Fig. 12, Generator speed shown in Fig. 13. From all these results it can be seen that the oscillations caused due to occurrence of SSR have been damped out by STATCOM with new controller in the system. The results shown in Fig.12 and Fig.13, are obtained when STATCOM is employed with SPWM and PI controller.

VII.   CONCLUSION

In this paper, the system studied connotes the radical impacts of SSR on the system. It is being observed that with no FACTS devices, the generator speed and different factors have oscillations that continue to amplify with time. The impacts of these oscillations can harm the shaft masses and also protective devices. The use of STATCOM in the transmission line enhances the condition of the system.

The STATCOM is fused with SPWM technique for switching and after that with PI controller. The aftereffects of both the procedures are examined. The PI controller has ended up being very powerful in damping out these oscillations. The proposed scheme for the alleviation of these oscillations is simple and effective.

REFERENCES

[1]   IEEE TF report, "Proposed terms and definitions for power system stability", IEEE Transactions on Power Apparatus and Systems, Vol. PAS-101, pp.1894-1897, July 1982.

[2]   IEEE/CIGRE Joint Task Force on Stability Terms and Definitions, "Definition and Classification of Power System Stability", IEEE Trans. on Power Syst., Vol. 19, No. 2, pp. 1387-1400, May 2004.

[3]   Hingorani N.G, Gyugyi L, Understanding FACTS: Concepts and Technology of Flexible AC Transmission Systems, IEEE Press, U.K., Wiley, 2000.

2nd IEEE International Conference on Power Electronics, Intelligent Control and Energy Systems (ICPEICES-2018)

[4] IEEE Committee Report, Terms, Definitions and Symbols for Sub-synchronous Oscillations, IEEE Transactions on Power Apparatus and Systems, Vol. PAS-104, pp. 1326-1334, June 1985.

[5] Sood V.K, HVDC and FACTS Controllers: Applications of Static Converters in Power Systems., Kluwer Academic Publishers, 2004.

[6] D. Suriyaarachchi, U. Annakkage, C. Karawita, and D. Jacobson, "A procedure to study sub-synchronous interactions in wind integrated power systems", IEEE Trans. Power Syst., Vol. 28, No. 1, pp. 377–384, Feb. 2013.

[7] X. Xie, L. Wang, Y. Han, 'Combined application of SEDC and GTSDC for SSR mitigation and its field tests", IEEE Trans. Power Syst., Vol. 31, No. 1, pp. 769-776, Jan. 2016.

[8] IEEE Subsynchronous resonance working group, Second benchmark model for computer simulation of subsynchronous resonance, IEEE Transactions on Power Apparatus and Systems, vol. 104, vol. 5, pp. 1057-1066, 1985.

[9] Y. Song, F. Blaabjerg,' Overview of DFIG-based wind power system resonances under weak networks", IEEE Trans. Power Electron., Vol. 32, No. 6, pp. 4370-4394, Jun. 2017.

[10] R.K. Varma, S. Auddy, Y Semsedini,' Mitigation of Subsynchronous Resonance in a Series-Compensated Wind Farm Using FACTS Controllers", IEEE Transaction on Power Delivery, Vol. 3, pp. 1645-1654, 2008

[11] D. H. R. Suriyaarachchi, U. D. Annakkage, C. Karawita, et al., "A procedure to study sub-synchronous interactions in wind integrated power systems", IEEE Transactions on Power Systems, Vol. 28, No. 1, pp. 377-384, Feb. 2013.

[12] IEEE Subsynchronous Resonance Working Group, Countermeasures to subsynchronous resonance problems, IEEE Transactions on Power Apparatus and Systems, Vol. PAS-99, No. 5, pp. 1810-1818, Sep. 1980.

[13] U. Karaagac, S. O. Faried, J. Mahseredjian, et al., Coordinated control of wind energy conversion systems for mitigating subsynchronous interaction in DFIG-based wind farms,"IEEE Transactions on Smart Grid, Vol. 5, No. 5, pp. 2440-2449, Sep. 2014.

[14] V. Yousuf, N. Yadav and N. Chopra, "Mitigation of subsynchronous resonance using UPFC with fuzzy logic control for power system stability," 2016 7th India International Conference on Power Electronics (IICPE), Patiala, 2016, pp. 1-6.

[15] A. E. Leon, J. A. Solsona," Sub-synchronous interaction damping control for DFIG wind turbines", IEEE Transactions on Power Apparatus and Systems, Vol. 30, No. 1, pp. 419-428, Jan. 2015.

[16] X. Xie, Ch. Zhang, H. Liu, Ch. Liu, D. Jiang and B. Zhou, "Continuous-Mass-Model-Based Mechanical and Electrical Co-Simulation of SSR and Its Application to a Practical Shaft Failure Event", IEEE Transactions on Power Systems, Vol.31, No. 6, pp. 5172-5180, Nov. 2016.

[17] Han Chen, Chunlin Guo, Jianting Xu, Pengxin Hou, Overview of Sub-Synchronous Oscillation in Wind Power System, Energy and Power Engineering, Vol. 5, No. 4B, pp.454-457, 2013.

[18] Wenjuan Du, Yang Wang, H.F. Wang, Strong sub-synchronous control interactions caused by open-loop modal resonance between the DFIG and series-compensated power system, International Journal of Electrical Power & Energy Systems, Volume 102, Pages 52-59, 2018.

[19] Wenjuan Du, Xubin Wang, Haifeng Wang, Sub-synchronous interactions caused by the PLL in the grid-connected PMSG for the wind power generation, International Journal of Electrical Power & Energy Systems, Volume 98, Pages 331-341, 2018.

978-1-5386-6624-1/18 $31.00 © 2018 IEEE

Analytical Approach for Multiple DSTATCOMs Allocation in Radial Distribution Systems for Enhancement of Energy, Cost and Emission Savings

Imran Ahmad Quadri, S.Bhowmick, D. Joshi

Electrical Engineering Department;
Delhi Technological University; Delhi,India imranquadri83@gmail.com

Abstract— **This paper presents an analytical approach for the allocation of DSTATCOMs in radial distribution systems (RDS). The proposed method is implemented for allocation of multiple DSTATCOMs considering realistic load profiles for yearly energy and cost reduction in 69-bus RDS. The analysis includes investigation on the optimal number of DSTATCOMs required for minimizing the operational costs. The performance of RDSs after DSTATCOMs allocation is demonstrated by several indices like qualified load index (QLI), voltage deviation index (VDI), voltage stability index (VSI) etc. Impact on environmental emission is analyzed, which are global concern for power utilities Comparative results of the proposed method show remarkable improvement in yearly energy savings and reduced operational costs over Immune algorithm (IA) and** Bat Algorithm(BA) **methods with substantial reduction in computational time (CT).**

Keywords— DSTATCOM; Analytical method; Radial Distribution system; Computational time; Emission Saving

I. INTRODUCTION

Enhancement in power demand and the need for cost-effective operation of electric power networks have motivated researchers to operate transmission and distribution networks with optimum efficiency and load management. Distribution system operation involves several issues like real power losses, voltage profile and stability [1]. These challenges are aggravated by high penetration of non-linear loads [2]. Highly resistive nature of distribution systems and unbalanced line loadings cause significant voltage drop in distribution networks. Advancement in technology has paved way for increased penetration of renewable-based electric sources. However, this affects the power flow through the network [3]. Since the last three decades, flexible alternating current transmission system (FACTS) and custom power (CP) technologies have enhanced the optimal utilization of distribution and their reliability [4]. Distribution Static compensator (DSTATCOM) is a shunt CP device which is used in distribution networks to enhance the power quality [5]. DSTATCOM constitutes a coupling transformer, an energy storage device and an inverter and has the capability to inject and absorb reactive power with a very fast, dynamical response. This makes it very useful in distribution systems. DSTATCOM installations in distribution networks enhance reactive power compensation [6], resulting in improved power factor [7], voltage profile [8] and line active power flows [9], while minimizing the impact of intermittency of renewable resources [10].

In order to render the distribution system operation more reliable and economically viable, DSTATCOMs of optimal capacity must be placed at suitable locations. Although several research works have been carried out for determining the optimal location and sizes of STATCOMs in power transmission systems, however, comparatively less literature is available on the optimal allocation of DSTATCOMs. In this respect, [8-23] present some comprehensive research works on the optimal allocation of DSTATCOMs/DGs in distribution networks based on analytical and evolutionary optimization algorithms by considering objective functions such as power loss minimization, energy savings, cost savings, power quality improvement etc. An analytical method has been suggested in [11] for determining the optimum location and capacity of a DSTATCOM in a radial distribution network. The proposed method was tested for the power loss reduction and the voltage profile improvement in the IEEE 33-bus RDS. It shows better results over genetic algorithm (GA) vis-a-vis voltage regulation, the number of buses that have an under/over voltage and DSTATCOM cost. [12] has reported optimal allocation of DSTATCOM in 85-bus RDSs based on reactive power stability index (QSI) for different load models. However, the analysis is limited to the placement of a single DSTATCOM. Nonetheless, this approach has not considered realistic loading of the distribution system. However, the analytical methods computational time adversely affected by complexities in the formulation of objective functions caused by multiple DSTATCOMs and multi-objective analyses. Development in soft computing techniques suggested several optimization algorithms for DSTATCOM placement in distribution systems. Many meta-heuristic methods are frequently used for siting and sizing of DSTATCOMs in distribution systems. Meta-heuristics methods are swarm intelligence-based optimization algorithms which are skilled in handling multimodal, multi-objective, discrete and constrained environments. Some notable meta-heuristic methods are detailed in [13, 14]. In [13], immune algorithm (IA) has been proposed for the optimal allocation of DSTATCOMs to improve energy losses, cost savings, power congestion management and voltage profile in 33-bus and 69-bus distribution systems. The effectiveness of IA in comparison to GA is validated by carrying out DSTATCOM allocation considering three different load levels viz. light, medium and peak. Although results demonstrate the superiority of IA over GA in DSTATCOM placement, however, the author has not considered an optimal number of DSTATCOM to maximize the cost benefits.

To prevent the premature convergence of several evolutionary algorithms, hybrid algorithms have been reported in the literature which yields improved results. [14] has used a hybrid of GA and ACO algorithm for allocation of multiple DSTATCOMs in the IEEE 30-bus distribution system. This proposed method has been used to minimize the power losses by considering the generation or absorption of optimal reactive power using three DSTATCOMs. While the proposed method results in a substantial reduction in the network power loss, optimal number of DSTATCOMs for cost-effective operation of the network has not been considered. Although the applicability of most of the meta-heuristic optimization techniques is solution oriented and easy to implement, they are not robust enough, resulting in premature convergence and non-optimal solutions.

This paper presents an analytical method for optimal allocation of multiple DSTATCOMs in distribution systems based on power loss sensitivity. This method does not involve the solution of differential equations and hence, requires significantly less computational time as compared to evolutionary algorithms, while maintaining the desired level of accuracy. The proposed work has considered realistic load profile for all three distribution networks based on daily loading and investigates the yearly energy loss and cost savings. The usefulness of the proposed method is tested on 69-bus RDSs while considering optimal number of DSTATCOMs for the network. Comparative results of the proposed method vis-a-vis IA and Bat Algorithm (BA)[19] methods show remarkable improvement in yearly cost savings with reduction in CT.

The organization of the paper as follows: Section II deals with the the problem formulation. Section III presents the simulation results and comparative analysis. Section IV presents the conclusions.

II. PROBLEM FORMULATION

The proposed technique is implemented on a deterministic problem of optimal DSTATCOM sizing and placement in RDSs addressing some core issues like active power losses, yearly energy loss, total cost saving.

The case studies assumptions are as following:

    a. All the RDS are balanced.
    b. The load shedding and line interruption are not considered.
    c. Daily load profile is considered for yearly energy loss, cost calculation and emission saving.
    d. DSTATCOMs are considered for injecting reactive power.

A. Optimum allocation of DSTATCOM

Radial distribution networks only supply from one side (sub-station). A section of a radial distribution network is shown in Fig. 1(a), where three phases are in balance condition. $R_j \& X_j$ is the resistance and reactance of the jth section line between bus m and m + 1,respectively. Load are connected at bus m and m + 1 are $P_{L,m} + jQ_{L,m}$ and $P_{L,m+1} + jQ_{L,m+1}$, respectively. V_m and V_{m+1} are voltages of these buses.

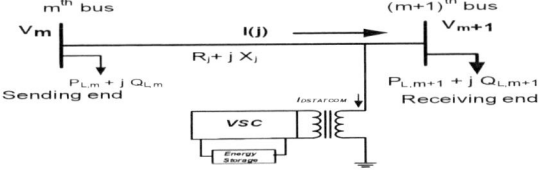

Fig. 1(a) Equivalent circuit of the jth branch of the network between buses 'm' and '(m+1)' with DSTATCOM

The proposed method for Optimum size and location of DSTATCOM is based on bus injection to branch-current

(BIBC) and Branch-current to Bus-voltage (BCBV) matrices. Development of BIBC and BCBV matrices are detailed in [19], uses the topological configuration of the distribution systems. The proposed method for multi-DSTATCOMs requires executing only load flow for the determination of the optimum size and location of DSTATCOM in the different RDS.

B. Theoretical analysis

The real power loss of the network is the primary concern while allocating DGs in RDS. Various formulations exist in the literature for load flow studies of RDS. The P_{loss} can be calculated as shown below [19].

$$P_{loss} = \sum_{j=1}^{nb} I_j^2 R_j \quad \text{Where} \quad I_j = \sum_{m=1}^{t} (P_{Lm}^2 + Q_{Lm}^2)/|V_m^2| \quad (1)$$

So, SOF to minimize real power loss is shown below

$$F_1 = Minimize\ P_{loss} \quad (2)$$

In Eqn. (1), 'nb' is the total number of branches in the network and 't' is the number of buses beyond branch 'j'.

Power loss reduction after DSTATCOM allocation

After DSTATCOM allocation, Q_{Lm} is modified according to eq. (3). Subsequently, the power losses (P_{loss}) calculated using modified Q_{Lm} using equation (1).

$$Q_{Lm} = Q_{Lm} - Q_{DSm} \quad (3)$$

Where $Q_{DS,m}$ is the reactive power injected by the DSTATCOM at the mth bus.

The loss sensitivity factor

The objective is to find the capacity of DSTATCOM at any bus 'm' to minimize total active power losses of the network. To find out the optimum capacity of DSTATCOM, the derivative of the total power losses per each bus injected reactive powers are equated to zero as

$$\frac{\partial P_{loss}}{\partial Q_m} = 0$$

The detailed description of Reactive power injected at bus m, i.e. Q_m is given in Appendix A.

$$Q_{DSTATCOM,m} = Q_m + Q_{load,m} \quad (4)$$

C. constraints

Following constraints are taken in consideration while allocating DSTATCOM in the radial distribution network.

a. Active and Reactive Power balance constraints

$$P_{sub} = P_{loss} + P_D \quad \&$$
$$Q_{sub} + Q_{DSTATCOM,m} = Q_{loss} + Q_D \quad (4)$$

Where P_{sub} and $Q\ sub$ are the substation active and reactive powers supplied to the load respectively and $Q_{DSTATCOM}$ injected reactive power by the DSTATCOM while P_D and Q_D are the total active and reactive load demands of the network.

b. Voltage constraint

$$0.90 \leq V_m \leq 1.05 \quad m = 1,2,3,4 \ldots \ldots n \quad (5)$$

c. Line Thermal limit [16]

$$I_j \leq I_j^{max} \quad (6)$$

Where I_j^{max} is the permissible loading of the branch 'j'.

d. Reactive power limit [17]

$$Q_m^{min} \leq Q_m \leq Q_m^{max} \quad (7)$$

Where Q_m^{min} / Q_m^{max} are the minimum / maximum limits, respectively, of the reactive power of the mth DSTATCOM.

978-1-5386-6624-1/18 $31.00 © 2018 IEEE

D. Procedure to find optimal size and location of multiple DSTATCOM

The optimum siting and sizing of DSTATCOM in the distribution system are determined as follows:

The objective is to minimize network active power losses(*Ploss*), in the system by injected or absorbed reactive power($Q_{DSTATCOM}$). The proposed method determines the optimal size and placement of multi-DSTATCOM is given below:

Step 1 Execute the base case power flow.

Step 2 Find the optimum capacity of new DSTATCOM for each bus except the substation bus using Eq. (4)

Step 3 Determine total power losses from Eq. (1) by placing the optimum capacity of DSTATCOM to each bus.

Step 4 Opt the bus which has the least network active power losses after allocating DSTATCOM in step 3, as optimum location and corresponding $Q_{DSTATCOM}$ capacity as optimum size.

Step 5 Check whether the bus voltages in the network are within the acceptable range.

Step 6 If the bus voltages are not within the acceptable range then omit DSTATCOM from the bus and return to Step 4.step 1 to step 4 determines the first optimal allocation of DSTATCOM in the network.

Step 7 Network Data is updated by placing $Q_{DSTATCOM}$ size at the optimum location for next location and size calculation of DSTATCOM.

Step 8 Step 1 to step 7 is repeated for finding multiple optimal location and size of DSTATCOM as shown in the flowchart.

E. Cost of DSTATCOM

Cost of investment can be extracted from the cost of DSTATCOM as shown below [8]

$$Cost_{DSTATCOM_{year,i}} = Cost_{DSTATCOM_i} \frac{(1+AR)^{n_{D-STATCOM}} * AR}{(1+AR)^{n_{D-STATCOM}} - 1} \quad (8)$$

Where $Cost_{DSTATCOM_{year,i}}$ the yearly is cost of DSTATCOM in the ith load level and $Cost_{DSTATCOM_i}$ is the cost of investment in the year of allocation in ith load level, $n_{DSTATCOM}$ is the durability of the DSTATCOM and AR is the asset rate of return.

This section deals with a computational procedure to utilize DSTATCOM units in the distribution systems for minimizing the energy loss and cost savings .

At first DSTATCOM size and location is calculated for minimizing the power loss at the peak load as previously mentioned. The DSTATCOM output is calculated as per the load demand curve. Finally, the yearly energy loss and cost saving is calculated based on the DSTATCOM output pattern. Following steps are included to find out for yearly energy saving for 24-h load demand.

Step 1 find out the optimal size and location of a DSTATCOM unit at the maximum load level only using the Procedure discussed earlier.

Step 2 Find the optimal output of the DSTATCOM unit at the optimal location only for period t as below, where p.u. demand (t) is the load demand (p.u.) at period t

$$Q_{DSTATCOM,m} = p.u.demand(t) \; x \; Q^{max}_{DSTATCOM,m} \quad (9)$$

Step 3 Run power flow with each DSTATCOM output i.e. $Q_{DSTATCOM,i}$ obtained in Step 2 for each period and calculate the total yearly energy loss as shown below:

The total yearly energy loss E_{loss} (MWh) in a RDS with time duration (Δt) of 1 h can be calculated as:

$$E_{loss} = 365. \sum_{t=1}^{24} P^t_{loss}.\Delta t \quad (9)$$

Where P^t_{loss} is the active power loss (APL) of the network at time t of the day

Step 4: calculate the total yearly cost saving shown below

F. Total Yearly Cost Saving (TYCS)

$$TYCS = 365. \left(K_e \sum_{i=1}^{24} T_i * P_{loss,i} - K_e \sum_{i=1}^{24} T_i * P^{with\,D-STATCOM}_{loss,i} \right) - \sum_{i=1}^{24} K_{ci} * \left(Cost_{D-STATCOM_{year,i}} \right) \quad (10)$$

Total yearly cost saving (TYCS) is the variation due to total energy loss cost prior to DSTATCOM installation and total energy loss cost after DSTATCOM placement. The yearly cost of DSTATCOM after installation in daily load levels is calculated using Eq. (10). TYCS data are given in Table 1.

G. Emission Saving (ES)

Air pollution is the global concern and every step is taken by several agencies to improve the air quality. Gases like CO_2, NOx and SO_2 are the major constituent of air pollutant associated with fossil fuel based power generation. DSTATCOM allocation in distribution network plays a vital role in reducing emission of such a pollutant. ES can be formulated as shown below,

$$ES = \left(P_{loss,base} - P_{loss,DSTATCOM} \right).ER.8760 \quad (11)$$

Where $P_{loss,base}$ and $P_{loss,DSTATCOM}$ are the distribution APL in the network without and with DSTATCOM respectively, ER is the emission rate of the grid. ES is the yearly emission (Ton) saving due to placement of DSTATCOM. The ER of grid is shown in Table 2 [43].

Table 1 *Data for cost calculation [13]*

$Cost_{DSTATCOM_i}$ (US\$/kVAr)	$n_{DSTATCOM}$ (year)	AR	K_e (US\$/kWh)
50	30	0.1	0.06

Table 2 *Emission Data [21]*

Source	Emission Rate (ER) (kg/MWh)		
	NOx	CO$_2$	SO$_2$
Grid	2.2952	921.25	3.5834

Table 3 *Performance Indices [22]*

Performance Indices	Formulation
Emission Saving (ES)(Ton)	$ES = E_b - E_{DSTATCOM}$
Active Power Loss Reduction (APLR) (%)	$APLR = \left(\frac{APL_b - APL_{DSTATCOM}}{APL_b} \right) X\,100$
Reactive Power Loss Reduction (RPLR) (%)	$RPLR = \left(\frac{RPL_b - RPL_{DSTATCOM}}{RPL_b} \right) X\,100$
Qualified Load-Index (QLI)	$QLI = \sum_{i=1}^{n} |V|_i.Load_i$
Voltage Deviation Index (VDI) (p.u.)	$VDI = \sum_{i=1}^{n} (1 - |V|_{i,DSTATCOM})^2$
Total Bus Violate Voltage Boundary (TBVVB)	$TBVVB=0$ $TBVVB = \sum_{i=1}^{n} if \; |V|_i < |V|_{min} \; or \; |V|_i > |V|_{max} \; TBVVB + 1 \; end$
Voltage Stability Index (VSI)	VSI_{m+1} $= |V_m|^4 - 4\{P_{m+1}X_j - Q_{m+1}R_j\}^2 - 4\{P_{m+1}R_j + Q_{m+1}X_j\}|V_m|^2$

The impact of DSTATCOM allocation in the distribution network is evaluated on the basis of indices detailed in Table 3.Voltage stability index indicate the vulnerability to voltage collapse in network, So VSI must be improved toward unity. Voltage deviation index represents the overall voltage fluctuation at all the buses in the network. Ideally VDI should be zero. Qualified load index (QLI) describe about the loadability of the network and it should be as high as possible.Total Bus violate voltage boundary indicate the number of buses in network goes beyond permissible limit.

III. SIMULATION OF CASE STUDIES AND RESULTS

The proposed method was primarily used to optimally allocated the single DSTATCOM in 69-bus RDS to illustrate the computational procedure involve in for finding optimum location and size of DSTATCOM under peak load condition and demonstrate the result subsequently.
Subsequently, the proposed method was used for multiple DSTATCOMs placement optimally in the IEEE 69-bus. Results are compared to the evolutionary algorithm to validate the effectiveness of proposed algorithm for multiple DSTATCOM placements in several distribution systems. The proposed analytical method was implemented in MATLAB R2015a environment on a Intel i5-4570, 3.2 GHz processor, 4GB RAM, desktop PC

a. Impact of Number of DSTATCOMs placement on 69-bus distribution network

The effect of DSTATCOM allocation in RDS APLand total yearly cost saving is investigated by increasing the number of DSTATCOM. From Fig. 2(a) it can be observed that although the system active power losses reduce when the number of DSTATCOMs is increased from 1 to 5, but yearly cost saving is maximum i.e. $ 21429.78 , while placing two DSTATCOMs. Hence, only 2 DSTACOMs placement is considered for 69-bus systems.

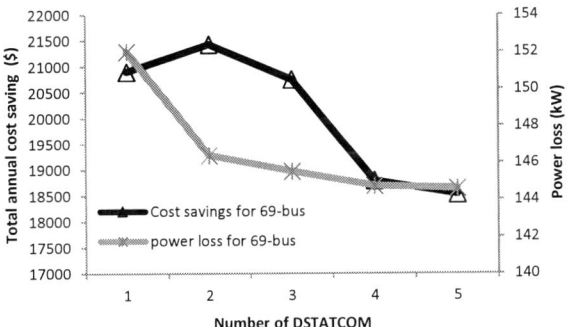

Fig. 2(a): Impact of number of DSTATCOMs on power loss and Total yearly cost saving for 69-bus RDS

b. 69-bus RDS case study

The detailed network data is given in [34]. It has 69 nodes, 7 laterals, 73 branches with 5 tie switches normally kept open. The Nominal voltage rating is 12.66 kV. Nominal load demand on the RDS is 3.8 MW and 2.69 MVAr. The base case real and reactive power losses are 224.9 kW and 102.13 kVAr, respectively. The base case VSI of this RDS is 0.6833 [15]. The Network base is 1000 KVA . DSTATCOM size at each bus is calculated using Eq. (4) and APL is calculated in the network after placing

calculated DSTATCOM size at respective bus using Eq.(1) It is observed from Fig. 3(a) that optimal location is Bus 61 and size is 1.2980 kVAr DSTATCOM where power loss is minimum i.e. 152.01 kW in the network. It is observed from Fig.3(b) that all the reactive power required by the load is fed from substation but DSTATCOM placement in network help to meet the reactive power demand by load locally results in reversal of reactive power flow in several lines in the network. This local reactive power management enhances the active power flow capability of the distribution network. Fig. 3 (c) and 3(d) shows that magnitude of branch current changes marginally but phase angle associated with branch current changes remarkable to reactive power compensation in the network. From Fig.3 (e) the minimum voltage in the network without DSTATCOM is 0.90919(p.u) at bus 65, while after one DSTATCOM placement minimum network voltage is 0.93016(p.u.) at 65. Multi-DSTATCOM placement i.e. 349.45 kVAr at bus 17 and 1291.80 kVAr at bus 61 results in power loss 136.51 kW whereas minimum voltage in the network is 0.93134 (p.u.) at bus 65. After placing DSTATCOM in the network at optimal location and size, the reactive power injected by the DSTATCOM is shown in Fig.4 (f) corresponding to the actual loading of the line for a day as shown in Fig. 3 (f). Fig.3 (g) shows the APL of the network without and with DSTATCOM for 24 hours. Energy loss and TYCS of the network is calculated using Eq.(9) and (10) for parameter provided in Table 1.It is observed from Table 4 that Cost saving is $ 20899.31 and 21429.78 for 1 and 2 DSTATCOM respectively. Fig.3 (e) depicts the voltage profile without, 1 and 2-DSTATCOM placed optimally in the distribution network. It is observed from Table 4 and 5 that all the indices have improvement except the TYCS, where two DSTATCOMs placement results in maximum cost saving i.e. $ 21429.7825. So optimal number of DSTATCOMs. 2 DSTATCOMs allocation result in 442.6704 tons of emission saving yearly. QLI have 1.21 % improvement, VDI increase by 10.21 %, VSI reduces to 0.05718 p.u., APLR and RPLR are 34.90 and 33.198 % respectively. It is observed from Table 6 that proposed method 22.64 % more yearly cost saving meanwhile computation time is only 1.41 sec instead of 32.30 sec. in comparison to the IA algorithm for 1 DSTATCOM placement. However, 2 DSTATCOM placement in network results in 9.23 % more yearly cost saving meanwhile computation time is only 1.86 sec instead of 12.83 sec. in comparison to BA.

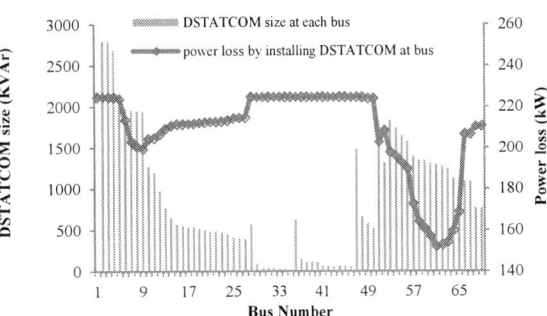

Fig. 3(a): The optimal size at each bus and total power loss for the corresponding DSTATCOM in 69-bus RDS

978-1-5386-6624-1/18 $31.00 © 2018 IEEE   157

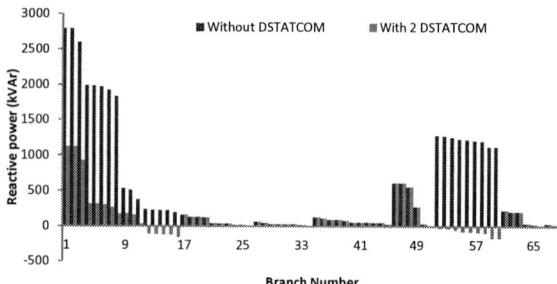

Fig. 3(b): Reactive power flow in each line without and with 2 DSTATCOM for 69-bus

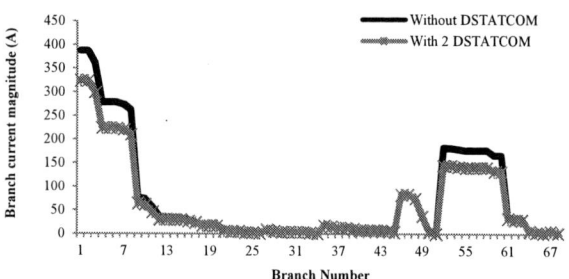

Fig. 3(c): Branch current magnitude without and with 2- DSTATCOMs for 69-bus

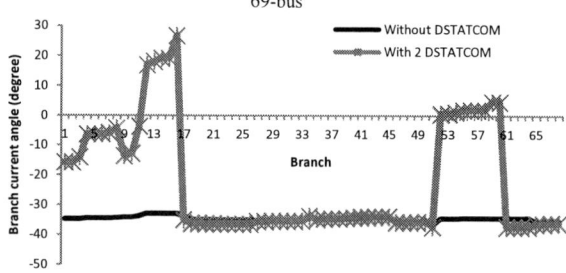

Fig. 4(d): Branch current angle without and with 2- DSTATCOMs for 69-bus

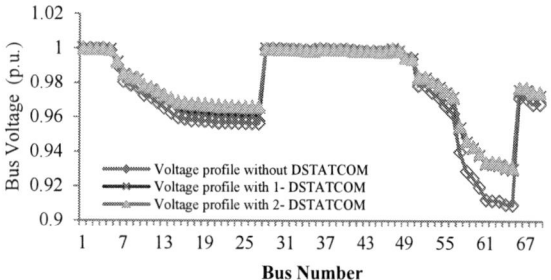

Fig. 3(e): Voltage profile of network without and with 1& 2 DSTATCOM for 69-bus

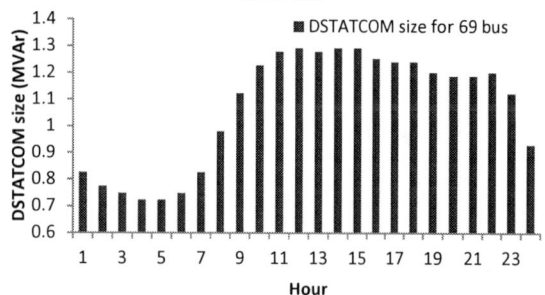

Fig. 3(f): Hourly Reactive power supplied by the optimally place 2-DSTATCOMs for 69-bus

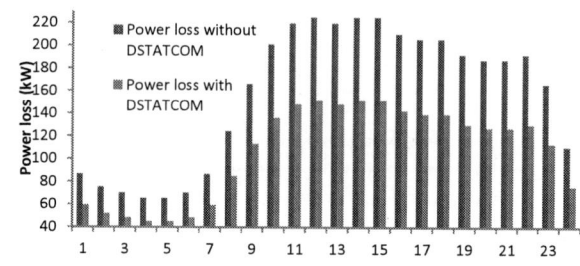

Fig.3(g): Hourly real power loss in the network without and with 2-DSTATCOMs for 69-bus

Table 4 Simulation results using proposed method for 69-bus RDS

Number of DSTATCOM	-	1	2
Size(Location) kVAr(Bus No.)		1291.8064 (61)	349.4544 (17) 1291.8064 (61)
APL (kW)	224.8974	152.0148	146.3932
RPL (kVAr)	102.115	70.528	68.2145
EL (kWh)	1381327.9423	938319.3798	903864.0621
ELC ($)	82879.6765	56299.1627	54231.8437
DSTATCOM cost ($)	-	5681.1954	7218.0503
TYCS ($)	-	20899.3183	21429.7825
Emission (Ton)	26879.983	26469.258	26437.313
ES (Ton)		410.7259	442.6704
QLI(p.u.)	3.6184	3.6575	3.6624
VSI (p.u.)	0.6833	0.6913	0.7531
VDI(p.u.)	0.0992	0.0648	0.05718
TBVVB	0	0	0
APLR (%)	-	32.406	34.9058
RPLR (%)	-	30.933	33.1987
MinVoltage (p.u.)	0.90919(65)	0.93017(65)	0.93135(65)
CT (Sec)	-	1.408053	1.860274

Table 5 Comparative results using proposed method for 69-bus RDS

69 Bus	IA [23]	Proposed Method	BA [41]	Proposed Method
Number of DSTATCOM	1	1	2	2
Size (Location) kVAr(Bus No.)	1702.42(61)	1291.81 (61)	480 (15) 1430 (61)	349.45 (17) 1291.81 (61)
PL (kW)	157.50	152.01	148.07	146.39
EL(kWh)	972395.33	938319.37	914371.81	903864.06
ELC($)	58343.71	56299.16	54862.30	54231.84
AC ($) DSTATCOM	7495.81	5681.19	8399.92	7218.05
TYCS($)	17040.14	20899.31	19617.4	21429.78
CT(Sec)	32.30	1.41	12.83	1.86

IV. CONCLUSION

This paper presents an analytical approach for optimal allocation of multiple DSTATCOMs in RDSs for yearly energy, cost savings and emission saving. This method is based on power loss sensitivity and does not involve the solution of differential equations thus requiring significantly less computational time as compared to IA and BA, while maintaining the desired level of accuracy. The proposed work has considered realistic load profile for all three distribution networks based on daily loading. The efficiency of the proposed method is tested on 69-bus RDSs while considering optimal number of DSTATCOMs for each network. Various performance indices indicate that increasing number of DSTATCOMs in RDS improves all the indices except cost saving. So economic aspect is considered for selecting optimal number of DSTATCOM placement in all the RDS. The several performance indicators clearly give insight of the network performance will helpful for the network operators. Comparative results of the proposed method show remarkable improvement in yearly cost savings with reduction in computational time over IA and BA methods.

REFERENCES

[1] Jordehi AR. Particle swarm optimisation (PSO) for allocation of facts devices in electric transmission systems: a review. Renew Sustain Energy Rev 2015; 52:1260–7.

[2] Prakash P, Khatod DK. Optimal sizing and siting techniques for distributed generation in distribution systems: a review. Renew Sustain Energy Rev 2016; 57:111–30.

[3] Sultana B, Mustafa M, Sultana U, et al. Review on reliability improvement and power loss reduction in distribution system via network reconfiguration. Renew Sustain Energy Rev 2016;66:297–310.

[4] Acha E, Fuerte-Esquivel CR, Ambriz-Perez H, et al. FACTS: modeling and simulation in power networks. John Wiley & Sons; 2004.

[5] Masdi H, Mariun N, Mahmud S et al., Design of a prototype DSTATCOM for voltage sag mitigation. p. 61–6.

[6] Majumder R. Reactive power compensation in single-phase operation of the microgrid.IEEE Trans Ind Electron 2013; 60(4):1403–16.

[7] Mahela OP, Shaik AG. A review of distribution static compensator. Renew Sustain Energy Rev 2015; 50:531–46.

[8] Mokhtari G, Nourbakhsh G, Zare F et al., A new distributed control strategy to coordinate multiple DSTATCOM in LV network. p. 1–5.

[9] Shahnia F, Chandrasena RP, Ghosh A, et al. Application of DSTATCOM for surplus power circulation in MV and LV distribution networks with single-phase distributed energy resources. Electr Power Syst Res 2014; 117:104–14.

[10] Yan R, Marais B, Saha TK. Impacts of residential photovoltaic power fluctuation on on-load tap changer operation and a solution using DSTATCOM. Electr Power Syst Res 2014; 111:185–93.

[11] Hussain SS, Subbaramiah M, An analytical approach for optimal location of DSTATCOM in the radial distribution system. p. 1365–9.

[12] Murthy V, Kumar A. Comparison of optimal DG allocation methods in radial distribution systems based on sensitivity approaches. Int J Electr Power Energy Syst 2013; 53:450–67.

[13] Taher SA, Afsari SA. Optimal location and sizing of DSTATCOM in distribution systems by the immune algorithm. Int J Electr Power Energy Syst 2014; 60:34–44.

[14] Bagherinasab A, Zadehbagheri M, Khalid SA, et al. Optimal placement of DSTATCOM Using hybrid genetic and ant colony algorithm to losses reduction. Int J Appl Power Eng (IJAPE) 2013; 2(2):53–60.

[15] Imran A. Quadri, S. Bhowmick, D. Joshi, A comprehensive technique for optimal allocation of distributed energy resources in radial distribution systems, Applied Energy, Volume 211, 2018, Pages 1245-1260

[16] Neeraj Kanwar, Nikhil Gupta, Khaleequr R. Niazi & Anil Swarnkar, "Optimal Allocation of DGs and Reconfiguration of Radial Distribution Systems Using an Intelligent Search-based TLBO, "Electric Power Components and Systems, Vol. 45 , Issue. 5, 2017.

[17] Z. Moravej and A. Akhlaghi, "A novel approach based on cuckoo search for DG allocation in distribution network," Int. J. Electr. Power Energy Syst., vol. 44, no. 1, pp. 672–679, 2013.

[18] H. Teng, A network-topology-based three-phase load flow for distribution systems, Proc Natl. Sci. Counc. ROC (A) 24 (4) (2000) 259–264.

[19] Yuvaraj, K. Ravi, K.R. Devabalaji, DSTATCOM allocation in distribution networks considering load variations using bat algorithm, Ain Shams Engineering Journal, Volume 8, Issue 3, 2017, Pages 391-403.

[20] T. Gözel and M. H. Hocaoglu, "An analytical method for the sizing and siting of distributed generators in radial systems," Electr. Power Syst. Res., vol. 79, no. 6, pp. 912–918, 2009.

[21] Mobin Esmaeili, Mostafa Sedighizadeh, Masoud Esmaili, "Multi-objective optimal reconfiguration and DG (Distributed Generation) power allocation in distribution networks using Big Bang-Big Crunch algorithm considering load uncertainty,"Energy,Volume 103, Pages 86-99, 2016.

[22] M.M. Aman, G.B. Jasmon, A.H.A. Bakar, H. Mokhlis, "A new approach for optimum simultaneous multi-DG distributed generation Units placement and sizing based on maximization of system loadability using HPSO (hybrid particle swarm optimization) algorithm,"Energy,Volume 66, Pages 202-215, 2014.

Appendix A

The loss sensitivity factor

The derivation of the jth branch power loss per k^{th} bus injected reative power i.e.$\partial Plossj/\partial Qk$, can be obtained from as shown below,

$$\frac{\partial P_{loss_j}}{\partial Q_m} = 2R_j . \sum_{u=2}^{n} \left(BIBC(j, u-1) \frac{P_u \cos(\theta_u) + Q_u \sin(\theta_u)}{|V_u|} \right) . BIBC(j, m-1) \frac{\sin(\theta_m)}{|V_m|}$$
$$+ 2R_j . \sum_{u=2}^{n} \left(BIBC(j, u-1) \frac{P_u \sin(\theta_u) - Q_u \cos(\theta_u)}{|V_u|} \right) . BIBC(j, m-1) . \frac{-\cos(\theta_m)}{|V_m|} \qquad A.1$$

Sum of the above expression leads to the derivation of the total power losses per k^{th} bus injected real power $\partial Ploss, j/\partial Qk$

$$\frac{\partial P_{loss}}{\partial Q_m} = 2 \sum_{j=1}^{nb} \left[R_j . \left(\sum_{u=2}^{n} BIBC(j, u-1) \frac{P_u \cos(\theta_u) + Q_u \sin(\theta_u)}{|V_u|} \right) . BIBC(j, m-1) \frac{\sin(\theta_m)}{|V_m|} \right]$$
$$+ 2 \sum_{j=1}^{nb} \left[R_j . \sum_{u=2}^{n} \left(BIBC(j, u-1) \frac{P_u \sin(\theta_u) - Q_u \cos(\theta_u)}{|V_u|} \right) . BIBC(j, m-1) . \frac{-\cos(\theta_m)}{|V_m|} \right] \qquad A.2$$

The objective is to find the size of DSTATCOM at any bus k to minimize total active power losses of the network. To determine the optimum size of DG, the derivative of the total power losses per each bus injected reactive powers are equated to zero as:

$$\frac{\partial P_{loss}}{\partial Q_m} = 0 \qquad A.3$$

$$Q_m = \frac{|V_m| . \sum_{j=1}^{nb} \left[R_j . \left\{ \left(\sum_{\substack{u=2 \\ u \neq m}}^{n} BIBC(j, u-1) . Im(I_u) \right) . BIBC(j, m-1) . \cos(\theta_m) - \left(\sum_{\substack{u=2 \\ u \neq m}}^{n} BIBC(j, u-1) . re(I_u) \right) . BIBC(j, m-1) \sin(\theta_m) \right\} \right]}{\sum_{j=1}^{nb} R_j . \left(BIBC(j, m-1) \right)^2} \qquad A.4$$

978-1-5386-6624-1/18 $31.00 © 2018 IEEE

Application of singular perturbation methodology for transient stability study of electric power system

Sunitha Anup*, T. S. Bhatti, Ashu Verma
Centre for Energy Studies
Indian Institute of Technology, Delhi
anup.sunitha@gmail.com

Abstract— Transient stability study is an important assessment in electric power system stability studies. An accurate methodology ensures precise boundary of stability after the occurrence of a fault in the power system. Singular perturbation is an effective methodology in non-linear dynamics which provides accurate results. The methodology is illustrated on a five generator interconnected power system and simulation results are compared with other methodologies namely step-by-step and catastrophe theory method.

Keywords—singularperturbation,synchronism,synchronous generator,critical clearing time,manifold,transient stability

I. INTRODUCTION

Electric power system transient stability phenomena is associated with synchronised operation of conventional and non-conventional generators in an interconnected electric power system, and becomes relevant with tranmission of reliable electric power to long distances. It is defined as the ability of an interconnected electric power system to remain in synchronism after the occurrence of a large disturbance in the electric power system [1-2]. In order to evaluate the stability of system in such a scenario, several numerical methods were resorted by earlier researchers[3-7]. Another approach of graphical method which used the idea of individualised energy, which eliminates the need of numerical integration. The graphical methodology is still considered useful for the assessment of stability limit. Introducing the concept of Lyapunov function to individualised energy was a major break through in the power system stability study[8-9]. However, the challenge of obtaining a suitable Lyapunov function made the researchers in the the energy and power sector, to resort to more impressive and effective methodgies.

Power system stability has a complicated nature as there are a number of factors involved in the study[10]. The non-linear modelling should not involve any linearisations and the time span of study should be suitably considered for large disturbance scenario. The occurrence of large disturbance may cause large excursion of the rotor angles of synchronous generator.In case, the excursion is not corrected in the correct time, then the synchronism of not only the affected synchronous generator , but also the entire interconnected power system may be affected[11]. If the system is stable, the swinging of generator rotor angle follows a decreasing path and maintains stabililty. If the system is unstable, the excursion of swinging follows an ever increasing path , which is a forerunner of 'out-of-sync' scenario and ultimately leads to grid collapse of electric power system.This detrimental nature of transient stability problem and non-linear behaviour of rotor angle swing necessitates an accurate nonlinear methodology . Singular perturbation is an important non-linear modelling tool ,widely used in control system domain. In this work, the application of singular perturbation methodology is employed to study the transient stability problem of an interconnected power system. It has a strong theoretical framework and at the same time provides accurate boundary of stability region[12].

In this work, the synchronous generator of electric power system is modelled in third order model. In most of the previous techniques for transient stability study, swing dynamics of rotor of the synchronous generator has not been studied in the detailed third order model,rather, these techniques have considered the variation of voltage behind transient reactance as constant[13-16]. The variation of rotor angle and speed are actually faster as compared to the variation of voltage behind transient reactance. For large disturbance study, the modelling of variation of voltage behind transient reactance is very relavant.The critical clearing time esimated from the study is the most important result of transient stability study. Singular perturbation methodology enables to model a dynamics of a non-linear system, such that the state variables may be from different time frames. In other words, all the state variables need not be in the same rate of variation and dynamics. Hence, singular perturbation methodolgy is an effective approach for studying transient stabiltiy of electric power system. The rotor angle and rotor speed of the synchronous generator are considered as fast variables and the voltage behind transient reactance is considered as slow variable. The direct axis transient reactance time constant of synchronous generator separates the singular perturbation model into slow and fast variables. If a three phase fault is occurred anywhere in the system, the asymptotic stability of these variables are evaluated in the right frame. If the fault is cleared within critical clearing time, the asymptotic stability of the variables are ensured. Also, it is possible to evaluate this critical clearing time mathematically using the methodology. The solution of stability is assumed to be lying in a region of stability and in the boundary of the stability region lies the critical clearing solution. In the terminology of singular perturbation, this boundary of stability region is termed as manifold, which can be algebraically and implicitly represented in terms of the state variables. Lyapunov function of the singular perturbation model is employed to obatain the location in this manifold, which is identified as the critical clearing of this large disturbance. Once, the fault is cleared within this critical clearing, large excursion of rotor angle of synchronous generator can be avoided and the synchronism of the electric power system is ensured.

The rest of the paper is arranged as follows. Section II gives the mathematical formulation of the problem. Section III illustrates the solution methodology for a 5-generator electric power system using singular perturbation method. Finally, the conclusion and future prospects of the study is given in Section IV.

II. MATHEMATICAL FORMULATION

The third order model of a synchronous generator can be mathematically represented as,

$$\frac{dE'_q}{dt} = \frac{1}{T'_{do}}(E_{fd} - E' - jX'_d I_t - jX_d I_d - jX_q I_q) \tag{1}$$

$$M\frac{d^2\delta}{dt^2} = P_m - P_e \tag{2}$$

Where, E'_q is the voltage behind transient reactance, T'_{do} is the direct axis transient reactance time constant, E_{fd} is the field excitation voltage, E' is the terminal voltage, X'_d is the transient reactance, I_t is the current flowing, X_d, X_q are the direct and quadrature synchronous reactances, I_d, I_q are the direct and quadrature components of current, M is the inertia constant, δ is the rotor angle, P_m is the mechanical input power P_e is the electrical power output. For a fault near the synchronous generator in the electric power system, the generator would be the most affected one among all the generators in the power system and it can be considered as a single machine connected to an infinite bus. Thus dynamics of the single machine would be studied. The above model is formulated using singular perturbation methodology as follows,

$$\left.\begin{array}{l} x' = f(t, x, y, \varepsilon); x(t_0) = \eta_0 \\ y' = g(t, x, y, \varepsilon); y(t_0) = \phi_0 \end{array}\right\} \tag{3}$$

Where, x is the slow variable which represents the dynamics of voltage behind transient reactance, y is the vector having two fast variables namely rotor angle and rotor speed of synchronous generator, ε is the singular perturbation parameter which is reciprocal of T'_{do}. η_0, ϕ_0 are initial conditions of the non-linear differential equations. The manifold of the singular perturbation model can be given as,

$$y = h(x) \tag{4}$$

Where, Eqn.(4) represents the simplest manifold with y implicitly represented only as function of x. Eqn.(4) is also called the slow manifold of the singular perturbation model .

If $h(x)$ is expanded using Taylor series as function of x and ε, then the manifold can be given as,

$$h(x, \varepsilon) = h_0(x) + \varepsilon h_1(x) + \varepsilon^2 h_2(x) + \dots \tag{5}$$

Eqn.(5) represents improved slow manifold of the singular perturbation model. The concept of boundary of stability region of a non-linear dynamical system using the singular perturbation methodology is illustrated in Fig. (1). The algebraic manifold implicitly represented as function of slow and fast variables. In either case of slow and improved slow manifolds, it is possible to identify the critical clearing in the boundary using Lyapunov function of the model. The Lyapunov function is given as,

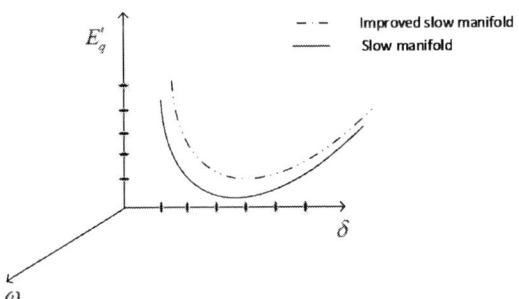

Fig.1 Concept of manifold stability region

$$E = (1-k)V(x) + kW(x, y) \tag{6}$$

$$V(x) = \int_0^x -f(x, y)dx \tag{7}$$

$$W(x, y) = \frac{D}{2M}(y_1 - g_1)^2 + y_2(y_1 - g_1) + \frac{M}{2D}y_2^2 + \tag{8}$$

$$\frac{M}{D}(1+x)\int_{g_1}^{y_1} \sin(y_1 + \overline{\delta}) - \sin(g_1 + \overline{\delta})d\delta$$

The Lyapunov function in Eqn. (6) is represented as weighted sum of slow component as well as fast component of singular perturbation model of a synchronous generator. V(x) is the slow component and W(x,y) is the fast component of Lyapunov function. For a three phase fault near one of the generators, that generator would be the most affected among other generators and that generator is modelled in singular perturbation methodology. The manifold is evaluated as slow manifold and improved slow manifold and Lyapunov functions are estimated. If the Lyapunov function is reached in first minima that solution is identified as critical clearing solution. The critical solution corresponds to critical clearing angle (CCA) of the generator. From the critical clearing angle, critical clearing time can be expressed as,

$$CCT = \sqrt{\frac{2M}{P_m}}(CCA - \delta_0) \tag{9}$$

The main advantage of the methodology is that the nonlinearity of the model is maintained and Lyapunov function of the model enables to identify the critical clearing solution in the manifold. From the critical clearing angle, critical clearing time can be obtained from Eqn.(9).

III. ILLUSTRATION ON 5-GENERATOR SYSTEM

The schematic diagram of a 5-generator system is given in Fig.2. The data of the system is given in reference [1]. It consists of five synchronous generators G1, G2, G3, G4 and G5 and fourteen interconnecting buses. The occurrence of a three phase fault can occur in any of these generator buses or interconnecting buses. The fault is required to be cleared within the critical clearing time. The proposed method of singular perturbation is employed to study the transient stability of the system. Third order model of synchronous generator is studied using the methodology. IEEE type 1 exciter is incorporated in each of the synchronous generator for the improvement of field voltage of synchronous generator of the system.

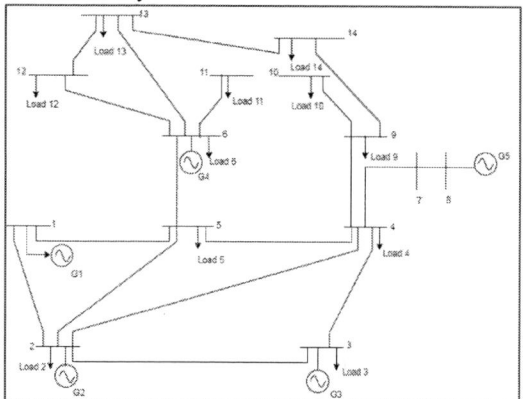

Fig.2 Schematic diagram

The swing dynamics of each synchronous generator characterizes the rotor angle oscillation of the generators. Due to the occurrence of the three phase fault, the oscillation of the rotor may be such that, it may accelerate and rotor angle may be different from the initial steady state value. Hence, post-fault scenario is studied in a region of attraction such that there is convergence of rotor angle to stable equilibrium point in this region. The energy calculated at each of the points in this region would be depending on the acceleration gained by a generator. It is advisable to have negative value of energy in order to remain in synchronism and the generator would not be further accelerated away from the stable condition. If the Lyapunov function is minima, it would correspond to critical point in the boundary of stability region and critical clearing angle and critical clearing time is evaluated at this point. The most accelerated machine is identified from the system and it corresponds to the most affected generator. For a three phase fault occurred at bus 2, generator G2 is likely to be the most affected generator due to proximity to the fault. G1 is taken as reference generator and rotor angles of remaining generators are obtained with respect to that of G1.If the fault is cleared within the critical clearing time, the rotor angle of the four generators with respect to rotor angle of G1 are as shown in Fig.3. As seen from the figure, the synchronism of the multi-machine system is ensured and the electric power system can be considered as stable system .If the fault is not cleared, the synchronism of the multi-machine power system is lost and the rotor angle of each generators would increase and move away from the steady state value as shown in Fig.4.It is also evident from the figure that the variation of the most affected machine G2 is maximum as compared to other machines. Thus, it is required to find out the critical clearing time of the fault occurred at bus 2 so that the synchronism is ensured for the multi-machine power system. Singular perturbation methodology enables to obtain the boundary of stability region and from the Lyapunov function of the model, critical clearing angle can be obtained from the slow manifold of stability region.

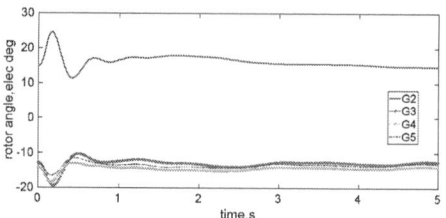

Fig.3 Swing curves of stable power system

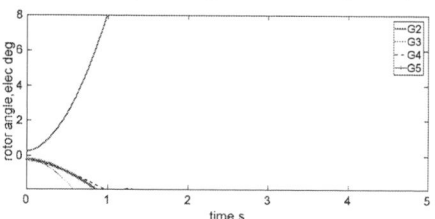

Fig.4 Swing curves of unstable power system

The voltage behind the transient reactance of the synchronous generators varies during this transient condition of fault as shown in Fig.5. The variation of E'_q is less compared to the variation of rotor angle variation. Singular perturbation modelling enables to study this variation of voltage behind transient reactance in slower time scale as compared to variation of rotor angle. The slow manifold obtained from the singular perturbation model enables to present a boundary of stability region. Critical clearing angle is identified in this slow manifold where the Lyapunov function attains minima.

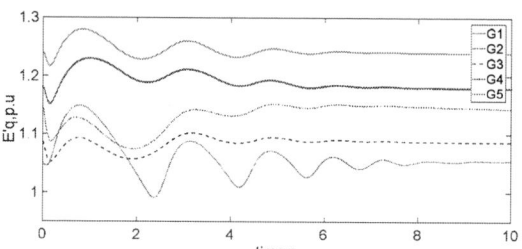

Fig.5 Variation of voltage behind transient reactance of five generators

The singular perturbation model of the generator G2 is studied for the 5-generator system and the state variables of the model, namely, x (internal voltage behind the transient reactance), y_1 (rotor angle, p.u.) and y_2 (speed, p.u.) are plotted in Fig.6.The state variables converge to the stable equilibrium point if the fault is cleared within the critical clearing time. With the slow manifold stability region of the singular perturbation model, it is possible to evaluate the exact value of critical clearing time from the Lyapunov function of the model.

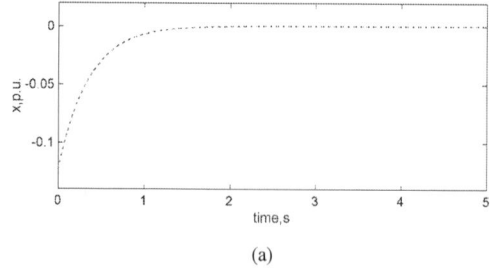

(a)

978-1-5386-6624-1/18 $31.00 © 2018 IEEE

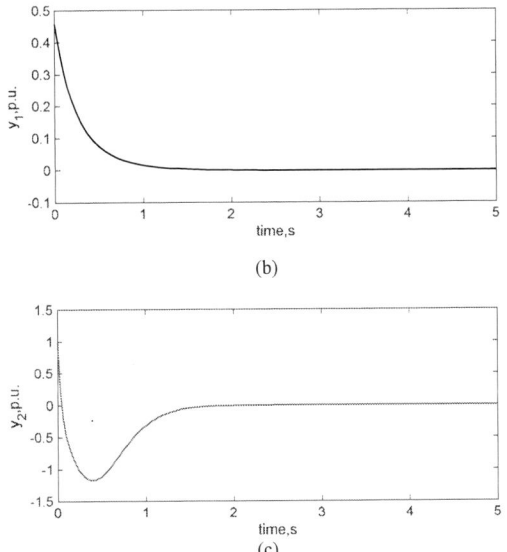

(b)

(c)

Fig.6 Singular perturbation model (**a**) Asymptotic stability of state variable x of G2, (**b**) Asymptotic stability of state variable y_1 of G2, (**c**) Asymptotic stability of state variable of state variable y_2 of G2

The improved slow manifold is obtained from the expanded Taylor series expansion of the algebraic manifold. The comparison of the slow and improved slow manifold of the singular perturbation model of G2 is shown in Fig.7. Both of them follow similar boundary characteristics. However, the critical clearing angles are different from both the manifolds. The Lyapunov function of the singular perturbation model vary as shown in due to the higher order terms in the Taylor series expansion. Thus, this variation is also visible in the identification of critical clearing angle in Fig.8

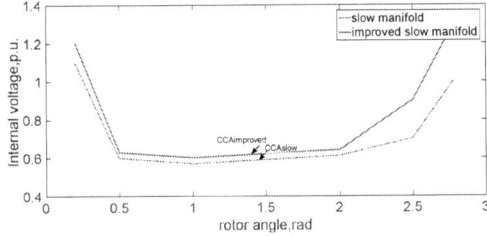

Fig 7 Comparison of slow manifold and improved slow manifold for critical clearing angle evaluation of singular perturbation model

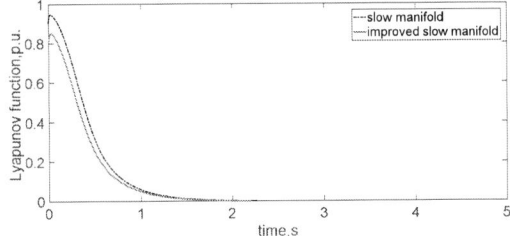

Fig 8 Comparison of Lyapunov function of slow manifold and improved slow manifold of singular perturbation model

Catastrophe theory method is a similar methodology which studies transient stability of power system from the stability region. The results of singular perturbation and catastrophe theory method are matching for the power system example. However, the improved slow manifold region has lesser value of critical clearing than slow manifold region, indicating that the rigorous modelling of the improved slow manifold than the slow manifold of the singularly perturbed model. The step-by-step method involves pre-determination of critical clearing time and there would a range of values inside which the critical clearing time is evaluated from trial and error. The comparison of critical clearing time of third order model of synchronous generator for using the singular perturbation method, catastrophe theory method and step-by-step method are shown in Table 1. The instability detection using the singular perturbation method in terms of the transient stability index for three phase faults in five bus locations are presented in Table 2.

Table 1. Comparison of critical clearing time of third order model

Sl no.	Fault near G1	Fault near G2	Fault near G3	Fault near G4	Fault near G5
Singular perturbation method	0.81s	0.99s	0.891s	0.79s	0.78s
Catastrophe theory method	0.85s	1.01s	1.00s	0.82s	0.81s
Step-by-step method	0.85-0.88s	1.00-1.03s	0.90-1.02s	0.78-0.82s	0.79-0.82s

Table 2. Singular perturbation method for instability detection

Fault location bus no	Critical clearing time (s)	Practical clearing time(s)	Stability detection
1	0.810	0.790	stable
		0.800	stable
		0.810	stable
		0.820	unstable
2	0.990	0.970	Stable
		0.980	Stable
		0.990	Stable
		1.000	Unstable
3	0.891	0.889	Stable
		0.890	Stable
		0.891	Stable
		0.900	Unstable
	0.790s	0.78	Stable
		0.79	Stable
		0.80	Unstable
		0.81	Unstable

		0.77	Stable
5		0.78	Stable
	0.780s	0.79	Unstable
		0.80	Unstable

IV. CONCLUSIONS

This paper presents the application of singular perturbation methodology for transient stability of power system. The methodology gives an effective approach for obtaining a definite stability boundary using manifold. The synchronous generator was modelled in third order modelling with the inclusion of variation of dynamics of voltage behind transient reactance of the generator. For a three fault scenario, Lyapunov function obtained from the methodology enables to identify critical clearing of the fault, with desired results. The Taylor series expansion of the manifold can be used to obtain more accurate boundary of stability region by taking one more term in the estimation of the manifold. Hence, improved slow manifold region gives more accurate value of critical clearing angle and critical clearing time of the model. The relevance of the improved manifold stability region is seen for the faults closer to the generator terminals as compared to faults far away from the generator terminals. The comparison of the results with catastrophe theory method and step-by-step method are also matching with the results of perturbation method. The critical clearing time of faults are reduced when the fault location is near the generator terminals. Hence, the protection devices closer to the generator terminals should ensure that the fault is cleared within the critical time so that the synchronism of the system is maintained. Singular perturbation methodology enables fast computation of critical fault clearing time as the number of calculations are few in comparison to the numerical methods. And the simplicity of this method is clear from the algebraic expression of the manifold stability region. The stability region of the methodology is unique and valid for any change in the system irrespective of the nature and location of the fault. Thus, singular perturbation methodology may be a promising technique for transient stability study of power system.

REFERENCES

[1] Padiyar KR. Power system dynamics. BS publications; 2008.

[2] IEEE/CIGRE Joint task force on stability terms and definitions: definition and classification of power system stability. *IEEE Transactions on Power Systems*, Vol 19, No 2, 2004

[3] Athay T, Podmore R, Virmani S. A practical method for the direct analysis of transient stability. IEEE Transactions on Power Apparatus and Systems. 1979 Mar(2):573-84

[4] Fouad AA, Vittal V, Oh TK. Critical energy for direct transient stability assessment of a multimachine power system. IEEE transactions on power apparatus and systems. 1984 Aug(8):2199-206.

[5] Bahbah AG, Girgis AA. New method for generators' angles and angular velocities prediction for transient stability assessment of multimachine power systems using recurrent artificial neural network. IEEE Transactions on Power Systems. 2004 May;19(2):1015-22.

[6] Pavella M, Ernst D, Ruiz-Vega D. Transient stability of power systems: a unified approach to assessment and control. Springer Science & Business Media; 2012 Dec 6.

[7] Vaahedi E, Li W, Chia T, Dommel H. Large scale probabilistic transient stability assessment using BC Hydro's on-line tool. IEEE Transactions on Power Systems. 2000 May;15(2):661-7

[8] Vu TL, Turitsyn K. Lyapunov functions family approach to transient stability assessment. IEEE Transactions on Power Systems. 2016 Mar;31(2):1269-77.

[9] Chiang HD, Wu FF, Varaiya PP. A BCU method for direct analysis of power system transient stability. IEEE Transactions on Power Systems. 1994 Aug;9(3):1194-208

[10] Dorfler F, Bullo F. Synchronization and transient stability in power networks and nonuniform Kuramoto oscillators. SIAM Journal on Control and Optimization. 2012 Jun 21;50(3):1616-42

[11] Weckesser T, Jóhannsson H, Østergaard J. Impact of model detail of synchronous machines on real-time transient stability assessment. InBulk Power System Dynamics and Control-IX Optimization, Security and Control of the Emerging Power Grid (IREP), 2013 IREP Symposium 2013 Aug 25 (pp. 1-9). IEEE

[12] Kokotovic P, Khali HK, O'reilly J. Singular perturbation methods in control: analysis and design. Siam; 1999

[13] Anup S. Comparative Fault Response study of Synchronous Generator in the presence of Wind Generator using Singular Perturbation based Transient Stability Index. International Journal of Renewable Energy Research (IJRER). 2018 Jun 26;8(2):994-1005.

[14] Kevorkian JK, Cole JD. Multiple scale and singular perturbation methods. Springer Science & Business Media; 2012 Dec 6.

[15] DeMarco CL, Bergen AR. Application of singular perturbation techniques to power system transient stability analysis. California Univ., Berkeley (USA); 1984 Jan 1

[16] Ayasun S, Nwankpa CO, Kwatny HG. Computation of singular and singularity induced bifurcation points of differential-algebraic power system model. IEEE Transactions on Circuits and Systems I: Regular Papers. 2004 Aug;51(8):1525-38

Study of P&O And INC PV MPPT Techniques For Different Environment Conditions

Bipin Singh, Bharat Verma and Prabin Kumar Padhy

Department of Electronics and Communication
Indian Institute of Information Technology Design and Manufacturing
Jabalpur, India

[1]1612205@iiitdmj.ac.in
[2]bharatbigj@gmail.com
[3]prabin16@iiitdmj.ac.in

Abstract— There are different maximum power point tracking techniques present in the literature, they vary according to the number of sensors required, speed of convergence, cost, effectiveness of techniques, easy hardware implementation and in other respects. In this paper comparison of two well-known conventional maximum power point tracking techniques, "perturb and observe" and "incremental conductance" is done for different environment condition. A DC-DC boost converter is also used for interfacing the load and photovoltaic module. Result is simulated in MATLAB/Simulink environment, and all the condition is kept same for both techniques in simulation for the best outcomes of this comparative study.

Keywords— DC-DC Boost Converter, MPPT, PV CELL, P&O, INC

I. INTRODUCTION

Electric Energy is necessary in our life, it plays important role to development of economies, and social growth. Conventional energy resources like fossil fuels are deleterious to the environment and also non-renewable. Combustion of fossils fuel produces the CO_2 and other air pollutant such as nitrogen oxide, sulphur dioxide and heavy metals. Environmentally unfriendly behaviour and unavailability of enough resources to fulfil the day-by-day increasing demand of electricity is taking the human to use alliterative energy resources, to fill this void there are different alternatives which is easily available. These energy resources also called renewable energy resources which are the prime energy resources are being used in this regard.

Sun is the massive source of energy which generate the more energy than anything else in our solar system. Human has figured out the way of harvesting the energy from sun which easily available around us. Harvesting of solar energy can be done by the Photovoltaics which is work on the principle of photovoltaic effect.

PV cell is made is semiconductor device which directly convert solar radiant light into the electrical energy. PV cell has low transforming efficiency about 17% (for low irradiance). Power conversion of solar PV cell is environment dependent (on solar irradiance and temperature of PV surface), high irradiance value gives the high power and more the temperature of PV panel surface gives less power. Low conversion efficiency and environment dependency of generated power conditions are the two basic problems associated with PV cell[1]. PV cell produce the nonlinear power at the output and it is desired that the operating point

coincide with MPP for all the time but due non-linearity of produced power operating point cannot stay at the maximum power point (MPP) for always, to tackle this problem a maximum power point tracking (MPPT) technique is required [1], [2][3] that can be able to make operating coincide with MPP regardless of any environment condition. There are different methods have proposed in literature for continuous extracting of maximum power. Most of them give good response when environment condition changes, and some them are suitable for constant environment condition[3]. below are the some MPPT techniques, Fuzzy logic control[4], Neural network [5], RCC [6], Perturb & observe (P&O) [7], [8][9]–[12], incremental conductance (INC) [13], [14], Fractional short-circuit current, Load current & load voltage maximisation, $dP/(dI)$ or dP/dV feedback control, Fractional open-circuit voltage are compared in[2], [3].

This comparative study is done between P&O and INC for performance analysis in constant and different weather conditions. Respective are the most popular techniques for tracking the maximum power from the PV system. Comparison is done in MATLAB SIMULATION and DC-DC boost converter is used as an interfacing device between the PV panel and load. All the condition is kept same in MATLAB for the both P&O and INC algorithm to distinguish the behaviour in constant and changing environment.

The paper is arranged in a following manner. Section II describes the mathematical model of a PV array, dynamics of boost converter in section III, Theory behind P&O & INC are detailed in IV, MATLAB/Simulink implementation and discussion are in section, and conclusion of the whole work is represented in VI.

II. PMODEING OF PV ARRAY IN MATHEMATICAL FORM

A. Ideal PV cell

Electrical circuit for ideal PV cell is shown in figure (1), basic equation of PV cell is given in equation (1),

$$I_{pv} = I_{ph} - I_0 \left[e^{\left(\frac{qV}{akT}\right)} - 1 \right], \tag{1}$$

where I_{pv} is the photo-current, I_0 is leakage or reverse saturation current of diode, q is charge of electron, k is Boltzmann constant T is temperature (in Kelvin) and a is the diode ideality factor[15].

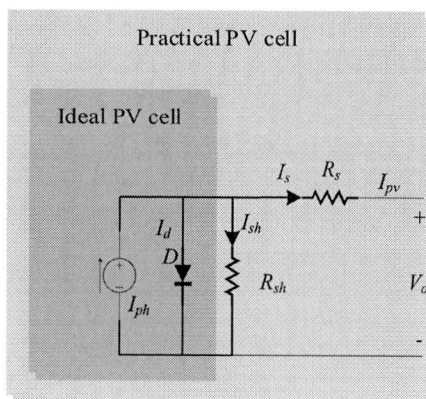

Figure 1: Single diode PV cell

B. Practical PV array

Practical circuit for the PV cell is shown in fig (1), in practical circuit, R_{sh} and R_s are extra element added, why these are added is studied in [15], by connecting the PV cells in series parallel combination an PV array can be formed. Let N_s is number of cells connected in series and N_p is number of PV cell connected in parallel. The PV current equation can be written as follows;

$$I_{pv} = N_P I_{ph} - N_P I_d - I_{sh}, \qquad (2)$$

where I_{sh} is current through the $\frac{N_s}{N_p} R_{sh}$ given by equation (3),

$$I_{sh} = N_P \left[\frac{\frac{V_{PV}}{N_s} + \frac{I_{PV} R_S}{N_P}}{R_{sh}} \right]. \qquad (3)$$

I_{ph} is photo-current and is given by (4),

$$I_{ph} = [I_{SC} + K_i(T - 298)] \, G/1000 \,. \qquad (4)$$

Here G is value of irradiance in W/m^2 .and K_i is short circuit current/temperature coefficient in A/K, and I_d in equation (2) is diode current given by following equation,

$$I_d = I_0 \left(\exp\left(q \frac{V_{PV}/N_s + I_{PV} R_S/N_P}{a \times K \times T} \right) - 1 \right), \qquad (5)$$

and I_0 is (diode leakage current) that has following expression,

$$I_0 = I_{rs} \left(\frac{T}{T_r} \right)^3 \exp\left(\left[\frac{qE_g}{a \times K} \left(\frac{1}{T_r} - \frac{1}{T} \right) \right] \right). \qquad (6)$$

In above equation, E_g is energy band gap (for silicon 1.13 eV at nominal temperature 25^0 C), T_r is reference temperature i.e 25^0 C, and I_{rs} is nominal saturation current of the diode and it is given by (7),

$$I_{rs} = \frac{I_{sc}}{\exp\left(\frac{q \times V_{OC}}{N_S \times a \times K \times T_r} \right)}, \qquad (7)$$

Here I_{SC} is the short circuit current and V_{OC} is the open circuit voltage of the PV array[15].

III. DYNAMICS OF BOOST CONVERTER

The Boost converter is used when the output voltage required to be higher than the input voltage, experimental analysis of the boost converter has been studied in [16]

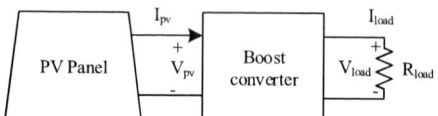

Figure 2 boost converter as an interfacing device

Boost converter can be described mathematically as follow, from figure (2),

$$\frac{V_{load}}{V_{pv}} = \frac{1}{1-D}, \qquad (8)$$

where, D denotes the duty ratio for the boost converter, and relation of current with the duty ratio is as follows ,

$$\frac{I_{Load}}{I_{PV}} = 1 - D, \qquad (9)$$

$$V_{Load} = I_{Load} \times R_{Load}, \qquad (10)$$

$$R_{eff} = (1 - D)^2 \times R_{Load}. \qquad (11)$$

R_{eff} is total effective resistance seen by the PV model. The duty cycle of the boost converter related to the PV array voltage[16][17] is given by (12)

$$V_{PV} = I_{PV} \times (1 - D)^2 \times R_{Load}, \qquad (12)$$

Equation (12) shows by decreasing the duty ratio (D) V_{PV} will increase and vice versa.

IV. THEORY OF P&O AND INC

A. Perturb and observe (P&O)

This is one of most famous technique for its simplicity and easy implementation. it has the good convergence property and it's also required the less hardware to implement. Characteristic curve for PV model is giver in figure (3)

This MPPT technique track the maximum power by continuous observation of change in power with respect to the voltage (dP/dV) and a small constant perturbation is provided according to the observation results, if Power increases with increasing the voltage or power increases with decreasing the voltage then next perturbation polarity need to be same as the last perturbation. If power decreased with the increasing the voltage or power decreased with decreasing the voltage then the next perturbation polarity need to be reversed [12].

$$\frac{dP}{dV} = 0, \quad \text{at MPP} \qquad (13)$$

$$\frac{dP}{dV} > 0, \quad \text{left of MPP} \qquad (14)$$

$$\frac{dP}{dV} < 0, \quad \text{right of MPP} \qquad (15)$$

For better understanding of algorithm, the explanation of algorithm is as follows,

Step 1: measure V and I of PV panel

Step 2: calculate power (P)

978-1-5386-6624-1/18 $31.00 © 2018 IEEE

Step 3: check, is power is bigger than the last time.

Step 4: check, is voltage is bigger than last time.

Step 5: if step 3 and step 4 both together are either true or false then increase the voltage otherwise decrease the voltage by giving a suitable perturbation size.

Environment condition is not always constant its vary time to time over the day, this techniques holds good response when environment condition is constant, but algorithm fails when environmental condition changes rapidly, operating point start oscillating around and diverge from maximum power point (MPP).

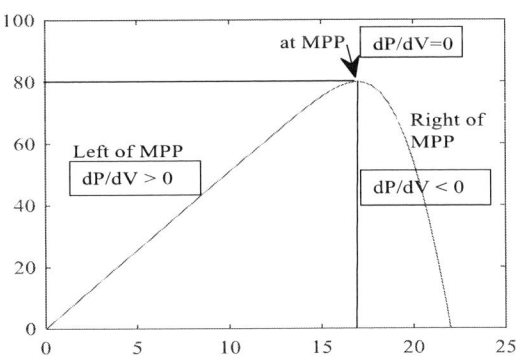

Figure 3 P-V characteristic of a PV system

B. Incremental conductance (INC)

Failure of P&O algorithms under rapidly changing environment can be overcome by the incremental conductance method, it works on the fact that slope of power-voltage (P-V) curve of the PV array (equation (13) (14) and (15)) Positive, negative and zero slope implies the position of operating point either lies at left of the MPP, at right of the MPP and at MPP respectively in the power-voltage (P-V) curve. It uses the constant perturbation size to increase and decrease the PV voltage.

Relation of instantaneous and incremented conductance can be evolved as follow

$$P = VI \qquad (16)$$

$$\frac{dP}{dV} = V\frac{dI}{dV} + I \cong I + V\frac{\Delta I}{\Delta V} \qquad (17)$$

Equations (13), (14), (15) can be written as,

$$\frac{\Delta I}{\Delta V} = -\frac{I}{V}, \quad \text{at MPP} \qquad (18)$$

$$\frac{\Delta I}{\Delta V} > -\frac{I}{V}, \quad \text{left of MPP} \qquad (19)$$

$$\frac{\Delta I}{\Delta V} < -\frac{I}{V}, \quad \text{right of MPP} \qquad (20)$$

where $\Delta I/\Delta V$ is incremented conductance and I/V is instantaneous conductance.

Once MPP achieved (i.e. $dP/dV = 0$), operating point force to remain on MPP until any change in ΔI is noted due to change in environmental condition. by following step explanation [13];

Step 1: measure $V(k)$ and $I(k)$ of the PV panel, where k denotes the present value of measured quantity.

Step 2: calculate $\Delta V = V(k) - V(k-1)$ and $\Delta I = I(k) - I(k-1)$, where, $(k-1)$ denoted the past value measured quantity.

Step 3: check, if $\Delta V = 0$ then go to step (4) otherwise follow the step (6).

Step 4: check whether $\Delta I = 0$ or not, after this follow the step5

Step 5: if $\Delta I = 0$ then go the step (11), if $\Delta I > 0$ first go to step (10) and then step (11). If $\Delta I < 0$ first go to step (9) and then step (11)

Step 6: check $\Delta I/\Delta V = -I/V$ or not, after this follow the step7

Step 7: If $\Delta I/\Delta V = -I/V$ then go to step (11) otherwise go to step (8).

Step 8: if $\Delta I/\Delta V > -I/V$, first go to step (10) and then step (11) otherwise first go step (9) and the step (11)

Step 9: $D = D + perturbation\ size$.

Step 10: $D = D - perturbation\ size$. Here D is the duty ratio of the converter.

Step 11: update $I(K-1) = I(k), and\ V(k-1) = V(k)$ and go to step (1)

TABLE 1 PARAMETER SPECIFICATION OF SOLAR ARRAY AT $25^o\ C$, $1000\ W/m^2$

Pmpp	80 W
Vmpp	16.8 V
Impp	4.8 A
Voc	22 V
Isc	5.1 A
Np	1
Ns	36
Ki	0.0017 A/K
Rsh	1000 Ω
Rs	0.0111 Ω
a	1.3

V. SIMULATION RESULT AND DISCUSSION

The PV model with the DC-DC boost converter as an interfacing device is simulated in MATLAB /Simulink software, to demonstrate the feature of the convention P&O technique in comparison with the INC MPPT method, and perturbation size for all simulation is kept 0.001, there is no change in perturbation size for any simulation.

CASE1 (Constant Environment Condition): In this case environmental condition is assumed to be constant at 1000 W/m^2 and 25^oC of temperature, perturbation is kept at 0.001 for this case. This simulation run is to verify the ability of tracking the maximum power of both techniques, Figure (4). It can be seen from the figure that INC gives good result in constant weather condition in transient state but in steady state there is no much difference between the both algorithms. P&O and INC both suffer from the same problem, that is, in steady state the operating point oscillates around the MPP and never achieve the true MPP.

CASE2 (Step Change in Irradiance) In this case the sudden change in the environment is assumed, happens when there is a step change in solar irradiance. Figure (5) shows the tracking behaviour of P&O and INC when irradiance changes suddenly from $500\ W/m^2$ to $800\ W/m^2$ @ 0.95 sec. It is clearly seen from figure (5) INC techniques gives good response towards sudden step changing behaviour of solar irradiance. INC give less deviation of operating and track new MPP faster than the P&O algorithm.

CASE3 (Continuously changing environment condition): This is case when a dark cloud passes over the surface of PV panel and the irradiance decreases and increases gradually. Figure (6) shows the change in irradiance gradually decreases from $700\ W/m^2$ to $400\ W/m^2$ within 3 seconds and starts increasing gradually from $400\ W/m^2$ to again $700\ W/m^2$ and stay $700\ W/m^2$. Figure (7) shows the tracking behaviour of techniques.

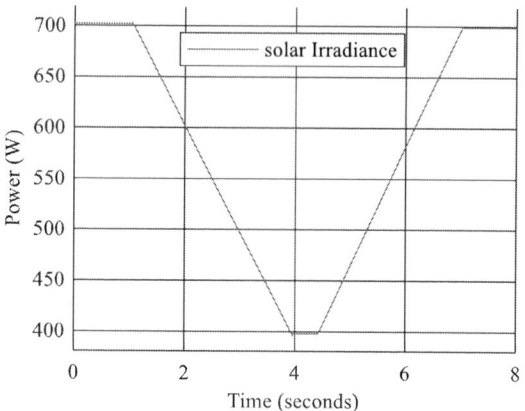

Figure 6 gradually changing irradiance for case 3

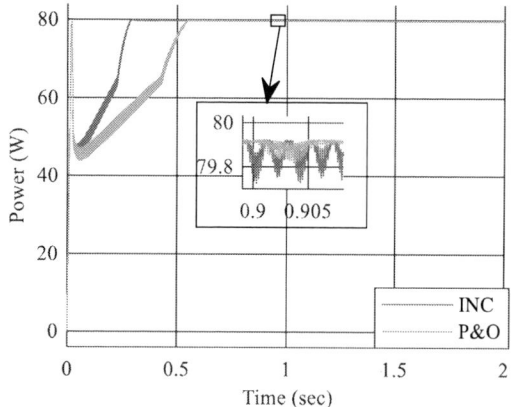

Figure 4 Power analysis of algorithms @1000 W/m² and Perturbation size is kept 0.001.

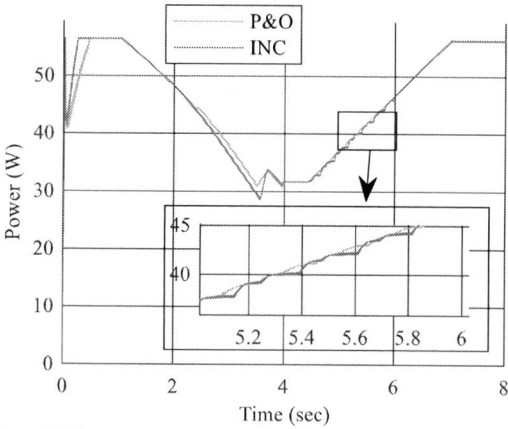

Figure 7 PV power output curve when irradiance increases and decreases gradually.

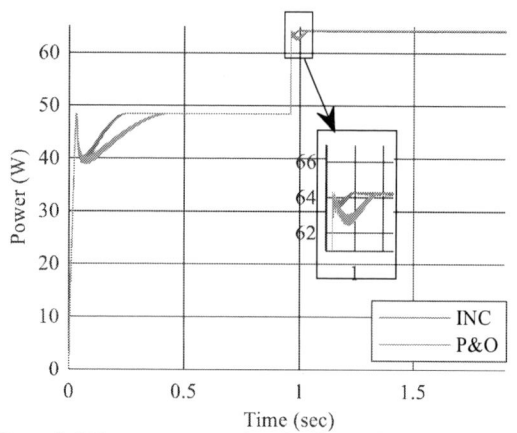

Figure 5 PV output power when irradiance changes sudden from 500 W/m² to 800 W/m²@ 0.95 sec and perturbation size is kept 0.001.

VI. CONCLUSION

In this paper, the DC-DC Boost converter is acting as an interfacing device between load the PV panel. Both the techniques required two sensors for implementation. From the above study it can be concluded that for constant irradiance condition both the techniques perform well but drawback of algorithms is oscillation in steady state, see figure (4). INC technique give good result than the P&O techniques under step change in irradiance condition, it tracks the new MPP faster than the P&O with less divergence of operating point from MPP see figure (5). Both the techniques give approximately same behavior under continuously changing environment condition see figure (7).

ACKNOLEDGEMENT

This publication is an outcome of the R&D work undertaken project under the Visvesvaraya PhD Scheme of Ministry of Electronics & Information Technology, Government of India, being implemented by Digital India Corporation.

References

[1] A. Q. Fertilizers, "A Survey of Maximum PPT techniques of PV Systems," 2012.

[2] T. Esram and P. L. Chapman, "Comparison of Photovoltaic Array Maximum Power Point Tracking Techniques," vol. 22, no. 2, pp. 439–449, 2007.

[3] B. Subudhi, S. Member, and R. Pradhan, "A Comparative Study on Maximum Power Point Tracking Techniques for Photovoltaic Power Systems," vol. 4, no. 1, pp. 89–98, 2013.

[4] A. I. Dounis, S. Stavrinidis, P. Kofinas, and D. Tseles, "Fuzzy-PID controller for MPPT of PV system optimized by Big Bang-Big Crunch algorithm," *IEEE Int. Conf. Fuzzy Syst.*, vol. 2015–Novem, 2015.

[5] L. M. Elobaid, A. K. Abdelsalam, and E. E. Zakzouk, "Artificial neural network-based photovoltaic maximum power point tracking techniques: a survey," *IET Renew. Power Gener.*, vol. 9, no. 8, pp. 1043–1063, 2015.

[6] T. Esram, J. W. Kimball, P. T. Krein, P. L. Chapman, and P. Midya, "'Dynamic maximum power point tracking of photovoltaic arrays using ripple correlation control,'" *IEEE Trans. Power Electron*, vol. 21, no. 217, pp. pp1282-1291, 2006.

[7] I. W. Christopher and R. Ramesh, "Open Access Comparative Study of P & O and InC MPPT Algorithms," vol. 408, no. 12, pp. 402–408, 2013.

[8] D. Sera, L. Mathe, T. Kerekes, S. V. Spataru, and R. Teodorescu, "On the perturb-and-observe and incremental conductance mppt methods for PV systems," *IEEE J. Photovoltaics*, vol. 3, no. 3, pp. 1070–1078, 2013.

[9] C. Y. Tan, N. A. Rahim, and J. Selvaraj, "Improvement of hill-climbing method by constraining voltage operating point," *CEAT 2013 - 2013 IEEE Conf. Clean Energy Technol.*, pp. 108–113, 2013.

[10] L. Piegari and R. Rizzo, "Adaptive perturb and observe algorithm for photovoltaic maximum power point tracking," *IET Renew. Power Gener.*, vol. 4, no. 4, p. 317, 2010.

[11] S. K. Kollimalla, S. Member, M. K. Mishra, and S. Member, "Variable Perturbation Size Adaptive P & O MPPT Algorithm for Sudden Changes in Irradiance," *IEEE Trans. Sustain. Energy*, vol. 5, no. 3, pp. 718–728, 2014.

[12] K. H. Hussein, I. Muta, T. Hoshino, and M. Osakada, "Maximum photovoltaic power tracking: an algorithm for rapidly changing atmospheric conditions," *IEE Proc. - Gener. Transm. Distrib.*, vol. 142, no. 1, p. 59, 1995.

[13] O. Wasynezuk, "Dynamic Behavior of a Class of Photovoltaic Power Systems," *IEEE Trans. Power Appar. Syst.*, vol. PAS-102, no. 9, pp. 3031–3037, 1983.

[14] and Y. K. F. Liu, S. Duan, Fei Liu, B. Liu, "A Variable Step Size INCMPPT Method for PV Systems," *IEEE Trans. Ind. Electron.*, vol. 55, no. 7, pp. 2622–2628, 2008.

[15] M. G. Villalva, J. R. Gazoli, and E. R. Filho, "Comprehensive Approach to Modeling and Simulation of Photovoltaic Arrays," vol. 24, no. 5, pp. 1198–1208, 2009.

[16] R. F. Coelho, F. M. Concer, and D. C. Martins, "Analytical and Experimental Analysis of DC-DC Converters in Photovoltaic Maximum Power Point Tracking Applications," no. 1, pp. 2778–2783, 2010.

[17] R. Khanna, Q. Zhang, W. E. Stanchina, and G. F. Reed, "Reference Adaptive Control," vol. 29, no. 3, pp. 1490–1499, 2014.

2nd IEEE International Conference on Power Electronics, Intelligent Control and Energy Systems (ICPEICES-2018)

A Unified Approach for Voltage Enhancement in Radial Distribution Network with Dynamic Loading

Swati Arora
Electrical Engineering Department
Punjab Engineering College
Chandigarh, India
aroras163@gmail.com

Sandeep Kaur
Electrical Engineering Department
Punjab Engineering College
Chandigarh, India
sandipsaroa@gmail.com

Rintu Khanna
Electrical Engineering Department
Punjab Engineering College
Chandigarh, India
rintukhanna1@rediffmail.com

Abstract—**Nowadays, distribution systems are widely into use at residential, commercial and industrial level leading to high load demand and thus losses in distribution lines are increasing and voltage of the system is getting affected which is order of main concern for network operators. In order to overcome this problem, firstly various energy storage devices like capacitor, DGs (wind and solar) of optimal rating are incorporated independently into the IEEE 13 bus radial distribution system at suitable locations to enhance the system voltage. Further impact of different loading cases have been seen to check the loading capacity of the network during peak load and contingency condition. To improve the voltage of the system during loading condition, different voltage improvement techniques have been implemented to reduce the system losses and to enhance the voltage profile of the network. A comparison is also shown between base case and different combination of loading cases to prove the significance of unified model approach.**

Keywords—Distribution System; Distribution Static Compensator (DSTATCOM); Photovoltaic (PV); Voltage Regulation.

I. INTRODUCTION

In our society usage of electricity is increasing day by day as many devices and loads require electrical energy. Aging assets and infrastructure causes energy losses hence more energy is required in transmission and distribution networks to meet load demand. To compensate losses and to fulfil energy demand, new power plants are needed to be installed either at source or at consumer end. Preferably consumer end is considered to avoid transmission losses. In addition, concerning about environment, renewable energy sources can be used as local power generators in distribution system[1].

Distribution network is more fluctuating than transmission network because all of the consumer loads (dynamic loads) are connected to distribution network. Therefore, analysis of distribution system is more important rather than transmission network. Distributed network has high R/X ratio and can be alienated in two types – Radial and Ring main distributed networks[2]. Distribution system can be further classified as – balanced and unbalanced distribution system[3]. Proper analysis of distribution system requires authenticate and appropriate modelling of loads and also depends upon its nature[4]. Inaccurate modelling of distribution network leads towards unnecessary huge losses in the system. Some of the algorithm like Voltage Control Algorithm (VCA) gives nature of accuracy or inaccuracy about distribution network. Monte Carlo (MC) is also a technique with VCA which gives acknowledgement of accuracy or inaccuracy of network[5]. For transmission networks having low R/X ratio, many conventional methods like Gauss Seidel, Newton Raphson,[6] and Fast Decoupled methods were used to give best result for load flow analysis whereas for radial distribution network, conventional methods faced difficulty in solving load flow of the system[3]. Some of the methods like backward forward sweep method and Axis Rotation Fast Decoupled Load Flow (ARFDLF) method are required for load flow calculation of distribution network as discussed in papers [2],[7],[8] and these methods require more than one iterative steps which can involve large computation time and memory storage. To minimise time and storage, a method Load Impedance Matrix [LIM] is discussed in paper [9]. This method provides an impedance matrix for distribution network which gives nodal voltage at any stage of the iteration.

Voltage profile improvement and stability of the system can be achieved by many methods which can be categorised in three categories – (a) By using self-regulating devices, (b) By using external energy devices, (c) By reconfiguration. Static VAR Compensator (SVC), On Load Tap Changer (OLTC)[10], Distribution Static Compensator (DSTATCOM)[11] and other FACTS devices work as energy storage device. FACTS devices store reactive power which is helpful in controlling the voltage of the system. These devices store energy when operating at off load period in distribution network and delivers energy to network during on load period so that flat voltage profile of the system can be maintained. In paper[10], SVC and OLTC were connected together into the system for improvement of voltage profile and two stage approach has been used for analysis purpose. Second category of energy devices are DGs like wind, photovoltaic cell, biogas etc. which can be used in distributed network for improvement of voltage regulation. Generally, these DGs are connected at consumer side or where local power is required so that transmission losses of DG can be minimised. Objective of these DGs are to improve several indices like reliability index, 3-ph unbalanced index, voltage stability index etc.[12]. It is noticed that size and placement of DGs or FACTS devices found to be important parameter for improvement of voltage profile and its stability in the system. Some objective functions like network losses, cost generation, and voltage profile are useful parameter to optimise size and location of DGs and FACTS devices in the system. Solution of these objective functions can be obtained by techniques like Particle Swarm Optimisation (PSO)[5], Cuckoo Search Algorithm(CSA)[13], and Genetic Algorithm(GA)[5].

Nowadays loading of the system found to be increased due to increase in population and therefore electricity requirement has also been raised, so distribution system need to be analysed on various levels of loading. In this paper,

978-1-5386-6624-1/18 $31.00 © 2018 IEEE

different types of spot loading (i.e. 150% and 180% loading on base load) in IEEE13 bus system is studied. Voltage profile and power losses are analysed at different buses in distribution network. After analysis of loading, various techniques discussed for the improvement of voltage regulation and minimisation of power losses are implemented on the radial distribution system. DGs like PV and wind along with capacitor and regulator is injected at different nodes and their effect has been observed from voltage profile and reduction in losses in the system. This paper aims to compare the effect of different voltage improvement techniques with DGs, regulators and capacitors in radial distribution network.

In this paper, modelling of distribution network and DGs is discussed in section II. Section III comprises of various case studies and their results. Section IV states about the conclusion.

II. MODELLING OF DISTRIBUTION SYSTEM

Modelling of distribution system components and their proper placement in the network is key factor of electricity used by consumers. Poor modelling affects power and voltage quality (waveforms) which in turn increases energy losses in the system. Size, location, selection of conductors, transformers, and source generator are extremely important for optimizing energy resources in the distribution system.

So, it would be necessary to properly model the components like line, load, shunt capacitor, and transformer in distribution system [14].

A. Line Modelling

Lines are one of the important components in distribution system modelling. It can be overhead or underground line. It becomes necessary to determine series and shunt impedance of the line before doing modelling of any distributed system. J. Carson developed several equations for computing self and mutual impedances of unbalance distributed line in his paper. Physical layout of Carson's concept is shown in Fig. 1.

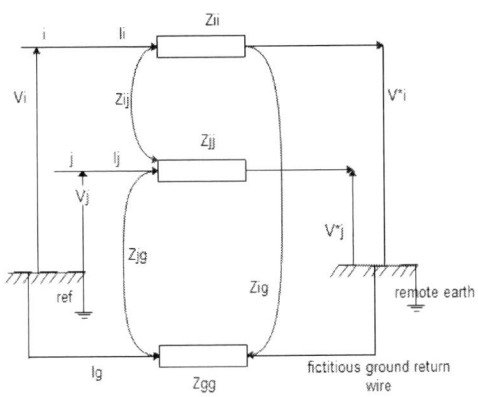

Fig. 1. Physical layout of carson's theory

He represented line with one end of conductor connected to source and other end grounded with earth. It is assumed that earth has uniform surface with constant resistivity therefore any effect from neutral grounding is neglected [14].

If frequency = 60 Hz and earth resistivity = 100 Ohm per meter. Approximated Carson's self (z_{ii}) and mutual (z_{ij}) impedances are-

$$z_{ii} = r_i + 0.09530 + j0.12134\left(\ln\frac{1}{gmr_i} + 7.93402\right)$$

Ohm/mile (1)

$$z_{ij} = 0.09530 + j0.12134\left(\ln\frac{1}{d_{ij}} + 7.93402\right)$$

Ohm/mile (2)

Where,

gmr_i = Geometrical mean radius of conductor

d_{ij} = Geometrical mean distance of conductor

After calculating impedances, impedance matrix $\lfloor z_{abc}\rfloor$ is designed and with help of this matrix, voltage and current matrices of line are given below.

Voltage matrix at node 'n',

$$\left[v\lg_{abc}\right]_n = [a]\cdot\left[v\lg_{abc}\right]_m + [b]\cdot\left[i_{abc}\right]_m \tag{3}$$

Where,

$$[a] = [U] + \frac{1}{2}\left[z_{abc}\right]\cdot\left[y_{abc}\right] \tag{4}$$

$$[b] = \lfloor z_{abc}\rfloor \tag{5}$$

Current matrix-

$$\left[i_{abc}\right]_n = [c]\cdot\left[v\lg_{abc}\right]_m + [d]\cdot\left[i_{abc}\right]_m \tag{6}$$

Where,

$$[c] = \left[y_{abc}\right] + \frac{1}{4}\left[y_{abc}\right]\cdot\left[z_{abc}\right]\cdot\left[y_{abc}\right] \tag{7}$$

$$[d] = [U] + \frac{1}{2}\left[z_{abc}\right]\cdot\left[y_{abc}\right] \tag{8}$$

Voltage required at node m is the function of voltage at node n and current at node m.

Voltage matrix at node 'm',

$$\left[v\ln_{abc}\right]_m = [A_t]\cdot\left[v\ln_{abc}\right]_n - [B_t]\cdot\left[i_{abc}\right]_m \tag{9}$$

Where,

$$[A_t] = [a_t]^{-1} \tag{10}$$

$$[B_t] = [a_t]^{-1}\cdot[b_t] \tag{11}$$

Shunt calculation requires conductance and susceptance. Conductance is neglected when it is compared with susceptance. Susceptance calculation is done in similar way that of inductance in series impedance section.

B. Load Modelling

There are various types of load combination which can be modelled in terms of constant ZIP (impedance, current and power) or different combination of constant ZIP. Current is controlling parameter because it can be controlled in effective and efficient way as compared to any other parameter. Modelling equations in current parameter terms for constant impedance, constant current, and constant power load models are given as-

1). Constant Impedance

$$z_{a,b,c} = \frac{\left|v_{a,b,c}\right|^2}{s_{a,b,c}^*} = \frac{\left|v_{a,b,c}\right|^2}{\left|s_{a,b,c}\right|} \angle \theta_{a,b,c} \tag{12}$$

$$= \left|z_{a,b,c}\right| \angle \theta_{a,b,c} \tag{13}$$

$$IL_{a,b,c} = \frac{v_{a,b,c}}{z_{a,b,c}} = \frac{\left|v_{a,b,c}\right|}{\left|z_{a,b,c}\right|} \angle (\delta_{a,b,c} - \theta_{a,b,c}) \tag{14}$$

$$= \left|IL_{a,b,c}\right| \angle \alpha_{a,b,c} \tag{15}$$

2). Constant Current

$$IL_{a,b,c} = \left|IL_{a,b,c}\right| \angle (\delta_{a,b,c} - \theta_{a,b,c}) \tag{16}$$

$$= \left|IL_{a,b,c}\right| \angle \alpha_{a,b,c} \tag{17}$$

3). Constant Power

$$\left|s_{a,b,c}\right| \angle \theta = P_{a,b,c} + jQ_{a,b,c} \tag{18}$$

$$IL_{a,b,c} = \left(\frac{s_{a,b,c}}{v_{a,b,c}}\right)^* = \frac{\left|s_{a,b,c}\right|}{\left|v_{a,b,c}\right|} \angle (\delta_{a,b,c} - \alpha_{a,b,c}) \tag{19}$$

$$= \left|IL_{a,b,c}\right| \angle \alpha_{a,b,c} \tag{20}$$

And

$$v_{a,b,c} = \left|v_{a,b,c}\right| \angle \delta_{a,b,c} \tag{21}$$

Where,

$v_{a,b,c}$ = voltage of three phase load,

$IL_{a,b,c}$ = line current of three phase load,

$s_{a,b,c}$ = complex power of load,

$P_{a,b,c}$ = active power,

$Q_{a,b,c}$ = reactive power,

$\delta_{a,b,c}$ = line-to-neutral voltage angles,

$\alpha_{a,b,c}$ = difference angle between line to neutral and power factor angle,

$\theta_{a,b,c}$ = power factor angles,

$z_{a,b,c}$ = impedances of load.

All of these load models are included in load file for calculation purpose.

C. Shunt Capacitor Modelling

Shunt capacitor provides reactive power in distribution system to improve voltage levels and to minimize requirement of reactive power demand. Modelling of Wye-Connected Capacitor bank is done by using following equations.

1). Wye- Connected Capacitor

Susceptance (B_c) for each unit is

$$B_c = \frac{k\,\text{var}}{k\,v_{ln}^2 \cdot 1000}\,s \tag{22}$$

And,

$$IC_a = jB_a \cdot v_{an} \tag{23}$$

$$IC_b = jB_b \cdot v_{bn} \tag{24}$$

$$IC_c = jB_c \cdot v_{cn} \tag{25}$$

Where, IC_a, v_{an} are phase current and phase voltage respectively. In our system, Wye connected capacitor bank is considered for load flow solution.

D. Transformer Modelling

Transformers are required to transfer voltage from transmission level to distribution feeder level. These are classified as stepping up, stepping down and phase shifting type transformers. There are different transformers combinations which can be used in radial distribution system as below-

1. D–Grounded Y
2. Open Y–Open D
3. Grounded Y–Grounded Y
4. Ungrounded Y–D
5. D–D

Where,

Y= star, and D= delta.

In our proposed system, type 1 and 3 are used for load flow solution purpose.

Transformer equations of voltage and current matrices are-

$$\left[v\lg_{abc}\right]_m = [A_t] \cdot \left[v\ln_{abc}\right]_n - [B_t] \cdot \left[i_{abc}\right]_m \tag{26}$$

$$\left[i_{abc}\right]_n = [C_t] \cdot \left[v\lg_{abc}\right]_m + [D_t] \cdot \left[i_{abc}\right]_m \tag{27}$$

Where, A_t, B_t, C_t, D_t are coefficients and their value depends on type of transformer connections used for calculation purpose [14].

E. Wind Modelling

Wind energy is a kind of solar energy, when atmospheric temperature and pressure changes of different places, low pressure and high pressure zones are generated. To compensate the differences, air flows from high pressure to low pressure zone and attains kinetic energy.

If mass of air is m and velocity v_w, then kinetic energy is given as-

$$KE = \frac{1}{2} m v_w^2 \tag{28}$$

Power $\left(P_w\right)$ is given as -

$$P_w = \frac{1}{2}\dot{m}v_w^2 \tag{29}$$

Where, mass rate flow $\dot{m} = \dfrac{dm}{dt}$

Now, let's define ρ is air density and A is rotor effective area, then -

$$\frac{dm}{dt} = \rho A\, v_w \tag{30}$$

Finally, equation of wind power is:

$$P_w = \frac{1}{2}\rho A\, v_w^3 \tag{31}$$

P_{tur} is extracted power from wind power P_w as:

$$P_{tur} = \frac{1}{2}\rho A v_w^3 C_p \tag{32}$$

F. PV (Photovoltaic) Modelling

The output power of solar depends on intensity of solar irradiance. Since solar irradiance changes throughout the day time hence output power also changes and its expression is given in the following equation.

$$Po_{pv} = \eta_{pv} \times A_{pv} \times \phi \tag{33}$$

Here, Po_{pv} is the total output power of PV unit in MW or kW.

η_{pv} is the efficiency of PV unit in (%),

A_{pv} is the surface area of PV array in m^2,

ϕ is the solar irradiance in W/m^2.

III. CASE STUDIES AND RESULTS

Three phase load flow analysis is done by interfacing MATLAB with Open DSS on IEEE 13 bus radial distribution network. Different case studies have been demonstrated for system voltage improvement. Base case and 150% and 180% loading cases are considered for examination purpose. Various energy storage devices like capacitor, DG's (solar and wind) and regulator of optimal rating are incorporated in the system at suitable locations for enhancing voltage profile of the distribution network. Their locations in the system are decided by iterative method for optimal results.

A. Case 1: Voltage enhancement with capacitor placement

Capacitors are used for low voltage profile improvement and their modeling is easy in comparison to other mentioned energy storage devices in the system.

Table I shows loss reduction to 11.94% with installation of 2 capacitors of optimal sizes in the system at specified locations. Where rating of capacitor is decided at consumer end and their locations are determined by iterative method.

TABLE I. CAPACITOR PLACEMENT ON BASE CASE

Capacitor	Bus no.	Size (kVAR)	System loss (kW)	Loss reduction (%)
Without cap	-	-	110.5	
One cap	8	600	101	8.59
Two cap	4,8	600,600	97.3	11.94

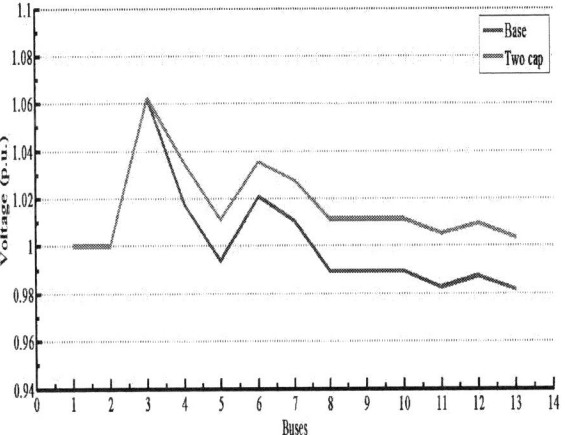

Fig. 2. Comparison of voltage profiles without and with 2 capacitors

Enhancement in voltage profile can be seen from Fig. 2. in comparison to base case.

B. Case 2: Voltage enhancement with DG placement

DG's are also called local power generators and are used for medium voltage profile improvement. Best DG locations and their sizes are given in table II for their different combinations employed in the system.

TABLE II. DIFFERENT TYPES OF DGs PLACEMENT AT OPTIMAL LOCATIONS

DGs	Bus no.	Size (kVA)	System loss (kW)	Loss reduction (%)
Without DG	-	-	110.5	-
One wind	11	1000	88.1	20.27
One wind+ One PV	11 (w)	1000	85.9	22.26
	10 (pv)	500		
Two wind	11	1000	78.5	28.96
	6	1000		
Two wind+ Two PV	6,11 (w)	1000,1000	74.6	32.48
	9,10 (pv)	500,500		
Best DG + 2 Cap	6,11 (w)	1000,1000	53.0	52.03
	9,10 (pv)	500,500		

It is found that percentage reduction in losses is 20.27, 22.26, 28.96, and 32.48 for 1 wind, 1 wind and 1 PV, 2 wind, and 2 wind and 2 PV cases respectively. Therefore, 2wind and 2PV combination is best case among all DG's for enhancing system performance in loss reduction and voltage enhancement terms. When best DG combination is employed with 2 capacitors in the system, huge reduction in losses occurs to 52.03% and enhancement in voltage profile is shown in Fig. 3. in comparison to base case.

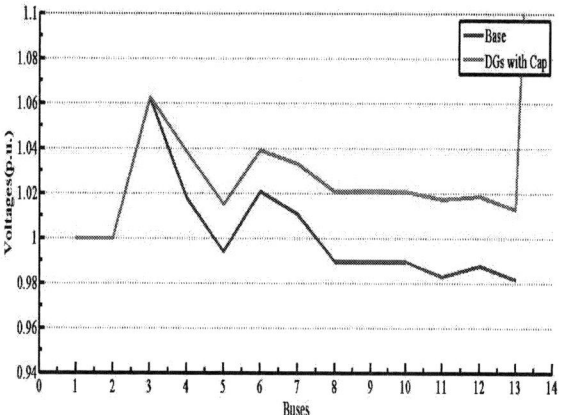

Fig. 3. Comaprison of voltage profiles without and with best DG and 2 capacitor combination

C. Case 3: Impact of loading on system profile

Loading of the system has been increased to check the voltage stability limit of the distribution network. Here, two loading cases are taken for examination purpose, a).150%, and b). 180%.

TABLE III.   COMPARISON OF BASE CASE WITH DYNAMIC LOADING CASES

Case	Loading	System loss (kW)	Min. bus voltage (p.u.) (Bus no.)
Base	-	110.5	0.9819 (13)
Loading	150%	282.2	0.9279 (13)
Loading	180%	436.9	0.8925 (13)

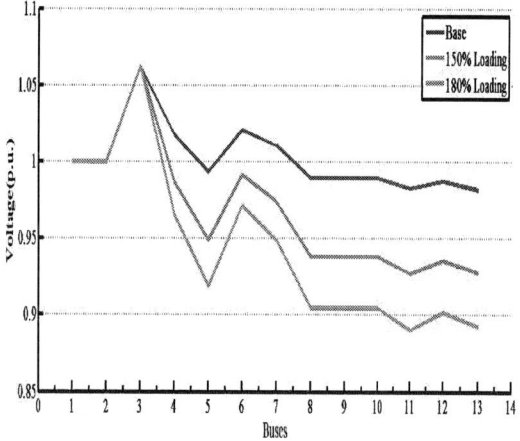

Fig. 4. Voltage profile comparison without and with 150% and 180% loading cases

It has been observed from table III that losses have increased to 282.2 kW and 436.9 kW from 110.5 kW for 150% and 180% loading respectively. Minimum bus voltage of the system has been fallen to 0.9279p.u. and 0.8925p.u. at 13th bus from 0.9819p.u. for 150% and 180% cases respectively. Their voltage profiles are also depicted in Fig. 4. for both loading cases in comparison to base case. Therefore, improvement of loading cases is discussed in case 4 by using different combinations of energy storage devices.

D. Case 4: Comparison of voltage profile improvement with all loading cases

The comparison for all of the configurations in terms of system losses, percentage loss reduction and minimum bus voltage is shown in table IV. System losses observed to be 110.5 kW for base case and minimum bus voltage was 0.9819p.u. at 13th bus. Then with 2 capacitor employment in the system, % loss reduction is 11.94 and minimum bus voltage improves to 1.004p.u. Further, with best DG combination (2 PV and 2 wind) in the network, % loss reduction increases to 32.48 and bus voltage improves to 1.006p.u. On simultaneous placement of DG and capacitor in the system, maximum reduction in losses occurs to 52.03% and minimum bus voltage boosts to 1.013p.u.

Impact of 150% and 180% loading and their improvement is also seen on distribution system. Initially, losses for 150% loading case have been increased to 282.2 kW from 110.5 kW in base case. Therefore, 282.2 kW is taken as basis for % loss reduction in 150% loading case. By employment of only capacitors in the system, % loss reduction is 12.57 and voltage increases to 0.9417p.u. from 0.9279p.u. Then, with only DG, and with simultaneous DG and capacitor placement in the system, % reduction in losses found to be 26.39%, and 41.42% respectively. Also, bus voltage improves to 0.9487p.u. and 0.9636p.u. for both cases respectively. Simultaneous implementation of DG, capacitor, and regulator in the system results into huge reduction in losses to 59.67% and enhances system bus voltage to 1.053p.u. Similarly, with 180% loading case, best results were found with combined incorporation of DG, capacitor, and regulator in the system, maximum reduction in losses occurs to 77.08% and bus voltage of the system boosts to 1.055p.u.

It can be concluded that as more no. of DG's and capacitors are connected in the system, system performance improves in terms of loss reduction and voltage enhancement. It is also found that DG's, capacitors, and regulator combination give best results at optimal rating and suitable location. These devices are rating and location sensitive. Voltage stability limit of the system exceeds when we increase loading beyond 180%.

TABLE IV.   COMPARISON IN LOSS REDUCTION AND VOLTAGE ENHANCEMENT OF DIFFERENT TECHNIQUES

Configuration	System loss (kW)	Loss reduction (%)	Min bus voltage (p.u.) (Bus no.)
Base	110.5	-	0.9819(13)
Only Capacitor (2 cap)	97.3	11.94	1.004(13)
Only DGs(2pv+2wind)	74.6	32.48	1.006(13)

Best DG+ 2 Cap	53.0	52.03	1.013(13)
150% Loading	282.2	-	0.9279(13)
150% Loading+ Capacitors	246.7	12.57	0.9417(13)
150% Loading +Best DGs(2pv+2wind)	207.7	26.39	0.9487(13)
150% Loading+ Capacitors+ Best DGs	165.3	41.42	0.9636(13)
150% Loading+ Capacitors+ Best DGs+ Regulator	113.8	59.67	1.053(13)
180% Loading	436.9	-	0.8925(13)
180% Loading+ Capacitors	385.9	11.67	0.9106(13)
180% Loading +Best DGs(2pv+2wind)	332.4	23.91	0.9126(13)
180% Loading+ Capacitors+ Best DGs	275.3	36.98	0.9318(13)
180% Loading+ Capacitors+ Best DGs +Regulator	100.1	77.08	1.055(13)

IV. CONCLUSION

In this paper, different voltage improvement techniques without and with loading impact have been examined on the proposed distribution system. From comparative study of various integrated schemes, it is found that simultaneous placement of DG's (wind and solar), capacitor and regulator at optimal location in distribution network leads to huge reduction in losses and better system voltage profile for both of the loading cases during peak load condition. By increasing the number of units of energy storage devices, performance of the system enhances. Optimal ratings and suitable location of energy devices also plays an important role for improving system voltage profile. It can also be concluded from results that when energy devices like capacitor, DGs, and regulator are altogether incorporated into the system for 180 % loading case, then voltage stability limit of the system has arrived after which the system may collapse therefore further loading of the network is not increased. The proposed model can be further extended to daily load demand for real time analysis by using various optimization techniques.

REFERENCES

[1] I. Ali and S. Kucuksari, "Voltage regulation of unbalanced distribution network with distributed generators," *2016 North Am. Power Symp.*, pp. 1–6, 2016.

[2] J. M. Rupa and S. Ganesh, "Power Flow Analysis for Radial Distribution System Using Backward / Forward Sweep Method," *Int. J. Electr. Comput. Energ. Electron. Commun. Eng.*, vol. 8, no. 10, pp. 1537–1541, 2014.

[3] F. Yang and Z. Li, "Effects of balanced and unbalanced distribution system modeling on power flow analysis," *2016 IEEE Power Energy Soc. Innov. Smart Grid Technol. Conf.*, pp. 1–5, 2016.

[4] A. Tyagi, A. Verma, and P. R. Bijwe, "Reconfiguration of balanced and unbalanced distribution systems for cost minimization," in *TENCON 2017 - 2017 IEEE Region 10 Conference*, 2017, pp. 2188–2192.

[5] B. B. Zad, J. Lobry, and F. Vallee, "Impacts of the load and line inaccurate models on the voltage control problem of the MV distribution systems," in *2017 52nd International Universities Power Engineering Conference (UPEC)*, 2017, no. Mc, pp. 1–6.

[6] R. G. Wasley and M. A. Shlash, "Newton-Raphson algorithm for 3-phase load flow," *Electr. Eng. Proc. Inst.*, vol. 121, no. 7, pp. 630–638, 1974.

[7] G. A. Setia, G. H. M. Sianipar, and R. T. Paribo, "The performance comparison between fast decoupled and backward-forward sweep in solving distribution systems," in *3rd IEEE Conference on Power Engineering and Renewable Energy, ICPERE 2016*, 2017, pp. 247–251.

[8] H. Liu, S. Cheng, C. Huang, and Y. Hou, "Unbalanced power flow calculation for low-voltage distribution systems including DGs," *2012 IEEE Innov. Smart Grid Technol. - Asia, ISGT Asia 2012*, pp. 1–5, 2012.

[9] U. Ghatak and V. Mukherjee, "A fast and efficient load flow technique for unbalanced distribution system," *Int. J. Electr. Power Energy Syst.*, vol. 84, pp. 99–110, 2017.

[10] N. Daratha, B. Das, and J. Sharma, "Coordination between OLTC and SVC for voltage regulation in unbalanced distribution system distributed generation," *IEEE Trans. Power Syst.*, vol. 29, no. 1, pp. 289–299, 2014.

[11] J. Sanam, S. Ganguly, and A. K. Panda, "Allocation of DSTATCOM and DG in distribution systems to reduce power loss using ESM algorithm," in *1st IEEE International Conference on Power Electronics, Intelligent Control and Energy Systems, ICPEICES 2016*, 2017, pp. 1–5.

[12] Y. Zheng, Z. Y. Dong, K. Meng, H. Yang, M. Lai, and K. P. Wong, "Multi-objective Distributed Wind Generation Planning in an Unbalanced Distribution System," *Csee J. Power Energy Syst.*, vol. 3, no. 2, pp. 186–195, 2017.

[13] T. Wang, M. Meskin, and I. Grinberg, "Enhancement of Voltage Profile in Unbalanced Distribution Systems with Variable Loads," *... Conf. (GreenTech), 2017*.

[14] W. H. Kersting, *Distribution system modelling and analysis*. 2002.

2nd IEEE International Conference on Power Electronics, Intelligent Control and Energy Systems (ICPEICES-2018)

Load Flow Solution for Radial Distribution Network

Om Pathak
Department Of Electrical Engineering
Delhi Technological university
Delhi,India
ompathak95@gmail.com

Prem Prakash
Department Of Electrical Engineering
Delhi Technological University
Delhi,India
ppyadav1974@gmail.com

Abstract—The paper proposes an elementary and straightforward approach for load flow evaluation for radial distribution networks. Two developed matrices – BIBC and BCBV (the bus injection to branch current matrix and the branch current to bus voltage matrix) and an easy matrix multiplication are applied to get load flow results. The developed matrices are formed with help of backward and forward sweep (BFS method) using KCL and KVL (Kirchhoff's current law and Kirchhoff's voltage law respectively). This approach can be enforced to the solution for three-phase and for single-phase configuration of the system. The considered method finds the array of branches for feeders, laterals and sub laterals. The presented algorithm has been tested for IEEE 85 bus system without tie lines. The algorithm converged well for constant power, constant current, constant impedance and composite load modeling.

Keywords — radial distribution system, load-flow, BIBC and BCBV matrix, backward forward sweep method, load models.

I. INTRODUCTION

The Load Flow Analysis is an essential requirement for power system studies as the flow of power from generators to loads should be balanced, reliable and economical. The Load flow information is useful for analysis of the normal operating mode and contingency analysis, outage security appraisal, dispatching and stability [1]-[2][14]. Various approaches for solution of load flow problem like Gauss-Seidel, Newton-Raphson and Fast Decoupled load flow have been studied and extensively used in power system operation, control, and planning. These techniques are not appropriate in case of distribution systems due to their special characteristics such as radial and weak structure, multiphase operation, unbalancing, continuous load fluctuation, extensive branch network and higher number of nodes, higher resistance and reactance values in comparison with transmission networks. Methods developed for the radial distribution systems may be studied into two sections, one is forward-backward sweep process to solve ladder networks and the other is by rectifying the existing methods such as Newton-Raphson.

BFS methodology is the most suitable process for calculations of distribution load flow. The first version of this method was developed using a radial system model and considering only constant complex power nodes. Several corrections are made since it was firstly proposed [3]; these modifications can solve systems with a weakly meshed topology [4], [5], voltage dependent loads systems [7], systems with DG (distributed generation) [15], three-phase systems [6], or three-phase star systems, including neutral grounding [10]. An approach for weakly meshed systems in which the loops are broken and the network is converted into radial was given in [4]; calculation of the breakpoint

currents with the help of multi-port compensation. The approach was later improved in [5]. A further expansion of this solution method included some important features, such as three-phase systems, PV node, distributed generation, distributed loads, or voltage regulators [6]. Reference [11] proposed an approach to solve radial and meshed systems in which only a backward sweep method is required. A thorough comparison and analysis of backward forward sweep methods is presented in [12]. Data for IEEE 85 bus are taken from [13]. The base values are considered as 100 MVA and 12.66 kV.

Conducting load flow evaluation for multiple scenarios ensures that the system is appropriately designed to meet the performance criteria. A perfect drafted system provides calculation for initial capital investments and future operating costs. Load flow analysis is required when:

- Significant plant or load expansion occurs
- Local generation is proposed to be added
- New utility feed installation
- Large motor is added to the system
- New transformer installation etc.

Normally prior to the analysis, we need three phase voltage at sub-station and complex power of all the loads and load models.

A power flow analysis of a feeder can determine the following by phase and total three phase:

- Magnitude of voltage and angles at each buses
- Power flow in each section having units as kW and kVAr, ampere and degrees.
- Line loss in all line section.
- Total feeder input in kW and kVAr.
- Total power loss in feeder.

II. .LOAD MODELING

Loads modeling developed is to be used in the process of iteration of a power flow program where the initial values of node voltage are assumed. Load on a feeder can be modeled as wye-connected or delta connected. The modeling of load can be done as :

- Constant complex power(constant PQ)
- Constant current (constant I)
- Constant impedance (constant Z)
- Composite (ZIP)

978-1-5386-6624-1/18 $31.00 © 2018 IEEE

All models are initially defined by a complex power per phase and an assumed line to neutral voltage (wye load) or an assumed line to line voltage (delta load).

The notation for the complex powers and voltage are as :

$$S = P + j * Q = |S| \angle \theta$$

$$V = |V| \angle \delta$$

i) CONSTANT PQ LOADS :-

The line currents for constant complex power loads are given by

$$IL_{PQ} = \left[\frac{|S|}{|V|} \right] \angle (\delta - \theta) \qquad (1)$$

In this modeling, the value of load voltage will change in each iteration to achieve convergence.

ii) CONSTANT IMPEDANCE (Z) LOADS :-

In this model, the constant load impedance is determined first. The calculation is based upon the complex power and assumed voltages, given as

$$Z = \left[\frac{|V|^2}{|S|} \right] \angle \theta \qquad (2)$$

Load currents as a function of the constant load impedance are given by

$$IL_Z = \left[\frac{|V|}{|Z|} \right] \angle (\delta - \theta) \qquad (3)$$

iii) CONSTANT CURRENT LOADS:-

In this modeling the magnitudes of the currents are calculated according to Equation (1) and are then held constant while the angle of the voltage (delta) changes, resulting in a changing angle of the current such that the load power factor remains constant.

$$IL_I = (S/V)^* = \left[\frac{|S|}{|V|} \right] \angle (\delta - \theta) \qquad (4)$$

Where δ = voltage angle

θ = power factor angle

iv) COMPOSITE LOAD (ZIP LOADS):-

ZIP loads are modeled by assigning a percentage of the total load to the three above load models. The total line current entering the load is the addition of the three load current components. The values of ZIP coefficients can be changed according to system specification.

ZIP COEFFICIENTS:-

- ZP = percentage of constant Z-type load
- IP = percentage of constant I-type load
- SP = percentage of constant PQ-type loads

Here we have taken, ZP=0.30, IP=0.10, SP=0.60

As $\quad ZP + IP + SP = 1$

$$IL_{ZIP} = I_Z + I_I + I_{PQ} \qquad (5)$$

Thus $\quad I_Z = ZP * IL_Z$

$$I_I = IP * IL_I$$

$$I_{PQ} = SP * IL_{PQ}$$

III. BIBC AND BCBV MATRIX.

A. BIBC MATRIX

It explains a direct correlation of the node current injection with branch current. The power injections at every node might be transformed into the equivalent current injection using (1), (3), (4) and (5). The relation of node current with branch current can be easily obtain by applying KCL (Kirchhoff's current law) with backward sweep. Now each the branch current can be shaped as a function of the equivalent current injection (ECI) [9].For example, in fig. 1, branch currents IB1, IB2 and IB3 can be expressed by ECI as

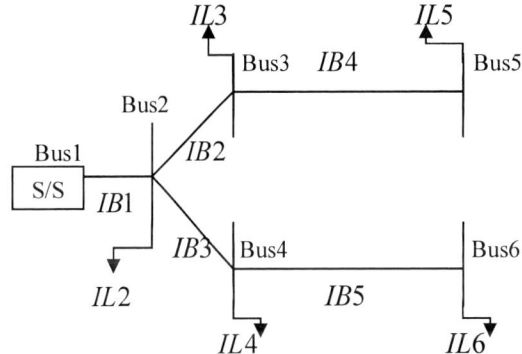

Fig.1. Simple distribution system

$$IB1 = IL2 + IL3 + IL4 + IL5 + IL6$$

$$IB2 = IL3 + IL5 \qquad (6)$$

$$IB3 = IL4 + IL6$$

$$
\begin{bmatrix} IB1 \\ IB2 \\ IB3 \\ IB4 \\ IB5 \end{bmatrix} =
\begin{bmatrix} 1 & 1 & 1 & 1 & 1 \\ 0 & 1 & 0 & 1 & 0 \\ 0 & 0 & 1 & 0 & 1 \\ 0 & 0 & 0 & 1 & 0 \\ 0 & 0 & 0 & 0 & 1 \end{bmatrix}
\begin{bmatrix} IL2 \\ IL3 \\ IL4 \\ IL5 \\ IL6 \end{bmatrix}
$$

$$[IB_i] = [BIBC][IL_{i+1}] \qquad (7)$$

B. BCBV MATRIX

It builds a direct correlation of branch current with node voltage. This relation can be achieved easily by applying

978-1-5386-6624-1/18 $31.00 © 2018 IEEE

KVL (Kirchhoff's voltage law) with forward sweep. It can be calculated using $[BIBC]$ and diagonal impedance matrix $[ZD]$ as given in step-IV of algorithm.

IV. ALGORITHM FOR LOAD FLOW OF DISTRIBUTION NETWORKS

- Step-I: Read the input line and bus data.

- Step-II: Determine the each node current or node current injection matrix using (1), (3), (4), (5), (6) and (7).

- Step-III: Calculate BIBC matrix as shown.

- Step-IV: Calculate the $[BCBV]$ matrix using $[BIBC]$ and diagonal impedance matrix $[ZD]$

$$[BCBV] = [BIBC]^T [ZD]$$

- Step-V: Calculate $[DLF]$ with simple multiplication of $[BCBV]$ and $[BIBC]$

$$[DLF] = [BCBV][BIBC]$$

- Step-VI: Evaluate the branch currents by using $[BIBC]$ matrix and equivalent current injections for their respective load models,

$$[IB] = [BIBC][IL]$$
$$[\nabla V] = [DLF][IL]$$

- Step-VII: Set iteration, $K = 0$

- Step-VIII: Iteration, $K = K + 1$

- Step-IX: Update voltage by using specified load models

$$[\nabla V^{K+1}] = [DLF][IL^K]$$
$$[V^{K+1}] = [V^0] + [\nabla V^{K+1}]$$

- Step-X: If max $[|V(K+1)| - |V(K)|] >$ tolerance, go to step VIII for next iteration.

- Step-XI: Calculate branch currents and losses from final node voltage.

- Step-XII: Display the magnitudes of node voltage and losses

- Step-XIII: Stop.

V. RESULT AND ANALYSIS

The bus voltage of 85 bus system is given in tables (1)-(4) for specific load modeling. Depending upon tolerance value, number of iterations may vary. The convergence tolerance is set at 0.0001. The proposed method took five iterations to give converged solution. The comparison of load models with minimum bus voltage value is given in table (5) that clearly shows that power loss decreases after including the voltage dependent loads than voltage independent loads in the system although the fraction of ZIP constant may vary according to given system.

Voltage profile for various load modeling is given in fig. (2),that clearly shows the effect of voltage dependency in voltage profile of system. The time consuming algorithms, such as the forward/backward substitution of the Jacobian matrix or Y matrix and LU factorization are not required in the proposed method. Thus, the proposed method is robust and efficient. As in the proposed method we require relation matrices (i.e. BIBC and BCBV) for load flow solutions.

Table 1: VOLTAGE FOR CONSTANT POWER LOAD

Bus Number	Voltage Magnitude	Bus Number	Voltage magnitude
1	1	43	0.913163473
2	0.996763383	44	0.909191564
3	0.991961686	45	0.90843817
4	0.985832424	46	0.907999055
5	0.982850117	47	0.907924347
6	0.972131486	48	0.907123738
7	0.965541085	49	0.906934619
8	0.937001954	50	0.906605549
9	0.935621435	51	0.906355812
10	0.932751977	52	0.905304842
11	0.930488718	53	0.904897
12	0.928573183	**54**	**0.904597046**
13	0.927849578	55	0.905067263
14	0.927630399	56	0.906859823
15	0.927499026	57	0.933794547
16	0.996534503	58	0.928748696
17	0.991505099	59	0.928651431
18	0.980813633	60	0.925682417
19	0.979555154	61	0.92489928
20	0.97911792	62	0.924360475
21	0.978593215	63	0.924955006
22	0.978097384	64	0.922138455
23	0.979462936	65	0.922015988
24	0.965245594	66	0.921918013
25	0.932915472	67	0.920685002
26	0.929771661	68	0.918580545
27	0.925645237	69	0.917022113
28	0.923729682	70	0.916619451
29	0.920193906	71	0.916432911
30	0.916954953	72	0.920586885
31	0.915429298	73	0.917534278
32	0.914475823	74	0.917386394
33	0.9137094	75	0.917191917
34	0.910365918	76	0.916323337
35	0.908634867	77	0.922003791
36	0.908572232	78	0.932412342
37	0.92957694	79	0.920499286
38	0.925106482	80	0.927242516
39	0.919898942	81	0.92677273
40	0.914007403	82	0.926724189
41	0.913319643	83	0.925950077
42	0.913226046	84	0.92553474
		85	0.927572884

Table 2: VOLTAGE FOR CONSTANT Z MODEL

Bus Number	Voltage Magnitude	Bus Number	Voltage magnitude
1	1	43	0.925561241
2	0.997155242	44	0.922257358
3	0.992941904	45	0.92162694
4	0.987591283	46	0.921259638
5	0.984994341	47	0.921197154
6	0.975777411	48	0 920540746
7	0.970119388	49	0.920383022
8	0.945651463	50	0.920108606
9	0.944458592	51	0.919900387
10	0.941961471	52	0.919028909
11	0.939994497	53	0.918689992
12	0.938331588	**54**	**0.918440786**
13	0.937702377	55	0.918831337
14	0.937511789	56	0.920320597
15	0.937397569	57	0.942887055
16	0.996927856	58	0.938554881
17	0.992492594	59	0.938470114
18	0.983036289	60	0.935925758
19	0.98182713	61	0.935248714
20	0.98140742	62	0.93478315
21	0.980904004	63	0.935304265
22	0.980428526	64	0.932898749
23	0.981738454	65	0.932793438
24	0.969842768	66	0.932709191
25	0.942186042	67	0.931659108
26	0.939523567	68	0.929867837
27	0.936044034	69	0.928541974
28	0.934431907	70	0.928199567
29	0.93146232	71	0.928040917
30	0.928746704	72	0.931574964
31	0.927468704	73	0.928976452
32	0.926670682	74	0.928850439
33	0.926030827	75	0.928684837
34	0.923240595	76	0.927947782
35	0.921798606	77	0.932782948
36	0.921746151	78	0.941663267
37	0.939353537	79	0.931499867
38	0.93557774	80	0.937177296
39	0.93120965	81	0.936770036
40	0.926274548	82	0.936727898
41	0.925693234	83	0.936057474
42	0.925614117	84	0.935697819
		85	0.937461774

Table 3: VOLTAGE FOR CONSTANT I LOAD MODEL

Bus Number	Voltage Magnitude	Bus Number	Voltage magnitude
1	1	43	0.920096843
2	0.996981778	44	0.91650148
3	0.992508005	45	0.915817279
4	0.986812819	46	0.915418574
5	0.984045379	47	0.915350745
6	0.974165187	48	0.914631755
7	0.968095437	49	0.914460319
8	0.941830056	50	0.914162034
9	0.940553976	51	0.913935684
10	0.937891249	52	0.91298599
11	0.93579254	53	0.912617013
12	0.934017341	**54**	**0.912345677**
13	0.93334615	55	0.912770965
14	0.933142847	56	0.91439249
15	0.933020999	**57**	0.938869411
16	0.996753692	58	0.934221503
17	0.992055296	59	0.934131178
18	0.982050816	60	0.931399053
19	0.980818685	61	0.930674948
20	0.980390808	62	0.930176901
21	0.979877466	63	0.930730773
22	0.979392495	64	0.928143713
23	0.98072836	65	0.928030806
24	0.967810216	66	0.927940481
25	0.938091854	67	0.926809678
26	0.935218053	68	0.924880234
27	0.931454905	69	0.923451781
28	0.929709834	70	0.923082799
29	0.926492345	71	0.922911847
30	0.923547725	72	0.926719353
31	0.922161388	73	0.923920494
32	0.921295385	74	0.923784828
33	0.920600232	75	0.923606484
34	0.917568318	76	0.922811463
35	0.916000144	77	0.928019561
36	0.915943236	78	0.93757457
37	0.935037045	79	0.926638727
38	0.930956502	80	0.932784671
39	0.926221009	81	0.93234963
40	0.920867505	82	0.932304646
41	0.920239452	83	0.931588166
42	0.920153977	84	0.931203778
		85	0.933089497

Table 4: VOLTAGE FOR COMPOSITE LOAD MODEL

Bus Number	Voltage Magnitude	Bus Number	Voltage Magnitude
1	1	43	0.9179885
2	0.99691558	44	0.91427772
3	0.992342407	45	0.91357234
4	0.986515611	46	0.91316127
5	0.983683012	47	0.91309134
6	0.97354827	48	0.91234722
7	0.967320415	49	0.91217037
8	0.940364377	50	0.91186266
9	0.939056653	51	0.91162914
10	0.936331455	52	0.91064837

11	0.934182994	53	0.91026747
12	0.932365344	**54**	**0.90998736**
13	0.931678308	55	0.91042643
14	0.931470205	56	0.91210041
15	0.931345477	57	0.9373289
16	0.996687266	58	0.93256017
17	0.991888579	59	0.93246775
18	0.981676186	60	0.92966364
19	0.980436342	61	0.92892166
20	0.980005722	62	0.92841127
21	0.979489048	63	0.92897738
22	0.979000891	64	0.9263205
23	0.980345463	65	0.92620469
24	0.96703218	66	0.92611204
25	0.936520065	67	0.92495011
26	0.933563965	68	0.92296735
27	0.929689949	69	0.9214993
28	0.927892816	70	0.92112005
29	0.924578067	71	0.92094435
30	0.921543454	72	0.92485742
31	0.920114508	73	0.92198126
32	0.919221744	74	0.92184187
33	0.91850477	75	0.92165862
34	0.915377445	76	0.92084116
35	0.913759363	77	0.92619315
36	0.913700703	78	0.93600787
37	0.933378822	79	0.92477467
38	0.929179342	80	0.93110302
39	0.924299559	81	0.93065747
40	0.918781504	82	0.93061141
41	0.918135243	83	0.92987748
42	0.918047292	84	0.92948372
		85	0.9314156

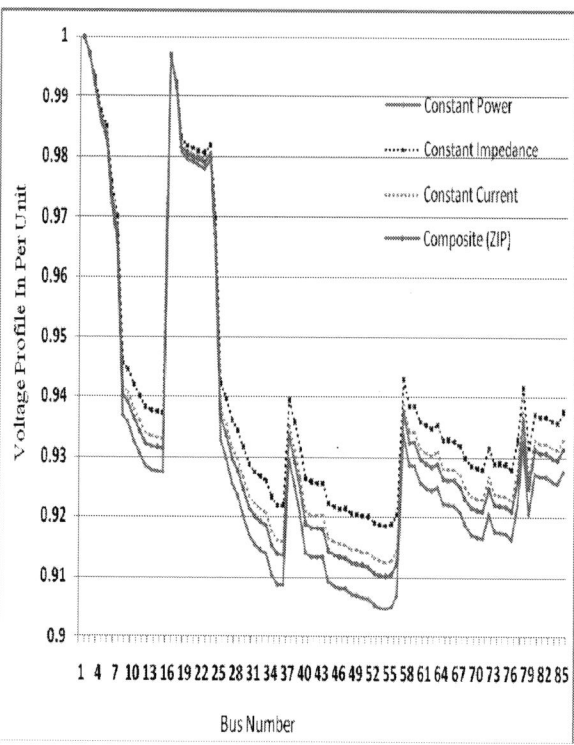

Figure 2: Voltage profile for 85 bus system for specified loads

VI. CONCLUSION

An efficient and decent method for load flow calculation is presented in the paper named forward backward sweep method with the help of BIBC matrix and BCBV matrix. The presented methodology converges for complex power modeling, constant impedance modeling, constant current modeling and composite load modeling. This methodology has been tested for IEEE 85 bus system without tie lines. From the results it is observed that the power loss decreases after including the voltage dependent loads than the voltage independent loads. After this loss evaluation, the various loss minimizing methods and improvement in voltage profile can be applied.

TABLE 5: TOTAL POWER LOSS AND MINIMUM OF BUS VOLTAGE VALUE

Types of Load	Total real Power loss(kW)	Total reactive Power loss (kVAr)	Node No.	Minimum Voltage value (per unit)
Constant Power	234.4670	147.8563	54	0.904597
Constant Z	173.9550	109.9251	54	0.918440
Constant I	199.5502	125.9818	54	0.912345
Composite (ZIP)	209.8584	132.4394	54	0.909987

REFERENCES

[1] IEEE Tutorial Course on 'Distribution Automation'.

[2] IEEE Tutorial Course on 'Power Distribution Planning'.

[3] R. Berg et al, "Mechanized calculation of unbalanced load flow on radial distribution circuits," IEEE Transactions On Power Apparatus and Systems, vol. 86, no.4, pp. 415-421, April 1967.

[4] D. Shirmohammadi et al, "A compensation based power flow method for weakly meshed distribution networks," IEEE Transaction on Power Systems, vol. 3, no. 2, pp. 753-762, May 1988.

[5] Luo and Semlyen, "Efficient load flow for large weakly meshed networks," IEEE Trans. on Power Systems, vol. 5, no. 4, pp. 1309-1316, November 1990.

[6] C.S. Cheng and Shirmohammadi, "A three phase power flow method for real time distribution system analysis," IEEE Trans. on Power Systems, vol. 10, no. 2, pp. 671–679, May 1995.

[7] M.H. Haque, "Load flow solution of distribution systems with voltage dependent load models," Electric Power System Research, vol. 36, pp. 151-156, 1996.

[8] Y. Zhu and Tomsovic, "Adaptive power flow method for distribution systems with dispersed generation," IEEE Transactions on Power Delivery, vol. 17, no. 3, pp. 822–827, July 2002.

[9] Jen-Hao Teng ," A direct approach for distribution system load flow solutions" IEEE transactions on power delivery, vol. 18, no. 3,pp. 882-887,July 2003.

[10] R.M. Ciric et al,, "Power flow in fourwire distribution networks – General approach," IEEE Trans. on Power Systems, vol. 18, no. 4, pp. 1283-1290, November 2003.

[11] A. Augugliaro et al., "A backward sweep method for power flow solution in distribution networks," Electrical Power and Energy System, vol. 32, pp. 271-280, 2010.

[12] Eminoglu and Hocaoglu, "Distribution systems forward/ backward sweep based power flow algorithms: A review and comparison study," *Electric Power Components and Systems*, vol. 37, pp. 91-110, 2009.

[13] D. Das et al. "Simple and efficient method for load flow solution of radial distribution system" Electrical Power and Energy Systems, vol. 17, no. 5, pp. 335-346, 1995.

[14] Kersting, William. (2007). Distribution System Modeling and Analysis. 10.1201/9781420009255.book

[15] Tanveer Husain, Muqueem Khan, Mujtahid Ansari, "Power flow analysis of radial distribution system" IJAREEIE, vol. 5, 2016.

An Exponential Smoothing Based Power Swing Detection Technique for Distance Protection

A.V. Koteswara Rao, K. M. Soni, S. K. Sinha
Department of Electrical & Electronics Engineering,
Amity School of Engineering & Technology,
Amity University Uttar Pradesh, Noida, India
koteswara.v@student.amity.edu, kmsoni@amity.edu,
sksinha6@amity.edu

Ibraheem Nasiruddin
Department of Electrical Engineering,
College of Engineering, Qassim University,
P.O.B 6677 Buraydah ,Saudi Arabia.
ibraheem_2k@yahoomail.com

Abstract— **Power system network is more susceptible to grid disturbances due to its frequent operation in stressed conditions to meet increase in energy demand. This may lead to power swings due to predominant changes in power flow measurements. If such disturbances are not distinguished from fault, then the distance relays may picks up and trip the healthy transmission lines. To mitigate this, it is essential to introduce an algorithm in numerical relay to block its operation during swing. If a fault occurs in a line which is under influence of swing, the relay must sense and trip the concerned breaker instantaneously to disconnect faulty line from the network. This paper introduces an exponential smoothing prediction based power swing detection technique to estimate impedance seen by the distance relay during power swing. The actual and estimated impedance values are examined to segregate a fault from swing in the transmission system. The performance of the model is evaluated statistically by measuring forecasting accuracy and model validation. The proposed method is tested for a 400 KV two-area power system network and compared with available methods. The method can initiate the blocking correctly during power swing and able to identify faults during power swing.**

Keywords— Exponential smoothing; Distance relaying; Power swing; Differential impedance; Forecasting accuracy

I. INTRODUCTION

Present global scenario is evident for the extreme energy demand which often compel the power system to
operate in reduced reliability margin. When the stressed power system is under the influence of large perturbations, the power swings will be generated. Due to significant changes in power flow measurements during swing, the impedance trajectory seen by the distance relay may enter into the relay operation characteristics. This can cause unintended tripping of transmission lines. The protection relays shall not trip the transmission line during swing so that the power system can regain its synchronous state. An unintended initiation of the protection relay during these swings may lead to catastrophic tripping of transmission lines and further blackout of the power system network. To mitigate this issue and to assure intensified power system security, the distance relay is

incorporated with power swing blocking (PSB) function [1]-[3].

The traditional methods are unsuccessful to initiate the blocking function before the impedance trajectory enter into the zone characteristics of the relay, which results into an unintended tripping of the lines. This urges a necessity to develop a new algorithm to track the impedance trajectory continuously and invoke the PSB function much before it enters into the protective zone region.

In [3], a technique to distinguish fault and swing, on the basis of rate-of-change of impedance is projected using blinder scheme. The method is complex for large network. Even though, the superimposed DC current characteristic component method and Swing-Center-Voltage method (SCV) [4], [5] are simple, they initiate the relay operation for all swings irrespective of the impedance perceived by distance relay. These methods fail to revoke the operation of distance relay in many stable swings. An adaptive based algorithm resistant to swing conditions is projected in [6]. A new technique is proposed in [7] to identify fault during swing by eliminating decaying DC offset and harmonics present in fault current. This method is prone to noise present in current signal. An error calculation method based on Taylor Series expansion is proposed in [8] for improving power swing detection and performance of the distance relay. The method requires about one and half cycles for identification. A technique using moving average of modulated current signals is illustrated in [9]. Different PSB techniques are compared in [10]-[11]. Impedance prediction using auto-regression is projected in [12].

In this paper, an exponential smoothing based PSB function is proposed to segregate fault and power swing. The proposed method uses past samples of impedance seen by relay to estimate accurate future samples. The differential impedance is computed using actual and predicted values of the impedance. During power swings, the future values can be predicted using the past values and hence the differential shall be negligible. In case of a fault, the bserved by the distance relay changes abruptly

978-1-5386-6624-1/18 $31.00 © 2018 IEEE

and it cannot be predicted using the past impedance trajectory. Therefore, the differential impedance (ΔZ) will be significant during fault. A 400 KV power system is simulated to verify the proposed algorithm using PSCAD/EMTDC software, and the results are evaluated in MATLAB.

II. PROPOSED METHOD

During power swing, the generators in one area oscillate with respect to the other area generators. This can be represented as two area power system where one area is considered as sending end and the other area as receiving end [13].

$$v_s(t) = V_{sm}\sin(2\pi f_s t + \phi_1) \qquad (1)$$
$$v_r(t) = V_{rm}\sin(2\pi f_r t + \phi_2) \qquad (2)$$

Where, V_{sm} and V_{rm} are amplitudes of voltage signals; ϕ_1, ϕ_2 are initial phase angle of voltage and f_s, f_r are frequencies at sending and receiving ends of power system network. Instantaneous current signal is given in [9] as,

$$i(t) = I_s \sin(2\pi f_s t + \phi_s) + I_r \sin(2\pi f_r t + \phi_r) \qquad (3)$$

Where, I_s and I_r are amplitudes; ϕ_s and ϕ_r are initial phase angles of current signal at sending and receiving end respectively. Equations (1) and (3) are represented in discrete form for sampling time T_s at k^{th} sample as,

$$v_s(k) = V_{sm}\sin(2\pi f_s(kT_s) + \phi_1) \qquad (4)$$
$$i(k) = I_s \sin(2\pi f_s(kT_s) + \phi_s) + I_r \sin(2\pi f_r(kT_s) + \phi_r) \qquad (5)$$

From equations (4) and (5), the impedance measured by the relay at sending end for k^{th} sample is calculated as,

$$Z_k = \frac{v_s(k)}{i(k)} \qquad (6)$$

A. Exponential smoothing prediction technique

Exponential smoothing is a quantitative forecasting technique suitable to estimate periodical signals based on past observations i.e. 'priori'. It is a particular case of weighted moving average (WMA) scheme, in which the weight is assigned to the most recent past observation. This method is very simple, easy and fast to compute as it considers most recent past observations. This method is highly accurate with acute Mean Square Error (MSE) in forecasting. In the proposed method, an exponential smoothing forecasting technique [13] is used to predict relay impedance characteristics. The future values of the impedance are estimated using the past trajectory. However, in case of a fault, the apparent impedance moves at a faster rate and hence not viable to estimate the future values precisely. The difference between estimated and actual samples is used to segregate fault from power swing.

The expression for the exponential smoothing weighted moving average model [14], [15] for n known previous observations is:

$$\hat{Z}_{t+1} = \beta[Z_t + (1-\beta)Z_{t-1} + \cdots + (1-\beta)^n Z_{t-n}] \qquad (7)$$

where Z_t, $\hat{Z}_t$ and $\hat{Z}_{t+1}$ are actual value of impedance at time t, estimated values of impedance at time t and $t+1$ respectively. β is a smoothing constant of range [0-1] and n indicates number of previous observations. The next impedance value can also be predicted using last observation and its corresponding forecast as follows,

$$\hat{Z}_{t+1} = \hat{Z}_t + \beta(Z_t - \hat{Z}_t) \qquad (8)$$

Re-arrange the equation (8) in terms of error correction form, we obtain,

$$\hat{Z}_{t+1} = \hat{Z}_t + \beta(e_t) \qquad (9)$$

Where, e_t is forecast error at time t. Eqn. (9) can also be represented in terms of an interpolation between previous observation and previous forecast as below,

$$\hat{Z}_{t+1} = \beta Z_t + (1-\beta)(\hat{Z}_t) \qquad (10)$$

It is evident from equation (9) that in this method, the forecast at period t is attuned the forecast error e_t by β times to obtain new forecast for a period $t+1$. The Performance of this method depends on choosing smoothing constant, β. The higher value of β is opted for greater responsiveness and lower value for greater stability. To obtain optimum performance and better estimation, a suitable value of β shall be chosen.

The signal model in (10) considers smoothing constant (β), actual impedance Z_t and its corresponding forecast value $\hat{Z}_t$, to predict relay impedance characteristics using MATLAB. The performance analysis of the proposed method is determined by measuring forecast accuracy using following statistical metrics.

1) Sum of forecasting error (SFE)

The forecast accuracy of the proposed model is determined by finding SFE. The test statistic for SFE is as follows:

$$SFE = \sum_{t=1}^{n} |Z_t - \hat{Z}_t| \qquad (11)$$

2) Mean absolute deviation (MAD)

It is a test to measure forecasting accuracy by considering absolute values of the forecast errors to avoid cancellation of positive and negative values while calculating mean. A low value of MAD indicates better forecasting accuracy. The test statistic for MAD is as follows:

$$MAD = \frac{1}{n}\sum_{t=1}^{n} |Z_t - \hat{Z}_t| \qquad (12)$$

978-1-5386-6624-1/18 $31.00 © 2018 IEEE

3) Mean square error (MSE)

It is the most significant test to find forecasting accuracy as it converts the negative signed observations to positive by squaring the forecast errors. A low value of MSE indicates better forecasting accuracy. The test statistic is as follows:

$$MSE = \frac{1}{n}\sum_{t=1}^{n}(Z_t - \hat{Z}_t)^2 \qquad (13)$$

B. Proposed power swing technique

The estimated impedance ($\hat{Z}_t$) is compared with the characteristic impedance of zone 3 (Z_o) as shown in the flow diagram Fig.1. If $\hat{Z}$ is greater than Z_o, then it infers normal operating state of system. Else, it indicates a disturbance such as a fault or swing in the system.

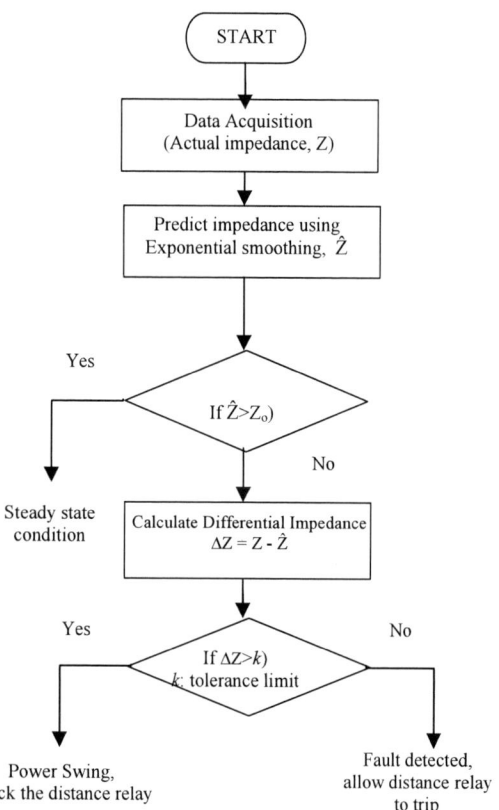

Fig. 1. Flow chart of proposed algorithm

For secure operation, the distance relay shall not trigger trip command. Contrary, for a fault condition, the relay shall initiate trip command. This can be achieved by implementing a concept of 'Differential impedance'.

Differential impedance (ΔZ), used to distinguish swing and fault conditions, is computed as follows,

$$\Delta Z = Z - \hat{Z} \qquad (14)$$

In case of power swing, differential impedance will be negligible as the proposed forecasting technique can estimate the future values using the past values precisely. However, for faults, a correct estimation of impedance could not be achieved and hence its value is very significant. Segregation of fault and swing can be done by choosing a tolerance limit (k) for ΔZ as follows,

$$\begin{aligned} &\text{If} \quad \Delta Z < k \qquad \text{Power swing is detected}\\ &\text{Else } \Delta Z > k \qquad \text{Fault is detected} \end{aligned}$$

$$(15)$$

The tolerance limit (k) is selected considering largest value of ΔZ recorded for fast power swings of slip frequency 5 Hz. For a power swing, the PSB function is invoked to block the operation of relay for tripping. Contrary, for faults, the PSB is revoked to unblock the operation of relay for tripping.

III. RESULTS AND DISCUSSION

A 400 KV power system [13] is simulated to verify the proposed algorithm using PSCAD/EMTDC software. A 3-phase fault is simulated in line 2 of the system in Fig. 2. It causes power oscillations in line 1. The severity of power swing depends on fault clearance time. The actual impedance observed by relay R1 is obtained from the above simulation network. It is used to predict the $\hat{Z}$ for the next time period using exponential smoothing technique. The Z_o is considered as zone-3 characteristics of the distance relay. Tolerance limit (k) is selected as 2 Ω considering ΔZ at slip frequency of 5 Hz.

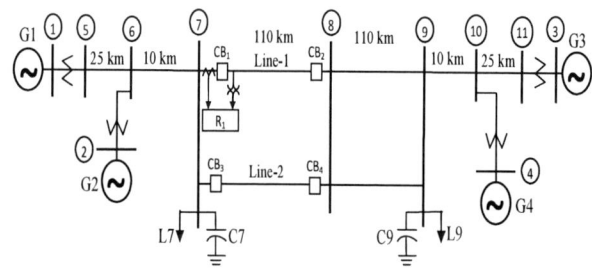

Fig. 2. 400 KV Two-area system

A. Choosing of an optimal value of smoothing constant

The impedance seen by relay R1 is recorded from simulation of system. It is used to estimate future values of impedance ($\hat{Z}$) using exponential smoothing technique. This estimation is carried out for various values of smoothing constant. SFE, MAD and MSE are calculated to measure forecasting accuracy. The results are mentioned in Table I.

978-1-5386-6624-1/18 $31.00 © 2018 IEEE

TABLE I
RESULTS OF FORECASTING ACCURACY

β value	SFE	MAD	MSE
0.0	1.6E+05	5.5E+01	3.6E+03
0.05	2.20E+03	0.7342	0.7955
0.09	4.45E+02	0.1484	0.0209
0.1	3.65E+02	0.1216	0.0040
0.105	**3.58E+02**	**0.1192**	**0.0021**
0.11	3.64E+02	0.1212	0.0037
0.2	9.93E+02	0.3310	0.1547
0.3	1.34E+03	0.4455	0.2879
0.4	1.51E+03	0.5033	0.3702
0.5	1.61E+03	0.5382	0.4246
0.6	1.68E+03	0.5614	0.4631
0.7	1.73E+03	0.5781	0.4917
0.8	1.77E+03	0.5906	0.5137
0.9	1.80E+03	0.6003	0.5312
0.95	1.81E+03	0.6044	0.5387
1.0	1.82E+03	0.6081	0.5455

In the proposed method, an optimal value of β is obtained from the plot between smoothing constant (β) and MSE [16]. It is evident from Fig. 3 that β=0.105 gives an optimal solution to predict impedance precisely with least MSE.

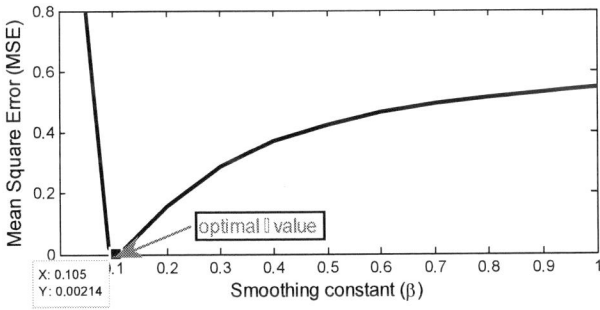

Fig. 3. Mean Square Error (MSE) for various values of smoothing constant (β).

B. Results for stable-swing case

The results of the proposed algorithm are compared with traditional blinder scheme for performance evaluation. In blinder scheme, for early detection of swing before entering into relay protective zones, two blinders are positioned outside zone 3 reach in the impedance plane. Outer blinder is placed with 20% margin w.r.t inner blinder. A reference cross-over time of two cycles (33.3 ms for 60 Hz power system) is taken to identify fault and power swing. From the impedance characteristics, the cross-over time is recorded as 46 ms which is more than reference. This confirms presence of power swing in line 1. Accordingly, the blinder method blocks distance relay.

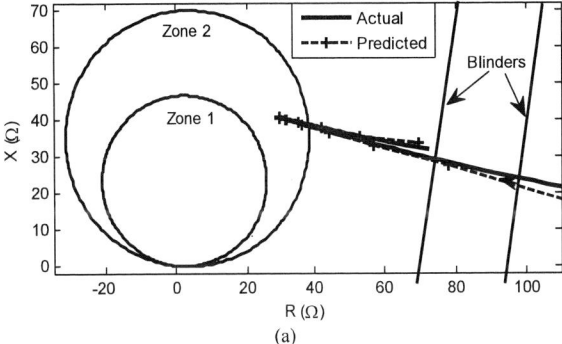

Fig. 4. Impedance trajectory during stable swing

In the proposed method, from results, maximum differential impedance is recorded as 0.6 Ω. This value is within the tolerance limit (k). Therefore, stable power swing is detected in line 1. Accordingly, the method enables the PSB. Both traditional blinder and proposed methods have succeeded to detect stable swings.

TABLE II.
RESULTS OF SFE, MAD AND MSE FOR STABLE SWING CASE

Type of method used	SFE	MAD	MSE
Auto-regression method	4.07e+03	1.358	2.746
Exponential smoothing (proposed method)	**3.58e+02**	**0.119**	**0.002**

The statistical metrics SFE, MAD and MSE of the proposed method are compared with auto-regression technique, as shown in Table II. The test results are better in the proposed method .The power system model is statistically validated using F-test. In the instant case, 3000 observations are considered for prediction. Hence, F-test is considered as appropriate [14].

1) Model validation:
The steps of validation of power system model using F-test are as follows:

a) Step1: Formulation of null hypothesis

Null Hypothesis (H_o): The prediction of impedance using exponential smoothing technique is inaccurate and the proposed model is invalid.

Alternate Hypothesis (H_a): *Alternate Hypothesis (Hₐ):* The prediction of impedance using exponential smoothing technique is accurate and the proposed model is valid.

b) Step 2: Choose 99% confidence level i.e. α= 0.99

c) Step 3: Selection of test static - F test
- The total no. of observations considered, n=3000
- The degree of freedom numerator, df_n =1
- The degree of freedom denominator, $df_d = n-2$ =2998
- The critical tabulated value as per F distribution at α, df_n, df_d is found of order of maximum in double digits from the statistic tables.

d) Step 4: Calculated value of F_{cal}

$$F_{cal} = \frac{MST}{MSE} - 1 = \frac{SST/df_n}{SSE/df_d} - 1 = \mathbf{5.8e+05}$$

Where MST, *MSE* are mean squares of total treatments or observations and errors respectively, which are derived from Table II [16].

e) Step5: Compare F_{cal} with F_{tab}

In this case, F_{cal}>>>F_{tab} . Hence, from the test of hypothesis, it is inferred that there is enough evidence against null hypothesis.

Therefore, Alternate hypothesis is accepted. i.e. the prediction of impedance using exponential smoothing technique is accurate and the proposed model is valid.

C. Results for an unstable-swing case

The impedance plot of unstable swing case is given Fig. 5. In blinder method, the cross-over time is recorded as 19 ms. which is less than reference value. Therefore, the blinder scheme is ineffective to identify swing. In the proposed method, it is obtained from results the estimated impedance plot entered into the zone 3 region with tolerable differential impedance of 1.1 Ω. Therefore, the method succeeded to detect unstable swing.

The statistical metrics MSE, MAD and SFE of the proposed method are compared with auto-regression technique, as shown in Table III.

TABLE III.
RESULTS OF SFE, MAD AND MSE FOR UNSTABLE SWING CASE

Type of method used	SFE	MAD	MSE
Autoregression method	5.48e+04	111.881	5.07e+04
Exponential smoothing	**3.098e+04**	**63.217**	**1.78 e+02**

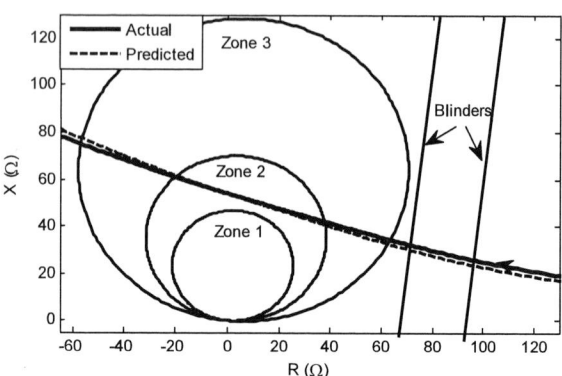

Fig. 5. Impedance trajectory of unstable swing case

D. Results for symmetrical fault during swing

It is difficult to identify a fault due to enormous variations in power flow measurements [17]. For secure operation, it is essential to disable PSB for a fault during swing. The conventional blinder method failed to identify the fault due to less cross-over time than reference value.

In the proposed method, this case is tested by simulating a 3-phase symmetrical fault in line1 (which is under influence of swing) at 6.01 s as shown in Fig. 6. The value of ΔZ is significant at the instant of fault. Accordingly, the distance relay tripped the faulty portion of the network.

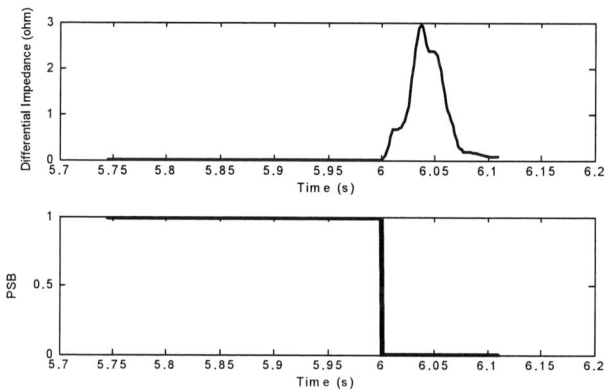

Fig. 6. Symmetrical fault occurred during power swing

E. Results for different swing frequencies

The proposed method is evaluated for various slip frequencies including a fault case during power swing. The results are compared with blinders method. It is clear from Table IV that traditional blinder scheme failed for slip frequencies more than 1.5 Hz.

TABLE III.
RESULTS FOR DIFFERENT SWING FREQUENCIES

Swing Frequency (Hz)	Conventional blinders method (Timer setting = 33 ms)		Proposed method
	Time (ms)	Decision	
0.28	182	Block	Block
1.0	90.0	Block	Block
1.5	30.3	-	Block
2	22.5	-	Block
2.5	15.4	-	Block
3.0	13.0	-	Block
3.5	10.0	-	Block
Fault during power swing	Can not be detected		Trip

The proposed method identified power swing in all cases and blocked the distance relay correctly. The proposed method is also successful to identify fault during power swing.

IV. CONCLUSION

In this paper, an algorithm for PSB function is developed using exponential smoothing forecasting method. The proposed method is used to predict impedance trajectory of relay. The segregation of swing and fault is carried out by measuring differential impedance. The model is validated statistically. The performance of the proposed PSB algorithm is tested for various slip frequencies and found to be superior to traditional methods.

REFERENCES

[1] D. Tziouvaras, "Relay performance during major system disturbances," 2006. Available: http://www.selinc.com.

[2] IEEE Power System Relaying Committee of the IEEE Power Eng. Soc., "Power swing and out-of-step considerations on transmission line," *Rep. PSRC WG D6., Jul. 2005.* Available: http://www.pes-psrc.org

[3] J. P. E. Mooney and N. Fischer, "Application guidelines for power swing detection on transmission systems," *Power system conference, PS-2006,* pp. 159–168, 2006

[4] G. Benmouyal, D. Hou and D. Tziouvaras, "Zero-setting power swing blocking protection," *The 31st Annual Western Protective Relay Conference, Washington,* 2004.

[5] A. P. Apostolov, D. Tholomier and S. H. Richards, "Superimposed components based sub-cycle protection of transmission lines," *Power Systems Conference and Exposition, IEEE PES,* vol.1, pp. 592 – 597, 2004.

[6] X. Lin, Z. Li, S. Ke and Y. Gao, "Theoretical fundamentals and implementation of novel self-adaptive distance protection resistant to power swings," *IEEE Trans. Power Del.,* vol.25, no. 3, pp. 1372-1383, July 2010.

[7] K. Seethalekshmi, S. N. Singh and S. C. Srivastava, "SVM based power swing identification scheme for distance relays," *IEEE Power and Energy Soc. Gen. Meet.,* pp. 1 – 8, July 2010.

[8] Ibrahim Gursu, Tekdemir, and Bora Alboyaci, , "A Novel Approach for Improvement of Power Swing Blocking and Deblocking Functions in Distance Relays", *IEEE Trans. Power Del.,* available in early access, DOI 10.1109/TPWRD.2016.2600638, 2016

[9] J. G. Raoand A. K. Pradhan, "Power-swing detection using moving window averaging of current signals," *IEEE Trans. Power Del.,* vol.30, no. 1, pp. 368-376, Feb. 2015.

[10] H. Khoradshadi-Zadeh, "Evaluation and performance comparison of power swing detection algorithms", *IEEE Power Eng. Soc. Gen. Meet.,* vol. 2, pp. 1842 – 1848, June 2005.

[11] A. Esmaeilian, A. Ghaderi, M. Tasdighi and A. Rouhani, "Evaluation and performance comparison of power swing detection algorithms in presence of series compensation on transmission lines," *Int. Conf. on Environment and Electrical Engg., IEEE,* pp. 1-4, May 2011.

[12] A.V. KoteswaraRao and Aziz Ahmad, "Power swing blocking (PSB) function for distance relay using prediction technique," *Springer: International Journal of System Assurance Engineering and Management, DOI 10.1007/s13198-016-0434-2,* Mar 2016.

[13] KundurP(1994) Power system stability and control. McGraw Hill, New York,NY,USA.

[14] Anderson Sweeney Williams (2008), "Statistics for Business and Economics",. Thomson South-Western, USA.

[15] Rob Hyndman, Anne Koehler, Keith Ord and Ralph Snyder (2008)." Forecasting with Simple Exponential Smoothing", Springer Series in Statistics, *DOI 10.1007/978-3-540-71918-2,* 2008

[16] Eva Ostertagova and Oskar Osterag, "Forecasting using Exponential Smoothing Method," *Journal of ActaElectrotechnia et Informatica, Vol. 12, No.39, pp. 62-66,DOI: 10.2478/v10198-012-0034-2,* Sep 2012.

[17] J. G. Rao and A. K. Pradhan, "Improved Transverse Current Differential Protection Resistant to Power Swing," Springer, INAE Letter (2016

Three inputs and Two Outputs Boost DC-DC Converter for DC Microgrid Applications

Kumaravel S.,
Department of Electrical Engineering
National Institute of Technology
Calicut, Kerala
kumaravel_s@nitc.ac.in

Arun Sarathi
Department of EEE
Vedavyasa Institute of Technology,
Calicut, Kerala
arunsarathi1991@gmail.com

Gangavarapu Guru Kumar
Department of Electrical Engineering
National Institute of Technology
Calicut, Kerala
gurukumar537@gmail.com

Sivaprasad A.
Department of Electrical Engineering
SRM Institute of Science and
Technology
Chennai, Tamilnadu
sivanuday@gmail.com

Abstract- **This paper introduces a new type of unidirectional, non-isolated, three inputs and two outputs boost DC-DC converter for DC microgrid applications. The proposed converter integrates three input sources such as solar panel, battery, and fuel cell, etc., and produces two outputs of different voltage levels using one inductor, two capacitors four MOSFETs and four diodes. The steady state output equation of the proposed converter is derived using the analytical waveforms obtained from different modes of operation. Small signal model of the converter is derived and the bode plot of significant transfer functions are also plotted. The simulation study is carried out in MATLAB/Simulink environment and the output waveforms are obtained. A laboratory prototype of the proposed converter is fabricated and tested successfully. The dSPACE 1104 real time digital controller is used to implement the control algorithm for the converter. Finally, the performance evaluation of the proposed converter is carried out and also compared with the existing power converter topologies reported in the literature.**

Keywords—DC-DC converter, multi-input multi-output, microgrid, small signal modelling

I. INTRODUCTION

Renewable energy sources are highly used to meet the local electricity demand in the form of distributed generators. Renewable energy sources such as solar and wind energy are significantly used for the electricity generation in the past two decades. A power electronic converter is always required to match the current-voltage characteristics between the renewable energy source and load. But these sources are small in ratings and the output voltages also vary due to the input sources. Usually, a boost converter is used between the photovoltaic panel and the DC micro-grid to match the dynamics present in solar-PV. The boost converters are mainly classified into isolated and non-isolated types. The isolated converter is also called galvanic isolated converter. The galvanic isolated converter includes a transformer which provides electrical isolation between the input and output side. However, the use of transformer in a DC-DC converter makes the overall system complex, bulky and results in high cost.

The isolated converters are classified into single input - single output, multiple inputs - single output, single input - multiple outputs and multiple inputs - multiple outputs converters. The topologies in [1-4] give an idea of combining

different DC sources in magnetic form by adding up the magnetic core of the coupled transformer. The concept of phase-shifted pulse width modulation (PWM) is discussed for the control of converter output voltage.

To overcome the demerits of isolated type converters, the idea of non-isolated type converters have been proposed. Due to the absence of the transformer to provide isolation, non-isolated type converters are simple in structure, compact and highly efficient. The nonisolated converters are classified into single input single output converter, multiple - inputs, single - output converter and multiple - inputs, multiple outputs converters. This converter doesn't have a transformer for electrical isolation, and each switching signals needs a delay for the conduction of power switches [5-12].

A new type of converter is proposed in this paper to integrate three input sources and two load resistances. The steady state analytical waveform for different modes of operations is obtained and based on the waveforms, the output voltage of the converter is derived. The dynamic analysis of the proposed converter is also carried out based on the small signal model. A detail simulation study and experimental validation of the converter also carried out. Different modes of the converter are discussed in the following section.

II. CONVERTER STRUCTURE AND MODES OF OPERATION

The diagram of proposed three inputs and two outputs boost DC-DC converter circuit is shown in Fig.1. The converter has input voltages of V_{in1}, V_{in2}, and V_{in3} and the voltage values are considered as $V_{in1} < V_{in2} < V_{in3}$. Similarly, two output voltages of V_{01} and V_{02} are considered. The value of output voltage V_{01} is greater than V_{02}. The proposed topology has four MOSFET switches S_1, S_2, S_3 and S_4 and also three diodes D_1, D_2 and D_3. One inductor (L) and two output capacitors (C_1, C_2) are used in the converter. The capacitors (C_1, C_2) supplies power to the load resistors (R_1, R_2).

The energy available with the input sources is transferred to the inductor L when the switch S_1 is turned on. The input source which has high voltage value i.e., V_{in3} first charges the inductor (S_1 and S_3 are on, S_2 and S_4 are off). Next, V_{in2} charges the inductor (S_1 and S_2 are on, S_3 and S_4

are off). Finally, V_{in1} charges the inductor (S_1 is on, S_2, S_3 and S_4 are off). When the switch S_4 is on and S_1 is off, the

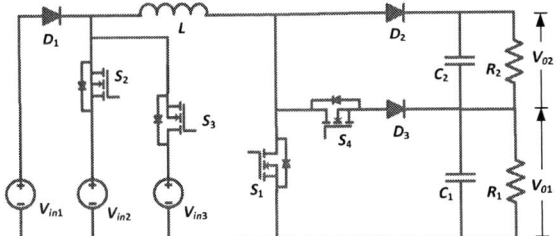

Fig. 1.Proposed converter

inductor delivers energy to the output capacitors. Totally five modes of operation exist in the proposed converter based on the on and off state of four switches present in the converter.

A.   Switching State 1:

In this switching state, S_1 and S_3 are turned on. Remaining switches and all diodes are in off state. The equivalent circuit of the proposed converter in this state is shown in Fig.3.*a* The voltage across the inductor is V_{in3}. The capacitors C_1 and C_2 discharges the energy to the load resistors R_1 and R_2 respectively.

B.   Switching State 2:

In this switching state, S_1 and S_2 are turned on while all other switches and diodes are in off condition. The equivalent circuit of the converter in this state is shown in Fig.3.b. The inductor is charged by a volatge of V_{in2} The inductor current further increases at a reduced slop of V_{in2}/L. The capacitors C_1 and C_2 discharges the energy to the load resistors R_1 and R_2 respectively.

C.   Switching State 3:

The power switch S_1 and diode D_1 are conducting in this switching state, whereas all the remaining power switches and diodes are in off state. The equivalent circuit of the converter for the given switching state is illustrated in Fig. 3.c. Here, the inductor is charged by the source voltage V_{in1}, while capacitors C_1 and C_2 discharges the energy to the load resistors R_1 and R_2 respectively.

D.   Switching State 4

In this switching state, the switch S_1 and diodes D_1 and D_3 are in conduction, while all other switches and diodes are in off state. The equivalent circuit of the converter in this state is shown in Fig.3.d. Here, the source V_{in1} simultaneously chariging the inductor and supplies the load R_1. Hence, during this interval voltage across inductor is $V_{in1} - V_{o1}$, which is of negative polarity (since $V_{in1} < V_{o1}$). Hence the current through the inductor gradually reduced with a slop of $(V_{in1} - V_{o1})/L$.

E.   Switching State 5

In this switching state, all the switching devices are in off state except diodes D_1 and D_2. The equivalent circuit of the converter in this state is shown in Fig.3.e. In this interval, the source V_{in1} simultaneously charging the inductor and supplies the entire load. Hence the output voltage is V_{oT}. So, the voltage across inductor is $V_{in1} - V_{oT}$, which is more negative

compared to the previous state (because $V_{oT} > V_{o1}$) . Hence the current through the inductor reduced further with a slop of $(V_{in1} - V_{oT})/L$.

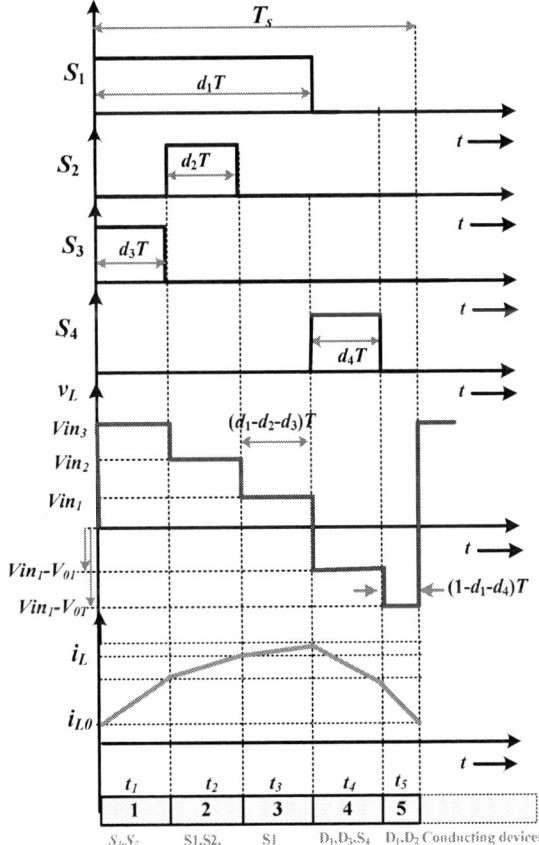

Fig. 2.Stady state waveform of proposed converter

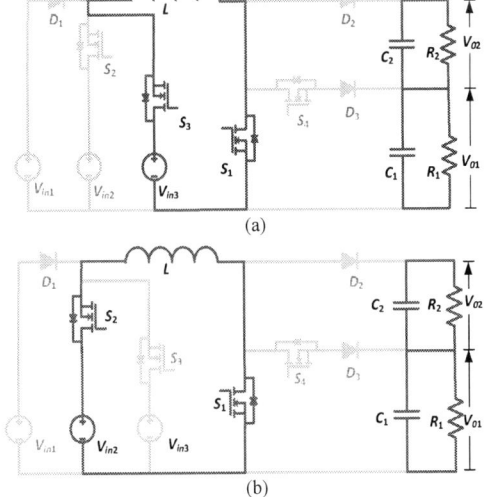

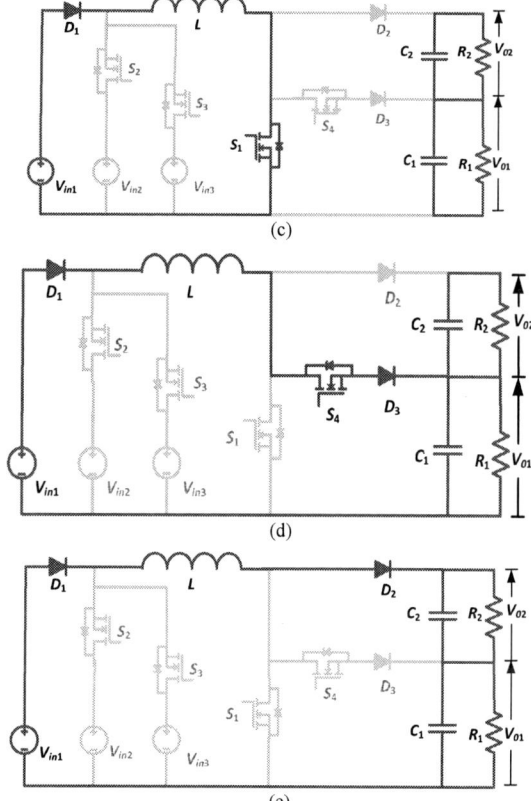

(c)

(d)

(e)

Fig.3 Equivalent circuit of state, (a) switching state 1, (b) switching state 2, (c) switching state 3,(d) switching state 4, (e) switching state 5

For a DC-DC converter, under steady state condition, the average value of the inductor voltage is zero according to the volt-second balance principle which can be expressed as given below.

$$V_{in3}d_3 + V_{in2}d_2 + V_{in1}(d_1 - d_2 - d_3) = (V_{o1} - V_{in1})d_4 + (V_{oT} - V_{in1})(1 - d_4 - d_1) \quad (1)$$

From the above equation, the output voltage (V_{o1}) of the converter is :

$$V_{o1} = \frac{V_{in1}(1 - d_2 - d_3) + V_{in2}d_2 + V_{in3}d_3 - V_{oT}(1 - d_1 - d_4)}{d_1 - d_4} \quad (2)$$

From the above equation, the output voltage (V_{o2}) of the converter is obtained as,

$$V_{o2} = \frac{V_{in1}(1 - d_2 - d_3) + V_{in2}d_2 + V_{in3}d_3 - V_{o1}(1 - d_1)}{1 - d_1 - d_4} \quad (3)$$

III. System Modelling and Design

Basically, the state space and transfer function are the two ways of representing the dynamic model of any converter which consists of linear circuit elements like R, L, C and also non-linear circuit elements, called semiconductor switches. But the circuit derived for each switching operation is a linear circuit. Hence it is viable to develop a dynamic and output equation for each switching operation. Since the energy storage elements (L/C) present in the converter are dynamic elements, a dynamic variable is linked to each element. Hence, the inductor current (i_L) and the capacitor voltage (v_c) are considered as the dynamic variables in this converter model The concept of small

signal model is very important to develop an effective dynamic model for an optimized converter. Therefore dynamic behaviour of this proposed converter can be described in a matrix form to perform its frequency domain based analysis based on the derived transfer functions. The dynamic and output equations of the converter for all three modes of operations are expressed below.

State 1:
By applying KVL to the loop containing source V_{in3}, switches S_1 and S_3 and the inductor L, the voltage across the inductor L is

$$L\frac{di_L}{dt} = V_{in3} \quad (4)$$

By applying KCL at the load side,

$$C_1\frac{dV_{o1}}{dt} = -\frac{V_{o1}}{R_1} \; ; C_2\frac{dV_{o2}}{dt} = -\frac{V_{o2}}{R_2} \quad (5)$$

State 2:
By applying KVL to the loop containing source V_{in2}, switches S_1 and S_2 and the inductor L, the voltage across the inductor L is

$$L\frac{di_L}{dt} = V_{in2} \quad (6)$$

By applying KCL at the load side,

$$C_1\frac{dV_{o1}}{dt} = -\frac{V_{o1}}{R_1} \; ; C_2\frac{dV_{o2}}{dt} = -\frac{V_{o2}}{R_2} \quad (7)$$

State 3:
By applying KVL to the loop containing source V_{in1}, switch S_1 and the inductor L, the voltage across the inductor L is

$$L\frac{di_L}{dt} = V_{in1} \quad (8)$$

By applying KCL at the load side,

$$C_1\frac{dV_{o1}}{dt} = -\frac{V_{o1}}{R_1} \; ; C_2\frac{dV_{o2}}{dt} = -\frac{V_{o2}}{R_2} \quad (9)$$

State 4:
By applying KVL to the loop containing source V_{in1}, switch S_4 and the inductor L, the voltage across the inductor L is

$$L\frac{di_L}{dt} = V_{in1} - V_{o1} \quad (10)$$

By applying KCL at the load side,

$$C_1\frac{dV_{o1}}{dt} = i_L - \frac{V_{o1}}{R_1} \; ; C_2\frac{dV_{o2}}{dt} = -\frac{V_{o2}}{R_2} \quad (11)$$

State 5:
By applying KVL to the loop containing source V_{in1}, diode D_3 and the inductor L, the voltage across the inductor is

$$L\frac{di_L}{dt} = V_{in1} - (V_{o1} + V_{o2}) \quad (12)$$

By applying KCL at the load side,

$$C_1 \frac{dV_{o1}}{dt} = i_L - \frac{V_{o1}}{R_1} \; ; C_2 \frac{dV_{o2}}{dt} = i_L - \frac{V_{o2}}{R_2} \tag{13}$$

By Comparing all the above state equations,

$$L\frac{di_L}{dt} = V_{in3}d_3 + V_{in2}d_2 + V_{in1}(d_1 - d_2 - d_3) + (V_{in1} - V_{o1})d_2 + (V_{in1} - V_{oT})(1 - d_4 - d_2) \tag{14}$$

$$C_1 \frac{dV_{o1}}{dt} = (d_4)i_L - \frac{V_{o1}}{R_1} \; ; \; C_2 \frac{dV_{o2}}{dt} = (1 - d_4 - d_2)i_L - \frac{V_{o2}}{R_2} \tag{15}$$

The above state variable equations have quantities which are time varying. This creates nonlinearity in the system. By adding small signal AC perturbations to the parameters like duty ratios, currents and voltages, it is possible to extract linearized state equations.

Applying local average,

$$\left. \begin{aligned} & i_L(t) = I_L + \Delta i_L, V_{o1}(t) = V_{o1} + \Delta V_{o1} \\ & V_{o2}(t) = V_{o2} + \Delta V_{o2}, d_1(t) = D_1 + \Delta d_1 \\ & d_2(t) = D_2 + \Delta d_2, d_3(t) = D_3 + \Delta d_3 \\ & d_4(t) = D_4 + \Delta d_4 \end{aligned} \right\} \tag{16}$$

$$L\frac{\Delta i_L}{dt} = (V_{in3} - V_{in2})\Delta d_3 + (V_{in2} - V_{o1})\Delta d_2 + V_{o1}\Delta d_4 + V_{o2}\Delta d_4 + V_{o1}D_4 - V_{o2}(1 - D_4 - D_2) \tag{17}$$

$$C_1 \frac{dV_{o1}}{dt} + C_1 \frac{\Delta dV_{o1}}{dt} = (1 - D_4 - \Delta d_4)(i_L + \Delta i_L) - \frac{(V_{o1} + \Delta V_{o1})}{R_1} \tag{18}$$

$$C_2 \frac{dV_{o2}}{dt} + C_2 \frac{\Delta dV_{o2}}{dt} = (1 - D_2 - \Delta d_2)(i_L + \Delta i_L) - \frac{(V_{o2} + \Delta V_{o2})}{R_2} \tag{19}$$

State space representation of any system can be expressed as,

$$\dot{x} = Ax + Bu$$
$$y = Cx + Du \tag{20}$$

From the above equations, the state space model of the proposed converter can be expressed as,

$$\begin{bmatrix} \frac{\Delta di_L}{dt} \\ \frac{d\Delta V_{o1}}{dt} \\ \frac{d\Delta V_{o2}}{dt} \end{bmatrix} = \begin{bmatrix} 0 & \frac{D_4}{L} & \frac{(1-D_1)}{L} \\ \frac{D_4}{C_1} & \frac{-1}{R_1 C_1} & 0 \\ \frac{(1-D_1)}{C_2} & 0 & \frac{-1}{R_2 C_2} \end{bmatrix} \begin{bmatrix} \Delta i_L \\ \Delta V_{o1} \\ \Delta V_{o2} \end{bmatrix} + \begin{bmatrix} \frac{V_{o2}}{L} & \frac{(V_{in3}-V_{in2})}{L} & \frac{(V_{in2}-V_{o1})}{L} & \frac{V_{o1}}{L} \\ 0 & 0 & 0 & \frac{-i_L}{C_1} \\ \frac{-i_L}{C_2} & 0 & 0 & 0 \end{bmatrix} \begin{bmatrix} \Delta d_4 \\ \Delta d_3 \\ \Delta d_2 \\ \Delta d_1 \end{bmatrix} \tag{21}$$

$$\begin{bmatrix} \Delta V_{o1} \\ \Delta V_{oT} \\ \Delta i_{b1} \\ \Delta i_{b2} \end{bmatrix} = \begin{bmatrix} 0 & 1 & 0 \\ 0 & 1 & 1 \\ D_3 & 0 & 0 \\ D_2 - D_3 & 0 & 0 \end{bmatrix} \begin{bmatrix} \Delta i_L \\ \Delta V_{o1} \\ \Delta V_{o2} \end{bmatrix} + \begin{bmatrix} 0 & 0 & 0 & 0 \\ 0 & 0 & 0 & 0 \\ 0 & 0 & I_L & 0 \\ 0 & 0 & 0 & I_L \end{bmatrix} \begin{bmatrix} \Delta d_4 \\ \Delta d_3 \\ \Delta d_2 \\ \Delta d_1 \end{bmatrix} \tag{22}$$

This topology has sixteen transfer functions. For each transfer function is derived from matrix A, B, C and D. Two transfer functions G, and F are selected out of eight transfer functions. These transfer functions are high stability, low steady state error and high bandwidth. The transfer function of G(s) and F(s) are shown in (23) and (24). Using MATLAB software, the bode diagrams of G(s) and F(s) are obtained as shown Fig.4. These are high stability, low steady state error and high bandwidth. The transfer function of G(s) and F(s) are shown in below. Using MATLAB software the bode diagrams of G(s) and F(s) shown Fig.4.

$$G(s) = \frac{\Delta V_{o1}}{\Delta d_4} = \frac{1.2 \times 10^8 S + 6 \times 10^9}{S^3 + 500 S^2 + 7.065 \times 10^5 + 1.61 \times 10^8} \tag{23}$$

$$F(s) = \frac{\Delta V_{oT}}{\Delta d_2} = \frac{4800 S^2 + 2.4 \times 10^6 S + 3 \times 10^8}{S^3 + 500 S^2 + 7.065 \times 10^5 + 1.61 \times 10^8} \tag{24}$$

IV. SIMULATION RESULTS

The performance evaluation of the proposed converter in simulation platform has been conducted using MATLAB/Simulink platform and the parameters used for simulation and experimental purpose are shown in Table 1. The input voltages are considered as $V_{in1} = 12$ V, $V_{in2} = 36$ V and $V_{in3} = 48$ V. The output voltage are $V_{01} = 80$ V, $V_{02} = 40$ V and total voltage is $V_{0T} = V_{01} + V_{02} = 120$ V. The simulation results are shown in Fig.5. The charging and

TABLE I
PARAMETERS OF THE PROPOSED CONVERTER

Sl. No	Parameters	Value
1	Swiching Frequency	20 kHz
2	Inductor L	5 mH
3	Capacitor C_1, C_2	100 μF
4	Resistors R_1, R_2	40 Ω
5	Voltage source 1	12 V
6	Voltage source 2	36 V
7	Voltage source 3	48 V

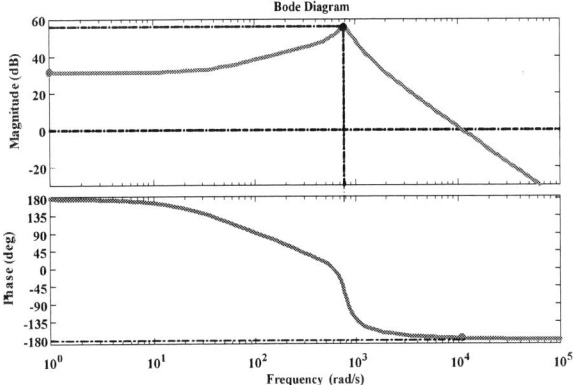

(a)

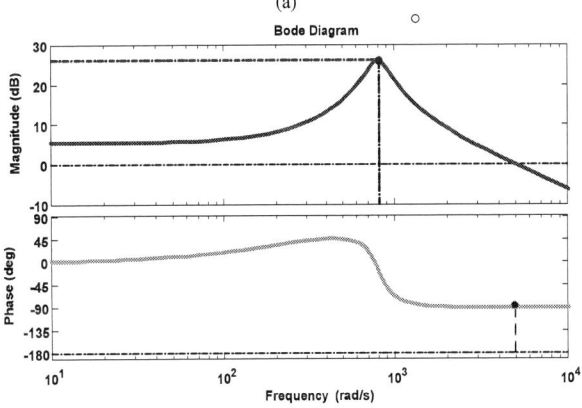

(b)

Fig.4.Bode plots of converter (a) G(s) and (b) F(s)

discharging of the inductor can be clearly observed from the inductor voltage and current waveform shown in Fig. 5. From the output voltage waveforms shown in Fig. 5, it is observed that the output voltages of 40 V (i.e., V_{02}), 80 V (i.e., V_{01}) and 120 V (i.e., V_{0T}) are obtained using simulation analysis. This clearly shows the validation of simulation results with the theoretical calculations.

V. EXPERIMENTAL RESULTS

To verify the feasibility of the proposed converter in experimental platform, a small scale laboratory prototype of the converter has been developed as shown in Fig. 6. The real time generation of the switching pulses has been performed using Dspace 1104 real time controller. The power switches are realized by IRFP 460 MOSFET. The input and output voltage specifications are already shown in Table 1. The experimental waveforms of the source voltages, output voltages, inductor voltage and inductor current are shown in Fig. 7 (a-d).

From the experimental waveforms shown in Fig. 7.b., it is clearly observed that the output voltages of 120 V (i.e., V_{0T}), 80 V (i.e., V_{01}), and 40 V (i.e., V_{02}), are obtained in experimental analysis. Also, the inductor voltage and current waveforms shown in Fig. 7.c, shows the performance of the converter in different switching states and the variation in the slope of inductor current according the variation in inductor voltages. All the experimental results are well matched with the simulation results which shows the efficient operation of the proposed converter in both simulation and experimental platforms.

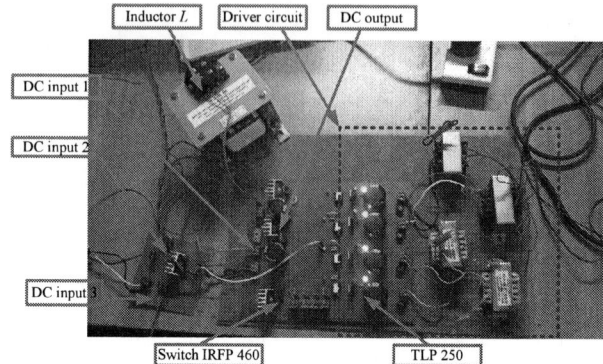

Fig.6. Experimental setup of the proposed converter

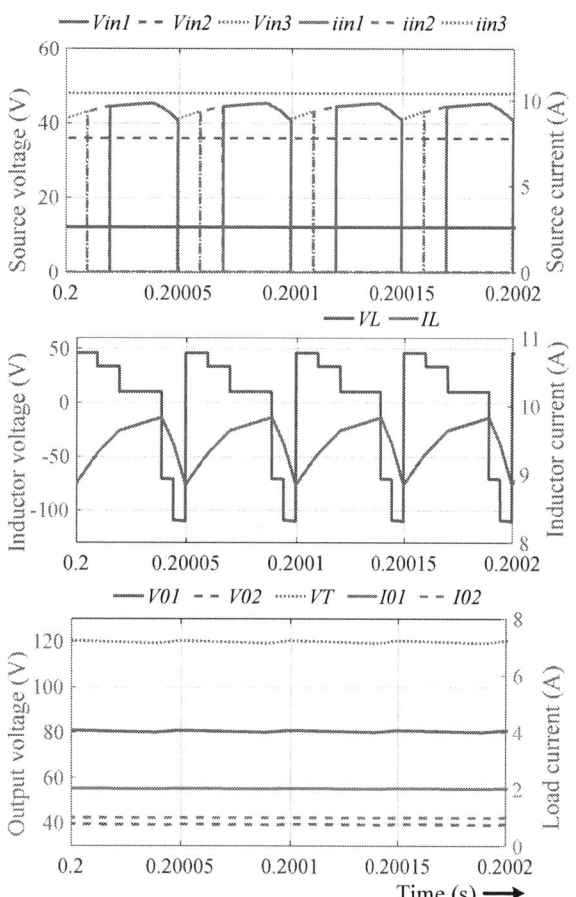

Fig.5. Simulation waveforms of the proposed converter

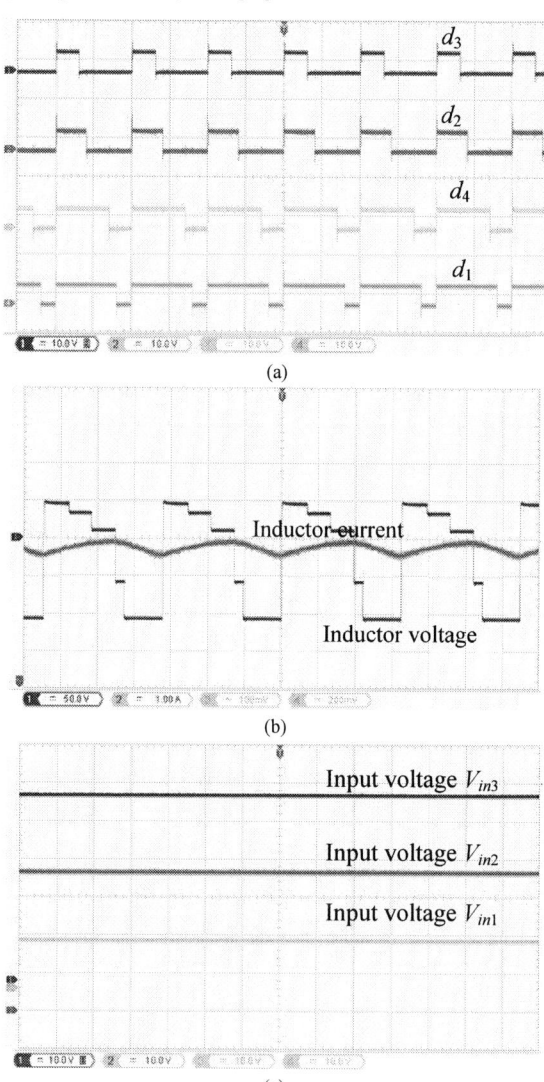

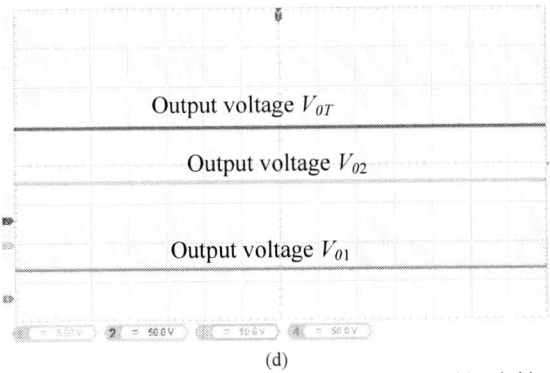

(d)

Fig. 7. Experimental waveforms of the proposed converter (a) switching pulses (b) inductor current and inductor voltage (c) Source voltages (d) output voltages

V. CONCLUSION

A three inputs and two outputs boost DC-DC converter for DC microgrid applications is proposed in this paper. The proposed converter has one inductor and can be used to integrate different energy resources such as solar PV, battery, fuel cell etc. The proposed DC-DC converter has two outputs such as low voltage DC and high voltage DC with different power levels. The low voltage DC can be used for lighting applications and other can be used for supplying electric motor in the domestic applications. Different switching states of the proposed converter, corresponding state equations and converter modeling have been carried out. Necessary transfer fucntions are derived and bode plot analysis of the converter has been conducted. Finally the feasibility of the porposed converter is verified in simulation and experimental platforms and the results obtained are well matched each other. The propsoed converter has less number of component count compared to other multi input multi output DC-DC converters which enhances the converter effficieny and also their use in various applications.

REFERENCES

[1] Yaow-Ming ChenYuan-Chuan Liu and Feng-Yu Wu, " Multi-Input DC/DC Converter Based on the Multi-winding Transformer for Renewable Energy Applications", IEEE transactions on industry applications, vol. 38, no. 4, August 2000

[2] Belloni, M., E. Bonizzoni, and F. Maloberti, "On the design of single-inductor double-output DC-DC buck, boost and buck-boost converters". Electronics, Circuits and Systems, International Conference ICECS 2008.

[3] Matsuo, H., et al, "Characteristics of the multiple-input DC-DC converter". IEEE Transactions on Industrial Electronics, 1993. 51(3): pp. 625-631.

[4] Wu, H., et al, "A family of three-port half-bridge converters for a stand-alone renewable power system". IEEE Transactions on Power Electronics, 2011. 26(9): pp. 2697-2706

[5] Lalit Kumar, "Multiple-input DC/DC converter topology for hybrid energy system", IET Power Electron., 2013, Vol. 6, Iss. 8, pp. 1483–1501.

[6] Akshatha N.M, "Dual Output DC-DC Converter for DC Micro-Grid Application", International Journal of Innovative Research in Electrical, Electronics, Instrumentation and Control Engineering. Vol. 4, Special Issue 2, April 2016

[7] Chiu, H.J., et al, "A multiple-input DC/DC converter for renewable energy systems", IEEE International Conference on Industrial Technology. 2006.

[8] Yang, D., et al. " Multiple-input full bridge dc/dc converter", Energy Conversion Congress and Exposition, 2009. ECCE. 2009.

[9] Wai, R.J., C.Y. Lin, and Y.R. Chang, "High Step-Up Bidirectional Isolated Converter With Two Input Power Sources", IEEE Transactions on Industrial Electronics, 2009. 56(7): pp. 2629-2643.

[10] Suetomi, M., et al. "A novel multi-input DC-DC converter with high power efficiency", Telecommunications Energy Conference. 2011.

[11] Sivaprasad A., G. G. Kumar, Kumaravel S. and Ashok S., "Performance Analysis of Novel Bridge Type Dual Input DC-DC Converters", in IEEE Access, 2017, Vol. 5, pp. 15340-15353.

[12] Sivaprasad A, Kumaravel S, Ashok S.: Design, Fabrication and Performance Analysis of a Two Input—Single Output DC-DC Converter, Energies, 2017, Vol. 10, pp. 1-18.

Novel intelligent controllers for Load Frequency Control of a Multi Area hybrid power System

Divyank Srivastava
Electrical Engineering Department
Delhi Technological University
New Delhi, India
divyanksri.srivastava@gmail.com

M. M. Tripathi
Electrical Engineering Department
Delhi Technological University
New Delhi, India
mmmtripathi@gmail.com

Abstract— **In this paper, an intelligent neuro-fuzzy based controller is proposed for a multi-area load frequency control with PV system, suitable for a restructured power system. The proposed controller has been applied in a multi area system with and without PV system. A comparative analysis of the hybrid neuro-fuzzy controller is done with different controllers such as PI, FGSPI and ANFIS. The proposed hybrid neuro-fuzzy controller evaluates the current system situation and regulates system characteristics efficiently. The functioning of proposed scheme has been tested on multi area AC network in which two areas are thermal generator based and one area is non-conventional PV array. The proposed control technique has been validated and found to be better than the conventional control techniques used in AGC designing in both the cases with and without PV system.**

Keywords—Load frequency control, PI controller, FGSPI, ANFIS controller, PV panel.

I. INTRODUCTION

One of the main motives of prevailing electric power system is to deliver secure, reliable and cost effective electrical energy sources to the consumers [1]. Nowadays, there is an increase in the demand of power supply but a long term planning needed to get that demand to be met is lacking. This insufficient planning facilitates the companies to supply unsecured and degraded quality of the power supply. Therefore, there is an immediate need of improving the power system quality.

There are several sources of electricity generation known to us. These sources can be classified as renewable and non-renewable sources of energy [1]. It has undoubtedly become essential to switch to renewable sources such as solar, wind, etc. to protect our environment of the ill effects caused by using non- renewable sources of energy such as thermal, diesel etc. Due to the aforementioned reasons it is critically important to use hybrid systems comprising a power grid employing a renewable source of energy. For example a hybrid system can be installed along with a backup power system of a thermal generator plant. Advantage of commissioning such a system is that during day time the majority of load is provided by the renewable source of energy such as a PV (Photovoltaic) Array and when the sunlight is insufficient, the electric power is provided by the thermal generator plant [2]-[3].

Employment of such a hybrid system comes with its own challenges. The main issue of concern is the maintenance

of constant frequency throughout. As the system frequency is reliant on the load changes, Load frequency control (LFC) or automatic generation control (AGC) is an important issue in the field of power system operation and control (PSOC). An interconnected power system with multi areas can be segregated into various control areas connected via tie-lines. Frequency of the system alters according to the generated power and load demand. A change in demand of active power at one place imitated in other places. Today's power system processes needs reconsideration for system instability and problems related to control [4]. Dynamic performance of all conventional standard controllers like Integral, Proportional, Proportional-Integral (PI), Proportional-Integral-Derivative (PID) controllers and intelligent controllers (Fuzzy-Tuned Controller) have been discussed in the literatures [5]-[6].

In this paper, an advanced and powerful controlling technique ANFIS (Adaptive neuro fuzzy interference system) is applied for optimization of different parameters for both primary and secondary control loops of the governor in a multi area hybrid power system having two DG based system and one solar PV array system. Both the conventional standard controller and proposed soft controller are implemented and their performance is compared so as to evaluate the best controller subjected to different environments. The proposed ANFIS controller being adaptive and intelligent has resulted in fast settling time of 4 sec. and reduced oscillations and overshoot compared to PI and FGSPI controller on different load change in different areas & types of power sources.

Rest of the paper is organized as follows. Section II discusses the modelling of the hybrid power system and section III presents the proposed control scheme. Simulation and results are discussed in section IV and conclusion is presented in section V.

II. MODELLING THE HYBRID POWER SYSTEM

A. PV Module

PV module consist of fixed number of solar cells. Solar cell is the basic unit of a PV module, each module is interconnect in series-parallel fashion. In solar cell a current source is connected in parallel with a diode, providing a constant current at the output as shown by equivalent circuit of solar cell in fig.1.

The output current equation in fig. 1 is given by eq. 1.

$$I = I_L - I_D - I_{sh} \tag{1}$$

978-1-5386-6624-1/18 $31.00 © 2018 IEEE

Using Shockley equation, the current equation I_D for a single diode can be given by eq. 2.

$$I_D = I_O \left[exp\left(\frac{V_S + IR_S}{nV_T}\right) - 1 \right] \tag{2}$$

Where n is the diode ideality factor (unit less, usually between 1 and 2 for a single cell).

The shunt current I_{sh} is given by eq. 3.

$$I_{sh} = (V_s + IR_s)/R_{sh} \tag{3}$$

Putting eq. 3 and eq. 2 in eq. 1 results in complete equation of a solar cell as given by eq. 4.

$$I = I_L - I_O \left[exp\left(\frac{V_S + IR_S}{nV_T}\right) - 1 \right] - \frac{V_S + IR_S}{.R_{sh}} \tag{4}$$

Where,

I_L = Ligth Current (A)

I_D = Voltage dependent Diode current (A)

I_O = Diode reverse saturation current (A)

R_S = Series resistance (Ω)

R_{sh} = Shunt resistance (Ω)

n = Diode ideality factor (unitless)

V_T = Thermal voltage (V)

V_s = Solar Cell Voltage

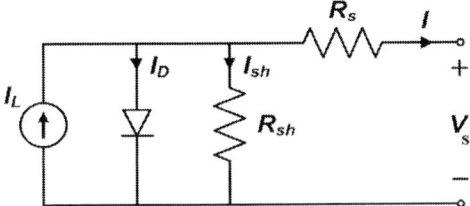

Fig. 1: Equivalent circuit of a solar cell

The output of PV panel had been controlled by MPPT technique for getting maximum output vltage [7]. After that a boost converter is implementedd to regulate the voltage level to desirable value i.e.230V. The equivalent circuit of boost converter is shown in fig. 2.

In fig. 2 when S=0, diode is on and when S=1, diode is off. Boost convertor works in two different modes [6]. First period occur when the switch is turned ON, i.e. $0 < t < T_D$. The source voltage for this mode is given by eq. 5.

$$V_S = L\frac{di(t)}{dt} \tag{5}$$

The output loop equation for this mode is given by eq. 6.

$$C\frac{dV_O}{dt} + \frac{V_O}{R} = 0 \tag{6}$$

Second period occur when the switch is turned off, i.e. $T_D \leq t \leq T_s$, output voltage V_o is given by eq.7

$$V_S - V_O = L\frac{di(t)}{dt} \tag{7}$$

In this mode inductor current is given by eq. 8.

$$i_L - C\frac{dV_o}{dt} + \frac{V_O}{R} = 0 \tag{8}$$

Parameters for the Boost Convertor are considered as given below.

f_s = 20 KHz (Switching Frequency)

V_s(min) = 46.6 V (Minium Output Voltage of PV Panel)

V_s(max) = 57.8 V (Maximum Output voltage of PV Panel)

V_O(max) = 230 V (Maximum Output voltage of PV system)

I_O = .98 A (Load Current)

L = 547 µH

C = 486 µF

Based on the above equations and data provided, transfer function of a PV module is given by eq. 9 [8]- [9].

$$G_{PV} = \frac{-18S + 900}{S^2 + 100S + 500} \tag{9}$$

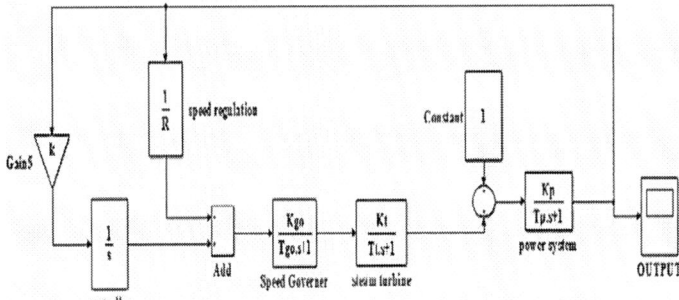

Fig. 2: Equivalent circuit diagram of boost converter

B. Thermal generator network

A single area power system consist of simple power system block which includes the generation,transmission, governer, prime-mover and its control stretagies. The load discrepancy has been consider as step change. The transfer function model has been developed using MATLAB simulink [10]. The different components used in transfer function model are interconnected to form a single area system as shown in fig. 3 [11]-[12]. The equations are used in s domain only.

Transfer functions used for speed governor, turbine, power system and speed regulation are as follows;

$$\text{Speed Governor Transfer Function} = \frac{K_{go}}{T_{go}s + 1} \tag{10}$$

$$\text{Turbine Transfer Function} = \frac{K_t}{T_t s + 1} \tag{11}$$

$$\text{Power System Transfer Function} = \frac{K_p}{T_p s + 1} \tag{12}$$

$$\text{Speed regulation} = \frac{1}{R} \tag{13}$$

Fig. 3: Transfer function Model of a single area LFC

978-1-5386-6624-1/18 $31.00 © 2018 IEEE

III. PROPOSED CONTROL SCHEME

It is important to design more reliable and robust controller for load frequency control to attain minimum overshoot as well as settling time. Here, we have suggested different controllers i.e. PI, FGSPI and ANFIS for AGC. Where PI is a conventional controller and FGSPI & ANFIS are intelligent controllers.

A. PI controller

There are various types of load-frequency controllers available but the PI controller is most widely used in speed-governor schemes for LFC schemes [13]. The PI controller has an upper hand on other controllers as it makes the steady-state error to zero. Mathematically it is given by eq. 14.

$$Up(s) = (K_p + K_i/s)E(s) \qquad (14)$$

B. FGSPI

With the help of gain scheduling technique we can design a controller which can be implemented on a system whose system dynamics have non-linear relation with operating condition [14]-[15]. In this paper, to determine different integral constant values (K_i) in PI controller, fuzzy controller output have been used.

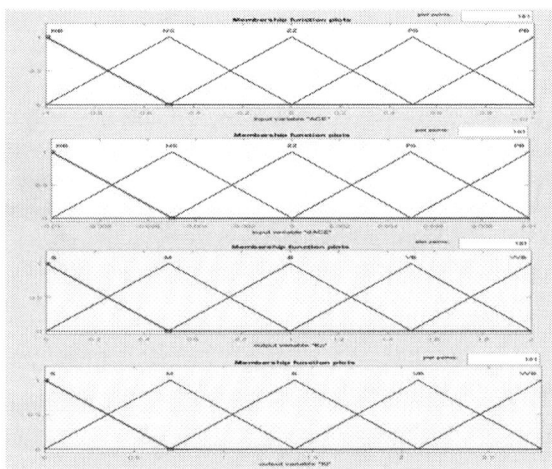

Fig. 4: Membership functions of input and output

Increasing the K_i value will bring out fluctuations in the response. Minimum deviation can be obtained with initial value of K_i. In contrast, when we decrease the value of K_i that will result in high damping with low overshoot but the study state error will increase [16]. In the same way value of proportional constant value (K_p) will also change the system dynamics. So to detect the exact value of Ki and K_p intelligent fuzzy controller is used that will follow the changes in the system and try to return set point to initial defined value.
In FGSPI, the inputs to fuzzy logic are area control error (ACE) and differential of ACE (i.e. dACE). The output of the fuzzy controller is K_i and K_p. In this work 2 inputs and 2 output variables and 5 triangular membership function for each variables are used as shown in fig.4. The rule base applied to the FGSPI controller is given in table 1. For defuzzification centroid method is used. Block diagram of FGSPI is shown in Fig:5

Table 1: Fuzzy rule base for FGSPI controller

ACE dACE / K_p K_i	NB	NS	Z	PS	PB
NB	S,S	S,S	M,M	M,M	B,B
NS	S,S	M,M	M,M	B,B	VB,VB
Z	M,M	M,M	B,B	VB,VB	VB,VB
PS	M,M	B,B	VB,VB	VB,VB	VVB,VVB
PB	B,B	VB,VB	VB,VB	VVB,VVB	VVB,VVB

NB: Negative Big, NS: Negative Small, Z: Zero, PB: Positive Big, PS: Positive Small, S: Small, M: Medium, B: Big, VB : Very Big, VVB: Very Very Big.

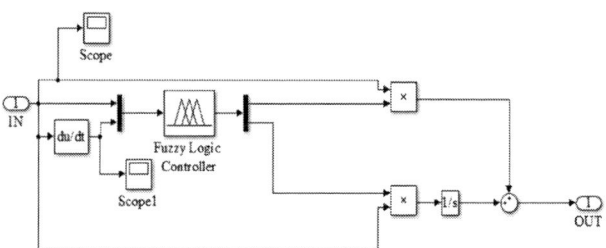

Fig. 5: Block diagram of FGSPI controller

C. ANFIS

Adaptive neuro fuzzy interface system (ANFIS) is like a neural network. It has no synaptic weights, adaptive and non-adaptive neurons are there. With the use of ANFIS controller the value of Ki can be calculate with more accuracy because ANFIS will train the fuzzy file with feed forward network and more we train our fuzzy file more accurate value of Ki will be achieved which reduces settling time [17]-[20]. In MATLAB for creating ANFIS based FIS file a datasheet with the above used FGSPI system is used in which ACE & dACE used as input and Ki as output. Then in ANFIS editor first datasheet is loaded and a FIS file is created and trained. The trained network is used for testing of a new data. The ANFIS architecture is shown in fig. 6, which is consisting of 5 layers. First layer and fourth layer shows the no. of input and output variables. Second layers shows the membership function of each input variables. In third layer rule base is defined.

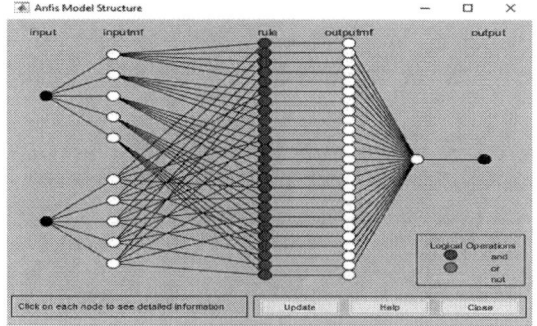

Fig. 6: ANFIS architecture

978-1-5386-6624-1/18 $31.00 © 2018 IEEE

IV. SIMULATION AND RESULTS

The MATLAB R16A was used for modelling and simulation of the controller for multi area hybrid power system. The fig. 7 shows the MATLAB Simulink model used for simulation, training and testing. Appropriate values of variable parameter of MATLAB Simulink model has been determined and applied. Fig. 8 shows a screenshot of training testing output.

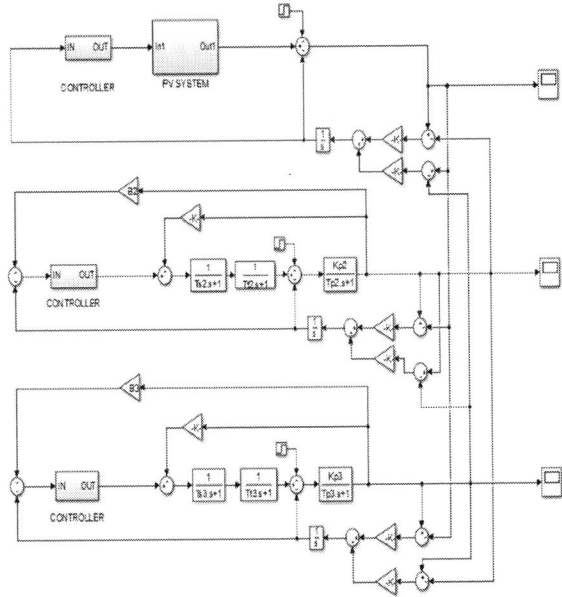

Fig. 7: MATLAB Simulink model of proposed Controller

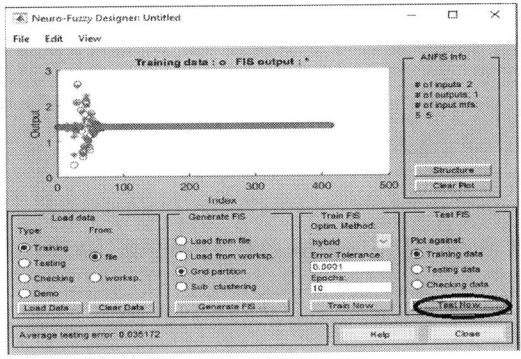

Fig. 8: Testing the training data in ANFIS

Simulation result of hybrid power system network (Thermal generator and PV) with different controllers are shown from fig. 9 to fig. 18. Table 2 presents the comparison of the performance of different controllers. Considering the hybrid system (DG set and solar PV), using PI controller, output of thermal generator is shown in fig.9 and PV system shown in fig.10. The settling time for thermal generator is 29 seconds and PV system is 19 second.

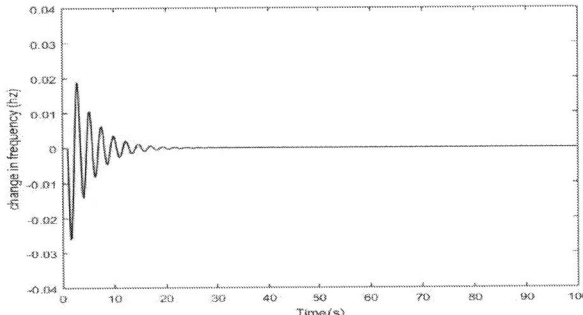

Fig. 9: Thermal generator output with PI controller

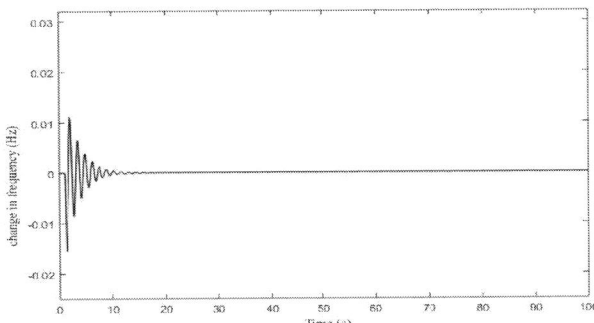

Fig. 10: PV system output with PI controller

Using FGSPI controller the output of hybrid system (DG set and solar PV) is shown in fig.11 and fig.12. The settling time for thermal generator is 15 second and for PV system is 7 second.

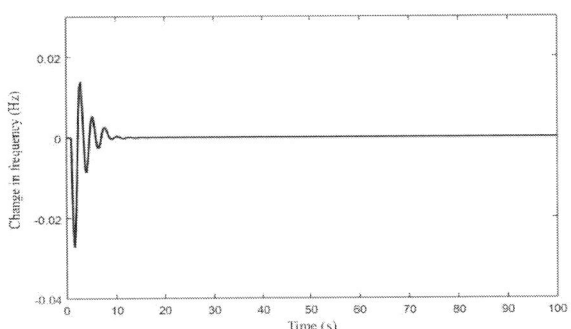

Fig. 11: Thermal generator output with FGSPI controller

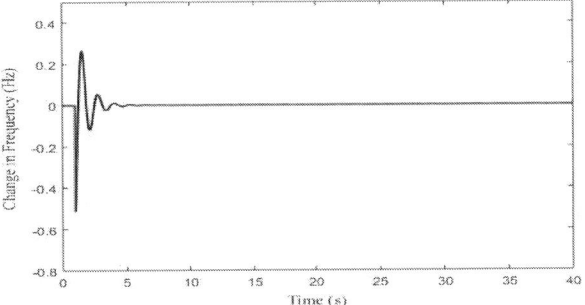

Fig. 12: PV system output with FGSPI controller

978-1-5386-6624-1/18 $31.00 © 2018 IEEE

Using ANFIS controller the output of hybrid system is shown in fig.13 and fig.14, in which the settling time for thermal generator is 11 second and for PV system is 4 second.

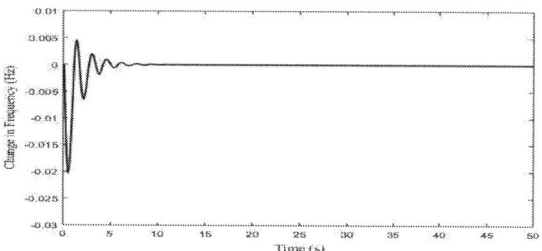

Fig. 13: Thermal generator output with ANFIS controller

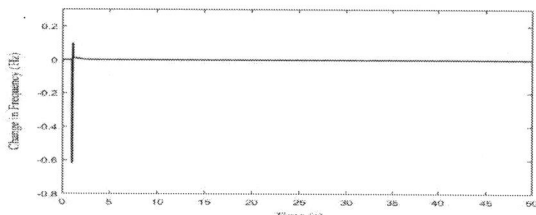

Fig. 14: PV system output with ANFIS controller

Tie-line power flow of hybrid system with PI, FGSPI and ANFIS controller are shown in fig. 15, fig.16 and fig.17 respectively. Settling time for ANFIS is better than other controllers.

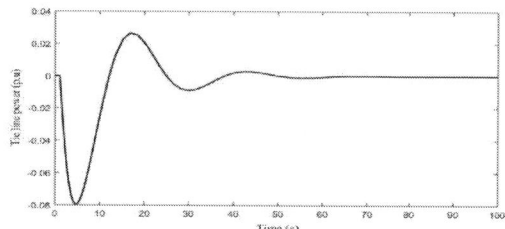

Fig. 15: Tie-line power with PI controller

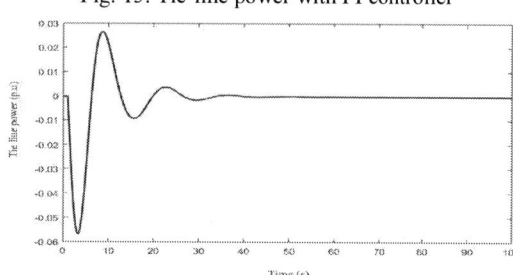

Fig. 16: Tie-line power with FGSPI controller

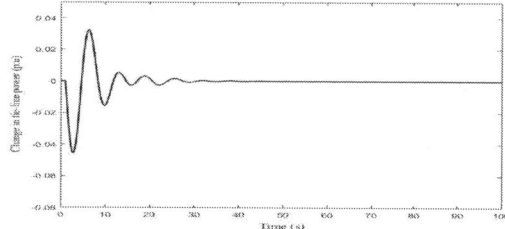

Fig. 17: Tie-line power with ANFIS controller

Table 2: Comparison of Different Controllers

Controllers	Thermal Generator System Settling Time (s)	Thermal generator system (Overshoot)	PV System Settling Time (s)	PV System (Overshoot)
PI	29	0.027	19.00	0.019
FGSPI	15	0.025	7.00	0.500
ANFIS	11	0.020	4.00	0.610

V. CONCLUSION

The controller being adaptive and intelligent are the most important factor for power system operation and control. The proposed intelligent controller is not only unique in design but also easy to implement. Simulation results obtained have demonstrated the usefulness of the proposed ANFIS controller for multi area hybrid power system. As it is seen from the results that FGSPI gives lesser settling time than PI controller and ANFIS gives least among all, it is expected that the proposed ANFIS controller will work effectively on different load change in different areas consisting of different types of power sources.

REFERENCES

[1] H K Yadav, Y Pal, M M Tripathi, "Photovoltaic power forecasting methods in smart power grid" Annual IEEE- India Conference (INDICON), 2015.

[2] G.Spagnuolo, "Renewable energy operation and conversion schemes," IEEE Industrial Electronics Magazine, Vol. 4, No. 2, pp. 38-51, March 2010.

[3] D. Sonnenenerggie, "Photovoltaic Basics, in Planning and installing photovoltaic systems: A guide for installers,architects,and engineers", 2nd., Deutsche Gesellschaft für Sonnenenergie, Ed. London: Earthscan Publications Ltd., x ch. 1, pp. 1-6, 2008.

[4] Sathans,Akhilesh Swarup "Intelligent Load Frequency of Two-Area Interconnected Power System and Comparative Analysis" International conference on Communication System and Neural Network Technologies,2011

[5] Y. Yorozu, M. Hirano, K. Oka, and Y. Tagawa, "Electron spectroscopy studies on magneto-optical media and plastic substrate interface," IEEE Trans. J. Magne. Japan, vol. 2, pp. 740–741, August 1987.

[6] Tridipta Kumar Pati, JyotiRanjanNayak, Binod Kumar Sahu, Sanjeeb Kumar Kar, "Automatic Generation Control of Multi-Area Thermal Power System using TLBO Algorithm Optimized Fuzzy-PID Controller", International Conference, 2015.

[7] A. J. Mahdi, W.H. Tang and Q.H. Wu " Improvement of a MPPT Algorithm for PV Systems and Its Experimental Validation", International Conference on Renewable Energies and Power Quality (ICREPQ'10), Granada, March.

[8] Seyed Ali Jeddi,Seyed Hamidreza Abbasi,Fereydoon Shabaninia "Load Frequency Control of Two Area Interconnected Power System (Thermal Generator and Solar PV) with PI and FGSPI Controller" The 16th CSI International Symposium on Artificial Intelligence and Signal Processing (AISP 2012).

[9] Tom santy, Mr.R.natesan "Load Frequency Control of a Two Area System Consisting of a Gri Connected PV System andThermal Generator" International Journal of Emerging Technology in Computer Science & Electronics (IJETCSE) ISSN: 0976-1353 Volume 13 Issue 1 – MARCH 2015.

[10] Shigeyuki Yanagawa, Takeyoshi Kato, Wu Kai Akimori Tabata, Yasunobu Yokomizu,Tatsuki Okamoto,Yasuo Suzuoki, "Load Frequency Control", IEEE Transactions on Energy Conversion, Vol. 11, No. 1, March 1996

[11] Miller N., Ye Z., "Report on Distributed Generation Penetration Study", NREL/SR-560347156, Golden, CO: National Renewable Energy Laboratory, 2003.

[12] Manoj Datta, Tomonobu Senjyu, Atsushi Yona, , Toshihisa Funabashi, and Chul-Hwan Kim, "A Coordinated Control Method for Levelling PV output Power fluctuations of PV-Thermal Hybrid System connected to 530 Isolated Power Utility", IEEE Transactions on Energy Conversion, Vol. 24, No.1, MARCH2009.

[13] Bhatti T S, Al–Ademi A. A. F et al, "Load Frequency control of isolated wind-thermal-micro hydro hybrid power systems. Elsevier-Energy, 22(5): 461-470, 1997.

[14] Soundarrajan. A et al, "Intelligent controllers for Automatic Generation Control", Proceedings of The International conference on Robotics, Vision, Information and signal processing", Malaysia, pp 307-311, 2003.

[15] Zhen-Yu Zhao, Masayoshi Tomizuka and Satoru Isaka, "Fuzzy gain scheduling of a PID controllers", IEEE Trans. Syst., Man Cyber. 23 (5) (1993).

[16] Kroposki, B.; Vaughn, "A. DG Power Quality, Protection, and Reliability Case Studies Report", NREL/SR-560-34645. Golden, CO: National Renewable Energy Laboratory. General Electric Corporate R&D, 2003.

[17] Chalmers S., Hitt, M., Underhill, J., Anderson, P., Vogt, P., Ingersoll, R. , "The Effect of Photovoltaic Power Generation on Utility Operation." IEEE Transactions on Power Apparatus and Systems; PAS-104, 1985, pp. 524–530.

[18] EPRI report EL-6754, Photovoltaic Generation Effects on Distribution Feeders, Volume 1: Description of the Gardner, Massachusetts, TwentyFirst Century PV Community and Research Program," 1990.

[19] A Demiroren , N S Sengor, H Zeynelgil, "Automatic generation control bynbusing ANN technique", Electr. Power Compon. Syst., 2001, 29, (10), pp. 883–896.

[20] C.S.Chain,K.T.K.Teo, "Fuzzy Logic Based MPPT for Photovoltaic Modules Influenced by Solar Irradiation and Cell Temperature" UKSim 13th International Conference on Modelling and Simulation, 2011.

A Comparative Study for Short Term Wind Speed Forecasting using Statistical and Machine Learning Approaches

Parul Arora
Department of Electrical Engineering
Indian Institute of Technology Delhi,
India
parularora@ee.iitd.ac.in

Himanshu Kumar
Department of Computer Science
SRM Institute of Science and Technology
Chennai, India
kumarkush36@gmail.com

B.K Panigrahi
Department of Electrical Engineering
Indian Institute of Technology
Delhi, India
bkpanigrahi@ee.iitd.ac.in

Abstract—**The paper presents a comparative study of statistical and machine learning based methods for wind speed forecasting. Effectively predicting wind speed forecasts plays a vital role in resource planning and managing production control. Already existing forecasting methods are compared with the new ones over two different wind farm zones for short term forecasting horizon. Three most accurate and accessible methods Support Vector Regression (SVR), Auto-Regressive Integrated Moving Average (ARIMA), and Recurrent Neural Network are discussed along with the procedure for selecting correct model parameters for fine-tuning.A novel work on Kernel variations in support vector regressors is done to improve forecast results.**
Index Terms—**Forecasting,Wind,ARIMA,SVR,LSTM,RNN**

I. INTRODUCTION

Wind Energy is the most important clean fuel source that has attracted many researchers over last decade towards bringing the advanced energy forecasting models, and advanced predictive capabilities have brought ease to operational management and short term planning for wind energy generation. Wind turbines require proper wind speed flow for generating electricity. Developed countries are ramping up their massive investment in renewable energy sources and emphasize on eliminating fossil fuel use within their borders. Wind power generation in a particular area is dependent on multiple variables that determine how much amount of power is going to be accumulated with more accurate forecasts available. One such variable that alters the production means is the wind speed, and thus it is required to develop accurate forecast models for the same. Various forecasting horizons are present for making the predictions such as short term forecasting relates to the prediction that is done on a microscopic scale and includes production control and planning and has range averaging from 10 min to few hours. Medium and long-term forecasting horizons are also very much critical as they determine the long-term availability of resources and planning of wind plants.

"Naive Predictors" uses the persistence method approach and still produce much better forecasts for very short term forecasting horizons [1]. Physical methods such as numerical weather predictors(NWP) can produce reasonable forecasts by working over a lot of complicated mathematical equations on weather data as temperature, pressure, precipitation and other variables[2] .These methods are generally deployed over an area where weather conditions are much more stable. Statistical methods work with a lot of measured training data and use the difference between actual value and predicted value to tune the model parameters. This study shows the comparative analysis of various statistical methods present for wind speed forecasting. Recent advancement in deep learning techniques has made it possible to implement deep neural architectures and new models which are being deployed for many applications majority in forecasting area. Various deep learning algorithms have been structured over to predict desired forecasts. These algorithms can extract complex features present in the series which cannot be easily extracted with conventional learning approaches. The use of deep learning techniques in time series has been implemented in many studies but rather less focused on selecting correct model parameters for fine-tuning.

The rest of the paper is organized in the following manner. Section II discusses various studies implemented over the goal of presenting the best forecasting results using different algorithms and approaches. The fundamentals of various forecasting methods are discussed in Section III. Section IV gives information about the experimental setup used in this study for implementing various model architectures. Finally, results are evaluated in Section V, and final remarks are presented in Section VI.

II. BACKGROUND

Statistical and machine learning approaches towards wind speed prediction have gained much focus in recent times, and therefore many researchers have been trying to implement new or hybrid approaches towards selecting the best model presenting their advantages. Some of the recent studies presenting their diverse approaches have motivated towards new research directions. J.L. Torres et al. [3]have presented a paper using wind speed data from North Dakota, USA in which persistence models and ARMA (autoregressive moving average process)

978-1-5386-6624-1/18 $31.00 © 2018 IEEE

are used for prediction up to 10 hours in advance. This method is applicable only for short-term forecasting as RMSE does not exceed 1.5 m/s. Errors with ARMA models are 12-20 % smaller than persistence models.

Rajesh G. Kavasseri and Krithika Seetharaman [4] have proposed wind forecasting model using fractionalARIMA model. This f-ARIMA model has been compared with Persistence and ARIMA models. The daily mean error of f-ARIMA is 33.18% whereas, for ARIMA and persistence models, it is 44.92% and 45.2% respectively-ARIMA model can capture time series measurements regarding both long-term as well as short-term.

Ergin Erdem and Jing Shi [5] have used ARMA based approaches for forecasting wind speed and direction. In the first method, wind speed in decomposed into longitudinal and lateral and components. In the second method, two independent traditional ARMA models for forecasting wind speed. The third model uses a Vector Autoregressive model to forecast using wind parameters. The fourth method uses a restricted version of the VAR.

Ladislav Zjavka and Stanislav Mik[6] have proposed direct wind power forecasting using the polynomial decomposition of the general differential equation. This method uses the inverse Laplace transform for representing local weather conditions up to 12 hours ahead. D-PNN forms a variety of PDE which represents the atmospheric dynamics. It is trained using some inputs are higher than outputs for half to 12 hours prediction

Yongning Zhao et al.[7]have proposed Sparsity-Controlled Vector Autoregressive (SC-VAR) model to get the sparse model structures in the space-time domain by redeveloping the original VAR model into Mixed Integer Non-Linear Programming (MINLP) problem. To reduce the complexity, SC-VAR method is modified to Correlation-Constrained (CCSC-VAR).CCSC-VAR is better because it allows inclusion of prior knowledge. This method is discussed using batch learning mode, and it could be extended to online adaptive version of CCSC-VAR to achieve better forecasting

Rajesh Karki et al. [8] has presented a method based on time series model to estimate the variation in wind and to calculate the possibility of wind power prediction. In this, a simple method is presented which is easy to apply in practical applications. The conditional probability distribution is used to quantify wind speed for wind power generation.

Tarek H.M.El-Fouly et al. [9]has proposed a linear time series based model. The proposed model is used for wind speed and direction prediction. It is better than a persistent model by 54.4% for MAE and by 55.3% for RMSE. This method has advantages over the persistent model regarding the higher scaling factor, reduced scatter and higher coefficients of correlation.

Jooyoung Jeon and James W. Taylor [10] have developed a method for wind power density forecasting using conditional kernel density estimation for 1-72 hours ahead. Uncertainty in wind speed and direction is taken by bivariate vector auto-regressive moving average-generalized autoregressive conditional heteroscedastic (VARMA-GARCH) model. Conditional kernel density (CKD) is used for nonparametric modeling of wind power density. The Monte-Carlo simulation of the VARMA-GARCH model and CKD enables wind power density forecasts. More work could be done by incorporating other parameters like air pressure, temperature and other meteorological parameters using these methods.

Yujie Lin et al. [11]have analyzed the existing models for short-term prediction of wind speed. It has been concluded in this paper that forecasting accuracy can be improved by assimilating seasonal effects, by incorporating daily measured variables like radiation and pressure and by independently measuring wind speed and direction. Auto-Regressive(AR) structures with external inputs (ARX) produce more accurate results than only AR structures. Moreover, the veracity of forecasting over a particular month, wind speed, and direction increase more if data for the same month is analyzed over many years.

III. METHODOLOGY FUNDAMENTALS

A. Autoregressive Integrated Moving Average Model (ARIMA)

ARIMA is one of the most prominent linear models that have been used in predicting future values based on its past observations or also known as time series forecasting. These models are utilized as a part of an extensive variety of applications ranging from engineering to economics. An ARIMA (p,d,q) is a mixture of Autoregressive(p) and Moving Average(q) models with a differencing operator as d and is widely implemented in time series forecasting domain. An AR(p) model can be represented mathematically[12] with the equation(1) as the linear regressive model for past p number of observations.

$$y_t = c + \sum_{i=1}^{p} \theta_i y_{t-i} + \delta_t = c + \theta_1 y_{t-1} + \ldots + \theta_p y_{t-p} + \delta_t \quad (1)$$

The above equation includes the y_t and δ_t as the actual and random error value at time t respectively,θ_i(i = 1,2,..p) are model parameters and c is defined to be a constant. Similarly,the MA(q) model is represented mathematically[12] with the equation(2) as the dependency of what happens at current time period t relates to randoms errors in past time periods.

$$y_t = \nu + \sum_{j=1}^{q} \phi_j \delta_{t-j} + \delta_t = \nu + \phi_1 \delta_{t-1} + \ldots \phi_q \delta_{t-q} + \delta_t \quad (2)$$

ν is called as the mean of the series under observation,ϕ_j(j=1,2..q) are model parameters and random errors are generally assumed to follow the normal distribution. ARIMA methodology combines both of these model parameters to produce the desired forecast that involves differentiating the original time series growing at a fairly constant rate d times to obtain stationarity. Finally,the equation(3) helps us to give the fairly accurate forecasts by observing the previous time steps with optimum AR and MA parameters.

$$y_t = c + \delta_t + \sum_{i=1}^{p} \theta_i y_{t-i} + \sum_{j=1}^{q} \phi_j \delta_{t-j} \qquad (3)$$

The principal issue in Box-Jenkins(1973) method is to choose which ARIMA parameter to take but in this paper Autocorrelation(ACF) and Partial Autocorrelation Function(PACF) plots are used which helps in finding AR/MA parameters .These statistical plots are generally done in reference with the time lags

B. Support Vector Regression(SVR)

Regression is defined to find a relationship amongst different input variables for predicting the target variable, and SVR is a support vector application seeking an optimal hyperplane such that the total distance to every sample point is minimal. Naturally, as all regressors, it endeavors to fit a line to information by minimizing the cost work. The deployment of a non-linear kernel makes it more advantageous over other regressors.

Fig. 1. Hyperplane

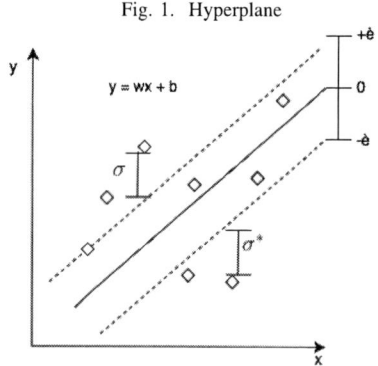

$$Minimize = 1/2\|w\|^2 + \sum_{i=1}^{N} (\sigma_i + \sigma_i^*) \qquad (4)$$

The main role is to minimize above equation as much as possible to get better prediction accuracy.

- Linear SVR

$$y = \sum_{i=1}^{N} (\beta_i - \beta_i^*).\langle x_i, x \rangle + \gamma \qquad (5)$$

- Non-Linear SVR

$$y = \sum_{i=1}^{N} (\beta_i - \beta_i^*).K\langle x_i, x \rangle + \gamma \qquad (6)$$

The $K\langle x_i, x \rangle$ represents different non-linear kernels that can be used in regression tasks.
- Kernel Functions
  $K\langle x_i, x \rangle$ can be replaced with following equations(7,8,9,10) depending upon the applications of kernel.

$$X_i.X_j - Linear \qquad (7)$$

$$(\lambda X_i.X_j+)^d - Polynomial \qquad (8)$$

$$exp(-\lambda|X_i - Xj|^2) - RBF \qquad (9)$$

$$tanh(\lambda X_i.X_j+) - Sigmoid \qquad (10)$$

C. Recurrent Neural Network(LSTM)

Recurrent neural network variations have been around for a long time and are used in various studies on different tasks related to forecasting domain[13] . Several LSTM Cells are connected to each other where each cell processes single time stamp individually and sending its output value to the next step. The data is fed to the network in the selected window size, and each window data is processed through the cell through multiple gates. The first gate in a cell is known to be the forget gate that decides what amount of information is going to be thrown away. Secondly, the next gate decides what piece of new information is going to merge with the earlier one. The next gate updates the old cell state with the new one, and finally, the output gate decodes what output is to be passed onto the next cell. The working mentioned above of the LSTM cell can be visualized through the figure given below.[14]

Fig. 2. LSTM Cell

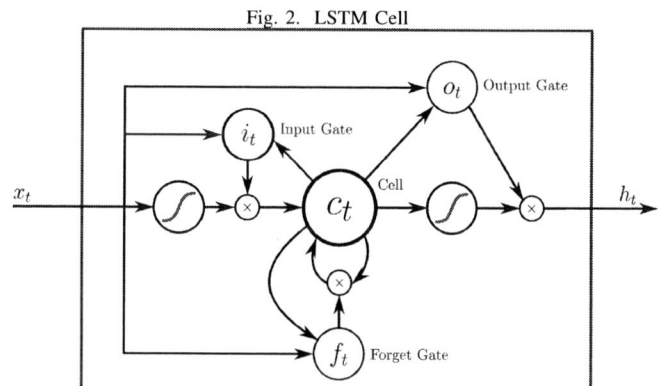

D. Error Metrics

Various method evaluation studies[15,16,17] have used different number of evaluation statistics to check how good the forecasting model is performing. Below are some of the metrics used in this study to compare different forecasting models where n is the number of samples ,$y_{(j)}$ and $\hat{y}_j$ are the actual and predicted values.

- Mean Absolute Error(MAE)

$$MAE = 1/n \sum_{j=1}^{n} |y_j - \hat{y}_j| \qquad (11)$$

2nd IEEE International conference on power Electronics, Intelligent Control and Energy systems (ICPEICES-2018)

- Mean Squared Error(MSE)

$$MSE = 1/n \sum_{i=1}^{n} (y_j - \widehat{y_j})^2 \qquad (12)$$

- Root Mean Squared Error(RMSE)

$$RMSE = \sqrt{1/n \sum_{i=1}^{n} (y_j - \widehat{y_j})^2} \qquad (13)$$

IV. EXPERIMENTAL ANALYSIS

A. Wind Data

The dataset in this paper is collected on an hourly basis over a one year period which includes u and v as zonal and meridional component that are transformed into wind speed using the (14) mentioned below. It involves two different wind farm zones of GEFCOM-14 at 10m height (above ground level). Also,the obtained data is found to be a univariate time series.

$$w = \sqrt{u^2 + v^2} \qquad (14)$$

Fig. 3. Wind Data (Zone 1)

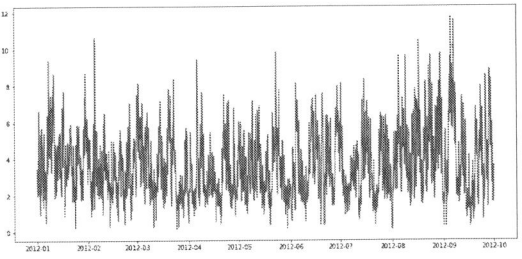

Fig. 4. Wind Data (Zone 2)

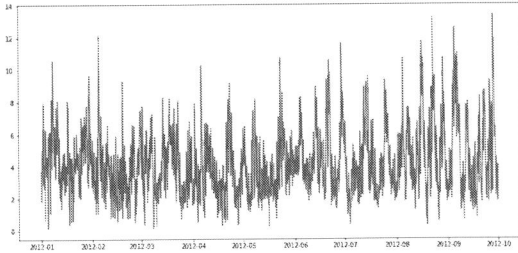

B. Experimental Setup

The whole task is performed on Amazon Web Service (Cloud) on p2.xlarge(GPU) instance that made the algorithms run much more faster on an average. Scikit-learn[18], keras[19], stastmodels[20] and other python based libraries are used in implementing the models.

C. ARIMA Plot Analysis

The wind data is not stationary initially and thus required to be differentiated to obtain stationarity. After taking the ARIMA parameter d = 1 as per Dickey-Fuller test by which the plots oscillate continuously around zero and the significant peaks in the ACF plot are dropped . After observing the plots , ARIMA(p,d,q) parameters for both the zones are selected. At the very beginning, the analysis did not help a lot in selecting the parameters, but the range is known, and thus, grid searching with different selected ARIMA models helped in finding parameters for both zones. However, the only exception is that the ARIMA model outperformed other methods in the first zone. This is also mentioned by Shi J et al.[21] in their hybrid forecasting study.

Fig. 5. ACF-PACF Plot (Zone 1)

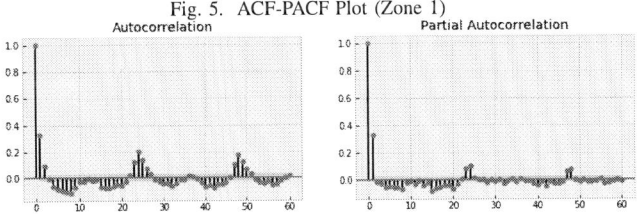

Fig. 6. ACF-PACF Plot (Zone 2)

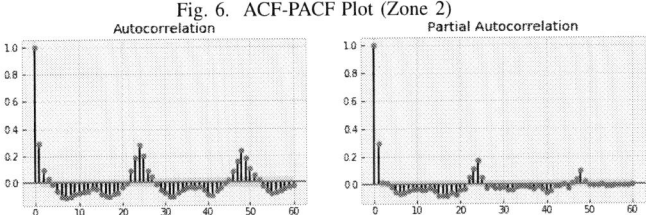

D. Wind Prediction by Optimized SVR

This section describes the steps for implementation of optimized support vector regression model for forecasting wind speeds at both zones.

- Step 1 (Transforming Data)
  The dataset in a total contains 6576 observations in both zones and is transformed into the supervised learning problem by shifting the dataset through the selected window size.
- Step 2 (Data Selection)
  The algorithm involves some amount of data to be selected as training and testing dataset, and by experimenting at specific divisions, it is found that 70/30(Training/Testing) is the best selection that showed reduced testing errors as compared to other splits. Time series involves the division of data into chronological order sets.
- Step 3 (Kernel Selection)
  Support Vector Regressors gives the ability to select different linear and non-linear kernels.Different non-linear kernels make it possible to transform the data into higher dimensional space and makes it easier for linear separation. Among different kernels available for

2nd IEEE International conference on power Electronics, Intelligent Control and Energy systems (ICPEICES-2018)

implementation, the performance is evaluated on each of the selected kernels separately on each dataset of different wind farm zones. Radial Basis Function(RBF) kernels are found to be performing much better on forecasting task as compared to others.

- Step 4 (Hyperparameter Optimization)
  The SVR function takes multiple parameters, and different combinations give different forecasting outputs. As to get the lowest metric error, multiple parameters are grid searched, and the best model is selected to give the predictions.

E. Recurrent Neural Network Forecasting Mechanism

Neural Networks resemble the working of neurons in human brain network, and a variant of recurrent neural networks known to be as Long Short Term Memory(LSTMs) are considered to be working well on memory-related tasks [22]. Thus, we chose to use these type of variants in our study and suitable look-backs are created on the dataset and found that seven look-backs are enough for the recurrent networks to tune the oscillations present in the series.

The essential task is to select the best model architecture that can suitably remember the events happening in previous time stamps and predict most accurate future forecasts. It has been shown that a one layer deep network architecture is suitable for time series prediction tasks [20]. Table I shows the best model architecture and training parameters used in this study to make forecasts.

TABLE I
PARAMETER SELECTION

Parameter	Selection
Layers	3
Loss	MAE
Optimizer	Adam
Batch Size	512
Epochs	500
Trainable Parameters	45,671

Using the above setup ,the forecast plots are generated as shown below.

Fig. 7. LSTM Forecast Plot(Zone 1)

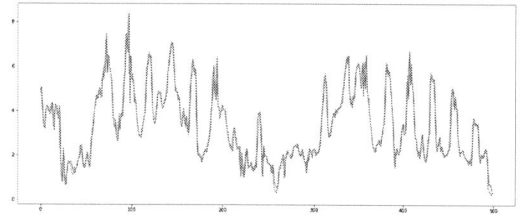

V. RESULT

This section evaluates results over different statistical methods including kernel variations in support vector regression.

Fig. 8. LSTM Forecast Plot(Zone 2)

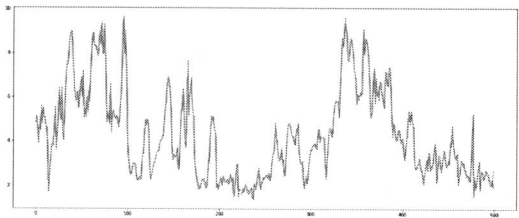

Results in each zone are calculated individually with the above discussed experimental setup.

- SVR(1) - Linear
- SVR(2) - Radial Basis Function
- SVR(3) - Polynomial

TABLE II
STUDY EVALUATION

Zone	Error	ARIMA	LSTM	SVR(1)	SVR(2)	SVR(3)
I	MAE	0.349	0.352	0.405	0.408	0.805
I	MSE	0.264	0.272	0.307	0.314	0.649
I	RMSE	0.514	0.522	0.554	0.560	0.805
II	MAE	0.372	0.358	0.408	0.403	0.625
II	MSE	0.311	0.294	0.322	0.325	0.700
II	RMSE	0.558	0.543	0.577	0.571	0.836

The above results showed that ARIMA outperformed RNN in the first zone as being the exception also discussed in the study by Shi J et al.[21] in their hybrid forecasting study. Comparing different kernels that can be implemented while using SVR for wind speed prediction showed that both linear and RBF kernels outperformed the polynomial one and produce many similar forecasts when compared to each other.

VI. CONCLUSION

In this paper,comparative study has been performed with different statistical and machine learning based algorithms like ARIMA,SVR and RNN(LSTM). Algorithms of each method is explained with their applications and advantages. Dataset from two different zones of GEFCOM-14 are taken for thorough study and wind speed prediction is done on this data. The study emphasizes more towards the fundamentals of implementing multiple statistical methods each having issues over selecting correct model parameters. Parameter selection is discussed and its effects on the predicted values. Wind prediction is done based on these parameters and comparative analysis of different algorithms is presented.

Future work aims to test the hybrid methods based on combination of statistical and machine learning parameters which could surpass the benchmark ARIMA and other individual models.

978-1-5386-6624-1/18 $31.00 © 2018 IEEE

2nd IEEE International conference on power Electronics, Intelligent Control and Energy systems (ICPEICES-2018)

ACKNOWLEDGMENT

This work acknowledges the financial a ssistance provided by the UI-ASSIST(US-India Collaborative for Smart Distribution System with Storage).

REFERENCES

[1] Bramer, Lisa, "Methods for modeling and forecasting wind characteristics" (2013). Graduate Theses and Dissertations. 13605.

[2] Yongqian Liu,Yimei Wang,Li Li1, Shuang Han, and David Infield,"Numerical weather prediction wind correction methods and its impact on computational fluid dynamics based wind power forecasting" Journal of Renewable and Sustainable Energy 8, 033302 (2016)

[3] J. L. Torres, A. Garca, M. De Blas, and A. De Francisco, Forecast of hourly average wind speed with ARMA models in Navarre (Spain), 2005 vol.79 (1) pp.65 - 77

[4] R. G. Kavasseri and K. Seetharaman, Day-ahead wind speed forecasting using f-ARIMA models, Renew. Energy, vol. 34, no. 5, pp. 13881393, May 2009.

[5] E. Erdem and J. Shi, ARMA based approaches for forecasting the tuple of wind speed and direction, Appl. Energy, 2011.

[6] L. Zjavka and S. Mik, Direct Wind Power Forecasting using a Polynomial Decomposition of the General Differential Equation, vol. 3029, no. c, 2018.

[7] Y. Zhao, L. Ye, P. Pinson, Y. Tang, and P. Lu, Correlation-Constrained and Sparsity-Controlled Vector Autoregressive Model for Spatio-Temporal Wind Power Forecasting, IEEE Trans. Power Syst., vol. XX, no. X, pp. 11, 2018.

[8] R. Karki, S. Thapa, and R. B. Billinton, A simplified risk-based method for short-term wind power commitment, IEEE Trans. Sustain. Energy, vol. 3, no. 3, pp. 498505, 2012.

[9] T. H. M. El-Fouly, E. F. El-Saadany, and M. M. A. Salama, One Day Ahead Prediction of Wind Speed and Direction, IEEE Trans. ENERGY Convers., vol. 23, no. 1, 2008.

[10] J. Jeon and J. W. Taylor, Using Conditional Kernel Density Estimation for Wind Power Density Forecasting, J. Am. Stat. Assoc., vol. 107, no. 497, pp. 6679, Mar. 2012.

[11] Y. Lin, U. Kruger, J. Zhang, Q. Wang, L. Lamont, and L. El Chaar, Seasonal Analysis and Prediction of Wind Energy Using Random Forests and ARX Model Structures, IEEE Trans. Control Syst. Technol., vol. 23, no. 5, pp. 19942002, Sep. 2015.

[12] I. Snchez,Short-term prediction of wind energy production International Journal of Forecasting, 22 (2006), pp. 43-56

[13] Ha Young Kim; Chang Hyun Won,"Forecasting the volatility of stock price index: A hybrid model integrating LSTM with multiple GARCH-type models",Expert Systems with Applications, ISSN: 0957-4174, Vol: 103, Page: 25-37

[14] (2018,Jul)https://en.wikipedia.org/wiki/Long/short-term/memory

[15] E. Cadenas, W. Rivera,Short term wind speed forecasting in La Venta, Oaxaca, Mexico, using artificial neural networks,Renew Energy, 34 (2009), pp. 274-278

[16] P. Ramasamy, S.S. Chandel, A.K. Yadav Wind speed prediction in the mountainous region of India using an artificial neural network model,Renew Energy, 80 (2015), pp. 338-347

[17] M.A. Mohandes, T.O. Halawani, S. Rehman, A.A. Hussain,Support vector machines for wind speed prediction,Renew Energy, 29 (6) (2004), pp. 939-947

[18] (2018,Jul) Github website https://github.com/scikit-learn/scikit-learn

[19] (2018,Jul) Github website https://github.com/keras-team/keras

[20] (2018,Jul) Github website https://github.com/statsmodels/statsmodels

[21] J. Shi, J. Guo, S. Zheng,Evaluation of hybrid forecasting approaches for wind speed and power generation time series,Renew Sustain Energy Rev, 16 (2012), pp. 3471-3480.

[22] J.S. Jung, R.P. Broadwater,Current status and future advances for wind speed and power forecasting,Renew Sustain Energy Rev, 31 (2014), pp. 762-777

978-1-5386-6624-1/18 $31.00 © 2018 IEEE

2nd IEEE International Conference on Power Electronics, Intelligent Control and Energy Systems (ICPEICES-2018)

Performance Analysis of a Distance Relay for Zone Identification

M. Kiruthika
Department of Electrical Engineering
Agnel charities 'Fr. Conceicao Rodrigues Institute of Technology,
Vashi .Navi Mumbai. India. m.kiruthika@fcrit.ac.in

Bindu S
Department of Electrical Engineering
Agnel charities 'Fr. Conceicao Rodrigues Institute of Technology,
Vashi .Navi Mumbai. India. bindu.s@fcrit.ac.in

Abstract—**Distance relay is a widely used protective scheme for transmission line protection due to their high speed fault clearance. A distance relay estimates the distance of the fault by calculating the impedance by measuring the current and voltage at the relay point and compares the result with a given threshold, which determines the correct zone of protection. This paper presents a distance relay model designed and modelled using MATLAB / SIMULINK which identifies the correct zone of the fault. The model is tested for various fault conditions. The results show that the relay is able to identify the fault zone and isolate the fault line as per the relay settings.**

Keywords—Distance relay, Zones of protection, relay settings.

I. INTRODUCTION

The demand of power is increasing day by day around the world. This lead to the development of proper infrastructure to transmit power from generation centres to the areas that are located at a large distance and is possible with long distance transmission lines .

Distance relays are generally used for primary and back-up protection of transmission lines. Under stressed conditions, the relay finds it difficult to differentiate a fault and a stress condition and consequently there is a possibility that relay may maloperate. Analysis of shape of the relay characteristic and choosing the settings of zone carefully are followed conventionally to avoid such maloperation. Sometimes, overreaching distance relays may trip under certain load conditions which have played a major role in cascading

blackouts. The most recent blackout in India in July 2012 was due to Zone3 relay tripping of 400kv Bina-Gwalior line due to load encroachment [3].

However, it is very difficult to prevent such situations under extreme conditions because it is not included in the regular setting process. Therefore, researchers are focusing on developing better methodologies to enhance the performance of distance relay. There are line protection schemes, protection schemes based on travelling wave, network based relay schemes, transient and adaptive based relay schemes developed for transmission line protection.

Zone3 is used in combination with the derivatives of voltage and current etc to prevent maloperation. However, to avoid this maloperation even impedance characteristics of the relay are used to analyse the problems. A novel zone3 scheme based

on combination of steady state components and transient components using a state diagram is proposed in [4]. S. Horowitz *et.al.* [5] felt the requirement to modify and reexamine the application of zone 3 when unexpected loading conditions occur. A table is presented by them which gives a clear understanding about where zone 3 should be used as a protection scheme and where it could be an option. Zone3 protection schemes to discriminate fault conditions and stress conditions using local measurements (impedance calculated from voltage and current signals) are discussed in [6-9]. Adaptive Zone 3 protection schemes developed to enhance the security of the protection of relay operation are discussed in [10-13].

Hence, the challenges available in this area are identifying approaches to reduce the unintended operation of the relay, assessing the behavior of relay during various events, proposing new relay principles, developing advanced fault analysis techniques.

Therefore, there is a need to revisit the transmission line protection. As a base to address the above issues, this paper presents a model designed in Simulink for distance relay. The relay model is designed and modelled with their settings. The values of current voltage and impedance seen by the relay when a fault occurs is observed. This value is compared with relay settings and a trip signal is sent as per the schedule to clear the fault. Impedance calculation for different types of faults is different and impedance algorithm is modelled. The model includes a subsystem for zone coordination which identifies the correct zone of fault and takes action as per settings. The model is simulated and tested for various fault conditions. The current and voltage waveforms are observed at the relay end and are presented in this paper.

II. DISTANCE RELAY:

Distance relays are usually implemented for protection of transmission lines. This operation is based on the impedance seen by the relay when a fault occurs at a specific location. Since, the impedance of a transmission line is proportional to its length, for distance measurement it is appropriate to use a relay that can measure the impedance of a line upto the reach point.

978-1-5386-6624-1/18 $31.00 © 2018 IEEE

A. Distance Relay Principle

Distance relay is the most widely used protective scheme for the protection of high and extra high voltage transmission and sub transmission lines.

B. Working of Distance relay

1) The voltage and current measured through voltage transformer and current transformer are the inputs of distance relay.

2) The distance relay algorithm now extracts the phasor of voltage and current signals.

3) Then, the apparent impedance which is seen by the relay is calculated and compared with preset settings.

4) If the impedance is falling into the protective zone, the relay is operated by assuming it as a fault and disconnects the faulty line.

5) The setting of the relay is done when it is installed in the system.

A backup scheme is also provided for each relay.

C. Zones of Protection

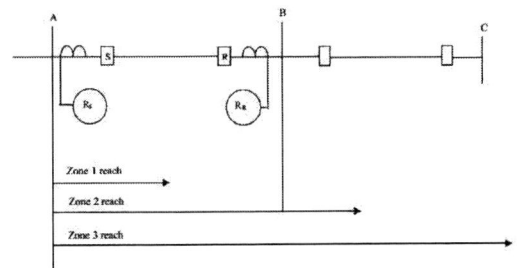

Fig. 1 Typical distance zone reaches

Zone 1 is an instantaneous (high-speed, no intentional delay) tripping zone that is normally set to provide less than 100% coverage of the protected line (AB as in Fig. 1).

Zone 2 is a time delayed tripping zone covering the protected line (AB) and a part of the next line (BC). It clears the faults in the protected line that is beyond zone 1 and also provides a backup for a failed zone 1 element in both the lines.

Zone 3 is also a time delayed tripping zone for backup protection. First two zones preserve continuity of service whereas zone 3 is a remote backup to clear a fault in the event of circuit breaker not tripping. Zone 3 is set to cover 100% of the next line (BC). Infeed can be ignored for zone 1 and zone 2. But, zone 3 should be set for maximum infeed conditions. This zone 3 is given a delay longer than that associated with zone 2 to achieve time coordination, and the time delay is typically in the range of 1-2 seconds [14].

III.   DESIGN OF A DISTANCE RELAY

The following flow diagram shows the design of a distance relay and the steps involved in identifying the fault conditions and isolating faulty line.

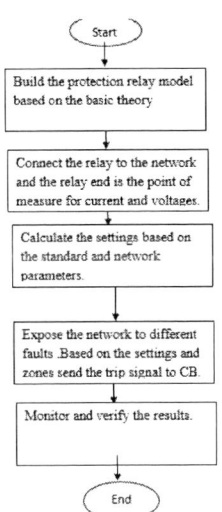

Fig. 2 Design of Distance Relay

IV.   MODELING AND SIMULATION OF A DISTANCE RELAY

Modeling and Simulation of a Distance Relay has been carried out using MATLAB / SIMULINK. The following steps have been followed to achieve the implementation.

The steps include :-

➢ Modeling of Transmission line
➢ Design of Relay Model
➢ Simulation of fault conditions

A. Modeling of Transmission Line and Relay:

Fig 3 shows the single line diagram of the modelled transmission line. The transmission line model is developed using MATLAB / SIMULINK. The developed model includes various components like three phase source, three phase load, transmission line, circuit breaker etc. which are readily available in the power system tool box of MATLAB/SIMULINK.

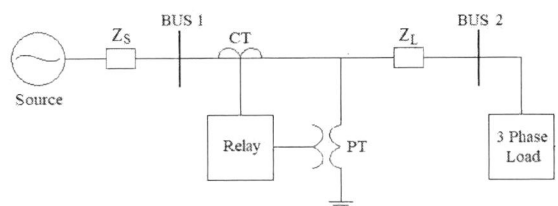

Fig. 3 Single line diagram of modeled transmission line

978-1-5386-6624-1/18 $31.00 © 2018 IEEE   207

The distance relay is modeled using MATLAB/SIMULINK and simulated for various fault conditions. As shown in the flow diagram in Fig 4 , the three phase voltage and current signal is given to the multiplexer from the measurement block to split into three single phase voltage and currents. Then, these quantities are passed through a low pass filter to eliminate the higher order harmonics. Fourier transform of the fundamental component eliminates the dc offset from the signal and provides magnitude and phase angle of the voltages and the current. Now, the values of voltage and current obtained are used to develop the mathematical model of impedance for various fault conditions. Impedance for various faults are calculated [14].

Zone coordination subsystem includes action blocks for each zone which gets activated based on relay settings given in table III [15].

Table III Relay settings

Zone	Setting	Time setting (in s)
Zone 1	80% 0f TL1	instantaneous
Zone 2	TL1+50%of TL2	0.3
Zone 3	TL1 +TL2+ 20% of TL3	0.5

The network under study consists of one three phase source supplying 400KV transmission line with three sections of 100km each. It is designed to deliver a power of 260 MVA to the load at the end of transmission line. The bus bars are equipped with current and voltage measurements.

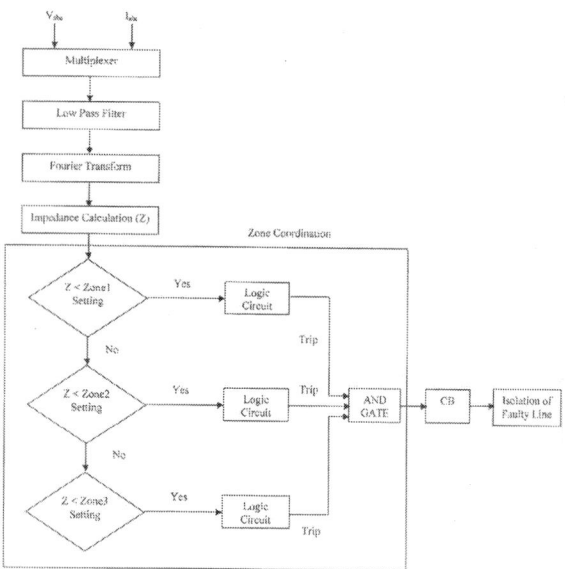

Fig. 4 Flow diagram of Relay model

IV.Simulation Results and Case studies:

Different case studies are performed using the developed relay model .

A. Performance of the relay without circuit breaker and relay settings.

The relay model is simulated tested for different types of faults like SLG, DLG and three phase faults. The results for three phase fault, line to ground faults(A-G) monitored at the relay end are shown in this paper.

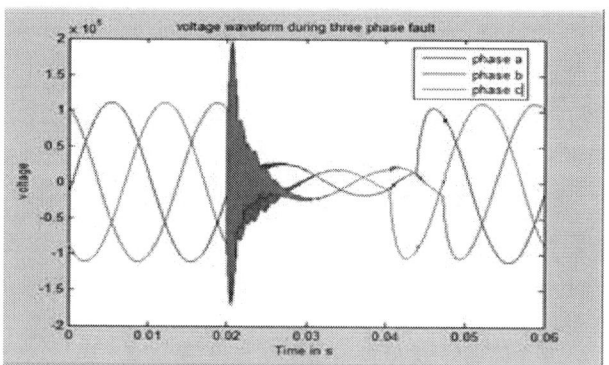

Fig 5: Voltage waveform for three phase fault

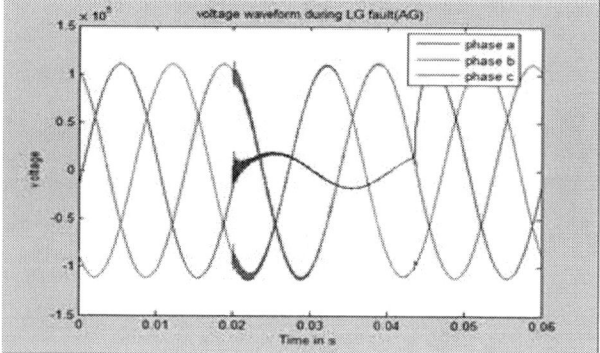

Fig 6: Voltage waveform for Line to Ground(AG) fault

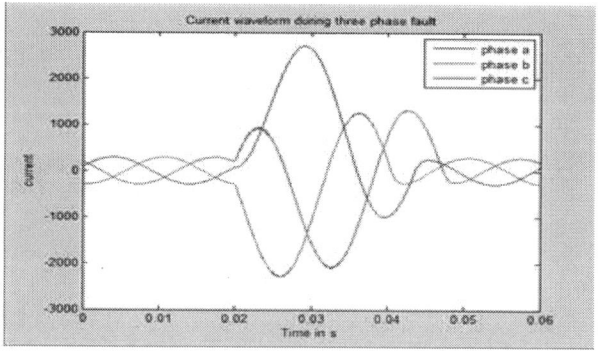

Fig 7: Current waveform for three phase fault

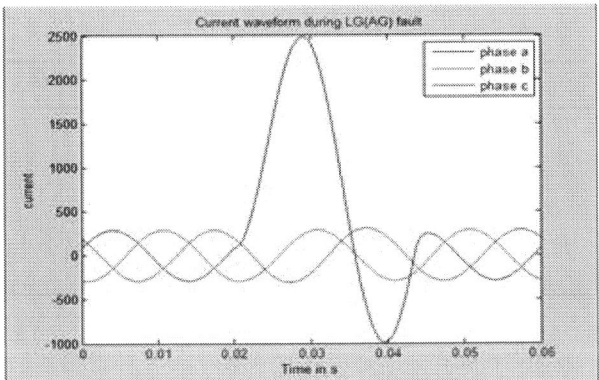

Fig 8: Current waveform for Line to Ground(AG) fault

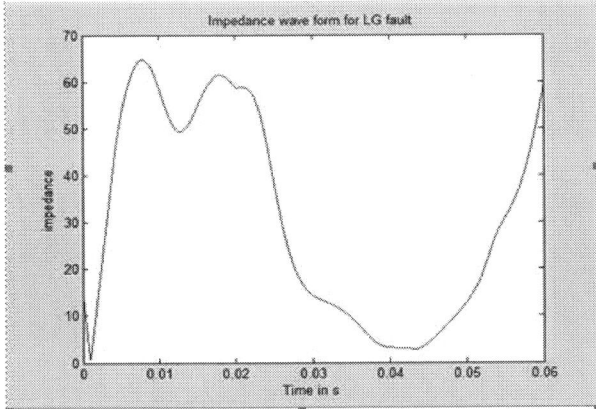

Fig 9 :Impedance waveform for Line to Ground (AG) fault

The simulation was carried out by assuming fault resistance 0.01 Ω and fault was initiated at 0.02 s and cleared at 0.04 s. From the waveform, it is observed that the impedance varies and drops during fault.

Fig 5, 6, 7, 8 shows the voltage and current waveforms observed at relay point for three phase and AG fault respectively without circuit breakers and relay settings. From the simulation, it has been observed that during fault ,voltage of the faulty phase drops drastically and the current increases. From Fig 9, it is observed that the impedance varies and drops during fault.

This concept is used for fault detection and fault isolation. Also, it is observed that these waveforms contain harmonics and dc offset and hence ,these quantities are taken during the relay design.

   B.   Performance of relay with circuit breaker and relay settings.

The relay model with circuit breaker and the relay settings is simulated for various fault conditions. Fault is applied to the network at 0.3 s and allowed to persist. Simulation results are observed by measuring current ,voltage at the relay end and impedance is calculated as per the algorithm given in table II. Relay settings are done as per the fault. In this, the impedance setting of distance relay is almost set as proportional to lengths.

   a)   Performance of relay for a three phase fault at zone1.

Fig. 10 and 11 shows the current and voltage waveforms for zone1. When a fault is applied at 80 km of length of transmission line which falls in zone 1 , then the relay operates by measuring the current and voltage at its end ,impedance gets calculated as per Impedance algorithm presented in Table: II. This impedance is then compared with the pre settings done in the relay. The impedance measured is within the reach of zone1 settings, hence the action block of zone1 gets activated. This action block in turn operates a logic circuit which issues a trip signal '0' as per the time set in the relay settings. At the same time, the action blocks of other zones would generate a signal '1' because fault has not occurred in their zone. These trip signals are further given to an AND gate which would give signal '0' to the breaker and the breaker opens at appropriate time set for zone 1. The time set for zone1 is 0.4 s (simulation time) ie. 0.1 s from the time of fault which is assumed as instantaneous clearance in this case. As per the settings the logic circuit sends a trip to the breaker for it to open. Therefore, it is observed from the waveform that the breaker opens the circuit at 0.4 s for zone1 fault.

   b)   Performance of relay for a three phase fault at zone 2.

When a fault is applied at 140 km of length of transmission line which falls in zone 2 at 0.3 s, the relay operates by measuring the current and voltage at its end. Since the fault occurred at zone2, the action block of zone2 gets activated and   a trip signal gets generated with some delay. The time set for zone2 is with a delay of 0.3 s from the time of occurrence of the fault. Hence, the circuit breaker opens at 0.6s which is observed from the waveforms. Fig 12,13,14 shows current, voltage and trip signal waveforms respectively. Similarly, the waveforms for AG fault at zone2 is shown in Fig 15,16.

   c)   Performance of relay for a three phase fault at zone3.

When a fault is applied at 200 km of length of transmission line which falls in zone 3 at 0.3 s , the relay operates by measuring the current and voltage at its end.Since the fault occurred at zone3, the action block of zone3 gets activated and   a trip signal gets generated with more delay. The time set for zone3 is with the delay of 0.5 s from the time of occurrence of the fault. Hence, the circuit breaker opens at 0.8s which is observed from the waveforms. Fig 17,18 shows current, voltage waveforms for zone3 respectively. Thus, the relay operates and sends a trip signal to the circuit breakers as per the fault location/zone.

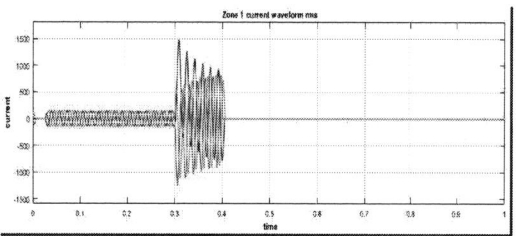

Fig 10. Zone 1current wave form

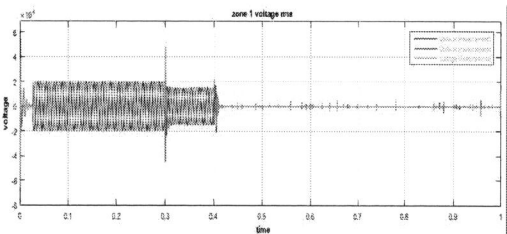

Fig 11 .Zone 1 Voltage Waveform

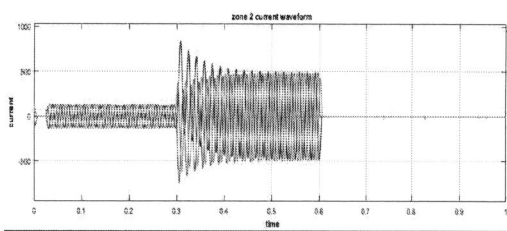

Fig 12. Zone 2 Current Waveform

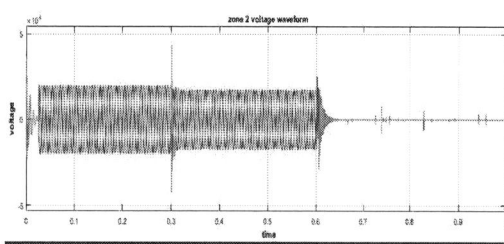

Fig 13. Zone 2 Voltage Waveform

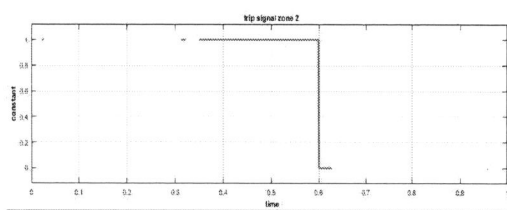

Fig 14 . Zone 2 Trip Signal

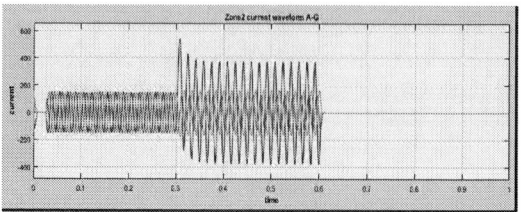

Fig 15. Zone 2 current wave form (A-G)

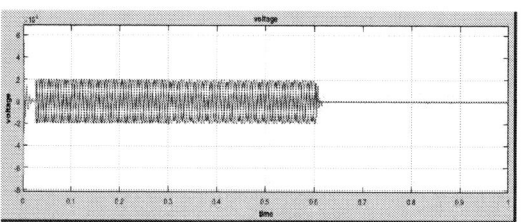

Fig 16 .Zone 2 Voltage wave form (A-G)

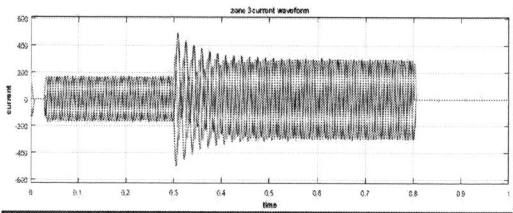

Fig 17 .Zone 3 Current Waveform

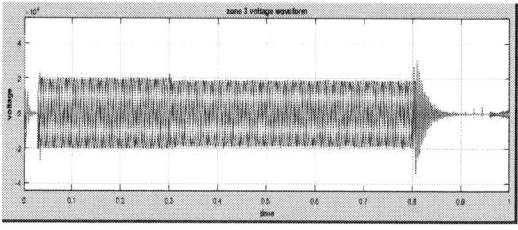

Fig 18. Zone 3 Voltage Waveform

978-1-5386-6624-1/18 $31.00 © 2018 IEEE   210

C. CONCLUSION

A Simulink model for a distant relay is designed. Current and voltage waveforms without and with fault conditions are observed. When a fault occurs it is observed that voltage decreases and high current flows in the line. The distance relay designed and modelled, operates for identifying the fault location/ fault zone and sends a trip signal as per the zone setting to the circuit breaker which opens and isolates the line. The designed model forms the basis for understanding the distance relay operation by measuring only local information. It is concluded that a distance zone protection scheme for identifying the correct zone of fault occurence is designed and the results are presented.

REFERENCES:

[1]. Hadisadat, "Power system analysis", 3rd edition, psapubishing, ISBN 0984543805, 9780984543809

[2]. August .14, 2003 Blackout: "NERC Actions to Prevent and Mitigate the Impacts of Future Cascading Blackouts," 2004 [Online]. Available: http://www.nerc.com/comm/PC/System Protection and Control Subcommittee SPCS DL/NERC Recommendations 2-10-04.pdf Accessed October 2015.

[3]. Report of the enquiry committee on grid disturbance in northern region on 30th July 2012 and in northern, eastern & northeastern region on 31st July. 2012. [On line].

[4]. Chul-Hwan, Jeong-Yong Heo and Raj K. Aggrawal, "An enhanced zone 3 algorithm of a distance relay using transient component and state diagram," IEEE Transaction on Power Delivery, Vol. 20, No. I, January 2005.

[5]. S. Horowitz and A. Phadke, "Third zone revisited," IEEE Transactions on Power Delivery, vol. 21, no. 1, pp. 23-29, 2006.

[6]. Seong-Il Lim, Member, IEEE, Chen-Ching Liu, Fellow, IEEE, Seung-Jae Lee, Member, IEEE, Myeon-Song Choi, Member, IEEE, and Seong-Jeong Rim, Member, IEEE, " Blocking of Zone 3 Relays to Prevent Cascaded Events," IEEE Transactions on Power Systems, vol. 23, no. 2, May 2008.

[7]. Amr El-Hadidy and Christian Rehtanz, "Mitigation of blackouts due to mal-operation of distance relays by using the fault resistance information," 25th International Conference on Electricity Distribution, June 2011.

[8]. Paresh Kumar Nayak, Ashok Kumar Pradhan and PrabodhBajpai, "Secured zone 3 protection during stressed condition," IEEE Transactions on Power Delivery, Vol. 30, No. 1, February 2015.

[9]. Prashant Gawande, Pallavi Bedekar, Vidyulata Joshit and Sanjay Dambhare, " An Algorithm to Secure the Zone 3 Operation of Distance Relay" 978-1-5090-2597-8/16/$31.00 ©2016 IEEE.

[10]. Jonsson, M.; Daalder, J.E., "An adaptive scheme to prevent undesirable distance protection operation during voltage instability," Power Delivery, IEEE Transactions on, vol.18, no.4, pp. 1174-1180, Oct. 2003.

[11]. M. Jin and T. S. Sidhu, "Adaption load encroachment prevention scheme for distance protection." Electric Power Systems Research, vol. 78. no. 10. pp. 1693-1700, 2008.

[12]. Avinash N. Sarwade, Pradeep K. Katti, Jayant G. Ghodekar, "A New Adaptive Technique for Enhancement of Zone-2 Settings of Distance Relay",Energy and Power Engineering, 2012, 4, 1-7 ,http://dx.doi.org/10.4236/epe.2012.41001 Published Online January 2012.

[13]. M. Ojaghi , K. Mazlumi, M. Azari , " Zone-3 Impadance Reach Setting of Distance Relays by Including In-feed Current Effects in an Adaptive Scheme ", IJE Transactions ,Vol. 27, No. 7, (July 2014) 1051-1060.

[14]. L. C. Wu, C. W. Liu and C. S. Chen, "Modeling and testing of a digital distance relay using Matlab/ Simulink", IEEE 2005.

[15]. S.Soman, A web course on digital protection,NPTEL,Accessed Jan 2015.

[16]. Christos A. Apostolopoulos.,y and George N. Korres " Real-time Implementation of digital relay models us-ing MATLAB/SIMULINK and RTDS" Euro. Trans. Electr. Power (2008).

[17]. M. H. Idris, S. Hardi and M. Z. Hassan, "Teaching Distance Relay Using Matlab/Simulink Graphical User Interface", Malaysian Technical Universities Con-ference on Engineering and Technology, November 2012.

978-1-5386-6624-1/18 $31.00 © 2018 IEEE

Artificial Neural Network based Model for Short Term Solar Radiation Forecasting considering Aerosol Index

Astitva Kumar, Mohammad Rizwan, and Uma Nangia

Depatment of Electrical Engineering

Delhi Technological University

Abstract— Solar photovoltaic (PV) power estimation is difficult due to intermittent meteorological parameters. Thus, it is utmost important to predict solar PV power for proper operation and control of renewable energy based power system. Before predicting solar PV system power, forecasting of solar irradiation is essential. In this paper an intelligent model based on ANN is developed to forecast the solar radiation using aerosol index as input data. Based on daily weather classification, the artificial neural network (ANN) using the Nonlinear Auto-Regressive for Exogenous inputs (NARX) approach is implemented to forecast the next day 3-hour solar radiation outputs. To find a precise prediction for 3-hour ahead PV power generation, earlier models were designed by considering temperature, humidity, wind direction and wind speed data. The predicted results of the proposed PV forecasting model are evaluated on the basis of statistical indicators Mean Square Error (MSE). The percentage MSE of the proposed model using Aerosol Index (AI) is 4.67%. Thus, showing that the proposed model is found to be accurate and may be used for PV Power forecasting and also in smart energy management system.

Keywords— Aerosol Index (AI), Solar Irradiance, Artificial Neural Network(ANN), Time Horizon, Photovoltaic Power Forecasting

I. INTRODUCTION

For providing reliability of operations and economic dispatch of modern power systems, the schedule for generation are usually planned for varying time horizons like real time, hour ahead, one-day-ahead, weekly, monthly, and yearly power system operations. Hence, making it almost essential for predicting power system load curves and generation outputs of renewable power plants, such as wind, and solar photovoltaic (PV) power stations. Forecasting the solar power output is needed for smooth operation of power grid and optimal management of the system [1]. One of the important challenge regarding future global energy supply is to have huge integration among renewable energy sources (particularly non-predictable ones as wind and solar) along with the existing or future energy supply structure. They would work in creating a sustainable balance between electricity production and its consumption at any given point of time.

Use of renewable energy into a predefined grid makes the system complex and the grid production and consumption continuity balance is also disturbed because of the non-linearity [2]. The studies have shown that PV power stations

generation are affected due to varying weather conditions. Thus to predict PV power has become a wide area of research of many scholars. PV forecasting can be done using two objectives [3-4]: a) predict environmental parameters related to the PV system, such as solar irradiation, then calculate the power, the other is b) predict the active solar power directly. Here, former technique is used for prediction. PV power forecasting can be broadly classified into two methods: a) cloud imagery with physical models, b) machine learning models. The choice of methods mainly depends on prediction horizon, objective selected. On reading various literature [5], these methods can be classified into two classes of techniques, firstly, Extrapolation and statistical process using satellite images. It is suitable for short-term forecast up to 6h. this can be further sub divided into very short-term (0-3h) forecast, and short term (3-6h). Secondly, Numerical Weather Prediction (NWP) is a technique suitable for forecasting up to two days ahead [6]. NWP models are coupled with post processing modules in combination with real time measurements.

Various PV power forecasting algorithms and methods have been proposed in the literature. To help in study of solar power output characteristics authors have used multiple linear regression analysis with various weather data and solar irradiance data. Some other machine learning algorithm like kNN regression algorithm, Support Vector Machine (SVM), Adaptive Fuzzy algorithm, Markov Chain, Cluster Evaluation and etc [7-11]. Among them one effective and efficient way to predict PV Power is using artificial neural network (ANN). Among various configuration of the neural networks, backpropagation (BP) is a good choice for mapping of non-linear function. ANN with BP has quite a few advantages such as strong learning ability, simplification of complex functions, large fault tolerance, good performance [12-13]. To develop the relationship between short wave irradiation or global horizontal irradiation (GHI) on solar PV panels located at any site with ambient parameters which include temperature (K), relative humidity (%), wind direction and wind speed (m/s) specifically for PV power calculation can be mathematically modelled [14-15]. Comprehensive way of calculation solar radiation on a site where PV is installed included knowing of sun-earth distance, latitude, and longitude of that location. So, to mathematically formulate a model to measure the irradiance is a tedious and complex task. Aerosols are mainly induced by power plant emission, desert dust, volcano smoke, biomass burning and so on. In fact, the optical depth of aerosol particles is strongly correlated with total solar radiation [16-17].

In this paper, an attempt has been made to develop correlation of AI along with other parameter, with GHI has been established. These suspended particles in the atmosphere along with reflecting down the sunlight also are the main reason of isotropic scattering which increases diffuse irradiance on earth's surface. Therefore, higher number of AI means greater attenuation of global solar irradiance which will reach earth's surface, thus, making solar PV panels even more less productive. The paper also aims in helping power utility staff, who do not have direct access to solar irradiance data as the meteorologists, geoscientists, or researchers do. So, it can be said that aerosols can be classified into a parameter affecting prediction of PV power, indirectly.

In this paper, a novel PV irradiation forecasting model is proposed, based on various environmental/meteorological data including AI, to get more accurate prediction. First, NARX ANN (Nonlinear Auto-Regressive Exogenous Artificial Neural Network) is trained by historical meteorological data of Delhi for the month of May, 2018 at an average of 3 hours, so, the temporal horizon considered is 3h for GHI forecasting in this paper. The efficiency of the proposed method is validated and tested by analyzing the mean square error (MSE) between predicted values and measured values.

The paper consists of following sections. Section I defined the introduction part. The short term forecasting model is presented and briefly described in section II. The proposed ANN forecasting model is defined in section III of the paper. The simulation results are given in section IV and concluded the paper in section V.

II. Short Term PV Forecasting Model

A. Artificial Neural Network

ANN are recently being used for nonlinear regression analysis in meteorology as they are very useful in data analysis and prediction. Mostly ANN is used for time series forecasting problems with nonlinear methods. This method is data driven and requires large amount of data as a prerequisite for effective prediction [18]. ANN method is defined by the following equation

$$y = y(x; w) = \sum_{j=1}^{m} w_i f(\sum_{i=1}^{d} w_{ij} x_j) \qquad (1)$$

ANN with d inputs, m hidden neurons and a single linear output unit defines a non-linear parameterized mapping from vector x to an output y is given in eq. 1. Every m hidden layer are tangent hyperbolic function $f(x) = \frac{e^x - e^{-x}}{e^x + e^{-x}}$. The parameter vector $w = (\{w_i\}, \{w_{ij}\})$ directs the nonlinear mapping of inputs to output during learning or training phase.

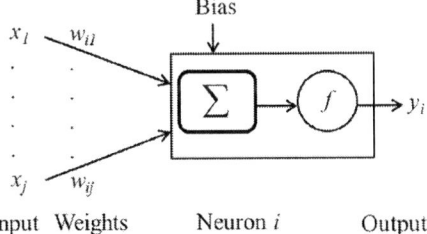

Fig. 1 Simple block diagram of ANN Structure

From the Fig. 1 and 2 shows that a classic Multi-Layer Perceptron (MLP) model constitutes three layers, namely, input, hidden, and output layers and comprises of weights, neurons and an activation function. The information flows in forward path from input to output layer. The eq. 1 is defined in fig. 1 above.

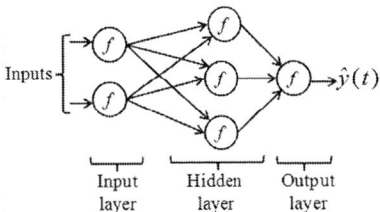

Fig. 2. Representation of a classic MLP Network

ANN model equivalent for 3h ahead forecasting of PV power is nonlinear autoregressive model with exogenous inputs (NARX) for time series [19]. The equation representing NARX model is

$$y(t) = f(y(t-1), y(t-2), ..., y(t-n_y), u(t-1),$$
$$u(t-2), ..., u(t-n_y) \qquad (2)$$

In eq. 2 the next value of the dependent output signal y(t) is regressed on previous values of the output signal and previous values of an independent (exogenous) input signal. Feedforward model is used for implementation of NARX model. One should be careful while building the model as a complex model will easily over fit when training the data. Techniques like Bayesian regularization, and Lavenberg-Marquardt learning algorithm can be implemented to control the complexity of ANN [20]. In the proposed model, latter is used as it stops training with a maximum fail parameter, this technique of training is based on approximation of Newton's method.

B. Aerosol Index (AI)

Aerosol Index (AI) is a measure of attenuation of the sunlight or solar radiation due to haze, dust and other suspended particles. Particle such as smoke, dust, cloud cover and etc. block the sunlight by absorbing, diffusing or scattering light. AI defines the amount of sunlight that is prevented from reaching the ground by aerosol particles. In other words, AI is the amount of UV-absorbing aerosols. It is noted that, GHI and PV power output at earth's surface is adversely affected by the concentration of aerosols, However, the relationship between AI values and radiation data is difficult to obtain, due to the complicated relationship of atmospheric parameters AI for the month of May,2018 at every 3 hours is used for training of ANN as one of the inputs.

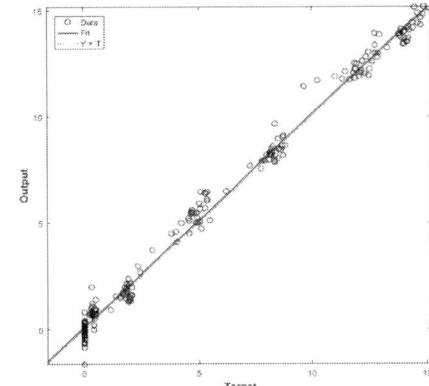

Fig. 3. Regression Line of AOD and irradiance (Wh/m²)

The above Fig. 3 shows the correlation of AOD to short wave radiation (Wh/m²) as the regression index comes to 0.89.

978-1-5386-6624-1/18 $31.00 © 2018 IEEE

C. Weather Parameters Considered for Forecasting

The accurate prediction and forecasting of GHI is difficult so it considers using AI, thus playing a significant role in prediction. So, a need to consider other physical parameters that effect the performance of the proposed forecasting model. The method proposed in this paper uses temperature, wind speed, wind direction and relative humidity as other parameters. Some papers have even used snowfall and precipitation but, the region considered for analysis is Delhi which has no snowfall and less rainfall when considered throughout the year.

As the value of the activation function is between 0-1, it is advisable to normalize the training data. This normalization is defined in eq. (3)

$$\dot{x}_i = \left. (x_i - x_{min}) \middle/ (x_{max} - x_{min}) \right. \tag{3}$$

Where x_i is the ith component in the original input data vector; x_{max} and x_{min} are the maximum and minimum values of the training data vector.

III. STRUCTURE OF THE PROPOSED MODEL

ANN is inspired from the human neural structure. It makes use of vast number of artificial neurons to imitate capability of living organisms. In recent years, backpropagation neural network is one of the most vastly used ANNs due to characteristic of making the system simple to understand and implement. The proposed ANN model is a feed forward multilayered BP network. Three layered Backpropagation network is shown in Fig. 4 where X7 is back propagated Y.

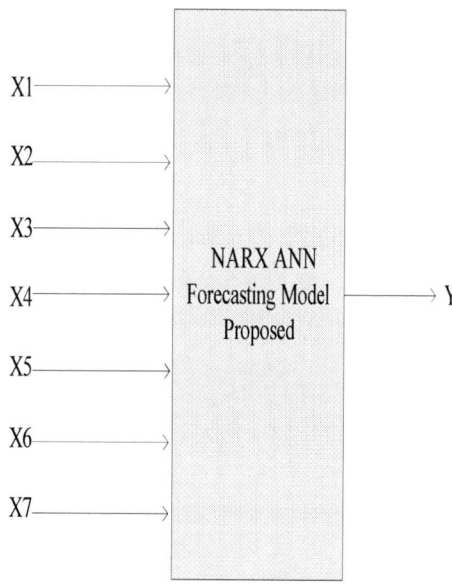

Fig. 4. Structure of Proposed NARX-ANN Forecasting Model

The above Fig. 4. shows the proposed prediction model using seven input variables to forecast the irradiance (Wh/m²) i.e. denoted by 'Y'. Here inputs are temperature (K), wind speed (m/s), wind direction, relative humidity (%), pressure (hPa), AI, and historical GHI data (Wh/m²). It is noted that the temperature and relative humidity is measured at 2 meters above ground, whereas, wind speed is at 10 meters above ground and pressure data is taken at ground level.

TABLE 1. Input data for ANN based Prediction Model

Input Variables	Meaning of Ambient Variables	Unit
X1	Ambient Temperature	Kelvin
X2	Wind speed	m/s
X3	Wind direction	Degree
X4	Relative humidity	%
X5	Pressure	hPa
X6	Aerosol Optical Depth (AOD)	550nm
X7=Y	Historical GHI Data	Wh/m²

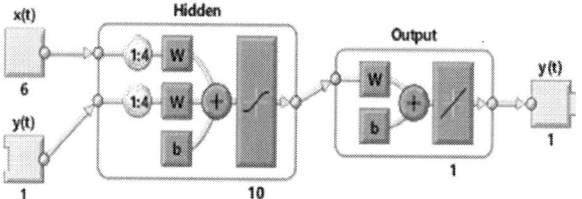

Fig. 5. Neural Network layout for PV Forecasting

The above Fig. 5 shows the six input along with an output fed to the input layer of the ANN. A total of 240 time steps in a set of 6 variables (mentioned before) are considered for training. The data is collected every minute and average is taken for every 3 hours of this each minute data. So for a day there are 8 data. The data is stored and analyzed for the month of May, 2018. The time horizon for prediction model is selected as 3 hours because AI doesn't change viably every minute or even every hour. The model is defined to use 10 hidden layers along with 4 delays applied to the inputs. The hidden layer and number of delays are chosen considering the performance of the system optimum, the mean square error and iterations are minimum for this selection. Generally, there is no apt method to decide the suitable number of hidden layers and delays. Simple "trial and error" have been used to find a suitable number of hidden layer neurons and delays with balanced precision and computational burden. Lavenberg-Marquardt algorithm is used for training as it is fast but is a bit complex and takes more memory. This is also known damped least squares method which is generally used in curve fitting applications. This technique interpolates between Gauss-Newton algorithm and gradient descent, thus making it to be more robust.

TABLE 2. Parameters for training of Neural Networks

Neural Network Variables	Numbers
Total neurons in input layer	7
Total neurons in hidden layer	10
Total neurons in ouput layer	1
Total Delays	4

In Fig. 6 clear sky GHI (Wh/m²) is shown for a day (24 hours) of the location which is Delhi Technological University. The graph plotted is showing GHI every minute of a day of May, 2018 i.e. 1440 minutes in total.

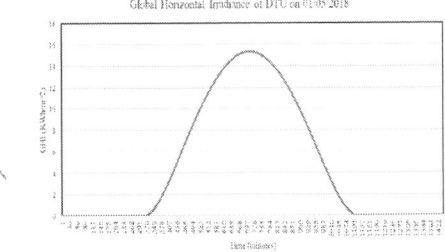

Fig. 6. Graph representing GHI of DTU on a single day

Mean Square Error (MSE) is selected as the performance indices for evaluation of the predicted result. It is also known as mean squared deviation. In the below given eq. 4 y is the vector of the observed values and $\dot{y}$ is vector of n predictions.

$$MSE = \frac{1}{n}\sum_{i=1}^{n}(y - \dot{y})^2 \qquad (4)$$

IV. SIMULATION RESULTS

The training data of 30 days is obtained for the site, Delhi Technological University, with latitude of 28.74☐ North and 77.11☐ East. Before prediction to be performed the historical data is analyzed. The meteorological data is collected every minute but, as discussed earlier the AOD doesn't vary much every minute or even every hour. Thus, a time horizon considered is 3 hours. Therefore, there are 240 data sets of inputs.

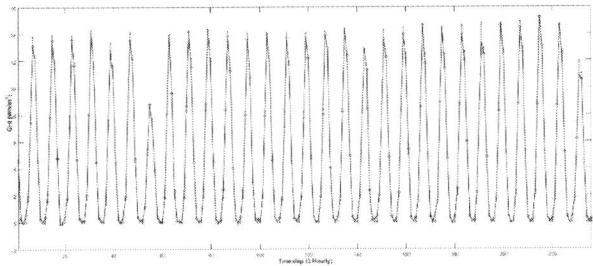

Fig. 7. Global Horizontal Irradiance of May 2018

Fig. 7 shows the 30 days GHI which is used for training of the proposed ANN predictor model. They can be classified into cloudy day and sunny day as overcast and rainy weather is not present in the month of May '18 for the site that is selected for test.

TABLE 3. Results of proposed ANN predictor technique

Phases of ANN model	Target Values	MSE	R
Training	192 (80%)	0.0607	0.998
Validation	24 (10%)	0.6910	0.989
Testing	24(10%)	0.1889	0.996

The above Table 3 shows the number of data sets used for different stages of modelling of the proposed ANN network to accurately predict and forecast the solar irradiance. For training of NN 80% of the data was used, whereas for validation and testing 10% each was selected. These values gave satisfactory results as MSE is limited to 0.691 (highest in validation phase) and regression factor (R) was always around 0.99. In the Fig. 8 the error is shown against number of epochs for the training and validation of the data. Best validation performance is noted at epoch 5 which corresponds to minimum error of 0.9021. Training of the proposed ANN

prediction model yields a Regression coefficient (R) of 0.994 considering the above given variables in Table 1. This coefficient defines the rate of change of independent variable (here it is GHI) for a unit change of dependent variables (here, these variables are wind speed, temperature, humidity, etc). The Fig. 9 below shows the regression line for the training of data.

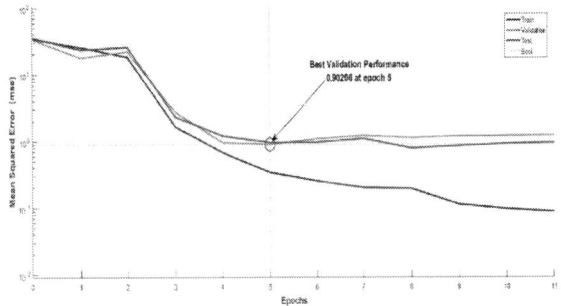

Fig. 8. Training performance of Proposed method

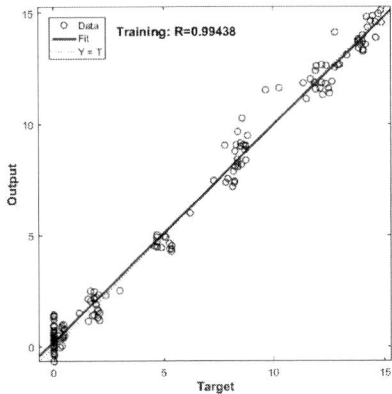

Fig. 9. Regression line of training phase of proposed model

A. Cloudy Day Test

The predicted PV irradiance for a cloudy day can be performed using the proposed model, the forecasted result is shown in Fig. 10. From the results it can be easily seen that error ranges from -0.2 to 0.5. The mean square error (MSE) is 2.66% for a cloudy day with 3 hours as the time horizon.

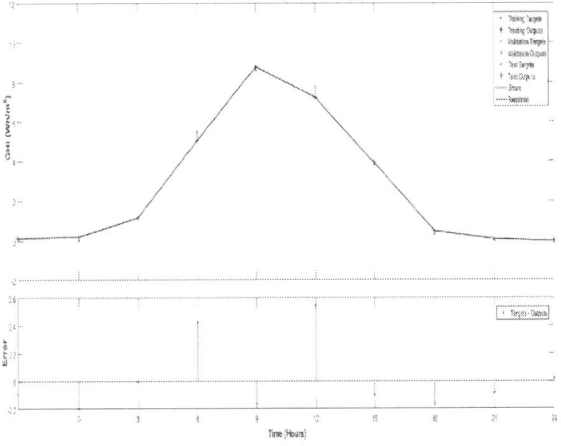

Fig. 10. Cloudy day prediction response of proposed model

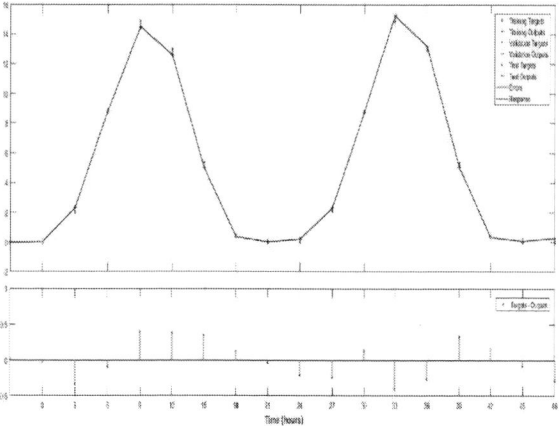

Fig. 11. Sunny Day prediction response of proposed model

B. Sunny Day Test

The forecasted values of 3 hourly PV irradiance for the sunny day weather are shown in Fig. 11. The results coincide well within the measured values for prediction with proposed model. In Fig. 11, shows two consecutive days with sunny weather conditions. The error ranges from -0.5 to 0.4 and MSE is around 2.43%.

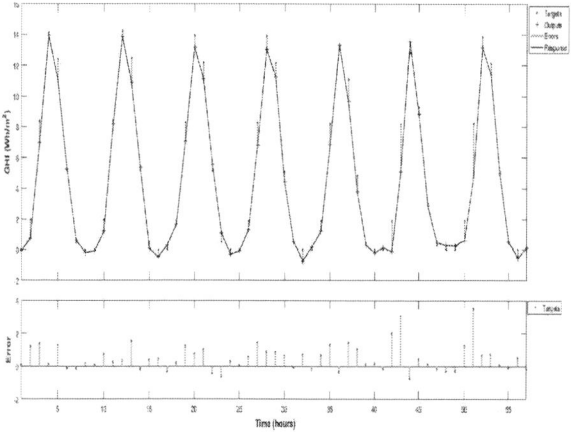

Fig. 12. Prediction Response of 7 days of proposed method

C. Testing results for 7 day period

The proposed NARX-ANN model is also tested for predicting next 7 days of GHI i.e. first week of June, 2018 as depicted in Fig. 12. MSE is 4.67% for these days of the forecasted data.

V. CONCLUSION

A novel model using NARX-ANN developed for irradiation forecasting based on Lavenberg-Marquardt training method. The proposed model uses aerosol index (AI) as a key input for effective and accurate prediction of GHI for every 3 hours. This model helps as a baseline in defining the meteorological conditions for PV generation more aptly by considering AI. Further, validation and tests have been performed by predicting the irradiance of 7 days on DTU (as the site) and the results when analyzed are within the parameters, as MSE is 4.67%. This forecasting accuracy will assist solar power operators as they can schedule PV power and also conventional power generation accordingly, also spare more reserve capacity for emergencies. Also, an added advantage of this method that it will be able to perform predictions even for smaller time horizons like 15 min and 30 min, if historical data of inputs are provided with the same time horizons. The developed model has shown that application of meteorological data from remote sensors will be an added advantage for renewable PV forecasting and scheduling.

REFERENCES

[1] H. Long, Z. Zhang, and Y. Su, "Analysis of daily solar power prediction with data-driven approaches," Application Energy, Vol. 126, pp. 29–37, 2014.

[2] M. Jamil, S. Kirmani, and M. Rizwan, "Techno-economic feasibility analysis of solar photovoltaic power generation: A review," Smart Grid and Renewable Energy, Vol. 3, pp. 266-274, 2012.

[3] Cyril Voyant, G. Notton, S. Kalogirou, M. Nivet, C. Paoli, F. Motte, and A. Fouilloy, "Machine learning methods for solar radiation forecasting: A review," Renewable Energy, Vol. 105, pp. 569-582, 2017.

[4] S. Kirmani, M. Jamil, and M. Rizwan, "Empirical correlation of estimating global solar radiation using meteorological parameters," International Journal of Sustainable Energy, Vol. 34, pp. 327-339, 2015

[5] M. David, F. Ramahatana, P.J. Trombe, and P. Lauret, "Probabilistic forecasting of the solar irradiance with recursive ARMA and GARCH models," Solar Energy, Vol. 133, pp. 58-67, 2016.

[6] H.-T. Yang, C.-M. Huang, Y.-C. Huang, and Y.-S. Pai "A weather-based hybrid method for 1-day ahead hourly forecasting of PV power output," IEEE Transaction on Sustainable Energy, Vol. 5, pp. 917-926, 2014.

[7] H. K. Elminir, Y. A. Azzam, and F. I. Younes, "Prediction of hourly and daily diffuse fraction using neural network, as compared to linear regression models," Energy, Vol. 32, pp. 1513–1523, 2007.

[8] S. I. Sulaiman et al., "Artificial neural network versus linear regression for predicting grid-connected photovoltaic system output," IEEE International Conference on Cyber Technology Automatic Control and Intelligent System, pp. 170–174, 2012.

[9] Y. Li and J. Niu, "Forecast of power generation for grid-connected photovoltaic system based on Markov chain," IEEE Asia-Pacific Power Energy Engineering Confeence, pp. 1–4, 2009.

[10] L. Ying-zi, L. Ru, and N. Jin-cang, "Forecast of power generation for grid-connected photovoltaic system based on grey model and Markov chain," IEEE Conference on Industrial Electronics Applications, pp. 1729–1733, 2008.

[11] S. Junseok et al., "Development of a Markov-chain-based energy storage model for power supply availability assessment of photovoltaic generation plants," IEEE Transaction on Sustainable Energy, Vol. 4, pp. 491–500, 2013.

[12] F. Wang, Z. Mi, S. Su, and C. Zhang, "A practical model for single-step power prediction of grid-connected PV plant using artificial neural network," IEEE PES Innovative Smart Grid Technology Asia (ISGT), pp. 1–4, 2011.

[13] Y. Ting-Chung and C. Hsiao-Tse, "The forecast of the electrical energy generated by photovoltaic systems using neural network method," International Conference on Electronics and Control Engineering, pp. 2758–2761, 2011.

[14] A. Moreno, M.A. Gilabert, and B. Martínez, "Mapping daily global solar irradiation over Spain: a comparative study of selected approaches", Solar Energy, Vol. 85, pp. 2072-2084, 2011.

[15] M. Rizwan, M. Jamil, and D.P. Kothari, "Solar energy estimation using REST model for PV-ECS based distributed power generating system," Solar Energy Materials and Solar Cells, Vol. 94, pp. 1324-1328, 2010

[16] J. Liu, W. Fang, X. Zhang, and C. Yang, "An Improved Photovoltaic Power Forecasting Model with the Assistance of Aerosol Index Data," IEEE Transactions On Sustainable Energy, Vol. 6, No. 2, pp. 434-442, 2015.

[17] H. Breitkreuz et al., "Short-range direct and diffuse irradiance forecasts for solar energy applications based on aerosol chemical transport and numerical weather modeling," Journal on Application of Meteorology and Climatol., vol. 48, pp. 1766–1779, 2009.

[18] D. Sánchez, P. Melin, and O. Castillo, "Optimization of modular granular neural networks using a firefly algorithm for human recognition," Engineering Application of Artificial Intelligence, Vol. 64, pp. 172–186, 2-17.

[19] E. Cadenas, W. Rivera, R. Campos-Amezcua, C. Heard, "Wind Speed Prediction Using a Univariate ARIMA Model and a Multivariate NARX Model," Energies, Vol. 9, 2016.

[20] T. Gong, T. Fan, J. Guo, and Z. Cai, "GPU-based parallel optimization of immune convolutional neural network and embedded system," Engineering Application of Artificial Intelligence, Vol. 62, pp. 384–395, 2017.

A Novel Scheme for Potential Identification of Customers for Demand Response

Pradeep Kumar Mallick
Centre for Energy Studies
Indian Institute of Technology Delhi
New Delhi-110016, India
e-mail: prdpkrmallick@gmail.com

Arjun Tyagi, *Student Member, IEEE*
School of Electrical Engineering
Shri Mata Vaishno Devi University
Katra-182320, India
e-mail: atiitd@hotmail.com

Ashu Verma, *Senior Member, IEEE*
Centre for Energy Studies
Indian Institute of Technology Delhi
New Delhi-110016, India
e-mail: averma@ces.iitd.ac.in

Abstract—Demand Response is one of the most promising technology to improve the financial benefits of utility and consumers, reliability of the grid and ancillary services to the power system etc. Due to the growing use of renewable energy and electric vehicles in the electricity structure, the provision of DR from loads like residential buildings or electric vehicles has become out-dare. There are numerous challenges in the real implementation of DR programs for such customers. Therefore, in this work a novel potential customer identification process has been developed for residential customers which uses only the load profile rather than conventional pilot project data. This method can be used in countries like India where there is lack of field study data available to support residential customer selection process for demand response. The effectiveness of the customer selection process has been verified by a DR program with customer benefit maximization objective.

Keywords—demand response, customer identification, load profiling.

I. INTRODUCTION

Energy supply and demand balancing is one of most challenging problem, as the storage is not sufficient and economically viable. Distribution utilities are facing a hard time to meet the peak demand both due to the high price of the spot market and congestion in the distribution system [1]. Therefore, regulatory authorities and policy makers worldwide are increasingly appreciating load flexibility, that is known as Demand Response (DR) [2, 3], to increase flexibility and enhance the system efficiency. Demand response (DR) is introduced, when the reliability of the system is under threat and market price for the short-term bidding to supply peak load is high [4]. Demand response (DR) is the process of modifying the electricity consumption pattern of the consumers by providing certain incentives to them. The advent of smart grid has enabled implementation of sophisticated DR techniques [5]. DR helps the integration of renewable distributed energy resources by modifying the load profiles to improve the energy efficiency and reliability [6, 7]. Demand Response is not only beneficial for distribution utilities but also to the to the consumers as they can lower their electricity bill by participating in DR programs [8-11].

According to the load profile, consumers can be classified as Residential, Commercial and Industrial. The quantum of the load is quite different among these sectors. Moreover, according to the type of control, DR can be classified into two categories, centralized control and distributed control. Communication infrastructure and decision center are the major difference between these two [12]. In centralized control, DR aggregator takes the decision and communicates it to the consumers individually. There is no communication link between the consumers. In decentralized control, only DR aggregator communicates the price signal to the consumers and the decision is taken by the consumer. There is a communication infrastructure between consumers. Decentralized control can have multiple objectives such as voltage balance, frequency regulation and dynamic stability etc. to improve the health of the grid.

However, power system has made significant technical advancement, still distribution system is struggling to address the growing demand. The energy demand is increasing in all the sectors as residential, industrial and commercial. The residential sector has significant energy conservation potential as averaged worldwide, the residential sector consumes approximately 30% of the total energy consumption [13]. Therefore, it is worthwhile to focus on techniques to decrease energy consumption in residential sector.

The basic idea behind designing a good Demand Response strategy is to give certain incentives to the electricity consumers to alter their consumption pattern which is beneficial not only to the consumers from the financial perspective but also to the electricity provider and overall reliability of the power system [14]. Various pricing mechanism for DR schemes such as Time of Use (TOI), Real Time Pricing (RTP) and Critical Peak Pricing (CPP) have been developed which addresses the above objectives [8].

Identification of target customer for DR and their DR potential is very crucial in designing DR scheme. Analysis of data generated from the Advanced Metering Infrastructure (AMI) is used for load profiling to identify target customers and their load flexibility. Clustering techniques such as K-means, Fuzzy K-means, Principal Component Analysis (PCA), Support Vector Machine (SVM) etc. gives better insight into the load profile data [15].

A stochastic knapsack formulation to framed the suitable customer identification has been proposed in [16]. Smith et al. [17] have used, only the peak timing as a parameter for identifying the target customers. The stability of load profile as reliability parameter is proposed in [18] and used it for customer selection.

A load modeling approach for residential customers has been developed in [19, 20], where they start form load curve of a large area and go deep into the load profile of individual customer by using the socioeconomic and geographical data of that region. In [21], a Monte Carlo Markov chain on time is used to model residential loads by considering price elastic loads, as well as controllable loads for demand response. A statistical model for residential demand and price sensitivity to design optimal real time tariff for utility is given in [22]. Kirschen et al. [23] have formulated load elasticity in the form of a matrix structure, where diagonal elements characterize self-elasticity and off-diagonal element characterizes cross elasticity of the load. Though some of the above work can be used to evaluate demand-price sensitivity, but they do not reflect the actual behavior of the consumer due to the inherent assumptions in the load modeling.

In the literature, broadly, two types of studies are done on the DR pilot data to identify the load flexibility and responsiveness as, technical modeling of the electricity demand of the consumer and econometric study of the consumers. Where, technical load models give insight into consumers behavior and scope for energy efficiency improvement program. An econometric study of the consumers requires pilot DR data for evaluation of load flexibility and incentive sensitivity. The objective of this work is the identification of the target residential customers prior to the implementation of actual DR program, while considering various technical parameters to maximize the overall benefit.

This paper is originated in five sections. In section II and III, the mathematical model of identification of target customers and formulation of DR are given. Section IV describes the results of different technical parameters and demand DR. The conclusions are described in section V.

II. IDENTIFICATION OF TARGET CUSTOMERS

The objective of this work is the identification of the target residential customers prior to the implementation of actual DR program. Three technical parameters (quantum of load, the flexibility of load and timing of occurrence of the peak) have been selected for the identification process. Load flexibility is further evaluated using the presence of non-essential appliances and presence of HVAC appliances. The workflow for identifying suitable residential customers for DR program is shown in Fig.1.

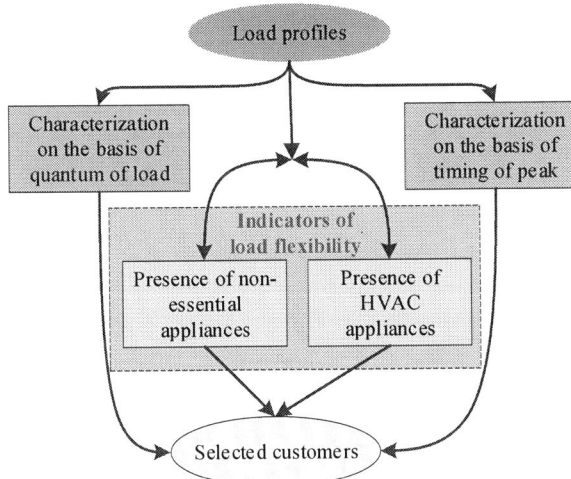

Fig. 1. Workflow for target customer Identification.

A. Quantum of Load

According to the load profile, consumers can be classified as residential, commercial and industrial. A good DR scheme should treat these sectors separately as the behavior and the quantum of the load is quite different among these sectors. Traditionally participation of residential consumer in DR program was highly ignored due to their low consumption, while energy intensity of a residential consumer is also one of the most important parameters to determine its effective DR. This indicates the importance of quantum of the residential load.

B. Timing of Peak:

In this process clusters are selected for a particular DR event based on their time of occurrence of the peak, this is important as consumers have the highest potential during their peak consumption. The k-means Clustering methods have been used to segment load profiles based on their peak occurrence time. The objective of the k-means algorithm is to decrease the sum of square error within the cluster. Given a d dimensional data $X_i = (x_{1i}, x_{2i}, x_{3i} \dots x_{di})$ with 'm' data points $S = \{X_1, X_2, X_3 \dots X_m\}$ k-means algorithm partitions the data into k clusters with an objective of decreasing inter-cluster distance and increasing intra-cluster distance. The algorithm for a simple k-means clustering with Euclidian distance measure is given below:

$$\text{Euclidian distance: } d(X, \mu) = (X - \mu)(X - \mu)' \quad (1)$$

C. Load Flexibility:

Presence of HVAC appliance gives an indication of load flexibility, as they have heat storage capacity. Energy consumption of HVAC appliance has a high correlation with temperature. In this part, we hypothesize that seasonal variability is a gauge of the presence of HVAC appliances large inter season temperature variation. Inter-season variance is calculated from the annual load profile matrix X as

$$X_{seasosnal\_mean} = \sum_{i \in season} X_{ij} \quad (2)$$

$$Var_{seasonal} = \sqrt{\frac{1}{3} \sum_{k \in season} |X_{seasonal\_mean} - \mu|} \quad (3)$$

$$\mu = \frac{1}{4} \sum_{k \in season} X_{seasonal\_mean} \quad (4)$$

III. FORMULATION OF DR

A DR program has been formulated to verify the performance of the selected consumers. The problem has been formulated with an objective of maximizing utility profit while satisfying the consumer constraints. The formulation is mathematically expressed as:

$$objective: minimise \sum_{t=1}^{N} \sum_{i=1}^{S} U(t) P_c^i(t) \times t \quad (5)$$

Subjected to

$$P_c^{min} \leq P_c(t) \leq P_c^{max} \quad (6)$$

$$P_c^{max} = P_{baseline} + \gamma^i C(t), \forall i \in S \quad (7)$$

$$P_c^{min} = P_{baseline} - \gamma^i C(t), \forall i \in S \qquad (8)$$

$$\sum_{t=1}^{N} P_c^i(t) \times t = E^i, \forall i \in S \qquad (9)$$

The above-formulated optimization problem takes into account maximum and minimum power consumption constraints P_c^{max} and P_c^{min} respectively. γ incorporates the flexibility associated selected house. Daily total energy consumption constraint has also been included in the above formulation.

IV. RESULT AND DISCUSSION

A. Identification of Target Customers

This work has revealed the load data of 200 residential customers for 365 days. The load is attained at 10 min. interval. For the identification process, the results of three technical parameters are as follows,

1) Quantum of load: Fig. 2 shows the probability density plot of consumers average daily consumption. A normal distribution with 29.24 kWh mean and 5.2 kWh standard deviation has been fitted in the distribution to show that load distribution across a region follows normal distribution. From the figure, it can be observed that the minimum average daily consumption of the sample residential customer is 16.55 kWh. Customers have been categorized into three load groups (16.55-24), (24-35) and more than 35 kWh represents the low, medium and high load group respectively. Fig. 3 shows variability of instantaneous power consumption associated with different load groups. It can be observed that variability is proportional to the energy intensity of the load group, as highest for high load consumers and lowest for low load consumers.

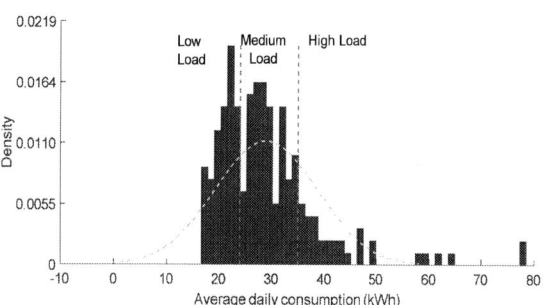

Fig. 2. Customer segmentation based on the quantum of load.

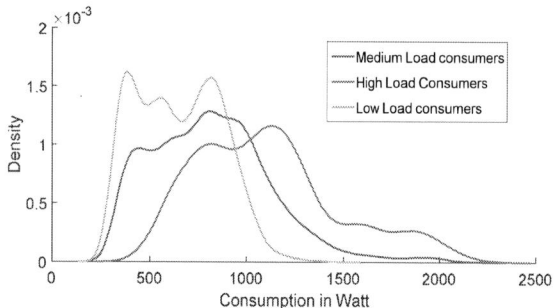

Fig. 3. Variability associated with load groups.

2) Timing of peak: Fig. 4 shows the cluster centroids of 6 clusters estimated by the k-means algorithm with Euclidian

distance metric. The timing of peak for different segmants are 2:49 PM (single peak), 2:32 PM (single peak), 9:00 AM & 8:00 PM (dual peak), 5:40 PM (single peak), 7:20 PM (single peak) and 8:20 PM (single peak).

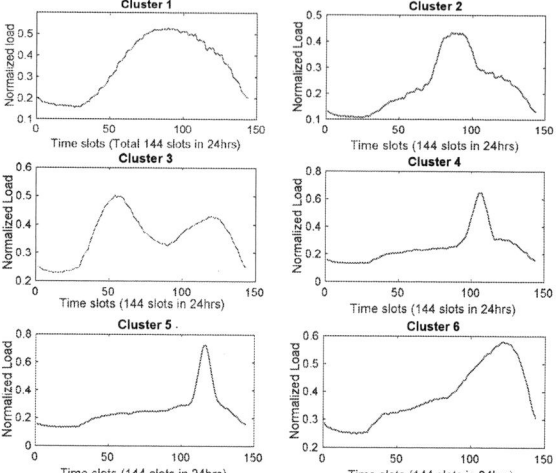

Fig. 4. Cluster centroids (K-means: Euclidian distance)

3) Load flexibility: As explained in section II, presence of HVAC appliance gives an indication of load flexibility and the energy consumption of HVAC appliance has a high correlation with temperature. The seasonal average consumption of a typical customer for different seasons is shown in Fig. 5. It can be observe form the figure that power consumption is higest in summer season and minimum in winter.

Fig. 6 shows the histogram plot of seasonal variance, houses lying on the right of red dotted line have potential HVAC presence.

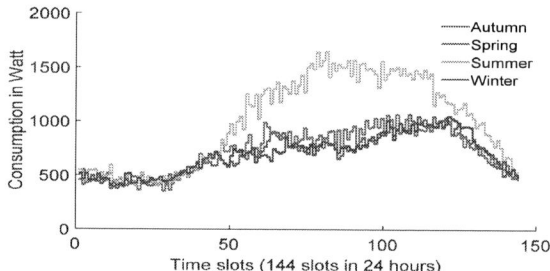

Fig. 5. Seasonal average consumption of a typical customer.

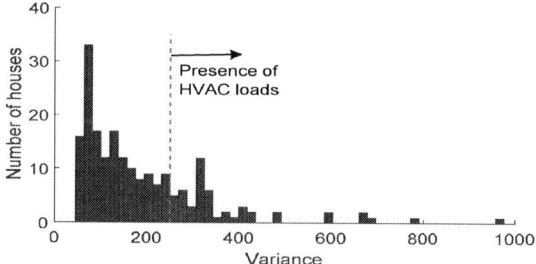

Fig. 6. Histogram plot of seasonal variance.

B. Formulation of DR

Utilities which implement DR program always try to maximize their profile while keeping consumer comfort and

978-1-5386-6624-1/18 $31.00 © 2018 IEEE

reliability of the grid in mind. Therefore, an optimization problem for DR has been solved through linear programming with an objective explained in equation 5.

Fig. 7 and Fig. 8 shows the electricity market price and time of use price structure respectively. Utility buys electricity from the spot market in a 15 min interval time slot, Time of Use (TOU) price structure is generally fixed for a particular season.

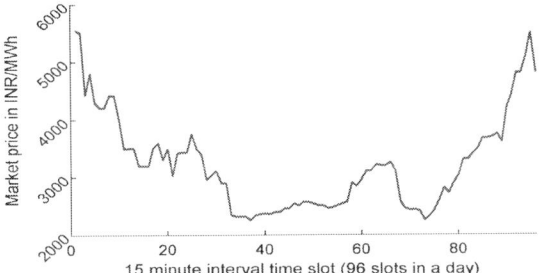

Fig. 7. Market price obtained from Indian Energy Exchange.

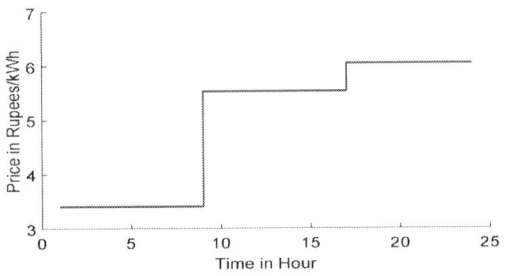

Fig. 8. TOU price structure.

Fig. 9 shows the before and after DR load profile of a residential consumer. Valley filling and load shifting due to TOU price structure can be clearly seen in the figure. Though this customer does not show its own peak clipping, it contributes towards system peak clipping. Fig. 10 shows the effect of DR with selected consumers on system load profile, valley filling and peak reduction of 6% can be clearly observed in it.

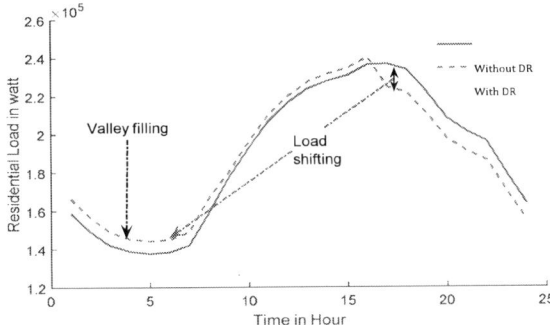

Fig. 9. Effect of DR on a selected residential consumer.

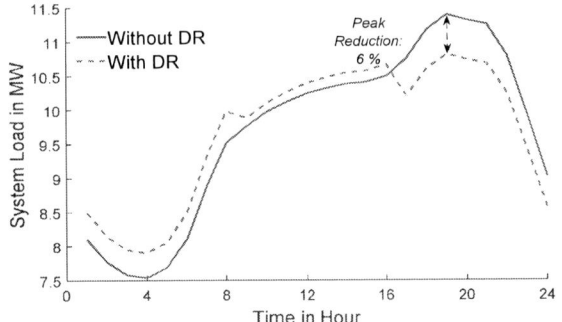

Fig. 10. Effect of DR on system load profile.

V. Conclusions

Application of load profiling information for designing the DR strategy can be a game changer for utility companies as they can design dedicated DR strategy for a specific group according to their behavior. Application of k-means algorithm revealed 6 natural clusters hiding inside the load profile data. Each cluster vary from other cluster on the basis of profile shape. Various critical information about the consumer such as demand variance, behavior in weekdays vs weekends, the sensitivity of demand to weather parameters and coincident peak load etc. have been visualized in a simplistic way to enhance consumers awareness about its energy consumption behavior. 107 potential customers out of 200 are selected by the use of the proposed methodology. These customers result in 6% system peak reduction.

References

[1] M. E. El Telbany, A. Youssef and A. A. Zekry, "Intelligent techniques for mppt control in photovoltaic systems: A comprehensive review," *4th International Conference on Artificial Intelligence with Applications in Engineering and Technology (ICAIET)*, Kota Kinabalu, Dec. 2014, pp. 17–22.

[2] "CEER Advice on ensuring market and regulatory arrangements help deliver Demand-Side flexibility," *Council of European Energy Regulators*, Brussels, Belgium, C14-NaN-40-3, 2014.

[3] C. Eid, E. Koliou, M. Valles, J. Reneses and R. Hakvoort, "Time-based pricing and electricity demand response: existing barriers and next steps," *Utilities Policy*, vol. 40, pp. 15-25, June 2016.

[4] D. Vercamer, B. Steurtewagen, D. Van Den Poel and F. Vermeulen, "Predicting Consumer Load Profiles Using Commercial and Open Data," *IEEE Transaction on Power System*, vol. 31, no. 5, pp. 3693–3701, Sept. 2016.

[5] J. S. Vardakas, N. Zorba and C. V. Verikoukis, "A survey on demand response programs in smart grids: Pricing methods and optimization algorithms," *IEEE Communications Surveys & Tutorials*, vol. 17, no. 1, pp. 152–178, July 2015.

[6] P. Palensky and D. Dietrich, "Demand side management: Demand response, intelligent energy systems, and smart loads," *IEEE Transactions on Industrial Informatics*, vol. 7, no. 3, pp. 381–388, Aug. 2011.

[7] O. Ma, N. Alkadi, P. Cappers, P. Denholm, J. Dudley, S. Goli, M. Hummon, S. Kiliccote, J. MacDonald, N. Matson, et al., "Demand response for ancillary services," *IEEE Transactions on Smart Grid*, vol. 4, no. 4, pp. 1988–1995, Dec. 2013.

[8] A. Brooks, E. Lu, D. Reicher, C. Spirakis and B. Weihl, "Demand Dispatch," *IEEE Power and Energy Magazine*, vol. 8, no. 3, pp. 20-29, June 2010.

[9] J. Medina, N. Muller and I. Roytelman, "Demand Response and Distribution Grid Operations: Opportunities and Challenges," *IEEE Transactions on Smart Grid*, vol. 1, no. 2, pp. 193-198, Sept. 2010.

[10] Q. Hu, F. Li, X. Fang and L. Bai, "A Framework of Residential Demand Aggregation With Financial Incentives," *IEEE Transactions on Smart Grid*, vol. 9, no. 1, pp. 497-505, Jan. 2018.

978-1-5386-6624-1/18 $31.00 © 2018 IEEE

[11] S. Mohagheghi, F. Yang and B. Falahati, "Impact of demand response on distribution system reliability," *IEEE Power and Energy Society General Meeting,* San Diego, CA, 2011, pp. 1-7.

[12] Y. Ota, H. Taniguchi, T. Nakajima, K. M. Liyanage, J. Baba and A. Yokoyama, "Autonomous Distributed V2G (Vehicle-to-Grid) Satisfying Scheduled Charging," *IEEE Transactions on Smart Grid,* vol. 3, no. 1, pp. 559-564, March 2012.

[13] L. G. Swan and V. I. Ugursal, "Modeling of end-use energy consumption in the residential sector: a review of modeling techniques" *Renewable and Sustainable Energy Reviews,* vol. 13, no. 8, pp. 1819–1835, Oct. 2009.

[14] S. Gottwalt, J. Gärttner, H. Schmeck, and C. Weinhardt, "Modeling and valuation of residential demand flexibility for renewable energy integration," *IEEE Transaction on Smart Grid,* vol. 8, no. 6, pp. 2565–2574, Nov. 2017.

[15] G. Chicco, R. Napoli, F. Piglione, P. Postolache, M. Scutariu, and C. Toader, "Load pattern-based classification of electricity customers," *IEEE Transactions on Power System,* vol. 19, no. 2, pp. 1232–1239, May 2004.

[16] J. Kwac and R. Rajagopal, "Data-driven targeting of customers for demand response," *IEEE Transactions on Smart Grid,* vol. 7, no. 5, pp. 2199–2207, Sept. 2016.

[17] B. A. Smith, J. Wong, and R. Rajagopal, "A Simple Way to Use Interval Data to Segment Residential Customers for Energy Efficiency and Demand Response Program Targeting Use of Segmentation Schemes within the Utility Industry : A Review," *ACEEE Summer Study Energy Efficiency Buildings,* pp. 374–386, 2012.

[18] H. Â. Cao, C. Beckel and T. Staake, "Are domestic load profiles stable over time? An attempt to identify target households for demand side management campaigns," *IECON 2013 - 39th Annual Conference of the IEEE Industrial Electronics Society,* Vienna, 2013, pp. 4733-4738.

[19] A. Capasso, W. Grattieri, R. Lamedica, and A. Prudenzi, "A bottom-up approach to residential load modeling," *IEEE Transactions on Power Systems,* vol. 9, no. 2, pp. 957–964, May 1994.

[20] A. Tyagi, A. Verma and P. R. Bijwe, "Reconfiguration of balanced and unbalanced distribution systems for cost minimization, *TENCON 2017 - 2017 IEEE Region 10 Conference,* Penang, 2017, pp. 2188-2192.

[21] K. McKenna and A. Keane, "Residential load modeling of price-based demand response for network impact studies," *IEEE Transactions on Smart Grid,* vol. 7, no. 5, pp. 2285–2294, Sept. 2016.

[22] R. Yu, W. Yang and S. Rahardja, "A statistical demand-price model with its application in optimal real-time price," *IEEE Transactions on Smart Grid,* vol. 3, no. 4, pp. 1734–1742, Dec. 2012.

[23] D. S. Kirschen, G. Strbac, P. Cumperayot and D. de Paiva Mendes, "Factoring the elasticity of demand in electricity prices," *IEEE Transactions on Power Systems,* vol. 15, no. 2, pp. 612–617, May 2000.

Phasor Measuring Units of Smart Grid for Uninterrupted and Stable Power Supply in India

Punit Mangal
Senior Analyst (Analytics)
EMERITUS Institute of Management
Mumbai, INDIA
punnumangal@yahoo.com

Abstract—I collected the data recorded by 7 Phasor Measuring Units (PMUs) for one hour (2 Lakh data points) from the North Regional Load Dispatch Centre, Delhi on 10th March 2016. In this research, I did a thorough analysis to leverage this big data of phasor measuring units to find any disturbance in the power grid and take immediate action to support stability, reliability, and resilience of the power system. Thus, this research was successful by recommending suitable actions at national and regional control rooms of power for better monitoring and supply of uninterrupted and stable power supply in India. I concluded the study in 2018.

Keywords—Phasor Measuring Units (PMU's), uninterrupted and stable power supply, smart Grid, Supervisory Control and Data Acquisition (SCADA), Rate Of Change Of Frequency (ROCOF), Adjusted Angle Difference (AAD)

I. INTRODUCTION

Average energy consumed per person has become the indicator to show the development of a country. Fossil fuels are a major source of energy presently but the pollution is associated with them. Presently major power is produced by coal but now the use of solar energy to produce power is increasing rapidly. Nowadays the most preferred source of energy is electric power. The Indian power grid is one of the largest grid in the world which covers 1.3 billion people and area of 3.2 million square kilometers.

The power supply has been monitored by the SCADA (Supervisory Control and Data Acquisition) system for many decades in the conventional grid system. The SCADA system measure voltage and frequency at 4-10 samples per second. The SCADA system does not measure voltage angle.

A phasor is a number that is having both voltage and voltage angle. The voltage is measured in volts and voltage angle is measured in degree. Phasor data of the same time are called synchrophasors.

The phasor measuring units are installed all over transmission lines and the phasor data recorded by them are transmitted to a central control room with a time stamp based on GPS time, on every data recorded. The phasor measuring units are recording samples at a high speed of 25 samples per second. They record the frequency, voltage, and voltage angle, all the three values of the transmission line, where they are installed, which is transmitted in almost real time to a central control room.

Actual data of PMU were obtained and several graphs were plotted for the fundamental power parameters in Excel and Tableau to gain critical insights about their behavior via visualization. A bad event was identified with a voltage dip observed for microseconds and a detailed event analysis was conducted.

II. SMART GRID TECHNOLOGY

There are three major systems, namely the smart infrastructure system, the smart management system, and the smart protection system in the Smart Grid. There are the smart energy subsystem, the smart information subsystem, and the smart communication subsystem for the smart infrastructure system. There are various management objectives, such as improving energy efficiency, profiling demand, maximizing utility, reducing cost, and controlling emission for the smart management system. There are security issues, privacy issues and various failure protection mechanisms which improve the reliability of the smart protection system. [1]

The time ahead for the smart grid is challenging but promising. The increasing cost of energy, the use of electricity by everyone, changes in climate, electrification of modern life etc. are the drivers to determine the speed at which these changes will take place. This is a different issue that how fast the implementation of the smart grid is done, but there is no doubt in mind of anybody that smart grid is to be implemented as fast as possible. It has been accepted by India, rather by various countries of the whole world, that implementation of the smart grid will bring improved electrical technology and business of country will make progress by the smart grid. [2]

A new synchronized technique software based IEEE 1588-2008 Global System for Smart Grid has been developed specifically for PMUs with synchronism needs up to the microsecond. The software base system provides, increased accuracy, higher time-stamp resolution, shorter synchronization intervals, correction field, new mappings, fault tolerance, unicast operation, and rapid re-configuration after network topology changes. [3]

The importance of renewable energy sources is increasing. However, different forms of renewable energy, being dependent on nature, keep on varying. Electrical storage is limited and very costly. Similarly, the demand for power is also fluctuating during different times of the day. It reaches a peak during the evening to midnight, minimum early in the morning and moderate during rest of the day. Further, electricity losses due to transmission are quite substantial. To address the existing limitations, the solution is moving towards smart grid which allows seamless integration of renewable energy sources with the non-renewable energy sources. [4]

Synchrophasor has many advantages over the SCADA system, exact information of less than a second, detect fault undetected by SCADA, increased system reliability, increased system efficiency, increased event analysis etc. [5]

The smart grid will be capable of handling quick recovery from grid disturbance, predict the future events, protect against internal and external threats uncertainties in schedules, transferring power across regions, accommodating renewable energy sources, optimizing the power transfer capability of the transmission and distribution networks and meeting the demand for increased quality and reliable supply, managing and resolving unpredictable events and uncertainties in operations, and planning more aggressively and smartly. [6]

In the present time, varying power price depending on the load on the grid has been widely accepted and has been started in many other countries [7]

Future is of smart control framework, where a few supporters share a typical energy source. This permits two-way correspondence among smart meters. [8]

The transmission of power is associated with risk. The generation of power is regarded as a heterogeneous product of discontinuous or stochastic power and uses data and control to configuration supporting strategies to deal with the danger of instability. [9]

In future, Vehicle-to-Grid (V2G) systems will be used where the battery of vehicle will be used to supply power into grid [10]

III. TRANSMISSION AND THEFT LOSSES

After a study of capacity & actual power supply data published by Central Electricity Authority of India as of 30.09.2015 following calculation had been done during the initial stage of research in 2015.

Monitored Capacity of power   =   241 GW

Planned Maintenance of power   = (-) 10 GW

The forced outage of power   = (-) 40 GW

Other reasons of power   = (-) 4 GW

The capacity of power online   = 187 GW

Actual power from Apr to Sep 2015 = 556 GU.

1 Unit power = 3.6 x 106 J

Days Apr to Sep 2015 = 183 days

187 GW = 187 x 109 Watt or J/Sec

$$= \frac{187 \times 109 \times 60 \times 60 \times 24 \times 183}{3.6 \times 106 \times 109}$$

= 821 GU

So, Transmission and theft losses from Apr to Sep-15 are 265 GU out of 821 GU i.e. 32.3% as calculated above.

This research revealed, how the smart grid can help in controlling theft.

IV. CONTROLLING THEFT LOSSES BY SMART GRID

Transmission and Theft losses are very high around 30% in India. The amount of systematic theft in places like Bihar, where even factories are run by illegal high-load siphon lines, boggles the mind. In India, the population is a problem, but in many of the heavily populated areas, a lot of the population isn't even on the grid. Therefore there is an urgent need to minimize transmission losses and control power theft. Smart Grid implementation will reduce these losses.

The smart grid is getting implemented with smart meters. Smart meters will be installed for every consumer to measure power consumed by that consumer and smart meter readings will be communicated to the control room where it will be saved in the computer installed at the control room. This computer can add the consumption by all the consumers.

The power supplied by the control room is also measured by smart meters which will be communicated to the same computer. In this way the computer will have both figures with him, the power supplied and the power consumed. The difference between these two figures can be calculated easily by the computer and displayed or report can be printed. This difference amount is the transmission and theft losses. As the difference can be calculated daily, when it increases suddenly, it will indicate some power theft is started on that date.

Further the report can be taken from the computer for all the areas to which this control room is supplying power, say Tilak Nagar, Chembur, and Ghatkopar, so the report will speak out in which area power theft has been started on that date on which difference has abruptly jumped, this will help in controlling power theft.

V. PMU DATA AND ANALYSIS

Data for seven Phasor Measuring Units in the Northern Region was collected to analyze and understand the variation and trends in the major parameters.

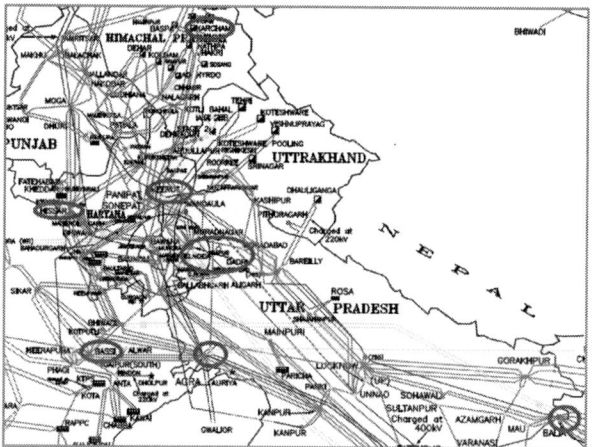

Fig. 1. Layout of seven PMUs.

A. Analysis 1 – Individual graphs of all PMUs (to effectively interpret trends of important power parameters)

Individual graph of each PMU has been plotted in Excel and Tableau to effectively interpret trends of important power parameters.

Fig. 2. Agra PMU Frequency Plot in Tableau.

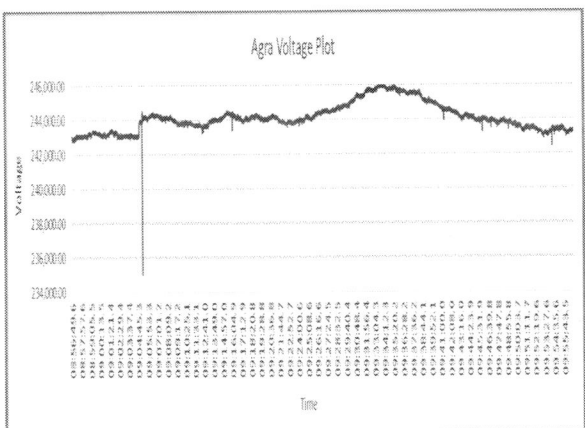

Fig. 3. Agra PMU Voltage Plot in Tableau.

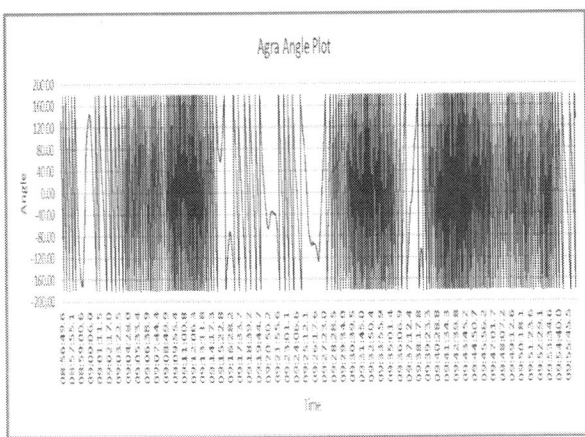

Fig. 4. Agra PMU Angle Plot in Tableau.

Key findings of Analysis 1: This analysis provided clarity on the maximum and minimum bounds, averages and standard deviation of frequency, voltage and angle.

B. Analysis 2 – Collaborated graph of all PMUs together (to compare trends closely across all PMUs)

Collaborated graphs of all PMUs were plotted together using Tableau to compare trends closely across all PMUs for Voltage, Frequency, and Angle. These graphs are as under.

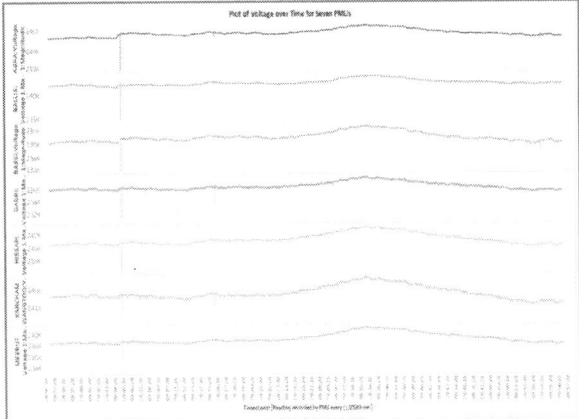

Fig. 5. Plot of Voltage for Seven PMUs.

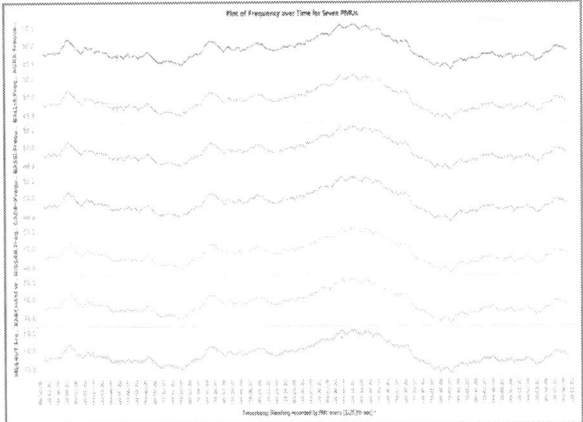

Fig. 6. Plot of Frequency for Seven PMUs.

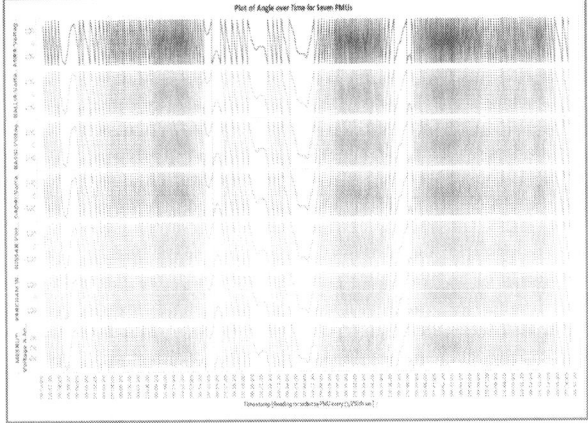

Fig. 7. Plot of Angle for Seven PMUs.

Key Findings of Analysis 2: It was observed that the frequency plots for all 7 PMUs follow the same trend. This indicates that if the frequency of a PMU fluctuates due to any disturbance/fault in the grid, then it will affect the frequency of other PMUs as well because currently the power lines are interconnected all over India and India has become "One Nation - One Grid" w.e.f. 31.12.2013. Fluctuation in one PMU will subsequently affect the entire grid.

C. Analysis 3 – Identification and analysis of bad event (to detect any disturbance to the grid immediately and take action quickly)

In analysis 3, a detailed analysis of bad event has been done which is shown below in Fig. 8 and Fig. 9 where graph for all seven PMUs is plotted in Tableau.

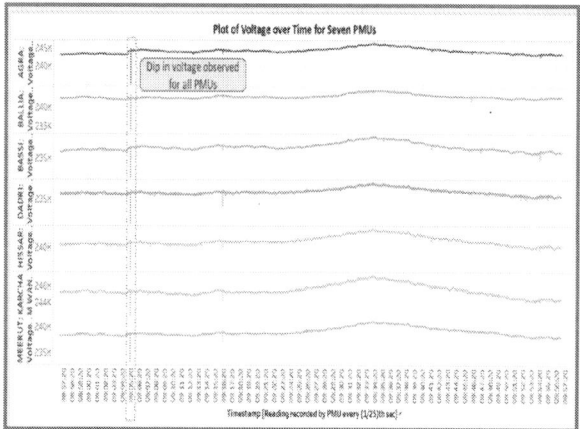

Fig. 8. Dip in voltage for Seven PMUs.

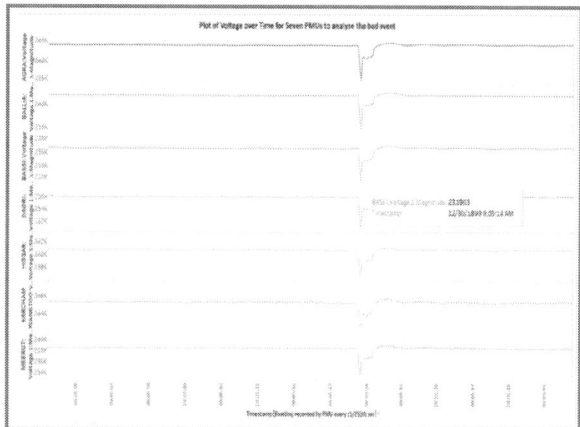

Fig. 9. Zoomed bad event voltage-seven PMUs.

Key Findings of Analysis 3: Synchrophasor data can be used to cause detailed event analysis for those bad events/disturbances which would have remained completely unnoticed with the traditional SCADA (Supervisory Control and Data Acquisition) system. As seen in the above analysis, the whole disturbance did not exist even for a complete second. In fact, the voltage drop persisted for only 10 readings (out of 25) i.e. (2/3) time of a second.

D. Analysis 4 – Identification and real-time tracking of additional critical index, Rate of Change of Frequency (to improve grid performance)

This phase of analysis focuses on identifying any additional variable in addition to the three important power

variables of frequency, voltage, and angle which can help a grid operator in early detection of any issue and troubleshoot it effectively. A thorough research was carried out to understand other calculated variables based on the three fundamental parameters.

An important variable identified is df/dt, Rate of Change of Frequency given short name ROCOF. df/dt is a key variable considered in emergency control and protection systems. It is used for fast load shedding by detecting the dip in frequency earlier and initiating load shedding to normalize frequency of the Grid.

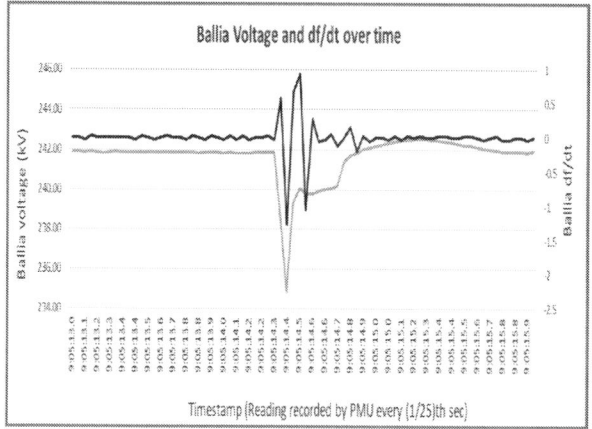

Fig. 10. Ballia Voltage and df/dt for bad event.

To further substantiate this analysis, the Rate of Change of Frequency (ROCOF) or df/dt plots were made for all the Phasor Measuring Units (PMUs) in Tableau.

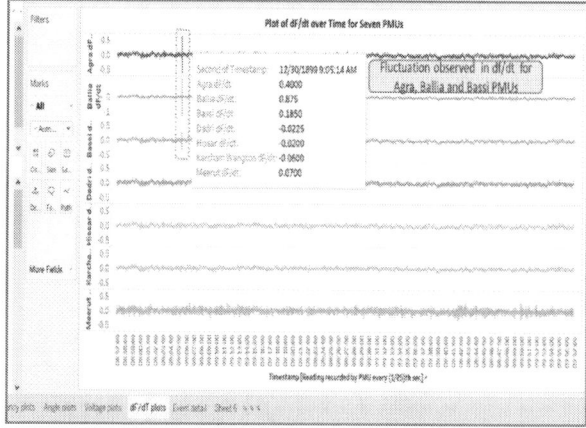

Fig. 11. df/dt shows three PMUs affected mainly.

This Fig. 11 clearly shows that a fluctuation in Rate of Change of Frequency (ROCOF) or df/dt was observed for Agra, Ballia and Bassi Phasor Measuring Units (PMUs), whereas the Rate of Change of Frequency (ROCOF) or df/dt trend was almost stable for the four other Phasor Measuring Units (including Dadri, Hissar, Karcham Wangtoo and Meerut).

It was observed that a fluctuation in the voltage has occurred around the timestamp of 09:05:14 AM. To see the fluctuation more clearly, the above plot (Fig. 11) is zoomed further around the timestamp of 09:05:14 AM and a graph has been plotted which is shown below (Fig. 12).

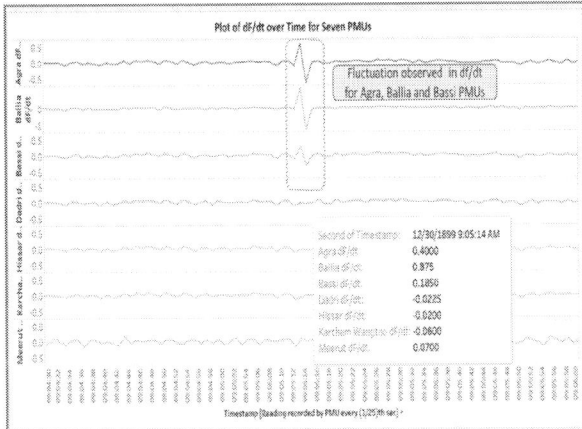

Fig. 12. Zoomed df/dt for PMUs affected.

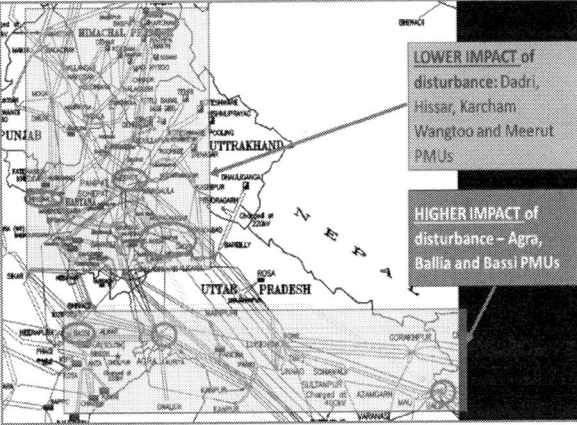

Fig. 13. Lower and Higher Impacted PMUs.

This graph of df/dt variation in Agra, Ballia and Bassi PMUs readings show that this three PMUs were impacted more due to the disturbance originated at 09:05:14 AM.

Key Findings of Analysis 4:

- This analysis helped in identifying an additional variable Rate of Change of Frequency in addition to the three important power variables of frequency, voltage, and angle to early detect any bad issue, gain critical insights and troubleshoot it effectively.

- The Rate of Change of Frequency showed variation during disturbance not captured by the frequency plots (when the frequency plot was almost stable).

- During the bad event found in this study, voltage plot showed a dip across all seven PMUs whereas out of that seven PMUs, only three PMUs (Agra, Ballia, and Bassi) had a fluctuation in df/dt. This indicates that the impact of the voltage spike was witnessed more at this PMUs.

- Thus, df/dt is indeed a key variable to detect early any issues in the grid which provides us with additional insights. It is a very useful tool for post-event analysis.

E. Analysis 5 – Identification and monitoring of another key variable - Adjusted Angle Difference (for better grid stability)

While plotting the graph of voltage angle, it was observed that the voltage angle fluctuates from +180 to -180. Another dimension to the analysis was to see if the difference between voltage angles of two nearby PMUs is plotted, it can lead to some conclusion and throw light on some crucial information. Thus, the angle difference data was calculated and graphs were plotted for angle difference between two nearby PMUs. One of them is as under.

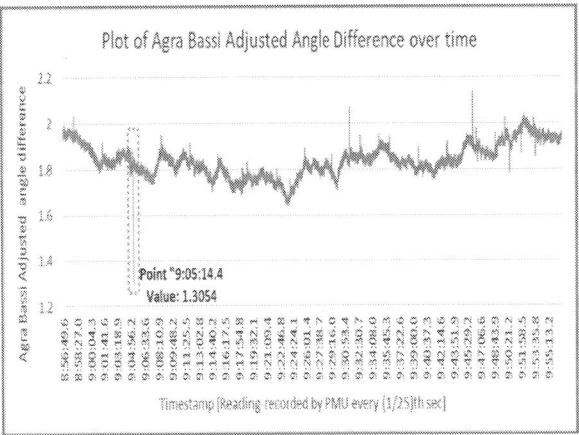

Fig. 14. Agra Bassi Adjusted Angle Difference Plot.

Key findings of Analysis 5:

- As the voltage angle of PMUs keeps on oscillating from +180 degrees to -180 degrees, it becomes difficult to detect any bad incident by monitoring the voltage angle. However, the adjusted angle difference between two PMUs is a key variable which can leverage PMU data to identify any defect/issue which arises in the grid.

- The Adjusted Angle Difference (AAD) is indeed a key variable to detect any issues in the grid at an early stage. The AAD provides us clear-cut insights to grid health. The AAD is another very useful tool for post-event analysis.

CONCLUSION

Transmission and Theft losses were calculated by capacity & actual power supply data published by Central Electricity Authority of India. A mechanism is suggested by which Transmission and Theft losses can be controlled by the smart grid and smart meters.

Two additional key indices were discovered to identify any voltage spike quickly and phasor measuring units (PMUs) which are at higher risk i.e. more impacted by the disturbance in comparison with other PMUs which are (i) Rate of Change of Frequency (ROCOF) and (ii) Adjusted Angle Difference (AAD). They can provide help in concluding post-event analysis.

An automated tool was developed by which e-mail alert notification is sent to the concerned executives, whenever voltage fluctuates beyond defined limits, for immediate remedial action to maintain the stability of the grid.

978-1-5386-6624-1/18 $31.00 © 2018 IEEE

REFERENCES

[1] Fang X, Misra S, Xue G (2012), Smart Grid, The New and Improved Power Grid, IEEE, ISSN 1553-877X, 14(4), Pages 944-980.

[2] Farhangi H (2010), The path of the smart grid, IEEE, ISSN 1540-7977, 8(1), Pages 1-6.

[3] Lopez VP, Munoz AM (2012), Synchrophasor for Smart Grid with IEEE 1588-2008 Synchronism, PRZEGLAD, ISSN 0033-2097, 88(1), Pages 31-36.

[4] Mangal Punit (2015), Intelligent Controller in Smart Grid towards National and Global development, International Journal of Renewable Energy Exchange, Rexjournal, ISSN 2321-1067, Vol 3, Issue 3, Pages 43-48.

[5] Mangal Punit, Vitkar Swati (July 2016), Uninterrupted & Stable power supply by Smart Grid in India & World, International conference on emerging trends in engineering and technology, IFERP, ISBN 97881-9295-8050, Pages 12-16.

[6] Momoh James (2012), Smart Grid: Fundamentals of Design & Analysis, WILEY, 978-0-470-88939-8, 16-27.

[7] Prasad Indrajeet (2014), Smart Grid Technology: Application and Control, IJAREEIE, ISSN 2278-8875, 3(5), Pages 9533-9542.

[8] Samadi P, Rad AHM, Schobar R (2010), Optimal Real-Time Pricing Algorithm Based on Utility Maximization for Smart Grid, IEEE, 978-1-4244-6512-5.

[9] Varaiya PP, Wu FF, Bialek JW (2010), Smart Operation of Smart Grid: Risk-Limiting Dispatch, IEEE, 1558-2256, 99(1), 40-57.

[10] Yang Z, Yu S, Lou W, Liu C (2011), Privacy preserving communication and precise reward architecture for V2G network in smart grid, IEEE, 1949-3053, 2(4), 697-706.

Optimal placement of Distributed Generation for Congestion Management: A Comparative Study

Md Sarwar[1], Anwar Shahzad Siddiqui[1], ZainulAbdin Jaffery[1], Imran Ahmad Quadri[2]
[1]Department of Electrical Engineering,
JamiaMilliaIslamia, New Delhi, India
[2]Department of Electrical Engineering, *Delhi Technological University*, Delhi, India
*sarwaramu@gmail.com

Abstract—An approach to manage congestion in an electricity market under deregulated environment has been presented. The objective is accomplished by optimally placing a DG in the power system network.Four different methods based on highest LMP, LMP difference, total congestion rent revenue, and total savings are used to find the potential locations for placing a DG in order to manage congestion. Using these methods, a priority ranking has been formed to get the optimal location such that the implantation of DG gives maximum benefit while fulfilling the objective of congestion mitigation. A comparative study of these methods has also been analysed to obtain the optimal location of placement of DG. The performanceofthe presented methodologies has been tested on IEEE 14-bus system.

Keywords— *Deregulation,locational marginal price, optimal power flow, distributed generation,transmission line congestion*

I. INTRODUCTION

To meet the increasing electricity demand at cheaper price, electricity market is becoming more competitive which endangers the security of the transmission system. This may cause the violation of the transfer limits of the transmission network. This condition is known as congestion in transmission line and is defined as the violation of any of the transfer limits of transmission lines such as stability limits, thermal limits voltage limits etc. The congestion of transmission line is an unwanted condition in power system operation as it deviates the power system operation from optimal point to a non-optimal or sub-optimal point thereby increasing the electricity price.

The adoption of competition in an electricity market often causes congestion in the transmission lines due to its inadequacy to accommodate all the committed power transactions to meet the demand. This results in reduction of social welfare. Therefore, to keep the equilibrium of competitive electricity market intact and to operate and maintain the power system reliably and securely, management of congestion plays a vital role.

Some of the techniques to manage congestion in a deregulated electricity market is presented in [1]-[2]. These methods include load shedding [6], generation rescheduling [3]-[5], reactive power management[7]-[8], zonal/cluster congestion management [8]-[9], voltage stability[11], use of FACTS devices [10] etc. Most of these methods make use of the generation side approach to mitigate congestion. However,with the advancement of technology and increasedcompetition in electricity market, congestion

management schemes involving demand side approach are being extensively used due to its ability toits high effectiveness in managing congestion thereby enhancing the security and reliability of the power system [12]. The load shedding or load curtailment is one such method to manage congestion. However keeping in mind to meet the electricity demand, the shedding of load does not seem viable. Therefore, another method which makes use ofallocation of distributed generations (DGs) is preferred over load shedding.The DG can be considered as negative power demandsand can be generally located in load pockets.Its inherent characteristic to respond quickly to the varying load conditions in electricity market makes it a viable choicefor power system planning and operation. Besides its benefit of easy installation and simple to operate, its optimizedallocationimproves voltage profile,reduces losses, defer system upgrades resulting in improvement in system reliability. In view of all these benefits,it is being extensively used for effectively and efficiently managing congestion in power system network. In [13], the authors have discussed the impact of DG in managing congestion. The authors have presented an approach to place DG optimally in order toalleviate congestion using realand reactive power flow along with voltage magnitude contribution factors. In [14], optimal allocation of DGs is performed based on LMP and congestion rent and benefit-to-cost ratio for the implementation of DG is also discussed. However the other potential location to optimally place a DG for congestion management can also be considered.

In this paper, different approaches to optimally place a DG to mitigate the congestion from a transmission network in deregulated environment has been presented.The method includes the DG placement based on total congestion rent revenue, total savings, highest LMP of buses as well as LMP difference between two busses in a system. A comparison between these methods for congestion management has been performed. Different probabilities of placement of DG based on bus priority ranking have been studied based on which the optimal method is identified.A priority list capturing the potential locations for placing a DG are formed using the presented methods based on which its optimal location is identified.

II. PROBLEM FORMULATION

A pool based market of electricity is adopted with the aim of maximizing the social welfare (or to minimize the total cost of generation in case there is no demand bid). The optical power flow (OPF) tool is utilized to evaluate the nodal prices of electricity (LMP) and to manage congestion in the system. The problem is formulated as:

$$\text{Minimize} \sum_{i=1}^{n_g} C_i(P_{G_i}) - \sum_{i=1}^{n_d} B_i(P_{D_i}) \qquad (1)$$

Subject to the following constraints:
1. Power balance constraint at each node

$$P_k - P_{G_k} + P_{D_k} = 0; \quad k = 1,2,\dots\dots,n_b \qquad (2)$$

$$Q_k - Q_{G_k} + Q_{D_k} = 0; \quad k = 1,2,\dots\dots,n_b \qquad (3)$$

2. Generator operating limit constraint

$$P_{G_m}^{min} \le P_{G_m} \le P_{G_m}^{max} ; \quad m = 1,2,\dots,n_g \qquad (4)$$

$$Q_{G_m}^{min} \le Q_{G_m} \le Q_{G_m}^{max} ; \quad m = 1,2,\dots,n_g \qquad (5)$$

3. Line flow constraints

$$F_{kl} \le F_{kl}^{max} ; \quad kl = 1,2,\dots,n_L \qquad (6)$$

4. Bus voltage limit

$$V_k^{min} \le V_k \le V_{mk}^{max} ; \quad k = 1,2,\dots,n_b \qquad (7)$$

where n_g and n_d are the number of generators and load respectively, $C_i(P_{Gi})$ and $B_i(P_{Di})$ are bid curve of ith generator and benefit curve for i^{th} load respectively, n_b is the number of system buses, P_{Gi}^{min} and Q_{Gi}^{min} are the minimum real and reactive power output limits of ith generator respectively, and P_{Gi}^{max} and Q_{Gi}^{max} are the maximum real and reactive power output limits of ith generator respectively, F_{kl} is the power flow through transmission line connected between bus-k and bus-l accomodating all contracts, F_{kl}^{max} is the limit of power flow through line connected between bus-k and bus-l, n_L is the number of lines, V_k^{min} and V_k^{max} are minimum and maximum voltage limits at k^{th} bus respectively.
The bid curve for i^{th} generator is given as quadratic cost function:

$$C_i(P_{G_i}) = a_i . (P_{G_i})^2 + b_i . (P_{G_i}) + c_i \qquad (8)$$

where P_{Gi} is the amount of electricity output from i^{th} generator and a_i, b_i and c_i are corresponding cost coefficients. The second term in equation (1) is equal to zero due to absence of demand bidding.

The objective function incorporating all the constraints of the optimization problem is formed as a Lagrangian function and is given as:

$$\mathcal{L} = \sum_{i=1}^{n_g} C_i(P_{G_i}) + \sum_{k=1}^{n_b} \lambda_{p_k} (P_k - P_{G_k} + P_{D_k})$$
$$+ \sum_{k=1}^{n_b} \lambda_{Q_k} (Q_k - Q_{G_k} + Q_{D_k}) + \sum_{kl=1}^{n_L} \mu_L(F_{kl} - F_{kl}^{max})$$

$$+ \sum_{i=1}^{n_g} \mu_{G_i}^- (P_{G_i}^{min} - P_{G_i}) + \sum_{i=1}^{n_g} \mu_{G_i}^+ (P_{G_i} - P_{G_i}^{max})$$

$$+ \sum_{i=1}^{n_g} \mu_{G_i}^- (Q_{G_i}^{min} - Q_{G_i}) + \sum_{i=1}^{n_g} \mu_{G_i}^+ (Q_{G_i} - Q_{G_i}^{max})$$

$$+ \sum_{k=1}^{n_b} \mu_{V_k}^- (V_k^{min} - V_k) + \sum_{k=1}^{n_b} \mu_{V_k}^+ (V_k - V_k^{max}) \qquad (9)$$

where λ_p and λ_Q, μ_L, μ_G and μ_V are Lagrangian multipliers vectors associated with equality constraints and inequality constraints respectively. The values of these multipliers are obtained by the OPF solution of equation (9). OPF solution is obtained using interior point method in MATLAB.

III. DG LOCATION

Since DG involves a huge investment cost, it must be located such that it provides maximum benefit. A number of methods have been reported in literature to optimally place a DG to manage congestion. However in this paper four different methods have been considered for the effective placement of a DG. These methods include total congestion rent revenue, total savings, highest LMP and LMP difference. Different probabilities of placement of DG based on the bus priority ranking are calculated depending on which the DG can be optimally located.

To reduce the solution space, the buses with large generation capacity than the demand connected to them are not considered for DG placement since they have low LMP. Therefore the buses with either low generating capacity than their demand or with no generating unit are considered for potential location of DG and can be mathematically given as:

$$P_{G_k} \le P_{D_k} ; \quad k = 1,2,\dots\dots,n_b \qquad (10)$$

The different performance indices or methods to find the optimal DG location for mitigating congestion are discussed as follows.

A. Highest LMP Method

For normal operation of a power system network, LMP at each bus is almost equal, but with the occurrence of congestion, the LMPs will become different at all buses. Thus it provides an indication of the congestion. Therefor LMP can be utilized to place DG for management of congestion. One such method is highest LMP method [14]. In this method buses which have high LMPs are placed on top of the priority ranking while the buses having low LMPs are ranked lower in the priority list for DG placement. The bus with highest rank in priority list is the optimal location of DG placement for congestion alleviation.

B. LMP Difference Method

Although, DG placement using highest LMP is simple, but sometimes it may increase the congestion in the network. Therefore, another method can be used to place the DG which is based on LMP difference of two buses across a line, known as "LMP difference method" and is mathematically given as:

$$\Delta LMP_{kl} = LMP_k - LMP_l \; ; \qquad kl = 1,2,\dots,n_L \quad (11)$$

where ΔLMP_{kl} is the difference of LMP across line-kl connected between bus-k and bus-l, LMP_k and LMP_l are the LMPs at bus-k and bus-l respectively.

This method is more reliable as compared to highest LMP method. This method has been proposed in [15] to optimally place TCSC for managing congestion. The congested line will have high LMP difference across it as compared to other lines. Therefore, a priority ranking of buseson the basis of LMP difference between them is calculated to find the potential locations for placing a DG in the network. The best possible bus location for DG to alleviate congestion is obtained with equation (10) as well as equation (11).

C. Total Congestion Rent Revenue

Congestion rent is the function of power flow through a line and LMP [16]. It is mathematically given as:

$$CR_{kl} = \Delta LMP_{kl}.P_{kl} \; ; \qquad kl = 1,2,\dots,n_L \quad (12)$$

where CR_{kl} is the congestion rent for line-kl connected between bus-k and bus-l, and P_{kl} is the power flow on line-kl.

The placement of DG is such that it would provide maximum benefit. Congestion rent is one of the methods of measuring the benefit obtained with the implementation of DG. With the placement of DG, the congestion rent of the line changes and accordingly the total congestion rent also changes, the total congestion rent being high without DG implementation in the system. The total congestion rent is mathematically given as:

$$TCR = \sum_{kl=1}^{n_L} \Delta LMP_{kl}.P_{kl} \quad (13)$$

The total congestion rent revenue is obtained by taking the difference of total congestion rent without DG and total congestion rent with implementing DG at a specific bus as given by:

$$TCRR = TCR_k - TCR_l \quad (14)$$

where TCRR is the total congestion rent revenue and TCR_k and TCR_l are total congestion rent when DG is placed at bus-k and bus-l respectively. The DG is placed at a bus for which the total congestion rent revenue is maximum.

D. Total Savings

Although the total congestion rent revenue gives a measure to analyse the benefit of DG implementation, if considered alone may give rise to a situation that the electricity generation cost is more. Therefore another method or performance index to optimally place a DG takes into consideration of generation cost and congestion rent revenue after implanting DG. This gives the total savings due to implementation of DG and is given as:

$$Savings = \left(C_{i_{BC}} - C_{i_{DG}}\right) + (TCR_{BC} - TCR_{DG}) \quad (15)$$

where $C_{i_{BC}}$ and $C_{i_{DG}}$ are generation cost without DG and with DG implementation respectively while TCR_{BC} and TCR_{DG} are total congestion rent without DG and with DG implementation respectively.

The placement of DG at a specific bus which gives maximum saving is ranked highest in the priority ranking while the location which gives minimum savings is ranked lowestin the priority ranking.

Thus all these four methods for placement of DG is utilized and analysed for their effectiveness.

IV. RESULTS AND DISCUSSIONS

The presentedmethods are tested and analysed on modified IEEE 14-bus system as shown in Fig.1, the network data of which is taken from[17] and data of generator is taken from [16]. The DG is considered to inject only real power of 5 MW.

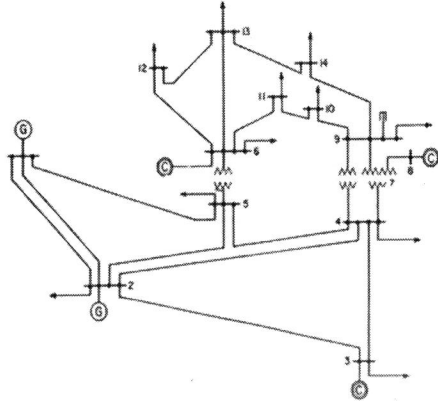

Fig. 1IEEE 14-bus system

TABLE I
DG BUS LOCATIONS PRIORITY BASED ON HIGHEST LMP METHOD

Priority Rank	LMP (\$/MWh)	Bus No.
1	35.22	14
2	35.20	4
3	34.99	7
4	34.89	9
5	34.77	10
6	34.00	11
7	33.91	13
8	33.58	12
9	32.59	5

The priority rank for DG locations based on highest LMP is shown in Table I. Taking into account equation (10), the LMPs for the load buses have been considered in Table I. which shows that bus-14 having highest LMP is ranked-1 in the priority list and thereafter bus-4 and so on. The generation cost without DG (base case) is 6353.65 \$/h while TCR is found to be 2119.85 \$/h.

Table II presents the priority ranking of the buses based on LMP difference across a line connected between two buses. It shows that the LMP difference across a line connected to bus-4 has highest LMP difference and is ranked-1 in the priority list for DG placement while bus-14 which is earlier ranked-1 in highest LMP method of DG placement is ranked-3 with this method.

TABLE II
DG Bus Locations Priority Based on LMP Difference Method

Priority Rank	LMP Difference (\$/MWh)	Bus No.
1	13.00	4
2	12.54	5
3	1.32	14
4	1.04	13
5	0.95	11
6	0.77	10
7	0.62	12
8	0.33	9
9	0.21	7

Similarly, Table III shows the priority ranking of the potential locations of DG placement based on total congestion rent revenue. With this method, it is found that bus-14 is ranked highest in the priority ranking while bus-10 is ranked next to bus-14 and so on.

The priority ranking of the potential locations for DG placement based on total savings due to implementation of DG at each bus is shown in Table IV. According to this method, placing a DG at bus-14 gives maximum savings while placing a DG at bus-5 gives the minimum savings.

TABLE III
DG Bus Locations Priority Based on TCRR

Priority Rank	TCRR (\$/MWh)	Bus No.
1	57.49	14
2	54.11	10
3	51.43	9
4	44.89	4
5	44.40	7
6	41.67	11
7	15.56	13
8	11.58	12
9	11.40	5

TABLE IV
DG Bus Locations Priority Based on Savings

Priority Rank	Savings (\$/MWh)	Bus No.
1	224.96	4
2	214.04	14
3	208.40	7
4	206.38	9
5	200.98	10
6	191.18	11
7	165.35	13
8	158.16	12
9	110.57	5

An analysis of the results obtained for potential locations of DG with the discussed methods reveals that with different methods, the priority rank of the buses changes as shown in Table V.

TABLE V
Comparison Table For DG Bus Locations Priority

Priority Rank	Bus No.			
	Highest LMP	LMP Difference	TCRR	Savings
1	14	4	14	4
2	4	5	10	14
3	7	14	9	7
4	9	13	4	9
5	10	11	7	10
6	11	10	11	11
7	13	12	13	13
8	12	9	12	12
9	5	7	5	5

The bus-14 which is ranked-1 in highest LMP method is ranked-3 in LMP difference, ranked-1 in TCRR method and ranked 2 in total savings while bus-4 which is ranked-2 in highest LMP method is ranked-1 in LMP difference method, ranked-4 in TCRR method and and again ranked-1 in total savings. Similarly the priority ranking of all other buses varies accordingly.

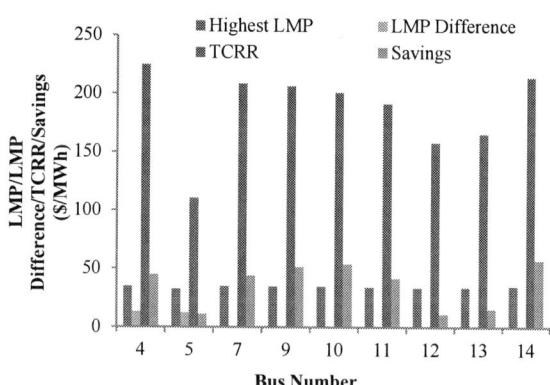

Fig. 2 LMP, LMP difference, TCRR and savings for different locations of DG

It can be inferred from Table V that the bus-4 and bus-14 are the most optimal locations for DG placement and are almost captured by each method of DG placement as discussed. The values obtained for LMP, LMP difference,

TCRR and savings for different bus locations of DG placement is illustrated in Fig. 1.

V. CONCLUSIONS

This paper presentsfour different methods to optimally place a DG for congestion management. Which includes highest LMP, LMP difference, TCRR and total savings.All these methods capture the potential location of DG placement, however these potential location differ in ranking with each method implemented. Even though the priority ranking of bus location differs in each method, the optimal potential locations in priority ranking is captured by each method. Since a DG allocation involves a heavy investment, its optimal location can be decided based on the priority in achieving the objective viz, minimum generation cost, total savings etc. in order to get the maximum benefit.

REFERENCES

[1] A. Kumar, S. C. Srivastava, and S. N. Singh, ''Congestion managementin competitive power market: A bibliographical survey,'' *Elect. PowerSyst. Res.*, vol. 76, pp. 153–164, 2005.

[2] R. D. Christie, B. Wollenberg, and I. Wangensteen, ''Transmission management in the deregulated environment,'' *Proc. IEEE*, vol. 88, no.2, pp. 170–195, Feb. 2000.

[3] A. Shandilya, H. Gupta, and J. Sharma, ''Method for generation rescheduling and load shedding to alleviate line overloads using local optimization'', in *Proc. Inst. Elect. Eng.*, vol. 140, pp. 337-342, 1993

[4] K. Talukdar, A. K. Sinha, S. Mukhopadhyay, and A. Bose, ''AComputationally simple method for cost-efficient generationrescheduling and load shedding for congestion management,'' *Int. J. Elect. PowerEnergy Syst.*, vol. 27, no. 5, pp. 379–388, Jun.–Jul. 2005.

[5] P. Boonyaritdachochai, C. Boonchuay, and W. Ongsakul, ''optimal congestion management in an electricity market usingparticle swarm optimization with time-varying acceleration coefficients'', *Computers and Mathematics with Applications*,vol. 60, pp. 1068-1077, 2010.

[6] S. Dutta, and S.P. Singh, Optimal rescheduling of generators for congestion management based on particle swarm optimization, *IEEE Transactions on Power Systems*, vol. 23, no. 4, pp. 1560-1569, 2008.

[7] S. Hao, and A. Papalexopoulos, ''Reactive power pricing and management'', *IEEE Trans. Power Syst.*, vol. 12, pp. 95–104, 1997.

[8] A. Kumar, S. C. Srivastava, and S. N. Singh, ''A zonal congestion management approach using real and reactive power rescheduling,,'' *IEEETrans. Power Syst.*, vol. 19, no. 1, pp. 554–562, Feb. 2004.

[9] A. Kumar, S. C. Srivastava, and S. N. Singh, ''A zonal congestion management approach using ac transmission congestion distribution factors,'' *Elect. Power Syst. Res.*, vol. 72, pp. 85–93, 2004.

[10] A. J. Conejo, F.Milano, and R. Bertrand, '' Congestion management ensuring voltage stability'', *IEEETrans. Power Syst.*, vol. 21, pp. 357–364, 2006.

[11] S. N. Singh, and A. K. David, ''Optimal location of FACTS devices for congestion management'', *Elect. Power Syst. Res.*, vol. 58, pp. 71–79, 2007.

[12] F. Rahimi, and A. Ipakchi, ''Demand response as a market resource under smart grid paradigm'', *IEEE Trans. Smart Grid*, vol. 1, pp. 82-88, 2010.

[13] J. Liu, M.M.A Salama, and R. R. Mansour, ''Identify the impact of distributed resources on congestion management'', *IEEETrans. Power Deliv.*, vol. 20, no. 3, pp. 1998–2005, Jul. 2005.

[14] M. Afkousi-Paqaleh, A. Abbaspour-TehraniFard, and M. Rashidinejad, ''Distributed generation placement for congestion management considering economical and financial issues'', *Elect. Eng.*, vol. 92, pp. 193-201, 2010.

[15] M. Sarwar, A. S. Siddiqui, '' Congestion management in deregulated electricity market using distributed generation'', in Proc. 12th IEEE India International Conference, pp. 1-4, New Delhi, Dec. 2015.

[16] N. Acharya, and N. Mithulananthan, ''Locating series FACTS devices for congestion management in deregulated electricity markets'', *Elect. Power Syst. Res.*, vol. 77, pp. 352–360, 2007.

[17] Power system test case archives (2004) [Online]. Available: http://www.ee.wasington.edu/research/pstca

978-1-5386-6624-1/18 $31.00 © 2018 IEEE

2nd IEEE International Conference on Power Electronics, Intelligent Control and Energy Systems (ICPEICES-2018)

Improvised Binary Sequence MPPT Method for Solar PV Applications

Swapnendu Narayan Ghosh
Department of Electrical Engineering
Jalpaiguri Government Engineering College
Jalpaiguri, India
E-mail: swapnendu99@gmail.com

Abstract—**With the increase in the use of solar power across the globe, it has become critical to extract the maximum power available from the photovoltaic(PV) modules. In order to extract maximum power from the modules, several algorithms have been proposed. Most of the commercial systems implement the Perturb & Observe(P&O) method and the Incremental Conductance(Inc) method. This paper proposes a novel MPPT algorithm for solar PV applications based on the binary sequence of a number which has been improvised for higher efficiency, accuracy and convergence speed along with a faster response to rapid change in operating circumstances. Also the algorithm converges within 18 iterations whatever be the change in the operating conditions. The proposed algorithm is discussed in detail, with simulation results verifying the viability algorithm using MATLAB/SIMULINK.**

Keywords—MPPT, Solar PV, Binary, P&O, Variable step

I. INTRODUCTION

Recently, in the past few years energy generation from renewable energy sources has gained a lot of importance as it is eco-friendly and abundantly available and more importantly the governments are encouraging their use and reduce dependence on traditional sources of energy. It was estimated that the solar energy reaching the earth's surface is a few times more than what the entire globe consumes at present. However, compared to other renewable energy sources, harnessing solar energy has captured the limelight as it is abundantly available, with advanced technology and reduced equipment cost. But the solar PV arrays cannot deliver maximum power on their own due to it's non-linear characteristic curve. Added to that, the maximum power point also varies on temperature and irradiance of the sun. The maximum power point(MPP) keeps on changing throughout the day and measuring the temperature and irradiance throughout the day may turn out to be a costly task. Therefore various maximum power point tracking algorithms are used to extract maximum power that is available. These algorithms are used to control the duty cycle of DC/DC converters, which are an integrated part of the PV systems, to dynamically change its operating condition so as to match the maximum power point of the PV curve.

There are many algorithms that exists today for finding the maximum power point of the PV. Each of the algorithms has an advantage but is also accompanied by a disadvantage. Some algorithms are simple and require very less calculations such as, the Fractional Open-Circuit Voltage [1-2], Perturb & Observe(P&O) [3-4], Incremental

Conductance(Inc) methods [5-6]. There are also a few complex methods which require complex or added complex circuitry for their operation such as, Fuzzy Logic control [7], Artificial Neural Network [8], Sliding mode control [9]. Complex calculations involved in these algorithms imply that they would require high end microprocessors, large memory to meet the demands of high computational load. Thereby this paper proposes a maximum power point tracking(MPPT) method which is much simpler to implement with a stable convergence point and high accuracy.

II. ELECTRICAL MODEL OF A PV MODULE

A PV cell is modeled using a current source, with a diode connected in anti-parallel across it. To account for the losses that take place in the PV cell a shunt and series resistances are used in the model. Equation (1) and (2) gives the current-voltage equation of a typical PV cell.

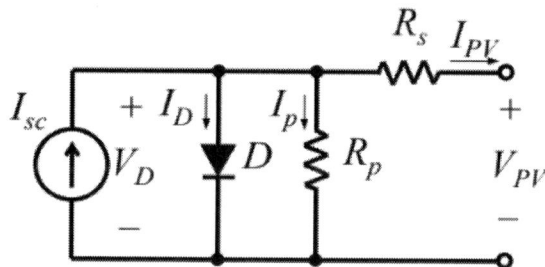

Fig. 1. Equivalent circuit model of a solar PV

$$I_{PV} = I_{SC} - I_D - I_P \qquad (1)$$

or

$$I_{PV} = I_{sc} - I_0(e^{\frac{V_{PV}+I_{PV}R_s}{mV_T}} - 1) - \frac{V_{PV} + I_{PV}R_s}{R_p} \qquad (2)$$

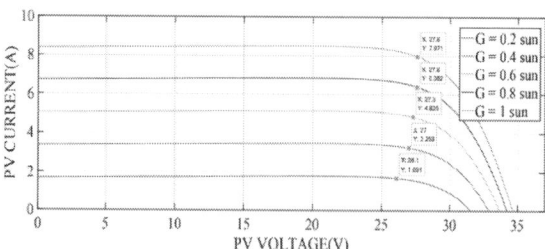

Fig. 2. IV characteristics with varying irradiance and constant temperature

978-1-5386-6624-1/18 $31.00 © 2018 IEEE

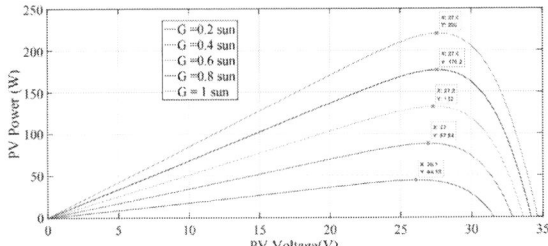

Fig. 3. PV characteristics with varying irradiance and constant temperature

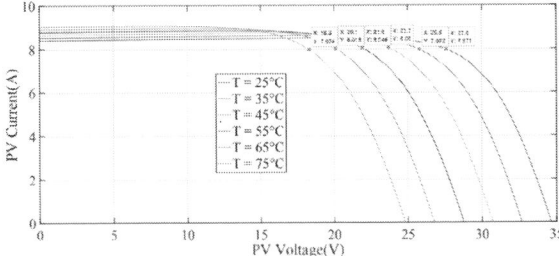

Fig. 4. IV characteristics with constant irradiance and varying temperature

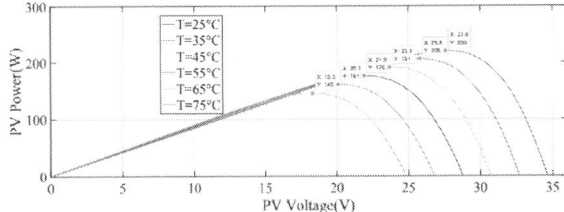

Fig. 5. PV characteristics with constant irradiance and varying temperature

III. BINARY SEQUENCE OF A NUMBER

A binary number is a sequence of 0s and 1s that can represent any whole number. A n digit binary number can be used to represent whole numbers in the range from 0 to 2^n - 1. In a binary number the multiplier of the right most digit is 1 and from there on multiplier of each of the digit to the left keeps on multiplying by 2. The left most digit is known as the most significant digit and the right most is known as the least significant digit. Considering any digit the sum of the multiplier to the digits on the right is exactly 1 less than its own multiplier.

To find the binary sequence of a number, N adequate length of a binary number is chosen. Then starting from the left each digit is made 1 and it is checked whether the whole number corresponding to the binary sequence exceeded N. If not the procedure is continued moving towards the digits on the right, if yes then the corresponding digit is turned to 0 and the process continues.

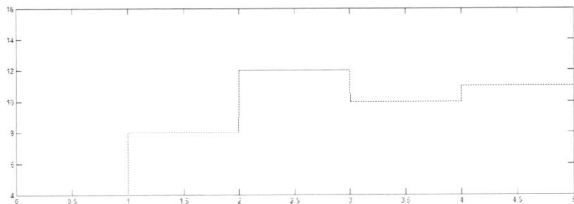

Fig. 6. Step wise approach to get binary sequence of a whole number

Let's suppose we have to find the binary sequence to the number 11. For that a 4 digit binary number is chosen, '0000'. At first the left most digit is made one, which results in '1000' which is equivalent to 8 < 11. Thus we move on to the second digit and flip it to 1. The result is '1100', equivalent 12 > 11. So in the next step, the second digit is overturned to 0 and the third digit is turned to 1. The advancement of the process with each step is graphically explained above in Fig. 6.

IV. THE P&O ALGORITHM

The P&O method is the most widely used algorithm for commercial applications and has been improvised over the years for better performance. A very small magnitude of perturbation is introduced in the system. The perturbation causes the power output from the solar cell to change immediately. If the power increases due to the perturbation, the process is continued in the same direction. The power decreases once the maximum power point is reached and the direction of perturbation is reversed. Thus the system begins to oscillate about the MPP point. The algorithm is shown below in Fig. 7.

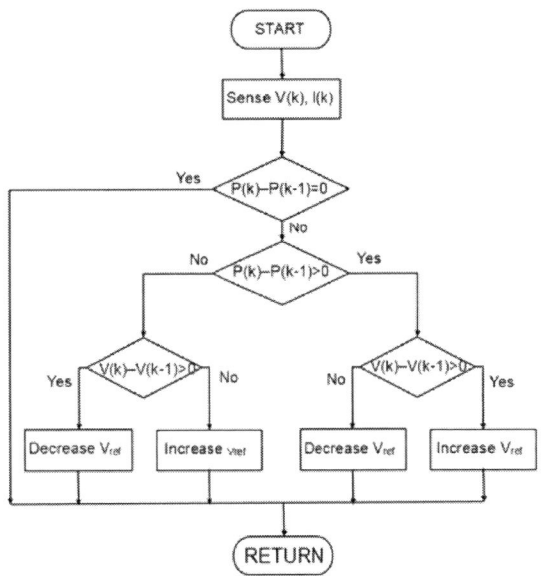

Fig. 7. Flow chart of the P&O Method

The oscillation of the algorithm about the MPP leads to significant amount of power loss. A PI controller is used to dampen the oscillations to an extent. The algorithm responds slowly under rapidly changing atmospheric conditions. If the perturbation size is reduced the system slows down while accuracy is increased and vice-versa. To improve the performance of the P&O method a modified P&O method has been developed. In an improved algorithm, proposed in [3], instead of comparing the average, the instantaneous values of i_{pv} and v_{pv} and the peak current control that presents one cycle speed of response for small variations in the reference current. Despite its shortcomings, the algorithm is very much popular due to its simplicity and requirement of less complex hardware.

It has also been discussed in [10] that the the P&O and the Inc methods have similar tracking performances under both static and dynamic conditions. Both of them work using the derivative of power with voltage. It has been recommended that the Inc method should not be treated as a separate MPPT method but can be considered to be a one of special cases of the P&O method.

V. PROPOSED IMPROVISED BINARY SEQUENCE ALGORITHM

The MPPT technique based on bisection search theorem, discussed in [11] and the binary search technique, discussed in [12] scans the whole range of the panel voltage starting from zero to V_{OC}. The total scanning of the range increases the time taken. If there is any disturbance in irradiance and temperature after the MPP point is reached the whole algorithm starts from the beginning. Also due to the above mentioned feature, the technique is not much preferable for use under fast changing irradiance conditions.

The binary search based P&O method, discussed in [13]. can give the MPP pretty fast under normal operating limits, as expected on a normal day, as their range of scan is confined to a limited space considering the speed. But the algorithm will take a lot more time if the conditions change abruptly, specially if the MPP point goes beyond the predetermined expected region which may happen in a few cases.

The proposed improvised binary sequence algorithm uses variable step duty cycle values for finding the MPP. The binary sequence is applied to the voltage range of the PV and not on the duty cycle. Thereby it can track the whole range of MPP points from zero to V_{OC} by reducing the area of search each time by half. But since in most of the regular cases the MPP point falls in the higher part of the voltage range, it skips scanning the lower half but if needed it has the capability to retrace back to the lower half. It reaches within tolerance limits within 10 iterations following which it continues on the process for maximum accuracy. The algorithm has also been designed for rapid changes in operating conditions and is also designed to give successful results in all cases possible. The algorithm of the MPPT has been shown above in Fig. 8.

The basic algorithm has been designed for use in a boost converter with $D_0 = 50$. The algorithm stops changing the duty cycle after the 12th stage as voltage changes of 0.009 V is not affecting the output power considerably. It is also seen after the fourth stage the power fluctuations can be limited in a particular range, which helps us to detect any abrupt change in conditions if power changes go beyond the band. This helps us to account for abrupt changes in operating conditions.

VI. SIMULATION

For the simulations the solar panel considered is the Mitsubishi Electric PV-TJ230GA6. At first, the most common algorithm, the P&O algorithm is tested followed by the proposed IBS algorithm. Both the algorithms were tested using simulation built on MATLAB & Simulink.

For simulation of the P&O algorithm a buck-boost converter was used to control the PV voltage. The power output from the PV is shown in Fig. 9.

Fig. 9. PV power output using the P&O algorithm

As is seen in Fig. 9, the PV power reaches the MPP quite soon(within 0.3 s). But the controller is kept busy for a longer period of time as concluded from the voltage reference curve from the controller as shown in Fig. 10.

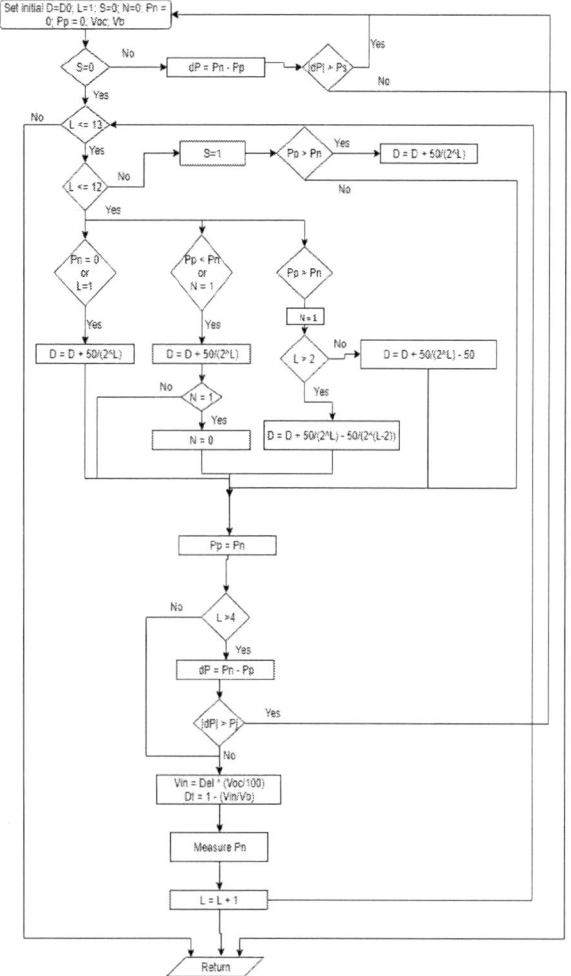

Fig. 8. Flowchart of the proposed Improvised Binary Sequence Method

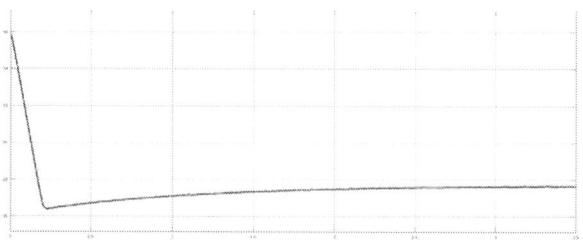

Fig. 10. Voltage reference as controller output

978-1-5386-6624-1/18 $31.00 © 2018 IEEE

As is seen in Fig. 10, the voltage reference reaches a stable value after around 3s. So, though it may appear(from the power output) that the controller has reached a stable operating point quite quickly, but the controller is kept busy for a longer period.

The proposed IBS algorithm was tested with the circuit diagram given in Fig. 11. A boost converter was used to transfer the power from the PV to the battery. A smoothing capacitor was connected across the PV terminals to reduce the oscillations. The components used: C = 5mF, L = 1mH ,f_{switch} = 1kHz.

For the MPPT, P_j = 15W, P_s = 0.5W, $V_{battery}$ = 48V, sampling time = 0.1s.

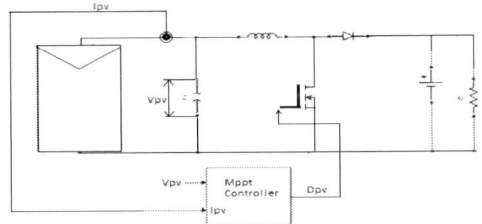

Fig. 11. Considered circuit model

Two tests were conducted. In the first test temperature was kept at 25°C and irradiance was 1000W/m² for the first 2 seconds and then changed to 500W/m². In the second test the reaction of the algorithm to rapid changing conditions were tested. The temperature was kept at 25°C and irradiance at 1000W/m² and was changed to 35°C and 500W/m² irradiance after 1 second.

The figures below show the results of the first test mentioned above when Improvised Binary Sequence method was used.

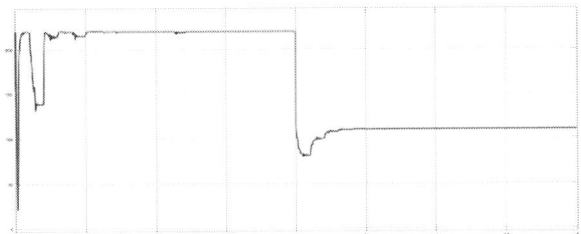

Fig. 12. PV power output of the first test

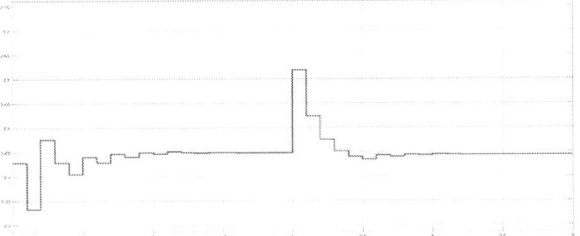

Fig. 13. Duty cycle output from the MPPT controller for the first test

The duty cycle output from the controller was passed through a first order transfer function to reduce the oscillations in power occurring due to sudden steep changes in duty cycle. The duty cycle fed to the PWM generator is given in Fig. 12.

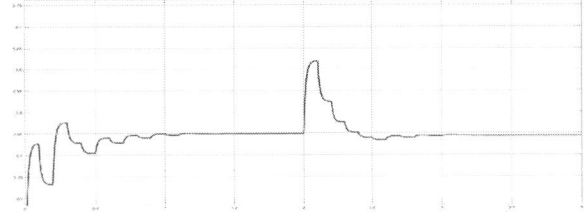

Fig. 14. Duty cycle waveform fed into the PWM generator for the first test

The results of the second test are given below.

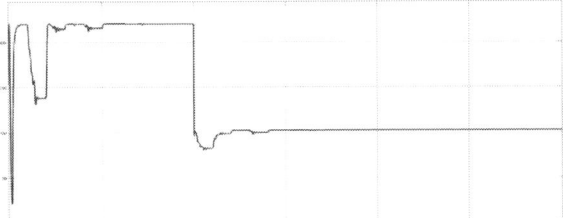

Fig. 15. PV power output of the second test

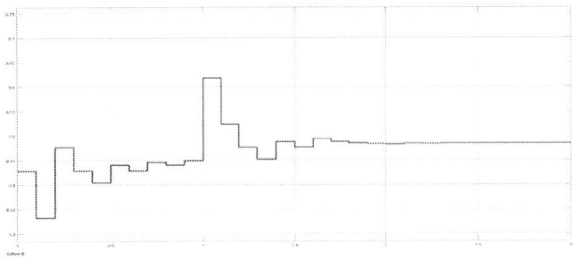

Fig. 16. Duty cycle output from the MPPT controller for the second test

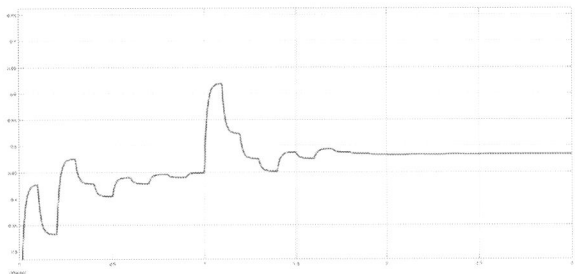

Fig. 17. Duty cycle waveform fed into the PWM generator for the second test

The algorithm takes anywhere between 1.3s to 1.8s to converge. From the results it can also be seen that the algorithm reaches the tolerance band within 0.5s following which it tries to accurately locate the MPP point. One of the main advantages of the proposed algorithm is that, once the maximum power is obtained from the PV implies that the controller has also achieved a stable operating point. The algorithm has also been designed to take care of the abrupt changes due to operating conditions while its converging. Since the algorithm is designed to take care of voltages from zero to V_{oc}, the initial fluctuation in power is large but it settles down soon.

During transition , unlike the BST method discussed in [11], the fluctuations are much less leading to increase in efficiency.

The algorithm was also tested under was tested with two conditions:

I) When the algorithm was tried out mathematically in MATLAB on the PV power voltage curves generated mathematically.

II) When the algorithm was tried out in simulation in SIMULINK according to the circuit in Fig. 11.

The results and the comparison are tabulated in Table 1.

TABLE I. COMPARISON OF RESULTS UNDER STATIC AND DYNAMIC CONDITIONS.

Operating Conditions		Power Output(W)	
Temperature(°C)	Irradiance (W/m²)	Case I	Case II
25	1000	220.01	219.5
20	800	181.9	181.9
35	500	102.26	102.2
15	900	210.85	210.9
30	300	63.42	63.4

On interpolating the results from the various tests done on the algorithm to test its utility, it has been found that the algorithm was accurate up to 99.9% at the best and 99.7% at the worst.

VII. CONCLUSION AND FUTURE WORK

The paper has presented a new MPPT technique based on the binary sequence of a whole number and was improvised for the task. The algorithm was tested on a Boost converter. The simulation results show that the proposed technique offers the combination of high accuracy with faster response and stability.

Future work on the algorithm may include the following: a way to make the algorithm converge faster. In the tests, the sampling time was taken as 0.1s for all the iterations, but it can be reduced for the later iterations as fluctuations keep decreasing to give an overall faster response. While continuous MPPT the Control starts from level zero when a shift in operating conditions take place. Instead if the control

can be shifted to some other level the speed of the algorithm will increase considerably.

REFERENCES

[1] Sivaramakrishnan S, "Linear extrapolated MPPT - an alternative to fractional open circuit voltage technique," 2016 Biennial International Conference on Power and Energy Systems: Towards Sustainable Energy (PESTSE), 2016, pp. 1-4.

[2] H. A. Sher, A. F. Murtaza, K. E. Addoweesh, M. Chiaberge, "A two stage hybrid maximum power point tracking technique for photovoltaic applications," 2014 IEEE PES General Meeting, Conference & Exposition, 2014, pp. 1-5.

[3] X. Liu, L. A. C. Lopes, "An improved perturbation and observation maximum power point tracking algorithm for PV arrays," 2004 IEEE 35th Annual Power Electronics Specialists Conference (IEEE Cat. No. 04CH37551), 2004, Vol. 3, pp. 2005 - 2010.

[4] N. Femia, G. Petrone, G. Spagnuolo, M. Vitelli, "Optimization of perturb and observe ,maximum power point tracking method," IEEE Transactions on Power Electronics, 2005, Vol. 20, pp. 963-973.

[5] A. Safari, S. Mekhilef, "Simulation and Hardware Implementation of Incremental Conductance MPPT With Direct Control Method Using Cuk Converter," IEEE Transactions on Industrial Electronics, 2011, Vol. 58, No. 4, pp. 1154-1161.

[6] F. Liu, S. Duan, Fei Liu, B. Liu, Y. Kang, "A Variable Step INC MPPT Method for PV Systems," IEEE Transactions on Industrial Electronics, 2008, Vol. 55, No. 7, pp. 2622-2628.

[7] B. L. Alajmi, K. H. Ahmed, S. J. Finney, B. W. Williams, "Fuzzy-Logic- Control Approach of a Modified Hill- Climbing Method for Maximum Power Point Tracking in Microgrid Standalone Photovoltaiv System," IEEE Transactions on power Electronics, 2011, Vol. 26, No. 4, pp. 1022-1030.

[8] H. S. Agha, Zafar-ullah Koreshi, M. B. Khan, "Artifitial neural network based maximum power point tracking for solar photovoltaics," 2017 International Conference on Information and Communication Technologies (ICICT), 2017, pp. 150-155.

[9] E. Bianconi, J. Calvente, R. Giral, E. Mamarelis, G. Petrone, C. A. Ramos-Paja, G. Spagnuolo, M. Vitelli, "A Fast Current-Based MPPT Technique Employing Sliding Mode Control," IEEE Transactions on Industrial Electronics, 2013, Vol. 60, No. 3, pp. 1168-1178.

[10] D. Sera, L. Mathe, T. Kerekes, S. V. Spataru, R. Teodorescu, "On the Perturb-and-Observe and Incremental Conductance MPPT Methods for PV Systems," IEEE Journal of Photovoltaics, 2013, Vol. 3, No. 3, pp. 1070-1078.

[11] P. Wang, H. Zhu, W. Shen, F. H. Choo, P. C. Loh, K. K. Tan, "A novel approach of maximizing energy harvesting in photovoltaic systems based on bisection search theorem," 2010 25th Annual IEEE Applied Power Electronics Conference and Exposition (APEC), 2010, pp. 2143-2148.

[12] N. Shiota, V. Phimmasone, T. Abe, M. Miyatake, "A MPPT algorithm based on the binary-search technique with ripples from a converter," 2013 International Conference on Electrical Machines and Systems (ICEMS), 2013, pp. 1718-1721.

[13] Chi-Thang Phan-Tan, Nam Nguyen-Quang, "A P&O MPPT Method for Photovoltaic Applications Based on Binary- Searching," 2016 IEEE International Conference on Sustainable Energy Technologies (ICSET), 2016, pp. 78- 82.

Placement of Distributed Generation in Radial Distribution System by Considering Uncertainty of Load Demand

Prem Prakash
Department of Electrical Engineering,
Delhi Technological University, Delhi- 110042
Email: ppyadav1974@gmail.com

Abstract - **The scope of the present study is to state the probabilistic load flow (PLF) for the distribution system. The PLF is extremely important in case of uncertain input variables. In real time the electric load demand and output power of wind and solar operated DGs are uncertain parameters. The uncertain input parameters give rise to uncertain output. To handle uncertain parameters the PLF provides accurate results while deterministic load flow (DLF) is not capable of providing satisfactory results. Therefore, the present paper aims to identify the buses for placement of Distributed Generation (DG) in the distribution by evaluating loss sensitivity factor with the help of PLF by considering the uncertainty of electric load demand. In the present paper, the electric load demand is considered as a normal probability distribution function (PDF). The three-point estimate method (PEM) based method is applied to get the solution of optimal probabilistic load flow (OPLF). The candidate buses are determined for placement of DG unit in distribution by PLF.**

Index Terms – Probabilistic Load Flow; Point Estimate Method; Load flow; Distributed Generation.

I. INTRODUCTION

Distributed generation (DG) technologies have become more significant in the power system since last decade because of its enumerable benefits over centralized power generating units. Nowadays, the efforts are being made to cut down the line losses, and emission of greenhouse gases along with maintains the system voltage profile within the prescribed limit. Therefore, the DG technologies which are powered by renewable energy sources (RES) such as wind turbine generators (WTG), solar photovoltaic (SPV), biomass, small hydro, geothermal are extremely useful. The above mentioned DG technologies are capable of reducing the emission of green houses gases (GHG). The other type of DG technologies which are powered by non-renewable energy sources is also existed such as internal combustion engines, fuel cells, etc. Furthermore, the optimal sizing and siting of DG units in the distribution system are of extremely important they reduce the system loss, improve voltage profile and system reliability. Furthermore, the placement of DG in distribution system becomes crucial when uncertainty imposed by the power generating units, electric load demand and electricity markets, fuel cost, etc. [1, 3, 5, 6, 7, 12]. Therefore, it significantly affects the operation and performance of the system. Moreover, non-optimal placement of DG units may lead to increase system losses, system cost and voltage in some load buses. Further, inefficiently and inadequately designed DG units also associated with some undesirable consequences on the operation of the system such as DG may lead to reverse power flow, enhancement of voltage leads towards voltage instability problem. Furthermore, non-optimal placement of DG units invites extra fault levels, increase total harmonic distortion as a consequence the performance of system gets deteriorates. Therefore, optimal placement of DG in distribution systems is not only beneficial for consumers but also for electric utilities.

Therefore, given the above facts, the optimum placement of DG (OPDG) is extremely important. Therefore, (OPDG) becomes more complex when uncertainties of load and source of generation especially when energized by RES like solar and wind is considered. Additionally, the intermittent nature of solar irradiance and wind speed lead towards uncertainties in the generation while system load is also an uncertain parameter. To handle the uncertainties associated with load and renewable resources, probabilistic techniques have been applied. The objectives of OPDG are to reduce the system losses and improve the power quality along with system voltage profile.

A. Literature Review

Evangelopoulos and Georgilakis [1] proposed a PEM based technique for probabilistic power flow (PPF) to consider the uncertainties of the power output of DG, load demand, fuel cost and electricity prices. Moreover, for the optimization GA based technique is applied for optimum sizing and placement of non-dispatchable DG. Aien et al. [2] applied Weibull probability distribution function has been considered for the uncertainty of wind generation. Additionally, for probabilistic load flow solutions are obtained to consider the uncertainty of load demand. Further, the point estimate based method has been applied to handle these uncertainties.

The probabilistic load flow was first introduced by [9]. Further, the authors of [1-8,10, 12, 36, 37], applied the point estimate based method for probabilistic load flow considering the uncertainty of load, generation, fuel prices and energy prices, etc. The author of [13] presents the probabilistic method for optimal placement of a wind turbine generator in the distribution network. [15-17, 19, 21, 22, 23, 24, 32], highlighted nature inspired meta-heuristic optimization algorithms for optimal placement of DG units in the distribution network. The authors of [16, 28] have presented an expansion planning of distribution system in the presence of DG units. The author of [18, 20] has analyzed the impact of

978-1-5386-6624-1/18 $31.00 © 2018 IEEE

the penetration level of DG units in the distribution system. Further, the authors of [25-27, 29, 30, and 31] have presented the optimal placement of DG units in the distribution system by considering the uncertainty of load and power output of DG which are operated by solar and wind.

Furthermore, for OPDG, the load flow solutions of the radial distribution system are extremely important to assess the power system status. The point estimate method [1-8, 12, 30, 36, and 37], have been applied for probabilistic power flow and genetic algorithm based approach to solve optimization problem by considering uncertainties of load growth, wind, and solar photovoltaic generation, fuel cost and electricity prices. Moussavi et al. [13] applied the probabilistic method for placement of wind turbine generator to assess the uncertainty of wind turbine generator and congestion management in the distribution system. Lyu et al. [14] described optimal probabilistic load flow for optimization of active and reactive power dispatch in the presence of a wind turbine generator.

B. Electric Load Demand Modelling

The normal distribution function mathematically is defined as follows

$$f(x) = \frac{1}{\sqrt{2\pi\sigma^2}} \exp\left(-\frac{(x-\mu)^2}{2\sigma^2}\right)$$; Where x is identified

uncertain variable μ is the mean of an uncertain variable, σ is the standard deviation. The variation of x and f(x) is presented here with for value of μ is zero and σ is 1. The Fig. 1 is applied to represent given uncertain function

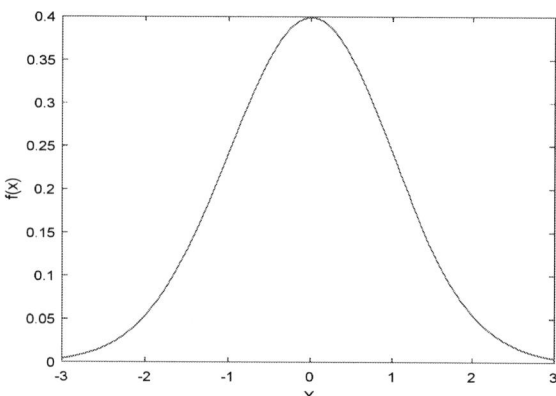

Fig. 1. Standard normal distribution function

In the present paper, load demand is considered as normal distribution probability density function (PDF). Peak load demand is considered as mean value (μ), and ±5 percent of the peak value of load demand is considered as standard deviation (σ) for the uncertainty of load.

The power output of the wind turbine is given in Fig. 2, the generated power is zero when the wind speed is less than the cut-in speed and generated power is linearly increased when

up to wind speed reached as nominal speed further the generated power remains constant when wind speed more than nominal speed and less than cut-out speed. In the present study, the cut-in speed is 3 m/s, nominal speed is 13 m/s, and cut-out speed is 25 m/s.

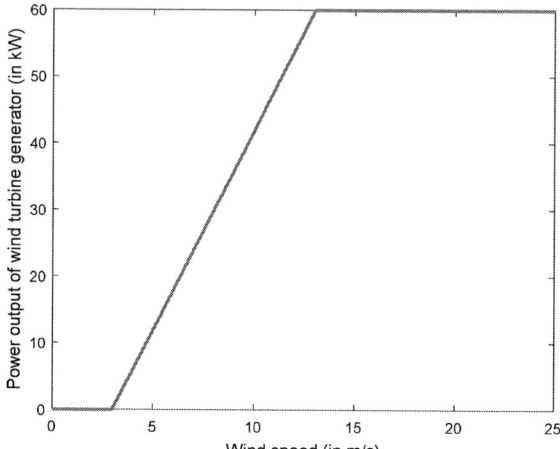

Fig. 2. Power output of wind turbine generator

C. Point Estimate Method

A point estimate method is an approximate method in which only a few points of the distribution function of a random variable are considered in calculating the statistical information or central moments of input random variable. These points are called concentrations. By using these points and the output-input relationship and the information of the output can be obtained [1-8].

Let's assume the function F is the set of non-linear load flow equations which relate the input and output stochastic variables. Let's assume Z (l, k) is the vector of stochastic output variables, and p_i is the ith random variables. Further, the set Z (l, k) of random output variables can be expressed as follows

$$Z (l, k) = F (p_1, p_2, \ldots, p_l, \ldots, p_m) \quad (1)$$

Further, PEM concentrates all the statistical information provided by the first central moments of the stochastic input variables and computes K points for each variable, named concentration. The kth concentration ($p_{l, k}$, $w_{l, k}$) of a random variable p_l can be defined as a pair of a location $p_{l, k}$ and a weight $w_{l, k}$. The location is the kth value of the variable p_l at which function F is evaluated and the weight $w_{l, k}$ is a weighting factor that accounts for the relative importance of this evaluation in the random output variable [6].

The new points or value of a random variable can be obtained as

$$p_{l,k} = \mu_l + \xi_{l,k}\sigma_l \text{ for } k = 1, 2 \ldots h$$

Further, the other parameters can be evaluated as

Procedure for determining various parameters such as $\lambda_{l,i}$

a) Find the standard central moments as:

978-1-5386-6624-1/18 $31.00 © 2018 IEEE

$$\lambda_{l,i} = \frac{E\left[(x_l - \mu_1)^i\right]}{\sigma_l^i}, \qquad i = 3\ldots\ldots 2m$$

It is to be noted that, $m = 2$ for three-point estimate method

b) Find the standard locations $\xi_{l,q}$, where q=1……m for 3 PEM these locations are calculated using (2)

$$\xi_{l,k} = \frac{\lambda_{l,3}}{2} + (-1)^{3-k} \times \sqrt{\lambda_{l,4} - \frac{3}{4}\lambda_{l,3}^2} \qquad \text{where}$$

$$\xi_{l,3} = 0, \quad k = 1,2 \tag{2}$$

c) After the calculation of standard locations $\xi_{l,k}$, obtain $x_{l,k}$ from (1) The weighting factors $w_{l,k}$ and $w_{l,k}$ 3for 3-PEM are computed from and the weighting factors equations

$$w_{l,k} = \frac{(-1)^{3-k}}{\xi_{l,k}\left(\xi_{l,1} - \xi_{l,2}\right)}, \qquad w_{l,3} = \frac{1}{n} - \frac{1}{\lambda_{l,4} - \lambda_{l,3}^2}, \qquad k = 1,2 \quad \text{and}$$

$$w_{\mu,1} = 1 - \sum_{l=1}^{n}\sum_{k=1}^{m} w_{l,k}$$

Finally, the output is given as follows

$$E(z) = \sum_{l=1}^{m}\sum_{k=1}^{2} w_{l,k} Z(l,k) + w_{\mu,1} Z(\mu,1) \tag{3}$$

$$E\left(z^j\right) = \sum_{l=1}^{m}\sum_{k=1}^{2} w_{l,k} Z(l,k)^j \tag{4}$$

D. Evaluation of Loss Sensitivity Factor for DG Placement

Loss Sensitivity Factor is applied to identify the candidate bus for placement of DG in the distribution system. Once the loss sensitivity factor is calculated for each bus, the buses are arranged in descending order of their sensitivities which is obtained by following formulae to get the most sensitive buses. Therefore, these sensitive buses serve as candidate buses for the DG placement.

The single line diagram of a distribution network having two nodes as i and j respectively is shown in Fig. 1. The active power loss of this line diagram is given by $I_k^2 R_k$ which can further be expressed as

$$P_L = \frac{P_i^2 + Q_i^2}{V_j^2} \times R_k, \quad Q_L = \frac{P_i^2 + Q_i^2}{V_j^2} \times X_k$$

The loss sensitivity factor can be determined as $\frac{\partial P_L}{\partial P_{eff}} = 0, \quad \frac{\partial Q_L}{\partial Q_{eff}} = 0$, differentiating above loss equations concerning P_i and Q_i respectively. Therefore, the active power loss and reactive power loss sensitivity is given by the following expressions as $\frac{\partial P_L}{\partial P_i} = \frac{2P_{eff}}{V_j^2} \times R_k$ and

$\frac{\partial Q_L}{\partial Q_i} = \frac{2Q_{eff}}{V_j^2} \times X_k$ respectively, where P_{eff} and Q_{eff} are the active and reactive power demand at bus i respectively.

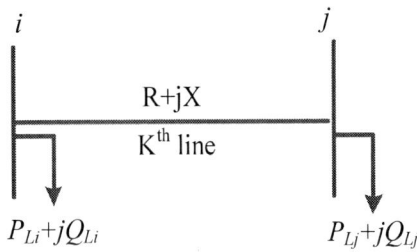

Fig. 3.   Single line diagram for general distribution system

The single line diagram of IEEE 33 bus radial test distribution system is shown in Fig. 4. Further, on the basis of the above formulations, the value of the loss sensitivity index (LSI) at different load buses has been shown in Fig. 3 for IEEE 33-bus test radial distribution system. From this figure, it is observed that active power loss sensitivity indices in their descending order on the bus number are 6, 3, 28, 4, 5, 9 and 24. Therefore, the candidate buses for DG placement are 6, 3, 28, 4, 5, 9 and 24

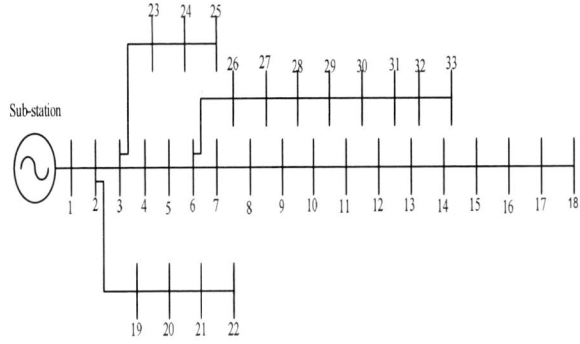

Fig. 4.   Single line diagram for 33-bus radial distribution network

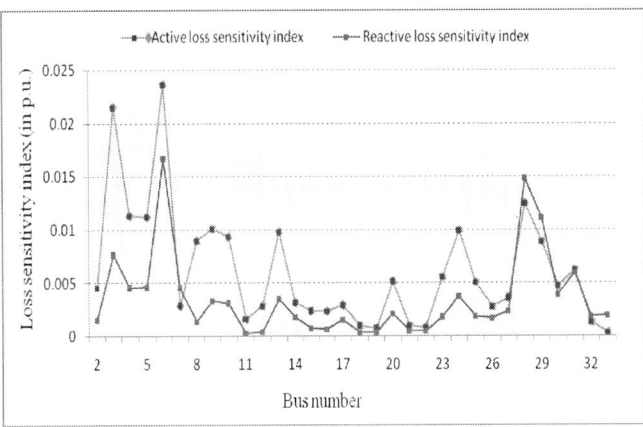

Fig. 5.   loss sensitivity index (in p.u.) for a 33-bus system for different buses

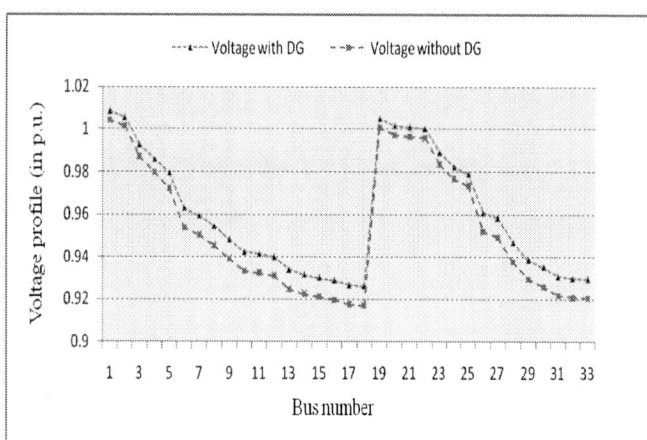

Fig. 6. System voltage profile with and without DG

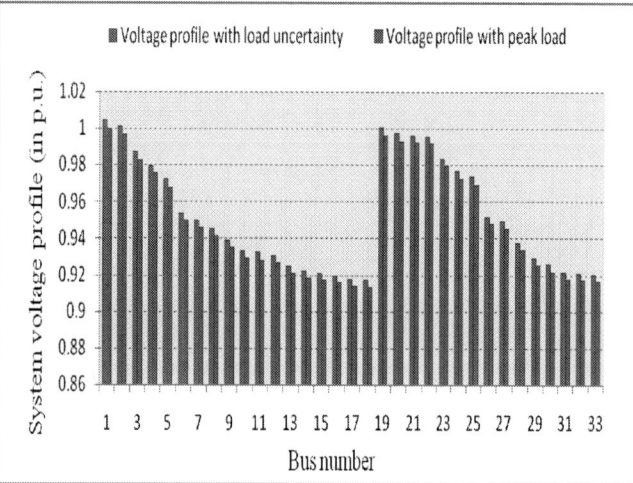

Fig. 7. System voltage profile with load uncertainty and at peak load

II. RESULT AND DISCUSSION

The developed algorithm has been applied to the IEEE 33 bus test radial distribution system. This system has a total load of (3.715+j2.300) MVA, and relevant programming has been developed in MATLAB. The proposed three points PEM based PLF solutions have been obtained with the help of a deterministic load flow (DLF) method. The active power loss of test system is 202.7 kW and 135.1 kVAR respectively in the base case, i.e., without DG. The Fig.5 provides the load sensitivity factor of the system at different buses according to this figure the maximum sensitivity occurs at bus six. Therefore, the bus six becomes the most sensitive bus, and it acts as a candidate bus for DG placement. When the DG unit of size 100 kW is placed at bus 6 the active, and reactive power loss of the system become 178.7 kW and 120.2 kVAR respectively. Therefore, system losses get reduced significantly. Further, system voltage profile with and without DG unit by considering the uncertainty of load is depicted in

Fig. 6. It is observed from this figure that the voltage profile of the system is also getting improved significantly.

Fig. 7 shows the system voltage profile. Here the profile has been evaluated by two different methods of load flow solutions. The first solution has been obtained with PLF by applying point estimate method in which the uncertainty of load demand has been considered, and another load flow solution has been obtained by deterministic load flow using backward and forward sweep method at peak load condition. Further, it is observed in Fig. 7 that the system voltage profile at each bus gets improved as compared to peak load condition.

CONCLUSIONS

In the proposed method probabilistic load flow solutions are obtained by three-point PEM to consider the uncertainty of a random variable. The PEM based method is seven-time faster as compare to the Monte Carlo simulation (MCS) technique which is the only available tool to solve the problem of the uncertain random variable. In the process of three PEM based method, only two points or two new values of the random variable are to be evaluated in addition to mean value. While in MCS five thousand or ten thousand samples of each random variable are generated. Therefore, it does not only increase computational bourdon but also increases complexity. Further, the results of MCS and point estimate are almost similar. The loss sensitivity factor is evaluated as per formulation described in section four of this study, and these factors are arranged in descending order to get the most sensitive buses for DG placement

REFERENCES

[1] Vasileios A. Evangelopoulos, Pavlos S. Georgilakis, "Optimal distributed generation placement under uncertainties based on point estimate method embedded genetic algorithm," *IET Generation, Transm. & Distrib.* vol. 8, no. 3, pp. 389-400, 2013.

[2] Morteza Aien, Reza Ramezani and S. Mohsen Ghavami, "Probabilistic Load Flow Considering Wind Generation Uncertainty," *Int. J. Engineering, Technology & Applied Science Research,* vol. 1, no. 5, pp. 126-132, 2011.

[3] Emilio Rosenblueth, "Point estimates for probability moments," *Proc. Nat. Acad. Sci. USA,* vol. 72, no.10, pp. 3812-3814, 1975.

[4] Christopher Scott Saunders, "Point Estimate Method Addressing Correlated Wind Power for Probabilistic Optimal Power Flow," *IEEE Trans. on Power Systems,* vol. 29, no. 3, pp. 1045-1054, 2014.

[5] C. Delgado and J.A. Domínguez-Navarro, "Point estimate method for probabilistic load flow of an unbalanced power distribution system with correlated wind and solar sources," *Int. J. Electrical Power and Energy Systems,* vol. 61, pp. 267-278, 2014.

[6] Juan. M. Morales and Juan. Pérez-Ruiz, "Point Estimate Schemes to Solve the Probabilistic Power Flow," *IEEE Trans. on Power Systems,* vol. 22, no. 4, pp. 1594-1601, 2007.

[7] Chun-Lien Su, "Probabilistic Load-Flow Computation Using Point Estimate Method," *IEEE Trans. on Power Systems,* vol. 20, no. 4, pp. 1843-1851, 2005.

[8] M. Mohammadi, A. Shayegani, and H. Adaminejad, "A new approach of point estimate method for probabilistic load flow," *Int. J. Electr. Power and Energy System,* vol. 51, pp. 54-60, 2013.

[9] Barbara Borkowska, "Probabilistic Load Flow," *IEEE Trans. On Power Apparatus and Systems,* vol. 93, no.3, pp. 752-759, 1974

[10] M. Hosseinzadeh and H. Afrakhte, "Probabilistic Optimal Allocation and Sizing of Distributed Generation," *Research Journal of Applied*

Sciences, Engineering and Technology, vol. 7, no. 3, pp. 430-437, 2014.

[11] Gregor Verbic, Antony Schellenberg, William Rosehart and Claudio A. Cañizares, "Probabilistic Optimal Power Flow Applications to Electricity Markets," Probabilistic Methods Applied to Power System, PMPAS, pp. 1-6, 2006

[12] Gregor Verbic and Claudio A. Cañizares, "Probabilistic Optimal Power Flow in Electricity Markets Based on a Two-Point Estimate Method," *IEEE Trans. on Power Systems,* vol. 21, no. 4, pp. 1883-1893, 2006.

[13] S. Z. Moussavi, A. Badri, and F. Rastegar Kashkooli, "Probabilistic Method of Wind Generation Placement for Congestion Management," *Int. J. Electrical, Computer, Energetic, Electronic and Communication Engineering,* vol. 5, no. 8, pp. 1102-1110, 2011.

[14] Jae-Kun Lyu, Jae-Haeng Heo, Jong-Keun Park, and Yong-Cheol Kang, " Probabilistic Approach to Optimizing Active and Reactive Power Flow in Wind Farms Considering Wake Effects," *Int. J. Energies,* vol. 6, pp. 5717-5737, 2013.

[15] D. K. Khatod, Vinay Pant and Jaydev Sharma, "Evolutionary Programming Based Optimal Placement of Renewable Distributed Generators," *IEEE Trans. on Power Systems,* vol. 28, no. 2, pp. 683-695, 2013.

[16] Naveen Jain, S. N. Singh and S. C. Srivastava, "A Generalized Approach for DG Planning and Viability Analysis Under Market Scenario," *IEEE Transactions on Industrial Electronics,* vol. 60, no. 11, pp. 5075-5085, 2013.

[17] Neeraj Kanwar, Nikhil Gupta, K.R. Niazi and Anil Swarnkar, "Improved meta-heuristic techniques for simultaneous capacitor and DG allocation in radial distribution networks," *Int. J. Electrical Power and Energy Systems,* vol. 73, pp. 653-664, 2015.

[18] H.M. Ayres, W. Freitas, M.C. De Almeida, and L.C.P. Da Silva, "Method for determining the maximum allowable penetration level of distributed generation without steady-state voltage violations," *IET Generation, Transm. & Distrib.* vol. 4, no. 4, pp. 495-508, 2009.

[19] Satish Kumar Injeti, Vinod Kumar Thunuguntla and Meera Shareef, "Optimal allocation of capacitor banks in radial distribution systems for minimization of real power loss and maximization of network savings using bio-inspired optimization algorithms," *Int. J. Electrical Power and Energy Systems,* vol. 69, pp. 441-455, 2015.

[20] Yasser Moustafa Atwa and E. F. El-Saadany, "Optimal Allocation of ESS in Distribution Systems With a High Penetration of Wind Energy," *IEEE Trans. on Power Systems,* vol. 25, no. 4, pp. 1815-1822, 2010

[21] Y. Mohamed Shuaib, M. Surya Kalavathi, C. Christober and Asir Rajan, "Optimal capacitor placement in radial distribution system using Gravitational Search Algorithm," *Int. J. Electrical Power and Energy Systems,* vol. 64, pp. 384-397, 2015.

[22] K.R. Devabalaji, K. Ravi and D.P. Kothari, "Optimal location and sizing of capacitor placement in radial distribution system using Bacterial Foraging Optimization Algorithm," *Int. J. Electrical Power and Energy Systems,* vol. 64, pp. 384-397, 2015.

[23] Arash Zeinalzadeh, Younes Mohammadi and Mohammad H. Moradi, "Optimal multi objective placement and sizing of multiple DGs and shunt capacitor banks simultaneously considering load uncertainty via MOPSO approach," *Int. J. Electrical Power and Energy Systems,* vol. 67, pp. 336-349, 2015.

[24] Ranjit Roy and H.T. Jadhav, "Optimal power flow solution of power system incorporating stochastic wind power using Gbest guided artificial bee colony algorithm," *Int. J. Electrical Power and Energy Systems,* vol. 64, pp. 562-578, 2015.

[25] Y. M. Atwa, E. F. El-Saadany and R. Seethapathy, "Optimal Renewable Resources Mix for Distribution System Energy Loss Minimization," *IEEE Trans. on Power Systems,* vol. 25, no. 1, pp. 360-370, 2010.

[26] Zhipeng Liu, Fushuan Wen and Gerard Ledwich, "Optimal Siting and Sizing of Distributed Generators in Distribution Systems Considering Uncertainties," *IEEE Trans on Power Delivery,* vol. 26, no. 4, pp. 2541-2551, 2011

[27] Ke-yan Liu, Wanxing Sheng, Yuan Liu, Xiaoli Meng, Yongmei Liu, "Optimal sitting and sizing of DGs in distribution system considering time sequence characteristics of loads and DGs," *Int. J. Electrical Power and Energy Systems,* vol. 69, pp. 430-440, 2015.

[28] Naveen Jain, S.N. Singh and S.C. Srivastava, "Planning and Impact Evaluation of Distributed Generators in Indian Context using Multi-Objective Particle Swarm Optimization," *IEEE Power and Energy Society General Meeting,* pp. 1-8, 2011

[29] Y.M. Atwa and E.F. El-Saadany, "Probabilistic approach for optimal allocation of wind based distributed generation in distribution systems," *IET Renew Power Gener.,* vol. 5, no. 1, pp. 79-88, 2009.

[30] Xue Li, Jia Cao and Dajun Du, "Probabilistic optimal power flow for power systems considering wind uncertainty and load correlation," *Int. J. Neuro computing,* vol. 148, pp. 240-247, 2015.

[31] T.B.M.J. Ouarda, C. Charron, J.-Y. Shin, P.R. Marpu, A.H. Al-Mandoos, M.H. Al-Tamimi, H. Ghedira and T.N. Al Hosary, "Probability distributions of wind speed in the UAE, *Int. J. Energy Conversion and Management,* vol. 92, pp. 414-434, 2015.

978-1-5386-6624-1/18 $31.00 © 2018 IEEE

Application of ALO to Economic Load Dispatch Without Network Losses for Different Conditions

A.V. Sudhakra Reddy
Dept. of Electrical & Electronics
Engineering,
C.B.I.T, Proddatur, A.P., India.
E-mail: sreddy4svu@gmail.com

Y. Praveen Kumar Reddy
Dept. of Electrical & Electronics
Engineering,
C.B.I.T, Proddatur, A.P., India.

B. Bhargava Reddy Reddy
Dept. of Electrical & Electronics
Engineering,
C.B.I.T, Proddatur, A.P., India.

Y. V. Krishna Reddy

*Dept. of Electrical & Electronics
Engineering,
S.V.University, Tirupath., India.*

Abstract— **This manuscript explores the novel Ant Lion Optimization (ALO) to resolve the load dispatch problem optimally. The primary role of Optimal Economic Load Dispatch (OELD) is to obtain an efficient and economical operation of a power system network. The main aim of performing the OELD is for minimizing the fuel cost of the real power generation when the losses are neglected. Ant lion optimization (ALO) is a narrative nature-inspired algorithm and apes the tracking mechanism of ant lions in life. The random walk of ants, building traps, entrapment of ants in traps, catching preys, and re-building traps are the five main steps involved in the hunting process. The principal objective of OELD is to minimize total cost of generation while honoring effective constraints of available generation resources. The anticipated ALO technique is utilized on three and six unit test system to various load demands for solving the load dispatch problem economically. A statistical result illustrates that the anticipated method has lowest fuel cost and superior in quality of solution than other optimization techniques accounted in most latest literature.**

Keywords—Ant Lion Optimization, Economic Load Dispatch, Fuel Cost, Different Load Demands, Real Power Generation, Coal Fired Station, Without Losses.

I. INTRODUCTION

Economic Dispatch forms the significant analysis functions dealing with Operation in a power transmission system. Economic Dispatch (ED) is defined as the route of distributing real power generation to the each generating unit, so that the total network load is supplied with most economical manner. In static economic dispatch, the point is to evaluate, the output power of all generating units so that all the loads are energized at minimum cost, while satisfying the various equal and inequality restrictions of the network and the generator limits.

Economic Dispatch models the electric power system and dispatch the essential load from the accessible, well, generating units for each control area in the most economic manner or in real-time situation. The point is to reduce the total generation cost (include fuel cost, but exclude the cost of network loss) by meeting the operational constraints.

In modern days, various evolutionary, heuristic and meta-heuristics optimization techniques have been developed simulating natural phenomena such as: Bat Algorithm (BA) [1], Particle Swarm (PSO) Optimization [2], Moth-Flame Optimization [3, 15], Cuckoo Search Algorithm (CSA) [4], FireFly Algorithm (FA) [5], Real Coded Genetic Algorithm

(RCGA) [6], DragonFly Algorithm (DFA) [7], Grey Wolf Optimizer (GWO) [8], Evolution Optimization Swarm Algorithm (EOSA) [9], Sine Cosine algorithm (SCA) [10], Glowwarm Swarm Algorithm (WSA) [11], Hybrid Big Bang-Big Crunch Algorithm (HBBBCA) [12], Ant Lion Algorithm (ALO) [13, 17], Novel Bat Algorithm (NBA) [14], stochastic whale optimization (SWO) [16, 18], Moderate Random Search PSO (MRS-PSO) [19], Multi-Verse Optimization (MVO) [20] and Sine Cosine Algorithm (SCA) [21]. Out of these heuristics evolutionary algorithms, some of these are used to solve economic load dispatch problem which is reported in published literature as Bat Algorithm (BA), Cuckoo Search Algorithm (CSA) [4], FireFly Algorithm (FA) [5], Real Coded Genetic Algorithm (RCGA) [6], Evolution Optimization Swarm Algorithm (EOSA) [9], Glowwarm Swarm Algorithm (WSA) [11], Hybrid Big Bang-Big Crunch Algorithm (HBBBCA) [12], Ant Lion Algorithm (ALO) [17], Novel Bat Algorithm (NBA) [14], stochastic whale optimization (SWO) [18], Moderate Random Search PSO (MRS-PSO) [19].

This manuscript recommends, ALO algorithm is utilized for the dolution of economic load dispatch to reduce the cost of fuel in steam power station for different load demands. The projected ALO has tested on simple 3-unit and 6-unit test systems.

II. PROBLEM FORMULATION

A. Objective Function

The fitness of the load dispatch hitch is to curtail the total generation cost while checking the limits of constraints, when the essential power demand of a network is being supplied. The main objective of thus function is to reducecost given by the subsequent equation.

$$\text{Minimize } F_T = F_i(P_{Gi}) = F_1(P_{G1}) + F_2(P_{G2}) + \text{------} + F_n(P_{Gn})$$

$$F_i(P_{Gi}) = \sum_{i=1}^{n} F_i(P_{Gi}) \qquad \text{......(1)}$$

$$F(P_G) = \sum_{i=1}^{n} \left(a_i P_{Gi}^2 + b_i P_{Gi} + c_i\right) \qquad \text{......(2)}$$

B. Equality Restraint

The total generation of all generators should be meet the required load demand. The power balance equation without losses is given by

$$\sum_{i=1}^{n} P_{Gi} = P_D \qquad \qquad ......(3)$$

C. Inequality Restraint

The active power of all generators in coal fired station is to be inhibited within its functional limits.

$$P_{Gi}^{min} \leq P_{Gi} \leq P_{Gi}^{max} \quad i=1,2,.......n \qquad ......(4)$$

III. ANT LION OPTIMIZATION

A. Introduction

This article suggested a novel environment motivated algorithm called Ant Lion Optimization [13, 17] as a different practice to crack optimization problems. As the name signifies, ALO algorithm emulates the sharp exploits of antlions in tracking ants in the atmosphere. The life series of AntLions is revealed in Fig. 1 and altered phases engaged while a corner and hunting the prey is publicized in Fig. 2.

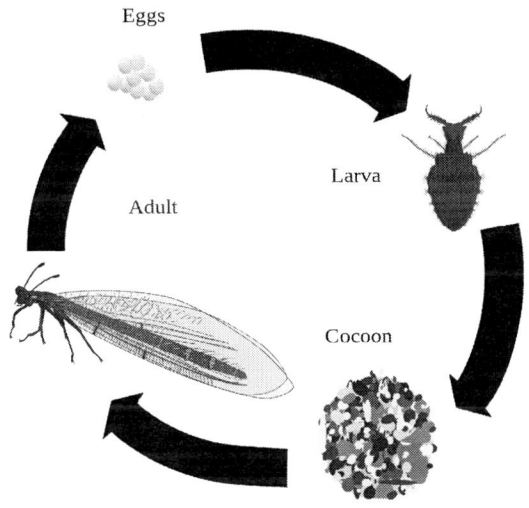

Fig. 1: Life Series of an Ant Lions

The life cycle of antlions comprises two phases: larvae and adult. A total lifespan can take up to 3 years, which mostly transpires in larvae. Antlions undergo metamorphosis in a cocoon to turn out to be an adult. They mostly pursue in larvae and the parenthood era is for reproduction.

The ALO mimics relations involving ant-lions and ants in the trap. To mould, such dealingss, ants are crucial to move over the search space, and antlions are authorized to hunt them and grow to be fitter using traps. Since ants voyage stochastically in life when penetrating for groceries, a random walk using roulette wheel is preferred for the modelling ant movement.

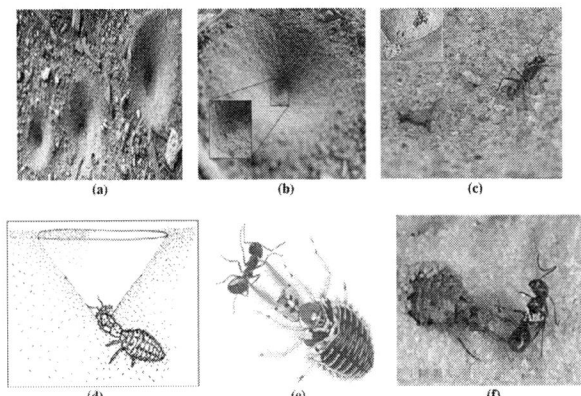

Fig. 2: a–c Building traps and snare of ants in traps; d–f contagious the prey and re-building traps

In order to keep the random walks within the search space, they are normalized using min–max normalization by the subsequent equation.

$$X = \left[0, CumSum\left(2*\left(rand\left(\max\_iter, 1 \right) > 0.5 \right) - 1 \right) \right]$$

$$RW(i) = X(i) = \frac{\left((X-a).*(d-c) \right)}{(b-a)} + c \qquad ......(5)$$

$$Where, \quad a = \min\left(Rand\_Walk \right)$$
$$b = \max\left(Rand\_Walk \right)$$
$$c = lb\left(dim \right)$$
$$d = ub\left(dim \right)$$

$$ant\_pos(i) = \left(ant\_pos(i).*\left(\sim \left(Flag4ub + Flag4lb \right) \right) \right)$$
$$+ ub.*Flag4ub + lb.*Flag4lb \qquad (6)$$

Pseudo Code
% Initialize input data, search agents and maximum iteration
% Initialize generator coeffecients, lower and upper limits each generator
% Initialize the positions ants
% Initially, evaluate the fitness
for $i = 1:n$
for $i = 1:dim$
$RandomWalk(i) = \left(d(i) - c(i) \right)*rand + c(i)$
Calculate $Ant\_Pos(i) = RA(i) + RE(i)/2$
end
end
% Boundar checking
% Update positions
% If any antlinons becomes fitter then Update elite position
% Update convergence criteria
display Best_score as Ant_Lion Fitness

B. Flow Chart for Optimal Economic Load Dispatch

The flow chart for Economic load dispatch problem by using ALO algorithm is presented in Fig. 3. The ants travel stochastically in nature, while searching for food.

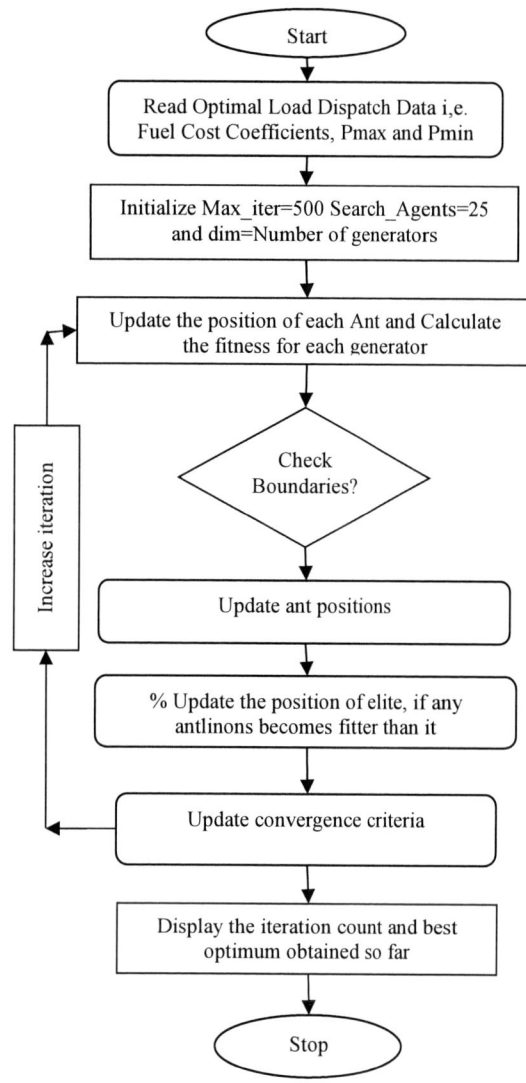

Fig. 3: Flow Chart for ELD Problem Using

Ant Lion Optimization

TABLE I. ALO PARAMETERS

Organize Parameters	
Search Agents	25
Maximum Iterations	500
Ub	Upper limit
Lb	Lower limit
Dim	3 (for 3-unit system)
	6 (for 6-unit system)

A. Three-Unit System

The generator coefficients and lower, upper boundaries for each generator in a thermal system data are revealed in Table. 2. The planned ALO is realized in the OELD problem when transmission losses are not included and end results of the fuel cost for altered load demands, i.e. 350, 400, 450, 500, 550, 600, 650, 700, 750 and 800 MW are offered in Table. 3.

TABLE II. GENERATOR COEFFFICINTS AND INEQUALITY CONSTRAINTS OF 3-GENERATOR SYSTEM

Unit	a	B	c	Pmin	Pmax
1	0.03546	38.30553	1243.5311	35	210
2	0.02111	36.32782	1658.5696	130	325
3	0.01799	38.27041	1356.6592	125	315

TABLE III. OPTIMUM LOAD DISPATCH RESULTS OF 3-GENERATOR SYSTEM

Power Demand, (MW)	P_1 (MW)	P_2 (MW)	P_3 (MW)	Fuel Cost (Rs./hr)
350	64.97302	155.9829	129.0441	18315.5651
400	75.7237	174.0416	150.2347	20480.2969
450	86.47439	192.1003	171.4253	22683.1507
500	97.22508	210.159	192.6159	24924.1263
550	107.9758	228.2177	213.8066	27203.2239
600	118.7264	246.2764	234.9972	29520.4433
650	129.4771	264.3351	256.1878	31875.7847
700	140.2278	282.3938	277.3784	34269.2481
750	150.9785	300.4525	298.569	36700.8333
800	163.5053	321.4947	315	39171.2478

IV. TEST RESULTS

The projected ALO is executed to decipher ELD problem with benchmark 3 and 6 thermal units. The MATLAB code tested on Version R2014a on an Intel, Core™ i3-3227U, CPU@ clock speed of 1.90 GHz, and 8GB RAM, 64-bit OS. The assessment results of the three and six unit systems are revealed. The capricious parameters of an ALO are offered in Table. 1.

B. Six-Unit System

The 6-unit test system generator coefficients, min and maximum limits for real power generation of data are exposed in the Table. 4.

The planned ALO is realized and the outcomes of the fuel cost for various load demands, i.e. 600, 700 and 800 MW are offered in Table. 5.

TABLE IV. GENERATOR COEFFFICINTS AND INEQUALITY CONSTRAINTS OF 6-GENERATOR SYSTEM

Unit	A	b	c	Pmin	Pmax
1	0.15240	38.53973	756.79886	10	125
2	0.10587	46.15916	451.32513	10	150
3	0.02803	40.39655	1049.9977	35	225
4	0.03546	38.30553	1243.5311	35	210
5	0.02111	36.32782	1658.5596	130	325
6	0.01799	38.27041	1356.6592	125	315

TABLE V. OPTIMUM LOAD DISPATCH RESULTS OF 3-GENERATOR SYSTEM

Power Demand (MW)	600	700	800
P1	21.18952	24.97371	28.75794
P2	10	10	10
P3	82.08603	102.6604	123.236
P4	94.37058	110.6347	126.8982
P5	205.3642	232.6838	260.0032
P6	186.9897	219.0473	251.1047
Cost(Rs./hr)	31445.6229	36003.1239	40675.968

C. Comparison of Six-Unit Test System Results

The test results of 6-unit system are compared with the other meta-heuristic optimization techniques which are previously published articles in the literature under the same operating conditions and presented in Table. 6.

TABLE VI. COMPARATIVE RESULTS OF 6-UNIT SYSTEM

Power Demand	FFA [5]	MFA [5]	VSMFA [5]	MFFA [5]	Proposed ALO
600	31489	31447	31576	31481	31446
700	36075	36006	36036	36021	36003
800	40739	40676	40701	40740	40676

From the above comparison, detect that ALO attain the least amount of fuel cost for different power demands as evaluated when the losses are neglected.

The graphical representation of comparative results for 600MW, 700MW and 800MWare exposed in Fig. 3, Fig. 4 and Fig. 5 correspondingly.

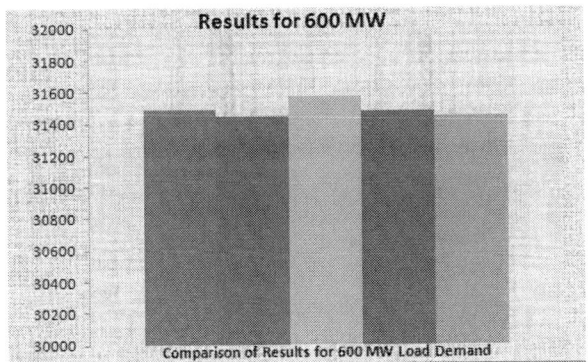

Fig. 4: Graphical representation For 600 MW Load Demand

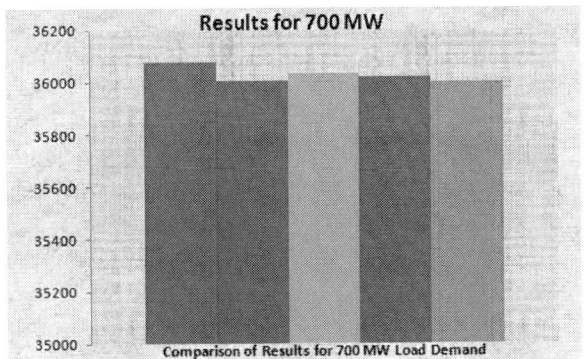

Fig. 5: Graphical Representation For 700 MW Load Demand

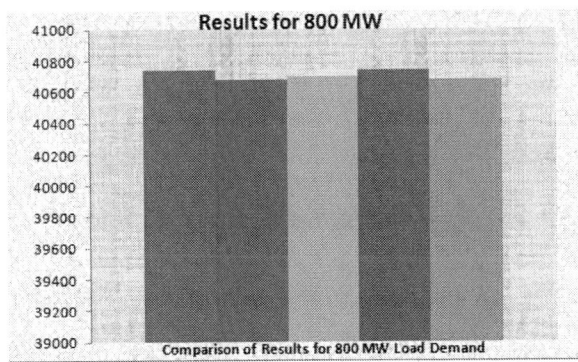

Fig. 6: Graphical Representation For 800 MW Load Demand

ACKNOWLEDGMENT

The investigators are delighted to the institute of C.B.I.T (Chaitanya Bharathi Institute of Technology) Vidya Nagar, Proddatur, India to grant financially sustain to do this research.

REFERENCES

[1] S. Gautham and J. Rajamohan, "Economic Load Dispatch Using Novel Bat Algorithm," *IEEE*, pp. 1–4, July 2016.

[2] Reddy AS, Reddy M.D., "Optimization of network reconfiguration by using Particle swarm optimization," *IEEE Int. Conference* on *in Power Electronics, Intelligent Control and Energy Systems (ICPEICES-2016)*, July 4, pp. 1-6, 2016.

[3] S.Bharathi, A.V.S. Reddy, Dr.M.D. Reddy, "Optimal Placement of UPFC and SVC using Moth-Flame Optimization Algorithm," *International Journal of Soft Computing and Artificial Intelligence*, vol.5, no.1, pp.41-45, May 2017.

[4] R. Chellappan and D. Kavitha "Economic And Emission Load Dispatch Using Cuckoo Search Algorithm," *IEEE*, pp. 1–7, 2017.

[5] F. S. Moustafa, A. El-Rafei, N.M. Badra and A. Y. Abdelaziz "Application and Performance Comparison of Variants of the Firefly Algorithm to the Economic Load Dispatch Problem," *IEEE*, pp. 1–4, Feb 2017.

[6] Dipayan De, D. Saha, T. Samanta, D. Jana, D. Palai, A. Maji, S.W. Ahmad, A. Poddar, P. Das, "Economic Economic Load Dispatch By Optimal Scheduling of Generating Units using Improved Real Coded Genetic Algorithm," *IEEE*, pp.305-308, 2017.

[7] Reddy, A.S. and Reddy, P.M.D., "Optimization of Distribution Network Reconfiguration Using Dragonfly Algorithm," *Journal of Electrical Engineering*, vol.16, no.4, pp.273-282, 2017.

[8] A.V.S.Reddy, M.D.Reddy, and M.S.K.Reddy "Network Reconfiguration of Primary Distribution System Using GWO Algorithm," *Int. J of Electrical and Computer Engineering (IJECE)*, vol.7, no.6, 2017.

[9] Y.Wu, B. Zhao and L. Liu "Glowworm Solving Economic Load Dispatch Problem with Valve Point Effect Using Mean Guiding Differential EvolutionSwarm Optimization Algorithm for Solving Non-Smooth and Non-Convex Economic Load Dispatch Problems," *IEEE,* pp.103-109, 2017.

[10] A.V.S.Reddy, M.D.Reddy, "Network Reconfiguration of Distribution System for Maximum Loss Reduction Using Sine Cosine Algorithm," *International Journal of Engineering Research and Applications (IJERA),* vol.7, no.10, pp.34-39, October 2017.

[11] H. Shahinzadeh, M. Moazzami, D. Fadaei and S. Rafiee-Rad "Glowworm Swarm Optimization Algorithm for Solving Non-Smooth and Non-Convex Economic Load Dispatch Problems," *IEEE,* pp.103-109, 2017.

[12] H. Shahinzadeh, M. Moazzami, S. Hamid Fathi and S. H. Hosseinian "Hybrid Big Bang-Big Crunch Algorithm for Solving Non-convex Economic Load Dispatch Problems," *IEEE,* pp.48-53, 2017.

[13] Reddy, A.S. and Reddy, M.D., "Optimal Capacitor Allocation for the Reconfigured Network Using Ant Lion Optimization Algorithm," *International Journal of Applied Engineering Research,* vol.12, no.12, pp.3084-3089, 2017.

[14] H. T. Ul Hassan, M. U. Asghar, M. Z. Zamir and H. M. Aamir Faiz "Economic Load Dispatch Using Novel Bat Algorithm With Quantum and Mechanical Behaviour," *IEEE,* pp.01-06, Feb 2018

[15] A.V.S.Reddy, M.D.Reddy, "Distribution Network Reconfiguration for Maximum Loss Reduction using Moth Flame Optimization," *International Journal of Emerging Technologies in Engineering Research (IJETER),* vol.6, no.1, pp.86-90, January 2018.

[16] A. V. S. Reddy, Dr. M. D. Reddy, "Application of Whale Optimization Algorithm for Distribution feeder reconfiguration," i-manager's *Journal on Electrical Engineering,* vol.11, no.3, pp.17-24, Jan-Mar 2018.

[17] Suharto, H. Sugiarto, Ruskardi and H. Moustafa "Ant Lion Optimization Algorithm For Environmental/Economic Dispatch Problem," *The International Journal of Engineering and Science,* vol.7, no.2, pp.01-06, Feb 2018.

[18] F. A. Mohamed, M. A. Nasser, K. Mahmoud and S. Kamel "Economic Dispatch Using Stochastic Whale Optimization Algorithm," *IEEE,* pp.19-24, Feb 2018.

[19] P. M. Dash, A. K. Baliarsingh, S. K. Mohapatra "Economic Load Dispatch using Moderate Random Search PSO with Ramp Rate Limit Constraints ," *IEEE,* pp.1-4, Mar 2018.

[20] A. V. S. Reddy, M. D. Reddy and Y. V. K. Reddy "Feeder Reconfiguration of Distribution Systems for Loss Reduction and Emissions Reduction using MVO Algorithm," *Majlesi Journal of Electrical Engineering,* vol.12, no.2, pp.1-8, June 2018.

[21] A.V.S. Reddy, M.D. Reddy, "Network Reconfiguration of Distribution System for Maximum Loss Reduction using Sine Cosine Algorithm, " *IJERA,* Vol.7, Issue.10, pp.34-39, October 2017

2nd IEEE International Conference on Power Electronics, Intelligent Control and Energy Systems (ICPEICES-2018)

Overvoltage and Undervoltage Protection of Load using GSM modem SMS Alert

Savita
Department of Electrical and
Electronics Engineering
Dr. Akhilesh Das Gupta Intitute of
Technology & Management
New Delhi, India
savita.singh@adgitmdelhi.ac.in

Sumit Shrivastava
Department of Electrical and
Electronics Engineering
Dr. Akhilesh Das Gupta Intitute of
Technology & Management
New Delhi, India
mrsumit187@gmail.com

Abhishek Arora
Department of Electrical and
Electronics Engineering
Dr. Akhilesh Das Gupta Intitute of
Technology & Management
New Delhi, India
abhishekarora801@gmail.com

Vikas Varshney
Department of Electrical and
Electronics Engineering
Dr. Akhilesh Das Gupta Intitute of
Technology & Management
New Delhi, India
vikas.varshney@adgitmdelhi.ac.in

Abstract— **In the paper, GSM Modem SMS Alert aims to build a system that monitors voltage and provides a breakpoint based low and high voltage tripping mechanism that avoids any damage to the load and sends a text message (SMS) to alert the user whenever the condition of under-voltage or overvoltage occurs. There is a chance of damaging electronic devices due to the fluctuations the AC mains supply. Therefore, a tripping system is required to avoid any damage to these loads. This system consists of a tripping mechanism that monitors the input voltage and trips according to limits provides. It uses 8-bit microcontroller ATmega328 with a GSM modem and 16*2 LCD attached to it externally. As soon as the input voltage falls out of the window range, it delivers an error on screen. Here, a dc motor is used as a load. This system is also configured with a buzzer that goes on as soon as tripping takes place. The states of the system are displayed on the LCD. Whenever fault occurs the microcontroller sends message to the GSM modem, then the GSM modem sends alert SMS to the user to protect their device as soon as possible [1].**

Keywords— *Overvoltage, Undervoltage, GSM modem, ATmega328, 16*2 LCD screen, Buzzer, SMS Alerts*

I. INTRODUCTION

The main motivation behind this paper was to get an elaborate and in depth knowledge on the protection of load with GSM alert and to use this method in an efficient way. There is a chance of damaging electronic devices due to fluctuations in ac mains supply. So a tripping system is required to avoid any damage to these loads. The main objectives of this paper to design a circuit that will protect the system from fluctuations.

Electrical power system can be divided into three sectors i.e. Generation, Transmission and Distribution. Electrical distribution system mainly comprises of electrical load. In this, electrical energy is distributed among different consumers. Consumers can be industrial, commercial or residential. Load forecasting can be done approximately but not accurately. Variation in load affects the voltage profile in power system. In power system, flat voltage profile i.e. constant voltage at every node is desirable. Overvoltage at any point can lead to insulation failure of power apparatus. It can even damage the power appliance. Under voltage is

similarly undesirable. Under voltage can increase the losses of induction motor. Full potential of electrical appliance at lesser voltage cannot be exploited. Small deviation from standard or normal voltage (usually 5%) causes under voltage or over voltage.

Sag and swell conditions in power system are more harmful than under voltage and over voltage. When the voltage falls below 90% of the rated normal voltage, then such condition is termed as voltage sag. Main reasons for voltage sag are heavy loading and shunt fault such as L-L fault, L-G fault etc. When the voltage rises above 110% of the rated normal voltage then this condition is called voltage swell.

Overvoltage and under voltage classification according to the IEEE 1159 standard are given in Table I & Table II. Also the variation in magnitude of overvoltage and undervoltage with different time durations are shown in Fig. 1 & Fig. 2.

TABLE I. CLASSIFICATION OF OVER VOLTAGE ACCORDING TO IEEE 1159 [2]

Types of over voltage	Duration	Magnitude
Instantaneous	0.5 - 30 cycles	1.1-1.8 pu
Momentary	30 cycles - 3 secs	1.1-1.4 pu
Temporaray	3 secs - 1 min	1.1-1.2 pu

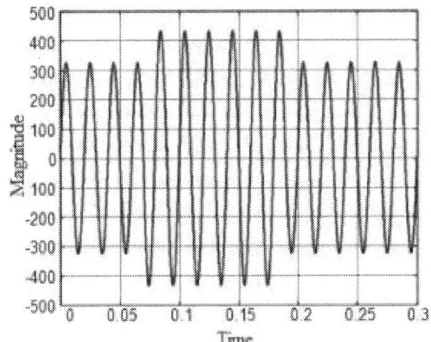

Fig. 1. Over voltage Waveform [3]

978-1-5386-6624-1/18 $31.00 © 2018 IEEE

TABLE II   Classification of Over voltage according to IEEE 1159
[2]

Types of under voltage	Duration	Magnitude
Instantaneous	0.5 - 30 cycles	0.1-0.9 pu
Momentary	30 cycles - 3 secs	0.1-0.9 pu
Temporaray	3 secs - 1 min	0.1-0.9 pu

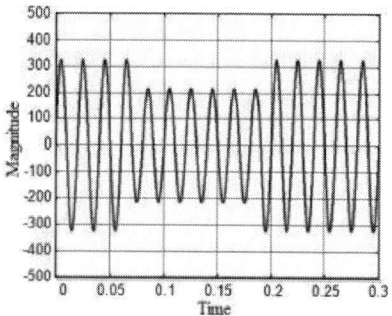

Fig. 2.   Under voltage Waveform [3]

II.   BLOCK DIAGRAM AND HARDWARE MODEL

The block diagram of U/O protection of load (motor) with GSM alert has been shown in Fig. 3. It comprises mainly of a transformer, GSM modem, self made Arduino board, ATmega328 microcontroller, motor and a tripping circuit. Here an IC LM324N is used to compare the voltage change and generates a signal to the relay to activate the buzzer and to trip the circuit [4].

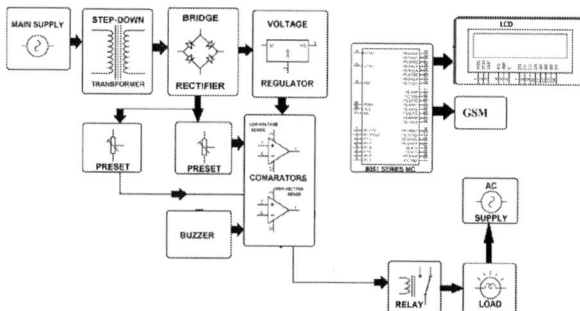

Fig. 3.   Block Diagram

Fig. 4.   Hardware Model

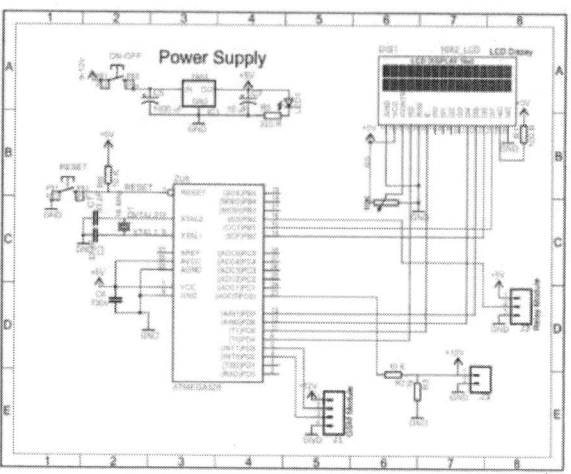

Fig. 5.   Interfacing of IC with LDC and GSM Modem

The hardware model of protection of load from overvoltage and undervoltage condition has been shown in Fig. 4.

III.   HARDWARE DESIGNING AND MAIN COMPONENT USED

In the hardware model a GSM alert system is used with the interfacing between the components. A GSM modem is used to send SMS to the user's mobile, Arduino basic circuit comprises of Atmega328 IC and LCD screen is used for the display purpose [5]. The interfacing among various components is shown in figure 5.

A.   GSM Modem

The SIM900A is shown in Fig. 8. It is a complete Quad-band GSM/GPRS solution in a SMT module which can be embedded in the customer applications. It delivers GSM/GPRS 850/900/1800/1900MHz performance for Data, voice, SMS, and Fax in a small form factor [6]. It has a tiny configuration therefore; it can fit in almost all space requirements. The GSM Modem unit is shown in Fig. 6.

B.   LCD Screen

LCD (Liquid Crystal Display) screen is an electronic display module and find a wide range of applications. A 16x2 LCD display is shown in Fig 7. It is a very basic module and is very commonly used in various devices and circuits [7].

Fig. 6.   GSM Modem

Fig. 7.   LCD Screen

Fig. 8.   Ardunio Hardware Model

C. Arduino Basic Circuit

Arduino hardware as shown in Fig.8 is an open-source circuit board. It has a microprocessor and input/output pins for communication and controlling physical objects. It is powered via USB or an external power supply. It is similar to C++ & it allows code writing, compiling, and then uploading it to Arduino for standalone use.

IV.   RESULT AND DISCUSSION

The circuit can be installed and used successfully for under-voltage and overvoltage protection of the load, i.e. DC motor with GSM Modem SMS Alert. Whenever there is any adverse condition which may cause damage to the system/load, the user will receive message instantly after these (overvoltage or undervoltage) conditions occurred, hence by this method the voltage can be adjusted within the safe range leads to load/system protection. The circuit generally works in four conditions. They are as follows:

A. Normally OFF

When there is no power supply to the circuit, then circuit is not in working mode i.e. it is the normally off condition and is shown in figure 9.

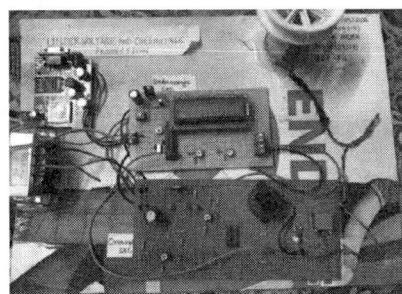

Fig. 9.   Normally OFF Mode

Fig. 10. Normally ON Mode

B. Normally ON

When the applied voltage is between the accepted values, i.e. between 6V and 10.5V, the motor will run. It is termed as normally on condition because no undervolage or overvoltage condition occurs. Fig. 10 shows normally on condition. A message 'Safe State' will be displayed on the LCD.

C. Under Voltage Condition

When the applied voltage is lower than the set value (6 v), i.e. the under-voltage condition, then the motor will not run, buzzer starts and the red LED will be turned off. A message 'Alert Undervoltage' will be displayed on the LCD and two SMS will be sent to the user [8]. This condition is shown in Fig. 11.

D. Over Voltage Condition

When the applied voltage is greater than the set value, i.e. above 10.5V, then it is referred as overvoltage condition. When the applied voltage is greater than 10.5V, the motor will not run. Buzzer starts and the red LED will be turned off. A message 'Alert Overvoltage' will be displayed on the LCD and SMS will be sent to the user [9].

Fig. 11. Under Voltage Condition and Displayed Message

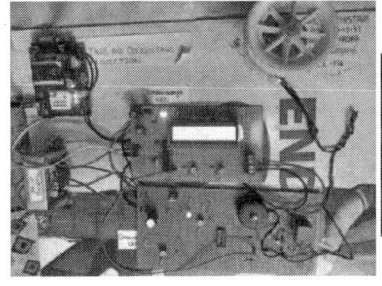

Fig. 12. Over Voltage Condition and Displayed Message

V. CONCLUSION

In this paper, a great way of protecting loads from voltage fluctuations either under voltage or overvoltage has been practically implemented. It proves to be cheap, requires very less maintenance and fully automatic. The GSM modem SMS technology alerts the user so that operator can respond as soon as possible to save load and other important equipments from getting damaged. The objective of this paper is to access the load in safe condition even when the operator is far away from the load side. The possibilities are numerous and can be explored further. Further modifications can be done to improve it in a better way.

REFERENCES

[1] F. Shawki, M. El-Shahat Dessouki, A. I. Elbasiouny, A.N. Almazroui, F. M. R. Albeladi, "Microcontroller based smart home with security using GSM technology" in "Internatinoal Journal of Research in Engineering & Technology," vol. 4, issue 6, June 2015.

[2] "IEEE Recommended Practice for Monitoring Electric Power Quality," IEEE Std. 1159-1995, June 1995.

[3] Manish Paul, Antara Chaudhury, Snigdha Saikia,"Hardware implementation of Overvoltage and Undervoltage Protection,"

"International journal of innovative research in Electrical, Electronics, Instrumentation and Control Engineering," Vol. 3, issue 6, June 2015.

[4] Electricity World.

Avalable at: http://electricity2all.blogspot.com/2013/05/over-voltage-and-voltage-swell.html

[5] Manish Paul, Antara Chaudhury, Snigdha Saikia,"Hardware implementation of Overvoltage and Undervoltage Protection," "International journal of innovative research in Electrical, Electronics, Instrumentation and Control Engineering," Vol. 3, issue 6, June 2015.

[6] Over & Under Voltage protection circuit . Available:https://www.elprocus.com/under-andovervoltage-protection-circuit/

[7] Interfacing 16*2 LCD With Arduino. Available: https://circuitdigest.com/microcontrollerprojects/ardu ino-lcd-interfacing-tutorial

[8] Over & Under Voltage protection circuit. Available: https://www.eleccircuit.com/over-voltage-and-lowvoltage-protection-circuit/

[9] Over/Under Voltage Protection of Electrical Appliances.Available:https://electronicsforu.com/ele ctronics-projects/under-over-voltage-protection.

Optimal Placement and Sizing of Distributed Generation using Flower Pollination Algorithm for Power Loss Reduction Maximization in Distribution Networks

Sabarinath.G
Dept. of electrical and electronics engineering
S.V.University College of Engineering
Tirupathi, India
sabarinath204@gmail.com

T.Gowri Manohar
Dept. of electrical and electronics engineering
S.V.University College of Engineering
Tirupathi, India
gsnphd@gmail.com

Abstract—In this article, power loss index (PLI) and flower pollination algorithm (FPA) are proposed for allocation and sizing of renewable energy resources based distributed generation (DG) in distribution network by considering power loss reduction maximization as an objective function. Here the most suitable locations for the placement of DG units are identified using PLI factor and then the FPA is applied to calculate the optimal power capacities of DG units. The effectiveness of introduced methodology is evaluated by testing with the standard IEEE 33 and 69 bus distribution networks (DN). The results produced with the FPA method are compared with the existing methods in the literature and show that the introduced method is capable of minimizing more power losses than the compared methods.

Keywords—Distributed generation, Flower pollination algorithm, Power loss index, Power loss minimization.

I. INTRODUCTION

Distribution network (DN) or Distribution system (DS) behaves like an interface which connects the distribution substation and the end user. Due to high current and low voltage levels, almost 70.00% of the overall power system losses are happening in the DN and the left over 30.00% of the power system losses happening in a transmission system. Distribution system losses are 15.50% of the total power production, but the expected is 7.50% of the total generation. Hence, many efforts have been performed on distribution system planning and design so as to minimize power loss and to enhance the voltage profile.

Several types of DGs and their definitions have been discoursed in [1]. DG is a small quantity power production which is directly interfaced to distribution network rather than transmission system. DG is mainly classified [2] into two categories based on type of fuel consumed
  ➢ Renewable energy resource based DGs
  ➢ Non-renewable energy resource based DGs

In recent years penetration of renewable energy resource based DG into the distribution system is expeditiously increasing because it has the following numerous technical, economic and environmental benefits [3]

Some important technical benefits of DG includes
  ❖ Reduced power loss
  ❖ Enhancement of overall system voltage profile
  ❖ Improved system reliability as well as security
  ❖ Better power quality
  ❖ Energy efficiency improvement

  ❖ Reduced burden on transmission network
  ❖ Less emission of pollutants
  ❖ Reducing peak demand
  ❖ Improving load factor

Below are the main economic benefits
  ✓ Reduction in cost for upgradation of system
  ✓ Less operation and maintenance cost
  ✓ Reduced health care cost due to improved environmental conditions
  ✓ Less fuel cost

Few years back DG is considered as active power source only but now with advances in technology DGs exists in different kinds [4] such as
  ⮱ DGs generating only active power. For example photo voltaic, fuel cells, micro turbines etc.
  ⮱ DGs generating only reactive power. For example DSTATCOM, synchronous condenser, capacitor etc.
  ⮱ DGs generating both active as well as reactive power such as synchronous generator.
  ⮱ DGs delivering active power and consuming reactive power such as induction generator.

The location and KVA capacity of a DG can greatly affect the power loss and overall voltage of a distribution network. Many studies have been carried out by the researchers and scientists to analyze the effect of DG on different aspects when it is connected to distribution network.

A technique for analyzing the influence of DG over power loss, voltage and system reliability of distribution network was proposed in reference [5]. Predefined loss reduction level based DG placement and sizing was proposed in [6]. So many methods such as numerical methods, analytical methods and artificial intelligence dependent methods have been used in the earlier research to suitably select the best location and optimal rating of DG.

A 2/3 analytical method has been used in [7]. According to this rule in a balanced DS, power loss minimization will be more if the DG is placed at a distance of 2/3 from the feeder with 2/3 size of DG. The main defect in this method is, it is suitable only for balanced distribution system but not suitable for unbalanced system.

An analytical approach was proposed by authors in [8] for optimal placement and sizing of time varying DGs in radial as well as networked DSs. An analytical method for minimization of power loss in DS by optimal placement and

978-1-5386-6624-1/18 $31.00 © 2018 IEEE

sizing using power stability index was proposed by authors in [9].

II. PROBLEM FORMULATION

A. Objective Function

To extract the extreme benefits from the DG the size and location of DG should be optimum. The main motive of this problem is to determine the suitable location, type and KVA rating of DG unit to minimize the power loss and to improve the voltage profile in the distribution network.

$$\text{Min } f = \min(\text{total power loss}) \tag{1}$$

B. Constraints

- Power balance constraints
$$P_{Loss} + P_d = P + \sum P_{DG} \tag{2}$$
- Voltage constraints
$$0.95 \leq V_i \leq 1.05 \tag{3}$$

III. OPTIMAL DG LOCATION

PLI [10] method is employed to pick the most effective buses to place DG. PLI values at each bus are evaluated using the below expressions (4)-(6). The buses with highest PLI value are assigned as candidate buses for DG sitting. Power loss reduction values at each individual bus excluding swing bus are calculated by injecting the reactive power equal to the load reactive power at each bus. Real power loss in m^{th} line is given by

$$PL(l) = \frac{(P^2[l] + Q^2[l]) + R_m}{(V[l])^2} \tag{4}$$

Reactive power loss in m^{th} line is given by

$$QL(l) = \frac{(P^2[l] + Q^2[l] + X_m}{(V[l])^2} \tag{5}$$

The power loss index is given as below

$$PLI(d) = \frac{LR(d) - LR_{min}}{LR_{max} - LR_{min}} \tag{6}$$

Here P and Q are net active power and net reactive power beyond bus l. $LR(d)$ is the value of loss reduction at bus d. LR_{min} and LR_{max} represents the least and highest loss reduction values.

IV. FLOWER POLLINATION ALGORITHM

A new meta-heuristic algorithm called flower pollination algorithm (FPA) motivated by the natural biological phenomenon of flowering plants was developed and introduced in [11] by Yang. Flower pollination (FP) is the process moving pollen to female flower from male flower through biotic or abiotic pollination. Almost around 90% of the pollination process is caused by animals and insects and the left over 10% is due to natural causes.

In cross pollination pollen moves from one flower to another flower where as in self-pollination pollen moves in the same flower itself. Biotic and cross pollination process generally takes place between flowers which are furthest from each other and hence they are considered as global optimization. Abiotic and self-pollination takes place in the same flower hence they are considered as local optimization.

The characteristics of FPA
1. Biotic and cross pollination are thought of global pollination because pollen follows Levy flight movement.
2. Abiotic and self-pollination are thought of local pollination.
3. Flower constancy can be developed by pollinators which is considered as production probability.
4. The interaction between local and global pollination process is managed by a switch probability $p \in [0, 1]$.

Couple of updating formulae is derived by changing the above said rules into updating equations.
First rule can be represented as

$$X_i^{t+1} = X_i^t + L(X_i^t - g_*) \tag{7}$$

Where X_i^t is the ith pollen, t is the iteration number g_* current best solution obtained and L is strength of pollination Because insects may travel over a long distance with different distance steps we can use a levy flight.

$$L = \frac{\lambda \Gamma \sin(\frac{\Pi \lambda}{2})}{\Pi} \frac{1}{s^{1+\lambda}} \tag{8}$$

Here $\Gamma(\lambda)$ is standard gamma function. This is valid for large steps S>0. In this $\lambda = 1.5$

Rule 2, 3 and local pollination can be represented as

$$X_i^{t+1} = X_i^t + \varepsilon(X_j^t - X_k^t) \tag{9}$$

Here X_j^t and X_k^t are pollen from separate flowers of the same plant species.

V. SIMULATION RESULTS AND ANALYSIS

Two standard radial distribution systems (RDS), IEEE 33-bus and IEEE 69-bus test systems have been taken to analyze the effectiveness of proposed FPA in determining the optimal DG size.

A. IEEE 33-bus test system

The detailed specifications of this test system are taken from [12] and are given in table 1.The single line diagrams of IEEE 33-bus test system is shown in figure 1. Before installation of DG unit into the first test system, the total active power loss (TAPL) and total reactive power loss (TRPL) are 210.99 KW and 143.03 KVAr, respectively. The minimum and maximum voltages before connection of DG are 0.9038 per unit and 1 per unit. The optimal location for DG placement for 33-bus test system is 30.

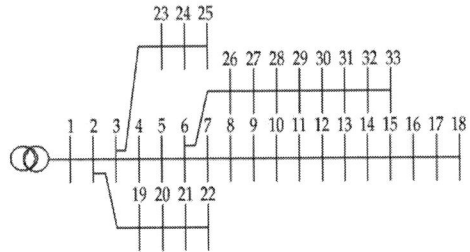

Fig. 1.   *Single line diagram of 33-bus system*

TABLE I.   SPECIFICATIONS OF 33- BUS SYSTEM

Test system	Base MVA	Base KV	Total connected load		Power loss	
			Active (KW)	Reactive (KVAr)	TAPL (KW)	TRPL (KVAr)
IEEE 33-bus	100	12.66	3715.00	2300.00	210.99	143.03

This section addresses the effectiveness of different types of DG units on the value of power loss reduction. Examinations have been conducted and the results are provided in table 2. The active power loss reduction (APLR) values for type-1, type-2, type-3 and type-4 DG units are 40.67%, 63.69%, 9.83% and 28.25%, respectively. Similarly reactive power loss reduction (RPLS) values are 37.58%, 61.07%, 6.99%, 27.42% respectively. From these values we can say that type-2 DG unit gives more power loss reduction and type-4 DG unit gives less power loss reduction as compared to other three types of DG units. Obtained results are compared with the existing method in table 3.

TABLE II.   RESULTS OF 33- BUS SYSTEM

DG Type	DG location	DG size (KVA)	APL (KW)	RPL (KVAr)	APLR (%)	RPLR (%)
Type-1	30	1554.40	125.16	89.27	40.67	37.58
Type-2	30	1975.86	73.59	55.68	63.69	61.07
Type-3	30	1481.50	190.24	133.02	9.83	6.99
Type-4	30	1257.82	151.37	103.81	28.25	27.42

TABLE III.   COMPARISION RESULTS OF 33- BUS SYSTEM

	Index vector method [13]		Proposed method (Type-2 DG)
DG location	15		30
Power factor	1	0.9	0.85
DG size(KVA)	1061	1255.89	1975.86
TLP (KW)	133.50	108.40	73.59
TLQ (KVAr)	90.79	74.77	55.68
Vmin (p.u)	0.9327	0.9390	0.9395

The effect of DG units on first test distribution system voltage profile is analyzed in this section. The minimum and maximum voltage levels obtained with the installation of different types of DG units are provided in table 4. From figure 2 it is evident that the overall system voltage profile is good when it is connected to type-2 DG unit as compared with all other three types of DG units.

TABLE IV.   MINIMUM AND MAXIMUM VOLTAGE LEVELS OF 33-BUS SYSTEM AFTER DG PLACEMENT

Test system	DG type	Minimum voltage(p.u)	Maximum voltage (p.u)
IEEE 33-bus	Type-1	0.9272	1.0000
	Type-2	0.9395	1.0044
	Type-3	0.9183	1.0000
	Type-4	0.9165	1.0000

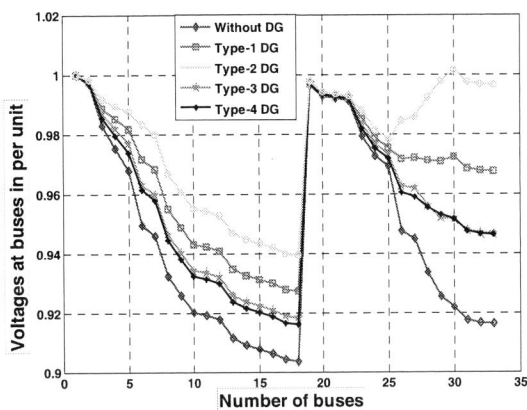

Fig. 2.   *Voltage profile of 33-bus system*

B.   IEEE 69-bus test system

The next test system considered is a commonly used 69-bus test system. The detailed specifications about this test system are taken from [12] and are given in table 5.The single line diagrams of IEEE 69-bus test system is illustrated in figure 3. Before integration of DG unit into this test system the TAPL and TRPL are 225 KW and 102.17 KVAr, respectively. The minimum and maximum voltage levels before connection of DG are 0.9092 per unit and 1 per unit. The optimal location for DG placement is 61.

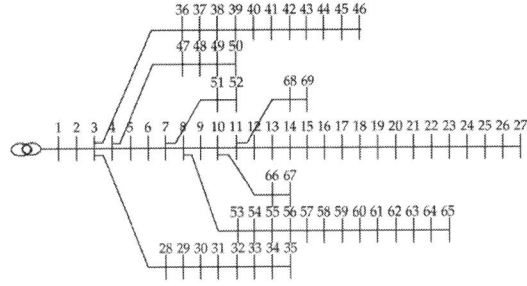

Fig. 3.   Single line diagram of 69-bus system

TABLE V.   SPECIFICATIONS OF 69-BUS SYSTEM

Test system	Base MVA	Base KV	Total connected load		Power loss	
			Active (KW)	Reactive (KVAr)	TAPL (KW)	TRPL (KVAr)
IEEE 69-bus	100	12.66	3802.19	2694.60	225.00	102.17

To validate the proposed methodology in allocating the different types of DG technologies it has been tested on 69-bus test system and the results are placed in table 6.From this results it is clear that type-2 DG unit operating at 0.8158 leading power factor minimizes the total APL by 89.71%.Similarly total RPL is reduced by 85.93 %. Comparative results of 69-bus test system are provided in table.

Figure 4 shows how the system voltage profile changes with the integration of different types of DG technologies into the second test system. Figure 4 show that system voltage profile is improved most when it is connected to type-2 DG unit and poor voltage profile is obtained with type-1 DG unit. The minimum and maximum voltage levels of this test system are provided in table 8.

TABLE VI.   RESULTS OF 69-BUS SYSTEM

DG Type	DG location	DG size (KVA)	APL (KW)	RPL (KVAr)	APLR (%)	RPLR (%)
Type-1	61	1871.92	83.180	40.52	63.03	60.34
Type-2	61	2243.35	23.140	14.37	89.71	85.93
Type-3	61	1780.70	157.96	72.44	29.79	29.09
Type-4	61	1330.41	152.00	70.48	32.44	31.01

TABLE VII.   COMPARISION RESULTS OF 69-BUS SYSTEM

	Index vector method [13]		Proposed method (Type-2 DG)
DG location	61		61
Power factor	1	0.9	0.8158
DG size(KVA)	1872.82	2217.39	2243.35
TLP(KW)	83.22	27.96	23.14
TLQ (KVAr)	40.53	16.46	14.37
Vmin (p.u)	0.9683	0.9724	0.9725

TABLE VIII.   MINIMUM AND MAXIMUM VOLTAGE
LEVELS OF 69-BUS SYSTEM AFTER DG PLACEMENT

Test system	DG type	Minimum voltage (p.u)	Maximum voltage (p.u)
IEEE 69-bus	Type-1	0.9683	1.00
	Type-2	0.9725	1.00
	Type-3	0.9597	1.00
	Type-4	0.9307	1.00

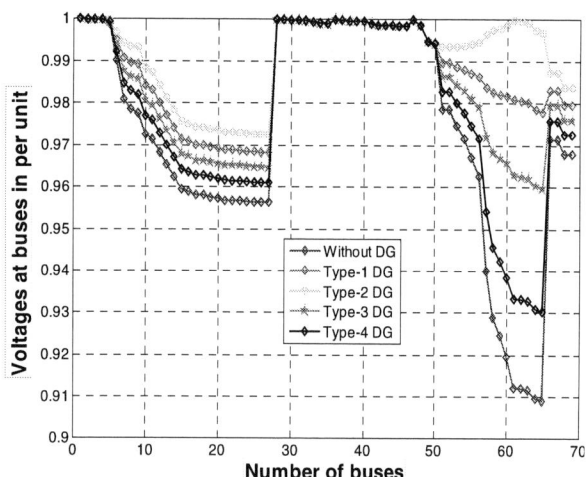

Fig. 4.   *Voltage profile of 69-bus system*

VI. CONCLUSIONS

In this paper the effect of four types of DG technologies on power loss reduction and voltage profile enhancement of a distribution system is analyzed. Here four types of DG technologies based on active power and reactive power consumption or injection have been considered. A novel method called power loss index is utilized to identify the best location for optimal DG sitting and flower pollination algorithm is applied to identify the optimal DG size. Two commonly used test systems, IEEE 33 and 69-bus systems are used to measure the effectiveness of proposed method and a detailed comparison has been carried out with existing methods in the literature. The results showed that the better performance has been gained with type -2 DG technology operating at power factor which is equal to total connected load power factor.

REFERENCES

[1]   T. Ackermann, G. Anderson, and L. Soder., "Distributed generation: a definition," Electric Power Systems Research, vol.57, pp.195–204, 2001.

[2]   Y.Xue,L.Chang,ang J.Meng, "Dispatchable distributed generation network-A new concept to advance DG technologies" in Proc.IEEE Power Eng. Soc. Gen.Meeting,pp.1-5,June 2007.

[3]   R.E.Brown and L.A.A.Freeman, "Analyzing the reliability impact on distributed generation," in IEEE power Engineering Society Summer Meeting, vol.2,pp.1013-1018,July 2001..

[4]   P.Mehta,P.Bhatt, and V.Pandya, "Optimal selection of distributed generating units and its placement for voltage stability enhancement and energy loss minimization"Ain Shams Eng.Jour.,vol.9,pp.187-201,2018.

[5]   C.L.T.Borges and D.M.Falcao, "Impacts of distributed generation allocation and sizing on reliability,losses and voltage profile" in Proc.IEEE Power Tech Conf.,Bologna,Italy,vol.2,pp1-5,2003.

[6]   A.D.T.Le,M.A.Kashem, M.Negnevitskv, and G.Ledwich, "Optimal distributed generation parameters for reducing losses with economic consideration," in Proc.IEEE Power Eng. Soc. Gen.Meeting,pp.1-8,2007.

[7]   V. A. Evangelopoulos and P. S. Georgilakis, , "Optimal distributed generation placement under uncertainties based on point estimate method embedded genetic algorithm," *IET Generation, Transmission & Distribution*, vol. **8**, pp. 389-400, (2014).

[8]   C.Wang, and M.H.Nehrir, "Analytical approaches for optimal placement of distributed generation sources in power systems,"IEEE Trans. on power systems,vol.19,pp.2068-2076,2004.

[9]   M.M.Aman,G.B.Jasmon,H.Mokhlis,and A.H.A.Bakar, "Optimal placement and sizing of a DG based on a new power stability index and line losses,"Electrical power and energy systems,vol.43,pp.1296-1304,2012.

[10]   P.D.P.Reddy,V.C.V.Reddy and T.G.Manohar, "Optimal renewable resources placement in distribution networks by combined power loss index and whale optimization algorithm," JEST,vol.53,pp.1-17,2017.

[11]   X.S.Yang,M.Karamanoglu,and X.He, "Flower pollination algorithm: A novel approach for multiobjective optimization"Talor & Francis.

[12]   M.E.Baran, and F.F.Wu, "Optimal sizing of capacitors placed on a radial distribution system'' IEEE Trans. on Power Delivery,Vol.4No.1,Jan 1989,pp.735-743.

[13]   P. D. P. Reddy, V.C. V. Reddy, T. G. Manohar, "Whale optimization algorithm for optimal sizing of renewable resources for loss reduction in distribution systems",Renewables: wind,water,and solar.,2017.

On-Line Monitoring and Simulation of Transmission Line Network Voltage Stability Using FVSI

Manish Kumar Meena, Narendra Kumar
Department of Electrical Engineering Delhi
Technological University
Delhi, India
manish.meena1986@gmail .com
dnk_1963@yahoo.com

Abstract— **Till now many of line indices have been proposed by the engineers for the assessment of voltage stability of a power transmission network. For the fast voltage stability assessment of a transmission network FVSI (Fast Voltage Stability Index) is used and assumed that the active and reactive power flow in the same direction in a line. The FVSI is helpful to bring down the computational effort for the stability assessment. In this paper FVSI is used to assess the voltage stability of a 14-bus transmission network based on IEEE-14 bus test data. This index depends on the type of transmission network and interconnection of buses. Here monitoring is done for the load and generation changes on complete network and individual bus.**

Keywords— Fast Voltage Stability Index (FVSI), load flow, Newton-Raphson, Voltage stability.

I. INTRODUCTION

Voltage stability is mainly load phenomenon whenever load changes up to critical level on a bus then abnormal flow of active and reactive power happened or reactive power demand increase which is not met by the available reactive power sources and voltage instability occurs on that bus. Voltage instability is the major problem [1] for the operation of a big power system network, if voltage level is not maintained at any bus and continuously changes right after the disturbances occurs then it may result into the major power blackout in the power system network and complete grid may collapse [2].

The increasing power demand forces the power systems to perform near to its stability limits. As electric utilities attempt to use their transmission system capacities to transmit real power maximally, voltage collapse acts as an extreme loading condition. Mostly voltage instability occurs due to heavy load conditions [3]. The voltage instability problem is one of the types of power system instability.

There are many of interconnection between the buses of a large power system network and it is difficult to identify the weakest line where the chances of voltage instability occur. It is observed from the previous researcher's papers and literature review [5, 6] that it is possible to identify the weakest line in a complex power system network by using Fast Voltage Stability Index (FVSI) with the Newton-Raphson load flow analysis.

In recent years a number of algorithms are developed to compute the voltage stability margin by performing load flow studies. Investigations were carried out [3, 4] to search for the voltage collapse point and determine stability. Continuation approaches could give more accurate results but the drawback was the slow convergence for on-line voltage stability assessment.

In this paper FVSI is used to assess the voltage stability of a 14-bus transmission network based on IEEE-14 bus test data. This index depends on the type of transmission network and interconnection of buses. Here monitoring is done for the load and generation changes on complete network and individual bus. In the MATLAB environment based on the standard IEEE-14 bus test data load flow analysis is done to find the line flow and implement the line flow data (reactive power flow and voltage at bus) on the FVSI formula to identify the weakest line in the 14-bus power system network.

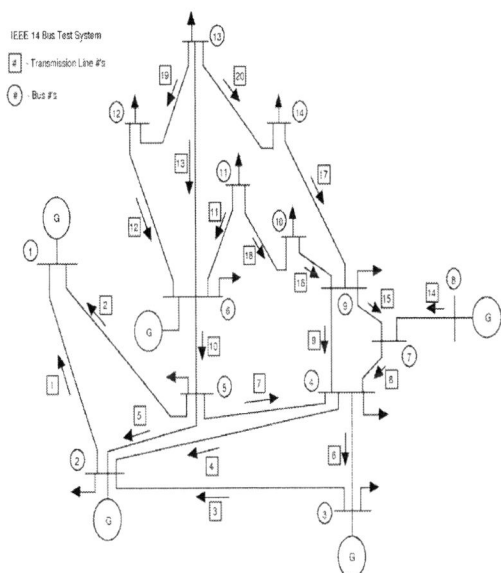

Fig.1: IEEE 14 Bus system

II. FAST VOLTAGE STABILITY INDEX (FVSI)

In order to assess the stability of a given power system an index known as Fast Voltage Stability Index (FVSI), which is used to specify the voltage stability state based on line or

978-1-5386-6624-1/18 $31.00 © 2018 IEEE

bus. The FVSI is resulting from the voltage quadratic equation at the receiving end bus. Any line in the transmission network that shows FVSI close to 1.0 indicates that the specific line is imminent to its instability point hence may lead to system violation. Therefore, FVSI has to be maintained less than unity in order to maintain a voltage stable system.

When the FVSI of a line approaches unity it means that the line is approaching its voltage stability limits. The FVSI of all the lines must be lower than 1 to assure the voltage stability of power system.

III. FORMULLATION OF FVSI

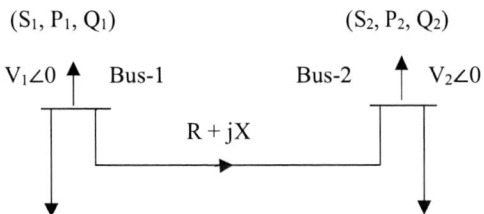

(S_1, P_1, Q_1) (S_2, P_2, Q_2)

$V_1 \angle 0$ Bus-1   Bus-2 $V_2 \angle 0$

$R + jX$

Fig.2. Two-bus power system model

V1, V2 = Voltages on sending and receiving buses
P1, Q1 = Active and reactive power on the sending bus.
P2, Q2 = Active and reactive power on the receiving bus.
S1, S2 = apparent power on the sending and receiving buses.
δ = angle difference between sending and receiving buses.

As we know that apparent power –

$$S_2^{*} = (P_2 + jQ_2)^{*} = V^{*}I$$

$$I = (P_2 + jQ_2)^{*} / V^{*}$$

$$V_2 \angle \delta + (P_2 + jQ_2)^{*} \frac{R + jX}{V_2 \angle -\delta} = V_1 \angle 0$$

$$V_2^2 + (P_2 - jQ_2)(R + jX) = V_1 V_2 \angle -\delta$$

$$V_2^2 + (P_2 R + jP_2 X - jQ_2 R + Q_2 X) =$$

$$V_1 V_2 \text{Cos } \delta - jV_1 V_2 \text{Sin } \delta$$

Equating real and imaginary parts-

$$V_2^2 + (P_2 R + Q_2 X) = V_1 V_2 \text{Cos } \delta \qquad (1)$$

&

$$P_2 X - Q_2 R = -V_1 V_2 \text{Sin } \delta \qquad (2)$$

So $$P_2 = \frac{Q_2 R}{X} - V_1 V_2 \frac{Sin\delta}{X}$$

From equation (1)

$$V_2^2 + R\left(\frac{Q_2 R}{X} - V_1 V_2 \frac{Sin\delta}{X}\right) + Q_2 X = V_1 V_2 \text{Cos } \delta$$

$$V_2^2 - V_1 V_2 \left(\frac{R}{X} Sin\delta + Cos\delta\right) + \left(X + \frac{R^2}{X}\right) Q_2 = 0$$

$$V_2 = V_1 \left(\frac{R}{X} Sin\delta + Cos\delta\right) \pm$$

$$\frac{\sqrt{\left[\left(V_1\left(\frac{R}{X} Sin\delta + Cos\delta\right)\right)^2 - 4\left(X + \frac{R^2}{X}\right)Q_2\right]}}{2}$$

For root must be real & discriminate in above equation

$$\left(V_1\left(\frac{R}{X} Sin\delta + Cos\delta\right)\right)^2 - 4\left(X + \frac{R^2}{X}\right)Q_2 \geq 0$$

Since $Sin\delta$ is very small so $Sin\delta = 0$ & $Cos\delta = 1$ then

$$V_1^2 - 4\left(X + \frac{R^2}{X}\right)Q_2 \geq 0$$

$$\frac{1 - 4\left(X + \frac{R^2}{X}\right)Q_2}{V_1^2} \geq 0$$

$$\frac{4\left(X + \frac{R^2}{X}\right)Q_2}{V_1^2} \leq 1$$

For generalized it taking 'j' as the receiving bus and 'i' as the sending bus, hence, the fast voltage stability index (FVSI) can be expressed as :

$$\text{FVSI}_{ij} = \frac{4Z^2_{ij}Q_j}{V^2_i X_{ij}} \qquad (3)$$

Where:
   Z_{ij} = line impedance
   X_{ij} = line reactance
   Q_j = reactive power at the receiving end
   V_i = sending end voltage

The objective is:
- To find the weakest line in a large power System network where there are chances of voltage instability.
- To quantitatively assess the voltage stability.
- To determine the voltage stability limit whenever the load changes.
- To determine the required reactive power compensation at each and every bus.

Methodology-

STEP-1: Taking IEEE-14 bus standard data for reference.

STEP-2: Create the Y bus matrix on MATLAB .

978-1-5386-6624-1/18 $31.00 © 2018 IEEE

STEP-3: perform load flow analysis using Newton-Raphson method on MATLAB platform by coding.

STEP-4: Analyze the FVSI code on MATLAB for voltage stability limit. For FVSI analysis input data is taking from N-R Load flow analysis. The bar chart will be prepared based on the FVSI data to indicate each line in this network which shows the voltage instability comparison at a glance.

STEP-5: Compare the results for different load and generation on different buses and predict the voltage stability limit and weakest line in 14 bus power system network.

IV. SIMULATION RESULTS

Case-I: INPUT (As per Standard IEEE-14 test bus) —

% change in active load on complete network: Nil
% change in reactive load on complete network: Nil
Change in load at bus no.: Nil
Change in active load in MW at this bus: 0
Change in reactive load in MVAR at this bus: 0
Add generation at bus no.: nil
Added real power generation in MW: 0
Added reactive power generation in MVAR: 0

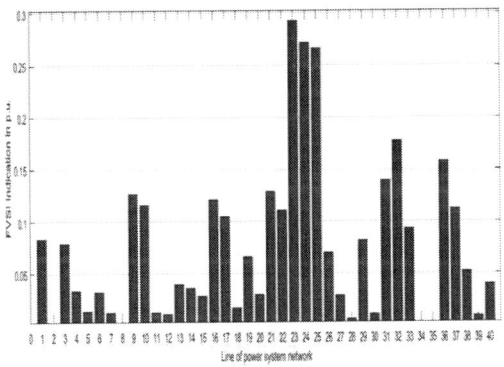

Fig.3: Output (FVSI BAR) of case I.

Remarks— From the above data and bar graph it seen that the complete system is stable and voltage is maintained within the limit, because of the FVSI value is less than 1 pu. For all connected lines in power system network

Table-1 : Case 1 FVSI output data

| From |	To |	FVSI |	From |	To |	FVSI |
| Bus |	Bus |	P.U. |	Bus |	Bus |	p.u. |
1	2	0.083	2	1	0.129
1	5	0.004	5	1	0.111
2	3	0.079	3	2	0.293
2	4	0.034	4	2	0.272
2	5	0.014	5	2	0.266
3	4	0.032	4	3	0.070
4	5	0.012	5	4	0.029
4	7	0.000	7	4	0.007
4	9	0.126	9	4	0.082
5	6	0.116	6	5	0.011
6	11	0.012	11	6	0.139
6	12	0.011	12	6	0.177
6	13	0.040	13	6	0.093
7	8	0.036	8	7	0.000
7	9	0.028	9	7	0.000
9	10	0.122	10	9	0.158
9	14	0.105	14	9	0.113
10	11	0.017	11	10	0.052
12	13	0.066	13	12	0.010
13	14	0.030	14	13	0.039

Case-II: INPUT (Faulty condition) —

% change in active load on complete network: Nil
% change in reactive load on complete network: Nil
Change in load at bus no. :5
Change in active load in MW at this bus: 100
Change in reactive load in MVAR at this bus: 150
add generation at bus no.: 2
added real power generation in MW: 20
added reactive power generation MVAR: 15

Table-2 : Case II FVSI output data

Fast Voltage Stability Index					
| From |	To |	FVSI |	From |	To |	FVSI |
| Bus |	Bus |	P.U. |	Bus |	Bus |	p.u. |
1	2	0.078	2	1	1.124
1	5	0.354	5	1	1.157
2	3	0.377	3	2	0.302
2	4	0.037	4	2	0.312
2	5	1.446	5	2	0.327
3	4	0.035	4	3	0.371
4	5	1.423	5	4	0.038
4	7	0.000	7	4	0.008
4	9	0.156	9	4	0.094
5	6	0.589	6	5	1.150
6	11	0.014	11	6	0.601
6	12	0.012	12	6	0.754
6	13	0.044	13	6	0.399
7	8	0.074	8	7	0.000
7	9	0.032	9	7	0.000
9	10	0.140	10	9	0.181
9	14	0.121	14	9	0.128
10	11	0.020	11	10	0.058
12	13	0.073	13	12	0.011
13	14	0.033	14	13	0.045

978-1-5386-6624-1/18 $31.00 © 2018 IEEE

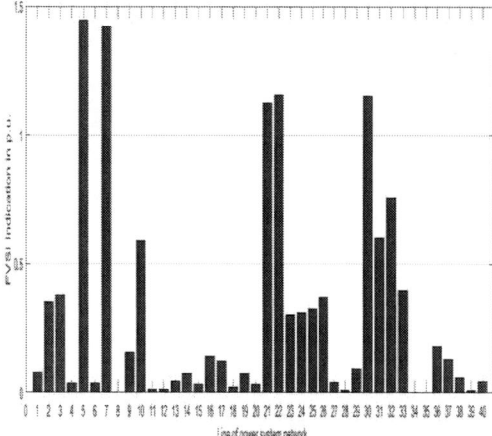

Fig.3: Output (FVSI BAR) of case II.

Remarks— From the above data and bar graph it shows that the complete system is not stable and voltage stability limit is exceeded at line 2-5, 4-5,2-1,5-1,6-5, because FVSI value comes more than 1 pu. For these lines. Hence, we have to take corrective action for these lines to maintain the voltage stability of this power network.

CONCLUSION

Fast voltage stability index (FVSI) is calculated from Newton-Raphson method to measure the voltage stability. The buses of connected lines with FVSI value closer or equal to 1 will be the most vulnerable buses in the system and helps in identifying weak area in the system which requires reactive power compensation. It can predict the voltage stability analysis for individual load pattern change on any bus also. It instantly shows the voltage stability index for complete network at a glance. The proposed approach provides fast computation of fast voltage stability index (FVSI) and can analyze any unknown load patterns. This powerful and versatile feature is useful for power system operation. The future work will focus on how to build early warning system based on ANN learning mechanism. Predict the exact voltage stability margin with change in load for individual bus and calculate the reactive power requirement for compensation purpose.

REFERENCES

[1] Abhijit Chakrabarti, D.P Kothari, A.K. Mukhopadhyay and Abhinandan De, " An Introduction to Reactive power control and voltage stability in power transmission system", PHI,

[2] Prabha Kundur, "Power system stability and control" NewYork : McGraw Hill.

[3] Swetha G C and H. R. Sudarshana Reddy "Voltage stability assessment in power network using artificial neural network", International Journal of Advanced Research in Electrical, Electronics and Instrumentation Engineering, Vol3, Issue 3, March 2014.

[4] Load flow analysis from NPTEL "www.nptel.ac.in/ courses/ Webcourse-contents /IIT KANPUK. Elissa, "Title of paper if known," unpublished.

[5] A. Ramasamy, R. Verayiah, H. I. Zainal Abidin and I.Musirin, "A study on FVSI index as an indicator for under voltage load shedding (uvls) Proceedings 2009 IEEE 3rd International Conference on Energy and Environment,7-8 December 2009, Malacca, Malaysia.

[6] N. Flatabo, R. Ognedal, and T. Carlsen," Voltage Stability Condition in a Power Transmission System Calculated by Sensitivity Methods", IEEE Trans. on Power Systems, Vol. 5, no. 4, Nov. 1990, pp. 1286-1293.

[7] M. M. Begovic and A. G. Phadke, "Control of Voltage Stability using Sensitivity Analysis", IEEE Trans. on Power Systems, Vol. 7, no. 1, Feb 1992, pp. 114-123. [5] Y. Lin Chen, C. Wei Chang,etc, " Efficient Methods For Identifying Weak Nodes In Electrical Power Network, " IEE Proc. - Gener. Transm. Distrib. ,Vol 142, No. 3 , May 1995.

[8] K. Sakameto, Y. Tamura, etc, " Voltage Instability Proximity Index (VIPI) Based On Multiple Load Flow Solutions In Ill-conditioned Power Systems, " Proceeding of the 27th Conference on Decision and Control, Austin, Texas, December 1988.

[9] F. Gutina and B. Strmcnik, " Voltage Collapse Proximity Index Determination Using Voltage Phasors Approach, " 94SM 510-8 PWRS, 1994.

Impact of Presence of Calender Feature in the Performance of Day-Ahead Electric Load Power Forecasting with New Framework Realized in Artificial Neural Network Tehinique using Matlab Programming

M.Vetri Selvi
Scientist 'D'
Centre for Development of Advanced Computing
Research Scholar
Indian Institute of Technology
Delhi, India
mvetriselvi@yahoo.com

Dr. Sukumar Mishra
Professor & IEEE Member
Electrical Engineering Department
Indian Institute of Technology
Hauz Khas, Delhi
India
sukumar@ee.iitd.ac.in

Abstract— In Smart Grid era, Electric Load Power Load Forecasting (ELPF) is an important process in any power system network to make it a customer friendly distribution network. Majority of literatures claims that an Artificial Neural Network (ANN) technique produces a better accuracy compared to any other technique. Hence ANN based model is chosen with a new architecture for this research work. This is a multiple input multiple output (MIMO) Feed Forward neural network with back propagation algorithm. The input features like climate, season, week end, week day, national and festival holiday factors are taken into due consideration in the development of the architecture. This research work is mainly focused on the impact of calendar variables in day-ahead hourly ELPF for the year 2017 by using data of the previous year 2016. The evaluation criteria are presented with determination of various statistical errors. The performance of each model is analyzed and the results are presented in this paper. These models developed based on ANN technique are implemented using Matlab programming. The proposed models are proven for their simplicity, less measurement requirement and easiness for implementation.

Keywords—load forecasting, day-ahead forecasting, artificial intelligence, ann, elpf, composite method

I. INTRODUCTION

Electric load power forecasting (ELPF) is the process used to forecast future electric load, based on given historical weather information, calendar information, previous day load information and current forecasted weather information. As per the literature survey, it is understood that continuous emphasis is being paid towards the problem of accuracy of electric load power forecasting techniques [1]-[2]. In the past few decades, several methodologies have been developed to forecast ELP more accurately. These methods can be classified into:

A. Conventional methods (regression models and statistical learning model) [3]–[5],
B. Alternative methods (Artificial Neural Network (ANN) and Expert Systems, among others) [6]–[8].

Most of the researchers asserted that the features such as air temperature and previous load play a major role in

predicting day-ahead ELP in addition to any other meteorological factors. After the analysis of data which is from Delhi Electricity Distribution Utility (DEDU) that Delhi is having extreme temperature variations of 4°C to 47°C throughout the year. Also, daily variation of temperature is high because of human activity. DEDU is loaded with more complex and combinational loads of residences, commercial complexes and industries. Because of high variations in air temperature and load in day-today activities, the electric utility managers here are facing problems in the accuracy of day-ahead ELPF with present available forecasters which ultimately affect the marketing and commercial activities of the power system network. Hence, this research study is aimed to develop new architecture which forecast the load more accurately. Here, this paper exhibits development of new architecture based on weather and load features as constant and adding calendar features one by one ("forward addition concept"). This framework is implemented using ANN based technique and the performance results of day-ahead ELPF are discussed. The paper is organized as following sections. Section I discusses the concept of the models developed. Section II focuses on the theoretical background of ANN architecture selection and usage. Section III discusses the development of models using ANN and its adoption for the load forecasting. Section IV is about implementation of models based on ANN in Matlab environment. Section V depicts the results of each model. Section VI provides conclusion and future work for the researchers working in this area.

II. CONCEPT BEHIND THE MODEL DEVELOPMENT

There are five different models built as shown in Table I with major three input features.

A. Calendar Variables(CV) Selected

- Season *(CV(1))*: It is classified into six different types as shown in Table II.
- Day of the Week *(CV(2))*: There are seven days in a week. Here, Sunday is considered as first day and hence assigned the value of '0' and Saturday as '6'.

- Week End Indicator *(CV(3))*: Saturday and Sunday are considered to be '1'and, other days as '0'.

TABLE I. DETAILS OF INPUT AND OUTPUT PARAMETERS SELECTED FOR CREATION OF DIFFERENT ARCHITECTURE OF MODELS

Model	Input Parameters*	Output Parameter*
Model 1	$T_a(24)$, (Day-1)$P_L(24)$	$P_L(24)$
Model 2	CV(1), $T_a(24)$, (Day-1)$P_L(24)$	$P_L(24)$
Model 3	CV(1), CV(2), $T_a(24)$, (Day-1) $P_L(24)$	$P_L(24)$
Model 4	CV(1),CV(2),CV(3),$T_a(24)$, (Day-1) $P_L(24)$	$P_L(24)$
Model 5	CV(1), CV(2), CV(3),CV(4), $T_a(24)$, (Day-1)$P_L(24)$	$P_L(24)$

*Hourly sampled data are used for the modeling.

TABLE II. CLASSIFICATION OF SEASONS

Season	Period	Value
Summer	16th Apr to 15th June	0
Monsoon	16th June to 16th Aug	1
Autumn	17th Aug to 16th Oct	2
Pre-Winter	17th Oct to 15th Dec	3
Winter	16th Dec to 14th Feb	4
Spring	15th Feb to 15th Apr	5

- Holiday Indicator *(CV(4))*: National holiday, Festival and adjacent day are considered as '1'; other days as '0'.The holidays selected for modeling are depicted in Table III. The days adjacent to national important days like Republic day and Independence Day are chosen as holiday. In addition to that major festival days like, 'Holi' and 'Diwali' in Delhi, the adjacent days to these days are also chosen as holiday in the model built-in. Mostly, day after the holiday is chosen for adjacent holiday criteria rather than prior days.

TABLE III. HOLIDAYS SELECTED

Republic Day	Good Friday
Independence Day	Buddha Purnima
Gandhi Jayanti	Janmastami
Christmas	Idul'l Filtr
Makar Sankranti	Dussehra
Milad-un-nabi	Idul Zuha(Bakrid)
Maha Sivratri	Maharishi Valmiki Jayanti
Holi	Diwali
Ram Navami	Muharram
Mahaveer Jayanti	Guru Nanak Birthday

B. Weather Variables Selected
- Only air temperature (T_a) is chosen for modeling because of very high variations in it throughout the day and major changes during seasons. Due to normal/abnormal variations in the temperature which ultimately affect human comfort leads to variations in the electric load power. And hence air temperature is taken for modeling.
  Where, T_a is Air Temperature in (°C)

C. Load Variable Selected
- As per electrical managers' opinion and claims of various literatures listed, the previous day hourly load power ((Day-1) P_L) is taken for modeling.
  Where, (Day-1)P_L – Previous day Load Power
  P_L is load power consumption in (kW)

The correlation analysis of the input variables with respect to output variable is studied and shown in Fig. 1 and

Fig. 2. Only the rho value and p-value of Model 5 is presented here. From the figures, it is clearly understood that previous day load has very strong positive correlation. Then, the air temperature has strong positive correlation. The seasonal effect has very strong negative correlation. The p-value of less than 0.05 shows the strong relationship between input and output variable. All other relationships are comparatively less significant. For more details, reader is directed to glance at reference given in [9].

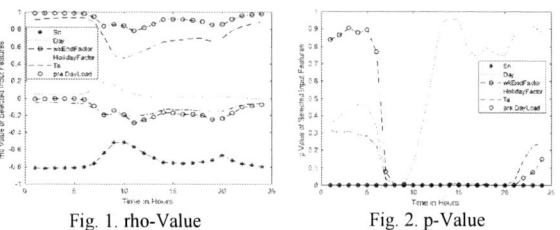

Fig. 1. rho-Value   Fig. 2. p-Value

III. USING ARTIFICIAL NEURAL NETWORKS; THEORETICAL BACKGROUND AND ITS ADOPTATION FOR THE LOAD FORECASTING APPLICATION

A. Theoretical Background

The ANN architecture proposed in this work has a Multi-layer perceptron (MLP) of feed forward back propagation network (BPN), which is commonly used structure in many such applications [10]. After a lot of trial and error analysis, it has been chosen to use only one hidden layer in order to have a good computational speed during training phase and not to complicate ANN architecture. The effectiveness of this choice (as evaluated) was confirmed in all simulation scenarios because the results have been satisfying. In the proposed technique, a two layer feed-forward network with sigmoid (logsig) activation function in the hidden layer neurons and linear (purelin) activation function in output layer neurons are selected. After defining the layers transfer functions, possible algorithms for training the neural network have been deeply analyzed to acquire the best forecast accuracy, and hence Levenberg-Marquadt back propagation algorithm is selected. This algorithm is designed to approach second order training speed without having to compute the hessian matrix. The training or learning of proposed model updates the ANN according to Levenberg-Marquadt optimization [11]. The feed forward back propagation network undergoes supervised training, with a finite number of pattern pair consisting of an input pattern and a desired pattern. In back propagation the mean squared error between calculated output and the desired value is back propagated into the previous layer to minimize the error. It is performed by adjusting the node weights and biases. For avoiding over training, cross validation is used. In this, data set is split into training and test set. The training set is again split into learning and validation sets. The ANN parameters are estimated on the learning set, and the performance of the model is tested every few iterations, on the validation set. When the performance starts to deteriorate (which means the ANN is over fitting the training data), the iterations are stopped, and the last set (test set) of parameters to be computed is used to generalize the ability of the model [12]-[14].

978-1-5386-6624-1/18 $31.00 © 2018 IEEE

B. Data Preprocessing

There are three important steps involved in the preprocessing stage.

a) Outliers prediction and removal: The actual data received from the Delhi Electricity Distribution utility was not having any outliers like spikes, zero crossings errors etc.,

b) Identification & Substitution of Missing Data: The actual data were having missing items. This was identified in MS Excel tool using filter function. And then the missed data was filled up using interpolation technique in Matlab.

c) Normalization: Data normalization means transforming all variables (features) in the data to a specific range. In ANN and other data mining approaches, there is a need of normalizing data (features) when they have different ranges. There are two important reasons for these; first to eliminate the influence of one factor over another (i.e. to give features equal chances), second reason is faster and stable convergence of weight and biases in back propagation with normalized data than with un-normalized data. So, if we have different features in same data range then there is no need for normalization. The formulae used for normalizing the data between 0 and 1 (p.u i.e. per unit) is given in (1).

$$X_p = \left(\frac{X - X_{Min}}{X_{Max} - X_{Min}} \right) \qquad (1)$$

Where,

X=raw data/ actual value of the data;
X_{Min}=minimum value of raw data;
X_{Max}=maximum value of raw data;
X_P=per unit value of raw data;

C. Performance Evaluation Criteria for Load Forecast

In most of the literatures for the evaluation of forecasting models, one or more errors listed in (2) to (4) were addressed. The reader is directed to look at reference given in [15]-[17] for detailed information. In an ideal case, all of the errors mentioned here must be zero. i.e. Forecasted value should be same as actual value.

$$P.E = P_{LActual} - P_{LForecasted} \qquad (2)$$

$$MAPE = \left(\frac{1}{n} \times \sum_{i=1}^{n} \left| \frac{\left(P_{LActual} - P_{LForecasted} \right)}{P_{LActual}} \right| \right) \times 100 \qquad (3)$$

$$MSE = \left(\frac{1}{n} \times \sum_{i=1}^{n} \left(P_{LActual} - P_{LForecasted} \right)^2 \right) \qquad (4)$$

Where,

'n' is Number of Samples
'$P_{LActual}$' is Actual Load Power in (kW)
'$P_{LForecasted}$' is Forecasted Load Power in (kW)
'PE' is Prediction Error or Residual in (p.u)
'MAPE' is Mean Absolute Percentage Error in (%)
'MSE' is Mean Squared Error in (p.u)

IV. IMPLEMENTATION OF MODELS USING ANN TECHNIQUE

There are five different models built to study the effect of calendar variables in the day-ahead load power consumption throughout the year. Here air temperature and previous day load are taken as constant variables. The calendar effects such as season, day of the week, week end factor and holiday factors are included one by one as "forward addition method" of forecaster development.

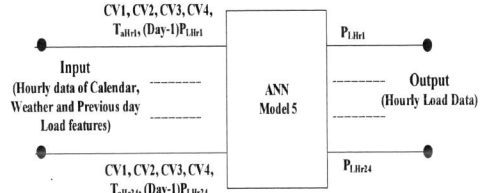

Fig. 3. Block diagram representation of Model 5

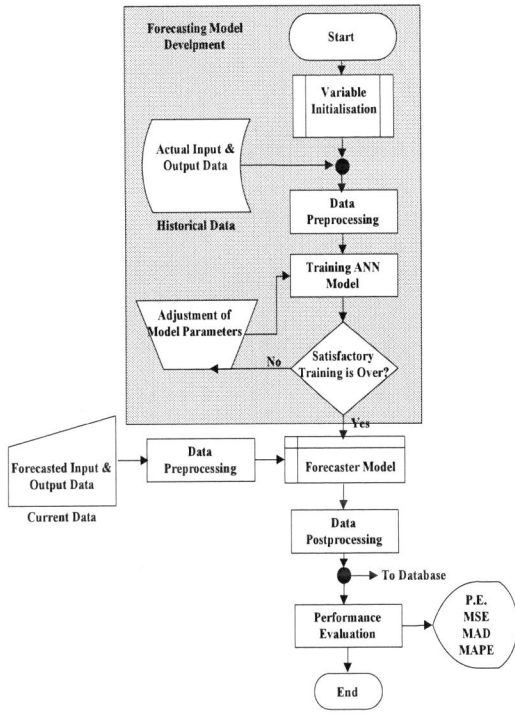

Fig. 4. Processes involved in Forecaster Model Development using ANN Technique

The block diagram of representation of Model 5 is shown in Fig. 3. There are five different models created using subset features of Model 5 (i.e. Main Model). There are six different features (CV1, CV2, CV3, CV4, Ta, (Day-1) PL) selected for new framework of modeling. The hourly data of year 2016 (i.e. total of 8760 observations in each selected feature during 1st January 2016 to 31st December 2016) were used as the training set data to create a forecaster model. The processes involved in forecaster model development are shown in Fig. 4. As discussed in section III, the forecaster models were developed and implemented in Matlab programming environment. The training of data is used only once in a year. There is no need of shifting and updating the training set either daily or weekly. It means there is no need of shifting the training data set till one year. The shifting of data with new data set did not turn out any satisfactory

978-1-5386-6624-1/18 $31.00 © 2018 IEEE

performance as compared with the results presented here. Hence, the idea of shifting data set was dropped out. The reason may be that while shifting data set with new day data, there is a chance of over/under frequency of the same number of patterns which results in lower accuracy of the ANN network that was observed in these models. Hence once in a year only ANN needs to be trained and the same trained ANN can be used for another one year of forecasting period (i.e. 1st January 2017 to 31st December 2017). This kind of development work makes a forecaster more reliable and very economical. Then, the forecasted models were tested with out-of-sample data of year 2017. Initially, for two weeks of April and two weeks of July 2017, the forecast value of weather was fed into trained ANN and tested. The results obtained were produced better as compared to sending actual input data of weather patterns. Due to the non-availability of forecasted weather data for the whole 2017 period, the models were tested with actual weather data. All of the obtained results are discussed and presented here in the following section.

V. RESULTS & DISCUSSIONS OF MODELS DEVELOPED

The comparison of obtained ANN training results of all five models are presented in Table IV. The performance goal of MSE of 0.001 (pre-defined) was achieved by all the models with less than 10 epochs. The regression values for all the models were approximately near 1 (above 0.95) achieved by all the models. Hence the performances of all models were equally good in all respects of parameters

TABLE IV.   COMPARISON OF OBTAINED ANN TRAINING RESULTS

Parameter	Model 1	Model 2	Model 3	Model 4	Model 5
Simulation Time in Seconds	7	12	13	21	15
Epoch	8	9	6	7	4
Performance goal achieved	0.0009	0.0007	0.0008	0.0008	0.0009
Best Validation Performance obtained	0.0018 at epoch 3	0.0019 at epoch 3	0.0012 at epoch 6	0.00105 at epoch 5	0.0013 at epoch 4
Regression_Training	0.9700	0.9782	0.9862	0.98263	0.9850
Regression_Validation	0.9712	0.9722	0.9749	0.98443	0.9795
Regression_Testing	0.9576	0.9850	0.9472	0.99021	0.9864
Regression_Overall	0.9697	0.9772	0.9826	0.98315	0.9839

listed in Table IV. The Model 5 has produced most accurate results i.e. close to least error space in all kind of errors addressed in Performance Evaluation criteria (section III.C) as compared to other Models presented here with. Hence, Model 5 is the best model in the list provided in Table I. Initially performance of each model is analyzed based on the prediction error (P.E).

The prediction error is calculated as per equation 2 given in section III.C. The variation of P.E corresponding to frequency of occurrences of samples in various ranges of error is shown as a histogram chart in Fig. 5. for Model 1 to understand the statistical fitness of the Model. The bell shaped and/or inverted 'l' shaped histogram charts of errors show that the model statistically proves to be normal distribution. The results of all five models are presented

TABLE V.   HOURLY VARIATION OF PREDICTION ERROR WITH RESPECT TO FREQUENCY OF OCCURANCE OF SAMPLES

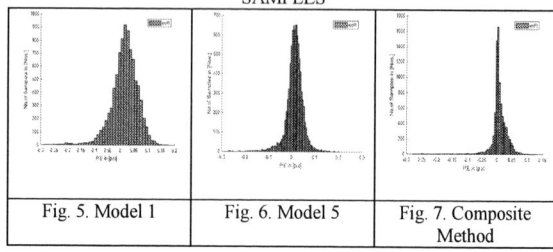

Fig. 5. Model 1	Fig. 6. Model 5	Fig. 7. Composite Method

TABLE VI.   HOURLY VARIATION OF PREDICTION ERROR WITH RESPECT TO TIME

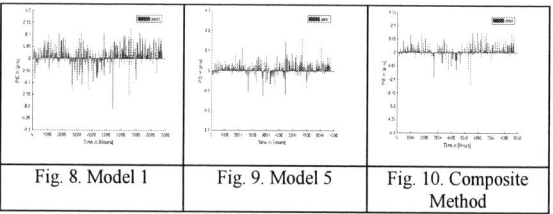

Fig. 8. Model 1	Fig. 9. Model 5	Fig. 10. Composite Method

TABLE VII.   GROUPING OF SAMPLE WITH RESPECT TO PREDICTION ERROR

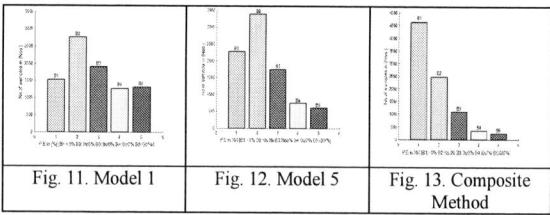

Fig. 11. Model 1	Fig. 12. Model 5	Fig. 13. Composite Method

TABLE VIII.   VARIATION OF MAPE WITH RESPECT TO DAYS IN THE YEAR 2017

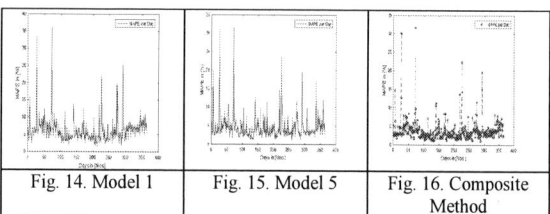

Fig. 14. Model 1	Fig. 15. Model 5	Fig. 16. Composite Method

TABLE IX.   VARIATION OF MSE WITH RESPECT TO DAYS IN THE YEAR 2017

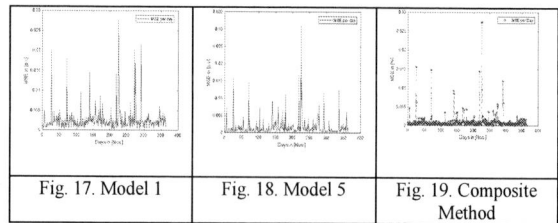

Fig. 17. Model 1	Fig. 18. Model 5	Fig. 19. Composite Method

here. The bar chart representation of P.E is shown in Fig. 8. It clearly depicts the over and under prediction error of each hourly sample for 365 days (8760 samples). The absolute value of prediction error under predefined error region is shown in Fig. 11. as bar chart. Model 1 has produced P.E of less than 0.01p.u (B1) for 1522 samples (out of total samples of 8760) hourly predictions. Similarly, it has produced 2763 hourly predictions having P.Es between 0.01p.u and 0.03p.u (B2). In the same manner 1907, 1264 and 1304 predictions are lying respectively between P.E.s of 0.03p.u and 0.05p.u (B3) and then between 0.05p.u and 0.07p.u (B4) and the remaining samples are above 0.07p.u (B5). It is predicted that almost 17% of hourly predictions are lying in less than 0.01p.u P.Es. 32% hourly predictions are in the range between 0.01p.u and 0.03p.u. Then 22%,

TABLE X.   ACTUAL & FORECASTED LOAD POWER VARIATION ON A MINIMUM MAPE DAY IN THE YEAR 2017

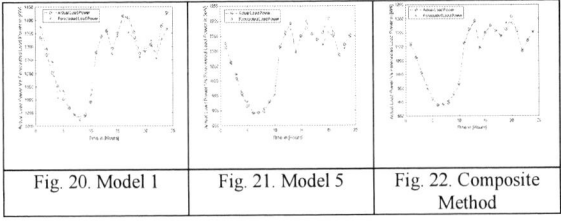

| Fig. 20. Model 1 | Fig. 21. Model 5 | Fig. 22. Composite Method |

TABLE XI.   AVERAGE ABSOLUTE PREDICTION ERROR (IN P.U) OBTAINED PER HOURLY SAMPLE FOR EVERY MONTH IN THE YEAR 2017

Month	Model 1			Model 5			Composite Method		
	Min.	Max.	Avg.	Min.	Max.	Avg.	Min.	Max.	Avg.
Jan	0.013	0.122	0.033	0.011	0.113	0.026	0.003	0.052	0.011
Feb	0.019	0.067	0.041	0.015	0.060	0.028	0.006	0.036	0.020
Mar	0.017	0.113	0.040	0.009	0.100	0.025	0.005	0.088	0.017
Apr	0.016	0.077	0.033	0.011	0.072	0.025	0.004	0.043	0.013
May	0.017	0.089	0.039	0.009	0.089	0.031	0.003	0.058	0.015
Jun	0.021	0.085	0.043	0.014	0.086	0.033	0.005	0.049	0.016
Jly	0.015	0.074	0.033	0.011	0.043	0.026	0.005	0.031	0.013
Aug	0.013	0.134	0.044	0.017	0.144	0.040	0.003	0.128	0.026
Sep	0.018	0.116	0.042	0.009	0.062	0.030	0.004	0.058	0.018
Oct	0.013	0.117	0.041	0.007	0.092	0.032	0.003	0.087	0.024
Nov	0.018	0.054	0.036	0.010	0.052	0.019	0.005	0.027	0.014
Dec	0.023	0.065	0.041	0.011	0.096	0.030	0.003	0.038	0.020
Avg. 2017	0.013	0.134	0.039	0.007	0.144	0.029	0.003	0.128	0.017

14% and 15% hourly predictions resulted in between 0.03p.u and 0.05p.u, 0.05p.u to 0.07p.u and above 0.07p.u ranges respectively. It means around 15% hourly predictions are above 7% P.Es in Model 1. The MAPE and MSE errors of Model 1 are shown in Fig. 14 and Fig. 17 respectively. Model 1 forecasted a minimum MAPE of 1.7% on 2nd Aug 2017 and is shown in Fig. 20. The maximum MAPE of 36% is forecasted on 13th March 2017. The maximum MAPE day, is a "Holi" festival day in Delhi. Then, Model 1 was completely analyzed for the Higher MAPE days; it has been understood that all the festival and adjacent festival days are lying in this region of higher errors that too most of the predictions were over forecasted (negative error). The reason for this may be that trained ANN has very lesser number of holidays and festival holiday pattern i.e. may be around 20 days in a year. The same higher MAPE issue was also observed in every model. So the ANN needs to have more number of festival days' pattern to get rid of higher errors for those kinds of inaccurate predictions. This will be taken care separately in future work. Model 2, Model 3, Model 4 and Model 5 have predicted a minimum MAPE of

1.49% on 25th July, 1.14% on 16th Apr, 1.05% on 15th Apr and 1.01% on 15th Oct 2017 respectively. Similarly all the results of Model 5 (best) are depicted in Fig. 6, Fig. 9, Fig. 12, Fig. 15, Fig. 18 and Fig. 21 respectively. The composite method has been created by comparing all the results of five models for each and every hourly prediction errors and daily errors. Then, the model which produces the least error has been chosen as the best model for that particular prediction time of forecast horizon either as sample wise or day wise. Table V shows the histogram charts of prediction errors of Model 1, Model 5 and Composite method. Sample wise P.Es of these models are shown in Table VI. Similarly, day wise MAPE and MSE predictions have also been observed and presented in Table VIII and Table IX. The actual and forecasted load power variation on a minimum MAPE day obtained in the year 2017 for Model 1, Model 5 and Composite Method is presented in Table X. The results of Composite Method are shown in Fig. 7, Fig. 10, Fig. 13, Fig. 16, Fig. 19 and Fig 22. The average absolute P.E.s obtained per sample for Model 1 (worst), Model 5 (best) and composite method respectively are presented in Table XI. Model 1 is forecasted with an average P.E of 3.9% per sample and Model 5 is forecasted with an error of 2.9% respectively. The Composite Method has an average P.E of 1.7% per sample throughout the year. Out of the total predictions of 8760 (for the whole year 2017), Model 1

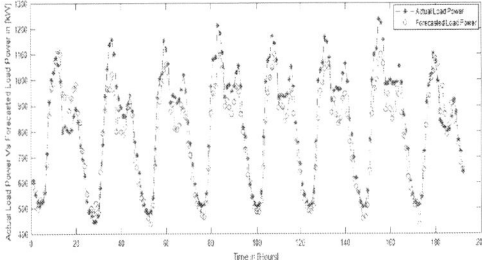

Fig. 23. Model 1: One Week Actual & Forecasted Load Power Variation in the Month of Feb 2017

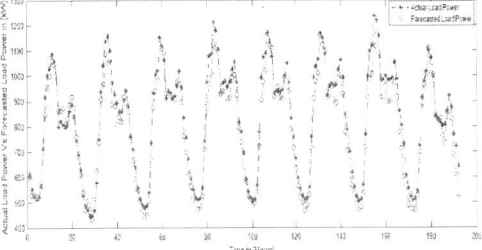

Fig. 24. Model 5: One Week Actual & Forecasted Load Power Variation in the Month of Feb 2017

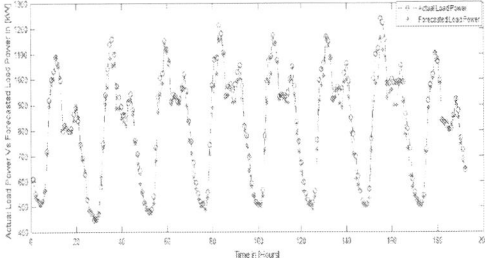

Fig. 25. Composite Method: One Week Actual & Forecasted Load Power Variation in the Month of Feb 2017

has produced 1307 times hourly least prediction errors. Model 2 and Model 3 have produced 1243 and 1597 times hourly least prediction errors respectively. The Model 4 and Model 5 have elected for 1924 and 2689 times for hourly least prediction errors. Based on least P.Es (per hourly sample), the percentage ratio of Models chosen (whole year) in the Composite Method is 15:14:18:22:31 (Model 1 to Model 5) respectively. The 'T' shaped and/or bell shaped histogram charts of individual and/or composite method reveal that the performance of each model is statistically proven to be the best model. Based on least MAPEs obtained per day, the percentage ratio of Models chosen for throughout 365days is 5:10:17:25:43 (Model 1 to Model 5) respectively. Hence it is understood that all the calendar variables (Model 5) are playing major role in the day-ahead ELPF which gives better accuracy compared to other with less number of calendar variables. The composite method of analysis reveals that even though Model 5 majorly produces high accuracy in most of the predictions, all other models are also play equally in different periods of the year. Hence, every model is equally good in the forecasting. The actual and forecasted load power variation for continuous seven days in the month of February 2017 for the Model 1, Model 5 and Composite Method is shown in Fig. 23, Fig. 24 and Fig. 25 respectively.

VI. CONCLUSION & FUTURE WORK

The new framework for day-ahead ELPF was developed using ANN technique and implemented in Matlab environment, with one year of historical data of year 2016 utilized for training. Out-of sample test was performed for the next year 2017(whole period) and the results are shown clearly in relevant histogram and bar charts. Model 5 has revealed an efficiency of above 95% at about 85% of sample predictions. The Composite Method is 95% efficient at 93% of hourly sample predictions. Model 1 is forecasted with an average error of 3.9% hourly and Model 5 has resulted in 2.9% respectively. The Composite Method has produced an average of 1.7% P.E per sample (hourly) throughout the year. All the models have less than 0.005p.u MSEs for above 90% sample predictions. The major drawback observed uniformly in all the models was obtainment of higher MAPE during festival and adjacent festival days. This will be a future research possibility to bring out a unique model for festival holidays separately. It may be because of very lesser number of festival days pattern available (around 25 days only) in the ANN training data. But ANN brings out all those days in the higher MAPE range i.e. it is understood that grouping is possible but prediction is inaccurate. This will be taken up for future possible research direction. Then, in general, it is understood that, this new frame work with MLP based feed forward back propagation neural network performs best using only temperature and load power data for Delhi Electricity Distribution utility which is always connected with high variation of loads (Residential, Commercial complexes and Industrial loads) with respect to times. Hence, the models discussed here, proved for their simplicity, less measurement requirement, easiness for implementation and most accurate method of forecasting day-ahead electric load power.

ACKNOWLEDGMENT

The authors express their sincere gratitude to Delhi Electricity Distribution Limited., for sharing the data required for carrying out this research successfully.

REFERENCES

[1] D.W.Bunn and Farmer, "Comparative Models for Electrical Load Forecasting", John Wiley &Sons, New York, USA, 1985.

[2] D.W.Bun, "Forecasting loads and prices in competitive power markets", Proceeding of IEEE, Vol. 88, pp.163-169, Feb 2000

[3] E.H. Barakat and M.A.M. Eissa, "Forecasting monthly peak demand in fast growing electric utility using a composite multi-regression decomposition model", in Proc. of. Second Regional Conf. CIGRE Committees Arab Countries, Amman, Jordan, May 1997.

[4] E. H. Barakat,J.M. Al-Qassim and S.A. Al Rashed, "New model for peak demand forecasting applied to highly complex load characteristics of a fast developing area", IEE Proceedings-C. Vol. 139, No. 2, March 1992.

[5] E. A. Feiberg,J.T. Hajagos and D. Genethliou, "Statistical load modelling, Power and Energy Systems", in Proc. of the Seventh IASTED International Multi-Conference, Palm Springs, CA, 2003.

[6] M. S. Kandil, S. M. El-Debeiky, and N. E. Hasanien, "Long- term forecasting for fast Developing Utility Using a knowledge-Based Expert System," IEEE Transaction on Power System, vol. 17, No. 2, pp. 491- 496, May 2000.

[7] M. Gavrilas, I. Ciutea and C. Tanasa, "Medium-term load forecasting with artificial neural network models", in Proc. of 16th CIRED Conference, Amsterdam, June 2001.

[8] M. Gravilas, "Neural Network Based Forecasting for Electricity Market", in Proc. of Technology Evolution for Future European Electricity Markets Discussion Forum, Kingstone upon Thames – London, September, 2002.

[9] https://www.mathworks.com/help/stats/corr.html

[10] A.Z.Chen and Y.Yang "Assessing Forecast Accuracy Measures", 2004.

[11] J.Nocedal and S.J.Wright, Numerical Optimization, Springer, 1999

[12] C.K.Goh, E.Teoh and K.C. Tan, "Hybrid Multi-objective Evolutionary design for artificial neural networks", IEEE Transactions on Neural Networks 19, 9(2008), 1531-1547

[13] Ludermir, T.B.Yamazaki and C. Zanchettin, "An optimization methodology for neural network weights and architectures", IEEE Transactions in Neural Networks 17, 6(2006), 1452-1457.

[14] E.Teoh, K.C.Tan and C.Xiang, "Estimating the number of hidden neurons in a feed forward network using the singular value decomposition", "IEEE Transactions on Neural Networks 17, 6(Nov.2006),1623-1629.

[15] Daniel T.Larose, Data Mining Methods and Models, A John Wiley & sons ,2006, pg.36

[16] J. Armstrong, F. Collopy, "Error measures for generalizing about forecasting methods: Empirical comparisons with discussion", International Journal of Forecasting, 8 (1992), pp. 69-80

[17] Lehmann, E. L, and Casella George, " Theory of Point Estimation (1998, 2nd Edison), New York, Springer

Voltage Profile Enhancement for IEEE-14 Bus System using UPFC, TCSC and SSSC

Gurudutt Bagha
Department of Electrical Engineering
NIT Kurukshetra
Haryana, India
gdbagha@gmail.com

Amit Kumar
Department of Electrical Engineering
NIT Kurukshetra
Haryana, India
amitkumar357@gmail.com

Abstract–Due to rapid growth in industrialization and urbanization much more capable transmission and distribution system is required to overcome the growing demand of electricity. So to meet up the demand of increasing power recent new technology, i.e., FACTS devices, has been developed in recent decades. The goal of this paper is to enhance the voltage profile of a test system using some FACTS devices, like UPFC, TCSC and SSSC. In this paper we have used an IEEE 14 bus as a test system and built it using PSAT software. First we find the critical bus having a minimum voltage profile and then by trial and error method we searched for appropriate FACTS device at appropriate location for maximum voltage enhancement of that critical bus.

Keywords: -IEEE-14 bus, UPFC, TCSC, SSSC, PSAT

I. INTRODUCTION

A major concern of a power engineer is the instability in the power system due to voltage collapse. Voltage fluctuation is a serious issue which is undesirable at the customer end. For enhancing voltage in a network stability analysis is very important. Voltage profile of a network is getting affected day by day due to increment in load demand. Load flow analysis is very important to tackle this problem. Some of the modern advanced devices like FACTS are very handful in voltage profile improvement of a system [1].

Continuous attention is required in the issued related to power quality. Due to the increment in the number of load which are sensitive to power quality it has created a problem in power quality in recent years. So the changes in user requirements and equipment have made a new goal for an electrical engineer to work in the area for providing a good power quality [1]. When the demand of reactive power is not met up then the problems related to voltage instability and collapse occurs in our power system. Continuous changes and restruction of transmission system is happening worldwide to make the transmission system much more flexible to load patterns and diverse power generation. For the purpose of generating employment and to support industry in developing industries, the investments over the transmission must be required to be used in optimized manner [2].

Operation and control of power system network is getting much more complex with the increment of power demand. The control of existing network is mostly mechanical. In recent times some devices like computers and high-speed communication devices have come into picture to be used for control and protection of transmission system. Although the control signal is sent from these electronic devices but still the final switching action is done from mechanical devise only. Since these mechanical devices are slow in action so a frequent control cannot be obtained from these mechanical devices. These slow action causes quick wear out of these devices as compare to static devices. Because of such reasons FACTS technology was developed. Introduction of Flexible AC Transmission System (FACTS) devices around three decades ago has led to a revolutionized power flow improvement in a transmission system [3-4].

Some of the FACTS devices which have been running in operation are TCSC, SSSC and UPFC comes under series and series-shunt family, respectively. Meikandasivam et. al [5] referred a brief study of a TCSC device about the operation, reactance characteristic and resonance and also degree of series compensation (K) brings an idea of selecting the TCSC capacitor. A controllable series compensation can be obtained using SSSC which is immune to network resonance [3]. Power flow reversement is possible in SSSC as mentioned in [6]. Resistive component of a line impedance can have its voltage drop be compensated in the presence of an external DC supply by SSSC [7]. System parameters like impedance, voltage and phase angle are controlled either simultaneously or individually by UPFC [8-9].

One of the open source software used in this paper is Matlab based PSAT. Some of the operations like power flow, continuation power flow etc. can be performed over a power system network in PSAT [10]. IEEE 14 bus used in this paper consists of five generators and four PV generators. Line, generator and PV generator data of IEEE 14 bus system used in this paper is given in [11].

The main aim of this paper is to analyze such FACTS device on an IEEE-14 bus system and there results are analyzed. The FACTS devices which are used for enhancement analysis are UPFC, TCSC and SSSC.

II. FACTS DEVICES

With the recent day by day increment in industrialization and modern urbanization of life style has led to major transformation in electricity industry worldwide. This huge transformation has led to a high complexity in existing power system. With this modernization it has created a huge task for a power engineer to come up with efficient ways to control the voltage and power profile in an efficient and economic manner. One of the methods is by installation of new

978-1-5386-6624-1/18 $31.00 © 2018 IEEE

transmission lines. But installing a new line is a very difficult task and high complexible. It is not economically and environmentally considered beneficial to install new transmission lines.

For much more economical and efficient transmission of power over a line a new technology called FACTS (Flexible Alternating Current Transmission System) devices have been introduced [5]. For the increment of controllability and transfer capability FACTS devices uses power electronic devices and controllers. With the introduction of FACTS it has become very easy to regulate phase angle, voltage, series impedance, current, shunt impedance and series impedance of a transmission line for voltage and power flow profile improvement [2].

For the voltage profile study of an IEEE-14 bus system following FACTS devices are used in this paper:

- Thyristor Controlled Series Capacitor(TCSC)
- Static Synchronous Series Capacitor(SSSC)
- Unified Power Flow Controller(UPFC)

A. TCSC

Thyristor Controlled Series Capacitor is used as a series compensator which is useful in governing the power flow in a transmission line by compensating the line reactance of the transmission line. In 1986, Vithayathil with others proposed the basic scheme of TCSC [3]. Fault current can also be limited by adjusting its impedance.

TCSC basically consists of three parts: series capacitor C, a bypass inductor L and a bi-direction thyristor SCR. TCSC can be made to act as an inductor or capacitor by modifying the reactance of TCSC.

Fig. 1 shows the basic circuit diagram of a TCSC which consists of a Thyristor Controlled Reactor (TCR) shunting a series compensating capacitor (C). Inductive reactance of a TCR is a function of firing angle (α). Thus the value of TCSC can be varied by changing the value of α. The formula for X_L as a function of α is [5]:-

$$X_L(\alpha) = X_L \frac{\pi}{\pi - \sin 2\alpha - 2\alpha} \qquad (1)$$

$$X_L = \omega L = 2\pi f L \qquad (2)$$

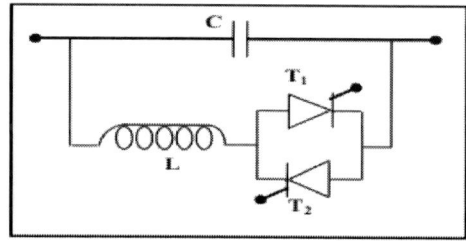

Fig 1. Schematic diagram of TCSC [5]

Depending upon the range of α TCSC can be made to work under inductive, capacitive and resonance region as shown in Fig. 2 and Table 1.

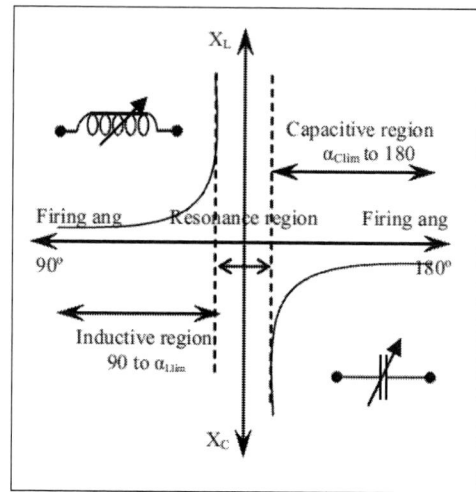

Fig 2. Characteristic curve of TCSC[5]

TABLE I. TCSC FIRING ANGLE (α) RANGE

Range of firing angle (α)	Region
$\alpha_{Llim} \leq \alpha \leq \alpha_{Clim}$	Resonance region
$90 \leq \alpha \leq \alpha_{Llim}$	Inductive region
$\alpha_{Clim} \leq \alpha \leq 180$	Capacitive region

B. SSSC

SSSC (Static Synchronous Series Compensator) being a FACTS device is a voltage source converter based device. The schematic diagram of SSSC is shown in Fig. 3. It is able to provide inductive and capacitive compensation which are independent of line current magnitudes [3].

Availability of any external energy source is not required for the operation of SSSC. Thus it is fully controllable and independent of transmission line current. Thus it is able to increase or decrease the overall voltage drop across the transmission line and thereby controlling the power flow in the transmission line.

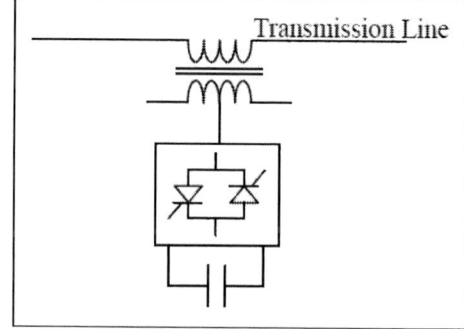

Fig 3. Schematic diagram of SSSC [6]

SSSC has got the capability to exchange both active and reactive power with the transmission line. Fig. 4 represents the performance curve of a SSSC.

978-1-5386-6624-1/18 $31.00 © 2018 IEEE

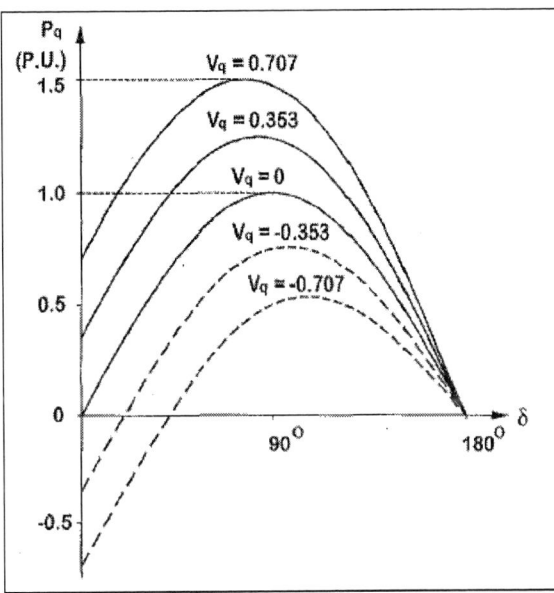

Fig 4. Power vs Angle curve of SSSC [7]

Power transmitted through the SSSC is given by:-

$$P_q = \frac{V^2 \sin \delta}{X} + \frac{V.Vq \cos \delta/2}{X}$$

P_q = Power transmitted through transmission line
V= magnitude of voltage
δ = angular difference between the voltage ends of a transmission line

C. *UPFC*

In 1991, GyuGyi proposed the idea of UPFC (Unified Power Flow Controller) [3]. On high-voltage transmission system UPFC can easily provide fast paced reactive power compensation. Many of the parameters (i.e. phase angle, voltage and impedance can be controlled with help of UPFC. Thus its capability to control all these parameters has got it the term "unified".

It is the latest in the FACTS technology. Features of two other FACTS device are being combined for the working of UPFC. Those two devices are STATCOM (Static Synchronous Compensator) and SSSC (Static Synchronous Series Compensator). These two FACTS devices are linked via a common DC voltage link. Transient stability of a power system network can be obtained with the help of UPFC, since; it is able to suppress the oscillation in power system network [8].

Fig. 5 represents the circuit diagram of UPFC. In this figure shunt converter corresponds to STATCOM and series converter corresponds to SSSC and they are linked by a common DC voltage link $V_{dc.}$

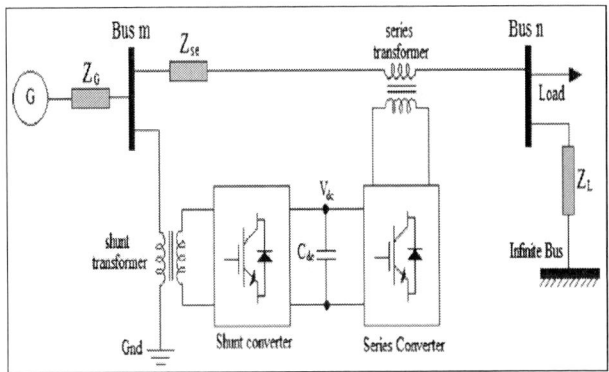

Fig 5. Circuit diagram of UPFC [9]

III. PSAT (POWER ANALYSIS AND TOOLBOX)

For the purpose of control and analysis of electrical power system MATLAB toolbox which is helpful is PSAT [10]. It is portable and open source. PSAT can run on some of the most common most operating system like Windows, Unix etc. Small signal stability, optimal power flow, power flow and continuous power flow are included in PSAT.

Most important in the PSAT is power flow algorithm. Once it is completed the user can perform further analysis. These analyses are:-

- Optimal power flow
- Continuation power flow
- Small-signal stability analysis
- Time domain simulations

Some of the other features which are included in PSAT:
- Simulink library to form a network
- User defined model installation and construction
- Filters for converting data
- Command logs

IV. POWER FLOW ANALYSIS OF IEEE-14 BUS SYTEM

For the purpose of load flow analysis we have taken a IEEE-14 bus system and simulated it in PSAT toolbox. This modified model was then compared with reference model as given in [10]. It was observed to follow the reference model. Block diagram of IEEE-14 bus in shown in Fig 6. Simulation model of IEEE-14bus system made in PSAT is shown in Fig 7.And also, all the input data ,i.e., PQ load data , branch data, generation data, PV generator data and transformer tap setting data, to the IEEE-14 bus PSAT simulink model are given in Tables 2 to 6 respectively [12].

Results of the IEEE-14 bus when run in PSAT are shown in Table 7.

10	0.09	0.058
11	0.035	0.018
12	0.061	0.016
13	0.135	0.058
14	0.149	0.05

TABLE 3. BRANCH DATA [11]

Line no	From bus	To bus	Line impedance		Half line charging susceptance (pu)
			R (pu)	X (pu)	
1	1	2	0.01938	0.05917	0.02640
2	2	3	0.04699	0.19797	0.02190
3	2	4	0.05811	0.17632	0.01870
4	1	5	0.05403	0.22304	0.02460
5	2	5	0.05695	0.17388	0.01700
6	3	4	0.06701	0.17103	0.01730
7	4	5	0.01335	0.04211	0.0064
8	5	6	0.0	0.25202	0.0
9	4	7	0.0	0.20912	0.0
10	7	8	0.0	0.17615	0.0
11	4	9	0.0	0.55618	0.0
12	7	9	0.0	0.11001	0.0
13	9	10	0.03181	0.08450	0.0
14	6	11	0.09498	0.19890	0.0
15	6	12	0.12291	0.25581	0.0
16	6	13	0.06615	0.13027	0.0
17	9	14	0.12711	0.27038	0.0
18	10	11	0.08205	0.19207	0.0
19	12	13	0.22092	0.19988	0.0
20	13	14	0.01709	0.34802	0.0

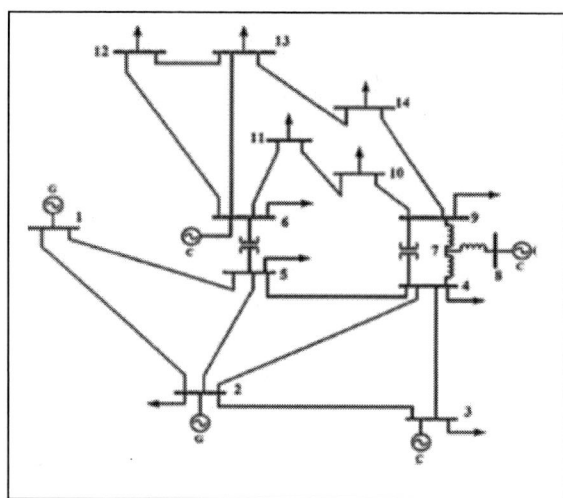

Fig 6. Basic block diagram of IEEE-14 bus system [12]

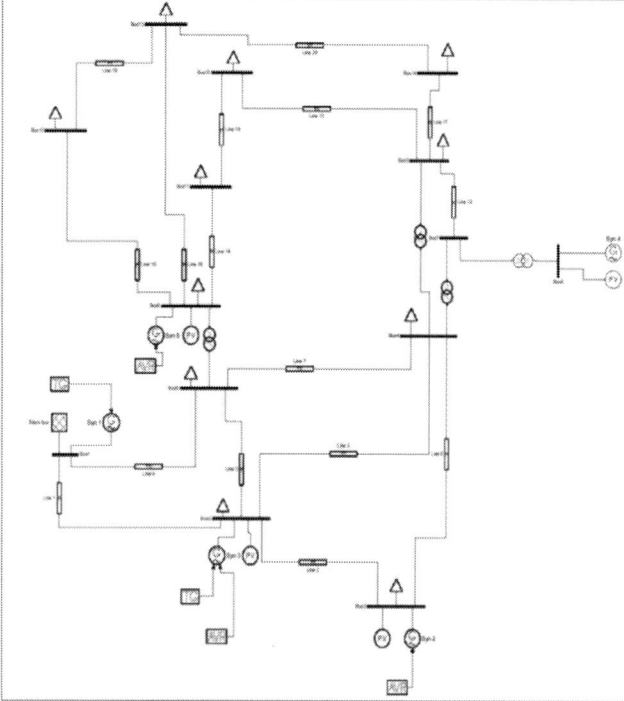

Fig 7. PSAT model of IEEE-14 bus system

TABLE 2. PQ LOAD DATA

Bus no.	Real power (pu)	Reactive power(pu)
1	0.0	0.0
2	0.217	0.127
3	0.942	0.19
4	0.478	0.04
5	0.076	0.016
6	0.112	0.075
7	0.0	0.0
8	0.0	0.0
9	0.295	0.166

TABLE 4. GENERATION DATA

Bus no	Real power (pu)	Reactive power (pu)
1	2.324	-0.169
2	0.4	0.424
3	0.0	0.234
4	0.0	0.0
5	0.0	0.0
6	0.0	0.122
7	0.0	0.0
8	0.0	0.174
9	0.0	0.0
10	0.0	0.0
11	0.0	0.0
12	0.0	0.0
13	0.0	0.0
14	0.0	0.0

TABLE 5. PV GENERATOR DATA

Bus no.	Voltage magnitude (pu)	Reactive power limit	
		Minimum MVAR (pu)	Maximum MVAR (pu)
2	1.045	-0.4	0.5
3	1.01	0.0	0.4
6	1.07	-0.06	0.24
8	1.09	-0.06	0.24

TABLE 6. TRANSFORMER TAP SETTING DATA

From bus	To bus	Tap setting value (pu)
4	7	0.978
4	9	0.969
5	6	0.932

TABLE 7. POWER FLOW RESULT OF IEEE-14 BUS WITHOUT ANY FACTS DEVICES

Bus no	Bus voltage (pu)	Phase angle(rad)
1	1.06	0
2	1.045	-0.0871
3	1.01	-0.22266
4	1.012	-0.17851
5	1.016	-0.15274
6	1.07	-0.25161
7	1.0494	-0.23092
8	1.09	-0.23092
9	1.0328	-0.25854
10	1.0318	-0.26224
11	1.0471	-0.25898
12	1.0534	-0.26645
13	1.047	-0.2671
14	1.0207	-0.28018

Voltage magnitude result of IEEE-14 bus in the absence of any FACTS device is shown in Fig 8. And also the impact of FACTS devices like UPFC, SSSC and TCSC are shown in Fig 9 to 11 respectively.

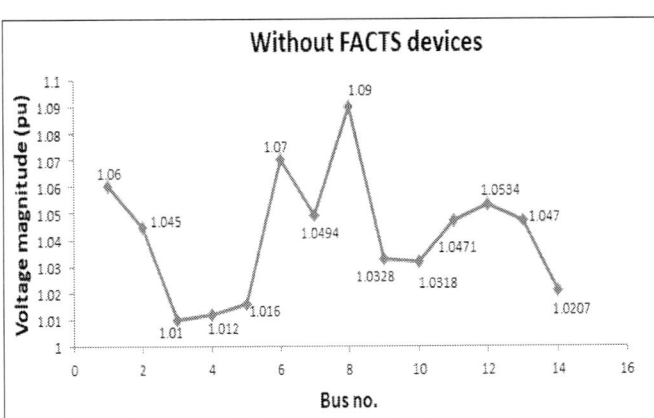

Fig 8. Graphical result of IEEE-14 bus without any FACTS devices

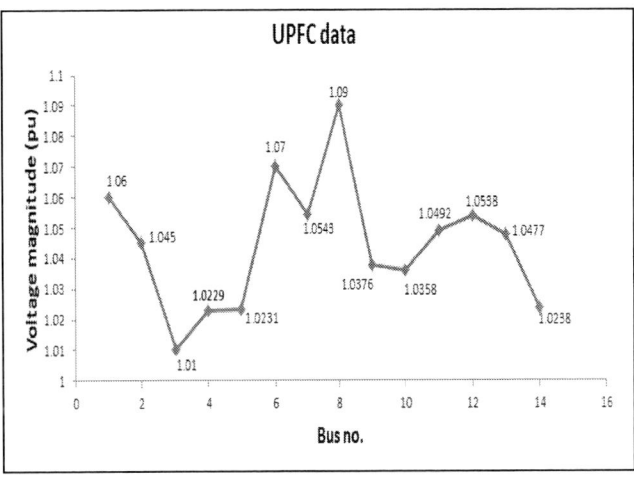

Fig 9. Graphical result of IEEE-14 bus in the presence of UPFC between bus no. 2 and 4

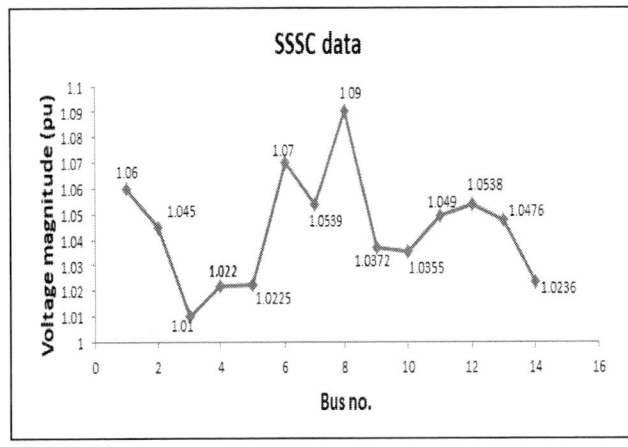

Fig 10. Graphical result of IEEE-14 bus in the presence of SSSC between bus no. 2 and 4

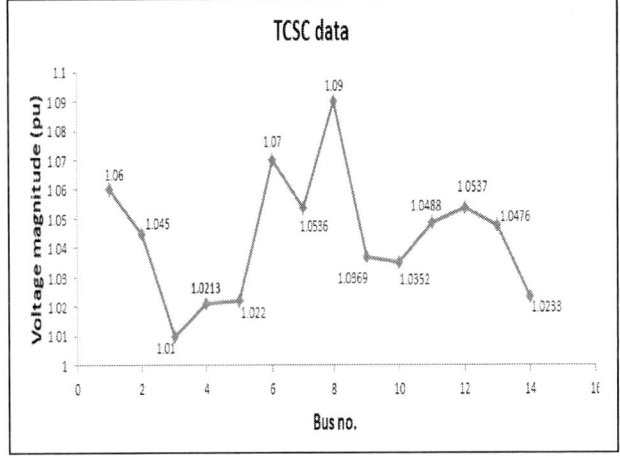

Fig 11. Graphical result of IEEE-14 bus in the presence of TCSC between bus no. 2 and 4

In the Table 7 voltage profile at bus 4 is found to be to be least. So in order to enhance its voltage profile all three FACTS devices, i.e. UPFC, SSSC and TCSC were placed at each transmission line simultaneously and by trial and error method voltage magnitude of bus 4 is recorded in Table 7.

TABLE 8. POWER FLOW RESULT OF BUS NO-4 WITH THE PRESENCE OF DIFFERENT FACTS DEVICES IN IEEE-14 BUS

From bus	To bus	UPFC	SSSC	TCSC
1	2	1.0120	1.0120	1.0120
2	3	1.0213	1.0123	1.0122
2	4	1.0229	1.0220	1.0213
1	5	1.0191	1.0184	1.0179
2	5	1.0172	1.0167	1.0163
3	4	1.0072	1.0072	1.0073
4	5	1.0147	1.0147	1.0147
7	9	1.0120	1.0120	1.0120
9	10	1.0120	1.0120	1.0120
6	11	1.0125	1.0125	1.0124
6	12	1.0121	1.0121	1.0121
6	13	1.0124	1.0124	1.0124
9	14	1.0118	1.0118	1.0118
10	11	1.0123	1.0123	1.0123
12	13	1.0121	1.0121	1.0121
13	14	1.0125	1.0123	1.0124

V. Conclusion

As it can be seen in Fig. 7 that IEEE-14 bus PSAT model is shown to have PV generator at buses 2, 3, 6 and 8. So it recorded that the voltage level at these buses won't change even with the addition of FACTS devices mentioned in this paper.

Power flow result of IEEE-14 bus is observed to have a least critical voltage of magnitude 1.012 pu at bus no. 4. So the voltage magnitude data of bus 4 is recorded with the addition of series compensation devices, i.e., UPFC, SSSC and TCSC, between every branch and results recorded in table 7 shows that the addition of UPFC between bus no 2 and 4 is giving a voltage magnitude of 1.0229 pu at bus no 4 which is greater than the voltage magnitude in the absence of any FACTS device.

It can also be noted that UPFC is giving much more satisfactory result than other FACTS devices, i.e., SSSC and TCSC. Thus it is concluded that UPFC gives much better satisfactory results than SSSC and TCSC for the enhancement of voltage profile in IEEE-14 bus system.

REFERENCES

[1] W. E. Kazibwe, R. 1. Ringlee, G. W. Woodzell and H. M. Sendaula, "Power quality: a review," Computer Applications in Power, IEEE, vol. 3, pp. 39- 42,1990.

[2] Anil Kumar, "A review: FACTS devices for modern power system", IJRMEE, vol. 2, issue 9, July 2008

[3] Hingorani, N.G. and Gyugyi, L., "Understanding FACTS", Concepts and Technology of Flexible AC Transmission System. New York: Ins. Elect. Electron. Eng., Inc., 2000

[4] Hussain Ashfaq, "Electrical Power Systems", CBS Publications, New Delhi, 2007.

[5] Meikandasivam S, Nema R.K and Jain S.K, "Selection of TCSC parameters: Capacitor and inductor," Proc. 2010 India International Conference Power Electronics (IICPE), pp. 1-5,2011

[6] Anwar S. Siddiqui, Naqui Anwer, Abdullah Umar, "Power flow management using FACTS controller for smart grid application," IJIRSET, vol. 2, issue 3, March 2014

[7] Lazlo Gyugyi, Colin D. Schander, Kalyan K. Sen, "Static synchronous series compensation: A solid state approach to the series compensation of transmission line," IEEE transactions on power delivery, vol. 12, issue:1, Jan 1997

[8] Sapna Khanchi, Vijay K. Garg, "Unified Power Flow Controller (FACTS device):A review," IJERA, vol. 3, issue 4, Jul-Aug 2013, pp. 1430-1435

[9] Vijay Kumara B, Srikanth NV, "Optimal location and sizing of Unified Power Flow Controller (UPFC) to improve dynamic stability: a hybrid technique," Int J Electr Power Energy Syst 2015; 64:429–38.

[10] F. Milano, "An Open Source Power System Analysis Toolbox," IEEE Transactions on Power Systems vol. 20, No.3, August 2005.

[11] P. Mishra, A. Choubey, V. Holkar," Enhancement of voltage profile for IEEE-14 bus system using STATIC-VAR compensation (SVC) when subjected to various changes in load," IJRSSET, vol. 1, issue 2, May 2014

[12] S. A. Taher, H. Mahmoodi, and H. Aghaamouei, "Optimal PMU location in power systems using MICA," Alex. Engr. J., Dec. 2015.

PSO Tuned AGC Strategy of Multi Area Multi-Source Power System Incorporating SMES

Nidhi Gupta[1], Narendra Kumar[2], Nisha Singh[3]

[1,3]*MSIT, GGSIPU, Delhi,* India

[2]*Delhi Technological University,* Delhi, India

[1]nidhi@msit.in, [2]dnk_1963@yahoo.com

Abstract—This paper investigates the effect of Super Conducting Magnetic Energy Storage System (SMES) in Automatic Generation Control (AGC) of multi area multi-source interconnected power systems. Primarily, the tuning of Integral controller (IC) in AGC strategy of three identical capacity control areas as non-reheat thermal, hydro and gas power system incorporating SMES is explored. The system dynamic stability is illustrated by comparing frequency responses, settling time, peak overshoot and performance index value of the system. For analysis, tuning of IC has been done by considering the performance index as Integral Square Error (ISE). The comparative analysis of different Artificial Intelligent techniques tuned AGC strategy with and without SMES on 1% step load perturbation in each area has been presented. The dominance of SMES with PSO (Particle Swarm Optimization) tuned integral controller has been established for the AGC strategy of thermal-hydro-gas interconnected power system. Moreover, this study shows that the sluggishness of dynamic response in case of participation of hydro power system is significantly improved with the application of SMES in multi area- multi source power system. Further, the comparative study is presented to demonstrate the effect of SMES on PSO tuned AGC strategy of multi area thermal-hydro-gas power system on 1% step load perturbation in different control areas. It has been analyzed that the studied system shows better performance in case of load disturbance in gas power system or thermal power system as a control area than the hydro power system as a control area.

Keywords—Thermal-Hydro-Gas Power System, Automatic Generation Control, Super Conducting Magnetic Energy Storage System, Particle Swarm Optimization, Area Control Error.

I. INTRODUCTION

The purpose of AGC is to keep the nominal value of frequency, to meet the interconnected tie line flows and to balance the active power generation with variable power demand at desired tolerance values [1-4]. Researchers and power engineers are very much inclined towards AGC studies [5-7]. Classical and optimal control strategies have been implemented in conventional AGC studies. An interconnected power system using a conventional Integral controller establishes that it is simple, economical and offered lesser calculations [8]. Thus, it is significant to consider the integral controller for the AGC strategies.

Since last many years, the conventional thermal, or hydrothermal power systems have been studied as one or two control areas. But in a realistic system, nowadays, multi area multi source power systems may simultaneously work in each control area [9,10]. Generally, optimizing the gains of supplementary controllers for each area in AGC of the interconnected power system reduces the effect of load disturbances. On the other hand, it takes longer duration to settle frequency and the tie line power deviations specifically when multi generating sources [11-12] are employed in the interconnected power system. To further improve the dynamic performance, addition of SMES unit has been incorporated in each control area. The effect of SMES has been studied for multi area multi source power systems by obtaining optimum values for the gain parameters of Integral controller.

Recent advancement in distributed generations, DC systems, superconducting technologies and power electronics spurred technical and economic growth of high temperature SMES in the demanding power system. The problems faced by power systems like low discharge rate, increased time required for power flow reversal and maintenance requirement has led to the evolution of SMES [13]. It is one of the most efficient energy storage technology, which is further explored for the AGC strategy. Effective use of SMES for sudden load changes includes multi area thermal-thermal [15], hydro-thermal [16], hydro–hydro [19] power systems.

Literature survey shows that very less work have been witnessed about the application of SMES in AGC of multi area interconnected thermal-hydro-gas power system. However, no literature is available to show the effect of SMES in three area thermal-hydro-gas power system on 1% step load perturbation in different control area. Moreover, such comparative study is not witnessed within multi area multi sources of thermal-hydro-gas making it further important to practice investigations on the same.

The main contribution of the present paper is to :

(a) Explore different artificial intelligence technique like PSO and GA algorithm for tuning integral controller for AGC strategy of multi area thermal-hydro-gas power system with/without SMES.

(b) Investigate the impact of SMES to PSO tuned integral controller for AGC strategy of multi area multi source power systems.

(c) Collate the system dynamic responses produced by assimilation of SMES in AGC strategy for multi area multi source power systems and to investigate their effectiveness with respect to wide change in load perturbation.

II. POWER SYSTEM INVESTIGATION

The power system under investigation is a multi-area multi source interconnected power systems incorporating SMES in each control area. The control areas of considered multi source power system model comprise of non-reheat thermal power system as first area, hydro power system as second area and gas power system as third area which are interconnected by tie lines as shown in Fig.1. For simulation study, transfer function block diagram of considered power system with integral controller is shown in Fig.2. The transfer function block diagram of SMES is given in Fig.4. Investigation is accomplished by performing the comparative analysis of the AGC strategy of thermal-hydro-gas interconnected power systems with other intelligent techniques incorporating and excluding SMES. The nominal parameters of the thermal, hydro and gas power systems are taken from [11] and given in Appendix B.

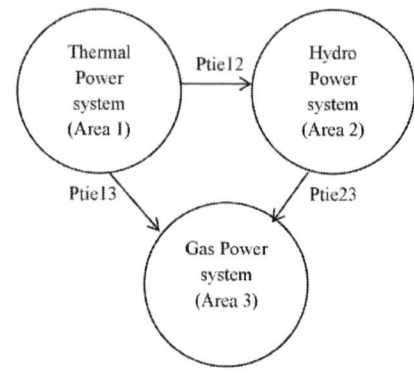

Fig. 1. Block diagram of multi area thermal-hydro-gas power system

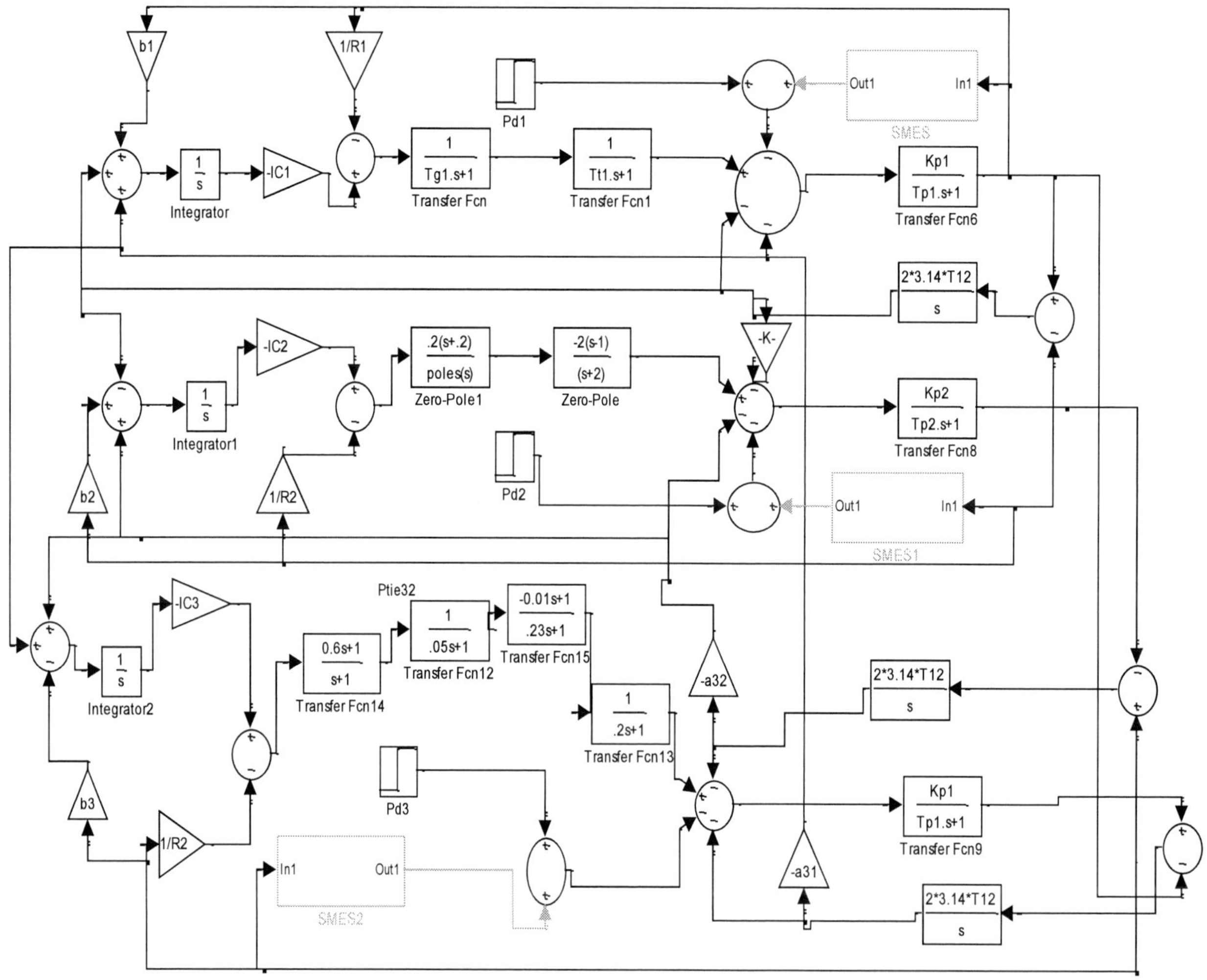

Fig. 2. Transfer function model of power system under exploration.

978-1-5386-6624-1/18 $31.00 © 2018 IEEE   274

Dynamic modeling of multi area thermal-hydro-gas power system is given as:

State space model:

$$X^. = A\,x + B\,U + \Gamma\,P_D \qquad (1)$$

$$Y = C\,x \qquad (2)$$

Where $X^.$, U, P_D and Y are the state, control, disturbance and system output vectors, respectively. A, B, C and Γ are real constant matrices of compatible dimensions.
The detailed dynamic model of thermal-hydro-gas is taken from reference [12].

System State vector:

$$[X]^T = [\Delta F_1\ \Delta P_{G1}\ \Delta X_E\ \Delta F_2\ \Delta P_{G2}\ \Delta X_h\ \Delta X_{RH}\ \Delta F_3\ \Delta P_{G3}\Delta P_{FC}\ \Delta P_V$$
$$\Delta X_g\ \Delta P_{tie12}\ \Delta P_{tie13}\ \Delta P_{tie23}\ \int ACE_1\ \int ACE_2\ \int ACE_3] \qquad (3)$$

System Control vector:

$$[U]^T = [\Delta P_{c1}\ \Delta P_{c2}\ \Delta P_{c3}] \qquad (4)$$

System Disturbance vector:

$$[\Delta P_D]^T = [\Delta P_{D1}\ \Delta P_{D2}\ \Delta P_{D3}] \qquad (5)$$

where the symbols are defined in Nomenclature.

The control signal will be of the form:

$$\Delta P_{ci}(t) = -\,IC_i \int (ACE_i)\,dt \qquad (6)$$

$$ACE_i = b_i\,\Delta F_i(t) + \sum \Delta P_{tie\,ij}(t) \qquad (7)$$

$$\Delta P_{tie\,ij}(t) = T_{ij}\,[\Delta F_i(t) - \Delta F_j(t)] \qquad (8)$$

Where $j = 1$ to 3, $j \neq i$

On each evaluation of the objective function, the model developed in Simulink is executed by considering a particular performance index. For finding the effect of SMES on multi area multi source of power system model, the main objective is to minimize the considered performance index of the system. The objective function for multi area multi source power system is formulated and it is based on the ACE of each control area. As ISE cogitate large error of system, it is considered to evaluate the performance index value.

System Performance Index:

$$J_0 = \int ACE_1^2(t)dt + \int ACE_2^2(t)dt + \int ACE_3^2(t)\,dt \qquad (9)$$

Intelligent techniques like PSO and GA are employed to tune integral controller for AGC strategy of multi area multi source power system under consideration.

III. MODELLING OF SUPERCONDUCTING MAGNETIC ENERGY STORAGE SYSTEM

An overview of SMES application in power system is given in [13]. A layout of the thyristor controlled SMES unit configuration has been shown in Fig.3. A complete unit of SMES is established with the help of superconducting magnetic coil, 12 phase bridge converter and star delta step down transformer with AC system [14]. The superconducting coil is placed within a helium vessel to maintain a cryogenic temperature. During normal operation of power system, the superconducting coil charges to a fixed value and then conducts high current in it's discharging mode. The increase in load results in the release of stored energy. Having a control on converter firing angle enables the DC voltage appearing across the inductor to be continuously varied between a wide range of positive and negative values. A low positive voltage is used to change the inductor initially to its rated current. When the current attain its rated value, it is kept constant by reducing the voltage across inductor to zero due to the super conducting nature of the coil. The expression for DC voltage after neglecting transformer and converter losses is given as:

$$E_d = 2\,E_{d0}\cos\alpha - 2\,I_d\,R_D \qquad (10)$$

Where E_d is the DC voltage directed to inductor (kV), I_d is the inductor current (kA), α is the firing angle, R_D is equivalent commutating resistance (kΩ), E_{do} is maximum open circuit bridge voltage (kV).
The block diagram representation of SMES unit is shown in Fig. 4. Parameter values for SMES unit applied to AGC strategy for multi area multi source power system are given in Appendix B.

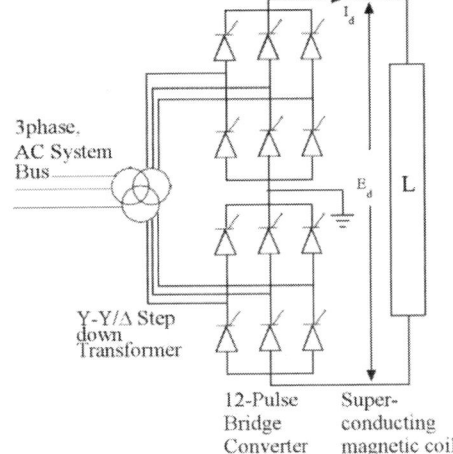

Fig. 3. Thyristor controlled SMES unit configuration

To improve the performance of controllers ACE is adopted as control signal. The DC voltage across superconducting inductor is continuously controlled in SMES power system. Moreover variation in inductor current helps in being utilized as negative feedback signal within the control loop of the SMES system.

978-1-5386-6624-1/18 $31.00 © 2018 IEEE

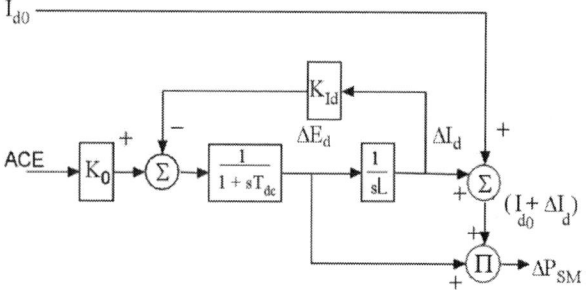

Fig. 4. Block diagram representation of SMES unit

The inductor current should regain its rated value very quickly when there is any load disturbances immediately. Therefore, inductor current deviation is sensed and fedback as negative signal in the SMES control loop to achieve quick restoration of current and SMES energy levels. As a result, the general equations for deviation in converter voltage and current are as follows:

$$\Delta E_d(s) = (1/(1+sT_{dc})) (K_o \, ACE(s) - K_{Id} \, \Delta I_d(s)) \quad (11)$$

$$\Delta I_d(s) = \Delta E_d(s)/ \, sL \quad (12)$$

Where ΔE_d and ΔI_d are incremental change in converter voltage (kV) and inductor current (kA), T_{dc} is converter time delay (s), K_o is the gain of the control loop (kV/unit ACE), L is the coil inductance (H), I_{do} is the nominal inductor current value (kA), E_{do} is the nominal inductor voltage value (kV), K_{Id} is the gain corresponding to negative current feedback (kV/kA). Initial power flow (P_{SM}) into the inductor coil without disturbance is given as:

$$P_{SM} = E_{do} \cdot I_{do} \quad (13)$$

The deviation in real power of inductor in SMES unit in response to the load disturbance is assumed positive for transfer from AC grid to DC grid which is given as:

$$\Delta P_{SM} = \Delta E_d (\, I_{d0} + \Delta I_d) \quad (14)$$

At any instant of time the amount of energy stored (W) in the inductor in time domain is given by

$$W = (LI_d^2)/2 \quad (15)$$

AGC strategy of thermal-hydro-gas interconnected power systems with IC incorporating SMES in first area, SMES1 in second area and SMES2 in third area which is shown in Fig. 2 are simulated to show the effect of SMES in considered thermal-hydro-gas power system.

IV. ARTIFICIAL INTELLIGENCE

Various AGC strategies have been proposed in literature which include conventional and artificial intelligence based techniques such as Genetic algorithm [12], Bacteria forging

optimization algorithm [12], Cuckoo search algorithm [15], BAT algorithm [17], Firefly algorithm [18], PSO [19-22] , Genetic algorithm [23] etc. These artificial techniques are usually developed using the knowledge extracted from the environment and available data. GA is the most popular and widely used artificial algorithm. The difficulties and problems associated with the use of GA have been overcome by the application of PSO technique. The PSO algorithm was found to be more robust, providing repeatable results unlike the GA which fails to converge sometimes, depending on the initial population used. Compared to GAs, the advantages of PSO are that PSO is easy to implement and there are few parameters to adjust. Kennedy defined PSO technique as a population (swarm) based stochastic optimization algorithm [19]. The quality attributes, principle of diverse response, principle of stability and the principle of adaptability of PSO algorithm make it a worthy choice for tuning controllers in AGC of power system. The parameters of PSO and GA tuned integral controller for AGC strategy of thermal hydro gas interconnected power system are given in Appendix C.

V. SIMULATION

In this study the simulation of AGC strategy of multi area multi source power system model incorporating SMES were developed in MATLAB software. The system data given in Appendix B has been used to study dynamic performance of the system on 1% step load perturbation for one or all the control areas. It has been observed that after the load disturbances, the area frequency response in multi area multi source power system is heavily perturbed. Hence, to suppress the oscillations and to have the optimal frequency responses, the impact of SMES in multi area thermal-hydro-gas power system has been explored and discussed. The transfer function model of multi area thermal-hydro-gas power system incorporating SMES has been shown in Figure 2. The gains of integral controller has been optimized by using different intelligent techniques like GA and PSO technique for 1% step load perturbation in each control areas, which has been given in Appendix C. The frequency responses of the considered system model with GA and PSO based AGC strategy incorporating and excluding SMES on 1% step load perturbation has been shown in Fig 5. ISE criterion has been considered as the performance index of system.

Table I COMPARATIVE ANALYSIS OF DIFFERENT ARTIFICIAL INTELLIGENCE TECHNIQUE APPLIED TO TUNING OF AGC STRATEGY POWER SYSTEM INCORPORATING AND EXCLUDING SMES

		PSO Tuned AGC with SMES	GA Tuned AGC with SMES	PSO Tuned AGC without SMES
Settling Time (s)	ΔF_1	30	46	71
	ΔF_2	41.7	44	62
	ΔF_3	30	46	60
	ΔP_{12}	46	66	73
	ΔP_{13}	46	58	74
	ΔP_{23}	5.9	44	27
Peak Overshoot Value×10⁻³	ΔF_1	5	5.1	37
	ΔF_2	5.6	5.7	78
	ΔF_3	5.3	5.3	42
	ΔP_{12}	0.7	0.7	10.2
	ΔP_{13}	0.7	0.8	9
	ΔP_{23}	0.2	0.5	1.9
P e a	ΔF_1	0.21	0.21	2.61
	ΔF_2	0.51	0.66	1.61

ΔF_3	0.21	0.21	1.12
ΔP_{12}	1.87	10.3	2.61
ΔP_{13}	2.55	2.54	2.77
ΔP_{23}	1.24	2.70	1.95

The dynamic performance of the system has been evaluated on the basis of frequency responses, settling time, peak overshoot as shown in Table I. The value of performance index for PSO tuned AGC with SMES is 0.000038 which is smaller than value of GA tuned SMES with SMES is 0.000057 and PSO tuned AGC without SMES is 0.00054. It has been investigated that PSO based AGC strategy with SMES is better than the GA based AGC strategy with SMES and PSO based AGC without SMES. Therefore, PSO tuned AGC strategy has been used to study system dynamic performances on 1% step load perturbation for different control areas.

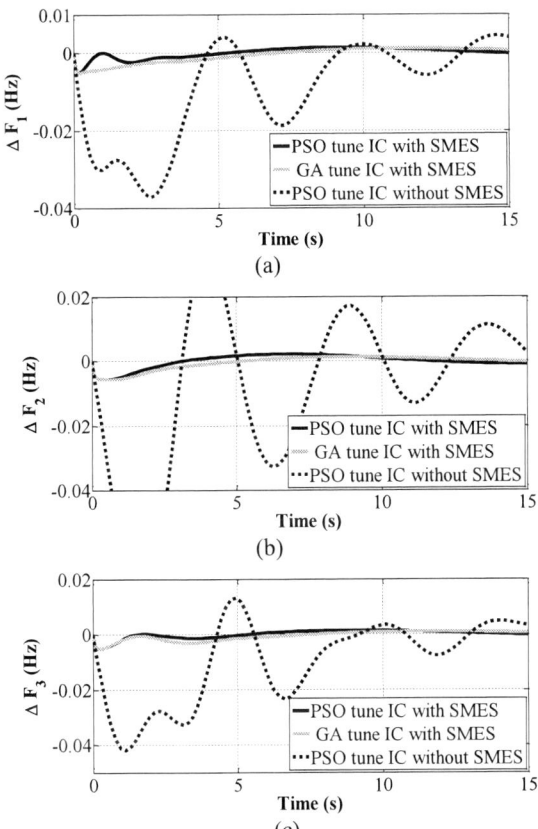

Fig.5. Frequency responses with and without SMES , Frequency deviation responses of (a) area-1 (b) area-2 (c) area-3, on 1% step load perturbation in the control area-1

Further, to demonstrate the effect of SMES for considered power system model, comparative analysis has been done for 1% step load perturbation in different control areas. As the multi area power system have been interconnected, the dynamics of other areas will be effected if there is disturbance in any of the area.

Therefore, the frequency responses of individual power system give transient even if there is no load perturbation in that control area.

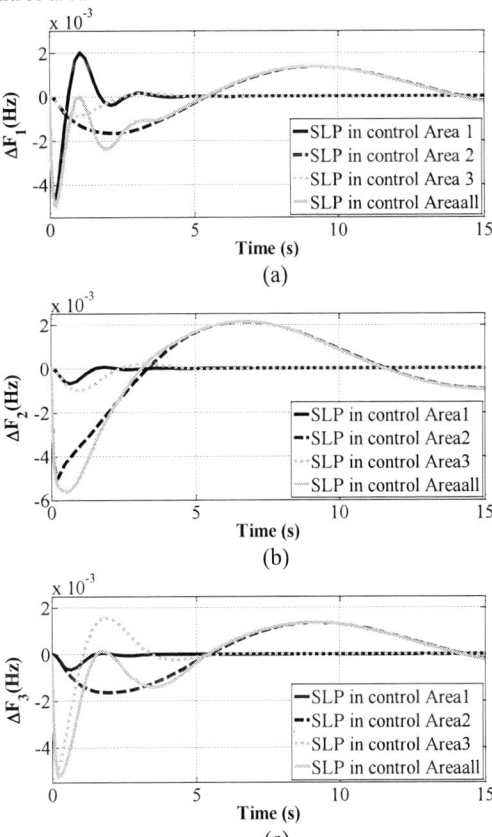

Fig.6. Frequency responses with SMES , Frequency deviation responses of (a) area-1 (b) area-2 (c) area-3, on 1% step load perturbation in different control areas

Table II COMPARATIVE STUDY OF AGC STRATEGY OF MULTI AREA THERMAL-HYDRO-GAS POWER SYSTEM INCORPORATING SMES ON 1% STEP LOAD PERTURBATION AT DIFFERENT CONTROL AREAS

	Settling Time (s)						Peak Overshoot Value ×(-)10⁻³ (Hz)						Peak Time Value (s)						Performance Index Value ×10⁻⁵
	ΔF_1	ΔF_2	ΔF_3	ΔP_{12}	ΔP_{13}	ΔP_{23}	ΔF_1	ΔF_2	ΔF_3	ΔP_{12}	ΔP_{13}	ΔP_{23}	ΔF_1	ΔF_2	ΔF_3	ΔP_{12}	ΔP_{13}	ΔP_{23}	ISE
Area1	3.42	3.24	3.24	3.24	48	3.3	4.8	0.68	0.65	0.23	0.006	0.23	0.15	0.68	0.68	0.52	1.5	0.52	0.18
Area2	47	42	47	47	47	36	1.7	5	1.6	0.69	0.69	0.006	2.07	0.22	1.91	2.26	2.26	3.91	3.3
Area3	4.7	4.7	5.1	48	6.4	5.8	0.89	0.99	4.9	0.01	0.34	0.35	0.88	0.88	0.15	1.9	0.88	0.88	0.34

Area all	30	41.7	30	46	46	5.9	5	5.6	5.3	0.72	0.73	0.27	0.21	0.51	0.21	1.87	2.55	1.24	3.9

The graph shown in Fig.6 represent frequency responses of the PSO tuned AGC strategy of multi area thermal-hydro-gas power system with SMES on 1% step load perturbation in different control areas. The investigation of Table II reveals that settling time, peak time, peak overshoot and performance index values of frequency and tie-line power for 1% step load perturbation in area 1 or area 3 are remarkably better when disturbance occurs at thermal or gas power system. However, the participation of hydro power system in multi area thermal-hydro-gas power system degrades the system response and makes it sluggish, which is highly suppressed by the application of SMES in AGC strategy.

VI. CONCLUSIONS

A maiden attempt has been made to study the effect of SMES in PSO tuned AGC strategy of multi area multi source power system on 1% step load perturbation in different control areas. In an interconnected multi area multi source power system, the control area 1 is non-reheat thermal power system, area 2 is hydro power system and area 3 is gas power system. The dynamic stability of system is illustrated by comparing frequency responses, settling time, peak overshoot and performance index value of the system. At first, the tuning of Integral controller (IC) with different intelligent techniques in AGC strategy of three identical capacity control areas incorporating and excluding SMES on 1% step load perturbation in each area are studied. Upon comparison of the result, it has been concluded that PSO tuned AGC for three area interconnected thermal-hydro-gas power system incorporating SMES provides better dynamic performances in terms of frequency response and least values of settling time, peak overshoot and performance index when compared to GA tuned AGC with SMES or PSO tuned AGC without SMES. The critical examination of the frequency performances reveals that the participation of hydro source in multi area multi source power system adversely affects dynamic stability of system. However, the sluggishness of hydro power system is significantly improved by the use of SMES in AGC strategy.

Moreover, the effect of SMES on PSO tuned AGC strategy incorporating SMES with 1% step increase of load on different control areas has been analyzed. The investigation of this analysis reveals that the system dynamic stability offered by load perturbation applied in thermal or gas as control area gives better result than load perturbation in hydro as a control area. In general, it has been observed that the dynamic performances of the system gets better with the incorporation of SMES in PSO tuned AGC strategy.

APPENDIX A

NOMENCLATURE

ΔF_i	System frequency deviation of ith area
R_i	Speed regulation of generator of ith area
P_{ri}	Rated power system capacity of ith area
T_g	Time constant of speed governor
T_t	Time constant of thermal turbine
T_P	Power system Time constant
T_W	Water starting turbine for hydraulic turbine
K_P	Power system gain
ΔP_{Gi}	Change in generated power of ith area
ΔP_{Di}	Change in load demand of ith area
ΔP_{Ci}	Change in speed governor of ith area
ΔX_{RH}	Incremental change in mechanical governor
ΔP_{tie}	Tie-line power deviation
T_{ij}	Synchronizing coefficient for tie-line
IC_i	Integral controller gain constant of ith area
ACE_i	Area Controller Error of ith area
b_i	Bias constant of ith area
T_s	Sampling period
s	Laplace operator
ΔX_E	Incremental change in governor valve position change.
ΔX_h	Incremental change in hydraulic governor
ΔP_{FC}	Incremental change in Fuel system and combustor
ΔP_V	Incremental change in Gas valve position
ΔX_g	Incremental change in Gas turbine speed governor

APPENDIX B - SYSTEM DATA

General Power System	Thermal Power System	Hydro Power System	Gas Power System	SMES
Pri=2000MW, f=60Hz, R_i= 2.4Hz/p.u. MW, b_i=0.425p.u.Mw/Hz, Kp=120, T_P=20s, P_{tie}=200MW	T_t=0.3s, T_g=0.08s	T_{RH}=41.6s, T_R=5s, T_{GH}=0.51s, T_W=1s	X=0.6, Y=1, b=0.05, c=1, T_F=0.23s, T_{CR}=0.01s, T_{CD}=0.2s	30MJ, L=2.65H, Tdc=0.03s, Ko=100kV/unit KW, I_{d0}= 4.5kA

APPENDIX C - PARAMETERS OF TUNED IC

	SMES WITH PSO TUNED IC	SMES WITH GA TUNED IC	WITHOUT SMES WITH PSO TUNED IC
IC1	17	0.9	0.4
IC2	15.5	12	0.3
IC3	8.8	11	0.1

REFERENCES

[1] IEEE Committee Report, "IEEE Standard Definition of Terms for Automatic Generation Control of Electric Power Systems", IEEE Trans. Power Apparatus and Systems, vol. PAS-89, Jul. 1970, pp. 1358-1364.

[2] J. Carpentier, "State of the art review" "To be or not to be modern, that is the question for automatic generation control point of view of a utility engineer", Elect. Press & Energy System,1985, 7, pp. 81-91.

[3] C. Concordia and L.K. Kirchmayer, "Tie line power and frequency control of electric power system: part-II," AISE Trans, IIIA, vol. 73, pp. 133-146, Apr. 1954.

[4] Nathan Cohn,"Some Aspects of Tie-Line Bias Control on Interconnected Power Systems", AIEE Trans. , vol. 75, Feb. 1957, pp. 1415-1436.

[5] Kumar, Prabhat, and Dwarka P. Kothari. "Recent philosophies of automatic generation control strategies in power systems." IEEE transactions on power systems 20, no. 1 (2005): 346-357.

[6] C. E. Fosha and O. I. Elgerd, "The Megawatt Frequency Control problem: A New Approach via Optimal Control Theory", IEEE Trans. on Power Apparatus and Systems, vol. PAS-89, no. 4, Apr. 1970, pp. 563-574.

[7] P. Kundur, "Power System Stability & Control," McGraw-Hill, New York, 1994, pp. 418-448.

[8] M. L. Kothari, B. L. Kaul, J. Nanda, "Automatic Generation Control of Hydro-Thermal System", Journals of Institute of Engineers (India), pt. EL-2, vol. 61, Oct. 1980, pp. 85-91.

[9] K.S.S. Ramakrishna amd T.S. Bhatti, "Automatic generation control of single area power system with multi-source power generation", Proc. IMechE vol. 222 Part A: J Power Energy,2008,pp.1-11.

[10] K.K.Challa and P.S.N.Rao" Analysis and design of controller for two area thermal-hydro-gas AGC System" Power Electronics Drives and Energy Systems (PEDES), 2010 Power India , pp. 1-4.

[11] K.S.S. Ramakrishna amd T.S. Bhatti, "Automatic generation control of single area power system with multi-source power generation", Proc. IMechE vol. 222 Part A: J Power Energy,2008,pp.1-11.

[12] Nasiruddin, Ibraheem, Terlochan S. Bhatti, and Nizamuddin Hakimuddin. "Automatic generation control in an interconnected power system incorporating diverse source power plants using bacteria foraging optimization technique." Electric Power Components and Systems 43, no. 2 (2015): 189-199.

[13] Ali, Mohd Hasan, Bin Wu, and Roger A. Dougal. "An overview of SMES applications in power and energy systems." *IEEE Transactions on Sustainable Energy* 1, no. 1 (2010): 38-47.

[14] Tripathy, S. C., M. Kalantar, and R. Balasubramanian. "Dynamics and stability of wind and diesel turbine generators with superconducting magnetic energy storage unit on an isolated power system." IEEE Transactions on Energy Conversion 6, no. 4 (1991): 579-585.

[15] Dash, Puja, Lalit Chandra Saikia, and Nidul Sinha. "Comparison of performances of several Cuckoo search algorithm based 2DOF controllers in AGC of multi-area thermal system." *International Journal of Electrical Power & Energy Systems* 55 (2014): 429-436.

[16] Abraham, Rajesh Joseph, D. Das, and Amit Patra. "Automatic generation control of an interconnected hydrothermal power system considering superconducting magnetic energy storage." *International Journal of Electrical Power & Energy Systems* 29, no. 8 (2007): 571-579.

[17] Dash, Puja, Lalit Chandra Saikia, and Nidul Sinha. "Automatic generation control of multi area thermal system using Bat algorithm optimized PD–PID cascade controller." International Journal of Electrical Power & Energy Systems 68 (2015): 364-372.

[18] Sahu, Rabindra Kumar, Sidhartha Panda, and Saroj Padhan. "A hybrid firefly algorithm and pattern search technique for automatic generation control of multi area power systems." *International Journal of Electrical Power & Energy Systems* 64 (2015): 9-23.

[19] Bhatt, Praghnesh, Ranjit Roy, and S. P. Ghoshal. "Comparative performance evaluation of SMES–SMES, TCPS–SMES and SSSC–SMES controllers in automatic generation control for a two-area hydro–hydro system." *International Journal of Electrical Power & Energy Systems* 33, no. 10 (2011): 1585-1597.

[20] J. Kennedy and R. Eberhart, "Particle swarm optimization" , Proc. 1995 IEEE Int. Conf. Neural Netw. , vol. IV , pp.1942 -1948 .

[21] P. Angeline, "Using selection to improve particle swarm optimization", Proc. IEEE Int. Conf. Evolutionary Computation , pp.84 -89 , 1998.

[22] J.B. Park, K.S. Lee, J.R. Shin, and K. Y. Lee, "A particle swarm optimization for economic dispatch with non-smooth cost functions" , IEEE Trans. Power Syst. , vol. 20 , no. 1 , pp.34 -42 , 2005.

[23] Lal, Deepak Kumar, and A. K. Barisal. "Comparative performances evaluation of FACTS devices on AGC with diverse sources of energy generation and SMES." Cogent Engineering 4, no. 1 (2017): 1318466.

978-1-5386-6624-1/18 $31.00 © 2018 IEEE

Bird Swarm Algorithm for Solving Multi-Objective Optimal Power Flow Problem

Saket Gupta, Narendra Kumar, Laxmi Srivastava[2]

Electrical Engineering Department
Delhi Technological University, Delhi, India
MITS, Gwalior, India
sguptamits@gmail.com[1], dnk_1963@yahoo.com[1], laxmigwl@gmail.com[2]

Abstract—**This paper applies bird swarm algorithm (BSA) to finding the solution of optimal power flow (OPF) problem. BSA is a recently developed bio-inspired evolutionary algorithm. It uses swarm intelligence derived from the social interactions and social behaviors in bird swarms for searching global optimal solution. The purpose of solving an OPF problem is to find the steady state operating point of a given network that optimizes a certain objective function. The BSA has been applied to carry out OPF for minimization of fuel cost, improvement of voltage profile, total emission cost minimization and power losses minimization. In order to show the efficacy of the proposed BSA, the standard IEEE 30-bus test systems has been selected to solve OPF problem with above mentioned objectives. The comparison of results obtained using BSA and other evolutionary computing based methods reported in the literature. It is clearly show that the proposed BSA based technique gives better result compare to other EC based techniques when solving the optimal power flow problem.**

Keywords—*Bird swarm algorithm; metaheuristic; optimal power flow; fuel cost minimization; voltage profile improvement; total emission cost minimization; minimization of power losses*

I. INTRODUCTION

The motive of solving an optimal power flow (OPF) problem is to determine the optimal setting of control variables in a given power system network that optimize one or more objective functions. The objective functions to carry out OPF are minimizes generation fuel cost, improvement of voltage profile, minimize real power losses and minimize emission cost via optimal adjustment of the power system independent variables, while at the same time it takes care of various network physical operating constraint. The control variable used for OPF problem are generator bus voltages, generators active power outputs (except slack bus), transformer tap-settings, phase shifters and other source of reactive power such as shunt capacitor or some shunt FACTS controllers. Some of them are discrete (e.g. reactive injections, phase shifters, and transformer tap setting) and others are continuous (e.g. generator real power outputs and generator voltages). Due to presence of the discrete nature of the variables providing challenge to optimization technique and makes OPF problem become a non-convex one. The power system equality and inequality constraint such as generator constraints shunt VAR constraints, transformer constraints, line flows and bus voltages

are effectively handled in OPF problem by implementing penalty factor approach. OPF problem formulation yields a highly non-linear, multi-modal, non-convex, non-differential objective function having discrete and continuous control variables and it has been introduced by Carpentier in the early1960's[1].

In the early decades, many classical method used to solved optimal power flow problem[2, 3]. Those are Linear Programming, Interior Point Method, Mixed Integer Programming, Non-Linear Programming, Quadratic Programming and Newton-Based Methods. Most of the classical methods have used slop of objective function with respect to optimization variables (gradient based/ derivative) and move the optimal solution. Therefore most of the time, they stuck local minima that prevents the algorithm for searching actual global optimal solution. They are facing the problem to deal with system having non convex, non-differentiable, multi-modal optimization function and constraint. Also, conventional optimization method required initial point so the conventional methods are not appropriate for solving optimal power flow problem. In order to overcome these drawbacks and handle such difficulties Evolutionary Computing (EC) technique comes into being as an alternative to the classical optimization methods. These techniques have been successfully applied to non-differentiable, non-smooth and non-convex optimization problem. The rapid developments of various branch of optimization method, EC methods have invited many researchers to apply their applications in solving OPF problem[4, 5] and other field of optimizations.

In recent years, various EC based techniques have been implemented to solve OPF problem. Some of these techniques are Dragonfly algorithm (DA) [6], black-hole-based optimization (BHBO) [7],Linear Adaptive Genetic Algorithm (LAGA) [8], Adaptive group search optimization algorithm (AGSOA) [9], Particle swarm optimization (PSO) [10], novel Sine-Cosine algorithm (NSCA) [11], Chaotic krill herd algorithm (CKHA) [12], Refined Genetic Algorithm (RGA) [13], Glowworm Swarm Optimization (GSO) [14], Tree-seed algorithm[15], Interior search algorithm (ISA) [16], Improved Differential Evolution Based approach (IDE) [17]and many others.

BSA has been successfully applied with other field of optimization application. Some of those are: an Improved Particle Filter in [18], author proposed Bird Swarm Algorithm with considered Particle Filter is well suited for strong nonlinear and non-Gaussian noise problem. In[19], authors proposed edge-based target detection for unmanned aerial vehicles using competitive Bird Swarm Algorithm. For optimization Benchmark Functions [20] authors present Bird Swarm Algorithm. To validate the effectiveness of the proposed method authors have been applied on various benchmark functions and economic load dispatch (ELD) problem. In [21], Bird Swarm Algorithm has been successfully applied for combinatorial testing data Generation. Combinatorial test data generation is a research hotspot in combination testing and many more.

The Bird Search algorithm was introduced by Xian-Bing Meng et.al in 2016. It is a population based bio-inspired algorithm. It works on the swarm intelligence and social interaction among the bird swarm. To validate and effectiveness of the BS algorithm has been applied on IEEE30-bus system with the different objectives function such as minimizing fuel cost and improving voltage profile, minimize real power losses and minimize emission cost.

The rest of the paper is structured as follows. In Section-2 optimal power flow problem modeling is described. The Section-3 presents the proposed bird swarm algorithm for solving OPF problem. In section-4 the numerical results are obtained and analyzed on the standard IEEE-30 bus system. Finally, conclusion is drawn in Section 5.

II.   PROBLEM FORMULATION

In general OPF problem can be mathematically written as follow:

Mathematically,

Min $F(a, b)$ (1)

Subject to: $\begin{cases} G(a,b) = 0 \\ H_l \leq H(a,b) \leq H_u \end{cases}$ (2)

Where,

F(a, b): objective function,

G(a, b), h(a, b) : equality and inequality constraints,

a,b: set of dependent and control variables,

H_l, H_u: lower and upper bounds of inequality constraint

Here, a can be expressed as

$$a = [P_{g_1}, V_1 \dots V_{NLB}, Q_{g1}, \dots Q_{gNG}, S_1, \dots S_{Ntl}]$$

while, b can be expressed as

$$b = [P_{g_2} \dots P_{g_{NGN}}, V_{g_1} \dots V_{g_{NGN}}, Q_{C_1} \dots Q_{C_{NC}}, T_1 \dots T_{NTR}]$$

Where, NLB, NGN: number of load and generator buses
Ntl: number of transmission lines
NC, NTR: numbers of shunt compensation and transformer

A. Constraints

The optimal power flow problem has two types of constraints namely equality and inequality constraints, as given below.

a) Equality Constraints

These constraints are the power balance equations and can be divided into real power and reactive power static load flow equations as:

$$P_{gi} - P_{di} - V_i \sum_{j=1}^{NB} V_j \left[G_{ij} \cos(\theta_{ij}) + B_{ij} \sin(\theta_{ij}) \right] = 0 \quad (3)$$

$$Q_{gi} - Q_{di} - V_i \sum_{j=1}^{NB} V_j \left[G_{ij} \sin(\theta_{ij}) - B_{ij} \cos(\theta_{ij}) \right] = 0 \quad (4)$$

Where, $\theta_{ij} = \theta_i - \theta_j$,

V_i, V_j: voltage magnitudes of bus i and bus j, respectively

P_{gi}, Q_{gi}: active and reactive power generation respectively at bus i,

NB is the number of buses,

P_{di}, Q_{di}: active and reactive load demand at bus i,

G_{ij}, B_{ij}: conductance and susceptance between bus i and bus j, respectively

b) Inequality Constraints

These constraints can be categorized into four types, namely, Generation constraints, shunt VAR compensation constraints, transformer constraints and security constraints.

- Generator Constraints:

$$V_{g_k}^{min} \leq V_{g_k} \leq V_{g_k}^{max} \quad , k = 1 \dots \dots NGN \quad (5)$$
$$P_{g_k}^{min} \leq P_{g_k} \leq P_{g_k}^{max} \quad , k = 1 \dots \dots NGN \quad (6)$$
$$Q_{g_k}^{min} \leq Q_{g_k} \leq Q_{g_k}^{max} \quad , k = 1 \dots \dots NGN \quad (7)$$

- Transformer Constraints:

$$T_k^{min} \leq T_k \leq T_k^{max} \quad k = 1 \dots \dots NTR \quad (8)$$

- Shunt VAR compensator constraints:

$$Q_{C_k}^{min} \leq Q_{C_k} \leq Q_{C_k}^{max} \quad k = 1 \dots \dots NGN \quad (9)$$

- Security Constraints:

$$V_{L_k}^{min} \leq V_{L_k} \leq V_{L_k}^{max} \quad k = 1 \dots \dots NLB \quad (10)$$

$$S_{l_k} \leq S_{l_k}^{max} \quad k = 1 \dots \dots ntl \quad (11)$$

The corresponding lower and upper limits are represented by scripts "min" and "max" in (5) to (11), respectively.

To obtain feasible solution, the inequalities are included into the objective function and the extended objective function can be formulated as:

$$F_{aug} = F + A_1 \cdot H(x_i) + A_2 \sum_{i=1}^{NGN} H(Q_{G_i}) + A_3 \sum_{i=1}^{NLB} H(V_{L_i})$$
$$+ A_4 \sum_{i=1}^{ntl} H(S_{l_i}) \quad (12)$$

Where A_1, A_2, A_3 and A_4 are the penalty factors corresponding to limit violations, and selected by the user.

B. Objective Function

The paper considers four objective functions for solving OPF problem. The objective functions are as follows:

1) Minimization of total Fuel cost

The first objective function $F_1(a, b)$ is represented as the summation of total fuel cost. It is mathematically expressed as below:

$$F_1(a,b) = \sum_{i=1}^{NG} f_i(P_{G_i})(\$/h) \tag{13}$$

These generators fuel cost characteristics are represented by the quadratic function as:

$$f_i(P_{G_i}) = C_i + B_i P_{G_i} + A_i P_{G_i}^2 (\$/h) \tag{14}$$

Where A_i, B_i and C_i are the fuel cost coefficients of the ith generating unit.

2) Voltage profile improvement

The second objective function $F_2(a, b)$ deals with the minimization of fuel cost and enrichment of voltage profile. It has been considered using two-fold objective function as given below:

$$F_2(a,b) = \sum_{i=1}^{NGN} f_i(.)\$/h + \alpha_{VD} \sum_{i \in NL} |V_i - 1| \tag{15}$$

Where α_{VD} is a weight factor and it is to be selected by the user.

3) Total emission cost minimization

The present case, two types of emission gasses mainly the oxides of SO_x (Sulphur) and NO_x (nitrogen) are taken as the main pollutant. The emission gasses are considered as combination of quadratic and an exponential function of the generator active power output. The total emission cost is defined as below:

$$F_3(a,b) = \sum_{i=1}^{NG} \alpha_i + \beta_i P_{G_i} + \gamma_i P_{G_i}^2 + \varepsilon_i \exp(\lambda_i P_{G_i}) \text{ ton/hr} \tag{16}$$

Where α_i, β_i, γ_i, ε_i and λ_i are the emission coefficients of the ith unit.

4) Minimization of real power losses

The system real power losses P_{loss} can be considered as follows:

$$F_4(a,b) = \sum_{i=1}^{NG} f_i(.) + \alpha_{PL} \sum_{k=1}^{NT} G_k |v_i^2 + v_j^2 + 2|V_i||V_j| \cos(\delta_i - \delta_j)| \tag{17}$$

Where α_{PL} as α_{VD} is a scaling factor, to be selected by user and G_k the conductance of kth line is connected between ith and jth buses: NT is the number of transmission lines: V_i are the voltage magnitudes at bus i: V_j are the magnitudes at bus j: δ_i is the voltage angles at bus i: δ_j is the voltage angles at bus j.

III. BIRD SWARM ALGORITHMS

Bird swarm algorithm is a recently developed bio-inspired algorithm. It is based on the swarm intelligence derived from the social interactions and social behaviors in bird swarms. Three kinds of behaviors are mainly possessed by all birds. Those are foraging behavior, vigilance behavior and flight behavior. For survival and searching good forage, birds use their own sense and move in flock. Each bird shares its experience with rest of birds in a flock. If it finds food, it frequently raises its head and scans near surrounding area which is called as vigilance behavior. All birds have competition to achieve their position at the center of the flock

because of more chances of being attacked at the periphery of the group. Birds may move from one location to another in order to escape from the hunter and to find their food. There are two different breeding group in flock birds exist namely producers and scroungers. The producers are actively searching their food and scroungers are those who follow the producers for their food.

All N virtual birds, depicted by their position $x_i^t (i \in [1, ... N])$ time step t, forage for food and fly in a D-dimensional space. The main steps used in bird swarm algorithm can be described as follows[22]:

3.1 Foraging behavior

All birds search for food as per their own and swarm experience. The mathematical expression of the foraging behavior can be written as below:

$$x_{i,j}^{t+1} = x_{i,j}^t + (p_{i,j} - x_{i,j}^t) \times C \times rand(0,1) + (g_j - x_{i,j}^t) \times S \, rand(0,1) \tag{18}$$

Where $(j \in [1, ... D])$, rand $(0, 1)$ represent a random number. It is uniformly distributed between $(0, 1)$.

S and C are two positive constraints, which can be respectively called as social and cognitive accelerated coefficients.

$p_{i,j}$ is represent the best j position of the bird b_i

g_j is the best previous position shared by the swarm.

2.2.2 Vigilance behavior

Each bird tries to attain the position at the center of the swarm to save themselves from the predators, and they inevitably compete with each other. Thus, each bird would not directly move towards the center of the swarm. These motions can be formulated as given below:

$$x_{i,j}^{t+1} = x_{i,j}^t + A1(mean_j - x_{i,j}^t) \times rand(0,1) + A2(p_{k,j} - x_{i,j}^t) \times rand(-1,1) \tag{19}$$

$$A1 = a1 \times \exp(-\frac{pFit_i}{sumFit+\epsilon} \times N) \tag{20}$$

$$A2 = a2 \times \exp((\frac{pFit_i - pFit_k}{|pFit_k - pFit_i + \epsilon|}) \frac{N \times pFit_k}{sumFit+\epsilon} \tag{21}$$

Where, k is a random positive integer, its value selected between 1 and N but not equal to i.

Both a1 and a2 are positive constants between 0 and 2.

$pFit_i$ is the best fitness value of the bird b_i and sumFit denote the sum of the best fitness value of the whole swarms.

ϵ is the smallest constraint in the computer which is used to avoid zero-division error.

$mean_j$ represent the jth element of the average position of the whole swarm.

2.2.3 Flight behavior

Birds fly to another location for searching the food and save themselves against the predators attack. A bird can be a producer or a scrounger and their behaviors can be expressed as follows respectively:

$$x_{i,j}^{t+1} = x_{i,j}^t + rand(0,1) \times x_{i,j}^t \tag{22}$$

$$x_{i,j}^{t+1} = x_{i,j}^t + (x_{k,j}^t - x_{i,j}^t) \times \times FL \times rand(0,1) \tag{23}$$

978-1-5386-6624-1/18 $31.00 © 2018 IEEE

Where, *randn* (0,1) is express as Gaussian distributed random number with μ=0 (mean) and σ = 1(standard deviation),

k∈[1,….N], (k≠ i).

FL (FL∈[0,2]) is a following factor which means scrounger would follow producer to search their food. For simplicity, we assume that each bird move to one site to another site every FQ unit interval.

The solution algorithm for solving OPF using BSA algorithm can be summarized in following steps:

Input: N: birds or population size, P: the probability of foraging for food, FQ: frequency of bird's flight behavior, C, S, a1, a2, FL: five constant parameters, M: the maximum number of iteration and the control /decision variables (*D*= 24 here). The minimum and maximum value of control variable in vector form $X_{Min} = [X_1^{min} ….. X_D^{min}]$ and $X_{Max} = [X_1^{max} ….. X_D^{max}]$. Initialise the load flow data.
Set count G=0 and define algorithm parameters, Population (P): Initialise the population randomly with uniformly distributed amongst $[X_{Min}, X_{Max}]$. For each bird, run load flow (e.g. NRLF) program and evaluate the fitness of each bird.
While (G < M)
If (t % FQ ≠ 0)
For i = 1: N
If rand (0, 1) < P
Birds forage for food (Equation 18)
Else
Birds keep vigilance (Equation 19)
End if End for
Else
Divide the swarm into two parts: producers and scroungers.
For i=1: N
If *i* is a producer
Producing (Equation 22)
Else
Scrounging (Equation 23)
End if End for
End if Evaluate new solutions
If the new solutions are better than their previous ones, update them
Find the current best solution
G=G+1; End while
Output: the individual with the best objective function value in the population
Stop
The basic flowchart of the BSA is shown in Figure 1.

I. RESULTS AND DISCUSSION

In this section, standard IEEE 30 bus is used to evaluate the effectiveness of proposed Bird Swarm algorithm with different objective function.

For the IEEE 30-bus system under study, the control variables limit, line data, bus data along with their initial settings are taken from [23]. The standard IEEE 30 bus system consist of 41 transmission line, six generator at bus nos. 1, 2, 5, 8, 11 and 13, reactive power source (shunt VAR compensation) at buses 10, 12, 15, 17, 20, 21, 23, 24 and 29 and four transformers with off-nominal tap settings in line nos. 6–9, 6–10, 4–12 and 27–28. For this system, 30 runs were taken using BSA to solve the OPF problem with different objective functions and best result are given here.

A. Case 1:Total Fuel cost Minimization

The BSA has been run for the total fuel cost minimization with considering the cost characteristics of all generator units are assuming quadratic nature. The control variable setting and fuel cost found by BSA are given in Table 2. For case 1 fuel cost obtain was 800.6293. The comparison of the results obtains by BSA with other EC based method are given in Table 1. From Table 1, it is clear that least generating fuel cost found by BSA compared to other EC based methods, this demonstrates the potential of the proposed BSA to solve OPF problem. Fig. 1 displays the convergence characteristic BS algorithm for minimization of total fuel cost.

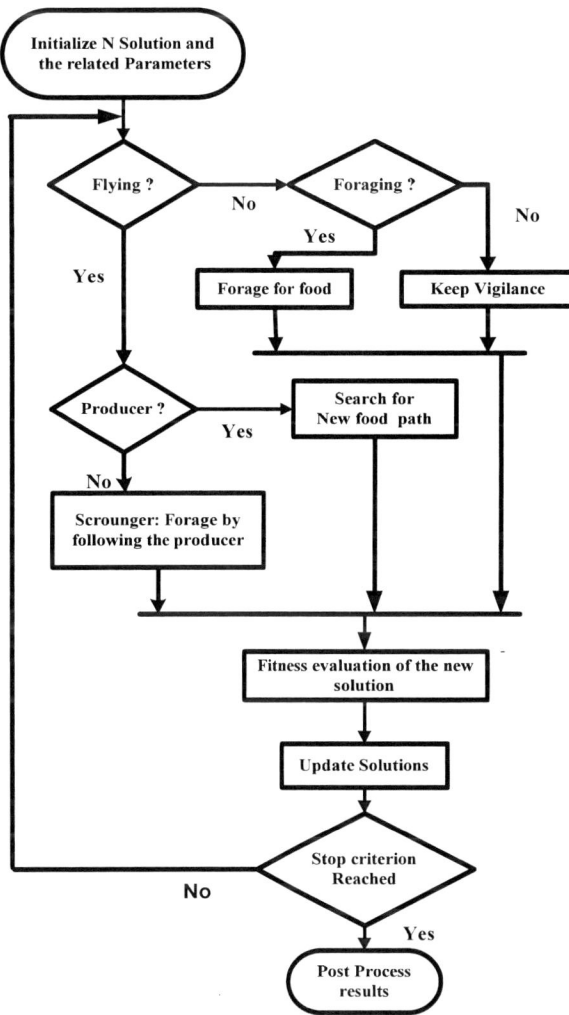

Fig. 1 Flowchart of Bird Swarm Algorithm

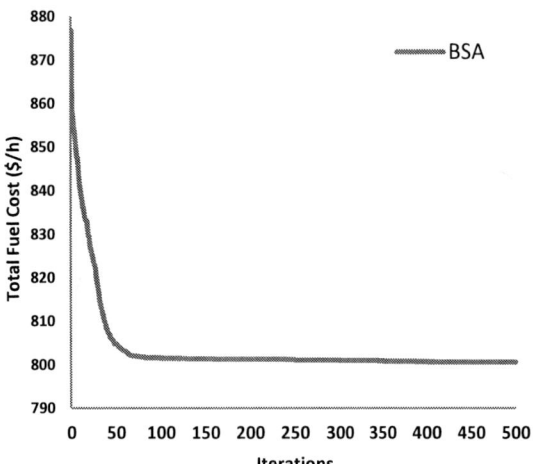

Fig.2 Convergence for fuel cost minimization

TABLE I.   COMPARISON OF RESULT: CASE 1

METHOD	FUEL COST
BSA	**800.6293**
BBO [24]	801.0562
GWO [25]	801.41
DE [25]	801.23
HSFLA-SA [26]	801.79
MSFLA [27]	802.287
SFLA [27]	802.5092
GPU-PSO [28]	800.53
ABC [29]	800.66
PSO [10]	800.41

B. Case 2:Voltage profile improvement

In this case, BSA find minimum total fuel cost and voltage deviation were 805.7171 $/h and 0.0786 p.u. respectively. The result obtained via BSA is compare with other method is present in Table 3. From the comparison of the result in table 3, it can be inferred that the total voltage deviation of 0.0786 p.u obtained by BSA, is 91.57% less than the compare to case 1 at the same time 0.6354% increased the fuel cost.The optimal setting of control variable is given in 3rd column of Table 2. Fig.3 is displays the total voltage deviation in case 1 and case 2.

C. Case 3: Emission cost minimization

For Case 3, only emission cost minimization has been selected as an objective function and it is defined in Eq. (16). The emission cost minimization obtains by proposed method was 0.2039. The comparison of emission cost minimization obtains by proposed algorithm and other methods are given in Table 4. It is found that the performance of proposed technique is superior as compared with the other algorithms available in the literature. In case 3 total emission cost is reduced to 0.2039 ton/h (39.07%) in comparison to 0.3347 ton/h in case 1. The control variable setting obtained by BSA is given in Table 2.

TABLE II.Optimum value of control variable for different case

ControlVariable	Case-1	Case -2	Case-3	Case-4
P_{G2}	0.48910	0.48970	0.66810	0.68934
P_{G5}	0.21410	0.21760	0.50000	0.50000
P_{G8}	0.21440	0.23530	0.35000	0.35000
P_{G11}	0.12300	0.12350	0.30000	0.30000
P_{G13}	0.11410	0.10490	0.40000	0.40000
V_{G1}	1.07910	1.02120	1.05410	1.03084
V_{G2}	1.05880	1.01220	1.05610	1.02600
V_{G5}	1.03140	1.01850	1.03880	1.01295
V_{G8}	1.03780	1.01150	1.04570	1.01590
V_{G11}	1.05130	0.98570	1.06760	1.01613
V_{G13}	1.04080	0.99830	1.00620	1.01860
T_{6-9}	1.12500	0.99690	1.06590	0.99407
T_{6-10}	0.90470	0.95860	1.02670	0.99401
T_{4-12}	0.98520	0.99000	1.01590	0.99310
T_{28-27}	0.97340	0.98090	1.05050	0.99658
Q_{C10}	0.07570	0.03830	0.04120	0.04143
Q_{C12}	0.05500	0.05990	0.03300	0.04022
Q_{C15}	0.02220	0.02250	0.04410	0.04109
Q_{C17}	0.04090	0.00690	0.04150	0.04018
Q_{C20}	0.02290	0.04110	0.04990	0.04060
Q_{C21}	0.06330	0.03320	0.05000	0.04133
Q_{C23}	0.01230	0.01850	0.04260	0.04256
Q_{C24}	0.03100	0.05410	0.05000	0.04236
Q_{C29}	0.01430	0.01820	0.04850	0.03584
Fuel cost($/h)	**800.6293**	805.7171	944.2058	947.4367
V.D (p.u)	0.9324	**0.0786**	0.6797	0.4724
Emission(ton\h)	0.3347	0.3342	**0.2039**	0.2039
P_{Loss}(M.W)	9.0549	10.4120	3.6573	**3.5517**

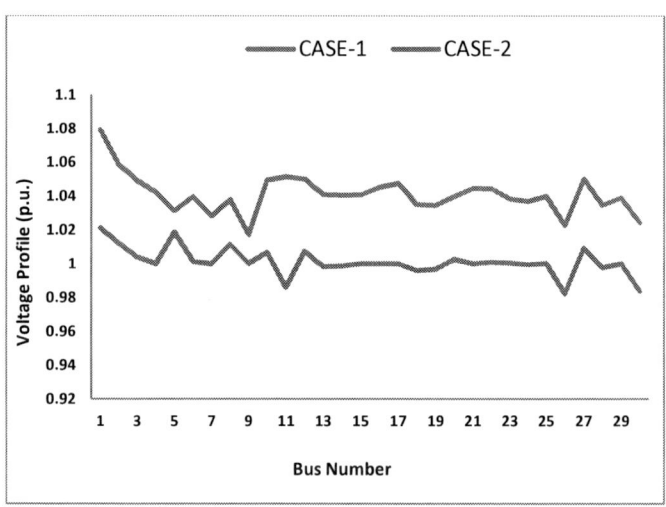

Fig.3 system voltage profile

TABLE III.COMPARISON OF RESULT: CASE 2

METHOD	FUEL COST	VOLTAGE DEVIATION
BS	805.7171	**0.0786**
MSA [30]	803.3125	0.10842
MPSO [30]	803.9787	0.1202
MDE [30]	803.2122	0.12646
MFO [30]	803.7911	0.10563
FPA [30]	803.6638	0.13659
DE [31]	805.2619	0.1357
PSO [10]	806.3800	0.0891

TABLE IV. COMPARISON OF RESULT: CASE 3

METHOD	ECOST
BSA	**0.2039**
MSA[30]	0.20482
MPSO[30]	0.23246
MDE[30]	0.20926
MFO[30]	0.20489
FPA[30]	0.20523
ARCBBO[24]	0.2048
ABC [29]	0.204826
MSFLA [27]	0.2247
SFLA [27]	0.2249

D. Case 4 Active power loss minimization

In case 4, BSA algorithm applied for minimization of active power loss for IEEE 30 bus test system. By applied of proposed BSA method, find the minimum active power loss was 3.5517 M.W.The results obtained BSAalgorithm and other EC based method are present in Table 5. From Table 5, it is clear that proposed method superior among compare to other EC method reported in this literature.

TABLE V. COMPARISON OF RESULT: CASE 4

METHOD	FUEL COST	Ploss
BSA	**947.4367**	**3.5517**
GWO [25]	968.38	3.41
DE [25]	968.23	3.38
MSA[30]	859.1915	4.5404
MPSO[30]	859.5841	4.5409
MDE[30]	868.7138	4.3891
MFO[30]	858.5812	4.5772
FPA[30]	855.2706	4.7981

CONCLUSION

In this paper, bird swarm algorithm (BSA) was suggested and applied to solve OPF problems with considering the different objective function such as to minimize of generation fuel cost, enhance the voltage profile, total emission cost minimization and minimization of real power losses.

The proposed BSA based algorithm has been successfully and effectively implemented in the IEEE 30-bus test systems. The numerically results obtained from the BSA method have been compared with other EC based methods in the literature. On comparison, this has been observed that BSA provides better results as compare with recently reported results in literature. This confirms the superiority and effectiveness of BSA algorithm over other EC based techniques to solve different cases of OPF problem with accurate and feasible results.

REFERENCES

[1] A. M. Shaheen, S. M. Farrag, and R. A. El-Sehiemy, "MOPF solution methodology," IET Gener. Transm. Distrib., vol. 11, no. 2, pp. 570–581, 2017.

[2] M. B. Maskar, "A Review on Optimal power flow problemand solution methodologies," International Conference on Data Management, Analytics and Innovation (ICDMAI) Zeal Education Society, Pune, India, Feb, 2017.

[3] J. a Momoh, "A Review of Selected Optimal Power Ftow Literature to 1993 Part I: NonLinear and Quadratic Programming Approaches IEEE Transactions on Power Systems, Vol. 14, No. 1, February 1999," Power, vol. 14, no. 1, pp. 96–104, 1999.

[4] M. NIU, C. WAN, and Z. XU, "A review on applications of heuristic optimization algorithms for optimal power flow in modern power systems," J. Mod. Power Syst. Clean Energy, vol. 2, no. 4, pp. 289–297, 2014.

[5] M. R. AlRashidi and M. E. El-Hawary, "Applications of computational intelligence techniques for solving the revived optimal power flow problem," Electr. Power Syst. Res., vol. 79, no. 4, pp. 694–702, 2009.

[6] T. K. Bashishtha, "Nature Inspired Meta-heuristic dragonfly Algorithms for Solving Optimal Power Flow Problem," Int. J. Electron. Electr. Comput. Syst., vol. 5, no. 5, pp. 111–120, 2016.

[7] H. R. E. H. Bouchekara, "Optimal power flow using black-hole-based optimization approach," Appl. Soft Comput. J., vol. 24, pp. 879–888, 2014.

[8] A. M. Abusorrah, "The Application of the Linear Adaptive Genetic Algorithm to Optimal Power Flow Problem," Arab. J. Sci. Eng., vol. 39, no. 6, pp. 4901–4909, 2014.

[9] N. Daryani, M. T. Hagh, and S. Teimourzadeh, "Adaptive group search optimization algorithm for multi-objective optimal power flow problem," vol. 38, pp. 1012–1024, 2016.

[10] M. A. Abido, "Optimal power flow using particle swarm optimization," Electrical Power and Energy Systems, vol. 24, pp. 563–571, 2002.

[11] A.F. Attia, R. A. El Sehiemy, and H. M. Hasanien, "Optimal power flow solution in power systems using a novel Sine-Cosine algorithm," Int. J. Electr. Power Energy Syst., vol. 99, no. July 2017, pp. 331–343, 2018.

[12] A. Mukherjee and V. Mukherjee, "Solution of optimal power flow using chaotic krill herd algorithm," Chaos, Solitons and Fractals, vol. 78, pp. 10–21, 2015.

[13] S. R. Paranjothi and K. Anburaja, "Optimal power flow using refined genetic algorithm," Electr. Power Components Syst., vol. 30, no. 10, pp. 1055–1063, 2002.

[14] S. Surender Reddy and C. Srinivasa Rathnam, "Optimal Power Flow using Glowworm Swarm Optimization," Int. J. Electr. Power Energy Syst., vol. 80, pp. 128–139, 2016.

[15] A. A. El-Fergany and H. M. Hasanien, "Tree-seed algorithm for solving optimal power flow problem in large-scale power systems incorporating validations and comparisons," Appl. Soft Comput. J., vol. 64, pp. 307–316, 2018.

[16] B. Bentouati, S. Chettih, and L. Chaib, "Interior search algorithm for optimal power flow with non-smooth cost functions," Cogent Eng., vol. 4, no. 1, pp. 1–17, 2017.

[17] D. Suganthi, S T, "An Improved Differential Evoluation Based Approach for Emission Constrained Optimal Power Flow," IEEE Conference 2013, pp. 1308–1314, 2013.

[18] L. Zhang, Q. Bao, W. Fan, K. Cui, H. Xu, and Y. Du, "An Improved Particle Filter Based on Bird Swarm Algorithm," 2017 10th Int. Symp. Comput. Intell. Des., pp. 198–203, 2017.

[19] X. Wang, Y. Deng, and H. Duan, "Edge-based target detection for unmanned aerial vehicles using competitive Bird Swarm Algorithm," Aerosp. Sci. Technol., vol. 78, pp. 708–720, 2018.

[20] M. Parashar, S. Rajput, H. M. Dubey, and M. Pandit, "Optimization of benchmark functions using a nature inspired bird swarm algorithm," 3rd IEEE Int. Conf., pp. 1–7, 2017.

[21] Y. Zhang, "Combinatorial Testing Data Generation Based on Bird Swarm Algorithm,"2nd International Conference on System Reliability and Safety pp. 491–499, 2017.

[22] X. B. Meng, X. Z. Gao, L. Lu, Y. Liu, and H. Zhang, "A new bio-inspired optimisation algorithm: Bird Swarm Algorithm," J. Exp. Theor. Artif. Intell., vol. 28, no. 4, pp. 673–687, 2016.

[23] K. Y. Lee, Y. M. Park, and J. L. Ortiz, "A United Approach to Optimal Real and Reactive Power Dispatch," IEEE Power App. ans Sys., vol. PAS-104, no. 5, pp. 1147–1153, May, 1985.

[24] A. R. Kumar and L. Premalatha, "Optimal power flow for a deregulated power system using adaptive real coded biogeography-

based optimization," Int. J. Electr. Power Energy Syst., vol. 73, pp. 393–399, 2015.

[25] A. A. El-Fergany and H. M. Hasanien, "Single and Multi-objective Optimal Power Flow Using Grey Wolf Optimizer and Differential Evolution Algorithms," Electr. Power Components Syst., vol. 43, no. 13, pp. 1548–1559, 2015.

[26] T. Niknam, M. R. Narimani, and R. Azizipanah-Abarghooee, "A new hybrid algorithm for optimal power flow considering prohibited zones and valve point effect," Energy Convers. Manag., vol. 58, pp. 197–206, 2012.

[27] T. Niknam, M. rasoul Narimani, M. Jabbari, and A. R. Malekpour, "A modified shuffle frog leaping algorithm for multi-objective optimal power flow," Energy, vol. 36, no. 11, pp. 6420–6432, 2011.

[28] V. Roberge, M. Tarbouchi, and F. Okou, "Optimal power flow based on parallel metaheuristics for graphics processing units," Electr. Power Syst. Res., vol. 140, pp. 344–353, 2016.

[29] M. Rezaei Adaryani and A. Karami, "Artificial bee colony algorithm for solving multi-objective optimal power flow problem," Int. J. Electr. Power Energy Syst., vol. 53, no. 1, pp. 219–230, 2013.

[30] A. A. A. Mohamed, Y. S. Mohamed, A. A. M. El-Gaafary, and A. M. Hemeida, "Optimal power flow using moth swarm algorithm," Electr. Power Syst. Res., vol. 142, pp. 190–206, 2017.

[31] A. A. Abou El Ela, M. A. Abido, and S. R. Spea, "Optimal power flow using differential evolution algorithm," Electr. Power Syst. Res., vol. 80, no. 7, pp. 878–885, 2010.

Design and Analysis of Renewable Energy Based Hybrid Model for RemoteApplications

P. Anand[a], A. H. Quadri[b], S. K. Bath[c], M. Rizwan[b], Narendra Kumar[b]

[a]Department of Electrical Engineering, I. K.Gujral Punjab Technical University, Kapurthala-144601,India
[b]Department of Electrical Engineering, Delhi Technological University, Delhi-110042, India [c]Department of Electrical Engineering, Giani Zail Singh Campus College of Engineering & Technology,, Bathinda-151001,Punjab, India
Email: [a]anand_priyanka10@yahoo.co.in, [b]arshadhusainquadri@gmail.com, [c]sjkbath77@gmail.com,
[b]rizwan@dce.ac.in, [b]dnk_1963@yahoo.com

Abstract— **India got independence in 1947, but after more than seven decades many of the villages are not either fully electrified or not connected with grid supply. Additionally, there are many rural households in the country still not have access to electricity. Further, In order to meet this challenge and fulfill the Government of India's mission "Power to All" by 2019, it is necessary to generate the electrical power by harnessing locally available green energy resources (GES).**

In the present study, design, techno-economical analysis and comparison of off grid and grid connected hybrid models comprising of GES viz. solar, biomass and biogas have been carried out to get feasible solution. Hybrid model comprising of solar, biomass, biogas along with battery in grid connected mode has been proposed for selected area due to least net present cost and cost of energy with minimum green house gas (GHG) emission.

Index Terms— Green energy sources (GES); Solar photovoltaic (SPV); biomass; biogas; green house gas (GHG).

I. INTRODUCTION

Rural development plays an important role to improve the social and economic status of rural people and electricity is one of most essential infrastructural input in development and growth of the Nation. In India, maximum population are residing in rural areas and the majority of them either getting limited grid power for eight to ten hours daily or do not have access of grid power at all. To provide electricity through diesel generator (DG) is not a feasible solution because of its high fuel and operation & maintenance cost and greenhouse gas emission. So, the power generation through GES like solar, wind, biomass, small hydro etc may be the best alternative solution to provide the energy security to rural people. Moreover, Government of India (GOI) is promoting people to use GES by giving various attractive schemes, incentives, etc. and also fixed a goal of adding up 175 GW power generations through RES by 2022 [1].

As on March 31, 2018, installed capacity of thermal power is 222907 MW i.e 65% of total capacity. 45293 MW (13%) of power is shared by large hydro and 6780 MW (2%) through Nuclear while 69022 MW (20%) of power is generated from RES respectively [2, 3]. The graphical representation of All India Installed capacity as on March 31, 2018 is depicted in Fig.1.

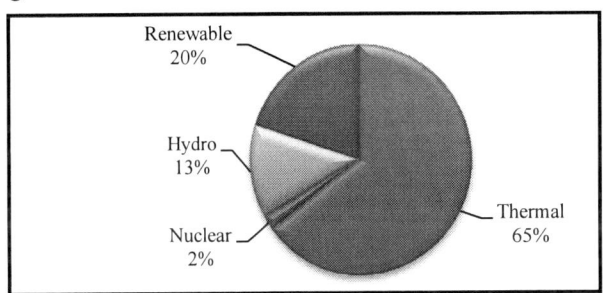

Fig.1: Installed Capacity of Electrical Power (MW) of India as on March 31, 2018

Further, the Pie- chart presentation of GES in Indian power sector as on March 31, 2018 is demonstrated in Fig.2.

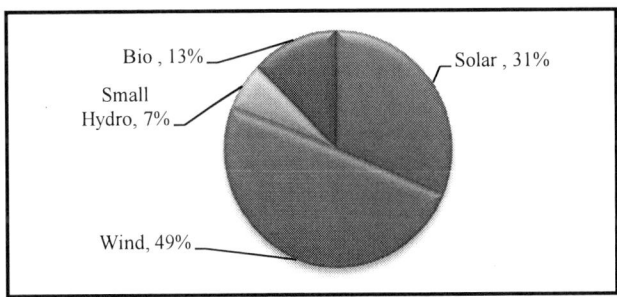

Fig.2: Contribution of GES in Indian Power Sector as on March 31, 2018

It is revealed from this chart that wind energy system is contributing half of the total renewable power whereas around one third power is shared by solar; however, small hydro and bio-mass/gas plants contribute as 7% and 13% of total power respectively.

However, due to intermittent nature of GES and cost effectiveness, integration of two or more energy sources called hybrid system is more reliable solution. Several hybrid systems using GES have been designed and developed for different locations by using various simulation and optimization software and techniques [4-29].

In light of above facts, main aim of the present study is to design and develop a hybrid power system by utilizing locally available GES viz. solar, biomass and biogas for rural villages located in Sonipat district, Haryana state, India. The proposed hybrid model will provide energy security to rural community. HOMER (Hybrid Optimization Model for Electrical Renewable) software has been employed to find most feasible system in terms of lowest net present cost and cost of energy. Additionally, techno-economic with GHG emission analysis has also been carried out.

II. HYBRID SYSTEM MODELLING METHODOLOGY

A systematic modelling methodology is used to design and develop a hybrid power system for group of rural villages in order to provide continuous and reliable power to the rural community. In the present study, the site identification, electrical hourly load estimation, GES potential evaluation and optimization methods are illustrated in the consecutive sections:

A. Description of study area

The four villages Khanpur, Kasanda, Kasandi and Sargathal situated in Tehsil Gohana of District Sonipat, Haryana (India) have been chosen in this proposed work. By carrying survey of this area, 533 un-electrified households have been found. Further, the general information of the selected area has been given in Table 1.

Table 1: Particulars of study area [30, 31]

Description	General particulars			
Country, State	India, Haryana			
District, Sub-division	Sonipat, Gohana			
Village	Khanpur Kalan	Kasanda	Kakana	Sargathal
Latitude	29.15° N	29.13° N	29.13° N	29.00° N
Longitude	76.75° E	76.83° E	76.80° E	77.01° E
Population	12544	2077	1455	3435
Total households	1987	399	261	634
Number of unelectrified households	424	64	24	21

B. Estimation of Electrical Hourly Load Demand

The electrical hourly load demand has been calculated considering present and future load demand of the rural community in this area. The load demand of the selected site has been classified as residential, commercial and community load. A residential load consists of LED, TV, fan, radio, mobile charger, water pump and heater. The commercial load includes shops and flour mill. School, veterinary hospital, health centre and street lights are taken in community load. In addition, two seasons' viz. summer and winter has been considered in the present study. The electrical load demand of selected area in summer and winter season has been estimated as 2658.85 kWh/day and 2183.86 kWh/day respectively. Further, the hourly load profile of both seasons has been shown in Fig.3 and Fig.4.

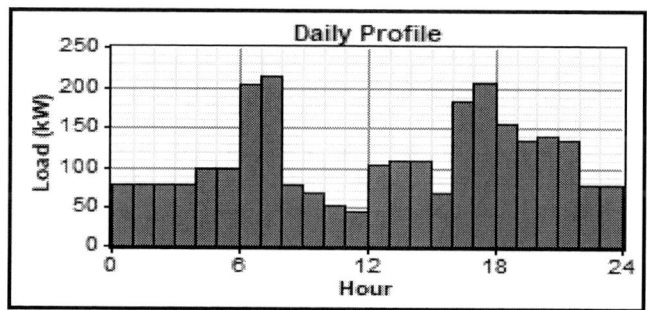

Fig. 3. Hourly load profile during summer season of selected site

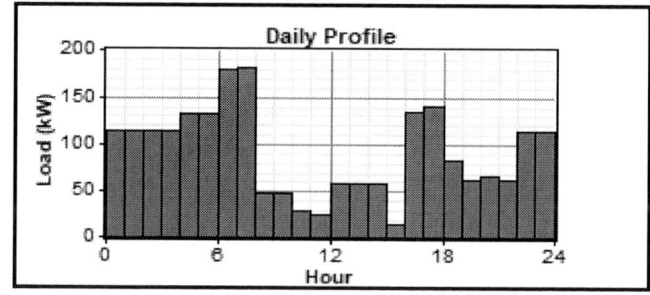

Fig.4. Hourly load profile during winter season of selected area

C. Potential Appraisal of Green Energy Resources

On the basis of extensive survey and data collection, the potential of GES viz. solar, biomass and biogas has been assessed and shown in Table 2.

Table 2: Potential of different GES at selected site

Sr. No.	Village name	Biomass (tons/year)	Biogas (m³/day)	Solar radiation (kWh/m²/day)
1	Khanpur Kalan	200.58	328.09	5.26
2	Kasanda	117.55	205.06	5.24
3	Kakana	58.02	123.04	5.24
4	Sargathal	94.04	164.05	5.14
Total		470.19	820.24	5.26

978-1-5386-6624-1/18 $31.00 © 2018 IEEE

A crop residue provides 470.19 tons/year biomass and 820.24 m³/day biogas from cattle dung has been computed in the proposed area. The average solar irradiance of 5.26 kWh/m²/day is available in the area. It has been found that a GES is available in the study area with huge potential that can be utilized to fulfil the energy requirements of the selected area.

In order to compute the power output of the solar photovoltaic (SPV) system (P_{PV} (t)), software uses the following formula [29]:

$$P_{PV} \ (t) = R_{PV} \times DF \times \frac{Q_{PV}(t)}{Q_{PV,STC}} \qquad (1)$$

Where, R_{PV} represents rated capacity of SPV array under standard test condition (STC); Q_{PV} (t) is radiation incident on SPV array in kW/m²; $Q_{PV,STC}$ represents the solar radiation at STC (1 kW/m²). DF depicts derating factor of SPV array to consider reduced output under real world conditions like shadow, dust, etc.

The output power from biomass generator system (P_M (t)) has been calculated by equation (2) [29]:

$$P_M(t) = \frac{Q_{AM} \times CV_M \times \eta_M \times 1000}{365 \times 860 \times H_M} \qquad (2)$$

Where, Q_{AM} depicts yearly available amount of biomass (tons/yr), CV_M represents calorific value of biomass in kcal/kg, η_M is conversion efficiency and H_M is operating hours per day of biomass generator system.

The power output (P_G (t)) of biogas generator system is computed as [29]

$$P_G(t) = \frac{Q_G \times F_G \times \eta_G}{860 \times H_G} \qquad (3)$$

Where, Q_G represents biogas availability per day (m³/d), F_G depicts calorific value of biogas (kcal/m³), η_G represents overall conversion efficiency from biogas to electrical power production. H_G is operating hours of biogas generator per day.

D. Simulation and Optimization Technique

In this case study, HOMER software is utilized for simulation and optimization purpose. Various possible configurations of the given model have been considered and out of which the best possible configurations is selected based on lowest NPC [32]. However, the technical and economical input parameters like size, cost, lifetime etc. of hybrid system components are given in Table 3.

III. SIMULATION RESULTS AND DISCUSSION

A hybrid model has been designed and analyzed in HOMER software for the proposed area by considering locally available GES, numerous input parameters and constraints etc. Further, three configurations in off grid mode are considered and compared based on NPC and COE. Finally, the best off grid configuration has been compared with grid connected and investigated most feasible.

Selected configurations in off grid mode are as follow:
    I. SPV/ Biogas /Biomass/Battery

    II. SPV/ Biogas /Battery
    III. SPV/Biomass/Battery

Further, the result of different configurations in off grid mode has been given in Table 4.

From Table 4, it is inferred that Hybrid model of SPV/biogas/ biomass with battery storage has smallest amount of NPC, COE and GHG emission.

Further, the best off-grid configuration has been compared with grid connected mode and it is found that grid connected hybrid model has lowest NPC and COE of $ 847365 and $ 0.114 respectively. The system rating and economical parameters of grid connected configuration is given in Table 5.

From Table 4 and Table 5, it is concluded that the grid connected hybrid model comprising of SPV/Biomass/Biogas/Battery has been proposed for the selected site. The proposed hybrid model is depicted in Fig.5. Further, the cash flow summery in terms of NPC and monthly average energy generation of proposed configuration is shown in Fig.6 and Fig.7.

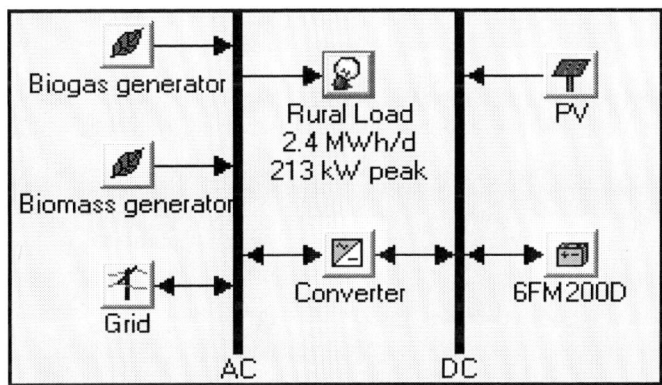

Fig.5. Schematic diagram of proposed hybrid SPV/ biogas/ biomass/battery system

IV. CONCLUSION

In the present investigation, optimal designing of hybrid model consisting of SPV/ biogas/ biomass/battery has been carried out for electrifying un-electrified households in order to get energy security with future needs of the rural areas of Haryana state in India. Further, different configurations of hybrid model in off grid and grid connected mode have also been presented and compared in view of techno-economic and greenhouse gas emissions analysis. The grid connected hybrid system comprising of 25 kW SPV, 40 kW biomass gasifier, and 150 kW biogas generator along with 72 kWh of battery has been proposed among all considered configurations at the study area due to smallest amount of NPC of $ 847365. The COE of the proposed system is 0.114 $/kWh with negligible GHG emissions. The proposed model would be very supportive in fulfilling the mission of rural electrification and "Power to all" in India.

978-1-5386-6624-1/18 $31.00 © 2018 IEEE

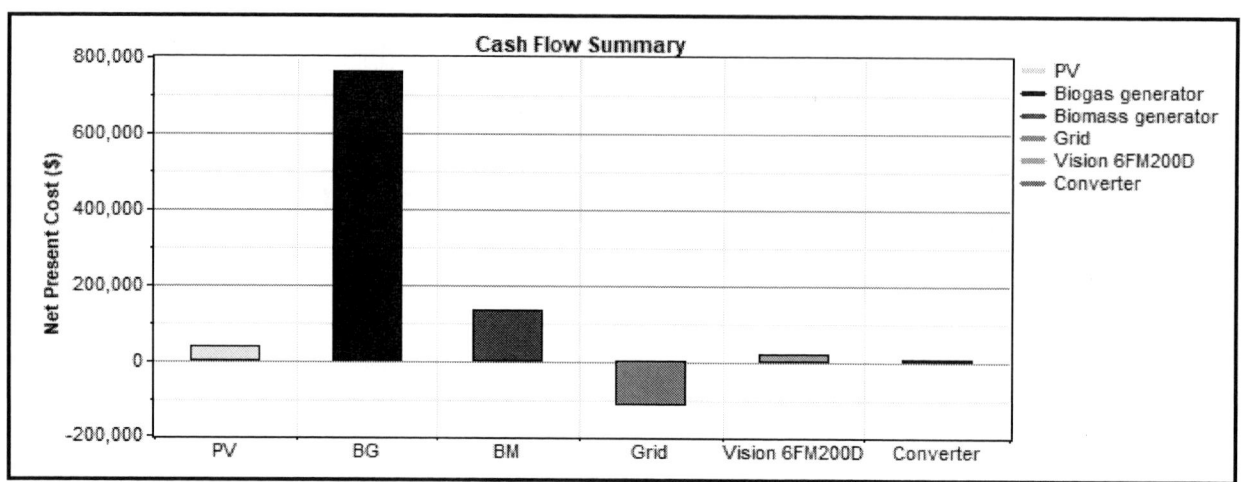

Fig.6. Cash flow summery in terms of NPC of proposed hybrid model

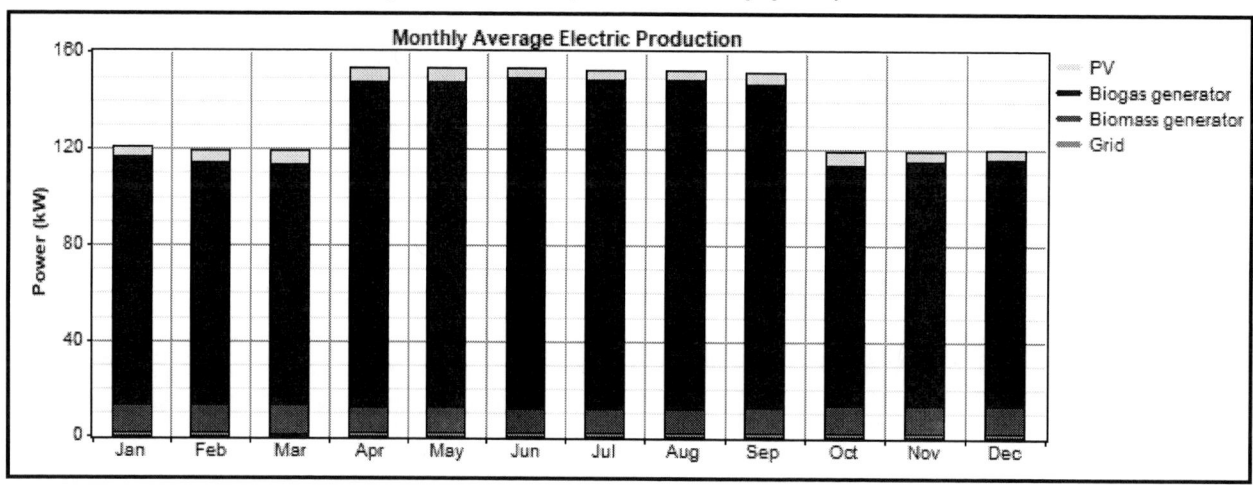

Fig.7. Monthly average energy generation of proposed grid connected hybrid model of study area

Table 3: Technical and economical input parameters of proposed hybrid system components [24]

Parameters	SPV System	Biomass generator	Biogas generator	Battery	Converter
Capital cost	1333 $/kW	1033 $/kW	660 $/kW	284 $/kW	117 $/kW
Replacement cost	1333 $/ kW	750 $/kW	450 $/kW	220 $/kW	117 $/kW
Operation and Maintenance (O &M) cost	26 $/kW/year	0.01 $/kW/hour	0.01 $/kW/hour	6 $/ year	4 $/year
Lifetime	25 years	15000 hours	20000 hours	5 years	10 years
Size to be considered	0-100	0-100	0-100	0-60	0-50
Project life	25 years				
Annual rate of interest	11 %				
Operating reserve as % of hourly load	10%				
Operating reserve as % of solar power output	25%				
Dispatch strategy	Cycle charging with 80% setpoint state of charge				

978-1-5386-6624-1/18 $31.00 © 2018 IEEE

2nd IEEE International conference on power Electronics, Intelligent Control and Energy systems (ICPEICES-2018)

Table 4: Results of feasible system configurations

Description	Parameter	SPV/biomass/ biogas generator with battery	SPV/ biogas generator with battery	SPV/biomass generator with battery
System rating	SPV (kW)	30	80	150
	Biogas generator (kW	150	170	--
	Biomass generator (kW)	50	--	140
	Battery (kWh)	50	200	400
	Converter (kW)	40	70	120
Economic	Total capital cost ($)	209520	283830	472210
	Total NPC ($)	926833	1038726	1347689
	Total annualized capital cost ($)	24878	33702	56070
	Total annual replacement cost ($)		264368	455071
	Total O & M cost ($)	105899	132130	129296
	Total fuel cost ($)	387873	358953	296963
	Total annualized cost ($)	110052	123339	160025
	Cost of energy ($/kWh)	0.124	0.140	0.181
Electrical power production	SPV array (kWh/year)	50953 (6 %)	135875 (15 %)	254766 (26%)
	Biomass generator (kWh/year)	720466 (81%)	--	716910 (74%)
	Biogas generator (kWh/year)	120494 (14%)	782134 (85%)	--
	Total electrical power production (kWh/year)	891914 (100%)	918009 (100 %)	971675 (100%)
	Renewable fraction (%)	100	100	100
	Unmet electric load (kWh/year)	0.000132 (0%)	0.00 (0%)	0.00 (0%)
	Capacity shortage (kWh/year)	0.00 (0%)	0.00 (0%)	0.00 (0%)
Green house gas emission	Carbon dioxide (CO_2) emission (tons/year)	0.531	0.492	0.407
	Carbon monoxide (CO) emission (tons/year)	0.020	0.0185	0.0153
	Unburned hydrocarbons (UHC) emission (tons/year)	0.00221	0.00205	0.00169
	Particular matter (PM) (tons/year)	0.0015	0.00139	0.00115
	Sulphur oxide (SO_x) (tons/year)	0	0	0
	Nitrogen oxide (NO_x) (tons/year)	0.178	0.165	0.0136

Table 5: Results of Grid Connected Hybrid model

Configuration	SPV (kW)	Biomass generator (kW)	Biogas generator (kW)	Battery (no.)	Converter (kW)	Grid (kW)	NPC ($)	COE ($/kWh)
SPV/ Biomass/ Biogas/Battery	25	40	150	30	40	10	847365	0.114

ACKNOWLEDGEMENT

The authors would like to acknowledge the support and facilities provided by the Delhi Technological University (DTU), Delhi and I. K. Gujral Punjab Technical University, Kapurthala (Punjab) to conduct the present research.

REFERENCES

[1] www.mnre.gov.in.

[2] Ministry of Power:
http://www.cea.nic.in/reports/monthly/installedcapacity/2018/installed_capacity-03.pdf

[3] Ministry of New and Renewable Energy:
http://www.mnre.gov.in/sites/default/file/uploads/mission/document_JNNSM.pdf

[4] M. Muselli, G. Notton, and A. Louche, "Design of hybrid-photovoltaic power generator, with optimization of energy management", Solar Energy, vol. 65, pp. 143-157, 1999.

[5] A. M. Elhadidy, and M. S. Shaahid, "Feasibility of hybrid (wind+solar) power systems for Dhahran, Saudi Arabia", Renewable Energy, vol.16, pp. 970-976, 1999.

[5] B. R. Bagen, "Evaluation of different operating strategies in small stand-alone power systems", IEEE Transactions on Energy Conversion, vol. 20, pp. 654-660, 2005.

[6] J.M. Khan, and T. M. Iqbal, "Pre-feasibility study of stand-alone hybrid energy systems for applications in Newfoundland" , Renewable Energy, vol. 30, pp. 835-854, 2005.

[7] S. Rehman, I. M. El-Amin, F. Ahmad, S. M. Shaahid, A. M. Al-Shehri, J. M. Bakhashwain, and A. Shash, "Feasibility study of hybrid retrofits to an isolated off-grid diesel power plant", Renewable and Sustainable Energy Reviews, vol. 11, pp. 635-653, 2007.

[8] Y. Himri, A. Boudghene Stambouli, B. Draoui, and S. Himri, "Techno-economical study of hybrid power system for a remote village in Algeria", Energy, vol. 33, pp. 1128-1136, 2008.

[9] G. Bekele, and B. Palm, "Feasibility study for a standalone solar–wind-based hybrid energy system for application in Ethiopia", Applied Energy, vol. 87, pp. 487-495, 2010.

[10] S. Kumaravel, and S. Ashok, "An optimal stand-alone biomass/solar-PV/pico-hydel hybrid energy system for remote rural area electrification of isolated village in western-ghats region of India", International Journal of Green Energy, vol. 9, pp. 398-408, 2012.

[11] S. Lal, and A. Raturi, "Techno-economic analysis of a hybrid mini-grid system for Fiji islands", International Journal of Energy and Environmental Engineering, vol. 3, pp. 1-10, 2012.

[12] A. H. Al-Badi, and H. Bourdoucen, "Study and design of hybrid diesel–wind standalone system for remote area in Oman", International Journal of Sustainable Energy, vol. 31, pp. 85-94, 2012.

[13] M. S. H. Lipu, M. S. Uddin, and M. A. R. Miah, "A feasibility study of solar-wind-diesel hybrid system in rural and remote areas of Bangladesh", International Journal of Renewable Energy Research, vol. 3, pp. 892-900, 2013.

[14] A. V. Anayochukwu, "Feasibility assessment of a PV-diesel hybrid power system for an isolated off-grid catholic church", Electronic Journal of Energy & Environment, vol. 1, pp. 49-63, 2013.

[15] M. Rizwan, R. Kumar, and D. Kumar, "Renewable energy based optimal hybrid system for distributed power generation", International Journal of Sustainable Development and Green Economics (IJSDGE), vol. 2, pp. 60-62, 2013.

[16] M. M. Rahman, M. M. Hasan, J. V. Paatero, and R. Lahdelma, "Hybrid application of biogas and solar resources to fulfill household energy needs: a potentially viable option in rural areas of developing countries", Renewable Energy, vol. 68, pp. 35-45, 2014.

[17] S. Upadhyay, and M. P. Sharma, "Development of hybrid energy system with cycle charging strategy using particle swarm optimization for a remote area in India", Renewable Energy, vol. 77, pp. 586-598, 2015.

[18] S. G. Sigarchian, R. Patela, A. Malmquist, and A. Pina, "Feasibility study of using a biogas engine as backup in a decentralized hybrid (PV/wind/battery) power generation system -case study Kenya", Energy, vol. 90, pp. 1830-1841, 2015.

[19] V. A. Ani, and B. Abubaka, "Feasibility analysis and simulation of integrated renewable energy system for power generation: a hypothetical study of rural health clinic", Journal of Energy, vol. 2015, pp. 1-7, 2015.

[20] S. Goel, and R. Sharma, "Feasibility study of hybrid energy system for off-grid rural water supply and sanitation system in Odisha, India", International Journal of Ambient Energy, vol. 37, pp. 314-320, 2016.

[21] P. Anand, S.K. Bath, and M. Rizwan, "Feasibility analysis of Solar-Biomass based standalone hybrid system for remote area", American Journal of Electrical Power and Energy Systems, vol.5, pp. 99-08, 2016.

[22] Priyanka, S.K. Bath, and M. Rizwan, "Design and Optimization of RES based Standalone Hybrid System for Remote Applications", Proceeding of 8th IEEE conference on Innovative Smart Grid Technologies (ISGT 2017) sponsored by IEEE Power and Energy Society (PES), Washington DC, USA, April 23-26, 2017.

[23] S. Bhardwaj, and S. K. Garg, "Rural electrification by effective mini hybrid PV solar, wind & biogas energy system for rural and remote areas of Uttar Pradesh", International Journal of Computer Science and Electronics Engineering (IJCSEE), vol. 2, pp. 178-181, 2014.

[24] A. Chauhan, and R.P. Saini, "Techno-economic feasibility study on integrated renewable energy system for an isolated community of India", Renewable and Sustainable Energy Reviews, vol. 59, pp. 388-405, 2016.

[25] M. J. Khan, A. K. Yadav, and L. Mathew, "Techno economic feasibility analysis of different combinations of PV-Wind-Diesel-Battery hybrid system for telecommunication applications in different cities of Punjab, India", Renewable and Sustainable Energy Reviews , vol. 76, pp. 577–607, 2017.

[26] R. Rajbongshi, D. Borgohain, and S. Mahapatra, "Optimization of PV-biomass-diesel and grid base hybrid energy systems for rural electrification by using HOMER", Energy, vol. 126, pp. 461-474, 2017.

[27] M. Rizwan, R. Kumar, and D. Kumar, "Renewable energy based optimal hybrid system for distributed power generation", International Journal of Sustainable Development and Green Economics (IJSDGE), vol. 2, pp. 60-62, 2013.

[28] P. Anand, S.K. Bath, and M. Rizwan, "Design of Solar-Biomass-Biogas based hybrid system for Rural electrification with environmental benefits", International Journal on Recent and Innovation Trends in Computing and Communication, vol.5 (6), pp. 450 – 456, 2017.

[29] P. Anand, S.K. Bath, and M. Rizwan, " Design and sizing of RES based hybrid system for rural applications of Haryana state in India", International Journal of Electronics Engineering, vol.9 (2), pp.79-85, 2017.

[30] http://www.censusindia.gov.in/2011census/dchb/DCHB.html.

[31] https://garv.gov.in/garv2/dashboard/main.

[32] www.nrel.gov/homer.

2nd IEEE International Conference on Power Electronics, Intelligent Control and Energy Systems (ICPEICES-2018)

Maximum Power Point Tracking Techniques under Partial Shading Condition- A Review

Md Reyaz Hussan
Department of Electrical Engineering
Aligarh Muslim University
Aligarh, India
Email: mreyazamu@gmail.com

Dr. Adil Sarwar
Department of Electrical Engineering
Aligarh Muslim University
Aligarh, India
Email: adilsarwar123@gmail.com

Abstract—**Partially shaded condition (PSC) is one of the major problems in large photovoltaic generation systems. It causes losses in output power and hot spot effects. Under PSC, PV characteristic curve exhibits multiple peaks having one global maximum power point and multiple local maximum power points. Tracking the global maximum power point is one of the main challenges the design engineers have to face. The paper presents the recent work done on the development of Global Maximum Power Point Tracking (GMPPT) algorithms under partial shading condition and their comparative analysis. To have focus on GMPPT techniques used in PSC, traditional MPPT techniques that cannot distinguish GMPP from local maximum power points have not been discussed.**

Keywords: Partial Shading Condition (PSC), Global Maximum Power Point Tracking (GMPPT), P-V characteristics.

I. INTRODUCTION

The problem of partial shading in the area of solar photovoltaic systems (SPV) has generated interest in the research community since 20 years. Partial shading occurs when the modules connected in series and parallel doesn't receive same illumination and the result is different power generation by the different modules for the same rating of the panel. And if same power doesn't flow through all the modules, the modules which are generating lower power will act as sink and the power will be absorbed from the modules which are generating more power leading to hotspot formation and consequent irrelevant damage of the PV module. To overcome this problem diodes are generally connected in parallel with the PV panel. Some manufacturers provide inbuilt bypass diodes with the PV module. Normally bypass diodes are provided in parallel with series connected PV cells as shown in figure 1. But it introduces complexity in the non-linear PV characteristic curve with multiple maxima under PSC. Even large PV plants are being built in a fixed series-parallel configuration and modules have bypass diode included in different configurations. To extract maximum power from the PV system MPPT is employed [1].

Various review papers for MPPT methods are available for solar PV power generating systems till date. Some papers have discussed both traditional methods as well as those suitable for partial shading condition (PSC) [2-5]. Some have discussed only MPPT methods for partial shading condition [6-8]. For further optimization of the PV system, partial shading has to be taken into consideration by the researchers. In this paper a comprehensive review of the papers on

maximum power point tracking under PSC has been presented. A comparative analysis based on the advantages and shortcomings has also been included.

II. PARTIAL SHADED CONDITION(PSC) PROBLEM

To enhance the power handling capacity of solar PV generating systems, the solar PV modules are connected in series (to enhance voltage level) and/or parallel (to enhance current level) with each other. Under uniform solar insolation on these modules the complete PV generating unit has a unique maximum power point that can be tracked using various gradient search based conventional methods. But under different insolation (due to clouding, tree or building shade etc.) the IV characteristics in series connected modules differ resulting in mismatch in the operating point. This causes some of the modules which are heavily shaded to act as load consuming power thereby giving rise to a situation called hotspot formation as shown in figure 1 which ultimately results in irreversibly damage of the module. To overcome this problem bypass diodes are connected across the module. The current thus bypasses the shaded module. Figure 1 shows a PV array composed of three modules in a string and its characteristic curves under uniform insolation and partial shading condition (with and without bypass diode). Two local MPPs can be seen on the PV characteristic curve other than the GMPP (global maximum power point). The popular gradient search or hill climbing method employing the logic of finding zero gradient (dp/dv=0) fails as the algorithm sticks and start operating in local maximum power point. Thus an efficient algorithm to track global maxima under PSC is required.

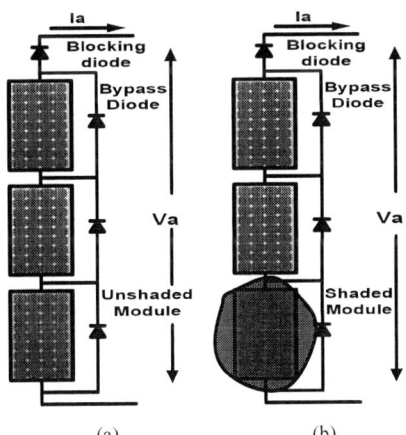

(a)   (b)

978-1-5386-6624-1/18 $31.00 © 2018 IEEE   293

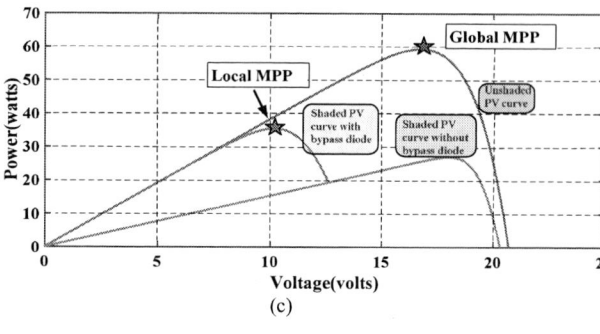

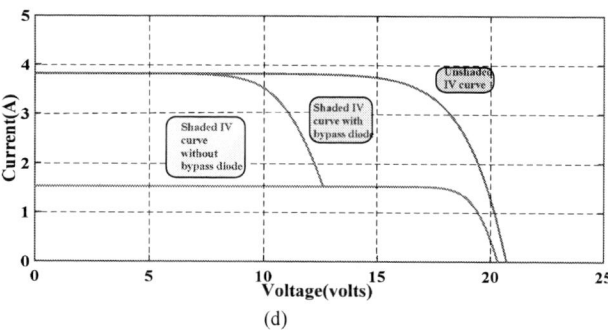

Fig1. (a) PV array under uniform insolation. (b) PV array under PSC.

(c) PV characteristic for (a) and (b). (d) IV characteristic for (a) and (b).

III. GLOBAL MAXIMUM POWER POINT TRACKING TECHNIQUES UNDER PSC

Recent algorithms for Global MPP tracking can be classified as follows.

I. Modified conventional MPPT
II. Utilizing the features of the characteristic PV curve
III. Metaheuristic nature inspired methods

I. Modified conventional MPPT

1) Two stage searching method: Search for the GMPP location interval is done in the first stage and then by using traditional MPPT methods in the second stage precise GMPP location is found out [6]. H. Patel and V. Agarwal [9] used this method with the basic search rules. They took 0.85 $V_{oc,all}$ (total O.C voltage of the system) as the P&O search starting point. Then they used the peak value found in step 1 as the basis to move the operating point one large step to the left. If the peak value obtained was greater than the previous one, step 2 was repeated, if it was smaller, then previous peak value was considered GMPP.

Advantage of this method is that its implementation is easy and it can be integrated into conventional Power Generating System. Its disadvantages are that it can fail to track GMPP in some cases. It is successful in cases when we move from start or from V_{OC} (as shown in figure 2). Also probability of the GMPP being tracked is dependent on value of the large step. Its tracking speed is also less as maximum power point of each curve has to be obtained using P&O method. Works in [10-12] has similar methods. Authors employed a large interval for the entire P–V characteristic curve and determined the largest peak value. Then a refined search is done near this peak value to locate the GMPP. Flowchart of the method is

given in figure 3.

2) System characteristic curve method: Works in [13-15] used a preset function (linear) to move the operating point near the GMPP. Flowchart in figure 4 describes the principle of this method. The linear function depends on various system parameters such as O.C voltage and S.C current. The tracking speed reported in the work is quite fast. But for obtaining O.C voltage and S.C current, open or short circuits are required which can lead to power loss or safety concerns. Moreover, this method fails under complex shading pattern.

3) Current sweeping method: In this method maximum power point is tracked using PV output current instead of PV output voltage. Most of the MPPT algorithms adjust the duty ratio of the DC to DC converter which indirectly adjusts the output voltage. However, changing the duty ratio affects both the output voltage and current. By utilizing the dynamics of the DC to DC converter, Tsang and Chan [16] used current sweeping method to develop a good current controller for the DC to DC converter by having a firm control on the output current. Then a prompt current sweeping signal can be sent for the converter making it possible to track the GMPP very quickly. Authors have included simulation & experimental results. The performance reported is satisfactory.

4) Distributed MPPT (DMPPT): The output of every PV module has its own DC to DC converter in this method. These DC to DC converters operate at low power levels since they supply only the equalization current. Thus power losses are significantly reduced [17-18]. For multiple strings in parallel without the loss of their MPPs, this current equalization topology is extended as shunt-series compensation in [19]. Each output of the DC to DC converter is connected in series to form the PV string. These strings are connected at the output of every module. Every DC to DC converter processes the whole power produced by the corresponding modules and then tracks the MPP for that individual module. The total available MPP power of the array is somewhat increased. But as separate DC to DC converter is there for every module, its implementation becomes complex.

5) Electrical PV array Reconfiguration: Works in [20] and [21] showed that immediate adjustment of the PV array configuration with respect to the pattern in shading can reduce the power losses caused by the partial shading condition. Although reconfiguration method can minimize effects of PSC, it requires use of a switching matrix to effect the changes in architecture. Therefore, the system becomes costly and the complexity of the controller design also increases. Moreover, reconfiguration technique can fail to track GMPP in some shading patterns.

II. Utilizing the features of the characteristic PV curve

1) Direct method: P–V characteristic curve resembles Lipschitz characteristics and hence peak values can be found by dividing rectangle (DIRECT) technique [22]. Here "**direct**" means to divide the searching area into three different areas with equal intervals and to find potentially optimal intervals using mathematical equations. Its merit is that it has a strong mathematical foundation and has good tracking speed. Its demerit is that it fails to track the GMPP under some shading pattern, and this method cannot be integrated directly into

978-1-5386-6624-1/18 $31.00 © 2018 IEEE

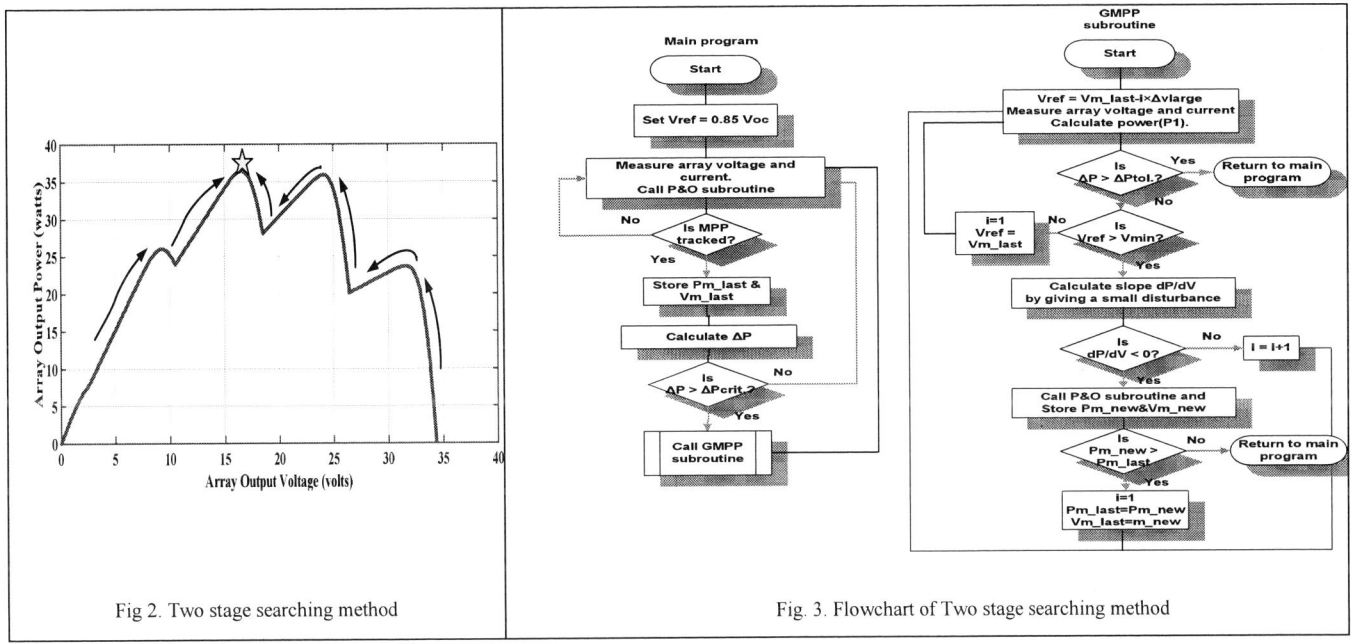

Fig 2. Two stage searching method

Fig. 3. Flowchart of Two stage searching method

conventional power generating system. Flowchart of the method is shown in figure 5.

2) Fibonacci methods: Fibonacci sequence which is used as the mathematical basis for segmentation is the main difference between Fibonacci method and DIRECT method. The Fibonacci search method search for a sorted array by using an algorithm of divide and conquer, narrowing down possible locations using Fibonacci numbers. It continuously narrows down the range having the optimal point always within that range. It has strong mathematical foundation and good tracking speed. Its disadvantage is that it fails to track the GMPP under some shading pattern, and it cannot be integrated directly into conventional power generating system. A random number method is used in [23] where power values have been randomly sampled under six well defined voltages, and then the new search range is obtained using that point which has the highest power value. Repeated random sampling was done with new search range for getting convergence.

III. Metaheuristic nature inspired methods

1) Fuzzy Logic Control (FLC): To understand the system to be controlled, traditional control system design uses accurate mathematical models for describing the system. But when the system to be controlled becomes very complex, system identification method cannot establish a system model. However, FLC converts the linguistic values into automatic control action with the help of expert knowledge using the fuzzy set theory. Its advantages are that there is no need of precise mathematical model. It is also suitable for systems which are nonlinear and vary with time. Systems whose complete model is unknown can also be dealt with this method. Therefore PSC (partially shading condition) problems can be dealt with FLC. Karatepe et al. [24] utilized FLC in place of traditional MPPT methods. Since each converter has an MPPT controller, tracking of GMPP is guaranteed.

3) Ant colony optimization: It is one of the evolutionary

computation methods and has been widely used in image processing [26], scheduling [27], power electronic circuit design [28] and many other fields. Its main advantage is that it adjusts the command values very fast according to environmental changes. Thus, this technique is suitable to track MPP under varying environmental condition. In this method a random path is selected by each agent at first. If the path chosen is short, the agent drops concentrated pheromone on that path. Then in the next iteration, the path is chosen on the basis of the concentration of pheromone on that path. The probability of the path to be chosen by the agent depends on the concentration of the pheromone. Jiang et al. [29] used this method to develop MPPT control scheme for PV systems under partial shading condition. Single current and voltage sensors were used which simplified the system and reduced its cost. It also provides fast convergence independent of the initial condition. But only the simulation results were provided in [29].

4) Differential Evolution (DE): It is a population based, stochastic evolutionary algorithm [30]. It differs from GA in the context that it relies on mutation process rather than crossover. Mutation operation is used as a search mechanism and selection operation for directing the search towards the prospective regions of the search space. This method has three advantages (1) locates accurate GMPP without depending on the initial values taken (2) convergence is fast (3) utilizes few control parameters, hence easy to use. [30] and [31] showed simulation results based on DE.

5) Particle swarm optimization (PSO): Among the various Evolutionary Algorithm techniques, PSO has one of the simplest structures that can be used to track MPP under PSC. The main advantage of the method proposed in [32] is the use of direct duty cycle control method by removing PI control loops. The PSO algorithm proposed in [33] took about 1 to 2 seconds to find the GMPP. More importantly, the response

978-1-5386-6624-1/18 $31.00 © 2018 IEEE

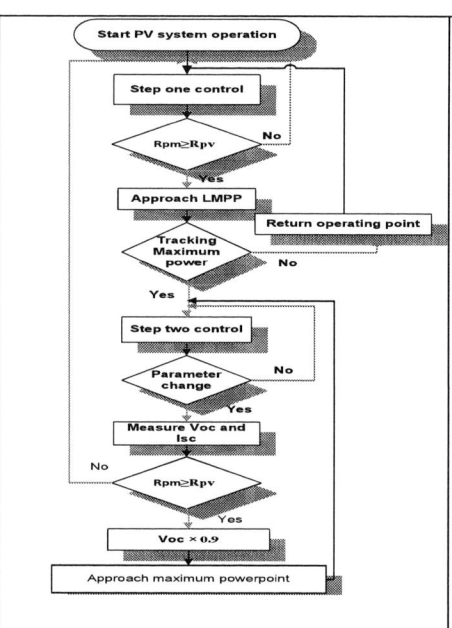

Fig. 4. Flowchart of System characteristic curve method

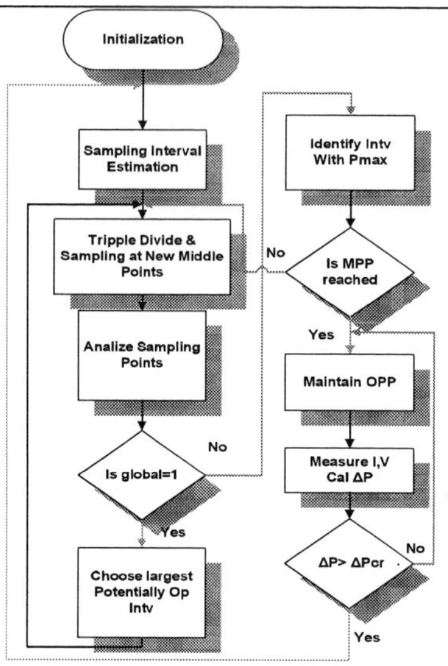

Fig. 5. Flowchart of DIRECT method

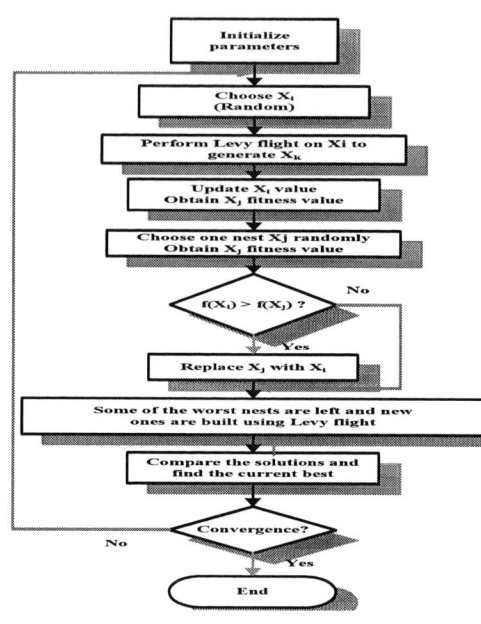

Fig. 6. Flowchart of Cuckoo search algorithm

time was almost independent of the search space dimensions taken and partial shading pattern.

6) Chaos search (CS) method: It is a stochastic search method which uses chaos theory as its basis. Its search results are much better than those of the search methods which use pure random numbers. In [34] control variables are randomly produced using dual carrier with the help of this method. Then output power is measured using the control variables which is used to find the GMPP. According to [34], dual- carrier Chaos Search can accurately track the GMPP under PSC, the search efficiency is improved, has good precision and system robustness and has a simple control mode.

7) Simulated Annealing method: In [35-36] a simulated annealing-based global MPPT method designed for PSC has been proposed. It is based on following the heating and cooling processes in metals to find the global optimum solution. The energy measured during the heating and cooling process is compared to the ongoing state and if the new operating point has more energy, the new working point is selected as the new one. The algorithm achieves objective but with some increment in computational complexity as compared to the P&O technique and it has also fewer parameters stored in the memory than the PSO technique. Starting value of the algorithm is not an issue for tracking the GMPP.

8) Grey- Wolf optimization: The Grey-Wolf optimization (GWO) method is proposed in [37-38] as an algorithm that overcomes problems such as steady-state oscillations, lower tracking efficiency, which have been seen in P&O and PSO methods. This method detects the shading pattern variations and is faster to converge to the global maximum, and has less steady-state oscillations. This algorithm has some

disadvantages also. It has a complex initialization part and there are lots of unknown parameters which have to be determined by the designer himself.

9) Artificial bee colony algorithm: An artificial bee colony (ABC) algorithm for global MPP has been proposed in [39-40]. The proposed method takes less tracking time for GMPP, compared with PSO and enhanced P&O (EPO). It has fewer control parameters and its convergence doesn't depend on initial conditions. The comparison results indicate that, ABC method is slightly better than PSO and EPO in efficiency, convergence and GMPP time parameters. But the implementation complexity of this method is quite high as compared to the other methods.

10) Cuckoo Search (CS) algorithm: It is an optimization algorithm based on parasitic reproduction strategies of cuckoo birds [41]. Yang and Deb introduced this method in 2009 [42]. Several birds of cuckoo family perform brood parasitism, i.e. Firstly, each cuckoo lays one egg in every iteration, and then a nest is randomly chosen by it to lay its egg in. Secondly, the best nest having the top quality solution will be carried forward to the next generation. Thirdly, there are fixed number of host nests and the probability of alien eggs being discovered by a host bird is pa $\in$ [0, 1]. It has several advantages such as higher efficiency, fast convergence and fewer tuning parameters [41]. Among them the most pronouncing is the dependence of PS size on random numbers. Its flowchart is shown in figure 7.

IV. CONCLUSION

Recent work on the topic of GMPPT tracking suggests it to be a popular & hot area of research. In this study various GMPPT methods have been discussed with their merits and demerits from diverse references. Table I sums up the advantages and disadvantages of these methods reported in literature.

Table I

Classification	Methods	Advantages	Disadvantages
Modified conventional MPPT	System characteristic curve method	Good tracking speed	Requirement of open or short circuits can cause power loss or safety concerns, method fails in some cases
	Two stage searching method	Its implementation is easy and it can be integrated into traditional PGS	It can fail to track GMPP in some cases
	Current sweeping method	Fast tracking speed	Requires periodical tracking of the MPP
	Electrical PV array Reconfiguration	Compensate the power losses caused by PSC	Expensive and the controller design is also complex, fail to track GMPP in some shading patterns
	Distributed MPPT (DMPPT)	total available MPP power of the PV array is increased	Implementation complexity is high
Methods based on utilizing the features of the characteristic PV curve	Direct method	Based on a strong mathematical foundation and good tracking speed	Cannot be directly integrated into conventional PGS
	Fibonacci methods	Based on a strong mathematical foundation	Fail to track GMPP in some cases and cannot be directly integrated into conventional PGS
Metaheuristic nature inspired methods	Fuzzy logic control	No need of precise mathematical model, it is very suitable for use in non-linear, time-varying and systems without complete models	high hardware cost
	Genetic Algorithm	Can optimize parameters of other algorithms such as FLC	Its implementation is complex and difficult to achieve using low cost microcontroller
	Ant colony optimization	Fast convergence and convergence independent of the initial condition	Implementation is difficult
	Differential Evolution	Fast convergence and convergence independent of the initial condition, easy to use	Some parameters may not guarantee optimal solution
	Particle swarm optimization	Simpler structure than other EA techniques	Optimization performance depends on parameter selection
	Chaos search method	Improved search efficiency, precision, and system robustness	High complexity
	Simulated annealing	Requires fewer parameters to be stored, independent of initial condition	Slight more complex
	Grey- Wolf optimization	Less oscillations, high efficiency, high speed	complex initialization part, more unknown parameters
	Artificial bee colony (ABC) algorithm	Less tracking time, fewer control parameters, independent of initial condition	Implementation complexity is high
	Cuckoo search	Fast convergence, high efficiency, fewer tuning parameters	Unknown parameters

REFERENCES

[1] A. Sarwar, M. Hasan and A. Q. Ansari, "Five parameter modelling and simulation of solar PV cell," *2015 International Conference on Energy Economics and Environment (ICEEE)*, Noida, pp. 1-5, 2015.

[2] K.Ishaque & Z.Salam, "A review of maximum power point tracking techniques of PV system for uniform insolation and partial shading condition", Renewable and Sustainable Energy Reviews, vol. 19, pp. 475-488,2013.

[3] M.Anwari, S. Twaha, K. Ishaque, A. Al-Turki Yusuf, "A review on maximum power point tracking for photovoltaic systems with and without shading conditions", Renewable and Sustainable Energy Reviews, vol. 67, 2017.

[4] D. Verma, S. Nema, A.M. Shandilya, S. Dash, "Comprehensive analysis of maximum power point tracking techniques in solar photovoltaic systems under uniform insolation and partial shaded condition", Journal of Renewable and Sustainable Energy. 7. 042701, 2015.

[5] A. Sayal, "MPPT techniques for photovoltaic system under uniform insolation and partial shading conditions", Proceedings of 2012 Students Conference on Engineering and Systems (SCES), pp. 1-6, 2012.

978-1-5386-6624-1/18 $31.00 © 2018 IEEE

[6] Y.H. Liu, J.H Chen, J.W. Huang, "A review of maximum power point tracking techniques for use in partially shaded conditions", Renewable and Sustainable Energy Reviews, vol. 41, pp. 436–453, 2015.

[7] A. Mohapatra, B. Nayak, P. Das, K. Mohanty, "A review on MPPT techniques of PV system under partial shading condition", Renewable and Sustainable Energy Reviews, vol. 80, pp. 854-867, 2017.

[8] A. Dadje, N. Djongyang, J.D. Kana, R. Tchinda, "Maximum power point tracking methods for photovoltaic systems operating under partially shaded or rapidly variable insolation conditions: a review paper", International Journal of Sustainable Engineering, 2016.

[9] H. Patel, V. Agarwal, "Maximum power point tracking scheme for PV systems operating under partially shaded conditions", IEEE Trans Ind Electron, vol. 55, pp. 1689–1698, 2008.

[10] H. Renaudineau, A. Houari, J. P. Martin, S. Pierfederici, F. Meibody-Tabar, B. Gerardin, "A new approach in tracking maximum power under partially shaded conditions with consideration of converter losses", Sol Energy, vol. 85, pp. 2580–2588, 2011.

[11] E. Koutroulis, F. Blaabjerg, "A new technique for tracking the global maximum power point of PV arrays operating under partial-shading conditions", IEEE J Photovolt, vol. 2, pp. 184–190, 2012.

[12] Kouchaki, H. Iman-Eini, B. Asaei, "A new maximum power point tracking strategy for PV arrays under uniform and non-uniform insolation conditions" Sol Energy, vol. 91, pp. 221–232, 2013.

[13] Kobayashi , Takano , Y. Sawada, "A study of a two stage maximum power point tracking control of a photovoltaic system under partially shaded insolation conditions", Sol Energy Mater Sol Cells, vol. 90, pp. 2975–2988, 2006.

[14] G. Carannante, C. Fraddanno, M. Pagano, L. Piegari, "Experimental performance of MPPT algorithm for photovoltaic sources subject to inhomogeneous insolation", IEEE Trans Ind Electron, vol. 56, pp. 4374–80, 2009.

[15] Y. H. Ji, D. Y. Jung, J. G. Kim, J. H. Kim, T. W. Lee, C. Y. Won, "A real maximum power point tracking method for mismatching compensation in PV array under partially shaded conditions", IEEE Trans Power Electron, vol. 26, no. 4, pp. 1001–9, April 2011.

[16] K. M. Tsang, W. L. Chan, "Maximum power point tracking for PV systems under partial shading conditions using current sweeping", Energy conversion and management, vol. 93, pp. 249-258, 2015.

[17] P. Sharma, P. K. Peter and V. Agarwal, "Exact maximum power point tracking of partially shaded PV strings based on current equalization concept," 38th IEEE Photovoltaic Specialists Conference, Austin, TX, pp. 001411-001416, 2012.

[18] P. Sharma and V. Agarwal, "Comparison of model based MPPT and exact MPPT for current equalization in partially shaded PV strings," IEEE 39th Photovoltaic Specialists Conference (PVSC), Tampa, FL, pp. 2948-2952, 2013.

[19] P. Sharma and V. Agarwal, "Maximum Power Extraction From a Partially Shaded PV Array Using Shunt-Series Compensation," in IEEE Journal of Photovoltaics, vol. 4, no. 4, pp. 1128-1137, July 2014.

[20] G. V. Quesada, F. G. Gispert, R. P. Lopez, M. R. Lumbreras, A. C. Roca ,"Electrical PV array reconfiguration strategy for energy extraction improvement in grid-connected PV systems", IEEE Trans Ind Electron, vol. 56, no. 11, pp. 4319–31, Nov 2009.

[21] D. Nguyen, B. Lehman, "An adaptive solar photovoltaic array using model-based reconfiguration algorithm", IEEE Trans Ind Electron, Vol. 55, No. 7, pp. 2644–54, July 2008.

[22] T. L. Nguyen, K. S. Low, "A global maximum power point tracking scheme employing DIRECT search algorithm for photovoltaic systems", IEEE Trans Ind Electron, vol. 57, no. 10, pp. 3456–67, Oct 2010.

[23] Y. Fan, C. Fang, Z. Liang, "New GMPPT algorithm for PV arrays under partial shading conditions", In: Proceedings of the 2012 IEEE international symposium on industrial electronics (ISIE), pp. 1046–51, 2012.

[24] E. Karatepe, T. Hiyama, M. Boztepe, M. Colak, "Voltage based power compensation system for photovoltaic generation system under partially shaded insolation conditions", Energy Convers Manag, vol. 49, pp. 2307–16, 2008.

[25] Y. Shaiek, M.B. Smida, A. Sakly, M.F. Mimouni, "Comparison between conventional methods and GA approach for maximum

power point tracking of shaded solar PV generators", Sol Energy, vol. 90, pp. 107–22, 2013.

[26] D. Picard, M. Cord, A. Revel, "Image retrieval over networks: active learning using ant algorithm", IEEE Transactions on Multimedia, vol. 10, no. 7, pp. 1356-1365, Nov 2008.

[27] D. Martens, M. De Backer, R. Haesen, J. Vanthienen, M. Snoeck, B. Baesens, "Classification with ant colony optimization", IEEE Transactions on Evolutionary Computation, vol. 11, no. 5, pp. 651-665, Oct 2007.

[28] J. Zhang, H. Chung, W.L. Lo, T. Huang, "Extended ant colony optimization algorithm for power electronic circuit design", IEEE Transactions on Power Electronics, vol. 24, no. 1, pp. 147-162, Jan 2009.

[29] L. L. Jiang, D. L. Maskell, J. C. Patra, " A novel ant colony optimization-based maximum power point tracking for photovoltaic systems under partially shaded condition", Energy and Buildings, vol. 58, pp. 227-236, 2013.

[30] H. Taheri, Z.Salam, K. Ishaque, A. Syafaruddin, "Novel maximum power point tracking control of photovoltaic system under partial and rapidly fluctuating shadow conditions using differential evolution", IEEE Symposium on Industrial Electronics & Applications (ISIEA), pp. 82–87, 2010.

[31] M. F. N. Tajuddin, S. M. Ayob, Z. Salam, "Tracking of maximum power point in partial shading condition using differential evolution (DE)", Proceedings of the IEEE International Conference on Power and Energy (PECon), pp. 384–89, 2012.

[32] K. Ishaque, Z. Salam, A. Shamsudin, M. Amjad, "A direct control based maximum power point tracking method for photovoltaic system under partial shading conditions using particle swarm optimization algorithm", Appl. Energy, vol. 99, pp. 414–22, 2012.

[33] M. Miyatake, M. Veerachary, F. Toriumi, N. Fujii, H. Ko, "Maximum power point tracking of multiple photovoltaic arrays: A PSO approach", IEEE Trans. on Aerosp. Electron Syst, vol. 47, no. 1, pp. 367–80, Jan 2011.

[34] L. Zhou, Y. Chen, K. Guo, F. Jia, "New approach for MPPT control of photovoltaic system with mutative-scale dual-carrier chaotic search", IEEE Trans Power Electron, vol. 26, no. 4, pp. 1038–48, April 2011.

[35] Lyden, Sarah & Haque, M.E, "A Simulated Annealing Global Maximum Power Point Tracking Approach for PV Modules Under Partial Shading Conditions", IEEE Transactions on Power Electronics, vol. 31, pp. 1-1, 2015.

[36] Guan, Tong & Zhuo, Fang, "An improved SA-PSO global maximum power point tracking method of photovoltaic system under partial shading conditions", pp. 1-5, 2017.

[37] S. Mohanty, B. Subudhi and P. K. Ray, "A New MPPT Design Using Grey Wolf Optimization Technique for Photovoltaic System Under Partial Shading Conditions," in IEEE Transactions on Sustainable Energy, vol. 7, no. 1, pp. 181-188, Jan. 2016.

[38] S. Mohanty, B. Subudhi and P. K. Ray, "A grey wolf optimization based MPPT for PV system under changing insolation level," 2016 IEEE Students' Technology Symposium (TechSym), Kharagpur, pp. 175-179, 2016.

[39] K. Sundareswaran, P. Sankar, P. S. R. Nayak, S. P. Simon and S. Palani, "Enhanced Energy Output From a PV System Under Partial Shaded Conditions Through Artificial Bee Colony," in IEEE Transactions on Sustainable Energy, vol. 6, no. 1, pp. 198-209, Jan. 2015.

[40] Abou soufyane Benyoucef, Aissa Chouder, Kamel Kara, Santiago Silvestre, Oussama Ait sahed, "Artificial bee colony based algorithm for maximum power point tracking (MPPT) for PV systems operating under partial shaded conditions", Applied Soft Computing, Vol. 32, pp. 38-48, 2015.

[41] Ahmed, Jubaer & Salam, Zainal, "A Maximum Power Point Tracking (MPPT) for PV system using Cuckoo Search with partial shading capability", Applied Energy, vol. 119, pp. 118–130, 2014.

[42] B. Peng, K. Ho and Y. Liu, "A Novel and Fast MPPT Method Suitable for Both Fast Changing and Partially Shaded Conditions," in IEEE Transactions on Industrial Electronics, vol. 65, no. 4, pp. 3240-3251, April 2018.

Power Quality Disturbance Prediction using PNN

Richa Sharma
Dept. of Electrical Engineering
MITS, Gwalior, India
richandsharma@gmail.com

Laxmi Srivastava
Dept. of Electrical Engineering
MITS, Gwalior, India
laxmigwl@gmail.com

Abstract— **In this paper the application of Probabilistic Neural Network for power quality disturbance prediction is given. Wavelet transform is used for the extraction of input patterns. The accuracy of the PNN is evaluated using Pattern Recognition Neural Network. The PNN has shown to be the faster predictor than PRNN.**

Keywords—power quality, wavelet transform, PQ disturbance, PNN, pattern recognition neural network

I. INTRODUCTION

With the increasing use of electronic components in modern electrical equipments and appliances, the issue of power quality has become prominent in present time. The use of electronic devices has greatly reduced the size of various electrical equipments and made them economically competitive but they have become more sensitive to power quality (PQ) anomalies [1]. In case the quality of power supplied to equipment is not up to the mark, the performance of the equipment degrades [2].

Quality power may be considered as uninterrupted power supply, at acceptable frequency and voltage magnitude with sinusoidal voltage and current waveform [3]. Due to transient and steady state disturbances various power quality issues like voltage sag, voltage swell, harmonics, flicker, notches, interruptions etc occur. Also, it is well known that be it industrial, commercial or domestic customers of power utilities all have experienced unhealthy power, in one way or another, due to such disruptions [2, 4].

The catastrophes related to power quality have overblown each of these three organs of consumer community either by affecting home appliances, industrial equipments or by causing power grid failures for instance 'the famous power blackouts of 2012' that concerned most of the Northern and Eastern India. This wreckage is possibly because the loads on consumer side has evolved from simple heaters, motors or incandescent lamps to advanced electronics dependent loads such as microwave Owens, computers or variable speed drives. The influence caused due to these non-linear loads brings in abrupt and deviated electrical characteristics from normal which signify PQ issues. Therefore, it is imperative to analyze these issues by monitoring and classification approaches [5].

Nowadays power quality is also used as a cadent for solving the power reliability anxiety owing to the conscious nature of present day customers. The heedfulness is due to the fact that many types of sensorial electronic appliances are strained by many power quality problems. These disturbances cause complications such as dimming of lights, tripping of electrical equipments, malfunctioning of protective instruments [6] inaccurate metering, instability, short life spans, overheating and the rest. Voltage sag can arise due to lightning, direct starting of large electric motors, either individually or in groups, capacitor switching, etc [7].Voltage swells, on the other hand, may materialize due to removal of large loads, by a damaged or loose neutral connection or due to unbalanced faults [8]. Harmonics are generated by saturation of the magnetic core of electric motors [9]. Interruptions, harmonics and notches are the responses of solid-state switching devices and non-linear power electronically switched loads. Transients are generally caused due to transformer energizing or by capacitor switching. Also, furnaces cause flicker and surges lead to spikes [10].

Aforementioned disturbances are inevitable in electrical systems hence it is paramount to detect, situate and classify them so that a system must promise a 'good power quality'. Alternatives are being investigated using various measurement and instrumentation techniques found in literature. The instrument units collect the data in bulk such as voltage and current waveforms related to power quality anomalies. However, for classification purpose, offline analysis is required from the recorded data. Fourier Transform (FT) has also been used for the harmonic and spectral analysis [11]. Nevertheless FT works satisfactorily only for stationary Power quality disturbances and gives inaccurate results for time-varying transient disturbances [20]. The Short-time Fourier transform (STFT) is discussed in [12] which overcame the demerits of FT. Although, fixed window size has imposed limitation on this approach [20].

The upgraded technique of STFT has also been used in power quality disturbance (PQD) detection [13]. It is known as ST. It utilizes a movable and scalable localizing Gaussian window. But detection quality of ST also deteriorates in examining some non-stationary PQ transients [10]. A probabilistic neural network for feature selection in combination with adaptive probabilistic neural network (APNN) is discussed in [14]. In [15] wavelet transform is used for feature extraction with radial basis function (RBF) neural network. Histogram based method combined with discrete wavelet transform (DWT) with extreme learning machine as a classifier is discussed in [16]. A fast fault detection approach applied in transmission line system using wavelet and Probabilistic neural network (PNN) is reviewed in [17].For sag and swell detection, wavelet transform (WT) is used in [18].

It is salient that WT has been progressively used in power quality analysis. It is because WT has outperformed other approaches with respect to time-frequency analysis. WT based approach such as Multi-Resolution Analysis (MRA) is extensively used for solving resolution based issues of other transforms [19]. The wavelet transform provides a long window for low frequency components and a short window for high frequency components, thereby providing outstanding time-frequency resolution. This helps in analyzing signals with stationary disturbance components and also for classifying low and high frequency PQ issues.

978-1-5386-6624-1/18 $31.00 © 2018 IEEE

II. WAVELET TRANSFORM

Principle of uncertainty proposed by Heisenberg says that the momentum and position of a moving particle cannot be determined simultaneously. This infers that we can only know the spectral components existing at an interval of time rather than knowing about the spectral component at any given time instant. This is the dispute related to resolution and is resolved by WT which provides a variable resolution and has replaced STFT in this regard. Higher frequencies are time domain resolved and lower ones are better resolved in frequency domain. The two types of wavelet transforms are explained in the following sections [20, 21].

A. Continuous Wavelet Transform

The concept of mother wavelet and its relation with the signal can be understood from the continuous wavelet transform(CWT).The following equation represents CWT:

$$CWT_x^\psi(\tau, S) = \varphi_x^\psi(\tau, S) = \frac{1}{\sqrt{|S|}} \int x(t) \, \psi^*\left(\frac{t-\tau}{s}\right) dt \qquad (1)$$

Points evident from the above equation are:
- The signal so transformed is a function of two main parameters τ stands for translation and refers to time information and S is scale, a mathematical operator which is the inverse of frequency. If the scale is large the signal is dilated and if the scale is small signal is compressed.
- Mother wavelet $\psi^*(t)$ is a transforming, compactly supported and oscillatory function. It begets the other functions with different support regions used in the process of transformation.
- CWT can be considered as the inner product of the signal under test with its basis function being $\psi_{(\tau,s)}(t)$. Where,

$$\psi_{(\tau,s)} = \frac{1}{\sqrt{s}} \, \psi\left(\frac{t-\tau}{s}\right) \qquad (2)$$

- So, the above equation specifies that there is some similar frequency content between basis function (wavelet) and the signals and CWT is the measure of it.
- CWT coefficients are the reference that the signal is close to the wavelet at a particular scale [20].

B. Discrete Wavelet Transform and Multi-Resolution Analysis

The intresting technique of signal decomposition and its separation into smoothed and detailed version can be better understood with the help of discrete wavelet transform (DWT).
- In discrete wavelet transform, to analyze the high frequencies and low frequencies the signal is passed through the series of high pass filters and low pass filters respectively.

- The resolution measures the details of the signal and is changed by filtering operations whereas scale is changed by upsampling and downsampling operations.
- Since CWT consists of the redundant information in large amounts it is better to translate and dilate the mother wavelet discretely. Therefore, when $S=S_0^m$ and $\tau = n\tau_0 S_0^m$ Where, $S_0 > 1$ and $\tau_0 > 0$ and m, n is an element of Z which is the set of positive integers. Therefore, discretized mother wavelet has the following equation:

$$\psi_{m,n}(t) = S_0^{\frac{-m}{2}} \, \psi\left(\frac{t - n\tau_0 S_0^m}{S_0^m}\right) \qquad (3)$$

Therefore, DWT becomes:

$$DWT_\psi \quad x(m,n) = \int_{-\infty}^{\infty} x(t) \, \psi_{m,n}^*(t) dt \qquad (4)$$

For the sake of simplicity let, $S_0 = 2$ and $\tau_0 = 1$. This choice leads to a refined algorithm called multi-resolution analysis (MRA). By selecting the above values, the wavelet transform is now called the dyadic-orthonormal wavelet transform. This states that there will be no redundant information.

C. Multi-Resolution Analysis

The multi-resolution analysis decomposes a given signal into its approximation and details as shown in "Fig.1". Approximation consists of a smoothed version of a signal and details consists of borders, changes and leaps in a signal, therefore, the original signal is distinguished from a disturbance by making separate analysis of a signal easy.

Let , $f_0(n)$ be a discrete time signal, the MRA is applied by decomposing the signal at scale 1 which yields A1(n) the approximated version and D1 (n) the detailed version of the original signal $f_0(n)$. Where,

$$A1 (n) = \sum_k h(k-2n) f_0(k) \qquad (5)$$

$$D1 (n) = \sum_k g(k-2n) f_0(k) \qquad (6)$$

h(n) and g(n) are associated filter coefficients that decompose $f_0(n)$ into A1(n) and D1(n) respectively. The next higher scale decomposition is now based on the signal A1 (n), the decomposed signal at scale 2 is given as [21]:

$$A2 (n) = \sum_k g(k-2n) A1(k) \qquad (7)$$

$$D2 (n) = \sum_k g(k-2n) A1(k) \qquad (8)$$

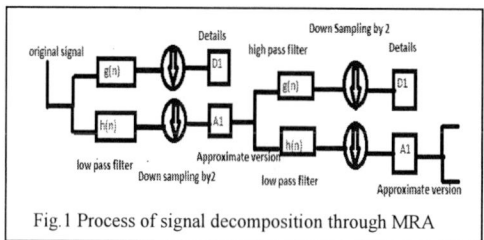

Fig.1 Process of signal decomposition through MRA

D. Mother Wavelet and Level of Decomposition

The literature contains many basis functions that may be utilized for wavelet transformation. The translation and dilation help in determining the characteristics of resulting wavelet transforms. The shape and the capability of the wavelet to analyze the signals are the two main parameters that result in the choice of wavelet to be used and it varies from application to application. In the case of power quality, literature and many simulation results have reviewed Daubechies wavelets to be the most suitable for feature extraction. It is well known that larger the energy of the wavelet, the more information is secured after the signal decomposition. This is due to the fact that energy related to a particular wavelet is the indicator of the energy concentration of wavelet coefficients.

In this work preserving energy in the analysis stage is important hence orthogonal wavelet is used. An orthogonal transform has a characteristic to preserve energy. The number of oscillations and the vanishing moments a wavelet has, have a loose relationship. The greater the number of vanishing moments a wavelet has, more the wavelet oscillates. Daubechies (Db4) wavelets used in this paper have maximal number of vanishing moments. A scaling function associated with it generates multi-resolution analysis. Many important features of a signal may get subtracted at much lower levels and many redundant are added at the higher ones so optimal level of decomposition is chosen as 8 according to the previous literatures [15].

III. PROBABLISTIC NEURAL NETWORK

Probabilistic neural network (PNN) is a supervised multilayered feed-forward neural network. It is found on the theory of classification given by Bayes and also is based on the valuation of probability density function. It has an ability of minimizing misclassification. The initial weights of PNN are adjusted by itself. It neither has a requirement of a learning process nor is any relationship needed between the processes of training and testing. Its four layers constitute input layer, pattern layer, summation layer and the output layer. Since it requires only a single iteration for classification it is considered as a fastest classifier [22, 23]

It is mandatory to arrange the input vectors into each of the two classes in a way in which the value of the inputs is not known but it is known that inputs are random variables and probability distribution of these variables is known. The probability density function (PDF) shall be integrating and the value of integral must be equal to 1 over all x. The pdf must also be positive over all.

The PNN uses the following evaluator to evaluate the pdf-

$$f_A (x) = \frac{1}{(2\pi)^{n/2}\sigma^n} - \frac{1}{m_n} \sum_{i=1}^{m_A} \exp\left[-2\frac{(x-x_A)^T(x-x_{Ai})}{\sigma^2}\right] \quad (9)$$

Where, x_{A_i} - i^{th} training pattern from Class A.

n - Dimension of the input vectors.

m_A - Number of training patterns in Class A.

σ - Smoothing parameter (relates to standard deviations of Gaussian distribution).

$f_A (x)$ is the sum of Gaussian distributions and can be considered as an evaluator until the probability density is continuous and smooth. Also, it tends to PDF as the number of data counts used for evaluation process is increased [24].

IV. PATTERN RECOGNITION NEURAL NETWORK

Neural networks have been applied majorly in Pattern recognition problems because it helps in making reasonable classification of patterns. The process of learning in pattern recognition network involves two main steps and these are up gradation of the architecture and weights connected in the network. This network is given desired supervised learning and is made to learn from a training set. It is a three layered feed forward network with size (13-30-8) which is trained to classify inputs in accordance to target classes.

The hidden layer consists of a sigmoid function and output is a softmax function. In the network each neuron in the input layer designates a feature set while in output layer each neuron designates a pattern which the network learns while being trained. The network is scaled conjugate gradient back propagation trained. It is called using nprtool in MATLAB or by employing a command line [25].

V. RESULTS AND DISCUSSION

A. Generation of Power Quality Disturbances

The power quality disturbances were generated using the parametric equations in MATLAB 2013a. Table I explain the notation of signals as:

TABLE I.   DESCRIPTION OF SIGNAL NOTATIONS

Signal Notation	Signal Description
s1	sag
s2	swell
s3	sag with harmonics
s4	swell with harmonics
s5	interruption
s6	harmonics
s7	transient
s8	flicker

Table II describes the mathematical model of the disturbance equations with the range of parameters.
The power quality disturbances so generated are shown in "Fig.2" to "Fig.9".

These disturbances are simulated by formulating the parametric equations given in table II in the form of MATLAB code. The parametric equations are selected because they represent the close approximation to real-time equations. The parametric equations for 8 types of power quality disturbances are shown. Out of these 8 equations, two are of complex power quality disturbances- sag with harmonics and swell with harmonics donated by s3 and s4.
The parameter variation is according to IEEE-1159[10].

978-1-5386-6624-1/18 $31.00 © 2018 IEEE

TABLE II. MATHEMATICAL MODEL OF PQ ISSUES

Signal	Parametric Equations	Range of Parameters
s1	$y(t)=A[1-\alpha(u(t-t_1)-u(t-t_2))]\sin(\omega t)$	$0.1\leq\alpha\leq0.9; T\leq t_2-t_1\leq9T$
s2	$y(t)=A[1+\alpha(u(t-t_1)-u(t-t_2))]\sin(\omega t)$	$0.1\leq\alpha\leq0.8; T\leq t_2-t_1\leq9T$
s3	$y(t)=A[1-\alpha(u(t-t_1)-u(t-t_2))][\alpha_1\sin(\omega t)+\alpha_3\sin(3\omega t)+\alpha_5\sin(5\omega t)]$	$0.1\leq\alpha\leq0.9; T\leq t_2-t_1\leq9T, 0.05\leq\alpha_3,\alpha_5,\alpha_7\leq0.15; \Sigma\,\alpha_i^2=1$
s4	$y(t)=A[1+\alpha(u(t-t_1)-u(t-t_2))][\alpha_1\sin(\omega t)+\alpha_3\sin(3\omega t)+\alpha_5\sin(5\omega t)]$	$0.1\leq\alpha\leq0.8; T\leq t_2-t_1\leq9T, 0.05\leq\alpha_3,\alpha_5,\alpha_7\leq0.15; \Sigma\,\alpha_i^2=1$
s5	$y(t)=A[1-\alpha(u(t-t_1)-u(t-t_2))]\sin(\omega t)$	$0.9\leq\alpha\leq1; T\leq t_2-t_1\leq9T$
s6	$y(t)=A[\sin(\omega t)+\alpha^{-\alpha(t-t_1)/\tau}\sin\omega_n(t-t_1)][u(t_2)-u(t_1)]$	$0.1\leq\alpha\leq0.8; 0.5T\leq t_2-t_1\leq3T, \quad 8ms\leq\tau\leq40ms; 300\leq f_n\leq900Hz$
s7	$y(t)=A[\alpha_1\sin(\omega t)+\alpha_3\sin(3\omega t)+\alpha_5\sin(5\omega t)+\alpha_7\sin(7\omega t)]$	$0.05\leq\alpha_3,\alpha_5,\alpha_7\leq0.15; \Sigma\,\alpha_i^2=1$
s8	$y(t)=A[1+\alpha_f\sin(\beta\omega t)]\sin(\omega t)$	$0.1\leq\alpha_f\leq2; 5\leq\beta\leq20\ Hz;$

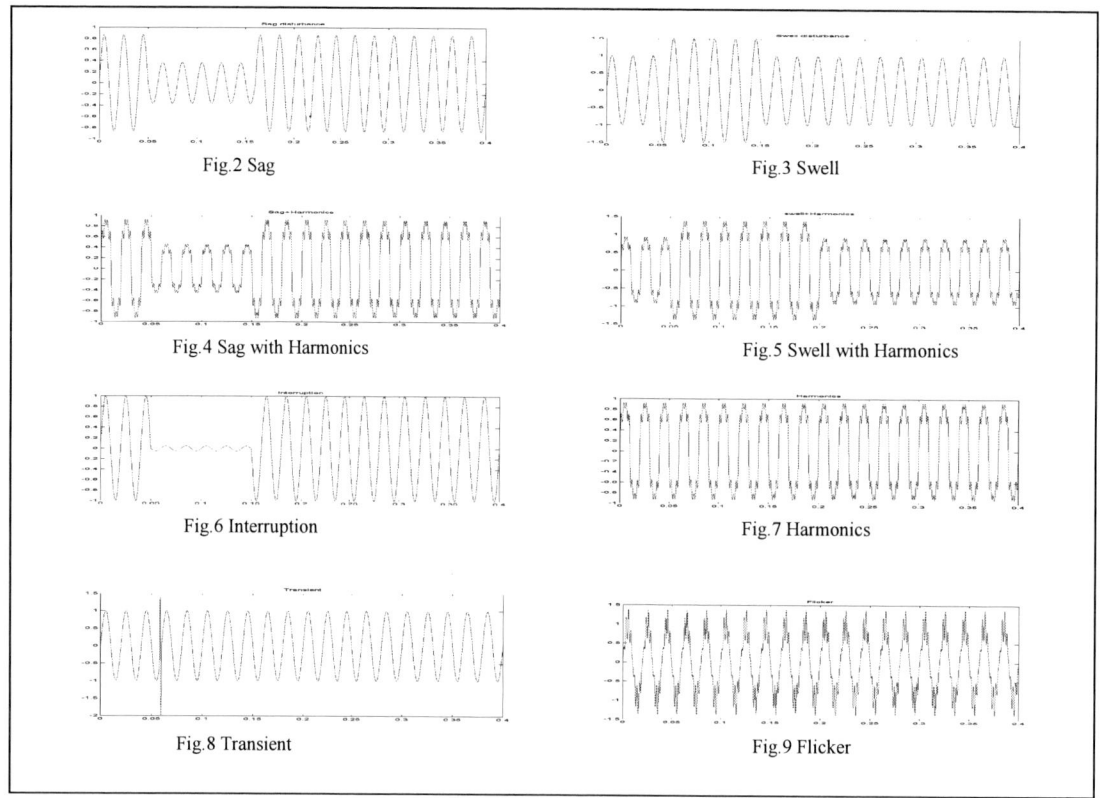

Fig.2 Sag

Fig.3 Swell

Fig.4 Sag with Harmonics

Fig.5 Swell with Harmonics

Fig.6 Interruption

Fig.7 Harmonics

Fig.8 Transient

Fig.9 Flicker

B. Feature extraction

The generated signals were analyzed using wavelets and were decomposed till 8 levels using wavelet analysis and synthesis toolbox in MATLAB 2013a. The decomposition of the sag signal is as shown in the "Fig.10". It is evident from the figure that the disturbance is occurred at 0.05 seconds and it lasted till 0.15 seconds which is resolved into the approximation and detail coefficients. The approximation coefficients were taken for feature extraction.

The features taken for the prediction task along with their values are shown in Table III. Since the features are statistical in nature therefore the difference between various PQDs is inevitable because of the difference in the statistical parameters of the signals. 20 orientations were formed for each of the signals which yielded 160 patterns, out of which 112 patterns were used for training, 24 sets were used for validation and the rest 24 for testing.

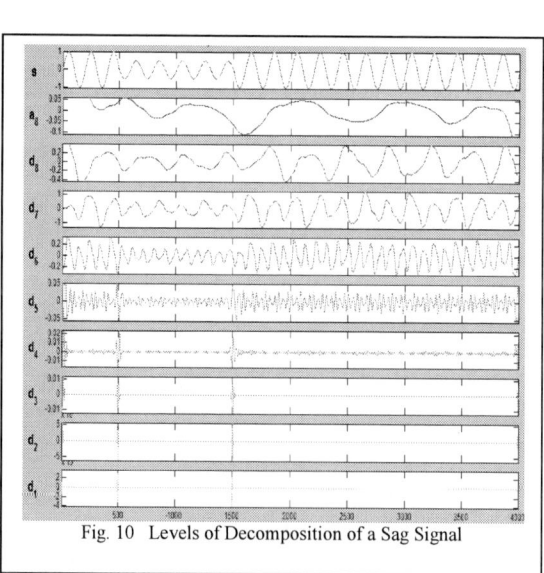

Fig. 10 Levels of Decomposition of a Sag Signal

TABLE III.   EXTRACTED FEATURES

Signal	Mean	Median	Sample maximum	Sample minimum	Range	Standard deviation	Median absolute deviation	Mean absolute deviation	L1-Norm	L2-Norm	Maximum norm	Energy	Amplitude
s1	0.5418	0.6476	2.356	-1.449	3.805	0.976	0.7071	0.7656	14.37	4.357	2.356	0.9775	1
s2	0.4801	0.8475	2.004	-1.237	3.241	0.9188	0.6646	0.898	16.75	4.572	2.004	0.9775	1
s3	0.3927	0.7533	1.712	-1.203	2.915	0.9218	0.594	0.8606	15.34	4.316	1.712	0.9771	0.866
s4	1.345	0.7551	3.853	-1.287	5.14	1.905	1.619	1.679	28.11	9.13	3.853	0.9685	1
s5	1.164	0.8449	3.337	-1.275	4.612	1.719	1.728	1.452	26.84	8.125	3.337	0.9687	0.866
s6	0.5121	0.6572	2.13	-0.7732	2.903	0.9553	0.7775	0.7977	14.76	4.229	2.13	0.9751	1
s7	1.092	0.5866	3.142	-0.6987	3.841	1.536	1.215	1.379	22.39	7.381	3.142	0.9694	0.866
s8	1.147	0.646	0.3308	-0.74655	4.071	1.1619	1.335	1.448	23.73	7.769	3.305	0.9138	0.866

C. Prediction Using Probabilistic Neural Network (PNN)

Probabilistic neural network (PNN) with spread value (δ) selected as 0.005 is used for prediction purpose.PNN classified all the training, validation and testing patterns correctly. The inputs for PNN are the generated patterns, the outputs are the PQ disturbances which are assigned as 1 in the case of a PQ event and 0 otherwise. The all confusion matrix including the classification of training, validation and testing patterns is shown in "Fig. 11".

The rows and columns correspond to the predicted class and true (target) class respectively. Correctly classified observations are shown by the diagonal. Misclassified observations if any are shown by off diagonal. The bottom right cell shows the overall accuracy. The figure hence depicts that all the patterns are correctly classified. The results are evaluated by PNN in a single iteration therefore the training time required is very less.

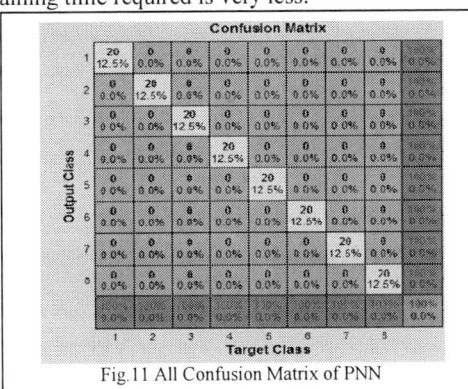

Fig.11 All Confusion Matrix of PNN

D. Comparison Using Pattern Recognition Neural Network (PRNN)

To evaluate the performance of the probabilistic neural network, the same set of training, validation and testing patterns was utilized for classification of various power quality issues using pattern recognition neural network.To develop the appropriate neural network various trials were taken by changing the number of neurons in hidden layer. Best results so,obtained are given in this paper and are with 30 neurons in hidden layer. Optimal training was provided to the Patternnet(13-30-8).Iteration wise training, validation and testing errors are shown in "Fig. 12". The classification although is accurate but it required many iterations as compared to the single iteration of PNN. The all confusion matrix of PRNN is also shown in "Fig.13" which represents that all the patterns are correctly classified.

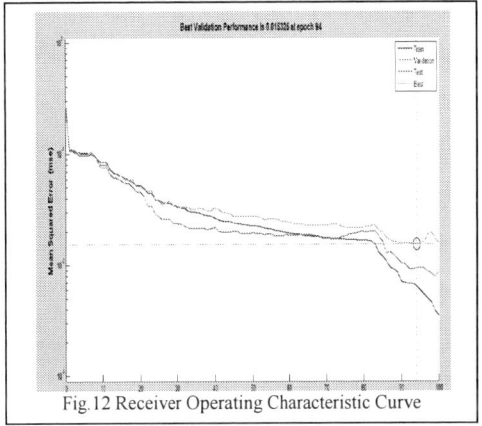

Fig.12 Receiver Operating Characteristic Curve

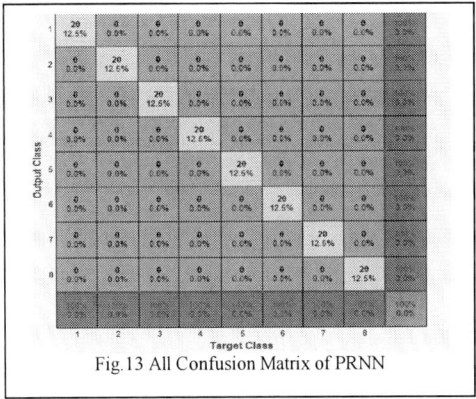

Fig.13 All Confusion Matrix of PRNN

VI. CONCLUSION

From the above discussion it is clear that the probabilistic neural network predicted the patterns correctly and without any misclassifications. The training time required by PNN is much lesser than that of PRNN as PNN provided correct results in a single iteration only. On the other hand, pattern recognition neural network, though, classified all the patterns accurately but it required too many iterations which indeed required too much of the time in its training so as to bring it to the point of 100 % classification.

Hence, PNN can be chosen as the faster and reliable classifier for the real time power quality disturbance detection.

REFERENCES

[1] D.Mittal, O. P. Mahela and R. Jain, "Classification of power quality disturbance in electric power system: A review," IOSR Journal of Electrical and Electronics Engineering, vol.3, pp 06-14, Nov-Dec 2012.

[2] R. C. Dugan, M. F. Granaghan, S Santoso, and H.W. Beaty, Electric Power systems quality, 2nd ed.,New York.:McGraw-Hill,2003,pp 4-9

[3] S.Chattopadhyay, M. Mitra and S.Sengupta, Electric Power Quality, Power Systems,1st ed.,Netherlands: Springer, 2011,pp5-12

[4] C. Sankaran, Power quality,Boca Raton:CRC press, 2002,pp 12-34

[5] A.K. Khan, "Monitoring power for the future," power engineering journal, IET, vol.15, pp 81-85,April 2001

[6] C. Bhartiraja , H. Chowdary V., "Real time Power quality phenomenon for various distribution feeders," Indonesian journal of Electrical Engineering and Computer Science,vol.3,pp 10-16,July 2016.

[7] M.Kezunovic and Y. Liao, "A new method for classification and characterization of voltage sags," Electric Power Systems Research,Elsevier,vol.58,pp 27-35,March 2001

[8] P.V. Dhote, B.T. Deshmukh, B.E.Kushare, "Generation of power quality disturbances using MATLAB- Simulink," IEEE International Conference on Computation of Power, Energy, Information and Communication, April 2015

[9] P.S. Bimbhra, Electrical Machinery,7th ed.,New Delhi:Khanna Publishers,2012, pp272

[10] S. Khokar, A. A. M. Zin, A.P. Memon and A.S. Mokhtar, "A new optimal feature selection algorithm for classification of power quality disturbances using DWT and PNN," Measurement, vol.95, pp 246-259, January 2017.

[11] G.T. Heydt, P.S. Fjeld, C.C. Liu, D. Pierce, L. Tu and G. Hensley, "Applications of the windowed FFT to electric power quality assessment," IEEE Trans. Power Delivery, vol.14, pp 1411-1416, October1999

[12] F. Jurado, J.R. Saenz, "Comparison between discrete STFT and wavelets for the analysis of power quality events," Electric Power Systems Research.,Elsevier,vol.62 ,pp183-190, July 2002

[13] S.Mishra, C.N. Bhende and B.K. Panigrahi, "Detection and classification of power quality disturbances using S-Transform and probabilistic neural networks," IEEE Trans. Power Delivery, vol.23, pp 280-286,January 2008.

[14] C.-Y. Lee and Y.-X. Shen, "Optimal feature selection for power-quality disturbances classification", IEEE Trans. Power Delivery, vol.26, pp 2342-2351, October 2011.

[15] P.Kanirajan, and V.S. Kumar, "Power Quality disturbance detection and classification using wavelet and RBFNN", Applied Soft Computing, vol.35, pp 470-481, October 2015.

[16] F.Ucar, O.F.Alcin, B.Dandil and F. Ata, "Power quality event detection using a fast extreme learning machine," Energies, MDPI, vol.11, pp 1-14, January 2018.

[17] M. F. Othman and H. A. Amari, "Online fault detection for power system using Wavelet and PNN," IEEE 2nd International Power and Energy Conference, December 2008.

[18] M.B. Latran and A. Teke,, "A novel wavelet transform based voltage sag/swell detection algorithm," Electrical power and energy systems,vol.71,pp-131-139, October 2015.

[19] S. G. Mallat, "A theory for multi-resolution signal decomposition: the wavelet representation", IEEE Trans.Pattern Analysis machine intelligence.,vol.II, pp 674-693, July 1989.

[20] R.Polikar, The wavelet tutorial,2nd ed., pp1-67,Available: web.iitd.ac.in/~sumeet/WaveletTutorial.pdf

[21] S.Santoso, Edward J. Powers, W.Mack Grady and Peter Hofmann, "Power quality assessment via wavelet transform analysis," IEEE Trans.Power Delivery.,vol.2, pp 924-930 ,April 1996.

[22] D.F. Specht, "Probabilistic neural networks", Neural Networks,vol.3,pp 109-118,1990.

[23] K.Z. Mao, K.-.C. Tan and W. Ser, "Probabilistic neural-network Structure determination for pattern classification," IEEE Trans. Neural Networks,vol.11,pp 1009-1016 ,July 2000

[24] S.N. Sivanandam, S .Sumathi and S.N. Deepa, Introduction to Neural Networks using MATLAB 6.0, 12th ed.,New Delhi: Tata McGraw-Hill, pp312-314.

[25] Mathworks, Available: https://www.mathworks.com › ... › Pattern Recognition

Intelligent Approach for Load Balancing on an 11 kV Feeder

Jagriti Surabhi , Priyanka Chaudhary, M. Rizwan
Department of Electrical Engineering
Delhi Technological University
New Delhi, India
gravity.jag@gmail.com

Abstract—The present electrical network is designed with the aim to deliver adequate power at the requisite voltages to the loads connected at the consumer end, both in retrospect and in the future. As per reports provided in January 2018, India's installed capacity is at 3,29,298 megawatt(MW) making it easily one of the forerunners in the Power sector ranked at 5th position after European Union, China, US and Japan. Handling such a large capacity must be consistent with ensuring good quality of supply, minimum losses along the line, reliability and most importantly, economy. Load balancing is a technique ensuring judicious supply in all conductors of the system henceforth minimizing losses and excessive voltage drops for a paradigmatic performance. In this paper, for the aforementioned real-time system, a load balancing technique using fuzzy logic is used for facilitating the load change decision. Fuzzy Logic has been implemented in this paper on an 11 KV Feeder named 6/18 s/s of the Rohini-5 grid in New Delhi, India under the DISCOM – TPDDL.

Keywords—- *Load Balancing, Feeder load balancing, Intelligent Techniques, Fuzzy Logic in Distribution system, Radial Distribution system*

I. INTRODUCTION

The Power System consists of three universal parts – Generation, Transmission and Distribution with the latter being that part of the system connecting the Transmission and Consumer service points. A genuine term for this is Feeder which can be defined as a Line directly feeding the consumer from a Distribution Substation. With dependence on electrical supply increasing manifold and automation making way for better control, intelligent techniques can be adopted for monitoring load variations and paving way for phase balancing [1].

The secondary side of a distribution transformer is usually configured in Wye as it comes with promising features, few being its ability to furnish both 3-phase and 1-phase loads, the latter even more so in a three phase 4 wire system for its grounding conductor along with supplanting the flow of dominant third harmonic current into supply lines by circulating it within the transformer itself. Single phase loads (Lighting) in a distribution system require a neutral as a return path and most efforts lie in ensuring even load dispensing in all three phases [2]. A Balanced System is one in which Impedance in all the three phases (RYB or ABC) are equal leading to equivalent current flow in the phases along with phase voltage drops. Hence, no Neutral current flows as opposed to the case of Imbalance in the phases of a system, which leads to violation of KCL at the star node and current flowing through the Neutral. An unwanted voltage

develops due to this at the neutral point resulting into unwanted waveforms and harmonic distortion. This imbalance also causes communication interference, over and under loading, malfunctioning of protective devices affecting the health of the system and compromising on the efficiency of the whole intended purpose.

Conventionally, this imbalance is addressed by physically changing the connection phases post analysis and field measurement. It still isn't an effective approach for it being time consuming and impractical in many cases. The redistribution of loads from heavily to lightly loaded area is a possible configuration to improve on most common cases of Imbalance. In contemporary power distribution systems, tie-switches are used for feeder reconditioning, the one demerit here being that it makes voltage and current imbalance worse in a few cases. The authors presented a real time glimpse of a distribution system in [3] and Neural networks based techniques has been implement to find a favorable rearranging condition of the loads connected to the individual phases of system in [5,6]. These results point towards Radial Topology in Feeders to satisfy the superlative condition.

For the problem undertaken, a mathematical model is formulated and solved using Fuzzy logic. This is to minimize the problem of phase imbalance on an existing 11 KV Feeder as a novel technique with few constraints defined.

II. PROBLEM DESCRIPTION

The proposed model is used for an existing feeder. Fuzzy logic, inputs and outputs are chosen suitably and a constraint algorithm is applied to depict model success along with its limitations.

A. Feeder Description

The distribution feeders are generally a three-phase, four-wire Wye system with an open loop or radial structure. It is designed considering the current capacity of conductors and without any tapping taken from it. The connection between a feeder in consideration and distribution transformer is suitably rearranged to facilitate a balance condition. There are four topologies to choose from, them being – Radial, Parallel, Ring Main and Meshed. From the construction and protection point of view, the Radial system is most simple and economic to work with and supports our proposed technique well. For the problem in this paper, an 11 KV Feeder is considered supplying to consumers in Sector -5 of Rohini suburb area. In Fig.1, an example distribution feeder is demonstrated load connected to any one of the three phases through the use of Tie-Switches.

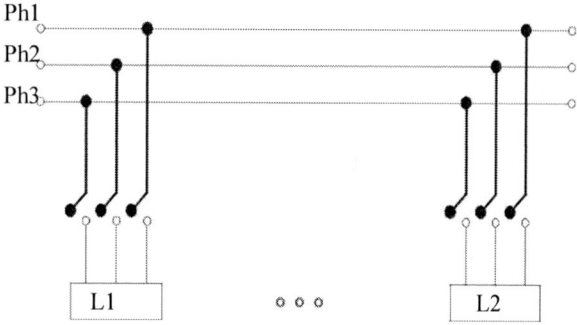

Fig. 1. A Prototype Distribution Feeder

B. Balancing of currents in each phase

Most of the domesticated areas in India use the average of five kilowatt power. The major use is by lighting and domestic loads. Any large single phase load, or a collection of small loads connected to only one phase result in relatively more current to flow in that phase hence causing voltage drop on that line. This phase will have poor voltage regulation. An instantaneous surge in power, like that arising by the use of heater etc, often introduces an unwanted component of harmonic power to flow in the distribution system. This ruptures the cables as well as affects the distribution transformers. The loss in power on a line is a function of real and reactive power flowing in the lines which is intrinsically based on the type of loads connected, real or reactive or both. The basic formulation in a phase balancing problem is done by keeping the load values as independent variables. Switch Breakers installed at each of the connections on the network feeders aspire to provide control for reduction in power loss.

C. Proposed Technique

In this paper, for the aforementioned real-time system, a proposed load balancing technique using fuzzy logic is used for facilitating the load change decision. The flowchart of this proposition is shown in Fig. 2 where the input fed to the system is the cumulative phase load (calculated for each of the phases). An upper limit is set for this model and the mean imbalance per phase is computed against it. If the mean imbalance per phase is below 10 kW, it can be deduced that the system is approximately balanced and will supplant any incoming load balancing.

It steers towards fuzzy logic-based balancing in an event of a case otherwise. The output is the change in load value per phase. The sign of this output indicates the nature of load, i.e. a positive output indicates that that specific phase is loaded less than its capacity and can accept that amount of load. Similarly, a negative output value is an indicator of that phase being over-loaded and can release that value of load.

This obtained change in load is fed as inputs to the intelligent system which aims at judiciously relocating the specific load points. But, sometimes the intelligent system may be unable to implement the definite value of change in load as obtained by the fuzzy rule. This happens because the actual load points for any phase might not lead to an adequate combination which totals up to the exact variation in value stated by the fuzzy rule.

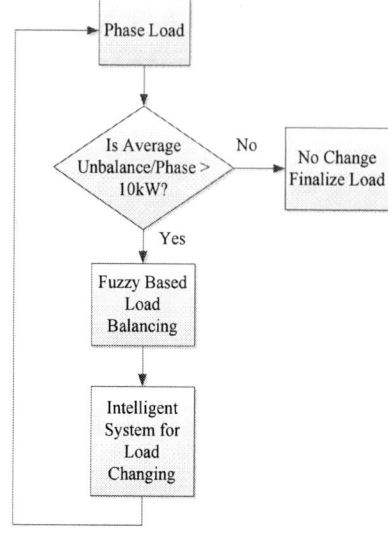

Fig. 2. Flowchart for applied Balancing in Load

So, it is executed from the best possible variation obtained from the intelligent system and it sequentially checks the system imbalance till the mean imbalance (MI) falls below 10 kW, if possible. Mean imbalance is calculated by the equation,

$$MI = |LoadPh1 - LoadPh2| + |LoadPh2 - LoadPh3| +$$

$$|LoadPh3 - LoadPh1| \quad (1)$$

$$\text{Mean Imbalance per Phase} = [\text{Mean Unbalance} / 3] \quad (2)$$

III. LOAD BALANCING USING FUZZY LOGIC

In this section, the fuzzy logic-based load balancing technique is explained in detail.

TABLE □. LOAD HANDELED ON THE 11 KV TDPPL FEEDER ON 18.07.2018

Time	R(KW)	Y(KW)	B(KW)
00:00	1750.342	1784.75	1696.211
01:00	1784.585	1965.612	1744.006
02:00	1740.695	1761.364	1652.695
03:00	1607.408	1652.827	1569.678
04:00	1133.935	1216.028	1095.028
05:00	992.838	1036.662	985.039
06:00	904.937	937.838	892.078
07:00	919.204	965.316	908.534
08:00	941.402	977.086	930.149
09:00	968.572	1022.461	921.283
10:00	978.505	1024.749	941.028
11:00	981.695	1026.146	951.819
12:00	1011.901	1074.986	940.819
13:00	1165.197	1341.197	1099.197
14:00	1399.794	1487.794	1377.794
15:00	1366.816	1443.816	1256.816
16:00	1008.194	1107.194	986.194
17:00	1025.838	1091.838	981.838
18:00	1042.151	1086.151	1020.151
19:00	1256.816	1333.937	1210.286
20:00	1319.835	1383.283	1306.151
21:00	1344.937	1453.056	1352.527
22:00	1605.538	1703.449	1571.625
23:00	1791.383	1969.627	1768.206
24:00	2004.816	2081.211	1976.194

As mentioned in section 2, the load balancing technique is applied on an existing 11KV feeder whose data is given in Table I as shown above.

For designing our Fuzzy Controller, an assumption is made that the upper limit of load capacity in any phase is 2800 kW. This is based on the load readings obtained from the Feeder. The Fuzzy Controller should be avoided beyond this limit for balancing the load. This is because, in practice, when a phase reaches 200% of its overload capacity, significant circulating current will flow within the transformer winding making it hotter than anticipated. If the overloading is instantaneous and high at that, there is a chance that the transformer may burn out.

A. Input & Output Design of the proposed Fuzzy Controller

The design sequence is initiated by choosing the variables for input and output. To balance the phase loads, the cumulative (kW) for each of the three phases is chosen as the input variables. Three inputs are designed for each phases.

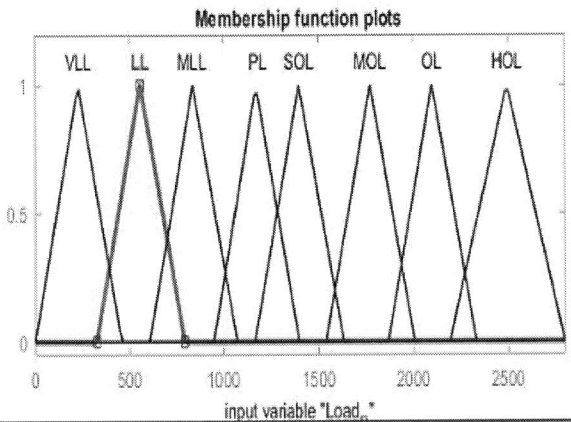

Fig.3. Triangular membership functions in Fuzzy for the input variables

TABLE □. FUZZY MNEMONICS FOR THE INPUT VARIABLES

Serial Number	Input (Load) character	Mnemonic	kW range
01	Very Less Loaded	'VLL'	[0] - [460]
02	Less Loaded	'LL'	[327] - [793]
03	Medium Less Loaded	'MLL'	[607] - [1073]
04	Perfectly Loaded	'PL'	[950] - [1400]
05	Slightly Overloaded	'SOL'	[1167] - [1633]
06	Medium Over Loaded	'MOL'	[1540] - [2007]
07	Over Loaded	'OL'	[1867] - [2333]
08	Heavily Overloaded	'HOL'	[2193] - [2800]

TABLE □. FUZZY MNEMONICS FOR THE OUTPUT VARIABLES

Serial Number	Output (Change) character	Mnemonic	kW range
01	High Subtraction	'HS'	[-1400] - [-793]
02	Subtraction	'S'	[-933] - [-467]
03	Medium Subtraction	'MS'	[-607] - [-140]
04	Slight Subtraction	'SS'	[-467] - [234]
05	Perfect Addition	'PA'	[0] - [467]
06	Medium Addition	'MA'	[327] - [794]
07	Large Addition	'LA'	[607] - [1073]
08	Very Large Addition	'VLA'	[934] - [1400]

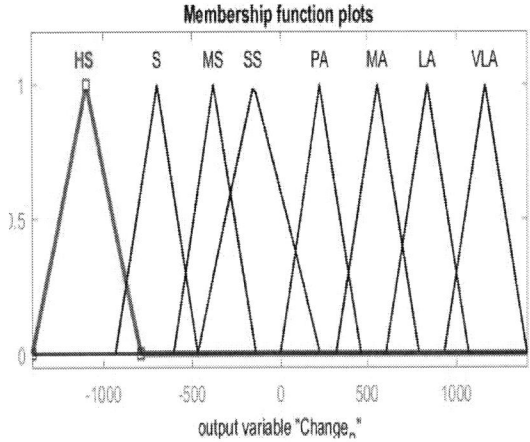

Fig. 4. Triangular membership functions in fuzzy for the output variables

Three sets of variables form the input representing each phase of the feeder is used. Similarly, three sets of output variables are defined to show the change in load in each phase possible to implement load balancing.

B. Fuzzy Design Rules & Non-Linear Surface Plot

Next, using the inbuilt fuzzy rule IF-THEN set [8], input and output variables are associated as stated in Table □.

In retrospect, Fig. 3 depicts the membership functions for the chosen input variables and Table II shows the fuzzy mnemonics.

978-1-5386-6624-1/18 $31.00 © 2018 IEEE

TABLE IV. FUZZY IF-THEN RULES FOR THE INPUT AND OUTPUT
VARIABLES

Rule No.	Rule Description
01	If Load is [VLL] then Change is [VLA]
02	If Load is [LL] then Change is [LA]
03	If Load is [MLL] then change is [MA]
04	If Load is [PL] then change is [PA]
05	If load is [SOL] then change is [SS]
06	If Load is [MOL] then change is [MS]
07	If Load is [OL] then change is [S]
08	If Load is [HOL] then change is [HS]

For the output variables, Table III shows the fuzzy glossary, and Fig. 4 the respective triangular fuzzy membership functions. Analogous to the fuzzy input, output load change variables and the associated rule set, the fuzzy surface is shown in Fig. 5, showing the non aligned relationship between the chosen input and output variables.

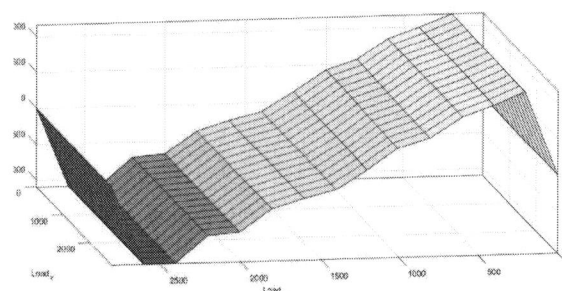

Fig. 5. Non aligned relationship between the input and the output variables considering two phases at a time

C. Simulation Results

This section shows the utilization results using the fuzzy logic-based balancing of load technique. Matlab® fuzzy toolbox [9] using the Mamdani fuzzy interfacing technique [10] is used for this simulation. To demonstrate, let us take a reading recorded at 22:00 hours as the input given by [1606 1703 1572] kW for the three phases. The aforementioned fuzzy controller is applied to it for balancing the load. The plot showing load change in all three phases corresponding to the input load fed to it using the designed fuzzy rules is shown in Fig. 5. The defuzzification is derived from the Mamdani (centroid) technique [10]. After approximating the change in load as output, for the input load :

$$P_{in} = [1606\ 1703\ 1572]^T \quad (kW) \tag{3}$$

the load change configuration (output) is:

$$\Delta P_{fuzzy} = [-263\ -373\ -165]^T\ (kW) \tag{4}$$

Notwithstanding, with this load change combination obtained, it exhibits an error. This is because the negative and positive sums are not equal, i.e. $\Sigma\Delta P_{fuzzy} \neq 0$ kW. So, if this load change combination is actualized, it will conclude in reduction of total load which is neither desirable nor our motive. The load balancing technique is about rearranging the load points between the three phases in a way that the total load same remains the same.

So, an error correction is put in order. The mean error (ME) is given as:

$$ME = round\ \frac{\Sigma\Delta Pfuzzy}{3} \tag{5}$$

An error matrix ΔP_{error} is constructed using this mean error by accomodating the computed ME judiciously among all three phases.

$$\Delta P_{error} = \begin{bmatrix} ME \\ ME \\ \Sigma P_{fuzzy} - 2*ME \end{bmatrix} \tag{6}$$

Using this, the load change array ΔP is obtained by subtracting ΔP_{error} array from the unchecked fuzzy output array ΔP_{fuzzy}.

$$\Delta P = \Delta P_{fuzzy} - \Delta P_{error} \tag{7}$$

and: $\Sigma\Delta P = 0 \tag{8}$

Applying equations (5) - (8) to our study case, the following criterion are:

$$ME = -267\ kW, \quad \Delta P_{error} = [-267\ -267\ -267]^T\ kW\ and:$$

$$\Delta P = \begin{bmatrix} -263 \\ -373 \\ -165 \end{bmatrix} - \begin{bmatrix} -267 \\ -267 \\ -267 \end{bmatrix} = \begin{bmatrix} 4 \\ -106 \\ 10 \end{bmatrix} \quad (kW) \tag{9}$$

The final output is :

$$P_{final} = P_{in} + \Delta P = \begin{bmatrix} 1606 \\ 1703 \\ 1572 \end{bmatrix} + \begin{bmatrix} 4 \\ 106 \\ 102 \end{bmatrix} = \begin{bmatrix} 1610 \\ 1597 \\ 167 \end{bmatrix} (kW) \tag{10}$$

Applying (8) on P_{in} and P_{final}, the Nascent Absolute Imbalance (NAIB)/Phase and Final Absolute Imbalance (FAIB)/Phase will be, respectively:

$$(NAIB)/Phase = 87.33\ kW \tag{11}$$

$$(FAIB)/Phase = 51.33\ kW \tag{12}$$

TABLE V. UTILIZATION RESULTS FOR DIFFERENT PHASE LOADING

Time	Nascent Load (kW)	IAIB/Ph (kW)	Nascent Fuzzy Change (kW)	Error (kW)	Final Fuzzy Change (kW)	Final Load (kW)	FAIB/ Ph (KW)
13: 00	1165.19	161.33	233	141.7	91.3	1256.3	59.933
	1341.19		-32.9	141.7	-174.6	1166.4	
	1099.19		233	141.7	91.3	1190.3	
15:00	1366.81	124.666	-84.1	-56.33	-27.467	1339.5	22.066
	1443.81		-128	-56.33	-71.367	1372.6	
	1256.81		42.2	-56.33	98.833	1355.8	
16:00	1008.19	80.533	403	364.6	38.33	1046.3	69.4733
	1107.19		233	364.6	-131.67	975.3	
	986.19		458	364.6	93.34	1079.5	
18:00	1042.15	44	318	308	10	1052	49
	1086.15		233	308	-75	1011	
	1020.15		373	308	65	1085	
19:00	1256.81	82.67	42.2	44.96	-2.767	1254.2	15.867
	1333.28		-20.3	44.96	-65.267	1268.7	
	1210.28		113	44.96	68.034	1278.0	
20:00	1319.83	51.33	-29	-47.54	18.54	1388.5	6.26
	1383.28		-100	-47.54	-52.46	1330.5	
	1306.15		-13.6	-47.54	33.93	1339.9	
21:00	1344.93	72	-57.9	-84.23	26.33	1371.3	25.867
	1453.05		-128	-84.23	-43.76	1409.2	
	1352.52		-66.8	-84.23	17.433	1370.4	
22:00	1606	87.33	-263	-267	4	1610	51.333
	1703		-373	-267	-106	1597	
	1572		-165	-267	102	1674	
23:00	1791.3	119.67	-373	-451	78	1869	36.67
	1969.6		-607	-451	-156	1814	
	1768.2		-373	-451	78	1846	
00:00	2005	70	-695	-672.6	-22.33	1982.6	47.33
	2081		-700	-672.6	-27.33	2053.6	
	1976		-623	-672.6	49.66	2016.6	

An improvement in phase balancing can be observed with the reduction in imbalance. Table V shows few application results for different hours of a day obtained from the feeder. A failure case has also been documented which violates our model's architecture.

The chosen inputs for this study were chronicled through a load survey on an 11 kV Feeder in Rohini area of New Delhi. It consists of current readings converted to load capacity on each phase taken over a period of 24 hours on 18.07.2018. The final fuzzy output change configuration is passed onto an expert system for initiating the load change and balance operation.

IV. CONCLUSION

A load balancing technique minimizing losses and excessive voltage drops using fuzzy logic based intelligent technique has been presented in this paper. The proposed technique has been applied on real-time system, which is 11 KV Feeder named 6/18 s/s of the Rohini-5 grid in New Delhi, India under the DISCOM – TPDDL.

REFERENCES

[1] M. W. Siti, A. A. Jimoh and D. V. Nicolae, "Phase load balancing in the secondary distributin network using fuzzy logic," AFRICON 2007, pp. 1.-7

[2] S. Cinvalar, J. J. Grainger, H. Yin and S. S. H. Lee, "Distribution feeder reconfiguration for loss reduction," IEEE Transactions on Power Delivery, vol. 3, no. 3, pp. 1217-1223, 1988.

[3] M.E. Baran and W. Kelly, "State estimation for real time monitoring of distribution systems," IEEE Transactions on Power Systems, vol. 9, no. 3, pp. 1601-1609, 1994.

[4] M. Siti, A. Jimoh and D. Nicolae, "Automatic load balancing in Distribution Feeeder," IECON 05, Raleigth North Caroline, USA..

[5] A. Ukil, W. Siti and J. Jordaan, "Feeder Load Balancing Using Neural Network," Lecture Notes in Computer Science 3972, pp. 1311-1316, 2006.

[6] M. A. Kashem, G.B. Jasmon and V. Ganapathy, "A new approach of distribution system reconfiguration for loss minimization," Electrical Power and Energy Systems, vol. 22, pp. 269-276, 2000.

[7] D. Das, "Reconfiguration of distribution system using fuzzy multi-objective approach," Electrical Power and Energy Systems, vol. 28, pp. 331-338, 2006.

[8] B. Kosko, Fuzzy Engineering, Prentice Hall, 1999.

[9] MATLAB® Documentation – Fuzzy Logic Toolbox, Version 6.5.0.180913a Release 13, The Mathworks Inc., Natick, MA.

[10] E. H. Mamdani, "Applications of fuzzy logic to approximate reasoning using linguistic synthesis," IEEE Transactions on Computers, vol. 26, no. 12, pp. 1182-1191, 1977.

[11] M. Rizwan, D. Kumar and R. Kumar, "Fuzzy Logic Approach for Short Term Electrical Load Forecasting," Electrical and Power Engineering Frontier, vol. 1, no. 1, pp. 8-12, 2012.

A Control Strategy for Subsynchronous Resonance Under Varying Degree of Series Compensation

Narendra Kumar, Shilpa Gupta, Nisha Singh
Dept. of Electrical Engg., *DTU*, Delhi-110042, India.
(e-mail: dnk_1963@yahoo.com, shilpagupta1807@gmail.com).

Abstract—A control strategy is obtained for restraining torsional oscillations owing to SSR over a varying degree of series compensation. A mid-point located SVS equipped with a joint supportive signal composed of line current and voltage deviations has been utilized to damp out torsional oscillations grown due to sub-synchronous resonance. The test system considered for analysis of the usefulness of the controller under consideration is built using IEEE first benchmark model having SVS connected at the mid-point of the transmission line. Eigenvalue analysis has been carried out for the linearized power system model to investigate the unstable torsional modes. To explicate the relative assessment of the system under consideration three secondary controllers are considered. To illustrate the robustness of proposed controller a comprehensive power system model is used which is described by various subsystems, their relations are been described in detail with accuracy. The eigen value analysis illustrates the efficacy of the Combined Line Current and Voltage (CLCV) SVS controller for different percentage of the series compensation as compared to the cases where no controller is used and case where only Line Current SVS controller, Voltage SVS controller are used. All results obtained at various levels of series compensation approves usefulness of the SVS controller under consideration.

Keywords—IEEE First Bench Mark Model, Series Compensation, SVS Controller, Eigenvalues.

I. INTRODUCTION

Series capacitor compensation is the most inexpensive solution to the problem of transferring large amount of power over a long distance without the construction of large transmission corridor. Though while decreasing the electrical length in between the generating station and load points series capacitive compensation can make power system susceptible to subsynchronous resonance event as detected at Mohave Generating Station in Southern Nevada in 1970 [1].

Numerous methods are proposed over years to protect the power system from the hostile effects of subsynchronous resonance since it was first observed in 1970.Looking into the severity of the problem in 1974 a distinct set of power system engineers were given responsibility to get insight of the SSR problem [2].

Pole face amortisseurs, SEDC, SMF relay, SSR monitor, static subsynchronous resonance filter (SRF) and static machine frequency (SMF) relay are few technologies proposed [3-5].

Various Flexible AC Transmission systems were also used for reducing oscillations owing to SSR. Rapidly acting Static Var Compensators have encouraged numerous researchers for developing new SVC controllers for improving steady state and transient performance. Many Artificial intelligence (AI) techniques are applied to design controller for SVC [6]. Many more SVC secondary signals such as Bus Frequency deviations, Voltage Angle deviations were also proposed.

Any modifications in the power system can also lead to SSR related problems even though projection scheme may exist. These modifications maybe in the form of variations in structure of network or degree of series capacitive compensation of transmission lines or infrequent switching state in the transmission network [7]. All this require re-modification of an existing SSR protection scheme which is a tedious process. In the given work a new control strategy is proposed and its effectiveness is tested over wide range of capacitive compensation so as to ensure stable operation under every condition.

In the existing work, the effectiveness of a novel secondary signal Combined Line Current and Voltage (CLCV) SVS controller is being established. The IEEE first benchmark model with SVS connected at the mid-point of transmission line is considered for this work. The study of given system is done by calculating the eigenvalue of the linearized model and undamped modes are examined over varying degree of series compensation. The unstable modes together with various system modes are alleviated using the CLCV controller. The Eigenvalue evaluated shows the success of the CLCV controller for the given study system for varying degree of series compensation.

II. SYSTEM MODELLING

IEEE first benchmark model for SSR is assessed as a system for analysis [8]. SVS is located at midpoint of the transmission line with capacitor bank located on the both sides of the Static Var System as shown in Fig.1.

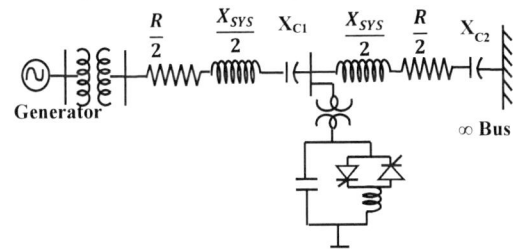

Fig. 1 IEEE First Benchmark Model for SSR study

A. Generator Flux

A detailed mathematical model from [9] is considered here for analysis. The various linearized state equations reflecting the modelling of generator are derived and compiled for further calculation under rotor circuit modelling and are presented as under:

$$\dot{x}_r = [A_r]x_r + [B_{r1}]u_{r1} + [B_{r2}]u_{r2} + [B_{r3}]u_{r3} \quad (1)$$

$$y_{r1} = [C_{r1}]x_r + [D_{r1}]u_{r1} \quad (2)$$

$$y_{r2} = [C_{r2}]x_r + [D_{r2}]u_{r1} + [D_{r3}]u_{r2} + [D_{r4}]u_{r3} \quad (3)$$

where

$$x_r = [\Delta\Psi_f \ \Delta\Psi_h \ \Delta\Psi_g \ \Delta\Psi_k]'$$

$$y_{r1} = [\Delta I_D \ \Delta I_Q]'$$

$$y_{r2} = [\Delta i_D \ \Delta i_Q]'$$

$$u_{r2} = [\Delta V_F]'$$

$$u_{r3} = [\Delta i_D \ \Delta i_Q]'$$

B. Mechanical Turbogenerator

The six-mass representation of the turbo generator shaft is modeled in Fig.2.

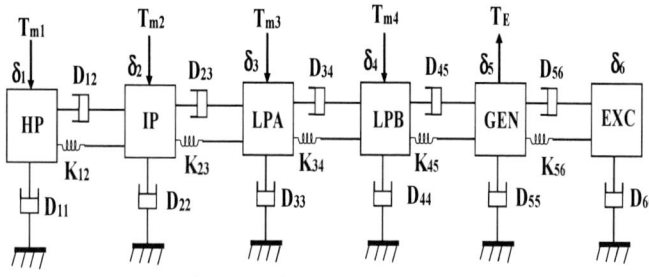

Fig. 2 Turbogenerator System

Turbine mechanical model from [10] is utilized for thorough analysis and linearized state space derivations of same is specified as:

$$\dot{x}_m = [A_m]x_m + [B_{m1}]u_{m1} + [B_{m2}]u_{m2} \quad (4)$$

$$y_m = [C_m]x_m \quad (5)$$

where

$$x_m = [\Delta\delta_1 \ \Delta\delta_2 \ \Delta\delta_3 \ \Delta\delta_4 \ \Delta\delta_5 \ \Delta\delta_6 \ \Delta\omega_1 \ \Delta\omega_2 \ \Delta\omega_3 \ \Delta\omega_4 \ \Delta\omega_5 \ \Delta\omega_6]'$$

$$u_{m1} = [\Delta I_D \ \Delta I_Q]'$$

$$u_{m2} = [\Delta i_D \ \Delta i_Q]'$$

$$y_m = [\Delta\delta_5 \ \Delta\omega_5]'$$

C. IEEE Exciter

The excitation system represented by IEEE type-1 from [11] is used for analysis as shown in Fig. 3.

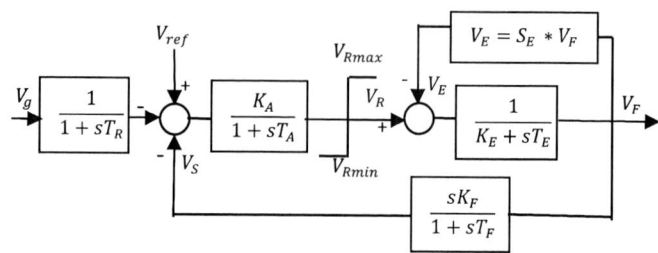

Fig. 3 IEEE Type I Exciter

The state eqs. representing above model is derived as:

$$\dot{x}_e = [A_e]x_e + [B_e]u_e \quad (6)$$

$$y_e = [C_e]x_e \quad (7)$$

where

$$x_e = [\Delta V_f \ \Delta V_s \ \Delta V_t]'$$

$$u_e = \Delta V_g$$

$$y_e = \Delta V_f$$

D. Transmission Line

The pi circuit model of transmission line as in [12] is used for analysis is shown in Fig.4.

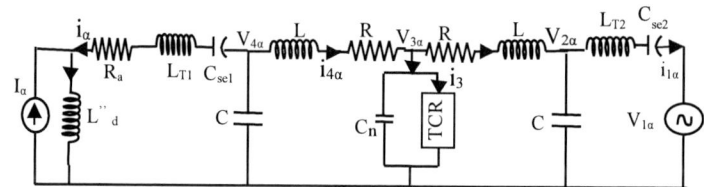

Fig 4 π-circuit representation of Transmission Line

The state eqs. representing the model are as follows:

$$\dot{x}_n = [A_n]x_n + [B_{n1}]u_{n1} + [B_{n2}]u_{n2} + [B_{n3}]u_{n3} \quad (8)$$

$$y_{n1} = [C_{n1}]x_n + [D_{n1}]u_{n1} + [D_{n2}]u_{n2} + [D_{n3}]u_{n3} \quad (9)$$

$$y_{n2} = [C_{n2}]x_n \quad (10)$$

$$y_{n3} = [C_{n3}]x_n \quad (11)$$

where

$$x_n = [\ \Delta i_{1D} \ \Delta i_{2D} \ \Delta i_{4D} \ \Delta i_D \ \Delta \dot{V}_{2D} \ \Delta \dot{V}_{3D} \ \Delta \dot{V}_{4D} \ \Delta \dot{V}_{5D}$$
$$\Delta i_{1Q} \ \Delta i_{2Q} \ \Delta i_{4Q} \ \Delta i_Q \ \Delta \dot{V}_{2Q} \ \Delta \dot{V}_{3Q} \ \Delta \dot{V}_{4Q} \ \Delta \dot{V}_{5Q}]$$

$$y_{n1} = [\Delta V_{gD} \ \Delta V_{gQ}]$$

$$y_{n2} = [\Delta i_D \ \Delta i_Q]$$

$$y_{n3} = [\Delta V_{3D} \ \Delta V_{3Q}]$$

$$u_{n1} = [\Delta i_{3D} \ \Delta i_{3Q}]$$

$$u_{n2} = [\Delta I_D \quad \Delta I_Q]$$
$$u_{n3} = [\Delta \dot{I}_D \quad \Delta \dot{I}_Q]$$

E. SVS System

The SVS control system with feedback is shown in Fig. 5.

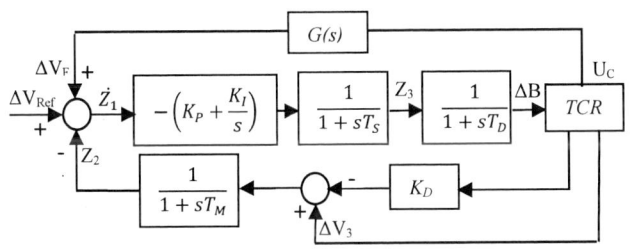

Fig.5 SVS Control System with Secondary Feedback

The state eqs. of the SVS system are specified as:
$$\dot{x}_s = [A_s]x_s + [B_{s1}]u_{s1} + [B_{s2}]u_{s2} + [B_{s3}]u_{s3} \tag{12}$$
$$Y_s = [C_s]x_s + [D_s]u_{s1} \tag{13}$$

where
$$x_S = [\Delta i_{2D} \quad \Delta i_{2Q} \quad Z_1 \quad Z_2 \quad Z_3 \quad \Delta B]'$$
$$u_{s1} = [\Delta V_{2D} \quad \Delta V_{2Q}]'$$
$$u_{s2} = [\Delta V_{ref}]$$
$$u_{s3} = [\Delta V_F]$$
$$y_s = [\Delta i_{2D} \quad \Delta i_{2Q}]'$$

III. Complete Model of the System

The state input and output eqs. of various above components are joined giving the complete model of the system is as follows:
$$\dot{x} = [A]x + [B]\Delta V_{ref} \tag{14}$$
where
$$x = [x_r \quad x_m \quad x_e \quad x_n \quad x_s]'$$
$$B = [0 \quad 0 \quad 0 \quad 0 \quad B_{S3} \quad 0]'$$
The system matrix $[A]$ having overall dimension of (43×43) is given by

$$[A]$$
$$= \begin{bmatrix} A_R & B_{R1}C_M & B_{R2}C_E & B_{R3}C_{N2} & 0 \\ B_{M1}C_{R1} & A_M + B_{M1}D_{R1}C_M & 0 & B_{M2}C_{N2} & 0 \\ B_E D_{N2}C_{R1} + B_E D_{N3}C_{R2} & B_E D_{N2}D_{R1}C_M + B_E D_{N3}D_{R2}C_M & A_E + B_E D_{N3}D_{R3}C_E & B_E C_{N1} + B_E D_{N1}D_S C_{N3} + B_E D_{N3}D_{R4}C_{N2} & 0 \\ B_{N2}C_{R1} + B_{N3}C_{R2} & B_{N2}D_{R1}C_M + B_{N3}D_{R2}C_M & B_{N3}D_{R3} & A_N + B_{N1}D_S C_{N3} + B_{N3}D_{R4}C_{N2} & B_{N1}C_S \\ 0 & 0 & 0 & B_{S1}C_{N3} & A_S \end{bmatrix}$$

$$I_3^2 = I_{3D}^2 + I_{3Q}^2$$

B. Line Voltage Secondary Controller

The SVS bus voltage is given by
$$V_3^2 = V_{3D}^2 + V_{3Q}^2$$

C. Combined Line Current and Line Voltage Controller

The Secondary signal in the study includes combined line current and voltage, the control scheme is depicted in Fig.7.

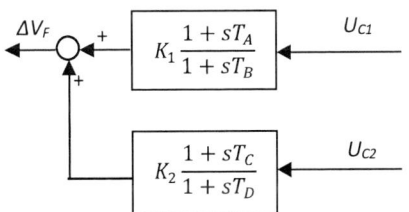

Fig. 7 Combined Secondary Controller

IV. Secondary Control of SVS

The state equations representing influence of the SVS control system, as shown in Fig. 6. are specified as:
$$\dot{x}_c = [A_c]x_c + [B_c]u_c \tag{15}$$
$$y_c = [C_c]x_c + [D_c]u_c \tag{16}$$
where
$x_C = Z_C$ and $y_C = \Delta V_F$

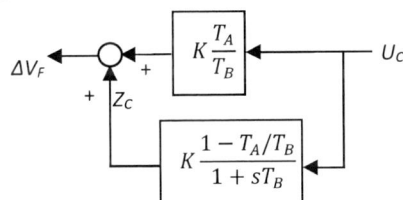

Fig. 6 General First order Secondary Controller

A. Line Current Secondary Controller:

The line current flowing to SVS bus specified as:

$$\begin{bmatrix} \dot{X}_{C1} \\ \dot{X}_{C2} \end{bmatrix} = \begin{bmatrix} A_{C1} & 0 \\ 0 & A_{C2} \end{bmatrix} \begin{bmatrix} X_{C1} \\ X_{C2} \end{bmatrix} + \begin{bmatrix} B_{C1} & 0 \\ 0 & B_{C2} \end{bmatrix} \begin{bmatrix} U_{C1} \\ U_{C2} \end{bmatrix}$$

U_{C1} is Secondary signal for line current and U_{C2} is Secondary signal for line voltage.

V. STUDY SYSTEM

The study system used for analysis is given in [13] The midpoint connected SVS with rating of 250 MVA leading to 150 MVA lagging.

This study evaluates the usefulness of Combined Line Current and Voltage (CLCV) SVS controller under varying 20%,

40%, 60% and 80% capacitive compensation for increasing the performance of capacitor connected transmission line.

Eigenvalues calculated for given study are for generation of 800MW for the cases of system without secondary controller, with Line Current SVS controller, Line Voltage SVS controller and Combined Line Current and Voltage (CLCV) SVS controller and are tabulated for 20% compensation in Table 1, for 40% compensation in Table 2, for 60% compensation in Table 3 and for 80% compensation in Table 4

TABLE 1
Eigenvalues at 20% Series Compensation

Controller Parameter / Modes	Without Secondary Controller	Line Current $K=-0.01; T_A=0.04; T_B=0.1$	Line Voltage $K=-0.5; T_A=0.0009; T_B=0.02$	Combined Line Current and Voltage $K_1=-0.01; K_2=-0.15; T_A=0.03; T_B=0.15; T_C=0.0075; T_D=0.0225$
5TH	$-0.0000168 \pm j298.1$	$-0.000035 \pm j298.1$	$-0.0000176 \pm j298.1$	$-0.00002146 \pm j298.1$
4TH	$0.09264 \pm j202.725$	$0.1121 \pm j202.728$	$0.1171317 \pm j202.669$	$-0.1034422 \pm j202.703$
3TH	$-0.02334 \pm j160.509$	$-0.00142 \pm j160.517$	$-0.01989 \pm j160.514$	$-0.0052099 \pm j160.513$
2TH	$-0.0033 \pm j126.962$	$-0.0049 \pm j126.969$	$-0.00708 \pm j126.965$	$-0.0039107 \pm j126.965$
1TH	$0.069157 \pm j98.666$	$-0.01471 \pm j98.675$	$-0.003921 \pm j98.609$	$-0.004246 \pm j98.679$
0TH	$-0.993099 \pm j5.873$	$-1.67769 \pm j2.972$	$-1.334299 \pm j4.359$	$-38.454446 \pm j4.877$

TABLE 2
Eigenvalues at 40% Series Compensation

Controller Parameter / Modes	Without Secondary Controller	Line Current $K=-0.002; T_A=0.01; T_B=0.03;$	Line Voltage $K=-0.2; T_A=0.009; T_B=0.04;$	Combined Line Current and Voltage $K_1=-0.2; T_A=0.0022; T_B=0.79; K_2=-0.9; T_C=0.27; T_D=0.08;$
5TH	$-0.000025 \pm j298.1$	$-0.0000276 \pm j298.1005$	$-0.0000263 \pm j298.1$	$-0.000000919 \pm j298.1006$
4TH	$-0.0013 \pm j202.689$	$-0.0001415 \pm j202.6876$	$-0.00077 \pm j202.686$	$-0.000259629 \pm j202.8433$
3TH	$0.1192 \pm j160.438$	$0.10780599 \pm j160.4456$	$0.12537 \pm j160.4517$	$-0.017493895 \pm j160.4858$
2TH	$0.00078 \pm j126.971$	$0.002448 \pm j126.97064$	$-0.000008 \pm j126.97$	$-0.012592 \pm j126.95983$
1TH	$-0.04085 \pm j98.677$	$-0.031319 \pm j98.690776$	$-0.04892 \pm j98.6738$	$-6.897956 \pm j98.667$
0TH	$0.000919 \pm j2.3627$	$-33.9021 \pm j2.3753528$	$-0.01417 \pm j2.4232$	$-4.014226 \pm j5.7229$

TABLE 3
Eigenvalues at 60% Series Compensation

Controller Parameter / Modes	Without Secondary Controller	Line Current $K=-0.03; T_A=0.08; T_B=0.05;$	Line Voltage $K=-0.4; T_A=0.07; T_B=0.09;$	Combined Line Current and Voltage $K_1=-0.2; T_A=0.01; T_B=0.09; K_2=-0.01; T_C=0.03; T_D=0.4;$
5TH	$0.0000019 \pm j298.1$	$-0.00000129 \pm j298.1$	$-0.000000024 \pm j298.1$	$-0.000000675 \pm j298.1005$

978-1-5386-6624-1/18 $31.00 © 2018 IEEE

4TH	-0.0163±j202.809	-0.009648±j202.813	-0.017015±j202.8104	-0.0105406±j202.80977
3TH	-0.0242±j160.503	-0.0165166±j160.5118	-0.024749±j160.503	-0.0158954±j160.506424
2TH	-0.0181±j126.985	-0.016924±j126.989	-0.01837±j126.98	-0.0146997±j126.986913
1TH	-0.9883±j99.114	-0.950024±j99.158575	-0.98629±j99.116	-0.9386611±j99.1404964
0TH	-20.5568±j12.944	-9.453285±j11.3804	-19.4445±j15.9728	-10.850897±j10.3245992

TABLE 4
Eigenvalues at 80% Series Compensation

Controller Parameter / Modes	Without Secondary Controller	Line Current $K_B=-0.01; T_A=0.08; T_2=0.09;$	Line Voltage $K_B=-0.03; T_A=0.08; T_2=0.01;$	Combined Line Current and Voltage $K_{B1}=-0.2; T_A=0.01; T_2=0.09; K_{B2}=-0.01; T_3=0.03; T_4=0.4;$
5TH	0.000001±j298.1	-0.0000000858±j298.1	-0.00000006±j298.1	-0.00000079±j298.10053599
4TH	-0.0119±j202.782	-0.01139526±j202.782	-0.01261±j202.782	-0.010436365±j202.78308
3TH	-0.0153±j160.474	-0.01427135±j160.474	-0.01569±j160.474	-0.012314359±j160.474
2TH	-0.0079±j126.966	-0.0074597±j126.966	-0.00804±j126.966	-0.006307629±j126.9674
1TH	-0.18981±j98.844	-0.18536±j98.858	-0.19114±j98.8436	-0.16517906±j98.88
0TH	-24.2936±j5.0567	-22.22356±j10.647	-24.5945±j4.5079	-14.57014±j10.1952

VI. CONCLUSION

In the present paper, the utility of Combined Line Current and Voltage (CLCV) SVS controller is been established over a wide range of series compensations. An exhaustive model based upon IEEE first benchmark model is used for the investigation which replicate the power structure dynamics very precisely. The series compensated transmission line model considered is an equivalent π-circuit. An analysis through calculations of eigenvalues of the linearized study system is tabulated in Tables 1-4 clearly shows that proposed CLCV SVS secondary signal steadies all the torsional frequency modes for 20%, 40%, 60% and 80% series capacitive compensations. Eigenvalues clearly shows that the CLCV SVS controller is able to stabilize all torsional modes as compared to the cases when no controller, Line Current SVS controller and Line Voltage SVS controllers are used. So, a conclusion is derived that the CLCV controller improves the performance of study system over SSR phenomenon by essentially stabilizing all the torsional frequency modes. Henceforth the investigation can be applied for further examination of the given system under transient conditions for varying series capacitor compensation by using proposed CLCV SVS controller.

REFERENCES

[1] D. N. Walker, C. E. J. Bowler, R. L. Jackson, and D. A. Hodges, "Results of subsynchronous resonance test at Mohave," IEEE Transactions on Power Apparatus and Systems, vol. PAS-94, no. 5, pp. 1874–1889, Sep. 1975.

[2] IEEE Subsynchronous Resonance Working Group, "Countermeasures to Subsynchronous Resonance Problems," IEEE Transactions on Power Apparatus and Systems, vol. PAS- 99, pp. 1810-1818, 1980.

[3] Z. Donghui, X. Xiaorong, L. Shiyu, Z. Shuqing, "Optimal design of SEDC for damping multi-mode SSR based on GASA", Power Tech, 2007 IEEE Lausanne.

[4] R. G. Farmer, A. L. Schwalb, Eli Katz, "Navajo project report on subsynchronous resonance analysis and solutions", IEEE Transactions on Power Apparatus and Systems, Vol. PAS-96, no. 4, July/August 1977.

[5] C.E.J. Bowler, J.A. Demcko, L. Mankoff, W.C. Kotheimer, D. Cordray, "The Navajo SMF Type Subsynchronous Resonance Relay", IEEE Transactions on Power Apparatus and Systems, Vol. PAS-97, No. 5, Sept/Oct 1978.

[6] Z. Xin, Y. Ruoying, G. Shan, "A Novel Objective Function and Analysis for Optimal SVC Parameters Design to Damping Subsynchronous Resonance", International Conference on Electricity Distribution (CICED 2016) Xi'an, 10-13 Aug, 2016.

[7] Atia Adrees, Jovica V. Milanović, "Optimal Compensation of Transmission Lines Based on Minimisation of the Risk of Subsynchronous

Resonance", IEEE Transactions on Power Systems, Vol. 31, No. 2, March 2016.

[8] IEEE Subsynchronous Resonance Working Group, "First Benchmark Model for Computer Simulation of Subsynchronous Resonance", IEEE Transactions on Power Apparatus and Systems, Vol. PAS-96, No. 5, September / October 1977.

[9] R.S. Ramsaw, K.R. Padiyar, "Generalized System Model for Slip Ring Machines", IEE Proc., Part C, 1973, Vol.120, No. 6, pp.647-658.

[10] Yan Guo, Boon-Teck, Ooi Howard, C. Lee, "Integration of turbo-generator modules in digital transient network analyzer", IEEE Transactions on Power Systems, 1993.

[11] Kundur, "Power System Stability and Control", McGraw Hill Inc, 1994.

[12] N. Kumar, S. Gupta, "Supplementary Signal of SVC for Damping Torsional Oscillation", International Conference CIPECH. 2014.

[13] N. Kumar. "Damping SSR in a Series Compensated Power System", 2006 IEEE Power India Conference, 2006

Development of Empirical Models for Forecasting Global Solar Energy

Gulnar Perveen
Department of Electronics and Communication Engineering
Delhi Technological University
Delhi-110042,India
gulegulnar@gmail.com

M. Rizwan
Department of ElectricalEngineering
Delhi Technological University
Delhi-110042,India
rizwan@dce.ac.in

Nidhi Goel
Department of Electronics and Communication Engineering
IGDTUW
Delhi-110006, India
drnidhigoel@gmail.com

Abstract—The present work involves two sections; in the first section, eight empirical sunshine-based models which include linear and non-linear correlations have been employed to forecast monthly average global solar energy on a horizontal surface using the relative duration of sunshine hours. In the second section, 35 empirical models have been presented correlating meteorological parameters such as the monthly average global solar energy with ambient temperature, duration of sunshine hours, wind speed, atmospheric pressure, relative humidity, rainfall, and cloudiness index. The data has been employed to generate multiple regression equations for five meteorological stations across India; exploiting the 15 years averaged recorded data. An exercise has been done for selecting the most suitable model/equation based on principal component analysis. Five statistical indicators have been used for measuring the performance of proposed models.

Further to check for accuracy of the proposed model, a comparison has been carried out with other models discussed in the literature i.e. Angstrom-Prescott, Rietveld, Page model, Akinoglu and Ecevit, Bahel, Newland and Abdalla model.

Keywords—climate zone, forecast, global solar energy, meteorological parameter, regression equations, statistical indicators

Abbreviations

a,b,c,d   regression coefficients of empirical models (dimensionless)
c_a averaged calculated value (dimensionless)
c_i i_{th} calculated value (dimensionless)
G_{sc} solar constant (W/m^2)
H_d diffuse solar radiation (MJ/m^2)
H_g global solar radiation (MJ/m^2)
H_o extra-terrestrial solar radiation on a horizontal surface (MJ/m^2)
m_a averaged measured value (dimensionless)
MBE   mean bias error (%)
m_i i_{th} measured value (dimensionless)
MPE   mean percentage error (%)
n   number of calculated and measured values (dimensionless)
P_m measured atmospheric pressure (hPa)
P_o maximum possible atmospheric pressure (hPa)
R^2 coefficient of determination (dimensionless)
r   correlation coefficient (dimensionless)
RF_m measured rainfall (mm)
RF_o maximum possible rainfall (mm)
R_m measured relative humidity (%)
R_o maximum possible relative humidity (%)
RMSE   root mean square error (%)

S   measured sunshine duration (hours)
S_o maximum possible sunshine duration (hours)
T_m measured ambient temperature ($^\circ$C)
T_o maximum possible ambient temperature ($^\circ$C)

I. Introduction

For the accurate design and development of solar energy devices, solar radiation resource data plays a significant role. Unfortunately, the measuring devices are not available at most of the developing countries because of the high cost of the instrument. Therefore, it is essential to develop methods that can forecast global solar energy based on more readily available data with reasonable accuracies.

Many previous researches were based on establishing model to forecast global solar energy utilizing meteorological parameters [1]. However, commonly used parameter is the sunshine hours [2]. Angstrom proposed the first theoretical model based on sunshine hours [3]. Further, Page [4] and Prescott [5] proposed the improved model so to determine the global solar energy from total solar insolation on an extra-terrestrial surface. Many authors have presented the models for Middle East and African countries [6]. However, such models are barely available for Indian climatic conditions. Kirmani *et al.* [7] developed regression models using five parameters but for a single location. Models taking into account weather factors have been proposed by Abdalla [8]. However, no recent finding is reported that uses different meteorological parameters for estimating global solar energy for distinct climate zone across India. Hence, there exists a clear scope, for developing an empirical model that can gauge the effect of adding meteorological parameters for forecasting global solar energy for Indian climatic conditions. The aim of this work is to establish models to forecast global solar energy using different meteorological parameters for five meteorological stations. The performance of the models has been evaluated by using statistical indicators. Principal component analysis has been carried out for selecting the most accurate model. Further, to check for model accuracy, comparison has been made with the well-established models from the literature. The paper is organized as follows: Section II presents the collection of data. Section III presents the methodology. Statistical performance tests have been discussed in Section IV. Results are discussions are carried out in Section V. The conclusions followed by references are presented in Section VI.

II. DATA COLLECTION

A. Climate zone

Within India, it is possible to define five distinct climate zones. The criteria of allocating location to these climate zones depend on weather conditions which prevail for six months or more. Based on this condition, Bansal and Minke in 1988 presented the climate zone by evaluating the average of the monthly data from 233 different meteorological stations, and then defining the five distinct climate zones as shown below in Table I [9].

B. Meteorological data

In this study, the long-term measured, 15 years (1986-2000) averaged, hourly data have been obtained from National Institute of Solar Energy (NISE) and IMD (Indian Meteorological Department) [10].

The data have been utilized for five meteorological stations across India and are presented below in Table II-VI.

TABLE I.   CLIMATE ZONE

Climate Zone	Meteorological Stations	Ambient temperature (°C)
Hot and dry	Jodhpur, Rajasthan	>30
Warm and humid	Chennai, Tamil Nadu	>30
Moderate	Pune, Maharashtra	25-30
Cold and cloudy	Shillong, Meghalaya	<25
Composite	New Delhi	This applies, when more than 6 months do not fall in above category.

TABLE II.   METOROLOGICAL DATA FOR CHENNAI, TAMIL NADU (WARM AND HUMID CLIMATE ZONE)

Month	Clearness Index (H_g/H_o)	S/S_o	Air Temp. (°C)	Relative Humidity (%)	Wind speed (km/hr)	Rainfall (mm)	Atmospheric pressure (hPa)	Cloudiness Index (H_d/H_g)
Jan	0.50	0.79	25.46	71.32	8.22	0.03	1012.84	0.44
Feb	0.54	0.84	26.61	76.29	9.06	0.09	1009.74	0.34
Mar	0.55	0.76	27.84	73.75	6.98	0.00	1008.99	0.32
Apr	0.52	0.76	30.53	71.26	8.78	0.00	1019.08	0.35
May	0.49	0.70	31.71	67.14	7.49	9.46	1033.84	0.40
Jun	0.46	0.58	30.73	64.04	8.48	0.57	1013.19	0.51
Jul	0.40	0.49	30.53	62.25	10.04	0.51	1014.78	0.79
Aug	0.41	0.38	29.07	71.30	8.85	1.55	1003.74	0.64
Sep	0.44	0.51	29.10	79.09	8.41	2.13	1014.47	0.53
Oct	0.43	0.56	27.75	80.06	6.25	1.39	1009.47	0.52
Nov	0.42	0.51	25.64	83.73	11.58	1.41	1011.95	0.59
Dec	0.41	0.63	26.09	78.13	9.49	0.31	1012.08	0.58

TABLE III.   METOROLOGICAL DATA FOR JODHPUR, RAJASTHAN (HOT AND DRY CLIMATE ZONE)

Month	Clearness Index (H_g/H_o)	S/S_o	Air Temp. (°C)	Relative Humidity (%)	Wind speed (km/hr)	Rainfall (mm)	Atmospheric pressure (hPa)	Cloudiness Index (H_d/H_g)
Jan	0.6	0.87	18.09	45.33	6.37	0.06	991.99	0.26
Feb	0.6	0.87	19.87	41.86	6.12	0.00	987.13	0.25
Mar	0.57	0.77	26.12	30.91	7.74	0.03	1009.11	0.29
Apr	0.52	0.78	32.91	23.98	5.7	0.02	979.69	0.35
May	0.53	0.77	34.88	35.82	8.65	0.05	978.01	0.30
Jun	0.47	0.66	33.52	45.91	14.04	0.10	974.44	0.50
Jul	0.38	0.6	31.52	60.78	14	0.27	974.42	0.60
Aug	0.43	0.63	31.44	62.33	5.99	0.97	976.47	0.51
Sep	0.56	0.8	29.7	59.59	6.75	0.15	980.46	0.24
Oct	0.58	0.86	28.46	42.03	4.38	0.00	993.23	0.25
Nov	0.61	0.87	22.08	42.01	3.3	0.00	988.79	0.21
Dec	0.57	0.82	18.64	48.58	3.21	0.00	991.78	0.30

TABLE IV.   METEROLOGICAL DATA FOR NEW DELHI, DELHI (COMPOSITE CLIMATE ZONE)

Month	Clearness Index (H_g/H_o)	S/S_o	Air Temp. (^{o}C)	Relative Humidity (%)	Wind speed (km/hr)	Rainfall (mm)	Atmospheric pressure (hPa)	Cloudiness Index (H_d/H_g)
Jan	0.39	0.75	14.11	65.48	5.15	0.17	990.03	0.39
Feb	0.38	0.72	18.64	59.49	7.71	0.43	986.33	0.38
Mar	0.55	0.74	22.73	53.30	7.34	0.02	983.27	0.25
Apr	0.56	0.72	30.03	36.19	8.42	0.08	979.30	0.22
May	0.49	0.62	34.14	34.30	9.52	0.03	976.22	0.22
Jun	0.41	0.72	33.40	52.56	10.59	0.28	972.74	0.27
Jul	0.39	0.66	30.48	70.64	10.40	0.31	984.73	0.28
Aug	0.40	0.51	29.14	79.36	9.57	0.82	974.82	0.31
Sep	0.34	0.32	29.73	69.28	9.43	0.43	979.55	0.40
Oct	0.55	0.42	26.18	64.52	6.34	0.00	983.97	0.29
Nov	0.54	0.52	20.92	49.80	6.53	0.00	986.67	0.37
Dec	0.50	0.85	16.00	65.68	5.93	0.00	991.57	0.46

TABLE V.   METEROLOGICAL DATA FOR PUNE, MAHARASHTRA (MODERATE CLIMATE ZONE)

Month	Clearness Index (H_g/H_o)	S/S_o	Air Temp. (^{o}C)	Relative Humidity (%)	Wind speed (km/hr)	Rainfall (mm)	Atmospheric pressure (hPa)	Cloudiness Index (H_d/H_g)
Jan	0.54	0.74	19.78	59.88	1.49	0.00	982.92	0.31
Feb	0.57	0.90	23.18	48.82	5.34	0.00	947.97	0.23
Mar	0.55	0.83	26.28	41.28	3.23	0.00	948.68	0.31
Apr	0.53	0.80	29.19	44.47	6.07	0.00	945.63	0.28
May	0.53	0.84	29.14	55.21	11.76	0.02	944.35	0.34
Jun	0.37	0.35	25.89	76.92	10.48	0.19	942.76	0.67
Jul	0.30	0.31	23.91	86.03	7.93	0.28	954.91	0.79
Aug	0.35	0.31	23.30	85.24	7.73	0.09	943.78	0.81
Sep	0.45	0.46	24.07	84.23	4.68	0.17	945.02	0.57
Oct	0.56	0.68	24.24	75.96	2.27	2.31	947.39	0.37
Nov	0.64	0.80	22.40	71.97	2.00	0.02	950.14	0.33
Dec	0.70	0.85	19.27	63.01	2.41	0.00	962.04	0.23

TABLE VI.   METEROLOGICAL DATA FOR SHILLONG, MEGHALAYA (COLD AND CLOUDY CLIMATE ZONE)

Month	Clearness Index (H_g/H_o)	S/S_o	Air Temp. (^{o}C)	Relative Humidity (%)	Wind speed (km/hr)	Rainfall (mm)	Atmospheric pressure (hPa)	Cloudiness Index (H_d/H_g)
Jan	0.52	0.67	9.95	75.58	3.61	0.22	840.99	0.34
Feb	0.51	0.56	10.24	71.84	3.80	0.68	838.66	0.42
Mar	0.54	0.61	15.54	59.65	5.65	1.87	838.72	0.41
Apr	0.45	0.30	18.26	63.53	7.63	5.61	839.83	0.47
May	0.38	0.37	20.69	80.29	4.29	3.53	838.13	0.60
Jun	0.33	0.26	21.04	85.50	3.63	15.68	834.75	0.70
Jul	0.33	0.20	21.21	87.52	3.15	14.15	835.36	0.80
Aug	0.31	0.17	20.64	89.23	1.23	12.92	836.08	0.77
Sep	0.34	0.23	20.00	85.92	0.87	7.53	838.44	0.73
Oct	0.43	0.52	18.39	80.74	2.29	1.53	842.01	0.48
Nov	0.54	0.66	15.27	75.60	2.65	0.33	843.31	0.43
Dec	0.60	0.73	11.89	74.40	0.84	0.01	841.67	0.22

III. METHODOLOGY

A. Empirical Models for Estimating Global Solar Energy

The main objective of the present work is to develop correlations for forecasting global solar energy using different meteorological parameters for five distinct climate zones. The input parameters are sunshine duration, wind speed, ambient temperature, relative humidity, rainfall, atmospheric pressure, and cloudiness index (H_d/H_g) whereas the output is the clearness index (H_g/H_o). The forecasted value of global solar energy H_g is obtained on multiplying the forecasted clearness index by H_o, where H_g is the monthly mean global solar radiation; and H_o is an extra-terrestrial solar radiation and can be calculated using standard geometric procedures [11]. In linear regression, the equation using one parameter is as follows:

$$y = a + bx, \qquad (1)$$

Since there is increase in the number of parameters so the correlation takes the form of multiple linear regressions as shown below:

$$y = a + bx_1 + cx_2 + dx_3 + ex_4 + fx_5 + \ldots + nx_n, \quad (2)$$

B. Sunshine-based Models for Estimating Global Solar Energy

In solar estimation models, the most widely used method for forecasting global solar energy is Angstrom [3] – Prescott [5] model which is based on the correlation of the ratio of the global solar radiation on a horizontal surface to an extra-terrestrial solar radiation with the ratio of relative sunshine duration. In the present study, eight sunshine-based models have been established and the performance is evaluated based on statistical error- tests.

- Model 1: Linear model
  Source - Angstrom-Prescott model [3,5]

978-1-5386-6624-1/18 $31.00 © 2018 IEEE

$$H/H_o = a + b(S/S_o) \tag{3}$$

- *Model 2: Quadratic model*
  Source -Akinoglu and Ecevit model [12]
  $$H/H_o = a + b(S/S_o) + c(S/S_o)^2 \tag{4}$$

- *Model 3: Cubic model*
  Source – Bahel *et al.* model [13]
  $$H/H_o = a + b(S/S_o) + c(S/S_o)^2 + d(S/S_o)^3 \tag{5}$$

- *Model 4: Linear logarithmic model*
  Source – Newland model [14]
  $$H/H_o = a + b(S/S_o) + c \log(S/S_o) \tag{6}$$

- *Model 5: Logarithmic model*
  Source - Ampratwum and Dorvlo model [15]
  $$H/H_o = a + b\log(S/S_o) \tag{7}$$

- *Model 6: Linear Exponential model*
  Source - Bakirci model [16]
  $$H/H_o = a + b(S/S_o) + c \exp(S/S_o) \tag{8}$$

- *Model 7: Exponential model*
  Source - Almorox and Hontoria model [17]
  $$H/H_o = a + b\exp(S/S_o) \tag{9}$$

- *Model 7: Exponent model*
  Source - Bakirci model [16]
  $$H/H_o = a + (S/S_o)^b \tag{10}$$

C. Principal Component Analysis

Principal component analysis has been performed for selecting the model that gives the best correlation. We have chosen coefficient of determination as closeness parameter. The high values of correlation of coefficient r and coefficient of determination R^2 are considered good for better analysis. The regression coefficients are obtained using curve fitting tool in MATLAB.

IV. STATISTICAL PERFORMANCE TESTS

To assess the performance of the models, five different statistical indicators such as MPE, MBE, RMSE, r, and R^2 are used for comparing the results statistically.

A. Mean Bias Error

It is defined by the following equation:-

$$\text{MBE} = \frac{1}{x}\sum_{i=1}^{x}(e_i - m_i) \tag{11}$$

where x is the number of observed data, m_i and e_i are the i_{th} measured and estimated data, respectively.

B. Mean Bias Error

It is the deviation of the estimated value with the measured value and this relationship is given by:-

$$\text{MPE} = \frac{1}{x}\sum_{i=1}^{x}\left(\frac{s_i - m_i}{m_i}\right) \tag{12}$$

C. Root Mean Square Error

It is described by the equation given by:-

$$\text{RMSE} = \sqrt{\frac{1}{x}\sum_{i=1}^{x}(e_i - m_i)^2} \tag{13}$$

where x is the number of observed data, m_i and e_i are the i_{th} measured and estimated data, respectively.

D. Coefficient of Determination (R^2)

It can be defined as:-

$$R^2 = \frac{\sum_{i=1}^{n}(c_i - c_a)(m_i - m_a)}{\sqrt{[\sum_{i=1}^{n}(c_i - c_a)^2][\sum_{i=1}^{n}(m_i - m_a)^2]}} \tag{14}$$

where c_a and m_a are the averaged calculated and measured values, respectively.

V. RESULTS AND DISCUSSIONS

A. Sunshine-based Model for Forecasting Global Solar Energy

In the first section of present work, models based on sunshine duration with linear and non-linear correlations have been established using multiple regression analysis. The regression coefficients a, b, c and d are obtained for eight models i.e. models 1-8 as shown in Eq. 3-10 of the sunshine-based models by using the measured data illustrated in Table II-VI for five meteorological stations across India that represents distinct climate zone. The performance has been evaluated by using statistical indicators and is presented in Table VII from which following can be briefly summarized as:-

1). Warm and humid climate zone: For this climate zone, it is observed that the dependencies are stronger for quadratic (16), cubic (17) and linear exponential term (20) with r = 0.80 and R^2 = 0.65 which means 65% of the clearness index can be accounted using relative sunshine duration.

2). Hot and dry climate zone: Similarly, for this climate zone, it is observed that the dependencies are stronger for logarithmic term (27) with r = 0.66 and R^2 = 0.47 which means 47% of the clearness index can be accounted using relative sunshine duration.

3). Composite climate zone: For this climate zone, it is observed that the dependencies are stronger for quadratic (32), cubic (33) and linear exponential term (36) with r = 0.66 and R^2 = 0.47 which means 47% of the clearness index can be accounted using relative sunshine duration.

4). Moderate climate zone: For this climate zone, it is observed that the dependencies are stronger for linear term (39) with r = 0.73 and R^2 = 0.57 which means 57% of the clearness index can be accounted using relative sunshine duration.

5). Cold and cloudy climate zone: For this climate zone, it is observed that the dependencies are stronger for linear (47), quadratic (48), cubic (49), and exponential term (53) with r = 0.72 and R^2 = 0.55 which means 55% of the clearness index can be accounted using relative sunshine duration.

For each of the climate zone, all the developed models have different determination coefficients. The largest difference between the determination coefficients of the best and worst model is only 0.23 for Chennai, 0.17 for Jodhpur, 0.07 for Delhi, 0.12 for Pune and 0.11 for Shillong station respectively. Further, the graphical representation of the monthly average global solar radiation has been shown in Fig. 1. for five meteorological stations across India.

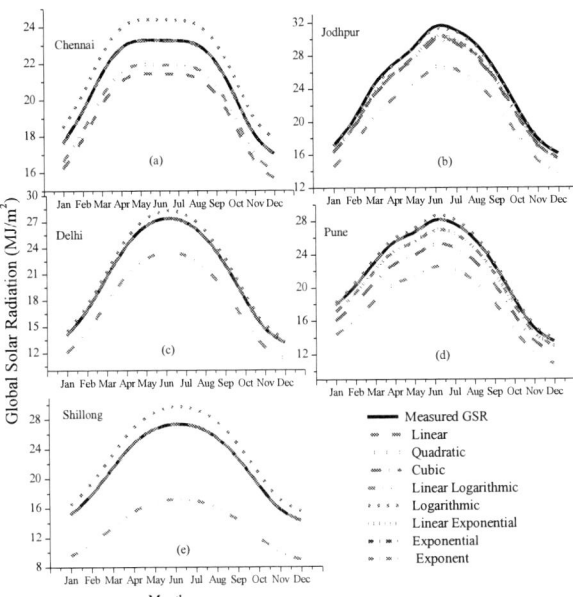

Fig.1. Graphical representation of measured and forecasted global solar energy for linear and non-linear sunshine based models for (a) Chennai, (b) Jodhpur, (c) Delhi, (d) Pune, and (e) Shillong

From Fig. 1, it may be noted that the weakest fit is obtained for Chennai station with the largest difference between the best and worst determination coefficients of 0.23. Similarly, the best fit is obtained for Delhi composite climate zone with the smallest difference between the best and worst determination coefficients of 0.07 as shown by the calculated data presented in the Table VII.

B. Empirical Models for Forecasting Global Solar energy

In the second section of the present work, 35 empirical models have been established using multiple regression analysis of parameters namely global solar radiation, sunshine hours, ambient temperature, wind speed, atmospheric pressure, rainfall and relative humidity for five meteorological stations across India that represents widely changing climatic conditions such as warm and humid climate, hot and dry climate, composite, moderate, and cold & cloudy climate conditions respectively. The developed models have been further processed based on principal component analysis to obtain the correlation with highest correlation coefficients. The performance of the models has been evaluated using statistical error-tests and is illustrated in Table VIII from which following can be summarized as:-

1). Warm and humid climate zone: The best result was obtained for correlation using seven parameters on the basis of closeness parameters with MPE=1.93%, RMSE=0.04%, r=0.93 and R^2=0.87 as shown in (61) which means 87% of the clearness index can be accounted using sunshine hours, ambient temperature, wind speed, atmospheric pressure, rainfall, and relative humidity.

2). Hot and dry climate zone: The best result was obtained for correlation using seven parameters on the basis of closeness parameters with MPE=0.36%, RMSE=0.03%,

r=0.79 and R^2=0.66 as shown in (68) which means 66% of the clearness index can be accounted using sunshine hours, ambient temperature, wind speed, atmospheric pressure, rainfall, cloudiness index, and relative humidity.

3). Composite climate zone: The best result was obtained for correlation using seven parameters on the basis of closeness parameters with MPE=0.42%, RMSE=0.03%, r=0.83 and R^2=0.81 as shown in (75) which means 81% of the clearness index can be accounted using sunshine hours, ambient temperature, wind speed, atmospheric pressure, rainfall, cloudiness index, and relative humidity.

4). Moderate climate zone: The best result was obtained for correlation using seven parameters on the basis of closeness parameters with MPE=0.25%, RMSE=0.02%, r=0.90 and R^2=0.81 as shown in (82) which means 81% of the clearness index can be accounted using sunshine hours, ambient temperature, wind speed, atmospheric pressure, rainfall, cloudiness index, and relative humidity.

5). Cold and cloudy climate zone: The best result was obtained for correlation using seven parameters on the basis of closeness parameters with MPE=0.69%, RMSE=0.03%, r=0.84 and R^2=0.71 as shown in (89) which means 71% of the clearness index can be accounted using sunshine hours, ambient temperature, wind speed, atmospheric pressure, rainfall, cloudiness index, and relative humidity.

Further, the graphical representation of the measured and forecasted data using one, two, three, four, five, six and seven parameter correlation have been shown in Fig. 2 for five meteorological stations.

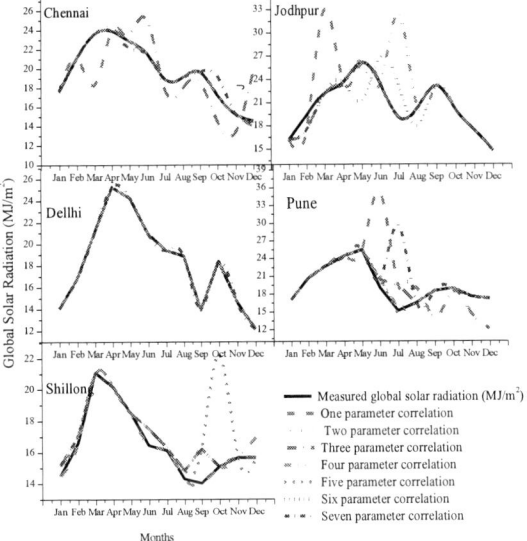

Fig. 2 Graphical representation of measured and forecasted global solar energy for empirical models for (a) Chennai, (b) Jodhpur, (c) Delhi, (d) Pune, and (e) Shillong.

An excellent match has been noticed between the measured and forecasted values for Delhi composite climate zone. The statistical analysis shows that the best estimates of global solar radiation are obtained from the proposed model using seven parameters correlation as compared to other proposed model. Thus in general, seven parameters correlation is found to be the most accurate model for each of the climate zone across the entire country.

978-1-5386-6624-1/18 $31.00 © 2018 IEEE

TABLE VII.   SUNSHINE-BASED MODELS WITH STATISTICAL ERRORS FOR FIVE METEOROLOGICAL STATIONS ACROSS INDIA

Stations	Model	Eq. No.	Equations	MPE (%)	MBE (%)	RMSE (%)	r	R²
Chennai (Warm and humid)	Linear	15	$H_g/H_o = 0.259 + 0.321(S/S_o)$	1.52	0.00	0.05	0.77	0.62
	Quadratic	16	$H_g/H_o = 0.289 + 0.252(S/S_o) + 0.041(S/S_o)^2$	1.79	0.00	0.05	0.78	0.63
	Cubic	17	$H_g/H_o = 0.426 - 0.232(S/S_o) + 0.505(S/S_o)^2 - 0.086(S/S_o)^3$	1.79	0.00	0.04	0.80	0.65
	Linear Logarithmic	18	$H_g/H_o = 0.134 + 0.434(S/S_o) + 0.011 \log(S/S_o)$	-10.35	-0.05	0.10	0.74	0.57
	Logarithmic	19	$H_g/H_o = 0.533 + 0.153 \log(S/S_o)$	15.60	0.04	0.08	0.62	0.42
	Linear Exponential	20	$H_g/H_o = 0.243 + 0.230(S/S_o) + 0.037 \exp(S/S_o)$	1.97	0.00	0.04	0.78	0.63
	Exponential	21	$H_g/H_o = 0.113 + 0.184 \exp(S/S_o)$	2.58	0.00	0.05	0.77	0.61
	Exponent	22	$H_g/H_o = 0.556(S/S_o)^{0.318}$	5.79	0.01	0.06	0.71	0.53
Jodhpur (Hot and dry)	Linear	23	$H_g/H_o = 0.375 + 0.214(S/S_o)$	-0.55	-0.01	0.05	0.51	0.33
	Quadratic	24	$H_g/H_o = 0.586 - 0.414(S/S_o) + 0.439(S/S_o)^2$	-0.44	-0.01	0.04	0.60	0.41
	Cubic	25	$H_g/H_o = -0.464 + 0.032(S/S_o) - 0.693(S/S_o)^2 + 0.769(S/S_o)^3$	1.18	0.00	0.04	0.50	0.40
	Linear Logarithmic	26	$H_g/H_o = -0.196 + 0.801(S/S_o) - 0.396 \log(S/S_o)$	-7.37	-0.05	0.13	0.53	0.35
	Logarithmic	27	$H_g/H_o = 0.570 + 0.125 \log(S/S_o)$	4.10	0.01	0.05	0.66	0.47
	Linear Exponential	28	$H_g/H_o = 0.114 - 0.673(S/S_o) + 0.309 \exp(S/S_o)$	-1.51	-0.02	0.05	0.60	0.41
	Exponential	29	$H_g/H_o = 0.297 + 0.111 \exp(S/S_o)$	-0.89	-0.01	0.05	0.52	0.34
	Exponent	30	$H_g/H_o = 0.569(S/S_o)^{0.234}$	-1.81	-0.02	0.05	0.49	0.30
Delhi (Composite)	Linear	31	$H_g/H_o = 0.367 + 0.205(S/S_o)$	1.24	0.00	0.04	0.59	0.40
	Quadratic	32	$H_g/H_o = 0.470 - 0.117(S/S_o) + 0.238(S/S_o)^2$	1.04	0.00	0.04	0.66	0.47
	Cubic	33	$H_g/H_o = 0.504 - 0.241(S/S_o) + 0.231(S/S_o)^2 + 0.067(S/S_o)^3$	1.44	0.00	0.04	0.66	0.46
	Linear Logarithmic	34	$H_g/H_o = 0.033 + 0.522(S/S_o) - 0.142 \log(S/S_o)$	-19.43	-0.09	0.12	0.54	0.40
	Logarithmic	35	$H_g/H_o = 0.543 + 0.098 \log(S/S_o)$	10.62	0.03	0.07	0.52	0.40
	Linear Exponential	36	$H_g/H_o = 0.208 - 0.282(S/S_o) + 0.245 \exp(S/S_o)$	1.74	0.00	0.04	0.65	0.46
	Exponential	37	$H_g/H_o = 0.261 + 0.124 \exp(S/S_o)$	1.61	0.00	0.04	0.61	0.40
	Exponent	38	$H_g/H_o = 0.543(S/S_o)^{0.185}$	2.31	0.00	0.05	0.54	0.40
Pune (Moderate)	Linear	39	$H_g/H_o = 0.156 + 0.281(S/S_o)$	3.10	0.01	0.04	0.73	0.57
	Quadratic	40	$H_g/H_o = 0.392 + 0.036(S/S_o) + 0.172(S/S_o)^2$	3.65	0.01	0.04	0.71	0.53
	Cubic	41	$H_g/H_o = 0.82 - 0.62(S/S_o) + 0.52(S/S_o)^2 - 0.6(S/S_o)^3$	2.46	0.01	0.04	0.70	0.53
	Linear Logarithmic	42	$H_g/H_o = 0.093 + 0.506(S/S_o) - 0.097 \log(S/S_o)$	5.19	0.00	0.08	0.70	0.52
	Logarithmic	43	$H_g/H_o = 0.567 + 0.193 \log(S/S_o)$	17.69	0.06	0.08	0.65	0.45
	Linear Exponential	44	$H_g/H_o = 0.230 + 0.006(S/S_o) + 0.133 \exp(S/S_o)$	3.19	0.01	0.04	0.72	0.56
	Exponential	45	$H_g/H_o = 0.119 + 0.194 \exp(S/S_o)$	3.15	0.01	0.04	0.72	0.54
	Exponent	46	$H_g/H_o = 0.590(S/S_o)^{0.445}$	3.16	0.01	0.04	0.72	0.54
Shillong (Cold and cloudy)	Linear	47	$H_g/H_o = 0.239 + 0.586(S/S_o)$	1.09	-0.01	0.07	0.70	0.51
	Quadratic	48	$H_g/H_o = 0.301 + 0.187(S/S_o) - 0.4(S/S_o)^2$	3.33	0.00	0.06	0.72	0.55
	Cubic	49	$H_g/H_o = 0.369 - 0.250(S/S_o) + 1.059(S/S_o)^2 - 0.601(S/S_o)^3$	3.15	0.00	0.06	0.70	0.52
	Linear Logarithmic	50	$H_g/H_o = 0.160 + 0.458(S/S_o) + 0.049 \log(S/S_o)$	0.11	-0.02	0.12	0.64	0.45
	Logarithmic	51	$H_g/H_o = 0.546 + 0.145 \log(S/S_o)$	18.88	0.06	0.10	0.64	0.44
	Linear Exponential	52	$H_g/H_o = 0.160 + 0.104(S/S_o) + 0.132 \exp(S/S_o)$	4.36	0.00	0.09	0.70	0.53
	Exponential	53	$H_g/H_o = 0.061 + 0.229 \exp(S/S_o)$	3.52	0.00	0.07	0.68	0.50
	Exponent	54	$H_g/H_o = 0.573(S/S_o)^{0.355}$	3.76	0.00	0.07	0.70	0.52

C. Comparison of Proposed Model with Other Models

The proposed model is further compared with other models as shown in Table IX. It can be observed that the proposed models presents the best result with mean percentage error of 1.93% for Chennai, 0.36% for Jodhpur, 0.42% for Delhi, 0.25% for Pune, and 0.69% for Shillong station which indicates a good agreement between the measured and forecasted data.

TABLE VIII. THE EMPIRICAL CORRELATIONS ALONG WITH STATISTICAL ERRORS FOR FIVE METEOROLOGICAL STATIONS ACROSS INDIA

Climate Zone	Eq. No.	Models/Equations	MPE (%)	MBE (%)	RMSE (%)	r	R²
Chennai (Warm and humid)	55	$H_g/H_o = 0.259 + 0.321(S/S_o)$	1.52	0.00	0.05	0.77	0.62
	56	$H_g/H_o = 0.550 + 0.107(S/S_o) - 0.265(H_d/H_g)$	1.51	0.00	0.04	0.89	0.79
	57	$H_g/H_o = 0.683 + 0.063(S/S_o) + 0.001(P_m/P_o) - 0.332(H_d/H_g)$	2.95	0.01	0.03	0.89	0.8
	58	$H_g/H_o = 0.530 + 0.048(S/S_o) + 0.004(T_m/T_o) - 0.001(R_m/R_o) - 0.330(H_d/H_g)$	0.48	0.00	0.03	0.9	0.82
	59	$H_g/H_o = 0.508 + 0.051(S/S_o) + 0.003(T_m/T_o) + 0.003(R_m/R_o) + 0.001(W_m/W_o) - 0.333(H_d/H_g)$	1.39	0.00	0.03	0.91	0.83
	60	$H_g/H_o = 0.938 + 0.073(S/S_o) + 0.002(T_m/T_o) + 0.003(W_m/W_o) + 0.001(RF_m/RF_o) + 0.001(P_m/P_o) - 0.304(H_d/H_g)$	0.85	0.00	0.03	0.92	0.86
	61	$H_g/H_o = 0.945 + 0.072(S/S_o) + 0.003(T_m/T_o) + 0.001(R_m/R_o) + 0.003(W_m/W_o) + 0.001(RF_m/RF_o) + 0.001(P_m/P_o) - 0.300(H_d/H_g)$	1.93	0.00	0.03	0.93	0.86
Jodhpur (Hot and dry)	62	$H_g/H_o = 0.656 - 0.352(H_d/H_g)$	-0.04	0.00	0.03	0.71	0.54
	63	$H_g/H_o = 0.587 + 0.069(S/S_o) - 0.307(H_d/H_g)$	-0.08	0.00	0.03	0.7	0.55
	64	$H_g/H_o = 0.541 + 0.073(S/S_o) + 0.001(T_m/T_o) - 0.293(H_d/H_g)$	-0.14	0.00	0.03	0.77	0.62
	65	$H_g/H_o = 0.543 + 0.067(S/S_o) + 0.001(T_m/T_o) + 0.005(W_m/W_o) - 0.305(H_d/H_g)$	-0.13	0.00	0.03	0.77	0.62
	66	$H_g/H_o = -0.789 + 0.051(S/S_o) + 0.001(T_m/T_o) + 0.004(W_m/W_o) + 0.002(P_m/P_o) - 0.299(H_d/H_g)$	-0.19	0.00	0.03	0.79	0.64
	67	$H_g/H_o = -0.087 + 0.036(S/S_o) - 0.001(T_m/T_o) - 0.001(R_m/R_o) - 0.011(RF_m/RF_o) + 0.002(P_m/P_o) - 0.299(H_d/H_g)$	-0.22	0.00	0.03	0.8	0.64
	68	$H_g/H_o = -0.087 + 0.034(S/S_o) - 0.002(T_m/T_o) - 0.001(R_m/R_o) - 0.001(W_m/W_o) - 0.011(RF_m/RF_o) + 0.002(P_m/P_o) - 0.305(H_d/H_g)$	-0.36	-0.01	0.03	0.79	0.66
Delhi (Composite)	69	$H_g/H_o = 0.670 - 0.405(H_d/H_g)$	0.67	0.00	0.03	0.78	0.62
	70	$H_g/H_o = 0.4597 + 0.625(S/S_o) + 0.055(H_d/H_g)$	0.49	0.59	0.07	3.63	0.71
	71	$H_g/H_o = 0.589 + 0.138(S/S_o) - 0.004(T_m/T_o) - 0.301(H_d/H_g)$	0.56	0.00	0.03	0.77	0.61
	72	$H_g/H_o = 0.723 + 0.121(S/S_o) - 0.006(T_m/T_o) - 0.001(R_m/R_o) - 0.282(H_d/H_g)$	0.52	0.00	0.03	0.8	0.65
	73	$H_g/H_o = 0.747 + 0.127(S/S_o) - 0.007(T_m/T_o) - 0.001(R_m/R_o) + 0.001(W_m/W_o) - 0.268(H_d/H_g)$	0.47	0.00	0.03	0.81	0.67
	74	$H_g/H_o = 0.735 + 0.123(S/S_o) - 0.006(T_m/T_o) - 0.001(R_m/R_o) - 0.030(RF_m/RF_o) + 0.001(P_m/P_o) - 0.274(H_d/H_g)$	0.46	0.00	0.03	0.82	0.69
	75	$H_g/H_o = 0.08 + 0.123(S/S_o) - 0.006(T_m/T_o) - 0.001(R_m/R_o) + 0.001(W_m/W_o) - 0.028(RF_m/RF_o) + 0.001(P_m/P_o) - 0.264(H_d/H_g)$	0.42	0.00	0.03	0.83	0.81
Pune (Moderate)	76	$H_g/H_o = 0.156 + 0.281(S/S_o)$	3.1	0.01	0.04	0.71	0.53
	77	$H_g/H_o = 0.426 + 0.159(S/S_o) - 0.178(H_d/H_g)$	-0.29	-0.03	0.07	0.58	0.41
	78	$H_g/H_o = 0.439 + 0.222(S/S_o) + 0.001(P_m/P_o) - 0.358(H_d/H_g)$	1.03	0.00	0.03	0.87	0.75
	79	$H_g/H_o = 0.586 + 0.168(S/S_o) - 0.001(R_m/R_o) + 0.003(W_m/W_o) - 0.252(H_d/H_g)$	-1.71	-0.02	0.04	0.86	0.75
	80	$H_g/H_o = 0.589 + 0.185(S/S_o) - 0.001(T_m/T_o) - 0.001(R_m/R_o) + 0.002(RF_m/RF_o) - 0.333(H_d/H_g)$	0.34	0.00	0.02	0.9	0.81
	81	$H_g/H_o = 0.9 + 0.210(S/S_o) + 0.002(T_m/T_o) + 0.004(W_m/W_o) + 0.002(RF_m/RF_o) + 0.001(P_m/P_o) - 0.340(H_d/H_g)$	0.33	0.00	0.05	0.89	0.8
	82	$H_g/H_o = 0.447 + 0.170(S/S_o) + 0.001(T_m/T_o) - 0.001(R_m/R_o) + 0.003(W_m/W_o) + 0.004(RF_m/RF_o) + 0.001(P_m/P_o) - 0.340(H_d/H_g)$	0.25	0.02	0.02	0.9	0.81
Shillong (Cold and cloudy)	83	$H_g/H_o = 0.729 - 0.563(H_d/H_g)$	4.75	0.00	0.07	0.7	0.77
	84	$H_g/H_o = 0.404 + 0.339(S/S_o) - 0.002(R_m/R_o)$	1.5	-0.01	0.06	0.73	0.56
	85	$H_g/H_o = 0.605 + 0.221(S/S_o) - 0.002(R_m/R_o) - 0.226(H_d/H_g)$	1.87	0.00	0.05	0.79	0.65
	86	$H_g/H_o = 0.390 + 0.247(S/S_o) + 0.004(T_m/T_o) + 0.004(W_m/W_o) - 0.263(H_d/H_g)$	2.23	0.00	0.05	0.8	0.66
	87	$H_g/H_o = 0.651 + 0.234(S/S_o) - 0.001(T_m/T_o) - 0.002(R_m/R_o) + 0.003(W_m/W_o) - 0.256(H_d/H_g)$	1.61	0.00	0.05	0.82	0.68
	88	$H_g/H_o = -0.718 + 0.246(S/S_o) - 0.004(T_m/T_o) - 0.002(R_m/R_o) + 0.001(RF_m/RF_o) + 0.001(P_m/P_o) + 0.002(H_d/H_g)$	1.62	0.00	0.05	0.83	0.69
	89	$H_g/H_o = -0.547 + 0.241(S/S_o) - 0.004(T_m/T_o) - 0.002(R_m/R_o) + 0.003(W_m/W_o) + 0.001(RF_m/RF_o) + 0.001(P_m/P_o) - 0.256(H_d/H_g)$	0.69	0.00	0.03	0.84	0.71

978-1-5386-6624-1/18 $31.00 © 2018 IEEE

TABLE IX. COMPARISON WITH OTHER MODELS

Model	Eq. No.	Equations	MPE (%)	MBE (%)	RMSE (%)
Proposed Model	61	$H_g/H_o = 0.945 + 0.072(S/S_o) + 0.003(T_m/T_o) + 0.001(R_m/R_o) + 0.003(W_m/W_o) + 0.001(RF_m/RF_o) + 0.001(P_m/P_o) - 0.300(H_d/H_g)$	1.93	0.00	0.04
	68	$H_g/H_o = -0.087 + 0.034(S/S_o) - 0.002(T_m/T_o) - 0.001(R_m/R_o) - 0.001(W_m/W_o) -0.011(RF_m/RF_o) + 0.002(P_m/P_o) -0.305(H_d/H_g)$	0.36	0.01	0.03
	75	$H_g/H_o = 0.08 + 0.123(S/S_o) -0.006(T_m/T_o) -0.001(R_m/R_o) +0.001(W_m/W_o) -0.028(RF_m/RF_o) + 0.001(P_m/P_o) -0.264(H_d/H_g)$	0.42	0.00	0.03
	82	$H_g/H_o = 0.447 + 0.170(S/S_o) +0.001(T_m/T_o) -0.001(R_m/R_o) +0.003(W_m/W_o) +0.004(RF_m/RF_o) + 0.001(P_m/P_o) -0.340(H_d/H_g)$	0.25	0.02	0.02
	89	$H_g/H_o = -0.547 + 0.241(S/S_o) -0.004(T_m/T_o) -0.002(R_m/R_o) +0.003(W_m/W_o) +0.001(RF_m/RF_o) + 0.001(P_m/P_o) -0.256(H_d/H_g)$	0.69	0.00	0.03
Angstrom- Prescott Model	90	$H_g/H_o = 0.0801 + 0.709(S/S_o)$	96.06	15.62	16.54
Rietveld Model	91	$H_g/H_o = 0.18 + 0.62(S/S_o)$	35.71	4.80	5.21
Page Model	92	$H_g/H_o = 0.23 + 0.48(S/S_o)$	25.90	3.32	3.52
Akinoglu and Ecevit Model	93	$H_g/H_o = 0.145 + 0.845(S/S_o) - 0.28(S/S_o)^2$	32.47	4.15	4.36
Bahel Model	94	$H_g/H_o = 0.16 + 0.87(S/S_o) - 0.16(S/S_o)^2 + 0.34(S/S_o)^3$	-53.24	-6.51	7.69
Newland Model	95	$H_g/H_o = 0.34 - 0.4(S/S_o) + 0.17 \log (S/S_o)$	88.87	16.72	18.54
Abdalla Model	96	$H_g/H_o = 0.5289 + 0.459(S/S_o) + 0.004073(T_m/T_o) - 0.006481 R_m/R_o)$	-93.19	-14.40	15.40

VI. CONCLUSION

In the present work, eight sunshine-based models and thirty five empirical models have been developed based on multiple regression analysis using seven meteorological parameters for five stations across India. The regression and correlation coefficients for each model is calculated and presented. Principle component analysis have been applied and five such empirical models have been proposed that incorporate seven parameters which is found to be accurate for estimating global solar energy for each of the climate zone across the entire country.

Good agreement is found between measured and estimated data based on seven parameter correlations, which makes it useful in estimating global solar energy. The best fit model is achieved by (61) with MPE=1.96% for warm and humid climate; (68) with MPE=0.36% for hot and dry climate; (75) with MPE=0.42% for composite climate; (82) with MPE=0.25% for moderate climate; and (89) with MPE=0.69% for cold and cloudy climate respectively. The models presented have reasonable estimation errors. Based on the overall analysis, it has been concluded that the parameters considered do have strong influence on estimating global solar energy. Therefore, the models being proposed could be successfully used to forecast global solar energy in distinct climate zone of India and in elsewhere with similar climatic conditions.

REFERENCES

[1] J. Hassan, "ARIMA and regression models for prediction of daily and monthly clearness index", Renewable Energy, vol. 68, pp. 421-427, 2014.

[2] Z. A. Al-Mostafa, A. H. Maghrabi, and S. M. Al-Shehri, "Sunshine-based global radiation models: A review and case study", Energy Conversion and Management, vol. 84, pp. 209-216, 2014.

[3] A. Angstrom, "Solar and terrestrial radiation", Quarterly Journal of Royal Meteorological Society, vol. 50(4), pp. 121-6, 1924.

[4] J. K. Page, "The estimation of monthly mean values of daily total short wave radiation on vertical and inclined surfaces from sunshine records for latitudes 40°N – 40°S", Proceedings of UN Conference on New Sources of Energy, vol. 4, pp. 378-90, 1961.

[5] J.A. Prescott, "Evaporation from water surface in relation to solar radiation", Transaction of the Royal Society of South Australia, vol. 64, pp. 114-118, 1940.

[6] A. Teke, and H. B. Yildrium, "Estimating the monthly global solar radiation for Eastern Mediterranean Region", Energy Conversion and Management, vol. 87, pp. 628-635, 2014.

[7] S. Kirmani, M. Jamil, and M. Rizwan, "Empirical correlation of estimating global solar radiation using meteorological parameters", International Journal of Sustainable Energy, Taylor and Francis, vol. 34(5), pp. 327-339, 2013.

[8] Y. A. G. Abdalla, "New correlation of global solar radiation with meteorological parameters for Bahrain", International Journal of Solar Energy, vol. 16, pp. 111-120, 1994.

[9] N. K. Bansal, and G. Minke, "Climatic zones and rural housing in India," Kernforschungsanlage Julich GmbH, Zentralbibliothek, Julich1988.

[10] P. T. Ajit, "Solar radiant energy over India," by Ministry of New and Renewable Energy and Indian Meteorological Department, 2009.

[11] J. A. Duffie, and W. A. Beckman, "Solar engineering of thermal process," New York: Wiley; 1991.

[12] B. G. Akinoglu, and A. Ecevit, "Construction of a quadratic model using modified Angstrom coefficients to estimate global solar radiation", Solar Energy, vol. 45, pp. 85-92, 1990.

[13] V. Bahel, H. Baksh, and R. Srinivasan, "A correlation for estimation of global solar radiation," Energy, vol. 12, pp. 131 - 135, 1987.

[14] F. J. Newland, "A study of solar radiation models for the coastal region of South China", Solar Energy, vol. 43(4), pp. 227-235, 1989.

[15] D. B. Ampratwum, and A. S. S. Dorvlo, "Estimation of solar radiation from the number of sunshine hours", Applied Energy, vol. 63, pp. 161-7, 1991.

[16] K. Bakirci, "Correlations for estimation of global solar radiations with hours of bright sunshine in Turkey", Energy, vol. 34, pp. 485-501, 2009.

[17] J. Almorox, C. Hontoria, and M. Benito, "Models for obtaining daily global solar radiation with measured air temperature data in Madrid (Spain)", Applied Energy, vol. 88, pp. 1703-1709, 2011.

A Comparative Analysis of AGC of Two-area Hydro-Thermal Power System Interconnected with AC-DC Parallel Link in Restructured Power System

Nisha Singh
Dept. of Electrical Engineering
Maharaja Surajmal Institute of Technology, Janakpuri,
Delhi, India
nisha.singh0261@gmail.com,

Narendra Kumar
Dept. of Electrical Engineering
Delhi Technological University Bawana Road,
Delhi, India
dnk_1963@yahoo.com

Yogendra Arya, Nidhi Gupta
Dept. of Electrical Engineering
Maharaja Surajmal Institute of Technology, Janakpuri, Delhi,
India
nidhi.msit@gmail.com

Abstract— The paper presents a comparative analysis of automatic generation control (AGC) of two-area deregulated power system linked by AC-DC parallel tie lines each area having hydro and thermal power sources. The state space model of the power system has been developed under restructured atmosphere have been developed and optimal PI regulators are designed to simulate all power market contracts which are possible in a restructured situation. The concept of DISCO participation matrix (DPM) is exploited to simulate the transactions. Eigenvalue analysis is carried out to assess the comparative performance of the power system with and without the effect of AC & DC parallel tie lines. Further, the dynamic responses of the hydro-thermal power system are compared and it has been established that inclusion of parallel DC link with AC improves the dynamic performance further.

Keywords—Hydro-Thermal Power System, Restructured Power System, AC-DC Parallel link.

I. INTRODUCTION

Automatic generation control (AGC) is one of the major control problems in interconnected power systems. It has mainly two objectives (a) maintain the frequency of each area within the specified limits and (b) controlling the inter area tie line power exchanges within the scheduled values [1,2,3]. In conventional power systems, the power generation, transmission, and distribution are governed by a single entity called vertically integrated utility (VIU). VIU supplies their consumers at specified rates. In the restructured power system VIU supplies power to their consumers at a specified rate after restructuring, the role of VIU is carried out by different market players like generating companies (GENCOs), transmission companies (TRANSCOs), distribution companies (DISCOs) and independent system operator (ISO). ISO is independent, disassociated agent for market participants who perform various ancillary services and among them is the AGC. These market players control the generation and load demand by keeping the entire power system stable under highly competitive and distributed control environment. Due to lack of proper controller design in deregulated power system, the instability may spread to other control areas

leading to severe system black out. In order to cope with these situations, a lot of work is being done on various control strategies in deregulated power systems notable among these are optimal control, sub-optimal control [4,14,15], self-tuning control, robust control, variable structure control and intelligent control strategies. In addition, various soft computing techniques such as artificial neural network (ANN)[6], fuzzy logic, genetic algorithm (GA)[7,3,9], particle swarm optimization[14], ANFIS controller, bacterial foragingalgorithm (BFA) [10,11], hybrid neuro fuzzy [8] Eigen structure assignment technique [4], active disturbance rejection [12], firefly algorithm[7] H_2/H_1 control [5], non-integer controller [11] etc are being applied.

Now a days flexible AC Transmission Systems devices such as thyristor controlled series compensator (TCSC), interline power flow controller (IPFC) and Thyristor controlled phase shifter (TCPS) are being applied in series with AC tie-line to alleviate the system frequency and tie-line power variations. One of the important utilities of HVDC lines is adding the DC link in parallel with AC link interlinking two control areas. Most of the researchers have studied the AGC of conventional power systems with diverse sources such as hydro, thermal, gas, and nuclear, working in control areas [1, 16, 17]. Optimal AGC controllers are developed and implemented on a two-area power system with hydro-thermal–gas generating units in each control area [1].

In this paper a comparative analysis of automatic generation control (AGC) of two-area restructured power system interlinked by AC-DC parallel tie lines each area having (a) hydro-hydro power sources (b) hydro and thermal power sources (c) two thermal power sources. The state space models of the power systems under restructured atmosphere have been developed and optimal proportional integral (PI) controllers are developed to simulate all power market contracts which are possible. The concept of DISCO participation matrix (DPM) is exploited to simulate the transactions. Eigenvalue analysis has been carried out to assess the comparative performance of each power system with and without the effect of AC-DC parallel links. Further,

978-1-5386-6624-1/18 $31.00 © 2018 IEEE

the dynamic responses of the hydro-hydro, hydro-thermal and thermal-thermal power systems are compared in restructured/ deregulated power system environment interconnected by (i) AC-DC link and (ii) AC link only. It has been established that the responses of hydro-hydro power system are sluggish with large overshoots as compared with hydrothermal and thermal-thermal systems. The inclusion of parallel AC-DC links improves the dynamic performance significantly.

II. MODELLING OF DREGULATED POWER SYSTEM

Deregulated power systems are generally planned to consider all power contracts which are likely poolco based, bi-lateral and a combination of these two. A GENCO in contract alters its power output till it does not exceed the contracted worth, while a DISCO is accountable for monitoring its demand to meet load as per requirements of contract [4]. To fulfill various transactions and to deliver information about the involvement of a DISCO in contract with a GENCO, a DISCO participation matrix (DPM) has been recommended [2]. The number of rows in DPM represent the number of GENCOs and the number of columns are equal to the number of DISCOs in the decentralized/ restructured power system [18]. Each element in DPM is designated as contract participation factor (cpf) which corresponds to a contracted load by a DISCO needed from the corresponding GENCO. The sum of the elements of a column of DPM is equal to one. In the proposed two-area power system, each control area has two DISCOs and two GENCOs. Assuming GENCO$_1$, GENCO$_2$, DISCO$_1$ and DISCO$_2$ in control area-1 and GENCO$_3$, GENCO$_4$, DISCO$_3$ and DISCO$_4$ in control area-2. The hydro-thermal system in control area-1 is constituted by one generating source which is single reheat thermal power plant (GENCO$_1$) and the other is mechanical governor based hydro power plant (GENCO$_2$). Similarly GENCO$_3$ is thermal and GENCO$_4$ is hydro source of the same configuration and in area-2. In the two area hydro-hydro configuration, each area is having two hydro sources GENCO$_1$, GENCO$_2$, and GENCO$_3$, GENCO$_4$ respectively. Each area interconnected with AC-DC matching links is considered to demonstrate the efficacy of the optimal proportional-integral controllers based on full state feedback control. Fig.1 shows complete transfer function model of the proposed power system. The following DISCO participation matrix (DPM) is formulated for the system under consideration is given as:

$$DPM = \begin{bmatrix} cpf_{11} & cpf_{12} & cpf_{13} & cpf_{14} \\ cpf_{21} & cpf_{22} & cpf_{23} & cpf_{24} \\ cpf_{31} & cpf_{32} & cpf_{33} & cpf_{34} \\ cpf_{41} & cpf_{42} & cpf_{43} & cpf_{44} \end{bmatrix} \quad (1)$$

The steady state power flow for a two-area power system linked via AC tie line only is given as:

$$\Delta Ptie_{actual} = \frac{2\pi T_{12}}{s} [\Delta F1 - \Delta F2] \quad (2)$$

The DC link is considered to be operated in constant current control mode. The incremental power flow ($\Delta Ptie_{dc}$) through DC link is modelled with incremental change in frequency at rectifier end. The representation of transfer function model

of DC link is taken from [1, 15, 18, 21] is shown in Fig.1. For small variations in power the DC tie-line flows can be modelled as:

$$\Delta Ptie_{dc} = \frac{K_{dc}}{1 + s\, T_{dc}} [\Delta F_1 - \Delta F_2] \quad (3)$$

For small load perturbation, the total tie-line power flows considering the presence of the DC link in parallel with the existing AC link can be given as:

$$\Delta Ptie_{mactual} = \Delta Ptie_{actual} + \Delta Ptie_{dc} \quad (4)$$

The scheduled steady state power flow on the tie-line can be given as:

$$\Delta Ptie_{scheduled} = \sum_{i=1}^{2} \sum_{j=3}^{4} cpf_{ij}\, \Delta P_{Lj} - \sum_{i=3}^{4} \sum_{j=1}^{2} cpf_{ij}\, \Delta P_{Lj} \quad (5)$$

$$\Delta Ptie_{scheduled} = [P_{exp1}] - [P_{imp1}]$$

Where P_{exp1}= total power exported from area-1 which is same as demand of DISCOs in area-2 from GENCOs in area-1. And P_{imp1}= total power imported in area-1 which is same as demand of DISCOs in area-1 from the GENCOs in area-2. The error in tie-line power flow can be written as:

$$\Delta Ptie_{error} = \Delta Ptie_{mactual} - \Delta Ptie_{scheduled} \quad (6)$$

In steady state $\Delta Ptie_{error} = 0$ as $\Delta Ptie_{mactual}$ attains $\Delta Ptie_{scheduled}$ tie-line flows. The area control error (ACE) is modified due to the presence of AC-DC parallel links and can be defined as:

$$ACE_{m1} = \beta_1 \Delta F_1 + \Delta Ptie_{error} \quad (7)$$

$$ACE_{m2} = \beta_2 \Delta F_2 + \alpha_{12} \Delta Ptie_{error} \quad (8)$$

Where α_{12} =area size ratio and β_1, β_2 = frequency bias constants of the respective area. Since there are usually more than one GENCO in each area, therefore new area control error (ACE) signal is to be shared by these GENCOs in proportion to their roles in AGC. The factors which denote this sharing, are called ACE participation factors (apfs) and it is assumed that all GENCOs contribute in the AGC according to their respective apfs. The equations for steady state produced power of GENCOs in contract with DISCOs for Hydro-thermal power system are:

$$\Delta P_{Gti} = cpf_{i1}\Delta P_{L1} + cpf_{i2}\Delta P_{L2} + cpf_{i3}\Delta P_{L3} + cpf_{i4}\Delta P_{L4};\ i=1,3 \quad (9)$$

$$\Delta P_{Ghi} = cpf_{i1}\Delta P_{L1} + cpf_{i2}\Delta P_{L2} + cpf_{i3}\Delta P_{L3} + cpf_{i4}\Delta P_{L4};\ i=2,4 \quad (10)$$

The market disturbance signal ($\Delta P_{L1;LOC} = \Delta P_{L1} + \Delta P_{L2}$) can be defined as the total local demand in area-1 and $\Delta P_{L2;LOC} = (\Delta P_{L3} + \Delta P_{L4})$ is the total local load demand in area-2. ΔP_{Li} represents the power demanded by DISCO$_i$. The disturbance signals ΔP_{UC1} and ΔP_{UC2} are the un-contracted power demanded by DISCOs in area-1 and area-2 respectively. The market signals ΔP_{UC1} and ΔP_{UC2} will be equal to zero, when un-contracted power demands of DISCOs are not present.

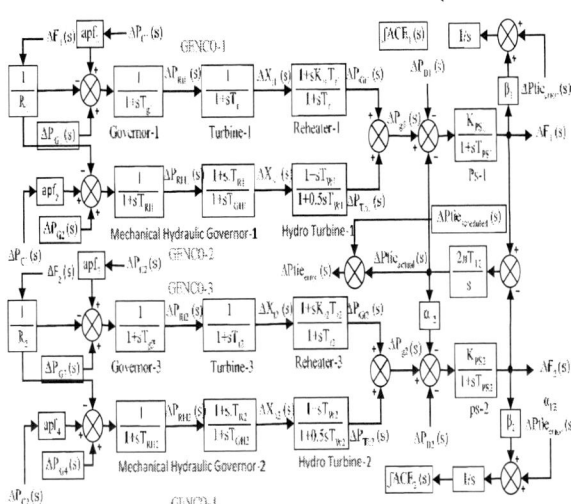

Fig.1. Hydro-thermal power system model

III. STATE SPACE MODEL

The power system model as given in Fig.1 is a linear continuous-time system and is $\dot{X}$ represented by the following state space equations

$$\dot{X}=AX+BU+\Gamma P_d \qquad (11)$$

$$X(0)=0 \qquad (12)$$

$$Y=CX$$

where X is the state vector of the dimension n×1, n is number of state variables = 18, U is control input vector of the dimension m×1, m is number of control variables = 2, P_d is disturbance vector of the dimension px1, p is number of disturbance variables = 6, Y is output vector of the dimension q×1, q is number of measured output variables = 18. A, B, C, and C are system, control, output, and disturbance matrices having dimensions of n × n, n × m, q × n, and n × p respectively, and are given in Appendix A. In optimal control theory application, the term P_d in Eq. (13) is eliminated in steady state occurring after a disturbance. So the Eq. (13) can be rephrased in a convenient form as

$$\dot{X}=AX+BU, \qquad (13)$$

$$X(0) = -X_{ss}$$

Where new state vector is equal to the old state vector minus its steady state value X_{ss} [11,12] State, control and disturbance vectors selected for the power system under study are given as follows:

State Vector for the hydro-thermal system:

$$X=[\Delta F_1 \Delta F_2\ \Delta Ptie_{mactual}\ \Delta P_{Gt1}\ \Delta P_{Gh1}\ \Delta P_{Gt2}\ \Delta P_{Gh2}\ \Delta X_{t1}\ \Delta X_{h1}$$
$$\Delta X_{t2}\ \Delta X_{h2}\ \Delta P_{Rt1}\ \Delta P_{RH1}\ \Delta P_{Rt2}\ \Delta P_{RH2}\ \int ACE_1 dt\ \int ACE_2 dt\ \ \Delta Ptie_{dc}]^T$$
$$(14)$$

Distribution Vector:

$$P_d=[\Delta P_{L1}\ \Delta P_{L2}\ \Delta P_{L3}\ \Delta P_{L4}\ \Delta P_{UC1}\ \Delta P_{UC2}]^T \qquad (15)$$

OPTIMAL PI CONTROLLER DESIGN

The design of optimal PI controllers has been re-counted in the literature [1, 15, 19, 21, 25]. This section is dedicated to derive optimal feedback gains K^* of optimal PI controllers. The continuous time dynamic model of power system in the state variable form is described by Eqns. (11)

& (12). The control vector U expresses the performance criterion to minimize the performance index J given by

$$J=\int_0^\infty \frac{1}{2}[X^T QX+U^T R U]dt \qquad (16)$$

Where Q is a positive semi-definite symmetric state cost weighting matrix and R is a positive definite symmetric control cost weighting matrix. The optimal control law is given [24] as:

$$U^*= -K^*X \qquad (17)$$

$$\text{Where } K^* = R^{-1}B^T P \qquad (18)$$

And P is the solution of algebraic matrix-Riccati equation

$$PA+A^T P-PBR^{-1}B^T P+Q=0 \qquad (19)$$

SIMULATION AND DISCUSSION

The state space modeling of the 2-area power system as shown in Fig. 1 with thermal and hydro sources in each area interconnected by parallel AC-DC link under restructured atmosphere is simulated with the optimal controller gains using MATLAB software. The system data are given in Appendix A. The optimal PI regulators are designed to simulate all types of possible power contracts like poolco, bilateral and a combination of the two contracts taking place in a restructured atmosphere. The optimal gains of optimal PI regulators are obtained for power system model considering (a) AC link and (b) AC-DC parallel tie lines along with the lowest performance index values J* as given in table.1. The open-loop and closed-loop system eigenvalues for the power system model are listed in table.2. For hydro-thermal system performance index J* is minimum (J* = 370.6373) when AC-DC parallel tie lines are used as interconnection than that obtained with only AC link (J* = 398.1168). In hydro-thermal power system with only AC tie lines eigenvalues of $\Delta F_1\ \Delta F_2\ \Delta Ptie_{12actual}$ are more negative in closed loop than open loop similarly all other states also are more negative in close loop than in open loop, in AC-DC tie line eigenvalues of closed loop area more negative than closed loop of AC Link so in close loop of AC/DC Link system stability improves significantly. The cases for different transactions are:

Case-A: Poolco based contracts

In poolco based contracts GENCOs contribute in AGC of their own control area, here area-1 only i.e., power is demanded only by DISCO-1 and DISCO-2. Let, the power demand of each DISCO be 0.04, i.e., $\Delta P_{L1,LOC} = 0.04$ pu and $\Delta P_{L2,LOC} = 0$ pu. Area control error (ACE) participation factor of thermal ($apf_1 = apf_2$) and hydro ($apf_3= apf_4$) are equal to 0.5. The DPM for this case is given as:

$$DPM = \begin{bmatrix} 0.5 & 0.5 & 0 & 0 \\ 0.5 & 0.5 & 0 & 0 \\ 0 & 0 & 0 & 0 \\ 0 & 0 & 0 & 0 \end{bmatrix}. \qquad (20)$$

Generations of GENCOs must match the demanded power of the DISCOs in contract with them, in the steady state. In steady state the theoretical values of generations are $\Delta P_{Gt1}=0.04$ pu, $\Delta P_{Gh1} = 0.04$ pu and $\Delta P_{Gt2} = \Delta P_{Gh2} = 0$ pu; for hydro-thermal. Also the scheduled power flow on the tie-line can be found by Eqn. (5), i.e., $\Delta Ptie_{scheduled} = 0$ pu. For Case-A, it is observed that in the steady state the deviations in

978-1-5386-6624-1/18 $31.00 © 2018 IEEE

each area, frequency deviation settle to zero under a step load change of DISCO-1 and DISCO-2, hence AGC requirement is fulfilled. The $\Delta Ptie_{scheduled} = \Delta Ptie_{actual} = \Delta Ptie_{mactual} = 0$ pu. Therefore, $\Delta Ptie_{error} = 0$ pu. The actual generations of all GENCOs attain the desired values in the steady state. As GENCOs situated in area-2 do not have contracts with any GENCO, $(\Delta P_{Gh3}=\Delta P_{Gh4}=0pu;$ for H-H),$(\Delta P_{Gt3}=\Delta P_{Gt4}=0pu;$for T-T), $(\Delta P_{Gt2}=\Delta P_{Gh2}=0$ pu; for H-T) in steady state.

Table 1

Optimal gain matrices of the optimal PI controllers for two-area single-source HYDRO-THERMAL restructured power system.

Type of area interconnections	Optimal feedback gain matrix [K*]									J*
AC link only	[0.6377	0.0998	-1.5770	5.1949	2.0432	0.4447	0.1477	-1.6747	3.8398	398.1168
	-0.2020	0.8697	0.4525	-0.0042	0.0015	-0.6063	1.0000	0.0000;		
	0.0998	0.6377	1.5770	0.4447	0.1477	5.1949	2.0432	-0.2020	0.8697	
	-1.6747	3.8398	0.0015	-0.6063	0.4525	-0.0042	-0.0000	1.0000]		
AC/DC links	[0.4054	0.3321	-0.1323	4.4442	1.3625	1.1955	0.8284	-1.5487	3.1050	370.6373
	-0.3280	1.6045	0.3925	0.5985	0.0616	-1.2091	1.0000	-0.0000	-0.0150;	
	0.3321	0.4054	-0.1323	1.1955	0.8284	4.4442	1.3625	-0.3280	1.6045	
	-1.5487	-3.1050	0.0616	-1.2091	0.3925	0.5985	0.0000	1.0000	0.0150]	

Table 2

Pattern of open-loop and closed-loop eigenvalues for two-area single-source HYDRO-THERMAL restructured power system.

S. No.	State variables	Eigenvalues with AC link		Eigenvalues with AC/DC links	
		Open-loop	Closed-loop	Open-loop	Closed-loop
1.	ΔF_1	0	-14.2014	0	-14.2014
2.	ΔF_2	0	-14.1913	0	-14.0921
3.	$\Delta Ptie_{actual}$	-12.9074	-0.5098 + 2.8635i	-12.7630i	-2.1269 + 7.8181i
4.	ΔP_{Gh1}	-12.9074	-0.5098 - 2.8635i	-2.0752 + 7.8229i	-2.1269 - 7.8181i
5.	ΔP_{Gh1}	-0.1628 + 2.9641i	-2.8423 + 0.8174i	-2.0752 - 7.8229i	-3.8493
6.	ΔP_{Gh3}	-0.1628 - 2.9641i	-2.8423 - 0.8174i	-12.9224	-2.823 + 0.8174i
7.	ΔP_{Gh4}	-2.6553 + 0.6186i	-2.8358 + 0.4309i	-3.4881	-2.823 - 0.8174i
8.	ΔX_{h1}	-2.6553 - 0.6186i	-2.8358 - 0.4309i	-1.9876 + 0.3891i	-1.1819 +1.2571i
9.	ΔX_{h2}	-1.2913	-1.1819 + 1.2571i	-1.9876 - 0.3891i	-1.1819 -1.2571i
10.	ΔX_{h3}	-2.6410 + 0.8493i	-1.1819 - 1.2571i	-0.4585	-1.9871 + 0.3153i
11.	ΔX_{h4}	-2.6410 - 0.8493i	-1.3844	-2.6410 + 0.8493i	-1.9871 - 0.3153i
12.	ΔP_{RH1}	-0.0204	-0.3358	-2.6410 - 0.8493i	-0.4905
13.	ΔP_{RH2}	-0.0980	-0.1971 + 0.0839i	-0.7514 + 1.3220i	-0.3358
14.	ΔP_{RH3}	-0.7514 + 1.3220i	-0.1971 - 0.0839i	-0.7514 - 1.3220i	-0.1944 + 0.0812i
15.	ΔP_{RH}	-0.7514 - 1.3220i	-0.0347	-0.0976	-0.1944 - 0.0812i
16.	$\int ACE_{m1} dt$	-0.2113	-0.0347	-0.0204	-0.1973
17.	$\int ACE_{m1} dt$	-0.0347	-0.0348	-0.2113	-0.0347
18.	$\Delta Ptie_{sc}$			-0.0347	-0.0348

Case - B Combination of poolco and bilateral transactions:

In this case, DISCOs have the freedom to have contracts with any GENCOs available in its own or other control areas, the DPM given by

$$DPM = \begin{pmatrix} 0.50 & 0.25 & 0.0 & 0.30 \\ 0.20 & 0.25 & 0.0 & 0.0 \\ 0.0 & 0.25 & 1.0 & 0.70 \\ 0.30 & 0.25 & 0.0 & 0.0 \end{pmatrix}$$

Let each DISCO demand 0.04 pu power from GENCOs as per the pattern of cpfs shown in the DPM, apfs: $a_{11} = 0.75, a_{12} = 0.25, a_{21} = a_{22} = 0.50$ Note that ACE participation factors (apfs) affect only the transient behaviour of the system and not the steady state behaviour when un-contracted loads are absent i.e., $\Delta P_{uc1} = \Delta P_{uc2} = 0.0$ pu. Using Eqn. (5), $\Delta Ptie_{scheduled} = \{(0.0 + 0.30 + 0.0 + 0.0) - (0.0 + 0.25 + 0.30 + 0.25)\}\ 0.04 = -0.02$ pu. For Case-B, the

simulation results for the system with AC/DC parallel links are also shown in Fig. 2. In steady state the deviation in frequency error settles to zero. The actual tie-line powers shown in Figs. 2 (c) settle to the desired value of −0.02 puMW, which is $\Delta Ptie_{scheduled}$ in the steady state. The steady state desired values of power generations are ($\Delta P_{Gh1}=0.042pu$, $\Delta P_{Gh2}=0.018pu$, $\Delta P_{Gh3}=0.078pu$, $\Delta P_{Gh4}=0.022pu;$for H-H),($\Delta P_{Gt1}=0.042$ pu,$\Delta P_{Gt2}=0.018$ pu, $\Delta P_{Gt3}=0.078$ pu, $\Delta P_{Gt4}=0.022pu;$ for T-T), ($\Delta P_{Gt1} = 0.042$ pu, $\Delta P_{Gh1} = 0.018$ pu, $\Delta P_{Gt2} = 0.078$ pu, $\Delta P_{Gh2} = 0.022$ pu; for H-T). These are verified in Figs. 5(d—i).

CASE B RESULT

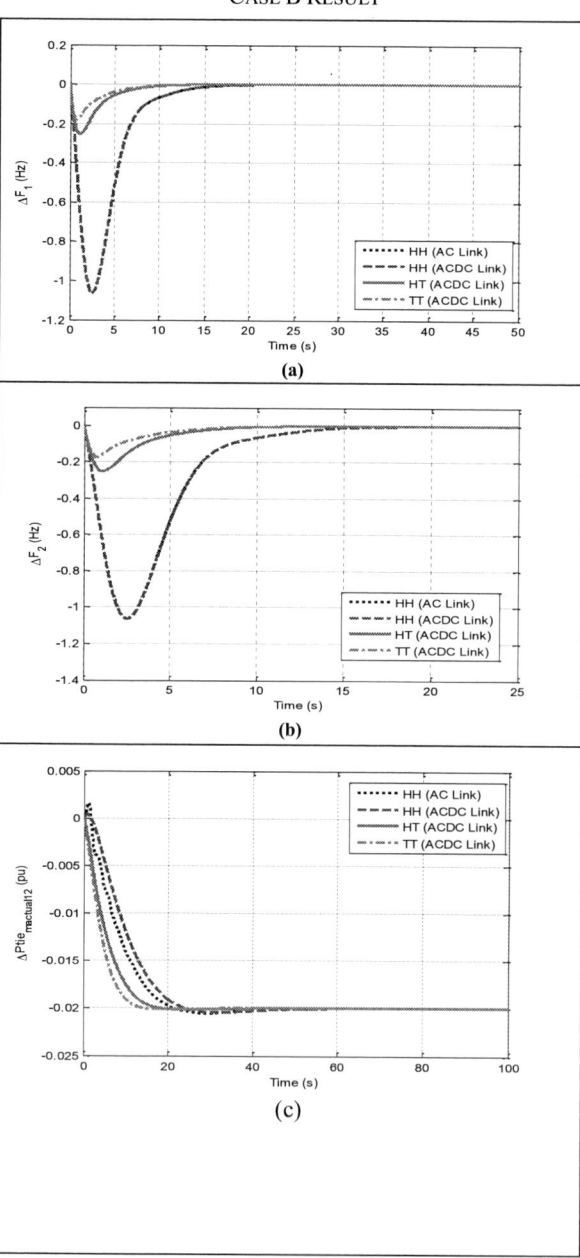

(a)

(b)

(c)

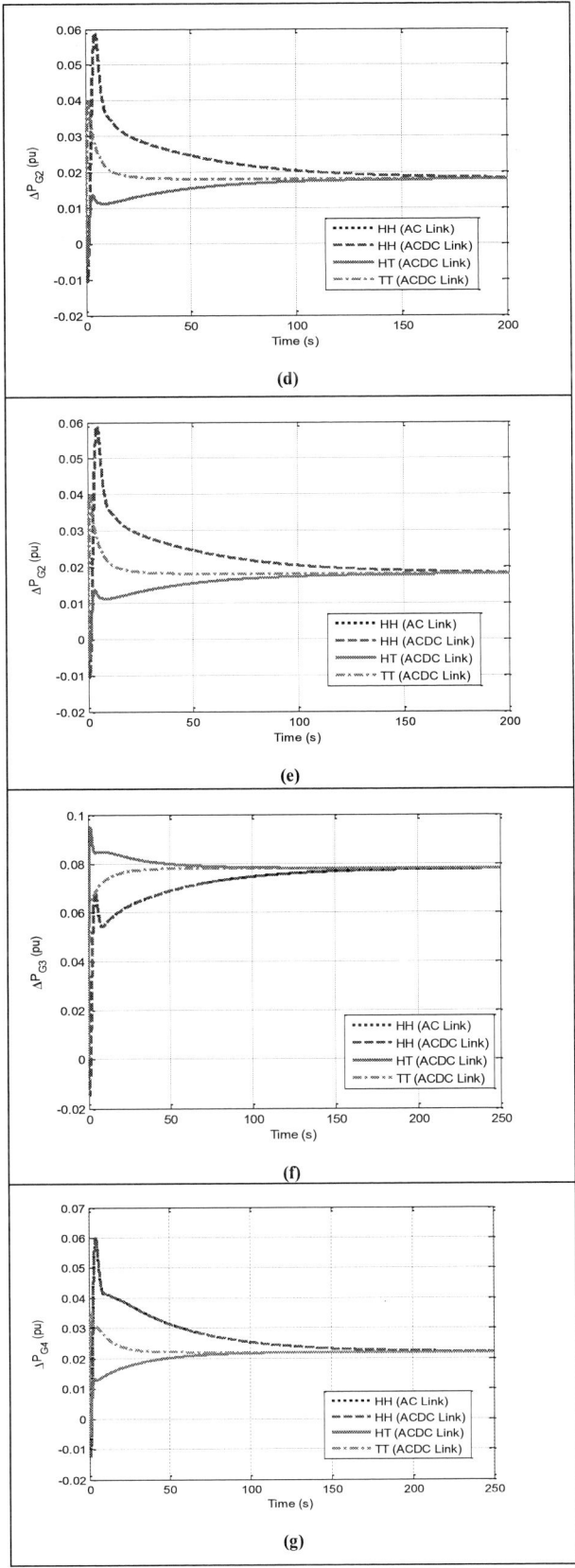

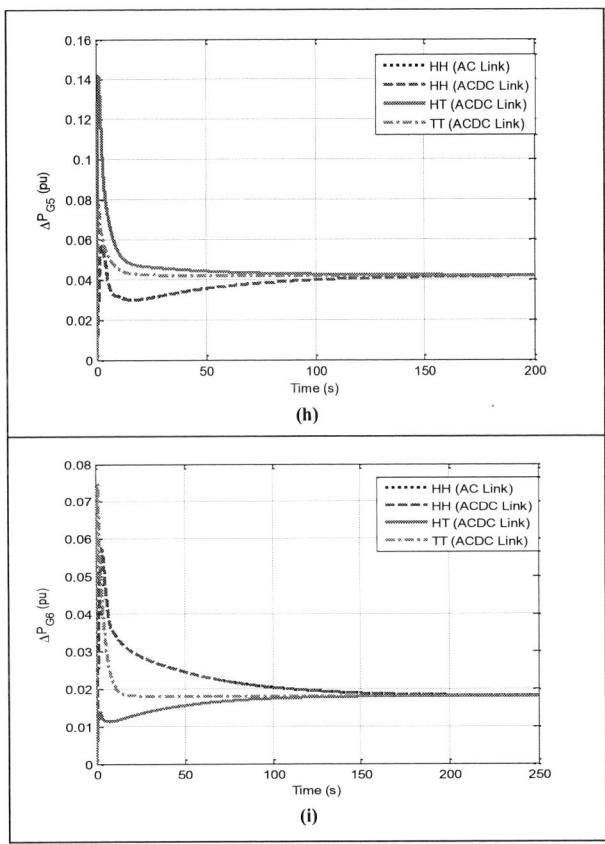

Fig.5 Dynamic performance of two-area single source and two-area multi source (a) F_1 (Hz) vs. time (b) F_2 (Hz) vs. time (c) $Ptie_{mactual}$ (pu) vs. time (d) P_{G1} (pu) vs. time (e) P_{G2} (pu) vs. time (f) P_{G3} (pu) vs. time (g) P_{G4} (pu) vs. time (h) P_{G5} (pu) vs. time (i) P_{G6} (pu) vs. time

CONCLUSION

The optimal AGC of a two-area Hydro-thermal, two-area Hydro-hydro and two-area thermal-thermal power system under restructured atmosphere has been analyzed and compared. In hydro-thermal two-area power system having hydro and single reheat thermal units in each area is interconnected with AC-DC parallel tie lines. The optimal proportional integral (PI) regulators are designed using full state feedback control strategy to implement all market transactions possible between GENCOs and DISCOs in a deregulated atmosphere. The analysis has been carried out for the AGC scheme under pool-co, bilateral and the combination of these two contracts. The MATLAB simulation results with optimal PI controllers for different generating sources and tie-line power flows of GENCOs are validated with the desired values. It has been observed that with both AC and AC-DC parallel tie lines the power system is marginally stable in open-loop and stable with good stability margins in closed-loop modes. In addition, dynamic stability margins appreciably improve further with AC-DC

978-1-5386-6624-1/18 $31.00 © 2018 IEEE   329

tie-lines. It is concluded that the system performance index is reduced and overall dynamic performance of the power system is enhanced considerably, with regards the reduced settling times, peak overshoots, under-shoots and oscillations with AC-DC parallel tie lines in comparison with AC link only.

REFERENCES

1. Ibraheem, Nizamuddin, T.S. Bhatti AGC of two area power system interconnected by AC/DC Links with diverse sourcs in each area .
2. Donde V, Pai MA, Hiskens IA. Simulation and optimization in an AGC system after deregulation. IEEE Trans Power Syst 2001;16(3):481–9
3. Demiroren A, Zeynelgil HL. GA application to optimization of AGC in three-area power system after deregulation. Int J Electr Power Energy Syst 2007;29 (3):230–40
4. Tyagi B, Srivastava SC. A decentralized automatic generation control scheme for competitive electricity markets. IEEE Trans Power Syst 2006;21(1): 312–20.
5. Shayeghi H. A robust decentralized power system load frequency control. J Electr Eng 2008;59(6):281–93.
6. Shayeghi H, Shayanfar HA, Malik OP. Robust decentralized neural networks based LFC in a deregulated power system. Elect Power Syst Res 2007;77(3– 4):241–51.
7. Sekhar GTC, Sahu RK, Baliarsingh AK, Panda S. Load frequency control of power system under deregulated environment using optimal firefly algorithm. Int J Electr Power Energy Syst 2016;74:195–211.
8. Shree SB, Kamaraj N. Hybrid neuro fuzzy approach for automatic generation control in restructured power system. Int J Electr Power Energy Syst 2016;74:274–85.
9. Deepak M, Abraham RJ. Load following in a deregulated power system with thyristor controlled series compensator. Int J Electr Power Energy Syst 2015;65:136–45.
10. Chidambaram IA, Paramasivam B. Optimized load–frequency simulation in restructured power system with redox flow batteries and interline power flow controller. Int J Electr Power Energy Syst 2013;50:9–24.
11. Debbarma S, Saikia LC, Sinha N. AGC of a multi-area thermal system under deregulated environment using a non-integer controller. Elect Power Syst Res 2013;95:175–83.
12. Tan W, Hao Y, Li D. Load frequency control in deregulated environments via active disturbance rejection. Int J Electr Power Energy Syst 2015;66: 166–77.
13. Rao CS, Nagaraju SS, Raju PS. Automatic generation control of TCPS based hydrothermal system under open market scenario: a fuzzy logic approach. Int J Electr Power Energy Syst 2009;31(7–8):315–22.
14. Sinha S, Patel R, Prasad R. Application of AI supported optimal controller for automatic generation control of a restructured power system with parallel AC–DC tie lines. Euro Trans Electr Power 2012;22(5):645–61.

15. Ibraheem, Kumar P, Hasan N, Singh Y. Optimal automatic generation control of interconnected power system with asynchronous tie-lines under deregulated environment. Elect Power Comp Syst 2012;40(10):1208–28.
16. Mohanty B. TLBO optimized sliding mode controller for multi-area multi-source nonlinear interconnected AGC system. Int J Electr Power Energy Syst 2015;73:872–81.
17. Nasiruddin I, Bhatti TS, Hakimuddin N. Automatic generation control in an interconnected power system incorporating diverse source power plants using bacteria foraging optimization technique. Elect Power Compon Syst 2015;43 (2):189–99.
18. Das D. Electrical power systems. 1st ed. Delhi: New Age International Publications; 2010.
19. Ibraheem, Kumar P. Current status of the Indian power system and dynamic performance enhancement of hydro power systems with asynchronous tie lines. Elect Power Comp Syst 2003;31(7):605–26.
20. Hasan N, Ibraheem, Kumar P. Optimal automatic generation control of interconnected power system considering new structures of matrix Q. Elect Power Comp Syst 2013;41(2):136–56.
21. Sharma G, Nasiruddin I, Niazi KR. Optimal automatic generation control of asynchronous power systems using output feedback control strategy with dynamic participation of wind turbines. Elect Power Comp Syst 2015;43 (4):384–98.
22. Hasan N, Ibraheem, Kumar P, Nizamuddin. Sub-optimal automatic generation control of interconnected power system using constrained feedback control strategy. Int J Electr Power Energy Syst 2012;43(1):295–303.
23. Mohanty B, Panda S, Hota PK. Controller parameters tuning of differential evolution algorithm and its application to load frequency control of multi-source power system. Int J Electr Power Energy Syst 2014;54:77–85.
24. Gupta SK. Power system engineering. 2nd ed. New Delhi: Umesh Publications; 2013.
25. Kothari ML, Nanda J. Application of optimal control strategy to automatic generation control of a hydrothermal system. IEE Proc Control Theory Appl 1988;135(D4):268–74.

APPENDIX

PARAMETERS

T_{gi}=0.08; T_{ti}=0.3; K_{ri}=0.5; T_{ri}=10; K_{PSi}=120; T_{PSi}=20; T_{12}=0.545; R_i=2.4; B_i=0.425; T_{RHi}=48.7; T_{Ri}=5; T_{GHi}=0.513; T_{Wi}= 1; a_{12}=-1; K_{Ii}=0.4; K_{dc}=1; T_{dc}=0.2;

Implementation of Grid Connected Wind Driven Induction Generator from MATLAB/Simulink to Real-Time

K.A. Naik

Dept. of Electrical Engineering
IIT, Roorkee, India
anilnaik205@gmail.com

C. P. Gupta

Dept. of Electrical Engineering
IIT Roorkee, India
cpg_umist@yahoo.co.in

E. Fernandez

Dept. of Electrical Engineering
IIT, Roorkee, India
eugenefdz@gmail.com

Abstract— Since the mid twentieth century the simulation tools are widely employed for design and improvement of electrical systems. Nowadays, most of the researchers want to develop and implement their model in real time platform. In this paper, therefore, an advanced distributed real-time simulator traditionally called as Real-Time Laboratory (RT-LAB) is employed to implement wind energy system (WES). Firstly, a simple grid connected WES model is developed in MATLAB/Simulink® according to the operating requirements of RT-LAB. This model is then transformed to be suitable for real-time operation mode in RT-LAB environment. Finally, the real time simulation results obtained for WES are presented and analyzed.

Index Terms— *Controller; Design; Fuzzy logic; MATLAB/Simulink®; Induction generator; Pitch angle; Real time laboratory (RT-LAB); Wind turbine.*

Abbreviations
RT LAB   - Real Time Laboratory
MATLAB   - Matrix laboratory
RTW   - Real Time Workshop
HIL   - Hardware-in-the-Loop
PSCAD   - Power Systems Computer Aided Design
PSASP   - Power System Analysis Software

I. INTRODUCTION

Taking consideration of the restraints due to the limited nonrenewable energy sources and with the ever increasing demand of electricity, it has become an absolute necessity to integrate the renewable energy systems in the grid [1]. Among the others, wind energy systems are most promising. Therefore, wind power market has become an important energy source in the global power industry. Today more than 341,320 wind turbines are spinning around the world [2].

The development of wind power technology during research, it is difficult to obtain physical experiments due to large and complexity of the system as well as a precise theoretical model. In such cases simulation is most suited and unique approach for study of wind power systems. Therefore, recently various research organizations are preferring simulations software for wind power systems. In this trend, various researchers have employed MATLAB/Simulink® as a platform to analyze the wind power systems [3-5]. In the same way different commercial simulation soft-wares such as

PSCAD/EMTDC™ [6], DigSILENT/Power Factory™ [7], PSASP™ [8] etc were employed to implement the wind energy systems. These existing simulation technologies mainly focus on off-line simulation. Although, these may be took great efforts to do the simulation. It is difficult to guaranty the accuracy of the simulation results. Thus, digital real time simulators are introduced to conduct the real-time simulation of studied systems. In this trend various real time digital simulators are developed [9].

The term "real time: is used by several research institutions as well as industries to describe time-critical technology [10]. However, the word "real time" mostly applied to OPAL-RT™ technology since it is reference to embedded systems. The RT-lab which is fully integrated with MATLAB/Simulink®, provides the flexibility and scalability to achieve the most complex real-time simulation applications in the power systems, power electronics, automotive, aerospace and industries [11-13]. For real time simulations of the stiff systems, in order to achieve strict constraints, the simulator uses advanced fixed time step solvers (ARTEMIS). RT-LAB can separate the complex systems into different subsystems such as SM_Master, SS_Master and SC_Console and do the parallel operations in multi processor. RT-LAB also employs the controller prototype and hardware in loop to connect physical devices to the simulation system to obtain the realistic results. Various authors were achieved real time simulation of wind power systems using RT-LAB [14-16]. However, still a detailed study is needed how to implement wind energy systems in real time using OPAL-RT™ Laboratory.

In this research article, at first a simple wind energy system integrated with grid has developed using MATLAB/Simulink® software. This model then exported to RT-LAB environment for real time operation according to operating requirements of RT-LAB. The studied system is modelled with ARTEMIS and Simulink block-sets for compensated fixed time step simulation. This type of work is suitable for the design and test of wind energy systems controllers as well as apparatus.

The organization of paper as follows: Section II provides the brief about the system configuration and modelling. The MATLAB/Simulator architecture of employed system is provided in section III. Section IV, discuss the RT-LAB real time simulation platform for WES. The real time results obtained are discussed in section V. Finally, important conclusions are drawn in section VI.

978-1-5386-6624-1/18 $31.00 © 2018 IEEE

II. STUDIED WIND ENERGY SYSTEM MODELLING

The employed one line diagram of simple grid connected wind energy system is as shown in Fig. 1. The induction generator (IG) stator winding is directly connected to the grid. The capacitor has connected to the low voltage bus of the WES in order to supply the needed amount of reactive power to the wind generator [17]. However, the employed wind turbine and generator design parameters are presented in Appendix-I. Before start of developing the MATLAB/Simulink model it is necessary to understand the operation of wind energy system. The following sub-sections briefly brush up the mathematical concept of studied system.

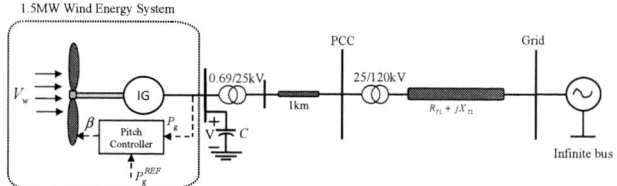

Fig. 1 The employed system single line diagram

A. Wind Turbine Modelling

The typical mathematical equation of wind turbine for extracted mechanical power from the wind is:

$$P_m = \frac{1}{2} \rho A_r C_p(\lambda, \beta) V_w^3 \qquad (1)$$

where, the air density is ρ (kg/m³), the power coefficient is A_r is C_p, the turbine swept area (m²) and V_w is the wind speed (m/s).

C_p can be determine the characteristics of the wind turbine where it depends on the pitch-angle (β) and rotor-tip speed ratio (λ). The mathematical expression of employed C_p curve can be illustrated in Eq. (2) [18]:

$$C_p(\lambda, \beta) = \left(c_2/\lambda_i - c_4 - c_3\beta\right)c_1 e^{-c_5/\lambda_i} + \lambda c_6 \qquad (2)$$

where

$$\frac{1}{\lambda_i} = \frac{1}{\lambda + 0.008\beta} - \frac{0.035}{\beta^3 + 1} \qquad (3)$$

The coefficients $c_1 - c_6$ are: $c_1 = 0.5176$, $c_6 = 0.0068$, $c_4 = 5$, $c_2 = 116$, $c_5 = 21$, $c_3 = 0.4$ and and the blade tip-speed (λ) can be defined as:

$$\lambda = \frac{\text{blade tip speed}}{\text{wind speed}} = \frac{\omega_r R}{V_w} \qquad (4)$$

where, ω_r is rotor speed [rad/s] and R is the radius of turbine rotor [m].

According to Eqns. (1) - (4), Fig. 2 obtained as the relation between the power coefficient and tip speed ratio ($C_p - \lambda$) with respect to different pitch angle beta (β).

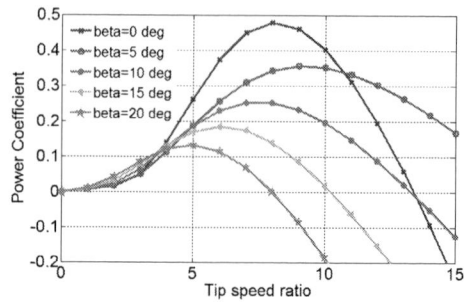

Fig. 2 Wind turbine $C_p - \lambda$ curve

B. Modelling of Induction Generator

The induction generator rotating reference frame in terms of d-q axis reference can be referred as [18].

$$p\varphi_{qr} = \omega_b(V_{qr} - R_r i_{qr}) - (\omega_b - \omega_r)\varphi_{dr} \qquad (5)$$

$$p\varphi_{dr} = \omega_b(V_{dr} - R_r i_{dr}) + (\omega_b - \omega_r)\varphi_{qr} \qquad (6)$$

$$p\varphi_{qs} = \omega_b(V_{qs} + R_s i_{qs} - \varphi_{ds}) \qquad (7)$$

$$p\varphi_{ds} = \omega_b(V_{ds} + R_s i_{ds} + \varphi_{qs}) \qquad (8)$$

All the variables of rotor have referred to the stator side. The electromagnetic torque expression for p.u is:

$$T_e = \varphi_{ds} i_{qs} - \varphi_{qs} i_{ds} \qquad (9)$$

Here the employed system is single mass system therefore; wind turbine and generator rotor can be represented as an equivalent mass. Finally, the dynamic equation of motion can be expressed as.

$$p\omega_m = \frac{\omega_b}{2H}(T_m - T_e) \qquad (10)$$

where, H is the equivalent inertia constant of both wind turbine and induction generator rotor.

C. Pitch Angle Controller

Primary function of the pitch-angle controller is to regulates the output power of wind turbine when the wind speed surpasses rated wind speed.

Fig. 3 shows the typical pitch-angle controller. The error obtained from reference power (P_g^{REF}) and the measured power (P_g) passes through the controller ($C(s)$), which regulates the output power accordingly by changing the pitch angle (β).

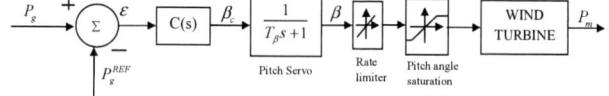

Fig. 3 Typical pitch angle controller

Beside this pitch angle controller also employed to smoothing the generator output power when the wind speed is below rated. A detailed concept and approach how the power smoothing can be done using pitch angle controller is presented in [5] and the same concept has been employed in this paper for WES.

III. MATLAB/SIMULINK MODEL OF THE SYSTEM

The simple grid connected wind energy system which is model using MATLAB/Simulink is shown in Fig. 4(a). The modules consist inside the WES system block are as shown in Fig. 4(b). These modules can be described as:

- Asynchronous machine module: this module consists of induction generator model, and it works as a conversion machine and convert mechanical energy into electrical energy.

- Wind Turbine module: this module related to wind turbine (WT) with one output (mechanical torque) and three inputs (rotor speed, pitch angle and wind speed data). Detailed mathematical equations of WT can be found in the sub-section above. The mechanical torque generated by the WT goes to generator as input.

- Pitch angle controller module: the pitch angle controller which is designed with PI/fuzzy is shown in Fig. 4(b). Here the difference between measured power and reference power defined as error (ε) goes to the controller as an input and accordingly it generate pitch angle (β).

- Breaker with SR-flip flop module: with this SR-flip flop and breaker one can able to disconnect/trip the WES when it run in abnormal condition.

- Load module: this module contains a load which can be varied from zero to 500kW and act as a frequency regulator to maintain the grid frequency constant.

- Data acquisition module: this module also called display module it consist of scopes to display the results of necessary system parameters.

- Capacitor module: here three phase delta connected capacitor bank of 200kvar is inculpated to supply the reactive power to the WES to maintain the power factor at reasonable value.

- Transformer module: in this study two step up transformers with different MVA rating are employed as shown in Fig. 4(a).

- Transmission line module: here also two transmission lines of pie type are employed, one is at local area of the WES and another is exporting generated power to the grid.

- Utility grid module: the grid modelled as a typical North American distribution grid. It includes 120kV transmission system.

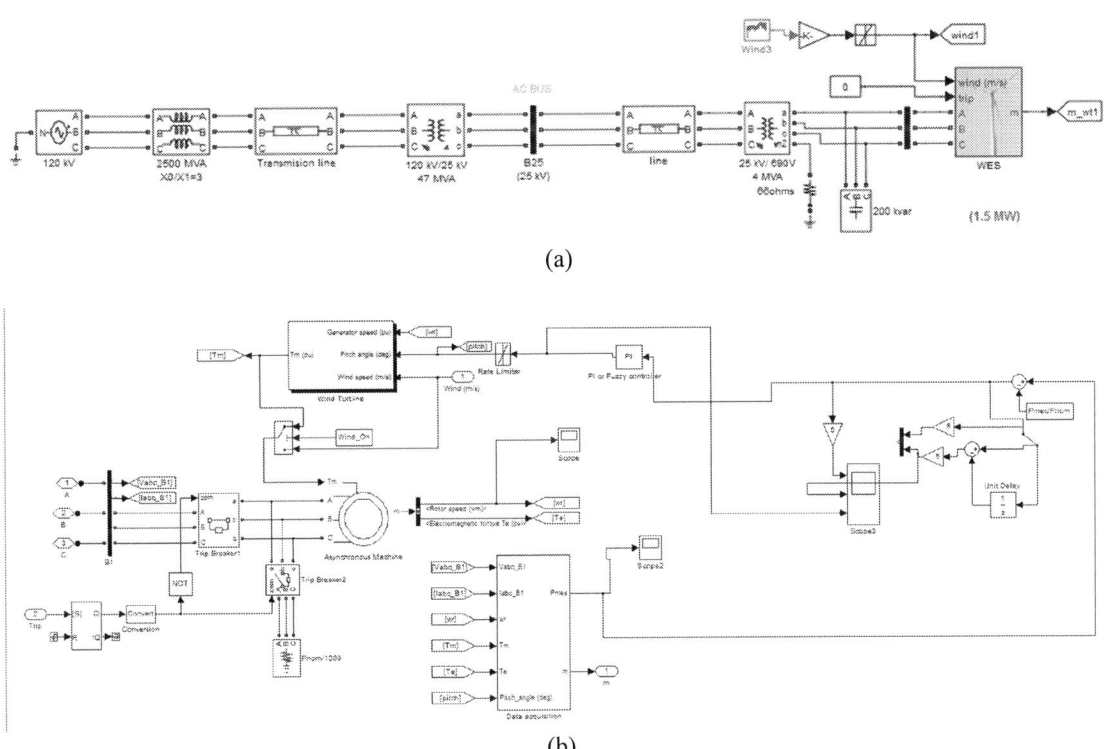

(a)

(b)

Fig. 4 Wind energy system integrated with grid model in MATLAB/Simulink platform (a) studied model (b) sub-system inside the WES block

IV. RT-LAB Environment for Studied System

The real time simulation of the studied system is carried on the OPAL-RT digital simulator. To run the models in real time simulation, the RT-lab allows user to readily converted simulink models through real time workshop (RTW). The RT-lab can separate a complex system into simple subsystems and performs parallel operation in multiple cores. Hence, one can needs to be separate the model into suitable subsystems for real time simulation, the studied WES as shown in Fig. 4(a) can be divided into two subsystems (SM_Mater and SC_Console)as shown in Fig. 5. The console subsystem denoted as SC_Console which contains parameters accessing and displaying blocks. This subsystem runs on the host PC, which can receive simulation results and display through the scope. Another subsystem is a computing subsystem named as SM_Master which contains all the calculation blocks (studied WES). This subsystem runs on the target machine with real time condition. The sampling time used to realize the system is $50\mu s$. The overview of the main control window is shown in Fig. 6. It mainly consists of preparing, compiling and execution buttons. In order to implement a model from MATLAB/Simulink to Real–Time using OPAL-RT simulator, the following eight main steps are required.

1. Open a model: in the main control window, the button open model allow user to open the model. When the model is opened a path is appear in the display.

2. Edit a model: the edit button from the main control view allow user to open a MATLAB/Simulink model and also can be drop required OPAL-RT blocks which are available for precise results. The model is distributed into two subsystems as shown in Fig. 5.

3. Compilation: the compilation button in the control window allow user to compile the model. If for any reason wants to stop the compilation while it launched, the button compile changed to abort to stop the process.

4. Assigning nodes: the assign nodes in the main control button allow user to assign the subsystems to targets. If right click on the assign button shows the options to assign nodes to subsystems.

5. Synchronization mode: this can be set at any time before model loading.

6. Load: the load button in the main control window allows user to load the Simulink model on the target(s). This is the final step before the model execution.

7. Execution: during the execution, model is on running state. User can pause or reset the model at any time. The results will display in Opcomm scopes.

8. Reset: this button from the main control window stop the software, hardware and console will close.

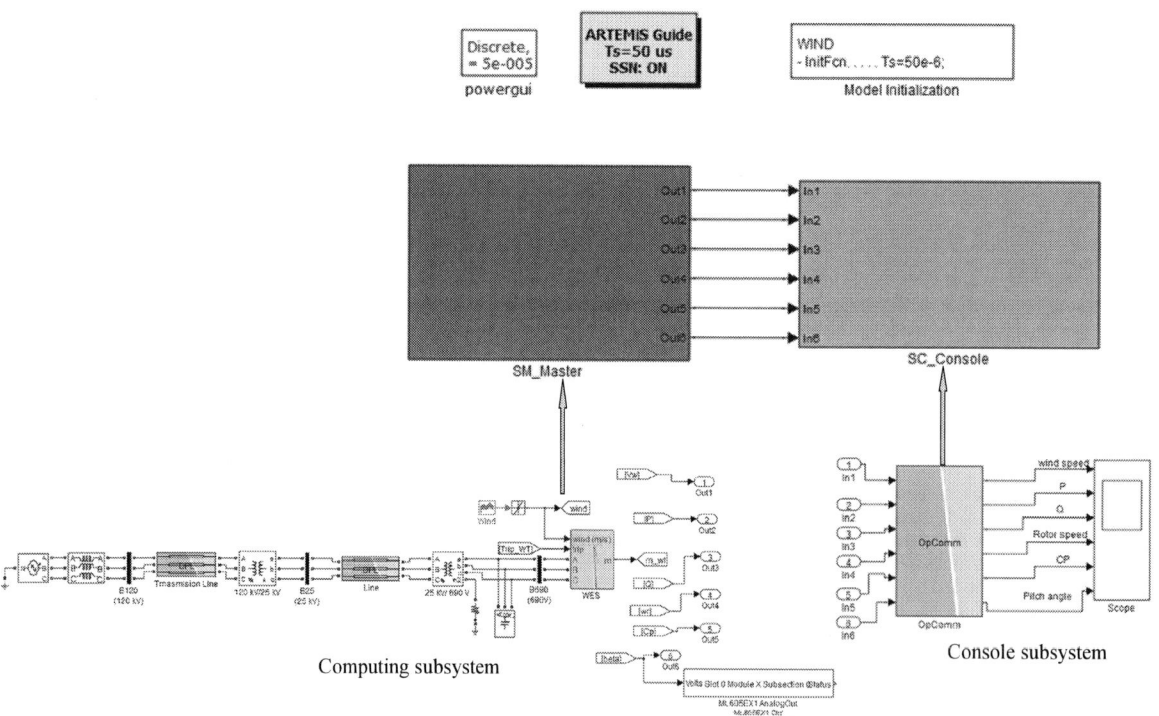

Fig. 5 Distributed model of the WES for real time simulation in OPAL-RT LAB

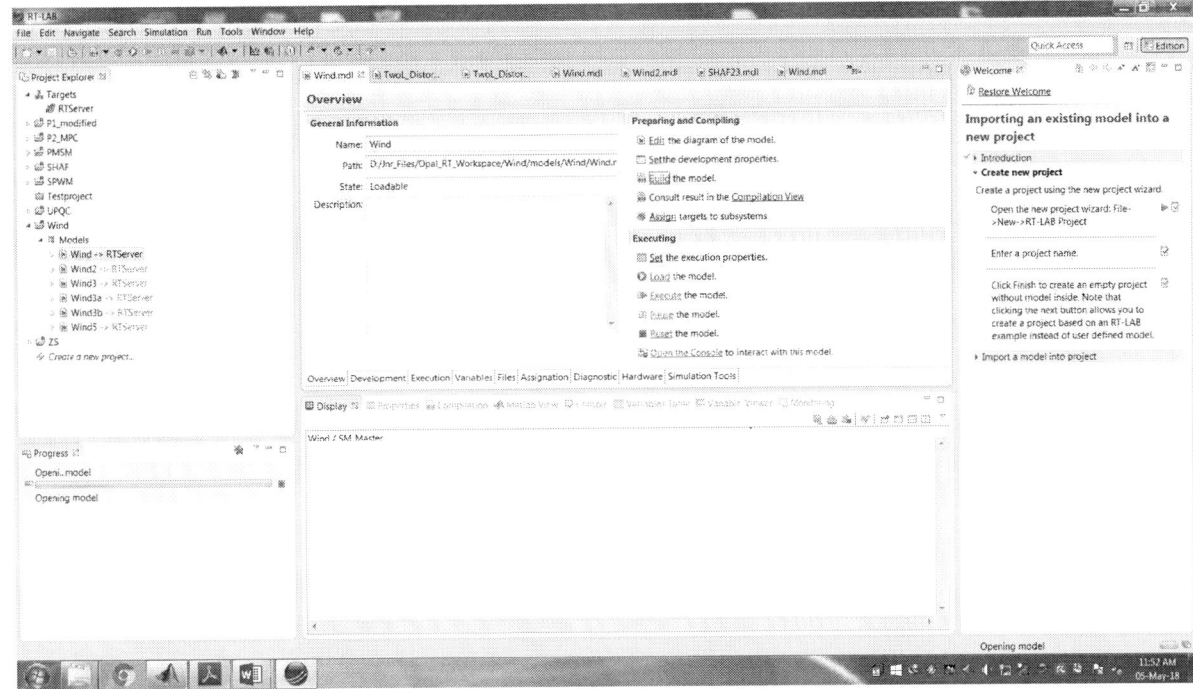

Fig. 6 over view of the main control window

The real time digital simulation laboratory setup is as shown in Fig. 7; it consists of host personal computer (Host-PC), target (real time digital simulator) and oscilloscope. The simulator employed here is OP5600 with one processor and four 3.33-GHz dedicated cores to perform parallel computations. The work station computer (Host-PC) executes the WES model and interacts with the real time digital simulator (RTDS) to produce the results of real time simulation. The digital oscilloscope is also employed to observe the real time results.

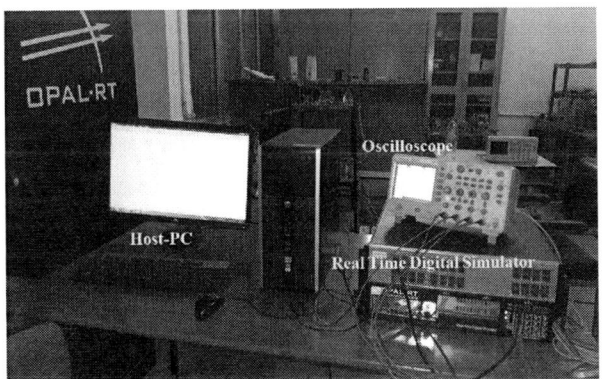

Fig. 7 Real time digital simulator laboratory setup

V. SIMULATION RESULTS IN OPAL-RT ENVIRONMENT

The real time results obtained using fuzzy logic and PI controller used, one at a time, for the pitch controller are shown in Figs. 8 and 9. However, the target of this paper is to implement MATLAB/Simulink model to real time. Therefore the real time results obtained are presented, but we are not going much discussion about them due to number of pages constraint. The wind speed shown in Fig. 7(a) is a below rated wind speed for which the PI based pitch angle controller did not generate any pitch angle and the output power follow the wind speed variations as shown in Fig. 8(b). While using fuzzy logic method the output power smoothing is obtained with proper generation of pitch angle as shown in Fig. 8(b).

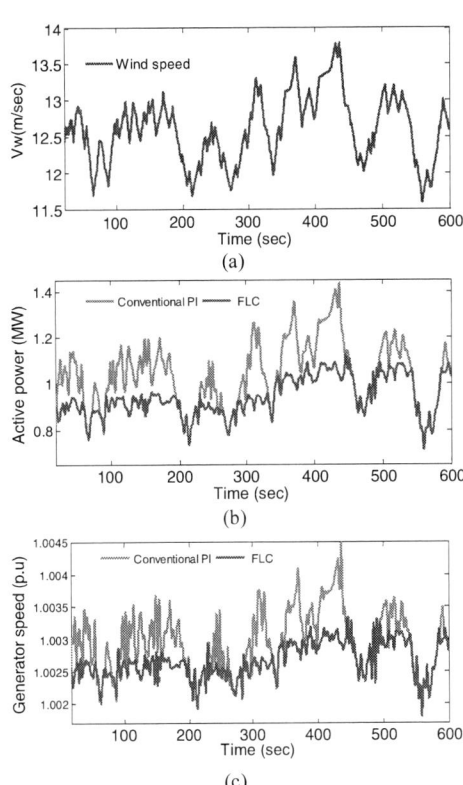

Fig. 8 Real time simulation results obtained using Opcomm scope (a) Wind speed (b) Generator active power (c) Generator rotor speed

978-1-5386-6624-1/18 $31.00 © 2018 IEEE   335

The generator rotor speed corresponding to both pitch-angle controllers are as shown in Fig 8(c). The pitch angle activities with PI and fuzzy logic controller (FLC) are recorded in the digital oscilloscope as shown in Fig. 9(a) and (b), respectively.

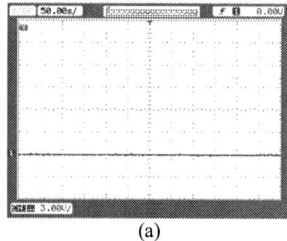 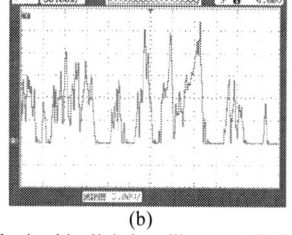

(a)   (b)

Fig. 9 Pitch angle profile obtained in digital oscilloscope (a) PI controller (b) Fuzzy logic controller (FLC)

VI. CONCLUSIONS

In this paper, a simple grid connected WES system has considered and implemented from MATLAB/Simulink to OPAL-RT simulator for real time environment. A detailed MATLAB/Simulink model is developed and demonstrated. Later on, how it can be carried out to the real time using RT-Lab is discussed with systematic procedure. The major applications of RT-Lab are in: power systems, power electronics and renewable energy resources. Moreover, it is particularly useful for hardware in loop (HIL) and rapid control prototyping. The work presented in this paper especially helpful for under graduate, post graduate and research students.

Appendix-I
SCIG generator parameters [20]

Parameters	Values
V_{rated}, P_{rated}	0.69 kV, 1.5 MW
L_r, L_s	0.1791 p.u, 0.1248 p.u.
R_r, R_s	0.004377 p.u, 0.004843 p.u
H, L_m	5.04 (s), 6.77 p.u
C	200 kVAR

Wind turbine specifications [19]

Parameters	Values
Rated power	1.5 MW
Number of blades	3
Rotor diameter	64m
Wind speed-rated (V_{wR})	14 m/s
Wind speed-cut-in (V_{wCI})	4 m/s
Wind speed-cut-out (V_R)	25 m/s
Generator	SCIG

REFERENCES

[1] K. A. Naik and C. P. Gupta, "Improved oscillatory behavior of a grid connected wind farm using IMC-PID based pitch angle controller," *2016 IEEE 6th Int. Conf. on Power Systems (ICPS)*, New Delhi, 2016, pp. 1-6.

[2] K. A. Naik and C. P. Gupta, "Transient stability enhancement of wind energy system using fuzzy logic based pitch angle controller," *2017 4th IEEE Uttar Pradesh Section International Conference on Electrical, Computer and Electronics (UPCON)*, Mathura, 2017, pp. 78-83.

[3] K. A. Naik and C. P. Gupta "Improved fluctuation behavior of SCIG based wind energy system using hybrid pitch angle controller", *IEEE Uttar Pradesh Section Int. Conf. on Elect., Comp. and Elect. Engg. (UPCON-2016)*, IIT-BHU, pp1-6,2016.

[4] K. A. Naik and C. P. Gupta, "Performance comparison of Type-1 and Type-2 fuzzy logic systems," *2017 4th International Conference on Signal Processing, Computing and Control (ISPCC)*, Solan, 2017, pp. 72-76.

[5] K. A. Naik and C. P. Gupta, "Output Power Smoothing and Voltage Regulation of a Fixed Speed Wind Generator in the Partial Load Region Using STATCOM and a Pitch Angle Controller," *Energies* 11, vol. 58, no. 1, pp. 1-18, 2017.

[6] S. Tao, C. Zhe, F. Blaabjerg, "Transient stability of DFIG wind turbines at an external short-circuit fault," Wind Energy, vol. 8, 2005.

[7] H.L. Guan, Y.N. Chi, H.Z. Dai, Y.H. Yang, "Small signal stability and control of wind turbine with asynchronous generator integration into power system," Automation of electric power systems, vol. 32, 2008.

[8] W. Wei, Y.H. Wang, X.Y. Li, et al, "Simulation of large scale wind farms integration into Gansu Jiajiu power system," Automation of electric power systems, vol. 33, 2009.

[9] M. D. Omar Faruque *et al.*, "Real-Time Simulation Technologies for Power Systems Design, Testing, and Analysis," *IEEE Power and Energy Technology Systems Journal*, vol. 2, no. 2, pp. 63-73, June 2015.

[10] RT-LAB Professional [online], available: http://www.opal-rt.com/product/rt-lab-professional.

[11] G. Sybille and Hoang Le-Huy, "Digital simulation of power systems and power electronics using the MATLAB/Simulink Power System Blockset," *2000 IEEE Power Engineering Society Winter Meeting. Conference Proceedings (Cat. No.00CH37077)*, 2000, pp. 2973-2981 vol.4.

[12] S. Mikkili and A.K. Panda, " FLC based shunt active filter (p–q and Id–Iq) control strategies for mitigation of harmonics with different fuzzy MFs using MATLAB and real-time digital simulator," *Int. J. Electr. Power Energy Syst. Elsevier*, vol. 47, pp. 313–336, 2013.

[13] S.Mikkili and A.K. Panda, "Types-1 and -2 fuzzy logic controllers-based shunt active filter Id–Iq control strategy with different fuzzy membership functions for power quality improvement using RTDS hardware," *IET Power Electron*, Vol. 6, no. 4, pp. 818–833, 2013.

[14] T. Wang, Y.J. Zou, X.H. Nian, Y. Hu, "Real-time Simulation of DFIGbased wind farm connected to power grid," Journal of system simulation, vol. 21, pp. 4306-4311, 2009.

[15] V.Jalili-Marandi, L.F. Pak, V. Dinavahi, "Real-Time simulation of gridconnected wind farms using physical aggregation," IEEE Transactions on industrial electronics, vol. 57, pp. 3010-3021, 2010.

[16] Y. Zhao, L. Shi, Y. Ni and L. Yao, "Modeling and Real-Time Simulation of Wind Farm," *2012 Asia-Pacific Power and Energy Engineering Conference*, Shanghai, 2012, pp. 1-4.

[17] K. A. Naik and C. P. Gupta, "Fuzzy logic based pitch angle controller for SCIG based wind energy system," *2017 Recent Developments in Control, Automation & Power Engineering (RDCAPE)*, NOIDA, India, 2017, pp. 60-65.

[18] T. Ackerman, *Wind power in power system*, John Wiley & Sons. Ltd, 2005.

[19] http://www.thewindpower.net/windfarm_en_86_challicum-hills.php.

[20] M. H. Haque "Evaluation of power flow solutions with fixed speed wind turbine generating systems, *Energy Conversion and Management*," vol 79, pp. 511-518, 2014.

2nd IEEE International conference on power Electronics, Intelligent Control and Energy systems(ICPEICES-2018)

A Modified Perturb & Observe Algorithm For Maximum Power Point Tracking With Zero Voltage Switching Buck Boost Converter in Photovoltaic System

Dubal Amol Jalindar
Dept. of Insrumentation & Control
Engineering, NIT Jalandhar, Punjab,India
amoldubal2411@gmail.com

Dr.Sheela Tiwari
Dept. of Insrumentation & Control
Engineering, NIT Jalandhar, Punjab,India
tiwaris@nitj.ac.in

Abstract—The conventional perturb and observe algorithm used for maximum power point tracking of PV system suffers from major drawbacks such as steady state oscillation around final maximum power point and loss of tracking direction i.e. divergence from actual maximum power point for rapid change in insolation. Both these problems result in significant power loss as the PV system operating point is getting diverted from actual maximum power point. As Buck-Boost converter plays an important role while changing the operating point of PV panel, it contributes to significant amount of power loss in the form of switching loss, conduction loss etc. This paper proposes a method to avoid the above given shortcomings of perturb & observe algorithm. Major focus is given on the designing of Buck-Boost converter as to date nobody has taken into consideration the losses occurring in the converter even though converter forms an integral part of the MPPT set up. The losses occurring in the conventional buck boost converter have been computed and its effect on the efficiency of MPPT system is investigated. The soft-switched buck boost converter for PV applications is designed so as to minimize the losses taking place in converter.

Keywords— Perturb and observe (P&O), Photo voltaic (PV), Zero voltage switching (ZVS), Soft switching, Buck-boost converter.

I. INTRODUCTION

The demand for electrical energy is growing day by day because of rapid increase in population and industry. Technological progress and improved life quality also played an important role in respect of increased energy demand. Shortage of current fossil fuels and their negative impact on environment resulted into exploration of renewable energy sources such as geothermal, wind, solar, tidal etc. Among all renewable energy resources, solar energy is forming the best suitable option as the energy obtained from solar is environment friendly and most importantly, available in huge amount. However, it suffers from some drawbacks such as high initial cost of installation, non-availability throughout the year, poor conversion efficiency etc.

Because of poor conversion efficiency of the PV modules, it is considered to be costlier compared to conventional source of energy. Therefore, tremendous work is going on in the field of solar PV to improve its conversion efficiency. The main reason behind having poor conversion efficiency is that the power obtained from the solar panel attains peak value for a particular operating voltage. Also the maximum power that a PV cell can attain is dependent on

solar irradiance and temperature. Therefore, in order to have maximum utilization of PV module one should extract maximum power from solar PV. Hence, Maximum power point tracking (MPPT) techniques came into existence [1]. These MPPT techniques help in tracking the unique maxima of power available for specific irradiance and temperature condition. An MPPT is a technique of getting maximum possible power over the period of time by changing the duty cycle of DC-DC converter so as to operate the PV panel at such a voltage which will correspond to maximum power on a Power versus Voltage (P-V) curve. The basic block diagram of MPPT system is shown in Fig.1.

As shown in Fig.1, the real time values of voltage and current are given to the control unit which consist of the MPPT algorithm along with controller. Controller provides the required duty cycle to the DC-DC converter so that the operating voltage of the PV panel corresponds to maximum power. To date, various MPPT techniques have been reported [1-2]. These techniques are classified into two categories as 1) conventional methods and 2) soft computing methods. MPPT techniques such as perturb & observe, incremental conductance and hill climbing algorithm fall under conventional approach.

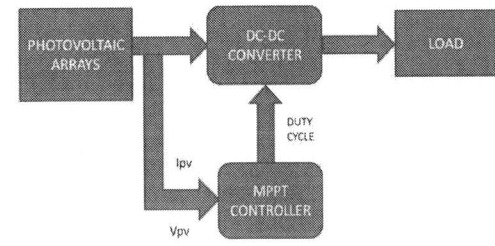

Fig. 1. Block diagram of the MPPT system [6].

On the other hand, the recently developed MPPT methods, based on fuzzy logic control, artificial neural network, genetic algorithm, particle swarm optimization, cuckoo based approach etc. come under soft computing approach [2].

Among all the conventional approaches of tracking maximum power, the perturb and observe algorithm forms the most suitable one as it is simple to implement and easy to understand. Also, it has good convergence towards maximum power point. But as stated earlier, it suffers from some drawbacks such as steady state oscillation and loss of

978-1-5386-6624-1/18 $31.00 © 2018 IEEE

tracking direction. It results in power loss and decreased conversion efficiency. It is to be noted that much work in this field has been done to address these issues, but both these problems are yet to be addressed in an effective manner [3-11].

Apart from these issues, the biggest gap lies in the field of dc-dc converter. To date, plenty of work has been done in the field of maximum power point tracking of PV system. But the losses that are occurring in the dc-dc converter have not been considered while calculating the efficiency of MPPT even though it is playing a vital role in tracking maximum power point of PV system.

Therefore, in this study, a modified perturb and observe method is implemented to eliminate the probability of having steady state oscillation and loss of tracking direction under sudden change of irradiance. Along with that, the losses occurring in the conventional buck-boost converter are computed and their effect on the efficiency of complete MPPT set up is shown [13]. A new buck-boost converter employing ZVS technique is designed and implemented [14]. Finally, the modified MPPT with soft switching buck-boost converter is simulated in MATLAB software.

I.   PRIOR WORK IN P&O ALGORITHM

A. Shortcomings of the conventional P&O algorithm

In the conventional perturb and observe method the power at any time instant is calculated by using photo voltaic current (Ipv) and voltage (Vpv) at that instant. The corresponding change in power is calculated by comparing power at any time 't' with its prior value at '(t-1)'. Accordingly, perturbation (ΔX) is provided as given below [12]. i.e.

$$Xnew = Xold + (\Delta X \times \varphi) \qquad (1)$$

In (1), 'X' is the variable which is to be controlled so as to achieve MPP. This controlled variable can be either duty cycle (D), current (I) or voltage (V) depending on which variable is to altered to track maximum power. 'Φ' is signed multiplication of perturbation (ΔX) and change in power (ΔP) whose value is either '1' or '-1'. The value of 'Φ' is computed using algorithm given below.

TABLE I.

Summary of P&O algorithm

Perturbation (ΔX)	Change in power (ΔP)	Next perturbation($\emptyset$)
Positive	Positive	Positive
Positive	Negative	Negative
Negative	Positive	Negative
Negative	Negative	Positive

As shown in table I, if the value of perturbation (ΔX) is resulting in increase in power then the polarity (Φ) of next perturbation is kept same as that of previous one(either positive(+1) or negative(-1)) until it reaches MPP. Once MPP is passed, power goes on decreasing as shown in Fig. 2.

Hence the polarity of next perturbation is reversed so that the algorithm again approaches the MPP. Because of this, PV panel operating point moves back and forth around final MPP resulting in loss of power. In this regard, selection of value of ΔX plays an important role. If the value of ΔX is very high then convergence speed is high but it also leads to large amount of oscillation once it passes through MPP, resulting in significant amount of power loss. On the other hand, if (ΔX) is set to have low value then the steady state oscillation gets reduced but convergence speed is greatly affected resulting in a sluggish response.

Also P&O algorithm may fail under rapidly changing atmospheric condition by diverging away from actual MPP as illustrated in the Fig.3. If one starts from an operating point A and atmospheric conditions stay approximately constant, a perturbation ΔV in the PV voltage V will bring

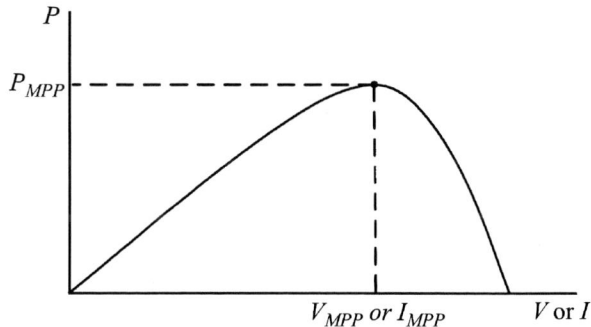

Fig. 2. P-V characteristics of Photo voltaic array [1].

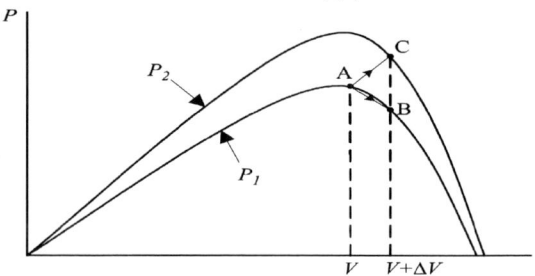

Fig. 3. Divergence of P&O algorithm from MPP [1].

the operating point to B and the perturbation will be reversed due to a decrease in power. However, if the irradiance increases and shifts the power curve from *P1* to *P2* within one sampling period, the operating point will move from A to C. This represents an increase in power and the perturbation is kept the same. Consequently, the operating point diverges from the MPP and will keep diverging if the irradiance steadily increases.

B. Mitigation methods using adaptive P&O algorithms & their drawbacks

Authors in [4,6-8] addressed the two most common issues i.e. steady state oscillation and loss of tracking direction from which P&O algorithm suffers using adaptive P&O approach. Adaptive term is used in relation with perturbation size. In adaptive P&O approach, the perturbation size is made adaptive taking into consideration operating point of PV module and the actual maximum power point. Normally, in most of the cases voltage is taken as perturbation variable. In [4, 15, and 16], the authors have

978-1-5386-6624-1/18 $31.00 © 2018 IEEE

proposed an adaptive perturbation size algorithm but the proposed algorithm loses its tracking direction whenever there is a sudden change in insolation.

In some techniques, value of duty cycle is perturbed in order to achieve MPP [6]. In duty cycle based adaptive P&O approach, duty cycle of dc-dc converter is varied to obtain the output voltage which will correspond to maximum power. But this technique is also unable to prevent the divergence from actual MPP. In [8] and [11], authors made use of the fact that change in radiation directly affects the value of current. Hence, use of current perturbation (ΔI) is done to reach the MPP. But these MPPT techniques are not affordable as they use expensive sensors to measure the value of insolation (G). Also this methodology did not address divergence issue.

Some authors comprehensively addressed both the issues of steady state oscillation and divergence away from MPP [7, 12]. Algorithm provided in [7] is limited for specific conditions. It does not considered the real time environmental behavior. Also the main shortcoming of this method is that voltage perturbation (ΔV) is reduced to zero once MPP is reached. As the oscillations completely drop down to zero, the algorithm does not provide any perturbation until a considerable change in photo voltaic current(I_{pv})occurs.Therefore, there will be MPPT mismatch which will result in power loss. On the other hand algorithm provided in [12] has made great efforts to eliminate the shortcomings of the conventional and adaptive P&O techniques discussed so far. But the algorithm provided is quite complex to understand and harder to implement. Moreover, the physical efficiency of dc-dc converter has not been considered even though converter is forming an integral part of whole MPPT set up.

So from this brief overview, it can be concluded that a research gap exists regarding consideration of dc-dc converter. Computation of the losses that are occurring in the converter and their effect on the efficiency of MPPT algorithm has not been considered. Therefore, in this work, the losses (such as switching loss, turn on loss, conduction loss etc.) which are taking place in conventional dc-dc buck-boost converter are computed and a thorough procedure is given to calculate it for any given converter. The physical efficiency of converter is computed and its effect on the efficiency of complete MPPT set up is shown. While doing so the work provided in the literature [12] is taken for reference. Finally zero voltage switching based soft switched dc-dc buck-boost converter without using auxiliary switches is designed to minimize the losses so as to improve the efficiency of MPPT algorithm. The entire procedure of designing a converter is provided.

II. PHOTOVOLTAIC SYSYTEM MODELLING

The equivalent circuit model of a PV cell is needed in order to simulate its real behavior. A cell can be modeled as a DC current source in parallel with two diodes that represent currents escaping due to diffusion and charge recombination mechanisms as shown in fig.1.4 (a). Assuming that the current due to charge recombination is small enough to be neglected, a simplified PV cell model can be reached as shown in fig.1.4 (b) known as single-diode model. Series resistance (R_s) is for representing hindrance in the path of flow of electrons from n to p junction and parallel resistance (R_p) is due to the leakage current.

The equivalent circuit of PV cell can be developed with a current source representing photo current(I_{ph}) & a diode. Its output current can be written as:

$$I = I_{ph} - I_{rs}\left(\exp\left(\frac{qV}{KT_cA}\right) - 1\right) \qquad (2)$$

$$I_{ph} = [I_{sc} - K_1(T_c - T_{ref})] * G \qquad (3)$$

Where$I_{rs}$is the cell reverse saturation current (of p-n junction), T_c is the cell's working temperature, A is the ideality factor, G is the irradiation, I_{sc} is the cell's short

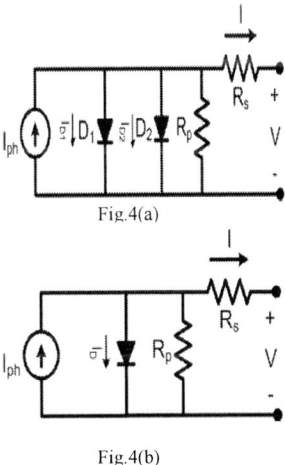

Fig.4(a)

Fig.4(b)

Fig. 4. Equivalent model of Photo voltaic system [11].

circuit current at reference temperature T_{ref} and K_1 is the short circuit current temperature coefficient of the cell. The net output PV current (I_{pv}) and power (P_{pv}) with N_p number of parallel cell and N_s number of series cell, and (V_{dc}) being output voltage is given by equation (4) & (5) respectively,

$$I_{pv} = N_p I_{ph} - N_p I_{rs}\left(\exp\left(\frac{q*V_{dc}}{KT_cA* N_s}\right) - 1\right) \qquad (4)$$

$$P_{pv} = V_{dc} * I_{pv} = f(V_{dc}, G, T_c) \qquad (5)$$

The specifications for the PV module (Trima solar TSM-250PA05-08) used in this paper are given in Table II.

TABLE II.

Specification of the PV module

Parameters	Label	Values
Open circuit voltage	V_{oc}	37.6 V
Short circuit current	I_{sc}	8.55 A
Voltage at maximum power	V_{mpp}	31.0 V
Current at maximum power	I_{mpp}	8.06 A
Maximum power	P_{max}	249.86 W
V_{oc} coefficient of temperature	K_V	- 0.35 % /°c
I_{sc} coefficient of temperature	K_1	0.06 % /°c
Cells in series per module	N_s	4
Cells in parallel per module	N_p	10

III. MODIFIED P&O ALGOROTHM

In the conventional P&O, even though adaptive perturbation size algorithms are used but they do not take into consideration the real time environment behavior. Basically in real time atmospheric conditions the irradiance (G) varies in two different ways as- 1) slow change in irradiance i.e. G < 10 w/m2 2) fast change in irradiance i.e. G > 10 w/m2 [12].The conventional adaptive P&O approach treats both these irradiance profiles in the same way as initial perturbation size is kept same irrespective of amount of change in irradiance.

But in this modified algorithm, both these irradiance profile have been given special treatment in the form of initial perturbation size (ΔV). If gradient of G is small, then conventional P&O approach manages to track MPP without divergence. But main problem arises when rapid change in the irradiance takes place, as conventional approach cannot differentiate between these two irradiance profiles providing the same perturbation size in both cases resulting in loss of tracking direction. This problem is completely eliminated by special algorithm given below:

If irradiance keeps on changing, MPPT algorithm takes two consecutive samples for the constant temperature. Then, from work reported in [12, 17], it can be said that

$$\frac{\Delta P}{P_1} = \frac{\Delta G}{G_1} \qquad (6)$$

Where ΔP and ΔG are change in power and irradiation respectively. P_1 and G_1 are the values of power and irradiation taken simultaneously at constant temperature. From (6), it can be concluded that normalized change in power is equal to the normalized change in irradiance. At standard test condition suppose that the value of $G_1 = 1000$ w/m^{2}and fast change in irradiance profile occurs i.e. G > 10 w/m^2.Then from above given equation, the threshold value of($\Delta P/\Delta P_1$) can be 0.01 (10/1000). So in proposed algorithm the value of ($\Delta P/\Delta P_1$) is continuously checked. If it exceeds its threshold value i.e. 0.01, then it can be treated as a sudden change of irradiance and perturbation size can be kept to initial set value otherwise (($\Delta P/\Delta P_1$) < 0.01) the perturbation size is kept to minimum possible value.

IV. ANALYSIS AND EVALUATION OF PRIOR ART

A. Computation of converter losses & its efficiency

The conventional buck-boost converter is as shown in Fig.5. The specification of the given converter is as given in Table III [12]

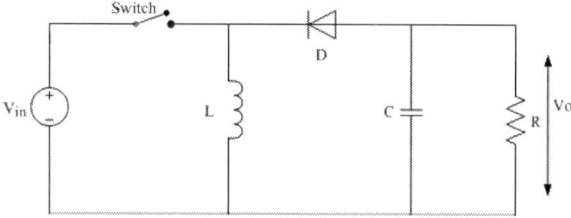

Fig. 5. dc-dc buck-boost converter [14].

Three main causes of power-dissipation in DC-DC converter are as given below.

1) Inductor conduction losses

2) MOSFET conduction losses

3) MOSFET switching losses

4) Miscellaneous losses

TABLE III

Specifications of converter component

Component	Specification
Switching frequency(f)	50 kHz
Inductor(L_1)	1 mH
Capacitor(C_1)	470 µF
Capacitor(C_2)	220 µF

The inductor conduction loss is given by:

$$P_L = I^2_{RMS(L)} * R_{DCR} \qquad (7)$$

(R_{DCR}: DC resistance of inductor)

The RMS inductor current is given by:

$$I^2_{RMS(L)} = I^2_0 + \left(\frac{\Delta I^2}{12}\right)$$

(ΔI=Ripple current & it is about 30% of output current)

In our case,

I_0=output current=3.5 Amp

ΔI=0.3*3.5=1.05 Amp &

R_{DCR}=0.104 Ω   (from Texas data sheet)

INDUCTOR CONDUCTION LOSS=1.28 Watt

The power dissipated in the MOSFET is given by:

$$P_Q = I^2_{RMS(Q)} * R_{DSON}$$

(R_{DSON}: on-time drain-to-source resistance of the MOSFET).

Substituting:-

$$P_Q = \frac{V_0}{V_{in}} * \left(I^2_0 + \frac{\Delta I^2}{12}\right) * R_{DSON} \qquad (8)$$

V_0 = Voltage at MPP=17.1

V_{in}= Open circuit voltage=21.1 V

I_0=3.5 A; ΔI=1.05 A;

R_{DSON} =0.026 Ω   (From Texas Data Sheet)

MOSFET CONDUCTION LOSS=0.2 Watt

MOSFET switching losses are directly proportional to frequency.

From Texas Data Sheet, Switching losses are 1.5 Watt for 76 KHz

In our case, Switching frequency (Fs) =50 kHz,

Therefore, we will take it as 1 Watt approx. Hence,

978-1-5386-6624-1/18 $31.00 © 2018 IEEE

MOSFET SWITCHING LOSS=1 Watt

The miscellaneous losses include losses in Input filter, Output Filter. Miscellaneous losses are approximated to be 10% of total losses. Therefore it is given as,

Misc. losses=0.1*(1.28+0.2+1) =0.25 Watt.

Hence,

MISCELLANEOUS LOSSES=0.25 Watt

Total power dissipated in the converter can be calculated by summing up all the losses.

Total Power loss in converter (P_d)= 1.28+0.2+1+0.25

$$= 2.13 \text{ Watt}$$

The efficiency of buck-boost converter is given by following equation [13]

$$\eta = \frac{P_0}{P_0 + P_d} \qquad (9)$$

As, $(P_0 = V_0 * I_0)$

In the current work, for standard test conditions the V_0 & I_0 is given as

V_0=17.1V, I_0 =3.5A (This V and I values corresponds to maximum power)

P_0=converter power loss=2.73 Watt.

Putting this values in above given equation (9), we can get the efficiency of buck-boost converter η_{conv} has been calculate as 95.63%.

B. Efficiency of complete MPPT set up

The efficiency of the complete MPPT set up can be calculated using the following formula:-

= (η of MPPT controller) * (η of Buck-Boost converter)

The efficiency of the complete MPPT set up was calculated for the following two conditions:-

1) Slow changing irradiation (G<10 W/m2)

2) Fast changing irradiation (G>10 W/m2)

The values of the efficiency obtained for the above two cases are,

For slow changing irradiation, η=95.02%

For fast changing irradiation, η=93.8%

V.   STATE OF THE ART DESIGN

A. Problems with conventional buck boost converter

The conventional buck-boost converter is as shown in fig.1.5. In the conventional converter, the switching is done as a hard switching. When the switch is in ON condition the voltage is not going to zero immediately and the current is increasing quickly but both are occurring at the same time, so it creates losses because both the voltage and current are

working at maximum ranges, it implies to get power loss. The problem of hard switching can be understood from the below given Fig.6.

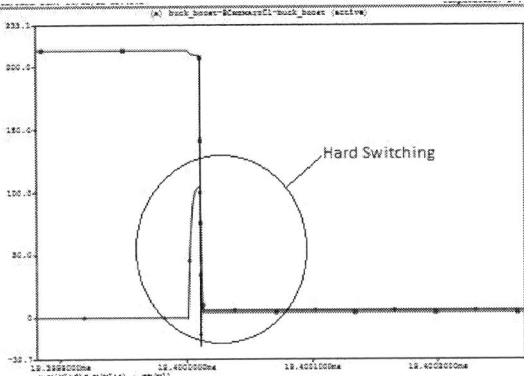

Fig. 6. Problem associated with hard switching [14]

B. Design of ZVS based soft swithed buck-boost converter

In order to mitigate the switching losses which were occurring in conventional hard switched buck boost converter, a ZVS (Zero Voltage Switching) converter reported in [14] is used in the current work. When switch is made ON, at first, voltage comes to zero then current starts increasing so as to have minimum switching loss. On the other hand, when the switch is OFF then the current goes to minimum and the voltage goes to maximum from the zero level. This is the ZVS converter, which is working by adding one inductor and the capacitor as shown in Fig.7 below,

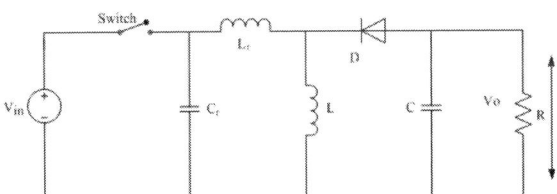

Fig. 7. ZVS based soft switched buck-boost converter [14]

L, C- Critical values of inductance & capacitance

L_r, C_r-Resonant inductance & capacitance respectively

R- Load resistance

D- Diode

V_0- Load/output voltage

V_{in}-Input voltage

F_s- Switching frequency

C. Circuit parameters of ZVS buck-boost converter

The circuit parameters of implemented ZVS based soft switched buck-boost converter in this proposed modified P&O algorithm is given in table IV.

TABLE IV.

ZVS converter circuit parameters

Parameter	Value
Input voltage (V_{in})	315.7 V
Output (V_{out})	310 V
Switching frequency(F_s)	50 kHz
Duty cycle(D)	0.495
Load(R)	10 Ω
Inductance(L)	50.5 μH
Capacitance(C)	495 ηF
Resonant inductance (L_r)	160 μH
Resonant capacitance (C_r)	1.55 μF

VI. RESULTS

In this work, the efficiency of conventional MPPT set up is computed by considering the losses that are occurring in DC-DC converter. It has been observed that, there is efficiency drop of almost 5% when losses of DC-DC converter are considered. In this study the new ZVS based soft switched converter is proposed so as to minimize losses and improve the efficiency. By inclusion of this ZVS converter, there has been efficiency improvement of 2.5% as compared to conventional MPPT technique. The work done so far in this study can be summarized as shown in table V.

TABLE V.

Efficiency comparison of conventional and proposed MPPT Techniques

Sr.No.	Irradiance profile(G)	% Efficiency		
		Conventional P&O without considering converter losses	Conventional P&O considering converter losses	Modified P&O with ZVS converter
1.	G<10 W/m2	99.4%	95.02%	97.5%
2.	G>10 W/m2	98.2%	93.80%	96.3%

VII. CONCLUSION

In the given work, modified adaptive P&O algorithm is developed in order to reduce the steady state oscillation to minimum value so as to have negligible power losses in a PV system. This paper mainly focuses on the losses which are taking place at converter side by taking the work presented in literature [12] as a reference. The impact of losses occurring in dc-dc converter on the performance of complete MPPT set up is shown. Finally the ZVS based soft switched dc-dc buck-boost converter is designed to minimize converter losses so as to improve efficiency of P&O MPPT system. The simulation results clearly demonstrate the effectiveness of ZVS converter in improving the efficiency of complete MPPT set up with losses under consideration.

REFERENCES

[1] T. Esram and P.L. Chapman, "Comparison of photovoltaic array maximum power point tracking techniques," IEEE Trans. Energy Convers., vol.22, no.2, pp. 439-449, jun. 2007.

[2] J. Prasanth Ram, T. Sudhakar Babu, and N. Rajasekar, "A comprehensive review on solar PV maximum power point tracking techniques," Renewable and sustainable energy reviews 67(2017) 826-847, Sept.2016

[3] M. A. Elgendy, B. Zahawi, and D. J. Atkinson, "Assessment of perturb and observe MPPT algorithm implementation techniques for PV pumping applications," IEEE Trans. Sustainable Energy, vol. 3, no. 1, pp. 21–33, Jan. 2012.

[4] F. Zhang, K. Thanapalan, A. Procter, S. Carr, and J. Maddy, "Adaptive hybrid maximum power point tracking method for a photovoltaic system," IEEE Trans. Energy Convers., vol.28, no. 2, pp. 353–360, Jun. 2013.

[5] R. Alonso, P. Ibaez, V. Martinez, E. Roman, and A. Sanz, "An innovative perturb, observe and check algorithm for partially shaded PV systems," in Proc. 13th Eur. Conf. Power Electron. Appl., 2009, pp. 1–8.

[6] A. Pandey, N. Dasgupta, and A. K. Mukerjee, "High-performance algorithms for drift avoidance and fast tracking in solar MPPT system," IEEE Trans. Energy Convers., vol. 23, no. 2, pp. 681–689, Jun. 2008.

[7] F. Paz and M. Ordonez, "Zero oscillation and irradiance slope tracking for photovoltaic MPPT," IEEE Trans. Ind. Electron., vol. 61, no. 11, pp. 6138–6147, Nov. 2014.

[8] S. K. Kollimalla and M. K. Mishra, "Variable perturbation size adaptive P&O MPPT algorithm for sudden changes in irradiance," IEEE Trans. Sustainable Energy, vol. 5, no. 3, pp. 718–728, Jul. 2014.

[9] N. Femia, G. Petrone, G. Spagnuolo, and M. Vitelli, "Optimization of perturb and observe maximum power point tracking method," IEEE Trans. Power Electron., vol. 20, no. 4, pp. 963–973, Jul. 2005.

[10] M. Killi and S. Samanta, "Modified perturb and observe MPPT algorithm for drift avoidance in photovoltaic systems," IEEE Trans. Ind. Electron., vol. 62, no. 9, pp. 5549–5559, Sep. 2015.

[11] S. K. Kollimalla and M. K. Mishra, "A novel adaptive P&O MPPT algorithm considering sudden changes in the irradiance," IEEE Trans. Energy Convers., vol. 29, no. 3, pp. 602–610, Sep. 2014.

[12] Jubaer Ahmed and Zainal Salam, "A modified P&O maximum power point tracking method with reduced steady-state oscillation and improved tracking efficiency" IEEE Trans. Sustainable Energy, vol. 7, no. 4, Oct. 2016.

[13] Arvind Raj, "calculating efficiency" An application report from Texas instruments-SLVA 390-Feb. 2010.

[14] B.Ram Kiran and G.A. Ezhilarasi, "Design and analysis of soft switched buck-boost converter for PV applications," India conference (INDICON), Mar.2016.

[15] L. Zhang, A. Al-Amoudi, and Y. Bai, "Real-time maximum power point tracking for grid-connected photovoltaic systems," in Proc. 8th Int. Conf. Electron. Variable Speed Drives 2000, pp. 124–129.

[16] Y. Yang and F. P. Zhao, "Adaptive perturb and observe MPPT technique for grid-connected photovoltaic inverters," Procedia Eng., vol. 23, pp. 468–473, 2011.

[17] M. Qiang, S. Mingwei, L. Liying, J.M. Guerrero, "A novel improved variable step size incremental-resistance MPPT method for PV systems," IEEE Trans. Ind. Electron., vol.58, no. 6, pp. 2427–2434, Jun. 2011.

2nd IEEE International conference on power Electronics, Intelligent Control and Energy systems (ICPEICES-2018)

Reduced Device Count Asymmetrical Multilevel Inverter Topology Using Different PWM Techniques

Shubham Kumar Gupta
Deparment of Electrical Engineering
MITS Gwalior, India
shubhamgsti@gmail.com

Praveen Bansal
Department of Electrical Engineering
MITS Gwalior, India
pbansal444@gmail.com

Abstract-- This paper presents a new topology in a multilevel inverter (MLI) field with reduce device count and less number of dc voltage source with different PWM technique. This Proposed topology offers high power capability associated with less commutation losses, less total harmonic distortion (THD). For a specific number of levels in output, proposed topology required less switching components and DC voltage sources in comparison of Conventional topologies. This topology used asymmetric voltage sources is in nature, that required three asymmetric voltage sources and eight switching devices for producing 15-Level single phase output voltage. This topology is simple and optimal, which can be easily extended for higher level in output voltage of multilevel inverter. This paper presents the comparison of total harmonic distortion, with different pulse width modulation technique, with different modulation index, and the simulation result of proposed topology for single phase 15-level multilevel inverter, which are carried out by using MATLAB/Simulink R2013a software version.

Keywords— Multilevel Inverter (MLI), Pulse Width Modulation (PWM), Total Harmonic Distortion (THD)

I. INTRODUCTION

Multilevel inverter (MLI) is an extended form of an inverter, which appears in power electronics systems in 1975, because of the requirement of the good quality power, with better system efficiency [1]. Multilevel inverters becomes more popular over the years in high power medium voltage application because of its capability of handling the high power with less harmonic distortion (THD), reduced switching losses and good power quality. As the number of levels in output voltage waveform increases, harmonics in the output voltage waveform decreases. In comparison with two level inverter, multilevel inverter has several advantages like lower commutation losses that's why efficiency is improved, less dv/dt stress across the switch, lower harmonic contents, lower electromagnetic interference (EMI), and reduced filter size [2][3][4]. In multilevel inverter (MLI) requirement of power electronics switching components is more, which makes configuration complex and costly [5], can be considered as a disadvantage of MLI.

Multilevel inverter is used to generate a multiple step output waveform, to achieve smoother and less distorted DC to AC power conversion at desired AC output voltage [6]. The general concept of multilevel inverter is, utilizing a more number of semiconductor switches to perform the power conversion in small voltage steps. There are mainly three types of conventional multilevel inverter, Diode clamped multilevel inverter (DCMLI) [7], Flying capacitor

multilevel inverter (FCMLI) [2], and cascaded H-bridge (CHB) multilevel inerter [8]. In DCMLI topology, as the number of levels in voltage increased, more number of clamping diodes required. In FCMLI topology, as the number of levels in voltage increased, more number of flying capacitor required. CHB multilevel inverter topology is the simplest topology among the all conventional topologies. In CHB multilevel inverter topology, clamping diodes and flying capacitors are not required, but as the number of levels in voltage increased, more number of DC voltage sources required.

Generally, CHB multilevel inverters can be classified in two types: Symmetric and Asymmetric multilevel inverters. When the value of DC voltage sources in H-bridges are equal, is called Symmetric multilevel inverters. When the value of DC voltage sources in H-bridges are unequal, is called Asymmetric multilevel inverter. Asymmetrical topologies required less switching devices and voltage sources as compared to symmetrical topologies for same levels in output voltage. This paper proposed a topology for Asymmetrical multilevel inverter, which required less number of switching devices in comparison with conventional topologies [4]. For 15-Level MLI total device count required in proposed topology is compared with conventional topologies, shown in Table III. This paper used different pulse width modulation techniques to analyze total harmonic distortion in the output voltage waveform.

II. PROPOSED TOPOLOGY

This paper proposed a topology based on asymmetrical multilevel inerter, shown in Fig. 1, which required N unequal voltage sources (V_1, V_2, V_3..... V_N) and (2N+2) unidirectional switches (S_1, S_2, S_3, S_4..... $S_{(2N+2)}$), for (2^N-1) levels in output voltage of proposed multilevel inverter. Generally, the proposed topology produces maximum voltage ($V_n + V_{n-1}$). The proposed topology starts with 7 level, for which two unequal voltage sources and six unidirectional switches, consists of IGBT with an antiparallel diode, required. For generating all positive and negative levels, magnitude of voltage sources (V_1:V_2) should be in the ratio of (1:2).

To generate 15-Level MLI, proposed topology required three unequal voltage sources and eight unidirectional switches, shown in Fig. 2, it is clear that switching combination of (S_1, S_2), (S_3, S_4), (S_5, S_6), and (S_7, S_8) should not be turned 'ON', simultaneously, to avoid

978-1-5386-6624-1/18 $31.00 © 2018 IEEE   343

short circuit across the voltages. To generate all the positive and negative levels in the output voltage, magnitude of the voltage sources (V_1:V_2:V_3) should be in the ratio of (1:2:5). For different levels of 15-Level MLI, different switching combinations are shown in Table I, in which, state 1 to state 7 present positive levels, state 8 present Zero level and state 9 to state 15 present negative levels in the output voltage waveform of proposed MLI. State condition "1" and "0" shows the corresponding switch is turn 'ON' or turn 'OFF', respectively.

In this paper for 15-Level MLI, V_1=50V, V_2=100V, V_3=250V, and Load R= 10Ω is used, and ±350V, ±300V, ±250V, ±200V, ±150V, ±100V, ±50V, and 0V, are the 15 levels in output voltage waveform generated, shown in Fig. 9.

Proposed topology required ten unidirectional switches and four unequal voltage sources to generate 31-Level MLI, which can be obtain by adding two switches and one voltage source in 15-Level MLI. Similarly, by connecting more switches and more voltage sources, more levels in output voltage of proposed MLI can be obtained.

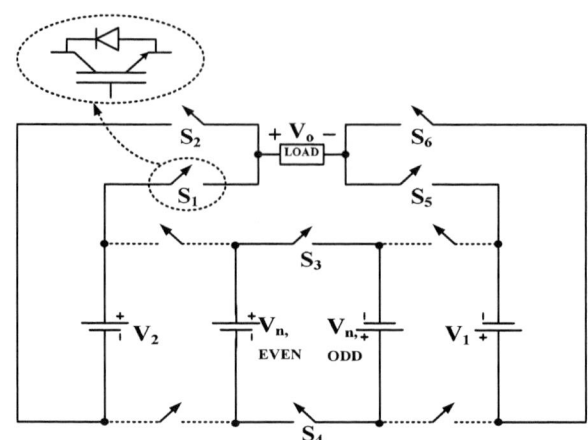

Fig. 1. General Structure of Proposed Topology

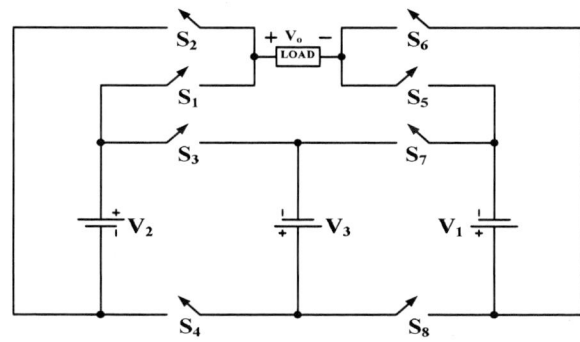

Fig. 2. Proposed Model of 15-Level Asymmetrical MLI

III. OPERATING MODES OF PROPOSED TOPOLOGY

TABLE I.   SWITCHING SCHEME FOR PROPOSED 15-LEVEL MLI

STATES	SWITCHING SCHEME								OUTPUT VOLTAGE
	S_1	S_2	S_3	S_4	S_5	S_6	S_7	S_8	
1	1	0	0	1	1	0	1	0	$V_3 + V_2$
2	1	0	0	1	0	1	1	0	$V_3 + V_2 - V_1$
3	0	1	0	1	1	0	1	0	V_3
4	0	1	0	1	0	1	1	0	$V_3 - V_1$
5	1	0	0	1	1	0	0	1	$V_2 + V_1$
6	1	0	0	1	0	1	0	1	V_2
7	0	1	0	1	1	0	0	1	V_1
8	0	1	0	1	0	1	0	1	0
	1	0	1	0	1	0	1	0	0
9	1	0	1	0	0	1	1	0	-V_1
10	0	1	1	0	1	0	1	0	-V_2
11	0	1	1	0	0	1	1	0	-($V_2 + V_1$)
12	1	0	1	0	1	0	0	1	-($V_3 - V_1$)
13	1	0	1	0	0	1	0	1	-V_3
14	0	1	1	0	1	0	0	1	-($V_3 + V_2 - V_1$)
15	0	1	1	0	0	1	0	1	-($V_3 + V_2$)

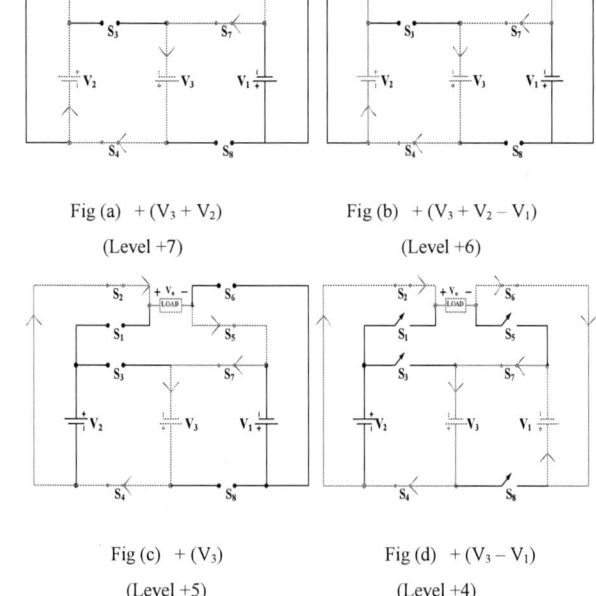

Fig (a)   + ($V_3 + V_2$)
(Level +7)

Fig (b)   + ($V_3 + V_2 - V_1$)
(Level +6)

Fig (c)   + (V_3)
(Level +5)

Fig (d)   + ($V_3 - V_1$)
(Level +4)

978-1-5386-6624-1/18 $31.00 © 2018 IEEE   344

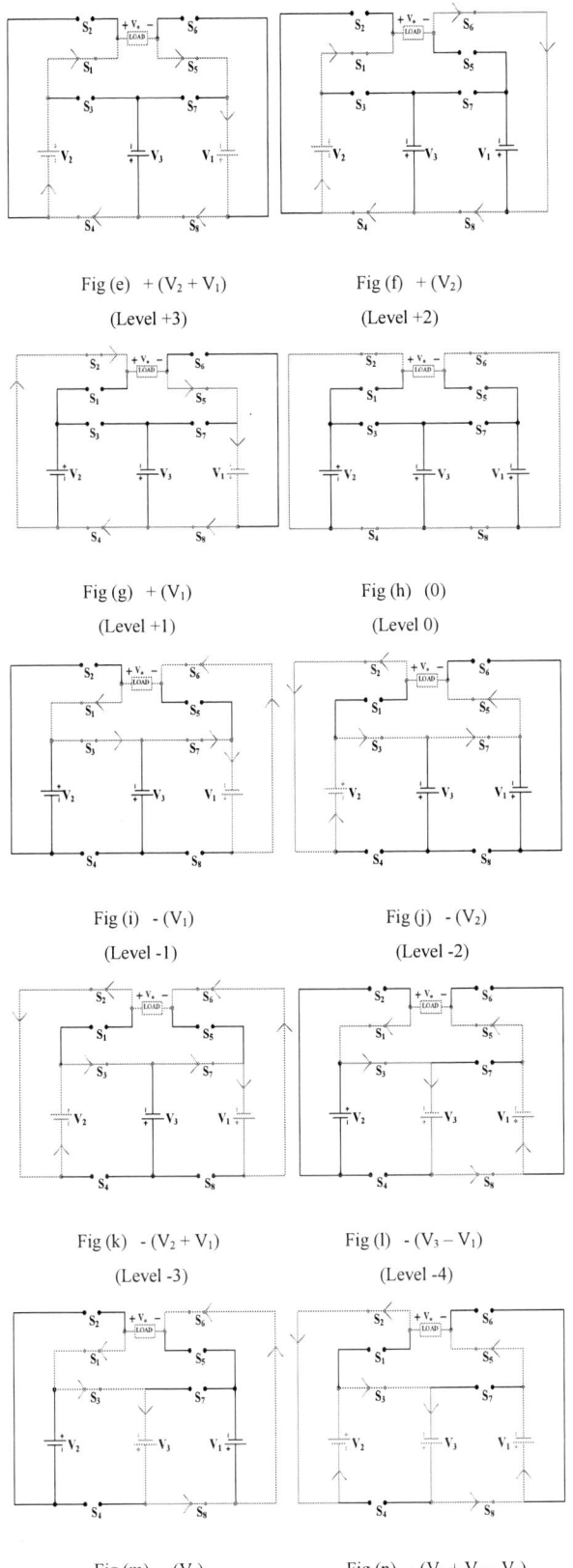

Fig (e) + (V₂ + V₁)

(Level +3)

Fig (f) + (V₂)

(Level +2)

Fig (g) + (V₁)

(Level +1)

Fig (h) (0)

(Level 0)

Fig (i) - (V₁)

(Level -1)

Fig (j) - (V₂)

(Level -2)

Fig (k) - (V₂ + V₁)

(Level -3)

Fig (l) - (V₃ – V₁)

(Level -4)

Fig (m) - (V₃)

(Level -5)

Fig (n) - (V₃ + V₂ – V₁)

(Level -6)

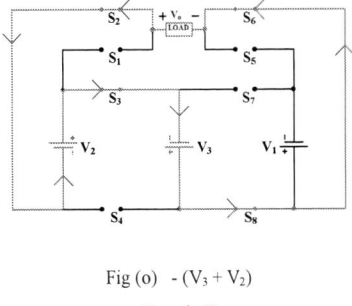

Fig (o) - (V₃ + V₂)

(Level -7)

Fig. 3. Fig (a-o) are the different Operating Modes of Proposed Topology for 15-Level MLI.

IV. MULTI CARRIER PWM TECHNIQUES

The main purpose of modulation techniques is to produce a train of switching pulses, which is responsible for generating sinusoidal waveform at output voltage in multilevel inverter (MLI). By proper use of these techniques THD in the output can be reduced. In modulation techniques Reference signal is compared with Carrier signal [9]. For N-level inverter (N-1) carrier waves are required. The most popular modulation technique used is, Multi-Carrier based Sinusoidal Pulse Width Modulation (MCSPWM) [10]. In this technique reference signal is sine wave of frequency 50 Hz and carrier signal is a triangular or inverted sine wave of higher frequency than reference signal frequency. At each and every point carrier signal is compared with the reference signal, when the reference signal is greater than carrier signal, switching pulse shows '1' otherwise '0'. Addition of the results of comparison between reference signal and carrier signal produces the switching pulse which generates levels in output voltage with different modulation index. Modulation Index (M_a) is the ration of peak magnitudes of the modulating waveform and the carrier waveform.

In this paper, PDPWM, PODPWM, and APODPWM ISCPWM, and VFISCPWM techniques are used for 15-Level MLI, shown in Fig. 4, Fig. 5, Fig. 6, Fig. 7, and Fig. 8 respectively.

- Phase Disposition Pulse Width Modulation (PDPWM) Technique :–

    In PDPWM technique, each and every carrier waveforms are in same phase with same amplitude and frequency.

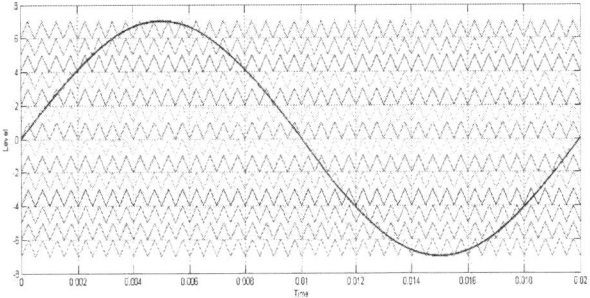

Fig. 4. Carrier Arrangement for PDPWM techniques.

- Phase Opposition Disposition Pulse Width Modulation (PODPWM) Technique :–
    In PODPWM technique, all carrier waveforms above zero reference line are in same phase with same amplitude and frequency, but carrier waveforms below zero reference line are 180° out of phase to above zero reference line carrier waveforms with same amplitude and frequency.

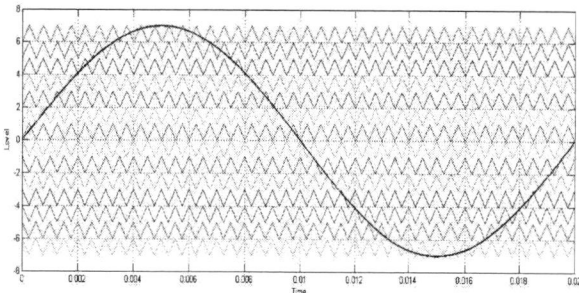

Fig. 5.   Carrier Arrangement for PODPWM techniques.

- Alternate Phase Opposition Disposition Pulse Width Modulation (APODPWM) Technique :–
    In APODPWM technique, all carrier waveforms has same amplitude and frequency, but all the carrier waveforms are 180° out of phase with by its neighboring carrier waveforms.

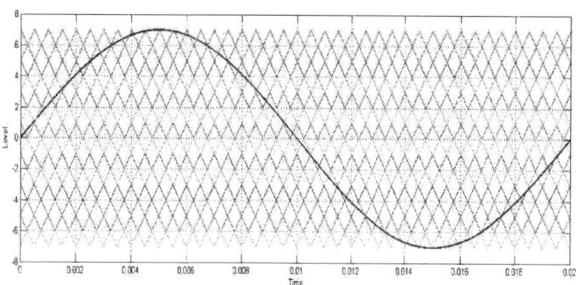

Fig. 6.   Carrier Arrangement for APODPWM techniques.

- Inverted Sine Carrier Pulse Width Modulation (ISCPWM) Technique :–
    In ISCPWM technique, reference signal is sinusoidal waveform and carrier signals are inverted sinusoidal waveforms. Frequency of carrier signals are higher than reference signal, which presents high fundamental output voltage [11].

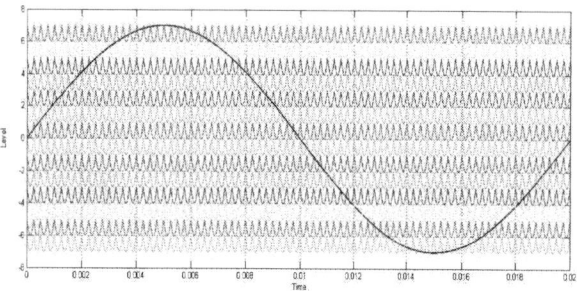

Fig. 7.   Carrier Arrangement for ISCPWM techniques.

- Variable Frequency Inverted Sine Carrier Pulse Width Modulation (VFISCPWM) Techniques :–
    In VFISCPWM technique, reference signal is sinusoidal waveform and carrier signals are inverted sinusoidal waveforms with variable frequencies. In this technique, frequencies of carrier signals are higher than reference signal. This technique provides an enhanced fundamental voltage and lower Total Harmonic Distortion (THD) [12].

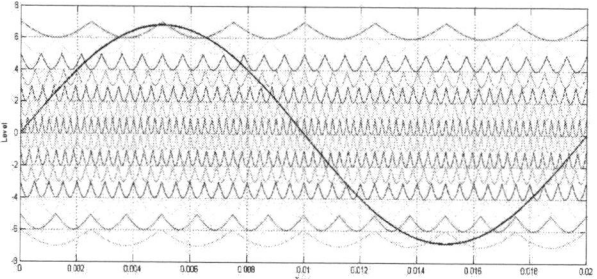

Fig. 8.   Carrier Arrangement for VFISCPWM techniques.

V.   SIMULATION RESULTS

The proposed topology for 15-Level asymmetric multilevel inverter, simulated in MATLAB. Table III shows the comparison of device count between proposed topology and conventional topologies for 15-Level MLI. Table I shows the switching schemes for 15-Level MLI. Table II shows comparison between Total Harmonic Distortion (THD) for different modulation technique at different modulation index. The Simulation parameters for 15-Level MLI are as following Resistance R=10Ω and DC Voltage sources are V_1=50V, V_2=100V, and V_3=250V. Output Voltage Waveform of 15-Level MLI shown in Fig. 9. This paper used five modulation techniques, PDPWM, PODPWM, APODPWM, ISCPWM, and VFISCPWM with different Modulation Index, at carrier signal frequency 2000Hz. Total Harmonic Distortion is carried out by using FFT analysis in MATLAB/Simulink R2013b, shown in Fig. 10, Fig. 11, Fig. 12, Fig. 13, and Fig. 14, respectively.

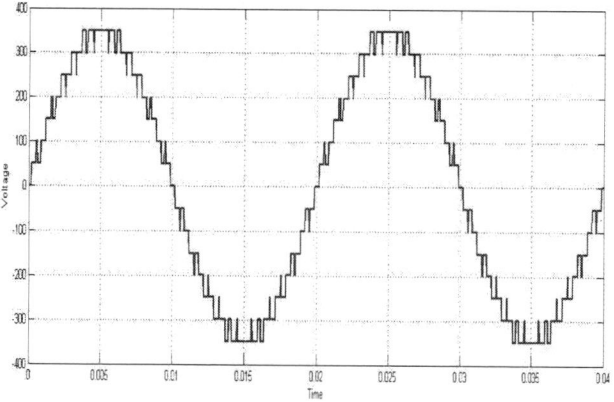

Fig. 9.   Output Voltage Waveform of 15-Level Proposed Asymmetrical Multilevel Inverter (ASMLI)

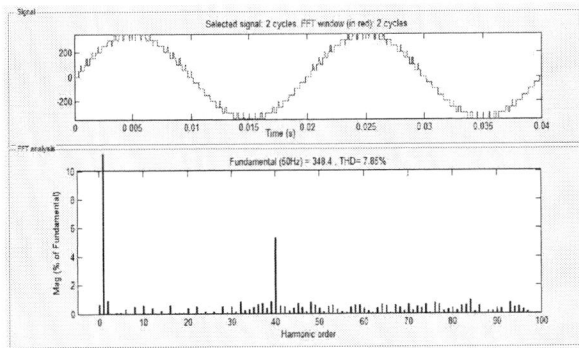

Fig. 10. %THD of 15-Level ASMLI with PDPWM technique
(M$_a$=1 and M$_f$=40)

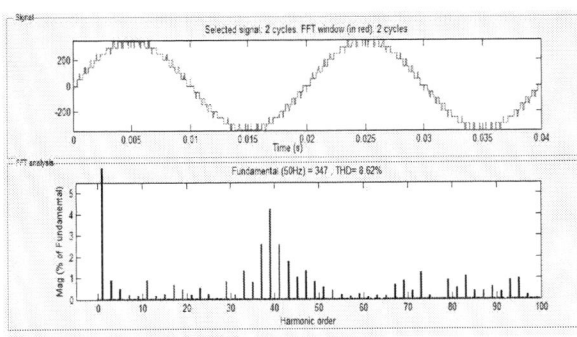

Fig. 11. %THD of 15-Level ASMLI with PODPWM technique
(M$_a$=1 and M$_f$=40)

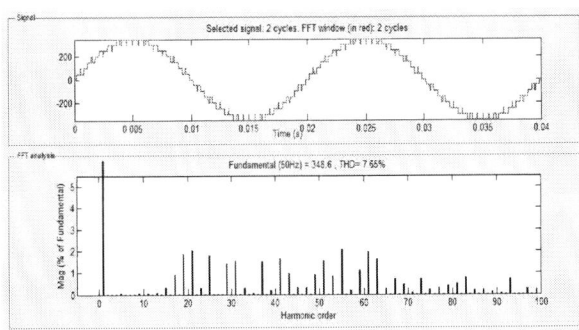

Fig. 12. %THD of 15-Level ASMLI with APODPWM technique
(M$_a$=1 and M$_f$=40)

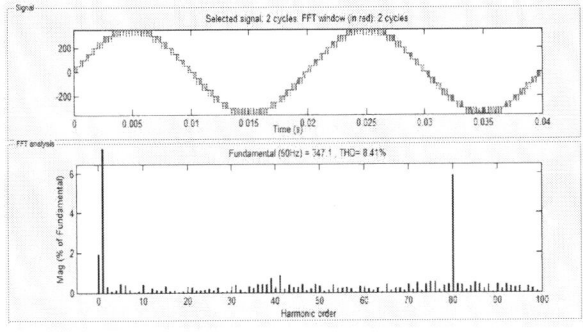

Fig. 13. %THD of 15-Level ASMLI with ISPWM technique
(M$_a$=1 and M$_f$=40)

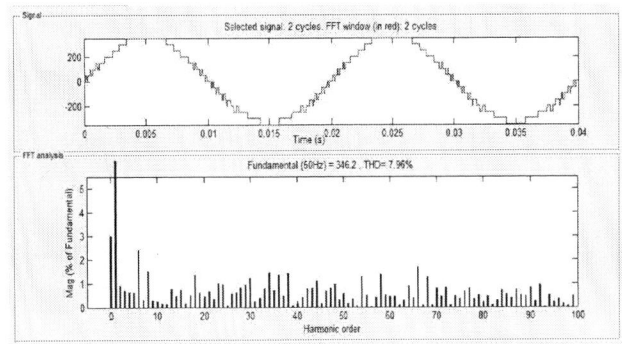

Fig. 14. %THD of 15-Level ASMLI with VFISCPWM technique
(M$_a$=1 and M$_f$=40)

TABLE II.   COMPARISON OF %THD FOR DIFFERENT MODULATION
TECHNIQUES WITH DIFFERENT MODULATION INDEX (M$_A$)

Level	Modulation Index	Modulation Techniques				
		PD	POD	APOD	ISC	VFISC
7 Level	1	18.23	18.18	18.39	19.67	19.97
	0.95	20.59	20.60	20.23	21.43	21.81
	0.90	22.40	22.11	21.98	22.77	22.65
15 Level	1	7.85	8.62	7.65	8.41	7.96
	0.95	8.56	8.53	8.89	9.12	8.68
	0.90	8.80	8.26	8.67	9.46	8.84

TABLE III.   COMPARISON OF COMPONENTS USED FOR 15-LEVEL MLI
BETWEEN PROPOSED TOPOLOGY AND CONVENTINAL TOPOLOGIES

MLI Component	Diode Clamped	Flying Capacitor	Cascaded H-Bridge	Proposed Topology
Switching Devices	28	28	28	8
Clamping Diode	182	0	0	0
DC Split Capacitor	14	14	0	0
Clamping Capacitors	0	91	0	0
DC Sources	1	1	7	3

978-1-5386-6624-1/18 $31.00 © 2018 IEEE

VI. CONCLUSION

This paper presents a new asymmetrical multilevel inverter (ASMLI) topology, which has greater performance than conventional topologies. The proposed ASMLI topology reduces total harmonic distortion (THD), shown in table. II, and modify the output voltage waveform. From table. III, the proposed topology required less switching components in comparison of conventional (DCMLI, FCMLI, and CHBMLI) topologies, so that the overall cost of switching components required, is reduced and efficiency improved. For 15-Level MLI, THD is 7.65% by using alternate phase opposition disposition pulse width modulation (APODPWM) technique with modulation index (Ma=1). The introduced topology was simulated and Harmonics analysis carried out by using MATLAB/Simulink R2013a version software.

VII. ACKNOWLEDGEMENT

The authors gratefully acknowledges the guidance and support of Electrical Engineering Department, Madhav Institute of Technology and Science for carrying out this work.

VIII. REFERENCES

[1] R. H. Baker, "High Voltage Converter Circuits", U. S. patent no. 4, 203, 1511 May 1980.

[2] J. Rodriguez, J. S. Lai, and F. Z. Peng, "Multilevel Inverters : A Survey of Topologies, Controls and Applications", IEEE Transactions Industrial Electronics, Vol. 49, no. 4, pp. 724-738, Aug 2002.

[3] J. Rodriguez, S. Bernet, B. Wu, J. Pontt, and S. Kouro, "Multilevel Voltage Source Converter Topologies for Industrial Medium Voltage Drives", IEEE Transactions Industrial Electronics, Vol. 54. no. 6, pp. 2930-2945, Dec 2007.

[4] J. Ebrahimi, E. Babaei, and G. B. Gharehpetian, "A New Multilevel Converter Topology with reduced number of Power Electronics Components", IEEE Transactions Industrial Electronics, Vol. 59, no. 2, pp.655-667, Feb 2012.

[5] G. Ceglia, V. Guzman, and C. Sanchez et al., "A New Simplified Multilevel Inverter Topology for DC-AC Conversion", IEEE Transactions Power Electronics, Vol. 21, no. 5, pp. 1311-1319, 2006.

[6] S. Daher, J. Schmind, and F. L. M. Antunes, "Multilevel Inverter Topologies for stand-alone PV Systems", IEEE Transactions Industrial Electronics, Vol. 55, no. 7, pp. 2703-2712, July 2008.

[7] X. Yuan, and I. Barbi, "Fundamentals of a New Diode Clamping Multilevel Inverters", IEEE Transactions Power Electronics, Vol. 15, no. 4, pp. 711-718, 2000.

[8] E. Babaei, "Optimal Topologies for Cascaded Sub-multlevel Converters", Power Electronics, Vol. 10, no. 3, pp. 251-261, May 2010.

[9] J. Holtz, "Pulse Width Modulation for Electronic Power conversion", proc. of IEEE, Vol. 82, no. 8, pp. 1194-1214, Aug 1994.

[10] J. Rushiraj, and P. N. Kapil, "Analysis of Different Modulation Techniques for Multilevel Inverters", ISTIEEE International Conference on Power Electronics, Intelligent Control and Energy System, 2016.

[11] Jeevananthan.S., Nandhakumar.R., and Dananjayan.P,"Inverted Sine Carrier for Fundamental Fortification in PWM Inverters and FPGA Based Implementations", Sebian Journal of Electrical Engineering, vol. 4, no. 2, pp.171-187, 2007.

[12] Seyezhai. R. And Mathur. B. L. "Implementation and Control of Variable Frequency Inverted Sine PWM Method for an Asymmetric Multilevel Inverter", European Journal of Scientific Research, Vol. 39, Issue 4, pp. 558-568.

2nd IEEE International conference on Power Electronics, Intelligent Control and Energy systems (ICPEICES-2018)

Implementation and Influence of C-SVPWM and BC-SVPWM Techniques using DSP for Voltage Oriented Control of Three Phase Active Front End Rectifier

Jaimin Mehta, Swathy Pillai, Atul Gupta, Rajamanickam R
Power Electronics & Automation Technology Centre
L & T Electrical & Automation
Navi-Mumbai, India
mehtajaimin2009@gmail.com, simplyswathy@gmail.com,
Atul.Gupta@lntebg.com, Raja.Manickam@lntebg.com

Abstract— **The Voltage Oriented Control method with Conventional Space Vector PWM (C-SVPWM) and Bus-clamp SVPWM (BC-SVPWM) for a Three Phase Active Front End rectifier which is used in 4-Quadrant drive has been designed, simulated and implemented. The system simulation has been done using Matlab/Simulink software. The CSVPWM and 60° Bus-clamped SVPWM techniques are implemented and compared experimentally using TI DSP TMS320F28335 as the controller platform. This paper investigates the effect of both PWM techniques on THD and power losses of three phase AFE. The measure of THD performance and switching loss are evaluated based on supply current ripple analysis and the variation of angle between line current and line voltage respectively. The switching loss mainly depends on the modulating wave shape and fundamental power factor in bus clamped PWM. Finally, the effectiveness of the proposed PWM technique is validate by theoretical and experimental results conducted on back to back converter loaded with induction motor. The result will be discussed and compared.**

Keywords— *Voltage Oriented Control, Space vector, Bus-clamp PWM, PWM Rectifier, unity power factor, Power Losses, THD.*

I. INTRODUCTION

The power quality issues are emerging in last two decades due to increasing usage of non-linear loads. These problems have been major concern and also created a specialized subject in power engineering. In industrial high power applications such as three phase ac motor drives (Fig 1.), uninterruptible power supplies, etc. it is convenient to employ three phase active front end PWM rectifier to reduce the harmonic content of supply current, dc-bus voltage regulation, bi-directional power flow and unity power factor [1]. Since the Three phase pwm rectifier always used for high and medium power applications with higher switching frequency its essential requirements are reduce input THD and power losses of the rectifier which are strongly depend on the PWM techniques [7].

Different PWM techniques are used to produce power converter output current and voltage with higher qualities for different types of loads. Several techniques have been proposed to meet the above mentioned objects by achieving a wide linear modulation range and improved overall power converter efficiency with minimum switching loss [11]. SVPWM is a popular PWM method because it simultaneously achieves a

higher DC bus utilization with lower switching loss and lower harmonic distortion.

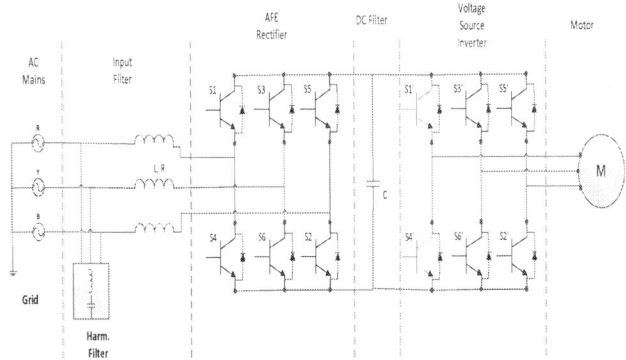

Fig 1. Three Phase PWM Rectifier with Inverter for 4Q Drive

There is classification of SVPWM techniques given in Fig 2. To reduce the current harmonics and switching losses, there is a degree of freedom in the placement of zero space vectors which can be either conventional SVPWM or Bus clamp PWM techniques. In a line cycle, the bus clamping techniques are clamp for 120° duration every phase to either the negative or positive DC rail. The orientation of the desired reference voltage will decide the clamped leg and the duration of the clamping. In BCSVPWM, 120°, 60° and 30° are the best known clamping methods. By clamping a phase to either negative or positive DC rail the BCSVPWM techniques can significantly reduce the switching losses and gives better spectral performance. Theoretically by clamping the phase which is carrying the highest current reduced the switching loss 33% to 50% compared to CSVPWM [7-11].

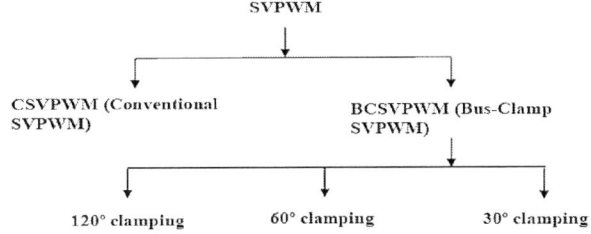

Fig 2. Classification of SVPWM Techniques

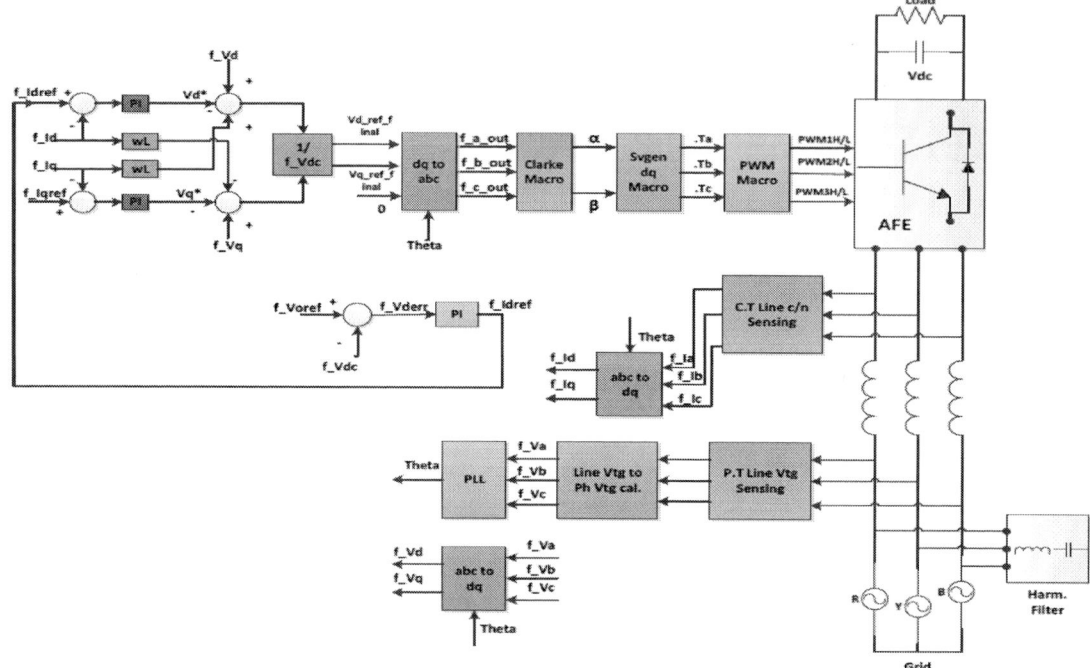

Fig 3. Voltage Oriented Control Diagram for Three Phase Active front End PWM Rectifier

In this paper, a Simulation, hardware implementation and performance evaluation of CSVPWM and 60° BCSVPWM techniques for three-phase unity power factor AFE PWM Rectifier which is used in 4Q Drive is presented. In Section 2, Design and analysis of VOC with both PWM techniques for three phase PWM rectifier are presented. In Section 3, Loss calculation for three phase PWM rectifier is presented. The simulation results, experimental results and conclusions are given in sections 4 and 5 respectively.

II. DESIGN AND ANALYSIS OF VOLTAGE ORIENTED CONTROL WITH CSVPWM AND BCSVPWM TECHNIQUES FOR THREE PHASE PWM RECTIFIER

Vector control is a popular method for control of three-phase induction motors. Similar control approach can be used for three phase AFE also as shown in Fig 3. Here, the three phase line currents and supply voltages are converted in to stationary reference frame which is also called as equivalent two phase system. These quantities are further transformed into a reference frame called synchronous reference frame, which revolves at the supply frequency. In synchronous reference frame, the components of current corresponding to reactive and active power are controlled in an independent manner similar to the flux and torque producing components in a motor drive [6].

In order to obtain a voltage oriented control (VOC) a well-known linear control technique such as the PI based controller could be applied both to the dc voltage control and to the ac current control. Since the main purpose of the active front end rectifier is to absorb or to generate sinusoidal currents. The reference current's components are dc quantities in dq frame, the reference current q-component f_iqref is controlled to obtain a unity power factor while he reference current d-

component f_idref is controlled to perform the dc voltage regulation. The dc voltage controller calculates the reference value for d-axis current controller, so this control structure defined as 'cascade'. There is three PI controller are used, two for d- and q-current controls and one for dc voltage control. In order to limit dangerous current overshoots all the controller adopt suitable limitations on the integrators as well as anti-windup devices.

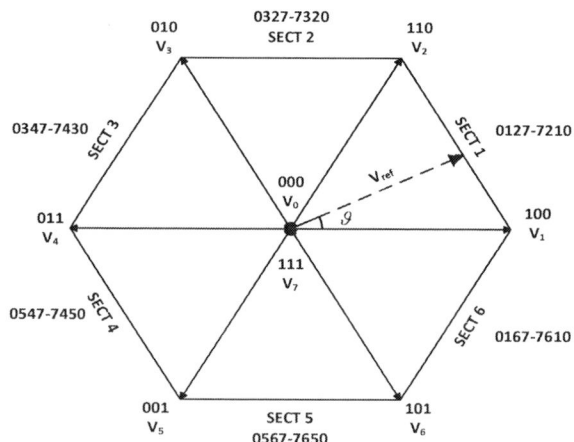

Fig 4. Voltage Vector Diagram for CSVPWM Technique

PWM waveform are usually produced either by comparing three phase modulating signal against a common triangular carrier or using the space vector approach [9]. In SVPWM technique, the reference voltage is provided by revolving reference vector which is sampled once every sub-

978-1-5386-6624-1/18 $31.00 © 2018 IEEE   350

cycle Ts. The reference vector is realized by the two zero vectors and the nearest two active vectors as shown in Fig. 4, for sector I based on the volt- second balance principle.

$$V_{ref} * T_s = V_1 * T_1 + V_2 * T_2 + V_0 * T_0 \qquad (1)$$

$$T_s = T_1 + T_2 + T_0 \qquad (2)$$

Where T_s and V_{ref} are the sampling period and the reference vector respectively. For the active vectors V_1 and V_2, T_1 and T_2 are the dwell times respectively. The same way for the zero vectors V_0 and V_7, the dwell time is T_0.

The dwell times T_1, T_2 and T_0 are defined as:

$$T_1 = mi * \frac{\sin(60^\circ - \theta)}{\sin(60^\circ)} * T_s \qquad (3)$$

$$T_2 = mi * \frac{\sin(\theta)}{\sin(60^\circ)} * T_s \qquad (4)$$

$$T_0 = T_s - T_1 - T_2 \qquad (5)$$

Where, mi is the modulation-index.

In the space vector domain there is flexibility for PWM strategies to divide dwell time T_0 between the two zero states. If dwell time T_0 divide equally between two zero states it will lead to lesser ripple current. In conventional space vector modulation technique, sequences 0127 and 7210 are divided equally and used to generate a given sample in sector I.

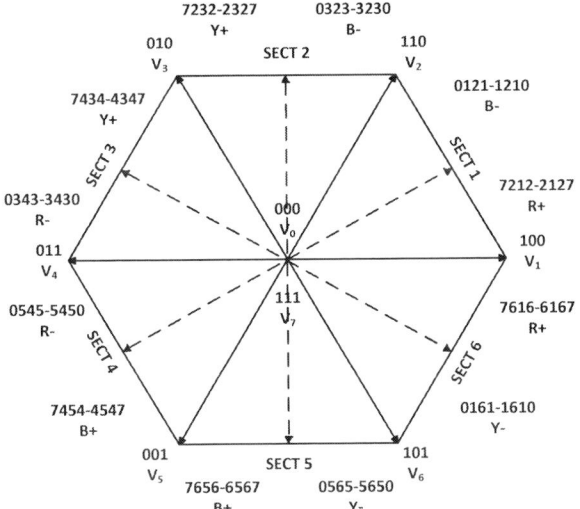

Fig 5. Voltage Vector Diagram for BCSVPWM Technique

Sequences like 012, 210, 721 or 127 also used to generate same sample using only one zero state. One another method is also possible which is proposed here. Divide one active state for time T_1 or T_2 into two equal intervals. It will give sequences 7212, 2127, 0121 or 1210 to generate the same sample. So, here either R-phase remains clamped to the positive bus or B-phase gets clamped to the negative bus, which is depend upon the choice of the zero state used. In bus-clamping

PWM, the two zero states are used alternately over 60° durations. For considerations of symmetry, at the middle of every sector zero state is usually changed. As shown in Fig 5 the zero state 7 is used from middle of the sector VI to sector I, which leads to 60° clamping. An improved discontinuous PWM modulation strategy that inherently balances the switching losses is to place the space vectors to alternately eliminate zero space vectors V0 and V7 for successive 60° segments of the fundamental cycle. Here each phase leg is now unmodulated for only 60° at a time (alternately switched to the upper or to the lower DC rail), so this this arrangement is called 60° discontinuous modulation and it has the particular benefit that the line-line switched voltages are now symmetrical. The space vector arrangement for this modulation strategy is shown in Fig 5.

There is the non-switching period for each phase leg in centre and symmetrical around the both positive and negative peaks of its fundamental reference voltage because of this method of discontinuous switching pattern. As we know that input voltage and current should be in phase for unity power factor in AFE, the line current's peak follow the fundamental voltage peak. So, this is one of the best clamping method for AFE rectifier because when current at its maximum value, each phase leg does not switch and minimizes switching losses.

III. POWER LOSS CALCULATION FOR THREE PHASE ACTIVE FRONT END RECTIFIER

Forward characteristics of the power devices are approximated by the following linear equations for IGBT and diode, respectively.

$$u_{CE} = u_{CE0} + r_c i_C \qquad (6)$$

$$u_D = u_{D0} + r_D i_D \qquad (7)$$

Assuming that the switching frequency is high, the input current of the rectifier can be assumed to be sinusoidal. That is,

$$i_C = I_M \cos(\theta - \emptyset) \qquad (8)$$

Where. I_M is a peak value of input current and $\emptyset$ is a phase angle difference between voltage and current. Using above equations. the conduction loss of one IGBT and diode can be obtained as

$$P_{CT} = \frac{1}{2\pi} \int_{-\frac{\pi}{2}+\emptyset}^{\frac{\pi}{2}+\emptyset} u_{CE} i \frac{1+V_{an}}{2} d\theta \qquad (9)$$

$$= \frac{u_{CE0}}{2\pi} \int_{-\frac{\pi}{2}+\emptyset}^{\frac{\pi}{2}+\emptyset} i_C \frac{1+V_{an}}{2} d\theta + \frac{r_c}{2\pi} \int_{-\frac{\pi}{2}+\emptyset}^{\frac{\pi}{2}+\emptyset} i_C^2 \frac{1+V_{an}}{2} d\theta \qquad (10)$$

$$P_{CD} = \frac{1}{2\pi} \int_{-\frac{\pi}{2}+\emptyset}^{\frac{\pi}{2}+\emptyset} u_D i_C \frac{1-V_{an}}{2} d\theta \qquad (11)$$

$$= \frac{u_{D0}}{2\pi} \int_{-\frac{\pi}{2}+\emptyset}^{\frac{\pi}{2}+\emptyset} i_C \frac{1-V_{an}}{2} d\theta + \frac{r_D}{2\pi} \int_{-\frac{\pi}{2}+\emptyset}^{\frac{\pi}{2}+\emptyset} i_C^2 \frac{1-V_{an}}{2} d\theta \qquad (12)$$

So, $P_{Cond} = P_{CT} + P_{CD} \qquad (13)$

$$= \frac{I_M}{2\pi}(u_{CE0} + u_{D0}) + \frac{I_M}{4\pi}(u_{CE0} - u_{D0})\left(M_i \frac{\pi \cos \emptyset}{2}\right)$$

978-1-5386-6624-1/18 $31.00 © 2018 IEEE

$$+\frac{I_M^2}{8}(r_C+r_D)+\frac{I_M^2}{4\pi}(r_C-r_D)\left(M_i\frac{4\cos\emptyset}{3}\right)+P_{PWM} \qquad (14)$$

Where, V_{an} is a pole voltage and M_i is a modulation index. P_{PWM} is adding losses due to Bus-clamp SVPWM techniques. Here, in our power stack we have used Fuji IGBT module part no 2MBI200N-120. So, based on that value of parameters from datasheet and condition $u_{CE0} = 1.3V$, $u_{D0} = 1.1V$, $I_M = 136A$, $r_C = 5m\Omega$, $r_D = 3.75m\Omega$, $M_i = 0.8$ and $\emptyset = 0°$ (at UPF) put all these values in above equations (14).

Here, we will get conduction losses for one IGBT and diode using CSVPWM technique. $P_{Cond} \cong 81W$. in the computer simulation, the difference of conduction loss (P_{PWM}) among CSVPWM and BCSVPWM scheme is less than 3% of total conduction loss. So, that more important it is to consider switching loss more than conduction loss. Conduction losses for one IGBT and diode using 60° BCSVPWM $P_{Cond} \cong 83W$.

Switching losses:

$$P_{sw} = \frac{1}{2\pi}\int_{-\frac{\pi}{2}+\emptyset}^{\frac{\pi}{2}+\emptyset}(E_{ONT}+E_{OFFT}+E_{OND})*i_C*f_{sw}\,d\theta \qquad (15)$$

$$P_{sw} = \frac{(E_{ONT}+E_{OFFT}+E_{OND})*f_{sw}}{\pi} \qquad (16)$$

We have taken turn on and turn off energy at peak value of current $E_{ONT} = 20mJ$, $E_{OFFT} = 15mJ$, $E_{OND} = 10mJ$ and $f_{sw} = 3.6$ kHz. Keep all these values in above equations we will get switching losses for one IGBT and diode using CSVPWM technique. $P_{sw} = 52W$. And in the case of 60° Bus-clamp PWM, following properties hold because of symmetry of pole voltage.

$$P_{sw}(-\emptyset) = P_{sw}(\emptyset) = P_{sw}(\pi-\emptyset) \text{ Where, } 0 < \emptyset < \Pi. \qquad (17)$$

Therefore, it is sufficient to consider the range of $0 < \emptyset < \Pi/2$ for the 60° Bus-clamp PWM schemes as follows.

$$P_{sw\_BC} = \begin{cases} P_{sw}*\left(1-\frac{1}{2}\cos(\emptyset)\right), 0<\emptyset<\frac{\pi}{3} \\ P_{sw}*\frac{\sqrt{3}\sin(\emptyset)}{2}, \frac{\pi}{3}<\emptyset<\frac{\pi}{2} \end{cases} \qquad (18)$$

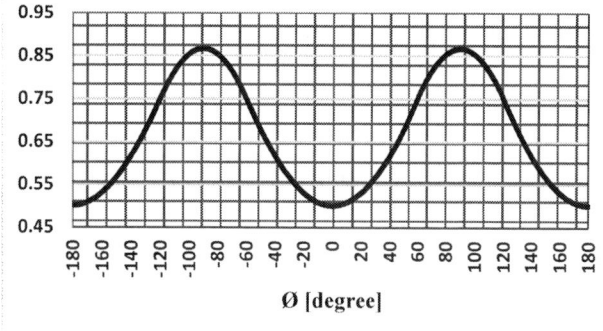

Ratio of Switching Losses

Fig 6. Ratio of Switching Losses of 60° BCSVPWM Technique Compared with CSVPWM Technique

We are working on three phase active front end pwm rectifier, so it will work at unity power factor and $\emptyset = 0°$. From the Fig 6. we can see the ratio of switching losses 60° BCSVPWM to the CSVPWM. And according to above equation, $P_{sw\_BC} \cong P_{sw}*0.5$. Switching losses for one IGBT and diode using 60 ° BCSVPWM $P_{sw\_BC} \cong 26W$.

Total Power losses for CSVPWM and 60° BCSVPWM:

$$P_{Total\_CSVPWM} = (P_{Cond}+P_{sw})\times 6 \cong 800 \text{ W}$$
$$P_{Total\_BCSVPWM} = (P_{Cond}+P_{sw})\times 6 \cong 650 \text{ W}$$

IV. THREE PHASE PWM RECTIFIER'S SIMULATION AND EXPERIMENTAL RESULTS

The proposed method has been simulated on 60kW AFE for three phase inverter drive. The system parameters used in the simulation are given in Table 1. The control strategy is also applied to DSP based 60kW experimental setup, showed in Fig. 12. Here Fig. 7-8 shows simulation results for input voltage and current of AFE in phase and Fig. 9-10 show FFT analysis of input current of AFE 3.06% and 4.37% with CSVPWM and BCSVPWM respectively. We have analyzed power quality of our converter using three phase power analyzer. The THD analysis and experimental results with various power ratings are shown in Fig. 18 and Table 2 respectively. Fig. 11 shows PLL output waveform with input phase voltage. Then Fig. 13 and 14 shows waveforms for AFE operation in motoring and regeneration mode respectively using CSVPWM technique.

Table 1 System Parameters for Three Phase PWM Rectifier

Symbol	Description	Value	Unit
Vin	Supply Input voltage	400	V
Vo	Output dc voltage	660	Vdc
Iin	Rated Input current (rms)	96	A
Io	Rated Output current	90	Adc
Po	Output power	60	kW
L_f	Input Inductance	1.46	mH
R_f	Internal Resistance	0.4	Ohm
R	Load resistor	7.34	Ohm
C_f	Dc link capacitor	3.9	mF
Fsw	Switching Frequency	3600	Hz
$\cos(\emptyset)$	Input power factor	0.95 or more	
η	Efficiency	95or more	%
THD	Total Harmonic Distortion	< 5	%

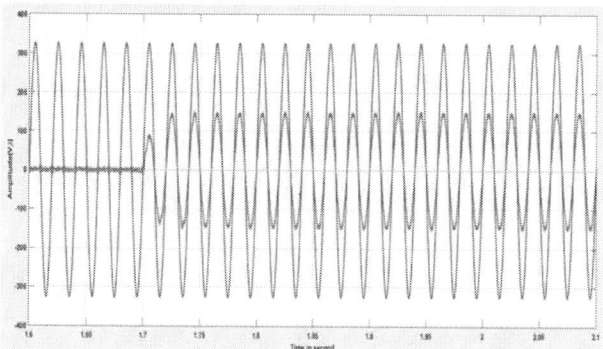

Fig 7. Input Voltage and Current of AFE with CSVPWM

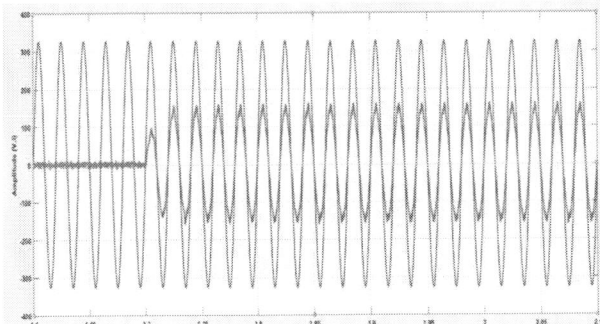

Fig 8. Input Voltage and Current of AFE with BCSVPWM

Here Fig. 15 shows the modulating signal generated by algorithm using DSP for 60° BCSVPWM. Then Fig. 16 and 17 shows waveforms for AFE operation in motoring and regeneration mode respectively using 60° BCSVPWM technique. We have measured THD for regeneration also, which are shown in Fig. 14 and 17. We are getting THD 4.3% to 9.9% and 5.5% to 8.5% for CSVPWM and 60° BCSVPWM respectively.

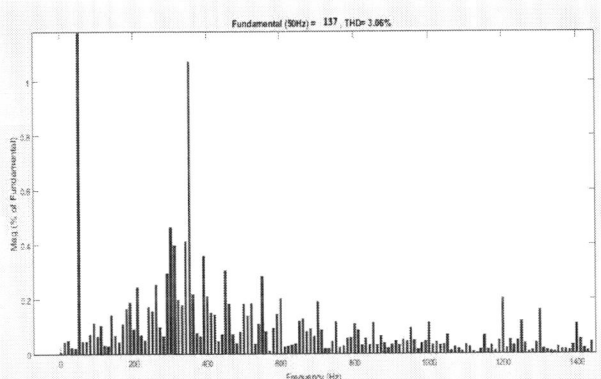

Fig 9. FFT Analysis of Input Current of AFE with CSVPWM

Fig 12. Experimental setup for Three Phase AFE Rectifier with Inverter for 4Q Drive

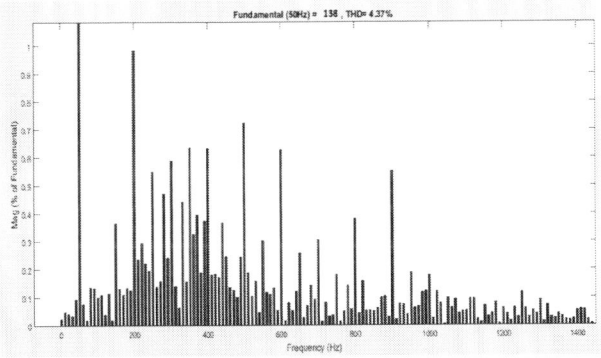

Fig 10. FFT Analysis of Input Current of AFE with BCSVPWM

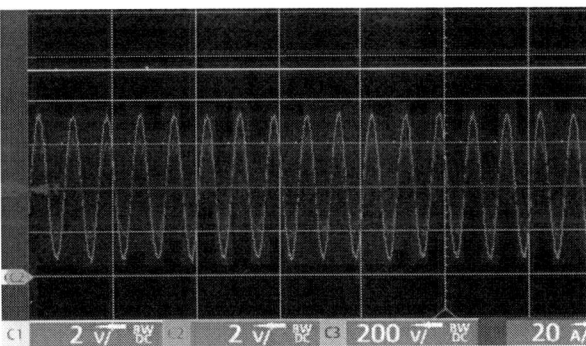

Fig 13. AFE operation in Motoring mode with CSVPWM Technique

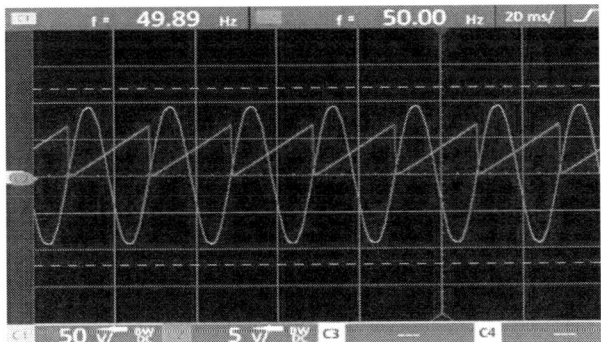

Fig 11. PLL output waveform with Input phase voltage

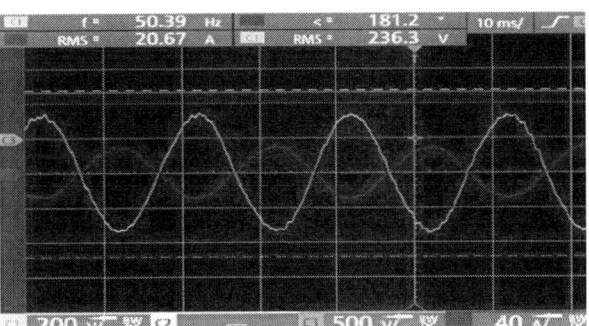

Fig 14. AFE operation in Regeneration Mode with CSVPWM

978-1-5386-6624-1/18 $31.00 © 2018 IEEE   353

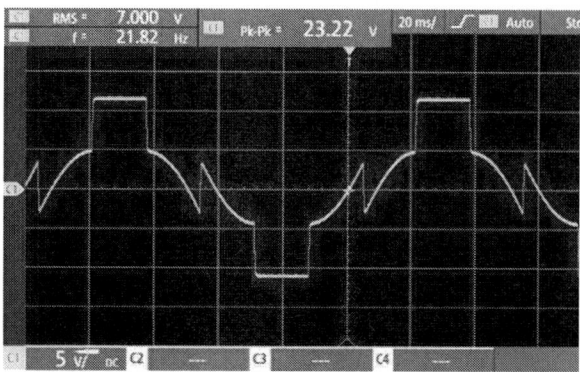

Fig 15. Modulating signal with 60° Bus Clamped SVPWM

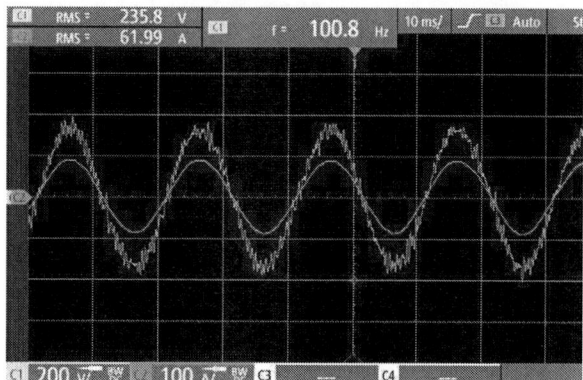

Fig 16. AFE operation in motoring mode with BCSVPWM Technique

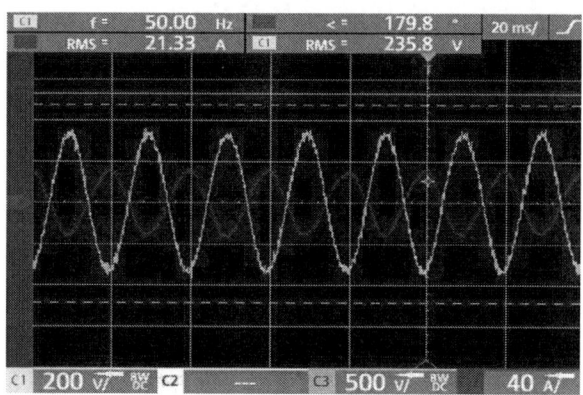

Fig 17. AFE operation in Regeneration Mode with BCSVPWM

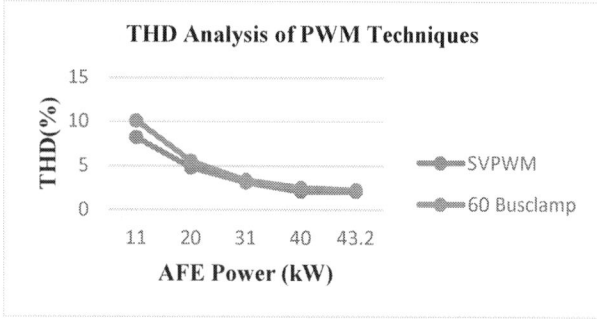

Fig 18. THD Analysis for CSVPWM and 60° BCSVPWM

Table 2 Power Analyzer Experimental Results with CSVPWM and 60° Bus-Clamped SVPWM

Power (in kW)	Current (in A) RMS	CSVPWM THD	60° BCSVPWM THD
11 kW	16.5 A	7.3 - 9 %	9.5 – 10.6 %
20 kW	28.7 A	4 – 5.5 %	4.98 – 6 %
31 kW	42.5 A	2.7 – 3.5 %	2.9 – 3.8 %
40 kW	57.2 A	1.9 – 2.2 %	2.2 – 2.8 %
43.2 kW	61.5 A	1.9 – 2.2 %	1.9 – 2.5 %

V. CONCLUSION

Hence, from the above theoretical and experimental analysis we can see that THD for both PWM techniques are almost same but with 60° Bus-clamped SVPWM total conduction loss are increase 2-3 % and at the same time total switching loss are reduce around 50 %. So, for our 60 kW prototype total power losses reduce from 800 W to 650 W. For high power converter and drive application, this PWM technique will be more desirable for highly efficient power conversion. It will reduce Total Power loss, Heat-sink requirement, temperature of power devices and improve overall performance of Drive.

REFERENCES

[1] J. Rodr'ıguez, J. Dixon, J. Espinoza, and P. Lezana, "PWM Regenerative Rectifiers: State of the Art," IEEE Transactions on Industrial Electronics, vol. 52, no. 1, pp. 5-22, February 2005.

[2] D. Grahame Holmes, Thomas A. Lipo, "Pulse Width Modulation for Power Converters: Principles and Practice," Wiley Text Books, First Edition, 2003, ISBN: 978-0-471-20814-3.

[3] S. L. Sanjuan, "Voltage oriented control of threephase boost pwm converters," Master's thesis, Chalmers University of Technology, Gothenburg,2010.

[4] M. Malinowski, M. P. Kazmierkowski, and A. M. Trzynadlowski, "A comparative study of control techniques for pwm rectifiers in ac adjustable speed drives," IEEE Transactions On Power Electronics, vol. 18, no. 5, pp. 1390–1396, May 2003.

[5] J. S. Siva Prasad, Tushar Bhavsar, Rajesh Ghosh, and G. Narayanan, "Vector control of three-phase AC/DC front-end converter," Sadhana, vol. 33, part 5, pp. 591-613. October 2008.

[6] Marco Liserre , Frede Blaabjerg & Antonio Dell'Aquila, "Step-by-step design procedure for a grid-connected three-phase PWM voltage source converter," International Journal of Electronics, vol. 91, no. 8, pp. 445-460, Aug—2004.

[7] Dae-Woong Chung, Seung-Ki Sul, "Minimum-Loss Strategy for Three-Phase PWM Rectifier," IEEE Transactions On Industrial Electronics, vol. 46, no. 3, pp. 517–526, June 1999.

[8] G. Narayanan, V. T. Ranganathan, "Two Novel Synchronized Bus-Clamping PWM Strategies Based on Space Vector Approach for High Power Drives," IEEE Transactions On Power Electronics, vol. 17, no. 1, pp. 84–93, Jan 2002.

[9] Di Zhao, V. S. S. Pavan Kumar Hari, Gopalaratnam Narayanan and Rajapandian Ayyanar, "Space-Vector-Based Hybrid Pulsewidth Modulation Techniques for Reduced Harmonic Distortion and Switching Loss," IEEE Transactions On Power Electronics, vol. 25, no. 3, pp. 760–774, March 2010.

[10] Olorunfemi Ojo, "The Generalized Discontinuous PWM Scheme for Three-Phase Voltage Source Inverters," IEEE Transactions On Industrial Electronics, vol. 51, no. 6, pp. 1280–1289, March 2004.

[11] Meenu D. Nair, Jayanta Biswas, G. Vivek and Mukti Barai, "Performance Evaluation of Various Bus Clamped Space Vector Pulse Width Modulation Techniques," Journal of Power Electronics, vol 17, no. 5, pp. 1244–1255, Sept. 2017.

2nd IEEE International conference on power Electronics, Intelligent Control and Energy systems (ICPEICES-2018)

A New 15-Level Asymmetrical Multilevel Inverter Topology with Reduced Number of Devices for Different PWM Techniques

Kuldeep Rathore
Department of Electrical Engineering
Madhav Institute of Technology & Science
Gwalior (M.P.), India
kuldeeprathore4545@gmail.com

Praveen Bansal
Department of Electrical Engineering
Madhav Institute of Technology & Science
Gwalior (M.P.), India
pbansal444@gmail.com

Abstract— **In this paper, a new 15-level asymmetrical multilevel inverter topology with reduced number of devices for different pulse width modulation (PWM) techniques introduces. It involves reduce number of IGBT switches, dc voltage sources, gate driver circuits used which reduces the complexity of circuit, area of installation and overall cost of system. This topology reduces total harmonic distortion (THD) of output across the load which leads to better operation of inverter and overall efficiency of system increases. The THD of the output waveform decreases as the levels of output voltage increases. Here different PWM techniques are used for multilevel inverter topology to generate 15-level output phase voltage. The output results and FFT analysis of 15-level asymmetrical multilevel inverter topology are ascertained by using MATLAB/SIMULINK software.**

Keywords—**Multilevel inverter (MLI), pulse width modulation (PWM) techniques, total harmonic distortion**

I. INTRODUCTION

Multilevel inverter is the power electronics converter that is widely used in high power applications. In the recent years, multilevel inverters have more attention and importance in the power industry because of their high frequency and high voltage operation, low electromagnetic interference (EMI) and high efficiency [1].The thought of multilevel inverter was introduced in 1975 [2]. It basically uses several lower dc sources and then adjust a near about sinusoidal waveform [3]. Multilevel inverter gives output voltage waveform with lower harmonic distortion and as we increase the number of levels then the total harmonic distortion reduces more [4].

A multilevel inverter has several advantages over a conventional two-level inverter i.e. it generate output voltage with extremely low distortion and lower dv/dt loss across the switch i.e. efficiency increases, electromagnetic interference (EMI) and size of filter reduces, draw input current with very low distortion, generate smaller common mode (CM) voltage, MLI can operate with a lower switching frequency [1-4].

Multilevel inverter has some disadvantages over two-level inverter i.e. power semiconductor (IGBT) switches used increases which increases gate drivers that makes the system complex and costly. These disadvantages minimized by using higher number of levels in output voltage waveform.

There are three main types of conventional multilevel inverters: cascaded H-bridge multilevel inverter (CHBMLI), flying capacitor multilevel inverter (FCMLI) and diode clamped multilevel inverter (DCMLI) or neutral point clamped multilevel inverter (NPCMLI) [5]. Firstly a series connection of H-bridge inverter is known as CHBMLI was introduced in 1960s [6].Then it was followed by FCMLI topology introduced in 1992 and DCMLI topology later on evolved into neutral point clamped multilevel inverter (NPCMLI) topology [7]. But these conventional topologies have some drawbacks, in case of DCMLI topology capacitor voltage unbalancing occurs and the number of clamping diode increases as we increase the number of levels. FCMLI topology balance capacitor voltages but as we increase the number of levels, number of balancing capacitors increases and it increases the circuit complexity. CHBMLI consist of series connection of several H-bridges as we increase the number of voltage levels number of switching devices given by 2(N+1) also increases. CHBMLI requires separate dc source, hence renewable energy sources can be utilized as input for this topology. But it reduces its area of applications [8].

Here a new 15-level asymmetrical multilevel inverter topology based on series connected dc sources has been proposed. It requires less number of switches, number of driver circuits and number of dc sources used. It also reduces total harmonic distortion (THD). The THD of proposed topology is compared with different PWM techniques.

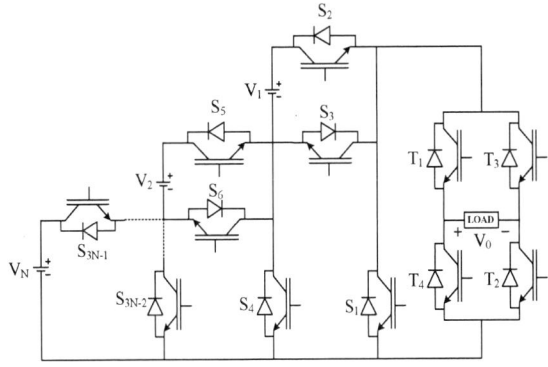

Fig.1. Proposed Topology.

978-1-5386-6624-1/18 $31.00 © 2018 IEEE   355

TABLE I. SWITCHING SCHEME OF PROPOSED 15-LEVEL MLI

GENERATION OF LEVEL	SWITCHING SCHEME												OUTPUT VOLTAGES
	S_1	S_2	S_3	S_4	S_5	S_6	S_7	S_8	T_1	T_2	T_3	T_4	
7	0	1	0	0	1	0	0	1	1	1	0	0	$V_1+ V_2+ V_3$
6	0	0	1	0	1	0	0	1	1	1	0	0	$V_2+ V_3$
5	0	1	0	0	0	1	0	1	1	1	0	0	$V_1 +V_3$
4	0	0	1	0	0	1	0	1	1	1	0	0	V_3
3	0	1	0	0	1	0	1	0	1	1	0	0	$V_1+ V_2$
2	0	0	1	0	1	0	1	0	1	1	0	0	V_2
1	0	1	0	1	0	0	0	0	1	1	0	0	V_1
0	1	0	0	0	0	0	0	0	1	1	0	0	0
-1	0	1	0	1	0	0	0	0	0	0	1	1	$- V_1$
-2	0	0	1	0	1	0	1	0	0	0	1	1	$- V_2$
-3	0	1	0	0	1	0	1	0	0	0	1	1	$-(V_1+ V_2)$
-4	0	0	1	0	0	1	0	1	0	0	1	1	$- V_3$
-5	0	1	0	0	0	1	0	1	0	0	1	1	$-(V_1 +V_3)$
-6	0	0	1	0	1	0	0	1	0	0	1	1	$-(V_2+ V_3)$
-7	0	1	0	0	1	0	0	1	0	0	1	1	$-(V_1+ V_2+ V_3)$

II. PROPOSED TOPOLOGY

Proposed topology for asymmetrical source configuration is shown in Fig.1 requires, N different value voltage sources, and $3(N+1)$ switches for generating $(2^{(N+1)}-1)$ number of levels in output voltage waveform. In the Fig.1 left side circuit generates the required output level hence it is called output level generator and right side circuit is called full bridge converter which reverses the voltage direction as per requirement. When positive levels required in output voltage waveform, switch T_1, T_2 will be ON, when negative levels required in output voltage waveform, switch T_3 and T_4 will be ON. Fig.2 shows the 15-level asymmetrical MLI, this topology used three voltage sources V_1, V_2, V_3 of values V_{dc}, $2V_{dc}$, $4V_{dc}$ and twelve switches, consists of IGBT with antiparallel Diode, which is less in comparison of conventional topologies. This topology can be used with both, symmetrical and asymmetrical type of sources. For symmetrical source configuration required, N equal voltage source, $3(N+1)$ switches for generating $(2N+1)$ number of levels in output voltage waveform. Advantages of asymmetrical source configuration over the symmetrical source configuration are that it requires less number of switches and driver circuits for same level of output voltage waveform. For 15-level asymmetrical MLI operating switching scheme is shown in Table I. There is short circuiting takes place across the voltage sources when the combination of switches (S_2, S_3), (S_5, S_6), and (S_7, S_8) are ON respectively.

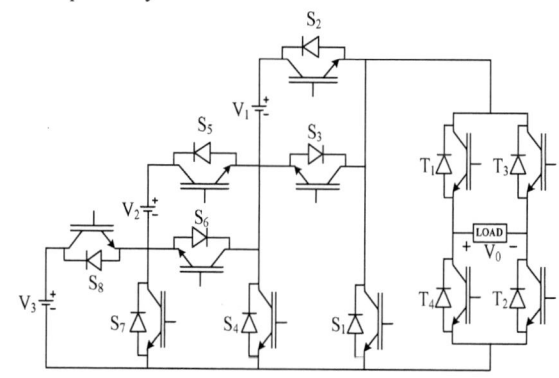

Fig.2. Proposed 15-Level Multilevel Inverter.

III. OPERATING MODES OF PROPOSED 15-LEVEL MLI

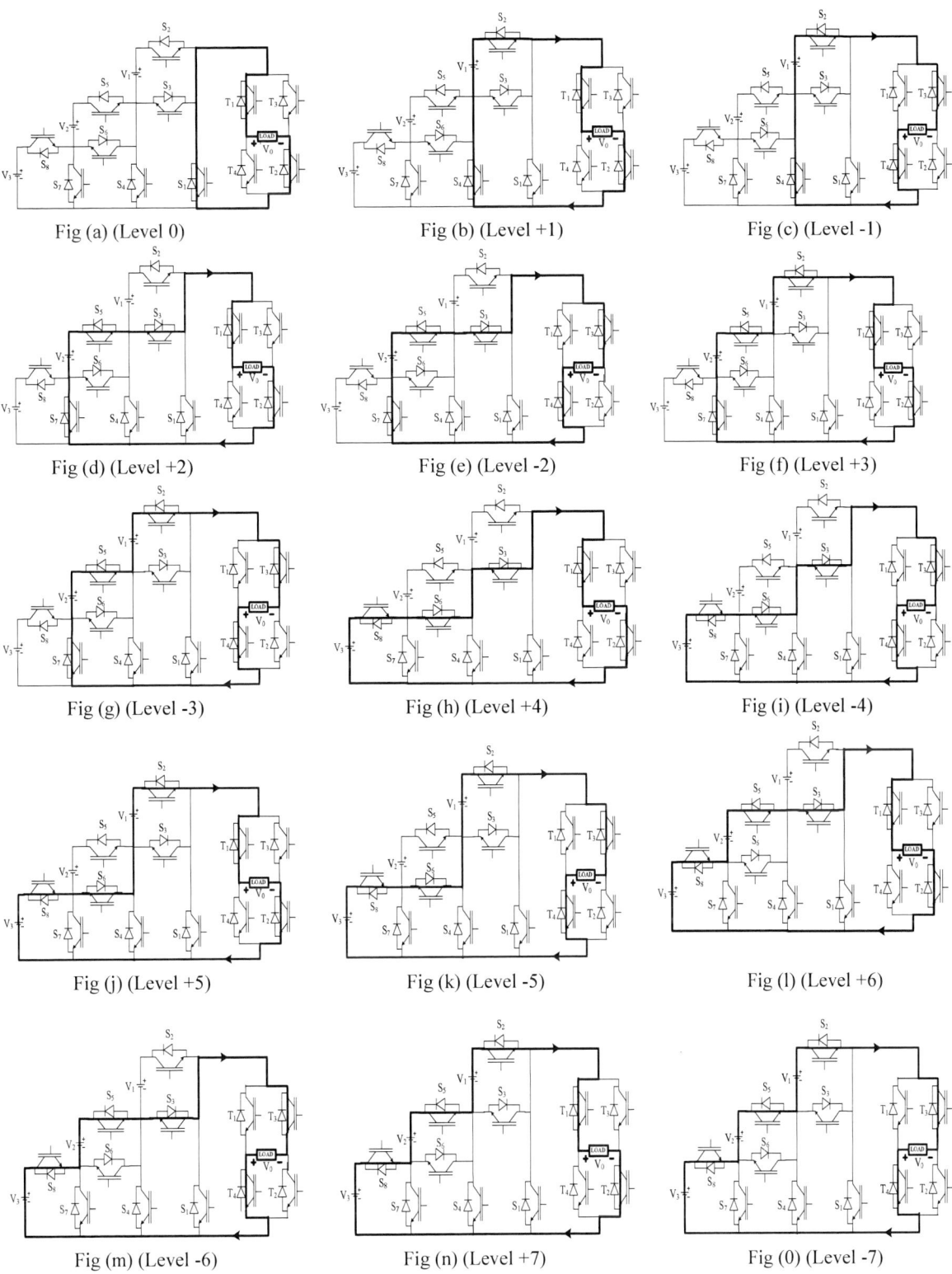

Fig. 3. Fig(a) to Fig(o) are the Different Operating Modes of Proposed 15-Level MLI.

IV. MULTICARRIER PWM TECHNIQUES

In multicarrier PWM techniques reference signal is compared with carrier signal. For M level inverter (M-1) carrier signals are occupied. In this paper sinusoidal reference signal of frequency 50 Hz are compared with carrier signal of higher frequency 2000 Hz. When reference signal is greater than carrier signal, the results in output is one otherwise zero. Here different PWM techniques are explained [9].

- Phase Disposition Pulse Width Modulation (PDPWM) Technique:

   In PDPWM technique has all carrier signals in same phase. For 15-level inverter the (15-1) =14 carrier signals required. All carrier signals are compared with reference signal of 50 Hz.

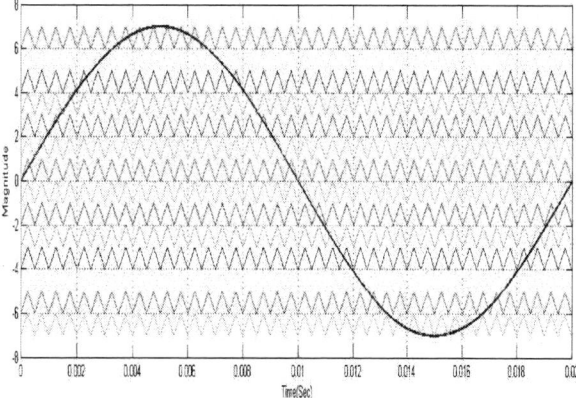

Fig.4. PDPWM Technique with modulation index (Ma=1)

- Phase Opposition Disposition Pulse Width Modulation (PODPWM) Technique:

   In the PODPWM technique, the career signal above zero reference are in same phase, and the carrier signal below zero reference are 180° out of phase with carrier signal above zero reference.

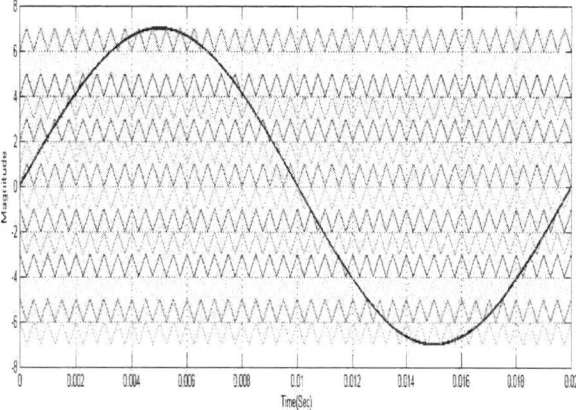

Fig.5. PODPWM Technique with modulation index (Ma=1)

- Alternate Phase Opposition Disposition Pulse Width Modulation (APODPWM) Technique:

   In the APODPWM technique all (M-1) carrier signal consist of same frequency and amplitude but 180° out of phase with its nearby carrier signals.

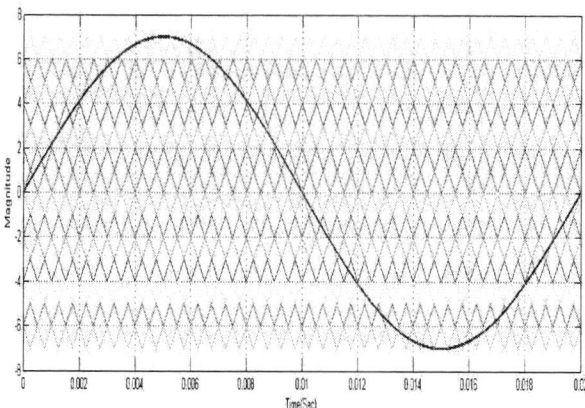

Fig.6. APODPWM Technique with modulation index (Ma=1)

- Inverted Sine Carrier Pulse Width Modulation (ISCPWM) Technique:

   In ISCPWM technique reference signal is taken as sine wave and carrier signals are the inverted sine wave of a high constant frequency of all signals. Here reference signal is compared with carrier signal at different modulation index.

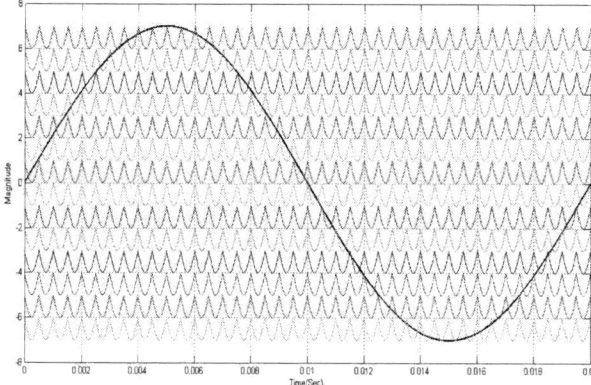

Fig.7. ISCPWM Technique with modulation index (Ma=1)

- Variable Frequency Inverted Sine Carrier Pulse Width Modulation (VFISCPWM) Technique:

   In VFISCPWM technique sine wave of 50 Hz is taken as reference signal and inverted sine wave of variable frequency is taken as carrier signal.

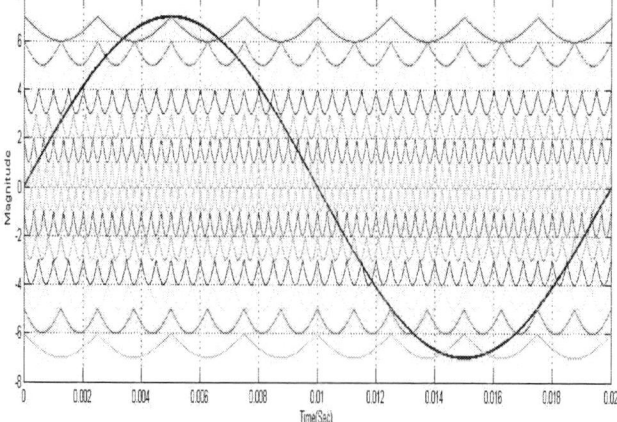

Fig.8. VFISCPWM Technique with modulation index (Ma=1)

V. SIMULATION RESULTS

The performance analysis of 15-level asymmetrical multilevel inverter is ascertained by MATLAB/SIMULINK R2013a software. In this topology three dc voltage source of values V_1=5V, V_2=10V, V_3=20V and switching frequency of carrier is 2000 Hz, frequency of reference signal is 50 Hz, Load resistance R=5Ω is taken. This paper presents five pulse width modulation (PWM) techniques to control the proposed multilevel inverter. Here some of the results are shown:-

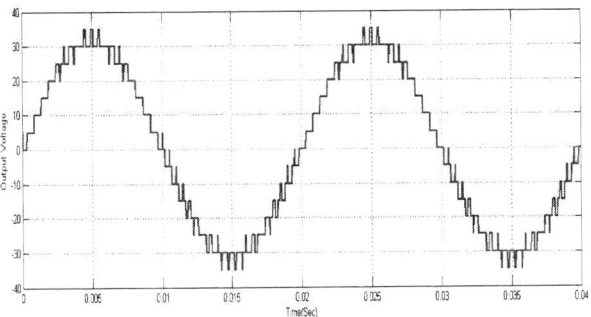

Fig.9. Output Voltage of Proposed 15-Level Multilevel Inverter with modulation index (Ma=1) and (Mf=40).

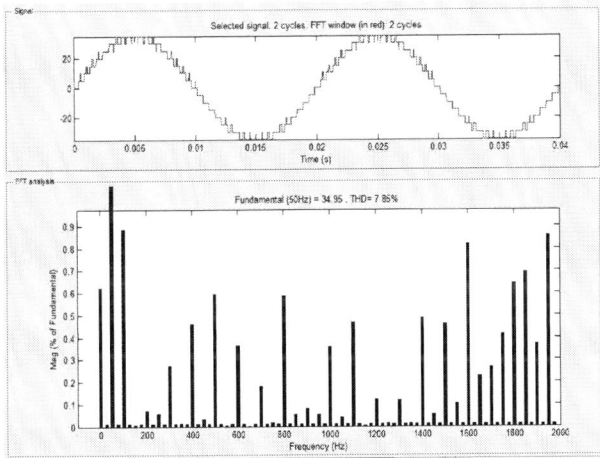

Fig.10. THD of 15-Level MLI with PDPWM technique (Ma=1).

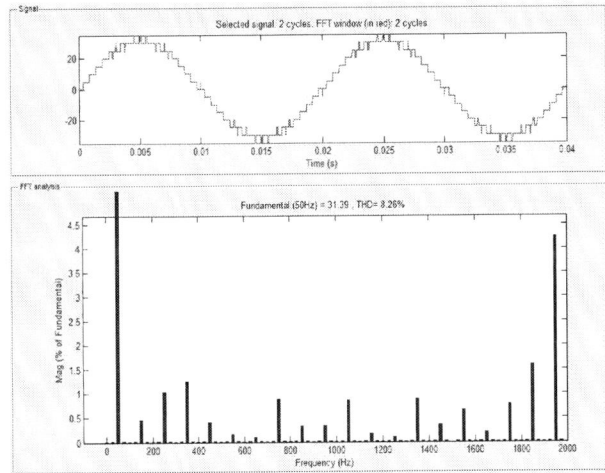

Fig.11. THD of 15-Level MLI with PODPWM technique (Ma=0.9).

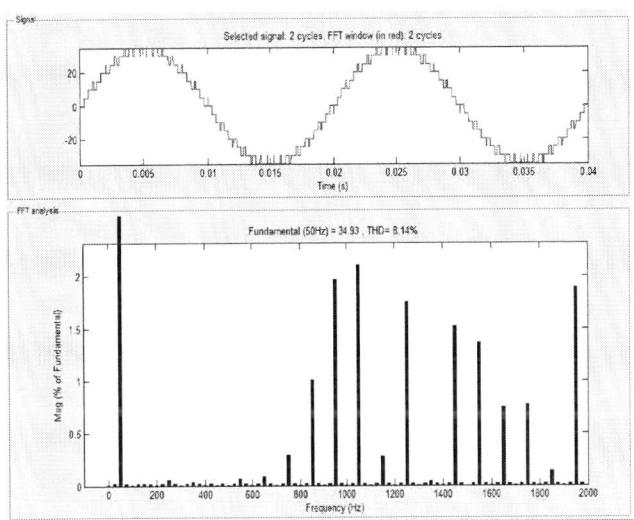

Fig.12. THD of 15-level MLI with APODPWM technique (Ma=1).

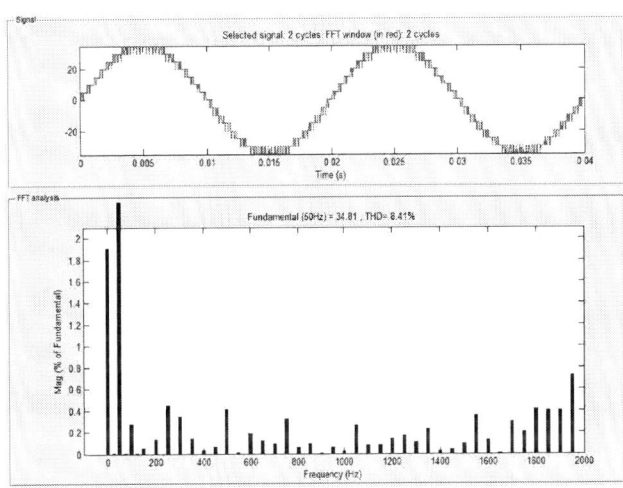

Fig.13. THD of 15-Level MLI with ISCPWM technique (Ma=1).

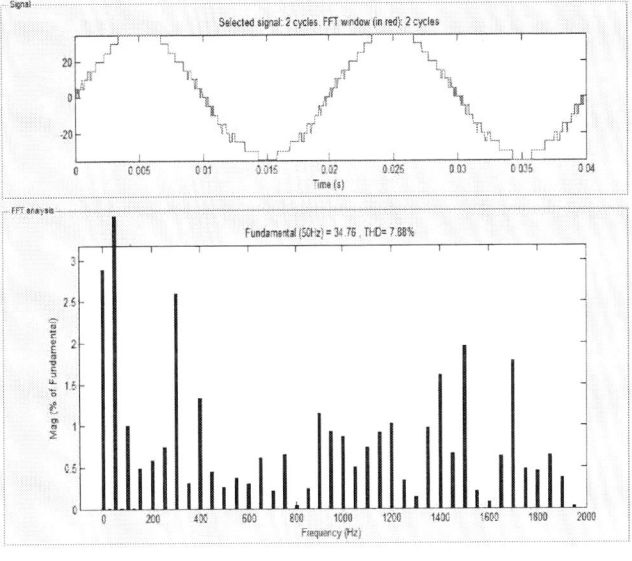

Fig.14. THD of 15-Level MLI with VFISCPWM technique (Ma=1).

TABLE II.   COMPARISON OF THD FOR DIFFERENT PWM TECHNIQUES
WITH DIFFERENT MODULATION INDEX

MODULATION INDEX	MODULATION TECHNIQUES				
	PD	POD	APOD	ISC	VFISC
1	7.85	8.62	8.14	8.41	7.88
0.95	8.57	8.53	8.39	9.12	8.90
0.9	8.81	8.26	9.01	9.46	9.10

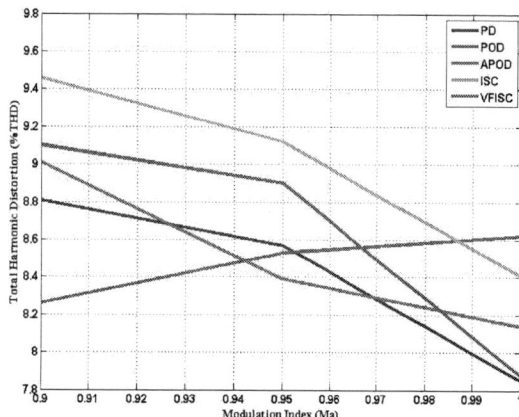

Fig.15. %THD at different modulation index and techniques

VI.   CONCLUSION

In this paper, a new 15-level asymmetrical multilevel inverter topology has been presented. It was shown that the proposed topology need less number of IGBTs and gate driver circuit as compared to conventional topologies. These factors reduce the installation area, circuit size, and cost, improves reliability and efficiency. The simulation results for a 15-level inverter based on the asymmetric source configuration have been presented. It was shown that by applying various PWM Techniques THD of the inverter is varying as shown in Table II; minimum THD is obtaining by Phase Disposition pulse width modulation (PDPWM) Technique which is 7.85% with modulation index 1.

VII.   ACKNOWLEDGEMENT

The authors would like to thanks the Department of Electrical Engineering, Madhav Institute of Technology and Science, Gwalior for guidance and support to carry out this work.

VIII.   REFERENCES

[1] J. Rodriguez, L. Jih-Sheng, P. Fang Zheng, "Multilevel inverters: a survey of topologies, controls, and applications", IEEE Transactions on Industrial Electronics, volume 49, pp. 724-738, 2002.

[2] R.H. Baker, L.H. Bannister, "Electric power converter," U.S. Patent 867 643, 1975.

[3] J. Ebrahimi, E. Babaei, G.B. Gharehpetian, "A new multilevel converter topology with reduced number of power electronics components", IEEE Transactions on Industrial Electronics, volume 59, pp. 655-667, February 2012.

[4] N.A. Rahim, K. Chaniago., J. Selvaraj, "Single-Phase Seven-Level Grid-Connected Inverter for Photovoltaic System," IEEE Trans. on Ind. Electron., Volume.58, Issue.6, pp.2435-2444, 2011.

[5] J. Rodriguez, S. Bernet, B. wu, J. pontt, and S. kouro, "Multilevel voltage-source converter topologies for industrial medium voltage drives", IEEE Transactions on Industrial Electronics, vol.54,issue.6, pp. 2930-2945, December 2007.

[6] Y. Hinago, H. Koizumi, "A single phase multilevel inverter using switched series/parallel DC voltage sources," Energy Conversion Congress and Exposition, 2009. IEEE, vol.57, no.8, pp. 1962-1967, 20-24 September 2009.

[7] Nabe, I. Takahashi and H. Akagi, "A new neutral point clamped PWM inverter", IEEE Transactions on Industrial Applications, vol. 1A-17, pp. 518-523, September 1981.

[8] E. Babaei, "Optimal topologies for cascade sub-multilevel converters", J. Power Electronics, vol. 10, no.10, pp.251-261, 2010.

[9] J. Holtz, "Pulse width modulation for electronic power conversion", proc. IEEE, vol. 82, no. 8, pp. 1194-1214, August 1994.

Generation of 13-Level Output Voltage from Single-Phase Multilevel Inverter Consisting of Cascaded Three H-Bridge Units

Tapan Kumar Chakraborty
Department of Electrical and Electronic Engineering
University of Asia Pacific
Dhaka, Bangladesh
tapan9550@gmail.com

Ashique Anan
Department of Electrical and Electronic Engineering
University of Asia Pacific
Dhaka, Bangladesh
abirashique@gmail.com

Sakhawat Hossen Rakib
Department of Electrical and Electronic Engineering
University of Asia Pacific
Dhaka, Bangladesh
rakib.h.eee@uap-bd.edu

Md. Imran Prodhan
Department of Electrical and Electronic Engineering
University of Asia Pacific
Dhaka, Bangladesh
imranpg.prodhan@gmail.com

Md. Mostofa Kamal
Department of Electrical and Electronic Engineering
University of Asia Pacific
Dhaka, Bangladesh
mkshobuj10@gmail.com

Md. Mahabubunnabi
Department of Electrical and Electronic Engineering
University of Asia Pacific
Dhaka, Bangladesh
kingraxapon@gmail.com

Abstract— This paper addresses the simulation and practical implementation of single-phase multilevel inverters consisting of three H-bridge units to generate maximum 13-level output voltage. The proposed system consists of a microcontroller, three separate input dc sources, isolating circuit and three cascaded H-bridge MOSFET-based voltage source inverters. The proper gate signals for MOSFETs of the three H-bridge inverters are generated by using ATmega 2560 microcontroller-based Arduino board. The complexity of generating gate drive signals for higher levels of inverter output voltage can be reduced dramatically using microcontroller. Different-level output voltage has been obtained from simulation and experimental works using three number of cascaded H-bridge inverters which require less power switching devices. The simulation and experimental results show that total harmonic distortion is reduced with increasing number of levels at the output voltage of the multilevel inverter. It is found that the experimental results are almost similar compared to the simulated results.

Keywords— total harmonics distortion, microcontroller, multilevel inverter, cascaded H-bridge units

I. INTRODUCTION

A multilevel inverter is a power electronics circuit which is capable of providing desired alternating voltage level at the output with variable voltage and frequency from single dc voltage or multiple lower level dc voltages as input [1].

Multilevel inverters have been proposed as the best choice in several medium and high voltage applications. In recent years, multilevel inverters have been attracting the attention of Electrical Engineers due to its high power quality, high voltage capability, low switching losses and low Electro-Magnetic Interference [2-5].

A conventional inverter has two possible voltage levels. Multilevel inverter can switch their outputs between many voltage or current levels and have multiple voltage or current sources. A multilevel inverter can be implemented in different topologies [6-10] along with its own advantages and limitations. The simplest technique adopted is parallel or series connection of conventional inverters to form the multilevel inverter. More complex structures involve, inserting inverter within inverter to form a multilevel inverter.

The main function of a multilevel inverter is to produce a desired ac voltage level from several dc voltage sources. This dc voltage source may or may not be equal to one another. The ac voltage produced from this dc voltage appears to be sinusoidal. One disadvantage of using multilevel inverter is to approximate sinusoidal waveforms concerned with harmonics. The staircase waveform produced by a multilevel inverter contains sharp transitions. The harmonics generated on the ac side greatly influence the power quality of the control system. The multilevel inverter improves the ac power quality by performing the power conversion in small voltage steps leading to lower

978-1-5386-6624-1/18 $31.00 © 2018 IEEE

harmonics. For this reason, researchers are doing considerable work on multilevel inverter in recent years.

A large number of research works [6-23] have been undertaken and various classifications of multilevel inverters have been made with different topologies, such as, diode clamped, flying capacitor, cascaded H-bridge, hybrid H-bridge and new hybrid H-bridge multilevel inverters. Out of these topologies, cascaded H-bridge multilevel inverter has been attracted much attention of the Electrical Engineers due to its simplicity. The topology created by serially connected H-bridge units with separate dc sources is called as cascaded H-bridge multilevel inverter. In this type of configuration, voltage on each dc source is same value [1,2]. However, the multilevel inverter has some drawbacks, such as, increased power switching devices for increasing the voltage levels in the output, power semiconductor switching losses, circuit complexity and economic aspects. Also, the overall reliability and efficiency of the power converter are reduced due to increasing number of switching devices.

The main objective of the present works is to design and implement cascaded H-bridge multilevel inverters with minimum number of switching devices in order to minimize the overall energy loss and total harmonic distortion.

II. DESCRIPTION OF PROPOSED MULTILEVEL INVERTER

The block diagram of the multilevel inverter circuit used in the present study is shown in Fig. 1. Three single-phase H-bridge inverter units are connected in series to form a cascaded multilevel inverter. The general function of this multilevel inverter is to obtain a desired voltage at the output from three separate dc sources, 6 v, 12 v and 18 v, which may be obtained from batteries, fuel cells, or solar cells. The multilevel inverter consists of three H-bridge inverters connected in series. One separate dc source is connected to an H-bridge inverter.

The ac terminal voltages of different level inverters are connected in series. The cascaded inverter does not require any voltage-clamping diodes or voltage-balancing capacitors like the diode-clamp or flying-capacitors inverter [1]. Each H-bridge consists of four IRF540 MOSFET switches. In H-bridge 1, MOSFET switches are denoted as M1, M2, M3 and M4. The upper two switches are indicated by M1 and M2, and the lower switches by M3 and M4. In H-bridge 2, MOSFET switches are denoted as M5, M6, M7 and M8. The upper two switches are indicated by M5 and M6, and the lower switches by M7 and M8. In H-bridge 3, MOSFET switches are indicated by M9, M10, M11 and M12. The upper two switches are indicated by M9 and M10, and the lower switches by M11 and M12. ATmega2560 microcontroller based Arduino has been used in the proposed work. It has 54 digital input/output pins, 16 analog inputs, a 16 MHz crystal oscillator, a USB connection, an on-board programmer chip, an ICSP header, and a reset button.

The microcontroller has been used to generate proper gate signals for MOSFETs. An optocoupler is connected at the input of each switch for isolating gate signal from the main power circuit. Generated pulses from microcontroller are applied to optocoupler through a resistor. The phase output voltage is synthesized by the sum of three inverter outputs as shown in Fig. 1. Each inverter level can generate three different voltage outputs, $+V_{dc}$, 0, and $-V_{dc}$, by connecting the dc source to the ac output side using different combinations of the four switches, M1, M2, M3 and M4. Switching off all switches yields zero output voltage. Similarly, the ac output voltage at each level can be obtained in the same manner.

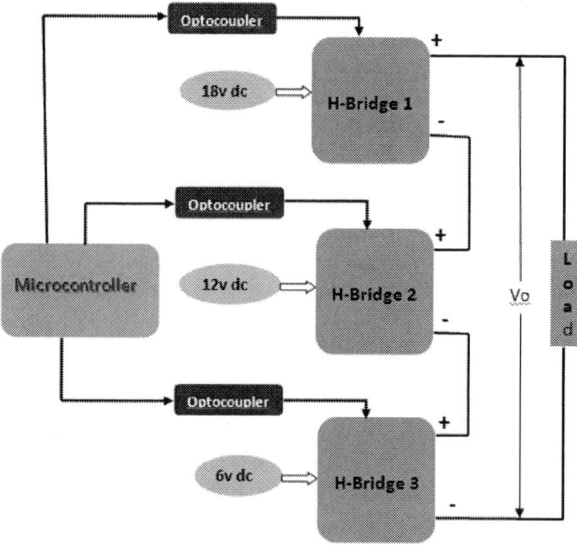

Fig. 1. Block diagram of the proposed cascaded H-bridge multilevel inverter

III. RESULTS AND DISCUSSION

A. Simulation Results

All simulation data were collected using Multisim simulation software. Figures 2-4 shows the simulation results obtained for 9-level, 11-level and 13-level output voltage by applying gate signals to the appropriate MOSFETs for on/off condition. Data collected from Multisim software were used in MATLAB-Simulink to calculate the total harmonic distortion.

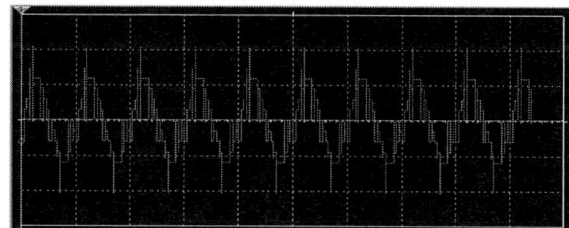

Fig. 2 Simulated 9-Level output voltage of multilevel inverter

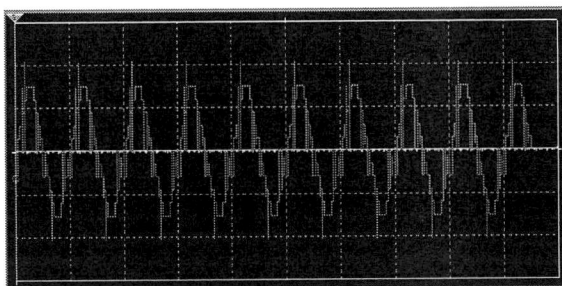

Fig. 3 Simulated 11-Level output voltage of multilevel inverter

Total harmonic distortion of the output voltage has been calculated from the simulation results for 9-level, 11-level and 13-level cascaded H-bridge multilevel inverters and they are found to be 22.68%, 15.68% and 11.48%, respectively.

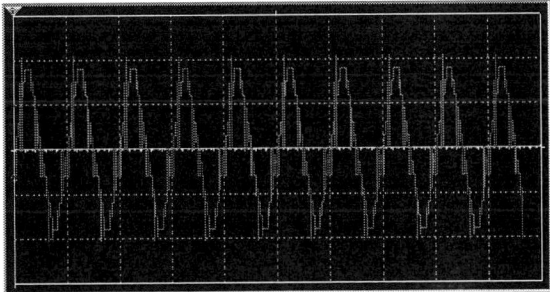

Fig. 4 Simulated 13-Level output voltage of multilevel inverter

B. Experimental Results

The experimental setup for studying the cascaded multilevel inverter is shown in Fig. 5. From the

Fig. 5. Experimental setup for multilevel inverter.

experimental study, the output voltage across the load for 9-level, 11-level and 13-level cascaded H-bridge multilevel inverters are shown in Figs. 6-8.

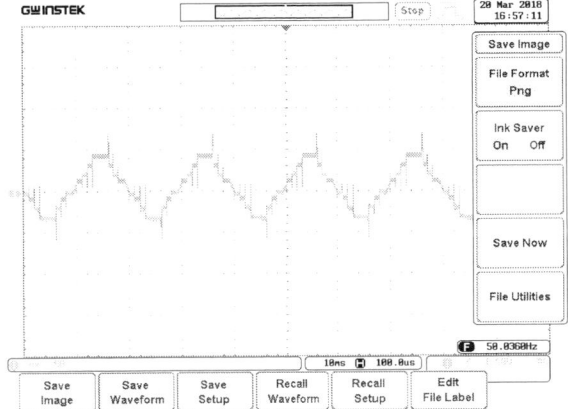

Fig. 6. Experimental 9-level output voltage across the load.

It is found that there are switching spikes in the output voltages as shown in the Figures. This switching spike can be minimized by using proper power switching devices and suitable gate drive circuits.

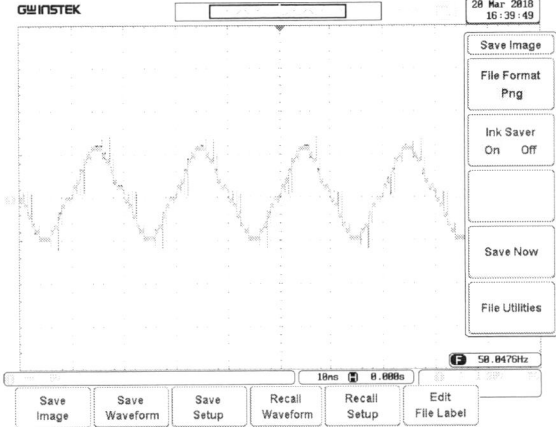

Fig. 7. Experimental 11-level output voltage across the load

From experimental results, total harmonic distortion in the output voltage has been calculated for 9-level, 11-level and 13-level cascaded H-bridge multilevel inverters and they are found to be 21.85%, 20.57% and 17.06%, respectively.

978-1-5386-6624-1/18 $31.00 © 2018 IEEE

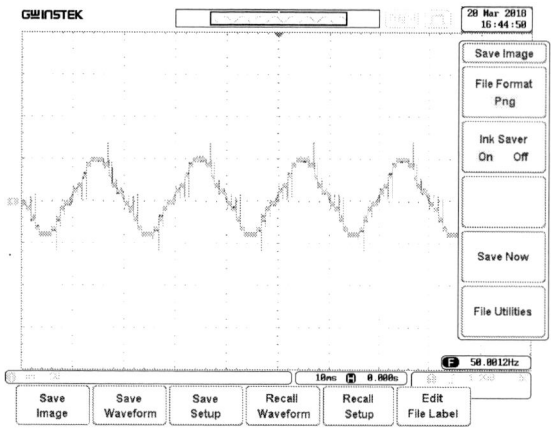

Fig. 8. Experimental 13-level output voltage across the load.

ACKNOWLEDGEMENT

The authors are grateful to the Institute of Energy, Environment, Research and Development of the University of Asia Pacific, Dhaka, Bangladesh for partial funding of this research works.

REFERENCES

[1] M. H. Rashid, " Power Electronics: Circuits, Devices and Applications", 3rd Edition, Pearson Education Inc, India, 2015.

[2] J. P. R. A. Méllo, C. B. Jacobina, "Single-Phase Converter With Shared Leg and Generalizations", IEEE Transactions on Power Electronics, vol. 33, pp. 4882-4893, 2018.

[3] R. S. Alishah, S. H. Hosseini, E. Babaei, M. Sabahi, "Optimization Assessment of a New Extended Multilevel Converter Topology", IEEE Transactions on Industrial Electronics, vol. 64, pp. 4530-4538, 2017.

[4] M. Aleenejad, H. Mahmoudi, R. Ahmadi, H. Iman-Eini, "A New High-Switching-Frequency Modulation Technique to Improve the DC-Link Voltage Utilization in Multilevel Converters", IEEE Transactions on Industrial Electronics, vol. 64, pp. 1807-1817, 2017.

[5] E. Babaei, S. Alilu, S. Laali, "A new general topology for cascaded multilevel inverters with reduced number of components based on developed H-bridge", IEEE Trans. Ind. Electron., vol. 61, no. 8, pp. 3932-3939, 2014

[6] E. Babaei, S. Laali, Z. Bayat, "A Single-Phase Cascaded Multilevel Inverter Based on a New Basic Unit With Reduced Number of Power Switches", IEEE Transactions on Industrial Electronics, vol. 62, no. 2, 2015.

[7] S. R. Alishah, D. Nazarpour, S. H. Hosseini, , M. Sabahi, "Reduction of Power Electronic Elements in Multilevel Converters Using a New Cascade tructure," IEEE Trans. Ind. electron., vol. 62, no. 1, pp.256-269, 2015.

[8] M. F. Kangarlu, E. Babaei, "A generalized cascaded multilevel inverter using series connection of sub-multilevel inverters", IEEE Trans. Power Electron., vol. 28, no. 2, pp. 625-636, 2013.

[9] E. Babaei, S. Sheermohammadzadeh, "Hybrid multilevel inverter using switched-capacitor units", IEEE Trans. Ind. Electron., vol. 61, no. 9, pp. 4614-4621, 2014.

[10] K. K. Gupta, S. Jain, "Topology for multilevel inverters to attain maximum number of levels from given DC sources", IET Power Electronics, vol. 5, no. 4, pp. 435-446, 2012.

[11] N. Farokhnia, S. H. Fathi, N. Yousefpoor, M. K. Bakhshizadeh, "Minimizations of total harmonic distortion in a cascaded multilevel inverter by regulating of voltages DC sources", IET Power Electronics., vol. 5, no. 1, pp. 106-114, 2012.

[12] V. Sridhar, S. Umashankar, "A comprehensive review on CHB MLI based PV inverter and feasibility study of CHB MLI based PV-STATCOM", Renewable and Sustainable Energy Reviews, vol. 78, pp. 138, 2017.

[13] R. S. Alishah, S. H. Hosseini, E. Babaei, M. Sabahi, G. B. Gharehpetian, "New High Step-Up Multilevel Converter Topology With Self-Voltage Balancing Ability and Its Optimization Analysis", IEEE Transactions on Industrial Electronics, vol. 64, pp. 7060-7070, 2017.

[14] Q. Xu, F. Ma, A. Luo, Y. Chen, Z. He, "Hierarchical Direct Power Control of Modular Multilevel Converter for Tundish Heating", IEEE Transactions on Industrial Electronics, vol. 63, pp. 7919-7929, 2016.

[15] M. A. Rao, K. Sivakumar, "A Fault-Tolerant Single-Phase Five-Level Inverter for Grid-Independent PV Systems", IEEE Transactions on Industrial Electronics, vol. 62, pp. 7569-7577, 2015.

[16] Bharti, K. Verma, A. Gupta, "A review on switching function of multi level inverter and applications", 2016 7th India International Conference on Power Electronics, pp. 1-5, 2016.

[17] B. P. McGrath, D. G. Holmes, "Multicarrier PWM Strategies for Multilevel Inverters", IEEE transaction on Industrial Electronics, vol. 49, no. 4, 2002.

[18] M. A. Rao, M. Sahoo, K. Sivakumar, "A three phase five-level inverter with fault tolerant and energy balancing capability for photovoltaic applications", 2016 IEEE International Conference on Power Electronics, Drives and Energy Systems, pp. 1-5, 2016.

[19] J. Rodriguez, J. S. Lai, F.Z. Peng, "Multilevel inverters: A survey of topologies controls and applications", IEEE transaction on Industrial Electronics, vol. 49, no. 4, pp. 724-738, Aug. 2002.

[20] Y. Suresh, A. K. Panda, "Investigation on hybrid cascaded multilevel inverter with reduced dc sources", Renewable and Sustainable Energy Reviews, vol. 26, pp. 49-59, 2013.

[21] J. P. R. A. Méllo, C. B. Jacobina, "DVR based on three-phase converter cascaded by transformers with only two pairs of windings", 2017 IEEE Energy Conversion Congress and Exposition, pp. 2192-2199, 2017.

[22] J. Rodriguez, J. S. Lai, F. Z. Peng, "Multilevel Inverters: Survey of Topologies Controls and Applications", IEEE Transactions on Industry Applications, vol. 49, no. 4, pp. 724-738, 2002.

[23] M. F. Kangarlu, E. Babaei, S. Laali, "Symmetric multilevel inverter with reduced components based on non-insulated DC voltage sources", IET Power Electronics, vol. 5, no. 5, pp. 571-581, 2012.

Power Loss Analysis in Multilevel Inverters using Multi-objective Optimization

Deepshikha Singla
Department of Electrical Engineering
YMCAUST
Faridabad, Haryana
deepshikha_16s@yahoo.com

Parsh Ram Sharma
Department of Electrical Engineering
YMCAUST
Faridabad, Haryana
prsharma1966@gmail.com

Abstract—**Power loss is the most significant parameter in power system analysis and its adequate calculation directly effects the economic and technical evaluation. This paper aims to propose a multi-objective optimization algorithm which optimizes dc source magnitudes and switching angles to yield minimum THD in cascaded multilevel inverters. The optimization algorithm uses metaheuristic approach, namely Harmony Search algorithm. The effectiveness of the multi-objective algorithm has been tested with 11-level Cascaded H-Bridge Inverter with optimized DC voltage sources using MATLAB/Simulink. As the main objective of this research paper is to analyze total power loss, calculations of power loss are simplified using approximation of curves from datasheet values and experimental measurements. The simulation results, obtained using multi-objective optimization method, have been compared with basic SPWM, optimal minimization of THD, and it is confirmed that the multilevel inverter fired using multi-objective optimization technique has reduced power loss and minimum THD for a wide operating range of multilevel inverter.**

Index Terms—**Multilevel inverter, Conduction Loss, Harmony Search Algorithm, Pulse Width Modulation, Total Harmonic Distortion, Modulation Index, Switching Loss.**

I. INTRODUCTION

The utilization of multilevel inverters at medium voltage and high power level has gained significance during previous years. Distinct combinations of power semiconductor switches can help in obtaining various different topologies of multilevel inverters for different applications [1], [2], [3] . Various literature works report efficient utilization of different topologies in various applications. However, out of three basic configurations, the cascaded H-bridge configuration has attracted the most reviewers due to its impressive features like modular structure, easy control and operation, adaptive to various modulation techniques [4], [5] . Cascaded H-Bridge Multilevel Inverter is basically a series/cascade combination of a number of H-bridge, that synthesize desired AC output voltage from several DC voltage connected at input of each H-bridge. As, each unit of H-bridge is supplied with its own isolated DC supply, the current in each unit and further in each power semiconductor switch of a particular unit is different from load current or source current. Thus, it becomes crucial to examine the behavior of power semiconductor switches and investigate the power losses.

Power loss is the most significant parameter in power system analysis and its adequate calculation directly effects the economic and technical evaluation [6], [7] . The power losses in a converter circuit comprises of conduction loss, switching loss, snubber loss, and off-state loss. But, as during the off

state of the device, leakage current is negligibly small, so off state power loss can also be neglected. Also, snubber losses are negligible in IGBT's. Therefore, only switching losses and conduction losses are necessary to be considered [8], [9] . It is a quite challenging to perform the power loss analysis in multilevel inverters. Equally important is to control the power quality while adopting the modulation method for minimiz- ing the power loss. Many reviewers have proposed SPWM based methods minimize harmonics and analyzed the total power loss in multilevel inverter. But as SPWM involves high switching frequency, thus power losses are high. Therefore, it is essential to optimize switching frequency for power loss reduction while minimizing THD [10] . Some methods in- volving optimization of switching angles for providing gating pulse to different switches of multilevel inverter are proposed which help in reducing the power losses to a greater extent, as the switching frequency is reduced significantly. Optimizing switching angles require use of some optimization algorithm, like Particle Swarming Optimization (PSO) [11], [12], [13] , Genetic Algorithm (GA) [14], [15] , BAT Inspired Algorithm, Evolution Algorithm, etc.

This research work presents a multi-objective optimization method which aims to minimize THD as well as optimize dc source voltages to give desired fundamental output voltage with better efficiency. Also, the optimized values of switching angles obtained from the proposed technique have values such that the power losses in the circuit are much more reduced than with single objective of minimizing THD. Optimization technique, Harmony Search is chosen as it is easy to implement and can be applied to higher order optimization problems. It has inherent feature of search efficiency and potential to avoid getting stuck at local minimum points. The algorithm efficiently reduces harmonic distortion and power loss analysis proves that the proposed method helps in significant reduction of power loss. Comparative analysis of power loss in MLI us- ing PWM, OTHD and multi-objective optimization technique has been performed. Simulation results proved that multi-objective optimization technique has minimum power loss as well as the THD has been minimized with the desired value of fundamental voltage.

II. LITERATURE REVIEW

Various techniques have been proposed in literature to evaluate the power loss in multilevel inverters. Some methods perform calculation along with simulation of circuit while some require intense mathematical analysis and evaluation. The choice of modulation technique affects the harmonics and power loss in multilevel inverters. Like, in MLIs controlled using the traditional modulation method, carrier based pulse width modulation (PWM), switching losses are most significant as they directly depends on switching frequency. But, reducing switching frequency is not possible without compromising the harmonic content of the voltage waveform, thus demanding the use of harmonic filters [16] . Space-vector modulation that operates with high switching frequency is also applied to multilevel inverter. They are efficient in reducing harmonics but cause high switching loss. A hybrid scheme [17] presents combination of carrier based space vector modulation and fundamental frequency modulation, to minimize switching losses with an improved harmonic performance. Compared to conventional CBSVM, it reduces 27% of power loss. In multilevel inverters operating at low switching frequency, conduction loss are a major part of the total power loss. But, as all semiconductor switches are gated with individual switching pulses, the amount of current flowing through them varies and hence the conduction loss in different switches also varies [18] . Also, estimation of switching and conduction times for each semiconductor switch is a slow and tedious process. Keeping numerous calculations in mind, some reviewers have suggested improving the power loss equations to simplify the task of power loss evaluation [19] .

Selective Harmonic Elimination (SHE) technique gives high output power quality because it uses low switching frequency, and thus gives reduced harmonic distortion. Many heuristic methods have been used to deal with the system of transcendental equations and optimize the switching instants so as to minimize harmonics while reducing the power loss. An optimization method has been proposed [20] using SHE where determination of switching angles is done using the application of Particle Swarming Optimization (PSO). But PSO tends to converge early at mid optimum points and takes time to converge in refined search stage. Another scheme where determination of switching angles is done using Genetic Algorithm (GA) is presented [21] . But the quality of solution deteriorates with increase in the level of inverter while using GA. A new heuristic method, Harmony Search is used in the multi-objective optimization because of its distinguishing features such as high rate of convergence and precision [22] . Also, the algorithm can be used with any level of multi-level inverter.

III. MULTI-OBJECTIVE OPTIMIZATION TECHNIQUE

Practically, MLIs are connected to solar power or photovoltaic panels for getting DC input. And if the magnitude of DC source is decided according to the required fundamental output value, the efficiency of inverter is enhanced. As reviewed in literature, the researchers mainly focus on minimization of harmonics. But here in this paper, an optimization function has been developed with an objective to optimize switching angles as well as the input dc source magnitudes   so as to achieve better power quality and efficiency. Some applications demand extending the operating range of multilevel inverter by varying the modulation index to get variable ac voltage. This method regulates the voltage levels of multilevel inverter by optimizing the dc source voltage magnitudes according to the desired modulation index.

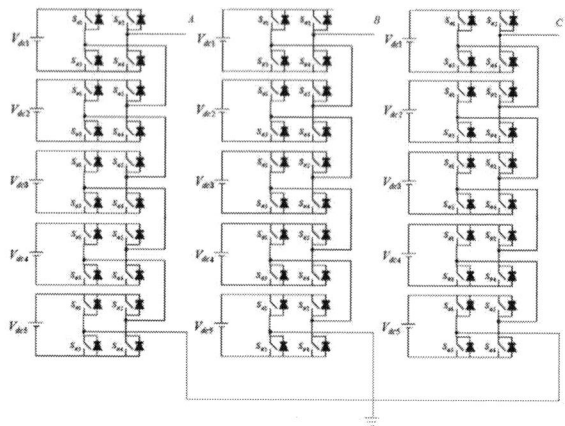

Fig. 1.   Three-Phase11- level Cascaded H Bridge Inverter

Now, for an 11-level Inverter shown in Figure 1 , switching angles α_1, α_2, α_3, α_4, α_5 and V_{dc1}, V_{dc2}, V_{dc3}, V_{dc4}, V_{dc5} are obtained for firing each H-bridge unit at desired switching instant with optimized output. As, the phase output voltage of multilevel inverter can be expressed in terms of Fourier series expansion:

$$V(t) = \sum_{n=1,3,5}^{\infty} \frac{4}{n\pi} V_n \sin n\alpha_n \qquad (1)$$

where Vn can be expressed as:

$$V_n = \frac{4(V_{dc1}\cos n\alpha_1 + V_{dc2}\cos n\alpha_2 + V_{dc3}\cos n\alpha_3 \, ...}{n\pi}$$

$$... + V_{dc4}\cos n\alpha_4 + V_{dc5}\cos n\alpha_5) \qquad (2)$$

where V_{dc1}, V_{dc2}, V_{dc3}, V_{dc4}, V_{dc5} are the output voltages of each H-bridge inverter units.

Since, we are considering line voltage and triplen harmonics are eliminated in line voltage. Thus, Total Harmonic Distortion of line voltage can be expressed as:

$$THD = \frac{\sqrt{\sum_{n=5,7,11,13,...}^{\infty} V_n^2}}{V_1} * 100 \qquad (3)$$

The fundamental component can be written as:

$$V_1 = \frac{4(V_{dc1}\cos \alpha_1 + V_{dc2}\cos \alpha_2 + V_{dc3}\cos \alpha_3 \, ...}{n\pi}$$

$$... + V_{dc4}\cos \alpha_4 + V_{dc5}\cos \alpha_5) \qquad (4)$$

Now, the multi-objective function is developed and coded in MATLAB to give the desired results in form of switching angles to fire the switches of multilevel inverter.

$$F(\alpha, V_{dc}) = W_1 * THD + W_2 * (V_1 - Desired) \qquad (5)$$

The condition for perfect waveform is that switching angles should be within zero and $\pi/2$ ($0 < \alpha_1 < \alpha_2 < \alpha_3 < \alpha_4 < \alpha_5 < \pi/2$).

Now, the Harmony Search Algorithm is implemented on

the developed objective function to obtain best results. It employs two probabilities Harmonic Memory Considering Rate (HMCR) and Pitch Adjustment Rate (PAR) to update the solution till requirement is fulfilled. The technique proves to be successful in avoiding the output from getting stuck in some neighboring values, and thus results in best optimum solution, also known as global best solution.

IV. CALCULATION OF POWER LOSS

During operation of any power circuit which involves switching of power components, there are mainly four types of power losses, namely, conduction loss, switching loss, gate loss and off-state loss. The Gate loss and Off-state loss are minimal and normally neglected. Also, considering the total of the conduction and switching loss gives a good estimation of the total power loss in any power circuit. Thus, only switching loss and conduction loss are investigated for power loss analysis. Loss to be calculated can be classified as given in Figure 2.

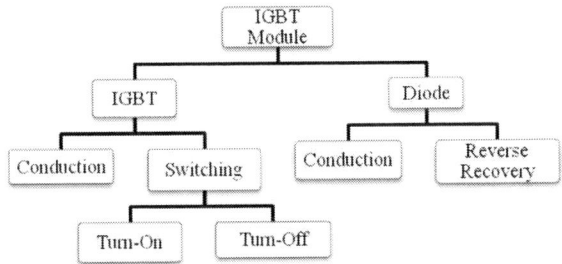

Fig. 2. Classification of Losses in IGBT Module

A. Conduction loss

The losses that appear during on-state or conduction of a device are termed as Conduction loss. Total conduction loss in the multilevel inverter is sum of real time loss of every component integrated over conduction period. So, the power loss can be represented by product of on-state saturation voltage V_{on} and on-state current I_{on}.

$$P_{Conduction} = P_{Cond(IGBT)} + P_{Cond(Diode)}$$

$$P_{Conduction} = (V_{CE} \times I_C) + (V_D \times I_D) \quad (6)$$

Here, V_{CE} and V_D are functions of current flowing through the IGBT and diode. And I_C and I_D are taken as mean values of current. Now the value of V_{CE} and V_D can obtained from the V-I characteristic curve of IGBT and Diode.

Now, by approximation of curves,

$$V_{CE} = -2 * 10^{-7} I_C^2 + 0.0018 I_C + 0.9661$$

$$V_D = -1 * 10^{-7} I_D^2 + 0.0012 I_D + 0.7796 \quad (7)$$

Conduction loss is also termed as on-state loss or steady state loss. They are independent of the switching frequency but depend on the duty cycle.

B. Switching loss

Switching loss appeared because the transitions from off-state to on-state and on-state to off-state do not occur immediately. During this transition, both current and voltage across the device are considerably high leading to high instantaneous power loss. These losses typically contribute a significant amount to the total system losses. As diode is considered to be an ideal switch, its turn-on loss is negligible. So, switching loss is total of IGBT turn-on loss, IGBT turn-off loss and Diode reverse recovery loss.

These losses can be computed from the switching energy equations which are function of switch current.

$$P_{Switching} = P_{Turn-ON} + P_{Turn-OFF} + P_{Rec.Diode}$$

$$P_{Switching} = (E_{Turn-ON} + E_{Turn-OFF}) * f_{sw(IGBT)} +$$

$$(E_{Rec.Diode}) * f_{sw(Diode)} \quad (8)$$

Now, by approximation of curves,

$$E_{Turn-ON} = 8 * 10^{-7} I_C^2 - 0.0023 I_C + 4.016$$

$$E_{Turn-OFF} = 3 * 10^{-7} I_C^2 - 0.0011 I_C + 3.1584$$

$$E_{Rec.Diode} = 7 * 10^{-7} I_D^2 - 0.0039 I_D + 6.6546 \quad (9)$$

Switching losses can be easily estimated using these equa- tions. These losses depend on the dc link voltage, load current, junction temperature and switching frequency. If the switching frequency is higher, then the losses will be higher.

V. SIMULATION AND RESULTS

For power loss investigation, an 11-level three-phase multi-level inverter in cascaded H-bridge configuration is modeled in MATLAB/Simulink environment. The H-bridges of multilevel inverter are supplied so as to give output of 430V, 10A when connected with a balanced load of 30Ω, 50mH as shown in Figure 3 .

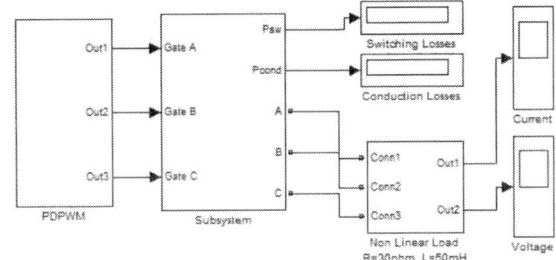

Fig. 3. MATLAB/Simulink of 11-level Cascaded H-Bridge Inverter

To evaluate the effectiveness of multi-objective optimization technique various simulations are carried out in MATLAB/Simulink. Power Loss Calculation and THD Analysis of 11-level Cascaded H-Bridge Inverter are done over a modulation range of 0.4 to 1.2 with an interval of 0.1 using SPWM and OTHD method. And the results are shown in Tables 1 and 2 . From the results, it is confirmed that switching losses are reduced by 54% approximately by using OTHD. It

978-1-5386-6624-1/18 $31.00 © 2018 IEEE

can also be observed that with SPWM, the desired fundamental output equals the desired voltage and the THD decreases with increase in modulation index but content of harmonics is not within acceptable range. However, using OTHD method decreases THD but fundamental is not equal to desired value. Power Losses have reduced to nearly half of the conventional technique. Also, conduction losses contribute only around 3-8% of total loss.

TABLE I: PERFORMANCE ANALYSIS OF MLI USING SPWM

M	Switching Loss (mW)	Conduction Loss (mW)	Total Loss (mW)	THD (%)	Fundamental Voltage (V)
0.4	2614	70.19	2684.2	17.79	169.1
0.5	2622	88.45	2710.5	13.38	211.3
0.6	2622	107.5	2729.5	11.26	254.8
0.7	2622	126.2	2748.2	9.09	297.0
0.8	2615	145.8	2760.8	8.34	340.8
0.9	2616	165.4	2781.4	7.61	383.7
1.0	2616	185.2	2801.2	7.01	426.8
1.1	2063	199.7	2262.7	6.96	456.2
1.2	1863	209.0	2072.0	7.06	472.7

TABLE II: PERFORMANCE ANALYSIS OF MLI USING OTHD

M	Switching Loss (mW)	Conduction Loss (mW)	Total Loss (mW)	THD (%)	Fundamental Voltage (V)
0.4	993.8	192.8	1186.6	5.89	442.8
0.5	993.8	192.8	1186.6	5.89	442.8
0.6	993.8	192.8	1186.6	5.89	442.8
0.7	993.8	192.8	1186.6	5.89	442.8
0.8	993.8	192.8	1186.6	5.89	442.8
0.9	993.8	192.8	1186.6	5.89	442.8
1.0	993.8	192.8	1186.6	5.89	442.8
1.1	1001	207.9	1208.9	6.13	471.4
1.2	1001	238.3	1239.3	6.66	519.1

When multi-objective optimization algorithm is executed, the optimized value of switching angles and magnitude of DC voltage input to be given to each H-bridge are obtained and shown in Table 3. Table 4 summarizes the performance of multilevel inverter using multi-objective optimization technique.

The simulation results obtained, using multi-objective optimization technique, are compared with those obtained using traditional PWM method, and optimal minimization of THD method. The graphical comparison is done for Total losses, THD and Fundamental value shown in Figures 4, 5 and 6.

From the graphical analysis, it can be clearly seen that the multi-objective optimization technique gives best results. Comparison shows that the power loss increase with increase in modulation index, and losses are further reduced by 2-10% with multi-objective optimization technique than OTHD

method.

Moreover, desired results can be obtained at any value of modulation index. By varying the value of modulation index, desired value of fundamental output can be achieved with THD in acceptable range.

TABLE III: PERFORMANCE OPTIMIZED SWITCHING ANGLES USING MULTI-OBJECTIVE OPTIMIZATION TECHNIQUE

M	α_1	α_2	α_3	α_4	α_5	V_D C_1	V_D C_2	V_D C_3	V_D C_4	V_D C_5
0.4	12.68	36.28	56.31	62.47	83.98	38.94	36.25	36.05	38.30	37.08
0.5	10.03	34.43	39.70	59.89	81.72	45.73	35.57	44.76	45.86	38.08
0.6	9.06	32.67	40.35	59.28	80.24	57.56	40.36	56.63	43.7	36.41
0.7	5.39	34.54	45.02	74.85	85.04	66.4	71.87	72.08	45.68	43.06
0.8	3.35	13.36	34.03	40.04	82.08	57.6	68.47	60.21	52.23	41.17
0.9	7.30	20.09	33.39	46.61	60.35	79.41	75.26	59.00	35.62	46.44
1.0	6.60	14.42	26.74	39.61	59.61	74.12	68.30	69.90	58.40	52.65
1.1	6.58	14.46	26.43	38.92	59.44	84.43	78.67	72.21	62.91	56.34
1.2	6.83	14.76	25.98	39.68	59.73	84.57	83.38	87.08	69.26	65.25

TABLE IV: PERFORMANCE ANALYSIS OF MLI USING MULTI-OBJECTIVE OPTIMIZATION

M	Switching Loss (mW)	Conduction Loss (mW)	Total Loss (mW)	THD (%)	Fundamental Voltage (V)
0.4	1000	69.7	1069.7	8.43	169.6
0.5	993.7	89.35	1083.05	5.57	214
0.6	993.6	106.5	1100.1	5.42	253.4
0.7	999.6	124.8	1124.4	5.12	298.6
0.8	1001	149.7	1150.7	4.92	342.4
0.9	993.9	167.5	1161.4	4.84	383.6
1.0	974.4	187.8	1162.2	4.35	426.5
1.1	974.3	207.9	1182.2	4.39	470.7
1.2	994.1	225.8	1219.9	4.28	512.2

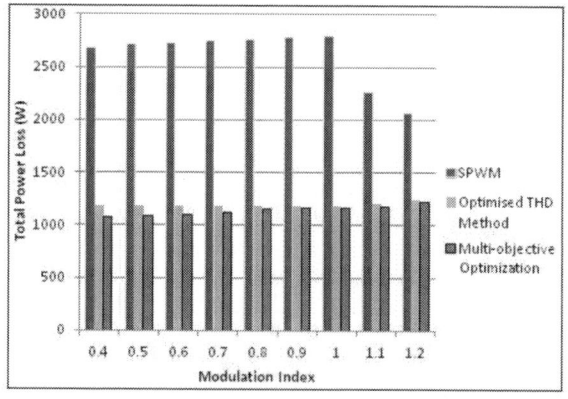

Fig. 4. Comparison of Total Power Loss at different modulation indices

978-1-5386-6624-1/18 $31.00 © 2018 IEEE

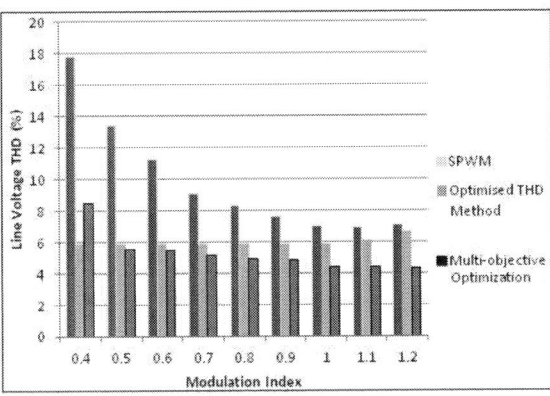

Fig. 5. Comparison of THD in Output Line Voltage at different modulation indices

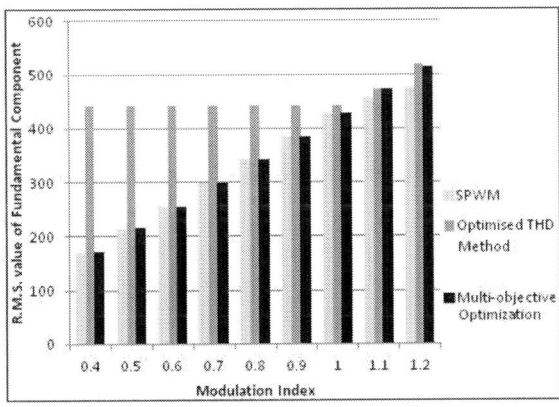

Fig. 6. Comparison of R.M.S. Value of fundamental component of Output Line Voltage at different modulation indices

VI. CONCLUSIONS

The multi-objective optimization efficiently solves for opti- mal switching angles for firing switches of multilevel inverter keeping the fundamental value of output voltage at desired value. Optimization is done using Harmony Search Algorithm, due to its distinguishing features like simplicity, better search efficiency and ability to avoid becoming trapped at local minima. The control signals given to multilevel inverter using optimized value of switching angles are proved to be very effective in reducing power loss and THD with optimized value of dc voltage sources at different value of modulation indexes. It was found that only about 3-8% of total power loss represent conduction loss. There is around 56 % reduction in power loss by using multi-objective optimization.

REFERENCES

[1] J.-S. Lai and F. Z. Peng, "Multilevel converters-a new breed of power converters," *IEEE Transactions on industry applications*, vol. 32, no. 3, pp. 509–517, 1996.

[2] J. Rodriguez, J.-S. Lai, and F. Z. Peng, "Multilevel inverters: a survey of topologies, controls, and applications," *IEEE Transactions on industrial electronics*, vol. 49, no. 4, pp. 724–738, 2002.

[3] P. Fang-Zen and Q. Zhao-ming, "Applications of cascade multilevel inverters," *Journal of Zhejiang University-SCIENCE A*, vol. 4, no. 6, pp. 658–665, 2003.

[4] M. Malinowski, K. Gopakumar, J. Rodriguez, and M. A. Perez, "A survey on cascaded multilevel inverters," *IEEE Transactions on industrial electronics*, vol. 57, no. 7, pp. 2197–2206, 2010.

[5] D. Singla, P. R. Sharma, and N. Hooda, "Power quality improvement using multilevel inverters – a review," *Int. J. of Engg. Sci. & Mgmt.(IJESM)*, vol. 1, no. 1, pp. 64–76, 2011.

[6] A. Farzaneh and J. Nazarzadeh, "Precise loss calculation in cascaded multilevel inverters," in *Computer and Electrical Engineering, 2009. ICCEE'09. Second International Conference on*, vol. 2, 2009, pp. 563–568.

[7] Y. Kashihara and J. ichi Itoh, "Power losses of multilevel converters in terms of the number of the output voltage levels," in *Power Electronics Conference (IPEC-Hiroshima 2014-ECCE-ASIA), 2014 International*, 2014, pp. 1943–1949.

[8] T.-J. Kim, D.-W. Kang, Y.-H. Lee, and D.-S. Hyun, "The analysis of conduction and switching losses in multi-level inverter system," in *Power Electronics Specialists Conference, 2001. PESC. 2001 IEEE 32nd Annual*, vol. 3, 2001, pp. 1363–1368.

[9] M. G. H. Aghdam, S. H. Fathi, and A. Ghasemi, "The analysis of conduction and switching losses in three-phase ohsw multilevel inverter using switching functions," in *Power Electronics and Drives Systems, 2005. PEDS 2005. International Conference on*, vol. 1, 2005, pp. 209–218.

[10] H. Toodeji, "An improved omthd technique for an $ n $-level cascaded multilevel inverter with adjustable dc sources," *Turkish Journal of Electrical Engineering & Computer Sciences*, vol. 25, no. 6, pp. 4841–4853, 2017.

[11] R. Mohanty, S. Rath, and S. P. Mishra, "A comparison study of harmonic elimination in cascade multilevel inverter using particle swarm optimization and genetic algorithm," *vol*, vol. 5, pp. 43–49, 2013.

[12] V. K. Gupta and R. Mahanty, "Optimized switching scheme of cascaded h-bridge multilevel inverter using pso," *International Journal of Electrical Power & Energy Systems*, vol. 64, pp. 699–707, 2015.

[13] D. Singla and P. Sharma, "Performance analysis of harmonic elimination in cascaded h-bridge multilevel inverters using constrained pso algorithm," *International Journal Series in Engineering Science (IJSES)(ISSN: 2455-3328)*, pp. 1–14, 2017.

[14] B. Ozpineci, L. M. Tolbert, and J. N. Chiasson, "Harmonic optimization of multilevel converters using genetic algorithms," in *Power Electronics Specialists Conference, 2004. PESC 04. 2004 IEEE 35th Annual*, vol. 5, 2004, pp. 3911–3916.

[15] A. Salami and B. Bayat, "Total harmonic distortion minimization of multilevel converters using genetic algorithms," *Applied Mathematics*, vol. 4, no. 07, p. 1023, 2013.

[16] P. K. Chaturvedi, S. Jain, P. Agrawal, R. K. Nema, and K. K. Sao, "Switching losses and harmonic investigations in multilevel inverters," *IETE Journal of research*, vol. 54, no. 4, pp. 297–307, 2008.

[17] C. Govindaraju and K. Baskaran, "Power loss minimizing control of cascaded multilevel inverter with efficient hybrid carrier based space vector modulation," *International journal of electrical and computer engineering systems*, vol. 1, no. 1, pp. 45–53, 2010.

[18] M. G. H. Aghdam, S. H. Fathi, and G. B. Gharehpetian, "A novel switching algorithm to balance conduction losses in power semiconductor devices of multi-level cascade inverters," *Electric Power Components and Systems*, vol. 36, no. 12, pp. 1253–1281, 2008.

[19] A. Babaie, B. Karami, and A. Abrishamifar, "Improved equations of switching loss and conduction loss in spwm multilevel inverters," in *Power Electronics and Drive Systems Technologies Conference (PEDSTC), 2016 7th*, 2016, pp. 559–564.

[20] S. T. PARVEEN and P. S. KRISHNA, "Simulation of three phase cascade h-bridge multilevel inverter with grid connected system modeling of switching and conduction losses," *International Journal of Research*, vol. 3, no. 12, pp. 828–833, 2016.

[21] B. Alamri and M. Darwish, "Precise modelling of switching and conduction losses in cascaded h-bridge multilevel inverters," in *Power Engineering Conference (UPEC), 2014 49th International Universities*, 2014, pp. 1–6.

[22] X. Z. Gao, V. Govindasamy, H. Xu, X. Wang, and K. Zenger, "Harmony search method: theory and applications," *Computational intelligence and neuroscience*, vol. 2015, p. 39, 2015.

Protection of Power Electronic Switches in a VSI against Over-Current and Shoot-Through Faults

V. K. Choubey
Electrical Engineering Department
Narula Institute of Technology
Kolkata, India
vkconline437@gmail.com

J. Simlai
Electrical Engineering Department
Narula Institute of Technology
Kolkata, India
jayatrasimlai@gmail.com

T. Ghosh
Electrical Engineering Department
Narula Institute of Technology
Kolkata, India
tarpanghosh003@gmail.com

A. Mitra
Electrical Engineering Department
Narula Institute of Technology
Kolkata, India
arkendu83@gmail.com

Abstract—**Voltage Source Inverter consists of power electronic switches, mostly MOSFET or IGBT operating at high frequency. These switches are costly and prone to various faults which if not prevented within a limited time, will result in their burning out and making the inverter inoperable. Further this would lead to potential halt of inverter based processes. Thus protection of these costly power electronic switches against such faults is of utmost importance. This experimental work investigates and prioritizes the protection of the power electronic switches of an inverter against over-current and shoot-through faults and also ensures automatic recovery of the inverter based system post the fault clearance.**

Keywords—*Inverter; IGBT; Opto-coupler; Over-current; Shoot-through*

I. INTRODUCTION

With recent development in power electronics, inverters are playing an important role in many of industries. There are inverters rated up to hundreds of kW and becoming prominent because of increased usage of DC power from AC mainly in power system control areas.

Inverters require power semiconductor switches and one of the widely used power switches for inverters is Insulated Gate Bipolar Transistors (IGBT), owing to their properties and advantages such as efficiency, versatility and reliability. During turn off of the IGBT, over-voltage may appear across the collector to emitter due to the parasitic inductance in the switching circuit. This may result the damage of the switch because of the over-voltage [1, 2].

As inverters carry heavy workload in industries, the switches within the inverters are prone to various common faults, mostly over-current and shoot-through faults. A typical requirement with regard to this, protection circuit suitable for these switches against these faults is highly essential. Due to an external short-circuit or phase fault, the switches within the inverter carry high current than rated. To protect the IGBT switches with fuse isn't possible. Thus an effective protection scheme of the switches against over-current fault is required. There are various works carried out by the researchers for protecting the device against over-current. Mainly the over-current protection of the switch can be done in two ways – a) by sensing the collector to emitter voltage of the IGBTs [3-5] and b) by inserting current transducers to sense the collector current of the IGBTs [6].

Some researchers proposed other developments to protect IGBT against over-current and short-circuit. [7-9].

The switches within the inverter can operate on very high frequency ranging upto few kHz to several MHz. At higher frequency, it may be possible that there happens a false triggering causing the IGBT switches in the same limb to turn on at same instant, thus causing severe current enter into the switches on that limb and this shoot-through may permanently damage the switches. An effective protection against shoot-through fault is very much vital when the inverter is operated from high switching frequency.

In this experimental work, how the inverter switches can be protected both from over-current shoot-through faults is described.

II. GATE DRIVING METHODS

An inverter is a controlled converter which converts DC supply to an utilizable AC using a suitable gate drive control scheme applied to the power electronic switches as per requirements of output voltage and frequency. A proper gate drive method is required for the optimum operation of the inverter. Also it has to be ensured that the gate driver circuit remains isolated from the power circuit to avoid high voltage puncture of the control circuit from the power circuit.

A. Conventional Gate Driving Method

Conventional method of gate drive circuit is shown in Fig. 1. When the output of the control circuit is low, the output transistor of the opto-coupler is ON. This will force the collector of the transistor at -9V so that the upper NPN transistor of totem-pole will OFF and the gate pulse will become low. Consequently the lower PNP transistor will set to discharge the gate capacitance of the switch.

Similarly, a high output from the control circuit drives the output transistor of the opto-coupler at OFF condition. This will make the collector of the transistor at +15V which will set the upper NPN transistor of the totem-pole circuit at ON position. The gate current flows from the positive supply through upper NPN transistor and the gate resistance and thus the switch at ON condition. Subsequently the lower PNP transistor of the totem-pole circuit will remain OFF to block discharging of the gate capacitance.

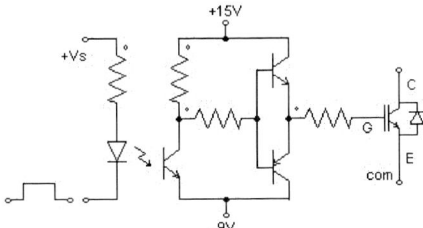

Fig. 1. Conventional gate driver circuit.

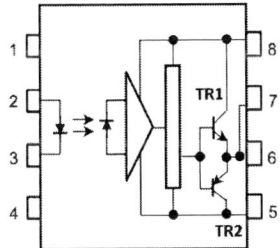

Fig. 2. On-chip gate driver module (TLP250 or equivalent).

Now-a-days, on chip gate driver module (TLP250 or equivalent) is available with totem-pole configuration as shown in Fig. 2. To avoid proper selection and connection of the different components, a single gate driver IC is highly appreciable in present days. It is clear from Fig. 2 that the connection of the gate driver IC with same positive and negative supply is very easier than connecting discrete components.

B. Over Current Protection Scheme

Over current protection of IGBT is a necessary feature that is to be incorporated in the gate drive circuit. Any IGBT has a maximum current rating under which the device operates safely. The IGBT exhibits de-saturation feature beyond a certain collector current level. In this condition, the collector to emitter voltage of the device increases very fast results in exessive power dissipation that may destroy the device. Thus, the rapid increase in collected emitter to voltage must be detected and the gate driver circuit guaranteed to turn off the switch as fast as possible to protect it.

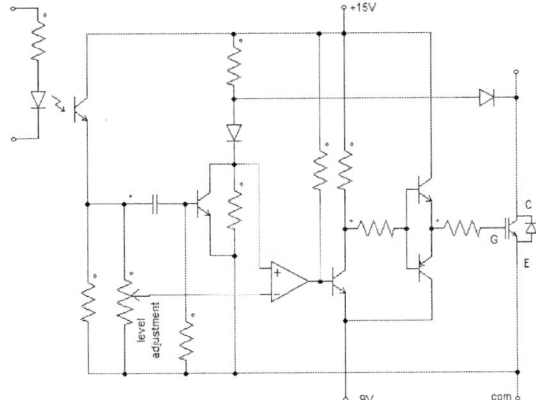

Fig. 3. Over current protection scheme.

A typical over current protection scheme for IGBR is shown in Fig. 3. When the opto-coupler in turned on, the pulse signal momentarily turns on the transistor connected to the non-inverting terminal of the comparator. This brings the non-inverting terminal at zero condition keeping output state of the comparator at low level. As the result, NPN transistor of the totem-pole circuit will be in conduction state and send the positve drive signal to IGBT gate. Under this condition, the collector to emitter voltage of the IGBT becomes low and te same signal is fed back to the non-inverting terminal of the comparator circuit. During steady-state operation of the drive, the transistor connected to the non-inverting terminal turns off and the output of the comparator remains off. When collector to emitter voltage of the IGBT increases, the same is reflected to the non-inverting terminal of the comparator. If this voltage exceeds the preset value of the inverting terminal, the output from the comparator swings to high value, results in cut off the drive signal from the IGBT gate irrespective of the status of the signals appearing from the opto-coupler.

With recent advancements, the over current protection is combined with gate drive circuit in a single chip. The block diagram of such IC is shown in Fig. 4. In this experimental work, instead of conventional driver circuit, a hybrid IC VLA542-01R is used which helps in protecting the power switches against over current alongwith proper gate driving.

When the drive carries an over-current, the collector to emitter voltage of IGBTs increases rapidly, this is sensed by the ultra-fast diode whose anode is connected to pin no. 1. This sensed voltage is compared with a reference value using a comparator. All the circuit is embedded within the single hybrid IC VLA542-01R and hence the protection of the IGBT against over current is carried out. Fig. 5 shows the typical connection diagram for VLA542-01R.

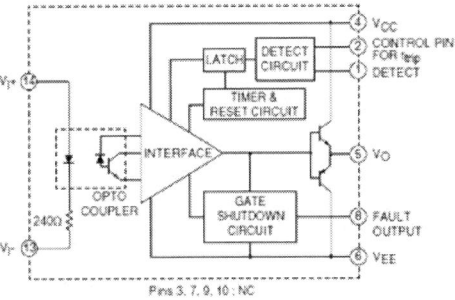

Fig. 4. Block diagram VLA542-01R or equivalent.

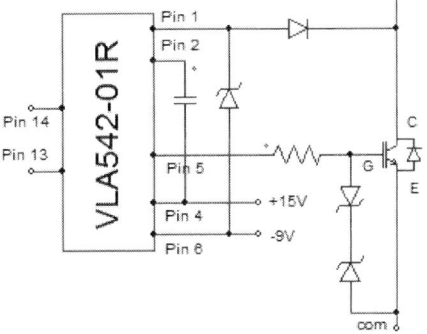

Fig. 5. Connection diagram of VLA542-01R.

C. Scheme for Shoot-through protection

There are several stray conditions which may cause two IGBTs of a limb in an inverter with bridge configuration to turn-on simultaneously. These results in short circuiting the dc bus followed by rupture of IGBTs. Thus protection of the IGBTs against shoot-through fault is very much essential from reliability aspect. An interlock is to be provided to protect the IGBTs against this fault. The protection scheme for the same is shown in Fig. 6.

When any one IGBT is turned on, it turns on the LED of the opto-coupler, thus turning on the output stage transistor of the device, which is directly connected to the base drive of the totem-pole circuit of the complementary IGBT. If any pulse tends to appear at the lower IGBT due to stray condition, the base drive for totem-pole circuit of the upper IGBT gets short-circuited through the transistor of the opto-coupler. In similar fashion pulse is forbidden to appear further in the lower IGBT if there is any stray condition appears in the upper IGBT gate drive. This ensures a proper protection against shoot-through fault.

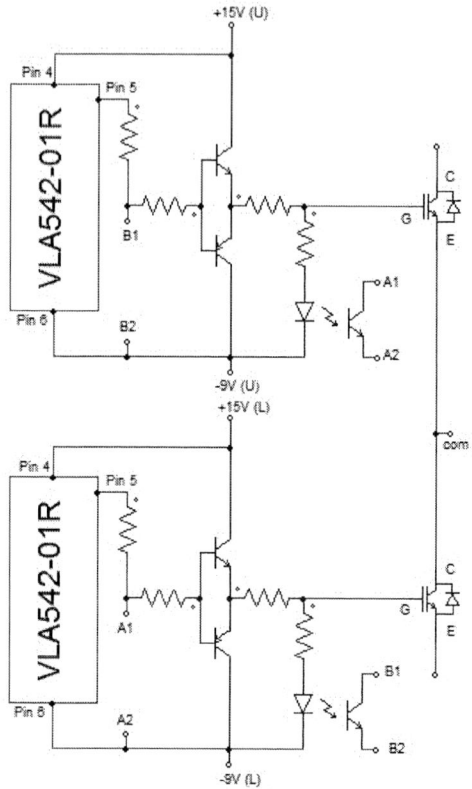

Fig. 7. Overall protection of IGBTs.

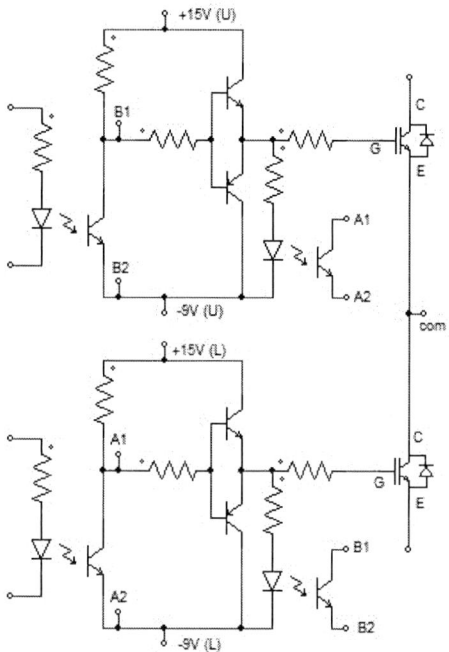

Fig. 6. Shoot-through protection scheme against IGBTs.

D. Combined Over-current Shoot-through protection

It is to be complied that the IGBTs connected in same leg of an inverter are to be protected both from over-current and shoot-through fault. During short-circuit condition of dc bus; an enormous current will flow through IGBTs, which cannot be protected by the over-current protection feature of the hybrid IC and so that an extra arrangement is to be provided to stop the shoot-through fault with immediate effect. This will ensure the overall protection of the inverter switches connected in bridge configuration. Fig. 7 shows the entire protection scheme of IGBTs connected in same leg of bridge configuration.

III. EXPERIMENTAL RESULTS

A microcontroller dsPIC30F2010 based inverter logic circuit has been established and as per requirement, the pulse output is fed to the VLA542-01R which has an inbuilt opto-coupler at its input side. This also ensures the over current protection of the drive. To prevent shoot-through, the circuit has been fabricated as shown in Fig. 7.

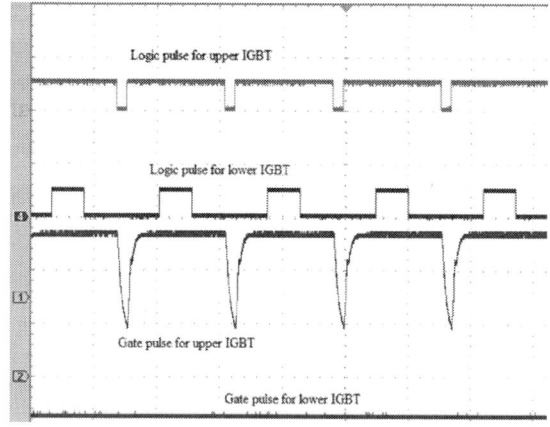

Fig. 8. Test result for shoot-through of IGBTs.

978-1-5386-6624-1/18 $31.00 © 2018 IEEE

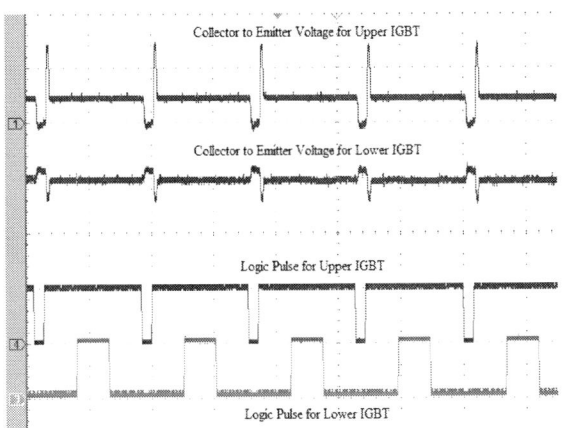

Fig. 9. IGBT collector-emitter voltages durig shoot-through.

For testing purpose of the shoot-through protection, a long duration pulse is along with a short duration pulse is programmed through the microcontroller to drive the gate circuit of the upper IGBT and lower IGBT respectively as shown in Fig. 8. This shows the output pulse from the gate driver is directly reflected to the gate of the upper IGBT. Whenever the driving pulse of the lower IGBT is going to turn on, it will immediately stop across the gate to emitter circuit of the lower IGBT and no pulse will appear to drive the lower switch. Fig. 9 shows the status of collector to emitter voltage for both the IGBTs has been checked with the DC voltage equal to 75V and the current through the limb is zero.

IV. CONCLUSION

The aim of this experimental work is to make arrangement for the protection of the costly power electronic switches used in the Voltage Source Inverters (VSI) for industrial drives against two major faults, over-current and shoot-through. IGBTs have been used as the power electronic switch for the inverter.

In the configuration of Voltage Source Inverter, two IGBTs are connected in same limb, which are generally driven by complementary gate pulses. A phase-to-phase or phase-to-ground fault in the output of such inverter causes the IGBTs to carry more current than the rated value. The shoot-through fault happens when an unexpected gate pulse appears on the gate terminal of an IGBT (upper or lower) while the other IGBT in the same limb is in ON (conducting) state. This unexpected gate drive causes both the IGBTs in the same limb to begin conduction simultaneously, thus causing immediate short-circuit of the DC bus. Therefore a robust, reliable and cost-effective method for integrated protection and prevention against both the faults has been carried out successfully in a prototype model of this experimental work.

REFERENCES

[1] N. Jovančić, N. Hadžimejlić, P. Ćeklić, "Efficient Control of IGBT Transistor as Part of Overvoltage Protection", International Journal of Electrical Engineering and Computing Vol. 1, No. 1 (2017), pp. 46-52.

[2] R. S. Chokhawala, "Switching Voltage Transient Protection Schemes for High-Current IGBT Modules", IEEE Transactions on Industry Applications, Vol. 33, No. 6, 1997, pp. 1601-1610.

[3] S. K. Biswas, B. Basak, K. S. Rajashekara, "Gate Drive Methods for IGBTs In Bridge Configurations", Proceedings of 1994 IEEE Industry Applications Society Annual Meeting, Vol.2, pp. 1310 - 1316.

[4] S. K. Biswas, B. Basak, K. S. Rajashekara, "A Modular Gate Drive Circuit for Insulated Gate Bipolar Transistor", Conference Record of the 1991 IEEE Industry Applications Society Annual Meeting, Vol.2, pp. 1490 – 1496.

[5] A. K. Jain, V. T. Ranganathan, "VCE Sensing for IGBT Protection in NPC Three Level Converters—Causes For Spurious Trippings and Their Elimination", IEEE Transactions on Power Electronics, Vol. 26, No. 1, 2011, pp. 298-307.

[6] Smart Gate Driver Coupler TLP5214 Application Note by Toshiba.

[7] Z. Wang, X. Shi, L. M. Tolbert, B. J. Blalock, M. Chinthavali, "A Fast Overcurrent Protection Scheme for IGBT Modules Through Dynamic Fault Current Evaluation", 2013 Twenty-Eighth Annual IEEE Applied Power Electronics Conference and Exposition (APEC), pp. 577 – 583.

[8] K. Yuasa, S. Nakamichi, I. Omura, "Ultra High Speed Short Circuit Protection for IGBT with Gate Charge Sensing", 2010 22nd International Symposium on Power Semiconductor Devices & IC's (ISPSD), pp. 37-40.

[9] R. Pagano, A. Raciti, "Evolution in IGBT's Protection against Short Circuit Behaviors by Gate-Side Circuitry", Proceedings of the IEEE International Symposium on Industrial Electronics, 2002, Vol.3, pp. 913 – 918.

[10] B. K. Bose, "Modern Power Electronics and AC Drives", Prentice Hall PTR.

[11] N. Mohan, T. M. Undeland, W. P. Robbins, "Power Electronics – Converters, Applications and Design", Second Edition, John Wiley & Sons, Inc.

Power Quality Improvement of Standalone Wind Energy Generation System for Non Linear Load

Seema Agrawal[*1], Vijay Kumar Gupta[2], D. K. Palwalia[3]
Department of Electrical Engineering
Rajasthan Technical University, Kota-324010, INDIA
[1]seema10dec@gmail.com, [2]vijaygupta2905@gmail.com, [3]dheerajpalwalia@gmail.com

R. K. Somani[4]
Department of Computer Science Engineering (BKIT)
Rajasthan Technical University, Kota-324010, INDIA
rksomani1@gmail.com

Abstract—The propose paper goal is to design hybrid shunt active power filter (HSAPF) for harmonic cancelation of standalone wind turbine (WT). The wind energy conversion system (WECS) has permanent magnet synchronous generator (PMSG) run by a fixed-speed 12 m/s wind turbine. The wind maximum power point tracking is obtained by switch mode rectifier (SMR) scheme. Voltage harmonic from the output of WEGS is eliminated by passive L_C filter. Current harmonics due to presence of nonlinear uncontrolled converter feeding with R-L load is compensated by shunt active power filter. Hybriding of active and passive gives cost effective solution of power quality. SAPF is consists of three phase insulated gate field effect transistor (IGBT) switches capacitor coupled voltage source inverter (VSI-CC). Reference current generation of SAPF is based on positive sequence detection method using PLL. DC bus voltage regulation is done using Fuzzy logic controller. Proposed system is simulated using MATLAB/Simulink (2015a). The demonstrated results show that suggested system successfully mitigates harmonics as per IEEE 519-1992 standards.

Keywords— Fuzzy Logic Controller (FLC), Wind Energy Generation System (WEGS), Permanent Magnet Synchronous Generator (PMSG), Hybrid Shunt Active Power Filter (HSAPF), PLL Synchronization, Capacitor Coupled Voltage Source Converter (CC-VSC).

I. INTRODUCTION

Global warming has been accredited to increase of atmospheric gases intensity produced by blaze of fossil fuel. Wind energy generation sources (WEGS) are most alternative to mitigate this problem above all due to its smaller green impact and its renewable quality that add for a sustainable growth [1].

It is gathering more consideration in entire globe due to less damaging environmental effect. WEGS are usually based on doubly-fed induction generators or permanent-magnet synchronous generators with power electronics interfaces. The heavy gearbox needs regular maintenance of WEGS with DFIG [20]. Therefore, WECS with PMSG are received much attention due to self excitation [2, 14, 17-19]. Optimum wind energy utilization is achieved by optimum tracking scheme such as maximum power/torque tracking, switch mode rectifier (SMR) etc. SMR is more preferable because feature of boosting [3].

The effectiveness and quality of conventional electrical networks mainly depend on the features of renewable energy generation sources (REGS) and loads type. Non-linear loads inject distorted current and voltage waveforms due to switching devices as power electronic converters [4, 13, 15].

In order to overcome power quality problems traditional passive harmonic filters can be used to remove higher order harmonics but due to sluggish response modern active harmonic filters (APFs) are investigated to eliminate random varying current[5-7, 16]. The significant cost reduction is obtained in both power semiconductor devices and signal processing devices by putting hybrid active power filters [8-9].

Performance of APF depends on accuracy and time taken in compensating harmonic components. Literature review revels include compensation techniques as d-q theory, instantaneous active reactive power p-q theory, synchronous reference frame method, notch filter method and self tuning filter [10]. Other method for reference current generation using unit template generation for SAPF is discussed in this article. The synchronization using PLL is explained.

In the ground of DC bus voltage regulation based on soft computing techniques such as artificial neural network (ANN), sliding mode control, fuzzy logic controller (FLC), genetic algorithm and evolutionary algorithm tuned conventional controller are focused for best PWM generation to minimize losses [11-12, 21].

In this paper, the WT with PMSG are used as power supply system. These sources provide AC variable power output, which may be converted in to DC supply. It is again converted into ac by PWM inverters at fundamental frequency. AC power is generated by WEGS is utilized to run non linear power electronic loads. Harmonics are introduced duo to non-linear loads. So hybrid filter is used to improve the power quality. Synchronization using PLL is applied for reference currents generation in SAPF and Fuzzy logic controller as DC bus voltage controller. To study performance of proposed system at fixed wind speed (12 m/s) simulation is carried out using MATLAB/Simulink (2015a) platform.

II. PROPOSED SYSTEM CONFIGURATION

The proposed system consists of wind energy generation system (WEGS) with HSAPF for harmonics mitigation due to presence of 3- phase uncontrolled rectifier non linear load. Series passive filter mitigates voltage harmonic present in wind AC output. SAPF improve the current quality by injecting equal and opposite current. Control unit of SAPF based on instantaneous unit template generation using PLL for reference current generation and hysteresis band controller for pulse generation fed VSI with DC link capacitor.

978-1-5386-6624-1/18 $31.00 © 2018 IEEE

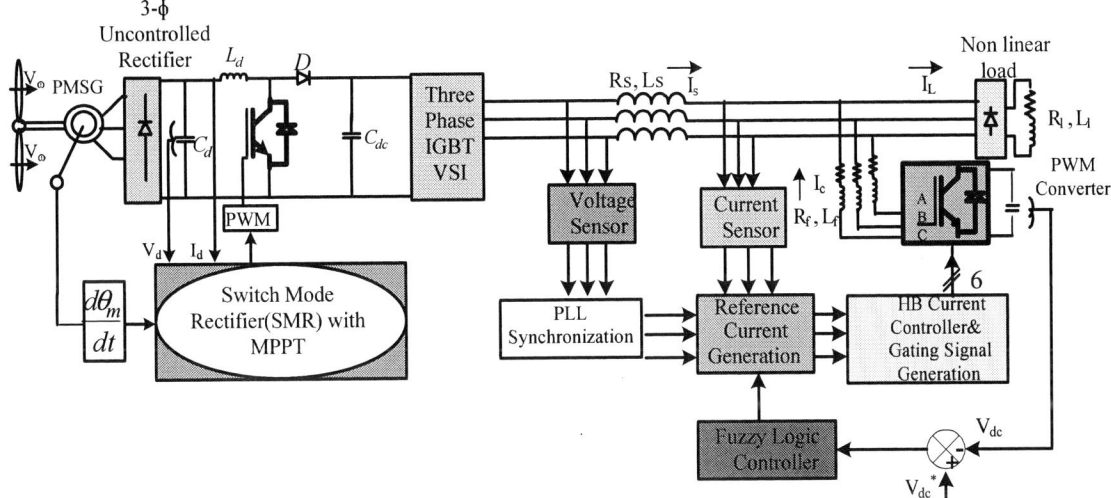

Fig. 1. Block diagram proposed architecture.

III. WIND ENERGY GENERATION SYSTEM

The proposed generation system model is developed using basic circuit equations of wind turbine and drive train.

A. Wind Turbine Modeling

The mechanical power is produced by turbine rotor given as:

$$P_m = 0.5 \rho A C_p(\lambda, \beta) v_w^3 \tag{1}$$

$$C_p(\lambda, \beta) = C_1 \left(C_2 \frac{1}{\beta} - C_3 \theta - C_4 \theta^x - C_5 \right) e^{-\left(\frac{C_6}{\beta} \right)} \tag{2}$$

$$\frac{1}{\beta} = \frac{1}{\lambda + 0.08\theta} - \frac{0.035}{1 + \theta^3} \tag{3}$$

$$\lambda = \frac{\omega_m R}{v_w} \tag{4}$$

$$T_m = \frac{P_m}{\omega_m} \tag{5}$$

Where,

P_m is mechanical power output from wind turbine(watt)

ρ is air density (kg/m³)

$C_p(\lambda, \beta)$ is power coefficient of wind turbine

λ is tip speed ratio and β is pitch angle taken zero

A is turbine rotor swept area ($A = \pi R^2$) in m²

B. Two-Mass Drive Train Modeling

Wind turbine dynamics is considered using more efficient two mass drive train models. Governing equations to this model are discussed as below:

$$2H_m \frac{d\omega_m}{dt} = T_m - T_{sh} \tag{6}$$

$$\frac{1}{\omega_{elb}} \frac{d\theta_{tw}}{dt} = \omega_m - \omega_r \tag{7}$$

$$2H_g \frac{d\omega_r}{dt} = T_{sh} - T_g \tag{8}$$

$$T_{sh} = K_{sh}\theta_{tw} + D_t \frac{d\theta_{tw}}{dt} \tag{9}$$

C. Switch-Mode Rectifier

Control block diagram of SMR for optimum torque tracking is illustrated as Fig 2. and optimum torque equation as given below:

$$T_{m\_opt} = K_{opt} \left(\omega_{m\_opt} \right)^2 \tag{10}$$

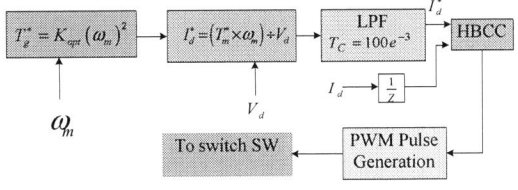

Fig. 2. Optimum torque tracking block

978-1-5386-6624-1/18 $31.00 © 2018 IEEE

IV. SAPF CONTROL STRATEGY

Reference current is generated using positive sequence detection method PLL and DC bus voltage signal is regulated with fuzzy logic controller.

A. Compensating Current Generation

Synchronizing unit template signals are measured by phase locked loop (PLL). FLC output is respected as amplitude 3-phase reference currents.

Three phase unit template signals are U_{sa}, U_{sb}, U_{sc} then unit vector (U) is expressed as

$$U = \begin{bmatrix} U_{sa} \\ U_{sb} \\ U_{sc} \end{bmatrix} = \begin{bmatrix} \sin(\omega t) \\ \sin(\omega t - 120^0) \\ \sin(\omega t + 120^0) \end{bmatrix} \quad (11)$$

Three phase reference source current is as follows:

$$\begin{bmatrix} i_{sa}^* \\ i_{sb}^* \\ i_{sc}^* \end{bmatrix} = \begin{bmatrix} U_{sa} \\ U_{sb} \\ U_{sc} \end{bmatrix} \begin{bmatrix} i_{sp} \end{bmatrix} \quad (12)$$

Switching signals for VSI are formed using difference of real and reference source currents. Errors are passed through hysteresis PWM to generate required switching signals. Switching (S) position is defined as below;

Here, HB limit is 0.01.

$$S = \begin{bmatrix} 1 & i_{sa} \le i_{sa}^* - HB; & i_{sb} \le i_{sb}^* - HB; & i_{sc} \le i_{sc}^* - HB; \\ 0 & i_{sa} \ge i_{sa}^* + HB; & i_{sb} \ge i_{sb}^* + HB; & i_{sc} \ge i_{sc}^* + HB; \end{bmatrix} (13)$$

B. FLC Controller

Artificial intelligent based fuzzy logic controller (FLC) is applied as DC side voltage regulation by mitigating real power losses. Fuzzy surface can be observed in Fig. 3.

Fuzzy logic rule is defined in Table I.

TABLE I. RULE MATRIX

E/CE	NB	NM	NS	ZE	PS	PM	PB
NB	NB	NB	NB	NB	NM	NS	ZE
NM	NB	NB	NB	NM	NS	ZE	PS
NS	N	NB	NM	NS	ZE	PS	PM
ZE	NB	NM	NS	ZE	PS	PM	PB
PS	NM	NS	ZE	PS	PM	PB	PB
PM	NS	ZE	PS	PM	PB	PB	PB
PB	ZE	PS	PM	PB	PB	PB	PB

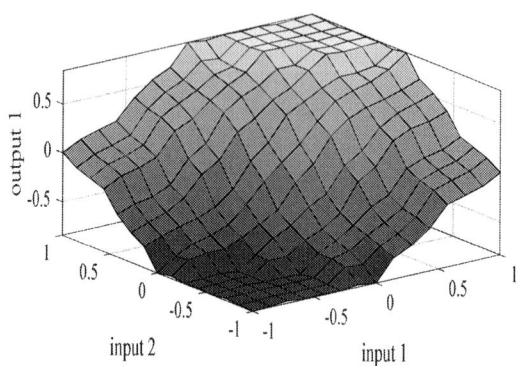

Fig. 3. Fuzzy surface.

V. SIMULATION REULTS AND DISCUSSION

Proposed stand alone wind energy generation system with HSAPF is illustrated as before in Fig. 1. System is examined with uncontrolled converter based non linear loads. The system data specifications for simulation study in MATLAB platform are listed in Table II. The simulation time is T=0 to T=1 s.

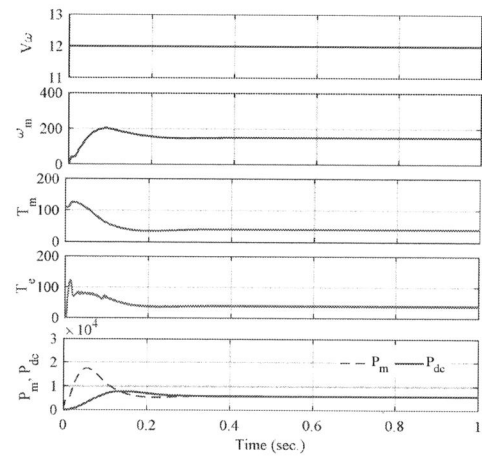

Fig.4. Wind speed, Generator speed, Turbine torque, Electromagnetic torque and Turbine mechanical input power & electrical output power from top to bottom respectively.

TABLE II. SYSTEM DESIGN PARAMERTERS

Source	Wind turbine	PMSG
	Air density=1.225 Kg/m3 Swept Area by turbine rotor A=1.67 m² R=0. 58m Base wind speed=12 m/sec K_{opt}= 1.67*10⁻³ Nm/(rad/s)²	No. of poles P=10, Rated speed=153 rad/sec Rated current = 12 A Stator resistance and inductance Rs=0.425Ω, Ls=8.4mH Magneticflux linkages=0.433 web Rated torque=40 Nm Rated power= 6 KW
APF	DC link capacitor, C_{dc} = 1500µF V_{dc}=700 Volts Filter with L_f= 3.35 mH, R_f= 0.5 Ω	
Loads	Three phase diode rectifier with L_l = 100 mH, R_l = 100 Ω load	

978-1-5386-6624-1/18 $31.00 © 2018 IEEE

Fig. 4 gives the simulation response of WEGS for wind speed 12 m/s. It can be seen that Te follows Tm quite well. Bottom of curve illustrates DC output power that shows DC current regulates turbine torque to extract peak power from wind turbine.

Three-phase output voltage signal of series passive filter is presented in Fig. 5(a) that is sinusoidal with 50 Hz. The non linear load current for phase 'a' is shown in Fig. 5(b). The shunt APF generates the harmonic current that is represented in Fig. 5(c). Compensated three phase current is depicted in Fig. 5(d) current compensation is obtained after 0.1 second switching pulse is supplied at 0.1 second. DC link voltage is regulated by artificial intelligence based fuzzy logic controller that is achieved constant value as at defined reference level shown in Fig. 5(e).

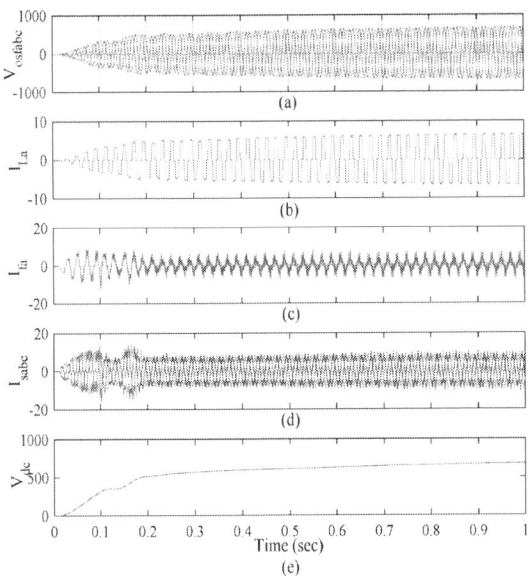

Fig.5. Performance parameter (a) Series passive filter output voltage signal (b) Load current (phase 'a') (c) filter current (phase 'a') (d) compensated current (e) DC link voltage respectively.

- **_Real Reactive Power and Power Factor Measurement_**

HSAPF with fuzzy logic controller based on PLL positive sequence detection method improves power quality of WEGS. The real (W) and reactive power (VAR) at 50 Hz are shown in Fig. 6.

Power factor improvement has been achieved approximate unity as shown at bottom of curve.

- **_Fast Fourier Transforms Analysis_**

The FFT analysis of AC generated current is carried out to find magnitudes of different harmonic components. Series passive filter mitigates voltage harmonics.

THD of voltage signal before and after series filter can be analyzed in Fig. 7. Fuzzy logic controller and PLL based SAPF is suppressing harmonic currents. THD analysis before and after compensation with HSAPF are plotted in Fig. 8 respectively.

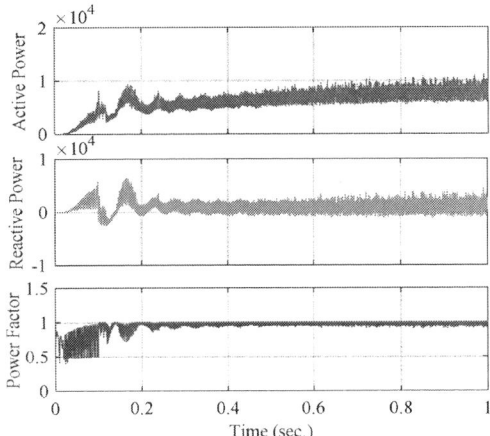

Fig. 6. Active, Reactive power and Power factor.

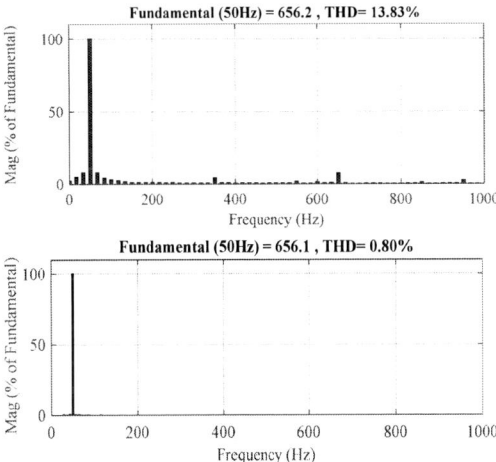

Fig.7. Voltage signal (phase a) harmonic spectrum before and after compensation with series passive filter.

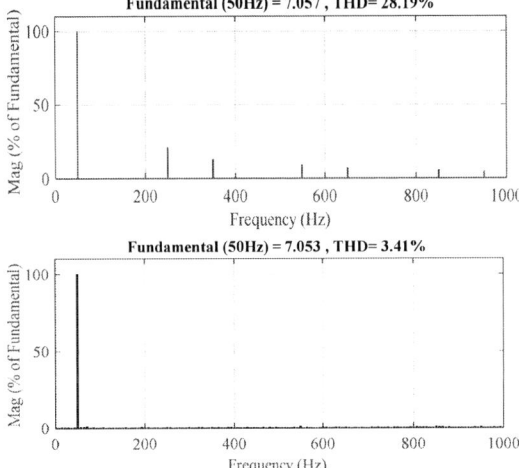

Fig. 8. Current signal (phase a) harmonic response before and after compensation with SAPF with uncontrolled converter non linear load.

VI. CONCLUSION

The performance of standalone wind energy generation system based on HSAPF has been investigated. Wind turbine and drive train modeling with wind speed 12 m/s as input parameter for electrical power generation has been considered. SMR algorithm based MPPT has been found to track peak power satisfactorily to maximize generated energy.

A single switch mode rectifier is applied to convert AC output voltage of generator to constant DC voltage then convert constant DC voltage to AC signal via three phase inverter. WEGS has been connected to uncontrolled non linear load. HSAPF is used to improve power quality generation system. HSAPF is based on phase locked loop positive sequence detection method and passive filter includes LC section. DC bus voltage has been regulated by fuzzy logic controller and its effectiveness is judged in term zero over shoot. The simulated response shows the effectiveness of HSAPF. The simulation results have clearly shown that series passive filter improve voltage quality and shunt active power filter (SAPF) improve current quality of given generation system. HSAPF has been provided cost effective solution of power quality issue such as harmonics. Performance index parameter %THDs have been obtained under limit prescribed by IEEE 519-1992 standard.

APPENDIX

H_t = Inertia constant of the turbine (4 s)

Hg = Inertia constant of the PMSG (0.4 s)

θ_{tw} = Shaft twist angle

ω_m = Angular speed of the wind turbine

ω_r = Rotor speed of the PMSG

ω_{elb} = Electrical base speed

T_{sh} = Shaft torque

K_{sh} = Shaft Stiffness (0.3 p.u./el.rad)

D_t = Damping coefficient (0.7 p.u.s/el.rad)

REFERENCES

[1] Global wind report: annual market update", Published by global wind energy council (GWEC), Accessible from http://www.gwec.net, accessed on May 2018.

[2] A.Beainy, C. Maatouk, N. Moubayed, and F. Kaddah, "Comparison of different types of generator for wind energy conversion system topologies", IEEE International Conference on Renewable Energies for Developing Countries (REDEC), Zouk Mosbeh, Lebanon , pp. 1-6, 2016.

[3] M.A. Abdullah, A.H.M. Yatim, C.W. Tan, and R. Saidur, "A review of maximum power point tracking algorithms for wind energy systems", Renewable and Sustainable Energy Reviews, vol. 16, no. 5, pp. 3220-3227, 2012.

[4] K. Singh, "Power system harmonics research: a survey," European Transaction on Electrical Power, vol.19, no. 2, pp. 151-172, August 2007.

[5] J. C. Das, "Passive filters-potentialities and limitations," IEEE Transactions on Industry Applications, vol. 40, no. 1, pp. 232-241, 2004.

[6] H. Akagi, "New trends in active filters for power conditioning," IEEE Transaction on Industrial Applications, vol. 32, no. 6, pp.1312-1402, November 1996.

[7] H. Rudnick, J. Dixon, and L. Morán, "Active power filters as a solution to power quality problems in distribution networks," IEEE Power and Energy Magazine, pp. 32-40, October 2003.

[8] S. Agrawal, and D. K. Palwalia, "Analysis of standalone hybrid PV-SOFC-battery generation system based on shunt hybrid active power filter for harmonics mitigation", Power India International Conference (PIICON), Bikaner, India, pp. 1-6, 2016.

[9] J. Tsai, and K. Tan, "APF harmonic mitigation technique for PMSG wind energy conversion system", Australasian Universities Power Engineering Conference (AUPEC), Perth, Australia, pp. 1-6, 2007.

[10] S. Chourasiya, and S. Agrawal, "A review: control techniques for shunt active power filter for power quality improvement from non-linear loads", International Electrical Engineering Journal, vol. 6, no.10, pp. 2028-2032, 2015.

[11] S. Agrawal, P. Kumar, and D. K. Palwalia, "Artificial neural network based three phase shunt active power filter", Power India International Conference (PIICON), Bikaner, India, pp. 1-6, 2016.

[12] S. Agrawal, B.K. sen, and D.K. Palwalia, "Performance analysis of shunt active power filter based on GSA tuned PI Controller", IEEE International Conference on Information Communication Instrumentation and Control (ICICIC-2017), 2017.

[13] M. A. M. Moftah, G. E. S. A. Taha, and E. N. A. Ibrahim, "Active power filter for variable-speed wind turbine PMSG interfaced to grid and non-linear load via three phase matrix converter", Eighteenth International Middle East Power Systems Conference (MEPCON),Cairo, Egypt, pp. 1013-1019, 2016.

[14] C. N. Bhende, S. Mishra, and S. G. Malla, " Permanent magnet synchronous generator-based standalone wind energy supply system", IEEE Transactions on Sustainable Energy, vol.2, no.7, pp. 361-373, 2011.

[15] Y. Oğuz, İ. Güney, and H. Çalık, "Power quality control and design of power converter for variable-speed wind energy conversion system with permanent-magnet synchronous generator", The Scientific World Journal, 2013.

[16] S. Nallusamy, D. Velayutham, and U. Govindaraja, "Design and implementation of a linear quadratic regulator controlled active power conditioner for effective source utilization and voltage regulation in low-power wind energy conversion systems", IET Power Electronics, vol. 8 , pp. 2145-2155, 2015.

[17] A.V. P. Kumar, A. M. Parimi, and K. U. Rao, "Investigation of small PMSG based wind turbine for variable wind speed", International Conference on Recent Developments in Control, Automation and Power Engineering (RDCAPE), pp. 107-112, 2015.

[18] N. A. Orlando, M. Liserre, and R. A. Mastromauro, and A. D. Aquila, "A survey of control issues in PMSG-based small wind-turbine systems", IEEE Transactions on Industrial Informatics, vol. 9, 2013.

[19] A. Patel, Sabha, and R. Arya, "Distributed power generation system using PMSG with power quality features", International Conference on Next Generation Intelligent Systems (ICNGIS), Kottayam, India ,pp. 1-8, 2016.

[20] A. Jain, C.T. Vijay, S. Shravanthi, and S. Gokul, "Comparative analysis of direct power control (dpc) and direct voltage control (DVC) for control of doubly fed induction generator (DFIG) connected to a variable speed wind turbine", IJCTA, vol. 9, pp. 8961-89717, 2016.

[21] S. Agrawal, D. Sharma, and D. K. Palwalia, "Performance analysis of SAPF based on self-tuned harmonic filter with fuzzy logic controller", IEEE Conference on Recent Developments in Control, Automation & Power Engineering (RDCAPE), pp. 487-492, 2017.

Harmony Search Algorithm Based Power Quality Improvement in Multilevel Inverter

Deepshikha Singla
Department of Electrical Engineering
YMCAUST
Faridabad, Haryana
deepshikha_16s@yahoo.com

Parsh Ram Sharma
Department of Electrical Engineering
YMCAUST
Faridabad, Haryana
prsharma1966@gmail.com

Abstract—**Cascaded Multilevel Inverters with isolate DC source have proven to be extremely beneficial in medium voltage and high-power applications. And DC normally has variations, which disturbs power quality of any system. The idea of this paper is to propose a Harmony Search method for deciding opti- mal switching angles for firing the switches of multilevel inverter to operate under all operating conditions. The HS Algorithm aims to lower the voltage deviation of fundamental component from desired value and diminish the harmonics. Along with THD minimization, the algorithm also focuses on decreasing even order harmonics and DC component to negligible. The proposed technique is tested with equal, unequal, equal and time-varying supply with non-linear and dynamic load. Analysis proves that the algorithm efficiently calculates the switching angles under all operating conditions. And more essentially, simulated output voltage and current equals calculated theoretical values, which mean the power quality is enhanced. For validation, the results are compared with the Sinusoidal Pulse Width Modulation (SPWM) results.**

Keywords—**Cascaded H-Bridge Multilevel inverter, Harmony Search Algorithm, Modulation Index, Selective Harmonic Elimination, Total Harmonic Distortion**

I. INTRODUCTION

Cascaded H-Bridge Multilevel Inverters (CHMLI) [1], [2], [3] are being increasingly used for utility applications including utility interface of renewable energy, voltage regulation, reactive power compensation, and harmonic filtering in power systems. CHMLIs have a characteristic feature of producing a medium voltage output from an arrangement association of (M-1)/2 control units provided from separate dc input, which enables them to use as an important alternative in the medium-voltage and high power applications [4], [5]. But, there is a limitation that this configuration requires separate DC voltage sources and practically, it is difficult to have sources that provide constant and equal dc voltage. Also, variation in supply voltage affects harmonics in power system. So, the main challenge in the design of an effective multilevel inverter is decision of modulation control [6], [7] such that the harmonics are minimized and power quality of is enhanced. The strategy should also be able to control multilevel inverter in practical conditions like unbalanced load, dynamic load [8], etc.

In order to minimize harmonics, most widely used methods are Sinusoidal-PWM (SPWM) [9], [10], [11] and Selective Harmonic Elimination Pulse Width Modulation (SHE-PWM) [12], [13]. There exist various SPWM methods

which are very effective for controlling the inverter output voltage but involve high switching frequency which causes switching loss. Selective Harmonic Elimination (SHE) technique is benefited from its low switching frequency and high output power quality, by using harmonic equations derived by applying Fourier Series Expansion. Newton-Raphson (NR) method is used to solve these non-linear transcendental harmonic equations but it also suffers from some drawbacks like divergence problems, require initial guess and gives no optimum solution. Nowadays, most of the reviewed research work discusses the application of several optimization algorithms like Genetic Algorithm [14], [15], [16], [17], Differential Evolution [18], Particle Swarming Optimization [19], [20], [21], [22], and numerous other algorithms. Here, the SHE equations are converted into the objective function and the algorithms are proposed for computing switching angles of multilevel inverter. But, in case of GA, the quality of solution deteriorates with increase in the level of inverter as this method requires proper selection of certain parameters such as crossover and mutation probability, number of generations, etc.

Heuristic algorithms have the ability to combat the above drawbacks. As an optimization technique, PSO is significantly more straightforward as it doesn't rely on the direction of the slope data. But, it has tendency to a fast and premature convergence in mid optimum points and slow convergence in refined search stage. Similar to other algorithms, Harmony Search [23] is a metaheuristic approach of solving an optimization problem that has achieved great success in many areas of research, particularly in control and electrical power systems, due to its ability to avoid becoming trapped at local minima. It has characteristic features of algorithm simplicity and search efficiency. In this paper, HS method is utilized to solve the SHE equations to obtain optimized switching angles which are then used to generate the gating pulses for switches of multilevel inverter.

II. SELECTIVE HARMONIC ELIMINATION TECHNIQUE

Selective Harmonic Elimination PWM refers to technique which aims at eliminating specific lower order harmonics, whereas all other harmonic components can be removed by filtering. The SHEPWM consists of non-linear transcendental equations, normally solved using numerical iterative methods. But these methods are complex and time consuming, computational algorithms based on Newton-

978-1-5386-6624-1/18 $31.00 © 2018 IEEE

Raphson and Resultant Theory can be developed.

In this paper, an 11-level H-Bridge Inverter is used for the study of power quality of multilevel inverter under various operating conditions. 11-level H-Bridge Inverter comprises of series combination of five H-bridges with separate dc inputs as V_{dc1}, V_{dc2}, V_{dc3}, V_{dc4}, and V_{dc5} as shown in Fig.1. The output phase voltage V_{an} of 11-level CH-MLI is the sum of individual H-bridge inverter unit outputs.

$$V_{an} = V_1 + V_2 + V_3 + V_4 + V_5 \qquad (1)$$

where V_1, V_2, V_3,....., $V_{(m-1)/2}$ are the output voltages of each H-bridge inverter units.

Considering that each H-Bridge is fired at switching angles $\alpha_1, \alpha_2, \alpha_3, \alpha_4$ and α_5, the Fourier series expansion of the output voltage is given as:

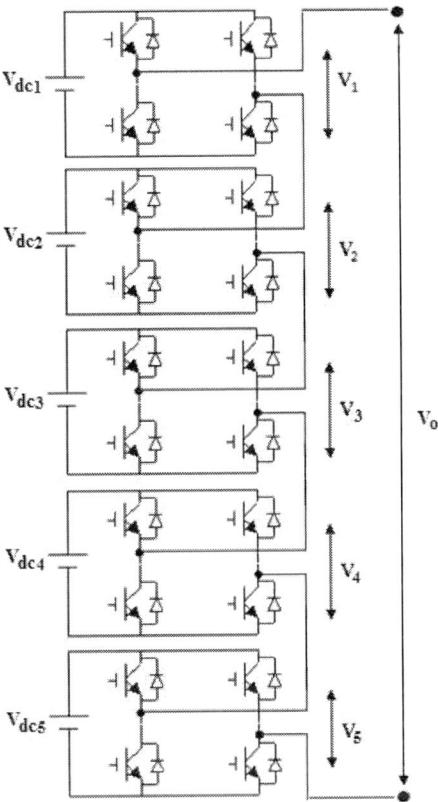

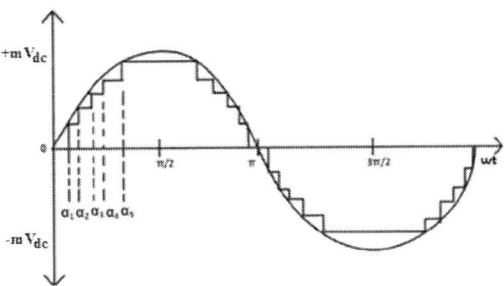

Fig.1 a) Single Phase 11- level Cascaded H Bridge Inverter, b) Output waveform

$$V(t) = \sum_{n=1,3,5}^{\infty} \frac{4}{n\pi} (V_{dc1}\cos n\alpha_1 + V_{dc2}\cos n\alpha_2 + V_{dc3}\cos n\alpha_3$$
$$+ V_{dc4}\cos n\alpha_4 + V_{dc5}\cos n\alpha_5)\sin n\alpha_n \qquad (2)$$

Because of odd quarter-wave symmetry, harmonics with even order become zero and triplen harmonics are eliminated in three phase balanced systems. Now, to obtain minimum harmonics, fundamental component of output voltage should be equal to desired voltage and lower order odd harmonics should be zero. Thus, the above equation reduces to Selective Harmonic Elimination equations or SHE equations:

$$\cos\alpha_1 + \cos\alpha_2 + \cos\alpha_3 + \cos\alpha_4 + \cos\alpha_5 = 5M$$

$$\cos 5\alpha_1 + \cos 5\alpha_2 + \cos 5\alpha_3 + \cos 5\alpha_4 + \cos 5\alpha_5 = 0$$

$$\cos 7\alpha_1 + \cos 7\alpha_2 + \cos 7\alpha_3 + \cos 7\alpha_4 + \cos 7\alpha_5 = 0, \quad (3)$$

and so on.Switching angles should be within zero and π/2. However, solving equations by NR method gets complicated when extended to higher levels.

III. PROPOSED OPTIMIZATION ALGORITHM: HARMONY SEARCH

Harmony search (HS) is a metaheuristic way to deal with various optimization problems. Basically, the metaheuristic approach avoids getting caught in local optimum solutions and aims at obtaining the global optimum solution. The HS method is presented by Geem in 2001, and is inspired by spontaneous act of music improvisation. It is an act of transforming the beauty and harmony of music into an optimization procedure through search for a perfect harmony, and thus named as Harmony Search Optimization. Here, the perfect condition of concordance in music is similar to best optimal solution of a problem. HS algorithm employs a Harmony Search Memory (HSM) of a specific size termed as Harmony memory size (HS), and two probabilities namely, Harmony Memory Considering Rate (HMCR) and Pitch Adjustment Rate (PAR), it keeps revising the solution till ideal solution/requirement is met. Harmony Memory Considering Rate is the probability of selecting a component from the HSM. Pitching Adjust Rate is the probability of a candidate from the HSM to be mutated. The spontaneous act of result optimization in Harmony Search is analogous to Genetic Algorithm. Where on one hand, the generation of offspring in GA is done using only single mutation or two crossovers of existing chromosomes, the modification of any arrangement in HS algorithm is done using all the existing HSM members.

Now, the aim here is to minimize the voltage deviation of fundamental voltage from the reference value and reduce the harmonic content of each order to a minimal value. So, the objective of optimization algorithm is given as:

978-1-5386-6624-1/18 $31.00 © 2018 IEEE

$$Objective = (A_1 - 5 * M) + \sum_{k=5,7,11}^{49} \frac{A_k}{k}$$

$$A_1 = (V_{dc1} \cos \alpha_1 + V_{dc2} \cos \alpha_2 + V_{dc3} \cos \alpha_3 +$$

$$V_{dc4} \cos \alpha_4 + V_{dc5} \cos \alpha_5)/ V_{dc(avg)}$$

$$A_k = (V_{dc1} \cos k\alpha_1 + V_{dc2} \cos k\alpha_2 + V_{dc3} \cos k\alpha_3 +$$

$$V_{dc4} \cos k\alpha_4 + V_{dc5} \cos k\alpha_5)/ V_{dc(avg)}$$

$$V_{dc(avg)} = \frac{V_{dc1} + V_{dc2} + V_{dc3} + V_{dc4} + V_{dc5}}{5} \qquad (4)$$

Where $\alpha_1, \alpha_2, \alpha_3, \alpha_4, \alpha_5$ are the switching angles of 11-level inverter, and M is the modulation index.
with constraint:

$$0 < \alpha_1 < \alpha_2 < \alpha_3 < \alpha_4 < \alpha_5 < \frac{\pi}{2}$$

The flowchart of HS Algorithm to optimize switching angles for multilevel inverter is shown in Fig.2.

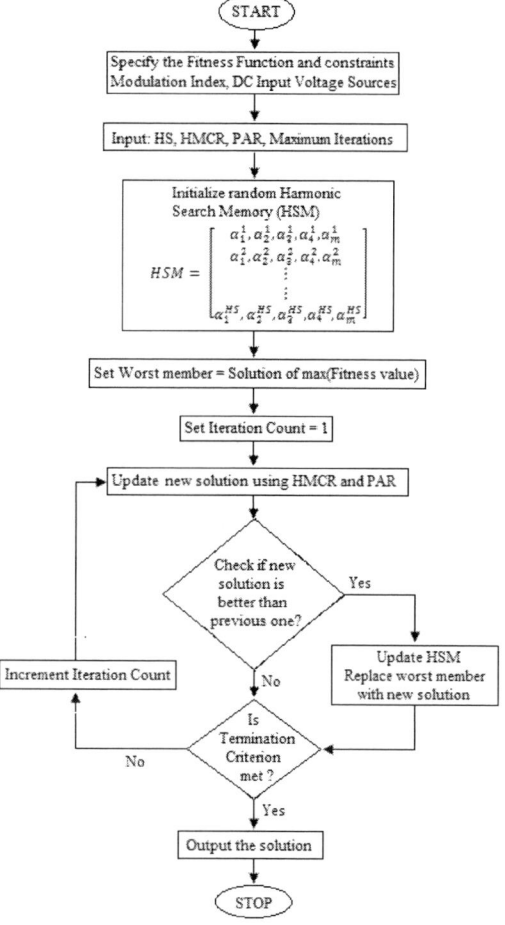

Fig. 2. Flowchart of Proposed Harmonic Search Algorithm

IV. SIMULATION ANALYSIS

For analysis of power quality of multilevel inverter, a three phase 11-level CHMLI is modelled in MATLAB/Simulink and simulations are carried out under different conditions. Harmony Search Algorithm is written in MATLAB and the size of Harmony Memory (HS) is taken as 200, HMCR is 0.95 and PAR is 0.1.

Normally, the MLIs are powered using photo-voltaic energy, solar power, wind power, etc and therefore there occurs variations in the dc supply voltage. So, we have considered different supply side variations of dc power with symmetric and asymmetric configurations. Additionally, load side variations (non-linear load as well as dynamic load) are considered. Comparative analysis of fundamental voltage deviation, harmonic spectrum, rms output voltage, output current and THD for different modulation indices is done.

A. Symmetrical system with Non-Linear Load

The 11-level Cascaded H-Bridge Symmetrical Inverter has equal and fixed dc sources (V_{dc1}= V_{dc2}= V_{dc3}= V_{dc4}= V_{dc5}= 10V) at the input of each H-bridge and balanced load comprises of three-phase nonlinear load with R=10ohm and L=10mH in all phases. The switching angles for generating gating pulse of multilevel inverter at different modulation indices are obtained using HS Algorithm, and are shown in Table I.

TABLE I: SWITCHING ANGLES FOR SYMMETRICAL 11-LEVEL INVERTER USING HS ALGORITHM

M	α_1 (degree)	α_2 (degree)	α_3 (degree)	α_4 (degree)	α_5 (degree)
0.85	7.42	13.20	20.88	33.26	59.13
0.875	6.27	13.34	25.64	39.13	58.79
0.90	3.25	10.42	21.91	28.29	44.04
0.925	3.46	9.70	15.38	26.29	38.67
0.95	3.00	10.03	15.44	26.06	38.20
0.975	2.90	9.09	16.80	27.44	37.47
1.00	3.05	9.34	15.58	26.81	37.00

These switching angles for a specific modulation index are used to generate gating pulse, and then multilevel inverter is simulated. And as the best results are obtained at modulation index m=0.925. Observation results summarizes that HS Algorithm based Harmonic Elimination reduces THD, while minimizing the fundamental voltage deviation at all modulation indices. The comparison is shown in Table II. The harmonic order chart concludes that HS algorithm based simulation reduces the even harmonics and dc component to almost negligible value. The output voltage and current are equal to the calculated theoretical values, thus power quality is enhanced.

TABLE II: COMPARISON OF OUTPUT LINE VOLTAGE AND OUTPUT CURRENT OF SYMMETRICAL CONFIGURATION

M		Vr.m.s (V)	Voltage THD	DC Component	Ir.m.s (A)	Current THD
0.85	SPWM	51.61	7.95	0.03739	4.92	0.65
	HS	66.11	4.83	4.33e-08	6.31	1.29
0.875	SPWM	53.43	7.79	0.1371	5.09	0.58
	HS	64.74	4.88	4.23e-08	6.18	0.79
0.90	SPWM	54.71	7.62	0.01246	5.22	0.51
	HS	70.06	4.45	4.94e-08	6.68	0.93
0.925	SPWM	56.57	6.95	0.1371	5.34	0.47
	HS	72.01	3.96	4.97e-08	6.87	0.63
0.95	SPWM	57.86	6.74	0.1122	5.52	0.56
	HS	72.02	4.04	0.004985	6.87	0.66
0.975	SPWM	59.61	6.81	0.03739	5.69	0.47
	HS	71.91	4.01	5.03e-08	6.86	0.60
1.0	SPWM	60.85	7.03	0.03739	5.81	0.56
	HS	72.12	4.08	4.99e-08	6.88	0.72

B. Asymmetrical system with Non-Linear Load

The 11-level Cascaded Asymmetrical Inverter has variable dc sources (V_{dc1}=8V, V_{dc2}=13V, V_{dc3}=11V, V_{dc4}=9V, and V_{dc5}=9V) for each H-bridge and balanced three-phase non- linear load with R=10ohm and L=10mH in all phases. The switching angles, obtained using HS algorithm, are shown in Table III.

The asymmetrical multilevel inverter is simulated for different values of modulation indices. Harmonics are reduced to a great level, shown in Table III and Fig. 3. As observed from graph, output voltage obtained using HS algorithm is approximately equal to theoretical values calculated using formulas, and thus the output power quality is much better.

TABLE III: SWITCHING ANGLES FOR ASYMMETRICAL 11-LEVEL INVERTER USING HS ALGORITHM

M	α_1 (degree)	α_2 (degree)	α_3 (degree)	α_4 (degree)	α_5 (degree)
0.85	5.98	13.44	25.97	39.35	59.36
0.875	2.64	14.56	20.28	33.85	53.23
0.90	2.20	10.32	22.18	32.30	44.02
0.925	2.99	9.45	17.18	28.32	37.68
0.95	2.85	9.50	17.19	28.0	37.15
0.975	2.73	9.44	17.23	28.53	37.07
1.00	2.87	9.43	17.27	28.39	37.44

TABLE III: COMPARISON OF VOLTAGE THD AT DIFFERENT MODULATION INDEX FOR ASYMMETRICAL CONFIGURATION

M		THD(%)	2nd	3rd	4th	5th	6th	7th	8th	9th	10th
0.85	SPWM	8.89	0.73	0.15	2.19	2.14	0.11	0.13	0.30	0.29	0.38
	HS	4.60	0.00	0.21	0.00	0.33	0.00	0.39	0.00	0.18	0.00
0.875	SPWM	8.33	1.00	0.43	1.78	1.33	0.46	0.42	0.41	0.22	0.16
	HS	5.18	0.00	0.11	0.00	0.11	0.00	0.09	0.00	0.15	0.00
0.90	SPWM	8.08	0.96	0.12	1.65	1.90	0.18	0.07	0.45	0.12	0.66
	HS	4.38	0.00	0.24	0.00	1.44	0.00	0.26	0.00	0.07	0.00
0.925	SPWM	7.32	0.82	0.25	1.21	1.81	0.08	0.49	0.23	0.33	0.51
	HS	4.10	0.00	0.05	0.00	0.54	0.00	0.79	0.00	0.11	0.00
0.95	SPWM	7.04	0.92	0.29	1.17	2.02	0.23	0.16	0.12	0.032	0.43
	HS	4.16	0.00	0.10	0.00	0.51	0.00	0.96	0.00	0.13	0.00
0.975	SPWM	7.10	1.02	0.09	0.74	1.80	0.14	0.5	0.26	0.24	0.30
	HS	4.16	0.00	0.10	0.00	0.51	0.00	0.96	0.00	0.13	0.00
1.0	SPWM	7.42	0.78	0.21	0.61	2.22	0.10	0.31	0.39	0.12	0.44
	HS	4.10	0.00	0.10	0.00	0.50	0.00	0.86	0.00	0.10	0.00

978-1-5386-6624-1/18 $31.00 © 2018 IEEE

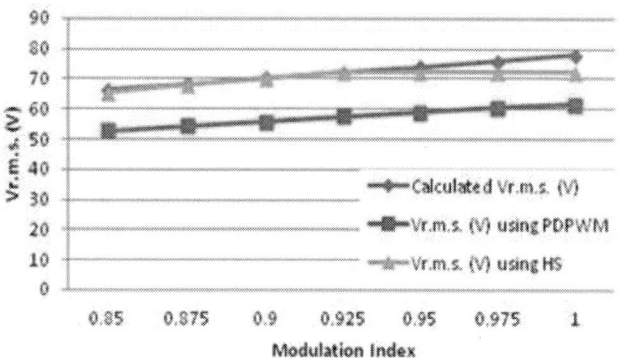

Fig.3. Performance Comparison of Output Line Voltage for Asymmetric system

C. Symmetrical system with Unbalanced Non-Linear Load

The symmetrical 11-level MLI is connected to unbalanced non-linear load with R=15ohm, L=10mH in first phase, R=5ohm, L=20mH in second phase, and R=10ohm, L=5mH in third phase. The system is simulated at m=0.925 and comparison is shown graphically in Fig.4.

The graphs indicate that harmonic distortion has reduced and the peak values of current and voltage have improved in all the three phases, thus improving power quality. The problem of voltage balancing in three phases during variations is also solved, since the voltage in three phases is same.

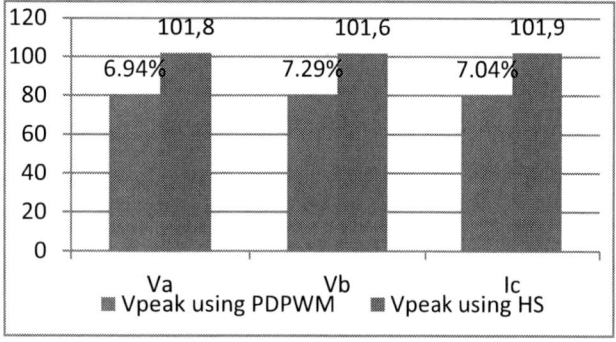

Fig.4. Graphical Comparison using SPWM and HS Algorithm of Symmetrical system with unbalanced load

D. Symmetrical system with Dynamic Load

The symmetrical system is connected to split-phase motor load in all phases as dynamic load. The simulation is done and results are shown graphically, shown in Fig. 5.

The output voltage and current quality has improved in case of dynamic load also. The rated value for voltage and current has been achieved, hence enhancing power quality.

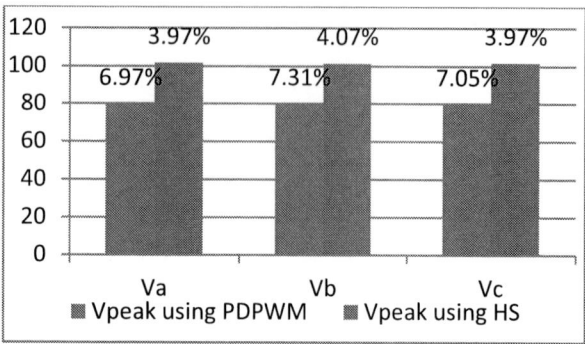

Fig.5. Graphical Comparison using SPWM and HS Algorithm of Symmetrical system with dynamic load

V. CONCLUSION

Harmony Search Technique based algorithm has been im- plemented to optimize switching angles for firing the switches of multilevel inverter. HS Algorithm is coded in m-file and control circuit for 11-level CHMLI is simulated in MAT- LAB/Simulink. The paper aims to control multilevel inverter to work efficiently under different unbalances on source as well as on load side. The suggested method has proved to   be efficient in optimizing switching angles for better power quality.

Minimum Harmonics content are obtained at modulation index m=0.925, hence all simulations are done at the same modulation index. A detailed analysis is performed in terms of RMS output voltage, peak output current, voltage and current harmonics, DC voltage component, and harmonic spectrum of output voltage. The simulation and comparison results proved that the switching angles obtained in both symmetric and asymmetric configuration using HS algorithm are the best optimal values for power quality enhancement.

Dynamic load and unbalances in the load are handled easily, so the inverter has proven to be independent of load. Also, a considerable amount of improvement in output line voltage and currents has been obtained and they have become approximately equal to those of theoretical values, resulting in minimum deviation and better power quality. Thus, the multilevel inverter controlled by switching angles obtained using Harmony Search Algorithm is efficient in all the operating conditions.

REFERENCES

[1] P. F. Lai JS, "Multilevel converters-a new breed of power converters," *IEEE Transactions on industry applications*, pp. 32–3, 1996.

[2] "A survey on cascaded multilevel inverters," *IEEE Transactions on industrial*, pp. 57–7, 2010.

[3] H. N. P. . S. D, "Quality improvement using multilevel inverters –," *A Review. Int. J. of Engg. Sci. & Mgmt.(IJESM)*, vol. 2011, pp. 1–1.

[4] P. F. Rodriguez J, Lai JS, "Multilevel inverters: a survey of topologies, controls, and applications," *IEEE Transactions on industrial electronics*, pp. 49–4, 2002.

[5] Fang-Zen P, "Applications of cascade multilevel inverters," *Journal of Zhejiang University-SCIENCE A*, pp. 1–4, 2003.

[6] P. F. W. J. [6]Corzine KA, Wielebski MW, "Control of cascaded multi- level inverters," in *Electric Machines and Drives Conference*, vol. 1. IEEE, June 2003, pp. 1549–1555.

[7] M. A. . K. J, "Analysis of various control schemes for minimal total harmonic distortion in cascaded h-bridge multilevel inverter," *Journal of Electrical Systems and Information Technology*, pp. 1– 3, 12 2016.

[8] V. K. Tamilvani P, "Harmonic mitigation in various levels of multilevel inverter with different loads," *International Journal of Innovative Re- search in Electrical Engineering*, vol. 2, p. 9, 2014.

[9] S. D. Urmila B, "Multilevel inverters: A comparative study of pulse width modulation techniques," *International Journal of Scientific & Engineering Research*, pp. 1–3, 2010.

[10] W. A. . J. FT, *The comparative analysis of multicarrier control tech- niques for SPWM controlled cascaded H-bridge multilevel inverter*, vol. 23.

[11] K. S. . M. D, "A comparative analysis of multi carrier spwm control strategies using fifteen level cascaded h-bridge multilevel inverter," *International Journal of Computer Applications*, pp. 1–41, 1 2012.

[12] A. V. Dahidah MS, "Selective harmonic elimination pwm control for cascaded multilevel voltage source converters: A generalized formula," IEEE Transactions, 2008, on power electronics. 2008 Jul;23(4):1620-30.

[13] T. H., "An improved omthd technique for an n-level cascaded multilevel inverter with adjustable dc sources," Turkish Journal of Electrical Engineering & Computer Sciences, pp. 3–25, 2017.

[14] C. J. . O. B, Harmonic optimization of multilevel converters using genetic algorithms. IEEE, 6 2004, vol. 20, no. 5.

[15] M. S. . M. R, "A comparison study of harmonic elimination in cas- cade multilevel inverter using particle swarm optimization and genetic algorithm," pp. 43–9, 2013.

[16] B. B. Salami A, "Total harmonic distortion minimization of multilevel converters using genetic algorithms," Applied Mathematics, pp. 28–4, 2013.

[17] K. N. . M. M, "Comparison of ga and pso algorithms in cascaded multilevel inverter using selective harmonic elimination pwm technique," Int J Adv Res Electr, vol. 2014, pp. 3–4.

[18] A. A. Salam Z, Majed A, "Design and implementation of 15-level cascaded multi-level voltage source inverter with harmonics elimination pulse-width modulation using differential evolution method. iet power electronics," IET Power Electronics, pp. 23–8, 2015.

[19] A. Al-Othman AK, "Thd elimination of harmonics in multilevel inverters with non-equal dc sources using pso. energy conversion and manage- ment," pp. 1–50, 2009.

[20] M. R. . G. VK, "Optimized switching scheme of cascaded h-bridge multilevel inverter using pso," International Journal of Electrical Power & Energy Systems, vol. 64, pp. 699–707, 1 2015.

[21] K. M. . A. SS, "Exploration of modulation index in multi-level inverter using particle swarm optimization algorithm," Procedia Computer Sci- ence, vol. 105, pp. 144–52, 1 2017.

[22] S. P. Singla D, "Performance analysis of harmonic elimination in cascaded h-bridge multilevel inverters using constrained pso algorithm," International Journal Series in Engineering Science (IJSES), vol. 27, no. 3328, pp. 1–4, 2017.

[23] X. H. W. X. Z. K. Gao XZ, Govindasamy V, "Harmony search method: theory and applications. computational intelligence and neuroscience," in Computational intelligence and neuroscience, 2015.

2nd IEEE International conference on power Electronics, Intelligent Control and Energy systems (ICPEICES-2018)

Power Quality Improvement of Grid Connected DFIG using Fuzzy Controller

<table>
<tr>
<td align="center">Nitin Goel
Department of Electrical Engineering
YMCA University of Sci. and Tech.
Faridabad, India
goel.nitin81@gmail.com</td>
<td align="center">Paras Ram Sharma
Department of Electrical Engineering
YMCA University of Sci. and Tech.
Faridabad, India
prsharma1966@gmail.com</td>
<td align="center">Dheeraj Joshi
Department of Electrical Engineering
Delhi Technological University
New Delhi, India
ee.dheeraj@gmail.com</td>
</tr>
</table>

Abstract— **Paper presents fuzzy based sensorless control strategy for grid connected doubly fed induction generator. In this system doubly fed induction generator connected with pulse width modulated voltage source converter to the rotor and stator is connected to utility grid. Rotor side voltage source converter connected with uncontrolled diode rectifier through dc link and battery storage system. A fuzzy based controller is used to improve the performance of rotor side controller. Comparatively reduced cost can be achieved by proposed control technique as a single voltage source converter and an uncontrolled diode rectifier is used in place of conventional back to back converter. Simulation studies were performed to analyse the performance of the proposed control technique. The result shows that the proposed strategy enhances the system stability, power factor and also improves the power quality with change in system parameters such as wind sped and connected load.**

Keywords— **Doubly fed induction generator (DFIG), wind energy, fuzzy controller, voltage control, total harmonic distortion**

I. INTRODUCTION

Today the renewable energy sources like hydro, solar, wind, geothermal, bio-power, ocean power are more preferred than the conventional energy sources for utilizing for the electrical power needs as they are less environment polluting and are in abundance [1]. Wind energy, as compare to other renewable energy sources is most economical energy source. The use of wind power generation has been started since the early 1980s. In recent years, wind energy conversion has implemented with the advanced technologies to extract maximum energy and supply to the grid. Conventional wind power production system is a fixed speed induction generator that represents a simple and robust solution. Now, variable speed turbine base induction generator has been adopted with the advantage of generating higher energy, increased active and reactive power control and fluctuation-free output.

In the wind energy conversion system DFIG mostly used [2], as the rating back-to-back converter does not exceed one third rating of machine. In this system the active and reactive power is controlled by rotor side converter and grid converter is synchronized with grid [3]. In standalone DFIG system voltage and frequency are controlled to the rated value. Mostly field oriented and direct power control techniques are used for regulation [3-4]. Different techniques have been introduced for replacing encoder by rotor position estimator [5]-[10].

The major problem in grid connected DFIG is voltage dips in stator voltage that creates oscillations in rotor over voltage and flux linkages in stator winding [11]-[12]. In [13]-[16] different control strategies have been developed to improve low-voltage ride-through capability. To overcome the effect of grid voltage dips, an extra grid series converter is attached on the stator side [13]. A dynamic programming power control is used for controlling oscillations for steady-state conditions and unbalanced voltage dips in [14]. To minimize the effect of unbalanced grid voltage on DC capacitor a novel dc-capacitor current control method is used in [16]. Stand-alone DFIG system with non linear loads also affected by these problems [6] and [17]. In [17] harmonic voltage components in stand-alone system are eliminating by proportional integral (PI) implementation.

In present study DFIG based wind plant is connected to grid with variable loading arrangement. DFIG is directly connected to grid at stator side and PWM voltage source inverter is connected at rotor side. Voltage source inverter is connected with diode rectifier via dc link capacitor. A battery is connected between the Diode rectifier and the Inverter so that the increased voltage in rectifier is utilized in charging the battery and voltage across the Inverter is constant. Fuzzy controller with Pulse generator is used to control the gate signals of the switching devices.

II. DFIG MODEL

The general model for wound rotor induction machine is similar to any fixed-speed induction generator as follows:

A. Voltage equations:

Stator Voltage Equations:

$$V_{qs} = R_s \times I_{qs} + p \times \lambda_{qs} + \omega_s \times \lambda_{ds} \qquad (1)$$

$$V_{ds} = R_s \times I_{ds} + p \times \lambda_{ds} - \omega_s \times \lambda_{qs} \qquad (2)$$

Rotor Voltage Equations:

$$V_{qr} = R_r \times I_{qr} + p \times \lambda_{qr} + \omega_r \times \lambda_{dr} \qquad (3)$$

$$V_{dr} = R_r \times I_{dr} + p \times \lambda_{dr} - \omega_r \times \lambda_{qr} \qquad (4)$$

B. Power Equations:

$$P_s = 3/2 \, (V_{ds} \times I_{ds} + V_{qs} \times I_{qs}) \qquad (5)$$

$$Q_s = 3/2 \, (V_{qs} \times I_{ds} - V_{ds} \times I_{ds}) \qquad (6)$$

C. Torque Equation:

$$T_s = -3P/4 \, (\lambda_{qs} \times I_{qs} - \lambda_{ds} \times I_{ds}) \qquad (7)$$

D. Flux Linkage Equations:

Stator Flux Equations:

$$\lambda_{qs} = L_s \times I_{ds} + L_m \times I_{dr} \qquad (8)$$

978-1-5386-6624-1/18 $31.00 © 2018 IEEE

$$\lambda_{ds} = L_s \times I_{qs} + L_m \times I_{qr} \qquad (9)$$

Rotor Flux Equations:

$$\lambda_{qr} = L_r \times I_{qr} + L_m \times I_{qs} \qquad (10)$$

$$\lambda_{dr} = L_r \times I_{dr} + L_m \times I_{ds} \qquad (11)$$

where λ_s and λ_r are the stator and rotor magnetic flux; L_s and L_r are the stator and rotor inductances; L_m is the magnetizing inductance ; V_s and V_r are the stator and rotor voltages; I_s and I_r are the stator and rotor currents; R_s and R_r are the stator and rotor resistances; ω_s and ω_r are the synchronous and rotating angular frequencies; P_s and Q_s are the active and reactive stator power; T_s is the torque; V_{dc} is the dc link voltage; p is the derivative function (d/dt); 'd' and 'q' represents the direct and quadrature axis quantities.

III. SIMULATION AND RESUTS

Fig-1 shows the proposed model. In the proposed circuit doubly fed induction generator is connected to grid with variable loading. The effect of variable loading on DFIG is being observed and controlling is done by the use of Fuzzy Logic control method. For that, single pulse width modulated voltage source converter is connected to the rotor terminals of DFIG and utility grid is connected to the stator side of the DFIG. It is found that there are many limitations in the conventional controlling techniques used for DFIG. Here propose a configuration with a battery connected storage system in between the Diode Rectifier and the Inverter dc link with a control strategy to maintain the grid power constant. This proposed control strategy is simulated in Matlab-Simulink (Fig-2).

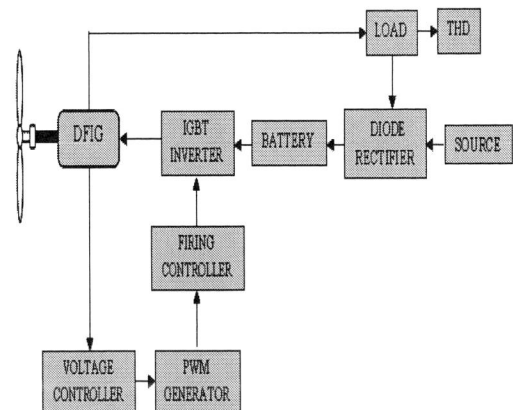

Fig.1. Circuit diagram of proposed DFIG model

Fig-3 and 4 shows the inputs to fuzzy controller and fig 5 and 6 shows the outputs of controller. Fig 7 represents fuzzy inference rules i.e. Negative large is N1, negative medium is N2, negative small is N3, zero, positive small is N4 and positive medium is N5. The FLC circuit is supplied with voltage feedback system which provide signal to the PWM generation system. The power factor is maximized by the circuit consists of inductor, capacitor, and one switch i.e., IGBT.

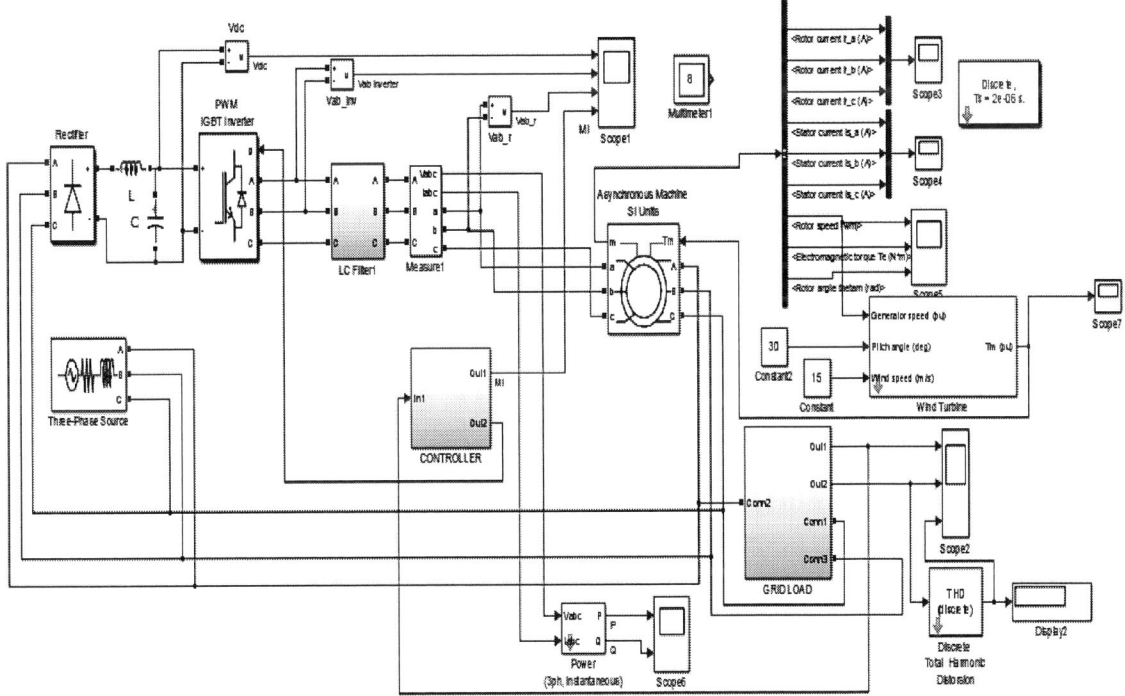

. Fig.2 Simulation of proposed circuit

978-1-5386-6624-1/18 $31.00 © 2018 IEEE

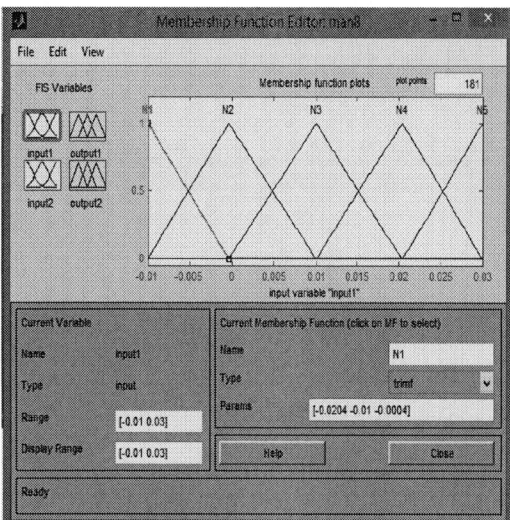

Fig.3. Input-1 of the fuzzy controller

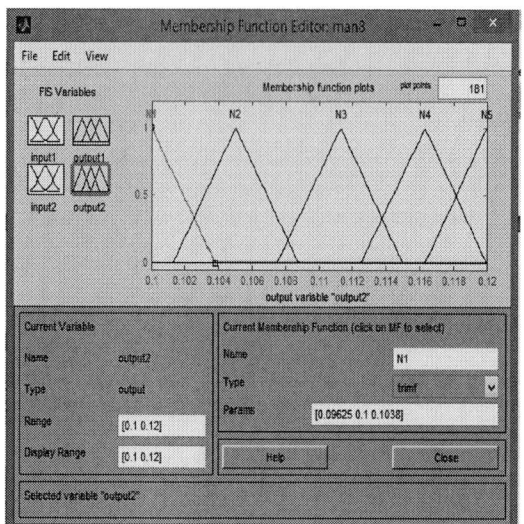

Fig.6. Output 2 of the Fuzzy controller

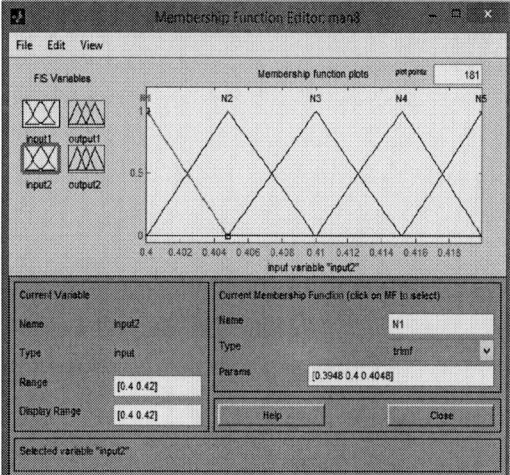

Fig-4. Input-2 of the fuzzy controller

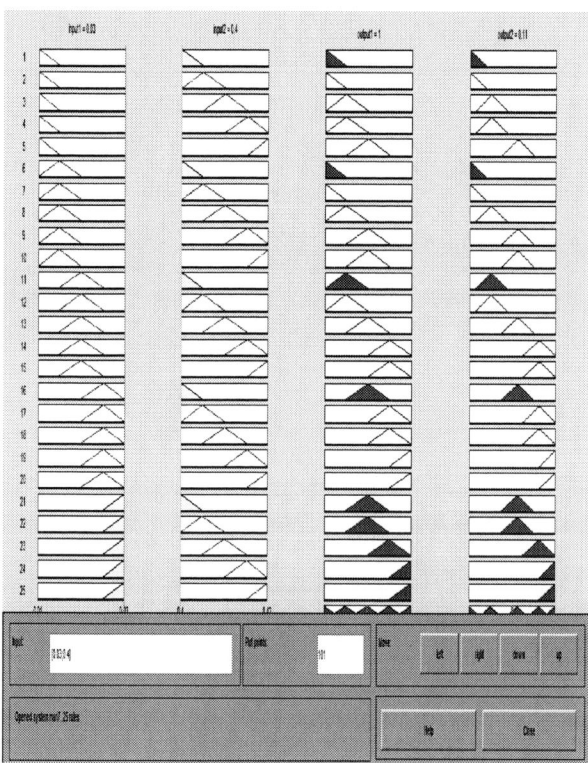

Fig.7. Fuzzy inference rules

Fig-8 to 13 are performance curves of propose DFIG model which present the supply voltage, dc bus voltage, stator and rotor current, inverter voltage, active and reactive power, stator and rotor current, rotor speed, electromagnetic torque and total harmonic distortion.

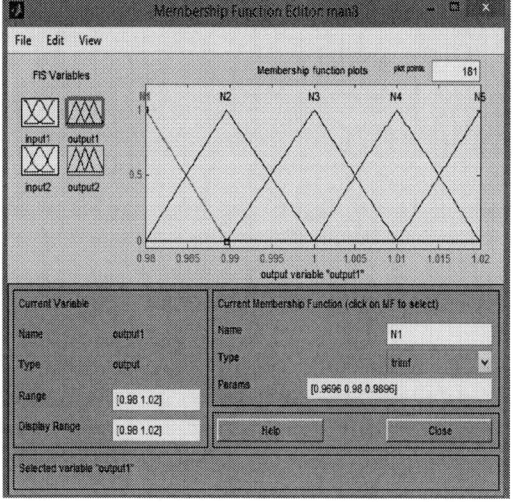

Fig.5. Output-1 of the Fuzzy controller

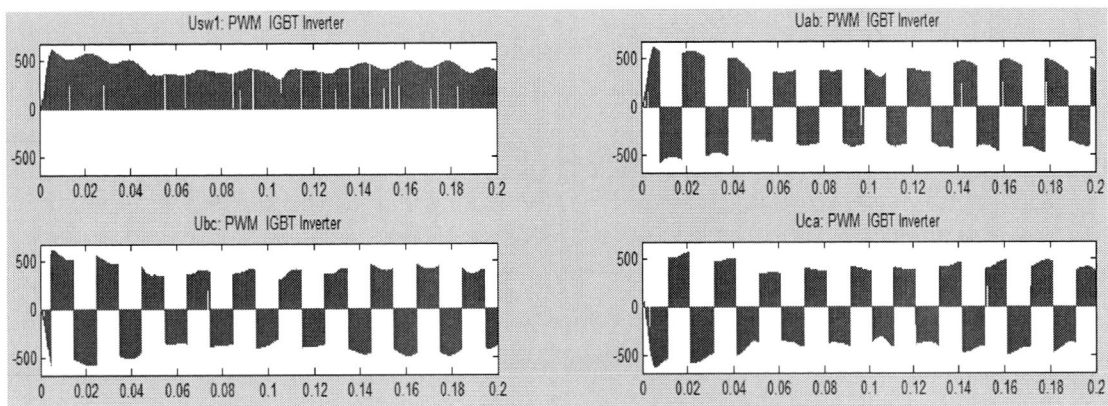

Fig.8. Converter bridge wave form for supply voltage

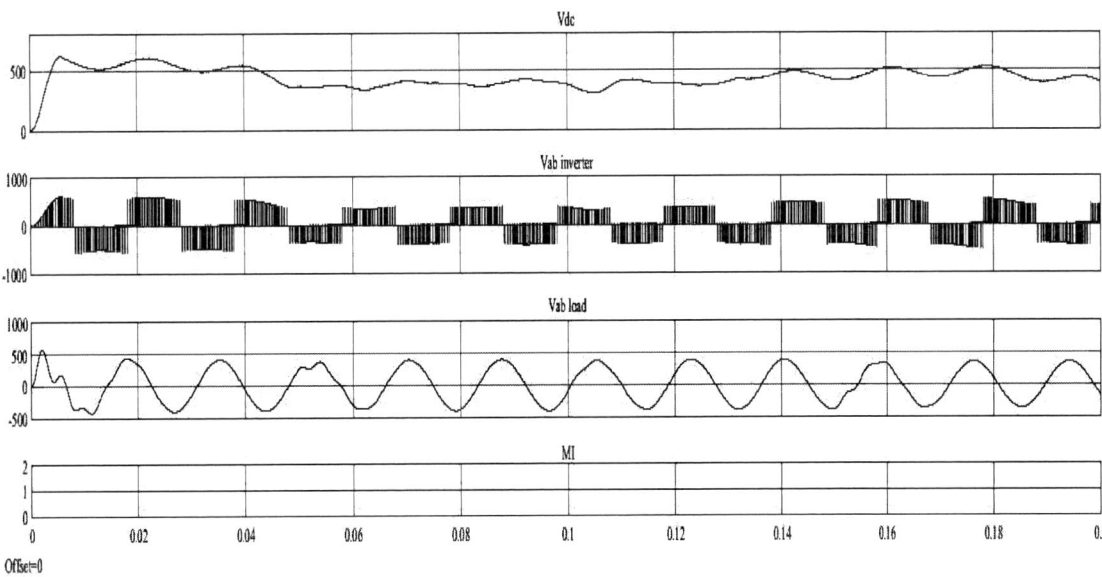

Fig.9. V_{dc} , V_{ab} inverter, V_{ab_r} (rotor voltage) and MI value of Convertor

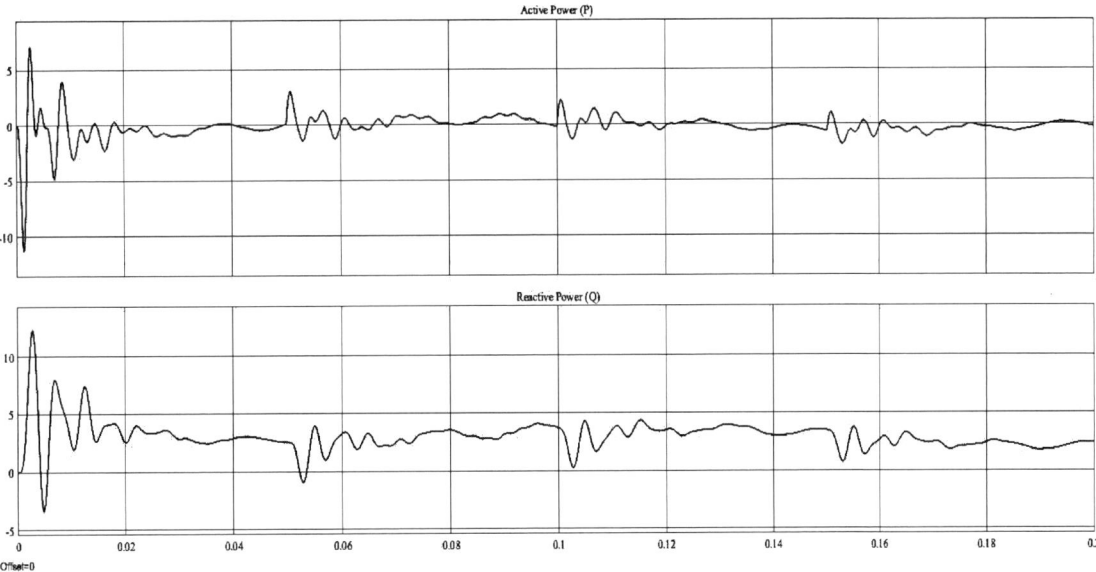

Fig.10. Active and Reactive Power

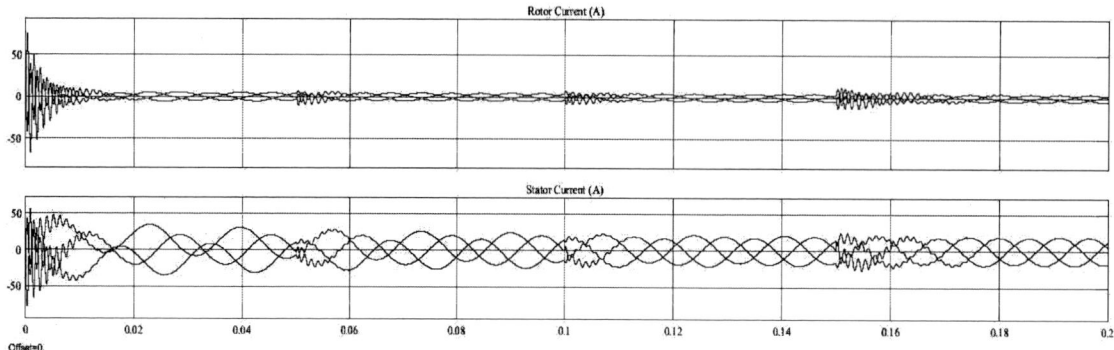

Fig.11. Rotor and stator currents waveform

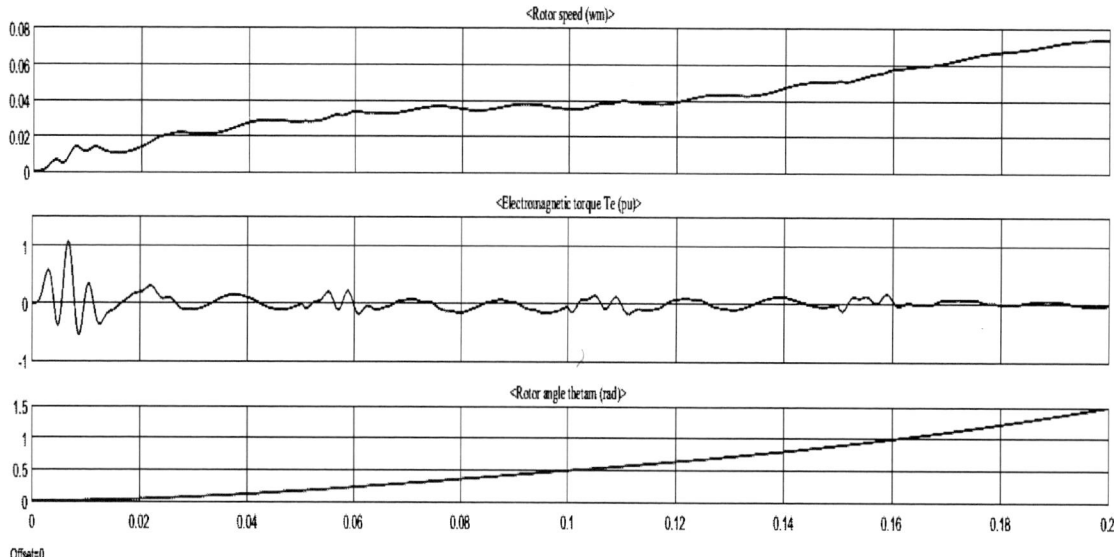

Fig.12. Rotor speed, Electromagnetic –Torque and Rotor angle wave form

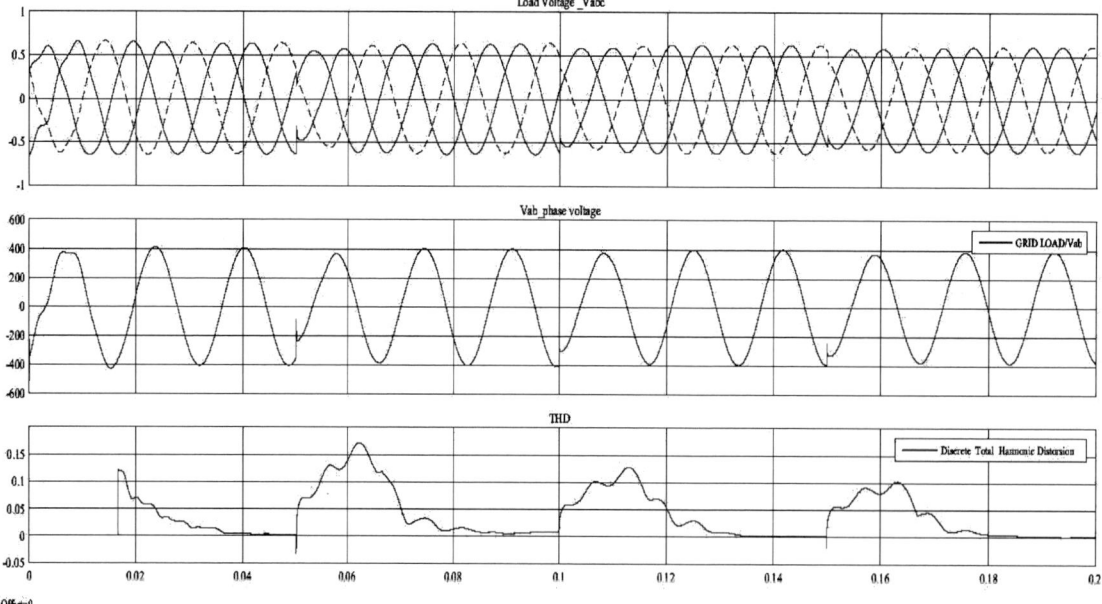

Fig.13. Grid load voltage V_{abc}, V_{ab} and THD wave form

Fig-12 shows rotor speed, electromagnetic torque and rotor angle waveform. Fig-13 present Grid load voltage (Vabc & Vab) and total harmonic distortion (THD) wave form. The variation of current, voltage and THD can be seen as the loading of the power grid is done at 0.05sec, 0.1sec and 0.15sec. The simulation result shows that the proposed strategy controls the power flowing in the grid system by enhancing the performance of the DFIG.

IV. CONCLUSION

The above study present a sensorless fuzzy based control strategy for a doubly fed induction generator connected to utility grid at stator side. Rotor side inverter is connected to dc link via diode rectifier and dc storage system. Here, conventional back to back converter is replaced with single voltage source inverter and a diode rectifier. The benefits of proposed scheme are robust control and reduced cost of converters. The loading of the power grid is done at 0.05, 0.1 and 0.15sec. The simulation results show that at variable loading the DFIG does not stop providing the power; also it controls the voltage stability and harmonics introduced by the loading. It also improves the power factor of system.

REFERENCES

[1] S. Muller, M. Deicke and R. W. De Doncker, "Doubly fed induction generator systems for wind turbine," IEEE Industry Applications Magazine, vol.3, pp 26-33, May/June 2002.

[2] R. Cárdenas, R. Pena, S. Alepuz, G. Asher, "Overview of control systems for the operation of DFIGs in wind energy applications," IEEE Trans. Ind. Electron., vol. 60, no 7, pp. 2776-2798, Jan 2013.

[3] H. Nian, Y. Song, "Direct power control of doubly fed induction generator under distorted grid voltage," IEEE Trans. Power Electron., vol. 29, no. 2, pp. 894–905, Feb 2014.

[4] R. Pena, J. C. Clare, and G. M. Asher, "Doubly fed induction generator using back-to-back PWM converters and its application to variable speed wind-energy generation", Proc. Inst. Elect. Eng. – Elect. Power Appl., vol. 143, pp. 231-241, May 1996.

[5] B. Shen, B. Mwinyiwiwa, Y. Zhang, B.T. Ooi, "Sensorless Maximum Power Point Tracking of Wind by DFIG Using Rotor Position Phase Lock Loop (PLL)," IEEE Trans. Power Electron., vol. 24, No.4, pp. 942–951, April 2009.

[6] A. Kumar Jain, V.T. Ranganathan, "Wound Rotor Induction Generator with Sensorless Control and Integrate Active Filter for Feeding Nonlinear Loads in a Stand-Alone Grid," IEEE Trans. on Ind. Electron., vol. 55, No.1pp. 218 – 228, Jan 2008.

[7] J. Morren and S. W. H. de Haan, "Ride through of wind turbines with doubly-fed induction generator during voltage dip," IEEE Trans. Energy Conversion, vol. 20, no. 2, pp. 435-441, June 2005.

[8] F. Castelli-Dezza, G. Foglia, M. F. Iacchetti, and R. Perini, "An MRAS observer for sensorless DFIM drives with direct estimation of the torque and flux rotor current components," IEEE Trans. Power Electron.,vol. 27, no. 5, pp. 2576–2584, May 2012.

[9] G. D. Marques, D. M. Sousa, "Air-gap power vector based sensorless method for DFIG control without flux estimator", IEEE Trans. Ind. Electron., vol. 58, no. 10, pp. 4717-4726, Oct 2011.

[10] T. Petru and T. Thiringer, "Modeling of wind turbines for power system studies," IEEE. Trans. Power Systems, vol. 17, no. 4, pp. 144-151, Nov 2002.

[11] F. K. A. Lima, A. Luna, P. Rodriguez, E. H. Watanabe, F. Blaabjerg, "Rotor voltage dynamics in the doubly fed induction generator during grid faults," IEEE Trans. Power Electron. , vol. 25, No.1, pp. 118–130, Jan 2010.

[12] G. D. Marques, D. M. Sousa, "Understanding the DFIG during Voltage Dips", IEEE Trans. En. Conv., vol. 27, no. 2, pp. 421-431, June 2012.

[13] J. G. Slootweg, S. W. H. de Hann, H. Polinder, and W. L. Kling, "General model for representing variable speed wind turbines in power system dynamic simulations," IEEE. Trans. Power Systems, vol. 18, no. 1, pp. 1132-1139, Feb 2003.

[14] P.S. Flannery, G. Venkataramanan, "A fault tolerant doubly fed induction generator wind turbine using a parallel grid side rectifier and a series grid side converter," IEEE Trans. Power Electron. , vol. 23, No.3, pp. 1126–1135, May 2008.

[15] D. Santos-Martin, J.L. Rodriguez,-Amenedo, S. Arnalte, "Providing ride-through capability to a doubly fed induction generator under unbalanced voltage dips," IEEE Trans. Power Electron. , vol. 24, No.7, pp. 1747–1757, July 2009.

[16] S. Hu, X. Lin, Y. Kang, and X. Zou, "An improved low-voltage ride through control strategy of doubly fed induction generator during grid faults," IEEE Trans. Power Electron., vol. 26, no. 12, pp. 3653–3665, Dec 2011.

[17] Changjin Liu, F. Blaabjerg, Wenjie Chen, Dehong Xu, "Stator current harmonic control with resonant controller for doubly fed induction generator," IEEE Trans. Power Electron, vol. 27, no. 7, pp. 3207-3220, July 2012.

2nd IEEE International conference on power Electronics, Intelligent Control and Energy systems (ICPEICES-2018)

A Robust Adaline Based Control of Shunt Active Power Filter without Load and Filter Current Measurement

Shivangni Sharma
Department of Electrical Engineering
National Institute of technology
Patna, India
shivangni.eepg15@nitp.ac.in

Vimlesh Verma
Department of Electrical Engineering
National Institute of technology
Patna, India
vimlesh.verma@nitp.ac.in

Abstract—**This paper presents a new robust control strategy which is based on Adaline approach. The purpose of the proposed technique is to eliminate load and filter current sensors, normally present in conventional Adaline based control of Shunt Active Power filter(SAPF) thus reducing its hardware cost. The proposed method also enables SAPF to work under highly dynamic load conditions where other similar approaches with reduced sensors ceases to perform. A comparative analysis with other similar classical approach under same load conditions is done to show the effectiveness of proposed method. MATLAB/SIMULINK results are shown under fast varying load conditions to validate the proposed work.**

Keywords—Sensorless control, harmonic, Active power filter, Adaline

I. INTRODUCTION

With the advent of Distributed energy resources, renewable energy resources which are highly dependent on power electronics devices, there is dire need of maintaining the power quality. Non-linearity present in power electronics-based devices is one of the basic reason behind power quality issues[1]. In view of increasing no. of problems caused by power quality in terms of loss of production, penalty imposed on industries, wastage of raw materials, Financial loss and so on, a wide variety of solutions and techniques has been proposed till now. These includes passive elements such as inductors, capacitors, various combination of these, custom power devices, matrix converters, improved power quality AC-DC converters and so on. Active Power Filter(APF)[2] is one of the dynamic solution to such power quality problem. SAPF (Shunt Active Power Filter) is one of the widely used topology among other in APF such as Series-APF and Unified Power quality controller (UPQC) as it deals with some of main power quality problems such as harmonics, reactive power, load unbalancing, voltage regulation, power factor improvement etc. and it can be optimized according to cost, size, and requirements. SAPF compensates harmonic current by introducing an equal and opposite current into the system.

SAPF requires four types of sensors i.e. Source voltage sensors for amplitude and phase information, Source current sensors for input current control, DC Voltage sensors for power flow control, Load current sensors for estimation of harmonics and Filter current (In case of direct current control). Failure and faults in sensors will lead to maloperation of SAPF[4] which will further aggravate harmonics in power system instead of compensating it. Reduction in number of

sensors presents clear advantages such as elimination of noise related to sensor, simplified calibration, increased robustness etc. Also, sensorless SAPF can be used as a backup in case of continuous operation of SAPF in case of sensor failure or fault. Most of the control strategies which deals with reduced number of current sensors (Load current sensors) are based on Indirect current control which directly imposes sinusoidal reference current on actual current. Here information about phase is obtained through source voltages and magnitude from DC voltage. Load current sensor reduction approach works well and it requires half no. sensors, it also present better harmonic results since the circuit is in closed loop. But these require extensive PI controller tuning which comes as a challenge in order to achieve optimum performance without load current sensors. First control strategy without using Load current sensor was based on basic PI controller [5][6]. Further the same technique was presented in [7] but with reduced number of inverter switches. These methods are only limited to low and medium power applications. In place of PI controller Fuzzy controller [8] and VS-APPC (Variable Structure Adaptive Pole Placement controller) was proposed in [9], remaining operation was same as in PI controller-based method discussed above. Both fuzzy and VS-APPC control have better dynamic response than PI controller. In [10] author proposed a control strategy by developing system model in d-q frame and controlling voltage across source inductance to be sinusoidal without use of any Load current sensor. Drawback of this control strategy is requirement of large source inductance.Conventional SAPF based on P-Q Theory based reference current generation techniques without sensing Load current was proposed in [11],[12]. Predictive current control-based control strategy without Load current sensors was developed in [13], which has limitation of having slow transient response. Voltage sensorless approach was given in [14] which works on the concept of Resistive-APF(R-APF). R-APF is controlled in such a way as to present infinite impedance to Harmonic frequencies. Drawback of this method is, it is based on impedance emulation accuracy of which depends on plant model. In a significant development a control strategy for selective harmonic compensation without using both Source voltages and load currents was proposed in [15], [16]. In [15] Virtual Flux Oriented Control (VFOC) was employed which uses virtual flux to determine grid synchronization angle. Concept of virtual flux is employed in [16] by using AF-PLL (Adaptive filter and Phase locked loop) to improve disturbance rejection and in these techniques, Sinusoidal reference current is directly imposed on Source current without employing separate harmonic rejection

978-1-5386-6624-1/18 $31.00 © 2018 IEEE

algorithm. Source current and DC voltage sensors are indispensable in SAPF due their use in overcurrent and over voltage monitoring, but in [16] control scheme for power flow control is presented which does not require DC Voltage sensor leaving SAPF with only source current measurement which is preferable as it is the desired output. Overall desired number of SAPF should be as less as possible. The present research challenge is to reduce sensors in SAPF as much as possible without compromising with its performance and without much complex circuitry which will be immune to variation of parameters and ensure stability. In recent years sensorless control of APF has been widely investigated to increase its economic viability at industrial level.

In proposed system, load current sensors has been reduced without compromising with SAPF performance, extensive PI controller tuning and complex circuitry. LMS (least Mean square) algorithm-based ADALINE control[17-19]is utilized to control SAPF which makes distorted source current to follow sinusoidal unit voltage vectors. Weights are adjusted to determine required sinusoidal reference current which is imposed on actual source current. This proposed method is based on conventional approach which works for all kind of load variations. This method also works for unbalanced voltage (in case of magnitude unbalance). MATLAB/ SIMULINK results are shown for different conditions in order to observe the performance of modified control strategy with reduced sensors. SECTION II shows overall block diagrams, Proposed Control strategy is discussed in SECTION III, SECTION IV deals with results and discussion and SECTION V concludes the work.

II. CONTROL ARCHITECTURE

Fig. 1 shows the basic block diagram of SAPF with all the sensors intact (highlighted in red). SAPF is connected near load end in parallel to source and load. Among various configurations, one of the most popular most popular 3 phase 3 wire Voltage Source Converter(VSC) based SAPF is utilized in this paper

Reference current generation techniques senses source voltages(v_{Sa}, v_{Sb}, v_{Sc}) and load currents(I_{La}, I_{Lb}, I_{Lc}) to generate reference currents(I^*_a, I^*_b, I^*_c). A capacitor is used at DC side of VSC to maintain real power flow between source and load. In case of Indirect current control [20] Source current (I_{Sa}, I_{Sb}, I_{Sc}) along with reference current (extracted Fundamental component of Load current) is fed to controller to generate pulses for VSC. In case of direct current control, Filter current along with compensating current (extracted Harmonic component((I_{Fa}, I_{Fb}, I_{Fc}) of Load current) is fed to controller to generate pulses for VSC. VSC thus generates equal and opposite harmonics into the system making source current sinusoidal.

III. PROPOSED CONTROL STRATEGY

ADALINE network works by continuously tracking desired output by adjusting weights. Least Mean Square (LMS) algorithm is used here to update weights. Block diagram of control strategy is shown in Fig 2. Control strategy is explained in steps as follows.

A. Estimation of weights

Load current (I_L) injects harmonics into the system, therefore same harmonics are reflected into Source current (I_S) as

well. I_L(t) and I_S(t) can be decomposed into Real fundamental Active (I_{Lp}(t), I_{Sp}(t)), reactive (I_{Lq}(t), I_{Lq}(t)) and harmonic component (I_{Lh}(t), I_{Sh}(t)) respectively.

$$I_L(t) = I_{Lp}(t) + I_{Lq}(t) + I_{Lh}(t) \qquad (1)$$

$$I_S(t) = I_{Sp}(t) + I_{Sq}(t) + I_{Sh}(t) \qquad (2)$$

Unit vector voltages(u) for each phase separately (u_a, u_b, u_c) are generated by expressing source voltages in per unit. Unit voltage can also be generated with the help of PLL, in this case algorithm will also work for unbalanced magnitude of voltages as it only senses phase angle of v_{Sa}, v_{Sb}, v_{Sc} (Fig.1) for sinusoidal unit voltage generation.

Initial estimate of real fundamental component $I_{Sp}(t)$ of I_S(t) is expressed as:

$$I_{Sp}(t) = w * u \qquad (3)$$

Where w represents weight in general. Error $e(k)$ at kth instant between desired and actual output can be expressed as:

$$e(k) = \{I_S(k) - w(k)u(k)\} \qquad (4)$$

Source current tracks Unit vector voltages by updating weights at kth instant as follows:

$$w(k + 1) = w(k) + \eta\{I_S(k) - w(k)u(k)\}u(k) \qquad (5)$$

η (0<η<1) is converging coefficient, weights are updated according to eqn. (5) for each phase (w_a, w_b, w_c) separately as shown in Fig.2 and average of weights(w_e) is done to eliminate unbalance in current components.

$$w_e = \frac{(w_a + w_b + w_c)}{3} \qquad (6)$$

B. DC Voltage Regulation

DC capacitor across VSC maintains real power flow between source and load. Voltage(V_{DC}) across DC capacitor is maintained by comparing it with reference voltage (V^*_{DC}) with the help of PI controller. w_{Loss} and V_{DE} represents Filter losses and DC Voltage error respectively. k_{pd} and k_{id} are proportional and integral gains.

$$V_{DE(k)} = V^*_{DC}(k) - V_{dc}(k) \qquad (7)$$

$$w_{Loss}(k + 1) = w_{Loss}(k) + k_{id} * \{V_{DE}(k + 1) - V_{DE}(k)\} + k_{pd} * V_{DE}(k + 1) \qquad (8)$$

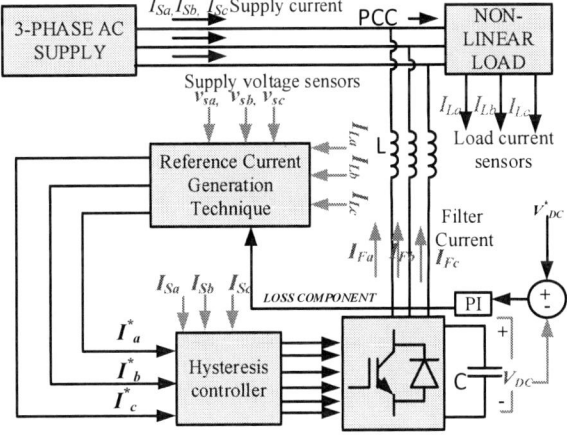

Fig. 1. Block diagram of SAPF showing all sensors

978-1-5386-6624-1/18 $31.00 © 2018 IEEE

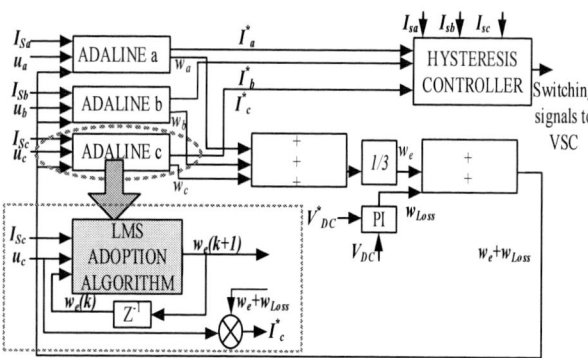

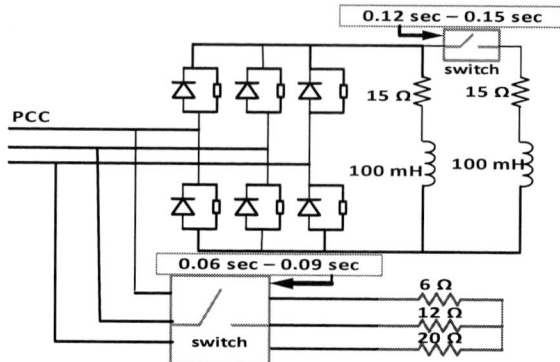

Fig. 4. Variations in Non-linear load

Fig. 2. Block diagram of Modified Adaline based control strategy

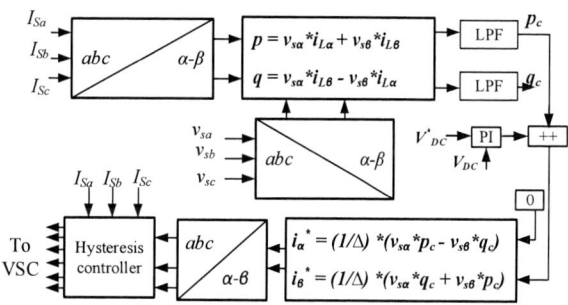

Fig. 3. Block diagram of Modified P-Q theory-based control of SAPF

C. Reference Current Generation

Reference Source current is calculated by multiplying total of loss component and equivalent weight to Unit vector voltages as shown in eqn. (9).

$$I_{Sa}^* = (w_e + w_{Loss}) * u_a \qquad (9)$$

Similarly reference currents are generated for phase b and c.

Reference current and actual source current are fed to Hysteresis controller to generate pulses for VSC. Pulses are generated for phase a as follows: If $I_a^* - I_{Sa} < HB$, upper switch is ON and lower switch is OFF and If $I_a^* - I_{Sa} > HB$, upper witch is OFF and lower switch is ON. Similarly, pulses are obtained for each phase separately. HB is hysteresis band.

IV. P-Q THEORY

Block diagram of modified P-Q theory without measurement of Load current is shown in Fig.3. Various equation related to modified P-Q theory are available in [11], [12].

V. RESULTS AND DISCUSSIONS

The performance of above mentioned techniques is observed with the help of MATLAB/SIMULINK results. System parameters: 3-phase ac supply 415V, Source resistances and inductances 0.02Ω and 0.1mH, coupling inductor 1mH , dc capacitor 3000µF, Reference DC Voltage (V_{DC}^*) 700 V and supply frequency 50 Hz respectively. 3 phase uncontrolled rectifier with R-L load at its side is taken as Non-Linear load. Non-linear and Unbalanced Load and its variations used for simulation is shown in Fig. 4.

A. Performance of Proposed control Strategy

Dynamic performance of SAPF is shown in Fig.5.Due to Non-Linear Load, Harmonics are injected into the system which makes Load Current Non-sinusoidal this in turn changes the source current to Non-sinusoidal having 27% THD (Fig.6(a)) as shown in Fig.5 Source (I_S) and Load current (I_L) waveforms. At instant t=0.03 sec, SAPF is introduced which improves Source current to 2.61 % THD(Fig.6(b)) for 0.03-0.06 sec. At 0.03 sec when SAPF is connected there is no delay in compensation which shows quick response of proposed control strategy. For 0.06-0.09 sec, unbalanced load is introduced, Source current quickly adjusted improving its THD to 2.66%(Fig.8(a)) from 24.44%. At t=0.09 sec, when unbalanced load is removed Source current converges fast to its original state without any delay. At 0.12 R-L load is changed by connecting an additional R-L branch to already connected R-L load at DC side of VSC as shown in Fig. 4. At this instant also, Source current quickly adjusted to according to changed R-L value improving THD from 32.40% to 4 %. DC Voltage regulation here is also satisfactory (V_{DC}, Fig. 5).

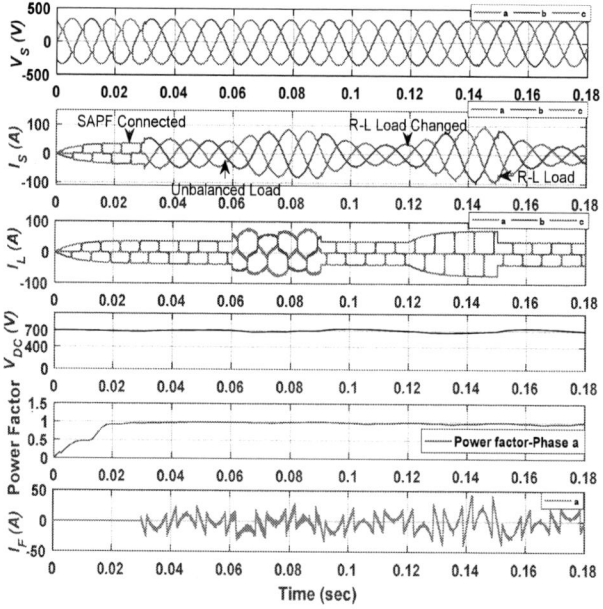

Fig.5.Performance of Proposed control strategy: Source voltage (V_S), Source current (I_S), Load current (I_L), DC Voltage (V_{DC}), Power factor and Compensating current (I_F) from top to bottom respectively

978-1-5386-6624-1/18 $31.00 © 2018 IEEE

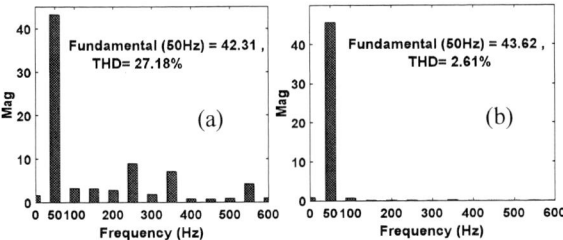

Fig.6.FFT Analysis of Source current (Phase a) in case of (a) Before SAPF connection (t=0 - 0.03 sec)) (b) Proposed control strategy after SAPF connection (t=0.03 sec)

Power factor improved after SAPF connection to almost unity (Power factor, Fig.5). Compensating current injected by SAPF is also shown (Fig. 5, I_F).

In order to demonstrate the effectiveness of Proposed ADALINE based control strategy, Performance of conventional ADALINE [17], [18] based control strategy under same parameter and load condition is also shown in Fig.7. After SAPF connection at t=0.03 sec, Source current improved to 1.59 % THD(Fig.8(b)) during normal R-L load, 2.56 % THD during Unbalanced load and 3.71 % THD during changed R-L load. Corresponding THD values are given in Table I.

B. Performance of Modified P-Q Theory

Performance of Modified P-Q theory is shown in Fig.9. SAPF is connected at t=0.03 sec after which source current (I_S) improves from THD of 27% to 2.12% (Fig.10(a)). At t= 0.06 sec Unbalanced load is connected, after delay of 1-2 cycle source current adjusted according to the load improving its THD from 24.44% to 3%. At t=0.09 sec when Unbalanced load is removed, then source current fails to resume back to its original state. From t=0.012 sec to 0.15 sec again R-L load is changed, here also Source current fails to respond. It has been observed at t=0.09, 0.12, 0.15 sec i.e. at points of disturbances system fails to respond which can be observed from waveforms of Source current (I_S) and compensating current (I_F, Fig.9). Therefore, for highly dynamic load Modified P-Q theory does not work effectively. It requires extensive PI controller tuning in order to maintain DC voltage (V_{DC}, Fig.9) constant across capacitor. Similar to proposed control strategy this method also follows indirect current control therefore Source current remains sinusoidal after SAPF connection but it fails to follow the change in magnitude according to variation in Load current (I_L, Fig.9). Power factor (Fig.9) is near unity after t=0.03 sec. In Modified P-Q theory in place of Load current, Source current is used to calculate instantaneous active and reactive power as shown in Fig.3.

Same system which is utilized for Modified P-Q theory is simulated using load current in place of Source current sensors (Fig.3) (same as Conventional P-Q theory [1], [3]) keeping other parameter and load same and the results are shown in Fig.11.When SAPF is connected at t=0.03 sec, Source current (I_S) improves from THD of 27% to 2.88%(Fig.10(b)). When Unbalanced load is connected from T=0.06 sec to 0.09 sec, after delay of 1-2 cycle, Source current adjusts itself according to the load improving its THD from 24.44% to 5.15%. When R-L load is changed from t=0.12 sec to 0.15 sec, here also source current waveform improved after delay of 1-2 cycle improving its THD from 32.40% to 4.09%. This delay of 1-2 cycle is usually present in conventional P-Q theory due to the

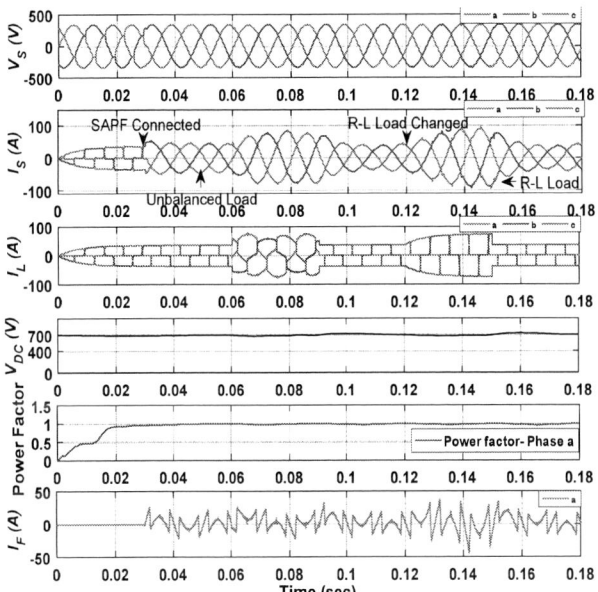

Fig.7.Performance of conventional ADALINE based control with all sensors intact. Source voltage (V_S), Source current (I_S) and Load current (I_L), DC Voltage (V_{DC}), Power factor and Compensating current (I_F) from top to bottom respectively

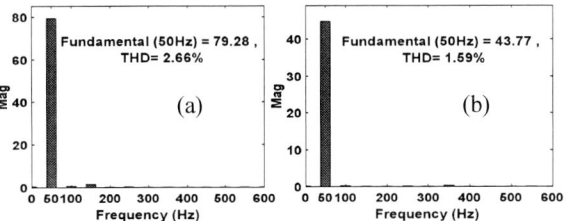

Fig.8.FFT Analysis of Source current (Phase a) in case of (a) Proposed control strategy during unbalanced load (t=0.06-0.09 sec) (b) Conventional ADALINE control strategy after SAPF connection (t=0.03 sec)

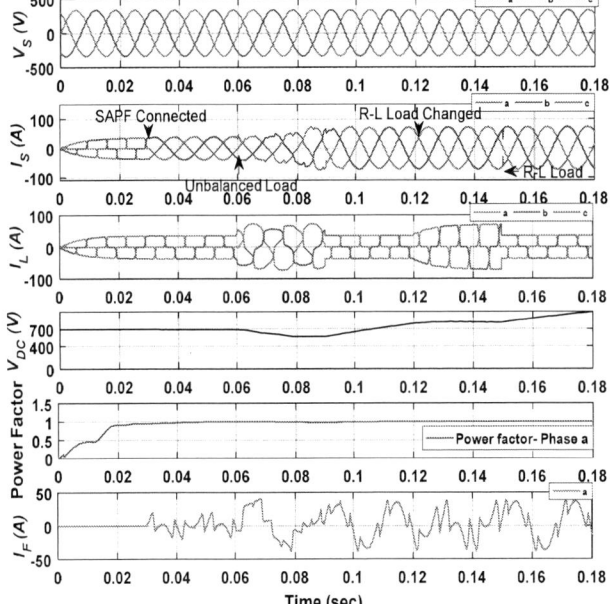

Fig.9.Performance of Modified PQ Theory without Load current sensor: Source voltage (V_S), Source current (I_S), Load current (I_L), DC Voltage (V_{DC}), Power factor and Compensating current (I_F) from top to bottom respectively

978-1-5386-6624-1/18 $31.00 © 2018 IEEE   395

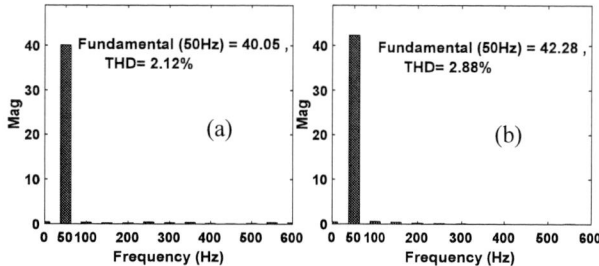

Fig.10.FFT Analysis of Source current (Phase a) in case of (a) Modified P-Q theory after SAPF connection (t=0.03 sec) (b) Conventional P-Q theory after SAPF connection (t=0.03 sec)

use of lot of transformations and Low Pass filter. DC Voltage regulation and Power factor can also be observed from Fig. 11. SAPF is able to adjust accordingly at instant of disturbance (i.e. t=0.06,0.09,0.12 and 0.15 sec respectively) which can be clearly observed from Source Current (I_S) and compensating current (I_F) waveforms (Fig.11). Therefore, Modified P-Q theory with reduced number of sensors does not performs same as conventional P-Q theory with all sensors.

VI. COMPARISON

Table I shows THD values of Phase a Source current for four types of control strategy discussed above. It can be clearly observed that Proposed ADALINE based control strategy performs effectively with reduced number of sensors without compromising with the SAPF performance. Modified P-Q theory failed after in case of certain load variations discussed earlier but THD remains 0.64% because of sinusoidal reference current generation in indirect current control.

Table II shows number of sensors utilized in various control techniques discussed. Proposed Adaline based control strategy does not involve load current measurement.

VII. CONCLUSION

Due to low cost and increased robustness, sensorless and fault tolerant operation of SAPF is utmost desirable. The paper presented a robust ADALINE based control strategy which enable SAPF to work without sensing Load current. While there are other similar methods available in literature but proposed method works without any complex circuitry and extensive PI controller tuning which makes it most suitable in comparison to other techniques. Also Proposed control strategy is based on ADALINE based control which has better performance in comparison to other techniques. Load current sensor is replaced by Source current sensor in control algorithm. Performance of SAPF under different conditions is shown with the help of simulation results. With reduced no. of sensors SAPF is able to reduce harmonics to considerable limit. Current state regarding reduction of sensors including most recent research and trends is also discussed. The proposed modification can be utilized for fault tolerant operation as it can be used as back up for continuous operation of SAPF in case of sensor failure. Further there is still need of in depth research for solving problem of high computational burden in sensorless techniques and parameter robustness. Paper suggested one such simplified and conventional approach to reduce sensors. Future work is to develop control strategy which require minimum number of sensors.

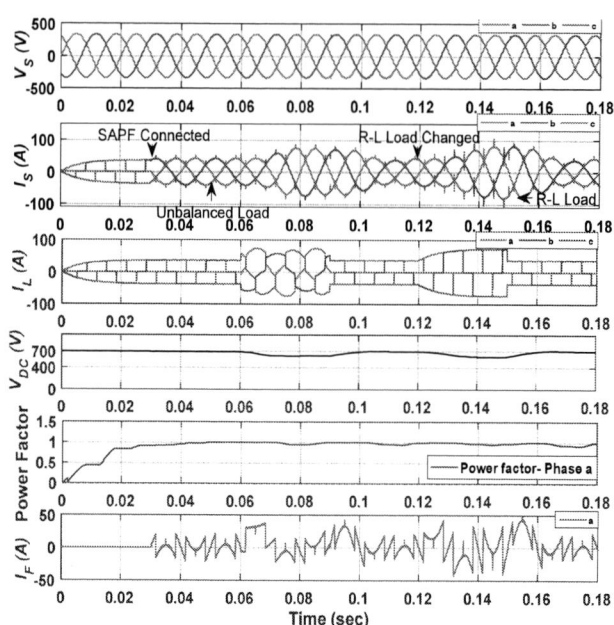

Fig.11.Performance of conventional PQ Theory with all sensors intact. Source voltage (V_S), Source current (I_S) and Load current (I_L), DC Voltage (V_{DC}), Power factor and Compensating current (I_F) from top to bottom respectively

TABLE I.   THD(%) OF SOURCE CURRENT IN ALL CASES

Load condition	Control Strategy			
	Proposed Modified Adaline control	Adaline with all sensors	Modified P-Q Theory	Classical P-Q Theory
Before Compensation	27.18	27.18	27.18	27.18
Non-linear Load	2.16	1.59	2.12	2.88
Non-linear and Unbalanced load	2.66	2.56	3	5.15
Change in R-L Load	4	3.71	0.64	4.09

TABLE II.   NO. OF SENSORS UTILISED

Control Strategy	Type of sensor			Total
	Source Voltage and DC Voltage	Source Current	Load current	
Proposed Modified Adaline control	4	3	0	7
Conventional Adaline control	4	3	3	10
Modified P-Q theory	4	3	0	7
Conventional P-Q theory	4	3	3	10

ACKNOWLEDGMENT

This work was supported by TEQUIP III towards Registration fee and Travelling Allowance.

REFERENCES

[1] Singh.B., Chandra.A, K.Al-haddad,"Power quality: problems and mitigation techniques",John Wiley and Sons, UK,2015.

[2] L.Gyugyi and Strycula," Active Power Filters," in *proc. Of IEEE Industrial Application Annual meeting*, vol.19-C, pp.529-535,1976.

[3] H. Akagi, Y. Kanazawa and A. Nabae, "Instantaneous Reactive Power Compensators Comprising Switching Devices without Energy Storage Components," in *IEEE Transactions on Industry Applications*, vol. IA-20,no. 3, pp. 625-630, May 1984.

[4] S.Sharma, V.Verma, "Performance of Shunt Active Power Filter Under Sensor Failure," 2017 *IEEE International WIE Conference on Electrical and Computer Engineering (WIECON-ECE)*, Dehradun, 2017, in press.

[5] H. L. Jou, "Performance comparison of the three-phase active-power-filter algorithms," in *IEE Proceedings - Generation, Transmission and Distribution*, vol. 142, no. 6, pp. 646-652, Nov 1995

[6] K. Chatterjee, B. G. Fernandes and G. K. Dubey, "An instantaneous reactive volt-ampere compensator and harmonic suppressor system," in *IEEE Transactions on Power Electronics*, vol. 14, no. 2, pp. 381-392, Mar 1999

[7] G. Joos, Su Chen and K. Haddad, "Four switch three phase active filter with reduced current sensors," *2000 IEEE 31st Annual Power Electronics Specialists Conference. Conference Proceedings (Cat. No.00CH37018)*, Galway, 2000, pp. 1318-1323 vol.3

[8] J. Dixon, J. Contardo and L. Moran, "DC link fuzzy control for an active power filter, sensing the line current only," *PESC97. Record 28th Annual IEEE Power Electronics Specialists Conference. Formerly Power Conditioning Specialists Conference 1970-71. Power Processing and Electronic Specialists Conference 1972*, St. Louis, MO, 1997, pp. 1109-1114 vol.2.

[9] R. L. A. Ribeiro, C. C. Azevedo and R. M. Sousa, "A non-standard adaptive control for shunt active power filter without current harmonic detection," *IECON 2010 - 36th Annual Conference on IEEE Industrial Electronics Society*, Glendale, AZ, 2010, pp. 2007-2012.

[10] M. Routimo, M. Salo and H. Tuusa, "Current sensorless control of a voltage-source active power filter," *Twentieth Annual IEEE Applied Power Electronics Conference and Exposition, 2005. APEC 2005.*, 2005, pp. 1696-1702 Vol. 3.

[11] E. Ozdemir, M. Ucar, M. Kesler and M. Kale, "A Simplified Control Algorithm for Shunt Active Power Filter Without Load and Filter Current Measurement," *IECON 2006 - 32nd Annual Conference on IEEE Industrial Electronics*, Paris, 2006, pp. 2599-2604

[12] N. Gupta, S. P. Dubey and S. P. Singh, "DSP based control algorithm for three-phase four-wire shunt active filter without load and filter current measurement," *2010 Annual IEEE India Conference (INDICON)*, Kolkata, 2010, pp. 1-6

[13] D. Nedeljkovic, M. Nemec, K. Drobnic and V. Ambrozic, "Active Power Filter with a Reduced Number of Current Sensors", *Electrotechnical review*, vol.76, no. 5, Sep. 2009, pp 275-280

[14] H. Bai, X. Wang and F. Blaabjerg, "A Grid-Voltage-Sensorless Resistive-Active Power Filter With Series LC-Filter," in *IEEE Transactions on Power Electronics*, vol. 33, no. 5, pp. 4429-4440, May 2018..

[15] M. B. Ketzer and C. B. Jacobina, "Nonlinear virtual flux oriented control for sensorless active filters," *2013 Brazilian Power Electronics Conference*, Gramado, 2013, pp. 393-398

[16] M. B. Ketzer and C. B. Jacobina, "Virtual Flux Sensorless Control for Shunt Active Power Filters With Quasi-Resonant Compensators," in *IEEE Transactions on Power Electronics*, vol. 31, no. 7, pp. 4818-4830, July 2016

[17] B.Singh and J. Solanki," An implementation of and Adaptive Control Algorithm for a Three-Phase Shunt Active Filter," *in IEEE Transactions on Industrial Electronics*, vol. 56, no.8, pp. 2811-2820, Aug.2009

[18] B. Singh, V. Verma and J. Solanki, "Neural Network-Based SelectiveCompensation of Current Quality Problems in Distribution System," in *IEEE Transactions on Industrial Electronics*, vol. 54, no. 1, pp. 53-60,Feb. 2007.

[19] A. Singh, M. Badoni and B. Singh, "Application of least means square algorithm to shunt compensator: An experimental investigation," *2014 IEEE International Conference on Power Electronics, Drives and Energy Systems (PEDES)*, Mumbai, 2014, pp. 1-6.

[20] P. Chittora, A. Singh and M. Singh, "Simple and efficient control of DSTATCOM in three-phase four-wire polluted grid system using MCCF-SOGI based controller," in *IET Generation, Transmission & Distribution*, vol. 12, no. 5, pp. 1213-1222,2018.

978-1-5386-6624-1/18 $31.00 © 2018 IEEE

2nd IEEE International conference on power Electronics, Intelligent Control and Energy systems (ICPEICES-2018)

THD Minimisation in 15- Level Hybrid Multilevel Inverter using Harmonic Minimization Technique

Abhishek Azad
Electrical Engineering Department
Punjab Engineering College
Chandigarh, India
Email: abhishekazad01@gmail.com

Jagdish Kumar
Electrical Engineering Department
Punjab Engineering College
Chandigarh, India
Email: jagdishkumar@pec.ac.in

Abstract— **Multilevel inverter technology is gaining more attention these days due to its various advantages over conventional 2/3-level inverter technologies. Researchers have proposed many topologies for MLI to reduce the switch and DC source counts. These topologies are known as 'Hybrid Multilevel Inverter topology'. The output waveform of multilevel inverter resembles with the sinusoidal wave but still has harmonic contents. THD of the output can be reduced further by applying any suitable modulation technique. A single-phase hybrid multilevel inverter topology with reduced switch count for fifteen levels is presented in this paper and total harmonics minimization modulation technique is applied to reduce THD of its output below 5% level as per IEEE 519 standards. The inverter is implemented & simulated in MATLAB/SIMULINK and output waveforms of voltage & current are shown along with their FFT analysis. Results from the simulation are also compared and verified with results obtained from experimental setup.**

Keywords—15-level, multilevel inverter, total harmonic minimization technique, THD minimization, Genetic algorithm, Hybrid topology.

I. INTRODUCTION

DC to AC conversion has now become an indispensable part of any modern power system. An Inverter system is utilised very frequently these days especially in static reactive power compensation, HVDC transmission, variable-speed drives and renewable sources & grid interfacing particularly for solar and fuel cells. For normal and efficient functioning of system, an advance, reliable and efficient inverter technology is needed to convert available DC source to an AC supply of desired voltage, frequency and quality. Multilevel inverter (MLI) technology has gained large attention of researcher community these days mainly for medium and high power applications due to its various advantages over available conventional 2/3-level inverters technologies. As conventional 2/3-Level inverters are not applicable in medium and high voltage applications due to its semiconductor switches blockage voltage limitation, the MLI has become the optimum choice for medium and high voltage grid interfacing. MLI synthesis waveform closer to sinusoidal wave having low harmonic contents. It reduces voltage stress on switches and can be operated at both high as well as fundamental switching frequencies. Also, MLIs have low interference with neighbouring instruments because of low dV/dt [1]. These all give MLI an edge over conventional inverters.

The main concept behind MLI is to synthesis more DC voltage levels in its output to fill the gap between pure sinusoidal and generated waveform. Voltage waveforms for 2-level, 3-level and & 7-level MLI are shown in Fig. 1. As output waveform of MLI is closer to sinusoidal wave, therefore, the harmonic content of output is quite low and can be reduced further by increasing number of voltage levels in output or by utilising appropriate modulation technique. This staircase like voltage waveform is achieved by complex networking of DC voltage sources and power electronic switching devices. A proper switching strategy is developed to control these switches to produce desired output voltage waveform. Levels in output voltage can be decided based on required voltage rating and quality. More levels in output led to high voltage rating and better power quality but also increases required number of DC voltage sources and power electronic switches making inverter more complex, bulky, costly and fault-prone. MLI requires many DC voltage sources and power electronic switches as compared to conventional inverter technologies and this is a major drawback of MLIs. Researchers are working continuously to reduce the number of DC sources & power switch count and proposed many MLI topologies with reduced DC source and switch count [1]-[8]. Some of traditional MLI topologies are flying capacitor (FC), neutral point clamped (NPC) and cascaded H-bridge (CHB) multilevel inverter. A new MLI category 'Hybrid multilevel inverter' is getting famous these days due to its flexible design and reduced components count.

In this paper, a 15-level hybrid MLI topology with reduced switch count is implemented in MATLAB/ SIMULINK for single phase and THD of its output is reduced below 5% level as per IEEE 519 standards using total harmonic minimization technique. GA toolbox in MATLAB is used to solve the set of non-linear equations. An experimental setup is developed to validate the results obtained through simulation. Various output waveforms from simulation and experimental setup are shown.

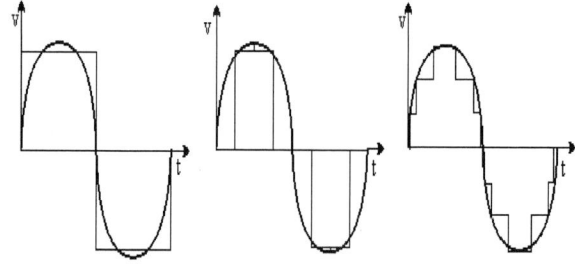

Fig. 1 Voltage waveform for 2-level, 3-level & MLI

978-1-5386-6624-1/18 $31.00 © 2018 IEEE 398

II. MULTILEVEL INVERTER TOPOLOGY WITH REDUCED SWITCH COUNT

A single-phase hybrid multilevel inverter topology with reduced switch count is shown in Fig. 2 for fifteen levels [8]. This topology consists of three DC voltage sources and ten power electronic switches which are lower than twelve switches required for CHB topology for implementing 15-level MLI with same source configuration. The whole topology can be divided into two parts i.e. Level generator and polarity generator. Level generator consists of all 'N' DC voltage sources and 2N power electronic bidirectional conducting unidirectional blocking switches where N depends on number of voltage level required in output and configuration of sources taken symmetrical or asymmetrical (N is three in this case with asymmetrical source configuration of $V_1 = 48V$, $V_2 = 96V$ & $V_3 = 192V$). Function of level generator is to synthesis different voltage levels using appropriate combination of sources and switches. Polarity generator consists of four power electronic bidirectional conducting unidirectional blocking switches irrespective of number of levels in output. Polarity generator reverses the polarity of levels generated by level generator alternately using proper switching combination to produce symmetrical waveform. Switching scheme for different voltage levels are tabulated in Table 1. Switches used in polarity generator are of high voltage and low frequency rating as these switches will be operated at fundamental frequency i.e. 50 Hz whereas switches used in level generator are of low voltage and high frequency rating.

III. TOTAL HARMONIC MINIMIZATION TECHNIQUE

Output of MLI resembles with pure sinusoidal wave and have low harmonic contains but it can be further reduced by applying appropriate modulation technique. Various modulation techniques are proposed by researchers to reduce harmonic contents in output [9]-[19]. Concept behind modulation is to adjust the switching angle between two DC levels such that output waveform become as sinusoidal as possible. For calculating exact switching angle, harmonic components are expressed as the functions of switching angles in trigonometric terms [19].

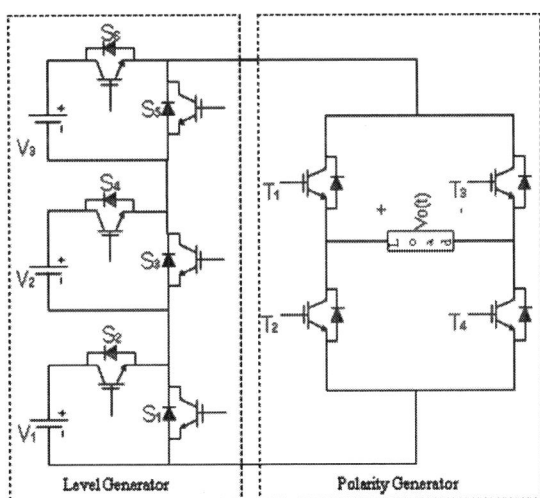

Fig. 2 Hybrid multilevel inverter topology with reduced switch count

Table 1
Switching table for different DC voltage level

State	Vo(t)	Voltage	Switches in ON State
1.	$V_1+V_2+V_3$	336V	S_2, S_4, S_6, T_1, T_4
2.	V_2+V_3	288V	S_1, S_4, S_6, T_1, T_4
3.	V_1+V_3	240V	S_2, S_3, S_6, T_1, T_4
4.	V_3	192V	S_1, S_3, S_6, T_1, T_4
5.	V_1+V_2	144V	S_2, S_4, S_5, T_1, T_4
6.	V_2	96V	S_1, S_4, S_5, T_1, T_4
7.	V_1	48V	S_2, S_3, S_5, T_1, T_4
8.	0	0V	S_1, S_3, S_5, T_1, T_4 S_1, S_3, S_5, T_2, T_3
9.	$-V_1$	-48V	S_2, S_3, S_5, T_2, T_3
10.	$-V_2$	-96V	S_1, S_4, S_5, T_2, T_3
11.	$-(V_1+V_2)$	-144V	S_2, S_4, S_5, T_2, T_3
12.	$-V_3$	-192V	S_1, S_3, S_6, T_2, T_3
13.	$-(V_1+V_3)$	-240V	S_2, S_3, S_6, T_2, T_3
14.	$-(V_2+V_3)$	-288V	S_1, S_4, S_6, T_2, T_3
15.	$-(V_1+V_3+V_3)$	-336V	S_2, S_4, S_6, T_2, T_3

$$V_1 = \frac{4V_{DC}}{\pi}\left(\cos(\emptyset_1) + \cos(\emptyset_2)\ldots + \cos(\emptyset_N)\right)\sin(\omega t) \quad (1)$$

$$V_3 = \frac{4V_{DC}}{3\pi}\left(\cos(3\emptyset_1) + \cos(3\emptyset_2)\ldots + \cos(3\emptyset_N)\right)\sin(3\omega t) \quad (2)$$

$$V_5 = \frac{4V_{DC}}{5\pi}\left(\cos(5\emptyset_1) + \cos(5\emptyset_2)\ldots + \cos(5\emptyset_N)\right)\sin(5\omega t) \quad (3)$$

$$V_n = \frac{4V_{DC}}{n\pi}\left(\cos(n\emptyset_1) + \cos(n\emptyset_2)\ldots + \cos(n\emptyset_N)\right)\sin(n\omega t) \quad (4)$$

$$THD = \frac{\sqrt{V_1^2 + V_3^2 + V_5^2 + \ldots + V_{49}^2}}{V_1} \quad (5)$$

$$\emptyset_1 \leq \emptyset_2 \leq \emptyset_3 \ldots \leq \emptyset_N \leq 90° \quad (6)$$

where V_n is the voltage of nth harmonic, n is order of harmonic, $\emptyset$ is the switching angle and N is the total number of switching transitions.

In total harmonic minimization technique, firing angle $\emptyset_1, \emptyset_2 \ldots \emptyset_N$ are calculated by solving nonlinear equation 5. Unlike selective harmonic elimination technique, no harmonic is eliminated in this technique, but overall harmonic contents are compressed to minimise the THD. Firing angles obtained by solving equation 5 gives minimum THD. The above nonlinear equation is solved using genetic algorithm (GA) toolbox in MATLAB. Genetic algorithm toolbox gives better results compared to Newton Raphson method and is quite easy to utilise.

Total harmonic minimization technique is quite effective technique and is better than SHE technique because it gives lower THD compared to SHE and also solution exists for each modulation index.

IV. GENETIC ALGORITHM OPTIMIZATION TECHNIQUE

Genetic algorithm is a stochastic global search method that is similar of natural biological evolution. Its operates on initial population utilising the principle of survival for the fittest to produce better and better solution. A new set of solutions are approximated in each generation by selecting individuals based on their level of fitness and are breed together to produce individual which are better suited to the environment. This process of evolution goes till satisfactory result obtained. This is quite easy optimisation technique as only a fitness equation and set of nonlinear constraints are needed to initiate the optimization [20].

V. SIMULATION RESULTS

Above presented 15- level MLI topology is implemented and simulated in MATLAB SIMULINK for single phase voltage & current waveforms for modulation index 0.8 and RL load of R=250Ω & L= 40mH are shown in Figs. 3 – 6 along with their FFT analysis. Total harmonic minimisation technique is utilised to estimate switching angles and calculations are done using 'Genetic Algorithm toolbox' in MATLAB.

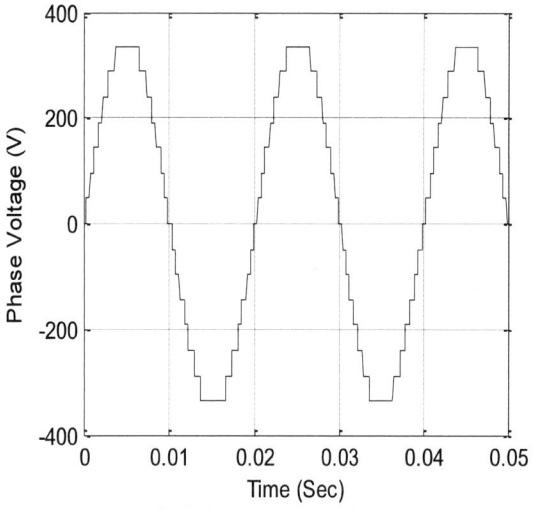

Fig. 3 Phase voltage waveform

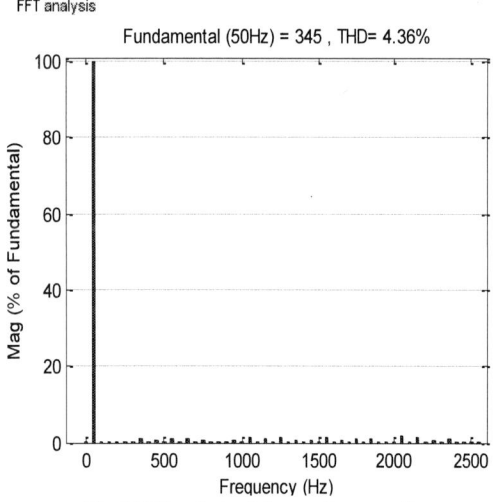

Fig. 4 FFT analysis of phase voltage waveform

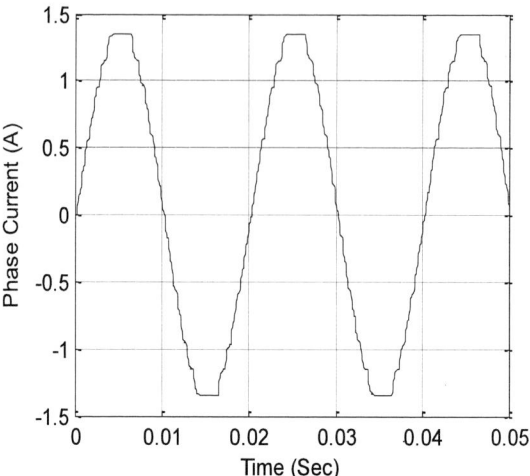

Fig. 5 Phase current waveform

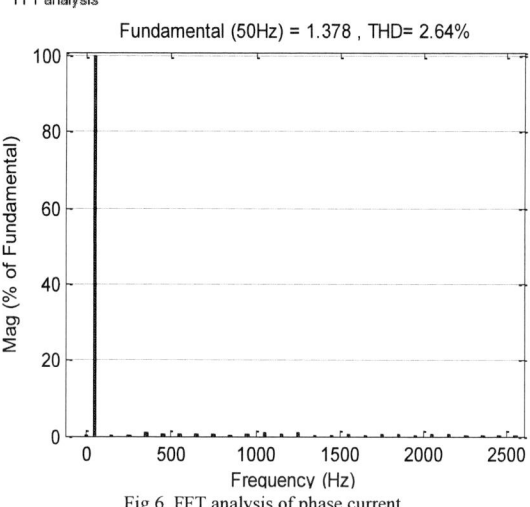

Fig.6 FFT analysis of phase current

VI. EXPERIMENTAL SETUP & RESULTS

Above discussed 15-level multilevel inverter is also implemented with hardware components to create an experimental setup and validate the results obtained through Simulations. Components used are DSP DS1104 of dSPACE for interfacing MATLAB and experimental setup. The DS1104 is specifically designed for the development of high-speed multivariable digital controllers and real-time simulations in various fields. It is a complete real-time control system based on a 603 Power PC floating-point processor running at 250 MHz. Other components used are 1200V/40A IGBT switches with anti-parallel power diode, driver & isolation circuit for hybrid MLI topology implementation and Autotransformers, isolation transformers & rectifier modules for creating stiff DC voltage sources. Experimental setup is shown in Fig. 7. Results obtained from experimental setup are also illustrated in Figs. 8 – 11. For waveform recording and purpose, Fluke power quality analyser and Tektronix TPS2000 Series TPS2014 Digital Oscilloscope are used.

978-1-5386-6624-1/18 $31.00 © 2018 IEEE

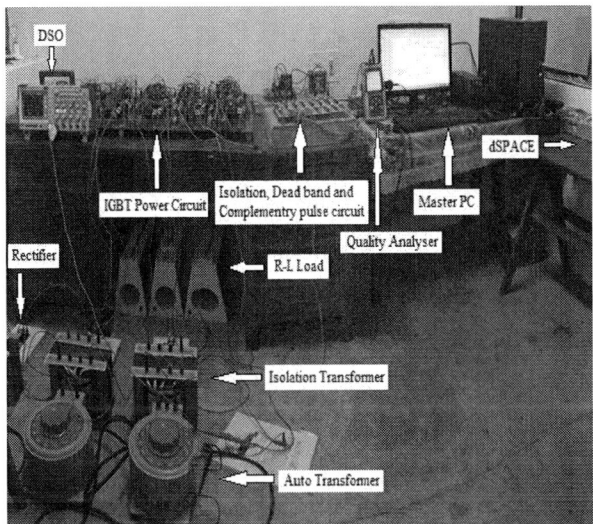

Fig. 7 Experimental Setup

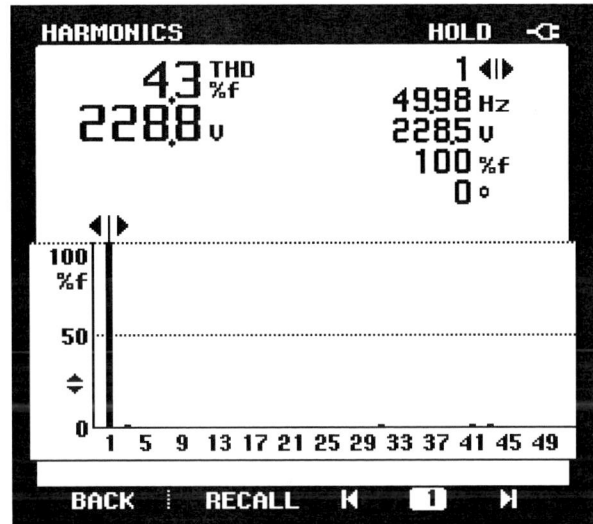

Fig. 10 Phase voltage THD in quality analyser

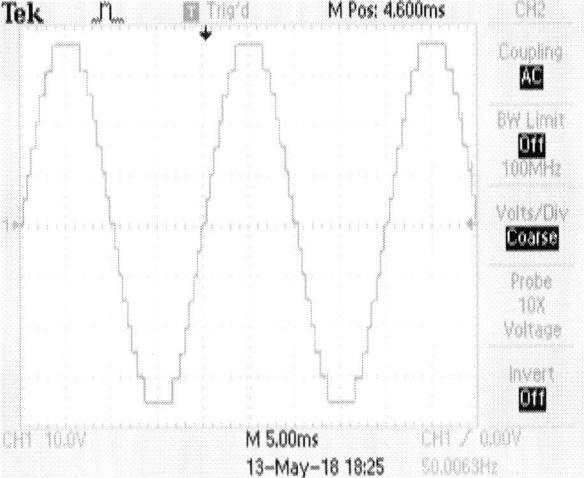

Fig. 8 Output waveform for 15-level MLI in DSO

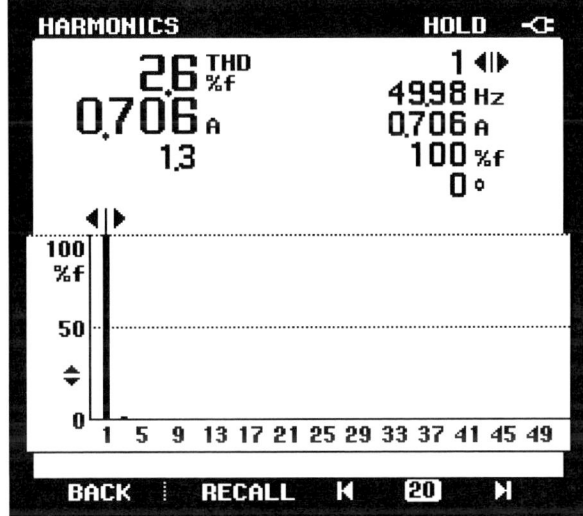

Fig. 11 Phase current THD in quality analyser

V. CONCLUSION

A 15-level hybrid multilevel inverter topology is simulated in MATLAB/SIMULINK and its phase voltage THD is minimised after calculating switching angles using total harmonic minimization modulation technique. It is found that THD of phase voltage is 4.36%and current THD is 2.64%. Obtained THDs of phase voltage and current are also verified with hardware implementation. Difference between simulation and hardware results founded negligible. Output THDs are below 5% level as per IEEE 519 standards therefore total harmonic minimization technique can be an effective method to minimise THD of multilevel inverters with large number of DC levels in its waveform.

VI. REFERENCES

[1] A. Krishna R and L. P. Suresh, "A brief review on multi-level inverter topologies," *2016 International Conference on Circuit, Power and Computing Technologies (ICCPCT), Nagercoil, 2016, pp. 1-6.*

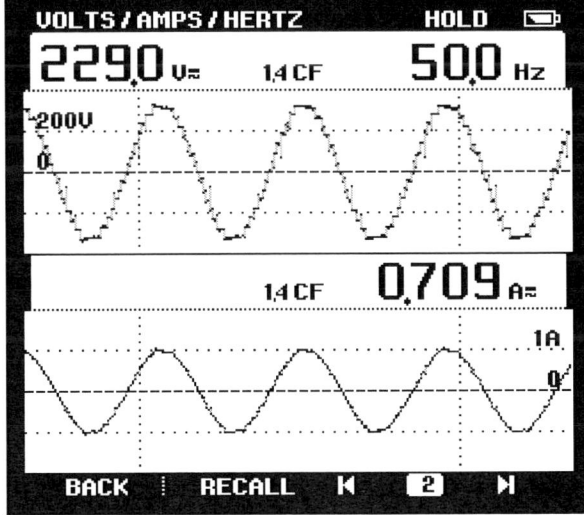

Fig. 9 Waveform of Phase voltage & current in Fluke quality analyser

978-1-5386-6624-1/18 $31.00 © 2018 IEEE   401

[2] K. K. Gupta, A. Ranjan, P. Bhatnagar, L. K. Sahu and S. Jain, "Multilevel Inverter Topologies with Reduced Device Count: A Review," *in IEEE Transactions on Power Electronics, vol.* 31, no. 1, pp. 135-151, Jan. 2016.

[3] Jih-Sheng Lai and Fang Zheng Peng, "Multilevel converters-a new breed of power converters," *in IEEE Transactions on Industry Applications, vol. 32, no. 3, pp. 509-517, May/Jun 1996.*

[4] S. Daher, J. Schmid and F. L. M. Antunes, "Multilevel Inverter Topologies for Stand-Alone PV Systems," in *IEEE Transactions on Industrial Electronics*, vol. 55, no. 7, pp. 2703-2712, July 2008.

[5] E. Babaei, M. F. Kangarlu, F. N. Mazgar, ''Symmetric and asymmetric multilevelinverter topologies with reduced switching devices'', Elsevier Electr. Power Syst.Res., 2012, 86, pp. 122–130

[6] J. Rodriguez, J. Lai, F. Z. Peng, ''Multilevel inverters: a survey of topologies, controls and applications'', IEEE Trans. Ind. Electron., 2002, 49, pp. 724–738

[7] M. Manjrekar, G. Venkataramanan, ''Advanced topologies and modulation strategies for multilevel converters''. Proc. IEEE Power Electron Spec. Conf., Baveno, Italy, 1996

[8] Gui-Jia Su, "Multilevel DC-link inverter," in *IEEE Transactions on Industry Applications*, vol. 41, no. 3, pp. 848-854, May-June 2005.

[9] M. S. A. Dahidah, V. G. Agelidis, ''Generalized formulation of multilevel selective harmonic elimination PWM: case I – non-equal DC sources''. 37th IEEE Power Electronics Specialists Conf., PESC '06, Jeju, South Korea, 2006, pp. 1–6

[10] P. N. Enjeti, P. D. Ziogas, J. F. Lindsay, ''Programmed PWM techniques to eliminate harmonics: a critical evaluation'', IEEE Trans. Ind. Appl., 1990, 26, pp. 302–316

[11] J. N. Chiasson, L. M. Tolbert, K. J. McKenzie, et al. ''Elimination of harmonics in a multilevel converter using the theory of symmetric polynomials and resultants'', IEEE Trans. Control Syst. Technol., 2005, 13, pp. 216–223

[12] J. N. Chiasson, L. M. Tolbert, Z. Du, et al. ''The use of power sums to solve the harmonic elimination equations for multilevel converters'', Eur. Power Electron. Drives., 2005, pp. 19–27

[13] K.J. McKenzie, ''Eliminating harmonics in a cascaded H-bridges multilevel inverter using resultant theory, symmetric polynomials, and power sums''. MSc thesis, University of Tennessee, Chattanooga, 2004

[14] Z. Du, M.L. Tolbert, J.N. Chiasson, ''Active harmonic elimination for multilevel converters'', IEEE Trans. Power Electron., 2006, 21, pp. 459–469

[15] H. Taghizadeh, M. TarafdarHagh ''Harmonic elimination of multilevel inverters using particle swarm optimization''. Proc. IEEE-ISIE, Cambridge, UK, 2008, pp. 393–397

[16] M. TarafdarHagh, H. Taghizadeh, K.Razi, ''Harmonic minimization in multilevel inverters using modified species-based particle swarm optimization'', IEEE Power Electron., 2009, 24, pp. 2259–2267

[17] D. Ahmadi,J.Wang,''Selective harmonic elimination for multilevel inverters with unbalanced DC inputs''. IEEE Conf., Dearborn, MI, USA, 2009.

[18] D. Ahmadi, K. Zou, C. Li, Y. Huang and J. Wang, "A Universal Selective Harmonic Elimination Method for High-Power Inverters," in *IEEE Transactions on Power Electronics*, vol. 26, no. 10, pp. 2743-2752, Oct. 2011.

[19] J. Kumar B. Das P. Agarwal "Selective harmonic elimination technique for a multilevel inverter" Proc. 15th NPSC pp. 608-613 Dec. 2008.

[20] A. Chipperfield, P. Fleming, H. Pohlheim and C. Fonseca, "Genetic Algorithm Toolbox" in Dept. of Automatic Control & System Engg., Version 1.2.

978-1-5386-6624-1/18 $31.00 © 2018 IEEE

Design, Simulation and Economical Analysis of Solar Powered Irrigation Water Pump

Archna Aggarwal
Deptt. of Electronics Engineering
YMCA Univ. of Science and Technology
Faridabad,INDIA
Email:archna.elect@gmail.com

Kriti Srivastava
Department of Electronics Engineering
YMCA University of Science and Technology
Faridabad,INDIA
Email: kriti.srivastava1@gmail.com

Abstract—**Irrigation is an age-old technique which is practiced on various levels on farms around the world. It allows variegation of crops while increasing crop yields. A typical irrigation system consume high amount of non-conventional energy by using generators to power the water pumps. Photo-voltaic energy can be used as a substitute of conventional energy. This paper deals with the simple yet effective design of a solar water pumping system in GHAZIABAD for irrigation purpose. In this work, the system for rice and wheat crop has been designed consisting of SPV modules along with maximum power point tracker (MPPT) to attain maximum power efficiency, DC-DC Boost Converter, permanent magnetic DC(PMDC) motor and Submersible pump. The simulation of the system has been done through MATLAB/SIMULINK which verifies the system's functionality along with its components. The study also focuses on the economic evaluation of renewable energy over traditionally used non-conventional source of energy**

Index Terms—**SPV-modules, Boost-Converter, PMDC motor, Submersible pump , Irrigation System, MATLAB/SIMULINK**

I. INTRODUCTION

Dwindling conventional energy resources and Ever increasing energy demand has created a huge gap betweem energy demand and supply.Energy scaracity and rising pollution levels worldwide have forced the researchers to look for options for a green energy production. Agriculture is the backbone of Indian economy. Reliable irrigation is so critical to farmer's livelihood that app. 25 million grid connected and diesel powered pump systems have been installed till date. Replacing these with Solar powered pumps will reduce air pollution, release of green house emissions and help in reducing energy scaracity. For developing countries like India, cost of the overall system plays a huge role in its practical feasibilty. Using plastic or metal tank for water storage purpose insted of using batteries for power storage [2]can help to reduce the cost of the Solar water pump.D.C. motors for Solar water pumps are most cost effective as PV modules produce d.c. which can be directly used by d.c. motors. But,A.C. motors require an inverter which adds to the cost of the whole system [3].Dynamic performance and efficiency of PMDC motor is better than A.C. motor [4].PV pumping system reduces the Net Present Cost (NPC) of the

system due to low operation & maintenance cost [8].Water conservation, water saving technology and precision farming can be used to improve the water use efficiency. Different irrigation techniques like drip irrigation [5] makes it further cost effective. Solar water pump gives better performance at MPPT point [6], but, variation in solar radiation makes the efficiency of pump dependent on motor parameters [7].

Based upon literature review, The author have designed a solar energy driving water pump system incorporating a PMDC motor for 20-Quintal of rice and wheat and simulated the design model in MATLAB/SIMULINK. The author have also done an economical analysis between different sources which is used to power the water-pump. The another aim of this paper is to observe the response of pump speed, discharge rate, armature current of motor and power received by the motor with the change in irradiance at constant temperature.

II. SITE LOCATION

Ghaziabad is a fast upcoming advanced industrial district of U.P which is located at 28.67deg N 77.42deg E. The solar radiations received by the district in 12 months has been shown in Table 1.

A. Agricultural Profile Of The Location

Ghaziabad is located in the western part of Uttar Pradesh covering an area of approx.1148 sq km. The normal rainfall in this Ghaziabad district is about 732 mm. The State has remarkable bearing on the Agricultural performance. The net sown area is 1.49 lakh ha with a net irrigated area of 1.38 lakh ha, with cropping intensity 172%. Wheat and Rice are the major crops grown in the district.

B. A. Calculation of Solar PV System for Water Pumping

- RICE

20quintal Of Rice Requires = 7000000 Litres For 150days
Load requirements = 47000 litres of water every-day from a depth of 50m

Amount of water to be pumped/day = $47m^3$
- WHEAT

20quintal Of Rice Requires = 1800000 litres for 150days

Load requirements = 12000 litres of water every-day from a depth of 50m

Amount of water to be pumped/day = $12m^3$

To decide the Total Dynamic Head (TDH)

Maximum elevation of piping unit inlet = 35m

Maximum head of running stream fluctuates = 15m

Total vertical lift = 35m + 15m = 50m

Frictional losses = 5% of total vertical lift = 5 % of TDH = 2.5m+Total vertical lift = 50m + 2.5m =52.5m

For estimating the load requirement with selected DC pump for RICE crop [Experimental]

Selected DC pump max head = 59m

Selected DC pump max flow = $6m^3$/hour

Supplied voltage = 24/36V

Power consumption, P = 750Watt

Required running hour/day = 8 hour/day

Required electrical energy/day

= Power consumption× Running hour/day

= 750Watt × 8hour/day = 6 kWh/day

For estimating the load requirement with selected DC pump for WHEAT crop [Experimental]

Selected DC pump max head = 59m

Selected DC pump max flow = $1.5m^3$/hour

Supplied voltage = 24/36V

Power consumption, P = 750Watt

Required running hour/day = 8 hour/day

Required electrical energy/day

= Power consumption × Running hour/day

= 750Watt × 8hour/day

= 6 kWh/day

To understand the Ampere hour requirement of DC load System for both crop

Voltage = 36V

Load Current = 20.83A

Required running hour/day = 8 hour/day

Required Ampere hour/day = Load current × Running hour /day

= 20.83× 8 hour/day

= 167Ah/day

Module efficiency due to temperature = 21.1 %

Total loss factor = 5

Estimated Ah requirements from PV module

= Possible max Load × Loss factor

= 167Ah/day × 5 = 835 Ah/day

Total Ampere requirement from a PV module = 282A

PV module system voltage = 36V

PV module system current = 282A

Required Power capacity of a PV module, P = 10.2kW

Selected PV module unit = 365W

Required numbers of PV modules

= Required PV power module/Selected module = 28 modules

Hence 28 modules can be used in around 800-1000sq.feet of

an area to run 1hp of a pump for the discharge of water at the rate of $6m^3$/h to the fields.

TABLE I
SOLAR RADIATION DATA OF GHAZIABAD

Months	Hours	Hours per Day	Kwh/m2/hour	Watt/m2
January	91.8	2.96	3.16	131.66
February	108	3.85	4.90	204.166
March	120.8	3.89	6.28	261.66
April	138.3	4.61	5.83	242.916
May	154.5	4.98	5.40	225
June	149	4.96	4.16	173.33
July	147.3	4.75	2.79	116.25
August	145	4.67	3.53	147.0833
September	111.8	3.72	4.74	197.5
October	98	3.16	4.28	178.33
November	89.8	2.99	3.41	142.0833
December	92.8	2.99	3.07	127.916

III. PROPOSED SYSTEM

The water pump system proposed in this work is a directly-coupled type, without battery backup. A storage tank can be used in place of batteries, to store the water. The SPV water pumping system is designed in this work for irrigation purpose, based on the actual requirement of the water for the crops(rice & wheat).

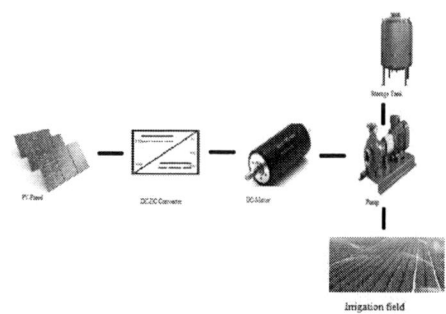

Fig. 1. Structure of SPV Pumping System

Figure 1 shows a very simple pumping system which consists of a photovoltaic module (attached to a maximum power point tracker (MPPT)), DC-DC converter, a DC submersible pump, and a storage tank. An actual size system is designed, which is required at present to fulfill the water needs of the crops for their harvesting period in a year. The theoretical study is done about the system and performance is analyzed in Matlab/Simulink.

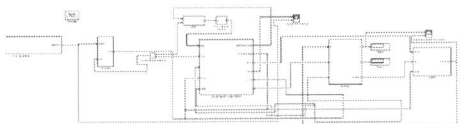

Fig. 2. Simulink Model for Solar Water Pumping System

Figure 2 shows the SIMULINK model of Solar Water pump in which PV-array is connected to a DC-DC Boost Converter along with permanent magnetic DC motor through a MPPT. Each subsystem in this model has its own characteristics and Simulink blocks as shown in APPENDIX(I-IV). The input is taken from spreadsheet which consists of 12 months radiation of the location at a constant temperature of 300K.

IV. RESULTS

The simulation result of solar water pump-system at different radiation and constant temperature has been shown All the results are based out of One Solar Panel. On a similar ground, the full system can be analyzed.

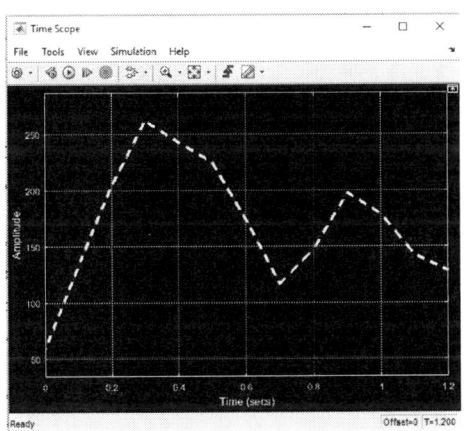

Fig. 3. Solar Radiation Data on the Scope

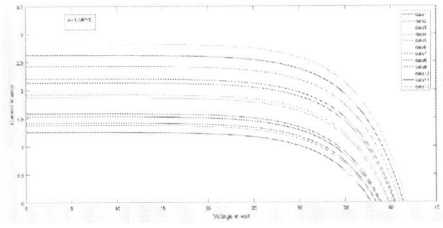

Fig. 4. I-V Curve

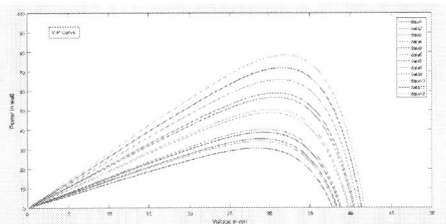

Fig. 5. V-P Curve

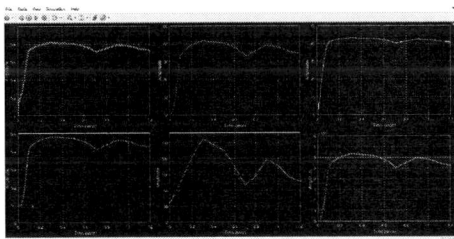

Fig. 6. Outputs After Simulation

Above figures show the results of PV-panels obtained after simulation for input radiations received by GHAZIABAD. Further, these output becomes the input for boost converter which will help to boost up the armature voltage for PMDC motor.

⋆ Solar irradiation is at its peak on 0.32 seconds, where the solar irradiance increases from 204W/m2 to 261W/m2 and correspondingly the input power to the PMDC motor is also increased.

⋆ Since the PV array is directly connected to the PMDC motor through MPPT, the power efficiency will be higher and will cause an increase in the mechanical output torque.

⋆ The current at 0.32 seconds is increased due to the increase of torque which is directly proportional to armature current as shown in Figure.7(a).

⋆ The speed of the rotor is increased proportionally with the increase in the solar irradiation as shown in Figure.7(c).

⋆ Now, with the increase of rotor speed, the flow rate of the submersible pump will start increasing. Figure.7(f) shows the increase of the flow rate when the solar irradiance is increased at 0.32 seconds.

⋆ The power attained by the PMDC motor is decreased when the solar irradiance decreases from 216W/m2 to 116W/m2 because they both are correlated to each other.

All the results show that solar irradiance is the prime concern for the system.

A. ECONOMICAL RESULTS OF SOLAR WATER PUMP

Current economics indicate an IRR of 10% for replacement of diesel pumps with solar pumps without factoring in crop yield improvement benefits due to water availability on demand. However, the capital cost of an SPV pump is about ten times of a conventional pump and hence it requires capital subsidy and financing support but looking at its positive side,

978-1-5386-6624-1/18 $31.00 © 2018 IEEE

it reduces the cost of the whole system in long term. The parameters of the cost analysis for the PV system are as follows; PV unit cost = Rs.50/PV Watt

PV structure = Rs.18/PV Watt

Cabinet & cables = Rs.7/PV Watt

Installation = 5% of the capital cost

O&M = Rs.6700/year (with interest rate of 6 %).

TABLE II
COST ANALYSIS OF PV PUMPING SYSTEM

PV Cost	Rs. 50000/-
PV Structure	Rs. 18000/-
Cabinet& Cables	Rs. 7000/-
Total Cost	Rs. 75000/-
Installation Cost	Rs. 3750/-(5%of the total)
NPC	Rs.78750/-

B. Diesel Systems

Diesel system uses the fuel to drive the water pump for operation. So the total cost of the system includes the cost of the diesel generator, fuel, and O&M. There are two systems one is only PV system and the other is only Diesel unit. Table-3.shows the comparison between the two of assumed 1KW systems.

TABLE III
COMPARISON OF COST BETWEEN PV AND DIESEL SYSTEM

PARAMETERS	PV	DIESEL
SYSTEM COST	Rs. 78750/-	Rs. 32000/
FUEL CONSUMPTION	00	0.7Litre/hr./
FUEL PRICE	00	Rs.0.67/Litre
ANNUAL REPAIR&MAINTENANCE	00(first year)	00(first year)
HOURS OF OPERATION	3600=1800	3600=1800
FIRST YEAR COST	Rs. 78,750/-	Rs. 1,16,420/-

It shows that for an assumed system of 1KW, system cost of PV system is higher than Diesel system but the total first-year cost of diesel system is higher than PV system. In next 5 years, PV system will be beneficial because with an inflation rate of approx. 2

Table.2 shows the results of the economic analysis of the 1KW[assumed] PV system

V. CONCLUSION

Based on the study main conclusions are as follows:

● 28 PV-Modules each of 365Watt with an efficiency of 21%, can be used in the system along with to provide the full power to the 1Hp of DC submersible pump at different solar radiation and water can be store in large storage tanks and can be utilized when no source of energy is available.

● The first year cost of PV system is Rs.78,750 whereas

the cost is high for traditionally used Diesel system, that is Rs.1,16,420. The O&M cost of PV system would cost around Rs.8734 whereas for diesel system it would be Rs.1lakh at the end of 5 years at 2% of inflation rate. It shows that Solar based water pump is a dependable alternative solution against diesel water pumps as it reduces the O&M cost of the system which seems to be a feasible solution in the long run.

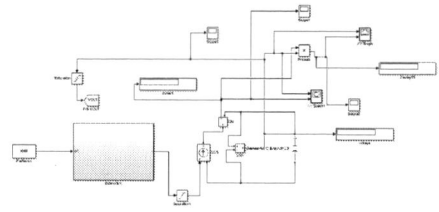

Fig. 7. Simulink diagram of PV Cell

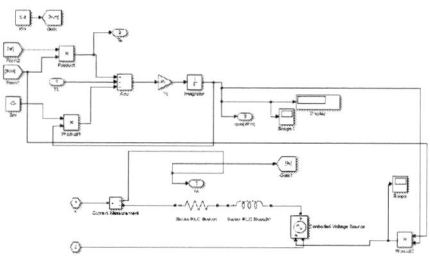

Fig. 8. PMDC Motor

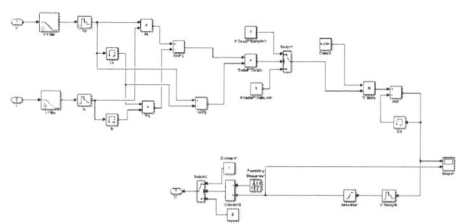

Fig. 9. P&O MPPT Algorithm

REFERENCES

[1] Bhong Sagar, Kale Madhav, Shinde Kishor, Bobade Rameshwar,Pandhi Tushar *International Research Journal of Engineering and Technology (IRJET)*,5,(2018),1324-1326

[2] M. Belarbi, K. Haddouche,A. Midoun,Modeling and simulation of photovoltaic solar panel,*WREEC. J. Clerk Maxwell, A Treatise on Electricity and Magnetism*,2,(2006),68-73

[3] Enas R. Shouman, E.T. El Shenawy, M.A. Badr,Economics Analysis of Diesel and Solar Water Pumping with Case Study Water Pumping for Irrigation in Egypt,,*International Journal of Applied Engineering Research*,11,(2016),950-954

[4] Chandrasekaran N, Thyagarajah K,Modeling and performance study of single phase induction motor in PV fed pumping system using MATLAB,*International journal of Electrical Engg.*,5,(2012),305-316

[5] Pande PC, Singh AK, Ansari S, Vyas SK, Dave BK,Design development and testing of a solar PV pump based drip system for orchards,*International journal of Renewable Energy*,2,(2003),385-396

[6] Katan RE, Agelidis VG, Nayar CV,Performance analysis of a solar water pumping system. *IEEE international conference on power electronics, drives, and energy systems for industrial growth (PEDES)*,(1996),81-87

[7] ,Muamer M. Shebani and Tariq Iqbal,Dynamic Modeling, Control, and Analysis of a Solar Water Pumping System for Libya.*Journal of Renewable Energy*,2,(2017),Article-ID 8504283

[8] Ibrahim Alkhubaizi,Solar Water Pump,*International Journal of Engineering Research and Application* ,7,(2017),1-5

2nd IEEE International conference on power Electronics, Intelligent Control and Energy systems (ICPEICES-2018)

Sizing and Financial Optimization of Hybrid PV/battery/Diesel System

Archna Aggarwal
Department of Electronics Engineering
YMCA University of Science and Technology
Faridabad,INDIA
Email:archna.elect@gmail.com

Divya Nagpal
Department of Electronics Engineering
YMCA University of Science and Technology
Faridabad,INDIA
Email: er.divyanagpal@gmail.com

Abstract—**Energy scarcity and environmental concerns are driving the World energy sector towards renewable energy sources. But,practical feasibility,reliability and high costs are the major hurdles. So,a judicious combination of all these factors and the various Renewable energy sources is the need of the hour. In this paper, a hybrid optimization technique for a residential building in Faridabad, Haryana, India has been proposed. A new scheme of utilizing battery costs as fixed deposits based on present market rates of installation makes the system financially viable. The proposed algorithm evaluates the system for desired LLP values & then economically optimizes it to provide a financially viable system with least total life cycle costs.**

Keywords—**Life Cycle Cost Analysis (LCCA), Levelized Cost of Energy (LCOE), Loss of Load probability (LLP), Financial optimization, standalone,**

I. INTRODUCTION

Various policies and renewable energy framework by the government has made possible large number of renewable energy deployments in India. In order to achieve grid parity by 2020, there are many schemes and subsidies by the government [1]. Low reliability power systems results in lower initial costs, however, that cost are served back in terms of outage costs. With the use of any auxiliary source of power like batteries, generators etc., economic evaluation of such systems becomes necessity to find the most optimal configuration for given parameters [2]. Parameters like LLP [4], tilt angle [5], [6], [23]have been used to ensure better performance and lower costs for PV systems. Energy resource provision is a process to assign battery and solar PV panels to a PV system. The costs incurred in doing so can be reduced by cost-efficient resource planning methods. Conventional methods results in time varying and node dependent results in WNS, which may be erroneous [7]. Load profiles are one of the major parameter for sizing and optimizing of PV/Wind/Diesel/Battery bank so as to minimize levelized cost of energy (LCE) and CO2 emissions via Multi-objectives genetic algorithm approach. It is observed that for project with lower LCE results in higher CO2. Apart from ANN and GA [20], Object Oriented Programming is used to design and sizing of PV-Wind hybrid system. It relies on various fundamentals like generator models, storage models, and loss of load probability algorithm to propose techno-economic analysis for the system [10]. To ensure better results, combination of two tools like HOMER

[17], [25],PV.MY and economic optimization [14], [15] along with mathematical modelling [16], [22], [26] are used based on real time statistics,market availability and maximum renewable penetration. For highly reliable system, many parameters like wasted energy, cost of energy and battery usage are evaluated for LLP = 0.01 [11], [12]. There are various limitations and challenges in optimization of HRES with storage elements. Future on RES lies upon grid parity process and its payback period of installations. Battery due to high initial and maintenance costs yields over fit or under fit sizing values. Therefore while modelling and designing standalone PV/battery systems battery constraints regarding its various parameters needs to be considered carefully [19]. Earlier residential loads in lower income countries like Nigeria, were dependent on diesel generators. However due to lower capital and maintenance costs of HRES [24], apart from rural sector, even residential buildings are promoting the use of standalone PV systems. These systems are not just environment friendly but also 30%cheaper [21]. Gradually PV is becoming economically viable and feasible due to capital costs associated with them are declining. The motivation behind presenting above case studies in form of literature summary is to provide methodology to be used in our study, fill out the gaps in studies and provide information for referencing and comparison. While global environment is facing a challenge from excessive use of fossil fuel based energy, a large population is still living under conditions that range from extreme to moderate energy deficit. Forming a concrete basis on these, it has been realized that given the state of the art in the technology and favorable financing terms in India, this energy deficiency can be decreased by optimal planning of PV installations. This study focuses to examine financial viability of a hybrid standalone system in urban environment and evaluation a sizing technique to optimize a hybrid standalone system economically with maximum reliability.

II. SITE ASSESMENT

In urban areas, the grid connectivity is considered to be most cost effective in terms of energy pay pack time and grid parity price during the lifetime of the PV plant. The site selected for this study is a 2-BHK residential building with daily load of 5.7 kWh. The load profile is collected

978-1-5386-6624-1/18 $31.00 © 2018 IEEE

from assessment of working loads in a day as in figure 1 given below. The site rooftop area is taken from google maps and end to end boundary is marked as shown in figure 2 below.

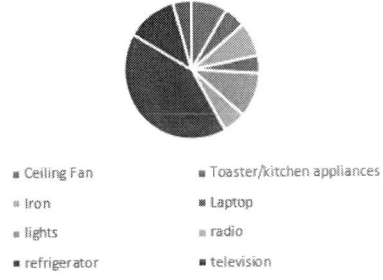

Fig. 1. Load Profile of the Selected Site Location

Fig. 2. Rooftop Area of Selected Site Location

A. Solar Radiation Assesment

Solar radiation is measured in term of intensity of radiation and number of hours that are received. TMY data which a typical meteorological year, lists hourly values of solar radiation and other weather data for selected coordinates location. This data is comprising of 8764 values from NREL database as shown in figure 3. The location coordinates of selected location are at 28.43 ° N 77.32° E.

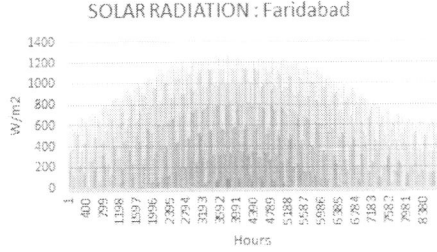

Fig. 3. Solar Radiation:Faridabad

III. DESIGNING AND FINANCIAL MODELLING

Any financial modelling for a renewable energy resource installation comprises of various steps as listed in figure 4.

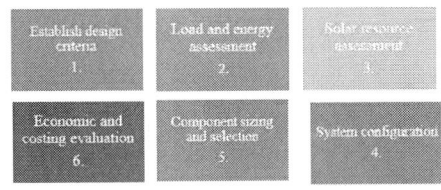

Fig. 4. Design Process of a Standalone System

The system configuration is evaluated on components available in market and their pricing variability in real times. Following sections define the entire system configuration for the chosen load consumption and their total installation costs.

A. Sizing and configuration of hybrid PV systems

After load profile assessment and solar radiation data available for site, it is concluded average annual solar radiation is 5.2 kWh/m2/day. Sizing is done via considering losses and efficiency at each conversion stage. Since the diesel generator is fixed for this space and its rating is taken as 2.2 kVA, the sizing for same is avoided here. 1.) Inverter capacity: Inductive loads produce surge power and may destroy appliances during on and off operation or voltage fluctuations. So considering it as reason for future defects, the load requirement is total amount of energy consumption per day and 3.5 times the amount of inductive load power so as to incorporate potential of inductive load surge effect. Moreover, if 25% load extension requirement is to be kept in mind for future scope, the total power is multiplied by 2.5 to fulfill the requirement. However for this study, surge protection demand is ignored for easy understanding as shown in table 1.

TABLE I
INVERTER CAPACITY

Total Power(TP)	2555W
Efficiency%	95%
Input Power to Inverter	2689W
Input DC Voltage	24V
Input DC Current	112.0416667A
Total Inverter Power	2555W
Power Factor	0.89
KVA Rating	2870.786517
Energy Produced by Inverter	5767Wh
Energy Input to Inverter	6070Wh
Output AC Voltage	230V
No. of Phase	Single
MPPT Voltage from PV	24V

2.) Battery Capacity: It is estimated from the calculations energy needed by the inverter is 6070 kWh. System DC bus line has voltage of 24 V DC. Days of autonomy are requisite set of information, but here since it is hybrid, autonomy days are not involved in battery calculation. The detailed battery configuration is shown in table 2, which is based on energy flow.

TABLE II
BATTERY CAPACITY

Autonomy	3 days
Battery Type	Lead Acid
Depth of Discharge	80%
Required Capacity of Battery Bank	316.145833 Ah
Battery Bank Voltage	24 V
Each Battery Capacity	150 Ah
Total no. of Battery Strings	2
No. of Batteries in a String	2
Total Batteries	4
C-Rating	C-10
Energy Required to Charge Batteries	75875 Kwh

3.) PV array Capacity: PV array is generally designed in by dicing sunshine hours with energy consumption requirement. However more accurate results are obtained when it takes into account power needed to charge the batteries. The array capacity calculation is shown in table 3.

TABLE III
PV ARRAY SIZING

Array Voltage Required	24V
Ah Required	351.2731481
Current Required	67.55252849 A
Panels Required	8
PV Capacity	9 250=2.2Kw
DC Wire Length	10m

B. Costing Summary of SPV system

Depending the market survey and various ratings of components available in market, and from the real stake holder in this sector, there is the cost break down of the initial capital spent as shown in table 4. Further the total cost derived is used for financial analysis to obtain CPBT i.e. cost payback time and NPV, IRR values from excel formulas based financial modelling as shown in table 5. It is observed that due to high initial capital invested in standalone SPV system, it is highly discourage in urban areas. However, by the methods adopted in this financial modelling, a lot of initial costs are saved due to banks FDs scheme. The total battery replacements costs after every 5 years for tenure of 25 years accounts to Rs. 16

lacs. If half the amounts is invested as fixed deposit initially, it results in acceptable NPV of system and IRR of 14%to make the system financial viable in case of urban areas as well.

TABLE IV
COSTING STRUCTURE OF 2 KW STANDALONE PV SYSTEM

Total Energy Consumption	5767 Kwh
Total Load Consumption	2 Kw
Required No. of Panels	8
Cost of Different components	
a. Cost of Each module	Rs. 8120/-
b.Cost of all modules	Rs. 64960/-
c.Cost of Inverter(3KVA)	Rs. 30000/-
d. Structure Cost	Rs. 6000/-
Battery Cost	Rs. 44140/-
Total cost(abcde	Rs. 145,400/-
Cost of wiring(4%of total)	Rs. 5804/-
Electrical hardware items	Rs. 3000/-
Supervision,design & installation	Rs. 3000/-
Total capital cost	Rs. 156,904/-
GST(5%)	5%of total
Total Expense	Rs.164,749/-

TABLE V
EXCEL BASED FINANCIAL MODELLING

NPV	Rs. 69453/-
IRR	14%
LCOE	Rs. 2.17 per watt
Fixed deposit in bank	Rs. 80,000.00/-
Total Investment	Rs.165000.00/-Rs.80,000.00/-
b.Savings	Rs. 2,76,908/- Rs.37,545/-

IV. OPTIMIZATION OF HYBRID SYSTEM

Remote areas are considered with the option to replace or completely offset diesel generator to economically fulfill electricity demands. A complete replacement is however not advisable because solar power cannot fulfill the entire needs on cloudy/rainy days and would require large battery banks. This would consequently increase the installation and maintenance costs of a system. However, a hybrid system can serve as a reliable and cost effective source, given the right conditions i.e. optimal sizing. A well-managed hybrid system can give lifetime of profitable results with reliable electricity supply. Also the consumption of diesel price is reduced hence, reduction in CO_2 remissions as well other pollutants that are harmful to health are reduced. This also provide economical option in places where grid extension is not possible in nearby future. The criteria for optimization of hybrid renewable energy systems is based on technical parameters like LLP , LPSP etc. and economical parameters like TLCC , NPV etc. In this study, only conventional methods are dealt with which forms concrete base on following ideas. 1.) The concept of energy balance: In this methodology, averaged solar radiation energy and averaged consumption demands are simultaneously balanced. The available energy from the sun can be determined from solar irradiance data. During this method the path losses and efficiencies of the source, converters and controllers at each stage is very well accounted. 2.) Reliability of electricity supply: Loss of load probability is taken as reliability factor which is ratio of energy deficit hours to energy demand hours during entire operational time , which is known as loss of load probability (LLP).

A. Proposed Optimization Technique

There are three methods to size a hybrid system: numerical, analytical and intuitive. Numerical methods energy balance equations form iterative loops, which are simple line codes logically arranged in an algorithm. The economical optimization is done on the system with LLP value between range 0-2 % as Haryana government accept at maximum 2%LLP in power transmission policy. Annualized total life cycle costs are evaluated for individual configuration with their specific number of PV panels and number of batteries for fixed diesel generator capacity already installed as 2.2 kVA.

B. Flowchart of Proposed Algorithm

The flowchart comprises of two three stages where first stage is approximating number of panels and batteries, Second step is run iterative loops as per the numerical method of sizing of hybrid PV system, and the third step comprises of financial optimization of the selected PV and battery capacity with LLP range in 0-2%. The algorithm is shown in figure 5.

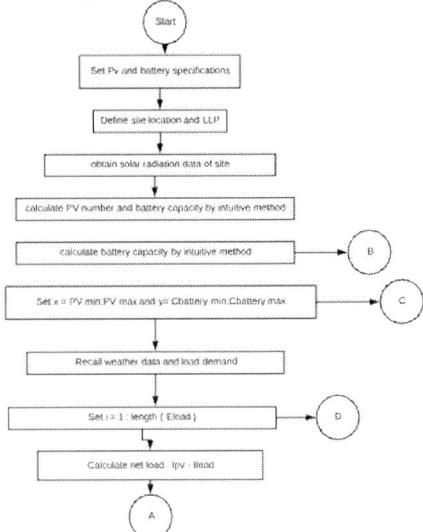

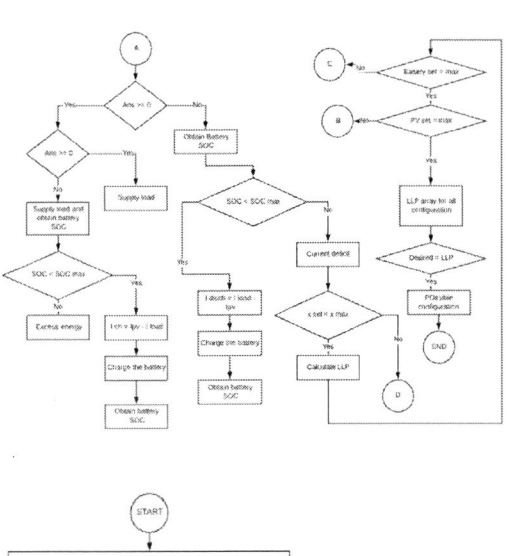

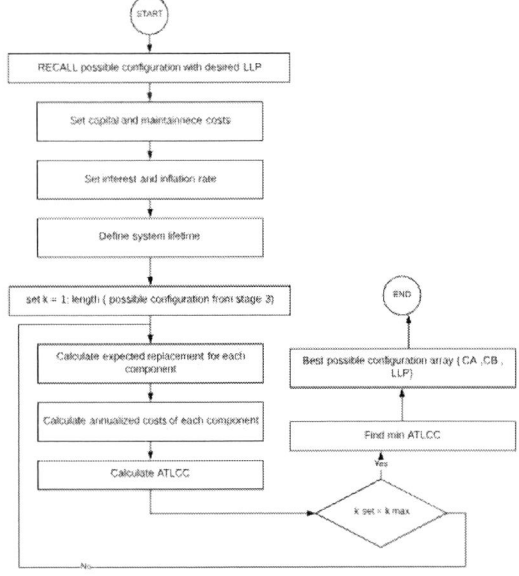

Fig. 5. Hybrid Optimization Algorithm

V. RESULTS

It is observed that excel based financial model yields LCOE turned out to be Rs. 2.2 /unit, and in due course of lifetime of SPV plant i.e. 25 years, this is beneficial in achieving grid parity. NPV of project is Rs. 69,453 and IRR as 14%, which states that this plant is viable and also, positive IRR value states that profitable income can be sourced from this system. NPV of savings in 25th year turns is evaluated as Rs. 2, 76,908. There is new scheme presented in which batteries are replaced every 5 years from the returns of fixed deposits done at the time of installation of the plant. The total costs of battery replacement, if cumulative, in every successive 5 years is Rs. 1, 64,670. If even half of this amount is deposited initially as FD in any bank at 8.75%, it will give beneficial returns as amount of Rs. 62,217.The ROI for this scheme is 13.1%. Cash flow is presented in figure 11 and comparison between COE of solar and grid supply is presented in figure 12. It is observed that grid electricity prices will increase approximately between 4-6% in coming years and the observed levellized cost of energy after installation of this hybrid system helps to achiever grid parity. Also, the payback period reduces 7.5 years, therefore post this period all the savings are turned into profit earned via this installation.

The proposed algorithm is executed in MATLAB line codes and results for Optimized model after executing proposed algorithm is number of PV panels is 7 and battery capacity is 2669 Wh with number of batteries of 150 Ah as 2 and loss of load probability as 0.007. As per the annual cost analysis, the annualize cost of total system after operations and maintenance is Rs. 66550/-. Capital recovery factor is CRF=0.09(1+0.09)25 (1+0.09)251. Therefore LCC = ATLCC* CRF. In a week generation = 50 units, therefore total energy = 5*4*12 approximately units and LCOE = LCC/Total Energy = Rs. 2.9 / watt. The selected LLP values as per the algorithm between 0-2 percent.

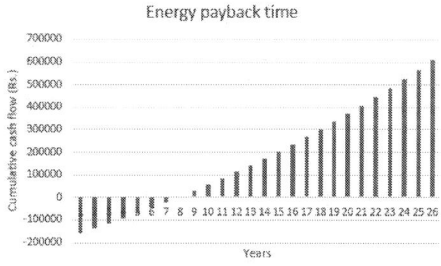

Fig. 6.   Energy Payback Time

VI.   Conclusion

Based on this case study for residential home in Faridabad for hybrid PV/diesel/battery system, main conclusions are as followed:

1.) The designed configuration for energy consumption of 2 kWh/day with 8 panels of 250 watts rating and battery capacity of 7200 Wh is evaluated for economic viability with FD scheme and it is observed that LCOE for the system is Rs. 2.17 / watt which is comparatively lower than present rate of per unit of electricity in residential homes of Faridabad.

2.) The same system is run for proposed algorithm gives optimized number of PV panels as 7 of 250 watts rating and battery capacity as 2996 Wh with LLP = 0.007% , which ensures system is reliable in a long term operation. LCOE for this optimized system turns out to be Rs. 2.9 / watt which is still lesser than unit electricity cost from grid. Due to lesser number of panels and battery capacity, capital costs for this optimized system is comparatively lesser than designed system. However the proposed optimization algorithm is more acceptable because it includes real time efficiencies, losses and optimal utilization of already installed diesel generator. The comparisons between both the methods on LCC, Annual production and LCOE is shown in figure 6, 7 and 8 respectively.

TABLE VI
COMPARISON BETWEEN EXCEL BASED FINANCIAL MODELLING AND PROPOSED OPTIMIZATION ALGORITHM.

	Excel based financial modellind	Proposed optimization Algorithm
LCC	Rs. 72345.00	Rs. 64550.00
Annual PV generation capacity	3350 units	2400 units
LCOE	Rs.2.17 per unit	Rs. 2.9 per unit

References

[1]] Arsalan Nisar.,Carlos Rodrguez Monroy,A review of the potential of renewable energy sources for the State of Jammu and Kashmir (India,*Energy Policy*,39,(2011),6667-71.

[2] France Lasnier, Wang Yaw Juen,The Sizing of Standalone Photovoltaic Systems using The Simulation Technique,*RERIC International Energy Journal*,12,(1990),21-39

[3] Tamer Khatib, Azah Mohamed, and K.Sopian K, A Software Tool for Optimal Sizing of PV Systems in Malaysia,*Modelling and Simulation in Engineering*,2,(2012),1-11

[4] Khaled Bataineh, Doraid Dalalah,Optimal Configuration for Design of Stand-Alone PV System,*Smart Grid and Renewable Energy*,3,(2012),139-147

[5] Bourrelle, J. Andersen, I Gustavsen, A,Energy payback: An attributional and environmentally focused approach to energy balance in net zero energy buildings. *International Journal of Energy and Buildings*,65,(2013),84-92.

[6] Hussein A. Kazem, T. Khatib, K.Sopian, Sizing of a standalone photovoltaic/battery system at minimum cost for remote housing electrification in Sohar,*International Journal of Energy and Buildings*,61,(2013),108-115.

[7] Saeid Lotfi Trazouei, Farid Lotfi Tarazouei, Mohammad Ghiamy,Optimal Design of a Hybrid Solar -Wind-Diesel Power System for Rural Electrification Using Imperialist Competitive Algorithm,*International Journal of Renewable Energy Research, ,3*,(2013),1-5

[8] B. Ould. Bilal, V. Sambou, P.A. Ndiaye, C.M.F. Kb, M. Ndongo,Study of the Influence of Load Profile Variation on the Optimal Sizing of a Standalone Hybrid PV/Wind/Battery/Diesel System *Energy Procedia*,36,(2013),1265-1275

[9] Mohammed T. Hussein, Shadi N. Albarqouni,Developing MATLAB software for PV and battery sizing for lighting projects in Gaza Strip, Palestine *Energy Procedia*,4,(2013),31-43

[10] Hocine Belmili, Mourad Haddadi, Seddik Bacha, Mohamed Fayal Almi, Boualem Bendib,Sizing stand-alone photovoltaicwind hybrid system: Techno-economic analysis and optimization *InternationalJournalof Renewable and Sustainable Energy Reviews*,30,(2014),821-832

[11] Tamer Khatib, Wilfried Elmenreic,Optimum Availability of Standalone Photovoltaic Power Systems for Remote Housing Electrification *International Journal of Photo energy*,10,(2014),1-5

[12] F. Z. Kadda, S. Zouggar, M. E. Hafyani and A. Rabhi,Contribution to the optimization of the electrical energy production from a Hybrid Renewable Energy system, *2014 5th International Renewable Energy Congress (IREC)*,(2014),1-6

[13] Mohd Shawal Jadin, Intan Zalika Mohd Nasiri, Syahierah Eliya Sabri and Ruhaizad Ishak,A Sizing Tool for PV standalone system , *ARPN Journal of Engineering and Applied Sciences*,10,(2015),10727-10732

[14] Fazia Baghdadi, Kamal Mohammedi, Said Diaf, Omar Behar,Feasibility study and energy conversion analysis of stand-alone hybrid renewable energy system *International Research Journal of Energy Conversion and Management*,1,(2015),471-479

[15] Vellanki Mehar Jyothi, T. Vijay Muni, S V N L Laltha,An Optimal Energy Management System for PV/Battery Standalone System *International Journal of Electrical and Computer Engineering (IJECE*,6,(2016),2538-2544

[16] Gupta A, Saini RP, Sharma MP,Steady-state modelling of hybrid energy system for o grid electrication of cluster of villages *International Research Journal of Renewable Energy*,35,(2010),520-35

[17] Ammar Mohammed Ameen, Jagadeesh Pasupuleti,Tamer Khatib,Simplified performance models of photovoltaic/diesel generator/battery system considering typical control strategies, *International Research Journal of Energy Conversion and Management*,99,(2015),313-325

[18] Rahul Rawat, S.C. Kaushik, Ravita Lamba,A review on modeling, design methodology and size optimization of photovoltaic based water pumping, standalone and grid connected system. *International Research Journal of Renewable and Sustainable Energy Reviews*,57,(2016),1506-1519

[19] Tamer Khatib, Ibrahim A. Ibrahim, Azah Mohamed,A review on sizing methodologies of photovoltaic array and storage battery in a standalone photovoltaic system, *International Research Journal of Energy Conversion and Management*,120,(2016),430-448

[20] Zaid Bin Siddique, Gagan Deep Yadav,Standalone Power Generation System using Renewable Energy Sources: A Review, *International Research Journa,l of Engineering and Technology (IRJET)*,3,(2016),1176-1182

[21] Chiemeka Onyeka Okoye, Ouz Solyali,Optimal sizing of stand-alone photovoltaic systems in residential buildings, *International Research Journal of Energy*,126,(2017),573-584

[22] Piyali Ganguly Akhtar Kalam, Aladin Zayegh,Optimum sizing of standalone PV/Wind power generating system with storage. *International Conference on Research in Education and Science (ICRES)*,(2017),345-354

[23] Ali Najah Al-Shamani,Mohd Yusof Hj Othman, Sohif Mat, M.H. Ruslan, Azher M. Abed,K. Sopian,Design Sizing of Stand-alone Solar Power Systems A house Iraq, *International Research Journal of Recent Advances in Renewable Energy Sources*,(2015),145-150

[24] Sonali Goel, Renu Sharma,Performance evaluation of standalone, grid connected and hybrid renewable energy systems for rural application:A comparative review, *International Research Journal of Renewable and Sustainable Energy Reviews*,78,(2017),1378-1389

[25] Saban Yilmaz, Furkan Dincerr,Optimal design of hybrid PV-Diesel-Battery systems for isolated lands: A case study for Kilis, Turkey, *International Research Journal of Renewable and Sustainable Energy Reviews*,77,(2017),344-352

[26] Razman Ayop, Normazlina Mat Isa, Chee Wei Tan,Components sizing of photovoltaic stand-alone system based on loss of power supply probability, *International Research Journal of Renewable and Sustainable Energy Reviews*,81,(2017),2731-2743

2nd IEEE International conference on power Electronics, Intelligent Control and Energy systems (ICPEICES-2018)

Power Quality Improvement using Fuzzy-PI Controlled D-STATCOM

Atma Ram
Department of Electrical Engineering
YMCA University of Sci. and Tech.
Faridabad, India
atma.ram12@gmail.com

Paras Ram Sharma
Department of Electrical Engineering
YMCA University of Sci. and Tech.
Faridabad, India
prsharma1966@gmail.com

Rajesh Kr. Ahuja
Department of Electrical Engineering
YMCA University of Sci. and Tech.
Faridabad, India
rajeshkrahuja@gmail.com

Abstract— **A novel method for controlling the current source converter (CSC) based D-STATCOM is presented in this paper using with Fuzzy-PI controller for mitigation of voltage sag, improvement of THD of source current , load current, load voltage and also .maintains DC link Voltage profile. The model has been tested and simulated in MATLAB/SIMULINK environment. The simulation result under normal and change in load conditions has been studied without and with D-STATCOM.**

Keywords— **Custom power Device, D-STATCOM, Fuzzy logic controller, Proportional Integral.**

I. Introduction

In distribution systems, power quality problems arise mainly due to different non-linear loads and unplanned expansion of distribution systems. Various power quality problems comprise harmonic currents, unbalanced load, high reactive power burden [1-8]. The application of power electronics technology used in power distribution led to design of new devices known as custom power devices (CD)[9].D-STATCOM is a shunt connected custom power device which is used to takes care of such type of problems in distribution systems. The CSC topology based D-STATCOM is used .The capacitor filter is used to AC side of D-STATCOM for improving the quality of output currents waveforms but it increase the cost of converter. It resonates with AC side inductance due to which some of harmonic present in output current might be amplified[10]when D-STATCOM is operated with sinusoidal pulse modulation technique(SPWM)[11] the magnitude of harmonic component is proportional to fundamental component of their output. The current injected under normal operating conditions by CSC based D-STATCOM is small percentage of line current, due to which current harmonics are small. Hence requirement of energy storage for CSC based topology is lesser when used for mitigation of voltage sag [10]. D-STATCOM can be used to prevent the rest of distribution system from polluting non linear loads. It can be used to offer continuous and dynamic   control of power supply, reactive power compensation, eliminations of harmonics ,and mitigation of voltage sag/swell[12]. Through investigation and performance analysis of CSC based D-STATCOM for mitigation of voltage sag which is the one of the power quality problems resulted from sudden change in load conditions connected to a distribution system [10]. The output voltage magnitude of inverter circuit used in CP device is proportional to their DC link voltage. Hence, it is essential to maintain DC link voltage at the time of designing of CSC based Custom power Device. An effort has also been made to keep DC link voltage by using FLC in control system of D-STATCOM.

II. CSC BASED D-STATCOM

It is a shunt connected device which able to inject unbalanced current to eliminate distortion in source current, load current and supply voltage. The D-STATCOM used for custom power application by using PWM switching control as oppose to fundamental frequency switching strategy which used in FACTS application as it is used at low power level.CSC based D-STATCOM can modify the basic configuration of VSC based one [12]. It consists of circuit which is interfaced with a coupling transformer & DC link reactor as shown in figure-1.

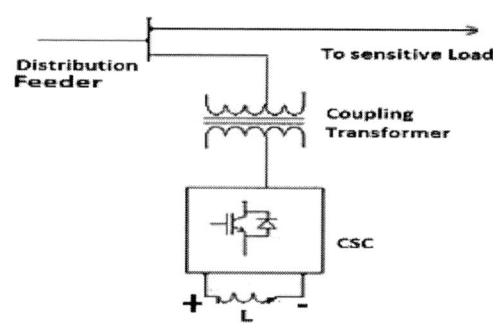

Fig 1.Schemetic Diagram of CSC based D-STATCOM

It consists of three phase CSC driven by SPWM,AC side low pass filter, coupling transformer & internal control system as proposed in figure 2.

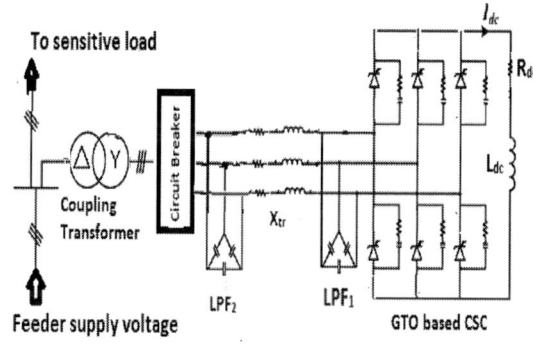

Fig 2.Current Source Converter based D-STATCOM setup

The choice of switching device & modulation scheme depend on application voltage & power rating[13]. A distribution level convertor topology for proposed model being CSC, GTO switch having sufficient reverse voltage with standing capability is selected. These switches are configured in bridge & fed from a DC link reactor act as a

978-1-5386-6624-1/18 $31.00 © 2018 IEEE

storage element. The reactive power to be generated by CSC can be calculated from the formula

$$Q= \sqrt{(3/2)}*V*M*Idc*cos\theta \qquad (1)$$

Where V= rms value of converter input, M= modulation index, I_{dc}= mean DC link current & θ = Phase shift [14].The output of CSC is filtered by three phase LPF comprising of three capacitor connected in shunt manner. These filters provide good sinusoidal output voltage & current waveforms of separation of high order harmonic components to coupling transformer. The external series reactor Xtr has been used on low voltage side of coupling transformer for adjusting corner frequency of input filter in implementation of proposed D-STATCOM.

III. CONTROL STRATEGY

The control system of the proposed D-STATCOM consists of a phase locked loop (PLL).proportional integral controller (PI).and Idqo transformation block, etc. Synchronous Reference Frame Theory (SRFT) is used for producing reference source currents for control of GTO based D-STATCOM [15].The block diagram is shown in figure3. The Point of Common Coupling (PCC) voltage Vt, Load currents (ILa, ILb, ILc), and Vdc voltage of D-STATCOM are taken as feedback signals.

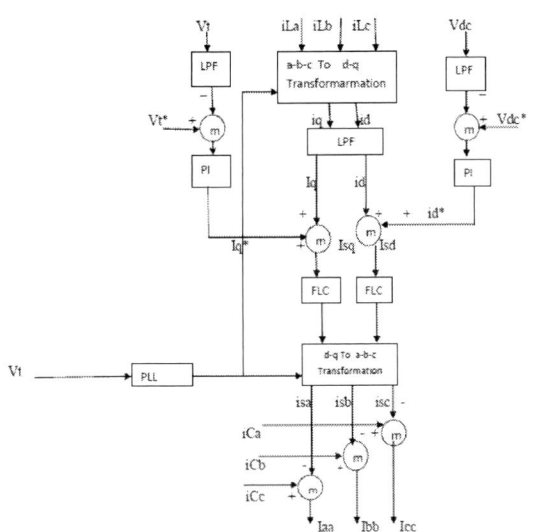

Fig 3.Block diagram of D-STATCOM control system

The load currents from the Ila,Ilb and Ilc are converted to d-q-0 frame by using park transformation relations as:

$$I_d= 2/3 \ [i_{La}sin\theta+i_{Lb}sin(\theta-2\pi/3)+i_{Lc}sin(\theta+2\pi/3)] \qquad (2)$$
$$I_q=2/3[i_{La}cos\theta+i_{Lb}cos(\theta-2\pi/3)+i_{Lc}cos(\theta+2\pi/3)] \qquad (3)$$
$$i_0= 1/3[i_{La}+i_{Lb}+i_{La}] \qquad (4)$$

The Phase Locked Loop (PLL) is used to match these signals with PCC voltage Vt. The DC component of ILd and ILq is extracted by passing the D-Q component through LPF. Error between V_{dcref} and V_{dc} sensed at the output terminal of D-STATCOM is given to a Proportional Integral (PI),output of which is taken as loss component of current and added to DC component I_d*. The error between V_{tref} and

V_t is given to second PI controller which is used to adjust the PCC voltage. The output of this controller I_q* is added to I_q.

A. FLC SECTION

FLC is a linguistic based control strategy; it is derived from knowledge and converted into an automatic control strategy. It doesn't need any difficult mathematical calculation like the others control system. While the others type of controller use complicated mathematical calculation to the control the plant, but the simple mathematical calculation is used in FLC to simulate the expert knowledge [16]. Without using any difficult calculation FLC gives result for control system. Thus it can be best available answers today for a broad class of challenging controls problems. A fuzzy logic control usually consists of the following:

- Fuzzification: It convert numerical variable into linguistic variable.
- Knowledge Base: It is the collection of rules to achieve the goal.
- Fuzzy Reasoning Mechanism: It performs the FLC operation and control the output according to fuzzy inputs.
- Defuzzification unit: It converts linguistic variable into numerical values.

FLC is very simple controller. It has an input stage, processing stage and output stage. The input stage combine the rules using linguistic values and membership function for each rules. The processing stage generates the output of each rule and combines the result of rules. Finally, the output stage converts this combine result into numerical values. Most common membership function used in FLC controller is triangular, and trapezoidal. But the curves of function are less important than the placement of fuzzy controller.FLC basic structure is shown in fig.4.

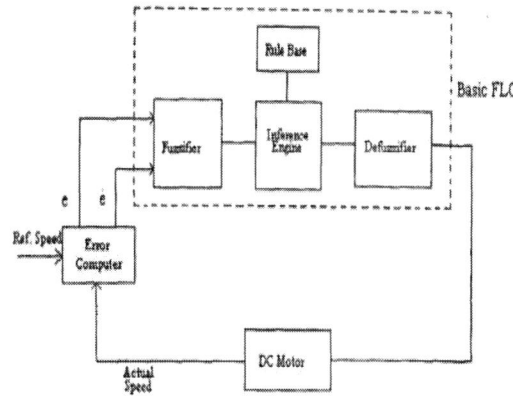

Fig 4. Fuzzy Logic controller

In proposed fuzzy PI controller two FLC block are used for error signal d & error signal q .Each controller has two inputs & one outputs. One input to FLCs is the derivative of output of PI controllers & abc to dq transformation is shown in fig 5.

978-1-5386-6624-1/18 $31.00 © 2018 IEEE

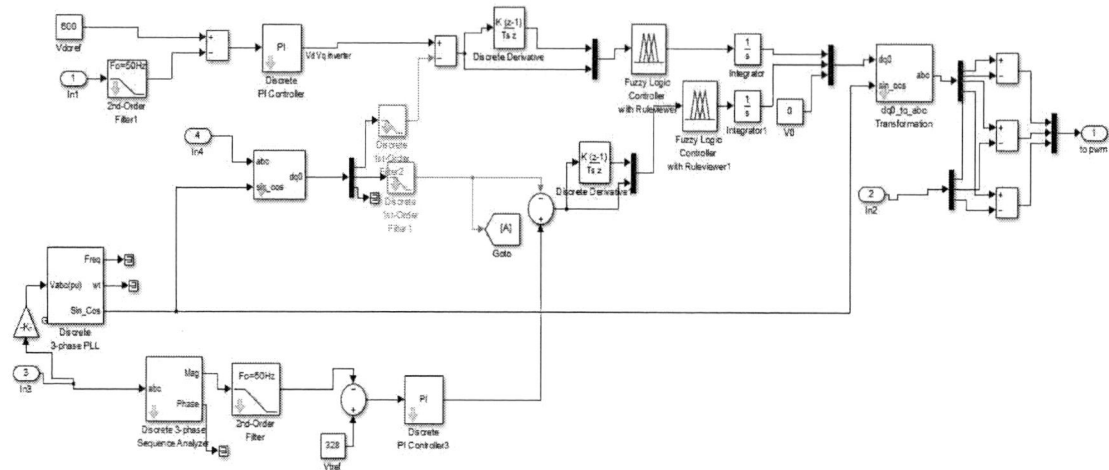

Fig 5. Simulink diagram for Fuzzy PI Controller

The other words input of FLCs are the error of d & q axis current & derivative of these error. An integrator is used to remove error in the output of FLC [16]. In this paper five membership are used for input variable and output variable namely representing negative big, negative small, zero, positive small & positive big respectively. A fizzy interface system is developed with triangular membership function with the help of FIS editor available in MATLAB. The 25 Fuzzy rules applied which are shown in Table 1. The basic fuzzy rules give the relationship between input and output [17].

Table1. Rule Base for FLC

de(i) \\ e(i)	NB	NS	Z	PS	PB
NB	NB	NB	NS	NS	Z
NS	NB	NB	NS	Z	Z
Z	NS	NS	Z	PS	PS
PS	Z	PS	PS	PB	PB
PB	Z	Z	PS	PB	PB

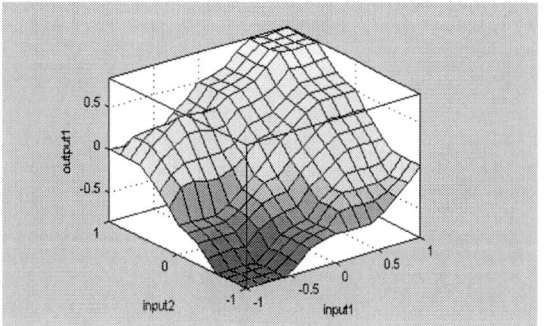

Fig 6.Surface Viewer of FLC

Fuzzy rules are framed using MATLAB and after investigation of these rules surface viewer is generated which is shown in fig.6. With this approach output coefficient of PI controller is tuned for the desired level there by achieving the advantage of improvement of sag voltage as well as the DC output of inverter at better level. FLC controller improves the inverter output voltage in better manner than PI controller.

IV. SIMULATION RESULT & DISCUSSION

Simulink model of CSC based D-STATCOM which is shown in Fig-7 has been simulated under steady state and dynamic change in load conditions. Initially simulation is done under normal load conditions without D-STATCOM and large inductive load2 by opening the circuit breaker1 and 2 respectively. Figure8 shows the waveforms of load voltage and load current under normal load condition. From the waveforms it is observed that there is no voltage sag taking place across the load.

Secondly this model is simulated without D-STATCOM under sudden change in load conditions. The D-STATCOM is disconnected by opening the circuit breaker1 and large inductive load is connected by closing the circuit breaker2 for transition time set from 0.4 to 0.6 second. Figure9 shows the waveforms of load voltage and load current and it is observed that the voltage sag is generated for time duration from 0.4 to 0.6 second. Figure 10 shows the waveforms of voltage and current at point of coupling without operating the D-STATCOM.

Thirdly this model has been simulated by using the D-STATCOM under sudden change in load conditions. To mitigate the voltage sag created by change in load conditions the D-STATCOM is brought to operation by operating the circuit breaker1 with its transition time setting from 0.4 to 0.6second. Figure11 shows the waveforms of load voltage and current. It is observed that in the load voltage during this period 0.4 to 0.6 second, sag has been mitigated by proposed D-STATCOM under this condition. Voltage profile can also be improved by operating the circuit breaker3.

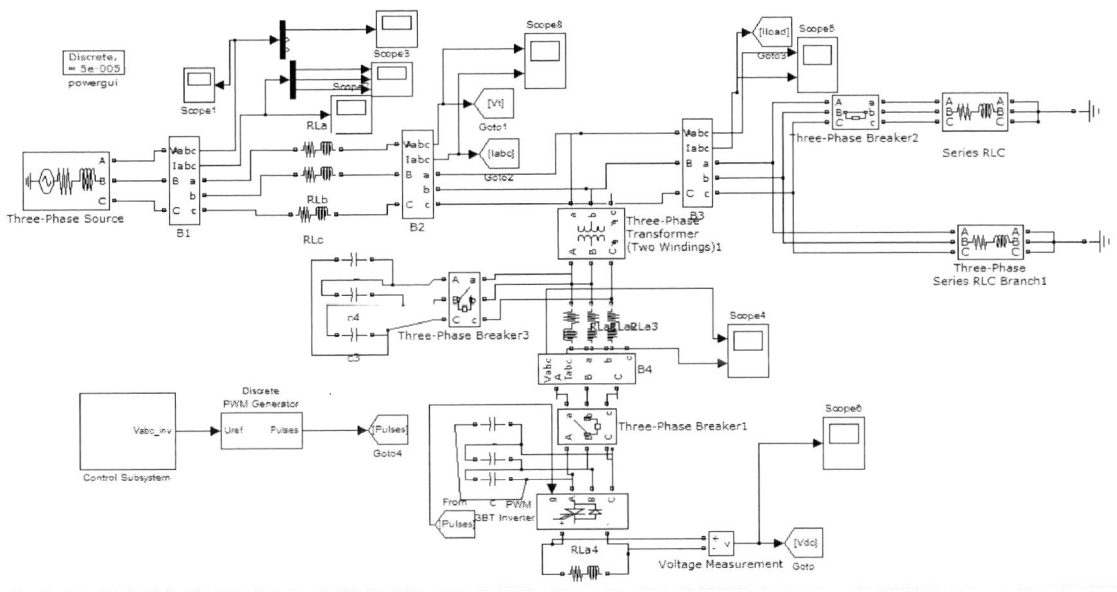

Fig 7.Simulink model of proposed D-STATCOM

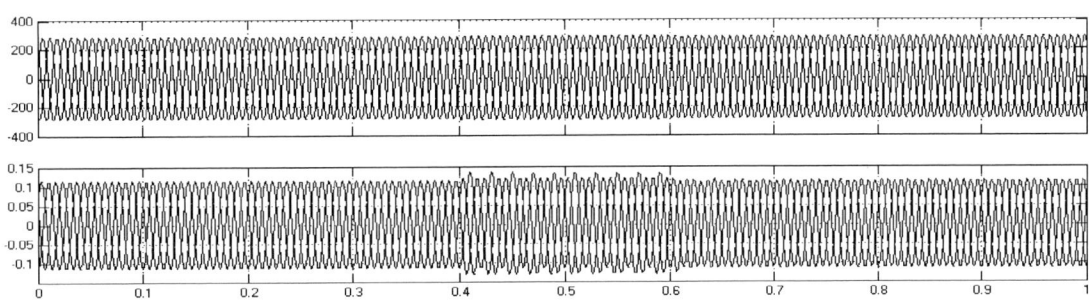

Fig 8. Load voltage and load current waveforms under normal load condition

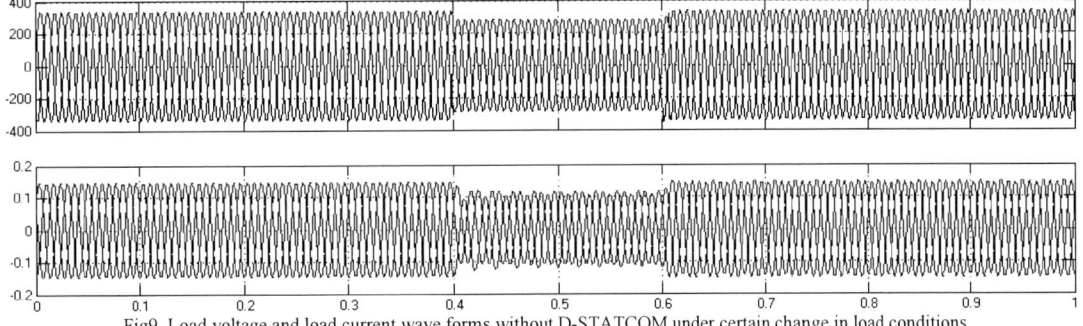

Fig9. Load voltage and load current wave forms without D-STATCOM under certain change in load conditions

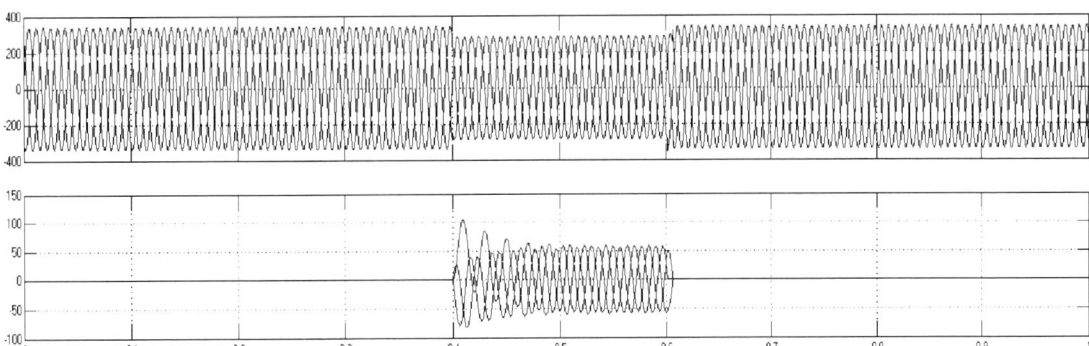

Fig10.Voltage and current waveforms without D-STATCOM under certain change in load conditions at point of common coupling

978-1-5386-6624-1/18 $31.00 © 2018 IEEE   417

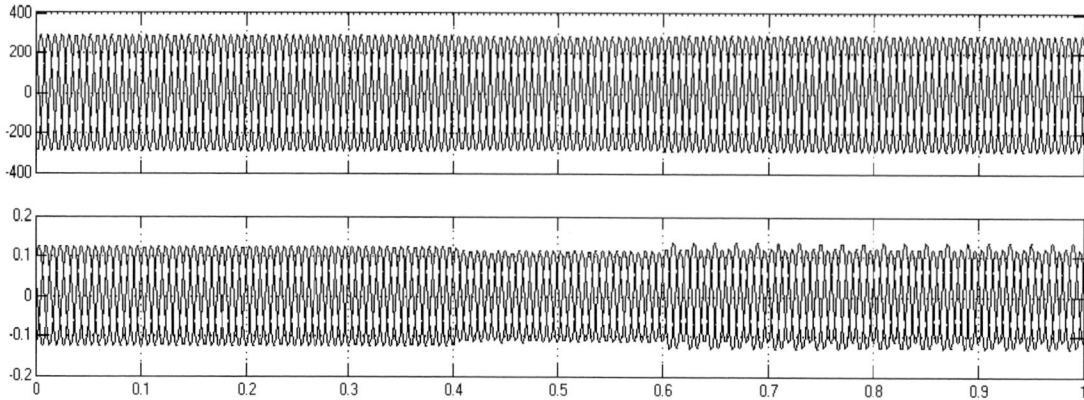

Fig11.Load voltage and load current waveforms with D-STATCOM under certain change in load condition

Lastly the model has been simulated with only PI controller for DC-link reactor voltage profile under small change in load condition which is shown in figure12. It is observed that the magnitude of DC link voltage is decreased from its initial value. It is learnt that with the use of Fuzzy PI controller the DC-link reactor profile voltage improved which is shown in figure13.

V. HARMONICS ANALYSIS

Fig 14 and 15 shows the simulation results of source current for compensation of harmonics by D-STATCOM system with non linear inductive load. The waveforms of source current without and with D-STATCOM are presented to demonstrate the performance of D-STATCOM. The corresponding THD of the source current is reduced from 71.20% to 0.98% with D-STATCOM. Figure16 and 17 shows the harmonics spectrum of source current without and with D-STATCOM respectively. Waveforms of Load voltage and Load current without and with D-STATCOM are shown in Figure9 and figure 11 respectively.Figure18 and figure19 shows the harmonics spectrum of phase a of load voltage without and with D-STATCOM respectively. The corresponding THD of load voltage is reduced from 1.78% to 0.03% with D-STATCOM. Figure20 and figure21 shows the harmonics spectrum of phase a of load current without D-STATCOM and with D-STATCOM respectively. The corresponding THD of load current is reduced from 5.47% to 2.35% with D-STATCOM.

VI. CONCLUSION

This paper presents the proposed Fuzzy-PI controller for CSC based D-STATCOM used for mitigation of voltage sag, improvement of voltage and current harmonics and to improve DC link voltage maintain DC-Link voltage profile under change in load condition in a distribution system. The voltage sag is created by switching large inductive load in distribution system for duration from 0.4 to 0.6 seconds. The load voltage shows sag during this time duration. This voltage sag is effectively mitigated by using a CSC based D-STATCOM during sudden change in load condition. THD of source current, load voltage and load current is also improved effectively and DC- link voltage becomes constant after the disturbance of   load variation thus minimizing the ripple injected into distribution system by D-STATCOM.

REFERENCES

[1] IEEE, "IEEE recommended practices and requirements for harmonics control in electric power systems," IEEE Std. 519, 1992.

[2] Ghosh, A., and Ledwich, G., Power Quality Enhancement Using Custom Power Devices, London: Kluwer Academic Publishers, 2002.

[3] Acha, E., Agelids, V. G., Anaya-Lara, O., and Miller, T. J. E., Power Electronic Control in Electric Systems, 1st ed., Oxford, UK: Newness Power Engineering Series, 2002.

[4] Dugan, R. C., McGranaghan,M. F., and Beaty, H.W., Electric Power Systems Quality, 2nd ed., New York: McGraw Hill, 2006.

[5] Akagi, H., Watanabe, E. H., and Aredes, M., Instantaneous Power Theory and Applications to Power Conditioning, NJ: John Wiley & Sons, 2007.

[6] Padiyar, K. R., FACTS Controllers in Transmission and Distribution, New Delhi: New Age International, 2007.

[7] Moreno-Munoz, A., Power Quality: Mitigation Technologies in a Distributed Environment, London: Springer-Verlag London Limited, 2007.

[8] Fuchs, E. F., and Mausoum, M. A. S., Power Quality in Power Systems and Electrical Machines, London: Elsevier Academic Press, 2008.

[9] Hingorani, N. G., &Gyugyi, L. (2013). Understanding FACTS:Concepts& technology of flexible AC transmission systems.London: Wiley.

[10] Singh, M. D., Mehta, R. K., & Singh, A. K. (2015). Current source converter based D-STATCOM for voltage sag mitigation. International Journal for Simulation and MultidisciplinaryDesign Optimization, 6, A5, 1–10. France. doi: 10.1051/smdo/2015005

[11] Mohan, N., Undeland, T. M., & Robbins, W. P. (1989). Power electronics: Converters, applications, and design. New York,NY: John Wiley & Sons.

[12] Acha, E., Agelidis, V. G., Anaya-Lara, O., & Miller, T. J. E. (2006). Power electronic control in electrical systems.Newnes power engineering series. Oxford: Elsevier.

[13] Han, B., Moon, S., &Karady, G. (2000). Static synchronous compensator using thyristor PWM current source inverter.Proceeding of IEEE Transactions on Power Delivery, 15, 1285–1290.

[14] Bilgin, H. F., &Ermis, M. (2010). Design and implementation of a current source converter for use in industry applications of D-STATCOM.Proceedings of IEEE Transactions on PowerElectronics, 25, 1943–2957.

[15] Bhim,S.,Jayaprakash,P.,&Kothari,D.P (2008).   A T-connected transformer and three leg VSC based D-STATCOM for power quality improvement. Proceeding of IEEE Transactions on power electronics 23, 2710-2718.

[16] Resul, C., Besir D., &Fikret A. (2011). Fuzzy-PI current controlled D-STATCOM. Gazi University Journal of Science, 24, 91–99.

[17] Deepa, S., &Ranjani, M.(2015). Dynamic voltage restorer controller using grade algorithm.Cogent Engineering, 2, 1017243, 1-11.doi:10.1080/23311916.2015.1017243.

978-1-5386-6624-1/18 $31.00 © 2018 IEEE   418

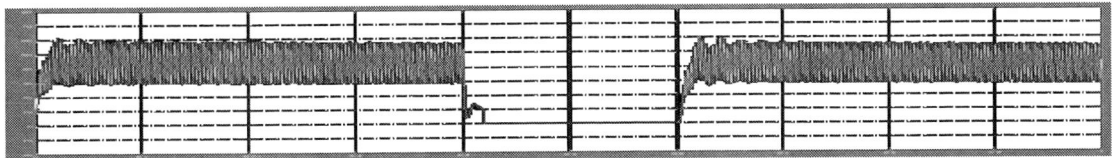

Fig 12 DC link rector voltage with PI controller alone

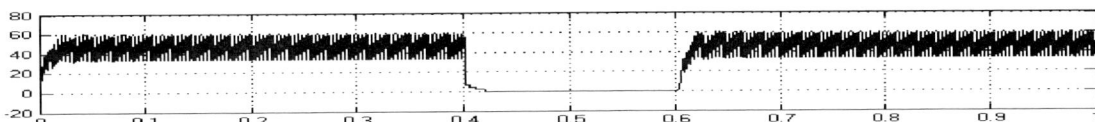

Fig13 DC link rector voltage with Fuzzy - PI controller

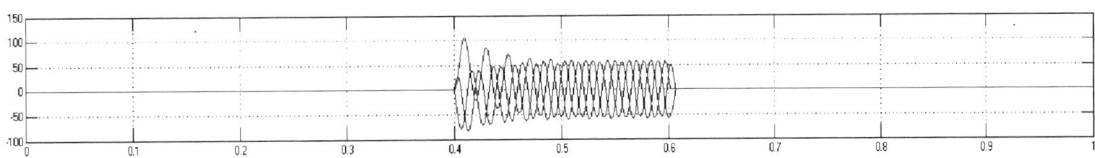

Fig14. Source current without D-STATCOM

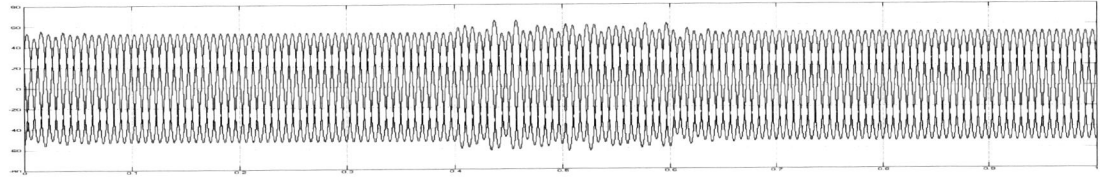

Fig15. Source current with D-STATCOM

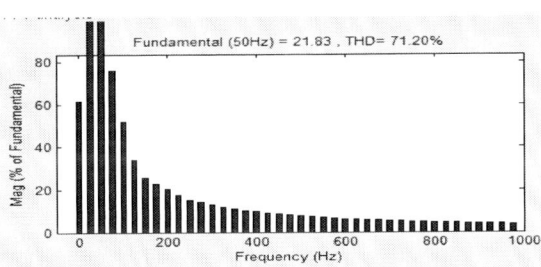

Fig16. THD of source current in phase1 without D-STATCOM

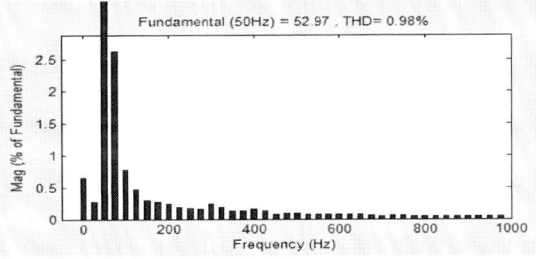

Fig17. THD of source current in phase1 with D-STATCOM

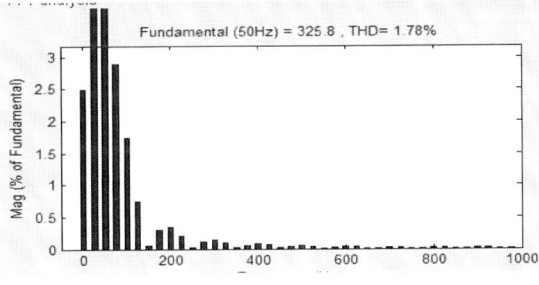

Fig18. THD of load voltage in phase a without D-STATCOM

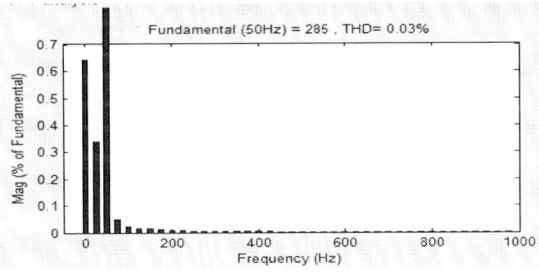

Fig19. THD of load voltage in phase a with D-STATCOM

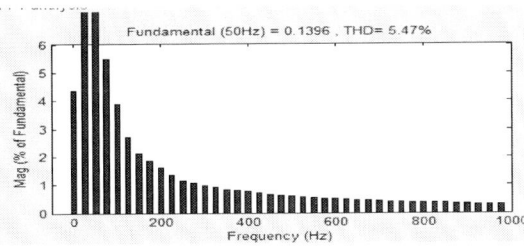

Fig20. THD of load current in phase a without D-STATCOM

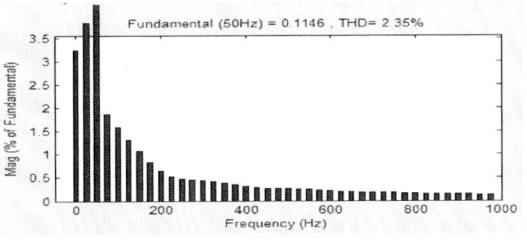

Fig21. THD of load current in phase a with D-STATCOM

Non-linear H-infinity Control for Grid-interactive Offshore Wind Farm and Marine Current Farm

Satendra Kr Singh Kushwaha Dept.
of Electrical Engineering *Motilal
Nehru National Institute of
Technology*
Allahabad-211004
ree1506@mnnit.ac.in

S. R. Mohanty
Dept. of Electrical Engineering
Motilal Nehru National Institute of
Technology
Allahabad-211004
soumya@mnnit.ac.in

Paulson Samuel
*Dept. of Electrical Engineering
Motilal Nehru National Institute of
Technology*
Allahabad-211004
paul@mnnit.ac.in

Abstract- **This paper presents the H-infinity loop-shaping robust control strategy for offshore wind farm (OWF) and marine current farm (MCF) under high stochasticity and disturbances. Offshore wind and marine current resource, being a fluctuating resource required robust control management that able to address the stability issues. The proposed H-infinity loop-shaping control objective is to optimize the control signal and enhanced the system stability. The performance of the proposed controller is evaluated under high stochastic wind and marine current speed. The simulation results and analysis conforms a significant improvement in the stability of the studied system with proposed controller under high stochasticity.**

Keywords- H-infinity robust control, offshore wind farm.

I. INTRODUCTION

In recent years the penetration of offshore wind farm in the existing system is increased. Currently the offshore wind farm (OWF) and marine current farm (MCF) is combined together in UK [1]. The ocean is a huge source of kinetic energy that is capable of contributes to our future energy needs. The high penetration of OWF and MCF in the national grid creates some critical issues of power quality in normal operating condition and stability issues [2-3] in transient condition. It can be recognized that the offshore wind farm and marine current farm controllers should play a critical role in the stability of the OWF and MCF connected to grid.

The DFIG is widely used in the offshore wind farms becoming the active and reactive power control ability. However the control of DFIG has proven to be difficult task due to non-linear system dynamics multivariable system. The variation of wind speed is more as compare to marine current speed. So DFIG based generator is used in for offshore wind farm and SCIG based generator is used in marine current farm in this study. The field oriented vector control technique has been used for the active and reactive power control of the DFIG based OWF. Traditionally the PI controller is used in the DFIG because of its simplicity and easy to implement. However due to nonlinear behavior of system dynamics, the PI controller is not incorporate the non-linear dynamics of the system the performance of the PI

controller is not satisfactory during high stochasticity and transient condition.

The commonly used controllers are adding more regulators to the existing PI controller such as proportional plus resonant (PIR) controller, proportional plus resonant (PR) controller [4-6]. These controllers are single input single output (SISO) controller and depend on certain model of plant. Therefore they give good performance and satisfactory results while a fixed operating point and uncertainty is not considered. The H-infinity loop-shaping based controller allows the uncertainty in wind and disturbances. Also it is a multi-input multi-output (MIMO) control strategy so it gives a satisfactory performance and robustness against high stochasticity and disturbance. Therefore the H-infinity [7-8] loop-shaping control strategy is adopted in this paper to ensure the robustness and stability of the studied system.

The aim of this paper is focused to model a DFIG based OWF; SCIG based MCF and develop an H-infinity loop-shaping controller that pertains the stability issues in the integrated system.

This paper is organized as follows: segregated modeling of system components such as DFIG based OWF and SCIF based MCF is given in section II. The modeling of proposed controller is given in section III. The simulation results and analysis is given in section IV. The conclusion and remarks is given in section V.

II. OWF AND MCF MODELING

In this section, modeling of DFIG based offshore wind farm and SCIG based marine current farm is discussed. The plant layout is shown in fig 1(a). The aggregated model of offshore wind farm is simulated by 2 MW DFIG based generator and the aggregated model of marine current generator is simulated by 2MW SCIG based generator. The modeling of OWF and MCF is illustrated in subsequent section.

A. MODELING OF MARINE CURRENT TURBINE

The modelling of MCT is expressed by the following eqn [9].

$$P_{mct} = P_{m0}.C_{p\_mct}(\lambda_{mr}, \beta_{mr}) \tag{1}$$

Where P_{m0} is the power available to marine current turbine and C_{p_mct} is the power coefficient of marine current turbine. P_{m0} and C_{p_mct} is expressed as [10]

$$P_{m0} = \frac{1}{2}\rho_{mr}.\pi R_{mr}^2.V_{mr}^3 \tag{2}$$

$$C_{p\_mct}(\psi_{mr},\beta_{mr}) = b_{c1}(\frac{b_{c2}}{\psi_{mr}} - b_{c3}.\beta_{mr} - b_{c4}.\beta_{mr}^{b_5} - b_{c6})e^{(-b_{c7}/\psi_{mr})} \tag{3}$$

in which

$$\frac{1}{\psi_{mr}} = \frac{1}{\lambda_{mr}+b_{c8}.\beta_{mr}} - \frac{b_{c9}}{\beta_{mr}^3+1}; \qquad \lambda_{mr} = \frac{R_{bmr}.\omega_{bmr}}{v_{mr}} \tag{4}$$

where ρ_{mr} is the density of water (kg/m3), V_w is the velocity of water (m/s), R_{mr} is the radius of marine current turbine (m), β_{mr} is the pitch angle of the turbine, ω_{mr} is the angular velocity of the turbine, λ_{mr} is the tip speed ratio of the turbine, and d_1–d_9 are the constant-coefficients of C_{p_mct}. The parameter of turbines and power coefficient is given in appendix.

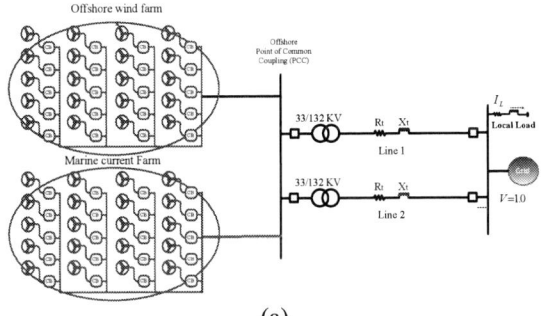

(a)

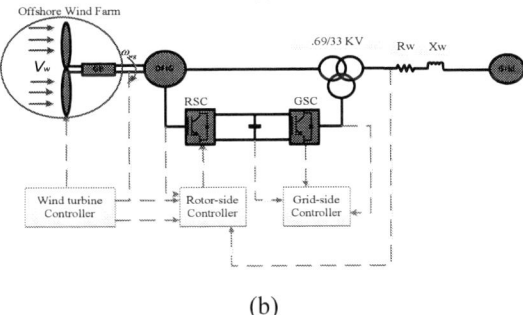

(b)

Fig. 1. (a) single line diagram of OWF & MCF connected to grid (b) Detailed diagram of DFIG based OWF

SCIG MODELING

The dynamic eqn of SCIG based marine current generator in d-q reference frame is defined as [9]

$$\begin{aligned} V_s &= r_s i_s + p\psi_s + \omega_{ref}\psi_s \\ 0 &= r_r i_r + (\omega_{ref} - \omega_r)\psi_r + p\psi_r \end{aligned} \tag{5}$$

where ψ_s is the flux linkage, ω_{ref} and ω_r are the angular velocity of rotating field and rotor of SCIG respectively, i_s and i_r are the stator and rotor current of SCIG respectively.

B. OFFSHORE WIND TURBINE

The modeling of variable speed offshore wind turbine is expressed by following eqn [9]

$$P_{owt} = P_{m0}.C_{pw}(\lambda_w, \beta_w) \tag{6}$$

Where P_{m0} is the power available to offshore wind turbine and C_{pw} is the power coefficient of wind turbine. P_{m0} and C_{pw} is expressed as

$$P_{m0} = \frac{1}{2}\rho_{owt}.\pi R_{owt}^2.V_w^3 \tag{7}$$

$$C_{p\_owt}(\psi_w,\beta_w) = c_{c1}(\frac{c_{c2}}{\psi_w} - c_{c3}.\beta_w - c_{c4}.\beta_w^{c5} - c_{c6})e^{(-c_{c7}/\psi_w)} \tag{8}$$

In which

$$\frac{1}{\psi_{ww}} = \frac{1}{\lambda_w+c_{c8}.\beta_w} - \frac{c_{c9}}{\beta_w^3+1}; \qquad \lambda_w = \frac{R_{bw}.\omega_{bw}}{v_w} \tag{9}$$

Where ρ_w the density of air, V_w is the velocity of wind, R_w is the radius of turbine, β_w is the pitch angle of the turbine, ω_w is the angular velocity of the turbine, λ_w is the tip speed ratio of the turbine, and c_1–c_9 are the constant-coefficients of C_{p_owt}. The parameter of turbines and power coefficient is given in appendix.

C. DFIG MODELING

The Dynamic eqn of DFIG based offshore wind generator in d-q reference frame is expressed as [10]

$$\begin{aligned} V_{dq0s} &= r_s i_{dqs} + \omega\lambda_{dqs} + p\lambda_{dq0s} \\ V'_{dq0r} &= r_s i_{dq0r} + (\omega-\omega_r)\lambda'_{dqr} + p\lambda'_{dq0r} \end{aligned} \tag{10}$$

$$i_{qs} = \left(-\frac{L_m}{L_{ls}+L_m}\right)i'_{qr}; \qquad i_{ds} = \left(\frac{\lambda_{ds}-L_m i'_{dr}}{L_{ls}+L_m}\right) \tag{11}$$

where λ is the flux linkage, L_{ls} is the magnetizing inductance and L_m is the mutual inductance of the DFIG. The active and reactive power of the DFIG is expressed as

$$P_s = \frac{3}{2}V_{qs}i_{qs}; \qquad Q_s = -\frac{3}{2}V_{qs}i_{ds} \tag{12}$$

The stator d and q axis current is controlled by the d and q axis current of the rotor side of the current so the P_s and Q_s is expressed as

$$P_s = -\frac{3}{2}\left(\frac{\omega\lambda_{ds}L_m}{L_{ls}+L_m}\right)i_{qs}; \qquad Q_s = \left(\frac{\omega\lambda_{ds}}{L_{ls}+L_m}\right)(\lambda_{ds}-L_m i'_{dr}) \tag{13}$$

Eq (13) presents the active and reactive power of DFIG which is independently controlled by the rotor current in dq frame of reference. The rotor winding is connected to grid through back to back rotor side and grid side converter, the control circuit of rotor side converter (RSC) and grid side converter (GSC) is shown in fig 2.

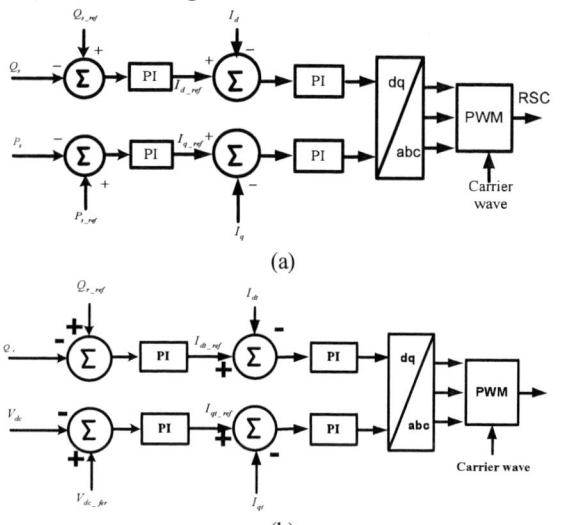

(a)

(b)

Fig. 2. (a) RSC controller of DFIG (b) GSC controller of DFIG

III. APPLICATION OF H-INFINITY CONTROL TO OWF & MCF

In this section, the design of loop-shaping H-infinity controller for DFIG based offshore wind farm is discussed. The objective of the controller design to optimize the controller performance subjected to uncertain wind speed and other external disturbances. The major concern to design a controller using H-infinity loop-shaping for integrated system is robust capability and ability of controller to perform satisfactory under various disturbances. Fig 3(a) shows the basic block diagram of H-infinity control problem. It is a multi-input multi-output (MIMO) controller it contains two sets of input and output signal. The signal w is an exogenous input, u is a control input, the output signal z denotes control error and output signal y is a feedback signal. The open loop transfer function G yields the following relations.

$$
\begin{bmatrix} z \\ y \end{bmatrix} = \begin{bmatrix} G_{11} & G_{12} \\ G_{21} & G_{22} \end{bmatrix} \begin{bmatrix} w \\ u \end{bmatrix}
$$
(14)

The state space of plant P

$$
\begin{bmatrix} \dot{x} \\ z \\ y \end{bmatrix} = \begin{bmatrix} A & B_1 & B_2 \\ C_1 & D_{11} & D_{12} \\ C_2 & D_{21} & D_{22} \end{bmatrix} \begin{bmatrix} x \\ w \\ u \end{bmatrix}
$$
(15)

Where (A, B_2, C_2) is stabilizeble and detectable to guarantee stability of controller K. D_{11} & D_{21} is full rank to ensure the controller are proper. The reference signal and disturbances

are marked as an exogenous input w to the system. The state space expression (17) is defined by following parameters.

$$
x = \begin{bmatrix} \psi_{ds} \\ \psi_{qs} \\ i_{qs} \\ i_{qr} \end{bmatrix}, \quad \omega = \begin{bmatrix} v_{ds} \\ v_{qs} \\ i_{dr}^{ref} \\ i_{qr}^{ref} \end{bmatrix}, \quad u = \begin{bmatrix} v_{dr} \\ v_{qr} \end{bmatrix}
$$

$$
y = [\psi_{ds} \ \psi_{qs} \ i_{dr} \ i_{qr}]^T, z = [i_{dr}^{ref} - i_{dr} \ i_{qr}^{ref} - i_{qr}]^T
$$
(16)

For the H-infinity control, to insure closed loop stability of plant 'P', a feedback controller 'K' is incorporated. Such that the H-infinity norm is smaller than the γ_{min} from disturbance signal 'w' to the output 'z'.

$$
\left\| T_{zw}(s) \right\|_{\infty} < \gamma
$$
(17)

The H-infinity control problem is to find the controller 'K' that minimizes and stabilizes the closed loop system.

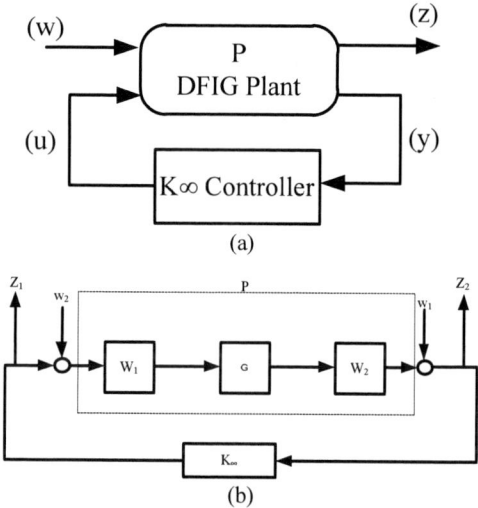

(a)

(b)

Fig. 3. (a) block diagram of the loop-shaping H-infinity control (b) H-infinity controller for shaped plant

The stabilizable controller is design using pre compensator W_1 and post compensator W_2. The controller design from shaped plant shown in fig 3(b) is

$$
K = W_1 K_{\infty} W_2
$$
(18)

Where $K\infty$ is the H-infinity controller. The weighting functions are chosen as [11]

$$
W_1 = K_w \left(\frac{S+a}{S+b} \right)
$$
(19)

Where 'Kw' is a gain, 'a' and 'b' are positive value. Usually the weighting function W_2 is considered as a unity. The algorithm of the K∞ controller design is shown in fig 4.

978-1-5386-6624-1/18 $31.00 © 2018 IEEE

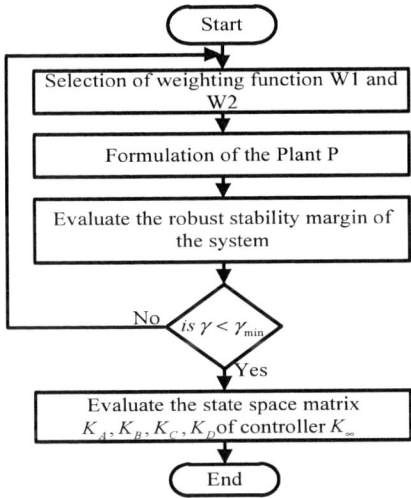

Fig. 4. Flow chart

IV. SIMULATION RESULTS AND DISCUSSION

In this section, the studied system is simulated and analyzed with H-infinity loop-shaping based proposed controller. The performance of the controller is evaluated for minimization of wind and marine current disturbances. The simulation results are analyzed in pu quantities except the wind and marine current velocity, which is in m/sec.

A. TIME DOMAIN SIMULATION

The time domain simulation in above studied system is evaluated under various disturbances.

Case 1: linear variation in wind speed and marine current speed: a linear variation in wind speed from 4 m/s to 25 m/s and marine current speed from 1 m/s to 5 m/s simultaneously applied to the studied system with proposed controller. The 3D plot of active and reactive power of MCF and OWF is shown in fig 5 (a) to fig 5(d) respectively. The total active and reactive power at grid is shown in fig 5(e) and fig 5(f) respectively. The terminal voltage at PCC is shown in fig 5(g). The fig 5(a) suggest that the active power variation of MCF is increase with speed up-to the rated speed, above the rated speed almost constant up-to the cutoff speed. Further the power variation of MCF is only depends on marine current speed and almost constant for wind speed variations. Fig 5(b) suggests that the reactive power variation of MCF is almost independent of the variation in marine current speed. Analogous to the Fig 5(a) and Fig 5(b), the Fig 5(c) and Fig 5(d) reflects the variation of active power of offshore wind generator w.r.t offshore wind speed and marine current speed respectively. The reactive power of DFIG based OWF is almost zero because of the unity power factor operation of converters. Fig 5(e) suggest that the total power export to the grid is depends on both marine current speed and wind speed up-to the rated speed, above the rated speed the total power is almost constant up-to the cut-off speed. The total reactive power to

grid shown in fig 5(f) is the sum of reactive power required to SCIG based MCF and reactive power developed by DFIG based OWF. Fig 5 (g) suggest that the flat voltage profile achieved at different speed of MCF and OWF with proposed controller.

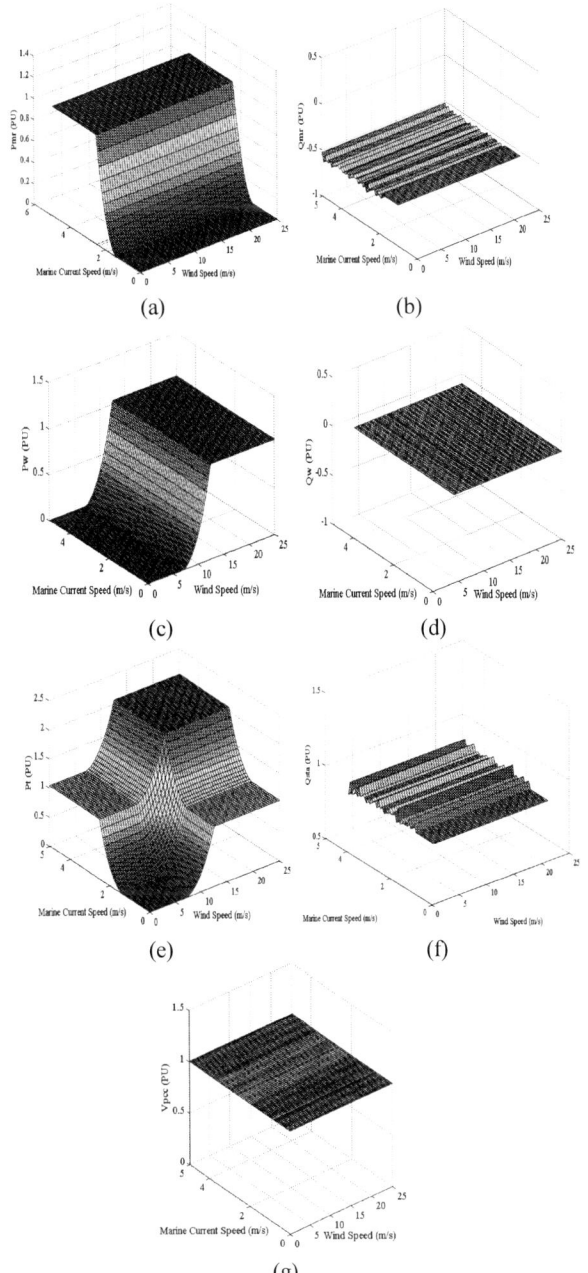

Fig. 5. (a) active power of MCF (b) reactive power of MCF (c) active power of OWF (d) reactive power of OWF (e) active power at grid (f) reactive power at grid (g) terminal voltage of PCC

Case 2: dynamic variation in wind speed and marine current speed: a random offshore wind speed and random marine current speed is simultaneously applied to the studied system to evaluate the performance of the studied system under dynamic condition.

The response of the system is compared between conventional PI controller and H-infinity loop-shaping based proposed controller. Fig 6(b) to fig 6(g) shows the dynamic response of the studied system. Fig 6(b) & fig 6(c) suggest that the active & reactive power deviation of OWF have less overshoot and settling time with H-infinity loop-shaping based proposed controller due to higher time constant. Analogous to the Fig 6(b) & fig 6(c) the fig 6(d) & fig 6(e) conforms that the active and reactive power deviation at grid is minimum under high stochasticity and it improves the time response of the system. The fig 6(f) shows the time response of grid voltage which gives a better voltage profile with proposed controller. Fig 6(g) shows the dc-link voltage of DFIG based OWF. The fluctuation in dc-link voltage is minimizing faster and less overshoot as compared to conventional PI controller. The overall performance and robustness of the studied system has been enhanced with proposed controller.

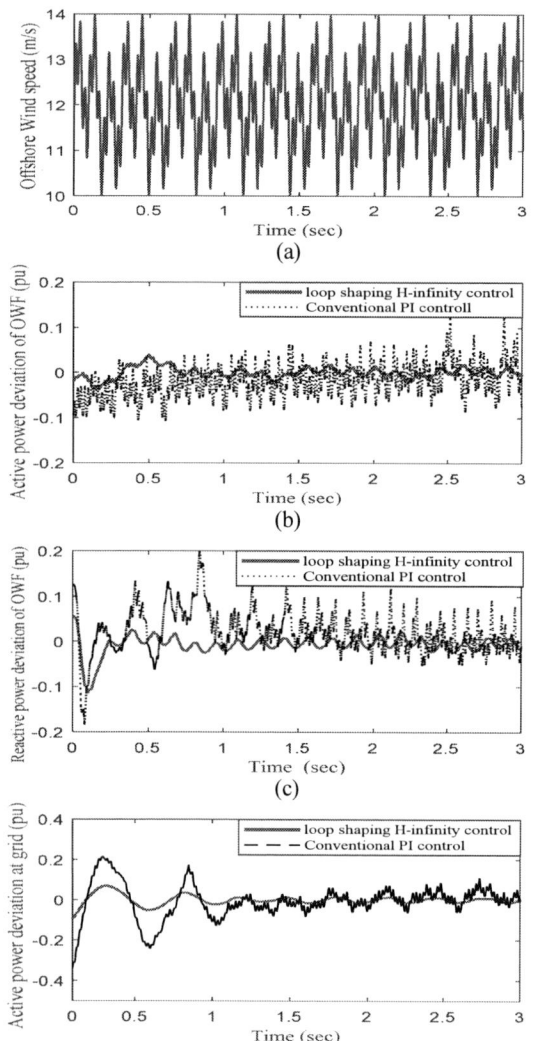

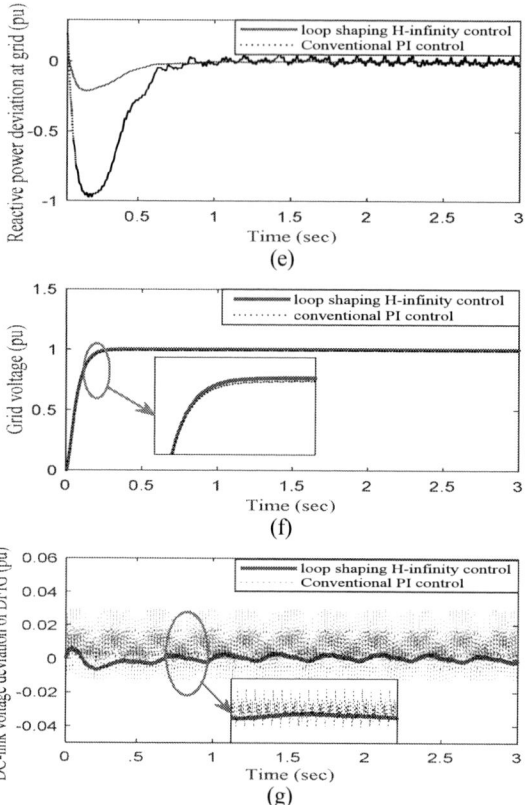

Fig. 6. (a) random wind speed (b) active power of OWF (c) reactive power of OWF (d) active power at grid (e) reactive power at grid (f) terminal voltage of grid (g) dc-link voltage of DFIG based OWF

V. CONCLUSION

This paper presents the stability issues in the grid integration of OWF and MCF under high stochastic offshore wind and marine current speed. It proposed a H-infinity loop-shaping based controller to improve the robustness and stability of the integrated system. The study has been performed to show the effectiveness of the proposed controller in terms of voltage and power at various point of the studied system. The damping of power oscillation and voltage fluctuation has been effectively achieved with proposed controller. Hence the stability and robustness of the integrated system is significantly enhanced with the proposed controller.

REFERENCES:

[1]. S. E. B. Elghali, R. Balme, K. L. Saux, M. E. H. Benbouzid, J. F. Charpentier, and F. Hauville,: A simulation model for the evaluation of the electrical power potential harnessed by a marine current turbine. *IEEE J. Ocean. Eng.*, vol. 32, no. 4, (2007) pp. 786–797

[2]. E. Nasr-Azadani, C. A. Canizares, D. E. Olivares, and K. Bhattacharya,: Stability Analysis of Unbalanced Distribution Systems With Synchronous Machine and DFIG Based Distributed Generators. *IEEE Trans. Smart Grid*, vol. 5, no. 5, (2014) pp. 2326–2338

[3]. H. Xu, J. Hu, H. Nian, and Y. He,: Dynamic modeling and improved control of DFIG under unbalanced and distorted grid voltage conditions *IEEE International Symposium on Industrial Electronics*, (2012) pp. 1579–1584.

[4]. E. Nasr-Azadani, C. A. Canizares, D. E. Olivares, and K. Bhattacharya,: Stability Analysis of Unbalanced Distribution Systems With Synchronous Machine and DFIG Based Distributed Generators. *IEEE Trans. Smart Grid*, vol. 5, no. 5, (2014) pp. 2326–2338

[5]. H. M. Jabr and N. C. Kar,: Effects of main and leakage flux saturation on the transient performances of doubly-fed wind driven induction generator. *Electr. Power Syst. Res.*, vol. 77, no. 8, (2007) pp. 1019–1027

[6]. R. Chattopadhyay, A. De, and S. Bhattacharya,: Comparison of PR controller and damped PR controller for grid current control of LCL filter based grid-tied inverter under frequency variation and grid distortion. *IEEE Energy Conversion Congress and Exposition (ECCE)*, (2014) pp. 3634–3641.

[7]. M. Djukanovic, M. Khammash, and V. Vittal,: Application of the structured singular value theory for robust stability and control analysis in multi-machine power systems. I. Framework development. *IEEE Trans. Power Syst.*, vol. 13, no. 4,(1998) pp. 1311–1316

[8]. J. C. Doyle, K. Glover, P. P. Khargonekar, and B. A. Francis,: State-space solutions to standard H/sub 2/ and H/sub infinity /control problems. *IEEE Trans. Automat. Contr.*, vol. 34, no. 8,(1989) pp. 831–847,

[9]. L. Wang and C. T. Hsiung,: Dynamic Stability Improvement of an Integrated Grid-Connected Offshore Wind Farm and Marine-Current Farm Using a STATCOM. *IEEE Transactions on Power Systems*, vol. 26, no. 2,(2011) pp. 690-698

[10]. K. Okedu, S. Muyeen, R. Takahashi, and J. Tamura,: Wind farms fault ride through using DFIG with new protection scheme. IEEE Trans. Sustain. Energy, vol. 3, no. 2,(2012) pp. 242–254

[11]. Ngamroo I,: An optimization of robust SMES with specified H-infinity controller for power system stabilization considering superconducting magnetic coil size. Energy Convers Manage (2011) 52:648–51.

Appendix

[12].SYSTEM PARAMETERS

Base parameters
$V_b = 0.69 Kv, S_b = 1 MW, \; f_b = 50 Hz$
DFIG based OWF parameters
$r_{sw} = r_{rw} = 0.005\,pu, x_{sw} = x_{rw} = 0.04\,pu, \; x_{mw} = 4\,pu,$ $\beta_{w\min} = 0^o, \beta_{w\max} = 30^o$ $v_{cut-in} = 4\,m/s, v_{rated} = 15\,m/s, v_{cut-off} = 25\,m/s$
SCIG based MCF parameters
$r_s = 0.05\,pu, X_s = 0.08\,pu, r_r = 0.005\,pu$ $X_r = 0.092\,pu, X_m = 4\,pu, \beta_{mr\min} = 0^o, \beta_{mr\max} = 30^o,$ $v_{cut-in} = 1\,m/s, v_{rated} = 2.5\,m/s, v_{cut-off} = 4\,m/s$
Constants of Cp in OWF and MCF
$C_{c1} = 0.33, C_{c2} = 125, C_{c3} = 0.3, C_{c4} = C_{c5} = 0, C_{c6} = 10.5$ $C_{c7} = 11, C_{c8} = 0.09, C_{c9} = 0.01$ $b_{c1} = 0.2, b_{c2} = 80.5, b_{c3} = 0.45, b_{c4} = 0.25, b_{c5} = 0.45$ $b_{c6} = 11.5, b_{c7} = 6.5, b_{c8} = .03, b_{c9} = -0.05$

978-1-5386-6624-1/18 $31.00 © 2018 IEEE

A Brief Review Regarding Sensor Reduction and Faults in Shunt Active Power Filter

Shivangni Sharma
Department of Electrical Engineering
National Institute of technology
Patna, India
shivangni.eepg15@nitp.ac.in

Vimlesh Verma
Department of Electrical Engineering
National Institute of technology
Patna, India
vimlesh.verma@nitp.ac.in

Abstract— Shunt Active Power Filter (SAPF) works basically with the help of two key operations i.e. Reference current generation technique and current controller. Both these operations require measurement of currents and voltages of the system which is done with the help of sensors. Failure due to faults in these sensors causes SAPF to aggravate power quality problems instead of compensating it. Also, the reduction in the number of sensors will make SAPF more economical and reliable. Therefore, sensorless and fault tolerant control of SAPF is a very attractive solution for its robust operation. The aim of this paper is to present and discuss details about the different solutions and development in sensor reduction and faults in SAPF. A literature review has been done in a classified manner based on the type of faults and the type of sensor reduced.

Keywords— Sensorless control, Shunt Active Power filter, faults, Estimation

I. INTRODUCTION

Active Power Filter (APF) was introduced by L. Gyugyi in 1976 [1], which provides a dynamic solution to major problems related to power quality such as harmonics, reactive power compensation, voltage regulation etc. APF provides fast, accurate, efficient and dynamic performance in comparison to other solutions present in the literature [2]. There are various advancements took place in APF since the 80th century. There are many configurations of APF available in the literature based on different type of supply, converter and topology. One of the most popular configurations is Three phase three wire Voltage Source Converter (VSC) based SAPF because of its efficient and economic operation. Primarily performance of SAPF is decided by the control strategy involved, which consists of a reference current generation technique and controller. One of the key research areas in SAPF is reference current generation technique. P-Q theory, SRF theory, Kalman filtering method, Adaline based Algorithms, ANN-based techniques [3], [4] etc. are some of the most popular techniques available in the literature. Different types of controller available in the literature include Hysteresis based controller, Predictive controller, PWM controller, Space vector-based controller, Pole shift cancellation controller, sliding mode controller etc. [4]. Another important area that plays a major role in the operation of SAPF is sensor failure and faults. For proper functioning of the control strategy, SAPF requires information of voltages and currents with the help of sensors. SAPF requires four types of sensors i.e. Source voltage sensors for amplitude and phase information, Source current sensors for input current control, DC voltage sensors for power flow control, load current sensors for estimation of harmonics and filter current in case of direct current control. Failure and faults in sensors will lead to maloperation of SAPF which will further aggravate harmonics in the power system instead of compensating it [5]. Apart from sensors and faults, some other concerns such as rating of inverter [6], switching losses [7] have also been discussed in the literature to increase economic viability at the industrial level. Reduction in number of sensors offers obvious advantages such as eliminating the noise associated with the sensor, simplified calibration, increased robustness etc. Thus, sensorless and fault tolerant operation of SAPF is much desired due to various technical and economical aspects. In this paper, a study related to sensor reduction and faults in SAPF is presented. A literature review is divided into two parts i.e. sensor reduction and faults. Further classification into a particular type of sensor reduction and faults is also done. Section II deals with the basic control strategy showing the role of all sensors. Section III deals with the reduction in the number of sensors, Section IV presents a discussion about various faults and its solution in SAPF and Section V presents the conclusion.

II. ROLE OF SENSORS IN SAPF

Fig. 1 shows the basic block diagram of SAPF with all the sensors intact (highlighted in red). SAPF is connected near the load end in parallel to source and load.

Reference current generation technique senses source voltages (v_{sa}, v_{sb}, v_{sc}) and load currents (I_{La}, I_{Lb}, I_{Lc}) to generate reference currents (I^*_a, I^*_b, I^*_c). A capacitor is used on the DC side of VSC to maintain an active power flow between the source and load. In case of indirect current control [8], source current along with reference current (a component of load current containing only 50 Hz signal) is sent to the current controller for gate pulse generation. In case of direct current control, filter current (I_{fa}, I_{fb}, I_{fc}) along with compensating current (extracted harmonic component of load current) is sent to the current controller [9]. Pulses generated controls VSC such that it generates equal and opposite harmonics into the system making source current sinusoidal.

III. SAPF WITH REDUCED NUMBER OF SENSOR

Classified discussion along with its limitation and technique involved is presented in following sections. Overall Table I shows number of sensors utilized for control

978-1-5386-6624-1/18 $31.00 © 2018 IEEE

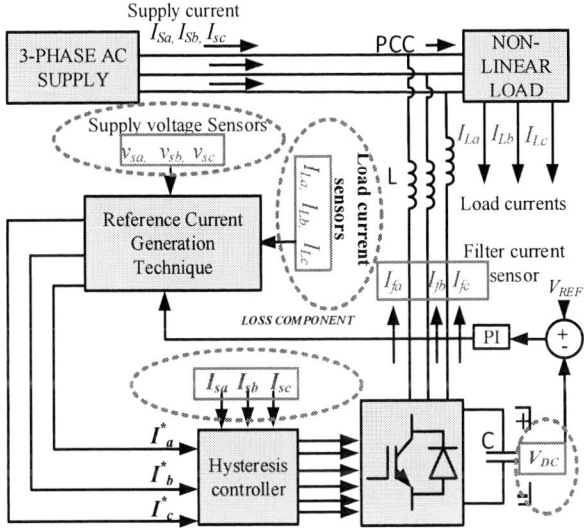

Fig. 1. Block diagram of SAPF showing all sensors

of SAPF in various papers discussed here, where sensor used is shown by "✓" and sensors not utilized is shown by "✗".

A. Load and Filter current sensor reduction

Most of the control strategies which deals with a reduced number of current sensors are based on indirect current control, so all papers discussed here employ indirect current control. First control strategy without using the load current sensor was based on the basic PI controller [10]. Phase and waveform information is obtained from mains and amplitude is obtained with the help of DC voltage. By detecting only one phase of the source voltage, the balanced sinusoidal wave is formed (In this case reactive power cannot be fully compensated). Voltage detector, sine wave generator and multiplier all are required for all phases separately in case of the unbalanced condition if reactive power compensation is also preferred. PI controller output provides information on the source current's peak value, which is required to be drawn by DC capacitor and reference current is obtained by multiplying the output of PI controller with sine wave generator whose phase is synchronized with source voltage with the help of PLL (Phase Locked Loop). In same PI controller-based method ripple in DC voltage was removed by sampling voltage at zero crossing thus eliminating the delay caused by Low Pass Filter for V_{DC} [11]. Further, the same technique was presented in [12] but with a reduced number of inverter switches. These methods are only limited to low and medium power applications. In place of the PI controller Fuzzy controller [13] and Variable Structure Adaptive Pole Placement controller(VSAPPC) was discussed in [14], remaining operation was the same as in PI controller-based method discussed above. Both fuzzy and VSAPPC control have a better dynamic response than the PI controller.

In [15] author proposed a control strategy by developing system model in the d-q frame and controlling voltage across source inductance to be sinusoidal without use of any Load current sensor. The drawback of this control strategy is a requirement of large source inductance.

Conventional SAPF based on PQ theory-based reference current generation techniques without sensing load current was proposed in [16]. Predictive current control-based

control strategy without load current sensors was developed in [17], which has the limitation of having a slow transient response.

An advance current control strategy was developed in [18] without load current sensors which include two control loops, Outer loop consist of PI controller which calculate fundamental reference current component in d-q frame (q axis component is set to 0) and inner control loop controls source current to be sinusoidal with the help of PI plus vector PI(VPI) control. VPI provides high gains at harmonic frequencies to regulate supply current. It also utilizes only four switches for VSC so overall cost is reduced.

Overall sensorless approaches which eliminate load and filter current sensors directly imposes a sinusoidal reference on source current instead of detecting harmonics from the load side and generating equal and opposite reference current.

B. Source Voltage sensor reduction

In case of voltage sensor reduction in conventional SAPF, it is observed that in frequency domain reference current generation technique eliminating voltage sensor requires complex circuitry [19]. But there are many other non-conventional approaches available in the literature which are discussed below.

Voltage sensors are mostly used for the purpose of synchronization hence a good sensorless estimator can be developed. Voltage sensorless techniques has been widely used in grid-connected VSC operations which includes VOC (Voltage oriented control) [20], Instantaneous power-based methods [21], Predictive control [22], state observer [23], current shaping techniques [24], virtual flux-based methods [25], [26] etc. Since controlled VSC when connected between source and load is responsible for the operation of SAPF, therefore, it is important to discuss some latest sensorless approaches related to VSC. In [24] the author presented a control technique for PWM rectifier which was based on current shaping strategy. This control strategy compensates selective harmonics from grid based on multiple resonant modes without voltage sensors. Whereas in [22] predictive control technique was presented for Active-Front-End rectifier which uses the predictive control to estimate source voltage with the help of DC link voltage, switching states and instantaneous and future values of active and reactive power. Similarly, the sliding mode state observer-based method was proposed in [23] to track source voltage which uses model predictive control for power regulation. Papers discussed above are independent of PLL. A grid voltage estimation technique is proposed in [27] hence, eliminating both voltage sensors and PLL. This alternative technique is based on MRAC (Model-Reference-Adaptive-Control) which basically utilized MRAC structures of active and reactive power to estimate grid voltage. MRAC based estimation techniques are widely employed in the estimation of speed and machine parameters in sensorless drives.

There are many other voltage sensorless control strategies present in the literature regarding grid connected VSC control but only a few latest techniques which have their application in Active Power Filtering operation are discussed. Voltage sensorless control strategies in SAPF are mostly derived from the methods discussed above.

978-1-5386-6624-1/18 $31.00 © 2018 IEEE   427

The model predictive voltage sensorless control algorithm was applied in [28], [29] both for reference current generation and current controller. Estimations in the algorithm are based on information about the previous and predicted value of equivalent electromotive-force (emf) of the grid. Dynamics in AC current shaping are only affected by system parameters. An algorithm using a single voltage sensor was proposed in [30]. This method was based on the assumption of having balanced supply voltages, therefore information of one phase was enough to obtain other phases information for control purposes. Also, harmonics are extracted by sensing load current by using Notch filter-based control algorithm. All these schemes require measurement of load current thus reducing only source voltages sensors.

Control based on current shaping technique was presented in [31] which is the extension of the same method as previously discussed in [24] for PWM rectifier, the method used here is based on impedance emulation. A discussion of multiple quasi-resonant schemes used for impedance emulation and generalization of current shaping control strategy is presented in [32]. Another voltage sensorless approach was given in [33] which works on the concept of Resistive-APF(R-APF). In R-APF harmonic frequencies are attenuated by an infinite impedance. It only utilizes DC voltages and filter current to develop a plant model in order to estimate impedance. The drawback of this method is its accuracy in impedance emulation which relies on the plant model.

C. Source Voltage and Load current sensor reduction

In a significant development, a control strategy without using both source voltages and load currents sensors for selective harmonic elimination was proposed in [34], [35], In [34] Virtual Flux Oriented Control (VFOC) was employed which uses virtual flux to determine grid synchronization angle. The concept of virtual flux is employed in [35] by using adaptive filter-phase locked loop (AF-PLL) for improving the process of rejecting disturbances and in these techniques, the sinusoidal reference current is directly imposed on source current without employing separate harmonic rejection algorithm. In addition to this, a separate control scheme for power flow control is presented which does not require DC voltage sensor leaving SAPF with only source current sensors.

D. DC Voltage sensor reduction

Although source current and DC voltage sensors are indispensable, few papers have discussed use of DC voltage sensors in over-voltage monitoring [24]. DC voltage regulation in SAPF directly depends on peak (maximum value) of the reference current, which is decided by the PI controller by adjusting quickly in reponse to any changes in the load. With the help of switching states and output currents i.e. Filter current (Fig.1) of VSC, A voltage estimation technique for DC voltage is presented in [36] for two-phase SAPF based on cacaded H-bridge topology. This method reduced each voltage sensors required for each capacitor in cacaded H-bridge topology, thus reducing the number of voltage sensors. Reduced DC voltage sensor based SAPF was also proposed in [35] as discussed in section III(C). In this technique , actual DC voltage is replaced by estimated DC voltage. This voltage is calculated

with the help of estimated source voltage and modulating signal used for PWM current control.

IV. FAULTS IN SHUNT ACTIVE POWER FILTER

A broad classification of faults in SAPF is shown in Fig.2.

There are very few papers available in literature related to sensor faults but they are also restricted for single sensor faults in a balanced condition. Aiming to the symmetry of load, a fault tolerant control strategy was proposed in [37] for sensor faults.

First paper which discussed faults in APF was for short circuit fault in the distribution side in case of series APF [38]. Fault protection technique presented in this paper involved use of varistor and an additional pair of antiparallel thyristors. Hence, the secondary side short circuit current will flow through the thyristors and varistors. This process will continue until the protection equipment in the power distribution system clears the fault.

In [39] the author developed a protection scheme to protect against ground faults, short circuit faults in DC link and output phases. This paper presented a unique arrangement of DC link to protect against possible DC link and short-circuit faults in output phases of VSC. A method was developed to determine the output current of VSC with the help of relation between the output current of VSC, DC link current and voltage vector.

Authors in [40] presented a fault tolerant technique against short circuit and open circuit faults of converter switches by isolating and reconfiguring the VSC circuit using the additional fourth leg. In addition, to compensate fault, fault detection scheme was also proposed in [41], both for sensor faults and switch faults. Fuzzy control-based fault tolerant four leg APF was also developed in [42]. Reference currents are obtained by fuzzy control and short circuit, open circuit and line to ground faults are compensated by reconfiguration of VSC. These solutions come with the increased cost. A similar approach of reconfiguring VSC circuit was presented in [43] but without using the additional fourth leg, instead fault leg is connected here between two capacitors in midpoint configuration type APF and a two-leg VSC topology is formed, accordingly a suitable control strategy for the disconnected leg is proposed. But this method is also expensive than normal VSC topology due to the requirement of bidirectional switches such as triacs. Reliability analysis of APF was done in [44] for both cases of post-fault reconfigured structures i.e. with or without an additional fourth leg.

TABLE I.   NO. OF SENSORS INVOLVED FOR CONTROL OF SAPF

Reference	Type of Sensor				
	Source voltage	*DC Voltage*	*Source current*	*Load current*	*Filter current*
[10-18]	✓	✓	✓	✗	✗
[28-29]	✗	✓	✓	✓	✗
[33]	✗	✓	✗	✗	✓
[34]	✗	✓	✓	✗	✗
[35]	✗	✗	✓	✗	✗
[36]	✓	✗	✓	✓	✓

978-1-5386-6624-1/18 $31.00 © 2018 IEEE   428

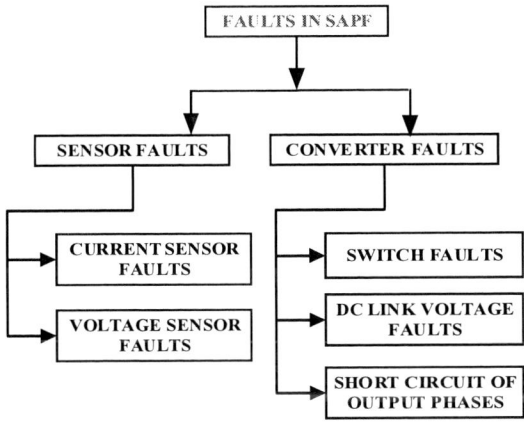

Fig. 2. Classification of Faults in SAPF

A new method to diagnose fault location in half period cycle is proposed in [45]. This method detects open circuit switch faults by comparing distortion in DC voltage with the normal value of DC voltage across the capacitor for Shunt Hybrid APF. An algorithm was developed in [46] with the help of control states and instant pole voltage in NPC (Neutral Point Clamped) based APF for open circuit fault diagnosis in IGBTs. An algorithm for identifying and locating open circuit faults of IGBT was presented in [47]. On the basis of a mathematical model of AC side VSC, an algorithm was developed with a different relationship between ac side output voltages of APF and turn-on duty cycle of the corresponding IGBT under different conditions. All these open circuit fault diagnosis methods did not require any external circuitry thus, providing an economic solution.

Apart from these VSC circuit faults, discussion about Line to ground and Line to Line faults was done in [48], Author compared the performance of Hysteresis and Space vector PWM current controllers in SAPF under both faults.

Overall there are a variety of effective and economical solutions available in the literature for switch faults. Discussion about sensors faults is not present in literature, very few papers have mentioned about sensors faults and the solution was limited to three phase balanced system in case of only one sensor fault.

V. CONCLUSION

Sensorless and fault tolerant operation of SAPF offers many advantages such as reduction in cost, increased reliability, robustness etc. The paper presented the current state regarding the reduction of sensors and faults in SAPF including most recent research and trends. Some paper dealt with reduction of voltage sensor or load current sensor and few papers with both kinds together. Elimination of DC voltage sensor was also presented in a paper. However, the presence of source current sensor is preferable than any other sensor present in SAPF as it is required output and is directly related with performance of SAPF. There is no much work found in the literature regarding sensor fault in APF especially for the unbalanced condition. But sensorless control of APF can be used as back up for continuous operation of SAPF. Further, there is still need of in-depth research for solving the problem of high computational burden in sensorless techniques and parameter robustness.

ACKNOWLEDGMENT

This work was supported by TEQUIP III towards Registration fee and Travelling Allowance.

REFERENCES

[1] L.Gyugyi and Strycula," Active Power Filters," in *proc. Of IEEE Industrial Application Annual meeting*, vol.19-C, pp.529-535,1976.

[2] H. Akagi, "Modern active filters and traditional passive filters,"*Bulletin of the Polish Academy of Sciences, Technical Sciences* , vol. 54, no. 3, pp. 255–269, 2006.

[3] A. Bhattacharya and C. Chakraborty, "A Shunt Active Power Filter With Enhanced Performance Using ANN-Based Predictive and Adaptive Controllers," in *IEEE Transactions on Industrial Electronics*, vol. 58, no. 2, pp. 421-428, Feb. 2011.

[4] Miska Prasad, A.K Akella, " Mitigation of power quality devices using custom power devices: A review," in *Indonesian Journal of Electrical Engineering and Informatics*, vol.5 , no.3 , pp.207-235 ,Sep 2017

[5] S.Sharma, V. Verma, "Performance of Shunt Active Power Filter Under Sensor Failure," 2017 *IEEE International WIE Conference on Electrical and Computer Engineering (WIECON-ECE)*, Dehradun, 2017, in press.

[6] P. Garanayak, G. Panda,"Harmonic Elimination and Reactive Power Compensation with a Novel Control Algorithm based Active Power Filter," in *Journal of Power Electronics*, vol. 15, no. 6, pp. 1619-1627, November 2015

[7] C. S. Lam, M. C. Wong, N. Y. Dai, W. H. Choi, X. X. Cui and C. Y. Chung, "Switching-loss reduction technique in active power filters without auxiliary circuits," in *IET Power Electronics*, vol. 9, no. 4, pp. 728-742, 3-30-2016.

[8] S.K Dash, G.Panda, P. K. Ray, S. S. Pujari, "Realization of active power filter based on indirect current control algorithm using Xilinx system generator for harmonic elimination, "*International Journal of Electrical Power & Energy Systems*, vol.74,pp 420-428,2016

[9] R. K. Patjoshi and K. K. Mahapatra, "Performance comparison of direct and indirect current control techniques applied to a sliding mode based shunt active power filter," *2013 Annual IEEE India Conference (INDICON)*, Mumbai, 2013, pp. 1-5.

[10] H. L. Jou, "Performance comparison of the three-phase active-power-filter algorithms," in *IEE Proceedings - Generation, Transmission and Distribution*, vol. 142, no. 6, pp. 646-652, Nov 1995

[11] K. Chatterjee, B. G. Fernandes and G. K. Dubey, "An instantaneous reactive volt-ampere compensator and harmonic suppressor system," in *IEEE Transactions on Power Electronics*, vol. 14, no. 2, pp. 381-392, Mar 1999

[12] G. Joos, Su Chen and K. Haddad, "Four switch three phase active filter with reduced current sensors," *2000 IEEE 31st Annual Power Electronics Specialists Conference. Conference Proceedings (Cat. No.00CH37018)*, Galway, 2000, pp. 1318-1323 vol.3

[13] J. Dixon, J. Contardo and L. Moran, "DC link fuzzy control for an active power filter, sensing the line current only," *PESC97. Record 28th Annual IEEE Power Electronics Specialists Conference. Formerly Power Conditioning Specialists Conference 1970-71. Power Processing and Electronic Specialists Conference 1972*, St. Louis, MO, 1997, pp. 1109-1114 vol.2.

[14] R. L. A. Ribeiro, C. C. Azevedo and R. M. Sousa, "A non-standard adaptive control for shunt active power filter without current harmonic detection," *IECON 2010 - 36th Annual Conference on IEEE Industrial Electronics Society*, Glendale, AZ, 2010, pp. 2007-2012.

[15] M. Routimo, M. Salo and H. Tuusa, "Current sensorless control of a voltage-source active power filter," *Twentieth Annual IEEE Applied Power Electronics Conference and Exposition, APEC*, pp. 1696-1702 vol. 3,2005

[16] N. Gupta, S. P. Dubey and S. P. Singh, "DSP based control algorithm for three-phase four-wire shunt active filter without load and filter current measurement," *2010 Annual IEEE India Conference (INDICON)*, Kolkata, 2010, pp. 1-6

[17] D. Nedeljkovic, M. Nemec, K. Drobnic and V. Ambrozic, "Active Power Filter with a Reduced Number of Current Sensors", *Electrotechnical review*, vol.76, no. 5, sep. 2009, pp 275-280

978-1-5386-6624-1/18 $31.00 © 2018 IEEE

[18] Q. N. Trinh and H. H. Lee, "An Advanced Current Control Strategy for Three-Phase Shunt Active Power Filters," in *IEEE Transactions on Industrial Electronics*, vol. 60, no. 12, pp. 5400-5410, Dec. 2013.

[19] S. M. Williams and R. G. Hoft, "Adaptive frequency domain control of PWM switched power line conditioner," *Fifth Annual Proceedings on Applied Power Electronics Conference and Exposition*, Los Angeles, CA, USA, 1990, pp. 505-511.

[20] S. Hansen, M. Malinowski, F. Blaabjerg, and M. Kazmierkowski, "Sensorless control strategies for PWM rectifier," in *Proc. 15th IEEE Appl. Power lectron. Conf. Expo.*, 2000, vol. 2, pp. 832–838.

[21] T. Noguchi, H. Tomiki, S. Kondo, and I. Takahashi, "Direct power control of PWM converter without power-source voltage sensors," *IEEE Trans. Ind. Appl.*, vol. 34, no. 3, pp. 473–479, May/Jun. 1998.

[22] M. Mehreganfar and S. A. Davari, "Sensorless predictive control method of three-phase AFE rectifier with MRAS observer for robust control," *2017 IEEE International Symposium on Predictive Control of Electrical Drives and Power Electronics (PRECEDE)*, Pilsen, 2017, pp. 107-112

[23] H. Yang, Y. Zhang, N. Zhang, P. D. Walker and J. Gao, "A voltage sensorless finite control set-model predictive control for three-phase voltage source PWM rectifiers," in *Chinese Journal of Electrical Engineering*, vol. 2, no. 2, pp. 52-59, Dec. 2016.

[24] M. B. Ketzer and C. B. Jacobina, "Sensorless Control Technique for PWM Rectifiers With Voltage Disturbance Rejection and Adaptive Power Factor," in *IEEE Transactions on Industrial Electronics*, vol. 62, no. 2, pp. 1140-1151, Feb. 2015.

[25] J.Suul, A. Luna, P. Rodriguez, and T. Undeland, "Virtual-flux-based voltage-sensor-less power control for unbalanced grid conditions," *IEEE Trans. Power Electron.*, vol. 27, no. 9, pp. 4071–4087, Sep. 2012

[26] J. G. Norniella, J. Cano, G. Orcajo, C. Rojas, J. Pedrayes, M. Cabanas, and M. Melero, "Improving the dynamics of virtual-flux-based control of three-phase active rectifiers," *IEEE Trans. Ind. Electron.*, vol. 61, no. 1, pp. 177–187, Jan. 2014.

[27] S. Mukherjee, V. R. Chowdhury, P. Shamsi and M. Ferdowsi, "Grid Voltage Estimation and Current Control of a Single-Phase Grid-Connected Converter Without Grid Voltage Sensor," in *IEEE Transactions on Power Electronics*, vol. 33, no. 5, pp. 4407-4418, May 2018

[28] D. Wojciechowski, "Novel estimator of distorted and unbalanced electromotive force of the grid for control system of PWM rectifier with active filtering," *2005 European Conference on Power Electronics and Applications*, Dresden, 2005, pp. 10 pp.-P.10

[29] D. Wojciechowski, "Sensorless predictive control of three-phase parallel active filter," in *Proc. AFRICON*, Sep. 2007, pp. 1–7.

[30] K. Kant, B. Singh and U. Kalla, "Single voltage measurement-based control algorithm for voltage source converter," *2016 IEEE 7th Power India International Conference (PIICON)*, Bikaner, 2016, pp. 1-4

[31] M. B. Ketzer and C. B. Jacobina, "Sensorless current shaping control technique for shunt active filters," *2014 11th IEEE/IAS International Conference on Industry Applications*, Juiz de Fora, 2014, pp. 1-7

[32] M. B. Ketzer, C. B. Jacobina and A. M. N. Lima, "Shaping control strategies for active power filters," in *IET Power Electronics*, vol. 11, no. 1, pp. 175-181, 1-12-2018

[33] H. Bai, X. Wang and F. Blaabjerg, "A Grid-Voltage-Sensorless Resistive Active Power Filter with Series LC-Filter," in *IEEE Transactions on Power Electronics*, vol. PP, no. 99, pp. 1-1.

[34] M. B. Ketzer and C. B. Jacobina, "Nonlinear virtual flux oriented control for sensorless active filters," *2013 Brazilian Power Electronics Conference*, Gramado, 2013, pp. 393-398

[35] M. B. Ketzer and C. B. Jacobina, "Virtual Flux Sensorless Control for Shunt Active Power Filters With Quasi-Resonant Compensators," in *IEEE Transactions on Power Electronics*, vol. 31, no. 7, pp. 4818-4830, July 2016

[36] M. Fajardo, J. Viola, J. Restrepo, F. Quizhpi and J. Aller, "Two-phase active power filter direct current control with capacitor voltages estimation and balance," *2015 IEEE Workshop on Power Electronics and Power Quality Applications (PEPQA)*, Bogota, 2015, pp. 1-6.

[37] J. Zhu and W. Yu, "The Study of Fault-Tolerant Control for Shunt Active Power Filter," *2007 IEEE International Conference on Control and Automation*, Guangzhou, 2007, pp. 1059-1062.

[38] L. A. Moran, I. Pastorini, J. Dixon and R. Wallace, "A fault protection scheme for series active power filters," in *IEEE Transactions on Power Electronics*, vol. 14, no. 5, pp. 928-938, Sep 1999.

[39] F. Blaabjerg, J. K. Pedersen, U. Jaeger and P. Thoegersen, "Single current sensor technique in the DC link of three-phase PWM-VS inverters: a review and a novel solution," in *IEEE Transactions on Industry Applications*, vol. 33, no. 5, pp. 1241-1253, Sep/Oct 1997.

[40] C. B. Jacobina, M. B. R. Correa, R. F. Pinheiro, A. M. N. Lima and E. R. C. da Silva, "Improved fault tolerance of active power filter system," *IEEE 32nd Annual Power Electronics Specialists Conference (IEEE Cat. No.01CH37230)*, Vancouver, BC, 2001, pp. 1635-1640 vol.3

[41] S. Karimi, P. Poure, S. Saadate and E. Gholipour, "Current sensors and power switches fault detection and compensation for shunt active power filters," *2007 IEEE International Symposium on Industrial Electronics*, Vigo, 2007, pp. 3157-3161

[42] N. Madhuri, S. R. Doradla and M. S. Kalavathi, "Fuzzy based Fault Tolerant shunt Active Power Filter," *2014 International Conference on Control, Instrumentation, Communication and Computational Technologies (ICCICCT)*, Kanyakumari, 2014, pp. 1334-1337.

[43] Q. Xu, Y. Wang, M. Wang, J. Zhan and G. Chen, "A novel fault-tolerant control scheme for Shunt Active Power Filter with high reliability," *2015 IEEE Energy Conversion Congress and Exposition (ECCE)*, Montreal, QC, 2015, pp. 5531-5537.

[44] P. Poure, P. Weber, D. Theilliol and S. Saadate, "Fault-tolerant Power Electronic Converters: Reliability Analysis of Active Power Filter," *2007 IEEE International Symposium on Industrial Electronics*, Vigo, 2007, pp. 3174-3179.

[45] Tao Peng, Shuai Zhao, Zhili Lin, Yao Sun and Yongmei Mao, "Fault diagnosis for shunt hybrid active power filter with open-circuit fault based on voltage distortion," *2015 Chinese Automation Congress (CAC)*, Wuhan, 2015, pp. 2163-2168.

[46] L. M. A. Caseiro, A. M. S. Mendes and P. M. A. F. Lopes, "Open-circuit fault diagnosis in Neutral-Point-Clamped active power filters based on instant voltage error with no additional sensors," *2015 IEEE Applied Power Electronics Conference and Exposition (APEC)*, Charlotte, NC, 2015, pp. 2217-2222.

[47] H. Zhang, C. da Sun, Z. x. Li, J. Liu, H. y. Cao and X. Zhang, "Voltage Vector Error Fault Diagnosis for Open-Circuit Faults of Three-Phase Four-Wire Active Power Filters," in *IEEE Transactions on Power Electronics*, vol. 32, no. 3, pp. 2215-2226, March 2017

[48] A. R. Mohanty and A. K. Kapoor, "Performance evaluation of HCC & SVPWM current controllers for shunt APF under fault conditions," *India International Conference on Power Electronics 2010 (IICPE2010)*, New Delhi, 2011, pp. 1-8.

978-1-5386-6624-1/18 $31.00 © 2018 IEEE

A Novel Active Anti-islanding Scheme for Inverter-Based Distributed Generation

Sheetal Chandak,
Electrical Engineering Department,
Siksha 'O' Anusandhan
(Deemed to be University)
Bhubaneswar, India
sheetalchandak91@gmail.com

Manohar Mishra, *Member*, IEEE
Electrical and Electronics Engineering
Department
Siksha 'O' Anusandhan
(Deemed to be University)
Bhubaneswar, India
manohar2006mishra@gmail.com

Pravat Kumar Rout
Electrical and Electronics Engineering
Department
Siksha 'O' Anusandhan
(Deemed to be University)
Bhubaneswar, India
pkrout_india@yahoo.com

Abstract—This study presents a harmonic distortion based active anti-islanding scheme for electrically integrated distributed generation system. The proposed method employs the injection of an intensified harmonic voltage signal at the distributed generator end so as to supervise the responses of the proposed islanding detection method. The detecting parameters considered in this work are voltage and frequency, measured at the point of common coupling (PCC). The threshold sets for the detecting parameters are according to the specific UL 1741 standards. The entire study is carried out with MATLAB/Simulink environment and the test outcomes obtained after extensive study show that the proposed technique does not possess a non-detection zone (NDZ) in distributed generation system.

Keywords: **Anti-islanding, distributed generation, harmonic distortion.**

I. INTRODUCTION

The encouragements and limitations on non-conventional and conventional power resources respectively have facilitated the setting up a grid coupled conservational power generation. The penetration of distributed power generation (DPG) within an electrical system employs a key idea of micro-grid. DPGs supply the local-loads and swiftly exchanges the energy with the interconnected AC network. Thus, it provides several advantages, such as reduced cost of storage implementation and transmission lines along with the minimum transmission losses, improved power quality and reliability with the enhanced environmental performance. Along with the mentioned benefits, there exist many operation challenges and among them one critical concern lies in the need to guard against the occurrence of islanding [1].

Unintentional islanding (UI) can be described as a state, when actively performing DPGs and local loads encounter a sudden cut-off of utility grid and the centralized control system of grid fails the capability to control the island. UI instances lead to several dangerous issues, like damaging the electrical equipment due to the uncontrolled frequency and voltage, increasing threat of electric shocks to the utility workers and a major menace is out-of-phase enclosure of the breakers connecting the DPGs and utility grid. Thus, UI is an undesirable event and immediate detection becomes important. Some standardized institute, UL-1741 standards [2], IEEE-1547 standards [3], IEEE 929 standards [4], recommend interconnection codes to ensure detection of UI within 2sec.

Numerous UI detection methods (DMs) projected till-date, can generally be segregated into two groups, such as remote and local techniques. Remote DMs implements the theory of communication between the DPG and the utility. Generally the remote methods are very reliable, but due to the higher execution cost of these methods are not compatible with the plan of micro-grid [5]-[6]. On the other hand, the local methods are executed with the control strategy of each DPGs. Local DMs can be categorized into passive and active detection methods.

The passive detection methods (PDMs) scrutinizes the changes and/or the rate-of-change of parameters at the target DPG ends, due to the disparity between the DPG generation and integrated load demands [7]–[8]. The fundamental concept of the PDM is very simple with the minimal implementation cost and technical installation complexity. In recent times, the PDMs mostly focuses on the application of intelligent digital signal processing techniques. For the extraction of useful signal parameters different methods are proposed such as, rotational invariance techniques [9], Fast Gauss-Newton Algorithm [10], Tufts–Kumaresan [11], Wavelet Transform [12], Wavelet Packet Transform [13], S-Transform [14], Hilbert-Haung Transform [15] are employed to select the significantly operating features under the noisy and distorted operational environment. Subsequently, the machine learning techniques such as decision tree [7], Naive-Bayes Classifier [16], Support Vector Machine [17], and Extreme Learning Machine [18] are applied for the detection of islanded events. Employing intelligent data processing algorithm on the computed variables, helps in reducing the non-detection zone (NDZ). However, the NDZ may still be present under the operating conditions of zero power mismatch between DG generation and local load demands. Thus, the complex data processing makes the PDMs less promising for distribution network.

Active detection methods (ADMs) are basically considered as an alternative to PDMs. ADMs are deliberately planned such that the DPGs are compelled to attain higher instability at the occurrence of an islanding event [19] and directly performs with the operation of the electric power system [20]. The major advantages of ADM over PDMs are the negligible NDZ and faster detection speed [21]. Several ADMs implements the concept of impedance measurement during the injection of a specified frequency signal [22], further certain ADM stimulates a cumulative feedback to attain a highly evident fluctuations and thus the corresponding deviations in the detection parameters [23] and a number of methods perturb the inverters by introducing the disturbance via phase locked loop [24]. The effectiveness of any active DM is generally measured based on the detection speed (i.e., the time difference between occurrence of islanding event and its detection time) and NDZ.

A novel active DM for islanding detection based on the injection of intensified voltage harmonic signal is proposed in this work. The key benefits of the proposed DM are insignificant adverse effects due to the introduced intensified harmonic voltage signal along with correct and quick islanding detection. To achieve a quick and efficient islanding DM, a VSC based photovoltaic (PV) system is considered in this study. The proposed DM constantly introduces a disturbing signal by means of a Direct-Quadrature (d-q) framed current-control loop. Further, to facilitate a precise measurement, an amplified disturbing signal is to be considered and thus, an intensified harmonic signal has been taken into account as an injecting parameter. The disturbance injected is calculated to be one percent of the extracted disturbing signal. During the grid following mode the impedance of the system is generally very less, therefore the voltage harmonic cannot be adversely affected due to the injected disturbance. Moreover, to achieve a proper detection, the voltage and frequency variations are monitored constantly at the point of common coupling. Several power mismatch conditions are simulated to validate the proposed DM for accurate and faster detection speed. The complete experiment is carried out with MATLAB/Simulink environment and the test outcomes obtained after extensive study show that the proposed technique results with a zero NDZ for all the islanding events in distributed generation system.

Rest part of the paper is structured as follows. Section II presents modeling of the test system considered for study and DPG control scheme. Section III portrays the proposed active anti-islanding scheme. Section IV provides the simulation result to validate the proposed approach. Lastly, Section V summarizes the proposed approach with few brief concluding statements.

II. MODELING OF PROPOSED TEST SYSTEM AND CONTROL SCHEME

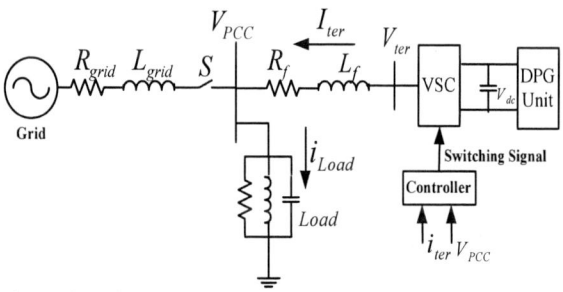

Fig. 1.Schematic presentation of a test system for a grid-connected inverter based DPG.

The proposed anti-islanding scheme is implemented on a simple grid-connected inverter-based DG. The single line diagram of the test system is shown in Fig. 1. The DPG interface controller for regulating the current using a control strategy based on self tuned proportional integral (STPI) is represented in Fig. 2. In grid following mode, voltage synchronization between the DPGs and grid is very essential. Therefore, a phase-locked loop (PLL) is equipped with an inverter controller, to extract the reference frequency and phase-angle at the point of common coupling. Thus, it assists the electric network to maintain the power factor nearly equals to one, by compelling the current signal to be in phase with the grid voltage. The control strategy, transforms the current at the point of common coupling into d-q frame by implementing park transformation. Additionally, the DC voltage signals are controlled and fed as a reference current signal ($I_{d_{ref}}$, $I_{q_{ref}} = 0$) in d-q frame. Afterwards, the error signals are generated by subtracting the obtained current signal from the reference. This error is summed up with an intentionally introduced disturbance and is made to pass through the STPI controller to generate reference voltage. With the intention of enhancing the sensitivity and dynamic response of the system, V_{dq} is feed-forward. The obtained voltage in the d-q axis frame are employed to extract the modulation index (m) and phase (ϕ), and are fed forward towards the pulse width modulated (PWM) signal generator [25].

The instantaneous current at the terminal of a voltage-source-converter (VSC) can be calculated as presented in the following equations [26]:

$$\frac{d}{dt} i_{ter_{abc}} = -\frac{R_f}{L_f} i_{ter_{abc}} + \frac{1}{L_f}\left[v_{ter_{abc}} - v_{PCC_{abc}}\right] \quad (1)$$

Here, R_f and L_f states the resistance and inductance respectively of the designed filter. i_{ter} and v_{ter} are the instantaneous voltage and current respectively at DPG terminal.And the instantaneous PCC voltage is represented as v_{PCC}. Applying park's transform on Eq.1 the rotating synchronous reference frame can be written as in Eq. 2 or Eq. 3:

$$\frac{d}{dt}\begin{bmatrix} i_{d_{ter}} \\ i_{q_{ter}} \end{bmatrix} = \begin{bmatrix} -\dfrac{R_f}{L_f} & \omega \\ -\omega & -\dfrac{R_f}{L_f} \end{bmatrix}\begin{bmatrix} i_{d_{ter}} \\ i_{q_{ter}} \end{bmatrix} + \frac{1}{L_f}\begin{bmatrix} v_{d_{ter}} - v_{d_{PCC}} \\ v_{q_{ter}} - v_{q_{PCC}} \end{bmatrix} \quad (2)$$

or

$$\frac{d}{dt}\begin{bmatrix} i_{d_{ter}} \\ i_{q_{ter}} \end{bmatrix} = \begin{bmatrix} -\dfrac{R_f}{L_f} & 0 \\ 0 & -\dfrac{R_f}{L_f} \end{bmatrix}\begin{bmatrix} i_{d_{ter}} \\ i_{q_{ter}} \end{bmatrix} + \frac{1}{L_f}\begin{bmatrix} u_d \\ u_q \end{bmatrix} \quad (3)$$

where,

$$u_d = v_{d_{ter}} - v_{d_{PCC}} + w L_f i_{q_{ter}} \quad (4)$$

$$u_q = v_{q_{ter}} - v_{q_{PCC}} - w L_f i_{d_{ter}} \quad (5)$$

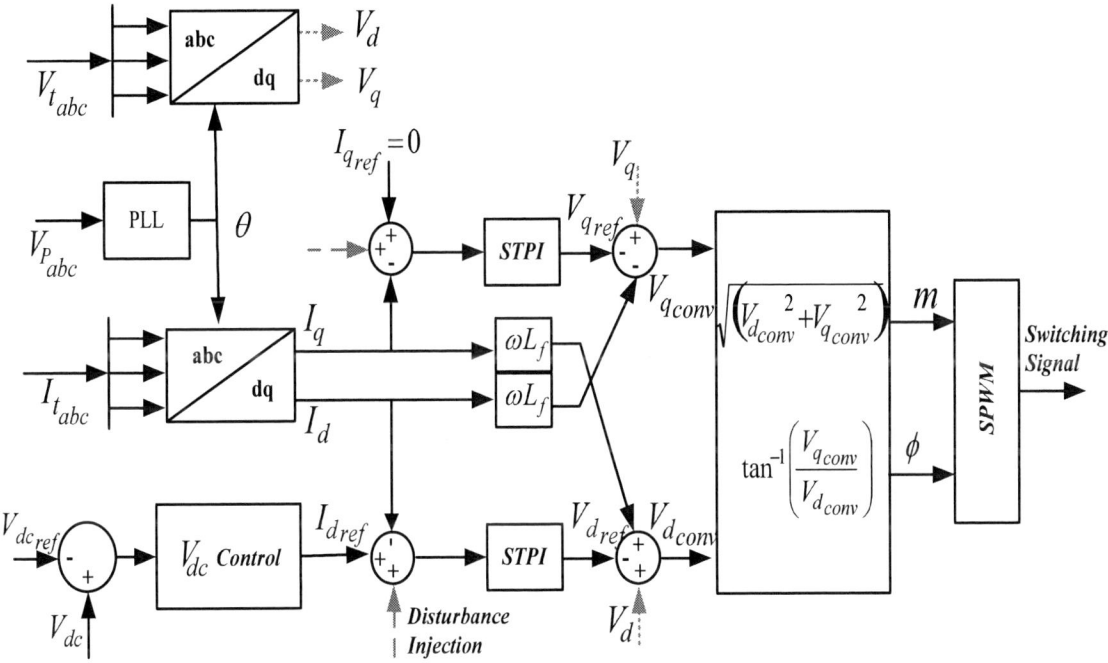

Fig. 2. DPG interfacing inverter controller.

The control strategy explained is implemented to design the controller of an inverter based DPG considered in this study.

The most important characteristic of the designed controller is the consideration of STPI controllers. The gains are regulated frequently corresponding to the changes in the operational circumstances. The proportional-gain parameter 'K_p' is in charge of enhancing the overshoot-time and the rise-time response, whereas, the integral-gain 'K_i' facilitates to reduce the steady-state error. Generally, the value considered for K_p and K_i are directly dependent on the error value (i.e. large for K_p and small for K_i) for better controllability and for reduction of steady-state error, respectively. Thus, the use of STPI controller enhances the performance to attain a smooth and quicker stability response under the steady-state and transient-state respectively. The error function implemented to tune the gain values are presented in Eq. 6 and Eq.7.

$$K_p(t) = K_{P_{\max}} - \left(K_{P_{\max}} - K_{P_{\min}}\right)e^{-[Ke(t)]} \qquad (6)$$

$$K_i(t) = \left(K_{i_{\max}}\right)e^{-[Ke(t)]} \qquad (7)$$

here, K symbolizes a constant, which decides the rate at which the $K_p(t)$ will vary between the predefined maximum $K_{P_{\max}}$ and minimum $K_{P_{\min}}$ values [27]. Table I enumerates the system and controller specifications considered while designing the test system.

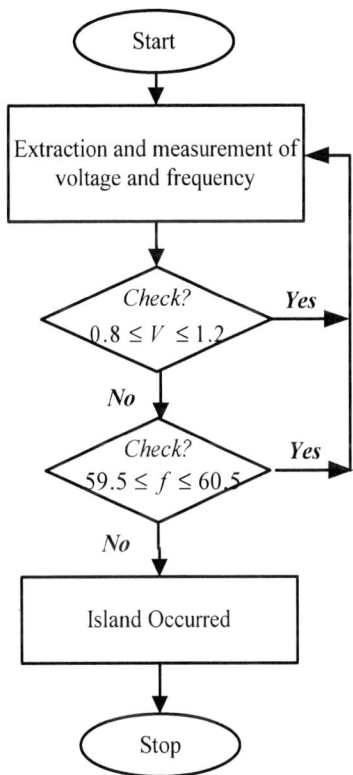

Fig. 3. Proposed anti-islanding algorithm.

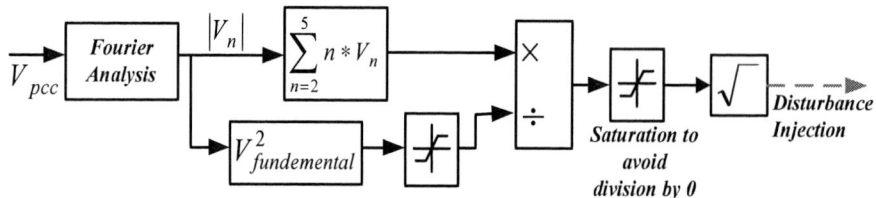

Fig. 4. Design to obtain the injecting signal.

TABLE I. SYSTEM SPECIFICATION

System Rating:	
DPG	100 KW
R_f	3.72e-3 Ω
L_f	375e-6 H
Load	100 KW
Controller Specification:	
STPI, K	0.084
Current Regulator	$K_{P_{max}} = 0.418$, $K_{P_{min}} = 0.237$ $K_{i_{max}} = 24.647$, $K_{i_{min}} = 17.500$
V_{dc} Regulator	$K_{P_{max}} = 9.848$, $K_{P_{min}} = 6.592$ $K_{i_{max}} = 1000$, $K_{i_{min}} = 575$

III. PROPOSED HARMONIC INJECTION BASED ISLANDING DETECTION SCHEME

The DPG may undergo serious transients, consequent to an unintentional islanding event. These transients might damage the energized DPGs and the loads. Thus designing a fast and efficient islanding DM turns out to be a major challenge. Therefore the proposed efficient anti-islanding scheme here, considers the two basic system parameters (i.e., voltage magnitude and frequency) to detect the islanding occurrence. Fig. 3. represents the steps followed to design the proposed anti-islanding scheme.

The parameter considered in the proposed active anti-islanding scheme is an intensified voltage harmonic signal. The simulink model designed to extract the injecting signal is shown in Fig. 4. The injecting parameter is evaluated as shown in Eq. 8.

$$IHS_v = \frac{1}{f_{fundamental}} \sqrt{\sum_{n \neq 1} (n * f_n)^2} \quad (8)$$

Here, IHS_v represents the intensified harmonic signal of the voltage. The nth frequency component is considered up to the 5th order harmonics. The injecting parameter enhances the sensitivity of the detecting parameters to a higher extent.

In this islanding DM based on harmonics, the concept of IHS_v injection as a feedback signal to the inverter controller is used. The detailed mathematical structure of the controller is depicted in Fig. 2. The amount of perturbation considered for the injection in the controller is 1% of the IHS_v signal. In grid following mode, the system impedance is too less and thus the perturbations rushes towards the grid without affecting the PCC voltage and current. However, on the incidence of an

islanding scenario, the harmonic disturbances flow into the connected loads and cause significant fluctuations in system parameters at PCC. Therefore, with an intention to achieve precise and faster islanding DM, the PCC voltage and frequency are monitored continuously along with the continues injection of perturbations in the controller. An appropriate threshold has been set for the voltage and frequency by considering the standards (i.e. ±0.2, ±0.5 for voltage and frequency respectively) [28]. On recognition of an islanding instance, the injected disturbance is disabled to avoidsystem damage.

IV. SIMULATION RESULTS AND ANALYSIS

This section of the manuscript provides a detailed analysis on the obtained simulation results of the suggested islanding DM. To validate the proposed islanding DM, UL1741 test standard is considered in this work for various power mismatch condition in between DPG and local loads. It states that the variation of active load power (P) can be +25%, ±50% and 100% of the rated power generation of an inverter with zero varying reactive loads. Moreover, the UL1741 standard also specify that the deviations of reactive power (Q) can be varied with 1% step size manner from -5% to +5% within the balanced operating conditions to test the system under study. Therefore, the proposed islanding DM test cases are designed referring to the UL1741 standard and are presented in TableII.

TABLE II. UL 1741 TESTING STANDARDS FOR LOADING CONDITIONS.

Cases	P(mismatch in %)	Q (mismatch in %))
C-1	100	95
C-2	100	105
C-3	50	100
C-4	100	100
C-5	125	100
C-6	150	100

Figure. 5 and Figure. 6 provides some of the time evolution simulated results based on the proposed islanding DM. The test system is equipped with the proposed DM and the evaluation of detection scheme is carried out by disconnecting the grid from the PCC at time, t = 1s. Fig. 5 shows the active and reactive power instability of the system at PCC subsequent to the occurrence of an islanding event. The change of frequency and d-axis voltage at PCC are the prime measurement index in this study, which are considered as detecting parameters for an islanding event. Fig. 6.(a) and Fig. 6.(b) presents the change in frequency and d-axis voltage magnitude respectively, at PCC due to the induction of islanding event.

Similarly, the change of frequency and direct-axis voltage magnitude at PCC is analyzed for some frequently occurring non-islanding events in Fig. 6.(c) and Fig. 6.(d). The non-islanding events simulated for verifying the effectiveness of the proposed islanding DM are capacitor switching, fault events and load switching scenarios etc. An appropriate threshold limit has been provided for these detecting parameter indices to attain an precise and rapid detection. The corresponding threshold limit for frequency and direct-axis voltage are $59.5 \geq f \geq 60.5$ and $0.8 \geq V \geq 1.2$ respectively. When both frequency and direct-axis voltage magnitude deviate from the approved threshold limits, the detection task of proposed islanding DM is accomplished as presented in Fig. 3.

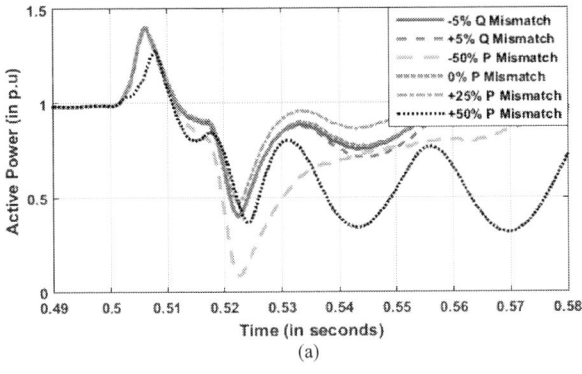

(a)

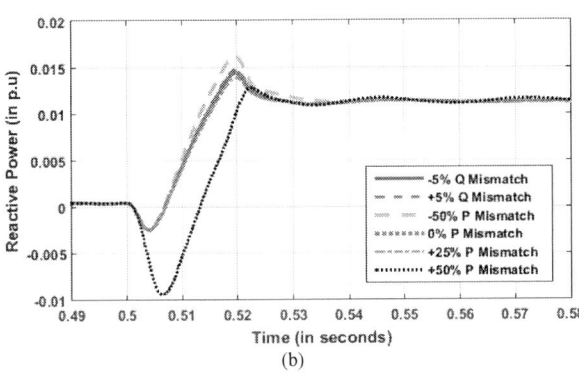

(b)

Fig. 5. Parameter variation analyzed in the proposed anti-islanding scheme (a) Variation of active power magnitude at the occurrence of an islanding events (b) Variation of reactive power magnitude at the occurrence of an islanding events.

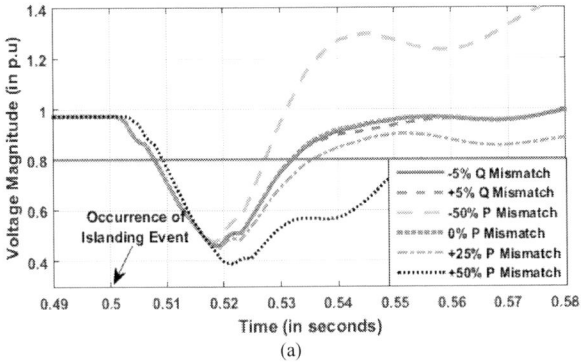

(a)

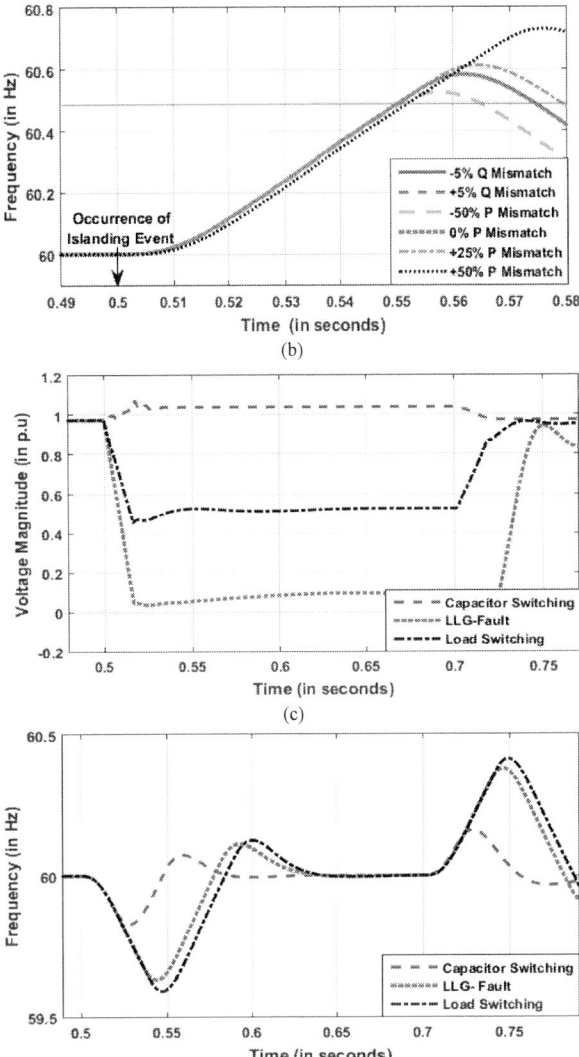

(b)

(c)

(d)

Fig. 6. Judging parameter and its variation studied for proposed approach. (a) Variation of direct-axis voltage magnitude at the occurrence of an islanding events (b) Variation of frequency magnitude at the occurrence of an islanding events (c) Variation of direct-axis voltage magnitude at the occurrence of an non-islanding events (d) Variation of frequency magnitude at the occurrence of an non-islanding events.

TABLE III. TIME TAKEN TO DETECT AN ISLANDING EVENT.

Cases Studied	Time taken to violate the voltage threshold	Time taken to violate the frequency threshold	Detection Time
C1	0.5082s	0.5452s	0.5452s
C2	0.5085s	0.5468s	0.5468s
C3	0.5098s	0.5494s	0.5494s
C4	0.5080s	0.5490s	0.5490s
C5	0.515s	0.554s	0.554s
C6	0.510s	0.558s	0.558s

It can be analyzed from Fig. 6.(c) and Fig. 6.(d) that the deviation of detecting indices of non-islanding cases, such as capacitor switching and load switching are satisfying the range of both the threshold limits and completely protected from nuisance tripping. However, for the fault condition (i.e., LLG-Fault), the magnitude of d-axis voltage reduces about to 0.1 p.u violating the voltage threshold limit. But, examining the other parameter i.e. frequency, it is clearly noticeable that magnitude of frequency deviations are significantly minor and does not violate the threshold. Hence, none of the non-islanding event gets wrongly detected as an islanded event.

Furthermore, fast detection is also one of the major concerns for islanding detection problem. Thus, this work also analyzed the detection time associated with this approach for accurate islanding detection and depicted in Table III. It can be clearly analyzed from TableIII that the max and min time undertaken by this approach are 558ms and 545ms respectively. Along with a faster detection of an islanding instance, the proposed method do not possess a NDZ, as the sensitive testing cases of power mismatch are simulated and are easily detected.

V. CONCLUSION

This manuscript proposed an active harmonic injection based islanding detection method for inverter based DPG. An intensified harmonic voltage signal is used as an injecting parameter to the controller interfaced between the utility grid and DPG. It has been analysed that an intensified harmonic voltage signal of 1% of the total extracted voltage harmonic value is sufficient for reliable detection, which results a smooth operation of DPG in grid-connected mode without affecting the PCC voltage harmonics considering very less value of grid impedance. The detection process has been based on the deviation of the frequency and direct-axis voltage at PCC. To validate the proposed islanding DM, UL1741 test standard is considered in this work for various power mismatch conditions in between DPG and local loads. The simulation results show the reliability of the proposed method regardless of operating condition and prove its effectiveness with zero NDZ.

REFERENCES

[1] Samui, A. and Samantaray, S.R.,"New active islanding detection scheme for constant power and constant current controlled inverter-based distributed generation," IET Generation, Transmission & Distribution, 7(7), pp.779-789,2013.

[2] Std, U.L.,1741. "Inverters, Converters, and Controllers for Use in Independent Power Systems," 2002.

[3] IEEE Standards Association, "IEEE Standard for Interconnecting Distributed Resources With Electric Power Systems," IEEE Std 1547, pp.1-16,2003.

[4] Hudson, R.M., Thorne, T., Mekanik, F., Behnke, M.R., Gonzalez, S. and Ginn, J., "Implementation and testing of anti-islanding algorithms for IEEE 929-2000 compliance of single phase photovoltaic inverters," In Photovoltaic Specialists Conference, Conference Record of the Twenty-Ninth IEEE (pp. 1414-1419),May,2002.

[5] Etxegarai, A., Eguía, P. and Zamora, I.," Analysis of remote islanding detection methods for distributed resources," In Int. Conf. Renew. Energies Power Quality,April, 2011.

[6] Akhlaghi, S., Ghadimi, A.A. and Akhlaghi, A.,"A novel hybrid islanding detection method combination of SMS and Qf for islanding detection of inverter-based DG," In Power and Energy Conference at Illinois (PECI), IEEE, pp.1-8, February, 2014.

[7] K. El-Arroudi and G. Joos, "Data mining approach to threshold settings of islanding relays in distributed generation," IEEE Transactions on power systems, 22(3), pp.1112-1119, 2007.

[8] A.H.K. Alaboudy and H.H. Zeineldin,"Islanding detection for inverter-based DG coupled with frequency-dependent static loads," IEEE Transactions on Power Delivery, 26(2), pp.1053-1063, 2011.

[9] W.K. Najy, H.H. Zeineldin, A.H.K. Alaboudy, and W.L. Woon, "A Bayesian passive islanding detection method for inverter-based distributed generation using ESPRIT," IEEE Transactions on Power Delivery, 26(4), pp.2687-2696, 2011.

[10] M. Padhee, P.K. Dash, K.R. Krishnanand, and P.K. Rout, "A fast Gauss-Newton algorithm for islanding detection in distributed generation," IEEE Transactions on Smart Grid, 3(3), pp.1181-1191, 2012.

[11] M. Bakhshi, R. Noroozian, and G.B. Gharehpetian, "Anti-islanding scheme for synchronous DG units based on Tufts–Kumaresan signal estimation method,"IEEE Transactions on Power Delivery, 28(4), pp.2185-2193, 2013.

[12] M. Mishra, R. Sahu, D. Ray, S. Swarup, and P.K. Rout, "Study of performance of pattern recognition techniques based on wavelet features for Islanding detection in distributed generation," In Power Electronics, Intelligent Control and Energy Systems (ICPEICES), IEEE International Conference on (pp. 1-6). IEEE., July 2016

[13] M. Mishra, P.K. Rout, and S. Patel,"A novel islanding detection technique based on wavelet packet transform," In Power, Communication and Information Technology Conference (PCITC), 2015 IEEE (pp. 697-702). IEEE, October 2015.

[14] M. Mishra and P.K. Rout, "Time-Frequency Analysis based Approach to Islanding Detection in Micro-grid System,"International Review of Electrical Engineering (IREE), 11(1), pp.116-129., 2016

[15] M. Mishra, M. Sahani, and P.K. Rout,"An islanding detection algorithm for distributed generation based on Hilbert–Huang transform and extreme learning machine," Sustainable Energy, Grids and Networks, 9, pp.13-26, 2017.

[16] O.N. Faqhruldin, E.F. El-Saadany, and H.H. Zeineldin,"A universal islanding detection technique for distributed generation using pattern recognition,"IEEE Transactions on Smart Grid, 5(4), pp.1985-1992, 2014

[17] M.R. Alam, K.M. Muttaqi, and A. Bouzerdoum,"A multifeature-based approach for islanding detection of DG in the subcritical region of vector surge relays," IEEE Transactions on Power Delivery, 29(5), pp.2349-2358, 2014.

[18] M. Mishra, P.K. Rout, R. Sahu, D. Ray and S. Swarup," December. Study the performance of S-transform based extreme learning Machine for islanding detection in distributed generation," In Power Systems Conference (NPSC), 2016 National (pp. 1-6). IEEE, 2016.

[19] H.H. Zeineldin, E.F. El-Saadany and M.M.A. Salama,"Impact of DG interface control on islanding detection and nondetection zones," IEEE Transactions on Power Delivery, 21(3), pp.1515-1523, 2006

[20] W. Xu, G. Zhang, C. Li, W. Wang, G. Wang, J. Kliber, "A power line signaling based technique for anti-islanding protection of distributed generators—Part I: Scheme and analysis," IEEE Transactions on Power Delivery, 22(3), pp.1758-1766, 2007

[21] H.H. Zeineldin, E.F. El-Saadany and M.M.A Salama,"Islanding detection of inverter-based distributed generation," IEE Proceedings-Generation, Transmission and Distribution, 153(6), pp.644-652, 2006.

[22] Reigosa, D., Briz, F., Charro, C.B., García, P. and Guerrero, J.M., "Active islanding detection using high-frequency signal injection," IEEE Transactions on Industry Applications, 48(5), pp.1588-1597, 2012.

[23] Chandak, S., Dhar, S. and Barik, S.K., "Islanding disclosure for grid interactive PV-VSC system using negative sequence voltage," In Power, Communication and Information Technology Conference (PCITC), 2015 IEEE, pp.497-504,October, 2015.

[24] Velasco, D., Trujillo, C., Garcera, G. and Figueres, E., "An active anti-islanding method based on phase-PLL perturbation," IEEE Transactions on Power Electronics, 26(4), pp.1056-1066, 2011.

[25] Balaguer, I.J., Lei, Q., Yang, S., Supatti, U. and Peng, F.Z.," Control for grid-connected and intentional islanding operations of distributed power generation," IEEE transactions on industrial electronics, 58(1), pp.147-157,2011.

[26] Vahedi, H., Noroozian, R., Jalilvand, A. and Gharehpetian, G.B.," A new method for islanding detection of inverter-based distributed generation using DC-link voltage control," IEEE transactions on power delivery, 26(2), pp.1176-1186,2011.

[27] Zaky, M.S., "A self-tuning PI controller for the speed control of electrical motor drives," Electric Power Systems Research, 119, pp.293-303,2015.

[28] "Charge Controllers for Use in Photovoltaic Systems Standard UL," Northbrook, IL: Underwriters Laboratories, 2001.

Selective Harmonic Elimination for Cascaded Three Phase Multilevel Inverter

Deepak Singh
Dept. of Electrical Engineering
North Eastern Regional Institute
of Science and Technology
Nirjuli, India
deepmzp30@gmail.com

Akhilesh Sharma
Dept. of Electrical Engineering
North Eastern Regional Institute
of Science and Technology
Nirjuli, India
as@nerist.ac.in

Pukhrambam Devachandra Singh
Dept. of Electrical Engineering
North Eastern Regional Institute
of Science and Technology
Nirjuli, India
pds@nerist.ac.in

S. Gao
Dept. of Electrical Engineering
North Eastern Regional Institute
of Science and Technology
Nirjuli, India
sg@nerist.ac.in

Abstract— **Frequent power failures, non-availability of electrical energy from conventional energy sources at the load center or insufficient generation of electrical energy are the major reasons of utilization of non-conventional resources to harness electrical energy. From these sources, usually direct current form of electrical energy is obtained which needs to be regulated for conversion into alternating form of voltages through inverter. Thus, the design of inverters which are widely used as standby power supply at domestic levels, ac drives control, distribution of power generation, play a vital role. The ease of utilization of inverter depends on the magnitude of ac output voltage and frequency which could be directly controlled by switching the controlled semi-conductor devices. However, this leads to injection of harmonics in the output waveform. The dominant order of harmonics needs to be eliminated which otherwise, causes power losses in the form of heat, damaging the inverter. Eliminating these harmonics is a big challenge. Therefore, selective harmonic elimination PWM (SHE-PWM) is one among many techniques which could be used for multi-level Cascaded H-bridge inverter. The optimized switching angles have been generated through Newton-Raphson by solving non-linear equations. This paper tries to bring comparative study of presence of harmonics for different types of load.**

Keywords— **CHB, NR, THD, SHE-PWM, FSF**

I. INTRODUCTION

In the present scenarios, when the demand of electrical energy is increasing day by day, the role of inverter, a device to converter a dc power into an ac power at desired output voltage and frequency has increased a lot, as, it is possible to directly converter direct current into alternating current. The renewable energy is spreading fast in numerous forms to supply electrical energy for domestic purposes, grid integration, power generation and hybrid vehicles hence it has become one of the essential devices. Single phase and three phase inverters with single voltage level can be used for low voltage and power applications. These inverters contain more harmonics thereby reducing the performance. To overcome some fraction of the harmonics, a multilevel inverter is better option.

In the design of inverter, gate pulse generation plays a vital role, if properly designed, it can reduce many numbers of harmonics order present in output signal [5]. The pulse

generation depends on many parameters. Modulation is one among many parameters. It is the process of switching power electronic device in power converter from one state to another.

The aim of modulation is to generate a step waveform which adjusts the reference signal with amplitude, frequency and fundamental component with best approximate form [19]. The reference signal is usually a sinusoidal wave in a steady state [7].

Different converters have different switching states configuration to achieve desire output voltage [3]. To keep all voltage sources balanced and to synthesis reference control signal, different modulation strategies are used. All Modulation techniques are not suited to each and every multilevel topology [9]. These techniques have their own advantages as well as disadvantages. The modulation techniques are classified on the bases of switching frequency i.e. fundamental switching frequency (FSF) and high switching frequency (HSF). Further, FSF is grouped into space vector control (SVC) and selective Harmonic Elimination (SHE), whereas high switching frequency is divided into Sinusoidal PWM (SPWM) and space vector PWM (SVPWM) [20] and [21]. Of the two types of techniques, high switching frequency is widely used in industrial application since switching devices conduct high commutation [24]. In low switching frequency, switching losses are low and efficiency is high but devices conduct less commutation [12].

II. CASCADED H-BRIDGE MULTILEVEL INVERTER

In a cascaded H- Bridge inverter, each single phase has a separate DC source (SDCS) V_{dc}. An N-level structure of Cascaded H-Bridge for leg A of three phase is shown in the Fig. 1. As seen from the figure, each H-Bridge can have three different output voltage level, namely, $+V_{dc}$, 0 and -V_{dc}, depending on the switching states. If two cascaded H-Bridges are considered then their corresponding voltages will be $+2V_{dc}$, $+V_{dc}$, 0, -V_{dc}, and -$2V_{dc}$ and so on [1] and [26], so for an N cascaded H – Bridge, the output voltage levels will be (2N+1).

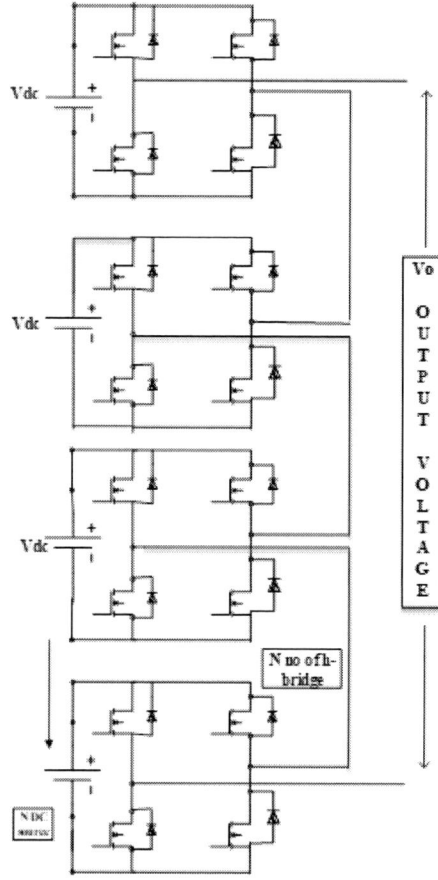

Fig. 1. N-level Cascaded H- Bridge Multilevel inverter for leg A

III. SELECTIVE HARMONIC ELIMINATION

In this method, a particular harmonic can be eliminated at fundamental frequency hence this is also called as fundamental switching frequency method [15]. This method provides high quality output among all PWM methods. It was first introduced by F.G. Turnball. Real time implementation of this method was done by H.S Patel and R.G. Hoft [10] in year 1970.

As per the Fourier Series representation, any periodic signal can be represented as sum of a dc, even and odd components. Since SHE-PWM shows a regular pattern with certain periodicity hence the concept Fourier Theory can be applied for decomposition of PWM voltage /current waveform [16]. It depends only on formulation of the given waveform and its properties [8]. To find analytical solutions of a SHE-PWM waveform, a suitable solving algorithm or method for formulation of the waveform is needed [18] and [22]. The Newton-Raphson (NR) technique is used to solve the non-linear equation for selecting switching angle for different SHE-PWM waveforms [28]. Initially, it was proposed for two and three level converters and later extended to multilevel and hybrid multilevel converters [23].

A. SHE-PWM Formulation

The waveform generated by a power electronic converter is periodic in nature. Any periodic signal can be

decomposed on the basis of Fourier – Series [11]. The generalized Fourier-Series equation can be represented as

$$F_n(t) = \frac{a_0}{2} + \sum_{n=1}^{N} \left(a_n \cos \frac{2n\pi t}{T} + b_n \sin \frac{2n\pi t}{T} \right) \quad (1)$$

In order to eliminate the dominant harmonics, the switching angle is calculated on the basis of equation (1). There are ways to define a SHE-PWM problem. Quarter-wave symmetry is simplest way for defining a problem in SHE-PWM method for both two level and multilevel [13] as it simplifies the formulation. The solution process becomes all DC components as even harmonic and sine coefficient of odd harmonic are equal to zero, resulting in the least number of equations to be solved [14] and [17].

Seven switching angles are obtained to reduce 3rd, 5th, 7th, 9th, 11th, 13th, and 15th order harmonics by optimizing the equation shown below.

$$\cos\theta_1 + \cos\theta_2 + \cos\theta_3 + \cos\theta_4 + \cos\theta_5 + \cos\theta_6$$
$$+ \cos\theta_7 = M \quad (2)$$

$$\cos 3\theta_1 + \cos 3\theta_2 + \cos 3\theta_3 + \cos 3\theta_4 + \cos 3\theta_5 +$$
$$\cos 3\theta_6 + \cos 3\theta_7 = 0 \quad (3)$$

$$\cos 5\theta_1 + \cos 5\theta_2 + \cos 5\theta_3 + \cos 5\theta_4 + \cos 5\theta_5 +$$
$$\cos 5\theta_6 + \cos 5\theta_7 = 0 \quad (4)$$

$$\cos 7\theta_1 + \cos 7\theta_2 + \cos 7\theta_3 + \cos 7\theta_4 + \cos 7\theta_5 +$$
$$\cos 7\theta_6 + \cos 7\theta_7 = 0 \quad (5)$$

$$\cos 9\theta_1 + \cos 9\theta_2 + \cos 9\theta_3 + \cos 9\theta_4 + \cos 9\theta_5 +$$
$$\cos 9\theta_6 + \cos 9\theta_7 = 0 \quad (6)$$

$$\cos 11\theta_1 + \cos 11\theta_2 + \cos 11\theta_3 + \cos 11\theta_4 +$$
$$\cos 11\theta_5 + \cos 11\theta_6 + \cos 11\theta_7 = 0 \quad (7)$$

$$\cos 13\theta_1 + \cos 13\theta_2 + \cos 13\theta_3 + \cos 13\theta_4 +$$
$$\cos 13\theta_5 + \cos 13\theta_6 + \cos 13\theta_7 = 0 \quad (8)$$

where, M is the modulation index.

$$M \triangleq \frac{\pi V_1}{4V_{dc}}$$

(9)t=a02+n=1Nancos2nπtT+bnsin2nπtT

B. Newton-Raphson Method

Newton-Raphson method is one of the traditionally preferred iterative methods to solve nonlinear transcendental equations [2]. This method, based on calculus approach, is a fast-iterative method with fast convergence to reach global minima. It begins with an initial guess and generally converges at zero. Numerical approach is one of the types of classical optimization techniques [27]. In these methods, the solution of equation is based on iterations [4]. The convergence of these techniques depends on the estimation of initial assigned values. If the initial guess is good, rate of convergence is fast and computational time is reduced. Providing good initial guess is greatly dependent on previous history. However, in most of the cases, probability of providing good initial values is quite low [6]. Hence, it is difficult to assess whether a solution of SHE problem exists or not. If a solution exists then what should be the proper initial guesses of such problems. Although, the predictions of initials values for a simple waveform is easier since very fewer switching angles are calculated [25]. This drawback can be overcome by random selection of initial guesses which could be considered to obtain analytical solution [29].

Algorithm used is as under and its corresponding flow chart is shown in Fig. 2:

1. Start
2. Initial Guesses of angles θ^o
3. Initialize the range of modulation index
4. Evaluate $F(\theta^o) = F^0$
5. Evaluate T by Linearizing the equation $F^o + [dF/d\theta] d\theta^o = T$
6. Solve for $d\theta^o$; $d\theta^o = INV[dF/d\theta]^o [T-F]$
7. Check whether $d\theta^o$ is within the tolerance range; if not, update the modulation index by 0.001 and increment the angle by $d\theta^{j+1} = \theta^j + d\theta^j$. If yes, solution found for particular MI and repeat step 4.
8. Check whether the calculated angle is greater than previous angle; if yes, store the results if not, pick the result as new initial angle and GOTO step 2 else ignore the angle and GOTO step 9.
9. END

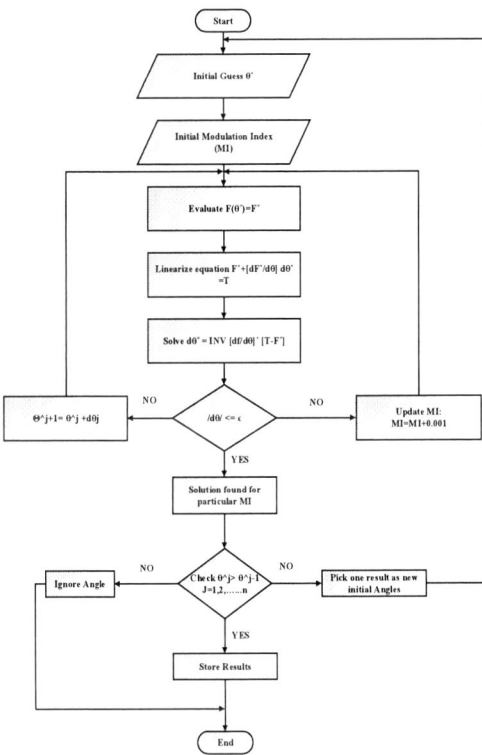

Fig. 2. Flowchart for calculation of switching angles

IV. SIMULATION RESULTS

The simulation has been carried out in MATLAB Simulink. Seven switching angles have been calculated using NR method. These angles have been used as gating pulse for cascaded multi-level inverter for removing dominant order of harmonics, namely 5th, 7th, 11th, and 13th. A Simulink model for three phase 15-level cascaded H-bridge inverter based on NR is shown in the Fig 3. The three-phase open circuit voltage waveform has been shown in Fig 4. Appendix I shows the parameters of inverter. It is

seen that the output voltage has 15 steps in each of the half cycle.

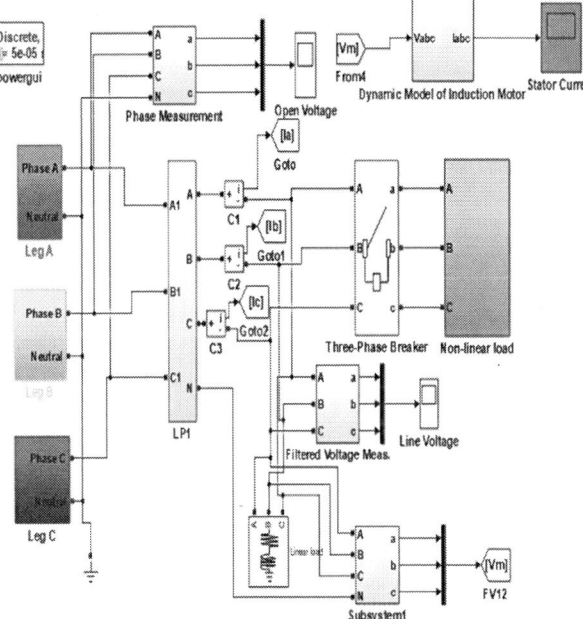

Fig. 3. Simulink Model Three Phase CHB Multilevel

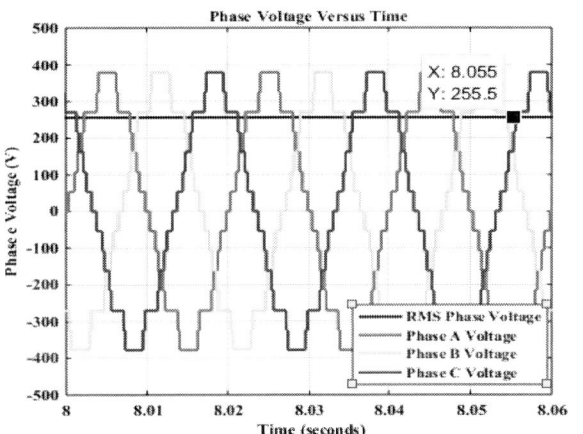

Fig. 4. Three Phase Voltages for H-bridge Inverter (15-level)

The FFT analysis of phase voltage is shown in Fig. 5. Its dominant orders are shown in Table I. It is seen form figures that the components of 3rd, 5th, 7th, 9th, 11th, 13th and 15th order of harmonics are 3.46%, 1.34%, 5.84%, 0.27%, 4.56%, 2.79%, and 0.46% respectively with THD of 10.40%.

With three different types of load, namely, linear, non-linear and a three-phase induction motor load, the load current, the three phase currents for linear and non-linear type of loads are shown in Fig. 6, Fig. 8 respectively whereas Fig. 10 indicates the THDs of 15.20 % for the load current with both linear and nonlinear loads connected together. Fig. 11 shows that an induction motor has a THD of 11.93%. their corresponding FFTs are shown in the Fig. 7, Fig. 9 and Fig. 11 respectively. These THDs have been tabulated in Table II. In case of three phase induction motor load.

978-1-5386-6624-1/18 $31.00 © 2018 IEEE   439

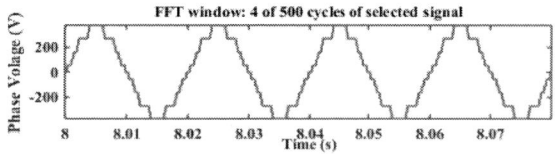

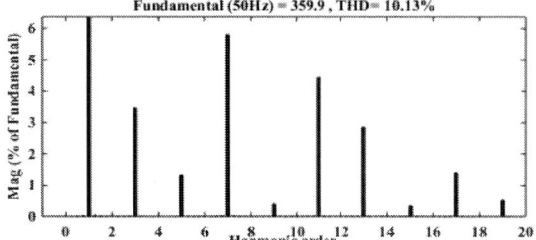

Fig. 5. FFT analysis of Harmonic Order

TABLE I: FFT ANALYSIS

FFT analysis

Sampling time = 3.90652e-05 s
Samples per cycle = 512
DC component = 0.003625
Fundamental = 360 peak (254.5 rms)
THD = 10.13%

0 Hz (DC):	0.00%	90.0°
50 Hz (Fnd):	100.00%	-1.4°
100 Hz (h2):	0.01%	179.5°
150 Hz (h3):	3.46%	176.7°
200 Hz (h4):	0.00%	175.0°
250 Hz (h5):	1.33%	-10.7°
300 Hz (h6):	0.00%	140.8°
350 Hz (h7):	5.78%	170.5°
400 Hz (h8):	0.00%	9.0°
450 Hz (h9):	0.41%	160.5°
500 Hz (h10):	0.00%	25.8°
550 Hz (h11):	4.44%	-15.6°
600 Hz (h12):	0.01%	160.4°
650 Hz (h13):	2.83%	163.5°
700 Hz (h14):	0.00%	-67.2°
750 Hz (h15):	0.34%	-36.3°
800 Hz (h16):	0.00%	-16.0°
850 Hz (h17):	1.41%	-26.1°
900 Hz (h18):	0.00%	143.9°
950 Hz (h19):	0.54%	-16.8°

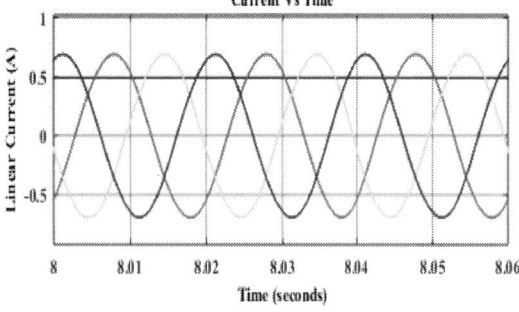

Fig. 6. Three Phase Line current for linear load

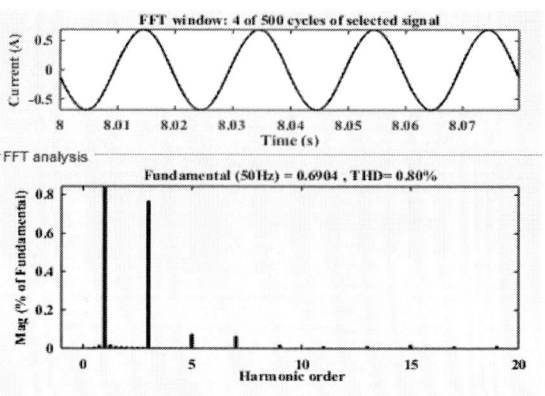

Fig. 7. FFT analysis of Line current for linear load

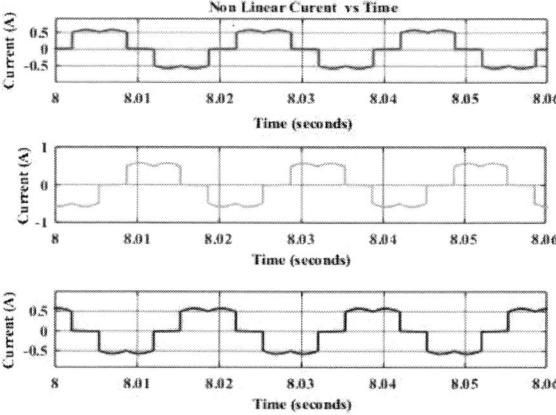

Fig. 8. Three Phase Line current for Non-Linear load

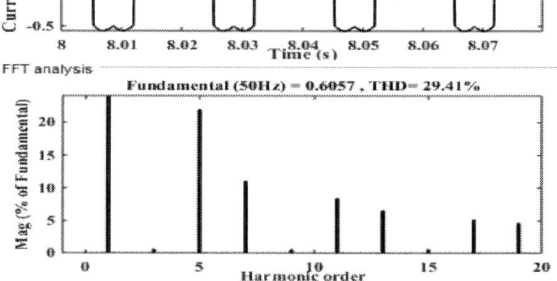

Fig. 9. FFT analysis of Line current for Non-Linear load

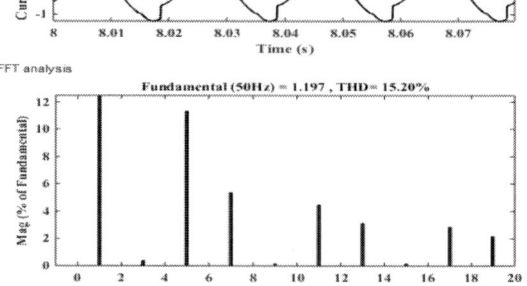

Fig. 10. FFT analysis of load current for both linear & non-linear load

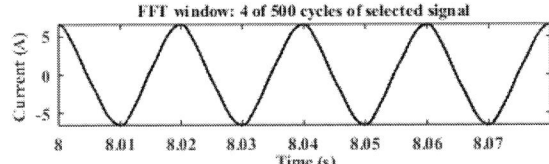

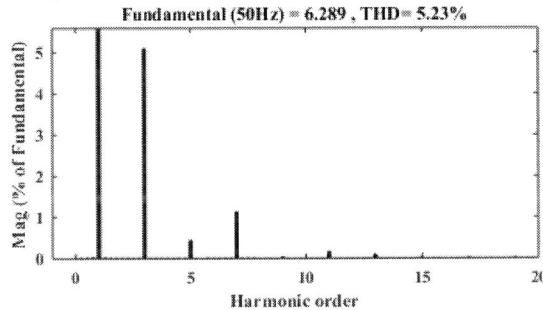

Fig. 11. FFT analysis of Stator current for Motor load

TABLE II: SIMULATION RESULTS

% THD of Phase Voltage	% THD in line current for different types of load			
	Linear Load	Non-linear Load	Both Linear & Non-linear Loads	Motor Load
10.13%	0.80%	29.41%	15.20%	5.23 %

V. CONCLUSION

It is seen that 3rd, 5th, 7th, 9th, 11th, 13th, and 15th order of harmonics has been successful suppressed using NR-method. As seen from the result, a 15-level voltage waveform is obtained for three phase cascaded H-bridge multilevel inverter at no-load and it's correspondingly THD is 10.13%. It is also seen that the THDs for linear, non-linear, combined load (linear and non-linear together) and induction motor loads are 0.80%, 29.41%, 15.20% and 5.23 % respectively. This clearly indicates that the THD depends on types of load. A linear load has least THD whereas non-linear load has maximum THD, about 82.22 % higher than a dynamic load. The performance of THD reduction is most satisfactory in linear and motor load hence it is better to use such load on inverter rather than non-linear loads.

APPENDIX -1: LIST OF PARAMETERS

Input Voltage for Three phase cascaded H-bridge inverter (15-level)	54 V for each H-bridge
Fundamental Frequency (F)	50 Hz
Parameters for Motor Load	
Input Voltage for Motor	412 V
Stator resistance	0.90 □
Rotor resistance	0.66 □
Stator inductance	0.00457 H
Rotor inductance	0.00457 H
Moment of Inertia	0.1384 kg m²
Number of Poles	4
Magnetizing Inductance	1 H

Parameters for Non-Linear Load	
Input voltage for three phase Rectifier	412 V
Number of diodes	6
Three phase ac R load Active power	650 W
Parameters for Motor Load	
Three phase ac RL load Active power for inverter	1000 W
Three phase ac RL load Reactive power for inverter	1 kVAR

REFERENCES

[1] Aniruddho Chatterjee, Adarsh Rastogi, Rajat Rastogi, Ajay Saini and Sarat Kumar Sahoo, "Selective Harmonic Elimination of Cascaded H-Bridge Multilevel Inverter using Genetic Algorithm", IEEE International Conference on Innovations in Power and Advanced Computing Technologies. pp1-4, 2017.

[2] Wahidah Abd Halim, Tengku Noor Ariana Tengku Azam, Komathi Applasamy and Auzani Jidin, "Selective Harmonic Elimination Based on Newton-raphson Method for Cascaded H-bridge Multilevel Inverter International Journal of Power Electronics and Drive System", IJPEDS Vol. 8, No. 3, pp. 1193-1202, September 2017, DOI: 10.11591/ijpeds. v8i3.pp1193-1202.

[3] K. Ravi1 and T.Rammohan "Dwindling Of Harmonics In CML Inverter Using Genetic Algorithm Optimization", International Journal of Pure and Applied Mathematics Volume 117 No. 16, pp-757-764. 2017

[4] E. Anandha Banu and D. Shalini Punithavathani, "Real Time GA And Ann Based Selective Harmonic Elimination In 9 Level Ups Inverter", ICTACT Journal On Soft Computing, Volume: 06, Issue: 04, July 2016, DOI: 10.21917/ijsc.2016.0174.

[5] Kotb. M. Kotb, Abd Elwahab Hassan and Essam M. Rashad, "Selective Harmonic Elimination using Genetic Algorithm for An Asymmetric Cascaded Multilevel Inverter", 17th International Middle East Power Systems Conference, Mansoura University, Egypt, December 15-17, 2015.

[6] Faouzi Armi, Lazhar Manai, And Mongi Besbes, "Newton Raphson algorithm for Selective Harmonic Elimination in Asymmetrical CHB *Multilevel Inverter using FPGA"*, Proceeding Of Engineering And Technology (PET) pp. 887-894, 2016.

[7] Zhang Wenyi, Meng Xiaodan, and Li Zhenhua, "The Simulation Research for Selective Harmonic Elimination Technique based on Genetic Algorithm", Proceedings of The 33rd Chinese Control Conference, Nanjing, China, pp- 8628-8632, July 28-30, 2014.

[8] Concettina Buccella, Carlo Cecati, Maria Gabriella Cimoroni, and Kaveh Razi, "Analytical Method for Pattern Generation in Five-Level Cascaded H-Bridge Inverter Using Selective Harmonic Elimination", IEEE Transactions on Industrial Electronics, Vol. 61, No. 11, pp-5811-5819, November 2014.

[9] Mohamed S. A. Dahidah, , Georgios Konstantinou, and Vassilios G. Agelidis, "A Review of Multilevel Selective Harmonic Elimination PWM:Formulations, Solving Algorithms, Implementation and Applications", IEEE Transactions On Power Electronics, Vol.99. Early Access, pp-1-16, 2014 DOI:http//dx.doi.org/10.1109/TPEL.2355226p.

[10] Kishor Bommassani, and P. Ram Prasad, "Harmonic Reduction In Multilevel Inverter Based On GA Optimization", International Journal Of Scientific & Technology Research Volume 3, Issue 5, May 2014 pp-316-321.

[11] V. Joshi Manohar, P. Sujatha and K. S. R. Anjaneyulu, "Lower Order Harmonics Minimisation in CHB Inverter Using GA and Decomposition by WT", World Academy of Science, Engineering and Technology International Journal of Electrical and Computer Engineering Vol:7, No.5, pp-542-547, 2013.

[12] Ayoub Kavousi, Behrooz Vahidi, Reza Salehi, Mohammad Kazem Bakhshizadeh, Naeem Farokhnia, and S. Hamid Fathi, "Application of the Bee Algorithm for Selective Harmonic Elimination Strategy in Multilevel Inverters", IEEE Transactions On Power Electronics, Vol. 27, No. 4, pp-1689-1696, April 2012.

[13] Madichetty Sreedhar, Navin Mani Upadhyay and Sambeet Mishra", Optimized Solutions for an Optimization Technique Based on

Minority Charge Carrier Inspired Algorithm Applied to Selective Harmonic Elimination in Induction Motor Drive", IEEE, 1st International Conference on Recent Advances in Information Technology, 2012.

[14] Wenyi Zhang, Qiang Zhang, Wensheng Chen and Liuzhong Zhang, "Analyzing of Voltage-Source Selective Harmonic Elimination Inverter", Proceedings IEEE International Conference on Mechatronics and Automation, Beijing, China, pp-1888-1892, 7-10 August 2011.

[15] Sangeetha.S and S. Jeevananthan, "A Software Tool for Selective Harmonic Elimination in Multilevel Inverters Using Mathematica and Visual C++", IEEE conference organised by Institute of Technology, Nirma University, Ahmedabad, pp-1-6, 08-10 Dec. 2011.

[16] Jin Wang and Damoun Ahmadi, "A Precise and Practical Harmonic Elimination Method for Multilevel Inverters", IEEE transactions on Industry Applications, Vol. 46, No. 2, pp-857-865, March/April 2010.

[17] Mohamed S. A. Dahidah and Vassilios G. Agelidis, "Selective Harmonic Elimination PWM Control for Cascaded Multilevel Voltage Source Converters: A Generalized Formula", IEEE Transactions on Power Electronics, Vol. 23, No. 4, pp-1620-1630, July 2008.

[18] Vassilios G. Agelidis, , Anastasios I. Balouktsis and Mohamed S. A. Dahidah, "A Five Level Symmetrically Defined Selective Harmonic Elimination PWM Strategy: Analysis and Experimental Validation", IEEE Transactions on Power Electronics, Vol. 23, No. 1, pp-16-26, January 2008.

[19] Jagdish Kumar, Biswarup Das and Pramod Agarwal, "Selective Harmonic Elimination Technique for a Multilevel Inverter", Fifteenth National Power Systems Conference (NPSC), IIT Bombay, pp-608-613, December 2008.

[20] Sule Ozdemirl, Engin Ozdemirl, Leon M. Tolbert and Surin Khomfoi, "Elimination of Harmonics in a Five-Level Diode-Clamped Multilevel Inverter Using Fundamental Modulation", IEEE conference, PEDS, , pp-850-854, 2007.

[21] K. Al-Othman, Nabil A. Ahmed, A. M. Al-Kandari and H. K. Ebraheem, "Selective Harmonic Elimination of PWM AC/AC Voltage Controller Using Hybrid RGAPS", Approach International Scholarly and Scientific Research & Innovation, World Academy of Science, Engineering and Technology International Journal of Electrical and Computer Engineering Vol.1, No.5, pp-746-752, 2007.

[22] Tianhao Tang, Jingang Han and Xinyuan Tan, "Selective Harmonic Elimination for a Cascade Multilevel Inverter", IEEE ISIE, Montreal, Quebec, Canada, pp-977- 96,. 9-12 July 2006.

[23] Zhong Du, Leon M. Tolbert and John N. Chiasson, "Active Harmonic Elimination for Multilevel Converters" IEEE Transactions on Power Electronics, Vol. 21, No. 2, pp-459- 469, March 2006.

[24] Eryong Guan, Pinggang Song, Manyuan Ye and Bin Wu, "Selective Harmonic Elimination Techniques for Multilevel Cascaded H-Bridge Inverters", IEEE conference, 2005 [PEDS, pp-1441- 1446, 2005].

[25] Jason R. Wells, Brett M. Nee, Patrick L. Chapman and Philip T. Krein, "Selective Harmonic Control: A General Problem Formulation and Selected Solutions", IEEE Transactions on Power Electronics, Vol. 20, No. 6, pp-1337- 1345, November 2005.

[26] John Chiasson, Leon M. Tolbert, Keith McKenzie and Zhong Du, "A Complete Solution to the Harmonic Elimination Problem", IEEE pp-596-602, 2003.

[27] Li Li, Dariusz Czarkowsk, Yaguang Liu and Pragasen Pillay, "Multilevel Selective Harmonic Elimination PWM Technique in Series-Connected Voltage Inverters", IEEE Transactions on Industry Applications, Vol. 36, No. 1, pp-160-170, January/February 2000.

[28] Hamid R. Karshenas, Hassan Ali Kojori and Shashi B. Dewan, "Generalized Techniques of Selective Harmonic Elimination and Current Control in Current Source Inverter/Converters", IEEE Transactions on Power Electronics, Vol. 10, No. 5, pp- 566- 573, September 1995.

[29] Jian Sun and Horst Grotstollen, "Solving Nonlinear Equations for Selective Harmonic Eliminated PWM Using Predicted Initial Values", IEEE, International Conference on Industrial Electronics, Control, Instrumentation, and Automationpp-259- 264, 1992.

978-1-5386-6624-1/18 $31.00 © 2018 IEEE

An ANN Based Intelligent MPPT Control for Wind Water Pumping System

Neeraj Priyadarshi
Department of Electrical Engg.
*Birsa Institute of Technology
(Trust) Ranchi, India*
neerajrjd@gmail.com

Vigna K. Ramachandaramurthy
Power Quality Research Group,
Dept. of Electrical Power Engg.
*UniversitiTenagaNasional,
Malaysia.*
vigna@uniten.edu.my ,

Sanjeevi kumar Padmanaban
Dept. of Energy Technology
*Aalborg University
Esbjerg, Denmark*
san@et.aau.dk

Farooque Azam
Department of Comp.Sc.&Engg.,
*Millia Institute of Technology
Purnea, India*
farooque53786@gmail.com

Amarjeet Kumar Sharma
Department of Electrical Engg.
*Birsa Institute of Technology
(Trust) Ranchi, India*
maxeramar@gmail.com

J.P Kesari
Department of Mechanical Engg.
*Delhi Technological University
Delhi, India*
drjpkesari@gmail.com

Abstract—**This paper explains the application of artificial neural network (ANN) method for wind energy conversion system with water pumping supplications. A back propagation supervised learning algorithm is employed for complete estimation. The induction generator linked wind turbine with induction motor supplied power from inverter controlled through fuzzy-dSPACE is employed in this research work. MATLAB-dSPACE environment has been provided for justification of ANN modeling for wind estimation. Optimal wind generated power is achieved and water pumping operation is performed accurately.**

Keywords—*ANN, back propagation, dSPACE, wind speed.*

I. INTRODUCTION &ANN COMPUTATIONAL ALGORITHM USING BACK-PROPAGATION FOR WIND ENERGY SYSTEM

Because of the varying rotor velocity, an induction generator fed wind turbine is employed to generate relevant power compared to different generators used [1-5]. Two types of power electronics circuitries are used for smooth operation of MPPT as well as running water pump, respectively. Several MPPT tracking methods have been discussed in the literature [6-9]. The artificial neural network as wind trackers with fuzzy-dSPACE hardware board [2] has been implemented. Through inverter supplied supply, the induction motor coupled centrifugal pump provides required pumping illustrated with Fig 1.

Induction generator utilizes the generated mechanical power obtained from converted K.E from wind velocity [7-15]. Wind turbine generated power is described mathematically as:

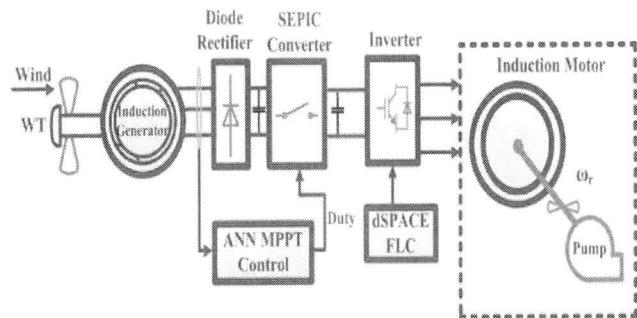

Fig. 1. Overall ANN MPPT wind scheme for water pumping

$$P_{windturbine} = \frac{1}{2}\rho_{air}V_{wind}^3\pi R_{Turbine}^2 C_{P.c}(1)$$

Where,

ρ_{air}= Density of air

$R_{Turbine}$= Radius of Turbine

$C_{P.C}$ = Coefficient of power

Also tip speed ratio is defined with relation:

$$\lambda_{tipspeedratio} = \frac{\omega_{Blade}*R_{Turbine}}{V_{wind}}(2)$$

Where,

$\lambda_{tipspeed\ ratio}$ = Tip speed ratio

ω_{Blade}= Angular blade velocity

V_{wind} = Wind velocity

978-1-5386-6624-1/18 $31.00 © 2018 IEEE

Power coefficient is expressed with relations who depend on classification of wind turbine.

$$C_{P.C} = C_A \left[\frac{c_B}{c_A} - C_C\beta - C_D\beta^y - C_E \right] e^{-C_F/\lambda_i} \quad (3)$$

Where,

$1/\lambda_i = (\lambda_K + 0.08)^{-1} - 0.035(1+\beta^3)^{-1}$

B = Pitch blade angle

A. Mathematical modeling of centrifugal pump

It is the most frequently used for the purpose of large water volume pumping which comprises better efficiency in which produced load torque is directly proportional to square of rotating speed of motor. Laws of similarity are employed for deducing the proper relation between pump flow, height and speed [10, 11]. Fig 2 depicts centrifugal pump model used for water pumping application.

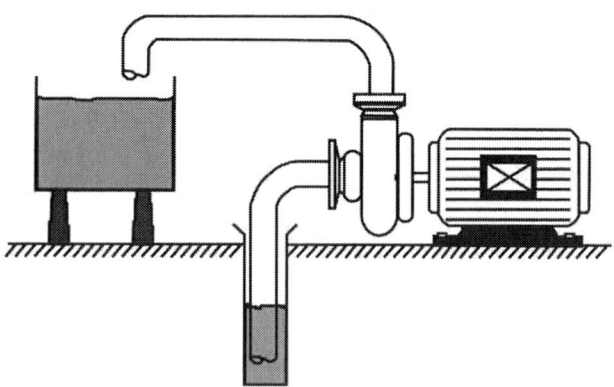

Fig. 2.Typical centrifugal pump

$$Q'_R = Q_R * \left(\frac{N'_R}{N_R} \right) \quad (4)$$

$$H'_{TT} = \left(\frac{N'_R}{N_R} \right)^2 * H_{TT} \quad (5)$$

$$P'_{TT} = \left(\frac{N'_R}{N_R} \right)^3 * P_{TT} \quad (6)$$

The sun insolation level and ambient temperature are the major factors on which the complete operation of induction motor depends. Based on equivalent circuit, the mathematical modeling of induction motor is carried out [10].

Electric power generated from motor pump can be expressed mathematically as:

$$P_{ELEC} = \frac{2.7250 * Q_R * H_T}{\eta_{Pump}} \quad (7)$$

Where,

η_{Pump} = Efficiency of motor pump

H_T = Nominal height

Q_R = Nominal pump flow

The centrifugal pump imposed load torque is matched with the motor torque [11]. The mathematical modeling of the system on rotor side can be expressed as

$$I_M \frac{d\omega}{dt} = (T_{EL} - T_{LOAD} - f * \omega) \quad (8)$$

Where,

I_M = Total machine inertia

T_{EL} = Electromagnetic torque

T_{LOAD} = Load torque

f = Friction coefficient

w = Machines mechanical speed

B. ANN Based Wind Energy Conversion System

The wind velocity is evaluated with ANN intelligent method. Rotor speed is calculated with the help of optimal tip speed ratio depicted by Fig 3. The turbine velocity and mechanical obtained power acts as inputs to the layer 1 in which back propagation feed forward ANN method for multilayer network is formed. These two inputs are evaluated for non-linear hidden layer which performs back computation in which weight moves opposite to forward gradient. MATLAB based training method is implemented with 20 neurons comprises hidden layer with rotor speed as only one output illustrated by Fig 4.

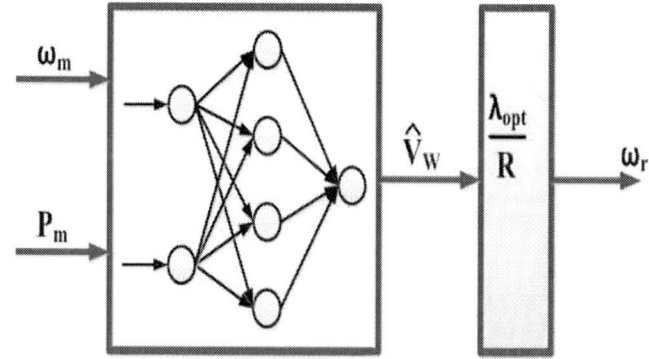

Fig. 3. ANN MPPT implementation

978-1-5386-6624-1/18 $31.00 © 2018 IEEE

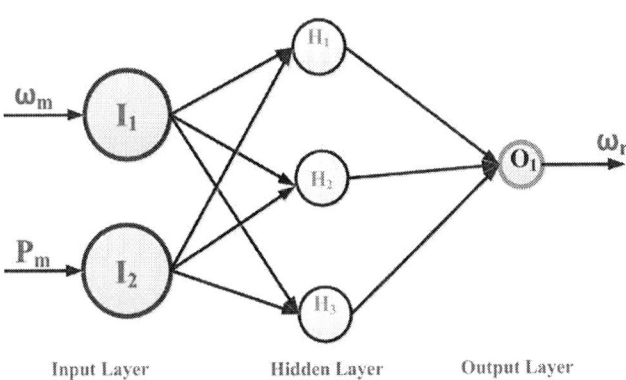

Fig. 4. ANN back propagation MPPT

The proper accuracy is improved by proper adjustment of weight of the layers and this modification is named training. Based on the minimized mean square error (MSE) achievement, the weights have been adjusted irrespective of variation of wind velocity. Fig 5 describes the supervised learning architecture in which global max. point and voltage at this point are adjusted according to changing sun variation to get optimal duty ratio. It lies in the 50 W/m^2 to 1200 W/m^2 and 5^0 C to 60^0 c sun insolation and temperature range, respectively.

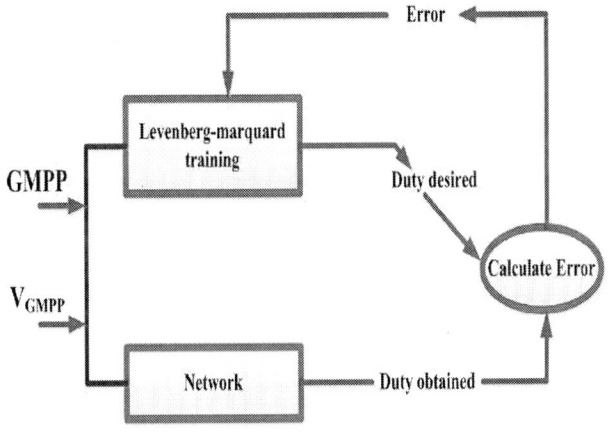

Fig. 5. ANN supervised learning architecture

Fig 6 illustrates the scheme of wind emulator controlled through ANN method. Rotor speed and generated mechanical power have been worked as input to ANN algorithm which produces reference rotor velocity and then translated to dSPACE platform to produce required control action to generator. The wind velocity is estimated easily with the application of ANN control from present metrological variables. It provides less sensor requirement for the proper placement of anemometer and performs better correlation between variables (input and output). In this work, sigmoid/hyperbolic gradient activation scheme is implementedfor modeling of neurons with 2 layersand 6neural arrangements. Hyperbolic gradient and linear neuron methods have been discussed for 1 and 2 layers, respectively and illustrated using Fig 7.

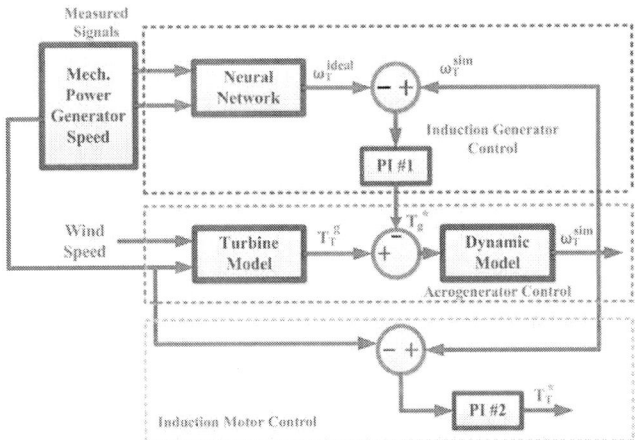

Fig. 6. Illustrates the scheme of wind emulator controlled through ANN method

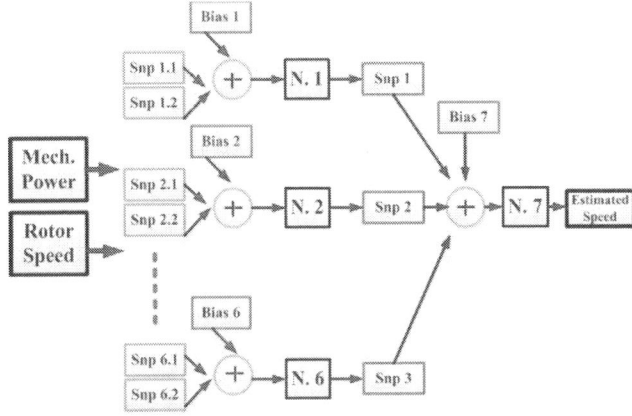

Fig. 7. Sigmoid/hyperbolic gradient activation scheme

Fig 9 demonstrates the dSPACE-MATLAB linked ANN implementation usingnntool which comprises 2 layers and 15 sigmoid neurons present in hidden layers. Back-propagation supervised feed forward method is employed for complete training process.Fig.8depicts the overall scheme of multilayer ANN architecture used for wind energy conversion system. Action function is calculated at T every step mathematically as:

$$Q_T = \sum_{j=1}^{N} W_{T,j}^{(2)} * D_{T,j} \ (9)$$

Where,

$$D_{T,j} = \frac{1}{1+e^{-H_{T,j}}} (10)$$

$$H_{T,j} = \sum_{i=1}^{M} W_{T,ji}^{(1)} * X_{T,i} (11)$$

Where,

N, M = Hidden input nodes

W= connection weight

978-1-5386-6624-1/18 $31.00 © 2018 IEEE   445

Updation in weight can be expressed as:

$$\Delta W_T = n_T[\gamma_{T+1} + \gamma_{A\epsilon a}^{max} Q_{T+1}(S_{T+1}, A) - Q_T] * \nabla_W Q_T (7)$$

Action vector gradient can be evaluated using back propagation training as:

$$\partial Q_T \big/ \partial W_{T,j}^{(2)} = D_{T,j} (12)$$

$$\partial Q_T \big/ \partial W_{T,ji}^{(1)} = W_{T,j}^{(2)} * D_{T,j}(1 - D_{T,j})X_{T,i} (13)$$

The eligibility trace matrix E_T can be expressed and updated as:

$$E_T = \nabla_W Q_T + \lambda_T \gamma E_{T-1} (14)$$

For implementation of wind maximum trackers, weights are initialized and online learning process has been started. Under wind stable condition, ANN based wind maximum trackers starts learning process. Current action and probability functions have been evaluated in every cycle. The induction generator based wind conversion scheme with ANN learning method is explained with Fig 10 which illustrates the overall online weight updation.

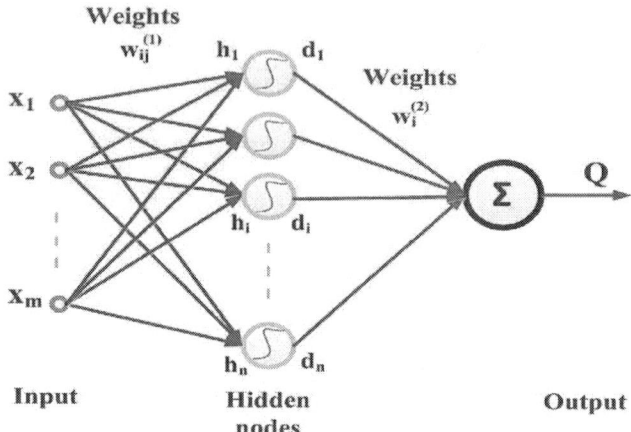

Fig. 8. Depicts the overall scheme of multilayer ANN architecture

Whenever the specifications of power characteristics changed by means of climate variation, the ANN online method started working and forces to new optimal power characteristics. highly fluctuating wind variations. The electrical generated wind power; power coefficient of wind turbine, rotor speed is validated under changing operated states. The sun insolation is also taken in consideration and centrifugal pump operation has been evaluated to calculate produced electromagnetic torque of induction motor with flow rate of coupled pump.

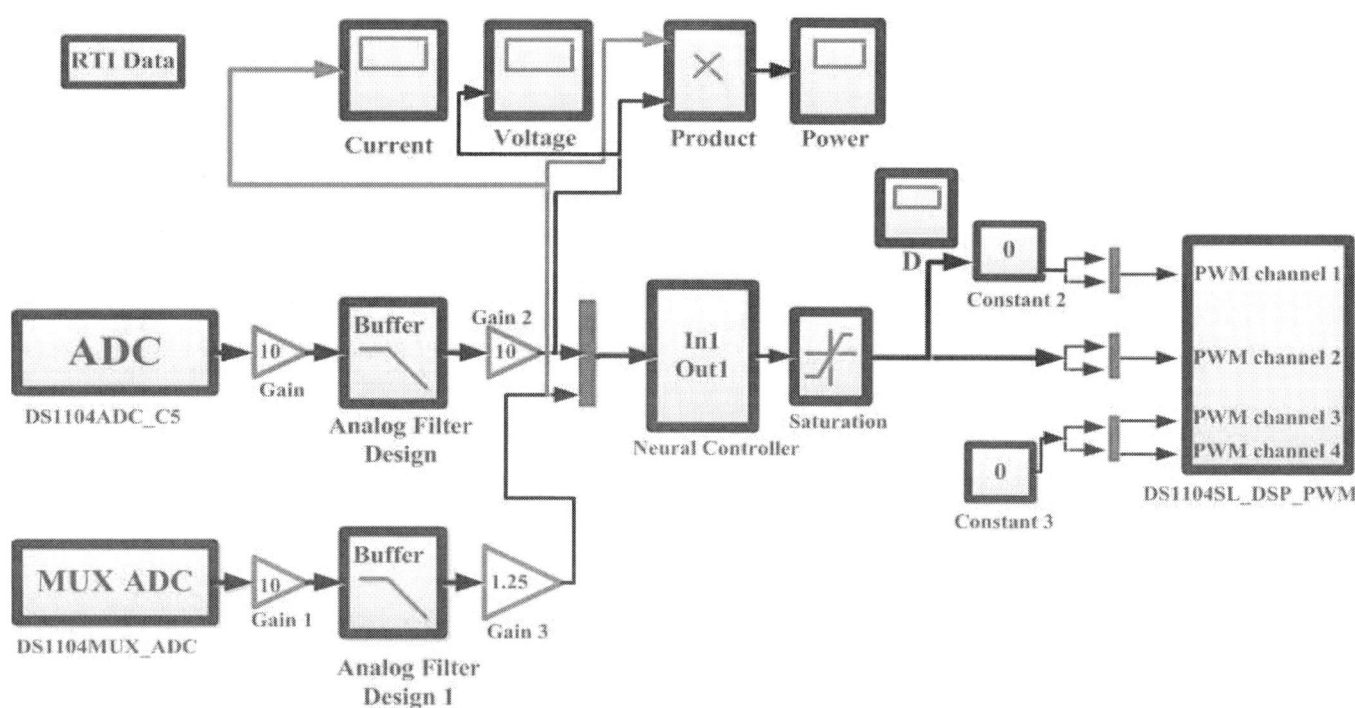

Fig. 9. Demonstrates the dSPACE-MATLAB linked ANN implementation using nntool

For implementation of wind maximum trackers, weights are initialized and online learning process has been started. Under

wind stable condition, ANN based wind maximum trackers starts learning process. Current action and probability functions have been evaluated in every cycle. The induction generator based wind conversion scheme with ANN learning method is explained with Fig 10 which illustrates the overall online weight updation. Whenever the specifications of power characteristics changed by means of climate variation, the ANN online method started working and forces to new optimal power characteristics. Complete wind energy power model has been designed by MATLAB simulation tool. The working model has been provided hardware interfacing to interpret proposed behavior with real time under step or highly fluctuating wind variations. The electrical generated wind power; power coefficient of wind turbine, rotor speed is validated under changing operated states. The sun insolation is also taken in consideration and centrifugal pump operation has been evaluated to calculate produced electromagnetic torque of induction motor with flow rate of coupled pump.

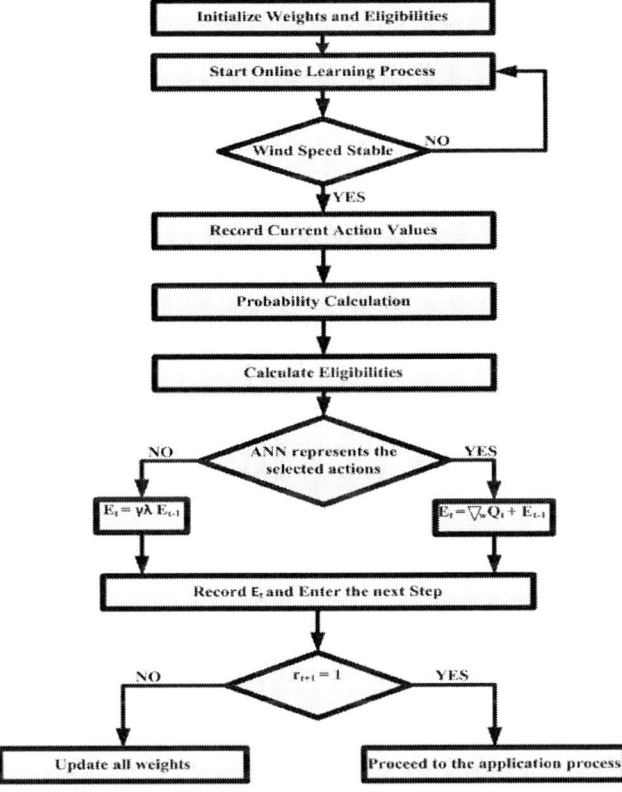

Fig. 10. Illustrates the overall online weight updation using ANN

II. MATLAB-dSPACE RESPONSES UNDER FLUCTUATING WIND VELOCITY WITH ANN CONTROLLER

Fig 11 presents the experimental responses of wind speed and maximum wind power extracted from turbine. The accurate and efficient wind power is found accordingly as the wind speed varies. The experimentation is performed to test the effectiveness of the proposed MPPT and inverter controller under different operating conditions. To check superiority of the proposed controller its responses are compared with conventional algorithms.

Under step change wind speed operating condition, the optimal convergence performance has been tested. The experimental wind speed, power coefficient, rotor speed and electrical power responses are presented in Fig.12 using proposed control.Here the wind speed varies between 6 m/s and 7 m/s for 30 seconds intervals.After 90 sec the C_P becomes close to the optimal value, since during this periodit gets

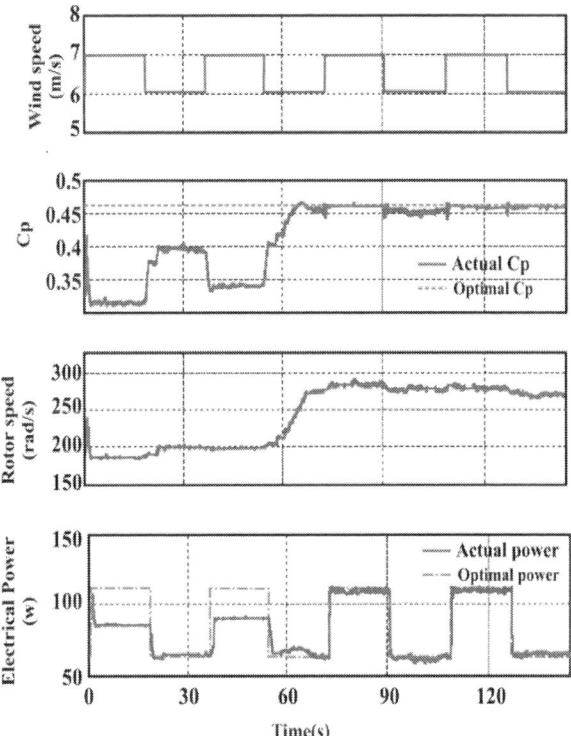

Fig. 11. Experimental responses of wind speed, C_P, Rotor speed and Electrical power with proposed control

experience as before 90 sec SCIG was very naïve. There is no absolute change in the rotor speed as controller has no idea where to go. Moreover, the actual and optimal power has large difference up to t=110 sec and after this period SCIG got experience and power coefficient reaches its optimal value. On the other hand, in case of conventional MPPT controller, the SCIG reaches MPP without preceding experience. There is a significant reduction in tracking efficiency and experimentally obtained output power using conventional control presented in Fig 12.

The performance of system has been investigated under varying sun insolation level. Fig 12 shows the variation of sun insolation with respect to time and its impact on induction motor driven centrifugal pump performance. It is found that the induction motor mechanical speed matched with its optimum speed. The experimental responses of electromagnetic torque explain the optimal behavior of IM which is depicted by Fig.12. Practical results also present the optimal flow rate of centrifugal pump. Presented characteristics explain the MPPT operation of centrifugal pump.

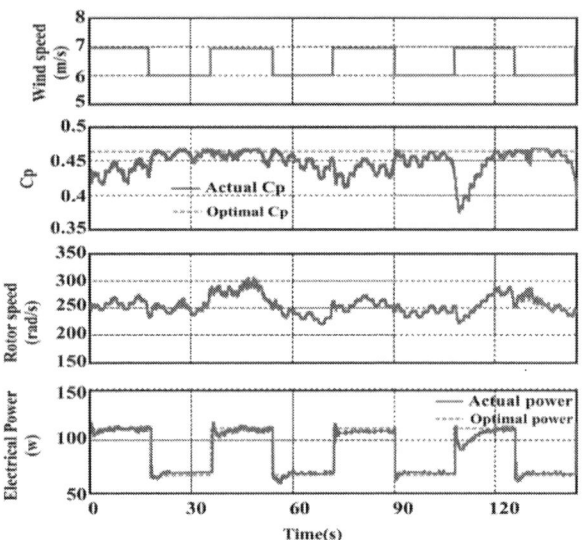

Fig. 12. Experimental responses of wind speed, C_P, Rotor speed and Electrical power with conventional control

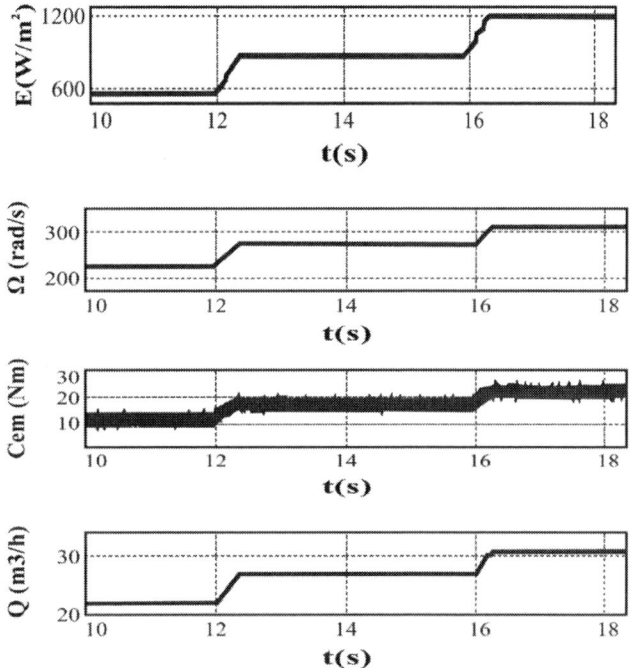

Fig. 12. Practical responses during varying sun insolation level: irradiance level, induction motor mechanical speed, electromagnetic torque and optimal flow rate of centrifugal pump

III. CONCLUSIONS

An ANN method for wind energy conversion system with water pump coupled induction motor is justified using MATLAB-dSPACE platform. SEPIC converter for impedance matched adapter has been implemented and required duty

ratio is generated under step wind velocity variation as well as fluctuation. Sensor less ANN control provides accurate influenced with acceptable estimation of induction motor coupled centrifugal water pumping supplications.

REFERENCES

[1] M. M. R. Singaravel and S. A. Daniel, "MPPT with Single DC-DC Converter and Inverter for Grid Connected Hybrid Wind-Driven PMSG-PV System," *IEEE Trans. on Ind. Electron.*,62.8(2015): 4849-4857.

[2] Y. -M. Chen, Y. –C. Liu, S. –C. Hung and C. –S. Cheng, "Multi-Input Inverter for Grid-Connected Hybrid PV/Wind Power System," *IEEE Trans. on Ind. Electron.*,22.3(2007): 1070-1077.

[3] R. G. Wandhare and V. Agarwal, "Novel Integration of PV-Wind Energy System with Enhanced Efficiency,*" IEEE Trans. on Power Electron.*,30.7(2015): 3638-3649.

[4] H. Geng, L. Liu and R. Li, "Synchronization and Reactive Current Support of PMSG based Wind Farm during Severe Grid Fault," *IEEE Transactions on Sustainable Energy*, 99(2018): 1-1.

[5] S. Das and B. Subudhi, "A H_{infty} Robust Active and Reactive Power Control Scheme for a PMSG based Wind Energy Conversion System," *IEEE Transactions on Energy Conversion,*99(2018): 1-8.

[6] J. Ahmed and Z. Salam, "An Enhanced Adaptive P&O MPPT for Fast and Efficient Tracking Under Varying Environmental Conditions," *IEEE Transactions on Sustainable Energy,* 99(2018): 1-10.

[7] X. Xiao, X. Huang and Q. Kang, "A Hill Climbing Method-Based Maximum Power Point-Tracking Strategy for Direct-Drive Wave Energy Converters," *IEEE Transactions on Industrial Electron.,*63.1(2016): 257-267.

[8] N. Kumar, I.Hussain, B. Singh, and B. K.Panigrahi, "Self-Adaptive Incremental Conductance Algorithm for Swift and Ripple Free Maximum Power Harvesting from PV Array," *IEEE Transactions on Industrial Informatics*, 99(2017): 1-10.

[9] N. Priyadarshi, A. K. Sharma, and F. Azam, "A Hybrid Firefly-Asymmetrical Fuzzy Logic Controller based MPPT for PV-Wind-Fuel Grid Integration," *International Journal of Renewable Energy Research,*7.4 (2017): 1546-1560.

[10] A. M. Kassem , "Modelling and robust control design of a standalone wind-based energy storage generation unit powering an induction motor variable-displacement pressure compensated pump ," IET Renewable Power Generation, IET Renew. Power Gener, pp. 275–286 Vol. 10, no. 3, 2016

[11] F. Mayssa, L. Sbita , " Advanced ANFIS-MPPT Control Algorithm for Sunshine Photovoltaic Pumping Systems ," First International Conference on Renewable Energies and Vehicular Technology, pp. 167-172, May. 2012.

2nd IEEE International conference on power Electronics, Intelligent Control and Energy systems (ICPEICES-2018)

Characterization Study of a Grid Connected DFIG Based WECS Under Variable Wind Speed and Loading Conditions

Amit Kumar Roy
Electrical and Instrumentation
Engineering
*Thapar Institue of Engineering
and Technology,* Patiala
Punjab, India
amit621988@gmail.com

Prasenjit Basak
Electrical and Instrumentation
Engineering
*Thapar Institue of Engineering
and Technology,* Patiala
Punjab, India
prasenjit@thapar.edu

Gyan Ranjan Biswal
Department of Electrical Engineering
and EEE
*Veer Surendra Sai University of
Technology,* Burla
Odisha, India
gyanbiswal@gmail.com

Abstract— Grid-connected doubly-fed induction generator (DFIG) based wind energy conversion system (WECS) is widely used in harnessing wind power. The paper attempts to characterize a 15- kW grid-connected DFIG based WECS operating under variable speed and loading conditions. Back to back converter topologies are utilized consisting of rotor side converter (RSC) and grid side converter (GSC) controlled in the synchronous reference frame coordinates. The design, modeling, and control of various system components are deliberated. The interaction of the grid with the proposed DFIG-WECS is analyzed where the regulation of the machine active power is achieved through the control of RSC. The current quality, voltage regulation is achieved by the control of GSC. Case studies involving constant wind speed, variable wind speed, and variable loading are undertaken for the characterization analysis. The performance characterizations of the entire system is validated in Matlab/Simulink simulation environment where the critical performance parameters like DC link voltage, generator torque, DFIG stator current, active and reactive power delivered by the DFIG to the grid, PCC side voltage/ current profile and the PCC current Total harmonic distortion (THD) are rigorously monitored, presented and discussed for each case studies.

Keywords—Wind energy conversion system, grid-connected, vector control, THD, DFIG.

I. INTRODUCTION

Replacement of conventional power plants with non-conventional counterpart is now an established trend. Non-conventional power plants have the advantages like high modularity, zero environmental hazards and low maintenance issues (in case of solar photovoltaic) [1]. This paper attempts to model a DFIG based wind energy conversion systems (WECS) and to explore its functionality under variety of possible system constraints. The related literature is deliberated covering the aspects of DFIG, its functional feature in grid-connected environment and the aspect of power quality induced by such systems. In reference [2], a DFIG-WECS composed of two back to back converter system connected by a capacitor is showcased. In [3], a control strategy using the combined reactive power compensation from both back-to-back power converters for their optimized lifetime distribution is presented. It is interesting to note that the power output of a conventional power plant can be controlled, whereas the power output of a WECS depends on the wind which attributes to a variable voltage and frequency at the output of the stator terminals [4]. As per the modern control strategies, a grid-connected DFIG-based WECS must remain connected to the grid

during faulty grid conditions. In [5], a modularized control strategy for providing fault ride through capability is presented where the back to back converters are controlled for reactive power compensation. Vector control remains a good choice for the control of WECS as the control scheme is flexible and provides multi-objective control functionality due to the provision of a transformation of the control signals in d-axis and q-axis respectively [6]. The vector control approach also has the advantage of low THD levels for the system current [11]. The control of WECS via direct torque control (DTC) proves flexible in the control of a PMSG based WECS where smooth control of stator flux and shaft torque [7].

DFIGs are preferred choice to be widely used in micro-grid environment [8], consensus-based algorithm proves effective in rendering load sharing and frequency regulation in a DFIG based micro-grid for the load sharing and frequency regulation in micro-grids [9]. The role of power converters is highly significant in a DFIG based WECS since wind speed is stochastic in nature which in turn affects the output voltage, power, and frequency in such systems. Hence, control of RSC is significantly important to capture the variable wind power and constantly track the maximum power possible from the available wind speed. The job of RSC also lies to maintain a constant frequency for the rotor injected voltages. The role of GSC is to propagate the generated power from the stator to grid at constant voltage amplitude and frequency [10].

An electrical power system contains a variety of loads such as balanced, unbalanced, linear, nonlinear etc; hence the aspect of good power quality is a prime concern in electric power systems. The same aspect is applicable for a DFIG-WECS. The aspect of power quality and its improvement in WECS supplied by DFIG is showcased in [11].

The main motivation of the paper is to characterize the functionality of DFIG- WECS when it is subjected to different operating scenarios by showcasing the various system parameters. The paper also attempts to investigate the following modalities; (a) to develop simulation model for the analysis of a DFIG based WECS, (b) to implement control techniques for GSC and RSC in order to regulate various system parameters, (c) to implement vector control scheme in order to achieve higher performance under variable loading and wind speed conditions. In Section II, the design of various system components is presented followed by the system modeling in detail has been deliberated in Section III. The control algorithm for RSC and GSC is highlighted in Section IV. Entire system performance is presented in Section V, followed by the conclusion in Section VII.

978-1-5386-6624-1/18 $31.00 © 2018 IEEE

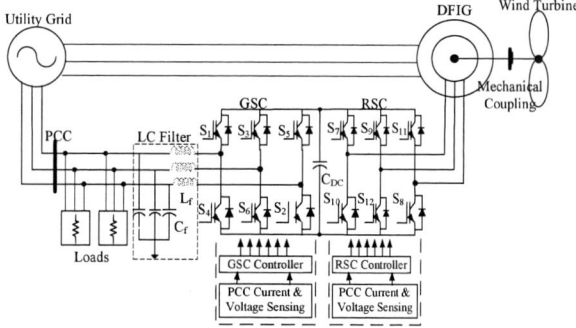

Fig. 1. Schematic of the DFIG based WECS.

II. System Architecture & Components

The system configuration on which the investigation to characterize the proposed DFIG based WECS is depicted in Fig. 1.

A. Wind Turbine

The ultimate function of the wind turbine is to capture the wind energy and provide mechanical output to the generator. Generally, a gearbox is used in between the coupling shaft connecting the turbine and generator whose function could be to provide stall control, pitch control and maximum power point tracking control. In this work a turbine rated a 15.5- kW is selected to render its nominal power at the base wind speed of 12 m/s. Pitch control is not performed, instead, a fixed pitch angle (β) of 0-degree is maintained throughout the operation.

B. Wind Generator

The function of a wind generator is to convert the mechanical input from the turbine to electrical output. A doubly fed machine proves to be an efficient choice for variable speed operations, simultaneously MPPT algorithms are well established in such generators. A machine of 15-kW is chosen for this work.

C. Controllers

A WECS has to perform the conditioning of AC power generated from the stator terminals to the electrical loads. However additional functionality such as maximum power point extraction, harmonic mitigations, grid synchronization, control of active and reactive powers are being performed by the converters namely RSC and GSC. The role of these converters changes intermittently depending on the speed of the wind turbine. When DFIG is run above synchronous speed GSC acts an inverter whereas the RSC acts as a rectifier. Below synchronous speed, the RSC converter the DC power to AC while, the functionality of the GSC is to convert the AC power to DC. The rating of the converters is determined from the rating of the DFIG-WECS, it is 25% of the DIFG rated power i.e, 3.75-kW.

D. DC Link Capacitor

A DC link is established at the intermediate junction of the GSC and RSC which is to be maintained at a fixed rated value with the help of capacitor known as DC link capacitor. The rated value of the DC link voltage is established as per

(1) where, the rated DC link voltage (V_{DC}) value is taken as 830V, modulation index is denoted by (m_f), and V_{RMS} denotes the AC side line- line voltage respectively. An important function of the DC Link capacitor is also to remove the chattering in the DC link voltage during operation. The value of the DC Link capacitor (C_{DC}) is determined on the basis rated DC power rating (P_{dc}), DC link voltage, allowable DC voltage ripple (ΔVdc) and the electrical angular frequency (ω).

$$V_{DC} = \frac{1}{m_f}\sqrt{\frac{2}{3}}V_{RMS}\;;\;C_{DC} = \frac{\left(\dfrac{P_{dc}}{V_{dc}}\right)}{\left(2\omega\Delta Vdc\right)} \qquad (1)$$

E. LC Filter

An LC filter is utilized to improve the power quality at the output of the GSC, such that quality power can be transmitted to the grid. The design of the LC filter is performed based on the equation (2) and (3).

$$\Delta I_{max} = 20\% * I_{rated} = \frac{1}{8} * V_{DC} * L_f * F_s \qquad (2)$$

$$\because Q = 15\% P_{rated}\;;\;C_f = \frac{Q_{rated}}{V^2(2\Pi f_{rated})} \qquad (3)$$

The derived values of L_f = 8.7mH and that of C_f is 3000μF respectively. The details of remaining system parameters are shown in Table. I.

III. System Modeling

A. Modeling of wind turbine

The mechanical power generated by wind turbine is given by (4)

$$P_m = \frac{1}{2}C_p(\lambda,\beta)\rho Av^3 \qquad (4)$$

Where, P_m is the mechanical power developed by the wind turbine, ρ is the air density in kilograms per cubic meter, A denotes blades swept area in meter square, v represents wind speed in meter per second, and C_p represents the power coefficient which is function of the tip speed ratio (λ) and pitch angle (β) [10].

TABLE I.   WECS COMPONENT RATINGS

Equipments/ Controller	Ratings/ Control Parameters
Wind Turbine	Pitch angle (β) = 0
	Wind Speed (V_w) = 12 m/s
	Rated Power = 15.5-kW
	Optimal Tip Speed Ratio (λ_{opt}) = 8.1
DFIG	Rated Power = 15-kW
	Voltage (L-L) = 415V
	Pole Pairs = 3
	Inertia Constant = 0.385
RSC	K_p = 0.6 , K_i = 8
GSC	K_p = 350, K_i = 3500
Grid	Voltage (L-L) = 415V, Frequency = 50Hz
DC Link Reference Voltage	830V
DC Link Capacitor	9000μF

978-1-5386-6624-1/18 $31.00 © 2018 IEEE

B. Modeling of DFIG

The purpose of modeling a DFIG system is to understand its various electrical parameters involved. In order to estimate the variations of numerous physical parameters like machine flux, stator and rotor currents, active and reactive power the modeling of DFIG is presented with the help of equivalent circuit model and various mathematical equations. The modeling is done in synchronously rotating reference frame where the synchronous reference frame quantities rotate synchronously with the stator flux. The stator and rotor voltages in (dq-axis) frame are given as per (5-8) respectively.

$$v_{ds} = r_s i_{ds} + \frac{d\lambda_{ds}}{dt} - \omega_e \lambda_{qs} \tag{5}$$

$$v_{qs} = r_s i_{qs} + \frac{d\lambda_{qs}}{dt} + \omega_e \lambda_{ds} \tag{6}$$

$$v_{dr} = r_r i_{dr} + \frac{d\lambda_{dr}}{dt} - (\omega_e - \omega_r)\lambda_{qr} \tag{7}$$

$$v_{qr} = r_r i_{qr} + \frac{d\lambda_{qr}}{dt} + (\omega_e - \omega_r)\lambda_{dr} \tag{8}$$

Where V_{ds}, V_{qs} are the stator voltage in synchronous rotating frame, i_{ds}, i_{qs} is the stator currents in dq frame. The supply and rotor angular frequency is represented by ω_e and ω_r respectively. λ_{ds} and λ_{qs} represents the stator flux linkage in dq- axis respectively. While in the rotor side, V_{dr}, V_{qr} are the rotor voltages in synchronous rotating frame. i_{dr}, i_{qr} are the rotor currents in dq frame. λ_{dr} and λ_{qr} represents the rotor flux linkage in dq- axis respectively. The stator and rotor resistances are represented by r_s and r_r respectively. An important aspect lies to control the active and reactive power from the DFIG stator terminals. This is conveniently done by the RSC and GSC controllers. As per the mathematical modeling the stator active (Ps) and reactive power (Qs) is given by (9) and (10) respectively.

$$P_s = \frac{3}{2}\left(v_{ds}i_{ds} + v_{qs}i_{qs}\right) \tag{9}$$

$$Q_s = \frac{3}{2}\left(v_{qs}i_{ds} - v_{ds}i_{qs}\right) \tag{10}$$

Even though the dependency of P_s and Q_s is seen between V_{ds},V_{qs}, i_{ds} and i_{qs} but it is indirectly related to the rotor currents i_{dr} and i_{qr} respectively. The same is enumerated in (15) and (16) respectively. Since the stator and rotor flux linkages in the dq- axis can be expressed as per (11) and (12) where, L_s and L_r denotes the stator inductance and rotor self inductances and L_m represents the mutual inductance respectively.

$$\lambda_{dqs} = L_s i_{dqs} + L_m i_{dqr} \tag{11}$$

$$\lambda_{dqr} = L_r i_{dqr} + L_m i_{dqs} \tag{12}$$

On fetching the stator dq-axis current from (11) and (12) and expressing them in terms of flux linkages, stator self inductance, mutual inductance and rotor currents in dq-axis coordinates leads to following expression as per (13)-(14) respectively.

$$i_{ds} = \frac{1}{L_s}\left[\lambda_{ds} - L_m i_{dr}\right] \tag{13}$$

$$i_{qs} = \frac{1}{L_s}\left[\lambda_{qs} - L_m i_{qr}\right] \tag{14}$$

It is interesting to note that upon substituting the values of i_{ds} and i_{qs} in the expression of P_s and Q_s gives the infernce about the controllabily aspect of the stator power with the control of rotor currents which is expedited as per (15) and (16) respectively.

$$P_s = \frac{3}{2}\left[\frac{1}{L_s}\left(v_{ds}\lambda_{ds} + v_{qs}\lambda_{qs}\right) - \frac{L_m}{L_s}\left(v_{ds}i_{dr} + v_{qs}i_{qr}\right)\right] \tag{15}$$

$$Q_s = \frac{3}{2}\left[\frac{1}{L_s}\left(v_{qs}\lambda_{ds} - v_{qs}\lambda_{qs}\right) - \frac{L_m}{L_s}\left(v_{qs}i_{dr} - v_{ds}i_{qr}\right)\right] \tag{16}$$

IV. CONTROL ALGORITHMS

This section showcases the major control algorithms associated for the operation of the DFIG - WSEG system. The working of various controllers namely maximum power point tracking (MPPT) controller, RSC and GSC controllers are briefly discussed with the supporting mathematical equations. They incorporated and validated in the Matlab/Simulink software and its operation is deliberated in section V.

A. Maximum Power Point Tracking

The MPPT control is embedded in the RSC which performs the tracking of reference mechanical turbine speed corresponding to the maximum power coefficient. From the general Cp vs λ curve [8], the maximum value of the power coefficient is achieved for β=0 corresponding to the value of λ=8.1, this value of λ is termed as optimal tip speed ratio (λopt).

B. RSC Control Algorithm

RSC control scheme is based on the vector control scheme [10]. Here, the control algorithm for RSC is implemented as per Fig. 2. The major agenda lies to fetch the MPPT from the turbine power and speed curve. The d-axis current reference controls the generator torque hence the power is indirectly controlled. The q-axis current reference for the rotor is determined from the stator's desired reactive power. In the present work unity power factor operation is intended; hence the reactive power reference is set to zero. The reference angular speed (ω_r^*) is generated from the optimal λ as per (17), which further generates the reference electromagnetic torque (T_e^*) as per (18) after the error between the reference and actual electromagnetic torque is processed by the PI controller.

$$\omega_r^* = \frac{V_w \cdot \lambda_{opt}}{r} \tag{17}$$

$$T_e^* = K_p(\omega_r^* - \omega_r) + K_I \int (\omega_r^* - \omega_r) \tag{18}$$

The reference torque helps in determining the reference d-axis rotor current (i_{dr}^*) provided the values of L_m, L_s and the q-axis flux (λ_{sq}) are known [3]; the corresponding mathematics is as per (19).

978-1-5386-6624-1/18 $31.00 © 2018 IEEE

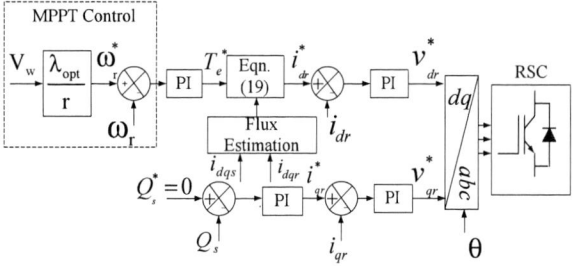

Fig. 2. Schematic of the RSC control system.

$$i_{dr}^* = \frac{T_e^*}{-\left(\dfrac{3}{2}\right)\left(\dfrac{L_m}{L_s}\right)\lambda_{sq}} \qquad (19)$$

The reference current for the q-axis is generated based on the knowledge of the reference reactive power. The error between the reference and actual reactive power is processed by the PI controller as per (20) which are further processed in the PI controller to generate the reference q-axis voltage.

$$i_{qr}^* = K_p(Q_s^* - Q_s) + K_I\int(Q_s^* - Q_s) \qquad (20)$$

C. GSC

The objective of the GSC is to maintain the DC link voltage at a constant value and to exchange the active and reactive power with the grid. The control algorithm of the GSC is implemented as per Fig. 3. The d-axis and q-axis stator voltages are regulated by their corresponding current references which are in turn derived by the reference DC link voltage (V_{DC}^*) and q-axis current reference (I_{qs}^*). The error between the reference and actual DC link voltage is processed by the PI controller which in turn generates the d-axis current reference (I_{ds}^*) expressed as per (21) and the value of I_{qs}^* set to zero for unity power factor operation. Further, the d-axis voltage reference (V_{ds}^*) and q-axis voltage reference (V_{qs}^*) for the GSC are generated as per (22) and (23) respectively.

$$I_{ds}^* = K_p(V_{DC}^* - V_{DC}) + K_i\int(V_{DC}^* - V_{DC}) \qquad (21)$$

$$V_{ds}^* = K_p(I_{ds}^* - I_{ds}) + K_i\int(I_{ds}^* - I_{ds}) - \omega L_f I_{qs} + V_{ds} \quad (22)$$

$$V_{qs}^* = K_p(I_{qs}^* - I_{qs}) + K_i\int(I_{qs}^* - I_{qs}) + \omega L_f I_{ds} + V_{qs} \quad (23)$$

Fig. 3. Schematic of the GSC control system.

V. RESULTS AND DISCUSSIONS

The modeling, control and design of the discussed DFIG-WECS are validated in Matlab/Simulink based simulation platform. The simulation schematic is presented as per Fig. 4. The system is being validated for the set following case studies namely Case-1, Case-2, Case-3 and Case-4 respectively. In case-1, system dynamics for constant wind speed and constant load is observed, in case-2 the system behavior for variable wind and constant load is performed. Case-3 deals with the system performance for variable wind and constant load. Lastly, WECS response for simultaneous variation in wind speed and load is dealt in case-4.

A. Case-1: Constant wind and constant load

The system performance of the DFIG-WECS for constant wind speed of 12- m/S and for the constant linear load of 10- kW is performed and various monitored parameters are shown as per Fig. 5. The velocity of wind, turbine torque, turbine speed, actual value of stator voltage, currents are presented as per Fig. 5 (a), (b), (c), (d), and (e) respectively. The profile of power delivered from the stator terminals, DC link voltage, PCC side voltage and current are illustrated as per Fig. 5 (f), (g), (h) and (i) respectively.

Since the wind speed is constant, the generator receives an uniform torque and experiences uniform angular speed. The stator terminals votage are maintained at the rated value while delivering a steady power of 11-kW as per Fig. 5 (f). The profile of DC link voltage is maintained constant at 830 V as per Fig. 5 (g) which justifies the apt working of the GSC. Simultaniously the PCC side voltage is maintained at rated value of 1-P.U. The aspect of power quality is enlightned where the THD analysis of the grid current is shown as per Fig. 6. As per the IEEE Std. 1547 the current THD at the PCC must be less than 5%. The proposed and designed DFIG- WECS is able to maintain its current THD at 4.39% which is well with in the acceptable limit.

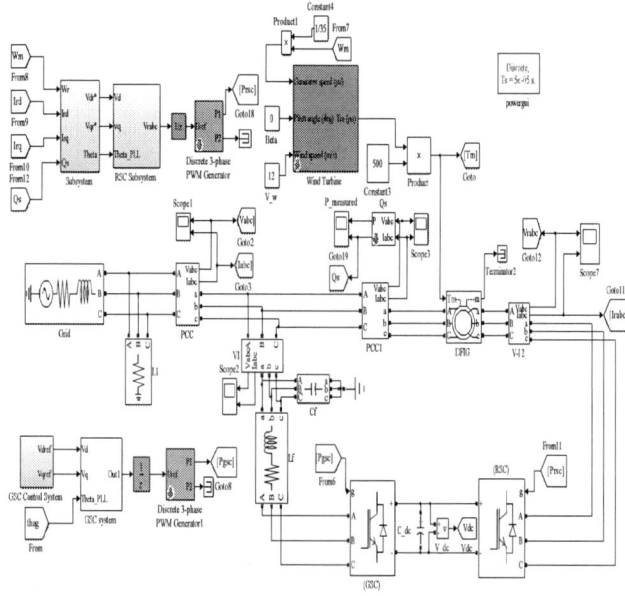

Fig. 4. Matlab/Simulink schematic of the proposed DFIG-WECS.

978-1-5386-6624-1/18 $31.00 © 2018 IEEE

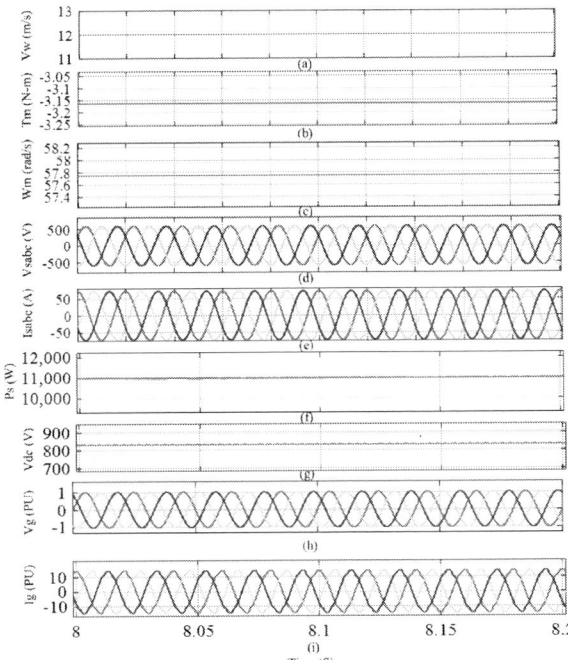

Fig. 5. DFIG-WECS performace under constant wind and load condition.

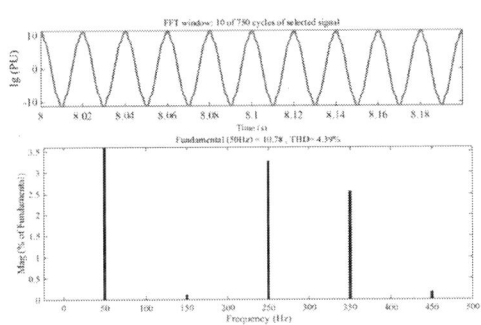

Fig. 6. THD analysis of the grid current.

B. Case-2: Constant wind and variable load

This subsection attempts to demonstrate the aspect of variable loading occurring at the PCC. It is interesting to observe the grid interaction with the DFIG- WECS when the DFIG- WECS is serving its power to the various loads. For the purpose of validation, the load connected at the DFIG terminals is kept more than the rating of DFIG. A loading up to 220-kW is performed in order to observe the self regulating capability of the utility grid to provide the excess power to the load. As per the profile of Fig. 7 (j), the machine is subjected to variable loading where the load power keeps changing in steps at the time instants 4-S, 6-S and 8-S respectively. The stator power from the machine is maintained constant due to the power control regulation performed by the RSC which maintain its power at 11-kW as per Fig. 7 (i). The power delivered by the grid is reduced according to the power demand of the load such that the power delivered by the DFIG and the grid is maintained equal to that of the load at every time instant; it is highlighted as per Fig. 7 (k).

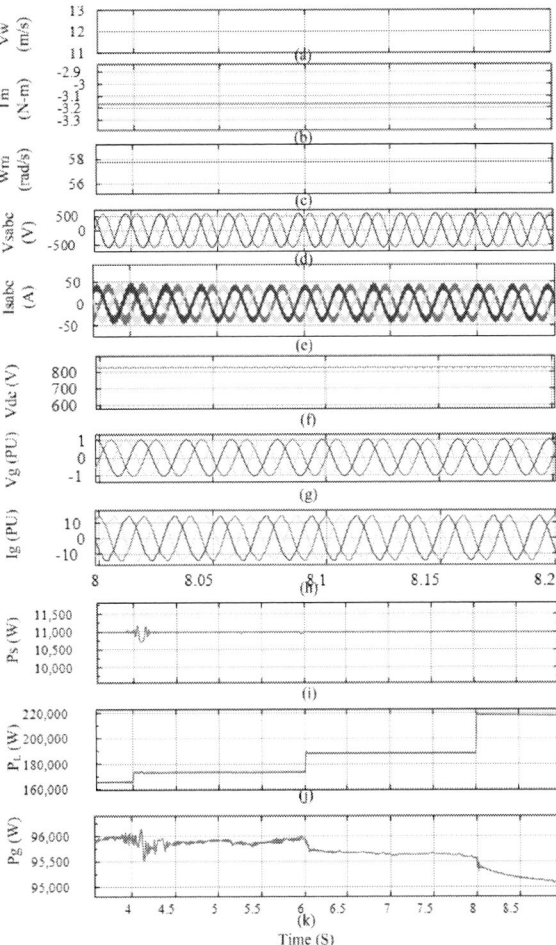

Fig. 7. DFIG-WECS performace under constant wind and variable load condition.

C. Case-3: Variable wind and constant load

Wind generation being sochastic in nature has the probabilty of vartions from its rated speed. A DFIG based WECS has the advantage of variable speed operation capabilty, this very aspect is present in this case study. The role of MPPT in the RSC becomes significant during variable wind speed operations. The preformnce of the discussed DFIG- WECS is illustrated as per Fig. 8 where the various system parameters for variable wind speed operations at 10 m/S, 12 m/S and 11 m/S are presented. Due to the wind speed variations, notches in the turbine machanical torque are prominent as per Fig. 8 (b). The tubine speed variation is seen during wind speed change instances as per Fig. 8 (c). The current at the stator terminals varies which in turn varies the stator terminal power. The machines stator terminal power is regulated as per the wind speed speed which is presented as per Fig. 8 (f), the DC link voltage is maintained at its reference value even during variable wind speed instances due to effective GSC control.

D. Case-4: Variable wind and variable load

Further charachterization of the DFIG based WECS is carried out in where the wind velocity and load changes are being performed simultaniously. The variation in wind prof-

978-1-5386-6624-1/18 $31.00 © 2018 IEEE   453

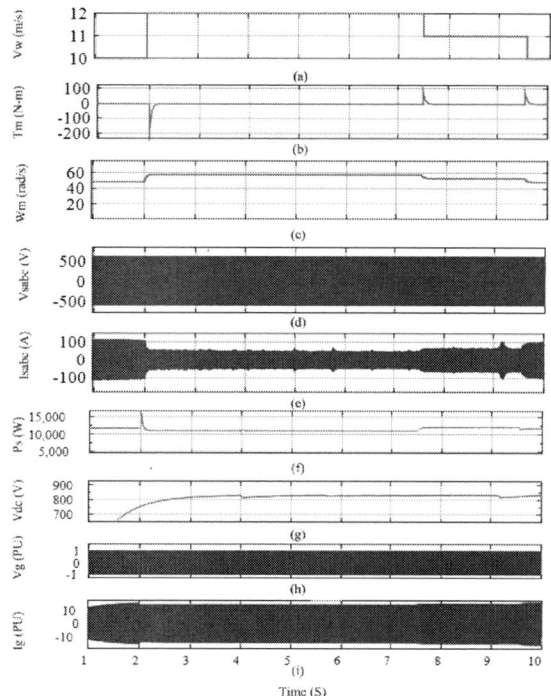

Fig. 8. DFIG-WECS performace under constant wind and variable load condition.

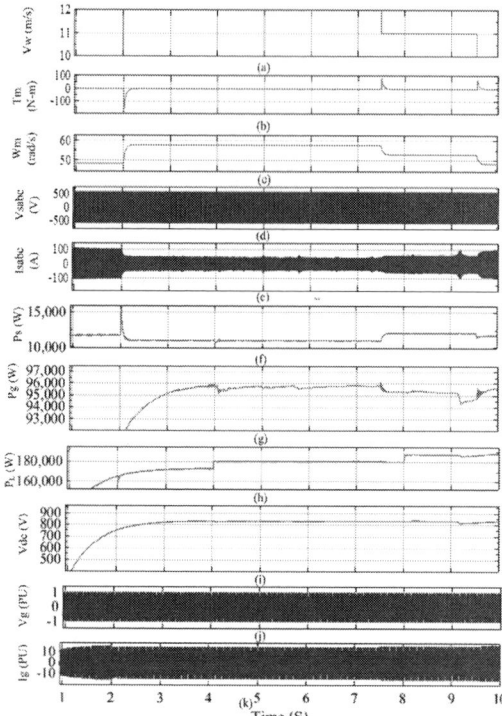

Fig. 9. DFIG-WECS performace under constant wind and variable load condition.

-ile is as per Fig. 9 (a), while the load switching instances are being performed at t=4-S and at t=8-S respectively. The variation in the load power (P_L) is as per Fig. 9 (h). It is compelling to note the power balancing aspect between the power delivered by the stator (P_S), power delivered by the grid (P_g) such that the load power is always balanced. The effective control action of the GSC is justified by visualizing the Vdc as per Fig. 9 (i), where it is controlled at its reference voltage. Also, the constant voltage profile maintained at the stator terminals and at the PCC terminals as per Fig. 9 (d), (j) justifies the efficacy of the GSC control.

VI. CONCLUSION

The paper highlighted the aspect of modeling, design and control of a 15-kW DFIG based WECS. The aspect and importance of MPPT control, RSC control and GSC control are showcased with the support of various mathematical equations. Vector control scheme is adopted for the rotor side controllers where the d-axis and q-axis currents controls the active and reactive power of the machine respectively. The system is operated in the grid connected mode where the power exchange between the machine and grid is done by the GSC, simultaniously the DC link voltage is maintained constant. Various case sudies involving, constant wind speed, variable wind speed, and variable loading conditons are performed where it is demonstrated that effective control of the DFIG-WECS is achieved. System parameters like turbine torque, stator terminal voltage and current, power delivered by the stator, DC link voltage are extensievly presented for each case studies. The systems grid interactive capability proves to be an added advantage when the system loading is beyond the DFIGs rated power rating. More over the apt design of LC filter maintains the grid current quality as per the IEEE reccomended practices.

REFERENCES

[1] S. Muller, M. Deicke, and R. W. De Doncker, "Doubly fed induction generator systems for wind turbines," *IEEE Ind. Appl. Mag.*, vol. 8, no. 3, pp. 26–33, May/Jun. 2002.

[2] A. Tapia, G. Tapia, J. X. Ostolaza and J. R. Saenz, "Modeling and control of a wind turbine driven doubly fed induction generator," *IEEE Trans. Energy Convers.*, vol. 18, no. 2, pp. 194-204, June 2003.

[3] B. Singh, S. K. Aggarwal and T. C. Kandpal, "Performance of Wind Energy Conversion System Using a Doubly Fed Induction Generator for Maximum Power Point Tracking," in *Proc. IEEE Industry Applications Society Annual Meeting*, Houston, TX, 2010, pp. 1-7.

[4] Y. Lei, A. Mullane, G.Lightbody, and R. Yacamini, "Modeling of the wind turbine with a doubly-fed induction generator for grid integration studies," IEEE Trans. Energy Conversion, vol. 21, no. 1, pp. 257-264, Mar. 2006.

[5] D. Zhou, Y. Song and F. Blaabjerg, "Modern control strategies of doubly-fed induction generator based wind turbine system," *Chinese Journal of Electrical Engineering*, vol. 2, no. 1, pp. 13-23, June 2016.

[6] E. Aydin, A. Polat and L. T. Ergene, "Vector control of DFIG in wind power applications," in *Proc IEEE Int. Conf. on Renewable Energy Research and Applications (ICRERA)*, Birmingham, 2016, pp. 478-483.

[7] S. Y. Yang, Y. K. Wu and H. J. Lin, "The direct torque control of the PMSG based windturbine with two level voltage source converter," 2014 CACS International Automatic Control Conference (CACS 2014), Kaohsiung, 2014, pp. 39-44.

[8] G. Ramtharan, J. B. Ekanayake, and N. Jenkins, "Frequency support from doubly fed induction generator wind turbines," *IET Renewable Power Generation*, vol. 1, no. 1, pp. 3–9, 2007.

[9] W. Zhang, Y. Xu, W. Liu, F. Ferrese and L. Liu, "Fully Distributed Coordination of Multiple DFIGs in a Microgrid for Load Sharing," *IEEE Trans. Smart Grid*, vol. 4, no. 2, pp. 806-815, June 2013.

[10] V. C. Ganti, B. Singh, S. K. Aggarwal and T. C. Kandpal, "DFIG-Based Wind Power Conversion With Grid Power Leveling for Reduced Gusts," *IEEE Trans. on Sustain. Energy*, vol. 3, no. 1, pp. 12-20, Jan. 2012.

[11] S. Mathur, R. Verma and B. Sharma, "Calculation of THD in vector controlled IGBT fed doubly fed induction generator," in *Proc Int. Conference on Electrical Power and Energy Systems (ICEPES)*, Bhopal, 2016, pp. 22-26.

Economic Scheduling of Hybrid Microgrid using planned management scheme

Iram Akhtar
Department of Electrical Engineering
Faculty of Engineering & Technology
Jamia Millia Islamia
New Delhi, India
iram1208@gmail.com

Sheeraz Kirmani
Department of Electrical Engineering
Faculty of Engineering & Technology
Jamia Millia Islamia New Delhi,
India sheerazkirmani@gmail.com

Majid Jamil
Department of Electrical Engineering
Faculty of Engineering & Technology
Jamia Millia Islamia
New Delhi, India
majidjamil@hotmail.com

Abstract- **Due to lack of fossil fuel, Microgrid with hybrid energy system and interconnection of different types of loads is rapidly developed in the recent years. In this respect, this work introduces the economic scheduling of Microgrid with renewable energy resources. The power level of different energy resources is measured and compared with the power which is needed for successful operation. In the case of an overvoltage, the system automatically operates the load according to the voltage capacity, but if any source is not available then the system automatically shifts on the AC source supply. This can be explained as the first load is satisfied by solar system till its maximum capacity is reached hence the load is shifted to wind system followed by the combination of both solar and wind then finally by the utility (main grid). If generating power from the renewable source is more than the demand then surplus power will be sold out to the utility. For the economical purpose, the single Lead acid battery is used which is charged by a solar panel and wind turbine. A microgrid which is integrated with the solar system, wind system, and main supply is investigated by PROTEUS Software. Therefore an idea of non-conventional energy generation and control the grid automation system has been given. So the whole system is designed in such a manner that it is economical as well as efficient.**

Keywords - **Solar energy, Wind Energy, Microcontroller, AC grid, Proteus Software**

I. INTRODUCTION

Nowadays, the economic scheduling of generation of power plays an important role because Coal cannot survive for long times to generate electrical power in Thermal Power Plant. A significant amount of money is needed for the transportation of coal and other requirements. It is very costly to keep it maintaining the whole system. Almost the same problems occur with Hydro Power Plant, here water, the primary source of generation is available free of cost but only at remote places. Whereas for transmitting the power to long transmission lines so there can be a large amount of power loss which is the main cause of catching fire of forests. It also increases the initial cost of installation of Hydro Power Plant. These are the main reasons for switching into the Solar Power Generation and Wind Power Generation. Because of different intermittent characteristics of the hybrid energy system, reliability is the big issue to cope up with this issue real-time energy management system constitutes with energy storage system. The total cost of energy is to be minimized by optimizing the generation of energy resources[1]. To enhance the economic operation of
a

microgrid, the dc control system is adopted. The result of dc microgrid is compared with a microgrid which shows that dc microgrid strategy for optimal operation of the hybrid energy system is an effective way of ensuring the response in a steady state [2]. Multi-period islanding constraints are very important for optimal scheduling and the main objective is to minimize the total operation cost which includes local resources generation cost and energy purchase cost from the main grid. If the sufficient generation is not got then the master problem is formed to examine the whole situation [3]. Recently, a hybrid energy system is intensively exploited because of high energy demand.AC/DC hybrid method enables the future integration of different energy resources for the smart grid. This needs an efficient power tool to determine the operation of such a hybrid energy system. The algorithm includes the specific characteristics of the hybrid system is to be developed [4]. A microgrid is very important for future of smart grid that includes all renewable energy resources. The major concerns for microgrids are frequency regulation and optimal scheduling. An integrated framework is developed to achieve minimum cost while considering the frequency regulation constraints [6]. Energy trading among microgrids is another matter of concerns so different distributed mechanism has to be developed. For this multiple microgrids have to use whereas some of them have the superfluous energy for storage or for sale and some other need to purchase some additional energy to satisfy the local demand. Buyers can submit independently unit price bid to the different sellers. Energy is distributed to the different buyers according to their bids [8]. It is very important to balance the generation and demand in the power system. The battery plays important role in the integration of renewable energy sources because it has the fast capability of storage of energy. Stochastic unit commitment and solution from this commitment is used for energy storage scheduling [9]. A control of grid voltage is necessary to know the grid power and instantaneous power sharing. The main aims to minimize the stored energy-related losses so the number of conversion steps is also reduced. A droop-voltage power-sharing scheme is employed in which the voltage of bus drops in retort to low supply and high demand [10].An introduction of distributed economic dispatch strategy with multiple energy storage systems for microgrids. It overcomes the tasks of dynamic couplings among all decision variables and stochastic variables. It can be used as a central controller as well as the local controller depending upon the needs [11]. The real-time energy management is needed for stand-alone microgrid which includes wing system, diesel generator, storage devices etc. This method is based on hour ahead wind speed forecast and this determines the reference output power of the wind turbine and the operation mode of the microgrid [12]. An online optimal energy and optimal power control method are used for energy storage operation in microgrid hybrid system. A mixed-integer linear program optimization is formulated for prediction of future electricity uses and renewable energy generation. Performance objectives comprise of electricity usage cost, operation costs of battery, and oriented goals of utility associated with the peak demand [13]. When the penetration of renewable energy is high then energy storage is very important to assure the stable operation in an isolated power grid or isolated microgrids. Each type of energy storage has different characteristics, which includes the round-trip efficiency, power, and energy rating etc [15].

In this paper, the Microgrid hybrid energy system scheme is described in Section II. In Section III, economic scheduling scheme, Automated Economic Scheduling test execution flow, and HIL simulator framework are presented. Results and discussion are described in Section IV. Finally, concluding statements are presented in Section V.

II. HYBRID ENERGY SYSTEM SCHEME

The electrical energy is generated from different energy resources such as solar energy and wind energy. This energy is stored in a battery bank and is applicable in industrial automation and load scheduling according to supply available. An idea of non-conventional energy generation and control the grid automation system has been given.Fig.1. Shows the hybrid energy system scheme in which algorithm is developed to decide which source would supply the loads.

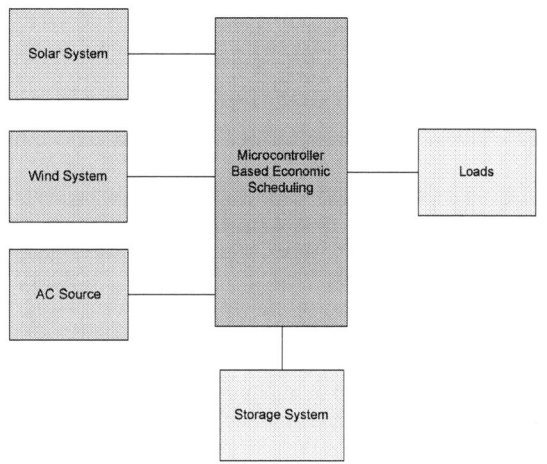

Fig.1. Block diagram of Hybrid Energy System Scheme

Solar, wind and AC source are connected to the Microcontroller for economic operation of the Microgrid hybrid energy system. In this system, an LCD has been installed which shows the status of the resources on which the system works at that moment along with the absence of resource. This system checks the phase and switches the load on the correct phase and operated device reaches in protection mode. At different moments the amount of generated energy with solar panel and wind generation may be different, if we use this energy directly along the load side the fluctuations in frequency and terminal voltage will be very high, hence to overcome with this problem battery bank has been used where the energy generated from these two sources is first stored and then transmitted to the load side by passing through the required conversion mechanisms. In this paper, Simulation of the microgrid which is integrated with a solar panel, wind turbine, and main supply is done by PROTEUS. The circuit is designed in such a manner that it is economical and efficient for e.g. wind turbine is represented by voltage supply, internal resistance and maximum load it can handle. This is done so that electrical characteristic remains the same. Similarly with a solar panel. The circuit is programmed in such a manner that economic scheduling is implemented. The solar system and wind system are complementary in nature. When solar supply is not available during the day then wind is available similarly if wind power is not available than solar is available. This nature emphasizes the use of renewable energy sources for energy generation. Power sharing between the load and sources increase the stability of the system and improve the efficiency also.

III. ECONOMIC SCHEDULING SCHEME BY PROTEUS SOFTWARE

Proteus 8 is a solo application which has many service modules proposing different functionality. Economic Scheduling is done by Microcontroller programming in solar and Wind is used to supply the loads if load demand is high then a combination of wind and solar are used along with a supply. An algorithm is used for Economic Scheduling of the hybrid energy system. A transformer has been taken having the turns ratio as 230/12V, to the transformer primary a 230V, 50Hz supply is given which gets converted into 12V on the secondary side. Now, this supply is rectified with the help of full wave bridge rectifier circuit and across the output of the rectifier. A capacitor is connected to reduce the ripples in the output if any. This output is now connected to the variable resistor using it the required value of output voltage can be achieved and again a capacitor has been installed to make the output ripple free as much as possible. Because these are harmful to the working of the circuit. A LED in series with a resistor is connected which is the indication of getting a fine amount of ripple-free supply. Along the input side of the microcontroller three sources namely Wind source, AC supply source, and the Solar System source. All are connected through the comparators to the required pins of the microcontroller. Along the output side of the microcontroller, different loads are connected. Along the input and load side, same color LEDs have been used to make sure about working with unique source. P_D is the surplus power , P_{gen} is the generating form from wind and solar sources , P_{load} is the load power and P_{grid} is the grid power.

$$P_D = P_{gen} - P_{load}$$

$$P_{gen} = P_{solar} + P_{wind}$$
$$P_D = P_{solar} + P_{wind} - P_{load}$$

When the main grid is used as backup & feedback the system
$$P_{grid} = k * P_D$$
$P_D > P_{load}$, if generation is more than load then surplus is sold out to the utility (Main Grid)
$P_D < P_{load}$, if the generation is less than load then load will be supplied by utility (Main Grid)
k=1; for lossless system

If any source is generating power more than the required by load then that energy is too sold back to the Utility using Smart Energy Meters. This can be clarified as the first load is fulfilled by solar system till its maximum capacity is achieved when the load is shifted to wind system followed by the combination of both solar system and wind system then finally by AC supply. This can be done by an algorithm. In the midnight, wind power is available. So during this time interval load is satisfied by wind source after this interval solar power is available, the load is satisfied by solar power. Smart energy meters are used to sell out the surplus power when surplus power is more than load demand.

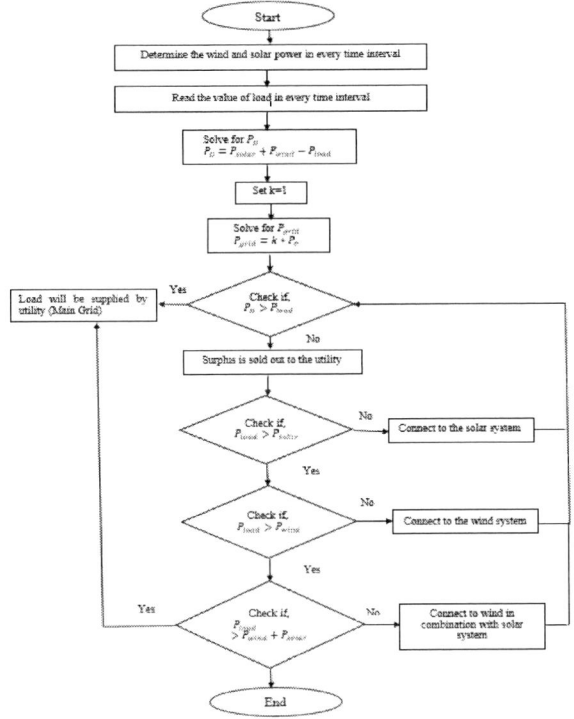

Fig.2. Flow Chart for Economic Scheduling

The process of Economic scheduling based on hardware in the loop is described as below.
Step 1- Monitor the outputs of solar and wind system.
Step 2- Monitor the load of the system as described in Section II.

Step 3- Calculate the Surplus Power by
$$P_D = P_{solar} + P_{wind} - P_{load}$$
Step 4-Set the value of k is equal to 1 for the lossless system.
Step 5- Calculate the grid power by
$$P_{grid} = k * P_D$$
Step 6- Check if, $P_D > P_{load}$ then load will be supplied by utility otherwise surplus is sold out to the utility.
Step 7- Check if, $P_{load} > P_{solar}$, then move to step 8 otherwise load will be supplied by solar system and move to step 6.
Step 8- Check if, $P_{load} > P_{wind}$, then move to step 9 otherwise load will be supplied by wind system and move to step 6.
Step 9- Check if, $P_{load} > P_{wind} + P_{solar}$, then again load will be supplied by utility (Main grid) otherwise load will be supplied by wind system and solar system and move to step 6. Finally, after the connection, the economic scheduling can be achieved in the hybrid energy system.
The flowchart of the above procedures is shown in fig 2.

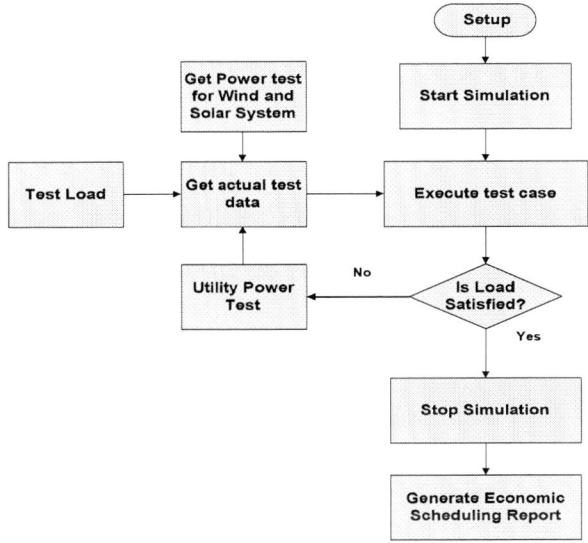

Fig.3. Automated Economic Scheduling test execution flow

Fig.3. shows the automated economic scheduling test execution flow. First Initialize the simulation environment and download the hybrid energy system model into a simulation environment and configure the processor. Then read the list of all possible test cases valid for execution under the given hybrid energy system i.e. for constant load and variable load. After that start simulation and then excite test case for the constant load. Check all the combination of the load as well as sources   Go to execution test case until all test cases in the test case list are considered. If all prerequisite variables are matched for economic scheduling than Stop simulation, reset the test environment and Generate economic scheduling report.
Monitoring of the load in the system is needed for unit commitment otherwise economic scheduling is not achieved.

978-1-5386-6624-1/18 $31.00 © 2018 IEEE   457

Surplus power can be sold out to the utility for economic operation. This power gives economic operation and stables the system also. When this power is sold out to utility then this power can be used when generated power is less than load demand. A separate meter is used for this purpose. The meter can handle the power and record the sold-out power as well as power taken by the utility. Real-time monitoring of different parameters is done by measurement devices and these parameters are compared with actual variable needed if there is any change between this, then load is shifted to another source. Economic Scheduling structure is shown in Fig. 4. Mainly Economic Scheduling execution is the real-time process with various input/output data and user interface. It can be implemented in hardware. There are so many modules available in the market for developing the real model for this. It can be expanded according to need. The model can be run as a real-time operating system with computer-based Simulink model. This is really easy to check costly system like microgrid based hybrid system. Microgrid based hybrid energy system is very costly because of the cost of the wind and solar system. So this provides an easy platform to verify the results. This needs a separate computer where Simulink model has to be run and interfaced with Microcontroller.

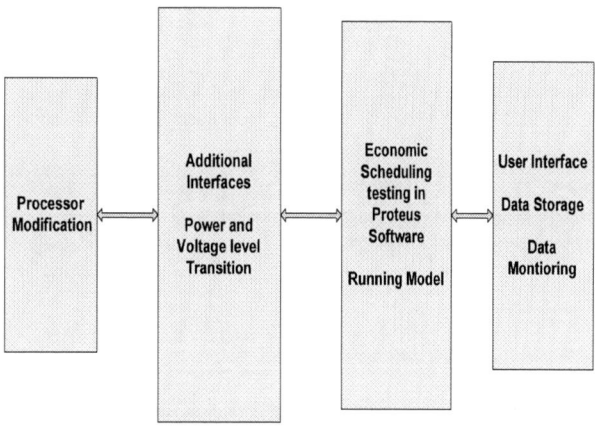

Fig.4. Economic Scheduling framework

Data monitoring and storage systems are needed for this because in every time interval load can be changed. Real-time monitoring is a must for this system otherwise economic scheduling cannot be implemented. Power and voltage level transition is necessary to satisfy the load demand according to the generation of the wind and solar system. If the solar system is not able to satisfy the load demand when the load will be shifted to another source. Similarly, if load demand is not satisfied by solar and wind sources than load will be satisfied by the main grid. So real-time monitoring is a must for unit commitment. For the lossless system, the value of k is 1. Real-time monitoring can easily satisfy the load demand so this system will be efficient. In the Proteus software, the microcontroller is programmed according to need. So the whole system is programmed in such a way that economic scheduling is executed. The function of the microcontroller can be changed according to need. Suppose if thermal unit also connects with this system than the function of microcontroller has to be changed. This is very economical to run this system and generate the result without using interface units.

IV. RESULTS AND DISCUSSIONS

A microgrid is an interconnection of different renewable energy sources and different loads. Renewable energy sources supply the different loads which are controlled by the Microcontroller. By algorithm, it can be decided which load is satisfied by which source. The load may be vary depending upon the demand. Surplus power varies with load changes. If Surplus power is greater than the load power then this surplus power is sold out to the utility. Similarly, if surplus power is less than load power. Than Main grid supplies the load. By this economic operation can be achieved.

Scenario I- Generation of solar and wind source with a constant load

TableI- shows the generation of solar and wind power with constant load.In this case, the load is constant say 10 MW and the wind and solar power vary. At 3.00 AM, solar power is 0 and wind power is 16,500 MW so surplus power is 6,500 MW similarly at 6.00 AM, solar power is 6,000 MW and wind power is 15,500 MW, so the surplus power is 11,500 MW which is more than load power. In this situation, this surplus power is given to the

Utility. Economic scheduling is done by a microcontroller based system. At 3.00 PM, surplus power is very high in this period this power is sold out to the utility. Wind power varies with time and solar power also depends on the sun. Multiplication factor k is taken to 1 for the lossless system. But if the system is not loss-free then the value of k is varied. Wind system is not available in some areas because for the production of electricity using wind turbine its speed must be more than 5 Km/hr as the power of wind turbine is directly proportional to the square of the speed of the wind. In India, very few places are there where the speed of wind is more than 5 Km/hr. So wind power is quite difficult to achieve in different areas.By a combination of solar and wind, economic scheduling can be done easily.

TABLE I. GENERATION OF SOLAR AND WIND POWER WITH CONSTANT LOAD

Time (Hrs)	P_{solar} (MW)	P_{wind} (MW)	P_{load} (MW)	$P_D = P_{solar} + P_{wind} - P_{load}$ (MW)	$P_{grid} = k * P_D$ (MW)
0000	0	16,500	10,000	6,500	6,500

03:00	0	14,000	10,000	4,000	4,000
06:00	6,000	15,500	10,000	11,500	11,500
09:00	15,500	16,000	10,000	21,500	21,500
12:00	19,000	16,500	10,.000	25,500	25,500
15:00	17,050	16,000	10,000	23,050	23,050
18:00	10,500	12,000	10,000	13,500	13,500
21:00	0	80	10,000	-9,920	-9,920
23:00	0	15,000	10,000	5,000	5,000

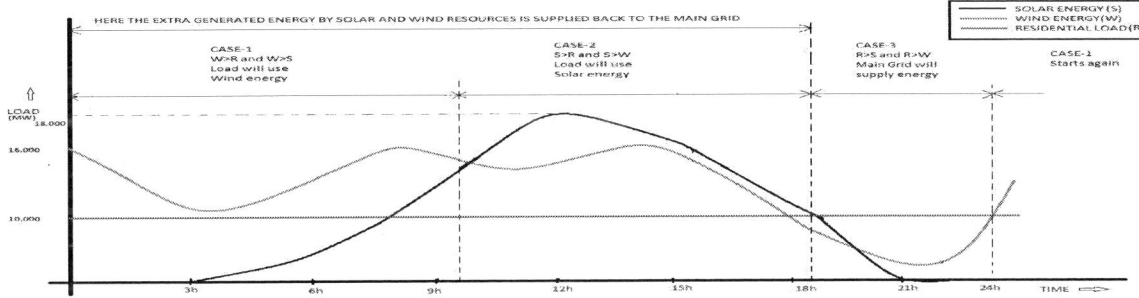

Fig.5.Solar and Wind Energy generation curves with a constant load

In different time interval, the wind and solar system supply the load. Before 18.00 PM, load uses solar energy because during this interval solar power is greater than wind power. After 18.00 PM, Main grid supplies the load. In starting time the wind and solar power supply the main grid also because load demand is less than a generation.

Scenario II- Generation of solar and wind source with a variable load. Table II- shows the generation of solar and wind power with a variable load. It is desirable that the solar panel is utilized to the maximum for a generation to be eco-friendly and economical. Since India is a tropical country sunlight is present in abundance. In short wind generation is eco-friendly but not accessible, unlike solar generation.so every source has some constraints. But when both sources are available than load is satisfied by anyone of this or combination of this. During some time interval surplus power is less than load power than AC source and the combination of the wind and solar power satisfy the load demand. At 3.00

AM, solar power is zero and wind power is 14,000 MW and load power in 12,000 MW so surplus power is equal to the 2000 MW. This power is sold out to the utility. At 9.00 AM, solar power is greater than wind power and the surplus power is 16,500 MW, this power is given to the main grid. At 23.00 PM at night, solar power is zero and wind power is 80MW. During this time interval, the main grid will supply energy. So load will use solar energy in morning time because at this time solar power is greater than wind power. By a combination of solar, wind and main grid, economic scheduling can be achieved. Fig 5. Shows the energy generation curves with a constant load. Fig 6. Shows the energy generation curves with variable loads. During different time interval, the load is satisfied by sources.It is designed in such a manner that it is economical and efficient for e.g. wind turbine is represented by voltage supply, internal resistance and maximum load it can handle. This is done so that electrical characteristic remains the same

TABLE II. GENERATION OF SOLAR AND WIND POWER WITH VARIABLE LOAD

Time (Hrs)	P_{solar} (MW)	P_{wind} (MW)	P_{load} (MW)	$P_D = P_{solar} + P_{wind} - P_{load}$ (MW)	$P_{grid} = k * P_D$ (MW)
0000	0	15,500	10,000	5,000	5,000
03:00	0	14,000	12,000	2,000	2,000
06:00	6,000	15,500	14,000	7,500	7,500
09:00	15,500	16,000	15,000	16,500	16,500
12:00	18,000	16,500	16,000	18,500	18,500
15:00	17,050	16,000	16,000	17,050	17,050
18:00	10,500	12,000	17,000	5,500	5,500
21:00	0	80	19,000	-18,920	-18,920
23:00	0	15,000	11,000	4,000	4,000

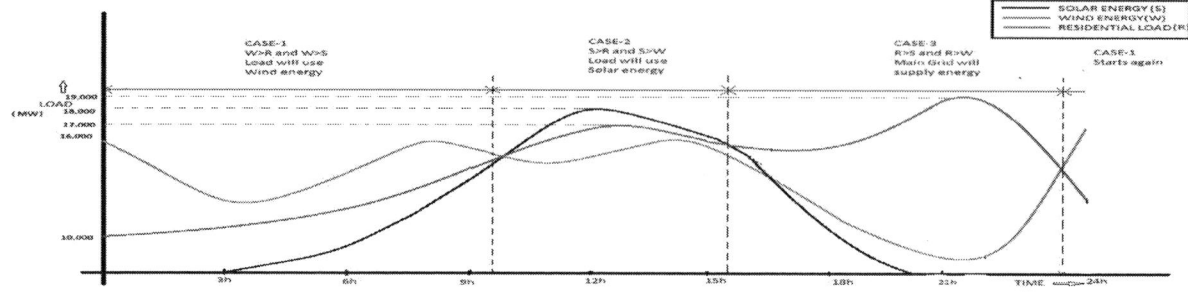

Fig.6. Solar and Wind Energy generation curves with variable load

Similarly with a solar panel. The whole system is programmed in such a manner that economic scheduling is implemented. This can be explained as the first load is satisfied by solar panel till its maximum capacity then the load is shifted to wind followed by a combination of both solar and wind then finally by AC supply. For economical purpose single Lead acid battery which is charged by the source.

V. CONCLUSION

In this paper, Simulation of the microgrid which is integrated with a solar panel, wind turbine, and main supply is done using planned management scheme. The power output of different energy resources is measured and compared with the load power. It will automatically connect the load to the source according to the power demand. If any source is not available then the system automatically shift on the main grid. In this respect, this paper implements the economic scheduling of Microgrid with renewable energy resources. This can be explained that the first load is satisfied by the solar system until its maximum capacity is attained when the load is shifted to the wind system followed by the combination of both the solar system and wind system finally load will be supplied by the main grid. Smart energy meters are employed to sell out the surplus power when surplus power is more than load demand. A microgrid which is integrated with the solar system, wind system, and main supply is observed by PROTEUS Software. An idea of renewable energy generation and unit commitment has been given. The hybrid energy system is designed in such a manner that it is economical and efficient.

REFERENCES

[1] O.O. Approach, K. Rahbar, S. Member, J. Xu, R. Zhang, Real-Time Energy Storage Management for Renewable Integration in Microgrid : An Off-Line Optimization Approach, IEEE Trans Smart Grid. 6 (2015) 124–134.

[2] L. Che, M. Shahidehpour, DC Microgrids : Economic Operation and Enhancement of Resilience by Hierarchical Control, IEEE Trans Smart Grid. 5 (2014) 2517–2526.

[3] M.I. Constraints, A. Khodaei, Microgrid Optimal Scheduling With Multi-Period Islanding Constraints, IEEE Trans Power Electron. 29 (2014) 1383–1392.

[4] A.A. Eajal, S. Member, M.A. Abdelwahed, S. Member, S. Member, K. Ponnambalam, A Unified Approach to the Power

Flow Analysis of AC / DC Hybrid Microgrids, IEEE Trans Sustain Energy. 7 (2016) 1145–1158.

[5] Y. Guo, J. Xiong, S. Xu, W. Su, Two-Stage Economic Operation of Microgrid-Like Electric Vehicle Parking Deck, IEEE Trans Smart Grid. 7 (2016) 1703–1712.

[6] A.H. Hajimiragha, M.R.D. Zadeh, S. Moazeni, Microgrids Frequency Control Considerations within the Framework of the Optimal Generation Scheduling Problem, IEEE Trans Smart Grid. 6 (2015) 534–547.

[7] M. Hong, X. Yu, N.-P. Yu, K.A. Loparo, An Energy Scheduling Algorithm Supporting Power Quality Management in Commercial Building Microgrids, IEEE Trans Smart Grid. 7 (2014) 1–1.

[8] J. Lee, J. Guo, J.K. Choi, S. Member, M. Zukerman, Distributed Energy Trading in Microgrids : A Game-Theoretic Model and Its Equilibrium Analysis, IEEE Trans Ind Electron. 62 (2015) 3524–3533.

[9] N. Li, S. Member, C. Uçkun, E.M. Constantinescu, J.R. Birge, K.W. Hedman, A. Botterud, Flexible Operation of Batteries in Power System Scheduling With Renewable Energy, IEEE Trans Sustain Energy. 7 (2016) 685–696.

[10] P.A. Madduri, S. Member, J. Poon, S. Member, J. Rosa, M. Podolsky, E.A. Brewer, S.R. Sanders, Scalable DC Microgrids for Rural Electrification in Emerging Regions, IEEE J Emerg Sel Top Power Electron. 4 (2016) 1195–1205.

[11] M. Mahmoodi, P. Shamsi, B. Fahimi, Economic Dispatch of a Hybrid Microgrid With Distributed Energy Storage, IEEE Trans Smart Grid. 6 (2015) 2607–2614.

[12] W. Microgrid, L. Guo, W. Liu, X. Li, Y. Liu, B. Jiao, W. Wang, C. Wang, S. Member, F. Li, S. Member, Energy Management System for Stand-Alone Wind-Powered-Desalination Microgrid, IEEE Trans Smart Grid. 7 (2016) 1079–1087.

[13] G. Microgrids, An Optimal Energy Storage Control Strategy for Grid-connected Microgrids, IEEE Trans Smart Grid. 5 (2014) 1785–1796.

[14] W. Pei, Y. Du, W. Deng, K. Sheng, H. Xiao, H. Qu, Optimal Bidding Strategy and Intramarket Mechanism of Microgrid Aggregator in Real-Time Balancing Market, IEEE Trans Ind Informatics. 12 (2016) 587–596.

[15] P. Yang, A. Nehorai, Joint Optimization of Hybrid Energy Storage and Generation CapacityWith Renewable Energy, IEEE Trans Smart Grid. 5 (2014) 1566–1574.

978-1-5386-6624-1/18 $31.00 © 2018 IEEE

2nd IEEE International Conference on Power Electronics, Intelligent Control and Energy Systems (ICPEICES-2018)

An Efficient Solar Energy Harvesting System for Wireless Sensor Nodes

Himanshu Sharma[1]
ECE Department,
KIET Group of Institutions,
Ghaziabad, U.P., India
himanshu.sharma@kiet.edu

Ahteshamul Haque[2],
Senior Member IEEE,
Electrical Engineering Department,
Jamia Millia Islamia University, New Delhi,
India

Zainul Abdin Jaffery[3],
Senior Member IEEE,
Electrical Engineering Department,
Jamia Millia Islamia University,
New Delhi, India

Abstract— The Wireless Sensor Networks (WSN) are the basic building blocks of today's modern internet of Things (IoT) infrastructure in smart buildings, smart parking, and smart cities. The WSN nodes suffer from a major design constraint that their battery energy is limited and can work only for a few days depending upon the duty cycle of operation. In this paper, we propose a new solution to this design problem by using ambient solar photovoltaic energy. Here, we propose a highly efficient and unique solar energy harvesting system for rechargeable battery based WSN nodes. Ideally, the optimized Solar Energy Harvesting Wireless Sensor Network (SEH-WSN) nodes should operate for infinite network lifetime (in years). In this paper, we propose a novel and efficient solar-powered battery-charging system with maximum power point tracking (MPPT) for WSN nodes. The research focus is on to increase the overall harvesting system efficiency, which depends upon Solar Panel Efficiency, MPPT controlled DC-DC converter efficiency and rechargeable battery efficiency. Several models for solar energy harvester system have been developed and iterative simulation was performed in MATLAB/SIMULINK for solar powered DC-DC converters with MPPT to achieve optimum results. From the simulation results, it is proved that our designed solar energy harvesting system has 96% efficiency(η_{sys}).

Keywords— *Smart Cities, Solar Energy Harvesting, DC-DC Converters, Maximum Power Point Tracking (MPPT), Battery Charging, Wireless Sensor Nodes*

I. INTRODUCTION

In the 21st century, the development of renewable energy harvesting systems is the most important technological design challenge due to the increase in global warming and other environmental issues. Recently, in August 2016, the Zigbee Alliance, USA has announced the new standard for Energy harvesting wireless sensor networks (EHWSNs) known as ZigBee Green Power (GP)[1]. The amendments in IEEE 802.15.4 communication standard protocol for low data rate wireless networks and the ZigBee Green Power (GP) standard for EHWSNs facilitates the use of the Green Power feature for ZigBee applications running on the low power wireless microcontrollers platform [2]. Now a day, the renewable energy harvesting based power management solutions are being proposed for wireless sensor networks by the commercial companies like Texas Instruments, ST Microelectronics and Linear Technology, USA. The design of an efficient solar energy harvesting systems is necessary for the proper planning of solar energy harvesting wireless sensor networks (SEH-WSN). The harvester system extracts the solar

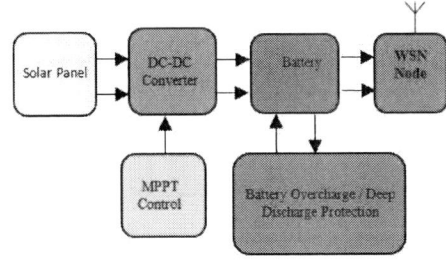

Fig.1. Block diagram of Solar Energy Harvesting System using MPPT Control

energy into the electrical form by using the Photovoltaic (PV) cells. Then, this electrical energy is used to charge the wireless sensor node battery. It reduces the human efforts to replace the battery of hundreds or thousands of sensor nodes by going out in the remote areas. Therefore, the design problem of limited energy availability of wireless sensor nodes is resolved and the human efforts to replace the battery periodically has been reduced. In the year 2008, Ref. [3] proposed Modelling and Optimization of a Solar Energy Harvester System for Self-Powered Wireless Sensor Networks. In 2009, Ref. [4] proposed Design of a Solar-Harvesting Circuit for Battery-less Embedded Systems. In this paper, the simulation results show that by using efficient solar energy harvester circuits the sensor network lifetime can be increased from few days to 20-30 years and higher. Section 1 provides an overview of a basic Solar Energy Harvesting System. Section 2 presents the operation of SEH-WSN Node. Section 3 provides two types of solar energy harvester system most commonly used in practice i.e pulse width modulation (PWM) controlled and MPPT controlled. Section 4 provides simulation parameters and section 5 provides simulation results, section 6 provides efficiency calculations and finally, section 7 provides the conclusion.

II. OPERATION OF AN SEH-WSN NODE

The operation of an SEH-WSN node is explained as follows: The solar energy-harvesting system provides a d.c. power supply (3.6 volts) to the WSN node. This voltage is harvested from the ambient sunlight by using the solar panels

978-1-5386-6624-1/18 $31.00 © 2018 IEEE

Solar Powered Boost Converter with MPPT for Battery Charging of a WSN Node

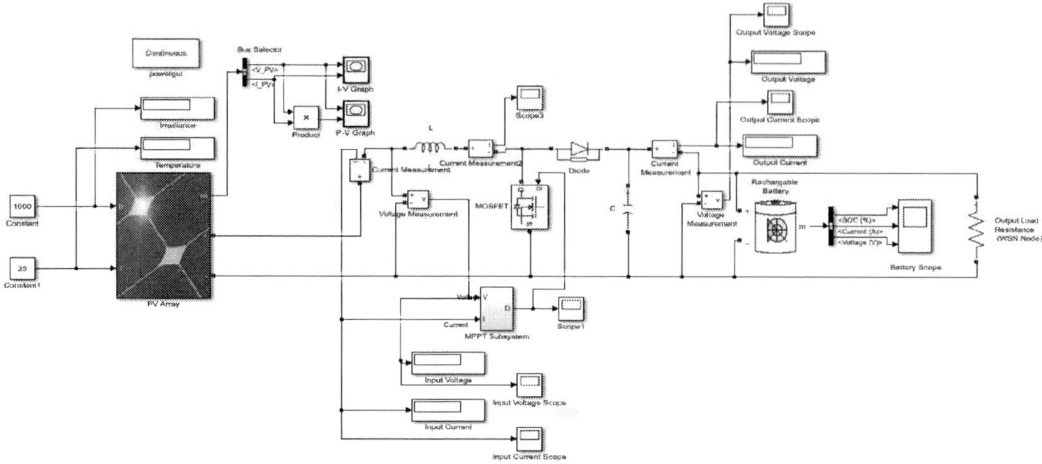

Fig.2. MATLAB/SIMULINK model for MPPT controlled SEH system for WSN Node

[5]. The solar panel converts light energy directly into the d.c. electrical energy. The DC-DC converter regulates this dc voltage to charge the battery. The rechargeable battery powers the WSN node. The WSN node measures the desired physical quantity (e.g. temp., light, humidity, and pressure) by using the sensor measurement unit. A microcontroller in computation unit processes this sensed data. The measured or sensed data is sent to the nearby network node wirelessly, in the form of data packets using transmitter unit. The information is sent to the USB gateway node via cluster head nodes [6]. Finally, the user can remotely monitor & control the application process e.g. temperature monitoring & control of an industrial boiler plant, Air conditioner cooling system control, Traffic light management in a smart city. In this paper, we will focus on modelling and optimization of solar energy harvesting system.

III. SOLAR ENERGY HARVESTING SYSTEM

The figure 1 shows a block diagram of a maximum power point tracking (MPPT) controlled solar energy harvester (SEH) system. The SEH system consists of a solar panel, a DC-DC buck converter, a rechargeable battery, a maximum power point (MPPT) controller, and a WSN sensor node connected as a d.c. load. The ambient solar light energy is harvested using the solar panel and converted into the electrical energy. The DC-DC Buck converter steps down and regulates the magnitude of this harvested voltage, and supplied to the rechargeable battery. The MPPT controller tracks the voltage and current from the solar panel and adjusts the duty cycle accordingly for the MOSFET of DC-DC Buck converter [7]. Finally, the battery voltage is utilized to operate the wireless sensor node. The WSN performs the function of sensing, computation, and communication with

other similar characteristics nodes. Thus, autonomous operation of monitoring and control of any physical phenomenon like temperature, humidity, pressure or acceleration can be achieved using SEH-WSN nodes. In this whole scenario, the efficiency of solar energy harvester circuit plays a very important role. If the efficiency of solar energy harvester system is poor, then the battery will not get recharge properly and hence the wireless sensor network lifetime will reduce.

IV. SIMULATION EXPERIMENT SETUP

The simulation parameters for a solar energy harvesting system are shown in Table 1. Figure 2 shows MATLAB Simulink model of solar energy harvester system using MPPT control. The solar irradiance of 1000 watts/cm^2 is incident on the solar panel with a constant temperature of 25-degree Celsius [8]. The Solar panel can extract only this solar energy into 15 mW/cm^2 with 15% efficiency [9]. For full irradiance on the simulated solar panel, the output voltage of solar panel is 6 volts, 500mA, and 3 watts. Now, this electrical energy from the solar cell is fed to the DC- DC boost converter [10], which increases the output voltage. The Boost converter output voltage is used to charge the rechargeable battery. The rechargeable battery is used to operate the WSN node. Here, the WSN load is modelled as output d.c. load resistance of 100 ohms.

V. SIMULATION RESULTS

The simulation results for the Battery State of Charge (SoC), battery Current (I_B) and battery voltage (V_B) as a function of time (seconds) are shown in figure 3 to figure 5.

978-1-5386-6624-1/18 $31.00 © 2018 IEEE

TABLE I. SIMULATION PARAMETERS

Parameters	Value	Parameters	Value
Irradiance (W/m²)	1000 Watts/m²	Capacitor (C)	100uF
Temperature (T)	25 degree Celsius	Inductor (L)	200uH
DC-DC Converter	Boost Converter	MOSFET Switching Frequency (f)	5KHz
Max. Solar Panel output voltage (V_m)	6 volts	Initial duty Cycle	0.5
Max. Solar Panel output current (I_m)	500mA	MOSFET Switching Power Losses (P_{sw})	0.5mW
Max. Power from Solar Cell (P_m)	3 watts	Switching Voltage Loss (V_{sw})	0.2 volts
Rechargeable Battery Type	NiCd	WSN Load Model	10-ohm resistor
Battery Voltage	3.6 volts	Inductor conduction Power Loss (P_L)	50 mW

A. Battery State of Charge (SoC), Voltage and Current during Charging using MPPT

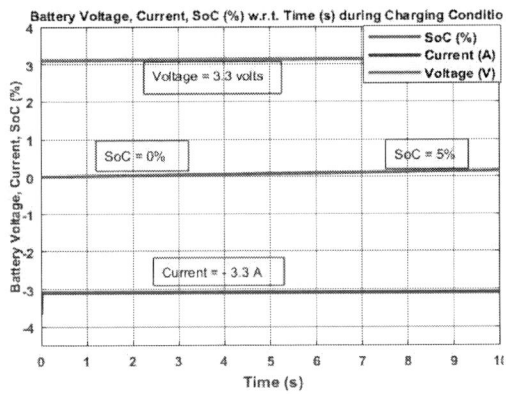

Fig. 3. Simulation results of MPPT controlled SEH system for 10 s.

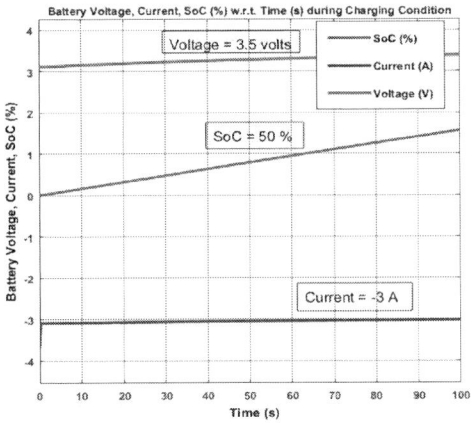

Fig. 4. Simulation results of MPPT controlled SEH system for 100 s.

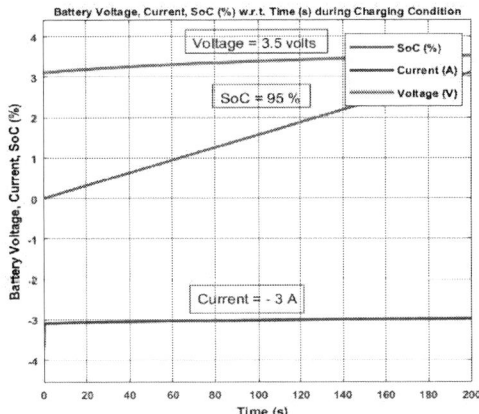

Fig. 5. Simulation results of MPPT controlled SEH system for 200 s.

In figure 3, MPPT controlled solar energy harvesting battery charger (i.e. Battery State of Charge (SoC), battery Current and Voltage) are shown for a simulation time of 10 seconds. Here, the battery SoC reaches from 0 to 5 % Similarly, in figure 4 the MPPT results for 100 s simulation time have battery SoC till 50 %. Finally, in figure 5 the battery SoC reaches till 95 % in just 200 s simulation time. Thus, the battery charging time is dynamically increased by using MPPT controlled solar energy harvesting systems for WSN nodes.

VI. ENERGY HARVESTER SYSTEM EFFICIENCY (η_{sys})

The energy harvester system efficiency is calculated for MPPT control methods. By using P&O MPPT the max. Power available from the solar panel is 2.8 watts. Now the MPPT efficiency is calculated as [11]:

$$MPPT \; Efficiency \; (\eta_{MPP}) = \frac{P_{MPP}}{P_m} \qquad (1)$$

From the simulation parameter table 1, the (P_{MPP}) is 2.8 watts and maximum theoretical power (P_m) is 3 watts. Thus MPPT efficiency is calculated as 2.8w / 3w = 93.33 %. Here, the P_{loss} also changes due to MPPT in DC-DC Buck Converter. The P_{loss} is the sum of MOSFET switching loss (P_{sw}) and Inductor conduction loss (P_L). From the simulation results table, the output power (P_o) is 1.8 W and MOSFET switching losses are 2 mW and inductor power loss is 20 mW. Thus buck converter efficiency is calculated as 1.8W /1.8 W+ 22 mW = 98.79%. Finally, the overall energy harvester circuit efficiency (η_{sys}) is the average of Buck converter efficiency and MPPT efficiency.

$$Harvester \; Systems \; Efficiecny \; (\eta_{sys})$$
$$= \frac{(\eta_{buck}) + (\eta_{MPP})}{2} \qquad (2)$$

TABLE 2. SIMULATION RESULTS FOR MPPT CONTROL SEH SYSTEM

Energy Harvester Parameters	Value
Max. Solar Panel output Power (P_m)	2.8 watts
Average Buck Converter Output Voltage(V_m)	3.6 volts
Average Buck Converter Output Current(I_m)	500mA
Buck Converter Output Power	1.8 watts
Inductor Loss	20mW
MOSFET Switching Loss	2mW
Harvester System Efficiency (η_{sys})	96 %

From the formula of eq.2, the calculated overall energy harvester system efficiency (η_{sys}) is (98.79%+93.33%) / 2 = 96.28%.

VII. CONCLUSION

In this paper, we have modeled and simulated an efficient solar energy harvester system for WSN nodes with MPPT in MATLAB/SIMULINK. The MPPT based harvesting system charges the battery using at a very fast rate. The battery SoC and Terminal voltage start increasing while charging. The overall energy harvester circuit efficiency (η_{sys}) is the sum of Boost converter efficiency and MPPT efficiency. From the simulation results, it is proved that the MPPT based Solar Energy Harvester system efficiency is 96.28%.

ACKNOWLEDGMENT

The authors are thankful for Advanced Power Electronics Lab, Electrical Engineering Department, Jamia Millia Islamia, (a Central Government University), New Delhi, India. This research work was supported by Ministry of New & Renewable Energy (MNRE), Government of India, New Delhi, India.

REFERENCES

[1]. IEEE Standard documents "802.15.4 for Low-Rate Wireless Networks, Amendment 2: Ultra-Low Power Physical Layer", IEEE Standards Association, IEEE Computer Society, 3 Park Avenue, New York, 10016-5997, USA, 2016.

[2]. "ZigBee Pro with Green Power User Guide", Revision 1.4, August 2016. [online]:www.nxp.com/documents/ user_manual/JN-UG-3095.pdf

[3]. Denis Dondi, Alessandro Bertacchini, Davide Brunelli, "Modelling and Optimization of a Solar Energy Harvester System for Self-Powered Wireless Sensor Networks", IEEE Transactions on Industrial Electronics, vol. 55, no. 7, July 2008.

[4]. Davide Brunelli, Clemens Moser, Lothar Thiele, "Design of a Solar-Harvesting Circuit for Batteryless Embedded Systems" IEEE Transactions on Circuits and Systems—I: regular papers, vol. 56, no. 11, November 2009.

[5]. Ian Mathews, Paul J. King, Frank Stafford, and Ronan Frizzell, "Performance of III–V Solar Cells as Indoor Light Energy Harvesters" IEEE Journal of Photovoltaics, vol. 6, no. 1, pp. 230-236, January 2016.

[6]. X. Liu and E. S_anchez-Sinencio, "A highly efficient ultralow photovoltaic power harvesting system with MPPT for internet of things smart nodes," IEEE Transections on VLSI System, vol. 23(12), pp. 3065–3075, 2015.

[7]. Jaw-Kuen Shiau and Chien-Wei Ma, "Li-Ion Battery Charging with a Buck-Boost Power Converter for a Solar Powered Battery Management System" Energies Journal, vol.6, MDPI, pp1669-1699, 2013.

[8]. S. Sivakumar et al., "An assessment on performance of DC-DC converters for renewable energy applications," Renewable Sustainable Energy Reviews, Springer, Elsevier, vol. 58, pp. 1475–1485 , 2016.

[9]. Minchul Shin, Inwhee Joe, "Energy management algorithm for solar-powered energy harvesting wireless sensor node for Internet of Things" , IEEE- IET Communications Journal, , volume: 10, Issue: 12, pp.1508-1521, 2016.

[10]. Himanshu Sharma, Ahteshamul Haque, and Zainul A. Jaffery "Solar energy harvesting wireless sensor network nodes: A survey", Journal of Renewable and Sustainable Energy, American Institute of Physics (AIP), USA, vol.10, no.2, pp 1-33, March 2018.

[11]. Himanshu Sharma, Ahteshamul Haque, and Zainul A. Jaffery "Modeling and Optimisation of a Solar Energy Harvesting System for Wireless Sensor Network Nodes", Journal of Sensor and Actuator Networks, MDPI, vol.7, Issue 40, pp.1-19, Sept. 2018.

2nd IEEE International conference on power Electronics, Intelligent Control and Energy systems (ICPEICES-2018)

High Voltage Gain Switched Capacitor Boost Converter with ANFIS Controller for Fuel Cell Electric Vehicle Applications

Jyotheeswara Reddy K
School of Electrical Engineering
Vellore Institute of Technology
Vellore, India
jyothireddy.kalvakurthi@gmail.com

Sudhakar N
School of Electrical Engineering
Vellore Institute of Technology
Vellore, India
nsudhakar@vit.ac.in

Abstract— **Designing of a high voltage gain or high step-up boost converter is the major challenge in the high power applications. In this work, a high voltage gain Switched Capacitor Boost Converter (SCBC) is designed for fuel cell electric vehicle (FCEV) applications. The designed converter achieves the high voltage gain with low duty cycle and reduces the voltage stress across the power semiconductor switches. The converter output voltage is given to the Brushless DC (BLDC) motor through a DC link and a Voltage Source Inverter (VSI) for the propulsion of the vehicle. An adaptive neuro-fuzzy inference system (ANFIS) based MPPT technique is adopted for the fuel cell system for extracting the maximum power at different temperature conditions. The performance analysis of fuel cell electric vehicle system with SCBC is done by using the MATLAB/Simulink environment.**

Keywords— *Fuel cell electric vehicle, Switched capacitor boost converter, BLDC Motor, MPPT, ANFIS.*

I. INTRODUCTION

From the past few years, fuel cells are becoming an alternative and attractive power source for electric vehicle applications because of their low noise, cleanness, high reliability and high efficiency [1]. Fuel cells are classified into different types based on the electrolyte material used [2].

Among all the available fuel cells, PEMFC is commonly used in electric vehicle applications because of its high energy density, low noise and no emissions [3].

The conventional powertrain architecture of the Fuel Cell Electric Vehicle (FCEV) with Brushless DC (BLDC) motor is as shown in the Fig. 1. It consists of a fuel cell engine, DC-DC boost converter, an inverter and a BLDC motor for the propulsion of vehicle wheels. PEMFC is a high current and low voltage power source. So, a high voltage gain DC-DC boost converter is required in fuel cell electric vehicle applications for providing an interface between the DC link and fuel cell [4, 5].

The duty cycle of the non-isolated boost converters is varied within a range of 0 to 1. For high voltage gain converters, the larger duty cycle is required. The use of larger duty cycle increases the conduction losses of the power switching devices [6]. In order to overcome this problem, a switched capacitor based high voltage gain DC-DC converter with smaller duty cycle is designed for high power applications. The proposed switched capacitor boost converter (SCBC) has the benefits as follows: less duty cycle, high voltage gain, simple structure and the voltage stress across the power semiconductor devices is very low.

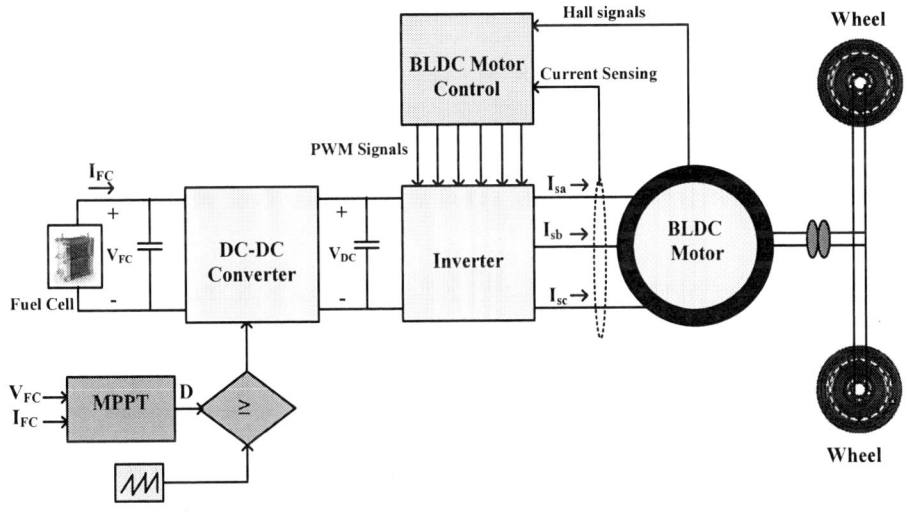

FIGURE 1. Traditional powertrain architecture of the FCEV.

978-1-5386-6624-1/18 $31.00 © 2018 IEEE

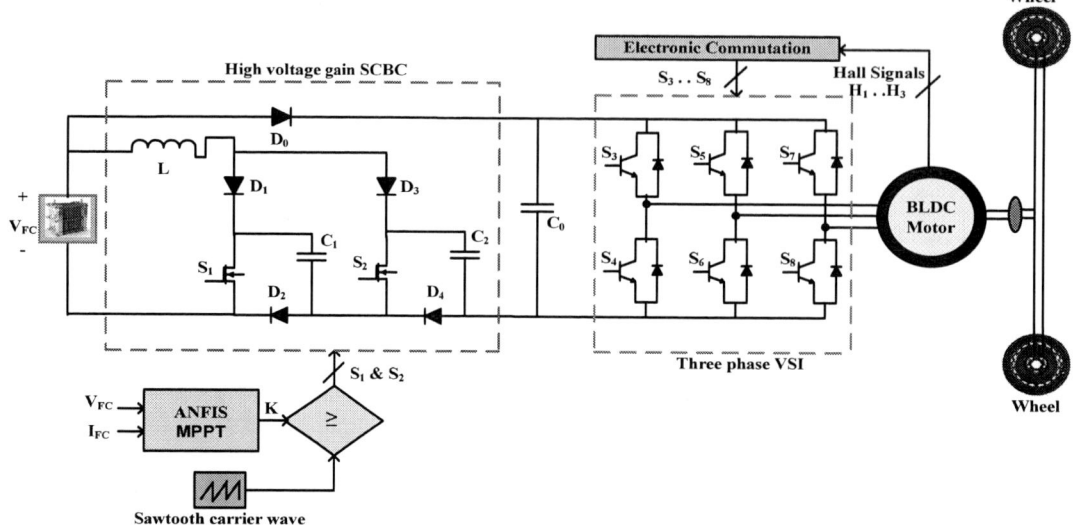

FIGURE 2. FCEV system with proposed SCBC.

The FCEV system with proposed high step-up SCBC is shown in Fig. 2. It consists of a fuel cell, high voltage gain SCBC, VSI and a BLDC motor. SCBC operates as an interface between the VSI and fuel cell. ANFIS based MPPT controller is executed to draw the maximum power from the PEMFC. The SCBC provides the required power to the BLDC motor via VSI and the VSI switches are controlled by electronic commutation of BLDC motor.

II. PEMFC MODELING

PEMFC is an electrochemical device converts chemical energy of the fuel into electrical energy. The inputs to the PEMFC are air and hydrogen, through chemical reaction air and hydrogen are converted into water and electrical energy. This electrical energy can be used to propel the BLDC motor drive through SCBC power conditioning circuit in electric vehicles. PEMFC consists of two electrodes (cathode and anode) and an electrolyte. The PEMFC output voltage is given as follows [7-11]:

$$V_{FC} = n_{FC} V_{cell} = n_{FC} \left(E_{Nernst} - V_{Ohm} - V_{Con} - V_{Act} \right) \quad (1)$$

where V_{cell} is fuel cell stack voltage; n_{FC} is number of cells in fuel cell stack; E_{Nernst} is open circuit thermodynamic voltage; V_{Ohm} is activation overvoltage; V_{Con} is concentration voltage; V_{Act} is activation overvoltage. Each term in Eq.(1) are determined by using the following expressions:

$$V_{Ohm} = R_{FC} . I_{FC} \quad (2)$$

$$V_{Con} = -0.016 \, ln \left(1 - \frac{I_{FC}}{25} \right) \quad (3)$$

$$V_{Act} = T_{FC} \left(x + y \, ln(I_{FC}) \right) \quad (4)$$

$$E_{Nernst} = 1.22 - 8.5 e^{-4} \left(T_{FC} - 298.15 \right) + \\ 4.308 \, e^{-5} \left[ln \left(P_{H2} \right) + 0.5 \, ln \left(P_{O2} \right) \right] \quad (5)$$

where TFC is fuel cell temperature; P_{H2} is Hydrogen gas partial pressure; P_{O2} is Oxygen gas partial pressure; x and y

are constants. The design specifications of 450W PEMFC are listed in the Table1.

TABLE 1. Design parameters of PEMFC.

Parameter Description	Rating
Maximum power (P_{max})	450 W
Maximum current (I_{max})	15 A
Maximum voltage (V_{max})	30V
Temperature (T)	55^0C
Nominal air flow rate	2400 lpm

III. HIGH VOLTAGE GAIN SCBC

The circuit configuration of high voltage gain SCBC is as shown in Fig. 3. It consists of two switches (S_1&S_2), five diodes (D_0-D_4), three capacitors (C_0-C_2) and one inductor (L). V_{FC} and V_{DC} are input and output voltages of the converter respectively. The gate pulses are given to the two switches S_1 and S_2 simultaneously. The steady-state waveforms of the SCBC in continuous conduction mode (CCM) are shown in Fig. 6. The modes of operation of the proposed converter are as follows:

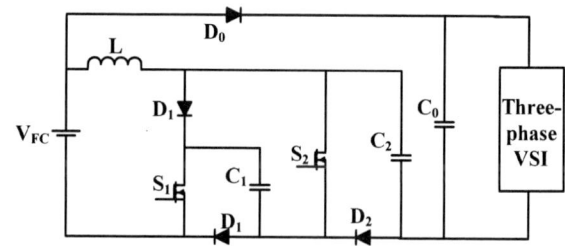

FIGIURE 3. Proposed high voltage gain SCBC.

Mode-1 (t_0-t_1): During this mode, both the switches S_1 and S_2 are conducting and the diodes D_1, D_2, D_3 and D_4 are in reverse biased condition as shown in Fig. 4. The current through the inductor increases linearly and the capacitors C_1

and C_2 are discharged. By applying KVL to the Fig. 4, we get

$$L\frac{dI_L}{dt} = V_{FC} + V_{C1} \quad (6)$$

$$V_{DC} = V_{FC} + V_{C1} + V_{C2} \quad (7)$$

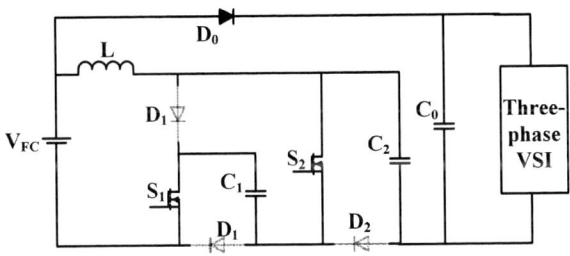

FIGURE 4. High voltage gain SCBC during mode-1.

Mode-2 (t_1-t_2): During this mode both the switches S_1 and S_2 are switched OFF. The diodes D_1, D_2, D_3 and D_4 are conducting and the diode D_0 is in reverse biased condition as shown in Fig. 5. The current through the inductor decreases linearly and the capacitors C_1 and C_2 are charged. By applying KVL to the Fig. 5, we get

$$L\frac{dI_L}{dt} = V_{FC} - V_{C1} \quad (8)$$

$$V_{C1} = V_{C2} \quad (9)$$

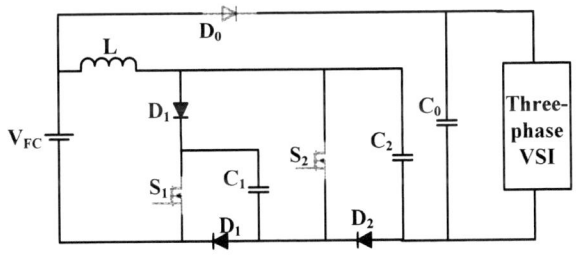

FIGURE 5. High voltage gain SCBC during mode-2.

The voltage gain (M) of the SCBC is derived by using volt-second balance technique. By using Eq. (6) to (9) we get

$$V_{C1} = V_{C2} = \frac{1}{1-2K} V_{FC} \quad (10)$$

where K is duty cycle. From Eq. (7) & (10), the voltage gain of the high voltage gain SCBC is obtained as follows:

$$M = \frac{V_{DC}}{V_{FC}} = \frac{3-2K}{1-2K} \quad (11)$$

The values of the inductor and output capacitor are calculated by using the Eq. (12) & (13) respectively.

$$L = \frac{(V_{FC} + V_{C1})K}{\Delta I_L F_S} \quad (12)$$

$$C_0 = \frac{I_0(1-K)}{\Delta V_0 F_S} \quad (13)$$

where, F_S is switching frequency, K is duty cycle, ΔI_L and ΔV_0 are input current ripple and output voltage ripple respectively. The converter parameter specifications are listed in Table 2.

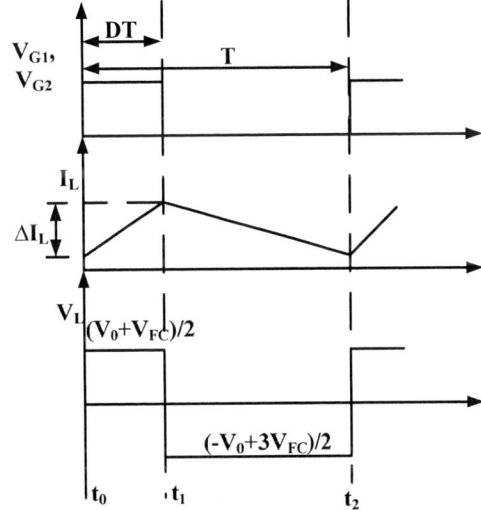

FIGURE 6. Steady-state waveforms of SCBC in CCM.

TABLE 2. High voltage gain SCBC Parameters.

Parameter	Value
Switching frequency (F_S)	50 kHz
Inductor (L)	0.3 mH
Capacitor ($C_1=C_2$)	28 μF
DC link capacitor (C_0)	153 μF
Duty cycle (K)	0.32

IV. CONTROLLING TECHNIQUES

Two control strategies are used in this paper. One is for to track the MPP of PEMFC and second is for BLDC motor operation.

A. ANFIS MPPT Controller

MPPT controller is required for PEMFC system for extracting the maximum power [12]. Among all the available MPPT techniques perturb and observe (P&O) and incremental conductance (INC) methods are more popular due to their simple structure and easy operation [13]. These techniques generate oscillations at steady state which may decrease the efficiency of the PEMFC [14]. To overcome these issues, artificial intelligence (AI) based MPPT controllers are developed to track the MPP. In this paper, an adaptive neuro-fuzzy inference system (ANFIS) based MPPT controller is designed for PEMFC. The designed MPPT controller is the combination of fuzzy inference system and neural networks. Hence, it provides the advantages of fuzzy inference mechanism and neural networks learning ability. The typical architecture of the ANFIS is illustrated in Fig. 7. It has two inputs (x and y) and single output (F) with 5 layers.

978-1-5386-6624-1/18 $31.00 © 2018 IEEE   467

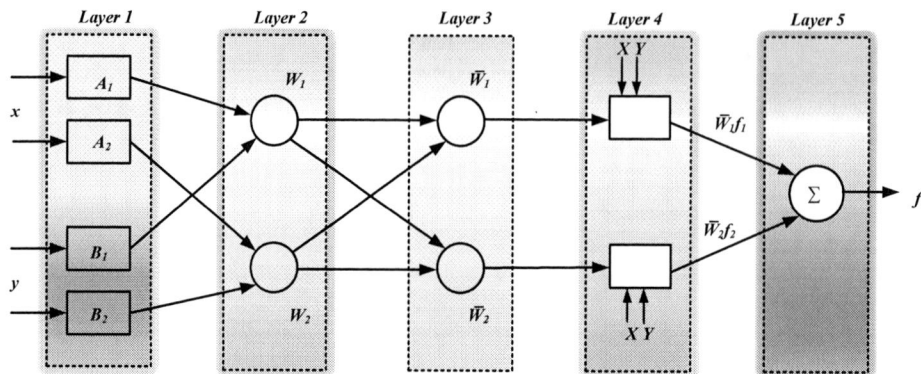

FIGURE 7. ANFIS Structure.

In the first layer the input values are converted to fuzzy values. Hence, it is also referred as fuzzification layer. Each node in the fuzzification layer is an adaptive node with a node function as

$$O_{1,i} = \mu_{A_i}(x) \qquad \text{for } i=1,2 \qquad (14)$$

$$O_{1,i} = \mu_{B_{i-2}}(y) \qquad \text{for } i=3,4 \qquad (15)$$

where, $O_{1,i}$ is membership value of x and y and μ is membership function.

The second layer in the ANFIS is known as product layer. All the nodes in the second layer are fixed nodes. The output of this layer is the multiplication of all receiving signals.

$$O_{2,i} = w_i = \mu_{Ai}(x)\mu_{Bi}(y) \qquad \text{for } i=1,2 \qquad (16)$$

The third layer is referred as normalization layer. Each node in this layer is fixed node. This layer normalizes the firing strength by dividing rules firing strength with sum of firing strengths of all rules.

$$O_{3,i} = \overline{w_i} = \frac{w_i}{w_1 + w_2} \qquad (17)$$

The fourth layer of ANFIS is defuzzification layer. Each node in this layer is an adaptive node with node function as

$$O_{4,i} = \overline{w_i} f_i = \overline{w_i}(p_i x + q_i y + r_i) \qquad \text{for } i=1,2 \qquad (18)$$

The final layer of the ANFIS is also called as output layer and it has only a fixed node. It calculates the overall output value by adding all the incoming signals.

$$O_5 = \sum_i \overline{w_i} f_i = \frac{\sum_i \overline{w_i} f_i}{\sum_i w_i} \qquad (19)$$

The input variables to the ANFIS MPPT controller are fuel cell voltage (V_{FC}) and current (I_{FC}) and the output variable is duty cycle (K).

B. BLDC Motor Control

BLDC motors are becoming more popular in adjustable speed drives over induction motor and wound rotor synchronous motor due to its significant advantages. BLDC motors have the advantages of high power factor, high torque to current ratio, silent operation and compact structure [15-17]. BLDC motors are widely used in all-electric vehicles, military and aeronautical applications. The brushless motors are generally controlled by three-phase VSI from the information provided by the position sensor. Three Hall Effect sensors are placed in 120^0 apart from each other mounted on the stator frame for sensing the rotor position. Based on the rotor position, the power switches are triggered sequentially for every 60 degrees electronically. These generated hall signals are converted into gate pulses to the switches of VSI by using a decoder circuit.

V. RESULTS DISCUSSION

The performance of the designed FCEV system with high voltage SCBC is analyzed in the MATLAB/Simulink environment. In order to examine the dynamic response of the proposed system, quick variations in the PEMFC temperature are considered and is shown in Fig. 8. The different temperature levels are given by using the signal builder block in MATLAB. The considered temperature levels are given as follows: at first from 0 to 0.3sec is 325 K, from 0.3sec to 0.6sec is 315 K and from 0.6sec to 0.9sec is considered as 335 K.

For the considered temperature levels; the output current, voltage and power waveforms of the 450W PEMFC are illustrated in Fig. 9. It generates a DC link power of 400W for 0 to 0.3 sec, 375W for 0.3sec to 0.6 sec and 446W for 0.6sec to 0.9sec.

The output current, voltage and power of the high voltage gain SCBC by using the fuzzy logic based MPPT controller are illustrated in Fig. 10. It generates a power of 390W for 0 to 0.3 sec, 360W for 0.3sec to 0.6 sec and 432W for 0.6sec to 0.9sec at the output of the converter. The output voltage of the SCBC is 190V for first 0.3 sec, 175V from 0.3 to 0.6sec and a voltage of 196V from 0.6 to 0.9sec.

978-1-5386-6624-1/18 $31.00 © 2018 IEEE   468

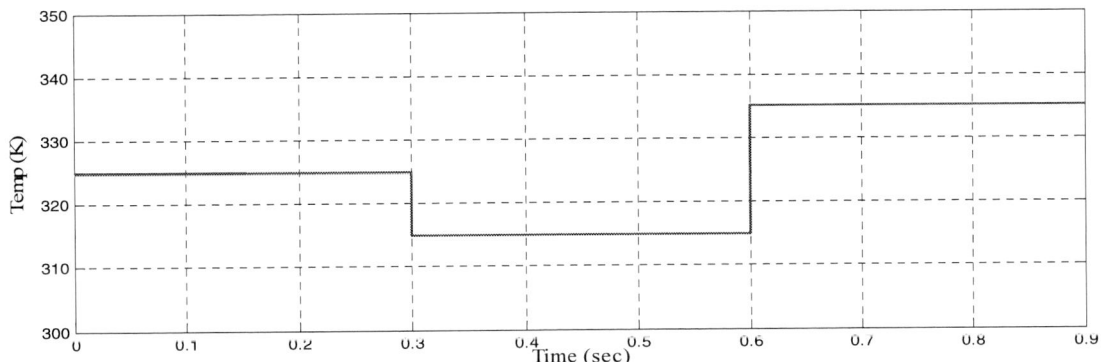

FIGURE 8. Different temperatures of PEMFC system.

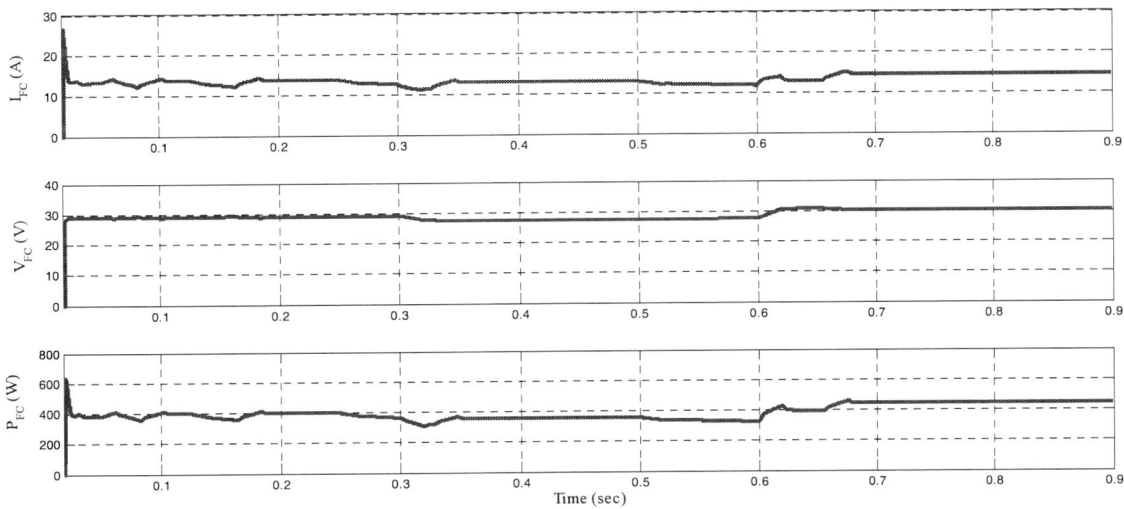

FIGURE 9. PEMFC output current, voltage and power.

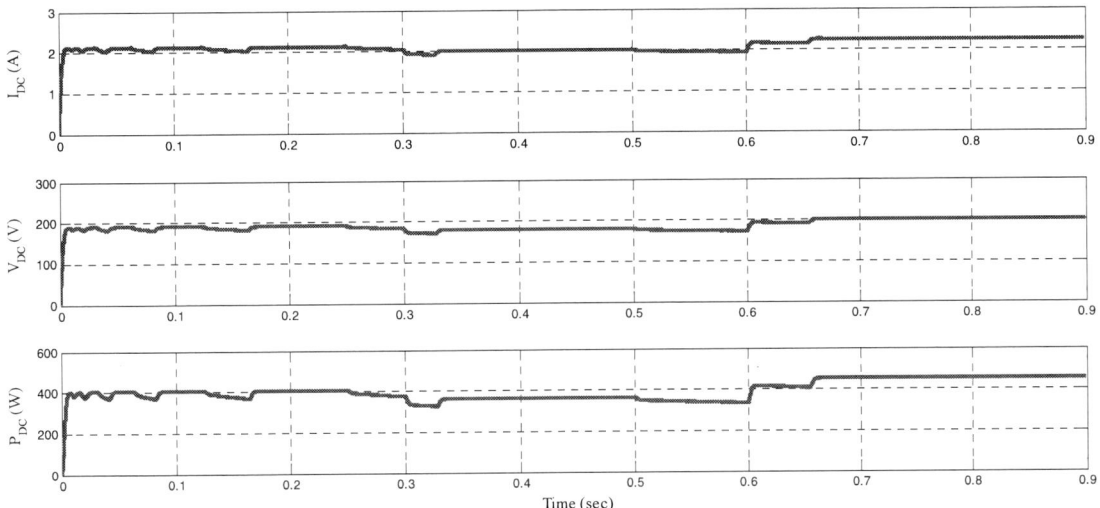

FIGURE 10. DC link output current, voltage and power.

978-1-5386-6624-1/18 $31.00 © 2018 IEEE

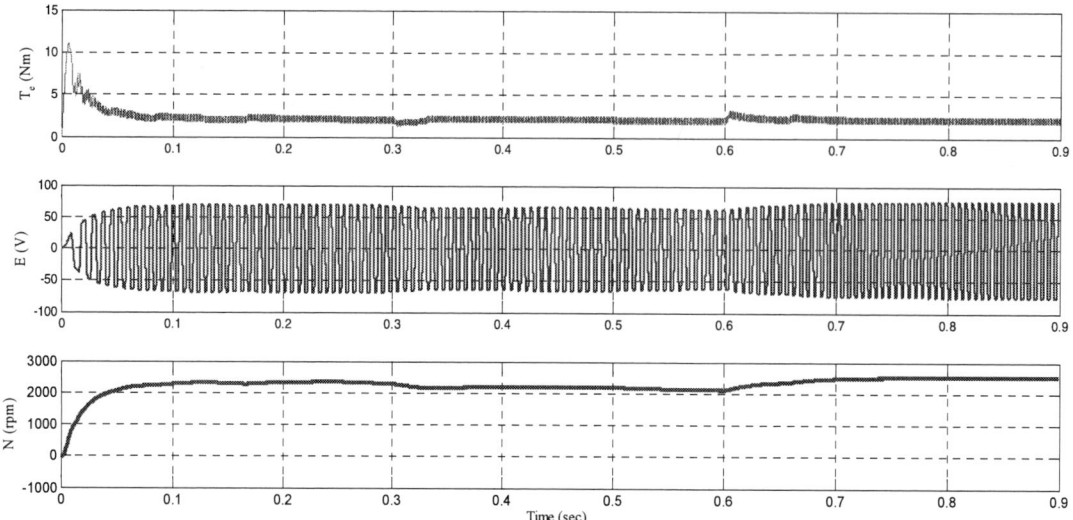

FIGURE 11. DC link output current, voltage and power.

The steady-state characteristics of the BLDC motor at various temperatures of PEMFC are shown in Fig. 11. The motor parameters such as back EMF (E), electromagnetic torque (Te) and speed (N) are presented at dynamic temperature conditions of the PEMFC. The fuel cell output power (PFC), DC link power (PDC) and motor speed (N) at different temperatures of the fuel cell are listed in Table 3.

TABLE 3. Comparison of different parameters

T (K)	P_{FC} (W)	P_{DC} (W)	N (rpm)
315	375	360	2170
325	400	390	2450
335	446	432	2730

VI. CONCLUSION

This paper presents a high voltage gain SCBC for BLDC motor driven FCEV system in MATLAB/Simulink platform. The designed SCBC has achieved the high voltage gain with low duty cycle and reduced the voltage stress across the power semiconductor devices. ANFIS based MPPT controller is designed for 450W fuel cell system for extracting maximum power at dynamic temperature conditions. The different characteristics of BLDC motor such as speed, electromagnetic torque, and back emf are analyzed at different PEMFC temperatures.

REFERENCES

[1] D. Gao, Z. Jin, J. Zhang, J. Li, and M. Ouyang, "Development and performance analysis of a hybrid fuel cell/battery bus with an axle integrated electric motor drive system", Int. J. of Hydrogen Energy, vol. 41(2), pp.1161-1169, 2016.

[2] K.J. Reddy, N. Sudhakar, N. Sarvanan and B. Chitti Babu "High Step-Up Boost Converter with Neural Network Based MPPT Controller for a PEMFC Power Source Used in Vehicular Applications", " Int. J. of Emerging Electric Power system., in press

[3] K.J. Reddy, and N. Sudhakar, "High Voltage Gain Interleaved Boost Converter With Neural Network Based MPPT Controller for Fuel Cell Based Electric Vehicle Applications", IEEE Access, 6, pp.3899-3908, 2018.

[4] A. Garrigos, F.Sobrino-Manzanares, "Interleaved multi-phase and multi-switch boost converter for fuel cell applications" Int. J. of Hydrogen Energy, vol. 40(26), pp.8419-8432, 2015.

[5] F. Sobrino-Manzanares, A. Garrigos, "An interleaved, FPGA-controlled, multi-phase and multi-switch synchronous boost converter for fuel cell applications." Int J of Hydrogen Energy, vol.40 (36), pp. 12447-56, 2015.

[6] M.K. Nguyen, T.D. Duong, Y.C. Lim, "Switched-Capacitor-Based Dual-Switch High-Boost DC-DC Converter", IEEE Trans. Power Electron., 2017.

[7] REDDY, Jyotheeswara, and Sudhakar NATARAJAN. "Control and Analysis of MPPT Techniques for Standalone PV System with High Voltage Gain Interleaved Boost Converter." Gazi University Journal of Science 31, no. 2: 515-530.

[8] K.J. Reddy, and N. Sudhakar, "A new RBFN based MPPT controller for grid-connected PEMFC system with high step-up three-phase IBC", Int. J. of Hydrogen Energy, vol. 43(37), pp.17835-17848, 2018.

[9] N. Mebarki, T. Rekioua, Z. Mokrani, D. Rekioua and S.Bacha, "PEM fuel cell/battery storage system supplying electric vehicle." Int J. of Hydrogen Energy, vol. 41(45), pp. 20993-21005. 2016.

[10] B. Allaoua, , B. Draoui, D. Belatrache, "Study of the energy performance of a PEM fuel cell vehicle", Int. J. Renewable Energy Research, vol.7(3), pp.1395-1402, 2017.

[11] K.J. Reddy, and N. Sudhakar, "Energy sources and multi-input DC-DC converters used in hybrid electric vehicle applications-A review", Int. J. of Hydrogen Energy, vol. 43(36), pp.17387-17408, 2018.

[12] M.H.Wang, H.T. Yau, T.Y. Wang, "Extension sliding mode controller for maximum power point tracking of hydrogen fuel cells", Abstract and Applied Analysis, Hindawi Publishing Corporation, 2013.

[13] K. Kumar, N. Ramesh Babu, K. R. Prabhu, "Design and Analysis of an Integrated Cuk-SEPIC Converter with MPPT for Standalone Wind/PV Hybrid System." Int. J. Renewable Energy Research, vol. 7, pp. 96-106, 2017.

[14] S.Narendiran, S.K. Sahoo, R.Das, "Control and Analysis of MPPT Techniques for Maximizing Power Extraction and Eliminating Oscillations in PV System", International Energy Journal, vol. 16(3), 2016.

[15] P. Yedamale, "Brushless DC (BLDC) motor fundamentals", Microchip Technology Inc, vol. 20, pp.3-15, 2003.

[16] B. Singh, R. Kumar "Simple brushless DC motor drive for solar photovoltaic array fed water pumping system." IET Power Electron., vol. 9(7), pp. 1487-1495, 2016.

[17] V. Bist, B. Singh "Reduced sensor configuration of brushless DC motor drive using a power factor correction-based modified-zeta converter." IET Power Electron., vol.7(9), pp. 2322-2335, 2014.

2nd IEEE International conference on power Electronics, Intelligent Control and Energy systems (ICPEICES-2018)

Internal Model Controller Design for HVAC System

Shahzad Hussain
Department of Electronics Engineering
Rajasthan Technical University
Kota, Rajasthan, India
shahzad.ais@rediffmail.com

Sapna Gupta
Department of Electronics Engineering
Rajasthan Technical University
Kota, Rajasthan, India
sepsa.gupta10@gmail.com

Rajeev Gupta
Department of Electronics Engineering
Rajasthan Technical University
Kota, Rajasthan, India
rajeev_eck@yahoo.com

Abstract— **This paper presents an Internal model controller (IMC) for temperature control of heating, ventilation and air conditioning (HVAC) system. Firstly, the temperature of an air-conditioning system of the test room is controlled by using PID controller. The performance and stability of the control process are degraded due to the significant time delay of air-conditioning process. So, smith-predictor scheme is used to compensate the time delay. But when the process has larger system parameters and longer time delay, smith-predictor does not provide better results. To overcome this problem, an internal model (IMC) controller is developed. Simulation results compare the performance of internal model (IMC) controller with conventional PID controller and smith predictor controller. Simulation result shows that an internal model controller gives satisfactory performance with less overshoot and less settling time.**

Keywords— **HVAC system; PID control; IMC control; smith- predictor control**

I. INTRODUCTION

Modeling of heat, ventilation and air conditioning (HVAC) system is very essential for the study of energy consumption and quality of indoor environment. Mainly HVAC system uses three types of modeling techniques such as data driven, physics based and gray box models. Various modeling techniques of the HVAC system are explained in [1]. The physics based techniques are classified under the deductive models whereas the data driven methods are classified as inductive models. The gray box models are said to be hybrid model because these models are categorized under both deductive and inductive models. Generally, HVAC systems are used in the rooms and industrial buildings. The main purpose of the HVAC system of a building is to give convenience to the people which are living in this environment [2], [3].

The environment of building is totally depends on the temperature and humidity. So, it is necessary to control the temperature and humidity of the HVAC system. Conventional PI/PID controllers are used to control the temperature of the HVAC system [4]. An adaptive self tuning PI controller is designed for the HVAC system in [5]. Various PID controller tuning methods are described in [6]. PID controller parameter's tuning for heat exchanger using optimization is discussed in [7]. PID controller is widely used in control applications due to its simplicity but it can't deal with uncertainties, time delay and external disturbances. The PI/PID controller is excellently works when the time delay is

small. Sometimes, this controller doesn't provide better performance specially when the time delay is large. To overcome such type of problems, smith predictor is used with significant delay [8]. Modified smith predictor controller design for integrating and unstable process is described in [9], [10]. The nominal and robust control performance analysis of modified smith predictor for first order plus time delay (FOPDT) and higher order processes is carried out in [11], [12], [13]. Smith predictor is not able to improve the performance of the system with large system parameters, longer time delay and process uncertainties. Smith predictor maintains the stability and performance of the process when the reference model is exactly matched with the delay free part of the actual process. But in the real situation, perfect matching of the actual plant and the reference model is unexpected in the HVAC system due to uncertainties and external disturbances. So the control performance is not as good as expected. Due to this reason, an internal model (IMC) controller is designed to control the temperature of heat ventilating and air-conditioning (HVAC) system. Internal model controller (IMC) provides an effective, simple and powerful framework for the analysis of control system performance [14]. The development and future aspects of the internal model (IMC) controller is discussed in [15]. The step by step design of IMC controller for heat exchanger system is explained in [16]. To improve the closed loop system, the performance of load disturbance rejection for the first order and second order system, modified internal model (IMC) controller is designed in [17].

This paper presents an internal model (IMC) controller to control the outlet temperature of heat ventilating and air conditioning (HVAC) system. Firstly, conventional PID controller and smith predictor controller is used to control the temperature of HVAC system, but these controllers don't provide better performance due to large system parameters, longer delay and uncertainties. To overcome this problem, an internal model (IMC) controller is designed. The performance of the controllers is analyzed using different transient criteria.

This paper is organized as follows. Section II provides the mathematical modeling of heat, ventilation and air conditioning (HVAC) system. Section III design the PID controller. Section IV design the smith predictor controller. Section V design the internal model (IMC) controller. Section VI provides the discussion of results. Section VII concludes the paper.

978-1-5386-6624-1/18 $31.00 © 2018 IEEE

II. MODELLING OF HVAC SYSTEM

The time response analysis of HVAC system is performed by obtaining the system models. Figure 1 shows the schematic diagram of an air-conditioning test platform of the HVAC system[5]. The main parts of the system are the cooling zone areas, heater, evaporator, cooling units, fan and dampers. Air conditioning test platform consists of two room, one is inner room and other is outer room. The temperature and humidity of the test rooms are kept constant when the performance of the room is tested.

Where, r(t), u(t) and y(t) represented as the set point temperature, output of the PID controller and the air room temperature. K_m is the proportional gain of the heater, C and R are the over all capacitor and resistor of the room. L_m is the time delay and $q_i(t)$ is the load disturbance.

The input and output of the system can be represented in equation 1:

$$Y(s) = \frac{K_m U(s) - q_i(s)}{sC + \dfrac{1}{R}} e^{-L_m s} \qquad (1)$$

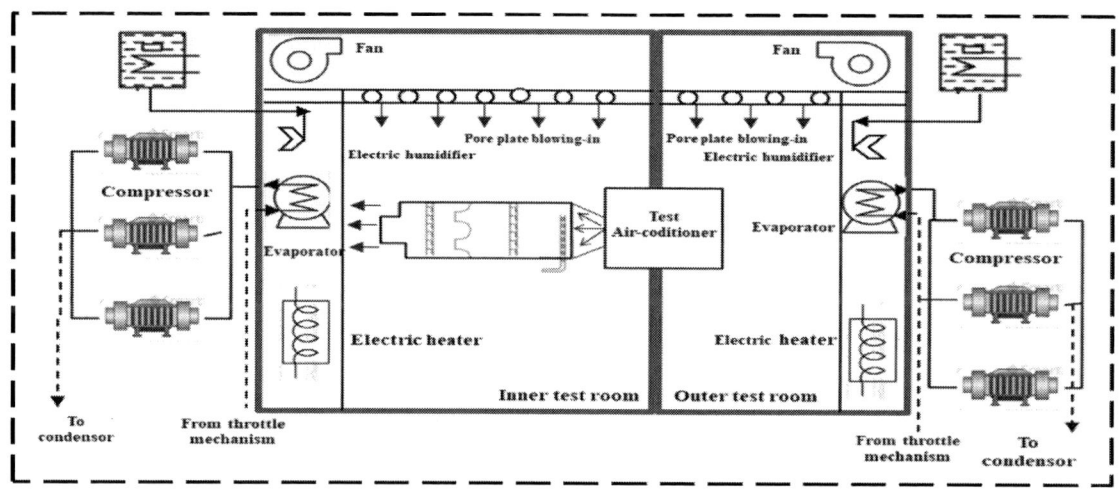

Fig. 1.   Schematic diagram representation of the air conditioner room

Both the air-conditioning room consist individual electric heater, fan, compressor and evaporator to manage the temperature in the room. In simple way, the temperature of the rooms is controlled by using heater. For the control purpose, HVAC control system is shown in figure 2.

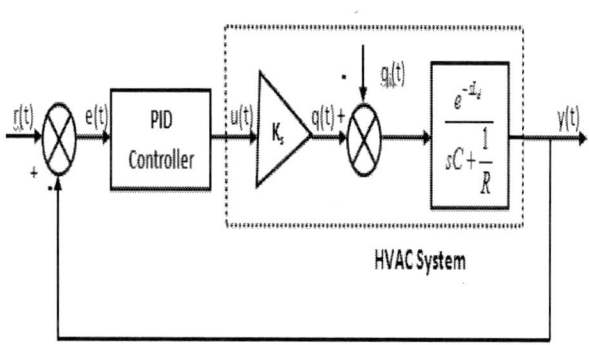

Fig. 2.   Block diagram of the HVAC system

This paper assume that the load disturbance is zero, then the equation (1) can be rewritten as

$$\frac{Y(s)}{U(s)} = \frac{K_m}{sC + \dfrac{1}{R}} e^{-L_m s} \qquad (2)$$

So, the equation 3 represents the HVAC system as first-order plus dead time model which has the gain $K_r = K_m R$, the time constant $T_r = CR$ and the time delay $L_p = L_m$.

Therefore, the transfer function of the HVAC model can be represented as

$$T_p(s) = \frac{K_r}{sT_r + 1} e^{-L_p s} \qquad (3)$$

The HVAC system can be represented by equation 4 [5].

$$T_p(s) = \frac{72}{60s + 1} e^{-5s} \qquad (4)$$

978-1-5386-6624-1/18 $31.00 © 2018 IEEE   472

III. PID CONTROLLER

Conventional proportional integral derivative (PID) controller is designed to control the temperature of the HVAC system. It is the most commonly used controller to find wide application in various areas of control engineering. This paper describes the set point regulation and load disturbance rejection feature of PID controller. There are different tuning methods of PID control such as Ziegler-Nichols, Cohen-Coon and Chien Hrones Reswick. In this paper, PID controller parameters are tuned by using Ziegler- Nichols and Cohen-Coon method. To find the best control output, researchers need to calculate the proper value of these parameters.

PID controller can be represented as:

$$G_c(s) = K_c\left(1 + \frac{1}{\tau_i s} + \tau_d s\right) \qquad (5)$$

Here, K_c is the proportional gain, τ_i is the integral time τ_d and is the derivative time.

IV. SMITH PREDICTOR CONTROLLER

Many HVAC systems have unknown time delay. The closed loop stability is reduced due to the system time delay specially when high feedback gains are used. Therefore, to manage the stability of the system and eliminate the adverse effect of the time delay, the smith predictor is the perfect and most widely used controller for the time delay compensation. Figure 3 represents the block diagram of smith predictor controller with the HVAC system.

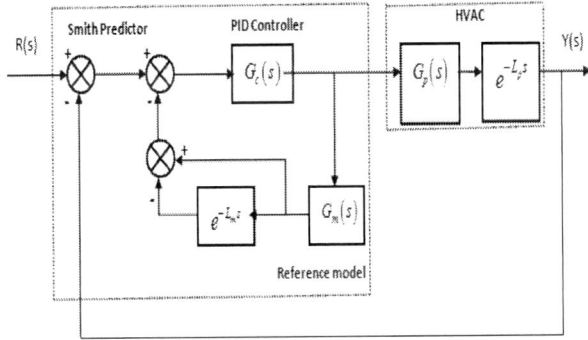

Fig. 3. Block diagram of smith predictor with HVAC system

Where, $G_c(s)$ is the PID controller. HVAC model is represented as $T_p(s) = G_p(s)e^{-L_p s}$, $G_p(s)$ is the process model and $e^{-L_p s}$ is the time delay of the HVAC system. The reference model is described as $T_m(s) = G_m(s)e^{-L_m s}$, $G_m(s)$ is the reference model and $e^{-L_m s}$ is the time delay of

the reference model. The complete system with controller can be represented by eq.6.

$$\frac{Y(s)}{R(s)} = \frac{G_c(s)T_p(s)}{1 + G_c(s)G_m(s) + G_c(s)\left(T_p(s) - T_m(s)\right)} \qquad (6)$$

$$= \frac{G_c(s)G_p(s)e^{-L_p s}}{1 + G_c(s)\left(1 - e^{-L_m s}\right)G_m(s) + G_c(s)G_p(s)e^{-L_p s}} \qquad (7)$$

This paper considered that, there is a perfect matching between the reference model and the actual process i.e. $G_m(s) = G_p(s)$ and $e^{-L_p s} = e^{-L_m s}$, so the overall transfer function can be represented as

$$\frac{Y(s)}{R(s)} = \frac{G_c(s)G_m(s)e^{-L_m s}}{1 + G_c(s)G_m(s)} \qquad (8)$$

From the above equation, it is clear that the time delay is removed from the control loop. So the controller $G_c(s)$ is designed with delay free part of the actual plant $G_p(s)$. This improves the system performance. But in the real situation, accurate matching of the actual plant and the reference model is unexpected in the HVAC system due to uncertainties and external disturbances. So the control performance is not as good as expected. Due to this reason, the internal model (IMC) controller is designed in figure 4.

V. INTERNAL MODEL CONTROLLER

PID controller is widely used for control purpose due to simplicity but it cannot deal with different type of uncertainties and disturbances. In this regards, Internal model controller (IMC) provides a sufficient, novel and simple framework for the study of control system performance mainly for robust properties [15].

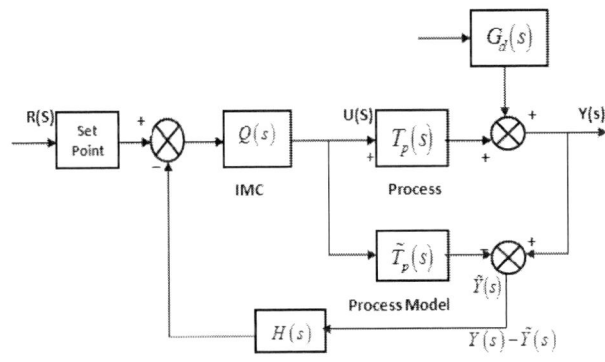

Fig. 4. Block diagram of IMC controller

978-1-5386-6624-1/18 $31.00 © 2018 IEEE

A. Design Procedure for IMC Controller

The transfer function of the process model is represented as

$$T_p(s) = \frac{72}{(60s+1)} e^{-5s} \qquad (9)$$

The process model $\tilde{T}_p(s)$ is split into two parts i.e invertible part $\tilde{T}_{p^+}(s)$ and non-invertible part $\tilde{T}_{p^-}(s)$. To make the internal model controller stable, factorization is performed.

$$\tilde{T}_p(s) = \tilde{T}_{p^+}(s)\tilde{T}_{p^-}(s) \qquad (10)$$

Where, $\tilde{T}_{p^-}(s) = \dfrac{72}{60s+1}$ and $\tilde{T}_{p^+}(s) = e^{-5s}$

The IMC controller $Q(s)$ can be represented as

$$Q(s) = \tilde{T}_{p^-}^{-1}(s) \qquad (11)$$

Where, $Q(s)$ is stable but not proper.

To make the controller proper and robust, a low pass filter $f(s)$ is augmented with the inverted controller $Q(s)$.

$$Q(s) = \tilde{T}_{p^-}^{-1}(s)\, f(s) \qquad (12)$$

$$Q(s) = \frac{(60s+1)}{72(\lambda s+1)} \qquad (13)$$

The tuning parameter λ can be adjusted by both online and offline methods. The online method is based on adaptive IMC scheme and the offline method is based on hit and trial method and various analytical approaches. In this paper, offline method is used to tune the parameter λ and the value is taken as $\lambda = 5, 10$ and 15.

VI. SIMULATION RESULTS

This section discusses the simulation results of different controllers which are designed to control the temperature of air- conditioning system. Firstly, the conventional PID controller is used to control the outlet temperature of the HVAC system. PID controller parameters are tuned by using Ziegler- Nichols method and Cohen- Coon method. Figure 5 and figure 6 shows the set point regulation and load disturbance rejection of Ziegler- Nichols method and Cohen – Coon method.

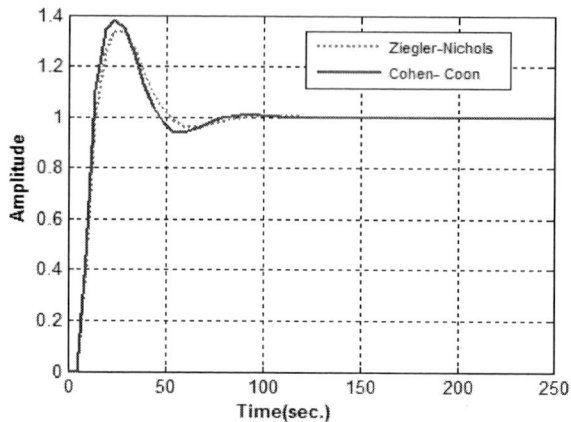

Fig. 5. Set point regulation of Z-N and Cohen-Coon method

In figure 6, A step load is given at 130 sec. So, there is a change in the transient state at 130sec. and then recover from the transient load and reach the steady state value.

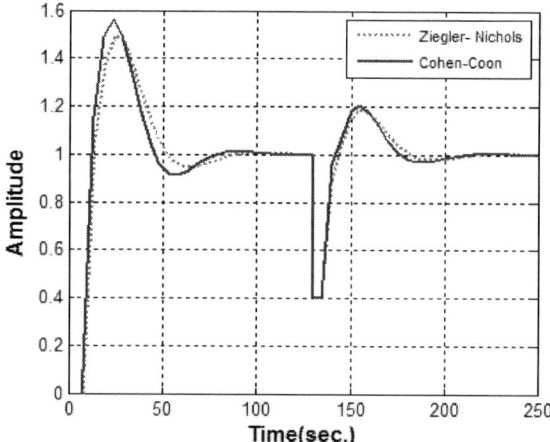

Fig. 6. Load disturbance rejection of Z-N and Cohen-Coon method

Table I and Table II shows the transient performance analysis of conventional PID controller for set point regulation and load disturbance rejection feature. Two different parameters i.e peak overshoot and settling time are evaluated for transient analysis.

TABLE I. TRANSIENT PERFORMANCE ANALYSIS FOR SET POINT REGULATION FEATURE

Tuning Methods	K_p	T_i	T_d	$(\%M_p)$	(t_s)
Ziegler- Nichols	0.10	14	2.50	33.99%	70.61
Cohen-Coon	0.21	12	1.82	38.19%	68.71

TABLE II. TRANSIENT PERFORMANCE FOR LOAD DISTURBANCE REJECTION FEATURE

Tuning Methods	K_p	T_i	T_d	(%M_p)	(t_s)
Ziegler- Nichols	0.10	14	2.50	48.99%	135.37
Cohen-Coon	0.21	12	1.82	56.33%	109.29

From Table I and Table II, it is clear that the Ziegler- Nichols tuning method gives less overshoot in both the conditions. PID controller doesn't deal with the large system parameters, longer time delay and different uncertainties. So, IMC controller is developed. The main difference in the PID controller and IMC controller is that PID controller having three tuning parameters whereas IMC controller consists only one parameter for tuning i.e. λ. Therefore, it is easy and quite simple to implement and give better performance over other controllers.

The step response of IMC controller for different values of λ is shown in figure 7.

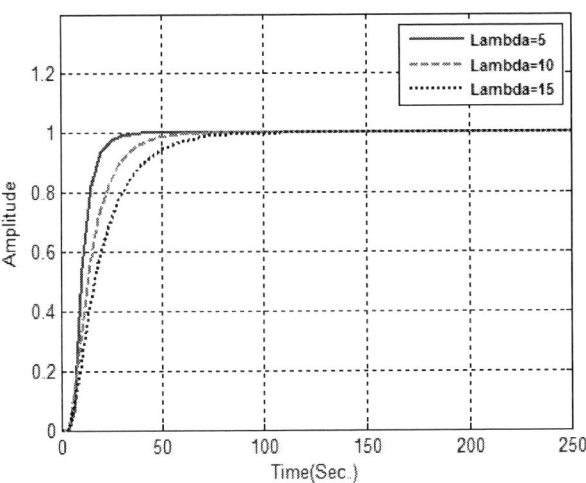

Fig. 7. Step response of IMC controller for different values of λ

From figure 7, it is observed that the overshoot and settling time is increased by increasing the value of λ. Figure 8 shows the step response of the HVAC system for different controllers and Table III provides the transient response analysis of the different controllers. From table III, it is clear that the IMC controller provides better performance with less overshoot i.e 0% and less settling time i.e. 9.21 sec.

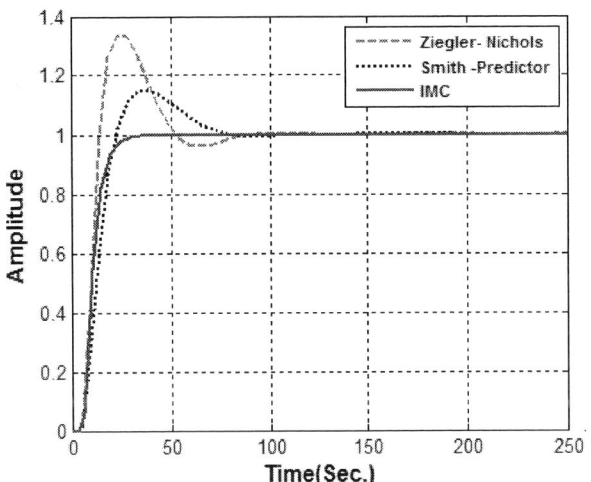

Fig. 8. Step response of HVAC system for different controllers

TABLE III. TRANSIENT PERFORMANCE ANALYSIS OF CONTROLLER

S.No	Controller	Peak Overshoot (%M_p)	Settling time (t_s)
1	PID Set point regulation (Z-N)	33.99%	70.61
2	PID Load disturbance rejection (Z-N)	48.99%	135.37
3	Smith Predictor	15.25%	68.04
4	IMC controller (λ =5)	0%	9.21

VII. CONCLUSION

This paper provides a comparative analysis of different controllers. These controllers are used to control the outlet temperature of air- conditioning system of the test room. Firstly, this paper presents the performance of PID controller using Ziegler-Nichols and Cohen-Coon method. Set point regulation and load disturbance rejection features of the controllers are simulated. Simulation result shows that the PID controller tuning using Z-N method gives good performance in set point regulation mode. PID controller is widely used due to its simplicity but it doesn't provide satisfactory performance when the process has longer time delay, different uncertainties and external disturbances. So, smith- predictor and internal model (IMC) controller is designed. The performance of the above controllers are evaluated by calculating two different parameters i.e. overshoot and settling time. The internal model (IMC) controller provides better performance with less overshoot and less settling time.

REFERENCES

[1] Abdul Afram, Farrokh Janabi-Sharifi, " Review of modeling methods for HVAC systems," Applied Thermal Engineering 67 (2014) , pp. 507-519.

[2] K.J. Chua, J.C. Ho, S.K. Chou, "A comparative study of different control strategies for indoor air humidity," Energy and Buildings 39 (2007), pp.537–545.

[3] J.A. Orosa, A. Baalina, "Passive climate control in spanish office buildings for long periods of time," Building and Environment 43 (12) (2008) 2005–2012.

[4] C.P. Underwood, HVAC Control Systems: Modelling, Analysis and Design, E & FN Spon, London and New York, 1999.

[5] Jianbo Bai, Shengwei Wang, Xiaosong Zhang, " Development of an adaptive smith predictor- based self- tuning PI controller for an HVAC system in a test room," Energy and Buildings 40 (2008) 2244–2252.

[6] S. B. Prusty, S. Padhee, U. C. Pati, and K. K. Mahapatra, "Comparative performance analysis of various tuning methods in the design of pid controller," in Proc. IET Michael Faraday International Summit, 2015, pp. 43–48.

[7] Sapna Gupta, Rajeev Gupta, " Controller parameter optimization using BFOA- PSO algorithm," 1st IEEE International Conference on Power Electronics. Intelligent Control and Energy Systems (ICPEICES-2016), 978-1-4673-8587-9/16/$31.00 ©2016 IEEE.

[8] J.M. Smith, "Closer control of loops with dead time," Chemical Engineering Process 53 (5) (1957), pp.217–219.

[9] T. Liu, Y.Z. Cai, D.Y. Gu and W.D. Zhang, " New modified smith predictor scheme for integrating and unstable processes with time delay," IEE Proc.-Control Theory Appl., Vol. 152, No. 2, March 2005.

[10] S.Maihi and D.P.Atherton, " Modified smith predictor and controller for processes with time delay," Electronics Letters 18th August 2011 Vol. 47 No. 17.

[11] K.V.L Narayana, Wendwosen Bellete Bedada and Kena Likassa Nefabas, " Enhanced modified smith predictor for higher order stable processes," IEEE Africon 2017 Proceedings 978-1-5386-2775-4/17/$31.00 ©2017 IEEE.

[12] Moina Ajmeri, Ahmad Ali, " Analytical design of modified smith predictor for unstable second order processes with time delay," International Journal Of Systems Science,2017, Vol.48,NO.8, 1671-1681.

[13] A. Seshagiri Rao, V. S. R. Rao, and M. Chidambaram," Simple Analytical Design of Modified Smith Predictor with Improved Performance for Unstable First-Order Plus Time Delay (FOPTD) Processes," Ind. Eng. Chem. Res. 2007, 46, 4561-4571.

[14] M. Morari, and E. Zafiriou, Robust Process Control, Prentice-Hall, Inc, New Jersey, 1989.

[15] Sahaj Saxenaa & Yogesh V. Hotea, "Advances in internal model control technique," IETE Technical Review , Vol 29, Issue 6, Nov-Dec 2012.

[16] Subhransu Padhee, Yuvraj Bhushan Khare, Yaduvir Singh, " Internal model based PID control of shell and tube heat exchanger system," proceeding of the 2011 IEEE Students' Technology Symposium, lIT Kharagpur, 978-1-4244-8943-5/11/$26.00 ©2011 IEEE.

[17] T. Liu, F. Gao, "New insight into internal model control filter design for load disturbance rejection," IET Control Theory Appl., 2010, Vol. 4, Iss. 3, pp. 448–460.

Adaptive Particle Swarm Optimization Based Maximum Power Point Tracking in Grid Connected PV System

Majid Jamil
Department of Electrical Engineering
Jamia Millia Islamia
New Delhi ,India
Majidjamil91@hotmail.com

M. Rizwan
Department of Electrical Engineering
Delhi Technological University
New Delhi, India
Rizwan@dce.ac.in

D.P. Kothari
Department of Electrixal Engineering
Wainganga Engineering College
Nagpur,India
dpkvits@gmail.com

Arjun Baliyan
Department of Electrical & Electronics Engineering
Inderprastha Engineering College
Ghaziabad, India
arjunbaliyaneee@gmail.com

Abstract—Maximum power point tracking (MPPT) approaches are exploited as a piece of Photovoltaic (PV) outlines to augment the PV array produce power through following reliably the maximum power point (MPP) which relies upon board's temperature and on irradiance circumstances.This paper deals with the improvement of MPPT approach by expanding the yield of the PV cluster under fractional shading circumstances. In our proposed strategy, the maximal power of the PV cluster will be tracked during partial shading conditions by using a novel procedure called Look-up table strategy and the optimization will be done by using the adaptive particle swarm optimization (APSO) algorithm. The resultant curves (*I-V* and *P-V* characteristic curves) are gotten because of using look-up table technique. The power produced by the traditional PSO algorithm and the suggested APSO based MPPT are especially analyzed under the partial shading states. The outcomes demonstrate that the APSO based MPPT can encourage the PV cluster to attain the global MPP and additionally aid the PV array to generate more stable yield control contrasted to the ordinary look-up technique. Asignificant benefit of the suggested methodology, measured up to the usual PSO algorithm is, it shouldn't get caught in the local optima and doesn't requires a higher number of function evaluations.

Keywords—MPPT; PV array; partially shaded conditions; adaptive particle swarm optimization (APSO); look-up table technique.

I.= INTRODUCTION

Owing to the speedy development in the electrical energy production and power electronics systems, the solar photovoltaic systems are outstanding amongst other regularly expanding renewable energy generation schemes since it offers an unconstrained, uncontaminated and environmentally friendly energy [1]. Photovoltaic (PV) energy is standout amongst the most significant power resources have a rising attention in electrical power claims that is spotless, pollution-free and inexhaustible and currently represents an extremely tiny share of electrical energy capability and fabrication [2]. One of the essential sorts of PV framework is known as Grid-connected photovoltaic power system in which the energy production depends on various factors. Grid-connected photovoltaic (PV) frameworks [3] is an electrical energyproducing solar PV power systems,which are associated with the utility grid that comprises of solar panels, one or more inverters, a power conditioning component and grid connection equipment. It is especially a low power, mostly single-phase PV ''rooftop'' systems [4] and their commitment to clean power generation are perceived increasinglyworldwide. An essential objective [5] of these frameworks is to expand theenergy infused to the grid by monitoring the most extreme PowerPoint of the panel, by decreasing the switching frequency, and by giving high dependability. Nowadays the grid-connected photovoltaic sources are utilized [6] in various applications for instance, home power supply,battery charging, swimming-pool warming frameworks, water pumping, satellite power frameworks and so forth. They have the favorable position [7] of being support and contamination-free, low conversion efficiency; owing to the low conversion proficiency, the general framework charge will be decreasedby means of high-efficiency power conditioners, moreover, are intended to remove the most extreme conceivable power from the PV module.

To boost the PV array yield, another strategy called MPPT has been used [8] by trailingpersistently the Maximum Power Point MPPthat reliesupon the panel's temperature and irradiance states, and the load electrical features. MPP is a unique point which has been used forshowing the current-voltage characteristic of the PV array under uniform irradiance. The MPPT controller [9] is normally executedin the dc/dc converter input phase of the solar unit. Essentially, the advanced voltage DC yield from solar panels has been converted to lower voltage required to control batteries [10].

This article is systematized such that, Section 2 examines existing work related to the PV model, Section 3 describes the individuality of the PV array beneaththe fractional shading conditions, Section 4 introduces our proposed optimization methodology called APSO algorithm, and includes the modeling of the proposed PV array using look-up table technique, Section 5 illustratesthe outcomes and the discussion, and Section 6 terminates the research study.

II. RELATED WORK

IulianMunteanu, AntonetaIulianaBratcu [11] came out with amaximumcontrol MPPT algorithm for grid-connected PV frameworkexclusive ofmediator DC-DC converter phase. The technique utilizedthe DC-connect voltage wave as irresolute signal thatnormallycomes about because of the framework inverter activity.

In [12] Ismail Hossain *et al.* proposed a keen strategy for MPPT of a PV system belowuneventemperature and insolation circumstances and transformation of thePV power into alleviated sine wave consumingtruncated distortion factor (DF), so that the PV power can be provided to grid and can workon the electrical and microelectronicstrategieseffectively.

A novel strategyfor MPPT by means of adaptive fuzzy logic control for grid-connected PVschemes had been suggested by NoppornPatcharaprakitiet al. in [13]. The framework was made from aboost converter and a single-phase inverter linked to a utility grid.

AymenChaouachiet al. [14] outlined a newstrategy for MPPT of a grid-connected PV framework via the neuro-fuzzy network. It anticipated the suggested PV voltage guarantying ideal power exchange among the PV originator and the primary utility grid. The neuro-fuzzy system was made from a fuzzy rule-based classifier and three multi-layered feed-forwarded Artificial Neural Networks (ANN).

Andrés Tobónet al. [15] deal with the optimization of MPPTwhilst a PVboard is demonstrated as two diodes. The received control is actualized using a Sliding Mode Control (SMC) and the enhancement was executed utilizing an upgradedpattern search method.

III. OPTIMAL POWER POINT TRACKING USING PSO

A. Modeling of photovoltaic system

PV cluster model [16] underfractionalshading is a $(m \times n)$ PV display model, where m representsthe sequentialseries in each division and n signifies the equivalentseries in the display.

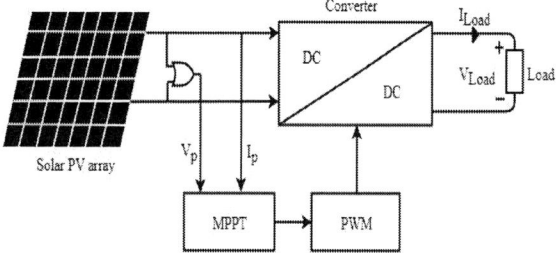

Figure 1: Block diagram of PV system with MPPT

Initially, amount of sequential and equivalentseries in the display is resolved. Also, modules with similar shading are viewed as a group. If nature stateretains consistent, the PV control yield is steady. On the off chance that the cluster is halfway shaded, radiation force to the exhibit diminishes, so the power yield diminishes and the PV yield trademark bends changes.

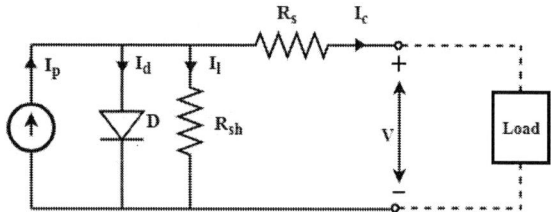

Figure 2 Circuit diagram of PV array

PV module comprises of various modeling methods. Among those various modeling methods here we are considering two types namely, (i) Single Diode PV model, (ii) Double Diode PV model.

a) Single Diode PV model

The single diode [17] comparable circuit of a solar unit is a presentresource in parallel with a solitary diode as

the shunt resistance and the series resistance. The design of this model is appeared in figure 3.

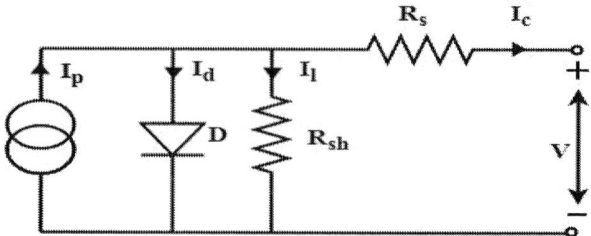

Figure 3: Circuit diagram of single diode PV model

The comparable representation has been unruffled of a current supply that can breed the photocurrent I_p and a solitary diode instead of the dispersionoccurrencepassed throughviathe current I_d. The spillage current from as of the shunt resistance R_{sh} is I_l that is origin by the circulated fabricatingdeformitywithin the solar cell configuration. The series resistance R_s reduces the solar cell effectivenessby means ofspreading the power in current form throughout the entireintersection substrates. Beneaththe light, the current I_c conveyed by the cell can be communicated in relationships of the photocurrent I_p the current I_d in the course of the diode and the leakage current I_l as indicated by the accompanying connection:

$$I_c = I_p - I_d - I_l \qquad (1)$$

The association of the current I_d and the leakage current I_l between the single voltage V across the cell possibly will be illustrated in (2) and (3) respectively.

$$I_d = I_{rs} \left[\exp\left(\frac{V + R_s I_c}{f_d V_t} \right) - 1 \right] \quad (2)$$

and

$$I_l = \frac{V + R_s I_c}{R_{sh}} \qquad (3)$$

Replace with the currents I_d and I_l, equation (1) becomes:

$$I_c = I_p - I_{rs} \left[\exp\left(\frac{V + R_s I_c}{f_d V_t} \right) - 1 \right] - \frac{V + R_s I_c}{R_{sh}} \quad (4)$$

$$V_t = \frac{T_{oc} k}{e} \qquad (5)$$

Where t is the temperature of the PV cell, I_{rs} is the reverse saturation current, f_d is the quality factor of the diode ranges between 1 and 2, T_{oc} is the temperature of the operating cell (°C), k denotes the Boltzmann constant and e is the charge of an electron.

B. Adaptive Particle Swarm Optimization (APSO) based MPPT

a) PSO Algorithm

PSO (Particle Swarm Optimization) [18] is a stochastic, populace-based inquiry strategy, displayed after the behavior of bird flocks. Itkeeps up a swarm of folks (particles), in which everyelementsymbolizes a candidate solution. Elementstake after aneffortless behavior: imitate the accomplishment of the nearest particles and its personalaccomplishedachievements. The position of a particle is, along these lines, impacted because ofthe preeminent particle in a region p_{best} in addition to the most excellent solution establishedwitheach of the elements in the overall population g_{best}. The position of the particle P_i has beenbalanced. In PSO algorithm, if N particles are assumedarbitrarily, every particle is a D-dimension vector. The ith particle is communicated as (6),

$$P_i = \left\{ P_{i_1}, P_{i_2}, \cdots, P_{i_D} \right\} \quad (6)$$

The ith particle knowledge's the present local optimal solution and present global optimal solution (best position), which can be denoted in Eqn. (7) and (8) respectively,

$$p_{best} = \left\{ p_{i_1}, p_{i_2}, \cdots, p_{i_D} \right\} \quad (7)$$

$$g_{best} = \left\{ g_1, g_2, \cdots, g_D \right\} \quad (8)$$

Flying velocity of the ith particle indicatedin (9)

$$v_i = \left\{ v_{i_1}, v_{i_2}, \cdots, v_{i_D} \right\} \quad (9)$$

The updationof the position and flying velocity of particles in the d-dimension $(1 \leq d \leq D)$, which can be represented as,

$$v_{ij}^{t+1} = w \cdot v_{ij}^t + \varphi_1 r_1 \left(p_{ij}^t - P_{ij}^t \right) + \varphi_2 r_2 \left(g_{ij}^t - P_{ij}^t \right) \quad (10)$$

$$P_{ij}^{t+1} = P_{ij}^t + v_{ij}^{t+1} \quad (11)$$

Where, w is an inertia weight coefficient that can be represented as $w = [0.4, 1]$, t denotes the iterations, φ_1 and φ_2 are denoted as learning factors to adjust learning step, r_1 and r_2 are two random numbers which are general as $r_1, r_2 \in [0, 1]$ to enlarge the ability of haphazardpenetrating. The value of φ_1 and φ_2 will be calculate by using the following equations

$$\varphi_1 = \frac{f_{max} + f_{min}}{2} \text{ and} \quad (12)$$

$$\varphi_1 = \frac{f_{max} - f_{min}}{2} \quad (13)$$

Where f_{max} denotes the maximum fitness vale (best fitness value) and f_{min} denotes the minimum (worst) fitness value. The inertia weight w can be represented as, in Eq. (7)

$$w = w_{max} - \left(w_{max} - w_{min} \right) \times {t_{current}}\Big/{G_{max}} \quad (14)$$

Where, $t_{current}$ is presentquantity of iterations and G_{max} is a predefined greatestamount of creations. A constriction factor C is used for the iterative progresses of PSO and the velocity is manipulated by,

$$v_{ij}^{t+1} = C \cdot \left[v_{ij}^t + \varphi_1 r_1 \left(p_{best_{ij}} - P_{ij}^t \right) + \varphi_2 r_2 \left(g_{best_{ij}} - P_{ij}^t \right) \right] \quad (15)$$

With

$$C = \frac{2}{\left| 2 - \theta - \sqrt{\theta^2 - 4\theta} \right|} \quad (16)$$

Where $\theta = \varphi_1 + \varphi_2$ and $\theta > 4$, the constriction factor C usually set to 2.05.

b) Adaptive PSO

The standard PSO algorithm has favorable position ofspeedycongregate rate yet owns a downsideto get caught in the local optima effortlessly when tackling complex multimodal issues and requires a higher number of capacity assessments. These deficiencies have confined the more extensive applications of the PSO. Consequently, accelerating convergence speed and maintaining a strategic distance from the local optima have turned into the two most vital and engaging objectives in the PSO research. To enhance this shortcoming numerousvariations of PSO have been proposed, yet it is discovered that accomplishing both goalsconcurrently is very difficult. To accomplishthose goals, Adaptive PSO (APSO) [19] has beenfigured. Numerous specialists proposedthe distinctive inertia weight procedures. They can be sorted as,

(i) Constant (Steady)

In thisprocedure, substantial estimation of inertia weight encouragesoverallstudy and a small value encouragesthe narrow search.

(ii) Random (Arbitrary)

In thisprocedure, thesystemsaugment the union of PSO in the beginning of the algorithm.

(iii) Time-varying

In time-varying inertia weight systems, it is characterizedthe same as a component of time.

In any case,every one of these systems has the shortcoming of untimelyconvergence to the local minimum. With the intention ofconquer the shortcoming, adaptive inertia weight strategies will be proposed.

(iv) Adaptive strategies

In thisprocedure, to enhance the searching ability and to keep up the assorted variety of the populace by modifying the inertia weight. During this investigation, fitness estimation of particles will be utilized for adjusting the inertia weight.

The proportion of the particle's fitness 'α' and the typical fitness of the swarm is figured by using the equations (17) and (18).

$$f_{avg} = \frac{\sum f_i}{n} \quad (17)$$

$$\alpha = \frac{f_i}{f_{avg}} \quad (18)$$

Where f_i is the fitness of ith particle, n is the quantity of the particles in the swarm. The term α is more than 1 for individuals whose present fitness is more than the average fitness of the swarm. Execution of theelements is better and they are nearer to the ideal solution. In this way, a smaller inertia weight ought to be used and computed by eqn. (19).

$$w_i = w_{\min} + \left(w_{\max} - w_{\min} \right) \times \left[\frac{\left(f_i - f_{avg} \right)}{\left(f_{\max} - f_{avg} \right)} \right] \quad (19)$$

The term α is less than 1 at the point when the fitness of the particleis not as much as the average fitness of the group. Expanding inertia weight know how tofortify the particle'sseeking ability and empowers it to hop out of local minimum. Along these lines, thehigh inertia weight ought to beused and computedvia eqn. (20).

$$w_i = w_{\min} + \left(w_{\max} - w_{\min} \right) \times \left[\frac{\left(1 + P \right)}{e^{\alpha}} \right] \quad (20)$$

Algorithm

1. Initialize the position $P_i = \left\{ P_{i_1}, P_{i_2}, \cdots, P_{i_D} \right\}$ and velocity $v_i = \left\{ v_{i_1}, v_{i_2}, \cdots, v_{i_D} \right\}$.
2. Set $p_{best_{ij}} = P_i$, $t_{current} = 0$, $w = 0.9$.
3. Calculate φ_1 and φ_2 using eqns (12) and (13).
4. Calculate g_{best}.
5. Estimate the transformative conditions of the algorithm and adaptively control the attributes.
6. Perform elitist learning operation in the union state.
7. Set $i = 1$.
8. **If**$\left(i \leq N \right)$
9.   Update the location and speedof particle i.
10.   **If**$\left(P_{\min} \leq P_i \leq P_{\max} \right)$
11.   Evaluate particle i and update $p_{best_{ij}}$ and g_{best}.
12.   Set $i = i + 1$ and go to step 7.
13.   **Else** set $i = i + 1$ and go to step 7.
14. **Else** assign $t_{current} = t_{current} + 1$
15.   **If**$\left(t_{current} < G_{\max} \right)$
16.   go to step 4.
17.   **Else** end If.

Algorithm 1: Proposed APSO algorithm

C. Modeling and Simulation

The modeling of MPPT systems is carried out using the look-up table technique. The look-up table block makes use ofa variety of information to outlinekeyprinciples to the yield values. Modeling carry outs a "look-up" task for theknownkey values, to recover the equivalentyield values from the table. If the key values are not characterized, look-up table block gauges the yield values in view of close-by values.

a) Look-Up Table technique

The look-up table technique [20] has been selected to follow the MPP of the PV cluster because it uses the basic technique, has ease of implementation and it is simple. The Lookup Table in MPPT method uses a preset table to determine the MPP which leads to very fast responses and do not require any forced biasing to find the MPP. It can predict the MPP for any number of conditions, however with each additional condition, the number of sensors increases proportionally, and the number of calculated values increases exponentially. This significantly limits the practicality of using such a method.

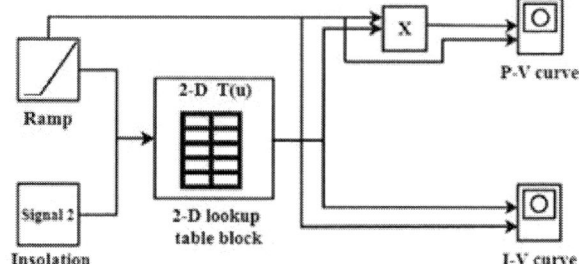

Figure 4: Circuit diagram of look-up table technique

In the look-up table techniques, the previous information of PV panel material [21], specialized information, panel attributes at different ecological conditions are required and are stored. At that point themeasured values of the PV paneyield current *I-P* and output voltage *V-P* and the detected power iscontrasted withthe accumulated values to trailthe MPP;the working point is shifted to new MPP and accumulated in the memory of MPPT's control framework. Throughout this task, the relating MPP for a constraint is chosen out of the memory and executed. In this way,alot of information is required to be stored before locating the MPP.

b) Lookup table technique in our proposed methodology

For our proposed MPPT strategy, a look-up table is detailed considering theexperimental evidenceacquired [22]. The electronic load is used for tracking the *I-V* and P-V qualities of the PV module which are then gotten from the stored data in digital storage oscilloscope. The MPP at different insolation levels is acquired from the *P-V* curve. The voltage and current information for different insolation levels are used to build up the lookup table. For any given insolation level, the created model imitates the behavior of the real PV module. Since the computations engaged with this model are straightforward.

So as to acquire the *I-V* and *P-V* characteristic curves by means of our proposed strategy, the insolation value is entered through a consistent block and the adjustment in voltage is specified by a ramp signal. The *I-V* and *P-V*attributes obtained by this strategy are appeared in figure-5 and figure-6, individually. The estimationsof V_{mp} at various insolation levels are then given to the lookup table in this way empowering it to compute the voltage V_{mp} which relates to the given insolation level without much calculation. A pilot board, which has comparative qualities as that of the alternate boards in the framework can be used to quantify the open circuit voltage and short circuit current. From this data, the insolation level can be resolved and corresponding V_{mp} can be gotten from the lookup table.

This approach decreases the complexity and tracks the V_{mp} quicker than the alternate algorithms.

IV. SIMULATION RESULTS

A. Implementation and Results

978-1-5386-6624-1/18 $31.00 © 2018 IEEE

The proposed strategy has been executed via MATLAB. The *I-V* and *P-V* characters of the PV cluster beneath incomplete shading circumstance are produced by using the look-up table technique are appeared in the figures. The yield power and the current spawned by the PV array are shown by the *P-V* and the *I-V* characteristics curves individually.

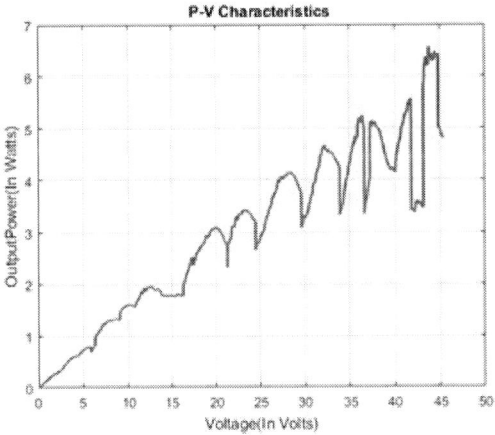

Figure 5: *P-V* characteristic curve of PV array

The above figure represents the *P-V* character of the PV cluster. The yield power of the PV cluster is measured based on the voltage of the PV cluster. According to the above figure, if the voltage of the PV array increases then the output power also rises simultaneosly. By using our proposed method, the maximum power reached by the PV array is 4.9W when the voltage is 45V.

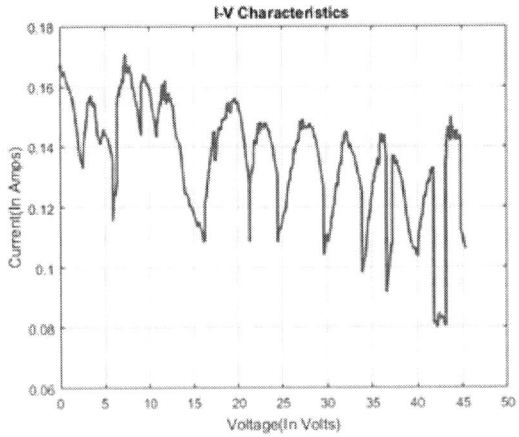

Figure 6: *I-V* characteristic curve of PV array

The current delivered by PV cluster in fractionalgloomingsurroundings is demonstrated in figure 6 by using the *I-V* characteristics curve. As same as the *P-V* curve, the current will be measured based on the amount of voltage. In the *I-V* characteristics, if the voltage is increased then the equivalent amount of the current will be decreased. The amount of current approximate 0.11A is produced besides the practical operating voltage from 0V to 50V.

B. Comparison

The power generation of the anticipated APSO-LUT procedure is associated to the standard PSO algorithm.

The examination has been made for various test conditions and from the outcomes; the LUT based MPPT reacts sooner i.e. optimal point locating time is slighter than the customary model. For compound interfacing frameworks like grid-connected PV classifications, the anticipated LUT-MPPT not just decreases the complexity of the entire system yet, in addition, condenses the locating time and simulation time.

The comparison will be performed for the yield control in view of the number of iterations of the suggested APSO and the conventional PSO algorithms. The delineated APSO based look-up table technique and the conventional PSO techniques are tested for two diverse shading designs, as portrayed in figure 7.

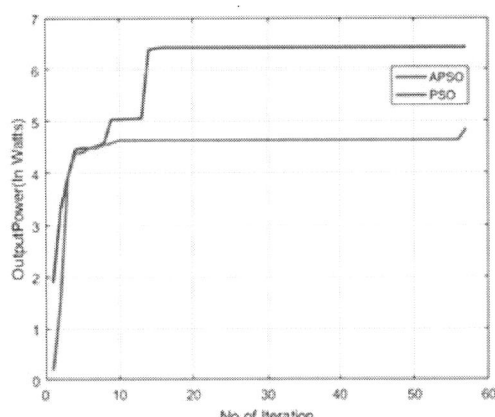

Figure 7: Comparison between PSO and APSO

In the conventional PSO algorithm it is shown that for the functional voltage from 0V to 60V, the maximum power produced is 4.5W. But in the proposed APSO algorithm for the same voltage range, the maximum output power reached is 7.5W. Thus, the above figure demonstrates that our proposed APSO technique has produced high output power than the conventional PSO algorithm.

V. CONCLUSION

An MPPT control procedure is pivotal for PV production frameworks, which can engender highest production power still if atmospheric conditions are distorted has been premeditated in this paper. This paper analyzes the PV array performance using Adaptive Particle Swarm Optimization (APSO) based Look-up technique for quick locating of MPP under varying lighting circumstances. The power generation of the proposed APSO based LUT are explored once the PV array presents under partially shaded environments. Furthermore, it decreases the number of computations of the PV interfacing scheme. For LUT method, the requisite data are extorted from the realistic characteristic using electronic load. The look-up table is developed in MATLAB and is tested for fluctuating illumination constraints. The output power generation of the proposed APSO based LUT has been balanced with the conventional PSO in terms of output power. It is found that the APSO based approach produces high power in ecological state of affairs and follows the MP more rapidly. This way of executing MPPT is very much constructive for simulation studies of superior PV supplied system but LUT

978-1-5386-6624-1/18 $31.00 © 2018 IEEE

requires real-time data to obtainappropriateoutcomes. In future, the accomplishment and the toughness of the suggested APSO are experiencedrealistically.

REFERENCES

[1] Enany, Mohamed A., Mohamed A. Farahat, and Ahmed Nasr. "Modeling and evaluation of main maximum power point tracking algorithms for photovoltaics systems." Renewable and Sustainable Energy Reviews, vol.58, pp.1578-1586, 2016.

[2] Kuo, Yeong-Chau, Tsorng-Juu Liang, and Jiann-Fuh Chen. "Novel maximum-power-point-tracking controller for photovoltaic energy conversion system." IEEE transactions on industrial electronics, vol.48, no.3, pp.594-601, 2001.

[3] Kadri, Riad, Jean-Paul Gaubert, and Gerard Champenois. "An improved maximum power point tracking for photovoltaic grid-connected inverter based on voltage-oriented control." IEEE transactions on industrial electronics, vol.58, no.1, pp.66-75, 2011.

[4] Villanueva, Elena, Pablo Correa, José Rodríguez, and Mario Pacas. "Control of a single-phase cascaded H-bridge multilevel inverter for grid-connected photovoltaic systems." IEEE Transactions on Industrial Electronics, vol.56, no.11, pp.4399-4406, 2009.

[5] Villanueva, Elena, Pablo Correa, José Rodríguez, and Mario Pacas. "Control of a single-phase cascaded H-bridge multilevel inverter for grid-connected photovoltaic systems." IEEE Transactions on Industrial Electronics, vol.56, no.11, pp.4399-4406, 2009.

[6] Selvaraj, Jeyraj, and Nasrudin A. Rahim. "Multilevel inverter for grid-connected PV system employing digital PI controller." IEEE Transactions on Industrial Electronics, vol.56, no.1, pp.149-158, 2009.

[7] Denholm, Paul, and Robert M. Margolis. "Evaluating the limits of solar photovoltaics (PV) in traditional electric power systems." Energy policy, vol.35, no.5, pp.2852-2861, 2007.

[8] Koutroulis, Eftichios, Kostas Kalaitzakis, and Nicholas C. Voulgaris. "Development of a microcontroller-based, photovoltaic maximum power point tracking control system." IEEE Transactions on power electronics, vol.16, no.1, pp.46-54, 2001.

[9] Femia, Nicola, Giovanni Petrone, Giovanni Spagnuolo, and Massimo Vitelli. "Optimization of perturb and observe maximum power point tracking method." IEEE transactions on power electronics, vol.20, no.4, pp.963-973, 2005.

[10] Rakhshan, Mohsen, NavidVafamand, Mohammad-Hassan Khooban, and FredeBlaabjerg. "Maximum power point tracking control of photovoltaic systems: a polynomial fuzzy model-based approach." IEEE Journal of Emerging and Selected Topics in Power Electronics, vol.6, no.1, pp.292-299, 2018.

[11] Munteanu, Iulian, and AntonetaIulianaBratcu. "MPPT for grid-connected photovoltaic systems using ripple-based

Extremum Seeking Control: Analysis and control design issues." Solar Energy, vol.111, pp.30-42, 2015.

[12] Hossain, Md Ismail, ShakilAhamed Khan, MdShafiullah, and Mohammad Jakir Hossain. "Design and implementation of MPPT controlled grid connected photovoltaic system." In IEEE Symposium on Computers & Informatics (ISCI), 2011, pp. 284-289. IEEE, 2011.

[13] Patcharaprakiti, Nopporn, SuttichaiPremrudeepreechacharn, and YosanaiSriuthaisiriwong "Maximum power point tracking using adaptive fuzzy logic control for grid-connected photovoltaic system." Renewable Energy, vol.30, no.11, pp.1771-1788, 2005.

[14] Chaouachi, Aymen, Rashad M. Kamel, and Ken Nagasaka. "A novel multi-model neuro-fuzzy-based MPPT for three-phase grid-connected photovoltaic system." Solar energy, vol.84, no.12, pp.2219-2229, 2010.

[15] Tobón, Andrés, JuliánPeláez-Restrepo, Juan Villegas-Ceballos, Sergio I. Serna-Garcés, Jorge Herrera, and AsierIbeas. "Maximum Power Point Tracking of Photovoltaic Panels by Using Improved Pattern Search Methods." Energies, vol.10, no.9, pp.1316, 2017.

[16] Sampaio, Priscila GonçalvesVasconcelos, and Mario Orestes Aguirre González. "Photovoltaic solar energy: Conceptual framework." Renewable and Sustainable Energy Reviews, vol.74, pp.590-601, 2017.

[17] Ashouri-Zadeh, Alireza, MohammadrezaToulabi, Ahmad SalehiDobakhshari, SiavashTaghipour-Broujeni, and Ali Mohammad Ranjbar. "A novel technique to extract the maximum power of photovoltaic array in partial shading conditions." International Journal of Electrical Power & Energy Systems, vol.101, pp.500-512, 2018.

[18] Ishaque, Kashif, Zainal Salam, Muhammad Amjad, and SaadMekhilef. "An improved particle swarm optimization (PSO)–based MPPT for PV with reduced steady-state oscillation." IEEE transactions on Power Electronics, vol.27, no.8, pp.3627-3638, 2012.

[19] Li, Hong, Duo Yang, Wenzhe Su, Jinhu Lu, and Xinghuo Yu. "An Overall Distribution Particle Swarm Optimization MPPT Algorithm for Photovoltaic System under Partial Shading." IEEE Transactions on Industrial Electronics, 2018.

[20] Kumar, Sumit, D. K. Yadav, and D. A. Khan. "An Adaptive PSO Algorithm Based Test Data Generator for Data-Flow Dependencies using Dominance Concepts." International Journal of Software Engineering and Its Applications, vol.10, no.11, pp.59-82, 2016.

[21] Salas, V., E. Olias, A. Barrado, and A. Lazaro. "Review of the maximum power point tracking algorithms for stand-alone photovoltaic systems." Solar energy materials and solar cells, vol.90, no.11, pp.1555-1578, 2006.

[22] Bhatnagar, Pallavee, and R. K. Nema. "Maximum power point tracking control techniques: State-of-the-art in photovoltaic applications." Renewable and Sustainable Energy Reviews, vol.23, pp.224-241, 2013.

Performance Evaluation of 3-Phase 4-Wire SAPF based on Synchronizing EPLL with Fuzzy Logic Controller

Seema Agrawal[*1], Deepika Sharma[2], Vijay Kumar Gupta[3]
Department of Electrical Engineering
Rajasthan Technical University, Kota-324010, INDIA
[1]seema10dec@gmail.com, [2]deepikabhardvaj@gmail.com,
[3]vijaygupta2905@gmail.com

R. K. Somani[4]
Department of Computer Science Engineering
(BKIT) Rajasthan Technical University,
Kota-324010, INDIA [4]rksomani1@gmail.com

Abstract— This paper work focuses on current harmonic cancelation, reactive power compensation, neutral current compensation and power factor adjustment in a 3 phase four wire (3-ph-4w) system using shunt active power filter (SAPF) by incorporating synchronizing EPLL with fuzzy logic controller. The proposed system designed with non-linear uncontrolled rectifier feeding with R-L load, eight IGBT switch for current controlled voltage source inverter, interfacing inductor and DC bus capacitor. Hysteresis controller generates pulses for CI-VSI. The synchronizing EPLL generate sine cosine signals exercised to generate unit vector for reference current generation in SAPF to suppress harmonics. EPLL has advantage over traditional PLL is to maintain same phase between input signal and output signal of voltage controlled oscillator (VCO). Artificial intelligence technique based fuzzy logic controller has incorporated for DC bus voltage regulation. Proposed system is designed under MATLAB/SIMULINK platform 2015(b). The simulation study shows that proposed system mitigate harmonics and improve power quality with efficacy.

Keywords— Power Quality, Capacitor Interfaced Voltage Source Converter (CI-VSI), Shunt Active Power Filter (SAPF), Enhanced Phase Locked Loop (EPLL), Fuzzy Logic Controller (FLC).

I. INTRODUCTION

Presently in distribution system networks has remarkable augmentation of non linear loads at industry as well as consumer level. Non linear load categorized of AC/DC or DC/DC converters, battery charger, electronic switches, adjustable speed drive etc. These loads accounts for non sinusoidal current and voltages at load side and results in increased losses poor quality of power, lower efficiency and high THD [1, 2].

In order to mitigate voltage as well as current harmonics several types of filters are reported as passive, active and hybrid filters. Conventional passive filter has shortcomings like bulkiness, sluggish response, resonance and ageing effect [3]. Active power filter gained more attention as their dynamic performance is fast, accurate and fine estimation of current reference [4].

Active filter are required to eradicate the unnecessary harmonic created in distribution system and compensate unsystematic varying currents. Performance of APF depends on accurateness and time taken in compensating harmonics constituent [5]. Reference current generation for active power filter has derived by incorporating various control strategies [6, 7].

Reference control / current signal for APF is generated by various time domain based control methods such as instantaneous active reactive power(P-Q) theory, instantaneous active reactive current (I_d-I_q) theory [8], modified P-Q theory [9], synchronous reference frame (SRF) theory [10], sliding mode control [11], self tuning filter (STF) [12] and positive sequence detection (PLL) as reported in various literatures [14]. All these methods have various rewards and drawbacks compared to each other but appropriate techniques for reference current extraction is selected by deciding working condition and extent up to its accuracy [15]. PLL has electronic circuitry incorporating a voltage and voltage controlled oscillator (VCO) that frequently regulate to match frequency of an input signal and creates an output signal whose phase is related to phase of an input signal [16].

The enhanced phase-locked loop (EPLL) overcomes ordinary PLL by removing its chief drawback, which is existence of double-frequency errors [17]. EPLL achieves this task by estimating amplitude of input signal and also provides a filtered version of input signal by removing error within a fresh loop. Thus, removing ripples by functioning as a filter as well as controller. This facilitates its application for phasor estimation and also for straightforward use in control mechanism. DC bus energy regulation is based on soft computing techniques [13, 18].

Hence, in this article, a new method for reference current generation using synchronizing EPLL with fuzzy logic controller is discussed. EPLL is used to extract fundamental component of source current and estimation of its frequency. The proposed control system with artificial fuzzy logic controller is simulated using MATLAB/Simulink platform. The efficacy of control strategy has been verified by simulation results.

II. SYSTEM CONFIGURATION

The suggested system structure is depicted in Fig 1. It comprises of 3-phase 4-wire AC source is attached with uncontrolled rectifier feeding with R-L load and shunt active power filter which is fabricated by incorporating eight IGBT switches and connected at PCC through filter resistance and inductance (R_f, L_f).

Capacitor is interfaced on DC link to make its self supportive. Reference current for SAPF is generated by unit template signals introduced by synchronizing EPLL to suppress harmonics.

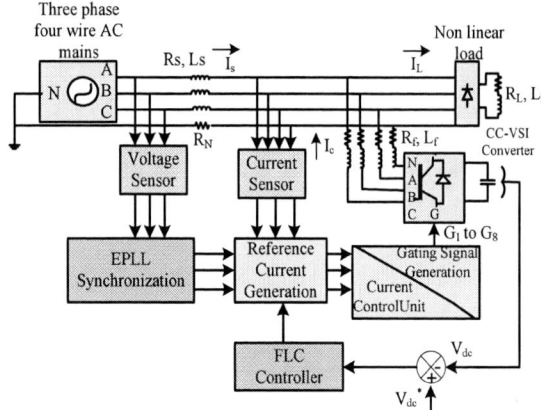

Fig.1. Proposed system block diagram with SAPF incorporating synchronizing EPLL with fuzzy logic controller.

Compensation ability of SAPF depends on accuracy of reference signal generation. SAPF introducing equal and opposite current as compared to harmonic current and compensate load current harmonics. Intelligent fuzzy logic controller is integrated with system to regulate error between actual DC link voltage (V_{dc}) and reference DC voltage (V_{dc}^*). Hysteresis controller generates pulses for CI-VSI. Capacitor interfaced VSI switching losses are provided by DC bus voltage controller. CC-VSI with eight IGBT switches starts conducting at 0.1sec before that harmonics are familiarized in network and source current waveforms turn out to be distorted due to non-linear load.

III. REFERENCE CURRENT PRODUCTION

Three phase reference currents ($I_{sa}^*, I_{sb}^*, I_{sc}^*$) for harmonic component mitigation has generated by using unit template generation with synchronizing EPLL strategy and fuzzy logic controller.

A. Enhanced phase locked loop

For unit template generation sine and cosine angle can generate by enhanced PLL. EPLL is a feedback control system as represented in Fig. 2.

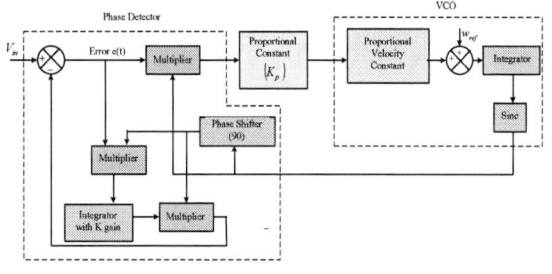

Fig. 2. Block diagram of enhanced phase locked loop (EPLL).

Phase detector is supplied with two inputs one from supply voltage while another from output of voltage source oscillator. The phase difference between these two signals can be made and maintained zero in EPLL.

Integration, three multiplication and subtraction operations are performed in suggested synchronizing PLL.

Let $e(t)$ be the phase detector output signal and $V_{in}(t)$, $V_{out}(t)$ are introduced and produced signals from EPLL respectively then mathematical equations governing its operation are listed as below:

$$V_{in}(t) = V\sin(\omega t + \phi) \tag{1}$$

$$V_{out}(t) = V\sin(\omega t + \bar{\phi}) \tag{2}$$

$$e(t) = V_{in}(t) - V_{out}(t) \tag{3}$$

Error has dependences on voltage amplitude (V), frequency (ω) and phase angle (φ).

$$e(V, \omega, \phi) = \left\| V_{in}(t) - V_{out}(t) \right\|^2 \tag{4}$$

The sine output signal of enhanced PLL is expressed as follows:

$$Y_s = \sin \omega t \tag{5}$$

Synchronizing unit vector signals are measured by EPLL shown in Fig. 3.

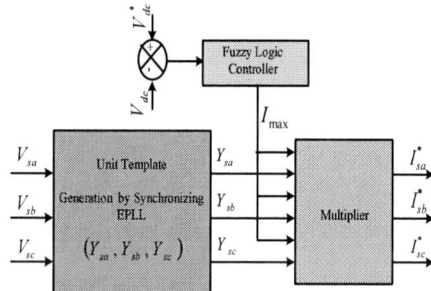

Fig. 3. Building block for unit template signal generation.

Three phase unit template signals are Y_{sa}, Y_{sb}, Y_{sc} then unit vector (U) is composed as:

$$\begin{bmatrix} Y_{sa} \\ Y_{sb} \\ Y_{sc} \end{bmatrix} \equiv \begin{bmatrix} \sin(\omega t) \\ \sin(\omega t - 120) \\ \sin(\omega t + 120) \end{bmatrix} \tag{6}$$

B. FLC Controller

Artificial intelligent based fuzzy logic controller (FLC) is applied as DC side voltage regulation by mitigating real power losses. The DC link voltage is sensed and compared with reference value of DC link voltage of VSI capacitor for getting error and change in error signal, as shown in (7)

$$e(k) = V_{dc}^* - V_{dc}(k)$$
$$ce(k) = e(k) - e(k-1) \qquad (7)$$

The subscript e(k) and ce(k) are error and change in error signal at k^{th} sampling time used as inputs for fuzzy processing.. The output of FLC is I_{max}^* and denoted as:

$$I_{max} = I_{max}^*(k-1) + \delta I_{max}^*(k) \qquad (8)$$

Above current I_{max} takes care of active power demand of load and losses in the converter system. Reference source and neutral currents are defined as follows:

$$\begin{bmatrix} I_{sa}^* \\ I_{sb}^* \\ I_{sc}^* \end{bmatrix} = \begin{bmatrix} Y_{sa} \\ Y_{sb} \\ Y_{sc} \end{bmatrix} [I_{max}] \qquad (9)$$

$$I_{sn}^* = I_{sa}^* + I_{sb}^* + I_{sc}^* \qquad (10)$$

Switching signals for VSI are formed using difference of real and reference source currents. Errors are passed through hysteresis PWM to generate required switching signals. Switching (S) position is defined as below:

0.01 is hysteresis band limit in equation 11.

$$S = \begin{bmatrix} 1 & i_{sa} \le i_{sa}^* - .01; & i_{sb} \le i_{sb}^* - .01; & i_{sc} \le i_{sc}^* - .01; \\ 0 & i_{sa} \ge i_{sa}^* + .01; & i_{sb} \ge i_{sb}^* + .01; & i_{sc} \ge i_{sc}^* + .01; \end{bmatrix} \qquad (11)$$

Fuzzy logic rules are defined in fuzzy rule matrix as Table I. Fig. 4 and Fig. 5 show degree of membership both inputs (e, ce) of Fuzzy controller. Degree of membership for output is displayed in Fig. 6 as per designed rule.

TABLE I. FUZZY RULE MATRIX

e/ce	NB	NM	NS	ZE	PS	PM	PB
NB	NB	NB	NB	NB	NM	NS	ZE
NM	NB	NB	NB	NM	NS	ZE	PS
NS	N	NB	NM	NS	ZE	PS	PM
ZE	NB	NM	NS	ZE	PS	PM	PB
PS	NM	NS	ZE	PS	PM	PB	PB
PM	NS	ZE	PS	PM	PB	PB	PB
PB	ZE	PS	PM	PB	PB	PB	PB

IV. SIMULATION RESULTS

MATLAB platform has been utilized to test compatibility of suggested system architecture as depicted in Fig. 1. Non linear diode rectifier is connected to introduce non linearity in system. it s efficacy is tested for T_s= 0 to 0.4 sec and pulse supplied by hysteresis PWM band controller after 0.1 sec and harmonic mitigation of source current is obtained.

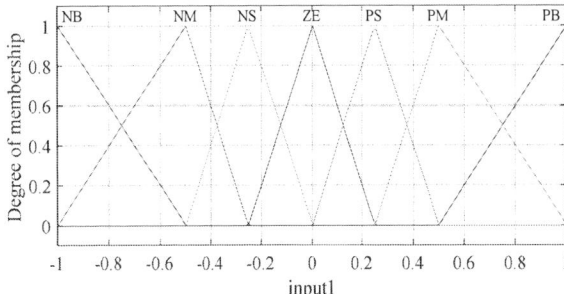

Fig. 4. Membership functions for input 1.

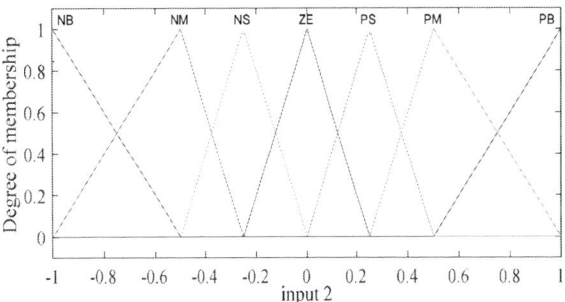

Fig. 5. Membership functions for input 2.

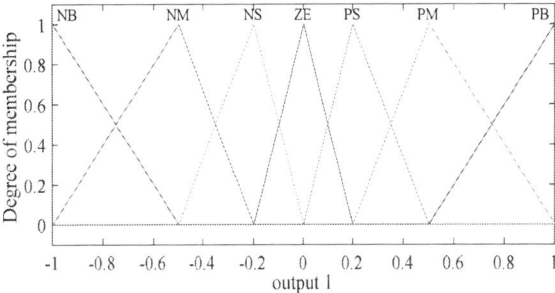

Fig. 6. Membership functions for output 1.

TABLE II. SIMULATION PARAMETERS

Source	Fundamental supply voltage 326V (peak phase-phase) Source frequency = 50 Hz, R_N =50ohm
SAPF	Reference DC link Voltage, V_{dc}^*=500V DC link Capacitor, C_{dc}=1600µF Filter inductor, L_f = 3.35mH, R_f = 0.4
EPLL	K_p=1.3424 and K=6.0781, K_v=.0000005 ω_{ref} = 314 rad/sec.
Loads	Three phase diode rectifier with L_L=100mH and R_L = 100Ω load. Simulation time 0 to 0.4 seconds

Before 0.1 step time, harmonic are inserted in system due to presence of uncontrolled converters, source current becomes distorted. At 0.1 sec., four leg SAPF come in picture through PWM pulse and proper sine shaped with Fig. 10. Active power, reactive power and power factor minimum permissible %THD source current signal I_{sabc} is produced.

Tracking phase for phase 'a' and sine cosine signal are

formed by EPLL as given away in Fig. 7.

Simulation study parameters such as source voltage (V_{sabc}), non linear load current (I_{labc}), compensated source current (I_{sabc}), harmonic current (I_{fabc}), regulate DC bus voltage (V_{dc}) from up to down in Fig. 8 using artificial intelligence based FLC controller.

Neutral current is illustrated in Fig. 9 which is also required compensation in case of three phase four wire grid structure.

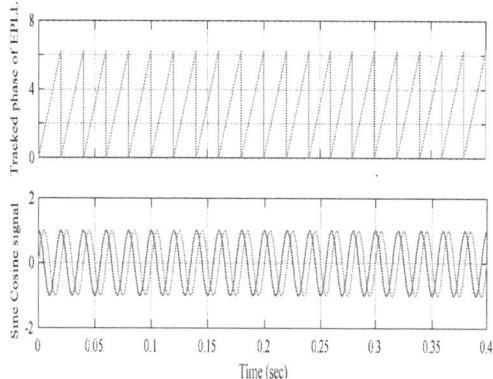

Fig.7. EPLL tracked phase / tacked signal for phase' a' and sine cosine angle measurement using of EPLL.

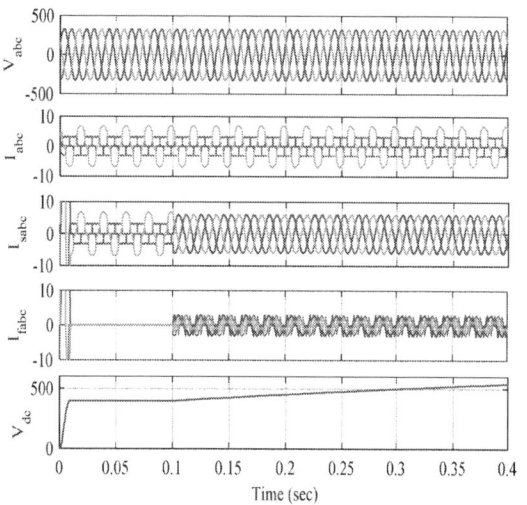

Fig. 8. Performance characteristics of SAPF (V_{sabc}, I_{labc}, I_{sabc}, I_{fabc}, V_{dc}).

Active power (kW), reactive power (KVAR) and Power factor are publicized in Fig. 10.

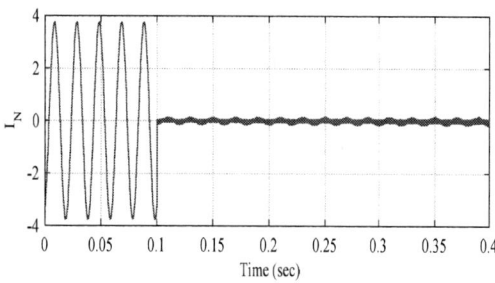

Fig. 9. Neutral current.

Improved power factor represents source current is in phase with source voltage after 0.1sec.

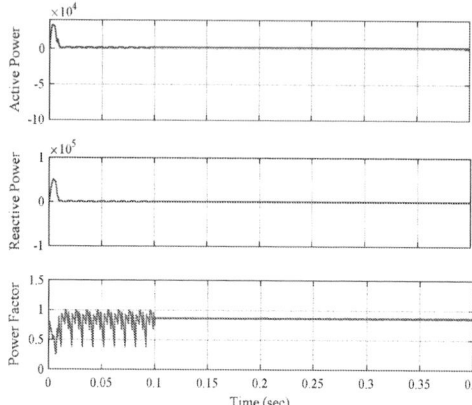

Fig. 10. Active power, reactive power and power factor

Performance index %THD of supply current before and after application of four leg SAPF are corresponding to Fig. 11 respectively.

After 0.1sec 99.82% reduction in total harmonic distortion is obtained with level of current 3.429 amperes for phase 'a'. This shows a effective elimination of THD under uncontrolled converter load.

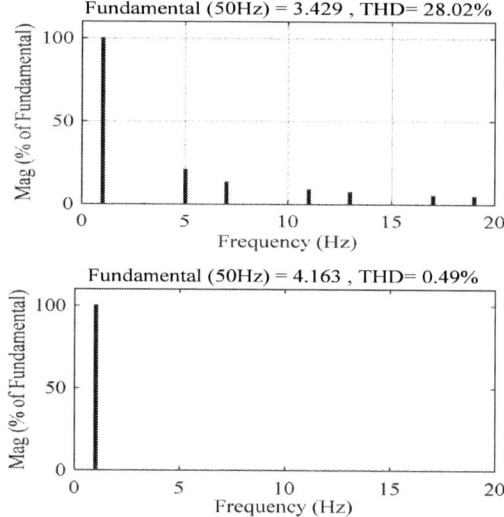

Fig. 11. Spectral analysis of uncompensated and compensated source current for phase 'a'.

V. CONCLUSION

The performance evaluation of three phase four wire (3-ph 4-wire) SAPF has been examined under uncontrolled converter feeding resistive inductive load for its ancillary applications as current harmonic cancelation, reactive power compensation, burden of neutral current compensation, power factor adjustment. Reference current extraction for shunt active power filter has been done by incorporating synchronizing EPLL and fuzzy logic controller as DC link energy regulation. Suggested control strategy has been proved efficacy for compensation of double frequency error generated in PLL. The EPLL has been achieved tracked phase and magnitude signal. The system performance with proposed FLC controller has been evaluated in terms of harmonic suppresion and DC bus energy regulation with no overshoot at transistion time. Source current has been made sinusoidal i.e. effective reduction of its THD from 28.02% to 0.05% (99.82%) with level of 3.429 amp. Hence, effectiveness of suggested system has been investigated to mitigate current harmonics and neutral current compensation.

REFERENCES

[1] K. Nikum, R. Saxena, and A. Wagh, "Power quality assessment of modern residential load", 7th India International Conference on Power Electronics, pp. 1-4, 2016.

[2] M. J. Ghorbani, and H. Mokhtari, "Impact of harmonics on power quality and losses in power distribution systems", International Journal of Electrical and Computer Engineering, vol. 5, no. 1, pp. 166-174, February 2015.

[3] A. Baitha, and N. Gupta, "A comparative analysis of passive filters for power quality improvement", IEEE International Conference on Advancements in Power and Energy, pp. 327-33, June 2015.

[4] L. B. G. Campanhol, S. A. O. da. Silva, and A. Goedtel, " Application of shunt active power filter for harmonic reduction and reactive power compensation in three-phase four-wire systems", IET Power Electronics, vol. 7, no.11, pp. 2825-2836, 2014.

[5] R. Dash, P. Paikray, and S. C. Swain, "Active power filter for harmonic mitigation in a distributed power generation system", IEEE International Conference on Power and Advanced Computing Technologies, pp. 1-6, April 2017.

[6] M. Aredes, J. Hafner, and K. Heumann, "Three-phase four-wire shunt active filter control strategies" ,IEEE Transactions on Power Electronics, vol. 12, no. 2, pp. 311-318, 1997.

[7] H. A. Moghaddam, A. Vahedi, and S. H. Ebrahimi, "Optimum control strategy for shunt active power filter under disturbed AC voltage and unbalanced load conditions in 4-wire power system", IEEE in Power Electronics, Drive Systems & Technologies Conference (PEDSTC), pp. 217-222, February 2017.

[8] S. Biricik, O. C. Ozerdem, S. Redif, and M. O. Kmail, "Performance improvement of active power filters based on p-q and d-q control methods under non-ideal supply voltage conditions", IEEE International Conference on Electrical and Electronics Engineering, pp. 312-316, 2011.

[10] K. Kelesidis, G. Adamidis, and G. Tsengenes, "Investigation of a control scheme based on modified p-q theory for single phase single stage grid connected PV system", International Conference on Clean Electrical Power, pp. 535-540, 2011

[11] M. C. Benhabib, and S. Saadate, "New control approach for four-wire active power filter based on the use of synchronous reference frame", Electric Power Systems Research, vol. 73, no. 3, pp. 353-362, 2005.

[12] S. Jain, S. Agarwal, A. Jain, and D. K. Palwalia, "Applied precise Multivariable control theory on shunt dynamic power filter using sliding mode controller", IEEE International Conference in Power Electronics, Intelligent Control and Energy Systems, pp. 1-4, July 2016.

[13] S. Agrawal, D. Sharma, and D. K. Palwalia, "Performance analysis of SAPF based on self tuned harmonic filter with fuzzy logic controller", IEEE in Control, Automation & Power Engineering, pp. 487-492, October 2017.

[14] P. Rodriguez, A. Luna, R. Teodorescu, F. Blaabjerg, and M. Liserre, "Control of a three-phase four-wire shunt-active power filter based on DC-bus energy regulation", IEEE International Conference on Optimization of Electrical and Electronic Equipment, pp. 227-234, May 2008.

[15] K. R. Patjoshi, and K. K. Mahapatra, "Performance analysis of shunt active power filter using PLL based control algorithms under distorted supply condition", IEEE Students Conference on Engineering and System, pp. 1-6, April 2013.

[16] M. I. M. Montero, E. R. Cadaval, and F. B. Gonzalez, "Comparison of control strategies for shunt active power filters in three-phase four-wire systems", IEEE Transactions on Power Electronics, vol. 22, no.1, pp. 229-236, 2007.

[17] K. G. Masoud, S. A. Khajehoddin, P. K. Jain, and A. Bakhshai, "Derivation and design of in-loop filters in phase-locked loop systems", IEEE Transactions on Instrumentation and Measurement, vol. 61, no. 4, pp. 930-940, April 2012.

[18] S. Agrawal, Y. K. Nagar, and D. K. Palwalia, "Analysis and implementation of shunt active power filter based on synchronizing enhanced PLL", IEEE International Conference on Information, Communication, Instrumentation and Control, pp. 1-5, August 2017.

[19] S. Chourasiya, and S. Agarwal, "PI, fuzzy and reduced FLC controllers act on shunt active power filters", International Electrical Engineering Journal, vol. 7, no.9, pp. 2391-2395, ISSN 2078-2365, 2017.

2nd IEEE International conference on power Electronics, Intelligent Control and Energy systems (ICPEICES-2018)

Event-Triggered Based Automatic Generation Control Considering Probabilistic Actuator Failure

Pankaj Dahiya, Pankaj Mukhija and Anmol Ratna Saxena
Department of Electrical and Electronics Engineering,
National Institute of Technology, Delhi-110040, India
pankajdahiya@nitdelhi.ac.in; pankajmukhija@nitdelhi.ac.in; anmolsaxena@nitdelhi.ac.in

Abstract—**The present work studies automatic generation control (AGC) of one-area thermal power system scheme, considering the effects of event based scheme and probabilistic actuator failure. The criteria for stability and stabilization of the AGC system is then derived in terms of linear matrix inequalities using Lyapunov theory. For effective communication, periodic event-triggered scheme is considered. The effects of delay introduced by the communication channel is also considered. The simulation results show that the designed state feedback controller has the ability to safely stabilize the AGC system under consideration, even with the larger release period.**

Keywords—*Probabilistic actuator failure, Linear matrix inequalities, Automatic generation control, Lyapunov theory.*

I. Introduction

Maintaining frequency to predefined range or active power balance in a power utility is recognized as automatic generation control (AGC) problem. Modern power system is a very complex structure consisting of distributed power generation sources. Considering distributed generation and deregulation of power grid, AGC needs an improved and open communication infrastructure. Where, communication channels will act as the backbone for the AGC scheme [1]. Dedicated communication lines are not suggested because of the disadvantages such as high-cost and rigidity. On the other hand, open channel is unreliable due to presence of delays, packet dropout and limited bandwidth [2]. For effective utilization of bandwidth, event-triggered control (ETC) has been given a greater attention in the recent past. In an ETC based scheme, the system states are measured using sensors and are transmitted through network to the controller, only when a given condition is violated. In comparison with sampling at regular interval, ETC schemes effectively utilize the network resources, which leads to increase sample release period and reduced usage of bandwidth.

Moreover, ETC has also been implemented for AGC scheme in [3]–[8]. In [3], an event based PI controller is implemented for the multi-area AGC scheme. Along with PI controller, supplementary adaptive dynamic programming is implemented for multi-area AGC scheme [4]. In [5], a sliding mode controller is implemented for multi-area AGC scheme. An event based state feedback controller is implemented for an isolated area consisting of wind-diesel generator [6]. A passivity based distributed ETC for two area AGC system is given in [7]. In [8], ETC and Periodic ETC has been designed for one-area thermal AGC system.

In networked control system, actuator failure and signal distortion may occur due to communication delay, actuator ageing etc. So, reliable controller design has been proposed [9] considering probabilistic actuator failure (PAF), delay and packet dropout. In [10], reliable controller design considering PAF with different failure rate has been studied. Recently, event-triggered based reliable controller considering PAF has been proposed [11], and is used in [12] for control of inverted pendulum and in [13] for control of a vehicle active suspension system problem. Also, similar algorithm is written for the autonomous platoon control, considering both actuator and sensor failure for designing output feedback controller [14]. But till date, no attempt has been made for designing event based controller considering effects of PAF for the AGC system. The algorithm of [11] looks promising for state feedback controller design for the AGC system. Hence, in present work, a full state feedback controller considering periodic event based communication and PAF is designed.

The organization of the paper is given as: in section II, description and state-space model (SSM) of a networked communication based, one-area, thermal power system considering PAF and periodic event-triggering with state feedback controller is done. In next section, criteria for the co-design of the event-trigger parameters and controller gain are derived using Lyapunov theory. Model is simulated for the obtained controller and results are presented and discussed in section IV. In the last section, conclusions are drawn.

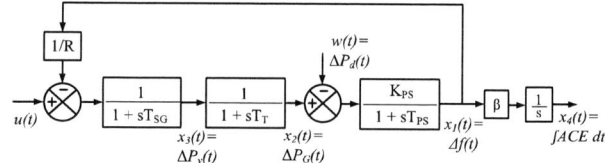

Fig. 1. AGC system under consideration.

II. AGC System Model with Event Based Scheme

The transfer function model describing one-area, state feedback, non-reheat thermal AGC system under consideration is shown in Fig. 1. SSM for the AGC system can be derived as given by Eq. (1). Where $x(t) \in \mathbb{R}^{4 \times 1}$ are the AGC system states, $u(t)$ is the controller output, $\Delta P_d(t)$ is deviation in load, A, B and F are AGC system, input, disturbance matrices, respectively:

$$\dot{x}(t) = Ax(t) + Bu(t) + F\Delta P_d(t) \qquad (1)$$

978-1-5386-6624-1/18 $31.00 © 2018 IEEE   488

where,

$$x(t) = \begin{bmatrix} x_1(t) & x_2(t) & x_3(t) & x_4(t) \end{bmatrix}^T,$$

$$A = \begin{bmatrix} -\dfrac{1}{T_{PS}} & \dfrac{K_{PS}}{T_{PS}} & 0 & 0 \\ 0 & -\dfrac{1}{T_T} & \dfrac{1}{T_T} & 0 \\ -\dfrac{1}{RT_{SG}} & 0 & -\dfrac{1}{T_{SG}} & 0 \\ \beta & 0 & 0 & 0 \end{bmatrix}$$

$$B = \begin{bmatrix} 0 & 0 & \dfrac{1}{T_{SG}} & 0 \end{bmatrix}^T$$

$$F = \begin{bmatrix} -\dfrac{K_{PS}}{T_{PS}} & 0 & 0 & 0 \end{bmatrix}^T$$

where the states $x_1(t)$-$x_4(t)$ are the deviations in frequency, generator power, steam output and time integration of area control error (ACE), respectively. Further, T_{SG}, T_T and T_{PS} denotes the time constants of speed governor, turbine and power system, respectively. Whereas, R, β and K_{PS} are, respectively, the speed drop of the governor, frequency bias factor and gain of the power system. Considering AGC scheme as shown in Fig. 2, with PAF.

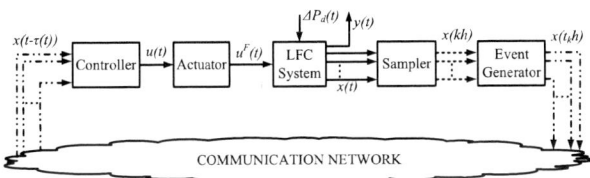

Fig. 2. AGC scheme under event-triggered control and probabilistic actuator failure

Taking full state feedback control, input to the AGC system, with and without actuator failure is taken as $u^F(t) = \Xi K x(t)$ and $u(t) = K x(t)$, respectively, where Ξ is a random variable having normal distribution with mean and variance as μ and σ^2. Also, $\Xi \in [0,\theta]$ with $\theta \geq 1$, $\Xi = 0$, $0 < \Xi < 1$, $\Xi = 1$ and $1 < \Xi \leq \theta$ represent the cases of actuator complete failure, partial failure, normal working and data distortion, respectively. From statistics, we have, $\mathbb{E}(\Xi) = \mu = \bar{\Xi}$, $\mathbb{E}(\Xi - \mu) = 0$ and $\mathbb{E}(\Xi - \bar{\Xi})^2 = \sigma^2$.

Again considering AGC scheme of Fig. 2, with actuator failure, event based sampling (sampling condition as given in Eq. (2)) and network delay 'τ', system Eq. (1) can now be written as Eq. (3), where $\tau(t)$ and $e_k(t)$ are as defined in [15].

$$[\mathbb{E}\{\Xi x((k+j)h)\} - \mathbb{E}\{\Xi x(kh)\}]^T \Omega [\mathbb{E}\{\Xi x((k+j)h)\} - \mathbb{E}\{\Xi x(kh)\}] \leq \rho [\mathbb{E}\{\Xi x((k+j)h)\}]^T \Omega [\mathbb{E}\{\Xi x((k+j)h)\}] \tag{2}$$

$$\dot{x}(t) = Ax(t) + B\bar{\Xi}K(x(t-\tau(t)) + e_k(t)) + B(\Xi - \bar{\Xi})K(x(t-\tau(t)) + e_k(t)) + F\Delta P_d(t) \tag{3}$$

Considering the exponential stability for the AGC system, SSM Eq. (3), the load disturbance '$\Delta P_d(t)$' is taken to be zero. The AGC system now can be rewritten as follows:

$$\dot{x}(t) = Ax(t) + B\bar{\Xi}K(x(t-\tau(t)) + e_k(t)) + B(\Xi - \bar{\Xi})K(x(t-\tau(t)) + e_k(t)) \tag{4}$$

III. Controller Designing Strategy

In this section, Stability and stabilization conditions in terms of LMIs proposed in [11,12] are used for finding exponential stability of the AGC scheme under consideration.

Theorem 1. The AGC scheme represented by Eq. (4) is exponential mean square stable, if for given μ, σ, ρ and state feedback gain 'K', there exists positive definite matrices P, Q, R, Ω and any appropriate dimensions matrices $N = \begin{bmatrix} N_1^T & N_2^T & N_3^T & N_4^T \end{bmatrix}^T$ and $M = \begin{bmatrix} M_1^T & M_2^T & M_3^T & M_4^T \end{bmatrix}^T$, such that following LMIs holds for $r = 1,2$:

$$\begin{bmatrix} \Psi_{11} & * & * \\ \Psi_{21} & \Psi_{22} & * \\ N_3 - M_1^T & M_3 - N_3 - M_2^T & -Q - M_3 - M_3^T \\ K^T \mu^T B^T P + N_4 & M_4 - N_4 & -M_4 \\ \sqrt{\tau_m} RA & \sqrt{\tau_m} RB\mu K & 0 \\ 0 & \sqrt{\tau_m}\sigma RBK & 0 \\ 0 & 0 & 0 \\ \phi_1^r & \phi_2^r & \phi_3^r \end{bmatrix}$$

$$\begin{bmatrix} * & * & * & * & * \\ * & * & * & * & * \\ * & * & * & * & * \\ -\Omega & * & * & * & * \\ \sqrt{\tau_m} RB\mu K & -R & * & * & * \\ 0 & 0 & -R & * & * \\ \sqrt{\tau_m}\sigma RBK & 0 & 0 & -R & * \\ \phi_4^r & 0 & 0 & 0 & -R \end{bmatrix} < 0 \tag{5}$$

where,

$\Psi_{11} = PA + A^T P + Q + N_1 + N_1^T$, $\Psi_{21} = K^T \mu^T B^T P + N_2 - N_1^T + M_1^T$,

$\Psi_{22} = \rho\Omega - N_2 + M_2 - N_2^T + M_2^T$, $\phi_i^1 = \sqrt{\tau_m} N_i^T$, $\phi_i^2 = \sqrt{\tau_m} M_i^T$,

where $i = \{x \in \mathbb{N} \mid x < 5\}$

Proof: Choose Lyapunov candidate as:

$$V(t) = x^T(t) Px(t) + \int_{t-\tau_m}^{t} x^T(s) Qx(s) ds + \int_{t-\tau_m}^{t}\int_{s}^{t} \dot{x}^T(v) R\dot{x}(v) dv ds \tag{6}$$

Applying infinitesimal operator 'Γ' on Eq. (6) as defined in [14,16] and then taking expectation, gives:

978-1-5386-6624-1/18 $31.00 © 2018 IEEE

$$\mathbb{E}\{\Gamma V(t) = 2x^T(t)P\gamma(t) + x^T(t)Qx(t) - x^T(t-\tau_m)Qx(t-\tau_m)$$

$$+ \mathbb{E}[\tau_m \dot{x}^T(t)R\dot{x}(t)] - \int_{t-\tau_m}^{t} \dot{x}^T(v)R\dot{x}(v)dv \qquad (7)$$

where $\gamma(t) = Ax(t) + B\mu Kx(t-\tau(t)) + B\mu Ke_k(t)$.
Introducing free weighting matrices in Eq. (8) and (9).

$$2\xi(t)^T N\left[-x(t-\tau(t)) - \int_{t-\tau(t)}^{t} \dot{x}(s)ds + x(t)\right] = 0 \qquad (8)$$

$$2\xi(t)^T M\left[-x(t-\tau_m) - \int_{t-\tau_m}^{t-\tau(t)} \dot{x}(s)ds + x(t-\tau(t))\right] = 0 \qquad (9)$$

where, $\xi(t)^T = \begin{bmatrix} x(t)^T & x(t-\tau(t))^T & x(t-\tau_m)^T & e_k(t)^T \end{bmatrix}$.

Now by using similar method as given in [11,12], one can conclude that the AGC system described by Eq. (4) is exponentially mean square stable. $\square$

Theorem 2: For given μ, σ, ρ and ε, the system Eq. (4) with control input $K = YX^{-1}$ is exponential mean square stable, If matrices $X > 0$, $\bar{Q} > 0$, $\bar{R} > 0$, $\bar{\Omega} > 0$ and any appropriate dimensions matrices $N^T = \begin{bmatrix} N_1^T & N_2^T & N_3^T & N_4^T \end{bmatrix}$ and $M^T = \begin{bmatrix} M_1^T & M_2^T & M_3^T & M_4^T \end{bmatrix}$ exist, such that following LMIs holds for $r = 1,2$:

$$\begin{bmatrix}
\bar{\Psi}_{11} & * & * \\
\bar{\Psi}_{21} & \bar{\Psi}_{22} & * \\
N_3 - M_1^T & M_3 - N_3 - M_2^T & -Q - M_3 - M_3^T \\
Y^T\mu^T B^T + N_4 & M_4 - N_4 & -M_4 \\
\sqrt{\tau_m}AX & \sqrt{\tau_m}B\mu Y & 0 \\
0 & \sqrt{\tau_m}\sigma BY & 0 \\
0 & 0 & 0 \\
\phi_1^r & \phi_2^r & \phi_3^r
\end{bmatrix}$$

$$\begin{bmatrix}
* & * & * & * & * \\
* & * & * & * & * \\
* & * & * & * & * \\
-\bar{\Omega} & * & * & * & * \\
\sqrt{\tau_m}B\mu Y & \Sigma & * & * & * \\
0 & 0 & \Sigma & * & * \\
\sqrt{\tau_m}\sigma BY & 0 & 0 & \Sigma & * \\
\phi_4^r & 0 & 0 & 0 & \Sigma
\end{bmatrix} < 0 \qquad (10)$$

where,
$\bar{\Psi}_{11} = AX + XA^T + \bar{Q} + N_1 + N_1^T$, $\bar{\Psi}_{21} = Y^T\mu^T B^T + N_2 - N_1^T + M_1^T$, $\bar{\Psi}_{22} = \rho\bar{\Omega} - N_2 + M_2 - N_2^T + M_2^T$ and $\Sigma = -2\varepsilon X + \varepsilon^2 \bar{R}$.

Proof: By applying Schur Complement lemma [17] to Eq. (5) and substituting $-PR^{-1}P$ with $-2\varepsilon P + \varepsilon^2 R$ then putting $X = P^{-1}$, $\bar{Q} = XQX$, $\bar{\Omega} = X\Omega X$, $\bar{R} = XRX$, $Y = KX$, $\bar{N} = XNX$, $\bar{M} = XMX$ and pre and post multiplying with $diag\{X,X,X,X,X,X,X,I\}$, we have Eq. (10), which completes the proof. $\square$

IV. CASE STUDY

In this section, simulation results obtained for the one-area, thermal power system are discussed. Table I presents the parameters for the power system under consideration and the corresponding model is shown in Fig. 1. Now we consider two cases.

TABLE I. LFC PARAMETERS

T_{SG}	T_T	K_{PS}	T_{PS}	R	β
0.1	0.3	1	10	0.05	21

A. Case I

In case I, the controller gain as obtained in [8], $K = \begin{bmatrix} 12.726 & -0.6844 & 0.997 & -0.2424 \end{bmatrix}$ is used. Then Theorem 1 is applied with $\mu = 1$, $\rho = 0$, $\sigma = 1$ and τ_m is found to be 0.03. For $\mu = 1$, $\tau_m = 0$, $\sigma = 1$ Theorem 1 gives ρ_{max} to be 0.07.

Now, the AGC system is simulated to check stability, taking $h = 30ms$ and various values of μ and σ. Fig. 3 represents the probabilistic actuator failure with various values of μ and σ. The time responses for 1% load disturbances are depicted in Fig. 34, clearly one can observe that frequency deviation for the AGC system under consideration is stable for the applied step disturbance. For effective utilization of bandwidth, ρ may also be varied up to ρ_{max} which results in decrease of the sampled data sent to stabilize the system. To show the effectiveness of ETC, the model is simulated for $\rho = 0$, 0.01 & 0.03. The corresponding frequency responses are shown in Fig. 5. The results show numbers of samples sent, decreases from 3,000 when $\rho = 0$, to 53 with $\rho = 0.01$ and only to 34 with $\rho = 0.03$.

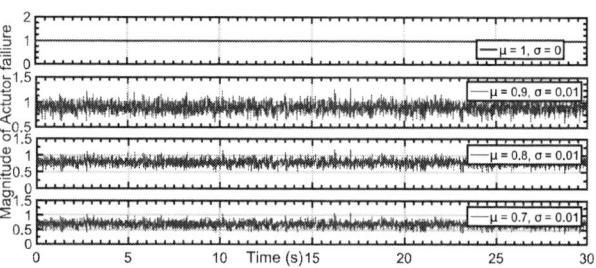

Fig. 3. Probabilistic actuator failures with different values of μ and σ.

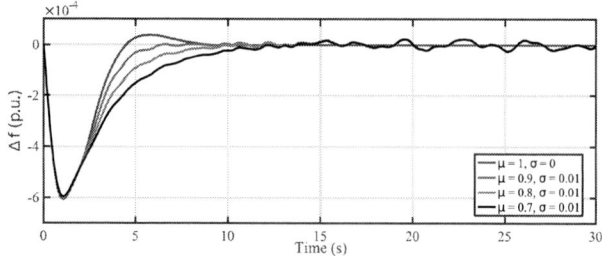

Fig. 4. Frequency response for 1% load disturbance at $h = 30ms$ and different values of μ and σ.

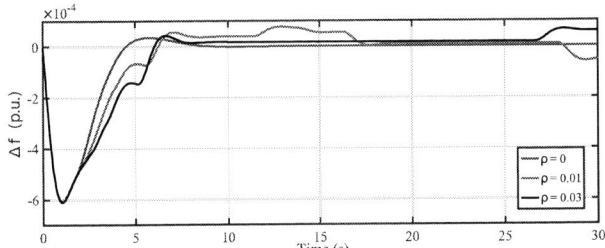

Fig. 5. Frequency response for 1% load disturbance at $h = 10ms$ and $\rho = 0$, 0.01 & 0.03.

B. Case II

Taking $\mu = 0.3$, $\sigma = 0.3162$, $\tau_m = 0$ and $\varepsilon = 0.53$ the obtained controller gain is K=[-20.5294 -1.1084 -0.3478 -1.2134]. Again, the system is simulated for stability, taking $h = 30ms$, various values of μ and σ. The time responses for 1% load disturbances are depicted in Fig. 56. The responses show that the system frequency deviation is stable for the applied step disturbance.

For effective utilization of bandwidth, the model is again simulated for $\rho = 0$, 0.03 & 0.09. The corresponding frequency deviations are shown in Fig. 7. The results show without much distortion in frequency deviation, numbers of samples sent, decreases to only 25 samples for $\rho = 0.03$ to only 16 samples for $\rho = 0.09$.

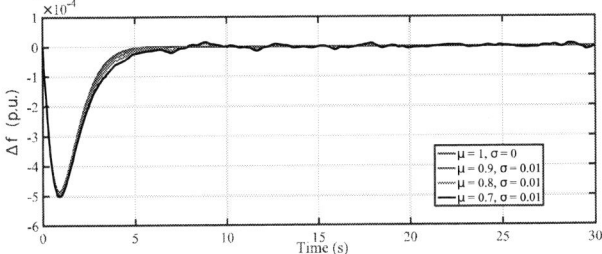

Fig. 6. Frequency response for 1% load disturbance at $h=30ms$ and different values of μ and σ.

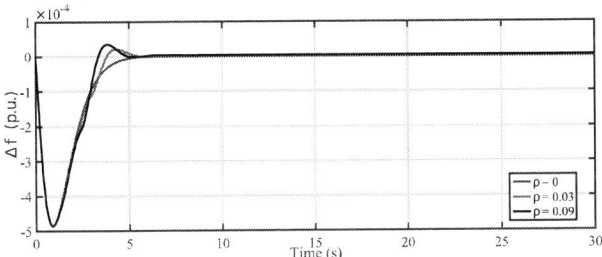

Fig. 7. Frequency response for 1% load disturbance at $h = 10ms$ and $\rho = 0$, 0.03 & 0.09.

V. CONCLUSION

In this article, state feedback controller is implemented, with event based triggering, for the frequency regulation of one-area, thermal power system. The controller is designed considering PAF along with periodic event based sampling. The results show, frequency deviation for the AGC problem can easily be stabilized using designed controller. Simulation studies also show, network bandwidth can be conserved without much deviation from desired response by using periodic event based sampling.

REFERENCES

[1] L. Jiang, W. Yao, Q. H. Wu, J. Y. Wen, and S. J. Cheng, "Delay-dependent stability for load frequency control with constant and time-varying delays," *IEEE Trans. Power Syst.*, vol. 27, no. 2, pp. 932–941, 2012.

[2] P. Dahiya, P. Mukhija, and A. R. Saxena, "Stability criteria for load frequency control systems with communication delays and packet dropout," in *1st IEEE International Conference on Power Electronics. Intelligent Control and Energy Systems (ICPEICES-2016)*, 2016, pp. 1–4.

[3] S. Wen, X. Yu, Z. Zeng, and J. Wang, "Event-Triggering Load Frequency Control for Multiarea Power Systems With Communication Delays," *IEEE Trans. Ind. Electron.*, vol. 63, no. 2, pp. 1308–1317, 2016.

[4] L. Dong, Y. Tang, H. He, and C. Sun, "An event-triggered approach for load frequency control with supplementary ADP," *IEEE Trans. Power Syst.*, vol. 32, no. 1, pp. 581–589, 2017.

[5] X. Su, X. Liu, and Y.-D. Song, "Event-Triggered Sliding Mode Control for Multi-Area Power Systems," *IEEE Trans. Ind. Electron.*, no. accepted for future issue, 2017.

[6] P. Dahiya, P. Mukhija, and A. R. Saxena, "Design of sampled data and event-triggered load frequency controller for isolated hybrid power system," *Int. J. Electr. Power Energy Syst.*, vol. 100, pp. 331–349, 2018.

[7] P. Bhui, "Passivity based distributed event-triggered load frequency control," in *2017 6th International Conference on Computer Applications In Electrical Engineering-Recent Advances (CERA)*, 2017, pp. 258–263.

[8] P. Dahiya, P. Mukhija, and A. R. Saxena, "Event-Triggered Control for Frequency Regulation of Single-Area Thermal Power System," in *6th IEEE International Conference on Computer Applications in Electrical Engineering-Recent Advances (CERA)*, 2017.

[9] E. Tian and D. Yue and C. Peng, "Brief paper: reliable control for networked control systems with probabilistic sensors and actuators faults," *IET Control Theory Appl.*, vol. 4, no. 8, pp. 1478–1488, Aug. 2010.

[10] E. Tian and D. Yue and C. Peng, "Reliable control for networked control systems with probabilistic actuator fault and random delays," *J. Franklin Inst.*, vol. 347, no. 10, pp. 1907–1926, 2010.

[11] J. Liu and D. Yue, "Event-triggering in networked systems with probabilistic sensor and actuator faults," *Inf. Sci. (Ny).*, vol. 240, pp. 145–160, 2013.

[12] Y. Zhai, R. Yan, H. Liu, and J. Liu, "Event-Triggered Reliable Control in Networked Control Systems with Probabilistic Actuator Faults," *Math. Probl. Eng.*, no. Article ID: 131942, p. 9, 2013.

[13] K. Bansal, P. Dahiya, and P. Mukhija, "Event-Triggered Based Reliable Control of Vehicle Active Suspension System Under Actuator Faults," *IFAC-PapersOnLine*, vol. 51, no. 1, pp. 196–201, 2018.

[14] W. Yue and L. Wang, "Event-triggered autonomous platoon control against probabilistic sensor and actuator failures," *Automatika*, vol. 58, no. 1, pp. 35–47, 2017.

[15] D. Yue, E. Tian, and Q. L. Han, "A delay system method for designing event-triggered controllers of networked control systems," *IEEE Transactions on Automatic Control*, vol. 58, no. 2, pp. 475–481, Feb. 2013.

[16] R. Khasminskii, *Stochastic stability of differential equations.* Springer Science & Business Media, 2011, vol. 66.

[17] M. Wu, Y. He, and J.H. She, *Stability Analysis and Robust Control of Time-Delay Systems.* New York: Springer-Verlag, 2010.

978-1-5386-6624-1/18 $31.00 © 2018 IEEE

2nd IEEE International Conference on Power Electronics, Intelligent Control and Energy systems (ICPEICES-2018)

Multi-Input Multi-Level Isolated DC-DC Converter for Enhanced Reliability of Renewable Sources

Ravi Vardhan Arya

Department of Electrical Engineering
Delhi Technological University, Delhi

Amritesh Kumar

Department of Electrical Engineering
National Institute of Technology, Silchar,

Aditya Narula

Department of Electrical Engineering
*Delhi Technological University,*Delhi

Abstract –This paper proposes an isolated multi-input multi-output power electronics converter for interfacing with multiple renewable energy sources. The converter provides the capability of multi-level output based upon the secondary side series-parallel configuration. The modular architecture of the converter enable plug and play operation. In Control algorithm proposed for the converter ensures equal load sharing between the inputs during continuous and discontinuous operating modes while maintaining the output voltage during load perturbations. The system is modeled and simulated under MATLAB Simulink environment with the presented results verifying the performance of the proposed converter.

Keywords - Multi-Input converter, Multi-Level converter, switched capacitor, close loop control.

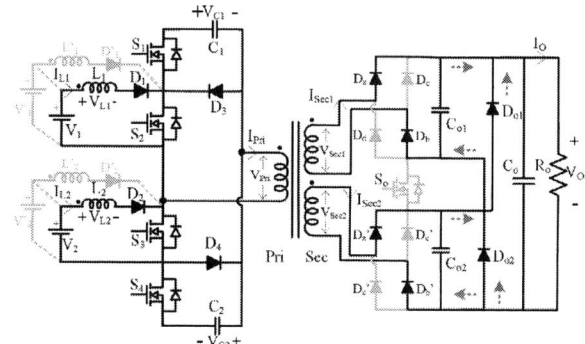

Fig. 1 Proposed Isolated Multi-Input Multi-Level Output DC-DC Converter

I. Introduction

Renewable energy sources (RES) have gained great attention due to its advantages of rich reserve and environment-friendly nature. It is regarded as the best alternative to fossil energy. However, these energy sources are easily gets perturbed by climatic conditions and these are unpredictable to provide continuous and reliable power supply.

In order to address the issue, it is preferred to integrate RES with conventional sources to form a hybrid configuration to get a stable and continuous power supply. To support such architecture Power Electronics based converter-inverter system becomes an integral for these RES for maximum power extraction and feeding power to load/grid. With this, the concept of Multi-Input Converter (MIC) has been proposed to accommodate multiple sources i.e. renewable and conventional sources [1-8]. Recently researchers have tried to integrate multiple batteries through a MIC to get a reliable output for Electric Vehicles [1]. Others have proposed PV-Wind-Battery based MIC to get a reliable dc supply which is further inverted to feed it to grid [2-3].

In literature, several topologies of MICs with resonant, non-resonant, isolated, non-isolated has been reported to combine different types of DC input sources to obtain single DC output [1-8]. In [4-5], a systematic approach for designing MIC is reported using a multi-winding transformer for renewable energy application.

In [1],[6], a non-isolated MIC with Modular Parallel topology is employed to get a single DC output. In this only one input transfers power to the load, at a time. If more than one sources are used to transfer power to the load at a time, then this increases the problem of circulating current. In [7], a non-

isolated MIC based on switched capacitor is proposed with very simple control circuitry. Author has employed the concept of LC oscillator, for improvement in output waveform and reducing the stress on the capacitor.

MIC provides a simple circuit, reduced number of components, centralized control, high reliability and decreased size/cost [8]. Most of these MIC topologies are confined to single output only. But in many application, variable output voltages are also requisite. This basically opens the avenue for isolated MIC topology capable of having multi-level output with multiple inputs to avoid multiplicity of converters. Though non-isolated topology is having low cost and less circuitry but lacks in safety and hazards protection. Isolated topology credits in terms of isolation, protection, and also avoid leakage/circulating current.

The authors have proposed single-input multi-level dc-dc converters for variable output [9]. But things are scarcely discussed on isolated multi-input multi-level dc-dc converter.

In this paper, an isolated multi-input multi-level dc-dc converter is proposed for variable output as shown in Fig. 1. For maintaining variable controlled output voltage irrespective of perturbation in loading, a close loop control strategy is designed. The proposed converter uses two distinct dc-dc converters on either side of a high-frequency transformer. Switching sequence of multiple switches is maintained in a way to provide continuous energizing to primary winding with proper boost. The primary side of the converter provides a base for integrating multiple inputs and secondary side of converter enables to get multi-level output. To verify the working operation of the proposed multi-input system, two uniform battery sources of 48V is used at each input. The effectiveness of the proposed topology and control logic is verified using MATLAB Simulink.

978-1-5386-6624-1/18 $31.00 © 2018 IEEE

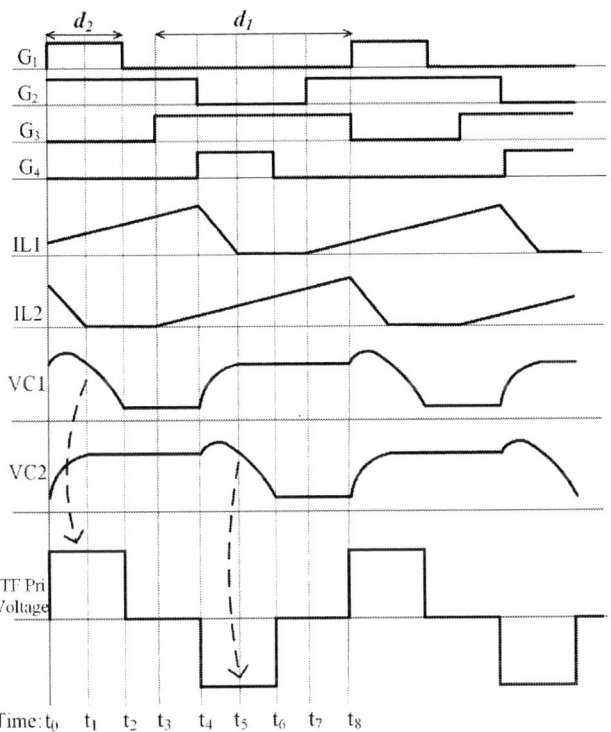

Fig.2: Gating Scheme and key waveforms

III. OPERATING PRINCIPLE

In this section operating modes of the converter is explained which are shown in Fig. 3 and Fig. 4. First, the operation of the secondary side is discussed. Owing the switching of S_o, the secondary side circuit form a multi-level dc output. When S_o is OFF, output terminals of both the full wave rectifiers which are connected to secondary windings of the transformer will make parallel connections through diodes D_{o1} and D_{o2} to feed the load. And when S_o is turned ON, the output terminals of both rectifiers will cascade (connect in series) as shown in Fig. 3(a-b).

Operation of primary side converter is divided into 8 various intervals discussed below without dead band consideration. All the operation interval of primary are shown in Fig. 2 & 4. Before the first interval, S_3 is ON and S_1 is OFF. Energy in L_2 stored to its peak. C_1 is fully charged and C_2 is discharged and L_1 starts storing energy from input-1.

Interval A (t_0-t_1): During this interval, S_3 is switched OFF and S_1 is Switched ON, due to which energy stored in L_2 and C_1 was transferred to the primary winding of transformer and excess energy get stored in C_2 and Inductor L_2 become fully discharged and enters DCM. And L_1 continues in storing energy.

Interval B (t_1-t_2): In this interval load is solely powered by C_1.

Interval C (t_2-t_3): In this interval S_1 is turned OFF, applying a zero voltage at the primary winding of Transformer, and a freewheeling path is made through diode D_3 and S_2.

Interval D (t_3-t_4): Till this interval, S_2 is turned ON and S_1, S_3, S_4 are turned OFF. Now S_3 is switched ON and L_2 starts storing energy from input 2. And Primary winding continues to freewheeling through D_3 and S_2. Also, the energy in L_1 is about to reaches its peak.

II. TOPOLOGY DESCRIPTION

Fig.1 shows the schematic of the proposed isolated Multi-input Multi-level (MI-ML) dc-dc converter. In the proposed structure two input sources in the form of battery stack of 48V each are integrated with the converter. The close loop operation of the converter is performed to have equal load sharing between the battery stack even in CCM. Switched capacitors (C_1, C_2) are also utilized in converter for efficiently absorbing and releasing excess energy from the input sources.

Input boost inductors (L_1, L_2) stores energy when switch S_2 and S_3 are turned ON respectively. When these switches are turned OFF the stored energy is transferred to the load and excess energy is stored in Capacitors C_1 and C_2. Switches S_1 and S_4 are switched to release or absorb excess energy in the capacitor. When S_1 (or S_4) is switched ON, then energy is released from Capacitor C_1 (or C_2). When these switches are turned OFF, input sources and stored energy in the inductor deliver power to the load and charges respective capacitor.

Gating scheme of all the switches is shown in Fig. 2. d_1 denotes the duty ratio of S_2 and S_3, and d_2 denotes the duty ratio of S_1 and S_4. Switching pulses of S_2 & S_3 are phase-shifted by 180° with respect to each other. Similarly, pulses of S1 & S4 are phase shifted by 180°. The duty ratio of middle switches (S_2 and S_3) is maintained greater than 0.5 to allow a freewheeling path for the transformer primary side current, and a limit is set around 0.72 for proper close loop operation under perturbing load condition. S_1 is switched ON just after the S_3 is OFF, and similarly, S_4 switched ON after the S_2 is OFF with proper dead band.

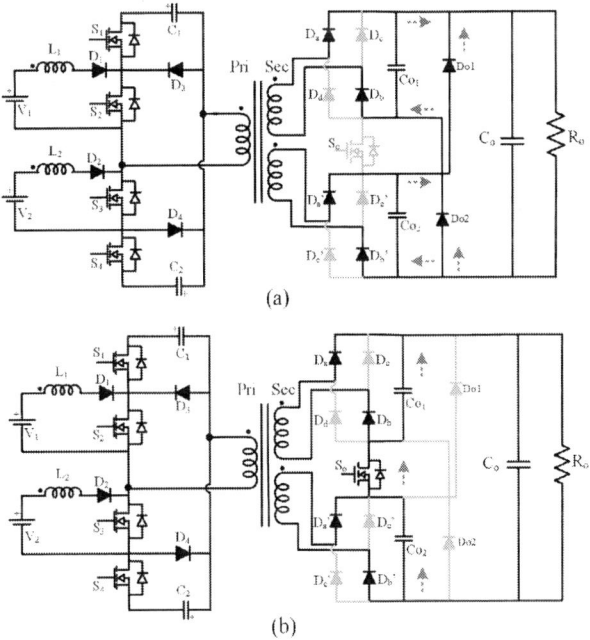

Fig.3 Secondary Windings connected in: (a) Parallel (S_o is OFF); (b) Series (S_o is ON)

978-1-5386-6624-1/18 $31.00 © 2018 IEEE   493

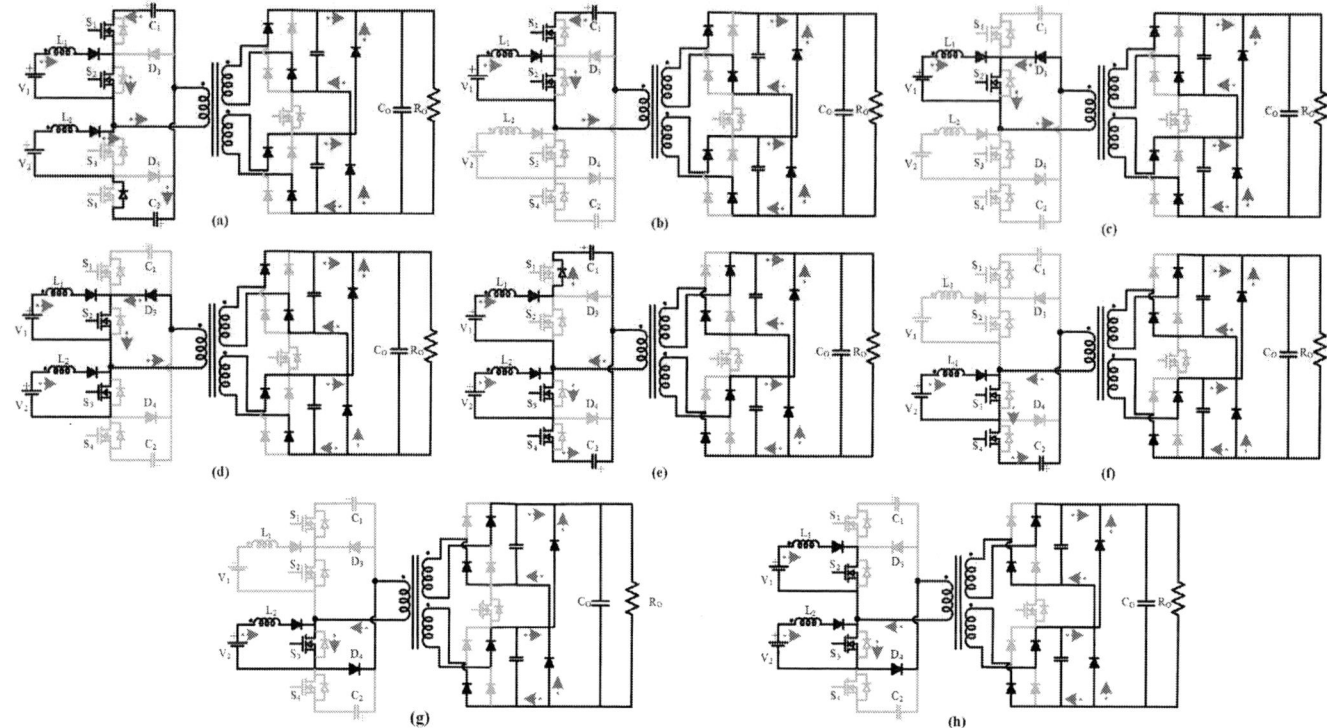

Fig.4: Operation intervals of converter: (a) Interval A(t_0-t_1), (b) Interval B(t_1-t_2), (c) Interval C(t_2-t_3), (d) Interval D(t_3-t_4), (e) Interval E(t_4-t_5), (f) Interval F(t_5-t_6), (g) Interval G(t_6-t_7), (h) Interval H(t_7-t_8)

Interval E (t_4-t_5): At this point of time S_2 is OFF and S_4 is ON. Stored energy in L_1 and C_2 is transferred to the load via primary having negative voltage and causing a reverse current in primary circuitry. Excess energy is absorbed by C_1. In this interval, L_1 is fully discharged and enters DCM. And L_2 continues to store energy form input-2.

Interval F (t_5-t_6): During this interval, the primary winding is solely energized by C_2.

Interval G (t_6-t_7): At this time, S_4 is OFF, applying a zero voltage at the primary winding of transformer, and a freewheeling path is made through diode D_4 and S_3.

Interval H (t_7-t_8): Before t_7, S_3 is ON and S_1, S_2, S_4 are turned OFF. Now at t_7 S_2 is ON, resulting in the increase in I_{L1} and L_1 starts storing energy from input-1. And primary winding continues to freewheeling through D_4 and S_3. Also, the energy in L_2 is about to reaches its peak.

IV. CONTROL STRUCTURE

A constant output voltage control is implemented in this converter, by regulating the duty cycle of S_1, S_2, S_3, and S_4. The close loop control strategy for the converter is shown in Fig. 5. Close loop control initializes with comparing of output DC-link voltage (V_o) with the reference voltage (V_{ref}) to generate the error signal. This error signal is fed to the PI Controller. Now from the output of the PI controller, 4 different gating pulses G_1, G_2, G_3 and G_4 for all the switches has to be generated by comparing it with Saw-tooth (carrier) signal. G_3 and G_2 are generated by directly comparing the PI output with 0° and 180°

phase shifted Saw-tooth signals of 20 kHz respectively as shown in fig.2. A D flip-flop is used in control circuitry to provide an initial delay for G_1 and G_4. On feeding the PI output, flip-flop result and both saw-tooth signals to Dead band controller (DBC), the dead band between the switching of S_1-S_3 and S_2-S_4 is being controlled and appropriate Saw-tooth signals are generated. K_a is used to provide dead band when middle switches (S_2, S_3) is switched OFF and outer switches (S_1, S_4) are turned ON. And similarly, K_b provide dead band when outer switches are switched OFF and middle switches are turned ON. After that, DBC output is compared with 0° and 180° phase shifted Saw-tooth signals to generate G_1 and G_4.

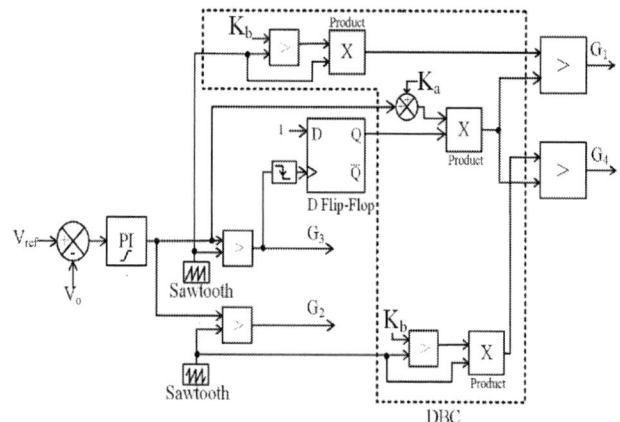

Fig.5: Close loop Control Logic for gating structure with proper dead band.

V. TOPOLOGY ANALYSIS

To understand the input-output relation of the proposed converter, it is necessary to analyze the steady state behavior of various storing elements like L_1, L_2, C_1, and C_2. Capacitors are switched at a proper time to power up the primary of the transformer and apply positive and negative voltages to get the required output. Similarly, L_1 and L_2 are properly switched ON and OFF for requisite power transfer in the secondary.

As discussed in the previous section, the duty of middle switches (S_2, S_3) is maintained greater than 0.5 to apply zero voltage at the primary of transformer. The relation between d_1 (duty ratio of G_2, G_3) and d_2 (duty ratio of G_1, G_4) is given as:

$$d_1 = 1 - d_2 - DB \qquad (1)$$

Where, DB is the dead band and must be considered between switching of S_1-S_3 and S_2-S_4 for providing proper freewheeling path and reversal of polarity without overshoot due to switching.

For the proposed isolated converter consisting of a high-frequency transformer having m no. of secondary windings, with turns ratio 1: n, then-

$$V_{Pri} = V_o/n \qquad \& \qquad I_{Pri} = nI_o \qquad (S_o\text{=OFF}), \qquad (2)$$
$$V_{Pri} = V_o/nm \qquad \& \qquad I_{Pri} = nmI_o \qquad (S_o\text{=ON}), \qquad (3)$$

For the steady-state operation of the converter, the net change in inductor current is zero, i.e. inductors should follow volt-second balance. Also, net change in capacitor voltage should be zero and hence it must follow ampere-second balance.

Considering the converter operating in continuous conduction mode, the following analysis is done.

Applying volt-sec balance across L_1 (as represented in Fig. 1-2):

$$V_1 d_1 + (1 - d_1)(V_1 - V_{Pri} - V_{c1}) = 0 \qquad (4)$$

i.e., $\quad V_{Pri} = \frac{V_1}{(1-d_1)} - V_{c1} = V_{c2} \qquad (5)$

Similarly, volt-sec balance across L_2 gives:

$$V_{Pri} = \frac{V_2}{(1-d_1)} - V_{c2} = V_{c1} \qquad (6)$$

Therefore, $\quad V_O = n \times V_{Pri} = \frac{n}{2}\left(\frac{V_1+V_2}{(1-d_1)} - (V_{c1} + V_{c2})\right) \qquad (7)$

And, $\quad d_1 = \left(1 - \frac{V_1}{(V_{Pri}+V_{c1})}\right) = \left(1 - \frac{V_2}{(V_{Pri}+V_{c2})}\right) \qquad (8)$

Now by Amp-sec balance in C_1:

$$-d_2(I_{Pri} - I_{L1}) + (1 - d_1)(I_{L2}) = 0 \qquad (9)$$

i.e., $\quad -d_2(I_{Pri} - I_{L1}) + (d_2 + DB)(I_{L2}) = 0 \qquad (10)$

$$I_{Pri} = \frac{I_{L2}(d_2+DB)}{d_2} + I_{L1} \qquad (11)$$

Similarly, Amp-sec balance in C_2:

$$I_{Pri} = \frac{I_{L1}(d_2+DB)}{d_2} + I_{L2} \qquad (12)$$

Therefore, $I_O = \frac{I_{Pri}}{n} = \frac{1}{2n}\left(\frac{(I_{L1}+I_{L2})(d_2+DB)}{d_2} + I_{L1} + I_{L2}\right) \qquad (13)$

And, $d_2 = \left(\frac{DB \times I_{L1}}{(I_{Pri}-I_{L1}-I_{L2})}\right) = \left(\frac{DB \times I_{L2}}{(I_{Pri}-I_{L1}-I_{L2})}\right) \qquad (14)$

From (11-12), it can be seen that input currents I_{L1} and I_{L2} are controlled by d_1 and d_2 in the PWM which is necessary for some applications such as renewable energy resources to extract maximum power.

Eq. (5-6) and (11-12) clearly shows that the load is equally power by both the input in DCM as well as CCM.

From the above analysis, the relation between Input voltages and Output voltage can be synthesized by averaging which is shown in Eq. (7) and (13).

According to (5), when source-1 is powering the load then at that point of time C_2 is also discharging and C_1 is getting charged by I_{L1}. And (6), clearly shows when source-2 is powering the load then at that time C_1 is discharging and C_2 is getting charged by I_{L2}.

Eq. (5-6) can be easily verified by the simulation results shown in Fig. 6. It can be seen clearly when C_1 is discharging then a positive voltage is applied on the primary winding of transformer (TF Voltage), I_{L2} is decreasing and C_2 is charging. And when C_2 is discharging then a negative voltage is applied on primary winding, I_{L1} is decreasing and C_1 is charging.

VI. SIMULATION RESULTS

The close loop operation of proposed topology is modeled and verified by simulation in MATLAB Simulink environment. Simulation results of both cascaded and parallel modes of the secondary side of transformer are observed under steady and dynamic conditions. Simulation results are obtained using two 48V battery sources as input to this MIC with the following parameter: $V_1=V_2=48V$, $L_1=L_2=200\mu H$, $C_1=C_2=100\mu F$, $f_{sw}=20$ kHz, 1:2 (4.5kVA) high-frequency transformer and $C_o=220\mu F$. Fig. 7 shows response of isolated MIC under the load variation from 130Ω to 90Ω with switch S_o OFF, i.e. secondary windings are connected in parallel, output voltage is maintained at 170V as shown in Fig. 7(b). To satisfy the load requirement, 1.3 A of

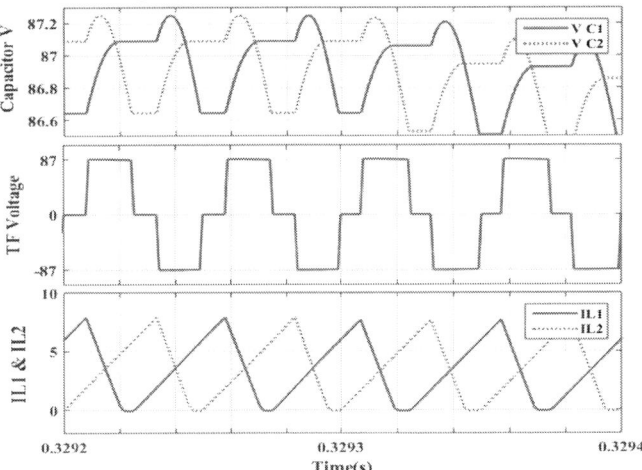

Fig.6: Capacitor Voltage Ripple and Transformer Primary Voltage

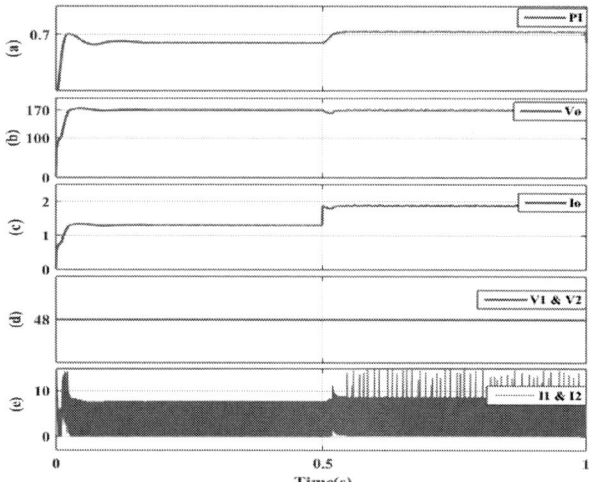

Fig. 7 Simulation Results with S_o is OFF showing: (a) PI output, (b)Output voltage, (c) Output current, (d) Input voltage V_1 and V_2, (e) Input inductor currents I_{L1} and I_{L2}

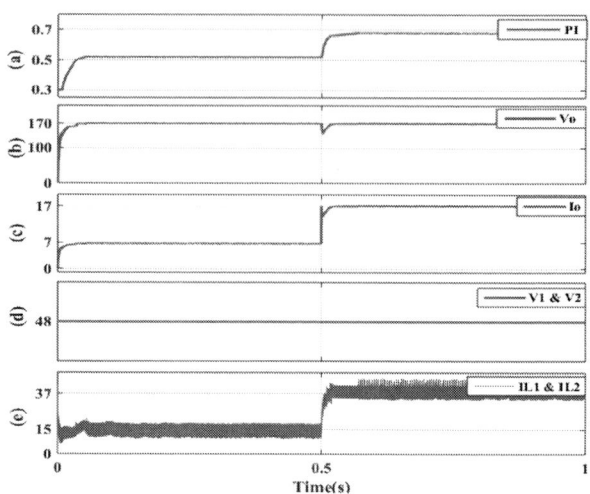

Fig. 9 Simulation Results with S_o is ON showing: (a) PI output, (b)Output voltage, (c) Output current, (d) Input voltage V_1 and V_2, (e) Input inductor currents I_{L1} and I_{L2}

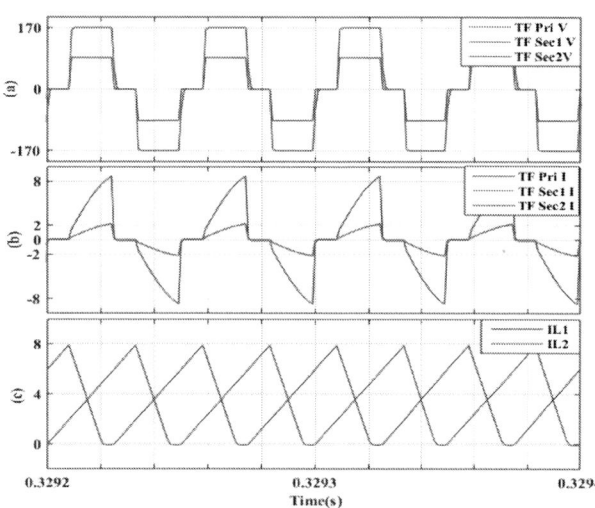

Fig. 8 Simulation Results with S_o is OFF showing: (a) Transformer Voltages across Primary and both secondary windings, (b) Transformer Current in Primary and both secondary windings, (c) Magnified waveforms of I_{L1} and I_{L2} in DCM

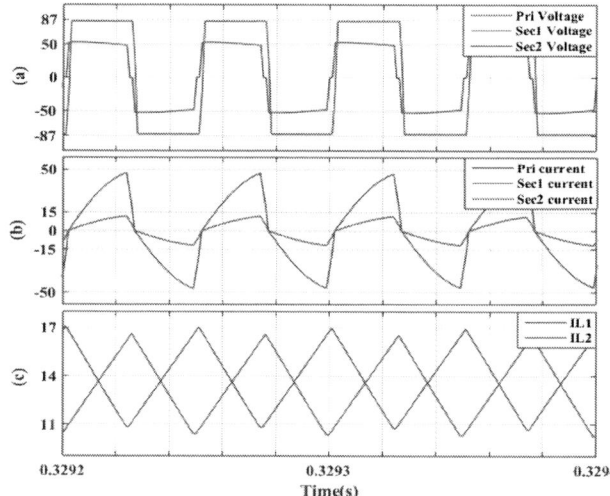

Fig. 10 Simulation Results with S_o is ON showing: (a) Transformer Voltages across Primary and both secondary windings, (b) Transformer Current in Primary and both secondary windings, (c) Magnified waveforms of I_{L1} and I_{L2} in CCM

current is drawn from the secondary terminals as shown in Fig. 7(c). At t=0.5s, a load perturbation command is initiated causing a step change in load current from 1.3A to 1.9A as shown in Fig. 7(c). With a change in load current output voltage is regulated at 170V as in Fig. 7(b). Variation in loading also changes the PI output from 0.65 to 0.71 as shown in Fig. 7(a), which control the duty of switches and stabilizes the output voltage to V_{ref}.

Fig. 8 shows the magnified waveform of transformer voltages, currents, and input inductor currents. Fig. 8(a) shows the transformer terminals input-output voltage resembling chopped square waveform. With the turns ratio of primary to secondary 1:2, for a peak primary voltage of 87V, a voltage 170V peak is observed at the secondary side. From Fig.8(c) it is clear that both I_{L1} and I_{L2} (input inductor current) operates in discontinuous mode (DCM).

Fig. 9 shows another mode of operation with S_o ON, giving a cascade connection of secondary windings. Dynamic response of the converter under the load perturbation from 25Ω to 10Ω is shown in Fig. 9. With the initial load of 25Ω connected on output, 7A of output current is drawn as in Fig. 9(c). With the same input voltage of 48V a load perturbation is initiated at 0.5s. It results in the increase in output current 7A to 17A, in a small time. Meanwhile, a steady reference voltage of 170V is maintained at the output as in Fig. 9(b).

Fig. 10(a-c) shows the magnified response of transformer winding currents, voltages and input inductor currents. Due to high current demanded at the output, input currents enter into CCM as clearly shown in Fig. 10(c).

978-1-5386-6624-1/18 $31.00 © 2018 IEEE   496

CONCLUSION

The isolated Multi-Input Multi-level output DC-DC Converter presented in this paper with close loop control strategy can be exploited as DC-link with the storage system for the generation of AC Power with various standard voltages. In this paper, the output voltage is referenced to 170V with two 48V batteries as input sources. Modular-structure enables multiplication of output voltages by adding the desired no. of inputs. Having the advantages of the centralized control scheme and high reliability the proposed converter credits with very good Voltage regulation.

REFERENCES

[1] Furkan Akar et al., ''A Bidirectional Non-isolated Multi-Input DC–DC Converter for Hybrid Energy Storage Systems in Electric Vehicles'', IEEE Trans. Veh. Tech., vol. 65, no. 10, pp.7944-7955, October 2016.

[2] V. Karthikeyan and Rajesh Gupta, "Multiple-Input Configuration of Isolated Bidirectional DC–DC Converter for Power Flow Control in Combinational Battery Storage", IEEE Trans Ind. Info, Vol. 14, No. 1, January 2018.

[3] B. Mangu et al. ,"Grid-Connected PV-Wind-Battery-Based Multi-Input Transformer-Coupled Bidirectional DC-DC Converter for Household Applications", IEEE Journal Of Emerging And Selected Topics In Power Electronics, Vol. 4, No. 3, September 2016

[4] Yuan-Chuan Liu and Yaow-Ming Chen,'' A Systematic Approach to Synthesizing Multi-Input DC–DC Converters'', IEEE Trans. Power Electron. , vol. 24, no. 1, pp.116-127, January 2009.

[5] Yuan-Chuan Liu and Yaow-Ming Chen, ''Multi-Input DC/DC Converter Based on the Multi winding Transformer for Renewable Energy Applications'', IEEE Trans. Ind. App, vol. 38, no. 4, pp.1096-1104, July/August 2002

[6] Ebrahim Babaei and Okhtay Abbasi, ''Structure for multi-input multi-output dc–dc boost converter'', IET Power Electron., vol. 9, no. 1, pp.1–11, May 2015.

[7] Ye Yuan-mao and Ka Wai Eric Cheng, ''Multi-input voltage-summation converter based on switched-capacitor'', IET Power Electron., 2013, Vol. 6, Iss. 9, pp. 1909–1916.

[8] Serkan Dusmez, IEEE, Xiong Li and Bilal Akin, ''A New Multi-Input Three-Level DC/DC Converter'', IEEE Trans Power Electron. , vol.31, no.7, pp. 1230-1240, 2016.

[9] Hurng-Liahng Jou et al., ''Novel Isolated Multi-level DC-DC Power Converter'', IEEE Trans. on Power Electron. , vol. 31, no. 4, pp. 2690-2694, 2016.

Grid Tied Hybrid Photovoltaic-Fuel Cell Power System for Residential Load

Mukul Chankaya
Department of Electrical Engineering
NIT Srinagar
Srinagar, India
Email: mukulchankaya@gmail.com

Aijaz Ahmad
Department of Electrical Engineering
NIT Srinagar
Srinagar, India
Email: aijaz54@nitsri.net

Abstract—**This paper presents the three-phase three wire grid-connected hybrid PV-PEMFC (Photovoltaic-Proton Exchange Membrane Fuel cell) system controlled by the Unit template and I-cos(phi) technique in a simulation environment. PV-PEMFC system can satisfy stable base load during odd weather conditions. During daytime, the PV system can simultaneously supply power and disintegrate hydrogen out of the water so that FC can work at night. Both the controlling techniques serve manifold objectives like harmonics suppression, load balancing, and voltage regulation. Simulation results show the exclusion of harmonics produced by power electronics devices and load itself. I-cos(phi) control technique has proved its superiority over Unit Template technique in reducing Total Harmonics Distortion.**

Keywords—Renewable Energy System (RES), Total Harmonics Distortion (THD), Proton Exchange Membrane (PEM)

I. INTRODUCTION

Renewable Energy Systems (RES) have proven themselves as a viable option for satisfying the power demands of various nature of consumer's load. The world is struggling with the effect of increasing greenhouse gases and RES have provided a constructive step towards a greener and cleaner future with no greenhouse gas emission. It is predicted that with the increasing penetration of RES in power sector, CO_2 emission will be reduced by 3.6 Gt in the year 2035, which would be about 10% of total greenhouse gas emission of the world. Photovoltaic (PV) and wind energy generation system are the main RES across the world. On account of latest breakthrough in solid state technology, decreasing cost and positively inclined government policies worldwide, PV power system would grow up to 26 times (i.e. 846Twh) that of the year 2010 (32Twh) [1]. Germany and Italy are the world pioneer in solar PV system and countries like US, India, and China are increasing their PV installed capacity and is expected to reach up to 68GW, 85GW, and 113GW in the year 2035 [2]. On the other hand, the fuel cell is currently breaking the commercial barrier with several existing power generating stations across the world. Several types of fuel cells are available with their distinct advantages, limitations and possible applications. Proton Exchange Membrane (PEM) fuel cell is simulated because it offers high energy density with low operating temperature and high responsiveness towards load variation [3]. It requires only hydrogen and oxygen. Alameda County Santa Rite Jail (California) and Hartford life center (Connecticut) uses hybrid PV-FC (Photovoltaic-Fuel Cell) system respectively of 1 MW and 300 MW capacity that proves its reliable and sustainable operation [4]. Fuel Cell (FC) stack has very less carbon

footprint as a PV system and provides energy without fuel and regular maintenance. It improves the whole system's fuel efficiency, improves battery performance and extends its lifetime. The fuel cell can provide full charging to the battery, unlike conventional generators do charging up to 80-85%, with the same amount of fuel consumption as it requires a normal operation. Its noise-free operation with the production of useful heat can be suitable for many applications.

In the presented work, the main emphasis is on creating a grid-connected hybrid PV-PEMFC (photovoltaic-Proton Exchange Membrane Fuel Cell) power system in the Simulink environment to satisfy the various requirement of linear and non-linear load and to provide adequate power quality in the distributed power system [5]. PV system and FC systems are designed with a 10 KW rating each. PEMFC is configured with the help of equations, while input pressure of H_2 and O_2 are maintained to provide an adequate power output. Voltage losses in PEMFC are a non-linear function of load, which are results of several factors like Oxygen reduction reaction, Hydrogen Oxidation reaction, mass transport limitation of electrons and ohmic losses due to membrane ionic conductivity are taken care of. Boost converter of FC rely on Proportional & Integral (PI) voltage controller for switching cycles, used to maintain DC bus voltage at 700 V. Whereas, PV system is designed with 10 modules in a series string and 5 strings in parallel connection of modules. Solar radiation and temperature inputs are provided externally by signal builder block. Maximum Power Point Tracking (MPPT) is developed by using the Perturb & Observe (P&O) technique, for extracting the maximum power out of the PV system. P&O has an advantage over other methods of being easy to understand and simple to simulate and operate. However, compromise has to be made between the steady state oscillation and the response speed by proper selection of step size. DC voltage boost converter for PV is implemented for maintaining the voltage at 700V. The boost converter is capable of providing a higher boost than buck-boost converter without deteriorating the efficiency. It has the same ground for driver and control circuit, that makes control circuit arrangement easier. It also maintains the continuity of input current, which is highly desirable for PV power system

A three-phase three wire grid integrated hybrid PV-FC system is simulated with the unit template and I-cos(phi) based controller [6]. The unit template is a simple algorithm, easy to implement, provide load balancing and voltage regulation at the point of common coupling (PCC) along with power factor correction and direct current control of voltage source converter (VSC). It uses two PI voltage regulators where one is used to regulating the DC bus voltage of VSC and the other one controls the amplitude of the voltage at PCC. With the help of both PI regulators, the amplitude of the in-phase and the quadrature components are estimated for generating a reference current. Whereas I-cos(phi) controlling algorithm uses load current, the voltage at PCC, source current and DC bus voltage as a feedback control signal. Active and apparent powers are measured separately with the subsequent division as required for determining power factor. Sample and hold logic and Zero crossing detector are used to derive the quadrature component of load current at zero crossing. In both the techniques reference current is generated by addition of both the in-phase and the quadrature component and then compared with the sensed supply current. Hysteresis current controller further processes these currents to generate the gate pulses for VSC. The proposed system is tuned in such a way so as to provide grid voltage and current THD within limit according to IEEE-519 and 929 standards [7]. Analysis of system is done in steady state and dynamic state by changing the solar insolation input to PV system.

II. PROPOSED TOPOLOGY

A. MPPT implemented by Perturb & Observe algorithm

PV system faces two main disadvantages like the low efficiency of the solar cell and high initial cost. For attaining higher efficiency its inevitable to operate PV system at the maximum power point. In P&O algorithm the sign of the last deflection and the last increment in power decide the next change. If perturbation shows an incremental change in power then the next perturbation should be in the same direction otherwise in opposite direction, if there is a decrement change in power. Hence the operation point oscillates across Maximum power point [8].

In this paper MPPT with boost converter is implemented. Boost converter maintains the voltage at a fixed level. The minimum DC voltage should be higher than twice the peak of the phase voltage of the distribution system.

B. Three-phase Three Wire grid-tied PV-PEMFC system

Three-phase three wire grid-connected PV-PEMFC system is simulated, which consist of Distribution Grid, interfacing inductor, ripple filter, load, VSC, DC link capacitor, Boost converter and PV-FC system is shown in figure 1. Distribution system was chosen with 415V AC, 50Hz system. 10 kW PV system is implemented by series-parallel connection of PV modules. Each module has a 200 W rating with 56 number of the cells connected in series, Voc=33.2, Isc=8.16, Vmp=26.6, and Imp=7.52 [9]. Voltage source converter used for DC to AC conversion receives the gating signal from controlling techniques [10].

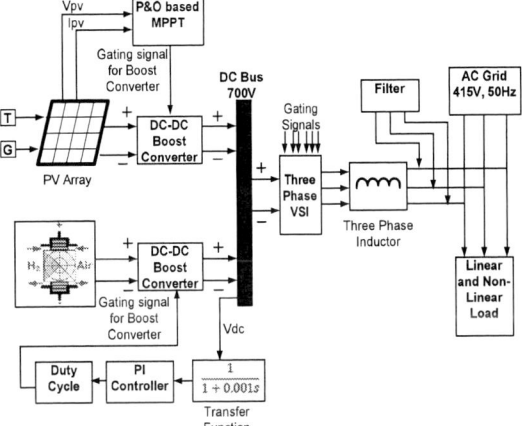

Fig 1. 3phase-3 wire grid-connected PV-PEMFC system

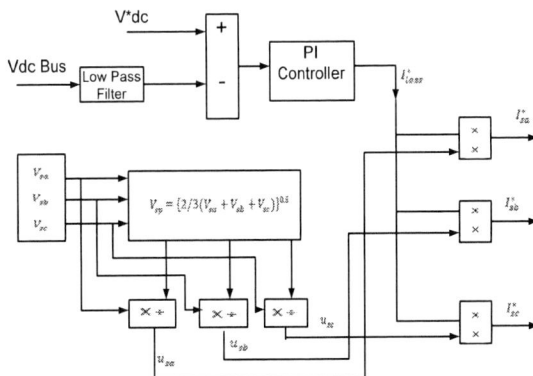

Fig. 2 Unit Template Controlling Technique

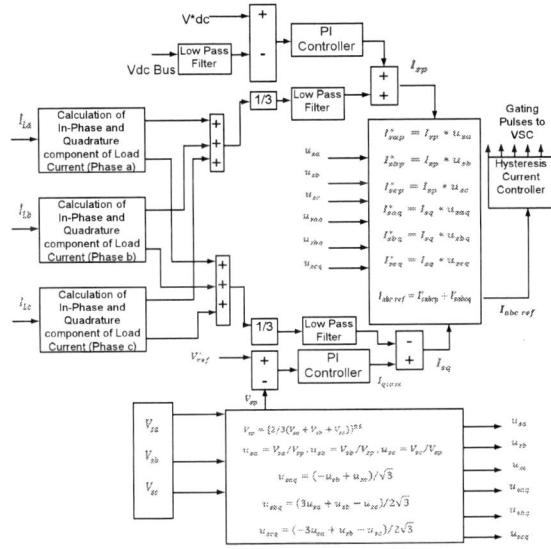

Fig.3. I-Cos (Phi) controlling Technique

978-1-5386-6624-1/18 $31.00 © 2018 IEEE

III. CONTROLLING TECHNIQUES

A. Unit Template controlling technique of VSC

The unit template is one of the time domain control techniques and it is very basic and easy to implement. Here, the DC bus voltage after passing from Low Pass Filter (LPF) is compared with the reference DC voltage and further delivered to the PI voltage controller for generating an active loss Component I_{loss}^{*} current.

$$V_{DC}(n) = V_{DC}^{*}(n)\text{-}V_{DC\,bus} \qquad (1)$$

LPF is used to extract DC component only. K_p and K_i are chosen to get the desired value of voltage response. V_{sp} is the amplitude of the voltage at the point of common coupling (PCC). The in-phase unit templates are calculated as shown in figure.2 [11].

$$V_{sp} = \{2/3(V_{sa} + V_{sb} + V_{sc})\}^{0.5} \qquad (2)$$

$$u_{sa} = V_{sa}/V_{sp}\,,u_{sb} = V_{sb}/V_{sp}\,,u_{sc} = V_{sc}/V_{sp} \qquad (3)$$

Further in-phase component of three-phase reference current can be computed by the following equation

$$I_{sa}^{*} = I_{loss}^{*} * u_{sa}, I_{sb}^{*} = I_{loss}^{*} * u_{sb}, I_{sc}^{*} = I_{loss}^{*} * u_{sc} \qquad (4)$$

The generated reference currents are compared with the sensed source current and fed to the hysteresis controller and triangular carrier signal of 10 KHz is used in Pulse width modulation technique (PWM).

B. I-Cos(phi) controlling technique of VSC

I-Cos(phi) technique for reference current generation requires sensed source and load currents, the voltage at the point of common coupling (PCC) and DC bus voltage. This technique requires the creation of both in-phase and the quadrature component of current [12]. In-phase and I_{loss}^{*} component of current is produced in the same manner as shown in equation 1,2,3.

The amplitude of active power component of the three-phase load current is extracted from the fundamental component of load current at zero crossing of the in-phase template of PCC voltage using Zero crossing detector and sample and hold circuit. Average of $I_{lpa}, I_{lpb}, I_{lpc}$ active power component of load currents is considered as the amplitude of the active power component under balanced system.

$$I_{Lp} = \left(\frac{I_{lpa}+I_{lpb}+I_{lpc}}{3}\right) \qquad (5)$$

For a self-supporting DC bus voltage, the I_{loss} component is added into amplitude the of active power components to create instantaneous active power component (I_{sp}) of reference supply current.

$$I_{sp} = I_{Lp} + I_{loss}^{*} \qquad (6)$$

The quadrature component is created as shown in the fig 3.

$$u_{saq} = (-u_{sb} + u_{sc})/\sqrt{3} \qquad (7\alpha)$$

$$u_{sbq} = (3u_{sa} + u_{sb} - u_{sc})/2\sqrt{3} \qquad (7\beta)$$

$$u_{scq} = (-3u_{sa} + u_{sb} - u_{sc})/2\sqrt{3} \qquad (7\chi)$$

The amplitude of reactive power components of the three-phase load current is extracted from the fundamental components of load current at zero crossing of the quadrature template of PCC voltage using Zero crossing detector and sample and hold circuit. Average of $I_{lqa}, I_{lqb}, I_{lqc}$ the reactive power components of load currents is considered as the amplitude of the reactive power component under the balanced system.

$$I_{LQ} = \left(\frac{I_{lqa}+I_{lqb}+I_{lqc}}{3}\right) \qquad (8)$$

The instantaneous value of the amplitude of the reactive power component of the reference supply current is calculated as

$$I_{sq} = I_{qloss} - I_{LQ} \qquad (9)$$

Where I_{LQ} is the reactive power component of VSC.

$$V_{ref}(n) = V_{ref}^{*}(n) - V_{sp} \qquad (10)$$

I_{qloss} is reactive loss component is derived by taking the difference of V_{ref}^{*} and V_{sp} and feeding into PI voltage controller.

In-phase components of the reference supply current are calculated as

$$I_{sap}^{*} = I_{sp} * u_{sa}, I_{sbp}^{*} = I_{sp} * u_{sb}, I_{scp}^{*} = I_{sp} * u_{sc} \qquad (11)$$

Whereas the reactive components of the reference supply current are calculated as

$$I_{saq}^{*} = I_{sq}*u_{saq}, I_{sbq}^{*} = I_{sq}*u_{sbq}, I_{scq}^{*} = I_{sq}*u_{scq} \qquad (12)$$

The In-phase components are

$$u_{sa} = \cos\emptyset_{pa}\,,u_{sb} = \cos\emptyset_{pb}, u_{sc} = \cos\emptyset_{pc} \qquad (13)$$

Likewise, the quadrature components are

$$u_{saq} = \sin\emptyset_{pa}\,,u_{sbq} = \sin\emptyset_{pb}, u_{sc} = \sin\emptyset_{pc} \qquad (14)$$

The overall reference current is derived as

$$I_{a\,ref} = I_{sap}^{*} + I_{saq}^{*}, I_{b\,ref} = I_{sbp}^{*} + I_{sbq}^{*}$$
$$I_{c\,ref} = I_{scp}^{*} + I_{scq}^{*} \qquad (15)$$

which is further compared with sensed source current and fed to the hysteresis controller for gate pulse generation by pulse-width modulation technique. Triangular carrier signal used with the frequency of 10 KHz.

C. Selection of AC Inductor

Three phase AC inductor value can be conceptualized by ripple current, DC bus voltage, and switching frequency.

$$L_{3\emptyset} = \sqrt{3} * V_{DC} * m \Big/ 12 * a * f_s * I_{ripp} = 1.35\,mH \qquad (16)$$

Where a is the diode ideality factor usually $1 \leq a \leq 2$, here it is a=1.3, switching frequency (f_s) is 10 KHz and I_{ripp} is the ripple current that varies about 10% - 40 % of output DC load current depending on the boost converter parameter, here 30 % of the load current is considered [13].

D. DC bus voltage and Capacitor

DC bus voltage is calculated as such, that it should be greater than twice the peak of the phase voltage as described in the following equation.

$$V_{DC} = \frac{2\sqrt{2} * V_{LL}}{\sqrt{3} * m} = 677.6\,V \qquad (17)$$

V_{LL} is the output phase voltage and m is the modulating index considered as unity (m=1). DC bus capacitor can be calculated as

$$C_{DC} = \frac{(6K_1 * V * a * I * t)}{(V_{DC}^2 - V_{DC1}^2)} = 13109.9\,\mu F \quad (18)$$

Where V_{DC} is the nominal voltage ($V_{DC} = 700$ V), V_{DC1} is the minimum voltage level shown in equation 17, 'V' is the phase voltage, 'I' is phase current, $K_1 = 0.1$ and 't' is the time within which DC bus voltage should be recovered.

E. Boost Converter

Boost converters are simulated to maintain DC bus voltage about 700 V. In a PV system, Boost converter compares the duty cycle generated by MPPT with 10 KHz of the triangular signal to provide a gating sequence to MOSFET. Whereas, in case of the fuel cell boost converter is using a PI controller to generate the duty cycle. Calculations of the inductor, DC capacitor, and the duty cycle is done as follows. Where D is the duty cycle, f_s is 10 KHz, V_{in} is the PV voltage input to boost converter, V_{out} is equivelent to DC bus voltage and ΔV is the acceptable voltage distortion content, chosen 1~2% of V_{out}. Inductor and capacitors are calculated to maintain continuous current mode of operation [14].

$$V_{out} = \frac{V_{in}}{1 - D} = 0.557 \qquad (19)$$

$$L = \frac{V_{in} * (V_{out} - V_{in})}{(I_{ripp} * f_s * V_{out})} = 10.188\,m\,H \quad (20)$$

$$C = \frac{(I_{out} * D)}{(f_s * \Delta V)} = 90.318\,\mu F \qquad (21)$$

F. PEM Fuel Cell Calculations

In PEMFC, anode and cathode are suplied by hydrated hydrogen through the air. At Anode Hydrogen gas is ionized in presence of a platinum catalyst [15].

$$H_2 = 2H^+ + 2e^- \qquad (22)$$

The variation of individual cell voltage can be found by decreasing various losses out of E_{Nernst}.

$$V_{fc} = E_{Nernst} - V_{act} - V_{ohm} - V_{conc} \qquad (23)$$

$$E_{cell} = n * V_{fc} \qquad (24)$$

$$E_{Nernst} = \frac{\Delta G}{2F} + \frac{\Delta S}{2F}(T - T_{ref}) + \frac{RT}{2F}\left[\ln(P_{H_2}) + \frac{1}{2}\ln(P_{O_2})\right] \quad (25)$$

For n number of cells, cell voltage can be calculated further. E_{Nernst} is the maximum voltage each cell can produce in the fuel cell.

$$V_{act} = -[\varepsilon_1 + \varepsilon_2 + \varepsilon_3 * T * \frac{1}{2}\ln(P_{O_2}) + \varepsilon_4 \ln(I_{stack}) \quad (26)$$

$$V_{ohm} = I_{stack}(R_m + R_c) \qquad (27)$$

$$V_{conc} = -\frac{RF}{n} * F * \ln\left(1 - \frac{i}{i_1}\right) \qquad (28)$$

V_{act} is the voltage drop due to the activation of the anode and cathode. V_{ohm} is a voltage drop due to the resistance offered to the flow of ions. Due to reduced partial pressure of reactant, a change in the concentration of reactant can be observed that results into voltage drop given as V_{conc}.

IV. RESULTS AND DISCUSSION

A. 10 KV Photovoltaic System

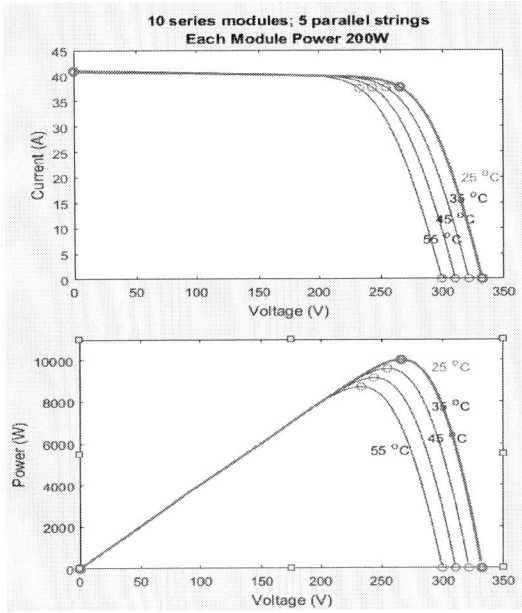

Fig.4. Power and PV terminal voltage with temperature variation

PV modules are connected as 10 modules in series and such 5 strings are connected in parallel to generate 10 KV power out of it. With the help of MPPT maximum power is achieved at 265 V and 25° C temperature. As the surroundings temperature increases, the power delivered by the PV system decreases

B. DC Bus voltage

DC bus voltage is calculated according to equation mentioned above and maintained at 700V after first over and undershoot. The capacity of the DC Bus Capacitor is also calculated by equation mentioned above. DC voltage is maintained at a specific voltage in case of both controlling strategies used. DC voltage at a certain level ensures the proper DC link between the PV system and VSI and at the same time, maximum power would be transferred to the AC system by proper charging of the capacitor.

978-1-5386-6624-1/18 $31.00 © 2018 IEEE

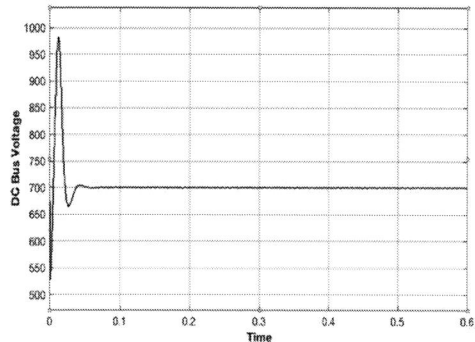

Fig 5. DC Bus Voltage

C. DC Power From Fuel Cell

PEMFC is simulated for generating 10 KW of power. By maintaining the adequate input hydrogen and oxygen pressure, a constant power, after a transient can be maintained at fuel cell end. Hydrogen disintegrated by PV system during daytime will be used in the fuel cell to provide essential power supply.

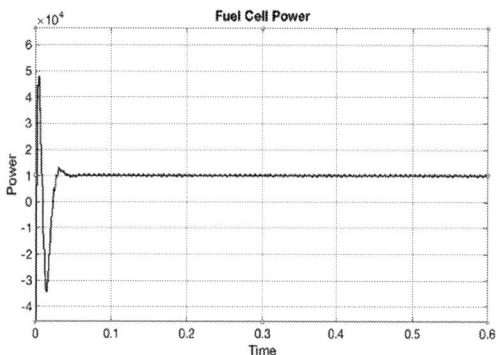

Fig 6. DC Power from Fuel Cell

D. Harmonics Analysis

The presented system is injected by harmonics from a combination of a linear load of 30 KW, a non-linear load of 5 KW and power electronics switches itself. The main emphasis of the presented work is to maintain the harmonics content within limit according to IEEE-519 and 929 standards. The harmonics content in current is calculated by fast fourier transformation at the source side and results show harmonic content is well below 5%, that ensures the satisfactory operation of the hybrid PV-PEMFC system under consideration [16]. With the application of both the schemes, power system will operate normally as harmonics content is well within the prescribed range [17]. I-Cos(phi) technique has shown its superiority over Unit template technique in harmonics elimination. I-Cos(phi) technique uses both direct and the quadrature component for creating the overall I_{ref} current for pulse generation for VSC.

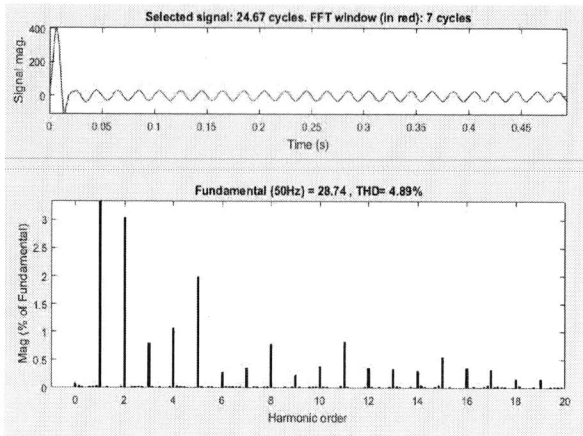

Fig.7. Current Harmonics content with Unit Template control scheme

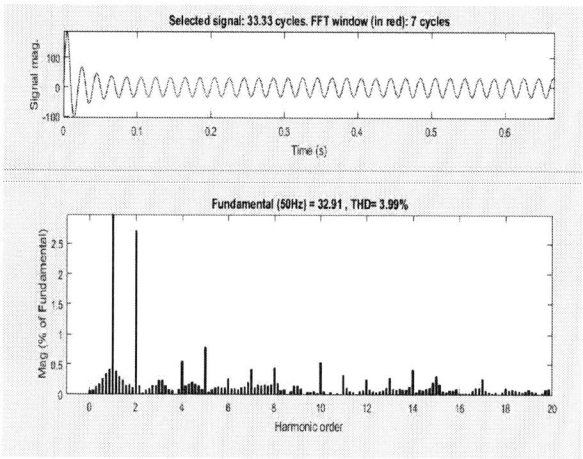

Fig.8. Current Harmonics with I-Cos(phi) control scheme

V. CONCLUSION

A Grid-Tied Hybrid system of PV-PEMFC is capable of providing the reliable power supply to the load. PV and PEMFC working together can be beneficial for the hilly area, where PV can operate in daytime and simultaneously disintegrate hydrogen out of water for PEMFC use during night time. Rest of the power system behaves as an infinite bus, which can help the hybrid PV-PEMFC system during fault and overload conditions. This would further reduce the need of installation of heavy transmission network resulting in the reduced installation cost of the transmission network and reactive power compensating devices. Chances of transmission line overloading get reduced and hence resulting in decreased maintenance cost. During low solar insolation state PEMFC can compensate the energy requirement needed to be satisfied by PV system load. Harmonics content are well within the acceptable limit set by the IEEE standard. Simulated system is capable of satisfying load while keeping the power quality of the system intact.

978-1-5386-6624-1/18 $31.00 © 2018 IEEE

REFERENCES

[1] World Energy Outlook 2017.

[2] Arnulf Jager-Waldau,"PV Status Report", JRC science for policy report of European Commission.

[3] M. Venkateshkumar, R. Raghavan, N. Kumarappan, "PEM fuel cell energy grid integration to electrical power system", Mechanical and Electrical Technology (ICMET) 2010 2nd International Conference on, *pp.* 678-681, 10-12 Sept. 2010.

[4] State of the States: Fuel Cell in America 2016, 7th Edition, Fuel cell technology office, November 2016.

[5] E. Özgirgin, Y. Devrim, A. Albostan, "Modeling and simulation of a hybrid photovoltaic (PV) module-electrolyzer- PEM fuel cell system for micro-cogeneration applications" Int. J Hydrogen Energy, 40 (2015), pp. 15336-15342

[6] B. Singh, A. Chandra, and K. Al-Haddad, "Power quality: problems and mitigation techniques", John Wiley & Sons Ltd., United Kingdom, 2015.

[7] IEEE Recommended Practice and Requirement for Harmonic Control on Electric Power System, IEEE Std. 519, 2014.

[8] S. Gupta, I. Hussain and B. Singh," A solar PV system controlled by least sum of exponentials algorithm", 2016 IEEE 7th Power India International Conference (PIICON).

[9] R. Agarwal, I. Hussain, B. Singh, "LMF based control algorithm for single-stage three-phase grid integrated solar PV system", IEEE Trans. Sustain. Energy, vol. 7, no. 4, pp. 1379-1387, 2016.

[10] V. N. Lal and S. N. Singh, "Control and Performance Analysis of a Single-Stage Utility-Scale Grid-Connected PV System," in IEEE Systems Journal, vol. 11, no.3, pp. 1601–1611, Sept.2017.

[11] A. Kumar and A. H. Bhat, "Fuzzy logic controlled hybrid filter for power quality improvement," 2016 International Conference on Electrical, Electronics, and Optimization Techniques (ICEEOT), Chennai, 2016, pp. 714-719.

[12] G. Bhubaneswari, M.G. Nair, " Design simulation and analog circuit implementation of a three-phase shunt active filter using I cos φ algorithm ", IEEE Trans. Power Deliv., *vol.* 23, no. 2, pp. 1222-1235, 2008.

[13] M. Kandpal, I. Hussain and B. Singh, "Grid integration of solar PV generating system using three-level voltage source converter", Annual IEEE India Conference (INDICON) 2015.

[14] T. H. Priya, A. M. Parimi, and U.M. Rao,"Performance evaluation of high voltage gain boost converters for DC grid integration" International Conference on Circuit, Power and Computing Technologies (ICCPCT) 2016.

[15] V. R. Kavya, K.S.Padmavathy, and M. Shaneeth," Steady-state analysis and control of PEM fuel cell power plant", International Conference on Control Communication and Computing (ICCC) 2013.

[16] B.H. Yong and V.K. Ramachandaramurthy," Harmonic Mitigation of Grid Connected 5MW Solar PV Using LCL Filter", 3rd IET International Conference on Clean Energy and Technology (CEAT) 2014.

[17] D. Mueller and E.H. Camm," Power quality standards for utility wind and solar power plants", PES T&D 2012.

978-1-5386-6624-1/18 $31.00 © 2018 IEEE

Variable Step HCS Based Sensorless Control of Wind Driven DFIG for Autonomous Operation

Bhim Singh, *Fellow, IEEE*, Sambasivaiah Puchalapalli, *Student Member, IEEE*, S.K. Tiwari, *Member, IEEE* and P.K. Goel

Department of Electrical Engineering, *Indian Institute of Technology Delhi*, Hauz Khas, New Delhi-110016, India
South MCD, Dr. SPM Civic Centre, New Delhi-110016, India

bsingh@ee.iitd.ac.in, sambasivaiah8888@gmail.com, shailendrakt@rediffmail.com and pkgoel3866@yahoo.com

Abstract—**This paper presents a sensorless scheme of a wind turbine driven doubly fed induction generator (DFIG) in standalone mode. The sensorless scheme is based on modified hill climb search (HCS) method and the system is immune to mechanical and electrical problems of anemometer. The scheme uses variable frequency sampling time for error free generation of speed set point for the wind turbine generator. The scheme is suitable for both standalone and grid connected modes of operation. The modeling and simulation of sensorless wind energy system, are carried out in MATLAB/Simulink environment using SimPowerSystems tool box for different scenarios like variable wind speeds, change in load and nonlinear unbalanced load connected at point of common coupling (PCC). The DFIG stator voltages and currents are found sinusoidal and balanced and, follow the IEEE 519 standard. A comparison of proposed maximum power point tracking scheme with tip speed ratio method incorporating wind speed measurement, is also presented. Finally, the presented scheme is tested on a developed prototype in the laboratory to validate its performance both in steady state and dynamic conditions.**

Keywords—*Doubly fed induction generator (DFIG), Wind energy, renewable energy system, sensorless control, vector control, power quality, battery energy storage (BES)*

I. INTRODUCTION

As a result of electrification combined with advent of modern gadgets and consumer goods, the per capita consumption of electricity of developing countries like India is continuously increasing. So there is a need to tap every bit of naturally available renewable energy (RE). Wind energy is technically matured and affordable RE source. Unlike solar energy, wind energy is available throughout the day with varying degree of strength. There are many locations where the wind power density is high enough to make the cost of energy comparable to the grid power. Wind energy generators require variable speed operation to realize maximum power point tracking (MPPT). There are many control techniques to achieve variable speed operation. Power electronics based control in addition to variable speed control also achieves reactive power control. As a result of this, power electronics based controller has gained popularity in multi-MW machine. MPPT is an important function of control as it can greatly affect the energy generation particularly small wind turbine, which are installed at lower height where wind variance is very high. In large commercially available wind turbine, since large revenue is at stake, procedure for power curve testing has been evolved and there are many accredited agencies undertaking the job as per standard [1]. For small wind turbine, there is no specific procedure and lower generation is attributed to low wind production during the period by the turbine supplier. MPPT operation of a wind turbine can be achieved by any of the three control namely, tip speed ratio (TSR) [2]-[4], power signal feedback (PSF) [5], hill-climb search (HCS) [6]. Tip

speed ratio (TSR) method may give best result but it needs wind speed data, which needs to be measured through anemometer [2] or estimated using adaptation method [4]. An anemometer used to measure wind speed, is exposed to extreme weather conditions and prone to problem of mechanical wear and tear. This method also requires optimum tip speed ratio of the wind turbine. The power signal feedback control approach necessitates the acquaintance of wind turbine's characteristics, that is to be traced by control process. This characteristic curve needs to be obtained by simulations or experimental tests, which increases the cost of implementation.

The hill climb search (HCS) method continuously searches for the optimum operating point through its algorithm. Normally such method uses gradient senor, which is prone to error due to presence of various noise signals because of nonlinear elements [7]. Dalala *et al.* [8] have presented a modified perturb and observe (P&O) based MPPT using wind sensorless method for PMSG based system. In this method, the DC bus current is taken as a perturbation variable to track MPPT. Hussain *et al.* [9] have presented an adaptive MPPT algorithm, which is a combination of HCS and TSR algorithms. In this method, a single-ended primary inductor converter is used, which has buck-boost feature and fast MPPT is achieved for wide range of wind speeds in a system of wind driven PMSG. But this scheme is complicated to incorporate in doubly fed induction generator (DFIG) based system. Chen *et al.* [10] have discussed an analytical approach to generate q and d-axis currents of rotor to maximize the output from the wind turbine by minimizing the total copper loss in DFIG. This algorithm is for grid connected system and its real time transient performance is not presented. In addition to MPPT control, its performance analysis in term of energy is also essential for commercial viability. This aspect has not been covered by many authors. Wind sensorless control for DFIG, has been covered by some authors. Tan *et al.* [11] have proposed suboptimal power point tracking (SOPPT) method to realize MPPT for DFIG based wind energy system for hybrid remote area power supply system. Vepa [12] has proposed a nonlinear, optimal control of DFIG for realizing MPPT. But there is no experimental validation of the system.

This paper presents a variable step hill climb search (HCS) method for doubly fed induction generator (DFIG) based sensorless wind energy system (SWES) for realizing MPPT. A comparison is also presented with normal HCS and TSR MPPT techniques. The proposed MPPT algorithm, is included in the rotor side converter (RSC) control to track maximum power from the wind, whereas load side converter (LSC) control looks after the maintaining of voltage and frequency in all cases.

The SWES is designed and simulated using SimPowerSystems tool box of MATLAB under variable wind speeds, change in load and nonlinear unbalanced load.

To validate the proposed SWES, tests are performed under steady state and dynamic conditions on developed prototype in the laboratory.

II. SYSTEM CONFIGURATION

The SWES in standalone mode with a battery energy storage (BES), is depicted in Fig. 1. The rotor windings are connected to an insulated gate bipolar transistor (IGBT) based voltage source converter (VSC) termed as RSC and the VSC, which is connected to DC bus of RSC is termed as LSC. Moreover, the BES is connected at DC link as shown in Fig. 1. The design of wind turbine driven DFIG, BES, transformer and other electrical parameters of SWES, is carried out based on the procedure reported in [3].

III. CONTROL PHILOSOPHY

The proposed SWES is effectively controlled using RSC control, an incremental speed set point estimator that is used to generate rotor speed reference and LSC control, which are illustrated as follows.

A. RSC Control Algorithm

The RSC control incorporates the MPPT function of wind turbine. The TSR based MPPT requires acquaintance of wind speed to determine the speed set point. In the presented scheme, HCS method is used, which requires DFIG speed and air gap power to determine speed set point.

The field oriented approach is used in RSC control to regulate the machine speed as depicted in Fig. 2. In field oriented control, the direct rotor component of current, is aligned toward the stator field and in such conditions, for constant stator flux, ϕ_{sd} the torque is proportional to the quadrature rotor component of current, I_{qr} as,

$$T_{em} = \left(\frac{p}{2} \times \frac{3}{2}\right) \phi_{sd} I_{qr} \qquad (1)$$

The quadrature reference rotor component of current, I^*_{qr} is derived from the speed controller as depicted in Fig. 2, as,

$$I^*_{qr(k)} = I^*_{qr(k-1)} + K_{p\omega}(\omega_{rerr(k)} - \omega_{rerr(k-1)}) + K_{i\omega}\omega_{rerr(k)} \qquad (2)$$

where, ω_{rerr} is the speed error, which is determined as,

$$\omega_{rerr(k)} = \omega^*_{r(k)} - \omega_{r(k)} \qquad (3)$$

$K_{p\omega}$ and $K_{i\omega}$ are the proportional and integral gains of PI speed controller, respectively.

In the proposed scheme, the speed of rotor (ω_r) is computed from the sensed rotor angle (θ_r). However, the reference speed, ω^*_r is derived as follows [2].

$$\omega^*_r = \frac{\lambda^* \times \eta_G \times V_w}{r} \qquad (4)$$

where λ^*, η_G, V_w and r represent the optimum tip speed ratio,

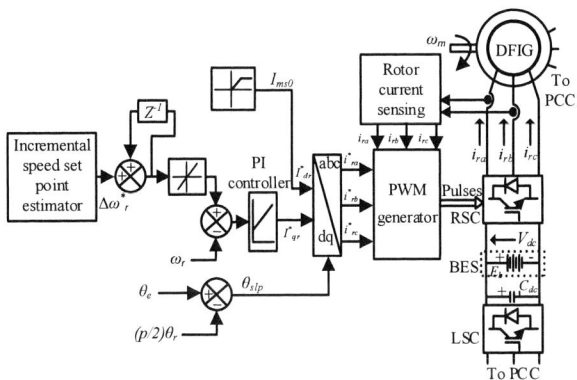

Fig. 2. Control diagram of RSC.

gear ratio, rated wind speed and turbine blade length, respectively.

The value of reference direct component of rotor current, I^*_{dr} is taken as machine magnetizing current and it depends on line rms voltage, V_{Lr} and magnetizing reactance, X_m as,

$$I^*_{dr} = I_{ms0} = \frac{\sqrt{2}V_{Lr}}{\sqrt{3}X_m} \qquad (5)$$

The derived reference quantities, I^*_{dr} and I^*_{qr} are passed through reverse Park's transformation in order to obtain reference three phase rotor currents, i^*_{ra}, i^*_{rb} and i^*_{rc} as,

$$\begin{bmatrix} i^*_{ra} \\ i^*_{rb} \\ i^*_{rc} \end{bmatrix} = \begin{bmatrix} \sin(\theta_{slp}) & \cos(\theta_{slp}) \\ \sin(\theta_{slp} - 2\pi/3) & \cos(\theta_{slp} - 2\pi/3) \\ \sin(\theta_{slp} + 2\pi/3) & \cos(\theta_{slp} + 2\pi/3) \end{bmatrix} \begin{bmatrix} I^*_{dr} \\ I^*_{qr} \end{bmatrix} \qquad (6)$$

where θ_{slp} refers to angle of transformation, which is computed by means of the following,

$$\theta_{slp} = \int_0^t (\omega_e - \omega_r) dt \qquad (7)$$

where ω_e represents rated angular stator field velocity, which is taken as 100π rad/s. The reference rotor currents derived from (6) and the sensed currents (i_{ra}, i_{rb} and i_{rc}) are compared and fed to the PWM current controller. The current controller generates switching signals for IGBTs of RSC.

B. Incremental Speed Set Point Estimator

This block estimates the incremental speed set point ($\Delta\omega^*_r$). The input for the estimator is the speed, ω_r and estimated air gap power, P_a. The power P_a is computed as,

$$P_a = (P_s + 3i_s^2 R_s) \times (1 - s_p) \qquad (8)$$

where P_s, i_s, R_s and s_p represent stator power, stator current, per phase stator resistance and operating slip of the machine.

The block diagram of $\Delta\omega^*_r$ estimator is shown in Fig. 3. As per Fig. 3, the estimator block consists of three sub-block, which are discussed as follows.

Energy Estimation: The total electrical power, P_a is averaged over a period of a specified time interval.

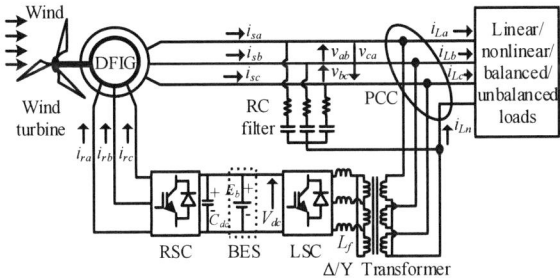

Fig. 1. Sensorless wind energy system with battery energy storage.

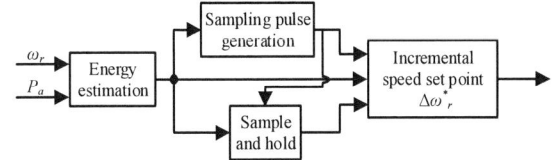

Fig. 3. Block diagram of incremental speed set point estimator.

Averaging of the signal for a specified time interval, removes the transients arising in the power output due to various nonlinear elements. The resultant signal becomes input to the second block to determine incremental change in power.

Sampling Pulse Generation: Sampling pulse is generated after meeting of the following two criteria.

- Time delay of 1.5 s is there from the generation of previous sampling pulse. This is the time taken by the machine to achieve steady state with maximum step change in speed output.

- Successive absolute difference of P_a over 0.25 s period is less than 100 W. The low value of difference can only be achieved when DFIG has attained steady state as shown in Fig. 4.

Fig. 4 illustrates the waveforms of airgap power (P_a), averaged airgap power (P_{avg}), change in airgap power (ΔP_a), sampling pulse ($u(t)$) and sampled airgap power (P_{sam}). The sampling pulse ($u(t)$) depicted in Fig. 4, confesses that the proposed MPPT method is a variable step. Fig. 5 depicts the flow diagram of sampling algorithm to achieve variable step. In the active MPPT zone, the power output is proportional to cube of the wind speed or rotational speed. Assuming slight change in λ on account of variation in wind speed, $\Delta\omega^*_r$ is estimated from sampled $P_{a(n)}$ and $P'_{a(n-1)}$, as depicted in Fig. 5. The change in airgap power (ΔP_a) is estimated as,

$$
\begin{aligned}
\Delta P_a &= P_a(n) - P'_a(n-1) \\
&= k_1(V_w + \Delta V_w)^3 - k_1(V_w)^3 \\
&= k_1(V_w^3 + 3V_w^2\Delta V_w + 3V_w\Delta V_w^2 + \Delta V_w^3) - V_w^3) \\
&\approx k_1'(V_w^2\Delta V_w)
\end{aligned}
\tag{9}
$$

P_a is expressed in terms of wind speed by ignoring mechanical losses as,

$$
P_a = k_2 V_w^3
\tag{10}
$$

$$
\left.
\begin{aligned}
\frac{\Delta P_a * V_w}{P_a} &= k \times \Delta V_w \\
\frac{\Delta V_w}{V_w} &= \frac{k \times \Delta P_a}{P_a}
\end{aligned}
\right\}
\tag{11}
$$

In (9-11), k_1, k_1', k_2 and k, are constant parameters. Moreover, $\Delta P_a/P_a$ gives information of per unit change in wind speed. Since in linear MPPT region, the ΔV_w is proportional to $\Delta\omega^*_r$, $\Delta P_a/P_a$ gives information of required per unit change in the generator speed as,

$$
\frac{\Delta\omega_r}{\omega_r} = \frac{K \times \Delta P_a}{P_a}
\tag{12}
$$

With the presented logic, the HCS algorithm functions to

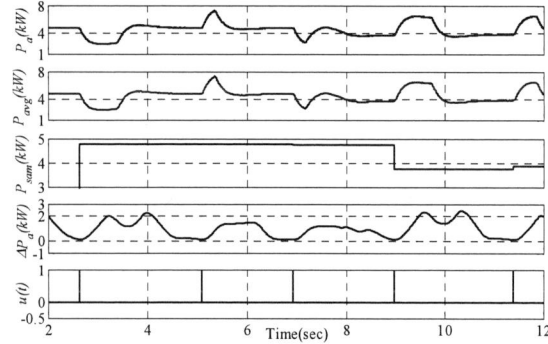

Fig. 4. Sampling of power in SWES.

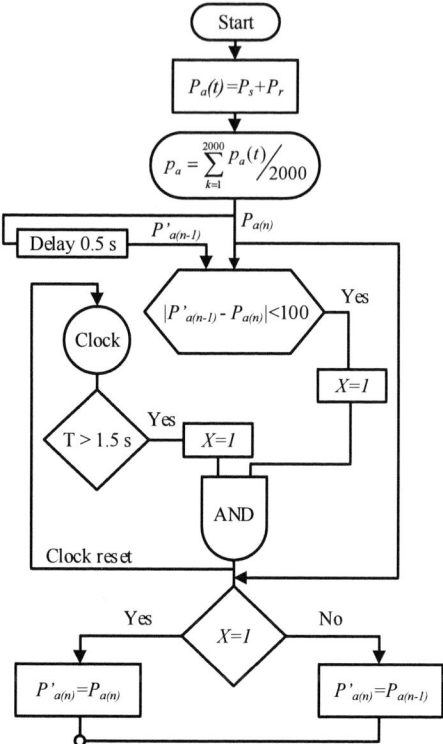

Fig. 5. Flow chart of the sampling algorithm.

achieve optimum tip speed ratio i.e. λ^*. The control logic detects the mode as shown in Fig. 6 and accordingly $\Delta\omega^*_r$ is generated as per Table-I. The constant factor K, as per (12) is chosen to arrive incremental per unit speed set point i.e. $\Delta\omega^*_r$. It is chosen such that the generator speed achieves MPP within three iterations. In the presented case, the value is chosen as 2. With this value, the maximum size of step speed demand is 6 rad/sec. It is to be noted that whenever there is a change in wind speed, incremental speed set point is generated. Figs. 7-8 depict the waveforms of V_w, ω_r, P_a and P_{sam} for normal HCS and modified HCS based SWES, respectively. It is observed that even though there is a reduction in wind speed at t=7.5 s, the normal HCS MPPT algorithm senses mode 'b' as per Table-I and increases DFIG speed set point, as depicted in Fig. 7. As a result of this, the

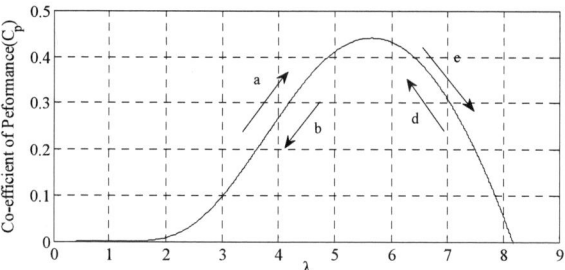

Fig. 6. C_p-λ characteristics with different modes of HCS algorithm.

TABLE I.   MODES OF OPEARATION OF HCS ALGORITHM

ΔP_a and $\Delta\omega_r$ Scenario	Operating Mode	Speed Set Point
$\Delta P_a>0$, $\Delta\omega_r>0$	a	$\omega^*_r + \Delta\omega^*_r$
$\Delta P_a<0$, $\Delta\omega_r<0$	b	$\omega^*_r + \Delta\omega^*_r$
$\Delta P_a<0$, $\Delta\omega_r>0$	c	$\omega^*_r - \Delta\omega^*_r$
$\Delta P_a>0$, $\Delta\omega_r<0$	d	$\omega^*_r - \Delta\omega^*_r$

978-1-5386-6624-1/18 $31.00 © 2018 IEEE

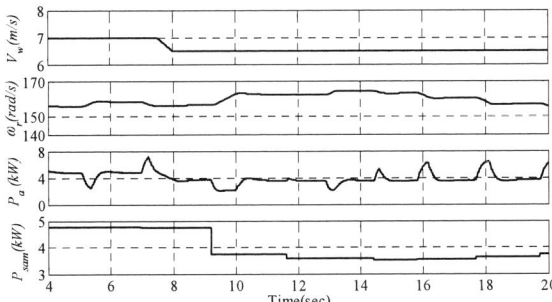

Fig. 7. Performance of SWES under normal HCS based control.

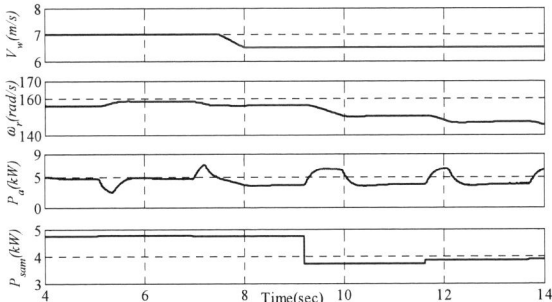

Fig. 8. Performance of SWES under HCS based with additional control loop.

controller takes more time in approaching λ^* and as a result, lesser energy is captured.

To exclude such operation, a control logic is incorporated in the incremental speed set point estimator block. It calculates the time interval, T_s from the previous sampling pulses and generates the incremental speed set point in case T_s exceeds threshold time (T_{TH}), as per the logic presented in Table-II. Based on the electrical and mechanical time constants, it is found that for a maximum step speed increase of 6 rad/s, the stabilization period remains within 1.5 s. Hence T_{TH} is chosen as 1.8 s. With the incorporation of the above said loop, the system progress toward λ^* in less iteration as shown in Fig. 8.

C. LSC Control Algorithm

The LSC objective is to maintain the rated voltage and frequency. It also provides path for differential power flow between stator-battery and battery-load in case of over-generation and under-generation, respectively. LSC through indirect vector control generates reference signals for stator currents. The total control loop as depicted in Fig. 9, consists of three sub-loops, described in the subsequent sections.

1) Reactive Reference Current Component (I^*_{ds}): I^*_{ds} is related to the reactive power flow through stator and it is obtained from the proportional-integral (PI) voltage controller output as,

$$I^*_{ds(k)} = I^*_{ds(k-1)} + K_{pv}(V_{tnerr(k)} - V_{tnerr(k-1)}) + K_{iv}V_{tnerr(k)} \quad (13)$$

where, V_{tnerr} is the voltage error and is derived as,

$$V_{tnerr(k)} = V^*_{tn} - V_{tn} \quad (14)$$

TABLE II. NOVEL CONTROL LOGIC TO ESTIMATE SPEED SET POINT

T_s	ΔP_a	Speed Set Point
$T_s > T_{TH}$	$\Delta P_a > 0,$	$\omega^*_r + \Delta\omega^*_r$
$T_s > T_{TH}$	$\Delta P_a < 0,$	$\omega^*_r - \Delta\omega^*_r$

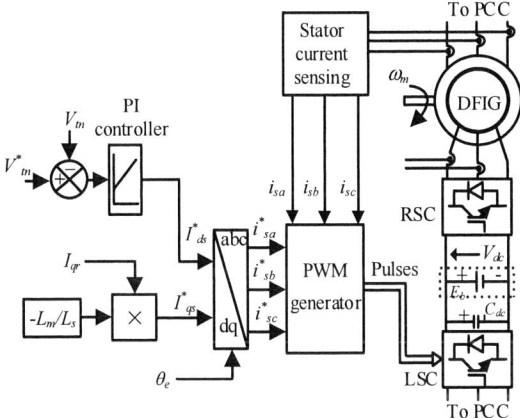

Fig. 9. Control diagram of LSC.

where V_{tn} is the peak of phase voltage at PCC and it is derived from the instantaneous phase voltages as,

$$V_{tn} = \{2(v^2_{an} + v^2_{bn} + v^2_{cn})/3\}^{1/2} \quad (15)$$

and V^*_{tn} is the reference voltage set point, which is equal to 338.84 V.

2) Active Reference Current Component (I^*_{qs}): I^*_{qs} is related to I_{qr}, which is torque component of rotor current as,

$$I^*_{qs(k)} = \frac{-L_m I_{qr(k)}}{L_s} \quad (16)$$

3) Transformation Angle Loop: Transformation angle, θ_e is computed by time integration of the reference load frequency f_L i.e. 50 Hz. The value of θ_e at kth sampling time is as,

$$\theta_e = 2\pi f_L k T_{samp} \quad (17)$$

T_{samp} is the sampling time period of the system, which is taken 10 μs. The I^*_{ds} and I^*_{qs} are transformed, to obtain stator reference currents (i^*_{sa}, i^*_{sb} and i^*_{sc}) using transformation angle θ_e as,

$$\begin{bmatrix} i^*_{sa} \\ i^*_{sb} \\ i^*_{sc} \end{bmatrix} = \begin{bmatrix} \sin(\theta_e) & \cos(\theta_e) \\ \sin(\theta_e - 2\pi/3) & \cos(\theta_e - 2\pi/3) \\ \sin(\theta_e + 2\pi/3) & \cos(\theta_e + 2\pi/3) \end{bmatrix} \begin{bmatrix} I^*_{ds} \\ I^*_{qs} \end{bmatrix} \quad (18)$$

The estimated reference currents together with sensed currents (i_{sa}, i_{sb} and i_{sc}) are passed through PWM generator to provide switching pulses for LSC.

IV. SIMULATION RESULTS AND DISCUSSION

A simulation model of the SWES is developed using SimPowerSystems tool box of MATLAB. The proposed variable step HCS block generates the reference generator speed set point to achieve MPPT based on DFIG speed and air gap power. The various signals used for performance demonstration, are wind speed (V_w), rotor speed (ω_r), power coefficient of turbine (C_p), stator frequency (f_s), rms value of PCC phase voltage (V_r), airgap power (P_a) and load power (P_L). The simulation of SWES is performed under varying wind speeds, varying load and nonlinear unbalanced load scenarios. The various parameters used for simulation are given in Appendix-A.

A. Dynamic Behavior of SWES at Variable Wind Speed

Dynamic behavior of SWES at varying wind speed is depicted in Fig. 10 (a). The system is initiated on connected load of 12 kW with 7 m/s wind speed. Consequently the control action takes the turbine speed at λ^* and wind speed is

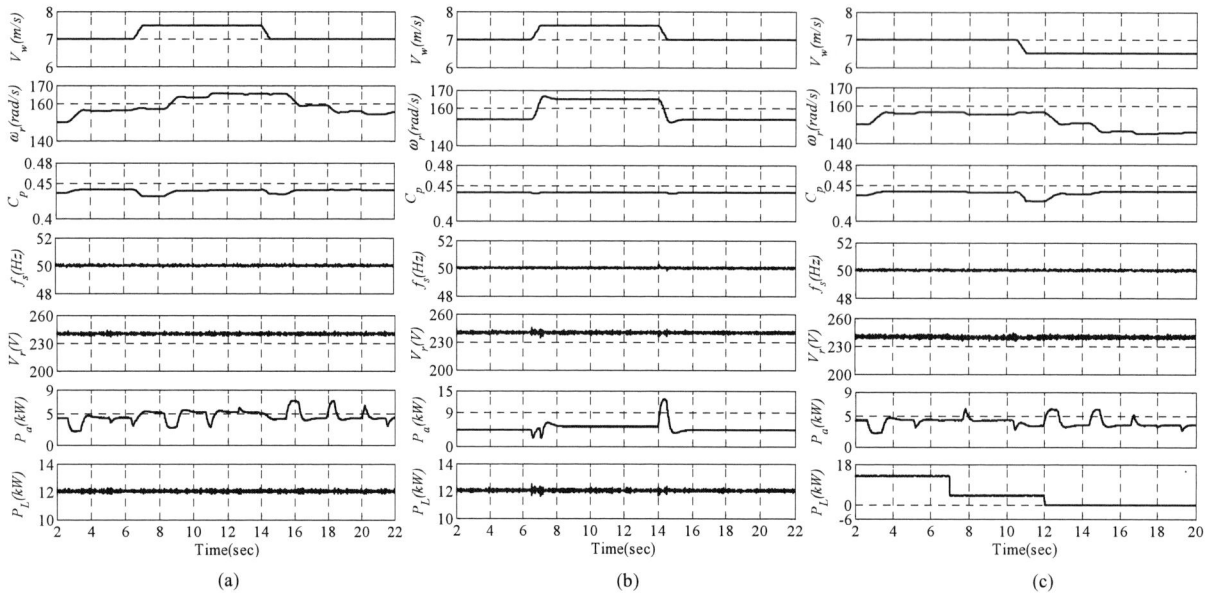

Fig. 10. Performance of SWES (a) during varying wind speed with variable step HCS MPPT method (b) during varying wind speed with TSR MPPT method (c) during change in load and wind speed.

raised to 7.5 m/s with an acceleration of 1 m/s². The control action senses change in power and gives incremental speed set point at regular interval to track MPPT. From t=14 s, wind speed starts reducing to reach its original value at same rate of change. It is found that the control action continues to take corrective action such that wind turbine operates at λ^*. Moreover, the performance of SWES with conventional TSR MPPT approach under varying wind speeds, is depicted in Fig. 10 (b), which is theoretically most efficient. Fig. 11 shows the energy diagram (Time vs C_p) of both scheme of MPPT. As per the result, the MPPT efficiency of presented scheme is more than 98%.

B. Dynamic Behavior of SWES at Change in Load and Nonlinear Unbalanced Load

Fig. 10 (c) and Fig. 12 depict the dynamic behavior of SWES during change in load and nonlinear unbalanced load, respectively. In Fig. 10 (c), phase 'a&b' is opened at t=7.0 s and again load of phase 'c' is also removed at t=12.0 s. Besides, wind speed pattern is also varied. However, the control action continues to take corrective action to run the turbine at MPPT irrespective of change in loading condition and varying wind speed, which is evident from C_p.

Fig. 12 shows the waveforms of DFIG stator voltages (v_{sabc}), load currents (i_{La}, i_{Lb}, i_{Lc}) and stator currents (i_{sabc}) at nonlinear load. To make the system unbalance, phase 'b' of load current is removed at 2.02 s and subsequently, phase 'c' is disconnected just after 2.06 s. However, it is evident that stator currents and voltages are sinusoidal and balanced.

V. HARDWARE IMPLEMENTATION

To validate the proposed DFIG based SWES in standalone with a variable step HCS MPPT technique, a prototype of 3.7 kW DFIG is developed in the laboratory. A wound rotor induction machine is used as a DFIG. The wind turbine characteristics are emulated with the help of DC shunt motor and a buck chopper. The BES placed at DC link is realized using 20 numbers of 12 V, 7 AH batteries to attain 240 V. Moreover, two sets of such battery banks are

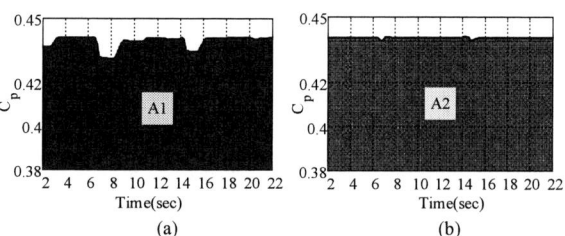

Fig. 11. Energy generation from MPPT method (a) HCS method (b) variable step HCS method.

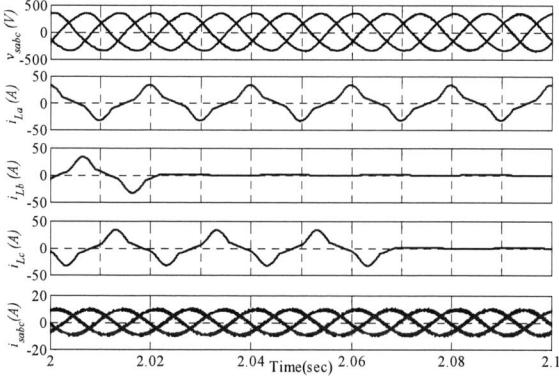

Fig. 12. Performance of SWES at nonlinear unbalanced load.

connected in parallel to achieve required AH capacity. Three phase two VSCs are utilized as RSC and LSC.

Moreover, the AC side of LSC, is connected to PCC through a delta/star transformer. The detailed parameters of experimental prototype are listed in Appendix-B.

Figs. 13 (a-b) show the performance of the system when the wind speed is varied and the SWES is producing generator set point to achieve MPPT. Fig. 13 (c) shows power variations of stator, load and LSC with respect to changes in wind speed. Fig. 13 (d) shows the dynamic performance of SWES during rotor speed transition from subsynchronous to supersynchronous region. Figs. 14 (a-f)

978-1-5386-6624-1/18 $31.00 © 2018 IEEE

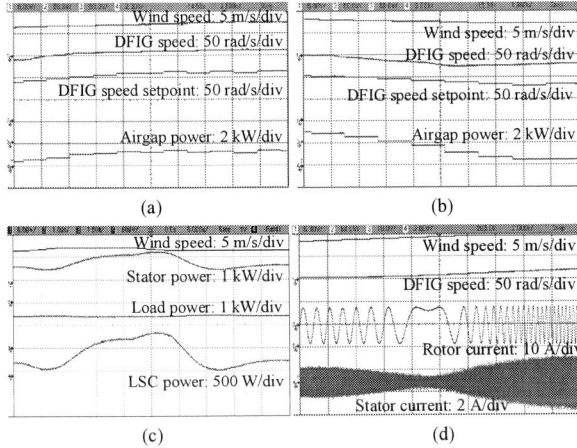

(a)   (b)

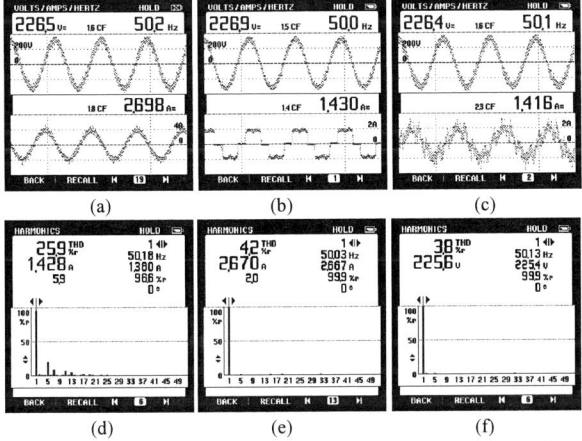

(c)   (d)

Fig. 13. Experimental results of SWES (a) increase of wind speed (b) decrease of wind speed (c) power response with change in wind speed (d) transition from subsynchrnous to supersynchronous mode.

Fig. 14. Experimental waveforms of SWES at steady state with nonlinear load (a-c) voltage and current waveforms of stator, load and LSC (d-f) harmonic spectrum of load current, stator current and PCC voltage.

show the steady state performance of SWES at nonlinear load. Figs. 14 (a-c) depict the voltage and current waveforms of stator, load and LSC. Figs. 14 (d-f) show the harmonic spectra of load current, stator current and PCC voltage, respectively. From Figs. 14 (e-f), it is clear that total harmonic distortion (THD) of DFIG stator currents and voltages are within the IEEE 519 standard.

VI. CONCLUSION

The proposed DFIG based SWES using a variable step HCS method works efficiently in tracking MPPT of the wind turbine. The proposed MPPT method, which is incorporated in the RSC control effectively provides the rotor speed set point to extract maximum power under varying wind speed conditions. The control of voltage and frequency, has been achieved by LSC through indirect vector control. Simulated results have shown the satisfactory performance of SWES under varying wind speeds, varying load and nonlinear unbalanced load connected at PCC. A comparison of proposed maximum power point tracking scheme with tip speed ratio method incorporating wind speed measurement has also been presented. Moreover, test results have shown the satisfactory performance of SWES under varying wind speeds and nonlinear load scenarios.

Moreover, the THDs of DFIG stator voltages and currents, are found as per the IEEE 519 standard.

APPENDICES

A. Parameters of Simulation

DFIG: Wound rotor induction machine of 415 V, 4 Pole, 15 kW, L_m=68.68 mH, L_{lr}=5.24 mH, R_r=0.236 Ω, L_{ls}=1.06 mH, R_s=0.2297 Ω; Wind turbine: 15 kW, Turbine rotor length, r=4.1 m, gear ratio η_G=15.93, λ^*=5.67, rotor power coefficient, C_p=0.441, rated wind speed=11 m/s, minimum and maximum rotor speeds of DFIG, ω_{rmin}=110 rad/s and ω_{rmax}=198 rad/s.

B. Experimental Parameters

DFIG: Make- McFEC Ltd, Rating: stator windings connection-Y, 4 pole, 50 Hz, 400 V, 3.7 kW, Inertia=0.1878 kg-m², L_{lr}=6.832 mH, L_{ls}=6.832 mH, L_m=219 mH, R_r=1.708 Ω, R_s=1.32 Ω; DC motor of 230 V, 5 kW, R_f = 220 Ω, R_a = 1.3 Ω, L_f = 7.5 H, L_a =7.2 mH, K_ϕ=1.3314.

REFERENCES

[1] International Electro-Technical Commission Std.-61400-12-1, "Power performance measurement of electricity producing wind turbine," Edition 2005.

[2] A. S. Satpathy, N. K. Kishore, D. Kastha, and N. C. Sahoo, "Control scheme for a stand-alone wind energy conversion system," *IEEE Trans. Energy Convers.*, vol. 29, no. 2, pp. 418-425, June 2014.

[3] S. K. Tiwari, B. Singh, and P. K. Goel, "Design and control of microgrid fed by renewable energy generating sources," *IEEE Trans. Ind. Appl.*, vol. 54, no. 3, pp. 2041-2050, May-June 2018.

[4] C. Shao, X. Chen, and Z. Lian, "Application research of maximum wind-energy tracing controller based adaptive control strategy in WECS," *IEEE 5th International Power Electronics and Motion Control Conference, 2006*, 14-16 August 2006.

[5] Eftichios Koutroulis and Kostas Kalaitzakis, "Design of a maximum power tracking system for wind-energy-conversion applications," *IEEE Trans. Ind. Electron.*, vol. 53, no. 2, April 2006.

[6] T. Pan, Z. Ji, and Z. Jiang, "Maximum power point tracking of wind energy conversion systems based on sliding mode extremum seeking control," *IEEE Conf. on Energy* 2030, 2008, 17-18 November 2008.

[7] I. K. Amin and M. N. Uddin, "MPPT based efficiently controlled DFIG for wind energy conversion system," *2017 IEEE International Electric Machines and Drives Conference (IEMDC)*, Miami, FL, 2017, pp. 1-6.

[8] Z. M. Dalala, Z. U. Zahid, and J. S. Lai, "New overall control strategy for small-scale WECS in MPPT and stall regions with mode transfer control," *IEEE Trans. Energy Convers.*, vol. 28, no. 4, pp. 1082-1092, Dec. 2013.

[9] J. Hussain and M. K. Mishra, "Adaptive maximum power point tracking control algorithm for wind energy conversion systems," *IEEE Trans. Energy Convers.*, vol. 31, no. 2, pp. 697-705, June 2016.

[10] B. A. Chen, T. K. Lu, Y. Y. Hsu, W. L. Chen, and Z. C. Lee, "An analytical approach to maximum power tracking and loss minimization of a doubly fed induction generator considering core loss," *IEEE Trans. Energy Convers.*, vol. 27, no. 2, pp. 449-456, June 2012.

[11] Y. Tan, L. Meegahapola, and K. M. Muttaqi, "A suboptimal power-point-tracking-based primary frequency response strategy for DFIGs in hybrid remote area power supply systems," *IEEE Trans. Energy Convers.*, vol. 31, no. 1, pp. 93-105, March 2016.

[12] R. Vepa, "Nonlinear, optimal control of a wind turbine generator," *IEEE Trans. Energy Convers.*, vol. 26, no. 2, pp. 468-478, June 2011.

2nd IEEE International conference on power Electronics, Intelligent Control and Energy systems (ICPEICES-2018)

An Intelligent MPPT Controller Based AC to DC Boost PFC Converter for Grid-Tied Wind Energy Conversion System

DamodharReddy
School of Electrical Engineering, *VIT,* India.
damodhar_reddy@ymail.com

Sudha Ramaramy
School of Electrical Engineering, *VIT,* India.
ishuma@gmail.com

Abstract - **This paper gives an intelligent MPPT controller based boost PFC converter topology for AC to DC conversion in the wind energy system. The single stage boost PFC circuit topology comprises of bridge rectifier plus boost converter, which is employed as AC to DC converter with wide conversion ratio, continuous input current, and low input current ripple. It can be widely used for AC to DC conversion with high efficiency and improved power factor. And a grid side converter is engaged for DC to AC conversion in the grid-connected mode, which includes the DC-link voltage control and grid side power control of the proposed system. In this paper, RBFN controller based single-stage PFC converter is designed for the grid-tied wind energy conversion system to supply the load demand of 1kW active power and 900VAr reactive power. The result analysis of both off-grid and grid-tied configurations is validated in MATLAB-Simulink.**

Keywords - **RBFN (Radial Basis Function Network) control algorithm, Wind Energy Conversion System (WECS), Single-stage PFC Converter, Grid Side Converter (GSC).**

I. INTRODUCTION

The wind energy system is the most popular and wide range installed renewable energy system in the power sector, due to rapid developments in low cost and highly efficient wind turbines design, turbine generators and power conversion units. In another hand, the price of fossil fuels is getting increased in the past decades which results in a heavy burden on the end-users. To overcome the deficiency of power from the conventional production units, the installation of renewable energy based alternative systems are essential to meet the load demand with effective utilization of natural resources. One of the clean energy productions is from [1] with different types of turbine configurations. There are mainly two types of wind systems widely used in the wind forms based on the operational conditions such as variable and fixed speed wind turbines.

Similarly, various types of turbine generators are used in wind energy system such as Induction Generators [2] (IGs), PMSGs (Permanent Magnet Synchronous Generators) and DFIGs (Double Fed Induction Generators) [3, 4]. The main aim of these turbine generators is to convert the mechanical energy into electrical energy. However, the PMSG is a preferable machine [5, 6] for low-speed wind turbine systems with high efficiency. Many MPPT control

techniques are proposed [7, 8] by the researchers to track the maximum power point such as INC (incremental conductance), P & O (Perturbation & Observation), fuzzy logic control and artificial neural network based control [9] etc. But the implementation of conventional MPPT controllers [10] like INC and P & O are very complex and also has low accuracy. Apart from all the control techniques, RBFN control technique is a fast and dynamic non-linear control method with minimum computational data [11]. This is one of the feed forward controllers of the artificial neural network (ANN) [12] with simple in design structure and fast convergence speed.

The two-stage converters [13, 14] are essential conversion units in the wind energy systems for AC to DC conversion with a two-stage conversion such as diode bridge rectifier + boost converter [15], boost PFC converter [16-18] and diode bridge rectifier + CUK converter [19, 20] etc. But these converter configurations have the drawbacks of low power density, high voltage stress across the switches and low efficiency. In another hand, the single stage PFC converters [21, 22] are preferable in AC to DC power conversion systems [23] where unidirectional power flow is adequate. Due to the multi-agent functionality, these are widely used in various applications like data centers, telecommunications, diesel turbine systems, and wind energy conversion systems and at AC mains for DC applications etc. The benefits of the proposed boost PFC topology [24, 25] are low voltage stress across the switches, high power density, low harmonic distortion, and high efficiency. In this paper, RBFN controller based single-stage PFC converter configuration is designed for both off-grid and grid-tied condition at variable wind speeds.

II. GRID-TIED WIND ENERGY CONVERSION SYSTEM

In this proposed system configuration as shown in Fig.1, the complete system operation involves in various stages: In the front-end, the kinetic energy is converted into mechanical energy at the wind turbine and fed to PMSG where mechanical energy is transformed to electrical energy. The generated alternative current (AC) from the wind turbine generator is converted into DC through the three-phase bridge rectifier with reduced input ripple content and the DC voltage is enhanced by boost converter for the desired voltage level. In the back-end, the DC-link voltage is converted into AC through the grid side converter (PWM inverter), which is fed to the load and as well as to the grid.

978-1-5386-6624-1/18 $31.00 © 2018 IEEE
510

The PWM (pulse width modulation) inverter has the capability to regulate the DC-link voltage and to control the active and reactive power based on the load on the requirement.

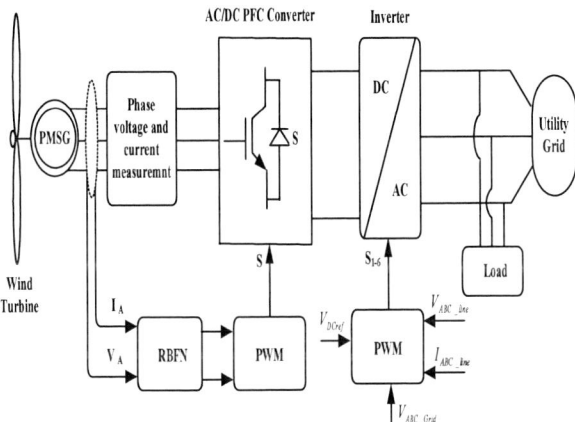

Fig.1 Design configuration of grid-tied wind energy conversion system

A. Wind Energy Conversion System

The wind energy conversion can be done in two stages; one is at the turbine blades where the kinetic energy is transformed into mechanical energy through a variable speed turbine. The second is at the turbine generator where the mechanical energy is transformed into electrical energy. The PMSG has widely used a generator in small-scale low-speed variable speed turbine systems due to due to high conversion ratio, maximum energy capture, and high efficiency.   It has many benefits such as fewer copper losses, low noise, and less weight etc. The PMSG has an ability to control a wide range of rotor speeds corresponding to the large variation in wind speeds and make it possible to obtain MPPT through various MPPT configurations. The relation between turbine output power and turbine speed of the wind system are shown in Fig.2 for different wind speeds at zero pitch angle.

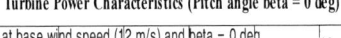

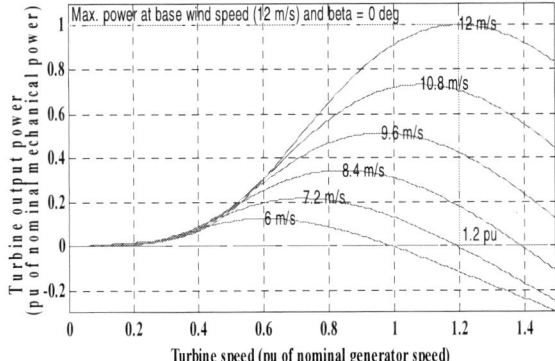

Fig.2 Turbine power and turbine speed characteristics

The turbine torque (T_m) can be written as,

$$T_m = \frac{1}{2}\rho A C_p(\lambda,\beta)v^3 \frac{1}{\omega_m} \qquad (1)$$

Where, ρ =Air density (kg/m³), C_p = Power Coefficient, A= Sweep area of turbine blades (m²), β = Pitch angle (deg), v = Wind speed (m/s), ω =Rotor angular velocity (rad/sec) and λ= Tip-ratio. And λ can be expressed as,

$$\lambda = \frac{\omega_m r}{v} \ , \text{ where r = turbine blade (rotor) radius (m) (2)}$$

Similarly, the turbine power can be written as,

$$P_m = \frac{1}{2}\rho A C_p(\lambda,\beta)v^3 \qquad (3)$$

The design parameters considered for wind turbine system simulation for the base speed are tabulated in Table.1.

TABLE.1: TURBINE AND PMSG PARAMETERS

S.No	Wind system parameters	Ratings
1	Wind base speed	12m/s
2	Pitch angle	0°
3	Torque constant	1.8
4	Flux linkage	1.2Wb-t
5	Stator phase resistance	3.07Ω
6	Armature inductance	6.57mH
7	Maximum voltage at base speed (peak)	230V
8	Maximum output power at base speed	1kW

B. Single Stage AC to DC Converter

Here the operation of the boost PFC converter is involved in a single stage as shown in Fig.3, which employed as AC to DC converter by the bridge rectifier and the DC to DC converter with the enhanced voltage level at the DC-link. The main function of the PFC boost topology is to convert AC to DC with enhanced input power factor. The converter is operated at a duty cycle of greater than 0.5 for boost/ step-up function. The operation of the converter can be expressed in two modes as follows,

Mode.1: When S is ON state, the input inductor L_i (L_i = $L_{A, B, C}$) start charging from input through the switch and the output capacitors C discharge through the load R, where diode D is inactive.

Mode.2: When S is in OFF state, the input inductor (L_i) energy releases through the load, capacitors C discharge through the load R, where diode D is active.

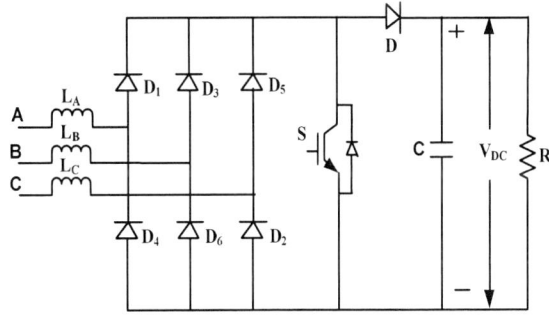

Fig.3 Circuit diagram of boost PFC converter

978-1-5386-6624-1/18 $31.00 © 2018 IEEE

The voltage equations for each mode of AC to DC bridge rectifier can be written as follows,

When diode D_1 and D_6 are in active state at $\dfrac{\pi}{3}, \dfrac{2\pi}{3}$

$$V_{AB} = i_{SA}R_A + L_A \frac{di_{SA}}{dt} - i_{SB}R_B - L_B \frac{di_{SB}}{dt} \quad (4)$$

When diode D_1 and D_2 are in active state at $\dfrac{2\pi}{3}, \pi$

$$V_{AC} = i_{SA}R_A + L_A \frac{di_{SA}}{dt} - i_{SC}R_C - L_C \frac{di_{SC}}{dt} \quad (5)$$

When diode D_3 and D_2 are in active state at $\pi, \dfrac{4\pi}{3}$

$$V_{BC} = i_{SB}R_B + L_B \frac{di_{SB}}{dt} - i_{SC}R_C - L_C \frac{di_{SC}}{dt} \quad (6)$$

When diode D_3 and D_4 are in active state at $\dfrac{4\pi}{3}, \dfrac{5\pi}{3}$

$$V_{BA} = i_{SB}R_B + L_B \frac{di_{SB}}{dt} - i_{SA}R_A - L_A \frac{di_{SA}}{dt} \quad (7)$$

When diode D_5 and D_6 are in active state at $0, \dfrac{\pi}{3}$

$$V_{CB} = i_{SC}R_C + L_C \frac{di_{SC}}{dt} - i_{SA}R_A - L_A \frac{di_{SA}}{dt} \quad (8)$$

When diode D_5 and D_4 are in active state at $0, \dfrac{\pi}{3}$

$$V_{CA} = i_{SC}R_C + L_C \frac{di_{SC}}{dt} - i_{SA}R_A - L_A \frac{di_{SA}}{dt} \quad (9)$$

Where, L_A, L_B, L_C are input inductors, R_A, R_B, R_C are input resistors and i_{SA}, i_{SB}, i_{SC} is input currents. From the above circuit diagram, the duty cycle D can be written for boost converter as,

$$D = \frac{V_o - V_{in}}{V_o} \quad (10)$$

Where V_o= DC output voltage, V_{in} = Vin = 0.86Vm = AC input voltage, Vm= Peak input voltage. The input inductance of L_i can be written as,

$$L_i = \frac{V_{in} . D_{max}}{(\Delta i_L . f_{s(min)})} \quad (11)$$

Where D_{max}=Maximum duty cycle, Δ_{iL}= inductor ripple current, and f_s=switching frequency.

And Δ_{iL} can be written as,

$$\Delta i_L = 10\% . \frac{i_{in}}{\eta} , \text{ where } \eta= \text{Efficiency} \quad (12)$$

The output capacitance C written as,

$$C = \frac{i_o . D_{max}}{(\Delta V_C . f_{s(min)})} \quad (13)$$

Where ΔV_C = Capacitor ripple voltage, i_o= DC load current.

C. RBFN Control Algorithm

In the present scenario, artificial intelligent MPPT techniques are more popular over the conventional control techniques due to the fast and dynamic response of the system with non-linear system control capability. Artificial Neural network based RBFN controller is easy to implement with very less computational input data even for wide variations in input. The maximum power tracking of the wind energy conversion system can be obtained by considering the PMSG voltage and current as an input variable to the RBFN controller. The input data is trained in such a way that the PFC converter is controlled to get the desired output voltage, where the output of the controller is a duty cycle.

It is a three layer network called the input layer, a hidden layer, and an output layer. The activation functions of the hidden layer are estimated by the distance between the input vector and the prototype vector. In the first step, the parameters which direct the basis function are estimated by unsupervised methods and in the second step, the final layer units are decided. The input variables (x_i^1) to an RBFN controller are voltage and current, and the output variable (y_k^3) is a duty cycle (D). The controller output as depicted in Fig.4 is given to the pulse generator which generates the switching signal for the PFC converter and the parameters considered for an RBFN configuration are shown in Table.2.

i) Input layer:

Here the estimated input values are transmitted to the hidden layer. The input & output of the first layer are written as,

$$net_i^1 = x_i^1(N) \quad (14)$$

$$y_i^1(N) = f_i^1(net_i^1(N)) = net_i^1(N) , \text{ where, i=1, 2..} \quad (15)$$

ii) Hidden layer:

The second layer of the proposed controller is known as a hidden layer. Here a Gaussian function is used for each and every node and the input and output variables of the hidden layer are as follows,

$$net_j^2(N) = (X - M_j)^T \sum_j (X - M_j) \quad (16)$$

$$y_j^2(N) = f_i^2(net_j^2(N)) = Exp(net_j^2(N)),$$
$$j=1, 2.....9 \quad (17)$$

Where Mean $= M_j = [m_{1j} m_{2j} .... m_{ij}]^T$ and

Standard deviation

$$= \sum_j = diag[\frac{1}{\sigma_{1j}^2} \frac{1}{\sigma_{2j}^2} \ldots \ldots \frac{1}{\sigma_{ij}^2}]^T$$

iii). Output layer:

The control output of the third layer is written as,

$$net_k^3 = \sum W_j y_j^2(N) \tag{18}$$

$$y_k^3(N) = f_k^3(net_k^3(N)) = net_k^3(N) = D \tag{19}$$

The input data of 617 values are considered from the output of PMSG to train the RBFN in order to generate the optimal control signal.

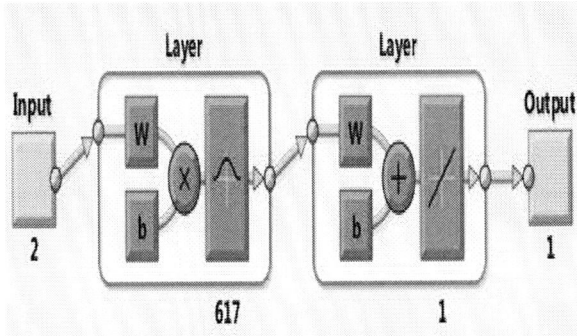

Fig.4 RBFN control structure

TABLE.2 PARAMETER CONFIGURATION OF AN RBFN

S.No	Controller parameters	Values/Methods
1	Input variables	$V_{A,B,C}$ and $I_{A,B,C}$
2	Output variables	D (Duty cycle)
3	Hidden neurons (Maximum limit)	617
4	Training algorithm	OLS (Ordinary Least Squares) method
5	Speed factor	0.03

III. SIMULATION RESULTS AND DISCUSSION

The proposed system is considered for the off-grid and grid-connected operation with an RBFN controller based single-stage boost PFC converter configuration. For the first case, the constant wind speed of 12 m/s is assumed for the wind turbine. The design parameters considered for the simulation of proposed topology are tabulated in Table.3.

TABLE.3: CIRCUIT PARAMETERS FOR THE BASE SPEED (12 m/s)

S.No	Circuit parameters	Ratings
1	Three phase input voltage from PMSG	230V
2	The maximum output voltage of Single stage boost PFC converter	400V
3	Input inductance ($L_A=L_B=L_C$)	10mH
4	Input filter capacitance ($C_A=C_B=C_C$)	100uF
5	DC-link capacitance (C)	200μF

6	Diode resistance (R_{ON})	0.001Ω
7	Load resistance (R)	160Ω
8	Maximum output power	1kW

A. An off-grid wind energy conversion system for variable wind speeds

In this case, the wind energy conversion system is operated for the stepwise wind velocity of 8m/s for 0 to 0.5sec, 11m/s for 0.5 to 1sec and 14m/s for 1 to 1.5sec as shown in Fig.5 with the same design parameters as mentioned in Table.3. The output variables of the PMSG and single stage boost PFC converter is changed with respect to the wind velocity as shown in Fig.6 and 7 respectively.

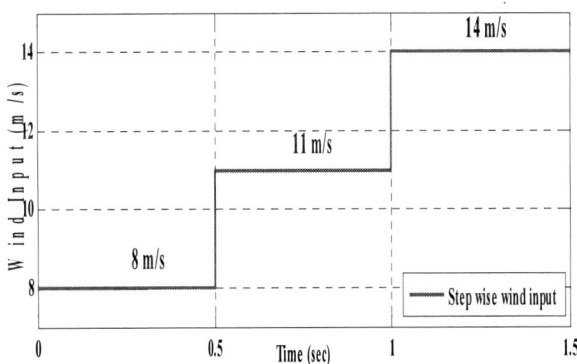

Fig.5 Different level of input wind velocity (m/s)

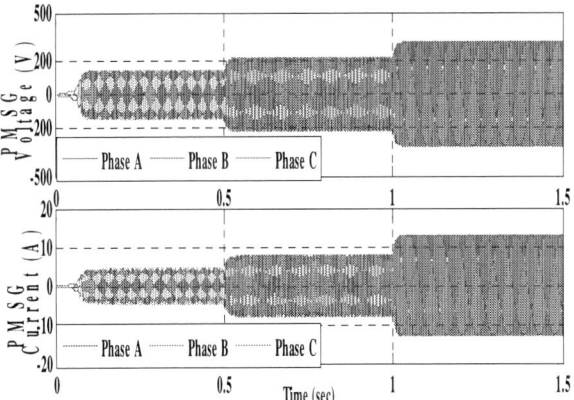

Fig.6 Output voltage and current of PMSG

The output voltage of the single stage converter is obtained as 283.3V for 0 to 0.5sec, 372V for 0.5 to 1sec and 460V for 1 to 1.5sec. The output current is obtained as 1.77A for 0 to 0.5sec, 2.32A for 0.5 to 1sec and 2.88A for 1 to 1.5sec. Similarly, the output power of the proposed system is obtained as 501.6W for 0 to 0.5sec, 864.8W for 0.5 to 1sec and 1332W for 1 to 1.5sec. The complete result analysis of the proposed converter is tabulated in Table.4 for different wind velocity and time intervals.

978-1-5386-6624-1/18 $31.00 © 2018 IEEE

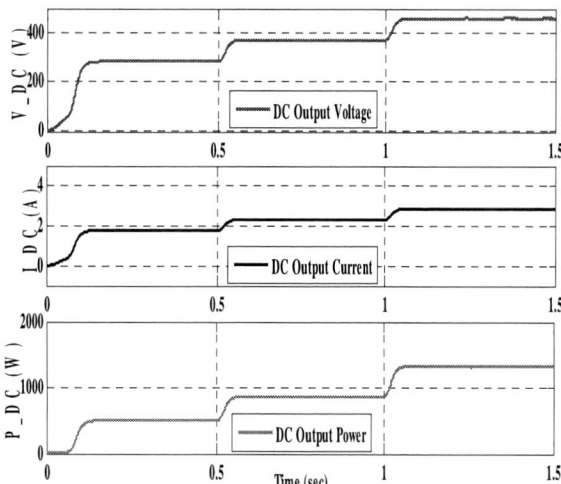

Fig.7 DC Output voltage, current, and power of single stage boost PFC converter

TABLE.4 OUTPUT PARAMETERS OF SINGLE STAGE BOOST PFC CONVERTER

Parameters	Output values of an RBFN controller based Single stage boost PFC converter		
Time Period (Sec)	0 to 0.5	0.5 to 1	1 to 1.5
Wind Velocity (m/s)	8	11	14
DC output voltage (V)	283.3	372	460
DC output current (A)	1.77	2.32	3.88
DC output power (W)	501.6	864.9	1332

B. A grid-tied wind energy conversion system for various wind speeds

The aim of the GSC is to feed the energy from a front-end conversion unit to the main grid and the load. Therefore, GSC acts as an AC to DC conversion unit at the back-end of the system in order to regulate the DC-link capacitor voltage and it can control the active and reactive power of the system. A direct power control method is considered in this system to regulates the instantaneous values of the power components which are decoupled into synchronous d-q reference frame. Here two control loops are enabled: the inner current loop is employed for the power control and outer loop is engaged for DC-link capacitor voltage control. The power contribution of the wind energy system and grid to the load is represented in Fig.8 and 9 respectively. The voltage balance equations are written based on the rotating reference frame as follows,

$$L_i \frac{di_{gd}}{dt} = E_{id} - R_i i_{gd} + \omega L_i i_{gq} - V \qquad (20)$$

$$L_i \frac{di_{gq}}{dt} = E_{iq} - R_i i_{gq} - \omega L_i i_{gd} \qquad (21)$$

Where i_{gd} and i_{gq} are the line current components of d-$axis$ and q-$axis$ respectively. Similarly, E_{id} and E_{iq} are the inverter voltage components of d-$axis$ and q-$axis$ respectively. R_i = line resistance and L_i = line resistance.

Therefore, active power (P) and reactive power (Q) are written in terms of grid voltage and line current components as follows,

$$P = \frac{3}{2} v_{gd} \, i_{gd} \qquad (22)$$

$$Q = \frac{3}{2} v_{gd} \, i_{gq} \qquad (23)$$

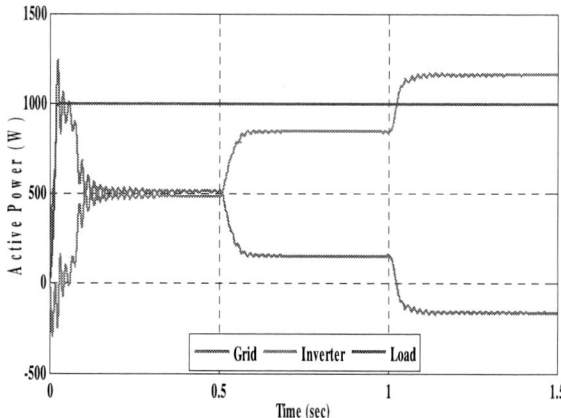

Fig.8 Active power output of the load, grid, and inverter

From the Fig.8, it is observed that the load demand of 1000W for a period of 0 to 0.5sec is supplied by both wind and grid sources with the contribution of 483.6W and 513.8W respectively. Similarly, for the period of 0.5 to 1sec, the load is supplied by both the wind energy system and grid with a contribution of 847.6W and 149.7Wrespectively. For the period of 1 to 1.5sec the power obtained from the wind energy system is 1159W which is more than the load demand, during this region excess of 157.8W wind-generated power is fed back to the grid.

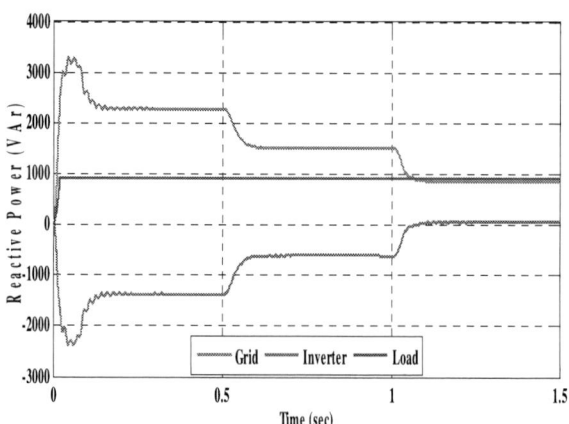

Fig.9 Reactive power output of the grid, inverter, and load

Similarly, from the Fig.9 it is observed that the load demand of reactive power is estimated for a time period of 0 to 0.5sec, 0.5 to 1sec and 1 to 1.5sec as tabulated in Table.5 for a combined contribution of the wind energy system and main grid.

978-1-5386-6624-1/18 $31.00 © 2018 IEEE

TABLE.5 OUTPUT PARAMETERS OF GRID-TIED WECS

Parameters	Active Power (W) =1000			Reactive Power (VAr) = 900		
Time Period (Sec)	0 to 0.5	0.5 to 1	1 to 1.5	0 to 0.5	0.5 to 1	1 to 1.5
Inverter output	483.6	847.6	1159	-1377	-611.6	51
Grid output	513.8	149.7	-157.8	2277	1511	848.9
Load output	997.4	997.4	997.4	899.6	899.6	899.6

IV. CONCLUSION

A grid-tied wind energy conversion system is carried for both off-grid and grid-tied configurations. The proposed system configuration is very efficient for the wind energy conversion at variable wind speeds with high conversion ratio. It is observed that the maximum power of 1332W obtained for the off-grid at the wind speed of 14 m/s in the time interval of 1 to 1.5 sec. Similarly, the power demand supplied by an inverter and the grid to the load for various time intervals is presented in Table.5. It infers that the preferred RBFN MPPT controller based converter topology is essential for AC to DC conversion in various systems such as renewable energy conversion systems (wind, tidal, ocean etc), diesel turbine systems and/or AC mains.

REFERENCES

[1] Oğuz, Y., Güney, İ. and Çalık, H., 2013. Power quality control and design of power converter for variable-speed wind energy conversion system with permanent-magnet synchronous generator. *The Scientific World Journal*, 2013.

[2] Nayanar, V., Kumaresan, N. and Gounden, N.A., 2016. A single-sensor-based MPPT controller for wind-driven induction generators supplying DC microgrid. *IEEE Transactions on Power Electronics*, 31(2), pp.1161-1172.

[3] Wang, X., Yuvarajan, S. and Fan, L., 2010, September. MPPT control for a PMSG-based grid-tied wind generation system. In *North American Power Symposium (NAPS), 2010* (pp. 1-7). IEEE.

[4] Benadja, M. and Chandra, A., 2012, December. A new MPPT algorithm for PMSG based grid connected wind energy system with power quality improvement features. In *Power India Conference, 2012 IEEE Fifth* (pp. 1-6). IEEE.

[5] Alizadeh, O. and Yazdani, A., 2013. A strategy for real power control in a direct-drive PMSG-based wind energy conversion system. *IEEE transactions on Power Delivery*, 28(3), pp.1297-1305.

[6] Chen, H., David, N. and Aliprantis, D.C., 2013. Analysis of permanent-magnet synchronous generator with Vienna rectifier for wind energy conversion system. *IEEE Transactions on Sustainable Energy*, 4(1), pp.154-163.

[7] Messalti, S., Harrag, A. and Loukriz, A., 2017. A new variable step size neural networks MPPT controller: Review, simulation and hardware implementation. *Renewable and Sustainable Energy Reviews*, 68, pp.221-233.

[8] Mendis, N., Muttaqi, K.M., Sayeef, S. and Perera, S., 2012. Standalone operation of wind turbine-based variable speed generators with maximum power extraction capability. *IEEE Transactions on Energy Conversion*, 27(4), pp.822-834.

[9] Xia, Y., Ahmed, K.H. and Williams, B.W., 2011. A new maximum power point tracking technique for permanent magnet synchronous generator based wind energy conversion system. *IEEE Transactions on Power Electronics*, 26(12), pp.3609-3620.

[10] Wei, C., Zhang, Z., Qiao, W. and Qu, L., 2016. An adaptive network-based reinforcement learning method for MPPT control of PMSG

wind energy conversion systems. *IEEE Transactions on Power Electronics*, 31(11), pp.7837-7848.

[11] Saravanan, S. and Babu, N.R., 2016. RBFN based MPPT algorithm for PV system with highstep up converter. *Energy Conversion and Management*, 122, pp.239-251.

[12] Lin, W.M., Hong, C.M., Ou, T.C. and Chiu, T.M., 2011. Hybrid intelligent control of PMSG wind generation system using pitch angle control with RBFN. *Energy conversion and management*, 52(2), pp.1244-1251.

[13] Dia, K.K.H., Islam, S. and Choudhury, M.A., 2016, September. A single phase input switched SEPIC PFC AC-DC converter. In *Electrical Engineering and Information Communication Technology (ICEEICT), 2016 3rd International Conference on*(pp. 1-6). IEEE.

[14] Reddy D, Ramasamy S. A fuzzy logic MPPT controller based three phase grid-tied solar PV system with improved CPI voltage. InPower and Advanced Computing Technologies (i-PACT), 2017 Innovations in 2017 Apr 21 (pp. 1-6). IEEE.

[15] Azazi, H.Z., Ahmed, S.M. and Lashine, A.E., 2017. Single-stage three-phase boost power factor correction circuit for AC–DC converter. *International Journal of Electronics*, pp.1-12.

[16] Kasa S, Ramanathan P, Ramasamy S, Kothari DP. Effective grid interfaced renewable sources with power quality improvement using dynamic active power filter. International Journal of Electrical Power & Energy Systems. 2016 Nov 1;82:150-60.

[17] Ramasamy S. Single Stage Energy Conversion through an RBFN Controller based Boost Type Vienna Rectifier in the Wind Turbine System. Gazi University Journal of Science.;30(4):253-66.

[18] Tiwari, R., Padmanaban, S. and Neelakandan, R.B., 2017. Coordinated Control Strategies for a Permanent Magnet Synchronous Generator Based Wind Energy Conversion System. *Energies*, 10(10), p.1493..

[19] Anand, A. and Singh, B., 2017. Power Factor Correction in Cuk-SEPIC Based Dual Output Converter fed SRM Drive. *IEEE Transactions on Industrial Electronics*.

[20] Kumar, K., Babu, N.R. and Prabhu, K.R., 2017. Design and Analysis of RBFN-Based Single MPPT Controller for Hybrid Solar and Wind Energy System. *IEEE Access*, 5, pp.15308-15317.

[21] Reddy D, Ramasamy S. Design of RBFN Controller Based Boost Type Vienna Rectifier for Grid-Tied Wind Energy Conversion System. IEEE Access. 2018;6:3167-75.

[22] Kumar, R. and Singh, N., 2017, April. Wind power grid interfacing using 3 level vs 5 level inverter with SEPIC converter. In *Electronics, Materials Engineering and Nano-Technology (IEMENTech), 2017 1st International Conference on* (pp. 1-6). IEEE.

[23] Yaramasu, V., Dekka, A., Durán, M.J., Kouro, S. and Wu, B., 2017. PMSG-based wind energy conversion systems: survey on power converters and controls. *IET Electric Power Applications*.

[24] RAMASAMY S, Reddy D. Design of a Three-phase Boost Type Vienna Rectifier for 1kW Wind Energy Conversion System. International Journal of Renewable Energy Research (IJRER). 2017 Dec 30;7(4):1909-18.

[25] RAMASAMY S, Sudheer K. Mitigating Voltage Imperfections with Photovoltaic fed ANFIS based ZSI-DVR in Three Phase System. International Journal of Renewable Energy Research (IJRER). 2017 Dec 30;7(4):2103-10.

Symmetrical Components Estimation of Unbalanced Three Phase Power System Using MO-ADALINE Structure and Hermite Polynomial Based Gauss Newton Algorithm

Swastik Acharya
Department of Electrical
and Electronics
Engineering,
*International Institute of
Information Technology
(IIIT)* Bhubaneswar,
India

Umamani Subudhi
Member, IEEE, Department
of Electrical and
Electronics Engineering
*International Institute of
Information Technology
(IIIT)* Bhubaneswar, India
umamani@iiit-bh.ac.in

Harish Kumar Sahoo
Senior Member, IEEE,
Department of Electronics
and Telecommunication
Engineering
*Veer Surendra Sai
University of Technology
(VSSUT),* Burla India
harish_etc@vssut.ac.in

Anuj Jena
Department of Electrical
and Electronics Engineering
*International Institute of
Information
Technology(IIIT)*
Bhubaneswar, India

Abstract— **The presence of different power quality (PQ) disturbances in the form of distorted voltage or current waveforms is an critical issue in modern power network. Due to increase use of power electronic devices and renewable sources of energy in unbalanced power networks, sequence or symmetrical components estimation across all the phases is really important. This paper proposes a MO-ADALINE (Multi Output Adaptive Linear Element) model based on Hermite Gauss-Newton algorithm for fast and accurate estimation of symmetrical components in terms of phase and amplitude across all the phases. The estimation performance of the model is compared on the basis of speed of convergence and steady state error. In addition to that the performance of the proposed model is also tested considering a fault condition generated using Matlab Simulink. The simulation results are used to validate the faster convergence and accuracy of Hermite Gauss Newton algorithm based estimation model.**

Keywords- Sequence Components, Hermite Polynomial , white Gaussian noise

I. INTRODUCTION

The unbalance condition in three phase system results in the development of negative sequence current flows through stator windings inducing negative sequence voltage in the rotor windings. As a result of which the rotor is short circuited and results in high current flow that damages the windings [1]. Hence faster and accurate detection of unbalance condition is required to protect the equipment. The method of symmetrical components is used for estimating unbalance condition in three phase network. There are many methods developed for the estimation of symmetrical components of current and voltage. One of the methods used for estimation is Instantaneous Reactive Power Theory, it works only if the supply voltage signal is ideal. Hence it is not suitable practical conditions where the

waveforms are distorted in nature [2]. An alternate method like Park Transformation with notch filters is also used to estimate the sequence components. But the method suffers from limitations in the form of low detection precision, sensitivity to change in the supply frequency and also to the change in parameters of the system [3]. To eliminate the limitations of the previously mentioned techniques, the technique of FFT (Fast Fourier Transform) was introduced which provides the spectral information[4]. Though it removes the problem of dealing with the distorted waveform (current or voltage), it depends on the width of the window used for analysis. FFT also shows inaccuracies in results in case of load changing in a dynamic manner. For estimation of time-varying parameters of the symmetrical components, Kalman Filter is popularly used as a recursive algorithm. Kalman filter is also capable of processing noisy measurements and the analysis results in the least square optimal form. The major limitation of Kalman Filter is the large number of calculations involved in the processing, which restricts its implementation in online measurements[5], [6]. In recent develop DLAVE (Dynamic Least Absolute Estimator) is introduced to estimate symmetrical components based on stochastic estimation theory. DLAVE algorithm is based on block processing algorithm and its accuracy is determined by the number of samples taken in a block which may not be easy to implement it in practical applications [7]. The ADALINE abbreviated as Adaptive Linear Element or Adaptive Linear Neuron is a single-layer artificial neural network used for tracking harmonics. ADALINE is generally used for single-output system and hence is not able to track the components of multi output systems [8]. Since in a three-phase system for tracking of symmetrical components all the three-phase currents or voltages are needed to be processed simultaneously, the ADALINE structure was modified into a multi-output system known as MO-ADALINE (MO stands for multi-output) [9]. The Hermite polynomial expansion and Gauss-Newton algorithm are used as the adaptation rule to update the weight vectors of adaline structure. The proposed estimation model for symmetrical components has advantage in terms of less computational complexity and high accuracy. The paper is divided into six sections. Section II presents the mathematical modeling for symmetrical components across three phases. Section III explains the MO-ADALINE structure generated from adaline network. Section IV describes the Hermite Gauss-Newton algorithm. Section V presents the simulation results using Matlab 2014a. The conclusion is presented in Section VI.

II. MODELING OF SYMMETRICAL COMPONENTS

The relationship between the unbalanced waveform and the symmetrical components is established using the state space representation and represented as

$$v_a = V_0 \sin(\omega t + \varphi_0) + V_1 \sin(\omega t + \varphi_1) + V_2 \sin(\omega t + \varphi_2)$$

$$v_b = V_0 \sin(\omega t + \varphi_0) + V_1 \sin(\omega t + \varphi_1 - 120°) + V_2 \sin(\omega t + \varphi_2 + 120°)$$

$$v_c = V_0 \sin(\omega t + \varphi_0) + V_1 \sin(\omega t + \varphi_1 + 120°) + V_2 \sin(\omega t + \varphi_2 - 120°)$$

$$(1)$$

where v_a, v_b, v_c are the three phases of the unbalanced waveform, φ_0 and V_0 are the phase angle and magnitude of zero sequence components respectively. Similarly, φ_1, V_1 and φ_2, V_2 are the phase angle and magnitude of positive and negative sequence components respectively. By the use of formula $sin\ (x+y) = sin\ (x)Xcos(y) + cos\ (x)Xsin(y)$ in (1), we get:

$$X(t) = \begin{bmatrix} \sin \omega t & \cos \omega t & \sin \omega t & \cos \omega t & \sin \omega t & \cos \omega t \\ \sin \omega t & \cos \omega t & \sin(\omega t - 120°) & \cos(\omega t - 120°) & \sin(\omega t + 120°) & \cos(\omega t + 120°) \\ \sin \omega t & \cos \omega t & \sin(\omega t + 120°) & \cos(\omega t + 120°) & \sin(\omega t - 120°) & \cos(\omega t - 120°) \end{bmatrix} \quad (8)$$

III. MO-ADALINE STRUCTURE

The conventional ADALINE structure is a powerful tool but its operation is limited to single output. The ADALINE in its conventional form cannot be used to estimate the symmetrical components as it is required to process the three-phase signals simultaneously. Therefore the ADALINE is modified into Multi-output ADALINE. It can be inferred from the symmetrical component model that for estimation of symmetrical components it is required to have three adaptive neurons with same weight vector W(n) and different input vector X(n) for the individual neuron. Hence it is needed to create an adaptive algorithm to update the weight vector of the MO-ADALINE. This is attained by repositioning the terms in the adaptive algorithm used in ADALINE and considering X(k) as a 'n × m' matrix (where n is the variables to be estimated and m is the outputs) in place of a vector as in conventional ADALINE [10,11]. The new structure known as MO-ADALINE can

$$v_a = V_0 \cos(\varphi_0) \sin(\omega t) + V_0 \sin(\varphi_0) \cos(\omega t) + V_1 \cos(\varphi_1) \sin(\omega t) \quad (2)$$
$$+ V_1 \sin(\varphi_1) \cos(\omega t) + V_2 \cos(\varphi_2) \sin(\omega t) + V_2 \sin(\varphi_2) \cos(\omega t)$$

$$v_b = V_0 \cos(\varphi_0) \sin(\omega t) + V_0 \sin(\varphi_0) \cos(\omega t) + V_1 \cos(\varphi_1) \sin(\omega t - 120°) \quad (3)$$
$$+ V_1 \sin(\varphi_1) \cos(\omega t - 120°) + V_2 \cos(\varphi_2) \sin(\omega t + 120°) +$$
$$V_2 \sin(\varphi_2) \cos(\omega t + 120°)$$

$$v_c = V_0 \cos(\varphi_0) \sin(\omega t) + V_0 \sin(\varphi_0) \cos(\omega t)$$
$$+ V_1 \cos(\varphi_1) \sin(\omega t + 120°) + V_1 \sin(\varphi_1) \cos(\omega t + 120°) \quad (4)$$
$$+ V_2 \cos(\varphi_2) \sin(\omega t - 120°) + V_2 \sin(\varphi_2) \cos(\omega t - 120°)$$

The above equation can be represented in matrix form as given in (5).

$$y(t) = X(t) \times W(t) \quad (5)$$

$$y(t) = \begin{bmatrix} v_a \\ v_b \\ v_c \end{bmatrix} \quad (6)$$

$$W(t) = \begin{bmatrix} V_0 \cos(\varphi_0) \\ V_0 \sin(\varphi_0) \\ V_1 \cos(\varphi_1) \\ V_1 \sin(\varphi_1) \\ V_2 \cos(\varphi_2) \\ V_2 \sin(\varphi_2) \end{bmatrix} \quad (7)$$

estimate the three phase of the unbalanced load simultaneously. This has the advantage of greater accuracy, less computation and easy to implement as compared to the methods mentioned in the literature review. In the proposed method m=3 and n=6. The working of the MO-ADALINE is shown in Fig. 1.

IV. HERMITE POLYNOMIAL BASED GAUSS-NEWTON ALGORITHM

The input vector X(n) is expanded using Hermite expansion up to second order as in (9).

$$X(n) = [h_{e0} \ h_{e1} \ h_{e2}] \quad (9)$$

where $h_{e0}=1$, $h_{e1}=X(n)$ and $h_{e2}=X(n)^2-1$. The unbalanced three phase load voltage is then represented as given in (10)

$$y(n) = X(n) * W(n-1) \quad (10)$$

where W(n) is the coefficient vector and X(n) is the expanded time varying input vector as in equation (7) and (8) respectively. The error is obtained as shown in (11).

$$e(n) = d(n) - y(n) \quad (11)$$

978-1-5386-6624-1/18 $31.00 © 2018 IEEE

where d(n) is the desired signal. Assuming that the signal frequency is known, a forgetting factor, λ based weighted error cost function is chosen as given in (12).

$$\xi(n) = \sum_{i=0}^{n} \lambda^{n-i} e^2(i) \qquad 0 < \lambda \le 1 \qquad (12)$$

The weight vector are updated as given in(13).

$$W(n) = W(n-1) - H^{-1}(n)\psi(n)e(n) \qquad (13)$$

where $H^{-1}(n)$ is the inverse Hessian matrix and $\psi(n)$ is the gradient vector and are calculated as

$$H(n) = \sum_{i=0}^{n} \lambda^{n-i} \psi(i)\psi^T(i) \qquad (14)$$

The inverse of the Hessian matrix can be calculated using matrix inverse lemma. But it is computationally very expensive. To reduce the complexity it is assumed that ω is not close to 0 or π, and the H(n) can be approximated as [12]

$$H^{-1}(n) = \begin{bmatrix} 6/c(n) & 0 & 0 & \cdots & & & \cdots & & 0 \\ 0 & 3/c(n) & 0 & \cdots & \cdots & & \cdots & & \vdots \\ 0 & \cdots & 3/c(n) & \cdots & & \cdots & & & \\ & \cdots & & 3/c(n) & \cdots & & & & \vdots \\ \vdots & & \cdots & & 3/c(n) & & & & \\ & & \cdots & & & 3/c(n) & & & \vdots \\ & & & \cdots & & & 3/c(n) & \cdots & \\ \vdots & \cdots & \cdots & & & & 3/2c(n) & \cdots & \vdots \\ \vdots & & \cdots & \cdots & & & & 3/2c(n) & \\ \vdots & & \cdots & \cdots & \cdots & & & 3/2c(n) & 0 \\ 0 & \cdots & \cdots & \cdots & & & & & 3/2c(n) \end{bmatrix} \quad (15)$$

where

$$c(n) = \frac{1 - \lambda^{n+1}}{2(1-\lambda)} \qquad (16)$$

The magnitude and phase of symmetrical components of the signal can be estimated using (17).

$$V_0(n) = \sqrt{W_2(n)^2 + W_3(n)^2}$$
$$V_1(n) = \sqrt{W_4(n)^2 + W_5(n)^2} \qquad (17)$$
$$V_2(n) = \sqrt{W_6(n)^2 + W_7(n)^2}$$
$$\phi_0(n) = \tan^{-1}(W_3(n) / W_2(n))$$
$$\phi_1(n) = \tan^{-1}(W_5(n) / W_4(n))$$
$$\phi_0(n) = \tan^{-1}(W_7(n) / W_6(n))$$

Pseudo Code for Hermite Gauss Newton Algorithm

Step I   Initialize MO-ADALINE with the expanded input samples $X(n)$ using (9) and weight vector $W(n)$ with random values and generate the estimated signal using (10).

Step II   The error was calculated using (11).

Step III   Calculate inverse Hessian matrix $H^{-1}(n)$ and parameter $c(n)$ using (15) and (16) respectively.

Step IV   Update the Weight Vector $W(n)$ using (13) till the maximum number of iterations are reached.

Step V   Updated weights are used to estimate the magnitude and phase of the unbalanced signal using (17).

Step VI   End

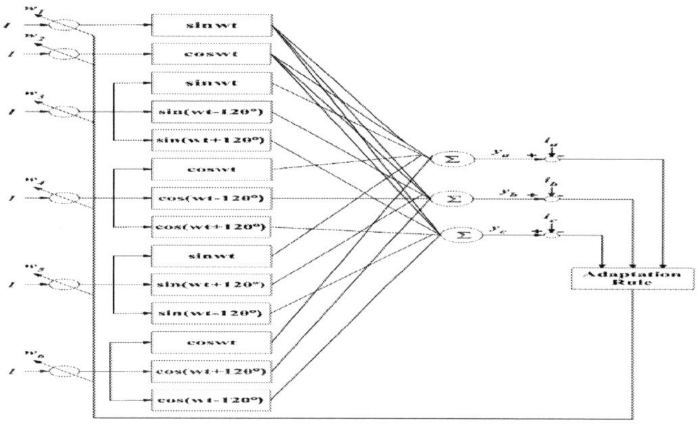

Fig. 1. Structure of MO-ADALINE

V. SIMLATION RESULTS AND DISCUSSIONS

For simulation purpose, a synthetic three-phase unbalanced signal was generated in MATLAB script using (18) for a

978-1-5386-6624-1/18 $31.00 © 2018 IEEE

time period of 0 to 0.1s. The input signal is as shown in Fig 2.

$$V_a = 300\sin(\omega t + 120°) = 300\angle 120°$$
$$V_b = 100\sin(\omega t + 30°) = 100\angle 30° \qquad (18)$$
$$V_c = 50\sin(\omega t + 60°) = 50\angle 60°$$

The fundamental frequency is taken as 50 Hz and the sampling Frequency as 10 kHz. The generated synthetic signals are made to pass through the MO-ADALINE unit for processing and the results were plotted.

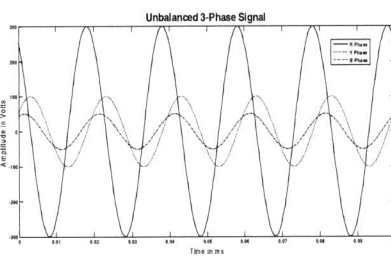

Fig. 2. Unbalanced Signal

From the figures i.e. from Fig. 3 – Fig. 8, it can be observed that the Hermite Gauss-Newton algorithm shows superior performance as compared to the other algorithms for estimation of magnitude and phase angles of symmetrical components.

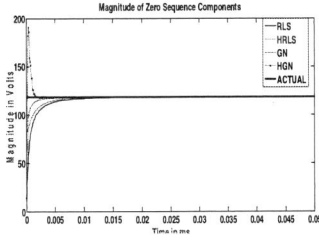

Fig. 3. Comparison Results of Zero Sequence Amplitude

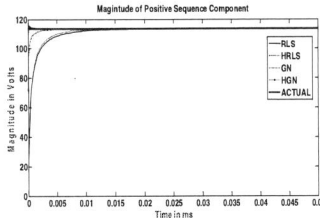

Fig. 4. Comparison Results of Positive Sequence Amplitude

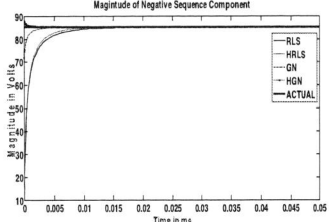

Fig. 5. Comparison Results of Negative Sequence Amplitude

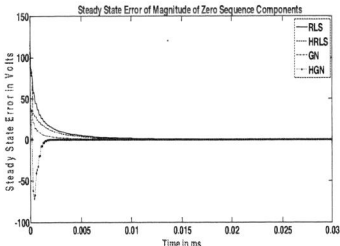

Fig. 6. Steady State Error of Zero Sequence Amplitude

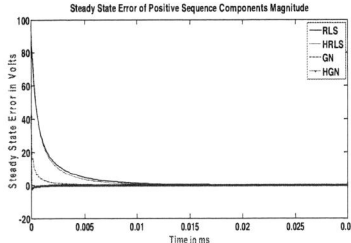

Fig. 7. Steady State Error of Positive Sequence Amplitude

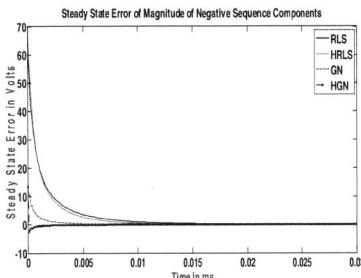

Fig. 8. Steady State Error of Negative Sequence Amplitude

Hermite Gauss-Newton algorithm takes less than 0.5ms to track the desired phase angle while, Gauss-Newton algorithm takes more than 1.5ms , RLS and Hermite RLS takes more than 2.5ms. From Table.1 it can be concluded that the Hermite Gauss-Newton gives minimum steady state error as compared to Gauss-Newton, RLS and Hermite RLS algorithms.

Case II: Sag in the Unbalanced Signal

A sag of 40% of the original magnitude is introduced in the synthetic signal generated in (18) during the time period 0.035s to 0.065s.

Table 1 Steady State Error of Magnitude of Symmetrical Components

Algorithm	Zero Sequence	Positive Sequence	Negative Sequence
RLS	$3.8210*10^{-4}$	$3.6522*10^{-4}$	$2.7433*10^{-4}$
Hermite RLS	$3.7779*10^{-4}$	$3.6090*10^{-4}$	$2.7042*10^{-4}$
Gauss Newton	$8.1730*10^{-5}$	$3.8738*10^{-5}$	$2.9119*10^{-5}$
Hermite Gauss Newton	$4.7203*10^{-6}$	$1.4872*10^{-6}$	$2.5291*10^{-6}$

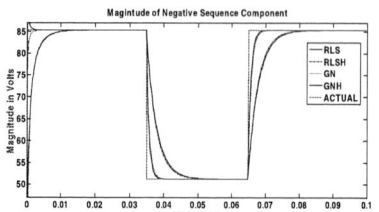

Fig. 11. Comparison Results of Negative Sequence Amplitude

From the figures i.e. from Fig. 9 – Fig. 11, it can be observed that the Hermite Gauss-Newton algorithm has better convergence when there is a sag in an unbalanced 3-phase signal, as compared to the other. Hermite Gauss-Newton algorithm takes around 5ms to track the path of the desired output while, Gauss-Newton algorithm takes more than 5ms , RLS and Hermite RLS takes more than 15ms.

Case III: Simulink Model

The simulation model was developed using MATLAB Simulink, which is used to simulate the 3-phase unbalanced load. Fig 12 shows the fault model. It consists of 11 kV, 30 MVA, 50 Hz three-phase source block feeding through 11 kV/420V, 3 MVA delta/wye transformers to a 10 kW resistive and 100VAR inductive load. The instantaneous waveform measurement scope is connected at both the load and source ends. A fault block is located at the source end to generate various types of faults. The line to ground (R-G) fault is introduced at 35ms till 65ms and the simulation runs for 100ms and is simulated using ode23tb solver [13]. From Fig. 13- Fig. 15 it can be observed that the Hermite Gauss-Newton algorithm has better tracking and speed of convergence as compared to RLS, H-RLS and GN.

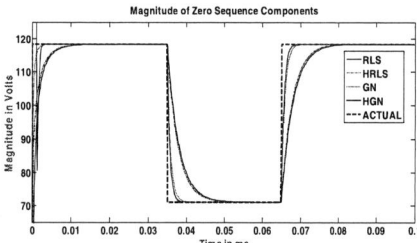

Fig. 9. Comparison Results of Zero Sequence Amplitude

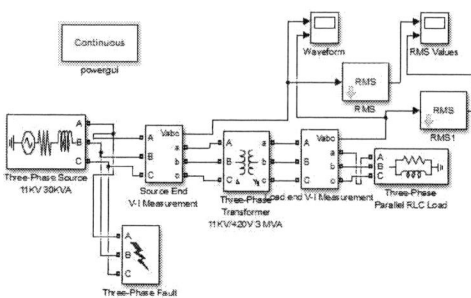

Fig. 12. Fault Model

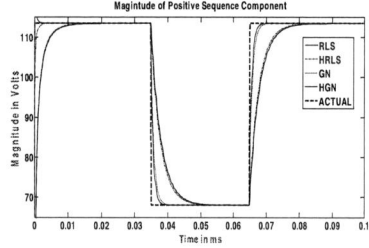

Fig. 10. Comparison Results of Positive Sequence Amplitude

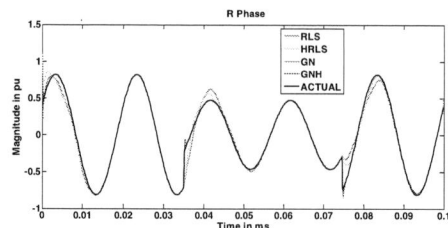

Fig. 13. Comparison Results of R Phase

978-1-5386-6624-1/18 $31.00 © 2018 IEEE

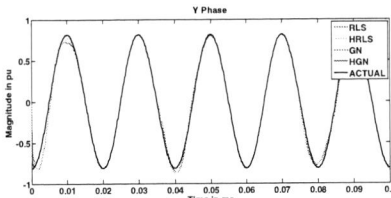

Fig. 14. Comparison Results of Y Phase

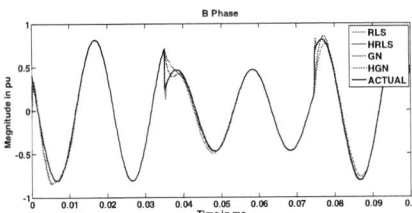

Fig. 15. Comparison Results of B Phase

VI. CONCLUSION

The unbalance in three phase system not only affects the power quality of the output but also it has an economic impact on the power industry. Hence it is necessary to track the sequence components through estimated amplitudes and phases which provide the correct information about the unbalanced condition. In this paper, proposed model is quite efficient to be used in three phase network. The four algorithms are compared on the basis of speed of convergence and steady-state error. It can be observed from the results that the Hermite polynomial based Gauss-Newton estimation model shows superior convergence and minimum steady state error. One of the most important advantages of this model is the independence of the MO-ADALINE structure from the varying system parameters and the type of waveform.

REFERENCES

[1] Andrea Bonetti, Romain Douib, "Test method for transformer differential relays based on symmetrical sequence components." ,IEEE International Symposium on Electrical Insulation Pg 1-5 , 2010.

[2] Yun Tengl, Chao Xiong, Chunlai Li, Zhongjie Zhang; Yutian Zhou; Danfeng Si, "Harmonic current detection based on the generalized

instantaneous reactive power theory" , IEEE International Conference on Progress in Informatics and Computing (PIC), Pg. 468-471, 2015.

[3] Kumarraja Andanapalli and B.R.K.Varma, "Park's Transformation based Symmetrical Fault Detection during Power Swing." Eighteenth National Power Systems Conference (NPSC), Pg. 1-5, 2014.

[4] Bilwal Rehman, Masood Ahmad, Jawad Hussain, "Analysis of power system harmonics using singular value decomposition, least square estimation and FFT" , International Conference on Energy Systems and Policies (ICESP), Pg 1-5, 2014.

[5] Anh Tuan Phan, Gilles Hermann, Patrice Wira, "Kalman filtering with a new state-space model for three-phase systems: Application to the identification of symmetrical components." IEEE International Conference on Evolving and Adaptive Intelligent Systems (EAIS) December 1-3, Pg. 1-6, 2015.

[6] Liming Shi, Jesper K. Nielsen, Jesper R. Jensen, Max A. Little, Mads G. Christensen, "A Kalman-based fundamental frequency estimation algorithm." , IEEE Workshop on Applications of Signal Processing to Audio and Acoustics (WASPAA), Pg 314-318, 2017.

[7] K. M. El-Naggar, "A fast method for identification of symmetrical components for power system protection," Int. J. Elect. Power Energy Syst., vol. 23, pp. 813–817, 2001.

[8] Priyabrat Garanayak, Gayadhar Panda, Pravat Kumar Ray, "Power System Harmonic Parameters Estimation Using ADALINE-VLLMS Algorithm." , International Conference on Energy, Power and Environment: Towards Sustainable Growth (ICEPE), Pg. 1-6, 2015.

[9] Mostafa I. Marei, Ehab F. El-Saadany, Magdy M. A. Salama, "A Processing Unit for Symmetrical Components and Harmonic Estimation Based on a New Adaptive Linear Combiner Structure", IEEE Transactions on Power Delivery, Vol. 19, No 3, Pg, 1245-1252, July 2004.

[10] Rodney Winter, Bernard Widrow, "MADALINE Rule II: A Training Algorithm for Neural Networks", IEEE Conference on Neural Networks, Pg, 401-408, 1988.

[11] Umamani Subudhi, Harish Kumar Sahoo, Sanjeev Kumar Mishra, "Harmonics and decaying DC Estimation using Volterra LMS/F aLgorithm", IEEE Transactions on Industry Applications, vol. 54, No. 2, march/april 2018

[12] P Stoica, R.L. Moses, B Friedlander, T,Soderstrom, "Maximum Likelihood Estimation of the Parameters of Multiple Sinusoids From Noisy Measurements.", IEEE Trans. ASSP, Vol 37, No. 3, pp. 378-392, 1989.

[13] Rodney H.G. Tan, Vigna K. Ramachandramurthy, "Chapter 3.A Comprehensive Modeling and Simulation of Power Quality Disturbances Using MATLAB/SIMULINK.", Power Quality Issues in Distributed Generation Jaroslaw Luszcz, IntechOpen, DOI: 10.5772/61209, Pg 83-107, Oct 2015.

A Normalized Adaptive Filter for Enhanced Optimal Operation of Grid Interfaced PV System

Vedantham Lakshmi Srinivas, *Member, IEEE*, Shailendra Kumar, *Member, IEEE*, Bhim Singh, *Fellow, IEEE*, Sukumar Mishra, *Senior Member, IEEE*,

Department of Electrical Engineering, *IIT Delhi*, New Delhi-110016, India.
vlsrinivas16@gmail.com, er.dwivedi88@gmail.com, bsingh@ee.iitd.ac.in, sukumariitdelhi@gmail.com

Abstract—This paper proposes a grid connected single-stage photovoltaic (GPV) system using an adaptive normalized notch filter (ANNF). An unknown time-varying fundamental frequency is precisely estimated from the normalized notch filter and consequently the fundamental load component is extracted using this algorithm. The convergence of the proposed algorithm is independent of both amplitude and the frequency of the fundamental component, and therefore, this algorithm is robust to these changes in the load current. The adaptive estimation of fundamental load current with perturbations on load/grid side system is well analyzed. The proposed ANNF control strategy is not only efficient in fundamental extraction, however, it also achieves the objectives of supplying real power to the grid, load current balancing and power factor correction. An experimental setup is developed to validate the proposed control under various perturbations. Test results are found satisfactory and the THDs in grid currents are achieved within the limits advised by IEEE 519 standard.

Keywords— Adaptive filter, Normalized frequency estimation, and Voltage Source Converter, power quality.

I. INTRODUCTION

The renewable energy sources for electricity production are abundant in nature, available worldwide and easily accessible. More than 13.5% of worldwide energy consumption is met in recent era through renewable power resources [1]. Moreover, feasibility of increased deployment of renewables worldwide has been increased through pro-active policy initiatives, which aims to assist the variegation of the electricity generation, to reduce the dependence on non-renewables and to help renewables a cost competitive counterpart for conventional energy sources [2][3]. Therefore, renewable energy sources are being developed and disseminated and the most significant role is expected from the solar energy [4]-[7]. The grid connected PV (photovoltaic) system proved to be one of the best options. The more convenient and practical to deploy is the GPV system, provided proper control over electrical parameters is established. In the integration of these renewables in distributed generation, the power electronic technology plays a very vital role [8]-[9]. Fig. 1 shows the typical configuration of the GPV system with local loads. The modeling methodologies for PV array are reported in the literature [9]-[11]. The single-stage GPV system is considered

here, where the buck-boost converter ensures the utmost power from the PV array and the VSC (Voltage Source Converter) accounts for proper control of active and reactive powers using the designed control strategy. Therefore, one of the key issues to be addressed is to improve the quality of the power conditioning stage.

The control strategies developed for the GPV system in the past which include SRFT and generalized integrator based approaches [12]-[13]. However, these algorithms developed suffer from their own demerits which may be stated as follows. The SRFT based approach suffers disadvantage of having cross-coupling blocks for reference current computation since the control structure is implemented in dq reference frame. Moreover, there are errors caused by the PLL in practical implementation, adding to its complexity. The essence of low-pass-filter in the algorithm also slows down the dynamics of the system.

The adaptive normalized notch filter based approach is implemented herein for the GPV system to address the aforementioned issues. The algorithm aims at fundamental and harmonic amplitude estimation along with accurate and fast tracking of fundamental frequency. The notch filter algorithms are proposed by authors in [14], nonetheless, there is always a constraint in selection of the adaptive gains of the filters for a trade-off situation for good steady-state and dynamic response. This increases the difficulty of tuning parameters and lowers the transient performance with time variant amplitudes. Moreover, an improved dynamic response and harmonic rejection are not attainable using a single notch filter [14].

The authors in [14] have presented adaptive notch filter based filtering to track the frequency of a sinusoidal signal analyzing both normalized and un-normalized adaptive filters. The authors in [15] have developed an un-normalized adaptive notch filter with necessary stability conditions and tracking performance. These are being employed for three phase systems by different authors [16]. However, it has been depicted in [14] that the input signal amplitude affects the convergence speed of a un-normalized algorithm in proportion to the square of input amplitude and value of input frequency, however, the normalized algorithm is least affected. A normalized adaptive notch filter [17] gets rid of the aforementioned difficulties. The main advantage of normalized adaptive notch filter is that the parameter update law gets decoupled from the filter itself and consequently the normalized approach is robust enough for the convergence speed to be unaffected by both amplitude and frequency of the input signal. The decoupled update law gives a representation of the filter as a feedback interconnection of the system with a nonlinear gain. The prominent contributions of the proposed work are depicted as- (i) Provision for precise frequency estimation and fundamental tracking with the convergence speed is unaffected by both amplitude and frequency of the input load current magnitude; (ii) The algorithm is simple and is advantageous for realization with embedded controllers and eliminated phase locked loop; (iii) The estimation of fundamental load components are independent of phase voltages and consequently the proposed scheme is independent from other phase.

The proposed control scheme is realized using DSP-dSPACE controller interfaced with the developed GPV laboratory prototype. The THD of the grid currents are attained within the standard limits.

II. GPV SYSTEM CONFIGURATION

The single-stage three phase GPV system is shown in Fig. 1. The single-stage topology includes the PV array, three phase VSC, ripple filter, nonlinear load, the boost converter, interfacing inductor and the DC link capacitor. The design, selection and modelling procedure for single-stage topology are presented in the literature [15-17]. The ripples in the grid current are eliminated using interfacing inductor and the VSC switching noises are absorbed by the VSC. The details of the system parameters are described in the Appendix.

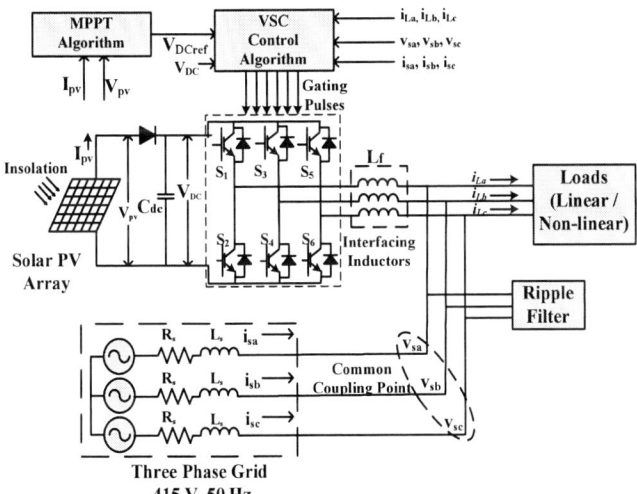

Fig. 1: Configuration of single-stage GPV system

III. CONTROL SCHEME

The overall control scheme for VSC is depicted in Fig. 2. The VSC switching scheme which comprises the substantial internal control signals of terminal voltage (V_z), unit templates (x_a, x_b and x_c), loss component (W_{DC}), feed-forward term of PV (W_{pvff}) and the reference currents ($i_a{}^{ref}$, $i_b{}^{ref}$ and $i_c{}^{ref}$). The voltage source converter (VSC) control strategy is depicted in section III.

The VSC control scheme can be presented in two divisions. The former presents the fundamental component from the load current which is estimated using notch filter based normalized adaptive algorithm. The latter includes estimation of the loss component (manifesting in the form of fluctuations in DC bus voltage), the PV feed-forward (PVFF) component, along with a control to incorporate adaptive DC bus voltage. Thereafter, the reference currents are evaluated and the VSC gating pulses are produced.

A. Load Current Fundamental and Harmonic content Extraction using Adaptive Normalized Notch Filter

The load current is contaminated with harmonic content. It can be represented in a generalized way as follows,

$$i_{La} = I_{La1}\sin(\omega_1 t + \delta_1) + I_{La2}\sin(\omega_2 t + \delta_2) +$$
$$...+ I_{LaN}\sin(\omega_N t + \delta_N) \qquad (1)$$
$$= \sum_{j=1}^{N} I_{Laj}\sin(\omega_j t + \delta_j)$$

Where, I_{Laj}, ω_j, δ_j are the amplitude, frequency and phase shift respectively of the jth component respectively. The fundamental frequency is represented by ω_1 and is an unknown and needs to be estimated. The jth component of signal in (1) is represented by,

$$i_{La}^{(j)} = I_{Laj}\sin(\omega_j t + \delta_j) \qquad (2)$$

For an input sinusoid signal ($i_{La}^{(j)}(t)$) given by a sinusoid ($=I_{Laj}\sin(\omega_j t + \delta_j)$), the structure of a notch filter, proposed by authors in [25], is characterized by the following differential equations,

$$\ddot{z}_j + \theta_j^2 z_j = 2\alpha_j(\theta_j^2 i_{La}^{(j)}(t) - \theta_j \dot{z}_j)$$
$$\dot{\theta}_j = -\beta_j z_j(\theta_j^2 i_{La}^{(j)}(t) - \theta_j \dot{z}_j) \qquad (3)$$

Where, 'θ_j' represents the estimated frequency corresponding to j^{th} harmonic content in load current signal, 'z_j' being an intermediate variable; 'α_j' and 'β_j' are both real and positive numbers which determine the behaviour i.e. accuracy and convergence speed of the algorithm. The equations given by (3) can be further modified by scaling the input signal by 'θ', and these modified signals are represented by (4). For a pure sinusoidal signal as input ($i_{La}^{(j)}(t)$), the solution of (4) boils down to a unique periodic orbit depicted in (5).

978-1-5386-6624-1/18 $31.00 © 2018 IEEE   523

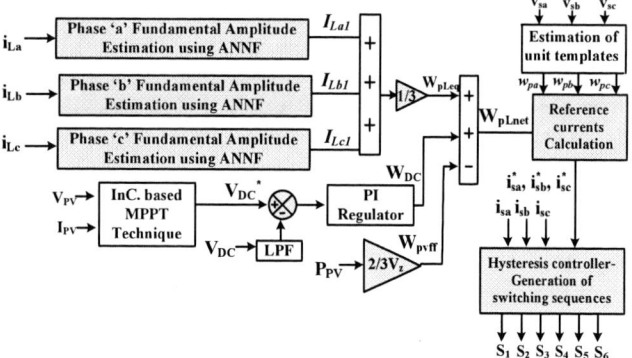

Fig. 2: Overall Control Schematic of ANNF based GPV System

$$\ddot{z}_j + \theta_j^2 z_j = 2\alpha_j \theta_j (i_{La}^{(j)}(t) - \dot{z}_j)$$
$$\dot{\theta}_j = -\beta_j z_j \theta_j (i_{La}^{(j)}(t) - \dot{z}_j) \quad (4)$$

$$\text{Periodic Orbit, } \Omega = \begin{pmatrix} z \\ \dot{z} \\ \theta \end{pmatrix} = \begin{pmatrix} -(I_{Laj}/\omega_j)\cos(\omega_j t + \delta_j) \\ I_{Laj}\sin(\omega_j t + \delta_j) \\ \omega_j \end{pmatrix} \quad (5)$$

Since '$\dot{z}$' in (5) exhibits a periodic orbit similar to the input signal $i_{La}^{(j)}(t)$, the system represented by (4) promises the desired frequency curve extraction from the input signal, if the variable '$\dot{z}$' for j^{th} component ($\dot{z}_j$) coincides with the input signal (i.e. $\dot{z}_j = i_{La}^{(j)}(t)$). The dynamics in (4) thus boil down to $\ddot{z} + \theta^2 z = 0$. Since '$z_j$' is an intermediate variable representing the dynamics, the '$\dot{z}_j$' may not always vary in the similar way as $I_{La}^{(j)}(t)$ in continuous time, and the corresponding error between the both is given by e(t)= $i_{La}^{(j)}(t)$- $\dot{z}_j(t)$, and the dynamic equation corresponding to (4) is modified as ' $\ddot{z}_j + \theta_j^2 z_j = 2\alpha_j \theta_j e_j(t)$ '. Extending this approach to the entire input signal given by (1), the error in the signal can be represented as,

$$e(t) = i_{La}(t) - \sum_{j=1}^{N} \dot{z}_j(t) \quad (6)$$

This is overall error in the input signal, and the error is manifested in every individual j^{th} component of load current. The filter dynamics to extract the load current components can therefore be formulated as follows,

$$\ddot{z}_j + \theta_j^2 z_j = 2\alpha_j \theta_j e(t) \quad j=1,2,...N$$
$$\text{and} \quad \dot{\theta}_1 = -\beta z_1 \theta e(t) \quad (7)$$

It may be noted in (7) that, in the latter equation, '$\dot{\theta}_1$' is dependent only on 'z_1', since the knowledge of fundamental frequency would be sufficient to know the harmonic frequencies and therefore update for 'θ' corresponding to fundamental component 'z_1' suffices the purpose. The unknown harmonic content in the load current can be extracted using this approach. The magnitudes of harmonic components along with the harmonic frequencies can be estimated from (7). However, (7) is non-normalized notch

filter based estimation. The amplitude of input sinusoid affects the convergence speed of non-normalized estimator in proportion of its square, whereas there is little effect on that of the normalized method. The notch filter equipped with normalized form of (7) has been discussed in [17] and for this purpose a new intermediate variable (y_j) is introduced and its corresponding update laws can be identified as,

$$\dot{z}_j = -\theta_j y_j + 2\alpha_j e(t), \quad j=1,2,.. N$$
$$\dot{\theta}_1 = \frac{2\alpha_j \beta_j y_1 e(t)}{\sqrt{z_1^2 + y_1^2}}$$
$$\text{and } \dot{y}_j = -\theta_j z_j \quad (8)$$

$$\text{Where, } e(t) = I_{La}(t) - \sum_{j=1}^{N} \dot{z}_j(t), \ \theta_j = j\theta_1$$

With the differential equations (8) base, the normalized frequency estimator for extraction of load current harmonics and corresponding fundamental frequency estimation is highlighted in Fig. 3. For the harmonic estimation purpose, the prominent harmonics in the GPV system, namely 5th to 23rd are accounted herein for the mitigation. The magnitude extraction of the corresponding harmonic component is also being highlighted in Fig. 3. These magnitudes are represented for three phases as I_{pa1}, I_{pb1} and I_{pc1}. The average value of of load current active components of each of phases provides the total magnitude of load current (W_{eq}). The fundamental and corresponding harmonic extraction capability of the proposed controller from the load current (i_L) is highlighted in Fig. 5. As depicted in the figure, the ANNF control scheme is highly efficient in its fundamental extraction and the dynamics are fast enough to settle down to zero under disconnection of the phase-'a' load. The corresponding amplitude estimation is therefore precise without any steady-state fluctuations, as shown in Fig. 4.

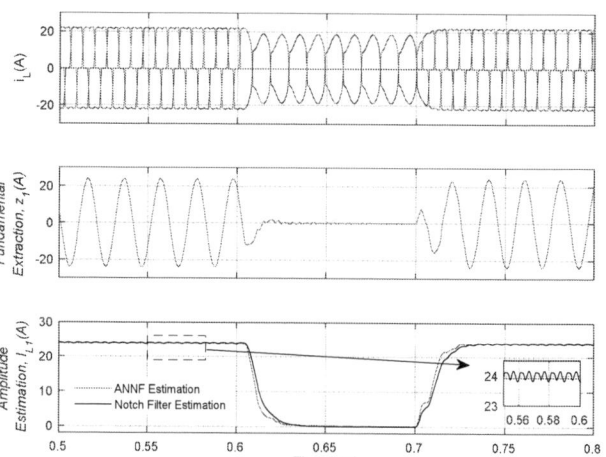

Fig. 4: Effectiveness in Fundamental and Amplitude Extraction of the Proposed Control Scheme

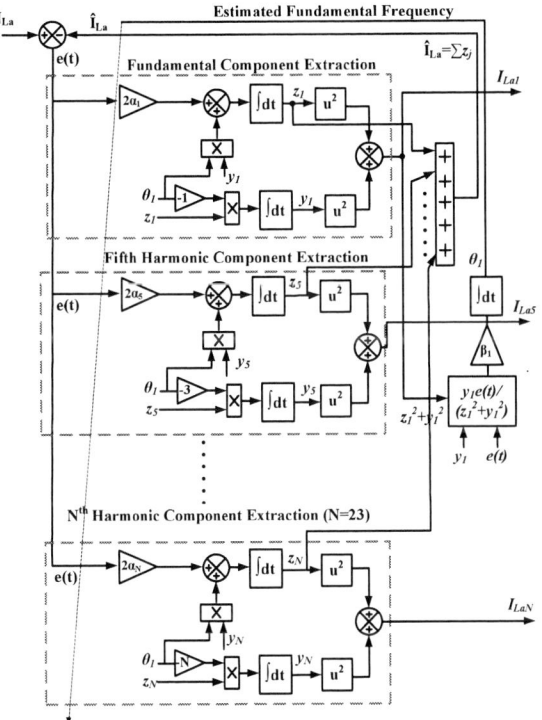

Fig. 3: Fundamental and Harmonic Components extraction using ANNF

B. DC Bus Voltage Regulation

The incremental conductance based MPPT (Maximum Power Point Tracking) algorithm is used to extract maximum power from the PV array, which outputs the reference DC bus voltage. The error between sensed and reference DC voltage (V_{dce}) is fed to a PI, whose output is governed by the following equation, (k_p being proportional gain and k_i being an integral gain).

$$W_{DC}(k) = W_{PI}(k-1) + k_p \left\{ V_{dce}(k) - V_{dce}(k-1) \right\} + k_i V_{dce}(k) \quad (9)$$

C. PV Feed Forward Compensation

An additional feed-forward unit has been incorporated to quicken the dynamic response under constant solar insolation fluctuations occurring in a practical photovoltaic system. Since the dynamics are directly associated with the change in PV array power, which is the effect of change in solar insolation, the feed-forward unit is given as,

$$W_{pvff} = \frac{(2P_{PV})}{(3V_z)} \quad (10)$$

D. Generation of Gating Signals

The net active load component is then given by (11) and the reference currents are generated using the in-phase voltage templates.

$$W_{pnet} = W_{peq} + W_{DC} - W_{pvff} \quad (11)$$

The computation of in-phase unit templates (x_{pa}, x_{pb}, x_{pc}) for the purpose of producing reference grid currents can be identified as follows,

$$\begin{bmatrix} x_{pa} & x_{pb} & x_{pc} \end{bmatrix} = \left(\frac{1}{V_z} \right) \begin{bmatrix} v_{sa} & v_{sb} & v_{sc} \end{bmatrix} \quad (12)$$

Where (v_{sa}, v_{sb}, v_{sc}) are the grid phase voltages and V_z is the terminal given by,

$$V_z = \sqrt{(2/3)\left(v_{sa}^2 + v_{sb}^2 + v_{sc}^2 \right)} \quad (13)$$

The reference currents are then extracted in the following way,

$$i_{sa}^* = W_{pnet}.x_{pa}, \quad i_{sb}^* = W_{pnet}.x_{pb}, \quad i_{sc}^* = W_{pnet}.x_{pc} \quad (14)$$

The error in between reference and sensed currents thus computed from (14) and the sensed grid current infiltrated through the fixed frequency current hysteresis controller producing firing signals (S1-S6) of the VSC.

IV. EXPERIMENTAL RESULTS

A prototype of GPV system is realized in the laboratory to validate control scheme. A solar PV array simulator is used to realization of solar power generation. The Hall-Effect based current sensors and voltage sensors are used for currents-voltages sensing, respectively, of loads solar PV and grid. An opto-coupler is employed to provide isolation between DSP signal and VSC. The proposed control algorithm using PV feed-forward term is implemented DSP-dSPACE controller. The behavior of GPV system is recorded and analyzed through a power analyzer (Fluke made: model 43-B) and four channel digital storage oscilloscope (Aglient make: model DSO714A). MPPT performance of the PV array simulator is demonstrated in Fig. 5.

The steady state and dynamic response of the proposed system are manifested with reference to the dynamic changes in grid and load side network. The dynamic behavior of GPV system is analyzed using in terms of grid line voltages (v_{sab}, v_{sbc}, v_{sca}) reference currents (i_{sa}^*, i_{sb}^*, i_{sc}^*) extraction, PV array power (P_{PV}), VSC currents (i_{VSC}), grid currents (i_s), PV feed forward component (W_{pvff}), load currents (i_L), amplitude of fundamental load currents (I_{La1}, I_{Lb1}, I_{Lc1}).

A. Experimental Behavior of under Load Unbalancing

To investigate the response of the control for GPV system, salient internal signals are recorded and the same are reported in Fig. 6. Figs. 6(a)-(b) shows the fundamental and the corresponding amplitude extraction capability of the proposed controller, along with precise frequency estimation.

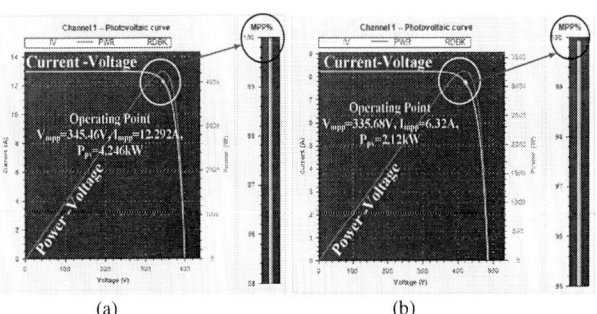

(a)   (b)

Fig. 5: MPPT response at (c) 1000 W/m² and (d) 500 W/m²

978-1-5386-6624-1/18 $31.00 © 2018 IEEE

The dynamics are observed to be very fast within less a half cycle, under disconnection of phase-'a' load. Figs. 6(c)-(d) depicts variation of the average weight component (W_{peq}) along with DC regulation (W_{DC}) component and the feed-forward component (W_{pvff}) under load disconnection. The grid current tracking capability with reference grid currents is well highlighted in Fig. 6(e). The current references raise as the surplus power due to load unbalancing being supplied to the distribution grid. Moreover, the perfect in-phase operation of extracted fundamental with the load current is depicted in Fig. 6(f). The amplitude extraction of the algorithm is observed to have fast dynamic response.

The corresponding system variables are demonstrated in Fig.7. The load currents (i_{La}, i_{Lb}, i_{Lc}), DC bus voltage (V_{DC}), the phase-'a' source current (i_{sa}) and source line voltage (v_{sab}), the PV array current (I_{PV}) and VSC current are presented in Figs. 7(a)-(d), with respect to unbalance realized by disconnection of phase-'a' loads. The increase in grid current is observed as a result of surplus power due to load reduction is being supplied to the grid.

B. Experimental Response under Balanced Nonlinear Load

Fig. 8 shows the performance of the system under balanced nonlinear load connect at PCC. Figs. 8(a)-(c) illustrate voltage and current waveforms of grid, VSC and load. Figs. 8(d)-(f) show the power flow through grid, power supplied by VSC and power consumption of load. The arithmetic sum of load power and grid power is equal to solar PV coupled VSC system which illustrates that PV coupled VSC supplies power to load as well as to the distribution network. The THD of grid current is achieved at 3.8% as shown in Figs. 8(g)-(i). The THD of grid voltage is 2.1% which is within limit to the prescribed standard. Fig. 8(j) depicts the V_{PV} and I_{PV}, whereas, Fig. 8(k) shows the power delivered by the PV array.

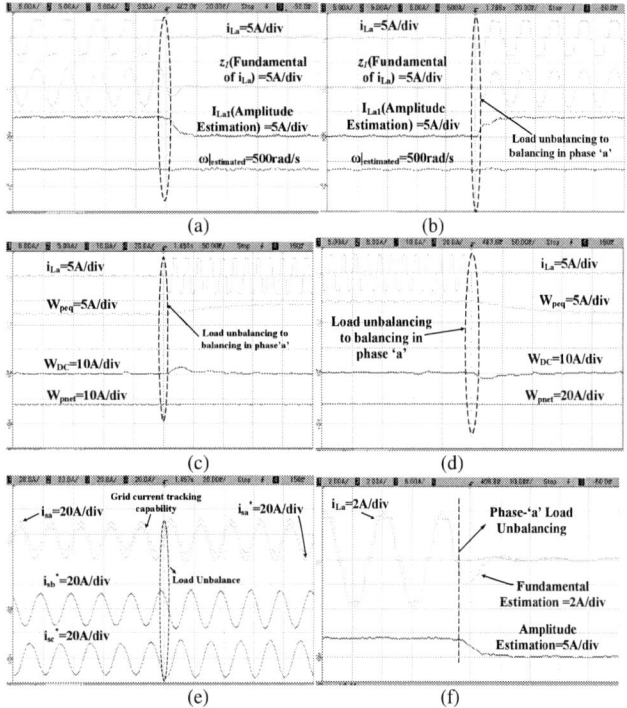

Fig.6: Salient internal signals under nonlinear load perturbation (a)–(b) I_{La} with z_l, I_{La1} and ω (c)–(d) I_{La} with W_{peq}, W_{DC} and W_{pnet} (e)–(f) i_{sa} with i_{sa}^*, i_{sb}^* and i_{sc}^*,

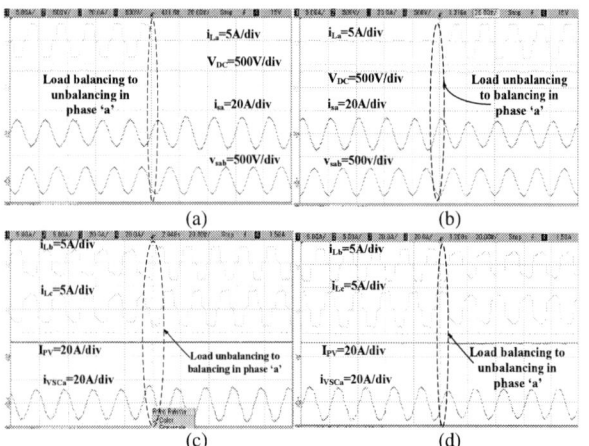

Fig. 7: Performance under nonlinear load perturbation (a)–(b) i_{La}, V_{dc}, i_{sa}, V_{sab} and (c)–(d) i_{Lb}, i_{Lc}, I_{pv} and i_{VSCa}.

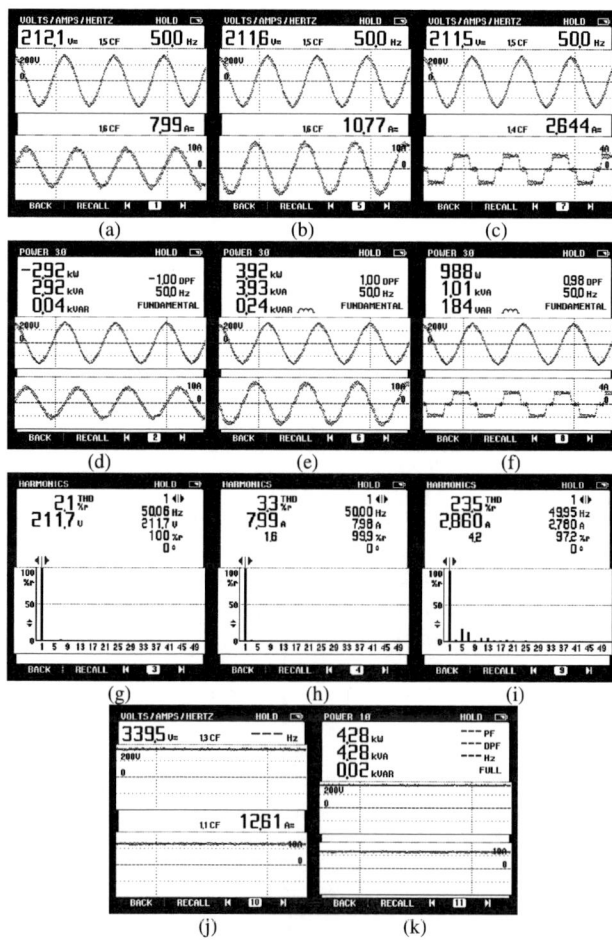

Fig. 8: Experimental response under balanced nonlinear loads (a-c) v_{sab}-i_{sa}-i_{VSCa}-i_{La} (d-f) power in grid-VSC-load (g-i) THDs of i_{sa}, v_{sa}, and i_{La}, (j)-(k) V_{Pv}-I_{PV}-P_{PV}

978-1-5386-6624-1/18 $31.00 © 2018 IEEE

C. Experimental Response under Variation in Solar Irradiations

The response of GPV system is analyzed as solar insolation varies from 1000W/m^2 to 500W/m^2 and vice versa. The behaviour in terms of V_{PV}, PV array current (I_{PV}), VSC current (i_{VSCa}) and source current (i_{sa}) is depicted in Figs. 9(a) and 9(b). The dynamics in grid and VSC currents are examined. The diminished real power being supplied to the grid at variable solar irradiations under fixed load power, resulted in reduced grid current. The solar PV array current and power are reduced as insolation is reduced. However, DC bus voltage is least affected even under variation of insolation from 1000W/m^2 to 500W/m^2 and vice versa.

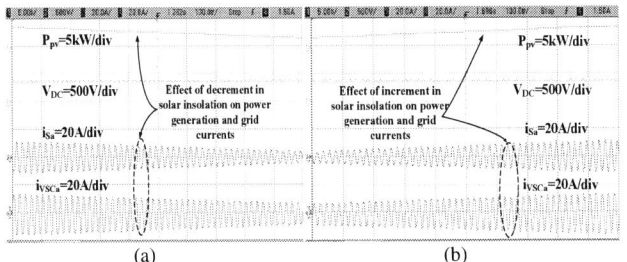

Fig 9: Performance under irradiations from (a) 1000-500 W/m^2, (b) 500-1000 W/m^2

V. CONCLUSION

An adaptive normalized estimator based notch filter has been proposed for single-stage GPV system. The algorithm is able to extract both fundamental and harmonic load components and is precise in its fundamental tracking with quick dynamics. It has been observed that the fundamental frequency is precisely estimated. Many fold objectives have been achieved with the proposed control scheme, namely, precise tracking of the fundamental frequency, feeding active power to the grid, load balancing and power factor correction. The proposed ANNF control strategy is capable to accomplish unity power factor even under steady-state/dynamic scenarios. The grid currents THDs are observed within the limits of IEEE-519 standard.

APPENDIX

Experimental Parameters

Parameter	Value	Parameter	Value
PV array current	I_{sc} = 13 A	PV array power	P_{PV} = 5.2 kW
Ripple filter	R_f = 6 Ω , C_f = 12μF	PV array voltage	V_{oc} = 400 V
DC link capacitor	2200μF	Nonlinear load	1 kVA
Interfacing inductor	L_f= 3.5 mH	VSC Rating	25kVA
DSP Sampling Time	30μs	Grid voltage	215 V
Control parameter	α= 220, β=80	DC voltage Controller,	k_p=0.6, k_i=0.2,

REFERENCES

[1] Solangi, K. H., "A review on global solar energy policy." *Renewable and sustainable energy reviews*, vol. 15, no. 4, pp. 2149-2163, 2011.

[2] Menz, Fredric C., and Stephan Vachon. "The effectiveness of different policy regimes for promoting wind power: Experiences from the states." *Energy policy*, vol. 34, no. 14, pp. 1786-1796, 2006.

[3] White, W., Lunnan, A., Nybakk, E. and Kulisic, B., "The role of governments in renewable energy: The importance of policy consistency" *Biomass and bioenergy*, vol. 57 pp. 97-105, 2013.

[4] Romero, Manuel, and Aldo Steinfeld. "Concentrating solar thermal power and thermochemical fuels." *Energy & Environmental Science*, vol. 5, no.11, pp. 9234-9245, 2012.

[5] V. L. Srinivas, S. Kumar, B. Singh and S. Mishra, "Partially Decoupled Adaptive Filter Based Multifunctional Three-Phase GPV System," in *IEEE Trans. on Sus. Energy*, vol. 9, no. 1, pp. 311-320, Jan. 2018.

[6] P. Shah, I. Hussain and B. Singh, "A Novel Fourth-Order Generalized Integrator Based Control Scheme for Multifunctional SECS in the Distribution System," in *IEEE Transactions on Energy Conversion*, vol. 33, no. 3, pp. 949-958, Sept. 2018.

[7] S. Vedantham, S. Kumar, B. Singh and S. Mishra, "Fuzzy logic gain-tuned adaptive second-order GI-based multi-objective control for reliable operation of grid-interfaced photovoltaic system," in *IET GTD*, vol. 12, no. 5, pp. 1153-1163, 13 3 2018.

[8] S. Vedantham, S. Kumar, B. Singh and S. Mishra, "RLMMN adaptive filtering based control scheme for multi-objective GPV system," *In proc.*, 6th *International Conference on Computer Applications In Electrical Engineering-Recent Advances (CERA)*, Roorkee , pp. 556-561, 2017

[9] V . L. Srinivas, S. Kumar, B. Singh and S. Mishra, "A Multifunctional GPV System Using Adaptive Observer Based Harmonic Cancellation Technique," in *IEEE Trans. on Ind. Electron.*, vol. 65, no. 2, pp. 1347-1357, Feb. 2018.

[10] L. H. I. Lim, Z. Ye, J. Ye, D. Yang and H. Du, "A Linear Identification of Diode Models from Single–Characteristics of PV Panels," *IEEE Trans. Indus. Electron.*, vol. 62, no. 7, pp. 4181-4193, July 2015.

[11] L. Hsu, R. Ortega and G. "Damm. A globally convergent frequency estimator," *IEEE Transactions on Automatic Control*, pp. 698– 713 April, 1999.

[12] A. K. Verma, B. Singh, and D. T. Shahani, "Grid interfaced solar photovoltaic power generating system with power quality improvement at AC mains," in Proc. IEEE 3rd Int. Conf. Sustain. Energy Technol., pp. 177–182, Sept. 2012.

[13] P. Shah, I. Hussain and B. Singh, "Fuzzy logic based FOGI-FLL algorithm for optimal operation of single stage three phase grid interfaced multifunctional SECS," *IEEE Transactions Industrial Informatics*, vol. 14, no. 8, pp. 3334-3346, Aug. 2018.

[14] X. He, H. Geng and G. Yang, "A generalized design framework of notch filter based frequency-locked loop for three-phase grid voltage," *IEEE Transactions Industrial Electronics*, vol. 65, no. 9, pp. 7072-7084, Sept. 2018.

[15] M. Mojiri and A. R. Bakhshai, "An adaptive notch filter for frequency estimation of a periodic signal," *IEEE Trans. Automatic Control*, vol. 49, no. 2, pp. 314-318, Feb. 2004.

[16] D. Yazdani, A. Bakhshai, P. K. Jain. "A three-phase adaptive notch filter-based approach to harmonic/reactive current extraction and harmonic decomposition," *IEEE Trans. Power Electron.*, pp. 914 – 923 April, 2010.

[17] Z. Chu, M. Ding, S. Du and X. Dong, "Normalized estimation of fundamental frequency and measurement of harmonics/interharmonics." *Journal of Control Theory and Appl.* , vol. 7, no. 1, pp. 10-17, 2013.

2nd IEEE International conference on power Electronics, Intelligent Control and Energy systems (ICPEICES-2018)

IFLL Based Control for PV System at Adverse Grid Conditions

Bhim Singh, Fellow IEEE
Department of Electrical Engineering
Indian Institute of Technology, Delhi
bhimsinghiitd61@gmail.com

Vandana Jain, Member IEEE
Department of Electrical Engineering
Indian Institute of Technology, Delhi
vandanadcrust@gmail.com

Abstract— **This work proposes a control algorithm based on improved frequency locked loop (IFLL) applied to single stage grid connected solar photovoltaic (PV) system at adverse grid situations like voltage distortion, voltage swell/sag etc. The proposed IFLL control algorithm extracts the grid voltage fundamental component, under various adverse grid conditions like voltage distortion and voltage swell/sag. For a single stage system, during dynamics, a fixed gain PI (Proportional-Integral) controller doesn't provide satisfactory performance. Thus, a shuffled frog leaping algorithm tuned PI (SFLAPI) controller is used to overcome the disadvantages of PI controller with fixed gains. The VSC (Voltage Source Converter) of PV system, supplies the PV energy to the grid even under the worst situations of voltage distortion and voltage swell/sag. An incremental conductance (INC) based MPPT (Maximum Power Point Tracking) is used for the peak power extraction from the PV array. Tests are conducted under various odd states like voltage distortion and voltage swell/sag, insolation change etc. and for the same, a prototype is built in the research laboratory. Performance of proposed system is found to be satisfactory, as recommended by the IEEE-519 and IEEE-1547 standards.**

Keywords— *SFLAPI, MPPT, IFLL, PV, Power Quality.*

I. INTRODUCTION

In the present scenario, renewable energy systems are emerging as alternatives to the conventional sources of energy for electric power production. Due to availability and clean nature of solar energy, it is the best alternative of renewable energy. However, the grid conditions in India, are very poor, erratic and degraded. It has the problems of poor power quality i.e. presence of harmonics, voltage distortion and voltage sag/swell etc. [1-2]. Power quality (PQ) problems in the grid, cause maloperation and undesired failure of the apparatus connected in the distribution network. Hence, it is a big task to integrate renewable energy systems due to its intermittent nature to the weak grid. These types of PQ problems are diminished using a VSC (Voltage Source Converter). This VSC works as a DSTATCOM (Distribution Static Compensator) as described in [1-2]. This VSC aids PV array in feeding the power to the grid and the load network at unity power factor with reduced harmonics. Various institutions have recommend benchmarks for the harmonics distortion in the distribution network [3-4].

A technique for PV array modelling is illustrated in [5]. A modified P&O (Perturb and Observe) based MPPT (Maximum Power Point Tracking) is demonstrated in [6] to overcome oscillations around MPPT during insolation change. A comparison of single stage and double stage topologies, is demonstrated in [7]. These topologies of grid interfaced PV systems, are analyzed for different modes of operations. Single stage topology is considered better than a

two stage topology due to high efficacy, less components, effective utilization of PV power and lower cost.

A grid interfaced solar PV system with VSC, provides an advantage that it improves power quality problems and directly transmits the power to the load and also to the grid, thereby reducing generation requirement from the non-renewable sources. During the absence of sunlight, VSC operates in DSTATCOM mode and supplies power to loads from the grid with an improved power quality. The improvement in power quality, depends upon the efficacy of the control algorithm, which is used for extraction of reference currents. Numerous control algorithms have been described in the literature. Some of them are synchronous reference frame (SRF), least mean square (LMS), least mean fourth (LMF), FFP (Fixed Forward Prediction), HTF (Hyperbolic Tangent Function) and MINF (Multiple Improved Notch Filter) [8-14]. The intelligent adaptive controls such as adaline-LMS, RLS (Recursive Least Squares), LMF, have provided an advantage of better dynamic response in comparison with conventional controls along with the ease of implementation.

An improved frequency locked loop (IFLL) algorithm is proposed in this work for extraction of fundamental component of voltage (FCV) during adverse grid circumstances of voltage distortion and voltage swell/ sag.

A conventional PI controller works well in steady state, however, during dynamics, performance of fixed gain PI controller is not satisfactory. Thus, a shuffled frog leaping algorithm tuned PI controller (SFLAPI) provides instant gains tuning of the PI controller during dynamics such as solar insolation variation.

In this work, the control algorithm for extraction of fundamental component of voltage (FCV) is applied on a single stage PV system connected to a three phase distribution grid, for an improvement of PQ during adverse grid circumstances of voltage distortion, voltage swell and voltage sag. For a single stage system, PV array is directly connected across DC-link capacitor. Thus, DC-link voltage, V_{dc} and PV voltage V_{pv} are same. Due to weak grid, there is continuous fluctuation in voltage, a PI controller is required to be tuned accordingly. This VSC supplies the PV energy to the distribution grid even under the worst operating conditions. The INC based MPPT is used on the PV array for peak power extraction from it. The system is tested at various adverse conditions like voltage distortion, voltage sag/swell, insolation change etc. and the system prototype is built in the laboratory and the performance of the system is found satisfactory, as recommended in the IEEE-519 standard and IEEE-1547 standard while injecting power in the grid [3-4].

978-1-5386-6624-1/18 $31.00 © 2018 IEEE   528

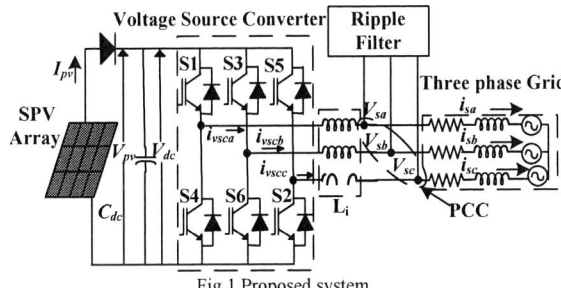

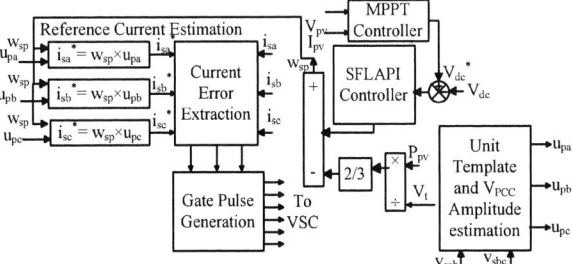

Fig.1 Proposed system   Fig.2 Control Algorithm

satisfactory, as recommended in the IEEE-519 standard and IEEE-1547 standard while injecting power in the grid [3-4].

The major contributions of this work, are as follows.

- The simplicity and robustness of this control algorithm are achieved with the introduction of a feed-forward term during dynamics such as variable solar irradiation.
- The grid dependence is also reduced and the peak power is also trimmed during the day time.
- The IFLL control extracts FCV effectively and decreases harmonics components. Thus providing superior performance in comparison to conventional FLL's.
- Single stage system provides better power extraction capability due to low loss and less complex in comparison with two stage system as demonstrated in [7].
- The PI controller with SFLAPI gains tuning, provides an improved performance during system dynamics.

To study the present system performance, a prototype is built in the laboratory and responses are verified at various insolation levels and adverse grid state of voltage distortion and voltage sag/swell.

II. SYSTEM DESCRIPTION

The structure of the system is depicted in Fig.1. The system comprises of a three phase grid, a ripple filter, interfacing inductors, a VSC, a PV array and the DC link capacitor. The PI controller regulates dynamically the voltage variation for the generation of reference DC bus voltage using a SFLAPI tuned PI controller.

III. CONTROL ALGORITHM

The proposed control algorithm is demonstrated in Fig. 2. This control algorithm is further classified into two subsections - MPPT control for the generation of reference DC bus voltage and VSC control. An INC based MPPT algorithm keeps a constant DC-link voltage for extraction of peak power from a PV array at fluctuating states using a PI controller. For a single stage system, during dynamics, fixed gain PI controller doesn't provide satisfactory performance. Thus, a PI controller using SFLAPI, is presented to overcome the limitations of PI controller with fixed gains. The PI controller adjusts DC bus voltage dynamically according to the voltage. The control scheme is described in the subsequent sections in details.

A. VSC Control Algorithm

An IFLL based control algorithm is used for the extraction of FCV for the estimation of reference grid currents for generation of gating pulses for the VSC, is

presented in Fig. 2. The fundamental aim of this method for VSC, is to adjust DC bus voltage dynamically while taking into account the grid voltage variations.

1) Unit Templates Estimation

The sensed PCC line voltages (v_{sab}, v_{sbc}) are used to estimate three phase voltages (v_{sa}, v_{sb} and v_{sc}). For the extraction of FCV from the unbalanced voltages, the PCC phase voltages (v_{sa}, v_{sb} and v_{sc}) are converted in α-β domain with Clark's transformation [1].

$$\begin{pmatrix} v_\alpha \\ v_\beta \end{pmatrix} = \frac{2}{3}\begin{pmatrix} 1 & -\frac{1}{2} & -\frac{1}{2} \\ 0 & \frac{\sqrt{3}}{2} & -\frac{\sqrt{3}}{2} \end{pmatrix} \times \begin{pmatrix} v_{sa} \\ v_{sb} \\ v_{sc} \end{pmatrix} \qquad (1)$$

These transformed voltages are then passed to IFLL based BPF's (Band-Pass Filters) for elimination of harmonics from them. This output is used for estimation of the positive sequence voltages (v_{pa}, v_{pb} and v_{pc}) by inverse Clark's transformation to minimize the voltage distortions.

$$\begin{pmatrix} v_{pa} \\ v_{pb} \\ v_{pc} \end{pmatrix} = \frac{3}{2}\begin{pmatrix} \frac{2}{3} & 0 \\ -\frac{1}{3} & \frac{\sqrt{3}}{3} \\ -\frac{1}{3} & -\frac{\sqrt{3}}{3} \end{pmatrix} \times \begin{pmatrix} v_\alpha \\ v_\beta \end{pmatrix} \qquad (2)$$

For grid adverse conditions of voltage swell, voltage sag, voltage distortion, DC offset etc., it is essential to extract FCV from the grid voltages. For the extraction of FCV at PCC, IFLL is proposed. Fig. 3 shows the structure of IFLL. IFLL is used for the elimination of harmonics distortion.

These PCC phase voltages (v_{pa}, v_{pb} and v_{pc}) are used in estimation of in- phase unit template (u_{pa}, u_{pb}, u_{pc}) as,

$$V_t = \sqrt{\frac{2}{3} \times \left(v_{pa}^2 + v_{pb}^2 + v_{pc}^2 \right)} \qquad (3)$$

where V_t is terminal PCC voltages.

Therefore, $u_{pa} = \dfrac{v_{pa}}{V_t}$, $u_{pb} = \dfrac{v_{pb}}{V_t}$, $u_{pc} = \dfrac{v_{pc}}{V_t}$ (4)

2) DC Bus PI Controller

To reduce the losses in VSC and its unwanted tripping, the DC link voltage has to be adaptable. For calculation of loss component, V^*_{dc} is compared with V_{dc} (sensed DC voltage) to obtain error. This error is given to PI controller and loss component is obtained. The subsequent equations demonstrate the above explanation as,

978-1-5386-6624-1/18 $31.00 © 2018 IEEE   529

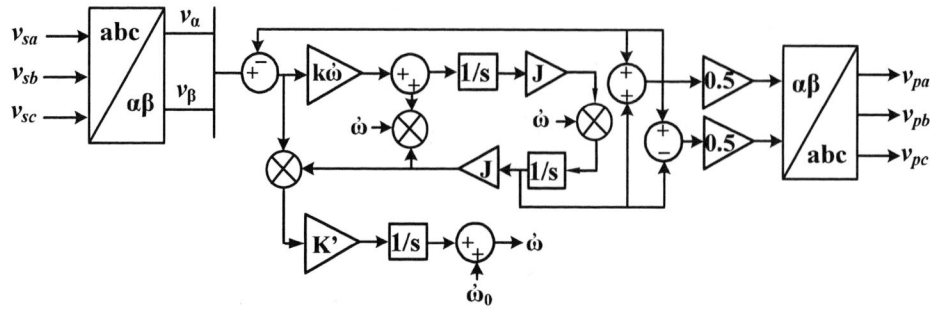

Fig. 3 Block diagram representation of IFLL

$$\Delta V_{dc}(n) = V_{dc}^{*}(n) - V_{dc}(n) \tag{5}$$

$$w_{cp}(n) = w_{cp}(n-1) + k_{pd}\left\{\Delta V_{dc}(n) - \Delta V_{dc}(n-1)\right\} \\ + k_{id}\Delta V_{dc}(n) \tag{6}$$

where $\Delta V_{dc}(n)$ and $\Delta V_{dc}(n-1)$ are errors of V_{dc}, k_{pd} is proportional and k_{id} is integral gain constant.

In case of a fixed gain PI controller, during system dynamics, PV array voltage, V_{pv} also fluctuates according to the variation in the grid voltage. In order to maintain PV array voltage constant, online gain tuning of PI controller is realized using SFLAPI. The gains of PI controller, are tuned for compensation of DC-link voltage error.

In SFLAPT algorithm, the population comprises of frogs, arranged in descending order according to their fitness. The frogs are divided into different groups (memplexes). The first frog (best) goes to first memplex, second best in second memplex and so on. Each memplex contains best and worst frog F_b and F_w respectively. Then worst frog position is updated as,

$$P_i = Rand(F_b - F_w) \tag{7}$$
$$F_w^{new} = F_w^{old} + P_i \quad (P_{il} < P_i < P_{ih}) \tag{8}$$

where P_i is updated position of the frog, Rand is between 0 to 1. P_{il} and P_{ih} are lowest and highest step size for frog position updation. If a better solution is obtained then worst frog is replaced else above process is repeated again unless the convergence criteria is met. Fig. 4 shows the flowchart of the SFLAPI algorithm.

3) Calculation of Weight of PV System

The disturbances in grid tied PV array, due to intermittent solar insolation, are compensated using feed-forward term. The value of this expression is estimated using P_{pv} and V_t. This reduces the grid current oscillations during varying conditions such as insolation change and voltage sag/swell. This term (w_{pv}) is achieved as,

$$w_{pv}(n) = \frac{2}{3} \times \frac{P_{pv}(n)}{V_t} \tag{9}$$

4) Grid Reference Currents Estimation

The gross weight for current component estimation of reference grid current, PV term is deducted from DC-link loss component. It is approximated from subsequent equation,

$$w_{sp} = w_{cp} - w_{pv} \tag{10}$$

The reference grid currents are estimated as,

$$i_{sa}^{*} = w_{sp}.u_{pa}, \ i_{sb}^{*} = w_{sp}.u_{pb}, i_{sc}^{*} = w_{sp}.u_{pc} \tag{11}$$

The current errors are generated by subtracting i_{sa}, i_{sb}, i_{sc} from i_{sa}^{*}, i_{sb}^{*}, i_{sc}^{*}. For generation of gate pulses for the VSC,

these current errors are passed to the hysteresis current controller.

IV.   SIMULATED PERFORMANCE

The system performance is examined by developing a model in MATLAB and testing it for several dynamic states such as load unbalance and erratic solar insolation. The response of grid voltages (v_{sabc}), grid currents (i_{sabc}), VSC currents (i_{vsc}), active power of grid (P_s), reactive power of grid (Q_s), DC bus voltage (V_{dc}) and solar insolation (G), are shown here for different conditions at t = 0.4 s. The specifications of system, are specified in Appendix.

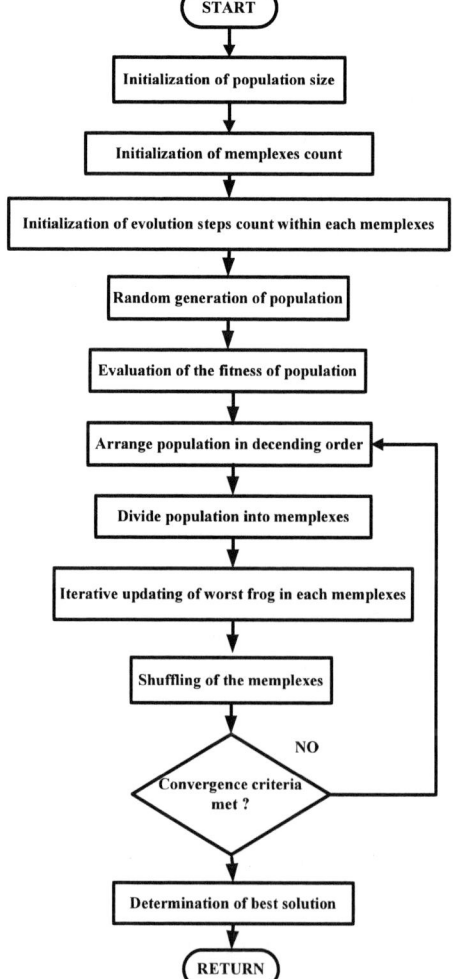

Fig. 4 Flowchart of SFLAPI

978-1-5386-6624-1/18 $31.00 © 2018 IEEE

A. Performance of System at Voltage Swell /Sag

The voltage sag/swell occurs due to abrupt removal or addition of a heavy load such as induction motors in a distribution network. Fig. 5 (a) shows waveforms of v_{sabc}, i_{sabc}, V_{dc}, i_{vsc}, P_s and Q_s at voltage sag occurred in the system. At 20% fall in the grid voltage of the distribution network, it is detected that grid currents are also increased by 20%. Similarly, during voltage swell, the grid current magnitude decreases by equivalent amount as depicted in Fig. 5 (b).

B. System Performance under Varying Solar Insolation

Fig. 5 (c) shows system performance at variable insolation. Waveforms of G, v_{sabc}, i_{sabc}, V_{dc}, i_{vsc} and P_s are shown in Fig. 5 (c). When solar irradiation is increased from 500 W/m² to 1000W/m², the PV array provides additional power to the grid. Thus, the grid power and grid currents, are increased. However, V_{dc} is constant and the PV system operates at MPPT.

C. Effectiveness of the proposed algorithm

The effectiveness of the IFLL is shown in Fig. 6. The system is tested in grid adverse conditions such as voltage sag, voltage swell and voltage distortion. Fig. 6 shows the waveforms of grid voltage, FCV of grid voltage, unit template and grid current. It is clear that even in the grid adverse conditions of voltage sag/swell and voltage distortion. in the grid, grid current is sinusoidal and balanced. FCV and unit templates are extracted accurately as shown in Fig. 6.

V. EXPERIMENTAL RESULTS

For the authentication of the system, a prototype is built in the research laboratory and the system responses are captured at different solar insolations and grid adverse situations of voltage swell/sag. The apparatus used for the development of the prototype are as described here.

- PV array simulator (ETS500x17DPVF from AMETEK)
- Power supplies (three phase AC and DC power supply). Recording instruments - DSO (Digital Storage Oscilloscope, DSO7014A) and power analyzer.
- Hall-Effect sensors, for sensing DC-link voltage, PCC voltages, grid current and PV current- current sensors (LA55-P), voltage sensors (LV25-P).
- DSP-dSPACE controller (dSPACE-1202 from MicroLabBox)

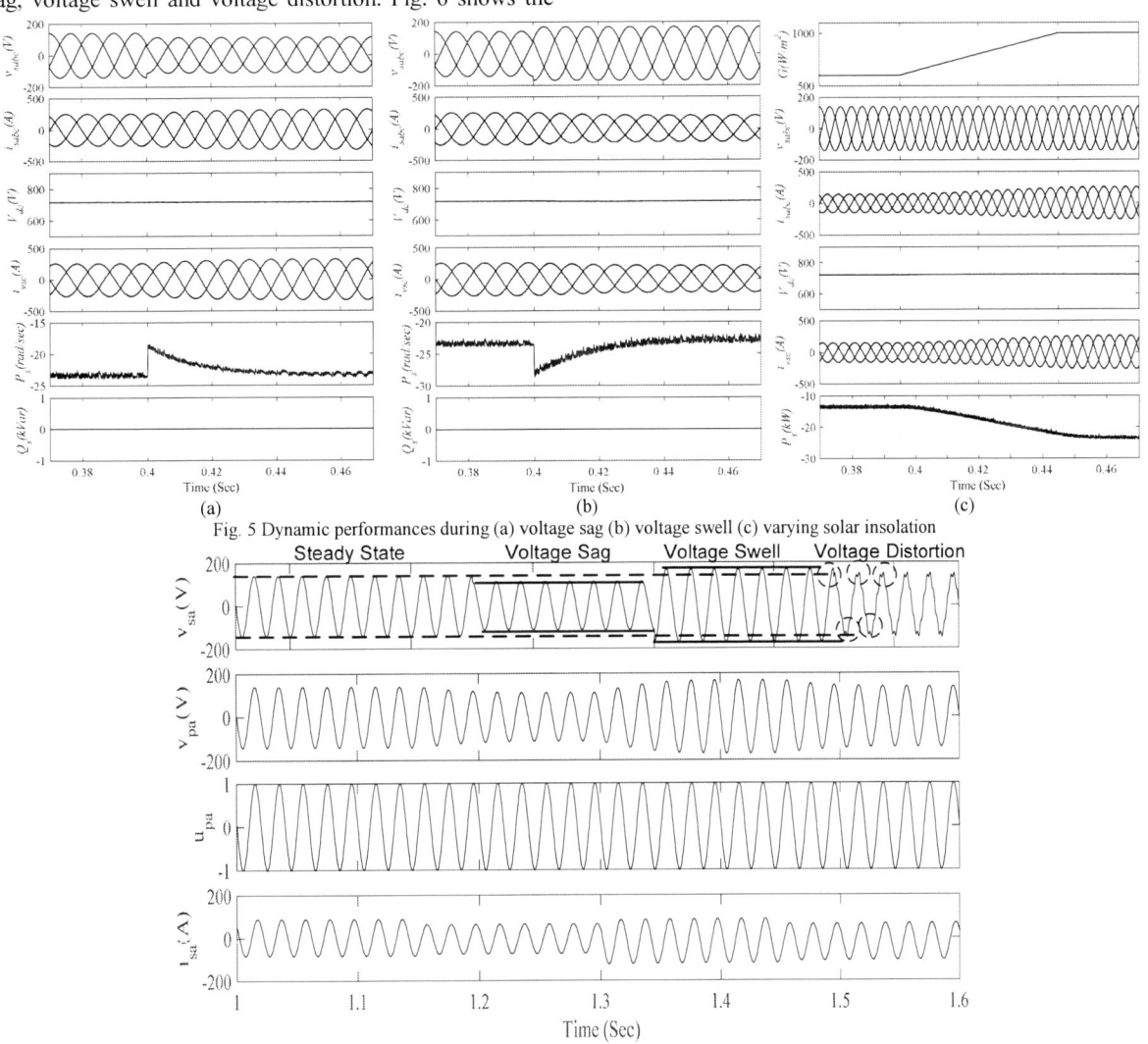

Fig. 5 Dynamic performances during (a) voltage sag (b) voltage swell (c) varying solar insolation

Fig. 6 Effectiveness of IFLL

- IGBT (Insulated Gate Bipolar Transistor) based VSC, DC link capacitor, AC Inductors, RC filters and opto-couplers, offers seclusion between VSC and DSP signals. The comprehensive information for prototype, are given in Appendix.

A. Steady State Performance of System

Figs. 7 (a-d) and Figs.8 (a-d) demonstrate steady state system performance at solar insolation of 1000W/m² and 500W/m², respectively. Fig. 7 (a) and Fig.8 (a) display grid voltage (v_{sab}), and current (i_{sa}). Fig. 7 (b) and Fig.8 (b) show the power supplied to the grid by the solar PV array. THD (Total Harmonic Distortion) values of voltage and current, are below 5% as defined in the IEEE-519 standards. Steady state waveforms of grid voltage and current, harmonics distortions of grid voltage are 4% and 2.3% as presented in Fig. 7 (c) and Fig. 8 (c), respectively. Fig. 7 (d) and Fig. 8 (d) demonstrates the THD's of grid current, which are 4.6% and 2.9 % respectively. THD's of grid voltage and current, are below 5%, which are in limit as recommended by the IEEE-519 standard.

B. Response during Voltage Sag/Swell

Fig. 9 (a) shows waveforms of v_{sab}, i_{sa}, i_{la} and i_{vsca} at voltage sag in the network. During 10% decrease in magnitude of three grid voltages, the magnitude of the grid currents, increases by 10%. Similarly, during voltage swell by 10%, the magnitude of the grid current, decreases by the equivalent factor. This phenomenon is depicted in Fig. 9 (b).

C. Response of System during Grid Voltage Distortion

Figs. 10 (a-c) show the response of the system under grid voltage distortions. Fig. 10 (a) shows the waveforms of v_{sab}, v_{pa}, u_{pa} and i_{sa}. The IFLL based positive sequence components are used for the generation of unit templates for the elimination of harmonics from the phase voltages. Thus the grid currents have low THD, less than 5% even though the PCC voltages are highly distorted as shown in Figs. 10 (b-c).

D. Operation of System during Solar Insolation Variation

Figs. 11 (a-d) depict the system performance at intermittent solar irradiations. Fig. 11 (a) demonstrates system performance during rise in solar irradiations from 500 W/m² to 1000 W/m². It is seen that with the rise in solar insolation, V_{dc} is marginally changed and there is an increase in the magnitude of i_{sa} and I_{pv}. The MPPT responses of the PV system at irradiation value of 500 W/m² and 1000 W/m² are presented in Figs. 11 (b-c). It illustrates that the MPPT is attained very near to 100%, i.e. peak power extraction from PV array even at wavering solar irradiation.

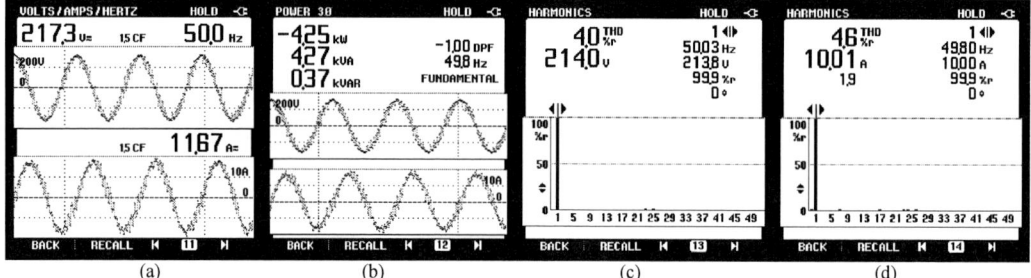

Fig. 7 Steady state system response at solar irradiation of 1000 W/m² (a)v_{sab} with i_{sa}(b) v_{sab}, i_{sa} (c-d) Harmonics distortion of phase 'b' (c) v_{sbc}(d) i_{sb}

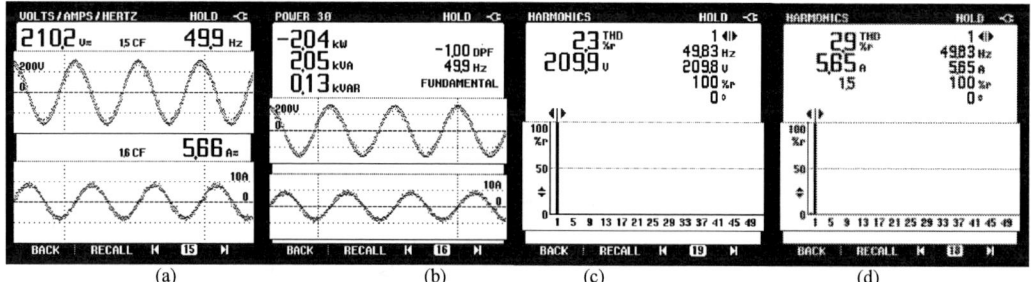

Fig. 8 Steady state system response at solar irradiation of 500 W/m² (a)v_{sab} with i_{sa}(b) v_{sab}, i_{sa} (c-d) Harmonics distortion of phase 'b' (c) v_{sbc} (d) i_{sb}

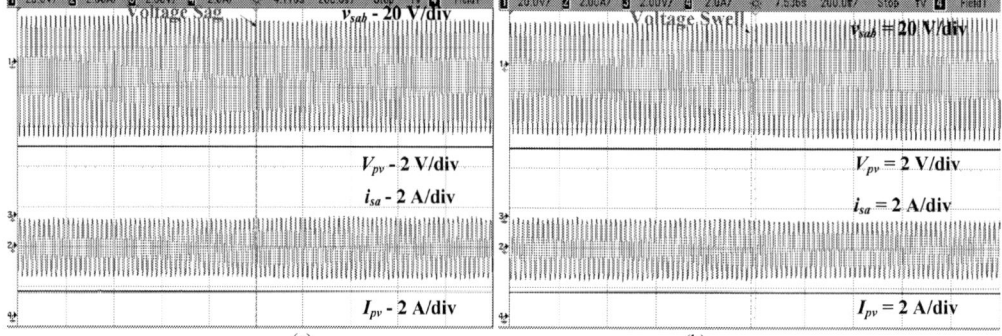

Fig. 9 Dynamic response at (a) voltage sag (b) voltage swell

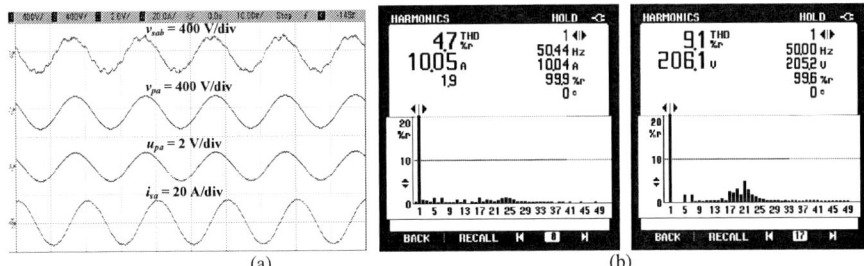

Fig. 10 System response during grid voltage distortions (a)v_{sab}, v_{pa}, u_{pa} and i_{sa} (b) Harmonic Spectra of i_{sa} (c)Harmonic Spectra of v_{sab}

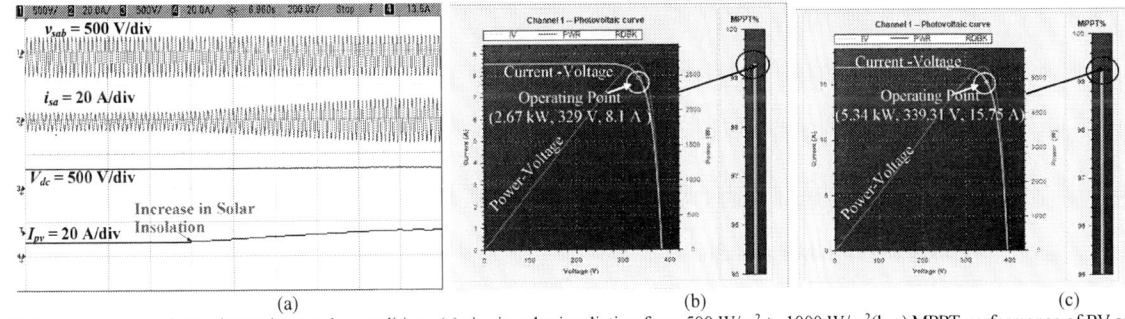

Fig.11 System response during intermittent solar conditions (a) rise in solar irradiation from 500 W/m² to 1000 W/m²(b-c) MPPT performance of PV array at 500 W/m² and 1000 W/m²

VI. CONCLUSION

A SFLAPI based PI controller and improved frequency locked loop algorithm, have been implemented on single stage three phase grid tied PV array system with an INC based MPPT algorithm. The SFLAPI based online gains tuning of PI controller has been presented to overcome the problems of PI controller with a fixed gains. Gain tuning of PI controller has been executed according to voltage variations. IFLL has extracted fundamental component of voltage efficiently during grid voltage disturbances. The response of the controller is robust and system performance is good in grid adverse situations. There is an enhancement in the power quality of the grid at insolation variation and also at grid adverse conditions of voltage swell/sag. Experimental results have demonstrated the system response to be satisfactory as recommended by the IEEE-519 and IEEE-1547 standards while delivering the PV power to the grid.

APPENDIX

A. *Simulation Data:* V_{MP} = 26.4 V; I_{MP} = 7.59 A; P_{MP} = 32.46 kW; N_p = 6; N_s = 27; V$_{dc}$ = 700 V; v_{sab} = 100 V (rms); L_f = 2.8 mH; T_s = 30µs; R_f = 5 Ω, C_f = 30 µF, k = 1.4; K' = -0.25; J = [0 -1; 1 0]; K_{pd} = 1 and K_{id} = 0.0001.

B. *System Parameters:* Solar simulator voltage, V_{MP} = 339.31 V; I_{MP} = 15.75 A; P_{MP} = 5.34 kW; V_{dc} = 339.31 V; T_s = 30 µs; V_L = 217 V (rms); ac inductor, L_f = 1.8mH.

ACKNOWLEDGMENT

The authors are highly thankful to MEITY and DST, Govt. of India, for supporting this project under Grant Number: RP03391.

REFERENCES

[1] B. Singh, A. Chandra and K. Al-Haddad, Power quality problems and mitigation techniques. John Wiley & Sons, 2015.

[2] X. Liang, "Emerging power quality challenges due to integration of renewable energy sources," *IEEE Transactions Industry Applications*, vol. 53, no. 2, pp. 855-866, March-April 2017.

[3] IEEE Recommended Practice and Requirement for Harmonic Control on Electric Power System, IEEE std. 519, 1992.

[4] IEEE Draft Recommended Practice for Establishing Methods and Procedures that Provide Supplemental Support for Implementation Strategies for Expanded Use of IEEE Standard 1547," IEEE P1547.8/D8, July 2014 , vol., no., pp.1-176, June 14 2014

[5] .W. Xiao, F. F. Edwin, G. Spagnuolo and J. Jatskevich, "Efficient Approaches for Modeling and Simulating Photovoltaic Power Systems," *IEEE Journal of Photovoltaics*, vol. 3, no. 1, pp. 500-508, Jan. 2013.

[6] M. A. Elgendy, B. Zahawi and D. J. Atkinson, "Assessment of Perturb and Observe MPPT Algorithm Implementation Techniques for PV Pumping Applications,"*IEEE Transactions Sustainable Energy*, vol. 3, no. 1, pp. 21-33, Jan. 2012.

[7] T. F. Wu, C. H. Chang, L. C. Lin and C. L. Kuo, "Power loss comparison of single- and two-stage grid-connected photovoltaic systems," *IEEE Transactions Energy Conversion*, vol. 26, no. 2, pp. 707-715, June 2011.

[8] S. Golestan, J. M. Guerrero and J. C. Vasquez, "Three-phase PLLs: A review of recent advances," *IEEE Transactions Power Electronics*, vol. 32, no. 3, pp. 1894-1907, March 2017.

[9] S. Golestan, J. M. Guerrero, A. M. Abusorrah and Y. Al-Turki, "Hybrid Synchronous/Stationary Reference-Frame-Filtering-Based PLL," *IEEE Transactions Industrial Electronics*, vol. 62, no. 8, pp. 5018-5022, Aug. 2015.

[10] N. J. Bershad, E. Eweda and J. C. M. Bermudez, "Stochastic analysis of the LMS and NLMS algorithms for cyclostationary white gaussian inputs," *IEEE Transactions Signal Processing*, vol. 62, no. 9, pp. 2238-2249, May1, 2014.

[11] Rahul Agarwal, Ikhlaq Hussain and Bhim Singh, "LMF based control algorithm for single stage three-phase grid integrated solar PV system," *IEEE Transactions Sustainable Energy*, vol. 7, no. 4, pp. 1379-1387, Oct. 2016.

[12] B. Singh, V. Jain and I. Hussain, "A FFP based adaptive control of SPV system tied to weak grid," *IEEE Transactions Industry Applications,* Early Access.

[13] B. Singh, V. Jain and I. Hussain,"A HTF based higher order adaptive control of single stage three phase grid integrated SPV system," *in Proc. India Int. Conf. on Power Electron.*, pp. 1-6, 2016.

[14] V. Jain and B. Singh, "A Multiple Improved Notch Filter Based Control for Single Stage PV System Tied to Weak Grid," *IEEE Transactions Sustainable Energy,* Early Access.

2nd IEEE International conference on power Electronics, Intelligent Control and Energy systems (ICPEICES-2018)

Internal Model Controller Design for Boost Converter by Stochastic Optimisation

Kelam Sudheer Kumar, Bharat Verma and Prabin Kumar Padhy

Department of Electronics and Communication,
Indian Institute of Information Technology Design and Manufacturing
Jabalpur, India

[1]`1612207@iiitdmj.ac.in`
[2]`bharatbigj@gmail.com`
[3]`prabin16@iiitdmj.ac.in`

Abstract— As the power electronics devices found in many constant voltage applications, control of these devices is essential for our day to day life. Of them the Boost converter control is a challenging task as it exhibits non-minimum phase behaviour when operated in continuous conduction mode (CCM). This paper presents a comparative analysis of controller design for Boost Converter with optimised Internal Model Control. The Internal Model Control only has single design variable, i.e., filter time constant (λ). The filter time constant is optimised with the help of the stochastic optimisation methods i.e., Genetic Algorithm and Particle Swarm Optimisation and performance of the both controllers are compared in terms of the transient performance parameters and Integral Square Error.

Keywords—DC-DC Boost converter, Non-minimum phase, IMC controller, Genetic Algorithms, Particle Swarm Optimisation.

I. INTRODUCTION

Power electronic devices play an important role in many constant voltage sources applications like aerospace, electronic gadgets communication etc., Among the power electronic devices Boost converter has found many applications in day to day life. Of all the different structures of DC-DC converters [1], Boost converters are the most economic ones having the standard efficiencies of about 70-95%. Boost converters are used for boosting up the input source (DC) voltage to higher output load (DC) voltage by varying the operating time of the switch in the converter. These are most commonly used in battery powered devices, in which the circuit needs more operating load voltage which is higher than the voltage supplied by the battery, e.g. Electronic gadgets, camera flashes & Electronic vehicles.

The typical open loop control of the DC-DC converter is having a lot of disadvantages when compared with the traditional closed-loop control which can easily regulate the line and voltage regulation. Closed loop control of the converters [2] helps in reducing the ripple content in the voltage or current profile of the converters. It is really crucial to know about the load variations occurring from time to time and countering them within a scheduled time is more required for the efficient converter. That's why the closed-loop control is more crucial one when compared with the open loop control.

Amid all the different converters, the controller design for the Boost converter is a troublesome as it exhibits non-minimum phase behavior. The boost converter is having an RHP Zero [3] in its transfer function model which shows a reverse response. Many kinds of literature have shown the problem of non-minimum phase behavior [4] in the Boost converter. Going with the conventional way of approach by designing and tuning of PID controller [5] [6], it does not reduce the effect of the RHP Zero in Boost converter and is also a difficult task.

In order to attain the robustness and better performance of the controller, the principle of Internal model control was introduced. It serves the many advantages in controlling the systems with uncertainties. Many control techniques like sliding mode control [7], Smith predictor control [8] also have used for improving the robustness and removing the effect of RHP Zero in the Boost converter. As they are having their own drawbacks in control of the converters the IMC control [9][10] approach is the best approach in improving the robustness and stability of the Boost converter.

In designing of the IMC Controller, the controller transfer function makes use of the low pass filter in order to become a proper function. All the parameters in designing of the controller are derived from the plant model except the low pass filter tuning parameter (λ) which must be tuned manually.

One must be more precise in the tuning of the low pass filter as the whole performance of the controller is dependent on the tuning parameter (λ). Here we have used the Particle Swarm (PSO) [11] based optimization method for tuning of the IMC Controller. There are several other optimization methods like Genetic Algorithms (GA) [12][13], Ant Colony Optimization (ACO) [14] which takes much time for the convergence. Differential evolution [15] is also one of the methods which are having the same problem of convergence. Artificial bee colony optimization [16] is having the advantages but is having the more computational cost. Among these optimization methods, PSO [17]–[19] is having more advantages in terms of convergence and accuracy in obtaining the optimal solution.

In this paper, firstly, we have designed the controller for the Boost converter using the IMC principle, as given in [20]. The filter time constant of designed Internal Model Control is then optimised, for the optimum Integral Square Error (ISE), with the help of Particle Swarm Optimization (PSO) and Genetic

978-1-5386-6624-1/18 $31.00 © 2018 IEEE

Algorithm, and their performance are compared on the basis of transient performance and Integral Square error.

II. DESIGNING OF IMC CONTROLLER

Internal model-based control is having the capability of controlling the systems with uncertainties in parameters. The systems with the time delays which can occur through the external conditions can also be easily compensated through this method for the proper functioning of the system.

The first systematic approach towards the designing of internal model control (IMC) [9][10] was first introduced by Daniel E. Rivera in 1986 and then later were advanced by M.Morari and C.E Garcia in 1989. Fundamentally, Internal Model Control principle formulates that the exact control can be obtained only if the control system recapitulates either implicitly or explicitly. This type of control method is having more advantages compared to conventional feedback controller in terms of internal stability and performance characteristics.

Although, the IMC design structure and open loop control design structure are equivalent to the implementation of IMC results in feedback control system. There are many advantages of Internal Model Control structure compared to the conventional feedback control structure.

- Designing of the internal model controller is relatively easy compared to the designing of the conventional feedback controller.
- The tuning parameters which are included in designing the IMC Controller are having the translucent relationship with the system performances like rise time, settling time, peak overshoot etc.
- The IMC control structure affords an accurate approach in the form of robust filter which can deal with the model plant mismatch and uncertainties in the system.
- Tuning of the robust filter is direct approach compared to the vapid procedures used in designing of the conventional feedback control systems for attaining the stability and robustness

A. Structure of Internal Model Control (IMC)

The internal model control structure comprises of the Plant model $G_p(s)$, process model $\tilde{G}_p(s)$, IMC controller $G_c(s)$ and the disturbance $d(s)$. In IMC scheme shown in the figure, the Internal Model Control loop calculates the difference between the outputs of the process and that of Internal Model.

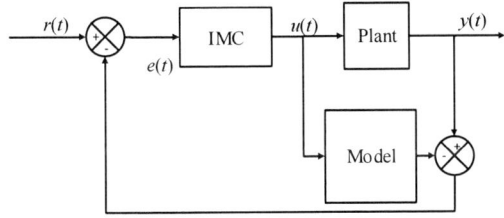

Figure 1: Basic structure of Internal Model Control

B. Designing of Internal Model Controller (IMC Controller)

In designing of the IMC Controller, the following steps have to be taken

- Firstly, the plant $G_p(s)$ has to be selected. Then the transfer function of the plant has to be determined.
- Now the process model $\tilde{G}_p(s)$ has to be selected.
- As the process model consists of RHP zeros, time delays the process model is factorized into non-minimum phase components and minimum phase components. The non-minimum phase part consists of RHP zeros, time delays etc. and the minimum phase part consists of LHP zeros or poles.
- For the perfect control, the product of $G_c(s)$ and $G_p(s)$ should be equal to one.

$$G_c(s) \times G_p(s) = 1 \qquad (1)$$

- From this, we can get that the transfer function of IMC Controller is the inverse of the process model

$$G_c(s) = \frac{1}{\tilde{G}^-(s)} \qquad (2)$$

The controller is then designed with the plant transfer function which is having invertible or minimum phase component and is cascaded with the low pass filter $F_r(s)$ to make the overall controller transfer function a proper.

$$C_1(s) = \frac{F(s)}{\tilde{G}^-(s)} \qquad (3)$$

where $F(s) = 1/(\lambda s + 1)^k$ here, k is the comparative order of the minimum phase component of the model and λ is the tuning constant.

TABLE I
PARAMETERS OF NON-IDEAL BOOST CONVERTER

Components	Values
Input Voltage (V_i)	10 V
Inductance (L)	3.1mH
Inductive resistance (R_l)	0.3Ω
Capacitance (C)	1930μF
Capacitive resistance (R_c)	0.08Ω
Load resistance(R)	90Ω
Duty ratio(D)	0.5
Output voltage (V_0)	20 V

By applying the state space averaging technique to the Boost converter, the transfer function of the Boost converter between the output load voltage to duty ratio is as follows [20],

$$\frac{\tilde{v}}{\tilde{d}} = \frac{V_0}{1-D} \frac{\left((1 + CR_c s)\left[R^2(1 - D)^2 - (R + R_c)\left(R_{eq} + Ls\right)\right]\right)}{Q} \qquad (4)$$

where,

978-1-5386-6624-1/18 $31.00 © 2018 IEEE   535

$$Q = R(1 - D)[(R(1 - D) + R_c(1 + C(R + R_c)s]$$
$$+ (R + R_c)(R_{eq} + Ls)(1 + C(R + R_c)s)$$

The following are the values of the components of the Non-Ideal Boost converter and the transfer function model are assumed from [4].

Substituting the values of the parameters in (4) we get the required transfer function as [20],

$$\frac{\tilde{v}}{\bar{d}} = \frac{22.0617(1.544 \times 10^{-4}s + 1)(-7.8287 \times 10^{-5}s + 1)}{1.3345 \times 10^{-5}s^2 + 1.8847 \times 10^{-3}s + 1} \quad (5)$$

This transfer function of the Boost converter is assumed as the plant transfer function in the IMC modeling procedure. By following the Internal model principle in designing the IMC Controller we got the final transfer function of the controller as follows,

$$\tilde{G}_P(s) = \frac{1.3345 \times 10^{-5}s^2 + 1.8847 \times 10^{-3}s + 1}{22.0617(1.544 \times 10^{-4}s + 1)(\lambda s + 1)} \quad (6)$$

This value of λ has to be manually tuned for getting the better output responses of the Boost converter. So, we go with the optimization methods for tuning of the lambda parameter.

III. Tuning of IMC Controller

In designing of IMC Controller, we have used the low pass filter in order to make the controller a stable one. As the low pass filter contains a tuning parameter (λ) it has to be properly tuned to get the desired and improved response of the system. Here we make use the optimization methods for tuning of the low pass filter. They are,

- Genetic Algorithms (GA)
- Particle Swarm Optimization (PSO)

These optimization methods make use of fitness function. In this paper, we have chosen as the integral square value of the error (ISE) function as,

$$\text{ISE} = \int_0^\infty e^2(t)dt. \quad (7)$$

A. Genetic Algorithms (GA)

Genetic Algorithms is the heuristic problem-solving method which was introduced in the mid-1970s by John Holland and his colleagues. Genetic Algorithms are based on the theory of biological evolution and principle of genetics and mimic the reproduction process in finding out the optimal solution for the system. This algorithm is based on the principle of "Survival of the fittest" which is employed in generating the individuals or solutions that are adopted to the environment.

Genetic algorithms use the following steps in generating the individuals or solutions and selecting them which are adaptive to the environment or the constraints of the problem.

1) Selection: In this, the individuals are selected based on the principle of the "Survival of the Fittest principle".

2) Cross over: This process mimics the process of biological mating in the production of population process. This process generates the biologically fittest solutions to the future population.

3) Mutation: This process helps in finding the global best solution in design space and helps not to trap in the local best and search for the best optimal individual or solution.

The optimization by GA includes a large population size which causes more number of generations which automatically increase the time of iterations for obtaining the better optimal solution.

B. Particle Swarm Optimisation (PSO)

Particle swarm optimization is one of the population-based heuristic optimization method introduced by Dr. Eberhart and Dr. Kennedy in the year 1995. Particle swarm optimization is inspired from the social behavior of bird flocking and fish schooling.to adapt to their environment in search of food and avoiding the predators by using the information sharing approach.

Particle swarm optimization also uses the following steps for finding the optimal solution to the problem.

1. Firstly, it generates the particles and their velocity and position in a random manner.

2. Then the velocity update of the particles.

3. Finally, the position update of the particles. The particles which are having the highest local best is assumed as the global best

In PSO the particle is defined as the point in the design space. The position of the particle in PSO is influenced by the best position of the neighbourhood. As discussed above the initial position of the particle is given as,

$$x_{k+1}^i = x_k^i + v_{k+1}^i \Delta t, \quad (8)$$

and the velocity of the particle is updated as follows,

$$v_{k+1}^i = w\, v_k^i + c_1 r_1 \left(P_{best}^i - x_k^i\right) + c_2 r_2 \left(G_{best}^i - x_k^i\right), \quad (9)$$

where, w is inertia weight factor, c_1 and c_2 is acceleration coefficients, r_1 and r_2 is random numbers between 0 to 1, P_{best}^i is particle best at i^{th} iteration, and G_{best}^i is global best of the group.

Once the velocity of the particle is calculated it is used in the updating the position of the particle. Now the position of each particle has calculated the particle which is having the highest best is updated as global best G_{best}^i. With P_{best}^i and G_{best}^i the velocity of the particle is again calculated and use in updating the position of the particle. In this way, the iterations are continued until we acquire the optimal position or solution to the system.

IV. Simulation Results

In this work firstly, the IMC controller has been modelled and the plant transfer function was derived from the state space technique. As the controller contains the tuning parameter (λ)., it was tuned using the optimization methods like Genetic Algorithms and Particle Swarm Optimization. These methods are simulated in the Matlab/Simulink and then the results are compared based on the performance characteristics.

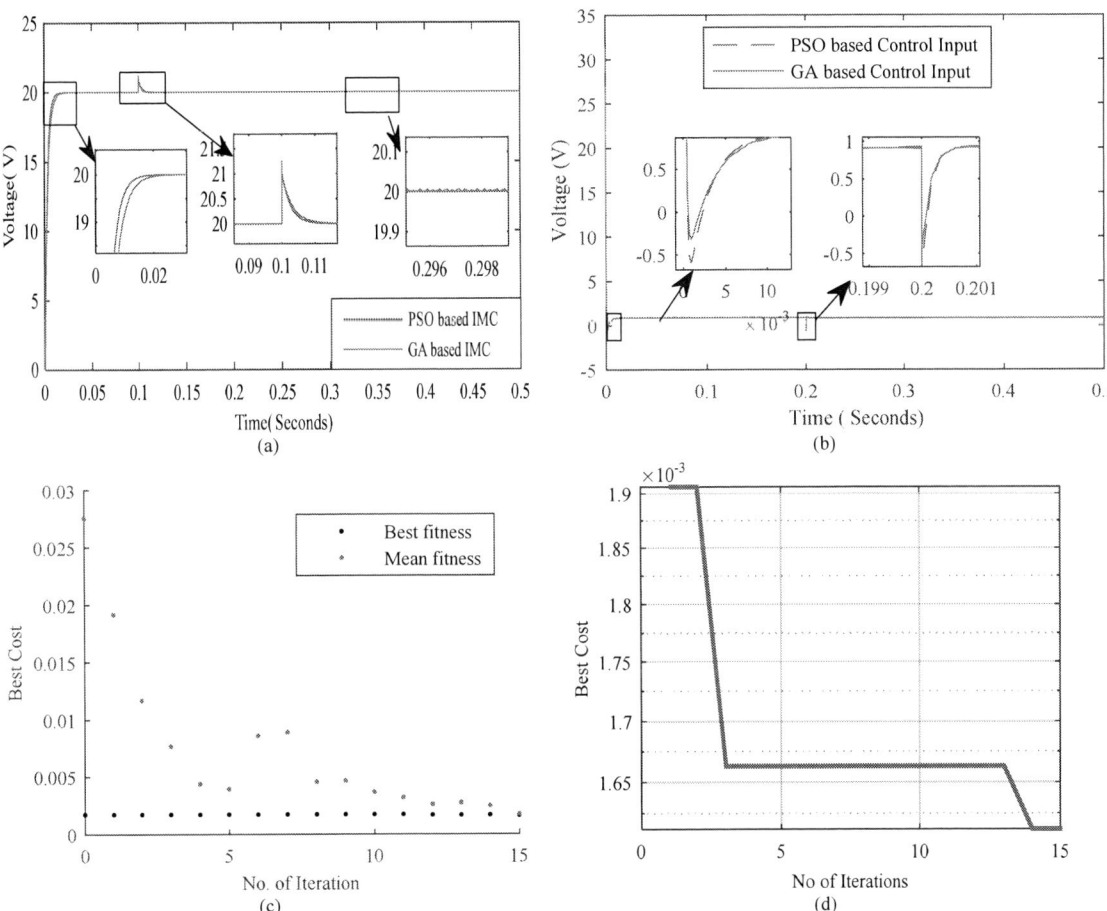

Figure 2: Simulation results of PSO and GA based IMC Controller for setpoint voltage of 20 V. a) Output Voltage b) Control input, c) Genetic algorithm Convergence, and d) PSO convergence.

From the above results shown in Figure.2a, the IMC Controller designed by the PSO algorithm is having less rise time and taking less time for settling compared with the GA. Overshoot and undershoot are also less compared with GA based IMC. The Integral Square Error (ISE) which is chosen as the performance index in the optimisation methods, PSO based controller is having the better performance compared to the GA based controller. The comparative analysis between the two stochastic methods are listed in the Table II.

To study the effect of disturbance in the boost converter, we have introduced a step disturbance of 1V at 0.1 sec in which PSO based controller is settling with the less time to the disturbance. Although the GA based IMC controller is having less overshoot shown in Figure.2b it is taking more time to get settled for the disturbance. GA based IMC controller is also having some ripple content in its final voltage profile which is unseen in PSO based IMC Controller.

In the Figure. 2d, we have compared the best cost function with the total number of iterations where we achieved $\lambda= 2.5$ ms

as the optimal solution in PSO. By GA from Figure.2c we have achieved $\lambda= 3$ms as the optimal solution. These results are compared with the performance characteristics.

TABLE II
COMPARATIVE RESULTS OF PSO AND GA FOR THE IMC CONTROLLER OF BOOST CONVERTER

Parameter	PSO ($\lambda=2.5$ ms)	GA ($\lambda=3$ ms)
Rise time (ms)	4.464 ms	5.144 ms
Settling time (ms)	16.624 ms	18.187 ms
Overshoot (%)	2.448%	2.535%
Undershoot (%)	2.018 %	2.376%
ISE	0.0016053	0.0016152

V. CONCLUSION

In this paper firstly, the IMC Controller for the DC-DC Boost converter have been modelled. The filter time constant λ is optimised with the help of stochastic optimisation methods i.e., Genetic algorithm and Particle Swarm Optimisation. From the results, we can conclude that PSO is having the better results compared to GA in terms of rise time and settling time. GA based controller is having a ripple content in its voltage profile which is clearly terminated in the PSO based IMC Controller. All the simulation results are carried out in the MATLAB/Simulink and the results are compared in terms of the system performances.

ACKNOLEDGEMENT

This publication is an outcome of the R&D work undertaken project under the Visvesvaraya PhD Scheme of Ministry of Electronics & Information Technology, Government of India, being implemented by Digital India Corporation.

REFERENCES

[1] S. ĆUK and R. D. Middlebrook, "Advances in Switched-Mode Power Conversion Part I," *IEEE Trans. Ind. Electron.*, vol. IE-30, no. 1, pp. 10–19, 1983.

[2] A. J. Forsyth and S. V. Mollov, "Modelling and control of DC-DC converters," *Power Eng. J.*, vol. 12, no. 5, pp. 229–236, 1998.

[3] S. Kapat, A. Patra, and S. Banerjee, "RHP zero elimination with near optimum regulation in a current controlled tri-state boost converter through inductor current filtering," in *Proceedings of the 2010 Power and Energy Conference at Illinois, PECI 2010*, 2010, pp. 83–87.

[4] T. Kobaku, S. C. Patwardhan, and V. Agarwal, "Experimental Evaluation of Internal Model Control Scheme on a DC-DC Boost Converter Exhibiting Nonminimum Phase Behavior," *IEEE Trans. Power Electron.*, vol. 32, no. 11, pp. 8880–8891, 2017.

[5] K. J. J. Åström and T. Hägglund, *PID controllers: theory, design and tuning*. International Society for Measurement and Control, 1995.

[6] M. I. Solihin, L. F. Tack, and M. L. Kean, "Tuning of PID Controller Using Particle Swarm Optimization (PSO)," *Proceeding Int. Conf. Adv. Sci. Eng. Inf. Technol.*, pp. 458–461, 2011.

[7] Y. B. Shtessel, A. S. I. Zinober, and I. A. Shkolnikov, "Sliding mode control of boost and buck-boost power converters using method of stable system centre," *Automatica*, vol. 39, no. 6, pp. 1061–1067, 2003.

[8] A. Visioli and Q. C. Zhong, "Smith-predictor-based control," in *Advances in Industrial Control*, no. 9780857290694, 2011, pp. 141–185.

[9] C. E. Garcia and M. Morari, "Internal model control. A unifying review and some new results," *Ind. Eng. Chem. Process Des. Dev.*, vol. 21, no. 2, pp. 308–323, Apr. 1982.

[10] C. E. Garcia and M. Morari, "Internal model control. 2. Design procedure for multivariable systems," *Ind. Eng. Chem. Process Des. Dev.*, vol. 24, no. 2, pp. 472–484, 1985.

[11] J. Kennedy and R. Eberhart, "Particle swarm optimization," in *IEEE International Conference on Neural Networks*, 1995, vol. 4, pp. 1942–1948.

[12] S. N. Sivanandam and S. N. Deepa, *Introduction to Genetic Algorithms*, no. Darwin. 2008.

[13] J. H. Holland, "Genetic Algorithms," *Sci. Am.*, vol. 267, no. 1, pp. 66–72, 1992.

[14] C. Blum, "Ant colony optimization: Introduction and recent trends," *Physics of Life Reviews*, vol. 2, no. 4. pp. 353–373, 2005.

[15] K. Fleetwood, "An Introduction to Differential Evolution," *New ideas Optim.*, pp. 79–108, 1999.

[16] D. Teodorović, "Bee Colony Optimization (BCO)," in *Optimization*, vol. 248, 2009, pp. 39–60.

[17] R. Hassan, B. Cohanim, and O. de Weck, "A comparison of particle swarm optimization and the genetic algorithm," in *1st AIAA multidisciplinary design optimization specialist conference*, 2005, pp. 1–13.

[18] B. Verma and P. K. Padhy, "Tuning of PID controller using sigmoidal weighted error function," in *2017 Innovations in Power and Advanced Computing Technologies (i-PACT)*, 2017, pp. 1–5.

[19] B. Verma and P. K. Padhy, "Optimal PID controller design with adjustable maximum sensitivity," *IET Control Theory Appl.*, vol. 12, no. 8, pp. 1156–1165, May 2018.

[20] S. Kelam, S. Sharma, and P. K. Padhy, "Voltage regulation of DC-DC Boost converter using Modified IMC controller," in *IEEE, 5th International Conference on Signal Processing and Integrated Networks*, 2018, pp. 1–6.

Output Voltage Control of DC-DC Boost Converter Using Model Predictive Control Approach

Kiran Vijay Khunte
Electrical and Electronics Engineering
National Institute of Technology
Karnataka,Surathkal
Surathkal, India
khuntekiran96@gmal.com

Sreedhar Madichetty
Electrical Engineering
Indian Institue of Technology,
Delhi Delhi, India
sreedhar.803@gmail.com

Sukumar Mishra
Electrical Engineering
Indian Institute of Technology, Delhi
Delhi,India
sukumariitdelhi@gmail.com

Abstract— This article contains an approach to control the output voltage of dc-dc boost converter using model predictive control algorithm. State space modeling of a boost converter is done for both continuous and discontinuous conduction mode. By implementing online optimization and constraint satisfaction it produces a control input which regulates the output terminal voltage to its given reference. Disturbances produced due to load variation at output terminal are taken care of by an observer which is implemented by using kalman state estimator. Simulations are carried out on Simulink-MATLAB platform to examine the merits of proposed strategy, which includes fast transient response and robustness.

Keywords— Model Predictive Control (MPC), Voltage control

I. INTRODUCTION

There has been a lot of development happened over the problems associated with dc – dc conversion power electronic topologies in past and their closed –loop controlled operation have been thoroughly analyzed and studied resulting in vast amount academic literature and industry application [1]. But still, with new requirements emerging from newer application dc-dc converters and the amount of progress done in the area control techniques and availability of cheap computational power nowadays have make it possible to use different control techniques to enhance the control operation of dc-dc converter. Taking the advantage of available options and providing another alternative for controlling dc-dc converters will only affect positive.

A. Motivation

Model Predictive Control schemes have been developed in late 70's and found wide application area in process industries as Dynamic Matrix Control (DMC) or Model

Algorithmic Control (MAC) [2]. The feature of MPC such as multivariable, non –square (no. of inputs not equal to no. of outputs) control system, process non-linearity, systematic and optimal handling are fascinating having the capability of application in wide range of area in electrical engineering. The difficulty of controlling of dc-dc boost converter due their switching characteristics, hybrid nature, non-minimum phase and very short time responses [1] needs the of attention of non-linear control strategies which needs to be implemented in real time. There have been reasonable effective control strategies available today but there are several challenges which have not meet yet. The objective of using MPC for controlling converter is not only about improving its closed-loop system performance but to provide a systematic design and implementation procedure [3].

B. Literature Review

Before looking into the literature available on MPC for dc-dc converters having clear understanding of some basic main components of MPC is must. A fundamental book on MPC by Liuping Wang [4] is very helpful. He has explained MPC for linear systems in easy manner in first three chapters and that is sufficient for one to understand the concepts used by authors of research paper on MPC for dc-dc converter. A. Pirooz and R. Noroozian in [5] used MPC for bidirectional dc-dc converter. Since, the state space models are different when switch is OFF and when is ON they define state equation for corresponding switch position for both buck and boost and use the basic philosophy of MPC predicting and optimizing is used to get the results. In [6] Andrea and Sebastien proposed the MPC with Extended Kalman Filter (EKF) for dc-dc converter, by having offline optimization they have reduced computational burden. In [7] Luca, Gionata suggest the use of MPC as an external loop for already installed dc-dc converter with linear controllers. In [8] Abu and Kang have proposed the combinatory optimization problem by simplifying the cost function for controller. J. Bonilla, R. D. keyser proposed the unconstrained non-linear MPC for dc-dc converter with EKF. Petros, Tobias and Stefanos in [3] proposed a MPC for dc-dc boost converter. They used basic philosophy of MPC

978-1-5386-6624-1/18 $31.00 © 2018 IEEE

to find out optimal control input with the help of hybrid model of converter.

C. Organization

The remaining section of the article is as follows: In Section-II we derive the discrete model of boost converter. Section-III defines the control objective for MPC. In Section-IV the all main components of MPC used in for controlling dc-dc converter have been explained. Section-V consists of all simulation results obtained on Simulink/MATLAB platform for designed controller. The conclusion has been drawn in Section VI.

II. BOOST CONVERTER MODEL

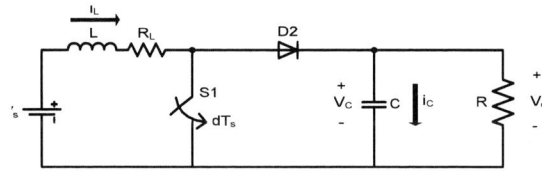

Fig. 1 Boost Converter

A. Continous- Time model

A dc-dc boost converter is a switching circuit (Fig. 1), which transforms the input (generally uncontrolled) dc voltage v_s to controlled high level output dc voltage v_o. The inductor L stores the energy which switch S is ON, and then push this energy to capacitor when switch S is OFF. We control this dynamics using a controlled switch which is a semiconductor device other than the diode D in the circuit. The load resistance R is attached at output terminal which receives the power at constant steady output voltage.

There can be three different modes of converter's dynamics corresponding to it switch and inductor current. These three different modes can be explained mathematically through linear state-space equation, which demonstrate the entire working of the boost converter.
1. When S=1 (ON) and $i_L(t)$ is increasing.
2. When S=0 (OFF) and $i_L(t)$ is decreasing.
3. When S=0 (OFF) and $i_L(t)$ is zero.

In first mode the topology of circuit is same as two different loops of *RLE* and *RC* circuit. In second mode the energy stored in inductor is supplied to output load and inductor current decreases, if this continues for longer time the inductor current reaches zeros and circuit become same as a single loop of capacitor and resistance.

The state-space representation of the converter can be given by following equations in continuous time domain [3]:

$$\frac{dx(t)}{dt} = \left(A_1 + A_2 * u(t)\right)x(t) + BV_s(t) \tag{1a}$$

$$y(t) = Cx(t) \tag{1b}$$

Where, $x(t) = [i_L(t) \; v_c(t)]^T$ contains inductor current and capacitor voltage as its state variables. $y(t) = v_o(t)$ is the output voltage.

$$A_1 = \begin{bmatrix} \dfrac{-d_{aux}R_L}{L} & \dfrac{-1}{L} \\ \dfrac{-d_{aux}}{C_o} & \dfrac{-d_{aux}}{RC_o} \end{bmatrix} \qquad A_2 = \begin{bmatrix} \dfrac{-1}{C_o} & 0 \\ 0 & \dfrac{1}{L} \end{bmatrix}$$

$$B = \begin{bmatrix} \dfrac{-d_{aux}}{L} & 0 \end{bmatrix}^T \quad \text{and} \quad C = \begin{bmatrix} 0 & 1 \end{bmatrix}.$$

The switch position is defined by value of u, which is either ON i.e. u = 1 or OFF i.e. u =0. The variable d_{aux} is used to present the conduction mode of the converter, which is an auxiliary binary variable.

$$d_{aux} = \begin{cases} 1 & when\; u = 0\; or\; u = 1\; with\; i_L(t) > 0 \\ 0 & when\; u = 0\; and\; i_L(t) = 0 \end{cases} \tag{2}$$

B. Discrete-Time model

MPC uses a mathematical model of the plant for the future prediction. This purpose will be served by a discrete-time model of the boost converter. Here we use a switch signal as a control input for the converter, since for each switch position there are different circuit topologies form from the converter circuit we use four different modes for the converter to demonstrate its full behavior [3].

$$x(k + 1) = \begin{cases} W_1 x(k) + E_1 v_s(k) & mode - 1 \\ W_2 x(k) + E_2 v_s(k) & mode - 2 \\ W_3 x(k) + E_3 v_s(k) & mode - 3 \\ W_4 x(k) + E_4 v_s(k) & mode - 4 \end{cases} \tag{3a}$$

$$y(k) = Vx(k) \tag{3b}$$

Modes are choose by position of switch and inductor current $i_L(t)$. When u = 0 and $i_L(k) > 0$ then mode-2 and if $i_L(k + 1) = 0$ with $i_L(k) > 0$ then mode-4. When u = 0 and $i_L(k) = 0$ then mode-3 and in other cases mode-1. Where, we assume $P_1 = A_1$ for $d_{aux} = 0$ and $P_2 = A_1$ for $d_{aux} = 1$, $P_3 = A_2$, Q = B for $d_{aux} = 1$. On the basis of continuous state space model (1), the state space matrices are obtained using forward Euler approximation approach.
$E_1 = E_2 = B.T_s.\; E_3 = [0 \quad 0]^T$
$W_1 = \mathbf{1} + (P_1 + P_2)T_s.$
$W_2 = \mathbf{1} + P_2 T_s.$
$W_3 = \mathbf{1} + P_1 T_s.$

Where, $\mathbf{1}$ is identity square matrix of dimension 2.

For mode-4, as we can look from the figure(2) the inherent incapability of determining sampling time for a signal changing its value from non-zero to zero value gives rise to this nonlinear equation as stated:

$$W_4 = (W_1\alpha_1 + W_2\alpha_2)/T_s \text{ and } E_4 = B\alpha_1.$$

978-1-5386-6624-1/18 $31.00 © 2018 IEEE   540

As, you can see the mode-4 is the average of mode-1 and mode-2 over sampling time. α_1 and α_2 are can be calculated using linear interpolation where α_1 is instant in between to sampling instant where inductor current $I_L(t)$ goes to zero and $\alpha_2 = T_s - \alpha_1$. The output matrix V = C.

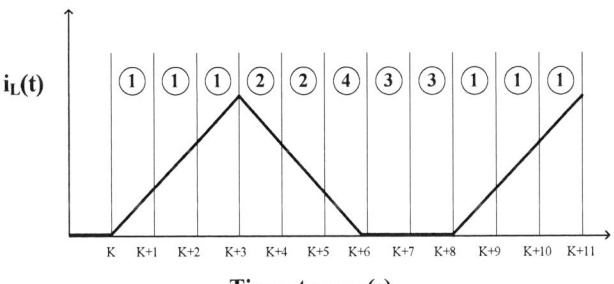

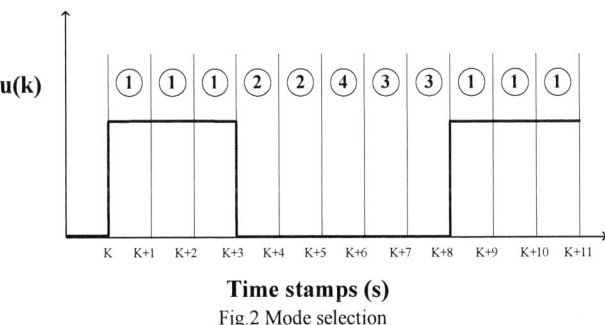

Fig.2 Mode selection

III. CONTROL OBJECTIVE

The dc-dc converter main objective is to track its reference voltage in spite of load variation on output terminals or change in reference or changes in its input voltage level. The required dynamics to be achieved by switching mosfet appropriately such as transient's should be as fast as possible with minimum overshoot.

IV. MODEL PREDICTIVE CONTROL

Here, we introduce the model predictive control approach to control the output voltage of dc-dc boost converter by directly controlling the switch position. The chosen state of switch is selected by minimizing the user defined objective function subjected to the constraints and model dynamics

The main components of control process are defined as below:

A. Prediction Horizon and Control Horizon

In MPC, we use the mathematical model of the plant to predict its future behavior using its current state variables. In this case dc-dc boost converter is our plant we have its discrete mathematical model (3). The presence of Right Half Plane Zeros (RHPZ) in transfer function of boost converter makes it nonminimum phase system. The nonminimum phase behavior of boost converter exhibits a negative gain relation of its switching action to the output at the start before going to its steady state value. The prediction horizon should be able to look beyond this dip in output voltage to

assist the correct decision of the controller. Here, we define the prediction horizon as N_p samples in future.

The control horizon can be defined as for how many next sampling instants N_c the control signals has been defined.

B. Receding Horizon

Every time we sampled the output we predict for next N_p samples in future and define our control action over the control horizon. We will only give the first control sample to the converter and then update our prediction horizon and control signal accounting for most recent measurement data which gives the receding horizon control law [4].

C. Modelling of Plant

The MPC used the mathematical model of converter for prediction, so converter model should be correct enough to give good prediction. The small modelling error will get modified by feedback loop.

D. Objective Function

The control action is taken in MPC by minimizing the objective function subjected to the given constraints. Generally, the objective function is called as cost function and denoted as J. The objective is related to an error function based on desired and actual response. Here, our cost function is defined as [3]

$$J(k) = \sum_{n=k}^{k+N_p-1} \left(\left| v_{o,err}(n+1|k) \right| + \lambda |\Delta u(n|k)| \right) \quad (4)$$

This takes into account of deviation of predicted output variable from desired value and also penalizes the change in the control input signal over the optimization window which is same as prediction horizon N_p here.

The first term takes in account of error in predicted output voltage from its reference value.

$$v_{o,err}(n+1|k) = v_{ref} - v_o(n+1|k) \quad (5)$$

By, penalizing the change in consecutive control input switching states, the second term tries to decrease the switching frequency [3].

$$\Delta u(n|k) = u(n) - u(n-1) \quad (6)$$

The weighting factor $\lambda > 0$, sets the trading factor between the output voltage error and change in switching frequency f_{sw}. Sampling interval T_s, hence, will always be same or greater than half of minimum of the switching period [3].

E. Optimization

In MPC, for given set-point signal $v_{ref}(k_i)$ at sampling time k_i, within prediction horizon the predictive control tries to bring predicted output near to the set point as much possible where we assume that the set-point signal remains constant.

Here, the optimization is carried out by minimizing the cost function J. The optimal control input sequence is chosen as following:

$$U^*(k) = \arg\min(J) \qquad (7)$$
Subject to (3).

Out of the taken input sequence we only apply the first sample to plant. The optimization procedure is repeated again at next sample $k + 1$. Due to this feedback is naturally incorporated in the control system design as per receding horizon principle.

The cost function minimization can be a difficult task depending upon the constraints applied, objective function itself and optimization window. Generally, the cost function is in quadratic form and constraints in linear form so to find optimum solution of problem becomes Non-linear Integer Programming (NMLP) for which finding a solution can be a tedious task. Here, the every possible input sequence is applied to the plant in prediction horizon and cost function is evaluated. The input sequence with minimum associated cost is chosen as optimal control input.

F. Input Blocking

As said earlier, due to the nonminimum phase behavior of dc-dc boost converter the prediction horizon should be as large as possible. The prediction horizon is same as N_pT_s, so by either increasing time stamp or increasing N_p we can achieve longer prediction horizon. But, time stamp should as small as possible to not to lost any significant dynamic of the plant and it may decrease the possible switching instants, since switching is possible at only sampling instants. Where, increasing the N_p will increase the computational power need exponentially.

The receding horizon policy allows us to give only first sample of control signal to the plant, so for predicting next input signal we need to know more about the dynamics in near future. Since, current input signal would not affect much the far in future dynamics due to receding horizon law. Here, we introduced the input blocking (IB) scheme [9]. Input blocking is one of the move blocking scheme and is standard practice to use it for optimization problems. In this scheme we will sample the near future dynamics of plant at sampling time T_s taking input constant over T_s and in far future of prediction horizon with N_sT_s sampling time in which input will be constant for longer period N_sT_s. Although this scheme does not provide any proof for stability or feasibility of the converter in future but its main objective of decreasing computational power need while having longer prediction horizon is fulfilled.

Taking N_p as prediction horizon, we sample first N_1 samples assuming control input constant for sampling time T_s and next N_2 samples assuming control input constant for with sampling time N_sT_s. So, the total prediction horizon length becomes $N_1T_s + N_2N_sT_s$ instead of N_pT_s.

In the fig (3), we can see advantages of using input blocking scheme. If we do not use the blocking scheme the prediction horizon length is not enough to look beyond the dip in output voltage. While in other case with the help of blocking scheme predictive model is able to look beyond the dip in output voltage and able to choose appropriate control input signal.

By applying input blocking scheme in other case the predictive model was able to look the rise in output voltage as expected and may produce optimal control input. If without the input blocking scheme we want predictive model to be able to look beyond the dip then we need to use $N_p = 8$, which increases the no. of possible input sequence by factor of 8 and need of computation power increases by 70%.

G. Load Uncertainties

The derived converter state space model assumes a constant load at its output terminals which is not the practical case. In most of the cases this load varies. This unknown variation in load can be considered as a disturbance in the state space model. We define two disturbance state variables i_d and v_d for inductor current and capacitor voltage respectively.

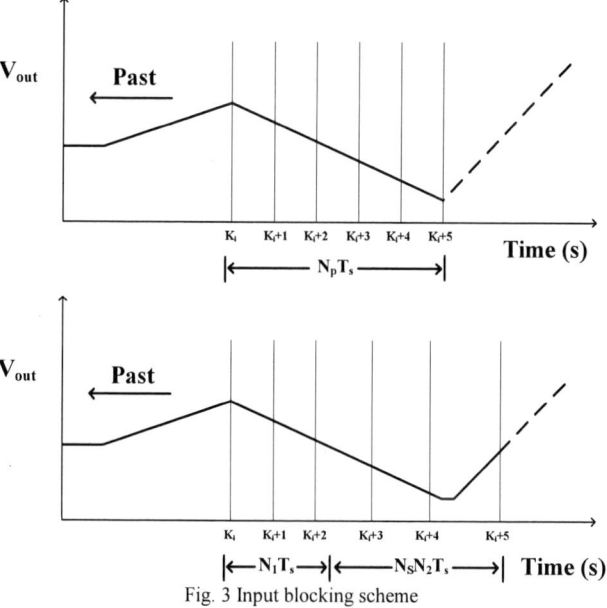

Fig. 3 Input blocking scheme

Disturbances drives the state variables away from its steady state values, so it's affect can be seen on state variables and it itself correspond as a state of the system. Our MPC controller do not have stochastic model of the converter which takes into account of measurement noise and process noise, so we will use an external loop to estimate a corrected current state of converter for the controller.

The state space equation of plant with noise is stated as:

$$x_a(k+1) = E_{ar}x_a(k) + F_{ar}V_s(k) + w(k) \qquad (8a)$$

$$y_a(k) = Gx_a(k) + v(k) \qquad (8b)$$

$x_a(k) = [i_L(k) \; v_o(k) \; i_d(k) \; v_d(k)]$, is augmented state model of converter. The augmented states are assumed to be slowly varying, almost constant.

$$i.e. \quad \begin{bmatrix} i_d(k+1) \\ v_d(k+1) \end{bmatrix} = \begin{bmatrix} i_d(k) \\ v_d(k) \end{bmatrix} \qquad (8c)$$

So,

$$E_{ar} = \begin{bmatrix} W_r & \mathbf{0} \\ \mathbf{0} & \mathbf{1} \end{bmatrix} \text{ and } F_{ar} = \begin{bmatrix} E_r \\ 0 \\ 0 \end{bmatrix}, G = \begin{bmatrix} \mathbf{1} & \mathbf{0} \\ \mathbf{0} & \mathbf{1} \end{bmatrix}$$

Where, $\mathbf{1}$ is identity square matrix of dimension 2 and $\mathbf{0}$ is zero square matrix of dimension 2 and the output variable is equal to sum of disturbance and state value. $w \in \mathbb{R}^n$, $v \in \mathbb{R}^q$ represent unmeasured disturbance (state noise) and measurement noise with n = no. of states, q = no. of outputs, respectively [2].

Here, $w(k)$ and $v(k)$ are assumed to be zero mean white Gaussian noise such that

$$Q = E[w(k)w(k)^T] \text{ and } R = E[v(k)v(k)^T].$$

A discrete Kalman filter is designed based on the augmented model of converter to be used as current state observer [10].

$$\widehat{x_p}(k) = E_{ar}\hat{x}(k-1) + F_{ar}v_s(k-1) \qquad (9)$$

$$\widehat{x_a}(k) = \widehat{x_p}(k) + K_r[y(k) - G\widehat{x_p}(k)] \qquad (10)$$

Where, $\widehat{x_p}(k)$ is called as priori estimate and $\hat{x}(k)$ as corrected current estimate. Four different Kalman gain K needs to be calculated, since there can be four different updates depending upon the switch position and inductor current value.

The noise covariance matrices Q and R need to be tuned according to your assumption of how much credibility you want to give to your measurements and your predicted values from mathematical model. The Kalman filter gives us the estimated values of disturbances. We can use these values to remove their influence from the output voltage.
So, the disturbance state $\widehat{v_d}$ is used to adjust the output voltage reference [3].

$$\widetilde{v_{ref}} = v_{ref} - \widehat{v_d} \qquad (11)$$

H. Control Algorithm

The proposed control algorithm for dc-dc boost converter can be termed as Finite Control-Set MPC [7]. At each iteration the predicted state variables are calculated over the prediction horizon and optimal input sequence is chosen corresponding to minimum deviation from set-point trajectory. The MPC control algorithm is explained in Fig.(5) and control scheme in block diagram shown in Fig.(4).

The control algorithm evaluates the predication horizon of $x(k)$ for every possible input sequence and check for the optimum solution simultaneously. The variable E makes sure that no redundant inner loop happens if its associated cost already increased than the one available. After all iteration we give the first sample of our optimum control sequence to the model and do all procedure once again for next control signal.

The MPC controller is consisting of the controller block and the observer. We sampled the current measurement of physical variable i_L and v_c and feed it to a current observer. The current observer which is implemented using Kalman state estimator give us the estimated augmented state variable vector which we can use in MPC control algorithm instead of direct measurements.

V. SIMULATION RESULTS

Here, we carried out the simulation of proposed control strategy on Simulink/MATLAB platform. The circuit parameters are chosen as L = 150µH, Co = 150µF, R_L = 3mΩ, R = 50Ω and v_s = 20V. The controller parameters are as follows N_p = 14, N_1 = 10, λ = 0.01, N_s = 6 and T_s= 15µS.

The input blocking scheme is used, so the last 4 samples are predicated using 90µS as sampling interval. We get prediction horizon of total 510µS. The noise covariance matrices are tuned so to match our expectation, $Q = diag([0.01 \quad 0.01 \quad 5000 \quad 5000])$ and $R = diag([0.1 \quad 0.1])$.

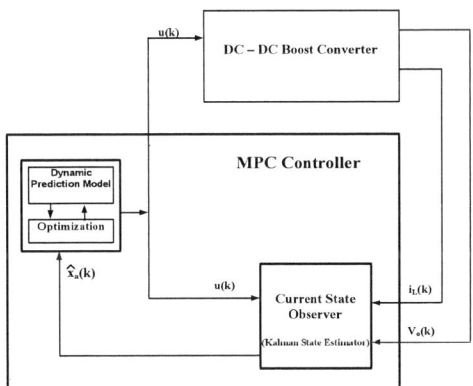

Fig. 4 Control architecture

978-1-5386-6624-1/18 $31.00 © 2018 IEEE

MPC Algorithm

```
function  u(kᵢ) = fcn(x(kᵢ), u(kᵢ-1))
Jₒ(kᵢ) = ∞, uₒ(kᵢ) = ∅, x(kᵢ) = x(kᵢ), E = ∞
        for all U over Nₚ  do
            J = 0
            for h = k to k + Nₚ -1
                if h <= N₁ then
                    x(h+1) = f1(x(h),u(h))
                else
                    x(h+1) = f2(x(h),u(h))
                end if
                v_{o,err}(h+1) = v_ref − v_o(h+1)
                Δu(h) = u(h) - u(h-1)
                J = J + abs(v_{o,err}(h+1)) + λ*abs(Δu(h))
                if E<J
                    break
                end if
            end for
            E = J
            if J<Jₒ(kᵢ)
                Jₒ(kᵢ) = J
                uₒ(kᵢ) = U(1)
            end if
        end for

end function
```

Fig. 5. MPC controller algorithm

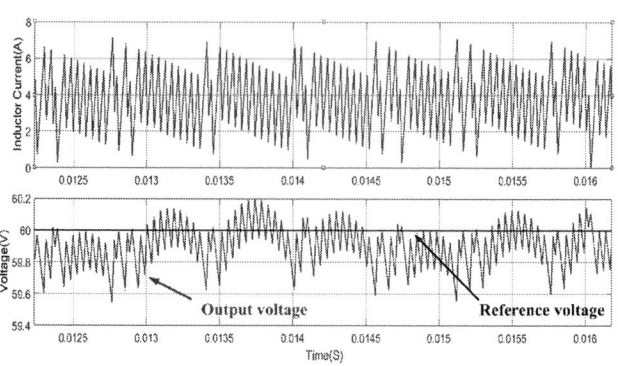

Fig. 7 Steady state

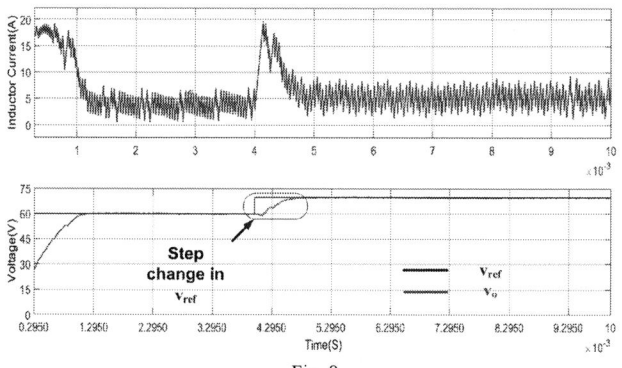

Fig. 8

A. Nominal Start-Up

For $v_{ref} = 60V$, simulation is carried out. There is no change in parameters while simulation is running. The resulting v_o and i_L are shown in Fig. 6. In 1.2ms v_o reached the steady state value without any overshoot. Fig. 7 shows the steady state waveform, which shows v_o ripple in between 59.5V to 60.2V voltage levels.

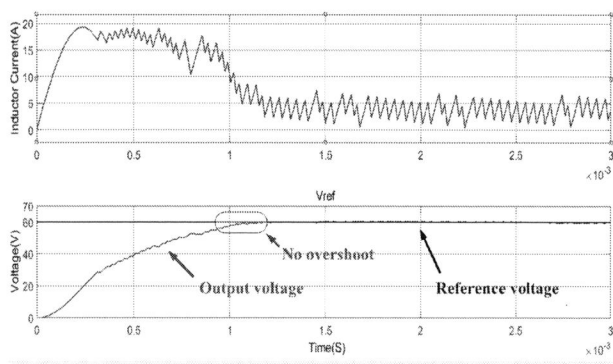

Fig. 6 Transients of v_o

B. Step-up v_{ref}

At t = 4 ms, v_{ref} changes from 60V to 72V. From Fig. 8, we can see that inductor rise up at that instant and then again goes to new steady state value once v_o settled at new reference voltage. It takes only 1ms for v_o to settle at new value.

C. Change in v_s

In this case we change v_s from its nominal value 20V to 16V at t = 4.25 ms and again to 19V at t = 15.7 ms and came back to its nominal value at $t = 2.75\ ms$. From the Fig. 9, we can see v_o track its steady state value as fast as possible despite change in its input voltage.

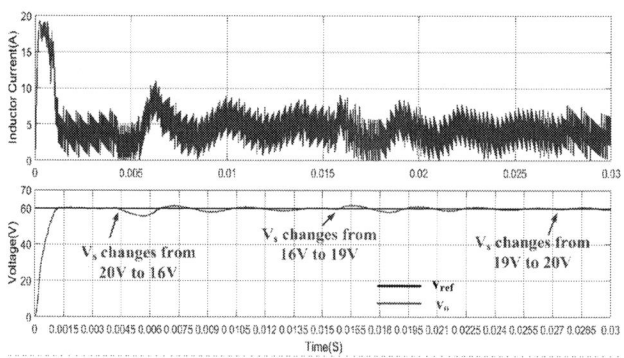

Fig. 9 Change in v_s

D. Change in load

At t = 5 ms, the load changes from 50Ω to 25Ω. The inductor current rises to keep the output voltage across the load same as before and then set at new value and then load changes from 25Ω to 100Ω at $t = 15\ ms$. When load changes, the v_o drop to 58V and 55V for first and second

978-1-5386-6624-1/18 $31.00 © 2018 IEEE

instant respectively, but soon come back to close proximity of its reference value.

Fig.11 shows the waveform of current and voltage over larger period of time, which shows output voltage hasn't deviated much from its reference values despite change in load resistance.

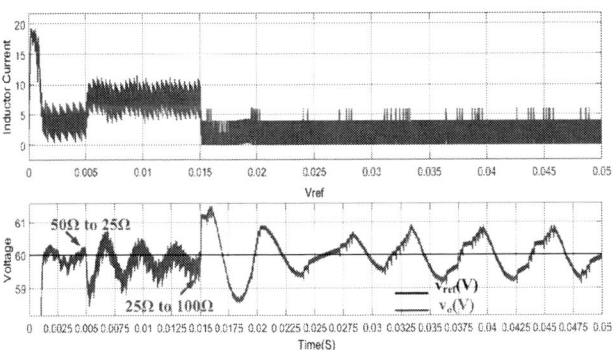

Fig. 10

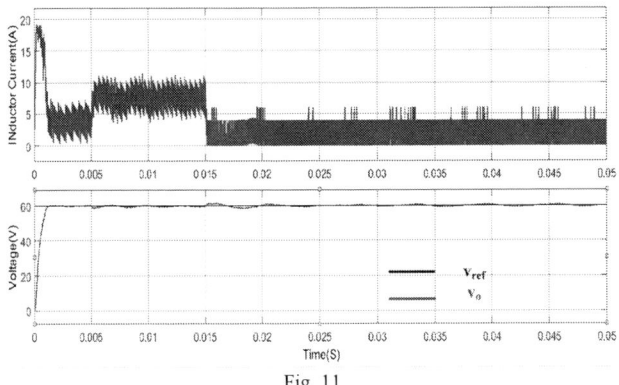

Fig. 11

VI. CONCLUSION

Here we have proposed MPC strategy for controlling output voltage of dc-dc boost converter with state estimator. The proposed control algorithm has any merits such fast response time, very little deviation from reference value despite change in other parameters of model. The proposed strategy is very easy to understand and gives a systematic approach for designing a controller. There is not requirement of tedious work of tuning the gains every time to get expected results, once set the model is capable of working over the wide range of values.

The drawback of need of very high computational power demand can be deal through by using different optimizing techniques in the controller. The stability check is also one of the factors we should look upon in future scope of this article. There are many other features of MPC which hasn't been explored here. In future there is chance of making a more robust algorithm and with maximum utilization of other aspects of MPC.

REFERENCES

[1] J. Bonila, R.D. Keyser, M.Diehl and J. Espinosa "Fast NMPC of a DC-DC converter: An exact Newton real-time iteration approach"

[2] Sachin C. Patwardhan "A Gentle Introduction to Model Predictive Control (MPC) Formulations based on Discrete Linear State Space Models"

[3] P. Karamanakos, T. Geyer and S. Manias, "Direct voltage control of DC-DC boost converters using model predictive control based on enumeration," *2012 15th International Power Electronics and Motion Control Conference (EPE/PEMC)*, Novi Sad, 2012, pp. DS2c.10-1-DS2c.10-8.

[4] Liuping Wang, "Model Predictive Control System Design and Implementation Using MATLAB®".

[5] A. Pirooz and R. Noroozian, "Model predictive control of classic bidirectional DC-DC converter for battery applications," *2016 7th Power Electronics and Drive Systems Technologies Conference (PEDSTC)*, Tehran, 2016, pp. 517-522

[6] A. G. Beccuti, S. Mariethoz, S. Cliquennois, S. Wang and M. Morari, "Explicit Model Predictive Control of DC–DC Switched-Mode Power Supplies With Extended Kalman Filtering," in *IEEE Transactions on Industrial Electronics*, vol. 56, no. 6, pp. 1864-1874, June 2009

[7] L. Cavanini, G. Cimini, G. Ippoliti and A. Bemporad, "Model predictive control for pre-compensated voltage mode controlled DC–DC converters," in *IET Control Theory & Applications*, vol. 11, no. 15, pp. 2514-2520, 10 13 2017

[8] A. Z. Ahmad and K. Z. Liu, "A new model predictive control approach to DC-DC converters based on combinatory optimization," *2008 34th Annual Conference of IEEE Industrial Electronics*, Orlando, FL, 2008, pp. 460-465.

[9] R. Cagienard, P. Grieder, E. C. Kerrigan and M. Morari, "Move blocking strategies in receding horizon control," *2004 43rd IEEE Conference on Decision and Control (CDC) (IEEE Cat. No.04CH37601)*, 2004, pp. 2023-2028 Vol.2.

[10] Piotr Tatjewski, "Disturbance modeling and state estimation for offset-free predictive control with state-space process model.

2nd IEEE International conference onn p power Electronics, Intelligent Control and Energy syysstems (ICPEICES-2018)

Design of an efficient Solar PV Battery Charge Controller under dynamic Motor load conditions

Sourish Ganguly
Dept. of Electrical Engineering
Institute of Engineering & Management
Kolkata, India
sourishganguly96@gmail.com

Subhrasish Pal
Dept. of Electrical Engineering
Institute of Engineering & Management
Kolkata, India
subhrasish1995@gmail.com

Imran Khan
Dept. of Electrical Engineering
Institute of Engineering & Management
Kolkata, India
khan.imran129@gmail.com

Debabrata Das
Dept. of Electrical Engineering
Institute of Engineering & Management
Kolkata, India
debabrata3130@gmail.com

Ahana Ghosh
Dept. of Electrical Engineering
Institute of Engineering & Management
Kolkata, India
ahanaghosh.16@gmail.com

Ankur Bhattacharjee
Dept. of Electrical Engineering
Institute of Engineering & Management
Kolkata, India
ankur.bhattacharjee@iemcal.com

Abstract—**This paper aims to develop an efficient solar charge controller which supplies power to both a DC motor and a rechargeable battery in an electric vehicle. The battery acts as a secondary source of power to drive the DC motor in case the PV module fails to provide sufficient power. The charge controller charges the battery at two modes: Constant Current (CC) charging mode and Maximum Power Point Tracking (MPPT) mode to efficiently charge the battery under dynamic load conditions. The modes of charging are decided on the basis of continuous monitoring of battery voltage. In CC charging mode, battery is charged at C/10 rate while in MPPT mode the charge controller aims to extract maximum power from the PV module under the impact of varying irradiance and varying load conditions. Model performance evaluation and analysis has been done through MATLAB/Simulink.**

Keywords—*Solar PV, Maximum Power Point Tracking (MPPT), Charge Controller, Buck Converter, Variable load.*

I. INTRODUCTION

The heavy dependence on limited energy sources, such as coal for production of electricity, has resulted in their rapid depletion, and has also affected the environment by increasing pollution levels. So, we need to shift our focus to renewable forms of energy which are pollution free and everlasting. Solar energy is the most promising alternative because of its ease of installation, less maintenance and its availability throughout the globe. Due to its poor energy conversion efficiency, it becomes necessary to operate the solar module at its maximum power point (MPP) region of its P-V characteristics. From literature, various designs of solar charge controllers and their analysis [3]-[9] have been proposed using popular MPPT algorithms, such as Incremental Conductance, Perturb & Observe etc. for load management and battery charging. In our work, a two-stage charging algorithm (Constant Current charging and MPPT) for a battery management system has been developed, intended for protection and longevity of a rechargeable battery. For our system, we have selected a lead-acid battery model. Additionally, we have analyzed the power of the solar PV module under variable DC motor load to verify the working of the MPPT algorithm at changing conditions. P&O algorithm has been used in our system for implementing MPPT, due to its simplicity and less computational demands. Also, no prior knowledge of the PV system is necessary for the algorithm to work. Rest of the paper contains the following sections; **Section II** contains the block diagram representation of the overall system. **Section III** describes the modeling and simulation of individual subsystem of the model. **Section IV** describes the

MPPT and CC charging algorithms, **Section V** discusses the simulation results. **Section VI** includes the conclusion of the work.

II. OVERALL SYSTEM DESCRIPTION

The proposed system, as shown in Fig. 1, consists of a photovoltaic module, a DC-DC Buck Converter, a controller for implementing the charging algorithms and providing input to the PWM Generator, a switching circuit, a 12V lead-acid battery, and a DC Motor as load. The closed-loop system with P&O algorithm automates the process of maximum electrical power transfer to the battery and the load from the solar PV-module, under dynamic load conditions. The switching circuit is meant for protection of the battery from overcharging and deep discharging, depicted by blocks 1 and 2 in Fig. 1. At highly discharged state, the Constant Current charging algorithm is implemented to ensure C/10 rate of charging, which effectively elongates battery-life. The load, used here, is a DC motor, to represent the actuator of an electric vehicle (EV). If the power from the solar-PV module is greater than the load requirement, then the PV module supplies power to the load, and the battery is charged automatically through appropriate algorithms at different charging stages of the battery. If the PV module power is equal to the load requirement, then, the entire power is transferred to the load, and the battery has no contribution to the system. On the other hand, if the solar power is less than the load requirements, then the battery and the PV module collectively supply the load, provided that the battery has not yet over-discharged. Lastly, if the battery is on the verge of over-discharge, the switching circuit connected to block-2 opens, and further discharge is disallowed. To prevent overcharging, the switch connected to block-1 opens, and charging is interrupted.

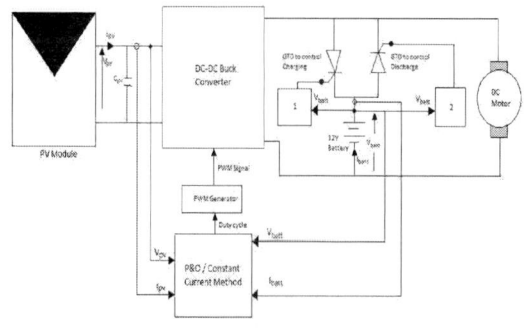

Fig. 1. Proposed System Topology.

978-1-5386-6624-1/18 $31.00 © 2018 IEEE

III. MODELLING AND SIMULATION OF VARIOUS COMPONENTS OF THE SYSTEM

A. Modeling of PV module

A solar cell is fundamentally a p-n junction diode which converts solar energy into electrical energy. Photons present in sunlight dislodge electrons and make them flow as DC current in the circuit connected to it. Single diode model of solar cell is shown below:

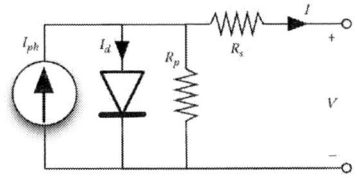

Fig. 2. Single Diode Electrical equivalent model of a solar cell [1].

The amount of current generated by the solar cell depends on its constructional features, ambient temperature and solar irradiance. This is in accordance with the following equation:

$$I_{ph} = \frac{I_{sc} \times S}{1000} + (T - T_i) \times K_i \qquad (1)$$

where, I_{ph} is the current generated by the solar cell
I_{sc} is the short circuit current of the PV module at S.T.C($S=1000$ W/m^2 and T=25°C)
S is the solar irradiance
T is the temperature of ambience
T_i is the reference temperature
K_i is temperature coefficient of solar cell

TABLE 1:SPECIFICATIONS OF PV MODULE

Parameter Name	Parameter Value
Module	SOLARSOVA 200P
Maximum Power	199.23 W
Cells per module	60
Open circuit voltage (V_{OC})	36V
Short circuit current (I_{sc})	7.67 A
Current at maximum power point (I_{mp})	6.87 A
Voltage at maximum power point (V_{mp})	29 V

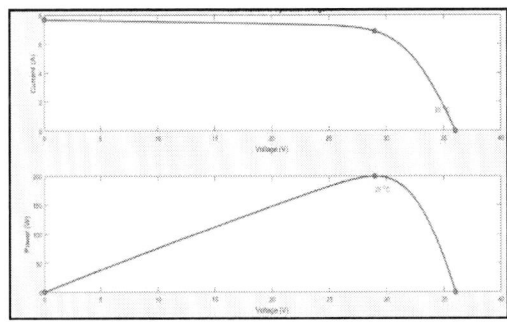

Fig. 3. I-V and P-V characteristics of the PV module used.

B. Buck Converter

A buck converter is a DC-DC step down converter. The operation of the converter is controlled by two switches mainly a transistor and diode. The freewheeling diode maintains the output current constant and the LC filter eliminates the harmonics.

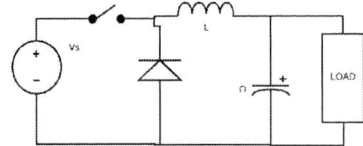

Fig. 4. Equivalent circuit of DC-DC BUCK converter.

When the switch is closed, the current flows from the supply to the load via the LC filter. The freewheeling diode is reverse biased. When the switch is opened, the stored energy in the inductor maintains the load current which freewheels through the diode.

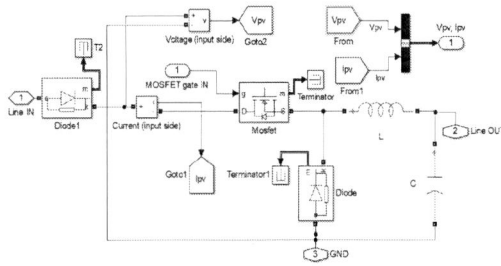

Fig. 5. Developed Simulink Model of DC-DC Buck Converter.

C. DC Motor

In order to test the system under varying load conditions, a permanent magnet dc motor model has been used as a load in this system. The induced e.m.f. generated in a dc machine is given by the following equation:

$$E = \frac{P\Phi ZN}{60A} \qquad (2)$$

where, E is the back e.m.f. of motor,
P is the no. of poles,
Φ is the flux per pole,
Z is the number of conductors,
N is the speed in r.p.m.,
A is the number of parallel paths.

Since in a permanent magnet DC motor, Φ, Z, P, A can be assumed to be constant, E can be written as follows:

$$E = KN \qquad (3)$$

where, $K = \frac{P\Phi Z}{60A}$

In the Simulink model, a permanent magnet DC motor with the following specifications has been used.

TABLE 2: SPECIFICATIONS OF DC MOTOR MODEL

Parameter Name	Parameter Value
Mechanical input	Torque TL

Field type	Permanent magnet
Armature resistance	2 Ohms
Armature inductance	0.1 H
Back E.M.F constant	0.012 V/rpm
Total inertia (J)	1 kg-m^2

D. Battery

A battery is a collection of electrochemical cells connected in series or parallel fashion. A lead-acid battery has been used to act as a secondary source of Power in case PV module fails to provide enough power output to drive the load.

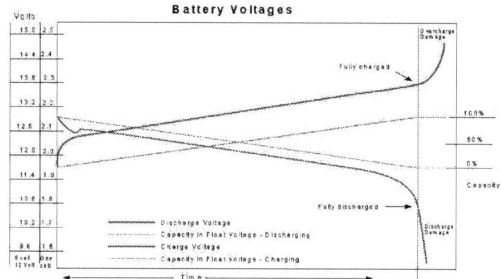

Fig. 6. Typical Charging Discharging graphs for a Lead acid battery [13].

TABLE 3: SPECIFICATIONS OF BATTERY MODEL

Parameter Name	Parameter Value
Nominal capacity	35 Ah
Nominal Voltage	12 V
Internal resistance	0.0034286 Ω

IV. CHARGING ALGORITHMS

In this paper a two-stage charging process is implemented: constant current (CC) and MPPT charging. When the battery voltage is less than 11.5V, CC charging is applied and when the voltage crosses the preset limit of 11.5V, MPPT mode comes into play. The flowchart for algorithm followed is shown below:

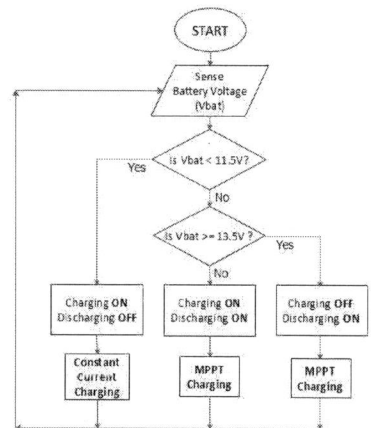

Fig. 7. Flowchart of Charging Algorithm followed.

Charging process is stopped at 13.5V to avoid overcharging of battery. While below 11.5V battery is not allowed to discharge to prevent low level discharge of battery.

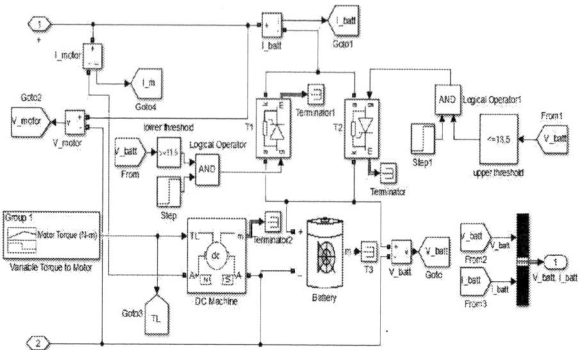

Fig. 8. Switching Circuit along with battery and DC motor.

A. Maximum Power Point Tracking (MPPT)

While charging the battery, MPPT algorithm is used in the linear region of the battery characteristic curve i.e. within 11.5V to 13.5V as shown in Fig. 6. Above 13.5V battery is disconnected but the PV module is still operated at MPPT to drive the DC motor. The MPPT algorithm implemented in this system is a standard P&O algorithm, as shown in the flow chart in Fig. 9. Fig. 10 shows the proposed model to implement the algorithm. The voltage perturbation is applied by changing the duty cycle of the DC-DC converter switch and the change in power is compared with the previous one in each step. If an increase in power is obtained then the direction of perturbation is kept unchanged. However, if the power decreases then the direction of perturbation is reversed in order to track the point of maximum power. These steps are repeated until no increase in power is possible.

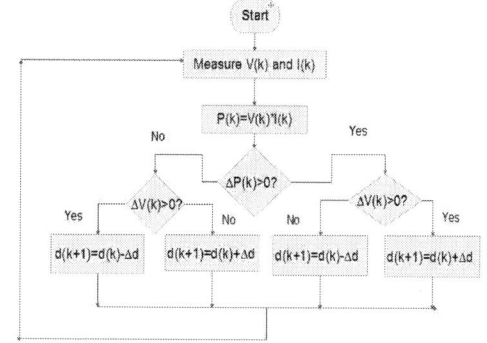

Fig. 9. Flowchart of P&O MPPT tracking algorithm.

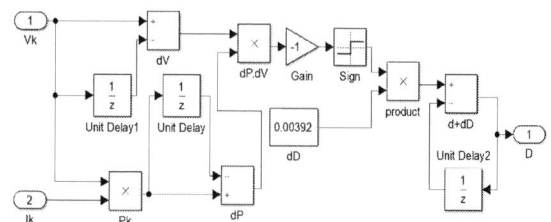

Fig. 10. Simulink Model of P&O algorithm.

B. Constant Current Charging

Constant current charging method charges the battery at C/10 rate, which is 10% of the nominal battery capacity. This mode is suitable for maintaining a relatively healthy battery life. In this paper, constant current charging is implemented as shown in Fig. 11 and Fig. 12.

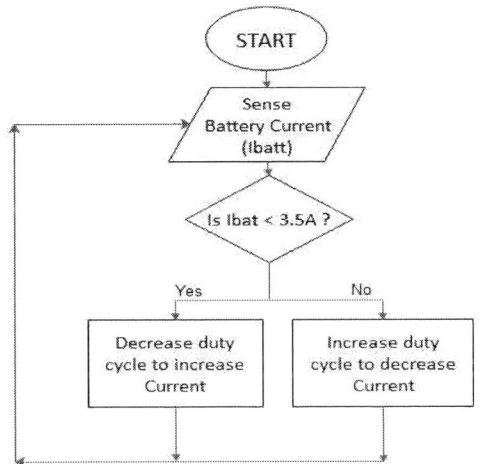

Fig. 11. Flow chart of Constant Current (CC) charging algorithm.

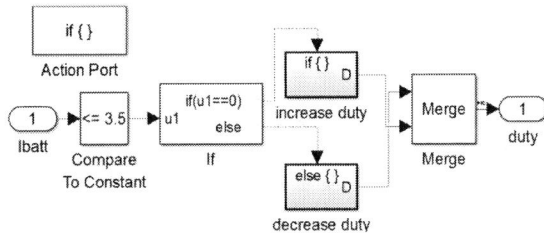

Fig. 12. Simulink Model of Constant Current (CC) charging.

In the model, battery current is measured and its value is compared with a constant value which is the C/10 rate of the battery. If the value of battery current is found to be less than the C/10 rate then the duty cycle is decreased to increase the current. Alternatively, if the battery current is more than the C/10 rate, then the reverse is followed. Thus, constant current is obtained.

V. RESULTS OBTAINED AND DISCUSSION

The resultant proposed SIMULINK model is implemented, as shown in Fig. 13, to implement the system topology shown in Fig. 1. As observed from the figure below, the solar module block is fed with an irradiance profile, and a value for the operating temperature of the model. To serve the purpose of dynamic irradiance values, a function generator block has been employed. As shown in Fig. 8, another function generator block has been implemented to feed the DC motor model with variable load torque. In the results obtained in Fig. 15 and Fig. 17, the system performances due to variable irradiance and dynamic torque have been evaluated by utilizing the above blocks.

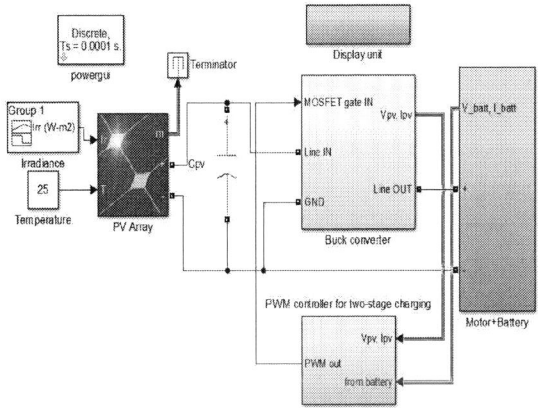

Fig. 13. MATLAB/Simulink model of the proposed overall system.

A. Shift from Constant Current charging to MPPT mode

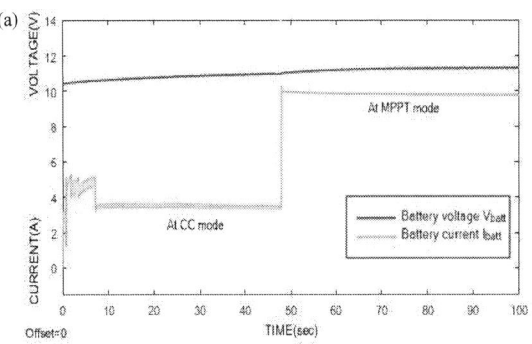

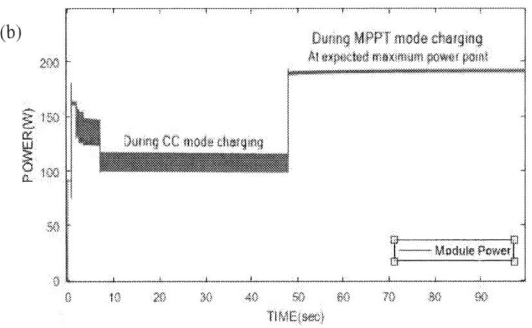

Fig. 14. (a) Current and Voltage characteristics during shift from CC charging to MPPT mode (b) Power output from PV module during the two modes.

Fig. 14(a) shows the battery current (I_{batt}) initially following the constant-current mode (at 3.5A). As the battery voltage (V_{batt}) crosses the threshold value (11.5V), the controller shifts to the MPPT mode. Fig. 14(b) shows the corresponding PV module-power during constant-current charging and at MPPT mode, which validates the working of the controller algorithms.

B. System Performance under varying irradiance

The peak power from the PV module changes under varying irradiance.

978-1-5386-6624-1/18 $31.00 © 2018 IEEE

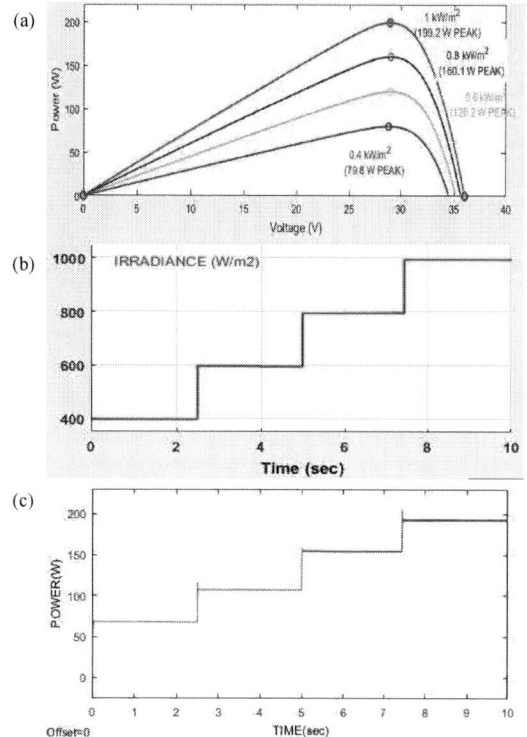

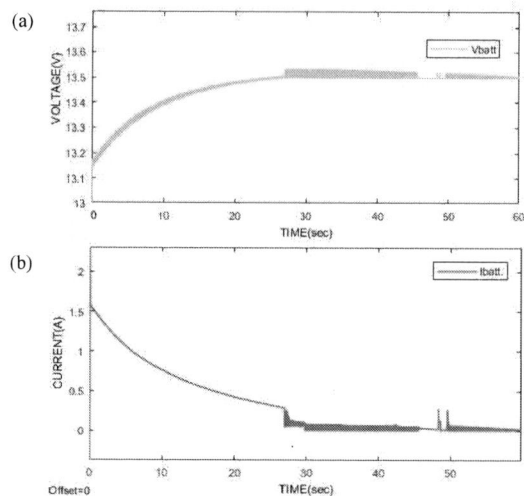

Fig. 16. System Performance at over-voltages (a) Battery voltage vs. time (b) Battery charging current vs. time.

Fig. 15. System Performance under varying irradiance at 25°C, full-load (a)Ideal P-V curves at varying irradiance (b) Irradiance vs. time (c) Power at MPP at respective irradiance.

Fig. 15(a), shows P-V curves of the given PV module under variable irradiance. A corresponding variable irradiance profile has been given as input to the solar PV model as shown in Fig. 15(b), Fig. 15(c), displaying the PV module output power, clearly shows that the power extracted from the PV module varies in accordance to the changing irradiance and is very close to ideal MPPs.

TABLE 4:   TRACKING EFFICIENCY UNDER VARYING IRRADIANCE

Irradiance (W/m²)	Expected Power (MPP) (W)	Observed Power (W)	Tracking Efficiency (%)
1000	199.2	191.5	96.13
800	160.1	160.0	99.94
600	120.2	118.0	98.17
400	79.8	76.1	95.36

C. System Performance at over-voltage of battery

Fig. 16(a) shows the terminal voltage of the battery during charging. Upon crossing the threshold value (13.5V), the GTO (gate turn-off thyristor) for charging is switched off. Fig. 16(b) shows the battery charging current which gradually decreases and after the battery voltage crosses the threshold, the switch prevents further flow of current to the battery. But at this stage the battery is allowed to discharge. Once it discharges below 13.5V, the charging switch (GTO) is turned ON. When, again the threshold value is reached, the GTO is turned off, and the cycle repeats, to maintain a constant voltage at 13.5V. This prevents over-charging.

D. System Performance under varying motor torque (Dynamic load)

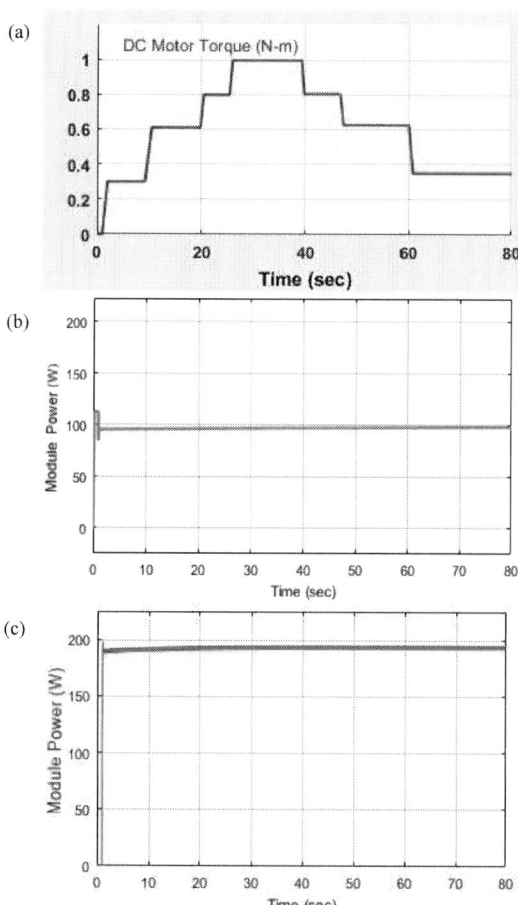

Fig. 17. System performance under varying torque at 1000 W/m²at 25°C (a) Dynamic Torque vs. time (b) Power from module without MPPT (c) Power from module with MPPT.

978-1-5386-6624-1/18 $31.00 © 2018 IEEE

Fig. 17(a) shows a variable torque vs. time curve input to the DC Motor to resemble sudden changes in motor load as observed in electric vehicles. As a result, Fig. 17(b) shows the solar module power, when the load is connected directly to the module, without closed-loop control. On modifying the system by adding the proposed power conditioning unit with implemented algorithms, Fig. 17(c) shows that the PV module always operates at the MPP (199.2 W at 1000 W/m^2), regardless of the varying load. The following table shows the results obtained on variation of motor torque:

TABLE 5: EFFICIENCY OF THE PROPOSED CHARGE CONTROLLER

Percentage Load (%)	Tracking Efficiency (%)
20	97.85
40	97.95
60	98.00
80	98.00
100	97.90

VI. CONCLUSION

In this paper, an optimized solar PV battery charge controller is designed and its performance is analyzed under dynamic load conditions of a DC motor. From the above results, it is observed that a high tracking efficiency up to 98% at 60%-80% load at $1kW/m^2$ irradiance (at 25°C) is obtained. From table 4, the highest tracking efficiency at full load, at an irradiance of 800 W/m^2 (at 25°C) is observed, at a maximum value of 99.94%. At a significantly discharged state (below 11.5V), it is observed that the controller operates at CC (C/10) mode, and thereafter the charging algorithm shifts successfully to MPPT mode and operates at this mode between 11.5V to 13.5V (13.5V is the specified overcharge limit of this battery). Also, the charge controller does not allow the battery to discharge below 11.5V, thereby preventing deep-discharge. The proposed charge controller topology can be used in Solar Electric Vehicle as it has been tested it under dynamic load conditions using a DC motor as load. Power conditioning unit (PCU) which is an essential part of an EV has also been integrated into the system.

Future research on implementation of 3-stage CC-CV charging shall be carried out as a continuation of work on the current system. The third stage involves float-charging algorithm, in which the battery is charged in Constant Voltage (CV) mode at a floating voltage point. This further ensures longevity of the battery after its sufficient cycle-use. By the use of P&O MPPT algorithm the tracking performance is affected by formation of local maxima in the I-V characteristic curve of solar module under dynamic operating conditions. This problem can be mitigated by using more robust MPPT algorithms, such as the Incremental Conductance (IC), fuzzy logic, Artificial Neural Networks (ANN) etc. The results have been observed by assuming an ambient temperature of 25°C. The effects of temperature change and ageing on battery, the solar PV module, and the DC motor are being analyzed for future work.

REFERENCES

[1] N. Femia, G.Petrone, G. Spagnuolo, M. Vitelli, "Power electronics and control techniques for Maximum Energy harvesting in photovoltaic system", *CRC Press,2013* Edition, Chapter 2,pp-43.

[2] Muhammad H. Rashid *"Power Electronics Circuit, Devices And Applications"*, Third Edition, 2016,pp 190.

[3] A. Durgadevi, S. Arulselvi and S. P. Natarajan, "Study and Implementation of Maximum Power Point Tracking Algorithm for Photovoltaic Systems." *1st International Conference on Electrical Energy Systems*, 2011.

[4] Md. Rokonuzzaman, Md. Hossam-E-Haider "Design and implementation of maximum power point tracking solar charge controller" , *Electrical Engineering and Information Communication Technology (ICEEICT), 2016 3rd International Conference* on 22-24 Sept,2016.

[5] B. Tar, A. Fayed "An overview of the fundamentals of battery chargers", *Circuits and Systems (MWSCAS),2016 IEEE 59th International Midwest Symposium* on 16-19 Oct,2016.

[6] M. Islam and Md. A. B. Sarkar "An Efficient Smart Solar Charge Controller for Standalone Energy Systems", *Electrical Drives and Power Electronics (EDPE), 2015 IEEE International Conference* on 21-23 Sept, 2015.

[7] D. Raveendhra, B. Kumar, D. Mishra and M. Mankotia "Design of FPGA based Open Circuit Voltage MPPT Charge Controller for Solar PV System", *IEEE International Conference on Circuit, Power and Computing Technologies (ICCPCT-2013)*, 21-22 March, 2013.

[8] T. Latif and S.R. Hussain "Design of a charge controller based on SEPIC and Buck topology using modified Incremental Conductance MPPT", *Electrical and Computer Engineering, 2014 IEEE International Conference* on 20-22 Dec. 2014.

[9] S. S. Valunjkar, S. D. Joshi and N. R. Kulkarni "Hardware and Simulation Study of MPPT Charge Controller for Non-Conventional Energy Sources", *Industrial Instrumentation and Control (ICIC), 2015 IEEE International Conference* on 28-30 May, 2015.

[10] Himanshu, R. Khanna "Various control methods for DC-DC buck converter", *Fifth IEEE Power India Conference*, 19-22 Dec, 2012.

[11] R.D Middlebrook and S Cuk, "A general unified approach to modelling switching Converter Power stages," in Proc. *IEEE PESC Rec.*, pp. 18-34, 1976.

[12] P. Vivek, R. Ayshwarya, S. JessibaAmali, A. S. NandiniSree, "A novel approach on MPPT algorithm for solar PV module using buck boost converter", *Energy Efficient Technologies for Sustainability (ICEETS),2016 International Conference* on 7-8 Apr,2016.

[13] Transtronics, Inc, 'Sealed Lead Acid Battery Applications'. [Online].Available:https://xtronics.com/wiki/Sealed_Lead_Acid _Battery_Applications.html [Accessed 1- July-2018]

978-1-5386-6624-1/18 $31.00 © 2018 IEEE

2nd IEEE International conference on power Electronics, Intelligent Control and Energy systems (ICPEICES-2018)

Hybrid Voltage Control for Stand Alone Transformerless Inverter

Mohammed Ali Khan
Department of Electrical Engineering,
Jamia Millia Islamia,
New Delhi, India.
mak1791@gmail.com

Ahteshamul Haque
Department of Electrical Engineering,
Jamia Millia Islamia,
New Delhi, India.
ahaque@jmi.ac.in

Kurukuru Varaha Satya Bharath
Department of Electrical Engineering,
Jamia Millia Islamia,
New Delhi, India.
kvsb272@gmail.com

Abstract- There has been a lot of progress going on in field of solar inverters. transformerless inverters has ben the most research area in recent time. For the operation of inverter in stand alone condition it is necessary to make sure that the control of the inverter is taken in account for maintaining the different operation parameters and making the system more safe and stable. This paper presents a comparative analysis of Highly Efficient and Reliable Inverter Concept (HERIC) inverter which is being controlled by hybrid controller. Stability testing of the HERIC inverter is performed based on load varying condition using Fuzzy- Proportional Integral Derivative (PID) and ANFIS-PID controller. Numerous different requirement is taken into consideration and analysis is carried out on basic of settling time and harmonic distortion present. At the end the experimental result conclude that system is running accurately with low settling time and harmonics distortion of less then 3.7%.

Keywords: HERIC transformerless inverter, control system, proportional integral derivative (PID), Settling time, Total Harmonic Distortion (THD)

I. INTRODUCTION

There has been a substituent decline in the use of non-renewable resources over the year. With rapid increase in demand of power, renewable source of power generation has been major focus area. Attempts have been made to obtain most out of the renewable source of energy over the years. In the field of inverters transformerless inverter has been most trending area of development as just by removing the transformer in the inverter there is a considerable increase in efficiency and even the size of the inverter has reduced. One concern that remains is the flow of leakage current from AC to DC half of the inverter due to the absence of galvanic isolation.

In the work presented by Vazquez et al. [1] a single phase grid connected inverter is linked with the grid for multilevel operation and the main aim is to reduce the leakage current of the inverter. Whereas Hu et al. [2] presented an enhanced HERIC inverter with focus to achieve excellent differential mode characteristics and reduce the leakage current. Patino et al. [3] performed a grid integration for HERIC inverter and controlled the performance by implementing second order generalized integrator- frequency lock loop controller. Yuksel et al. [4] proposed a control of single phase stand alone transformerless inverter. PI controller along with LCL filter is implemented to control the output. Somani et al. [5] work aimed on design a highly efficient solar inverter with control of power which is injected by the inverter into the grid. Dzung et al. [6] presented simulation of HERIC topology of an inverter along with PSIM simulator Hardware in loop technology is used for the purpose. Zaid et al. [7]

worked on a modified HERIC topology with the aim to improve the utilization factor of grid connected system and a review of previous work is presented. Xiao et al. [8] in his work proposed a zero voltage transition HERIC topology, by the integration of resonant tank and freewheeling switch. ZVT has an advantage over diode reverse recover, lower dv/dt and constant common mode voltage. Sun et al. [9] designed a new topology for transformerless inverter which freewheeling loop separation which provide isolation from harm caused by leakage current. Xiao et al. [10] in his another work presented unipolar sinusoidal pulse width modulation for switching of full bridge inverter keeping in mind the issue regarding reduction of leakage current. Kerekes et al. [11] in his work presented a AC bypass circuit with diode rectifier and switch which tends to clamp the midpoint. Koutroulis et al. [12] proposed a method of systematic design process which is capable to impact the cost and power loss of PV inverter.

In this paper an overview of HERIC topology is presented in section 2, whereas different control techniques are discussed in section 3, designing aspect and results are presented in section 4 and 5 respectively. In section 6 the work is concluded.

II. TRANSFORMERLESS INVERTER HERIC TOPOLOGY

Amid various transformerless inverter topologies, HERIC is considered as one of the most important topologies. This topology deals with H bridge inverter and a parallel branch connected alongside the H bridge. The back to back arrangement of switches is observed in the topology as shown in figure 2 which are operated at grid frequency. The additional parallel branch in the topology encourages the formation of third voltage level and isolates the DC end to that of the AC side of the inverter. The configuration and arrangement of switches in HERIC achieves low leakage current and high efficiency making it the most widely utilized topology in Photovoltaic applications.

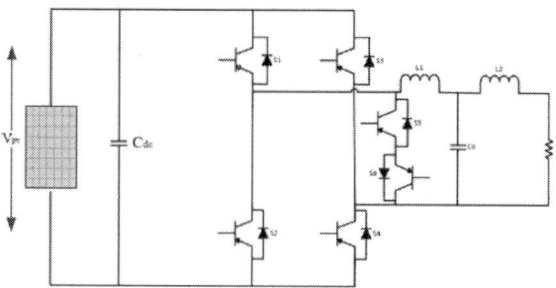

978-1-5386-6624-1/18 $31.00 © 2018 IEEE

Fig 1. HERIC topology

A. MODE 1&3

In this mode the switching frequency of S5 & S6 is observed, switch S5 is kept ON during positive half cycle, and S6 is ON continuously for negative half cycle. The positive voltage vector is obtained when high-frequency switching takes place instantaneously between S5 and S4 switches. When Switch S5 is ON the current flow is observed from S1-S5 and a return path is created through S4, as shown in fig. 2(a). When Switch S6 is ON the current flow is observed from S3-S6 a return path is created through S2, as depicted in fig. 2(c).

B. MODE 2&4

During Zero voltage vector, depending on the sign of the reference voltage switches S5 or S6 is turned on. The switches S1 to S4 are in OFF state and sources is disconnected as depicted in fig. 2(b) and 2(d). The output voltage tends to be unipolar, maintain the switching frequency, and achieving the Zero- voltage vector. The high frequency fluctuation will not be present at DC terminal of the PV array. As per freewheeling period, the inverter efficiency is kept higher, short circuiting the load current through S5 and S6 which is dependent on the grid current sign.

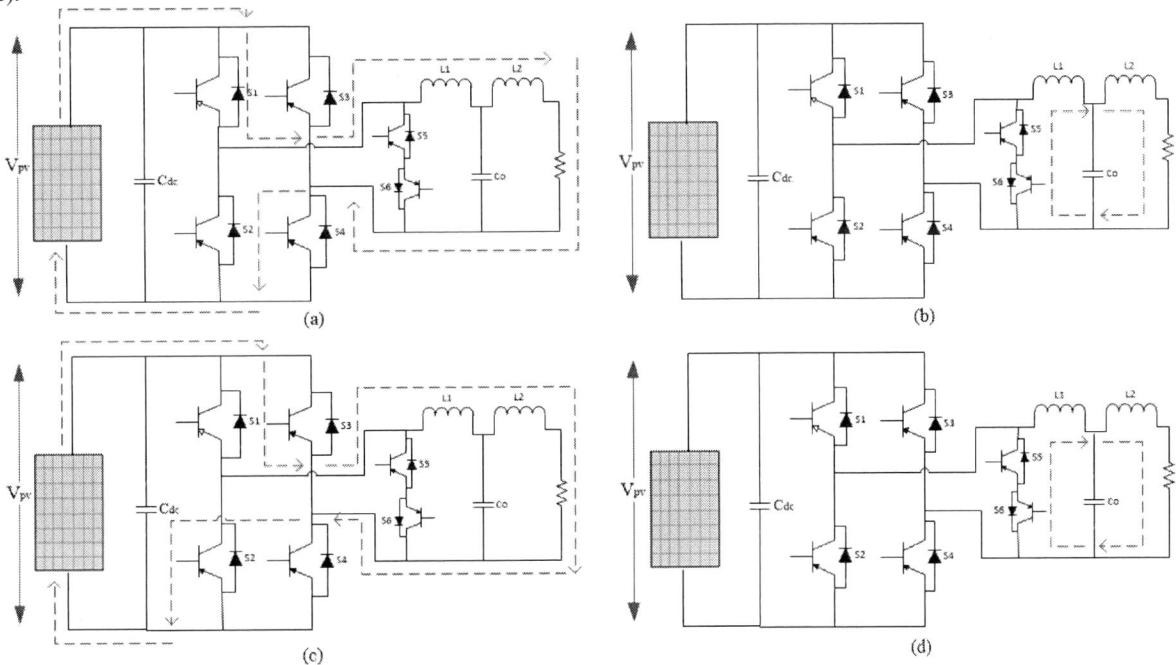

Fig 2. Modes of operation for HERIC topology

III. CONTROL ALGORITHM

To obtain voltage stability different control techniques are tested with the inverter topology. Explanation of different techniques is presented in the section below.

A. Fuzzy-PID Controller

Fuzzy tuned PID is a supervisory control in which PID controller is tuned with the help of fuzzy rules set. It helps system to be stable and get best performance with reliable output from the system. Input to the fuzzy are provided in for of error and change in error and rules are trained for controlling the gain value of the PID controller which is operating in close loop. The membership function and rules for fuzzy controller are illustrated in figure 3.

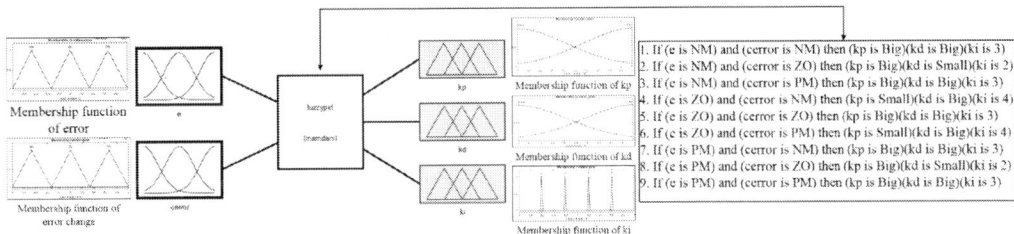

Fig. 3. Fuzzy rules and Membership function

Input for PID controller is obtained by the defuzzied output of fuzzy logic controller. The output obtained by PID is fed to the PWM generator where a comparative analysis is carried out with the carrier wave for providing a triggering pulse to different switches for the inverter.

B. ANFIS-PID Controller

To control the reliability of the system for error minimization, membership function for fuzzy are required to be defined whereas implementation of ANFIS is performed

978-1-5386-6624-1/18 $31.00 © 2018 IEEE

in parallel structure and implanted in control system. Not a lot of research is present in ANFIS-PID implementation for control of voltage regulation of inverter. Error is utilized to compute the ANFIS for each of the different PID gain values of K_p, K_i and K_d. In figure 4 a block representation of ANFIS PID is illustrated.

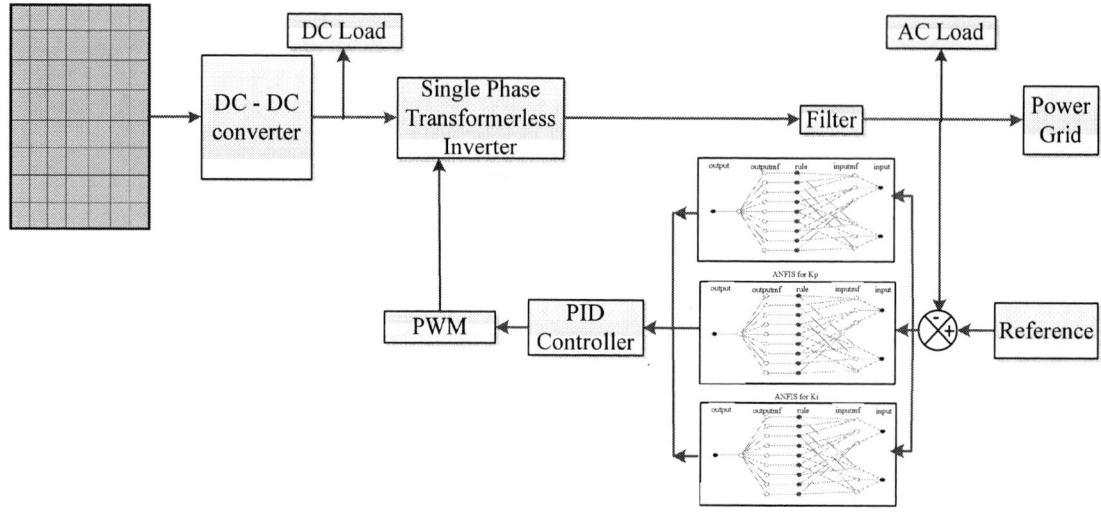

Fig. 4. Block representation of ANFIS-PID controller

IV. SIMULATION AND DISCUSSION

Simulation was performed on MATLAB 2017a regarding the study of HERIC topology and different controlling technique was implemented for obtaining stable output of the system. Different component is modelled such as panel and filter for implementation of the system. System components designing is explained in the section below: -

A. Designing of Solar Panel

For designing a 2kW Sanyo panel was considered. Various are presented in the value of irradiance and the temperature of the atmosphere by keeping the ambient condition to be 100 W/m² and 25°C respectively. In table 1 other parameter which needed while designing are mentioned.

Table 1 Solar panel parameters

Specification	Value
Current- Short Circuit value (I_{Shr})	5.57A
Power- Maximum value ($P_{\_maxmum}$)	210W
Voltage- Open Circuit value (V_{opn})	50.0V

B. Designing of Filter

In order to remove the harmonics and noise from the system output, filter is implimented on the output end of the system. A T shaped LCL filter is implemented which consist of a capacitor along with two inductor arranged in T shape. Following formulas are implimented for the calcuation purpose.

$$c = \frac{1}{2 \text{X} \pi \text{X} f \text{X} z} \quad (1)$$
$$L = CZ^2 \quad (2)$$

Where,
f= cut-off frequency
Z = impedance

V. RESULTS

An ideal range of voltage for operation of inverter is 230 V. The focus while designing the control algorithm for inverter was to maintain a stable 230 V while the load and other external condition such as irradiance and temperature is varied. As per the National Electricity Code published by bureau of Indian standard [13] up to 6% error is acceptable. For studying the system stability even runtime load variation was performed and settling time and error after settling was noticed. It was found that the system settling time is very small with a very small error and making system work almost accurately. Figure 5(a) it can be implied that the system is stable near to almost 230 V with some 3-4 volt of error which is less than 6% of permissible limit. Figure 5(b) depicts time required for the system to settle down when it is turned on at different load values for HERIC topology. Depending on the response time, HERIC topology settling time for ANFIS PID controllers is faster then Fuzzy PID controller hence it is fast in achieving stability.

978-1-5386-6624-1/18 $31.00 © 2018 IEEE

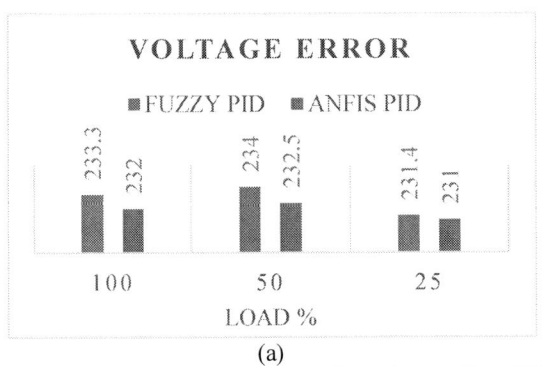

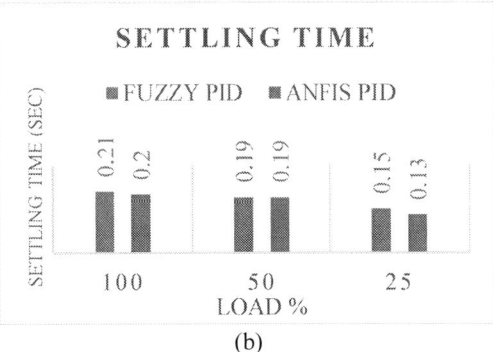

(a)　　　　　　　　　　　　　　　　(b)

Fig. 5. (a) HERIC topology voltage profile at different load, (b) HERIC topology settling time at initial loading

Different error values are presented in table 2 by assuming 230 V to be the reference voltage value. the value of error is large in case of Fuzzy PID when operating at 100% and 50 % of the load. Best results were obtained for ANFIS PID operating at all the various load with very less variation error.

Table 2 HERIC topology error calculation

Load %	HERIC	
	Fuzzy PID	ANFIS PID
100	1.434%	0.86%
50	1.739%	1.08%
25	0.608%	0.43%

A complete analysis of HERIC topology with variable load is presented in figure 6. The response on load change is illustrated at different control algorithm. For ANFIS PID most satisfactory output is obtained where as in case of Fuzzy PID even some variations are visible at 75 to 100 % load variation. System get settled faster at higher load variation.

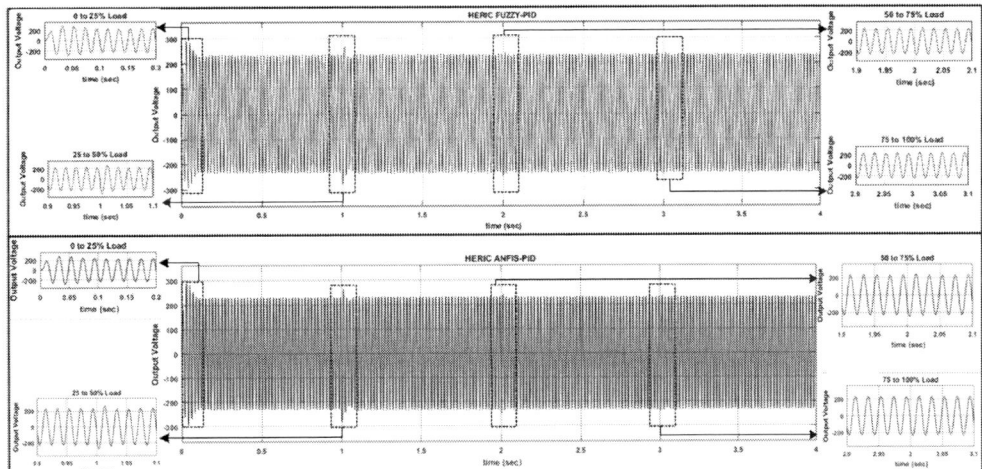

Fig. 6. Runtime load variation under operation implemented in HERIC topology

Analysis of Total Harmonic Distortion is also carried out and as per EN50160 [14] and IEEE 516 [15], the acceptable range for THD was kept to be 5%. THD characterize the waveform in such a way that higher load with low THD has more impact then the high THD with lower load. FFT analysis to obtain THD is performed in the HERIC topology at both the controller are illustrated in figure 7 and it can be observed that ANFIS-PID has a THD of 3.2% which is far better than fuzzy PID 3.46%.

978-1-5386-6624-1/18 $31.00 © 2018 IEEE

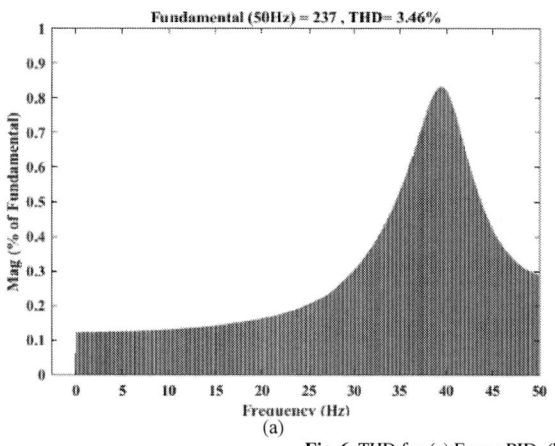
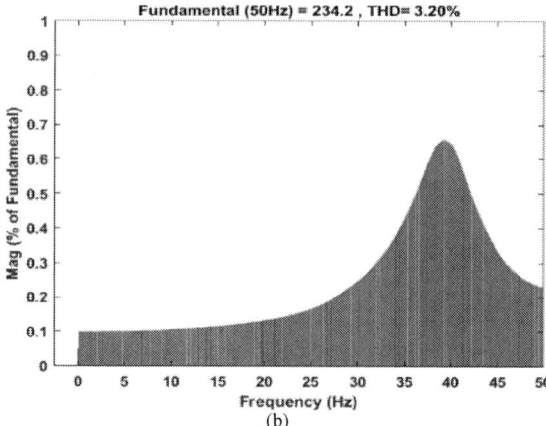

Fig. 6. THD for (a) Fuzzy-PID, (b) ANFIS-PID in HERIC topology

HERIC topology stability analysis is performed in over the transformerless inverter system and bode plot was obtained in the simulation which is illustrated in figure 7. It was observed that the gain and phase margin of the system was in the same half hence the system is stable. The gain and phase margin for both the controller is depicted in table 3.

Table 3 Bode plot characteristics of HERIC topology

	HERIC Topology	
	Fuzzy- PID	**ANFIS- PID**
Gain margin (G_m)	1.1198e+05	7.2476e+04
Phase Margin (P_m)	90.0005	90.0008
Gain Crossover Frequency (W_{gm})	1.3	1
Phase Crossover Frequency (W_{pm})	2.2393e-11	3.4606e-11

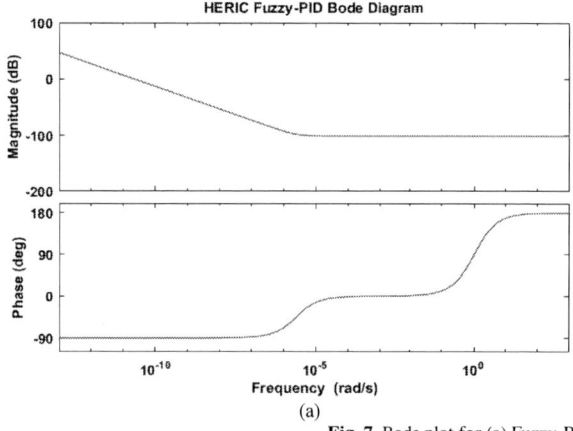
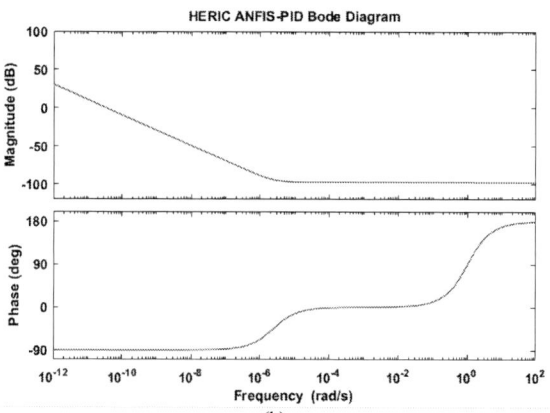

Fig. 7. Bode plot for (a) Fuzzy-PID, (b) ANFIS-PID in HERIC topology

VI. CONCLUSION

The work presented in the paper is regarding the comparative analysis of fuzzy PID and ANFIS PID for the HERC topology of transformerless inverter. Switching techniques are implemented using PWM. Control stability analysis and load variation controller for voltage was simulated using MATLAB. Run time load variation results was simulated to find the stability of the system and response time of controller for obtaining the stability.

The performance parameters like settling time, THD etc are evaluated. It was found that error for ANFIS PID was less in compare to Fuzzy PID and even system with ANFIS PID settled faster than Fuzzy PID system. Even in terms of THD ANFIS PID Controller presented a better result.

Work related to hardware implementation can be focused as future of the research along with more stability analysis and study of steady state error can be performed.

REFERENCE

[1] J. M. Sosa, G. Escobar, M. A. Juarez, and A. A. Valdez, "H5-HERIC Based Transformerless Multilevel Inverter for Single-Phase Grid Connected PV Systems," in IECON 2015, 2015, pp. 1026–1031.

[2] S. Hu, C. Li, and W. Li, "Enhanced HERIC Based Transformerless Inverter with Hybrid Clamping Cell for Leakage Current Elimination," in Energy Conversion Congress and Exposition (ECCE), 2015, pp. 5337–5341.

[3] G. David, "Implementation a HERIC inverter prototype connected to the grid controlled by SOGI-FLL," no. Dc, 2015.

[4] A. Yüksel, E. Özkop, and E. M. Bölümü, "A Single Phase Standalone Photovoltaic System with HERIC Inverter Control," pp. 373–376.

[5] P. Somani, "Design of HERIC Configuration Based Grid

Connected Single Phase Transformer less Photovoltaic Inverter," in International Conference on Electrical, Electronics, and Optimization Techniques (ICEEOT), 2016, pp. 892–896.

[6] H. I. L. Concept, P. Q. Dzung, D. N. Dat, N. B. Anh, L. C. Hiep, and H. Lee, "Design of HERIC Inverter for PV Systems by Using," pp. 2035–2040, 2014.

[7] S. A. Zaid and A. M. Kassem, "Review , analysis and improving the utilization factor of a PV-grid connected system via HERIC transformerless approach," Renew. Sustain. Energy Rev., vol. 73, no. December 2016, pp. 1061–1069, 2017.

[8] L. Xiao, Hua F.; Zhang, "A Zero-Voltage-Transition HERIC-Type Transformerless Photovoltaic Grid-Connected Inverter," IEEE Trans. Ind. Electron., vol. 64, no. 2, pp. 1222–1232, 2017.

[9] M. Sun, J. Zhao, K. Qu, F. Li, L. Mao, and M. Feng, "Novel Single-Phase Transformerless Inverter Based on The Freewheeling Loop Separation for PV systems," in IEEE 8th International Power Electronics and Motion Control Conference, 2016.

[10] H. Xiao, S. Xie, Y. Chen, and R. Huang, "An optimized transformerless photovoltaic grid-connected inverter," IEEE Trans. Ind. Electron., vol. 58, no. 5, pp. 1887–1895, 2011.

[11] T. Kerekes, R. Teodorescu, P. Rodríguez, G. Vázquez, and E. Aldabas, "A New High-Efficiency Single-Phase Transformerless PV Inverter Topology," IEEE Trans. Ind. Electron., vol. 58, no. 1, pp. 184–191, 2011.

[12] E. Koutroulis and F. Blaabjerg, "Methodology for the optimal design of transformerless grid-connected PV inverters," IET Power Electron., vol. 5, no. 8, p. 1491, 2012.

[13] National electrical code. 2011.

[14] H. (Cooper D. A. Markiewicz and A. (Wroclaw U. of T. Klajn, "Voltage Disturbances," 2004.

[15] "IEEE Guide for Maintenance Methods on Energized Power Lines," IEEE Std 516-2003 (Revision IEEE Std 516-1995), 2003.

978-1-5386-6624-1/18 $31.00 © 2018 IEEE

2nd IEEE International conference on power Electronics, Intelligent Control and Energy systems (ICPEICES-2018)

Performance Analysis of Different Modulating Techniques for Multilevel Inverter

Allu Bhargav
Dept. of Electrical Engineering
Delhi Technological University
Delhi, India
allubhargav@gmail.com

Alka Singh
Dept. of Electrical Engineering
Delhi Technological University
Delhi, India
alkasingh.dr@gmail.ac.in

Ankita arora
Dept. of Electrical Engineering
Delhi Technological University
Delhi, India
arora.ankita09@gmail.com

Abstract—**Multilevel converters are getting increasingly popular these days due to their high-power capability under medium voltage applications. In this paper, the modelling and analysis of a seven-level cascaded H-bridge multilevel inverter connected with photovoltaic (PV) cells is presented. Conventional modulating techniques give large total harmonic distortion (THD), so the focus of this paper is on the analysis and study of different modulating techniques. The presented techniques viz. phase disposition (PD) carrier based PWM, phase opposition disposition (POD) carrier based PWM, alternate phase opposition disposition (APOD) carrier based PWM and phase shifting (PS) carrier based PWM are discussed and compared in terms of total harmonic distortion (THD). It is observed that phase shifting modulating technique results in lowest THD as compared to other techniques.**

Keywords— Photovoltaic, P&O, Boost, phase disposition, phase shifting, THD

I. INTRODUCTION

Focus on renewable energy sources in place of conventional sources has increased even in developing nations of the world. Limited resources and environmental pollution due to high emissions has led scientists and engineers search for new and reliable energy sources. Many non-conventional energy sources are available today, but solar photovoltaic (PV) and wind energy capture the market today. The installation of solar energy has been increasing at a fast pace and there is a growth potential of 20-25% over the last 10 years. PV source is pollution free and continuous energy supply during day light adds to its advantages [1].

Maximum power point tracking (MPPT) control plays an vital role to extract the maximum available power from the PV array. Different MPPT techniques are presented in the literature [2]. Integration of PV to the grid can be achieved using a boost converter. It is implemented to boost the input voltage and its duty cycle is regulated by MPPT control [3]. In general, to transform DC to AC, we use a conventional two level inverter, which gives high THD. Decreasing the THD to permissible levels as per IEEE standards requires the use of additional AC filters which have large size and cost.

The use of multilevel inverters is preferred to get high power from medium voltages. Though it has been introduced lately, it has gained a lot of attention nowadays. Many advantages result with multilevel converters such as low total harmonic distortion which decrease the size of filters, low switching losses, reduced electromagnetic interference and lower voltage stresses. Multilevel configurations are broadly categorized into three types viz. Cascaded H-bridge, Neutral

point clamped, Flying capacitor multilevel inverter [4]. In this paper, cascaded H-bridge multilevel inverter [5] is implemented because it is the easiest to implement as compared to others.

This paper discusses the design of seven level cascaded H-bridge multilevel inverter [6] with PV connected at the DC link of each H-bridge. For MPPT control, perturb and observe (P&O) algorithm is used and boost converter is used at the DC link in multilevel inverter. Different modulation techniques [7]- [11] used in multilevel inverter are discussed in further sections.

II. SYSTEM DESCRIPTION

Seven level Cascaded H-bridge multilevel inverter connected with PV modules is illustrated in Fig. 1. Each H-bridge is a single-phase converter with four IGBTs and diode connected in reverse polarity. Three such bridges are connected in cascade configuration as illustrated in Fig.1.

A PV array is integrated at the DC link of each H-bridge. The design and modelling of PV array, MPPT and design of boost converter are discussed briefly.

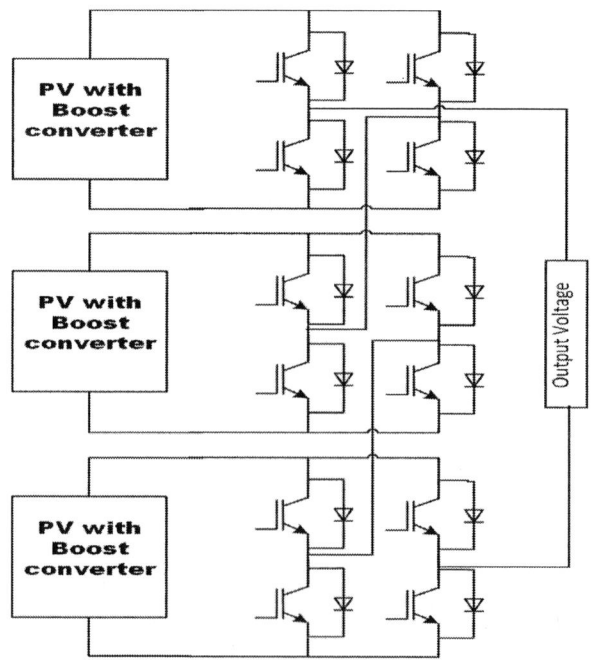

Fig. 1. Seven level Cascaded H-bridge based multilevel inverter connected with three PV modules

978-1-5386-6624-1/18 $31.00 © 2018 IEEE

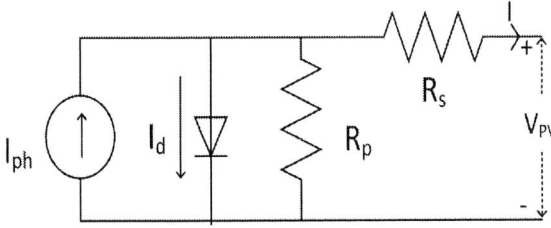

Fig. 2. Electrical equivalent network of PV cell

A. Modelling of PV Module

The equivalent circuit of PV cell is shown in Fig. 2. It shows that a solar cell can be modelled by current source and a diode in parallel with it and having a shunt (R_p) and series resistance (R_s) due to leakage current.

The modelling design of PV module is described below [12]:

The output PV module current is given in Eq. (1)

$$I = N_p I_{ph} - N_p I_d - I_p \qquad (1)$$

where I_{ph}= photo current in Amperes

I_d = diode current in Amperes

I_p = leakage current in Amperes in shunt resistance R_p

The Photo current (I_{ph}) can be computed from Eq. (2)

$$I_{ph} = [I_s + K_s(T_e - 298)] \times I_r/1000 \qquad (2)$$

where I_s = short circuit current in Amperes

K_s = short circuit current constant

T_e= operating temperature in kelvin

I_r = solar irradiation in W/m^2

The diode current (I_d) is computed from Eq. (3)

$$I_d = I_O\left[\exp\left(\frac{V_{oc}}{A.NV_T}\right) - 1\right] \qquad (3)$$

where I_O = saturation current

V_{oc} = open circuit voltage in volts

A = ideality factor

N = Number of series connected solar cells

V_T = thermal voltage

The thermal voltage V_T is given in eq. (4)

$$V_T = \frac{kT_e}{q} \qquad (4)$$

where k   = Boltzmann's constant

q   = charge of electron in coulombs

The saturation current (I_o) is given in eq. (5) below

$$I_O = I_{rs}\left(\frac{T_e}{T_r}\right)^3 \times exp\left[\frac{qE_g}{Ak}\left(\frac{1}{T_r} - \frac{1}{T_e}\right)\right] \qquad (5)$$

where I_{rs} = reverse saturation current

T_r = nominal temperature in kelvin

E_g= band gap energy in eV

The reverse saturation current (I_{rs}) is given in eq. (6) below

$$I_{rs} = \frac{I_s}{\exp\left(\frac{qV_{oc}}{NkAT_e}\right)-1} \qquad (6)$$

The leakage current (Ip) is given in eq. (7) below

$$I_p = \frac{V \times \frac{N_p}{N_s} + I \times R_s}{R_p} \qquad (7)$$

where N_p = number of parallel connected modules

N_s = number of series connected modules

TABLE I. PV CHARACTERISTICS

Parameters	Value
V_{oc}	32.9 V
I_s	8.21 A
P_{max}	200 W
A	1.3417
N	54
N_s	1
N_p	1
R_{sh}	951.9317 Ω
R_s	0.2127 Ω
I_r	1000 W/m^2
K_s	0.387×10^{-3}
E_g	1.1 Ev
V_{mp}	27.2 V
I_{mp}	7.35 A

The P-V characteristics and I-V characteristics of PV array is illustrated in Fig. 3 and Fig. 4.

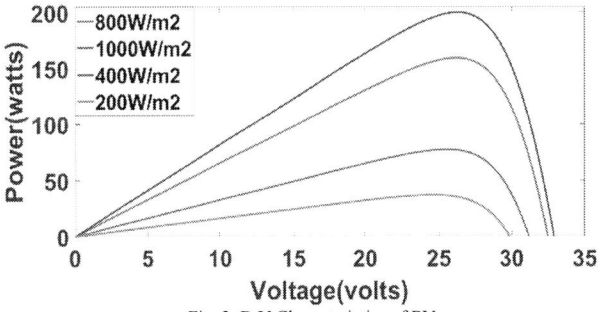

Fig. 3. P-V Characteristics of PV array

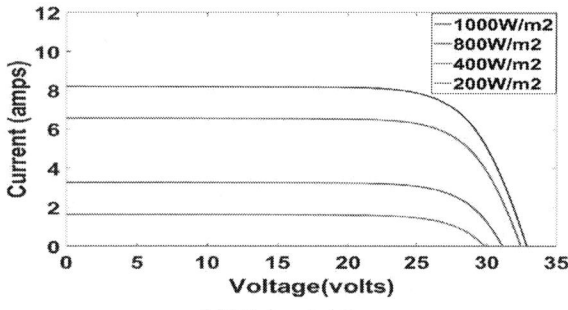

(a) I-V characteristics

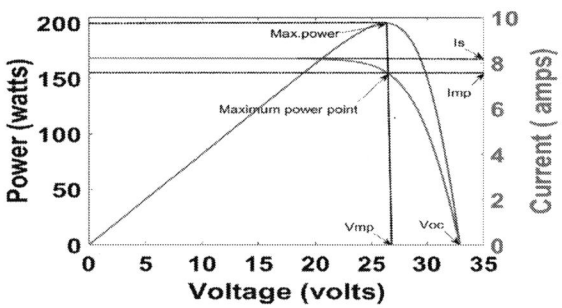

(b) Characteristics showing maximum power point

Fig. 4. Characteristics of PV array (a) I-V (b) showing maximum power point

B. Maximum power point tracking(MPPT)

Maximum power point tracking control is used to obtain maximum power from solar PV module. It is also used to generate the duty cycle required for DC-DC converter. In this paper, Perturb and Observe algorithm has been used. It is a conventional algorithm which is easy to implement in real time as compared to other methods and gives fairly accurate results. The P&O algorithm can be mathematically represented by

$$\frac{dP}{dV_{pv}}(i) = \frac{P(i) - P(i-1)}{V_{pv}(i) - V_{pv}(i-1)} \qquad (8)$$

where P(i), P(i-1) are present and previous power and $V_{pv}(i), V_{pv}(i-1)$ are present and previous voltage.

When $\frac{dP}{dV_{pv}}$ >0 then direction moves forward towards maximum power point, voltage must be incremented or increased. If $\frac{dP}{dV_{pv}}$ <0 then direction moves backward towards maximum power point, voltage must be decremented.

C. Boost converter

A boost converter is a step-up converter and plays a vital role in PV applications because low voltage is obtained from PV module which is not sufficient to drive the load. Boost converter is used to step-up the PV voltage and then connected to the DC link of each of the H-bridge inverter. The required duty cycle is generated by MPPT controller. The illustrative figure of boost converter is drawn below in Fig. 5.

The designing of boost converter and its parameters are calculated with the help of equations 9-11.

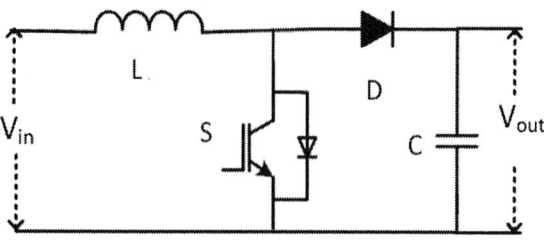

Fig. 5. Diagram of Boost Converter

$$\text{Duty cycle (D)} = \frac{V_o - V_i}{V_o} \qquad (9)$$

where V_o = output voltage

V_i = input voltage

$$\text{Inductor (L)} = \frac{V_i \times D \times T}{\Delta i_l} \qquad (10)$$

where T = time period of switching

Δi_l = inductor ripple current

$$\text{Capacitor (C)} = \frac{V_o \times D \times T}{R \times \Delta V_o} \qquad (11)$$

where R   = resistance

ΔV_o = capacitor ripple voltage

Where the Δi_l is the 3-5% of output current and ΔV_o is the 3-5% of output voltage. The values of inductor and capacitor and duty cycle and resistance are values are given in Table II.

Table II. Design values of boost converter

Parameters	Value
L	2.4mH
C	87.77μF
D	0.395
R	10Ω

D. Seven level Cascaded H-bridge based multilevel inverter

A cascaded H- bridge multilevel inverter is a popular configuration used in medium voltage applications. In this topology, cascade connection of three H-bridges is carried out so that the output voltage of whole system will be combination of the output of each H-bridge to form different level voltages. If we cascade 'n' number of H-bridges, then we get 2n+1 level of voltages in output voltage wave form. Each H bridge requires separate dc source which can be provided by either battery or PV cell. Cascaded H-bridge can be operated in symmetric as well as asymmetric conditions. Under symmetric condition, all the dc level voltages are the same and under asymmetric condition, different dc level voltages are used. In this paper, seven level cascaded H-bridge multilevel inverter has been realized in symmetric condition so that all the dc voltages are equal as $V_{dc1} = V_{dc2} = V_{dc3} = V_{dc}$ and output voltage will be in form of $V_o = V_{dc1} + V_{dc2} + V_{dc3}$.

978-1-5386-6624-1/18 $31.00 © 2018 IEEE

III. MODULATING TECHNIQUES

Several modulation techniques have been developed for multilevel inverter. As described in Fig. 6, modulating techniques can be broadly categorized into two viz. operation at fundamental switching frequency and high switching frequency PWM. The main focus of this paper is on multilevel carrier based PWM. The principle of carrier based PWM method is to use the i-1 carrier signals with reference signal for i level inverter. Difference of carriers will be as follows

- carrier frequency
- carrier amplitude
- carrier phase
- carrier dc offset

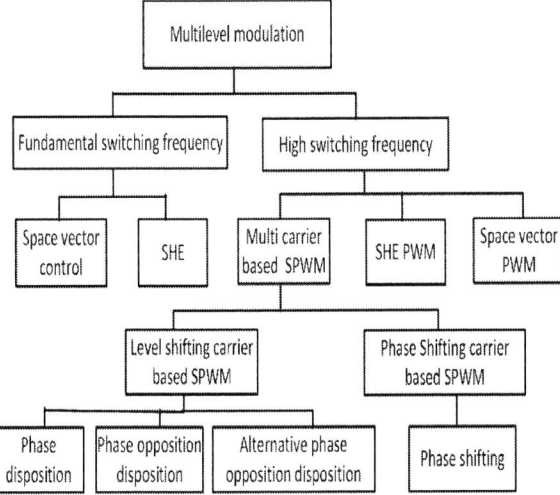

Fig. 6. Classification of multilevel modulation techniques

Multilevel carrier based PWM are classified as follows

- phase disposition (PD) carrier-based PWM
- phase opposition disposition (POD) carrier based PWM
- alternate phase opposition disposition (APOD) carrier based PWM
- phase shifting (PS) carrier based PWM

These techniques are discussed below and developed for the system represented in Fig.1.

A. Phase disposition carrier-based PWM

In phase disposition (PD) carrier-based PWM, all the carriers will be in same phase angle, same frequency and equal magnitude, but the difference will be in dc offsets so as to occupy contiguous bands. This is illustrated in Fig.7. In this, maximum harmonic energy is found at the carrier wave frequency and all other harmonic components are centered as side band harmonics on carrier wave frequency. In cascaded H-bridge seven level multilevel inverter, six carrier waveforms (of frequency 1KHz) have been used along with a sinusoidal reference waveform of 50Hz. The gate control is obtained when each of carrier waveforms with sinusoidal reference signal are compared, which in turn controls respective gate.

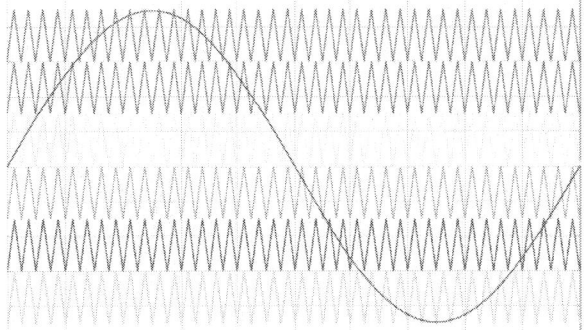

Fig. 7. PD modulation carrier waveforms

B. Phase opposition disposition carrier-based PWM

In this modulating technique as shown in Fig.8, phase opposition disposition (POD) carrier-based PWM, the carriers are having equal magnitude and same frequency, but the carriers have difference in dc offset and the carriers just above the zero reference will be in same phase angle and just below the zero reference carriers are in same phase angle and the phase angle difference between the above zero reference carriers and below zero reference carriers will have the 180 degrees phase shift. The maximum harmonic energy is found around the carrier wave frequency as side band harmonics

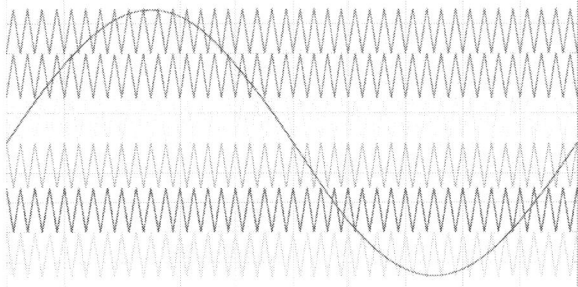

Fig. 8. POD modulation carrier wave forms

C. Alternate phase opposition disposition carrier-based PWM

In alternate phase opposition disposition (APOD) carrier-based PWM, all the carriers will have equal magnitude and same frequency, but the carriers have difference in dc offset and phase shifted by 180˚ alternately. The maximum harmonic energy is centered around carrier wave frequency as side band harmonics and no harmonics at carrier wave frequency. This is represented in Fig.9.

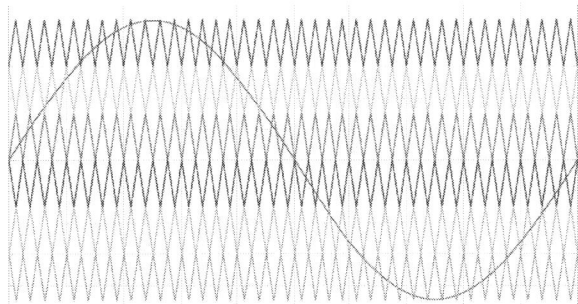

Fig. 9. APOD modulation carrier wave forms

D. Phase shifting carrier-based PWM

In phase shifting (PS) carrier-based PWM, all the carriers will have equal magnitude and same frequency and dc offset but the difference is that, each carrier will have certain phase shift angle π/m where, m= number of cascaded H bridges that are connected in series. This technique gives rise to cancellation of all the carrier and associated sideband harmonics. In cascaded H bridge seven level inverter, the phase shifting angle is 60˚.

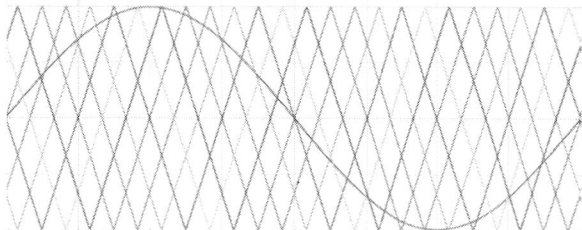

Fig. 10. PS modulation carrier wave forms

IV. SIMULATION RESULTS

Fig. 11. gives the output voltage results for the PV integrated with boost converter at the DC link of three H bridges. All PV modules are identical and operated under same irradiance and same temperature condition. For an input voltage of 27.2 V, the output voltage is boosted to 45V.

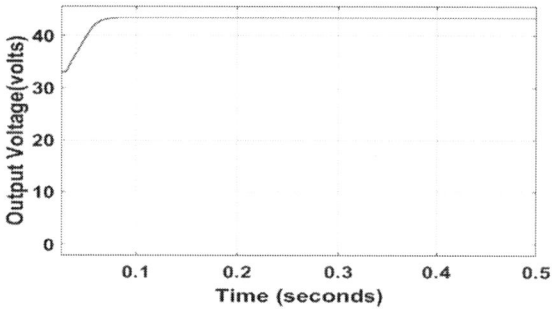

Fig. 11. output voltage of PV with boost converter and MPPT

Fig. 12. gives the output voltages with three cascaded H - bridge multilevel inverter. It can be observed that with each of the modulating techniques, the levels are obtained as follows +45, +90, +135, 0, -45, -90, -135.

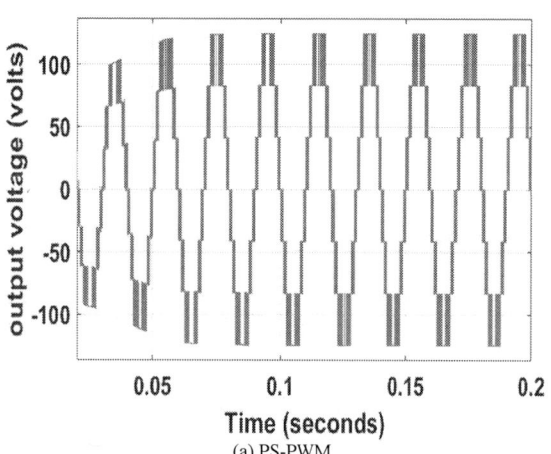

(a) PS-PWM

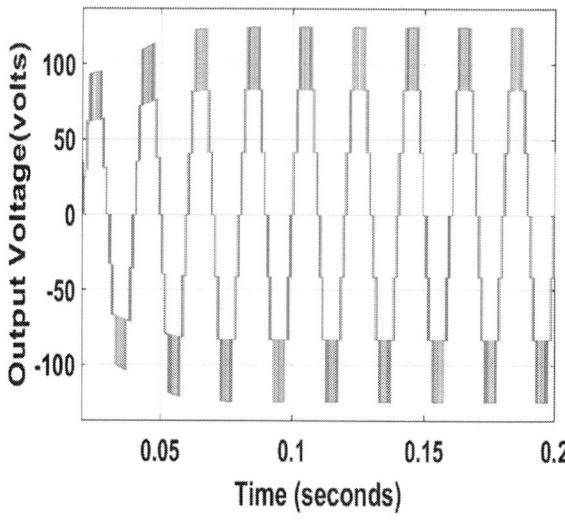

(b) PD-PWM

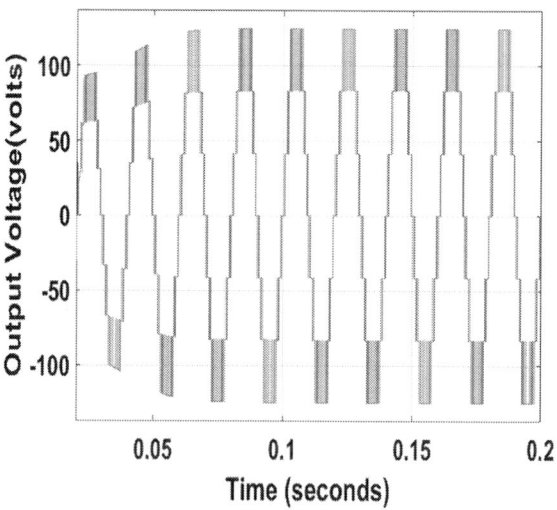

(c) POD-PWM

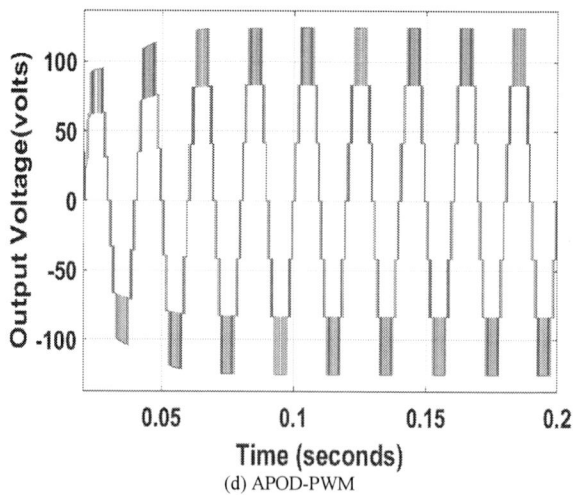

(d) APOD-PWM

Fig. 12. Output voltage waveforms of cascaded H bridge inverter with different modulating techniques

TABLE III. Comparison of different Modulating Techniques

Modulation technique	Output Voltage(rms)	%THD
PD	88.83	18.73
POD	88.73	18.93
APOD	88.72	19.22
PS	89.24	17.98

V. CONCLUSION

This paper has discussed the design of PV modules which are integrated at each of the H-bridge of the multilevel converter. The three H-bridges are connected in cascade configuration. Simulation model has been developed using MATLAB and Simpower system. Performance results with four different PWM techniques like PD, POD, APOD and PS techniques have been discussed. It can be inferred from the results of Table III, that Phase Shifting technique gives less percentage THD as compared to the other techniques. It is preferable to use PS-PWM technique for seven level cascaded H bridge multilevel inverter when integrated with PV modules.

ACKNOWLEDGMENT

The authors are thankful to the DST, SERB for the Sponsored project (EMR/2016/001874).

REFERENCES

[1] G. Eas Kannan N, Vakeesan D. Solar energy for future world: – a review. renew sustain energy rev,vol. 62,pp. 1092–105,2016.

[2] Liu L, Meng X, Liu C. "A review of maximum power point tracking methods of PVpower system at uniform and partial shading"Renew Sustain Energy Rev,vol. 53,pp. 1500–1507,2016.

[3] N. prabaharan, K. palanisamy, "Analysis and integration of multilevel inverter configuration with boost converters in photovoltaic system" Energy conversion and management, vol. 128, pp. 327-342, 2016.

[4] Jose´ Rodrı´guez, Leopoldo G. Franquelo, Samir Kouro, Ramo´n C. Portillo, Ma A´ngeles Martı´n Prats, Marcelo A. Pe´rez" Multilevel converters: an enabling technology for high-power applications" Proc. IEEE,Vol. 97,pp. 1786-1817, 2009.

[5] Natarajan Prabaharan, Kaliannan Palanisamy, "Analysis of cascaded H-bridge multilevel inverter configuration with double level circuit", IET power electron., vol. 10, no. 9, pp. 1023-1033, 2017.

[6] Prachi Salodkar, Sandeep. N, P. S. Kulkarni, Udaykumar R. Y" A comparison of seven-level inverter topologies for multilevel DC-AC power conversion" IEEE Int. Conf. on Power Electronics, Drives and Energy Systems (PEDES), 2014.

[7] Jose I. Leon, Sergio Vazquez, Leopoldo G. Franquelo" Multilevel converters: control and modulation techniques for their operation and industrial applications" Proc. IEEE,Vol. 105, pp. 2066-2081, 2017.

[8] Abuzaid Saeed Gadalla, Xiangwu Yan, Sara Yahia Altahir, Hashim Hasabelrasul" Evaluating the capacity of power and energy balance for cascaded H-bridge multilevel inverter using different PWM techniques" The sixth IEEE Int. Conf. on Renewable Power Generation (RPG) ,2017.

[9] P. Palanivel, S.S. Dash, "Analysis of THD and output voltage performance for cascaded multilevelinverter using carrier pulse width modulation technique", IET Power Electron., vol. 4, no. 8, pp. 951-958, 2010.

[10] Christopher D. Townsend, Terrence J. Summers, Robert E. Betz" Phase-shifted carrier modulation techniques for cascaded H-bridge multilevel converters" IEEE Trans. Ind. Electron.,vol. 62, pp. 6684-6696, 2015.

[11] R. Naderi, A. Rahmati, "Phase-shifted carrier PWM technique for general cascaded inverters", IEEE Trans. Power Electron.,vol. 23, no. 3, pp. 1257-1269, 2008.

[12] Xaun Hieu Nyuyen, Minh Phuong Nguyen "Mathematical modelling of photovoltaic cell/module/arrays withs tags in Matlab/simulink" a springer open journal , environmental system research,vol. 4, no. 24,2015.

[13] Janardhan Kavali, Arvind Mittal "Analysis of various control schemes for minimal total harmonic distortion in cascaded H-bridge multilevel inverters" Journal of Electrical Systems and Information Technology ,vol. 3 , pp. 428–441,2016.

2nd IEEE International conference on power Electronics, Intelligent Control and Energy systems (ICPEICES-2018)

Autonomous Operation of Single Stage Solar PV-BES Based Microgrid

Shubhra, *Member, IEEE*
Department of Electrical Engineering
Indian Institute of Technology Delhi
New Delhi, India
shubhra72@gmail.com

Bhim Singh, *Fellow, IEEE*
Department of Electrical Engineering
Indian Institute of Technology Delhi
New Delhi, India
bsingh@ee.iitd.ac.in

Abstract—**This paper deals with the autonomous operation of solar photovoltaic (PV) and battery energy storage (BES) based microgrid. The single stage PV system is incorporated to the DC link of voltage source converter (VSC). The BES is incorporated across the DC link through a bidirectional converter for load management. VSC operates on a voltage control algorithm in standalone mode. A perturb and observe approach is implemented to achieve the peak power from a solar PV array. The charging and discharging currents of battery, are controlled by the bidirectional converter. The solar PV-BES microgrid in standalone mode, controls the power requirement by the loads and maintains the frequency and voltage within stipulated limits. The implementation of bidirectional converter has reduced the battery rating as compared to the battery incorporation across the DC link of VSC directly. The operation of proposed microgrid is observed satisfactory in steady state and dynamic conditions and validated through test results on a developed prototype.**

Keywords—*Bidirectional Converter, Microgrid, MPPT, PV Array, Standalone, Power quality, Voltage Control, VSC.*

I. INTRODUCTION

The fossil fuel emits greenhouse gases, which has severe environment and health effects. The solar energy as a renewable source, is economical, easy to design and install as compared to non-renewable energy resources [1]. The extraction of maximum power using various maximum power point tracking (MPPT) controllers for solar photovoltaic (PV) applications, is given in [2]. A perturb and observe (P&O) algorithm used for solar PV system, is reported in [3]. The microgrid autonomous operation, has gained considerations for the benefits of the steadiness of service, higher reliability and good power quality. The autonomous operation of a microgrid is required, when the main grid is disconnected from the microgrid by the power electronics switch. The islanded operation of microgrid could be unintentional, when the grid outage or a fault is detected within the microgrid or intentional, when a preplanned maintenance is required for the main grid. The development of autonomous microgrid is considered for the rural areas, where the electrical power generation from main grid is not economical. Balancing of active power in microgrid, is proposed for the DC link voltage control of voltage source converter (VSC) based distributed generations (DGs) in [4]. The analysis of the operating modes and control methods for different power converters; grid-feeding, grid-forming and grid-supporting configurations, are given in [5]. An AC/DC hybrid microgrid control with balanced operation of all the converters in various load and source conditions, is

depicted in [6]. The implementation of renewable resources with the battery based microgrid for different operating conditions by control algorithm, is shown in [7]. A modified voltage control technique for power management between solar PV array, BES and loads, is depicted in [8]. The voltage regulation by control algorithm to stabilize voltage fluctuations due to variations in solar power, are described in [9]. The modelling, design and stability analysis of wind-solar PV-battery based system with modified P&O algorithm, are shown in [10]. The dynamic performance and stability analysis of solar PV–BES based system, are reported in [11]. The different power quality problems surfaced by incorporation of renewable sources, are described in [12].

The proposed single stage solar PV-BES microgrid system uses a voltage control technique for VSC in autonomous mode. The P&O method is utilized to obtain peak solar PV power. The microgrid works in standalone mode and the battery energy storage (BES) releases the power during peak load condition and stores the surplus power during light load condition. The power flow of solar PV-BES microgrid in standalone mode regulates the voltage and frequency within the allowable limits. The charging and discharging of battery are as per the load variations. The voltage rating of BES, decreases as the battery is not incorporated directly across the DC link, but it is connected through DC to DC buck boost converter. The experimental results of the system reveal satisfactory behaviour under perturbation in environmental conditions and load. The total harmonic distortion (THD) of load voltage is observed within a range (below 5%) as per the IEEE-519 standard [13].

II. SYSTEM CONFIGURATION

The proposed single stage PV-BES based microgrid configuration is depicted in Fig. 1. The microgrid has a battery, solar PV array, and three phase two level voltage source converter. The solar PV array is incorporated across the DC link of VSC. The VSC is made of three insulated gate bipolar transistor (IGBT) legs utilized for DC-AC conversion. In the proposed microgrid system, there is no essential requirement of extra DC to DC converter for peak power extraction from solar PV array, therefore, a single stage topology is chosen. Therefore, the cost and the losses are minimized in single stage system. The BES is integrated across the DC link by a bidirectional converter. The BES is utilized for the load management by dissipating power during peak load conditions and storing excess power during light load conditions. The output terminals of VSC, are connected

978-1-5386-6624-1/18 $31.00 © 2018 IEEE

564

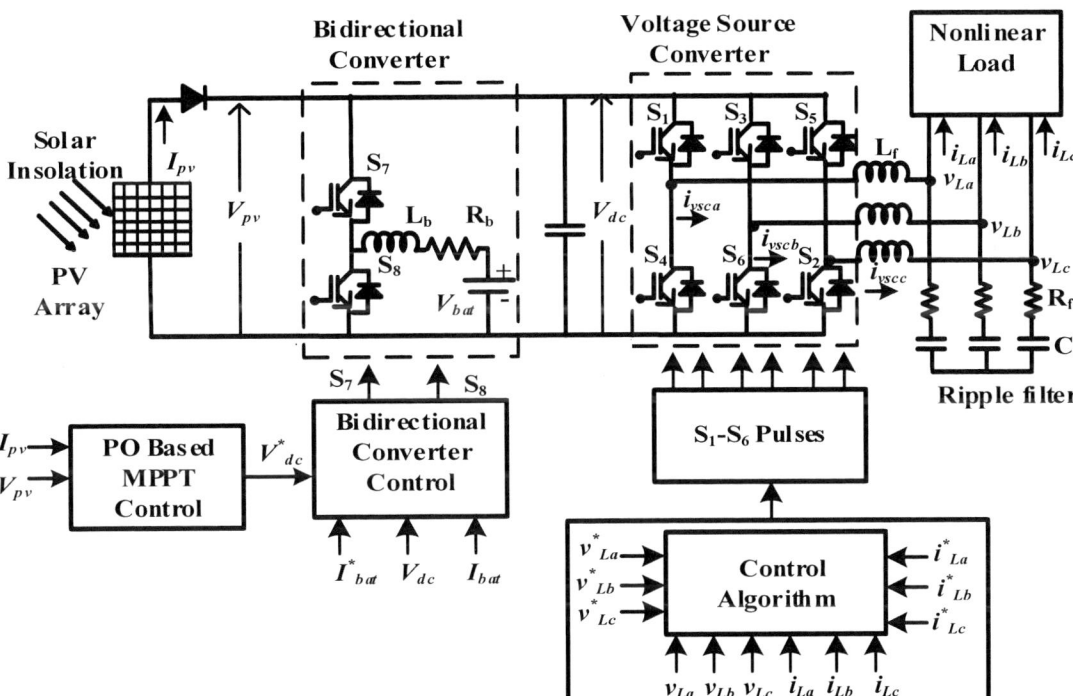

Fig.1 Proposed system configuration

to point of common coupling (PCC) by interfacing inductors. A nonlinear load is connected at PCC, wherein a ripple filter is also connected to absorb the switching ripples created by the VSC.

A. Selecting Selection of DC Link Voltage

The DC link voltage (V_{dc}) of VSC at PCC is calculated as,

$$V_{dc} = \frac{2*\sqrt{2}*V_{LL}}{m*\sqrt{3}} = \frac{2*\sqrt{2}*220}{1*\sqrt{3}} = 359V \quad (1)$$

where, V_{LL} is the PCC line voltage, m is the modulation index, which are considered as 220 V and 1respectively.

B. Selection of DC Link Capacitor

The capacitance of DC link is evaluated as,

$$C_{dc} = \frac{P_{pv}}{V_{dc}*w*2*V_{dccr}}$$

$$= \frac{1400}{360*(2*pi*50)*2*0.01*360} = 1720\mu F \quad (2)$$

where, P_{pv} is the solar PV power, w is the angular frequency and V_{dcrr} is % voltage ripples in V_{dc} and its value is taken as 1 % of the DC link voltage.

C. Selection of Interfacing Inductor

The value of interfacing inductor is evaluated as [14],

$$L_f = \frac{V_{dc}*m*\sqrt{3}}{12*h*f_{sw}*\Delta i} = \frac{360*1*\sqrt{3}}{12*1.2*10000*0.13*5.5} = 6mH \quad (3)$$

where, the value of switching frequency (f_{sw}) of VSC is taken

as 10 kHz, h (overloading factor) and m (modulation index) are constants, which are taken as 1.2 and 1 respectively. The Δ_i is % ripple current and its value is taken as 13 % of the peak current.

D. Bidirectional Converter Design

There are two operating modes for the bidirectional DC to DC converter. The converter works in boost mode, when the battery is in discharging mode and the converter works in buck mode, when the battery is in charging mode. The duty cycle (D_{buck}) for the buck mode is calculated as,

$$D_{buck} = \frac{V_{bat}}{V_{dc}} = \frac{240}{360} = 0.66 \quad (4)$$

where, the value of battery voltage (V_{bat}) is taken as 240 V. The value of inductor in buck mode is calculated as,

$$L_{buck} = \frac{(V_{dc} - V_{bat})*D_{buck}}{f_{sw}*\Delta I_l} = \frac{(360-240)*.66}{10000*0.2*5.5} = 6mH \quad (5)$$

where, the value of f_{sw} for bidirectional converter is taken as 10 kHz. The Δ_l in the buck mode is taken as 20 % of discharging current. The duty cycle (D_{boost}) for the boost mode is calculated as,

$$D_{boost} = \frac{V_{dc} - V_{bat}}{V_{dc}} = \frac{360-240}{360} = 0.333 \quad (6)$$

The value of inductor in boost mode is calculated as,

$$L_{boost} = \frac{D_{boost}*V_{bat}}{f_{sw}*\Delta I_l} = \frac{0.333*240}{10000*0.2*5.5} = 6mH \quad (7)$$

The Δ_l in the boost mode is taken as 20 % of charging current

978-1-5386-6624-1/18 $31.00 © 2018 IEEE

Therefore, the value of inductor for bidirectional converter (L_b) considered as 6 mH.

III. CONTROL ALGORITHM

The performance of control strategy of solar PV-BES based microgrid system in standalone mode has three main parts; MPPT control for solar PV array, voltage control for VSC and the battery with bidirectional converter to control V_{dc} of VSC.

A. MPPT Controller

To achieve the peak power from a solar PV array, the P&O MPPT method is utilized. The derivative of power to voltage at MPPT is zero. The solar PV current (Ipv) and voltage (Vpv) are given as inputs to MPPT controller. The principle of P & O method depends on the solar PV power versus voltage (Ppv-Vpv) curve. If the perturbation in voltage at any point on Ppv - Vpv curve, is such that the change in power (dPpv) is positive, then the operating point of solar PV array shifts towards the new MPP but the perturbation remains in the same direction. If the change in voltage is such that dPpv is negative, then the operating point of solar PV array shifts away from the MPP and the direction of perturbation is opposite. The reference DC link voltage V^*dc is output of the MPPT controller.

B. Control Technique for VSC

In standalone mode, the voltage control technique is utilized to produce the switching pulses for VSC as depicted in Fig. 2. The reference load voltages (v^*_{La}, v^*_{Lb}, v^*_{Lc}) are generated as,

$$v^*_{La} = V_{pr}\sin(w_r t), v^*_{Lb} = V_{mr}\sin\left(w_r t - \frac{2\pi}{3}\right),$$
$$v^*_{Lc} = V_{pr}\sin\left(w_r t + \frac{2\pi}{3}\right) \tag{8}$$

where, Vpr and w_r are reference peak voltage amplitude and where, and are reference peak voltage frequency, respectively. The sensed load voltages (v_{La}, v_{Lb}, v_{Lc}) are compared with v^*_{La}, v^*_{Lb}, v^*_{Lc} and the voltage errors are generated as,

$$v_{Laer}(q) = v^*_{La} - v_{La}, v_{Lber}(q) = v^*_{Lb} - v_{Lb}, v_{Lcer}(q) = v^*_{Lc} - v_{Lc} \tag{9}$$

The v_{Laer}, v_{Lber} and v_{Lcer} are given to PI regulators to give reference load currents voltages (i^*_{La}, i^*_{Lb}, i^*_{Lc}) as,

$$i^*_{La}(q+1) = i^*_{La}(q) + k_{pLaer} v_{Laer}(q+1) + k_{iLaer}\{v_{Laer}(q+1) - v_{Laer}(q)\} \tag{10}$$
$$i^*_{Lb}(q+1) = i^*_{Lb}(q) + k_{pLber} v_{Lber}(q+1) + k_{iLber}\{v_{Lber}(q+1) - v_{Lber}(q)\} \tag{11}$$
$$i^*_{Lc}(q+1) = i^*_{Lc}(q) + k_{pLcer} v_{Lcer}(q+1) + k_{iLcer}\{v_{Lcer}(q+1) - v_{Lcer}(q)\} \tag{12}$$
$$k_{pLaer} = k_{pLber} = k_{pLcer} = k_{pv} \tag{13}$$
$$k_{iLaer} = k_{iLber} = k_{iLcer} = k_{iv} \tag{14}$$

where, k_{pv} and k_{iv} are the proportional and integral constants of PI regulator to regulate the load voltages, respectively. The i^*_{La}, i^*_{Lb} and i^*_{Lc} are compared with sensed load currents (i_{La}, i_{Lb}, i_{Lc}) and current errors are generated, which are given to the hysteresis controller to produce the switching pulses for VSC in voltage control mode and are described as,

$$i_{erLa} = i^*_{La} - i_{La}, i_{erLb} = i^*_{Lb} - i_{Lb}, i_{erLc} = i^*_{Lc} - i_{Lc} \tag{15}$$

C. Bidirectional Converter Controller

The bidirectional controller as depicted in Fig. 3 consists of two main parts; the voltage control of DC link of VSC and the current control of the battery. The bidirectional converter is used to control the sensed voltage V_{dc} near to the reference DC link voltage (V^*_{dc}) according to MPPT in standalone mode. The V^*_{dc} and V_{dc} are compared and voltage error (V_{dcer}) is given as input to a PI regulator.

$$V_{dcer}(n) = V^*_{dc}(n) - V_{dc}(n) \tag{16}$$

The reference battery current (I^*_{bat}) is the output of PI regulator and calculated as,

$$I^*_{bat}(n+1) = I^*_{bat}(n) + k_{pdc} V_{dcer}(n+1) + k_{idc}\{V_{dcer}(n+1) - V_{dcer}(n)\} \tag{17}$$

where, k_{pdc} and k_{idc} are the proportional and integral constants of PI regulator to regulate V_{dc}, respectively. The sensed battery current (I_{bat}) is subtracted from I^*_{bat}, a battery current error (I_{bater}) is calculated as,

$$I_{bater} = I^*_{bat} - I_{bat} \tag{18}$$

The I_{bater} is given as an input to the PI regulator. The output of PI regulator current (I^*_{er}) is evaluated as,

$$I^*_{er}(n+1) = I^*_{er}(n) + k_{pbat} I_{bater}(n+1) + k_{ibat}\{I_{bater}(n+1) - I_{bater}(n)\} \tag{19}$$

where, k_{pbat} and k_{ibat} are the proportional and integral constants of PI regulator to control I_{bat}, respectively.

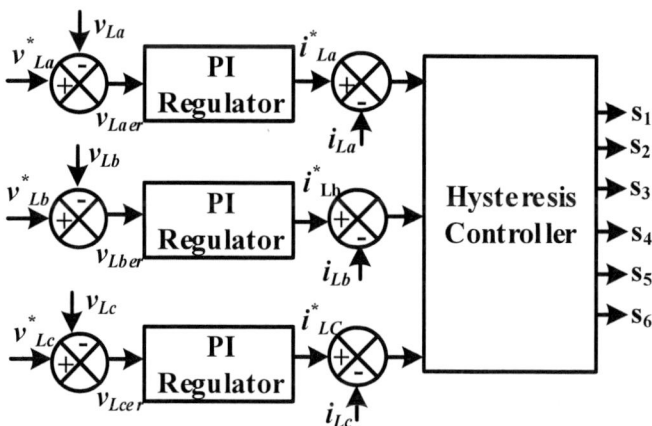

Fig. 2 Voltage control algorithm for VSC in standalone mode

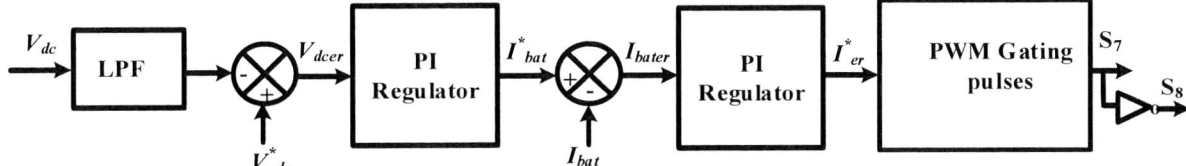

Fig. 3 Bidirectional converter control

IV. RESULTS AND DISCUSSION

A prototype is developed for a 1.4 kW microgrid in the laboratory, to perform the tests to validate the feasibility of the proposed solar PV-BES based microgrid in standalone mode. The Hall-Effect voltage sensors (LV-25) and current sensors (LA-55P) are utilized for sensing the signals v_{Lab}, v_{Lbc}, V_{dc}, i_{La}, i_{Lb}, I_{pv} and I_{bat}. The solar PV simulator (AMETEK, ETS600/17) is utilized to realize solar PV system. The control algorithm is loaded into OPAL-RT. The OP4510 is a real time simulator with 4 Intel cores 3.30-GHz processor. The proposed system is implemented through OPAL-RT (OP4510) with a sampling time of 20 μs. The dynamics of the system is recorded using a digital storage oscilloscope (DSO). The output pulses from OPAL-RT are isolated optically from the power circuit by opto-couplers. The parameters for the proposed system, are depicted in Appendix. The dynamic behaviour of system is analysed for perturbation in load and solar irradiance. The behaviour of system is deliberated here as follows.

A. Steady State Behaviour of System

The steady state behaviour of system is depicted in Figs. 4 (a)-(d) and Figs. 5 (a)-(h). The waveforms of v_{Lab}, v_{La}, v^*_{La} i^*_{vsc}, v_{Lb}, v^*_{Lb} v_{Lc}, v^*_{Lc} i_{La}, i_{Lb}, i_{Lc} and V_{dc} are shown in Figs. 4 (a)-(c). The V_{dc} is maintained constant as depicted in Fig. 4 (c). The waveforms of V_{bat}, I_{pv} ,V_{pv}, and I_{bat} are depicted in Fig. 4(d). Figs. 5 (a)-(c) present the current of phase 'c', power and THD of load current. The THD of i_{Lc} is 28.5 %. Figs. 5 (d)-(e) present current of phase 'c' and power of VSC. The load voltage THD is 3.7 % as shown in Fig 5 (f). The battery voltage (V_{bat}) with I_{bat} and battery power (P_{bat}) are shown in Figs. 5 (g)-(h).

B. Dynamic Response of System under Perturbation in Load

Figs. 6 (a)-(d) depict the response of system under load variation. Figs. 6 (a)-(b) show the waveforms of v_{Lab}, i_{La}, i_{vsca} and V_{dc} system for a decrease and an increase in the load current. The magnitude of i_{vsca} decreases and increases with the corresponding change in i_{La}. There is no variation in the

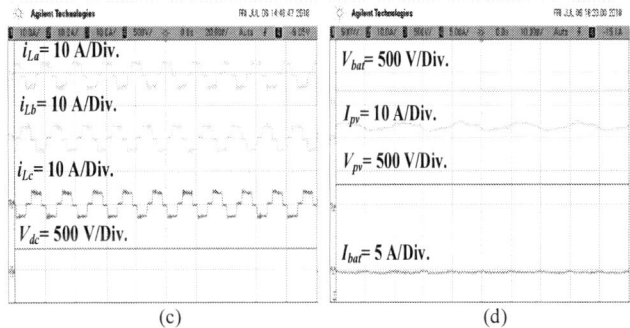

Fig. 4 (a)-(d) Steady state response of system

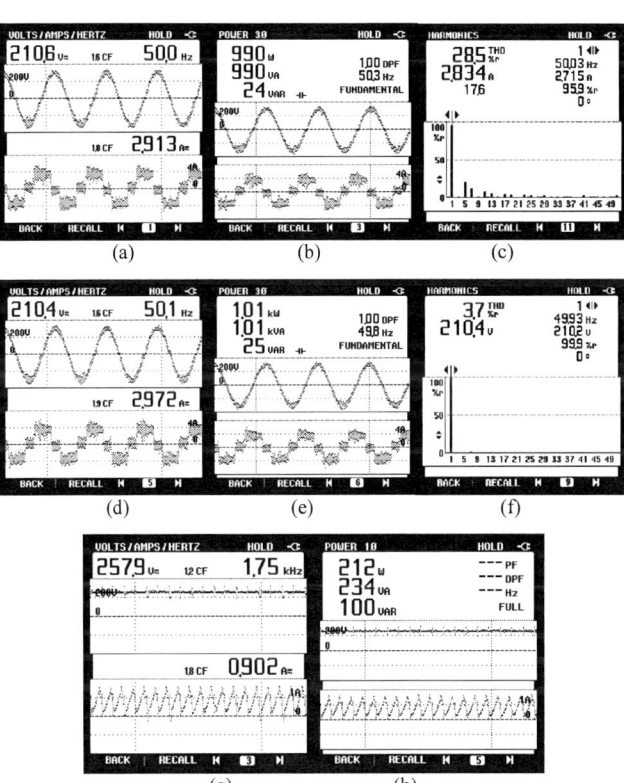

Fig. 5 Response of microgrid system (a)-(c) v_{Lab} with i_{Lc}, load power and THD of i_{Lc}, (d)-(f) v_{Lab} with i_{vscc}, VSC power and THD of v_{Lab}, (g)-(h) BES voltage with current and BES power

magnitudes of v_{Lab} and V_{dc}. Figs. 6 (c)-(d) illustrate the waveforms of V_{bat}, I_{pv} ,V_{pv}, and I_{bat} under change in the load. There is no variation in V_{pv} and. I_{pv}. During the light load condition, the battery current is negative, therefore, the BES is in charging mode as depicted in Fig. 6 (c). When the load

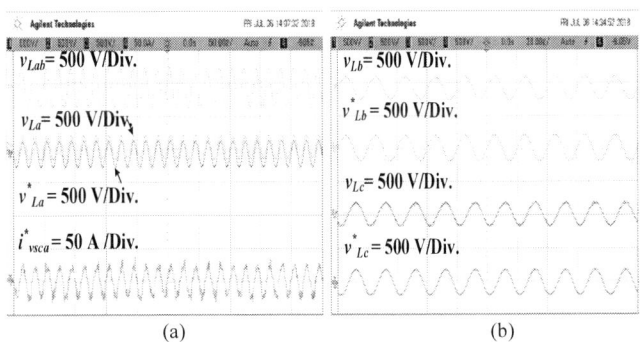

(a)　　　　　　　　(b)

978-1-5386-6624-1/18 $31.00 © 2018 IEEE　　　567

demand is increased, the BES is in discharging mode as shown in Fig. 6 (d).

C. Dynamic Behaviour of System Under Perturbation in Solar Irradiance

Figs. 7 (a)-(d) present the system behavior in case of solar irradiance change. Figs. 7 (a)-(b) show the waveforms of V_{bat}, I_{pv}, V_{pv}, and I_{bat}. Fig. 7 (a) depicts the solar irradiance decrease from 1000 W/m² to 500 W/m². The magnitude of I_{pv} is decreased and BES comes into discharging mode from the charging mode, to fulfil the load demand. The battery current is positive (discharging mode). Fig. 7 (b) shows increased solar PV power from 500 W/m² to 1000 W/m². The magnitude of I_{pv} is increased and the battery shifts from discharging mode to the charging mode. The battery current is negative (charging mode). The V_{pv} under solar insolation change remains at its required MPP value. However, during these transitions v_{Lab}, i_{La}, i_{vsca} and V_{dc} are unaffected as depicted in Figs. 7 (c)-(d), because the load currents are constant during this change. The V_{dc} maintains to its actual magnitude during these changes. The performance of MPPT algorithm is represented in Figs. 8 (a)-(b) at 500 W/m² and 1000 W/m², which shows almost 100 % MPPT of a PV array.

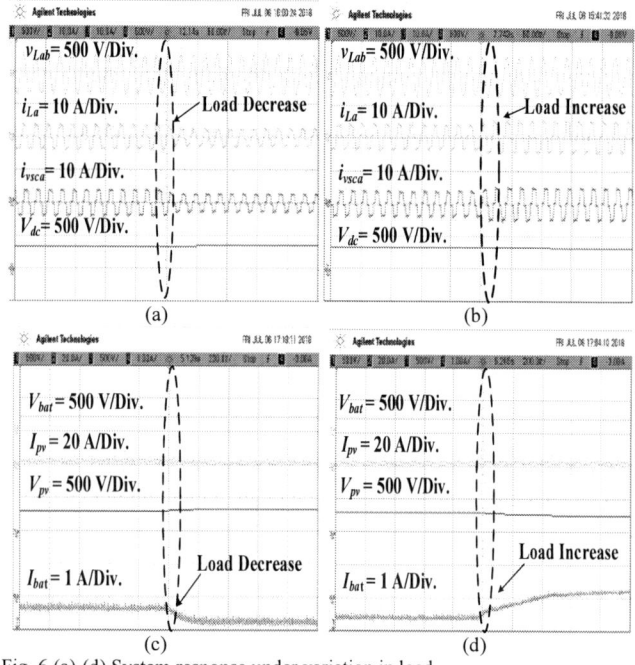

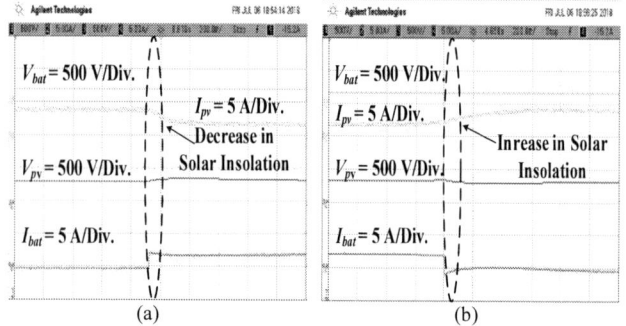

Fig. 6 (a)-(d) System response under variation in load

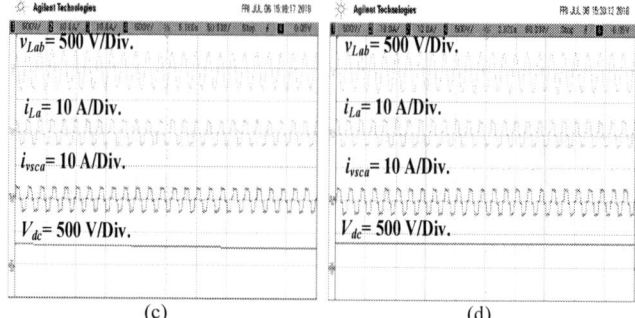

Fig. 7 System response (a)-(d) Under variation in solar insolation

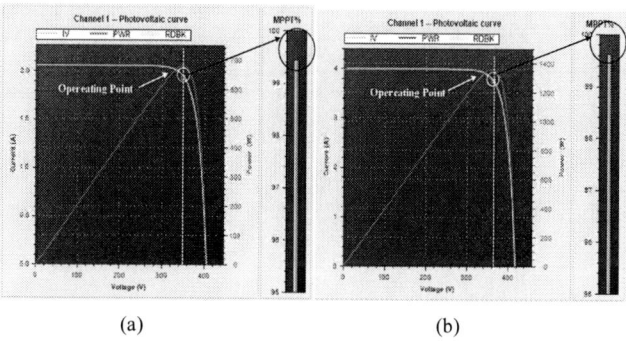

Fig. 8 Response of MPPT method at (a) 500 W/m² (b) 1000 W/m²

V. CONCLUSION

The voltage control of single stage solar PV-BES based microgrid system in standalone mode, has been demonstrated. The microgrid feeds the load in standalone mode. The solar PV voltage is maintained to required MPP value under perturbation in insolation level. The battery incorporated through bidirectional converter has maintained MPPT voltage to the appropriate value across the DC link. The charging mode and discharging mode of battery, under change in solar irradiance and load perturbation have been maintained with a bidirectional controller. The voltage control provides voltage and frequency regulation for the loads in autonomous mode of the microgrid. The system mitigates load voltage harmonics. The solar PV-BES microgrid system has satisfactory dynamic and steady state performances. The experimental results of the proposed system, have been validated by the voltage regulation technique for remote islands.

ACKNOWLEDGMENT

Authors are thankful to OPAL-RT for funding this project under Grant Number: RP03253.

APPENDIX

Parameters for Proposed System: P_{pv} = 1.4 kW, V_{oc} = 418 V, I_{sc} = 4 A, V_{mp} = 360 V, I_{mp} = 3.79 A, Nonlinear load: Diode bridge rectifier (3 phase) load with R = 60 Ω, L = 0.2 H, RC filter; C_f = 10 μF and R_f = 5 Ω, Inductor (interfacing) = 6 mH, f = 50 Hz; V_{dc} = 360 V, C_{dc} = 1720 μF. Switching frequency =

10 kHz, Voltage Controller PI gains, $k_{pv} = 0.95$, $k_{iv} = 0.001$; BES Rating, $V_{bat} = 240$ V and Ampere hour (Ah) = 7; Bidirectional DC-DC converter, $k_{pdc} = 0.14$, $k_{idc} = 0.002$, $k_{pbat} = 0.1$, $k_{ibat} = 0.0001$ and inductance $L_b = 6$ mH.

REFERENCES

[1] M. Malinowski, J. I. Leon and H. Abu-Rub, "Solar photovoltaic and thermal energy systems: current technology and future trends," *Proc. IEEE*, vol. 105, no. 11, pp. 2132-2146, Nov. 2017.

[2] M. A. G. de Brito, L. Galotto, L. P. Sampaio, G. d. A. e Melo and C. A. Canesin, "Evaluation of the Main MPPT techniques for photovoltaic applications," *IEEE Trans. Ind. Electron*, vol. 60, no. 3, pp. 1156-1167, March 2013.

[3] G. Petrone, G. Spagnuolo and M. Vitelli, "A multivariable perturb-and-observe maximum power point tracking technique applied to a single-stage photovoltaic inverter," *IEEE Trans. Ind. Electron.*, vol. 58, no. 1, pp. 76-84, Jan. 2011.

[4] P. H. Divshali, A. Alimardani, M. Abedi and S. H. Hosseinian, "Decentralized cooperative control strategy of micro sources for stabilizing autonomous VSC-based microgrids," *IEEE Trans. Power Syst.*, vol. 27, no. 4, pp. 1949–1959, Nov. 2012.

[5] J. Rocabert, P. Rodriguez, F. Blaabjerg and Alvaro Luna, "Control of power converters in ac microgrids," *IEEE Trans. Power Electron*, vol. 27, no. 11, pp. 4734-4749, Nov. 2012.

[6] X. Liu, P. Chiang Loh and P. Wang, "A hybrid ac/dc microgrid and its coordination control," *IEEE Trans. Smart Grid*, vol. 2, no. 2, pp. 278-286, June 2011.

[7] S. S. Thale, R. Wandhare and V. Agarwal, "A novel reconfigurable microgrid architecture with renewable energy sources and storage," *IEEE Trans. Ind. Appl.*, vol. 51, no. 2, pp. 1805–1816, Mar./Apr. 2015.

[8] I. Anand, S. Subramaniam, D. Biswas and M. Kaliamoorthy, "Dynamic power management system employing single stage power converter for standalone solar PV applications," *IEEE Trans. Power Electron.* Early Access.

[9] J.Krata and T. K. Saha, "Real-Time coordinated voltage support with battery energy storage in a distribution grid equipped with medium-scale PV generation," *IEEE Trans. Smart Grid*, Early Access.

[10] M. Rezkallah, A. Hamadi, A. Chandra and B. Singh, "Design and implementation of active power control with improved P&O method for wind-PV-battery-based standalone generation system," *IEEE Trans. Ind. Electron.*, vol. 65, no. 7, pp. 5590-5600, July 2018.

[11] X. Xiong, C. K. Tse and X. Ruan, "Bifurcation analysis of standalone photovoltaic-battery hybrid power system," *IEEE Trans. Circuits and Systems I: Regular Papers*, vol. 60, no. 5, pp. 1354-1365, May 2013.

[12] X. Liang, "Emerging power quality challenges due to integration of renewable energy sources," *IEEE Trans. Ind. Appl.*, vol. 53, no. 2, pp. 855-866, Mar. /Apr. 2017.

[13] IEEE Recommended Practices and Requirements for Harmonic Control in Electrical Power Systems, IEEE Standard 519, 1992.

[14] B. Singh, A. Chandra, and K. Al-Haddad, 'Power quality: problems and mitigation techniques' John Wiley & Sons Ltd., United Kingdom, 2015.

Wavelet Based Control of Shunt Compensator for Power Quality Enhancement

Masood Anzar
Department of Electrical Engineering
Delhi Technological University
New Delhi, India
masoodanzar@gmail.com

Narendra Kumar
Department of Electrical Engineering
Delhi Technological University
New Delhi, India
dnk_1963@yahoo.com

M. Rizwan
Department of Electrical Engineering
Delhi Technological University
New Delhi, India
rizwan@dce.ac.in

Abstract— **Wavelet based control is proposed in the paper for the removal of harmonics and power factor improvement. The non-linear balanced or unbalanced load current is divided into direct and quadrature axis. The active and reactive current weights are extracted by the wavelet transform. The technique is capable to eliminate of harmonics, compensate reactive power and load balancing. The approach is finds to be among the fastest weight learning approach.**

Keywords—Harmonics Elimination, Reactive Power Compensation, Wavelets Control, Shunt Compensator

I. INTRODUCTION

The disturbances in the power system are increasing with the increase in the usage of electronic equipment. As a result, many researchers are focused to find the possible solution on the enhancement of reliability and power quality of distribution system.

Non-linear reactive and unbalanced loads have adversely affected power quality supplied by the source. Some of the problems include the temperature rise of electrical equipments, neutral current, voltage between ground and neutral and supply rating. Compensators are proposed as a solution of these power quality (PQ) issues. The shunt compensator provides possible solution for harmonics current, low power factor and load unbalancing, which are the major issues of power quality.

There are two operational process of a shunt compensator. One includes the identifying current harmonics while the other governs the injection of the compensated current into power network. The algorithms based on accuracy, the speed, stability, computation complexity and implementation cost are proposed in due time. The authors have proposed various solutions based on time domain like IRP, SRF, and symmetrical component theory [1]. But such time domain techniques have some limitations of their own. The performance is degraded in case of unequal power disturbances in all the phases. Periodic signals are required for frequency domain techniques [2]. Fourier transform estimates the frequencies present in the signal over the entire time. It fails to find the time of the transient. Also, the Fourier analysis fails to consider non stationary signals where frequencies evolve with time [3]. Fourier techniques also have some additional issues like aliasing with non-stationary signals and Gibbs' phenomena. STFT (Short Time Fourier Transform) was also proposed [3] but it does not provide good resolution both in time and frequency due to fixed window size. Due to this

the signal transient analysis was still a hindrance in the STFT.

Wavelet Transform based method provides an optimized resolution of frequency and time, overcoming the above discussed drawbacks. Thus wavelets transform is a proper tool to deal with real life signals, i.e., low duration, high frequency disturbances and signals of low frequencies with longer time duration [4]. Thus wavelets are best suited for the power system signals analysis. As wavelet transform based method extract a band of frequency, the technique is also found to be less sensitive for frequency oscillation in power system. The time domain based algorithm for the estimation of phase-angle and amplitude can also be synchronously applied with other frequency estimation algorithms for frequency variation detection [5]. But these techniques are complex and not as fast as wavelet transform based algorithm in harmonic conditions. *Musab.A et al* [6] used wavelet transform for active component estimation from the load current. The computation complexity increases because it requires wavelet transform analysis of the currents of all the three phases. Also, the absence reactive power compensation and ZVR mode provided us a research gap. The paper introduces the wavelet control for the removal of harmonics and compensation of reactive power and load unbalancing.

II. WAVELET TRANSFORM

The signal is decomposed into oscillatory mathematical functions called as wavelets in wavelet transforms. It provides "variable time frequency" resolution in comparison to Fourier transform which provide only frequency resolution. The wavelets oscillations make it highly useful in non-stationary signals. It is classified into continuous and discrete form

A. Continuous Wavelet Transform

The time and scale parameters are continuous in continuous wavelet transform (CWT). The transform of a time function $X(t)$ is given as:

$$F(s,\tau) = \frac{1}{\sqrt{s}} \int X(t) \Phi^*(\frac{t-\tau}{s}) dt \qquad (1)$$

Where, "s" and "τ" are the scale and translation of the wavelet with $\Phi(t)$ as a mother wavelet of zero mean value.

$$\int_{-\infty}^{\infty} \Phi(t) dt = 0 \qquad (2)$$

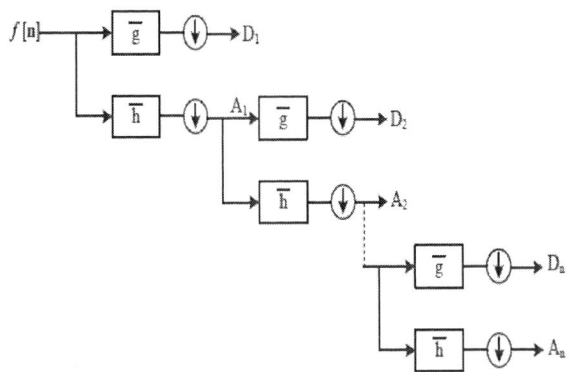

Fig (1): Multi Resolution Analysis of a Signal

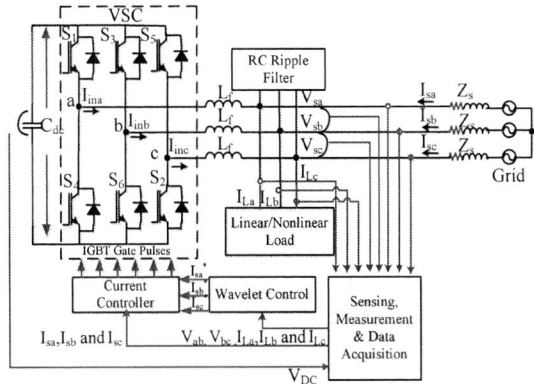

Fig (2): Schematic of Shunt Compensator

B. Discrete Wavelet Transform

In Discrete Wavelet Transform (DWT) scale and translation is incorporated in the wavelet function by providing binary dilations (2j) and dyadic translations (k/2j). The DWT of a signal is defined as:

$$DWT(m,k) = \frac{1}{\sqrt{s_0^j}} \sum_{n=0}^{N-1} X[n]\Phi(\frac{k-na_o^j}{s_o^j}) \qquad (3)$$

where, j,k are integers

The multi-resolution decomposition of the signal is very useful in signal analysis of the real world. The signal is converted into a series of $[A_1, D_1, A_2, D_2, ..., A_n, D_n]$ of coefficients at different level, where n is the number of level. The low pass coefficients $[A_1, A_2, ..., A_n]$ are known as approximations and high pass coefficient $[D_1, D_2, ..., D_n]$ are known as details.

Various discrete wavelets available to us are Haar, Daubechies, and Symlet etc. Daubechies wavelets represent the wavelet with adjoin, i.e. orthogonal, wavelet and inverse wavelet transform. The vanishing moments are maximum in Daubechies wavelets and hence represent polynomial functions of higher degree. It results in progressively finer discrete samplings. We have used Daubechies wavelet (db1) to filter a high frequency component from the direct axis current (I_d).

III. CONTROL SCHEME

The wavelet based proposed control scheme requires the source voltages (V_s), source current (I_s), load currents (I_L) and dc link voltage (V_{dc}) for the generation of reference currents $(Ia_{ref}, Ib_{ref}, Ic_{ref})$ as shown in Figure 2and 3. The generated reference current are compared with the source current (I_s) and produce the proper firing sequence for the voltage source converter (VSC) to produce compensating currents (I_c). These compensating currents are able to eliminate harmonics, maintaining the unity power factor under different load.

A. Unit Template Estimation of Load Voltages

The amplitude (V_m) of three phase voltages at PCC is estimated by the instantaneous value (V_{sa}, V_{sb}, V_{sc}) of the three sensed phases.

$$V_m = \sqrt{\frac{2(V_{sa}^2 + V_{sb}^2 + V_{sc}^2)}{3}} \qquad (4)$$

The unit templates in phase (v_{pa}, v_{pb}, v_{pc}) and quadrature vectors (v_{qa}, v_{qb}, v_{qc}) with voltages are calculated as given in [8-10].

$$v_{pa} = \frac{V_{sa}}{V_m}, \ v_{pb} = \frac{V_{sb}}{V_m}, \ v_{pc} = \frac{V_{sc}}{V_m} \qquad (5)$$

$$v_{qa} = \frac{1}{\sqrt{3}}(v_{pc} - v_{pb}), v_{qb} = \frac{\sqrt{3}v_{pa}}{2} + \frac{(v_{pb} - v_{pc})}{2\sqrt{3}}$$

$$v_{qc} = -\frac{\sqrt{3}v_{pa}}{2} + \frac{(v_{pb} - v_{pc})}{2\sqrt{3}} \qquad (6)$$

B. Average Active Power Weights Estimation

To control the dc link voltage, a proportional-integral controller based (PI) is used in the shunt compensator. The measured dc link voltage (V_c) is first passed through a low with the reference value of dc link voltage (V_c^{ref}) to calculate an error (V_{ce}). This error gives us the loss component when passed through a proportional integral (PI) controller. The controller output (w_{dc}^p) meets the compensator losses and regulates the dc link capacitor voltages. The output is given by:

$$w_{dc}^p(n) = w_{dc}^p(n-1) + k_p^p[V_c(n) - V_c(n-1)] + k_i^p V_c(n) \qquad (7)$$

where, the proportional and integral constants are k_p^p and k_i^p of controller.

Load currents are passed through Clark's transformation to obtain I_α and I_β

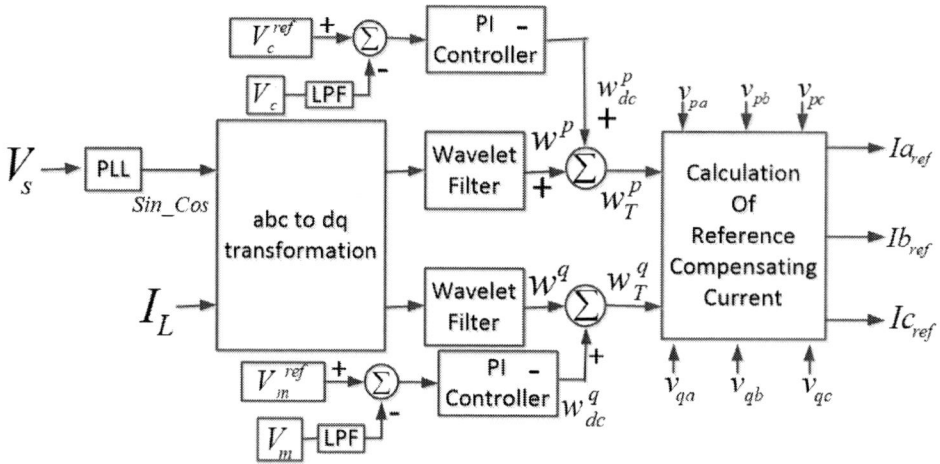

Figure (3): Block Diagram of Wavelet Based Algorithm

$$\begin{bmatrix} I_\alpha \\ I_\beta \end{bmatrix} = \sqrt{\frac{2}{3}} \begin{bmatrix} 1 & -1/2 & -1/2 \\ 0 & \sqrt{3}/2 & -\sqrt{3}/2 \end{bmatrix} \begin{bmatrix} iL_a \\ iL_b \\ iL_c \end{bmatrix} \qquad (8)$$

The obtained I_α and I_β are passed through Park's Transformation to get I_d and I_q.

$$\begin{bmatrix} I_d \\ I_q \end{bmatrix} = \begin{bmatrix} \cos\theta & \sin\theta \\ -\sin\theta & \cos\theta \end{bmatrix} \begin{bmatrix} I_\alpha \\ I_\beta \end{bmatrix} \qquad (9)$$

The direct axis current (I_d) is processed with DWT for the fundamental current component estimation. The direct axis current (I_d) is buffered and passed through the MATLAB function to calculate the wavelet decomposition of the signal. The signal is subdivided into two coefficients as approximation and detail with wavelets "db1" at level 10. The approximate coefficient (A_{10}) is our filtered signal which is further smoothed out by taking the windowing-mean some of the samples of A_{10} to find the fundamental active weight (w_{lpav}). The total active power weight (w_T^p) is calculated by adding the loss component (w_{dc}^p) to the average fundamental active power weight (w^p)

$$w_T^p = w^p + w_{dc}^p \qquad (10)$$

C. Estimation of Average Reactive Power Weights

The amplitude of sensed voltage (V_m) is compared with the reference value (V_m^{ref}) and the output is passed through PI controller. The PI controller output (w_{dc}^q) regulates the magnitude of phase voltage at its reference value and is:

$$w_{dc}^q(n) = w_{dc}^q(n-1) + k_p^q[V_m(n) - V_m(n-1)] + \\ k_i^q V_m(n) \qquad (11)$$

where, k_p^q and k_i^q are proportional and integral constants of ac bus controller.

This reactive power component (w^q) compensates the change in load current and the drop in voltage because of the impedance of the source. The reference reactive power (w_T^q) is obtained by subtraction of the average reactive weight (w_{dc}^q) from (w^q)

$$w_T^q = w^q - w_{dc}^q \qquad (12)$$

D. Estimation of Reference Currents

The active current component can be calculated by multiplying the active power unit templates with the active power weights. Similarly, reactive power current component can be obtained by multiplying reactive power unit templates with the reactive power weights. The active and reactive currents are combined to estimate the total reference current as:

$$Ia_{ref} = (w_T^p \times v_{pa}) + (w_T^q \times v_{qa})$$
$$Ib_{ref} = (w_T^p \times v_{pb}) + (w_T^q \times v_{qb})$$
$$Ic_{ref} = (w_T^p \times v_{pc}) + (w_T^q \times v_{qc}) \qquad (13)$$

IV. SIMULATION RESULTS

Simulation model based on wavelet control for the shunt compensator is developed in MATLAB/Simulink environment. Simulated results are presented to verify the wavelet control for power factor correction (PFC) and voltage regulation (VR) mode with nonlinear balanced/unbalanced loads.

A. Performance under load change for PFC

The dynamic performance of shunt compensator connected across the nonlinear load for PFC (power factor correction) mode can be seen in Figure 4. The waveforms of supply voltages (V_s), supply currents (I_s), load currents (I_L), compensator currents (I_f) and dc link voltages (V_{dc}) are shown in this figure. For time $t < 0.2$ sec the dc link voltage is maintained at its reference value and the source is delivering balanced sinusoidal currents in steady state.

978-1-5386-6624-1/18 $31.00 © 2018 IEEE   572

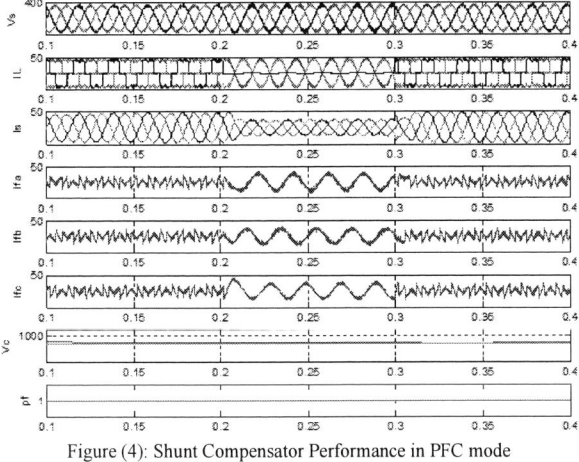

Figure (4): Shunt Compensator Performance in PFC mode

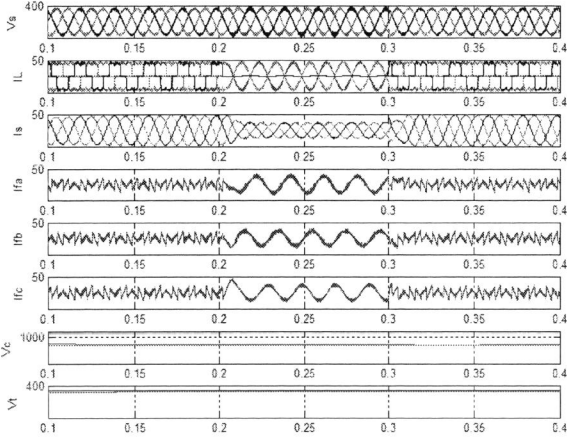

Figure (6): Shunt Compensator Performance in VR mode

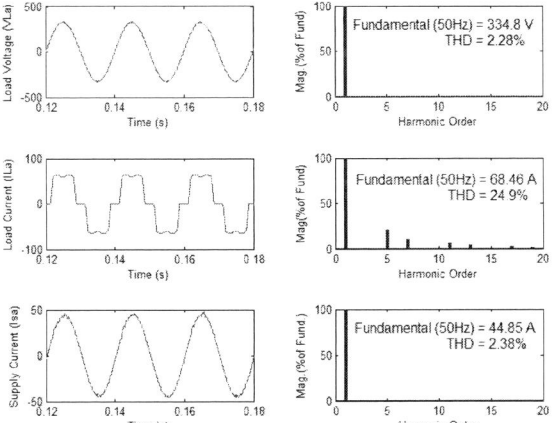

Figure (5): Supply Voltage, supply current and Load Current Harmonics in PFC Mode.

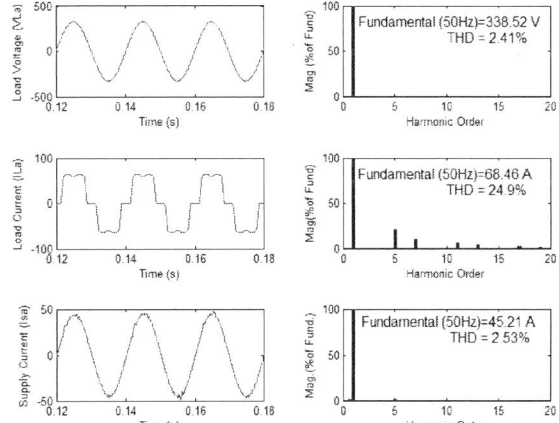

Figure (7): Supply Voltage, supply current, Load Current in VR Mode.

During time 0.2 < t < 0.3 sec, phase "c" is switched disconnected to create unbalancing in the load currents. It can be observed that the control continues to provide balanced supply currents during load unbalance.

The harmonic analysis of phase "a" for source voltage (V_s), source current (I_s) and load current is shown in figure 5. The total harmonic distortions (THD) observed are 2.28% in supply voltage (V_s), 2.38% in supply current (I_s) and 24.9% in load current (I_L). The satisfactory performance of compensator with wavelet control can be observed from the fact that the supply current THD is reduced to 2.28% even if the load requires the current of THD 24.95%. The obtained THD values of supply voltage (V_s) and supply current (I_s) fulfills the IEEE-519 standard.

B. Performance under load change for VR

The shunt compensator performances are shown in Figure 6. The supply voltages (V_s), supply currents (I_s), load currents (I_L), compensator currents (I_c), dc link capacitor voltage (V_c) and Voltage amplitude (V_m) are plotted in the figure 6.

The steady state condition results are observed for time $t \leq 0.2 \sec$. The source is providing sinusoidal balanced currents, with the dc link capacitor voltages and terminal voltage maintained at their reference values, i.e., $V_{dc}^* = 700V$, $V_t^* = 338.8V$.

The dynamic performance of control is also tested with load unbalancing by disconnecting phase "c" during $0.2 \leq t \leq 0.3 \sec$. The harmonic content of supply voltage (V_{sa}), supply current (I_{sa}) and non linear load current (I_{La}) for phase-A in voltage regulation are shown in Figure 7. The total harmonic distortions (THDs) for V_{sa}, I_{sa} and I_{La} are 2.41%, 2.53% and 24.9% respectively.

. The THD values of V_{sa} and I_{sa} are satisfying the permitted IEEE-519 standards [7]. It can be observed that the control is able to provide balanced sinusoidal supply currents under maintained values of dc link voltages and PCC voltages for harmonics PFC and VR.

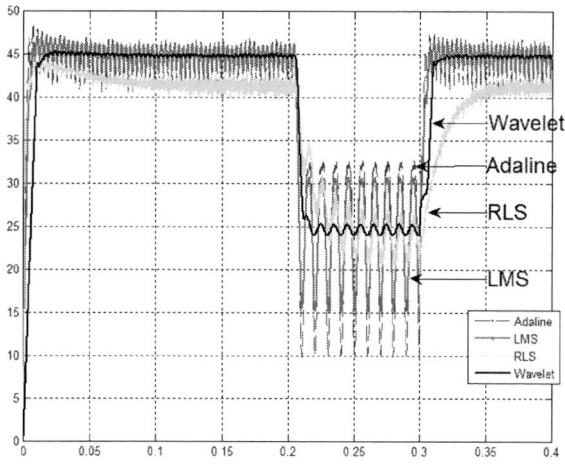

Figure (8): Comparison of Adaline, LMS, RLS and Wavelet Control

C. Comparative Analysis of Wavelet Based Control

The parameters of wavelet based control are compared with Adaline, LMS and RLS algorithm. The convergence of active weight of load current is compared with all the four algorithms in Figure 8. Result shows that wavelet control converge the weight faster in few cycles as compared to other controls algorithms. The estimated weight fluctuates around the mean value with Adaline, LMS and RLS control algorithm. But wavelet control provides more stable and fast weight update as compared with RLS algorithm. Adaline, LMS and RLS algorithm when implemented need a low pass filter after the weigh extraction, adding more delay into the system. But with wavelet control, no such low pass filter is required.

It can be also observed from these results that wavelet control provides best convergence of active weights under unbalanced load also. The wavelet control is seen to perform better in weight estimation and system delays.

TABLE I.   PARAMETERS IN DIFFERENT MODES

Modes	Parameters	Magnitude	THD (%)
Power Factor Correction	Load Voltage	334.8 V	2.28
	Source Current	44.85A	2.38
	Load Current	68.46 A	24.9
Voltage Regulation	Load Voltage	338.52 V	2.41
	Source Current	45.21 A	2.53
	Load Current	68.46 A,	24.9

[a.)]

V. CONCLUSION

The proposed wavelet control is applied for reference current determination from non-linear load currents. The control is also compared with other algorithms such as Adaline, LMS and RLS. The proposed wavelet based control algorithm is found to offer fast convergence and robustness as compared to other algorithms. The working is found satisfactory in both power factor correction (PFC) and voltage regulation modes under balanced and unbalanced loading conditions. The elimination of harmonic, compensation of reactive power and balance source currents are achieved with dc link voltage regulation. The supply

current THD is reduced from 24.9% to 2.38% and 2.53% in PFC and VR mode respectively, satisfying IEEE-519 standards

APPENDIX

3-ph, 50Hz, 415V AC Supply with source resistance $R_s = 0.01\Omega$ and source inductance $L_s = 2mH$, DC link voltage $V_c = 700V$, Interfacing Inductor: $L_f = 2.7mH$, Passive filter: Resistance $R_f = 5\Omega$, capacitance .DC Link Controller: $k_p^p = 0.1$, $k_i^p = 0.001$ Terminal Voltage PI controller: $k_p^q = 0.05, k_i^q = 0.001$, Nonlinear load: 3-phase bridge rectifier (uncontrolled) with $R = 13\Omega$.

REFERENCES

[1] U. Rao, Mahesh. K. Mishra, and A. Ghosh, Control strategies for load compensation using instantaneous symmetrical component theory under different supply voltages, IEEE Transactions on Power Delivery, vol. 23, no. 4, pp. 23102317, 2008.

[2] L. Asiminoaei, F. Blaabjerg, S. Hansen, Evaluation of harmonic detection methods for active power filter applications, in: Proceedings of APEC'05, vol. 1, March 2005, pp. 635–641.

[3] A. Galli, G. Heydt, and P. Ribeiro, Exploring the power of wavelet analysis, Computer Applications in Power, IEEE, vol. 9, no. 4, pp. 3741, Oct 1996.

[4] Shyh-Jier Huang and Cheng-Tao Hsieh, "Computation of continuous wavelet transform via a new wavelet function for visualization of power system disturbances," 2000 Power Engineering Society Summer Meeting (Cat. No.00CH37134), Seattle, WA, 2000, pp. 951-955 vol. 2.

[5] C. H. Kim and R. Aggarwal, Wavelet transforms in power systems. i. general introduction to the wavelet transforms, Power Engineering Journal, vol. 14, no. 2, pp. 8187, April 2000.

[6] Mus-ab.A, Mahesh K. Mishra, "Wavelet Transform Based Algorithms for Load Compensation Using DSTATCOM", IEEE PES Asia-Pacific Power and Energy Engineering Conference (APPEEC) 8-10 Nov. 2017

[7] A. Mus-ab and M. K. Mishra, "Wavelet transform based algorithms for load compensation using DSTATCOM," *2017 IEEE PES Asia-Pacific Power and Energy Engineering Conference (APPEEC)*, Bangalore, 2017, pp. 1-6.

[8] R. Kumar, B. Singh, D. T. Shahani and C. Jain, "Dual-Tree Complex Wavelet Transform-Based Control Algorithm for Power Quality Improvement in a Distribution System," in *IEEE Transactions on Industrial Electronics*, vol. 64, no. 1, pp. 764-772, Jan. 2017.

[9] W. Li, B. Rahmani and G. Liu, "A Wavelet-Based Shunt Active Power Filter to Integrate a Photovoltaic System to Power Grid," *2016 Sixth International Conference on Instrumentation & Measurement, Computer, Communication and Control (IMCCC)*, Harbin, 2016, pp. 482-485.

[10] B. Singh and S. Arya, "Composite observer-based control algorithm for distribution static compensator in four-wire supply system," IET Power Electron., vol. 6, Iss. 2, pp. 251–260, 2013.

[11] B. Singh and S. Arya, "Implementation of single-phase enhanced phase-locked loop-based control algorithm for three-phase DSTATCOM," IEEE Trans. Power Del., vol. 28, no. 3, pp. 1516-1524, July 2013.

[12] I. D. Landau, R. Lozano, M. M"Saad, and A. Karimi, Adaptive Control: Algorithms, Analysis and Applications. New York, NY, USA: Springer- Verlag, 2011.

[13] M. I. M. Montero, E. R. Cadaval, and F. B. Gonzalez, ''Comparison of control strategies for shunt active power filters in three phase four wire systems,'' IEEE Trans. Power Electron., vol. 22, no. 1, pp. 229–236, Jan. 2007

[14] M. Forghani and S. Afsharnia, ''Online wavelet transform-based control strategy for UPQC control system,'' IEEE Trans. Power Delivery, vol. 22, pp. 481–491, Jan. 2007.

[15] J. Barros and R. I. Diego, ''Analysis of harmonics in power systems using the wavelet-packet transform,'' IEEE Trans. Instrum Meas., vol. 57, pp. 63–69, Jan. 2008.

[16] G. Escobar, P. R. Martinez, J. Leyva Ramos, and P. Mattavelli, ''A negative feedback repetitive control scheme for harmonic compensation,'' IEEE Trans. Ind. Electron., vol. 53, no. 4, pp. 1383–1386, 2006.

[17] B. Singh, V. Verma, and J. Solanki, ''Neural network-based selective compensation of current quality problems in distribution system,'' IEEE Trans. Ind. Electron., vol. 54, no. 1, pp. 53–60, Feb. 2007.

[18] D. O. Abdeslam, P. Wira, J. Merckle, D. Flieller, and Y.-A. Chapuis, ''A unified artificial neural network architecture for active power filters,'' IEEE Trans. Ind. Electron., vol. 54, no. 1, pp. 61–75, Feb. 2007.

2nd IEEE International conference on power Electronics, Intelligent Control and Energy systems (ICPEICES-2018)

Analysis of Different MPPT Techniques Under Partial and Un-shaded Condition for Solar PV System

Anirudh Dube
Dept. of Electrical Engineering
Jamia Millia Islamia
Delhi, India
anirudhdube12@gmail.com

Majid Jamil
Dept. of Electrical Engineering
Jamia Millia Islamia
Delhi, India
majidjamil@hotmail.com

M Rizwan
Dept. of Electrical Engineering
Delhi Technological University
Delhi, India
rizwan@dce.ac.in

Abstract—**This paper focuses on the analysis of two different Maximum Power Point Tracking (MPPT) techniques for the solar photovoltaic (PV) system under partial shaded and un-shaded condition. Perturb & Observe (P&O) and Particle Swarm Optimization (PSO) method of MPPT tracking have been incorporated in the solar PV system and the results have been compared among each other. The main aim of this paper to apply the MPPT technique which can track global maxima and not the local maxima under the partial shaded condition of the PV module. The above two MPPT techniques have been tested under partial shaded condition and uniformly shaded condition of PV module. The test system is modeled and simulated using MATLAB/Simulink.**

Keywords—SPV, MPPT, P&O, PSO, and DC-DC converter

I. INTRODUCTION

In today's scenario when our society is moving towards the advancement in the technology, the usage of energy is increasing. With the constant depletion of the conventional sources of energy and steep enhancement in the demand of energy, the requirement of renewable energy sources to meet out the current electricity demand are becoming essential. To cater the demand of the power, it is essential to generate optimum power from renewable sources of energy. Being a well established technology and ample availability, solar power popularity is more among the other renewable energy sources. The maximum energy from the solar photovoltaic (SPV) module can be track using maximum power point tacking (MPPT) methods. The review on different MPPT techniques and its classification are given in [1]. The examination of cost and energy performance of MPPT techniques for solar PV applications is being carried out in [2].

Many researchers have worked out in the field of maximum power point tracking (MPPT) algorithms and come out with their results. [3-6] has presented the comparison of various MPPT techniques for uniform and partial shaded condition for solar PV system, while [7] focuses particularly on the MPPT techniques for the partial shaded conditions. Further, [8] proposes a precise detection scheme to determine the occurrence of partial shading whereas, a novel and fast MPPT method for both partial shaded condition as well as fast charging is proposed in [9].

The researches have tried to modify the conventional MPPT algorithms by combining them with advance strategies. [10] has proposed the modified perturb and observe MPPT algorithm for fast response in PV system while [11] has developed the novel adaptive MPPT for

sudden changes in the irradiance and [12] has proposed an improved P&O algorithm with soft computing techniques. Further researches have been made for tracking MPPT using soft computing techniques. [13] and [14] have proposed the particle swam optimization techniques for tracking the MPPT even under variation of irradiance. The comparison between the conventional methods and PSO based MPPT algorithm is proposed in [16]. Some authors have proposed new schemes for tracking the MPPT [17-18]. [20] proposes the novel techniques for tracking the MPPT under partial shaded condition for the grid connected solar PV system whereas [21] has come out with the MPPT problem as an optimization problem and used heuristic algorithm. Significant research has been done in the field of tracking MPPT for solar PV system either standalone or grid connected PV system and its continuing going on.

In this work, a comparative analysis between the two MPPT algorithms have been proposed, one the conventional P&O algorithm and other the advance PSO method. Both the MPPT techniques have been applied under partial and un-shaded condition to determine their responses under sudden changes in the environmental parameters like irradiance. Mathematical modeling of solar PV module is presented in the part II of this paper whereas, proposed system with MPPT techniques have been explained in section III. Section IV shows the simulation results of the proposed algorithm with different irradiance conditions.

II. MODELLING OF SOLAR PV MODULE

The equivalent circuit of the PV cell with single diode model is shown figure 1 wherein, one series and one parallel resistance is connected with diode and current source.

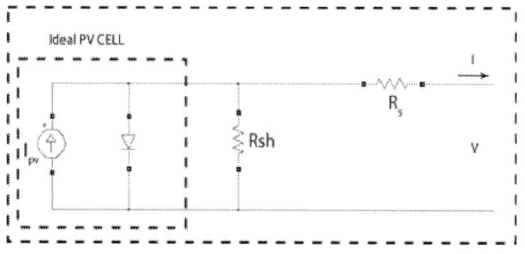

Fig. 1. Single Diode Model of PV cell

The photo current generated by solar irradiance at p-n junction at any instant k+1 as proposed in [19] may be given as:

978-1-5386-6624-1/18 $31.00 © 2018 IEEE

$$I_{pv}^{k+1} = N_p \left[I_{ph} - I_{d1} - \left(\frac{V_{pv} - I_{pv}^k R_s}{R_{sh}} \right) \right] \qquad (1)$$

where, N_p is the number of strings connected in parallel; I_{ph} is the photo current generated by solar irradiance to the p-n junction cell at any interval k; I_{d1} is the current flowing through diode; R_s is the resistance connected in series whereas R_{sh} is the resistance in shunt.

Photo current generated is represented as:

$$I_{ph} = \left(I_{phSTC} + k_i \Delta T \right) \frac{G}{G_{STC}} \qquad (2)$$

where, G is the actual irradiance and G_{STC} is the irradiance at standard test condition (STC) (1000W/m², 25°C and air mass 1.5), k_i is the short circuit current coefficient and ΔT is the difference between actual temperature and STC temperature.

An improved diode current is represented as equation 3 below:

$$I_{d1} = I_{01} \left[\exp \left(\frac{V_{pv} - I_{pv}^k R_s}{\alpha V_t} \right) \right]; \quad V_t = \frac{N_s kT}{q} \qquad (3)$$

I_{01} is the reserve saturation current of the p-n diode when not exposed to solar radiation; α is the diode ideality factor; V_t is the thermal voltage; k is the Boltzmann constant (1.38065×10^{23}J/K); and q is the electron charge (1.6021×10^{19}C).

The reverse saturation current I_{01} is expressed as:

$$I_{01} = \frac{I_{sc\_STC} + K_i \Delta T}{exp \left[\left(V_{oc\_STC} + K_v \Delta T \right) / \alpha V_t \right] - 1} \qquad (4)$$

The generation from the solar PV module is dependent on the value of the irradiance received. For an ideal case, the value of irradiance is 1000 W/m² but in actual it depends on the site. The generation also varies with the temperature which is not considered in the test system.

III. TEST SYSTEM WITH MPPT TECHNIQUES

Figure 2 presents the test solar PV system on DC side which is common for both grid connected and off grid system. The solar modules joined in series or parallel according to the current and voltage requirement resulting in formation of an array. It is connected with the DC-DC boost converter and the DC link capacitor before connecting to the voltage source converter (VSC) or DC load.

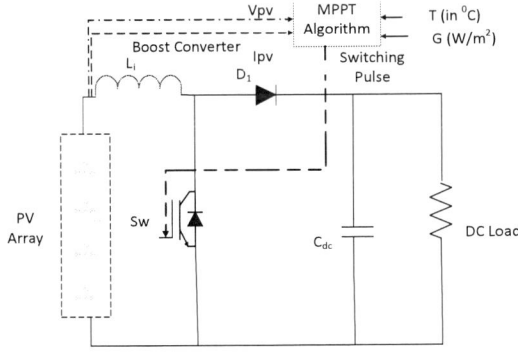

Fig. 2. Test system configuraton

The MPPT controller is used to generate pulses for the IGBT/MOSFET switch of DC-DC converter. This MPPT controller takes voltage and current of the PV system as an input and also the environmental factors like irradiance and temperature and accordingly based on algorithm being implemented, tracts the maximum power point. This maximum power point tracked by MPPT controller is given to the DC-DC converter in the form of pulse to control its switch to extract the maximum power for the solar PV system. Also, DC-DC converter converts the variable DC from the solar PV modules or arrays into constant DC for charging batteries or input the DC-AC converter for further applications.

Here, the test system considered with two environmental conditions which is partial shading condition where the irradiance of 1000 W/m², 800 W/m², 600 W/m² and 400 W/m² have been considered in different solar PV module and un-shaded condition. Figure 3 shows the difference in the above mentioned two cases. In the partial shading, few solar cells within the module is shaded resulting in the multiple maximum power points and the other is un-shaded condition where, all solar PV cells are without shading resulting in only one maximum power point.

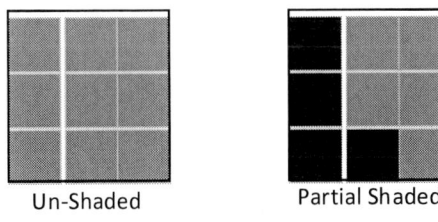

Un-Shaded   Partial Shaded

Fig. 3. Un-shaded vs Partial shaded condion of PV Module

Here, two algorithms have been compared, one which is the most common i.e. Perturb & Observe (P&O) algorithm and other the Particle Swarm Optimization (PSO). The reason for choosing these algorithms is their area of applications. During the partial shaded condition, as explained above, there are multiple maximum power points; out of which one is global maxima and rest are local maximas. The P&O algorithm fails to detect the global maxima and track the first maxima point whereas in PSO, the algorithm tracks the whole path and then tracks the global maxima among the various maxima point. Whereas, in the un-shaded condition, only one maxima is present and hence the time taken to track the maxima point is less in P&O algorithm as compared to PSO. PSO being the complex algorithm as compared to the P&O takes more time to track the maxima point in the un-shaded condition. The two algorithm are defined below.

A. Perturb & Observe (P&O) Algorithm

In this MPPT technique, the current and voltage of the solar PV system is measured to calculate the power at any instant say k. Then this instant power is compared with the previous power to decide the further operation. It notifies whether to increase or decrease the duty cycle. With the comparison from previous measured power, it determines whether to increase or decrease the duty cycle of the converter in order to maximize the power. If the input voltage is greater than the previous value, then the duty cycle (D) is decreased. If the input voltage was less than previously measured then increase the duty cycle to track the maximum power point. If the (ΔD) is very large then it will oscillate infinitely and it would never settle down at maximum power point. It is oscillatory in nature and its oscillations increase very large when the environmental or atmospheric conditions are changing thus making it not suitable for partial shaded condition.

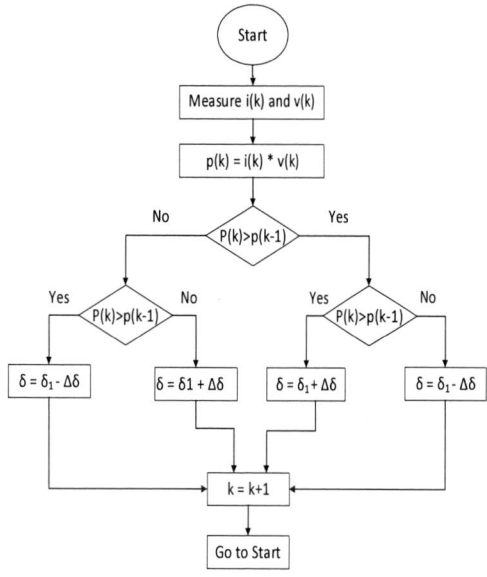

Fig. 4. P&O Algorithm for Solar PV system

B. Particle Swarm Optimization (PSO) Algorithm

The particle swarm optimization techniques is an intelligent techniques which optimizes the problem by evaluating the fitness function and then updating the local best (LB) and global best (GB) point thus resulting in tracking the global maxima in the case of partial shading condition. The flow diagram of the PSO algorithm is illustrated in figure 5. After updating the values of LB and GB, the fitness of the particle is being evaluated.

The standard PSO method may be characterized using the equation 5 & 6 given below:

$$v_{i(k+1)} = w(k)v_i(k) + c_1 r_1 \big(l_{best} - x_i(k)\big) + c_2 r_2 \big(g_{best} - x_i(k)\big) \quad (5)$$

$$x_i(k+1) = x_i(k) + v_i(k+1) \quad (6)$$

The position of the particle is $x_i(k)$, The velocity of the particle is given by $v_i(k)$ and the iteration number is denoted by the value k. The inertia weight is represented by w (k),r_1,r_2 are variables that are distributed randomly [0,1] . The c_1 and c_2 are represented the cognitive and the social coefficient. The l_{best} represent the local best or personal best position and is found by the particles so far and the g_{best} is used to get the global best position.

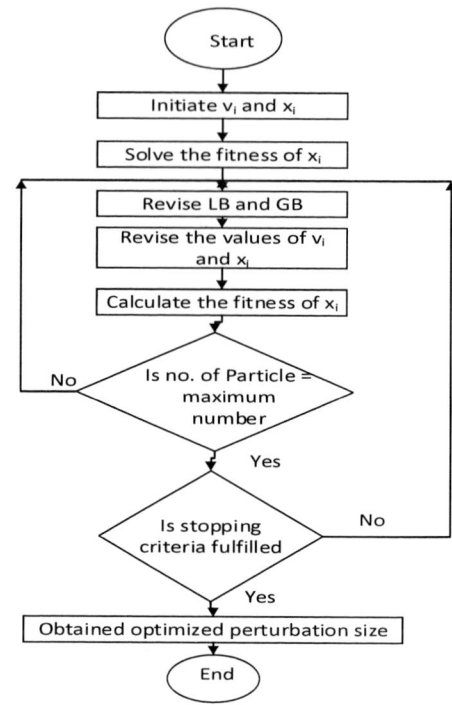

Fig. 5. PSO Algorithm for Solar PV system

The PSO can be combined to other conventional MPPT methods to give improve results [15] and also improves the efficiency.

IV. SIMULATION AND RESULTS

A test system of 400 W has been created in MATLAB/Simulink using Simscape toolbox as shown in figure 6.

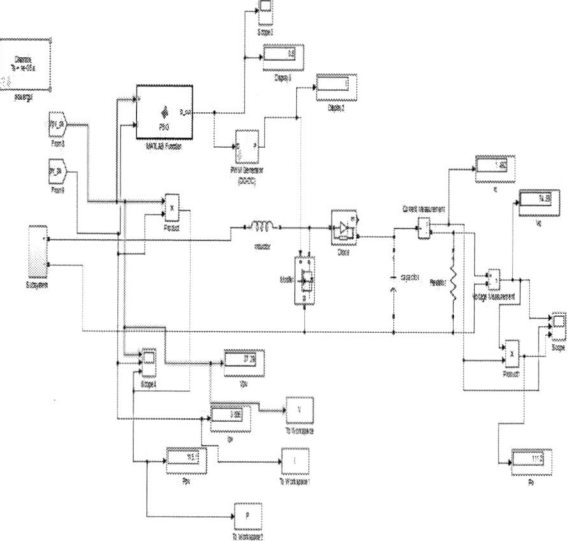

Fig. 6. Test System in MATLAB/Simulink

The values of the elements and parameters considered in the test system are mentioned in table I below:

978-1-5386-6624-1/18 $31.00 © 2018 IEEE   578

TABLE I.   SELECTED VALUES OF ELEMENTS AND PARAMETERS

Element	Formula	Selected Value
MPP Voltage of Module (V_{mpp})	$V_{mpp} = 0.85 \times V_{oc}$ V_{oc} is the open circuit voltage of module	17.3 V
MPP Current of Module (I_{mpp})	$I_{mpp} = 0.85 \times I_{oc}$ I_{oc} is the open circuit current of module	5.79 A
Power at maximum point (P_{mp})	$P_{mp} = (n_s*V_{mpp})*(n_p*I_{mpp})$ n_s and n_p are the number of modules in series and parallel respectively.	400.6 W
Duty cycly, D of DC-DC Converter	$D = 1 - \dfrac{V_{in}}{V_{out}}$	$D = 1 - \dfrac{37.29}{74.58} = 0.5$
Inductor , L_i	$L_i = \dfrac{RD(1-D)^2}{2f}$	5 mH
DC Link capacitor, C_{dc}	$C_{dc} = \dfrac{D*V}{R_{crit}.f.\Delta v}$	1200 µF

The simulation is done in the MATLAB/Simulink and the responses obtained are shown in below figure. Figure 7 shows the response curve of voltage, current and power of the solar modules. Figure 8 shows the on load response of I-V and P-V curve of solar array. The response curve of duty cycle generated by the MPPT controller is shown in figure 9 and figure 10 shows the photovoltaic voltage (V_{pv}), current (I_{pv}) and power (P_{pv}) of the PV array.

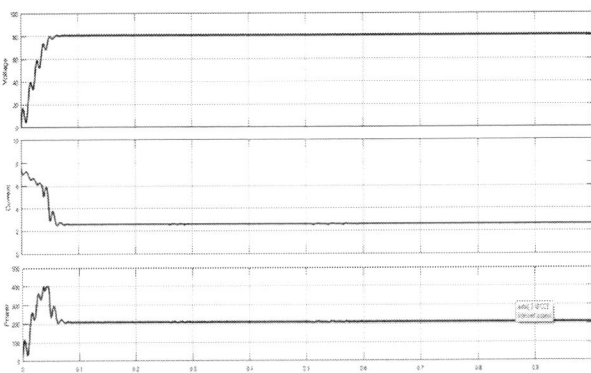

Fig. 7.   Voltage, Current and Power from the solar PV module

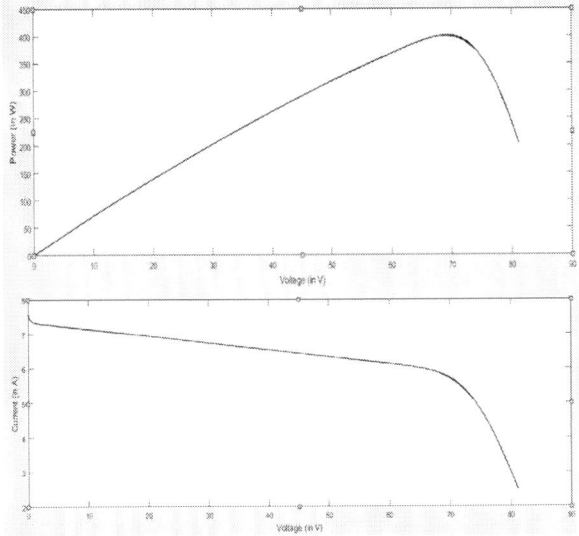

Fig. 8.   P-V and I-V curve of the PV array on load condition

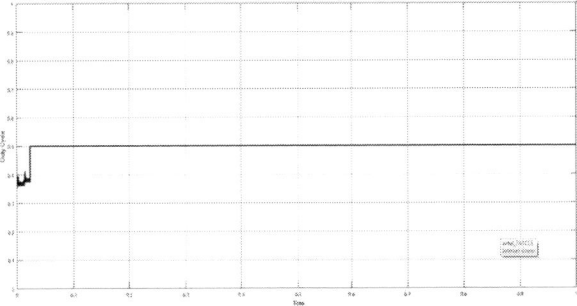

Fig. 9.   Duty Cycle generated by the MPPT controller

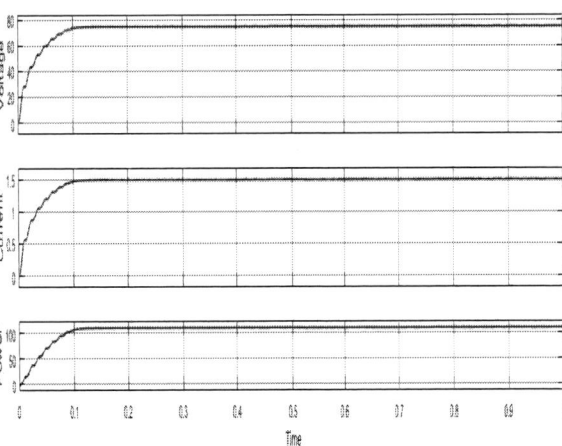

Fig. 10. Voltage (V_{pv}), Current (I_{pv}) and Power (P_{pv}) response of the PV array

The system is tested with the two MPPT techniques, one is P&O algorithm and second is PSO algorithm. The responses obtained after implementing these algorithms is discussed in details below.

A. System with P&O Algorithm

As explained in the algorithm shown in figure 4, the P&O algorithm is implemented in the test system. The MPPT technique is tested under partial shaded and un-shaded condition of the solar PV module. The response under un-shaded and partial shaded is shown in figure 11 and 12 respectively.

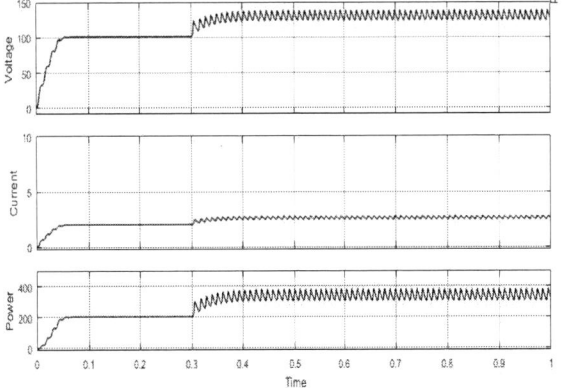

Fig. 11.   Voltage, Current and Power output after tracking by P&O MPPT under un-shaded condition

978-1-5386-6624-1/18 $31.00 © 2018 IEEE

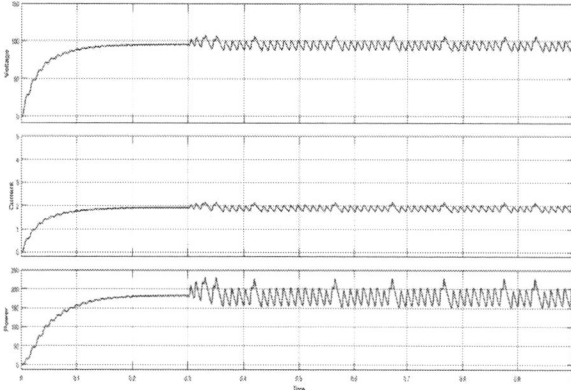

Fig. 12. Voltage, Current and Power output after tracking by P&O MPPT under partial shaded condition

It has been observed that the MPPT tracks in 0.3 seconds but after tracking the MPPT, oscillations in the response of current, voltage and power has been observed.

B. System with PSO Algorithm

The PSO algorithm explained in figure 5 is implemented as a MPPT technique in the test system. It has been observed that using PSO algorithm, the maximum power point is tracked in less than 0.1 second which is faster as compared to P&O technique. The respose curve obtained under un-shaded and partial shaded condition is shown in figure 13 and 14 respectively.

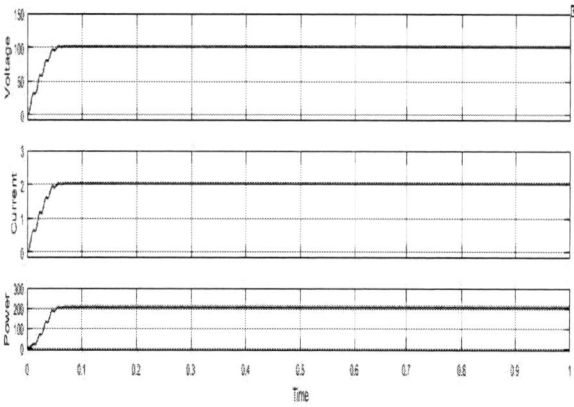

Fig. 13. Voltage, Current and Power output after tracking by PSO MPPT under un-shaded condition

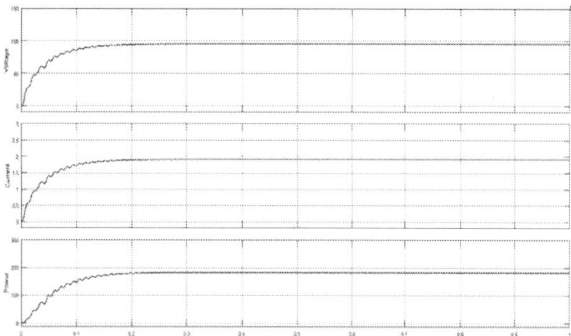

Fig. 14. Voltage, Current and Power output after tracking by PSO MPPT under partial shaded condition

It is observed that no oscillations are present after tracking the maximum power point in partial as well as un-shaded condition of PV module.

CONCLUSION

A test system of 400 W has been developed in the MATLAB/Simulink wherein, modules of 100 W each have been connected in series. Two situations have been created for PV module, one where no shading effect was considered and other where partial shading was considered. Irradiance of 1000 W/m^2, 800 W/ m^2, 600 W/ m^2 and 400 W/ m^2 have been considered in the partial shading condition. However, in the present research temperature variation has not been considered. Two algorithms have been applied in the test system to track maximum power point. Firstly, P&O algorithm has been considered which is the conventional method for tracking the MPPT algorithm. The other MPPT algorithm considered for the proposed system is the soft computing based technique i.e. Particle Swam Optimization techniques (PSO). The test system is designed for two conditions which are un-shaded condition where all PV modules were without shading (i.e. value of irradiance considered is 1000 W/m^2) and partial shaded condition where PV modules are imposed with different irradiance values. The results were obtained using MATLAB/Simulink. The result shows that, in the conventional method i.e. P&O, the tracking time is 0.3 second and after tracking the oscillations were present. The oscillations were less in case of un-shaded condition where as much more oscillations were present during the partial shaded condition. When MPPT was tracked using PSO algorithm, the tracking time was less than 0.1 second and there were no oscillations present even in partial shaded condition. Hence, this proves the novelty and fastness of the PSO algorithm as compared to the conventional P&O algorithm and can be implemented practically where the variation in irradiances is a common practice.

REFERENCES

[1] Nabil Karami, Nazih Moubayed, Rachid Outbib, "General review and classification of different MPPT Techniques",Renewable and Sustainable Energy Reviews,volume 68, Part 1, 2017, pp. 1-18.

[2] Osisioma Ezinwanne, Fu Zhongwen, Li Zhijun, "Energy Performance and Cost Comparison of MPPT Techniques for Photovoltaics and other Applications", Energy Procedia,Volume 107,2017,pp. 297-303.

[3] T. Logeswaran, A. SenthilKumar,"A Review of Maximum Power Point Tracking Algorithms for Photovoltaic Systems under Uniform and Non-uniform Irradiances", Energy Procedia,Volume 54,2014,pp. 228-235.

[4] B. Subudhi and R. Pradhan, "A Comparative Study on Maximum Power Point Tracking Techniques for Photovoltaic Power Systems," in *IEEE Transactions on Sustainable Energy*, vol. 4, no. 1, pp. 89-98, Jan. 2013.

[5] L. Gil-Antonio et al., "Maximum power point tracking techniques in photovoltaic systems: A brief review", *13th International Conference on Power Electronics*, 20-23 June 2016.

[6] R. Rawat, S. S. Chandel, "Review of maximum-power-point tracking techniques for solar-photovoltaic systems", *Energy Technology Elsevier*, vol. 1, no. 8, pp. 438-448, 2013.

[7] Y. Yorozu, M. Hirano, K. Oka, and Y. Tagawa, "Electron spectroscopy studies on magneto-optical media and plastic substrate interface," IEEE Transl. J. Magn. Japan, vol. 2, pp. 740–741, August 1987 [Digests 9th Annual Conf. Magnetics Japan, p. 301, 1982].

[8] J. Ahmed and Z. Salam, "An Accurate Method for MPPT to Detect the Partial Shading Occurrence in a PV System," in *IEEE Transactions on Industrial Informatics*, vol. 13, no. 5, pp. 2151-2161, Oct. 2017.

[9] B. R. Peng, K. C. Ho and Y. H. Liu, "A Novel and Fast MPPT Method Suitable for Both Fast Changing and Partially Shaded Conditions," in *IEEE Transactions on Industrial Electronics*, vol. 65, no. 4, pp. 3240-3251, April 2018.

[10] M. Killi, and S. Samanta, "Modified perturb and observe MPPT algorithm for drift avoidance in photovoltaic systems," IEEE Transactions on Industrial Electronics, vol. 62, no. 9, pp. 5549-5559, 2015.

[11] S. K. Kollimalla, and M. K. Mishra, "A novel adaptive P&O MPPT algorithm considering sudden changes in the irradiance," IEEE Transactions on Energy Conversion, vol. 29, no. 3, pp. 602-610, 2014.

[12] K. Sundareswaran, V. Vigneshkumar, P. Sankar, S. P. Simon, P. Srinivasa Rao Nayak, and S. Palani, "Development of an improved P&O algorithm assisted through a colony of foraging ants for MPPT in PV system," IEEE Transactions on Industrial Informatics, vol. 12, no. 1, pp. 187-200, 2016.

[13] R. B. A. Koad, A. F. Zobaa and A. El-Shahat, "A Novel MPPT Algorithm Based on Particle Swarm Optimization for Photovoltaic Systems," in *IEEE Transactions on Sustainable Energy*, vol. 8, no. 2, pp. 468-476, April 2017.

[14] K. Ishaque, Z. Salam, "A deterministic particle swarm optimization maximum power point tracker for photovoltaic system under partial shading condition", *IEEE Trans. Ind. Electron.*, vol. 60, no. 8, pp. 3195-3206, Aug. 2013.

[15] B Ishaque, K.; Salam, Z. "A Deterministic Particle Swarm Optimization Maximum Power Point Tracker for Photovoltaic System under Partial Shading Condition". IEEE Trans. Ind. Electron. 2013, pp. 3195–3206.

[16] R. B. A. Koad, A. F. Zobaa, "Comparison between the conventional methods and PSO based MPPT algorithm for photovoltaic systems", *World Acad. Sci. Eng. Technol. Int. Sci. Index 88 Int. J. Elect. Comput. Energetic Electron. Commun. Eng.*, vol. 8, no. 4, pp. 673-678, 2014.

[17] C. Kai , T. Shulin, C. Yuhua, and B. Libing , "An improved MPPT controller for photovoltaic system under partial shading condition," IEEE Trans. Sustain. Energy , vol. 5, no. 3, pp. 978–985 , Jul. 2014.

[18] Y. Hu , W. Cao, J. Wu, B. Ji , and D. Holliday , "Thermography-based virtual MPPT scheme for improving PV energy efficiency under partial shading conditions," IEEE Trans. Power Electron, vol. 29, no. 11, pp. 5667–5672, Nov. 2014.

[19] S. Dhar, R. Sridhar and G. Mathew, "Implementation of PV cell based standalone solar power system employing incremental conductance MPPT algorithm," *2013 International Conference on Circuits, Power and Computing Technologies (ICCPCT)*, Nagercoil, 2013, pp. 356-361.

[20] F. Rong , X. Gong, and H. Shoudao, "A novel grid-connected PV system based on MMC to get the maximum power under partial shading conditions ," IEEE Trans. Power Electron. vol. 32, no. 6, pp. 4320–4333, Jun. 2017.

[21] S. Lyden and M. E. Haque, "A simulated annealing global maximum power point tracking approach for PV modules under partial shading conditions," IEEE Trans. Power Electron., vol. 31, no. 6, pp. 4171–4181, Jun. 2016.

Third Order Sinusoidal Integrator Control of PV-Hydro-BES Based Isolated Micro-grid

Vineet P. Chandran
Dept. of Electrical Engineering
Indian Institute of Technology, Delhi
vineetp.chandran@gmail.com

Shadab Murshid
Dept. of Electrical Engineering
Indian Institute of Technology, Delhi
mail2smurshid@gmail.com

Bhim Singh, *Fellow IEEE*
Dept. of Electrical Engineering
Indian Institute of Technology, Delhi
bsingh@ee.iitd.ac.in

Abstract— Power integration and voltage frequency control (VFC) of a microgrid (MG) with two energy sources, permanent magnet synchronous generator (PMSG) based hydro and PV (photovoltaic) array, is dealt in this work. A battery energy storage (BES) is used to provide load levelling in the system. A constant DC-link voltage is achieved using a BES connected to DC-link of the VSC. An incremental conductance (INC) based control strategy with DC-DC boost converter is used for maximum power point tracking (MPPT) operation of the PV array. The VSC is used to maintain voltage and frequency of the MG and it also improves power quality by providing reactive power compensation, harmonics elimination and load balancing. A third order sinusoidal integrator (TOSI) based control algorithm is used to estimate fundamental component of load current for VSC control. Experimental results using TOSI based control strategy, are demonstrated to validate the performance of the proposed MG.

Keywords- Hydro-PV based micro-grid, Voltage source converter (VSC), MPPT, Power Quality.

I. INTRODUCTION

The energy demand is increasing day by day due to increase in population, so energy production must be increased to fulfill that demand. The non-renewable energy resources are major cause to the increase in pollution. Hence the renewable energy sources like solar, wind, hydro are the alternative energy resources available in future and less affects the environment. There are many advantages of using renewable energy sources (RES) but irregular energy supply is the disadvantage because this type of energy is dependent on weather conditions. So, depending on one type of RES for power generation is not reliable. With the integration of small capacity renewable energy generations, the existence of micro-grid has been evolved [1].

The solar PV array and hydro based green energy sources are gaining popularity due to their cost effectiveness. The intermittent power supply of PV array is a concern as it is not able to deliver energy during night or with substantial clouds present in the day [2]. A maximum power point tracking (MPPT) technique is necessary for harnessing maximum power from this system. The P&O (perturb and observe) and incremental conductance (INC) method have been used for MPPT tracking due to their ease of implementation. The INC method is used in this work for MPPT tracking as P&O method during large fluctuation of insolation, is confused and loses its direction while MPPT tracking [3]. The boost converter connected to the PV array increases the output voltage [4]. The most commonly used generators for hydro systems, are synchronous and induction generators. Lately due to development of high field intensity permanent magnets, permanent magnet synchronous generator (PMSG) has found its importance in many applications like wind and hydro based energy conversion systems. In this work, PMSG is used in

hydro based energy conversion system due to its high efficiency, less losses, high power coefficient and reduced size [5]. In the presence of multiple renewable sources of a MG, simultaneous variation in generation and loads, are observed, which results into frequent voltage and frequency fluctuations [6]. A battery energy storage (BES) connected to voltage source converter (VSC) is a viable solution to maintain power balance in the system. Moreover, it can ameliorate the conversion efficiency, reduces noise and vibrations in such MG [7]. The power quality issue is another great challenge because to stabilize the generated power from renewable energy resources (RES), power electronics converter should be used. Since, it introduces harmonics and poor power factor, which increases losses in the system. A VSC with efficient control algorithm is necessary to overcome this problem [8].

Several harmonics current extraction techniques are used to eliminate harmonics from distorted currents and to extract fundamental component from load currents. These techniques can be grouped into time domain and frequency domain. In this work, a time domain-based control approach is used as frequency domain control requires large computation burden, they are less accurate and more complicated [9]. Moreover, the memory requirement of time domain algorithms are minimal compared to frequency domain approach. The most popularly used time domain control techniques are instantaneous reactive power theory (IRPT) and synchronous reference frame (SRF). These techniques involve abc-to-αβ0 transformation in a fixed reference frame or abc-to-dq0 transformation in a rotating reference frame. Due to involvement of multiple transformations in fundamental current extraction, these algorithms suffer an error due to time delay [10]. The second order generalized integrator (SOGI) based control algorithm has capability of eliminating such delay. However, these algorithms fail to remove DC offset component in load currents introduced during dynamic load changes [11]. A third order generalized integrator (TOGI) based control algorithm can be used to filter out harmonics components and remove DC offset presence in load currents. However, it has the characteristics of a notch filter which has very limited attenuation for DC offset current component [12]. The forth order generalized integrator (SO-SOGI) is an effective solution as it can attenuate DC offset component but its performance is compromised for higher order harmonics [13].

In this work, a third order sinusoidal integrator-based control strategy (TOSI) is proposed for extraction of fundamental component of load current. This algorithm has superior performance for both higher and lower order harmonics. The performance results of the MG for different dynamic operating conditions, are presented in this work. The major contribution of the presented work is as follows.

1. TOSI based control strategy for fundamental current extraction and to provide gating pulses to VSC.
2. Stability analysis and parameters selection for proposed TOSI based control approach is presented to demonstrate its robust behavior.
3. Development of a prototype in laboratory with power management for PMSG based hydro-PV-BES supported MG.

II. PROPOSED SYSTEM CONFIGURATION

Micro-grid (MG) presented in this work, deals with the integration of PMSG based hydro and a boost converter supported PV array with the battery energy storage (BES), as shown in Fig 1. The BES is used to maintain active power balance in the MG. Moreover, the battery also maintains DC-link voltage by charging and discarding action. The boost converter is used to extract maximum power from the PV array. An RC filter connected at point of common interface (PCI) is used to reduce ripples in the voltage. The interfacing inductor L_f is connected between VSC and PCI to reduce ripples from converter currents. All parameters of the PMSG based generator, prime mover, PV array, BES are given in Appendix.

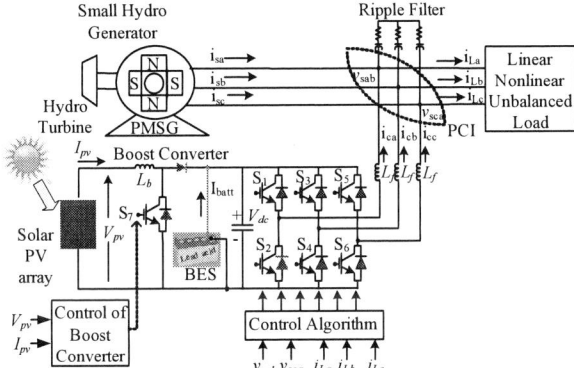

Fig. 1 PMSG based hydro–PV–BES based MG.

III. CONTROL STATERGY

In this work, a TOSI based control algorithm is used to evaluate active and reactive power components of load currents as shown in Fig. 2. This algorithm has superior performance in removing DC offset and higher order harmonics. The stability analysis of this algorithms is exclusively discussed in next section.

A. Estimation of Reference Source Currents

The phase voltages are estimated from sensed line voltages v_{sab} and v_{sbc} as [14],

$$\begin{bmatrix} v_{sa} \\ v_{sb} \\ v_{sc} \end{bmatrix} = \frac{1}{3} \begin{bmatrix} 2 & 1 \\ -1 & 1 \\ -1 & -2 \end{bmatrix} \begin{bmatrix} v_{sab} \\ v_{sbc} \end{bmatrix} \tag{1}$$

The amplitude of terminal voltage is determined from phase voltages as,

$$V_t = \sqrt{\frac{2}{3}(v_{sa}^2 + v_{sb}^2 + v_{sc}^2)} \tag{2}$$

The in-phase voltage template is obtained from instantaneous values of phase voltages as [14],

$$w_{pa} = \frac{v_{sa}}{v_t}, w_{pb} = \frac{v_{sb}}{v_t}, w_{pc} = \frac{v_{sc}}{v_t} \tag{3}$$

The quadrature voltage template of grid phase voltages is then estimated from in-phase templates w_{pa}, w_{pb} and w_{pc} as,

$$w_{qa} = \frac{\left(-w_{pb} + w_{pc}\right)}{\sqrt{3}} \tag{4}$$

$$w_{qb} = \frac{\left(3w_{pa} + w_{pb} - w_{pc}\right)}{2\sqrt{3}} \tag{5}$$

$$w_{qc} = \frac{\left(-3w_{pa} + w_{pb} - w_{pc}\right)}{2\sqrt{3}} \tag{6}$$

The reference value of terminal voltage $V_t{}^*$ is compared with calculated terminal voltage V_t to create an error voltage V_{te} as,

$$V_{te}(k) = V_t{}^*(k) - V_t(k) \tag{7}$$

This error signal is sent through the voltage PI controller to estimate the I_{qq} component for terminal voltage control of PCC. The PI controller output, I_{qq} is formulated as,

$$I_{qq}(k) = I_{qq}(k-1) + k_{pv}\{V_{te}(k) - V_{te}(k-1)\} + k_{iv}(k) \tag{8}$$

Similarly, the reference value of frequency, f^* is compared with system frequency, f to create an error value, f_e.

$$f_e(k) = f^*(k) - f(k) \tag{9}$$

This frequency error signal is sent to PI controller to estimate I_{df} component for the frequency control. Where I_{df} component is formulated as,

$$I_{df}(k) = I_{df}(k-1) + k_{pf}\{f_e(k) - f_e(k-1)\} + k_{if}(k) \tag{10}$$

The individual in-phase and quadrature components of load currents obtained from all three phases are added individually and averaged. The average of in-phase current component is subtracted from I_{df} component generated from frequency error (f_e) to obtain net in-phase template (I_{pT}) as,

$$I_{pT} = I_{df} - \left(\frac{I_{pL1} + I_{pL2} + I_{pL3}}{3}\right) \tag{11}$$

Similarly, the average quadrature current component is subtracted from I_{qq} component to obtain net quadrature template (I_{qT}) as,

$$I_{qT} = I_{qq} - \left(\frac{I_{qL1} + I_{qL2} + I_{qL3}}{3}\right) \tag{12}$$

These net in-phase and quadrature components are then multiplied with their corresponding voltage templates and added together to generate reference source currents $i_{sa}{}^*$, $i_{sb}{}^*$ and $i_{sc}{}^*$ as,

$$i_{sa}{}^* = I_{pLT} * w_{pa} + I_{qLT} * w_{qa} \tag{13}$$

$$i_{sb}{}^* = I_{pLT} * w_{pb} + I_{qLT} * w_{qb} \tag{14}$$

$$i_{sc}{}^* = I_{pLT} * w_{pc} + I_{qLT} * w_{qc} \tag{15}$$

978-1-5386-6624-1/18 $31.00 © 2018 IEEE

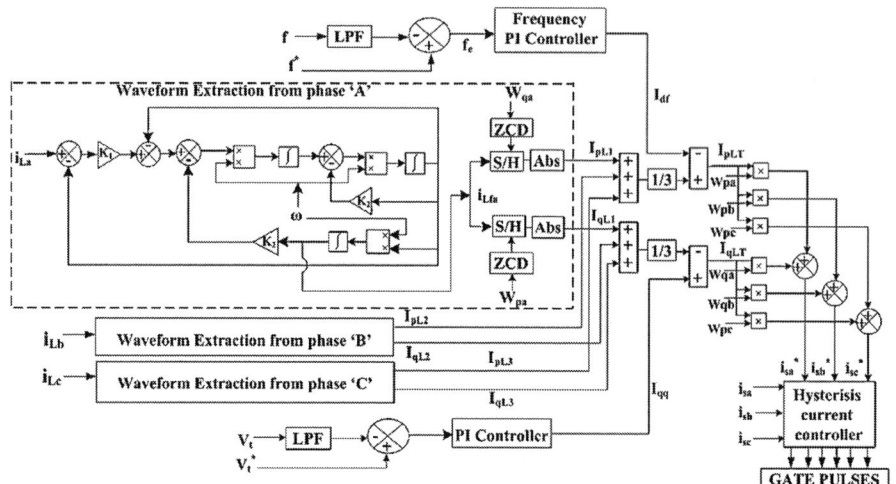

Fig. 2 TOSI based control algorithm.

These generated reference source currents are compared with generator source currents i_{sa}, i_{sb} and i_{sc} using hysteresis current controller for current tracking and generating pulses to VSC.

B. Structure and Stability Analysis of TOSI

The structure of TOSI control consists of three integrators and two parameters K_1 and K_2 as shown in Fig. 2. The transfer function of this control structure is $G_1(s)$, is given as the ratio of input signal i_{La} to output signal i_{Laf}, and is given as,

$$G_1(s) = \frac{i_{Lfa}(s)}{i_{La}(s)} = \frac{K_1\omega^3}{s^3 + K_2\omega s^2 + (K_1+1)\omega^2 s + K_2\omega^3} \quad (16)$$

The main attractive feature of this algorithm is its capability to eliminate higher and lower order harmonics including DC offset and sub harmonics. The capability of TOSI control and its comparison with SOGI [11], TOGI [12] and forth order based SO-SOGI [13] control are analyzed through Bode diagram as shown in Fig. 3. The value of parameters used, is also shown in this Fig. 3, for these controls for obtaining critical damping and good dynamic performance.

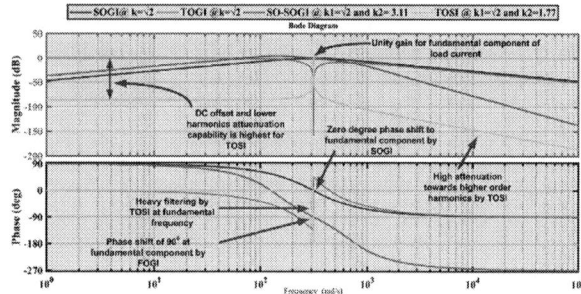

Fig. 3 Bode diagrams of conventional SOGI, TOGI, FOGI and TOSI control

It is clearly visible from Fig. 3, that compared to all Bode diagrams, the bode of TOSI (yellow line) is superior for removing DC offset and lower order harmonics. It offers highest and constant attenuation to all lower order harmonics nearer to the fundamental frequency ω, which is 314.16 rad/sec for this system. Moreover, it is seen that TOSI offers unity gain for fundamental frequency component of load current in the magnitude plot. The good filtering capability of the TOSI is shown in Fig. 3. It also shows that the TOSI based

control has 90^0 phase shift at the fundamental frequency. The TOGI (green line) has characteristics of a notch filter and its attenuation for lower order harmonics below fundamental is very minimal close to 0 dB. The SOGI (blue line) has a LPF characteristics, but its attenuation towards higher order harmonics is poor. The SO-SOGI (redline) has characteristics for eliminating both higher and lower order harmonics, but for distortion near fundamental frequency, like third and fifth the attenuations are less compared to TOSI. Hence TOSI has superior performance for attenuating higher and lower order frequencies plus prominent third and fifth order harmonic frequency.

C. Parameter Selection of TOSI

A simple procedure based on SOGI filter is followed for parameters tuning of K_1 and K_2 of TOSI, algorithm. In this procedure, the K parameter is selected as $\sqrt{2}$, and it is selected based on the damping factor, $\xi = 0.707$, which gives critical level of damping and preferred filtering response. The same value is used for TOSI parameter, K_1 for achieving same filtering performance as SOGI. Hence, in this work, K_1 value is set as $\sqrt{2}$ and once this value is fixed then K_2 value is adjusted to get desired dynamic response. To get the better understanding for selection of K_2 a pole-zero plot is used as shown in Fig. 4. The value of K_2 is varied from 0 to 5 and it is clearly observed that as the value of K_2 is increased above 2 of or reduced below 1.5 the conjugate poles move toward imaginary axes, which is an unstable zone. So, an optimum value of $K_2=1.77$ is selected in this work to get desirable filtering and good dynamic performance.

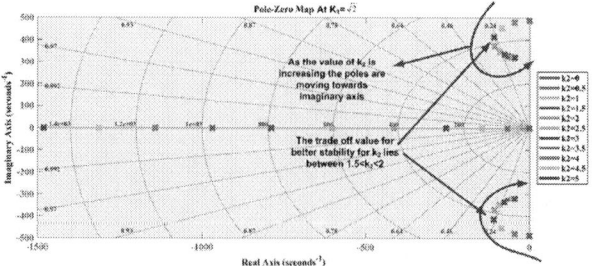

Fig. 4 Pole-Zero plot of TOSI to decide trade-off value for damping factor K_2.

978-1-5386-6624-1/18 $31.00 © 2018 IEEE

D. MPPT Control Statergy for PV

To extract the maximum power from a PV array, the MPPT technique is used. ☐There are lots of MPPT techniques developed by many researchers, but incremental conductance (INC) and perturb and observe (P&O) are popular, and these techniques are easy to implement and best for tracking the MPP of a PV array. A fixed step size is chosen for perturbation in PV voltage point where maximum power is available. The duty ratio changes, according to the maximum power point. Moreover, the duty ratio controls the output voltage of the DC-DC boost converter. Thus, the INC based algorithm accurately predicts this MPP point. The flowchart of INC method is given in Fig 5. In this technique, change in PV voltage is applied and the power before and after perturbations is measured. The MPP point is obtained at the point where the derivative of the power-vs-voltage curve is zero. This point is the MPP and is defined according to the comparison between the instantaneous conductance and the differential conductance after perturbation. The sign of the sum of conductance is the basic parameter to locate the MPP.

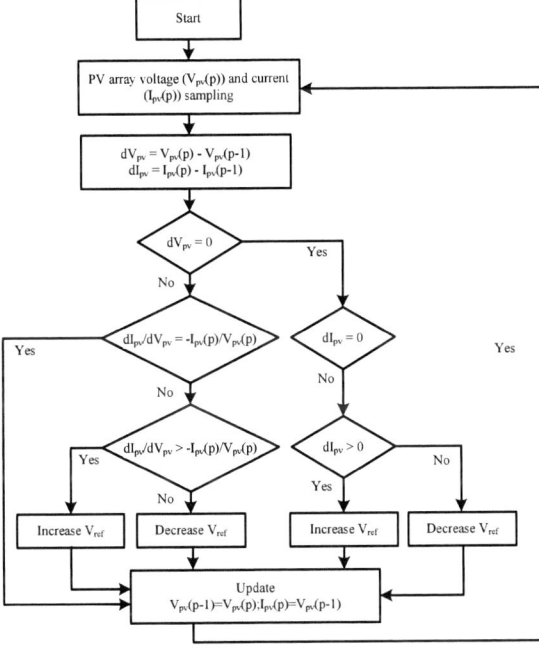

Fig. 5 Flow chart for INC algorithm.

E. DC-DC Boost Converter Control Scheme

In this control scheme as shown in Fig. 6, an INC based control algorithm is used for MPPT. It generates a reference PV voltage (V_{PVref}). According to V_{PVref}, an optimal duty cycle (d) for the peak voltage is generated, by keeping battery volatage (V_{bat}) constant. The duty ratio is compared with saw-tooth wave of fixed frequency to generate gate pulse to IGBT of a boost converter

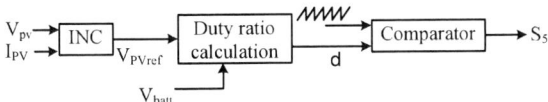

Fig. 6 Controller of bidirectional converter (BDC).

IV. RESULTS AND DISCUSSION

The hardware implementation of isolated MG for a 3.7 kW, PMSG based hydro with PV array simulator supported with BES is carried out. A 220 V, 1500 rpm and 3.7 kW DC motor is used as prime mover to run the PMSG based hydro system. A PV array simulator is used to realize a 1.7 kW solar PV array system, whose MPPT voltage (V_{PV}) and current (I_{PV}) are 311 V and 5.6 A, respectively. A total of ten Hall-Effect sensors are used for sensing voltages and currents. Two source currents (i_{sa}, i_{sb}), two load currents (i_{La}, i_{Lb}), two line voltages (V_{sab}, V_{sbc}) at PCC, PV voltage (V_{pv}), PV current (I_{pv}), battery voltage (V_{batt}) and battery current (I_{batt}) are sensed using these sensors. Test results of the MG system under various loading conditions are demonstrated in Figs. 7- 13.The performance is observed for system parameters like generator voltage and current (v_{sab} and i_{sa}), generated powers (P_g and Q_g), total harmonic distortion (THD) of source current ($i_{s THD}$), THD of source voltage ($V_{sab THD}$), load current(i_{La}), load power (P_L), THD of load current ($i_{La THD}$), converter current (i_{ca}), converter active and reactive power (P_{con},Q_{con}), THD of converter current ($i_{ca THD}$), BES voltage and current (V_{bat},I_{batt}), BES power (P_{bt}), PV power (P_{pv}).

A. Performance of Proposed MG at Unbalanced Nonlinear Load

The performance of the system at unbalanced nonlinear load is shown in Fig. 7. The PV array connected to the MG, is always generating, 1.7 kW at 1000 W/m². From, Figs 7 (b) the PMSG based hydro is generating 1.87 kW. So, the combined power generation in MG from hydro and PV array is 3.57 kW. Moreover, the load connected is consuming only 691 W, due to unbalancing condition as seen in Fig. 7(e). Hence, the excess power of 2.879 kW should be stored in BES. It is clearly evident from Fig. 7 (h), 2.77 kW power is fed to BES after some losses in the VSC. In this unbalanced load condition, the voltage and frequency across the generator terminals are not affected. Moreover, the THD's of the current ($i_{sa THD}$) and voltage ($V_{sab THD}$) are 4.4 % and 3.1 % respectively, which are well within the IEEE-519 standard [17] as shown in Figs. 7 (c) and 7(i).

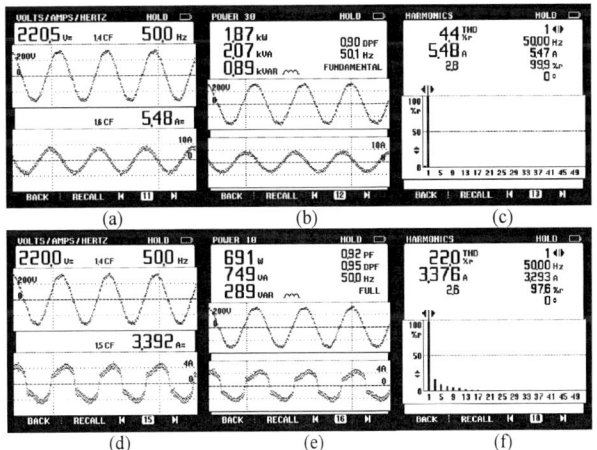

978-1-5386-6624-1/18 $31.00 © 2018 IEEE

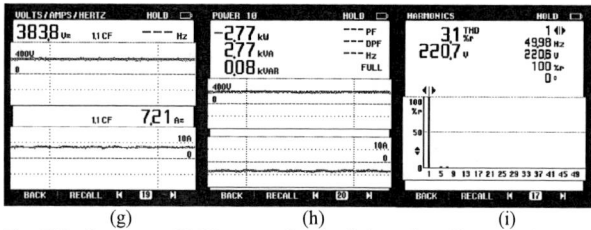

(g)   (h)   (i)

Fig. 7 Performance of MG system feeding balanced nonlinear load (a) V_{sab} and I_{sa} (b) P_g and Q_g (c) $i_{s\ THD}$ (d) V_{sab} and I_{La} (e) P_L and Q_L (f) $i_{s\ THD}$ (g) V_{batt} and I_{batt} (h) P_{batt} (i) $V_{sab\ THD}$.

B. Performance of Proposed MG When Connceted Load is more than Generation in MG.

When the power generated from connected sources is less than the connected load in MG due to fall in insolation of a PV array or when PV array is not generating. Then the BES should discharge power to meet the extra load demand. In this situation, the PMSG based hydro generation is generating reduced power of 1.17 kW as seen in Fig. 8(a). The PV array is disconnected, and it is not generating power. So, the total power available in the MG is only 1.17 kW, from hydro generator as seen in Fig. 8(b). However, the load connected is 1.50 kW, which is more than the generation seen in Fig. 8(e). Hence the BES should discharge the excess power to meet the excess load, so it is observed that the battery is providing rest of the power required by the load to maintain active power balance in the system as shown in Fig. 8 (g). The THD's of the current ($i_{sa\ THD}$) and voltage ($V_{sab\ THD}$) are 4.7 % and 3.2 %, respectively.

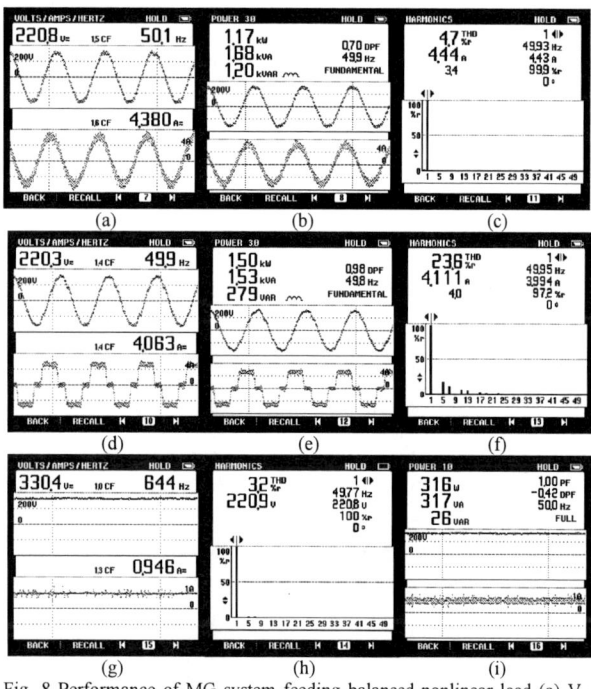

Fig. 8 Performance of MG system feeding balanced nonlinear load (a) V_{sab} and I_{sa} (b) P_g and Q_g (c) $i_{s\ THD}$ (d) V_{sab} and I_{La} (e) P_L and Q_L (f) $i_{s\ THD}$ (g) V_{batt} and I_{batt} (h) P_{batt} (i) $V_{sab\ THD}$.

C. Dynamic Response of Proposed MG at Unbalanced Nonlinear Load

The dynamics of system during load unbalancing is shown in Fig. 9. The 'a' phase of the load is cutoff and reconnected to observe the performance. Despite the load being unbalanced,

the PMSG currents are still sinusoidal as seen in Fig. 9(a). Moreover, due to reduction in the load the charging current of battery (I_{batt}), is increased and excess power (generated-consumed), is stored in the battery. The terminal voltage (V_t) and frequency (f) of MG, are quickly settled during this dynamic change to stable values as seen in Fig. 9 (b). During the load unbalancing, the battery charging power is increased to store excess power generated in the MG.

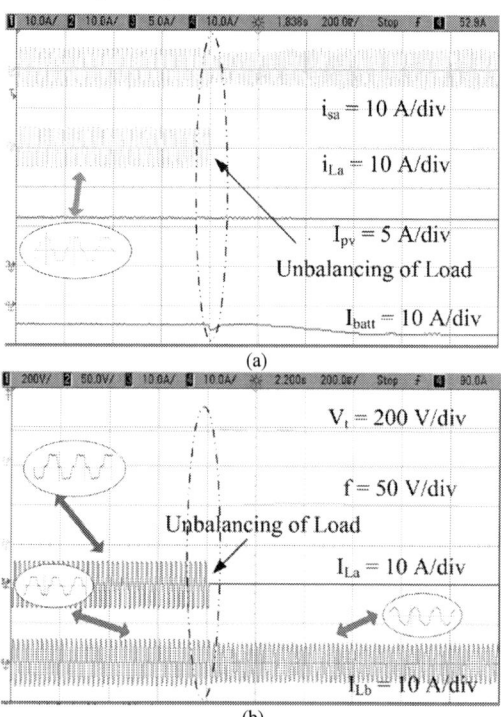

Fig. 9 Dynamic performance of proposed system under unbalanced nonlinear loading.

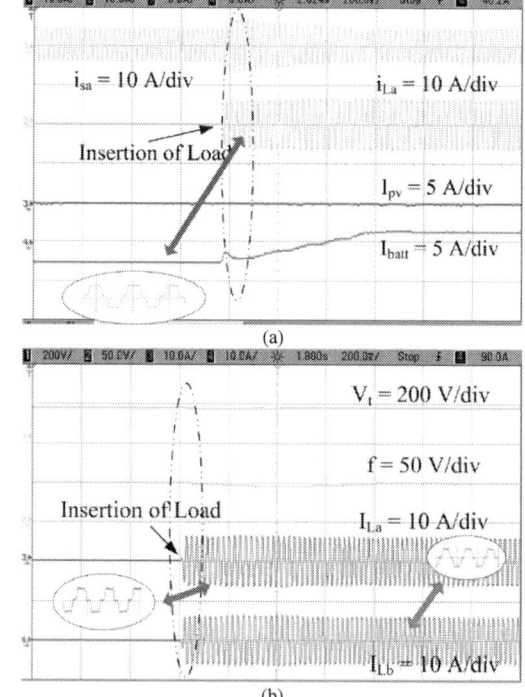

Fig. 10. Dynamic performance of proposed MG for change in BES mode.

D. Dynamic Response of Proposed MG For Charging to Discharging Mode of BES

Dynamic performance of the system for insertion of nonlinear load greater than the generation of PV array and hydro sources, is shown in Fig. 10. The PV is operated at a reduced insulation of 500 W/m². From Fig. 10 (a) it is evident that when there is sudden increase in load, more than the generation in MG, then the BES charging current quickly changes its mode to discharging for feeding the excess load. The terminal voltage (V_t) and frequency (f) of the MG are also not affected during this operation as seen from Fig. 10 (b). It can also be observed that the battery current quickly changes from discharging mode to charging mode.

E. Dynamic Response of BES in MG During Change in Insolation

The dynamic response of the BES during change in insolation from 1000 W/m² to 500 W/m² and back to 1000 W/m² of PV array is seen in Fig. 11 (a). During this operation, the PV current changes, consecutively the BES charging current changes accordingly to balance the active power in the proposed MG. The MPPT performance of the PV array system is achieved as seen in Fig. 11(a), despite the change in insolation the PV voltage (V_{pv}) remains unchanged during this operation. In Fig 11 (b), the MPPT performance of the PV is shown at insolation levels of 1000 W/m² and 500 W/m². The MPPT algorithms is tracking the maximum MPP voltage at both insulation levels. Moreover, when the insolation is reduced from 1000 W/m² to 500 W/m² the power has reduced from 1.761 kW to 877.99 kW.

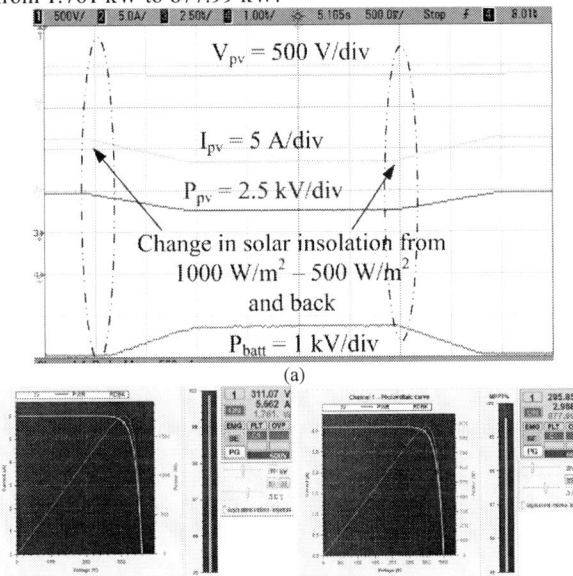

(a)

(b)

Fig. 11. Dynamic performance of MG during change in insolation of PV

V. CONCLUSION

The prototype in the laboratory of a MG by integrating PMSG based hydro power with solar PV array using BES is implemented. The performance of the MG under different nonlinear loading conditions has been demonstrated. It has been observed that during different loading conditions, the voltage and frequency of the (MG) are maintained constant. The dynamic power sharing capability of proposed MG with using BES is clearly achieved. The THDs of source voltage

and currents are always maintained within limits below 5% as per the IEEE-519 standard [15]. The proposed TOSI based control algorithm has been found to have superior lower, higher harmonics elimination capability with good dynamic response.

APPENDIX

PMSG: Three- phase, 3.7 kW, 220V Star-connected, 1500 rpm, 50 Hz. DC Machine (Prime mover): 230 V, 1500 rpm, 3.7 kW. BDC: L_b = 4 mH .BES rating: V_{bat}=380 V, 7 Ah. VSC Parameters: V_{dc}=380 V, L_f = 3.5 mH, C_{dc}= 2000 µF, K_{pv}= 0.010 , K_{iv} =0.18, K_{pf}=0.13, K_{if}=.035 Controller parameters: K_1=$\sqrt{2}$, K_2=1.77 : ω = 314.16 rad/sec,Ts=50 µs. Ripple Filters : R_f =5 Ω, C_f =10 µF. PV parameters @ insolation of 1000W/m², P_{pv} =1.7 kW, V_{pv}= 311 V, I_{pv}=5.6 A.

ACKNOWLEDGEMENT

The authors are grateful to the Department of Science and Technology, Govt. of India, the project entitled with an Indo-UK Project-UKCERI-I, SERI-II and J. C. Bose Fellowship for funding this project under grant number: RP03391, RP03357 and RP03128, respectively.

REFERENCES

[1] N. Lopez and J. F. Espiritu, "An approach to hybrid power systems integration considering different renewable energy technologies," *Procedia Computer Science, Elsevier*, vol. 6, pp. 463–468, 2011.

[2] C. Manickam, G. P. Raman, G. R. Raman, S. I. Ganesan and N. Chilakapati, "Fireworks Enriched P&O Algorithm for GMPPT and Detection of Partial Shading in PV Systems," *IEEE Transactions on Power Electronics*, vol. 32, no. 6, pp. 4432-4443, June 2017.

[3] S. Sreejith and K. S. Smitha, "A high efficiency resonant DC-DC power converter-analysis and comparison of P&O and INC algorithms," *in Proc. of International Conference on Control Communication & Computing India (ICCC)*, Trivandrum, pp. 182-187, 2015.

[4] N. Kumar, I. Hussain, B. Singh and B. K. Panigrahi, "Normal Harmonic Search Algorithm Based MPPT for Solar PV System and Integrated with Grid using Reduced Sensor Approach and PNKLMS Algorithm," *IEEE Transactions on Industry Applications*. Early Access.

[5] T. F. Chan, L.T. Yan and L. L. Lai, "Permanent-magnet synchronous generator with inset rotor for autonomous power-system applications," *Proc. Inst. Elect. Eng.–Gen., Trans. and Distrib.*, vol. 151, no. 5, pp. 597–603, Sept. 2004.

[6] H. Ahmed and A. Bhattacharya, "PMSG based constant power delivery standalone WECS using SST with bidirectional buck-boost BESS," *in Proc. of IEEE 7th Power India International Conference (PIICON)*, Bikaner, pp. 1-6, 2016.

[7] S. Kewat, B. Singh and I. Hussain, "Power management in PV-battery-hydro based standalone microgrid," *IET Renewable Power Generation*, vol. 12, no. 4, pp. 391-398, 2018.

[8] Seema and B. Singh, "Intelligent control of SPV-battery-hydro based microgrid," *in Proc. of IEEE International Conference on Power Electronics, Drives and Energy Systems (PEDES)*, Trivandrum, pp. 1-6, 2016.

[9] K. Mathuria, I. Hussain and B. Singh, "Power improvement using hartley S-transform for solar energy conversion system," *in Proc. of 4th IEEE Uttar Pradesh Section International Conference on Electrical, Computer and Electronics (UPCON)*, Mathura, pp. 427-433, 2017.

[10] B. Singh and J. Solanki, "A Comparison of Control Algorithms for DSTATCOM," *IEEE Transactions on Industrial Electronics*, vol. 56, no. 7, pp. 2738-2745, July 2009.

[11] B. Singh, S. Kumar and C. Jain, "Damped-SOGI-Based Control Algorithm for Solar PV Power Generating System," *IEEE Trans. on Industry Applications*, vol. 53, no. 3, pp. 1780-1788, May-June 2017.

[12] G. Fedele and A. Ferrise, "Biased sinusoidal disturbance rejection with plant uncertainty via an adaptive third-order generalized integrator," in *Proc.of 20th Mediterranean Conference on Control & Automation (MED)*, Barcelona, pp. 253-258, 2012.

[13] H. K. Yada and A. S. Kumar, "An SO-SOGI based control for a three-phase DVR under distorted grid conditions including DC offset," in *Proc. of TENCON IEEE Region 10 Conference*, Penang, pp. 3000-3005, 2017.

[14] B. Singh, A. Chandra and K. Al-Haddad, *Power quality problems and mitigation techniques*, Chichester, West Sussex United Kingdom: John Wiley & Sons Inc, 2015.

[15] IEEE Recommended Practice and Requirement for Harmonic Control on Electric Power System, IEEE std. 519, 2014.

Frequency Estimation of Unbalanced Three Phase Power System Using Legendre Polynomial Based Gauss Newton Algorithm

Shreeva Pattanaik
Department of Electrical and Electronics Engineering
International Institute of Information Technology
Bhubaneswar, India

Umamani Subudhi
Member, IEEE, Department of Electrical and Electronics Engineering *International Institute of Information Technology* Bhubaneswar, India
umamani@iiit-bh.ac.in

Harish Kumar Sahoo
Senior Member, IEEE, Department of Electronics and Telecommunication Engineering
Veer Surendra Sai University of Technology ,Burla India
harish_etc@vssut.ac.in

Saswat Panigrahi
Department of Electrical and Electronics Engineering,
International Institute of Information Technology
Bhubaneswar, India

Abstract— **Frequency is a vital parameter to analyze unbalance condition in three phase power system. Frequency variations need to be tracked and mitigated to ensure protection of power system. In this paper Legendre Polynomial based Gauss Newton Method is proposed to determine the unknown frequency of the three phase unbalanced voltage signal. By utilizing Lengendre polynomial based expansions in Gauss Newton algorithm results in a powerful adaptive estimation model to estimate the frequency in case of unbalanced three phase system. The performance of the algorithm is tested considering the real time signal from IEEE database. It is observed from the simulation results that the proposed model for estimation of frequency in a distorted power system signal exhibits superior estimation performance in terms of tracking time and speed of convergence as compared to performances of some of the existing techniques such as Least Mean Square(LMS), Recursive Least Square(RLS) and Gauss Newton algorithm.**

Keywords— *Clarke's transformation, Gauss Newton algorithm, Legendre Polynomial*

I. INTRODUCTION

Problems related to power quality are one of the greatest worry these days. The utilization of electronic equipments like equipment related to Data Technology sector, drives with adjustable speed, energy-conserving lighting resulted in a total conversion of nature of loads from electrical point of view. The above mentioned categories of loads are significant causes of power quality issues. All these loads are non-linear, therefore these causes interferences in the voltage waveform. The enhanced sensitivity of the processes like residential and industrial services to power quality problems has diminished the performance of these equipments, therefore increasing the demand of better electric power quality. The most critical areas are electronic devices plants, pharmaceutical plants, automotive production units, glass plants and textile industries. Power frequency variations are basically the differences of the electric power system fundamental frequency from its defined standard value (50Hz or 60Hz).

The frequency of power system is related undeviatingly to rotation speed of the electric generators providing supply to the utility system. Few variations in the value of frequency can occur because of the variation in synchronism between load connected to the utility and generator. The magnitude of frequency deviation is dependent on the properties of the load and impact on the control system of the generation to changes in load. Power frequency variations are generally caused due to (1) generators mal-operation (2) Variation in the source of electric power generation. Frequency variations of the power system affect the power system in many ways. Slight frequency variations will cause self executing load shutting down or other controlling activities to restore frequency of the system. Variation in system frequency will cause variation in the speed of the electric motors, change in magnetizing currents in transformers and induction motors etc. A huge increase in frequency will cause an increase in the harmonic currents and it will result in heating of the

978-1-5386-6624-1/18 $31.00 © 2018 IEEE

system which may lead to insulation failure. Frequency of a power system also decides the real power balance in the system as line parameters will change with variation in frequency.

Various adaptive filtering techniques have been proposed for the estimation of frequency for balanced three phase power system. Linear adaptive estimation method [1], least error square [2], Novel Kalman filtering [3], Extended complex Kalman filtering [4] , Clarke's Transformation[5,6], Adaptive frequency estimation techniques [7], demodulation of complex signals [8], grid voltage parameters estimation [9,10], Recursive algorithm[11], Least mean square technique [12]   and iterative area. Other soft computing techniques that have been developed, like neural network and genetic algorithms (GA)s are also being used for frequency measurement in power system. In the proposed method, Legendre polynomial based Gauss-Newton Method [13] is applied for frequency estimation to reduce the tracking time with increased speed of convergence. Here, a weighted error cost function is considered as an objective function which is minimized to obtain the frequency value. The forgetting factor is updated recursively to increase the robustness of the algorithm. The performance of the proposed algorithm is established through comparison results with that of LMS, RLS and Gauss-Newton method in MATLAB(ver. 2014 a) environment. Section II describes the frequency estimation model. In section III, the Legendre Polynomial based Gauss-Newton Algorithm to estimate the frequency value of unknown signal is described. In Section IV simulation and experimental results are presented for evaluating its accuracy and efficiency in the presence of signal distortion. Finally the conclusion is drawn in section V.

II. FREQUENCY ESTIMATION MODEL

Discrete form of three phase unbalanced signal is expressed as given in (1).

$$
\begin{aligned}
v_a &= V_a \cos(\omega k T_s + \varphi_0) \\
v_b &= V_b \cos(\omega k T_s + \varphi_0 - 2\pi / 3) \\
v_c &= V_c \sin(\omega k T_s + \varphi_0 + 2\pi / 3)
\end{aligned}
\tag{1}
$$

where Va is the maximum voltage in phase a, V$_b$ is the peak voltage value of phase b, and Vc is the voltage value of phase c. k is sampling instant, φ is the fundamental component phase angle and ω is frequency of voltage signal (ω= 2πf where f is frequency of the system). The task is to evaluate the unknown value of frequency f. To analyze three phase signal like single phase approach, the Clarke's transform or also known as αβ-transform is applied. Using this technique three phase voltage signal is converted to two phase voltage signal and then unknown frequency value is estimated using the proposed approach. The αβ-transformed signal is represented as in (2).

$$
\begin{bmatrix} V_{\alpha k} \\ V_{\beta k} \end{bmatrix} = \begin{bmatrix} 1 & \dfrac{-1}{2} & \dfrac{-1}{2} \\ 0 & \dfrac{\sqrt{3}}{2} & \dfrac{-\sqrt{3}}{2} \end{bmatrix} \begin{bmatrix} V_{ak} & V_{bk} & V_{ck} \end{bmatrix}^T
\tag{2}
$$

This transformation results in a complex-valued voltage that can show the 3-phase power system. The noiseless αβ signal can be evaluated as in (3).

$$
\begin{aligned}
y_k &= V_{\alpha k} + j V_{\beta k} \\
&= (A + jB)\cos(2\pi f k\tau + \theta) + (B + jC)\sin(2\pi f k\tau + \theta)
\end{aligned}
\tag{3}
$$

where

$$
A = \sqrt{\frac{2}{3}} V_a + \frac{1}{2\sqrt{6}} (V_b + V_c)
\tag{4}
$$

$$
B = \frac{1}{2\sqrt{2}} (V_b - V_c)
\tag{5}
$$

$$
C = \frac{1}{2}\sqrt{\frac{3}{2}} (V_b + V_c)
\tag{6}
$$

Now, the task is to correctly evaluate ω which is explained in the subsequent section.

The signal expressed in (3) is represented in vector form as given in (7).

$$
y_k = X_k \times W_k
\tag{7}
$$

where X_k is the input vector and is expressed as given in Eq.(8)

$$
X_k = \begin{bmatrix} e^{j\omega k\tau} & 0 \end{bmatrix}
\tag{8}
$$

The weight vector is considered as shown in (9)

$$
W_k = \begin{bmatrix} A & 2\cos(\omega k\tau) \end{bmatrix}
\tag{9}
$$

III. FREQUENCY ESTIMATION MODEL USING LEGENDRE GAUSS NEWTON METHOD

For time-varying power signals, a cost function proposed in [13] for calculating the parameters of the weight vector is expressed as (10)

$$
\varepsilon(k) = \sum_{i=0}^{k} \lambda^{k-i} e^2(i)
\tag{10}
$$

where λ is a forgetting factor with limits $0 < \lambda < 1$ and e(i) is an error function of the form given in (11).

$$
\begin{aligned}
e(k) &= y(k) - y_d(k) = (A)\cos(2\pi f k\tau + \theta) \\
&+ (B)\sin(2\pi f k\tau + \theta) - A e^{j\omega k\tau}
\end{aligned}
\tag{11}
$$

The gradient vector is obtained as (12)

$$\psi(k) = \frac{\partial e(k)}{\partial W(k)} \tag{12}$$

Deriving the gradient vector using equations we obtain

$$\psi(k) = \frac{\partial e(k)}{\partial W(k)} = \left[2\cos(\omega)\sin(\omega) \quad \sin(\omega)\right] \tag{13}$$

The Hessian matrix is obtained as (14)

$$H(k) = \sum_{i=0}^{k} \lambda^{k-i} \psi(i).\psi^{T}(i) \tag{14}$$

where $\psi^{T}(i)$ is the transpose of the gradient vector.

Deriving the Hessian matrix using (13) and (14), the hessian matrix for this model is obtained as (15)

$$H(k) = \frac{1-\lambda^{k+1}}{1-\lambda^{k}}\begin{bmatrix} 4\cos^2(\omega)\sin^2(\omega) & 0 \\ 0 & \sin^2(\omega) \end{bmatrix} \tag{15}$$

A recursive Gauss-Newton algorithm is formulated as follows

$$W(k) = W(k-1) - H^{-1}(k).\psi(k).e(k) \tag{16}$$

$$H(k) = \lambda H(k-1) + \psi(k)\psi^{T}(k) \tag{17}$$

Where the weight vector is updated using (16) and the Hessian matrix is updated using (17).

The forgetting factor is updated as (18)

$$\lambda(k) = \lambda(k-1) + (1-\lambda(k-1)).e^{\left(-|e(k).e(k-1)|\right)} \tag{18}$$

Applying this above mentioned algorithm to the model the fundamental frequency is obtained from the weight vector as in (20).

$$\omega_k dt = \cos^{-1}(W_k(2)/2) \tag{19}$$

$$f_k = \frac{\cos^{-1}(W_k(2)/2)}{2\pi dt} \tag{20}$$

Steps for Frequency Estimation using Legendre Gauss Newton Method

Step I) Initialize the weight vector using (9).

Step II) Expand the input vector through Legendre polynomial up to second order to generate the estimated signal.

$$X = \begin{bmatrix} 1 & X & \dfrac{(3X^2-1)}{2} \end{bmatrix}$$

where X denotes the input measurement (10).

Step III) Calculate error using equation (11).
Step IV) Calculate the gradient vector using (13).

Step V) Evaluate the hessian matrix using (14).

Step VI) Update the weight function in each step using (16).

Step VII) Update the Hessian matrix in each step using (17).

Step VII. Go to step V and VI until final iteration is reached.

Step VIII. Estimate the value of fundamental frequency using (20).

IV. SIMULATION RESULTS AND DISCUSSION

Extensive computer simulations have been carried out for the evaluation of the proposed adaptive algorithms to track the power system frequency under generated synthetic and real time conditions, including a practical test system [14]. For every test case, white Gaussian noise with 40 dB SNR is considered as additive noise. The parameters in the process are set as follows: The initial forgetting factor being set as 0.997, the fundamental frequency is set to 50 cycles/s. The signal is sampled with a sampling frequency of 200Hz. The performances of the adaptive algorithms are tested for the following conditions.

Case 1-Sinusoidal signal in presence of noise

A static three-phase signal is considered as given in (24) for observing the performance of the adaptive algorithms where the properties of the signal are not being changed .

$$V_{ak} = V_a \sin(\omega kdt + \phi)$$
$$V_{bk} = V_b \sin((\omega kdt + \phi) + \frac{2\pi}{3}) \tag{24}$$
$$V_{ck} = V_c \sin((\omega kdt + \phi) - \frac{2\pi}{3})$$

where V_a=0.6p.u., V_b=0.8p.u. and V_c=0.9p.u. and $\phi = \pi/3$ radians per second and ω is the angular frequency of the three-phase signals. A white Gaussian noise of SNR 40dB is added to analyze the signal.

The case is studied using Least Mean Square (LMS) technique, Recursive Least Square (RLS), Gauss Newton method, Legendre polynomial based Gauss Newton method. Figure 1 indicates the significantly increased tracking time of different proposed adaptive algorithms. TABLE 1 shows the tracking time to reach the steady state value by the different algorithms. It is observed that the Legendre Polynomial based Gauss Newton Method takes the least tracking time and LMS takes the longest tracking time.

TABLE 1. Tracking Time Comparison for Frequency Estimation

978-1-5386-6624-1/18 $31.00 © 2018 IEEE

Algorithms	Tracking time(in ms)
Legendre Gauss Newton	0.10
Gauss Newton	0.20
RLS	0.21
LMS	1.80

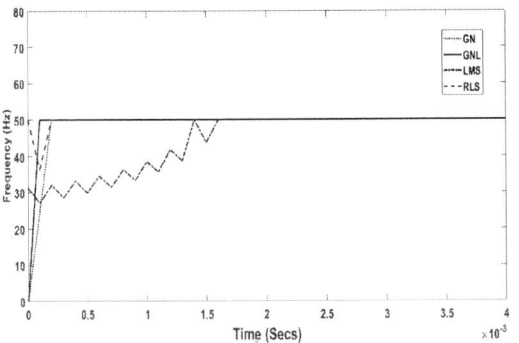

Figure 1. Comparative frequency estimation figures of different algorithm

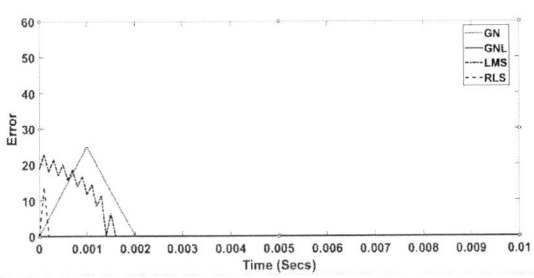

Figure 2. Steady-state error plot

Figure 2 shows the comparative analysis of the steady state error plots of the four adaptive algorithms. It is quite apparent from the plot that Legendre polynomial based Gauss Newton method shows superior performance in terms of convergence speed as compared to other algorithms.

Case 2-Data Collected From IEEE database

The adaptive algorithms are tested with the data collected from the IEEE database [14]. The three phase unbalanced signal considered is as shown in Figure 3.

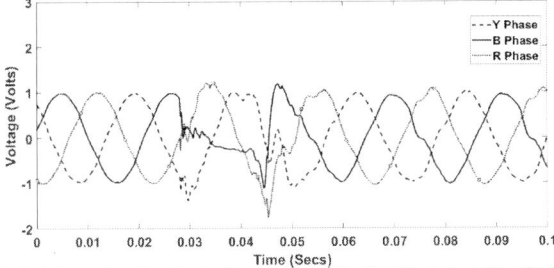

Figure 3. Unbalanced three-phase signal

Frequency estimation performance using different algorithms shown in Figure 4 shows that all of them track the frequency, but the Legendre Polynomial based Gauss Newton method tracks the frequency in the least time. The Least Mean Square (LMS) algorithm takes the maximum time to track the frequency accurately.

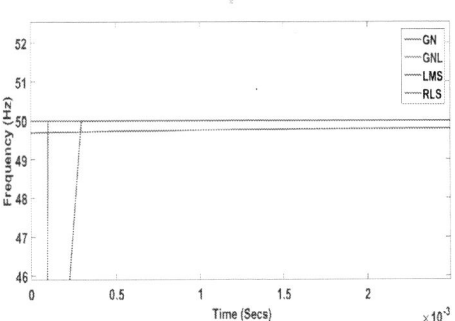

Figure 4. Comparative analysis of frequency estimation of real voltage data case.

V. CONCLUSION

Gauss Newton method being a derivative based optimization method, the estimation may diverge with increase order of system nonlinearity. But the proposed model utilizes the convergence property of Legendre polynomial which can effectively track the time-varying frequency across all the phases in case of three phase unbalanced system. Clarke transform is used to generate a complex model using three phase voltages which is helpful to obtain the Hessian matrix for estimation. A thorough compassion results are presented to show the estimation accuracy of the proposed model.

REFERENCES

[1] Yili Xia*, Lulu Qiao*, Qi Yang*, Wenjiang Pei*, and Danilo P. Mandic, "Widely Linear Adaptive Frequency Estimation for Unbalanced Three-Phase Power Systems with Multiple Noisy Measurement", 22ND) International Conference, IEEE, Digital Signal Processing (DSP) 2017

[2] M. S. Sachdev and M. M. Giray, "A least square technique for determining power system frequency," IEEE Trans. Power App. Syst., vol. PAS-104, no. 2, pp. 437–443, 1985.

[3] A. Routray, A.K Pradhan, K. Prahallad Rao,"A novel Kalman filter for frequency estimation of distorted signals in power system," IEEE Trans. Instrumentation and measurement, vol. 51, no. 3,june 2002.

[4] P. K. Dash, R. K. Jena, G. Panda, and A. Routray, "An extended complex Kalman filter for frequency measurement of distorted signals," IEEE Trans. Instrum. Meas., vol. 49, pp. 746–753, 2000.

[5] Elias Aboutanios,"An Adaptive Clarke Transform Based Estimator for the Frequency of Balanced and Unbalanced Three-Phase Power System", IEEE 25th European Signal Processing Conference (EUSIPCO) 2017 .

[6] Mario Mañana Canteli, Alfredo Ortiz Fernandez, Luis Ignacio Eguíluz, Carlos Renedo Estébanez "Three-Phase Adaptive Frequency

Measurement Based on Clarke's Transformation", IEEE Transactions On Power Delivery, Vol. 21, No. 3, July 2006.

[7] Reza Arablouei, Stefan Wernert and Kutluyull Dogamayt, "Adaptive Frequency Estimation Of Three-Phase Power Systems With Noisy Measurements", IEEE Conference ICASSP 2013.

[8] M. Akke, "Frequency estimation by demodulation of two complex signals," IEEE Trans. Power Del., vol. 12, no. 1, pp. 157–163, Jan. 1997.

[9] Zhiyong Dai and Wei Lin, "A Dynamic System Method for Estimation of Grid Voltage Parameters under Unbalance and Harmonic Distortion Operation", 2017 American Control Conference, May 24-26 2017.

[10] Zhiyong Dai and Wei Lin, "A Dynamic System Method for Estimation of Grid Voltage Parameters under Unbalance and Harmonic Distortion Operation", 2017 American Control Conference, May 24-26 2017.

[11] A Fast Recursive Algorithm for the Estimation of Frequency, Amplitude, and Phase of Noisy Sinusoid - P. K. Dash, Shazia Hasan- IEEE Transactions On Industrial Electronics, Vol. 58, No. 10, October 2011

[12] Power System Frequency Estimation Using Least Mean Square Technique – A.K. Pradhan, A. Routray, Abir Basak - IEEE Transactions On Power Delivery, Vol. 20, No. 3, July 2005

[13] A Gauss–Newton ADALINE for Dynamic Phasor Estimation of Power Signals and Its FPGA Implementation - Sarita Nanda, P.K. Dash - IEEE Transactions on Instrumentation and Measurements, Vol. 67, No. 1, January 2018

[14]]IEEE1159.2, Power quality event characterization, IEEE power quality standards. http://www.grouper.ieee.org/groups/1159/2/testwave.html

2nd IEEE International conference on power Electronics, Intelligent Control and Energy systems (ICPEICES-2018)

Power Quality Improvement Using Multilayer Gamma Filter Based Control for DSTATCOM Under Nonideal Distribution System

Pavitra Shukl, *Member, IEEE*
Department of Electrical Engineering
IIT, Delhi
New Delhi, India
pavitra.shukl@gmail.com

Bhim Singh, *Fellow, IEEE*
Department of Electrical Engineering
IIT, Delhi
New Delhi, India
bsingh@ee.iitd.ac.in

Abstract— The control technique utilizing multilayer gamma filter control algorithm is presented and analysed in the paper for the grid connected three phase distribution static compensator (DSTATCOM) system. For improving the power quality, DSTATCOM developed using a voltage source converter (VSC) is utilized. A PI (Proportional Integral) controller maintains DC link voltage of VSC in accordance to the reference value. An adaptive gamma filter based control technique is used for the control of VSC. The fourth order multilayer gamma filter is used for extracting the load current fundamental component and generating the pulses for switching of VSC. This system is developed under nonideal conditions where with vast variations in voltages and loads it performs satisfactorily.

Keywords— *Distribution Static Compensator, Gamma Filter, Power Quality, Voltage Source Converter (VSC)*

I. INTRODUCTION

The Power quality has always been a major concern in working of the distribution system and various endeavours have been in place to tackle this issue [1]. Improved power quality has numerous benefits including enhanced loading capability, maximum utilization of various electrical equipment, zero voltage regulation etc. There are majorly two broad categories on which the poor power quality sources can be based upon – distribution system including their subsystem and the consumer loads. These loads can be classified as linear loads and nonlinear loads [2]. Nowadays with the ever increasing usage of these nonlinear loads in distribution systems like rectifiers, computers and switched mode power supply (SMPS), the power quality problems of the electric distribution system, are gradually increasing. Moreover, the weak conditions of the grid, present in the electric distribution system in the developing countries, make the power quality problems even more severe [3]. These power quality problems basically arise due to current harmonics and voltage harmonics generated in the system. The harmonics occur due to nonlinear loads, unbalanced loads, voltage sag/swell conditions, addition/reduction of loads and voltage distortion. International standards have also been formulated for determining the acceptable level of power quality, such as IEEE-519 standard [4].

For the mitigation of these harmonics and improving power quality, power electronics based shunt custom power devices (CPDs) are used to increase reliability of the distribution systems. The most commonly used CPDs [5] are the distribution static compensators (DSTATCOMs), which are connected at the point of interconnection (PIC) in the system. They are used for producing compensating currents for

balancing of the grid currents. These are also suitable in handling different order of harmonics present in the system, in correspondence to distortion in voltage and therefore reducing harmonics resonance in the system. There are various methods of implementing the DSTATCOM.

However, for its optimal use, it is implemented by connecting a voltage source converter (VSC) in shunt to the electric distribution system at the point of interconnection [6]. A DC link capacitor is connected across the VSC for reducing ripples in the DC voltage. This configuration ensures a cost effective solution. The DSTATCOM can perform the functions of balancing the reactive power, regulating the voltage at PIC and improving the power quality [7]. The control algorithm ensures the proper operation of DSTATCOM in the electric distribution system. The prominence of control algorithm lies in the unerring estimation of fundamental load current component which is utilized to generate switching pulses of VSC for balancing of grid currents in case of abnormal conditions. It is also used for generating the pulses for switching of the VSC, for harmonics reduction and power quality improvement [8].

There are various control algorithms existing in the literature for the control of DSTATCOM. However, the utilization of an adaptive control enables an adjustment in the parameters and gains of controller in the real time, which helps in achieving the required performance [10]. This also accounts for the fast operation of the filters. The control algorithms utilize filters for fundamental component estimation of the load currents. The control algorithm with leaky LMS control is shown in [11], and neural network based control algorithm is shown in [12].

In this paper, the gamma filter is used for producing reference grid currents and gating pulses of VSC. It has combined properties of infinite impulse response system (IIR) and finite impulse response system (FIR). The gamma filter is adaptive in nature and has less complexity [13]. With the use of this filter efficient operation of DSTATCOM is performed by generating switching pulses of VSC, which improves the power quality of the system [14].

II. SYSTEM CONFIGURATION

The configuration of proposed system is represented in Fig. 1. An AC grid interfaced three phase DSTATCOM system is used. This system consists of a three phase AC grid to which nonlinear loads are connected. The DSTATCOM based on VSC, is used at the point of interconnection (PIC), to inject compensating currents so that the harmonics of load currents are cancelled, which improves the power quality of the system. The desired voltage and current rating of the proposed

978-1-5386-6624-1/18 $31.00 © 2018 IEEE

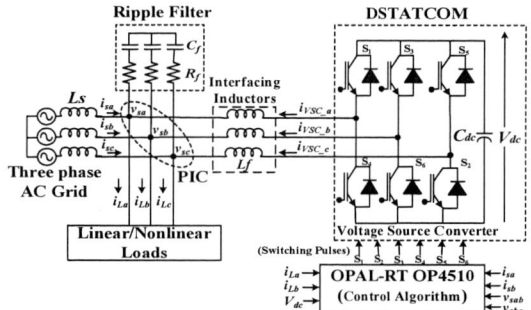

Fig. 1 Proposed System Configuration

system determines the insulated gate bipolar transistors (IGBTs) rating of the system.

For reducing the current ripples in the compensating currents, interfacing inductors (L_f) are utilized. The employment of ripple filters alleviates the switching ripples in the system, which are a combination of resistors (R_f) and capacitors (C_f) connected in shunt at PIC. Appendix constitutes of the parameters utilized in the operation of the proposed system.

For proper injection of compensating currents, is based on the control technique for the proper estimation of load current fundamental component so that the reference grid currents are evaluated.

The control algorithm requires load currents (i_{La}, i_{Lb}), grid currents (i_{sa}, i_{sb}), grid voltages (v_{sab}, v_{sbc}) and voltage of DC link (V_{dc}), which are sensed using current and voltage sensors for the generation of pulses for switching of the IGBTs in VSC. Fig. 2(a) shows the control structure, which consists of fourth order gamma filter for the estimation of load current fundamental component and harmonics mitigation of the load current.

III. CONTROL STRUCTURE

Fig. 2 (a) represents the proposed control structure of the system. In Fig. 2 (b) control algorithm is presented using multilayer gamma filter for acquiring fundamental load current component (I_{fla}). With combined features of finite impulse response (FIR) and infinite impulse response (IIR), gamma filter is developed which is a type of digital filter. It is a generalized feed-forward filter with gamma delay operator ($\mu / (z-(1-\mu))$). The gamma delay operator has a gain of (1- μ) in the feedback loop. With the utilization of multilayer gamma filter, the noise from the signals is alleviated with reduced parameters. The detailed control algorithm is explained as follows.

A. Evaluation of Load Currents Active Components

The determination of phase voltages of the grid (v_{sa}, v_{sb}, v_{sc}) is carried out using sensed grid line voltages (v_{ab}, v_{bc}) as [17],

$$v_{sa} = \frac{1}{3}(2v_{ab} + v_{bc}), \ v_{sb} = \frac{1}{3}(-v_{ab} + v_{bc}), \ v_{sc} = \frac{1}{3}(-v_{ab} - 2v_{bc}) \quad (1)$$

The grid phase voltages are used to calculate the amplitude of terminal voltage (V_t) as [14],

$$V_t = \sqrt{\frac{2(v_{sa}^2 + v_{sb}^2 + v_{sc}^2)}{3}} \quad (2)$$

Using these calculated grid phase voltages and terminal voltage, the in-phase unit templates are calculated as [14],

$$u_{pa} = \frac{v_{sa}}{V_t}, \ u_{pb} = \frac{v_{sb}}{V_t}, \ u_{pc} = \frac{v_{sc}}{V_t} \quad (3)$$

The in-phase unit templates are employed for the evaluation of quadrature unit templates as [14],

$$u_{qa} = \frac{(-u_{pb} + u_{pc})}{\sqrt{3}}, u_{qb} = \frac{(3u_{pa} + u_{pb} - u_{pc})}{2\sqrt{3}}, u_{qc} = \frac{(-3u_{pa} + u_{pb} - u_{pc})}{2\sqrt{3}} \quad (4)$$

With the proper estimation of fundamental load current component, the proficiency of control algorithm is realized in the system. Accordingly, due to the continuous variation of weights, according to the system requirements the gamma filter is termed as an adaptive filter. The fourth order multilayer gamma filter is utilized for the complete removal of harmonics and sub-harmonics in the system. In Fig. 2(b) for layer-1, the intermediate signal i_{L6} is calculated from its input signal i_{L5} as,

$$i_{L6} = \frac{\mu z^{-4}}{1 + (1-\mu)z^{-4}} \times i_{L5} \quad (5)$$

Similarly, the intermediate signal of i_{L5} is calculated from the obtained signals of Layer-2 and Layer- 3 are calculated as,

$$i_{L5} = i_{L3} - i_{L4}; i_{L3} = i_{La} - i_{L2} \quad (6)$$

Here, i_{L2} is the output obtained from Layer-2 and i_{L4} is the output obtained from Layer-3. From these, the transfer function between the input (i_{La}) and output (i_{Lfa}) of control algorithm is obtained as,

$$\frac{i_{Lfa}}{i_{La}} = k_1 \left\{ \left(1 - k_2 \frac{2\mu^2 z^{-8}}{1 + (1-\mu)z^{-4}}\right) \left(1 - k_3 \frac{2\mu^2 z^{-8}}{1 + (1-\mu)z^{-4}}\right) \left(\frac{\mu^2 z^{-8}}{1 + (1-\mu)z^{-4}}\right) \right\} \quad (7)$$

Assuming k_1, k_2 and k_3 equal to unity for simplicity in calculation of the transfer function, it can be written as,

$$\frac{i_{Lfa}}{i_{La}} = \frac{\mu^2 z^{-8} + \mu^2 z^{-12} - \mu^3 z^{-12} - 2\mu^4 z^{-16}}{\left[1 + (1-\mu)z^{-4}\right]^2} \quad (8)$$

The stability of gamma filter is dependent upon the proper selection of parameter μ, it is found from [13], that $0 < \mu < 2$ for stable operation of the filter. The multilayer gamma filter is developed to enhance the harmonics mitigation in the extracted load currents fundamental components (i_{Lfa}). The i_{Lfa} is procured free from harmonics and therefore, is sinusoidal in nature.

The zero crossing detector (ZCD) is supplied with i_{Lfa} generated from the output of multilayer gamma filter alongwith u_{qa}. The obtained signal is passed through sample and hold (S&H) logic for determination of load currents active power components. Therefore, after processing of the output signal through the absolute block, the generated signal is termed as load current active power component (I_{fpa}). The load current equivalent weight is calculated by the average of the obtained active components generated from all the remaining phases (I_{fpb}, I_{fpc}) as [14],

$$I_{pLavg} = \frac{I_{fpa} + I_{fpb} + I_{fpc}}{3} \quad (9)$$

B. Generation of switching pulses of VSC

The voltage error component of DC link is evaluated as [14],

$$V_{dce} = V_{dc}^*(m) - V_{dc}(m) \quad (10)$$

978-1-5386-6624-1/18 $31.00 © 2018 IEEE
594

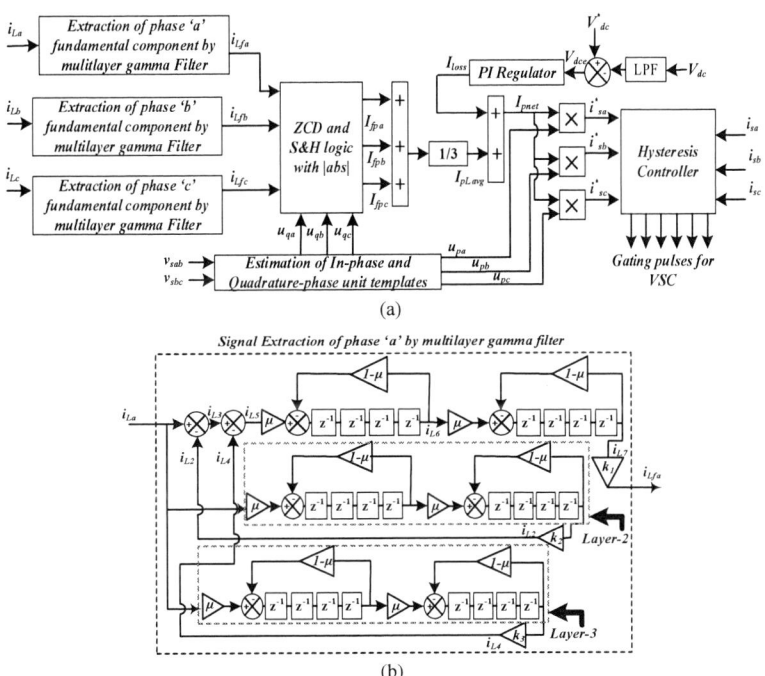

(a)

Signal Extraction of phase 'a' by multilayer gamma filter

(b)

Fig. 2 (a) Block Diagram of Fourth Order Multilayer Gamma Filter based Control Structure (b) Control Algorithm for 'a' phase

Here the V^*_{dc} is the reference voltage level of DC link and V_{dc} is the sensed voltage of DC link obtained of the VSC. For the estimation of loss component (I_{loss}), the V_{dce} signal obtained is passed through controller based on proportional-intergral (PI) mechanism.

$$I_{loss}(m) = I_{loss}(m-1) + K_p \{V_{dce}(m)\} + K_i \{V_{dce}(m) - V_{dce}(m-1)\} \quad (11)$$

Here, the PI controller gains of the DC link voltage are termed as K_i (integral gain) and K_p (proportional gain).

Therefore, the load current active component (I_{pnet}) is derived from the addition of loss (I_{loss}) and I_{pLavg} component,

$$I_{pnet} = I_{pLavg} + I_{loss} \quad (12)$$

The reference grid currents are evaluated by calculating the product of active power component with the unit vectors, which are in-phase with load current fundamental component as,

$$i^*_{sa} = I_{pnet} \times u_{pa}, \quad i^*_{sb} = I_{pnet} \times u_{pb}, \quad i^*_{sc} = I_{pnet} \times u_{pc} \quad (13)$$

The estimation of currents errors is performed as,

$$i_{esa} = i^*_{sa} - i_{sa}, \quad i_{esb} = i^*_{sb} - i_{sb}, \quad i_{esc} = i^*_{sc} - i_{sc} \quad (14)$$

The VSC switching pulses are generated by passing the calculated current errors through hysteresis controller.

IV. EXPERIMENTAL RESULTS

In the laboratory, by utilizing current sensors (LA55-P), voltage sensors (LV25-P), voltage source converter (VSC) and grid a prototype is developed. The OPAL-RT controller (OP4510) is fed with the signals sensed from these sensors for the operation of system. The control algorithm manipulates these signals and the output is obtained by digital signal oscilloscope (DSO). and power quality analyser. The control structure is presented in Fig. 2 (a). The multilayer gamma filter performs the operation of generating reference currents of grid by proper estimation of load current active component. For the functioning of the system the parameters utilized by the system are presented in Appendix.

A. Steady-State Response of System at Nonlinear Load

The analysis of grid current (i_{sa}), load current (i_{La}) and VSC current (i_{VSC_a}) is performed as per Figs. 3 (a-c). At steady state the power delivered by grid to load and DSTATCOM is observed in Figs. 3 (d-i). It is observed that the power factor correction (PFC) mode is implemented as power factor of grid current and voltage is obtained around unity. The load current total harmonic distortion (THD) is 28 % as shown in Fig. 3 (i). The waveform of load current is quasi-square in shape due to nonlinear load characteristics. The THD of grid current obtained is 2.9 % in Fig. 3 (h), which is within acceptable limit of the IEEE-519 standard and is sinusoidal in shape.

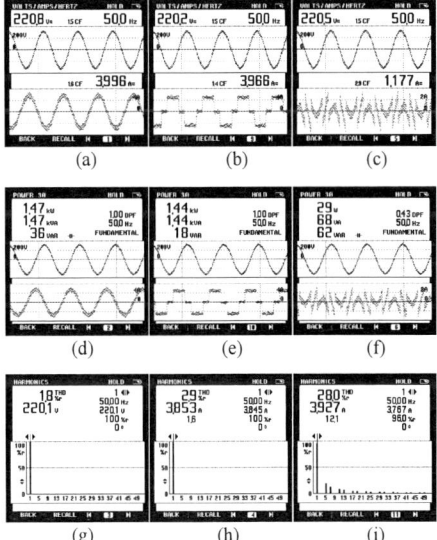

Fig. 3 Test results of balanced nonlinear loads connected to grid with DSTATCOM (a–c) v_{sab} alongwith i_{sa}, i_{La} and i_{VSC_a}, (d) three phase grid power (e) three phase load power (f) DSTATCOM power (g) harmonic spectrum of v_{sab} (h) harmonic spectrum of i_{sa} (i) harmonic spectrum of i_{La}

B. Behavior of System in Steady-State under Load Removal

For observing the steady-state behaviour of system under load unbalancing the load of phase 'a' is disconnected as shown in Figs. 4 (a-n). The waveforms in Figs. 4 (a-c), show the grid currents (i_{sa}, i_{sb}, i_{sc}), Figs. 4 (e-g) show load currents (i_{La}, i_{Lb}, i_{Lc}) and Figs. 4(i-k) show VSC currents (i_{VSC_a}, i_{VSC_b}, i_{VSC_c}). The unbalanced condition is realised by removing the phase 'a' load as observed in Fig. 4 (e).The THD of grid voltage is 1.8 % and grid current is 4.7 % as shown in Figs. 4 (m-n), which shows that the system performance is within the IEEE-519 standard. The waveforms of grid currents are maintained sinusoidal in shape, ensuring the standard power quality of the system.

C. Dynamic Behavior under Nonlinear Load Removal

Figs. 5 (a-i) show the performance of the system at unbalanced nonlinear load condition. The voltage of DC link (V_{dc}) is maintained at the reference value without any variation in Fig. 5 (c). It is observed that the grid currents remain sinusoidal when load of phase 'a' nonlinear load current is removed (Figs. 5 (e-f)). The waveform of VSC currents change shape when the load is removed as shown in Fig. 5 (f), as they provide compensation to the load currents for maintaining sinusoidal grid currents. In Figs. 5 (g-i), the internal signals of control algorithm during the unbalancing of load are presented. As observed, during the unbalancing of load, the magnitude of grid currents is decreased in but retain their sinusoidal nature hence showing satisfactory performance according to IEEE-519 standard with THD less than 5 %, as shown in Figs. 5 (d-e).

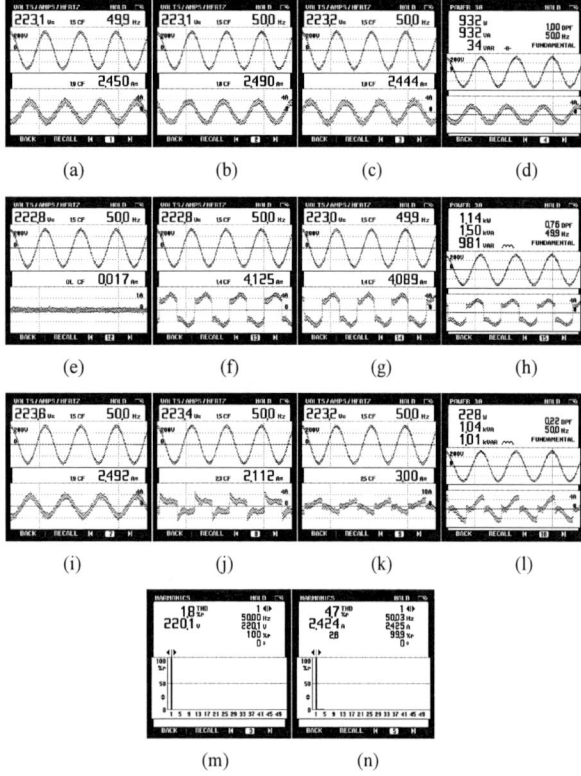

Fig. 4 Test results of unbalanced nonlinear loads connected to grid with DSTATCOM (a–c) v_{sab} alongwith i_{sa} , i_{sb} and i_{sc} (d) Grid power (e–g) v_{sab} alongwith i_{La} , i_{Lb} and i_{Lc} (h) Load power (i-k) v_{sab} alongwith i_{VSC_a}, i_{VSC_b} and i_{VSC_c} (l) DSTATCOM power (m–n) harmonic spectra of v_{sab} and i_{sa}

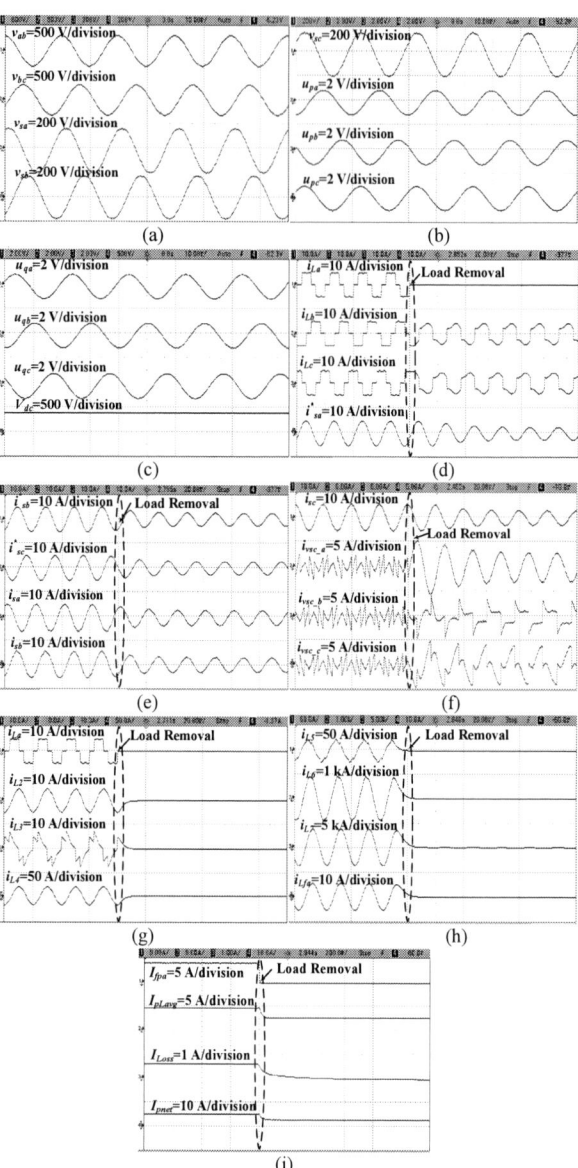

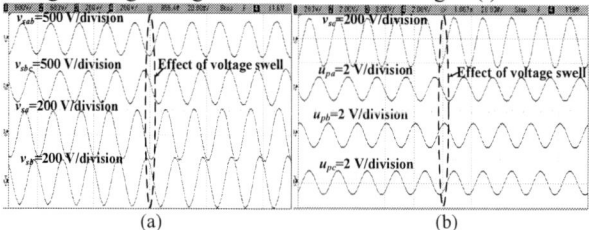

Fig. 5 (a-f) Dynamic behavior under load removal of phase 'a' for DSTATCOM connected to grid (g-i) control algorithm signals

D. Dynamic Behavior under Voltage Swell

The system performance is observed satisfactory under voltage swell for 20 % increase in grid voltages.. Figs. 6 (a-i) show the waveforms of the system under voltage swell. An increase in grid voltages can be seen in Fig. 6 (a). This has led to a increase in grid currents as shown in Figs. 6 (e-f). The DC link is sustained to the reference voltage level as shown in Fig. 6 (c). The internal signals are shown in Figs. 6 (g-i) which shows the fundamental load current obtained in Fig. 6 (h) is sinusoidal in nature. The loss component undergoes change during voltage swell as obtained in Fig. 6 (i).

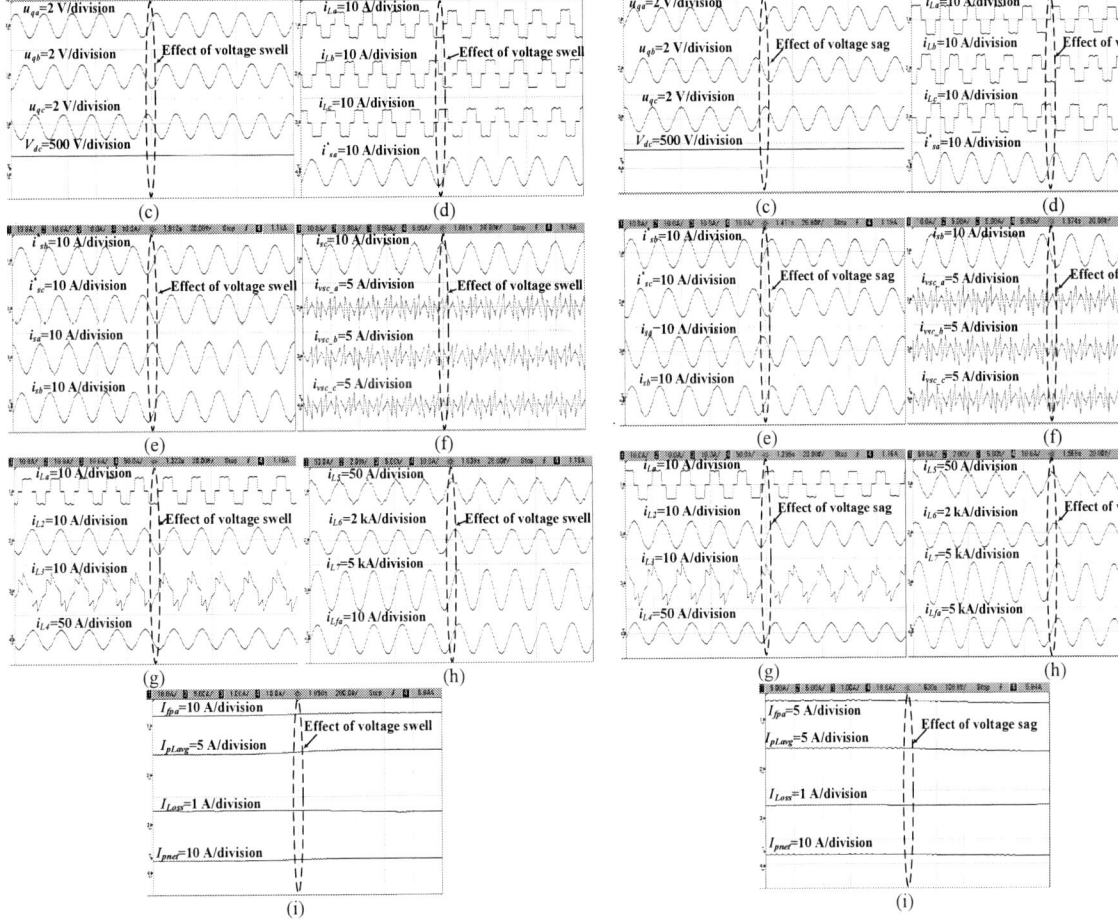

Fig. 6 (a-f) Response of grid connected DSTATCOM under voltage swell (g-i) internal signals

Fig. 7 (a-f) Dynamic behavior under voltage sag for DSTATCOM connected to grid (g-i) control algorithm signals

E. Dynamic Behavior under Voltage Sag at Nonlinear Load

The voltage sag conditions is shown under the decreasing grid voltage by 20%. The performance under a decrease in grid voltage is shown in Fig. 7. The grid currents are reduced in Fig. 7 (e) due to the voltage sag condition but the system performs satisfactorily as the grid currents are sinusoidal in nature. The various internal signals of Figs. 7 (g-i) show the performance of the control algorithm under the voltage sag conditions. The load current fundamental component extracted, is shown in Fig. 7 (h). The current loss component obtained is shown in Fig. 7 (i), which shows that the value is less than 1A. The reference currents obtained are sinusoidal in nature and hence the harmonics have been reduced as shown in Figs. 7 (d-e).

F. Dynamic Behavior under Voltage Distortion

The grid voltage distortion is observed in Figs. 8 (a-i), where the performance of the system is presented. The voltage distortion in grid line and phase voltages are shown in Figs. 8 (a-b). The in-phase and quadrature components obtained are shown in Figs. 8 (b-c). The DC link voltage is maintained to the reference value of DC link voltage in Fig. 8 (c). The reference grid currents in Fig. 8 (d), obtained from the control algorithm show the satisfactory performance of the system under voltage distortion condition. The VSC currents obtained are shown in Fig. 8 (f). The grid currents obtained in Fig. 8 (e), are sinusoidal in nature and are within the acceptable limit of the IEEE-519 standard with THD less than 5%. The internal signals obtained from control algorithm are shown in Fig. 8 (i).

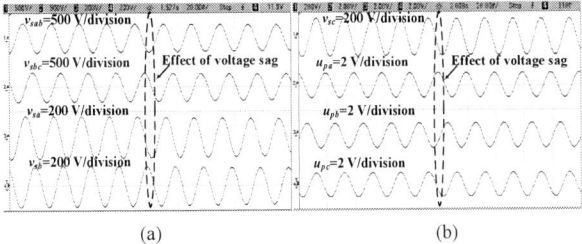

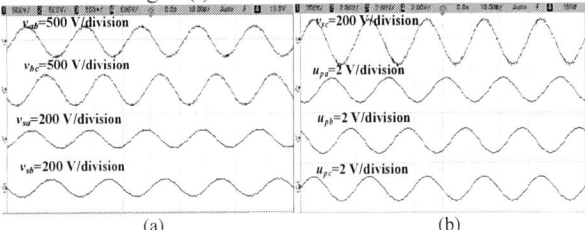

978-1-5386-6624-1/18 $31.00 © 2018 IEEE

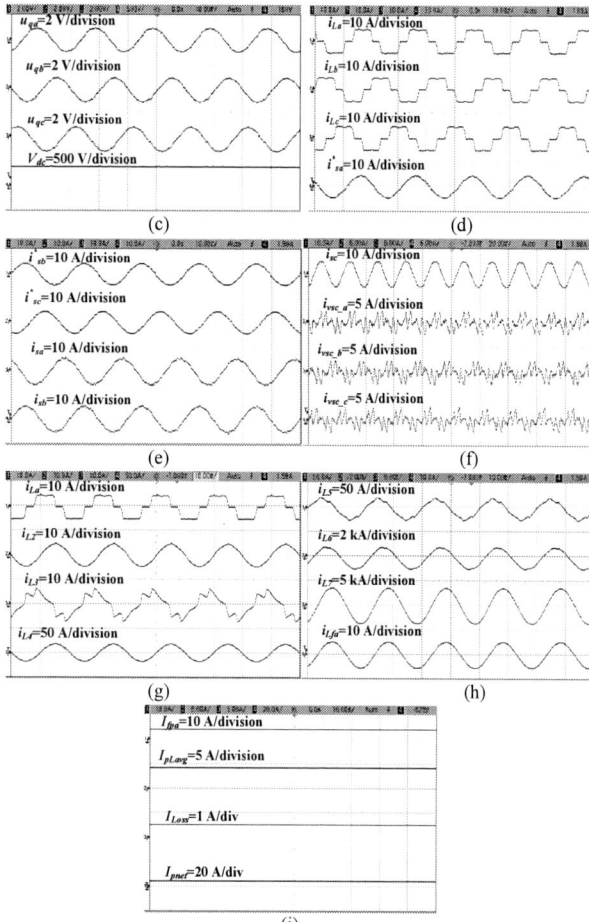

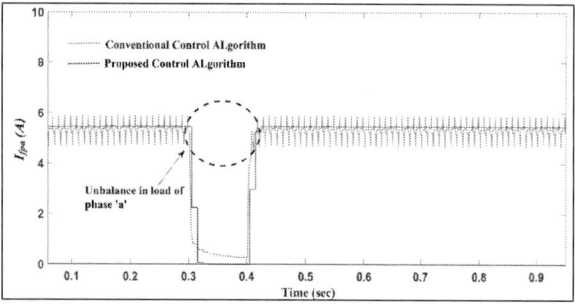

Fig. 8 (a-f) Response of grid connected DSTATCOM under voltage distortion (g-i) internal signals

G. Comparison of Proposed Multilayer Gamma Filter with Conventional Control Algorithm

The proposed multilayer gamma filter is compared with the least mean fourth (LMF) algorithm, during removal and injection of 'a' phase load current. The comparison in the extraction of amplitude of load current fundamental component (I_{fpa}) is presented in Fig. 9. As observed, the performance of the proposed multilayer gamma filter is fast during the load removal/injection conditions with reduced distortions in the extracted amplitude of load current fundamental.

Fig. 9 Comparison of multilayer gamma filter with conventional control algorithm during removal and injection of 'a' phase load current

V. CONCLUSION

The performance of grid interfaced three phase DSTATCOM has been demonstrated under nonlinear load, load removal/injection conditions, voltage swell/sag conditions, addition/reduction of loads and voltage distortion conditions. The utilization of multilayer gamma filter alleviates the computational burden due to its adaptive nature and noise can be easily eliminated from the signal. With the help of the filter switching pulses of VSC are generated by estimating the reference grid currents and hence the control algorithm is implemented for the DSTATCOM. According to a limit of IEEE-519 standard, THD of grid current is satisfactory, which is within the range of 5 %. The performance under voltage swell/sag conditions, load unbalancing and voltage distortion of the system is tested and the system is observed to be satisfactory.

ACKNOWLEDGMENT

The authors are thankful to OPAL-RT for providing financial support under project no. RP03253.

APPENDIX

System parameters: v_{ab} = 220V, f_0 = 50Hz; Impedance of grid: 0.1mH, 0.01Ω; V_{dc} = 360V; C_{dc} = 2200 µF; Nonlinear load: R = 50Ω, L = 100mH; L_f = 4mH, Ripple filter: C_f =10µF, R_f = 5Ω; Gains of multilayer gamma filter: μ = 1.96, k_1 = 0.00015, k_2 = 0.0005, k_3 = 0.002, PI controller gains K_p = 1.2, K_I = 0.001, Sytem processing: T_s = 20e-6

REFERENCES

[1] S. R. Arya, B. Singh, R. Niwas, A. Chandra and K. Al-Haddad, "Power quality enhancement using DSTATCOM in distributed power generation System," *IEEE Trans. Indus. App.*, vol. 52, no. 6, pp. 5203-5212, Nov.-Dec. 2016.

[2] L. Sainz and J. Balcells, "Harmonic interaction influence due to current source shunt filters in networks supplying nonlinear loads," *IEEE Trans. Power Deliv.*, vol. 27, no. 3, pp. 1385-1393, July 2012.

[3] R. T. Hock, Y. R. de Novaes and A. L. Batschauer, "A voltage regulator for power quality improvement in low-voltage distribution grids," *IEEE Trans. Power Electron.*, vol. 33, no. 3, pp. 2050-2060, March 2018.

[4] IEEE Recommended Practices and Requirements for Harmonic Control on Electric Power System, IEEE Standard 519, 1992.

[5] A. Domijan, A. Montenegro, A. J. F. Keri and K. E. Mattern, "Custom power devices: an interaction study," *IEEE Trans. Power Sys.*, vol. 20, no. 2, pp. 1111-1118, May 2005.

[6] C. Kumar and M.K. Mishra, "A voltage-controlled DSTATCOM for power-quality improvement," *IEEE Trans. Power Deliv.*, vol. 29, no. 3, pp. 1499-1507, June 2014.

[7] C. Kumar and M. K. Mishra, "An improved hybrid DSTATCOM topology to compensate reactive and nonlinear loads," *IEEE Trans. Indus. Electron.*, vol. 61, no. 12, pp. 6517-6527, Dec. 2014.

[8] M. Kumar, A. Swarnkar, N. Gupta and K. R. Niazi, "Design and operation of DSTATCOM for power quality improvement in distribution systems," *J. Engineer.*, vol. 2017, no. 13, pp. 2328-2333, 2017.

[9] B. Singh and J. Solanki, "A comparison of control algorithms for DSTATCOM," *IEEE Trans. Indus. Electron.*, vol. 56, no. 7, pp. 2738-2745, July 2009.

[10] P. Shukl and B. Singh, "Grid integration of three-phase single-stage PV system using adaptive laguerre filter based control algorithm under non-ideal distribution system," *IEEMA Engineer Infinite Conference (eTechNxT)*, pp. 1-6, 2018,

[11] S. R. Arya and B. Singh, "Performance of DSTATCOM using leaky LMS control algorithm," *IEEE J. Emer. Sel. Topics Power Electron.*, vol. 1, no. 2, pp. 104-113, June 2013.

[12] M. T. Ahmad, N. Kumar and B. Singh, "Generalised neural network-based control algorithm for DSTATCOM in distribution systems," *IET Power Electron.*, vol. 10, no. 12, pp. 1529-1538, 10 6 2017.

[13] J. C. Principe, B. de Vries and P. G. de Oliveira, "The gamma-filter-a new class of adaptive IIR filters with restricted feedback," *IEEE Trans. Sig. Process.*, vol. 41, no. 2, pp. 649-656, Feb 1993.

[14] B. Singh, Chandra, and K. Al-Hadad, '*Power Quality: Problems and Mitigation Techniques*', John Wiley & Sons Ltd., U. K., 2015.

2nd IEEE International conference on Power Electronics, Intelligent Control and Energy systems (ICPEICES-2018)

A Hysteresis Current Controlled Grid connected Full Bridge Inverter With Zero Current Switching

Shashank Kurm
Department of Electrical
Engineering Indian Institite of
Technology, Bombay Mumbai, India
shashank.sk51191@gmail.com

Vivek Agarwal
Department of Electrical
Engineering Indian Institite of
Technology, Bombay Mumbai, India
agarwal@ee.iitb.ac.in

Abstract—Grid connected solar inverters are operated in current controlled mode. Hysteresis current control is one of the easiest ways to control current to track any reference, but it requires a high switching frequency for power devices for good tracking of reference, which leads to a high switching loss. This paper proposes a full bridge inverter, operated with hysteresis current control, with zero current switching. Thus it employs the advantages of hysteresis control, but without the penalty of high switching loss. The proposed converter is an H-bridge inverter in which two of the power devices operate at line frequency, and the other two are switched at high frequency, while employing zero current switching. Since the inverter is hysteresis current controlled, it can accurately track the reference current easily. The ZCS technique employed in the proposed converter is similar to a PWM soft switched converter, where resonance is activated only during switching transition. For rest of the duration in a switching cycle, the converter behaves just like a PWM converter. The increased current stress on the main power device is not significant. The operation of the proposed inverter has been verified by simulation. Hardware/experimental results will be included in the final paper.

Keywords—Hysteresis current control, Zero current switching, Grid connected inverter, Hybrid PWM

I. INTRODUCTION

Due to the fast depleting fossil fuels, installation of renewable energy sources such as solar PV plants has become necessary to keep up with the electric energy demand. However, the installation of PV plants needs a lot of land space which is expensive. Hence, it has become common for residential and commercial buildings to install rooftop solar PV plants, which feed the generated power directly to the grid. These generators are connected to the grid through power electronic converters/inverters.

Since almost all of the power supply network across the world are AC systems, solar inverters are used to interface PV modules to the grid. These solar inverters are operated in current control mode, injecting power to the grid at near unity power factor [1]. The voltage reference is obtained from the grid voltage. The solar inverters have to comply with the harmonics requirements of the grid i.e. they can only inject current having very low THD (< 5%) [15]. So the injected current has to closely follow a sinusoidal reference, which is in phase with the grid voltage. The easiest way to generate a current waveform following any reference is to use hysteresis

current control. It is easy to implement hysteresis control using analog controller, leading to a low cost implementation. Also it offers features like unconditional stability, fast response and excellent tracking capability [2-7]. However, the problem with using hysteresis control is that the switching frequency for the power devices may go very high for a close tracking. This leads to high switching loss, which severely deteriorates efficiency. To obtain a low switching loss, while employing hysteresis current control, it may be a good option to implement soft switching i.e. zero voltage or zero current switching. The problem of a variable switching frequency in the hysteresis control may be solved by employing fixed switching frequency hysteresis control [3, 6].

By employing soft switching in an inverter, the reference current waveform can be closely tracked, without having to incur high switching loss. Also, by employing soft switching technique, the switching frequency (for constant switching frequency control schemes) can be made to obtain large bandwidth and reduced filter size.

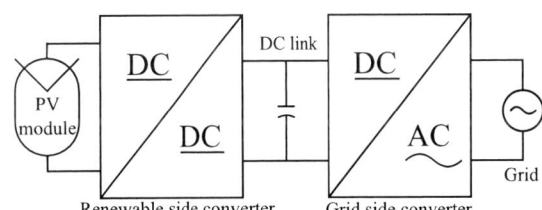

Fig 1. Basic structure of a grid connected solar PV system

Additionally, by reducing the device loss, cooling requirement can also be reduced. Both of these factors contribute to increase in power density, which is a critical requirement for the current trend of module integrated inverters. PWM soft switching technique must be used instead of resonance or quasi resonance soft switching for controlling an inverter to follow a reference, since the inverter is expected to behave like a hard switched inverter for most of the time. Resonance should be activated only during switching transition to minimize switching loss [8-12]. Addition of the soft switching feature to any converter requires additional resonant components to achieve ZVS/ZCS condition(s). This resonant network is active only during the switching transitions.

978-1-5386-6624-1/18 $31.00 © 2018 IEEE
599

While imparting the soft switching feature to the inverter, following points should be taken care of to keep the efficiency and cost of the converter under check:
1. The addition of auxiliary components does not alter the basic working of the circuit.
2. The circulating resonance current does not stress the main power devices to a large extent. The increase in device stress should be minimum.
3. The stress on the auxiliary devices should not be more, and their required ratings should be lesser compared to the main power devices to keep cost in check.
4. Gate pulse for the auxiliary devices should be derived from the gate pulses for main device itself, without additional computation burden.

Section II describes the proposed inverter, and the control strategy adopted. Section III describes the different operating modes, that occur during the inverter operation. Section IV describes the design procedure for auxiliary components for soft switching. Section V explains the logic for switching signal generation for the main and auxiliary switches. Section VI presents the simulation results, which is followed by major conclusions of the work in section VII.

II. PROPOSED INVERTER TOPOLOGY

In this paper, a zero current switching inverter is presented, which employs hysteresis current control. The basic circuit topology is a full bridge inverter, which is operated using the hysteresis current control technique. Two of the switches of the bridge are switched at line frequency, according to the grid voltage waveform, and the remaining two switches are switched to track the reference current waveform. Other than the main power devices, for creating ZCS condition, there are auxiliary components, which are rated much lower than the main components.

Fig. 1 shows the circuit diagram of the proposed inverter circuit. S_1, S_2 are the two high frequency switching devices, and S_3, S_4 are line frequency switching devices. L_f is the filter inductor.

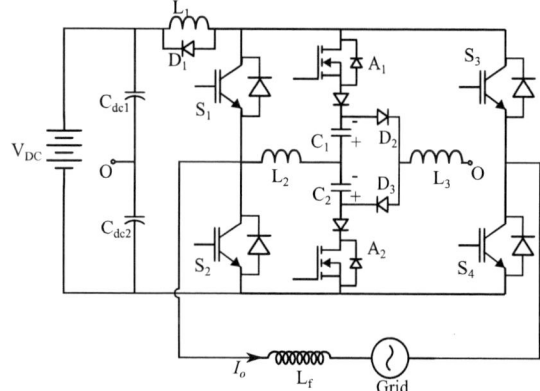

Fig. 2. Proposed inverter circuit

Two auxiliary switches A_1 and A_2 correspond to the turn off switches for S_1 and S_2 respectively. The turn on command for S_1, S_2 is generated by comparing the output current to the reference through a hysteresis comparator. The hysteresis width is 2h. When the Io exceeds $I_{ref}+h$, S_1 is turned on. When I_o goes below $I_{ref}-h$, turn off process of S_1 is initiated by turning on A_1. Similarly, S_2 is controlled for the opposite current direction. S_3 and S_4 are operated as per the polarity of grid voltage. When V_g is positive, S_4 is on and vice versa.

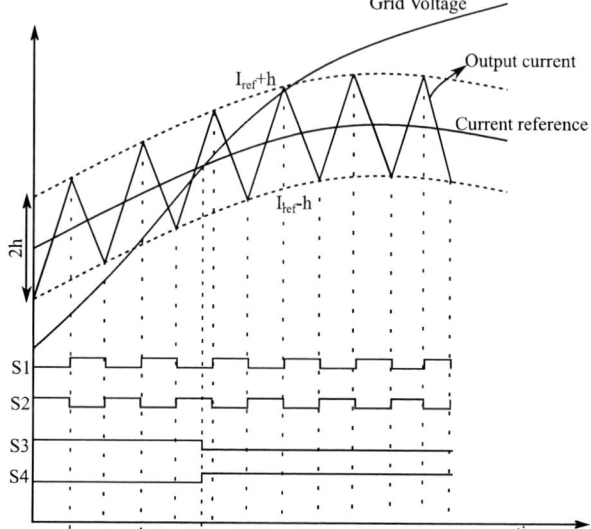

Fig. 3. Switching logic for the proposed inverter.

C_1 and C_2 are resonant capacitors for achieving zero current condition through the main switches S_1 and S_2. L_2 is the resonant inductor. D_2 (D_3) and L_3 are used to charge C_1 (C_2) when S_1 (S_2) is on, to initiate resonance for achieving ZCS condition for positive (negative) current polarity from the half DC bus C_{dc1} (C_{dc2}).

L_1 slows down the turning on of S_1 and S_2 to achieve ZCS turn on. D_1 is used for the current in L_1 to freewheel when S_1 (S_2) are turned off.

III. CIRCUIT OPERATING MODES

This section explains the different modes of operation of the proposed inverter, when the grid voltage is positive and the output current is controlled to track a sinusoidal reference in phase with the grid voltage.

Since the grid voltage is positive, S_4 is turned on. S_1 and S_2 are operated to track the reference current I_{ref}. The operation of the inverter circuit for either current direction can be divided into 8 distinct modes

A. Mode 0:

Mode 0 is the interval prior to turning on of S_1. I_o was freewheeling through the antiparallel diode of S_2. Resonance takes place in the network S_2-L_2-C_2-D_2-L_3-C_{dc2}. Mode 0 ends when I_o goes below $I_{ref}-I_h$, and the hysteresis comparator signals to turn on S_1 at $t=t_0$.

978-1-5386-6624-1/18 $31.00 © 2018 IEEE

B. Mode 1:

When S_1 is turned on, Mode 1 starts. Current in S_1, which is same as that in L_1 starts from zero and thus zero current turn on is achieved for S_1. In mode 1, both the IGBT S_1 and the anti-parallel diode of S_2 are conducting. Mode 1 ends when current in S_1 becomes equal to I_o (at $t=t_1$). Resonance still continues as in mode 1.

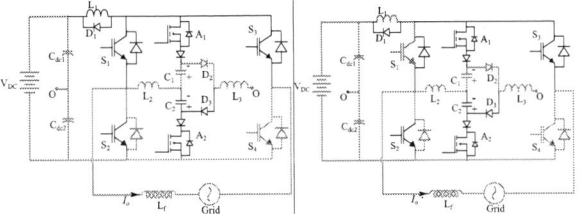

Fig 4. Circuit operation in Mode 0 and Mode 1

C. Mode 2:

When current in L_1 becomes equal to I_o, current in the anti-parallel diode of S_2 becomes zero, and S_2 turns off at $t=t_2$. Resonance path is now O-C_{dc1}-L_1-S_1-L_2-C_2-D_2-L_3-O. Mode 2 ends when current in D_2 becomes zero. At the end of mode 2, C_1 is charged to a voltage V_{c1}^0, with the polarity as shown in the figure. This is required so that the current in S_1 reduces when turn off process is initiated

D. Mode 3:

When D_2 turns off by natural commutation, mode 3 starts. Mode 3 continues till turn off is initiated at $t=t_3$ by turning on A_1. Voltage across C_1 is constant at V_{C1}^0 over mode 3.

Modes 2 and 3 correspond to the on-duration of a PWM converter. There is a slight increase in the switch S_1 current as compared to that in a hard switched converter due to charging of capacitor C_2 in mode 2.

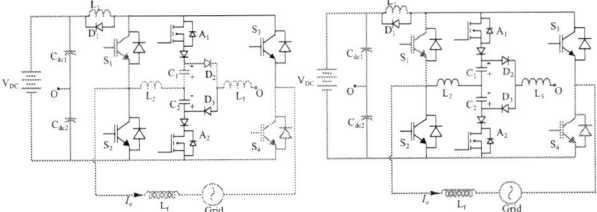

Fig 5. Circuit operation in Mode 2 and Mode 3

E. Mode 4:

When A_1 is turned on, resonance starts in S_1-A_1-C_2-L_2. Current in S_1 in mode 4 is given by:

$$i_{S1} = I_O - \frac{V_{C1}^0}{Z_r}\sin(\omega_r t) \qquad (1)$$

Where,

$$Z_r = \sqrt{\frac{L_2}{C_2}}, \omega_r = \frac{1}{\sqrt{L_2 C_2}}$$

Mode 4 ends when current in S_1 become zero (at $t=t_4$), and goes negative. At $t=t_4$, gate pulse for S1 can be turned off.

F. Mode 5:

When current in S_1 goes negative, the anti-parallel diode of S_1 starts conducting. The voltage across S_1 goes negative. There is a comparator to detect the voltage across S_1, and when this voltage goes negative, the comparator signals to turn off the gate pulse for S_1. Mode 5 continues till the antiparallel diode of S_1 is conducting, i.e. when the current in S_1 goes to zero at $t=t_5$, and tends to go positive. At this point current in L_2 also becomes zero.

Current in L_3 rises linearly in modes 4 and 5 due to the voltage of C_{dc1}. At the end of mode 5, the voltage of C_1 is V_{c1}^1.

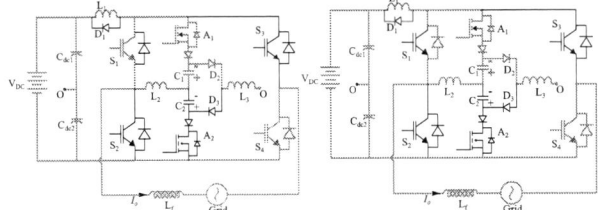

Fig 6. Circuit operation in Mode 4 and Mode 5

G. Mode 6:

At $t=t_5$, mode 6 begins and the turn off process of S_1 is complete. Current I_o is carried by the diode of S_2, current in L_1 freewheels through D_1, and decays by the forward voltage drop of D_1. Resonance takes place in the network S_2-L_2-C_1-D_2-L_3-O-V_{dc}. Current in A_1 during mode 6 is given by:

$$i_{A1} = \frac{V_{dc} - V_{C1}^1}{Z_r}\sin \omega t - \frac{V_{dc}}{L_3}t \qquad (2)$$

When i_{A1} becomes zero (at $t=t_7$), A_1 commutates naturally, since it is a unidirectional switch. At $t=t_7$, mode 6 ends.

H. Mode 7:

At $t=t_7$, mode 7 begins, and continues till the current in L_1 decays to zero (at $t=t_8$). This completes the total turn on and off cycle. After $t=t_8$, mode 0 begins.

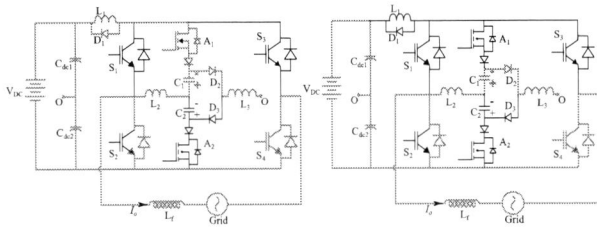

Fig 7. Circuit operation in Mode 6 and Mode 7

Fig. 8 shows the different current and voltage waveforms for the various modes during the positive current.

For the opposite current polarity, the main power devices conducting are S_2 and the anti-parallel diode of S_1. The corresponding auxiliary switch is A_2, and the resonant capacitor is C_2. The circuit's operating modes are same as

978-1-5386-6624-1/18 $31.00 © 2018 IEEE

described above. Assuming unity power factor operation, S_3 will be on for negative grid voltage.

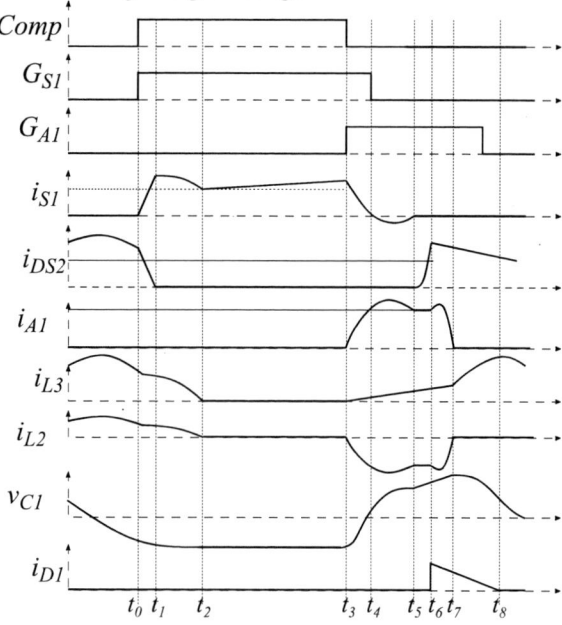

Fig 8. Waveforms for positive current direction

IV. DESIGN CONSIDERATIONS

One of the major challenges while designing any converter for hysteresis control is the variable switching frequency. To design the auxiliary components, that point in the reference current waveform where the switching frequency is highest is considered, i.e. near the zero crossing of grid voltage. By using the soft switching technique described earlier in this paper, the range of duty cycle that can be achieved is reduced because the turn off process has been lengthened. So the DC bus voltage requirement is increased.

1. The values of L_2, C_1 must be selected such that the magnitude of resonance current due to voltage V_{c1}^0 is higher than the peak value of I_o.

$$\sqrt{\frac{L_2}{C_2}} < \frac{V_{C1}^0}{I_o} \qquad (3)$$

2. The time period of one resonance cycle corresponding to $(L_1+L_2+L_3)$ and C_2 must be greater than the time period corresponding to the maximum switching frequency. This takes place near the zero crossing of grid voltage for unity power factor operation.

$$\pi\sqrt{(L_1 + L_2 + L_3)C_1} < 1/F_{sw,max} \qquad (4)$$

Maximum switching frequency can be calculated from [5]:

$$F_{sw,max} = \frac{V_{dc}}{4hL_f} \qquad (5)$$

Where, $2h$ is hysteresis width, and L_f is the value of filter inductor [5].

3. The value of L_1 must be such that the entire current in L_1 decays to zero in the least possible off duration. Otherwise, zero current turn on will not be achieved. However, since the turn on loss of an IGBT is much smaller than the turn off loss, this constraint can be overlooked.

4. The voltage rating of C_1 and C_2 must be at least 1.5 times the DC bus voltage.

V. CONTROL SCHEME FOR SWITCHES

The gate pulse for the low frequency switches is generated by comparing the grid voltage to zero. When the grid voltage is positive, S_4 is turned on and vice versa.

To generate the gate pulse for S_1 and S_2, a bistable multivibrator is used, which is set and reset as per the logic given below:

1. Set by the switching command given by the hysteresis comparator.
2. When the diode starts conduction (in mode 5), the gate signal for the IGBT is turned off. This is indicated by a diode conduction sense circuit. This circuit detects if the IGBT or the antiparallel diode is conducting on the basis of the voltage across the switch [13].

The gate pulse generation circuit logic and diode conduction sense circuit have been shown in Fig. 9 and 10 respectively.

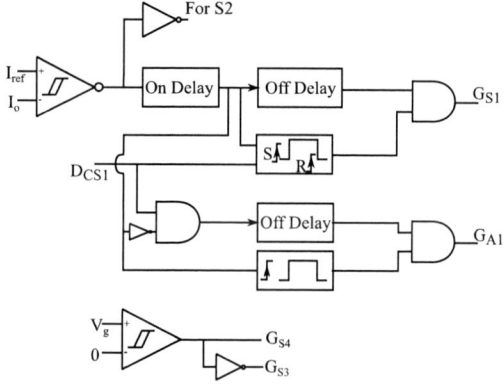

Fig 9. Gate signal generation for the main and auxiliary switches

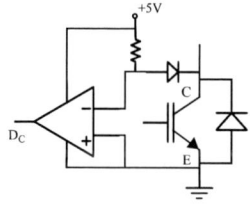

Fig 10. Diode conduction sense circuit for generating Dc

The gate command for S_2 is generated in a similar manner. There is a dead time between the input gate commands for the two switches. The dead time must be higher than the total turn off interval i.e. mode (4+5+6). Also, the maximum time for which the turn off of S_1 can be delayed has been limited, so that even if the reset signal does not occur, the current does not become very high.

The gate command for A_1 and A_2 is generated by ANDing of the following two signals:

1. A constant width pulse, starting at the falling edge of the input gate command which is given by the hysteresis comparator. This pulse duration is kept higher than the duration of the mode (4+5+6)

2. A high to low going pulse starting at the falling edge of (input AND DC_{S1}). This is to make sure that A_2 is not triggered for positive current direction and vice versa.

VI. SIMULATION RESULTS

The proposed inverter circuit has been simulated in MATLAB/SIMULINK. The circuit parameters, auxiliary components values, and the timing parameters are given in Table 1.

Vdc	350 V	Vgrid	230V (rms)
I_o (peak)	10 A	L_f	5 mH
Hysteresis band	±0.4 A	Fundamental Frequency	50 Hz
L_1	1μH	L_3	200 μH
L_2	2 μH	C_1, C_2	10 nF

Table 1. Simulation conditions

The timing parameters used in the control circuit are given in Table 2:

Dead time between S_1/S_2	2 μs
Maximum turn off delay for S_1/S2	1.8 μs
Width of gate pulse for A_1/A_2	1.5 μs

Table 2. Timing parameters for signal generation

Fig. 11 shows the simulated waveforms of gate pulses for S_1, A_1, and currents through S_1 and A_1. It can be seen from the simulation waveforms that the circuit works as per the analysis given above.

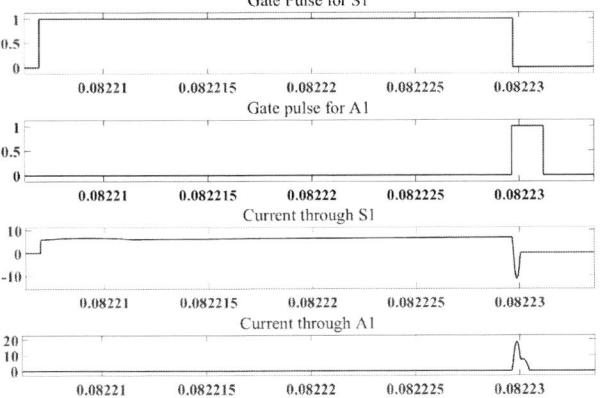

Fig. 11. Simulated waveform showing gate pulse and current for main and auxiliary switches.

Fig. 12 shows the simulated waveforms of grid voltage, output current and gate pulses for S_1 and S_4.

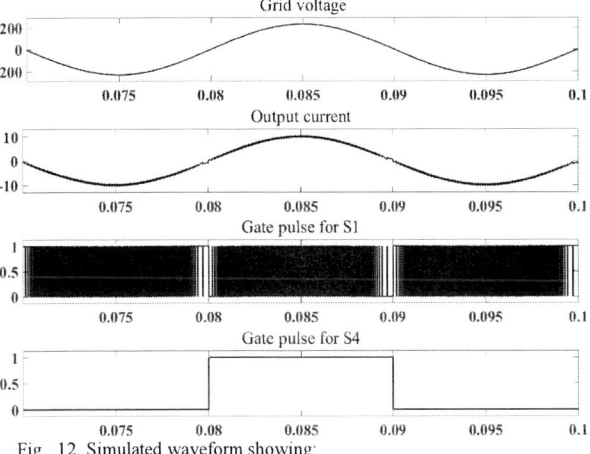

Fig. 12. Simulated waveform showing:

Fig. 13 shows simulated waveforms of voltage across S_1 and current through it. It can be seen that zero current switching is takes place-leading to ideally zero switching loss.

Fig. 14 shows the harmonic components in the output current. THD in the output current is 3.75%, which is pretty low. The switching frequency harmonics are spread around 15-20 kHz, using an output filter inductor of 5 mH. THD can be further improved by reducing the hysteresis band width which is ±0.4A used in simulation.

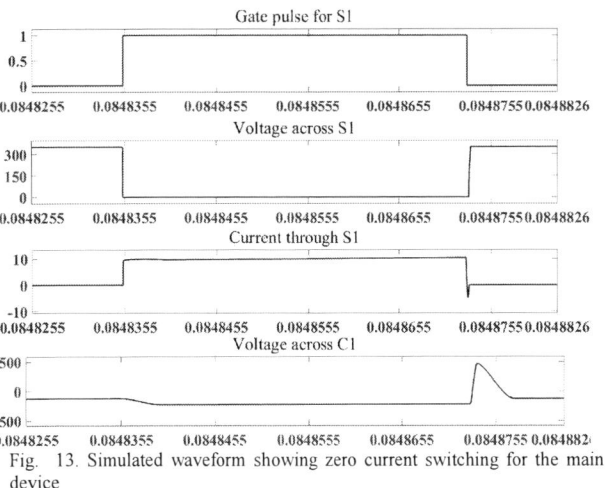

Fig. 13. Simulated waveform showing zero current switching for the main device

The proposed inverter circuit can also work with non-unity power factor, thus supplying/absorbing reactive power from the grid. This enables the inverter to take part in ancillary services such as delivering reactive power for grid support. This can be seen in the Fig. 14 and Fig. 15 where the reference current leads and lags the grid voltage waveform respectively.

978-1-5386-6624-1/18 $31.00 © 2018 IEEE

VII. CONCLUSION

A hysteresis current controlled inverter has been presented in this paper and the operation has been verified by simulation. To eliminate the high switching loss resulting from the hysteresis control, soft switching has been implemented Of the 4 main power devices, 2 are switched at line frequency and the other two are zero current switched at high frequency, thus reducing switching loss.. The soft switching technique does not increase the current stress on the main power switch significantly

. Simulations results validate the operation of the proposed inverter at leading, lagging and unity power factor. The THD in the output current is 3.75% which is within the limit.

The auxiliary components added to achieve soft switching are all having low ratings. Only L_1 has a rating that is similar to the main power circuit components. But its value itself is very less (1 µH). Experimental results will be presented in the final paper.

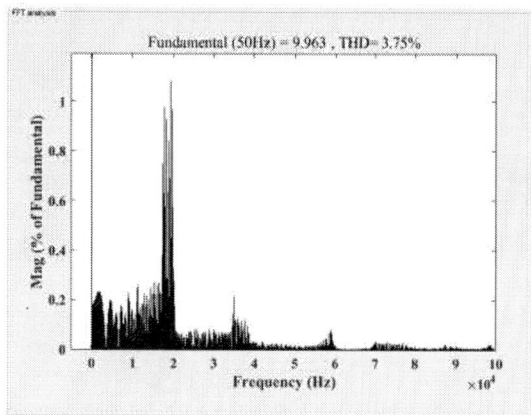

Fig. 14. FFT of output current showing THD within limits

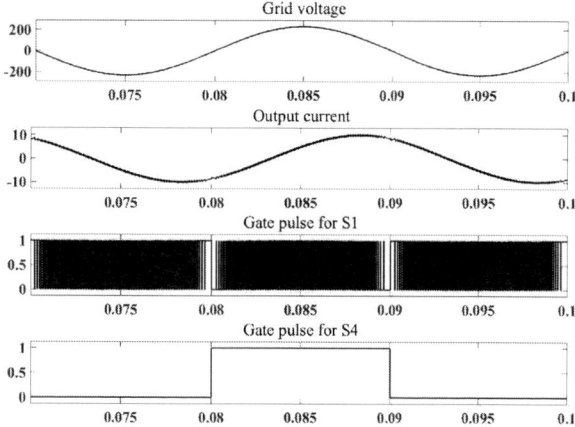

Fig 14. Simulated waveforms when the reference current lags the grid voltage.

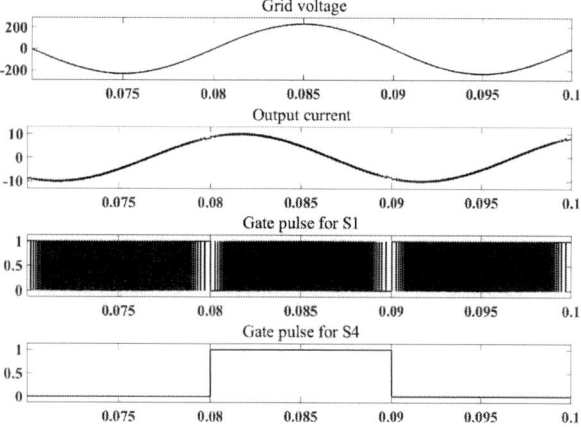

Fig 15. Simulated waveforms when the reference current leads the grid voltage:

REFERENCES

[1] S. B. Kjaer, J. K. Pedersen and F. Blaabjerg, "A review of single-phase grid-connected inverters for photovoltaic modules," in *IEEE Transactions on Industry Applications*, vol. 41, no. 5, pp. 1292-1306, Sept.-Oct. 2005.

[2] S. Buso, L. Malesani and P. Mattavelli, "Comparison of current control techniques for active filter applications," in *IEEE Transactions on Industrial Electronics*, vol. 45, no. 5, pp. 722-729, Oct 1998.

[3] L. Malesani and P. Tenti, "A novel hysteresis control method for current-controlled voltage-source PWM inverters with constant modulation frequency," in *IEEE Transactions on Industry Applications*, vol. 26, no. 1, pp. 88-92, Jan/Feb 1990

[4] B. K. Bose, "An adaptive hysteresis-band current control technique of a voltage-fed PWM inverter for machine drive system," in *IEEE Transactions on Industrial Electronics*, vol. 37, no. 5, pp. 402-408, Oct 1990.

[5] P. A. Dahono, "New hysteresis current controller for single-phase full-bridge inverters," in *IET Power Electronics*, vol. 2, no. 5, pp. 585-594, Sept. 2009.

[6] C. N. M. Ho, V. S. P. Cheung and H. S. H. Chung, "Constant-Frequency Hysteresis Current Control of Grid-Connected VSI Without Bandwidth Control," in *IEEE Transactions on Power Electronics*, vol. 24, no. 11, pp. 2484-2495, Nov. 2009.

[7] C. Attaianese, M. Di Monaco and G. Tomasso, "High Performance Digital Hysteresis Control for Single Source Cascaded Inverters," in *IEEE Transactions on Industrial Informatics*, vol. 9, no. 2, pp. 620-629, May 2013.

[8] G. Hua and F. C. Lee, "Soft-switching techniques in PWM converters," in *IEEE Transactions on Industrial Electronics*, vol. 42, no. 6, pp. 595-603, Dec 1995.

[9] A. Jain and V. Agarwal, "Design and fabrication of quasi resonant converter", *Journal of Indian Institute of Science*, Sept.-Oct. 1097, 77, 457-468

[10] Jih-Sheng Lai, "Resonant snubber-based soft-switching inverters for electric propulsion drives," in *IEEE Transactions on Industrial Electronics*, vol. 44, no. 1, pp. 71-80, Feb 1997.

[11] Y. S. Lee and G. T. Cheng, "Quasi-Resonant Zero-Current-Switching Bidirectional Converter for Battery Equalization Applications," in *IEEE Transactions on Power Electronics*, vol. 21, no. 5, pp. 1213-1224, Sept. 2006.

[12] P. Cancelliere, V. D. Colli, R. Di Stefano and F. Marignetti, "Modeling and Control of a Zero-Current-Switching DC/AC Current-Source Inverter," in *IEEE Transactions on Industrial Electronics*, vol. 54, no. 4, pp. 2106-2119, Aug. 2007.

[13] A. Datta, A. Guha and G. Narayanan, "An advanced gate driver for insulated gate bipolar transistors to eliminate dead-time induced distortions in inverter output," *2014 IEEE International Conference on Power Electronics, Drives and Energy Systems (PEDES)*, Mumbai, 2014, pp. 1-6.

[14] IEEE Recommended Practice and Requirements for Harmonic Control in Electric Power Systems," in *IEEE Std 519-2014 (Revision of IEEE Std 519-1992)*, vol., no., pp.1-29, June 11 2014.

Implementation of PSO based Selective Harmonic Elimination Technique in Multilevel Inverters

Peeyush Kala
Department of Electrical & Electronics Engineering,
Shivalik College of Engineering
Dehradun, India
peeyush.kala@gmail.com

Sudha Arora
Department of Electrical Engineering
G.B.P.U.A.&T. Pantnagar
Pantnagar, India
arora.sudha@gmail.com

Abstract— In multilevel inverters (MLIs), the selective harmonic elimination (SHE) technique is becoming popular owing to its twofold prime advantages. The first advantage is its ability to compute the desired switching angles such that the actual fundamental output voltage is maximized to desired voltage level while simultaneously minimizing the dominant low order harmonics. The second advantage of this scheme is that it can operate in low switching frequency mode which reduces switching losses in medium voltage and high power inverters. However, SHE method requires solution of transcendental nonlinear equations using conventional means such as algebraic and iterative methods which is quite cumbersome task. In order to alleviate this problem, a heuristic based optimization technique particle swarm optimization (PSO) is being used in this paper. Performance of PSO based SHE technique has been compared with continuous genetic algorithm (CGA) based SHE method which shows the superiority of proposed PSO based SHE technique. Simulations and programming have been performed using MATLAB/Simulink software environment for an eleven-level inverter.

Keywords— *selective harmonic elimination, particle swarm optimization, multilevel inverters, continuous genetic algorithm*

I. Introduction

Multilevel inverters (MLIs) play a vital role in various medium voltage and high power dc to ac conversion applications. MLIs possess several advantages over PWM inverters such as improved power quality, economical structure, low dv/dt stress, transformerless operation, and low filter requirement [1]. In industrial applications, mainly three conventional MLI topologies Neutral Point Clamped (NPC) MLI, Flying Capacitor (FC) MLI, and Cascaded H-bridge (CHB) MLI are widely used [2]. NPC MLI and FC MLI topologies are used in various applications in which high voltage single dc-bus is available whereas CHB MLI topology is suitable for applications having distributed dc sources such as fuel cells, photovoltaic arrays [3].

Over the years, there has been a lot of topological advancement seen in MLIs [4]. Apart from these topological advancements, various researchers have put a lot of thrust in the development and selection of suitable switching techniques for MLIs [5]. In the literature, for medium voltage and high power applications, use of low switching frequency modulation schemes is preferred over high switching frequency modulation schemes as former can reduce the switching losses

significantly [6]. One of the popular modulation techniques is known as selective harmonic elimination (SHE) technique. The popularity of this technique is attributed to its manifold advantages. The prime benefit of this scheme is that it eliminates or minimizes the low order dominant harmonics such as 5th, 7th, 11th and 13th for 11-level inverter while simultaneously maximizing the fundamental component of voltage to desired level. Other benefits of SHE technique are: reduced switching losses, control of fundamental voltage, elimination of non-triplen odd order harmonics so as to avail the advantage of 3-phase system, real-time and offline control, and ease of hardware implementation [7], [8].

The origin of SHE technique was documented in [9] where the output voltage waveform of an inverter was transformed into set of transcendental and nonlinear equations and then iterative method was applied for the solution of these set of equations to find switching angles. This technique was later adopted and implemented by various researchers on CHB MLIs [10]-[12]. Solution of SHE equations is a very tedious task. There are several methods proposed by some of researchers to solve these SHE equations. [13] proposed a method based on the resultant theory of polynomials in which a set of trigonometric equations was converted into polynomial equations. The disadvantages of this method were high computational burden and complexity. In [14], Newton-Raphson (NR) method was used for the solution of SHE equations in 11-level inverter. In this method, initial guess for switching angles is taken in the range of 0 to $\pi/2$ for range of modulation indices o to 1. [15] presented a programmed PWM method for the elimination of specific higher order harmonics in MLIs. A new optimum pulse width modulation (OPWM) method was implemented in two-level and MLI with unbalanced dc sources in [16]. [17] reviewed various SHE PWM techniques for MLIs and classified these techniques into five categories which are as follows: numerical approaches, algorithms with multiple solutions capabilities, theory of resultants, optimization based techniques, and methods for on-line implementation.

The conventional numerical methods are based on the proper selection of initial guess. These methods sometimes fail to find the solution of SHE problem for some modulation indices. Theory of resultants and symmetric polynomials increases the computational burden when these methods are applied for solving the SHE equations in high-level inverters. Therefore, in order to overcome these limitations, researchers

978-1-5386-6624-1/18 $31.00 © 2018 IEEE

have implemented the various optimization techniques for the solution of SHE problem. [18] implemented a genetic algorithm (GA) based SHE technique for determining the switching angles in a CHB MLI which reduces specified lower order harmonics while obtaining the desired fundamental voltage for different values of modulation indices. [19] proposed a modified optimization technique and implemented it on 11-level inverter for finding solutions of SHE equations. This modification has improved the ability of the algorithm in finding the global solution in the search space. Similarly other researchers have also tested other optimization techniques which provide solution of SHE problem with faster rate of convergence. In this way, we can confidently hope that other metaheuristic approaches can be used to solve the SHE problem in MLIs.

In this paper, a PSO based SHE technique is implemented in eleven level inverter. Performance of the proposed technique is evaluated with respect to the performance of continuous genetic algorithm (CGA) on the basis of minimum value of fitness function and rate of convergence. Comparative study shows the superiority of PSO based SHE technique over CGA based SHE technique in finding the precise solution at faster convergence rate. Simulation results along with the harmonic analysis of phase voltage waveform are also presented for eleven-level inverter. The paper organization is as follows. Section II describes the CHB MLI topology and mathematical formulation of SHE problem. It is followed by a brief description about the working of PSO in section III. Section IV describes the execution of proposed technique for finding solution of SHE problem. Comparative and simulation results are discussed in section V, whereas paper concludes with section VI.

II. MATHEMATICAL FORMULATION

A. CHB MLI Topology

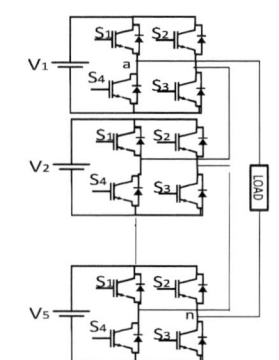

Fig. 1. Eleven-level cascade inverter

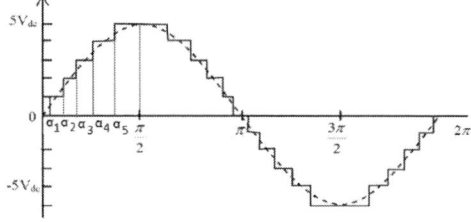

Fig. 2. Eleven-level staircase voltage waveform

Fig. 1 shows the CHB MLI topology for 11-level configuration. Here, five H-bridge cells are connected using cascade connection. Each H-bridge consists of a dc source and four switching devices. A H-bridge inverter produces three voltage levels: -V, 0, +V and a quassi-square waveform is obtained. In CHB MLI, phase shifted modulation technique is used for the generation of stepped AC voltage waveform. Fig. 2 shows the 11-level stepped single-phase voltage waveform. This waveform exhibits quarter wave symmetry. Hence, calculation of only five switching angles α_1 to α_5 is required for the SHE implementation in CHB MLI.

B. SHE Problem

The Fourier series analysis of staircase AC voltage waveform shown in Fig.2 can be mathematically expressed as in (1).

$$V(\omega t) = \sum_{n=1,3,5,\ldots}^{\infty} V_n \sin(n\omega t) \tag{1}$$

where, V_n represents the amplitude of the nth harmonic. Even harmonics are absent due to odd quarter-wave symmetry of stepped AC waveform. Equation (2) gives the expression of V_n.

$$V_n = \frac{4V}{n\pi} \sum_{k=1}^{5} \cos n\alpha_k \; ; \text{ for odd values of n} \tag{2}$$

here, nominal dc source voltage is denoted by V. kth switching angle α_k satisfies the condition shown in (3).

$$0 \le \alpha_k \le \frac{\pi}{2} \; ; \text{ where k} = 1,2,3,4,5 \tag{3}$$

In 3-phase systems, line voltage does not contain triplen harmonics, so our objective is to eliminate non-triplen odd order harmonics from the phase voltage. Therefore, nonlinear transcendental equations for eleven-level inverter can be written as follows in (4)-(9).

$$V_1 = \frac{4V}{\pi}[\cos\alpha_1 + \cos\alpha_2 + \cos\alpha_3 + \cos\alpha_4 + \cos\alpha_5] \tag{4}$$

$$V_5 = \frac{4V}{5\pi}[\cos(5\alpha_1) + \cos(5\alpha_2) + \cdots + \cos(5\alpha_5)] \tag{5}$$

$$V_7 = \frac{4V}{7\pi}[\cos(7\alpha_1) + \cos(7\alpha_2) + \cdots + \cos(7\alpha_5)] \tag{6}$$

$$V_{11} = \frac{4V}{11\pi}[\cos(11\alpha_1) + \cos(11\alpha_2) + \cdots + \cos(11\alpha_5)] \tag{7}$$

$$V_{13} = \frac{4V}{13\pi}[\cos(13\alpha_1) + \cos(13\alpha_2) + \cdots + \cos(13\alpha_5)] \tag{8}$$

$$M = \frac{V_1\pi}{4 \times 5 \times V} \; ; \text{ where } 0 \le M \le 1 \tag{9}$$

To achieve objective of SHE problem, (4) to (8) can be written as in (10) to (14).

$$\cos\alpha_1 + \cos\alpha_2 + \cos\alpha_3 + \cos\alpha_4 + \cos\alpha_5 = 5M \qquad (10)$$

$$\cos(5\alpha_1) + \cos(5\alpha_2) + \cdots + \cos(5\alpha_5) = 0 \qquad (11)$$

$$\cos(7\alpha_1) + \cos(7\alpha_2) + \cdots + \cos(7\alpha_5) = 0 \qquad (12)$$

$$\cos(11\alpha_1) + \cos(11\alpha_2) + \cdots + \cos(11\alpha_5) = 0 \qquad (13)$$

$$\cos(13\alpha_1) + \cos(13\alpha_2) + \cdots + \cos(13\alpha_5) = 0 \qquad (14)$$

Now, solution of (10) to (14) give the 5 switching angles for the range of modulation index M between 0 to 1.

III. PARTICLE SWARM OPTIMIZATION

Kennedy and Eberhart proposed a population based stochastic optimization technique which was motivated by the social behavior of animals such as fish schooling or bird flocking [20]. This technique is based on the intelligence and movement of swarms searching for food. The algorithm begins with a number of particles with random velocity and position (same as the chromosomes of GA) constituting a swarm which moves around the search space of problem to find the best solution. Here, each particle adjusts its position and velocity according to its own flying experience as well as the flying experience of other particles in the search space. The velocities and positions of particles are updated on the basis of their local and global best solutions which enable the individuals to move towards the better solution. Position and velocity updation of an ith particle in dth dimension after the "t" number of iterations are expressed as in (15) and (16).

$$x_i^d(t+1) = x_i^d(t) + v_i^d(t+1) \qquad (15)$$

$$v_i^d(t+1) = C\{w(t)v_i^d(t) + C_1 r_{i1}(p_{besti}^{\ d}(t) - x_i^d(t)) + C_2 r_{i2}(g_{best}^{\ d}(t) - x_i^d(t))\} \qquad (16)$$

where, C_1 and C_2 are cognitive and social coefficients that weights the stochastic terms r_{i1} and r_{i2}. $p_{besti}^{\ d}(t)$ and $g_{best}^d(t)$ are the particle's best position and the best position of the swarm after "t" iterations, respectively. Inertia weight, w and constriction factor, C are represented by (17) and (18), respectively. Both of these parameters control the rate of convergence.

$$C = \frac{2}{\left| 2 - (C_1 + C_2) - \sqrt{((C_1 + C_2)^2 - 4(C_1 + C_2))} \right|} \qquad (17)$$

$$w = (t_{max} - t)/t_{max} \qquad (18)$$

where, t_{max} is the maximum number of iteration. Fig. 3 shows the working of PSO algorithm and steps involved in this algorithm are as follows:

Step 1. Initialize swarm (a set of particles) in the search space of SHE problem. The position and velocity of the particles are randomly initialized as given in (19) and (20).

$$X_i = (x_i^1, x_i^2, \ldots, x_i^d, \ldots, x_i^n) \qquad (19)$$

$$V_i = (v_i^1, v_i^2, \ldots, v_i^d, \ldots, v_i^n) \qquad (20)$$

Step 2. Compute the fitness of all the particles in each iteration and for a minimization problem the best fitness is the minimum value of the function.

Step 3. Compute the pbest and gbest value corresponding to all the particles in the population.

Step 4. Position and velocity updation of all the particles in each dimension using (15) and (16).

Step 5. Update pbest and gbest if the current fitness value is smaller than the previous fitness value.

Step 6. Go to step 2 until the convergence criteria is met. The algorithm terminates when the minimum error criteria or maximum number of iterations is attained.

The algorithm is successfully used for solving various engineering optimization problems. In this work, we have used PSO for the computation of optimum switching angles.

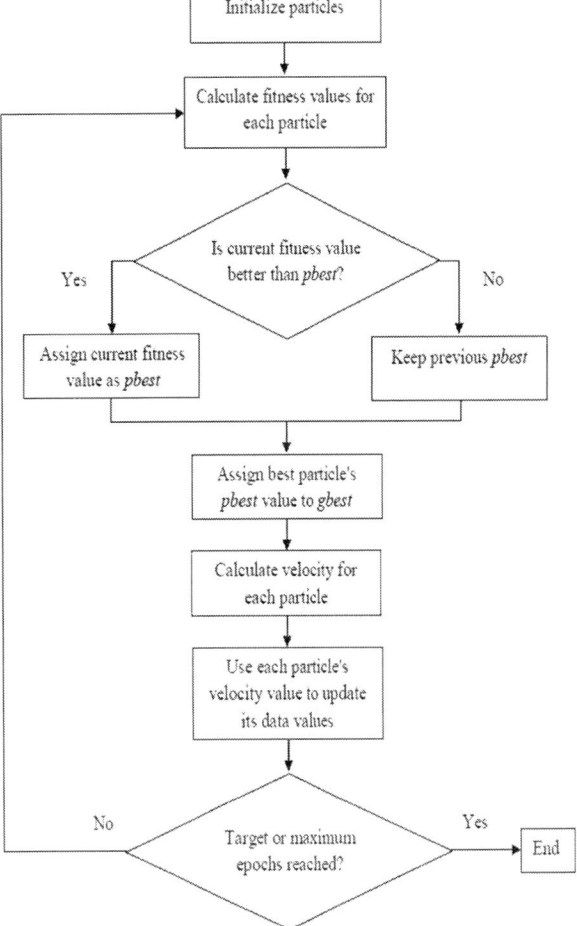

Fig. 3. Working of PSO algorithm.

IV. IMPLEMENTATION OF PSO BASED SHE TECHNIQE

In this section, implementation of PSO based SHE technique is described for 11-level inverter. A computer program was written in MATLAB/Simulink environment. The population size is taken 30 for the simulation. The stopping criterion is chosen as maximum number of iterations. For the SHE problem, fitness function is defined as in (21).

$$f = \min_{\alpha_i} \left\{ \left(100 \frac{\left(V_1^d - V_1 \right)}{V_1^d} \right)^4 + \sum_{i=2}^{5} \frac{1}{4h_i} \left(100 \frac{V_{hi}}{V_1} \right)^2 \right\}$$

where, $0 \leq \alpha_i \leq \dfrac{\pi}{2}$ and i = 1,2,…,5. (21)

V_1^d is the desirable fundamental voltage, i denotes the number of switching angles in a quarter cycle, and h_i is the ith order non-triplen odd harmonic present at output, e.g., in 11-level inverter: $h_2 = 5$; $h_3 = 7$; $h_4 = 11$; $h_5 = 13$. The first term of (21) represents the minimization of percentage error in fundamental voltage which implies the negligible value of fitness for error less than 1%. The second term of equation denotes the minimization of percentage harmonics and it indicates that harmonics having magnitude less than 2% of fundamental will be insignificant. It is also to be noticed that each harmonic ratio is scaled by its order which further reduces the magnitude of lower order harmonics significantly. Now, PSO technique can be applied to minimize the fitness function.

Step 1: Initialization of the population of 30 particles in the search space [0, π/2] radians having random velocities and positions in 5 dimensions.

Step 2: Select M in the range (0, 1) and calculate the fitness value of each particle using (21).

Step 3: Apply the PSO algorithm and update the velocity and position of the particles using (15) and (16), respectively. This completes an iteration of program.

Step 4: The process in step 3 is repeated until stopping criterion is met. At this stage, the positions of particles in five dimensions corresponding to low fitness values are achieved. In this manner, optimum switching angles for SHE problem are found for a particular M.

Step 5: Select another value of M by increasing its value by 0.01 and repeat this process from steps 2 and 4. Again obtain the switching angles which satisfy (21). This process works in a loop until M = 1.

V. COMPARATIVE STUDY AND SIMULATION RESULTS

In this section, simulated results obtained from the comparison between PSO based SHE technique and continuous genetic algorithm (CGA) based SHE technique are presented. The performance indices for comparison between these two techniques are: rate of convergence of algorithms and optimum fitness value for a given M. Simulation parameters are presented in Table I. Here, N denotes the size of the population or swarm, t is the maximum count of iterations, w represents the inertial weight (initial value = 0.9), N_{par} represents the number of switching angles, and M_{rate} is the mutation rate.

TABLE I.   SIMULATION PARAMETERS

Method	Simulation parameters						
	N	t	w	C_1, C_2	N_{par}	M_{rate}	Count (runs)
PSO	30	500	0.9 - 0.4	2	5	NA	15
CGA	30	500	NA	NA	5	0.15	15

Fig. 4 shows the comparison between the PSO based SHE technique and CGA based SHE technique from the point of view of rate of convergence. Performance of both techniques has been observed at M = 0.78 for 50 iterations. From the figure, it is quite clear that PSO based SHE algorithm has faster rate of convergence as compared to the CGA based SHE technique.

In Fig. 5, PSO and CGA based SHE techniques are compared on the basis of minimum fitness value with respect to M ($0.3 \leq M \leq 1$) is shown. It is evident that PSO based SHE technique shows its superiority over CGA based SHE technique. Fig. 6 shows the simulated optimum values of switching angles α_1 to α_5 for 11-level inverter. It can be also observed that PSO based SHE technique is able to find the multiple solutions for a particular value of M. Table II shows the multiple solutions obtained from the PSO based SHE technique at M=0.7.

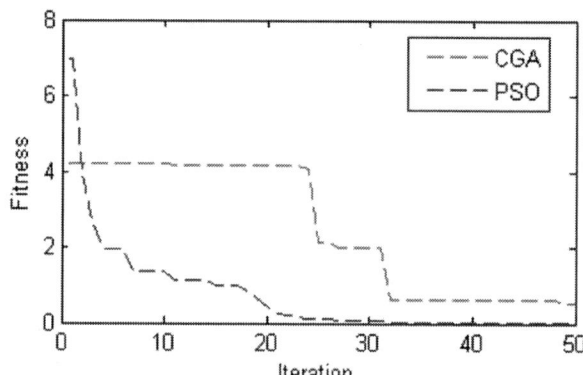

Fig. 4. Fitness value comparison between PSO and CGA.

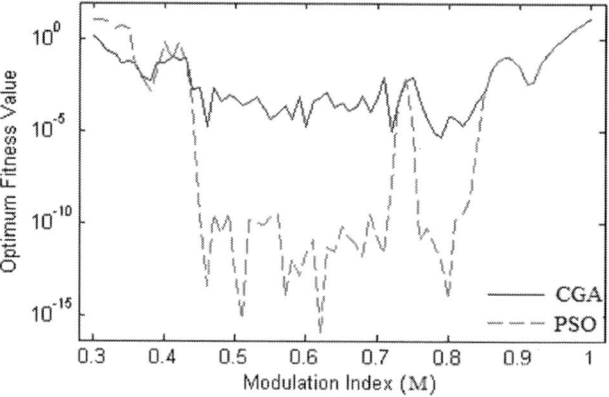

Fig. 5. Optimum fitness value versus M.

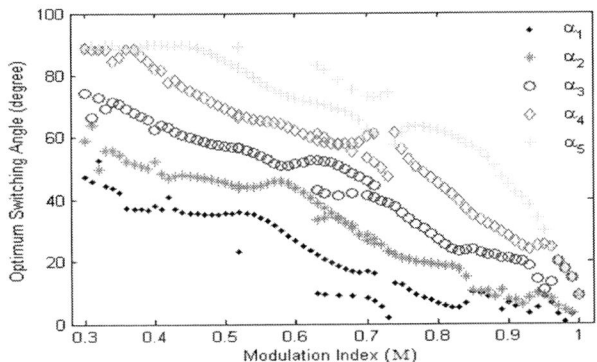

Fig. 6. Switching angles for M (0.3 to 1).

TABLE II. DIFFERENT SOLUTIONS AT M=0.7

Cost Function value	Switching Angle (degree)				
	α_1	α_2	α_3	α_4	α_5
3.6e-11	9.84	20.38	38.4	54.43	60.41
7.5e-11	9.84	20.38	38.4	46.29	60.41
1.5e-11	9.84	20.38	23.6	38.4	60.42
3.1e-11	9.84	20.38	38.4	60.4	70.75
3.8e-11	9.84	20.38	38.4	43.47	60.41

In Fig. 7, variation in percentage harmonic content and line voltage THD is shown with respect to M. It can be seen clearly that line THD and harmonics of order 5th, 7th, 11th, and 13th are minimized for M = [0.42, 0.855] as shown in the figure.

Fig. 8 illustrates the simulated phase voltage waveform of an 11-level inverter for M=0.78. Here, magnitude of each dc source is taken 12V for the simulation. Switching angles were computed by using PSO based SHE technique which are as follows.

$$\alpha_1 = 8.36^\circ ; \ \alpha_2 = 19.6^\circ ; \ \alpha_3 = 30.55^\circ ; \ \alpha_4 = 48.64^\circ ; \ \alpha_5 = 63.45^\circ .$$

FFT analysis of phase voltage waveform has been performed in MATLAB/Simulink as shown in Fig. 9.

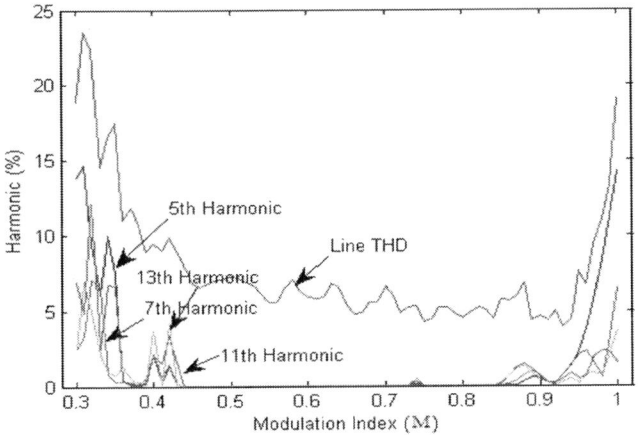

Fig. 7. Harmonics content versus M.

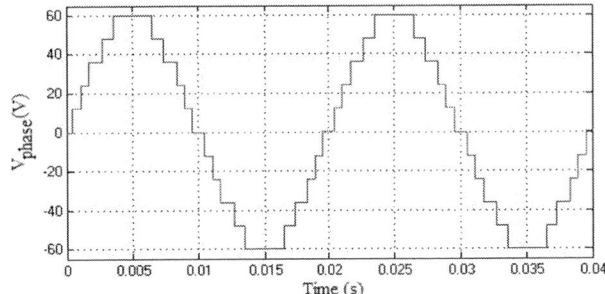

Fig. 8. Waveform of phase voltage.

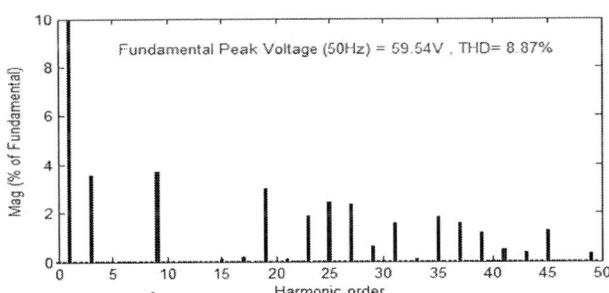

Fig. 9. FFT analysis of phase voltage waveform.

It clearly shows the minimization of non-triplen odd harmonics of order 5th, 7th, 11th, and 13th significantly. It also shows that actual peak fundamental voltage is obtained as given in (9) which prove the efficacy of the PSO based SHE technique.

VI. CONCLUSION

This paper explains the methodology of PSO based SHE technique in 11-level inverter. This technique has been used for the computation of optimum switching angles. From the comparative studies and simulation results, it was shown that the PSO based SHE technique offers advantages such as faster rate of convergence and minimum value of fitness function as compared with CGA based SHE technique. This is primarily due to the less computational requirement in PSO as only previous local and global positions are required to calculate the current position of particle whereas in CGA various steps such as encoding and decoding of chromosomes, crossover, selection, and mutation are required for updating the solution which offer more computational burden as compared to PSO algorithm. Therefore, PSO has faster rate of convergence as compared to CGA. The proposed scheme not only reduces the non-triplen odd order harmonics and THD but it also helps in obtaining the desired fundamental voltage. The superiority of the proposed scheme was also proved, as it is able to find the multiple solutions for some modulation indices. Hence, it can be concluded that PSO based SHE technique is simple, fast, and efficient modulation method for MLIs.

REFERENCES

[1] J. Rodriguez; J.S. Lai and F.L. Peng, "Multilevel inverters: A survey of topologies, controls and applications," *IEEE Trans. Ind. Electron.*, vol. 49, no. 4, pp. 724-738, Aug. 2002.

[2] I. Colak, E. Kabalci, and R. Bayindir, "Review of multilevel voltage source inverter topologies and control schemes," *Energy Conversion Manag.*, vol. 52, no. 2, pp. 1114- 1128, 2011.

[3] M. Malinowski, K. Gopakumar, J. Rodriguez, and M.A. Perez, "A survey on cascaded multilevel inverters," *IEEE Trans. Ind. Electron*, vol. 57, no. 7, p p. 2197-2206, July, 2010.

[4] D.A.B. Zambra; C. Rech and J.R. Pinheiro, "comparison of neutral-poi nt-clamped, symmetrical and hybrid asymmetrical multilevel inverters," *IEEE Trans. Ind. Electron.*,vol. 57,no. 7,pp. 2297-2306, July 2010.

[5] M.A. Memon, S. Mekhilef, M. Mubin, and M. Aamir, "Selective harmonic elimination in inverters using bio-inspired intelligent algorithms for renewable energy conversion applications: A review," *Renewable and Sustainable Energy Reviews*, vol. 82, no. 3, pp. 2235-2253, Feb. 2018.

[6] N. Yousefpoor, S.H. Fathi, N. Farokhnia, H.A. Abyaneh, "THD minimization applied directly on the line-to-line voltage of multilevel inverters," *IEEE Transactions on industrial electronics*, vol. 59, no. 1, pp. 373-80, Jan., 2012.

[7] W. Fei, B. Wu, and Y. Huang, "Half-wave symmetry selective harmonic elimination method for multilevel voltage source inverters," *IET Power Electronics*, vol. 4, no. 3, pp. 342 – 351, March 2011.

[8] A. Sanchez-Ruiz, G. Abad, I. Echeverria, I. Torre, and I. Atutxa, "Continuous phase-shifted selective harmonic elimination and dc-link voltage balance solution for h-bridge multilevel configurations, applied to 5l hnpc," *IEEE Transactions on Power Electronics*, vol. 32, no. 4, pp. 2533–2545, April 2017.

[9] H. S. Patel and R. G. Hoft, "Generalized techniques of harmonic elimination and voltage control in thyristor inverters: Part I—harmonic elimination," *IEEE Trans. Ind. Appl.*, vol. 9, no. 3, pp. 310–317, May/Jun, 1973.

[10] W. Fei, X. Du, and B. Wu, "A generalized half-wave symmetry SHE-PWM formulation for multilevel voltage inverters," *IEEE Trans. Ind. Electron.* vol. 57, pp. 3030–3038, 2010

[11] P. Kala and S. Arora,"Selective harmonic elimination in multilevel inverters using gravitational search algorithm," in *proc. 6th IEEE conf. CERA*, 6-8 oct, Roorkee, India, 2017, pp. 545-550.

[12] R. L. Haupt and S. E. Haupt, "More Natural Optimization Algorithms," in *practical genetic algorithms*, 2nd ed., New Jersey, USA: John Wiley & Sons, 2004, pp. 187-203.

[13] K. Yang, Q. Zhang, R. Yuan, W. Yu, J. Yuan, and J. Wang, "Selective Harmonic Elimination With Groebner Bases and Symmetric Polynomials*," IEEE Transactions on Power Electronics*," vol. 31, no. 4, pp. 2742-2752, 2016.

[14] J. Kumar, B. Das, and P. Agarwal, "Selective harmonic elimination technique for a multilevel inverter" in *proc. NPSC*, IIT Bombay, 2008, dec., pp. 608-613.

[15] Z. Du, L.M. Tolbert, J.N. Chiasson, "Harmonic elimination for multilevel converter with programmed PWM method," in *proc. 39th IAS Annual Meeting Industry Applications Conference*, 2004, 3- 4oct., pp. 2210-2215.

[16] L. Li, D. Czarkowski, Y. Liu and P. Pillay, "Multilevel selective harmonic elimination PWM technique in series-connected voltage inverters," *IEEE Transactions on Industry Applications*, 2000, Jan., vol. 36, no. 1, pp. 160-170.

[17] M.S. Dahidah, G. Konstantinou, and V.G. Agelidis, "A review of multilevel selective harmonic elimination PWM: formulations, solving algorithms, implementation and applications," *IEEE Transactions on Power Electronics*, vol. 30, no. 8, pp. 4091-4106, 2015.

[18] B. Ozpineci, L.M. Tolbert, and J.N. Chiasson, "Harmonic optimization of multilevel converters using genetic algorithms," in *power electronics specialists conference IEEE 35th annual*, 2004, Jun 20, vol. 5, pp. 3911-3916.

[19] A. Kavousi, B. Vahidi, R. Salehi, M.K. Bakhshizadeh, N. Farokhnia, S.H. Fathi, "Application of the bee algorithm for selective harmonic elimination strategy in multilevel inverters," *IEEE Transactions on power electronics*, vol. 27, no. 4, pp. 1689-1696, 2012 .

[20] J. Kennedy. and R .Eberhart,"Particle swarm optimization" in *Proc. IEEE international conference on neural networks*, Perth, Australia, 1995, vol. 4, pp. 1942–1948.

978-1-5386-6624-1/18 $31.00 © 2018 IEEE

Multi-Objective Control Algorithm for Solar PV-Battery based Microgrid

Yashi Singh, Bhim Singh, *Fellow, IEEE, and* Sukumar Mishra, *Senior Member, IEEE*
Department of Electrical Engineering, *IIT Delhi*, New Delhi-110016, India
iyashi.singh@gmail.com, bhimsinghiitd60@gmail.com and sukumariitdelhi@gmail.com

Abstract— This paper deals with a versatile control strategy for voltage source converter (VSC) in grid tied mode (GTM) and standalone voltage control mode (SVCM) using mixed order sinusoidal integrator phase locked loop (MOSSI-PLL) based control. The microgrid is based on voltage source converters (VSC) acts as an active power filter, harmonics and achieves the harmonic elimination compensation and reactive power compensation. Under normal condition, the grid tied solar photovoltaic (PV)-battery hybrid system works in current control mode (CCM) so as to provide the generated power to the grid and the nonlinear load. When solar PV-battery microgrid hybrid system is cut off from the main grid then proposed control detects the islanding condition and transfers the VSC control to SVCM to provide the constant voltage across the load. High order adaptive filter based synchronization controller is implemented for smooth switching to standalone mode and vice versa with high reliability and droop-less control. A MOSSI based control achieves grid current, PCC (Point of Common Coupling) voltage, and battery current under defined modes and helps in harmonics mitigation and enhancing the power quality of the single phase grid system though extracting power from the solar PV array. The effectiveness of the proposed structure of MOSSI-PLL is evaluated through test results under different modes and conditions on the developed prototype in the laboratory.

Keywords: Solar PV Array, Microgrid, Power Quality.

I. INTRODUCTION

In coming years, the need of solar energy (SE) is critically high because of reduced dependence on fossil fuels in order to maintain the pollution-free climate. The solar photovoltaic systems (PVSs) transform SE into electrical energy (EE) directly, without polluting the environment. PVSs are easy to operate and reliable, and their maintenance costs are relatively low. The research aims on renewable energy system, are to convert green energy into EE to facilitate the demand of consumer loads or the utility [1-3].

A well-defined way for realization of the developing potential of distributed generation (DG) is carried as a system, which assessments power generation and connected loads defined as a "microgrid" or subsystem [4-5]. The microgrid can be well-defined as locally generated DG consisting a group of micro sources such as micro turbines, solar photovoltaic (PV) generation system, wind turbines, energy storage, for example batteries, ultra-capacitors, and flywheels etc. It can be operated independently and interlinked to the grid (grid mode) or when disconnected from the grid (island mode) during climate disturbances and faults, therefore improving the quality of supply, consumers can attain a high the efficient, low-cost

and clean energy [6]. Moreover, microgrid can increase local dependability, deliver low investment, less emissions, enhance power quality and decrease losses of distribution system [7-8]. The main challenge for renewable energy generation system (REGS) is to deal with the irregular energy generation with varying load demand. For mitigating this issue, an additional energy storage system (ESS) is used in the REGS. In the power distribution based system consists of distributed energy resources (PV, wind etc.) containing battery ESS and load management. This system can work as independently from the main utility or as grid tied system [9].

The PV generation system (PGS) depicts the non-linear characteristic so to overcome this issue, researchers have proposed various maximum power tracking (MPT) algorithms for extracting maximum power. Many MPT techniques have been proposed by researchers, which have been broadly described in the literature [10-11]. The furthermost extensively implemented perturb and observe (P&O) algorithm comprises on-going perturbation till it reaches the peak point and estimated the reference voltage at varying insolation.

The utility grid is disconnected from the system due to the event of a fault happened either due to overloading of transmission network or natural climate condition impacts on the stability of the grid ties solar PV system. The numerous research has been carried out to maintain an uninterruptible power to the load, particularly critical loads such as major computer installation, safety monitors, telecommunication network, hospital intensive care unit, etc. However to take over this, a rigid and transient free synchronization control is needed between the utility and load voltage. So many phase-locked loop (PLL) based algorithms have been offered in the literature [12].

Specific switching controls are developed for re-connecting the utility grid to the distribution system when the fault is removed. The PLL-based seamless transfer control in a three phase grid between grid-connected and standalone modes is presented in [13]. The convention PLL is used for synchronization and islanding in both modes at liner loads. A single phase based dual mode system has been presented in [14] which consists of seamless control switches between the defined modes at the zero crossing of voltage and current. An intentional islanding and synchronization based control for the distribution network has been proposed for a three-phase in [15].

In this paper, a grid-tied solar PV-battery based dual mode system is presented with multi-objective functionalities and flexible control. The main contributions of the presented work, are given as follows,

- Extraction of maximum power of PGS under varying solar insolation employing P&O algorithm.
- PV-VSC (Voltage Source Converter)-battery microgrid system is capable to maintain uninterpretable supply to the load in islanding mode and grid tied mode (GTM).
- In GTM, PV interfaced VSC functions in current control mode (CCM). It provides the features of stable grid synchronization and current limitation.
- In standalone mode, voltage controller (VC) is used for PV-VSC, because no stiff grid source is available. However VC-VSC is not good choice for grid tied system as it does not provide high performance grid synchronization and stability.
- The higher order PLL based controller is used for proposed system thus there is no requirement for additional PLL and synchronization unit. The higher order current controller provides high degree of freedom (load fundament current components and grid and load voltage angle estimation) and better disturbance rejection capabilities.

In proposed topology, mixed (second and third) order based PLL control [16] is implemented for effective synchronization of the distribution system under the presence of nonlinear load. The MOSSI filter has capabilities along with less computation burden and less complexity. Moreover, the control algorithm has ability to inject reference active power to the grid and also extracts fundamental orthogonal components, which provide the power quality ancillary services. The passive method [15] is used to detect the islanding condition of the grid under grid outage condition. Test results present the satisfactory operation of the proposed topology when exposed to fluctuating disturbances.

II. SYSTEM TOPOLOGY

The topology of the proposed PV-batter microgrid system is illustrated in Fig. 1. The components of proposed system are PV array operating at maximum power point, the battery connected at DC link, boost converter and filters. The PV array connected to the DC bus through a boost converter. The PV array generation is modeled for maximum power rating of 1.9 kW. The battery storage is modeled according to the average load connected to the system and they are stacked in series and parallel to according to the required voltage and current

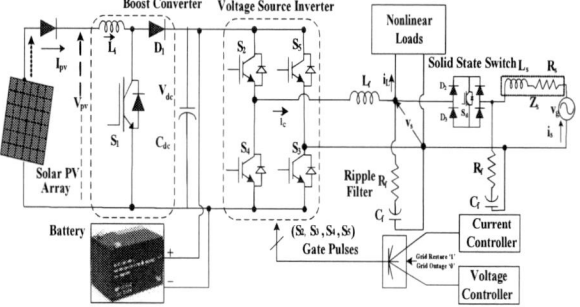

Fig 1 Proposed topology

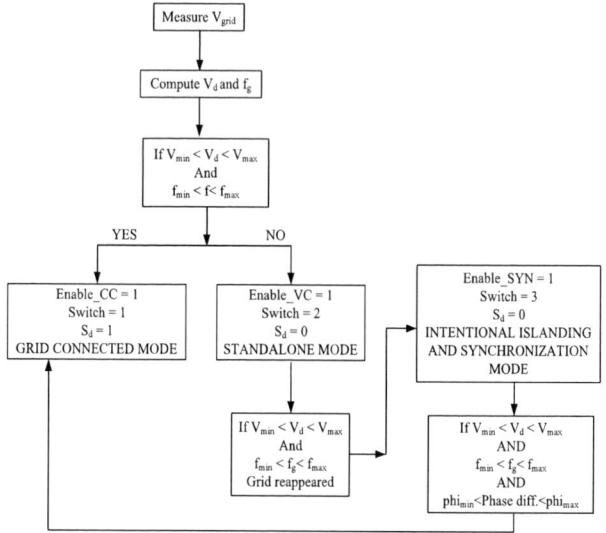

Fig.2 Master Control for modes transition

by using series and parallel combination of each battery of rating 12 V, 7 Ah. The VSC is interfaced to the utility with an interfacing inductor and parallel a ripple filter (R-C). It is realized in the system to absorb the high frequency switching harmonics from the VSC switching. The PV-batter microgrid system is continuously supplying power to nonlinear load represented through diode bridge rectifiers (DBR) connected to RL load. For synchronization and re- synchronization to the grid, a solid-state switch (SSS) is used in series with the grid and operates as per the system operating condition. It allows a bidirectional flow of current which consists of an IGBT switch and four diode operate using master control (MS) as shown in Fig 2.

III. SYSTEM CONTROL

The prime aim of the proposed control scheme of PV-batter microgrid system is feeding high quality and an uninterruptable power to the loads, mitigating the power quality (PQ) issues in the grid and extracting MPP from PV generation array as shown in Fig. 3-6. The control scheme of PV-batter microgrid system has two parts namely achieving MPP for PV array, VSC control for operating reconfigurable system in both modes standalone voltage control mode (SVCM) and grid tied mode (GTM) to improve PQ in the grid current and grid voltage.

A. MPT Control of a Boost Converter

The solar generation use P&O algorithm due to its simplicity and simple implementation compared to other MPP tracking algorithm. To achieve MPPT, gating pulses generation for the DC–DC boost converter is shown in Fig. 3.

B. VSC Control

The VSC control has distinguished feature of automatic switching between grid current control (GCC) in GTM, and a voltage-frequency control (VFC) in SVCM and resynchronization control during emergency as shown in Fig 3. The GCC produces switching logic for VSC in GTM. The

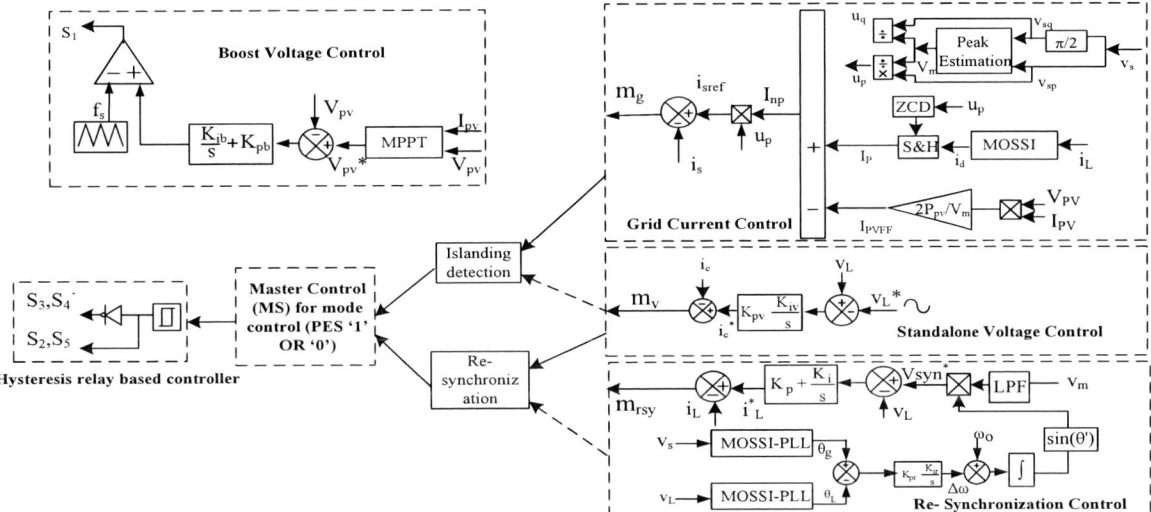

Fig 3. Proposed system control

VFC generates switching logic for VSC in SM. The resynchronizing control is used for smooth transition of VSC from GTM to SVCM and vice-versa. The master control checks the system parameters of the grid and sends ON/OFF signal to SSS depending on grid status and synchronization status.

C. Grid Current Control of VSC

In GTM, the grid voltage and frequency are maintained at PCC and GCC controls the power flow in the grid tied solar PV-battery microgrid system. At the same time, GCC eliminates the harmonic present in the grid current due to the nonlinear load at the PCC. The generation of reference grid current (i_{sref}) for GCC is implemented through the calculation of fundamental load component and PV feed-forward term (I_{PVFF}) as shown in Fig. 3.

The MOSSI filter is implemented for extraction of fundamental component (FC) of the load current (i_L) under nonlinear load as shown in Fig. 4. The MOSSI is a filter by adding an extra filtering branch to mitigate the DC offset of input signal in SOGI (Second Order Generalized Integrator) based filter . The output of the filter is the two orthogonal components i_d and i_q at fundamental frequency (ω_n). The two transfer functions (TF) of MOSSI-filter are defined as,

$$H_d(s) = \frac{i_d}{i_L}(s) = \frac{k\omega_n s(\omega_n - s)}{(s + \omega_n)(s^2 + k\omega_n s + \omega_n^2)} \tag{1}$$

$$H_q(s) = \frac{i_q}{i_L}(s) = \frac{k\omega_n(\omega_n - s)}{(s + \omega_n)(s^2 + k\omega_n s + \omega_n^2)} \tag{2}$$

Where ω_n, k are the natural frequency in rad/sec and filter parameter respectively of the MOSSI filter. The appropriate value of ω_n is chosen to extract the harmonic component of load current. The in-phase FC (i_d) of MOSSI, is used for extracting the amplitude of FC (I_P) by using zero crossing detectors (ZCD), sample and hold logics (S&H).

The effect of dynamic change in PV array power on i_s is calculated by using feed-forward component (I_{PVFF}). It is proportional to PV power (P_{PV}). The I_{PVFF} is estimated for GTPGS as,

$$I_{PVFF} = \frac{2P_{PV}}{V_m} \tag{5}$$

The amplitude of i_{sref} is estimated by adding the amplitude of FC, PI controller output as I_{Loss} and I_{PVFF}, respectively. The amplitude of reference grid current (I_{np}) is generated as,

$$I_{np} = I_P - I_{PVFF} \tag{6}$$

The generated I_{np} is multiplied by sin for calculating reference grid current i_{sref}, given as,

$$i_{sref} = I_{np} * \sin\theta \tag{7}$$

This i_{sref} is compared i_s and estimated error is given to the hysteresis switching controller (HSC). The output of the HSC is given to the switching elements using the bipolar modulation technique.

D. Grid Frequency Estimation

The structure of the MOSSI-PLL is shown in the Fig 5. For proper extraction of FC under varying grid frequency, it is required to use closed-loop system that forms the system adaptive to the varying utility grid condition. This adaptability is achieved by employing the PLL to calculate the grid frequency. The input to the PLL is fundamental components of

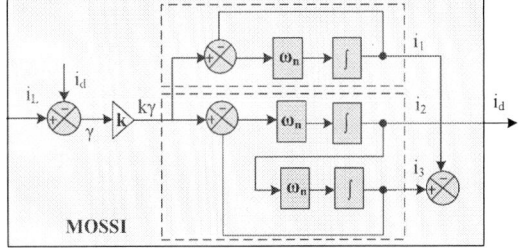

Fig 4 Block diagram of MOSSI for fundamental extractor of load current

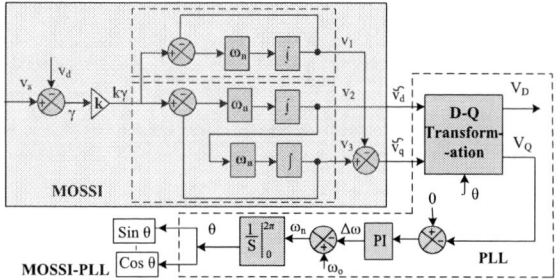

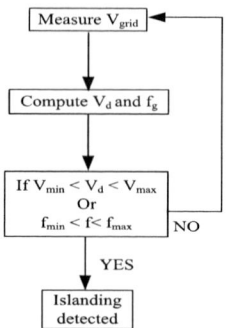

Fig 5 Block diagram of MOSSI-PLL for grid frequency estimation

the sensed voltage that transformed into D-Q frame are applied to the PI controller for approximation of the frequency. The D-Q components of the voltage are estimated by using Park transformation as given as,

$$\begin{bmatrix} V_D \\ V_Q \end{bmatrix} = \frac{1}{2}\begin{bmatrix} 1 & -q \\ q & 1 \end{bmatrix}\begin{bmatrix} \tilde{v}_d \\ \tilde{v}_q \end{bmatrix} \tag{8}$$

Where $q = e^{-j2\pi/2}$ is the phase operator. The frequency estimated by PLL is used in the MOSSI-filter for fundamental estimation, so the dynamic of the MOSSI filter depends on the dynamic of PLL. However to select the proper gain (proportion (K_p) and integral (K_i)) of PI for PLL structure.

E. Islanding and Standalone Voltage Control (SVC) Mode

Under grid failure condition, microgrid is cut-off from the main grid. At that instant islanding operation is detect in order to change the distribution system between GTM and SVCM. The islanding operation is required to detected abnormal conditions, protects from damage to the critical loads and detects variations in voltage and frequency. In this work passive islanding operation is implemented. The passive islanding method based on the instantaneous measuring of the grid voltage magnitude, grid frequency and phase. When the main grid is in fault state, then all measuring grid parameters cross the limit demarcated by the standard, the SSS is tripped, it disconnects the main grid from the PV-batter microgrid system. A passive islanding control algorithm for islanding operation is shown in Fig. 6. The PV-VSC is operated in SVCM. For voltage control, the reference voltage are calculated at 50 Hz in order to control the load voltage across the load using PI controller in. The governing equation of PI controller is given as,

$$i_c^* = \left(K_{pv} + \frac{K_{iv}}{s}\right)\left(v_L^* - v_L\right) \tag{9}$$

This output of the PI controller is compared with VSC current (i_c) and the error is fed to a HSC to generate the high frequency gating signal for VSC in SVCM.

F. Resynchronization Control:

After fault removed, main grid is regain to within the nominal voltage range, the grid-tied VSC is transferred to GCC for grid tied operation. The load voltage, v_L should match the grid voltage (v_s) in terms of magnitude and phase in order to avoid the high voltage spike across the inductor, causing failure in GCC. The resynchronization control is used to achieve a smooth

Fig. 6 Passive islanding algorithm

transition of the VSC from SVCM to the GTM. As grid recovers, instead of the directly transfer to the GTM, it transfers to the resynchronization mode. As shown in Fig. 4, MOSSI-PLL is used for determine the phase angles of the grid voltage (θ_g) and load voltage (θ_L) are compared and generated error is compensated to zero using a PI compensator. It is given as,

$$e_p = \left[\left(\theta_g - \theta_L\right)\right] \tag{10}$$

$$\Delta\omega = \left(K_{pr} + \frac{K_{ir}}{s}\right)e_p \tag{11}$$

$$v_{syn}^* = V_m\sin\left(w_o + \Delta w\right)t \tag{12}$$

The reference voltage v_{syn}^* and v_L are compared and the error is given to the PI controller for reducing the error. If the system is synchronized, the sensed quantities grid voltage amplitude (V_m), phase error (θ_g-θ_L) and f_g (grid frequency) are within the prescribed limit (grid recover signal is '1') and well within the IEEE 1547 standard.

IV. RESULTS AND DISCUSSION

To perform the validity of the grid tied PV-battery microgrid system, a prototype is developed in the laboratory of 1.9 kW. A solar PV array simulator (AMETEK make ETS600x17DPVF), a boost converter, a VSC are controlled by a digital signal processor (DSP dSPACE-1202). The voltage sensors (LEM25P) based on Hall Effect, are used to detect the PV voltage (P_{PV}), DC link voltage (V_{dc}) and grid voltage (v_s) of GTPGS. The current sensors (LV-25) are used to sense the PV current (I_{PV}), load current (i_L) and grid current (i_s). These sensed signals are used to control the developed prototype in the DSP (dSPACE-1202 based controller) through ADCs (analog to digital convertor).

A. Steady-State Behaviour of the PV Battery Microgrid under Various Modes

Steady-state behavior of solar PV battery hybrid system in GTM and SVCM at the presence of nonlinear mode are described as below.

1. Performance of the PV battery Microgrid under GTM at Presence Nonlinear Load

In GTM, the PV battery hybrid system is supplying power to the grid when generated solar PV power is more than load

978-1-5386-6624-1/18 $31.00 © 2018 IEEE

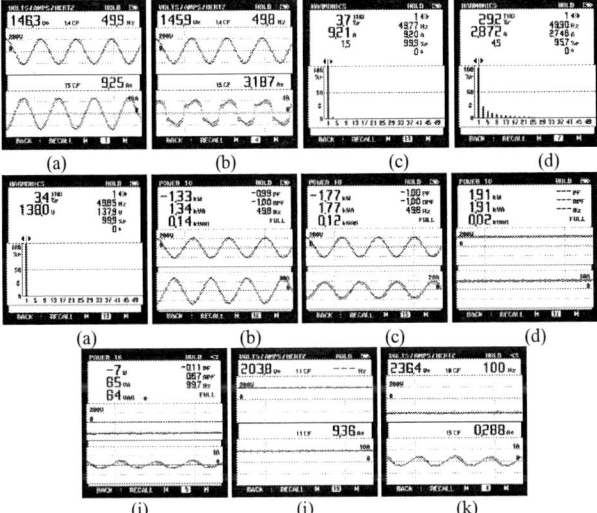

(a) (b) (c) (d)

(a) (b) (c) (d)

(i) (j) (k)

Fig 6 Steady state performance of the system under GTM. (a) v_s with i_s (b) v_s with i_L (c-e) Harmonic content of i_s, i_L, vs (f) P_L (g) P_g (h) solar PV (PPV). (i) battery power (P_b) (j) Solar PV voltage with current (k) battery current and voltage

power (P_L). The PV-VSC is operated in the CCM, and experimental results are presented in Fig. 7, at nonlinear load. Fig. 7 (a)–(b) illustrate the grid voltage (v_s) with grid current (i_s) and v_s and load current (i_L) respectively. The THD of v_s, i_s and i_L are 3.4%, 3.7% and 29.2% respectively which lies under the IEEE standard 519 as shown in Fig. 7 (c) and (e). The power scenario of proposed system is shown in Fig. 7 (g)–(i). The grid is fed with a P_g = 1.7 kW, and P_L is 1.3 kW. The total solar power is (P_{PV} = P_g+ P_L = 1.7 + 1.3 = 1.9 kW) and the battery power (P_b = 7 W). Fig 7 (j) shows the Vpv (PV voltage) and I_{PV} (PV current) which are 203 V, 9.3A respectively, accordingly DC link V_{DC} is maintained 240 V and the battery voltage is operating at 237 V.

2. Performance of the PV Battery Microgrid under SVCM at Nonlinear Load

In islanding mode, PV battery microgrid system is supplying power to the nonlinear load and excess power delivered to the battery for storage at with sufficient at grid outage condition is illustrated in Fig. 7. The load current (V_L) and i_L are shown in

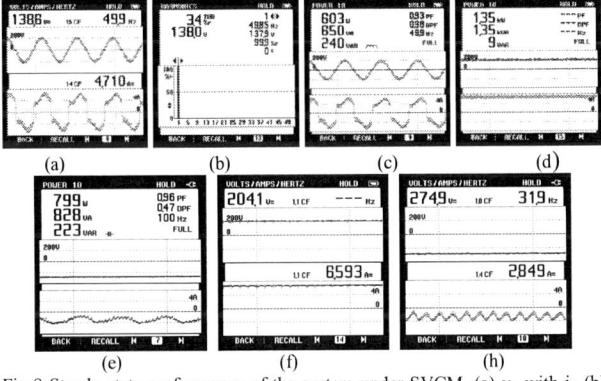

(a) (b) (c) (d)

(e) (f) (g) (h)

Fig 8 Steady state performance of the system under SVCM. (a) v_L with i_L (b) Harmonic content v_L(c) P_L (d) P_{PV} (e) battery power (P_b) (f) Solar PV voltage with current (h) battery current and voltage

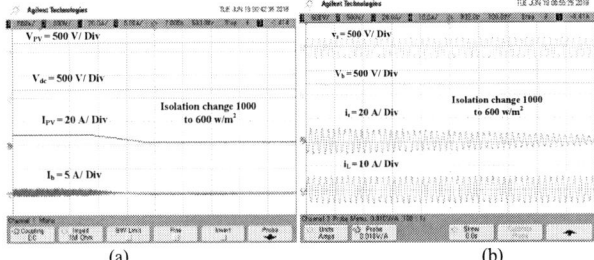

Fig 9 (a-b) Performance of the system under insolation change

Fig. 8 (a). The load voltage is sinusoidal in nature and its total harmonic distortion (THD) is 3.4%, as shown in Fig. 8 (b). The solar PV power (P_{PV}), load power (P_L) and battery power (P_b) are depicted in Figs. 8 (c) and (e). It is demonstrated that the load power demand is fulfilled via the PV-VSC power by transforming PV power into AC power. The excess PV power (P_{PV} >P_L) is the stored into the battery after feeding the load as shown in Fig. 8 (e).

B. Performance of PV Battery Microgrid Under Varying Insolation

The dynamic response of the PV-battery hybrid system for changing insolation (1000 W/m² to 700 W/m²) is illustrated in Fig. 9 (a)–(b). The insolation is deceased from 1000 to 700 W/m2 resulting in an equal change in I_{PV} and V_{PV}. However, due to variable grid power control, grid current is changed and no change is observed in battery voltage and battery current as PV power is large than the load power (P_{PV} > PL) shown in Fig. 9 (b). The load power is fixed due this load current is unchanged.

C. Performance of the PV Battery Microgrid at Under Load change

The response of the PV-battery hybrid system for load disconnected and reconnection, is shown in Figs. 10 (a) and (b). The system intermediate signals (I_p, I_{PVFF}, I_{np} and I_{Loss}) are

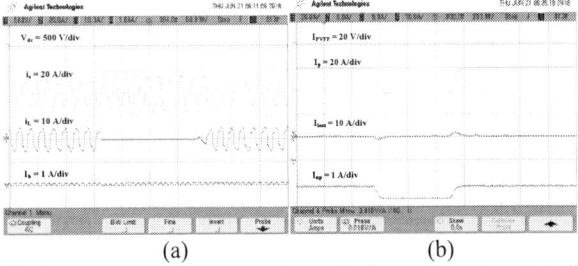

(a) (b)

Fig 10 (a) Performance of the system under load change (b) intermediate signal of the proposed algorithm

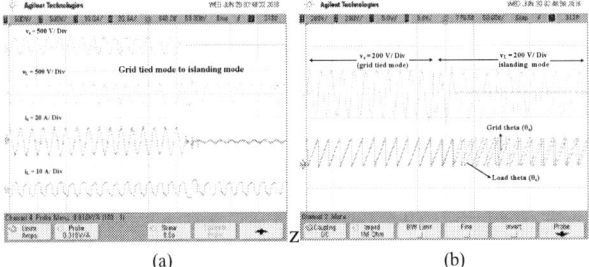

(a) (b)

Fig. 11 De-synchronization processes of the PV-battery hybrid system at nonlinear load: (a) v_s, v_L , i_s , i_L. (b) v_s, v_L, θ_g and θ_L.

978-1-5386-6624-1/18 $31.00 © 2018 IEEE   615

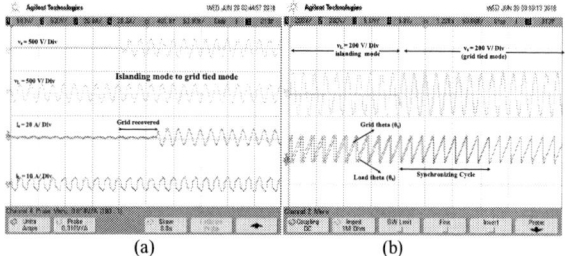

(a)   (b)

Fig. 12 Re-synchronization processes (islanding mode to grid tied mode) of the PV-battery hybrid system at nonlinear load: (a) v_s, v_L, i_s, i_L. (b) v_s, v_L, θ_g and θ_L.

demonstrated at load disconnection and reconnection in Figs. 9 (b-c). The P_{PV} remains unaffected during load disconnection and reconnection condition so the PV array feed-forward term (I_{PVFF}) remains constant. The amplitude of fundamental component (i_d) of i_L is reduced to zero and regains its value within couple of cycles due no oscillations in steady state and it is shown in Fig. 10 (b). The amplitude of FC load current is zero, therefore, i_s is enlarged in magnitude due to total generated power is fed to the grid at load disconnection. The generated P_{PV} always greater than load power ($P_{PV} > P_L$) so no variation is observed in the battery current. The battery operates in floating mode at this condition.

D. System Performance for De-synchronization from GTM to Islanding Mode

The response of the PV-battery microgrid system when subjected to grid disconnection is illustrated in Fig. 11. The i_s is zero at the point of the grid outage due to fault and no significant change on the load current (i_L) is shown in Figs 11 (a-b). The phase angles θ_g and θ_L are synchronized before the grid failure, so no phase difference is shown between v_s and v_L. However, after the grid failure, both θ_g and θ_L are out of synchronism and grid voltage goes to zero. The proposed control is capable of maintaining the load voltage within the prescribed limit in islanding mode.

E. System Performance of System Under Grid Resynchronization

The dynamic behavior of the PV-battery hybrid system under mode transition SVCM to GTM is presented in Fig. 12 (a) and (b). In islanding mode, the grid voltage and grid current are zero. After the grid fault is recovered, grid voltage regained but, i_s is still zero representing the disconnection between the grid and the system. At the beginning, synchronization controller takes couple of cycles to synchronization the v_s and v_L however initially both voltages are out of phase. A proposed algorithm successfully facilities the load voltage to track the voltage at grid until synchronization operation is completed as illustrated in Fig. 12 (b). The intermediate signals of the controller showing the phase angle θ_g and θ_L and the process of synchronization.

V. CONCLUSION

The proposed control structure is presented with three interfaced control: current control for GTM, SVCM for islanding and resynchronization for mode transition. A resynchronization control is used for switching between the two controllers. Moreover, it is depicted that the performance of the PV-battery microgrid under the proposed control schemes is proficient of sustaining the voltage and current within the prescribed limit during GTM and SVCM. Tests results have illustrated that proposed control scheme has ability to transfer the GTM to SVCM smoothly. The VSC has ability to eliminate the harmonics in grid current and load voltage resulting in <5% THD as defined by the IEEE 519 standard.

REFERENCES

[1] A. Aggarwal, A. Singhal and S. J. Darak, "Clean and Green India: Is Solar Energy the Answer?," *IEEE Potentials*, vol. 37, no. 1, pp. 40-46, Jan.-Feb. 2018.

[2] N. J. Guilar, T. J. Kleeburg, A. Chen, D. R. Yankelevich and R. Amirtharajah, "Integrated Solar Energy Harvesting and Storage," *IEEE Transactions on Very Large Scale Integration (VLSI) Systems*, vol. 17, no. 5, pp. 627-637, May 2009.

[3] F. Katiraei and J. R. Aguero, "Solar PV Integration Challenges," *IEEE Power and Energy Magazine*, vol. 9, no. 3, pp. 62-71, May-June 2011.

[4] G. K. Venayagamoorthy, R. K. Sharma, P. K. Gautam and A. Ahmadi, "Dynamic Energy Management System for a Smart Microgrid," *IEEE Trans. on Neural Networks and Learning Systems*, vol. 27, no. 8, pp. 1643-1656, Aug. 2016.

[5] H. Kanchev, D. Lu, F. Colas, V. Lazarov and B. Francois, "Energy Management and Operational Planning of a Microgrid With a PV-Based Active Generator for Smart Grid Applications," *IEEE Trans. on Industrial Electronics*, vol. 58, no. 10, pp. 4583-4592, Oct. 2011.

[6] V. A. Suryad, S. Doolla and M. Chandorkar, "Microgrids in India: Possibilities and Challenges," *IEEE Electrification Magazine*, vol. 5, no. 2, pp. 47-55, June 2017.

[7] D. Li and Z. Q. Zhu, "A Novel Integrated Power Quality Controller for Microgrid," *IEEE Tran. on Industrial Electronics*, vol. 62, no. 5, pp. 2848-2858, May 2015.

[8] H. Dong, S. Yuan, Z. Han, X. Ding, S. Ma and X. Han, "A Comprehensive Strategy for Power Quality Improvement of Multi-Inverter-Based Microgrid With Mixed Loads," *IEEE Access*, vol. 6, pp. 30903-30916, 2018.

[9] L. G. Meegahapola, D. Robinson, A. P. Agalgaonkar, S. Perera and P. Ciufo, "Microgrids of Commercial Buildings: Strategies to Manage Mode Transfer From Grid Connected to Islanded Mode," *IEEE Trans. on Sustainable Energy*, vol. 5, no. 4, pp. 1337-1347, Oct. 2014.

[10] D. Sera, L. Mathe, T. Kerekes, S. V. Spataru and R. Teodorescu, "On the Perturb-and-Observe and Incremental Conductance MPPT Methods for PV Systems," *IEEE Journal of Photovoltaics*, vol. 3, no. 3, pp. 1070-1078, July 2013.

[11] J. Ahmed and Z. Salam, "A Modified P&O Maximum Power Point Tracking Method With Reduced Steady-State Oscillation and Improved Tracking Efficiency," *IEEE Transactions on Sustainable Energy*, vol. 7, no. 4, pp. 1506-1515, Oct. 2016.

[12] M. S. Padua, S. M. Deckmann, G. S. Sperandio, F. P. Marafao and D. Colon, "Comparative analysis of Synchronization Algorithms based on PLL, RDFT and Kalman Filter," *in Proc. IEEE International Symposium on Industrial Electronics*, Vigo, 2007, pp. 964-970.

[13] T. V. Tran, T. W. Chun, H. H. Lee, H. G. Kim and E. C. Nho, "PLL-Based Seamless Transfer Control Between Grid-Connected and Islanding Modes in Grid-Connected Inverters," *IEEE Transactions on Power Electronics*, vol. 29, no. 10, pp. 5218-5228, Oct. 2014.

[14] Z. Yao, L. Xiao, and Y. Yan, "Seamless transfer of single-phase grid interactive inverters between grid-connected and stand-alone modes," IEEE Trans. Power Electronics., vol. 25, no. 6, pp. 1597–1603, Jun. 2010.

[15] I. J. Balaguer, Q. Lei, S. Yang, U. Supatti and F. Z. Peng, "Control for Grid-Connected and Intentional Islanding Operations of Distributed Power Generation," *IEEE Transactions on Industrial Electronics*, vol. 58, no. 1, pp. 147-157, Jan. 2011

[16] C. Zhang, X. Zhao, X. Wang, X. Chai, Z. Zhang and X. Guo, "A Grid Synchronization PLL Method Based on Mixed Second- and Third-Order Generalized Integrator for DC-Offset Elimination and Frequency Adaptability," *IEEE Journal of Emerging and Selected Topics in Power Electronics, Early Access, 2018*.

[17] R. Chilipi, N. Al Sayari, K. Al Hosani and A. R. Beig, "Control scheme for grid-tied distributed generation inverter under unbalanced and distorted utility conditions with power quality ancillary services," *IET Renewable Power Generation*, vol. 10, no. 2, pp. 140-149, 2 2016.

AUTHOR INDEX

Aarfin, Shamshul 1087
Abedi, Mehradad 915
Acharya, Swastik 516
Agarwal, Bhavna 1154
Agarwal, Pramod 809
Agarwal, Ravi 1232
Agarwal, Siddhartha 944
Agarwal, Tanuj 976
Agarwal, Vijyant 1036
Agarwal, Vivek 599
Aggarwal, Archna 403, 408
Aggarwal, Gianeshwar 944
Agrawal, Deepak 1093
Agrawal, Seema 374, 483
Ahmad, Aijaz 149, 498
Ahmad, Samiuddin 143
Ahuja, Rajesh Kr. 414
Akhtar, Iram 455
Allahloh, Ali S. 815
Anan, Ashique 361
Anand, P. 287
Ankarao, Mogili 892
Anup, Sunitha 160
Anuranjana 958
Anzar, Masood 570
Aree, Pichai 837
Arora, Abhishek 249
Arora, Ankita 558
Arora, Parul 200
Arora, Sudha 605
Arora, Swati 170
Arya, Ravi Vardhan 492
Arya, Yogendra 325
Azad, Abhishek 398
Azam, Farooque 102, 443
Azeem, Abdul 636
Bagai, Komal 78
Bagha, Gurudutt 267
Balguvhar, Sumit 972
Baliyan, Arjun 477
Ballal, Makarand S. 754
Banerjee, Indranil 1041
Banerjee, Subrata 760
Bansal, Praveen 343, 355
Bansal, Rahul 1219
Basak, Prasenjit 449
Bath, S. K. 287
Bazaz, Mohammad Abid 108
Bhadouria, Vivek Pratap Singh 23
Bhalla, Suresh 972
Bhandari, Manisha 1154
Bhandari, Shruti 1115, 1199
Bharath, Kurukuru Varaha Satya 552, 870
Bharatiraja, C. 636
Bhargav, Allu 558

Bhargava, Vishal 734
Bharti, Roshan 864
Bhaskar, D. R. 1237
Bhaskar, Mukesh Kumar 132
Bhattacharjee, Ankur 546
Bhattacharjee, Shayari 876
Bhattacharya, Chitra 886
Bhattacharyya, Sudip 659
Bhatti, T. S. 160, 783
Bhosle, Abhijit B. 1067
Bhowmick, S. 154
Bindu, S 206
Biswal, Gyan Ranjan 449
Blaabjerg, Frede 689
Bonev, Boncho 1178, 1193
Celeita, David 33
Chaithra, A 39
Chakraborty, Tapan Kumar 361
Chandak, Sheetal 431
Chandra, Kanchan 1121
Chandran, Vineet P. 582
Chankaya, Mukul 498
Chari, Shrivatsan K 841
Chaturvedi, Devendra Kumar 1046
Chaudhari, Saurabh 1056
Chaudhary, Priyanka 305
Choubey, V. K. 370
Choudekar, Pallavi 44
Choudhury, Anwesha 743
Choudhury, Shreeram 114
Dabas, Shivam 78
Daftari, Mohammad Ali 1062
Dahiya, Pankaj 488
Dalal, Sahil 982, 988
Dang, Radhika 927
Das, Debabrata 546
Das, P. Vipin 689
Das, S K 1041
Debnath, Manoj Kumar 114, 1142
Dehghani, Majid 915
Deshmukh, Rohit R. 754
Deshpande, A. P. 1056
Dey, Anamika 743
Dhembare, Sushil B. 1067
Dhiman, Rishav 841
Dimri, Ashish 948
Dixit, T. V. 909
Dmesh, Panditi 892
Dobhal, Daksh 1228
Dogra, Priyanka 743
Dua, Vaibhav 73
Dube, Anirudh 576
Dubey, Garima 809
Dwivedi, Nikhil 85
Fatima, Mehtab 44

AUTHOR INDEX

Fernandez, E. ...331
Gangoli, Hirdesh ...1121
Ganguly, Sourish ...546
Gao, S. ...437
Garai, Rabindranath ...743
Garg, Rachana ...96, 671, 717
Gaur, Aditi ...1199
Gaur, Prerna ...1036
Gautam, Amit Kumar ...1172
Ghose, Udayan ...827
Ghosh, Ahana ...546
Ghosh, Swapnendu Narayan ...234
Ghosh, T. ...370
Giri, Ravi ...1056
Goel, Arnav ...1228
Goel, Nidhi ...317
Goel, Nitin ...386
Goel, P. K. ...504
Gupta, Abhishek Kumar ...92
Gupta, Akhilesh Kumar ...881
Gupta, Anshul ...953, 1228
Gupta, Archit ...944
Gupta, Atul ...349
Gupta, C. P. ...331
Gupta, Ekta ...1213
Gupta, Himanshu ...29
Gupta, Maneesha ...1016, 1022, 1052
Gupta, Nidhi ...273, 325, 1148
Gupta, Nitesh ...967
Gupta, Rajeev ...471
Gupta, Ritesh K. ...729
Gupta, Ritesh Kant ...743
Gupta, S. K. ...50
Gupta, Saket ...280
Gupta, Sapna ...471
Gupta, Shilpa ...311
Gupta, Shubham Kumar ...29, 343
Gupta, Srishti ...8
Gupta, Tripurari Nath ...630
Gupta, Vijay Kumar ...374, 483
Gurumurthy, S. R. ...948
Handa, Himesh ...967
Haque, Ahteshamul ...461, 552, 870
Hasan, Naimul ...61, 143, 1046
Hussain, Ikhlaq ...683
Hussain, Shahzad ...471
Hussan, Reyaz ...293
Ibraheem ...61, 1109
Indu, S. ...1183
Istiyaque, Md. ...1046
Iyer, Parameswar K. ...729
Iyer, Parameswar Krishnan ...743
Jaffery, Zainul Abdin ...229, 461, 1109
Jain, Akansha ...1010
Jain, Jyoti ...921

Jain, N. K. ...921
Jain, Pranav ...85
Jain, R. K. ...1213
Jain, Vandana ...528
Jain, Vertika ...96
Jaint, Bhavnesh ...1183
Jalindar, Dubal Amol ...337
Jamil, Majid ...455, 477, 576, 1083
Jaroli, Shweta ...1199
Jena, Anuj ...516
Joshi, D. ...154
Joshi, Dheeraj ...386, 809
Joshi, Sunil ...1115, 1199
Juneja, Kapil ...962
Kala, Peeyush ...605
Kamal, Md. Mostofa ...361
Kamra, Rakhi ...8
Kanaujia, Anoop Kumar ...831
Kandpal, Tara C. ...777
Kansal, Veenu ...1188
Kansara, H R ...739
Kant, Piyush ...771
Karandikar, P. B. ...1056, 1067
Karimipour, Hadis ...915
Karimyan, Peyman ...915
Karthikeyan, V. ...689
Kartikeya, Kumar ...1228
Kashif, Mohd. ...765
Kataria, Gaurav ...1136
Kaur, Daman Preet ...1097
Kaur, Harpreet ...1188
Kaur, Mandeep ...1005
Kaur, Manpreet ...958, 1228
Kaur, Sandeep ...170
Kaur, Sanmukh ...958
Kaur, Simar Preet ...792
Kesari, J. P ...102, 443
Kesarwani, Aditya ...958
Khan, Imran ...546
Khan, Kuhsro ...897
Khan, Mohammed Ali ...552, 870
Khanna, Rintu ...170
Khare, Sudhir ...1121
Khatoon, Shahida ...1046
Khondokar, Shihab ...1132
Khunte, Kiran Vijay ...539
Kirmani, Sheeraz ...127, 455
Kiruthika, M. ...206
Kothari, D. P. ...477
Kulkarni, N. R. ...1067
Kulkarni, R. D. ...948
Kumar, Ajay ...1232
Kumar, Akshiv ...56
Kumar, Amit ...267
Kumar, Amritesh ...492

AUTHOR INDEX

Kumar, Aniket 1213
Kumar, Arvind 1073
Kumar, Astitva 212
Kumar, Bhavnesh 56
Kumar, Deepak 881, 1103
Kumar, Gangavarapu Guru 188
Kumar, Himanshu 200
Kumar, Jagdish 398
Kumar, Jitender 65
Kumar, Kelam Sudheer 534
Kumar, M. Vijaya 892
Kumar, Manoj 120
Kumar, Mukesh 50
Kumar, Narendra 65, 257, 273, 280, 287, 311, 325, 570, 623
Kumar, Pragati 1237
Kumar, Prakash 1032
Kumar, Punit 617
Kumar, Ravinder 1178, 1193
Kumar, Sanjiv 92, 831, 1232
Kumar, Shailendra 522, 659, 665, 677, 706, 777, 783
Kumar, Sujay 976
Kumar, Sushil 788, 1224
Kumaravel, S. 188
Kumari, Shweta 1016
Kundu, Sourabh 760
Kurm, Shashank 599
Kushwaha, Satendra Kr Singh 420
Lakshamanan, M 976
Lau, Shreya 876
Lodi, Kaif Ahmed 636
Lone, Ashiq Hussain 108
Madichetty, Sreedhar 539
Mahabubunnabi, Md. 361
Mahajan, Priya 671, 717
Maheshwari, Sudhanshu 1093
Majumdar, Sudipta 1172, 1219
Mallick, Pradeep Kumar 218
Mandal, Ranabir 1121
Mangal, Punit 223
Mangal, Shubham Kumar 976
Manohar, T. Gowri 253
Maurya, Satvik 927
Meena, Duli Chand 821
Meena, Manish Kumar 257
Meena, Rakesh K. 729
Mehta, Jaimin 349
Mishra, A 809
Mishra, Ashish 648
Mishra, Heena 909
Mishra, Manohar 431
Mishra, Sukumar 261, 522, 539, 611, 642
Mishra, Utkarsh 1056
Mitra, A. 370
Mittal, Ayush 748

Mittal, Prag 1087
Modi, Gaurav 665
Modi, Sangeeta 39
Mohammad, Sarfraz 815
Mohanty, Pradeep Kumar 1142
Mohanty, S. R. 420
Mukhija, Pankaj 488
Murshid, Shadab 582, 630, 765
Nagarajan, S. T. 96
Nagpal, Divya 408
Naik, K. A. 331
Nair, Manu J 1205
Nangia, Uma 212, 921
Naqvi, Syed Bilal Qaiser 706
Narayana, O. V. L. 798
Narayanan, Vivek 722
Narula, Aditya 492
Nasiruddin, Ibraheem 143, 182
Nataraj, J. 948
Nath, Shabari 729
Nekoui, Mohammad Ali 1062
Nirmal, A V 739
Padhy, Prabin Kumar 165, 534, 858, 864
Padmanaban, Sanjeevikumar 102, 443
Padmini, N 44
Pahuja, Vaibhav 1205
Pal, Kanwar 777
Pal, Nidhi Singh 132
Pal, Subhrasish 546
Palwalia, D. K. 374
Panchal, Tejas H 804
Panda, Jeebananda 927, 1165
Pandey, Anand K 127
Pandey, Karnika 876
Pandey, Neeta 1183
Pandey, Vishal Kumar 1078
Pandey, Yudhishthir 61
Panigrahi, B. K 200
Panigrahi, Saswat 588
Pant, B. K. 13
Pant, Millie 1121
Pant, Peeyush 700
Parthasarathy, Harish 1036, 1078
Patel, Amit N. 804
Patel, Nimai Charan 114
Patel, Rajesh M 804
Pathak, Om 176
Pattanaik, Shreeva 588
Perveen, Gulnar 317
Petkov, Peter 1178, 1193
Pillai, Swathy 349
Prakash, Prem 176, 239
Prakash, Surya 108
Pranith, Sai 783
Prasad, Dinanath 623

AUTHOR INDEX

Prasad, Jeetendra................................712
Priya, B. K. 798
Priyadarshi, Neeraj........................102, 443
Prodhan, Md. Imran..........................361
Puchalapalli, Sambasivaiah................504
Purwar, Ravindra Kr. 1160
Pushkar, Kanhaiya Lal.............788, 1224
Quadri, A. H................................287
Quadri, Imran Ahmad...............154, 229
Raghava, N. S. 734, 1000, 1126
Rahman, Obaidur 717
Rahul, K..729
Rai, J. N. 85, 993
Rai, Jitendra Nath...........................73
Raj, Ajishek..................................1237
Rajamanickam, R. 349
Rakib, Sakhawat Hossen....................361
Ram, Atma......................................414
Ramachandaramurthy, Vigna K. 102, 443
Ramachandran, T. 1232
Ramaramy, Sudha...........................510
Ramesh, T. K 798
Ramola, Ayushman.........................1087
Ramos, Gustavo 33
Rana, Chhavi..................................962
Rana, Kailash 821
Rana, Rohit 1036
Rao, A. V. Koteswara.......................182
Rathore, Kuldeep.......................29, 355
Ray, Shashwati...............................909
Reddy, A. V. Sudhakra......................244
Reddy, B. Bhargava Reddy 244
Reddy, Damodhar..............................510
Reddy, K. Jyotheeswara.....................465
Reddy, Madhav...............................798
Reddy, Y. Praveen Kumar 244
Reddy, Y. V. Krishna.........................244
Rizwan, M........... 1, 287, 305, 317, 477, 570, 576, 1083
Rizwan, Mohammad...........................212
Rohilla, Komal 1224
Rout, Pravat Kumar 431
Roy, Amit Kumar 449
Roy, Sumit 944, 1205
Rudaba, Quazi Nasrul 1132
Sabarinath., G 253
Sachan, Ankit 1103
Sadiq, Md. Zyed Ibn 1132
Sahoo, Harish Kumar...................516, 588
Sahoo, Sandeep Kumar 677
Saini, R. P 120
Sampath, N. 798
Samuel, Paulson 420, 881, 1103
Sarathi, Arun 188
Sarkar, Ujjawal 976
Sarwar, Adil 293

Sarwar, Md 229
Satapathy, Priyambada......................1142
Savita 249
Sawant, Kunal 1087
Saxena, A. K. 13
Saxena, Anmol Ratna 488
Saxena, Ravi 92
Saxena, Rohit 841
Sayed, Abdullah Abu 1132
Seema 722
Sehwag, Vinay 73
Selvi, M. Vetri 261
Shaik, Abdul Gafoor 17
Shaik, Mahmood 17
Shakya, Amit Kumar 1087
Sharma, Abhishek 8
Sharma, Akhilesh 437
Sharma, Amarjeet Kumar 102, 443
Sharma, Ankit 976
Sharma, Deepika 483
Sharma, Himanshu 461
Sharma, Paras Ram 386, 414
Sharma, Parsh Ram 365, 380
Sharma, Rakhi 623
Sharma, Richa 299
Sharma, S V 739
Sharma, Sahil 748, 1056
Sharma, Shailendra Kumar 886
Sharma, Shivangni 392, 426
Sharma, Sudeep 858
Shatil, Abu Hena Md. 1132
Shekher, Vineet 73, 85, 993
Shivangi 13
Shobhit, R. P. Agarwal 1213
Shrivastava, Sumit 249
Shrivastava, Tanmay 29
Shrujan, M 1041
Shubhra 564
Shukl, Pavitra 593
Shukla, Saurabh 897
Siddiqui, Anwar Shahzad 229
Siddiqui, M. Shadab 13
Simlai, J.......................................370
Singh, Ajendra 73, 85, 993
Singh, Alka 558, 792
Singh, Amresh Kumar 683
Singh, Ashish 743
Singh, Bhim...............504, 522, 528, 564, 582, 593, 611,
630, 653, 659, 665, 677, 683, 706, 722, 765,
771, 777, 783, 897
Singh, Bipin 165
Singh, Chhabindra Nath 881, 1103
Singh, Deepak 437
Singh, Ghanshyam 788
Singh, Himanshu 1121

AUTHOR INDEX

Singh, Hitesh 1178, 1193
Singh, Indu Prabha 648
Singh, Jaspreet 671
Singh, Jyotsna 1078
Singh, Kailash 1136
Singh, Kamaljeet 739
Singh, Krishna 1148
Singh, Kunwar 1032
Singh, Lokesh Shankar 967
Singh, Madhusudan 847, 953, 1027, 1228
Singh, Manminder 1005
Singh, Mayank 1027
Singh, Nisha 273, 311, 325
Singh, Preeti 1022
Singh, Pukhrambam Devachandra 437
Singh, Rajveer 92
Singh, S 1041
Singh, Simranjit 1188
Singh, Upma 1
Singh, Vinay Kumar 858
Singh, Vishwamitra 1183
Singh, Yashi 611
Singhal, Sidhartha 972
Singla, Deepshikha 365, 380
Sinha, S. K. 182
Sisaudia, Varsha 982, 988
Sivaprasad, A. 188
Somani, R. K. 374, 483
Soni, Ashu 1052
Soni, K. M. 182
Sreejeth, Mini 938
Sriavstava, S P 809
Srikakulapu, Ramu 138
Srinivas, Vedantham Lakshmi 522
Srivastava, Ajay 1073
Srivastava, Divyank 194, 903
Srivastava, Kriti 403
Srivastava, Laxmi 23, 280, 299
Srivastava, Neha 827
Subudhi, Umamani 516, 588
Sudhakar, N. 465
Sulochana, V. 1097
Sumedh, N. 798
Surabhi, Jagriti 305
Suryawanshi, H. M. 754
Suthar, Bhavik N. 804
Suvesh, P. 1205
Swaroopa, Manuhar 8
Talapur, Girish G. 754
Tanwar, Lalit Kumar 1183
Tanwar, Lavi 927, 1165
Taparia, Rajashree 1010
Tariq, Mohd 636
Tej, M. Arun 729
Teron, Biki 743

Tharani, Kusum 78
Tigade, Chinmay 938
Tiwari, S. K. 504
Tiwari, Shalini 1087
Tiwari, Sheela 337
Tomar, Anuradha 642, 748
Torres, Esperanza S. 33
Tripathi, M. M. 194, 903
Tripathi, Ramesh Kumar 712
Trivedi, Rishika 864
Tushir, Meena 876
Tyagi, Arjun 218
Valluru, Sudarshan K. 847, 953, 1027, 1228
Varma, Mukesh Kumar 1109
Varshney, Vikas 249
Veerachary, M. 694
Veerachary, Mummadi 617
Venkatesh, V. 739
Verma, Aditya 953, 1228
Verma, Akhilesh 1126
Verma, Anjeet 653
Verma, Ashu 160, 218
Verma, Bharat 165, 534
Verma, H. K. 886
Verma, Vimlesh 392, 426
Verma, Vishal 700
Vinatha, U. 138
Vishwakarma, Virendra P. 827, 852, 982, 988
Viswanadha, Karteek 1000
Vohra, Pawan Singh 958
Yadav, Madhuri 1160
Yadav, Mohini 1083
Yadav, N. K. 50
Yadav, Sandeep Kumar 17
Yadav, Sudesh 852
Yadav, Vinod Kumar 132
Yashwante, Meghna R. 1067
Yousuf, Viqar 108, 149

IEEE
445 Hoes Lane
Piscataway, NJ 08854-4141

ISBN 978-1-5386-6624-1

2018 2nd IEEE International Conference on Power Electronics, Intelligent Control and Energy Systems (ICPEICES 2018)

Delhi, India
22 – 24 October 2018

Pages 617-1240

IEEE Catalog Number: CFP18E68-POD
ISBN: 978-1-5386-6624-1

2018 2nd IEEE International Conference on Power Electronics, Intelligent Control and Energy Systems (ICPEICES 2018)

Delhi, India
22 – 24 October 2018

Pages 617-1240

IEEE Catalog Number: CFP18E68-POD
ISBN: 978-1-5386-6624-1

**Copyright © 2018 by the Institute of Electrical and Electronics Engineers, Inc.
All Rights Reserved**

Copyright and Reprint Permissions: Abstracting is permitted with credit to the source. Libraries are permitted to photocopy beyond the limit of U.S. copyright law for private use of patrons those articles in this volume that carry a code at the bottom of the first page, provided the per-copy fee indicated in the code is paid through Copyright Clearance Center, 222 Rosewood Drive, Danvers, MA 01923.

For other copying, reprint or republication permission, write to IEEE Copyrights Manager, IEEE Service Center, 445 Hoes Lane, Piscataway, NJ 08854. All rights reserved.

*** This is a print representation of what appears in the IEEE Digital Library. Some format issues inherent in the e-media version may also appear in this print version.

IEEE Catalog Number: CFP18E68-POD
ISBN (Print-On-Demand): 978-1-5386-6624-1
ISBN (Online):   978-1-5386-6625-8

Additional Copies of This Publication Are Available From:

Curran Associates, Inc
57 Morehouse Lane
Red Hook, NY 12571 USA
Phone: (845) 758-0400
Fax: (845) 758-2633
E-mail: curran@proceedings.com
Web: www.proceedings.com

TABLE OF CONTENTS

ANALYSIS OF FUZZY LOGIC, ANN AND ANFIS BASED MODELS FOR THE FORECASTING OF WIND POWER 1
Upma Singh ; M. Rizwan

ENERGY AUDIT AND ENERGY CONSERVATION FOR A HOSTEL OF AN ENGINEERING INSTITUTE 8
Srishti Gupta ; Rakhi Kamra ; Manuhar Swaroopa ; Abhishek Sharma

DEVELOPMENT OF A CHARACTERIZATION TOOL FOR DETERMINATION OF SHEET RESISTANCE AT 9 PLACES OF LARGE AREA 156 MM BY 156 MM DIFFUSED SILICON WAFER IN LESS THAN 5 SECONDS AND ITS APPLICATION IN MAKING HIGH EFFICIENCY SILICON SOLAR CELLS 13
M. Shadab Siddiqui ; A. K. Saxena ; Shivangi ; B. K. Pant

AN EMD BASED PROTECTION SCHEME FOR DISTRIBUTION SYSTEM 17
Mahmood Shaik ; Sandeep Kumar Yadav ; Abdul Gafoor Shaik

GREY WOLF OPTIMIZATION ALGORITHM FOR OPTIMAL SITING AND SIZING OF CAPACITORS 23
Vivek Pratap Singh Bhadouria ; Laxmi Srivastava

ORDER REDUCTION OF LINEAR TIME INVARIANT SYSTEM USING PARTICLE SWARM OPTIMIZATION 29
Shubham Kumar Gupta ; Himanshu Gupta ; Tanmay Shrivastava ; Kuldeep Rathore

STATE OF THE ART OF HUMAN FACTORS ANALYSIS APPLIED TO INDUSTRIAL AND COMMERCIAL POWER SYSTEMS 33
Esperanza S. Torres ; David Celeita ; Gustavo Ramos

POWER SYSTEM VULNERABILITY ASSESSMENT USING VOLTAGE COLLAPSE PROXIMITY INDEX 39
A Chaithra ; Sangeeta Modi

TRANSMISSION CONGESTION MANAGEMENT OF IEEE 24-BUS TEST SYSTEM BY OPTIMAL PLACEMENT OF TCSC 44
N Padmini ; Pallavi Choudekar ; Mehtab Fatima

EFFECT OF FACTS DEVICES ON CONGESTION MANAGEMENT USING ACTIVE & REACTIVE POWER RESCHEDULING 50
S. K. Gupta ; N. K. Yadav ; Mukesh Kumar

PITCH ANGLE CONTROL OF VARIABLE SPEED WIND TURBINE BY USING FUZZY LOGIC 56
Akshiv Kumar ; Bhavnesh Kumar

TRANSMISSION LOSS ALLOCATION IN RESTRUCTURED POWER SYSTEM WITH OPTIMIZATION LOSS CRITERION 61
Naimul Hasan ; Ibraheem ; Yudhishthir Pandey

DETERMINATION AND ALLOCATION OF OPTIMAL SIZE OF TCSC AND STATCOM FOR OBSTRUCTION ALLEVIATION IN POWER SYSTEM 65
Jitender Kumar ; Narendra Kumar

POWER FACTOR CORRECTION USING APFC PANEL ON DIFFERENT LOADS 73
Vinay Sehwag ; Vaibhav Dua ; Ajendra Singh ; Jitendra Nath Rai ; Vineet Shekher

SELECTION OF RENEWABLE ENERGY SOURCES FOR OFF-GRID ELECTRIFICATION OF NORTH-EASTERN STATES OF INDIA 78
Kusum Tharani ; Komal Bagai ; Shivam Dabas

INVESTIGATIONS ON EFFECT OF FREQUENCY VARIATION ON THE PERFORMANCE OF COMBINED CYCLE GAS TURBINE 85
J. N. Rai ; Ajendra Singh ; Vineet Shekher ; Pranav Jain ; Nikhil Dwivedi

A TECHNIQUE FOR POWER FACTOR MEASUREMENT OF HOUSEHOLD AND INDUSTRIAL LOAD USING LABVIEW 92
Abhishek Kumar Gupta ; Ravi Saxena ; Rajveer Singh ; Sanjiv Kumar

STUDY OF FORCED OSCILLATIONS IN TWO AREA POWER SYSTEM 96
Vertika Jain ; S. T Nagarajan ; Rachana Garg

AN ANFIS ARTIFICIAL TECHNIQUE BASED MAXIMUM POWER TRACKER FOR STANDALONE PHOTOVOLTAIC POWER GENERATION 102
Neeraj Priyadarshi ; Vigna K. Ramachandaramurthy ; Sanjeevikumar Padmanaban ; Farooque Azam ; Amarjeet Kumar Sharma ; J. P Kesari

LOAD FREQUENCY CONTROL OF TWO AREA INTERCONNECTED POWER SYSTEM USING SSSC WITH PID, FUZZY AND NEURAL NETWORK BASED CONTROLLERS 108

Ashiq Hussain Lone ; Viqar Yousuf ; Surya Prakash ; Mohammad Abid Bazaz

LOAD FREQUENCY CONTROL BASED ON NON-INTEGER TYPE PID CONTROLLER TUNED BY WOA 114

Manoj Kumar Debnath ; Nimai Charan Patel ; Shreeram Choudhury

ENERGY AND EXERGY ANALYSIS FOR HELIOSTAT BASED SOLAR THERMAL POWER PLANT 120

Manoj Kumar ; R. P Saini

EFFICIENT APPROACH FOR DG PLACEMENT AND SIZE IN MEDIUM VOLTAGE DISTRIBUTION SYSTEMS 127

Anand K Pandey ; Sheeraz Kirmani

A COMPARATIVE PERFORMANCE ANALYSIS OF AUTOMATIC GENERATION CONTROL OF MULTI-AREA POWER SYSTEM USING PID, FUZZY AND ANFIS CONTROLLERS 132

Mukesh Kumar Bhaskar ; Nidhi Singh Pal ; Vinod Kumar Yadav

DESIGN OF A HYBRID CONTROLLER BASED ON GA-SMC FOR THE MULTI-TERMINAL VSC-HVDC TRANSMISSION SYSTEM 138

Ramu Srikakulapu ; U. Vinatha

MARKET-BASED LOAD FREQUENCY CONTROL OF INTERCONNECTED POWER SYSTEM 143

Naimul Hasan ; Ibraheem Nasiruddin ; Samiuddin Ahmad

A CONTROL STRATEGY FOR STATCOM IN ALLEVIATION OF SUBSYNCHRONOUS RESONANCE IN POWER SYSTEMS 149

Viqar Yousuf ; Aijaz Ahmad

ANALYTICAL APPROACH FOR MULTIPLE DSTATCOMS ALLOCATION IN RADIAL DISTRIBUTION SYSTEMS FOR ENHANCEMENT OF ENERGY, COST AND EMISSION SAVINGS 154

Imran Ahmad Quadri ; S. Bhowmick ; D. Joshi

APPLICATION OF SINGULAR PERTURBATION METHODOLOGY FOR TRANSIENT STABILITY STUDY OF ELECTRIC POWER SYSTEM 160

Sunitha Anup ; T. S. Bhatti ; Ashu Verma

STUDY OF P&O AND INC PV MPPT TECHNIQUES FOR DIFFERENT ENVIRONMENT CONDITIONS 165

Bipin Singh ; Bharat Verma ; Prabin Kumar Padhy

A UNIFIED APPROACH FOR VOLTAGE ENHANCEMENT IN RADIAL DISTRIBUTION NETWORK WITH DYNAMIC LOADING 170

Swati Arora ; Sandeep Kaur ; Rintu Khanna

LOAD FLOW SOLUTION FOR RADIAL DISTRIBUTION NETWORK 176

Om Pathak ; Prem Prakash

AN EXPONENTIAL SMOOTHING BASED POWER SWING DETECTION TECHNIQUE FOR DISTANCE PROTECTION 182

A. V. Koteswara Rao ; K. M. Soni ; S. K. Sinha ; Ibraheem Nasiruddin

THREE INPUTS AND TWO OUTPUTS BOOST DC-DC CONVERTER FOR DC MICROGRID APPLICATIONS 188

Kumaravel S. ; Arun Sarathi ; Gangavarapu Guru Kumar ; Sivaprasad A.

NOVEL INTELLIGENT CONTROLLERS FOR LOAD FREQUENCY CONTROL OF A MULTI AREA HYBRID POWER SYSTEM 194

Divyank Srivastava ; M. M. Tripathi

A COMPARATIVE STUDY FOR SHORT TERM WIND SPEED FORECASTING USING STATISTICAL AND MACHINE LEARNING APPROACHES 200

Parul Arora ; Himanshu Kumar ; B. K Panigrahi

PERFORMANCE ANALYSIS OF A DISTANCE RELAY FOR ZONE IDENTIFICATION 206

M. Kiruthika ; S Bindu

ARTIFICIAL NEURAL NETWORK BASED MODEL FOR SHORT TERM SOLAR RADIATION FORECASTING CONSIDERING AEROSOL INDEX 212

Astitva Kumar ; Mohammad Rizwan ; Uma Nangia

A NOVEL SCHEME FOR POTENTIAL IDENTIFICATION OF CUSTOMERS FOR DEMAND RESPONSE 218

Pradeep Kumar Mallick ; Arjun Tyagi ; Ashu Verma

PHASOR MEASURING UNITS OF SMART GRID FOR UNINTERRUPTED AND STABLE POWER SUPPLY IN INDIA 223

Punit Mangal

OPTIMAL PLACEMENT OF DISTRIBUTED GENERATION FOR CONGESTION MANAGEMENT: A COMPARATIVE STUDY 229

Md Sarwar ; Anwar Shahzad Siddiqui ; Zainul Abdin Jaffery ; Imran Ahmad Quadri

IMPROVISED BINARY SEQUENCE MPPT METHOD FOR SOLAR PV APPLICATIONS 234

Swapnendu Narayan Ghosh

PLACEMENT OF DISTRIBUTED GENERATION IN RADIAL DISTRIBUTION SYSTEM BY CONSIDERING UNCERTAINTY OF LOAD DEMAND 239

Prem Prakash

APPLICATION OF ALO TO ECONOMIC LOAD DISPATCH WITHOUT NETWORK LOSSES FOR DIFFERENT CONDITIONS 244

A. V. Sudhakra Reddy ; Y. Praveen Kumar Reddy ; B. Bhargava Reddy Reddy ; Y. V. Krishna Reddy

OVERVOLTAGE AND UNDERVOLTAGE PROTECTION OF LOAD USING GSM MODEM SMS ALERT 249

Savita ; Sumit Shrivastava ; Abhishek Arora ; Vikas Varshney

OPTIMAL PLACEMENT AND SIZING OF DISTRIBUTED GENERATION USING FLOWER POLLINATION ALGORITHM FOR POWER LOSS REDUCTION MAXIMIZATION IN DISTRIBUTION NETWORKS 253

G Sabarinath. ; T. Gowri Manohar

ON-LINE MONITORING AND SIMULATION OF TRANSMISSION LINE NETWORK VOLTAGE STABILITY USING FVSI 257

Manish Kumar Meena ; Narendra Kumar

IMPACT OF PRESENCE OF CALENDER FEATURE IN THE PERFORMANCE OF DAY-AHEAD ELECTRIC LOAD POWER FORECASTING WITH NEW FRAMEWORK REALIZED IN ARTIFICIAL NEURAL NETWORK TEHINIQUE USING MATLAB PROGRAMMING 261

M. Vetri Selvi ; Sukumar Mishra

VOLTAGE PROFILE ENHANCEMENT FOR IEEE-14 BUS SYSTEM USING UPFC, TCSC AND SSSC 267

Gurudutt Bagha ; Amit Kumar

PSO TUNED AGC STRATEGY OF MULTI AREA MULTI-SOURCE POWER SYSTEM INCORPORATING SMES 273

Nidhi Gupta ; Narendra Kumar ; Nisha Singh

BIRD SWARM ALGORITHM FOR SOLVING MULTI-OBJECTIVE OPTIMAL POWER FLOW PROBLEM 280

Saket Gupta ; Narendra Kumar ; Laxmi Srivastava

DESIGN AND ANALYSIS OF RENEWABLE ENERGY BASED HYBRID MODEL FOR REMOTE APPLICATIONS 287

P. Anand ; A. H. Quadri ; S. K. Bath ; M. Rizwan ; Narendra Kumar

MAXIMUM POWER POINT TRACKING TECHNIQUES UNDER PARTIAL SHADING CONDITION - A REVIEW 293

Reyaz Hussan ; Adil Sarwar

POWER QUALITY DISTURBANCE PREDICTION USING PNN 299

Richa Sharma ; Laxmi Srivastava

INTELLIGENT APPROACH FOR LOAD BALANCING ON AN 11 KV FEEDER 305

Jagriti Surabhi ; Priyanka Chaudhary ; M. Rizwan

A CONTROL STRATEGY FOR SUBSYNCHRONOUS RESONANCE UNDER VARYING DEGREE OF SERIES COMPENSATION 311

Narendra Kumar ; Shilpa Gupta ; Nisha Singh

DEVELOPMENT OF EMPIRICAL MODELS FOR FORECASTING GLOBAL SOLAR ENERGY 317

Gulnar Perveen ; M. Rizwan ; Nidhi Goel

A COMPARATIVE ANALYSIS OF AGC OF TWO-AREA HYDRO-THERMAL POWER SYSTEM INTERCONNECTED WITH AC-DC PARALLEL LINK IN RESTRUCTURED POWER SYSTEM 325

Nisha Singh ; Narendra Kumar ; Yogendra Arya ; Nidhi Gupta

IMPLEMENTATION OF GRID CONNECTED WIND DRIVEN INDUCTION GENERATOR FROM MATLAB/SIMULINK TO REAL-TIME 331

K. A. Naik ; C. P. Gupta ; E. Fernandez

A MODIFIED PERTURB & OBSERVE ALGORITHM FOR MAXIMUM POWER POINT TRACKING WITH ZERO VOLTAGE SWITCHING BUCK BOOST CONVERTER IN PHOTOVOLTAIC SYSTEM 337

Dubal Amol Jalindar ; Sheela Tiwari

REDUCED DEVICE COUNT ASYMMETRICAL MULTILEVEL INVERTER TOPOLOGY USING DIFFERENT PWM TECHNIQUES 343

Shubham Kumar Gupta ; Praveen Bansal

IMPLEMENTATION AND INFLUENCE OF C-SVPWM AND BC-SVPWM TECHNIQUES USING DSP FOR VOLTAGE ORIENTED CONTROL OF THREE PHASE ACTIVE FRONT END RECTIFIER 349
Jaimin Mehta ; Swathy Pillai ; Atul Gupta ; Rajamanickam R.

A NEW 15-LEVEL ASYMMETRICAL MULTILEVEL INVERTER TOPOLOGY WITH REDUCED NUMBER OF DEVICES FOR DIFFERENT PWM TECHNIQUES 355
Kuldeep Rathore ; Praveen Bansal

GENERATION OF 13-LEVEL OUTPUT VOLTAGE FROM SINGLE-PHASE MULTILEVEL INVERTER CONSISTING OF CASCADED THREE H-BRIDGE UNITS 361
Tapan Kumar Chakraborty ; Ashique Anan ; Sakhawat Hossen Rakib ; Md. Imran Prodhan ; Md. Mostofa Kamal ; Md. Mahabubunnabi

POWER LOSS ANALYSIS IN MULTILEVEL INVERTERS USING MULTI-OBJECTIVE OPTIMIZATION 365
Deepshikha Singla ; Parsh Ram Sharma

PROTECTION OF POWER ELECTRONIC SWITCHES IN A VSI AGAINST OVER-CURRENT AND SHOOT-THROUGH FAULTS 370
V. K. Choubey ; J. Simlai ; T. Ghosh ; A. Mitra

POWER QUALITY IMPROVEMENT OF STANDALONE WIND ENERGY GENERATION SYSTEM FOR NON LINEAR LOAD 374
Seema Agrawal ; Vijay Kumar Gupta ; D. K. Palwalia ; R. K. Somani

HARMONY SEARCH ALGORITHM BASED POWER QUALITY IMPROVEMENT IN MULTILEVEL INVERTER 380
Deepshikha Singla ; Parsh Ram Sharma

POWER QUALITY IMPROVEMENT OF GRID CONNECTED DFIG USING FUZZY CONTROLLER 386
Nitin Goel ; Paras Ram Sharma ; Dheeraj Joshi

A ROBUST ADALINE BASED CONTROL OF SHUNT ACTIVE POWER FILTER WITHOUT LOAD AND FILTER CURRENT MEASUREMENT 392
Shivangni Sharma ; Vimlesh Verma

THD MINIMISATION IN 15- LEVEL HYBRID MULTILEVEL INVERTER USING HARMONIC MINIMIZATION TECHNIQUE 398
Abhishek Azad ; Jagdish Kumar

DESIGN, SIMULATION AND ECONOMICAL ANALYSIS OF SOLAR POWERED IRRIGATION WATER PUMP 403
Archna Aggarwal ; Kriti Srivastava

SIZING AND FINANCIAL OPTIMIZATION OF HYBRID PV/BATTERY/DIESEL SYSTEM 408
Archna Aggarwal ; Divya Nagpal

POWER QUALITY IMPROVEMENT USING FUZZY-PI CONTROLLED D-STATCOM 414
Atma Ram ; Paras Ram Sharma ; Rajesh Kr. Ahuja

NON-LINEAR H-INFINITY CONTROL FOR GRID-INTERACTIVE OFFSHORE WIND FARM AND MARINE CURRENT FARM 420
Satendra Kr Singh Kushwaha ; S. R. Mohanty ; Paulson Samuel

A BRIEF REVIEW REGARDING SENSOR REDUCTION AND FAULTS IN SHUNT ACTIVE POWER FILTER 426
Shivangni Sharma ; Vimlesh Verma

A NOVEL ACTIVE ANTI-ISLANDING SCHEME FOR INVERTER-BASED DISTRIBUTED GENERATION 431
Sheetal Chandak ; Manohar Mishra ; Pravat Kumar Rout

SELECTIVE HARMONIC ELIMINATION FOR CASCADED THREE PHASE MULTILEVEL INVERTER 437
Deepak Singh ; Akhilesh Sharma ; Pukhrambam Devachandra Singh ; S. Gao

AN ANN BASED INTELLIGENT MPPT CONTROL FOR WIND WATER PUMPING SYSTEM 443
Neeraj Priyadarshi ; Vigna K. Ramachandaramurthy ; Sanjeevikumar Padmanaban ; Farooque Azam ; Amarjeet Kumar Sharma ; J. P Kesari

CHARACTERIZATION STUDY OF A GRID CONNECTED DFIG BASED WECS UNDER VARIABLE WIND SPEED AND LOADING CONDITIONS 449
Amit Kumar Roy ; Prasenjit Basak ; Gyan Ranjan Biswal

ECONOMIC SCHEDULING OF HYBRID MICROGRID USING PLANNED MANAGEMENT SCHEME 455
Iram Akhtar ; Sheeraz Kirmani ; Majid Jamil

AN EFFICIENT SOLAR ENERGY HARVESTING SYSTEM FOR WIRELESS SENSOR NODES 461
Himanshu Sharma ; Ahteshamul Haque ; Zainul Abdin Jaffery

HIGH VOLTAGE GAIN SWITCHED CAPACITOR BOOST CONVERTER WITH ANFIS CONTROLLER FOR FUEL CELL ELECTRIC VEHICLE APPLICATIONS..465

K. Jyotheeswara Reddy ; N. Sudhakar

INTERNAL MODEL CONTROLLER DESIGN FOR HVAC SYSTEM..471

Shahzad Hussain ; Sapna Gupta ; Rajeev Gupta

ADAPTIVE PARTICLE SWARM OPTIMIZATION BASED MAXIMUM POWER POINT TRACKING IN GRID CONNECTED PV SYSTEM..477

Majid Jamil ; M. Rizwan ; D. P. Kothari ; Arjun Baliyan

PERFORMANCE EVALUATION OF 3-PHASE 4-WIRE SAPF BASED ON SYNCHRONIZING EPLL WITH FUZZY LOGIC CONTROLLER...483

Seema Agrawal ; Deepika Sharma ; Vijay Kumar Gupta ; R. K. Somani

EVENT-TRIGGERED BASED AUTOMATIC GENERATION CONTROL CONSIDERING PROBABILISTIC ACTUATOR FAILURE...488

Pankaj Dahiya ; Pankaj Mukhija ; Anmol Ratna Saxena

MULTI-INPUT MULTI-LEVEL ISOLATED DC-DC CONVERTER FOR ENHANCED RELIABILITY OF RENEWABLE SOURCES..492

Ravi Vardhan Arya ; Amritesh Kumar ; Aditya Narula

GRID TIED HYBRID PHOTOVOLTAIC-FUEL CELL POWER SYSTEM FOR RESIDENTIAL LOAD..498

Mukul Chankaya ; Aijaz Ahmad

VARIABLE STEP HCS BASED SENSORLESS CONTROL OF WIND DRIVEN DFIG FOR AUTONOMOUS OPERATION...504

Bhim Singh ; Sambasivaiah Puchalapalli ; S. K. Tiwari ; P. K. Goel

AN INTELLIGENT MPPT CONTROLLER BASED AC TO DC BOOST PFC CONVERTER FOR GRID-TIED WIND ENERGY CONVERSION SYSTEM...510

Damodhar Reddy ; Sudha Ramaramy

SYMMETRICAL COMPONENTS ESTIMATION OF UNBALANCED THREE PHASE POWER SYSTEM USING MO- ADALINE STRUCTURE AND HERMITE POLYNOMIAL BASED GAUSS NEWTON ALGORITHM...516

Swastik Acharya ; Umamani Subudhi ; Harish Kumar Sahoo ; Anuj Jena

A NORMALIZED ADAPTIVE FILTER FOR ENHANCED OPTIMAL OPERATION OF GRID INTERFACED PV SYSTEM...522

Vedantham Lakshmi Srinivas ; Shailendra Kumar ; Bhim Singh ; Sukumar Mishra

IFLL BASED CONTROL FOR PV SYSTEM AT ADVERSE GRID CONDITIONS.........................528

Bhim Singh ; Vandana Jain

INTERNAL MODEL CONTROLLER DESIGN FOR BOOST CONVERTER BY STOCHASTIC OPTIMISATION...534

Kelam Sudheer Kumar ; Bharat Verma ; Prabin Kumar Padhy

OUTPUT VOLTAGE CONTROL OF DC-DC BOOST CONVERTER USING MODEL PREDICTIVE CONTROL APPROACH..539

Kiran Vijay Khunte ; Sreedhar Madichetty ; Sukumar Mishra

DESIGN OF AN EFFICIENT SOLAR PV BATTERY CHARGE CONTROLLER UNDER DYNAMIC MOTOR LOAD CONDITIONS...546

Sourish Ganguly ; Subhrasish Pal ; Imran Khan ; Debabrata Das ; Ahana Ghosh ; Ankur Bhattacharjee

HYBRID VOLTAGE CONTROL FOR STAND ALONE TRANSFORMERLESS INVERTER..........552

Mohammed Ali Khan ; Ahteshamul Haque ; Kurukuru Varaha Satya Bharath

PERFORMANCE ANALYSIS OF DIFFERENT MODULATING TECHNIQUES FOR MULTILEVEL INVERTER...558

Allu Bhargav ; Alka Singh ; Ankita Arora

AUTONOMOUS OPERATION OF SINGLE STAGE SOLAR PV-BES BASED MICROGRID...........564

Shubhra ; Bhim Singh

WAVELET BASED CONTROL OF SHUNT COMPENSATOR FOR POWER QUALITY ENHANCEMENT..570

Masood Anzar ; Narendra Kumar ; M. Rizwan

ANALYSIS OF DIFFERENT MPPT TECHNIQUES UNDER PARTIAL AND UN-SHADED CONDITION FOR SOLAR PV SYSTEM..576

Anirudh Dube ; Majid Jamil ; M Rizwan

THIRD ORDER SINUSOIDAL INTEGRATOR CONTROL OF PV-HYDRO-BES BASED ISOLATED MICRO-GRID..582

Vineet P. Chandran ; Shadab Murshid ; Bhim Singh

FREQUENCY ESTIMATION OF UNBALANCED THREE PHASE POWER SYSTEM USING LEGENDRE POLYNOMIAL BASED GAUSS NEWTON ALGORITHM 588
Shreeva Pattanaik ; Umamani Subudhi ; Harish Kumar Sahoo ; Saswat Panigrahi

POWER QUALITY IMPROVEMENT USING MULTILAYER GAMMA FILTER BASED CONTROL FOR DSTATCOM UNDER NONIDEAL DISTRIBUTION SYSTEM 593
Pavitra Shukl ; Bhim Singh

A HYSTERESIS CURRENT CONTROLLED GRID CONNECTED FULL BRIDGE INVERTER WITH ZERO CURRENT SWITCHING 599
Shashank Kurm ; Vivek Agarwal

IMPLEMENTATION OF PSO BASED SELECTIVE HARMONIC ELIMINATION TECHNIQUE IN MULTILEVEL INVERTERS 605
Peeyush Kala ; Sudha Arora

MULTI-OBJECTIVE CONTROL ALGORITHM FOR SOLAR PV-BATTERY BASED MICROGRID 611
Yashi Singh ; Bhim Singh ; Sukumar Mishra

HYBRID SWITCHED INDUCTOR / SWITCHED CAPACITOR BASED QUASI-Z-SOURCE DC-DC BOOST CONVERTER 617
Punit Kumar ; Mummadi Veerachary

MODELING AND SIMULATION OF MICROGRID SOLAR PHOTOVOLTAIC SYSTEM WITH ENERGY STORAGE 623
Dinanath Prasad ; Narendra Kumar ; Rakhi Sharma

SINGLE-PHASE GRID INTERFACED WEGS USING FREQUENCY ADAPTIVE NOTCH FILTER FOR POWER QUALITY IMPROVEMENT 630
Tripurari Nath Gupta ; Shadab Murshid ; Bhim Singh

PERFORMANCE ANALYSIS OF DISCONTINUOUS PULSE WIDTH MODULATION SCHEMES ON PUC-5 INVERTER 636
Abdul Azeem ; Mohd Tariq ; Kaif Ahmed Lodi ; C. Bharatiraja

GRID INTERACTIVE MISO CONVERTER BASED PV SYSTEM 642
Anuradha Tomar ; Sukumar Mishra

PERFORMANCE COMPARISON OF DIFFERENT INVERTER STAGES CS-VCO IN 0.18μM CMOS TECHNOLOGY 648
Ashish Mishra ; Indu Prabha Singh

REWEIGHTED SPARSE- LEAST-MEAN MIXED-NORM ADAPTIVE CONTROL FOR SOLAR PV INTEGRATED EV CHARGING STATION 653
Anjeet Verma ; Bhim Singh

ADAPTIVE FREQUENCY ESTIMATION TECHNIQUE FOR GRID CONNECTED PHOTOVOLTAIC SYSTEM 659
Sudip Bhattacharyya ; Shailendra Kumar ; Bhim Singh

ACOUSTIC ECHO CANCELLATION BASED ADAPTIVE CONTROL ALGORITHM FOR GRID INTEGRATED SECS 665
Gaurav Modi ; Shailendra Kumar ; Bhim Singh

DESIGN & IMPLEMENTATION OF SOLAR FED INTENSITY CONTROLLED STREETLIGHT 671
Jaspreet Singh ; Priya Mahajan ; Rachana Garg

MODIFIED GRADIENT SPECTRAL VARIANCE SMOOTHING ADAPTIVE FILTER CONTROL FOR GRID CONNECTED PV SYSTEM 677
Sandeep Kumar Sahoo ; Shailendra Kumar ; Bhim Singh

MULTIFUNCTIONAL ADAPTIVE RZA-NLMF CONTROL TECHNIQUE FOR DOUBLE STAGE SEGS INTERFACED WITH THREE-PHASE DISTRIBUTION FEEDER 683
Amresh Kumar Singh ; Ikhlaq Hussain ; Bhim Singh

IMPLEMENTATION OF MPPT CONTROL IN FUEL CELL FED HIGH STEP UP RATIO DC-DC CONVERTER 689
V. Karthikeyan ; P. Vipin Das ; Frede Blaabjerg

DESIGN AND ANALYSIS OF FIFTH-ORDER BUCK-BOOST CONVERTER 694
M. Veerachary

SOURCE SCHEDULING FOR POWER MATCHING OF CLUSTERED INDUCTION GENERATORS FED VSC SUPPORTED MICRO GRID 700
Peeyush Pant ; Vishal Verma

IMPLEMENTATION OF A MODIFIED DISTRIBUTED NORMALIZED LEAST MEAN SQUARE CONTROL FOR A MULTI-OBJECTIVE SINGLE STAGE SECS 706
Syed Bilal Qaiser Naqvi ; Shailendra Kumar ; Bhim Singh

A DC-DC BOOST CONVERTER FOR SEDIMENT MICROBIAL FUEL CELL ENERGY HARVESTING 712

Jeetendra Prasad ; Ramesh Kumar Tripathi

MODELING OF LED USING PIECEWISE LINEAR APPROXIMATION AND MACLAURIN SERIES EXPANSION 717

Obaidur Rahman ; Priya Mahajan ; Rachana Garg

SOLAR PV-BES BASED MICROGRID SYSTEM WITH SEAMLESS TRANSITION CAPABILITY 722

Vivek Narayanan ; Seema ; Bhim Singh

BATTERY CHARGING OF SMART PHONES USING ORGANIC SOLAR CELLS 729

K. Rahul ; Rakesh K. Meena ; Ritesh K. Gupta ; Parameswar K. Iyer ; M. Arun Tej ; Shabari Nath

IMPROVE COLLISION IN HIGHLY DENSE WIFI ENVIRONMENT 734

Vishal Bhargava ; N. S. Raghava

MICROWAVE CIRCUITS CHARACTERIZATION ON CARBON FIBRE (CFRP) BASED CARRIER PLATES 739

Kamaljeet Singh ; H R Kansara ; V. Venkatesh ; A V Nirmal ; S V Sharma

MORPHOLOGY CONTROL OF MIXED HALIDE PEROVSKITE FOR ITS APPLICATION IN LOW-COST THIN FILM TRANSISTOR 743

Anwesha Choudhury ; Priyanka Dogra ; Biki Teron ; Ritesh Kant Gupta ; Anamika Dey ; Ashish Singh ; Rabindranath Garai ; Parameswar Krishnan Iyer

PV-PIEZO HYBRID GRID CONNECTED SYSTEM 748

Anuradha Tomar ; Ayush Mittal ; Sahil Sharma

DISTRIBUTED CONTROL FOR POWER MANAGEMENT BASED ON FUZZY LOGIC IN DC MICROGRID 754

Rohit R. Deshmukh ; Makarand S. Ballal ; Girish G. Talapur ; H. M. Suryawanshi

SELECTIVE HARMONIC MINIMIZATION SCHEME APPLIED TO CASCADED H-BRIDGE INVERTER FOR SATISFYING CIGRE WG 36-05 AND EN 50160 GRID CODES 760

Sourabh Kundu ; Subrata Banerjee

ADAPTIVE SMC BASED DTC OF POSITION SENSORLESS PMSM DRIVEN SOLAR PV WATER PUMPING SYSTEM 765

Mohd. Kashif ; Shadab Murshid ; Bhim Singh

MULTIPULSE AC-DC CONVERSION FED 3RD HARMONIC INJECTION BASED SPWM CONTROLLED CASCADED MLI DRIVEN VCIMD 771

Piyush Kant ; Bhim Singh

ADAPTIVE NEURAL NETWORK BASED CONTROL OF PV CONNECTED DISTRIBUTION SYSTEM 777

Kanwar Pal ; Shailendra Kumar ; Bhim Singh ; Tara C. Kandpal

IMPROVED LAPLACIAN KERNEL FILTER BASED CONTROL OF MULTIFUNCTIONAL PV SYSTEM WITH ENHANCED POWER QUALITY 783

Sai Pranith ; Shailendra Kumar ; Bhim Singh ; T. S. Bhatti

SINGLE VDBA-BASED VOLTAGE-MODE UNIVERSAL BIQUADRATIC FILTER 788

Kanhaiya Lal Pushkar ; Ghanshyam Singh ; Sushil Kumar

DSOGI BASED GRID SYNCHRONIZATION UNDER ADVERSE GRID CONDITIONS 792

Simar Preet Kaur ; Alka Singh

MODULAR ASSEMBLY SYSTEMS IN INDUSTRY 4.0 MILIEU 798

N. Sumedh ; O. V. L. Narayana ; Madhav Reddy ; N. Sampath ; B. K. Priya ; T. K Ramesh

COMPARATIVE PERFORMANCE ANALYSIS OF RADIAL FLUX AND DUAL AIR-GAP AXIAL FLUX PERMANENT MAGNET BRUSHLESS DC MOTORS FOR ELECTRIC VEHICLE APPLICATION 804

Amit N. Patel ; Bhavik N. Suthar ; Tejas H. Panchal ; Rajesh M. Patel

A COMPLETE FUZZY LOGIC BASED REAL-TIME SIMULATION OF VECTOR CONTROLLED PMSM DRIVE 809

A Mishra ; Garima Dubey ; Dheeraj Joshi ; Pramod Agarwal ; S P Sriavstava

DEVELOPMENT OF THE INTELLIGENT OIL FIELD WITH MANAGEMENT AND CONTROL USING IIOT (INDUSTRIAL INTERNET OF THINGS) 815

Ali S. Allahloh ; Sarfraz Mohammad

SELF EXCITED INDUCTION GENERATOR FOR ISOLATED PICO HYDRO STATION IN REMOTE AREAS 821

Kailash Rana ; Duli Chand Meena

PREDICTING THE POPULARITY OF WEBSITES USING MULTILAYER PERCEPTRON AND EXTREME LEARNING MACHINE 827

Neha Srivastava ; Virendra P. Vishwakarma ; Udayan Ghose

A HYBRID TWENTY FIVE-LEVEL INVERTER FOR AN OPEN-END WINDING INDUCTION MOTOR (OEWIM) DRIVE 831
Anoop Kumar Kanaujia ; Sanjiv Kumar

EFFECTS OF NEUTRAL CONDUCTOR ON INDUCTION MOTOR STEADY-STATE PERFORMANCE UNDER LOSS OF ONE PHASE OF SUPPLY VOLTAGES 837
Pichai Aree

NOVEL AND ROBUST HYSTERESIS CURRENT CONTROL STRATEGIES FOR A BLDC MOTOR: A SIMULATION STUDY AND INVERTER DESIGN 841
Shrivatsan K Chari ; Rishav Dhiman ; Rohit Saxena

MULTI-OBJECTIVE GENETIC AND ADAPTIVE PARTICLE SWARM OPTIMIZATION ALGORITHMS: A PERFORMANCE ANALYSIS WITH BENCHMARK FUNCTIONS 847
Sudarshan K. Valluru ; Madhusudan Singh

FUZZY QUATERNION-BASED PIXEL WISE INFORMATION EXTRACTION FOR FACE RECOGNITION 852
Sudesh Yadav ; Virendra P. Vishwakarma

CONTROLLING OF AVR VOLTAGE AND SPEED OF DC MOTOR USING MODIFIED PI-PD CONTROLLER 858
Vinay Kumar Singh ; Sudeep Sharma ; Prabin Kumar Padhy

DESIGN OF FPI-PD CONTROLLER FOR BRUSHLESS DC MOTOR 864
Roshan Bharti ; Rishika Trivedi ; Prabin Kumar Padhy

CONDITION MONITORING OF PHOTOVOLTAIC SYSTEMS USING MACHINE LEAMING TECHNIQUES 870
Kurukuru Varaha Satya Bharath ; Ahteshamul Haque ; Mohammed Ali Khan

A COMPARATIVE STUDY OF FUZZY SYSTEMS AND NEURAL NETWORKS FOR SYSTEM MODELING AND IDENTIFICATION 876
Karnika Pandey ; Shayari Bhattacharjee ; Shreya Lau ; Meena Tushir

FUZZY C-MEANS BASED MODEL SIMPLIFICATION USING JAYA OPTIMIZATION ALGORITHM 881
Chhabindra Nath Singh ; Akhilesh Kumar Gupta ; Deepak Kumar ; Paulson Samuel

A VOLTAGE CONTROLLED SENSORLESS SPEED CONTROL OF PMBLDC MOTOR DRIVE FOR AN ELECTRIC TWO WHEELER 886
Chitra Bhattacharya ; Shailendra Kumar Sharma ; H. K. Verma

DYNAMIC PERFORMANCE ANALYSIS OF REACTIVE POWER AND IMPROVED ROTOR FLUX BASED MRAS FOR INDUCTION MOTOR DRIVES EMPLOYING PI AND FUZZY CONTROLLER 892
Mogili Ankarao ; M. Vijaya Kumar ; Panditi Dmesh

NEURO-ELO BASED SPEED ESTIMATION OF IMPROVED DESIGNED INDUCTION MOTOR DRIVE FOR SINGLE STAGE PHOTOVOLTAIC FED WATER PUMPING 897
Kuhsro Khan ; Saurabh Shukla ; Bhim Singh

TRANSFORMER HEALTH MONITORING SYSTEM USING INTERNET OF THINGS 903
Divyank Srivastava ; M. M. Tripathi

PARAMETRIC OPTIMIZATION OF PROTON EXCHANGE MEMBRANE FUEL CELL USING SUCCESSIVE TRUST REGION ALGORITHM 909
Shashwati Ray ; Heena Mishra ; T. V. Dixit

DYNAMIC BEHAVIOR CONTROL OF INDUCTION MOTOR WITH STATCOM 915
Majid Dehghani ; Peyman Karimyan ; Mehradad Abedi ; Hadis Karimipour

IMPACT OF PARTICLE SWARM OPTIMIZATION PARAMETERS ON ITS CONVERGENCE 921
N. K. Jain ; Uma Nangia ; Jyoti Jain

AUDIO WATERMARKING USING BEAT DETECTION AND PITCH ESTIMATION 927
Lavi Tanwar ; Radhika Dang ; Satvik Maurya ; Jeebananda Panda

IMPLEMENTATION OF V/F ADJUSTABLE SPEED DRIVE FOR INDUCTION MOTOR USING DSPACE DS1104 938
Chinmay Tigade ; Mini Sreejeth

EVALUATION OF EMISSION AND PERFORMANCE CHARACTERISTICS OF DIFFERENT BIODIESEL BLENDS WITH VARYING FIP ON A SINGLE CYLINDER CI ENGINE 944
Archit Gupta ; Siddhartha Agarwal ; Gianeshwar Aggarwal ; Sumit Roy

DESIGN AND SIMULATION OF SENSORLESS CONTROL ALGORITHMS OF BRUSHLESS DC MOTOR: A REVIEW 948
Ashish Dimri ; R. D. Kulkarni ; S. R. Gurumurthy ; J. Nataraj

BIFURCATION CURVES FOR ELECTRICAL DC MOTORS 953
Sudarshan K Valluru ; Madhusudan Singh ; Aditya Verma ; Anshul Gupta

PERFORMANCE ANALYSIS OF FSO LINK UNDER DIFFERENT CONDITIONS OF FOG IN DELHI, INDIA 958

Aditya Kesarwani ; Anuranjana ; Sanmukh Kaur ; Manpreet Kaur ; Pawan Singh Vohra

FEATURE EXPANDED AND WEIGHT SELECTIVE MODEL TO CLASSIFY THE HEART DISEASE PATIENTS 962

Kapil Juneja ; Chhavi Rana

SYNCHRONIZATION AND ANTI SYNCHRONIZATION OF FRACTIONAL ORDER SYSTEM 967

Lokesh Shankar Singh ; Himesh Handa ; Nitesh Gupta

ENERGY HARVESTING USING D_{33} MODE BY INSOLE EMBEDDED LOW COST PIEZO-SENSORS 972

Sumit Balguvhar ; Sidhartha Singhal ; Suresh Bhalla

GAIN ENHANCEMENT OF CIRCULARLY POLARISED MICROSTRIP PATCH ANTENNA USING METASURFACE 976

Shubham Kumar Mangal ; Sujay Kumar ; Tanuj Agarwal ; Ujjawal Sarkar ; Ankit Sharma ; M Lakshamanan

EFFICIENT FEATURE EXTRACTION USING DWT-DCT FOR ROBUST FACE RECOGNITION UNDER VARYING ILLUMINATIONS 982

Virendra P. Vishwakarma ; Sahil Dalal ; Varsha Sisaudia

ECG CLASSIFICATION USING KERNEL EXTREME LEARNING MACHINE 988

Sahil Dalal ; Virendra P. Vishwakarma ; Varsha Sisaudia

DESIGN AND EXPERIMENTAL ANALYSIS OF INTEGER ORDER PID CONTROLLER FOR CERAMIC IR HEATING SYSTEM 993

Vineet Shekher ; Ajendra Singh ; J. N. Rai

DESIGN AND MATHEMATICAL ANALYSIS OF MICROSTRIP BANDPASS POWER DIVIDER FOR KU BAND COMMUNICATIONS 1000

Karteek Viswanadha ; N. S. Raghava

CONTRAST ENHANCEMENT AND PSEUDO COLORING TECHNIQUES FOR INFRARED THERMAL IMAGES 1005

Mandeep Kaur ; Manminder Singh

LAGUERRE FUNCTION BASED MODEL PREDICTIVE CONTROL FOR VAN-DE-VUSSE REACTOR 1010

Akansha Jain ; Rajashree Taparia

HIGHLY LINEAR CURRENT FOLLOWER TRANSCONDUCTANCE AMPLIFIER (CFTA) DESIGN AND ITS FILTER APPLICATION 1016

Shweta Kumari ; Maneesha Gupta

HIGH GAIN TRANSIMPEDANCE AMPLIFIER USING SELF CASCODE STRUCTURE 1022

Preeti Singh ; Maneesha Gupta

APPLICATION OF LINEAR QUADRATIC METHODS TO STABILIZE CART INVERTED PENDULUM SYSTEMS 1027

Sudarshan K Valluru ; Madhusudan Singh ; Mayank Singh

IMPLEMENTATION OF HIGH PERFORMANCE CLOCK-GATED FLIP-FLOPS 1032

Prakash Kumar ; Kunwar Singh

AN EFFICIENT UNSCENTED KALMAN FILTER FOR JOINT ANGLES ESTIMATION AND CONTROL OF OMNI BUNDLE WITH NOISE 1036

Rohit Rana ; Prerna Gaur ; Vijyant Agarwal ; Harish Parthasarathy

APPLICATION OF WIRELESS TECHNOLOGY TO ENHANCE SAFETY AND PRODUCTIVITY IN STEEL PLANT 1041

Indranil Banerjee ; M Shrujan ; S K Das ; S Singh

OBSERVER BASED CONTROLLER DESIGN FOR INVERTED PENDULUM SYSTEM 1046

Shahida Khatoon ; Devendra Kumar Chaturvedi ; Naimul Hasan ; Md Istiyaque

DESIGN OF FRACTIONAL ORDER BUTTERWORTH FILTER USING GENETIC ALGORITHM 1052

Ashu Soni ; Maneesha Gupta

ANALYSIS OF AQUEOUS SUPERCAPACITOR FOR VARIATION IN SEPARATOR WETNESS. 1056

Ravi Giri ; Saurabh Chaudhari ; Utkarsh Mishra ; A. P. Deshpande ; P. B. Karandikar ; Sahil Sharma

ANALYSING STABILITY OF TIME DELAYED SYNCHRONOUS GENERATOR AND DESIGNING OPTIMAL STABILIZER FRACTIONAL ORDER PID CONTROLLER USING PARTICAL SWARM OPTIMIZATION TECHNIQUE 1062

Mohammad Ali Daftari ; Mohammad Ali Nekoui

INVESTIGATION OF EFFECT OF CHARCOAL PARTICLE SIZE ON EARTH'S RESISTANCE 1067

Meghna R. Yashwante ; P. B. Karandikar ; N. R. Kulkarni ; Sushil B. Dhembare ; Abhijit B. Bhosle

DESIGN AND ANALYSIS OF HIGH PERFORMANCE LINE STARTED PERMANENT MAGNET SYNCHRONOUS MOTOR 1073
Arvind Kumar ; Ajay Srivastava

FRACTIONAL ORDER TCHEBICHEF MOMENT AND ITS INVARIANTS 1078
Vishal Kumar Pandey ; Jyotsna Singh ; Harish Parthasarathy

MICROCONTROLLER BASED LOAD PRIORITIZATION TECHNIQUE IN RESIDENTIAL SECTOR 1083
Mohini Yadav ; Majid Jamil ; M. Rizwan

VISUAL REPRESENTATION OF CHANGE IN VEGETATION AREA OF DEHRADUN, UTTARAKHAND, INDIA USING NORMALIZED DIFFERENCE VEGETATION INDEX (NDVI) 1087
Amit Kumar Shakya ; Ayushman Ramola ; Kunal Sawant ; Shalini Tiwari ; Shamshul Aarfin ; Prag Mittal

ELECTRONICALLY TUNABLE CURRENT MODE UNIVERSAL FILTER USING A SINGLE MX CCCII 1093
Deepak Agrawal ; Sudhanshu Maheshwari

DESIGN AND IMPLEMENTATION OF CACHE COHERENCE PROTOCOL FOR HIGH-SPEED MULTIPROCESSOR SYSTEM 1097
Daman Preet Kaur ; V. Sulochana

MODEL REDUCTION OF CONTINUOUS-TIME INTERVAL SYSTEMS USING EIGEN SPECTRUM ANALYSIS 1103
Chhabindra Nath Singh ; Deepak Kumar ; Paulson Samuel ; Ankit Sachan

COMPARATIVE STUDY AND INVESTIGATION OF THE BROADBAND POWERLINE CHANNEL MODEL 1109
Zainul Abdin Jaffery ; Ibraheem ; Mukesh Kumar Varma

COGNITIVE RADIO TECHNOLOGY IN 5G WIRELESS COMMUNICATIONS 1115
Shruti Bhandari ; Sunil Joshi

DESIGN AND DEVELOPMENT OF DIGITAL SIGNAL CONTROLLER BASED MOTORIZED ZOOM CONTROLLER FOR 16X ZOOM THERMAL IMAGER 1121
Himanshu Singh ; Millie Pant ; Sudhir Khare ; Ranabir Mandal ; Kanchan Chandra ; Hirdesh Gangoli

MILLIMETER WAVE RECONFIGURABLE VIVALDI ANTENNA USING POWER DIVIDER FOR 5G APPLICATIONS 1126
Akhilesh Verma ; N. S. Raghava

GENERATING ELECTRICITY ON ROADSIDE USING INVELOX 1132
Abdullah Abu Sayed ; Md. Zyed Ibn Sadiq ; Quazi Nasrul Rudaba ; Shihab Khondokar ; Abu Hena Md. Shatil

COMPENSATING A THIRD ORDER PROCESS HAVING INVERSE RESPONSE 1136
Gaurav Kataria ; Kailash Singh

DESIGN OF PD-PID CONTROLLER WITH DOUBLE DERIVATIVE FILTER FOR FREQUENCY REGULATION 1142
Priyambada Satapathy ; Manoj Kumar Debnath ; Pradeep Kumar Mohanty

DESIGN AND SIMULATION OF LOW-POWER CONDITIONAL-DISCHARGING FLIP FLOP 1148
Nidhi Gupta ; Krishna Singh

COMPOSITE NONLINEAR FEEDBACK CONTROL FOR INVERTED PENDULUM WITH INPUT SATURATION 1154
Bhavna Agarwal ; Manisha Bhandari

INTEGRATING WAVELET COEFFICIENTS AND CNN FOR RECOGNIZING HANDWRITTEN CHARACTERS 1160
Madhuri Yadav ; Ravindra Kr. Purwar

REVIEW OF DIFFERENT TRANSFORMS USED IN DIGITAL IMAGE WATERMARKING 1165
Lavi Tanwar ; Jeebananda Panda

STATE ESTIMATION OF SINGLE-PHASE RECTIFIER BASED LOAD CIRCUIT USING UNSCENTED KALMAN FILTER 1172
Amit Kumar Gautam ; Sudipta Majumdar

A NOVEL METHOD FOR PREDICTING ATTENUATION CAUSED BY CLOUDS FOR HIGHER FREQUENCY BANDS 1178
Hitesh Singh ; Boncho Bonev ; Peter Petkov ; Ravinder Kumar

AN EFFICIENT WEIGHTED TRUST METHOD FOR MALICIOUS NODE DETECTION IN CLUSTERED WIRELESS SENSOR NETWORKS 1183
Bhavnesh Jaint ; Vishwamitra Singh ; Lalit Kumar Tanwar ; S. Indu ; Neeta Pandey

PERFORMANCE ANALYSIS OF NON-COHERENT MODULATIONS OVER FTR FADING MODEL 1188
Veenu Kansal ; Harpreet Kaur ; Simranjit Singh

A NOVEL APPROACH FOR PREDICTING ATTENUATION OF RADIO WAVES CAUSED BY RAIN...1193

Hitesh Singh ; Boncho Bonev ; Peter Petkov ; Ravinder Kumar

REAL TIME ANALYSIS OF MAC BASED AND LEVEL BASED ROUTING PROTOCOL FOR WIRELESS SENSOR NETWORK..1199

Aditi Gaur ; Shweta Jaroli ; Sunil Joshi ; Shruti Bhandari

EVALUATION OF EMISSION CHARACTERISTICS OF GREEN DIESEL IN A SINGLE CYLINDER CI ENGINE...1205

Manu J Nair ; Vaibhav Pahuja ; P. Suvesh ; Sumit Roy

COMPARATIVE RESEARCH FOR MANAGING DELAY IN SIGNAL PROCESSING VIA MULTIPLIERS...1213

Aniket Kumar ; Ekta Gupta ; R. P. Agarwal Shobhit ; R. K. Jain

PERTURBATION BASED NONLINEAR ANALYSIS OF MOSFET CIRCUIT....................................1219

Rahul Bansal ; Sudipta Majumdar

RESISTORLESS ELECTRONICALLY CONTROLLABLE QUADRATURE SINUSOIDAL OSCILLATOR EMPLOYING VDIBA...1224

Kanhaiya Lal Pushkar ; Komal Rohilla ; Sushil Kumar

DESIGN OF MULTI-LOOP L-PID AND NL-PID CONTROLLERS: AN EXPERIMENTAL VALIDATION..1228

Sudarshan K. Valluru ; Madhusudan Singh ; Arnav Goel ; Manpreet Kaur ; Daksh Dobhal ; Kumar Kartikeya ; Aditya Verma ; Anshul Gupta

RADIO FREQUENCY BASED (RF) CONTROL & OPERATION OF ELECTRICAL/ELECTRONIC APPLIANCES IN HOME/OFFICES...1232

T. Ramachandran ; Sanjiv Kumar ; Ajay Kumar ; Ravi Agarwal

MULTIPLE-INPUT SINGLE-OUTPUT UNIVERSAL BIQUAD FILTER USING SINGLE OUTPUT OTAS..1237

Ajishek Raj ; D. R. Bhaskar ; Pragati Kumar

Author Index

2nd IEEE International Conference on Power Electronics, Intelligent Control and Energy Systems (ICPEICES-2018)

Hybrid Switched Inductor / Switched Capacitor based Quasi-Z-source DC-DC Boost Converter

Punit Kumar, Mummadi Veerachary

Dept. of Electrical Engineering, Indian Institute of Technology, New Delhi

E-mail: mvchary@ee.iitd.ac.in

Abstract- **A hybrid switched inductor quasi-Z-source dc-dc converter with cascaded switched capacitor arrangement to increase the voltage gain is proposed in this paper. The proposed converter has higher voltage gain at lower duty ratio, reduced voltage stress across all the components compared to conventional impedance source converter configurations. The voltage gain is almost two times the two switched inductor quasi-Z-source inverter arrangement. The operation principle and detailed mathematical analysis is presented. The design equations for the passive components are formulated. The proposed converter is then compared with other existing converters reported in literature. Simulation results verifies the mathematical analysis of proposed converter.**

Keywords- Z-source, quasi-Z-source, switched capacitor, voltage stress, boost, high-gain.

I. INTRODUCTION

A Z-source inverter (ZSI) presented in [1] eliminates the limitations of the conventional voltage-source inverter and current-source inverter, i.e. shoot-through problem. A family of quasi-ZSI (QZSI) is presented in [2] with continuous input current (CIC) and discontinuous input current (DIC), with voltage gain similar to ZSI. The inductors of ZSI is replaced by switched inductor (SL) cells in [3] to increase the voltage gain, thereby meeting the load demand. The SLQZSI is introduced in [4], in which one inductor of QZSI is replaced by SL cell, to have a higher voltage gain than QZSI with continuous input current. Then two SLQZSI is introduced in [5], in which both inductors of QZSI is replaced by SL cells, having same voltage gain as SLZSI in [3].

Z-source concept is then implemented in dc-dc power conversion [6]-[16]. The inductor of basic boost converter is replaced by Z-network in [7] to realize a higher voltage gain Z-source dc-dc converter (ZSC). The steady-state analysis of ZSC in continuous current mode (CCM) is explained in [8]. To increase the voltage gain of ZSC a switched capacitor (SC) cell is cascaded in [9]. A quasi-ZSC (QZSC) with voltage-lift technique is presented in [10] to increase the voltage gain, but it suffers higher voltage stress across devices. A modified ZSC is presented in [11] which has same voltage gain as ZSC but with a lower voltage stress on Z-network capacitors is presented in Fig. 1. A QZSC is presented in [12] with a slight modification in converter of [11], i.e. replacement of output side diode with an inductor to reduce output voltage ripple. A 3-Z-network boost converter is introduced in [13] with very high gain, which is basically a SLZSC with one Z-network capacitor replaced by a diode. A common grounded ZSC is presented in [14] with higher voltage gain than ZSC and lower voltage stress

on all its components. An impedance network boost converter is presented in [15] which has identical behavior to SLZSC, i.e. high voltage gain with higher voltage stress across devices. A QZSC with switched capacitor is presented in [16], which has nearly twice the voltage gain than converter in [11], but with a higher capacitor voltage stress. A hybrid QZSC with high voltage gain is presented in [17], but with higher system order. Hybrid ZSC are presented in [18] which has high voltage gain with higher number of inductors and capacitors. A hybrid quasi-Z-source based dc-dc converter is presented in [19] and [20] which has higher voltage gain than other existing converters. The voltage gain is still not large, as per requirements of modern day PV power systems. A SLSCQZSC is presented in this paper which has higher voltage gain than all the reported converters so far, with lower voltage stress on its components.

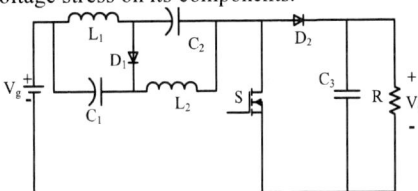

Fig. 1. Modified Z-source DC-DC converter [11].

The detailed mathematical analysis and operation principle of proposed converter is presented in section II, section III describes the detailed comparison with existing converters followed by parameter design in section IV, simulation results are presented in section V followed by conclusion of the paper in section VI.

II. ANALYSIS OF PROPOSED CONVERTER

The proposed converter is shown in Fig. 2. The converter structure is very simple that consists of a quasi-Z-network with two SL cells (inductors L_1, L_2, L_3 and L_4, capacitors C_1 and C_2 and seven diodes D_1-D_7), a SC arrangement (C_3, C_4, D_8 and D_9), filter capacitor C_5, switch S and diode D_{10}. The following conditions are assumed for the analysis: (1) $L_1 = L_2$ and $L_3 = L_4$, (2) The converter is assumed to be operated in CCM. This converter operates in two different modes i.e. switch S is ON or OFF, whose equivalent circuits are depicted in Fig. 2. The detailed steady-state and small-signal analysis is presented in this section. Fig. 3 represents the key waveforms of and Table I shows the switching pattern of the proposed converter.

TABLE I
SWITCHING PATTERN OF PROPOSED CONVERTER

	S	D_1	D_2	D_3	D_4	D_5	D_6	D_7	D_8	D_9	D_{10}
Mode-1	√	√	×	√	×	√	×	√	×	√	×
Mode-2	×	×	√	×	√	×	√	×	√	×	√

978-1-5386-6624-1/18 $31.00 © 2018 IEEE   617

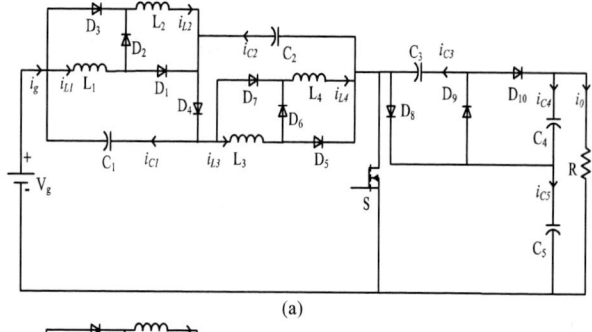

(a)

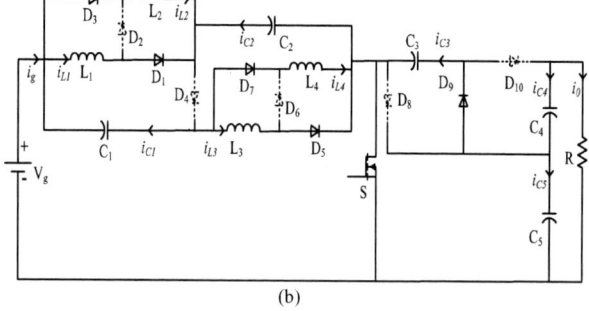

(b)

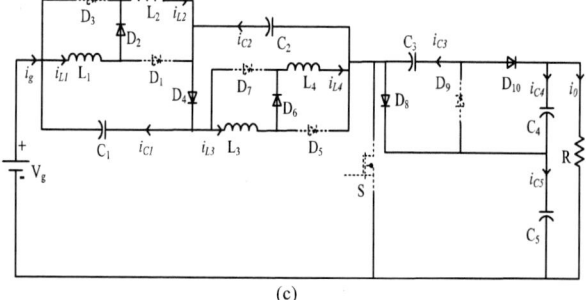

(c)

Fig.2.(a) Proposed converter (b) Equivalent circuit of Mode-1 (c) Equivalent circuit of Mode-2.

1. Steady-State Analysis

Mode-1: Fig. 2 (b) depicts the equivalent circuit for this mode of operation. The switch S is turned-ON during this mode of operation as identified from Table I. The inductor currents keeps on increasing and voltage across capacitors C_1 and C_2 starts decreasing. D is the duty ratio for this mode of operation. The voltage expressions obtained using KVL are:

$$\begin{cases} v_{L1} = v_{C2} + V_g \\ v_{L1} = v_{L2} \\ v_{L3} = v_{L4} \\ v_{L3} = v_{C1} + V_g \\ v_{C3} = v_{C5} \\ V_0 = v_{C4} + v_{C5} \end{cases} \quad (1)$$

Mode-2: Fig. 2 (c) depicts the equivalent circuit for this mode of operation. The switch S is turned-OFF during this mode of operation. The inductor currents starts decreasing immediately during start of this mode. The voltage across capacitors C_1 and C_2 starts increasing. The voltage expressions obtained using KVL are:

$$\begin{cases} v_{L1} + v_{L2} = -v_{C1} \\ v_{L3} + v_{L4} = -v_{C2} \\ v_{C3} = v_{C4} \\ V_0 = v_{C4} + v_{C5} \\ V_0 = v_{C1} + v_{C2} + v_{C3} + V_g \end{cases} \quad (2)$$

By applying the volt-sec balance to the inductors, the voltage across capacitor C_1 is obtained as

$$V_{C1} = \left(\frac{2D}{1-3D}\right)V_g \quad (3)$$

The voltage across Z-network capacitor C_2 is obtained as

$$V_{C2} = \left(\frac{2D}{1-3D}\right)V_g \quad (4)$$

The voltage across capacitors C_3, C_4 and C_5 is obtained as

$$V_{C3} = V_{C4} = V_{C5} = \left(\frac{1+D}{1-3D}\right)V_g \quad (5)$$

The voltage gain is obtained as

$$\frac{V_0}{V_g} = \left(\frac{2(1+D)}{1-3D}\right) \quad (6)$$

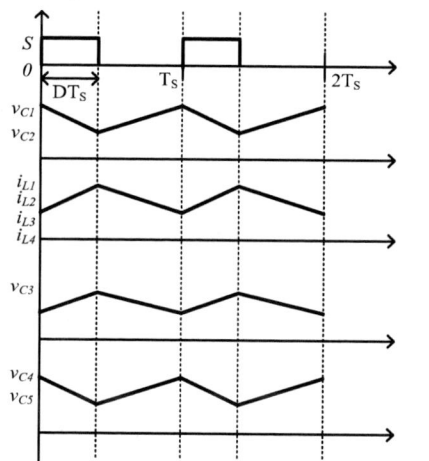

Fig. 3. Key waveforms for proposed converter.

2. Small-signal Analysis

The dynamic performance of converter is predicted by its state-space model. The converter operates in two different modes in CCM. State-space averaging technique [21],[22] is implemented and its model is as follows.

$$[\dot{x}] = [A_k][x] + [B_k][u] \qquad 7(a)$$

$$[y] = [E_k][x] + [F_k][u] \qquad 7(b)$$

where k = 1, 2 denotes the operating mode of the converter.

$[x] = [i_{L1} \quad i_{L2} \quad i_{L3} \quad i_{L4} \quad v_{C1} \quad v_{C2} \quad v_{C3} \quad v_{C4} \quad v_{C5}]^T$,

$[u] = [V_g]$ and $[y] = [V_0]$

$$A = A_1 D + A_2(1-D) \qquad 8(a)$$

$$B = B_1 D + B_2(1-D) \qquad 8(b)$$

$$E = E_1 D + E_2(1-D) \qquad 8(c)$$

$$F = F_1 D + F_2(1-D) \qquad 8(d)$$

978-1-5386-6624-1/18 $31.00 © 2018 IEEE

[A] is the state matrix, [B] is the input matrix, [E] is the output matrix and [F] is the feed forward matrix, [x] is the state vector, [y] is the output vector and [u] is the forcing function vector. The steady-state solution of the converter is expressed as

$$X_{ss} = -A^{-1}BV_g \tag{9}$$

The non-ideal behavior of the converter is also considered in the model. Using the state-space model described in eqn (7), the control-to-output transfer function is formulated as follows:

$$G_{vd}(s) = \frac{\hat{v}_o(s)}{\hat{d}(s)}\bigg|_{\hat{v}_g(s)=0} = \left(E(sI-A)^{-1}(h_1)\right) + h_2 \tag{10}$$

$$h_1 = (A_1 - A_2)X_{ss} + (B_1 - B_2)V_g$$

$$h_2 = (E_1 - E_2)X_{ss} + (F_1 - F_2)V_g$$

$$[A_1] = \begin{bmatrix} \frac{-(r_1+r_{C2})}{L_1} & \frac{-r_{C2}}{L_1} & 0 & 0 & 0 & \frac{1}{L_1} & 0 & 0 & 0 \\ \frac{-r_{C2}}{L_2} & \frac{-(r_2+r_{C2})}{L_2} & 0 & 0 & 0 & \frac{1}{L_2} & 0 & 0 & 0 \\ 0 & 0 & \frac{-(r_3+r_{C1})}{L_3} & \frac{-r_{C1}}{L_3} & \frac{1}{L_3} & 0 & 0 & 0 & 0 \\ 0 & 0 & \frac{-r_{C1}}{L_4} & \frac{-(r_4+r_{C1})}{L_4} & 0 & 0 & 0 & 0 & 0 \\ 0 & 0 & \frac{-1}{C_1} & \frac{-1}{C_1} & 0 & 0 & 0 & 0 & 0 \\ \frac{-1}{C_2} & \frac{-1}{C_2} & 0 & 0 & 0 & 0 & 0 & 0 & 0 \\ 0 & 0 & 0 & 0 & 0 & \frac{k_5}{k_1 C_3} & \frac{-r_{C5}}{k_1 C_3} & \frac{k_4}{k_1 C_3} \\ 0 & 0 & 0 & 0 & 0 & \frac{k_3+k_5}{k_1 C_4} & \frac{-(r_{C3}+r_{C5})}{k_1 C_4} & \frac{k_4-k_2}{k_1 C_4} \\ 0 & 0 & 0 & 0 & 0 & \frac{k_3}{k_1 C_5} & \frac{-r_{C3}}{k_1 C_5} & \frac{-k_2}{k_1 C_5} \end{bmatrix}$$

$$[B_1] = [1/L_1 \ 1/L_2 \ 1/L_3 \ 1/L_4 \ 0 \ 0 \ 0 \ 0 \ 0]^T \ ; \ [F_1] = [0]$$

$$[E_1] = [0 \ 0 \ 0 \ 0 \ 0 \ 0 \ k_6 \ -k_7 \ k_8]$$

$$[A_2] = \begin{bmatrix} \frac{-(r_1+r_2+r_{C1}k_{31})}{2L_1} & 0 & \frac{-r_{C1}k_{32}}{2L_1} & 0 & \frac{-(r_{C1}k_{27}+1)}{2L_1} & \frac{-r_{C1}k_{27}}{2L_1} & \frac{-r_{C1}k_{28}}{2L_1} & \frac{-r_{C1}k_{29}}{2L_1} & \frac{-r_{C1}k_{30}}{2L_1} \\ \frac{-(r_1+r_2+r_{C1}k_{31})}{2L_2} & 0 & \frac{-r_{C1}k_{32}}{2L_2} & 0 & \frac{-(r_{C1}k_{27}+1)}{2L_2} & \frac{-r_{C1}k_{27}}{2L_2} & \frac{-r_{C1}k_{28}}{2L_2} & \frac{-r_{C1}k_{29}}{2L_2} & \frac{-r_{C1}k_{30}}{2L_2} \\ \frac{-r_{C2}k_{25}}{2L_3} & 0 & \frac{-(r_3+r_4+r_{C2}k_{26})}{2L_3} & 0 & \frac{-r_{C2}k_{27}}{2L_3} & \frac{-(r_{C2}k_{27}+1)}{2L_3} & \frac{-r_{C2}k_{28}}{2L_3} & \frac{-r_{C2}k_{29}}{2L_3} & \frac{-r_{C2}k_{30}}{2L_3} \\ \frac{-r_{C2}k_{25}}{2L_4} & 0 & \frac{-(r_3+r_4+r_{C2}k_{26})}{2L_4} & 0 & \frac{-r_{C2}k_{27}}{2L_4} & \frac{-(r_{C2}k_{27}+1)}{2L_4} & \frac{-r_{C2}k_{28}}{2L_4} & \frac{-r_{C2}k_{29}}{2L_4} & \frac{-r_{C2}k_{30}}{2L_4} \\ \frac{k_{31}}{C_1} & 0 & \frac{k_{32}}{C_1} & 0 & \frac{k_{27}}{C_1} & \frac{k_{27}}{C_1} & \frac{k_{28}}{C_1} & \frac{k_{29}}{C_1} & \frac{k_{30}}{C_1} \\ \frac{k_{25}}{C_2} & 0 & \frac{k_{26}}{C_2} & 0 & \frac{k_{27}}{C_2} & \frac{k_{27}}{C_2} & \frac{k_{28}}{C_2} & \frac{k_{29}}{C_2} & \frac{k_{30}}{C_2} \\ \frac{k_{13}}{C_3} & 0 & \frac{k_{14}}{C_3} & 0 & \frac{k_{15}}{C_3} & \frac{k_{16}}{C_3} & \frac{k_{16}}{C_3} & \frac{k_{17}}{C_3} & \frac{k_{18}}{C_3} \\ \frac{-r_{C1}}{k_9 C_4} & 0 & \frac{-r_{C2}}{k_9 C_4} & 0 & \frac{-1}{k_9 C_4} & \frac{-1}{k_9 C_4} & \frac{k_{10}}{k_9 C_4} & \frac{-k_{11}}{k_9 C_4} & \frac{-k_{12}}{k_9 C_4} \\ \frac{k_{19}}{C_5} & 0 & \frac{k_{20}}{C_5} & 0 & \frac{k_{21}}{C_5} & \frac{k_{21}}{C_5} & \frac{k_{32}}{C_5} & \frac{k_{33}}{C_5} & \frac{k_{34}}{C_5} \end{bmatrix}$$

$$[B_2] = \left[\frac{-r_{C1}k_{27}}{2L_1} \ \frac{-r_{C1}k_{27}}{2L_2} \ \frac{-r_{C2}k_{27}}{2L_3} \ \frac{-r_{C2}k_{27}}{2L_3} \ \frac{k_{27}}{C_1} \ \frac{k_{27}}{C_2} \ \frac{k_{15}}{C_3} \ \frac{-1}{k_9 C_4} \ \frac{k_{31}}{C_5}\right]^T \ ; \ [F_2] = [k_{35}]$$

$$[E_2] = [k_{33} \ 0 \ k_{34} \ 0 \ k_{35} \ k_{35} \ k_{36} \ k_{37} \ k_{38}]$$

With the simulation parameters presented in Table V and duty ratio 0.25, MATLAB [23] simulation studies are made. AC-sweep analysis of PSIM simulator [24] is used for the circuit depicted in Fig. 2 (a) to validate the frequency response plot, obtained from transfer function of (10) as depicted in Fig. 4. The MATLAB model-based frequency response and PSIM simulator frequency response are in close agreement for both magnitude and phase responses.

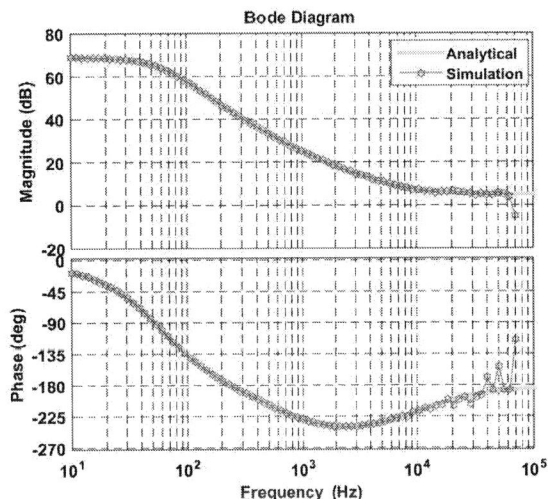

Fig. 4. Bode plot of control-to-output transfer function.

III. COMPARISON WITH OTHER CONVERTERS

This section shows the comparison of proposed converter with converters in [15] and [18]. To make a fair comparison, the converters considered for comparison are of same duty ratio range as their denominator is same. Table II shows the detailed comparison between voltage gain, the voltage across capacitors, switch and diodes. The voltage gain comparison is shown in Fig. 5, which shows that proposed converter has higher voltage gain at all duty ratio. The voltage gain is nearly two times that of converter in [15]. The voltage stress across capacitors, diodes and switch is lesser than other converters at same input and output power conditions, or at same load voltage. The voltage stress across switch and diodes is presented in terms of load voltage for a better understanding. As can be seen from Table II, the voltage stress across switch and diodes in proposed converter is nearly half of the converters in [15] and [18] respectively. The proposed converter also has common ground between its input and output terminals.

TABLE II
COMPARISON WITH EXISTING CONVERTERS

	Proposed	Converter in [15]	Converter in [18]
Voltage Gain	$\left(\frac{2(1+D)}{1-3D}\right)$	$\left(\frac{1+D}{1-3D}\right)$	$\frac{1}{1-3D}$
V_{C1}	$\left(\frac{2D}{1-3D}\right)V_g$	$\left(\frac{2D}{1-3D}\right)V_g$	$\left(\frac{1-2D}{1-3D}\right)V_g$
V_{C2}	$\left(\frac{2D}{1-3D}\right)V_g$	$\left(\frac{1-D}{1-3D}\right)V_g$	$\left(\frac{2D}{1-3D}\right)V_g$
V_{C3}	$\frac{V_o}{2}$	V_o	$\frac{V_o}{D}$
V_{C4}	$\frac{V_o}{2}$	NA	$\frac{V_o}{D}$
V_{C5}	$\frac{V_o}{2}$	NA	V_o
V_S	$\frac{V_o}{2}$	V_o	V_o
V_D	$\frac{V_o}{2}$	V_o	V_o
Common Ground	Yes	No	Yes

978-1-5386-6624-1/18 $31.00 © 2018 IEEE

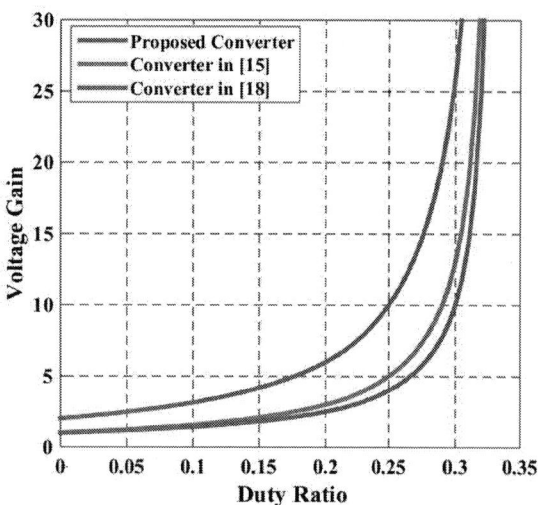

Fig. 5. Voltage gain comparison.

IV. PARAMETER DESIGN

This section determines the design equations for inductors and capacitors of the proposed converter. The rated voltage across each component calculated in section II is presented in Table III. By applying Kirchoff's current law (KCL) in the equivalent circuits of Fig. 2 and key waveforms in Fig. 3, the rated current through each component is calculated and listed in Table IV. Using the equations listed in Table III and IV, the design equations (11)-(15) for inductors and capacitors of proposed converter are formulated. The design parameters considered in simulation are listed in Table V.

TABLE III
VOLTAGE ACROSS EACH COMPONENT

	Mode-1	Mode-2
V_{C1}, V_{C2}	$\left(\dfrac{2D}{1-3D}\right)V_g$	$\left(\dfrac{2D}{1-3D}\right)V_g$
V_{C3}, V_{C4}, V_{C5}	$\left(\dfrac{1+D}{1-3D}\right)V_g$	$\left(\dfrac{1+D}{1-3D}\right)V_g$
$V_{L1} \& V_{L2}$	$\left(\dfrac{1-D}{1-3D}\right)V_g$	$\left(\dfrac{-D}{1-3D}\right)V_g$
V_{L3}, V_{L4}	$\left(\dfrac{1-D}{1-3D}\right)V_g$	$\left(\dfrac{-D}{1-3D}\right)V_g$
V_S	0	$\left(\dfrac{1+D}{1-3D}\right)V_g$
$V_{D1}, V_{D3}, V_{D5}, V_{D7}$	0	$\left(\dfrac{-D}{1-3D}\right)V_g$
V_{D2}, V_{D6}	$\left(\dfrac{D-1}{1-3D}\right)V_g$	0
V_{D4}, V_{D8}, V_{D10}	$\left(\dfrac{-(1+D)}{1-3D}\right)V_g$	0
V_{D9}	0	$\left(\dfrac{-(1+D)}{1-3D}\right)V_g$

$$L_1 = L_2 = L_3 = L_4 = \left[D(1-D)V_g\right]\big/\left[\Delta i_L f_s(1-3D)\right] \quad (11)$$

$$C_1 = C_2 = \left[4DI_0\right]\big/\left[\Delta v_{C1} f_s(1-3D)\right] \quad (12)$$

$$C_3 = \left[I_0\right]\big/\left[\Delta v_{C3} f_s\right] \quad (13)$$

$$C_4 = \left[DI_0\right]\big/\left[\Delta v_{C4} f_s\right] \quad (14)$$

$$C_5 = \left[(1+D)I_0\right]\big/\left[\Delta v_{C5} f_s\right] \quad (15)$$

TABLE IV
CURRENT THROUGH EACH COMPONENT

	Mode-1	Mode-2
$I_{L1}, I_{L2}, I_{L3}, I_{L4}$	$\left(\dfrac{2}{1-3D}\right)I_0$	$\left(\dfrac{2}{1-3D}\right)I_0$
I_{C1}, I_{C2}	$\left(\dfrac{-4}{1-3D}\right)I_0$	$\left(\dfrac{4D}{(1-D)(1-3D)}\right)I_0$
I_{C3}	$\dfrac{I_0}{D}$	$\dfrac{-I_0}{1-D}$
I_{C4}	$-I_0$	$\dfrac{D}{1-D}I_0$
I_{C5}	$\dfrac{-(1+D)}{D}I_0$	$\left(\dfrac{1+D}{1-D}\right)I_0$
I_S	$\left(\dfrac{1+5D}{D(1-3D)}\right)I_0$	0
$I_{D1}, I_{D3}, I_{D5}, I_{D7}$	$\left(\dfrac{2}{1-3D}\right)I_0$	0
I_{D2}, I_{D6}	0	$\left(\dfrac{2}{1-3D}\right)I_0$
I_{D4}	0	$\left(\dfrac{2(1+D)}{(1-D)(1-3D)}\right)I_0$
I_{D8}, I_{D10}	0	$\dfrac{I_0}{1-D}$
I_{D9}	$\dfrac{I_0}{D}$	0

TABLE V
SYSTEM PARAMETERS

Parameter	Value
V_g	36 V
$C_1 = C_2 = C_3 = C_4 = C_5$	47 μF
$r_{C1} = r_{C2} = r_{C3} = r_{C4} = r_{C5}$	0.5 Ω
$L_1 = L_2 = L_3 = L_4$	900 μH
$r_1 = r_2 = r_3 = r_4$	0.1 Ω
f_s	100 kHz
R	700 Ω

V. SIMULATION RESULTS

To verify the theoretical analysis of proposed converter, PSIM [24] is used for simulation. The simulation parameters are presented in Table V. The simulation results for presented in Fig. 6 are consistent with the key waveforms of proposed converter. In Fig. 6 (a), from top to bottom are gate to source voltage for switch S (V_{gs}), current through diodes D_1-D_{10} for D = 0.2. The switching pattern of the converter can be identified from this waveform. In Fig. 6 (b), from top to bottom are gate to source voltage for switch S (V_{gs}), current through inductors L_1, L_2, L_3, L_4 and diodes, voltage across capacitors C_1, C_2, C_3, C_4, C_5 and load voltage. According to the simulation parameters of Table IV, at D = 0.2, current through inductors L_1, L_2, L_3 and L_4 is 1.45 A. The voltage across capacitors C_1, C_2, C_3, C_4 and C_5 is 33.09, 33.09, 101.09, 100.87 and 102.69 V respectively whereas the load voltage is 203 V which is clearly indicated in Fig. 6

(b). In Fig. 6 (c), from top to bottom are gate to source voltage for switch S (V_{gs}), current through diodes D_1-D_{10} for D = 0.25. The current through inductors L_1, L_2, L_3 and L_4 is 3.52 A. The voltage across capacitors C_1, C_2, C_3, C_4 and C_5 is 58.61, 58.61, 153.24, 152.87 and 155.22 V respectively whereas the load voltage is 308 V which is clearly indicated in Fig. 6 (d).

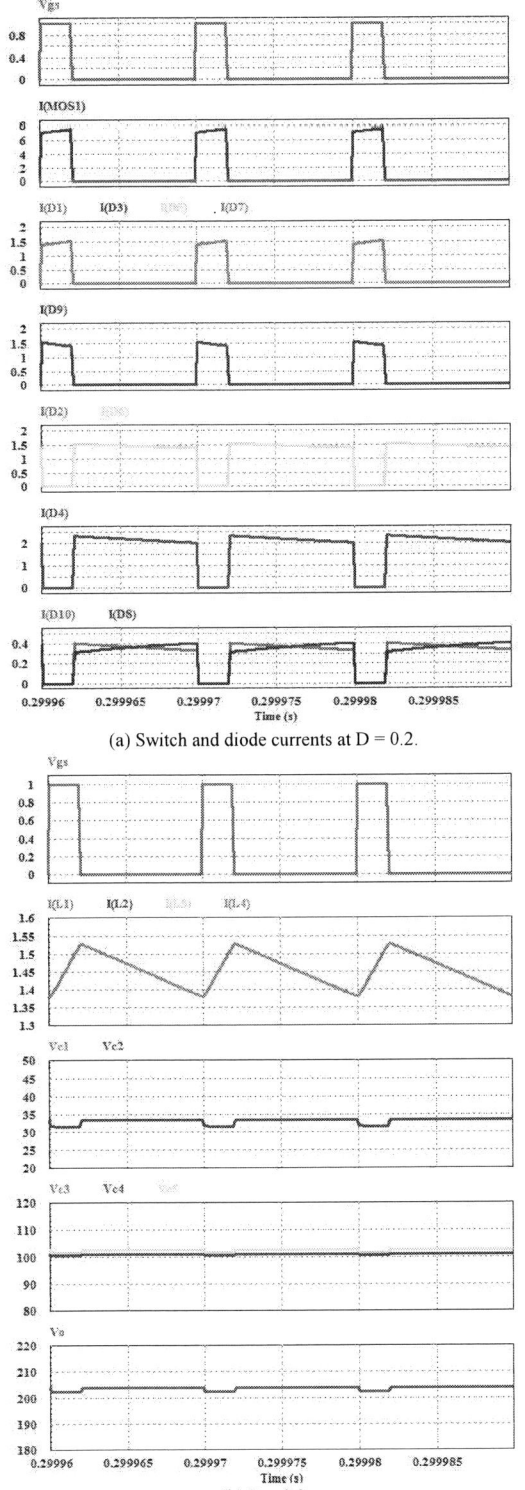

(a) Switch and diode currents at D = 0.2.

(b) D = 0.2.

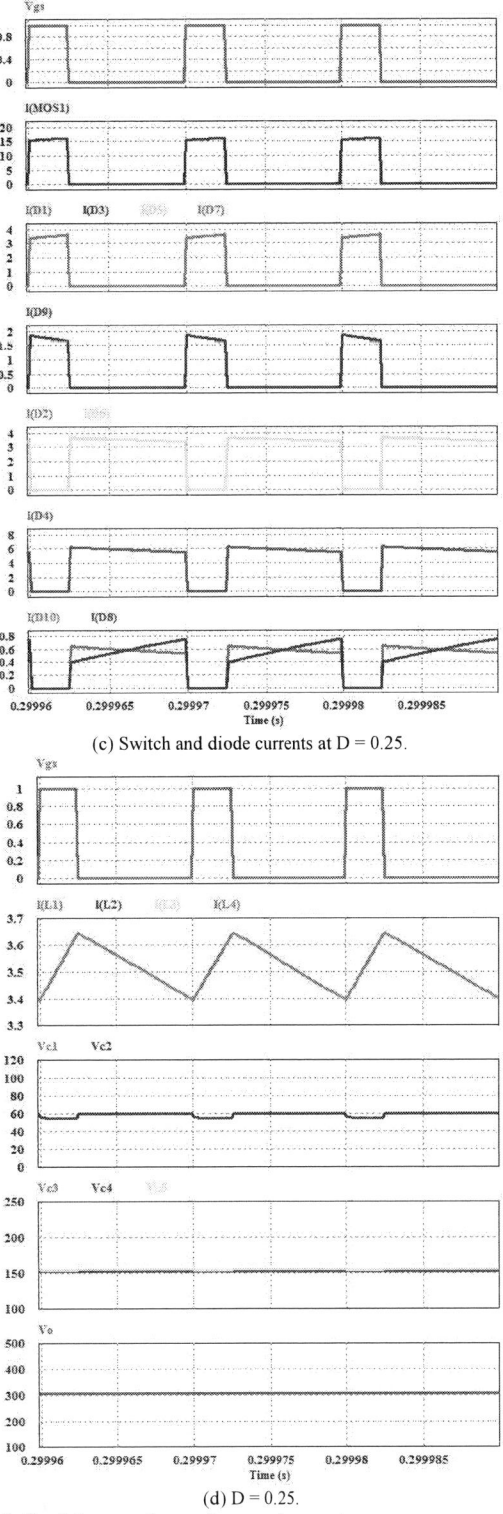

(c) Switch and diode currents at D = 0.25.

(d) D = 0.25.

Fig. 6. Simulation waveforms of proposed converter.

VI. CONCLUSION

An improved voltage gain hybrid quasi-Z-source dc-dc boost converter based on switched inductors and switched capacitor arrangement has been introduced in this paper. The proposed converter has higher voltage gain and reduced voltage stress on all the components than conventional converters. The detailed steady-state and small-signal analysis, parameters design and comparison with other reported converters is presented in detail. Simulation results verified the theoretical analysis.

REFERENCES

[1] F. Z. Peng, "Z-source inverter," in *IEEE Trans. Ind. Appl.*, vol. 39, no. 2, pp. 504-510, Mar/Apr. 2003.

[2] J. Anderson and F. Z. Peng, "Four quasi-Z-Source inverters," in Proc. *IEEE PESC*, pp. 2743-2749, 2008.

[3] M. Zhu, K. Yu and F. L. Luo, "Switched inductor Z-source inverter," in *IEEE Trans. on Power Electron.*, vol. 25, no. 8, pp. 2150-2158, Aug. 2010.

[4] M. K. Nguyen, Y. C. Lim and G. B. Cho, "Switched-inductor quasi-Z-source inverter," in *IEEE Trans. on Power Electron.*, vol. 26, no. 11, pp. 3183-3191, Nov. 2011.

[5] M. K. Nguyen, Y. C. Lim and J. H. Choi, "Two switched-inductor quasi-Z-source inverters," in *IET Power Electronics*, vol. 5, no. 7, pp. 1017-1025, August 2012.

[6] Xupeng Fang, "A novel Z-source dc-dc converter," in Proc. *IEEE ICIT*, 2008, pp. 1-4.

[7] J. Zhang and J. Ge, "Analysis of Z-source DC-DC Converter in Discontinuous Current Mode," in *Proc. IEEE APPEEC*, 2010, pp. 1-4.

[8] V. P. Galigekere, M. K. Kazimierczuk, "Analysis of PWM Z-source dc–dc converter in CCM for steady state", *IEEE Trans. Circuits Syst. I. Reg. Papers*, vol. 59, no. 4, pp. 854-863, Apr. 2012.

[9] Y. Shindo, M. Yamanaka and H. Koizumi, "Z-source DC-DC converter with cascade switched capacitor," in Proc. *IEEE IECON*, pp. 1665-1670, 2011.

[10] T. Takiguchi and H. Koizumi, "Quasi-Z-source dc-dc converter with voltage-lift technique," in Proc. *IEEE IECON*, pp. 1191-1196, 2013.

[11] L. Yang, D. Qiu, B. Zhang, G. Zhang and W. Xiao, "A modified Z-source DC-DC converter," in Proc. *IEEE EPE*, pp. 1-9, 2014.

[12] L. Yang, D. Qiu, B. Zhang, G. Zhang and W. Xiao, "A quasi-Z-source DC-DC converter," in Proc. *IEEE ECCE*, pp. 941-947, 2014.

[13] G. Zhang, B. Zhang, Z. Li, D. Qiu, L. Yang and W. A. Halang, "A 3-Z-network boost converter," in *IEEE Trans. on Ind. Electron.*, vol. 62, no. 1, pp. 278-288, Jan. 2015.

[14] H. Shen, B. Zhang, D. Qiu, L. Zhou, "A common grounded Z-Source DC–DC converter with high voltage gain," in *IEEE Trans. on Ind. Electron.*, vol. 63, no. 5, pp. 2925-2935, May. 2016.

[15] G. Zhang *et al.*, "An impedance network boost converter with a high-voltage gain," in *IEEE Trans. on Power Electron.*, vol. 32, no. 9, pp. 6661-6665, Sept. 2017.

[16] Y. Zhang, C. Fu, M. Sumner and P. Wang, "A wide input-voltage range quasi-Z source boost DC-DC converter with high voltage-gain for fuel cell vehicles," in *IEEE Trans. on Ind. Electron.*, vol. 65, no. 6, pp. 5201-5212, June 2018.

[17] M. M. Haji-Esmaeili, E. Babaei and M. Sabahi, "High step-up quasi-Z source DC-DC converter," in *IEEE Trans. on Power Electron.*, vol. PP, no. 99, pp. 1-1.

[18] H. Shen, B. Zhang and D. Qiu, "Hybrid Z-source boost DC–DC converters," in *IEEE Trans. on Ind. Electron.*, vol. 64, no. 1, pp. 310-319, Jan. 2017.

[19] P. Kumar, M. Veerachary, "Analysis and design of improved voltage gain impedance source DC-DC boost converter," in Proc. *IEEE PESTSE*, pp. 1-6, 2018.

[20] Y. Wang, Q. Bian, X. Hu, Y. Guan and D. G. Xu, "A high performance impedance-source converter with switched inductor," in *IEEE Trans. on Power Electron.*, vol. PP, no. 99, pp. 1-1.

[21] R. W. Erickson and D. Maksimovic, *Fundamentals of Power Electronics*, 2nd ed. Norwell, MA, USA: Kluwer, Jan. 2001.

[22] P. Kumar and M. Veerachary, "Analysis of modified impedance-network half-bridge converter," in Proc. *IEEE CERA*, pp. 415-420, 2017.

[23] MATLAB user manual, Mathworks, 2015.

[24] PSIM, Powersimtech, 2009.

APPENDIX

Coefficients for proposed converter

$$p = \frac{(R + r_{C4})}{R} \; ; \; p_1 = \frac{(pr_{C3} + r_{C4})}{r_{C3}} \; ; \; p_2 = \frac{(R + r_{C3})}{Rr_{C3}}$$

$$k_1 = r_{C4}r_{C5} + r_{C4}r_{C3} + r_{C3}r_{C5} + Rr_{C5} + Rr_{C3} \; ; \; k_2 = R + r_{C4} + r_{C3} \; ; \; k_3 = R + r_{C4}$$

$$k_4 = \frac{k_1 - k_2 r_{C5}}{r_{C3}} \; ; \; k_5 = \frac{k_3 r_{C5} - k_1}{r_{C3}} \; ; \; k_6 = \frac{k_5 r_{C4} + k_3 r_{C4} + k_3 r_{C5}}{k_1}$$

$$k_7 = \frac{-k_1 + r_{C4}r_{C5} + r_{C4}r_{C3} + r_{C3}r_{C5}}{k_1} \; ; \; k_8 = \frac{k_1 + k_4 r_{C4} - k_2 r_{C4} - k_2 r_{C5}}{k_1}$$

$$k_9 = \left[\frac{r_{C1}r_{C4} + r_{C2}r_{C4} + r_{C1}r_{C3} + r_{C2}r_{C3}}{r_{C3}} + \frac{r_{C1}Rp_1 + r_{C2}Rp_1 + r_{C5}Rp_1}{r_{C5}} \right]$$

$$k_{10} = \frac{r_{C1}r_{C5} + r_{C2}r_{C5} + Rr_{C1} + Rr_{C2} + Rr_{C5}}{r_{C3}r_{C5}}$$

$$k_{11} = \left[\frac{r_{C1} + r_{C2}}{r_{C3}} + \frac{(r_{C1} + r_{C2} + r_{C5})Rp_2}{r_{C5}} \right] \; ; \; k_{12} = \frac{r_{C1} + r_{C2}}{r_{C5}} \; ; \; k_{13} = \frac{-r_{C4}r_{C1}}{k_9 r_{C3}}$$

$$k_{14} = \frac{-r_{C4}r_{C2}}{k_9 r_{C3}} \; ; \; k_{15} = \frac{-r_{C4}}{k_9 r_{C3}} \; ; \; k_{16} = \frac{k_{10}r_{C4} - k_9}{k_9 r_{C3}} \; ; \; k_{17} = \frac{-k_{11}r_{C4} + k_9}{k_9 r_{C3}}$$

$$k_{18} = \frac{-k_{12}r_{C4}}{k_9 r_{C3}} \; ; \; k_{19} = \frac{Rp_1 r_{C1}}{k_9 r_{C5}} \; ; \; k_{20} = \frac{Rp_1 r_{C2}}{k_9 r_{C5}} \; ; \; k_{21} = \frac{Rp_1}{k_9 r_{C5}}$$

$$k_{22} = \frac{-r_{C3}Rp_1 k_{10} + Rk_9}{k_9 r_{C3}r_{C5}} \; ; \; k_{23} = \frac{Rp_1 k_{11} - Rp_2 k_9}{k_9 r_{C5}} \; ; \; k_{24} = \frac{Rp_1 k_{12} - k_9}{k_9 r_{C5}}$$

$$k_{25} = k_{13} - \frac{r_{C1}}{k_9} - k_{19} \; ; \; k_{26} = k_{14} - \frac{r_{C2}}{k_9} - k_{20} + 1 \; ; \; k_{27} = k_{15} - \frac{1}{k_9} - k_{21}$$

$$k_{28} = k_{16} + \frac{k_{10}}{k_9} - k_{22} \; ; \; k_{29} = k_{17} - \frac{k_{11}}{k_9} - k_{23} \; ; \; k_{30} = k_{18} - \frac{k_{12}}{k_9} - k_{24}$$

$$k_{31} = k_{13} - \frac{r_{C1}}{k_9} - k_{19} + 1 \; ; \; k_{32} = k_{14} - \frac{r_{C2}}{k_9} - k_{20} \; ; \; k_{33} = r_{C3}k_{13} + r_{C5}k_{19}$$

$$k_{34} = r_{C3}k_{14} + r_{C5}k_{20} \; ; \; k_{35} = r_{C3}k_{15} + r_{C5}k_{21} \; ; \; k_{36} = r_{C3}k_{16} + r_{C5}k_{22} + 1$$

$$k_{37} = r_{C3}k_{17} + r_{C5}k_{23} \; ; \; k_{38} = r_{C3}k_{18} + r_{C5}k_{24} + 1$$

978-1-5386-6624-1/18 $31.00 © 2018 IEEE

2nd IEEE International conference on power Electronics, Intelligent Control and Energy systems (ICPEICES-2018)

Modeling and Simulation of Microgrid Solar Photovoltaic System with Energy Storage

Dinanath Prasad
Dept. of Electrical Engineering
Delhi Technological University
Delhi, India
dinanath6@gmail.com

Narendra Kumar
Dept. of Electrical Engineering
Delhi Technological University
Delhi, India
dnk_1963@yahoo.com

Rakhi Sharma
Dept. of Electrical Engineering
Indira Gandhi National Open
University Delhi, India
rakhis_ignou@rediffmail.com

Abstract— **This paper highlights the integrated operations of the photovoltaic system with energy storage device. The variations in the energy produced and the variations in the load, the DC link voltage does vary, which has to be tackled using an energy storage device so as to maintain the DC link voltage. The battery is used as a storage device which is connected to a bidirectional converter and switching of the bidirectional converter has to be done in such a way that it either charges or discharges the battery depending upon the load conditions. The DC bus is connected to the utility grid, the power sharing and transfer should be in such a way to maximize the efficiency. An efficient algorithm is to be used to attain maximum efficiency through an interactive inverter. The SPV generating system with MPPT and boost converter is accomplished with a storage unit, and simulations are performed and the results are analyzed.**

Keywords— *Microgrid, SPV, MPPT, PVSC, Grid integration, SOC*

I. INTRODUCTION

With technological advancement and extensive R & D in the power electronics field very sophisticated and advanced microgrid have been developed. As the conventional sources are running out at a very fast rate distributed generators (DG) including renewable sources, within microgrids can help to overcome the power system limitations to efficiency improvement, reduction in emissions and manage the variability of renewable sources [1]. Nowadays small generators and the energy storage devices are being used with power electronics to control the flow of energy within the system.The non conventional energy sources with these devices can be efficiently used to meet the increase in load demand. The use of Microgrid with solar array acts as an energy source has also emerged as an efficient alternative to meet the increase in load demand. Since the solar energy is the best and cheapest source of energy, we can use their energy for power generation when they connected to grid. This incredible renewable source of energy having numbers of advantages like it neither pollutes the environment not destroy the forest cover, harnessing of energy is absolutely free, there is no need of maintenance, there are multiple other uses such as heating of water,

keeping the house warm, also used in many more applications [1, 2].

The main problem is to design a suitable energy balance of microgrid and its integration with the utility grid. Energy generation by the solar photovoltaic source is highly fluctuating and the maintenance of the DC bus voltage at a constant value is one of the major problems. With the variety of loads the maintenance of the voltage is much difficult. The effective utilization of the renewable energy by the loads and its integration to the utility grid is the major concerns.

This paper presents modeling of the PV array, with MPPT to utilize the maximum power from the PV array. Energy storage unit is used with PV array for the maintaining the DC link voltage using a control strategy. The importance of the storage unit in Microgrid emphasized where the power is either injected or absorbed by the storage unit in order to maintain the voltage at a constant value [3]. To connect the microgrid to the utility grid an energy management algorithm is proposed. The simulation is carried out and the results are analyzed in Matlab/ Simulink environment.

II. GRID-CONNECTED MICROGRID SYSTEM

Fig.1, shows the main circuit topology for grid-connected microgrid system. This system consist of the PV array and the battery storage system which is connected to the DC bus and then utility grid using an inverter. The DC bus is connected to the utility grid, for the power sharing and transfer that should be in such a way to maximize the efficiency. An efficient algorithm is to be used to attain maximum efficiency through interactive inverter. The grid-connected mode is carried out using a 3 leg 3 phase inverter and the switching to the inverter is done with the help of a droop control technique to generate reference current. PV array is connected through boost converter are used, to fulfill the objective is to produce reference current by MPPT algorithm. The battery is connected with the help of bi-directional, which can either charge the battery by absorbing the energy from the grid or by discharging the battery by injecting the power into the grid. Bi-directional converter is used for achieving the following, for which the switching has to be in such a way that it either charges or discharges the battery depending on the load condition. From the fig.1, we can observe that solar as a renewable energy source and

978-1-5386-6624-1/18 $31.00 © 2018 IEEE   623

"2nd IEEE International conference on power Electronics, Intelligent Control and Energy systems (ICPEICES-2018)"

battery as an energy storage device is connected to the DC bus by which, they are connected to the DC to AC converter [4-5].

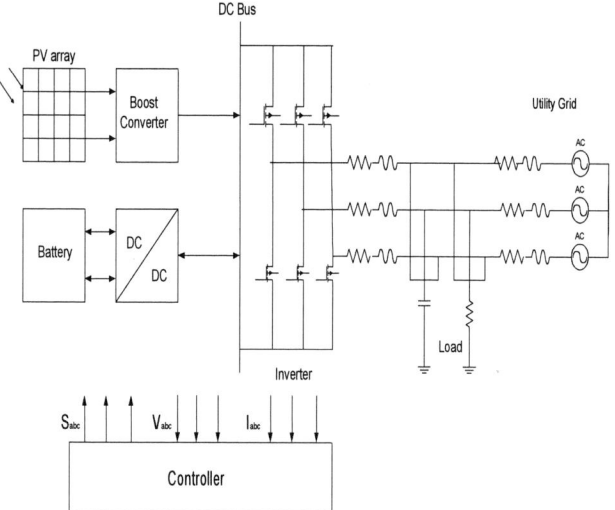

Fig.1. Schematic diagram of grid connected system

III. MODELING OF MICROGRID COMPONENTS

A. Modeling and Analysis of PV cell

PV cell is a solid state device which converts solar energy into electrical energy using photovoltaic effect. Model of Photo Voltaic cell is developed by analyzing the properties using Matlab/Simulink.

B. Equivalent circuit of PV cell

Fig. 2, shows the equivalent circuit of a PV cell, a current source represented by I_{ph} indicates the current caused by movement of electron and holes due to the incident light called photons. A diode current I_d , is represented parallel to the current source, a shunt resistance R_{sh} ,represents the leakage of current due to the presence of impurities in the junction diode, and series resistance R_s , due to the metallic contacts of the semiconductor [6, 7].

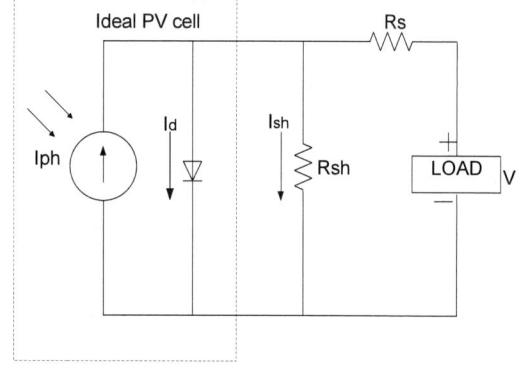

Fig.2. Equivalent circuit diagram of PV cell

V_{oc} is the open circuit voltage which is the voltage produced by the P-N junction diode. I_{sc} is the short circuit current, which is the photon current during short circuit condition. From the figure (2) ,using Kirchhoff's law

$$I_{pv} = I_{ph} - I_d \qquad (1)$$

The photon current (I_{ph}) is,

$$I_{ph} = [(I_{sc} + \alpha(T - T_{ref})]\frac{\beta}{\beta_{ref}} \qquad (2)$$

Where,

I_{ph} is the photon current (A)

I_{sc} Is the photon current at standard conditions (=298K,

$\beta_{ref} = 1000W/m^2$)

α is temperature coefficient

The diode current $I_d = I_s(e^{\frac{qV}{nkT}} - 1)$ (3)

Where I_s represents saturation current of the diode (A), q represents charge of an electron ($1.6*10^{-19}$ C), n represents the ideality factor, k is Boltzmann constant ($1.381*10^{-23}$ J/K) and T is the absolute temperature (K).

The saturation current of the diode varies with temperature as

$$I_s = I_0(\frac{T}{T_{ref}})^3 e^{\frac{qv_g}{nk}}(\frac{1}{T_{ref}} - \frac{1}{T}) \qquad (4)$$

Where I_0 is the reverse saturation current, T_{ref} is the reference Temperature (298 K), v_g is the bandgap voltage of the semiconductor (V).

By using initial conditions, I_{pv} =0 on the open circuit condition and the constant temperature. Equation (1) can be transformed into

$$0 = I_{sc} - I_d$$

$$I_{sc} = I_s(e^{\frac{qV}{nkT}} - 1)$$

$$I_s = I_0 \, at \, T = T_{ref}$$

$$I_{sc} = I_0(e^{\frac{qV}{nkT}} - 1)$$

$$I_0 = \frac{I_{sc}}{(e^{\frac{qV}{nkT}} - 1)}$$

From equation (1) the PV current can be expressed by

$$I_{pv} = I_{ph} - I_s(e^{\frac{qV}{nkT}} - 1)$$

Since the voltage across the diode itself is the voltage across the PV cell $V_d = V_{pv}$

$$I_{pv} = I_{ph} - I_s(e^{\frac{qV_{pv}}{nkT}} - 1) \qquad (5)$$

So, the output current can be expressed as

978-1-5386-6624-1/18 $31.00 © 2018 IEEE

$$I_{pv} = [I_{sc} + \alpha(T - T_{ref})]\frac{\beta}{\beta_{ref}} - I_s(e^{\frac{qV_{pv}}{nkT}} - 1) \qquad (6)$$

By using the above equations a model which fits the configuration is simulated in the Matlab Simulink environment. The following parameters are used for the simulation is tabulated in the table.

Table I. PARAMETERS FOR PV CELL SIMULATION

Parameters	Value
Open circuit Voltage (Voc)	0.6 V
Short circuit Current (Isc)	1.8 A
Band Gap Voltage (VG)	1.12 V
Diode Ideality Factor	1.24
Short circuit current temeperature coefficient	2.47*10^3

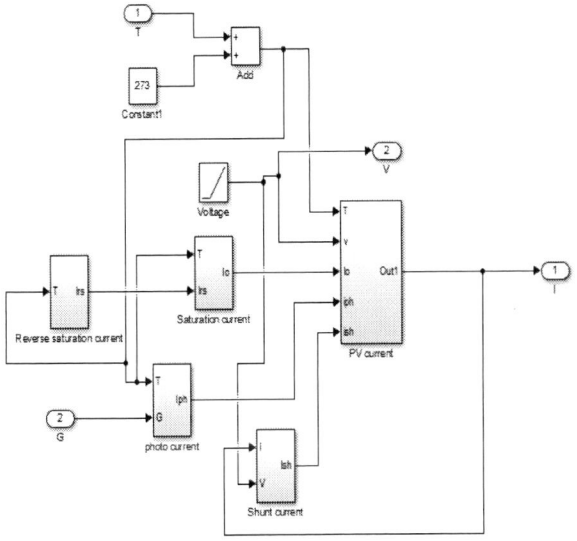

Fig.3. PV cell simulink model

The simulation of the PV array is similar to that of the PV cell, but instead of current being calculated, from the equation (6) and operating at reference temperature

$$Ipv - I_{sc}\frac{\beta}{\beta_{ref}} - I_s(e^{\frac{qVpv}{nkT}} - 1) = 0 \qquad (7)$$

Solving the following equation, Vpv can be calculated, which should be the voltage across the PV cell. For the consideration of an array, the equation (7) can be transformed in to equation (8)

$$\frac{Ipv}{N_p} - I_{sc}\frac{\beta}{\beta_{ref}} - I_s(e^{\frac{qVpv}{nkTNs}} - 1) = 0 \qquad (8)$$

The following simulations are carried out for Ns=100 and Np=10.

IV. MAXIMUM POWER POINT TRACKING TECHNIQUES

Maximum power point tracking (MPPT) plays a significant role in renewable energy sources, in this paper focuses on PV system as a source because of its intermittent nature with the changes in solar irradiance level, ambient temperature, and load the output of the PV array is nonlinear. For the determination of efficient PV output continuously adjust the operating point of the PV array to be around the maximum power point. This process is called MPPT. The available MPPT techniques are perturbed and observe (P&O), Incremental conductance (IC), fractional

Voltage, fractional current, fuzzy logic, feedback control, neural network. The MPPT classifications could be made on the basis of control strategies. Direct, indirect and probabilistic are three different types that are in use. Perturb and observe (P&O), incremental conductance (IC), fuzzy logic, and neural network are considered as direct methods. For estimating the Maximum Power Point (MPP), the indirect method is also used. Indirect methods include the look-up table, curve fitting, open circuit and short circuit methods [8].

In this paper perturb and observe (P&O) / Hill-climbing/ true seeking method is modeled for the determination of MPP. Because this algorithm could be easily implemented, and the accuracy of the sensors is not needed very high. The disadvantage of this method is that it may result in oscillations of power output near the MPP. The perturbance steps are used for the determination of the oscillation amplitude. The larger the perturbance step , faster the tracking is, but it serves the fluctuations. Also, if the solar irradiance changes suddenly, the P&O method may become invalid.

V. PV SIDE CONVERTER

For the PV side converter (PVSC) a boost converter is used. The main objective is that the I_{ref} (reference current) produced by the MPPT algorithm is tracked, for that by varying the duty cycle of the boost converter, the maximum power is transmitted and I_{ref} is to be tracked by the PV array.

TABLE II. PARAMETERS OF BOOST CONVERTER

Parameter	Value
Input value	ꭥ 200 V
Output value	600 V
Frequency	20 KHZ
Maximum Power	ꭥ 3 KW

The value of the inductor and capacitor are calculated from the following equations (9) and (10). The value of ripple is taken as 5% of the nominal value.

$$L = \frac{V_{pv} * d}{\Delta I * f} \tag{9}$$

$$C = \frac{I_o * d}{\Delta V * f} \tag{10}$$

The simulation is carried out with the following values, for the duty cycle input is analyzed in the next section.

The control circuit of the boost converter is shown below

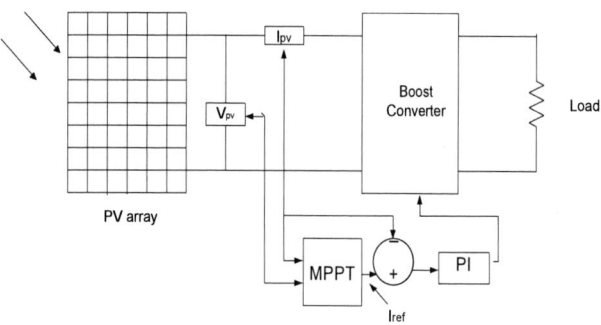

Fig.4. Control of boost converter

The control of boost converter is done for the tracking the reference current. The reference current is compared in between the MPPT with the actual current into the boost converter; the error is processed by PI controller. The output of the PI controller is further processed to generate PWM signals to the boost converter, by which the current is tracked [9].

VI. MICROGRID WITH STORAGE

It is essential for maintaining the DC grid voltage constant by injecting power if required in the system or conserving the energy by storing the excess amount of energy produced by the PV array. In this section, the operation of the Microgrid is carried out with a storage unit and simulations are carried out and analyzed. Since the battery is to be charged or discharged, depending on whether the power is to be injected or ejected from the DC grid, an algorithm is proposed for effective maintenance of DC grid voltage.

A. Storage connected to DC grid

In DC grid storage i.e. battery is not directly connected. The function of the battery is to inject the power or absorb the power from the grid. So, the battery is connected via bidirectional converter which can either be injecting the power into the grid by discharging the battery or either absorbing the energy from the grid to charge the battery. To fulfill the above requirements, switching of bidirectional converter has to be in such a way that it either charges or discharges the battery depending upon the load condition. Figure (5) shows the block diagram of the Microgrid with storage connected to the DC bus. The PV array is connected to the grid directly using the boost converter and the battery is connected to the grid using the bidirectional converter. The major bidirectional converter which is commonly used is the Buck-Boost converter. There are many bi-directional converters, but buck-boost is usually selected with the simplicity of its operation

implementation. Using the buck-boost converter, the battery is either charged or discharged, depending on the switching to the converter.

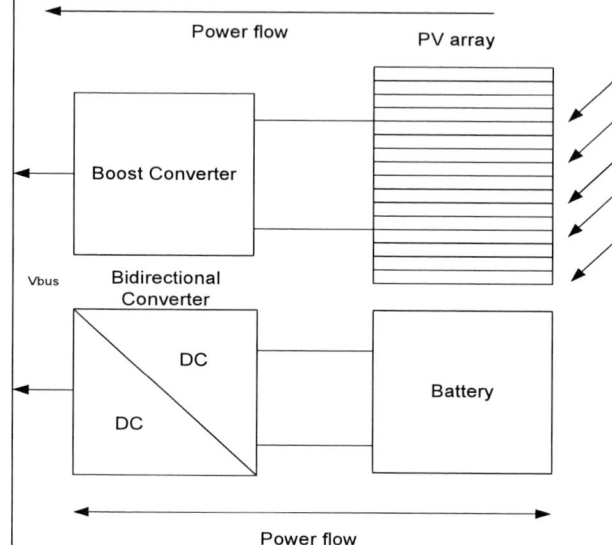

Fig.5. Microgrid with storage

B. Buck-Boost Converter

The Buck-Boost converter is very useful in real-time battery powered applications, as the name suggests it can act as a buck converter or it can act as a boost converter, that all depends upon the switched to the converter. For the buck operation or the step-down operation the power flow from the high voltage side to the low voltage side and for the boost operation or the step-up operation the power flow from the low voltage side to the high voltage side. The power flow can be controlled by suitably choosing the switching to either charge or discharge the battery [10].

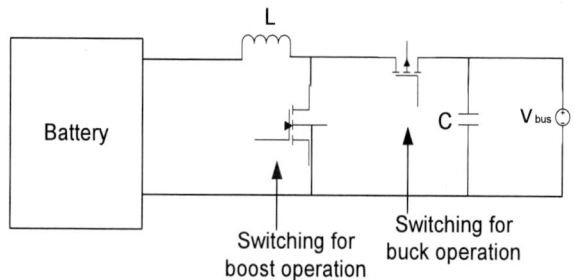

Fig.6. Buck-Boost circuit diagram

In the specified region of operation, the buck-boost converter parameters are so chosen in such a way that it can get both as buck and boost converter. The analysis of buck-boost converter can be done into buck operation and boost operation as the value of inductor and capacitor are calculated and selected for the lowest ripple value.

"2nd IEEE International conference on power Electronics, Intelligent Control and Energy systems (ICPEICES-2018)"

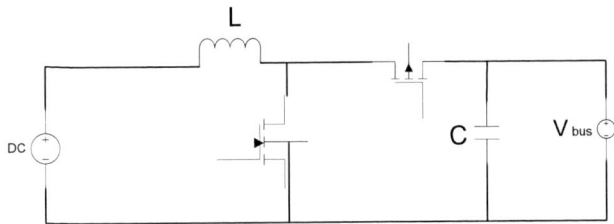

Fig.7. Bi-directional converter

$$(V_{bus} - V_{Battery})T_{ON} = V_{Battery}(T - T'_{ON}) \qquad (11)$$

$$D_1 = \frac{V_{Battery}}{V_{bus}} \qquad (12)$$

$$D_1 = \frac{T_{ON}}{T} \qquad (13)$$

The value of inductance (L) can be calculated as follows

$$L = \frac{(V_{bus} - V_{Battery})D_1}{\Delta I * f} \qquad (14)$$

The value of capacitance (C) can be calculated as follows

$$C = \frac{\Delta I}{8 * \Delta V * f} \qquad (15)$$

For good estimation the ripple current is 20% to 40% of the output current or $0.2 < \Delta I < 0.4$. The parameters of the boost converter are calculated from the equation, the value for which the lowest ripple of current and voltage is obtained is selected.

TABLE III. PARAMETERS OF THE BUCK-BOOST CONVERTER

PARAMETERS	VALUE
V_{BUS}	600 V
$V_{BATTERY}$	240 V
SWITCHING FREQUENCY	20 KHZ
INDUCTANCE (L)	14 MH
CAPACITANCE (C)	100 μF

The power is to be injected or to be absorbed from the grid is decided from the selection of Buck/Boost converter which is very important for the following a selection mechanism is required. If it is to be noted that solar PV power is greater than the load, then the voltage of the DC bus increases above the desired value, hence to maintain the DC voltage steady at required voltage should be absorbed. And

when solar PV power is less than the load, then the voltage of the DC bus decreases from the required voltage, hence to maintain the DC voltage steady at required voltage power should be injected [11- 14] .

C. Control of switching for the Buck-Boost converter

The switching for the buck-boost converter should be done so that the voltage reference is maintained at the DC voltage bus. To acquire the above, the voltage of the DC bus is sensed and the error in voltage between the DC grid voltage and the reference voltage is fed to a PI controller from which a reference current (Iref) is generated.

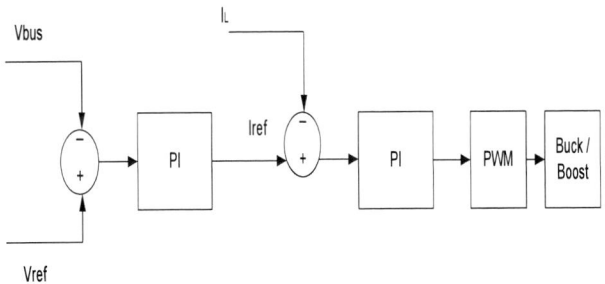

Fig.8. Control diagram for a buck - boost converter

VII. SIMULATION RESULTS

The Simulations are carried out for different cases when microgrid system connected to the DC link with and without storage unit. Microgrid without storage unit can also be classified into two different sections; either the load power at 600V is greater than the power produced by the PV array or when the load power at 600V is less than the PV array. So for the case when power required by the load at 600V is more than power by PV array with the maximum rated power at around 3KW, the load resistance is taken 120 ohms, for that load resistance at which the nominal value of about 600V is reached. Now for the load less than 120 ohms, to maintain the voltage of 600V, additional power is to be injected into the DC grid, but due to the lack of storage unit connected to the grid, the bus lack power and the voltage drops from the nominal value. The following simulation is carried out a load of 60 ohms for which to maintain the 600V nominal value of the DC bus, 6KW of power should be transmitted by the PV array. The maximum power from the PV array is around 3KW, so the DC bus voltage drops below 600V mark. It settles around 400V. This is the very unsuitable case without the storage unit system in the DC bus.

978-1-5386-6624-1/18 $31.00 © 2018 IEEE

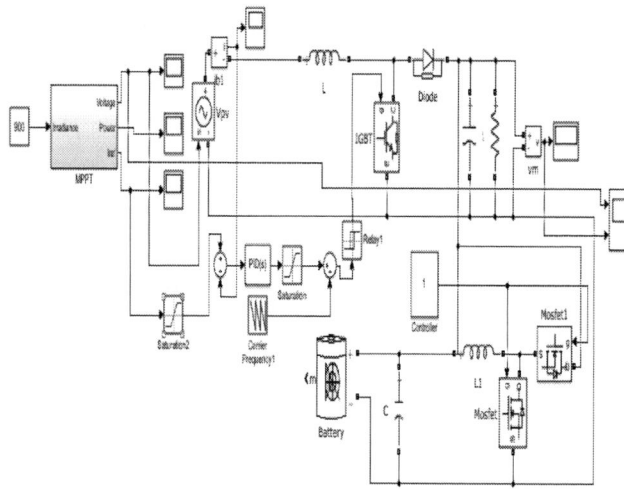

Fig.9. Matlab/Simulink diagram for PV, battery connected to DC link

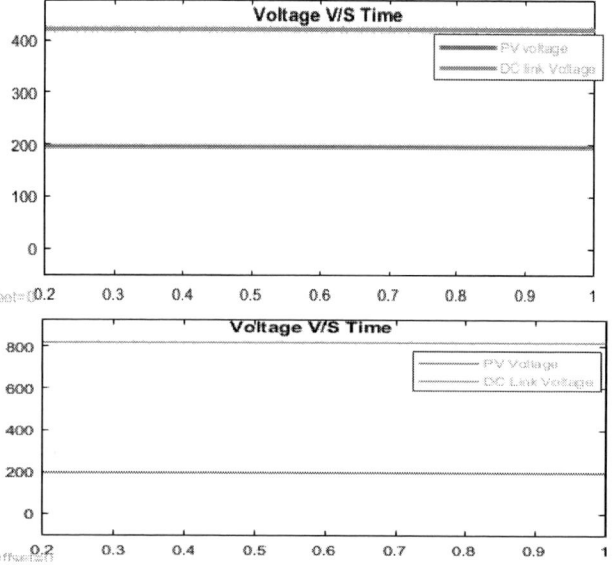

Fig.10(a). PV voltage and DC link Voltage when R_{load} = 60 ohm (b) PV voltage and DC link Voltage when R_{load} = 240ohm.

The DC link grid voltage has to be maintained at about 600V, but from the figure 10(a), the DC grid voltage is approx. 400V. The same situation is carried out a load of 240 ohms, and since the load power at 600V is 1.5KW but the power generated by the PV array is around 3KW. Since there is an excess generation of power than the load power, the voltage instead of maintaining at 600V will increase and reach a voltage around 850V which is shown in figure 10(b).

From the following, we can conclude that an energy source is essential for the maintenance of the DC link grid at a constant voltage. If the following DC bus is connected to the energy source, which can inject/eject the remaining power to maintain the voltage level at 600V. The other simulation is carried out where the PV array and the storage device is

connected to the utility grid using an inverter and the switching to the inverter is executed by droop control method. The inverter is connected to the grid and the required power is transmitted to the grid.

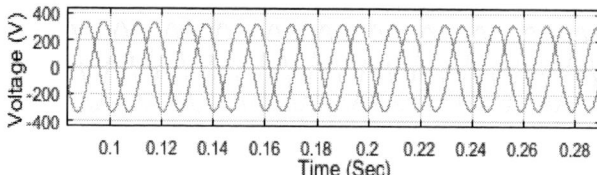

Fig.11. Grid Voltage

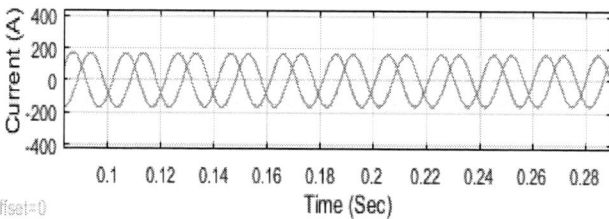

Fig.12. Grid Current

The following system was tested under the following conditions: Vs= 415 V, switching frequency = 10 kHz, Vdc= 600 V, line impedance Ls= 2.6 MH, Rs= 0.6 ohms, filter inductance L_L = 0.6 MH, filter capacitance = 1500µF, a linear load of 75 KW.

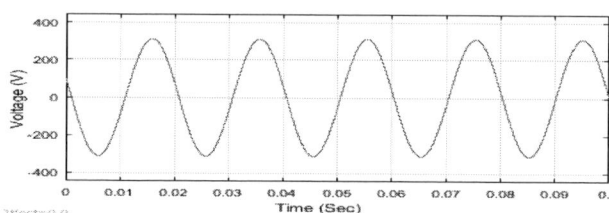

Fig.13. Line to line voltage during grid connected mode

Grid voltage and grid current in the respected figure 11 and 12 demonstrate system performance, and figure 13 shows the line to line voltage during grid-connected mode.

VIII. CONCLUSION

The Microgrid integration with the utility grid will help in the effective utilization of the renewable sources for improving the quality of the grid. Through this paper, explains the modeling of the components of the PV array, and uses Maximum Power Point Tracking to utilize the maximum power from the PV array and the is connected to an energy storage so as to maintain the DC link voltage using a control strategy. Algorithm for energy management is proposed to connect the Microgrid to the utility grid. The simulations are carried out, and the results are analyzed in Matlab/Simulink environment.

978-1-5386-6624-1/18 $31.00 © 2018 IEEE

REFERENCES

[1] K. Venkateswarlu and J. Krishna Kishore, "Modeling and simulation of micro grid system based on renewable power generation units by using multilevel converter," in International Journal of Engineering Research & Technology (IJERT), ISSN: 2278-0181, Vol. 1 Issue 6, August – 2012.

[2] Juan Manuel Carrasco, Leopoldo Garcia Franquelo, Jan T. Bialasiewicz, Eduardo Galván, Ramón C. Portillo Guisado, Ma. Ángeles Martín Prats, José Ignacio León and Narciso Moreno-Alfonso, "Power-Electronic Systems for the Grid Integration of Renewable Energy Sources: A Survey," IEEE transactions on industrial electronics, vol. 53, no. 4, august 2006.

[3] Hamid R. Teymour, Danny Sutanto, Kashem M. Muttaqi and P. Ciufo, "Solar PV and battery storage integration using a new configuration of a three-level NPC inverter with advanced control strategy," IEEE transactions on energy conversion, 2014.

[4] Irvin J. Balaguer, Qin Lei, Shuitao Yang and Athena Supatti, "Control for grid-connected and intentional islanding operations of distributed power generation," IEEE transactions on industrial electronics, vol. 58, no. 1, January 2011.

[5] Amir Khaledian and Masoud Aliakbar Golkar, "Analysis of droop control method in an autonomous microgrid," Journal of Applied Research and Technology 15 (2017) 371–377.

[6] Ba'amani beams, Dr. Iliya Tithe Thuku and Simon Samuel, "Design of a grid connected solar system for yola north and south, adamawa state ," International Journal of Science and Engineering Applications, Volume 6 Issue 10, 2017, ISSN-2319-7560.

[7] Cemal Keles, B. Baykant Alagoz, Murat Akcin, Asim Kaygusuz and Abdulkerim Karabiber, "A photovoltaic system model for Matlab/Simulink simulations," 4th International Conference on Power Engineering, Energy and Electrical Drives, Istanbul, Turkey, 13-17 May 2013.

[8] Bidyadhar Subudhi and Raseswari Pradhan, "A comparative study on maximum power point tracking techniques for photovoltaic power systems," IEEE transactions on sustainable energy, vol. 4, no. 1, January 2013.

[9] Bhim Singh, Shailendra Dwivedi, Ikhlaq Hussain and Arun Kumar Verma, "Grid integration of solar pv power generating system using a couple based control algorithm," IEEE. India International Conference (PIICON), Dec. 2014.

[10] Nupur Saxena, Ikhlaq Hussain, Member, Bhim Singh and A. L. Vyas, "Implementation of grid integrated pv-battery system for residential and electrical vehicle applications," IEEE transactions on industrial electronics, Volume: 65, Issue: 8, Aug. 2018.

[11] David Velasco de la Fuente, César L. Trujillo Rodríguez, Gabriel Garcerá, Emilio Figueres and Rubén Ortega González, "Photovoltaic power system with battery backup grid-connected and island operation capabilities," IEEE transactions on industrial electronics, vol. 60, no. 4, April 2013.

[12] Ya-Xiong Wang, Fei-Fei Qin, and Young-Bae Kim, "Bidirectional DC-DC converter design and implementation for lithium-ion battery application," IEEE PES Asia-Pacific Power and Energy Engineering Conference (APPEEC), Dec. 2014.

[13] Cody A. Hill, Matthew Clayton Such, Dongmei Chen, Juan Gonzalez, Student Member and W.Mack Grady, "Battery energy storage for enabling integration of distributed solar power generation" IEEE transactions on smart grid, vol. 3, no. 2, June 2012.

[14] Sathish Kumar Kollimalla, Mahesh Kumar Mishra, and N. Lakshmi Narasamma, "Design and analysis of novel control strategy for battery and supercapacitor storage system," IEEE transactions on sustainable energy, vol. 5, no. 4, October 2014.

2nd IEEE International conference on power Electronics, Intelligent Control and Energy systems (ICPEICES-2018)

Single-Phase Grid Interfaced WEGS using Frequency Adaptive Notch Filter for Power Quality Improvement

Tripurari Nath Gupta
Dept. of Electrical Engineering
Indian Institute of Technology
Delhi, New Delhi-110016, India
tnakgec@gmail.com

Shadab Murshid, *Member, IEEE*
Dept. of Electrical Engineering,
Indian Institute of Technology
Delhi, New Delhi-110016, India
mail2smurshid@gmail.com

Bhim Singh, *Fellow, IEEE* Dept. of
Electrical Engineering
Indian Institute of Technology Delhi,
New Delhi-110016, India
bsingh@ee.iitd.ac.in

Abstract—In this paper, frequency adaptive notch filter (FANF) based control algorithm is proposed for power quality improvement in single phase utility grid integrated wind energy generation system (WEGS). The FANF control is designed for extraction of fundamental component of load current. The main objective of the system, is to supply active power to loads as well as single phase grid connected at PCC (Point of Common Coupling) meanwhile mitigating the power quality issues like current harmonics distortion and load reactive power compensation. To achieve maximum power, a P&O (Perturb and Observe) based MPPT (Maximum Power Point Tracking) algorithm is used. A feed-forward term for the wind contribution is introduced for improving the dynamic response for change in wind speeds. The implementation of proposed control is easy and the response of the system is fast. The system behavior is tested under dynamic loading and changing wind speed conditions on a developed prototype in the laboratory. The performance of proposed system is found to be satisfactory and the grid current total harmonic distortion (THD) is found well within the IEEE 519 standard.

Keywords—*GSC, WEGS, power quality improvement and MPPT.*

I. INTRODUCTION

The renewable energy is inexhaustible and free in nature such as wind energy, solar energy, solar thermal energy, fuel cell energy and bio-mass energy etc., and these renewable energy sources for generating electrical energy, are preferred by most of the government in the world. Government of India is investing in renewable energy sources for production of power and has taken several initiatives to promote and boost rural electrification. The increasing demand of energy, has forced researchers to think about alternate sources of energy to reduce the pollution created by the use of fossil fuels for electricity generation [1-2]. As the demand of electricity is increasing exponentially and majority of the power plants are based on the fossil fuels, which are limited in nature and costing more per unit of electricity generation. Therefore, the need is look forward for renewable energy sources, which are free in nature. As mentioned above, the wind energy is one of the best energy sources for production of electricity. The wind energy shares a large part in energy production [1-4].

The WEGS (Wind Energy Generation System) requires an electric generator for generation of electricity. There are electric machines such as SCIG (Squirrel Cage Induction Generator), DFIG (Doubly Fed Induction Generator), PMSG (Permanent Synchronous Generator), and high-temperature superconducting SG (Synchronous Generator), which have been used for harvesting the wind energy. However, PMSG

is reasonable preferable as other machines require reactive power for their operation [5-8]. A maximum power point tracking (MPPT) algorithm is required for extraction of optimum power as the wind turbine characteristic is non-linear. Many MPPT techniques are already discussed in the existing literature, however, perturb and observe (P&O) algorithm is simple in implementation and does not depend on the system parameters. P&O algorithm involves the perturbation of the generator speed in the direction of maximum turbine output power [1].

Extensive use of linear and nonlinear loads at the PCC (Point of Common Coupling) of the grid connected WEGS system, faces several power quality issues. The harmonics created and reactive power drawn by these loads, result in decrease in efficiency and voltage distortion and so the power factor of the system. Various techniques such as synchronous reference frame (SRF) and extended Kalman filter (EKF), have been used in past for mitigation of power quality issues. EKF involves complex computation and it requires high end processors for its implementation [9]. SRF technique based on PLL is widely used. However, it requires multiple reference transformations [2,9]. Moreover, PLL deteriorates the dynamic response of the system due to inherent dependency on derivative term. It is also verified that the conventional PLL does not give satisfactory results under DC offset condition and distorted waveform condition. Therefore, the system robustness is compromised.

In order to solve the above mentioned issues, an attempt is made to develop a PLL-less system in this work. A frequency adaptive notch filter (FANF) based control algorithm is proposed, which estimates the active power component of the load current. A proportional-integral (PI) controller is used to estimate the loss component of voltage source converter (VSC) used in this system. A feed-forward component is introduced, which adapts itself in accordance with the variable wind speeds. The advantages of the proposed system, lie in its simplicity, fast convergence and ease of tuning. The performance of the system is studied for varying load conditions. The system behavior is also analyzed during variable wind speed conditions. It can be visualized from the acquired results that the system performs satisfactorily under all operating conditions and the total harmonic distortion (THD) of the grid current is found well within 5% as stated under the IEEE-519 standard.

II. SYSTEM CONFIGURATION

The proposed topology of Fig. 1, consists of a wind turbine, permanent magnet synchronous generator (PMSG), two voltage source converters, DC link capacitor (C_d), RC filter (R_f, C_f), the single phase grid (grid voltage, v_s and grid

978-1-5386-6624-1/18 $31.00 © 2018 IEEE

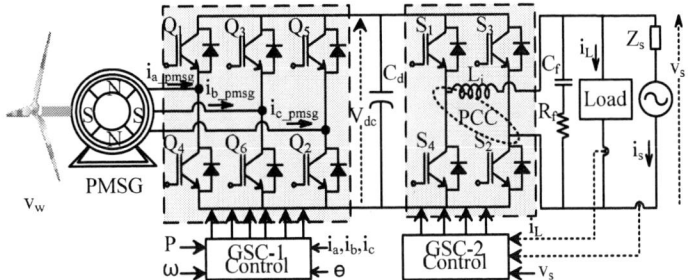

Fig. 1 Block diagram of the single phase grid connected WEGS system

impedance, Z_s) and connected linear and nonlinear loads. Two converters are connected at a common DC link (V_{dc}). The converter, which is connected to PMSG is termed as generator side converter (GSC-1) and the grid connected converter is termed as grid side converter (GSC-2). The system assures the maximum power extraction from the wind energy system and feeds it to the grid and nonlinear loads. The wind turbine driven PMSG system, is interfaced to GSC-1, which ensures MPPT by using P&O algorithm. GSC-2 is interfaced to the single phase grid and a nonlinear load, which improves the power quality of the grid. A coupling inductor (L_f) is used for reducing the harmonics in the injected grid current (i_s). A ripple filter is used for elimination of higher order switching harmonics.

III. CONTROL ALGORITHM

The proposed system mitigates power quality issues arising due to presence of nonlinear load at PCC as primary objective, while feeding peak power to the grid even during variable wind speed conditions. The converter GSC-1 is controlled using vector control and P&O algorithm is used to extract maximum power from WEGS. The control block diagram is shown in Fig. 2. The GSC-2 is controlled in such a way that it is capable of bidirectional power flow. The excess energy after feeding the load, is fed to the grid. Moreover, when the energy generated from the generator is not enough to feed the load, the utility grid provides the remaining energy. The control algorithm also takes care of the power quality issues arising due to nonlinear load integration. The control strategy of GSC-2, is shown in Fig. 4.

A. MPPT Algorithm

The detailed study and description of several MPPT algorithms, are listed in the literature. The P&O technique is simple to implement because of less complexity. The dynamic response of the P&O technique is good under varying wind speed conditions for WEGS. In this MPPT algorithm, the operating point is randomly perturbed in one direction and the power is calculated at new operating point. In case the power generated from wind generator increases, the operating point is perturbed in one direction and perturbed in other direction when the power generated from the wind generator decreases. The output of the MPPT algorithm estimates the reference generator speed. The governing equation of the algorithm is given as,

$$w_{ref}(k) = w_{ref}(k-1) + \Delta w \; if \; dP_w > 0 \, and \, dw > 0 \qquad (1a)$$
$$or \; dP_w < 0 \, and \, dw < 0$$

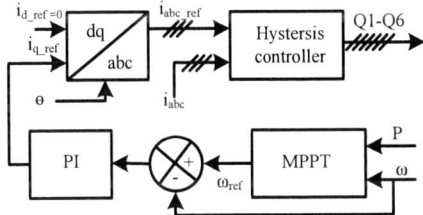

Fig. 2 Block diagram of the GSC-1 control block

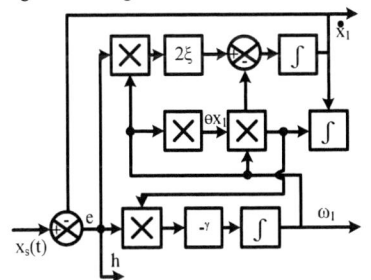

Fig. 3 Block diagram of frequency adaptive notch filtering scheme

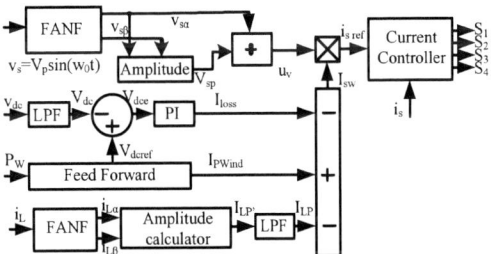

Fig. 4 Block diagram of the GSC-2 control

$$w_{ref}(k) = w_{ref}(k-1) - \Delta w \; if \; dP_w > 0 \, and \, dw < 0 \qquad (1b)$$
$$or \; dP_w < 0 \, and \, dw > 0$$

B. VSC (GSC-1) Control Block

GSC-1 is interfaced to PMSG driven by a wind turbine. The main objective of this converter, is to extract the maximum power from the available wind by controlling the rotational speed of the wind turbine. The maximum power is extracted using a well-known P&O based MPPT technique. The inputs are power and torque to the MPPT controller. The rotor position and rotational speed of the generator, are estimated using back emf based technique. The output of the P&O algorithm based MPPT controller is the reference generator speed. An error between the reference speed and the estimated speed, is input to the PI controller. The PI controller outputs the reference generator q-axis current. The

978-1-5386-6624-1/18 $31.00 © 2018 IEEE   631

d-axis reference current is set to zero for fast dynamic response. Using dq to abc conversion the reference generator currents are converted into three phase reference generator currents. The errors between reference currents and sensed currents are input to the hysteresis current controller. The output of the hysteresis current controller is the switching pulses of GSC-1.

C. VSC (GSC-2) Control Block

Fig. 4 shows the block diagram of GSC-2 control. The switching control of GSC-2 is obtained by estimating: (1) phase and magnitude of active power component of the load current, i_L; (2) fundamental in-phase and quadrature components at the point of common coupling voltage; (3) peak of the PCC voltage (V_{sp}); (4) feed-forward component of the wind power; and (5) switching losses of VSC. The active power component of load current, i_L is obtained by the designed frequency adaptive notch filter shown in Fig. 3, which is tuned to mitigate the harmonics, represented by $i_{L\alpha}$. The different components of current, are obtained by using FANF scheme.

In general, input to a synchronization algorithm is in the form defined in (2) can be voltage or current signal, in which the amplitudes A_i, the frequencies ω_i and the phase φ_i, are unknown parameters.

$$x_s(t) = \sum_{i=1}^{n} A_i \sin \phi_i \quad \text{where} \quad \phi_i = \omega_i t + \varphi_i \qquad (2)$$

The FANF scheme can accurately estimate the unknown parameters of such a measureable signal, $x_s(t)$. Thus, compensating signal is extracted by the scheme. The dynamic behavior of the proposed scheme FANF is realized for a single frequency n=1, by following equations:

$$\ddot{x} + \theta^2 x = 2\zeta\theta e(t)$$
$$e(t) = x_s(t) - \dot{x} \qquad (3)$$
$$\dot{\theta} = -\gamma\theta x e(t)$$

$x_s(t)$, Θ represent the input signal and estimated frequency, respectively. ξ and Υ which determine the FNAF performance in terms of convergence speed and accuracy, are both real and positive. For sinusoidal signal, $x_s(t)=A_1\sin(\omega_1 t+\delta)$, the dynamical system of (3), has a sole periodic orbit located at,

$$F(t) = \begin{pmatrix} x_1 \\ \dot{x}_1 \\ \theta \end{pmatrix} = \begin{pmatrix} -\dfrac{A_1}{\omega_1}\cos(\omega_1 t + \delta) \\ A_1 \sin(\omega_1 t + \delta) \\ \omega_1 \end{pmatrix} \qquad (4)$$

The estimated frequency is the last entry of F(t), and it is equal to its correct value, ω_1. Fig. 3 shows the detailed implementation of the FANF structure. The FNAF provides an online estimate of fundamental and quadrature component $x_s(t)=A_1\sin(\omega_1 t+\delta)$ and $x_{90}^0(t)=A_1\cos(\omega_1 t+\delta)$, respectively shown in (4). The outputs of FNAF are $\dot{x}_1$ and θx_1 as fundamental and quadrature signals.

The magnitude of the fundamental signal can simply be computed from $A_1 = (\theta^2 x_1^2 + \dot{x}_1^2)^{1/2}$. The fundamental frequency of the input signal, ω_1 is obtained from the output

Θ, and the FANF is composed of adders, integrators, and simple multipliers.

The values of in-phase and quadrature component of voltages, are also calculated by the notch filtering scheme. The unit template, u_v is calculated for synchronizing the reference current and the grid voltage. The governing equation is as follows,

$$V_{sp}(k) = \sqrt{(v_{s\alpha}^2(k) + v_{s\beta}^2(k))} \qquad (5)$$

V_{sp}, $v_{s\alpha}$ and $v_{s\beta}$ are the peak value, in-phase and quadrature component of the grid voltage. The unit template for the in-phase component of the voltage is given as,

$$u_v(k) = \frac{v_{s\alpha}(k)}{V_{sp}(k)} \qquad (6)$$

The voltage of the DC link is maintained by using a PI controller at a predefined value. This helps in determination of losses in VSC. The PI controller generates compensating currents, which is as follows,

$$i_{loss}(k) = i_{loss}(k-1) + k_i[e_{dc}(k) - e_{dc}(k-1)] + k_p[e_{dc}(k)] \quad (7)$$

$$e_{dc}(k) = V_{dc}^*(k) - V_{dc}(k) \qquad (8)$$

where, the gains of PI controller are represented by k_p and k_i, and the reference voltage input for the PI controller is V_{dc}^*. The calculation of the reference current is shown in Fig. 3.

1) Feed-Forward Term of Wind Component Calculation

The weight of wind component (I_{Pwind}) is estimated as,

$$I_{Pwind} = \frac{2P_w}{V_{sp}} \qquad (9)$$

2) Weight of Fundamental Grid Current

The estimated weight of the active grid component (I_{SW}) is the difference between the estimated weight of the wind component (I_{Pwind}) and sum of weight of average active weight of fundamental load current (I_{LP}) component and DC loss component (I_{loss}). The reference weight is as follows,

$$I_{sw} = I_{Pwind} - I_{loss} - I_{LP} \qquad (10)$$

$$I_{LP} = \sqrt{(i_{L\alpha}^2 + i_{L\beta}^2)} \qquad (11)$$

where, $i_{L\alpha}$ and $i_{L\beta}$ are fundamental in-phase and quadrature components of load current, , respectively.

As the grid frequency is constant, an adaptive notch filter is used. The proposed filtering scheme uses three integrators for extracting the fundamental component. The gain ζ is tuned to regulate the settling time and bandwidth of the filter. Since, the bandwidth is inversely proportional to settling time; a trade-off is made between these two. The open loop response of the system is studied and by inspection ζ= 0.55 and Υ= 0.15 are selected, which gives satisfactory response of the system while running in transient as well as steady state conditions.

IV. RESULTS AND DISCUSSION

The proposed system constitutes a 3.7 kW PMSG, which is operated under varying load conditions on the developed prototype in the laboratory. The steady state performance is recorded on a power analyzer (Fluke 43B), whereas dynamic response is recorded using the digital signal oscilloscope (Agilent DSO 6014A).

A) Normal Operating Condition

The system is operated under loaded condition and the load rating is 1.0 kW. Under normal operating condition, the wind speed, V_w is kept at 12 m/sec. The electrical output power generated by the PMSG under this condition is 2.21 kW. The estimated rotor speed (w_{est}), generator current (i_{a_pmsg}) and rotor position (Θ_{est}) are shown in Fig. 5(a). The DC link voltage is maintained at 400V and can be seen in Fig. 5(b). The generator currents are sinusoidal in nature. The grid current THD under normal operating condition is 4.1% and the load current THD is 25.1% as, shown in Figs. 6(c) and 6(f), respectively. THD of the grid current is less than 5% even if the load current THD is 25.1%. The control does it work satisfactorily.

B) Steady State Performance under Variable Loading Condition

Fig. 6 shows the steady state results recorded on power analyzer under different load conditions. Figs. 6(a) and 6(b) show the load voltage and load current waveforms when the load is connected at the system. Since, the load is non-linear, the current drawn is distorted in nature. It is clear from Fig. 6(c) that the load current THD is 25.1%. Since the proposed control is self-sufficient to maintain the grid current to follow the grid voltage, therefore, unity power factor is maintained as shown in Fig. 6(e). The grid current THD is 4.1% and 2.5%, which can be seen from Fig. 6(f) and Fig. 6(h), where the system is operating under load and no load conditions, respectively. The grid current THD in both the conditions are well within the IEEE-519 standard.

C) Dynamic Response of System under Variable Wind Speed Condition

The system performance is observed under the variable wind speed conditions and also with the change in the load on the wind energy generation system. The variations of various indices for change in wind speed from 7.2 m/sec to 12 m/sec and from 12m/sec to 7.2 m/sec, are shown in Figs. 7 (a) and 7(b). In the first case the wind speed is changed from 7.2 m/sec to 12 m/sec, which results in an increase of electrical power output generated. The MPPT algorithm tracks the maximum power point and makes the converter to control the generator speed. The increase in the wind speed, results in the increase in the rotor speed and the generator current as shown in Fig. 7(a) and vice versa is shown in Fig. 7(b). Fig. 7(c) shows the variation of grid power with the change in wind speed/power.

D) Dynamic Response of the System at Variable Load Condition

The load is changed from 1.0 kW to no load at a constant wind speed of 12 m/sec and the effect on different indices can be seen in Figs. 8(a) and 8(b). The time for which, the wind speed is constant and the load is connected to the system, the remaining power generated is fed into the grid. When the load is disconnected from the system, all the generated power is fed into the grid. Since, the wind power is constant there is no change in the generator current. The grid takes power when the generated power is more than the required power by the load, this can be seen from Fig. 8(a)

and it feeds the load when there is deficiency or no power available from the generator. Fig. 8(a) shows the effect of load change from 1.0 kW to 0.0 kW on load current (5.08 A to 0.0A), the grid current (5.52 A to 9.96 A) and no change in the generator current as the wind power input is constant. Fig. 8(b) shows the variation of power fed to the grid under load on and off conditions. The estimated weight components used for the control of GSC-2, are shown in Fig. 8(c), under this condition.

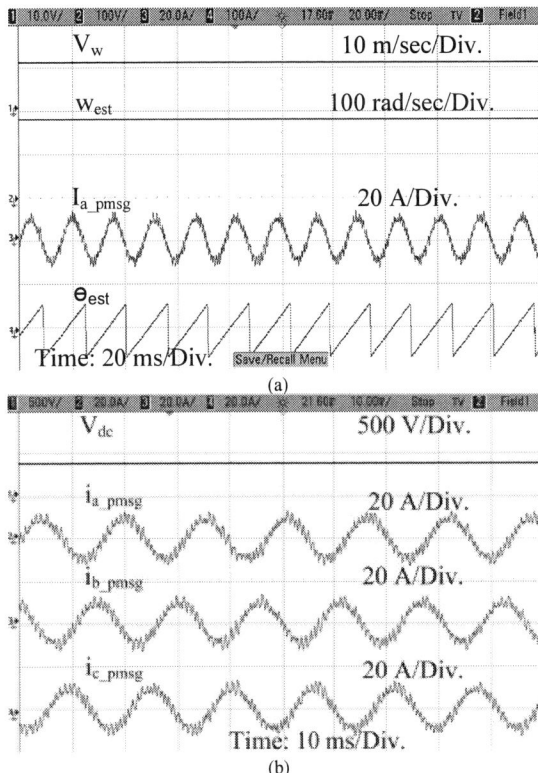

Fig. 5 Steady state behavior of the system under normal operating condition
(a) V_w, w_{est}, i_{a_pmsg}, Θ_{est} (b) V_{dc}, i_{a_pmsg}, i_{b_pmsg}, i_{c_pmsg}

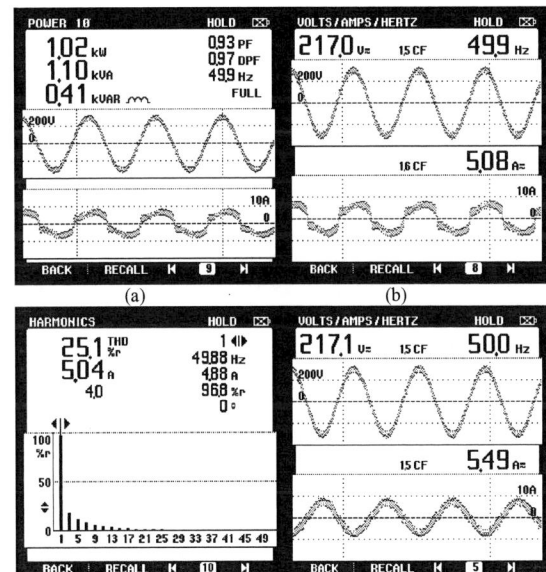

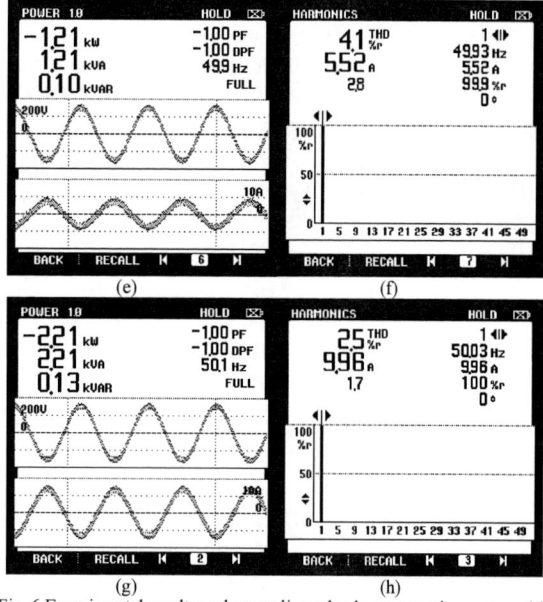

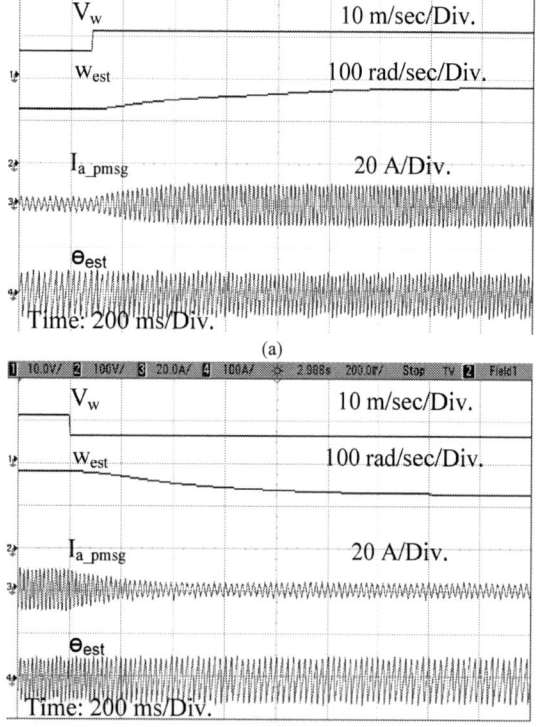

Fig. 6 Experimental results under non-linear load compensating system, (a), (b) v_L and i_L, (c) load current THD, (d), (e) v_s and i_s, (f) grid current THD, (g) v_s and i_s, (h) grid current THD

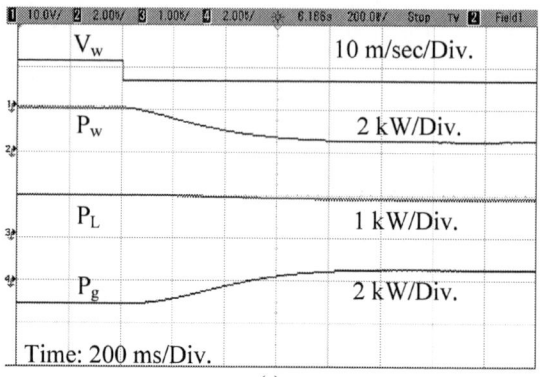

Fig. 7 Dynamic behavior of the proposed system at variable wind condition (a) V_w, w_{est}, i_{a_pmsg}, Θ_{est} (7.2 m/sec to 12 m/sec), (b) V_w, w_{est}, i_{a_pmsg}, Θ_{est} (12 m/sec to 7.2 m/sec) and (c) V_w, P_w, P_L, P_g

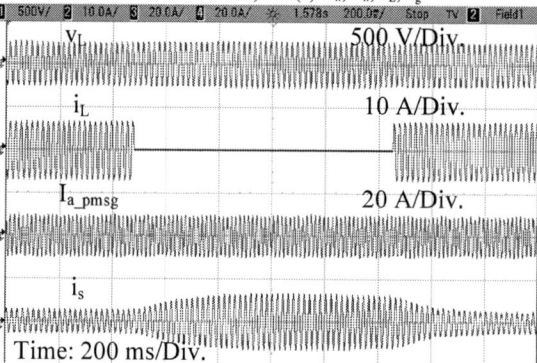

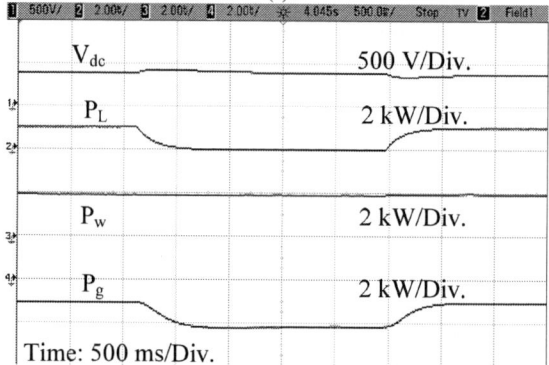

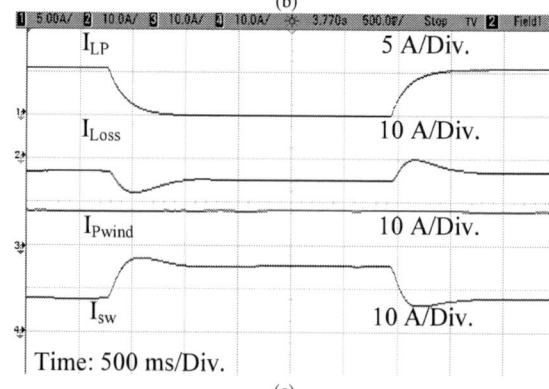

Fig. 8 Dynamic behavior of the proposed system at varying load condition (a) v_L, i_L, i_{a_pmsg}, i_s (b) V_{dc}, P_L, P_w, P_g, and (c) I_{loss}, I_{LP}, I_{Pwind} and I_{sw}.

V. CONCLUSION

The performance of proposed WEGS is observed at steady state and dynamic conditions and has been realized on a developed prototype in the laboratory. The control has been validated through DC-link voltage regulation, extraction of real component of the load current and reference current generation by extracting unit template of the PCC voltage. In addition, the proposed frequency

978-1-5386-6624-1/18 $31.00 © 2018 IEEE

adaptive notch filter technique has not used PLL for its operations. The system performance is depicted under variable wind speed conditions and also variable load conditions. The grid current THD is maintained below 5% even if the load current THD is high. The power factor of the grid is maintained at unity, under load and no-load conditions. The reactive power demand of the load, has also been compensated. In each case, the various parameters are observed and found satisfactory. It has been observed that the grid current THD is always under 5% even though the load is nonlinear and thereby satisfies the IEEE 519-standard. Therefore, the proposed system gives a feasible solution for effective wind energy harvesting.

APPENDICES

PMSG Rating: 3- phase, 230V, 3.7kW, Wind turbine rating: DC motor drive: Armature parameters- 220V, 20A, Field parameter- 220V, 1.65A, Ripple filter Capacitance C_f = 5μF and resistance R_f = 5Ω, Interfacing Inductor (L_i) : 4mH.

ACKNOWLEDGMENT

The authors are highly thankful to DST, Govt. of India, for supporting project, UKCERI (an Indo-UK Project), Clean Energy Project (SERI-2) and J. C. Bose Fellowship under Grant Number: RP03391, RP03357 and RP03128, respectively.

REFERENCES

[1] A. Shahi and C. Bhattacharjee, "A study & analysis of fuzzy based P&O MPPT scheme in PMSG based wind turbine," Technologies for Smart-City Energy Security and Power (ICSESP), *Bhubaneswar, India, 2018, pp. 1-4.*

[2] F. Shaikh and B. Joseph, "Simulation of synchronous reference frame PLL for grid synchronization using Simulink," *Int. Conf. Advances in Computing, Communication and Control (ICAC3)*, Mumbai, 2017, pp. 1-6.

[3] M. Singh, V. Khadkikar and A. Chandra, "Grid synchronisation with harmonics and reactive power compensation capability of a permanent magnet synchronous generator-based variable speed wind energy conversion system," *IET Power Electron.*, vol. 4, no. 1, pp. 122-130, January 2011.

[4] R. K. Agarwal, I. Hussain and B. Singh, "Dual-function PV-ECS integrated to 3P4W distribution grid using 3M-PLL control for active power transfer and power quality improvement," *IET Renewable Power Generation*, vol. 12, no. 8, pp. 920-927, 6 11 2018.

[5] M. Biweta and M. Mamo, "Closed loop control strategy of back to back PWM converter fed by PMSG using PLECS toolbox on Matlab/Simulink for wind energy application," *Proc. of 2017 IEEE AFRICON*, Cape Town, pp. 1313-1318, 2017.

[6] C. Khomsi, M. Bouzid, K. Jelassi and G. Champenois, "Harmonic current compensation in a single-phase grid connected photovoltaic system supplying nonlinear load," *Proc of 9th International Renewable Energy Congress (IREC)* Hammamet, Tunisia, pp. 1-6, 2018.

[7] M. Tarafdar Hagh, M. Jadidbonab and M. Jedari, "Control strategy for reactive power and harmonic compensation of three-phase grid-connected photovoltaic system," *CIRED - Open Access Proc Journal*, vol. 2017, no. 1, pp. 559-563, 2017.

[8] O. Alizadeh and A. Yazdani, "A Strategy for Real Power Control in a Direct-Drive PMSG-Based Wind Energy Conversion System," *IEEE Trans. on Pow. Delivery*, vol. 28, no. 3, pp. 1297-1305, July 2013.

[9] A. Bagheri, M. Mardaneh, A. Rajaei and A. Rahideh, "Detection of Grid Voltage Fundamental and Harmonic Components Using Kalman Filter and Generalized Averaging Method," *IEEE Trans. on Pow. Electron.*, vol. 31, no. 2, pp. 1064-1073, Feb. 2016.

[10] M. N. Geethanjali and M. H. Sidram, "Performance evaluation and hardware implementation of MPPT based photovoltaic system using DC-DC converters," *Int. Conf. Technological Advancements in Power and Energy (TAP Energy)*, Kollam, 2017, pp. 1-6.

978-1-5386-6624-1/18 $31.00 © 2018 IEEE

2nd IEEE International conference on power Electronics, Intelligent Control and Energy systems (ICPEICES-2018)

Performance Analysis of Discontinuous Pulse Width Modulation Schemes on PUC-5 Inverter

Abdul Azeem[1], Mohd Tariq[2*], Kaif Ahmed Lodi[3] and C. Bharatiraja[4]

[1,2,3] Department of Electrical Engineering, Aligarh Muslim University, Aligarh, India
[4] Department of Electrical and Electronics Engineering, SRM Institute of Science and Technology, India.

Abstract: - **Considerable research is being carried out on the effect of the choice of PWM scheme on inverter performance. Discontinuous PWM schemes, initially developed by "Depenbrock" in 1977, has become one of the most promising modern PWM methods. This modulation scheme is also known as "Depenbrock PWM" scheme. In this paper the discontinuous modulation schemes are applied on the recently introduced five level Packed U-Cell (PUC-5) inverter. The THD index, common mode voltage (CMV) and utilization of DC link voltage are analyzed as they are important parameters of a typical inverter. To control theses parameters and output voltage, PWM is a best option for it. A comprehensive analysis of these modulation schemes are shown on the basis of the output THD and common mode voltage applied for PUC-5 inverter. A single phase prototype model is also presented which is used for validation of the concept.**

Keywords: Packed U-Cell (PUC) inverter, Pulse Width Modulation (PWM), Total Harmonic Distortion (THD), Common Mode Voltage (CMV), Multilevel Inverter (MLI)

I. INTRODUCTION

In the variable speed drives (VSD) application systems, to control the output variables (voltage and frequency) in the conventional inverter (two level inverter) various pulse width modulation schemes are used. The multilevel inverter provides many advantages over the conventional two inverter such as high power and medium voltage application, lesser requirement of bulky filters, good quality of output etc. [1]. As the number of pole voltage increases, the THD index at line voltages are reduced. But it leads to an increase in the number of active and passive device counts. Thus making the inverter complex and less reliable.

There are many topologies of multilevel inverter which are found in the literature such as Cascaded H-Bridge (CHB), Neutral Point Clamped (NPC) and Flying Capacitor (FC) multilevel inverter. Selection among these inverter depends upon the area of application [2]. NPC inverter suffers the problem of neutral point voltage balancing as it requires complex control for balancing capacitor voltage for greater than three levels [3]. Flying capacitor can be used for both active and reactive power supply [4] but have same problem of capacitor balancing as in NPC inverter when used in reactive power compensation. CHB inverter topology is used when multiple independent DC source are available especially when it is integrated with the solar, fuel cell.

Due to the continuous research in optimization of the device counts in multilevel inverter, many new topology of MLI has been invented. One of the recently introduced topology is known as Packed U-Cell (PUC) inverter topology [5]. This inverter topology was invented in 2008 by deduction of CHB inverter. This is one of the most reduced device inverter available in the literature for five

levels. It has only six switches, one DC source and one auxiliary capacitor to produce the five level output [6].

The performance of the inverter depends on the form of pulse width modulation schemes [7]. Continuous research is going on to introduce the high efficient PWM schemes. These are continuous PWM (CPWM) and discontinuous PWM (DPWM). In continuous mode of PWM, the modulating signal is continuously time varying signal like the sinusoidal PWM (SPWM), third harmonic injection PWM (THIPWM) and space vector PWM (SVPWM). Among the continuous PWM, space vector PWM was invented for DC bus utilization range 0%-100%, in case of third harmonic injection PWM (THIPWM) this range is 0%-115.4% [8]. Discontinuous PWM also provide same DC bus utilization range (0%-115.4%) in linear range of modulation range along with other features like low switching loss and less common mode voltage (CMV) [9]. The common mode leakage current can be limited by using common mode filters. In this paper, a detailed analysis of the three phase five level PUC inverter is done by using different discontinuous PWM schemes.

Rest part of the paper is organized as: in section II, a detail modeling of the PUC inverter with auto capacitor balancing feature is presented, modulation schemes and their performance is presented in section III. Simulation result of PUC-5 inverter is presented in section IV and a single phase prototype model is discussed in section V. Finally conclusion is given in section VI.

II. THREE PHASE FIVE LEVEL PUC INVERTER

A power circuit model of the three phase five level PUC inverter is presented in the Fig.1.

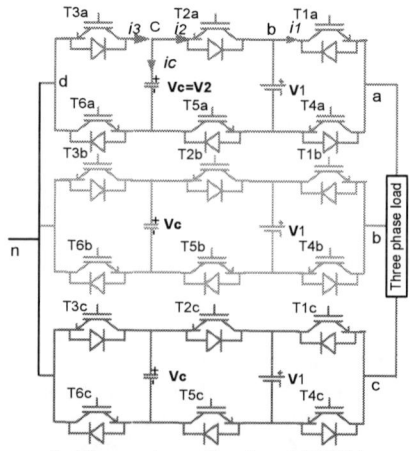

Figure 1: Three phase five level PUC inverter

978-1-5386-6624-1/18 $31.00 © 2018 IEEE

Table 1: Switching table of PUC inverter

States	T1	T2	T3	Output
1	1	0	0	V_1
2	1	0	1	V_1-V_2
3	1	1	0	V_2
4	1	1	1	0
5	0	0	0	0
6	0	0	1	$-V_2$
7	0	1	0	V_2-V_1
8	0	1	1	$-V_1$

It comprises six active switches in three pairs and each pair of switches (T1 & T4, T2 & T5, and T3 &T6) is working complementarily. The secondary DC source is the capacitor which has to be controlled at particular value to get desired levels at output waveform [10].

Table 1 shows the switching table for PUC inverter. It contains the eight switching states which gives the different operating states. For five level operation assuming $V_1= 2V_2=2V$, the 5-level output can be 0, $\pm E$, $\pm 2E$. For three phase operation three similar module of PUC inverter as shown in Fig. 1 is connected to get three phase five level output voltage [11].

The switching function of the PUC inverter is defined as:

$$Ti = 0 \;\; if \;\; T_i \;\; is \; off$$
$$= 1 \;\; if \;\; T_i \;\; is \; on \tag{1}$$

The inverter output voltage can be written as Eqn. (2), from the circuit diagram Fig. 1

$$Vad = Vab + Vbc + Vcd \tag{2}$$

Where the node a, b, c and d are shown in Fig.1 and voltage across the point can be calculated according to switching function:

$$Vab = (T_1 - 1)V_1$$
$$Vbc = (1 - T_2)(V_1 - V_2) \tag{3}$$
$$Vcd = (1 - T_3)V_2$$

Where, V_1 and V_2 are the main DC link voltage and capacitor voltage respectively.

By substituting the equation (3) into equation (2), we get Eqn. (4) as:

$$Vad = (T_1 - 1)V_1 + (1 - T_2)(V_1 - V_2) + (1 - T_3)V_2$$
$$= V_1T_1 - V_1 + V_1 - V_2 - V_1T_2 + V_2T_2 + V_2 - V_2T_3$$
$$= (T_1 - T_2)V_1 + (T_2 - T_3)V_2 \tag{4}$$

Since one of switches in pairs of T_1 & T_4, T_2 & T_5, and T_3 & T_6 are turned ON, the switches current can be expressed as function of load current in Eqn. (5):

$$i_1 = T_1 i_l$$
$$i_2 = T_2 i_l \tag{5}$$
$$i_3 = T_3 i_l$$

By applying KCL at node c, we get the following:

$$i_3 = i_c + i_2 \tag{6}$$

And

$$i_c = (T_3 - T_2)i_l \tag{7}$$
$$C\frac{dV_2}{dt} = (T_3 - T_2)i_l \tag{8}$$

We can write the equation (8) as:

$$\frac{dV_2}{dt} = \frac{(T_3-T_2)i_l}{C} = -\frac{i_l}{C}T_2 + \frac{i_l}{C}T_3 \tag{9}$$

As well, for the voltage and load current the KVL law is written as below:

$$V_l = Vad - i_l R_f - L_f\frac{di_l}{dt} \tag{10}$$

Where R_f and L_f be the filter resistance and inductance respectively. Calculating the equation (10) with the help of the Eqn. (4), the following relation will be obtained:

$$\frac{di_l}{dt} = \frac{V_1}{L_f}T_1 + \frac{V_1-V_2}{L_f}T_2 - \frac{V_2}{L_f}T_3 - \frac{R_f}{L_f}i_l - \frac{V_L}{L_f} \tag{11}$$

Considering the average model based on the following equation, the average model of the PUC inverter can be attained using relations (9) and (11):

$$\frac{dX}{dt} = A(x,t) + B(x,t)U + C(t) \tag{12}$$

By choosing the state variables as $x_1 = i_1$ and $x_2 = V_2$ and using duty cycles (d_1, d_2, d_3) of switch (T_1, T_2, T_3) as input matrix, the following state space average model of the PUC inverter is derived.

$$A(x,t) = \begin{bmatrix} \frac{-R_f}{L_f} \\ 0 \end{bmatrix}, \qquad B(x,t) = \begin{bmatrix} \frac{-V_1}{L_f} & -\frac{V_1-V_2}{L_f} & \frac{-V_2}{L_f} \\ 0 & \frac{-i_l}{C} & \frac{i_l}{C} \end{bmatrix}$$

$$d = \begin{bmatrix} d_1 \\ d_2 \\ d_3 \end{bmatrix}, \qquad C(t) = \begin{bmatrix} \frac{-V_l}{L_f} \\ 0 \end{bmatrix}$$

The above state space model can be used to design the controller to regulate the capacitor voltage to a particular value to get the desired multilevel output voltage [12]. This state space model has been used to design the voltage controller in seven level and fifteen level PUC inverter [13].

For five level operation, the capacitor voltage has to be maintained at ½ of the main DC link voltage, but fortunately five level PUC has self-balancing capacitor voltage feature. This balancing phenomenon can be proved by energy storing in positive and negative cycle of PUC inverter operation [14]. Five

from table 2. When the capacitor is connected to the source it get charging and the capacitor is charged (states 2 & 7). When the capacitor is connected to the load it feeds the power, means it is discharging (states 3 & 6).

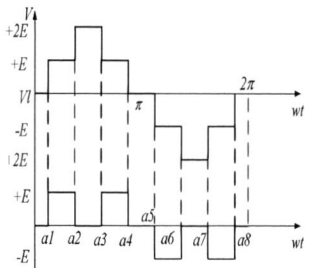

Figure 2: Output voltage wave for PUC-5 inverter

Table 2: Capacitor states

State	Cap. Voltage
1	No effect
2	charging
3	discharging
4	No effect
5	No effect
6	discharging
7	charging
8	No effect

Fig. 2 shows one cycle of the output voltage of PUC-5 waveform. V_E is a part of output voltage generated by the capacitor $(+E \; or - E)$) whether connected to load with series of main DC link voltage in charging or alone with the capacitor with load in discharging. It is assumed that output voltage and current is a sine function:

$$V_l = V_m \sin(\omega t) \tag{13}$$

$$i_l = I_m \sin(\omega t) \tag{14}$$

Where, Vm and Im are the peak value of output voltage and current. And ωt is the phase difference between output voltage and current, energy storing or releasing to the load by DC capacitor is given by Eqn. (15):

$$I = \frac{dq}{dt}$$

$$\rightarrow dU = Vdq = VIdt \tag{15}$$

Where, I, q, V and U shows the current flowing through capacitor, electronic charge voltage and energy stored by capacitor, respectively, substituting Eqn. (13) and (14) into Eqn. (15), energy sored in positive half cycle we get,

$$U^+ = \int_0^\pi V_E I_m \sin(\omega t - \theta_0) \, d(\omega t)$$

$$= I_m \int_0^\pi V_E \sin(\omega t - \theta_0) d(\omega t)$$

$$= I_m[\int_0^{a1} 0 * \sin(\omega t - \theta_0)d(\omega t) + \int_{a1}^{a2} E * \sin(\omega t - \theta_0) d(\omega t) + \int_{a2}^{a3} 0 * \sin(\omega t - \theta_0) d(\omega t) + \int_{a3}^{a4} E * \sin(\omega t - \theta_0) d(\omega t) + \int_{a4}^{\pi} 0 * \sin(\omega t - \theta_0)d(\omega t)]$$

$$= -EI_m \cos(\omega t - \theta_0) \begin{Bmatrix} a2 \\ a1 \end{Bmatrix} - EI_m \cos(\omega t - \theta_0) \, \mathrm{I}_{a3}^{a2}$$

$$= EI_m[\cos(a_1 - \theta_0) - \cos(a_2 - \theta_0) + \cos(a_3 - \theta_0) - \cos(a_4 - \theta_0)] \tag{16}$$

Similarly in negative half cycle

$$U^- = \int_\pi^{2\pi} V_E I_m \sin(\omega t - \theta_0) d(\omega t)$$

$$= I_m \int_\pi^{2\pi} V_E \sin(\omega t - \theta_0) d(\omega t)$$

$$= EI_m \cos(\omega t - \theta_0) \, \mathrm{I}_{a5}^{a6} \; + EI_m \cos(\omega t - \theta_0) \, \mathrm{I}_{a7}^{a8}$$

$$= [\cos(a_6 - \theta_0) - \cos(a_5 - \theta_0) + \cos(a_8 - \theta_0) + \cos(a_8 - \theta_0)] \tag{17}$$

It is cleared form Eqn. (16) and Eqn. (17) that, both equations are same but have opposite sign this means in positive and negative cycle, storing and releasing energy is same.

$$\left. \begin{aligned} a_5 &= \pi + a_1 \\ a_6 &= \pi + a_2 \\ a_7 &= \pi + a_3 \\ a_8 &= \pi + a_4 \end{aligned} \right\} \tag{18}$$

Hence,

$$U^+ = U^- \tag{19}$$

This means that the capacitor stored energy in a complete cycle is balanced and constant which keeps the capacitor voltage fixed at the 1/2 of the DC link voltage level in all conditions.

III. Modulation Schemes

In discontinuous PWM, the modulating signal is the sum of the sinusoidal fundamental and zero sequence injection signals, the later comprises a portion of the difference between the dc rails and envelop of the three phase sinusoidal references [15]. The zero sequence is combined to each three references signals to generate discontinuous PWM modulating signals as shown in Fig.3.

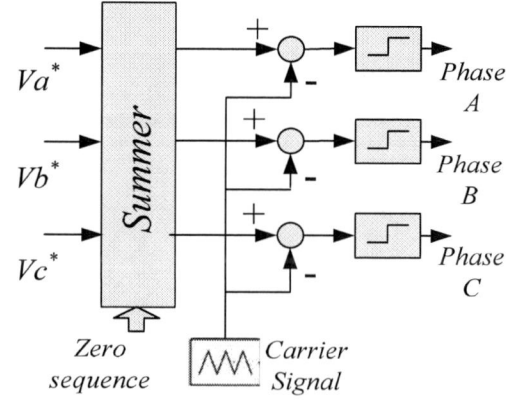

Figure: 3: General block diagram of DPWM generation

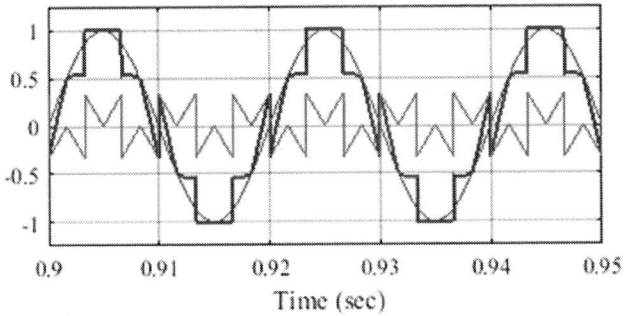

Figure 4: DPWM level-1 scheme

The generation principle of level-1 DPWM is shown in the Fig. 4. The dark blue line shows the resultant modulating signal.

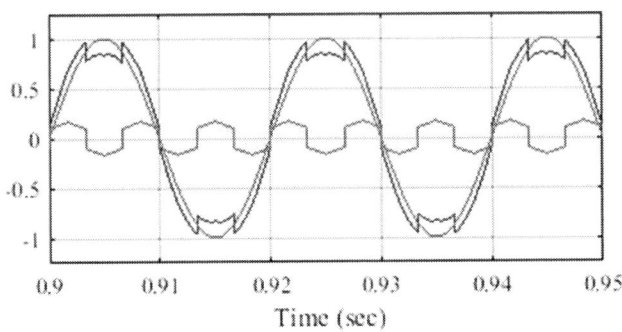

Figure 5: DPWM level-3 scheme

Level-3 DPWM scheme is shown in the Fig. 5. The modulating signal is compared with the triangular carrier wave and resulting switching signals are applied to the IGBT switches.

IV.   SIMULATION RESULTS

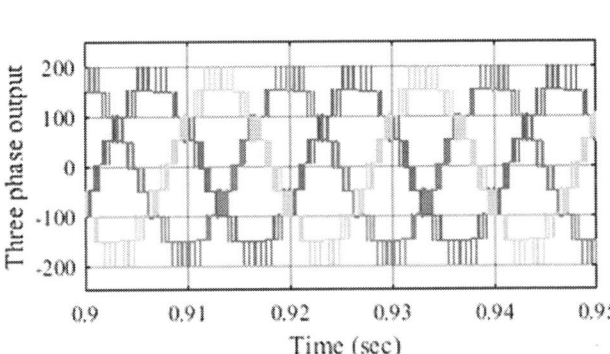

Figure 6: Three phase output in DPWM level-1 scheme

Fig. 6 shows the three phase output line voltage waveform of PUC five level inverter under DPWM level-1 scheme, line voltage has 9-level output. THD analysis shows in Fig. 7 which is 17.70% in line voltage. The peak amplitude of the rms value of the fundamental is 160.5 V. Common mode voltage is the voltage difference between the neutral point power source and the neutral point of a three-phase load. For a balanced source the common mode voltage is zero. But voltage source inverter does not constitute the ideal balanced source. This can be defined by:

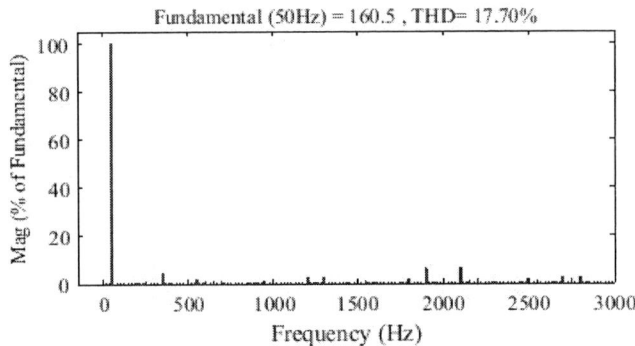

Figure 7: FFT analysis of THD in DPWM level-1 scheme

$$V_{cm} = \frac{V_{an}+V_{bn}+V_{cn}}{3} \qquad (20)$$

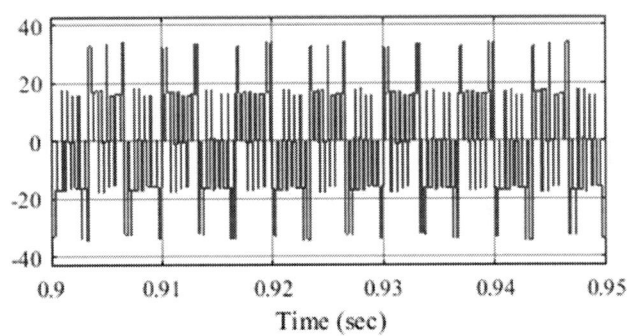

Figure 8: Common mode voltage in DPWM level-1 scheme

Fig. 8 shows the common mode voltage. The peak amplitude of the CMV is the 34 Volt approximately.

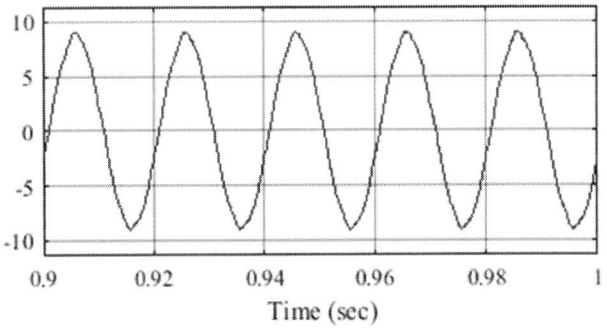

Figure 9: Load current in DPWM level-1 scheme

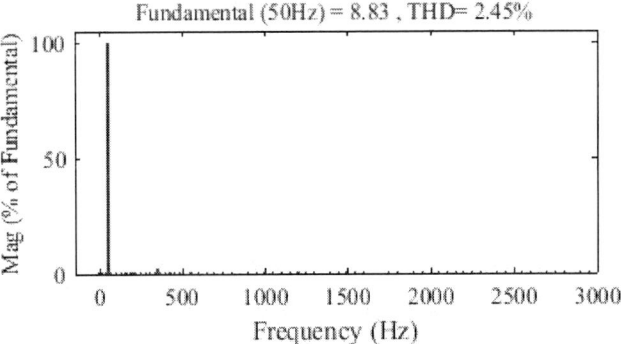

978-1-5386-6624-1/18 $31.00 © 2018 IEEE   639

Fig. 9 and Fig. 10 shows the load current and its THD. Load current is 6.25 A rms and THD is 2.45%.

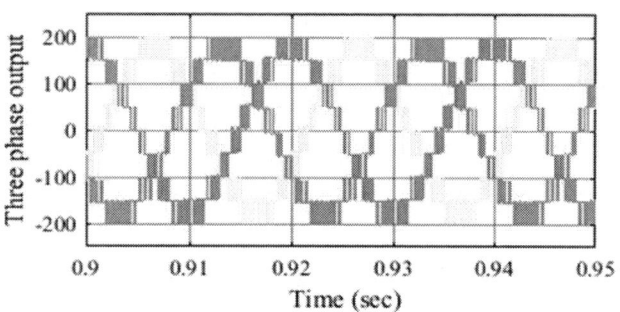

Figure 11: Three phase output in DPWM level-3 scheme

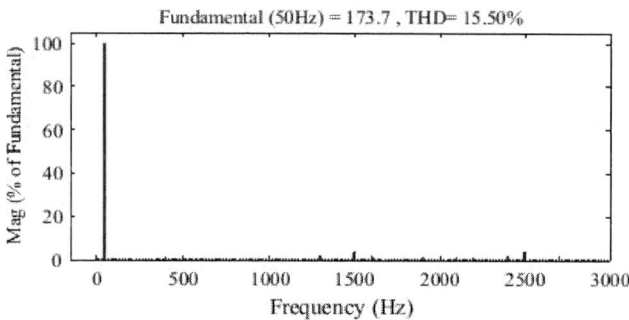

Figure 12: FFT analysis of THD in DPWM level-3 scheme

Fig. 11 shows the three phase line voltage of the PUC five level inverter under DPWM level-3 scheme. THD analysis of the above modulation scheme is shown in the Fig. 12. THD analysis shows 15.50% in line voltage. The peak value of the rms value of the fundamental value is 173.3 V.

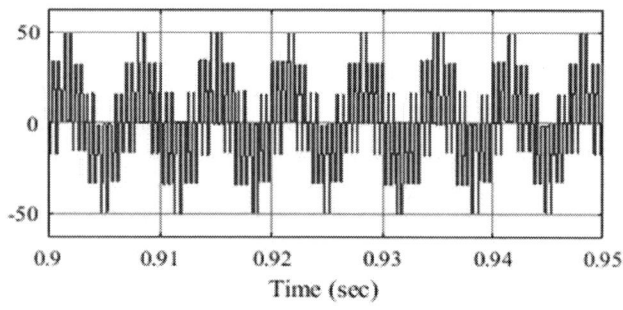

Figure 13: Common mode voltage in DPWM level-3 scheme

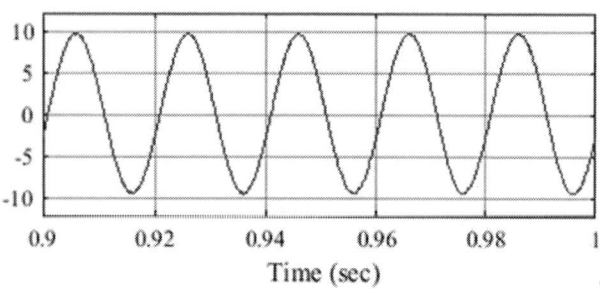

Figure 14: Load current in DPWM level-3 scheme

Figure 15: FFT analysis of load current

Fig. 13 shows the common mode voltage. The peak amplitude of the CMV is the 50 Volt approximately. Fig. 14 and Fig. 15 shows the load current and THD respectively.

Table 3: Simulation parameters

Particulars	Values
DC bus voltage	100 V
Auxiliary capacitor	4.7 mF
Switching frequency	2 kHz
Fundamental frequency	50 Hz
AC load	10 Ω, 10 mH

Discussion on results:

Table 4: Result table

Particulars	DPWM level-1	DPWM level-3
% THD in voltage	17.70%	15.50%
% THD in current	2.45%	0.98%
Peak value of CMV	34 V	50 V

Fig. 6-15 shows the results of an RL load fed by three phase five level PUC inverter. From the results it can be observed that the THD in line voltage is reduced in Discontinuous PWM level-3 but the peak value of common mode voltage is 50 V which is higher than the DPWM level-1.

V. HARDWARE MODEL AND FUTURE WORK

A prototype hardware model is developed to validate the single phase PUC-5 inverter. Three phase inverter hardware designing is under progress. A prototype circuit model in shown in the Fig. 16. The hardware constitute the following elements:

Table 5: Hardware specification

S. No.	Particulars	Model No.	Ratings
1	IGBT	FGA25N120ANTD	1200V, 25 A
2	Optocoupler	TLP250	
3	Adapter	-	12V, 1 A
4	DC source	RIGOL DP-832	30V, 3A
5	Capacitor	-	4.7 mF
6	DSP board	TMS320F28335	150 MHz

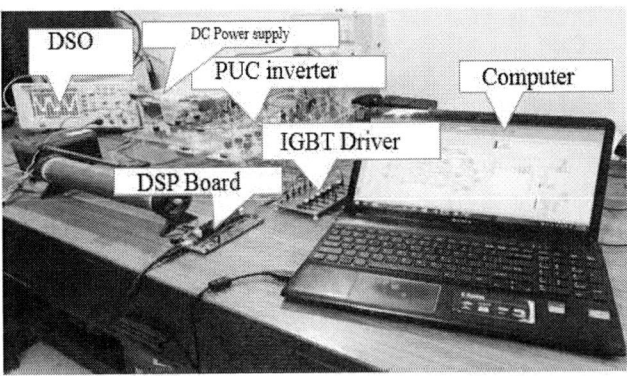

Figure 16: Experimental prototype setup

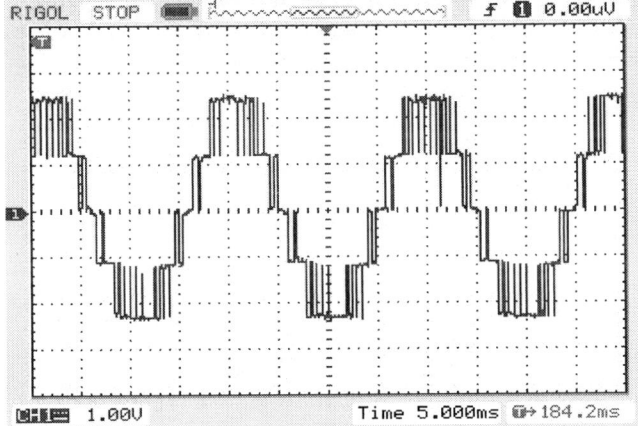

Figure 17: Output result (waveform is scaled by 10)

Experimental result for single phase PUC-5 inverter is shown in Fig. 17, the waveform is scaled by a factor of 10.

VI. CONCLUSION

The aim of this paper was to analyze the performance of discontinuous PWM scheme on PUC-5 inverter. In case of DPWM level-3 line voltage THD was 15.50% and 0.98 load current which was lower than the DPWM level-1 scheme, which was 17.70 % in line voltage and 2.45 % in load current. DPWM level-3 has given better performance in terms of power quality. Common mode voltage (peak value) was 50 V in level-3 scheme which was higher than the level-1 DPWM scheme.

REFERENCES

[1] J. Rodríguez, J. S. Lai, and F. Z. Peng, "Multilevel inverters: A survey of topologies, controls, and applications," *IEEE Trans. Ind. Electron.*, vol. 49, no. 4, pp. 724–738, 2002.

[2] K. K. Gupta and S. Jain, "Comprehensive review of a recently proposed multilevel inverter," *IET Power Electron.*, vol. 7, no. 3, pp. 467–479, 2014.

[3] J. Rodriguez, S. Bernet, P. K. Steimer, and I. E. Lizama, "A survey on neutral-point-clamped inverters," *IEEE Trans. Ind. Electron.*, vol. 57, no. 7, pp. 2219–2230, 2010.

M. Prats, "The Age of Multilevel Converters Arrives," no. June, pp. 28–39, 2008.

[5] H. Vahedi, H. Y. Kanaan, and K. Al-Haddad, "PUC converter review: Topology, control and applications," *IECON 2015 - 41st Annu. Conf. IEEE Ind. Electron. Soc.*, pp. 4334–4339, 2015.

[6] H. Vahedi, A. A. Shojaei, L. Dessaint, and K. Al-haddad, "Reduced DC Link Voltage Active Power Filter Using Modified PUC5 Converter," *IEEE Trans. Power Electron.*, vol. PP, no. 99, 2017.

[7] H. B. Chandwani and M. K. Matnani, "A review of modulation techniques for hybrid multilevel inverter," *Proc. 2012 1st Int. Conf. Emerg. Technol. Trends Electron. Commun. Networking, ET2ECN 2012*, 2012.

[8] B. Wu, *High-Power Converters and ac Drives*, 2006th ed. Hoboken, NJ, USA: John Wiley & Sons, Inc., 2006.

[9] S. Arun Rahul, S. Pramanick, M. Boby, K. Gopakumar, and L. G. Franquelo, "Extended Linear Modulation Operation of a Common-Mode-Voltage-Eliminated Cascaded Multilevel Inverter with a Single DC Supply," *IEEE Trans. Ind. Electron.*, vol. 63, no. 12, pp. 7372–7380, 2016.

[10] M. T. Chebbah, H. Vahedi, and K. Al-haddad, "Real-Time Simulation of 7-Level Packed U-Cell Shunt Active Power Filter," pp. 1386–1391, 2015.

[11] M. Tariq, M. T. Iqbal, M. Meraj, A. Iqbal, A. I. Maswood, and C. Bharatiraja, "Design of a Proportional Resonant Controller for Packed U Cell 5 Level Inverter for Grid-Connected Applications," pp. 3–8, 2016.

[12] Y. Ounejjar, K. Al-Haddad, and L.-A. Gregoire, "Packed U Cells Multilevel Converter Topology: Theoretical Study and Experimental Validation," *IEEE Trans. Ind. Electron.*, vol. 58, no. 4, pp. 1294–1306, 2011.

[13] Y. Ounejjar and K. Al-Haddad, "Fourteen-band hysteresis controller of the fifteen-level packed U cells converter," *IECON Proc. (Industrial Electron. Conf.*, pp. 475–480, 2010.

[14] H. Vahedi, P. A. Labbé, and K. Al-Haddad, "Sensor-Less Five-Level Packed U-Cell (PUC5) Inverter Operating in Stand-Alone and Grid-Connected Modes," *IEEE Trans. Ind. Informatics*, vol. 12, no. 1, pp. 361–370, 2016.

[15] K. Li, M. Wei, C. Xie, F. Deng, J. M. Guerrero, and J. C. Vasquez, "Triangle Carrier based Discontinuous PWM for Three-Level NPC Inverters," *IEEE J. Emerg. Sel. Top. Power Electron.*, pp. 1–1, 2018.

2nd IEEE International conference on power Electronics, Intelligent Control and Energy systems (ICPEICES-2018)

Grid Interactive MISO Converter Based PV System

Anuradha Tomar
Electrical Engineering Department
JSS Academy of Technical Education,
Noida, India

Sukumar Mishra
Electrical Engineering Department
Indian Institute of Technology,
New Delhi, India.

Abstract— **Standalone PV microgrids requires either alternative energy source or Energy storage system which can act as an alternative to overcome the limitation of PV due to its intermittent nature. Further, when various PV modules are connected in series and/or parallel configurations to fulfill the required voltage/current levels, resultant PV system suffers from the problem of mismatching/partial shading. To address these limitations, in this work, a Multi-input Single-output (MISO) DC-DC converter based grid interactive PV microgrid is proposed. As the proposed system is grid interactive in nature so, it does not required any battery storage system and thus reduce the cost of installation and maintenance cost of PV system. Application of MISO converter enables proposed system to solve the problem of partial shading and thus enables the system to operate at MPPT. Performance of proposed system is verified through MATLAB simulation, however an experimental verification is required further in order to derive on a strong conclusion.**

Keywords— *Grid interactive, Multi-input single-output DC-DC converter (MISO), Microgrid, Photovoltaic (PV) system, Offgrid Standalone PV system.*

I. INTRODUCTION

Now a day's Photovoltaic (PV) systems are being considered as feasible solutions to fill the demand supply gap of electricity that too with green and clean energy. However, its wide practical implementation is limited by its high installation cost. Therefore, efficient design of PV system is an important concern. Further, PV systems suffers from the problems like partial shading/mismatching which may occur due to various reasons like clouds, trees/buildings, dust, ageing, inclination angle, variations in specifications and types of PV modules connected together.

To limit the dependency of PV systems on storage devices and avoiding requirement of additional margin in capacity of PV installations, grid interactive nature of PV systems is desirable. Reactive power variation method was applied to determine the value of reactive power required for injection in PV based system and effect of variation in frequency for different operational modes were analyzed [1]. With the objective to improve system stability, performance and reliability in [2], an optimized reactive power compensation algorithm (RPCA) was proposed. A hybrid multilevel inverter topology based on neutral point clamp and cascaded H-Bridge topologies was presented in [5]. To optimize the selection of

DC bus capacitor rating with less ripples, highest possible voltage is chosen at DC bus. An energy management system for rooftop PV is developed in [7] by integrating uninterruptible power supply. Based on available PV power and load demand, proposed control is able to select different modes of operation and thus maintaining desired system parameters.

A sliding mode control (SMC) was proposed for a grid interactive PV system consisting of boost converter control to maintain maximum power point tracking (MPPT) operation and inverter control to maintain desired voltage, frequency was presented in [3]. Cascaded control was applied keeping current and voltage in inner loop and outer loop respectively. Further, response of the proposed system under partial shading scenario is need to explore [3]. To enhance the performance of the system, a fuzzy logic based MPPT control is proposed for a grid interactive single phase PV system. Proposed system is capable of supplying both active, reactive power during day and only reactive power during night hours [6].

To detect islanding mode of operation a new technique based on negative sequence voltage variations at point of common coupling (PCC) was proposed in [4]. However, for voltage source converter P-Q control is applied. Proposed method is able to maintain DC link voltage constant, but implementation of proposed method for application in distributed generation is required further.

In a grid interactive PV microgrid, various PV modules are required to be connected in series and/or parallel configurations depending upon the desired level of current and voltage. For extraction of maximum possible PV power MPPT operation is must and this will get halted if any kind of mismatching/partial shading occurs in PV modules. Therefore there is a need of mechanism which enables to operate system always as MPPT. Application of Multi-input Single-output (MISO) DC-DC converter enables to address this limitation [8]. Techno-economical analysis in carried out in [9] proves that addition of MISO DC-DC converter is feasible.

This work made the following contributions to address above discussed limitations:

- Proposed system is capable of handling partial shading/mismatching issues and thus ensures MPPT operation.

978-1-5386-6624-1/18 $31.00 © 2018 IEEE   642

2nd IEEE International conference on power Electronics, Intelligent Control and Energy systems (ICPEICES-2018)

- Grid interactive control eliminates the need of Battery energy storage (BES). Thus, reduced installation and maintenance system cost.
- Elimination of bypass and blocking diodes along with application of MISO converter; ensures extraction of best possible power from PV modules under partial shading/mismatching scenarios.
- Simplified inverter control enables switching of PV system from off-grid to on-grid mode and vice-versa, according to the availability of PV power and load demand.

This paper is structured as follows: System description and operation of proposed system along system specifications is included in Section II. Control of MISO DC-DC converter for MPPT operation along with inverter control for selection operational mode of proposed system is presented in Section III. In order to verify the performance of PV system, proposed system is simulated in MATLAB/Simulink environment and results thus obtained are included in section IV. At last section V concludes the main findings of the work with a light on future perspectives.

II. SYSTEM DESCRIPTION & OPERATION

Proposed grid interactive MISO converter based PV standalone system is represented in Fig. 1. Proposed system consists of PV source, MISO DC-DC converter and 3-phase inverter, filter and 3-phase load along with two controllers to maintain flow of power from PV arrays to DC bus and from DC bus to 3-phase load. Three PV modules as per specifications mentioned in Table I are connected in series to form one PV array and such three PV arrays are connected in parallel to form one PV unit (2.745 kW). Total PV rating of considered microgrid is 10.980 kW (2.745*4) at STC. These four PV units are connected to the DC bus through MISO converter to the DC bus.

Connection of PV arrays through MISO converter helps in eliminating mismatching losses that may cause if PV arrays were connected to the DC bus directly. MISO DC-DC converter ensures operation of PV modules at MPPT under all working conditions by controlling the switching of MISO converter switches (S_{w1}-S_{w6}). A 3-phase inverter is connected with the DC bus to the input side and its output is connected to the load though filter. This PV standalone system is connected to the utility grid at the point of common coupling. Inverter control enables transfer of power from DC bus to 3-phase load maintaining a 3-phase sinusoidal voltage with desired magnitude and frequency.

PV modules needs to be operate at MPPT for extraction of maximum possible power. However, MPPT operation fails to continue under mismatching/partial shading scenario and this non-MPPT operation results as an additional power loss to the system. Various complex algorithms are available in literature which ensures MPPT operation under all operating conditions,

but practical implementation of these algorithms is still limited.

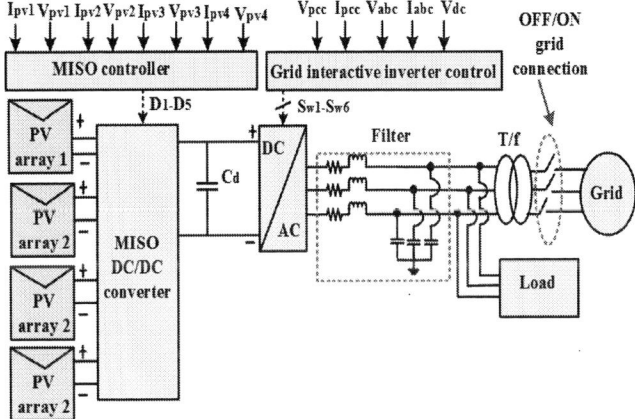

Fig. 1 Schematic for Grid Interactive PV system with Moisture Control

TABLE I. SPECIFICATION OF EACH PV MODULE

MPPT Power, P_{mppt} (W)	305
MPPT voltage, V_{mppt} (V)	54.7
Open circuited voltage V_{oc} (V)	64.2
MPPT Current, I_{mppt} (A)	5.58
Short circuit current, I_{sc} (A)	5.96
Maximum power of each PV array; P_{pv_mppt} (W)	305*9= 2745
Rated MPPT PV power of system at STC (kW)	2.745*4 = 10.980

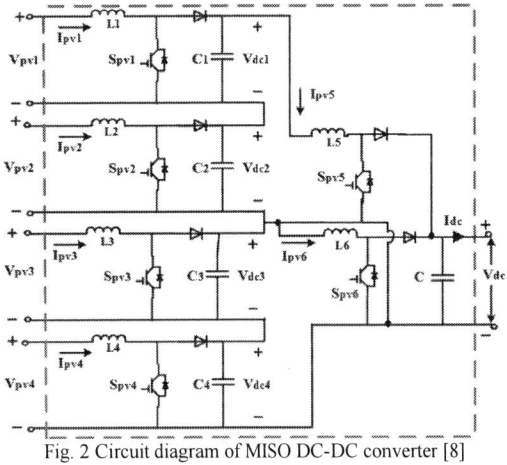

Fig. 2 Circuit diagram of MISO DC-DC converter [8]

TABLE II. PARAMETER VALUES OF MISO DC-DC CONVERTER [8]

Inductor, L_1= L_2= L_3= L_4 (mH)	5
Inductor, L_5 (µH)	312
Internal resistance of L_5 (Ω)	0.0005
Inductor, L_6 (µH)	700
Internal resistance of L_6 (Ω)	0.0005
Capacitor, C_1= C_2=C_3=C_4 (µF)	800
Capacitor, C (µF)	5000
I_{pv1}= I_{pv2}= I_{pv3}= I_{pv4} (A) (at Standard test condition)	16.74
Switching frequency, f_{sw} (kHz)	10

978-1-5386-6624-1/18 $31.00 © 2018 IEEE

2nd IEEE International conference on power Electronics, Intelligent Control and Energy systems (ICPEICES-2018)

In this work implementation of MISO DC-DC converter is considered. MISO DC-DC converter eliminates the impact of partial shading and ensures MPPT operation under all operating conditions [8]. Phenomena of partial shading may take place due to various factors like variation in PV module specifications, partial shading, difference in inclination angle, manufacturing issue etc. Configuration of the proposed MISO converter is given in Fig. 3. V_{pv1}, V_{pv2}, V_{pv3}, V_{pv4} and I_{pv1}, I_{pv2}, I_{pv3}, I_{pv4} are the voltage across each PV unit and current through the each PV unit respectively. Four PV units are connected to the MISO converter and single output of MISO converter is connected to the common DC bus, thus resulting in MISO DC-DC converter.

Output terminals of the MISO DC-DC converter is connected to the input of 3-phase inverter, to fed the PV power to the load. Perturb & Observe (P&O) based MPPT algorithm is implemented at MISO converter to ensure extraction of maximum power. Duty ratios D_1, D_2, D_3, D_4 are the outcome of P&O algorithm and with the help of Pulse width modulation (PWM), switching signals for S_{w1}, S_{w2}, S_{w3}, S_{w4} S_{w5} and S_{w6} are further obtained.

III. PROPOSED CONTROL STRUCTURE

This section consists of description of control structure for the proposed system. Proposed system's control architecture consist of two types of control i.e. First is MISO DC bus control (Fig. 3) to ensure MPPT operation for extraction of maximum power from PV modules and second is Inverter control (Fig. 4) to maintain desired 3-phase voltage and frequency. MISO converter control implements P&O algorithm for MPPT operation. In Fig. 3 V_{pv1}, V_{pv2}, V_{pv3}, V_{pv4}, V_{pv5}, V_{pv6} and I_{pv1}, I_{pv2}, I_{pv3}, I_{pv4}, I_{pv5}, I_{pv6} represents voltage and current for PV units 1, 2, 3 and 4 respectively. I_5 and I_6 are the intermediate currents as shown in Fig. 3.

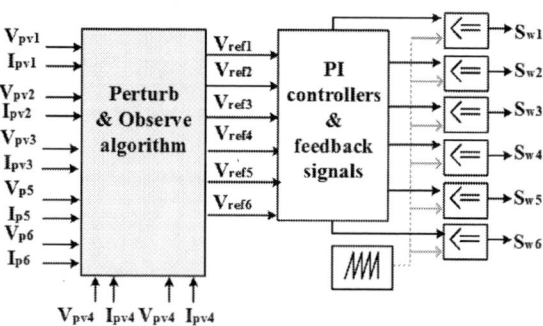

Fig. 3 MISO DC-DC converter control

The inverter control (Fig. 4) is designed considering the objectives of maintaining the system operation at desired voltage and frequency in both modes i.e. standalone and grid connected. Inverter Control has following objectives: (i) first to maintain desired voltage and frequency in standalone mode

and second (ii) is to maintain flow of active and reactive power in grid connected mode. Fig. 5 represents the flowchart for selection of operational mode of system i.e. grid connected mode or standalone mode. Here P_{vest} in Fig. 5 represents estimated PV power and is obtained using irradiance values at that time instant or one can use look up table as provided by manufacturer. If $P_{vest} > P_{load}$ then system will work in standalone mode or vice-versa. Accordingly status to flay (S_W) will be assigned.

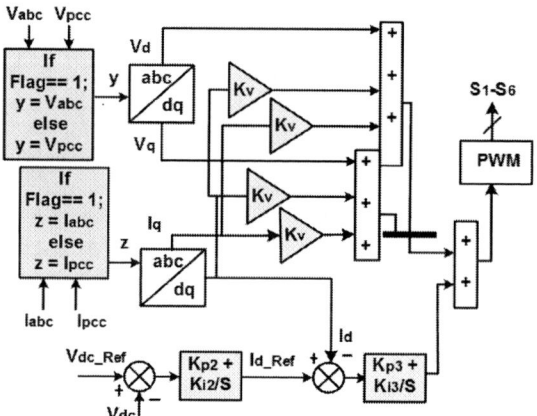

Fig. 4 Grid interactive inverter control

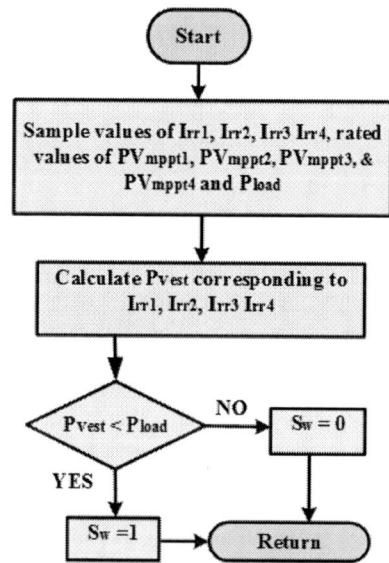

Fig. 5 Flowchart for selection of mode of operation

IV. RESULT & ANALYSIS

Proposed grid interactive MISO converter based PV microgrid scheme as per Fig. 1, has been simulated in MATLAB/Simulink environment. Two conditions are considered to analyze the system performance in the steady

2nd IEEE International conference on power Electronics, Intelligent Control and Energy systems (ICPEICES-2018)

state and dynamic conditions and responses has been recorded (Fig. 6-18).

It is assumed that initially all PV units are exposed to different irradiance as shown in Fig. 6 and at t = 12 S, irradiance suddenly rises to 1000 W/m². Considered values of irradiance in Fig 6 represents a scenario of mismatch/partial shading i.e. before t = 12 S PV unit 1, PV unit 2, PV unit 3 and PV unit 4 are exposed to 200, 250, 280, and 230 respectively. Connected 3-phase load is considered as 10.9 kW and is assumed as constant throughout the simulation.

PV power being extracted through all four PV units is depicted in Fig.7. All PV units are operating at their respective maximum power point in both situations i.e. under non-uniform irradiance (as per Fig. 6) and uniform irradiance of 1000 W/m². Thus, application of MISO DC-DC converter results in increased extraction of power as it enables PV modules to operate at MPPT.

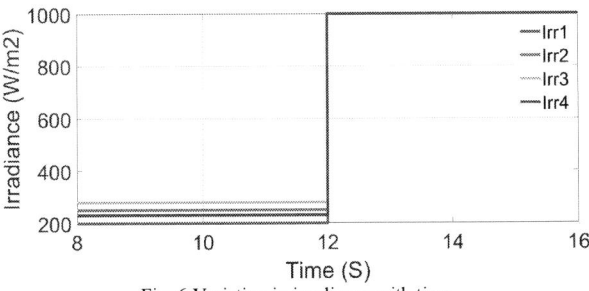
Fig. 6 Variation in irradiance with time

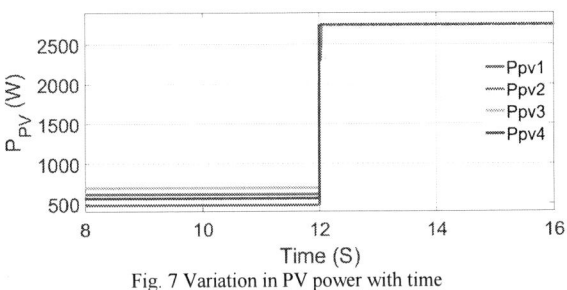
Fig. 7 Variation in PV power with time

Fig. 8 shows that before t = 12 S, system is operating in grid connected mode as generated PV power at DC bus (Fig. 11) is insufficient to fulfill the load demand (load is considered as 10.9 kW). This imbalance of demand and supply power is detected by the system control and as a result status of flag becomes one (Fig. 8). After t = 12 S, available PV power increases with increase in irradiance and becomes sufficient to fulfill the load demand. At this instant, flag status becomes zero which means that system is now operating in standalone mode (Flag = '1' for grid connected and '0' for standalone mode). Response of voltage and current at DC bus. PCC and grid to this change in flag status at t = 12 S is shown

in Fig. 9-11 and Fig. 12-14 respectively. Fig. 15 represents the frequency response for the proposed system.

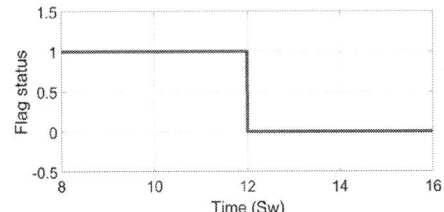
Fig. 8 Variation in flag status with time

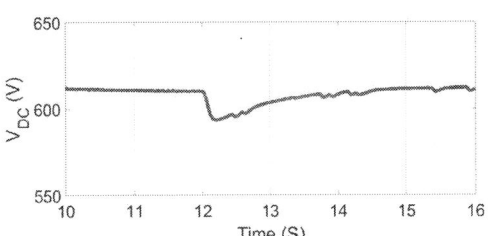
Fig. 9 Variation in DC bus voltage with time

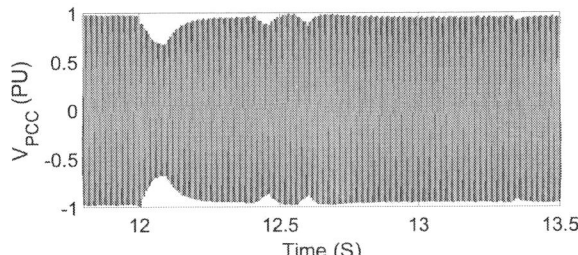
Fig. 10 Variation in voltage at PCC with time

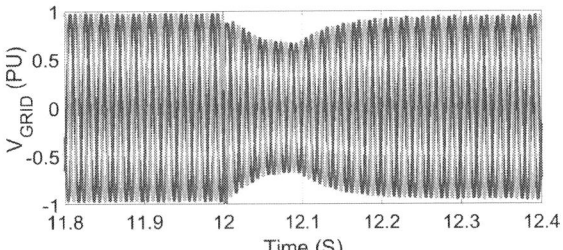
Fig. 11 Variation in grid voltage with time

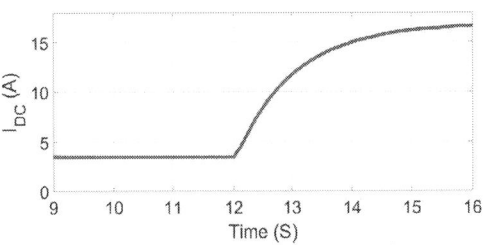
Fig. 12 Variation in DC bus current with time

Variation in power PV power along with power available at PCC and grid is depicted in Fig. 16. Before t = 12 S, as available PV power is less then load demand, additional

978-1-5386-6624-1/18 $31.00 © 2018 IEEE   645

2nd IEEE International conference on power Electronics, Intelligent Control and Energy systems (ICPEICES-2018)

required power is provided by Grid and thus able to cater the load demand. After t = 12 S, due increase in irradiance value, now available PV power is sufficient for meet the load demand and thus grid contribution becomes zero and system goes in standalone mode of operation. Fig. 17 and Fig. 18 shows the Total harmonic distortion (THD) values of voltage and current at PCC which are well within the limits.

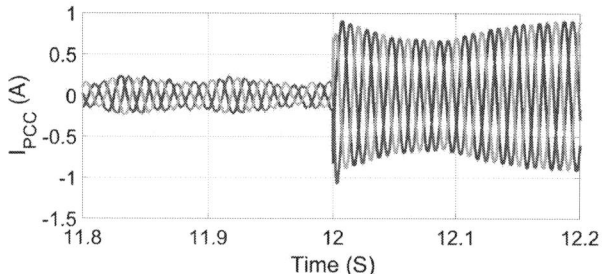

Fig. 13 Variation in current at PCC with time

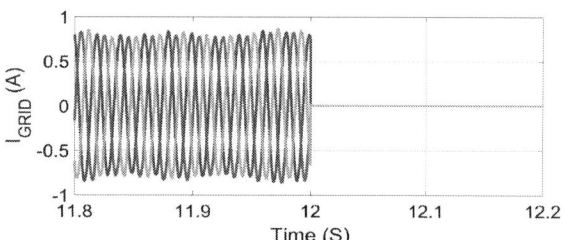

Fig. 14 Variation in grid current with time

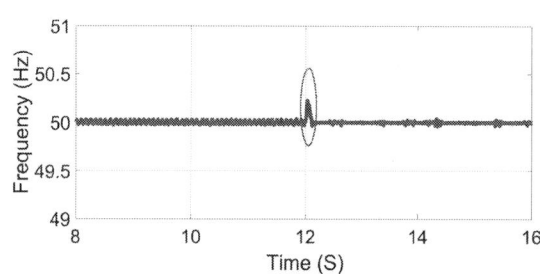

Fig. 15 Variation in frequency with time

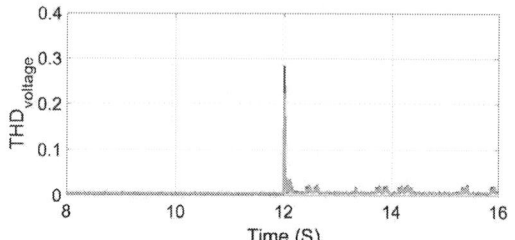

Fig. 17 THD values for voltage at PCC

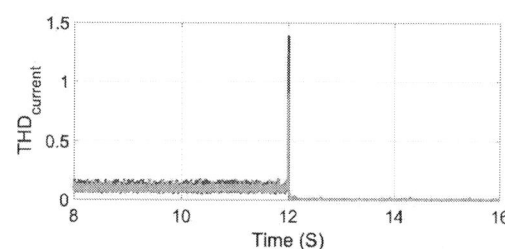

Fig. 18 THD values for current at PCC

V. CONCLUSION

In this paper a grid interactive MISO DC-DC converter based PV system is proposed. Application of MISO DC-DC converter in proposed system addressed the problem of mismatching/partial shading and thus increases the extracted power by enabling MPPT operation of the system. Grid interactive control proposed the work eliminated the need of BES and thus reduces the system implementation and maintenance cost. Verification of the proposed system is done using MATLAB/Simulink environment and satisfactory results are obtained. Further, an experimental verification is required in order to reach on a strong conclusion.

REFERENCES

[1] Gyu-Ha Choe, Hong-Sung Kim, Han-Goo Kim, Young-Ho Choi, Jae-Chul Kim, "The Characteristic Analysis of Grid Frequency Variation under Islanding Mode for Utility Interactive PV System with Reactive Power Variation Scheme for Anti-Islanding", 37th IEEE Power Electronics Specialists Conference, June 2006.

[2] Liming Liu, Hui Li, Yaosuo Xue, Wenxin Liu, "Reactive Power Compensation and Optimization Strategy for Grid-Interactive Cascaded Photovoltaic Systems", IEEE Transactions on Power Electronics, IEEE Transactions on Power Electronics, Vol. 30, Iss. 1, Pp. 188-202, Jan. 2015

[3] Ridha Benadli, Brahim Khiari, Anis Sellami, "Three-Phase Grid-Connected Photovoltaic System with Maximum Power Point Tracking Technique Based On Voltage-Oriented Control and Using Sliding Mode Controller", 6th International Renewable Energy Congress (IREC), 2015.

[4] Sheetal Chandak, Snehamoy Dhar, S K Barik, "Islanding Disclosure for Grid Interactive PV-VSC System using Negative Sequence Voltage", IEEE Power, Communication and Information Technology Conference (PCITC), Bhubaneswar, India, 2015.

[5] Sumit K. Chattopadhyay, Chandan Chakraborty and Bikash C Pal, "Cascaded H-Bridge & Neutral Point Clamped Hybrid Asymmetric Multilevel Inverter Topology for Grid Interactive Transformerless

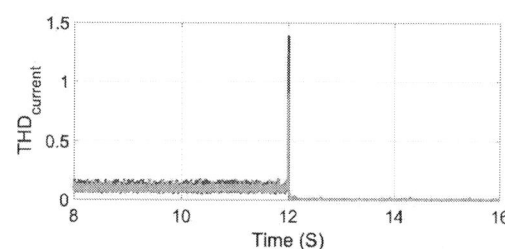

Fig. 16 Variation in power at PCC, load and PV

978-1-5386-6624-1/18 $31.00 © 2018 IEEE

2nd IEEE International conference on power Electronics, Intelligent Control and Energy systems (ICPEICES-2018)

Photovoltaic Power Plant", IECON 2012 - 38th Annual Conference on IEEE Industrial Electronics Society.

[6] Nilesh Shah, R. Chudamani, "Single-Stage Grid Interactive PV System Using Novel Fuzzy Logic Based MPPT with Active and Reactive Power Control", 2012 7th IEEE Conference on Industrial Electronics and Applications (ICIEA), Pp. 1667, 1972, July 2012.

[7] P. C. Kushwaha, C. N. Bhende, "Single–Phase Rooftop Photovoltaic based Grid–Interactive Electricity System", 2016 IEEE Annual India Conference (INDICON), 2016.

[8] Anuradha Tomar, Sukumar Mishra, Chandrashekhar N. Bhende, "Modified MISO DC-DC Converter Based PV Water Pumping System", DOI: 10.1109/POWERI.2016.8077193, IEEE PIICON-2016.

[9] Anuradha Tomar, Sukumar Mishra, Chandrashekhar N. Bhende, "Techno-Economical Analysis for PV based Water Pumping System under Partial Shading/Mismatching Phenomena", DOI: 10.1109/POWERI.2016.8077391, IEEE PIICON-2016, Nov. 2016.

2nd IEEE International conference on power Electronics, Intelligent Control and Energy systems (ICPEICES-2018)

Performance Comparison of Different Inverter Stages CS-VCO in 0.18µm CMOS Technology

Ashish Mishra
E.C.E. Department
S.R.M.G.P.C., Lucknow, India
ashishrohit08@gmail.com

Indu Prabha Singh
E.C.E. Department
S.R.M.G.P.C., Lucknow,
India
induprabhasingh@gmail.com

Abstract—**This paper presents a comparative study of two different topologies of Current Starved Oscillator (CS-VCO) i.e., (3-stage & 5-stage) on the basis of phase noise, power dissipation and centre frequency of oscillation parameters. With the observation, it is measured that the power dissipation can be scaled down to 28.53% for 3-stage w.r.t. 5-stage CS-VCO topology. In such manner, a design of optimal and robust current starved-VCO is achieved. The performance of circuit is simulated on 0.18µm CMOS Technology.**

Keywords—PLL, CMOS, CSVCO, P-Noise, PSS, LPF

I. INTRODUCTION

VCO is an important block for communication system. The designing of VCO incorporate several trade-offs among frequency of oscillation, power dissipation, phase noise and operating speed. So for a PLL design, it is important to focus upon these parameters. To design IC's, it is essential to obtain the VCO for low power dissipation. In this approach, an effective approach is considered to scale down the power dissipation by decreasing the supply voltage.

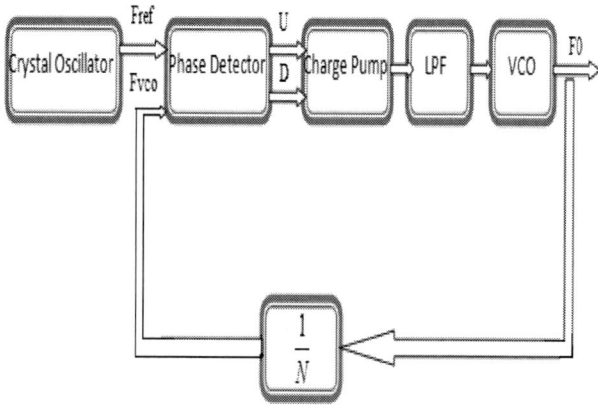

Fig.1. Basic Block Diagram of PLL

Phase Locked Loop (PLL) plays a vital role in many electronic communication based devices which consists of phase frequency detector (PFD), charge-pump (CP), low pass filter (LPF), voltage controlled oscillator (VCO) and a divide- by-N circuit as shown in fig.1 [1], [4].

VCO is the heart of PLL system which operates at high oscillation frequency. Proper design of oscillator is important for complete PLL design with reduced power. There are generally two types of voltage controlled oscillator (VCO) which is mostly used in recent technology: LC Tuned oscillator, ring oscillator or current starved VCO. Tuned oscillator results improved phase noise performance but covers large chip area because of an spiral winding of inductor which is exactly undesirable for design value. Also the frequency range is limited for LC-VCO while in case of CS-VCO, the phase noise result is not up to the mark compared to LC-VCO but the oscillation frequency range is much high and also the area consumed by CS-VCO is very small [3], [4]. The wide frequency range obtained from CS-VCO design is suitable to handle the variation in the process easily. The main aim to design of several stages CS-VCO is to scale down the power consumption and area occupied by the complete PLL and also to reduce the phase noise performance. The noise is also present in two form to a system: One is long term jitter which occurs due to deviation over time in the output clock edge to that of an ideal one which is exactly periodic one and another is periodic type jitter which is available due to variation over time in the output clock period [9], [5]. VCO is an essential block for several RF transceivers for frequency selection and generation of signal. PLL for RF transceivers demands programmable carrier frequency. PLL normally available with less precised oscillator in feedback which is controlled by the reference input voltage. For an ideal VCO, there is always a linear relationship between oscillation frequency and controlled voltage. Most of the application are based on tuned oscillator [12].

In this work, 3-stage and 5-stage CS-VCO is designed for two inverter stages. It is found that 3-stage CS-VCO occupies less power for its circuit operation as compared to other configuration. According to the relation (6), the oscillation frequency is inversely proportional to the number of inverter stages, it means as the number of inverter stage increases, oscillation frequency reduces. For 3-stage CS-VCO, the oscillation frequency is high as compared to 5-stage CS-VCO configuration. It is suitable to achieve high oscillation frequency with the reduction of inverter stages and also improved phase noise performance is obtained [15].

978-1-5386-6624-1/18 $31.00 © 2018 IEEE

648

This paper is divided into four sections: Section I: This section covers the abstract and introduction part which is already discussed, Section II: details about schematic of 3-stage and 5-stage, Section III: details about design equations used for sizing of transistors, In Section IV: simulation results are explained and finally the paper is concluded in last section V.

II. CIRCUIT DESCRIPTION

A. *Three-Stage Current-Starved VCO*

The schematic for 3-stage Current-Starved VCO is shown in fig.2. The working is almost similar to ring VCO. In such configuration, the MOSFETs ordered in the middle stage are responsible for several inverter stages, while top-most and bottom-most MOSFETs (M1) and (M4) of the oscillator circuit serves as current sources which limit the current available to the inverter or in other manner we can say that the inverter is starved for current. The drain current for MOSFETs (M0) and (M3) is same and is controlled by the input reference voltage. Currents in MOSFETs (M0) and (M3) are mirror image in each inverter/current source stages [4], [5], [15].

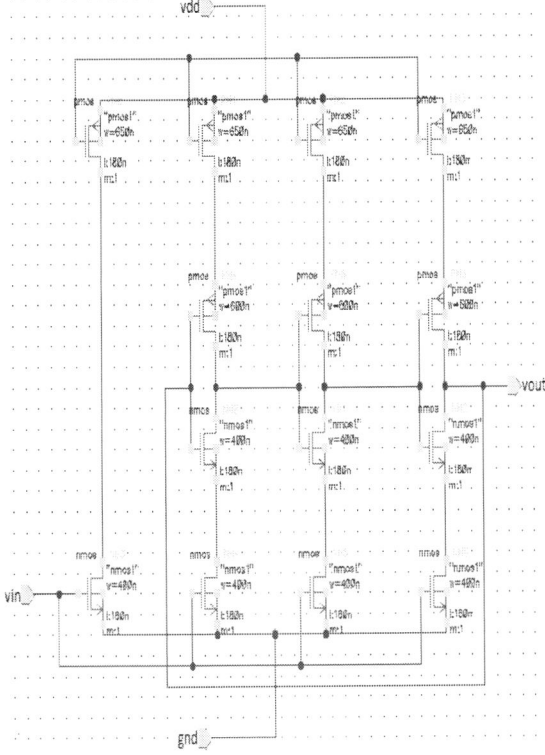

Fig.2. Designed Three-Stage Current-Starved VCO

Fig.2 depicts three-inverter stages that are cascaded. The applied input signal is Pulse-Wise Linear (PWL) to the test circuit of same.

B. *Five-Stage Current-Starved VCO*

Fig.3 is the schematic of five-stage CS-VCO. In this designed VCO, 5-stages of inverter are cascaded. The applied input is again same i.e., Pulse-Wise Linear (PWL) signal to the test circuit of same [4].

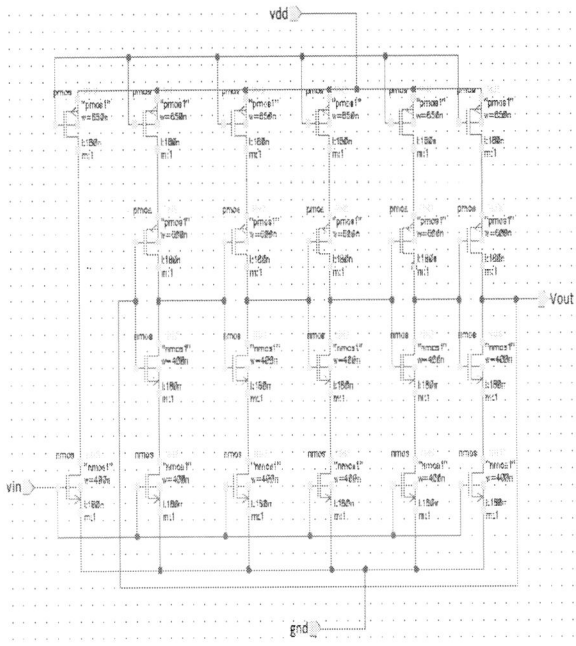

Fig.3. Designed Five-Stage Current-Starved VCO

III. DESIGN EQUATIONS FOR VCO

To evaluate the design parameters, transistor sizing is required initially. So W/L ratio for each CMOS transistor is calculated by the design equations as given below. For a simple single inverter stage circuit, the resultant capacitance on the drain terminal is calculated by [1], [6]:

$$C_{tot} = C_{out} + C_{in} = \frac{5}{2} C_{ox}(W_P L_P + W_n L_n) \qquad (1)$$

- Where (C_{ox}) is Gate Oxide Capacitance, can be calculated as,

$$C_{OX} = \frac{\in_r \, * \in_0}{t_{ox}} \qquad (2)$$

- Charging time acquired for capacitance (C_{tot}) to charge from 0 to V_{sp} with a constant current I_{D4} can be obtained as:

$$t_1 = C_{tot} * \frac{V_{SP}}{I_{D4}} \qquad (3)$$

- And time acquired for discharging (C_{tot}) from V_{DD} to V_{sp} with a constant current I_{D1} can be obtained as:

$$t_2 = C_{tot} * \frac{(V_{DD} - V_{SP})}{I_{D1}} \tag{4}$$

- Now total time for ($I_{D4} = I_{D1} = I_D$):

$$(t_1 + t_2) = C_{tot} * \frac{V_{DD}}{I_D} = \frac{1}{T_d} \tag{5}$$

- Oscillation frequency can be found as [8]:

$$f_{osc} = \frac{1 * T_d}{N} = \frac{I_D}{N * C_{tot} * V_{DD}} = \frac{V_{DD}}{2} \tag{6}$$

- Where time delay is denoted by (T_d). Here $f_{osc} = f_{center}$ if

$$V_{inVCO} = \frac{V_{DD}}{2}, I_D = I_{D\,center}$$

- Here (N) is inverter stage counting. Drain Current can be found as:

$$I_{Dcenter} = N * C_{tot} * V_{DD} * f_{center} \tag{7}$$

Also, $$I_{Dcenter} = \frac{\beta (V_{gs} - V_{thn})^2}{2} \tag{8}$$

- (β) can be calculated as:

$$\beta = \frac{K_p * W}{L} = \frac{\mu * C_{ox} * W}{L} \tag{9}$$

- The average power dissipation can be calculated as:

$$P_{avg} = V_{DD} * I_{avg} = V_{DD} * I_D \tag{10}$$

TABLE I. Sizing of the transistors for 3-stage and 5-stage CS-VCO

Devices	3-Stage CS-VCO (W/L)	5-Stage CS-VCO (W/L)
Current Starved (PMOS)	20	0.78
Current Starved (NMOS)	6.66	1.91
Inverter Stage (PMOS)	22.22	6.67
Inverter Stage (NMOS)	5.55	2.22

IV. SIMULATION RESULTS

To accomplish optimal performance for the circuit, it is required to observe different characteristics of VCO and parameters under different varying conditions. To obtain the characteristics, analysis as like Periodic Steady State (PSS), Phase Noise (P-NOISE) and Transient analysis are essential.

A. Transient Analysis

This analysis is done to perform time domain behavior of the signal.

B. PSS Analysis

PSS analysis is performed to observe the fundamental frequency, output power, tuning range, power dissipation etc. In this analysis, voltage of particular nodes is observed periodically so that time period can be calculated.

C. P-Noise Analysis

Phase noise is a form of flicker noise. It is measured in dBc/Hz. In oscillators, phase noise is the actual source of noise [10].

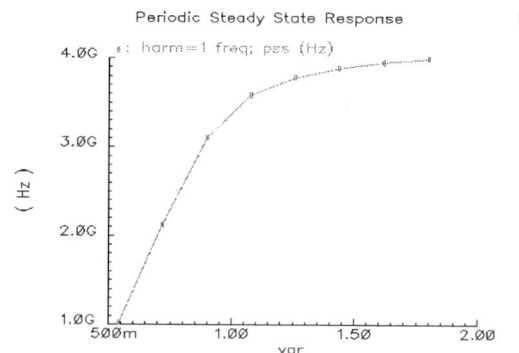

Fig.4. Tuning Frequency Vs Control Voltage Plot for 3-Stage CS-VCO

Fig.4 shows the plot of oscillation frequency variation w.r.t. the control voltage applied at input of 3-stage CS-VCO. It helps to find the range of frequency up-to which the plot is almost linear to change of controlled input voltage and the linearity is achieved in the range of (1.0229 GHz-3.1071 GHz).

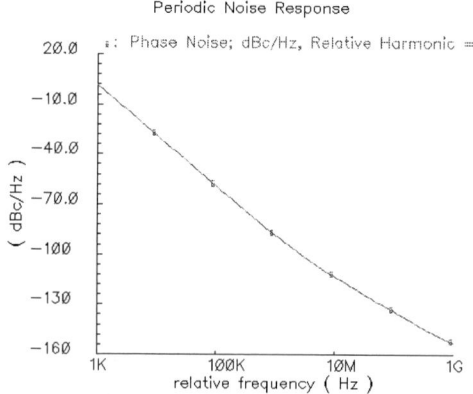

Fig.5. P-Noise Vs Offset Frequency for 5-Stage CS-VCO

Fig.5 represent the response of phase noise in dbc/Hz w.r.t. relative offset frequency for five-stage CS-VCO. The measurement result observe that the phase noise response is (-88.16)dBc/Hz @1MHz and (-113.01)dbc/Hz @10MHz offset frequency.

978-1-5386-6624-1/18 $31.00 © 2018 IEEE   650

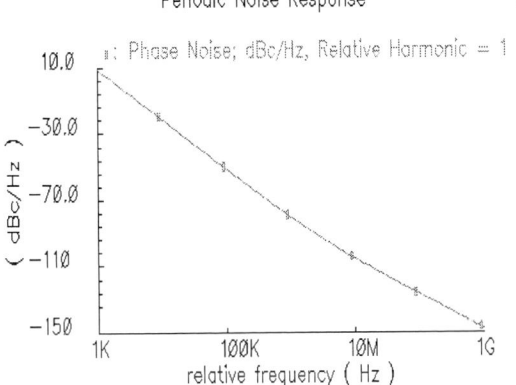

Fig.6. P-Noise Vs Offset frequency for 3-Stage CS-VCO

Fig.6 shows the phase noise plot for three-stage CS-VCO. The plot observes that the phase noise is (-80.17)dbc/Hz @1MHz offset and (-105.31)dbc/Hz @100MHz offset frequency.

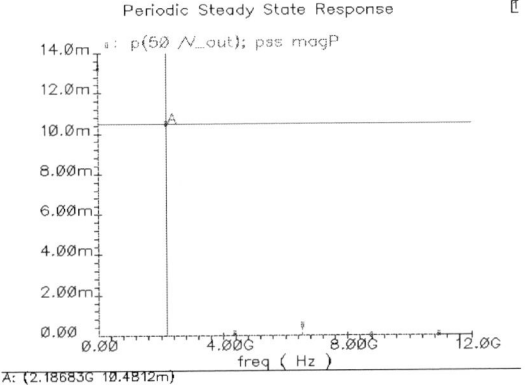

Fig.7. Power Dissipation Vs Frequency Plot for 5-Stage CS-VCO

Fig.7 is the plot for power dissipation in (mW) for 5-stage CS-VCO. The power dissipation observed is 10.4812 mW for the fundamental frequency of 2.1868 GHz.

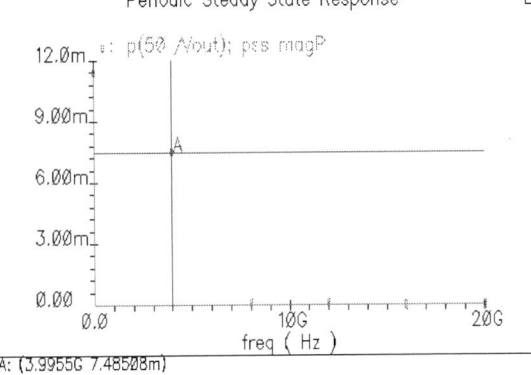

Fig.8. Power Dissipation Vs Frequency Plot for 3-Stage CS-VCO

Fig.8 is the plot for power dissipation in (mW) for 3-stage CS-VCO. By the plot, the dissipated power is 7.48508 mW for the fundamental frequency of 3.9955 GHz.

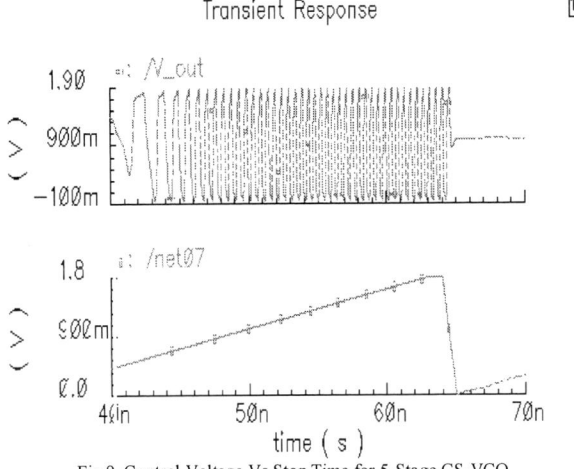

Fig.9. Control Voltage Vs Stop Time for 5-Stage CS-VCO

Fig.9 displays the transient response between control voltage and time for five-stage CS-VCO. The oscillations are performed for 100ns time period. The input applied is Pulse-Wise Linear (PWL). The plot shows that oscillations are initialized from 40.06ns.

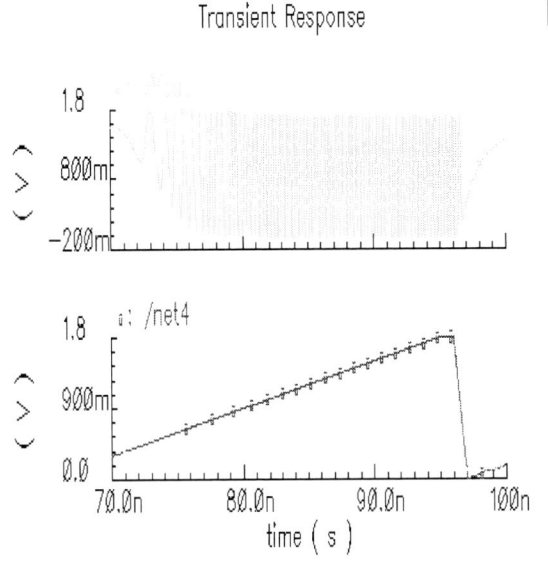

Fig.10. Control Voltage Vs Stop Time for 3-Stage CS-VCO

Fig.10 shows the transient response between control voltage Vs time for three-stage CS-VCO. The oscillations are observed for 100ns time period. The input is Pulse-Wise Linear (PWL). The plot depicts that oscillations are initialized from 72.213ns.

978-1-5386-6624-1/18 $31.00 © 2018 IEEE

TABLE II. Comparative study of different types of VCO design on different CMOS TECHNOLOGY.

Topology	Centre Frequency/ Operating Frequency Range (GHz)	VDD (Volt)	Power Dissipation (mW)	Phase Noise (dbc/Hz)
Ring VCO [2]	1.0	2.5	15.5	-106 @600 KHz, 900 MHz
Ring VCO [8]	866 MHz	3.3	24.5	-106@ 500KH
Ring VCO [3]	3.125	1.8	12.6	-91@ 1MHz
[10]	5.79	1.8	------	-99.5@ 1MHz
Differential Ring VCO [14]	1.41	1.8	12.5	-89.8@ 600KHz
CS-VCO (3-stage)	(1.0229-3.995)	1.8	7.49	-80.17@ 1MHz and -105.31@ 10MHz
CS-VCO (5-stage)	(437.47-2.187)	1.8	10.48	-88.16@ 1MHz and -113.01@ 10MHz

V. CONCLUSION

In this paper, the performance comparison of two different inverter stages CS-VCO is done evolved. The analysis concludes that 5-stage current starved VCO has better phase noise performance as compared to 3-stage CS-VCO while the power dissipation for 3-stage CS-VCO is less (7.49mW) as compared to 5-stage i.e. (10.48mW) CS-VCO. So the analysis depicts that 3-stage CS-VCO can be utilized for the purpose where less power dissipation is main aspect to the design while 5-stage CS-VCO can be adopted in the applications where better phase noise performance is the main concern. One more advantage with the 3-stage CS-VCO is that, it can be used for the applications works upon high oscillation frequency range such as Bluetooth, GSM and so on.

REFERENCES

[1] T.H. Lee and A. Hajimiri, " Oscillator Phase noise: A tutorial," IEEE J. Solid-State Circuits, vol. 35, pp. 326-336, March 2000.

[2] W. Shing, T. Yan and H.C. Luong, "A 900-MHz CMOS low-phase noise Voltage Controlled Ring Oscillator," IEEE Transactions on Circuits and System II: Analog and Digital Signal Processing, vol. 48, pp. 216-221, February 2001.

[3] A. Maxim, S. Baker, E. M. Schneider, M.L. Hagge, S. Chacko and D. Stiurca, "A low-jitter 125-1250 MHz process-independent and ripple-poleless 0.18um CMOS PLL based on a sample-reset loop filter," In IEEE J. Solid-State Circuits, vol. 41, pp. 1673-1683, Nov. 2001.

[4] R. J. Baker and D. E. Boyce, "CMOS Circuit Design, Layout, and Simulation," IEEE Press Series on Microelectronic Systems, 2002.

[5] B. Razavi, T.C. Lee, "A stablization technique for Phase-Locked Frequency Synthesizers," IEEE J. Solid State Circuits, vol. 38, pp. 888-894, June 2003.

[6] S. Kang and Y. Leblebici, "CMOS Digital Integrated Circuits: Analysis and Design," Tata McGraw-Hill Edition, 2003.

[7] D. P. Bautista and M. L. Arnada, "A low power and high speed VCO Ring-Oscillator," In Circuits and Systems, ISCAS' 04, vol. 4, pp. IV-752, May 2004.

[8] W. Xin, Y. Dunshan and S. Sheng, "A full swing and low power Voltage-Controlled Ring Oscillator," In Electron Devices and Solid-State Circuits, pp. 141-143, December 2005.

[9] H. Janardhan, M.F. Wagdy, "Design of a 1.0 GHz digital PLL using 0.18µm CMOS Technology" In null, pp. 599-600, April 2006.

[10] A. Bansal, Y. Zheng and C.H. Heng, "2.0 GHz CMOS noise-cancellation VCO," IEEE Asian Solid- State Circuits Conference, pp. 461-464, Nov 2008.

[11] F. Aznar, S. Celma and B. Calvo, "Inductorless AGC amplifier for 10G Base-LX4 Ethernet in 0.18µm CMOS," Electronics Letter, vol. 44, pp. 409-410, March 2008.

[12] L. S. Paula, S. Banpi, E. Fabris and A. A. Susin, "A wide band CMOS Differential Voltage-Controlled Ring Oscillator," Proc. of 2Ft Annual Symposium on Integrated Circuits and System Design, pp. 85-89, Sep. 2008.

[13] F. Aznar, S. Celma and B. Calvo, "A 0.18-µm CMOS 1.25 Gbps front end receiver for low-cost short reaches optical communications," Proceedings of the 36th European Solid-State Circuits Conference, pp.554-557, Sep. 2010.

[14] C. S. Azqueta, S. Celma and F. Aznar, "A 3.125-GHz, four stage Ring VCO in 0.18µm CMOS," In Circuits and Systems (ISCAS), vol. 44, pp. 1137-1140, May 2011.

[15] A. Mishra, G.K. Sharma, and D. Boolchandani, "Performance Analysis of Power Optimal PLL Design Using Five-Stage CS-VCO In 180nm," In Signal Propagation and Computer Technology (ICSPCT), pp. 764-768, July 2014.

Reweighted Sparse- Least-Mean Mixed-Norm Adaptive Control for Solar PV Integrated EV Charging Station

Anjeet Verma, *Member, IEEE* and Bhim Singh, *Fellow, IEEE*
Indian Institute of Technology, New Delhi, India
anjeet15@gmail.com

Abstract— Connecting a large number of electric vehicle (EV) to the distribution system, causes the unbalanced loading of the three phases and thereby resulting in the excessive neutral current to flow in the neutral wire of the transformer. To address these issues, in this paper, a solar photovoltaic (PV) array powered multifunctional EV charging station is proposed. The charging station provides the facility to connect multiple single phase EV chargers. Moreover, the charging station employs a voltage source converter (VSC) to interface the PV array and storage battery. The VSC of the charging station performs the multiple tasks such as, (1) extraction of maximum power of the PV array, (2) compensation of neutral current, (3) compensation of harmonics current and reactive power requirement of the EVs, (4) making the three phase source currents balanced and sinusoidal, (5) improvement in the power quality of source voltage and current, and (6) dynamic power management. To achieve all these functionalities, the VSC of the charging station, is controlled using the reweighted sparse least mean mixed norm (RS-LMMN) based adaptive control algorithm. The proposed control algorithm utilizes the benefits of both LMS, and LMF based adaptive control algorithm on minimizing the steady-state error, and improving the convergence rate. Moreover, the proposed control algorithm also adds the correction factor for sparsity in the input matrix. This further improves the tracking capability, convergence rate, and reduces the steady state error. The prototype of the charging station, is used to validate the control algorithm and the various claimed functionalities experimentally.

Keywords— Electric Vehicle, Charging Station, SPV Generation, Multifunctional VSC, Power Quality.

I. INTRODUCTION

Since last decade, the automotive industry is undergoing a paradigm shift from conventional internal combustion engine based vehicle to the electric vehicle (EV). As a result, more than two million people are already using EVs by the end of 2016 [1]. Since the electric energy is the spine of the EVs, many residential and public charging stations are installed. Most of the EVs are charged using the single-phase onboard charger [2]. However, connecting a large number of EVs with different power level, creates an unequal loading of the three phase of the grid. Because of this, huge harmonics current flows through the secondary winding of the transformer. Moreover, a large neutral current also flows through the neutral wire of the transformer. This may lead to the bursting of the neutral conductor [3]. Therefore, in this paper, a charging station is proposed that provides the facility to connect the single phase charger while keeping the neutral current zero even with the unequal loading of the three phases of the supply system.

Another paradigm shift, the world is undergoing through, is the generation of electricity from the renewable energy sources [4]. Following it, the proposed charging station, integrates the photovoltaic (PV) array for charging the EVs. Verma et al. [5], Abeywardana et al. [6], and Yan et al. [7] have suggested the PV array integrated EV charger for self-sustained, reliable, clean and cost-effective charging. Choudhari et al. [8] have proposed the SPV integrated smart charging for PHEV, in which a DC-DC boost converter is utilized to interface the PV array and the charger. Monteiro et al. [9] have also integrated the PV array to EV charger using a DC-DC boost converter. However, in this paper, the DC-DC conversion stage is eliminated, and the PV array is directly interfaced to the charging station using a diode.

Various publications have been reported the control the PV array assisted charging station. Tazay et al. [10] have discussed the d-q theory based control of the hybrid converter for an EV charging station. However, the control algorithm uses the phase locked loop (PLL), and the dq-to-abc transformation, which requires a lot of computation. Moreover, the control does not consider the nonlinearity of the EV charger. Because of it, the charger injects the harmonics into the grid. Therefore, in the present work, RS-LMMN based control algorithm is used, that takes the nonlinearity of the EV chargers into account.

Out of various adaptive filters discussed in the literatures [11], the least mean squares (LMS) based adaptive filter is used widely owing to its simple structure, and easy implementation. However, the convergence speed of LMS algorithm has strong dependency on the eigenvalue distribution of the input-signal autocorrelation matrix. Another disadvantage of the LMS algorithm is that, it utilizes fixed step size for updating the filter adaptation coefficients, which requires the tradeoff between the convergence speed and maladjustment error [12]. Moreover, the LMS algorithm [13] exhibits some stability issue due to appearance of some undamped response as the sampling time increases. Another famous adaptive algorithm is the recursive least squares (RLS) algorithm [14]. The RLS algorithm utilizes the best

978-1-5386-6624-1/18 $31.00 © 2018 IEEE

approximation of Wiener solution to find the least square error at each iteration based on all the previous data. Due to it, the RLS algorithm converges quickly. However, the RLS algorithm is computationally expensive, and numerically unstable for ill input conditions.

The data-reusing LMS (DR-LMS) algorithm [15] utilizes the received signal and current desired signal several times in each iteration to converge quickly. The performance can be enhanced using normalized and unnormalized new data-reusing LMS (NNDR-LMS and UNDR-LMS) where the past data is also used in each iteration along with the current desired output and the current input [16].

In this paper, RS-LMMN based adaptive control algorithm is used for implementation of PV array integrated multiport multifunctional charging station. The proposed control algorithm utilizes the features of both LMS and LMF algorithm for performance improvement. The RS-LMMN algorithm also uses a penalty factor for the sparsity of input signal matrix to improve the convergence speed.

II. System Description

The proposed charging station is designed to charge the electric vehicles using single phase onboard charger in three phase four wire configuration of the distribution system as shown in Fig. 1. Each EV is connected between the phase, and the neutral wire. A four leg VSC is coupled to the point of common coupling (PCC) of charging station using a coupling inductor. A PV array is interfaced to the DC bus (DCB) of the charging station through a diode. This diode stops the reverse power flow from DCB to the PV array. A storage battery connected at charging station using a DC-DC converter ensures the uninterruptible charging operation of the charging station under energy deficit condition. A RC ripple filter is utilized to eliminate the high-frequency noise caused by the switching of VSC.

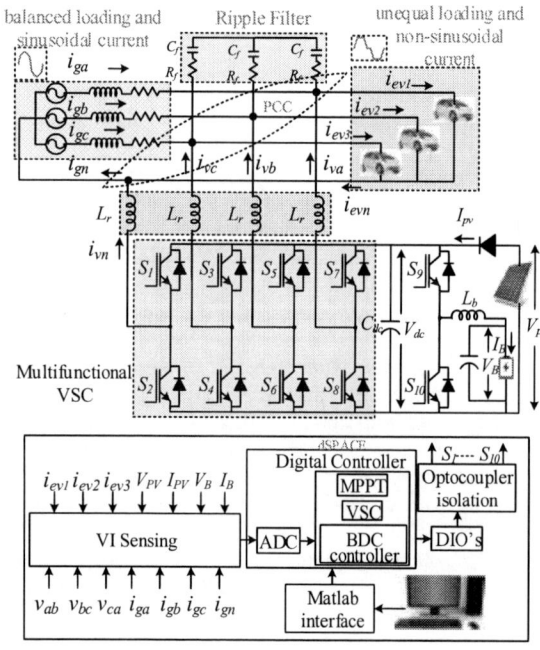

Fig. 1 Circuit diagram of proposed charging station

III. Control Algorithm

The proposed charging station is designed to charge the multiple EVs simultaneously using the SPV array power. Therefore, the control algorithm aims to utilize the PV array power maximally without much disturbing the grid. The control algorithm is designed to perform the following tasks.

- Estimation of maximum power point (MPP) voltage of the SPV array, and regulation of the DC link voltage equal to the MPP voltage for maximum power extraction.
- Control the flow of PV array power through VSC to the EVs for charging.
- Control the power exchange with the source depending upon the EV charging requirement, storage battery requirement and PV generation.
- Ensure the unity power factor operation (UPF) of the source. This means that no harmonics current or reactive power exchange with the source. Moreover, make the source currents balanced and sinusoidal all the time, even the different loading of the three phases.
- The zero neutral current to flow into the source.
- Manage the power flow balance among different components of the charging station even under transient condition.
- Control the charging/discharging current of the storage battery.

Based on these tasks, the control of the charging station is divided into the subsections such as (1) MPPT and DCB voltage control, (2) Control of multifunctional VSC, and (3) control of bi-directional DC-DC converter.

A. MPPT and DC Bus Voltage Control

In the proposed charging station, VSC harnesses the maximum possible power of the PV array by regulating the DCB voltage (V_{dc}) to such a value at which the impedance seen by the PV array equals the internal impedance of PV array itself. However, the voltage corresponding to the maximum power (V_{dc}^*) is estimated using the maximum power point tracking (MPPT) algorithm. An incremental conductance (INC) [17] based MPPT algorithm is used here. However, in case of unavailable PV power generation, the DCB voltage (V_{dc}^*) is regulated at fixed value of 400V. A proportional integral (PI) controller is utilized to maintain the DCB voltage equal to the reference DCB voltage (V_{dc}^*). The mathematical formulation of PI controller response (I_d) is given as,

$$I_d(s) = I_d(s-1) + k_{pd}\{V_{de}(s) - V_{de}(s-1)\} + k_{id}V_{de}(s) \quad (1)$$

Where V_{de} is the voltage error and k_{pd}, k_{id} are the gains of the PI controller.

B. Control of Multifunctional VSC

The control diagram of VSC is exhibited in Fig. 2. The purpose of the VSC control algorithm is to estimate the active (I_p) and a reactive component (I_q) of the charging current. For it, the RS-LMMN based adaptive control algorithm is used. Using I_p, I_q currents, the reference sinusoidal source currents (i_{ga}^*, i_{gb}^*, i_{gc}^*) are obtained, and the gate pulses of the VSC are generated. The detail description of the control algorithm is discussed as follows.

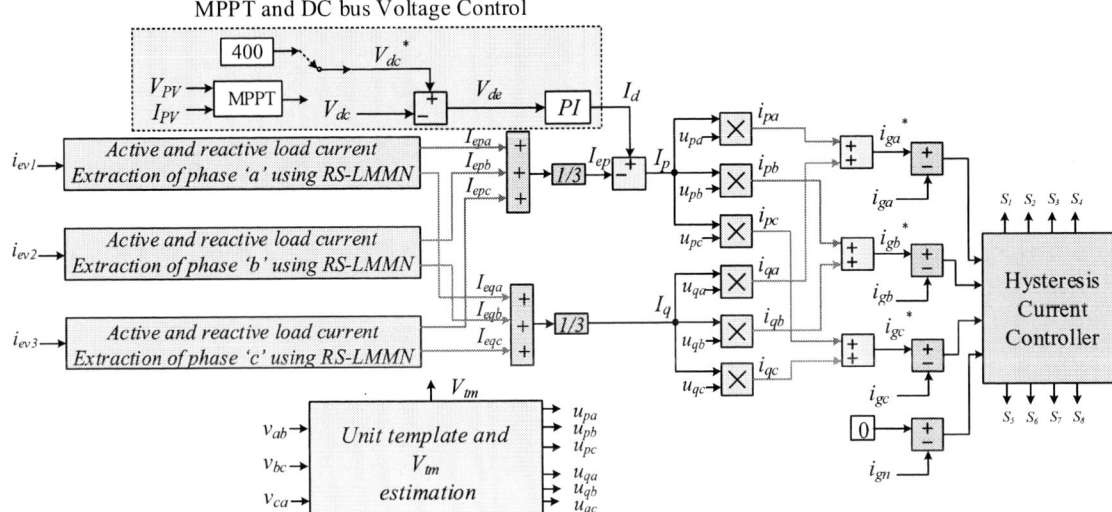

Fig. 2 Control algorithm schematic

1) Estimation of PCC Voltage Amplitude, In-phase and Quadrature-phase Unit Templates

The amplitude of the PCC voltage amplitude is estimated as [3].

$$V_{tm} = \sqrt{\frac{2}{3}} \times \sqrt{v_{ga}^2 + v_{gb}^2 + v_{gc}^2} \tag{2}$$

Where, v_{ga}, v_{gb}, and v_{gc} are the phase to neutral voltage of the source. The phase to neutral voltages are calculated using the line voltages (v_{ab} and v_{bc}) at PCC as,

$$v_{ga} = \frac{1}{3}\left(2v_{ab} + v_{bc}\right), v_{gb} = \frac{1}{3}\left(-v_{ab} + v_{bc}\right), v_{gc} = \frac{1}{3}\left(-v_{ab} - 2v_{bc}\right) \tag{3}$$

The in-phase unit (u_{pa}, u_{pb}, u_{pc}) and quadrature phase (u_{qa}, u_{qb}, u_{qc}) unit templates are estimated as,

$$u_{pa} = \frac{v_{ga}}{V_{tm}}, u_{pb} = \frac{v_{gb}}{V_{tm}}, u_{pc} = \frac{v_{gc}}{V_{tm}} \tag{4}$$

$$u_{qa} = -\frac{u_{pa}}{\sqrt{3}} + \frac{u_{pc}}{\sqrt{3}}, u_{qb} = \frac{\sqrt{3}u_{pa}}{2} + \frac{(u_{pb} - u_{pc})}{2\sqrt{3}} \tag{5}$$

$$u_{qc} = -\frac{\sqrt{3}u_{pa}}{2} + \frac{(u_{pb} - u_{pc})}{2\sqrt{3}}$$

2) Estimation of Weight of Real Component of Desired Source Currents

The weight of real component of charging current for phase 'a' (I_{epa}) is estimated as,

$$I_{epa}(r+1) = \underbrace{I_{epa}(r) + me_a(r)\left\{d + (1-d)e_a^2(r)\right\}u_{pa}(r)}_{LMMN\ algorithm} \tag{6}$$

$$\underbrace{-r\frac{\mathrm{sgn}\left[I_{epa}(r)\right]}{1 + z\left|I_{epa}(r)\right|}}_{Sparse\ penalty}$$

Where, μ is the step-size, $\rho = \mu\gamma\zeta$, is regularization factor, which decides the level of trade-off between the estimation error and sparsity strength. ζ controls reweighting factor, and its value is defined as $\zeta < 0$, whereas γ value is defined as $\gamma > 0$. δ is the constant, which decides the nature of the adaptive algorithm, and it is defined as $0 \le \delta \le 1$. For $\delta = 1$, the LMMN algorithm poses the feature of LMS. However, for $\delta = 0$, the

LMMN algorithm behaves as LMF algorithm. Therefore, δ value is selected between 0 and 1 to take the benefits of both the LMS and LMF algorithms. The schematic of RS-LMMN [18] algorithm is shown in Fig. 3.

The $\mathrm{sgn}[I_{epa}]$ is the component-wise function, which is defined as,

$$\mathrm{sgn}\left[I_{epa}\right] = \begin{cases} \dfrac{I_{epa}}{\left|I_{epa}\right|} & I_{epa} \neq 0 \\ 0 & I_{epa} = 0 \end{cases} \tag{7}$$

The addition of the sparse penalty speeds up the convergence of the proposed RS-LMMN based control algorithm.

Similarly, the real weights component of charging current for phase 'b' and 'c', are estimated as,

$$I_{epb}(r+1) = I_{epb}(r) + \mu e_b(r)\left\{\delta + (1-\delta)e_b^2(r)\right\}u_{pb}(r) -\rho\frac{\mathrm{sgn}\left[I_{epb}(r)\right]}{1 + \zeta\left|I_{epb}(r)\right|} \tag{8}$$

$$I_{epc}(r+1) = I_{epc}(r) + \mu e_c(r)\left\{\delta + (1-\delta)e_c^2(r)\right\}u_{pc}(r) -\rho\frac{\mathrm{sgn}\left[I_{epc}(r)\right]}{1 + \zeta\left|I_{epc}(r)\right|} \tag{9}$$

Using (6), (8), and (9) the total weight of the real components of the charging current, is obtained as,

$$I_{ep} = \frac{I_{epa} + I_{epb} + I_{epc}}{3} \tag{10}$$

The total weight of the real component of the desired source current, is obtained as,

$$I_p = I_d - I_{ep} \tag{11}$$

3) Estimation of Weight of Reactive Component of Desired Source Currents

Using (6), the weight of reactive components (I_{eqa}, I_{eqb}, and I_{eqc}) of charging currents, are given as,

$$I_{eqa}(r+1) = I_{eqa}(r) + \mu e_a(r)\left\{\delta + (1-\delta)e_a^2(r)\right\}u_{qa}(r) -\rho\frac{\mathrm{sgn}\left[I_{eqa}(r)\right]}{1 + \zeta\left|I_{eqa}(r)\right|} \tag{12}$$

978-1-5386-6624-1/18 $31.00 © 2018 IEEE

$$I_{eqb}(r+1) = \underbrace{I_{eqb}(r) + \mu e_b(r)\left\{\delta + (1-\delta)e_b^{\,2}(r)\right\}u_{qb}(r)}_{LMMN\ algorithm}$$

$$\underbrace{-\rho\frac{\mathrm{sgn}\left[I_{eqb}(r)\right]}{1+\zeta\left|I_{eqb}(r)\right|}}_{Sparse\ penalty} \qquad (13)$$

$$I_{eqc}(r+1) = I_{eqc}(r) + \mu e_c(r)\left\{\delta + (1-\delta)e_c^{\,2}(r)\right\}u_{qc}(r)$$

$$-\rho\frac{\mathrm{sgn}\left[I_{eqc}(r)\right]}{1+\zeta\left|I_{eqc}(r)\right|} \qquad (14)$$

The total weight of the reactive component of the desired source current, is obtained as,

$$I_q = \frac{I_{eqa} + I_{eqb} + I_{eqc}}{3} \qquad (15)$$

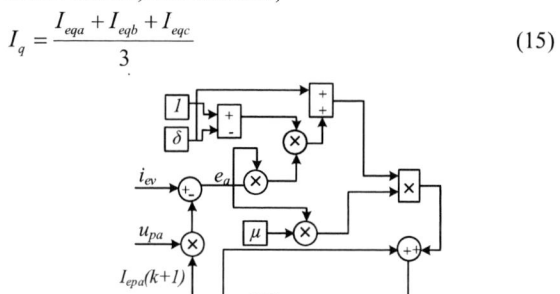

Fig. 3 Schematic of RS-LMMN algorithm

4) Estimation of Sinusoidal Desired Source Currents

Using (11) and (15), the sinusoidal desired real (i_{pa}, i_{pb}, i_{pc}) and sinusoidal desired reactive (i_{qa}, i_{qb}, i_{qc}) source currents, are estimated as,

$$i_{pa} = I_p \times u_{pa}, i_{pb} = I_p \times u_{pb}, i_{pc} = I_p \times u_{pc} \qquad (16)$$

$$i_{qa} = I_q \times u_{qa}, i_{qb} = I_q \times u_{qb}, i_{qc} = I_q \times u_{qc} \qquad (17)$$

Finally, using (16) and (17), the desired source currents (i_{ga}^*, i_{gb}^*, i_{gc}^*) are obtained as,

$$i_{ga}^* = i_{pa} + i_{qa}, i_{gb}^* = i_{pb} + i_{qb}, i_{gc}^* = i_{pc} + i_{qc} \qquad (18)$$

The VSC gate pulses are generated through the hysteresis current controller.

C. Bi-directional DC-DC Converter Control

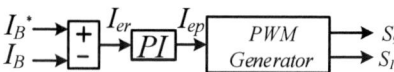

Fig. 4 DC-DC converter control

The control of the storage battery connected at DCB using a bi-directional DC-DC converter is given in Fig. 4. The EV charges/discharges at constant current. Based on the charging/discharging, the desired current can be positive or negative. The expression of PI controller for constant current mode is given as,

$$I_{ep}(k) = I_{ep}(k-1) + k_{ep}\{I_{er}(k) - I_{er}(k-1)\} + k_{ei}I_{er}(k) \qquad (19)$$

Where I_{er} is the error, and k_{ep} and k_{ei} are proportional and integral gains of the PI controller. Now the gating pulses are generated using the PWM generator.

IV. RESULTS AND DISCUSSION

The experimental performance of the proposed charging station, and its control algorithm under steady state condition and dynamic state condition, are shown in Figs. 5-9. A PV array of 5.6kW is used in the implementation of the charging station. The PV array is emulated using the Ametek TerraSAS SPV emulator. The open circuit voltage and short circuit current of the PV array, are 460V and 15A, respectively. However, the maximum power point (MPP) voltage and current of the PV array, are 396V and 14.2A, respectively. Since the EV chargers draw the non-sinusoidal current from the source, a diode bridge rectifier followed by the resistive-inductive load is used to emulate the EV charger. The insulated gate bipolar junction transistors (IGBTs) are used to design the multifunction VSC and the DC-DC converter for a storage battery. A 240V, 35Ah lead acid battery is used as a storage battery. The 5Ω resistor and 10μF electrolytic capacitor are used to design the ripple filter. Various Hall Effect based voltage (LEM make LV-25) and current sensor (LEM make LA-55) are used for sensing the system parameters. The sensed voltage and current signals are converted into a digital signal using analog to digital converters. Using these digital signals, the control algorithm of the charging station is implemented in the digital controller (dSPACE 1006), and the control signals are generated for VSC, and DC-DC converter.

A. Steady State Performance

The steady state performance of the charging station when the charging station is fully powered by the solar PV array, and three EVs, and a storage battery are charged using the PV power, is exhibited in Fig. 5. The excess power is also supplied to the source at a unity power factor (UPF) as justified by the unity displacement power factor (DPF) in Figs. 5 (j)-(o). The PV array is generating 5.6 kW of electrical energy as given in Fig. 5(a). Out of the 5.6kW, 2.77kW (EV1+EV2+EV3=0.93kW+0.89kW+0.95kW) is consumed by the three EVs, the storage battery takes 550W, and remaining 2.21kW power (phase 'a'+phase 'b' +phase 'c'= 0.73kW+0.75kW+0.73kW=2.21kW) is fed back to the source at UPF. Figs. 5 (b)-(g) show the voltages (v_{ga}, v_{gb}, v_{gc}), currents (i_{ev1}, i_{ev2}, i_{ev3}), and power of the EVs. Figs. 5 (h)-(i) show the voltage (V_B), current (I_B) and power of the storage battery. Similarly, Figs. 5 (j)-(o) show the per phase voltages (v_{ga}, v_{gb}, v_{gc}), currents (i_{ga}, i_{gb}, i_{gc}), and power of the source. While supplying power to the source, the total harmonic distortion (THD) of the source line voltage (v_{ab}) and line current (i_{ga}) is also observed at 1.8% and 4.3%, respectively as shown in Fig. 5 (p)-(q). Whereas, the THD of the current drawn by the EV1 (i_{ev1}) is 28% as shown in Fig. 5(r). Since the EVs connected at charging station, draw the unequal and non-sinusoidal current, the neutral current (i_{evn}) is significant as shown in Fig. 5(s) However, the source neutral current (i_{gn}) is almost reduced to very small value (Fig. 5(t)) because the VSC of the charging station, is supplying the neutral current (i_{vn}) as in Fig. 5(u). Moreover, the unbalanced phase currents are also mitigated by the VSC. Due to it, the source currents become balanced and sinusoidal.

978-1-5386-6624-1/18 $31.00 © 2018 IEEE

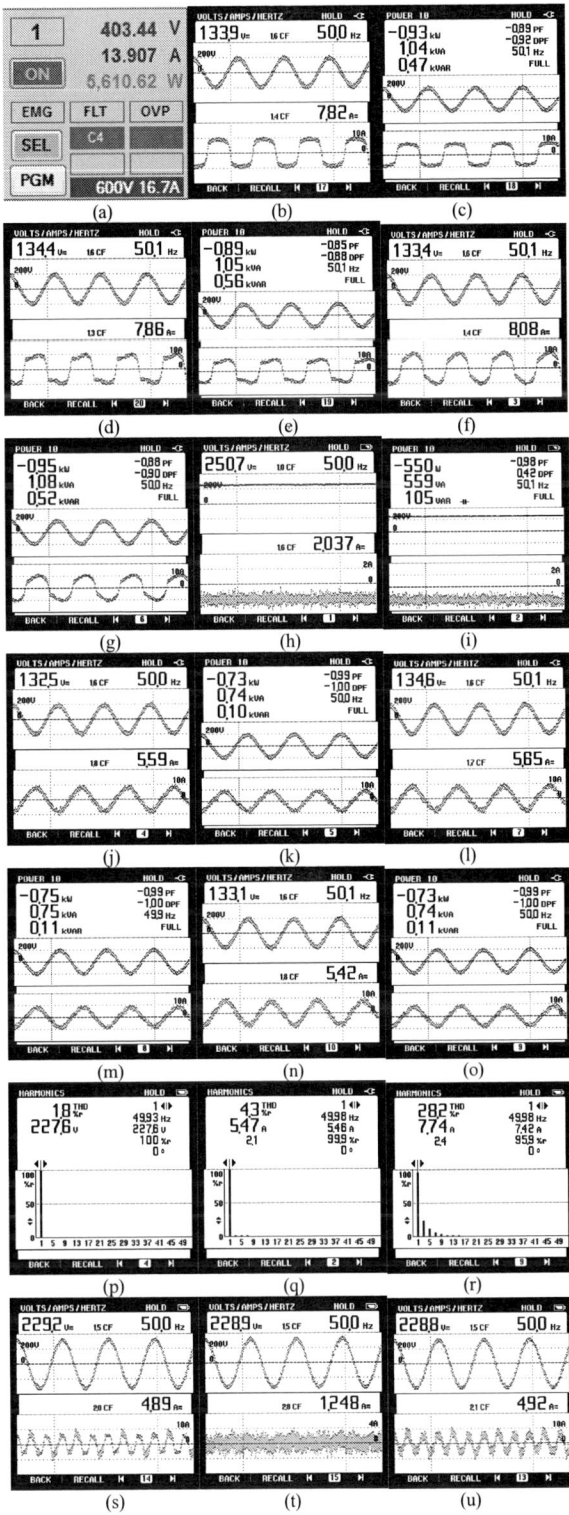

Fig. 5 Steady state performance, (a) V_{pv}, I_{pv}, and power, (b)-(c) v_{ga}, i_{ev1}, and EV1 power, (d)-(e) v_{gb}, i_{ev3}, and EV2 power, (f)-(g) v_{gc}, i_{ev3}, and EV3 power, (h)-(i) V_b, I_b, and power, (j)-(k) v_{ga}, i_{ga}, and power, (l)-(m) v_{gb}, i_{gb}, and power, (n)-(o) v_{gc}, i_{gc}, and power, (p) harmonic spectra of v_{ab}, (q) harmonic spectra of i_{ga}, (r) harmonic spectra of harmonic spectra of i_{ev1}, (s) i_{evn}, (t) i_{gn}, (u)i_{vn}

B. Dynamic Performance

During the operation, the charging station undergoes the transient caused by the change in solar irradiance,

connection/disconnection of the EVs, and a change in the storage battery charging/discharging. The performance under these transients, is shown in Figs. 6-8.

1) Performance Under Solar Irradiance Change

Since the solar irradiance changes throughout the day, therefore, the controller of the charging station should be fast enough to observe the solar irradiance change, and it should be able to harness the maximum power under all irradiance condition. Moreover, the charging of the other EVs should not be affected, and the power equilibrium in the system should not be affected. Usually, the solar irradiance changes gradually, however, in this implementation, the worst case is considered, and the solar irradiance is changed in step from 1000W/m^2-400W/m^2. Due to change in the solar irradiance, the PV generated power reduces as the PV current reduces as shown in Figs. 6(a)-(b). Since the PV power is inadequate to meet the charging demand of the EVs, the charging station starts drawing power from the source. On the contrary, the power is being fed into the source before the change in solar irradiance change. The charging of the EV and charging of the storage battery, do not get affected by the solar irradiance change.

2) Performance Under Connection/Disconnection of EVs and Change in Charging Current of Storage Battery

Throughout the day, the EVs are connected/disconnected at the charging station, and their charging currents change dynamically. Under this condition, the charging of the other EVs should not be affected, and power equilibrium should be achieved. Figs. 7(a)-(c) show the performance under a sudden disconnection of EV1. Due to disconnection of the EV1, the power required for charging the EVs reduces. As a result, the power supplied to the source increases since the PV generation is not affected by these transients. The disconnection of EV1 does not affect the charging of other EVs as shown in Fig. 7 (b). Moreover, to balance the power, the source phase 'a' current increases as shown in Fig. 7(b). The disconnection does not affect the source's neutral current, and it remains zero. Moreover, after the disconnection of the EV1, the VSC current becomes equal to the source phase 'a' current. Fig. 7(d) shows the performance of the charging station when the storage battery is changed from charging to discharging.

3) Multifunctional Operation of VSC

Figs. 8 (a)-(b) exhibit the performance of the charging station with or without VSC in operation. From Fig. 8(a), it is understood that without VSC in operation, the source current is not sinusoidal because the harmonics current required by the EVs are being drawn from the source. Moreover, due to unbalanced charging, the neutral current is also significant. After switching 'ON' VSC, the source current becomes sinusoidal and neutral current also becomes zero as VSC is supplying the harmonics current and the neutral current. From Fig. 8 (b) it is observed that the VSC neutral current is equal to the neutral current caused by the EVs. Fig. 8(b) also exhibits that the performance of the VSC under disconnection of EV1. Fig. 8(c) exhibits the voltage correction capability of VSC under distorted source voltage condition. After switching 'ON' VSC, the source voltage becomes sinusoidal.

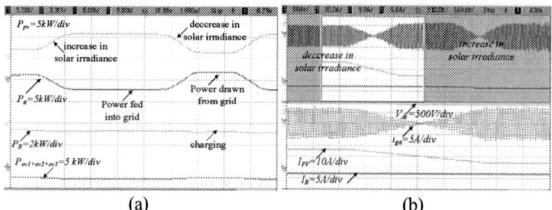

Fig. 6 Dynamic performance under solar irradiance change

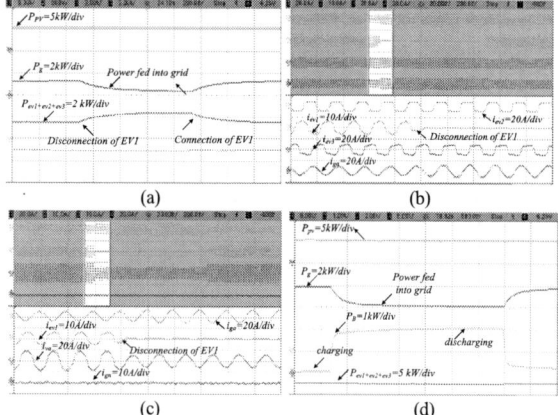

Fig. 7 Dynamic performance under connection/disconnection of EVs, and change in charging/discharging current of storage battery

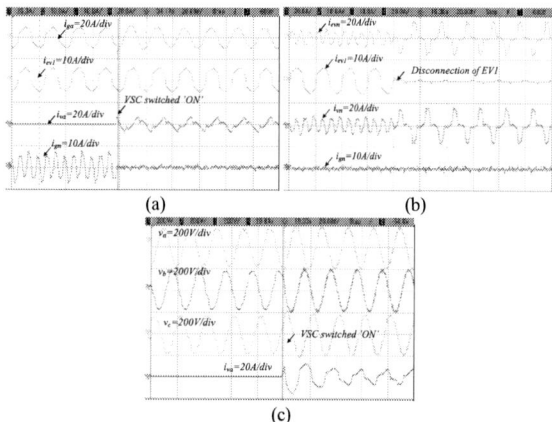

Fig. 8 Multifunctional operation of VSC, (a) neutral current compensation, (b) under EV connection/disconnection, (c) voltage correction capability

V. CONCLUSION

The experimental validation of a PV array powered multifunctional charging station and RS-LMMN based adaptive control algorithm, has been presented in this work. The presented results show that the three EVs multiple EVs can be charged simultaneously at the charging station with zero neutral current in the fourth wire of the distribution system at the unbalanced and nonlinear load. Moreover, the three phase's source currents are also maintained balanced and sinusoidal. While charging the EVs, the power quality of the source voltage and current, is also improved, and the THD is always maintained less than 5% as required by the IEEE 519 standard. The removal of the DC-DC converter power stage of the PV array has not affected the performance of the charging station, and presented results show that the maximum PV power is extracted under all operating conditions. The presented results have also verified the controller ability to manage the power flow balance under

different transients caused by the solar irradiance change, EVs connection and disconnection and change in charging/discharging current of the storage battery. The comprehensive evaluation of the presented results have proved the competency of the proposed topology and its control algorithm for the development of charging station.

ACKNOWLEDGEMENT

The authors are very grateful to the Renew Power Ventures Pvt. Ltd. for funding this project under Grant Number: RP03461 (Charging Station for Electric Vehicles).

APPENDICES

SPV array: V_{oc}=460V, I_{sc}=15A, V_{MPP}=396V, I_{MPP}=14.2A
Source Supply: Three phase 230V (L-L RMS), 50Hz
Storage Battery: 240V 35Ah Lead Acid

REFERENCES

[1] International Energy Agency-Global EV Outlook 2017-Two Million and Counting. [Online].Available: https://www.iea.org/publications/freepublications/publication/Global EVOutlook2017.pdf
[2] International Energy Agency-World Energy Outlook 2017. [Online]. Available: https://www.iea.org/weo2017/
[3] Bhim Singh, Ambrish Chandra and Kamal Al-Haddad, Power Quality: Problems and Mitigation Techniques. John Wiley & Sons, 2015.
[4] National Renewable Energy Agency- Distributed Solar Photo Voltaics For Electric Vehicle Charging [Online]. Available: https://www.nrel.gov/docs/fy14osti/62366.pdf
[5] A. Verma and B. Singh, "Multi-objective reconfigurable three phase off-board charger for EV," in *IEEE Transport. Electrificat. Conf. (ITEC-India)*, Pune, 2017, pp. 1-6.
[6] D. B. Wickramasinghe Abeywardana, P. Acuna, B. Hredzak, R. P. Aguilera and V. G. Agelidis, "Single-Phase Boost Inverter-Based Electric Vehicle Charger With Integrated Vehicle to Grid Reactive Power Compensation," *IEEE Trans. Power Electron.*, vol. 33, no. 4, pp. 3462-3471, 2018.
[7] Q. Yan, B. Zhang and M. Kezunovic, "Optimized Operational Cost Reduction for an EV Charging Station Integrated with Battery Energy Storage and PV generation," *IEEE Trans. Smart Grid*, Early Access.
[8] K. Chaudhari, A. Ukil, K. N. Kumar, U. Manandhar and S. K. Kollimalla, "Hybrid Optimization for Economic Deployment of ESS in PV-Integrated EV Charging Stations," *IEEE Trans. Ind. Informat*, vol. 14, no. 1, pp. 106-116, Jan. 2018.
[9] V. Monteiro, J. G. Pinto and J. L. Afonso, "Experimental Validation of a Three-Port Integrated Topology to Interface Electric Vehicles and Renewables With the Electrical Grid," *IEEE Trans. Ind. Informat.*, vol. 14, no. 6, pp. 2364-2374, June 2018.
[10] A. Tazay and Z. Miao, "Control of a Three-Phase Hybrid Converter for a PV Charging Station," *IEEE Trans. Energy Convers.*, Early Access.
[11] Ali H. Sayed, Robust *Adaptive Filters*, 1, Wiley-IEEE Press, 2008.
[12] B. Widrow; J.M. McCool; M. Larimore and C.R. Johnson; "Stationary and non-stationary learning characteristics of the LMS adaptive filter;" in *Proc. of IEEE*, vol. 64, no. 8, pp. 1151-1162, Aug. 1976.
[13] N. Beniwal, I. Hussain and B. Singh, "Implementation of DSTATCOM with i-PNLMS Based Control Algorithm under Abnormal Grid Conditions," *IEEE Trans. Ind. App.*, Early Access.
[14] M. Han, S. Zhang, M. Xu, T. Qiu and N. Wang, "Multivariate Chaotic Time Series Online Prediction Based on Improved Kernel Recursive Least Squares Algorithm," *IEEE Trans. Cybernetics*, Early Access.
[15] A. Rathore and D. K. Panda, "Performance analysis of data reusing least mean square algorithm for smart antenna system," in *IEEE Int. Conf. Comput., Communicat. and Automat.*, 2017, pp. 1391-1394.
[16] M. T. H. Alouane, "A square root normalized LMS algorithm for adaptive identification with non-stationary inputs," in *J. of Communicat. and Netw.*, vol. 9, no. 1, pp. 18-27, March 2007.
[17] J. Liu, J. Li, J. Wu and W. Zhou, "Global MPPT algorithm with coordinated control of PSO and INC for rooftop PV array," in *The J. of Eng.*, vol. 2017, no. 13, pp. 778-782, 2017.
[18] Y. Li, Y. Wang and F. Albu, "Sparse channel estimation based on a reweighted least-mean mixed-norm adaptive filter algorithm," *24th Euro. Sig. Process. Conf.*, 2016, pp. 2380-2384.

2nd IEEE International conference on power Electronics, Intelligent Control and Energy systems (ICPEICES-2018)

Adaptive Frequency Estimation Technique for Grid Connected Photovoltaic System

Sudip Bhattacharyya, Shailendra Kumar, *Member, IEEE* and Bhim Singh, *Fellow, IEEE*
Department of Electrical Engineering, *IIT Delhi,* New Delhi-110016, India
Emails: sudipiitd2017@gmail.com, er.dwivedi88@gmail.com and bsingh@ee.iitd.ac.in

Abstract— **This paper deals with the adaptive estimation of frequency observation (AEFO) based technique for compensating the harmonics component of two-stage grid-connected solar photovoltaic (PV) system. The maximum PV power is extracted using maximum power point tracking (MPPT) algorithm. Moreover, the input DC voltage of the voltage source converter (VSC) is maintained at desired level using a proportional integral (PI) controller. The AEFO control injects the required reactive component, which shapes the balanced grid currents. Moreover, the active power is injected in to the grid, which eliminates the harmonics in the grid currents. Using this controller, the fundamental reference currents are estimated using nonlinear load currents. This system maintains the grid current total harmonic distortion (THD) in accordance to the IEEE 519 standard. The dynamic response is improved by adding the PV feed-forward term with the estimated magnitude of $i_{e\alpha}$ and $i_{e\beta}$ current quantities. This control improves the dynamic response when insolation is changed and also provided a good response when switched from PV array to DSTATCOM and vice versa. Moreover, the power quality issues such as load unbalancing, power factor correction and reduced the grid side currents harmonics. Test results has shown the reactive power compensation of balanced and unbalanced nonlinear loads and the steady-state and dynamic performances of the system and PV array to DSTATCOM and vice-versa.**

Index Terms— *Power quality, AEFO, MPP, Solar PV Array.*

I. INTRODUCTION

This is the time to take attention towards the alternative resources due to limitation of fossil fuels. The renewable energy is one of the way to overcome the problem and thus is free from any conversion processes like fossil fuels and therefore, it does not hamper the ecology [1]. The electrical energy produced by renewable sources, is clean and green way, which protects the environment from air pollution, noise pollution and all kind of greenhouse gasses [2].

There are various types of non-conventional resources available in the nature like solar photovoltaic (PV) energy, wind energy, fuel cell, tidal power etc. All others renewable sources depend upon various parameters. Among of them the solar photovoltaic (PV) is widely used due to its simplicity and cost efficiency. In this paper, a PV array is used as a non-conventional energy source and the peak power is transferred through the VSC (Voltage Source Converter) to the grid [3]-[5]. Ideally the solar PV array power basically depends on three factors (i) solar irradiation in W/m², (ii) ambient temperature ^0C and (iii) air mass in AM. Therefore, the main limitation of output power totally depends on whether condition. It works at high efficiency around its ideal conditions.

Though the output of all PV arrays, is highly depending upon the atmospheric condition. Therefore, when the use of solar PV energy as a source to the system, it is unable to produce the maximum peak power due to uncertainty of whether condition. The solar PV array has a serious limitation of unavailability of solar energy during night time, which directly reduces the overall energy yield by 50%. The efficiency is further reduced due to non-ideal condition of solar energy in the atmosphere. There are several methods to extract the maximum power at different atmospheric conditions. In the literature [6]-[8], the algorithms like classical perturb and observe (P&O), drift free improved P&O, incremental conduction method etc., popularly track the peak power from solar panel. The P&O algorithm is executed by varying the perturbation size. The MPP algorithm tracks the slope of the PV array curve to get maximum power according to its parameters. There are two techniques for solar PV system, (i) DC link voltage control by generating reference voltage form MPP algorithm, which tracks the MPPT without a boost converter, (ii) direct duty ratio control is a double stage conversion, where it needs one additional DC-DC converter. The maximum PV array power is fed to the voltage source converter (VSC) using a feed-forward loop.

The grid-connected three phase VSC based single stage and double stage topologies, are presented in the literature [9]-[10], which has been reported only active power control. The additional DC-DC converter is used to extract the peak power from the PV array with the help of MPPT algorithm. The MPPT algorithm senses the voltage (V_{pv}) and current (I_{pv}) of a PV array system and estimates the reference duty ratio. The reference duty ratio is compared with high frequency carrier signal to generate the desired gate pulses of the DC-DC boost converter. It has a wide scope to develop simple control with high accuracy depending on load and grid conditions, which are reported in the literature [11]-[12]. The three-phase rotating reference frame to two-phase rotating reference frame, is used in the literature [13]. This SRF (Synchronous Reference Frame) based theory gives the idea to make unity power factor or zero voltage regulation by extracting the reference grid currents. Besides, it suffers from very poor dynamic response and 2nd harmonic component is more pronounced in d-axis component. The low pass filter increased the response under a change from unbalanced load condition to healthy condition and vice-versa. It takes large time to settle because of low cut-off frequency. On the other hand, the IRPT (Instantaneous Reactive Power Theory) algorithm [14] has better response than the SRF algorithm. However, at unbalanced or large load condition, the values of PI (Proportional Integral) controller gains, are extremely large in IRPT algorithm, which reduces the settling time of output of PI (Proportional Integral) controller. The high precision quadrature control [15] slightly improves the dynamic response but the computational burden is

978-1-5386-6624-1/18 $31.00 © 2018 IEEE

very high as it is very complex as compared to proposed adaptive controller. It needs a large memory to store the all previous samples. The performance is sluggish in the grid during the dynamic of the system. The adaptive notch type filter is reported in [16] with fast response but the input to cost function is instantaneous. Therefore, it compromises between steady and dynamic responses. The problem is easily overcome and also improved the power quality [17] by using adaptive detection based method with selective harmonic cancelation technique.

The proposed two-stage PV system deals with AEFO (Adaptive Estimation of Frequency Observation) based current controller in this work. It injects active power near load side, which also helps to reduce the losses in the distribution system. Moreover, it mitigates the harmonics current and maintains the grid currents balanced with unity power factor at even unbalanced load condition.

II. SYSTEM CONFIGURATION

Fig.1 shows a two-stage PV grid interfaced system feeding active power to a three-phase distribution network. The proposed system includes a PV array with a blocking diode, a boost converter, with an inductor L, 3-leg VSC with six power electronics switches $S_1 \ldots S_6$. The VSC is connected in between the grid and PV array via interfacing inductors L_f. The three-phase grid voltages are denoted by v_{ga}, v_{gb}, v_{gc} with source impedances Z_{sa}, Z_{sb}, Z_{sc}. The grid currents are represented by i_{sa}, i_{sb}, i_{sc}.

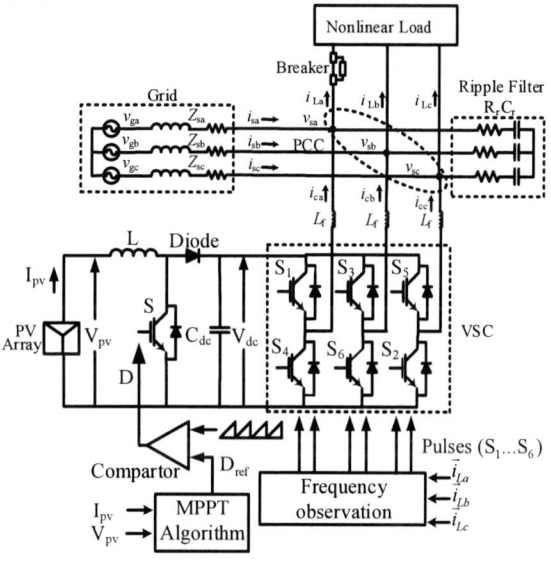

Fig. 1 Three phase Grid connected PV system configuration

The ripple filters R_r, C_r, and nonlinear load (i_{La}, i_{Lb} and i_{Lc}), are connected to point of common coupling (PCC). The PCC voltages (v_s) are denoted by v_{sa}, v_{sb}, v_{sc}, and the VSC produces compensating currents i_{ca}, i_{cb}, i_{cc} to the PCC.

III. CONTROL ALGORITHM

The maximum solar power is delivered to the grid using the MPP voltage ($V_{mp}=V_{dc}$) and the MPP current (I_{mp}). The MPPT algorithm generates the duty ratio D_{ref} which compares with the high frequency repetitive triangular wave to produce the gate pulses for the DC-DC boost converter in two stage grid connected system. The DC link voltage (V_{dc}) of the VSC, is separately maintained by the boost converter. AEFO (adaptive estimation of frequency observation) based current controller is used for switching the VSC. To observed the unbalanced load condition in the system, a breaker is used in phase 'a'for some time interval of phase 'a' to verify the response of the AEFO control. The data of the system configuration, are given in Appendix.

A. Maximum Power Point Tracking Using P&O Algorithm

There are huge number of algorithms available to track the maximum power which are described in the literature [7], [8]. The most regularly used technique, is the P&O (Perturb and Observe) technique due to its simplicity. The V_{pv} and I_{pv} are sensed and the reference duty ratio D_{ref} is produced. The extraction of peak power is based on the slope of p-v curve (dP_{pv}/dV_{pv}) of PV system. When the change of PV power dP_{pv} is less then zero but change of voltage dV_{pv} is greater than zero, then the reference duty ratio, D_{ref} decreases or if dV_{pv} is less than zero then the reference duty ratio D_{ref} voltage is increased and when the ratio of dP_{pv}/dV_{pv}, is greater than the zero then there is an increment in the reference duty ratio D_{ref}, otherwise there is a decrease in reference duty ratio D_{ref}.

B. Adaptive Feed-Forward Harmonic Cancellation Method

The generation of estimated frequency from the estimated load current in αβ axis is represented by adaptive estimation of frequency observer (AEFO) control technique in Fig. 2.

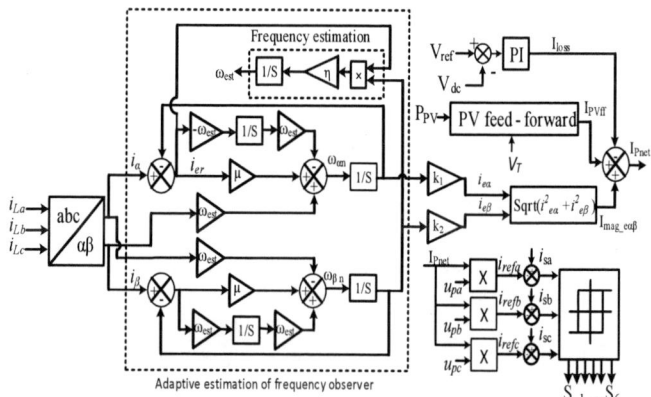

Fig. 2 Adaptive estimation of frequency ovserver control

The nonlinear load current in stationary reference frame, is represented as shown,

$$i_\alpha(t) = \sum_{l=1,3\ldots} I_{kl} \sin(l\omega t + \phi_l)$$

$$i_\beta(t) = \sum_{l=1,5\ldots} I_{kl} \cos(l\omega t + \phi_l) \qquad (1)$$

978-1-5386-6624-1/18 $31.00 © 2018 IEEE

Implementation of MPPT Control in Fuel Cell Fed High Step up Ratio DC-DC Converter

V. Karthikeyan, *Member, IEEE*
Dept. of Electrical Engineering
National Institute of Technology
Calicut, Kerala, India
karthikeyan@nitc.ac.in

Vipin Das P, Student *Member, IEEE*
Dept. of Electronics and
Communication Engineering
MNNIT Allahabad, UP, India
rel1512@mnnit.ac.in

Frede Blaabjerg, Fellow *IEEE*
Department of Energy Technology
Aalborg University,
Aalborg, Denmark
fbl@et.aau.dk

Abstract— **This paper presents an alternative approach towards implementation of maximum power point tracking (MPPT) control in fuel cell based power systems utilizing a high step up ratio DC-DC converter using single state PWM for DC Micro grid based applications. The P & O algorithm is implemented for the full bridge isolated forward converter to track the maximum power from fuel cell. In order to integrate the low voltage producing fuel cell with any existing DC micro grid, a high step up converter is required to boost the voltage and ensure proper regulation at the load side. The high frequency conversion link produces lower secondary voltage ripple. A capacitive filter is appended at the secondary side in order to impart smoother waveform characteristics to the output dc voltage. The design of C-filter is dealt with in the present paper using specified and acceptable ripple limits on the voltage and current. The theoretical analysis is verified through simulation studies using PSCAD/EMDTC software.**

Keywords— Fuel cell, Full Bridge Converter, Maximum power point tracking (MPPT), Single State PWM.

I. INTRODUCTION

Due to environmental and economic reasons, renewable energy penetration is increasing in modern power systems. Renewable energy sources are better alternatives to the fossil fuel based conventional power systems. Fuel cell (FC) is the best choice among the renewable energy sources due to its low emissions, noise free operation and high output efficiency. FC is now used for portable electronics, electric vehicles and distributed generation power plants due to its eco-friendly nature .According to usage of fuel types, it can be classified as i) polymer electrolyte membrane (PEM) fuel cells or PEMFCs (also called PEFCs), ii) solid oxide fuel cells (SOFCs), iii) alkaline fuel cells (AFCs), iv) phosphoric acid fuel cells (PAFCs), and v) molten carbonate fuel cells (MCFCs). The maximum electrical conversion efficiency of FCs are as high as 60% and by considering the co-generation it can be increased up to 80% with 90% reduction in major pollutants like CO2, CO, etc [1-3].

Due to the chemical reaction of hydrogen with anode, electrons are produced which moves toward cathode through an external electric circuit [4, 5]. The power generation ability of FCs is limited due to the internal barriers and losses. It is necessary that the FCs has to be forced to operate around the point which corresponds to the Maximum Power Point (MPP). In general, a fuel cell system (FCS) is operated in conjunction with a dc–dc power converter whose duty cycle is modulated in order to track the instantaneous MPP of the FCS [6]. It varies with temperature. In order to operate at MPPT, several methods have been implemented to extract the maximum power in the literature [7-9].

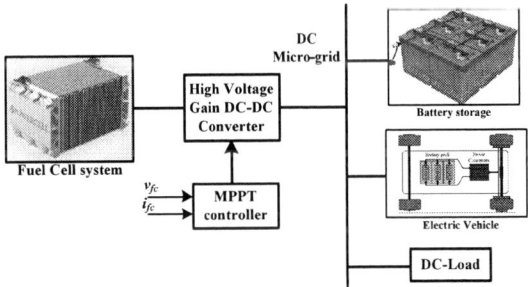

Fig. 1 Schematic of proposed layout

Boost type dc-dc converter is commonly used to achieve higher voltage ratio. However, due to saturation boost inductor core, the switch duty cycle is limited, thus voltage conversion ratio is limited to the range of 4-5 [8]. Usually, fuel cell operating voltage falls due to ohmic losses. It is hardly about 1 to 1.2 volts/cell. It is necessary to increase the voltage to higher levels to utilize the fuel cell based electrical energy. FCs is generally accumulated in stacks to improve the voltage. However, it is not advisable to accumulate many fuel cells in a single stack. It is a strict requirement that a high step up ratio converter boosts the voltage to the desired level. Therefore, a full bridge forward converter is the most suitable converter for fuel cell based power system application. In isolated dc-dc converter, many basic and derived configurations are available in the literature. Using a full bridge forward converter giver greater control flexibility over the power and helps achieve a higher conversion ratio to perform MPPT. The proposed layout is depicted in Fig. 1.

This paper particularly focuses on the MPPT operation in a full bridge dc-dc converter using single state PWM technique. The P&O algorithm approach has been implemented to perform MPPT for forward converter. Design consideration for proper functioning of the LC filter have been looked into and implemented. Section 2 delves into the fuel cell V-I characteristics and associated mathematical modelling of the losses incurred during its MPPT operation. Section 3 is dedicated to the topology, closed loop control and filter considerations of the full bridge forward converter employed in this study. MPPT implementation via the P&O algorithm is discussed in Section 4. Section 5 presents the simulation studies for the proposed system.

II. FUEL V-I CHARACTERISTICS

Fig.2 shows the generalized V-I characteristics of a fuel cell. The characteristic is non-linear in nature due to internal, physical and chemical barriers. The following

978-1-5386-6624-1/18 $31.00 © 2018 IEEE

$$I_{Pnet} = I_{mag\_ea\beta} + I_{loss} - I_{PVff} \tag{15}$$

Where, $I_{mag\_ea\beta}$ the magnitude of load current component, I_{loss} is the loss current component and I_{PVff} is the PV array feed-forward term.

The active reference grid currents components are represented as,

$$i_{refa} = I_{Pnet} \times u_{pa}; i_{refb} = I_{Pnet} \times u_{pb}; i_{refc} = I_{Pnet} \times u_{pc} \tag{16}$$

The error between sensed grid currents and the reference grid currents, is an input of the hysteresis current controller, and the switching pulses are the output of the VSC.

IV. SYSTEM PERFORMANCE

A prototype of a two stage grid connected PV system is developed by using PV simulator (Tera SAS ETS 600/17), which follows the PV array characteristics under ideal and dynamic condition. The Hall-Effect sensors (LV-25) and (LA- 55p) are used for sensing the voltages and currents, respectively. The proposed control is implemented in a real-time controller based DSP (Digital Signal Processor dSPACE1103). The switching pulses are collected from I/Os (Input/Outputs) of DSP. Moreover, a digital oscilloscope (DSO-DSO6010A) and three-phase power quality analyzer, are used to record the test results. MPPT performance of PV array at two different insolations, is shown in Figs. 3(a)-(b), which represent the MPPT as well as the V_{oc} and I_{sc} of the PV array simulator at two different irradiation levels (1000W/m² and 500W/m²).

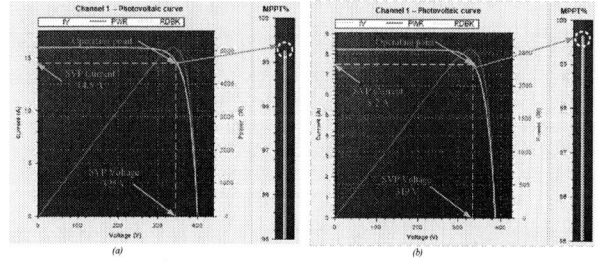

Figs. 3(a)-(b) MPPT performance at 1000W/m² and 500W/m² respectively

The MPPT is tracking nearly 99.95% when solar insolation falls from 1000W/m² to 500W/m². The V_{mp} and I_{mp} at 1000 W/m² are 325 V and 14.5 A, whereas at 500 W/m² V_{mp} and I_{mp} are 319 V and 8.2 A, respectively.

It implies that the PV system operating at maximum power point of the PV array. In this condition, various steady-state and dynamic performances of the solar tied distribution system, are discussed. The experimental data are reported in Appendix.

A. Performance at Non Linear Load Condition

Figs. 4(a)-(e) show the internal signals of the control under sudden disconnection of load. Fig. 4(a) shows the operation along with load current i_{La} and it's $\alpha\beta$ axis i_α and i_β current and the exact error current i_{er} at load disconnection. Fig. 4 (b) shows the load current with the estimated frequency component and the frequency component in $\alpha\beta$ reference frame when the load is disconnected.

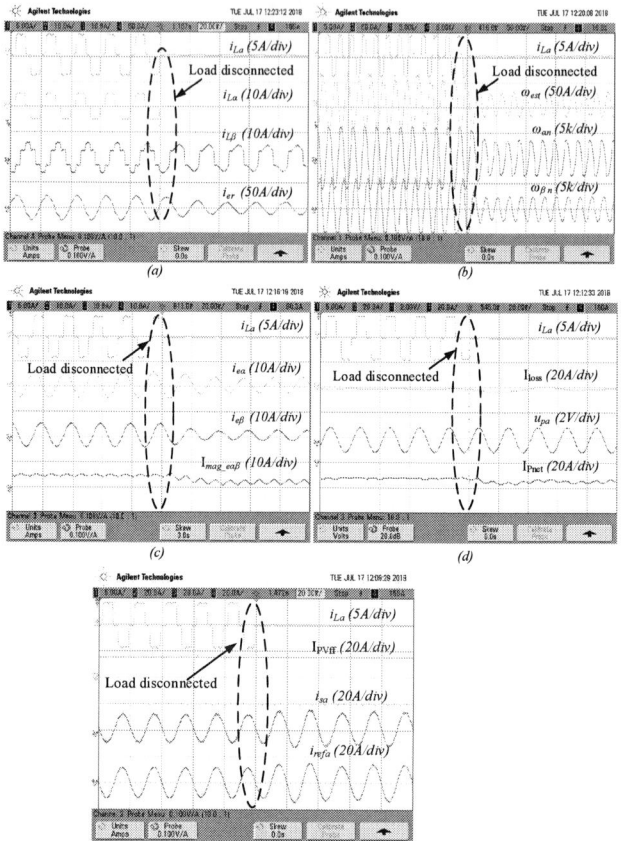

Figs. 4(a)-(e) Internal signals of the adaptive frequency observation control in load disconnected to the system

Fig. 4(c) shows the load current, i_{La} along with the estimated load current in $\alpha\beta$ reference $i_{e\alpha}$ and $i_{e\beta}$ and its magnitude $I_{mag\_ea\beta}$ at load disconnection. The magnitudes of $i_{e\alpha}$, $i_{e\beta}$ are reduced but it has maintained the fundamental frequency. The magnitude of $I_{mag\_ea\beta}$ is reduced because one of the phase become disconnected. Fig. 4(d) presents the load current i_{La}, the loss current component I_{loss}, the unit templet component u_{pa} and the net current I_{Pnet} at load disconnection. Because of load disconnection, the grid current is increased corresponding to the reference current and the PV feed-forward term is maintained and improves the responds of the system as shown in Fig. 4(e).

B. Performance at Load Unbalanced Condition

Figs. 5(a)-(e) show the load connected to a phase of the system. Fig. 5(a) shows the load connection of load current i_{La} and it's $\alpha\beta$ axis i_α and i_β current and the exact error current i_{er}. In Fig. 5(b), the frequency component are reached to the desired value by connected the phase a. in Fig. 5(c) shown the estimated quantities of load currents and it magnitude in $\alpha\beta$ reference frame along with the load current are shown. The loss current, unit templet voltage quantities and net current component are shown in Fig. 5(d) with load current at load connection. Fig. 5(e) shows the PV feed-forward term, source current and the reference current.

978-1-5386-6624-1/18 $31.00 © 2018 IEEE

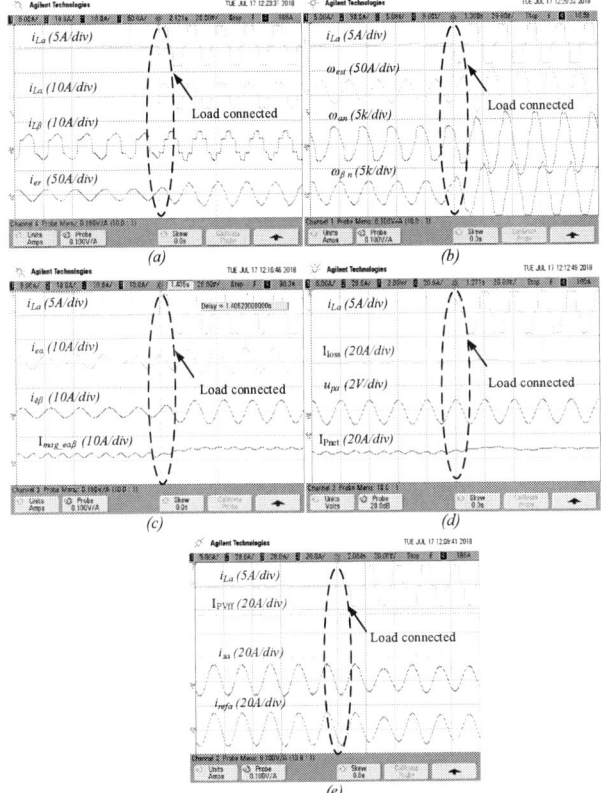

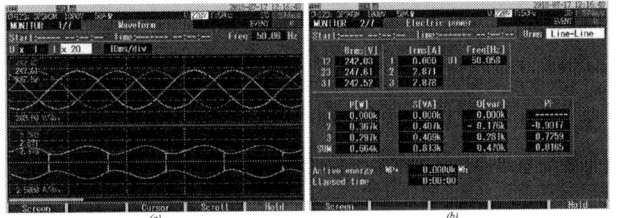

Figs. 5(a)-(e) Internal signals of the adaptive frequency observation control in load connected to the system

Figs. 6(a)-(b) show the load currents at discontinuous condition. Fig. 6(b) shows the load at discontinuous condition,

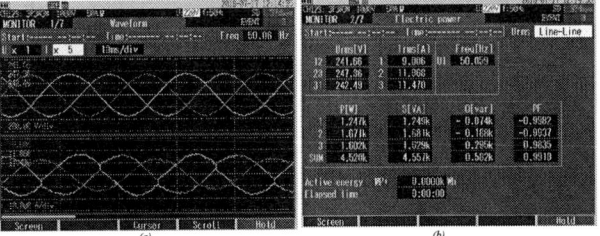

Figs. 7(a)-(b) The VSC current under unbalanced load condition with grid voltage.

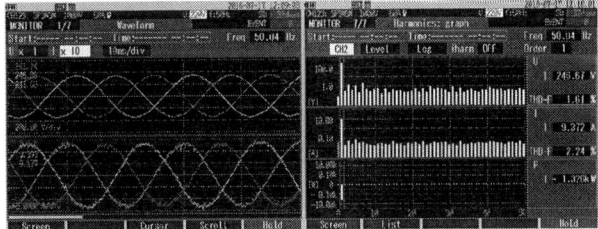

Figs. 8(a)-(b) The grid currents under unbalanced load condition with grid voltages.

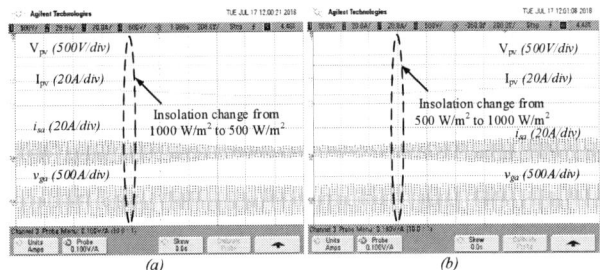

Figs. 9(a)-(b) Steady and dynamic responses of a PV system at insolation change from 1000 W/m² to 500 W/m² and vise-versa.

In Fig. 9(a), the PV voltage V_{pv} is slightly reduced over the dynamic condition because of insolation reduced to 50%. Though, the V_{pv} is logarithmically varied with I_{PV}, therefore, it does not change to very low values. However, the PV current, I_{PV} is directly varied with solar irradiation and, therefore, it is reduced nearly 50% of it original value. Since, the PV power is reduced, the magnitude of grid current i_{sa} is also lower as the injected power is reduced. The grid current is maintained the THD as per the IEEE 519 standard in both dynamic and steady state conditions.

The dynamic response of the control under insolation rise from 500 W/m² to 1000W/m², are shown in Fig. 9(b) where it has maintained the PV voltage V_{pv}, the PV current I_{pv}, grid current i_{sa} and grid voltage at desired limit. Therefore, the PV power, P_{pv} is also changed and reach rated value.

D. Prerformance of DSTATCOM Mode to PV Mode and vise-versa

Figs. 10(a)-(b) show the performance of DSTATCOM mode to PV mode and vise-versa. Fig. 10(a) shows the grid current i_{sa}, VSC current i_{VSC}, load current i_{La} and the PV current I_{pv}. The compensator VSC current, i_{VSC} compensates at the PCC when PV

Figs. 6(a)-(b) The load current under unbalanced load condition corresponding to grid voltage.

which is 664 W with power factor 0.8165 lag. Under this condition, the in phase, a VSC current is sinusoidal because no need of compensation. Moreover, the others two phases are compensated the reactive component to the PCC shown in Figs. 7(a)-(b). Figs. 8(a)-(b) show that the grid currents are increased in the load unbalanced condition, therefore, the extra power is fed to the grid and with grid current THD (Total Harmonic Distortion) of 2.24%.

C. Dynamic Performance of PV System

The dynamic performance of grid connected PV system, is shown in Figs. 9(a)-(b) when, insolation is changed from 1000 W/m² to 500 W/m² and vise-versa.

978-1-5386-6624-1/18 $31.00 © 2018 IEEE   663

array generation is off and the PV current, I_{pv} is zero in that instant.

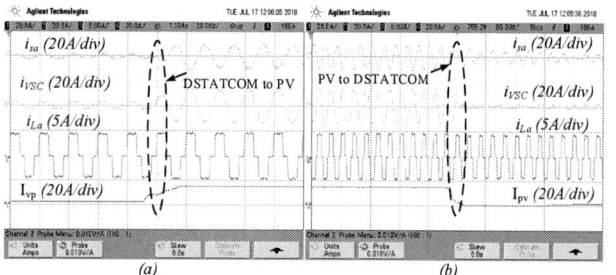

(a)   (b)

Figs. 10(a)-(b) Performance of DSTATCOM mode to PV mode and vise-versa.

However, the PV array is on the PV current is increased from zero to its original value and, therefore, reduced requirement of compensation at PCC. In this, the load is unaffected throughout the condition. The reverse is occoured when changed from PV mode to DSTATCOM mode. Though the PV current is reduced to zero yet the load current is not hampered, due to DSTATCOM action.

V.   CONCLUSION

A grid-connected two stage PV system with AEFO based control has been used for load balancing, reducing the harmonics in the grid currents at unity power factor and voltages as well as injection of the PV power to the grid. The maximum power is delivered to the grid using MPPT algorithm. The AEFO based control has provided good dynamic response under disturbances in the load and insolation. The VSC injects the compensating currents to maintain the grid current THD according to the IEEE 519 standard under unbalanced condition. The performance under unbalanced load condition has shown the fast dynamic response of the control. Although, the switching actions of DSTATCOM mode to PV mode and vise-versa are very fast due to the inborn notch filter behavior of the controller.

ACKNOWLEDGMENTS

The authors are thankful to Department of Science and Technology, Govt. of India, for supporting this research work under Grant Number: RP03357.

APPENDIX

PV array parameters: PV cell voltage V_{oc} = 32.9 V, PV cell current I_{sc} = 8.21 A, V_{mp} = 27.8 V, I_{mp} = 7.5 A, V_{dc}= 400 V, C_{dc} = 2210 μF, L_f = 5.8 mH, PI Controller: k_p = 0.96, k_i =0.32, utility

parameter: L_s = 2.98 mH, R_s = 0.08 Ω, nonlinear load: DBR and 155 mH, 65 Ω, R-C filter: C_r = 4 μF, R_r = 8 Ω, controller parameter: k_1=0.258, k_2=0.271, η=6.77

REFERENCES

[1]   J. Qi and D. Wu, "Green Energy Management of the Energy Internet Based on Service Composition Quality," *IEEE Acces.*, vol. 6, pp. 15723-15732, 2018.

[2]   D. Li, P. Deng, Y. Xu, L. Gao and G. Zhang, "Joint Access Spectrum and Backhaul Energy Allocation for Green Cognitive Heterogeneous Networks," *IEEE Acces.*, vol. 6, pp. 24793-24808, 2018.

[3]   T. Kerslake, F. Haraburda and J. Riehl, "Solar power system options for the Radiation and Technology Demonstration spacecraft," *IEEE Aerosp. and Electron. Systems* , vol. 16, no. 2, pp. 9-19, Feb 2001.

[4]   Y. Tsuchiya, "A photovoltaic AC fusion converter," *IEEE Trans. Energy Convers.*, vol. 14, no. 3, pp. 849-854, Sep 1999.

[5]   K. Emery, "The rating of photovoltaic performance," *IEEE Trans. Electron Dev.*, vol. 46, no. 10, pp. 1928-1931, Oct 1999.

[6]   M. Metry, M. B. Shadmand, R. S. Balog and H. Abu-Rub, "MPPT of Photovoltaic Systems Using Sensorless Current-Based Model Predictive Control," *IEEE Trans. on Ind. Applica.*, vol. 53, no. 2, pp. 1157-1167, March-April 2017.

[7]   J. Ahmed and Z. Salam, "An Enhanced Adaptive P&O MPPT for Fast and Efficient Tracking Under Varying Environmental Conditions," IEEE Trans. Sustan. Energy, vol. 9, no. 3, pp. 1487-1496, July 2018.

[8]   T. M. Chung, H. Daniyal, M. Sulaiman and M. Bakar, "Comparative study of P&O and modified incremental conductance algorithm in solar maximum power point tracking," *4th IET Clean Energy and Technol. Conf. (CEAT 2016)*, Kuala Lumpur, 2016, pp. 1-6.

[9]   B. Singh, S. Kumar and C. Jain, "Damped-SOGI-Based Control Algorithm for Solar PV Power Generating System," *IEEE Trans. on Ind. Appli.*, vol. 53, no. 3, pp. 1780-1788, May-June 2017.

[10] N. Kinhekar, N. P. Padhy, F. Li and H. O. Gupta, "Utility Oriented Demand Side Management Using Smart AC and Micro DC Grid Cooperative," *IEEE Trans. on Pow. Syst.*, vol. 31, no. 2, pp. 1151-1160, March 2016.

[11] S. Buso, T. Caldognetto and Q. Liu, "Analysis and Experimental Characterization of a Large-Bandwidth Triple-Loop Controller for Grid-Tied Inverters," *IEEE Trans. on Pow. Electron.* Early Access.

[12] S. Kayalvizhi and D. M. Vinod Kumar, "Load Frequency Control of an Isolated Micro Grid Using Fuzzy Adaptive Model Predictive Control," IEEE Acces., vol. 5, pp. 16241-16251, 2017.

[13] S. Vedantham, S. Kumar, B. Singh and S. Mishra, "Fuzzy logic gain-tuned adaptive second-order GI-based multi-objective control for reliable operation of grid-interfaced photovoltaic system," *IET Gener. Trans. & Distri.*, vol. 12, no. 5, pp. 1153-1163, 2018.

[14] S. Peng, A. Luo, Z. Lv, J. Wu and L. Yu, "Power control for single-phase micro-grid based on the PQ theory," *6th IEEE Conf. Ind. Electron. and Appl.*, Beijing, 2011, pp. 1274-1277.

[15] S. Kumar, I. Hussain, B. Singh, A. Chandra and K. Al-Haddad, "An Adaptive Control Scheme of SPV System Integrated to AC Distribution System," *IEEE Trans. on Ind. Appli.*, vol. 53, no. 6, pp. 5173-5181, Nov.-Dec. 2017.

[16] R. El Shatshat, M. Kazerani, and M. M. A. Salama, "Modular active power-line conditioner," IEEE Trans. Power Del., vol. 16, no. 4, pp. 700–709, Oct. 2001.

[17] B.Singh, A.Chandra and K Al-Haddad, Power Quality Problems And Mitigation Techniques. Wiley, Jan. 2015.

2nd IEEE International conference on power Electronics, Intelligent Control and Energy systems (ICPEICES-2018)

Acoustic Echo Cancellation Based Adaptive Control Algorithm for Grid Integrated SECS

Gaurav Modi
Electrical Engineering Department
Indian Institute of Technology
New Delhi-110016, India
gaurav07modi@gmail.com

Shailendra Kumar, Member, IEEE
Electrical Engineering Department
Indian Institute of Technology
New Delhi-110016, India
er.dwivedi88@gmail.com

Bhim Singh, Fellow, IEEE
Electrical Engineering Department
Indian Institute of Technology
New Delhi-110016, India
bsingh@ee.iitd.ac.in

Abstract— This paper presents an acoustic echo cancellation (AEC) based adaptive control for grid integrated solar energy conversion system (SECS) feeding power to the grid. It contains a three phase two level PWM voltage source converter (VSC) connected at the point of common coupling (PCC) to the three phase ac distribution system and three phase linear/nonlinear load. The VSC system performs multi-functions like delivering photovoltaic (PV) array power to the grid and the load, improving gird power factor to unity by compensating the reactive power requirement of the load, improving the grid current power quality by supplying the harmonics power to the load and load balancing. It uses an AEC based adaptive control algorithm to control the VSC and incremental conductance (INC) method to take care of the maximum power available across the PV array. The system performance is validated through simulation and test results.

Keywords- SECS, AEC, INC, power quality.

I. INTRODUCTION

Approximately 75-78% of world energy demand is met by conventional energy resources like oil, gas, and coal, which put an adverse impact on the environment like greenhouse gases, global warming and cause subsequently health hazards. As energy demand is increasing, the reserve of fossil fuel is decreasing with tremendous rate, therefore, the cost of energy production increases. In contrast, the renewable energy resources like solar and wind are found abundantly in nature. These resources have a very little impact on nature and can be utilized for the energy production [1]. The problems with renewable energy resources are their intermittent nature and high initial investment but continuous development in technologies, government incentives, and policies have opened the new possibility for the nonconventional energy resources in the energy sector. All these developments help in the cost reduction of photovoltaic (PV) system which makes them economically attractive [2].

Solar energy conversion system (SECS) can be utilized either in grid connected mode or standalone mode. In [3], various grid connected SECS topologies are discussed. In this paper, single stage topology is used for grid integrated SECS, has its own advantages. Due to the absence of DC-DC conversion in single stage topology, its efficiency is higher and the cost is lesser as compared to the double stage topology. In [4-5], modelling of PV system in ideal atmospheric condition and partial shading conditions is discussed. There are many techniques for peak power extraction from the PV panel in which, P&O method [6] and INC method [7] are common because of their easiness. In [8], a comparison of various MPPT techniques is given. In this paper, an INC method is incorporated as it provides good tracking accuracy, dynamic behavior and ideally no fluctuation at the peak of the PV curve in steady state condition.

The single stage grid integrated SECS can perform multifunction tasks depending upon the solar energy availability. When PV array output is available than it works as a SECS and transfers the PV array power to the grid and the load and provides filtering activity like power factor improvement, load balancing, and harmonics mitigation and when PV array output is not available (during the night hours, cloudy and rainy season) than it works as DSTACOM and provides filtering activity alone. In this way, the VSC can be utilized more efficiently and economically. In [9-14], various control technologies are discussed for the control of SECS. In [9] a modified controller is suggested, which has used feedback linearization (FBL) for the dc link voltage control in single stage PV system but it does not provide any filtering activity to improve the current power quality. In [10], an adaptive controller based algorithm using a fixed step size parameter is suggested to control the SPV system, while in [11] a variable step size parameter based least mean square (VSS-LMS) controller is suggested to control the VSC. In [12], an enhanced phase-locked loop (EPLL) based control algorithm is suggested for VSC control and to decrease the dc offset in the system. In [13], a digital multistage (DMSI) based scheme is suggested for the control of VSC. In [14], a least mean forth based control scheme is suggested to calculate the grid reference current to control the VSC. However, it used a fixed step size parameter, which causes the poor dynamics of the system. In [15], a least mean square based concept is suggested for cancellation of acoustic echo in the speech.

In this paper, an acoustic echo cancellation (AEC) based adaptive controller is proposed for a grid integrated solar energy conversion system (SECS). The proposed controller adjusts its step size parameter according to the error in the system and hence have good dynamic accuracy and response.

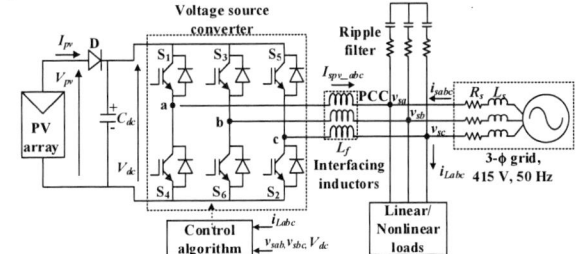

Fig 1. System configuration

978-1-5386-6624-1/18 $31.00 © 2018 IEEE 665

II. System Configuration

The schematic representation of grid integrated SECS is shown in Fig. 1. The PV array is directly connected across the dc link capacitor (C_{dc}) through a blocking diode (D). AC distribution system, VSC and load, are connected to each other at PCC. In the presence of solar irradiation, VSC feeds the active power to the distribution system and the load through interfacing inductor (L_f). The value of the interfacing inductor decides by the permissible current ripples in VSC current and VSC switching frequency. Switching of VSC causes the ripples in PCC voltage, are minimized by the ripple filter (r_f, c_f) connected at PCC.

III. Control Algorithm

The control scheme is separated into two parts: One is to calculate the PV voltage at peak power point (V_{PPP}) of PV curve and another is to generate the reference grid currents to control the VSC. An INC method is incorporated to calculate the V_{PPP} and VSC is controlled by using acoustic echo cancellation based adaptive control algorithm.

A. INC Based MPPT Method

The slope of the P-V curve (m_{pv}) at peak power point (PPP) is zero. On the left side of this PPP, it is positive then and on the right side, it is negative. It can be expressed as

$$m_{pv} = dP_{pv}/dV_{pv}, \text{ and } P_{pv} = V_{pv}.I_{pv} \quad (1)$$

$$\text{So } m_{pv} = I_{pv} + V_{pv}*(dI_{pv}/dV_{pv}) \quad (2)$$

An INC method utilizes the above equations to calculate the peak power point voltage (V_{PPP}) by sensing the PV terminal voltage (V_{pv}) and current (I_{pv}) flowing through it. This method reduces the fluctuations in V_{PPP} around the peak power point thereby minimizing the system steady state losses.

At MPP, $m_{pv} = 0$,

$$\Delta I_{pv}/\Delta V_{pv} = -I_{pv}/V_{pv} \text{ and } V_{PPP} = V_{PPP\_old} \quad (3)$$

At the left of MPP, $m_{pv} > 0$,

$$\Delta I_{pv}/\Delta V_{pv} > -I_{pv}/V_{pv} \text{ and } V_{PPP} = V_{PPP\_old} + \Delta V_{PPP} \quad (4)$$

At the right of MPP, $m_{pv} < 0$,

$$\Delta I_{pv}/\Delta V_{pv} < -I_{pv}/V_{pv} \text{ and } V_{PPP} = V_{PPP\_old} - \Delta V_{PPP} \quad (5)$$

Where ΔV_{PPP} is the step size of V_{PPP}.

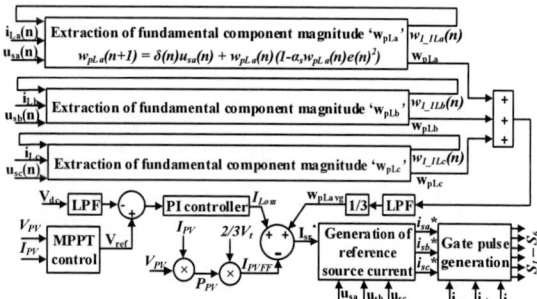

Fig 2. Control algorithm block diagram for AEC control

B. Acoustic Echo Cancellation Based Adaptive Control

The block diagram of AEC control is presented in Fig. 2. It involves the estimation of unit templates (u_{sa}, u_{sb}, u_{sc}) and

loads current fundamental active component magnitudes (w_{pLa}, w_{pLb}, w_{pLc}) by using AEC based adaptive control. The Feedforward term of PV current (I_{PVFF}) is estimated to minimize the effect of PV array power variation on system dynamics. The active loss component of current (w_{Loss}), which represents the VSC losses is calculated by comparing V_{ppp} and dc link voltage V_{dc}. The scheme estimates the grid reference current ($i_{sa}{}^*$, $i_{sb}{}^*$, $i_{sc}{}^*$) to transfer the PV array active power to the load and the grid and finally, it generates the switching pulses using the hysteresis controller.

1) Estimation of Unit Templates (u_{sa}, u_{sb}, u_{sc})

Unit templates are estimated by sensing any two PCC line voltage (v_{sab}, v_{sca}) by voltage sensors. These voltages are used to estimate the grid phase voltages as.

$$v_{sa} = (2vs_{ab} - vs_{ac})/3, \ v_{sb} = (-2vs_{ab} + vs_{ac})/3, \ v_{sc} = (vs_{ab} + 2vs_{ac})/3 \quad (6)$$

$$\text{And } u_{sa} = v_{sa}/V_t, \ u_{sb} = v_{sb}/V_t, \ u_{sc} = v_{sc}/V_t \quad (7)$$

Where V_t is the peak magnitude of grid phase voltages which is estimated as

$$V_t = \sqrt{2/3(v_{sa}{}^2 + v_{sa}{}^2 + v_{sa}{}^2)} \quad (8)$$

2) Load Current Fundamental Active Component Magnitudes (w_{pLa}, w_{pLb}, w_{pLc})

The magnitudes of load current fundamental active components (w_{pLa}, w_{pLb}, w_{pLc}) for the nth sample are estimated by using acoustic echo cancellation (AEC) based adaptive algorithm. For the phase 'a' load current, w_{pLa} is calculated as

$$w_{pLa}(n+1) = \delta(n)u_{sa}(n) + w_{pLa}(n)(1 - \eta w_{pLa}(n)e_{La}(n)^2) \quad (9)$$

Where $e_{La}(n)$ is the error input, can be estimated as

$$e_{La}(n) = i_{La}(n) - w_{pLa}(n).u_{sa}(n) \quad (10)$$

And δ is the constant learning rate and η is the variable leakage factor, is given as

$$\eta = 0.0001 \text{ if } |e(n).w(n)| < \zeta \text{ otherwise } \eta = 0 \quad (11)$$

Similarly w_{pLb}, w_{pLc} are estimated for phase 'b' and phase 'c' as

$$w_{pLb}(n+1) = \delta(n)u_{sb}(n) + w_{pLb}(n)(1 - \eta w_{pLb}(n)e_{Lb}(n)^2) \quad (12)$$

$$w_{pLc}(n+1) = \delta(n)u_{sc}(n) + w_{pLc}(n)(1 - \eta w_{pLc}(n)e_{Lc}(n)^2) \quad (13)$$

3) Estimation of Feed-Forward Term of PV Current (I_{PVFF}), Active Loss Component of Current (w_{Loss})

As atmospheric parameters like temperature and solar irradiation, fluctuate, the PV array output fluctuates. Therefore the grid current fluctuates. To reduce these fluctuations, a feed-forward term of PV power (I_{PVFF}) is incorporated to estimate the grid reference current. For the nth sample, I_{PVFF} is estimated as

$$I_{PVFF}(n) = 2P_{pv}(n)/3V_t \quad (14)$$

The active loss component of current is calculated by comparing the peak power voltage (V_{PPP}) of PV curve and sensed dc link voltage (V_{dc}). For the calculation of VSC losses and to take out the peak power from PV terminal, the error $v_{de}(n)$ between reference voltage (V_{PPP}) generated by the MPPT algorithm is compared with the sensed dc link voltage (V_{dc}) as,

$$v_{de}(n) = V_{PPP}(n) - V_{dc}(n) \quad (15)$$

The compared signal is fed to the PI controller, which reduces the error. Its output serves as w_{Loss} given as,

$$w_{Loss}(n+1) = w_{Loss}(n) + K_p[v_{de}(n+1) - v_{de}(n)] + K_i v_{de}(n+1) \quad (16)$$

In this way, the PI controller regulates V_{dc} equal to the V_{PPP} and thus the PV system works at peak power point.

    4)   Estimation of Reference Gird Current (i_{sa})*

The following equation estimates the reference grid current magnitude (I_{sr}*),

$$I_{sr}^*(n) = w_{pLavg}(n) - w_{Loss}(n) + I_{PVFF}(n) \qquad (17)$$

where $w_{pLavg}(n)$ is the average magnitude of load current fundamental active component and calculated as,

$$w_{pLavg}(p) = (w_{pLa} + w_{pLb} + w_{pLc})/3 \qquad (18)$$

The reference grid currents are obtained as

$$i_{sa}^*(n) = I_{sr}^*(n).u_{sa}(n), \; i_{sb}^*(n) = I_{sr}^*(n).u_{sb}(n), \; i_{sc}^*(n) = I_{sr}^*(n).u_{sc}(n) \qquad (19)$$

The error between sensed grid current (i_s) and reference grid current (i_s*), is obtained and this error is utilized to produce the gating pulses for the VSC.

IV. SIMULATION RESULTS

The grid integrated SECS is modelled and its performance is simulated for a three phase 415 V, 50 Hz grid system and 15 kW photovoltaic array. On the load side, a balanced three phase nonlinear load is connected, which is also realized as unbalanced by opening the circuit breaker of the phase 'a'. The system reaction is inspected in steady state condition and in perturbed conditions of load unbalancing and changing PV output conditions. The various simulation parameters are provided in Appendix A.

A. Dynamic Response at Nonlinear Load

The system response at dynamic nonlinear load condition is given in Figs. 3-4, which is examined at constant PV array output condition (at standard test condition). The dynamic loading condition is achieved by opening the breaker of phase 'a' at 0.5 s and reclosing of the breaker at 0.6 s. In this way, the load becomes balanced to unbalanced at 0.5 s and unbalanced to balanced at 0.6 s. From these results Figs. 3(a)-(b) it is evident that in both conditions, the grid current remains sinusoidal and balanced with unity power factor. Form these results it can also be seen that when the load becomes balanced to unbalanced than the grid current magnitude increases and when the load becomes balanced to unbalanced than the grid current magnitude decreases. From Figs. 4(a)-(b) it is evident that i_s is out of phase with v_s, which showcases that the grid is accepting power.

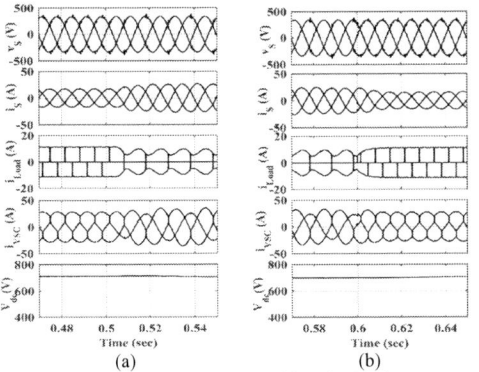

Fig. 3(a-b) System response under sudden load changes condition (balanced to unbalanced and unbalanced to balanced load condition)

B. Response of Acoustic Echo Cancellation Based Adaptive Controller

Figs. 5(a)-(b) show the response of control algorithm in dynamic loading condition and in dynamic PV array

conditions. Fig. 5(a) shows the internal signals of AEC control in dynamic nonlinear loading condition, while the Fig. 5(b) shows the internal signal of control in variable PV array output conditions. Dynamic loading condition is realized by opening the breaker of phase 'a' at 0.5 s and closing the breaker at 0.55 s. While the variable PV array condition is implemented by reducing the solar irradiation form 1000 W/m2 to 500 W/m2 at 0.5 s. From the result system quick response can be observed.

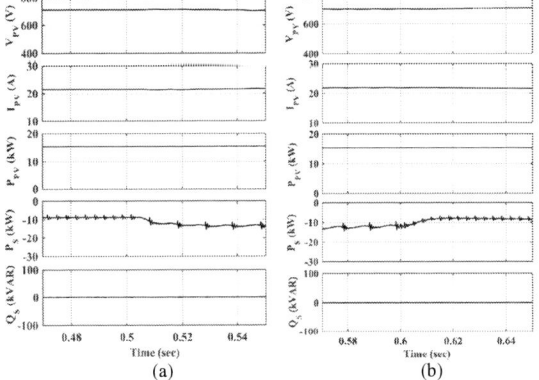

Fig. 4(a-b) System response under sudden load changes condition (balanced to unbalanced and unbalanced to balanced load condition)

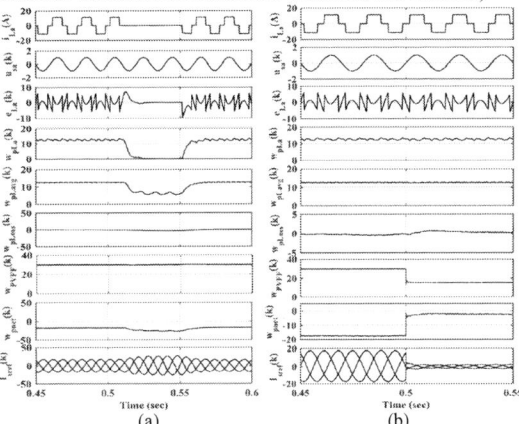

Fig. 5(a-b) Various internal signal of control signal (a) under dynamic load condition, (b) under varying PV output condition

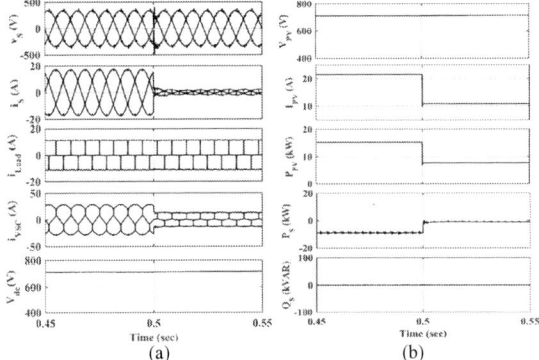

Fig. 6(a-b) System response under varying PV output condition

C. Response of System Under Inconsistent PV Output Condition

Figs. 6(a)-b) exhibit the system behavior under inconsistent PV array output condition. The PV array output is varied by varying the solar irradiation form 1000 W/m² to

500 W/m² at 0.5 s while the load remains constant. From these results, it can be observed that when PV array output is decreased the grid currents magnitude is also decreased. In low solar irradiation value, the system behaves like a DSTATCOM and improves the grid currents power quality.

V. EXPERIMENTAL RESULTS

A prototype of grid integrated SECS is made in the laboratory for 5.14 kW$_p$ PV system. The PV system is realized using PV array simulator which circuit parameters are given in appendix B. The system consists of three phase two level VSC, dc link capacitor (C_{dc}), interfacing inductor (L_f) and a ripple filter (r_f, c_f) and a balanced three phase nonlinear load of 1.5 kW which can be made unbalanced by opening the circuit breaker of phase 'a'. Test results are obtained by using a three phase power quality analyzer (HIOKI, model: PQ3100) and DSO (Agilent make, model: DSO7014A) in steady state and dynamics condition. The PV array output is varied by varying solar irradiation form 1000 W/m² to 600 W/m² and vice versa. MPPT tracking performance is also validated in the test results.

A. Response of AEC control Under Dynamic Loading Condition

The various internal signals of AEC based adaptive control are shown in Figs. 7(a-j). When the breaker of phase 'a' is opened then the load current, i_{La} becomes zero instantaneously. Corresponding, w_{pLa} becomes zero within one cycle. When phase 'a' breaker recloses again then w_{pLa} settles within one cycle. As the load is connected or disconnected, the w_{pLavg} decreases and increases respectively. On the other hand, net magnitude of reference grid current (I_{sr}^*) and reference grid current (i_{sr}^*) are increased and decreased respectively.

B. System Response at Balanced Nonlinear Load

The system response at balanced nonlinear load is shown in Figs. 8-10. Test results of grid side parameters are exhibited in the Figs. 8(a-d). Test results of load side parameters are exhibited in Figs. 9(a-d). Test results of VSC side parameters are exhibited in Figs. 10(a-d). Form these test results it can be noticed that grid current remains sinusoidal with THD 3.39% even the load current THD is 28.94%, which is under the limit recommended by the IEEE standards [16].

C. System response at Unbalanced Nonlinear Load

The system response under unbalanced nonlinear load is exhibited in Figs. 11-13. The load unbalanced condition is realized by opening the circuit breaker of the phase 'a'. These results exhibit that the grid currents remain sinusoidal and balanced even load is unbalanced and nonlinear in nature. Since the load is reduced as phase 'a' is cut off, the PV array delivers more power leading to larger amplitude of the grid currents. Moreover, the grid current THD obtained as 3.04% is better as compared to the previous case.

D. System Response at Inconsistent PV Output Condition

The system response under inconsistent PV array output condition is exhibited in Figs. 14(a-f). The PV array output is varied by varying the solar irradiation. In Figs. 14(a-b),

solar irradiation is varying from 1000 W/m² to 600 W/m² and vice versa while in Figs. 14(c-f) the solar irradiation is varying from 1000 W/m² to 50 W/m² and vice versa. These test results of MPPT tracking is validated by Figs. 15(a)-(b) which is shown that MPPT tracking accuracy of more than 99%, is achieved. From these results, it can be observed, as PV array output is varied, the grid currents also change. At higher solar irradiation, the PV array generates more power and delivers it to the grid and the load. Hence i_s becomes out of phase with v_s and VSC works as SECS. In the absence of solar irradiation, PV generates no power, therefore the load is powered only through the grid. Thus, i_s becomes in phase with v_s and VSC work as a DSTATCOM.

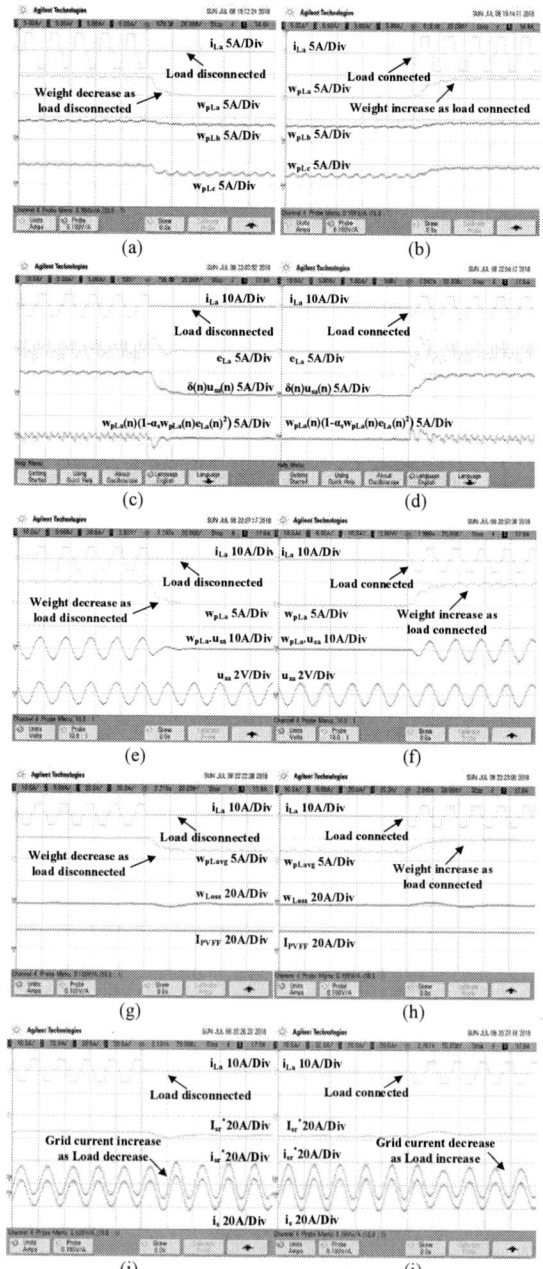

Fig. 7(a-j) various internal signal of control algorithm

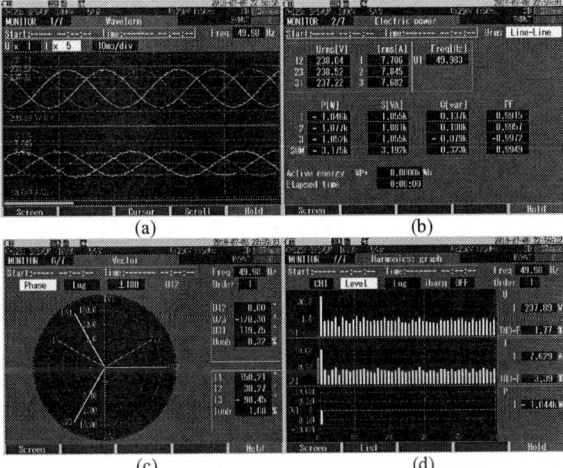

Fig. 8 System performance under balanced non-linear load condition. (a) Grid voltage-current (v_s, i_s), (b) Grid power, (c) Grid voltage- current phasor, (d) gird voltage-current THD

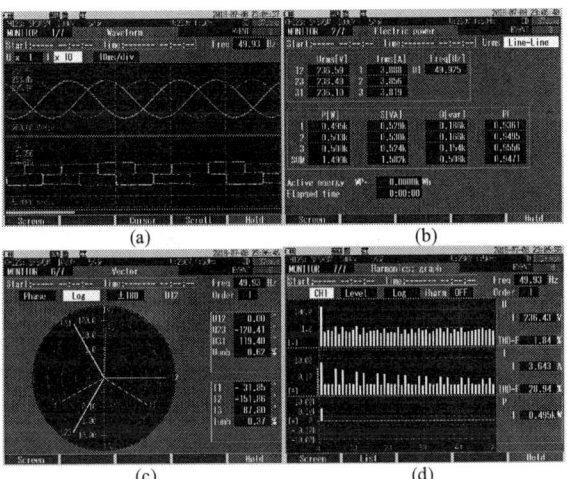

Fig. 9 (a) Load voltage-current (v_L, i_L), (b) Load power, (c) Load voltage-current phasor, (d) Load voltage-current THD

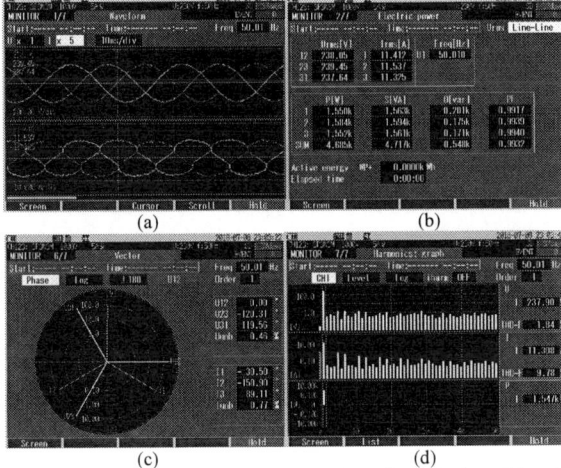

Fig. 10 System performance under balanced non-linear load condition (a) VSC voltage-current (v_{VSC}, i_{VSC}), (b) VSC power, (c) VSC voltage-current phasor

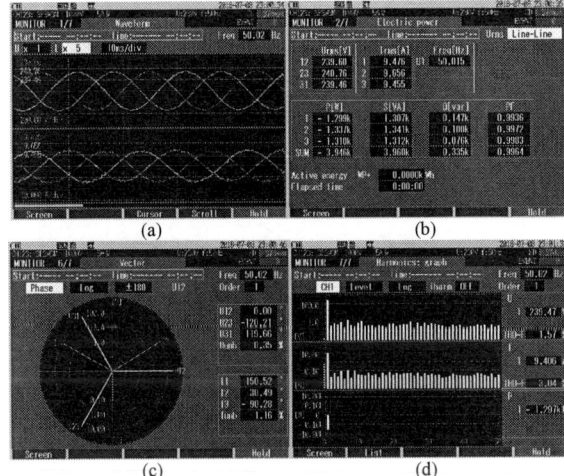

Fig. 11 System performance under unbalanced non-linear load condition. (a) Grid voltage-current (v_s, i_s), (b) Grid power, (c) Grid voltage- current phasor, (d) gird voltage-current THD

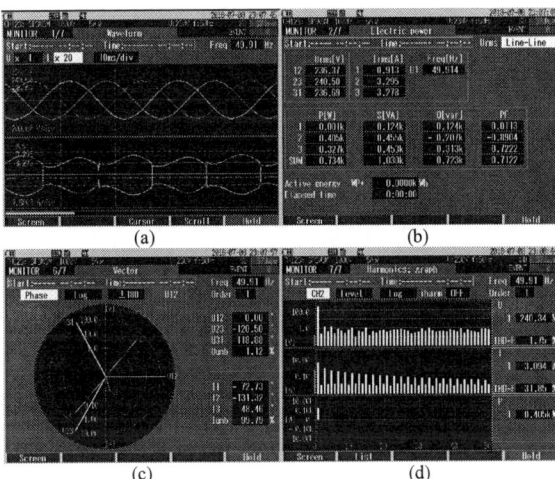

Fig. 12 (a) Load voltage-current (v_L, i_L), (b) Load power, (c) Load voltage-current phasor, (d) Load voltage-current THD

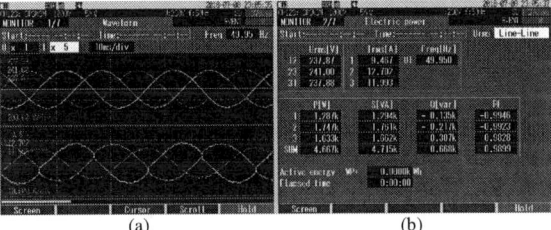

Fig. 13 (a) VSC voltage-current (v_{VSC}, i_{VSC}), (b) VSC power

VI. CONCLUSION

An adaptive control based on acoustic echo cancellation (AEC) has been used for a single stage SECS and the INC MPPT method is incorporated for drawing out the peak power from the PV array terminal. The system aims to transfer the PV array power to the grid and the load and minimizes the distribution system losses by improving the power factor, power quality, and load balancing. System performance is validated through simulation and test result in the various steady state and dynamic conditions, which validates the robustness and reliability of control algorithm.

978-1-5386-6624-1/18 $31.00 © 2018 IEEE

MPPT tracking test results show that an INC has method achieved more than 99% tracking accuracy. The grid currents remain balanced and sinusoidal at a unity power factor in both balanced-unbalanced nonlinear loading conditions. The harmonic content in the grid current is observed below 5% which complied with the requirement of the IEEE-1547 and IEEE-519 standards.

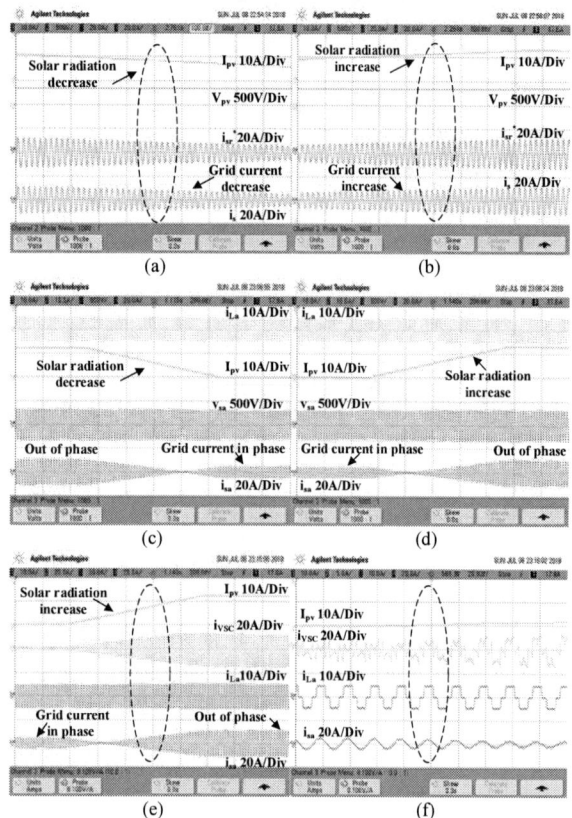

Fig. 14(a-h) System response in varying PV output condition

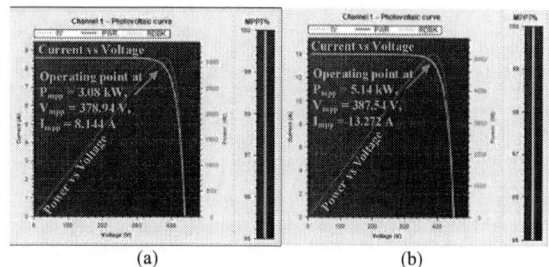

Fig. 15(a-b) MPPT performance of solar simulator (a) at 600 W/m² solar irradiance (b) at 1000 W/m² solar irradiance

ACKNOWLEDGMENT

The authors would express their gratitude towards the Department of Science and Technology, Govt. of India, for sponsoring this work under grant number: RP03391G.

APPENDIX

A. Simulation modelling Parameters of System

Grid parameters v_s = 415 V, 50 Hz r_s = 0.1Ω, L_s = 3.5 mH, PV system parameters V_{oc} = 871.2 V, I_{sc} − 23.52 A, P_{pv} = 15.1 kWp, C_{dc} = 4000 μF, k_p = 0.09, k_i = 0.005 interfacing inductance L_f = 4 mH, load parameters R= 50 Ω, L = 150 mH, ripple filter parameters r_f = 5 Ω, c_f = 5 μF, δ = 0.0025, ζ = 0.00004.

B. Experimental Parameters of System

Grid parameters v_s = 235 V, 50 Hz r_s = 0.2Ω, L_s = 0.1 mH, PV system parameters V_{oc} = 450 V, I_{sc} = 14A, P_{pv} = 5.143 kWp, dc link parameters, C_{dc} = 3500 μF, k_p = 0.2, k_i = 0.02, interfacing inductance L_f = 5 mH, load parameters R= 60 Ω, L = 150 mH, ripple filter parameters r_f = 5 Ω, c_f = 5 μF, δ = 0.09, ζ = 0.0001.

REFERENCES

[1]. A. Aggarwal, A. Singhal and S. J. Darak, "Clean and Green India: Is Solar Energy the Answer?," *IEEE Potentials*, vol. 37, no. 1, pp. 40-46, Jan.-Feb. 2018.

[2]. P. Vithayasrichareon and I. F. MacGill, "Valuing large-scale solar photovoltaics in future electricity generation portfolios and its implications for energy and climate policies," *IET Renew. Power Gen.*, vol. 10, no. 1, pp. 79-87, 2016.

[3]. Tsai-Fu Wu, Chih-Hao Chang, Li-Chiun Lin and Chia-Ling Kuo, "Power loss comparison of single- and two-stage grid-connected photovoltaic systems," *IEEE Trans. on Energy Conver.*, vol. 26, no. 2, pp. 707- 715, June 2011.

[4]. M. G. Villalva, J. R. Gazoli, and E. R. Filho, "Modeling and circuit-based simulation of photovoltaic arrays," in *Proc. Brazilian Power Electronics Conference*, Bonito-Mato Grosso do Sul, 2009, pp. 1244-1254.

[5]. A. Xenophontos and A. M. Bazzi, "Model-Based Maximum Power Curves of Solar Photovoltaic Panels Under Partial Shading Conditions," *IEEE Journal of Photovoltaics*, vol. 8, no. 1, pp. 233-238, Jan. 2018.

[6]. N. Femia, G. Petrone, G. Spagnuolo and M. Vitelli, "Optimization of perturb and observe maximum power point tracking method," *IEEE Trans. Power Electron.*, vol. 20, no. 4, pp. 963-973, July 200

[7]. N. E. Zakzouk, M. A. Elsaharty, A. K. Abdelsalam, A. A. Helal and B. W. Williams, "Improved performance low-cost incremental conductance PV MPPT technique," *IET Renew. Power Gener*, vol. 10, no. 4, pp. 561-574, April 2016.

[8]. T. Esram and P. L. Chapman, "Comparison of photovoltaic array maximum power point tracking techniques," *IEEE Trans. Energy Conversion*, vol. 22, no. 2, pp. 439-449, June 2007

[9]. V. N. Lal and S. N. Singh, "Control and Performance Analysis of a Single-Stage Utility-Scale Grid-Connected PV System," *IEEE Sys. Journal*, vol. 11, no. 3, pp. 1601-1611, Sept. 2017.

[10]. S. Kumar, I. Hussain, B. Singh, A. Chandra and K. Al-Haddad, "An Adaptive Control Scheme of SPV System Integrated to AC Distribution System," *IEEE Trans. Ind. Appl.*, vol. 53, no. 6, pp. 5173-5181, Nov.-Dec. 2017

[11]. S. Pradhan, I. Hussain, B. Singh and B. K. Panigrahi, "Modified VSS-LMS-based adaptive control for improving the performance of a single-stage PV-integrated grid system," IET Science, Measure. & Techno., vol. 11, no. 4, pp. 388-399, Jul. 2017.

[12]. S. Kumar, A. K. Verma, I. Hussain, B. Singh, and C. Jain, "Better control for a solar energy system: using improved enhanced phase-locked loop-based control under variable solar intensity," IEEE Ind. Appl. Magazine, vol. 23, no. 2, pp. 24-36, March-April 2017.

[13]. S. Kumar and B. Singh, "Multi-Objective Single-Stage SPV System Integrated to 3P4W Distribution Network Using DMSI-Based Control Technique," IEEE Trans. Ind. Appl., vol. 54, no. 3, pp. 2656-2664, May-June 2018.

[14]. R. K. Agarwal, I. Hussain, and B. Singh, "LMF-based control algorithm for single stage three-phase grid integrated solar PV system," IEEE Trans. Sustainable Energy, vol. 7, no. 4, pp. 1379-1387, Oct. 2016.

[15]. M. S. Kumar, "Variable Step Size (VSS) control for Circular Leaky Normalized Least Mean Square (CLNLMS) algorithm used in AEC," 2008 IEEE TENCON Region 10 Confer., Hyderabad, 2008, pp. 1-6.

[16]. IEEE Recommended Practices and requirement for Harmonic Control on Electric Power System, IEEE Std.519, 1992.

2nd IEEE International Conference on Power Electronics, Intelligent Control and Energy Systems (ICPEICES-2018)

Design & Implementation of Solar Fed Intensity Controlled Streetlight

Jaspreet Singh[1], Prof. Priya Mahajan[2] and Prof. Rachana Garg[3]

[1,2,3]Department Of Electrical Engineering, *Delhi Technological University,* Delhi

E-mail: [1]jaspreet741@yahoo.co.in, [2]priyamahajan.eed@gmail.com,

[3]rachana16100@yahoo.co.in

Abstract— **SPV based LED Streetlight has advantages over other conventional lighting systems as no power conversion is needed. LED work on DC and energy optimization is possible by controlling the duty cycle of the LED driver. The components of Solar Fed LED Street lighting system are SPV array, MPPT, dc-dc converter and battery unit. In this paper, the intensity of solar fed Street light is controlled from traffic hours to non-traffic hours which results in saving the electricity consumption. A hybrid street light model is also designed and developed. The simulation studies are performed in Matlab-Simulink environment.**

Keywords— *PV array ; DC-DC converter; Li-ion battery; LED array; MPPT; AC/DC module*

I. INTRODUCTION

Solar PV is a renewable source of energy accessible all over the world but in Direct Current (DC) form. Standalone SPV system is very advantageous as compared to conventional grids especially for dc loads since there is no need to convert from ac to dc, which results in an increase in efficiency of the system. SPV based streetlight is one of the example of Stand-alone PV system [1]. Light emitting diodes (LEDs) are becoming more popular in lighting applications because they exhibit high intensity or luminous with low power consumption and also, they work on DC effectively [2].This makes PV based street lighting using LEDs a very attractive and viable possibility.

The quantity of power developed by a SPV system is dependent on the solar irradiance so by sensing irradiance, energy management between SPV, battery and load can be controlled.

This work highlights the energy efficient SPV street lighting system using LEDs through intelligent algorithm interface for energy management between SPV panel, Battery and Load. Also the amount of energy used by LED load is controlled by controlling its intensity during traffic hours and low traffic hours. So, the proposed SPV based LED streetlight can be operated with adequate solar charging with optimum energy consumption. Further, a PV hybrid LED street lighting system is also designed in this paper.

II. SYSTEM DESCRIPTION

A SPV system of a suitable capacity is being utilized throughout the day to charge a battery i.e. Li-ion type through the buck converter with MPPT. At night, the stored energy of battery is utilized by the Boost converter to power the Load i.e.; LEDs lamp. The description of the proposed system is shown in Fig.1.

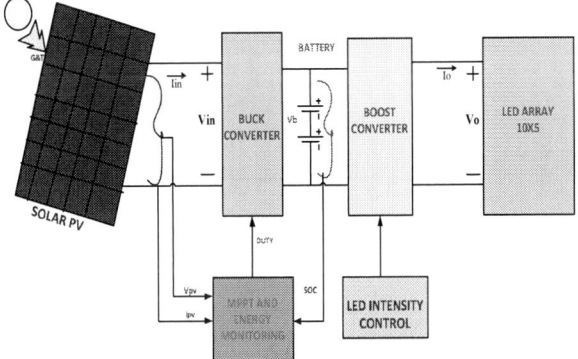

Fig.1 Solar Street Light Block Diagram

A SPV system of a capacity of a 200W peak with an output voltage and current of 39V & 5.20A is source of supply for the lighting system. Each street light is made of an array of 10x5 LEDs. Each LED is of 1W rating and has forward voltage of 2.9V.

Solar PV is designed to produce the output voltage of 39V; a buck converter with MPPT is used to step down the SPV voltage to 24V for charging the battery. A Li-ion battery having a capacity of 24V has been used for storing the energy of SPV system during daytime. The battery is charged in daytime and discharged during night through the LED array. The battery capacity is selected such that even if there is cloudy and rainy climate for consecutive three days, the battery will supply electricity to streetlight.

During high traffic hours i.e. from evening dusk to midnight 12:00 AM, the LED lamp is planned to glow bright for 100% intensity and during non-traffic hours i.e. after midnight the LED lamp brightness will be reduced down to 50%. The closed loop boost converter has been utilized to control the brightness of the LED lamp. When LED lamp is at the full brightness, the dc-dc boost converter will boost the battery output voltage

978-1-5386-6624-1/18 $31.00 © 2018 IEEE   671

of 24 V to 29V. During non-traffic hours (12:00AM to 6AM) the intensity is reduced to 50% by varying the current to save the energy consumption. This is done by varying the duty of boost converter.

The different components of the systems are as explained below:

A. PV Array

A SPV system of a capacity of 200W peak is used to fed LED Streetlight.

The maximum power is produced at 39V i.e. 200W under the STC conditions. Similarly, 100W at Solar Irradiance of 500W/m^2 as shown in fig.2

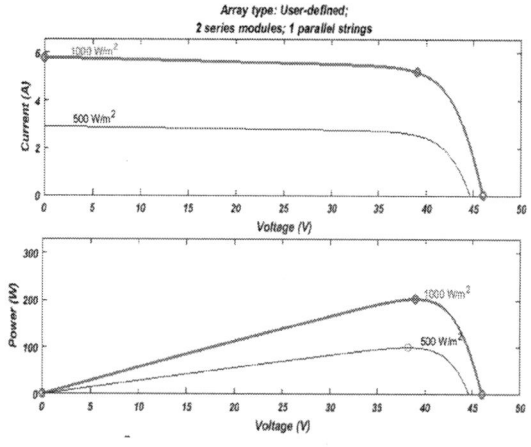

Fig.2 V-I and P-V Characteristics

B. DC-DC converter with MPPT

The duty cycle in buck converter is controlled by MPPT controller. The MPPT algorithm used for extracting the maximum power for charging the battery is Incremental conductance method(INC) as shown in fig.3 [3].The INC is used with buck converter for switching the MOSFET at 125KHz. Buck converter is used for stepping down the PV output voltage of 39 V to 24V which is required by battery.

The buck converter has been designed by using given equations [4]

$$\text{Inductor } (L_f) = \frac{D(1-D)Vs}{f_s.\Delta I_L} = 0.8488 \ \mu H \qquad (1)$$

where V_s is input voltage, D is Duty cycle of the converter, ΔI_L is Inductor ripple current, f_s is switching frequency, 125KHz

$$\text{Capacitor } (C_f) = \frac{D(1-D).Vs}{8.f_s^2 L \Delta V_C} = 0.73 \ \mu F \qquad (2)$$

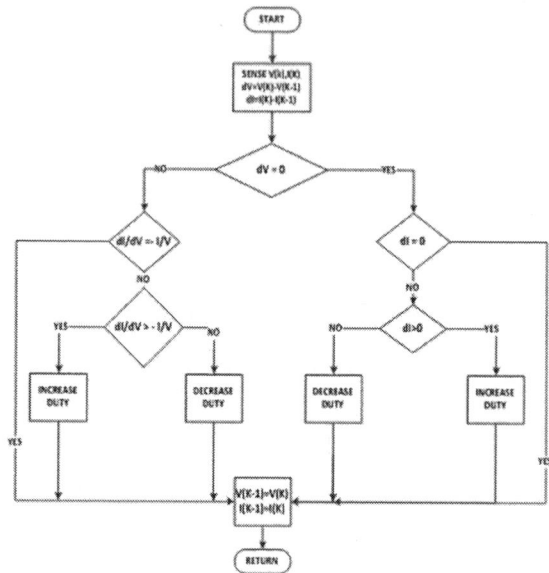

Fig.3 Incremental Conductance Algorithm

Where L_f is Inductance value, ΔV_c is Capacitor ripple voltage.

C. Solar Battery Storage

The energy provided by the PV array in daytime is utilized by 24V battery to charge it. For this purpose Li-ion generic parameterized battery model is used which can be accessible in Matlab-Simulink. The battery capacity is selected such that even if there is cloudy and rainy climate i.e. overcastting for consecutive three days, battery will give supply without any interruption.

D. DC-DC boost converter

A boost converter is used for intensity control of LED and stepping up the battery voltage i.e. 24V to the desired output voltage. The converter is working in continuous conduction mode (CCM) and value of inductor and capacitor is calculated from the following equation.

$$\Delta I = \left(\frac{(current tolerance)*(I)}{100}\right); \qquad (3)$$

$$\Delta V = \left(\frac{(voltage tolerance)*(V_s)}{100}\right); \qquad (4)$$

$$L_{boost} = \left(\frac{duty*Vs}{(f_s*\Delta I)}\right); \ = 0.9 * 10^{-4} \ H \qquad (5)$$

Where V_s is input voltage, f_s is switching frequency 13KHz, ΔI is Inductor ripple current

$$C_{boost} = \left(\frac{duty*I}{(f_s*\Delta v)}\right); \ = \ 3 * 10^{-5} F \qquad (6)$$

Where L_{boost} is Inductance value, ΔV is Capacitor ripple voltage

E. LED control and description

A 10X5 array of LED is used to design 50W Streetlight[5,6]. Each LED used have a forward voltage(V_f) of 2.9V and a forward current(I_f) of 350mA as shown in fig.4 at 25⁰C. The Resistance of the LED can be calculated from the slope of the tangent to the exponential I-V curve at the operating point Q(V_f, I_f) is given as,

$$\frac{1}{R_{LED}} = \frac{I_f - I_{f1}}{V_f - V_{f1}} \qquad (7)$$

Here, I_f is 0.35 A, I_{f1} is 0 A, V_f is 2.9V & V_{f1} is 2.6V from fig. 4.7.

Putting all values of I_f, I_{f1}, V_f and V_{f1} in above equations, the value of R_{LED} is,

$$\frac{1}{R_{LED}} = \frac{1}{1.16}$$

That result as R_{LED} = 0.823Ω

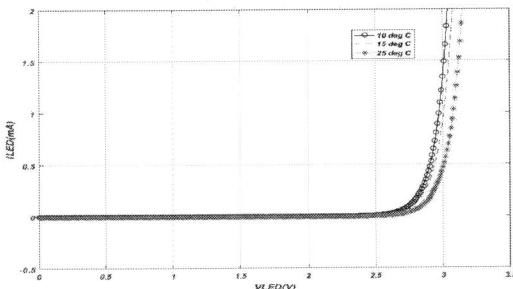

Fig.4 LED characteristics at different temperature

This data's is taken from Cree® XLamp® XPE2 LED data sheet [7]. The intensity of LED is controlled by changing the duty of the boost converter after 2.5sec in simulation.

III. SYSTEM DESIGNING & IMPLEMENTATION

Battery Charge capacity Calculation:
Power rating of LED = 1W
The energy utilized by the LEDs for operating for a duration of 12 hours ($6 * 50W + 6 * 25$) =450Wh
The energy required for 3 days ($450 * 3$) = 1350 Wh.
For 24V battery, Required Charge Capacity is 1350/24 = 56.25Ah.
Deep discharge batteries are used here with DoD in the range of 60% to 80%, Taking DoD of 70%, Required Charge Capacity = *54/0.7* = 80.35 Ah

PV Array rating calculation:
Consider the efficiency of the Battery as 90% and efficiency of SPV Array and Converter, 90%
Energy required at the battery $= \frac{1350}{0.9} = 1500 Wh$
Energy that should be supplied by Buck Converter $= \frac{1500}{0.9} = 1666.67 Wh$

Total Ah generated $= \frac{1666.67}{24} = 69.44 Ah$

The SPV module's Power capacity is varied with respect to Solar Irradiation. In India average of 1000W/m² irradiation is for 6hrs & 500W/m² irradiation is for 4hrs, So total hrs for Solar irradiation of 1000W/m² can be taken as 8hrs in a day. So, total current generated at buck converter output is 8.68A and correspondingly solar array should produce 5.2A short circuit current
So, solar PV array rating will be 200W at STC and 2 Nos. of Module of 100W are connected in series of rating 5.2A I_{mp} and 19.5V V_{mp} each.

A. Simulation Result & Discussion

The developed scheme has been modelled using MATLAB-SIMULINK software. All the parameters have been designed using the above calculated values. The LED lamp load is modelled as simple series resistance of 0.823Ω with a battery.

As during the day the solar irradiance varies, so to analyze the performance of the system, the battery charging is performed for two solar irradiations i.e. 1000W/m² and 500W/m². The Power, Voltage and Current graph of solar panel for both irradiance is shown in fig.5, 6 & 7 respectively. As is clear from fig, that at 1000W/m² the maximum power is 200W and for decreased irradiance of 500W/m² i.e. after 2.5 sec of simulation time the maximum power is reduced to 100W. The battery parameters i.e. SOC, voltage and current are shown in fig.8, 9 & 10 respectively.

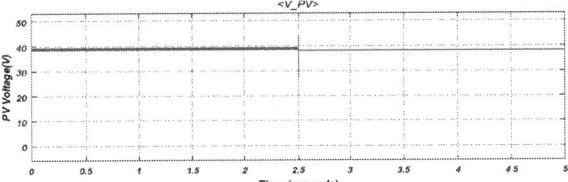

Fig.5 PV voltage at two Irradiation (1000W/m², 500W/m²)

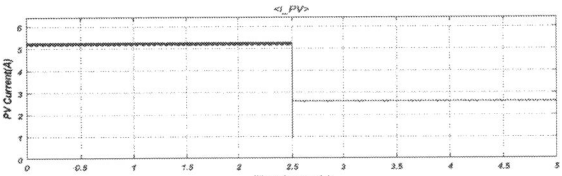

Fig.6 PV current at two irradiation (1000W/m², 500W/m²)

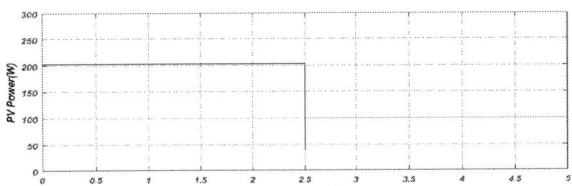

Fig.7 PV power at two irradiation (1000W/m², 500W/m²)

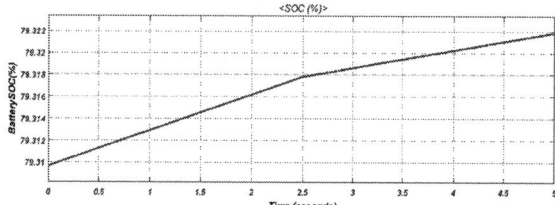

Fig.8 Battery SOC (%) at two irradiation (1000W/m², 500W/m²)

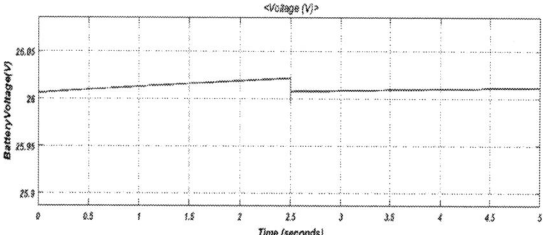

Fig.9 Battery Voltage (V) at two irradiation (1000W/m², 500W/m²)

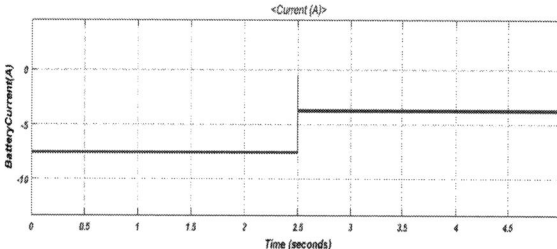

Fig.10 Battery Current (V) at two irradiation (1000W/m², 500W/m²)

When solar is not available, the battery is supplying power to the LED lamps as shown in fig.11,12. From dusk to midnight, the LED output current and voltage is 1.75A and 29 V respectively, so the power yield at the output is approx 50W. After midnight, as the traffic becomes very less so intensity of LED light can be decreased by controlling duty cycle of boost converter to save electrical energy i.e. from midnight to 6:00AM. It is shown in simulation after 2.5 sec correspondingly the LED output current and voltage is change to 0.86A and 27.8 V respectively, so the power yield at the output is approx 24W. As it is clear from the results that the approximately 50 % of the energy is saved during the non traffic hours.

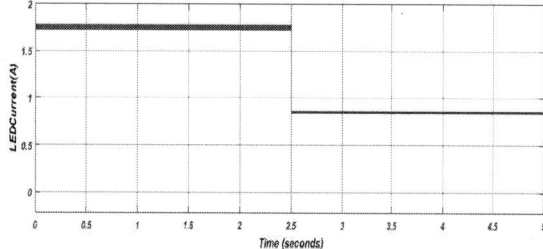

Fig.11 Output current (A) at different intensities

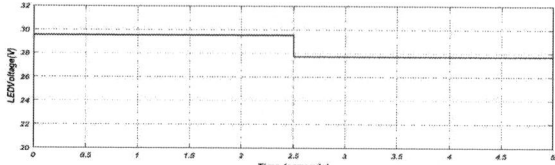

Fig.12 Output voltage (V) at different intensities

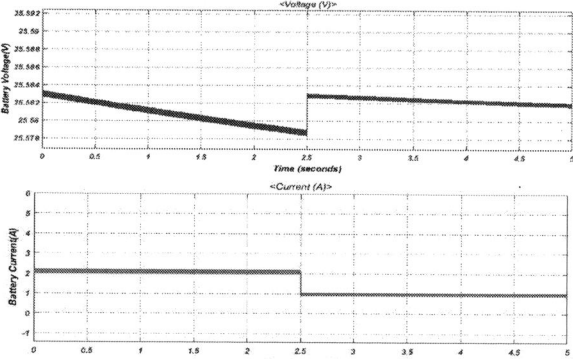

Fig.13 Battery Voltage (V), Current (A) at different intensities

Fig.13 shows battery parameters i.e. battery current & voltage. And it can be seen that the battery is discharging or supplying power.

IV. PV HYBRID LED STREET LIGHTING SYSTEM

The output power of PV panel used in the street lighting system is usually limited to 200W because of the limited size of a mounting structure [8]. So, there is a need for another energy supplying element that can be Wind, Grid (Utility) etc.

The structural block diagram of PV Hybrid system is shown in fig.14 and the working of the system has been explained with the help of vitality control algorithm as shown in fig.15. The framework comprises of SPV array of 200W, a lithium-ion battery of 24V/72Ah, INC based MPPT, dc-dc Buck converter, a 230V AC/DC module [9, 10] and LED load of 50W.

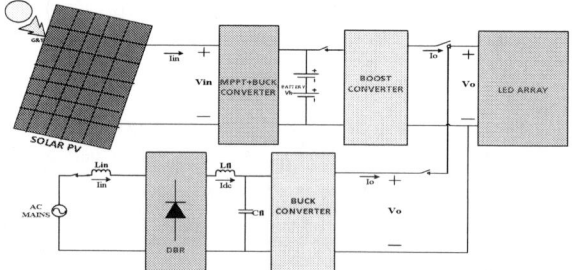

Fig.14 Block Diagram of PV Hybrid LED Street Lighting System

During daytime, the battery will be charged by P-V array. During night, the battery discharges to light up LED array. During nonstop cloudy or rainy climate, the battery may not have any electric charge left. Under such conditions, AC/DC converter is used to light the LED array. The power exchange switch is utilized to choose either the battery or AC/DC module to light up LED array from dusk to dawn, while in the daytime the switch is in open mode because no power is needed by load as it is in OFF state.

A. System Designing & Implementation

Battery Charge capacity Calculation:
Power rating of LED = 1W
The energy utilized by the LEDs for operating for a duration of 12 hours (50 ∗ 12) =600Wh
The energy required for 2 days (600 ∗ 2) = 1200 Wh.
For 24V battery, Required Charge Capacity is 1200/24 = 50 Ah.
Deep discharge batteries are used here with DoD in the range of 60% to 80%, taking DoD of 70%, Required Charge Capacity = 50/0.7 = 71.42 Ah

PV Array rating calculation:
Consider the efficiency of the Battery, 90% and efficiency of SPV Array and Converter, 90%
Energy required at the battery $= \frac{1200}{0.90} = 1333.33Wh$
Energy that should be supplied by Buck Converter $= \frac{1333.33}{0.9} = 1481.47Wh$
Total Ah generated $= \frac{1481.47}{24} = 61.72Ah$
The SPV module's Power capacity is varied with respect to Solar Irradiation. In India average of 1000W/m² irradiation is for 6hrs & 500W/m² irradiation is for 4hrs, So total hrs for Solar irradiation of 1000W/m² is 8hrs a day. So, total current generated at buck converter output is 7.71A and correspondingly solar array should produce 5.24A short circuit current (Duty of converter = 0.65)
So, solar PV array rating will be approx 200 W at STC and 2 Nos. of Module of 100W are connected in series

Vitality Control Algorithm: This algorithm is used for maintaining the energy balance between PV Array, Battery, and Utility (AC/DC Module). The algorithm only sense voltage (V) & current (I) of PV and SOC of the battery, if Power (P = V*I) is less than 40W and battery SOC is between 30 to 80% the battery will supply the load. If SOC is not within the given limit the grid will be connected from dusk till dawn. If power of PV is greater than 40W then battery will be charged and other sources and load will be in OFF condition.

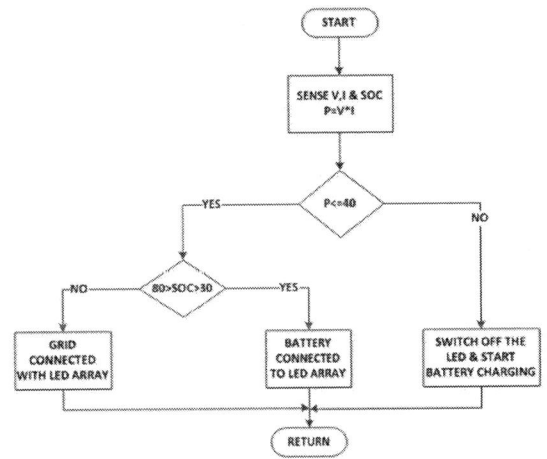

Fig.15 System Control Algorithm

B. Simulation Result & Discussion

The developed scheme has been modelled & simulated using MATLAB/SIMULINK software. All the parameters have been designed using the above calculated values; The LED lamp load is modelled as simple series resistance with battery as described in previous section and 10x5 array of LED is used as a Load.
When solar is not available, the battery is supplying power to the LED lamp as shown in fig.16. From dusk to dawn the LED output current and voltage is 1.73A and 29.5V respectively and the current from utility is zero as shown in fig.17.

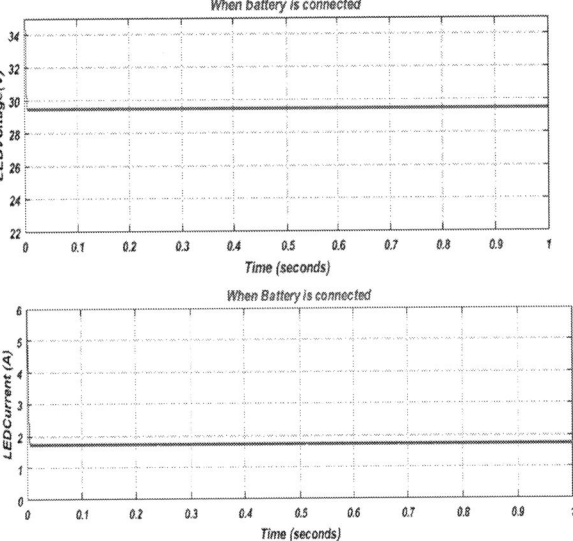

Fig.16 Voltage and current of LED when fed from battery

978-1-5386-6624-1/18 $31.00 © 2018 IEEE

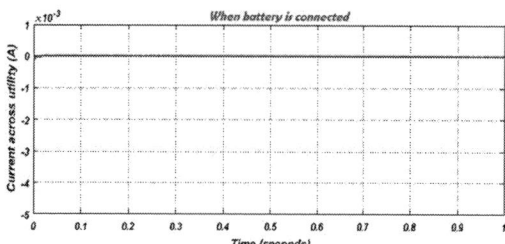

Fig.17 Output current from utility

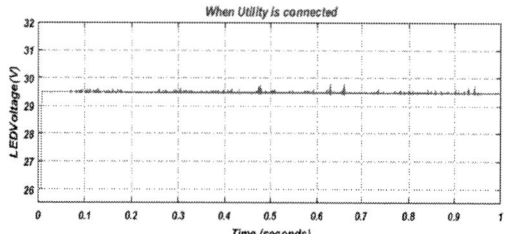

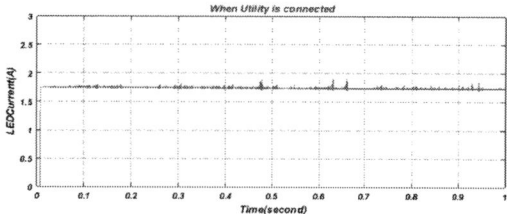

Fig.18 Voltage and current of LED when fed from utility

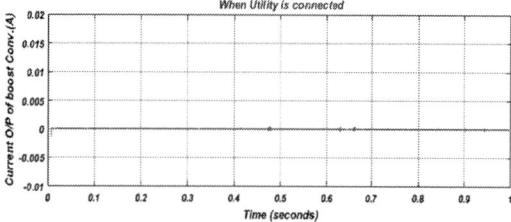

Fig.19 Output current of boost converter

If there is continuous overcast situation, the battery may fail to charge under those conditions to maintain constant supply to LED, it is lit by AC/DC module. The corresponding values of output voltage and output current of LED and boost converter are shown in fig. 18 and fig.19 respectively. It has been observed that the voltage and current remain same, whether the LED load is lit up by battery or utility. So the system is working very efficiently and accurately

V. CONCLUSION

The two models of smart solar LED street lighting are successfully designed and implemented. Based on the Model calculation and simulink results it has been observed that intensity control of LED has benefits such as less energy consumption and battery can work more efficiently giving higher duration of backup than compared to the streetlight which is working without any intensity control. Further a PV hybrid system is designed which can continuously under any weather conditions.

VI. REFERENCES

[1] Vieira, J.A.B.; Mota, A.M., "Implementation of a stand-alone photovoltaic lighting system with MPPT battery charging and LED current control," *Control Applications (CCA), 2010 IEEE International Conference on* , vol., no., pp.185,190, 8-10 Sept. 2010.

[2] Jaspreet Singh, Prof. Priya Mahajan, Prof. Rachana Garg "Techno Economic Analysis Of Different Lighting Schemes: A Case Study" presented in *IEEE International Conference on Power Energy, Environment & Intellegent Control, PEEIC* , 2018 to be published on IEEE Xplorer.

[3] R chafle, srushti & B vaidya, uttam "Incremental Conductance MPPT Technique FOR PV System," *International Journal of Advanced Research in Electrical, Electronics and Instrumentation Engineering*, 2(6), 2719-2726,2013

[4] Saurabh Thakran ,Jaspreet Singh, Prof. Priya Mahajan, Prof. Rachana Garg "Implementation of P&O Algorithm for MPPT in SPV System" presented in *IEEE International Conference on Power Energy, Environment & Intellegent Control, PEEIC* , 2018 to be published on IEEE Xplorer.

[5] Ray-Lee Lin, Jhong-Yan Tsai,"Four-Parameter Taylor Series-Based Light-Emitting-Diode Model," *IEEE journal of emerging and selected topics in power electronics, VOL. 3, NO. 3*, SEPT 2015

[6] Ray-Lee Lin, Yi Fan Chen, "Equivalent Circuit Model of Light-Emitting-Diode for System Analyses of Lighting Drivers," *IEEE Industry Applications Society Annual Meeting*, Houston, Oct 2009.

[7] http://www.cree.com/~/media/Files/Cree/LED%20Compon ents%20and%20Modules/XLamp/Data%20and%20Binnin g/XLampXPE2.pdf. [Access - 5-March-2018]

[8] Liuyi Ling, Xiaoliang Wu, Mengyuan Liu, Zhiqiang Zhu ,Van Li, Benben Shang, "Development of Photo voltaic Hybrid LED Street Lighting System," *IEEE Advanced Information Management, Communicates, Electronic and Automation Control Conference (IMCEC)*,China, 2016

[9] Ramanjaneya Reddy U and Narasimharaju B. L, "Unity power factor buck-boost LED driver for wide range of input voltage application," *2015 Annual IEEE India Conference (INDICON)*, pp. 1-6, New Delhi, 2015.

[10] Gacio, D., Alonso, J.M., Calleja, A.J., Garcia, J., Rico-Secades, M.:'A universal-input single-stage high power-factor power supply for HB-LEDs based on integrated buck-flyback converter', *IEEE Trans. Ind. Electron.*, 58, (2), pp. 589–599, 2011.

Modified Gradient Spectral Variance Smoothing Adaptive Filter Control for Grid Connected PV System

Sandeep Kumar Sahoo, Shailendra Kumar, *Member, IEEE* and Bhim Singh, *Fellow, IEEE*
Department of Electrical Engineering, *IIT Delhi*, New Delhi-110016, India
Emails: sandeepksahoo257@gmail.com, er.dwivedi88@gmail.com, bsingh@ee.iitd.ac.in

Abstract- **The modified gradient spectral variance smoothing (MGSVS) based control is used for a grid integrated two stage solar photovoltaic (PV) system. The boost converter is used as the first stage for tracking maximum power through the perturb and observe (P & O) control technique. A voltage source converter (VSC) is placed at the next stage, which handshakes between the PV array and the grid. MGSVS based control is used to provide gating pulses for the VSC. This PV system operates at unity power factor, compensates reactive power, reduces the grid harmonics current. The fast convergence is achieved by providing an optimal step size in the algorithm. This system is modeled and its performance is simulated in MATLAB-SIMULINK platform under different operating conditions and these simulated results are validated on an experimental prototype. Test results obtained, are reasonable with enhanced steady state and dynamic performances. The total harmonic distortion (THD) of grid current follows the IEEE-519 standard.**

Keywords- **MPPT, VSC, PV system, MGSVS, THD, power quality.**

I. INTRODUCTION

From ancient times, electricity generation is based on fossil fuels, whose reserves are limited and their amount is decreasing day by day. They cause environmental hazards due to CO_2 emission, which results in global warming. The pollution caused by these sources, creates severe health problems. Because of the above-said problems, the entire world is looking for renewable energy. These problems motivate researchers to go for clean energy alternatives like solar photovoltaic (PV), wind, hydro etc. Among several clean energy sources, PV energy is noiseless, pollution free, available in plenty. Due to advancement of technology, the efficiency of PV panels, have been increasing and the cost has been reducing day by day. Government has been providing financial support for installing rooftop PV panels. So, PV array has been a reliable alternative of traditional fossil fuels and hence demand of solar PV system has been increasing [1]. The PV system may be single stage or double stage. It may be connected to the utility grid or in standalone mode. Battery banks are used in standalone mode for maintaining continuity of the supply. Moreover in grid connected system, grid safeguards continuity of supply i.e. if

the PV system is unable to address the load demand then that extra power is supplied by the grid. Among single stage and double stage configurations, the controller design is easy in case of the later one, so double stage configuration is used here. Moreover, a double stage configuration has the flexibility in operation and improved performance for wide variations in PV array power generations and grid voltages.

Performances of the PV system, are dependent upon temperature and solar insolations. So for effective utilization of the system, the panel must run at maximum power point. That's why maximum power point tracking (MPPT) algorithms are used. There are several MPPT techniques like perturb and observe (P & O), incremental conductance, fractional open circuit voltage, fractional short circuit current, fuzzy logic control and neural network, which are available in the literature [2]. In this work, for tracking MPP, P & O algorithm is used because of its simplicity. A voltage source converter (VSC) is used for connecting the PV system to the grid. Several control algorithms like instantaneous reactive power theory (IRP Theory), synchronous reference frame theory (SRF Theory) [3], damped-SOGI based control algorithm [4], adaptive control [5]-[8] etc. are available in the literature for VSC to interface PV system to the grid. So, several control techniques are being developed for gating the switches of VSC to have fast, reliable and efficient operation of grid interfaced PV system.

Several filters using adaptive techniques have been used to track environmental condition changes. The least mean squares (LMS) algorithm is a primitive member of adaptive family [9]. LMS has better convergence rate but it is not effective in low signal-to-noise (SNR) ratio. However, it has disagreement between rate of convergence and steady state error. The proposed algorithm has improved the error exponent to enhance steady state performance using optimal step size. The proposed grid hand-shaked PV system is modeled and its performance is simulated in MATLAB/Simulink platform. Moreover, experiments are conducted on a developed prototype to validate the authenticity of the PV system.

II. SYSTEM CONFIGURATION

The proposed PV system consists of two stages as shown in Fig. 1. A boost converter is there, which tracks MPPT

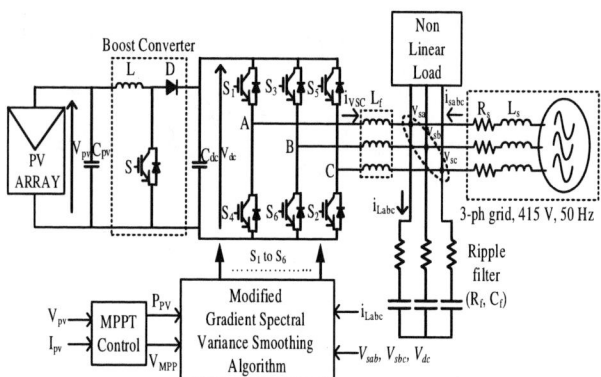

Fig. 1 System configurations

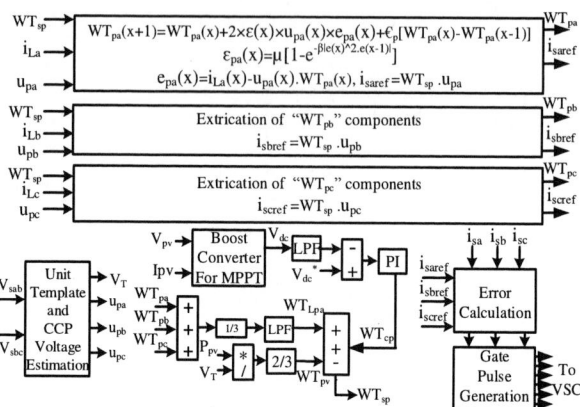

Fig.2 MGSVS based control algorithm

using the P & O control algorithm. After this stage, a DC link capacitor is there followed by a voltage source converter, which feeds active power to the grid and simultaneously fulfills the load demand. A combination of diode bridge rectifier with RL load, is used as a nonlinear load. For reducing switching ripple, a ripple filter (RC type) is used at the common connection point (CCP).

III. CONTROL ALGORITHM

The control algorithm comprises of couple of stages. First stage of it, consists of perturb and observe (P & O) control method for getting peak power from a PV array. The second stage consists of modified gradient spectral variance smoothing (MGSVS) based adaptive current control algorithm for generating gating pulses for the voltage source converter (VSC) switches to get the features like load balancing, elimination of harmonics, grid currents balancing and power factor correction ensuring improvement in power quality [10].

The output of the MPP tracking algorithm is the voltage across the DC link capacitor, which is compared with the reference DC link voltage value and the generated error is fed to the PI controller to get active power loss component. The MGSVS control algorithm is shown in Fig. 2, which is used to get in phase unit templates (u_{pa} , u_{pb} , u_{pc}) of sensed voltage of CCP, estimation of peak value of CCP voltage (v_T), extrication of weights of fundamental active components ($WT_{pa}, WT_{pb}, WT_{pc}$) of three phase load currents, active loss components (wt_{Lpa}), generation of reference grid currents (i_{saref} , i_{sbref} , i_{scref}). By comparing the reference grid currents with the sensed grid currents switching signals are generated for VSC to get the desired features for power quality improvement.

A. Maximum Power Point Tracking Control

The main aim of the MPPT algorithm is to maximize the obtained power from PV array in varying atmosphere conditions and to get DC link voltage for VSC. In the literature [2], several MPPT algorithms are available for MPPT tracking. Among them, the P & O technique is the simplest one to implement. So, this is used for MPP tracking. The boost converter operates in such a way that it extricates peak power from PV array. Here, the required duty cycle is generated by P & O based MPPT algorithm. This is a hill climbing technique. In the power versus voltage curve, by giving a perturbation to voltage, the changes in voltage and power, are calculated. If the product of these couple of quantities, is positive, then the reference duty ratio is reduced to move towards MPP. Similarly after perturbation, if the product is negative, then the reference duty ratio is increased to attain MPP. If the magnitude of power is unchanged, then the PV array is running at MPP. Inputs to the control of the boost converter, are PV array output voltage and current and voltage across the capacitor is the output.

A. MGSVS Based Adaptive Current Control Algorithm

The schematic diagram of the MGSVS based current control topology is shown in Fig. 2. Followings are shown in this algorithm.

1) *Unit Template Calculation and Amplitude of Peak Terminal Voltage*

From the CCP, sensed line voltages (v_{sab}, v_{sbc}) are used to get instantaneous phase voltage for three phases as,

$$v_{sa} = (2v_{sab} + v_{sbc})/3; v_{sb} = (-v_{sab} + v_{sbc})/3;$$
$$v_{sc} = (-v_{sab} - 2v_{sbc})/3 \tag{1}$$

The peak of CCP voltage is calculated as,

$$V_T = (2/3(v_{sa}^2 + v_{sb}^2 + v_{sc}^2))^{1/2} \tag{2}$$

u_{pa} , u_{pb} , u_{pc} are in-phase unit templates, which are eestimated as,

$$u_{pa} = v_{sa}/V_T; u_{pb} = v_{sb}/V_T; u_{pc} = v_{sc}/V_T \tag{3}$$

978-1-5386-6624-1/18 $31.00 © 2018 IEEE   678

2) Calculation of Loss Components

The DC link voltage across the capacitor is sensed and the error found at 'x' th sampling point is,

$$V_{der}(x) = V_{dc}^{*}(x) - V_{dc}(x) \tag{4}$$

where, sensed DC link voltage and reference DC link voltage, are V_{dc} and V_{dc}^{*}, respectively. The extracted error is fed to PI controller to get WT_{cp} as,

$$WT_{cp}(x+1) = WT_{cp}(x) + K_{pa}(V_{der}(x+1) - V_{der}(x)) + K_{ia}V_{der}(x+1) \tag{5}$$

Where, K_{ia}, K_{pa} are integral, proportional gains of the PI controller.
The PV array feed forward component is found as follows,

$$WT_{pv}(x) = (2P_{pv}(x)/3V_T) \tag{6}$$

where, P_{pv} is the obtained PV power by MPPT action.

3) Fundamental Weight Extrication of The Load Currents

MGSVS based adaptive control algorithm [11] is used to extricate the fundamental components of load currents. This algorithm enhances the speed of extrication of weights maintaining proper accuracy. The weight of the fundamental active component of load current of phase 'a' (i_{La}) is calculated as,

$$WT_{pa}(x+1) = WT_{pa}(x) + 2\varepsilon_{pa}(x) \times u_{pa}(x) \times e_{Pa}(x) + \textit{€}_p[WT_{pa}(x) - WT_{pa}(x-1)] \tag{7}$$

$$\varepsilon_{pa}(x) = \mu[1 - e^{-\beta|e(x)|^2 \cdot e(x-1)|}] \tag{8}$$

Where, μ, β and $\textit{€}_p$ are adaptive gains of the controller. ε_{pa} is the step factor for the algorithm.

$$e_{pa}(x) = i_{La}(x) - u_{pa}(x).WT_{pa}(x) \tag{9}$$

where, $u_{pa}(x), i_{La}(x)$ and $WT_{pa}(x)$ are the in phase unit template of phase 'a', load current of phase 'a' and weight of active reference component. $e_{pa}(x)$ is the adaptive component error. Likely, the weights for the fundamental active components of load currents of phase 'b' (i_{Lb}) and 'c' (i_{Lc}) are;

$$WT_{pb}(x+1) = WT_{pb}(x) + 2\varepsilon_{pb}(x) \times u_{pb}(x) \times e_{pb}(x) + \textit{€}_p[WT_{pb}(x) - WT_{pb}(x-1)] \tag{10}$$

$$WT_{pc}(x+1) = WT_{pc}(x) + 2\varepsilon_{pc}(x) \times u_{pc}(x) \times e_{pc}(x) + \textit{€}_p[WT_{pc}(x) - WT_{pc}(x-1)] \tag{11}$$

4) Reference grid currents generation

The net active weight component (WT_{sp}) of reference grid current is calculated by summing the average fundamental active weight (WT_{cp}) with DC loss component (WT_{Lpa}) and subtracting PV array feed-forward component (WT_{pv}).

$$WT_{sp} = WT_{cp} + WT_{Lpa} - WT_{pv} \tag{12}$$

where,

$$WT_{Lpa} = (WT_{pa} + WT_{pb} + WT_{pc})/3 \tag{13}$$

The reference grid currents are found as,

$$i_{saref} = WT_{sp}.u_{pa}; i_{sbref} = WT_{sp}.u_{pb}; i_{scref} = WT_{sp}.u_{pc} \tag{14}$$

5) Generation of Gating Pulses for VSC

To generate gating pulses for VSC, current error signals are estimated from the variation between reference grid currents ($i_{saref}, i_{sbref}, i_{scref}$) and sensed grid currents (i_{sa}, i_{sb}, i_{sc}). After this, the generated error signals are fed to hysteresis based current controller to gate the switches of VSC.

IV. EXPERIMENTAL RESULTS

The prototype of a PV system is developed by using a PV simulator (Terra SAS ETS 600/17), which replicates the PV array characteristics. A PV array simulator of 4.245 kW (V_{MPP}= 344.48 V and I_{MPP}= 12.324 A) is used to study the performance of the PV system. DSP (dSPACE1103) is used to implement the proposed controller. Hall Effect voltage sensors (LV-25) and current sensors (LA-55p) are used to sense voltage and current signals. An optical isolator is used to give isolation between power level and control level. A four channel digital storage oscilloscope (Agilent, DSO-DSO6010A) and three phase power quality analyzer (HIOKI) are used to record the test results. MPPT performance at different insolations (1000 W/m² and 500 W/m²), is shown in Figs. 3(a)-(b). They represent MPP along with V_{oc} and I_{sc}. In these conditions, several features of a three phase three wire (3P3W) grid interfaced PV system, are analyzed.

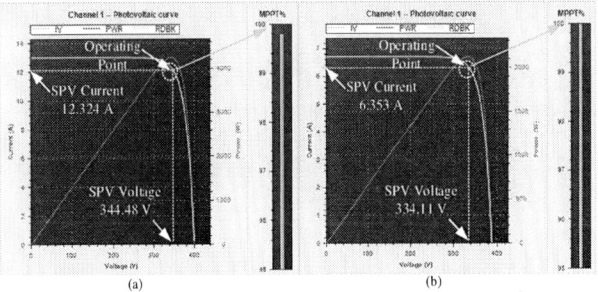

Figs. 3(a)-(b) MPP Tracking features at 1000 W/m² and 500 W/m²

A. System Performance at Nonlinear Load

Steady state performance at nonlinear load, is observed from Figs. 4(a)-(h). Grid voltage and current in steady state, are shown in Fig. 4(a). Fig. 4(b) shows power delivered to the grid in steady state i.e. 2.408 kW. All total 2.418 KVA is delivered to the grid at 0.9955 lagging power factor. Total harmonic distortion (THD) of grid voltage and current, are observed 1.7 % and 4.03 %, respectively as shown in Fig. 4(c). These waveforms of supply voltage and compensating current provided by VSC, are viewed in Fig. 4(d). Fig. 4(e) shows the THD of compensating current, which is 11.27 %. The load current profile is nonlinear, which is viewed in Fig. 4(f).

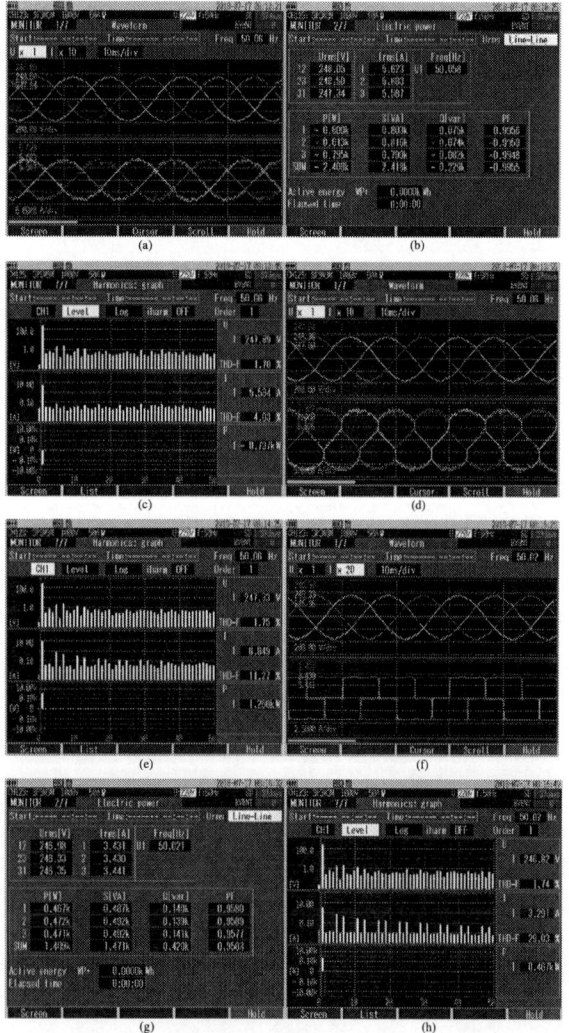

Figs. 4(a)-(h) Steady state features with nonlinear load

From Figs. 4(g)-(h), it can be viewed that 1.409 kW is delivered to the nonlinear load at 0.9583 lagging power factor.

B. System Performance at Unbalanced Loading Condition

Figs. 5(a)-(g) show the internal signals of the control algorithm with sudden disconnection of the load from phase 'a'. Fig. 5(a) shows the weight of phase 'a' (WT_{pa}) is reduced to zero, when load from phase 'a' is removed. The net power fed to the load is reduced and hence WT_{pb} and WT_{pc} are also reduced. Fig. 5(b) shows the extracted error

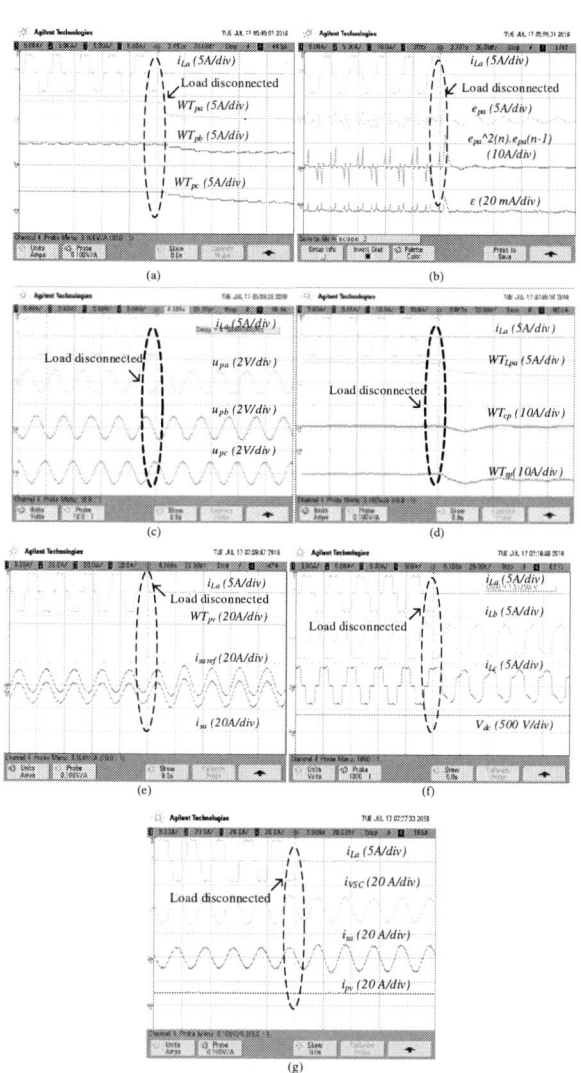

Figs. 5(a)-(g) Internal signals of the MGSVS based adaptive controller when load is disconnected

terms of the proposed control. Fig. 5(c) shows the in phase unit templates of the grid voltages, which are not disturbed though there is load unbalancing. Fig. 5(d) shows the average weight of the active component of the load currents (WT_{Lpa}), DC loss component (WT_{cp}) and weight of the net current WT_{sp}. The total weight of the current is reduced because of removal of load from one phase but due to proper control action that weight is increased instantaneously. Fig. 5(e) shows the weight of PV feed-forward component (WT_{pv}), reference current (i_{saref}) and sensed grid current (i_{sa}). i_{saref} and i_{sa} are in phase. Fig. 5(f) shows load currents of phase 'a' (i_{La}), phase 'b' (i_{Lb}) and phase 'c' (i_{Lc}) and input voltage of the VSC (V_{dc}). V_{dc} is maintained though there is sudden disconnection of the load. Fig. 5(g) shows VSC current (i_{VSC}), sensed current of phase 'a' and PV current (i_{pv}). Since the load from phase 'a' is removed no compensation is needed for phase 'a', so VSC current is sinusoidal for phase 'a'.

Figs. 6(a)-(d) show test results obtained from power analyzer. Fig. 6(a) shows that when the load from phase 'a' is removed, the power delivered to the load is reduced to 679 W, which is shown in Fig. 6(b). The phasor diagrams are shown in Fig. 6(c). VSC current is shown in Fig. 6(d)

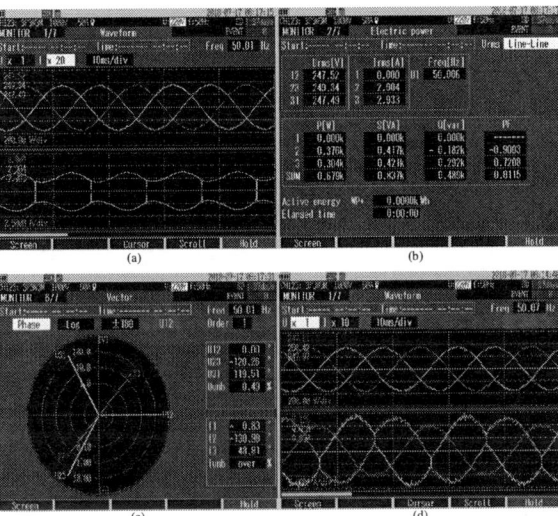

Figs. 6(a)-(d) Analysis of power quality when load is unbalanced

which, indicates that, as there is no load in phase 'a', no compensation is needed for that phase.

C. *Performance at Change in Insolations*

The PV system performance under variable insolation is shown in Fig. 7. Fig. 7(a) indicates that there is a decrement in solar insolation from 1000 W/m² to 500 W/m². V_{dc} is maintained as that of the previous case because of the double stage system. I_{pv} is reduced because of fall in solar insolation. PV power i.e. P_{PV} is also reduced. Since the generated power from the PV array is reduced, the power fed to the grid, reduces but the THD for grid current is maintained according to the IEEE-519 standard.

Fig. 7(b) shows the case when the solar insolation is raised from 500 W/m² to 1000 W/m². V_{dc} is maintained as earlier. I_{pv} is increased and hence the PV power.

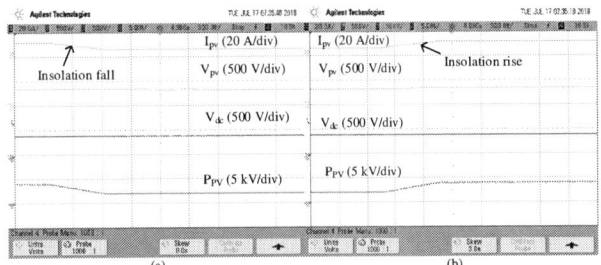

Figs. 7(a)-(b) Performance analysis during variable insolation

D. *Performance Analysis of PV Mode to DSTATCOM Mode and Vice Versa*

When PV array power generation is reduced to zero, then distribution static compensator (DSTATCOM) action takes place. The compensating currents are provided by the VSC. THDs of grid current and voltage are maintained as per the IEEE-519 standard. When PV array operates, at that time, grid voltage (V_{sa}) and grid current (i_{sa}) are 180⁰ out of phase, that implies power is delivered to the grid but when PV power generation reduces to zero, at that time V_{sa} and i_{sa} are in phase i.e. power is supplied by the grid. These results are shown in Fig. 8.

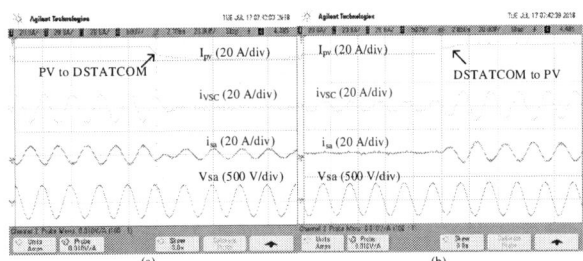

Figs. 8(a)-(b) Performance analysis during PV to DSTATCOM and vice versa

V. CONCLUSION

A grid connected two stage PV system with MGSVS based control is used for feeding power to the grid, to reduce harmonics in the grid current and voltage, to have load balancing and to correct the power factor. The peak power is extracted from the PV system by P & O control algorithm. The efficacy of MGSVS control is observed from its accurate evaluation of weights of phases. The proposed PV system has shown fast performance under various dynamic conditions like sudden change of solar insolation, sudden change of load and ensures the THD of grid voltage and current according to the IEEE-519 standard, thereby improving the power quality.

ACKNOWLEDGEMENT

The authors are highly obliged to Dept. of Science and Technology (DST), Government. of India, for the monetary cooperation under Grant No.: RP03357.

APPENDIX

SPV array parameters: For each panel-V_{oc}= 32.8 V, I_{sc}= 8.2 A, V_{MPP}= 27.8 V, I_{MPP}= 7.5 A, V_{dc}= 400 V, C_{dc}= 3500 μF, C_{pv}= 50 μF; grid parameters: R_s= 0.1 Ω and L_s= 2 mH; RC filter: R_f= 6 Ω, C_f= 10 μF; PI controller: K_{pa}= 0.8, K_{qa}= 0.8, nolinear load: diode bridge rectifier with 150 mH and 50 Ω; Controller parameters: β= 0.01, μ= 0.1, $€_p$= 0.1.

REFERENCES

[1] C. S. Solanki, Solar Photovoltaic Technologies and Systems. PHI Learning Pvt. Ltd., 2013.

[2] N. Kumar, I. Hussain, B. Singh and B. K. Panigrahi, "Normal Harmonic Search Algorithm Based MPPT for Solar PV System and Integrated with Grid using Reduced Sensor Approach and PNKLMS Algorithm," *IEEE Trans. on Indus. Appli.*, Early Access.

[3] B. Singh and J. Solanki, "A Comparison of Control Algorithms for DSTATCOM," *IEEE Trans. on Indus. Electron.*, vol. 56, no. 7, pp. 2738-2745, July 2009.S.

[4] B. Singh, S. Kumar and C. Jain, "Damped-SOGI-Based Control Algorithm for Solar PV Power Generating System," *IEEE Trans. on Indus. Appli.*, vol. 53, no. 3, pp. 1780-1788, May-June 2017.

[5] B. Singh and S. Kumar, "Grid integration of 3P4W solar PV system using M-LWDF-based control technique," *IET Renew. Pow. Gener.*, vol. 11, no. 8, pp. 1174-1181, 6 28 2017.

[6] V. L. Srinivas, S. Kumar, B. Singh and S. Mishra, "Partially Decoupled Adaptive Filter Based Multifunctional Three-Phase GPV System," *IEEE Trans. on Sustain. Energy*, vol. 9, no. 1, pp. 311-320, Jan. 2018.

[7] S. Kumar, I. Hussain, B. Singh, A. Chandra and K. Al-Haddad, "An Adaptive Control Scheme of SPV System Integrated to AC Distribution System," in *IEEE Trans. on Indus. Appli.*, vol. 53, no. 6, pp. 5173-5181, Nov.-Dec. 2017.

[8] S. Kumar and B. Singh, "Multi-Objective Single-Stage SPV System Integrated to 3P4W Distribution Network Using DMSI-Based Control Technique," in *IEEE Trans. on Indus. Appli.*, vol. 54, no. 3, pp. 2656-2664, May-June 2018.

[9] B. Widrow, M.E. Hoff, Adaptive switching circuits, STANFORD UNIV CA STANFORD ELECTRONICS LABS, 1960

[10] B. Singh, A. Chandra, and K. Al-Haddad, *Power Quality: Problems and Mitigation Techniques*. Hoboken, NJ, USA: Wiley, Jan. 2015

[11] Q. Niu and T. Chen, "A new variable step size LMS adaptive algorithm," 2018 Chinese Control And Decision Conference (CCDC), Shenyang, China, 2018, pp. 1-4.

2nd IEEE International conference on power Electronics, Intelligent Control and Energy systems (ICPEICES-2018)

Multifunctional Adaptive RZA-NLMF Control Technique for Double Stage SEGS Interfaced With Three-Phase Distribution Feeder

Amresh Kumar Singh, *Member, IEEE*, Ikhlaq Hussain, *Member, IEEE* and Bhim Singh, *Fellow, IEEE*
Department of Electrical Engineering, *IIT Delhi*, New Delhi-110016, India
Email id: amreshkumar98@gmail.com, ikhlaqiitd@gmail.com and bhimsinghiitd61@gmail.com

Abstract— This work present an intelligent adaptive technique is used for control a two stage solar power generating system (SEGS) feeding to a three phase distribution network. This system utilizes PV (Photovoltaic) modules, VSI (voltage Source Converter), linear and nonlinear loads. A RZA-NLMF (Reweighted Zero Attracting-Normalized Least Mean Fourth) adaptive approach is utilized to extricate the fundamental component of polluted load currents and for multiple purpose, such as improvement of quality of power at PCI (Point of Common Interconnection). For extracting, the extent amount of available power from the given SEGS at different climatically conditions, this system uses a simple P&O (Perturb and Observe) method. It also offers its capability of harmonics reduction and ensures the power quality norms. The proposed grid integrated SEGS is modeled in the MATLAB/Simulink. For check the performance of proposed approach at varying nonlinear loads on different environment conditions, the proposed control technique is examined through various recorded results on a hardware setup in the laboratory.

Keywords— *RZA-NLMF, PV, SEGS, MPPT, VSI, Power Quality and Three phase Distribution feeder.*

I. INTRODUCTION

There is a global issue of shortage of energy in the world because of fast depletion of conventional energy sources. Moreover, the degrading environment is one of the reason for moving towards the non-traditional eco-friendly green energy sources. Solar energy is most promising feasible cost effective option [1-2]. There is broadly two ways of generating solar PV (Photovoltaic) energy which are standalone and grid connected systems. In [3-4], especially for rural electrification, few standalone systems are considered for PV power generation. These PV systems are used with three port converters in battery energy management and battery charging stations. The batteries are the major requirement of standalone PV systems. However, the major demerits of these batteries are that these need periodic maintenance and timely replacement. These batteries are not needed in the grid connected solar power generation systems.

In [5-6], many harvesting methods are reported to extricate the peak power from solar energy generating systems (SEGSs). The basic philosophy of MPPT algorithm used in PV system has been given in [6]. In [7], the authors have reported VI characteristics of PV module with power analysis. Moreover, there are different method of harnessing maximum power from source such as P&O algorithm or hill climbing, fuzzy logic control technique, neural network method, incremental conductance method (INC) which are compared in term of its performance in the literature [7]. Because of easy and simple implementation, the P&O method is widely utilized in domestic as well as industrial applications. Two approaches are present for execution of P&O method in solar

PV energy generation system, which depends on selection of control variables like reference voltage or duty ratio as given in the literature [8-9].

Now a days, power electronics converters are used in majority of loads in distribution networks. These power converter based loads draw non-sinusoidal current which results in high distribution losses, voltage distortion and harmonics generation etc. Many researchers have proposed to mitigate these power quality issues through the use of DSTATCOM (Distribution Static Compensator), active filters and passive filter etc. [10-11].

In [11], the authors have presented SRFT (Synchronous Reference Frame Theory) technique for interfacing renewable source into distribution feeder network. The merits of SRF based control technique is that it proceeds simultaneously for all the three phase load currents. The direct axis DC component current is equally distributed average power absorbing component of load currents. So, there is a need of one low pass filter for calculation of direct axis DC component current. Moreover, the second harmonic content is a dominating in direct axis component in load unbalanced condition. So, to solve this issue and for achieving the good steady state response in load unbalanced condition, very low cut-off frequency is to be set for the low pass filter. The demerit of this low cut-off frequency, create a poor dynamic behaviour. Therefore, this is one of the major drawback of SRFT based control technique. Moreover, there are some other control algorithms for interfacing solar PV systems into the grid which are presented in the literature such as SRFT [12], Damped SOGI (Second Order Generalised Integrator) [13], adjustable step adaptive neuron based Control [14]. In [15], the authors have presented the neural network based control approach. A control technique based on LLMS (Leaky Least Mean Square) method is reported in [16]. In many of the control techniques, which are explained above, the calculation of average power absorbing component of load currents is achieved through filtering of load currents. Therefore, in the literature [12-16], many of the researchers have proposed various types of control techniques for integrating solar PV systems into the three phase distribution grid. Hence, there is a need of developing new control methods for appropriate switching of voltage source inverter [12-16] to get a fast, reliable and flexible interfacing of solar PV array into the three phase grid which also serves multifunctions such as active power filtering and mitigation power quality issues.

In last few decades, an adaptive filter control approach is also used which has shown its potential to track variation in the environment and characteristics. In varying environmental conditions, the parameters of the filters are self-tuned that the behaviour of the filter is changed according to the change in environment conditions. The RZA-NLMF (Reweighted Zero

978-1-5386-6624-1/18 $31.00 © 2018 IEEE   683

Attracting-Normalized Least Mean Fourth) control technique is one of technique belonging from the family of adaptive filters. An adaptive control algorithm using LMS (Least Mean Square) technique has a gap between convergence and steady state performance. However, the conventional adaptive filter technique has stability problem that may put a penance to its use. In [17, 18], the authors have reported that the stability of these methods decided by the input which is given to the adaptive control and through the noise power. Unlike, the LMF technique, the convergence behaviour of RZA-NLMF is independent of the input data correlation. Moreover, the initial settings of adaptive filters weight have also an influence on the stability of algorithm. In [17], the author has reported that the mean fourth error estimation based adaptive algorithm has relatively less noise compared to mean square technique for the same speed of convergence. In [18], the RZA-NLMF has been used with fast convergence speed and relatively less noise in weight and also stable in low SNR's (signal to noise ratios). Moreover, it also maintains the balance between convergence and steady state performances [18]. In literature [18], it is found that adaptive filter technique like RZA-NLMF with high order of error, provides less MSE (Mean Square Error) than traditional adaptive techniques like LMS. Moreover, the merits of this control technique over other existing control technique are its better DSP (Digital Signal Processing) speed, less complexity, better accuracy response, less computation burden, better MSE, fast convergence, can make it better performance in all tested condition than existing SRF and LMF technique. Furthermore, the proposed control technique is belonging from adaptive family, so it easily follows the changes in input signals and simply implemented in real time through a DSP or another embedded controller circuit.

The proposed RZA-NLMF technique is implemented for integrating a solar PV system into the three phase distribution network. In comparison of conventional control techniques like SRFT, IRPT etc., it involves simple computation and its implementation is easy as it utilizes simple mathematical blocks for estimation, while SRFT method and IRPT method utilizes complicated blocks like PLL (Phase Locked Loop). Moreover, the proposed technique takes a less time to settle.

The proposed distribution feeder interfaced SEGS is designed, modelled and simulate on MATLAB/Simulink platform. The real-time hardware validation has been carried out for UPF (Unity Power Factor) mode with harmonics mitigation and grid current balancing. The proposed grid interfaced SEGS uses a P&O method to extricate the extent amount of power from double stage solar PV grid tied system.

II. SYSTEM CONFIGURATION

A 1.954 kW two stage grid tied PV system is designed in proposed system as depicts in Fig. 1. This system comprises of a various components which are designed such as PV array, DC link capacitances voltage, DC link capacitance, voltage source inverter utilizes IGBT (Insulated Gate Bipolar Transistor) rating and other components. In [8-9], the designed, selection and modelling are given in detail and selected data are shown in Appendices.

III. CONTROL ARCHITECTURE

The Control algorithm comprises of a P&O based MPPT method for harnessing of an extent amount of power from solar PV system by generating duty cycle of DC-DC converter

and RZA-NLMF adaptive control technique for fed that power into the three phase distribution network and connected linear

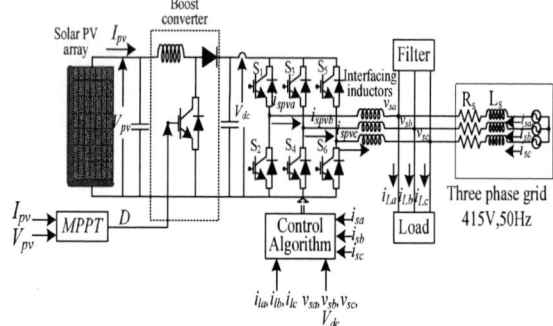

Fig.1 Schematic digram of PV power generation system.

and nonlinear loads through generating switching pulses of voltage source inverter . Therefore the control section is divided in two part. In which first part is implemented for harnessing extent amount of power from the source and second part is employed for fed that power in efficient way into the local distribution feeder as well as linear and non-linear loads. Here, in this section, the brief detail of the proposed control architecture is explain in detail.

A. MPPT Approach

The Control algorithm comprises of an improved P&O scheme for harnessing of an extent amount of power from solar PV system integrated with distribution network and connected nonlinear loads. The principle aim of P&O scheme is to extricate the extent amount of energy from SEGS at regularly changing climatic and loading conditions.

Moreover, there are two methods of implementation of P&O MPPT algorithm to a SEGS such as control of reference voltage for single stage and control of switching pulses of boost converter in a double stage SEGS. In [8-9], the detail of implementation for P&O scheme is explained. A code for P&O algorithm is developed in embedded Matlab function and executed on dSPACE based controller. The P&O based MPPT algorithm is implemented by perturbation of solar PV array voltage and so solar power.

B. Control Approach for Grid Interfaced Voltage Source Converter

The RZA-NLMF algorithm is depicted in Fig. 2. Fig. 2 depicts the calculation of weight in RZA-NLMF control algorithm which is obtained through input signal u(k) whereas u(k) is estimated from voltages which is sensed at PCI (Point of Common Interconnection) and the error signal e(k). This e(k) waveform is found from load current signal i_L(n) and the output component w(k). The weights obtained by the adaptive method are upgraded and converged to reduce the amount of error.

By using sensors and data acquisition devices two PCI line voltages (v_{sab} and v_{sac}) are recorded. The phase voltages at PCI and their amplitude peak are estimated as in [27],

$$V_{sa} = \frac{2V_{sab} + V_{sbc}}{3} \; ; V_{sb} = \frac{-V_{sab} + V_{sbc}}{3} \; ; V_{sc} = \frac{-V_{sab} - 2V_{sbc}}{3}$$

(16)

$$V_t = \sqrt{\frac{2}{3}(v_{sa}^2 + v_{sb}^2 + v_{sc}^2)}$$

(17)

The estimation of in-phase unit templates (u_{pa}, u_{pb} and u_{pc}) by phase voltages have been obtained as,

978-1-5386-6624-1/18 $31.00 © 2018 IEEE

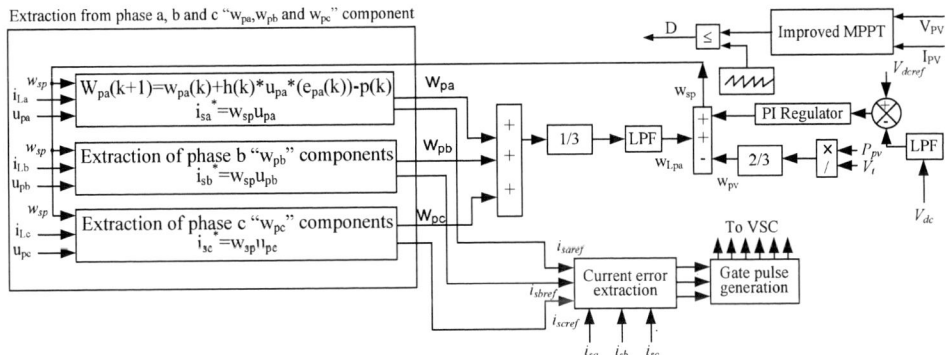

Fig.2 Schmatic diagram for control Scheme for SEGS

$$u_{pa} = \frac{v_{sa}}{V_t}; u_{pb} = \frac{v_{sb}}{V_t}; u_{pc} = \frac{v_{sc}}{V_t} \qquad (18)$$

For adjusting the DC-link capacitance voltage at set value, calculate the DC link capacitance voltage error $V_{dcerror}$ (k) which is the difference between the VSC reference DC link voltage V_{dcref} (k) and the recorded DC link capacitance voltage V_{dc} (k). The calculation of error at nth moment is obtained as,

$$V_{dcerror}(k) = V_{dcref}(k) - V_{dc}(k) \qquad (19)$$

This error function is send to PI (Proportional-Integral) controller and the output of this controller is considered as active loss factor (w_{loss}) which is utilized to adjust the DC link capacitance voltage of VSC as,

$$w_{loss}(k+1) = w_{loss}(k) + K_P\{V_{dcerror}(k+1) \\ -V_{dcerror}(k)\} + K_I V_{dcerror}(k+1) \qquad (20)$$

Where, K_P and K_I represents the value of gain for PI (Proportional-Integral) control.

The solar PV contribution term for SEGS is calculated as,

$$W_{pv}(k) = \frac{2P_{pv}(k)}{3V_t} \qquad (21)$$

Where, P_{pv} denotes the extracted solar power.

The fundamental weight component (W_{pa}) for the load current of phase 'a' is obtained as,

$$W_{pa}(k+1) = W_{pa}(k) + h_{pa}(k)e_{pa}(k)u_{pa}(k) - p_{pa}(k) \qquad (22)$$

Where,

$$h_{pa}(k) = \frac{\mu e_{pa}^2(k)}{\|u_{pa}\|_2^2 \left(\|u_{pa}\|_2^2 + e_{pa}^2(k)\right)} \qquad (23)$$

And

$$p_{pa}(k) = \frac{\rho_{RZA} sign_{pa} W_{pa}(k)}{1 + \varepsilon_{RZA}\left|W_{pa}(k)\right|} \qquad (24)$$

Here, re-weighted zero attracting factor is represented byρ_{RZA}, signpa denotes the sign function and ε_{RZA} re-weighted factor e_{pa} denotes the error and h(n) represents the variable step size, μ is a constant which represents the initial step size and it is choose appropriatly to achive the minimum value of epa, which is calculted as,

$$e_{pa}(k) = i_{La}(k) - u_{pa}(k) * W_{pa}(k) \qquad (25)$$

Likewise, the calcultion of fundamental weight component for the current of load for phase 'b' and 'c' is calculated as,

$$W_{pb}(k+1) = W_{pb}(k) + h_{pb}(k)e_{pb}(k)u_{pb}(k) - p_{pb}(k) \qquad (26)$$

$$W_{pc}(k+1) = W_{pc}(k) + h_{pc}(k)e_{pc}(k)u_{pc}(k) - p_{pc}(k) \qquad (27)$$

The total active weight component of reference distribution feeder currents is obtained as,

$$W_{sp} = W_{Lpa} + W_{loss} - W_{pv} \qquad (28)$$

Where,

$$W_{Lpa} = \frac{\left(w_{pa} + w_{pb} + w_{pc}\right)}{3} \qquad (29)$$

The reference distribution feeder currents are obtained as,

$$i_{saref} = w_{sp} * u_{pa}; \ i_{sbref} = w_{sp} * u_{pb}; \ i_{scref} = w_{sp} * u_{pc} \qquad (30)$$

The current error signals (which are obtained from the comparision of the reference grid currents (i_{saref}, i_{sbref}, i_{scref})generated through the control technique to the grid currents (i_{sa}, i_{sb} and i_{sc}) which are recorded and send to the VSC for produceing switching pulses.

IV. EXPERIMENTAL RESULTS

A hardware prototype of double stage, g integrated solar PV system is established in the laboratory. This prototype comprises of VSC, ripple filter, nonlinear load, interfacing inductors, and a solar PV array. For realize the function of solar PV system, 1.954 kW (V_{mp}=180.85V I_{mp}=10.87A) solar photovoltaic array simulator (AMETEK made ETS600x17DPVF) is deployed. Hall-Effect current recorders (LA55-P) have been utilized for record the distribution feeder currents, PV converter currents, load currents and PV array current. Similarly, the Hall-Effect voltage sensors (LV25- P) are deployed for sensing the two PCI voltages and DC-link capacitance voltage.

A. Experimental Result Under Nonlinear Load
The steady state behaviour of proposed SEGS system is depicts in Figs. 3-4. In Fig. 3 (a), the distribution feeder voltage v_{sbc} is shown with i_{sa}. The distribution system current (i_{sa}) with total harmonic distortion (THD) value is depicts in Fig. 3 (b). The distribution network current THD is 3.8 % which is an acceptable range of [29]. The load voltage with load current (i_{la}) waveforms are given in Fig. 3 (c). The load current (i_{la}) THD is nearly 27.6 % as it is depicts in Fig. 3 (d). Fig. 4 (a) reveals the voltage at common coupling point (v'_{sbc}) with VSC current (i_{vsca}). Finally, Figs. 4 (b-d) show the

978-1-5386-6624-1/18 $31.00 © 2018 IEEE

distribution feeder power, power consumed by load and PV converter power, respectively. The negative power of distribution feeder means, that the power is send to the distribution system. It can be found from experimental results that the distribution feeder power factor is sustained at unity. Moreover, from test results, it is found that overall performance is satisfactory and the system works well in steady state condition.

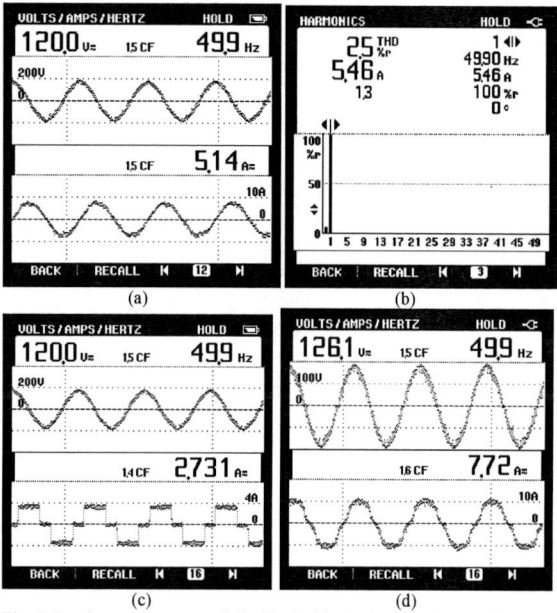

Fig. 3 Steady-state response of distribution feeder interfaced SEGS (a-b) v_{sbc} with i_{sa} and Harmonics of distribution system current (c) v_{sbc} with i_{la} (d) Harmonics spectrum of i_{la}

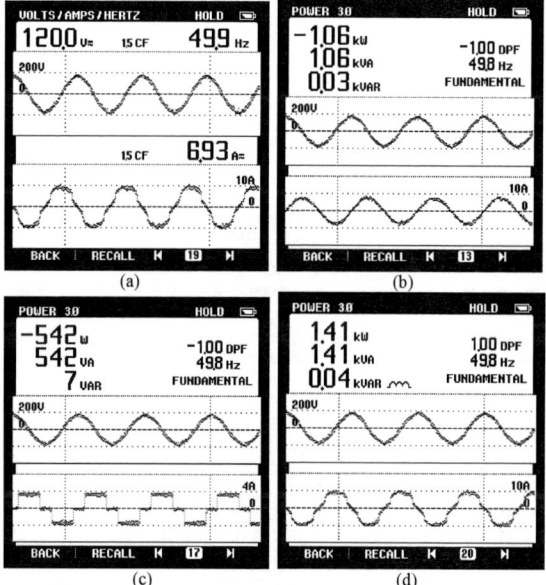

Fig. 4 Steady-state behaviour of distribution system interfaced SEGS (a) v_{sbc} with i_{vsc_a} (b) Distribution feeder power (c) load power (d) VSC power

B. Dynamic Response Under Unbalance Load

The dynamic response of proposed SEGS integrated to the three phase distribution system is given in Figs. 5 (a-d), when a load is disconnected from phase 'a'. The VSC-DC link voltage (V_{dc}), load current (i_{La}), PV converter current (i_{vsca}), and sensed grid currents (i_{sa}) of phase 'a' are depicted in Fig. 5 (a), respectively. The waveforms for reference distribution feeder current of phase 'a' (i_{saref}), grid current (i_{sa}), for phase 'b' distribution feeder current (i_{sbref}) and distribution feeder current (i_{sb}) are given in Figs. 5 (b). For examining the response of algorithm at unbalanced load, the intermediate signals are given in Figs. 5 (c) which are the weight of active loss factor (W_{loss}), weight factor of feed-forward path of SEGS (W_{pv}), weight of the average fundamental load component (W_{Lpa}) and weight of total PV power which is fed to the grid (W_{sp}). At last, the changes of intermediate waveforms such as adaptive error (e_{pa}), cube of error ($e_{pa}3$) the sign function of phase 'a' ($sign_{pa}$), and weight of active power component of phase 'a' (W_{pa}) are obtained, which are shown in Figs. 5 (d). The response with unbalanced loading condition is observed in which distribution feeder currents are balanced. Moreover, the intermediate signals are varying according to variation in the load. Therefore, it is found that overall performance of proposed configuration is satisfactory during load unbalancing.

C. Dynamic Response Under Varying Insolation Situation

The proposed system response with varying insolation from 1000 W/m2 to 700 W/m2 is shown in Figs. 6(a-b). The DC link voltage (V_{dc}), PV current (I_{pv}), converter current ivsca, and sensed grid current (i_{sa}) are given in Figs. 6(a). Fig. 6 (b) depicts the variation of the loss factor of PI controller (W_{loss}), the changing in intermediate variables such as weight of SEGS power component (W_{pv}), weight of the average fundamental active load component (W_{Lpa}) and weight of solar power which is fed from solar source to grid (W_{sp}) on the decrement of solar insolation. Figs. 6 (c-d), show MPPT response of PV array at (c) 1000 W/m^2 (d) 700 W/m^2. From which, it is clear that extraction of PV power from PV system is almost 100%. Moreover, these responses have clearly proved that control algorithm and MPPT works well under varying insolation.

V. CONCLUSION

A double stage PV system interfaced to three phase distribution feeder with multifunction RZA-NLMF control algorithm has been proposed here. The proposed algorithm not only extract peak power from solar array and feed that power to the distribution feeder. It also enhances the quality of power at PCC. Which is one of the significant issue when the power is fed from individual domestic consumer to the distribution grid.The control technique is deployed for obtaining the reference distribution feeder currents. The reference distribution feeder currents are obtained by using load currents, feed forward loop, and loss factor of VSC, which is further utilized to produce the switching signals of VSC. The response of system has been examined to show the harmonics reduction, load balancing and maintained power factor at unity.

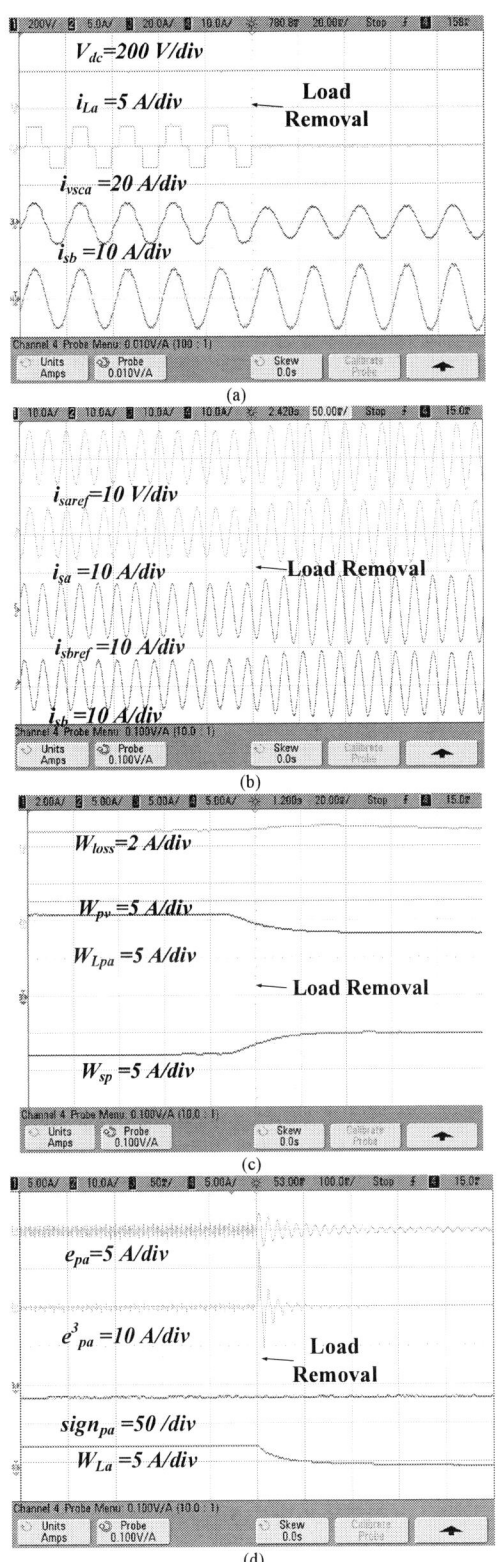

Fig. 5 Dynamic response of grid interfaced SEGS for non-linear load(a) V_{dc}, i_{la}, i_{vsc_a} and i_{sa} (b V_{dc}, i_{lb}, i_{vsc_b} and i_{sb} (c) w_{loss}, w_{pv}, w_{Lpa} and w_{sp} (d) e_{pa}, e_{pa}^3. $sign_{pa}$ and w_{pa}.

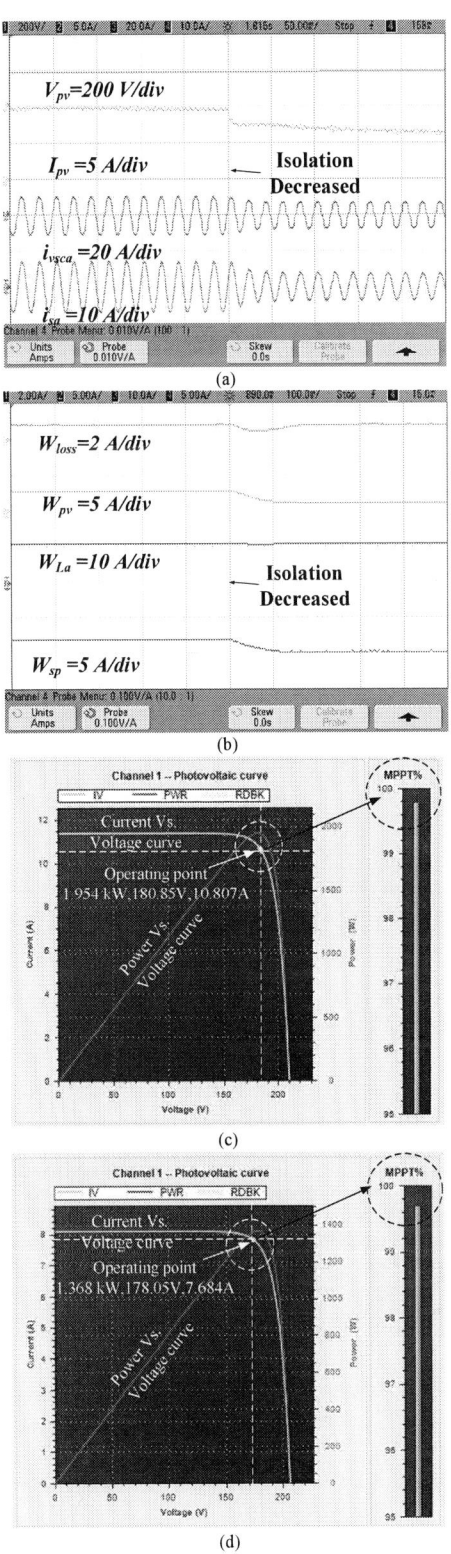

Fig. 6 Dynamic behaviour of SPV system under varying insolation 1000 W/m^2 to 700 W/m^2 (a) V_{pv}, I_{pv}, i_{vsc_a} and i_{sa} (b) w_{loss}, w_{pv}, w_{Lpa} and w_{sp} (c-d) SPV performance of solar array (c) 1000 W/m^2 (d) 700 W/m^2

978-1-5386-6624-1/18 $31.00 © 2018 IEEE   687

The THDs of grid side currents follow an IEEE-519 standard. For validating the performance of control algorithm, an experimental responses are examined for different test conditions, which clearly shown the potential of proposed technique under various operating conditions. Moreover, from test results, it is clear that the system performance is satisfactory in both steady-state and different transient situations.

APPENDICES

Experimental Prototype Parameters: PV array voltage, V_{MPP} = 180.85 V; I_{MPP} = 10.807 A; P_{MPP} = 1.954 kW; V_{dc} = 200 V; L_f = 2.4 mH; T_s = 30 μs; VSC rating = 25 kVA; didtribution system voltage VLL = 127.7 V (rms); R_f=5 Ω and C_f=10 μF, proportional controller K_{pd}=0.04 and integral controller K_{id}= 0.01, *adaptive constant, μ = 0.0001 and ρza=.00001, LPF cut-off frequency fc= 10 Hz.*

REFERENCES

[1] Nicholas Etherden, and Math HJ Bollen. "Increasing the hosting capacity of distribution networks by curtailment of renewable energy resources" in IEEE Trondheim PowerTech, Jun 2011, pp. 1-7.

[2] P. Mints, "The history and future of incentives and the photovoltaic industry and how demand is driven", Progress in Photovoltaic: Research and Applications, vol. 20, no. 6, pp. 711-716, 2012.

[3] A. Shakya., "Solar Irradiance Forecasting in Remote Microgrids Using Markov Switching Model," IEEE Transactions on Sustainable Energy, vol. 8, no. 3, pp. 895-905, July 2017.

[4] [115]N. Saxena, B. Singh and A. L. Vyas, "Single-phase solar PV system with battery and exchange of power in grid-connected and standalone modes," *IET Renewable Power Generation*, vol. 11, no. 2, pp. 325-333, 2 8 2017.

[5] Chee Wei Tan, T. C. Green, C. A Hernandez-Aramburo, "Analysis of perturb and observe maximum power point tracking algorithm for photovoltaic applications," in Proc. 2nd International Power and Energy Conference (PECon), 2008, pp.237-242.

[6] A. Ahmed, R. Li and J. Bumby, "Perturbation parameters design for hill climbing MPPT techniques," in IEEE International Symposium on Ind. Electron., May 2012, pp. 1819-1824.

[7] M. A. Elgendy, B. Zahawi, and D. J. Atkinson, "Assessment of the incremental conductance maximum power point tracking algorithm," *IEEE Trans. Sust. Energy*, vol. 4, no. 1, pp. 108–117, Jan. 2013.

[8] R. K. Agarwal, I. Hussain and B. Singh, "LMF-Based Control Algorithm for Single Stage Three-Phase Grid Integrated Solar PV System," *EEE Transactions on Sustainable Energy*, vol. 7, no. 4, pp. 1379-1387, Oct. 2016..

[9] A. K. Singh, I. Hussain and B. Singh, "Double-Stage Three-Phase Grid-Integrated Solar PV System With Fast Zero Attracting Normalized Least Mean Fourth Based Adaptive Control," *IEEE Transactions on Industrial Electronics*, vol. 65, no. 5, pp. 3921-3931, May 2018.

[10] B. Singh, P. Jayaprakash, D. P. Kothari, A. Chandra, and K. Al Haddad, "Comprehensive study of DSTATCOM configurations," *IEEE Trans. Ind. Inform.*, vol. 10, no. 2, pp. 854-870, May 2014.

[11] C. Kumar and M. K. Mishra, "An improved hybrid DSTATCOM topology to compensate reactive and nonlinear loads," IEEE Trans. Ind. Elect., vol. 61, no. 12, pp. 6517-6527, Dec. 2014.

[12] B. Singh and J. Solanki, "A Comparison of Control Algorithms for DSTATCOM," in IEEE Transactions on Industrial Electronics, vol. 56, no. 7, pp. 2738-2745, July 2009.

[13] B. Singh, S. Kumar and C. Jain, "Damped-SOGI-Based Control Algorithm for Solar PV Power Generating System," in IEEE Transactions on Industry Applications, vol. 53, no. 3, pp. 1780-1788, May-June 2017.

[14] B. Singh, C. Jain and A. Bansal, "An Improved Adjustable Step Adaptive Neuron Based Control Approach for Grid Supportive SPV System," in IEEE Transactions on Industry Applications, vol. PP, no. 99, pp. 1-1.

[15] G. Franceschini, E. Lorenzani, C. Tassoni and A. Bellini, "Synchronous Reference Frame Grid Current Control for Single-Phase Photovoltaic

Converters," Industry Applications Society Annual Meeting, 2008. IAS '08. pp.1,7, 5-9 Oct. 2008..

[16] B. Singh, S. Kumar and C. Jain, "Damped-SOGI-Based Control Algorithm for Solar PV Power Generating System," *IEEE Transactions on Industry Applications*, vol. 53, no. 3, pp. 1780-1788, May-June 2017.

[17] E. J. Candès, M. B. Wakin, and S. P. Boyd, "Enhancing sparsity by reweighted L1 minimization," *Journal of Fourier Analysis and Applications*, vol. 14, no. 5–6, pp. 877–905, Oct. 2008.

[18] E. Eweda and N. J. Bershad, "Stochastic analysis of a stable normalized least mean fourth algorithm for adaptive noise canceling with a white Gaussian reference," *IEEE Transactions on Signal Processing*, vol. 60, no. 12, pp. 6235–6244, Dec. 2012.

2nd IEEE International conference on power Electronics, Intelligent Control and Energy systems

Implementation of MPPT Control in Fuel Cell Fed High Step up Ratio DC-DC Converter

V. Karthikeyan, *Member, IEEE*
Dept. of Electrical Engineering
National Institute of Technology
Calicut, Kerala, India
karthikeyan@nitc.ac.in

Vipin Das P, Student *Member, IEEE*
Dept. of Electronics and
Communication Engineering
MNNIT Allahabad, UP, India
rel1512@mnnit.ac.in

Frede Blaabjerg, Fellow *IEEE*
Department of Energy Technology
Aalborg University,
Aalborg, Denmark
fbl@et.aau.dk

Abstract— **This paper presents an alternative approach towards implementation of maximum power point tracking (MPPT) control in fuel cell based power systems utilizing a high step up ratio DC-DC converter using single state PWM for DC Micro grid based applications. The P & O algorithm is implemented for the full bridge isolated forward converter to track the maximum power from fuel cell. In order to integrate the low voltage producing fuel cell with any existing DC micro grid, a high step up converter is required to boost the voltage and ensure proper regulation at the load side. The high frequency conversion link produces lower secondary voltage ripple. A capacitive filter is appended at the secondary side in order to impart smoother waveform characteristics to the output dc voltage. The design of C-filter is dealt with in the present paper using specified and acceptable ripple limits on the voltage and current. The theoretical analysis is verified through simulation studies using PSCAD/EMDTC software.**

Keywords— Fuel cell, Full Bridge Converter, Maximum power point tracking (MPPT), Single State PWM.

I. INTRODUCTION

Due to environmental and economic reasons, renewable energy penetration is increasing in modern power systems. Renewable energy sources are better alternatives to the fossil fuel based conventional power systems. Fuel cell (FC) is the best choice among the renewable energy sources due to its low emissions, noise free operation and high output efficiency. FC is now used for portable electronics, electric vehicles and distributed generation power plants due to its eco-friendly nature .According to usage of fuel types, it can be classified as i) polymer electrolyte membrane (PEM) fuel cells or PEMFCs (also called PEFCs), ii) solid oxide fuel cells (SOFCs), iii) alkaline fuel cells (AFCs), iv) phosphoric acid fuel cells (PAFCs), and v) molten carbonate fuel cells (MCFCs). The maximum electrical conversion efficiency of FCs are as high as 60% and by considering the co-generation it can be increased up to 80% with 90% reduction in major pollutants like CO2, CO, etc [1-3].

Due to the chemical reaction of hydrogen with anode, electrons are produced which moves toward cathode through an external electric circuit [4, 5]. The power generation ability of FCs is limited due to the internal barriers and losses. It is necessary that the FCs has to be forced to operate around the point which corresponds to the Maximum Power Point (MPP). In general, a fuel cell system (FCS) is operated in conjunction with a dc–dc power converter whose duty cycle is modulated in order to track the instantaneous MPP of the FCS [6]. It varies with temperature. In order to operate at MPPT, several methods have been implemented to extract the maximum power in the literature [7-9].

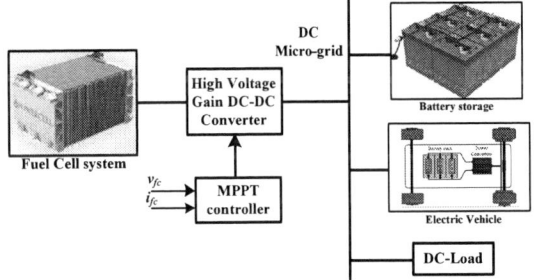

Fig. 1 Schematic of proposed layout

Boost type dc-dc converter is commonly used to achieve higher voltage ratio. However, due to saturation boost inductor core, the switch duty cycle is limited, thus voltage conversion ratio is limited to the range of 4-5 [8]. Usually, fuel cell operating voltage falls due to ohmic losses. It is hardly about 1 to 1.2 volts/cell. It is necessary to increase the voltage to higher levels to utilize the fuel cell based electrical energy. FCs is generally accumulated in stacks to improve the voltage. However, it is not advisable to accumulate many fuel cells in a single stack. It is a strict requirement that a high step up ratio converter boosts the voltage to the desired level. Therefore, a full bridge forward converter is the most suitable converter for fuel cell based power system application. In isolated dc-dc converter, many basic and derived configurations are available in the literature. Using a full bridge forward converter giver greater control flexibility over the power and helps achieve a higher conversion ratio to perform MPPT. The proposed layout is depicted in Fig. 1.

This paper particularly focuses on the MPPT operation in a full bridge dc-dc converter using single state PWM technique. The P&O algorithm approach has been implemented to perform MPPT for forward converter. Design consideration for proper functioning of the LC filter have been looked into and implemented. Section 2 delves into the fuel cell V-I characteristics and associated mathematical modelling of the losses incurred during its MPPT operation. Section 3 is dedicated to the topology, closed loop control and filter considerations of the full bridge forward converter employed in this study. MPPT implementation via the P&O algorithm is discussed in Section 4. Section 5 presents the simulation studies for the proposed system.

II. FUEL V-I CHARACTERISTICS

Fig.2 shows the generalized V-I characteristics of a fuel cell. The characteristic is non-linear in nature due to internal, physical and chemical barriers. The following

978-1-5386-6624-1/18 $31.00 © 2018 IEEE   689

points can be derived from the given characteristics:

- The theoretical voltage is slightly greater than the open circuit voltage.
- The voltage fall in the initial stage is very small and the graph is more linear.
- The voltage falls rapidly at higher current densities.

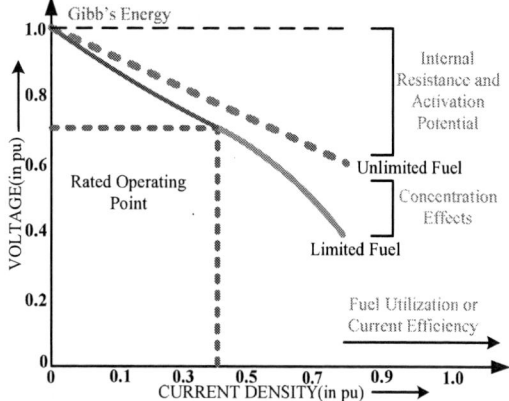

Fig. 2 V-I Fuel cell V-I characteristics

The operational voltage (V_{FC}) of fuel cell is given by equation (1).

$$V_{FC} = E_{Nernst} - \left(\eta_{act} + \eta_{Ohm} + \eta_{con} \right) \quad (1)$$

Here E_{Nernst} is the Nernst thermodynamic voltage, η_{act} are the activation losses, η_{Ohm} are the ohmic losses and η_{con} are the concentration losses [10-14].

A. Nernst reversible voltage (E_{Nernst})

Nernst reversible voltage is also known as thermodynamic potential. It is calculated as the open circuit voltage when the current density of fuel cell is zero. Thermodynamic potential is related to the partial pressure of hydrogen, oxygen and water as given in the equation (2).

$$E_{FC} = N_0 \left(E_0 + \frac{RT}{2F} \left[\ln \frac{p_{H_2} p_{O_2}^{0.5}}{p_{H_2O}} \right] \right) \quad (2)$$

Here, p_{H2}, p_{O2} and p_{H2O} are the partial pressure of hydrogen, oxygen and water, respectively. These partial pressures can be determined by the following equations,

$$p_{H_2} = \frac{1/K_{H_2}}{1 + \tau_{H_2}s} \left(q_{H_2}^{in} - 2K_r I_{fc} \right) \quad (3)$$

$$p_{O_2} = \frac{1/K_{O_2}}{1 + \tau_{O_2}s} \left(q_{O_2}^{in} - K_r I_{fc} \right) \quad (4)$$

$$p_{H_2O} = \frac{1/K_{H_2O}}{1 + \tau_{H_2O}s} \left(2K_r I_{fc} \right) \quad (5)$$

Here $K_r = N_o/4F$ is the model constant of fuel cell, q^{in}_{H2} and q^{in}_{O2} are the flow rates of hydrogen and oxygen, respectively as defined by the following equation;

$$q_{H_2}^{in} = \frac{1}{1 + \tau_f s} \left(\frac{2K_r I_{fc}}{U_f} \right) \quad (6)$$

$$q_{O_2}^{in} = \frac{q_{H_2}}{r_{H-O}} \quad (7)$$

The ratio of hydrogen and oxygen rH-O is about 1.168. The fuel cell utilization factor, U_f is the amount of hydrogen (mf, mass flow rate, Kg/s) utilized for reaction,

$$U_f = m_f q_{H_2}^{r} / m_f q_{H_2}^{in} \quad (8)$$

Typically for the fuel cell modelling, 80-90 % of fuel cell is utilized. The amount of Hydrogen reacts with Oxygen can be given by,

$$q_{H_2}^{r} = \frac{N_0 I_{fc}}{2F} = 2K_r I_{fc} \quad (9)$$

By considering the equation (8), when $U_f > 0.9$, it leads to fuel starvation which can cause permanent damage to the fuel cell. When $U_f < 0.7$, it can lead to higher fuel cell voltage. The current demand of a fuel cell for a fixed input fuel flow is constrained by the following equation.

$$\frac{0.8q_{H_2}^{in}}{2K_r} \leq I_{fc}^{in} \leq \frac{0.9q_{H_2}^{in}}{2K_r} \quad (10)$$

By adjusting the input fuel flow the fuel cell utilization factor can be regulated on the basis of output current from the fuel cell. The optimum utilization factor U_f is assumed in this model as 85%. Therefore, the value of fuel input flow can be written as,

$$q_{H_2}^{in} = \frac{2K_r I_{fc}^{r}}{0.85} \quad (11)$$

It can be seen from (11), the fuel input flow depends on fuel cell current

B. Activation loss (η_{act})

Activation loss is caused due to the control action of slow electrode kinetic on the electrochemical reaction. The η_{act} can be described by the general form of the Tafel equation as follows,

$$\eta_{act} = \frac{RT}{\alpha n F} \log \left(\frac{I_{fc}}{I_O} \right) \quad (12)$$

C. Ohmic loss (η_{Ohm})

Ohmic loss can occur due to resistance of the flow of ions in the electrolyte and resistance of flow of electrons through the electrode materials. This resistance depends on cell temperature which is obtained by,

$$\eta_{Ohm} = I \alpha \exp \left(\beta \left(\frac{1}{T_0} - \frac{1}{T} \right) \right) \quad (13)$$

D. Concentration loss (η_{conc})

Concentration loss occurs due to the change in concentration of the reactant, which can be written as,

978-1-5386-6624-1/18 $31.00 © 2018 IEEE

$$\eta_{con} = \frac{RT}{n_a F} \ln\left(1 - \frac{I_{fc}}{I_L}\right) \tag{14}$$

The expression of fuel cell current can be written as,

$$I_{fc} = \frac{P_{ref}}{V_{FC}}\left(\frac{1}{1+\tau_e s}\right) \tag{15}$$

The total power generated by the fuel cell stack can be given as,

$$P_{FC} = N_0 V_{FC} I_{fc} \tag{16}$$

From the fuel cell modeling equations it is clear that the fuel cell output is depending on various physical parameters such as temperature, pressure and humidity. The maximum power operating point of fuel cell varies with these parameters. In order to operate the fuel cell efficiently a MPPT controller is required.

III. High Step Up Ratio DC-DC Converter

Fig. 3 demonstrates a full-bridge topology setup with MPPT operation in fuel cell module. When one of the primary switches is active for the half-bridge topology, the voltage over the primary winding is $V_{in}/2$. For the full bridge topology, the switches are operated diagonally. At the point when diagonally opposite switches are active, the voltage over the primary winding is the full input voltage V_{in}. In this manner for a given power, the primary current will be half for a full-bridge topology when contrasted with half-bridge. The full bridge topology provides higher efficiency at high load as compared with half bridge topology. Full bridge converter operates at higher frequency with controlled output voltage, thereafter it provides DC output. The power can be transferred via high frequency link conversion. It supports high voltage conversion ratio with smooth power transfer. It also separates the low input and high output voltage ground.

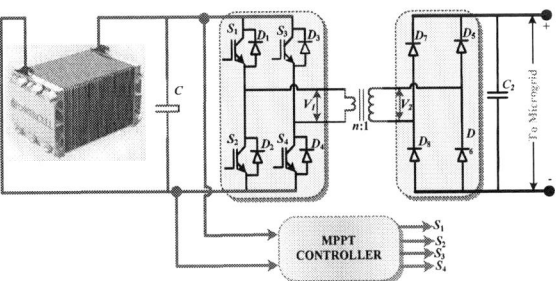

Fig. 3 Generalized block diagram of proposed system

A. Closed Loop Controller

The proposed control scheme achieves MPPT operation in full bridge converter with low voltage ripple. Fig. 4 shows the complete control block diagram of the control strategy. Four gating signals are generated and FCs voltage is held at constant value at the maximum power point by adjusting the width of pulse signal. Since a single error function is compared with triangular signal to produce desired single state PWM signals, the number of PI controller is reduced. Therefore, full bridge converter is used to extract maximum energy from fuel cell system.

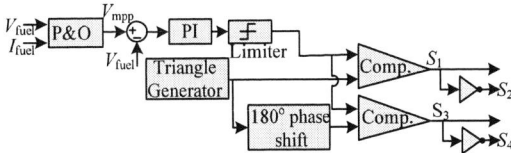

Fig 4. Block Diagram of proposed control scheme

B. Design of Capacitive Filter

Sizing of the output capacitor is an important aspect of design topology. Selection of the capacitor value depends on desired voltage ripple, ESR and RMS ripple current. The physical volume of the converter is affected by the capacitor rating, thereby affecting its performance indirectly.

For a closed loop operation, any load transient in the output causes variation in output current (ΔI) also leads to change in the voltage (dV). The response time of the closed loop (dt) to the change in output voltage and current. The value of output capacitance is taken as C. In order to simplify calculations the effect of ESR and ESL has not been considered.

$$dV = \frac{\Delta I}{C} \times dt \tag{17}$$

It is implied that the minimum capacitor value depends on specified output voltage ripple and the response time of the closed loop controller as given below:

$$C = \Delta I \times \frac{dt}{dV} \tag{18}$$

The maximum allowed variation on the output voltage of V is 2%.

$$\begin{aligned} dV &= V_{OUT} \times \%V_{OUT} \\ &= 400 \times 2\% = 8V \end{aligned} \tag{19}$$

In practical situations, considering the effect of delay loop crossover becomes equal to $f_{sw}/10$ which comes out to be 0.5 ms for operating switching frequency equal to 20 kHz. Current variation is supposed to be equal to 1 A.

$$\begin{aligned} C &= \Delta I \times \frac{dt}{dV} = 1 \times \frac{0.5 \times 10^{-3}}{8} \\ &= 62.5\mu F \end{aligned} \tag{20}$$

As the frequency of operation of the high gain DC-DC converter is high, small value of capacitor is sufficient to filter the ripple contents.

IV. Implementation Of MPPT In Forward Converter

P&O MPPT technique is most commonly used technique among the various available MPPT technologies. The implementation of P&O is easy as compared with other techniques; this makes the use of P&O MPPT common. The P&O MPPT operates in such a way that the change in power ΔP becomes zero. The load changes in grid connected mode of operation will oscillate the operating point around MPP due to large disturbances. The P&O provides faster response during these disturbances. The main task for designing P&O MPPT is to achieve high accuracy with small settling time. [15-17].

978-1-5386-6624-1/18 $31.00 © 2018 IEEE

The voltage and current of FCs are required to calculate the point at which the operating point should move in fuel cell V-I curve so that it approaches the MPP. Utilizing the equations mentioned in section 2, we compute the instantaneous current and voltage of the fuel cell. The MPPT operation is performed at the same conditions throughout. The flow chart for the above mentioned P&O is shown in Fig.5. In this method the instantaneous power is compared with the previous power by applying small perturbation. This process will continue until the change in power becomes zero.

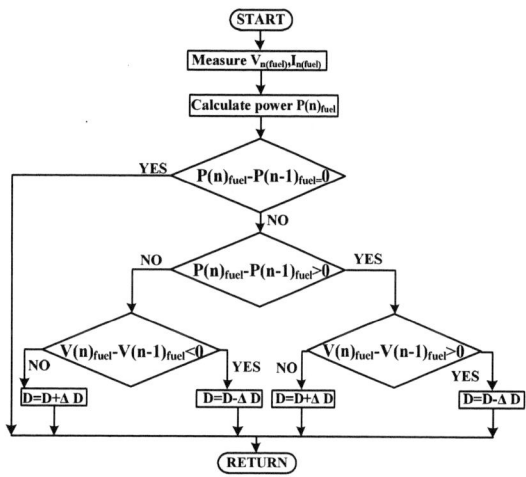

Fig. 5 Flow chart of MPPT controller

Fig.6 shows the polarization curve of a typical fuel cell. It can be seen from the figure that the MPP point has ΔP equals to zero.

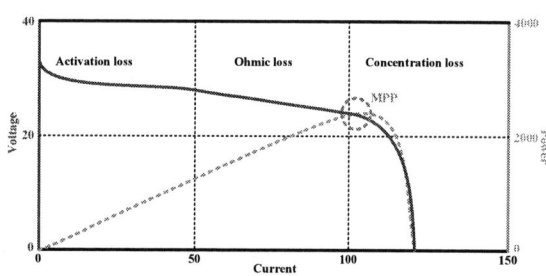

Fig. 6 Polarization curve of fuel cell

V. SIMULATION RESULTS

The simulation results are obtained using PSCAD/EMDTC software. The temperature of the fuel cell is changed from 700, 900 and 600 at a time period of 2.05s, 2.1s and 2.15s respectively as shown in Fig. 7a. The PWM switching frequency moves towards 20 kHz by comparing two reference signals with high frequency carrier triangular signal. The modulating index decides the shape of the output voltage of the full bridge converter. The P&O algorithm tracks the operating point rapidly and precisely in both conditions which is clearly shown in Fig. 7b. In order to verify the control loop design, a measurement of the fuel voltage and reference voltage generated using closed loop controller are connected in same scope and shown in Fig. 7b.

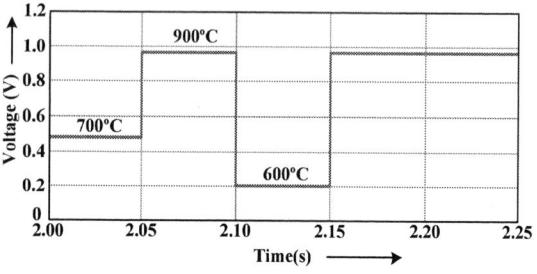

(b)

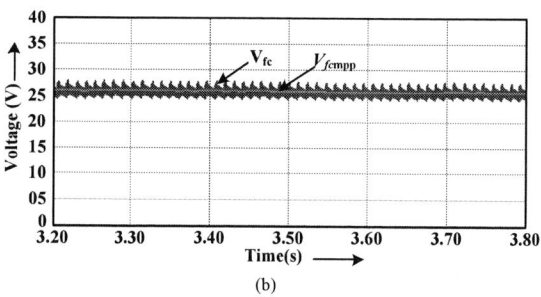

(b)

Fig.7 (a) Changes in output voltage of Fuel cell respect to temperature, (b) V_{FCmpp} and VFC tracking

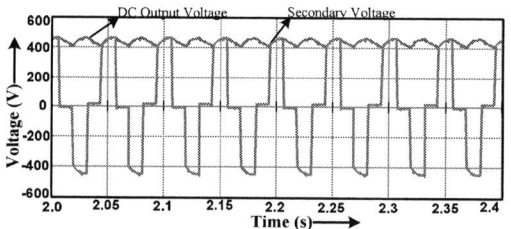

(a)

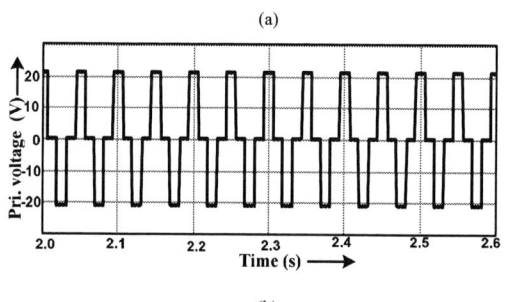

(b)

Fig. 8(a) Voltage of high voltage conversion converter (a) Secondary end voltage, (b) Primary end voltage

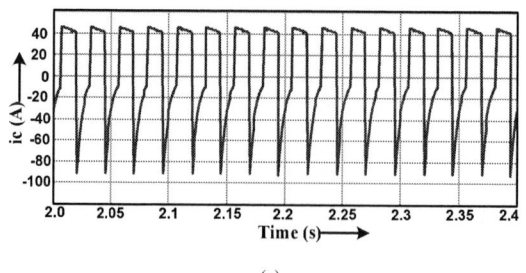

(a)

978-1-5386-6624-1/18 $31.00 © 2018 IEEE

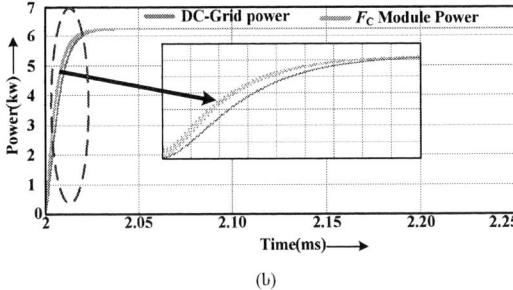

(b)

Fig. 9(a) Output capacitor current, (b) Fuel cell and DC micro grid power

Since increasing output current decreases the fuel cell voltage, in order to track the maximum power from fuel cell, modulation index is increased in such a way so that the fuel current will decrease. Therefore, as per algorithm, modulation index can be reduced to a certain value since power remains constant. Using this approach the current will be affected and this process is repeated until power generation becomes equal to maximum power point. Fig. 8(a) shows the results of dc output and secondary voltage of transformer. Choosing appropriate capacitor value, the output voltage of full bridge forward converter maintains uniform voltage with lesser ripple which is shown in Fig. 8 (a). Fig. 8(b) depicts the voltage on the primary side. DC output voltage contains more ripples. The DC link voltage reduces when the capacitor is trying to supply the voltage due to this effect capacitor current is increasing randomly to supply when lack of voltage in output side. The capacitor current is shown in Fig. 9(a). The power generated from fuel cell and power fed to DC micro grid is almost constant when negligible losses. The power generation and injecting to the micro grid is shown in Fig. 9(b).

TABLE I. SIMULATION SETUP PARAMETERS

Parameter	Simulation
Fuel cell capacity	6 kW
Switching Frequency	20 kHz
Fuel cell stack rated voltage	32 V
Output filter capacitor	62.5μF
Transformer rating	20 kHz, 10 KVA
Transformer Turn ratio	1:18

VI. CONCLUSION

Ease of Use In this paper, a P&O based maximum power point tracking approach for PEM fuel cell is presented and its performance is investigated through simulation. The analysis and simulation studies are performed on a system including of a PEMFC, and a high gain isolated boost DC/DC converter for both normal and time varying fuel cell temperature operating conditions. A single state pulse width modulation technique is implemented with a closed loop controller design which decreases the complexity of the control scheme. The performance of the high step up ratio DC-DC converter is also analyzed and the results are found

to be satisfactory. A capacitive filter is designed for the application of high gain DC-DC converter due to which the ripple content in the output voltage is reduced considerably. The analysis of output voltage and capacitor currents is conducted and is found to be adequate. The results are indicative of the performance of the proposed MPPT method.

REFERENCES

[1] Wilberforce Tabbi, Alaswad A, Palumbo A, Dassisti M, Olabi AG. Advances in stationary and portable fuel cell applications. Int J Hydrogen Energy 2016;1 14.

[2] Nguyen HQ, Aris AM, Shabani B. PEM fuel cell heat recovery for preheating inlet air in standalone solar-hydrogen systems for telecommunication applications: an exergy analysis. Int J Hydrogen Energy 2016; 41:2987-3003

[3] Garcia, CA, et al. Improving voltage harmonic compensation of a single phase inverted-based PEM fuel cell for stand-alone applications. Int J Hydrogen Energy 2014, 39: 4483-4492.

[4] Huang, B., Yutong Q., and Murshed M. Solid oxide fuel cell: Perspective of dynamic modeling and control. Journal of Process Control.2011; 21.1426-1437.

[5] Nguyen, G., et al. Dynamic modeling and experimental investigation of a high temperature PEM fuel cell stack. Int J Hydrogen Energy 2016; 41. 4729-4739.

[6] Benyahia, N., et al. "MPPT controller for an interleaved boost dc–dc converter used in fuel cell electric vehicles."International Journal of Hydrogen Energy 2014; 39.15196-15205.

[7] Hellman HL, van den Hoed L. Characterising fuel cell technology: challenges of the commercialization process. International Journal of Hydrogen Energy 2007; 32:305-15.

[8] Yu X, Starke M, Tolbert L, Ozpineci B. Fuel cell power conditioning for electric power applications: a summary. IET Electric Power Appl 2007; 1:643–56.

[9] Cardenas, A., Agbossou, K., and Henao, N. Development of power interface with FPGA-based adaptive control for PEM-FC system. IEEE Transactions on Energy Conversion 2015; 30.296-306.

[10] Wu, HW. A review of recent development: Transport and performance modeling of PEM fuel cells. Applied Energy 2016; 165. 81-106.

[11] Friede, W., Raël, S. and Davat, B. Mathematical model and characterization of the transient behavior of a PEM fuel cell. IEEE Transactions on Power Electronics 2004; 19.1234-1241.

[12] Gao, F., et al. PEM fuel cells stack modeling for real-time emulation in hardware-in-the-loop applications. IEEE Transactions on Energy Conversion 2011; 26.184-194.

[13] Puranik, SV. Keyhani A., and Khorrami, F. State-space modeling of proton exchange membrane fuel cell. IEEE Transactions on Energy Conversion 2010; 25. 804-813.

[14] Xue XD, Cheng KWE, Sutanto D. Unified mathematical modelling of steady-state and dynamic voltage-current characteristics for PEM fuel cells. ElectrochimActa 2006; 52:1135-44.

[15] Ahmed J, Salam Z. An improved perturb and observe (P&O) maximum power point tracking (MPPT) algorithm for higher efficiency. Appl Energy 2015; 150:97–108

[16] Benyahia N, et al. MPPT controller for an interleaved boost DC-DC converter used in fuel cell electric vehicles. Int J Hydrogen Energy 2014; 39:15196-205.

[17] Becherif M, Hissel D. MPPT of a PEMFC based on air supply control of the moto compressor group. Int J Hydrogen Energy 2010; 35:12521-30.

2nd IEEE International Conference on Power Electronics, Intelligent Control and Energy Systems (ICPEICES-2018)

Design and Analysis of Fifth-order Buck-Boost Converter

M. Veerachary

Dept. of Electrical Engineering, *Indian Institute of Technology Delhi,* New Delhi, INDIA

mvchary@ee.iitd.ac.in

Abstract— **A fifth-order buck-boost converter suitable to point of load applications is presented in this paper. It offers more buck-boost conversion ratio together with less ripple content in the source current and thus reducing the filtering requirements on the source side. A detailed time-domain and steady-state analysis is presented to establish the buck-boost features. Voltage transformation characteristic features are formulated for continuous inductor current mode of operation and equations defining L and C components are evolved. The state-space models are derived and small-signal analysis is performed to obtain the relevant transfer functions. Later, these are used in the controller design. The proposed circuit is able to perform bucking as well as boosting of the load voltage, similar to the conventional buck-boost converter and has no interaction issues as it has common ground. A 50** *Watt,* **100** *kHz* **prototype fifth-order converter is built to supply power at constant load voltage of either 60 or 15 V. A 24** *V* **dc-battery is used for powering the prototype converter operation both in simulation and experimentation. The proposed converters' effectiveness is demonstrated, both in simulation and experimentation, in terms of lower ripple content as well as in buck-boosting over wide ranges.**

Keywords— **Buck-boost converter, Point of load converter, Two-switch buck-boost converter, State-space model.**

I. INTRODUCTION

With technological advancement, the requirement of power conversion at high frequencies is soaring and is dominant in applications requiring low power for their operation. Converters for point of load applications are being developed by the designers while laying special emphasis on achieving higher conversion efficiency at full-load, increased power density, and lower radiation. Use of several point of load converters is common in low power dc system wherein many design challenges [1]-[9] must be resolved by the application engineer so as to ensure reliable power distribution. Some of them are: (i) formulation of transformerless non-isolated topologies as minimization or elimination of transformer leakage inductance is a difficult task, (ii) achieving reduced ripples with minimal L, C component requirements, and (iii) reduction in size and weight of the filtering components which will result in enhanced power densities. Many dc-dc conversion circuits are present in the literature which produce stable voltages to drive the dc-loads and is briefly classified as: (i) bucking based circuits, (ii) boosting circuits, and (iii) buck-boost and other higher-order or derived converter circuits. These converters find broad application in areas pertaining to controlled power such as: (i) customized low-power integrated

circuits, (ii) powering compact and tiny automotive loads, (iii) sophisticated loads such as bio-medical equipment, (iv) internet, wide and local area network services, and (v) telecommunication power supply systems, on-board spacecraft power systems, and defence equipment, etc.

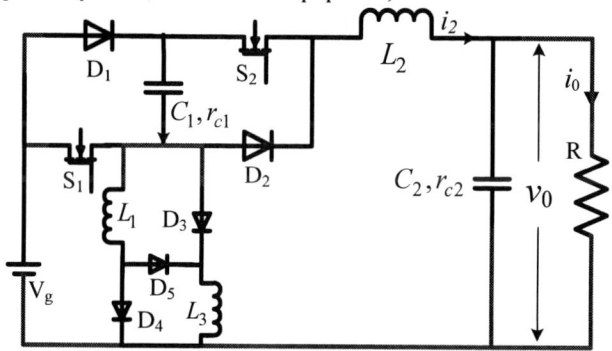

(a) Proposed topology: Fifth-order buck-boost converter

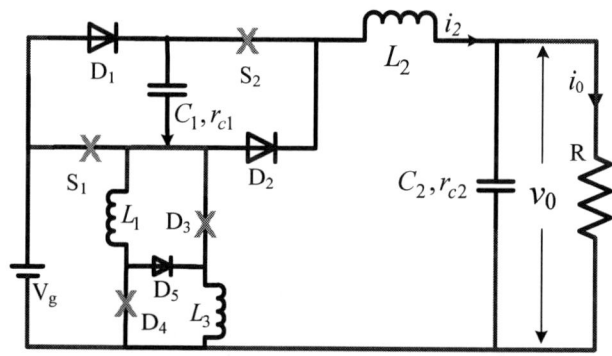

(b) Operation of fifth-order buck-boost converter (mode-1)

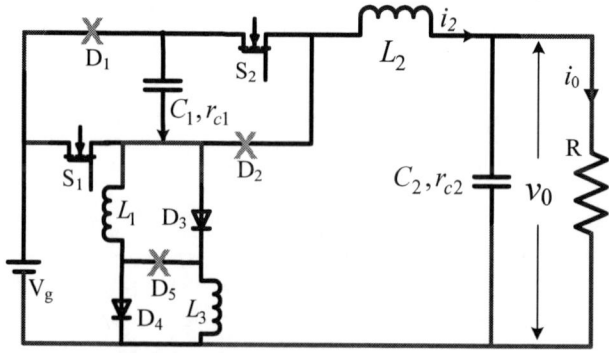

(c) Operation of fifth-order buck-boost converter (mode-2)

Fig. 1. Fifth-order buck-boost converter and its equivalent circuits.

Generally, the use of buck-boost circuits is in back-end power processing in which wide variations occur in the battery

978-1-5386-6624-1/18 $31.00 © 2018 IEEE

voltage. They are also used in point of load converter back-end applications where low voltage dc batteries are used to drive high voltage rating loads or vice-versa. One such converter is the single switch buck-boost dc-dc converter (SSBBC) which is broadly used as it is capable of smooth changeover from buck to boost and vice-versa. However, this comes with the problem of higher ripples on source side which affects the battery life and reliability. The buck-boost converter with input filter (BBCIF), addresses the high ripple problem to some extent but at the expense of increased system order [7]. However, addition of only the input filter is not recommended because at times it induces unwanted oscillations and may force the system into instability [8]. The insertion of the damping network in the L-C filter may help in stabilizing the system by reducing the oscillations.

The higher-order topologies [1] like SEPIC and CUK converters are reported in the literature which limits the ripple content while ensuring buck-boost voltage transformation. These are higher-order systems and may pose control issues due to right half of s-plane zeros in the control-to-output transfer functions. Cascading of buck followed by boost results in a two-switch buck-boost converter (TSBBC) [9] which also exhibits voltage gain same as that of the SSBBC. However, in this converter the degree of freedom is more and thus offers flexibility in the selection of controlling schemes. Although the TSBBC offers more flexibility but it is at the cost of high ripple content. Many fourth-order buck-boost topologies have also been reported [2]-[4] recently to supply power to loads at constant voltage. In literature, a robust controller design method for higher-order boost converters is reported [7]-[8]. Features like lower ripple content with better voltage buck-boost ratio is the main motive in the formulation of the proposed fifth-order buck-boost topology. This topology primarily uses a switched inductor and switching capacitor. The following paragraph pay attention to the steady-state analysis, state-space modeling and control aspects. Transfer function formulations reported in [7] are adopted in the closed-loop controller design.

II. MODELING OF FIFTH-ORDER BUCK-BOOST CONVERTER

A fifth-order buck-boost converter (FBBC) is evolved in this paper and shown Fig. 1. It comprises a switching capacitor C_1, located in the central arm of bridge, and a split inductor. In the bridge there are two uncontrollable diodes (D_1, D_2) and two controllable switching devices (S_1, S_2). The capacitor C_1 connects to source V_g or to the inductor L_2 through the controlled switches S_1, and S_2 ON-OFF operation. Energy transfer from the source to the load takes place due to this changeover of capacitor C_1, from source V_g to the inductor L_2. This interconnection of C_1 and L_1 also keeps continuity in the source current and thus the ripple content in source current is low. Further, the converter exhibits better buck-boost voltage gain features than the TSBBC and conventional SEPIC converter. The reported TSBBC is a second-order, while the proposed FBBC is fifth-order and the increased order is responsible in mitigating the source current ripples together with better voltage transformation features. The buck-boost converter with input filter is fourth-order while the proposed

converter is fifth-order system. Although the order has increased by one but it provides additional benefits of better buck-boost features. Due to the distributed nature of L, C elements the proposed FBBC exhibits the following salient features: (i) better buck-boost features than conventional TSBBC, (ii) lesser ripple content in source current both in bucking and boosting operations, and (iii) positive output with common ground between the source and load.

A. Steady-state Performance Analysis of Fifth-order Buck-Boost Converter

A time-domain analysis is presented in this section to determine the (i) steady-state performance features, (ii) voltage and current transformation ratios, and (iii) voltage and current ripple along with L, C energy storage elements design expressions. The proposed FBBC has two controllable switches, and hence four different possibilities for driving the switches. Among these only one control scheme is analyzed in this paper, i.e. S_1 and S_2 both in pulse width modulation (PWM) simultaneously. There are two types of devices in this circuit operation, controlled switches and uncontrolled diodes. Due to the circuit configuration, the simultaneous switching of S_1 and S_2 automatically synchronize the complementary operation of diodes D_1, D_2 and D_5. An application of volt-sec balance to the inductors L_1 and L_2 is useful in establishing buck-boost features of the proposed FBBC. Voltage across inductive components L_1, L_2 is given by (i) in mode-1: $v_{L1} = V_g$, $v_{L2} = (V_g - v_{c1} + v_0)$; $v_{L3} = (V_g - v_{c1} + v_0)$; $v_{L2} = V_g$, (ii) in mode-2: $v_{L1} = (V_g - v_{c1})$; $v_{L2} = (V_g - v_{c1} - v_0)/2$, $v_{L3} = (V_g - v_{c1} - v_0)/2$, Applying volt-sec balance to the inductor L_1 gives the identity defined by eqn. 1.

$$(D)(V_g - v_{c1}) + (1-D)(V_g - v_{c1}) = 0 \qquad (1)$$

Upon simplification, the above equation results in $v_{c1} = V_g/(1-D)$, which means that the capacitor C_1 charging voltage is following the boost conversion. Similar approach is extended to the inductors L_2 and L_3 resulting in eqn. 2.

$$(1+D)(V_g) + (3D-1)v_{c1} = (1+D)v_0 \qquad (2)$$

Incorporating the identity obtained from eqn. 1 and then simplification of eqn. 2 results in the voltage gain of the FBBC and it is given by eqn. 3. This voltage gain is entirely different from TSBBC and its variation is highly non-linear. Furthermore, the resultant load voltage is positive output polarity and there is no common ground issue with this proposed topology.

$$\frac{v_0}{V_g} = \frac{(3D - D^2)}{(1 - D^2)} \qquad (3)$$

For the given input/output specifications, the energy storage elements need to be properly designed to meet various ripple standards. The ripple currents and voltage expressions can easily be derived using network equations along with simple time-domain steady-state analysis. In steady-state the ripple current in inductor L_1 and L_3 is primarily decided by the impressed source voltage. The voltage impressed across L_2 is dependent on the charging voltage of C_1 and load voltage. Based on this analogy the design expressions in terms of ripple

978-1-5386-6624-1/18 $31.00 © 2018 IEEE

current/ voltage are given below.

$$L_1 > V_g D / (f_s \Delta i_1) \tag{4}$$

$$L_2 > 2V_g D(D^2 - D - 1) \big/ [(1-D)f_s \Delta i_2] \tag{5}$$

$$L_3 > 2V_g D(D^2 - D - 1) \big/ [(1-D)f_s \Delta i_3] \tag{6}$$

$$C_1 \geq (I_{L1} + I_{L2})(1-D) / (f_s \Delta v_{c1}) \tag{7}$$

$$C_2 > DI_0 / (f_s \Delta v_0) \tag{8}$$

B. Small-signal Analysis and Transfer Functions of the FBBC

Fig. 1b and 1c represents the FBBC operating modes wherein in first mode energy stored in the energy storage elements L_1, L_3 and C_1 while L_2 and C_2 will supply the energy to the load. In the second mode, the stored energy is transferred to load via L_2 and also charge C_2. Using network mesh analysis, the behaviour of these two circuits can easily be analysed independently. Here, the circuit behaviour essentially defined by the inductor current and capacitive elements voltages, which leads to the formulation of a set of first-order differential equations through mesh analysis and then represented using matrix notation as stated in eqn. 9.

$$\left. \begin{array}{c} [\dot{x}] = [A_k][x] + [B_k][u] \\ [y] = [E_k][x] + [F_k][u] \end{array} \right\} \quad t_k < t < t_{(k+1)} \tag{9}$$

here $[A_k]$ is the system state matrix, $[B_k]$ the input matrix, $[E_k]$ the output matrix, $[x]$ the state vector, $[y]$ the output vector, and $[u]$ is the input forcing function vector.

$$[A_1] = \begin{bmatrix} \dfrac{-r_1}{L_1} & 0 & 0 & 0 & 0 \\ 0 & \dfrac{(r_2 + r_{c1} + a)}{-L_2} & \dfrac{(r_1 + a)}{L_2} & \dfrac{1}{L_2} & \dfrac{-b}{L_2} \\ 0 & \dfrac{(r_1 + a)}{L_3} & \dfrac{(r_3 + r_{c1} + a)}{-L_3} & \dfrac{1}{L_3} & \dfrac{-b}{L_3} \\ 0 & \dfrac{1}{C_1} & \dfrac{-1}{C_1} & 0 & 0 \\ 0 & \dfrac{b}{C_2} & \dfrac{b}{C_2} & 0 & \dfrac{-b}{RC_2} \end{bmatrix}; \tag{10}$$

$$B_1 = B_2 = [1/L_1 \quad 1/2L_2 \quad 1/2L_3 \quad 0 \quad 0]^T;$$

$$E_1 = [0 \quad a \quad a \quad 0 \quad b]; \quad E_2 = [0 \quad 0 \quad a \quad 0 \quad b]$$

$$[A_2] = \begin{bmatrix} \dfrac{-(r_1 + r_{c1})}{L_1} & \dfrac{-r_{c1}}{L_1} & 0 & \dfrac{1}{L_1} & 0 \\ \dfrac{-r_{c1}}{2L_2} & \dfrac{(r_2 + r_{c1} + a)}{-2L_2} & \dfrac{-a}{2L_2} & \dfrac{-1}{2L_2} & \dfrac{-b}{2L_2} \\ \dfrac{-r_{c1}}{2L_3} & \dfrac{(r_2 + r_{c1} + a)}{-2L_3} & \dfrac{-a}{2L_3} & \dfrac{-1}{2L_3} & \dfrac{-b}{2L_3} \\ \dfrac{1}{C_1} & \dfrac{1}{C_1} & 0 & 0 & 0 \\ 0 & 0 & \dfrac{b}{C_2} & 0 & \dfrac{-b}{RC_2} \end{bmatrix}; \tag{11}$$

The small-signal transfer functions, both in s and z-domain, can easily be obtained from eqn. 8 after its linearization. The state-space model in discrete-time domain, in terms of ϕ and Γ, is defined as:

$$\hat{x}[NT_s] = \phi \hat{x}[(N-1)T_s] + \Gamma \hat{d}[(N-1)T_s] \tag{12}$$

where $\alpha = [(A_1 - A_2)X + (B_1 - B_2)V_g]$, $\phi_1 = e^{A_1 t_d}$, $\phi_2 = e^{A_2 D_2 T_s}$,

$\phi = e^{A_1(DT_s - t_d)} \phi_1 \phi_2$, $\Gamma = e^{A_1(DT_s - t_d)} \alpha \phi_2$. These 'phi' and 'gamma' matrices are used in formulating the z-transfer functions as listed in Table II.

TABLE -I. COMPONENT STRESS OF FBBC OVER THE REPORTED BUCK-BOOST CONVERTERS

Quantity	SSBBC	TSBBC	Proposed FBBC
Voltage Gain	$\dfrac{-D}{(1-D)}$	$\dfrac{D}{(1-D)}$	$\dfrac{(3D - D^2)}{(1 - D^2)}$
PVS	$(v_0 - V_g)$	v_0	v_{C1}
PDS	$(v_0 - V_g)$	$(v_0 - V_g)$	v_{C1}
SCR	High	High	Low
OCPCS	Identical	Identical	Low

PVS: Switch peak voltage stress, PDS: Diode peak voltage stress, SCR: Source current ripple, OCPCS: Peak current stress in the output capacitor.

TABLE II. Z-TRANSFER FUNCTIONS FORMULATION[8]

Transfer Function	Formulation
Control-to-Output	$\hat{v}_0(z) / \hat{d}(z) = E'(zI - \phi)^{-1}\Gamma$
Audio Susceptibility	$\hat{v}_0(z) / \hat{v}_g(z) = E'(zI - \phi)^{-1}\Gamma + F'$
Output Impedance	$\hat{v}_0(z) / \hat{i}_0(z) = [E'(zI - \phi)^{-1}\Gamma + J']$
Input Admittance	$\hat{i}_{in}(z) / \hat{v}_g(z) = P'[(zI - \phi)^{-1}\Gamma]$

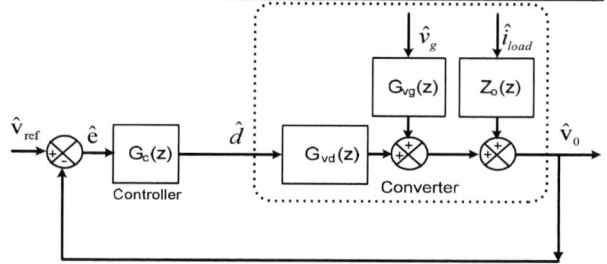

Fig. 2. Block-diagram: Closed-loop controlled FBBC.

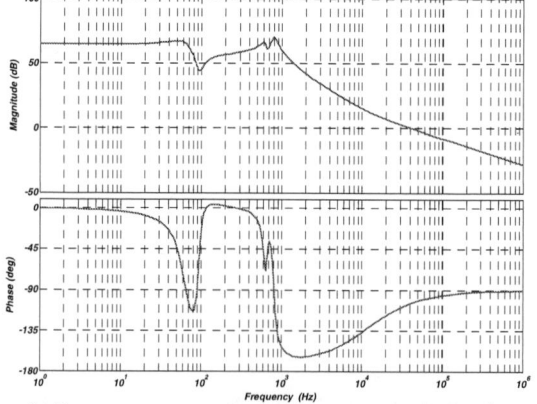

(a) Frequency response of control-to-output transfer function

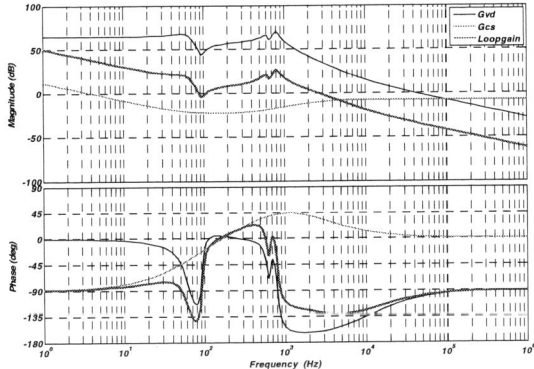

(b) Frequency response of controller and loopgain transfer functions

Fig. 3. Frequency response plots.

III. SYSTEM DESIGN

Linear control theory is extensively used in the design of controllers for power electronic systems [10]. Majority of these designs are evolved in the literature are on the basis of frequency domain models. Since the power electronic converters are switching systems, they constitute two or more different circuit systems in one cycle of operation. Although it is possible to apply the non-linear system theory for such systems but complexity involved for higher-order system increases tremendously. In view of this limitation, state-space average modelling is widely used so that extension of linear system theory becomes easier task. From control point of view, the FBBC along with its controller is shown in the block-diagram representation depicted in Fig. 2. Since the FBBC is regulating the load voltage through application of suitable PWM signal, its impact is represented in the block diagram by the duty ratio -to- output load voltage transfer function. Here, the controller included in the feed-forward path must be yielding sufficient relative stability margins while ensuring the system absolute stability. The remaining two transfer functions in the block diagram, output–to-input voltage and output impedance, reflects the impact of disturbances on the load voltage. In the design of controller the feed-forward path gain transfer function, indicated in Fig. 2, plays an important role. For the controller selection and design single-input single output tool of Matlab [10] is used. This tool offers placing of controller poles and zero in the desired locations such that the resulting feed-forward path gain transfer function exhibits absolute stability. Once it satisfies the absolute stability criterion, it becomes necessary to choose a controller which will ensure the relative stability specifications.

Disturbances are more common in the point of load converter applications and their rejection is decided by the relative stability specifications. The minimum gain margin (GM) should be at least 6 dB (higher value is recommended), the phase margin (PM) somewhere in between (45°~75°) with sufficient bandwidth which is decided by a combination of controller and FBBC system. In the controller design stage, firstly a pole at the origin assumed in the controller so as to reduce the steady-state error. A second pole is introduced to attain sufficient disturbance rejection capability. To ensure flat-top frequency response in the high frequency region, two zeros are added to the controller transfer function. In the closed-loop

realization process the high voltage levels can not be handled and hence the load voltage is sensed and scaled-down to less than 5 V range. In view of this, controller gain is reduced and pole-zero locations get modified to meet the relative stability margins as discussed above. The final controller form in s-domain is

$$G_c(s) = \frac{k(s+a_1)(s+a_2)}{(s+b_1)(s+b_2)} \qquad (13)$$

For experimentation, the controller defined in eqn. 13 is realized on a digital platform and the s-domain controller is then transformed into the digital domain. The digital equivalent of the s-domain controller is defined in eqn. 14.

$$G_c(z) = \frac{k_{11}(z+a_{11})(z+a_{22})}{(z+b_{11})(z+b_{22})} \qquad (14)$$

where gain k_{11}=0.405, a_{11}=-0.9806, a_{22}=-0.949, b_{11}=-1.0, b_{22}=-0.1983. Frequency response plots of the control-to-output and feed-forward path transfer functions are given in Fig. 3. These frequency response plots exhibiting sufficient relative stability margins and thus the closed-loop FBBC maintain voltage regulation even if there are any variations in the supply voltage or in the load. To confirm the effectiveness of frequency responses in the system stability quantification, a step response plot is generated in MATLAB platform and it is shown in Fig. 4. Since the closed-loop FBBC constitutes a higher-order system, the step response is slowly increasing with time and it reaches the reference in nearly 30 ms. It is thus evident that the proposed closed-loop system is stable and also possesses sufficient disturbances rejection features.

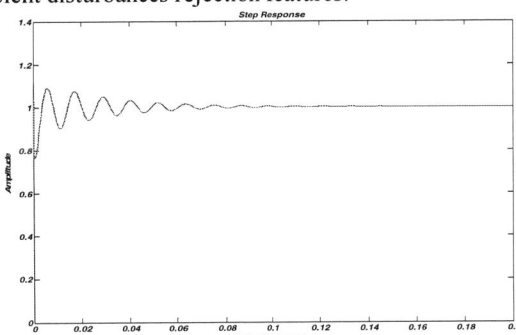

Fig. 4. Step response plots of FBBC.

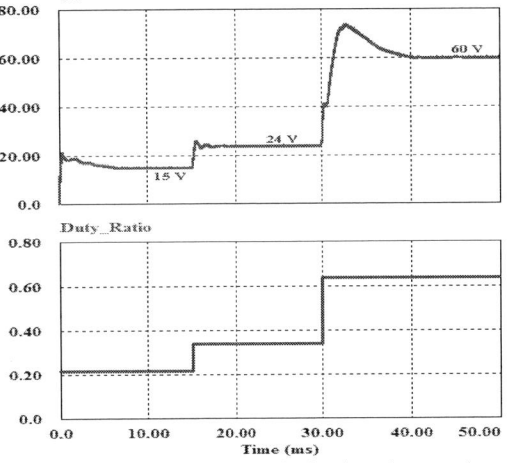

(a) Load voltage transition from buck to boosting operations

978-1-5386-6624-1/18 $31.00 © 2018 IEEE

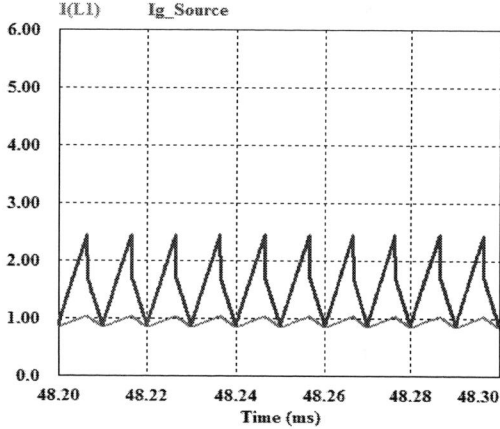

(b) Enlarged steady-state waveforms

Fig. 5. Step response and steady-state waveforms of FBBC.

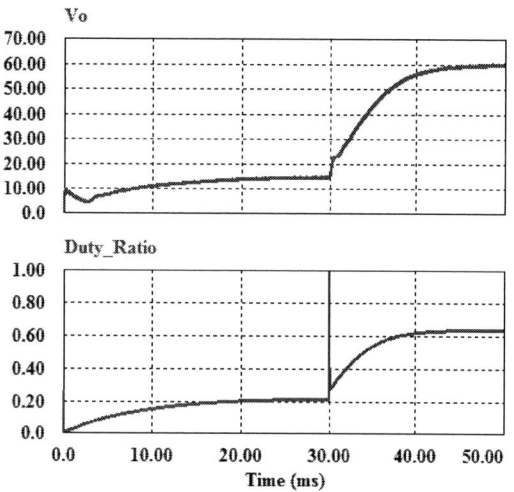

Fig. 6. Response of load voltage (V_{o_ref}: 15 to 60 V; Vg: 24 V; R: 30 to 15 Ω)

IV. EXPERIMENTAL RESULTS

A 50 Watt, 24 to 15/ 60 V FBBC prototype is built for demonstrating its performance both in simulation and experiments. Table III is included with the parameters used in simulation as well as in experimentation. Initially, the PSIM software simulator [11] tool is used to predetermine the steady-state and dynamic performance of the proposed FBBC. At first, different duty ratios (D=0.22, 0.33, and 0.62) have been set and the load voltage is monitored as shown in Fig. 5a. At D=0.22 the load voltage 15 V while it is equal to source voltage (V_o=24 V) at a duty ratio of 0.33. For D>0.33 the FBBC is boosting the load voltage and as indicated in Fig. 5a for D=0.62. Due to better buck-boost features, as indicated in eqn. 3, the buck to boost mode transition is taking place at low value of duty ratio. Due to this feature more range of duty ratio available for the boosting operation. The steady-state waveforms indicating the inductor and source current are shown in Fig. 5b. The source current waveform is smooth and its ripple content is also lower. The controller, eqn. 14, is now used to test the FBBC performance in closed-loop condition. To begin with, the reference voltage is set to 15 V, for bucking operation, and a changeover (reference voltage is changed to 60 V) is initiated at

t=30 ms and corresponding response characteristics are shown in Fig. 6. This simulation result clearly shows the bucking (V_g=24 V, V_o=15 V) and boosting (V_g=24 V, V_o=60 V) feature. To demonstrate the controller effectiveness in terms of voltage regulation and disturbance rejection, sample simulation results are illustrated in Fig. 7. Fig. 7a shows the dynamic response of load voltage against step change in load (R: 100 to 50 Ω) while Fig. 7b against supply voltage change (V_g: 24 to 36 V). Here, the controller effectively suppressing the disturbances. The maximum transient response settling time is about 20.0 ms.

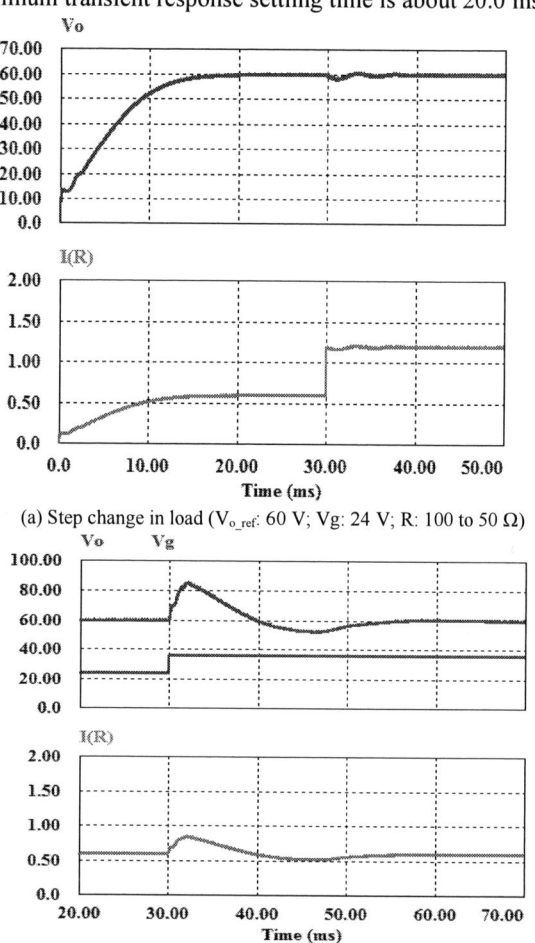

(a) Step change in load (V_{o_ref}: 60 V; Vg: 24 V; R: 100 to 50 Ω)

(b) Step change in source voltage (V_{o_ref}: 60 V; Vg: 24 to 36 V; R: 100 Ω)

Fig. 7. Dynamic response of FBBC load voltage against load perturbation (boost mode of operation).

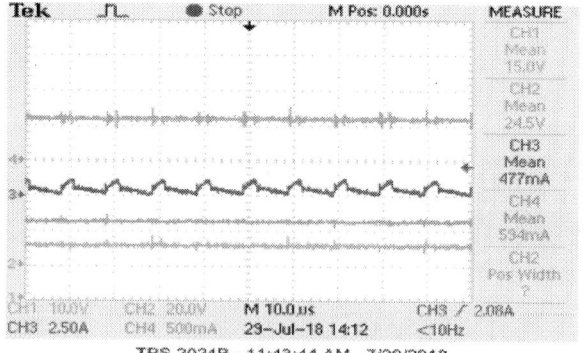

(a) Buck mode of operation

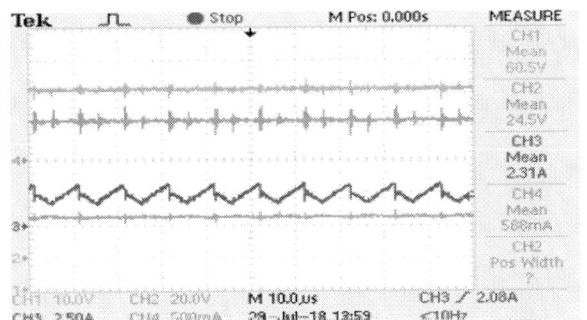

Ch-1:(V_0):10 V/div;Ch-2:(V_g):20 V/div;Ch-3:(i_g):2.5 A/div; Ch-4:(i_o):0.5 A/div

(b) Boost mode of operation

Fig. 8. Steady-state waveforms of FBBC.

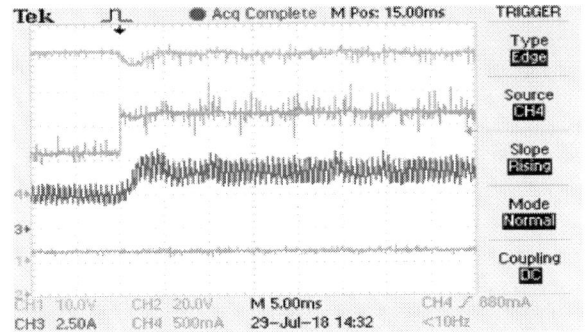

Ch-1:(V_0):10 V/div;Ch-2:(V_g):20 V/div;Ch-3:(i_g):2.5 A/div; Ch-4:(i_o):0.5 A/div

(a) Load perturbation

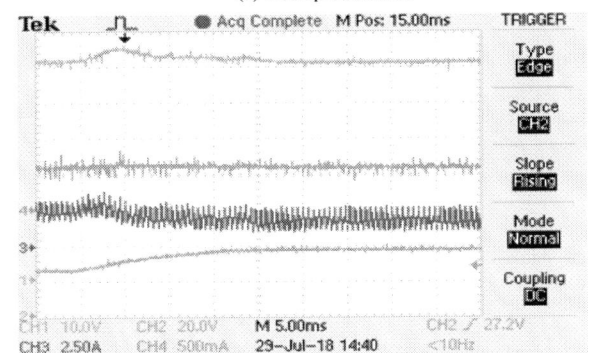

Ch-1:(V_0):10 V/div;Ch-2:(V_g):20 V/div;Ch-3:(i_g):2.5 A/div; Ch-4:(i_o):0.5 A/div

(b) Source voltage perturbation

Fig. 9. Dynamic response of load voltage (in boost operation).

To demonstrate the proposed FBBC's merits, design and its operation principles with the controller, discussed in Section III, experiments [12] are performed. Parameters used for experimentation circuit are listed in Table III. The measured steady-state waveforms are plotted in Fig. 8. The measured source current both in bucking and boost modes is smooth enough to exhibit low ripple content. The measured average load voltage is 15 V in bucking mode while in boosting mode it 60 V. The dynamic response of FBBC load voltage against load and source perturbations is also measured and illustration of voltage regulation measurement, for boosting operation, is shown in Fig. 9. Against step load perturbation the FBBC is responding quickly (response time close to 5.0 ms) while in boosting operation it is a bit slow (response time close to 15.0 ms). The discrepancies and a slight mismatch in the waveforms, in the simulation and experiments, are due to: (i) nature of

simulation platform, step size and simulation timings adopted, (ii) difficulty in estimating the time delays involved in the actual experimentation and their inclusion during simulation, (iii) mismatch in accounting the circuit non-ideal parameters of measurement set-up while performing simulations, and (iv) adopted step size in the solution of differential equations in simulation platform.

TABLE III. PROPOSED FBBC PARAMETERS

Parameter	Value
Vg	24 V
$L_1 - L_3$	300 μH
L_2	600 μH
C_1	47 μF
C_2	100 μF
f_s	100 kHz

V. CONCLUSION

A fifth-order buck-boost converter exhibiting better buck-boost features than the conventional buck-boost/SEPIC converter was proposed. The steady-state analysis revealed that buck to boost mode transition takes place at lower duty ratio and thus leaving more range of duty ratios for boosting operation. Detailed small-signal analysis was established and accordingly a controller was designed. The effectiveness of the controller was demonstrated for both bucking as well as boosting operations. The analytics of the proposed FBBC, its effectiveness in terms of load voltage regulation were simulated and supported with experimental measurements. Simulation and measurement results were in close agreement.

REFERENCES

[1] K. C. Daly, "Ripple determination for switch-mode dc-dc converters," *IET Proc.*, vol. 129, Pt. G, no. 5, pp. 229-234, Oct. 1982.

[2] Jingquan Chen, D. Maksimovic, Robert Ericksom, "Buck-boost PWM converters having two independently controlled switches," Proc. of *IEEE APEC.*, vol. 2, 2001, pp. 736-741.

[3] Kerui Li, Andrian Ioinovici, "Large DC gain nonisolated converter based on a new L-C-D step-up switching cell," *Proc. of IEEE APCCAS*, 2014, pp. 284-287.

[4] Shiyu Zhang, Jianping Xu, Ping Yang, "A single-switch high-gain quadratic boost converter based on voltage-lift-technique", *Proc. of IEEE IPEC*, 2012, pp. 71-75.

[5] K. I. Hwu, Y. T. Yau, "A KY boost converter," *IEEE Trans. on Power Electron.*, vol. 25, no. 11, pp. 2699-2703, Nov. 2010.

[6] K. I. Hwu, T. J. Peng, "A novel buck-boost converter combining KY and buck converters," *IEEE Trans. on Power Electron.*, vol. 27, no. 5, pp. 2236-2241, May. 2012.

[7] M. U. Iftikhar, D. Sadarnac, C. Karimi, "Input filter damping design for control-loop stability of dc-dc converters," in *Proc. of IEEE, ISIE*, 2007, pp. 353-358.

[8] M. Veerachary, Anmol Ratna Saxena, "Design of Robust Digital Stabilizing Controller for Fourth-Order Boost DC–DC Converter: A Quantitative Feedback Theory Approach," *IEEE Trans. on Ind. Electron.*, vol. 59, no. 2, pp. 952-963, Feb. 2012.

[9] Chuan Yao, Xinbo Ruan, Weijie Cao, Peilin Chen, "A two-mode control scheme with input voltage feed-forward for the two-switch buck-boost dc-dc converter," *IEEE Trans. on Power Electron.*, vol. 29, no. 4, pp. 2037-2048, Apr. 2014.

[10] MATLAB, user manual, 2005.

[11] PSIM, user manual, 2005.

[12] dsPIC30f6010, user manual, 2010

Source Scheduling for Power Matching of Clustered Induction Generators fed VSC supported Micro grid

Peeyush Pant
Department of Electrical Engineering
Delhi College of Engineering (Now DTU) Delhi
110 042, India
pantpeeyush@gmail.com

Vishal Verma
Department of Electrical Engineering
Delhi Technological University, Delhi 110 042,
India vishalverma@dce.ac.in

Abstract—This paper proposes an autonomous off-grid generation system fed through clustered cage-rotor induction generators (IGs) acting in parallel and forming a micro-grid. The soul of the control for proposed generation system is rested in the selection of the capacity of participating generators in binary weighted IG unit. The arrangement provides optimal capacity utilization of each source, power matching during load perturbations acting together with very small capacity storage system and easy management of the source leading to regulated voltage and frequency at point of common coupling (PCC). The aforesaid system is supported with current-controlled voltage source converter (VSC) with smaller capacity battery energy storage (BES) to substantiate the deficit real power matching during disturbances. The VSC also supplies reactive power both for generator excitation and voltage regulation at PCC. The current control of VSC is actuated in tandem with source scheduling routine during perturbations of the load. The proposed scheme offers optimal head conversion opportunities for the micro-hydro turbines with improved efficiency and reduced cost. The performance of the system is evaluated in Matlab/ Simulink and its effectiveness is gauged through simulation results.

Keywords— Induction generator, Dynamic excitation, Micro-grid, Source scheduling, Current control, Voltage source converter

I. INTRODUCTION

Fast depletion of fossil fuel has encouraged the researchers to investigate the possibilities of electricity generation using renewable energy resources [1]. Utilizing solar, wind and/or micro/pico-hydro resources seems attractive option to generate electricity effectively without carbon footprints [2]-[6]. Selecting micro-hydro system with cage rotor induction generators offers brushless, rugged, maintenance-free structure and fault immunity features for smaller capacity generation in remote locations since it experiences lesser perturbations on input power side, besides near constant head [7]-[11]. This arrangement of electricity generation utilizing micro hydro source integrated with small units of induction generator(IG) in parallel and forming a micro-grid advocates attainment of improved performance with efficient outcome and fault immune generation features utilizing near constant head siphon turbines[12],[13].

Parallel operation of induction generator for attaining improved reliability, stability and loading capacity is rarely investigated. The reported schemes do not addresses exact compensation of reactive power when dynamic reactive power consuming loads are connected at PCC making it unsuitable for off-grid application[14]-[16]. The deployment of fixed capacitors for IG excitation thus also takes short leading unregulated voltage/ fluctuations during load perturbations [17], [18]. Recently, IG fed micro-grid for optimal source utilization of the source is reported [19]-[20]. Since, IG is perceived as a weak source before PCC the issues of unregulated voltage and frequency amid reactive power consuming loads connected at PCC become prominent and is remediation is yet to be addressed. Hybrid combination of capacitors and VSC supported autonomous distribution system is also reported [21]-[25]. Large capacitor often provokes over excitation of generator during light load at PCC and even causes severe voltage spike during declutching of the source from the micro-grid and thus cannot confirm to capacity utilization of the source [26]. Dynamic excitation control of individual induction generator with current controlled VSC presents attractive option to attain precise voltage control even during declutching of generator [27]-[31]. Forming a micro-grid by integrating cluster of induction generators operating in parallel with dynamic excitation from a single VSC offers an attractive option. However, reported literature indicates large capacity battery storage source for maintaining frequency constant by matching the load. Thus investigation with small capacity battery storage acting with scheduling of IG source with constant head siphon turbine is yet to be explored for operations with optimal head converters.

In this paper a micro-grid fed through the binary weighted clustered parallel connected IGs with a current controlled VSC with storage is presented. Each CRIG is coupled to a separate micro-hydro siphon turbine of respective capacity. The capacity of each participating IG in the formation of the micro-grid is binary weighted so that rated capacity of next higher IG is nearly double of the former unit. This arrangement enables the capacity ratio as 2^n (for n=0 to 3) for optimal source scheduling amid harnessing the energy when generation system caters perturbing load in off-grid system. The micro-grid is supported by a VSC for adequate supply of reactive power both for IG excitation and reactive power hungry loads at PCC. The control action acts to regulate the voltage and frequency at PCC of micro-grid by source scheduling and battery storage support deployed at the dc-bus of VSC. The battery storage also emulates as a load controller when operating under battery charging mode and a real power source to releases the part/full real power demanded during load perturbations. The proposed current controlled scheme is demonstrated to present simple, cost-effective, efficient and responsive hardware. The scheme effectively and efficiently converts the head of micro-hydro system leading to improved reliability of the overall system. The proposed system is realized in Matlab/ Ps-Block set and the effectiveness of the system is gauged through simulation results.

II. SYSTEM CONFIGURATION

978-1-5386-6624-1/18 $31.00 © 2018 IEEE

The block diagram of the proposed system is shown in Fig.1. The system integrates clustered parallel connected induction generators along with a current controlled VSC forming a micro-grid. The capacity ratio of the generators are selected as 2^n (n=0 to 3) that of the former. The proposed system considers four IGs of rated power capacities 2.2kW, 5.5kW, 14kW and 30kW each of 415V, 50Hz rating. Small capacitors (600V, 1μF each) are connected at stator terminals of each IG for harmonics filtering. This IG and capacitor combination is hence forth

considered as source. A current controlled VSC with a capacitor(3000μF, 1200V) at dc-bus is integrated at PCC which supplies adequate reactive power both for generators excitation and voltage regulation at PCC. The dc-bus of VSC is also supported by a small battery storage source of capacity 135Ah. The dc-bus voltage is maintained in self-support mode and transacts requisite amount of real-power by absorbing or delivering current from the battery source. This arrangement eliminates the necessity of installing costlier governor and/or complex load controllers.

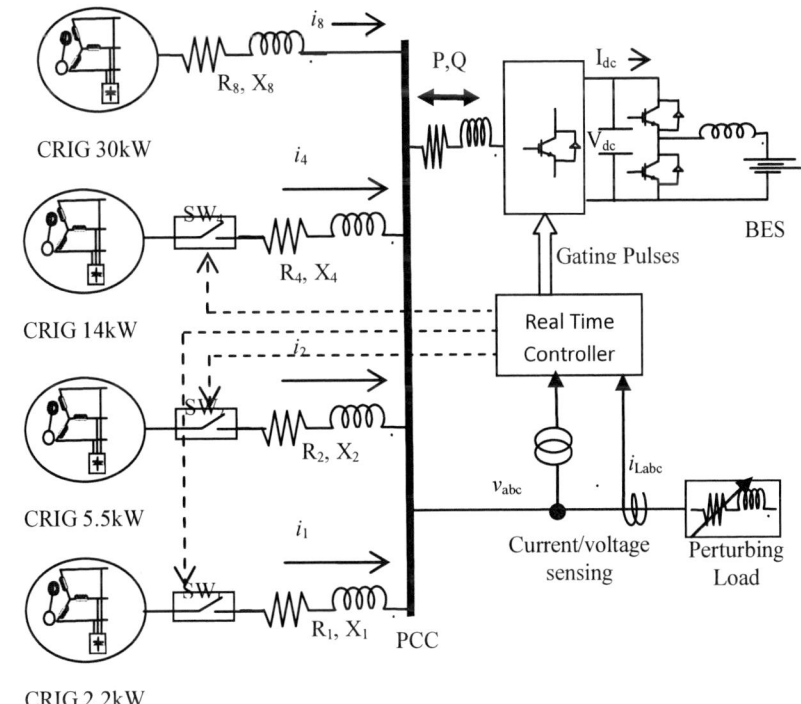

Fig.1 Block diagram of the proposed system depicting cluster of IGs operation in parallel and forming a micro-grid.

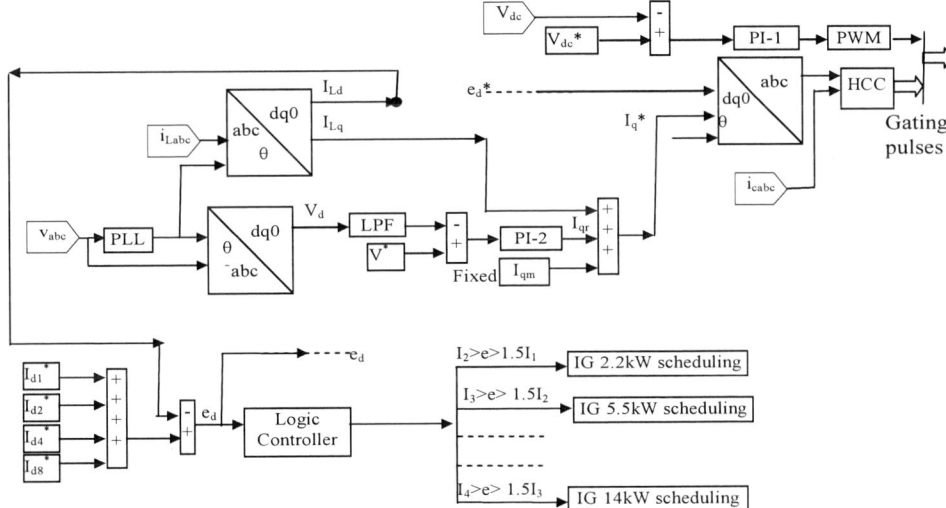

Fig.2 Control scheme pertaining dynamic excitation control of IG and source scheduling during load/ source disturbances.

Regulation of the input side potential energy utilization is made through adequate selection of binary weighted CRIGs providing the small power for load matching by the battery storage. A three-phase reactive power consuming load is connected at PCC to gauge the effectiveness of the control under adverse conditions. The loading perturbations enables battery charging/ discharging operation in tandem with source scheduling to offset the incremental load difference leading to the condition of load matching enabling near constant frequency operation. The system offers head conversion opportunities along with improved efficiency at reduced cost of electricity generation. An indirect current control of VSC is implemented in the synchronous reference frame and the same is explained in the next section.

III. CONTROL STRATEGY

The block diagram of the control algorithm to regulate voltage and frequency at PCC of the clustered IGs fed micro-grid with current controlled VSC and a small battery energy storage support at dc-bus is depicted in Fig.2. The VSC supplies reactive power both for generator excitation and voltage regulation at PCC. The control algorithm is realized in synchronous reference frame. To implement the control action the load current (i_{Labc}), VSC current (i_{Cabc}), phase voltage (v_{abc}) and dc-bus voltage (V_{dc}) are sensed. The proposed control scheme is oriented in the following sub-sections.

A. Regulation of dc-bus voltage and estimation of reference current I_d

Source scheduling/ load perturbation at PCC of micro-grid may lead to mismatch in real power pertaining unregulated voltage at dc-bus of VSC. Eventually larger loading may lead to voltage sag at dc-bus of VSC. The control action pertaining dc-bus voltage regulation is realized by estimating the error between the sensed voltage in dc-bus and the reference voltage ($V_{dc}*$). The error is passed through PI controller (PI-1) which estimates the current corresponding to real power drawn from the battery storage for dc bus voltage regulation. Similarly during deficit load the deviation of loading from rated value on generator simultaneously affect the dc-bus voltage and the energy is transferred across the battery storage and maintaining dc-bus voltage regulated.

The current corresponding to real power drawn by the load (I_{Ld}) is compared with the cumulative current equal to the rated capacity of all integrated IG sources. The error estimates the reference current I_d for real power equalization by extracting rated capacity power from the IG sources connected at PCC.

B. Voltage regulation and estimation of reference current I_q

The estimation of reference current corresponding to reactive power (I_q) is made by accumulating the current required for IG magnetization (I_{qm}) along with the current corresponding to reactive power (I_{qr}) required for voltage regulation at PCC. The estimation of the former component is derived by comparing per unit rms value of the sensed phase voltage (v_{abc}) with the unit reference voltage ($v*$) through PI controller (PI-2). This estimates the requisite current (I_{qr}) for voltage regulation.

C. IG source scheduling

The current corresponding to real power drawn by the load (I_{Ld}) is compared with the current corresponding to cumulative rated capacity of all the sources (i.e. $I_{d1} + I_{d2} + I_{d4} + I_{d8}$). The estimated error($e_d$) is passed through the logic controller circuit which leads to the scheduling action of suitable capacity IG source. This presents real power matching of the source and load. The cumulative action of source scheduling and battery storage in turn pivots each IG source to contribute its rated capacity real power at PCC.

A digital PLL is implemented to realize the fast control action in turn producing synchronization signals via Park/Clark transformation. The estimated values of reference real and reactive currents in dq0 frame are reverse transformed and compared with the sensed VSC currents (i_{Cabc}). The error is fed through a hysteresis current controller (HCC) which forces the VSC currents to follow the reference currents and maintains requisite real and reactive power transaction at PCC. Matlab based simulations are performed and the system performance is presented in the forthcoming section of the paper. The attained results validate that the scheme offers fast, easy and efficient control for effective system operation.

IV. SIMULATION SCHEME

The proposed system is simulated in Matlab/Simulink. The induction generator and the system parameters are kept close to the standard values enabling correct assessment of the modeling and control algorithms. The rated capacities of considered IGs are selected as 2.2kW, 5.5kW, 14kW and 30kW each of 415V, 50 Hz rating. The micro-grid is supported by a VSC and small battery storage at dc-bus. The three-phase reactive power hungry load perturbing in nature is connected at PCC. Each IG of the system is mechanically coupled to the respective low head constant power prime mover (siphon turbine). Three capacitors (each 1µF, 600V) are connected in star configuration across the stator terminals of each generator for harmonic filtering. A VSC with a tank capacitor at dc-bus has been used in tandem with a small battery at PCC and regulates the exchange of both real and reactive power. The tank capacitor voltage is maintained at 600V. A hysteresis band of 0.1A is considered to realize carrier-less PWM signals operating in current control mode. The system is operated at 12.8 kHz switching frequency. The simulation model of the system is executed with fixed-step solver using ode3 (Bogacki-Shampine) for estimating both dynamic and steady state simulation based performance of the system. The performance of the proposed system is evaluated by appropriate tuning of both simulation and control parameters of the system. The simulation scheme is realized by triggering the VSC when self-excitation of the IG is established leading stabilized voltage at PCC. During steady state operation the reactive power supplied by the VSC is sufficient to compensate the magnetization current of IG sources and also establishes voltage regulation at PCC. To demonstrate the potential of the scheme under adverse load condition load perturbations are made at PCC and the onus is transacted via battery source and source scheduling bringing the system to equilibrium again. The voltage across the dc-bus has been maintained by BES charging/ discharging action. This ensures compensation of real and

978-1-5386-6624-1/18 $31.00 © 2018 IEEE

reactive powers through the VSC in controlled manner. The next section demonstrates the attained results and performance of the system for validating successful implementation of the proposed control algorithm.

V. PERFORMANCE OF THE SYSTEM

Matlab based simulations are performed to evaluate the performance of the proposed system. The performance is validated in terms of the following (Refer Table I) illustrations.

1) Steady-state operation of the system.
2) Dynamic operation of the system during load perturbation at PCC.
3) Power matching by source scheduling and small battery support.

The performance of the system based on simulation results validates the effectiveness of the proposed micro grid which integrates cluster of IGs in parallel with VSC support. The performance of system is evaluated during steady- state operation and dynamic load perturbations. The attained waveforms are depicted in Fig.3 which presents PCC voltage (v_{abc}), source currents (i_{abc}), VSC current (i_{cabc}), load current (i_{Labc}), and the dc bus voltage (V_{dc}). The results are analyzed in the following three time sections.

(i) t=0.5s-2.0s: Steady-state operation of system with VSC supplies reactive power in dynamic manner
(ii) t=2s - 4s: Load perturbation at PCC and
(iii) t=4.5s onwards: Re-insertion of IG source at PCC of micro grid.

At t=0.5s the VSC is connected at PCC of micro-grid.

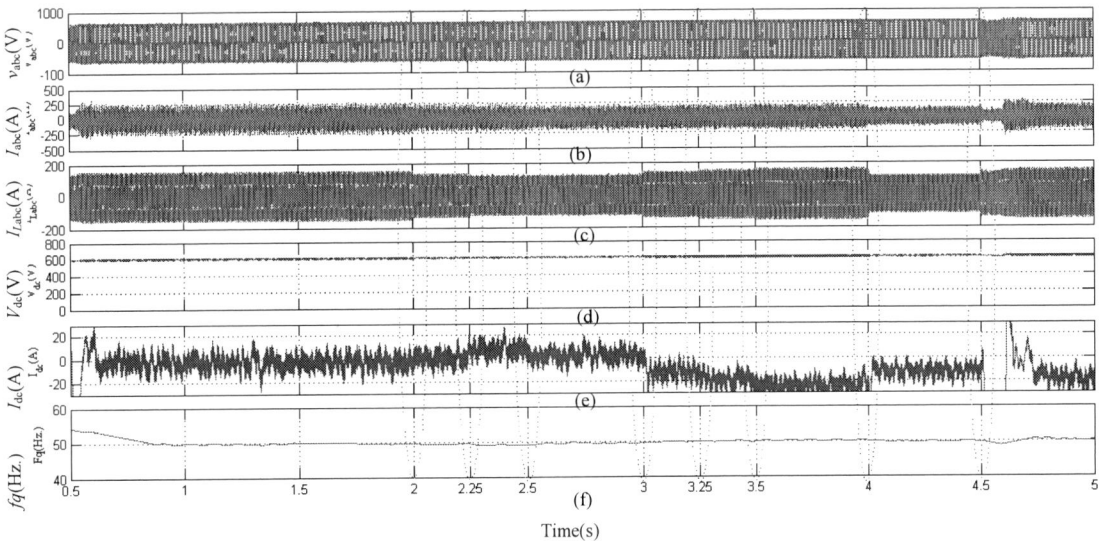

Fig.3 Performances of the proposed system during load perturbations at PCC of micro grid.

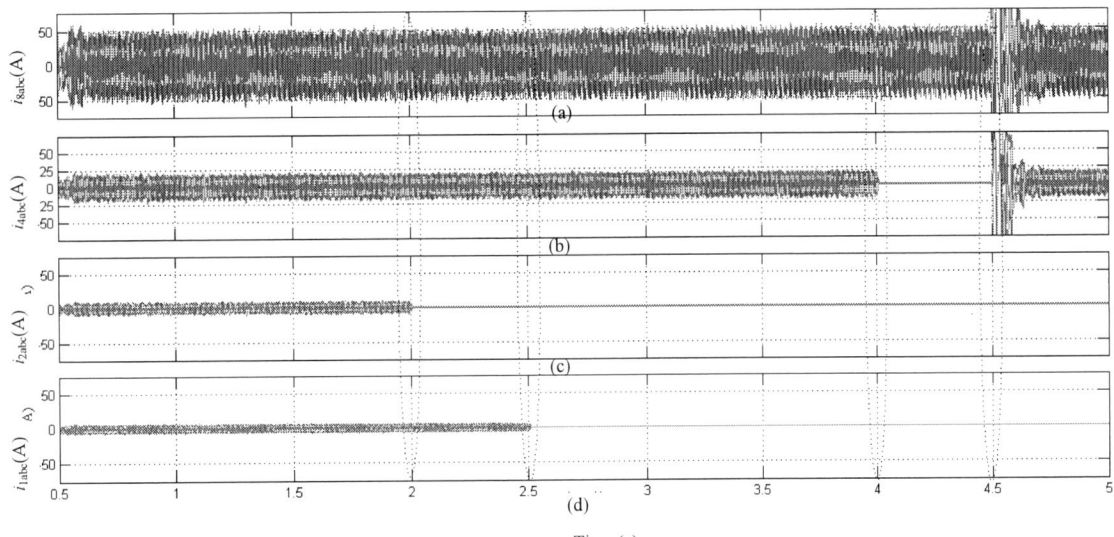

Fig.4. Simulation results depicting stator current of each IG sources connected at PCC

As shown in Fig.3(a,b) current transient and glitch in PCC voltage is observed. Small transient is also observed in the current transacted through dc-bus and is shown in Fig. 3(e). Further, the system comes to steady state within few cycles and no abnormal state is observed in the system.

At t=2s-5s the system experiences load perturbations in steps and are depicted in Fig.3(c). During perturbations the power matching is made fast by real power transaction to/from the battery and IG source scheduling and is presented in Table-I. At t=2.25s and 2.5s load at PCC is reduced to 46kW and 44kW respectively in steps. For real-power matching the logic controller schedules 5.5kW IG source from the system and is shown in Fig. 4(c). Effective power matching is maintained in the system vis-a- vis the VSC regulates reactive power and presents regulated voltage at PCC. The battery inacts in charging mode and the same may be observed in Fig.3(e). Further, as shown in Fig.3(c) at t=2.5s, 3s and 3.25s, the load changes to 43kW, 46kW and 48kW respectively. The logic controller matches the real-power by scheduling 2.2kW IG source from PCC and the same is shown in Fig. 4(d). Also it is observed at 2.5s-3s in Fig.3(e) that battery operates in charging mode and than inacts in discharging mode during t=3s-4s. At t=4s shown in Fig. 3(c) the loading at PCC changes to 32kW. To match the real power the 14kW IG is scheduled by the controller action (Refer Fig. 4(b)) and only 30kW IG source integrated in the system caters the load at PCC. Moreover the transient load matching is made by battery operation inacting in discharging mode and is clearly seen in Fig.3(e). At t= 4.5s - 5s the loading at PCC rises to 44kW. In this condition the real power matching is made by resynchronization action of 14kW IG source. As shown in Fig. 4(a, b) the synchronization of IG at PCC may results the current transients under acceptable limits. It is observed in Fig. 3(e) that requisite current is drawn by the battery storage to meet the loading perturbations. The battery turns idle at t=5s.

It may also be clearly observed in Fig.3 that controller action for dynamic compensation of reactive power and real-power matching operates the system well under synchronism. The PCC voltage and frequency are depicted in Fig.3 (a,f) are well regulated and pivoted under limits. Also the dc-bus voltage shown in Fig. 3(d) is well regulated at 600V throughout the system operation. This validates the effectiveness of the proposed control.

The results depicting the stator current of each IG is shown in Fig.4(a-d). It may be clearly demonstrated that the proposed scheme maintains rated extraction of real power from each IG source even under sudden load changes, scheduling and reinsertion of sources. The overall performance of the system presents fast compensation of both real and reactive power and effective load matching.

VI. CONCLUSION

The performance of the VSC supported autonomous generation system integrating cluster of IGs and small battery to cater reactive power consuming load at PCC is successfully demonstrated. The system successfully works well under load perturbations in steps and source scheduling operations. The PCC voltage and frequency is regulated by current controlled VSC by reactive power compensation and exact power matching. Employing VSC enables the scheme

to respect dynamic excitation control of induction generator. It is successfully demonstrated that even under extreme load perturbation the voltage at dc-bus of VSC is maintained. The scheme involves lesser computation effort in deriving current capacity of IGs sources from their rated capacity. The scheme is simple, fast acting and maintains healthy voltage profile of the connected loads. The proposed system effectively fulfills the requirement of utilizing renewable energy resources for electricity generation and forming a micro-grid.

TABLE I.   SOURCE SCHEDULING AND BATTERY OPERATION DURING LOAD PERTURBATION

Time Range	Load at PCC of micro-grid	IGs delivering power at PCC	Scheduled IGs for real power matching	Battery operation
0-2s	52kW	2.2kW, 5.5kW, 14kW and 30kW (51.7kW)	nil	Discharge mode for real power topping
2s-2.25s	46 kW	2.2kW, 14kW and 30kW (46.2kW)	5.5kW	Charging mode (real power matching)
2.25s-2.5s	44kW	2.2kW, 14kW and 30kW (46.2kW)	5.5kW	Trigger charging (prior to source scheduling)
2.5s-3.0s	43kW	14kW and 30kW (44kW)	5.5kW and 2.2kW	Charging mode (Real power matching)
3s-3.25s	46 kW	14kW and 30kW (44kW)	5.5kW and 2.2kW	Discharging mode (real power supply)
3.25s-4.0s	48kW	14kW and 30kW (44kW)	5.5kW and 2.2kW	Discharge mode to compensate transient overloads
4.0s-4.5s	32kW	30kW (30kW)	14kW, 5.5kW, 2.2kW	Discharging mode
4.5s-5.0s	44kW	30kW and 14kW IG resynchronized at 4.5s (44kW)	5.5kW and 2.2kW	Battery idle after stabilization

REFERENCES

[1] I.R. Pillai and R. Banerjee, "Renewable energy in India: status and potential," Energy, Vol.34, No. 8, pp.970-980, August 2009.

[2] M.A. Elhadidy and S. M.Shaahid, "Parametric study of hybrid (wind +solar+ diesel) power generating systems," Renewable Energy, vol. 21, no.2, pp.129-139, October 2000.

[3] Y.Jiang, L.Lu, H.Lu, "A novel model to estimate the cleaning frequency for dirty solar photovoltaic (PV) modules in desert environment," Solar Energy, vol. 140, pp. 236-40, December 2016.

[4] S.Adhikari, F. Li, "Coordinated V/f and PQ control of solar photovoltaic generators with MPPT and battery storage in micro-grids," IEEE Transactions on Smart Grid, vol.5, no.3, pp.1270-81, May 2014.

[5] M.Motevasel, A.R.Seifi, "Expert energy management of a micro-grid considering wind energy uncertainty," Energy Conversion and Management, Vol. 83, pp. 58-72, July 2014.

[6] Thomsen B, J.M.Guerrero, P.B.Thøgersen, "Faroe islands wind-powered space heating micro-grid using self-excited 220-kW induction generator," IEEE Transactions on Sustainable Energy, vol.5, no.4, pp.1361-66, October 2014.

[7] L.Wang, S.J.Chen, S.R.Jan, and H.W.Li, "Design and implementation of a prototype underwater turbine generator system for renewable microhydro power energy," IEEE Transactions on Industry Applications, vol. 49, no.6, pp.2753-2760, June 2013.

[8] A.S. Ridwan, S.Anjar, and I.Pudji, "Design and analysis of the prototype of pico hydro scale submersible type turbine-generator for flat flow river application," Teknologi Indonesia, vol. 35, no. 3, pp. 1–8, February 2015.

[9] A.O.Edeoja, J.S.Ibrahim, and E.I. Kucha, "Investigation of the Effect of Penstock Configuration on the Performance of a Simplified Pico-hydro System," British Journal of Applied Science & Technology, vol.14, no.5, pp. 1-11, January 2016.

[10] A.Alidai and I.W. Pothof, "Hydraulic performance of siphonic turbine in low head sites," Renewable Energy, vol. 75, pp. 505-511, March 2015.

[11] N. P. Smith, "Induction generator for stand-alone micro-hydro systems," in Institute of Electrical and Electronics Engineers, Inc., Piscataway, NJ (United States), 1995.

[12] D.Zhou, and Z. D. Deng, "Ultra-low-head hydroelectric technology: A review," Renewable and Sustainable Energy Reviews, vol.31, no.78, pp.23-30, October 2017.

[13] V.Z. Silva, A.J. Rezek, and R.D. Corrêa, "Analysis of synchronous and induction generators in parallel operation mode in an isolated electric system," IEEE 8th Intl Symposium on Power Electronics for Distributed Generation Systems (PEDG), 2017, Apr. 17, pp. 1-8.

[14] D.B.Watson, and I.P.Milner, "Autonomous and parallel operation of self-excited induction generators," International Journal of Electrical Engineering Education, vol.22, no.4, pp.365-374, Oct 1985.

[15] L.Wang, and C.H.Lee, "Dynamic analyses of parallel operated self-excited induction generators feeding an induction motor load," IEEE Transactions on Energy Conversion, vol. 14, no.3, pp. 479-485, September 1999.

[16] I.Tamrakar, L.B. Shilpakar, B.G.Fernandes, and R. Nilsen, "Voltage and frequency control of parallel operated synchronous generator and induction generator with STATCOM in micro hydro scheme," IET Generation, Transmission & Distribution, Voo.1, no.5, pp. 743-750, September 2007.

[17] Y.K. Chauhan, V.K.Yadav, and B. Singh, "Optimum utilisation of self-excited induction generator," IET Elect. Power Appl., vol. 7, no.9, pp.680-692, November 2013.

[18] Salimikordkandi and T. Surgevil, "Modeling and analysis of self-excited induction generator with fixed capacitor excitation and shunt voltage regulation," In IEEE 16th Power Electronics and Motion Control Conference and Exposition, September 21, 2014, pp. 149-155.

[19] M.Jain, S.Gupta, D.Masand, D.Agnihotri and S.Jain, "Real-Time Implementation of Islanded Micro-grid for Remote Areas. Journal of Control Science and Engineering," Journal of computer science and Engineering, April 2016.

[20] L.Che, X.Zhang, M.Shahidehpour, A.Alabdulwahab and A.Abusorrah, "Optimal interconnection planning of community micro-grids with renewable energy sources," IEEE Transactions on Smart Grid, vol.8, no.3, pp.1054-63, May 2017.

[21] N. Bottrell, M. Prodanovic, T.C. Green, "Dynamic stability of a micro-grid with an active load," IEEE Transactions on Power Electronics, vol. 28, no.11, pp.5107-19, November 2013.

[22] G. K. Kasal and Bhim Singh, "Decoupled voltage and frequency controller for isolated asynchronous generators feeding three-phase four-wire loads," IEEE Transactions on Power Delivery, vol. 23, no. 2, April 2008.

[23] L.G. Scherer, R.V. Tambara, and R.F. de Camargo, "Voltage and frequency regulation of standalone self-excited induction generator for micro-hydro power generation using discrete-time adaptive control," IET Renewable Power Generation, vol.10, no.4, pp. 531-540, January 2016.

[24] B.Singh and G.K.Kasal, "Neural network-based voltage regulator for an isolated asynchronous generator supplying three-phase four-wire loads," Electric Power Systems Research, vol.78, no. 6, pp. 985-994, June 2008.

[25] L.Wang and D.J. Lee, "Coordination control of an AC-to-DC converter and a switched excitation capacitor bank for an autonomous self-excited induction generator in renewable-energy systems," IEEE Transactions on Industry Applications, vol.50, no. 4, pp.2828-2836, July-August 2014.

[26] Li Wang, and Dong-Jing Lee, "Coordination Control of an AC-to-DC Converter and a switched excitation capacitor bank for an autonomous self-excited induction generator in renewable-energy systems," IEEE Trans. Ind. Appl., Vol.50, No. 4, pp.2828-2836, Aug. 2014.

[27] N. Bottrell, M. Prodanovic, T.C. Green, "Dynamic stability of a micro-grid with an active load," IEEE Transactions on Power Electronics, vol. 28, no.11, pp.5107-19, November 2013.

[28] M. Srinivasan and A.Kwasinski, "Autonomous hierarchical control of DC micro-grids with constant-power loads," IEEE Applied Power Electronics Conference and Exposition APEC, Charlotte NC, March 15, 2015, pp. 2808-2815.

[29] Alireza Kahrobaeian, and Yasser Abdel-Rady I. Mohamed, "Analysis and mitigation of low-frequency instabilities in autonomous medium-voltage converter-based micro-grids with dynamic loads," IEEE Transactions on Industrial Electronics, vol. 61, no. 4, pp. 1643- 1658, April 2014.

[30] R. Zhu, Z.Chen, X. Wu, and F.Deng, "Virtual damping flux-based LVRT control for DFIG-based wind turbine" IEEE Transactions on Energy Conversion, vol. 30, no. 2, pp.714-25, June 2015.

[31] P.T. Manditereza and R. Bansal, "Renewable distributed generation: The hidden challenges–A review from the protection perspective," Renewable and Sustainable Energy Reviews, vol.31, no.58, pp. 1457-65, May. 2016.

Implementation of a Modified Distributed Normalized Least Mean Square Control for a Multi-Objective Single Stage SECS

Syed Bilal Qaiser Naqvi, *Member, IEEE*
Department of Electrical Engineering,
Indian Institute of Technology Delhi,
New Delhi - 110016, India
bilalnaqviee@gmail.com

Shailendra Kumar, *Member, IEEE*
Department of Electrical Engineering,
Indian Institute of Technology Delhi,
New Delhi - 110016, India
er.dwivedi88@gmail.com

Bhim Singh, *Fellow, IEEE* Department
of Electrical Engineering, Indian
I*nstitute of Technology Delhi,*
New Delhi - 110016, India
bsingh@ee.iitd.ac.in

Abstract—**This work presents a single stage multi-objective solar energy conversion system (SECS), which integrates a photovoltaic (PV) array to a three-phase three wire distribution system. To harvest maximum power from the PV array, the reference DC bus voltage is calculated using an MPPT algorithm based on an incremental conductance (INC) technique. A voltage source converter (VSC), which is controlled by a modified distributed normalized least mean square (dNLMS) based control technique integrates the PV array to the grid. The SECS has the capabilities of a distribution static compensator (DSTATCOM) such as power factor correction, grid currents balancing, and harmonics currents reduction to meet the IEEE 519 standard. Moreover, it can continue operating as a DSTATCOM, in the absence of sufficient solar irradiance. The control technique is tested under various dynamic conditions using test results on a prototype developed in the laboratory.**

Keywords— *Modified Distributed Normalized Least Mean Square, Power Quality, INC Based MPPT, Solar PV System, DSTATCOM, Multi-Objective SECS.*

I. INTRODUCTION

The electricity demand is increasing worldwide. The current electricity demand is met using fossil fuels. However, such widespread use of fossil fuels, is not only leading to their hastened depletion, but it is also causing damage to the environment. Solar photovoltaic (PV) energy has emerged as a viable energy resource [1]. Falling prices of solar panels and increasing awareness about harms of fossil fuels, have further accelerated its adoption [2]. Grid connected PV systems are becoming popular as they don't require an expensive battery bank, which is required by stand-alone PV systems for reliability of supply. The grid acts as an energy storage, supplying power when the load demand surpasses generated PV power, and accepting power when the PV power exceeds load demand [3].

A single stage solar energy conversion system (SECS) is used to interface the PV array to the grid. The PV array is directly coupled to the DC bus of the voltage source converter (VSC), thus a DC-DC converter is not required for operating the array at its maximum power point (MPP), as required in a two stage SECS [3-4]. Single stage and two stage PV systems, are compared in [5]. The elimination of DC-DC converter stage in single stage SECS, leads to reduced system cost, size and complexity. Moreover, the losses in the DC-DC converter, are avoided, leading to reduced system losses, and enhanced system efficiency. The operating point of a PV array, shifts with the variation of temperature, solar irradiance, or load. Since the installation cost of a PV system is high, it is imperative that it be operated at its MPP. In a single stage SECS, the VSC DC link voltage is continuously tracking the MPP voltage, so that the maximum power is harvested. A comparison of various MPPT techniques, is given in [6]. An MPPT technique based on an incremental conductance (INC) method [7] is used in this work, owing to its ease of implementation and fast tracking performance.

Due to reduced size, cost and maintenance, power electronic devices have become very common for domestic, and industrial applications [8]. However, they cause power quality issues such as drawing excessive reactive power, harmonics currents, and cause load unbalancing. These power quality issues are harmful for the other equipment connected to the grid. Distribution static compensators (DSTATCOMS) have been used in [9,10] for mitigating such issues in the distribution network. To control the VSC switching, several control techniques are reported in the literature [11,12]. A control algorithm to control the three phase VSC using high precision quadrature, is demonstrated in [11], whereas a multilayer perceptron based control technique is used in [12]. The least mean square (LMS) based control techniques have been widely reported in the literature. A generalized subband decomposition normalized LMS algorithm is reported in [13], whereas a selective partial update normalized LMS algorithm (SPU-NLMS) is presented in [14].

A modified distributed normalized least mean square (dNLMS) algorithm is used in this work to extract fundamental weights of each phase independently, from distorted load currents. For improving dynamic response under changing environmental conditions, and grid voltage variations, a PV feed forward term is integrated in the control, which reduces the computational drain on the proportional integral (PI) controller. The dNLMS controlled SECS, has added power conditioning capabilities, such as power factor correction, grid currents balancing, and harmonics currents reduction within the IEEE 519 standard [15]. In the absence of sufficient insolation, the SECS functions as a DSTATCOM, which increases the utilization of the installed equipment. The modified dNLMS control is tested on an experimental prototype developed in the laboratory and evaluated for dynamic conditions such as sudden load unbalancing, and abrupt irradiance variations.

II. System Configuration

Fig. 1 portrays the system topology. The maximum power is harvested from the PV array and fed to a capacitor at the DC bus of the VSC. A three phase VSC injects the generated PV power into the grid via interfacing inductors, and also supplies the reactive power demand and the harmonics currents demand of the nonlinear load connected at the point of common coupling (PCC). An R-C filter is coupled at PCC, to suppress voltage ripples due to VSC switching. Hall Effect based current and voltage sensors, provide the required input signals to a digital signal processor, which executes the dNLMS based control algorithm, and the output switching pulses are given to the VSC.

III. Control Algorithm

A modified distributed normalized least mean square (dNLMS) algorithm is utilized to obtain the fundamental weights from distorted load currents and it generates the required switching pulses for the VSC. To adjust the DC bus voltage to the MPP voltage of the PV array, an INC based algorithm is utilized.

A. MPPT Control Algorithm

The power output of the PV array keeps changing with variation in solar irradiance or temperature. In order to harvest maximum power, under varying environmental and load conditions, the PV array should be operated at its MPP. In a single stage SECS, this is achieved by regulating the DC bus voltage to the MPP voltage of the PV array. Thus, an INC based method is used to generate an appropriate reference DC bus voltage. The PV power is calculated by sensing the PV array current (I_{PV}) and voltage (V_{PV}) as,

$$P_{PV} = V_{PV} * I_{PV} \tag{1}$$

The gradient of the P_{PV}-V_{PV} curve is zero, at the MPP. Thus, an incremental conductance is calculated as,

$$\frac{dP_{max}}{dV_{PV}} = I_{PV} + V_{PV} * \frac{dI_{PV}}{dV_{PV}} = 0 \Rightarrow \frac{dI_{PV}}{dV_{PV}} = -\frac{I_{PV}}{V_{PV}} \tag{2}$$

Using following equations, the MPP is tracked as,

$$\text{If } \frac{dI_{PV}}{dV_{PV}} > -\frac{I_{PV}}{V_{PV}} \Rightarrow V_{dc}^* = V_{dc\ old}^* + \Delta V_{dc} \tag{3}$$

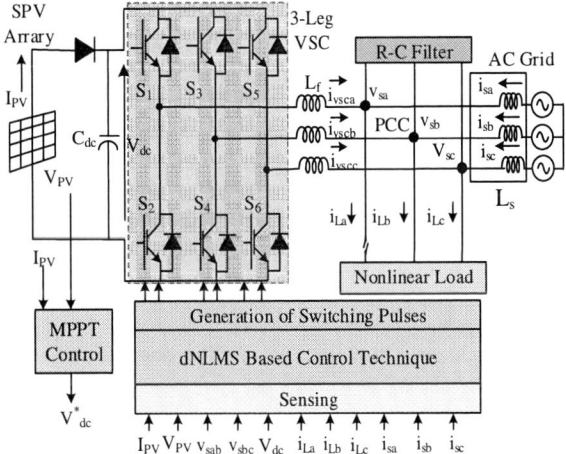

Fig. 1. System configuration

$$\text{If } \frac{dI_{PV}}{dV_{PV}} = -\frac{I_{PV}}{V_{PV}} \Rightarrow V_{dc}^* = V_{dc\ old}^* \tag{4}$$

$$\text{If } \frac{dI_{PV}}{dV_{PV}} < -\frac{I_{PV}}{V_{PV}} \Rightarrow V_{dc}^* = V_{dc\ old}^* - \Delta V_{dc} \tag{5}$$

Where V_{dc}^* is the reference DC bus voltage, $V_{dc\ old}^*$ is the previous reference DC bus voltage and ΔV_{dc} is the defined step change in voltage.

B. Modified dNLMS Based Control Technique

The structure of modified *dNLMS* based control is presented in Fig. 2. For operating the SECS at unity power factor, the unit templates of grid voltages are required. The three phase voltages are estimated by sensing two line voltages as [8],

$$v_{sa} = \frac{2v_{sab} + v_{sbc}}{3}; v_{sb} = \frac{-v_{sab} + v_{sbc}}{3}; v_{sc} = \frac{-v_{sab} - 2v_{sbc}}{3} \tag{6}$$

The amplitude of PCC voltages is assessed as,

$$V_t = \sqrt{2(v_{sa}^2 + v_{sb}^2 + v_{sc}^2)/3} \tag{7}$$

The in-phase unit templates are computed as,

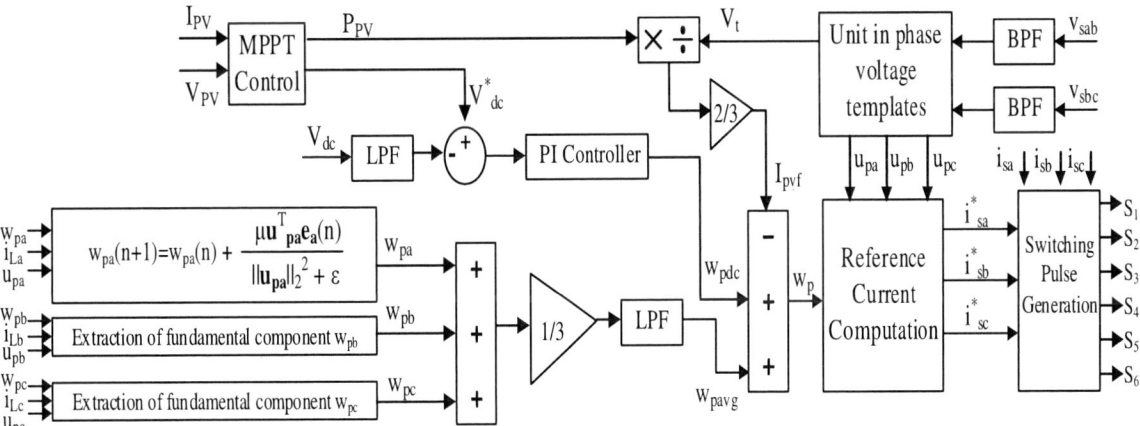

Fig. 2. Modified distributed normalized least mean square based control algorithm

978-1-5386-6624-1/18 $31.00 © 2018 IEEE

$$u_{pa} = v_{sa}/V_t, \quad u_{pb} = v_{sb}/V_t, \quad u_{pc} = v_{sc}/V_t \qquad (8)$$

In a single stage SECS, the reference DC bus voltage, V^*_{dc} is dictated by the MPPT algorithm. The DC bus voltage V_{dc} is sensed, error is calculated and given to a PI controller as,

$$V_e(n) = V^*_{dc}(n) - V_{dc}(n) \qquad (9)$$

$$w_{pdc}(n) = w_{pdc}(n-1) + k_{pdc}\{V_e(n) - V_e(n-1)\} + k_{idc}\{V_e(n)\} \quad (10)$$

Where k_{pdc} is the proportional gain and k_{idc} is the integral gain of the PI controller. In the absence of insolation, a fixed DC voltage is taken as a reference DC link voltage, and the SECS functions as a DSTATCOM.

For fast system dynamic response under changes in irradiance and grid voltage variations, a PV feed-forward term (I_{pvf}) is included in the controller. The PV feed-forward term accounts for such variations instantaneously, thus improving system response and reducing burden on the PI controller. It is computed from PV array power and amplitude of phase voltages as,

$$I_{pvf} = \frac{2 V_{PV} I_{PV}}{3 V_t} \qquad (11)$$

For obtaining the fundamental weights from distorted load currents, modified dNLMS based control is used. The adaptive weight update equation for phase 'a' is,

$$w_{pa}(n+1) = w_{pa}(n) + \frac{\mu u_{pa}^T e_a(n)}{\left\| u_{pa} \right\|_2^2 + \varepsilon} \qquad (12)$$

Where $w_{pa}(n)$ is the obtained weight of phase 'a' load current, μ is the learning rate, ε is a constant added to prevent division by zero, u_{pa} and e_a are vectors defined as,

$$u_{pa} = [u_{pa}(n)\ u_{pa}(n-1)\ ......u_{pa}(n-m+1)]^T \qquad (13)$$

$$e_a(n) = [e_a(n)\ e_a(n-1)\ ......e_a(n-m+1)]^T \qquad (14)$$

The vector e_a consists of 'm' previous filter output errors computed as,

$$e_a(n) = i_{La}(n) - u_{pa}(n) w_{pa}(n) \qquad (15)$$

Here, m is taken as 2 i.e. last two calculated errors are used in updating the fundamental weights. Similarly, phase 'b' (w_{pb}), and phase 'c' (w_{pc}) weights are calculated. The equivalent per phase fundamental weight is calculated as,

$$w_{pavg} = (w_{pa} + w_{pb} + w_{pc})/3 \qquad (16)$$

The net reference weight is obtained from the average fundamental weight, loss component, and PV feed-forward term as,

$$w_p = w_{pavg} + w_{pdc} - I_{pvf} \qquad (17)$$

The reference grid currents are obtained by multiplying the net reference weight with unit templates as follows,

$$i^*_{sa} = w_p u_{pa}, \quad i^*_{sb} = w_p u_{pb}, \quad i^*_{sc} = w_p u_{pc} \qquad (18)$$

For generation of gating pulses for switches of the VSC, a hysteresis controller is used. The sensed grid currents are compared with these reference currents, and the error is maintained within a hysteresis band, for indirect control of grid currents.

IV. EXPERIMENTAL RESULTS

The modified dNLMS control technique is validated on a developed prototype in the laboratory. It comprises of a solar PV array simulator, having characteristics similar to a rooftop PV array, a VSC to interface the array to the grid via interfacing inductors, an R-C ripple filter, and a nonlinear load. Hall Effect based current and voltage sensors are used to sense input signals to digital signal processor (DSP), for execution of control algorithm. The generated switching signals are amplified via an optocoupler circuit, and provided to the gate terminals of the VSC switches. The system parameters are provided in Appendix.

The performance of an INC based MPPT technique is demonstrated, for solar irradiance of 600 W/m² and 1000 W/m² in Fig. 3. It can be observed that MPP is tracked with an efficiency greater than 99%.

A. Dynamic Response of Modified dNLMS Based Control Technique at Load Change

The phase 'a' load is abruptly disconnected and reconnected, and the response of the control technique is evaluated in Fig. 4 using various intermediate signals. Various signals are quickly converging to their changed steady state values. Fig. 4(a) demonstrates the load current (i_{La}), error (e_a), $\left\| u_{pa} \right\|_2^2 + \varepsilon$, and incremental update in weight (Δw_{pa}). As the load is disconnected, load current reduces to zero, and error increases. Since $\left\| u_{pa} \right\|_2^2 + \varepsilon$ depends on grid voltage only, it does not vary with load changes. Fig. 4(b) exhibits the fundamental weights w_{pa}, w_{pb} and w_{pc}. The equivalent per phase weight magnitude (w_{pavg}), output of the PI controller (w_{pdc}), and net weight of reference grid currents (w_p) are presented in Fig. 4(c). The reference phase 'a' grid current (i^*_{sa}), grid current (i_{sa}), and computed unit template (u_{pa}), are exhibited in Fig. 4(d). When i_{La} falls to zero, w_{pavg} is reduced, which increases magnitude of w_p. As a result, the grid current increases. Moreover, the load current in phase 'a' is zero, but the grid current in phase 'a' is sinusoidal and out of phase with u_{pa}, indicating that the grid is accepting power at unity power factor. Figs. 4(e-h) exhibit the above signals at abrupt load reconnection in phase 'a'. When the load is reconnected, w_{pa} increases, and the current fed into the grid reduces.

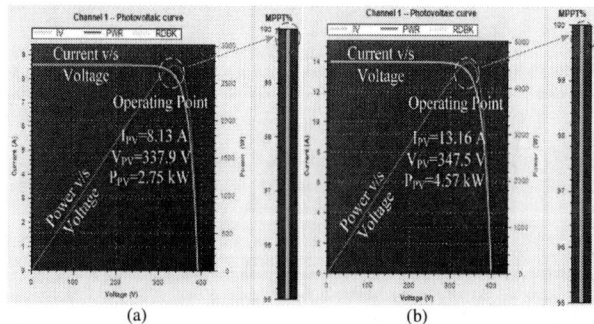

Fig. 3. Performance of MPPT for irradiance of (a) 600 W/m² (b) 1000 W/m²

978-1-5386-6624-1/18 $31.00 © 2018 IEEE

B. Dynamic Response of SECS Under Sudden Load Unbalance

The phase 'a' load is removed and inserted, and the response of SECS is demonstrated in Figs. 5(a-b) for sudden load removal, and Figs. 5(c-d) for sudden load insertion. The load currents of three phases (i_{La}, i_{Lb} and i_{Lc}), and the PV array voltage (V_{PV}) are exhibited in Fig. 5(a), whereas the grid voltage (v_{sa}), VSC current (i_{VSCa}), and the DC bus voltage (V_{dc}) are depicted in Fig. 5(h). Due to fast control action, V_{dc} and V_{PV} remain stable. After disconnection of nonlinear load in phase 'a', i_{VSCa} becomes sinusoidal, as it does not have to compensate harmonics currents and reactive power of phase 'a' load. The grid voltage is unaffected by sudden load changes.

C. Steady State Behaviour of SECS at Unbalanced Nonlinear Load

The SECS performance when load of phase 'a' is disconnected is demonstrated in Fig. 6. The phase 'a' load current is zero, as depicted in Fig. 6(c). However, the grid currents in all three phases, are sinusoidal and balanced, as shown in Figs. 6(a,g), demonstrating grid currents balancing capability of the control technique. From Fig. 6(k), the current THD for load is 37.16%. However, the THD for the grid current is 2.65%, as shown in Fig. 6(j), which is meeting the IEEE 519 standard [15]. Since the phase 'a' load is removed, the VSC currents are unbalanced, as exhibited in Figs. 6(b,h,l).

D. Response of SECS at Fall and Rise of Irradiance

Figs. 7(a-b) demonstrate the system response under decreasing, and increasing irradiance. The grid voltage (v_{sa}), PV feed-forward term (I_{pvf}), the grid current (i_{sa}), and DC bus voltage (V_{dc}), are exhibited in Fig. 7(a). On decreasing the irradiance from 1000 W/m² to 600 W/m², the grid voltage is unchanged, and I_{pvf} reduces. The reduction of I_{pvf} leads to decrease in the grid current. When irradiance is increased from 600 W/m² up to 1000 W/m², I_{pvf} increases, thereby increasing grid current and power fed into the grid. The DC link voltage remains stable at the voltage corresponding to MPP of the PV array.

E. Transition from SECS to DSTATCOM and DSTATCOM to SECS

In the absence of sufficient solar irradiance, the SECS transitions to DSTATCOM mode, and continues to perform the functions of power factor correction, harmonics currents reduction, and balancing of grid currents, thereby leading to efficient utilization of installed equipment. The transition from SECS to DSTATCOM operation is demonstrated in Fig. 8(a).

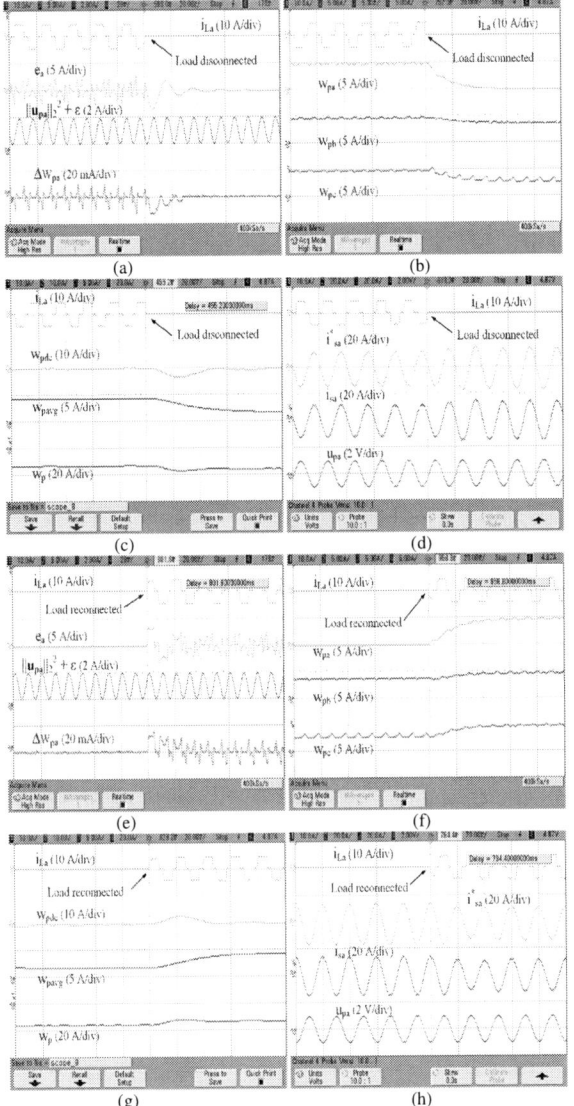

Fig. 4. Intermediate signals of dNLMS control technique, (a)-(d) Abrupt load disconnection and, (e)-(h) Abrupt load reconnection

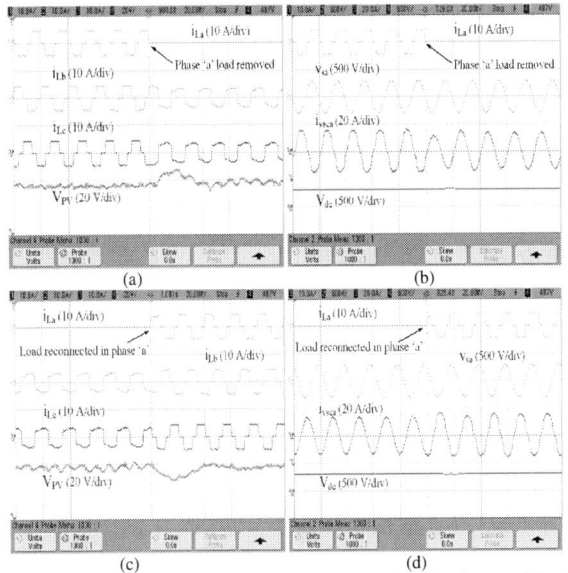

Fig. 5. Dynamic response of SECS under (a)-(b) sudden load removal in 'a' phase, and (c)-(d) Sudden load injection in 'a' phase

978-1-5386-6624-1/18 $31.00 © 2018 IEEE

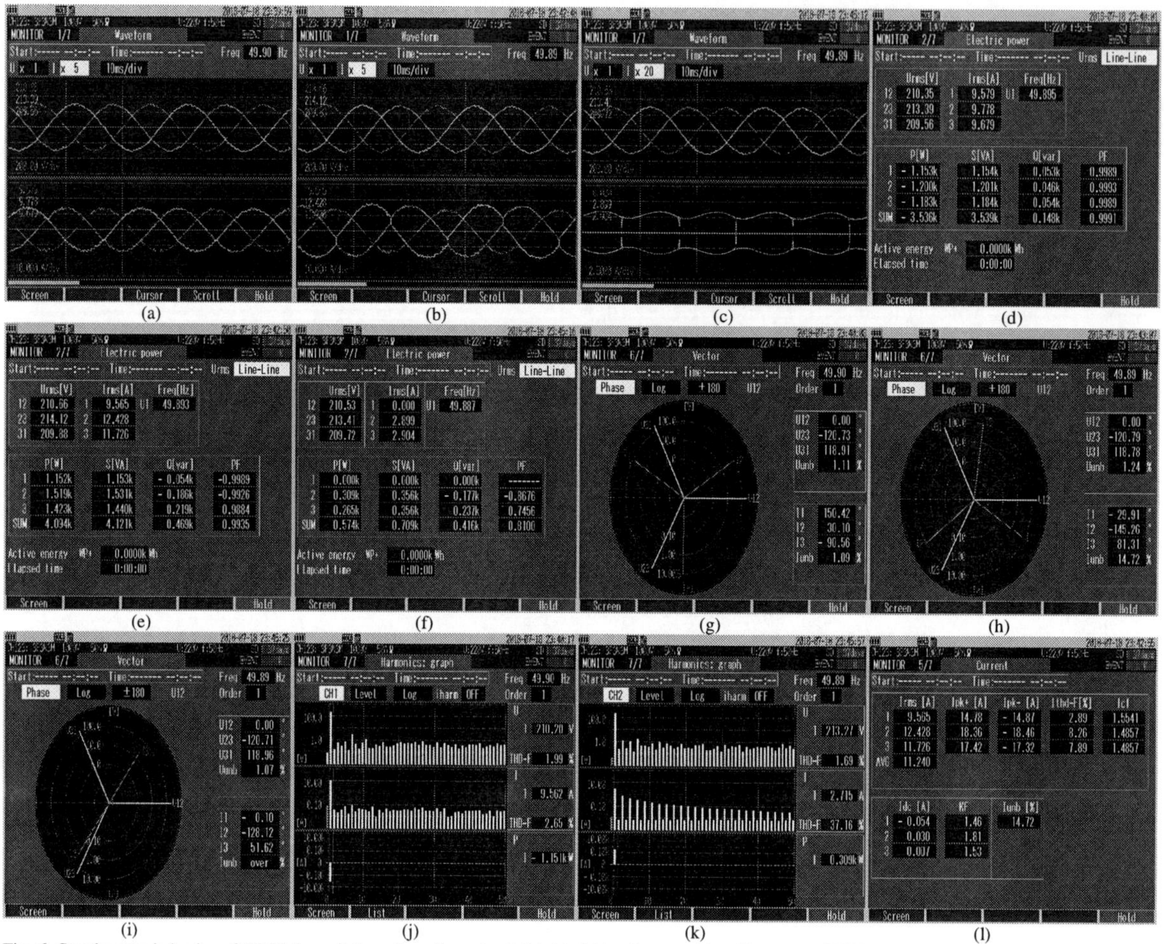

Fig. 6. Steady state behavior of SECS for unbalanced nonlinear load: (a)-(c) Grid voltages with grid currents, VSC currents, and load currents, (d)-(f) Grid power, VSC power and Load power, (g)-(i) Phasor diagram of Grid voltages with grid currents, VSC currents, and load currents (j)-(k) Harmonic spectrum of grid voltage with i_{sa}, and i_{Lb} (l) THD in VSC currents i_{VSCa}, i_{VSCb}, and i_{VSCc}

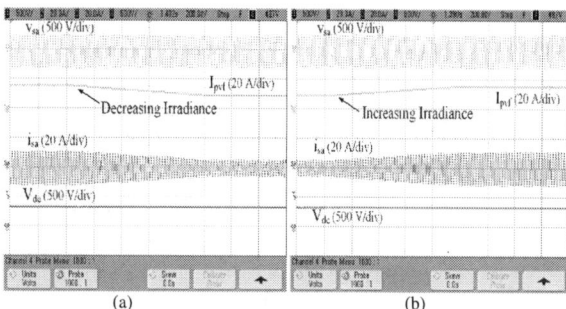

Fig. 7. System response under, (a) Decreasing Irradiance and (b) Increasing irradiance

As the solar irradiance decreases, the PV current (I_{PV}) decreases, while the grid voltage (v_{sa}) is unchanged. As the irradiance decreases, there comes a point when the generated PV power becomes lesser than the load power. Therefore, the grid current reverses, and the deficit power is obtained from the grid. On further reduction of irradiance, the entire power demanded by the load, is fed from the grid, with the VSC supplying the harmonics currents required by the load, and VSC current transitions to DSTATCOM compensating current.

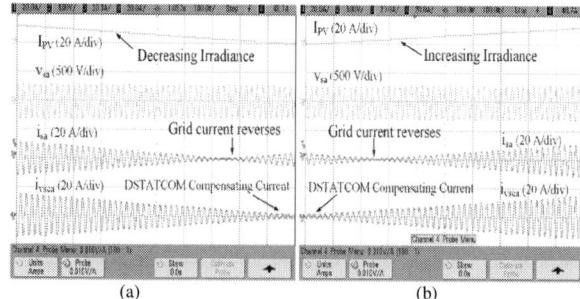

Fig. 8. Response of SECS under transition from (a) SECS to DSTATCOM operation mode and (b) DSTATCOM to SECS operation mode

V. CONCLUSION

A single stage multi-objective SECS is successfully implemented with a modified dNLMS based control technique. The PV array is directly coupled to the DC bus capacitor, thereby decreasing the cost, size and system complexity. The MPPT function is successfully carried out using an INC based technique. The performance of SECS has been evaluated on a developed prototype, under dynamic conditions of sudden load unbalancing and swift changes in irradiance. The response of the control is found to be reasonably fast, and it successfully extracts fundamental weights from distorted load currents. The multi-objective SECS has also demonstrated the features of grid currents balancing, power factor correction, and harmonics reduction to meet the IEEE 519 standard. These functions are also carried out by the system even in the absence of solar irradiance, and the SECS operates as a DSTATCOM.

APPENDIX

Solar PV array parameters: V_{oc} = 400 V, I_{sc} = 14 A, C_{dc} = 2200 μF, L_f = 5 mH, Controller parameters: μ=0.005, ε=0.5, PI Controller: k_{pdc}=0.4 and k_{idc}=0.01, Utility parameters: 200 V (L-L), 50 Hz, L_s = 3 mH, R_s = 0.06 Ω, R-C filter: C_f = 5 μF, R_f = 9 Ω, Nonlinear load: DBR and 200 mH, 66.5 Ω.

ACKNOWLEDGMENT

The authors are grateful to DST, Govt. of India, for financial assistance under the Project RP03391G.

REFERENCES

[1] C. S. Solanki: *Solar Photovoltaics: Fundamentals, Technologies and Applications*, Prentice Hall of India, 2015.

[2] S. Comello, S. Reichelstein and A. Sahoo, "The road ahead for solar PV power", *Renew. and Sust. Energy Reviews*, vol. 92, pp. 744-756, Sept 2018.

[3] S. B. Q. Naqvi, S. Kumar and B. Singh, " Implementation of Recurrent Neurocontrol Algorithm for Two Stage Solar Energy Conversion System," *IEEE Engineer Infinite, e-TechNxt-2018*, Greater Noida, March 2018.

[4] V. L. Srinivas, S. Kumar, B. Singh and S. Mishra, "A Multifunctional GPV System Using Adaptive Observer Based Harmonic Cancellation Technique," *IEEE Trans. Ind. Electron.*, vol. 65, no. 2, pp. 1347-1357, Feb. 2018.

[5] T. F. Wu, C. H. Chang, L. C. Lin and C. L. Kuo, "Power Loss Comparison of Single- and Two-Stage Grid-Connected Photovoltaic Systems," *IEEE Trans. Energy Conv.*, vol. 26, no. 2, pp. 707-715, June 2011.

[6] S. Jain and V. Agarwal, "Comparison of the performance of maximum power point tracking schemes applied to single-stage grid-connected photovoltaic systems," *IET Electric Power Appl.*, vol. 1, no. 5, pp. 753-762, Sept. 2007.

[7] N. A. Rahim, A. Amir, A. Amir and J. Selvaraj, "Modified incremental conductance MPPT with direct control and dual scaled adaptive step-size method," *IET Clean Energy Tech. Conf. (CEAT 2016)*, Kuala Lumpur, pp. 1-8, 2016.

[8] B. Singh, A. Chandra, and K. Al-Haddad: *Power Quality: Problems and Mitigation Techniques*, Wiley, 2015.

[9] R. T. Hock, Y. R. de Novaes and A. L. Batschauer, "A Voltage Regulator for Power Quality Improvement in Low-Voltage Distribution Grids," *IEEE Trans. Power Electron.*, vol. 33, no. 3, pp. 2050-2060, March 2018.

[10] M. Mangaraj and A. K. Panda, "Performance analysis of DSTATCOM employing various control algorithms," *IET Gen., Trans. & Distr.*, vol. 11, no. 10, pp. 2643-2653, July 2017.

[11] S. Kumar and B. Singh, "Linear coefficient function based control approach for single stage SPV system integrated to three phase distribution system," *IET Gen., Trans. & Distr.*, vol. 11, no. 3, pp. 676-684, Feb 2017.

[12] I. Hussain, R. K. Agarwal and B. Singh, "MLP Control Algorithm for Adaptable Dual-Mode Single-Stage Solar PV System Tied to Three-Phase Voltage-Weak Distribution Grid," *IEEE Trans. Industr. Inform.*, vol. 14, no. 6, pp. 2530-2538, June 2018.

[13] J. E. Kolodziej, O. J. Tobias and R. Seara, "Stochastic Model for the Generalized Subband Decomposition ε NLMS Algorithm with Gaussian Data," *IEEE Int. Conf. Acoustics Speech and Signal Proc. Proceedings*, Toulouse, pp. 764-767, 2006.

[14] M. S. E. Abadi and A. R. Danaee, "Selective partial update normalized least mean square algorithms for distributed estimation over an adaptive incremental network," *Iranian Conf. Electrical Eng. (ICEE2012)*, Tehran, pp. 1329-1333, 2012.

[15] IEEE Recommended Practice and Requirement for Harmonic Control on Electric Power System, *IEEE Standard* 519, 2014.

2nd IEEE International conference on Power Electronics, Intelligent Control and Energy systems

A Dc-Dc Boost Converter for Sediment Microbial Fuel Cell Energy Harvesting

Jeetendra Prasad[1], Student Member, IEEE and Ramesh Kumar Tripathi[2], Senior Member, IEEE

Department of Electrical Engineering
Motilal Nehru National Institute of Technology
Allahabad, Uttar Pradesh 211004, India
[1]ree1555@mnnit.ac.in, [2]rktripathi@mnnit.ac.in

Abstract—This paper presents sediment microbial fuel cell (SMFC) and a Dc-Dc boost converter. In which, an SMFC generates a maximum voltage of 1.16V, which is not sufficient to drive the low power electronic device. Therefore, a self-sustainable Dc-Dc boost converter is designed to boost up the SMFC voltage from 1.161V to 3.289V. This is simulated in the LTspice software by using an LTC3108 integrated circuit (IC). The performance of this proposed boost converter is analyzed at different load resistance. During the analysis, we found that output voltage and current of the boost converter are almost constant when the load resistance made equal to the internal resistance of the SMFC. The energy harvester Dc-Dc boost converter can deliver upto 3.289V, which is sufficient to power sensor devices, remote water sensing and other remote broadcasts etc. The proposed Dc-Dc boost converter is self-sustainable because it is powered entirely from harvested energy without requiring extra external power sources.

Keywords— Sediment microbial fuel cell; Dc-Dc boost converter; load resistance; voltage

I. INTRODUCTION

Sediment microbial fuel cell (SMFCs) is a bio-electrochemical device which generates the electrical energy through redox reactions catalyzed by microbes present in sediments [1]. In the process of degradation of organic matter in which presents a special type of microbes such as Shewanella which produced protons and electrons. The protons are exchanged to the cathode internally from the anode compartment to the cathode compartment, and electrons transfer from anode compartment to the cathode through the external circuit. The direct current will flow by movement of the electrons in the circuit. Previously, different researches were developed microbial fuel cell using the different electrode materials and different electrolytes. Researchers have high expectations that SMFC will prove to be a sustainable energy source. A Terrestrial microbial fuel cell was presented using the carbon cloth electrode which generates the voltage of 0.75V [2]. The maximum voltage generated from the sediment microbial fuel cell was 0.931V with graphite and zinc electrode and 1.16V with the copper and zinc electrodes [3]. Unluckily, single SMFC cannot generate the maximum output voltage of a 1.16V and power less than 1.0mW, which cannot drive the electronic devices continuously [4-5]. For example, the lowest operating voltage is 1.6 to 3.5V and the lowest operating power is 30mW require for a single light emitting diodes (LEDs) [6]. Most of the sensors require a minimum voltage of 3.3V and mill watt power level for humidity monitoring, temperature and pressure [7]. So as to get the

application in the industrial production early, different procedures and techniques have previously been employed to improve the voltage and power of the SMFC. The main focus of the researcher's on substrate concentration, the cathode or anode materials, reactor structure and the enhancement of the bio-electrochemical system and so on [2, 4, 8]. Electrical point of view, the output load is another essential parameter that specifically impacts power generation and which is often neglected. Most of the researchers used fixed resistor as a load in SMFC designs [6-8]. On the other hand, SMFC is a complex dynamic system, and the voltage generation and internal resistance change with the changes in microbial biomass, temperature, pH and substrate concentration. The aim of this study is to boost the voltage level of SMFC and evaluate the performance of the boost converter at different load resistance. The maximum output voltage is found when the external load resistance made equal to the internal resistance of the power source. Schematic of an energy harvesting system for SMFC shows in Fig.1. Dc-Dc boost converter extracts the energy from SMFC and boosts up its. The output voltage of the Dc-Dc boost converter is sufficient for driving the electronic device, soil moisture monitoring, light emitting diodes, wireless sensors etc. This study gives the direction with a larger application perspective in the process of using SMFC as a renewable energy source.

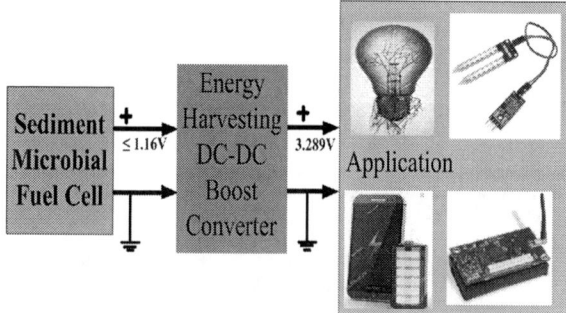

Fig. 1. Schematic of energy harvesting system for SMFC.

II. MATERIALS AND METHODS

A. Sediment microbial fuel cell

Sediment was collected from water bottom at about 5cm from Ganga River, Allahabad. The sediment used as electrolyte in the SMFC reactor. SMFC was designed in a 1L plastic cylindrical reactor. A copper anode and Zinc cathode were used as electrodes both with a thickness of 2mm, 7.5cm

978-1-5386-6624-1/18 $31.00 © 2018 IEEE

height and 2.5cm width (purchased from Naveen Scientific Industries). The copper anode emerged into sediment and zinc cathode placed in 400ml of water. Microbes present in the organic matter of sediment oxidized at the anode and produce the electron, proton and carbon dioxide as shown in Fig.2(a). These electron and proton produce the water at the cathode surface in the presence of oxygen. Fig.2(b) represents the SMFC rector developed in the laboratory. Electrode was connected by copper wire diameter 1.5mm^2 through an external load. The voltage across the electrode was monitored by a digital multi-meters (Agilent U1232A).

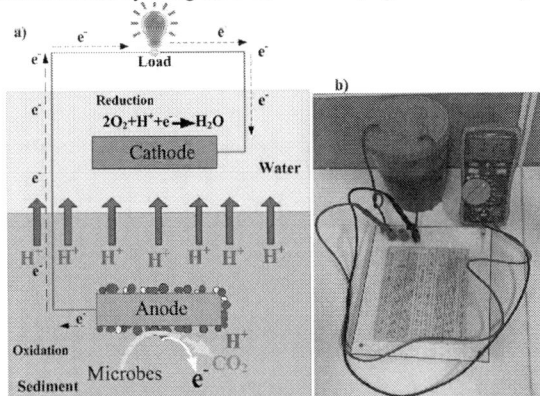

Fig.2.(a) Mechanism of power generation by SMFC. (b) A developed SMFC reactor in the laboratory.

B. Electrical Equivalent Circuit of SMFC

The electrical equivalent circuit of SMFC is a first-order approximation in which represent as a voltage source with a high series resistance, which limits the amount of available current to be supplied [9]. Fig.3 indicates an electrical equivalent circuit of an SMFC. V_{smfc} represents steady-state (dc) voltage generated by the SMFC and V_{int} represents the dynamic voltage. The value of R_{smfc} is reflected to be the internal resistance of SMFC composed of different components (cathodic resistance, electrolyte resistance and anodic resistance) [10]. Resistance (R_{smfc}) measured by a digital multi-meters (Agilent U1232A). The value V_{smfc} depends on various variables, such as temperature, pH, and electrolyte concentration in the anode compartment and varies nonlinearly [11].

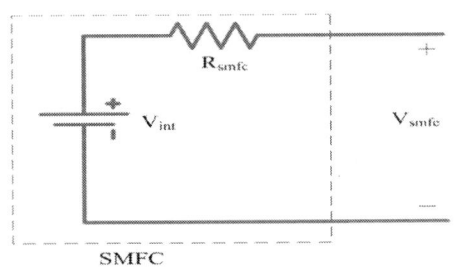

Fig.3. Electrical equivalent circuit of SMFC.

C. Praposed DC–DC Boost Converter Circuit

Sediment microbial fuel cell has a weak voltage source and its output voltage does not almost constant when the load was connected across it. So this type of voltage source cannot be used directly to power the electronic devices. In this paper, an ultra-low voltage boost converter proposed

which step up and regulate the voltage of SMFC. The boost converter was simulated on the LTspice software using LTC3108IC. The proposed circuit diagram of boost converter has shown in Fig.4. A step-up transformer with turn's ratio of 1:100 was used to form a resonant step-up oscillator, which allows the voltage of SMFC to boost up. The primary winding of the transformer connected between the SMFC model and Internal N-channel switch (SW). The secondary winding of transformer was connected between the ground and capacitors (C$_1$ of 1nF and C$_2$ of 1pF).

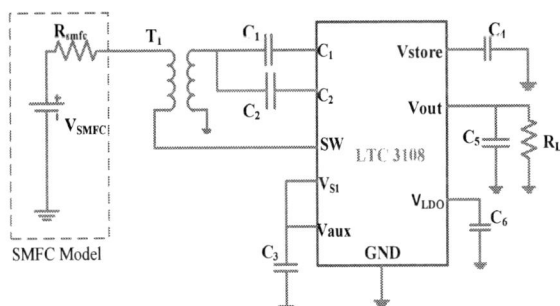

Fig.4. Proposed Dc-Dc boost converter for sediment microbial fuel cell.

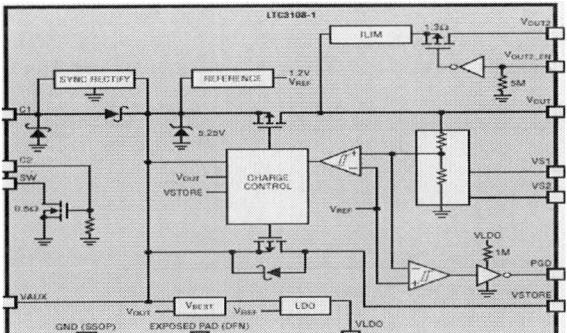

Fig.5. Block diagram of the LTC3108IC.

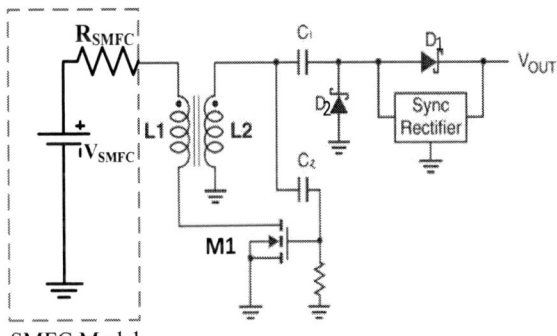

Fig.6. The circuit diagram of the resonant boost oscillator.

The main components of block diagram of the LTC3108IC are a synchronous reciter, a step-up oscillator, a low dropout linear regulator module, a power switching unit and a charging control unit as shown in Fig.5. The resonant boost oscillator is obtained by using a MOSFET (M1), coupling capacitors and a step-up transformer as shown in Fig.6. The output of the resonant boost oscillator used to power the IC (via the V$_{aux}$ pin) and charge the capacitor (C$_3$ of 1μF) connected between V$_{S1}$ and V$_{aux}$ pin. The resonant boost oscillator consists of the two active diodes which are

978-1-5386-6624-1/18 $31.00 © 2018 IEEE   713

made up of MOS switch and comparators. The comparators were energized by V_{aux} pin. A capacitor C_6 of 2.2µF was connected at dropout linear regulator (LDO) pin which aims to provide in regulation and to power low power processors or other low power electronic ICs. The output capacitor (C_{OUT}) was charged from the Vaux pin and provides regulated output voltage V_{out}. When Vout has reached maximum voltage, the output of V_{store} pin used to charge a storage capacitor (C_5) of 1mF. Once V_{out} has achieved the regulated voltage, the V_{store} pin allows charging the V_{aux} voltage. A storage capacitor C_5 of 470µF was connected to V_{out} pin which provides the regulated voltage 3.289V when input was applied on the primary side of the transformer. The load was connected across the capacitor (C_5). The proposed boost converter work well above the input voltage of 20mV.

III. RESULT AND DISCUSSION

A. SMFC Start- up and Operation

Microorganism presents within the SMFC required an adaptation method and want some time for the formation of the biofilm on the anode surface. Consequently, an anaerobic fluidized bed SMFC startup required some time. The experiment process for 30 days, the voltage of SMFC increases until it reaches a stable state value while no additional substrate has been ever provided to the SMFC during the experiment. At the beginning of start-up, open circuit voltage (OCV) of SMFC was 0.644V and continuously increasing. After 20 days, the stable OCV of SMFC appeared 1.161V across the electrode, and OCV of SMFC was stable for 30 days shown in Fig.7.

The maximum generated current and power of SMFC was 0.301mA and 3.491mW for the steady-state operating condition. Other researcher examined duel Chamber microbial fuel cell 5.780mA and 6.046mW were obtained using copper cathode as electron an acceptor and 1.515mA and 1.944mW were obtained using Potassium Permanganate as electron acceptor [12].

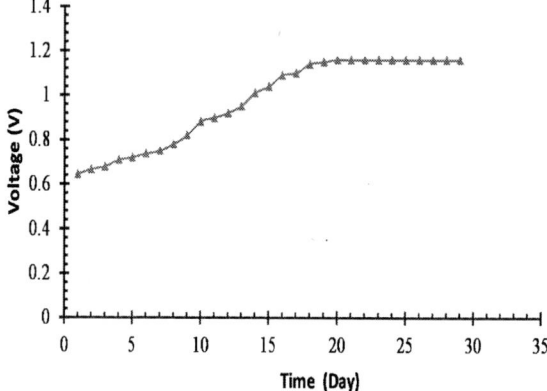

Fig.7. Experimental SMFC generated voltage of 30 days investigation.

B. Start- up and Operation of Dc-Dc boost converter

SMFC voltage is too low for powering the wireless sensors, led lighting etc. SMFC voltage can be increased by series and parallel connection but there required a large number of SMFC [13] which cover the large area. The dc to dc boost converter helps to increase the voltage of a single SMFC. This SMFC's voltage (1.16) is able to provide the sufficient power for the proposed boost converter.

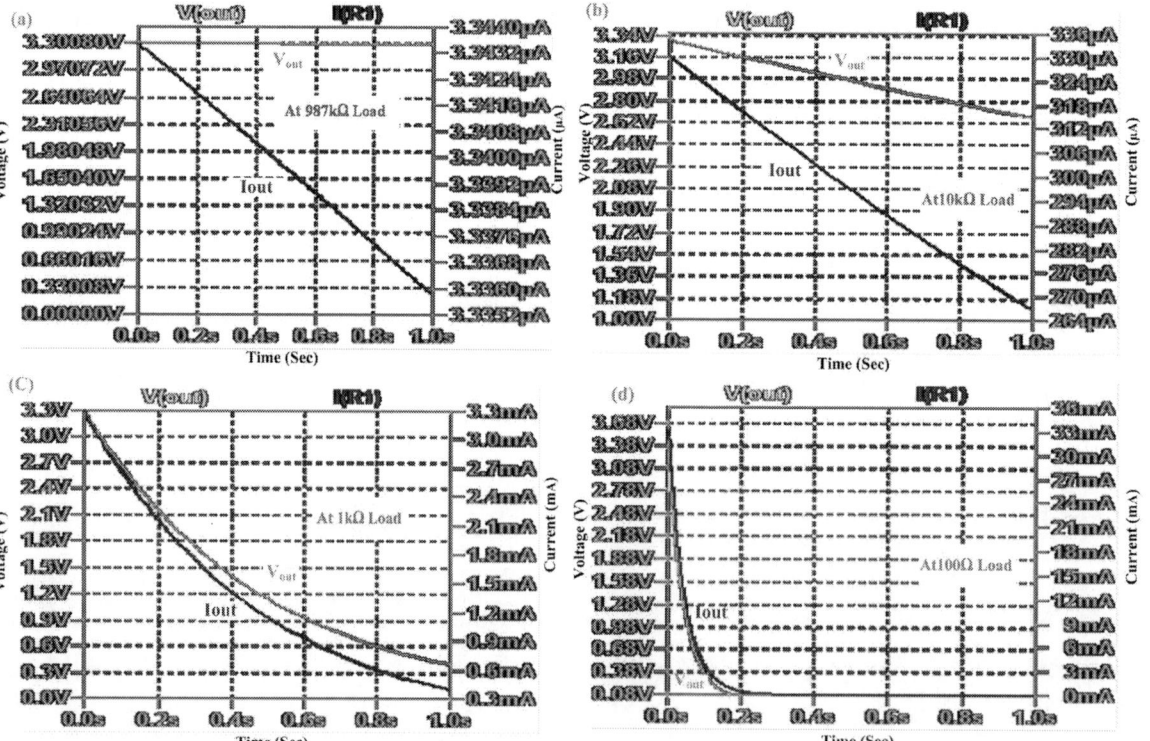

Fig.8. Simulation results of output voltage and output current at different load resistance (a) Boost converter output voltage at 987kΩ load resistance. (b) Boost converter output voltage and output current at 10kΩ load resistance. (c) Boost converter output voltage and output current at 1kΩ load resistance. (d) Boost converter output voltage and output current at 100Ω load resistance.

In this paper LTC3108 boost converter simulated with the different load resistance. The simulation result is shown in Fig.8 using the liner technology software LTspice. The output voltage of the boost converter was almost constant of 3.289V if load resistance (987 kΩ) made by equal to source resistance shown in Fig.8a. The load current was also almost constant of 3.343µA with 987 kΩ load resistance. The load resistance changed by 10kΩ, output voltage and current both were decreased and its final voltage was 2.752V from 3.289V and the current from 330µA to 264µA shown in Fig.8b. The voltage of SMFC step up by boost converter from 1.161V to 3.289V at a load resistance of 1kΩ but the output voltage continuous decreases and reach a final value of voltage 0.425V and the current decreased from 3.30mA to 0.40mA shown in Fig.8c. If the load resistance decreases to 100Ω, the output voltage and current of the boost converter decreases exponentially from 3.68V to 0.08V and current decreased from 33mA to 0mA shown in Fig.8d. The output current of the boost converter also decreases with the load decrease. This behavior of the output voltage due to source impedance is too high.

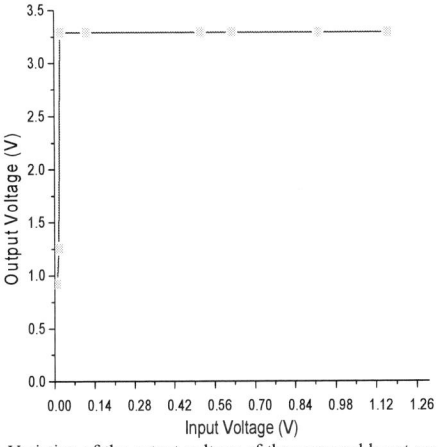

Fig.9. Variation of the output voltage of the proposed boost converter with respect to input voltage.

Fig.9 shows the variation of the output voltage of the boost converter with respect to the input voltage. The output voltage of the proposed boost converter was 3.289V with the input voltage above the 20mV. When the input voltage was less than 20mV, the output voltage of the boost converter was decreased instantaneously. SMFC is a weak power source which terminal voltage continuously decreases with the load. So this boost converter is suitable for the SMFC energy harvesting.

TABLE I. PERFORMANCE SUMMARY OF DIFFERENT CONVERTERS

Specification	[14]	[15]	[16]	[17]	This work
Input voltage (V)	0.30–0.72	<0.24	>0.6 (start-up)	0.315	1.161
Output voltage (V)	2.5	0.6–1.3	0.9–1.2	1.1	3.289
Extra power source	Yes	Yes	No	No	No

This work compares with the other converters topology for microbial fuel cell given by other researchers in Table I. in the previous research, the maximum voltage was boosted up 2.5V with the input voltage range of 0.30V to 0.72V. Some researcher uses the extra power source in the operation of switches [14-15]. Other researcher was reported different

boost converter topology without using external power supply but not stepped up the sufficient voltage to power the electronic device [16-17]. In this paper, SMFC voltage stepped up from 1.16V to 3.289V without any external power source. The proposed boost converter works at the input voltage of 20mV and above so it is the advantages for application of sediment microbial fuel cell energy harvesting. The application of this research is to provide the power for the wireless "smart" MEMS sensor networks. The wireless "smart" MEMS sensor networks are used by the military which needs to 3V input power for data transmitting [18].

IV. CONCLUSIONS

Sediment microbial fuel cell is a sustainable power source. A sediment microbial fuel cell developed in the laboratory which generates the maximum voltage and current of 1.16V and 0.301mA respectively. However, the SMFC has a weak power source so it cannot directly connect with the low power electronic device. Therefore, a Dc-Dc boost converter is proposed to step up and regulate the output voltage of SMFC. Performance of the boost converter also examined at the different load and maximum stable output voltage (3.289V) was obtained when the load resistance was matched with the SMFC internal resistance. The proposed boost converter is self-sufficient do not require the extra power supply. The proposed boost converter is suitable for applications that require self- sufficient, renewable, intermittent and maintenance-free power source in the remote area.

ACKNOWLEDGMENT

The work is done at Department of Electrical Engineering, MNNIT Allahabad, UP, India by the support of "Media Lab Asia Visvesvaraya Ph.D. Scheme".

REFERENCES

[1] Junyeong An, Jonghyun Nam, Bongkyu Kim, Hyung-Sool Lee, Byung Hong Kim, In Seop Chang, "Performance variation according to anode embedded orientation in a sediment microbial fuel cell employing a chessboard-like hundred-piece anode," Bioresource Technology, 2015, 190:175-181.

[2] D. Zhang, Y. Ge and W. Wang, "Study of a terrestrial microbial fuel cell and the effects of its power generation performance by environmental factors," International Conference on Advanced Mechatronic Systems, Luoyang, 2013, pp. 445-448.

[3] J.Prasad and R. K. Tripathi, "Energy Harvesting from Sediment Microbial Fuel Cell Using Different Electrodes", International Journal of ChemTech Research,2018, 7(1), Vol.11 No.07, pp 219-225.

[4] Ioannis A.Ieropoulos, John Greenman and Chris Melhuish, "Miniature microbial fuel cells and stacks for urine utilisation," International journal of hydrogen energy, 2013, 38(1) :492-496.

[5] Daxing Zhang, Fan Yang, Tsutomu Shimotori, Kuang-Ching Wang,Yong Huang, "Performance evaluation of power management systems in microbial fuel cell-based energy harvesting applications for driving small electronic devices," Journal of Power Sources, 2012, 217 (11) :65-71.

[6] Dermot Diamond, Shirley Coyle, Silvia Scarmagnani, and Jer Hayes, "Wireless sensor networks and chemo- biosensimg," Chemical Reviews, 2008, 108(2):652-679.

[7] Avinash Shantaram, Haluk Beyenal, Raaja Raajan Angathevar Veluchamy, and Zbigniew Lewandowski, "Wireless sensors powered by microbial fuel cells," Environmental Science & Technology, 2005, 39 (13): 5037- 5042.

[8] N. Degrennea, F. Bureta, B. Allardb and P. Bevilacqua, "Electrical energy generation from a large number of microbial fuel cells operating at maximum power point electrical load," Journal of Power Sources, 2012, 205 (2):188-193.

[9] Jae-Do Park and Zhiyong Ren, "Hysteresis-Controller-Based Energy Harvesting Scheme for Microbial Fuel Cells With Parallel Operation Capability," IEEE Transactions on Energy Conversion, vol. 27, no. 3, pp. 715-724, 2012.

[10] Y. Fan, E. Sharbrough, and H. Liu, "Quantification of the internal resistance distribution of microbial fuel cells," Environmental Science & Technology, vol. 42,pp. 8101–8107, 2008.

[11] G. S. Jadhav and M. M. Ghangrekar, "Performance of microbial fuel cell subjected to variation in pH, temperature, external load and substrate concentration," Bioresource Technology, vol. 100, pp. 717–723, 2009.

[12] S.S.Uddin, K. S. Roni, A. H.M. Shatil, S. Ahmed, "Using Copper Sulphate and Potassium Permanganate as Electron Acceptor in a Duel Chamber Microbial Fuel Cell," American Journal of Engineering & Natural Sciences (AJENS), 2016, Volume 1.Issue 1.

[13] Jeetendra Prasad and Ramesh Kumar Tripathi, "Scale up sediment microbial fuel cell for powering Led lighting", International Journal of Renewable Energy Development,2018, 7(1), 53-58.

[14] S. Carreon-Bautista, C. Erbay, A. Han and E. Sánchez-Sinencio, "Power Management System With Integrated Maximum Power Extraction Algorithm for Microbial Fuel Cells," IEEE Transactions on Energy Conversion, vol. 30, no. 1, pp. 262-272, 2015.

[15] E. J. Carlson, K. Strunz, and B. P. Otis, "A 20 mV input boost converterwith efficient digital control for thermoelectric energy harvesting," IEEE Journal of Solid-State Circuits, vol. 45, no. 4, pp. 741–750, 2010.

[16] X. Zhang, H. Ren, S. Pyo, J. I. Lee, J. Kim and J. Chae, "A High-Efficiency DC–DC Boost Converter for a Miniaturized Microbial Fuel Cell," IEEE Transactions on Power Electronics, vol. 30, no. 4, pp. 2041-2049, April 2015.

[17] J.-D. Park and Z. Ren, "High efficiency energy harvesting from microbial fuel cells using a synchronous boost converter," Journal of Power Sources, vol. 208, pp. 322–327, Jun. 2012.

[18] https://www.eol.ucar.edu/isf/facilities/isa/internal/CrossBow/DataSheets/mica2.pdf.

Modeling of LED using Piecewise Linear Approximation and Maclaurin series Expansion

Obaidur Rahman, *student*, Priya Mahajan, *Member, IEEE*, Rachana Garg, *Member, IEEE*
Dept. of Electrical Technology, Delhi Technological University
New Delhi, Delhi -110042

Abstract—**LED is a light emitting semiconductor device which is nowadays, swiftly replacing conventional lighting system, owing to its high intensity and low energy consumption. In order to understand the operating characteristics of LED, its modeling is required. The very basic modeling of LED can be done by considering it as a resistor. But this has some serious limitations as the non linear I-V characteristics of the LED cannot be modeled using a resistor. To model the non linear IV characteristics of LED, some advanced modeling techniques viz. Piecewise Linear Approximation and Maclaurin series expansion based modeling have been taken up in this paper. Finally, the I-V curves are plotted and compared using MATLAB, implementing both the techniques.**

Keywords—**LED, I-V characteristics, PLA, Maclaurin series expansion.**

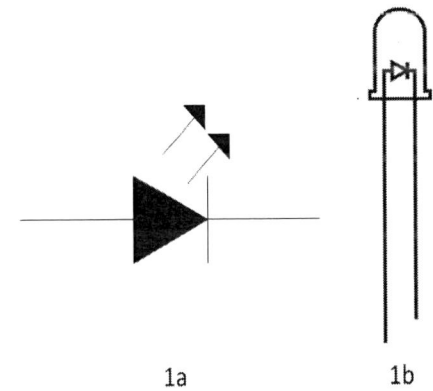

Fig. 1   LED symbol and its 2 pin variant.

I. INTRODUCTION

Modern day lighting technology mainly revolves around LEDs. LED or light emitting diode is a semiconductor device which, as the name suggests, emits light when current is passed through it. When the LED is forward biased, the holes and electrons recombine. As a result, photons are generated in the visible spectrum which reach the human eye and create the sensation of vision through light. In the last few years, LEDs have gained widespread popularity in a wide range of applications ranging from tiny little indicators to big video screens. LEDs are preferred over other sources of light because of availability in compact size, less consumption of energy and longer shelf life [1]-[6].To utilize the LED in the most efficient manner, it becomes imperative to understand its voltage and current relationships. Fig. 1 shows a 2 pin variant of LED and the general LED symbol. As can be seen from Fig.1b, the longer leg always denotes the anode and the shorter one denotes cathode. This indifference in length is done for the sake of remembrance. In this paper, an LED sample LNL-190UW-4H [7] has been taken up for study. The focus here is mainly on the I-V characteristics of the sample and not on its optical properties. An LED is basically a diode which has a non linear IV characteristic. Thus, it is not a good idea to model it using just a resistor which has a linear I-V characteristic. To tackle this problem, an approximate linear model [8] was developed as can be seen in Fig. 2. In this model, the diode remains in OFF state until the applied voltage reaches a particular voltage known as knee voltage. The diode remaining in OFF state simply means that no current flows through it during this state.

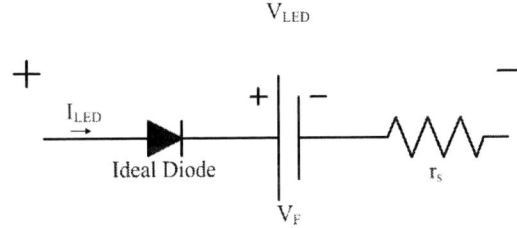

Fig. 2.   Approximate linear model.

Taking the above model as the basis, a new modeling technique has been presented in this paper, namely Piecewise Linear Approximation (PLA) technique. In this technique, several diodes are connected in parallel. Each branch has different knee voltages and gets turned ON at different values of forward voltage applied. Thus, as the applied voltage increases, the branches turn on one by one, making the resistance change each time a new branch conducts. The slope of the characteristics changes for different pieces of curve and hence the technique gets its name Piecewise Linear Approximation (PLA).

The PLA technique becomes more efficient as the number of branches increases. At the same time, it becomes very complicated too. So, to handle this issue, a better technique has been introduced in this paper viz Maclaurin series expansion based modeling as the measured I-V curve on inspection reveals that its non linear nature is quite similar to that of an exponential function. The Taylor series expression based modeling of LED as presented in [9] had been done using SIMPLIS and convergence to the measured sample curve was achieved after taking 23 terms of the

978-1-5386-6624-1/18 $31.00 © 2018 IEEE

series. The Maclaurin series model is a simplified version of Taylor series model and has been modeled using MATLAB with results converging by taking lesser number of terms.

II. MODELING OF LED

The measured I–V curve of the LED sample is shown in Fig. 3. As can be seen from Fig. 3, the nature of the curve is non linear.

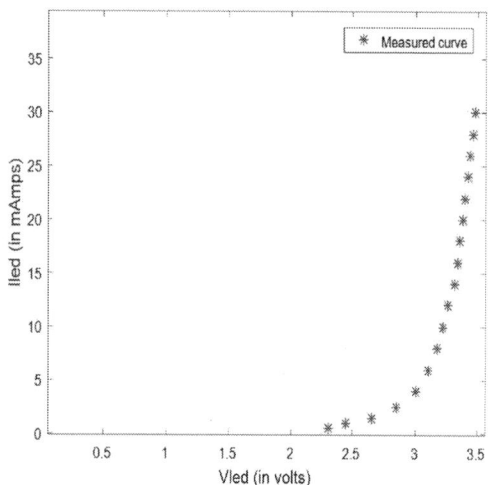

Fig. 3 Measured I-V curve of LED sample, LNL-190UW-4H [7].

In this paper, the I-V characteristic of the given LED sample has been proposed and modeled using two different modeling techniques, namely:
1 Piecewise linear approximation
2 Maclaurin series expansion based modeling.

A. Modeling using Piecewise Linear Approximation

The modeling of LED I-V characteristics using PLA technique involves multiple branches of series connected diode connected in parallel. More is the number of branches more is the accuracy of the plotted curve. Fig. 4 shows the generalized circuit of PLA technique. In this paper, five series connected diode branched circuit has been used as can be seen from Fig. 5. The first branch has only one diode, second branch has two diodes, branch three has three diodes connected in series and so on. It is clear from Fig 5 that as the applied voltage increases branches get switched ON in serial fashion with branch one getting switched ON first, then second and so on. Since in parallel connection, the equivalent resistance is smaller than the smallest branch resistance, therefore, the equivalent resistance of the circuit decreases as the number of branches getting switched ON increases. Further, the slope of the I-V curve increases with decrease in equivalent resistance. In this technique, 5 points from the datasheet have been taken to achieve the resulting curve. The choice of selection has been made such that the modeled curve best traces the plot of discrete datasheet points.

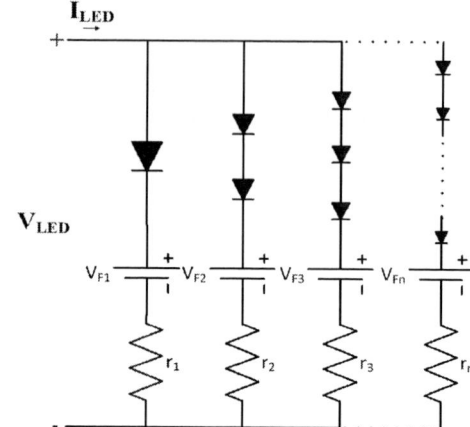

Fig. 4 Generalized PLA model

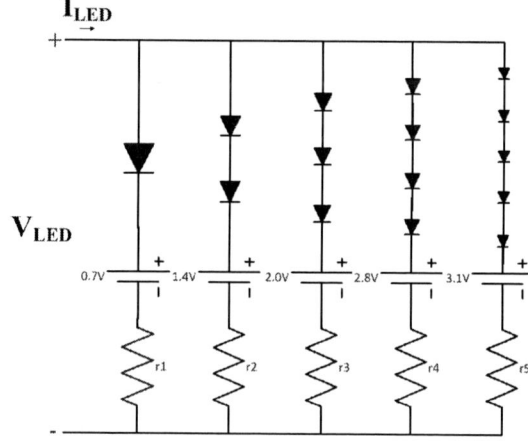

Fig. 5. Five branched PLA model.

B. Modeling using Maclaurin Series Expansion

The modeling of LED I-V characteristics using Maclaurin series expansion technique involves the use of the famous Shockley equation as given below:

$$I_{LED} = I_{SAT}(e^{(qV_F)/(\eta kT)} - 1) \qquad (1)$$

Where,

V_F	Voltage Drop across the LED
I_{LED}	Current passing through the LED
I_{SAT}	Saturation current of LED
η	Ideality factor
q	Magnitude of the Electronic Charge (1.602×10^{-19} C)
k	Boltzmann's constant (1.38×10^{-23} J/K)
T	Absolute temperature ($273 + T_a$) in Kelvin
T_a	Ambient temperature (in ˚C)

In the forward biased region, beyond the knee voltage, the exponential term predominates and the equation of LED I-V characteristics can be modified and rewritten as follows:

$$I_{LED} = I_{SAT}(e^{(V_F)/(\eta VT)}) \qquad (2)$$

Where

$$V_T = kT/q \qquad (3)$$

V_T is thermal voltage.

The Maclaurin series expansion of a function is given by the following equation:

$$f(x) = f(0) + \frac{f'(0).x}{1!} + \frac{f''(0).x^2}{2!} + \cdots \qquad (4)$$

The use of Maclaurin series expansion is preferred over the conventional exponential function because it saves a valuable amount of computational time. As can be seen from equation (1), the LED current is a function of internal temperature also. Thus for the sake of simplicity, ambient temperature of LED sample is taken to be 25 °C. Fig. 6 shows the equivalent circuit of LED used during the I-V characteristics analysis using the Maclaurin series expansion of the Shockley equation. The equation of the LED forward voltage V_F can be expressed in (5)

$$V_F = V_{LED} - r_s.I_{LED} \qquad (5)$$

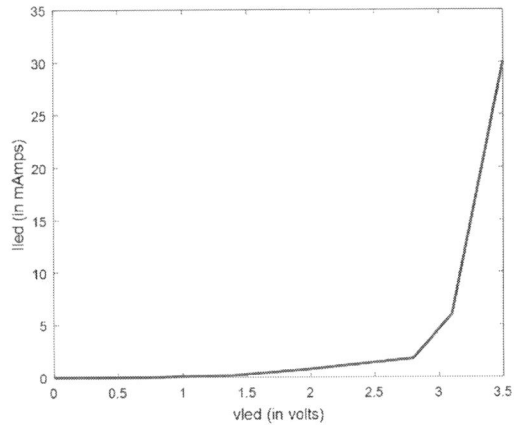

Fig. 6 Equivalent circuit for Maclaurin series based model.

By using equation (2) (3) and (5), the expression for LED forward voltage can be written as follows:

$$V_{LED} = \eta.V_T \ln\left(\frac{I_{LED}}{I_{SAT}}\right) + r_s.I_{LED} \qquad (6)$$

III. Implementation

In this paper, the I-V characteristics of the given LED sample have been modeled and implemented using two different modeling techniques, namely:
1 Piecewise linear approximation (PLA).
2 Maclaurin series expansion based modeling.

To implement the modeling techniques, certain parameters need to be calculated. The calculation method and the implementation of the techniques for both the methods are discussed below.

A. Calculation of parameters for piecewise model

The voltage drop for each diode (considering it to be silicon based) in the branch has been taken to be the standard 0.7 V, when forward biased. Then, a few key points have been taken from the datasheet such that the voltage co-ordinate

matches with the knee voltage drop of the branches. The internal equivalent resistance for each voltage range is then calculated using the ohm's law equation given as follows:

$$R_{eq} = V_{LED}/I_{LED} \qquad (7)$$

TABLE I
PARAMETERS FOR PIECEWISE MODEL

S no.	Voltage Range (volts)	V_t(volts)	r_s(kΩ)	Equivalent resistance R_{eq}(kΩ)
-	0-0.7	-	-	Infinity
1	0.7-1.4	0.7	3.5	3.5
2	1.4-2.0	1.4	1.4	1
3	2.0-2.8	2.0	4	0.8
4	2.8-3.1	2.8	0.0784	0.0714
5	Above 3.1	3.1	0.0216	0.0166

Using parameters listed in Table I, the I-V characteristics using PLA technique have been plotted as can be seen from Fig.7.

Fig. 7 I-V characteristics using PLA technique.

B. Calculation of parameters for Maclaurin series approximation

In the datasheet of LED LNL-190UW-4H [7], the parameters η, I_{Sat}, and r_s are not mentioned explicitly. However, in order to accurately determine the I-V characteristic curve of LEDs, the values of the unknown parameters η, I_{Sat}, and r_s are required as can be seen in (6). These can be obtained by following the steps mentioned below.

STEPS TO CALCULATE r_s, η AND I_{SAT}

1.) Locating The Maximum Operating Point (V_M, I_M): Fig. 8 shows the section of the curve plotted using the datasheet where the maximum operating point can be easily located. Above this point, the loading of LED should not be done so as to avoid overheating problems.

2.) Locating The Knee Point (V_K, I_K): Fig. 8 also mentions the information about the knee point. This is the point on the characteristics where the diode starts to conduct and after this point the curve almost becomes linear.

3.) Calculating The Unknown parameters: In Fig. 8, the slope of the curve between the rated operating point (V_R, I_R) and the maximum operating point (V_M, I_M) can be easily calculated due to the linear nature of the curve in this region. Since the value of internal resistance here will be the inverse of the magnitude of the slope of the curve, therefore, the equation to calculate r_s can be obtained as shown in equation (8).

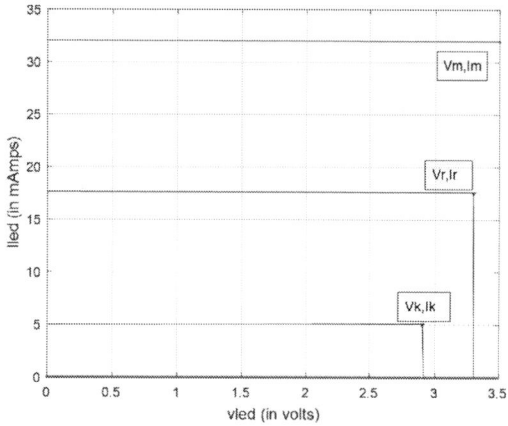

Fig. 8. Section of LED *I-V* curve from datasheet locating key points [7].

$$r_s = \frac{V_M - V_K}{I_M - I_K} \tag{8}$$

Following the above mentioned steps and applying equation (8) to equation (6), η and I_{Sat} can be calculated as per the following derived equations

$$\eta = \frac{V_R - V_M - r_s.(I_R - I_M)}{V_T.\ln(\frac{I_R}{I_M})} \tag{9}$$

$$I_{SAT} = \frac{I_K}{e^{(\frac{V_K - r_s.I_K}{\eta V_T})}} \tag{10}$$

TABLE II
PARAMETERS FOR MACLAURIN MODEL

Key point	Co ordinate value
(V_M,I_M)	(3.5,32.5)
(V_R,I_R)	(3.3,17.5)
(V_K,I_K)	(2.9,5)

Now, having calculated the essential parameters required the Maclaurin series expansion method can be applied to the Shockley equation as per the following expression:

$$I_{LED} = I_{SAT}\sum_{j=0}^{\infty} \frac{(\frac{V_{LED} - r_s.I_{LED}}{\eta V_T})^{\wedge}j}{j!} \tag{11}$$

Using the calculated parameters, the I-V characteristics using Maclaurin series expansion based modeling technique have been plotted as can be seen from Fig. 9.

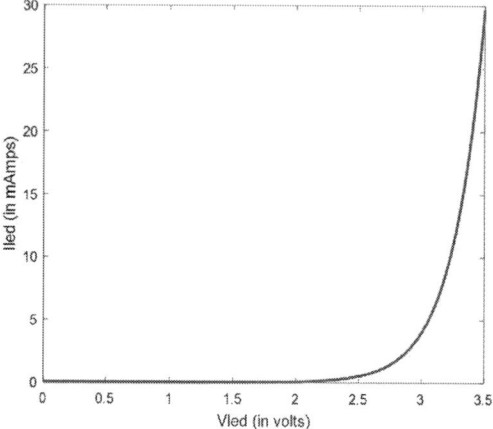

Fig. 9 I-V characteristics using Maclaurin series expansion based modeling technique.

IV. RESULTS AND DISCUSSONS

The measured curve has been plotted using the (V, I) points of the datasheet. The piecewise Linear approximation has been done with the help of data from Table 1. The Maclaurin series expansion based curve has been plotted using the Shockley equation and the parameters calculated in section 3.

From Fig.10, it can be seen that the Maclaurin series Expansion based modeled curve resembles the measured characteristics in much more appreciable way as compared to the Piecewise Linear Approximation model. The error between the measured curve and Maclaurin series Expansion based modeled curve is negligible while that for Piecewise Linear Approximation model is a little bit on the higher side.

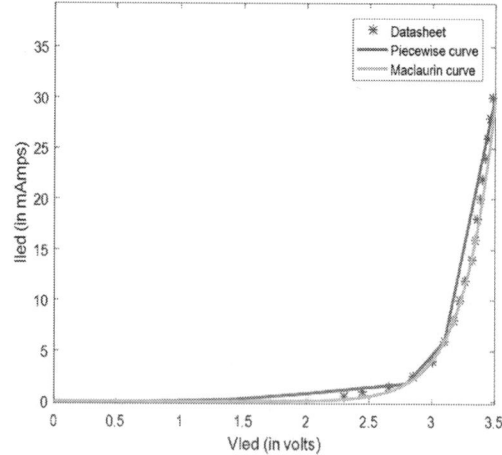

Fig. 10 Comparison of Measured curve with Piecewise and Maclaurin model curve done in MATLAB 2016a.

V. Conclusion And Future Scope Of Work

In this paper, two methods have been employed to model the I-V characteristics of a LED sample. The PLA technique is the reprise version of the Piecewise linear approximation technique that has been proposed in [9]. It has been observed that the proposed PLA technique is more simplified as compared to Piecewise linear approximation technique. The second method namely, Maclaurin series expansion also has the advantage of saving processor time. The characteristics shown above have been plotted by rigorous coding done in MATLAB software. The results started to converge with the original datasheet plot after merely eight iterations. Moreover, the PLA technique has the disadvantage that to achieve a better accuracy, it has to rely more on the datasheet which is not desirable. The Maclaurin series method, on the other hand, relies only on three datasheet points to get the whole characteristics plotted. The Maclaurin series expansion based model can be used to predict the current flowing through the LED at a particular voltage without actually applying voltage to it. The above work can be extended to model the larger arrays of LED connected in series and parallel and design its power ratings. It can further be used to develop the small signal model of LED which helps in determining the dynamic internal resistance of the LED. It can be of immense help in designing the wireless LED lighting system or designing the LED panels for artificial horticulture which are the two main future prospects of the above carried out work.

References

[1] Jaspreet Singh, Prof. Priya Mahajan, Prof. Rachana Garg "Techno Economic Analysis Of Different Lighting Schemes: A Case Study" presented in *IEEE International Conference on Power Energy, Environment & Intelligent Control, PEEIC* , 2018 to be published on IEEE Xplore.

[2] M. Jordan, *Single Inductor, Tiny Buck–Boost Converter Provides 95% Ef-ficiency in Lithium-Ion to 3.3V Applications*. Milpitas, CA, USA: LinearTechnol. Corp., 2002, Design Notes.

[3] Tomasz Torzewicz, Marcin Janicki and Andrzej Napieralski, "Experimental Determination of Junction-to-Case Thermal Resistance in LED Compact Thermal Models." *17th IEEE Intersociety Conference on Thermal and Thermomechanical Phenomena in Electronic Systems (ITherm) 2018,* pp. 768-772.

[4] Riccardo Pittini, Thomas Andersen, Toke M. Andersen, Martin Rodgaard, Jakob Monster and Mickey Madsen, "Multi-MHz LED Drivers: Design for Lifetime and Reliability.", *PCIM Europe 2018; International Exhibition and Conference for Power Electronics, Intelligent Motion, Renewable Energy and Energy Management* 2018, pp. 1 – 7.

[5] Jui-Hung Cheng and Hsin-Hung Lin, "Design and development of interactive LED display for fan applications.", *IEEE International Conference on Applied System Invention (ICASI) 2018,* pp. 393 – 396.

[6] Dipika Sadhvani and Mohamed Al Musleh, "Moving from CFL to LED lighting: Case study: University laboratory in Dubai". *International Conference on Intelligent Sustainable Systems (ICISS) 2017,* pp. 725 – 730.

[7] LNL-190UW-4H: SMD LED, Datasheet, LIGHTOP Technology Co.,Taipei County, Taiwan, 2007.[Online]. Available: http://www.lightop. com.tw

[8] A. S. Sedra and K.C. Smith, *Microelectronic Circuits*. London, U.K.: Oxford Univ. Press, 2004.

[9] R.-L. Lin and Y.-F. Chen, "Taylor series Expression based Equivalent circuit model of LED for analysis of LED driver System," in *Conf. Rec. IEEE IAS Annu. Meeting*, Houston, TX, USA, Oct. 4–8, 2009, pp. 1–5.

2nd IEEE International conference on power Electronics, Intelligent Control and Energy systems (ICPEICES-2018)

Solar PV-BES Based Microgrid System with Seamless Transition Capability

Vivek Narayanan, *Member IEEE*
Electrical Engineering
IIT Delhi
viveksw.narayanan7@gmail.com

Seema, *Member IEEE,*
Electrical Engineering
IIT Delhi
seemajmi2013@gmail.com

Bhim Singh, *Fellow IEEE*
Electrical Engineering
IIT Delhi
bhimsinghiitd62@gmail.com

Abstract— This paper demonstrates a solar photovoltaic (PV)-battery energy storage (BES) based microgrid system with multifunctional voltage source converter (VSC). It dealts with maximum power extraction from a PV array, reactive power compensation, harmonics elimination and seamless transition from the grid connected mode (GCM) to standalone mode (SAM) and vice versa. The maximum power extraction from a PV array, is achieved by using a DC-DC converter. A bidirectional DC-DC converter (BDDC) is used for regulating the DC link voltage. Whenever the BES is not connected then the VSC performs the regulation of the DC link voltage. The system behavior is studied on a prototype of the microgrid system under various operating conditions.

Keywords— *Solar PV array, MPPT, DC-DC converter, BES, BDDC, VSC, SAM and GCM, power quality.*

I. INTRODUCTION

The increasing population and industrialization, demands huge amount of energy. It is not possible to meet the energy demands by the conventional sources alone due to their depleting nature. So the renewable energy sources are becoming important. The energy production from the renewable sources is not causing any environmental pollutions hence it becomes popular for generating electrical energy [1-2]. Due to the ease of availability, environment friendly scheme and the absence of rotating parts in the energy conversion, makes the solar energy as prior choice among other renewable energy sources.

The main drawback of the solar energy, is its intermittent nature. So the battery energy storage (BES) is provided to meet the load demand without causing any interruption. Solar photovoltaic (PV)-BES based microgrid system is reported in the literature [3-4]. The extraction of maximum power from the solar PV array is achieved by using maximum power point tracking (MPPT) algorithm. Some of the MPPT control algorithms are reported in the literature [5-6]. In addition to the maximum power extraction, the improvement of power quality becomes an important consideration in the microgrid system due to the use of nonlinear loads at the point of common coupling (PCC). This task is achieved by controlling the voltage source converter (VSC). Various control algorithms are developed for improving the power quality. Some of the control algorithms are discussed in the literature [7-8].

The sudden failure of the grid is usually happened due to the occurrence of fault in the system. So it needs to operate the system in standalone mode (SAM), in order to ensure the continuity in supplying the load. Thus, the system needs to transfer its operation from grid connected mode (GCM) to SAM and vice versa without causing any oscillations or overshoots. The seamless transition can be achieved by using proper control techniques. Some control techniques are reported in the literature that presents seamless transition between GCM to SAM and from SAM to GCM [9-10].

This paper dealts with a solar PV-BES based microgrid system. Here the DC-DC converter performs the extraction of maximum power from a PV array. The DC bus voltage regulation is performed by the bidirectional DC-DC converter (BDDC), by operating as in boost or buck mode depending upon the mode of operation. The VSC is controlled such that multifunctional feature is achieved, such as reactive power compensation, harmonics elimination, balancing of grid currents, operation in SAM and seamless transition from the GCM to SAM and vice versa. Whenever the BES is present in the system then the control of VSC is such that the constant power is fed to the grid. This avoids the power fluctuations in the grid that arises due to the continuous variations in the solar irradiance.

II. SYSTEM DESCRIPTION

Fig. 1 shows the PV-BES based microgrid configuration with multifunctional VSC. The DC-DC boost converter extracts the peak power from a PV array. The BDDC regulates the DC link voltage to the desired DC voltage. The VSC converts the DC power into AC power and it feeds to the PCC through interfacing inductors. The system operates in GCM when the grid is available and it operates in SAM when the grid fails. It also performs seamless transition from the GCM to SAM and vice versa. The ripples in grid and load voltages, are eliminated by using ripple filters in the grid and load sides.

III. CONTROL STRATEGY

The control of proposed microgrid configuration is composed of MPPT control and the VSC control.

A. MPPT Control

The MPPT control is used for extracting the maximum power from the solar PV array in all operating conditions. Here, the perturb & observe (P&O) algorithm is used for harvesting the peak power. It gives a reference voltage (V_{pv}^*), which corresponds to the maximum power point (MPP) voltage.

B. VSC Control

This proposed configuration works in both GCM and SAM. So, the control for the VSC, is different in these operating conditions. The control of the VSC in GCM and SAM, are discussed in the following sections.

1) Control of VSC in GCM of Operation

In the GCM of operation, different cases are considered, mainly a fixed power mode (FPM) and variable power mode (VPM). Whenever, the BES is present in the system, then the FPM of operation takes place that is a fixed power is fed to the utility grid irrespective of the solar PV generated power. When the BES is absent, then the VPM of operation takes place that is the power feeding to the grid is depending on the solar PV generating power. The control of VSC in GCM of operation, is

978-1-5386-6624-1/18 $31.00 © 2018 IEEE

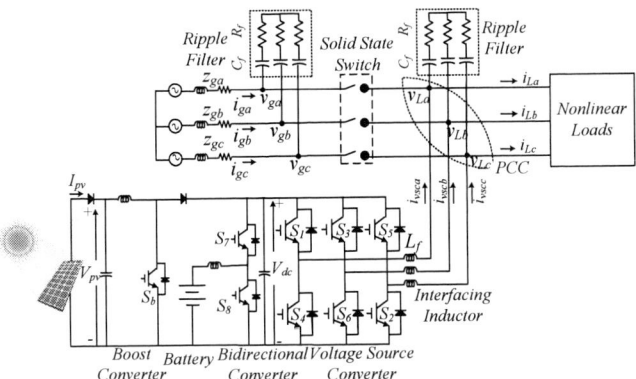

Fig. 1 Proposed PV-BES microgrid configuration

shown in Fig. 2. The main steps of operation in VSC control in the GCM, are discussed here.

a) Calculation of Unit Templates

The grid phase voltages are estimated as,

$$v_{ga} = \frac{2v_{gab} + v_{gbc}}{3}, v_{gb} = \frac{-v_{gab} + v_{gbc}}{3}, v_{gc} = \frac{-v_{gab} - 2v_{gbc}}{3} \quad (1)$$

Where, v_{gab}, v_{gbc} are the sensed grid line voltages.
The amplitude (V_t) of the grid voltage is given as,

$$V_t = \sqrt{\frac{2}{3}(v_{ga}^2 + v_{gb}^2 + v_{gc}^2)} \quad (2)$$

The in-phase unit templates are generated as,

$$u_{ia} = \frac{v_{ga}}{v_t}, u_{ib} = \frac{v_{gb}}{v_t}, u_{ic} = \frac{v_{gc}}{v_t} \quad (3)$$

b) Calculation of Active Loss Components

The DC voltage regulation is performed by the BDDC when the BES is present in the system otherwise the regulation of the DC link voltage is done by the VSC. The reference DC voltage (V_{dc}^*) is compared with the sensed DC link voltage (V_{dc}). At the jth instant it is given as,

$$V_{de}(j) = V_{dc}^*(j) - V_{dc}(j) \quad (4)$$

Where, V_{dc}^* is the reference DC link voltage (380V).
This error voltage V_{de} is fed to the proportional-integral (PI) controller, which is responsible for the DC link voltage regulation. The output of the PI controller as,

$$w_{ci}(j+1) = w_{ci}(j) + K_{pd}(V_{de}(j+1) - V_{de}(j)) + K_{id}V_{de}(j+1) \quad (5)$$

Where, w_{ci} is the active loss component and K_{id}, K_{pd} are the gains of the PI controller.

c) Calculation of Feed-Forward Components

Whenever, the BES is present in the system then the FPM operation becomes active. The feed-forward component for FPM operation, is given as,

$$w_{ff} = \frac{2P_{fixed}}{3V_t} \quad (6)$$

Where, P_{fixed} is the fixed power supplied to the grid.
The feed-forward weight of the PV power is given as,

$$w_{pv}(j+1) = \frac{2P_{pv}(j)}{3V_t} \quad (7)$$

Where, P_{pv} is the power output from the PV array.

d) Calculation of Fundamental Active Weights of Load Currents

The fundamental active weight of load current of phase 'a' using robust mixed-norm (RMN) control is given as [11],

Where, $e_{ia}(j)$ is the error of adaptive component, μ and λ are step size and mixing parameter, respectively.

$$e_{ia}(j) = i_{La}(j) - u_{ia}(j)w_{ia}(j) \quad (9)$$

Where, $w_{ia}(j)$, $i_{La}(j)$ and $u_{ia}(j)$ are the active weight, load current and the in-phase unit template of 'a' phase respectively at the jth instant.
Similarly, the phase 'b' and 'c' reference active current components are estimated as,

$$w_{ib}(j+1) = w_{ib}(j) + \mu u_{ib}(j)\{2\lambda e_{ib}(j) + (1-\lambda)sign(e_{ib}(j))\} \quad (10)$$

$$w_{ic}(j+1) = w_{ic}(j) + \mu u_{ic}(j)\{2\lambda e_{ic}(j) + (1-\lambda)sign(e_{ic}(j))\} \quad (11)$$

e) Generation of Reference Grid Currents

The performance is carried out at unity power factor (UPF) operation at the grid side. In FPM operation, the grid active weight w_{gi} is given as,

$$w_{gi} = w_{ff} \quad (12)$$

Where, w_{ff} is the fixed weight that is fed to the utility grid.
In VPM operation, the total active weight (w_{gi}) of grid currents is given as,

$$w_{gi} = w_{Lavg} + w_{ci} - w_{pv} \quad (13)$$

Where,
w_{Lavg} is the average active weight of load currents and is given as,

$$w_{Lavg} = \frac{(w_{ia} + w_{ib} + w_{ic})}{3} \quad (14)$$

The weight w_{ci} is the active loss component, which is utilised for regulating the DC link voltage and w_{pv} is the weight of PV power.
The reference grid currents are obtained as,

$$i_{ga}^* = w_{gi}u_{ia}, i_{gb}^* = w_{gi}u_{ib}, i_{gc}^* = w_{gi}u_{ic} \quad (15)$$

The switching pulses for the VSC, are generated by using the hysteresis controller. The error signals obtained by comparing the reference grid currents and the sensed grid currents, are fed to the hysteresis controller, which gives the switching pulses.

2) Control of VSC in SAM of Operation

The reference load voltages are given as,

$$v_{La}^* = V_t^* \sin(w_o t), v_{Lb}^* = V_t^* \sin(w_o t - \frac{2\pi}{3}), v_{Lc}^* = V_t^* \sin(w_o t + \frac{2\pi}{3}) \quad (16)$$

Where, V_t^* and w_o are the amplitude and frequency (314 rad/sec) of the voltages.
These reference load voltages are compared with the sensed load voltages. The error signals are fed to PI controller, which gives the reference load currents. These reference load currents are compared with the sensed load currents and the hysteresis controller gives the switching signals for the VSC in SAM. The control of VSC in SAM is depicted in Fig. 3.

C. Synchronization Control

The synchronization control is depicted in Fig. 4. When the grid voltage magnitude (V_t), grid frequency (f_g) and the difference between the grid phase angle (θ_g) and load phase angle (θ_L) become in the specified range then the main grid is connected to the system by controlling the solid state switch. Otherwise, the system operates in the SAM.

D. DC-DC Boost Converter Control

It tracks the PV array at its MPP. The duty cycle for the boost converter switch is generated as,

$$D = 1 - \frac{V_{pv}^*}{} \quad (17)$$

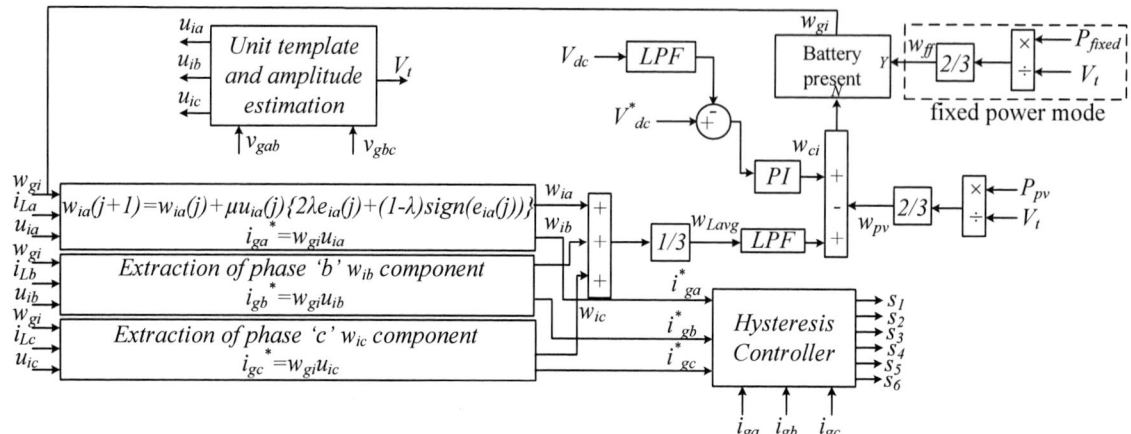

Fig. 2 VSC control algorithm in GCM

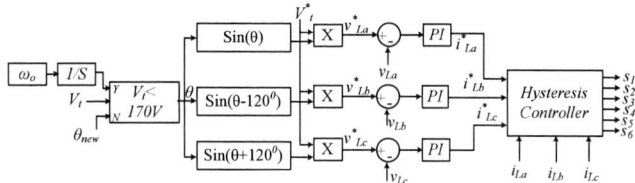

Fig. 3 VSC control algorithm in SAM

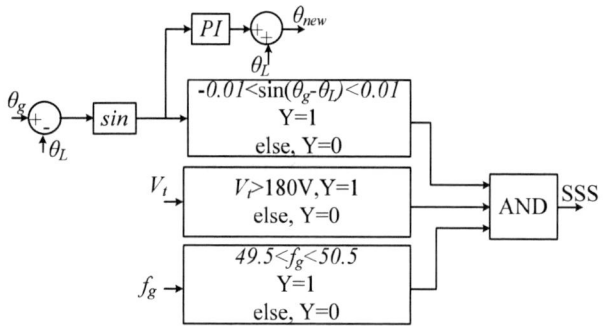

Fig. 4 Synchronization control

Where, V_{pv}^* is the reference voltage output from the MPPT controller and V_{dc} is the sensed DC link voltage.

E. BDDC Control

The BDDC regulates the DC link voltage to the desired DC voltage when the BES is present in the system. The PI controllers are used for generating the duty signals for the switches S_7 and S_8 as shown in Fig. 5. The comparison between the reference DC link voltage and sensed DC link voltage yields error signal, which is given to a PI controller. The PI controller output expression is given as,

$$I_b^*(j+1) = I_b^*(j) + K_{pd}(V_{de}(j+1) - V_{de}(j)) + K_{id}V_{de}(j+1) \quad (18)$$

Where, K_{id}, K_{pd} are gains of the PI controller.

$$V_{de}(j) = V_{dc}^*(j) - V_{dc}(j) \quad (19)$$

The reference battery current (I_b^*) from the PI controller is compared with the sensed battery current (I_b). The compared output signal is given to another PI controller, which gives the duty signals for the switches. The PI controller output expression is given as,

$$D(j+1) = D(j) + K_{pb}(I_{be}(j+1) - I_{be}(j)) + K_{ib}I_{be}(j+1) \quad (20)$$

Where,

$$I_{be}(j) = I_b^*(j) - I_b(j) \quad (21)$$

Fig. 5 BDDC control

IV. TEST RESULTS

The system response under different operating modes, is analysed by conducting experiments on the system prototype.

A. Performance of System in GCM of Operation

The GCM of operation of proposed system in different operating scenarios, is discussed in the following sections.

1) Steady State Response of Solar PV-BES

Fig. 6 shows the steady state behavior of the system at 1000W/m² solar irradiation. Fig. 6 (a) shows the grid voltages and grid currents waveforms. Figs. 6 (b)-(c) show the total harmonic distortions (THDs) of grid voltages and grid currents, it shows the THDs of grid voltages and grid currents are within the recommended values. The performance is studied while feeding a fixed power of 1.7kW to the utility grid as shown in Fig. 6 (d). Figs. 6 (e)-(f) show the load current waveforms and load power. Fig. 6 (g) shows the load current THD. It is 24.3%, since the load is nonlinear consisting of diode bridge rectifier with RL load. Figs. 6 (h)-(i) show the VSC current and the VSC power. Fig. 6 (j) shows the battery voltage and battery current. Here, the BES is operated in charging mode since the solar PV array generating power is higher than the load demand and the fixed grid power. The solar PV array voltage, current and power at 1000W/m² irradiance, are depicted in Fig. 9 (a). From this, it is clear that the maximum power is well extracted from the PV array at 1000W/m² irradiance. Fig. 6 (k) shows the DC link voltage and DC link current. It gives, the BDDC regulates the DC link voltage to the desired value.

2) Dynamic Response under Solar PV Array Disconnection

Fig. 7 shows the dynamic response of the system under the disconnection of the PV array. Fig. 7 (a) shows the grid current, load current, battery current and the PV array current. Fig. 7 (b) shows the grid power, load power, battery power and the PV array power. Here, the performance is studied in FPM operation. Initially, the irradiance is 1000W/m², so the excess power is available after meeting the fixed grid power (1.7kW) and the load demand, which is used for charging the BES.

978-1-5386-6624-1/18 $31.00 © 2018 IEEE   724

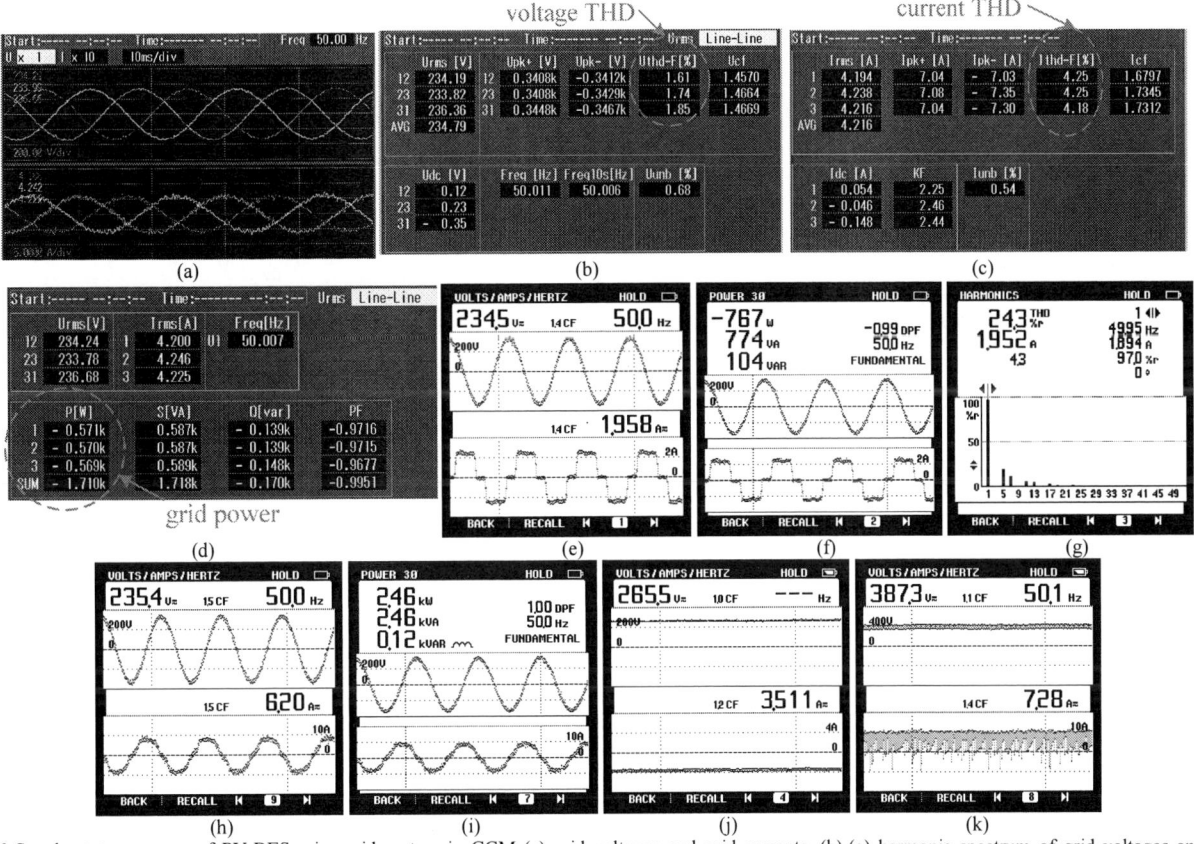

Fig. 6 Steady state response of PV-BES microgrid system in GCM (a) grid voltages and grid currents, (b)-(c) harmonic spectrum of grid voltages and grid currents, (d) grid power, (e) load current, (f) load power, (g) harmonic spectrum of load current, (h) VSC current, (i) VSC power, (j) battery voltage and battery current, (k) DC link voltage and DC link current

When the solar PV array is disconnected then the BES changes from the charging mode to discharging mode in order to keep the power that is fed to the grid as constant and for supplying the load demand as shown in Fig. 7 (b). Fig. 7 (c) shows the grid voltage, DC link voltage, PV array voltage and PV array current. It shows that the DC link voltage is maintained to the desired voltage even though during the disturbance in the solar irradiance.

3) Dynamic Response under Varying Irradiance

Fig. 8 shows the VPM operation of the microgrid configuration under varying irradiance. In this mode of operation, the BES is not present. Here the performance is analysed by changing the solar irradiance from $1000\,W/m^2$ to $500\,W/m^2$ and after some time the PV array is disconnected. Fig. 8 (a) shows the grid power, load power, PV array power and the PV array current. The performance is studied at constant load power.

When the irradiance is reduced from $1000\,W/m^2$ to $500\,W/m^2$, the PV array power is reduced, thereby the power that is fed to the grid, is also reduced and it is depicted in Fig. 8 (a). After the solar PV array disconnection, the grid power becomes positive. That is now the grid is supplying the load demand. Fig. 8 (b) shows the grid voltage, grid current, VSC current and the PV array current during the irradiation change from $1000\,W/m^2$ to $500\,W/m^2$ and during the solar PV array disconnection. Fig. 8 (c) shows the grid voltage, load current, DC link voltage and the PV array current. Even though during the disturbances in the solar irradiation, the DC link voltage is maintained to the desired voltage by the VSC. The variations of internal signals such as total grid active weight, average active weight of load currents and PV feed forward weight with respect to the PV array current, are depicted in Fig. 8 (d). After the PV array disconnection, the system is working as distribution static compensator (DSTATCOM).

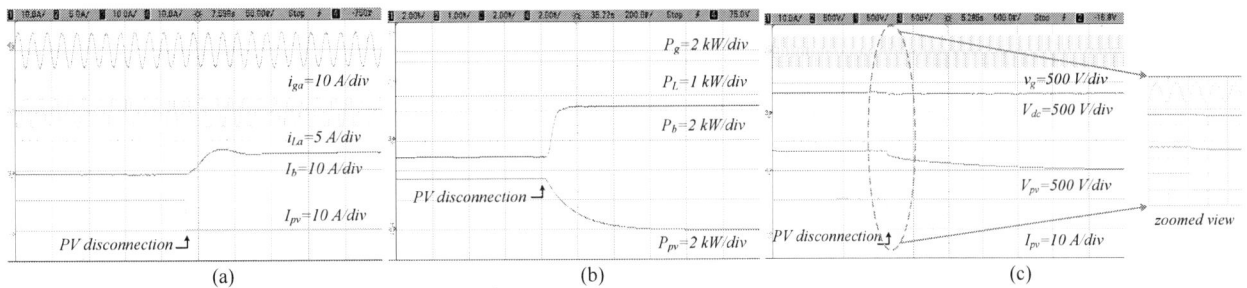

Fig. 7 (a)-(c) Dynamic response under solar PV array disconnection

978-1-5386-6624-1/18 $31.00 © 2018 IEEE

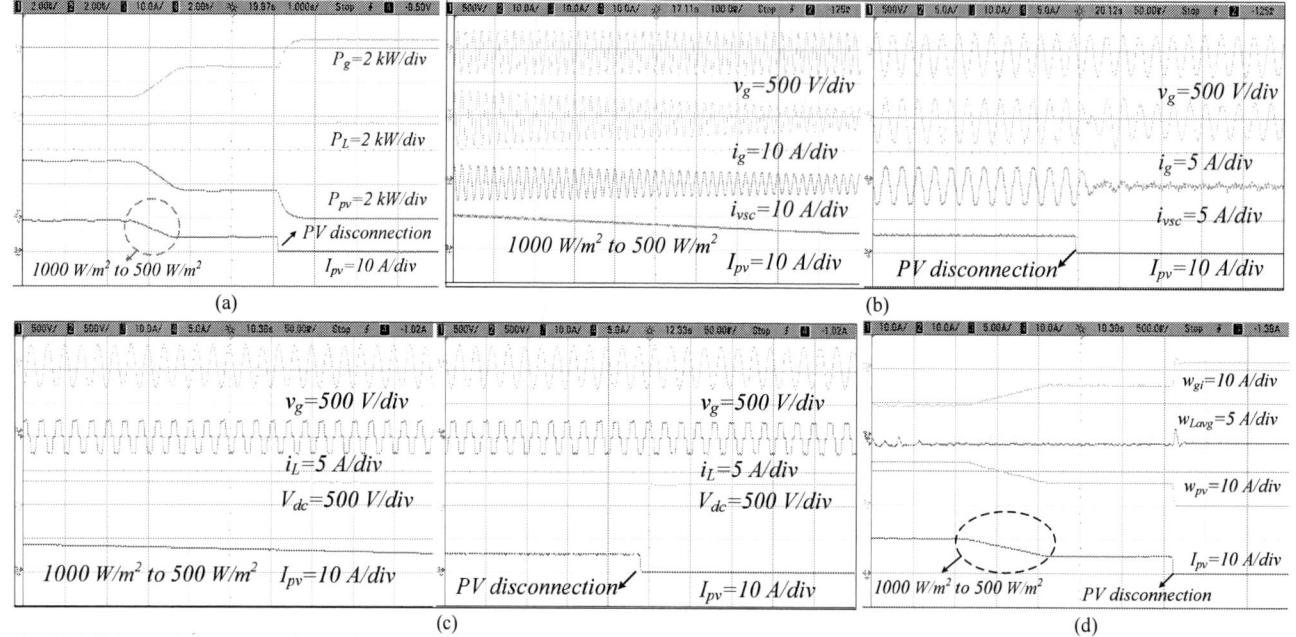

Fig. 8 (a)-(f) Dynamic response under varying irradiance

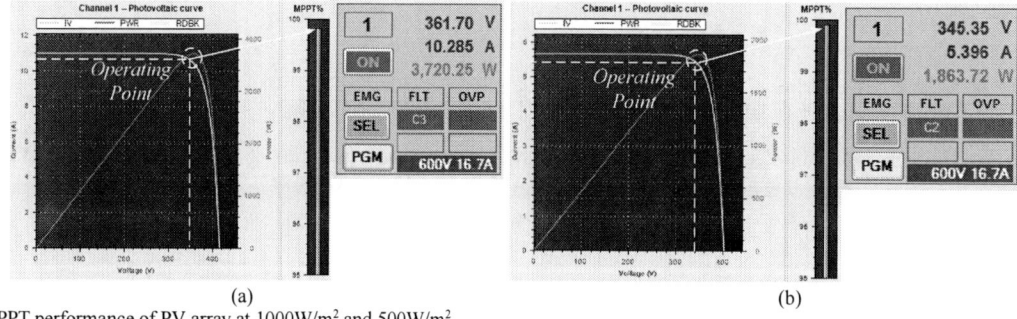

Fig. 9 (a)-(b) MPPT performance of PV array at 1000W/m² and 500W/m²

Figs. 9 (a)-(b) show the performance of the PV array at 1000W/m² and 500W/m² solar irradiations. It depicts that the maximum power is effectively extracted from the PV array.

4) Dynamic Response under Unbalanced Nonlinear Load

Fig. 10 shows the response of the system under unbalanced load condition. Here the performance is studied at 1000W/m² irradiance and without BES. This mode of operation is dealt with VPM. Fig. 10 (a) shows grid voltage, grid current, VSC current variations with respect to the load current. The unbalance is created by removing the 'a' phase load for a particular duration of time. During the load removal, the current fed to the grid is increased. Fig. 10 (b) shows the grid power, PV array power, load power and the load current. It shows that during the removal of load current in 'a' phase the power fed to the grid is increased. Here the PV array power is unaltered. Fig. 10 (c) shows the DC link voltage, PV array voltage, PV array current and the load current. It shows that the DC link voltage is regulated to desired voltage by VSC even during the load disturbances. Figs. 10 (d)–(e) show the variations in the internal signals during the load unbalance. Fig. 10 (d) shows the total grid active weight, 'a' phase unit template and 'a' phase reference grid current with respect to the load current. Fig. 10 (e) shows the average active weight of load currents, error signal, and active weight of 'a' phase load current with respect to the

phase becomes zero and hence the average active weight of load currents, is reduced. Figs. 10 (f)-(g) show the waveforms of grid currents with grid voltages and the grid current THDs, respectively before the load unbalancing. Figs. 10 (h)-(i) show the grid current waveforms with grid voltages and the grid current THDs at load unbalancing. It is clear that the current THDs are within the recommended value.

B. Performance of System in SAM of Operation

The performance of the microgrid configuration in SAM of operation is depicted in Fig. 11. Fig. 11 (a) shows the load voltages and the load currents. Fig. 11 (b) shows the THDs of the load voltages. It is clear that the THD of load voltages are within the level recommended by the IEEE 519 standard. Here, the performance is studied for a decrease in solar irradiance from 1000W/m² to 500W/m². Fig. 11 (c) shows DC link voltage, battery voltage, battery current and the PV array current. The DC link voltage is regulated to desired voltage by the BDDC. During the reduction in the irradiance, the battery changes its mode from charging to discharging in order to meet the load demand.

C. Seamless Transition of Microgrid from GCM to SAM and Vice Versa

Figs. 12 (a)-(b) show the seamless transition capability of the

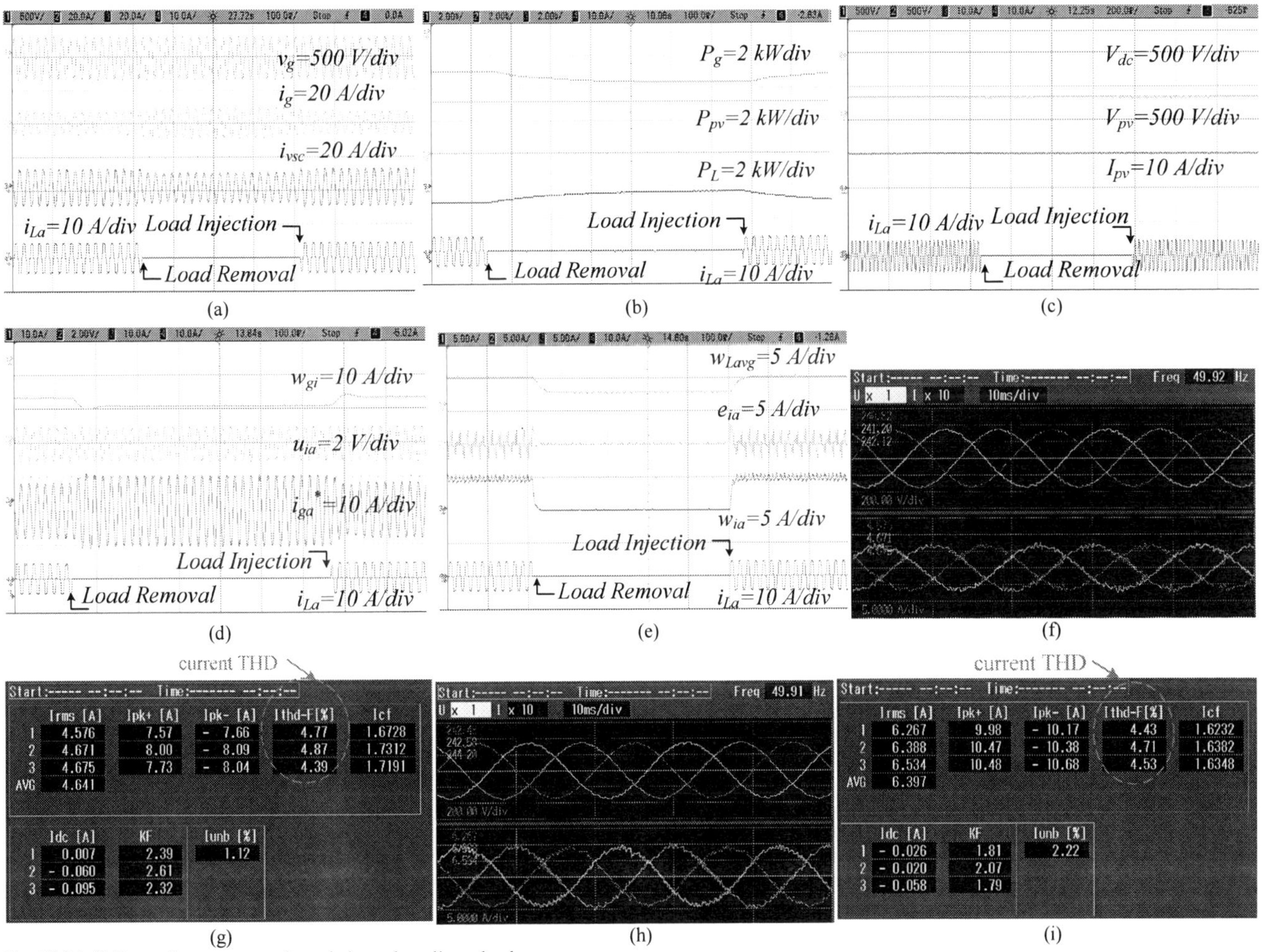

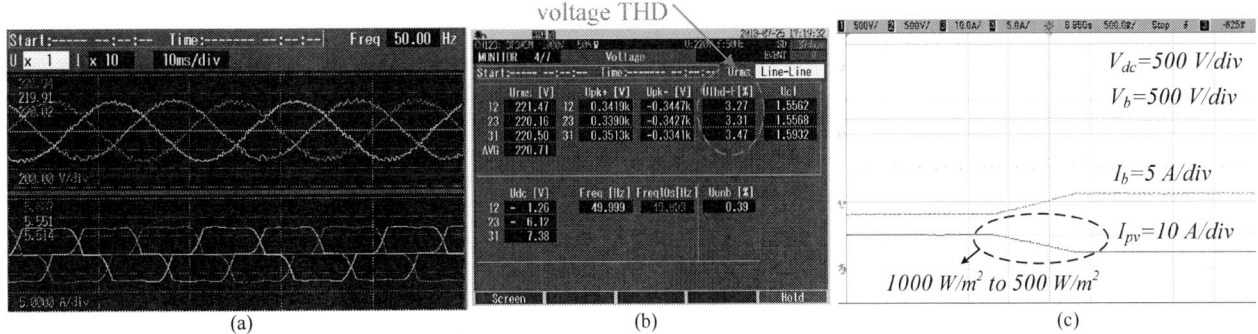

Fig. 10 (a)-(i) Dynamic response under unbalanced nonlinear load

Fig. 12 (a) shows grid voltage (v_g), load voltage (v_L), grid phase angle (θ_g) and load phase angle (θ_L). When the grid fails, the grid voltage and the grid phase angle, become zero. When the grid is restored, θ_L tracks θ_g with synchronization control. Fig. 12 (b) shows grid voltage (v_g), load voltage (v_L), grid current (i_g) and the load current (i_L). Here the performance is studied at 500W/m² irradiance and FPM operation. When the grid fails, the grid voltage becomes zero then the VSC control is shifted to the SAM control hence the load voltages are maintained to the desired voltages. The main grid is connected to the system when the grid voltage magnitude (V_l), sine of the phase angle difference (θ_g-θ_L) and the grid frequency (f_g) are become within the specified limits.

D. Comparison of Proposed Control with Conventional Controls

The proposed control is compared with the conventional least mean fourth (LMF) and least mean square (LMS) controls under the unbalanced load condition. Figs. 13 (a)-(b) show the variations of the active weight of the load current of phase 'a' (w_{ia}) and the average active weight of the load currents (w_{Lavg}) under the load unbalance. It shows that the oscillation associated in the weight with the proposed RMN control, is lesser as compared to those with the conventional LMF and LMS controls.

Fig. 11 (a)-(c) Performance of the system in SAM of operation

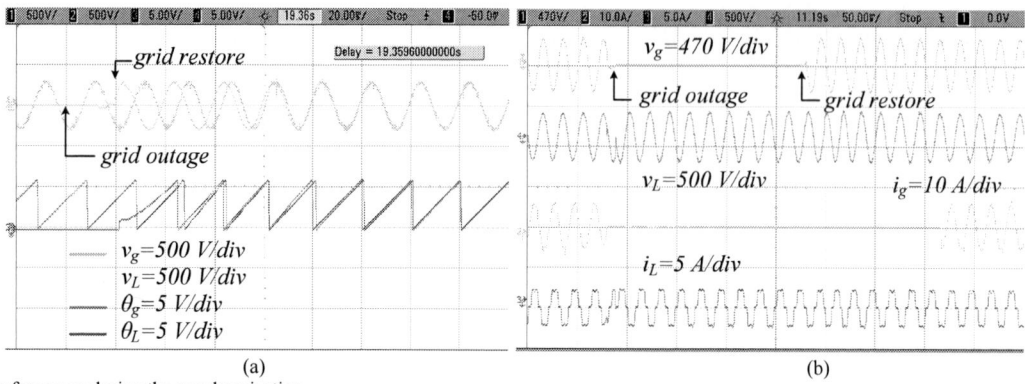

(a)   (b)

Fig. 12 (a)-(b) Performance during the synchronization

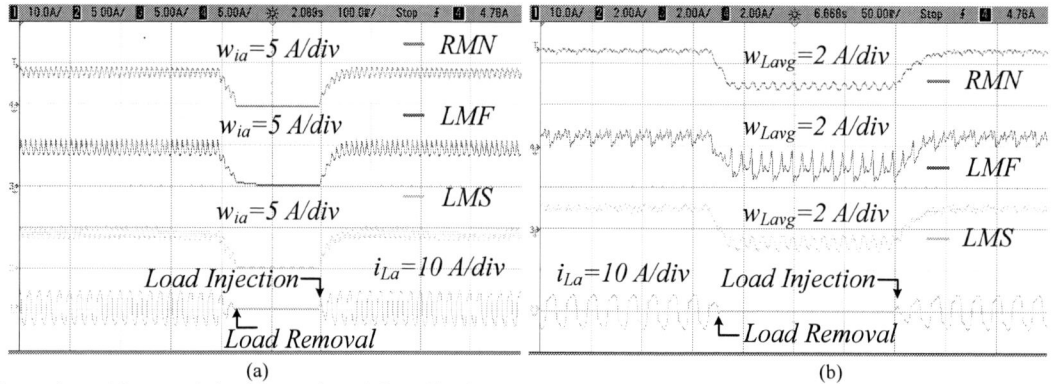

(a)   (b)

Fig. 13 (a)-(b) Comparison of the control algorithms under unbalanced load

V. CONCLUSION

The seamless transition of PV-BES microgrid from GCM to SAM and vice versa with compensation of reactive power and harmonics, balancing of grid currents and maximum power extraction from the PV array are demonstrated in this work. The PV array maximum power is harvested by controlling the DC-DC converter. The BDDC functions the regulation of the DC link voltage to the reference voltage. Whenever the BES is absent, then the control for regulating the DC link voltage, is automatically shift to VSC. The comparative analysis depicts that the proposed RMN control gives performance better than the conventional LMF and LMS controls.

ACKNOWLEDGEMENT

The authors are very grateful to the Dept. of Science and Tech., Govt. of India, for funding this project under Grant Number: RP03195 (FIST Proj.) and RP03443G (INDO-US Proj.).

APPENDICES

	PV array	V_{mpp}= 360 V, I_{mpp}= 10.3 A, P_{mpp}= 3.7kW
Test parameters	DC-link voltage	V_{dc}= 380 V
	DC link capacitor	C_{dc}=3mF
	Battery	240 V, 49 Ah
	Grid voltage	V_{LL}=230V (rms)
	Interfacing inductor	L_f= 4mH
	Step size	μ = 0.0156
	Mixing parameter	δ = 0.5

REFERENCES

[1] B. S. Pali and S. Vadhera, "Renewable energy systems for generating electric power: A review," *IEEE Inter. Conf. on Pow. Elect., Intel. Con. and Ene. Sys.*, 2016, pp. 1-6.

[2] E. Goutard, "Renewable energy resources in energy management systems," *IEEE PES Innov. Smart Grid Tech. Conf. Europe*, 2010, pp. 1-6.

[3] M. D. Vijay, I. Hussain, B. Singh and G. Bhuvaneswari, "Energy management and control of SECS and BESS integrated AC microgrid," *IEEE Inter. Symp. on Ind. Elect.*, 2017, pp. 975-980.

[4] N. Beniwal, I. Hussain and B. Singh, "Control and operation of a solar PV-battery-grid-tied system in fixed and variable power mode," *IET Gen., Transm. & Distri.*, vol. 12, no. 11, pp. 2633-2641, 2018.

[5] S. Amin, S. Khan and A. Qayoom, "Comparative analysis about the study of maximum power point tracking algorithms: A review," *Inter. Conf. on Comp., Math. and Eng. Tech.*, 2018, pp. 1-8.

[6] O. Singh and S. K. Gupta, "A review on recent Mppt techniques for photovoltaic system," *IEEMA Eng. Inf. Conf.*, 2018, pp. 1-6.

[7] V. Narayanan and P. Jayaprakash, "Performance comparison of three-phase four-wire grid integrated solar photovoltaic system with various control algorithms," *Inter. Conf. on Intel. Comp., Instr. and Cont. Tech.*, 2017, pp. 406-414.

[8] V. L. Srinivas, S. Kumar, B. Singh and S. Mishra, "A Multifunctional GPV System Using Adaptive Observer Based Harmonic Cancellation Technique," *IEEE Trans. on Ind. Elect.*, vol. 65, no. 2, pp. 1347-1357, 2018.

[9] S. Kumar and B. Singh, "Seamless transition of three phase microgrid with load compensation capabilities," *IEEE Ind. Appl. Soc. Annual Meet.*, 2017, pp. 1-9.

[10] T. H. Nguyen, K. Al Hosani, N. Al Sayari and A. R. Beig, "Seamless transition scheme between grid-tied and stand-alone modes of distributed generation inverters," *IEEE Inter. Fut. Ene. Elect. Conf. and ECCE Asia*, 2017, pp. 344-349.

[11] J. Chambers and A. Avlonitis, "A robust mixed-norm adaptive filter algorithm," *IEEE Sig. Proce. Letters*, vol. 4, no. 2, pp. 46-48, 1997.

978-1-5386-6624-1/18 $31.00 © 2018 IEEE

Battery Charging of Smart Phones using Organic Solar Cells

Rahul K.*, *Rakesh K. Meena*, Ritesh K. Gupta†, Parameswar K. Iyer‡, Arun Tej M.*, Shabari Nath*

Dept. of Electronics and Electrical Engineering, † Center for Nanotechnology, ‡ Dept. of Chemistry
Indian Institute of Technology Guwahati, Assam, India
Email: {riteshkantgupta, aruntej, snath}@iitg.ac.in

Abstract—Organic solar PV (OPV) cells are gaining attention due to their light weight, flexibility and low cost. Charging of smart phones using OPV modules is a prime application area. However, the chargers developed using organic PV modules are bulky as they require carrying long rolls of PV arrays. This paper proposes an integrated charger for smart phones, which includes an OPV module and a charging circuit with an op-amp based precision diode. The proposed charger uses an OPV module of size equal to that of the smart phone display. The proposed design reduces the size and weight of charger significantly. The experimental results of OPV module characterization, battery characteristics, blocking diode performance, and increase in state of charge (SOC) of the smart phone using proposed charger are presented.

Keywords— *Blocking diode, charging circuit, flexible organic PV module, precision diode, size of OPV, smart phone*

I. INTRODUCTION

The energy production from fossil fuels not only has an adverse impact on global ecology and human health, but will be insufficient for after a few decades. Therefore, there is a strong need to develop highly efficient and alternate energy source which is cheap and eco-friendly. Trapping energy from the sun is one of the best methods to answer all the energy demands of the globe. Inorganic solar cells involving silicon, selenium, copper indium gallium selenide (CIGS) etc. as active materials have been well demonstrated for energy applications. But, the inorganic solar cells suffer from the disadvantage of huge processing cost and require a lot of energy during production. Also, the materials used in the active layer of inorganic solar are toxic and scarce and do not decompose easily. Thus, the demand for organic solar cell (OSC) has increased as it contains organic materials in the active area which are available in plenty and are easy to synthesize.

Organic solar cells use organic semiconductors instead of silicon to convert photons into electricity. The attractive features of organic solar cells are lightweight and flexibility. So they can be used for various applications that are previously inhibited due to bulkiness, inflexibility and heavy weight of silicon based solar cells [1]. In comparison to silicon solar cells, organic solar cells have lesser efficiency and lower lifetime but are significantly cheaper and environmental friendly.

Charging of smart phones by various means such as solar photovoltaics (PV), wireless charging etc. have been of interest for many researchers. Charging of smart phones using organic solar PV modules has also been developed [2]. However, they all require of carrying long rolls of flexible PV modules due to lesser efficiencies of organic PVs. Moreover, the blocking diode used in charging circuits reduces the net output voltage. Blocking diode is a diode used in series with a PV module to block the reverse flow of current from battery to PV module. The voltage drop across the blocking diode reduces the performance of organic solar PV module.

This paper proposes a charger for smart phone, using - (i) flexible organic solar PV (OPV) module of size of the smart phone and (ii) precision diode in charging circuit. Because of the size as chosen in proposed charger, use of long rolls of OPV modules has been avoided. Long rolls of OPV modules increases the total weight and size of the charger and needs to be carried separately in addition to the phone. In addition, precision diode has been used in place of signal diode as blocking diode. The voltage drop across blocking diode is nullified with use of precision diode. The charger proposed in this paper, aims to (i) reduce the size (ii) keep on charging the smart phone slowly whenever internal circuit of the phone allows, and hence (iii) to increase the discharging time, as the proposed charger can be connected to phone always.

Rest of the paper is organized as follows. Section II reviews the developments in OSCs and charging circuits. The characteristics of selected OPV module, obtained using the standard solar simulator, are also shown. The experimental results are presented in Section III. The charging circuit details and experimental results are explained in Section IV. The proposed charger has been developed and its performance is presented in Section V.

II. STATE OF ART OF ORGANIC SOLAR CELLS

A. Organic solar cells

Organic solar cells (OSCs) have gathered greater attention of scientists and researchers in the last couple of decades for their ability to be fabricated on large area flexible substrates using very cheap solution-processing techniques [3,4]. The first OSC was demonstrated in 1986 with power conversion efficiency (PCE) of 1 % [5] and recently a PCE of over 14 % has been reported, which is almost comparable with that of some of the commercial inorganic solar cells available in the market [6]. Numerous molecular and device engineering techniques have been employed to increase the PCE of the OSCs. Basically, a bulk heterojunction OSC involves the use of donor and acceptor materials which are blended together to form nano-sized domains (approx. 20 nm) for efficient charge transport across the active material. The most commonly used donor polymers are P3HT, PTB7-Th, etc. and the small molecule acceptors are fullerene based

PCBM and nonfullerene based ITIC. The optimization of device PCE includes not only controlling the domain size but also tuning the crystallinity, molecular orientation and proper phase separation of the donor and acceptor material [7,8]. To achieve these targets, methods such as thermal annealing, solvent annealing and incorporation of additives have been established [9-11]. Continuous efforts are being made for the development of new donor and acceptor molecules for increasing the PCE. Other than the active layer material, water/alcohol soluble interfacial layers are also being designed and developed for improving the PCE and stability [12,13]. Various device architecture of OSC such as conventional, inverted, tandem and ternary has been demonstrated for improving the PCE further.

B. Existing charging techniques using organic solar cells

F. C. Krebs et al [14] have demonstrated an application of OSCs for lighting purposes. Flexible and semi-transparent polymer solar cell modules, fabricated using roll-to-roll process, were used to charge a lithium ion battery which, in turn, supplies current to light emitting diodes. The charging circuit was simple and only consisted of a commercially available blocking diode (1N4148). The purpose of such a diode is to block any current getting sourced from the battery into the solar cell. Using a 320 cm^2 active area module with an efficiency of 1 %, the charging time for a 105 mAh battery was found to be close to 2 hours. Recently, M. Song et. al. [15] have reported on the charging of lithium ion battery using P3HT and PC61BM bulk heterojunction solar cell semi-modules. They showed that voltage-time ($V-t$) characteristics of these OSCs followed a power law equation $V = \alpha t\beta$ where α and β are constants that depend on the number of solar cells in the module and their method of connection. The charging circuit consisted of only a diode. In a report on the industrialization of polymer solar cells, H. Lauritzen et al [16] have focused on two types of charging circuits for powering light sources. One of the circuits consisted of only a blocking Schottky diode (MMSD701) and the other consisted of a Zener diode. The report mentions that because of the low current production by the solar cells, the battery never get overcharged and this avoids the need for additional battery protection circuits. As can be seen from these examples, there is no standard circuit for charging of batteries using OSCs, especially for application in smart phones.

III. ORGANIC SOLAR CELL

The custom designed flexible solar cell has been purchased from [2] and used as received for developing the mobile battery charger. The solar module used in this work has ten cells connected in series having a total area of 112.5 cm^2 (Figure 1). The current density versus voltage (JV) characteristic curves were measured using Keithley 2400 source meter under argon atmosphere by illuminating the module with a solar simulator (AM 1.5 G, 100 mW/cm^2, Oriel Sol 3 A solar simulator, Newport). Short circuit current density (J_{sc}), open circuit voltage (V_{oc}), fill factor (FF) and PCE are calculated using the JV curve.

Figure 2 shows the experimental setup for characterizing organic solar PV module. The JV curve is given in Figure 3. Power versus voltage curve (Figure 4) has also been calculated from the JV curve and it was found that maximum power of approximately 50 mW has been observed at 3 V . A

V_{oc} of 5.43 V has been obtained. This data is estimated conservatively as approximately 80 % area of the module was illuminated under the solar simulator. The specifications of the organic solar module are summarized in Table I. This module has an efficiency of 2.22 %. Since the lamp output has a beam size that is lower than the module size, only a section of the module is illuminated. In such a case the J_{sc} and FF values will be grossly underestimated. Nevertheless the V_{oc} is close to the manufacturer data. This is because unlike J_{sc} and FF, V_{oc} is relatively is less affected the series resistance of the dark regions in the module.

Development of the charger using OPV module, has been divided into two parts - first the study of the smart phone's charging characteristics and second the external charging circuit required. It can be noted that smart phone has an internal charging circuit inside the phone. This circuit is different for different manufacturers and its design is usually unknown. The external charging circuit is the circuit that needs to be connected from outside the smart phone, to be able to charge it using the OPV module.

Fig. 1. Picture of the organic solar PV module

Fig. 2. Experimental setup for characterizing organic solar PV module

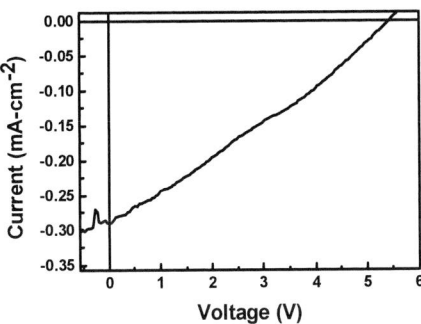

Fig. 3. *JV* characteristics at 1 sun intensity from partially illuminated module.

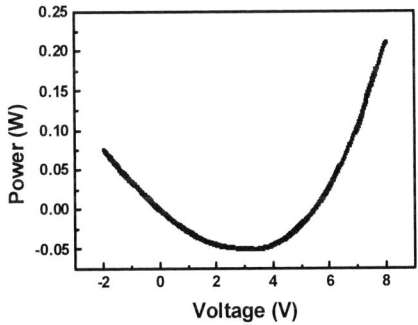

Fig. 4. *PV* characteristics at 1 sun intensity from partially illuminated module

TABLE I
SOLAR CELL SPECIFICATIONS

Area of solar cell	Output current	Output voltage	Output power
112.5 cm^2	70−90 mA	5.5 V	250 mW

IV. CHARGING

Development of the charger using OPV module, has been divided into two parts - first the study of the smart phone's charging characteristics and second the external charging circuit required. It can be noted that smart phone has an internal charging circuit inside the phone. This circuit is different for different manufacturers and its design is usually unknown. The external charging circuit is the circuit that needs to be connected from outside the smart phone, to be able to charge it using the OPV module.

A. Battery Characteristics

A smart phone by Lenovo has been used for developing the charger. Table II shows the specifications of the battery present inside the smart phone.

The internal charging circuit of smart phones can allow batteries to be charged either at constant current - constant current (CC) mode or with variable charging current - constant voltage (CV) mode. The current in CC mode is usually in the order of an ampere. From the results obtained in section III, it can be seen that the chosen PV module can charge the smart phone only in CV mode. The current

provided by PV module is in range on mA whereas CC mode requires current in range of A.

Figure 5 and Figure 6 show the change in battery voltage with time and charging current with time, respectively, during charging. The charging curve of battery shows that most of time battery is charging in constant current (CC) mode and only when phone is nearly fully charge, charging mode is switched to constant voltage (CV) mode. So, the PV module used in this paper can charge the smart phone only when state of charge (SOC) has reached to significant value.

TABLE II
BATTERY SPECIFICATIONS

Battery Technology	Nominal voltage	Limited charge voltage	Rated capacity
Li-ion polymer	3.8 V	4.35 V	2900 mAH

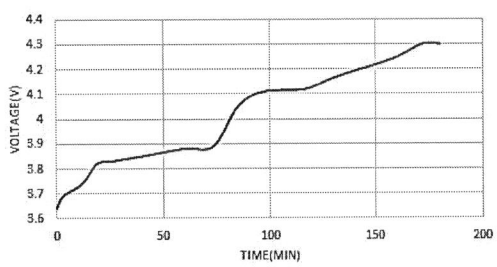

Fig. 5. Battery voltage vs time

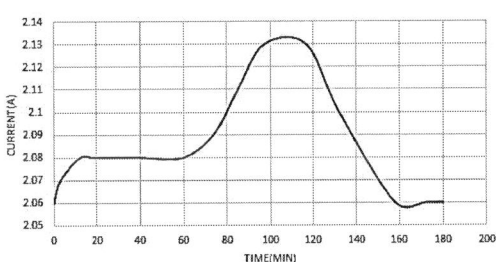

Fig. 6. Charging current vs time

B. Blocking diode

Blocking diode is used in series with solar PV modules to block the reverse flow of current - from battery to PV module. The reverse current may damage the PV module. For organic solar PV module, it is very important to choose appropriate blocking diode as the current provided by PV module is in range on mA. Figure 7(a) shows the circuit for charging smart phone using signal diode. The voltage drop v_D for signal diodes is usually 0.7 V . So, the output voltage v_o obtained is lesser than v_{PV} by 0.7 V .

Figure 7(b) shows a precision diode/rectifier as blocking diode. A precision diode/rectifier is formed by connecting an operational amplifier. It nullifies the voltage drop v_D across the diode and improves the diode characteristics.

978-1-5386-6624-1/18 $31.00 © 2018 IEEE

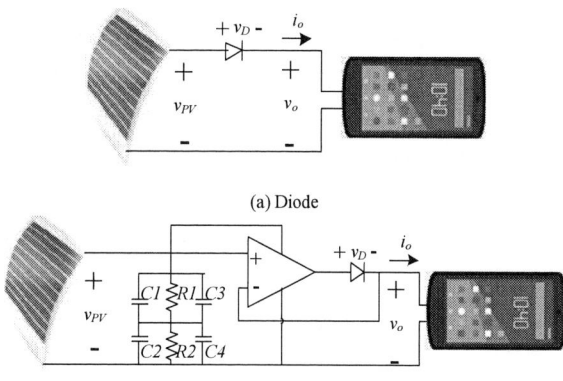

(a) Diode

(b) Precision Diode

Fig. 7. Blocking diode circuits: Use of (a) Signal diode (b) precision diode circuit using operation amplifier

The V_{oc} of OPV module has been measured for charging circuits formed using both signal diode and precision diode. The results obtained are shown in Table III. It can be seen that precision diode improved the output voltage by 0.7 V and due to this, the output current is changed from 10 mA to 9.1 mA.

TABLE III
BLOCKING DIODE RESULTS

Diode technology	Solar module voltage	Output voltage	Output current
Signal diode	6.1 V	5.4 V	10 mA
Precision diode	6.1 V	6.1 V	9.1 mA

V. DEVELOPED ORGANIC SOLAR PV CHARGER

Based on the experimental study and measurements obtained in Sections III and IV, a charging circuit with precision diode is used to develop the charger for smart phone using OPV module. Figure 8 shows the picture of the OPV module with the cover of smart phone. It can be seen that the chosen OPV module is of same size as the cover of the smart phone. The benefits of the proposed design are (i) reduced size of charger (ii) continued charging of the smart phone slowly whenever its internal circuit allows, and (iii) increase of the discharging time as the proposed charger can be connected to phone always.

Figure 9 shows the prototype of the proposed charger using phone sized OPV module and precision diode. Table IV presents the experimental measurements on the change in state of charge (SOC) when the developed charger is connected to the smart phone at different levels of SOC. It can be seen that when the SOC is 5 % then the OPV module is unable to charge the phone. This is because the internal phone circuit is in CC mode and does not allow to charge using mA of current. When the SOC is 60 % then the OPV module is able to charge the phone as the phone is in CV mode. It is observed that in an interval of 90 minutes the smart phone's SOC increased by 2 %. The use of precision diode improves the output power by 1.58 mW.

Fig. 8. Smart phone sized organic solar PV module

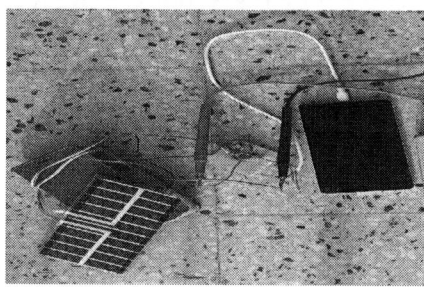

Fig. 9. Picture of charging the smart phone using organic solar PV module

TABLE IV
STATE OF CHARGING RESULTS

Initial SOC (%)	Charging time (minutes)	Change in SOC (%)
5	90	0
60	90	2

VI. CONCLUSION

A charger for smart phone, using OPV module of size of the smart phone and precision diode in charging circuit, has been proposed and developed in this paper. The JV and PV characteristics of the chosen OPV module have been obtained. It is found that the module gives an open circuit voltage of 6.1 V and a short circuit current of 28 mA under ambient conditions. The study of phone charging characteristics show that the phone charges in constant current mode till the state of charge becomes significant. After that the phone allows variable charging currents. So, organic PV modules can be used for charging smart phones only when SOC is significant. The voltage drop of blocking diode has been nullified by use of precision diode. The developed charger is able to charge the smart phone by 2 % in an interval of 90 minutes under daylight available during experiments.

REFERENCES

[1] S. Rafique, S. M. Abdullaha, K. Sulaimana and M. Iwamoto, "Fundamentals of bulk heterojunction organic solar cells: An overview of stability/degradation issues and strategies for improvement," Renew. Sust. Energ. Rev., vol. 84, pp. 43-53, March 2018.

[2] Organic solar PV module. [online] available at: https://infinitypv.com

[3] D. Gust, T. A. Moore, and A. L. Moore, "Solar fuels via artificial photosynthesis", Acc. Chem. Res., vol. 42 (12), pp 1890-98, 2009.

[4] N. S. Lewis, "Toward cost-effective solar energy use", Science, vol. 315 (5813), pp 798-801, 2007.

[5] L. T. Dou, Y. S. Liu, Z. R. Hong, G. Li, and Y Yang, "Low-Bandgap Near-IR Conjugated Polymers/Molecules for Organic Electronics", Chem. Rev., vol. 115, pp 12633, 2015.

[6] K. Zhang, Z. Hu, C. Sun, Z. Wu, F. Huang, and Y. Cao, "Toward Solution-Processed High-Performance Polymer Solar Cells: from Material Design to Device Engineering", Chem. Mater., 2017, vol. 29, pp 141, 2017.

[7] C.W. Tang, "Two-layer organic photovoltaic cell", Appl. Phys. Lett., vol. 48 (2), pp 183–185. 1986.

[8] S. Li, L. Ye, W. Zhao, H. Yan, B. Yang, D. Liu, W. Li, H. Ade, and J. Hou, "A Wide Band Gap Polymer with a Deep Highest Occupied Molecular Orbital Level Enables 14.2% Efficiency in Polymer Solar Cells", J. Am. Chem. Soc. vol. 140 (23), pp 7159-7167, 2018.

[9] L. Lu, T. Zheng, Q. Wu, A. M. Schneider, D. Zhao, and L. Yu, "Recent Advances in Bulk Heterojunction Polymer Solar Cells, Chem. Rev., vol. 115, pp 12666−12731, 2015.

[10] G. J. Hedley, A. Ruziicka's, and I. D. W. Samuel, "Light Harvesting for Organic Photovoltaics", Chem. Rev., vol. 117, pp 796−837, 2017.

[11] G. Li, V. Shrotriya, J. Huang, Y. Yao, T. Moriarty, K. Emery, and Y. Yang, "High-efficiency solution processable polymer photovoltaic cells by self-organization of polymer blends" Nat. Mater., vol. 4, pp 864−868, 2005.

[12] C. Duan, K. Zhang, C. Zhong, F. Huang and Y. Cao, "Recent advances in water/alcohol-soluble p-conjugated materials: new materials and growing applications in solar cells", Chem. Soc. Rev., vol. 42, pp 9071–9104, 2013.

[13] H.-L. Yip and A. K. Y. Jen, "Recent advances in solution-processed interfacial materials for efficient and stable polymer solar cells", Energy Environ. Sci., vol. 5, pp 5994–6011, 2012.

[14] F. C. Krebs, T. D. Nielsen, J. Fyenbo, M. Wadstrm and M. S. Pedersen, Manufacture, integration and demonstration of polymer solar cells in a lamp for the Lighting Africa initiative, Energy Environ. Sci., vol. 3, pp 512525, 2010.

[15] M. Song, S. Lee, D. Kim, C. Lee, J. Jeong, J. Seo, H. Kim, D. I. Song, D. Kim, and Y. Kim, Charging characteristics of lithium ion battery using semi-solar modules of polymer:fullerene solar cells, Energies, vol. 10, pp 1886, 2017.

[16] H. Lauritzen, J. Bork, R. B. Andersen, B. Bentzen, and F. C. Krebs, Industrialization of polymer solar cells phase 1, Technical University of Denmark, EUDP project 64009-0050, 2012.

[17] G. Gabian, J. Gamble, B. Blalock and D. Costinett, "Hybrid buck converter optimization and comparison for smart phone integrated battery chargers," 2018 IEEE Applied Power Electronics Conference and Exposition (APEC), San Antonio, TX, 2018, pp. 2148-2154.

[18] P. G. Horkos, E. Yammine and N. Karami, "Review on different charging techniques of lead-acid batteries," 2015 Third International Conference on Technological Advances in Electrical, Electronics and Computer Engineering (TAEECE), Beirut, 2015, pp. 27-32.

[19] H. A. Serhan and E. M. Ahmed, "Effect of the different charging techniques on battery life-time: Review," 2018 International Conference on Innovative Trends in Computer Engineering (ITCE), Aswan, 2018, pp. 421-426.

[20] Y. Kanesaka, K. Iida and J. Kondoh, "Analyses of string current and loss in blocking diodes in PV power system," 2016 19th International Conference on Electrical Machines and Systems (ICEMS), Chiba, 2016, pp. 1-4.

[21] J. C. Wiles and D. L. King, "Blocking diodes and fuses in low-voltage PV systems," Conference Record of the Twenty Sixth IEEE Photovoltaic Specialists Conference - 1997, Anaheim, CA, USA, 1997, pp. 11051108.

[22] A. A. D. T. Adikaari, D. M. N. M. Dissanayake and S. R. P. Silva, "OrganicInorganic Solar Cells: Recent Developments and Outlook," in IEEE Journal of Selected Topics in Quantum Electronics, vol. 16, no. 6, pp. 1595-1606, Nov.-Dec. 2010.

[23] S. Adami, N. Degrenne, C. Vollaire, B. Allard and F. Costa, "Ultralow power, low voltage, autonomous resonant DC-DC converter for low power applications," 4th International Conference on Power Engineering, Energy and Electrical Drives, Istanbul, 2013, pp. 1222-1228.

[24] S. Dietrich, L. Liao, F. Vanselow, R. Wunderlich and S. Heinen, "A 1mV voltage ripple 0.97mm2fully integrated low-power hybrid buck converter," 2013 Proceedings of the ESSCIRC (ESSCIRC), Bucharest, 2013, pp. 395-398.

[25] S. Adami, V. Marian, N. Degrenne, C. Vollaire, B. Allard and F. Costa, "Self-powered ultra-low power DC-DC converter for RF energy harvesting," 2012 IEEE Faible Tension Faible Consommation, Paris, 2012, pp. 1-4.

[26] S. G. Nassif-Khalil, S. Honarkhah and C. A. T. Salama, "Low voltage CMOS compatible power MOSFET for on-chip DC/DC converters," 12th International Symposium on Power Semiconductor Devices & ICs. Proceedings (Cat. No.00CH37094), Toulouse, France, 2000, pp. 43-46.

978-1-5386-6624-1/18 $31.00 © 2018 IEEE

2nd IEEE International conference on power Electronics, Intelligent Control and Energy systems (ICPEICES-2018)

Improve Collision in Highly Dense WiFi Environment

Vishal Bhargava
Department of Computer Science & Engineering
Delhi Technological University
Delhi, India
vishalbharg@gmail.com

N.S. Raghava
Department of Electronics & Communciation
Delhi Technological University
Delhi, India
nsraghava@dce.ac.in

Abstract— **Wi-Fi works on the principle of IEEE 802.11 based carrier sense multiple access –collision avoidance (CSMA/CA) to transmit packets and distributed coordination function (DCF) protocol based Inter-frame spacing uses to make the gap between two frames. DCF protocol functioning & its performance depends on the number of participating stations and contention window (CW) size. Several algorithms talk about contention windows effect on the wireless performance like G. Bianchi models suggest in WLAN network the optimal value of the minimum CW is N √ 2Tc, where N is the number of participating station in the network and Tc is the collision time [1]. the network also gives the best performance in terms of throughput when optimal contention windows select [2]. In this paper, we describe the simulation of contention windows size effect on throughput with the number of station & in case of high dense overlap basic service set (OBSS) how does can improve the performance of the overall network and reduce collision.**

Keywords—Contention, 802.11e, QOS, OBSS, Contention windows.

I. INTRODUCTION

802.11 uses inter-frame spacing for all high-level frames to avoid the collision and a device access medium randomly, CW (contention window) mechanism introduce into the distributed wireless medium. Inter-frame spacing provides priority access for a pick transmission of frames necessary for correct network operation. 802.11 defined several types of Interframe spacing. DCF based Interframe spacing (IFS) discussed here:

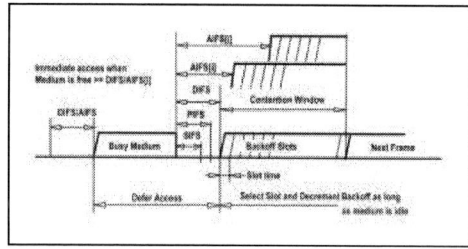

Fig. 1: IFS Relationship [17]

The Short InterFrame Spacing (SIFS) worth is employed for control frame (like ACK & CTS) for high priority ack based data; DCF InterFrame Spacing (DIFS) is employed for non-QoS data frames; Arbitrated InterFrame Spacing (AIFS) is employed for QoS information frames and is variable supported the WMM Access class (AC) to that the frame is allotted & these change according to modulation.

Before each frame transmission, Wi-Fi stations wait for a random timer which varies according to the contention window. As we discussed Wi-Fi works on CSMA/CD. A different device can also send data at the same time, this is called an external collision. When an external collision happens station do random back-off & increase contention windows (Below figure):

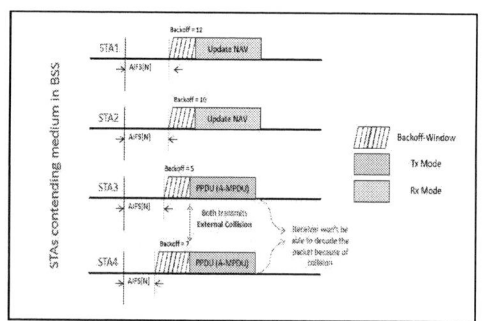

Fig. 2: External Collision

If a collision happens (as the station does not receive acknowledgment frame in decided duration), then the transmission stations double the contention window size and if again sequentially collision happens it's done till contention window maximum size. This can be known as Truncated Binary Exponential Backoff. The initial tiny contention window size is marked as contention Window Minimum (CWMin) and most boundary size is marked as contention Window maximum (CWMax). Once WMM QoS is in use, a mathematical advantage provides to frame in forms of inter-frame spacing and contention window size which varies according to frame category. This technique of probability-based medium competition introduces an outsized quantity of network overhead to reduce the likelihood of a frame collision.

The 802.11e specification defines a priority scheme for the different type of traffic. Wi-Fi have 4 access categories:

1. Background data

2. Best efforts

3. Voice

4. Video

978-1-5386-6624-1/18 $31.00 © 2018 IEEE   734

Actually, this priority queue creates to prioritize data frames of every station. And these queues named with above 4 access categories according to data type. So, depends the type of data is placed in the respective queue as if it is video data it will be placed in 4th queue & categorized as video access categories.

If a device has a certain type of data which is going to transmit at the same time [14] – an internal collision can occur.

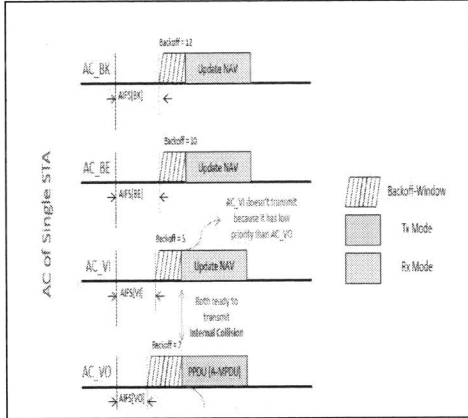

Fig 3: Internal Collison

The amended IEEE 802.11 wireless protocol specification suggest using a priority theme that may give up to eight priority categories for wi-fi traffic. This amendment added in IEEE 802.11e wi-fi specification to support QoS. Wi-Fi Alliance also provides certification test to test this type of device which supports 802.11e through the Wi-Fi transmission (WMM) certification method.

To support QoS (Quality of Service) in wi-fi data 802.11e wi-fi spec defines eight user priorities. These eight user priorities are divided into four specific Access classes (AC) referred as Voice, Video, Best Effort, and Background category and every access category contains two totally different user priorities of the same type of class.

Some major changes done in operation of 802.11 to improve the operation of initial 802.11 Distributed Coordination operate (DCF) which is now currently referred as EDCA (Enhanced Distributed Coordination Access). The major changes are:

- These four priority queues are subject to per station, every queue treated as station.

- AIFS i.e. Arbitrated Inter-Frame Spacing values which are introduced to give priority to one access category over another one and it is applicable for all information and management frames and it's enhanced the DCF capabilities.

- Random Backoff time also different for every pruority queue which is a drive from CWmin and CWmax values.

The performance of IEEE 802.11 distributed coordination operates (DCF) protocol directly proportional to a number of challenger wi-fi stations and also the contention window size [2] (CW).

First, inter-frame spacing establishes baseline intervals that sure forms of frames are needed to attend before having the ability to transmit.

The Backoff time & IFS (inter-frame spacing) calculation done independently for each queue and parallely decremented while system wait for ideal environment. If two queue want to transmit at same time an internal collision occurs, in this case, higher priority will granted access by MAC (like Voice over Best effort) and another queue will take this as a physical collision, stop transmission & it will increment its counter of retry, and according to binary exponential backoff increase their CW values. In this manner a queue work as a self-oriented station. The following figure illustrates these priority queues:

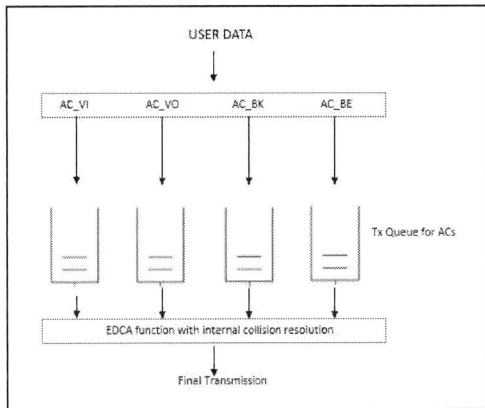

Fig. 4: Frame with the priority queue

This organization of this paper is as follows: Section II presents related work and research objectives. Our Simulation and Test Results & it's conclusion discussed in section III. Proposed work & feedback described in Section IV and finally, conclusions and future work are drawn in section V.

II. RELATED WORK AND RESEARCH OBJECTIVES

Several paper talks performance of different data transmission to support the QoS (define in IEEE 802.11e MAC protocol [15]) in WLANs and their results show that the how different QoS parameter can improve the throughput & quality of service transmission in the Wi-Fi network [2,3,4,12].

Several new techniques and algorithm proposed to minimize collision rate in the noisy environment, whether it's new backoff algorithm [13] or accurate delay distribution of

DCF functionality [10]. The researcher also proposed new QoS architecture (using 802.11e EDCA) in which adaptive CW used to increase Performance in the noisy environment. All the above method consider exiting Wi-Fi protocol & do amendments in it. Inspired from the above, motivation was drawn to propose an addition of Wi-Fi frame which easies to deploy in vendor's network and increase throughput in the high noise environment.

III. SIMULATION AND RESULTS

In this section, we show our simulator setup and simulation results. We used MATLAB WLAN toolbox to simulate the results. We assume all the STA's in the BSS have frames to transmit all the time and we have taken AIFSN value for each access category:

TABLE I.   ACCESS CATEGORY AIFSN DATA

Access Category	AIFSN Value
Background data	9
Best effort data	6
Video data	3
Voice data	2

We used below formula to calculate collision rate:

Collision rate = external collision / total transmission

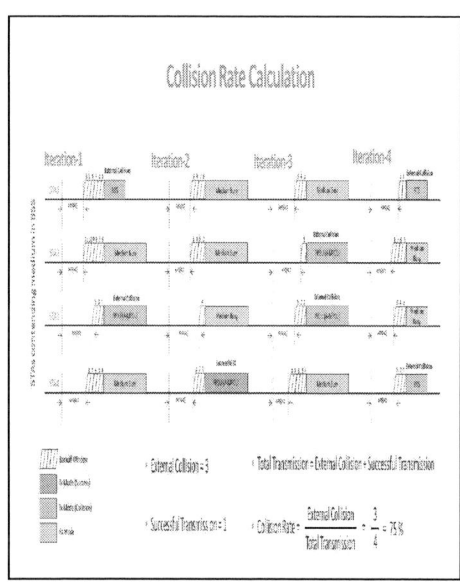

Fig. 5: Collision Rate

A. Collision Rate Vs [CWmin, CWmax]

We have simulated Collision rate according to a number of station & with different contention windows on the basis of access categories. Below graph showing the relationship between them.

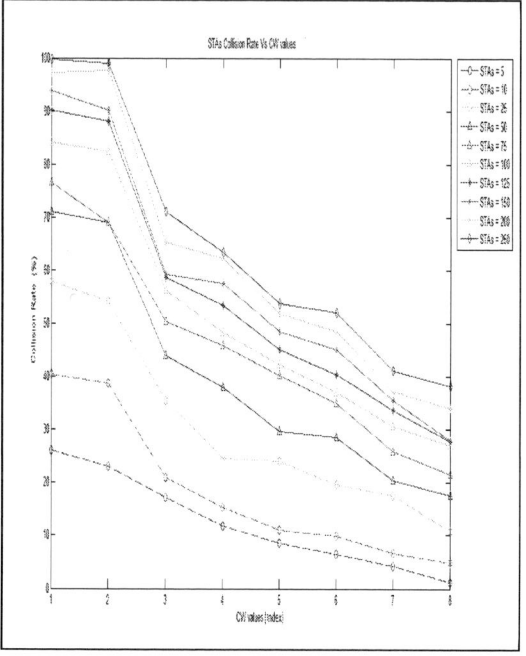

Fig 6 : STAs Collision Rate vs CW values

B. Conclusion derive from the above test

- As the number of STAs associated with the BSS increases, the collision rate of the BSS decreases, as [CWmin, CWmax] value increases.
- Thus, the AP should increase the [CWmin, CWmax] values to avoid a high collision rate in the highly congested environment.
- As per the Simulation results, the [CWmin, CWmax] values to have collision rate less than 40 % is:

TABLE II.   CW WITH DIFFERENT ACCESS CATEGORY

Cwmin, Cwmax	AC_BK	AC_BE	AC_VI	AC_VO
Index 1	15,1023	15,1023	7,15	3,7
Index 2	15,1023	15,1023	7,31	7,7
Index 3	31,1023	31,1023	15,31	7,15
Index 4	31,1023	31,1023	15,63	15,15
Index 5	63,1023	63,1023	15,63	15,31
Index 6	63,1023	63,1023	31,127	15,31
Index 7	63,1023	63,1023	31,127	31,63
Index 8	127,1023	127,1023	63,255	31,63

TABLE III.   CW FOR DIFFERENT ACCESS CATEGORY WITH THE NUMBER OF STATIONS

Number of Stations	Cwmin, Cwmax			
	AC_BK	AC_BE	AC_VI	AC_VO
5	15,1023	15,1023	15,1023	3,7
10	31,1023	31,1023	31,1023	7,15
25	31,1023	31,1023	31,1023	15,15
50	63,1023	63,1023	63,1023	15,31

75	63,1023	63,1023	63,1023	31,63
100	63,1023	63,1023	63,1023	31,63
125	127,1023	127,1023	63,255	31,63
150	127,1023	127,1023	63,255	31,63
200	127,1023	127,1023	63,255	31,63
250	127,1023	127,1023	63,255	31,63

This is the feedback we have derived from our test:

- No matter single BSS or multiple OBSS, more STAs want to TX at the same time will induce more collisions. The material is still correct for MBSS.

- Most accurate indicator to apply adequate CW setting:
  - Collision rate estimation

  - STA number linked to AP

- Seems too rough, STAs connected to AP did not means it always has traffic to content w/ other STAs.
- It cannot reflect the OBSS traffic contention.
- RTS retry statistics
  - Usually, we use a more robust rate to send RTS. The RTS retry statistics might be more close to "collision rate".
  - The "collision rate" estimation proposal in your material seems to be "STA number linked to AP".
  - Some STAs (which already connected to AP) might not adjust their EDCA parameters when AP announces new setting in Beacon after their connection.

IV. PROPOSED WORK

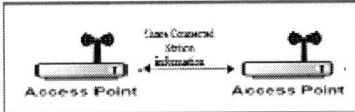

Fig. 6: Access Point Sharing information with each other

We want Access Point share information about a number of connected stations to itself, so we have added an IE (Information Element) in the beacon of the access point, which shows currently connected the station to AP. Current IEEE 802.11 specification, does not have such type of IE.

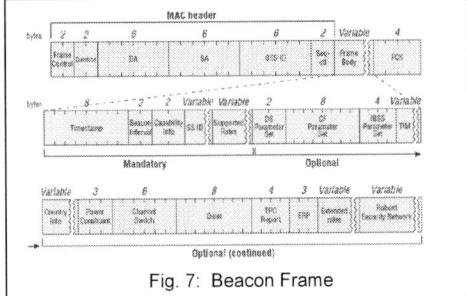

Fig. 7: Beacon Frame

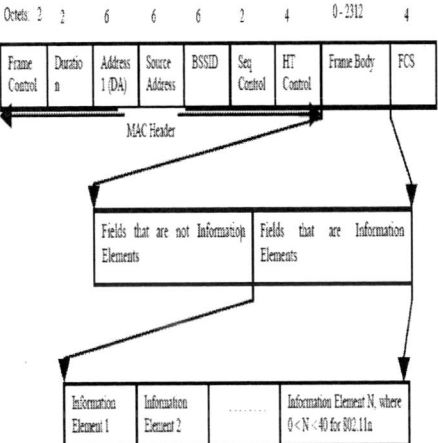

Fig. 8: Information Element (IE) in Beacon Frame

Information element shows the number of the station connected to AP and this beacon can listen via another APs in the environment, so they can set their contention window according to this information.

As this IE is not defined in IEEE 802.11 specification, here is our assumption "We should have all APs in the enterprise network, then, we can have proprietary IE in Beacon/Probe Response to indicate a number of STAs present in each BSS." Using this approach all AP in the dense environment can set CW windows as we discussed in "conclusion driver from our test" and this approach is applicable for BSS and MBSS.

V. FUTURE WORK & CONCLUSION

A. Future Work

- The current approach does not consider station power policy; whether AP should indicate only awake station, not the station which is in power save mode.
- And if suppose 75% station is only awakened at the time than can AP announce less number of the station to other AP & set TIM bit for only that number of station.
- Consider a BSS with around 128 STAs out of which 100 are in PS, and there are data buffered at AP for 75 STAs. Then the AP need to set TIM bit in the Beacon frames for all the 75 STAs.
- Since all 75 STAs tries to contend the channel at the same time, causing high collision rate.
- Thus, if Access Point sets TIM bit only for few STAs, then the collision rate can be reduced.

B. Conclusion

In this paper, from the simulation result, the effect of CW value on overall Wi-Fi throughput is shown. It is shown that how to optimize CW value can work well in the high

978-1-5386-6624-1/18 $31.00 © 2018 IEEE   737

dense environment. And Access point self is not capable enough to do that without understanding the whole environment. So AP also needs to talk to each other as they are sharing a common environment.

it is described that how an AP shared information about the connected station with other AP in OBSS so a collision can reduce in the highly dense environment. Currently, IEEE 802.11 specification does not define any IE so an AP can talk to other AP.

REFERENCES

[1] G. Bianchi, "Performance Analysis of the IEEE 802.11 Distributed Coordination Function," IEEE Journal on Selected Areas in Communications, vol. 18, vo. 3, March 2000.

[2] M.A. Parvej, S. Chowdhury, N. I. Hia, and M. F. Uddin, "Capture effect on the optimal contention window in IEEE 802.11 based WLANs" in Proc. IEEE ICIEV, 2013.

[3] Riyadh Qashi, Martin Bogdan & Klaus Hiinssgen "Case Study: The Effect of Variable Priority Parameters on the QoS of WLANs IEEE 802.11e EDCF" Communication Software and Networks (ICCSN), 2011 IEEE 3rd International Conference.

[4] R. Qashi M. Bogdan and K. Hanssgen "Performance analysis of WLANs for the IEEE 802.11 EDCF in real-time applications " in International Academic Conference of Young Scientists "Computer Science and Engineering 2010.

[5] P. Pocta P. Kortis and M. Vaculik "Impact of background traffic on speech quality inVoWLAN " Adv. Multimedia vol. 2007 no. 1 pp. 1-9 2007. [6] S. Sehrawat R. P. Bora and D. Harihar "Performance analysis of QoS supported by enhanced distributed channel access (edca) mechanism in IEEE 802.11e " in IMECS 2006 pp. 943-948.

[6] M.-S. Kim I-P. Ryu T. Byun and K.-J. Han "Throughput analysis of IEEE 802.11e edca protocol " in High-Speed Networks and Multimedia Communications vol. 3079. Springer Berlin-Heidelberg 20 04 pp. 579-588.

[7] S.-C. Wang and A. Helmy "Performance limits and analysis of contention-based IEEE 802.11 mac " Local Computer Networks Annual IEEE Conference on vol. 0 pp. 418-425 2006.

[8] I. Inan F. Keceli and E. Ayanoglu "Analysis of the 802.11e enhanced distributed channel access function " CoRR vol. abs/0704.1833 2007.

[9] S. Choi "On the performance characteristics of WLANs: revisited " in Proc. ACM SIGMETRICS05 2005 pp. 97-108.

[10] Shao-Cheng Wang and A. Helmy. "Performance Limits and Analysis of Contention-based IEEE 802.11 MAC". Local Computer Networks Annual IEEE Conference on. Vol. 0 pp. 418-425. 2006. USA .

[11] H. L. Vu and T. Sakurai "Accurate delay distribution for IEEE 802.11 DCF" IEEE Comms. Letters vol. 10 no. 4 pp. 317-319 Apr. 2006.

[12] J. Sengupta and G. S. Grewal "Performance evaluation of IEEE 802.11 mac layer in supporting delay sensitive services " International Journal of Wireless and Mobile Networks no. 1 2010.

[13] T. Nadeem and A. Agrawala "IEEE 802.11 DCF enhancements for noisy environments" in Proceedings of the 15th IEEE International Symposium on Personal Indoor andMobile Radio Communications (PIMRC 2004) vol. 1 pp. 93-97 Sept. 2004.

[14] A. Iyer, C. Rosenberg and A. Karnik, "What is the right model for Wireless channel Interference?" IEEE/ACM Trans. on Networking, vol. 8, no. 5, May 2009

[15] IEEE standard for Wireless LAN Medium Access Control (MAC) and Physical Layer (PHY) Specifications, ISO/IEC 8802-11:2012(E), 2012.

[16] R. Achary, V. Vaityanathan, P.R. Chellaih, N. Srinivasan, "A new QoS architecture for performance enhancement of IEEE 802.11e EDCA by contention window adaption", Proc. of the Fourth International Conference on Computational Intelligence and Communication Networks, pp. 74-78, Nov. 2012.

[17] Chong Han, Student Member, IEEE, Mehrdad Dianati, Member, IEEE, Rahim Tafazolli, Senior Member, IEEE, Ralf Kernchen, and Xuemin (Sherman) Shen, Fellow, IEEE, "Analytical Study of the IEEE 802.11p MAC Sub-layer in Vehicular Networks", Repository from University of Surry, U.K.

2nd IEEE International conference on power Electronics, Intelligent Control and Energy systems (ICPEICES-2018)

Microwave Circuits Characterization on Carbon Fibre (CFRP) based Carrier plates

Kamaljeet Singh[1], H R Kansara[2], V.Venkatesh[1], A V Nirmal[1], S V Sharma[1]

[1] U R Rao Satellite Centre, Bangalore

[2] Space Application Centre, Ahemdabad

kamaljs@isac.gov.in

Abstract— **This article details the indigenous development of CFRP based carrier plates for possible replacement of standard Kovar based carrier plates employed in front-end RF circuits at microwave frequencies. The development encompasses fabrication of CFRP based carrier plates, plating of the assembly, temperature trials, attachment and assembly of the substrates along with the RF characterization. The front-end of RF circuits consists amplifier, filter which are realized on alumina substrate. The assembly is attached to the developed gold plated CFRP carrier plates. The resulting structure achieved six times weight reduction compared to standard topology without any performance degradation. The development encompasses process details, comparative analysis, various trials along with performance characterization.**

Keywords- Carbon fibre reinforced plastic (CFRP),thin film process, radio frequency, filter

I. INTRODUCTION

The scope for continual miniaturization and weight reduction are important consideration for the electronic systems. Ease of processing, repeatability, cost reduction are added considerations which are to be addressed for newly proposed system in the era of 5G and SOC. CFRP (carbon fibre reinforced plastic) based material are the strong contender to replace mechanical housing traditionally using alumina, kovar or stainless steel [1]. This technology is blend of carbon fibre with the resin material which is much lighter and robust. Also mechanical and thermal considerations such as yield strength, stress, fracture, density, poisson ratio, thermal expansion are some of the important parameters. In satellite operations, apart from environmental conditions, the electrical performance is prime consideration before substitution of any new material or process. These aspects become critical at radio frequency as any thermal mismatch, vibration, shock should not result in stress or crack which is detrimental to the overall chain failure. Front end of TT&C receiver consists of LNA and filter topologies which are realized on alumina substrate [2]. Realized topologies are attached to the Kovar carrier plate to ensure proper

grounding and ensuing TCE match. The CFRP based topologies are reported to be used in detectors and antenna applications [3].In the present development CFRP based carrier plate is developed for the RF circuits attachment. Apart from attachment, other processes such as ribbon and wire bond, gap cap, resistors are important processes due to thermal considerations at various stages.

This article is based on the investigation of the alternative material which can replace the existing approach without deviating process or compromising electrical performances. In this article C-band front end realization and implementation on CFRP is taken as a test case. The circuit consists of extension line, two stage LNA and band pass filter topologies which needs three carrier plates of varies sizes. Key parameter associated with new material is the associated thermal stress due to CTE mismatch. This aspect is taken into account and the realized filter assembly on CFRP is presently taken for testing and environmental test performed on the same which is detailed in this article. The development is segregated into various steps as below:

- CFRP based carrier plates realization
- Gold plating of the carrier plates
- Process for substrates attachment
- Ribbon / wire bond and connector soldering
- Electrical testing
- Environmental testing

This article details the methodology adopted for realization of CFRP material, attachment processes, filter characterization and performance analysis with space borne environmental conditions.

II. CHARACTERIZATION

Standard Kovar based plates are the important module for attaching the microstrip based circuits to the package. The main consideration is to provide rigidity apart from the electrical grounding without propagating the stress to the package. The main consideration of alternative material is to posses similar CTE of the attached substrate so as to avoids stress related failures.

CFRP based carrier plates is having comparable CTE to the standard Kovar based carrier plate.It can be the possible replacement due to light weight and ease of processing. Table-1 provides the basic comparison

978-1-5386-6624-1/18 $31.00 © 2018 IEEE   739

Table1: Comparative table of various properties

Parameters	CFRP	Kovar
Material	Carbon fibre	Fe-Co-Ni
CTE(inch/inch °C)	2-3	5
Thermal cond.(W/m K)	24	~50
Plating	Ni-Au	Ni-Au

Exposure of CFRP material to various temperature are carried out in-house to check the feasibility of enduring temperature.

A. Thermal Trials

In the present trials two different types of CFRP materials are developed and are subjected to heat treatment. It was found out that continual exposure 250°C temperatures for 5 min may degrade the fibre composition. Fig 1 shows the material before and after thermal treatment.

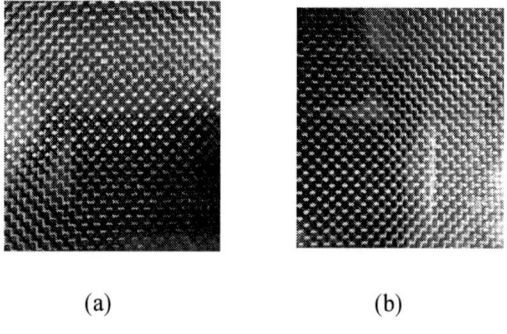

(a)   (b)

Fig1: CFRP material (a) before heat treatment (b) after heat treatment (250C,> 5 min)

Two different variants of the CFRP material is fabricated: standard topology and chopped fibre topology for better thermal conductivity. Sample of developed carrier plate on CFRP is shown in Fig 2which is coated with Ni-Au (300nm-5 um) using the RF sputtering unit.

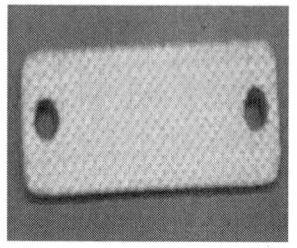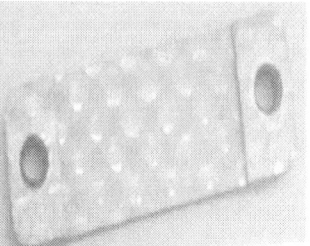

Fig2 Realized gold plated CFRP based carrier plate showing top and bottom side

After carrying out the temperature trials (Table-2) the further process of attachment is carried out on the CFRP based carrier plate. Both hot gun and oven are employed to deduce the effect of temperature and time exposure. The trials are briefed as shown in Table-2

Table2: Heater exposure trials on CFRP

Temperature	Timings	Remarks
100°C	5-10 min	No degradation
200°C	5-10 min	No degradation
220°C	5 min	No degradation
250°C	1-2 min	No degradation
250°C	>5 min	Degradation starts

Standard processes are around 200°C with exposure duration of less than 5 min of duration, so the same is implemented for the attachment of various components as standard processes. The fabricated gold plated carrier plates on CFRP are compared with the standard kovar based carrier plates. Table-3 shows the comparative analysis of the same.

Table3: Comparative analysis

Nomenclature	Kovar plate	CFRP
NR001(extension card) **5mm x 20 mm**	3.53 gms	**0.49 gms**
NR102,103,104 (LNA & filter) **30 mm x 40 mm**	8.59 gms	**1.49** gms

A comparative analysis of the standard and proposed material is being carried out and as per Table-3, the weight is considerably reduced using CFRP without compromising other parameters. This is due to the specific gravity which is at much lesser compared to Kovar (8.3 g/cc).

B. Assembly Trials

The attachments of the patterned alumina substrate (tanδ<0.0005) carried out using the solder perform (Sn:Pb) method on to the developed gold plated CFRP based carrier plate. The attachments process is carried out in two steps: tinning and simultaneously heating the perform along with etched substrate on the hot plate for 2 min at 200° C. Further components attachments namely FET device, wire bonding, chip capacitors, resistors are mounted as per standard processes. Standard amplifier and filter at microwave frequency are taken for establishment of mounting and attachment process. Fig 4 shows the complete assembly of 2-stage amplifier circuit where device, gap caps, resistors are attached on CFRP based carrier plate apart from wire and ribbon bond attachments to CFRP body [5].

 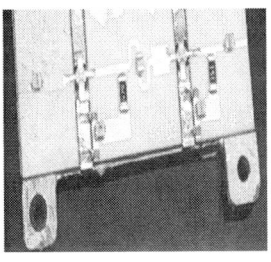

(a) bare CFRP   (b) assembled structure

Fig4: Amplifier assembly attachment on gold plated CFPR carrier plate   (courtesy: sub-system)

III. EXPERIMENTAL RESULTS AND DISCUSSION

The edge coupled filter topology [6] at C-band is fabricated on alumina substrate and attached to standard as well as on CFRP based carrier plate using In:Pb solder instead of gold germanium solder[7]. The attached filter along with connectors assembly on CFRP are shown in Fig 5.

Fig5: Filter assembly on CFRP

Figure 6 shows the comparative analysis of RF performance

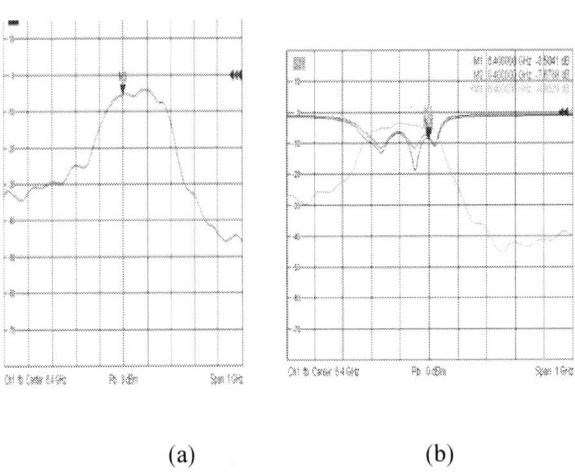

(a)   (b)

Fig6: Filter characterization at microwave (a) standard Kovar based carrier plate(b) CFRP based carrier plate

The realized filter is having the same signature as actual filter and no deviation in RF performance noticed. Further the filter is subjected to the temperature cycles without any performance degradation. Amplifier is under characterization and results will be presented in the conference.

The CFRP material is used for the complete front end package realization using 3-D printing techniques and tight tolerances are achieved which are shown in Fig 7.

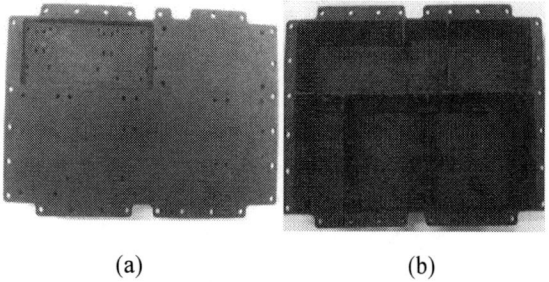

(a)	(b)

Fig7: CFRP based RF package top deck (a) Bottom and (b) Top view

The package weight is reduced by more than 60% with the proposed material and further trials related to plating and achieving tight tolerances are to be taken up. Slight increase in the warpage and dimensional inaccuracies are being looked into so as to address in future trials.

IV.   DISCUSSIONS

The indigenous developments of CFRP based carrier plate are demonstrated to be six times lighter than the existing topology. Very high stiffness and strength, tailorable thermal properties, low density, reduced thermal stress, warpage and distortion associated with this development results in increased reliability. The development encompasses all indigenous processes for the same and can be easily reproducible. The added advantage is the faster turnaround time, ease of re-work, corrosion resistance and can reduce E/M emissions because of electrical resistivity of CFRP. As the RF performance and processes are not altered so this can be replicated in all kovar based carrier plate for microwave circuits. Authors believe that this development is having potential to replace all the kovar based carrier plates presently employed in C-,X-,Ku-bands at transponders and payloads applications resulting in considerable weight reduction.

ACKNOWLEDGMENT

The authors hereby acknowledge the support and encouragement from various entities of URSC,SAC and industry partner. Special thanks to sub-system and TnFD in supporting the activities.

References:
[1]A.Ahmed,BTavakol,R.Das,R.Doven,P.Roozbehjavan,B.Minaie, " Study of thermal expansion in carbon fibre reinforced polymer composites,"

[2] Kamaljeet Singh, A V Nirmal," Ku-band low noise amplifier design approach using HEMT device,"IJRECE, Vol4,Issue4,2016

[3] *State of the art practices & Technology trends*, SAC,Document Pt-9/10,2013

[4]EECON 2017 Presentation Compilation, ISAC Digital Library,Bangalore-2017

[5] D M Pozar,Microwave Engineering ,Wiley Interscience 2010

[6] J G Hon,M J Lancaster', Microtrip Filters for RF/Microwave Applications, John Wiley & Sons 2001

[7] D.Matsuoka,M.Smith,O S Es-said," The attachement of thin alumina ceramic plates to Kovar using gold germanium solder," Engineering Failure Analysis,Vol 5,No1,pp 69-75,1998

Morphology Control of Mixed Halide Perovskite for its Application in Low-cost Thin Film Transistor

Anwesha Choudhury
Centre for Nanotechnology
Indian Institute of Technology Guwahati
Guwahati, India
anwes176153007@iitg.ac.in

Priyanka Dogra
Department of Electronics and Communication
Engineering Maharaja Surajmal Institute of
Technology, Delhi
Delhi, India
priyankamsit96@gmail.com

Biki Teron
Centre for Nanotechnology
Indian Institute of Technology Guwahati

Guwahati, India
bikiteron@iitg.ac.in

Ritesh Kant Gupta
Centre for Nanotechnology
Indian Institute of Technology Guwahati
Guwahati, India
riteshkantgupta@iitg.ac.in

Anamika Dey
Centre for Nanotechnology
Indian Institute of Technology Guwahati
Guwahati, India
d.anamika2012@gmail.com

Ashish Singh
Centre for Nanotechnology
Indian Institute of Technology Guwahati
Guwahati, India
ashish.iitg17@gmail.com

Rabindranath Garai
Department of Chemistry
Indian Institute of Technology Guwahati
Guwahati, India
rgarai02@gmail.com

Parameswar Krishnan Iyer
Centre for Nanotechnology,
Department of Chemistry
Indian Institute of Technology Guwahati
pki@iitg.ac.in

Abstract—**Organic-inorganic perovskite has gained popularity in recent times because of its high carrier mobility, easy and low cost processing for various optoelectronic application. In this paper we have demonstrated the variation of the growth and nucleation of perovskite thin film on various dielectric layers. Thereafter we report the fabrication and characterization of perovskite FET. We propose that growth of the perovskite depends on the dielectric layer on which it is coated. MACl and PbI$_2$ based perovskite forms a stable film and the nucleation depends on the aging of the precursor solution. Hence by controlling of the perovskite film morphology, a high performing device can be obtained. Hence a low cost and well performing TFT is reported.**

Index Terms—**nucleation, perovskite, dielectric, annealing, morphology.**

I. INTRODUCTION

Organometal halide perovskite are categorized as rising materials for optoelectronics device applications owing to their excellent photon absorption and better luminescence. Perovskite materials show their fascinating properties such as tunable bandgap [1], by changing the halide composition, high photoluminescence quantum yield [2], low cost fabrication [3], high performance carrier mobility and low temperature processing are achieved [4]. Perovskite material has great

advantages due to presence and combination of its inorganic and organic parts. The carrier mobility of perovskite materials is found to be high owing to its inorganic part present in the perovskite material as well as easily processable in solution process and low temperature owing to its organic part. It has now become an emerging material in solar cell owing to its weak bound exciton formation and as a result the charge separation process can easily take place and collected over in both the electrodes. Perovskite solar cell has achieved best performance in terms of its efficiency as compared to conventional organic materials used [5-7].

Similarly perovskite based materials as a semiconducting channel has found application in field effect transistors (FETs) or thin film transistors (TFTs) [8-10]. Though the highest carrier mobility achieved in perovskite based FET is 396.2 cm^{2}V^{-1}s^{-1} [11] using methyl ammonium lead iodide (CH$_3$NH$_3$PbI$_3$) as active layer but mixed halide pervoskite are very less explored as an active layer for transistor.

Till date it is a great challenge to the researchers to obtained superior perovskite film morphology In this paper, we are reporting the variation of morphology and nucleation of perovskite on different dielectric layers, different aging time, annealing time and annealing temperature.

II. EXPERIMENTAL

A. Thin film Morphology Study

First, Glass slides were cut into dimensions of approximately (1 × 0.5 inch) and were cleaned sequentially by ultrasonic bath of soap solution, DI water, acetone, ethanol, isopropanol and then dried. Then dielectric layer was spin coated on the cleaned glass substrates. On the top of this dielectric layer, perovskite precursor solution D1 (14 mg of MAI and 100 mg of PbI$_2$ were added in 1 mL of Dimethylformamide solvent (DMF) and kept on stirring for 24 hrs at 70°C.) were spin coated. To investigate the role of various dielectrics, different precursor composition, the effect of aging, annealing time, temperature in the growth of perovskite film and the TFT device performance; perovskite thin films were made in different condition.

1) Perovskite growth on different dielectric layers: A solution of 3% PVA in DI water, PS in toluene and PMMA in Anisole were made and spin coated at 3000 rpm on different glass substrates. Above which the perovskite precursor solution were spin coated at 4000 rpm for 30 s and annealed for 30 mins at 70°C.

2) Film preparation at different annealing temperature: Same D1 precursor solution was prepared again and spin coated on PMMA dielectric layer at 4000 rpm for 30 s and annealed at 60°C, 90°C, 120°C respectively for 1 hour.

3) Film preparation at different annealing time: D1 precursor solution was spin coated on PMMA coated substrates and annealed at 70°C for 30 mins and 1 hr respectively.

4) Film preparation with different precursor composition: Three solutions were prepared of ratio 1:1, 1:2 and 2:1 of MACl and PbI$_2$. These Solution were then spin coated on PMMA coated substrates at 4000 rpm for 30 s and annealed for 30 mins at 70°C.

   (a) Solution R1: 14.6 mg of MACl and 100 mg of PbI$_2$ were added in 1 mL of Dimethylformamide solvent (DMF) and kept on stirring for 24 hrs at 70°C.

   (b) Solution R2: 7.3 mg of MACl and 100 mg of PbI$_2$ were added in 1 mL of Dimethylformamide solvent (DMF) and kept on stirring for 24 hrs at 70°C.

   (c) Solution R3: 14.6 mg of MACl and 100 mg of PbI$_2$ were added in 1 mL of Dimethylformamide solvent (DMF) and kept on stirring for 24 hrs at 70°C

5) Film preparation with different precursor solution composition: Three different solutions were made spin coated on dielectric coated substrates at 4000 rpm for 30 s and annealed for 30 mins at 70°C.

   (a) Solution C1: 14 mg of MACl and 100 mg of PbI$_2$ were added in 1 mL of Dimethylformamide solvent (DMF) and kept on stirring for 24 hrs at 70°C.

   (b) Solution C2: 21.96 mg of MACl and 150 mg of PbI$_2$ were added in 1 mL of Dimethylformamide solvent (DMF) and kept on stirring for 24 hrs at 70°C.

   (c) Solution C3: 29.29 mg of MACl and 200 mg of PbI$_2$ were added in 1 mL of Dimethylformamide solvent (DMF) and kept on stirring for 24 hrs at 70°C.

B. Thin Film Characterization

Images of the perovskite film were captured using Optical Microscope (Leica DM 2500P with QJ CAM FAST1394 camera)and field-emission scanning electron microscopy(JSM-7610F, JEOL) to study the surface morphology.

C. TFT Fabrication

Top contact bottom gate structure of TFT was fabricated. Fig. 1 shows the fabricated TFT architecture. Aluminium gate was thermally deposited under vacuum at 10-6 mbar on glass substrate. Al$_2$O$_3$ was grown on Al gate by anodization. Polymer dielectric PVA 3% was spin coated to form a 100nm thick dielectric layer and dried at 100°C. PMMA 1% was spin coated at 3000 rpm over PVA for 1 min to get a thickness of 50 nm and dried for 30 mins at 100°C. Above that perovskite precursor solution was spin coated and annealed for 1 hr at 80°C. Shadow masking was used and copper source-drain was thermally deposited under vacuum at 10-6 mbar. Spin coating of perovskite precursor, annealing and shadow masking for source drain was done inside glovebox under inert atmosphere.

D. Device Characterization

Keithley 4200SCS semiconductor parameter analyzer was used for measurement of all the electrical properties of the fabricated OFETs.

E. Equations

The field-effect mobility (μ) in the saturation regime was extracted using the equation:

$$I_{DS} = \frac{\mu W C_i}{2L}(V_{GS} - V_{TH})^2 \qquad (1)$$

under the conditions of $V_{DS} \geq (V_{GS} - V_{TH})$, The linear region mobility was calculated using the equation:

$$\mu = \frac{dI_{ds}}{dV_G} \frac{L}{V_d C_i W} \qquad (2)$$

where μ is the field-effect mobility, W is the channel width, L is the channel length, C_i is the capacitance per unit area of the gate dielectric layer, and V_{GS}, V_{TH} and V_{DS} are the gate, threshold and source-drain voltages respectively [13]. The Sub threshold swing was calculated as the minimum value of the inverse of the slope of log I_D Vs V_G curve.

$$SS = \frac{dV_G}{d\log I_{DS}} \qquad (3)$$

III. RESULTS AND DISCUSSION

A. Perovskite Thin Film Morphology

The formation of pinhole free, lesser grain boundary, smooth is very crucial for the carrier transport in TFT. Hence, to obtain good device performance the morphology of the active layer film is studied by capturing optical microscopic images. Fig. 2(a)shows the optical microscope image of the perovskite formed on dielectric PMMA, (b) on dielectric PS and (c) on dielectric PVA. From Fig. 2(a), (b) and (c), we

978-1-5386-6624-1/18 $31.00 © 2018 IEEE

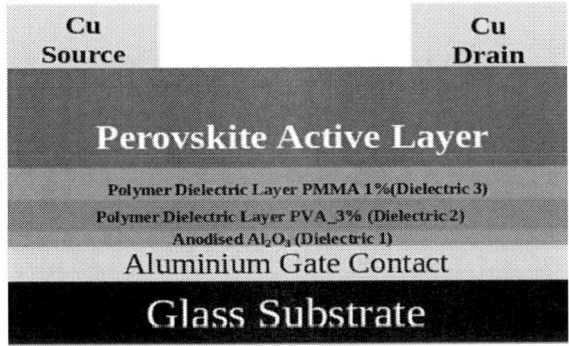

Fig. 1. Perovskite TFT structure. (Schematic view of the transistor with Cu source drain)

can conclude that larger domains and better film coverage are observed in (a) PMMA. Fig. 2(d), (e) and (f) shows the ratio variation. The best result is obtained for (f) 1:1 ratio of PbI_2:MACl, the film completely degraded for (d)1:2 whereas (e) 2:1 shows better coverage, however perovskite domain formation is not confirmed. Fig. 2(g) shows the effect aging of 15 mins and (h) shows the effect of aging of 48 hrs of the precursor solution. After 48 hrs of aging, large domains of about 60μm are formed but the boundaries are also large which effect adversely the transport of carriers. With aging of (g)15 mins, perovskite is not formed.

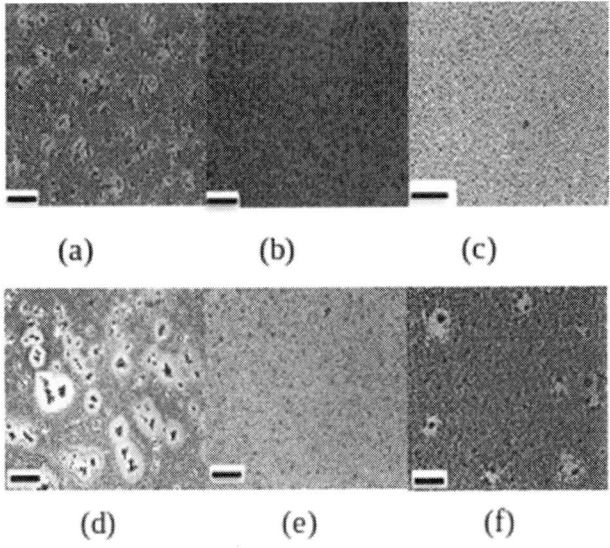

This concludes that the nucleation depends on the aging of the precursor solution. Fig. 2(i) shows the perovskite film after annealing at 60°C, (j) 90°C, (k) 120°C for 1 hr. At (k)120°C the film degraded, whereas grains started forming at 60°C. From this we can conclude that the optimum temperature is between 60°C and 90°C.

Concentration optimization is shown in Fig. 2 (l)100 mg of PbI_2, (m)150 mg of PbI_2, (n)200 mg PbI_2. Better domain

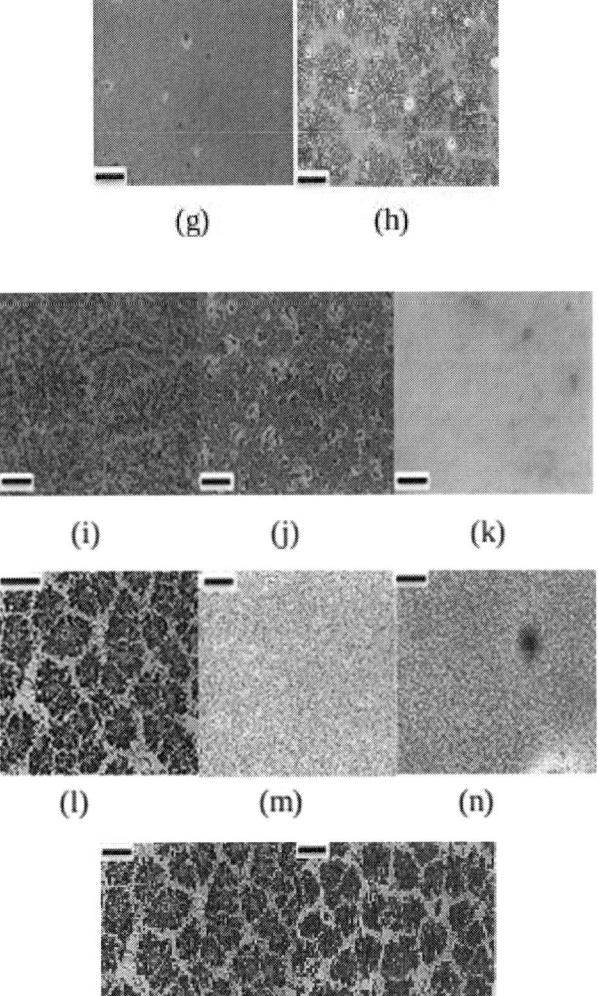

Fig. 2. Optical microscope images of the perovskite film on a dielectric layer. Perovskite coated on (a)PMMA (b)PS (c)PVA. Perovskite of ratio(d)1:2 (e)2:1 (f)1:1 coated on PMMA. Aging of (g)15 mins (h)48 hrs. Annealing temperature variation of (i) 60°C (j) 90°C (k) 120°C. Concentration variation of (l)100 mg (m) 150 mg (n)200 mg of PbI_2. (o)just after anneling (p) after 72 hrs of anneling kept at ambient atmosphere. The scale bar corresponds to 20 μm

formation is observed at 100 mg of PbI_2 here a ratio of 1:1 is maintained for all three concentrations. Fig. 2(o) shows the film just after annealing whereas (p) shows the film after 72 hrs of keeping in ambient condition. From this we can say that the film is stable even after keeping it at ambient atmosphere for 72 hrs. Further to verify the morphology study FESEM was done. Fig3 shows the FESEM image, from which a good film coverage is concluded.

B. Device Parameter

Device parameters obtained after fabricating the device with

978-1-5386-6624-1/18 $31.00 © 2018 IEEE   745

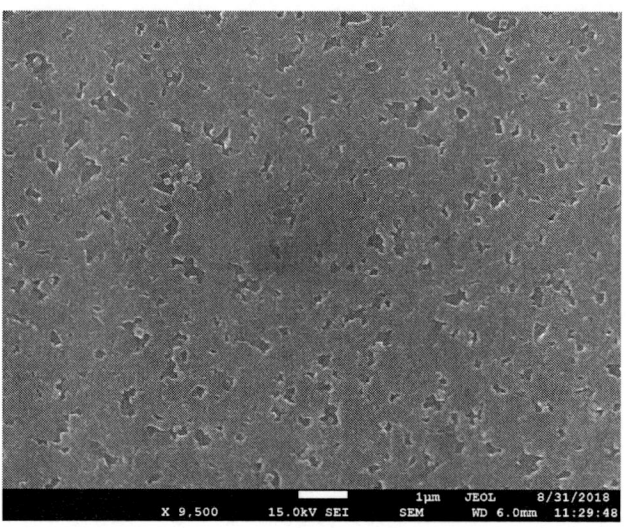

Fig. 3. FESEM Image of the Perovskite film

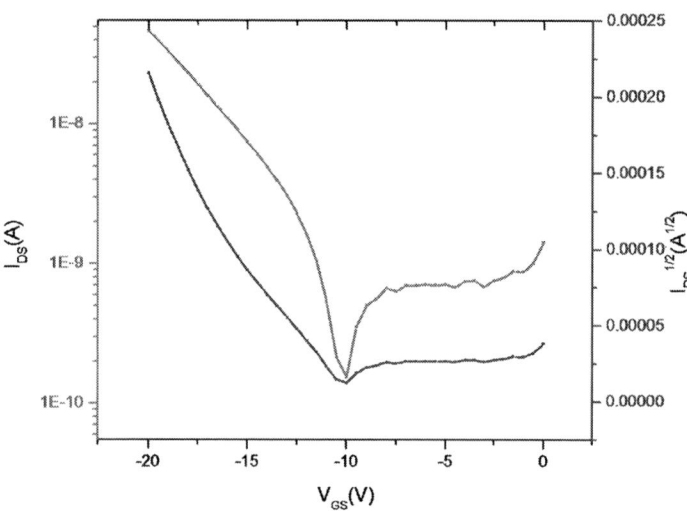

Fig. 4. Transfer Characteristic curve

TFT showing P-type characteristics. The device parameters are derived from transfer characteristic graph. At V_{DS} = 1V, V_{GS} was swept from 0 to -20 V. The linear regime mobility was calculated as 0.064 cm²V^{-1}s^{-1}, threshold voltage as -10V and current on/off ratio as 10^3.

TABLE I
SUMMARY OF DEVICE PARAMETERS

Mobility (cm²V^{-1}s^{-1})	Subthreshold Swing (V/decade)	Threshold Voltage (V)	Current ON/OFF ratio
0.064	3.75	-10	10³

IV. CONCLUSION

In summary, we have successfully demonstrated the different morphology of the mixed halide based perovskite thin films in various conditions and a very low cost device is fabricated with it. Device fabrication cost is very less because instead of costly Si substrate and gold electrodes, glass substrate and Cu electrodes were used.

By analyzing the morphology of the perovskite thin film formed on various dielectrics, it is confirmed that the growth and nucleation of the perovskite depends on the interfacial properties that also governs the accumulation of the charge carriers and hence formation of channels and carrier transport.

ACKNOWLEDGMENT

Financial grants from Department of Science and Technology(DST), New Delhi, India through the projects DST/TSG/PT/2009/23, DST/SB/S1/PC-020/2014 and Max-Planck-Gesellschaft IGSTC/MPG/PG(PKI)/2011A/48 are acknowledged and Centre for Nanotechnology, Department of Chemistry, Central Instruments Facility IIT Guwahati are acknowledged for providing various instrument facilities.

REFERENCES

[1] D. Ju, Y. Dang, Z. Zhu, H. Liu, C. Chueh, X. Li, L. Wang, X. Hu, A. Jen and X. Tao, "Tunable Band Gap and Long Carrier Recombination Lifetime of Stable Mixed CH₃NH₃PbₓSn₁₋ₓBr₃ Single Crystals", Chemistry of Materials, vol. 30, no. 5, pp. 1556-1565, 2018.

[2] Z. Tan, R. Moghaddam, M. Lai, P. Docampo, R. Higler, F. Deschler, M. Price, A. Sadhanala, L. Pazos, D. Credgington, F. Hanusch, T. Bein, H. Snaith and R. Friend, "Bright light-emitting diodes based on organometal halide perovskite", Nature Nanotechnology, vol. 9, no. 9, pp. 687-692, 2014.

[3] C. Kagan, "Organic-Inorganic Hybrid Materials as Semiconducting Channels in Thin-Film Field-Effect Transistors", Science, vol. 286, no. 5441, pp. 945-947, 1999.

[4] Y. Wu, J. Li, J. Xu, Y. Du, L. Huang, J. Ni, H. Cai and J. Zhang, "Organic–inorganic hybrid CH₃NH₃PbI₃ perovskite materials as channels in thin-film field-effect transistors", RSC Advances, vol. 6, no. 20, pp. 16243-16249, 2016.

[5] M. Lee, J. Teuscher, T. Miyasaka, T. Murakami and H. Snaith, "Efficient Hybrid Solar Cells Based on Meso-Superstructured Organometal Halide Perovskites", Science, vol. 338, no. 6107, pp. 643-647, 2012.

[6] Q. Zhang, F. Hao, J. Li, Y. Zhou, Y. Wei and H. Lin, "Perovskite solar cells: must lead be replaced – and can it be done?", Science and Technology of Advanced Materials, vol. 19, no. 1, pp. 425-442, 2018.

[7] H. Zhang and J. Toudert, "Optical management for efficiency enhancement in hybrid organic-inorganic lead halide perovskite solar cells", Science and Technology of Advanced Materials, vol. 19, no. 1, pp. 411-424, 2018.

[8] I. Chin, D. Cortecchia, J. Yin, A. Bruno and C. Soci, "Lead iodide perovskite light-emitting field-effect transistor", Nature Communications, vol. 6, no. 1, 2015.

[9] F. Li, C. Ma, H. Wang, W. Hu, W. Yu, A. Sheikh and T. Wu, "Ambipolar solution-processed hybrid perovskite phototransistors", Nature Communications, vol. 6, no. 1, 2015.

[10] F. Li, C. Ma, H. Wang, W. Hu, W. Yu, A. Sheikh and T. Wu, "Ambipolar solution-processed hybrid perovskite phototransistors", Nature Communications, vol. 6, no. 1, 2015.

[11] T. Matsushima, S. Hwang, A. Sandanayaka, C. Qin, S. Terakawa, T. Fujihara, M. Yahiro and C. Adachi, "Solution-Processed Organic-Inorganic Perovskite Field-Effect Transistors with High Hole Mobilities", Advanced Materials, vol. 28, no. 46, pp. 10275-10281, 2016.

[12] Y. Wu, J. Li, J. Xu, Y. Du, L. Huang, J. Ni, H. Cai and J. Zhang, "Organic–inorganic hybrid $CH_3NH_3PbI_3$ perovskite materials as channels in thin-film field-effect transistors", RSC Advances, vol. 6, no. 20, pp. 16243-16249, 2016.

[13] Babu Krishna Moorthy, S. (2016). Thin film structures in energy applications. : Springer International Pu, p.112.

PV-Piezo Hybrid Grid Connected System

Anuradha Tomar
Electrical Engineering
JSSATE, Noida
Uttar Pradesh, India
eranu28@gmail.com

Ayush Mittal
Electrical & Electronic Engineering
HMRITM, New Delhi
Delhi, India
am74333@gmail.com

Sahil Sharma
Electrical & Electronic Engineering
HMRITM, New Delhi
Delhi, India
sharmasahil328@gmail.com

Abstract—Increasing population, with never ending but rising electrical energy demand, emphasis towards the need of more focused research in the domain of renewable/alternate sources for energy. Efficient use of installed capacity of existing renewable/non-renewable energy resources along with addition in capacity is required to curtail the future needs. Due to continuous motivation towards promotion & installation of renewable energy resources by Government sector, in our developing Nation; Grid connected PV systems are being installed and continue to install on large scale. In this paper, integration of Piezo-electric crystals along with installed grid connected PV system is proposed. Proposed models may find suitable for public places, metro stations, road underpasses etc. But alternate sources are alone not able to match with the energy requirements of increasing population. In proposed system, I&C MPPT controller is used to ensure extraction of maximum possible energy from PV source and inverter control is implemented to keep 3-phase voltage and frequency constant as per desired value. Piezo electric stack after rectification process is hooked to the DC bus through a DC-DC converter. To validate the performance of the desired system, proposed model is simulated in MATLAB/Simulink environment and satisfactory results are obtained. However, a detailed analysis of performance is required for understanding the techno-economic benefits of the designed system.

Keywords—*Grid connected, piezostack, piezo electricity, photovoltaic (PV), renewable energy/sources.*

I. INTRODUCTION

One of the most optimistic implementation of the renewable energy is the fact, that – it can be optimized & combined with various sources of energy to compose a hybrid grid of generation. In the review paper [1], the combination of the Photovoltaic (PV) along with wind energy is studied and mentioned the modelling of PV system and wind energy system along with the hybrid control for the energy flow & management. Modelling & use of diesel generator along with PV/Wind energy is there to attenuate the shortfalls in power during no wind or poor sunlight.

In the technical paper [2], hybrid system having fuel cell in addition with PV/Wind energy is proposed along with the fuzzy logic control for the optimized energy generation from PV/Wind energy. The fuel cell is used as a compensator, to supply the full load in case the renewable is not able to meet

the required demand. The following paper [3] deals with the simulation of grid connected hybrid system of solar and wind energy using PV and Wind turbines, without any secondary/backup source, for 3 different locations in Iraq which have the capability to power a small village or as a source during total shutdown. The paper [4] shows the combination of PV along with a small hydro-power generator as primary sources. In addition to this, Diesel generator and battery is used as secondary sources, to attenuate the unstable power supply via the primary sources. In the paper [5], biomass is used as a combination of hybrid system along with PV/Wind system as a measure to electrify the rural/remote areas.

After visualizing and studying various combinations of Hybrid system with PV and their proposed control system for efficient power flow and management; In this paper, the following – Proposed model of grid connected combination of PV and Piezo along with the control of the power flow – is proposed, performed and implemented using MATLAB.

The paper starts with the fundamental of piezoelectricity and Photovoltaic required for understanding the proposed & simulated model; followed by proposed system description and operation of various controllers – I&C MPPT Controller, DC-DC and Inverter controller – that are connected in the system for making it more stable and efficient; followed by simulation results/graphs and draining the subject with various real life applications and conclusion regarding the proposed model with outputs and limitations, that are present in the proposed system.

II. FUNDAMENTAL OF PV & PIEZO-ELECTRICITY

Photovoltaic (PV), as the name suggests is somehow electrically related to Photons and Voltage. PV is the technology which convert the radiant (photon) energy from the sunlight to the DC power [7]. But, the electrical energy (or power) generated using PVs is much smaller in magnitude and with the rising demand for renewable energy, the output from PV array is desired to be as large as possible under the same irradiance and temperature conditions, because PV array accepts irradiance and temperature as variable parameters which help in generation of I-V and P-V characteristics. Continuous efforts are taken to maintain the power output constant throughout the period using MPPT controllers (P&O, INC, CVT, etc.) which is proposed for the Maximum Power Point Tracking of PV arrays for working in optimal conditions with minimal fluctuations. [11] An equivalent circuit for the single solar cell is shown in Fig. 1, which can be represented by equation (1) –

$$I = I_{ph} - I_O \left[\exp\left(\frac{V+R_S I}{V_T \alpha}\right) - 1 \right] - \frac{V+R_S I}{R_P} \quad (1)$$

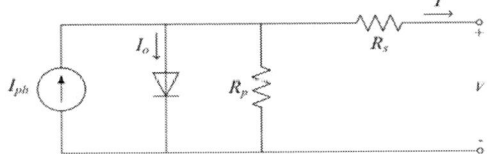

Fig. 1: Equivalent Circuit of a Solar Cell

Where: -

$$V_T = N_s \left[\frac{k \times T}{q} \right] \quad (2)$$

I_{ph} is the solar-generated current; I_O is the diode saturation current; V_T is thermal voltage of the array; N_s is the number of cells connected in series; α is the diode ideality constant; R_S is series resistance; R_p is parallel resistance; k is Boltzmann's constant; q is the electron charge; T is the temperature of PV array in kelvin.

The current generated with the help of photons (or radiant energy) is given by equation 3 –

$$I_{ph} = \frac{G}{G_n} \left[I_{ph,n} + K_i(T - T_n) \right] \quad (3)$$

Where: -

$I_{ph,n}$ is the solar generated current at the nominal condition (25° C and 1000W/m²); G is the irradiance; G_n is the nominal irradiance; T is the cell temperature; T_n is the nominal cell temperature; K_i is the short-circuit current/temperature coefficient.

However, a practical PV array consists of N_s number of solar cells connected in parallel or in series to work at optimum condition. Therefore, equation 1 can be amended to form an equation (4) –

$$I = I_{ph} - I_O \left[exp\left(\frac{V+IR_S\left(\frac{N_{SS}}{N_{pp}}\right)}{V_T \alpha \times N_{SS}}\right) - 1 \right] - \left(\frac{1}{N_{pp}}\right) \frac{V+R_S I\left(\frac{N_{SS}}{N_{pp}}\right)}{R_P\left(\frac{N_{SS}}{N_{pp}}\right)} \quad (4)$$

Where: -
N_{pp} = no. of PV modules connected in parallel; N_{SS} = no. of PV modules connected in series.

Finally, according to equation – 1 to equation – 4, PV array can be established with any arbitrary constants [14] [15].

The universal law of conservation energy states that energy can neither be created nor be destroyed but converted from one form of energy into another form of energy [12]. Piezoelectric stack is similar to transducer, which utilizes and convert the strain over the piezo element (usually, ceramic or crystals) and kinetic energy or vibrations due to mechanical force into electrical energy [13]. Therefore, Piezoelectricity can be defined as – the property of crystal to get polarized when subjected to pressure. [16] The word piezoelectricity is derived from two Greek words, "piezein" or "piezo" meaning "to press" or "to squeeze" and "electric" or "electron", which means production of electricity from pressure [6]. Piezoelectricity means production of electricity in a solid material due to the mechanical press applied on that material. The generated electricity in piezoelectricity can be calculated using formulas:

$$D = d\sigma + \epsilon^\sigma E \quad (5)$$

$$S = s^E + dE \quad (6)$$

Where D is Electric displacement/polarization [C/m²]; d is the piezoelectric charge coefficient [m/V or C/N]; S is the strain [m/m]; σ is the stress [N/m²]; E is the electric field [N/C or V/m]; ϵ^σ is the dielectric constant (permittivity) under constant stress [F/m]; s^E is the compliance when the electric field is constant [m²/N} (the superscript E denotes that the electric field is constant).

III. PROPOSED MODEL

In this paper, a PV-Piezo hybrid grid connected system as shown in Fig. 2 is proposed. Proposed model is designed for locations/places which daily large number of people walks in/out daily (like metro stations, foot over bridges, under passes) where grid connected PV system is already installed and addition of piezo stacks will further lead to increment in generated green energy. PV arrays are connected to the common DC bus through a DC-DC boost converter which operates according to the pulses received by maximum power point tracking (MPPT) controller to enable extraction of maximum possible PV power. Further, piezostack, which generates AC with the applied pressure, is connected with an AC/DC converter which is further connected to DC/DC converter for boosting the converted DC voltage and this system is controlled by the DC/DC controller (Fig. 5). Operation of this second DC-DC converter is maintained by piezoelectric controlled. As shown in Fig. 1, common DC bus is connected to the utility grid through a sequence of devices i.e. 3-phase inverter, RLC filter and transformer.

Overall installed capacity of proposed system is 53.5 kW, which can be divided as 50 kW and 3.5 kW for PV arrays and piezo electric crystals respectively. To avoid losses due to mismatch, all PV modules are considered similar and specification of each PV module is given in Table I. Further,

5 PV modules are connected in series to form one PV array and such 33 arrays are connected in parallel to make 50 kW PV system. Two Piezo Electric stacks are connected in series to make 3.5 kW piezo power contribution.

conductance of PV system is compared and depending on the result the final array terminal voltage is adjusted [9]. Block diagram for MPPT control unit for PV grid that we have used in this model is shown in Fig. 4. PV voltage (V_{pv}) and current (I_{pv}) are the inputs to the MPPT controller as depicted in Fig. 4. Output from the MPPT Controller is fed into the DC/DC$_{PV}$ which is synchronized to the bus connecting the output from DC/DC$_{Piezostack}$.

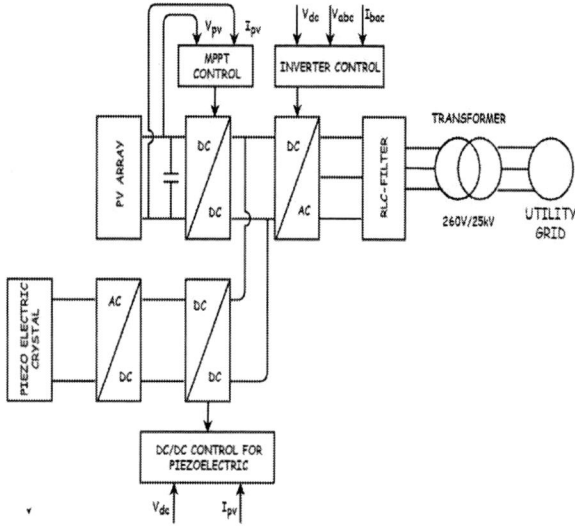

Fig. 2: Proposed combination of PV grid and piezo-electricity.

TABLE I. SPECIFICATION OF EACH MODULE

Maximum power, P_{MPPT} (W)	305
Voltage at Maximum Power Point, V_{MPPT} (V)	54.7
Open Circuit Voltage, V_{OC} (V)	64.2
Current at Maximum Power Point, I_{MPPT} (A)	5.58
Short Circuit Current, I_{SC} (A)	5.96
Number of PV Cells connected in series, N_S	96
Series Resistance, R_S (Ω)	0.037998
Parallel Resistance, R_P (Ω)	993.51
Saturation Current, I_{SAT} (μA)	0.011753

The Maximum Power Point Tracking (MPPT) is the process of achieving maximum power from the PV cells by varying the ratio of the voltage and current [7]. A number of algorithms are proposed for achieving MPPT such as – Fuzzy Logic, Incremental & Conduction (I&C), Perturbation and Observation (P&O), etc. [8]. In the proposed system, extraction of maximum power is assured by implementing I&C algorithm through MPPT controller. In I&C algorithm (shown in Fig. 3), the incremental and instantaneous

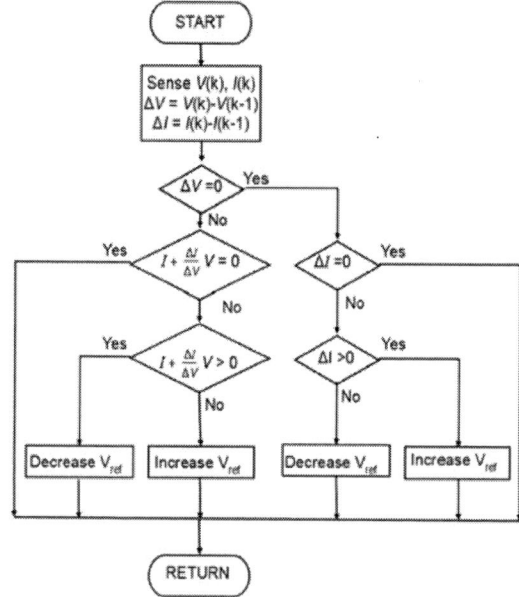

Fig. 3: Incremental & Conduction algorithm [10]

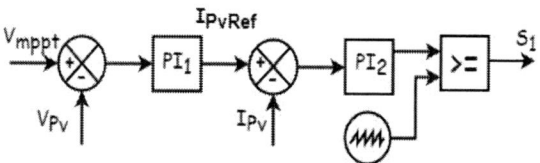

Fig. 4: MPPT Control Unit

Rectified output piezo stack voltage needs to be boost up, in order to match with common DC bus requirement (500V). Operation of DC-DC converter associated with piezo stack is controlled by control logic as shown in Fig. 5. Error in common DC bus voltage decides amount of piezo energy to be transferred to the 3-phase inverter through Dc-DC converter. The objective of this control is to maintain common DC bus voltage close to the reference voltage while ensuring transfer of maximum piezo stack power to the inverter. 3-phase inverter control is depicted in Fig. 6. Two control loops are used in this control i.e. outer loop is voltage loop and its output acts as a reference for inner current loop. Feed-forward control is used as shown in Fig. 6.

978-1-5386-6624-1/18 $31.00 © 2018 IEEE

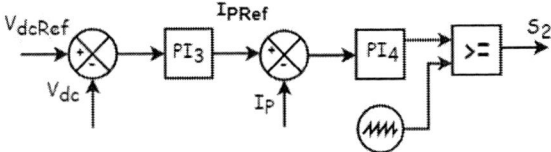

Fig. 5: DC/DC Converter Control for Piezoelectric

This output from DC/DC converters and MPPT control units is passed to an inverter control unit.

Inverter Controller: This inverter control unit is used to get a constant supply as an output. The inverter control unit schematics are shown in Fig. 6.

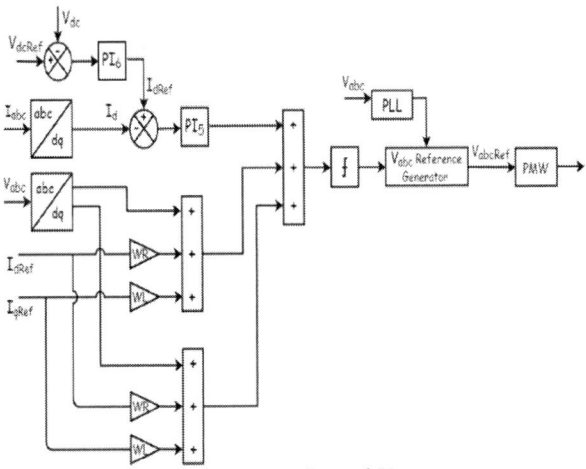

Fig. 6: Inverter Control Unit

IV. RESULT & ANALYSIS

The proposed PV-Piezo hybrid grid connected system is simulated in MATLAB/Simulink in order to verify its steady state and dynamic performance. Irradiance pattern for simulation is considered as per Fig. 7 i.e. initially irradiation is considered as 1000 W/m^2 and at t = 3 sec. irradiation suddenly falls to 300 W/m^2.

As shown in Fig. 8, before t = 3 s PV system is operating at MPPT and as expected at t = 3 s a sudden drop in extracted PV power is observed with fall in irradiation. However, MPPT controller ensures MPPT operation of PV modules in both the conditions. Fig. 9 represents the voltage across parallel connected PV arrays.

Variation in the reference DC bus voltage and measured DC bus voltage is presented in Fig. 10. It is observed that on decreasing the irradiance level from (100%) 1000 Watt/m^2 (in 0-3 seconds) to almost (30%) 300 Watt/m^2 (in 3-6 seconds), as shown in Fig.6, the output DC voltage remains constant for both the time periods with minimal fluctuations during the switching of irradiance from 1000 to 300 Watt/m^2 and that for only very few milliseconds and system again gets constant on 500 Volts (as shown in Fig. 10).

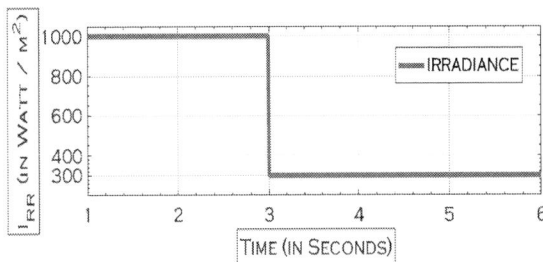

Fig 7: Variation in irradiance with time

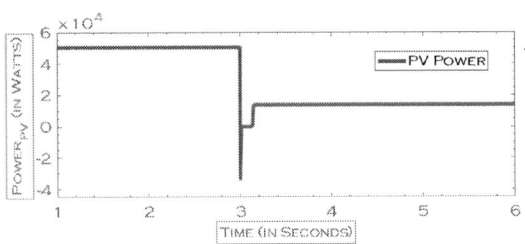

Fig 8: Variation in PV power with time

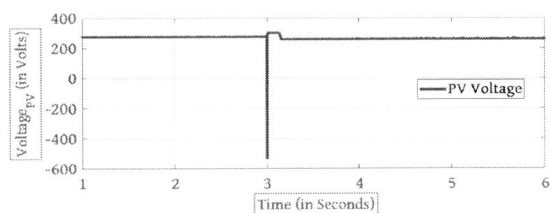

Fig. 9: Variation in PV voltage with time

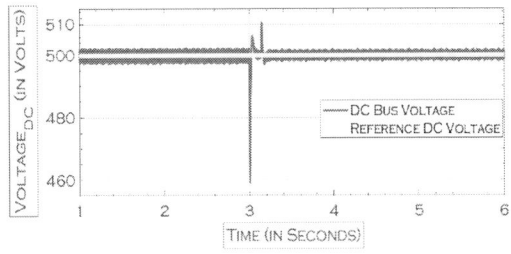

Fig. 10: Variation in common DC Bus voltage & reference DC bus voltage with Time

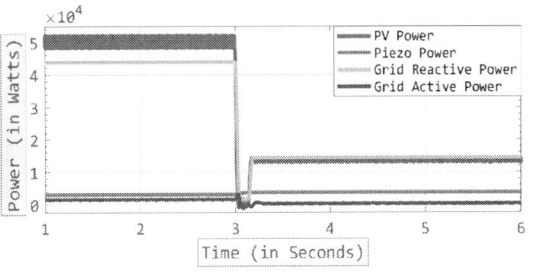

Fig. 11: Variation in PV, piezostack and grid power with time

978-1-5386-6624-1/18 $31.00 © 2018 IEEE

As shown in Fig. 11 before t = 3 sec, PV is operating at MPPT and is generating 50 kW and piezo stack is generating approximately 2.5 kW, however at t = 3 sec as PV power reduces, due to fall in irradiance, piezo power *increased* to 3.5 kW. This increment in piezo power is due to available potential of piezo power which was not utilized properly before t = 3 sec due to the fact that before t = 3 sec PV is generating at full capacity and to maintain V_{dc} voltage at 500 Volts, piezo power is re-drated; as in absence of any storage device.

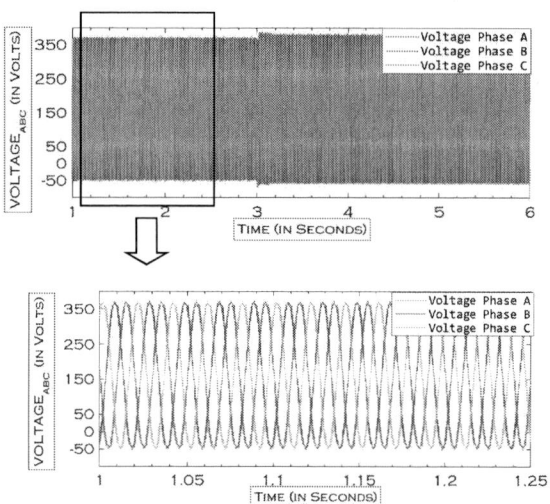

Fig. 12: 3-phase voltage vs Time plot

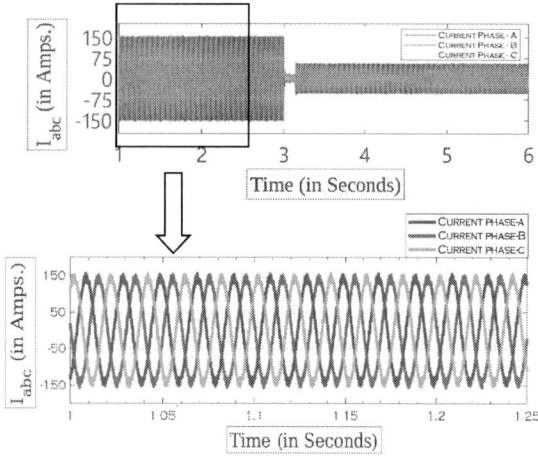

Fig. 13: 3-phase Current vs Time plot

3-phase voltage at point of common coupling (PCC), remains constant for the complete simulation time period (as shown in Fig. 12). At t = 3 sec, a small dip in voltage is observed with fall in irradiation, but voltage is recovered within a fraction of micro second duration. However, for

3-phase current at PCC response of current is depicted in Fig. 13. At t = 3 sec, 3-phase current reduced with fall in irradiance as overall power being transferred to the grid will be less as compared to the scenario before t = 3 sec. The PV voltage remained constant throughout the system with minimal dip at the time of switching but the Photovoltaic power got reduced from 50 kW (0-3 seconds) to 15kW (3-6 seconds) with fall in irradiance.

V. CONCLUSION

A good amount of green electricity can be generated from the combination of Photovoltaic Array and Piezoelectric crystal/stack as demonstrated with the above proposed model along with simulation. Proposed hybrid PV-piezo grid connected system has a wide range of applications in the crowded areas where piezo stacks can be mounted beneath the road crust & a good amount of pressure due to moving vehicles, humans, etc. on the piezo stack can generate much higher power at places such as – metro stations, on roads close to red light, market area, parking lots, etc. And also the generated electricity is not only eco-friendly but also is independent of any burning of fossil fuel and don't need any additional area for settling of apparatus and equipment as required in hydro-power plants or wind energy or solar cells. This is a cost-effective, reliable and significant alternative to non-renewable source of energy and is also one of the best element for any of the hybrid systems. There are still number of improvements such as – number of controllers, converters, etc. that can be implemented to generate stable output with high magnitude.

REFERENCES

[1] Pragya Nema, R.K. Nema, Saroj Rangnekar, "A current and future state of art development of hybrid energy system using wind and PV-solar: A review", Renewable and Sustainable Energy Reviews, Volume 13, Issue 8, 2009, Pages 2096-2103, ISSN 1364-0321, https://doi.org/10.1016/j.rser.2008.10.006.

[2] Thanaa F. El-Shatter, Mona N. Eskander, Mohsen T. El-Hagry, "Energy flow and management of a hybrid wind/PV/fuel cell generation system, Energy Conversion and Management", Volume 47, Issues 9–10, 2006, Pages 1264-1280, ISSN 0196-8904, https://doi.org/10.1016/j.enconman.2005.06.022.

[3] Salwan S. Dihrab, K. Sopian, "Electricity generation of hybrid PV/wind systems in Iraq", Renewable Energy, Volume 35, Issue 6, 2010, Pages 1303-1307, ISSN 0960-1481, https://doi.org/10.1016/j.renene.2009.12.010.

[4] Joseph Kenfack, François Pascal Neirac, Thomas Tamo Tatietse, Didier Mayer, Médard Fogue, André Lejeune, "Microhydro-PV-hybrid system: Sizing a small hydro-PV-hybrid system for rural electrification in developing countries", Renewable Energy, Volume 34, Issue 10, 2009, Pages 2259-2263, ISSN 0960-1481, https://doi.org/10.1016/j.renene.2008.12.038

[5] P. Balamurugan, S. Ashok & T. L Jose (2009) Optimal Operation of Biomass/Wind/PV Hybrid Energy System for Rural Areas, International Journal of Green Energy, 6:1, 104-116, DOI: 10.1080/15435070802701892

[6] Lumbumba Taty-Etienne Nyamayoka, Lijun Zhang, Xiaohua Xia, "Optimal Power Management for Grid connected piezoelectric Energy Harvesting System with Battery", IFAC (International Federation of Automatic Control), 2017.

[7] Selvan D. Saravana, "Modeling and Simulation of Incremental Conductance MPPT Algorithm for Photovoltaic Applications", International Journal of Scientific Engineering and Technology, Volume 2, Issue 7, 2013, Pages 681-685, ISSN 2277-1581

[8] M. Praful Raj and A. M. Joshua, "Design, implementation and performance analysis of a LabVIEW based fuzzy logic MPPT controller for stand-alone PV systems," *2017 IEEE International Conference on Power, Control, Signals and Instrumentation Engineering (ICPCSI)*, Chennai, 2017, pp. 1012-1017, doi: 10.1109/ICPCSI.2017.8391863

[9] S. Khadidja, M. Mountassar and B. M'hamed, "Comparative study of incremental conductance and perturb & observe MPPT methods for photovoltaic system," 2017 International Conference on Green Energy Conversion Systems (GECS), Hammamet, 2017, pp. 1-6, doi: 10.1109/GECS.2017.8066230

[10] https://in.mathworks.com/discovery/mppt-algorithm.html

[11] A. Aggarwal, A. Singhal and S. J. Darak, "Clean and Green India: Is Solar Energy the Answer?," in *IEEE Potentials*, vol. 37, no. 1, pp. 40-46, Jan.-Feb. 2018, doi: 10.1109/MPOT.2016.2593024

[12] C. Maxwell, "Law of conservation of energy", http://www.britannica.com/science/conservation-of-energy.

[13] Hiba Najini and Senthil Arumugam Muthukumaraswamy, "Piezoelectric Energy Generation from Vehicle Traffic with Technoeconomic Analysis," Journal of Renewable Energy, vol. 2017, Article ID 9643858, 16 pages, 2017. https://doi.org/10.1155/2017/9643858

2nd IEEE International conference on power Electronics, Intelligent Control and Energy systems (ICPEICES-2018)

Distributed Control for Power Management Based on Fuzzy Logic in DC Microgrid

Rohit R. Deshmukh
Department of Electrical Engineering
Visvesvaraya National Institute of Technology
Nagpur, India
deshmukh.rohit55@gmail.com

Makarand S. Ballal, *Senior Member, IEEE*
Department of Electrical Engineering
Visvesvaraya National Institute of Technology
Nagpur, India
drmsballal@rediffmail.com

Girish G. Talapur
Department of Electrical Engineering
Visvesvaraya National Institute of Technology
Nagpur, India
girish223@gmail.com

H. M. Suryawanshi, *Senior Member, IEEE*
Department of Electrical Engineering
Visvesvaraya National Institute of Technology
Nagpur, India
hms_1963@rediffmail.com

Abstract—**A standalone DC microgrid is consist of renewable energy sources (*RES*), energy storage system (*ESS*) and loads. This paper proposes a distributed control algorithm for adequate power sharing between the sources. The mismatch in power sharing between sources is overcome by modifying the droop control based on fuzzy logic. This assures the adequate power sharing between the sources depending on available power in the source. This maintains the stable power flow within microgrid. It also maintains the system bus voltage at set value. The fuzzy logic is used to enhance or to curtail the power of individual source considering the availability of power in source. The proposed control algorithm is decentralized and does not require communication network. The proposed modified control is verified on DC microgrid in real time simulator.**

Keywords—*DC microgrid; droop control; fuzzy logic controller; adequate power sharing; energy storage system.*

I. INTRODUCTION

The increase in power demand leads to use more renewable sources (*RES*). This arise the problem of integration of these *RES* in stable operation. The microgrid is an emerging solution to integrate *RES*. The microgrid system is generally based on *RES* such as solar *PV*, Wind. However these sources are usually stochastic and intermittent in nature. Therefore, it is essential to operate *RES* in effective way while keeping microgrid stable, reliable and economical. The DC microgrid reduces the conversion stages, less complex control and do not have issues like harmonic current, reactive power, etc. The DC microgrid consists of *RES*, *ESS* and local loads in defined boundary [1-3].

The power balance between stochastic *RES* and load is overcome by *ESS*. It can provide the power to essential loads in the absence of source. The energy management is required for optimal operation of the *ESS* [4]. Therefore the state of charge (SoC) balancing is achieved in [5] with energy

management algorithm. But, the problems like number of operation of *ESS*, avoiding over charging and deep discharging, etc are still needs more advance and complex control algorithm.

The power management within the microgrid is possible by sharing proportional power among the sources. The power sharing could be achieved by using droop control and master-slave control. In master-slave control the power sharing is possible by giving reference power to individual controllers. In this one of the source is voltage controlled source and other are power sources [6]. It needs the communication links, so that it could define the new power reference for slave units [7-8]. In this method if the master voltage source fails the whole system shut down. To overcome this new voltage source is formed but it requires communication and shifting of control mode. The slave source injects the constant power into the system irrespective of available power in source [9].

In droop control all the sources are voltage controlled sources therefore failure of system because of voltage source is not possible. In this method the control scheme is distributed and decentralized which does not require communication network. The proportional power sharing between sources is achieved by inserting virtual resistance in the source [10]. The power sharing error reduces with increase in droop constant, but this will increases the voltage deviation [11]. The voltage deviation is corrected by modifying droop control along with line resistance drop [12-13]. However the unpredictable nature of available source power creates mismatch in power sharing between the sources.

With the aim to overcome aforesaid issues in power sharing in the microgrid, this paper proposes a modified droop control along with fuzzy logic controller (FLC). The conventional droop control is modified to overcome e the power sharing error because of change in available source power. A FLC is proposed which corrects the droop voltage according to available power in individual source. Therefore,

978-1-5386-6624-1/18 $31.00 © 2018 IEEE

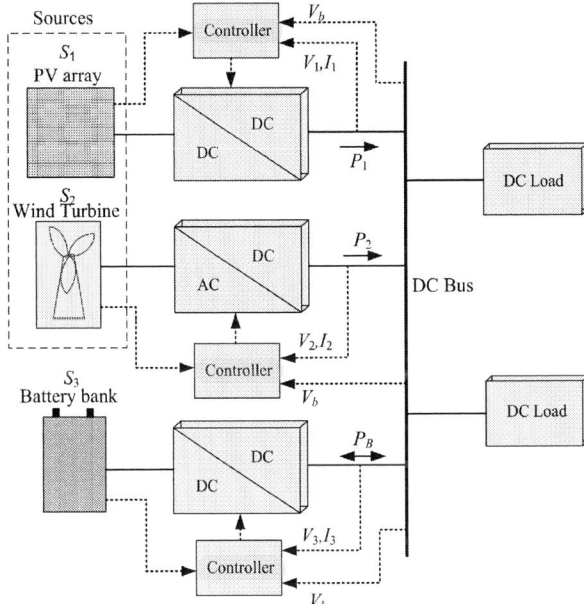

Fig.1. DC microgrid system under consideration

the optimal use of source power could be possible. It reduces the stress on *ESS* as well as interconnecting converters. It maintains the system bus voltage at its nominal value. The overall control scheme is decentralized which does not require communication network. The proposed control algorithm modifies the droop control without affecting the stable operation of DC microgrid. The considered DC microgrid system is explained in next section.

II. SYSTEM DESCRIPTION

The DC microgrid system is composed of two sources and an energy storage system. The considered DC microgrid is shown in Fig. 1. It consists of two sources photovoltaic (*PV*) and wind turbine generator (WTG). These sources are operated in voltage control mode. The *ESS* is operated as source in voltage control mode in discharging mode and as load in charging mode. Apart from that, each source is worked with sensor which provides the information about the available power in the source. Each source power is controlled by DC/DC converters.

The output variables voltages and currents are (V_1, I_1), (V_2, I_2) and (V_3, I_3) for source S_1, S_2 and S_3 respectively. Each source is controlled by decentralized controller. The bus voltage for the DC microgrid is taken as 48V [2], which is suitable voltage for low voltage DC distribution system.

III. PROBLEM CONSIDERATION AND PROPOSED CONTROL TECHNIQUE

To achieve power management in DC microgrid a distributed control scheme is used in this study. This does not require communication network. The power flow from the individual source is controlled by interconnecting DC/DC converters. These converters are operated in droop control.

A. Conventional droop control

The voltage droop control is applied for sources to have proportional power sharing among the sources. The decentralized control for each source without communication could be achieved with droop control. The droop control is implemented by adding virtual resistance in the source. The conventional droop control for two sources is given as [11]

$$V_1 = V_{ref} - I_1 R_{d1}$$
$$V_2 = V_{ref} - I_2 R_{d2} \tag{1}$$

where, V_1, I_1 and V_2, I_2 are the voltage and current outputs of source S_1 and S_2 respectively. V_{ref} is the reference voltage, R_{d1} and R_{d2} are droop constants of source S_1 and S_2.

Therefore power sharing between sources is depends on droop constant as well as new voltage from the droop equation given in (1). The sources having small droop constant shares more power and vice versa. However, in droop control the accuracy of power sharing could be improves by large droop constants. But, it increases the system bus voltage deviation which results in poor voltage regulation.

Considering that the power available in one of the source is reduced, this could be reflected in error of power sharing and bus voltage. To overcome the error in adequate power sharing between the sources as well as to improve the system bus voltage regulation, the conventional droop control is modified with the fuzzy logic controller (FLC). The FLC modifies the control to reduce the error in power sharing and improve the voltage deviation at DC bus.

For change in power ΔP from the source by keeping system bus voltage constant, hence current output from the source should be change, this leads to

$$P \pm \Delta P = V(I \pm \Delta I) \tag{2}$$

where the ΔI is the change in current, the '+' sign indicates the increase in power and '−'sign indicates the decrease in the power. Therefore the conventional droop equation is expressed as

$$V_1 = V_{ref} - R_{d1}(I_1 \pm \Delta I_1) \tag{3}$$

Considering (3) for two sources, (1) could be written as

$$V_1 = V_{ref} - I_1 R_{d1} \pm \alpha_{V1} \tag{4}$$

$$V_2 = V_{ref} - I_2 R_{d2} \pm \alpha_{V2} \tag{5}$$

where α_{V1} and α_{V2} are the voltage correction factors added in droop control. These correction factors shifts the conventional droop characteristics according to available power in source. The shifting of droop characteristics is shown in Fig. 2.

The correction factor in (4) and (5) are the output of FLC. The correction factor is decided by the rules of FLC, these are

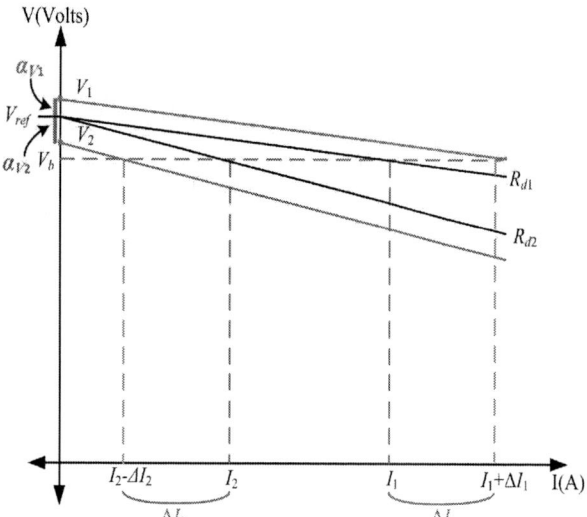

.Fig. 2. Shifting of droop characteristics for adequate power sharing between sources.

based on input parameters to fuzzy. The rules for FLC are defined based on system dynamics

B. Proposed Fuzzy Logic Control

A fuzzy logic is developed to mitigate the power sharing mismatch between the sources. It also maintains the system bus voltage at its nominal value. The block diagram of the proposed fuzzy logic control is shown in Fig.3. It includes two input to FLC and voltage correction factor as a output variable. This output minimizes the power sharing mismatch error as well as voltage deviation error. The reference voltage generated by the droop control is adjusted by fuzzy logic controller. The power variation is expressed as

$$\Delta P_a = P_{avi} - P_o \tag{6}$$

$$\Delta P_{min} < \Delta P_a \leq \Delta P_{max} \tag{7}$$

where P_{avi} is available power and P_o is the required output power. The maximum range of power variation for individual source is given by (7). The voltage variation at DC bus is given as

$$\Delta V_b = V_{bref} - V_b \tag{8}$$

$$\Delta V_{b min} < \Delta V_b < \Delta V_{b max} \tag{9}$$

where V_{bref} is the reference bus voltage and V_b is the actual bus voltage. The maximum range is mentioned by (9).

The FLC consists of 9 rules which operate on input and output [14]. The available power variation and voltage deviation error is taken as input to the FLC. The output of the FLC shifts the conventional droop characteristic which adjusts the power sharing among the sources depending upon available power in individual source. The 9 rules are extensively defined by studying the system dynamics. For each input variable three membership function (MFs) are considered. Each MFs is defined by low (L), medium (M) and high (H). The range for the input MFs is defined by the

Table I
Rules for FLC

ΔP_a \ ΔV_b	L	M	H
L	L	L	M
M	L	M	H
H	M	M	H

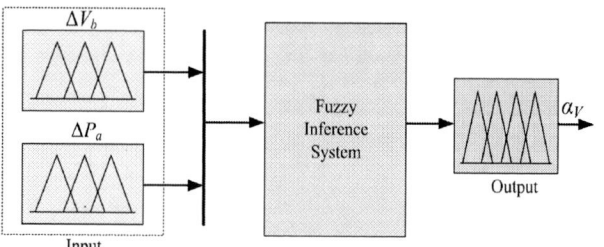

Fig. 3. Fuzzy logic system.

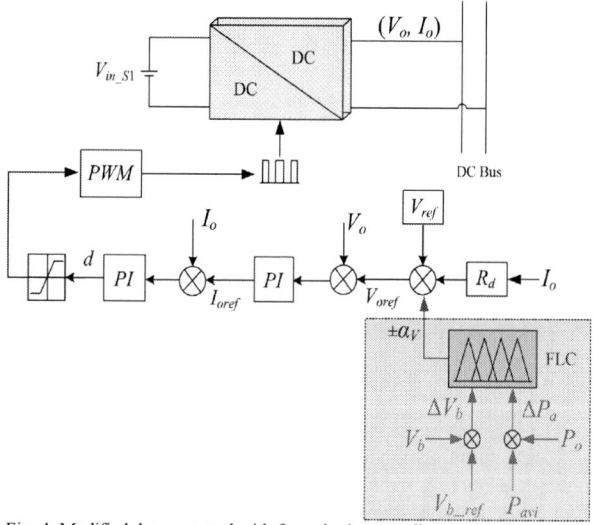

Fig. 4. Modified droop control with fuzzy logic controller.

Table II
Range for MFs in FLC

Parameter	Range
Input: Power variation (ΔP_a)	(-100 to 70) W for S_1 (-50 to 30) W for S_2
Input: Bus voltage variation (ΔV_b)	(-1.2 to 1.2) V for S_1 and S_2
Output: Correction factor (α_v)	(-0.7 to 1.2) for S_1 (- 1.4 to 2.4) for S_2

maximum available power variation in individual source and maximum voltage deviation at bus. The rules for FLC are mentioned in table I.

- If available power variation in low and bus voltage variation is high then output correction factor is medium.

The range of MFs for FLC is given in Table II. The FLC consists of Mamdani inference system with defuzzification of Center of Gravity is considered in this proposed control scheme.

The output of FLC is given to droop control which modifies the conventional droop control to minimize the adequate power sharing error. The modified droop control block diagram is shown in Fig. 4. The modified new voltage error is given to *PI* controller. This gives the new current reference as

$$I_{ref} = (K_{pv} + \frac{K_{iv}}{s})((V_{ref} - IR_d \pm \alpha_V) - V_o) \quad (10)$$

In (10) K_{Pv} and K_{iv} are the proportional and integral gains of *PI* controller. The new reference current I_{ref} is compared with actual current I_o and the respective error is given to *PI* controller. The output of this controller is modulation signal for PWM signals.

$$d = (K_{pi} + \frac{K_{ii}}{s})(I_{ref} - I_o) \quad (11)$$

where K_{ip} and K_{ii} are the proportional and integral gains for *PI* controller. The modified control scheme is shown in Fig. 4. The control is tested with different power scenarios discussed in next section.

IV. RESULT AND DISCUSSION

The proposed control algorithm is tested and verified on DC microgrid shown in Fig.1 in real time simulator (OPAL-RT). The parameters for considered DC microgrid are given in Table III. In this the FLC modifies the conventional droop control to achieve adequate power sharing among sources without communication signals. The different power variation conditions are discussed below.

In Fig. 5(a) the total load demand is 200W, initially the source S_1 shares 133.5W and source S_2 shares 66.6W based on droop control to provide to load demand. At $t = 3$s available power in the S_1 reduced by 60W, therefore the FLC modifies the droop voltage and new voltage reference is given to converter 2. To provide power to the required load demand the power output from source S_2, the correction factor from FLC increases voltage reference as per (4-5). Now source S_1 share 73.6W and source S_2 share 126.9W, the load demand is 200W. The power curves are shown in Fig. 5(a) and the new voltage references are shown in Fig.5 (b), the converters settled to new voltage.

Table III
DC microgrid parameters

Parameter	Symbol	Value
Source S_1 power rating	P_{S1}	300W
Droop constant source S_1	R_{d1}	0.25
Source S_2 power rating	P_{S2}	150W
Droop constant source S_2	R_{d2}	0.5
ESS	P_B	10Ah
System Bus Voltage	V_b	48V

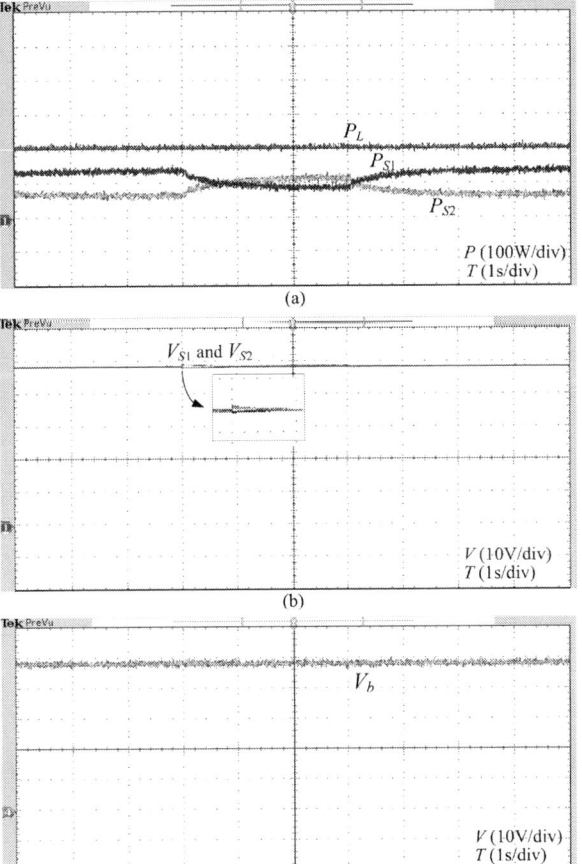

Fig. 5. (a) Power variation of sources. (b) Individual source voltage. (c) System bus voltage.

At $t = 6$s power availability of source S_1 restored to normal value. Therefore from $t = 6$s the source S_1 share 133.2W and source S_2 share 66.87W proportional to their individual power ratings. The system bus voltage is constant at 47.9V and is shown in Fig. 5(c). In the case shown in Fig. 6, *ESS* plays a significant role. Up to $t = 3$s the load sharing between sources is normal, source S_1 shares 200.3W and source S_2 shares 99.87W. The total load demand is 300W and the system bus voltage is maintained at $V_b = 47.8$V. At $t = 3$s availability of source S_2 is reduced by 50W. Therefore source S_1 shares the extra power of 50W to fulfil the load demand by modifying the voltage reference with FLC. Now the sources S_1 share 250.1W and S_2 shares 49.88W. The different power sharing between sources is made possible by modified droop control, the new voltage references for sources are different and decided by FLC depend on availability of source power and bus voltage deviation. At $t = 6$s the load demand increases it is change from 300W to 350W, however the available sources power is not sufficient to fulfil the required load demand. The MFs in FLC are its maximum range. Therefore, at $t = 6$s the *ESS* starts discharging and *ESS* shares the

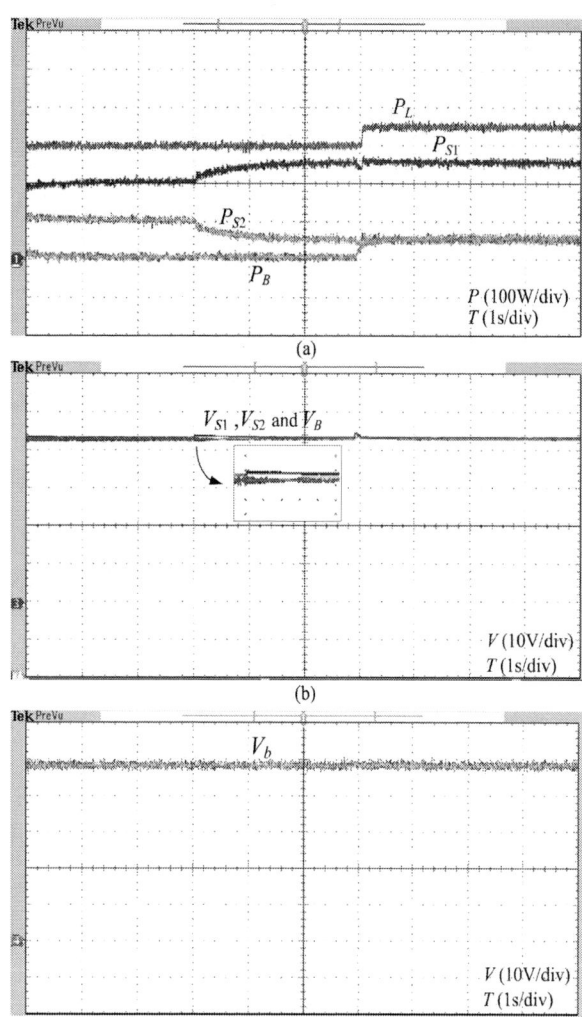

(a)

(b)

(c)

Fig.6. (a) Power variation of sources with *ESS* operation to compensate load demand. (b) Individual source voltage. (c) System bus voltage.

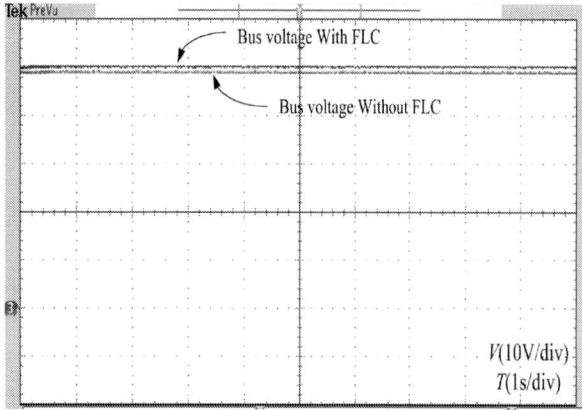

Fig. 7. System bus voltage variation with and without FLC.

remaining load demand. Initially system bus voltage is constant at $V_b = 47.8$ V. The power variation of sources is as shown in Fig. 6(a) along with new voltage references are shown in Fig. 6(b). The references voltage change for sources is shown in Fig. 6(c) this case the total load demand is 350W, source S_1 shares 250.1W and source S_2 shares 49.88W. The remaining 50.1W is provided by *ESS* to the required load demand. The system bus voltage changed to 47.93V from 47.8V as the new source is also sharing the required load demand. Therefore the FLC reduces the adequate power sharing error between sources. In Fig. 7 the system bus voltage comparison with FLC and without FLC is shown. The system bus voltage is maintained at its nominal value. It also changes the adequate power sharing between the sources based on available power in individual source. The designed FLC does not require any special system parameters and designed knowledge. It does not require communication system which reduces the reliability of overall system.

V. Conclusion

A modified control algorithm with fuzzy logic controller is proposed to achieve adequate power sharing and maintains the system bus voltage at its nominal value. The main objective of proposed control is to minimize the power sharing error between the sources because of stochastic nature of *RES*. It is overcome by designed FLC along with droop control. The FLC considers the available power in individual source and system bus voltage deviation. Based on this FLC, the droop control is modified. It helps to utilize the source power optimally without affecting the operation of DC microgrid. It also reduces the stress on *ESS* and stress on interconnected converters. The designed control algorithm is decentralized and does not require communication signals. The proposed designed controllers are distributed in operation. The performance of proposed control scheme with different power conditions is evaluated in real time simulator.

References

[1] B. Nordman and K. Christensen, "DC Local Power Distribution: Technology, Deployment, and Pathways to Success," in IEEE Electrification Magazine, vol. 4, no. 2, pp. 29-36, June 2016.

[2] T. Hakala, T. Lähdeaho and P. Järventausta, "Low-Voltage DC Distribution—Utilization Potential in a Large Distribution Network Company," in IEEE Transactions on Power Delivery, vol. 30, no. 4, pp. 1694-1701, Aug. 2015.

[3] P. A. Madduri, J. Poon, J. Rosa, M. Podolsky, E. A. Brewer and S. R. Sanders, "Scalable DC Microgrids for Rural Electrification in Emerging Regions," in IEEE Journal of Emerging and Selected Topics in Power Electronics, vol. 4, no. 4, pp. 1195-1205, Dec. 2016.

[4] D. Wu, F. Tang, T. Dragicevic, J. C. Vasquez and J. M. Guerrero, "A Control Architecture to Coordinate Renewable Energy Sources and Energy Storage Systems in Islanded Microgrids," in IEEE Transactions on Smart Grid, vol. 6, no. 3, pp. 1156-1166, May 2015.

[5] T. R. Oliveira; W. W. A. G. Silva; P. F. Donoso-Garcia, "Distributed Secondary Level Control for Energy Storage Management in DC Microgrids," in IEEE Transactions on Smart Grid , vol.PP, no.99, pp.1-11,

[6] M.S.Ballal, et al, "A Control and Protection Model for the Distributed

978-1-5386-6624-1/18 $31.00 © 2018 IEEE

Generation and Energy Storage Systems in Microgrids", Journal of Power Electronics, Vol. 16, No. 2, pp. 748-759, March 2016.

[7] N. Korada and M. K. Mishra, "Grid Adaptive Power Management Strategy for an Integrated Microgrid With Hybrid Energy Storage," in IEEE Transactions on Industrial Electronics, vol. 64, no. 4, pp. 2884-2892, April 2017.

[8] R. R. Deshmukh, M. S. Ballal, H. M. Suryawanshi and G. G. Talapur, "A control algorithm for energy management and transient mitigation in DC microgrid," 2017 National Power Electronics Conference (NPEC), Pune, 2017, pp. 270-275.

[9] G. G. Talapur, H. Suryawanshi, L. Xu and A. Shitole, "A Reliable Micro-grid with Seamless Transition between Grid Connected and Islanded Mode for Residential Community with Enhanced Power Quality," in IEEE Transactions on Industry Applications.

[10] A. Khorsandi, M. Ashourloo, H. Mokhtari and R. Iravani, "Automatic droop control for a low voltage DC microgrid," in IET Generation, Transmission & Distribution, vol. 10, no. 1, pp. 41-47, 1 7 2016.

[11] J. Xiao, P. Wang and L. Setyawan, "Multilevel Energy Management System for Hybridization of Energy Storages in DC Microgrids," in IEEE Transactions on Smart Grid, vol. 7, no. 2, pp. 847-856, March 2016.

[12] X. Zhao, Y. W. Li, H. Tian and X. Wu, "Energy Management Strategy of Multiple Supercapacitors in a DC Microgrid Using Adaptive Virtual Impedance," in IEEE Journal of Emerging and Selected Topics in Power Electronics, vol. 4, no. 4, pp. 1174-1185, Dec. 2016.

[13] S. Augustine, N. Lakshminarasamma and M. K. Mishra, "Control of photovoltaic-based low-voltage dc microgrid system for power sharing with modified droop algorithm," in IET Power Electronics, vol. 9, no. 6, pp. 1132-1143, 5 18 2016.

[14] D. Arcos-Aviles, J. Pascual, L. Marroyo, P. Sanchis and F. Guinjoan, "Fuzzy Logic-Based Energy Management System Design for Residential Grid-Connected Microgrids," in IEEE Transactions on Smart Grid, vol. 9, no. 2, pp. 530-543, March 2018.

Selective Harmonic Minimization Scheme Applied to Cascaded H-Bridge Inverter for Satisfying CIGRE WG 36-05 and EN 50160 Grid Codes

Sourabh Kundu*, Student Member, IEEE, Subrata Banerjee*, Senior Member, IEEE

*Department of Electrical Engineering, National Institute of Technology Durgapur, Durgapur, 713209, India
E-mails: sourabhkundu38@gmail.com, bansub2004@yahoo.com

Abstract—In this paper, a selective harmonic minimization-PWM (SHM-PWM) scheme for three-phase five-level cascaded H-bridge (CHB) inverter has been presented with having four number of switching transitions to meet the gird code limits such as CIGRE WG 36-05 and EN 50160, for the medium-voltage level applications. Here, the DC-link voltages has been taken to vary equally because it helps to minimize more no. of higher order harmonics with a reduced no. of switching transitions for the full modulation region. The variable DC-link voltages and switching angles have been determined by applying particle swarm optimization (PSO) technique. The efficacy of the proposed SHM-PWM scheme has been studied through simulation and compared with the SHE-PWM scheme. The proposed PWM scheme has been validated by hardware results extracted from the experimentation on a three-phase five-level CHB inverter.

Keywords—Cascaded H-bridge (CHB); CIGRE WG 36-05; EN 50160; Grid code limits; Selective harmonic minimization-pulse width modulation (SHM-PWM).

I. INTRODUCTION

In recent days, CHB inverter (Fig. 1) draws great attention towards medium-voltage high-power grid application due to its attractive features [1-3]. But it is challenging to restrict the harmonics and %THD of the output voltage to meet some specific grid code limits [4-5]. This can be accomplished by employing a suitable filtering system or an efficient pulse width modulation scheme. The first option makes the system bulky, heavy, costly, and inefficient. So, researchers go for the second option i.e. development of some efficient low-frequency switching based PWM schemes such as selective harmonic elimination-pulse width modulation (SHE-PWM) [6-10], selective harmonic mitigation-pulse width modulation (SHM-PWM) [11-14] and selective harmonic minimization-pulse width modulation (SHM-PWM) [15]. The well-established SHE-PWM scheme is used to remove 'n' no. of low-order harmonics by using '(n+1)' no. of switching instances in a quarter cycle of output voltage [6-8]. But it cannot remove or control other higher order harmonics. So, it requires more number of switching transitions for eliminating all individual harmonics and to satisfy standard grid code limits [9-10]. Some recently presented SHM-PWM schemes [11-15] focus on the improvement of the voltage waveform quality by fulfilling some standard grid code limits. In these schemes, the main aim is to reduce or mitigate each harmonic component below its certain grid code limits, rather than making it to zero.

SHM-PWM was first introduced by Napoles et al. [11] for medium power three-phase three-level diode-clamped inverter application with having 750 Hz switching frequency which

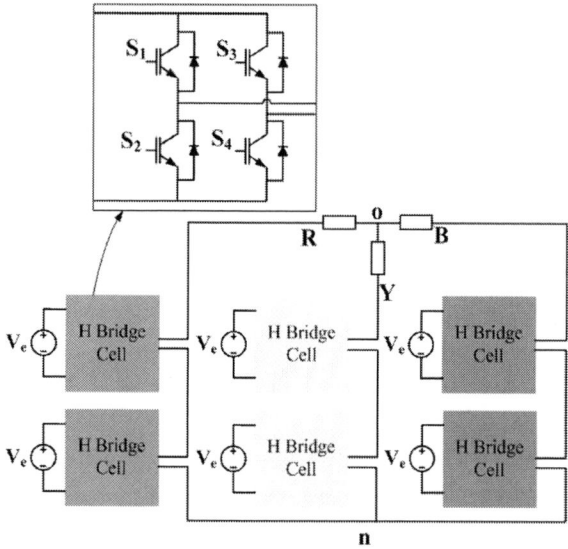

Fig. 1. Structure of three-phase five-level CHB inverter

mitigates all non-triplen odd harmonics up to 50th order satisfying grid codes CIGRE JWG C4.07, EN 50160. An optimal selective harmonic mitigation scheme [12] has been implemented for a constant DC link-voltage fed 7-level CHB inverter with having 150Hz switching frequency fulfilling some standard grid codes CIGRE JWG C4.07, EN 50160 and IEC 61000-3-6 for a wide modulation index range. Ref. [13] presents an equally variable DC-link voltage based optimal selective harmonic mitigation PWM scheme to fulfill some standard grid code limits for both three and single-phase seven-level CHB inverter for the entire modulation region. In [14] a non-equal DC-link voltage based SHM-PWM scheme has been presented to meet some grid code limits for both single and three-phase nine-level CHB inverter using 50 Hz and 150 Hz switching frequency respectively. In [15] a non-equal DC-link voltage based modified selective harmonic minimization scheme for seven-level CHB inverter has been introduced to meet EN50160 and CIGRE WG 36-05 by using 50 Hz switching frequency.

The main contribution of this work is to develop an improved SHM-PWM scheme in a three-phase five-level CHB inverter with four no. of switching instances to meet the voltage harmonic limits and %THD value satisfying CIGRE WG 36-05 [4] and EN 50160 [5] grid codes, for medium-voltage level applications. Here, the DC-link voltages have been considered to vary linearly with the

978-1-5386-6624-1/18 $31.00 © 2018 IEEE

modulation index. The proposed work has been studied in simulation and verified through hardware results. Additionally, a comparative simulation study has been performed between the proposed SHM-PWM and conventional SHE-PWM schemes.

The rest of the paper is summarized as follows: the brief discussion of the proposed SHM-PWM scheme for five-level CHB inverter in three-phase application and their optimization solutions are reported in section II. The simulation and hardware results are detailed in section III. Section IV concludes the manuscript.

II. PROPOSED SHM-PWM SCHEME

SHM-PWM scheme is extensively used for the improvement of voltage waveform quality with satisfying some of the standard grid code limits. The main objective of the proposed SHM-PWM scheme is to satisfy some standard grid code limits with a minimum number of switching instances for three-phase five-level CHB inverter. Here, two grid code limits CIGRE WG 36-05 [4] and EN 50160 [5] have been considered and its individual harmonic voltage components and %THD limits have been listed in Table 1. In case of a three-phase three-wire system, all triplen order harmonics are automatically disappeared. Therefore, the odd non-triplen harmonics up to 25th order have to be minimized to fulfill CIGRE WG 36-05 and EN 50160.

For the fulfillment of these standards, a predefined output voltage waveform (Fig. 2) in a quarter of a period for a five-level CHB inverter with having four switching transitions is considered. It shows that the first cell of the CHB inverter has one switching transition i.e. 50Hz switching frequency, while the second cell has three switching transitions i.e. 150Hz switching frequency.

The fundamental voltage component (H_1) of the predefined waveform in Fig. 2 can be deduced as:

$$H_1 = \frac{4V_{dc}}{\pi}[v_e\{\cos(\delta_{11}) + \cos(\delta_{21}) - \cos(\delta_{22}) + \cos(\delta_{23})\}]$$

$$v_e = \frac{V_e}{V_{dc}}, \quad m_a = \frac{\pi H_1}{4V_{dc}}, \quad 0 \le m_a \le 2 \qquad (1)$$

where 'V_{dc}' denotes the nominal DC-link voltage, 'V_e' implies equally variable DC-link voltages, 'v_e' indicates the

TABLE I.   SUMMARY OF INDIVIDUAL HARMONIC COMPONENT OF DIFFERENT NATIONAL (REGIONAL) STANDARDS

Harmonics order (n)	% of individual harmonic voltage components	
	CIGRE WG 36-05 (1-50kV) [4]	EN50160:1999 (1-35kV) [5]
3	5	5
5	6	6
7	5	5
9	1.5	1.5
11	3.5	3.5
13	3	3
15	0.5	0.5
17	2	2
19	1.5	1.5
21	0.5	0.5
23	1.5	1.5
25	1.5	1.5
>25	N/A	N/A
%THD	8% up to 25th	8% up to 25th

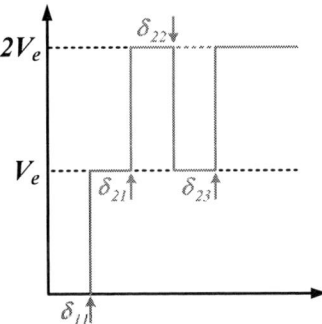

Fig. 2. Predefined waveform in a quarter of period for a five-level CHB inverter

equally variable normalized DC-link voltages, and 'm_a' signifies the modulation index. From Eq. (1), the acceptable modulation index range is in between 0-2. The other non-triplen odd n^{th} order harmonic component (H_n) of the predefined waveform in Fig. 2, can be deduced as,

$$H_n = \frac{4V_{dc}}{n\pi}[v_e\{\cos(n\delta_{11}) + \cos(n\delta_{21}) - \cos(n\delta_{22}) + \cos(n\delta_{23})\}]$$
$$n = 5,7,11, \ldots, 25 \qquad (2)$$

The value of H_n should be minimized such that each individual harmonics should be within the limit (L_i) defined by the grid code standards.

For the proposed PWM problem, the cost function (CF) can be formulated as follows,

$$F_{SHM} = \min_{(\delta_{11}, \delta_{21}, \delta_{22}, \delta_{23}, v_e)} \left\{ \left(100\frac{H' - H_1}{H'}\right)^2 + \sum_{n=1}^{2}\left(50\frac{H_{(6n\pm1)}}{H_1}\right)^2 + \sum_{n=3}^{4}\left(100\frac{H_{(6n\pm1)}}{H_1}\right)^2 \right\}$$

Subject to $0 \le \delta_{11} \le \delta_{21} \le \delta_{22} \le \delta_{23} \le \frac{\pi}{2}$

$$0 \le v_e \le 1.0518 \text{ p. u. and } H_n \le L_n \qquad (3)$$

The first term in (3) implies the percentage change in actual (H_1) fundamental voltage component with respect to desired (H') value. The other terms denote odd non-triplen harmonics up to 25th order with the appropriate weighting factors to make each harmonic value below the limits stated in Table 1. All terms of (3) fines by a power of 2.

Here, the SHE-PWM scheme has been taken into account considering the same predefined voltage waveform pattern as demonstrated in Fig. 2 to compare with the proposed scheme. So, for SHE-PWM scheme the cost function (CF) can be calculated as follows,

$$F_{SHE} = \min_{(\delta_{11}, \delta_{21}, \delta_{22}, \delta_{23})} \left\{ \left(100\frac{H' - H_1}{H'}\right)^2 + \sum_{n=5,7,11}\frac{1}{n}\left(50\frac{H_n}{H_1}\right)^2 \right\}$$

Subject to $0 \le \delta_{11} \le \delta_{21} \le \delta_{22} \le \delta_{23} \le \frac{\pi}{2}$ (4)

The first term in (4) implies the percentage change in actual (H_1) fundamental voltage component with respect to desired (H') value. The other term assigned in (4), responsible for the removal of 5th, 7th, and 11th order harmonics.

Fig. 3(a) demonstrates the solution of optimized switching angles with the change of modulation index. Fig.

978-1-5386-6624-1/18 $31.00 © 2018 IEEE

3(b) demonstrates the solution of optimized equally variable normalized DC-link voltages versus modulation index. It varies in between 0.0053 and 1.0518 per unit (p.u.) linearly for the entire modulation region. Based on the %THD up to 25th order harmonics, the minimum and maximum amount of %THD values are obtained at 0.92 and 1.79 modulation indices respectively. For the case of minimum THD, the switching angles are $\delta_{11} = 6.592°$, $\delta_{21} = 20.666°$, $\delta_{22} = 24.688°$, $\delta_{11} = 28.614°$, and the equally variable normalized DC-link voltages in p.u. is $v_e = 0.4839$. For the case of maximum THD, the switching angles are $\delta_{11} = 6.312°$, $\delta_{21} = 20.429°$, $\delta_{22} = 24.410°$, $\delta_{11} = 28.379°$, and the equally variable normalized DC-link voltages in p.u. is $v_e = 0.9388$. Furthermore, the optimized switching angles for conventional SHE-PWM scheme at 1.67 modulation index are $\delta_{11} = 13.123°$, $\delta_{21} = 37.673°$, $\delta_{22} = 46.478°$, $\delta_{11} = 53.613°$, and the DC-link voltages (v_e) is set at a constant value of 1 p.u.

III. SIMULATION AND HARDWARE RESULTS:

To validate the proposed SHM-PWM scheme, this section presents both hardware and simulation results for a three-phase five-level CHB inverter and it has also been compared with the conventional SHE-PWM scheme through simulation. The simulation has been performed in SIMULINK/MATLAB environment and the experimentation has been carried out on a practical setup of a three-phase five-level CHB inverter. The operating frequency of output voltage is set at 50Hz and the nominal DC-link voltage (V_{dc}) is taken as 30volts. Fig. 4 and Fig. 5 illustrate the phase voltage, line voltage and its harmonic spectrum of the proposed SHM-PWM scheme obtained through both simulation and experimentation while achieving minimum THD respectively. Fig. 6 illustrates the bar graph plot of individual harmonics and %THD up to 25th order harmonics of line voltage obtained by simulation

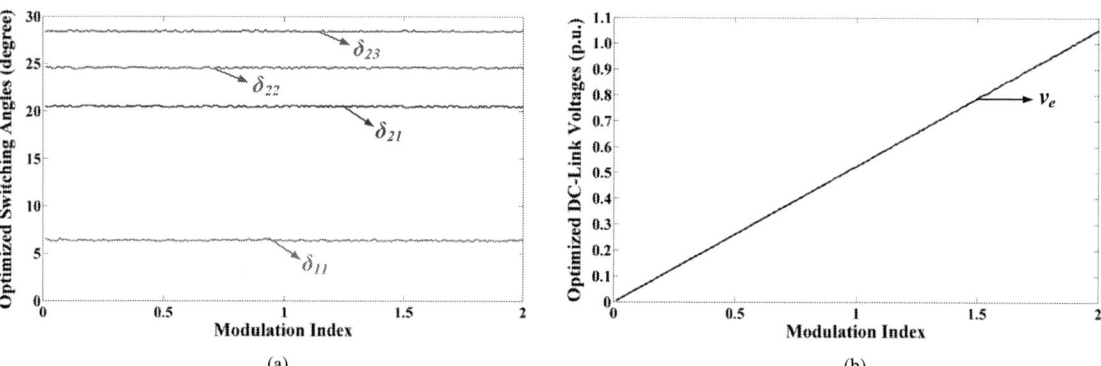

Fig. 3. Optimization results: (a) Optimized switching angles versus modulation index, (b) Optimized equally variable normalized DC-link voltages versus modulation index

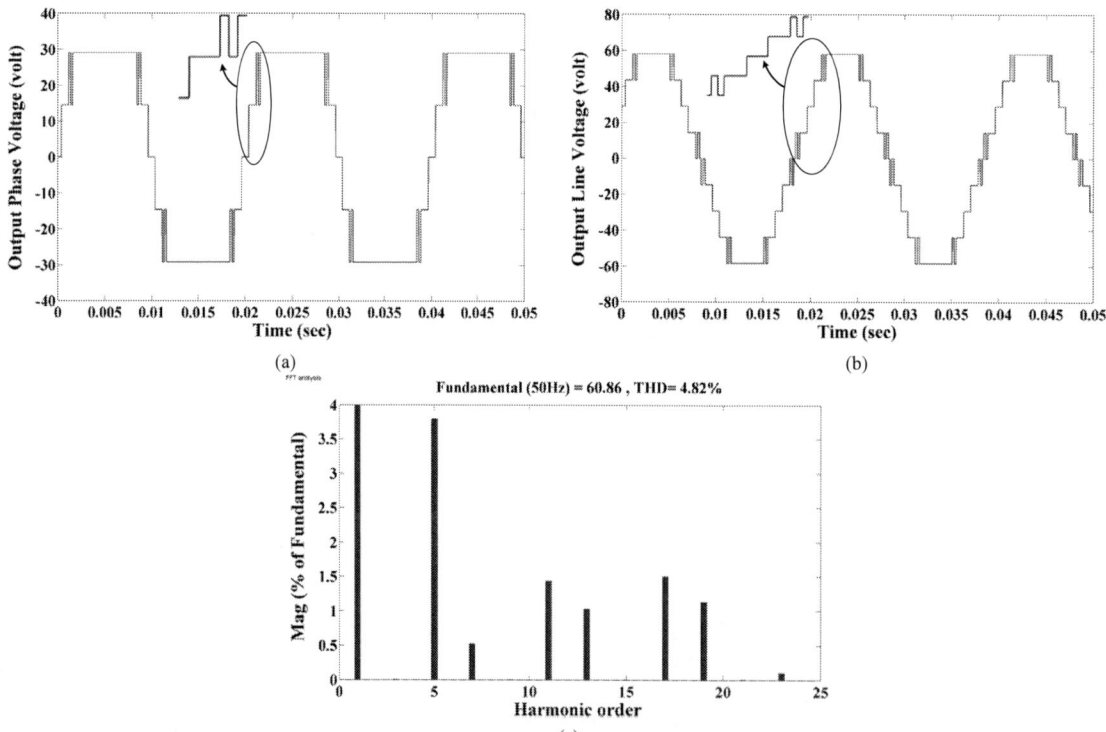

Fig. 4. Simulation result: (a) Phase voltage, (b) Line voltage, and (c) Harmonic spectrum of line voltage for proposed SHM-PWM scheme at best THD value.

result of SHE-PWM scheme, simulation result of the proposed SHM-PWM scheme at both best and worst %THD case, hardware result of the proposed SHM-PWM scheme at best %THD case and compared with the grid-code standards. Both simulation (for minimum and maximum THD case) and hardware (for minimum THD case) results show that the proposed PWM scheme can minimize each individual non-triplen harmonics and %THD up to 25th order harmonic such that it satisfies CIGRE WG 36-05 and EN 50160 grid codes. On the other hand, the standard SHE-PWM scheme eliminates the 5th, 7th, and 11th order harmonics, but the other high-order harmonics are totally uncontrolled and %THD cannot meet the above-mentioned grid code limits.

IV. CONCLUSION

This paper presents an equally variable DC link-voltage based SHM-PWM scheme in a three-phase five-level CHB inverter to fulfill CIGRE WG 36-05 and EN 50160 grid code limits for the medium-voltage level applications. PSO algorithm has been employed to evaluate the optimized switching transitions and variable DC-link voltages. The simulation and hardware results established the efficacy of the proposed SHM-PWM scheme by minimizing each individual non-triplen harmonics and %THD up to 25th order below the certain limit specified by the grid code standards. Additionally, a comparative simulation study between the proposed SHM-PWM and SHE-PWM schemes have been carried out with having same i.e. four number of switching

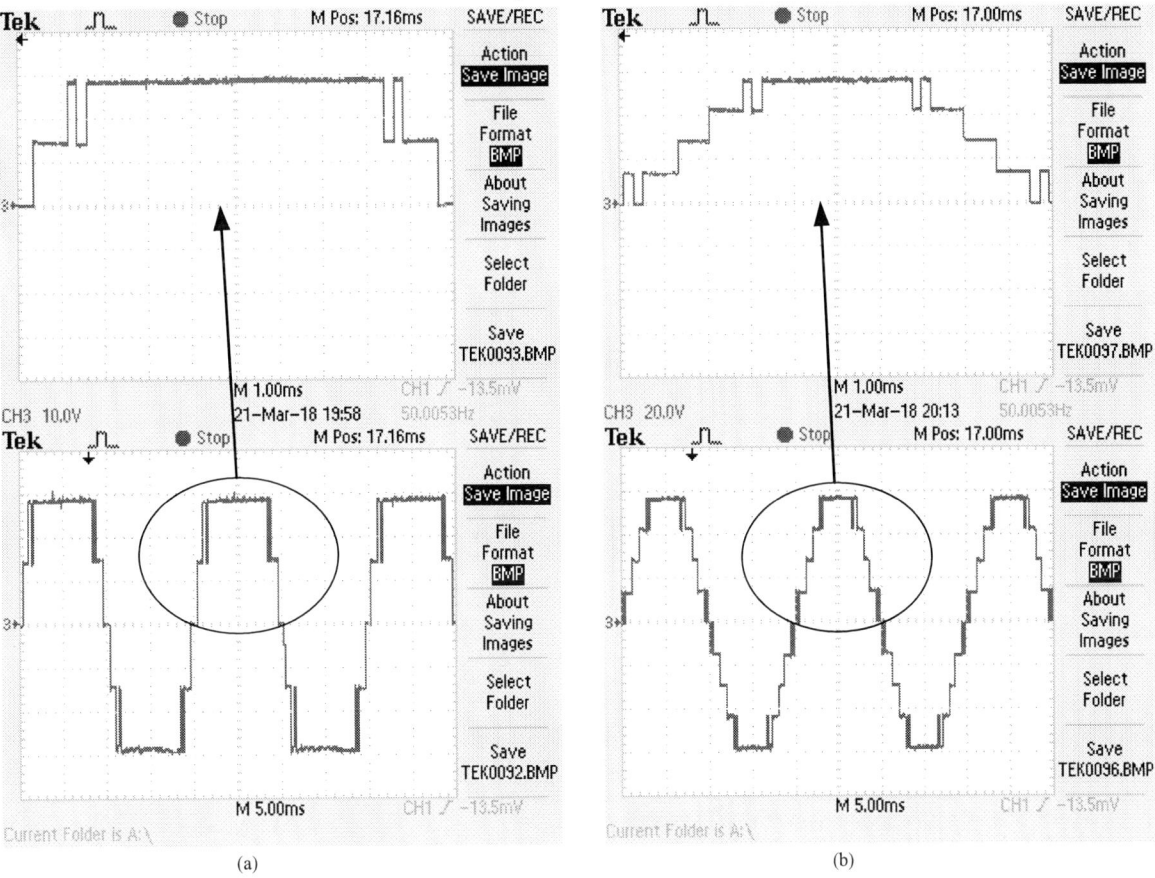

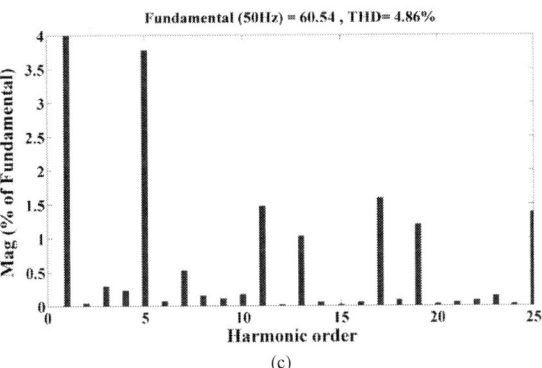

Fig. 5. Hardware result: (a) Phase voltage, (b) Line voltage, and (c) Harmonic spectrum of line voltage for proposed SHM-PWM scheme at best THD value.

978-1-5386-6624-1/18 $31.00 © 2018 IEEE

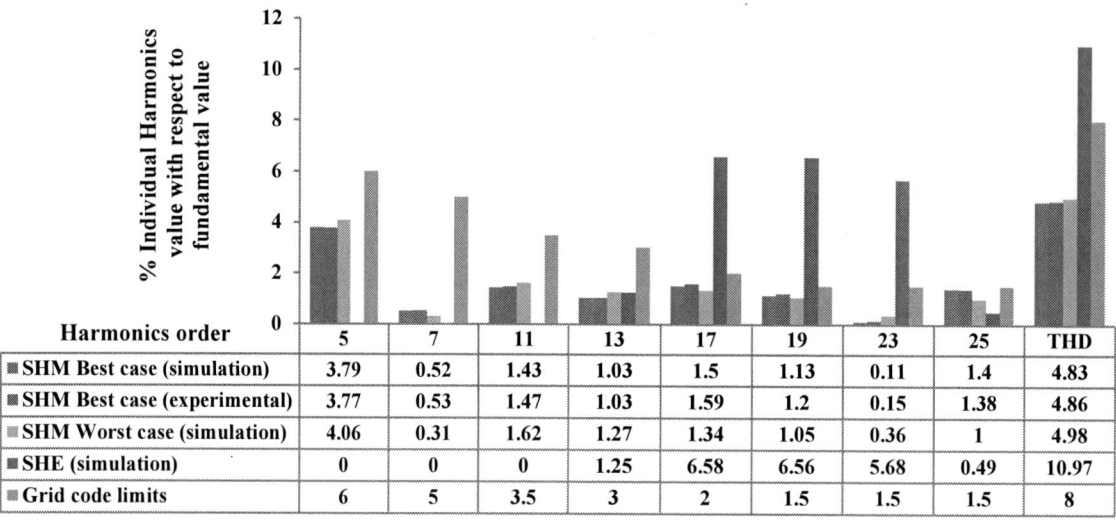

Harmonics order	5	7	11	13	17	19	23	25	THD
■ SHM Best case (simulation)	3.79	0.52	1.43	1.03	1.5	1.13	0.11	1.4	4.83
■ SHM Best case (experimental)	3.77	0.53	1.47	1.03	1.59	1.2	0.15	1.38	4.86
■ SHM Worst case (simulation)	4.06	0.31	1.62	1.27	1.34	1.05	0.36	1	4.98
■ SHE (simulation)	0	0	0	1.25	6.58	6.56	5.68	0.49	10.97
■ Grid code limits	6	5	3.5	3	2	1.5	1.5	1.5	8

Fig. 6. Bar graph plot of individual harmonic and % THD up to 25th order harmonics of line voltage obtained by simulation result of SHE-PWM scheme, simulation result of proposed SHM-PWM scheme at best case, worst case, and hardware result of the proposed SHM-PWM scheme at best case and compared with the grid-code standards

instances. The result shows that the SHE-PWM scheme eliminates only three low-order harmonics (i.e. 5th, 7th, 11th order harmonic) and other high-order harmonics are totally uncontrolled and %THD does not meet the grid code standards.

References

[1] J. I. Leon, S. Vazquez, and L. G. Franquelo, "Multilevel converters: control and modulation techniques for their operation and industrial applications," *Proc. IEEE*, vol. 105, no. 11, pp . 2066–2081, Nov. 2017.

[2] Y. Li et al., "Distributed generation grid-connected converter testing device based on cascaded H-bridge topology," *IEEE Trans. Ind. Electron.*, vol. 63, no. 4, pp. 2143–2154, Apr. 2016.

[3] H. D. Tafti, A. I. Maswood, G. Konstantinou, C. D. Townsend, P. Acuna, and J. Pou "Flexible control of photovoltaic grid-connected cascaded H-bridge converters during unbalanced voltage sags," *IEEE Trans. Ind. Electron.*, vol. 65, no. 8, pp. 6229–6238, Aug. 2018.

[4] *Power Quality Indices and Objectives*, Final Rep. CIGRE JWG C4.07, 2004.

[5] *"Voltage Characteristics of Electricity supplied by Public Distribution Systems,* European standard CENELEC EN 50160:1999," 1994.

[6] H. Taghizadeh, and M. T. Hagh, "Harmonic elimination of cascade multilevel inverters with nonequal dc sources using particle swarm optimization," *IEEE Trans. Ind. Electron.*, vol. 57, no. 11, pp. 3678–3684, Nov. 2010.

[7] A. Kavousi, B. Vahidi, R. Salehi, M. K. Bakhshizadeh, N. Farokhnia, and S. H. Fathi "Application of the bee algorithm for selective harmonic elimination strategy in multilevel inverters," *IEEE Trans. Power Electron.*, vol. 27, no. 4, pp. 1689–1696, Apr. 2012.

[8] S. Kundu, A. D. Burman, S. Giri, S. Mukherjee, and S. Banerjee, "Comparative study between different optimisation techniques for

finding precise switching angle for SHE-PWM of three-phase seven-level cascaded H-bridge inverter," *IET Power Electronics,* vol. 11, no. 3, pp. 600–609, Mar. 2018.

[9] M. S. A. Dahidah, G. S. Konstantinou, and V. G. Agelidis, "Selective harmonic elimination pulse-width modulation seven-level cascaded H-bridge converter with optimised DC voltage levels," *IET Power Electron.*, vol. 5, no. 6, pp. 852–862, Sep. 2012.

[10] M. Najjar, H. Iman-Eini, and A. Moeini, "Increasing the range of modulation indices with the polarities of cells and switching constraint reliefs for the selective harmonic elimination pulse width modulation technique," *J. Power Electron.*, vol. 17, no. 4, pp. 933–941, Jul. 2017.

[11] L. G. Franquelo, J. Nápoles, R. C. P. Guisado, J. I. Leon, and M.A. Aguirre, "A flexible selective harmonic mitigation technique to meet grid codes in three-level PWM," *IEEE Trans. Ind. Electron.*, vol. 54, no. 6, pp. 3022–3029, Dec. 2007.

[12] A. Marzoughi, H. Imaneini, and A. Moeini, "An optimal selective harmonic mitigation technique for high power converters," *Int. J. Elect. Power Energy Syst.*, vol. 49, pp. 34–39, Jul. 2013.

[13] M. Najjar, A. Moeini, M. K. Bakhshizadeh, F. Blaabjerg, and S. Farhangi, "Optimal selective harmonic mitigation technique on variable dc link cascaded H-bridge converter to meet power quality standards," *IEEE J. Emerg. Sel. Topics Power Electron.*, vol. 4, no. 3, pp. 1107–1116, Sep. 2016.

[14] A. Moeini, H. Iman-Eini, and M. Najjar, "Non-equal DC link voltages in a cascaded H-bridge with a selective harmonic mitigation-PWM technique based on the fundamental switching frequency," *J. Power Electron.*, vol. 17, no.1, pp. 106–114, Jan. 2017.

[15] S. Kundu, A. D. Burman, S. Giri, S. Mukherjee, and S. Banerjee, "Implementation of selective harmonic minimization PWM on seven-level cascaded H-bridge inverter with improved inverter performances and equal power sharing among cells," in *Proc. IEEE Int. Conf. Power and Embedded Drive Control (ICPEDC 2017)*, Mar.2017, pp. 348–353.

Adaptive SMC Based DTC of Position Sensorless PMSM Driven Solar PV Water Pumping System

Mohd. Kashif, Shabad Murshid and Bhim Singh, *Fellow IEEE*
Department of Electrical Engineering, *Indian Institute of Technology, Delhi,*
New Delhi-110016, India
kashifeed123@gmail.com, mail2smurshid@gmail.com, and bsingh@ee.iitd.ac.in

Abstract— In this paper, an adaptive sliding mode controller (ASMC) based direct torque control of position sensorless permanent magnet synchronous motor (PMSM) driven single stage solar photovoltaic (PV) water pumping system is proposed. An ASMC based DTC is used as it reduces the influence of external disturbances and parameter variations on the proposed system since the conventional SMC is robust only in the course of sliding mode. To operate solar PV array at its maximum point, a modified perturb and observe (PO) maximum power point tracking (MPPT) algorithm is used. The MPPT algorithm is based on fractional short-circuit current (FSCC) calculation. It is used due to its better tracking performance under rapidly changing solar irradiation and low-power oscillations around MPP. This increases the efficiency of FSCC-PO algorithm in comparison to conventional PO-MPPT algorithm. The angular rotor position is estimated using stator flux linkage based method, thus eliminating the mechanical rotor position sensors. This improves the reliability and decreases overall cost of the system. The 3-level and 2-level hysteresis controllers are used for processing torque error and flux error respectively. The appropriate switching states based on torque and flux error are, applied to the inverter. A laboratory prototype of the proposed system is made. The experimental results are obtained for steady-state, starting and dynamic conditions under variable solar irradiation.

Keywords—Adaptive sliding mode controller (ASMC), direct torque controller, PV array, PMSM, VSI, water pump

I. INTRODUCTION

Renewable energy sources based systems are perceived as the sustainable source of power due to increasing price of fossil fuel based energy resources and environmental degradation caused by consumption of fossil fuels for generation of electric power, transportation, and industrial purposes. Solar photovoltaic (PV) systems are proving their credibility as one of the most potential renewable energy source. This is the result of considerable drop in cost of solar PV panels. By the end of 2018, the cost of monocrystalline silicon based PV modules would be only USD 0.24 per watt [1]. The other factors that give lead to solar PV systems over other renewable energy based systems are their modularity, scalability, sparse environmental footprints and versatile location. In addition to this, the peak power demand hours meet the peak generating hours of solar PV systems. In view of this, a large number of countries, have established polices to support solar PV based power generating systems. Every continent and at least 29 countries have solar PV installations of at least 1 GW or more [2].

Water pumping systems (WPSs) are one of the potential applications of solar PV system due to the following reasons:

- The cost of supplying utility power to remote and off-grid areas is very high. The initial cost for one grid connection in such areas is between USD 400 to USD 1200 excluding the cost of wiring and the monthly payments for consumed power thereafter; whereas the cost of 4.8 kWh energy from solar PV system is less than USD 2 [3]. Moreover, in [4] a decline of 34% in the price of multicrystalline solar PV panels is predicted by the end of the year 2018.

- Diesel generators (DGs) based WPS, have higher cost than solar water pumping systems (SWPSs) over an operating period of 20 years. The per kWh price, operation and maintenance cost of DGs operated water pumps over a year, are three times more than SWPSs [5]. Moreover, DGs based WPSs cause air and sound pollution.

- Hand-held water pumps are unfeasible for high pumping heads and large water requirements. Thus, they are not suitable for deserts, mountainous regions and for applications like irrigation, sericulture, de-icing, municipal engineering and others.

- The high water discharge period from SWPS, matches with the peak water demand period for water pumping needs like aquaculture, irrigation, animal husbandry and a like.

Thus, solar PV based water pumps are the economic, environmentally pleasant, application-suitable and natural solution for water pumping needs. The Government of India (GoI) has initiated a program on solar water pumps that is to increase the solar PV installed capacity by 150 GW [6].

The DC motor powered SWPS may be a good alternative for water pumping since DC motors are directly fed by solar PV array and PV array is operated at its maximum power point (MPP) through armature voltage and field current variations. However, larger size, high cost, more weight and lower efficiency of DC motor in comparison to AC motors limits their use in SWPSs. Moreover, due to sparking at commutator and brush interface, DC motor based SWPSs are impractical for submersible water pumping requirements. The induction motors (IMs) based SWPSs are cheap, reliable, robust and require less maintenance than the DC motor based SWPS. However, less efficiency of IMs especially under light solar irradiations, reduces water discharge drastically from IM operated SWPS. This limits the use of IMs for SWPSs. The recent advances in permanent magnet (PM) technology, have made permanent magnet synchronous motors (PMSMs)

978-1-5386-6624-1/18 $31.00 © 2018 IEEE

a competitive contender for SWPSs. Now PMSMs have higher efficiency, power density, torque-to-inertia ratio and unity power factor (upf) operation as compared to IMs. The ratings of voltage source inverter (VSI) and solar PV array, are less for PMSM based SWPS due to their upf operation and higher efficiency. The high power density makes PMSMs suitable for submersible water pumping. Thus, PMSMs based SWPSs are suitable for both surface and submersible water pumping applications.

This paper proposes an adaptive sliding mode control (ASMC) based direct torque control (DTC) of position sensorless PMSM-drive used for water pumping and fed by single-stage solar PV array. Fig. 1 shows the presented system. An ASMC minimizes influence of external disturbances and parameter variations on the system as the conventional first order SMC is robust only in the course of sliding mode [7]. In addition to this, an ASMC maintains the tracking performance, steady-state behavior, and robustness of the system as by the first order sliding mode control, while reducing the chattering [8]. The ASMC generates the torque reference. The motor is operated below rated speed since it is couples to a water pump. Therefore, the reference flux linkage is kept constant. The motor torque and flux are estimated using winding currents. The three-level and two-level hysteresis controllers process motor torque and stator flux linkage errors respectively, which are then used to impress switching states on voltage source inverter (VSI) through proper control. The angular rotor position signal is generated through estimation of stator flux linkages. This eliminates the mechanical rotor position sensor reducing system cost and increasing reliability of the SWPS. Apart from this, a mechanical position sensorless system is best suited for submersible water pumping applications. A fractional short-circuit current (FSCC) based perturb and observe (FSCC-PO) MPP tracking (MPPT) control is used for MPP operation of solar PV array. The FSCC-PO control gives better tracking performance under dynamic solar irradiations and reduces oscillation around MPP [9]. An experimental prototype of the proposed system is developed. Test results are obtained for steady-state, starting and dynamic conditions under variable solar insolation.

II. Proposed System

The components and operation of the proposed SWPS, are discussed in this section.

A. Proposed System

The presented ASMC based DTC of position sensorless PMSM driven single-stage solar PV water pumping system is shown in Fig. 1. The system has a PV array, a VSI, a surface mounted PMSM and a water pump. The VSI operates the solar PV array at MPP and also controls the motor winding currents. The reference DC link voltage is generated by MPPT controller. The DC link ASMC controller processes the DC voltage error to generate reference speed command. The ASMC generates the reference torque command. The PMSM is run below rated speed, therefore the reference stator flux linkage is kept constant. The motor developed torque and stator flux linkages, are estimated by using three phase motor winding currents. The table of switching (TOS) is used to generate inverter switching states. The angular rotor position

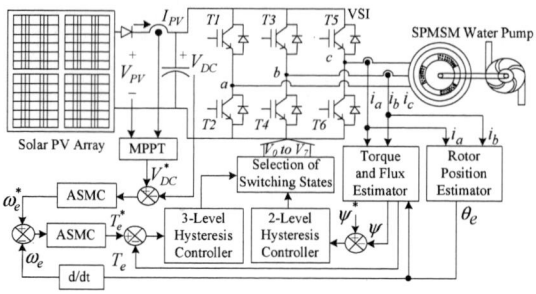

Fig. 1 Adaptive SMC based DTC of position sensorless PMSM driven single-stage solar PV array powered water pumping system

signal is generated through estimation of stator flux linkages.

B. Operation

The water pump is run by the PMSM. The solar PV array feeds power to PMSM through VSI. The VSI controls the motor winding currents and operation of PV array at MPP.

The MPPT controller generates the reference DC link voltage, V_{DC}^* using FSCC-PO MPPT technique. The sensed DC link voltage, V_{DC} and V_{DC}^* are used to generate an error signal, which is then processed by DC link ASMC controller to produce reference speed signal, $\dot{\omega}_e^*$. The $\dot{\omega}_e^*$ and estimated rotor speed ω_e are compared to generate a speed error. The speed error is processed by ASMC to generate reference torque command. The reference stator flux linkage is kept constant since the motor is operated below rated speed. The motor torque and stator flux linkage, are estimated through motor winding currents. The motor torque and stator flux linkage errors are generated using reference and estimated torque and stator flux linkages, respectively. The three-level and two-level hysteresis controllers generate appropriate control signals for impressing switching states on VSI through table of switching (TOS) using torque and stator flux linkage errors. The rotor position, θ_e is estimated using d- and q-axes stator flux linkages.

III. Design of Proposed System

A 2.1 kW, three phase surface mounted PMSM is used in the proposed system. Table I lists the specifications of PMSM. The designs of solar PV array and water pump are presented in the following subsections.

A. Design of Solar PV Array

Taking into account, losses in VSI and PMSM, a 2.37 kW peak power solar PV array is designed. The ratings of PV module and PV array are given in Table II under standard condition (1000 W/m² and 25 ˚C).

B. Design of Water Pump

Since input power to a water pump is in proportion to cube of its rotational speed, therefore the pump proportionality constant, K for input power of 2.1 kW and speed of 1500 rpm is [10],

$$K = \frac{P}{\left(\frac{\pi}{30}.N\right)^3} = \frac{2.1 \times 10^3}{\left(\frac{\pi}{30} \times 1500\right)^3} = 0.541\, mW.\left(sec/\, rad\right)^3 \qquad (1)$$

978-1-5386-6624-1/18 $31.00 © 2018 IEEE

TABLE I
RATING OF SURFACE MOUNTED PMSM

Rated Quantities	Symbol	Magnitude	Unit
Power	P_r	2.1	kW
Speed	N_r	1500	rpm
Voltage	V_r	415	V
Pole Pairs	N_P	2	----
Phase-phase resistance	$R_{ph\text{-}ph}$	0.8	Ohm
Torque	T	13	Nm
Frequency	f	50	Hz

TABLE II
RATINGS OF SOLAR PV MODULE / ARRAY AT 1 KW/M^2, 25 °C

Rated Quantities		Symbol	Magnitude	Unit
Solar PV Module	MPP Power	P_m	5.00	W
	MPP Voltage	V_m	6.00	V
	MPP Current	I_m	0.83	A
	Open circuit voltage	V_o	7.20	V
	Short circuit current	I_s	0.99	A
	No. of solar cells	N_{SC}	12	----
Solar PV Array	MPP Power	P_{PV}	2.38	kW
	MPP Voltage	V_{PV}	408	V
	MPP Current	I_{PV}	29	A
	Open circuit voltage	V_{OC}	489	V
	Short circuit current	I_{SC}	34.65	A
	Modules in series	N_S	68	----
	Modules in parallel	N_P	35	----

IV. CONTROL OF PROPOSED SYSTEM

The MPP operation of solar PV array is done through modified PO MPPT control based on calculation of fractional short-circuit current (FSCC). The FSCC based PO algorithm is used as it gives good tracking response and less power oscillations around MPP [9]. The rotor position command is estimated through stator flux linkage methods [11-12]. The ASMC based DTC is used for controlling PMSM drive due to its advantages [13-14]. The aforementioned control techniques used in the proposed system, are discussed in this section.

A. FSCC Based PO MPPT Algorithm

The FSCC-PO MPPT technique is a modified PO algorithm. The algorithm begins by selecting the initial operating point using fractional short-circuit current and then shifts to classical PO algorithm. This increases the MPPT efficiency under dynamic solar irradiations and reduces power oscillations around MPP [9], thus extracting more power than conventional PO method [15]. The algorithm is explained in forthcoming paragraphs.

The short circuit current, I_{SC} is measured and stored by disconnecting solar PV panels from the system. The maximum power point (MPP) is then, calculated as

$$I_{MPP} = c_1 I_{SC} \qquad (2)$$

where c_1 depends on the type of the PV panel. Its value is between 0.72 and 0.92. The PV current error is calculated as,

$$\Delta I = I_{MPP} - I_{PV} \qquad (3)$$

The PV current error generates the duty ratio, D. When the error is reduced below a tolerance, ξ_1, the PV power and

duty ratio are stored. Now, PO part of algorithm begins. The PV voltage, V_{PV} is decreased first and updated PV power, P_{PV} is calculated. If the power error, ΔP_{PV} is positive, a subroutine is called, which decides the time of measurements of I_{SC} for calculation of I_{MPP}, otherwise V_{PV} is increased after calling the subroutine. The ΔP_{PV} is calculated again, and if $\Delta P_{PV} > 0$, subroutine is called and V_{PV} is increased, otherwise ΔV_{PV} is decreased after calling subroutine. The flowchart of FSCC-PO algorithm is shown in Fig. 2.

B. Stator Flux Linkages Based Rotor Position Estimation

The stator voltages and stator flux linkages of PMSM in α-β frame respectively, are as follows [16]

$$\begin{bmatrix} v_\alpha \\ v_\beta \end{bmatrix} = \begin{bmatrix} R_{ph} & 0 \\ 0 & R_{ph} \end{bmatrix} \cdot \begin{bmatrix} i_\alpha \\ i_\beta \end{bmatrix} + \begin{bmatrix} d\psi_\alpha / dt \\ d\psi_\beta / dt \end{bmatrix} \qquad (4)$$

and,

$$\begin{bmatrix} \psi_\alpha \\ \psi_\beta \end{bmatrix} = \begin{bmatrix} L_\alpha(\theta_e) & 0 \\ 0 & L_\beta(\theta_e) \end{bmatrix} \cdot \begin{bmatrix} i_\alpha \\ i_\beta \end{bmatrix} + \begin{bmatrix} \psi_{PM} \cos\theta_e \\ \psi_{PM} \sin\theta_e \end{bmatrix} \qquad (5)$$

where, v_α, v_β, i_α, i_β, ψ_α, ψ_β, L_α, L_β are stator voltages, currents, stator flux linkages and inductances respectively in α-β frame; ψ_{PM}, θ_e and R_{ph} are PM flux linkage, estimated rotor position and per phase stator winding resistance respectively. The i_α and i_β are obtained from sensed currents (i_a, i_b, i_c) using

$$\begin{bmatrix} i_\alpha \\ i_\beta \end{bmatrix} = \frac{2}{3} \begin{bmatrix} 1 & -1/2 & -1/2 \\ 0 & \sqrt{3}/2 & -\sqrt{3}/2 \end{bmatrix} \begin{bmatrix} i_a \\ i_b \\ i_c \end{bmatrix} \qquad (6)$$

Using (4), the stator flux linkages are, then estimated as,

$$\begin{bmatrix} \hat{\psi}_\alpha \\ \hat{\psi}_\beta \end{bmatrix} = \int \left(\begin{bmatrix} v_\alpha(\tau) \\ v_\beta(\tau) \end{bmatrix} - \begin{bmatrix} R_{ph} & 0 \\ 0 & R_{ph} \end{bmatrix} \cdot \begin{bmatrix} i_\alpha(\tau) \\ i_\beta(\tau) \end{bmatrix} \right) d\tau \qquad (7)$$

The stator currents in α-β frame, are estimated using (5) as

$$\begin{bmatrix} \hat{i}_\alpha \\ \hat{i}_\beta \end{bmatrix} = \frac{1}{L} \begin{bmatrix} \hat{\psi}_\alpha \\ \hat{\psi}_\beta \end{bmatrix} + \begin{bmatrix} \psi_{PM} & 0 \\ 0 & \psi_{PM} \end{bmatrix} \cdot \begin{bmatrix} \cos\hat{\theta}_i \\ \sin\hat{\theta}_i \end{bmatrix} \qquad (8)$$

where, $\hat{\theta}_i$ is the initial rotor position, obtained in [12] using the following torque equation as

$$T_e = J \frac{d^2\theta_i}{dt^2} + B \frac{d\theta_i}{dt} + T_l \qquad (9)$$

where, T_e, T_l, J and B are motor torque, load torque, moment of rotor inertia and damping coefficient respectively. In order to update $\hat{\theta}_i$, the current and stator flux linkage errors are calculated respectively as,

$$\begin{bmatrix} \Delta i_\alpha \\ \Delta i_\beta \end{bmatrix} = \begin{bmatrix} i_\alpha \\ i_\beta \end{bmatrix} - \begin{bmatrix} \hat{i}_\alpha \\ \hat{i}_\beta \end{bmatrix} \qquad (10)$$

and,

$$\Delta\psi_e = \left(\frac{\partial\psi_e}{\partial i} \right) \Delta i + \left(\frac{\partial\psi_e}{\partial\theta} \right) \Delta\theta_e \qquad (11)$$

This expression is obtained through linearization of (5). The initial estimation of rotor position ensures that linearization errors are small. A position correction $\Delta\hat{\theta}_e$ reduces the flux

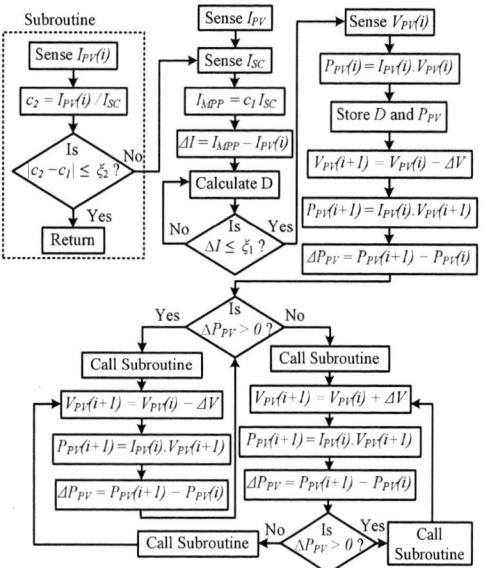

Fig. 2 Flowchart of fractional short-circuit current based perturb and observe (FSCC-PO) MPPT algorithm

error $\Delta\psi_e$ to zero. Therefore, $\Delta\hat{\theta}_e$ becomes

$$\Delta\hat{\theta}_e = -\Delta i . \frac{\partial\psi}{\partial i} . \frac{\partial\theta_e}{\partial\psi} \tag{12}$$

Thus, the updated rotor position becomes

$$\hat{\theta}_e(i+1) = \theta_e(i) + \Delta\hat{\theta}_e \tag{13}$$

The updated value of rotor position is used to calculate flux linkage and current errors, which are then used to calculate correction in rotor position, $\Delta\hat{\theta}_e$. The use of $\Delta\hat{\theta}_e$ in (13) gives estimated rotor position. The time derivative of $\hat{\theta}_e$ gives estimated rotor speed, $\hat{\omega}_e$.

C. ASMC Based DTC of PMSM Drive

The controlled system is subjected to parameter variations over time and system uncertainties; and the conventional first order SMC is robust during the course of sliding mode only. The performance of the system is improved further, in ASMC an observer, is designed for online estimation of system uncertainties Moreover, an ASMC also reduces chattering without affecting tracking performance, steady-state behavior, and robustness of the system.

The speed error, e_s and its derivative $\dot{e}_s$, are selected as the system states. The estimated speed is given by [7],

$$\ddot{\hat{\omega}}_e = u - d(t) \tag{14}$$

where, u and d are control input and external disturbances, respectively. The objective of the control, is to design an adaptive SMC for speed regulation of the motor. The sliding surface is designed as [17]

$$s = \dot{e}_s + n_1 \dot{e}_s + n_2 \int_0^t |\dot{e}_s|^\tau \operatorname{sgn}(\dot{e}_s) dx \tag{15}$$

where, $n_1 > 0$, $n_2 > 0$, and $0 < \tau < 1$

The derivative of s becomes

$$\dot{s} = \ddot{e}_s + n_1 \dot{e}_s + n_2 |e_s|^\tau \operatorname{sgn}(e_s) \tag{16}$$

Using (14) in (16) gives,

$$\dot{s} = \ddot{\hat{\omega}}_e^* - u + d(t) + n_1 \dot{e}_s + n_2 |e_s|^\tau \operatorname{sgn}(e_s) \tag{17}$$

Substitution of $\dot{s} = 0$ in (17), gives the equivalent control law, which causes the speed error to remain on sliding surface. Thus, equivalent control law becomes

$$u_{eq} = \ddot{\hat{\omega}}_e^* + d(t) + n_1 \dot{e}_s + n_2 |e_s|^\tau \operatorname{sgn}(e_s) \tag{18}$$

To push the convergence of system states to zero robustly in limited time, a ST technique is used for designing the switching control [7]. Thus, the switching control law becomes,

$$u_{sw} = n_3 |s|^{0.5} \operatorname{sgn}(s) + n_4 \int_0^t |\operatorname{sgn}(s)|^\tau dx \tag{19}$$

where, n_3 and n_4 are constants such that $n_3 > 0$, $n_4 > 0$. Thus, the sliding mode control input, u becomes

$$u = u_{eq} + u_{sw} \tag{20}$$

The values of the controller constants, are given in Appendices. Since the system is subjected to external disturbances and parameter variations; and the classical first order SMC is robust merely during sliding mode. Thus, to improve response of the controller and for estimation of system uncertainties, $d(t)$ online, a disturbance observer is designed in [7] as

$$\left.\begin{array}{l} \dot{z}_o = v_o + \ddot{\hat{\omega}}_e^* - u \\ v_o = z_1 - \lambda_o |z_o - \dot{e}_s|^{1/3} L^{1/3} \operatorname{sgn}(z_o - \dot{e}_s) \\ \dot{z}_1 = v_1 \\ v_1 = z_2 - \lambda_1 |z_1 - v_o|^{1/2} L^{1/2} \operatorname{sgn}(z_1 - v_o) \\ \dot{z}_2 = -L\lambda_2 \operatorname{sgn}(z_2 - v_1) \end{array}\right\} \tag{21}$$

where, λ_o, λ_1, λ_2, are observer coefficients; v_o, v_1, z_o, z_1, z_2 are observer variables; L is the Lipchitz constant. For the observer in (21), z_o, z_1, z_2 converge to $\dot{e}_s$, $d(t)$ and $\dot{d}(t)$ respectively in some limited time. Thus,

$$\left.\begin{array}{l} \varepsilon_o = z_o - \dot{e}_s \\ \varepsilon_1 = z_1 - d(t) \\ \varepsilon_2 = z_2 - \dot{d}(t) \end{array}\right\} \tag{22}$$

also tend to zero in some limited time. Therefore, the equivalent control law, then becomes

$$\hat{u}_{eq} = \ddot{\hat{\omega}}_e^* + \hat{d}(t) + n_1 \dot{e}_s + n_2 |e_s|^\tau \operatorname{sgn}(e_s) \tag{23}$$

Thus, the overall control input becomes

$$\hat{u} = \hat{u}_{eq} + u_{sw} \tag{24}$$

The integration of disturbance observer and switching control law, gives an ASMC. The ASMC is robust against parameter variations and external disturbances. To show the stability of speed control system, following Lyapunov function is selected

$$V = 0.5\varepsilon_1^2 + 0.5s^2 \tag{25}$$

Its derivative becomes

$$\dot{V} = s\dot{s} + \varepsilon\dot{\varepsilon} \tag{26}$$

or,

$$\dot{V} = s\left(\ddot{\omega}_e^* - \hat{u} + d(t) + n_1\dot{e}_s + n_2\left|e_s\right|^\tau \mathrm{sgn}(e_s)\right) + \varepsilon\dot{\varepsilon} \tag{27}$$

This gives

$$\dot{V} = s\left(\varepsilon_1 - n_1\left|s\right|^{0.5}\mathrm{sgn}(s) - n_2\int_0^t\left|\mathrm{sgn}(s)\right|^\tau dx\right) \tag{28}$$

or, $\dot{V} = s\varepsilon_1 + \varepsilon_1\dot{\varepsilon}_1 - n_1\left|s\right|^{3/2}\mathrm{sgn}(s) - n_2\int_0^t\left|s\right|^\tau dx$ (29)

Consider that $n_2\int_0^t\left|s\right|dx \geq 0$ holds for all time, ε_1 tends to zero in small time then

$$\dot{V} \leq -k_1\left|s\right|^{3/2} \leq 0 \tag{30}$$

As $\dot{V}$ is negative semi-definite, the system with the controller and the observer is stable.

The torque equation of a surface PMSM is rewritten as,

$$\dot{\omega}_e = \frac{1}{J}\left(T_e - B\omega - T_L\right) \tag{31}$$

where T_e, J, B, and T_L are motor torque, moment of rotor inertia, damping coefficient, and load torque respectively. The relationship between reference torque command, T_e^* and controller input, $\hat{U}(s)$ is [7] as,

$$T_e^* = J.\frac{n_2}{(n_1 + n_2.s)}.\hat{U}(s) \tag{32}$$

Thus, the output of ASMC controller is the reference torque command, T_e^*, which is generated from control input, $\hat{u}$ to the ASMC.

The PMSM is operated below speed as the torque required by the water pump increases in proportion to the square of its rotational speed. Thus, the reference stator flux linkage, ψ^* is kept constant. The stator flux linkage, ψ and motor torque, T_e are estimated using (5) and (6), respectively, as

$$\left.\begin{array}{l}\psi = \sqrt{\left(\psi_\alpha^2 + \psi_\beta^2\right)/2} \\ T_e = 1.5N_p\left(\lambda_\alpha i_\beta - \lambda_\beta i_\alpha\right)\end{array}\right\} \tag{33}$$

The flux error, $\Delta\psi$ and torque error, ΔT_e generate two (0,1) and three levels (-1,0,1) of output of flux hysteresis, $\delta\psi$ and torque hysteresis, δT_e controller, respectively. On the basis of $\delta\psi$ and δT_e, the active (V_i, $i = 1$ to 6) and zero (V_k; $k = 0,7$) switching states are applied to the VSI, where V_1 to V_7 switching states are defined in [18]. In other words, $\delta\psi$ and δT_e assist in selection of six sectors for inverter switching. Table III shows the selection of sectors.

V. EXPERIMENTAL PERFORMANCE

A laboratory prototype of proposed system is developed on which, test results are obtained. The experimental prototype comprises of a PMSM, a VSI, a solar PV simulator and a DC generator. The DC generator feeding resistive load realizes a water pump. The DSP-dSPACE microlab box generates inverter switching states. The rotor speed and position signals are generated through sensed motor currents

TABLE III
SELECTION OF VSI SWITCHING STATES

$\delta\psi$	δT_e	θ_1	θ_2	θ_3	θ_4	θ_5	θ_6
0	0	V_7	V_0	V_7	V_0	V_7	V_0
0	1	V_3	V_4	V_5	V_6	V_1	V_2
0	-1	V_5	V_6	V_1	V_2	V_3	V_4
1	0	V_0	V_7	V_0	V_7	V_0	V_7
1	1	V_2	V_3	V_4	V_5	V_6	V_1
1	-1	V_6	V_1	V_2	V_3	V_4	V_5

using aforementioned methods. The ratings of the PV emulator is given in Appendix. The hardware performance of the proposed system is discussed in this section.

A. MPP Operation

The solar PV simulator curves at 1000 W/m² and 500 W/m² are shown in Fig. 3(a) and Fig. 3(b), respectively. The temperature is set to 25°C. The simulator works at MPP with a efficiency close to 100%. The PV simulator power, voltage, current at MPP are 2.37 kW, 404 V, 5.87 A and 1.19 kW, 392 V, 3.03 A, respectively at 1000 W/m² and 500 W/m².

B. Starting and Steady-State Performances

The response at starting of the proposed system is shown in Fig. 4. The motor reaches a steady-state speed of 157 rad/sec and 100 rad/sec at 1000 W/m² and 500 W/m², respectively. The smooth and soft starting of the motor is also observed from Fig. 4.

The steady-state response of the proposed system is shown in Fig. 5. The steady-state values of T_e, I_a, ψ are 7.5 Nm, 7.07 A, 0.177 mWb-t and 5 Nm, 5.3 A, 0.177 mWb-t respectively at 1000 W/m² and 500 W/m². The system operates satisfactorily in steady-state as observed in Fig. 5.

C. Dynamic Performance

The response of the proposed system for solar irradiation change from 1000 W/m² to 500 W/m² and 500 W/m² to 1000 W/m² are shown in Figs. 6(a)-(c) and Figs. 6(b)-(d), respectively. All system parameters approach their steady state values after change of solar insolation without any abnormalities. The operation of the system is stable throughout the dynamic conditions as observed from Fig. 6.

VI. CONCLUSION

The ASMC based DTC of position sensorless PMSM drive operating water pump and fed by single stage solar PV array, has been proposed in this paper. The performances of the presented SWPS during starting, steady-state and for irradiation change, are obtained on a prototype of the system. The use of ASMC reduces influence of parameter variations and external disturbances on the system. Moreover ASMC also reduces chattering while maintaining the same tracking performance and robustness as with the conventional SMC. The angular rotor position has been estimated using stator flux linkage method. It has been observed that the proposed system gives quite a satisfactorily performance under all operating conditions.

APPENDICES

A. *ASMC Parameters*: n_1=50, n_2=625, n_3=20000, n_4= 20000

B. *Solar PV Simulator Parameters*: $P_{PV} = 4.68$ kW; $V_{PV} = 430$ V; $I_{PV} = 10$ A; $V_{OC} = 537$ V; $I_{SC} = 12.5$ A

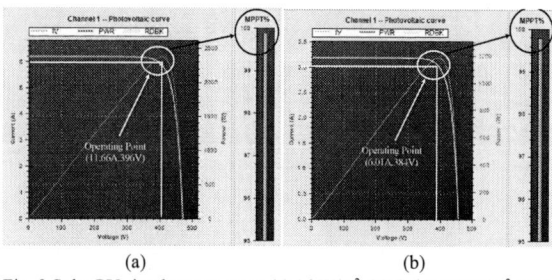

(a) (b)

Fig. 3 Solar PV simulator curves at (a) 1 kW/m², 25°C (b) 500 W/m², 25°C

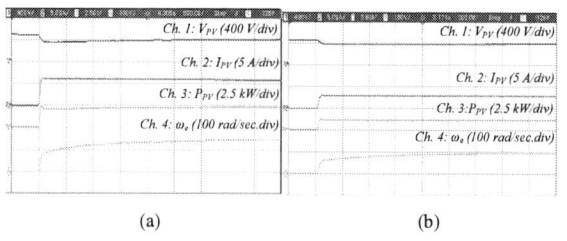

(a) (b)

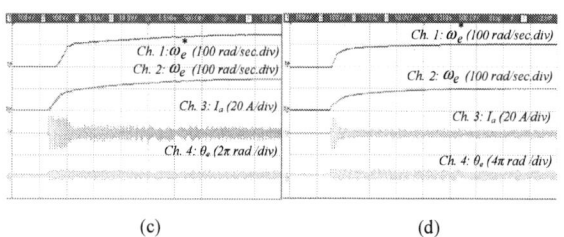

(c) (d)

Fig. 4 Starting performance at (a), (c) 1000 W/m², and (b), (d) 500 W/m²

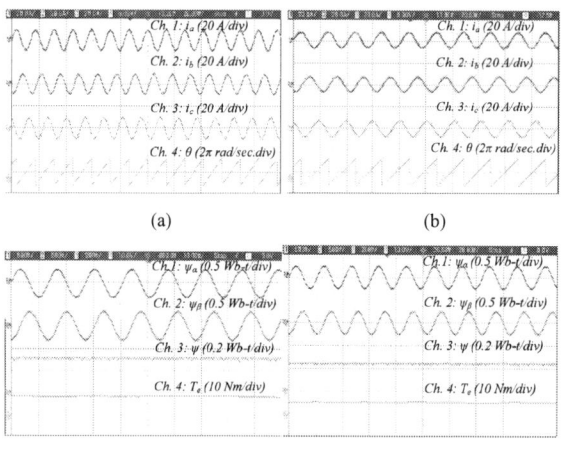

(a) (b)

(c) (d)

Fig. 5 Steady-state performance at (a), (c) 1 kW/m², and (b), (d) 500 W/m²

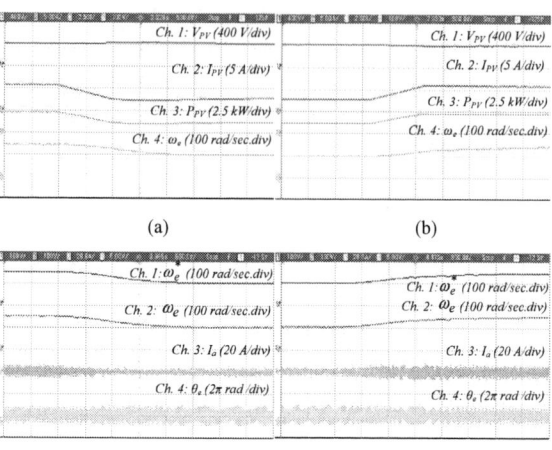

(a) (b)

(c) (d)

Fig. 6 Dynamic performance of proposed system at 25°C for solar irradiation change from (a), (c) 1000 W/m² to 500 W/m², and (b), (d) 500 W/m² to 1000 W/m²

ACKNOWLEDGMENT

This work is supported under the realm of Uchhatar Avishkar Yojana by Ministry of Human Resource and Development, Govt. of India and Shakti Pumps (India) with the Grant Number: RP03222.

REFERENCES

[1] C. Roselund, "Global PV module prices collapse," *PV Magazine International*, 21 June, 2018.

[2] REN21 2018, "Renewables 2018 Global Status Report," REN21 Community, Paris, France, 2018.

[3] A. M. Herscowitz, "Rethinking the Cost of Oh-Grid Power: Let's Do the Math", Power Africa, 2017. [Online] Available: https://medium.com/power-africa/rethinking-the-cost-of-off-grid-power-lets-do-the-math-1e31bddb1240 [Accessed: 8 Aug. 2018]

[4] C. Roselund, "BNEF expects 34% fall in PV module prices in 2018," *PV Magazine International*, 5 June, 2018.

[5] E. R. Shouman, E. T. El Shenawy and M. A. Badr, "Economics Analysis of Diesel and Solar Water Pumping with Case Study Water Pumping for Irrigation in Egypt," *Int. J. of Applied Eng. Research*, vol. 11, no. 2, pp. 950-954, 2016.

[6] M. Willuhn, "Indian Solar Water Pump Scheme Could Add 150 GW," *PV Magazine India*, 1 Aug. 2018.

[7] C. Mu, W. Xu and C. Sun, "Adaptive sliding mode control for the speed regulation of PMSMs with load change," in Proc. *17th Int. Conf. on Elect. Mach. and Syst.*, 2014, pp. 549-555.

[8] A. Kareem, "Fuzzy Logic Based Super-Twisting Sliding Mode Controllers for Dynamic Uncertain Systems", Ph.D. Thesis, Dept. of Electron. & Commun. Eng., St. Peter's Univ., Chennai, India, 2015

[9] H. A. Sher, A. F. Murtaza, A. Noman, K. E. Addoweesh, K. Al-Haddad and M. Chiaberge, "A New Sensorless Hybrid MPPT Algorithm Based on Fractional Short-Circuit Current Measurement and P&O MPPT," *IEEE Trans. Sustain. Energy*, vol. 6, no. 4, pp. 1426-1434, Oct. 2015.

[10] P. Girdhar, O. Moniz, *Practical Centrifugal Pumps Design, Operation and Maintenance*, Elsevier-Newnes, MA, USA, 2005.

[11] C. French and P. Acarnley, "Control of permanent magnet motor drives using a new position estimation technique," *IEEE Trans. on Ind. Appl.*, vol. 32, no. 5, pp. 1089-1097, Sept.-Oct. 1996.

[12] O. C. Kivanc and S. B. Ozturk, "Sensorless PMSM Drive Based on Stator Feedforward Voltage Estimation Improved With MRAS Multiparameter Estimation," *IEEE/ASME Trans. on Mechatronics*, vol. 23, no. 3, pp. 1326-1337, June 2018.

[13] D. Bao, X. Pan, Y. Wang, X. Wang and K. Li, "Adaptive Synchronous-Frequency Tracking-Mode Observer for the Sensorless Control of a Surface PMSM," *IEEE Trans. on Ind. Appl.*, to be published.

[14] J. Liu, H. Li and Y. Deng, "Torque Ripple Minimization of PMSM Based on Robust ILC Via Adaptive Sliding Mode Control," *IEEE Trans. on Power Electron.*, vol. 33, no. 4, pp. 3655-3671, April 2018.

[15] M. Kashif, S. Murshid and B. Singh, "Standalone solar PV array fed SMC based PMSM driven water pumping system," in Proc. *IEEMA Engineer Infinite Conference*, 2018, pp. 1-6.

[16] R. Krishnan, *Permanent Magnet Synchronous and Brushless DC Motor Drives*, CRC Press, New York, 2010.

[17] D. Liang, J. Li and R. Qu, "Super-twisting algorithm based sliding mode observer for wide-speed range PMSM sensorless control considering VSI nonlinearity," in *Proc. IEEE Int. Elect. Mach. and Drives Conf.*, 2017, pp. 1-7

[18] P. Vas, *Sensor less Vector and Direct Torque Control*, Oxford University Press, Oxford, 1998.

2nd IEEE International conference on power Electronics, Intelligent Control and Energy systems (ICPEICES-2018)

Multipulse AC-DC Conversion Fed 3rd Harmonic Injection Based SPWM Controlled Cascaded MLI Driven VCIMD

Piyush Kant, *Member, IEEE*
Dept. of Electrical Engineering
Indian Institute of Technology Delhi
New Delhi-110016, India
piyushkant357@gmail.com

Bhim Singh, *Fellow, IEEE*
Dept. of Electrical Engineering
Indian Institute of Technology Delhi
New Delhi-110016, India
bsingh@ee.iitd.ac.in

Abstract—**This paper present a 36-pulse AC-DC converter fed 5-level cascaded multilevel inverter (MLI) driven vector controlled induction motor drive (VCIMD). A modular transformer is utilized to make the grid current nearly sinusoidal and the transformer is connected to achieve a 36-pulse AC-DC converter. A symmetrical 5-level cascaded MLI is utilized on the motor side, which offers the good performances of the proposed VCIMD. A 3rd harmonics injection based sinusoidal pulse width modulation (SPWM) is proposed in this paper, for variable speed induction motor drive to achieve the performances similar to space vector modulation (SVM). A 5-level cascaded MLI has less switching loss. It is because, it is operated at low switching frequency. The Simulink/MATLAB base simulations results are verified by the experimental results and both the simulation and experimental results are given in this paper.**

Keywords—Induction motor control, multipulse AC-DC Converter, MLI, modulation technique and power quality.

I. INTRODUCTION

A conventional variable frequency induction motor drive (VFIMD) has a 3-phase diode bridge rectifier (DBR) at the grid side and a 2-level voltage source inverter (VSI) at the motor side. For medium power application, a step down transformer is required at the grid side to meet the required voltage of IMD. This 3-phase DBR introduces high quantity of current harmonics on grid side. Apart from this, a 2-level VSI has high voltage and current total harmonics distortion (THD), more common mode voltage and electro-magnetic interference [1]-[4]. To limit the above discussed problem of conventional IMD, a multi-winding transformer based multi-pulse AC-DC conversion is utilized on grid side to make the grid current nearly sinusoidal and reduce its THD to an acceptable level. The problem of conventional IMD on motor side is limited to an acceptable by utilizing a multilevel inverter (MLI) at the motor side. It is because, a MLI has improved power quality, lower switching losses, better electromagnetic compatibility, and high voltage capability than 2-level VSI [5]-[7].

In this paper, two-separate isolated transformers are utilized. These two transformers, are configured as an 18 pulse AC-DC converters. Due to the phase-shift angles in primary of these transformers, it is re-configured as 36-pulse AC-DC converter. The conventional MLIs (NPC, flying capacitor and cascaded) are mainly utilized in medium power application. A NPC MLI inverter offers better THD performances than other MLIs. However, a NPC inverter has DC-link capacitor voltage balancing problem and a very complex control algorithm is

required to maintain the DC-link capacitor voltages at the required values [8]-[9]. The flying capacitor MLI has flying capacitors and DC-link capacitors voltage balancing problems. The voltages across each flying capacitor should be sensed, in order to balance these capacitor voltages at the required voltage. Therefore, the requirement of voltage sensors, is more in case of flying capacitor MLI. Apart from this, the control algorithm for this MLI is quite complex [10]-[11]. The cascaded MLI is more popular for medium power application. It is because of it easy control and structure. Moreover, the power rating in cascaded MLI, can be easily scaled. Therefore, a 5-level cascaded MLI is utilized here to drive an induction motor [12]-[13].

One of the main objective of this paper, is that to propose an appropriate modulation technique for a 5-level cascaded MLI driven VFIMD. There are many modulation techniques, which have been reported in the literature to control a cascaded MLI [14]-[16]. A fundamental switching technique for 5-level cascaded MLI driven IMD has more ripple in stator current and hence more ripple in torque [17]. Therefore, a pulse width modulation technique (PWM) is utilized in this work. A simple sinusoidal pulse width modulation (SPWM) technique like carrier shifted and phase shifted are not much suitable for VFIMD. It is because, by utilizing this technique, one cannot achieve the maximum utilization of DC-link capacitor voltages. As compared to SPWM, a space vector modulation (SVM) has 15 % more fundamental voltage with same DC-link voltage as SPWM has [18]. However, a SVM technique is quite complex for MLIs and its complexity increases as the level increases. Therefore, a 3rd harmonic injection based SPWM technique is utilized in this paper. With a proper selection of the amplitude of 3rd harmonic component, one can achieve the similar performances as SVM technique has. However, the implementation of 3rd harmonic injection technique for constant fundamental frequency operation is quite easy. Whereas in case of VFIMD, implementation of 3rd harmonic injection based SPWM technique is quite complex. It is because, in case of VFIMD, the amplitude and frequency of the modulating signal generated from motor control is directly proportional to speed of the motor and IMD can have any speed as per the requirement. Therefore, in this paper, an implementation of 3rd harmonic injection based SPWM for 36-pulse AC-DC conversion and 5-level cascaded MLI driven VCIMD, is proposed for medium power industrial application. The extensive simulation and experimental results are given here to verify the usefulness of proposed IMD for medium power application.

978-1-5386-6624-1/18 $31.00 © 2018 IEEE

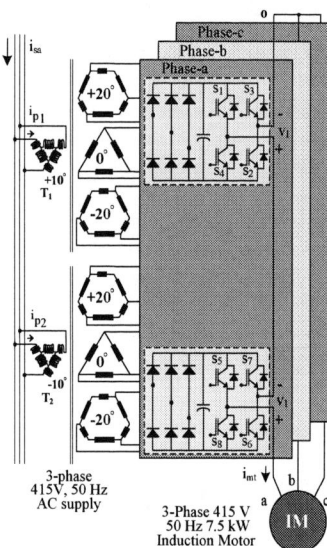

Fig.1 Schematic diagram of proposed VCIMD

II. Design of Proposed VCIMD

The structure of presented VCIMD, is illustrated in Fig. 1. This VCIMD has a 36-pulse AC-DC conversion at grid side. Whereas, on the motor side, it has a 5-level cascaded MLI. A 415 V, 50 Hz AC supply is utilized to drive a 7.5 kW (10-hp), 415 V, 50 Hz induction motor.

A. Calculation of DC-Link Capacitor Voltages

In this paper, a symmetrical 5-level cascaded MLI is utilized to drive an induction motor. The DC-link capacitors voltages are obtained by utilizing step by step procedure given in [19]. The DC-link capacitor voltages of each power cell is calculated as,

$$V_{ab1}=0.612m_a\left(K\text{-}1\right)V_{DC}=415\text{ V}$$

$$V_{DC}=415/(0.612\times0.92\times4)=184.27\text{ V}$$

The DC-link capacitor voltage of each power cell is taken as 185 V.

B. A 36-Pulse AC-DC Conversion

In this paper, two separate isolated transformers (T_1 and T_2), are utilized on grid side of proposed IMD. Secondary of each transformer is configured as an 18-pulse AC-DC converter and their primary has $+10^0$ and -10^0 angles. Because of these primary angles, these transformers are reconfigured as a 36-pulse AC-DC converter. The input and output winding voltages are obtained from step by step procedure given in [19]. The line voltage of a 3-phase DBR is calculated as,

$$V_{LL}=\pi V_{DC}/3\sqrt{2}=\left(\pi\times185\right)/\left(3\sqrt{2}\right)=136.99\text{ V}$$

The small (V_{DS}) and large (V_{LS}) secondary winding voltages are calculated as [19],

$$V_{DS}=\sin20^0/\left(\sin120^0\times136.99\right)=54.102\text{ V}$$

$$V_{DL}=\sin40^0/\left(\sin120^0\times136.99\right)=101.68\text{ V}$$

The input winding of T_1 and T_2 are connected in extended delta style the small (V_{EDS}) and large (V_{EDL}) winding voltages are calculated as [19],

$$V_{EDS}=N_{ps}V_{pan}=0.50771\times239.6=121.647\text{ V}$$

$$V_{EDL}=N_{pL}V_{pan}=0.87938\times239.6=210.699\text{ V}$$

III. Control Algorithms

In this paper, an indirect vector control (IVC) is utilized to control an IM and a 3rd harmonic injection based SPWM technique is utilized to operate a 5-level cascaded MLI. The complete control algorithm of proposed VCIMD is given here in detail.

A. Indirect Vector Control

In this paper, the motor speed is sensed and compared with reference speed as illustrated in Fig. 2. The output of the comparator is, the input for Proportional Integral (PI) Controller. The output of the PI controller is the reference torque. This reference torque is utilized to estimate the reference quadrature axis current and field weakening controller is utilized to estimate the reference direct axis current. The two-phase stator winding currents, are sensed and third phase current is estimated as $i_c=-(i_a+i_b)$. These stator winding currents, are converted into dq0-frame by utilizing abc to dqo conversions. These reference and sensed currents are compared and errors are passes through two PI controllers. The output of these two PI controllers, gives the direct and quadrature axes voltages. These dq0 axes voltages are converted in to abc co-ordinates. These v_a, v_b and v_c are the modulating signals for the proposed 3rd harmonics injection based sinusoidal pulse width modulation (SPWM) [4] and [7].

B. 3rd Harmonic Injection Based SPWM

In this paper, a 3rd harmonic injection based SPWM is proposed to control a 5-level cascaded MLI. A 15.5 % more fundamental voltage it achieved without changing the DC-link voltage by adding a 3rd harmonic component into the modulating signal. If the load fundamental frequency is fixed, then the frequency of 3rd harmonic component is 3-times the fundamental frequency. Moreover, if the modulating signal has fixed unity magnitude, then the magnitude of the 3rd harmonic current is equal to 1/6 [17]. However, in case of variable frequency induction motor drive (VFIMD), the magnitude and frequency of the modulating signals generated from indirect vector control, are not fixed and it is directly proportional to the speed of the motor. In case of VFIMD, the motor can have any speed as per the requirement. Hence, the modulating signals have varying magnitude and frequency. Therefore, in order to add a 3rd harmonic component into the modulating signal, one should require the information of magnitude and fundamental frequency.

In this paper, a MATLAB code is utilized to collect the magnitude information over wide range of operation of proposed VCIMD. An indirect vector control is utilized to control the speed of an IM. In indirect vector control, in order to find the rotor position angle, the synchronous speed is estimated by adding slip speed with sensed speed. The synchronous speed is utilized to estimate the fundamental frequency. From above discussion, it is observed that by doing this, one has both magnitude and fundamental frequency information. A 3rd harmonic has 3-times the fundamental frequency and its magnitude is equal to 1/6 time the estimated magnitude. The flowchart for an implementation of proposed 3rd harmonic injection based SPWM technique is illustrated in Fig. 3. From this figure, it is observed that, first magnitude (A_m) and frequency ($\omega_e t$) of the reference voltage signals, are estimated. Then (A_m/6)×$\sin(3\omega_e t)$ is added in each reference voltage signal. After this, the final modulating signal is equated with the 4-level shifted carrier to get the required PWM pulses.

978-1-5386-6624-1/18 $31.00 © 2018 IEEE

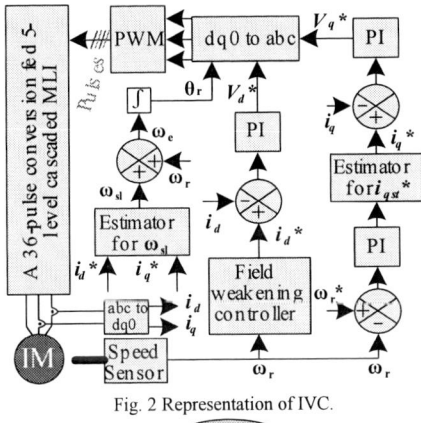

Fig. 2 Representation of IVC.

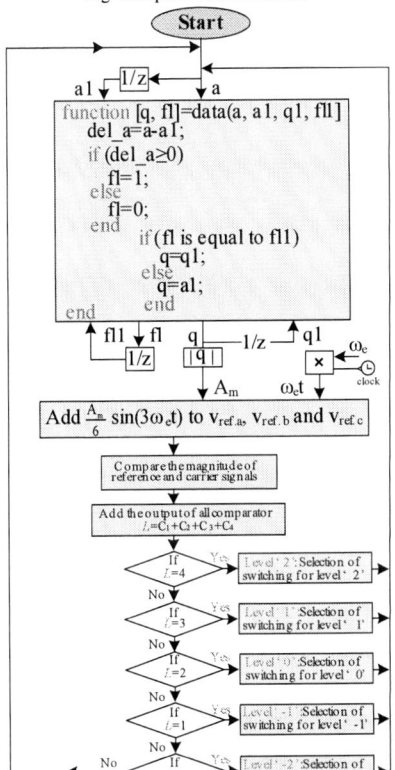

Fig. 3 Diagram of proposed 3rd harmonic injection based SPWM.

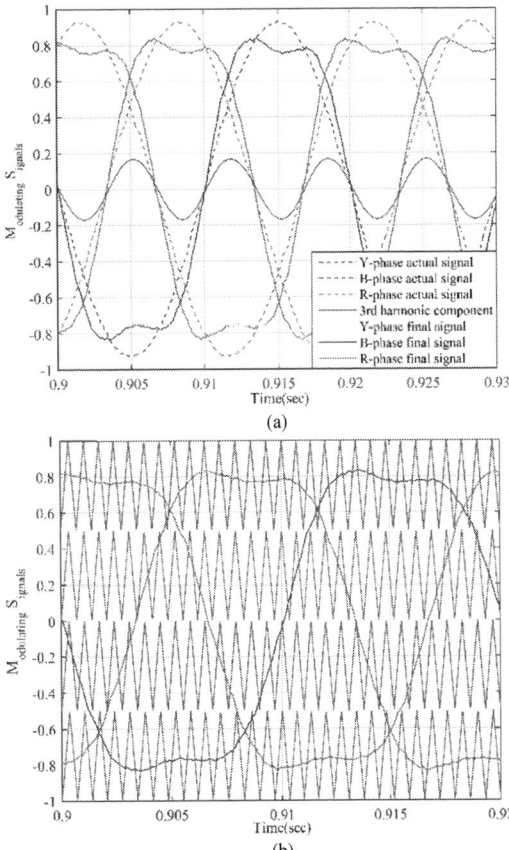

Fig. 4 Modulation signals (a) plot of modulating signals before and after addition of 3rd harmonic component (b) comparison of modulating signal with carrier signals.

The steady state response of proposed VCIMD is illustrated in Fig. 5(b). The starting response of proposed VCIMD is illustrated in Fig. 5(c). From above simulation results, it is witnessed that presented VCIMD has needed response during steady state and starting condition.

IV. SIMULATION RESULTS

The proposed VCIMD is simulated using Simulink/MATLAB based platform and its response is presented in this paper. The parameters of IM are mentioned in Appendix. Simulated results, are illustrated in Figs. 4-5. This figures depicts supply voltage (v_s), supply current (i_s), direct axis current (i_d), quadrature axis current (i_q), motor line voltage (v_{mt}), motor current (i_{mt}), speed (ω_m) and torque (T_e).

The modulation signals generated from indirect vector control, 3rd harmonic component and final modulating signals after adding 3rd harmonic component, are illustrated in Figs. 4(a)-4(b). The secondary (i_{s1}, i_{s2} and i_{s3}) and primary (i_{p1}) winding currents of transformer T_1, secondary (i_{s4}, i_{s5} and i_{s6}) and primary (i_{p2}) winding currents of transformer T_2 and resultant supply current, are illustrated in Fig. 5(a).

V. EXPERIMENTAL RESULTS

The proposed VCIMD is implemented in hardware by utilizing DSP-dSPACE (DS1006) controller. Two-current sensors are utilized to sense the two phase stator currents and third phase current is estimated as $i_{mc}=(-i_{ma}-i_{mb})$. A speed sensor is utilized to sense the motor speed. Experimental results are illustrated here in all the required operation of IMD. In this, experimental results in terms of supply line voltage (v_{s_ab}), supply current (i_{s_ab}), primary (i_{p1}) and secondary winding currents (i_{s1}, i_{s2}, i_{s3}) of transformer T_1, primary (i_{p2}) and secondary winding currents (i_{s4}, i_{s5}, i_{s6}) of transformer T_2, reference direct (i_d^*) and quadrature (i_q^*) axes currents, sensed direct (i_d) and quadrature (i_q) axes currents, motor line voltages (v_{mtab}, v_{mtbc}, v_{mtca}), motor line current (i_{mtab}, i_{mtbc}, i_{mtca}) are illustrated in detail.

A. Steady State Response

In order to examine the presented VCIMD during rated condition, the experimental results are illustrated if Fig 6. During rated speed and rated torque, the primary and secondary currents of transformer T_1 is illustrated in Fig. 6(a). The primary and secondary currents of transformer T_2 are illustrated in Fig. 6(b). The secondary currents of these two

978-1-5386-6624-1/18 $31.00 © 2018 IEEE

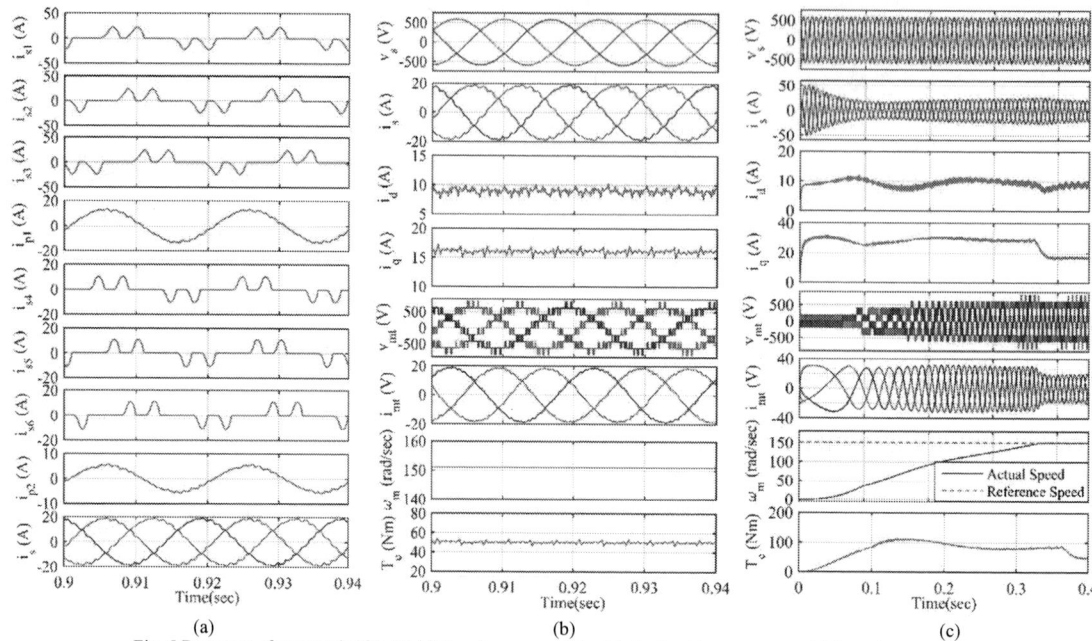

(a)   (b)   (c)

Fig. 5 Response of proposed VCIMD (a) transformer currents, (b) steady state response and (c) starting response.

transformers are seen almost similar. However, there is time delay between them. It is because, the voltage applied across the DBR has $+20^0$, 0^0 and -20^0 phase angles with respect to each other. The supply line voltage, primary currents of transformer T_1, T_2 and supply line current, are illustrated in Fig. 6(c), respectively. The second plot of this figure represents the 18-pluse primary current of transformer T_1 and third plot of this figure represents the 18-pluse primary current of transformer T_2. Whereas, the last plot of this figure, represents the supply current and it represent the 36-pulse current waveform. From these test results illustrated in Figs. 6(a)-6(c), it is witnessed that the proposed 36-pulse AC-DC converter has satisfactory and desired response.

The characteristic of proposed VCIMD during rated condition is illustrated in Fig. 6(d). Three-phase stator line voltages and stator line currents during rated condition are illustrated in Fig. 6(e). The internal signals of an IVC like reference and sensed value of direct and quadrature axes currents during rated condition are illustrated in Fig. 6(f). From the experimental results illustrated in Figs. 6(d)-6(f), it is witnessed that presented VCIMD has satisfactory response during rated condition.

B. Dynamic Response

In order to examine the dynamic response of proposed VCIMD, the starting characteristic of presented VCIMD is illustrated in Fig. 7(a). During starting response, the reference speed of the motor is set to 1000 rpm and speed PI controller output is limited to 1.3 times the rated torque. From Fig. 7(a), it is observed that the stator line current is 1.3 times the rated value during starting of the motor and it reduces to no load value, when the speed PI controller error becomes zero. From these experimental results, it is witnessed that the proposed VCIMD has satisfactory starting response. The speed perturbation characteristic of presented VCIMD is illustrated in Fig. 7(b). During speed control, initially motor is running at 1000 rpm and then speed of the motor is changed from

1000 to 1440 rpm. The response of supply line voltage, current, motor line voltage and current during speed control, is illustrated in Fig. 7(b).

The load change characteristic of presented VCIMD is illustrated in Fig. 7(c). During load control, initially motor is running at 40 % of rated load and then the load of the motor is changed from 40 % to 80 %. The response of supply line voltage, current, motor line voltage and current during load control, is illustrated in Fig. 7(c). From these experimental results illustrated in Figs. 7(a)-7(c), it is witnessed that the proposed VCIMD has satisfactory response during staring, speed and load control.

C. Power Quality Performance

The power quality performances of proposed VCIMD at the grid side, are illustrated in Figs. 8(a)-8(d). The waveforms of supply voltage and current, are illustrated in Fig. 8(a). The input power of multi-winding transformer is measured and depicts in Fig. 8(b). From this figure, it is observed that the 8.01 kVA power is drawn from the grid to drive a 7.5 kW. The input current waveform and its fast Fourier transform (FFT) analysis, are illustrated in Fig. 8(c). It depicts 3.5 % THD. The FFT of supply voltage is illustrated in Fig. 8(d). It depicts 1.8 % THD. From these experimental results illustrated in Figs. 8(a)-8(d), it is witnessed that the proposed VCIMD has satisfactory power quality performances at grid side.

The power quality performances of proposed VCIMD at the motor side in illustrated in Figs. 9(a)-9(d). The waveforms of motor line voltage and current, are illustrated in Fig. 9(a). The input power at the motor terminals, is measured and depicts in Fig. 9(b). From this figure, it is observed that the 6.81 kW power at the motor terminals. The motor current waveform and its fast Fourier transform (FFT) analysis, are illustrated in Fig. 9(c). It depicts 2.6 % THD. The FFT of motor line voltage is illustrated in Fig. 9(d). It depicts 8.6 % THD.

978-1-5386-6624-1/18 $31.00 © 2018 IEEE   774

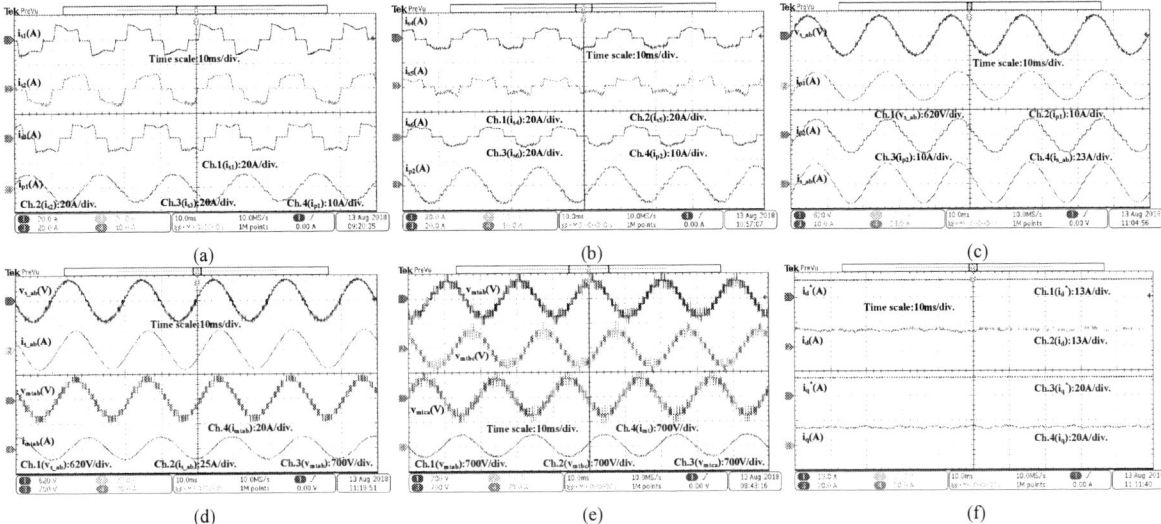

(a)   (b)   (c)

(d)   (e)   (f)

Fig. 6 Steady state response of proposed VCIMD (a) primary and secondary currents of transformer T_1, (b) primary and secondary currents of transformer T_2, (c) primary currents of transformer T_1, T_2 and resultant supply currents, (d) steady state response, (e) motor line voltages and (f) reference and sensed value of direct and quadrature axes currents.

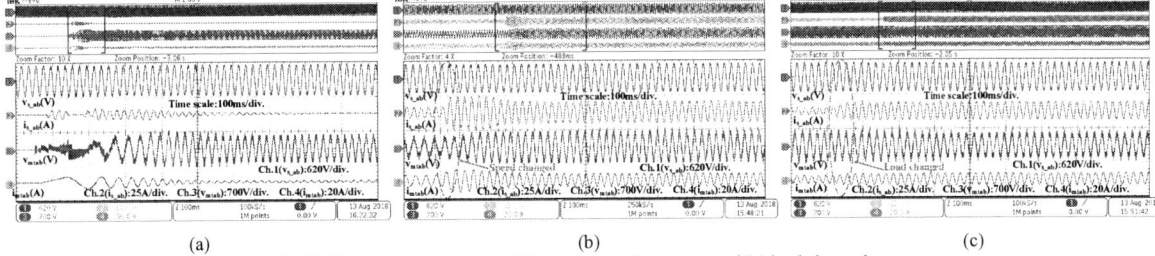

(a)   (b)   (c)

Fig. 7 Transient response of proposed VCIMD (a) starting response, (b) speed control response and (c) load changed response.

From these experimental results illustrated in Figs. 9(a)-9(d), it is witnessed that the proposed VCIMD has desired power quality performances at motor side.

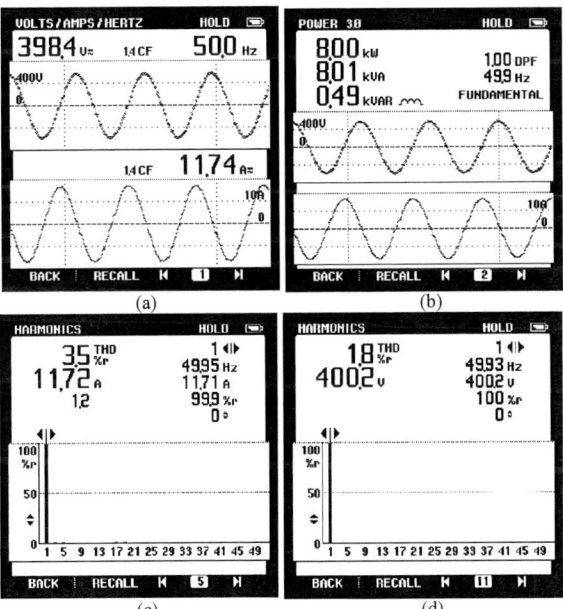

(a)   (b)

(c)   (d)

Fig. 8 Power quality at grid side (a) supply voltage and current waveform, (b) input power of multi-winding transformer, (c) supply current THD and (d) supply voltage THD.

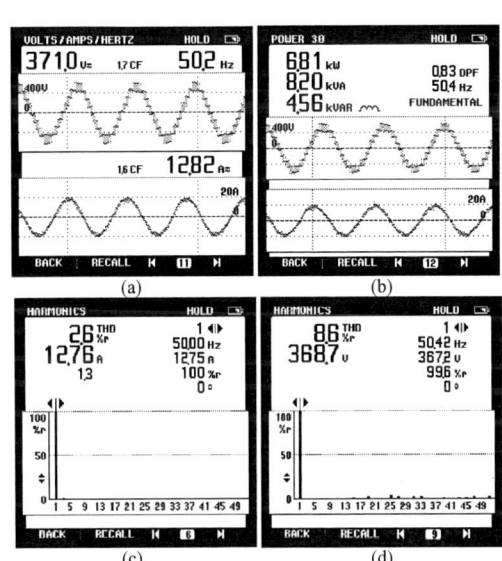

(a)   (b)

(c)   (d)

Fig. 9 Power quality at motor side (a) motor voltage and current waveform, (b) input power of IM (c) motor current THD and (d) motor voltage THD.

In this paper, a 36-pulse AC-DC converter fed 3rd harmonic injection based SPWM controlled symmetrical 5-level cascaded MLI driven VCIMD is proposed for medium rating application. From all the above simulation and experimental results, it is witnessed that the proposed VCIMD has desired response during wide range of operations.

978-1-5386-6624-1/18 $31.00 © 2018 IEEE   775

VI. Conclusions

A 36-pulse AC-DC converter fed 3rd harmonic injection based sinusoidal pulse width modulation controlled symmetrical 5-level cascaded MLI driven VCIMD has been proposed for medium power application. A 36-pulse converter makes the supply current nearly sinusoidal and it has reduced its THD to an acceptable level. A symmetrical 5-level cascaded MLI enhances the required performances of the proposed IMD. A 3rd harmonic injection based SPWM technique has been proposed for VFIMD to achieve MLI performance similar to SVM technique with simple and easy control. An indirect vector control has been utilized to achieve the good dynamic performances of IMD. All the required performances of IMD has been simulated in Simulink/MATLAB and simulated results has been verified from the experimental results to show the importance of proposed VCIMD aimed on medium power industrial applications.

Appendix

An IMD 7.5 kW (10 hp), 4 pole, 415 V, 50Hz, R_s=1 Ω, R_r=0.76 Ω, L_{ls}=0.00486 H, L_{lr}=0.00486 H, L_m=0.12 H and J=0.1 kg-m^2.

Acknowledgment

The authors are extremely grateful to Govt. of India, for funding the FIST project (RP03195) and J.C. Bose fellowship (RP03128).

References

[1] ABB/ BU Machines / HV Induction motors IEC catalogue / Standard motors EN 12-2007.

[2] A. Salem, E. M. Ahmed, M. Orabi and M. Ahmed, "New Three-Phase Symmetrical Multilevel Voltage Source Inverter," *IEEE Journ. Emerg. Selected Topics in Cirs. and Sys.,* vol. 5, no. 3, pp. 430-442, Sept. 2015.

[3] J. Venkataramanaiah, Y. Suresh and A. K. Panda, "Design and Development of a Novel 19-Level Inverter Using an Effective Fundamental Switching Strategy," *IEEE Journal of Emerging and Selected Topics in Power Electronics.* Early Access.

[4] B. Singh and P. Kant, "A 54-pulse AC-DC converter fed 15-level inverter based vector controlled induction motor drive,"*2017 IEEE Industry Applications Society Annual Meeting,* Cincinnati, OH, 2017.

[5] S. K. Chattopadhyay and C. Chakraborty, "Three-Phase Hybrid Cascaded Multilevel Inverter Using Topological Modules with 1:7 Ratio of Asymmetry," *IEEE Journal of Emerging and Selected Topics in Power Electronics.* Early Access.

[6] H. Khounjahan, M. Abapour and K. Zare, "Switched-Capacitor based Single source Cascaded H-bridge multilevel inverter featuring boosting ability," *IEEE Trans. Power Electrons.,* Early Access.

[7] P. Kant, B. Singh, A. Chandra and K. Al-Haddad, "Twenty pulse AC-DC converter fed 3-level inverter based vector controlled induction motor drive," *IECON 2017 - 43rd Annual Conference of the IEEE Industrial Electronics Society,* Beijing, 2017, pp. 2225-2230.

[8] N. Susheela, P. Satish Kumar and S. K. Sharma, "Generalized Algorithm of Reverse Mapping based SVPWM Strategy for Diode Clamped Multilevel Inverters," *IEEE Trans. Indust. Appls.,* Early Access.

[9] D. Cui and Q. Ge, "A Novel Hybrid Voltage Balance Method for Five-Level Diode-Clamped Converters," *IEEE Trans. Indust. Electrons.,* Early Access.

[10] J. Ebrahimi and H. Karshenas, "A New Single dc Source Six-Level Flying Capacitor Based Converter with Wide Operating Range," *IEEE Trans. Power Electrons.* Early Access.

[11] L. He and C. Chen, "A Bridge Modular Switched-Capacitor-Based Multilevel Inverter With Optimized SPWM Control Method And

Enhanced Power-Decoupling Ability," *IEEE Trans. Industr. Electrons.,* Early Access.

[12] P. W. Hammond, "A new approach to enhance power quality for medium voltage AC drives," *IEEE Trans. Indust. Appls.,* vol. 33, no. 1, pp. 202-208, Jan/Feb 1997.

[13] T. Parreiras, A. A. P. Machado, F. Amaral, G. Lobato, J. A. S. Brito and B. Cardoso, "Forward Dual-Active-Bridge Solid State Transformer for a SiC-Based Cascaded Multilevel Converter Cell in Solar Applications," *IEEE Trans. Indust. Appls.* Early Access.

[14] B. P. McGrath and D. G. Holmes, "Multicarrier PWM strategies for multilevel inverters," *IEEE Trans. Indust. Electrons.,* vol. 49, no. 4, pp. 858-867, Aug 2002.

[15] J. Weidong, L. Wang, J. Wang, X. Zhang and P. Wang, "An Carrier-Based Virtual Space Vector Modulation with Active Neutral Point Voltage Control for Neutral Point Clamped Three-Level Inverter," *IEEE Trans. Indust. Electrons.* Early Access.

[16] Y. Wang, C. Hu, R. Ding, L. Xu, C. Fu and E. Yang, "A Nearest Level PWM Method for the MMC in DC Distribution Grids," *IEEE Trans. Power Electronics,* Early Access.

[17] S. Bifaretti, A. Lidozzi, L. Solero and F. Crescimbini, "Modulation With Sinusoidal Third-Harmonic Injection for Active Split DC-Bus Four-Leg Inverters," *IEEE Trans. Power Electrons.,* vol. 31, no. 9, pp. 6226-6236, Sept. 2016.

[18] Md. R. Islam, Y. Guo and J. Zhu, "Power Converters for Medium Voltage Networks", Springer Heidelberg New York Dordrecht London, 2014, chapt. 3.

[19] P. Kant and B. Singh, "A multi-pulse AC-DC converter fed 5-level NPC inverter based VCIMD," *2018 IEEMA Engineer Infinite Conference (eTechNxT),* New Delhi, India, 2018, pp. 1-6.

2nd IEEE International conference on power Electronics, Intelligent Control and Energy systems (ICPEICES-2018)

Adaptive Neural Network Based Control of PV Connected Distribution System

Kanwar Pal, Shailendra Kumar, *Member,IEEE*, Bhim Singh, *Fellow, IEEE* and Tara C. Kandpal, *Senior Member, IEEE,*
Department of Electrical Engineering, IIT Delhi, New Delhi-110016, India

Abstract— **This work presents an adaptive neural network-based control approach for maximum power extraction and injection of active power in a photo-voltaic (PV) array connected distribution system. The control algorithm based on adaptive neural network (ANN), improves the performance of a PV connected distribution system. The ANN based control algorithm comprises of the sigmoidal and Gaussian functions for linearizing the load currents within a unity band to improve the performance of PV connected distribution system. This control algorithm calculates the amplitude of active and reactive components of nonlinear load currents to mitigate harmonics and to compensate reactive power for improving the power quality of proposed system. The ANN-based control algorithm has good dynamic response as compared to other ANN-based control algorithm. This algorithm requires only few training layers and unknown weights, which simplifies the control algorithm and reduces its execution time. It also enhances the performance of the system by estimating real time weight and specifies its merits.**

Keywords—PVarray, power quality, ANN Control Grid and MPPT.

I. INTRODUCTION

The PV array connected distribution systems, are on prime focus of researchers to meet the growing power demand of developing world community [1]. A PV connected distribution system consists of two sections: the first is solar PV array-based generation and the second is voltage source converter (VSC) tied to the distribution system [2]. The major issues in that PV array connected distribution system, are extraction of maximum power, an efficient utilization of this power and good stability of the system. Corresponding to each level of solar irradiation and PV array temperature, there is a unique point on PV characteristic, which represents efficient operating point. Thus, control algorithms are required to locate maximum power operating point accurately corresponding to each solar irradiation level and PV array temperature and it maintains the PV array and PCC voltage constant to preserve continuous functioning of the system in efficient and accurate manner.

The power distribution system is required to distribute the clean power to the connected utilities at fundamental frequency [3-4]. Besides the nonlinear loads, the switching of VSC also generates high order harmonics, which deteriorates the power quality of the distribution system [6]. The power quality problem is also solved by using UPQC (Unified Power Quality conditioners), DVR (Dynamic Voltage Restorer) and DSTATCOM [7]. However, harmonics in the distribution system, are produced by nonlinear load connected to it. These harmonics are mitigated by injecting same negative amplitude components of current at common integration point of the DSTATCOM [8].

The solar PV array integrates the distribution system through a voltage source converter (VSC), which is controlled to maintain required DC-link voltage, to get maximum power extraction from PV array and to improve power quality of the distribution system. The various control techniques, vector control [9], proportional resonant (PR) [10] and sliding mode control (SMC) [11] have been applied with hysteresis to work on high switching frequency, which requires expensive VSC. Recently researchers have focused on advanced control schemes to improve dynamic performance of the PV array connected distribution system, such as a model reference predictive controller and second order controllers [12]. These controllers are used to extract active component of nonlinear load current. The synchronous reference frame theory, to control grid interfaced VSC is reported in [13]. Wang et.al [14] have presented a multi-functional design of SVC for hybrid system to mitigate the harmonics. A controller for efficient power flow control from PV array to the distribution system and a multi-function VSC for power quality improvement in a three phase four wire system is described in [15]. Still there is a scope for researchers to develop a fast response control algorithm to improve stability, efficiency, dynamic response and accuracy of PV connected distribution system.

II. SYSTEM TOPOLOGY

Fig. 1 presents a single stage PV connected distribution system with DSP controller (dSPACE-1103). The configuration of the system consists of a VSC, opto-coupler based isolation circuit, a diode bridge rectifier connected at point of common coupling (PCC) to feed a highly inductive load and R-C filters for voltage ripple mitigation. The interfacing inductors are used to reduce the ripple in the grid currents.

III. CONTROL TECHNIQUES

The PV connected distribution system is required to utilize maximum power from the PV array without introducing harmonics in the distribution system. The control algorithm is required to generate a sequence of the switching pulses for multifunctional VSC. The adaptive controller based on neural network, is used to extract fundamental active component of load current by using a summation and a product neuron with adaptive weights. The output of these neurons, is summed up, which gives the resultant output of ANN based algorithm. The number of weights depend on number of input pulses and it is taken more than twice of input pulses.

978-1-5386-6624-1/18 $31.00 © 2018 IEEE

2nd IEEE International conference on power Electronics, Intelligent Control and Energy systems (ICPEICES-2018)

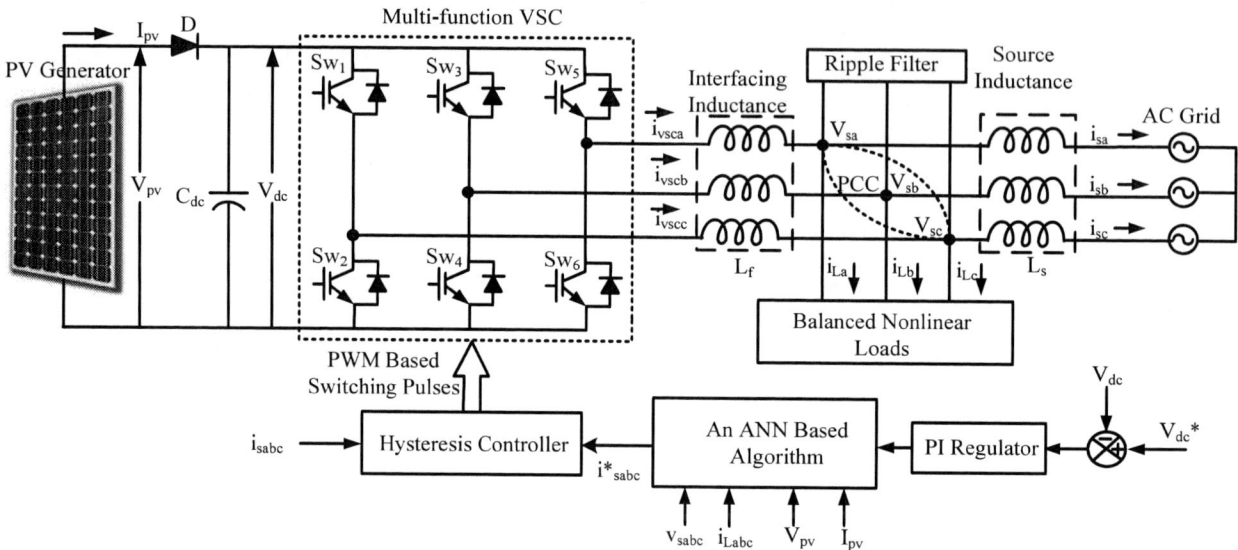

Fig. 1 Configuration of PV connected distribution system

A. Incremental Conductance Approach for MPPT

The maximum PV power is a function of solar insolation's and ambient temperature. In this work, an incremental conductance (INC) approach is used for extraction of maximum power from the PV array. In this technique, the P_{pv} versus V_{pv} and V_{pv} versus I_{pv} characteristics are used to decide maximum power and slope of the characteristics curve is analyzed. The maximum PV power is corresponding to zero slope of the characteristics and every point on the left present's negative slope and on the right positive slope. This INC based approach is simple, efficient and easy to implement. The change in voltage (ΔV_{pv}) and current (ΔI_{pv}), is observed at every instant in this approach to reach MPPT.

The INC based MPPT approach is explained in detail as,

$$P_{max} = V_{pv}*I_{pv} \qquad (1)$$

$$\frac{dP_{pv}}{dV_{pv}} = I_{pv} + V_{pv} \times \frac{dI_{pv}}{dV_{pv}} = 0 \qquad \text{(at MPPT)} \qquad (2)$$

From (2), if

$$\frac{I_{pv}}{V_{pv}} \rangle - \frac{dI_{pv}}{dV_{pv}} \text{ Then } \rho_{new} = \rho_{old} + \Delta\rho \text{ (positive slope point)}$$

If $\quad \dfrac{dI_{pv}}{dV_{pv}} = -\dfrac{I_{pv}}{V_{pv}} \quad$ then $\quad \rho_{new} = \rho_{old} \quad$ (zero slope point)

and $\dfrac{dI_{pv}}{dV_{pv}} \langle -\dfrac{I_{pv}}{V_{pv}}$ at that time $\rho_{new} = \rho_{old} - \Delta\rho$ (negative slope point)

In above expression, I_{pv} and V_{pv} are sensed current and voltage and ρ is the conductance of PV array.

B. ANN Based Control Algorithm

The proposed control algorithm is used to evaluate reference currents for PV connected distribution system. Both active and reactive components of load currents, are estimated with the use of a summing neuron and a product neuron. The output of summing neuron is used as an input to sigmoidal transfer function and product neuron is used as an input to the Gaussian transfer function. Three phase voltages (v_{sa}, v_{sb}, v_{sc}), DC link voltage (V_{dc}) of VSC and load currents (i_{La}, i_{Lb}, i_{Lc}), are sensed for this ANN based control algorithm.

The in-phase unit templates of three phase voltages are estimated as,

$$W_{pn} = \frac{v_{sn}}{V_t} \quad (n = a, b, c) \qquad (3)$$

Where w_{pa}, w_{pb} and w_{pc} are in phase unit templates of three phases.

The quadrature unit templates are derived using in phase templates as,

$$w_{qa} = -\frac{w_{pb}}{\sqrt{3}} + \frac{w_{pc}}{\sqrt{3}}, \ w_{qb} = \frac{\sqrt{3}w_{pa}}{2} + \frac{(w_{pb} - w_{pc})}{2\sqrt{3}},$$

$$w_{qc} = \frac{-\sqrt{3}w_{pa}}{2} + \frac{(w_{pb} - w_{pc})}{2\sqrt{3}} \qquad (4)$$

where w_{qa}, w_{qb} and w_{qc} are the quadrature unit templates of three phases.

The value of V_t is calculated as,

$$V_t = \sqrt{\frac{2}{3}(v_{sa}^2 + v_{sb}^2 + v_{sc}^2)} \qquad (5)$$

Fig. 2 shows ANN based control algorithm, which consists of one summation and one product neurons, f_Σ, $f_{\Sigma 1}$ and f_Π, $f_{\Pi 1}$

2nd IEEE International conference on power Electronics, Intelligent Control and Energy systems (ICPEICES-2018)

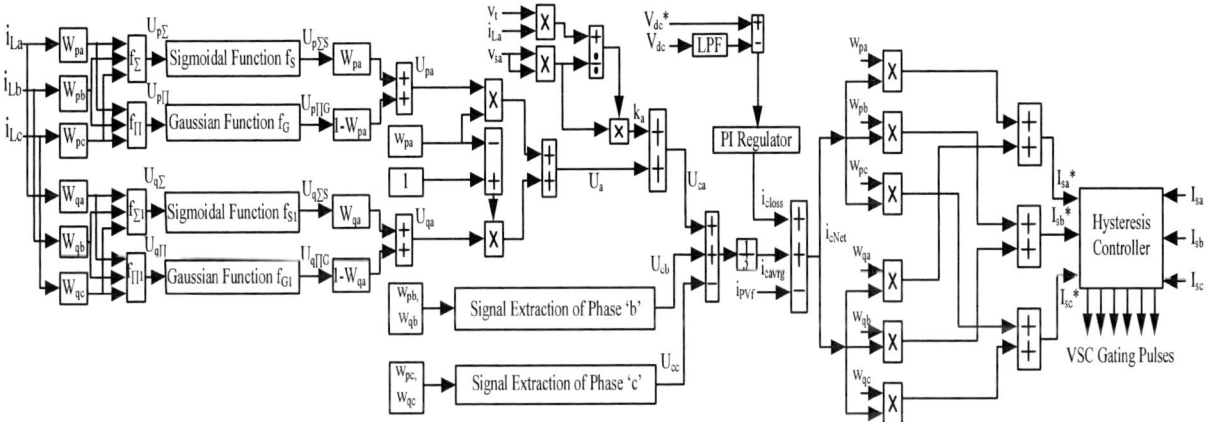

Fig. 2 ANN based control algorithm

respectively with adaptive weights w_Σ (w_{pa}, w_{pb} and w_{pc}) and w_Π (w_{qa}, w_{qb} and w_{qc}) for filtering higher order harmonics from load currents. Outputs of these neurons, are summed with these weights w_Σ and w_Π respectively. One additional weight is required for on-line training.

The summation function (f_Σ) output is expressed as,

$$U_{P\Sigma} = \sum_{j=a,b,c} i_{Lj} w_{pj} + w_{\Sigma bais} \qquad (6)$$

where $p_{\Sigma bias}$ is initial bias of summation neuron (f_Σ). The f_Σ function transforms load currents into internal signal, $U_{P\Sigma}$ which is transformed within unity magnitude by sigmoidal transfer function. The sigmoidal function (f_s) is expressed as,

$$f_s = \frac{1}{1 + e^{-\lambda_\Sigma U_{P\Sigma}}} \qquad (7)$$

where λ_Σ is a gain factor. The f_s transforms $U_{P\Sigma}$ into in phase internal signal, $U_{P\Sigma S}$. Similarly, sigmoidal function (f_{s1}) transforms internal signal ($U_{q\Sigma}$), using summation neuron ($f_{\Sigma 1}$) into $U_{q\Sigma s}$.

The output of product neuron (f_Π) is presented as,

$$U_{P\Pi} = \prod_{j=a,b,c} i_{Lj} w_{pj} \times w_{\Pi bais} \qquad (8)$$

Where $w_{\Pi bias}$ is initial bias of product neuron (f_Π). f_Π results in three phase load currents into internal signal, $U_{P\Pi}$. This is further processed by Gaussian transfer function (f_G), which results into internal signal, $U_{P\Pi G}$ within unity amplitude. Similarly, product neuron ($f_{\Pi 1}$) results into internal signal ($U_{q\Pi}$), which is transformed by Gaussian function (f_{G1}), into internal signal ($U_{q\Pi G}$). The Gaussian function is expressed as,

$$f_G = e^{-\lambda_\Pi \times U_{P\Pi}^2} \qquad (9)$$

Where λ_Π is Gaussian gain factor. The (8) and (9) are the output equations of summation function and product function. The ANN based control algorithm results in linear equations for three in phase components of load currents which are expressed as,

$$U_{pa} = U_{P\Sigma s} * w_{pa} + U_{P\Pi G}(1 - w_{pa}) \qquad (10)$$

$$U_{pb} = U_{P\Sigma s} * w_{pb} + U_{P\Pi G}(1 - w_{pb}) \qquad (11)$$

$$U_{pc} = U_{P\Sigma s} * w_{pc} + U_{P\Pi G}(1 - w_{pc}) \qquad (12)$$

Similarly, this ANN based control algorithm results in also linear equations for three quadrature phase components of load currents are expressed as,

$$U_{qa} = U_{q\Sigma s} * w_{qa} + U_{q\Pi G}(1 - w_{qa}) \qquad (13)$$

$$U_{qb} = U_{q\Sigma s} * w_{qb} + U_{q\Pi G}(1 - w_{qb}) \qquad (14)$$

$$U_{qc} = U_{q\Sigma s} * w_{qc} + U_{q\Pi G}(1 - w_{qc}) \qquad (15)$$

The control algorithm transforms U_{pa} and U_{qa} using in phase unit template of phase 'a' into lossless unit power component (U_a) of i_{La}. Similarly, the lossless unity power components (U_b and U_c) of i_{Lb} and i_{Lc} are estimated respectively. The U_a, U_b and U_c result in with load summing factors (k_a, k_b and k_c) into internal signals U_{ca}, U_{cb} and U_{cc} respectively. The average of U_{ca}, U_{cb} and U_{cc} is estimated as,

$$I_{cavrg} = \frac{U_{ca} + U_{cb} + U_{cc}}{3} \qquad (16)$$

The performance of PV connected distribution system under unbalanced load is improved by this average component (I_{cavrg}) for generating the reference currents. The voltage (V_{dc}) at input of VSC is controlled by proportional-integral (PI) controller. The PI controller is used to reduce the error between sensed V_{dc} and reference $V_{dc}*$. The output of PI controller gives loss component (I_{closs}) of the current. The power injected by PV array to the utility is given as,

$$I_{PVf} = \frac{2P_{PV}}{3V_t} \qquad (17)$$

The net in phase component of load current is estimated as,

$$I_{cnet} = I_{cavrg} + I_{closs} - I_{PVf} \qquad (18)$$

The reference current of three-phase for producing the control signals for VSC are estimated as,

$$i_{sa}* = I_{cnet} \times w_{pa} + I_{cnet} \times W_{qa} \qquad (19)$$

$$i_{sb}* = I_{cnet} \times w_{pb} + I_{cnet} \times W_{qb} \qquad (20)$$

$$i_{sc}* = I_{cnet} \times w_{pc} + I_{cnet} \times W_{qc} \qquad (21)$$

The sensed grid currents (i_{sa}, i_{sb} and i_{sc}) and estimated

978-1-5386-6624-1/18 $31.00 © 2018 IEEE   779

reference currents are compared and fed to the hysteresis controller to generate switching sequence for VSC.

IV. EXPRIMENTAL RESULTS

A prototype of this PV connected distribution system is developed in the laboratory that includes a PV array simulator, which behaves exactly as PV array roof top system, a ripple filter, a VSC with interfacing inductors, Hall Effect sensors for sensing the voltages and currents (EM010 BB and EL50P1BB), a three-phase diode bridge rectifier connected with highly inductive load. The developed PV connected distribution system is controlled with the help of digital signal processor (DSP: dSPACEl103). The performance characteristics of system, are recorded using a power analyzer (Fluke 43-B) and a 4-channel digital storage oscilloscope. The voltages and currents signals are sensed and sent to the ADC channel of DSP controller. Figs. 3 (a)-(b) show the performance characteristics of PV array for executing peak power at 1000W/m² and 500W/m², respectively. The peak power point is represented by a small dote on the graph, which presents maximum operating performance of PV connected distribution system. It has been observed that power of PV array corresponding to this dote point is 99% of maximum power. Therefore, PV array simulator in PV connected distribution system, is feeding maximum power to the system. The parameters of developed hardware prototype, are shown in the Appendix.

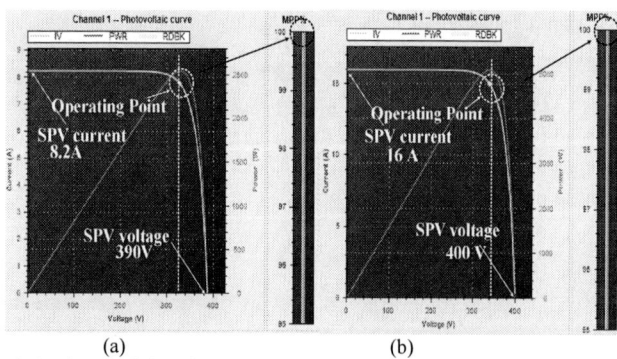

Fig.3 characteristic of PV indicating maximum power point at varying insolations

A. Response of ANN based control algorithm

Figs.4 (a)-(b) present load current (i_{La}) of phase 'a' and summation function output signals, $U_{p\Sigma}$, extracted component (U_{pa}) and quadrature component (U_{qa}) of ANN algorithm. It has been observed that sudden removal and connection of load of phase 'a'; the amplitude of $U_{p\Sigma}$, U_{qa} and U_{pa} is reduced.

Figs.5 (a)-(b) show i_{La} and power components (U_{ca}, U_{cb} and U_{cc}) load current. It has been observed that due to load removal and load connection of phase 'a', the amplitude of active component of particular phase reduced to zero.

Figs.6 (a)-(b) show i_{La} and average power component of load currents (I_{cavrg}), power loss component of PI controller (I_{closs}) and net power component of load currents (I_{cnet}). It has been observed that on sudden removal phase 'a', I_{cavrg} decreases, I_{closs} remains constant and I_{cnet}, has increased.

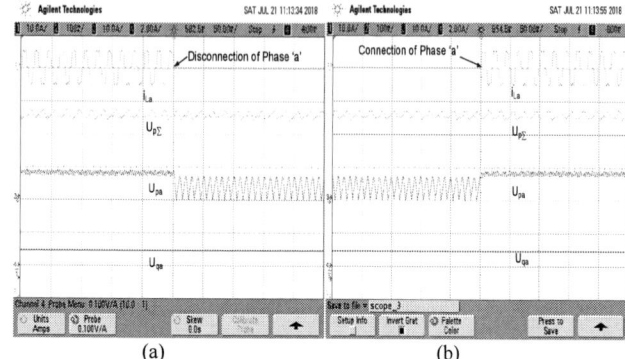

Figs. 4 (a)-(b) various signals i_{La}, $U_{p\Sigma}$, U_{pa} and U_{qa} of ANN at sudden disconnection and connection of phase 'a'

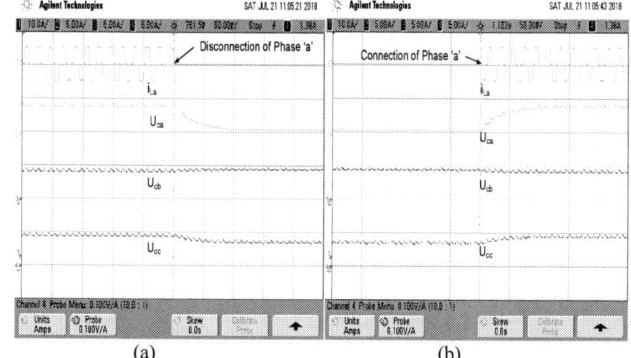

Figs. 5 (a)-(b) Various signals, i_{La}, U_{ca}, U_{cb} and U_{cc} of ANN based control algorithm at sudden load disconnection and connection on phase 'a'

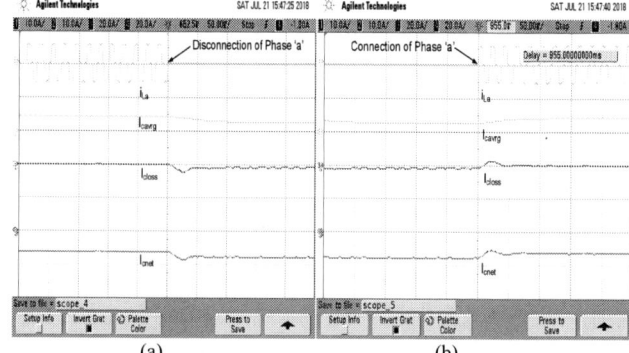

Figs. 6 (a)-(b) various signals, i_{La}, I_{cavrg}, I_{closs} and I_{cnet} of ANN based controller in case of sudden disconnection and connection of load current of phase 'a'

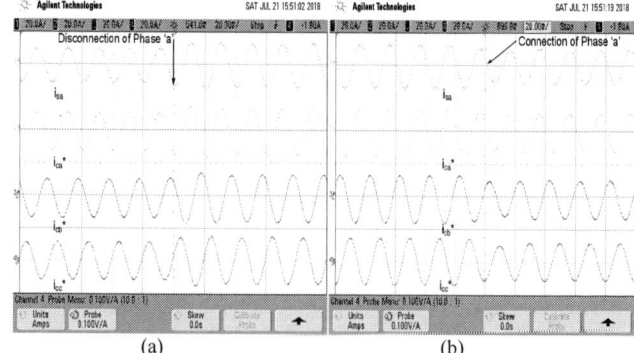

Figs.7 (a)-(b) Various signals, i_{sa}, i_{ca*}, i_{cb*} and i_{cc*} at disconnection and connection of i_{La}.

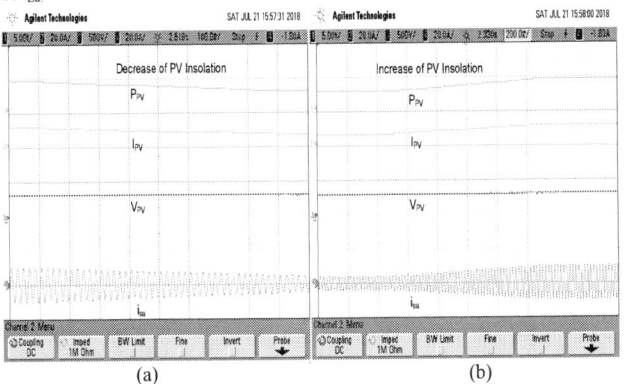

(a)   (b)

Figs. 8 (a)-(b) various signals, P_{pv}, I_{pv}, V_{pv} and i_{sa} in the case of decrease and increase of PV insolation

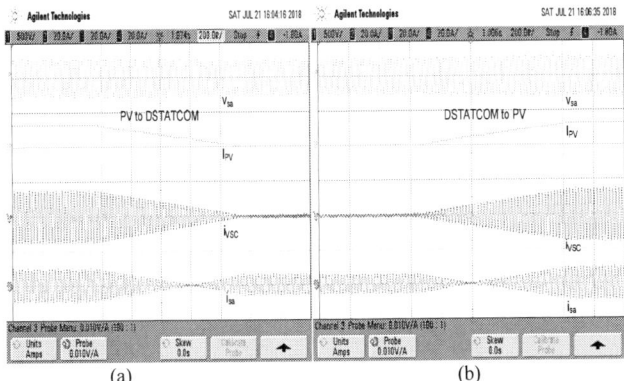

(a)   (b)

Figs.9 (a)-(b) Various signals v_{sa}, I_{pv}, i_{vsc} and i_{sa} from PV to DSTATCOM and DSTATCOM to PV

Figs.7 (a)-(b) show distribution system phase 'a' current (i_{sa}) and three phase reference currents $i_{ca}*$, $i_{cb}*$ and $i_{cc}*$. It has been observed that on sudden removal of phase 'a', the reference currents and i_{sa} are increased and at connection of load these are reduced.

Figs. 8 (a)-(b) exhibit various signals, PV power (P_{pv}), PV current (I_{pv}), PV voltage (V_{pv}) and distribution system phase 'a' current (i_{sa}) under decrease and increase of solar insolation. It has been observed that P_{pv}, I_{pv} and i_{sa} are decreased and increased, respectively. However, V_{pv} remains almost constant. Figs. 9 (a)-(b) show distribution system voltage (v_{sa}) of phase 'a', PV current (I_{pv}), VSC current (i_{VSC}) and distribution system current (i_{sa}) of phase 'a' under operation of PV to DSTATCOM mode. It has been observed that i_{VSC} and I_{pv} are reached zero and the compensator is supplying only reactive current. The direction of grid current is reversed and now the load is fed by the distribution system. A vice-versa is observed in Fig. 9 (b).

B. Test of PV Connected Distribution System under Balanced Load

Fig. 10 (a) visualizes distribution system voltages and currents and Fig. 10 (b) shows active and reactive power fed to the distribution system under balanced load condition. Fig. 11 (a) illustrates the phasor diagram of distribution system and Fig. 11 (b) shows the THD of the distribution system voltage (i.e.

2.02%) and current (i.e. 3.55%). Fig.12 (a) shows voltages and currents of VSC and Fig. 12 (b) shows the power fed by the VSC to the distribution system and to balance load. Fig.13 (a) illustrates phasor diagram of VSC and Fig. 13 (b) shows the THD of VSC voltage (i.e. 2.03%) and current (i.e. 9.75%). Figs.14 (a) shows distribution system voltages and load currents and Fig. 14 (b) shows load power under balanced condition. Figs.15 (a) presents phasor diagram of voltages and currents at PCC and Fig. 15 (b) shows the THD of PCC voltage (i.e. 2.07%) and load current (i.e. 28.07%) under balanced load. Fig.16 (a) shows THD and harmonics component of distribution system current and Fig. 16 (b) shows THD and harmonics component of load current at balanced loads. It is observed that THD of current fed to the load is 28.87% and THD of grid current is improved to 3.55%.

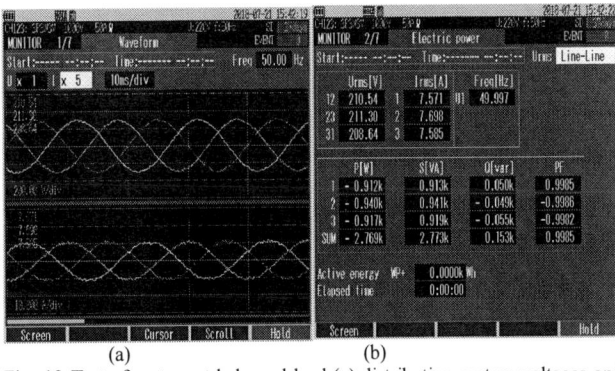

(a)   (b)

Figs.10 Test of system at balanced load (a) distribution system voltages and currents and (b) active and reactive power feed to the distribution system

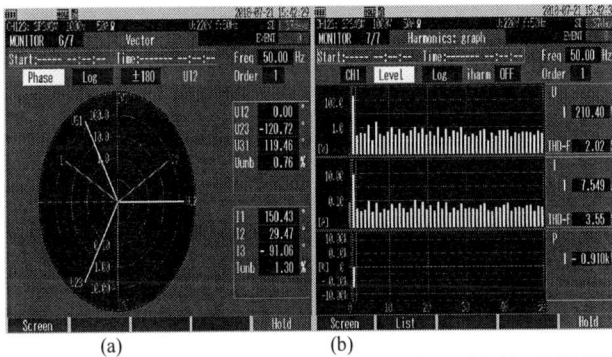

(a)   (b)

Figs.11 Test of system at balanced load (a) Phaser diag. of Grid and (b) THD of the the distribution system voltage and current

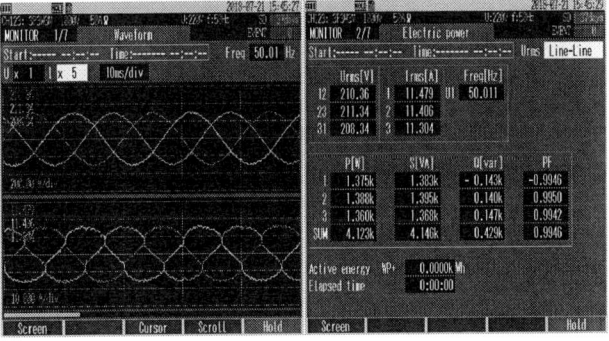

2nd IEEE International conference on power Electronics, Intelligent Control and Energy systems (ICPEICES-2018)

(a)   (b)

Fig.12 Test of the system at balanced load (a) VSC voltages and currents and (b) power fed by VSC

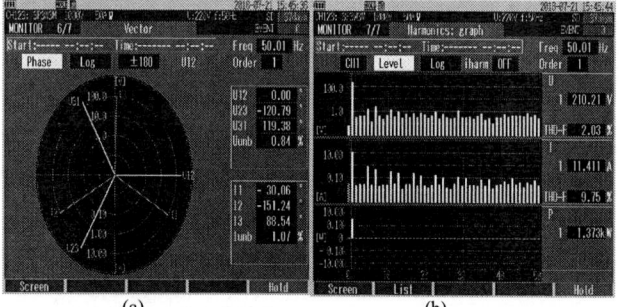

(a)   (b)

Figs.13 Test of system at balanced load (a) phaser diag. of VSC and (b) THD of VSC voltage and current

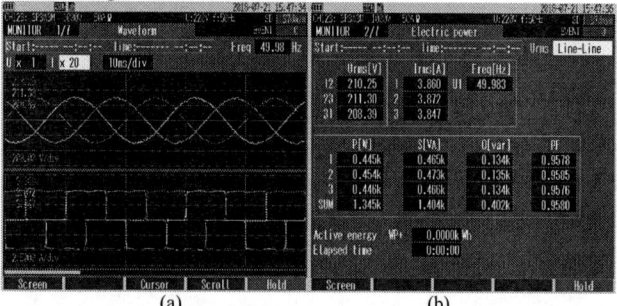

(a)   (b)

Figs.14 Test of system at balanced load (a) Grid voltages and load currents and (b) load power

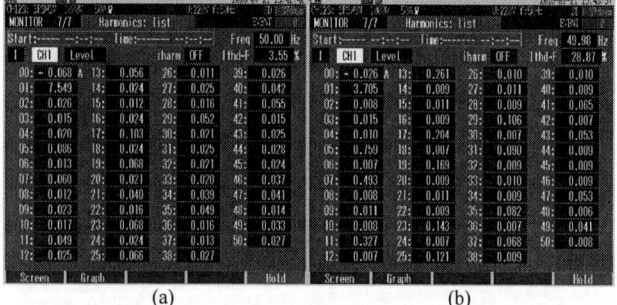

(a)   (b)

Figs.15 Test of system at balanced load (a) phaser diag. of PCC and (b) THD of PCC voltage and load current

(a)   (b)

Figs.16 Test of the system at balanced load (a) THD and harmonics components of the distribution system current at different loads and (b) THD and harmonics component of load current at different loads

V. CONCLUSION

An ANN based control algorithm for a PV connected distribution system has been proposed in this work. The performance of the proposed control algorithm is validated for balanced and unbalanced nonlinear loads and varying solar insolation. Test results are demonstrated to realize performance of ANN based control algorithm. The proposed control algorithm has been designed by using a product and a summing neuron with a single layer, which also improves the dynamics of this system under sudden change in working conditions. The extraction of maximum power from the PV array and regulation of DC voltage of VSC to its reference value demonstrate the efficacy of ANN based control algorithm. The power quality performance of PV connected distribution system has been found satisfactory as harmonics spectra and THD are in acceptable range (‹5%) according to IEEE-519 standard.

APPENDIX

Open Circuit voltage and current: 400 V, 16 A, VSC capacitor=2300 μF, Coupling Inductors= 4.5 mH, R-C filter =5 , 10 μF, Nonlinear load: 1.345 kW, Grid parameters: 230 V, 50 Hz, Controller parameters: K_p = 9.4 and K_i = .84.

REFERENCES

[1] Renewable EnergyWorld Editors. Residential Solar Energy Storage Market Could Approach 1 GW by 2018.

[2] R. Mastromauro, M. Liserre, and A. Aquila, "Control issues in single-stage photovoltaic systems: MPPT, current and voltage control," *IEEE Trans. Ind. Informatics,* vol. 8, no. 2, pp. 241–254, May 2012.

[3] IEEE Recommended practices and requirement for harmonic control on electric power system, IEEE Std.519.

[4] S. Arya, R. Niwas, K. Bhalla, B. Singh, A. Chandra and K. Al-Haddad, "Power Quality Improvement in Isolated Distributed Power Generating System Using DSTATCOM," *IEEE Tran. on Ind. Appl.,* vol. 51, no. 6, pp. 4766-4774, Nov.-Dec. 2015.

[5] S. Arya and B. Singh, "Neural Network Based Conductance Estimation Control Algorithm for Shunt Compensation," *IEEE Tran. on Indust. Inform.,* vol. 10, no. 1, pp. 569-577, Feb.2014.

[6] M. Qasim and V. Khadkikar, "Application of Artificial Neural Networks for Shunt Active Power Filter Control," *IEEE Trans. on Ind. Inform.,* vol. 10, no. 3, pp. 1765-1774, Aug. 2014.

[7] S. Kumar and B. Singh, "A Multipurpose PV System Integrated to a Three-Phase Distribution System Using an LWDF-Based Approach," *IEEE Trans. Power Electronics,* vol. 33, no. 1, pp. 739-748, Jan. 2018.

[8] B. Singh and S. R. Arya, "Back-Propagation Control Algorithm for Power Quality Improvement Using DSTATCOM," *IEEE Tran. on Ind. Electron.,* vol. 61, no. 3, pp. 1204-1212, March 2014.

[9] A. Abrishamifar, A. Ahmad and M. Mohamadian, "Fixed Switching Frequency Sliding Mode Control for Single-Phase Unipolar Inverters," *IEEE Trans. Power Electron.,* vol. 27, no. 5, pp. 2507-2514, May 2012.

[10] B. Singh, S. Kumar and C. Jain, "Damped-SOGI-Based Control Algorithm for Solar PV Power Generating System," *IEEE Trans. on Ind. Appl.,* vol. 53, no. 3, pp. 1780-1788, May-June 2017.

[11] P. Li, X. Yu and B. Xiao, "Adaptive Quasi-Optimal Higher Order Sliding Mode Control Without Gain Overestimation," in *IEEE Transactions on Industrial Informatics (Early Access),* 2018.

[12] M. Ahmad, N. Kumar and B. Singh, "Fast multilayer perceptron neural network-based control algorithm for shunt compensator in distribution systems," *IET Gen., Trans. & Dist.,* vol. 10, no. 15, pp. 3824-3833, 17 11 2016.

[13] S. Chauhan, M. Shah, R. Tiwari and P. Tekwani, "Analysis, design and digital implementation of a shunt active power filter with different schemes of reference current generation," *IET Power Electrn.,* vol. 7, no. 3, pp. 627-639, March 2014.

[14] L. Wang, C. Lam and M. W. Wong, "Multi-Functional Hybrid Structure of SVC and Capacitive Grid Connected Inverter (SVC//CGCI) for Active Power Injection and Non-Active Power Compensation," *IEEE Trans. Ind. Electron. (Early Access),* 2018.

[15] S. Kumar and B. Singh, "Multi-Objective Single-Stage SPV System Integrated to 3P4W Distribution Network Using DMSI-Based Control Technique," *IEEE Trans. on Ind. Appl.,* vol. 54, no. 3, pp. 2656-2664, May-June 2018.

2nd IEEE International conference on Power Electronics, Intelligent Control and Energy systems (ICPEICES-2018)

Improved Laplacian Kernel Filter based Control of Multifunctional PV System with Enhanced Power Quality

Sai Pranith[1], Shailendra Kumar[2], Bhim Singh[2], *Fellow, IEEE* and T.S. Bhatti[1]
[1]Centre for Energy Studies, IIT Delhi, New Delhi 110016, India
[2]Department of Electrical Engineering, *IIT Delhi*, New Delhi 110016, India
Emails: saipranith.eee@gmail.com, cr.dwivedi88@gmail.com, bsingh@ee.iitd.ac.in, tsh@ces.iitd.ac.in

Abstract—This paper proposes an improved Laplacian Kernel filter based control of a grid-connected photovoltaic (PV) system. This algorithm improves the power quality by mitigating harmonics distortion, provides power factor correction and also enables injection of maximum power from the PV array to the utility. The system under study, uses a PV array, a boost converter for maximum power point tracking and a voltage source converter (VSC) for connecting the PV boost converter to the grid. The proposed control algorithm is tested by varying irradiance, and under load disturbances. The performance of the control is found satisfactory under test conditions and the obtained total harmonic distortion of grid current satisfies the criteria set by the IEEE 519 standard.

Keywords—Laplacian Kernel, photovoltaic (PV) array, power quality, voltage source converter (VSC), Boost converter

I. INTRODUCTION

Traditional fossil fuels have hazardous effects on the environment in the form of greenhouse gases and global warming [1]. Renewable energy sources like Photovoltaics (PV) can play a key role in countering the harmful effects of fossil fuels. PV array is used in various kinds of applications ranging from powering loads and the grid [2], water pumping [3] etc. The PV systems can be designed to operate in grid-connected applications or standalone systems. For a standalone system, the battery is essential to make it reliable. However, in the grid-connected scenario, the grid takes care of the PV array power intermittency, ensuring continuous supply to the load. Especially, in rural India, the agricultural activity happens mostly during daytime and there are 300 sunny days available every year, which makes battery storage redundant. These factors also make PV array, a very good option for rural electrification.

However, the benefits have some challenges like the power quality issues, harmonics, changes in feeder power factor etc [1, 4]. Further, the PV array has nonlinear I-V characteristics, which makes it necessary to regulate performance at the proper operating point. For the optimal utilization of the PV array, maximum power point tracking (MPPT) algorithms are implemented by a power converter to control the input voltage [5]. A boost converter is used to implement the incremental conductance (INC) MPPT in this work.

The PV array with a boost converter is connected to the utility through a voltage source converter (VSC). There are various control techniques available in the literature like synchronous reference frame (SRF) theory, unit template based control, frequency observer based control [2] to make the VSC function like a DSTATCOM [6-8], thus enabling multiple functionalities of the grid-PV system. Adaptive filters can also be used for controlling the VSC [9-10]. These filters adapt their transfer function to the changes in the sensed signals over time by minimizing the error or loss function defined for the filter. The learning for the adaptation process can be done using an online algorithm and employing intelligent techniques like neural networks (NN) and Kernel filters.

Unlike NN methods, the Kernel filter has convex optimization [10]. The traditional non-linear methods like Hammerstein and Wiener models do not possess universal approximation property. The Kernel methods have the universal approximation property and are also easy to compute. In this work, an improved Laplacian Kernel-based method is proposed to control the grid-PV system to obtain multiple functionalities like reactive and harmonics compensation and load balancing along with extracting maximum power. The control has shown satisfactory performance under various test conditions as per the IEEE 519 standard.

II. SYSTEM CONFIGURATION

The proposed system comprises of a PV array with a boost converter feeding a nonlinear load and utility through a VSC as shown in Fig. 1. To eliminate the voltage ripples caused by converter switching, an RC ripple filter is connected at the point of common coupling (PCC).

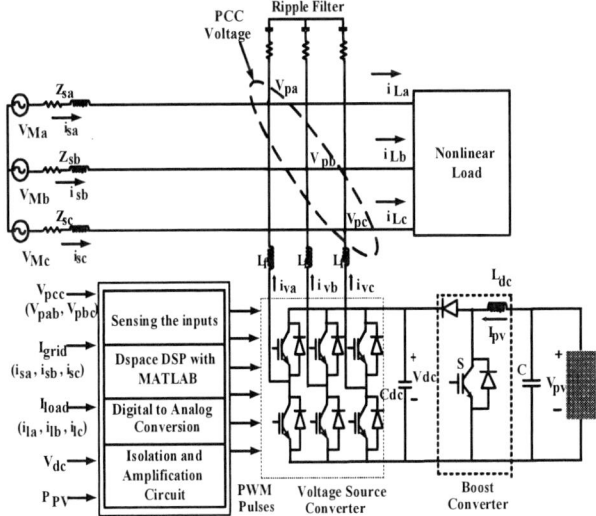

Fig. 1. Schematic of Grid-PV system

978-1-5386-6624-1/18 $31.00 © 2018 IEEE

III. CONTROL ALGORITHMS

The system has a two-stage configuration for interfacing PV system to the utility. A boost converter is used to implement the MPPT and a VSC is used to connect the PV array and a boost converter to the grid.

A. Incremental Conductance based MPPT

The control algorithm used in this work, is the incremental conductance method [5]. This algorithm differentiates the PV power output to voltage and bringing the result to zero. The relationships are represented as follows:

$$\frac{dI_{pv}}{dV_{pv}} > -\frac{I_{pv}}{V_{pv}}; \left(\frac{dP_{pv}}{dV_{pv}} > 0\right) \quad towards \ the \ left \ of \ MPP \tag{1}$$

$$\frac{dI_{pv}}{dV_{pv}} < -\frac{I_{pv}}{V_{pv}}; \left(\frac{dP_{pv}}{dV_{pv}} < 0\right) \quad towards \ the \ right \ of \ MPP \tag{2}$$

The overall MPPT algorithm is shown in Fig. 2.

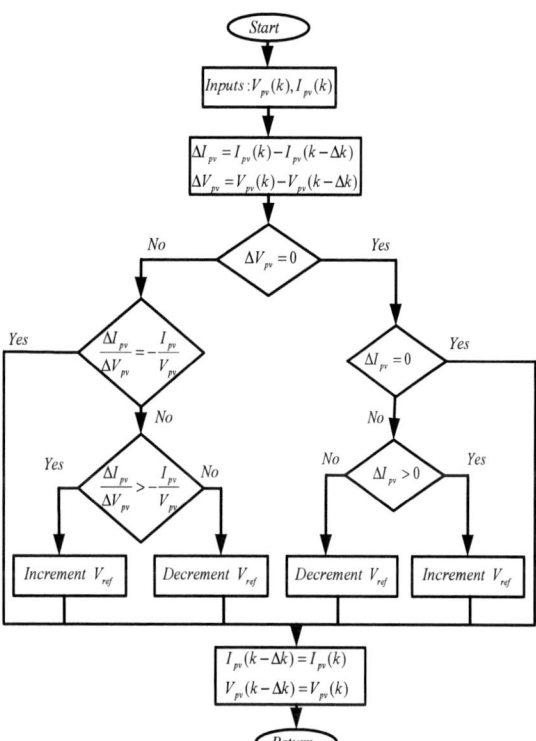

Fig. 2. Schematic of Incremental Conductance MPPT algorithm

B. Improved Kernel-based Control

In this work, a grid current control is used for the VSC and the desired grid currents references for the control, are obtained using the proposed improved Laplacian Kernel-based control method. This algorithm is used to control the VSC to obtain multiple functionalities like reactive power and harmonics compensation, load balancing along with injecting the obtained power from the PV array. The step-by-step procedure for implementing this algorithm proceeds with calculating the unit templates, weights for the fundamental component, DC link loss component and the PV feed-forward component and estimating the reference grid currents as explained here.

1) Calculation of peak voltage amplitude and unit templates

The line voltages (v_{pab}, v_{pbc}) are utilized to obtain the phase voltages (v_{pa}, v_{pb}, v_{pc}) as,

$$v_{pa} = \frac{2v_{pab} + v_{pbc}}{3}; v_{pb} = \frac{-v_{pab} + v_{pbc}}{3}; v_{pc} = \frac{-v_{pab} - 2v_{pbc}}{3} \tag{3}$$

A bandpass filter is used to eliminate the voltage harmonics. The amplitude of the terminal voltage V_t is calculated as,

$$V_t = \sqrt{\frac{2}{3}\left(v_{pa}{}^2 + v_{pb}{}^2 + v_{pc}{}^2\right)} \tag{4}$$

In-phase unit templates (u_{pa}, u_{pb}, and u_{pc}) are estimated as,

$$u_{pa} = \frac{v_{pa}}{V_t}; u_{pb} = \frac{v_{pb}}{V_t}; u_{pc} = \frac{v_{pc}}{V_t}. \tag{5}$$

2) Calculation of weight of DC loss component

For maintaining a constant DC link voltage, the voltage error is obtained as shown in Fig. 3.

$$V_{dc\_error}(g) = V_{dc\_ref}(g) - V_{dc}(g) \tag{6}$$

where
V_{dc_ref} = Reference DC link voltage, V_{dc} = Sensed DC link voltage
A PI controller is used to eliminate the voltage error. To maintain the set DC link voltage, the output W_{loss} is calculated as:

$$W_{loss}(g+1) = W_{loss}(g) + K_P\{V_{dc\_error}(g+1) - V_{dc\_error}(g)\} + K_I V_{dc\_error}(g+1) \tag{7}$$

where K_P and K_I are the proportional and integral gains of the PI controller

3) Calculation of weight of feed-forward term

The feed forward term (W_{pv}) of PV array is estimated as,

$$W_{pv} = \frac{2P_{PV}}{3V_T} \tag{8}$$

where P_{PV} = extracted PV power

4) Calculation of weight of fundamental Load Current:

Prediction error, $e_{pa}(g+1)$ of Phase 'a' as,

$$e_{pa}(g+1) = i_a(g+1) - u_a \times w_{pa}(g) \tag{9}$$

where $e_{pa}(g+1)$ is minimized through proper updating of $w_{pa}(g+1)$ as,

$$w_{pa}(g+1) = w_{pa}(g) + \alpha_a \times u_a(g+1) \times e_{pa}(g+1) \tag{10}$$

In (10), the learning during weight updating process is enhanced by using learning factor (Ω). Therefore improved version of Eq. (11) as,

$$w_{pa}(g+1) = w_{pa}(g) \times (1 - \Omega \times \alpha_a) + \alpha_a \times u_a(g+1) \times e_{pa}(g+1) \tag{11}$$

In (11), α_a is active weight constant. The value of α_a is also improved by using autocorrelation factor $\sigma_a(g)$, which is calculated as,

$$\sigma_a(g+1) = \beta \times \sigma_a(g) + (1-\beta) \times e_{pa}(g+1) \times e_{pa}(g) \tag{12}$$

$$\alpha_a(g+1) = \gamma \times \alpha_a(g) + \tau \times (\sigma_a(g+1)^2) \qquad (13)$$

where γ & τ are accelerating parameters, β is the auto-correlation parameters. Therefore improved equation is described as,

$$w_{pa}(g+1) = (1-\Omega \times \alpha_a(g+1)) \times w_{pa}(g) + \alpha_a(g+1) \times U_a(g+1) \times e_{pa}(g+1) \quad (14)$$

The accuracy of pattern recognition of (14) is further improved by using Laplacian Kernel Function (K_{pa}) [10] which is described as,

$$K_{pa} = e^{-\Omega \times |\mu_{pa} - \sigma_{pa}|} = e^{-sy*|A_{pa1} - P_{pa1}|} \qquad (15)$$

$$K_{pa}(\alpha_a(g+1), \sigma_a(g+1)) = e^{-\tau|\alpha_a(g+1) - \sigma_a(g+1)|} \quad (16)$$

Therefore final equation of $w_{pa}(g+1)$ is described as,

$$W_{pa}(g+1) = (1-\Omega \times \alpha_a(g+1)) \times W_{pa}(g) + \left(\frac{\alpha_a(g+1) \times u_a(g+1) \times e_{pa}(g+1)}{e^{-\tau|\alpha_a(g+1)-\sigma_a(g+1)|}} \right) \quad (17)$$

$$W_{pb}(g+1) = (1-\Omega \times \alpha_b(g+1)) \times W_{pb}(g) + \left(\frac{\alpha_b(g+1) \times u_b(g+1) \times e_{pb}(g+1)}{e^{-\tau|\alpha_b(g+1)-\sigma_b(g+1)|}} \right) \quad (18)$$

$$W_{pc}(g+1) = (1-\Omega \times \alpha_c(g+1)) \times W_{pc}(g) + \left(\frac{\alpha_c(g+1) \times u_c(g+1) \times e_{pc}(g+1)}{e^{-\tau|\alpha_c(g+1)-\sigma_c(g+1)|}} \right) \quad (19)$$

5) Calculation of the reference Grid Currents

The weight of active component (W_{As}) is calculated as,

$$W_{As} = W_{LAa} + W_{loss} - W_{PV} \qquad (20)$$

where
W_{loss} = DC loss component,
W_{pv} = PV feed-forward term, and
W_{LAa} = the average of the load fundamental active component calculated as,

$$W_{LAa} = \left(\frac{W_{pa} + W_{pb} + W_{pc}}{3} \right) \qquad (21)$$

The reference currents are calculated as,

$$i^*_{sa} = W_{As} u_{pa}, \; i^*_{sb} = W_{As} u_{pb}, \; i^*_{sc} = W_{As} u_{pc} \quad (22)$$

6) Switching Pulse generation for VSC:

The difference between reference grid currents and sensed grid currents is calculated to get the error and a hysteresis controller is used to generate the pulses in the grid current control method.

IV. EXPERIMENTAL RESULTS

An experimental prototype comprises of a PV array simulator, VSC, ripple filter and nonlinear load to verify the performance of the improved Kernel control. The parameters used in the experimentation are given in Appendix. Test results under various conditions are obtained using a power analyzer and a digital signal oscilloscope.

A. Performance of the Grid-PV System in steady state

Figs. 4(a)-(h) show the performance of the grid-PV system under normal conditions at a nonlinear load. The grid voltages and currents are shown in Fig. 4(a). The active power injected to the grid along with the reactive power with power factor, are shown in Fig. 4(b). It is observed that the grid currents are sinusoidal and balanced at UPF. Figs. 4(c)-(d) represent the voltage, currents, and powers of VSC. The voltage at the PCC, load currents and powers are shown in Figs. 4(e)-(f). The power quality aspects are studied, by comparing the harmonics spectra of load and grid currents shown in Figs. 4(g)-(h). The grid current THD is 2.97% in spite of having a nonlinear load current with THD 28.54%. This shows the ability of the proposed Kernel-based control in maintaining the grid current THD within the IEEE 519 standard ($i_{Grid\_THD} < 5\%$).

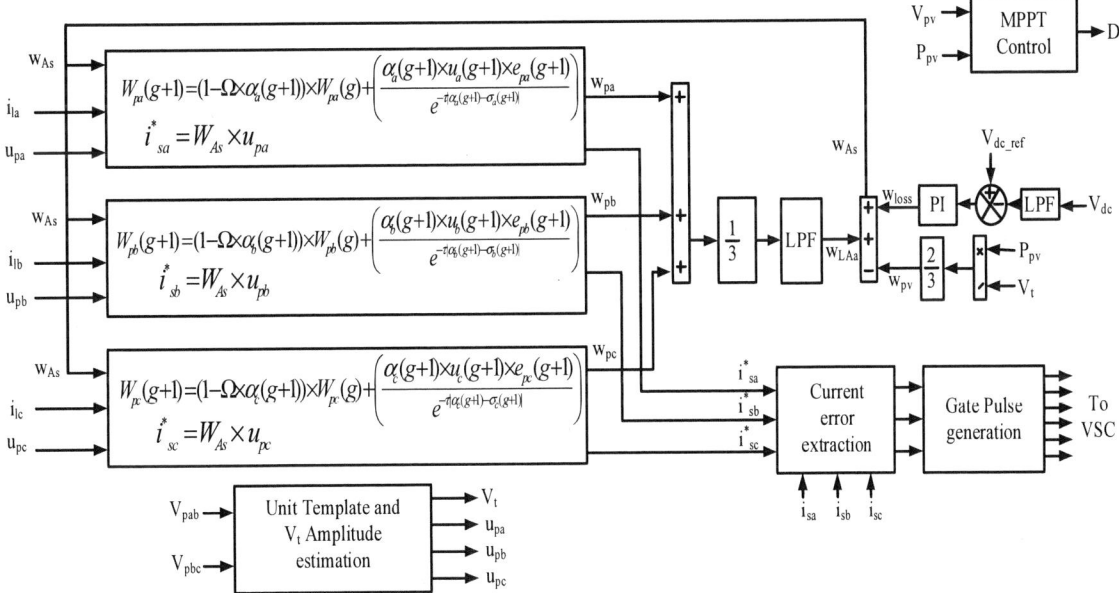

Fig. 3. Schematic of control algorithm for the grid-PV system

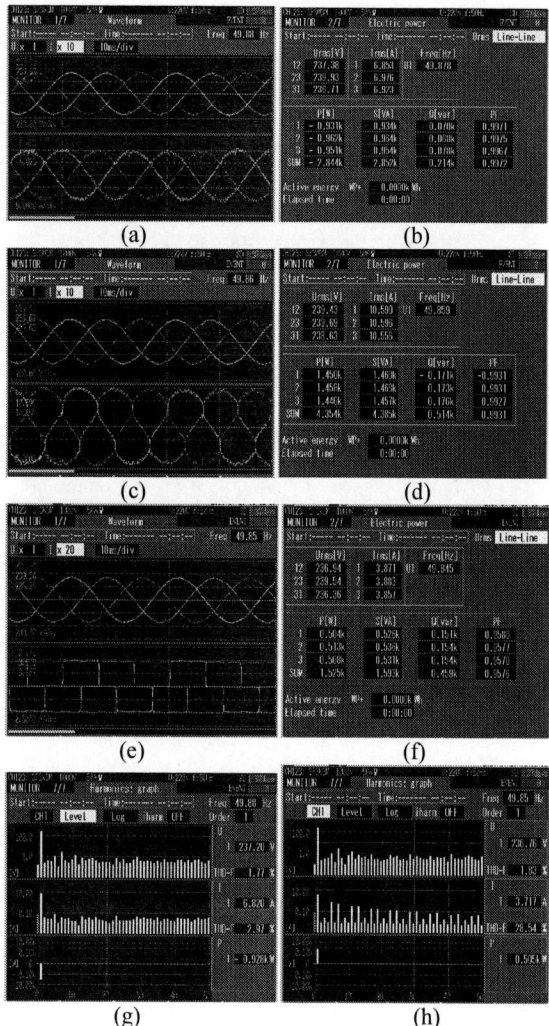

Fig. 4. Steady-state behavior of SPV system under non-linear load (a) grid voltage and grid current (b) Active power injected to the grid along with reactive power and power factor (c) Grid voltages and VSC currents (d) power carried by the VSC (e) PCC voltages and load currents (f) load power (g) THD of load current (h) THDs of grid voltage and current

B. Dynamic Performance of Grid-PV System

Figs. 5(a)-(j) show the dynamic performance of the grid-PV system during load disturbance. Figs. 5(a)-(b) show the reference currents generated as a result of load phase disconnection and reconnection. Figs. 5(c)-(d) show the load phase currents and PV output current on disconnection and reconnection of the load. Figs. 5(e)-(f) show the weights obtained for the phases on disconnection and reconnection of the load Phase 'a'. Figs. 5(g)-(h) show the total average net component, loss component, and the total net weight component. The control algorithm is able to update the weights to respond to the load unbalance showing satisfactory dynamic performance. Figs. 5(i)-(j) show the Phase 'a' currents of load, VSC, and grid and the DC link voltage on disconnection and reconnection. These responses show that the DC link voltage is maintained during dynamic conditions.

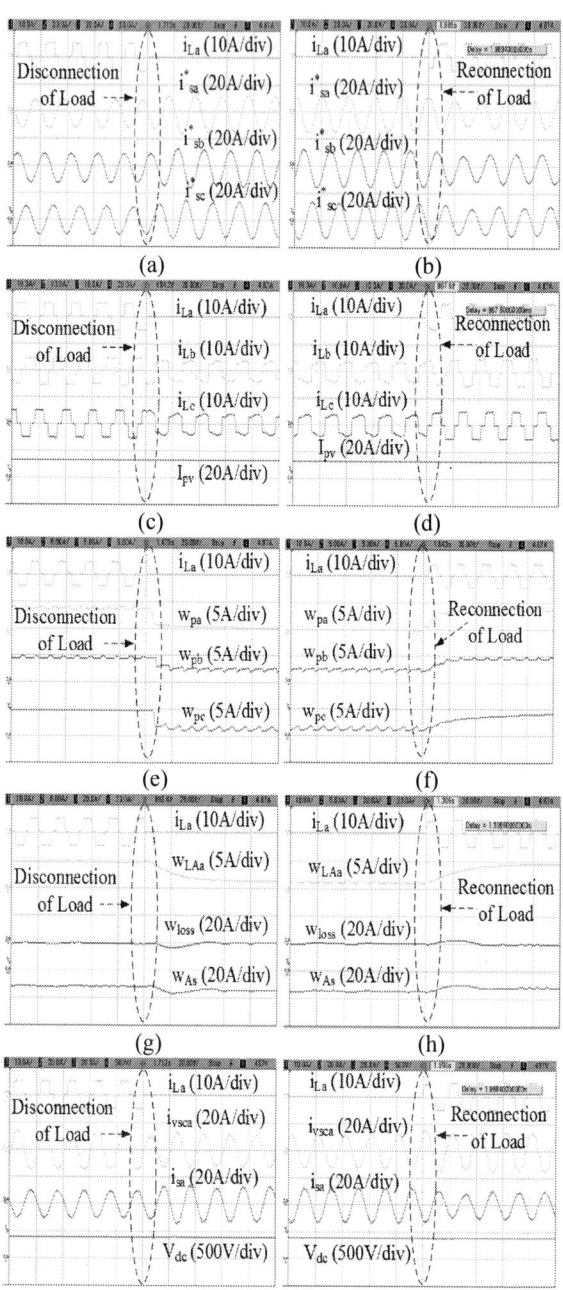

Fig. 5. Dynamic behavior of SPV under non-linear load (a)-(b) i_{La}, W_{pLa}, W_{pLb}, W_{pLc} (c)-(d) i_{La}, W_{pLavg}, W_{loss}, W_{pLnet}, (e)-(f) i_{La}, reference grid currents (i_{sa}*, i_{sb}*, i_{sc}*) (g)-(h) Load current i_{La}, i_{Lb}, i_{Lc} and I_{pv} (i)-(j) Load current of phase 'a', vsc current, grid current, dc voltage

C. Dynamic Performance During Mode Transition Between PV, Grid and DSTATCOM

Figs. 6(a)-(b) depict the performance during mode transition between PV, grid and DSTATCOM. It is observed that when PV generation is zero, the power supplied to the grid has stopped and the grid starts feeding the load. The VSC stops injecting power into the grid and starts working as a DSTATCOM performing only power quality improvement.

978-1-5386-6624-1/18 $31.00 © 2018 IEEE

On returning the power generation of the PV array, the VSC moves from DSTATCOM mode to renewable power injection mode of operation and feeds the PV power to the load and the grid.

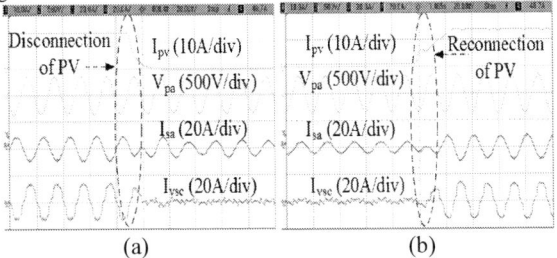

(a)   (b)

Fig. 6. Dynamic behavior of SPV under mode transitions. (a)-(b) PV output current I_{pv}, grid voltage of phase 'A', grid current and VSC current

D. Dynamic Performance at Varying Solar Irradiance

Figs. 7(a)-(b) depict the performance of the grid-PV system when irradiance varies from 1000W/m² to 600 W/m² and vice-versa. As the irradiance decreases, the I_{pv} reduces. Thus the power generated by the PV array decreases. But the load supply remains unaffected. The power injected into the grid is reduced which is reflected from the reduced I_{sa} values. Figs. 8(a)-(b) show the performance of the MPPT at varying irradiance. Fig. 8(a) shows that 5kW is produced at 1000W/m² and Fig. 8(b) shows the power reduction to 3kW when irradiance reduced to 600W/m². In both the conditions, the MPPT is found to operate at more than 99% efficiency. These results indicate the ability of the proposed Kernel-based control to learn quickly and adapt to the changes in the system operating conditions.

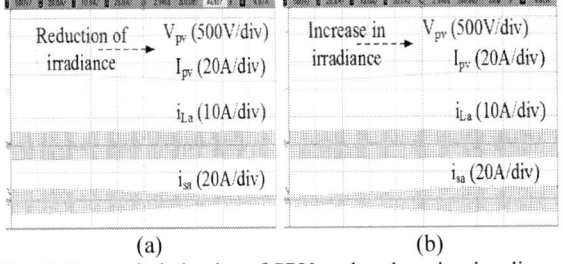

(a)   (b)

Fig. 7. Dynamic behavior of SPV under changing irradiance (a)-(b) PV Voltage and Current (Vpv, I_{pv}), Phase 'A' curents of Load and Grid (i_{La}, i_{sa})

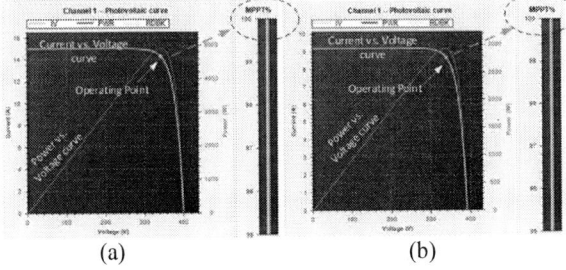

(a)   (b)

Fig. 8. Dynamic performance of MPPT under changing irradiance (a)-(b) solar array at 1000 w/m² and 600w/m²

V. CONCLUSION

The design and control of multifunctional grid-PV system have been proposed as illustrated in this work. The improved adaptive Kernel-based control has enabled the VSC to provide the required functionalities. The proposed system has been tested under various scenarios and the system has performed satisfactorily. It has proved to be a good option for grid integration of PV array and also providing auxiliary functions like reactive power and harmonics compensation. The dynamic performance is also found to be satisfactory. Test results show that the PV array is successfully operated at MPP and the grid current THD is found to be satisfying the IEEE 519 standard.

APPENDIX

PV array values: PV: V_{oc} = 400 V, I_{sc} = 15 A, L_b = 5 mH, F_{sw} = 10 kHz, V_{dc}= 400 V, C_{dc} = 2200 μF, L_f = 5 mH, Controller: k_{pd}= 0.2, k_{id}= .55, Grid values: 230 V, 50 Hz, L_s = 3 mH, R_s = 0.06 Ω, load: diode bridge rectifier, 160 mH, 65 Ω, Ripple filter: R_f = 10 Ω, C_f = 5 μF; Controller parameter: α = 0.02, u = 0.001.

REFERENCES

[1] Bo Zhao, C. Wang, and Xuesong Zhang, "Grid-Integrated and Standalone Photovoltaic Distributed Generation Systems: Analysis, Design, and Control," Wiley Publishers, 2017.

[2] Shailendra Kumar and Bhim Singh, "A frequency observer based control for solar energy conversion system," *IECON 2017 - 43rd Annual Conference of the IEEE Industrial Electronics Society*, Beijing, pp. 2321-2325, 2017.

[3] Utkarsh Sharma, Shailendra Kumar and Bhim Singh, "Solar array fed water pumping system using induction motor drive," *2016 IEEE 1st International Conference on Power Electronics, Intelligent Control and Energy Systems (ICPEICES)*, Delhi, pp. 1-6, 2016.

[4] Manasseh Obi, and Robert Bass. "Trends and challenges of gridconnected photovoltaic systems–A review," Renewable and Sustainable Energy Reviews 58, pp. 1082-1094, 2016.

[5] B. Subudhi and R. Pradhan, "A Comparative Study on Maximum Power Point Tracking Techniques for Photovoltaic Power Systems," *IEEE Transactions on Sustainable Energy*, vol. 4, no. 1, pp. 89-98, 2013.

[6] Shailendra Kumar and Bhim Singh, "Multi-Objective Single-Stage SPV System Integrated to 3P4W Distribution Network Using DMSI-Based Control Technique," *IEEE Transactions on Industry Applications*, vol. 54, no. 3, pp. 2656-2664, 2018.

[7] Bhim Singh and J. Solanki, "A Comparison of Control Algorithms for DSTATCOM," IEEE Trans. Ind. Electr., vol. 56, no. 7, pp. 2738- 2745, 2009.

[8] Bhim Singh, A. Chandra, and K. Al-Haddad: "Power Quality: Problems and Mitigation Techniques," Wiley, 2015.

[9] S. Vazquez, J. Sanchez, M. Reyes, J. Leon and J. Carrasco, "Adaptive Vectorial Filter for Grid Synchronization of Power Converters Under Unbalanced and/or Distorted Grid Conditions," *IEEE Trans. Ind. Electron.*, vol. 61, no. 3, pp. 1355-1367, 2014.

[10] Weifeng Liu, Jose C. Principe and Simon Haykin, "Kernel Adaptive Filtering," Wiley Publishers, 2010.

Single VDBA-Based Voltage-Mode Universal Biquadratic Filter

Kanhaiya Lal Pushkar
Department of Electronics &
Communication Engineering, *Maharaja Agrasen Institute of Technology*
Rohini, New Delhi-11086, India
klpushkar17@gmail.com

Ghanshyam Singh
Department of Electronics & Communication Engineering, Vaagdevi College of Engineering JNT University Hyderabad, India
ghanshyamsingh_09@rediffmail.com

Sushil Kumar
School of Engineering & Technology, Noida International University
Gautam Budh Nagar, Uttar Pradesh-2030201, India
sushilkumar0108@gmail.com

Abstract—This paper demonstrates a voltage-mode universal biquadratic filter employing a voltage differencing buffered amplifier (VDBA), one resistor and two capacitors. The proposed filter design has the ability of realizing all the five standard biquadratic filter functions i.e. band pass, high pass, low pass, band reject and all pass without changing the circuit topology. The proposed design of the same offers many benefits like; no requirement(s) of component(s) matching, low active and passive sensitivities. The theoretical analysis has been verified by the PSPICE simulation using 0.18μm Taiwan Semiconductor Manufacturing Company (TSMC) process technology.

Keywords—voltage-mode, universal biquadratic filter, voltage differencing buffered amplifier

I. INTRODUCTION

A number of single input multiple-output (SIMO) or multiple-input single-output (MISO) type current-mode (CM) or voltage-mode (VM) universal active filters have been described in the literature because of their flexibility and versatility for practical applications. Among these, MISO type filters are particularly attractive because the same filter structure can be used for different filter functions. Of late, proper attention has been given to the realization of VM or CM filter structure designs using different types of single active building blocks [1-10]. Although, these filter configurations have the drawback of passive component matching requirements or inversion of input signals. VDBA is one of the new active building blocks proposed in [11] and has been used in the realization of filters [12], [13] (ii) simulated inductance circuits [14], and (iii) electronically controllable sinusoidal oscillator [15-16]. The purpose of this paper is to propose a new MISO-type VM universal biquad employing a single VDBA, two capacitors, and a resistor to realize (by appropriate selection of input signal voltages) all the five standard filter functions, namely, low pass, high pass, band pass, band stop and all pass without any passive component matching requirements from the same circuit topology. The presented filter configuration also provides firstly, independent electronic control of angular frequency (ω_0) and bandwidth (BW) and secondly, low active and passive sensitivities. The validity of the proposed configuration has been confirmed by PSPICE simulation with TSMC 0.18 μm process parameters.

II. PROPOSED BIQUAD CONFIGURATION

The symbolic notation and equivalent model of the VDBA are shown in Fig. 2(a) and 2(b) respectively [12]. Using standard notations, the ports' voltage-current relations of VDBA can be described by the following matrix.

$$
\begin{pmatrix} I_p \\ I_n \\ I_z \\ V_w \end{pmatrix} = \begin{pmatrix} 0 & 0 & 0 & 0 \\ 0 & 0 & 0 & 0 \\ g_m & -g_m & 0 & 0 \\ 0 & 0 & \beta & 0 \end{pmatrix} \begin{pmatrix} V_p \\ V_n \\ V_z \\ I_w \end{pmatrix} \tag{1}
$$

Where β is a non-ideal voltage gain of VDBA. The value of β in an ideal VDBA is unity and g_m is transconductance VDBA.

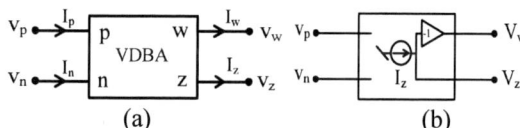

Fig.1. (a): Symbolic notation (b) equivalent model of VDBA

The proposed structure of universal biquad is given in Fig. 2. A routine circuit analysis of proposed biquad universal filter shown in Fig. 2, expression for voltage output in terms of input voltages is given below

$$
V_o = \frac{V_1 s^2 + V_3 s \dfrac{1}{C_1 R_0} + V_2 \dfrac{g_m}{R_0 C_1 C_2}}{s^2 + s \dfrac{1}{C_1 R_0} + \dfrac{g_m}{R_0 C_1 C_2}} \tag{2}
$$

From eq. (2) various filter functions can be realized as follows:

1. $T(s)\big|_{LP} = \dfrac{\dfrac{g_m}{C_1 C_2 R_0}}{D(s)}$; If $V_2 = V_{in}$, $V_3 = V_1 = 0$ (grounded), a low pass filter can be realized.

2. $T(s)|_{HP} = \dfrac{s^2}{D(s)}$; If $V_1 = V_{in}$ and $V_2 = V_3 = 0$

(grounded), a high pass filter can be realized.

3. $T(s)|_{BP} = \dfrac{\dfrac{s}{C_1 R_0}}{D(s)}$; If $V_3 = V_{in}$ and $V_1 = V_2 = 0$

(grounded), a band pass filter can be realized.

4. $T(s)|_{BR} = \dfrac{s^2 + \dfrac{g_m}{R_0 C_1 C_2}}{D(s)}$; If $V_1 = V_2 = V_{in}$ and $V_3 = 0$, a

band reject filter can be realized.

5. $T(s)|_{AP} = \dfrac{s^2 - s\dfrac{1}{C_1 R_0} + \dfrac{g_m}{R_0 C_1 C_2}}{D(s)}$; If

$V_1 = V_2 = -V_3 = V_{in}$, an all pass filter can be realized.

Where: $D(s) = s^2 + s\dfrac{1}{C_1 R_0} + \dfrac{g_m}{R_0 C_1 C_2}$

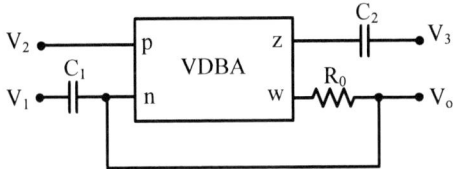

Fig. 2. Proposed universal filter structure employing a single VDBA

Expressions for natural frequency (ω_0) and bandwidth (BW) are given by

$$\omega_0 = \sqrt{\frac{g_m}{R_0 C_1 C_2}} \qquad (3)$$

$$BW = \frac{1}{C_1 R_0} \qquad (4)$$

From (3) and (4), it is seen after fixing bandwidth; ω_0 can be independently and electronically controlled through g_m. Further, it is clear that component matching is not required in any of the five filters realization. Only in APF, sign inversion is required. V_2 does have ideally infinite input impedance while the remaining two do not have ideally infinite input impedance while the output node does not have required low output impedance. So, in some cases additional voltage followers may be required to be employed for facilitating cascading of several such filters as for example in case of higher filter design.

III. NON- IDEAL AND SENSITIVITY PERFORMANCE

Let R_Z and C_Z denote the parasitic capacitance of the Z-terminal and R_W denotes the parasitic resistance of the W-terminal of VDBA. Taking the non-idealities into account, namely, the voltage of W-terminal $V_{W-} = \left(-\beta^+ V_Z + I_w R_w\right)$

where $\beta^+ = 1 - \varepsilon_p$ ($\varepsilon_p \ll 1$) denotes the voltage tracking error of Z-terminal of Z-terminal of VDBA, the expression output voltage in terms of input voltages is written below:

$$V_0 = \frac{V_1 s^2 + s\left\{\dfrac{V_1}{R_z(C_2 + C_z)} + \dfrac{V_3 C_2 \beta^+}{C_1(C_2 + C_z)R_0}\right\} + V_2 \dfrac{g_m \beta^+}{C_1(C_2 + C_z)R_0}}{s^2 + \left\{\dfrac{1}{R_z(C_2 + C_z)} + \dfrac{1}{C_1 R_0}\right\} + \dfrac{g_m \beta^+}{C_1(C_2 + C_z)R_0}} \qquad (5)$$

$$\omega_0 = \sqrt{\frac{\beta^+ g_m}{R_0 C_1 (C_2 + C_z)}} \qquad (6)$$

$$BW = \left\{\frac{1}{R_z(C_2 + C_z)} + \frac{1}{C_1 R_0}\right\} \qquad (7)$$

IV. SENSITIVITY ANALYSIS

The sensitivity is an important performance criterion of any circuit structure. The sensitivities of ω_0 with respect to active and passive elements are given by

$$S_{C_1}^{\omega_0} = S_{R_0}^{\omega_0} = -\frac{1}{2}, \quad S_{g_m}^{\omega_0} = S_{\beta^+}^{\omega_0} = \frac{1}{2},$$

$$S_{C_2}^{\omega_0} = -\frac{1}{2}\frac{C_2}{C_2 + C_z}, \quad S_{C_z}^{\omega_0} = -\frac{1}{2}\frac{C_z}{C_2 + C_z} \qquad (8)$$

It may be easily observed from eq. (8) that all sensitivities are lower than unity in magnitude, for the proposed universal biquad filter. It ensures that the sensitivity performance is good.

V. SIMULATION RESULTS AND DISCUSSION

The circuit was simulated using CMOS VDBA (as shown in Fig. 3) to confirm workability of the proposed universal biquad. The CMOS VDBA is implemented using 0.18 µm TSMC transistors process parameters [17]. The supply and bias voltages are given by $V_{DD} = -V_{SS} = 1.5$ V and $V_{B1} = -0.44$ V, $V_{B2} = -0.9$ V. The aspect ratios of the transistors are shown in Table 1. The simulation results show that this choice yields the trans-conductance value of $g_m = 748$ µA/V for the VDBA and parasitic impedances of $R_Z = 315$ kΩ, $C_Z = 0.35$pF and $R_W = 21$ Ω, parasitic parallel resistances and capacitances at Z-terminal and parasitic series resistances at W-terminal respectively. The power consumption of the proposed VDBA is 0.97mW [12]. For this purpose the passive elements were selected as $C_1 = C_2 = 0.01$nF, $R_0 = 10.57$KΩ. The transconductance of VDBA was controlled by bias voltage V_{B1}. The PSPICE simulated frequency responses

978-1-5386-6624-1/18 $31.00 © 2018 IEEE

of the proposed biquad are shown in Fig. 4. The Fig. 5 and Fig. 6 represent phase and magnitude plots of APF respectively. Therefore, these simulation results confirm the validity of the proposed biquad filter.

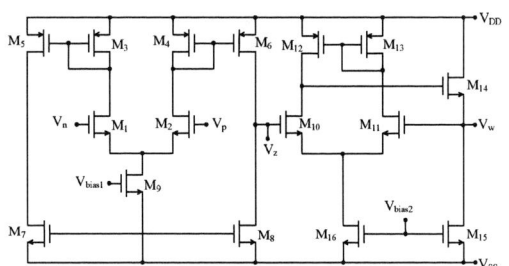

Fig.3. CMOS implementation of VDBA [12], $V_{DD} = V_{SS} = 0.9V$

Table 1: The aspect ratios of transistors used in Fig.3

Transistor	W (μm)	L (μm)
M1-M4, M10, M11, M15, M16	7	0.35
M5, M6	21	0.7
M7, M8	7	0.7
M9	3.5	0.7
M12-M14	14	0.35

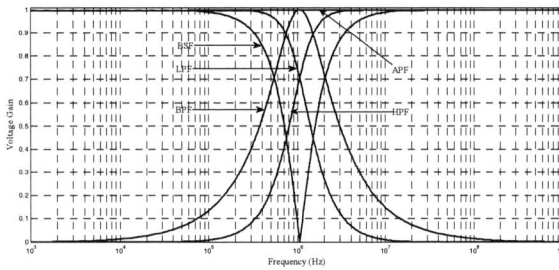

Fig.4. Frequency Response of Proposed Universal Biquadratic Filter

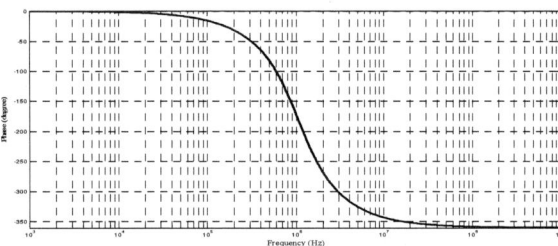

Fig.5. Phase Plot of APF

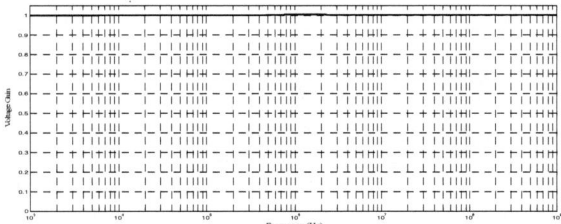

Fig.6. Frequency Response of APF

VI. CONCLUSION

The biquad filter structure proposed in this paper employs basically minimum number of active and passive elements (one VDBA, two capacitors and a resistor) to realize all the standard second order five filter responses. The proposed filter design configuration also provides the following advantages which are not available simultaneously in any of the single active device or element based (with VDBA) universal MISO type VM biquad realizing all the five filter functions, firstly, no requirement of any passive component(s) matching condition, secondly, independent electronic control of angular frequency (ω_0) and bandwidth (BW) and thirdly, low active and passive sensitivities. Simulation results using 0.18μm TSMC process technology has been included to confirm the feasibility and workability of the new proposed biquad filter design.

REFERENCES

[1] K. L. Pushkar and K. Gupta, "MISO-type voltage-mode universal biquadratic filter using single universal voltage conveyor," *Circuits and Systems*, vol.8, no.9, pp.227-236, 2017.

[2] J. Sirirat, W. Tangsrirat and W. Surakampontorn, "Voltage-mode electronically tunable universal filter employing single CFTA," In*Proceedings of the 7th Annual International Conference on Electrical Engineering/Electronics, Computer, Telecommunications and Information Technology (ECTI-CON '10)*, pp.759–763, 2010.

[3] D. Prasad, D. R. Bhaskar and A. K. Singh, "Multi-function biquad using single current differencing transconductanceamplifier," *Analog Integrated Circuits and Signal Processing*, vol.61, no.3, pp.309-313, 2009.

[4] J. W. Horng, "Voltage/current-mode universal biquadratic filter using single CCII+," *Indian Journal of Pure & Applied Physics*, vol.48, no.10, pp.749-756, 2010.

[5] A. U. Keskin, "Multi-function biquad using single CDBA," *Electrical Engineering*, vol.88, no.5, pp. 353-356, 2006

[6] S. A. Bashir and N. A. Shah, "Voltage mode universal filter using current differencing buffered amplifier as an active device," *Circuits and Systems*, vol.3, pp.1-4, 2012.

[7] N. Herencsar, J. Koton, K. Vrbaand, O. Cicekoglu, "Single UCC-N1B 0520 device as a modified CFOA and its application to voltage-mode and current-mode universal filters," *Applied Electronics*, pp.127-130, 2009.

[8] N. A. Shah, M. F. Rather and S. Z. Iqbal, "A novel voltage-mode universal filter using a single CFA," *Active and Passive Electronic Devices*, vol.1, pp.1183–188, 2005.

[9] J. W. Horng, C. K. Chang and J. M. Chu, "Voltage-mode universal biquadratic filter using single current-feedback amplifier," *IEICE Transactions on Fundamentals of electronics, communications and Computer Sciences*, vol.85, no.8, pp.70–1973, 2002.

[10] K. L. Pushkar, D. R. Bhaskar and D. Prasad, "Voltage-mode universal biquad filter employing single voltage differencing differential input buffered amplifier," *Circuits and Systems*, vol.4, no.1, pp.44-48, 2012.

[11] D. Biolek, R. Senani, V. Biolkova and Z. Kolka, "Active elements for analog signal processing: classification, review, and new proposals," *Radioengineering*, vol.17, no.4, pp. 15–32, 2008.

[12] F. Kacar, A. Yeil, and A. Noori, "New CMOS realization of voltage differencing buffered amplifier and its biquad filter applications," *Radioengineering*, vol.21, no.1, pp. 333–379, 2012.

[13] P. Gupta, and R. Pandey, "Single VDBA based multifunction filter," *International Journal of Control Theory and Applications*, vol.10, no.6, pp.651-661, 2017.

[14] F. Kacar and A. Yesil, "VDBA-based lossless and lossy inductance simulators and its filter applications," 24th *Signal Processing and Communication Application Conference (SIU)*, pp. 909-912, 2016.

[15] C. Malhotra, V. V. Ahalawat,V. Kumar, R. Pandey and N. Pandey, "Voltage differencing buffered amplifier based quadrature oscillator," *1st International Conference on Power Electronics, Intelligent Control and Energy Systems* (ICPEICES-2016), pp.1-4, 2016.

[16] K. L. Pushkar, "Voltage-mode third-order quadrature sinusoidal oscillator using VDBAs," *Circuits and Systems*, vol.8, no.12, pp. 285-292, 2017.

[17] S. Minaei and E. Yuce, "Novel voltage-mode all-pass filter based on using DVCCs," *Circuits System and Signal Processing*, vol.29, no.3, pp. 391-402, 2010.

2nd IEEE International conference on power Electronics, Intelligent Control and Energy systems (ICPEICES-2018)

DSOGI based Grid Synchronization under Adverse Grid Conditions

Simar Preet Kaur
Electrical Engineering Department
Delhi Technological University,
Delhi, India
simarpreet9211@gmail.com

Alka Singh
Electrical EngineeringDepartment
Delhi Technological University,
Delhi, India
alkasingh.eed@gmail.com

Abstract—**This paper discusses the design together with the application of a Double Second Order Genralized Integrator (DSOGI) for grid integration. A moderately distorted supply (having a THD equal to 8.49%) feeds a non - linear load and a Shunt Active Power Filter (SAPF) is used for as a compensator. In this paper, two different control approaches have been used to derive reference currents for the SAPF. These algorithms are the unit template (UT) based controller and synchronous reference frame (SRF) theory based approach. Both the algorithms fail to achieve less than 5% distortion in supply currents when used under distorted/ polluting grid conditions. Hence a DSOGI based controller has been realized to generate perfectly sinusoidal reference currents even under adverse grid conditions. With the proposed modification in the controller design using the DSOGI block, perfectly sinusoidal grid currents can be generated. System has been simulated in MATLAB environment using SIMULINK and PSB toolboxes. Performance results demonstrate the effectiveness of DSOGI based non - phase locked loop approach for grid synchronization.**

Keywords—*SAPF, unit template, SRFT, DSOGI, distorted supply*

I. INTRODUCTION

Distributed Generation (DG) systems based on renewable energy sources such as hydro power, wind energy conversion system (WECS), photovoltaic (PV) power, biofuels is getting popular to meet the growing power demands. For a grid connected system, these resources need to be connected in proper synchronization. These techniques estimate and track the magnitude, frequency as well as phase angle of the input voltage signals. Both Phase Locked Loop (PLL) and non PLL techniques are in use today. Researchers have focused on a no. of 1φ and 3φ PLLs.

An APF is a power electronic converter - based equipment which is intended to reduce the power quality issues induced by nonlinear loads.

A number of strategies for active power filter have been recommended, with the most extensively used being the shunt APFs [1] – [3]. A SAPF is meant for compensation of reactive power as well as to improve the supply current power factor, voltage regulation, load balancing, harmonic reduction etc. [4]. For this purpose, multiple control strategies have been documented, for instance, strategy using neural network techniques, Power Balance Theory (PBT) – based control technique, computation based on per phase basis, I cos φ – based control technique, current compensation with the help of dc bus regulation, SRFT, Current Synchronous Detection (CSD) control technique, Instantaneous Symmetrical Components Theory (ISCT) – based control technique, Instantaneous Reactive Power Theory (IRPT) theory, single – φ PQ theory – based control technique, single – φ DQ theory – based control technique [5] – [13].

Phase Locked Loops (PLLs) are used as a significant method aimed at grid synchronization besides tracking the phase angle. Single - φ PLL based on Transport delay, adaptive transport delay, power - PLL, inverse Park PLL etc. are popular and commonly employed.

A popular non - PLL based syncronization technique is Second Order Generalized Integrator (SOGI) and can be realized easily in single - φ systems. An extension of two

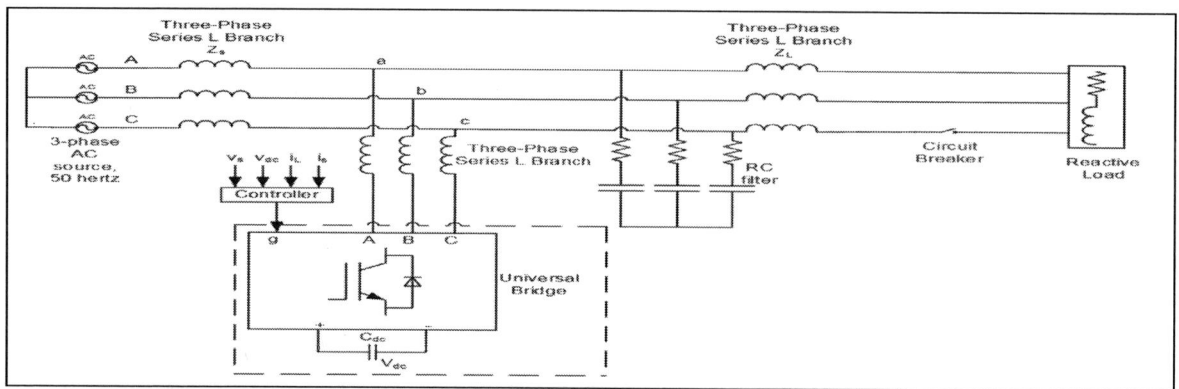

Fig. 1. Basic circuit diagram of the SAPF system

978-1-5386-6624-1/18 $31.00 © 2018 IEEE

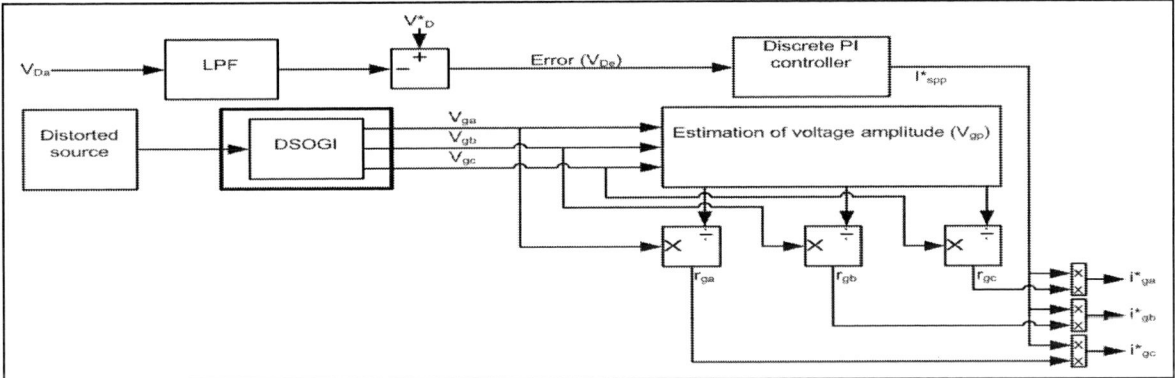

Fig. 2. UT-based control technique of SAPFs (with DSOGI technique)

SOGI circuits can be realized for three - φ systems and is termed as DSOGI based synchronization. In this paper, a DSOGI based synchronization method is designed for effective filtering of voltage distortions in polluted grid. Two algorithms based on unit template - based theory and SRFT are used for the control of SAPF. The grid is considered to be moderately polluted with THD of 8.49%. The effect of SAPF performance without/with DSOGI synchronization is investigated.

Fig. 1. shows a SAPF connected in parallel with the non – linear load, and controlled using a control algorithm to inject or draw iC, compensating current, to or from the supply so that at the point of common coupling (PCC) iG, the grid current, is a sinusoidal wave in - phase with the grid voltage, vG. The non - linear load is connected in the form of diode rectifier. A three - φ, three leg insulated gate bipolar thyristor (IGBT) based bridge is modelled as a SAPF.

Undoubtedly the most significant step in the control process of a APF is the reference compensating current generation. In literature many algorithms for reference current generation (RCG) have been cited. The RCG techniques can broadly be categorized into frequency and time domain algorithms. In case of frequency domain the most commonly used techniques are Recursive Digital Fourier Transform (RDFT) or Sliding DFT (SDFT), Fast Fourier Transform (FFT) and Digital Fourier Transform (DFT).

Two control algorithms viz. unit template - based theory and SRFT are described in detail for the control of SAPF. The performance is studied under the distorted grid conditions.

II. UNIT TEMPLATE-(UT-) BASED CONTROLLER

The UT based controlling system is a basic technique which is meant for load compensation. Besides this, it can also used for balancing and equalization of unbalanced loads. The above mentioned algorithm for the control of SAPF can be adapted according to the need either for the improvement of the power factor to maintain a Unity Power Factor (UPF) or for the control of voltage in order to get Zero Voltage Regulation (ZVR) by employing reactive power compensation at the PCC. A self - sufficient DC bus of the Voltage Source Converter (VSC) is used as a SAPF in this control process. This technique generates the reference grid currents and in addition to this, this technique can also be employed for direct current control of VSC currents of SAPF.

However, in order to get Pulse Width Modulated (PWM) switching pulses for the devices employed in Current Controlled - VSC employed as a SAPF, an indirect grid current control is favored. The benefits such as the characteristic removal of sharp peaks in currents, decreased computational load on the Digital Signal Processor and fast switching are some of the benefits of the indirect current control of the SAPF. For the pulse generation, three - φ reference grid currents are generated by using three - φ AC voltages which are sensed at PCC. In addition the DC bus voltage (V_D) of the SAPF is taken as an input for the control algorithm.

Fig. 2 demonstrates the UT - based control logic of SAPF developed for power factor correction (PFC) of the grid

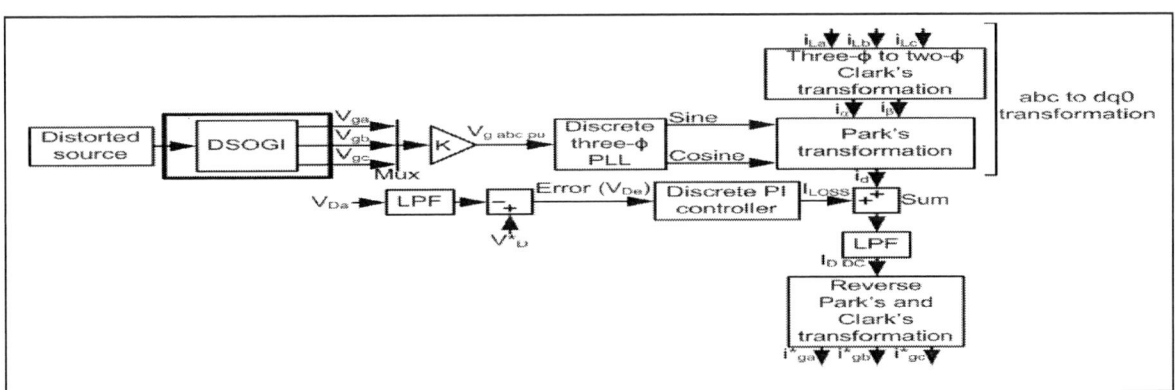

Fig. 3. SRFT based control algorithm (with DSOGI technique)

978-1-5386-6624-1/18 $31.00 © 2018 IEEE   793

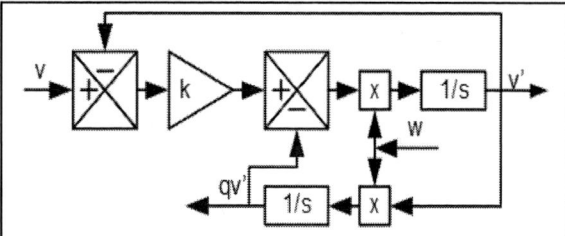

Fig. 4. QSG based on SOGI

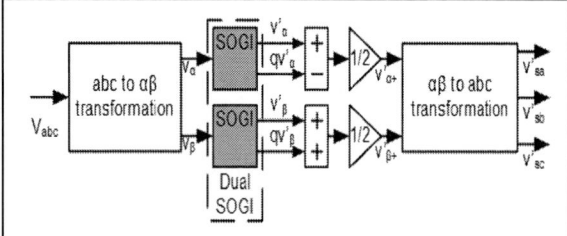

Fig. 5. Basic scheme of suggested Reference Current Generation (RCG) technique

currents at PCC. The Fig. shows three - φ distorted grid voltage signals as inputs and three - φ conditioned voltage signals (v_{ga}, v_{gb}, v_{gc}) are outputs after processing by the DSOGI block. Discrete PI controller is used to control the DC link voltage (V_{Da}) to its reference value (V^*_D). The output of the discrete PI controller on the DC bus voltage of the SAPF is assumed to be the magnitude I^*_{spp}. The in - phase reference grid current components i.e. i^*_{ga}, i^*_{gb} and i^*_{gc} are obtained using three - φ unit current templates i.e. r_{ga}, r_{gb} and r_{gc} are in - phase with the conditioned grid voltages i.e. v_{ga}, v_{gb} and v_{gc}. The product of the generated in - phase unit templates and the in-phase magnitude gives the three-φ reference grid currents i.e. i^*_{ga}, i^*_{gb} and i^*_{gc}. Therefore, for fundamental UPF grid currents, the in-phase reference grid currents, which are calculated in the above-described way, turn out to be the reference grid currents. A discrete PI controller is used to calculate the magnitude of reference grid currents by using V^*_D and average value V^*_{Da} of the SAPF. The difference in V^*_{Da} and V^*_D voltage values of the SAPF results in a voltage error (V_{De}), which is further fed into a discrete PI regulator as shown in the equation below:

$$V_{De}(l)=V^*_D(l)-V_{Da}(l) \qquad (1)$$

The voltage error that is generated as above is given as an input to the discrete PI regulator, further the discrete PI regulator output is shown in the equation below:

$$I_{spp}(l)=I_{spp}(l\text{-}1)+K_1\{V_{De}(l)\text{-}V_{De}(l\text{-}1)\}+K_2V_{De}(l) \qquad (2)$$

in which K_1 is the integral gain constant and K_2 is the proportional gain constant of the DC voltage discrete PI regulator, respectively and V_{De} (l) = V^*_D (l) - V_{Da} (l) is the difference or the error between V^*_D and V_{Da} DC voltages at the lth sample instant.

The magnitude of the voltage denoted by V_{gp} at the PCC can be calculated as per the following equation:

$$V_{gp}=\sqrt{\frac{2}{3}(v_{ga}^2+v_{gb}^2+v_{gc}^2)} \qquad (3)$$

The in-phase unit templates r_{ga}, r_{gb} and r_{gc} can be derived as per the following equation:

$$r_{ga}=\frac{v_{ga}}{V_{gp}},r_{gb}=\frac{v_{gb}}{V_{gp}},r_{gc}=\frac{v_{gc}}{V_{gp}} \qquad (4)$$

Now, in order to attain a desired DC voltage output proportional gain constant (K_2) and integral gain constant (K_1) is selected accordingly. Finally, the three-φ reference grid current in-phase components are calculated by the in-phase unit templates which are derived in phase with the PCC voltages in addition to the above mentioned magnitude:

$$i^*_{ga}=I^*_{spp}r_{ga}, \ i^*_{gb}=I^*_{spp}r_{gb}, \ i^*_{gc}=I^*_{spp}r_{gc} \qquad (5)$$

In case of distorted grid conditions, this technique fails to generate perfectly sinusoidal reference grid currents. Hence, DSOGI based UT controller is developed.

III. SYNCHRONOUS REFERENCE FRAME CONTROL ALGORITHM OF SAPF

The principle on which the SRF model is developed is based on the transformation of currents in synchronously rotating d–q frame. Fig. 3 demonstrates the elementary block diagram on which this concept is based.

As shown in Fig. 3, the input to the controller has three-φ voltages viz. v_a, v_b in addition to v_c and also three-φ load currents viz. i_{La}, i_{Lb} in addition to i_{Lc} along with DC link voltage V_{Da} given as an input to the controller. The unit voltage templates viz. sine and cosine waveforms are produced using a PLL to which the normalized grid voltage signals in per unit are fed as inputs. The three-φ current signals on the other hand are transformed into direct-quadrature (d-q) frame current signals (i_d and i_q) using the SRF theory. Since these are DC signals now, these current signals are filtered using a Low Pass Filter (LPF). The I_{LOSS} component generated using the DC link voltage controller is added to the d component of the current signal. Thereafter the signals are reconverted to again get i_{ga}, i_{gb} and i_{gc}. These a-b-c frame current signals are the reference grid signals and given as input to the Hysteresis Current Controller (HCC) which is used in order to produce final switching pulses which are fed to the SAPF for its controlling.

A detailed explanation of the SRFT technique is as follows. Like the p–q model, current components in α–β frame are produced. Moreover, by means of Park's transformation, these currents are changed from α–β frame to d–q frame i.e. by using a transformation angle viz. θ as shown in matrix equation (6):

$$\begin{bmatrix} i_d \\ i_q \end{bmatrix} = \begin{bmatrix} cos\theta & sin\theta \\ -sin\theta & cos\theta \end{bmatrix} \begin{bmatrix} i_\alpha \\ i_\beta \end{bmatrix} \qquad (6)$$

Now, with the help of reverse Park's transformation the calculated DC current components $i_{d, dc}$ and $i_{q, dc}$ are changed

978-1-5386-6624-1/18 $31.00 © 2018 IEEE

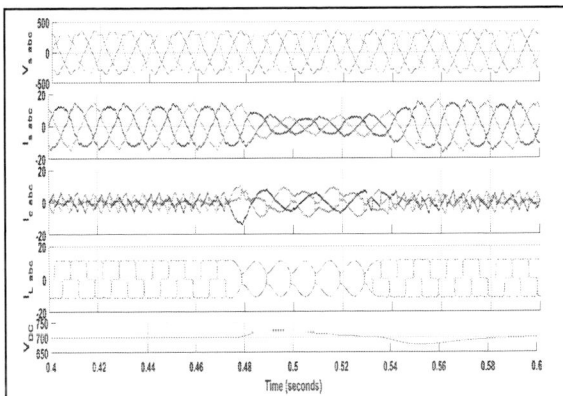

Fig. 6. UT-based control algorithm outcome (without DSOGI technique)

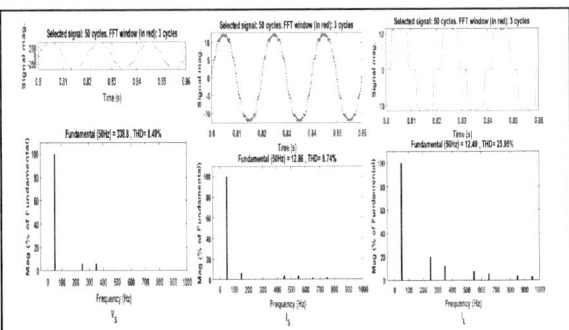

Fig. 7. THD values for UT-based control algorithm (without DSOGI technique)

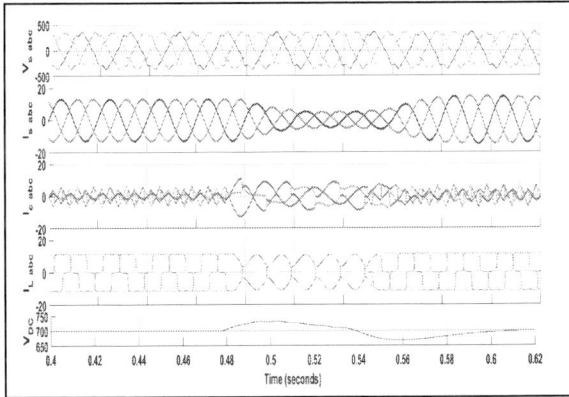

Fig. 8. UT-based control algorithm outcome (with DSOGI technique)

back into α–β frame signals as illustrated in matrix equation (7):

$$\begin{bmatrix} i_{\alpha,DC} \\ i_{\beta,DC} \end{bmatrix} = \begin{bmatrix} cos\theta & -sin\theta \\ sin\theta & cos\theta \end{bmatrix} \begin{bmatrix} i_{d,DC} \\ i_{q,DC} \end{bmatrix} \quad (7)$$

By using the above derived α-β frame current signals, the conversion is made for deriving three-φ reference grid currents in a-b-c frame by means of matrix equation (8):

$$\begin{bmatrix} i_{ga}^* \\ i_{gb}^* \\ i_{gc}^* \end{bmatrix} = \sqrt{\frac{2}{3}} \begin{bmatrix} \frac{1}{\sqrt{2}} & 1 & 0 \\ \frac{1}{\sqrt{2}} & -\frac{1}{2} & \frac{\sqrt{3}}{2} \\ \frac{1}{\sqrt{2}} & -\frac{1}{2} & -\frac{\sqrt{3}}{2} \end{bmatrix} \begin{bmatrix} i_0^* \\ i_{g\alpha}^* \\ i_{g\beta}^* \end{bmatrix} \quad (8)$$

By keeping the i_q current component nil in order to calculate the reference grid current components, the reactive power compensation can as well be provided.

IV. DESIGN OF DUAL SOGI BASED ALGORITHM

The SOGI based PLL is used to faultlessly detect and also separate the phase angle of positive in addition to negative sequence components with a good dynamic response.

The SOGI block open loop transfer function is shown as in equation (9):

$$T(s) = \frac{Kw_r s}{s^2 + w_r^2} \quad (9)$$

The closed loop transfer functions of v'$_\alpha$ and qv'$_\alpha$ of the SOGI block with error signal (V*$_{De}$) are given as:

$$\frac{v'_\alpha(s)}{v} = \frac{Kw_r s}{s^2 + Kw_r s + w_r^2} \quad (10)$$

$$\frac{qv'_\alpha(s)}{v} = \frac{w_r}{s} * \frac{v'_\alpha(s)}{v} = \frac{Kw_r s}{s^2 + Kw_r s + w_r^2} \quad (11)$$

It is essential to have a proper tuning of the gain K and the angular frequency, ω. So, in this paper K and ω values are chosen as 0.5 and 314 respectively.

Dual QSGs with the use of SOGI are used in order to get the d-q components of the α axis voltages (which are denoted by v'$_\alpha$ and qv'$_\alpha$ correspondingly) in addition to β axis voltages (denoted by v'$_\beta$ and qv'$_\beta$ correspondingly).

The subsequent comparisons shown in equations (12) and (13) are used to calculate the α and also β axis voltages for positive and negative sequence components:

$$\begin{cases} v_{\alpha+} = \frac{1}{2} * (v'_\alpha - qv'_\beta) \\ v_{\beta+} = \frac{1}{2} * (qv'_\alpha + v'_\beta) \end{cases} \quad (12)$$

$$\begin{cases} v_{\alpha-} = \frac{1}{2} * (v'_\alpha - qv'_\beta) \\ v_{\beta-} = \frac{1}{2} * (-qv'_\alpha + v'_\beta) \end{cases} \quad (13)$$

V. OUTCOMES AND DISCUSSION

The subsequent inferences are observed based on the outcomes of the performance of SAPF considered without/with DSOGI control.

978-1-5386-6624-1/18 $31.00 © 2018 IEEE

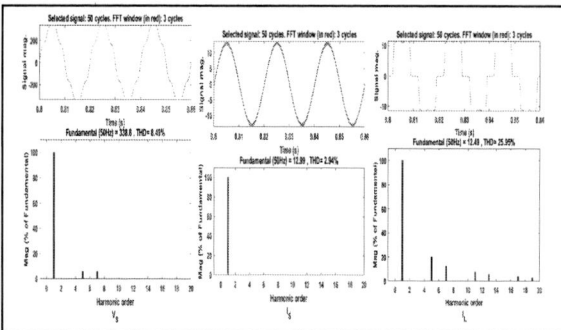

Fig. 9. THD values for UT-based control algorithm (with DSOGI technique)

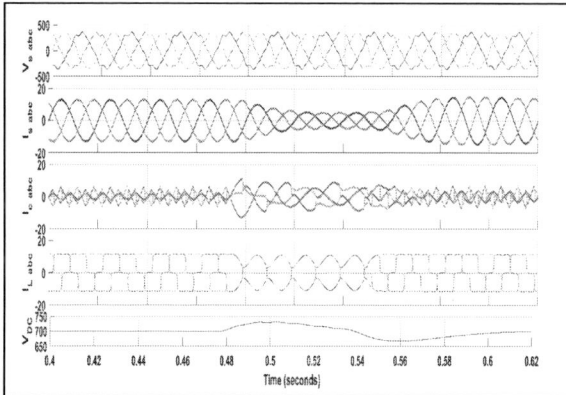

Fig. 10. SRFT based control algorithm outcome (without DSOGI technique)

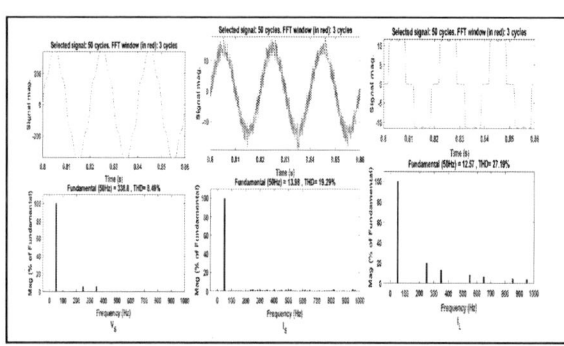

Fig. 11. THD values for SRFT based control algorithm (without DSOGI technique)

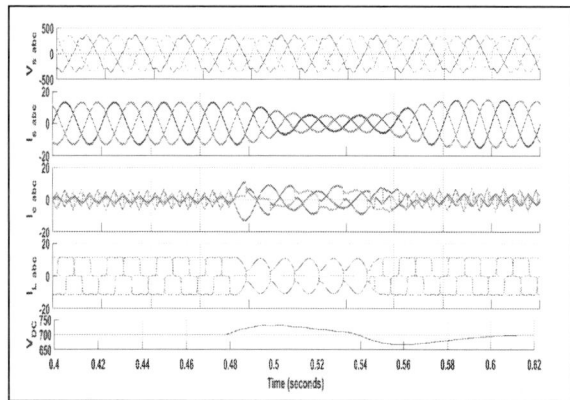

Fig. 12. SRFT based control algorithm outcome (with DSOGI technique)

A. Control of SAPF with the use of unit template-based theory without use of DSOGI

Figs. 6. and 8. demonstrate the dynamic performance of a SAPF with the unit template theory-based current extractor without and with using DSOGI technique respectively.

Fig. 6. shows the plots of grid voltage $V_{g, abc}$, grid current $I_{g, abc}$, compensator current $I_{c, abc}$, load current $I_{L, abc}$, and DC link voltage V_{Da} without DSOGI technique. Voltage and current inputs are distorted. An unbalance is created at 0.47 seconds wherein one of the phases is open circuited at the load end. Again, at 0.53 seconds all the three phases are normally connected. It is seen that V_{Da} settles to the 700 V value at 0.59 seconds. 700 V is the reference value taken during simulation.

Fig. 7. shows the corresponding THD values for V_g, I_g, and I_L. The THD value for I_g is 8.74% which does not meet the IEEE 519 standard. Hence, this technique fails with distorted source.

B. Control of SAPF by means of unit template-based theory with use of DSOGI

Fig. 8. shows the plots of grid voltage $V_{g, abc}$, grid current $I_{g, abc}$, compensator current $I_{c, abc}$, load current $I_{L, abc}$, and DC link voltage V_{Da} with DSOGI technique. Voltage and current inputs are no more distorted as they have been made perfectly sinusoidal by the DSOGI technique. The model is tested for the same unbalance conditions as in case I. It is observed that V_{Da} settles to the 700 V value at 0.59 seconds.

Fig. 9. shows the corresponding THD values for V_g, I_g, and I_L. The THD value for I_g now is 2.94% which meets the IEEE 519 standard. Hence, the proposed DSOGI technique gives the desired results. The considered load is resistive–reactive connected at the end of a diode rectifier. The unbalance is introduced at 0.5 s by using a circuit breaker in one of the three phases.

C. Control of SAPF by means of SRFT based algorithm without use of DSOGI

Figs. 10. and 12. show the SAPF performance controlled with the help of SRFT-based scheme without and with using DSOGI technique respectively.

Simulation is carried out for alike unstable situation as of the preceding situation. Fig. 10. shows the plots of grid voltage $V_{g, abc}$, grid current $I_{g, abc}$, compensator current $I_{c, abc}$, load current $I_{L, abc}$, and DC link voltage V_{Da} without DSOGI technique. Voltage and current inputs are distorted. Model is tested for similar unbalance condition as in case I. 700 V is the reference value taken during simulation. It is seen that V_{Da} settles to the 700 V value at 0.57 seconds. This shows that SRFT based technique is better as compared to UT-based control theory as far as settling of the V_{Da} back to the set V^*_D is concerned.

Fig. 11. shows the corresponding THD values for V_g, I_g, and I_L. The THD value for I_g is 19.29% which does not meet the IEEE 519 standard. Hence, this technique also fails with distorted source.

The templates are created by means of PLL, and when the grid is distorted, the references are also affected. Perfect sinusoial reference currents cannot be generated under adverse grid conditions.

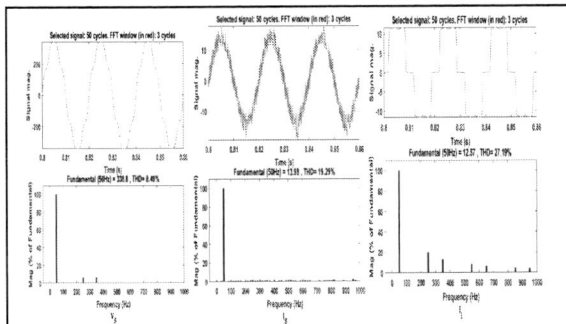

Fig. 13. THD values for SRFT based control algorithm (with DSOGI technique)

Fig. 12. shows the plots of grid voltage $V_{g,\,abc}$, grid current $I_{g,\,abc}$, compensator current $I_{c,\,abc}$, load current $I_L,\,_{abc}$, and DC link voltage V_{Da} with DSOGI technique. Voltage and current inputs are no more distorted as they have been made perfectly sinusoidal by the DSOGI technique. The model is tested for the same unbalance conditions as in case I. It is seen that V_{Da} settles to the 700 V value at 0.57 seconds.

Fig. 13. shows the corresponding THD values for V_g, I_g, and I_L. The THD value for I_g now is 2.88% which meets the IEEE 519 standard. Hence, the proposed DSOGI technique gives the desired results. THD value for SRFT based control algorithm is less than the one obtained with UT-based control algorithm when both these techniques are fed with undistorted input obtained via the DSOGI technique.

Table I shows the comparison of THD values for grid and load currents and grid voltage using both the control algorithms used for controlling SAPF under adverse grid conditions.

TABLE I.   COMPARISON OF TOTAL HARMONIC DISTORTION (THD)

Case			THD%
Unit template- or PI controller- based control algorithm	With distorted grid	V_S	8.49
		I_S	8.74
		I_L	25.95
	Using DSOGI technique	V_S	8.49
		I_S	2.94
		I_L	25.95
SRFT based control algorithm	With distorted grid	V_S	8.49
		I_S	19.29
		I_L	27.19
	Using DSOGI technique	V_S	8.49
		I_S	2.88
		I_L	27.23

A comparison of the THD values shows a significant drip in the distortion values of the grid current (I_g) i.e. now within 5% when using DSOGI. Without the DSOGI block, the two algorithms fail to achieve <5% THD in grid currents. The resaon is that reference currents are not perfectly generated as the voltage is itself distorted. DSOGI helps to achieve perfect grid synchronization and compute the positive sequence component of grid voltages.

VI. CONCLUSION

Two simple and effective algorithms based on UT-based and SRF theory for controlling the SAPF have been studied in this paper under distorted grid conditions. It was found that both the control schemes failed to achieve less than 5% THD in grid currents even under moderately distorted grid conditions. DSOGI block has been proposed to filter the distortions and produce perfect sinusoidal voltage templates. With a slight modification in control algorithms, the two approaches yielded perfectly sinusoidal grid currents under adverse grid conditions also. Simulation results under load changes and unbalanced load conditions gave satisfactory results.

ACKNOWLEDGMENT

The authors are thankful to the DST, SERB for the sponsored project (EMR/2016/001874).

REFERENCES

[1] A. Moreno-Munoz, Power Quality: Mitigation Technologies in a Distributed Environment. London, U.K.: Springer-Verlag, 2007.

[2] J. Matas, L. G. de Vicuna, J. Miret, J. M. Guerrero, and M. Castilla, "Feedback linearization of a single-phase active power filter via sliding mode control," IEEE Trans. Power Electron., vol. 23, no. 1, pp. 116-125, Jan. 2008.

[3] J. M. Maza-Ortega, J. A. Rosendo-Macias, A. Gomez-Exposito, S. Ceballos-Mannozzi, and M. Barragan-Villarejo, "Reference current computation for active power filters by running DFT techniques," IEEE Trans. Power Del., vol. 25, no. 3, pp. 446-456, Jul. 2010.

[4] B.-S. Chen and Y.-Y. Hsu, "A minimal harmonic controller for a STATCOM," IEEE Trans. Ind. Electron., vol. 55, no. 2, pp. 655–664, Feb. 2008.

[5] H. Akagi, E. H. Watanabe, and M. Aredes, Instantaneous Power Theory and Applications to Power Conditioning. Hoboken, NJ: Wiley, 2007.

[6] R. S. Herrera, P. Salmeron, and H. Kim, "Instantaneous reactive power theory applied to active power filter compensation: Different approaches, assessment, and experimental results," IEEE Trans. Ind. Electron., vol.55, no. 1, pp. 184–196, Jan. 2008.

[7] D. M. Divan, S. Bhattacharya, and B. Banerjee, "Synchronous frame harmonic isolator using active series filter," in Proc. Eur. Power Electron. Conf., 1991, pp. 3030–3035.

[8] B. Singh and V. Verma, "Selective compensation of power-quality problems through active power filter by current decomposition," IEEE Trans. Power Del., vol. 23, no. 2, pp. 792–799, Apr. 2008.

[9] C. Lascu, L. Asiminoaei, I. Boldea, and F. Blaabjerg, "Frequency response analysis of current controllers for selective harmonic compensation in active power filters," IEEE Trans. Ind. Electron., vol. 56, no. 2, pp. 337– 347, Feb. 2009.

[10] A. Luo, Z. Shuai, W. Zhu, and Z. J. Shen, "Combined system for harmonic suppression and reactive power compensation," IEEE Trans. Ind. Electron., vol. 56, no. 2, pp. 418–428, Feb. 2009.

[11] K.-K. Shyu, M.-J. Yang, Y.-M. Chen, and Y.-F. Lin, "Model reference adaptive control design for a shunt active-power-filter system," IEEE Trans. Ind. Electron., vol. 55, no. 1, pp. 97–106, Jan. 2008.

[12] S. Mohagheghi, Y. Valle, G. K. Venayagamoorthy, and R. G. Harley, "A proportional-integrator type adaptive critic design-based neurocontroller for a static compensator in a multimachine power system," IEEE Trans. Ind. Electron., vol. 54, no. 1, pp. 86–96, Feb. 2007.

[13] Z. Shu, Y. Guo, and J. Lian, "Steady-state and dynamic study of active power filter with efficient FPGA-based control algorithm," IEEE Trans. Ind. Electron., vol. 55, no. 4, pp. 1527–1536, Apr. 2008.

[14] C. Lascu, L. Asiminoaei, I. Boldea, and F. Blaabjerg, "High performance current controller for selective harmonic compensation in active power filters," IEEE Trans. Power Electron., vol. 22, no. 5, pp. 1826-1835, Sep. 2007. (Pubitemid 47423057).

[15] D. Yazdani, A. Bakhshai, G. Joos, and M. Mojiri, "A real-time three-phase selective-harmonic-extraction approach for grid-connected converters," IEEE Trans. Ind. Electron., vol. 56, no. 10, pp. 4097-4106, Oct. 2009.

978-1-5386-6624-1/18 $31.00 © 2018 IEEE

Modular Assembly Systems in Industry 4.0 milieu

N. Sumedh, O.V.L Narayana, Madhav Reddy, N.Sampath, Priya B.K, Dr. T.K Ramesh
Department of Electronics and Communication Engineering
Amrita School of Engineering, Bengaluru
Amrita Vishwa Vidyapeetham
India
sumedhn97, ovlnarayana@gmail.com, bk priya@blr.amrita.edu

-

Abstract—**High industrial productivity is a necessity for any aspiring economic society in the 21st century. The current panacea for consistent industrial growth seems to hover around the complete realization of automation and computer vision. Taking into consideration the recent ad- vances in Industry 4.0 paradigm, flexibility and scalability have become pre-requisite for the manufacturing sector. This study proposes a flexible, power effective and envi- ronment - adaptive assembly system for usage in contem- porary Industry 4.0 manufacturing milieu. Categorically, it is an attempt at exploring novel and innovative methods for developing a procedural setting to provide maximum customization in a batch processing assembly unit frame- work. Furthermore, the discussed design schematics have been proved capable of minimizing power consumption with the use of a prototype. The necessary calculations and mathematical formulations for increasing performance and efficiency have been provided.**

Keywords - Assembly System, Industry 4.0, Power Efficient Manufacturing, Superscalar Manufacturing.

I. INTRODUCTION

Due to the cut throat competition and numerous socio-economic constraints, present day manufacturing sector is universally and unanimously intensifying its effort to embed and integrate principle enabling technologies such as Internet of Things (IoT), Wireless Sensor Network (WSN), Machine Learning (ML), Cloud Computing, Big Data Analytics to increase the productivity and output yield. To this effort, the German government instituted a working group which proposed and launched Industry 4.0 paradigm in early 2013 [1]. This newfound interconnection among the most recent disruptive technologies has helped to solve or at least add a new dimension of thought into solution - oriented design.

The scope of such a model is quite comprehensive and global. In this paper, the base principles of Industry 4.0 have been incorporated to provide a full-fledged assembly unit, the schematics or which are provided in

A major pattern has been analysed in the emerging trends with respect to assembly systems. Automation and control are the basic pre-requisite in any design but modern day systems have been formulated to provide modularity within the system. This intra - system modularity has been conceived by using asynchronous signalling techniques. The use of SS7 [2] and more recently packet switched networking is evidence to modular designs. Despite improved efficiency and control, a basic draw-back to this mechanism is the increase of overhead power consumption. The primary focus of this research is to create a power efficient and highly decentralized auto-mated system for the purposes of component assembling. Moreover, the deliberations made in paper are grouped in this order. Section II outlines the prior work in the field of assembly line robotics and conveyor belts in a systematic way. Section III confers the methodological overview. Section IV examines and analysis the results of this research. Section V establishes a the scope of the current work with all relevant references for further analysis being provided for in a sequential manner in the bibliography section. Section VI acknowledges the support and motivation provided by Amrita Vishwa Vidyapeetham for this research.

II. RELATED WORK

Being an area of active interest, huge amounts of qual- ity research is available for detailed analysis in the field of assembly robotics and conveyor belt mechanisms. With the advent of industrial robotics, all kinds of mun- dane and iterative work has been automated, resulting in efficiency and faster processing without any change in quality of the output. The most recent development in industrial robots has been self-learning and automation. In [3 - 4], self-learning paradigm has been elucidated upon whereas in [5], high level of automation was

TABLE I
COMPARATIVE STUDY AMONG DIFFERENT CONTROLLERS

Parameters	Programmable Logic Controller (PLC)	Micro-Controller	Micro-Processor
Speed of operation	Real Time	Fast	Slowest
Clock Frequency	Real Time	<50 MHz	>1GHz
Power Consumption	High	Low	Low
Input Interfaces	High	Low	Low
Memory Type	RAM MROM	RAM EEPROM	RAM EEPROM
Memory Architecture	Von- Neumann	Harvard	Harvard
Initial Cost	High	Low	Low
Re-Programming Capability	Yes	Yes	Yes
Area Consumption	High	Least	Low

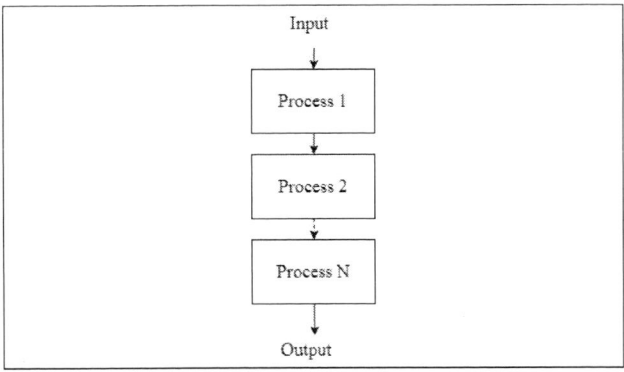

Fig. 1. Proposed Execution of N processes

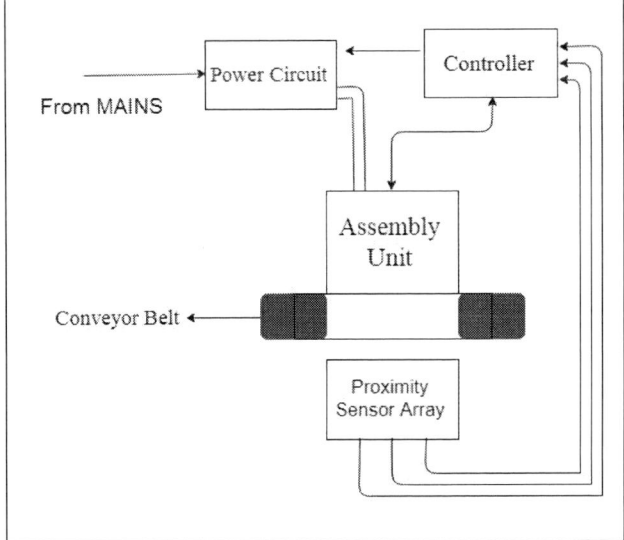

Fig. 2. Proposed Assembly Unit

achieved. Few case studies have been analysed for in-depth understanding of the underlying issues with the above mentioned technologies in [6 - 10]. In today's factory environment, it is found that all processing equipment are static except for a conveyor belt. The idea is to encapsulate procedures layer by layer on a raw product and transform it to a complete product while it navigates around the conveyor belt. Examples of such machines and their applications have been well documented in [11 - 12]. The proposed automation is provided for by a logic controller. Different controllers have been analysed with respect to many performance parameters. The general- ized results of this study have been tabulated in Table I. 3 major categories of industrial controllers have been examined, each providing different features. The se- lection of hardware is left to implementer's discretion for customization purposes. On a general basis, current technologies have seen a lot of progress, but the practical implementation of the above technologies in the form of a comprehensive mechanism was found missing. It is this void which the proposed schematic would adhere to.

III. METHODOLOGICAL OVERVIEW

The objective of the proposed design is to implement any procedural algorithm as shown in Fig.1. Fig.1 has N distinct processes which are to be executed sequentially.

Design schematic for the proposed model is shown in Fig.2. The primary focus in formulation of an assembly unit is based on 4 sub-schematics, namely:

A. Conveyor Belt Design
B. Power System Design
C. Controller and Feedback Design
D. Branching and Pipeline

A. Conveyor Belt Design:

In the design of conveyor belt, connected and modular belts have been considered instead of single monolithic belts for optimal power consumption as discussed in the next section. Each process is assigned to an individual conveyor belt. The implication being that a N process procedure will contain N individual belts connected by connectors as

shown   in   Fig.3.
The equivalent graph representation is shown in Fig.4.
Therefore, the logical equivalent for the system can be
analysed using concepts of graph theory for stability,
loop elimination and timing analysis.

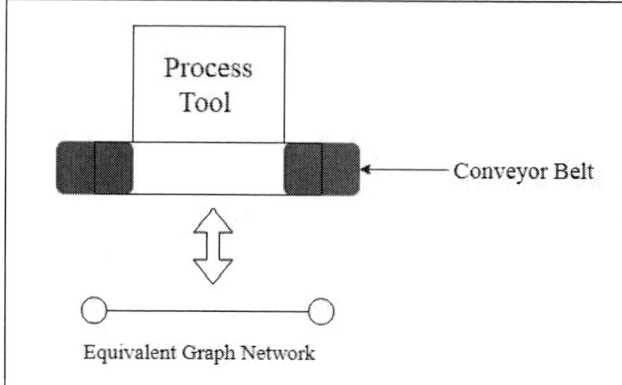

Fig. 4. Proposed Graph Theory Representation

all conveyor belts are active all the time as not all process
are happening simultaneously.

Therefore, this system is responsible to power all the
necessary processes while cutting off power supply to
inactive process states. The schematic for the power
switching system is shown in Fig.5.

The instruction decoder relies on the sensor feedback
given to it as input to generate signals to activate the
control signal generator, which acts like a control store.
The control signal generator activates the required nodes
of the crossbar switch which initiates power transmission. As the control signal generator varies the control
signal every clock cycle, a closed loop automated system
is achieved.

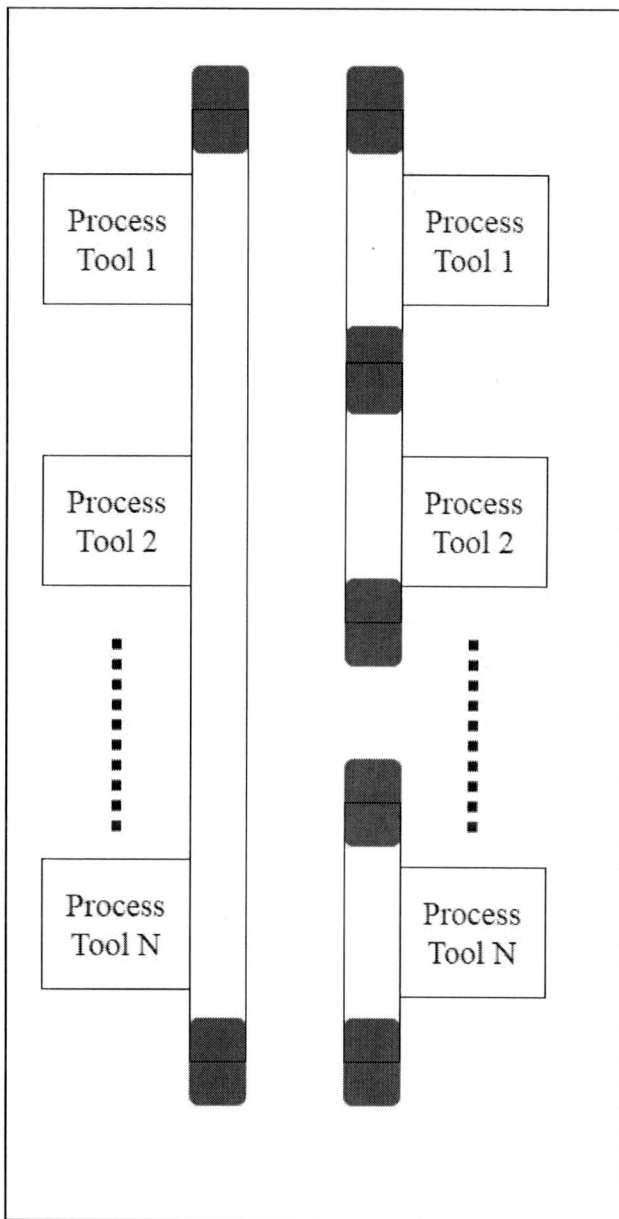

Fig. 3. Modular Belt Structure: Intuition

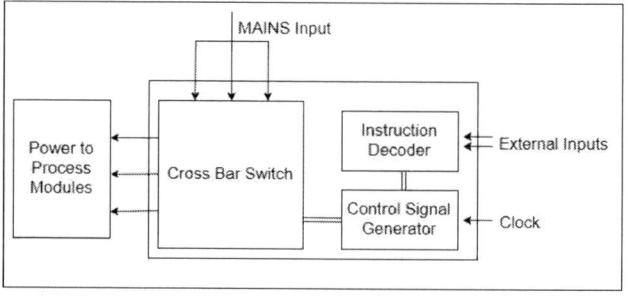

Fig. 5. Proposed Power Switching Circuit

B. Power System Design:

An automated power switching system has been designed to reduce power consumption. Considering the
processing of discrete objects, it can be inferred that not

C. Controller and Feedback Design:

Any type of controller tabulated in Table I can be
used in the development of controller subsystem. This
choice is left to the implementer's discretion. Regardless, a
feedback from the sensor array present in the process tools
is arranged as an input for the controller. This allows the
controller to update itself with process status and initiate a
subsequent response with respect to the

process completion.
To test and validate this mechanism, a hardware prototype based on ATmega328 microcontroller (shown in Fig.9) was designed and assembled. An array of proximity sensors were put up parallel to the conveyor belt to track the object along discrete length intervals. Based on the feedback from the sensors, the microcontroller would initiate (or) stop the movement of the conveyor belt.

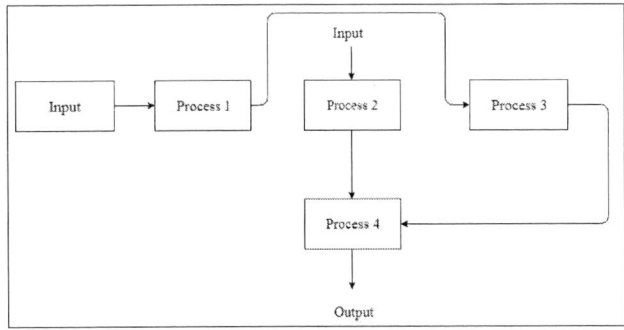

Fig. 7. Pipelining: Intuition

D. Branching and Pipeline:

The major motivation for the use of a small, modular conveyor belts ahead of single, monolith conveyor belt was to incorporate the features of branching and pipelining. If the procedure (as defined by Fig.1) has a branching operation, the appropriate process is executed next. A Boolean switching example has been depicted in Fig.6.

2) Pipelining
3) Decrease in power consumption

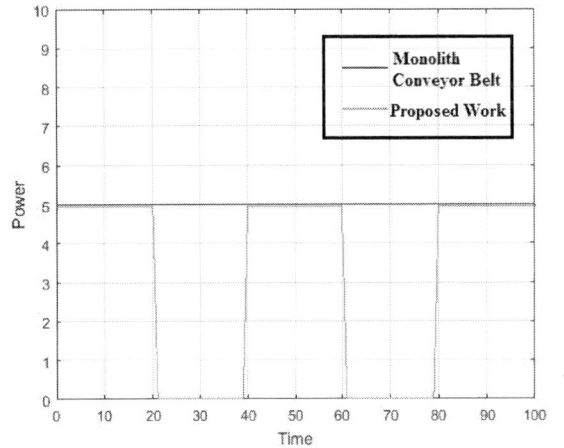

Fig. 8. Power vs. Time

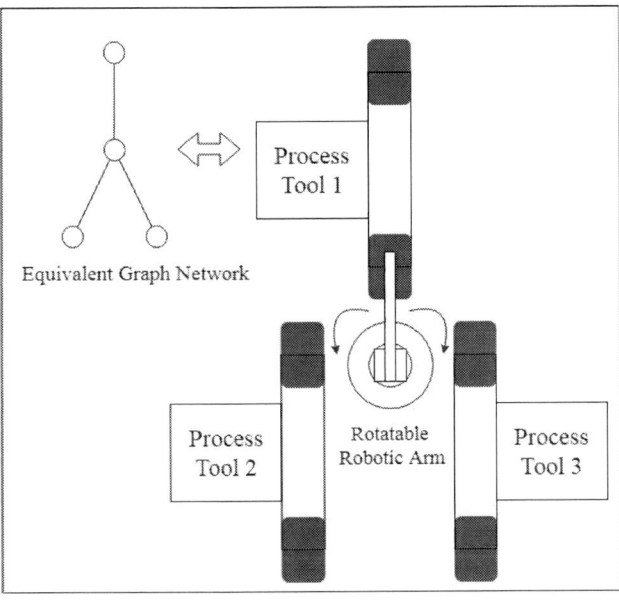

Fig. 6. Branching: Intuition

As all branches (and in general all conveyor belts) are uncorrelated, they can be used to execute a pipeline like paradigm as shown in Fig.7.

IV. DISCUSSION AND RESULTS

An interconnected chain of conveyor belts have been proposed as a replacement to widely used single mono- lith belt. The advantages of process oriented conveyor belts are:
1) Branching

A power vs. time graph (shown in Fig.8) obtained from hardware prototype verifies the third point. In Fig. 8, the green line indicates power consumption by the prototype whereas the red line represents the power consumed using single conveyor belt.
The general formulae to calculate power saved is given by:

$$P = P_c \times T_{off} \times N \qquad (1)$$

Where P is the total power saved, P_c is the power consumed in a unit time, T_{off} is the time period for which the process is inactive and N is the number of processes.
This implies that as time interval among conveyor belt objects increases, relative power consumption decreases. It has also been observed that the proposed schematic follows a logical star topology which relieves designers of physical constraints.

978-1-5386-6624-1/18 $31.00 © 2018 IEEE

The crossbar switch used in the power switching system can act as a smart grid if the corresponding controller is well programmed. This would lead to further increase in power efficiency.

Fig. 9. Hardware Prototype

Modularity has been achieved using the proposed design. This has been possible because all the processes have been considered independent. Therefore, a designer may group all correlated process into a single process to use the proposed schema.

V. CONCLUSION AND SCOPE

This document highlights a modular approach to designing of an automated assembly unit based on Industry 4.0 standards. In addition to a flexible solution, new methods have been proposed to inculcate branching and pipelining while reducing overhead power consumption. Due to a large use of independent processes among procedural paradigms, the concept of process oriented conveyor belt has been conceived.

Future work in the domain of smart grid and intelligent controller systems would make the current work more proficient. Any effort directed towards developing real time wireless feedback mechanism would lead to a subsequent reduction in hardware costs.

All the relevant design schematics and computations are provided, while the results are proved using graphical plots. A hardware prototype was developed for in-depth understanding of the underlying issues. In the case of power switching systems, a cross bar switch was implemented and the resultant outputs verified. The hardware schematics of all the devices for the implementation of

Assembly unit have been shown in section III, Fig [1 - 6]. In this paper, Industry 4.0 standards have been used as a pillar for building a power efficient and cost effective system rather than deploying it as the primary technology. This seems the way forward for the future development of intelligent and adaptive systems.

VI. ACKNOWLEDGMENT

The current research is a result of constant motivation and support granted to the authors by Amrita Vishwa Vidyapeetham. We thank our colleagues from Amrita School of Engineering, Bangalore for contributing and assisting the authors in terms of formatting and proof-reading.

REFERENCES

[1] D. Trotta and P. Garengo, "Industry 4.0 key research topics: A bibliometric review," 2018 7th International Conference on Industrial Technology and Management (ICITM), Oxford, 2018, pp. 113 – 117.

[2] A. A. El-Toumi and M. A. McGrew, "Interconnecting SS7 signaling networks," IEEE International Conference on Communications, Including Supercomm Technical Sessions, Atlanta, GA, 1990, pp. 589 – 593 vol.2.

[3] A. A. Al Sallab and M. A. Rashwan, "Self learning machines using Deep Networks," 2011 International Conference of Soft Computing and Pattern Recognition (SoCPaR), Dalian, 2011, pp. 21 – 26.

[4] M. Al Karim, J. Currie and T. T. Lie, "Dynamic Event Detection Using a Distributed Feature Selection based Machine Learning Approach in a Self Healing Microgrid," in IEEE Transactions on Power Systems.

[5] Ahmed Asif Shaik and G. Bright, "Closed loop sensor system for automated machines," AFRICON 2007, Windhoek, 2007, pp. 1 – 6.

[6] Y. Liu, "Learning self-awareness in committee machines," 2016 International Conference on Machine Learning and Cybernetics (ICMLC), Jeju, 2016, pp. 888 – 893.

[7] A. Ozadowicz and J. Grela, "Impact of building automation control systems on energy efficiency University building case study," 2017 22nd IEEE International Conference on Emerging Technologies and Factory Automation (ETFA), Limassol, 2017, pp. 1 – 8.

[8] L. Guerra, S. D. Sousa and E. P. Nunes, "Statistical process control automation in the final inspection process: An industrial case study," 2016 IEEE International Conference on Industrial Engineering and Engineering Management (IEEM), Bali, 2016, pp. 876 – 880.

[9] R. V. Arvind, Raj, R. R., and Prakash, N. K., "Industrial automation using wireless sensor networks, Indian Journal of Science and Technology, vol. 9, 2016.

[10] Rajeswari P., Shekar G., Devi S., and Dr. Purushothaman A., "Geometric Programming-Based Power Optimization and Design Automation for a Digitally Controlled Pulse Width Modulator, Circuits, Systems, and Signal Processing, 2017.

[11] Y. Zhang, L. Li, M. Ripperger, J. Nicho, M. Veeraraghavan and A. Fumagalli, "Gilbreth: A Conveyor-Belt Based Pick- and-Sort Industrial Robotics Application," 2018 Second IEEE International Conference on Robotic Computing (IRC), Laguna Hills, CA, 2018, pp. 17 – 24.

[12] J. Schrimpf, "Automated sewing using conveyor belts," 2016 IEEE 21st International Conference on Emerging Technologies and Factory Automation (ETFA), Berlin, 2016, pp. 1 – 4.

2nd IEEE International Conference on Power Electronics, Intelligent Control and Energy Systems (ICPEICES - 2018)

Comparative Performance Analysis of Radial Flux and Dual Air-Gap Axial Flux Permanent Magnet Brushless DC Motors for Electric Vehicle Application

Amit N. Patel
Electrical Engineering Department
Institute of Technology, Nirma University
Ahmedabad, India
amit.patel@nirmauni.ac.in

Bhavik N. Suthar
Electrical Engineering Department
Government Engineering Department
Bhuj, India
bhavikiitd@gmail.com

Tejas H. Panchal
Electrical Engineering Department
Institute of Technology, Nirma University
Ahmedabad, India
tejas.panchal@nirmauni.ac.in

Rajesh M. Patel
Electrical Engineering Department
MEFGI's Faculty of PG Studies
Rajkot, India
r_mpatel77@hotmail.com

Abstract — This paper presents comparative performance analysis between radial flux permanent magnet brushless DC motor and axial flux permanent magnet brushless DC motor for electric vehicle application. Initially motor rating is calculated based on requirements of electric vehicle application and vehicular dynamics. Radial flux motor and axial flux motors are designed for that calculated rating based on sizing equations and assumed design variables. Dual air-gap sandwiched stator type topology is selected for axial flux motor. Important design aspects of radial flux motor and axial flux motor are discussed. Comparative analysis is done for performance parameters like requirement of copper, iron and permanent magnet materials, various losses, efficiency, weight and cogging torque. Based on comparative analysis it is concluded that axial flux motor is the most compatible motor for electric vehicle application.

Keywords — *Radial flux permanent magnet brushless DC motor, axial flux permanent magnet brushless DC motor, electric vehicle, performance analysis, FE analysis*

I. INTRODUCTION

Cost of fossil fuel increases day by day due to its limited resources. Electric Vehicle (EV) technology is witnessing immense growth due to less running cost and environment friendly features. Many researchers are working to establish reliable technologies in the field of electrical vehicle. Overall efficiency and performance of electric drive is governed by electric motor. High efficiency, high power density, high torque to current ratio and flat shape are basic required characteristics of an electrical machine in EV drive system [1]. Researchers are contributing a lot in development of motor comprising majority of required characteristics for EV. The development in field of permanent magnets and semiconductor devices boost up EV technology. Permanent magnet (PM) motors are inherently efficient and compact. According to direction of magnetic flux and current, PM motors are classified in two categories like radial flux motor (RFM) and axial flux motor (AFM). In radial flux motor flux travels radially and current travels axially whereas in axial flux motors flux travels axially and current travels radially

[2]. Axial flux motors are further classified based on numbers and relative position of stator and rotors. Main classifications are single rotor single stator, double rotor single stator, single rotor double stator and multi disc. Dual rotor single stator axial flux motor is the best suited motor for vehicle application.

The rating of motor is calculated based on vehicle dynamics and requirements of application. Electric Vehicle has laden weight of 250 kg, velocity 25 km/hr and acceleration of 0-25 km/hr in 09 sec. Assumed parameters are coefficient of rolling resistance 0.011, drag coefficient 0.7 and frontal area 0.9 m². Calculated motor power and torque ratings for this application are 250 W and 15.9 N.m. respectively. Performance equations for emf, torque, inductance, etc. of the brushless permanent magnet motors are available in the literature [3]. Design procedure for radial flux surface mounted PM brushless dc motor and sandwiched stator dual rotor surface mounted axial flux PM motor is explored and finalized. Computer Aided Design (CAD) program with two decision making loops is finalized. Both motors are designed as per calculated rating and subsequently parametric analysis is carried out. Performance comparison is done for both motors. In order to establish correctness of CAD program and comparative analysis, finite element analysis (FEA) is carried out. In FEA 2-D model of radial flux motor and 3-D model of axial flux motor are prepared based on CAD output. Comparative performance analysis and compatibility of motor for EV application is discussed in result section.

II. INTRODUCTION AND SIZING EQUAQTIONS OF RADIAL FLUX PMBLDC MOTOR

Fig. 1 shows cross sectional diagram of radial flux permanent magnet motor. NdFeb sintered magnets are used to achieve high torque density. Sintered PM has high residual flux density and high conductivity [4]. M19 silicon steel type ferromagnetic material is used in core. Rotor outer diameter, stator outer diameter and axial length are main dimensions of radial flux PM motor, depending on assumed specific magnetic and electric loadings.

978-1-5386-6624-1/18 $31.00 © 2018 IEEE

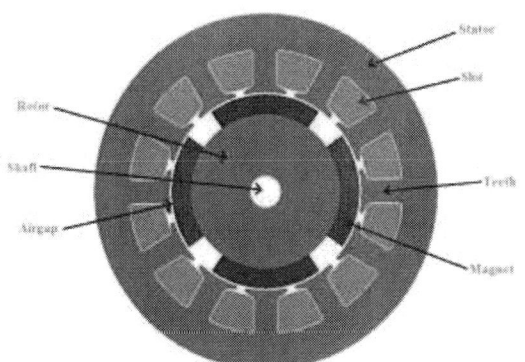

Fig. 1. Sectional diagram of radial flux permanent magnet motor

Rotor outer diameter, stator outer diameter of the motor are calculated using following equations [5].

Rotor outer diameter, $D_{ro} = \sqrt{\dfrac{2T}{\pi B_g K_w K_s (ac) L_i}}$ (1)

Stator outer diameter, $D_{so} = \dfrac{D_{ro}}{\text{Split ratio}}$ (2)

Where, T is torque, B_g magnetic loading, K_w winding factor, K_s skew factor, ac specific electric loading and L_i axial length of motor.

Width of stator back iron, width of stator teeth and width of rotor back iron are calculated by assuming suitable flux densities in these parts.

Width of stator back iron, $W_{sy} = \dfrac{\pi R_{ro} B_g}{N_m K_{st} B_{sy}}$ (3)

Width of stator teeth, $W_{st} = \dfrac{2\pi R_{ro} B_g}{N_s K_{st} B_{st}}$ (4)

Width of rotor back iron, $W_{ry} = \dfrac{\pi R_{ro} B_g}{N_m K_{st} B_{ry}}$ (5)

Where B_{sy}, B_{st} and B_{ry} are flux densities in stator back iron, stator teeth and rotor back iron respectively. R_{ro} is rotor outer radius, K_{st} stacking factor, N_m number of poles and N_s number of stator slots.

Stator back iron radius is given by, $R_{sb} = R_{so} - W_{sy}$ (6)

From which stator slot depth is calculated as,

$d_s = R_{sb} - R_{ro} - l_g$ (7)

where R_{so} is stator outer radius, l_g is length of airgap.

Efficiency of the motor is calculated using,

$\eta = \dfrac{P}{P + P_L} \times 100\%$ (8)

$P_L = P_r + P_c + P_s$ (9)

where, P is output power, P_L total losses in motor, P_r resistive loss, P_c core loss and P_s stray loss.

III. INTRODUCTION AND SIZING EQUATION OF AXIAL FLUX PMBLDC MOTOR

Dual rotor sandwiched stator axial flux permanent motor is shown in Fig. 2.

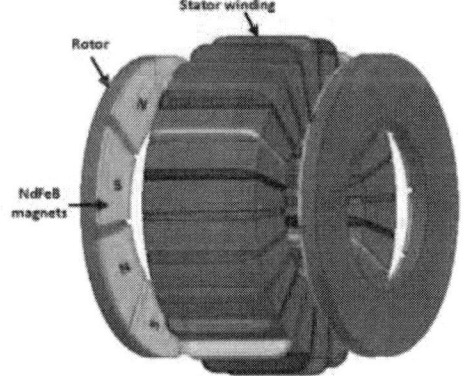

Fig. 2. Double rotor axial flux permanent magnet motor

Outer radius and inner radius are main dimensions of axial flux motor. Main dimensions can be calculated form following equations [6].

Outer radius,

$R_0 = \sqrt{\dfrac{3 * T}{2 * \eta * N_c * N_m * N_{spp} * K_w * B_g * I_s}}$ (10)

and inner radius, $R_i = \dfrac{R_0}{K_r}$ (11)

where T torque, η efficiency, N_c number of coils conducting at a time, N_m number of poles, N_{spp} number of slots per pole per phase, K_w winding factor, B_g magnetic loading, K_r diametric ratio and I_s electric slot loading.

Width of stator teeth can be calculated from following equation.

$W_{steeth} = \dfrac{B_g * \tau_{pi}}{N_{sm} * B_{max} * K_{st}}$ (12)

where τ_{pi} pole pitch, N_{sm} number of slots per pole, B_{max} maximum permissible flux density of core material and K_{st} stacking factor.

Slot dimensions can be calculated from following equations.

Width of slot base, $W_{sb} = (Z_{si} - W_{steeth})$ (13)

Depth of slot, $d_s = \dfrac{I_s}{K_{cp} * W_{sb} * J_{max}}$ (14)

where Z_{si} slot pitch, W_{steeth} width of stator teeth, K_{cp} slot packing factor and J_{max} maximum current density in conductor.

Number of conductors depend on voltage rating and flux in magnetic circuit. Sectional area of conductor depends on motor current and assumed current density[7]. Efficiency of motor can be estimated form following equation.

Efficiency, $\eta = \dfrac{P}{P + P_L} \times 100\%$ (15)

$P_L = P_r + P_c + P_s$ (16)

where, P output power and P_L total losses in motor, P_r resistive loss, P_c core loss and P_s stray loss.

IV. COMPUTER AIDED DESIGN AND PARAMETRIC ANALYSIS

Many design variables are assumed like average magnetic loading, slot loading, packing factor, winding factor, stator current density, length of air-gap, diametric ratio, carter's coefficient and stacking factor. Availability of material, motor specifications, efficiency and aforesaid assumed data for the design are provided as the input for the CAD program. Other dimensions of magnetic circuit and electric circuit are calculated based on assumed flux density and current density. The developed CAD program consists two loops. The first loop is used to calculate magnet thickness to establish required air-gap flux density. The second loop is used to check estimated efficiency with assumed efficiency of analytical design. If the estimated efficiency does not match with the assumed efficiency, assumed design variables are changed to get desired efficiency.

In parametric analysis out of all design variables one design variable is changed at a time and performance estimation is done. For both type of motors number of poles are decided considering its influence on weight of iron & overhang, switching losses and iron losses. As number of poles increase weight of iron & overhang decrease but switching frequency and iron loss increase. Selection of soft magnetic material also influence operational efficiency of motor. Necessity of PM material is greatly affected by selection of length of air-gap. Selection of stator current density influence requirement of copper and efficiency. In case of high stator current density copper requirement decreases with reduction in efficiency.

V. FEA FOR DESIGN VARIFICATION

Finite element analysis (FEA) is carried out for verification of CAD program and comparison between cogging torque, average torque of radial flux motor and axial flux motor. FEA is computer simulation technique used to solve engineering problems accurately. It uses a numerical method called Finite Element Method (FEM) for analysis. FEM divides the problem into number of small finite elements and solve them individually. It integrates solution of each finite element to obtain a particular solution. The

FEA comprises of three major steps called pre-processing, solver and post-processing.

Model is prepared according to design information obtained and FEA is carried out. Design information is shown in Table I for radial flux motor and axial flux motor. Design and comparative performance analysis for both motors is carried out with following considerations:

Equal number of phases, stator slots, rotor poles, same type of permanent magnet material, same type of core material and equal length of air gap.

Table I. Design information

Particulars	Value	
	Radial Flux Motor	*Axial Flux Motor*
Stator outer radius	94.5 mm	91 mm
Stator inner radius	48 mm	52 mm
Axial length of motor	84.6 mm	90 mm
Number of phase	3	3
Number of stator slots	48	48
Number of poles	16	16
Type of PM	NdFeb	NdFeb
Type of core material	M19	M19
Length of air-gap	0.5 mm	0.5 mm

Evolution of magnetic flux density in various parts is essential. If the flux density in core or stator teeth goes into saturation, the efficiency reduces affecting performance of motor. Flux density plot and torque profile of radial flux motor obtained from FEA are shown in Fig. 3 and Fig. 4 respectively.

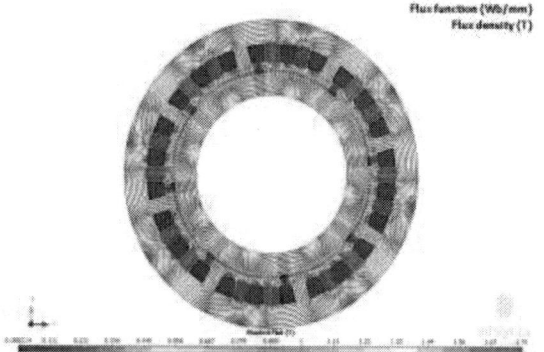

Fig. 3. Flux density plot of radial flux PM motor

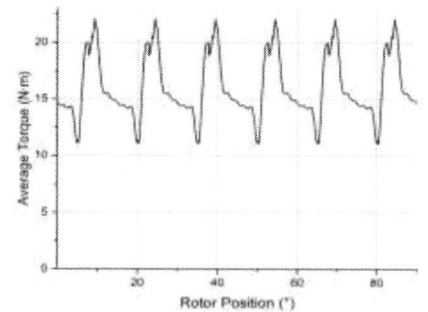

Fig. 4. Torque profile of radial flux PM motor

Flux density plot and torque profile of axial flux motor obtained from FEA are shown in Fig. 5 and Fig. 6 respectively.

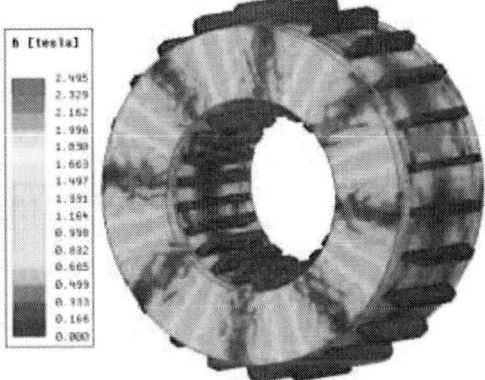

Fig. 5. Flux density plot of radial flux PM motor

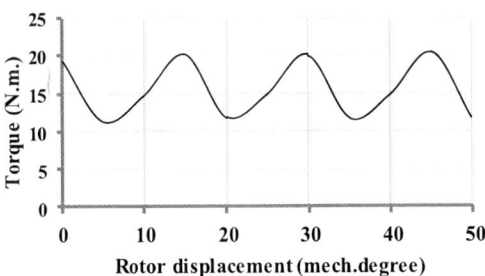

Fig. 6. Torque profile of axial flux PM motor

Flux densities in various sections of radial flux motor and axial flux motors are according to assumed flux densities. Cogging torque profiles of radial flux motor and axial flux motor are shown in Fig. 7 and Fig. 8 respectively.

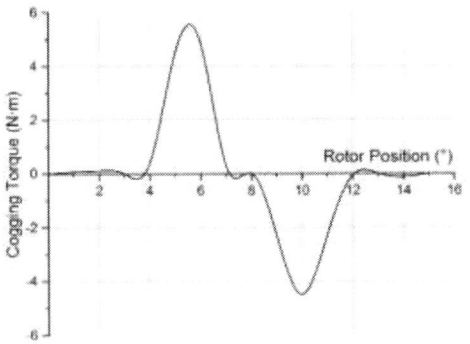

Fig. 7. Cogging torque profile of radial flux PM motor

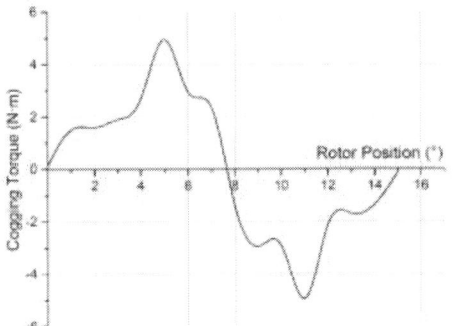

Fig. 8. Cogging torque profile of axial flux PM motor

Analysis reveals that peak to peak cogging torque is slightly less in radial flux motor than axial flux motor. Peak to peak cogging torque is 9.9 N.m. in axial flux motor and 9.8 N.m. in radial flux motor.

VI. COMPARATIVE PERFORMANCE ANALYSIS

According to CAD and FEA, It is analyzed that axial flux motor imparts efficiency of 88.76 % while radial flux motor imparts efficiency of 83.53 % for same PM material usage of 0.64 kg in both motors. Performance parameters of radial flux motor and axial flux motor are shown in Table II.

TABLE II. PERFORMANCE PARAMETERS OF RADIAL AND AXIAL FLUX MOTORS

Performance Parameter	Radial Flux Motor	Axial Flux Motor
Efficiency	88.76 %	83.53 %
Weight	9.9 kg.	12.15 kg.
Cogging Torque (peak-peak)	9.9 N.m.	9.86 N.m.

Utilization of copper is better in axial flux motor as its overhang is short. In this analysis copper requirement in axial flux motor and radial flux motors are 1.27 kg and 1.71 kg respectively. Stator core weight is 6.35 kg in axial flux motor and 6.82 kg in radial flux motor. Rotor core weight is 2.44 kg in axial flux motor and 2.98 kg in radial flux motor. Total weight of axial flux motor is 9.9 kg and radial flux motor is 12.15 kg.

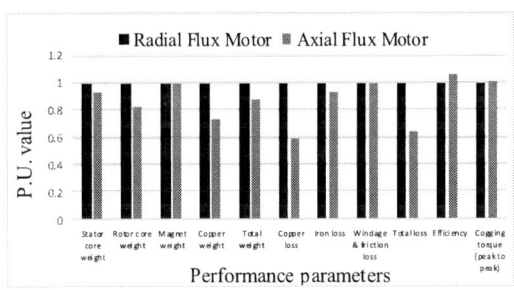

Fig. 9. Performance comparison on p.u. basis

Per unit comparison between performance parameters of axial flux motor and radial flux motor considering radial flux motor value as base is shown in Fig. 9. It has been observed that power per weight ratio is more in axial flux PM motor compared to radial flux PM motor. This feature of axial flux motor makes it an obvious selection in applications where compactness is important performance parameter.

VII. CONCLUSION

Rating of electric motor is determined based on vehicle dynamics and subsequently radial flux & dual air-gap axial flux PM motors are designed. Axial flux motor imparts 6.26 % high efficiency and 11.85 % less weight than radial flux motor of same rating using same amount of PM materials. Axial flux motor offers high efficiency and power density compared to radial flux motor. Peak cogging torque in axial flux motor is marginally more than radial flux motor. Axial flux topology is more advantageous in flat shape due to short overhang compared to radial flux motor. It is concluded that dual air-gap axial flux motor is the best suited motor for electric vehicle application due to its superior performance and flat shape.

REFERENCES

[1] N A Rahim, Hew Wooi Ping, M Tadjuddin, "Design of Axial Flux Permanent Magnet Brushless DC Motor for Direct Drive of Electric Vehicle", Power Engineering Society General Meeting, IEEE, pp. 1-6, July 2007.

[2] C C Chan, "Axial Field Electrical Machines-Design and Applications", IEEE Transactions on Energy Conversion, vol. PER-7, no. 6, pp. 49-50, June 1987.

[3] Ajay Nair, K R Rajagopal, "Generic model of an electric vehicle for dynamic simulation and performance prediction", IEEE International Conference on Electrical Machines and Systems, pp.753-757, December 2010.

[4] R. Tsunata, M. Takemoto, S. Ogasawara, A. Watanabe, T. Ueno and K. Yamada, "Development and Evaluation of an Axial Gap Motor Using Neodymium Bonded Magnet," in IEEE Transactions on Industry Applications, vol. 54, no. 1, pp. 254-262, Jan.-Feb. 2018.

[5] B. Singh and D. Goyal, "Computer Aided Design of Permanent Magnet Brushless DC Motor for Hybrid Electric Vehicle Application," 2006 International Conference on Power Electronic, Drives and Energy Systems, New Delhi, 2006, pp. 1-6.

[6] Parag R. Upadhyay, K.R.Rajagopal, "FE Analysis and Computer-Aided Design of a Sandwiched Axial-Flux Permanent Magnet Brushless DC Motor" in IEEE Transactions on Magnetics, vol. 42, no. 10, pp. 3401-3403, October 2006.

[7] D C Hanselman, "Brushless Permanent Magnet Motor Design", New York, McGraw-Hill, 1994.

A Complete Fuzzy Logic Based Real-Time Simulation of Vector Controlled PMSM Drive

A Mishra
EE Deptt, NIT Patna
Patna, India
ambrishee@gmail.com

Garima Dubey
EC Deptt
Rajkiya Plytechnic
Kanpur, India
garidubey2012@gmail.com

Dheeraj Joshi
EE Deptt
DTU Delhi
India
joshidheeraj@dce.ac.in

Pramod Agarwal
EE Deptt
IIT Roorkee
Roorkee, India
pramgfee@iitr.ac.in

S P Sriavstava
EE Deptt
IIT Roorkee
Roorkee, India
satyafee@iitr.ac.in

Abstract—The replacement of speed proportional plus integral (PI) controller by fuzzy logic controller is a most promising application of artificial intelligence in motion control application. The stator current divided in two components- d-axis and q-axis currents. The mathematical model of PMSM in per unit values is presented. The whole vector controlled drive has two loops; inner currents and outer speed loop. In this paper fuzzy-logic is employed for all these controllers. To execute the control rules of fuzzy inference engine, the actual error inputs to controller have been normalized by using input scaling factors. To assure that output of fuzzy logic controller is appropriate for actual system being controlled, output scaling factors are used. Each controller has different parameters like range of MFs, input-output range, so the tuning of three controllers are also different. Tuning of individual controllers is done by investigating the performance with PI controller. The exhaustive simulation is done in different operating conditions, and implemented in real time at FPGA based RT-LAB. The performance of complete fuzzy controller is compared with PI controller for PMSM drive. Simulation and experimental results validate the efficacy of proposed controllers.

Keywords—FPGA, FLC, Modeling, PI-controller, PMSM, Per Unit, RT-LAB, SVM

I. INTRODUCTION

The Permanent Magnet Synchronous Motor (PMSM) is a motor that uses permanent magnets to produce the air gap magnetic field rather than using electromagnets. These motors have significant advantages, attracting the interest of researchers and industry for use in many applications. The benefits of PMSM on dc motor are; less audible noise, longer life, sparkless operation, higher speed, good heat transfer, higher smaller size. Advantages of PMSM on induction motors are; better efficiency, better power factor, higher power density and smaller size, and better heat transfer. Moreover popularity of PMSM comes from their desirable features; compact size, high efficiency, low noise and robustness, high torque to inertia ratio, high torque to current ratio, high air gap flux density, and high acceleration and deceleration rate.

With the recent developments in digital electronics, DSPs and ASICs PMSMs are gradually replacing the DC motors in wide range of drive applications. The development in high energy permanent magnets like NdBFe PMSM is becoming more popular in adjustable speed drive

applications[1]. Due to inherent coupling effect the torque control becomes complicated. In vector controlled PMSM drives decoupled torque, flux makes the control quite easier. Speed controller used in PMSM drive plays an important role in achieving high performance[2]. In motion control applications speed controller plays an important role as it affects the efficiency, dynamic response etc of motor [3, 4]. In high performance drive (HPD) the controller exhibits a important role so it is affecting the supply of current to motor and improves the dynamic performance of motor [3, 5] [6].

One of the inherent problems with PI controller is sluggish response. The improved performance with d-q axis current controller in synchronous reference frame using PI controller are presented in [7, 8]. Both methods utilizes the complex vector to reduce complication of system to be implemented and designed. These techniques are principally move plant pole towards zero of controller or zero of controller towards pole of plant.

II. BACKGROUND THEORY

Scalar control is based on relationships valid in steady state. It is simple but due to the inherent coupling effect (i.e., torque and flux are proportional to the voltage or current and frequency) gives sluggish response and the system can be easily prone to instability. Vector control clearly requires instantaneous control of stator current[9]. Vector control usually realized with digital PWM controller in rotating (d-q) reference frame[10]. Aim of vector control is to control flux and torque of machine to drive the motor to accurately trace the reference command value irrespective of load, machine parameter and any external environmental changes. In vector control stator current is controlled instantaneously which reduces the torque ripples and improves overall performance of machine[11].

Among artificial intelligent techniques[12] (ANN, Fuzzy, GA), the FLC is less complex as compared to other techniques for achieving the desired performance. To improve the robustness of system, the complication of neural network controller increases and implementation becomes challenging with limited processing speed.

To reduce the torque ripples, Ref. [13]proposed a fuzzy

logic algorithm to refine the selection of the voltage vectors. By using the space vector modulation (SVM), the torque and flux ripples can be more significantly reduced [14].

A fuzzy logic controller is a non-linear controller which provides good performance with robustness for linear and non-linear system [15]. Fuzzy logic is a mathematics based logic combination, artificial intelligence, and probability algorithms to implement the mankind way for solving problems by reasoning to combine various data and to achieve desired result.

Three FLCs are used in this paper for vector controlled PMSM drive and investigated. The FLCs are used with scaling factors. The values of these scaling factors are distinct for speed and current controllers based on their input error range, required output range, and interval of membership function defined for that individual controller. The tuning of controller is with system parameter to achieve better performance. The fuzzy based drive is implemented in Simulink. Simulation and experimental results validate the effectiveness of the controllers.

III. MATHEMATICAL MODEL OF PMSM IN P.U.

Mathematical modeling of motor is presented and simulation and analysis of drive system is done. The equations of PMSM are given d-q reference frame[16].
It is assumed that nominal power of motor as base power. Peak per phase voltage is assumed as base voltage. Peak value of instantaneous stall current is base current. Base voltage is derived from nominal power and continuous stall current. Here the motor equations are derived in p.u. Base values of PMSM is based on,

$$V_b = \sqrt{2}V_{rms} ; ib = \sqrt{2}i_{rms} ; S_b = \frac{3}{2}V_b i_b ; T_b = \frac{S_b}{\omega_{bm}}$$

ω_{bm} = Base mechanical speed in Rad. /sec. $\omega_{be} = \frac{P}{2}\omega_{bm}$

ω_{be} = Base electrical speed in Rad. /sec.
Motor d-q axis rotor reference frame equations are,

$$V_{ds} = i_{ds}r + l_d\frac{di_d}{dt} - \omega_s l_q i_q \qquad (1)$$

$$V_{qs} = i_{qs}r + l_q\frac{di_q}{dt} + \omega_s l_d i_d + \lambda_{af}\omega_s \qquad (2)$$

Equation (1) and (2) are the voltage equations of PMSM used in vector control, based on that equivalent circuit of PMSM is obtained as shown in figure 1.

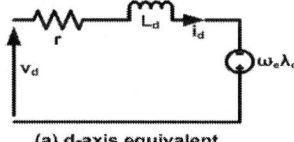
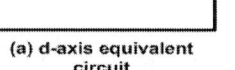
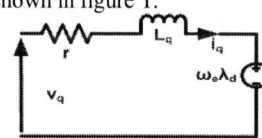

(a) d-axis equivalent circuit   **(b) q-axis equivalent circuit**

Fig.1.PMSM equivalent circuit

Dividing both sides of equations (1) and (2) by base voltage V_b

$$\frac{V_{ds}}{V_b} = \frac{ri_{ds}}{V_b\frac{i_b}{i_b}} + \frac{l_d}{V_b\frac{i_b}{i_b}\frac{\omega_{be}}{\omega_{be}}}\frac{di_d}{dt} - \frac{l_q}{V_b\frac{i_b}{i_b}\frac{\omega_{be}}{\omega_{be}}}\omega_s i_q \qquad (3)$$

$$\frac{V_{ds}}{V_b} = \frac{ri_{ds}}{V_b\frac{i_b}{i_b}} + \frac{l_d}{V_b\frac{i_b}{i_b}\frac{\omega_{be}}{\omega_{be}}}\frac{di_d}{dt} - \frac{l_q}{V_b\frac{i_b}{i_b}\frac{\omega_{be}}{\omega_{be}}}\omega_s i_q \qquad (4)$$

Hence, equations in P.U. (taking time in actual values) are;

$$V_{ds\_p.u.} = i_{ds\_p.u.}r_{\_p.u.} + l_{d\_p.u.}\frac{di_{d\_p.u.}}{\omega_{be}dt} - \omega_{s\_p.u.}l_{q\_p.u.}i_{q\_p.u.} \quad (5)$$

$$V_{qs\_p.u.} = i_{qs\_p.u.}r_{\_p.u.} + l_{q\_p.u.}\frac{di_{q\_p.u.}}{\omega_{be}dt} +$$
$$\omega_{s\_p.u.}l_{d\_p.u.}i_{d\_p.u.} + \lambda_{af\_p.u.}\omega_{s\_p.u.} \qquad (6)$$

Torque equation of motor is given as;

$$T_e = T_L + B\frac{2}{P}\omega_s + J_m\frac{2}{P}\frac{d\omega_s}{dt} \qquad (7)$$

Dividing by base torque in both sides

$$\frac{T_e}{T_b} = \frac{T_L}{T_b} + \frac{2}{P}\frac{B}{T_b\frac{\omega_{be}}{\omega_{be}}}\omega_s + \frac{2}{P}\frac{J_m}{T_b\frac{\omega_{be}^2}{\omega_{be}^2}}\frac{d\omega_s}{dt} \quad (8)$$

Torque equation in p.u. (taking time in actual values)

$$T_{e\_p.u.} = T_{L\_p.u.} + \frac{2}{P}B_{\_p.u.}\omega_{s\_p.u.} + \frac{2}{P}J_{m\_p.u.}\frac{d\omega_{s\_p.u.}}{dt} \quad (9)$$

The torque developed by motor is;

$$T_e = \frac{3}{2}\frac{P}{2}\left[\lambda_{af}i_{qs} + \left(l_q - l_d\right)i_{qs}i_{ds}\right] \qquad (10)$$

Dividing by base torque on both sides

$$\frac{T_e}{T_b} = \frac{3}{2}\frac{P}{2}\left[\frac{\lambda_{af}i_{qs}}{T_b\frac{i_b}{i_b}} + \frac{\left(l_q - l_d\right)i_{qs}i_{ds}}{T_b\frac{i_b^2}{i_b^2}}\right] \qquad (11)$$

Equation in p.u. (time in actual value) is;

$$Te = \left(\frac{P}{2}\right)\left[\lambda_{af\_p.u.}i_{qs\_p.u.} + \left(l_{q\_p.u.} - l_{d\_p.u.}\right)i_{qs\_p.u.}i_{ds\_p.u.}\right](12)$$

Electrical rotor Angle can be calculated from;

$$\theta_{re} = \int \omega_s dt \qquad (13)$$

$$\theta_{re} = \int \omega_{s\_p.u.}\omega_{be}dt \qquad (14)$$

Normally the drive is considered with standard PI controllers. For the research in area of drives to design most of the controllers, first of all the performance of the drive system is analyzed with PI controllers. Then based on the performance of drive, speed error, current error, speed of motor, torque generated by motor etc. provides a basic understanding of the required control actions. By using this knowledge as base criteria the controller design is being started.
Equation of PI controller;

$$T^* = K_p d\omega_s + K_i \int d\omega_s dt \qquad (15)$$

Dividing by base torque on both sides

$$\frac{T^*}{T_b} = \frac{K_p}{T_b \dfrac{\omega_{be}}{\omega_{be}}} d\omega_s + \frac{K_i}{T_b \dfrac{\omega_{be}}{\omega_{be}}} \int d\omega_s dt \qquad (16)$$

The equation in p.u. is

$$T^*_{\_p.u.} = K_{p\_p.u.} d\omega_{s\_p.u.} + K_{i\_p.u.} \int d\omega_{s\_p.u.} dt \qquad (17)$$

For the simulation of the motor with actual parameter values Eq. (1), (2), (7), (10), (13) and (15) are used. Generally to perform more accurate calculations normalized or p.u. values of parameter are used for this Eq. (5), (6), (9), (12), (14), and (17) are used.

IV. SYNTHESIS OF FUZZY LOGIC CONTROLLER

Fig. 2 shows the block diagram of FLC. The design steps for fuzzy logic controller for speed and current control of PMSM are as follows

1. Find out input and output variables.
2. Select membership functions and specify control rules.
3. Specify possible inference with membership functions and control rules.
4. Translate the fuzzy set into crisp set.
5. Tune the input and output gains appropriately to get the desired performance.

To control the speed of PMSM using FLC in this paper error (e) and change in error (ce) in corresponding variables are considered as input crisp variables and are defined as[17]

Error in speed is defined as

$$\Delta\omega_r(n) = \omega^*_r(n) - \omega_r(n) \qquad (18)$$

Error in d-axis current is defined as

$$\Delta i_d(n) = i^*_d(n) - i_d(n) \qquad (19)$$

Error in q-axis is defined as

$$\Delta i_q(n) = i^*_q(n) - i_q(n) \qquad (20)$$

Change in error for all three variables is defined as

$$\Delta e(n) = e(n) - e(n-1) \qquad (21)$$

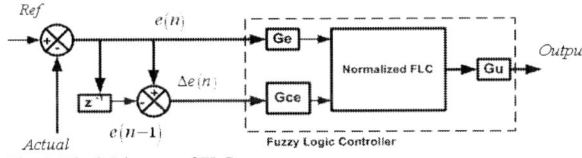

Fig. 2: Block Diagram of FLC.

Based on the prior experience of the system, the membership functions and control rules defined are defined. The normalized inputs and output are obtained for fuzzy

logic controller by using gain blocks as scaling factors G_e, G_{ce} and G_u [18]as shown in fig. 2.

In this paper for current controller only five variables are chosen 1) negative medium (NM); 2) negative small (NS); 3) zero (Z); 4) positive small (PS); 5) positive medium (PM); as shown in figure 3. For speed controller 7 variable are considered: 1) Negative Large (NL); 2) Negative Medium (NM); 3) Negative Small (NS); 4) Zero (Z); 5) Positive Small (PS); 6) Positive Medium (PM); 7) Positive Large (PL) as shown in figure 4.In case of speed controller more variables chosen because the variation and range of speed error is more as compared to current error.

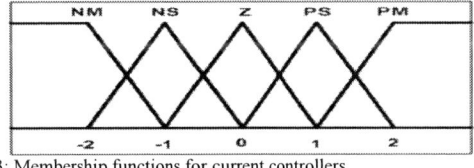

Fig. 3: Membership functions for current controllers

Fig. 4: Membership functions for speed controller

In the second stage of FLC, variables E and CE executed by inference engine of FLC that executes control rules stored in (7×7) rule base for speed controller and (5×5) rule base for current controller. Every rule is expressed in form of
Rule: IF 'E' is A and 'CE' is B, THEN 'Output' is C
Where 'E', 'CE', and 'Output' are fuzzy subsets. The control rules are formulated using behavior of PMSM. Derivation of control rules are based on following criteria for PMSM.

- When speed error is more positive then to catch up the reference speed, current reference has to be more.
- When speed error is small positive and change in speed error is large then current reference has to be kept constant to avoid overshoot
- At zero speed error current reference has to be unchanged.
- At negative speed error current reference has to be negative.

A robust controller with FLC requires tuning of parameter [19]. As shown in block diagram of FLC, there are three (two input and one output) scaling factors G_e, G_{ce} and G_u used. Depending upon the parameter of PMSM, inverter, load, and reference speed these scaling factors are tuned.

The current controller is same as speed controller shown in fig.2. Scaling factors for different controller is different, because input and output requirement of every controller is different. For example fuzzy speed controller has speed error as input and, q-axis current controller has q-axis current error (difference of i^*_q and i_q), and d-axis current controller has d-axis current error (difference of i^*_d and i_d). So the gains have to be different to normalize them into range specified by respective MFs. In this paper for the linguistic variable of current controller triangular MFs are selected in the interval [-2 2], for speed controller the interval is [-15 15]. To obtain the normalized error and

978-1-5386-6624-1/18 $31.00 © 2018 IEEE

change in the error in range defined for corresponding controller gains blocks have to be used. The control rules will be executed only when the inputs E and CE are normalized using gain blocks, in the range specified for the corresponding controller. In this paper max-min algorithm is implemented to obtain output from inputs executed by control rules. In centroid algorithm of defuzzification crisp values are achieved from center of gravity of MFs.

V.   OPAL-RT LAB TECHNOLOGY

Real time simulation of full fuzzy controlled PMSM drive on FPGA based real time simulator which produce the actual results. RT-LAB, from Opal-RT Technologies, is a real-time simulation platform that enables real time and HIL (hardware in loop) simulation of controllers, electric plants or both, through automatic code generation methods. The entire process occurs without the need for handwritten 'C' code, enabling very rapid deployment of prototyped controllers or HIL-simulated plants. The process is notably very efficient when applied to I/O code because RT-LAB provides a set of simulink blocks that automatically configure common I/O functions, like analog input/outputs and time-stamping capable digital I/Os, with a 10 nanosecond resolution. Special interpolating models use this timing information to greatly increase simulation accuracy [23]. RT-LAB simulator is equipped with a user-programmable FPGA card. The FPGA card can be programmed with the Xilinx System Generator blockset for Simulink enabling implementation of complex sensor models like resolvers, Resolver-To-Digital and FM resolvers or even complex motor drives [24]. RT-Lab is used as real time hardware-in-loop controller in this implementation for easy and flexibility [20]. Table I summarizes the characteristics of FPGA board used in this paper.

TABLE I.
RECONFIGURABLE FPGA BOARDS

Model Name	FPGA	Bus type	Gate	I/O lines	Logic cells	FPGA clock	
OP5142	Xilinx Spartan 3 XG3S500	PCI-Express 1x	5M	296	74.880 ((74k)	100 MHz	

VI.   PERFORMANCE EVALUATION

The vector control of PMSM is applied to obtain the performance as of dc-machine. The stator current is devided in d-axis and q-axis currents[3]. The voltage reference is generated by current controllers for both axis. The schematic diagram of PMSM drive is shown as in fig. 5.

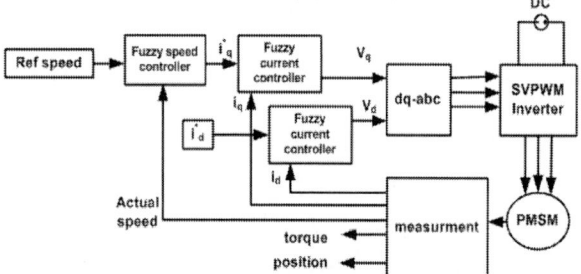

Fig. 5: Block diagram of PMSM drive with complete fuzzy control

As explained in previous section scaling factors initially estimated by using PI controller for selected input and output interval for individual controllers [21]. SVPWM inverter is feeding power to PMSM[22]. The speed control loop gives the reference torque, voltage reference for SVM inverter is generated by current controllers. The dc link voltage used here is 250V.

Table II: Parameters of PMSM

Motor Parameters		
Symbol	Parameter	Value
R_s	Resistance	2.0 ohm
L_d	d-axis inductance	4.5 mH
L_q	q-axis inductance	4.5 mH
J	Inertia	$0.8 \times 10\text{-}3$ kgm2
P	No of Poles	8

A.   Simulation Performance

The vector controlled PMSM drive simulated with PI, speed fuzzy controller, and full fuzzy (fuzzy speed and fuzzy current controller). The parameter of PMSM is given in table II. For a step change in load torque (0-5 Nm) the drive with PI speed and PI controller, fuzzy speed and current PI, and full fuzzy logic takes different times to generate desired load torque as given in table III.

Table III.

Controller	Time (sec)
PI speed +PI current	0.25
Speed fuzzy +PI current	0.17
Full fuzzy	0.11

The waveforms of electromagnetic torque for a step change of 0-5 Nm, in load torque with these three controllers based PMSM drive are shown in figure 6, figure 7, and figure 8 respectively.

Torque response of PMSM with PI controllers is shown in fig 6. During zero load torque ripples are more. Figure 7 shows the torque performance with fuzzy speed and PI current controllers. The ripples in torque are less with fuzzy speed controller as compared to PI speed controller. Torque response with full fuzzy logic controller is shown in figure 8. It has less torque ripples and takes less time to reach load torque as compared to other two controllers.

The torque response of drive is governed by q-axis current controller, and speed response is governed by d-axis current controller. The speed controller just generates the q-axis current reference for q-axis current controller. Speed response of full fuzzy logic controlled PMSM drive with 760-500 rad/sec step change in speed is shown in figure 9.

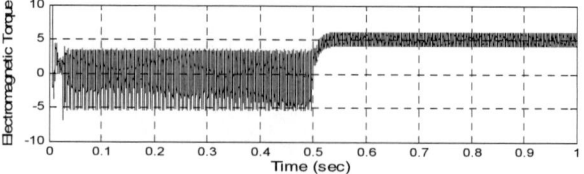

Fig. 6: Torque response with PI speed and PI current controller

978-1-5386-6624-1/18 $31.00 © 2018 IEEE

Fig. 7: Torque response with fuzzy speed and PI current controller

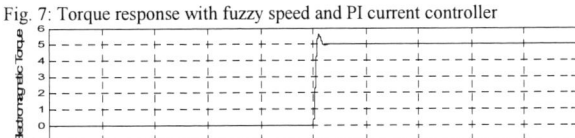

Fig8: Torque response with full fuzzy logic controller

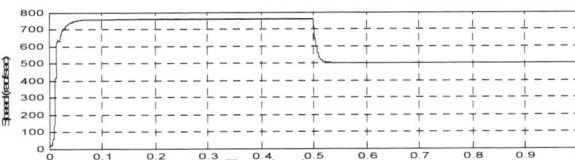

Fig. 9: Speed response of full fuzzy controlled PMSM with step change (760-500 rad/sec) in speed.

B. Real time implementation

The full fuzzy logic based PMSM drive is simulated in RT-LAB of OPAL-RT. The implementation of computer controlled RT-LAB based on MALAB is shown in Figure 10.

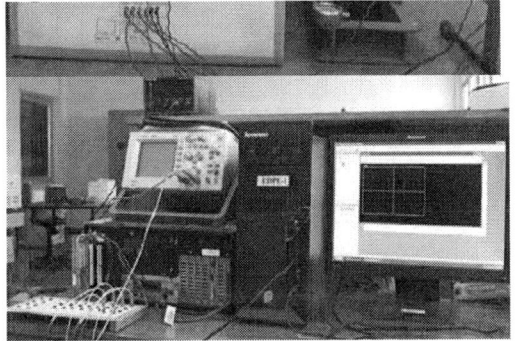

Fig. 10 Experimental setup for hardware implementation

Figure 11 shows the speed and torque response wit first speed is changed from 0-500 rad/sec, then after some time 0-4 Nm load torque is applied. Figure 12 shows the speed and torque response when speed is changed in step of 0-200-500 rad/sec at no load. At the transition moment of speed peak is generated in torque and vice versa. At positive step change in speed a positive peak is appearing in torque and at positive step change in torque causes negative peak in speed as shown in figure 12.

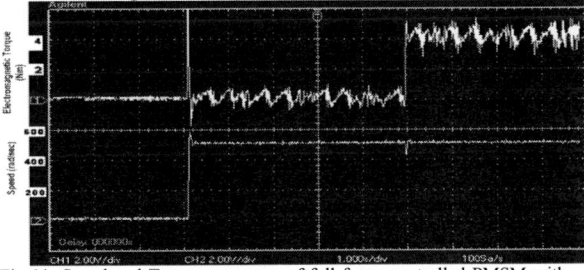

Fig 11: Speed and Torque responses of full fuzzy controlled PMSM with step change (0-500 rad/sec) in speed and (0-4 Nm) in torque at different instant.

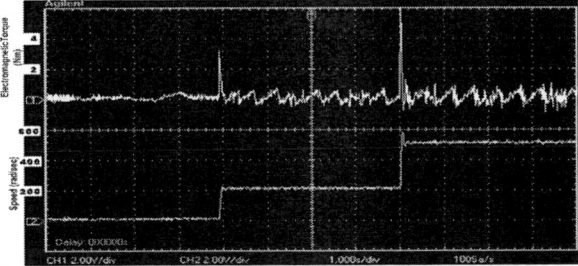

Fig. 12: Speed and Torque responses of full fuzzy controlled PMSM with step change (0-200-500 rad/sec) in speed at no load.

VII. CONCLUSION

To evade the enslavement of PI controllers on machine parameter and operating conditions a full fuzzy logic for PMSM drive is proposed with adequate gain tuning for a robust performance. Simulation performance is observed in MATLAB software, further the controller is implemented in real time using FPGA based RT-LAB system, which provides the real time performance of full fuzzy logic controlled drive.

REFERENCES

[1] P. Pillay and R. Krishnan, "Modeling of permanent magnet motor drives," *IEEE Transactions on Control Systems Technology,* vol. 35, pp. 537-541, 1988.

[2] B. K. Bose, *Modern power electronics and AC drives*: Prentice Hall PTR USA, 2002.

[3] M. Kadjoudj, M. E. H. Benbouzid, C. Ghennai, and D. Diallo, "A robust hybrid current control for permanent-magnet synchronous motor drive," *IEEE Transactions on Energy Conversion,* vol. 19, pp. 109-115, 2004.

[4] Z. Ibrahim and E. Levi, "A comparative analysis of fuzzy logic and PI speed control in high-performance AC drives using experimental approach," *IEEE Transactions on Control Systems Technology,* vol. 38, pp. 1210-1218, 2002.

[5] A. Mishra, V. Mahajan, P. Agarwal, and S. P. Srivastava, "Fuzzy logic based speed and current control of vector controlled PMSM drive," in *2nd International Conference on Power, Control and Embedded Systems (ICPCES), 2012* pp. 1-6.

[6] F. B. del Blanco, M. W. Degner, and R. D. Lorenz, "Dynamic 0analysis of current regulators for AC motors using complex vectors," *Industry Applications, IEEE Transactions on,* vol. 35, pp. 1424-1432, 1999.

[7] W. Kaewjindam and M. Konghirun, "A DSP - Based Vector Control of PMSM Servo Drive Using Resolver Sensor," in *IEEE Region 10 Conference TENCON,* 2006, pp. 1-4.

[8] F. Briz, M. W. Degner, and R. D. Lorenz, "Dynamic analysis of current regulators for AC motors using complex vectors," in *Industry Applications Conference, 1998. Thirty-Third IAS Annual Meeting. The 1998 IEEE,* 1998, pp. 1253-1260 vol.2.

[9] A. Mishra, J. A. Makwana, P. Agarwal, and S. P. Srivastava, "Modeling and implementation of vector control for PM synchronous motor drive," in *International Conference on Advances in Engineering, Science and Management (ICAESM), 2012,* pp. 582-585.

[10] Z. Q. Zhu and D. Howe, "Influence of design parameters on cogging torque in permanent magnet machines," *IEEE Transactions on Energy Conversion,* vol. 15, pp. 407-412, 2000.

[11] Y.-z. Li and K.-g. Zhao, "Speed sensorless control of the permanent-magnet synchronous motor based on wavelet neural networks," in *8th IEEE International Conference on Control and Automation (ICCA),* 2010, pp. 2073-2076.

[12] D. Sun, Y. He, and J. G. Zhu, "Fuzzy logic direct torque control for permanent magnet synchronous motors," 2004, pp. 4401-4405 Vol. 5.

[13]   D. Sun, J. G. Zhu, and Y. K. He, "A space vector modulation direct torque control for permanent magnet synchronous motor drive systems," 2003, pp. 692-697 Vol. 1.

[14]   L. Zhen and L. Xu, "Fuzzy learning enhanced speed control of an indirect field-oriented induction machine drive," *IEEE Transactions on Control Systems Technology,* vol. 8, pp. 270-278, 2000.

[15]   [17]   M. N. Uddin, T. S. Radwan, and M. A. Rahman, "Performances of fuzzy-logic-based indirect vector control for induction motor drive," *IEEE Transactions on Control Systems Technology,* vol. 38, pp. 1219-1225, 2002.

[16]   K. Hakiki, A. Meroufel, V. Cocquempot, and M. Chenafa, "A new adaptive fuzzy vector control for permanent magnet synchronous motor drive," in *18th Mediterranean Conference on Control & Automation (MED)* 2010, pp. 922-927.

[17]   F. Chen, X. Jiang, X. Ding, and C. Lin, "FPGA-based sensorless PMSM speed control using adaptive sliding mode observer," in *43rd Annual Conference of the IEEE Industrial Electronics SocietyI ECON 2017*, pp. 4150-4154.

[18]   A. Mishra, J. Makwana, P. Agarwal, and S. P. Srivastava, "Mathematical modeling and fuzzy based speed control of permanent magnet synchronous motor drive," in *7th IEEE Conference on Industrial Electronics and Applications (ICIEA), 2012* pp. 2034-2038.

[19]   A. Lidozzi, L. Solero, F. Crescimbini, and A. Di Napoli, "SVM PMSM Drive With Low Resolution Hall-Effect Sensors," *IEEE Transactions on Power Electronics,* vol. 22, pp. 282-290, 2007.

[20]   M. N. Uddin and R. S. Rebeiro, "Online Efficiency Optimization of a Fuzzy-Logic-Controller-Based IPMSM Drive," *IEEE Transactions on Control Systems Technology,* vol. 47, pp. 1043-1050, 2011.

[21]   P. Vas, *Sensorless vector and direct torque control* vol. 729: Oxford university press Oxford, UK, 1998.

2nd IEEE International Conference on Power Electronics, Intelligent Control and Energy Systems (ICPEICES-2018)

Development of The Intelligent Oil Field With Management and Control using IIoT (Industrial Internet of Things)

Ali S. Allahloh, Sarfraz Mohammad
Department of Electrical Engineering
Aligarh Muslim University, Aligarh, India
eng a.allahloh@yahoo.com, msarfraz@zhcet.ac.in

Abstract—In the past decade there is a huge development in artificial intelligence technologies for various applications. Credit goes to recent researches which have increased the computing capabilities manifold particularly of the workstation and microcontrollers it has opened the way to use these technologies in various fields o f i ndustry a nd l ed t o t he f ourth industrial revolution in which we are living today. The recent addition to this is Internet of Things technology that has made way for the entry artificial i ntelligence a ll a round u s. T he u se of this technology in the process of oil and gas production will increase the efficiency o f t he p roduct a nd r educe t he c ost of production because it will replace the current costly systems such as PLCs, DCS. etc. to build intelligent wireless system enables the application of intelligent management and control systems for oil and gas fields. This paper focuses on how to build a sophisticated industrial management and control system to manage oil and gas production based on the Internet things by exploiting the technologies currently available such as SCADA and LabVIEW on the workstations and microcontrollers connected to Wi-Fi networks on the equipments and the role of the OPC server to connect all the equipment to the single window system for the purposes of monitoring and control and the exploitation of large computing capabilities of workstations for the implementation of complex control algorithms such as Neural Networks and Fuzzy logic and even a Neuro-Fuzzy hybrid, while the microcontrollers in turn control the entire equipment on its own IoT Device The necessary parameters are obtained and updated continuously from the workstation via the wireless network to make the most of all the equipments and find the required harmony between them for smooth and continuous production and integration with ERP system modules such as SAP.

Keywords—*PIC microcontrollers, LabVIEW, MPLAB, TCP/IP, Wireless Ethernet, SBC65EC board, OPC Server, Kepware, Neural Networks, Fuzzy Logic, Neuro-Fuzzy, Industrial Internet of Things, IIoT, IoT.*

I. INTRODUCTION

The complexity of industrial automation and process control because the non-linearity of the process make the Neuro-Fuzzy technique ideal to solve the control problem. Neuro-Fuzzy is a hybrid system of two intelligent techniques Neural Networks and Fuzzy Logic. The Neural Networks is process information system based on human brain capability of processing the information. A Fuzzy Logic MVL (Multi Variable Logic) is the people thinking way. While the fuzzy logic is rule based system and each member in fuzzy set have degree of membership make it useful for capture acknowledge of experienced operator. However, fuzzy logic does not have known method to do the process of transformation the practical experienced in to fuzzy inference system based on if-then rules in addition to the long time needed to adjust the membership functions, So the needed to Neuro-Fuzzy hybrid system because the higher capability of learning process of Neural Networks the system can adjust the membership function automatically and adapt itself to environment. The exploration and innovation in industrial automation was an exciting in last two years where is the new technology such as Industrial Internet of Things (IoT) introducing the new and different products. The impacting of these technologies in industry become more and more intense, with related innovations including industrial wireless, new communications protocols, sensors, actuators, high-performance microprocessors and cloud computing. Logically these technologies will cut the cost of automation and expand the options of design. The emergence of these new technologies and the popularity of artificial intelligence technology make it imperative for us to think seriously about how we can employ the Neuro-Fuzzy systems, industrial IoT and industrial wireless to optimize the operations and process that will lead to improve the products quality, efficiency and safety environment in industrial fields. And lead to the integration of Systems, Devices and Applications. The powerful developments already done in last two years such as Intel Compute Card a full computer elements platform including wireless connectivity, Intel SoC, Memory, storage, flexible I/O and 7th generation Intel processor with footprint of credit card size will bring the Intelligent Facility. Beside the advantages of Neuro-Fuzzy system, it is difficult to implement real time system because the complex operations of algorithm regard to time constrains. So, the sequential implementations cannot work in real time in most of the cases. Different techniques used to implement the Neuro-Fuzzy controller models including parallel architecture FPGA, this paper focuses on how to use new technologies to implement hardware and software online Neuro-Fuzzy controller.

A. Literature Survey

[1], [2], [3] The authors discuss and implement the Neuro-Fuzzy and Fuzzy-Sliding Controllers for waste water treatment and twin rotor MIMO system for helicopter control also discuss on board fast controller learning. They investigates

978-1-5386-6624-1/18 $31.00 © 2018 IEEE   815

an adaptive Neuro-Fuzzy control system. [4], [5], [6] The authors present various types of intelligent controllers these controllers are predictive controller based on deep learning Neural Network, indirect type 2 Neuro-Fuzzy controller and Neuro-Fuzzy-PID controller. [7], [8], [9] The authors design self constructing Neuro-Fuzzy controller, self learning disturbance observer for nonlinear systems in feedback-error learning scheme and Neuro-Fuzzy inference system. [10], [11], [12] The authors implement Observer-Based adaptive Neuro-Fuzzy controller for non-linear systems with input delay, also discuss the application of Neuro-Fuzzy interface in modeling CO2-crude oil minimum miscibility pressure and discuss how to generate fuzzy rules using Quaternion neuro-fuzzy learning algorithm. [13], [14], [15] The Authors introduce the Artificial Intelligence-Based Semantic IoT in a User-Centric Smart City, A Resource Service Model in the Industrial IoT and Hy-LP: A novel IIoT protocol, evaluated on a wind park. [16], [17], [18] The authors present IIoT monitoring solution for advanced predictive maintenance applications, Development of the IIoT Competences in the Areas of Interaction, Process, and Organization Based on the Learning Factory Concept and Enabling distributed intelligence IIoT Controller. [19], [20], [21] The authors discuss using existing communication technology for Networking for IoT and applications, an exploratory case study how the IoT powered servitization of manufacturing and Efficient certificateless access control for IIoT. [22], [23], [24] The authors present modeling and synthesis of smart sensor networks for IIoT, IIoT-Based Framework and Research Challenges for Collaborative Sensing Intelligence and using IIoT for Adapting an industrial automation protocol to remote monitoring of mobile agricultural machinery. [25], [26], [27] The authors discuss Automation and Control Systems based on Application of selected supervised learning methods for time series classification, improvement the performance of Fuzzy Logic Controller Using Neural Network and using Neuro-Fuzzy system for Determination of minimum miscibility pressure in CO2-IOR projects. [28], [29] The authors present Social IIoT for Industrial and Manufacturing Assets and A Role to Play in Wireless Sensor Networks open hardware.

II. OIL PRODUCTION PROCESS

The new production and processing facility need to be design for independent treating of produced fluid coming from production well. The crude oil need to be prepared on the quality of the export specification, stored and loaded over metering to transport through pipe or trucks. Pneumatic, electrical instruments and local gauges are used for the measuring. Control valves are used for Regulatory Functions. The Control valves are used with Pneumatic positioner for continuous control. Limit Switches are used for direct interlock function. The electrical instruments shall be connected into the PLC cabinet via junction boxes and multi-core cables to measure Pressure, Temperature, Level and Flow, also to control by actuators such as Pneumatic control valves, Self-regulating valves and ESD On-Off valve. The Facility Control System (FCS) realizes the safe and smooth operation of the processing facility process. The FCS covers two main systems: Basic Process Control System (BPCS), Emergency Shut Down System (ESD). The control, measuring, SCADA functions are realized by the BPCS. The BPCS includes two types of control: pneumatic level control of the separators, and the

electronic control. The pneumatic control system is separated from each other, they are autonomous systems. The electronic control system is based on a Siemens S7-300 type PLC. The electronic control-, measuring functions are implemented on this PLC. The Emergency Shutdown (ESD) functions (Pre-Alarms, Interlocks) are implemented in the Siemens S7-300 Failsafe CPU. The PLC will be installed in an indoor cabinet at a non-hazardous area. This cabinet receives the cables from the field instruments, Fire and Gas System and RC cabinet. The RC cabinet is the interface between the instrumentation and electrical. Individual marshalling cabinet is not required. A new operator workstation with WinCC software will be installed to execute the SCADA function for the FCS. The indoor cabinet and the operator workstation will be in the operator cabin. This paper discussed how the new IIoT technology will replace this costly and traditionally system with better performance and less cost.

III. DESIGN AND CONFIGURATION

A. Proposed System

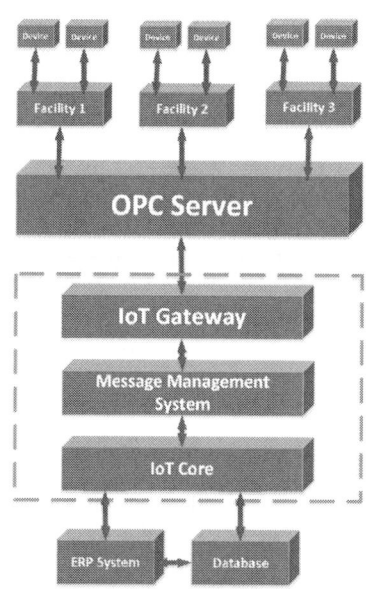

Fig. 1. Proposed System

B. Industrial Internet Of Things

The emergence of Internet stuff leads to a radical change that would reshape the industry in a revolutionary way to reach the so-called Fourth Industrial Revolution after the previous three of the steam engine, production lines and industrial automation respectively.

C. OPC-UA Server

The development of the OPC server technology allowed the connection of the various devices produced by different companies and used different communication protocols with each other using the one-window principle for the purposes of monitoring, control and automation. This window is SCADA, LabVIEW, MATLAB ... etc. In this paper the Kepware OPC

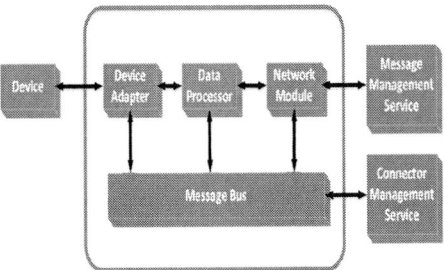

Fig. 2.   Required IIoT Device.

Server support IoT Protocols such as MQTT used to connect different PLCs and IIoT devices to Citect SCADA Software for monitoring and control purpose Also connect to LabVIEW to apply complex control algorithm such as Fuzzy, Neural or NeuroFuzzy Control systems.

finally, The Kepware OPC Server make the merging process of IIoT Technology with the present Technologies easier to activate the AI in the production facilities

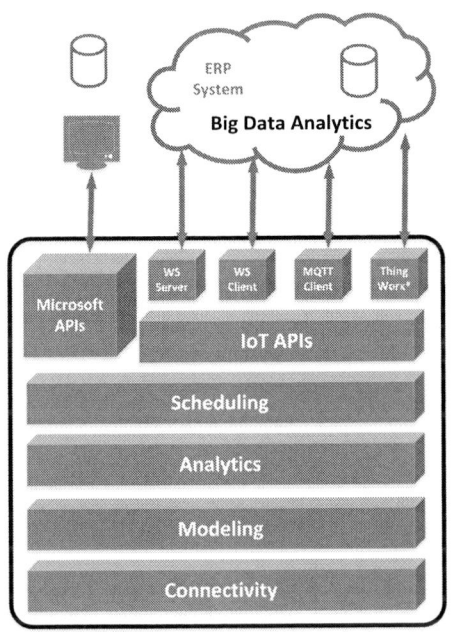

Fig. 3.   Kepware OPC Server.

D. Intelligent Control System

Neural networks and fuzzy logic introduce the new method to build intelligent control systems, this paper focuses on how this Technologies used to identify model and control industrial processes. System identification is a way to understand and investigate the world around by the deriving mathematical model of the system. While the adaptive control system provide approach for adjusting the controller parameters automatically in real time. It is using inputs, outputs, states and known disturbance to measure the performance index of a controller. The important feature of intelligent neuro control is online adaptation of parameters to tune a fuzzy control using different

strategies.

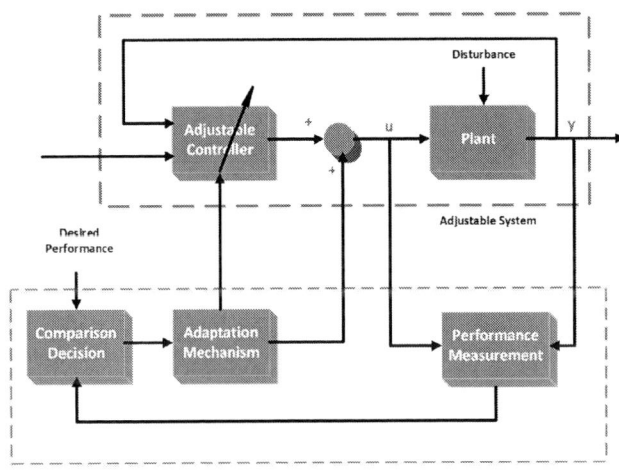

Fig. 4.   Adaptive control system configuration.

IV.   MONITOR AND CONTROL TEST

Through the pursuit of intelligent management and control this paper presented two important experiments first how to connect two PLCs from Siemens and SCADA system from Schneider Electric through the system of OPC Server from Kepware that shown in fig.7.

The second experiment how to connect the device we developed based on IoT technology to communicate over the same Wi-Fi network that Kepware OPC Server uses a system designed using the LabVIEW to implement a Level Loop control system based on Neuro-Fuzzy technology for the calculation the parameters of the PID system designed on the PIC Microcontroller, which is the heart of the IoT device and constantly updated. Now replace the PLC by the IIoT device

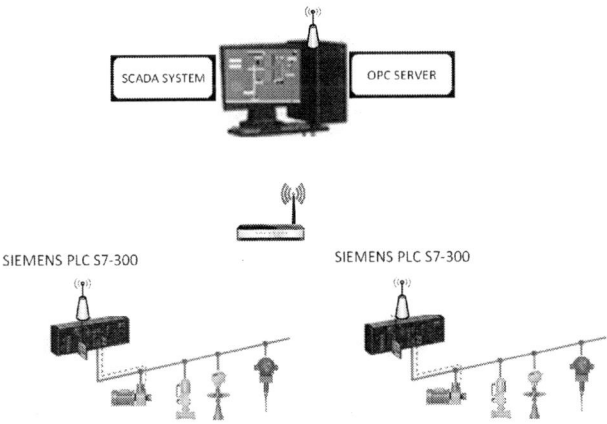

Fig. 5.   Citect SCADA to Siemens PLCS.

shown in fig.8 developed using PIC18F6627 microcontroller and Microchip TCP/IP stack connected to LabVIEW software. This device will accomplish all PLC functions rather than apply the intelligent control algorithm to the facility, this part

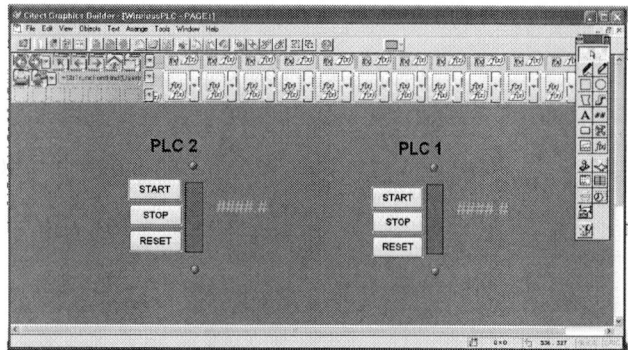

Fig. 6.   Citect SCADA GUI for Test Two Siemens PLCS.

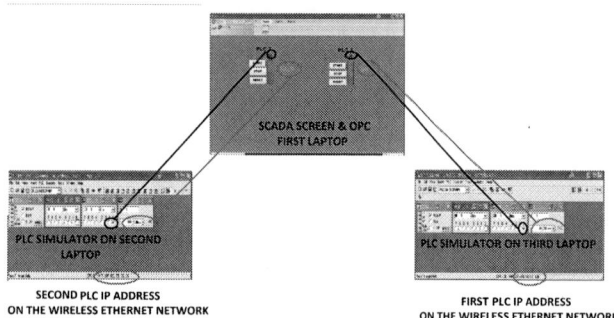

Fig. 7.   Citect SCADA to Siemens PLCS Test.

discusses the result of pneumatic control valve response that used as final control element in the field where the transfer function of that valve is:

$$\frac{X(s)}{P(s)} = \frac{1}{\tau^2 s^2 + 2\tau\zeta s + 1}$$

where τ is time constant and ζ is damping ratio. Using the proteus software shown in fig.9 the simulation of IIoT device connected to valve process developed and connected through virtual Ethernet network to LabVIEW software were the intelligent control system developed.

The step response of pneumatic valve that manipulate the level in the horizontal tank without any controller shown in fig.10 after that PD controller built on PIC18F6627 microcontroller and the step response of the valve shown in fig.11

$$U_n = K_p \left(e_n + \frac{1}{T_i} \sum_{j=1}^{n} e_j T_s + T_d \frac{e_n - e_{n-1}}{T_s} \right)$$

$$\Delta U_n = K_p \left(e_n - e_{n-1} + \frac{1}{T_i} e_n T_s \right)$$

$$U_n = U_{n-1} + \Delta U_n$$

Finally, the Neuro-Fuzzy-PD fig.12 Controller applied to tune the K_p and T_d online and update them on PIC18F6627 microcontroller

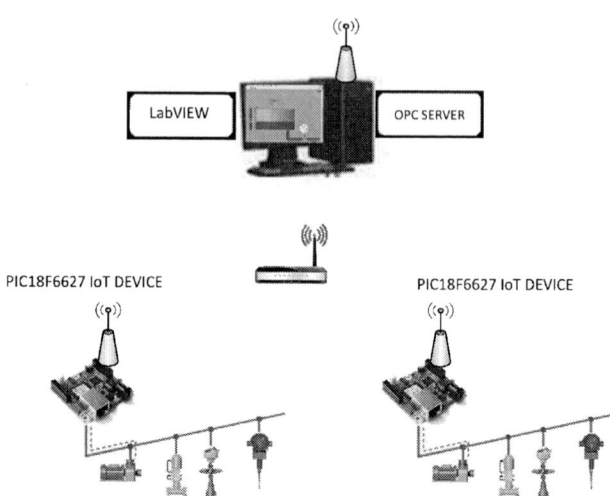

Fig. 8.   LabVIEW to industrial IoT Device.

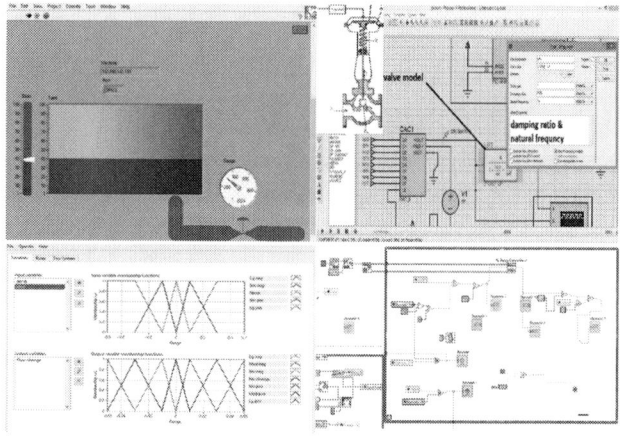

Fig. 9.   LabVIEW implementation and Proteus simulation of IoT Device.

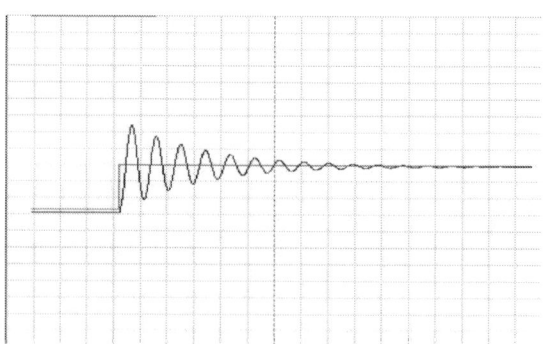

Fig. 10.   Control Valve Step Response Without any Controller.

978-1-5386-6624-1/18 $31.00 © 2018 IEEE   818

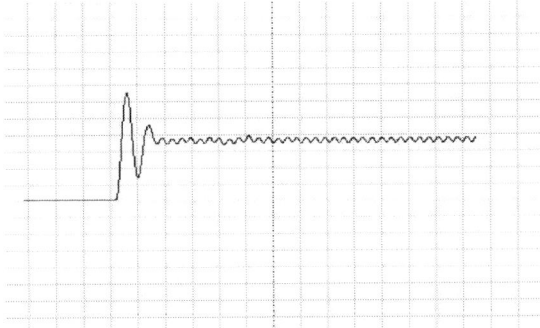

Fig. 11. Control Valve Step Response Using PD controller.

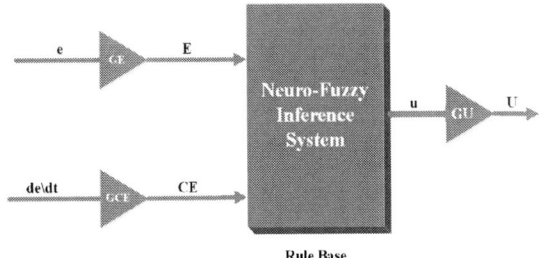

Fig. 12. Neuro-fuzzy-PD Control System.

$$U_n = f\big(GE * e(n), GCE * e^\cdot(n)\big) * GU$$

$$f\big(GE * e(n), GCE * e^\cdot(n)\big) = GE * e(n) + GCE * e^\cdot(n)$$

$$K_p = GE * GU, T_d = \frac{GCE}{GE}$$

The Neuro-Fuzzy inference system and common rule base of Sugeno fuzzy model shown in the fig13 is:

$$if \ \ x \ \ is \ \ A_1 \ \ And \ \ y \ \ is \ \ B_1 \ \ Then \ \ f_1 = p_1 x + q_1 y + r_1$$
$$if \ \ x \ \ is \ \ A_2 \ \ And \ \ y \ \ is \ \ B_2 \ \ Then \ \ f_2 = p_2 x + q_2 y + r_2$$

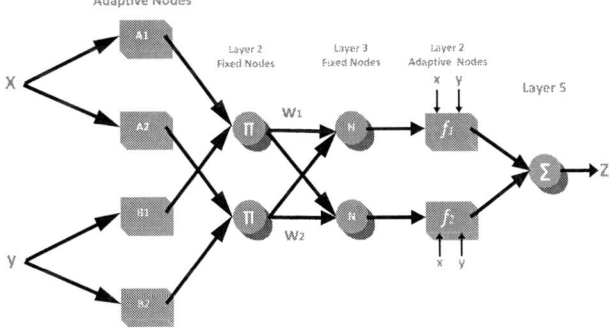

Fig. 13. Adaptive Neuro-Fuzzy Inference System for two inputs Sugeno fuzzy model.

And the layers outputs are:

- Layer1 output:

$$\mu A_i(x) = \frac{1}{1 + e^{a_i(x - c_i)}} \tag{1}$$

The membership function shape varies change according to the values of a_i and c_i Parameters.

- Layer2 output:

$$w_i = \mu A_i(x) * \mu B_i(y) \tag{2}$$

- Layer3 output:

$$\overline{w_i} = \frac{w_i}{\sum w_i} = \frac{w_i}{w_1 + w_2} \tag{3}$$

- Layer4 output:

$$\overline{w_i} f_i = \overline{w_i}(p_i x + q_i y + r_i) \tag{4}$$

- Layer5 output:

$$Z = \sum \overline{w_i} f_i = \frac{\sum w_i f_i}{\sum w_i} \tag{5}$$

finally, the step response of pneumatic control valve after applied Neuro-Fuzzy-PD controller shown in fig.14 and fig.15

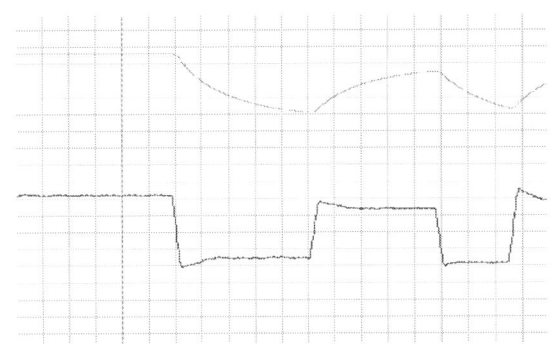

Fig. 14. Level and Flow step response with Fuzzy-PD controller when the setpoint is changed by less than 50% of full scale.

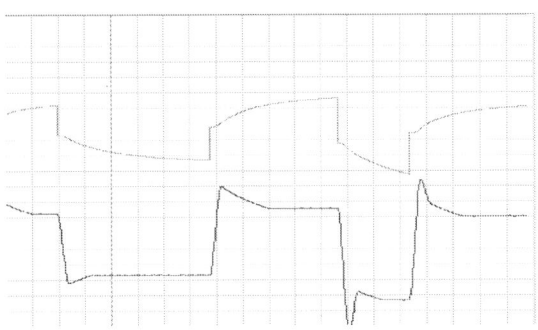

Fig. 15. Level and Flow step response with Fuzzy-PD controller when the setpoint is changed by more than 50% of full scale.

978-1-5386-6624-1/18 $31.00 © 2018 IEEE

V. CONCLUSION

This paper focused on how to integrate the industrial Internet of things technology within the oil and gas fields to obtain a high-quality product at a lower cost considering the increasing rise in global demand for energy and proved that the time has come to take advantage of this technology, especially considering the great development in technology and availability of elements Needed to replace costly technologies currently in use. This paper provides an overview of how to use the Kepware OPC Server, Citect SCADA, Microchip PIC microcontrollers, Wi-Fi and LabVIEW software to integrate with the SAP ERP system, which is used by most of the oil and gas companies around the world to manage the oil and gas fields, especially with the addition of SAP IoT module that support of the communication protocols currently used by IoT technology. These new technologies will lead to make the Plant Introduce Itself to the System in other word realize (Plug and Play Plant) where the AI system on workstation will use common database to visualize, manage and control the plant equipment in harmony mode to reach the production target. With the new IoT protocols such MQTT and publish subscribe method the researchers will continue to develop IIoT devices support this protocol also develop new and more easier protocols to realize great communications between equipments each others and using common database that used by ERP system to integrate all management modules of organization rather than help to convert traditional devices to smart device without replace them.

REFERENCES

[1] J.-F. Qiao, Y. Hou, L. Zhang, and H.-G. Han, "Adaptive fuzzy neural network control of wastewater treatment process with multiobjective operation," *Neurocomputing*, vol. 275, pp. 383–393, 2018.

[2] S. Zeghlache and N. Amardjia, "Real time implementation of non linear observer-based fuzzy sliding mode controller for a twin rotor multi-input multi-output system (trms)," *Optik-International Journal for Light and Electron Optics*, vol. 156, pp. 391–407, 2018.

[3] Y. Bodyanskiy, O. Vynokurova, G. Setlak, D. Peleshko, and P. Mulesa, "Adaptive multivariate hybrid neuro-fuzzy system and its on-board fast learning," *Neurocomputing*, vol. 230, pp. 409–416, 2017.

[4] S. Deepa and I. Baranilingesan, "Optimized deep learning neural network predictive controller for continuous stirred tank reactor," *Computers & Electrical Engineering*, 2017.

[5] K. Sabahi, S. Ghaemi, J. Liu, and M. A. Badamchizadeh, "Indirect predictive type-2 fuzzy neural network controller for a class of nonlinear input-delay systems," *ISA transactions*, vol. 71, pp. 185–195, 2017.

[6] R. Patel and V. Kumar, "Artificial neuro fuzzy logic pid controller based on bf-pso algorithm," *Procedia Computer Science*, vol. 54, pp. 463–471, 2015.

[7] A. R. Tavakoli, A. R. Seifi, and M. M. Arefi, "Designing a self-constructing fuzzy neural network controller for damping power system oscillations," *Fuzzy Sets and Systems*, 2018.

[8] E. Kayacan, J. M. Peschel, and G. Chowdhary, "A self-learning disturbance observer for nonlinear systems in feedback-error learning scheme," *Engineering Applications of Artificial Intelligence*, vol. 62, pp. 276–285, 2017.

[9] J. de Jesús Rubio, "Usnfis: uniform stable neuro fuzzy inference system," *Neurocomputing*, vol. 262, pp. 57–66, 2017.

[10] M. K. Talkhoncheh, M. Shahrokhi, and M. R. Askari, "Observer-based adaptive neural network controller for uncertain nonlinear systems with unknown control directions subject to input time delay and saturation," *Information Sciences*, vol. 418, pp. 717–737, 2017.

[11] A. Karkevandi-Talkhooncheh, S. Hajirezaie, A. Hemmati-Sarapardeh, M. M. Husein, K. Karan, and M. Sharifi, "Application of adaptive neuro fuzzy interface system optimized with evolutionary algorithms for modeling co2-crude oil minimum miscibility pressure," *Fuel*, vol. 205, pp. 34–45, 2017.

[12] R. Hata, M. M. Islam, and K. Murase, "Quaternion neuro-fuzzy learning algorithm for generation of fuzzy rules," *Neurocomputing*, vol. 216, pp. 638–648, 2016.

[13] Guo, Kun and Lu, Yueming and Gao, Hui and Cao, Ruohan, "Artificial Intelligence-Based Semantic Internet of Things in a User-Centric Smart City." *Sensors (Basel, Switzerland)*, vol. 18, no. 5, 2018.

[14] W. Li, B. Wang, J. Sheng, K. Dong, Z. Li, and Y. Hu, "A resource service model in the industrial iot system based on transparent computing," *Sensors*, vol. 18, no. 4, p. 981, 2018.

[15] G. Hatzivasilis, K. Fysarakis, O. Soultatos, I. Askoxylakis, I. Papaefstathiou, and G. Demetriou, "The industrial internet of things as an enabler for a circular economy hy-lp: A novel iiot protocol, evaluated on a wind parks sdn/nfv-enabled 5g industrial network," *Computer Communications*, vol. 119, pp. 127–137, 2018.

[16] F. Civerchia, S. Bocchino, C. Salvadori, E. Rossi, L. Maggiani, and M. Petracca, "Industrial internet of things monitoring solution for advanced predictive maintenance applications," *Journal of Industrial Information Integration*, vol. 7, pp. 4–12, 2017.

[17] N. Gronau, A. Ullrich, and M. Teichmann, "Development of the industrial iot competences in the areas of organization, process, and interaction based on the learning factory concept," *Procedia Manufacturing*, vol. 9, pp. 254–261, 2017.

[18] H. Rahman and R. Rahmani, "Enabling distributed intelligence assisted future internet of things controller (fitc)," *Applied Computing and Informatics*, vol. 14, no. 1, pp. 73–87, 2018.

[19] S. Mukherjee and G. Biswas, "Networking for iot and applications using existing communication technology," *Egyptian Informatics Journal*, 2017.

[20] A. Rymaszewska, P. Helo, and A. Gunasekaran, "Iot powered servitization of manufacturing–an exploratory case study," *International Journal of Production Economics*, vol. 192, pp. 92–105, 2017.

[21] F. Li, J. Hong, and A. A. Omala, "Efficient certificateless access control for industrial internet of things," *Future Generation Computer Systems*, vol. 76, pp. 285–292, 2017.

[22] C.-H. Chen, M.-Y. Lin, and X.-C. Guo, "High-level modeling and synthesis of smart sensor networks for industrial internet of things," *Computers & Electrical Engineering*, vol. 61, pp. 48–66, 2017.

[23] Y. Chen, G. M. Lee, L. Shu, and N. Crespi, "Industrial internet of things-based collaborative sensing intelligence: framework and research challenges," *Sensors*, vol. 16, no. 2, p. 215, 2016.

[24] T. Oksanen, R. Linkolehto, and I. Seilonen, "Adapting an industrial automation protocol to remote monitoring of mobile agricultural machinery: a combine harvester with iot," *IFAC-PapersOnLine*, vol. 49, no. 16, pp. 127–131, 2016.

[25] J. Fütterer, M. Kochanski, and D. Müller, "Application of selected supervised learning methods for time series classification in building automation and control systems," *Energy Procedia*, vol. 122, pp. 943–948, 2017.

[26] S. Rajan and S. Sahadev, "Performance improvement of fuzzy logic controller using neural network," *Procedia Technology*, vol. 24, pp. 704–714, 2016.

[27] S. Mollaiy-Berneti, "Determination of minimum miscibility pressure in co2-ior projects with the aid of hybrid neuro-fuzzy system," *Fuel*, vol. 180, pp. 59–70, 2016.

[28] H. Li and A. K. Parlikad, "Social internet of industrial things for industrial and manufacturing assets," 2016.

[29] R. Fisher, L. Ledwaba, G. Hancke, and C. Kruger, "Open hardware: A role to play in wireless sensor networks?" *Sensors*, vol. 15, no. 3, pp. 6818–6844, 2015.

Self Excited Induction Generator for Isolated Pico Hydro Station in Remote Areas

Kailash Rana
Department of Electrical Engineering
Delhi Technologcal University,
Delhi, India

Duli Chand Meena
Department of Electrical Engineering
lDelhi Technological University,
Delhi, India

Abstract—**This paper presents analysis of 3-phase Self excited induction generator (SEIG) which is most suitable for stand-alone Pico-hydroelectric power generation with constant input power. The increasing use of non conventional/renewable energy sources such as bio gas, wind, solar and hydro potential because of its low cost generating system, which are capable to operate in the remote rural areas or where supply of power from grid is not possible or not available. Induction generator is a good choice over well developed synchronous generator for alternative sources of energy like wind based turbine or micro/mini hydro generator because of its low price, simplicity, little maintenance, ruggedness, brushless (for squirrel cage), absence of dc source, protection against short circuit and overload. When driven by fixed speed prime mover or fixed head hydro turbine the SEIG faces the poor voltage regulation which is the main drawback of SEIG. In this paper the simulation using MATLAB /Simulink environment have been carried out to evaluate the performance of isolated 3-phase SEIG in micro/Pico hydropower plant using different load with Electronic Load Controller technique used to improve the power quality and maintaining voltage and frequency at desired level.**

Keywords—**self excited induction generator (SEIG), MATLAB/Simulink, Electronic Load Controller, excitation capacitor, micro/pico hydro power plant.**

I. INTRODUCTION

In remote locations, harnessing of electrical energy from such local resources can be cheaper and easier in comparison to grid connection, which involves long transmission lines and associated losses. A micro/Pico-hydro system using natural hydro potential with minimal civil works is a strong and good candidate in this race. However, the system must be economical, rugged, and user friendly since local communities are usually not well informed about modern technologies. The wind energy sources and small hydro sources has potential and seems to be good candidate for delivering energy to underdeveloped, rural, remote or isolated areas . The objective of the harnessing of such non-conventional energy sources could be achieved in a big way by development of the suitable low cost generating systems.[1],[2]. The generated electric power from these non conventional sources will not only supply the energy to the remote and isolated rural areas, but

can also supply the power demands to other interconnected systems.

However, these systems will become more viable if their cost is reduced to the minimum.

A self excited induction generator (SEIG) is an ideally suited electricity generating system for the renewable energy conversion [3],[4]. A standalone SEIG coupled by small hydro or wind turbines, biogas engines has ability of supplying electric power to domestic and agricultural loads particularly in the hilly and remote areas, where the grid supply is not available or difficult. The SEIG over the years has emerged as an alternative to the conventional synchronous generator for such applications [5].

Previously synchronous generator is widely used and dominate the generation of electricity but now induction generator become more applicable in the field of renewable energy because of their many advantages over the synchronous generator, the induction generator is rugged can with stand rough condition, brushless, low cost, and low maintenance and operation cost, self-protected against short-circuit, and it is capable of generating power at various range of speed, it require external source for excitation to generate a rotating magnetic field, the required reactive power for excitation can be supplied from the grid in this mode known as grid connected induction generator, or it can be supplied by a capacitor bank connected to the stator terminal in this mode it is known as self-excited induction generator[6]-[8].

Despite of offering so many advantages, SEIG's suffers poor voltage regulation. As load keeps varying, the voltage regulation gets poorer. Its inability to maintain a good voltage regulation is its prime disadvantage. Various methods were proposed over the years to overcome this advantage. Static VAR compensators (SVC), which use thyristors or IGBTs, STATCOMs, which are based on DC-AC converters, are used for this purpose[9][10]. But the main disadvantage of these methods is their complexity and low reliability because of power electronic circuits which inject harmonics in the line current.

In micro/pico hydro power plant where generating power capacity is less than 100KW which are not connected to grid a suitable control is necessary to maintain the voltage & frequency within permissible limits under varying load condition. This is achieved by using Electronic Load

Controller along with dump load[11] [12]. The Electronic Load Controller feeds a dump load, enabling the total power supplied by the generator to match the sum between the consumer's load and dump load and maintain voltage and frequency within desired level[16]-[20].

Power generated (P_g) = Power consumed by load (P_L) + Power consumed by ELC (P_D). This ELC dump power (P_D) may be used for space heating, water heater, battery charging, for cooking, baking etc.

II. Hydro Power Energy

Hydropower works on the principle of energy conversion, it convert kinetic energy available in water flow from higher elevation to lower elevation, to mechanical energy through hydro turbine, then utilize this mechanical energy to drive the generator in power house and generate electricity, the difference between higher level and lower level is known as head which can be naturally as in rivers and waterfalls or it can be artificially by building dams and canals.

The potential of hydro power depends upon the availability of head (H) and the flowing discharge (Q), following terminology related to estimation of hydropower potential.

The hydro potential of turbine shaft can be calculated from the following equation.

$$P = \eta \, \rho \, g \, Q \, H \qquad (1)$$

Where, P = mechanical power produced (kW),

η = turbine hydraulic efficiency,

ρ = the density of water (kg/ m^3),

g = the acceleration due to gravity (m/s^2),

Q = the quantity of water flowing through hydraulic turbine (m^3/s),

H = Net available head in meters (m).

While, input torque can be calculated using equation,

$$T = \frac{P}{\omega} \qquad (2)$$

where, ω = angular velocity of turbine runner (rad/s)

III. Self Excited Induction Generator

When three phase supply connected to stator winding of three phase induction motor, current flow through the winding and it will cause a rotating magnetic field of constant magnitude to set up. This rotating magnetic will rotate with synchronous speed which is given in equation (2)

$$Ns = 120 \, f / P \qquad (3)$$

Where, f = supply frequency (Hz),

P = number of pair poles of stator winding.

The difference between the rotating magnetic field and speed of the rotor is known as slip.

$$\text{Slip speed} = Ns - Nr \qquad (4)$$

Where, Ns = synchronous speed rpm,

Nr = actual rotor speed rpm.

The slip express as fractional slip

$$S = Ns - Nr / Ns \qquad (5)$$

The induction machine will work as motor when the speed of rotor is below synchronous speed at positive slip (0<S<1), and it will works as generator when the rotor speed is above synchronous speed at negative slip (S<0). And the induction machine will work as a break when the slip in more than one (S>1).

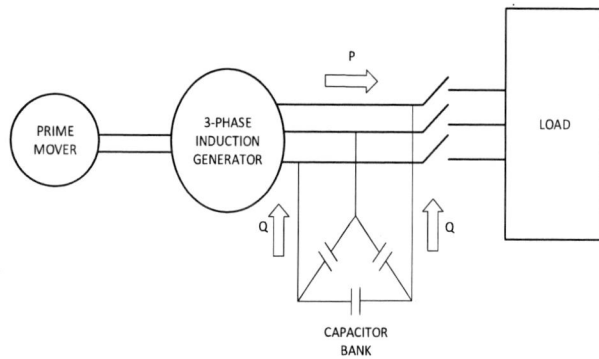

Fig 1. Self excited induction generator.

Here, the three-phase induction machine with excitation capacitors connected in delta for three-phase output is used as shown in Fig 1. Normal 1- Φ Induction motor not be able to use as single-phase Self Excited Induction Generator (SEIG) because it requires some additions or modification to work as SEIG. Single-phase IM are high in cost, which is compared with 3-phase induction machine of same size. For supplying three It is possible to use 3- Φ SEIG for supplying single phase load. The motor's Output Power and Voltage rating is considered. The specifications of the machine are shown in Table-1.

TABLE-1

SPECIFICATION OF INDUCTION MACHINE

Parameters	Value
Power	4KW
Voltage	400v
Frequency	50Hz
Stator- Resistance	2.976 ohm
Stator- Inductance	0.002882 H
Rotor- Resistance	1.408 ohm
Rotor- Inductance	0.002891 H

Mutual Inductance	0.137 H
J Moment of Inertia	0.0131 J

A   Design of Excitation Capacitor

For the proper operation a suitable rating excitation capacitor is is required. Capacitor can be connected in Delta or Star form. The rating of the capacitor is chosen as per

$$\text{Apparent power} \qquad S = \sqrt{3}. V_L . I_L \qquad (6)$$

$$\text{Active power} \qquad P = S\cos\theta \qquad (7)$$

$$\text{Reactive power absorbed,} \qquad Q = \sqrt{S^2 - P^2} \qquad (8)$$

$$\text{Reactive Power needed (per phase),} \quad q = \frac{Q}{3} \qquad (9)$$

$$\text{Per phase Voltage,} \qquad V_P = \frac{V_L}{\sqrt{3}} \qquad (10)$$

$$\text{Capacitive current,} \qquad I_C = \frac{Q}{V_P} \qquad (11)$$

$$\text{Capacitive reactance(per phase),} \quad X_C = \frac{V_P}{I_C} \qquad (12)$$

$$\text{Capacitance (per phase),} \qquad C = \frac{1}{2\pi f X_C} \qquad (13)$$

B . Electronic Load Controller.

Self-Excited Induction Generator (SEIG) for isolated /standalone mode of operation require Electronic Load Controller (ELC) with proper auxiliary dump load where ELC is used to connect and disconnect the dumpload during the working of the system. The Isolated SEIG with ELC is Shown below in fig (2).

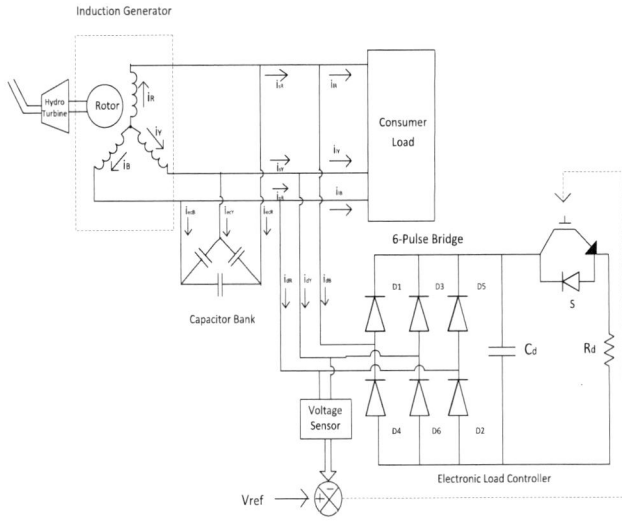

Fig 2.   Self excited induction generator with ELC.

ELC consists of, control circuit, an uncontrolled diode rectifier bridge and IGBT based chopper.
The rating of the different components of the Electronic Load Controller is given as follows:

V_{LL} is the RMS value of the line-to-line voltage of Generator. 4 KW SEIG has 400V as line voltage and the value of V_{DC} is given by

$$V_{DC} = (1.35) \times 400 = 540 \text{ V} \qquad (15)$$

For transient condition an overvoltage of 10% of the rated voltage is considered; so the Root Mean Square AC input voltage which will be obtained with a peak value is

$$V_{peak} = \sqrt{2} \times 440 \text{ V} = 622.25 \text{ V} \qquad (16)$$

This peak voltage will be shown by the components of ELC during the operation of the system. Uncontrolled rectifier and Chopper switch rating is decided by active component of input AC current and obtained by using the formula:

$$I_{AC} = \frac{P}{\sqrt{3}V_{LL}} \qquad (17)$$

where P is the power rating of SEIG. The current I_{AC} of generator is calculated using

$$I_{AC} = \frac{4000}{\sqrt{3} \times 400} = 5.77 \text{ A} \qquad (18)$$

The 3-Φ uncontrolled rectifier draws quasi-square current with distortion factor of 0.955.The calculation of input AC current of ELC is obtained using

$$I_{DAC} = \frac{I_{AC}}{0.955} = \frac{5.77}{0.955} = 6.041 \text{ A} \qquad (19)$$

The variation of Crest Factor (CF) is from 1.4 to 2.0 of the AC current drawn by an uncontrolled rectifier with a capacitive filter ; therefore, the AC input peak current can be calculated using

$$I_{peak} = 2I_{DAC} = 2 \times 6.041 = 12.083 \text{ A} \qquad (20)$$

So the maximum voltage and peak current in the uncontrolled Rectifier is 622.25 volts and 12.083 amperes, respectively. The rating of an uncontrolled rectifier and chopper switch is 700V and 15A, which is more than the calculated values, respectively.
The Auxiliary dump-load resistance rating is obtained by using

$$R_D = \frac{V_{DC}^2}{P_{rated}} = \frac{540^2}{4000} = 72.9\Omega \qquad (21)$$

$$V_{DC} = \frac{3\sqrt{2}V_{LL}}{\pi} = (1.35)\,V_{LL} \qquad (14)$$

The value of the DC link capacitance of the ELC is chosen on the basis of the ripple factor.

The Ripple Factor (RF) and the value of DC-link capacitance for a 3-phase uncontrolled rectifier is related by

$$C = \left(\frac{1}{12\,f\,R_D}\right)\left(1 + \frac{1}{\sqrt{2}\,RF}\right) \qquad (22)$$

Normally 5% ripple factor is permitted in average value of DC link voltage. So the capacitance value is calculated using the previous formula which is given by

$$C = \left(\frac{1}{12\times50\times72.9}\right)\left(1 + \frac{1}{\sqrt{2}\times0.05}\right) = 333.1\ \mu F \qquad (23)$$

The integral part of control mechanism involves the use of ELC which is basically a switching circuit using fast electronic switching devices such as IGBT or MOSFET to connect or disconnect dump load according to consumer load

across the generator. As the consumer load increases, the corresponding decrease in DC link voltage forces the control circuit to decrease the on time of IGBT. The simulink model of SEIG with ELC is shown in Fig. 3.

IV. RESULTS AND DISCUSSION

The performance of SEIG with ELC is carried out using MATLAB/Simulink environment to observe the load dynamic behavior. The ELC is to design so as to maintain the desired load level which enables the voltage and frequency within the limits. The output waveforms are shown in fig (4, 5). It is observed that the initial load of 3.5kW is changed to 1.05kW (70% decrease in load) at 2 second the variation in voltage and frequency is within limits.

When the load is decreased by 70% at 2 sec then most of the current flow through electronic load controller circuit and power consumed by dump load. Due to the presence of rectifier circuit certain harmonics are present in the circuit which results in slight distortion in voltage and current waves.

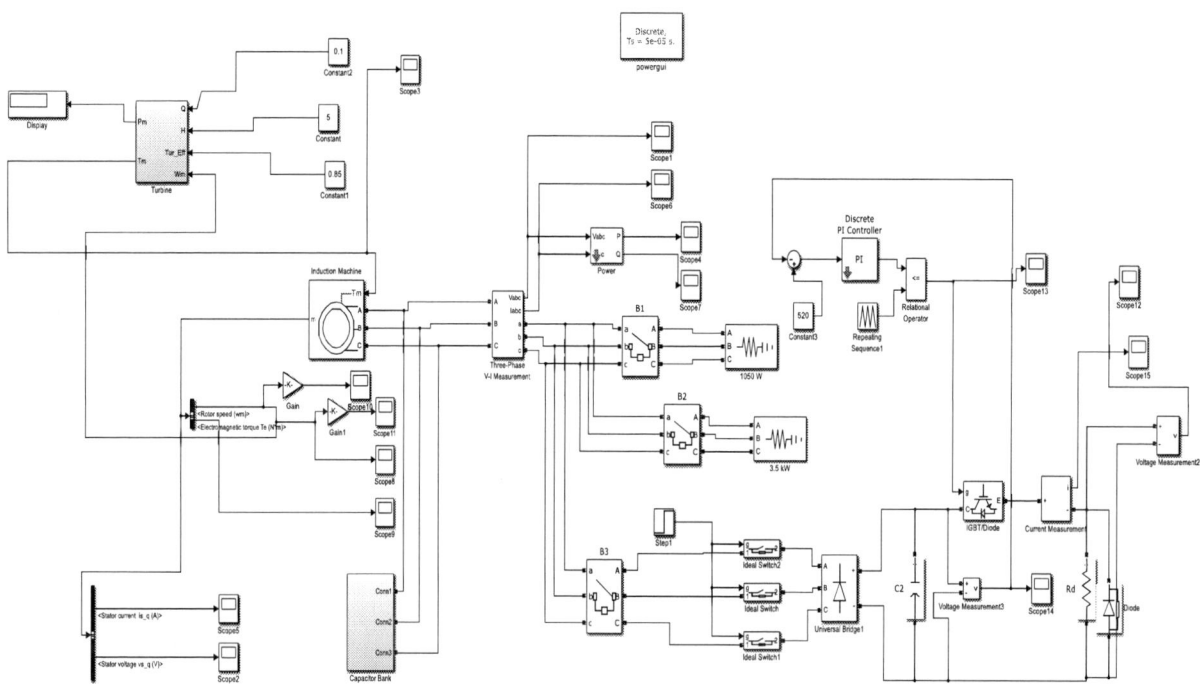

Fig 3. MATLAB/Simulink model of SEIG with ELC

2nd IEEE International conference on power Electronics, Intelligent Control and Energy systems

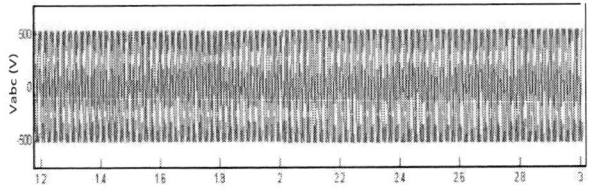
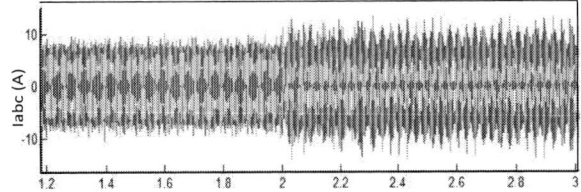

Fig 4. Generated Voltage and Current of SEIG

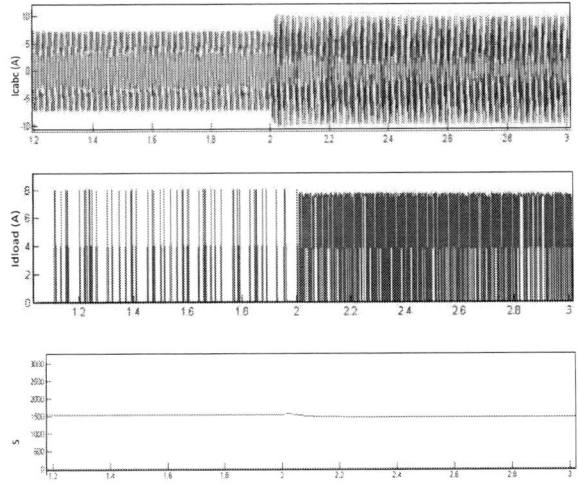
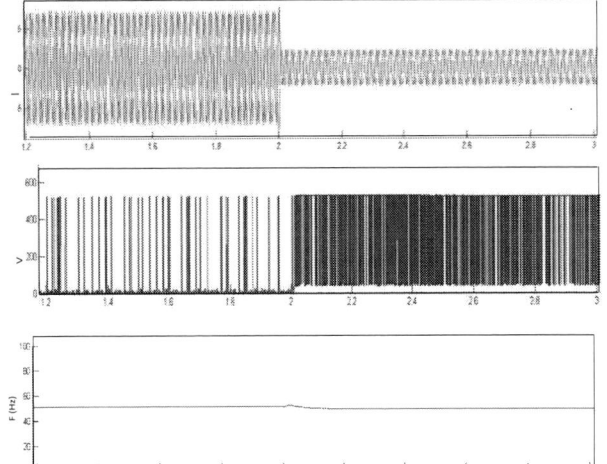

Fig.5 Capacitor Current (I_{cabc}), Load current (I_{Labc}), Dump load current (I_{dload}), Dump load Voltage (V_{dload}),

Speed in rpm of SEIG and Frequency

V. CONCLUSION

A MATLAB/Simulink based dynamic model of the SEIG system has been discussed and the simulated performance characteristic of 3-phase self excited induction with Electronic Load Controller shows that the discussed system can be used in Pico/micro hydro application. The simulated results reveals that the voltage and frequency remain within limit during the load changes A Pico/micro hydro system can be installed easily and economically in remote location/ rural area/ hill regions. The SEIG with ELC is reliable, simple and excellent option for micro/Pico hydro application.

REFERENCES

[1]. D. P. Kothari, R. C. Bansal, T. S. Bhatti, "Bibliography on the Application of Induction Generators in Nonconventional Energy Systems," IEEE Transactions. Energy Conversions., vol. 18, no.3, pp. 433 –439, 2003

[2]. G. K. Singh, "Self-excited induction generator research -A survey," ELSEIVER Electr. Power Syst. Res., vol. 69, no. 2–3, pp. 107–114, 2004

[3]. F. M. Potter and E. D. Bassett, "Capacitive excitation forinduction generators," Electrical Engineering, vol. 54, no. 5, pp. 540–545, 1935.

[4]. J.Arrillaga and D.B.Watson, "Static power conversion from selfexcited induction generators," Proceedings of the Institution of Electrical Engineers—Generation Transmission Distribution, vol. 125, no. 8, pp. 743–746, 1978.

[5]. Nigel P.A. smith "Induction Generators for Stand –alone Micro-Hydro system" Proc. Of international conference on power electronics,drivesand energy systems for industrial growth,V01.2,pp.669.Jan1996

[6]. T. F. Chan, "Capacitance requirements of selfexcited induction generators," IEEE Transctions, on Energy Conversion, Vol. 8, No, 2, pp. 304 to 311, June 1993.

[7]. R. J. Harrington and F. M. M. Bassiouny, "New approach to determine the critical capacitance for self-excited induction generators," IEEE Transations, on Energy Conver, Vol. 13, No. 3, pp. 244-249, Sept. 1998.

[8]. T. F. chan and L. L. Lai, "Capacitance requirements of a three-phase induction generator se!f-excited with a single capacitance and supplying a single-phase load," IEEE Transactions, on Energy Convers, Vol. 17, No. 1, pp. 90-94, Mar 2002.

[9]. Y.K. Chauhan, S.K. Jain, and B.Singh, "A prospective on voltage regulation of self-excited induction generators for industry applications", IEEE Transaction's on Industry Applications, Vol.46, No.2, pp 720-730, 2010.

[10]. H. C. Rai, A. K. Tandan, S. S.Murthy, B. Singh, and B. P. Singh, "Voltage regulation of self-excited induction generator using passive elements," in Proceedings of the 6th IEEE International Conference on Electrical Machines and Drives, Conference.

[11]. Bhim Singh, senior member, IEEE, Madhusudan Singh, member, EEE, and A.K. Tandon, Senior Member, IEEE, 'IEEE, "Transient Performance of series compensated three- Phase self –excited Induction generator feeding Dymanic loads", IEEE Transaction on Industry Application.

[12]. Bhim Singh, Madhusudan, and A.K. Tandon "Dynamic Modelling and Analysis of Three Phase Self Excited Induction Generator using Matlab," NATIONAL POWER SYSTEMS CONFERENCE, NPSC 2002.

[13]. A.Harvey, Micro-hydro Design Manual. Rugby: Intermediate Technology Publications, 2006.

978-1-5386-6624-1/18 $31.00 © 2018 IEEE   825

[14]. Shridhar L. Murthy S. S. Singh B. Jha, C. and B. P. Singh. Selection of capacitors for the self regulated short shunt self-excitet generator. IEEE, 3 1995.

[15]. M. H. Haque. "Characteristics of shunt, short-shunt and long shunt single-phase induction generators," IEEE, 06 2009. Publication no. 376, pp. 240–245, IET, Oxford, UK, September 1993.

[16]. B. Singh, S. S. Murthy, and S. Gupta, "Analysis and design of electronic load controller for self-excited induction generators," IEEE Transactions on Energy Conversion, vol. 21, no. 1, pp. 285–293, 2006.

[17]. E.Mishra and S. Tiwari, "Proportional Integral based Electronic load controller for Self excited induction generator," International Journal of Innovative Engineering Research, vol. 6, no. 2,2016.

[18]. B. Singh, S. S. Murthy and S. Gupta, "Transient analysis of self-excited induction generator with electronic load controller (ELC) supplying static and dynamic loads," IEEE Transactions on Industry Applications, vol. 41, no. 5, pp. 1194-1204, Sept./Oct. 2005.

[19]. B.Singh, S.S. Murthy and S. Gupta, "Analysis and Implementation of an Electronic Load Controller for a Self-Excited Induction Generator", IEE Proc.-Gener. Transm. Distrib., Vol. 151, No. 1, pp. 51-60, Jan. 2004.

[20]. Bhim Singh, S. S. Murthy and Sushma Gupta, "An Improved Electronic Load Controller for Self-Excited Induction Generator in Micro-Hydel Applications", IEE proceeding 2003, pp 2741 -2746.

[21]. C.Kathirel, K.Porkumaran and S.Jaganathan, " Design and Implementation of Improved Electronic Load Controller for Self-Excited Induction Generator for Rural Electrification", Hindawi Publishing Corporation, The Scientific World Journal Vol 2015. Art ID 340619.

Predicting the Popularity of Websites using Multilayer Perceptron and Extreme Learning Machine

Neha Srivastava[1], Virendra P. Vishwakarma[2] and Udayan Ghose[3]

University School of Information, Communication & Technology,

Guru Gobind Singh Indraprastha University, Sector 16- New Delhi, India

[1]srivastava.neha46@yahoo.in,

[2]virendravishwa@rediffmail.com,[3]udayan@ipu.ac.in

Abstract— **Due to increasing usage of internet, the prediction of the popularity of a website is becoming a popular topic of research. There are millions of users who are interacting every day. In this paper, we predict the popularity of website using classification and regression method based 3on multiple attributes. Multilayer perceptron and extreme learning machine have been used for predicting the popularity of the website.**

Keywords— **Websites popularity prediction, Multilaye perceptron, Extreme learning Machine**

I. INTRODUCTION

In the current era, World Wide Web (WWW) is well known which provides an interactive medium to circulate the information. The content of web data is enormous, distinct and dynamic. At present we are facing the problem of the drowning of web information and overloading of web data [1]. Web users face this problem when they are interacting with the websites. They either use search engine services or they browse the web when they are searching for any information. When a web user uses a query based search engine, they usually input a query to the search engine interface. Based on the query, a response is created using the list of page ranks. Because of the low accuracy of many search results, finding relevant information is difficult. Another issue is low recall rate that arises because of incapability to index all the information which are available on the web. This results in a difficulty in searching the unindexed information which is relevant. Therefore, techniques of web mining come into the picture to resolve this specific problem of information overloading directly or indirectly [1].

Due to increasing usage of internet, the prediction of the popularity of a website is becoming a popular topic of research [2] [3] [4] [5]. There are millions of users who are interacting with search engine every day. It is important to predict which webpage is more relevant for the user. Kelwin

et.al had proposed that popularity of online news can be measured based on the number of shares, number of likes or number of comments [6]. Some authors presented a method for popularity prediction of online videos, they used deep recurrent neural network [7]. First, they explained popularity prediction problem of online videos as a classification task after that they specified the deep neural network model which tries to resolve their task. For popularity measure of online videos, a regression method was given in which the popularity was measured by a number of views on videos. They used support vector regression technique including Gaussian Radial basis functions. They also used visual features, for the outputs of deep neural networks which were feasible for predicting the popularity before publication of content [8].

Gitte and Meert described the prediction of web page popularity based on time series data from the Chartbeat web analytics engine. They proposed a random forest regression technique to predict the page visit on the particular website as well as they also predicted the number of likes on facebook, number of visitors, number of tweets on twitter [9].

A simple linear regression model was presented in which the comments on news articles were used for popularity prediction [10].

For predicting the popularity of the website, the prediction module includes features for input and for output number of shares on that article. Therefore, for popularity prediction module we are using neural network supervised learning algorithm (classification, regression) to train our model. Here we are testing on two techniques multilayered perceptron and extreme learning machine. The dataset from UCI Machine learning repository is collected for present investigation.

In this paper, it is classified as given under. In Section 2 description of multilayered feedforward perceptron and extreme learning machine. Section 3 includes regression and classification model which have been used. In Section 4

Implementation and analysis of result have been described. Section 5 includes Conclusion and Future Work.

II. PRELIMINARIES

FeedForward Neural Network consists of one or more hidden layers and participates in computation. As the neurons of this layer can't be directly seen from input or output side hence it is called hidden layer by stacking one or more layers. From the input, the network is able to extract higher order of statistics. The learning process will make a decision of which features have to be taken for input which will correspond to hidden neurons. It is hard for the learning process to search for larger possible functions and selection is to be created between the alternative representative of input patterns. A common method is used for training an MLP is the back-propagation algorithm. The training can be done in two stages:

1. In the first stage which is a forward phase, the network weights are static and within the network, the input signal is circulated layer by layer until it reaches the output. Therefore, in this stage modification is restricted in a network for the activation function and output of the neuron.
2. In the second stage which is backward phase of a network, outcome is an error signal after examining the difference between the output of the neural network and desired response. The error signal is propagated back, layer by layer and adjustment of weights is done until the error is minimized [11].

Another technique which we have used is extreme learning machine(ELM) which is feed forward structure of single layer or multiple layers of hidden nodes, here it is not necessary to tune the parameters of the hidden nodes. It is simple to implement without iterative calculations because inputs weights are randomly generated and it does not depend on any specific applications. Extreme learning machine is capable of classifying newly observed objects [12]. Here, [13][14] single layer feedforward neural networks with N hidden nodes where weights of the input layer and hidden layer biases are randomly chosen and corresponding hidden nodes also known as random hidden nodes which can accurately learn N different observations. It is not necessary that we alter the weights of the input as well as hidden layer biases in the application. When the weights of the input along with hidden layer biases are selected randomly, the single layer feed forward neural network can be considered as a linear system and the output weights could be decided by simple generalized inverse operation of the hidden layer output matrices.

III. PROPOSED APPROACH

For predicting the popularity of websites, the dataset has been collected from the UCI Machine learning repository. First, we read the dataset and analyse what attributes are required for pre-processing and then we define features and labels and divide the dataset into two parts for training and testing. And then we train our model and calculate the RMSE value and we repeat this process until the loss is minimized, after that we make a prediction on the test dataset. Here we have extracted the 11 features for input which are predictive attributes and goal is to predict the number of shares.

As the number of shares is continuous value, regression models are used here to predict the continuous valued function. In the phase of regression, the target is to map data items with a real-valued prediction variable [15].

Classification models are used to map data into pre-defined groups or classes [13]. For classification model here we are assuming a binary classification. If the number of shares on a particular website is greater than a threshold value then it is considered to be "popular" otherwise, if it is less than decision threshold value then it is considered to be "unpopular". For prediction model, we are predicting the popularity of the website, which website is popular by continuous valued function in label field. Multilayer feed forward back propagation and extreme learning machine have been tested for the prediction model.

IV. IMPLEMENTATION AND ANALYSIS OF RESULT

For Predicting the popularity of websites, the dataset has been taken from the UCI Machine learning repository. The main goal is to predict the popularity of website. First, we have to select the parameters for the features for input and label for output. List of attributes which is taken for feature extraction:- Number of words in the title, Number of words in the article, Average word length, Rate of non-stop words, Rate of non-stop words in the content, Number of links, Number of self-reference links, Number of images, Number of videos, Average token length, Number of keywords in the metadata. The goal is to predict the number of shares. The graph which is presented below is showing the non-linear behaviour of all attributes.

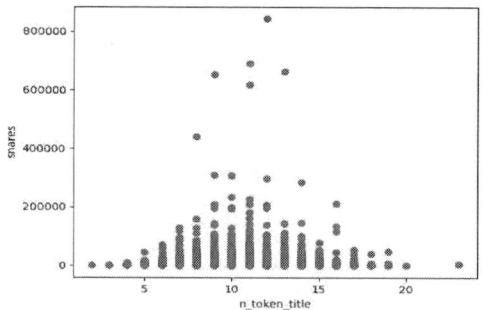

Figure1 no. of shares v/s words in the title

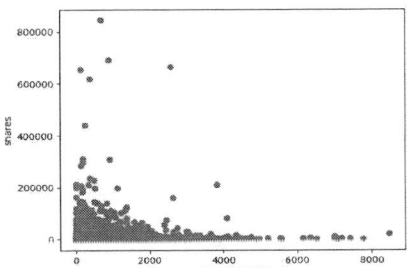

Figure2 no. of shares v/s words in the content

Here the graph is shown for two attributes. It can also be shown that all other attributes also have non-linear characteristics. Therefore, for prediction we have to do non-linear regression. After defining the parameters we have normalized the dataset by calculating the mean and standard deviation of parameters, and then standardization has been done on the dataset. Then we have defined a network parameters for MultilayerPerceptron, a standard training algorithm feedforward backpropagation and extreme learning machine is modeled here.

NON-LINEAR REGRESSION USING MLP

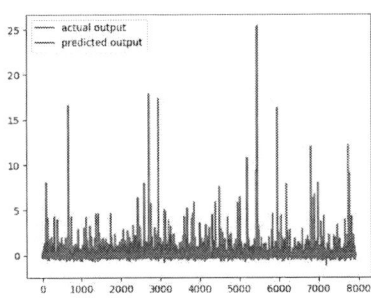

Figure3 Non-linear regression(multilayer perceptron)

80% and 20% of the dataset have been used for training and testing purpose respectively. The logistic activation function is used here. Figure3 shows non-linear regression where red line denotes actual output and green line denotes predicted output. Here the RMSE value for test data is 0.80.

NON-LINEAR REGRESSION USING ELM

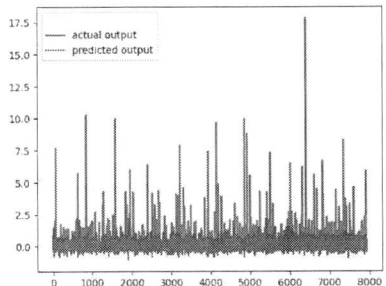

Figure4 Non-linear regression(extreme learning machine)

Here also, 80% and 20% of the dataset have been used for training and testing purpose respectively. The gaussian activation function is used at hidden layer. Figure4 shows non-linear regression where red line denotes actual output and green line denotes predicted output. RMSE value for test data comes out to be 0.60.

CLASSIFICATION USING MLP AND ELM

For classification modeling, the desired output correspond to popular and non-popular has been modeled using a threshold value. Implementation is done on a common binary classification. We tested on two methods Multilayer perceptron and Extreme learning machine.

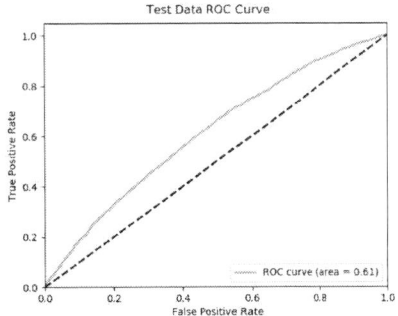

Figure5 ROC curve on Test Data

75% and 25% of the dataset have been used for training and testing purpose with relu activation function. Hidden layer sizes are 2500. Figure5 shows ROC curve for binary classification to study the classifier output of Multilayer perceptron.
ELM and MLP testing accuracy is same for classification.

ELM Training Accuracy	0.64
ELM Testing Accuracy	0.57
MLP Training Accuracy	0.60
MLP Testing Accuracy	0.57

Training Data Performance Metrics

Accuracy	F1	Precision	Recall	AUC
0.61	0.6	0.61	0.61	0.65

Testing Data Performance Metrics

Accuracy	F1	Precision	Recall	AUC
0.58	0.58	0.58	0.58	0.61

Performance of Training and Testing data metrics has been shown. In which accuracy, F1, precision, recall, AUC has been tabulated.

V. CONCLUSION AND FUTURE WORK

Here, we predicted the popularity of the website in which set of features has been extracted and optimize the result based on the number of shares which will determine the relevance of web pages. We are applying feedforward back propagation and extreme learning machine for predicting the popularity of website. In Future work we will use kernel extreme learning machine for predicting the popularity of websites for better accuracy.

REFERENCES

[1] Raymond and Blockeel, Hendrik Kosala, "Web mining research: A survey," ACM Sigkdd Explorations Newsletter, vol. 2, pp. 1--15, 2000.

[2] Mohamed and Spagna, Stella and Huici, Felipe and Niccolini, Saverio Ahmed, "A peek into the future: Predicting the evolution of popularity in user generated content," ACM, pp. 607--616, 2013.

[3] Roja and Asur, Sitaram and Huberman, Bernardo A Bandari, "The pulse of news in social media: Forecasting popularity," ICWSM, vol. 12, pp. 26--33, 2012.

[4] Gabor and Huberman, Bernardo A Szabo, "Predicting the popularity of online content," Communications of the ACM, vol. 53, pp. 80--88, 2010.

[5] Alexandru and Antoniadis, Panayotis and De Amorim, Marcelo Dias and Fdida, Serge Tatar, "From popularity prediction to ranking online news," Social Network Analysis and Mining, vol. 4, p. 174, 2014.

[6] Kelwin and Vinagre, Pedro and Cortez, Paulo Fernandes, "A proactive intelligent decision support system for predicting the popularity of online news," pp. 535--546, 2015.

[7] Tomasz and Andruszkiewicz, Pawe and Boche, "Recurrent Neural Networks for Online Video Popularity Prediction," pp. 146--153, 2017.

[8] Tomasz and Rokita, Przemys, "Predicting popularity of online videos using support vector regression," IEEE Transactions on Multimedia, vol. 19, pp. 2561--2570, 2017.

[9] Gitte and Meert, Wannes Vanwinckelen, "Predicting the popularity of online articles with random forests," 2014.

[10] Alexandru and Leguay, Jrmie and Antoniadis, Panayotis and Limbourg, Arnaud and de Amorim, Marcelo Dias and Fdida, Serge Tatar, "Predicting the popularity of online articles based on user comments," p. 67, 2011.

[11] Simon Haykin, "Neural Networks and learning machines," 2009.

[12] Guang-Bin and Zhu, Qin-Yu and Siew, Chee-Kheong Huang, "Extreme learning machine: theory and applications," Neurocomputing, vol. 70, pp. 489--501, 2006.

[13] Shin'ichi and Tateishi, Masahiko Tamura, "Capabilities of a four-layered feedforward neural network: four layers versus three," IEEE Transactions on Neural Networks, vol. 8, pp. 251--255, 1997.

[14] Guang-Bin Huang, "Learning capability and storage capacity of two-hidden-layer feedforward networks," IEEE Transactions on Neural Networks, vol. 14, pp. 274--281, 2003.

[15] Mittal C and Panchal, Mahesh Patel, "A review on ensemble of diverse artificial neural networks," International Journal of Advanced Research in Computer Engineering \& Technology (IJARCET), vol. 1, pp. pp--63, 2012.

2nd IEEE International Conference on Power Electronics, Intelligent Control and Energy Systems (ICPEICES-2018)

A Hybrid Twenty Five-level Inverter for an Open-End Winding Induction Motor (OEWIM) Drive

Anoop Kumar Kanaujia
Department of Electrical Engineering
Harcourt Butler Technical University
Kanpur, Uttar Pradesh, India
akk.nsi@gmail.com

Sanjiv Kumar
Department of Electrical Engineering
Harcourt Butler Technical University
Kanpur, Uttar Pradesh, India
sanjiv.iitr@gmail.com

Abstract—In this paper, a hybrid twenty five-level inverter topology for an OEWIM (Open-End Winding Induction Motor) drive is proposed. In this topology, one end of OEWIM is fed by a FC based hybrid five-level inverter and other end is connected to a symmetrical five-level inverter. The hybrid five-level inverter is achieved by cascading a three-level flying capacitor inverter with capacitor fed H-bridge. The symmetrical five-level inverter is obtained by cascade connection of four 2-level inverters with equal DC-link voltages. The combined effect of five-level hybrid and five-level symmetrical inverters generate twenty five distinct levels in the phase voltage. In this proposed topology, a total of 3,43,000 space phasor combinations are accessible; hence using switching redundancies, the capacitor voltages can be maintained at desired levels. The proposed topology requires only 48 switching devices, as compared to conventional topologies that require 144 switches and fully eliminates the requirement of clamping diodes, which are required in NPC inverter. It, also, requires only 6 capacitors unlike to 828 capacitors required in twenty five-level FC inverter and 5 DC power supply sources as compared to twenty five-level CHB inverter that require 36 DC power supply sources. Simulation study is carried out to evaluate the performance of the proposed inverter and results are presented.

Keywords— Induction Motor Drive; Multi-level Inverter; Open-End Winding; NPC Inverter; FC Inverter; CHB Inverter.

I. INTRODUCTION

A conventional 2-level inverter is generally used in low voltage and low power applications due to the restricted availability of voltage & power rating of power semiconductor switching devices. A conventional 2-level inverter can be used in high voltage and high power applications by the series-parallel connection of power semiconductor switches. However, such inverters experience with high dv/dt, large EMI (Electromagnetic Interference), CMV (Common Mode Voltage), low efficiency and high switching loss because of high switching frequency [1].

The various problems associated with the conventional 2-level inverter in the field of high power applications were overcome by the introduction of the Multi-level inverter (MLI) based induction motor drive. The term multi-level commenced with the 3-level inverter and are articulated by more than two voltage levels of motor phase voltage as compared to conventional 2-level inverter [2], [3].

Abundant MLI topologies have been proposed during the past few decades. Moreover, three main topologies of MLI for high power applications are diode clamped (neutral-point clamped) inverters [4], cascaded H-bridges inverters [5] and

flying capacitor (capacitor clamped) inverters [6]. The first NPC (neutral-point clamped), fundamentally a 3-level inverter was proposed by Nabae, Takahashi, and Akagi in 1981 [4]. Meynard and Foch introduced a FC (flying capacitor) based inverter in 1992 [7]. The structure of FC inverter is analogous to that of the NPC inverter except that the FC inverter uses capacitors instead of clamping diodes. A number of articles have been published comparing these three MLI topologies i.e. NPC, CHB & FC in the field of high power induction motor drive applications [8]–[12].

However, these MLI topologies face some problems in their uses. NPC inverter experiences neutral point fluctuations as the DC-link capacitors carry the load current. CHB (cascaded H-bridges) inverter requires separate DC power supply sources for all the three phases which increases the cost and complexity of power circuit. FC inverter does not require the clamping diodes but needs many capacitors due to floating DC voltage sources. Also, as the output levels in MLI configuration increases, the cost and complexity of the circuit enhances due to induction of large number of power electronics components.

The problem faced by these above MLIs was then overcome by the introduction of OEW topology. The concept of open-end winding structure for induction motor drive was first proposed by H. Stemmler and P. Guggenbach in 1993 [13], by feeding an open-end winding induction motor (OEWIM) with 2-level inverters from both of its ends. The OEWIM is achieved by detaching the star-point of an induction motor stator winding [14]. Further, if this motor is connected with two dissimilar or same type of inverters on both side of the stator winding, then the new arrangement is termed as OEWIM drive.

Open-end winding structure MLI topologies for induction motor drive are becoming more and more adoptable due to its numerous advantages over the conventional MLI topologies i.e. NPC, CHB and FC inverters. The OEW scheme is free from capacitor voltage unbalance and does not require any clamping diodes as in the case of NPC inverter topology. It requires lesser number of DC power supply sources unlike CHB inverter topology and additional capacitor banks are also not needed as in the case of FC inverter topology [15].

A hybrid seven-level [16] and nine-level inverter topology [17], [18] for an OEWIM were proposed by the researchers with the advantage that the inverter can still be operated at reduced voltage level if any bridge fails, thus enhances the reliability of circuit and available redundant states are utilized for capacitor balancing as well.

978-1-5386-6624-1/18 $31.00 © 2018 IEEE

In this paper, a hybrid twenty five-level inverter topology is proposed for an OEWIM. One end of OEWIM is fed by a FC based hybrid five-level inverter and other end is connected to a symmetrical five-level inverter. The capacitor voltages are maintained at the desired level by use of the redundancy switching available for generating different voltage levels. The combined effect of five-level hybrid and five-level symmetrical inverters generate twenty five distinct levels in the phase voltage.

II. PROPOSED TWENTY FIVE-LEVEL INVERTER TOPOLOGY

A. Power Circuit

The proposed circuit diagram to realize twenty five-levels in the phase voltage is shown in Fig. 1. In this circuit topology, an OEWIM is fed from a hybrid five-level inverter-A from one end of open-end stator winding (A_1, B_1, C_1) and a symmetrical five-level inverter-B from other end of the motor winding (A_2, B_2, C_2). Both the inverter A and B are connected by isolated DC power supply sources. The DC link voltage of inverter-A and inverter-B are $20V_{DC}/24$ and $V_{DC}/24$ of each sub-inverter respectively, where V_{DC} is the DC link voltage of an equivalent conventional two-level inverter drive. A three-level FC inverter is cascaded with a capacitor fed H-bridge to realize hybrid five-level inverter. Capacitors C_{A1}, C_{B1} and C_{C1} are maintained at voltage level of $10V_{DC}/24$, whereas capacitors C_{A2}, C_{B2} and C_{C2} are kept at voltage level of $5V_{DC}/24$. Similarly, four 2-level inverters are cascaded to realize the symmetrical five-level inverter.

The pole voltages of inverter-A; for example, phase A pole voltage (V_{A1O}) of inverter-A can have five voltage levels viz. 0, $5V_{DC}/24$, $10V_{DC}/24$, $15V_{DC}/24$ and $20V_{DC}/24$. Similarly,

phase B (V_{B1O}) and phase C (V_{C1O}) pole voltages can also have these voltages levels depending upon their switching strategies with respect to its reference point O.

The pole voltages of inverter-B; for example, phase A pole voltage ($V_{A2O'}$) of inverter-B can have five voltage levels viz. 0, $V_{DC}/24$, $2V_{DC}/24$, $3V_{DC}/24$ and $4V_{DC}/24$. Similarly, phase B ($V_{B2O'}$) and phase C ($V_{C2O'}$) pole voltages can also have these voltages levels depending upon its switching strategies with respect to its reference point O'.

Twenty five-levels, generated for phase A, when these hybrid five-level inverter-A and symmetrical five-level inverter-B are switched with different pole voltage levels V_{A1O} and $V_{A2O'}$ of the OEWIM drive, are shown in Table I with few switching states. Total 70 redundant switching states are possible for phase A voltage in both the direction of current (+/-) to realize twenty five-level operation.

B. Generation of Different Phase Voltage

The topology proposed here is capable of generating 25 voltage levels: $-(4/24)V_{DC}$, $-(3/24)V_{DC}$, $-(2/24)V_{DC}$, $-(1/24)V_{DC}$, 0, $(1/24)V_{DC}$, $(2/24)V_{DC}$, $(3/24)V_{DC}$, $(4/24)V_{DC}$, $(5/24)V_{DC}$, $(6/24)V_{DC}$, $(7/24)V_{DC}$, $(8/24)V_{DC}$, $(9/24)V_{DC}$, $(10/24)V_{DC}$, $(11/24)V_{DC}$, $(12/24)V_{DC}$, $(13/24)V_{DC}$, $(14/24)V_{DC}$, $(15/24)V_{DC}$, $(16/24)V_{DC}$, $(17/24)V_{DC}$, $(18/24)V_{DC}$, $(19/24)V_{DC}$ and $(20/24)V_{DC}$ in phase voltage, as shown in Table I. Other voltage levels do not have redundancy switching states to balance the capacitors voltages; hence, they are not used in proposed topology. Few switching states to generate twenty five-level phase voltage (V_{A1A2}) in the motor stator winding of phase A with its effect on capacitor voltages based on the direction of current are shown in Table I.

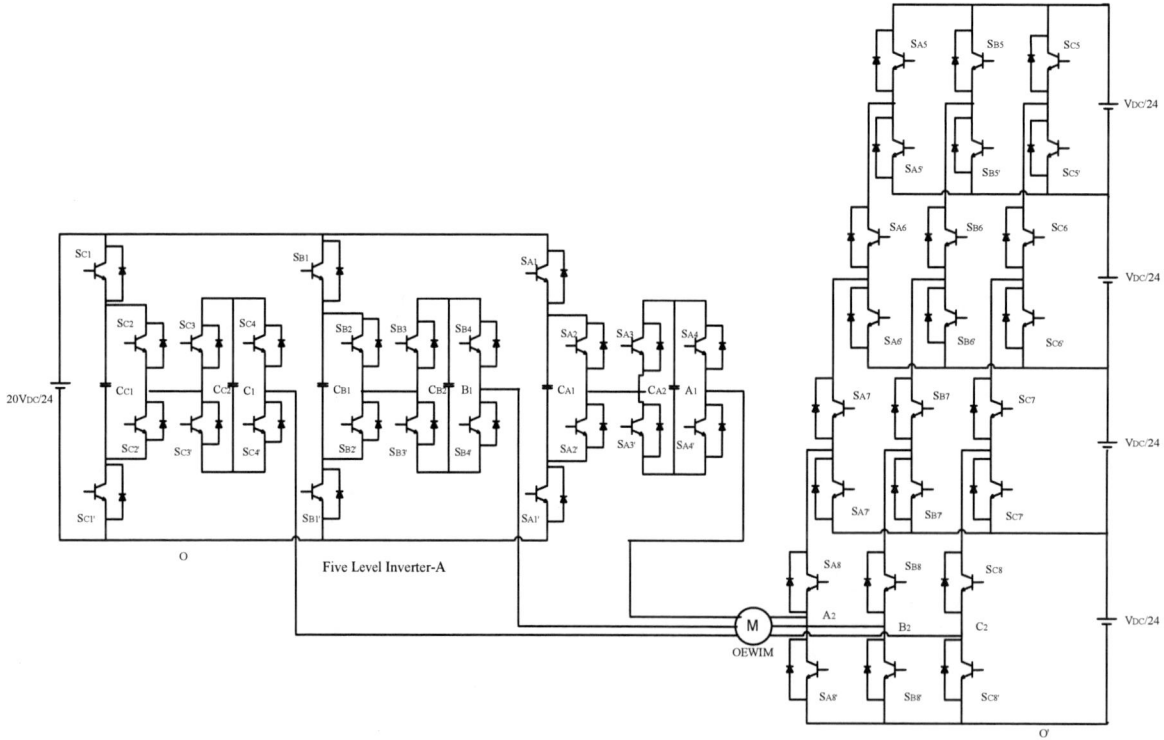

Fig. 1. Proposed twenty five-level inverter topology with OEWIM

978-1-5386-6624-1/18 $31.00 © 2018 IEEE

Voltage levels $-(4/24)V_{DC}$, $-(3/24)V_{DC}$, $-(2/24)V_{DC}$, $-(1/24)V_{DC}$, 0, $(1/24)V_{DC}$, $(2/24)V_{DC}$, $(3/24)V_{DC}$, $(4/24)V_{DC}$, $(5/24)V_{DC}$, $(6/24)V_{DC}$, $(7/24)V_{DC}$, $(8/24)V_{DC}$, $(9/24)V_{DC}$, $(10/24)V_{DC}$, $(11/24)V_{DC}$, $(12/24)V_{DC}$, $(13/24)V_{DC}$, $(14/24)V_{DC}$, $(15/24)V_{DC}$, $(16/24)V_{DC}$, $(17/24)V_{DC}$, $(18/24)V_{DC}$, $(19/24)V_{DC}$ and $(20/24)V_{DC}$ correspond to levels L_1 to L_{25}. These voltage levels are generated by proper combination of the pole voltages of hybrid five-level inverter-A with the pole voltages of symmetrical five-level inverter-B.

There are two redundant switching states to generate voltage levels of $-(4/24)V_{DC}$, $-(3/24)V_{DC}$, $-(2/24)V_{DC}$, $-(1/24)V_{DC}$, 0, $(16/24)V_{DC}$, $(17/24)V_{DC}$, $(18/24)V_{DC}$, $(19/24)V_{DC}$ and $(20/24)V_{DC}$; for voltage levels $(1/24)V_{DC}$, $(2/24)V_{DC}$, $(3/24)V_{DC}$, $(4/24)V_{DC}$, $(5/24)V_{DC}$, $(11/24)V_{DC}$, $(12/24)V_{DC}$, $(13/24)V_{DC}$, $(14/24)V_{DC}$ and $(15/24)V_{DC}$, there are three redundant switching states, and to generate $(6/24)V_{DC}$, $(7/24)V_{DC}$, $(8/24)V_{DC}$, $(9/24)V_{DC}$ and $(10/24)V_{DC}$, there are four redundant switching states available. The capacitors (C_{A1} and

C_{A2}) voltage remain unchanged for the voltage level of $-(4/24)V_{DC}$, $-(3/24)V_{DC}$, $-(2/24)V_{DC}$, $-(1/24)V_{DC}$, 0, $(16/24)V_{DC}$, $(17/24)V_{DC}$, $(18/24)V_{DC}$, $(19/24)V_{DC}$ and $(20/24)V_{DC}$, but for other voltage levels, capacitors are either charges or discharges, depending upon the direction of current I_A. Similar logic can be implemented for other two phases of the motor as well.

The capacitors (C_{A1}, C_{B1} & C_{C1} and C_{A2}, C_{B2} & C_{C2}) voltages are sensed & compared with reference voltages $10V_{DC}/24$ and $5V_{DC}/24$ respectively in every switching period and any error is passed through controller. The controller selects the appropriate switching from available redundant switching states. The capacitor voltages are rapidly balanced with reference voltages as corrective action is taking place in every switching period. Thus, the capacitors may be either charged or discharged in any direction of current I_A by using the available redundant switching states.

TABLE I. LEVEL REALIZATION IN PHASE A

Pole Voltage V_{A1O}	Pole Voltage $V_{A2O'}$	Motor Phase A Voltage Level V_{A1A2}	Level of Operation	Direction of Current I_A [#]	Switching States* S_{A1}, S_{A2}, S_{A3}, S_{A4}, S_{A5}, S_{A6}, S_{A7}, S_{A8}	Effect on Capacitor C_{A1}	Effect on Capacitor C_{A2}
0	$(4/24)V_{DC}$	$-(4/24)V_{DC}$	L_1	+	00001111	No effect	No effect
0	$(3/24)V_{DC}$	$-(3/24)V_{DC}$	L_2	+	00000111	No effect	No effect
0	$(2/24)V_{DC}$	$-(2/24)V_{DC}$	L_3	+	00000011	No effect	No effect
0	$(1/24)V_{DC}$	$-(1/24)V_{DC}$	L_4	+	00000001	No effect	No effect
0	0	0	L_5	+	00000000	No effect	No effect
$(5/24)V_{DC}$	$(4/24)V_{DC}$	$(1/24)V_{DC}$	L_6	+	00011111	No effect	Discharge
$(5/24)V_{DC}$	$(3/24)V_{DC}$	$(2/24)V_{DC}$	L_7	+	00010111	No effect	Discharge
$(5/24)V_{DC}$	$(2/24)V_{DC}$	$(3/24)V_{DC}$	L_8	+	00010011	No effect	Discharge
$(5/24)V_{DC}$	$(1/24)V_{DC}$	$(4/24)V_{DC}$	L_9	+	00010001	No effect	Discharge
$(5/24)V_{DC}$	0	$(5/24)V_{DC}$	L_{10}	+	00010000	No effect	Discharge
$(10/24)V_{DC}$	$(4/24)V_{DC}$	$(6/24)V_{DC}$	L_{11}	+	10001111	Charge	No effect
$(10/24)V_{DC}$	$(3/24)V_{DC}$	$(7/24)V_{DC}$	L_{12}	+	10000111	Charge	No effect
$(10/24)V_{DC}$	$(2/24)V_{DC}$	$(8/24)V_{DC}$	L_{13}	+	10000011	Charge	No effect
$(10/24)V_{DC}$	$(1/24)V_{DC}$	$(9/24)V_{DC}$	L_{14}	+	10000001	Charge	No effect
$(10/24)V_{DC}$	0	$(10/24)V_{DC}$	L_{15}	+	10000000	Charge	No effect
$(15/24)V_{DC}$	$(4/24)V_{DC}$	$(11/24)V_{DC}$	L_{16}	+	01011111	Discharge	Discharge
$(15/24)V_{DC}$	$(3/24)V_{DC}$	$(12/24)V_{DC}$	L_{17}	+	01010111	Discharge	Discharge
$(15/24)V_{DC}$	$(2/24)V_{DC}$	$(13/24)V_{DC}$	L_{18}	+	01010011	Discharge	Discharge
$(15/24)V_{DC}$	$(1/24)V_{DC}$	$(14/24)V_{DC}$	L_{19}	+	01010001	Discharge	Discharge
$(15/24)V_{DC}$	0	$(15/24)V_{DC}$	L_{20}	+	01010000	Discharge	Discharge
$(20/24)V_{DC}$	$(4/24)V_{DC}$	$(16/24)V_{DC}$	L_{21}	+	11111111	No effect	No effect
$(20/24)V_{DC}$	$(3/24)V_{DC}$	$(17/24)V_{DC}$	L_{22}	+	11110111	No effect	No effect
$(20/24)V_{DC}$	$(2/24)V_{DC}$	$(18/24)V_{DC}$	L_{23}	+	11110011	No effect	No effect
$(20/24)V_{DC}$	$(1/24)V_{DC}$	$(19/24)V_{DC}$	L_{24}	+	11110001	No effect	No effect
$(20/24)V_{DC}$	0	$(20/24)V_{DC}$	L_{25}	+	11110000	No effect	No effect

[#] Direction of current in OEWIM from terminal A_1 to A_2 is considered positive (+) and from A_2 to A_1 is negative (-).

* Switching state '1' denotes switch is ON and '0' denotes switch is OFF.

III. VOLTAGE SPACE PHASOR & CONTROL LOGIC

The three phase voltage equations in term of their pole voltages may be written as;

$$V_{A1A2} = V_{A1O} - V_{A2O},\qquad(1)$$

$$V_{B1B2} = V_{B1O} - V_{B2O},\qquad(2)$$

$$V_{C1C2} = V_{C1O} - V_{C2O},\qquad(3)$$

where V_{A1A2}, V_{B1B2} and V_{C1C2} are the phase voltages of phases A, B and C respectively. Voltage space vector V_S formed by three-phase voltages is given by

$$V_S = V_{A1A2} + V_{B1B2}\, e^{j(2\pi/3)} + V_{C1C2}\, e^{j(4\pi/3)}\qquad(4)$$

There are total 70 possible switching combinations to generate twenty five different levels in motor phase A. Therefore, a total of 3,43,000 (70 X 70 X 70) switching states are possible for the proposed three phase twenty five-level inverter.

A modified level shifted pulse width modulation (LSPWM) technique is considered here to realize the SVPWM based control strategy. Depending upon the reference speed, the required reference voltage phasor V_S* is generated using constant V/f control. The three reference phase voltages ($V*_{A1A2}$, $V*_{B1B2}$, $V*_{C1C2}$) are generated by relocating the orthogonal components of reference voltage phasor V_S* into the three phase quantity, given by equations as follows;

$$V^*_{A1A2} = V_m^* Sin\omega t\qquad(5)$$

$$V^*_{B1B2} = V_m^* Sin(\omega t - 2\pi/3)\qquad(6)$$

$$V^*_{C1C2} = V_m^* Sin(\omega t + 2\pi/3)\qquad(7)$$

The developed SVPWM model utilizes twenty four level shifted triangular carrier waves (C_1-C_{24}) of magnitude V_C to generate twenty five-levels (L_1-L_{25}) in the phase voltages.

The modulating reference waves are placed at the center of first triangular carrier for lower modulation indices i.e. two-level operation. With the increase in modulation index such that $V_{C2} < V^*_m < V_C$, a DC voltage of magnitude $V_C/2$ is added to the reference wave and the operation is shifted from two-level to three-level operation. In same way, as the inverter operation is shifted from three-level to four-level, then from four-level to five-level and so on, up to twenty five-level operation, DC voltage of magnitude $V_C/2$ is added to the reference wave in each step. Triangular carrier wave are shown in Fig. 2 when motor operation is performed at modulation index 0.2, 0.4, 0.6 and 0.8 corresponding to base frequencies of 10, 20, 30 and 40 Hz respectively. The ripple in capacitor voltages considerably affects the performance of proposed twenty five-level inverter. The capacitors are designed in such a way that the ripple would not exceed the allowable limit.

IV. SALIENT FEATURES

The proposed hybrid twenty five-level inverter topology requires only 48 semiconductor power switching devices unlike to conventional topologies (NPC, FC, CHB) wherein, twenty five-level inverter require 144 switches. The proposed topology needs only 5 isolated DC power supply sources, whereas the conventional CHB twenty five-level inverter requires 36 isolated DC supply sources. The proposed topology employs three capacitors C_{A1}, C_{B1} and C_{C1} of voltage rating $10V_{DC}/24$ and three capacitors C_{A2}, C_{B2} and C_{C2} of voltage rating $5V_{DC}/24$, whereas the conventional FC twenty five-level inverter requires 828 balancing capacitors. The proposed topology completely removes the necessity of 1656 clamping diodes, which are required in the NPC twenty five-level inverter. A comparison of the proposed topology with the other twenty five-level inverter topologies, in terms of the number of components used, is given in Table II. The cost, complexity of the control circuit and conduction losses is reduced with the reduction in number of components.

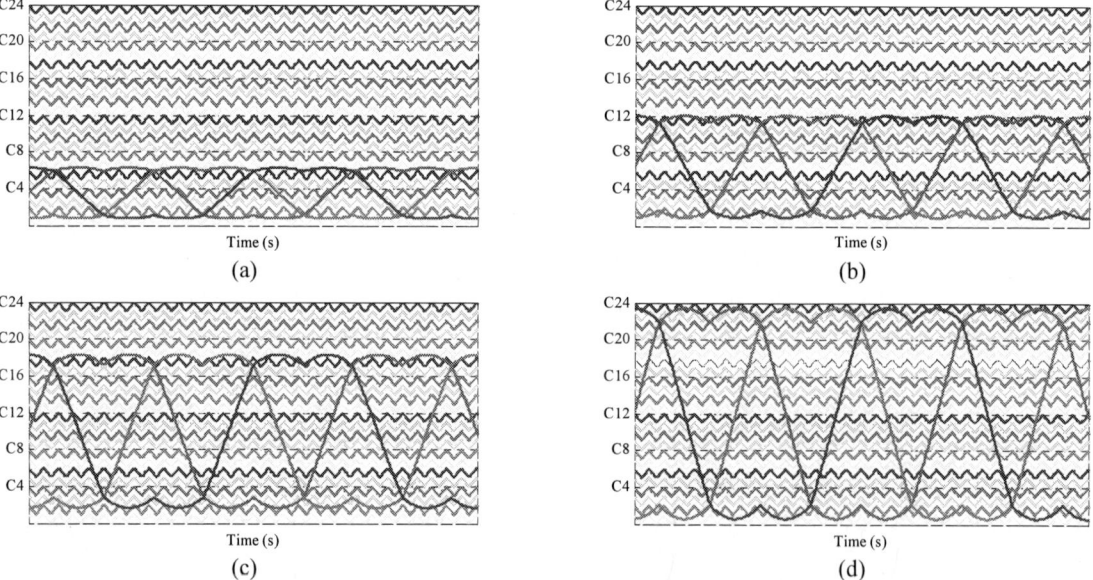

Fig. 2. Triangular carrier wave when motor operated in: (a) eight-level mode $M=0.2$, (b) thirteen-level mode $M=0.4$, (c) twenty-level mode $M=0.6$, (d) twenty five-level mode $M=0.8$

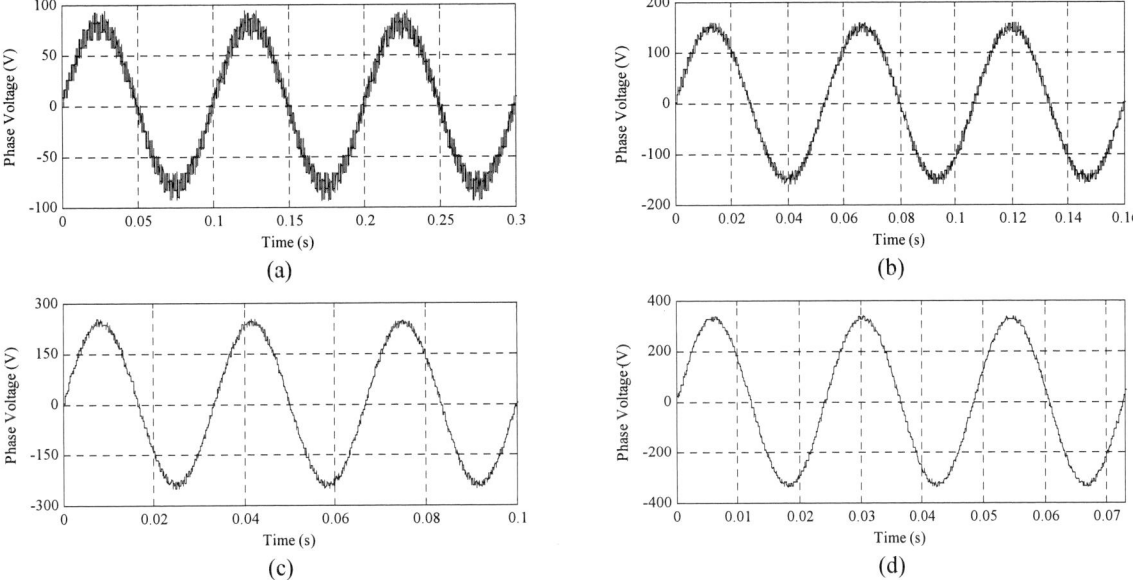

(a)

(b)

(c)

(d)

Fig. 3. Phase *A* voltage (*V_A1A2*) when motor operated in: (a) eight-level mode *M=0.2*, (b) thirteen-level mode *M=0.4*, (c) twenty-level mode *M=0.6*, (d) twenty five-level mode *M=0.8*

In the proposed inverter topology, the switching status of the switches activated at a higher DC-link voltage does not change with the change in direction of current in any level of inverter operation; this results in reduced switching losses and improved efficiency.

TABLE II. COMPARISON OF TWENTY FIVE-LEVEL TOPOLOGIES

Component required	NPC	FC	CHB	Proposed inverter
Main switches	144	144	144	48
Main diodes	144	144	144	48
Clamping diodes	1656	0	0	0
Capacitors	0	828	0	6
DC sources	1	1	36	5

Reliability of the system is key factor of any industrial application. In the proposed topology, if any one of the H-bridge fails, the inverter can still be operated as a fifteen-level inverter in the entire modulation range at full load by directing the current through the devices in the complementary pathway. This feature of the proposed configuration makes the inverter more reliable.

Though, the proposed inverter topology employs five isolated DC power supply sources as compared with the conventional NPC and FC topologies, it prevents the flow of triplen harmonic current (common mode voltage) through the switches and motor stator winding.

V. RESULTS AND DISCUSSION

In simulation DC-link voltage is considered as 600 volts. Inverter is operated for complete modulation range and corresponding waveforms are shown in Fig. 3 for motor phase *A* voltage *V_A1A2*. The modulation index is taken as the ratio of the magnitude of the voltage space phasor (*Vs**) to the maximum magnitude of the voltage (*V_DC*). The

modulation index (*M*) is the ratio of actual reference voltage magnitude to the maximum magnitude of voltage space phasor (*V_DC*). The limit of modulation index (*M*) for linear modulation is 0.866. The motor operation is performed at modulation index 0.2, 0.4, 0.6 and 0.8 corresponding to base frequencies of 10, 20, 30 and 40 Hz respectively and results are shown in Fig. 3.

Fig. 3 (d) shows the waveform of phase *A* voltage (*V_A1A2*) pertaining to twenty five-level mode of operation at modulation index of 0.8. The fundamental frequency in this mode of operation is 40 Hz. It can also be observed from the Fig. 4 that the capacitor voltages (*V_CA1* and *V_CA2*) are well balanced.

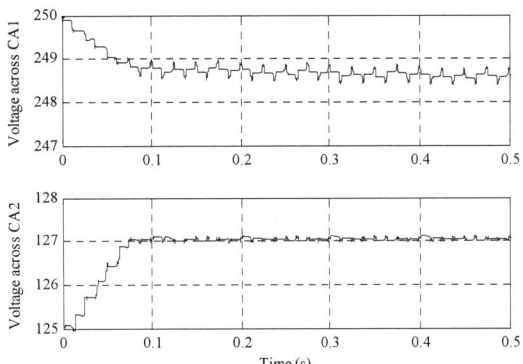

Fig. 4. Phase *A* capacitor voltage (*V_CA1* & *V_CA2*) for twenty five-level operation *M=0.8*

The performance under eight-level, thirteen-level and twenty-level mode of operations corresponding to a modulation index of 0.2, 0.4 and 0.6 respectively are shown in Fig. 3 (a), (b) and (c). In eight, thirteen and twenty-level mode of operations, the fundamental frequency of the motor-phase voltage is attained as 10, 20 and 30 Hz respectively. The switching frequency is kept at 1 kHz for entire range of speed and allowable ripple in capacitor voltages is 4 volt for motor peak load current 8 A. It is also apparent that the capacitor voltages (V_{CA1} and V_{CA2}) are well balanced and peak to peak ripple is less than 2 volt in all the operating conditions.

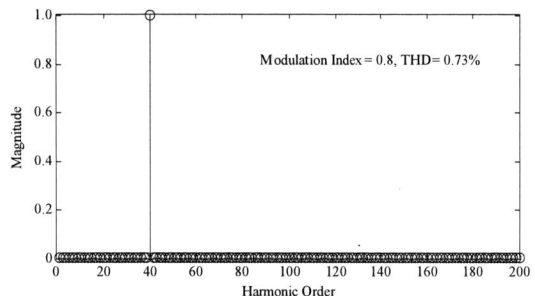

Fig. 5. Normalized harmonic spectrum of phase *A* voltage for twenty five-level operation *M=0.8*

The normalized harmonic spectrum of phase *A* voltage for modulation index 0.8 with respect to its fundamental frequency 40 Hz is shown in Fig. 5. The THD of phase A voltage pertaining to eight, thirteen, twenty and twenty five-level mode of operation at modulation index M=0.2, 0.4, 0.6 and 0.8 is found to be 3.22%, 2.11% and 1.27% and 0.73% respectively. It can be seen that the THD reduces as number of levels in phase voltage waveform increases.

VI. CONCLUSION

A hybrid twenty five-level topology for an OEWIM using a hybrid five-level and a symmetrical five-level inverter has been proposed in this paper. The hybrid five-level inverter is obtained by cascading of a FC inverter with capacitor fed H-bridges and the symmetrical five-level inverter is achieved by cascading of four conventional 2-level inverters. The proposed inverter topology requires lesser components as compared to conventional twenty five-level MLI topologies. The proposed topology reduces the switches losses results in improved overall efficiency of the system. It is pertinent to mention that if any of the H-bridges fails, it can be operated in lower level mode by bypassing the H-bridges in the system; thus the reliability of the system get enhanced. The proposed topology is capable to eliminates the triplen harmonic current in motor stator winding as well as switching devices as both the inverters are connected by isolated DC power supply sources. In this proposed topology, the capacitor voltages are well balanced and peak to peak ripple is within the range of allowable limit in all operating conditions. It is obvious and also achieved that the THD reduces as number of levels in phase voltage waveform increases. The simulation results indicate the potential of the proposed inverter topology for high voltage and high power drive applications.

REFERENCES

[1] L. M. Tolbert and F. Z. Peng, "Multilevel Converters for Large Electric Drives," *IEEE Transactions on Industry Applications*, vol. 35, no. 1, pp. 36–44, 1999.

[2] J. Rodriguez, L. Jih-Sheng, and F. Z. Peng, "Multilevel Inverters : A Survey of Topologies , Controls , and Applications," *IEEE Transactions on Industrial Electronics*, vol. 49, no. 4, pp. 724–738, 2002.

[3] B. K. Bose, "Multi-Level Converters," in *International Conference on Power Electronics*, 2015, vol. 4, pp. 582–585.

[4] A. Nabae, I. Takahashi, and H. Akagi, "A New Neutral Point Clamped PWM Inverter," *IEEE Transactions on Industry Applications*, vol. IA-17, no. 5, pp. 518–523, 1981.

[5] M. Malinowski, K. Gopakumar, J. Rodriguez, and M. A. Pérez, "A Survey on Cascaded Multilevel Inverters," *IEEE Transactions on Industrial Electronics*, vol. 57, no. 7, pp. 2197–2206, 2010.

[6] J. Rodríguez, S. Bernet, B. Wu, J. O. Pontt, and S. Kouro, "Multilevel Voltage-Source-Converter Topologies for Industrial Medium-Voltage Drives," *IEEE Transactions on Industrial Electronics*, vol. 54, no. 6, pp. 2930–2945, 2007.

[7] T. A. Meynard and H. Foch, "Multi-level Conversion: High Voltage Choppers and Voltage-Source Inverters," in *23rd Annual IEEE Power Electronics Specialists Conference (PESC,'92)*, 1992, pp. 397–403.

[8] J. A. Alves, G. Da Cunha, and P. Torr, "Medium Voltage Industrial Variable Speed Drives," in *International Power Electronics Conference*, 2014, pp. 3476–3481.

[9] S. Kouro, M. Malinowski, K. Gopalkumar, J. Pou, L. G. Franquelo, and B. Wu, "Recent Advances and Industrial Applications of Multilevel Converters," *IEEE Transactions on Industrial Electronics*, vol. 57, no. 8, pp. 2553–2580, 2010.

[10] N. P. Schibli, T. Nguyen, and A. C. Rufer, "A Three-Phase Multilevel Converter for High-Power Induction Motors," *IEEE Transactions on Power Electronics*, vol. 13, no. 5, pp. 978–986, 1998.

[11] S. Tamai, "High Power Converter Technologies for Saving and Sustaining Energy," in *IEEE International Symposium on Power Semiconductor Devices & IC's (ISPSD)*, 2014, pp. 12–18.

[12] P. K. Steimer and M. D. Manjrekar, "Practical Medium Voltage Converter Topologies for High Power Applications," in *IEEE Industry Applications Conference*, 2001, vol. 3, pp. 1723–1730.

[13] H. Stemmler and P. Guggenbach, "Configurations of High-Power Voltage Source Inverter Drives," in *European Conference on Power Electronics and Applications*, 1993, pp. 7–14.

[14] E. G. Shivakumar, K. Gopakumar, S. K. Sinha, A. Pittet, and V. T. Ranganathan, "Space vector PWM control of dual inverter fed open-end winding induction motor drive," in *Sixteenth Annual IEEE Applied Power Electronics Conference and Exposition (APEC)*, 2001, vol. 1, pp. 399–405.

[15] S. Kumar and P. Agarwal, "A Novel Eighteen-Level Inverter for an Open-End Winding Induction Motor," in *IEEE India International Conference on Power Electronics (IICPE)*, 2014, pp. 1–6.

[16] P. P. Rajeevan, K. Sivakumar, C. Patel, R. Ramchand, and K. Gopakumar, "A Seven-Level Inverter Topology for Induction Motor Drive Using Two-Level Inverters and Floating Capacitor Fed H-Bridges," *IEEE Transactions on Power Electronics*, vol. 26, no. 6, pp. 1733–1740, 2011.

[17] S. Kumar and P. Agarwal, "A Hybrid Nine-level Inverter Topology for an Open- end Stator Winding Induction Motor," *Taylor & Francis "Electric Power Components and Systems,"* vol. 44(16), no. 1532–5008, pp. 1801–1814, 2016.

[18] P. P. Rajeevan, K. Sivakumar, K. Gopakumar, C. Patel, and H. Abu-Rub, "A nine-level inverter topology for medium-voltage induction motor drive with open-end stator winding," *IEEE Transactions on Industrial Electronics*, vol. 60, no. 9, pp. 3627–3636, 2013.

2nd IEEE International conference on power Electronics, Intelligent Control and Energy systems (ICPEICES-2018)

Effects of Neutral Conductor on Induction Motor Steady-State Performance Under Loss of One Phase of Supply Voltages

Pichai Aree
Department of Electrical Engineering,
Thammasat University, Pathumthani,
Thailand, apichai@engr.tu.ac.th

Abstract—**Induction motors are widely used in industry. Their steady-state performances are significantly influenced by voltage unbalance. With advent of smart grid concept, there is a reliability benefit that could be achieved through implementations of single-phase tripping. In this case, it is possible that industrial induction motors can experience a single phasing, where one of their three-phase voltages is lost. Hence, this paper presents an experimental investigation of induction motor's performance under an occurrence of phase loss. Effects of neutral wire that connects the star points of supply source and motor together on steady-state torque and current characteristics are investigated. The study results show that the neutral conductor plays an important role in keep the motor running in two-phase operation when one phase of a three-phase voltage is lost. Hence, maintaining the motor output torque and reducing in pulsation torque.**

Keywords—*induction motor performance, single-phasing, neutral lead, speed-torque curve, speed-current curve.*

I. INTRODUCTION

Induction motor is the horse power of the industry because its cost and maintenance are cheap [1]. It comes from a fractional horse power up to thousands of horse power. The majority of induction motors are designed for three-phase balanced supply. Often, the system voltages differ excessively from the nominal value. The motor steady-state performances will be degraded [2-3]. The unbalanced voltages results in unbalanced current in the motor stator windings, causing overheating. Increased temperature leads to a reduction in motor's lifetime.

A phase unbalance may be caused by unstable utility supply, unbalanced transformer bank, blown fuse, thermal overload, broken wire, worn contact, or mechanical failure [4-8]. Moreover, single phasing (phase loss) can be caused by an interrupting a fault of automatic circuit reclosers. These devices are widely used by utilities on distribution network. With an advent of smart grid concept for reliability improvement, there is an increasing tendency to interrupt and reclose only lines that are faulted to ensure that the non-faulted phase remains intact [9-10]. In this situation, the induction motor load will experience single phasing.

It is generally known that three-phase induction motors will continue to operate when one of the three-phase supply voltages is lost. At this time, the motors remarkably behave like a single-phase motor because their star points are not connected to the star point of the power supply (floating) via the neutral lead. The motor's torque is rapidly fallen. The motors may be damaged due to overheating [11]. Often,

running the motors with two remaining phase voltages is greatly needed at least for some more time because very critical industrial process cannot be simply interrupted due to damage. To improve the motor's performance, this paper demonstrates an importance of the neutral lead that can be immediately switched to connect the star points of the supply source and motor when one phase of the supply voltages is lost. With the neutral wire in service, the motor will fully work as two-phase motor, in which the remaining torque can be improved, keeping it running without rapid stopping. To investigate the motor performances, the torque versus speed with and without the neutral load are measured and compared with those of single phasing operation. The discussions are made to point out a need of the neutral lead in providing a current return path for improving the motor performance.

II. TEST SYSTEM

To investigate motor's steady-state performance under loss of one phase of supply voltages, the test system in Fig. 1 is used.

Fig. 1. Test system

The system consists of Y-connected induction motor, whose nameplate is shown in Appendix A. It is rated 2.2kW, 4.7A, 1420rpm. Its shaft is mounted with torque sensor rated 200Nm and speed sensor. The input currents of motor are detected using the hall-effect sensors. An amount of motor output torque can be adjusted by controlling the amount of force applied to the break lever. The brake lever produces the input force needed to push hydraulic brake fluid to the slave cylinder or caliper and causes the brake pads to clamp the rotor disk. During the experiment, the motor speed is initially started with direct-on-line method. Then, the force is applied by the brake lever to control the motor from no-load speed until it is at rest. Torque, speed,

978-1-5386-6624-1/18 $31.00 © 2018 IEEE

and current signals are then delivered to data acquisition (DAQ) card, interfaced with LabVIEW software. In the experiment, these signals are recorded with sampling frequency of 20k samples/sec. The noise reductions in these recorded signals are performed by the smoothing operation using moving average filters.

III. STEADY-STATE PERFORMANCES OF INDUCTION MOTOR

In this section, the steady-state performances of induction motor are discussed. Before exploring the torque-speed curve under phase loss, let us exam its torque characteristics under full voltages. In this case, the tested motor is connected to the power supply voltage according to the schematic diagram as shown in Fig. 2a and 2b. In Fig. 2a, the three-phase motor is connected in wye with floating neutral.

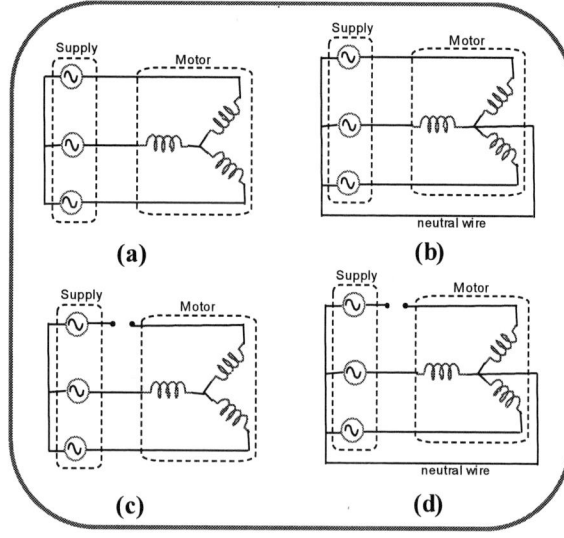

(a) **(b)** **(c)** **(d)**

Fig. 2. Circuit diagram of the motor under test

After conducting the experiments in the laboratory, the torque-speed curves are obtained as plotted in Fig. 3, respectively. It can be seen that the torque-speed curves are quite linear in the speed region ranging from no load to full load. The zoomed in version in the torque curves is given in Fig. 4. The corresponding input current is also plotted as displayed in Fig. 5. It can be seen that the motor's torque and current curves under three-phase balanced conditions are relatively closed to each other regardless of whether the neutral cable is connected or disconnected. It is apparent that the motor consumes current at no load about 3.6A. As speed drops, the current is increased. The current reaches the rated value (4.7A) around 1449rpm. So, this speed may be regarded as lowest (rated) speed, in which the motor can drive its mechanical load safely and continuously. The motor cannot be run at 1420rpm as stated in the nameplate since the input current exceeds limits. However, it is found that the motor only produces net output torque of 12.7Nm (Fig. 4). So, the actual kW output is approximately 1.93kW, which is less than the specified value as given from the nameplate (2.2kW) by 0.27kW.

Next, the investigation is moved to the case where one of the supply voltages (phase C) is lost as shown in Fig. 2c. The motor then continues running with the remaining two phase

(A and B) voltages. In the experiment, the motor is initially set up to start with three-phase supply voltages. It can be seen from Fig. 5 that when one of the supply voltages is suddenly removed, the no-load current of the remaining phase A and B equally jumps up to 4.7A. The motor takes nearby 6.3A at 1449rpm, which is 1.34 times greater than the rated value. Referred to Fig. 4, it is evident that the motor shaft torque is significantly reduced from 12.7Nm to 8.6Nm. It becomes 0.68 times (32% less) as compared to the case of full running three-phase voltages. If overload devices permit up to 125 percent overload for the full load current, the maximum output torque that motor can deliver is about 58% (7.4Nm) of its rating at 1457rpm.

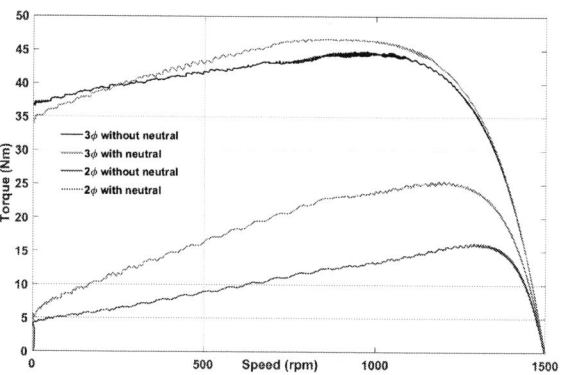

Fig. 3. Torque-speed curves

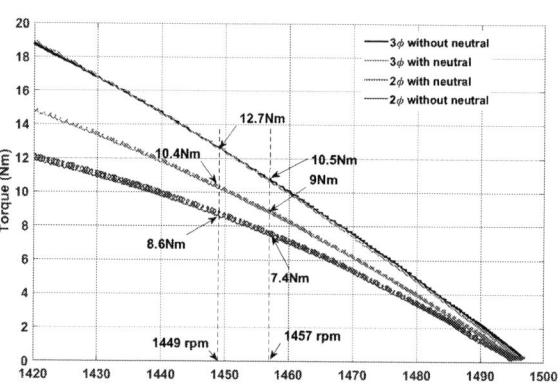

Fig. 4. Zoomed version of torque-speed curve in Fig. 3

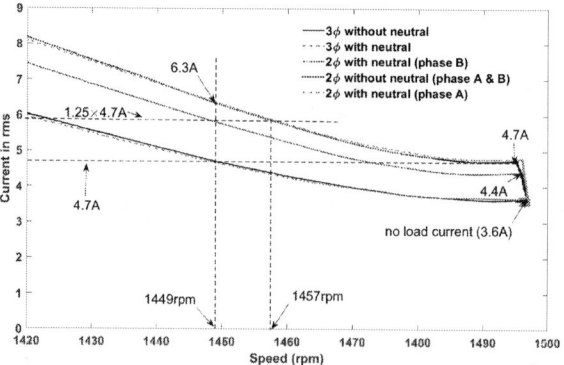

Fig. 5. Current-speed curves

The effect of neutral wire on steady-state torque performance is demonstrated by joining the star points between motor and supply source as indicated by Fig. 2d. As compared to the previous case, the measured current in rms in Fig. 5 reveals that the no load currents between phase A and B are no longer equal. One of them is reduced, while the other is almost the same as previous case. If the same degree of overcurrent (125%) is allowed, the corresponding output torque (at 5.8A) is about 9Nm, which becomes 0.71 times, i.e. only 29% decreases from the rated value (12.7Nm). Therefore, the motor output torque is improved when the neutral conductor is introduced to serve as a return path for the motor current. Utilization of neutral conductor during phase loss increases the motor's capability to handle the mechanical load.

It is interesting to explore the torque waveform of motor under phase loss. The tested motor is initially run as three-phase operation. The recorded torque signal is plotted against time as shown in Fig. 6. It can be seen that there is a small magnitude in torque pulsations when the motor runs under three-phase voltages. When one phase of the supply voltage is lost, a great increases in the undesired pulsating torque due to unbalanced current is observed if the star points between the motor and the supply source is not joined. The motor surely experiences vibrations and mechanical stresses. However, if the star points is joined by the neutral wire, the significant reduction in torque ripple can be noticed as shown in Fig. 6. It should be noted that the frequency signature being observed under loss of one phase is 250Hz.

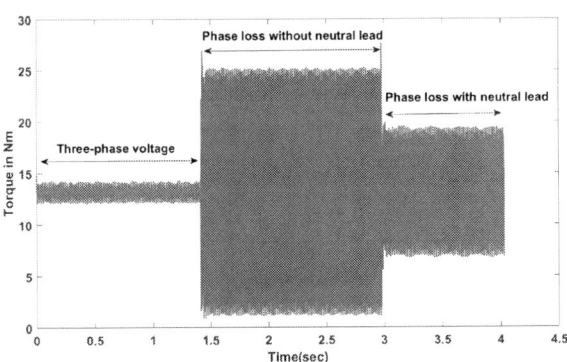

Fig. 6. Measured torque waveform of induction motor with and without neutral wires.

The torque improvement under phase loss due to making use of the neutral conductor can be explained from the waveforms of motor input current. The plots of currents with and without neutral wire are shown in Fig. 7 and 8. It is apparent that the three-phase currents are not equal in magnitudes. When the phase C voltage is cut off (point a), no current and power is flowing through it. The currents in the remaining phase A and B are immediately out of phase for the case where the motor star point is floating. This result confirms that the motor runs in single phase condition with line-to-line voltage across two windings. That is why a dramatic fall in the running torque is observed through Fig. 4. Moreover, because the motor works in single phasing manner, it fails to develop the starting torque as seen from a

rapid fall of the torque curve as the motor speed nearly approaches zero (Fig. 3). On the other hand, if the neutral conductor is employed, the phase displacement of motor's currents in phases A and B are maintained at 120 degrees apart as displayed in Fig. 8. Certainly, the motor works under two phase conditions. As expected, a significant improvement in the running torque is fully obtained as confirmed via Fig. 4.

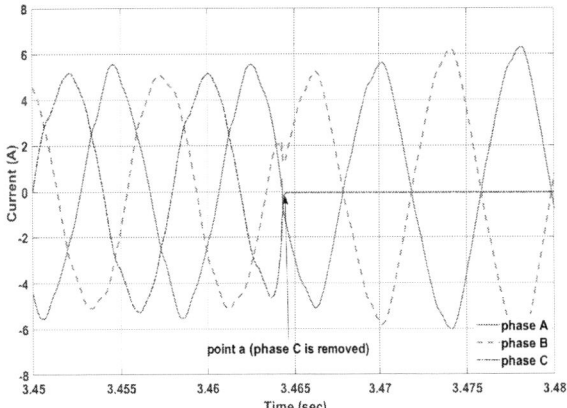

Fig. 7. Current waveforms under loss of one phase without neutral lead

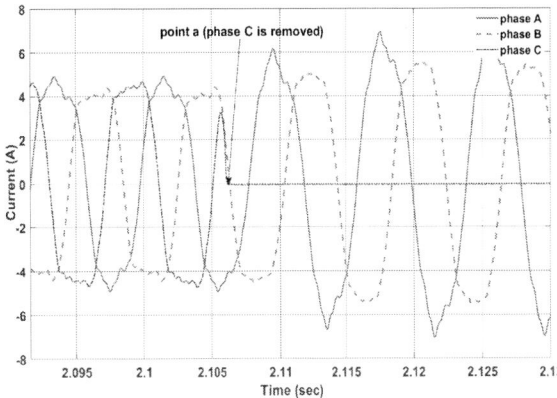

Fig. 8. Current waveforms under loss of one phase without neutral lead

The effect of neutral current is further studied. Commonly, three-phase wye-connected induction motor doesn't utilize the neutral conductor because there is no net neutral current. Another possible reason for absence of neutral wire is get rid of unbalance current from the motor winding. However, when one of the three-phase voltages is suddenly lost, the neutral lead may be switched in service, resulting in a higher torque production to continue driving the mechanical load. In this case, the neutral current will flow. According to the experimental result in Fig. 9, it is surprisingly found that the magnitude of neutral current is more than that of the phase currents. Thus, the minimum allowable size neutral conductor must have sufficient ampere capacity to carry the maximum unbalanced current.

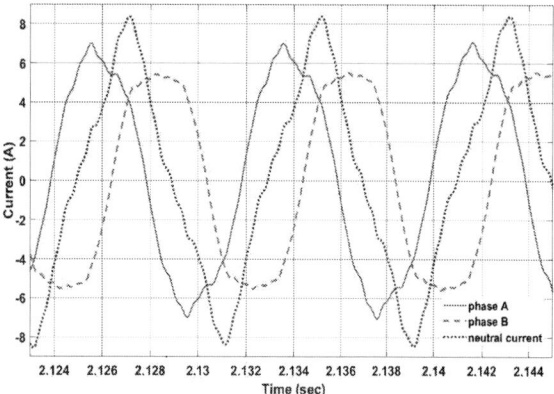

Fig. 9. Current waveform under loss of one phase without neutral lead

IV. CONCLUSION

In this paper, the torque-speed and current-speed curves of three-phase induction motor are measured under steady-state conditions to highlight a need of connecting the neutral conductor in service when the phase loss occurs. It is found that the neutral lead connected between the star points of the motor and power supply is necessary to prevent the motor running in the single-phasing but rather two-phasing conditions when one of the phase voltages is lost. The measurement results confirm that the two-phase operation gives much higher torque production and reduces unwanted pulsating torque in comparison with the single-phasing operation. Thus, utilization of neutral conductor during phase loss may be considered for increasing the motor's torque capability to continue driving the critical industrial process for a short period of time.

V. APPENDIX A

Appendix A provides a detail of the motor nameplate shown in Fig. 1.

Kw	V	I	PF	N	Pole	f (Hz)
2.2	400	4.7A	0.82	1420	4	50

REFERENCES

[1] A. Siddique, G. S. Yadava, B. Singh, "A review of stator fault monitoring techniques for induction motors", *IEEE Trans. Energy Convers.*, Vol. 20, pp. l06-114, 2005.

[2] P. Pillay, P. Hoffmann and M. Manyage, "Derating of induction motors operating with a combination of unbalanced voltages and over or under voltages", IEEE Trans. on Energy Conversion, Vol. 17, No. 4, pp. 485-491, 2002.

[3] J. Faiz, H. Ebrahimpour, and P. Pillay, "Influence of unbalanced voltage on the steady-state performance of a three-phase squirrel-cage induction motor," IEEE Trans. Energy Convers., Vol. 19, No. 4, pp. 657–662, 2004.

[4] W. H. Kersting, "Causes and effects of unbalanced voltages serving an induction motor," *IEEE Trans. Ind. Appl.*, Vol. 37, pp. 165–170, 2001.

[5] R. A. Walling, R. Saint, R. C. Dugan, J. Burke, and L. A. Kojovic, "Summary of distributed resources impact on power delivery systems", *IEEE Trans. Power Del.*, Vol. 23, pp.1636–1644, 2008.

[6] P.E. Sutherland, T.A. Short, "Effect of single-phase reclosing on industrial loads", 2006 IEEE IAS Annual Meeting, Vol. 5, pp. 2636-2644, Tampa, Florida, Oct. 2006

[7] M. Sudha and P. Anbalagan, "A novel protecting method for induction motor against faults due to voltage unbalance and single phasing", *33rd Annual conference of the IEEE*, pp.1144-1148, 2007.

[8] W. H. Kersting, "Causes and effects of single-phasing induction motors", *IEEE Trans. Ind. Appl.*, Vol. 41, pp. 1499-1505, 2005.

[9] E. Agamloh, S. Peele, and J. Grappe, "Induction motor single-phasing performance under distribution feeder recloser operations," IEEE Trans. Ind. Appl., Vol. 50, No. 2, pp. 1568–1576, Mar./Apr. 2014.

[10] E. Agamloh, S. Peele, and J. Grappe, "Response of motor thermal overload relays and phase monitors to power quality events," IEEE Trans. Ind. Appl., Vol. 52, No. 6, pp. 5336–5344, Nov./Dec. 2016.

[11] F. J. T. E. Ferreira, and A. M. Silva, "Single-phase protection of line-operated motor of different effficiency classes," IEEE Trans. Ind. Appl., vol. 54, no. 3, pp. 2071–2084, May./June. 2018

2nd IEEE International conference on power Electronics, Intelligent Control and Energy systems (ICPEICES-2018)

Novel and Robust Hysteresis Current Control Strategies For a BLDC Motor : A Simulation Study and Inverter Design

Shrivatsan K Chari
Department of Elecctrical Engineering
Delhi Technological University New
Delhi, India
shrivatsan3@gmail.com

Rishav Dhiman
Department of Electrical Engineering
Delhi Technological University,
New Delhi, India
dhiman.rishav97@gmail.com

Rohit Saxena
Department of Elecctrical Engineering
Delhi Technological University, New
Delhi, India
rohit.saxena17.1.96@gmail.com

Abstract – **The major objective of the present study is to control the torque of the BLDC motor through hysteresis current control for electric vehicle applications. In this regard, we have proposed two strategies. In the case of small machines due to their low inductance values, a buck-boost circuit is implemented using a three-phase inverter configuration with a novel sensing algorithm. Speed control is implemented using a PI controller which generates the reference current signal. For machines with higher inductance value, a novel topology has been suggested to control the current using a single current sensor, thereby decreasing sensor costs by one third. Also the strategies presented are robust as they are minimally dependent on motor parameters.**

Keywords— BLDC motor, Buck-Boost circuits, Hysteresis current control, torque control, speed control, PI controller, Current sensor

I. INTRODUCTION

Products powered by alternative energy are being thrust into the market because of the ill effects of non-renewable energy sources. Due to latest developments in Electrical Engineering technologies, reliability and sustainability of renewable sources such as solar power have improved leaps and bounds [1]. Electrical Vehicles have no gaseous emissions and when combined with a grid powered by renewable energy, have a negligible carbon footprint. Hence, there has been an avid interest in electric vehicle research and development to have a greener future. Brushless Direct Current motors are used in low power electric vehicles because of their reliability, efficiency and high power-to-volume ratio [2]. They have similar torque and speed characteristics to a traditional brushed DC motor. Because of the electronic commutation and absence of brushes, BLDC motors require lower maintenance and have longer lifetimes than brushed DC motors. Hall sensor feedback is often employed to govern the commutation. When the accelerator pedal is pressed in a gasoline powered vehicle, reference acceleration is commanded by the driver. To get the same feel and ride in an electric car, it requires torque control. This method has the advantages of being able to scale inclines from a stationary position and start the motor with higher payload (passengers). Also, the ride would be smoother. Hysteresis current control keeps the current within a tolerance band, and the average value of current is used to control the torque [3]. In a BLDC motor, the torque developed is directly proportional to the phase current [4].

II. CONTROL STRATEGIES

We propose two different methods to carry out the intended result. First method is suitable for smaller machines with lower inductances. The second method utilizes reversal of voltage polarity to reduce current in machines with large inductances. Both strategies utilize a 3-phase inverter bridge.

A. For smaller machines

Consider the inverter structure Fig 1.a. The leg which has to be fired is determined through the hall sensor data coming from the motor. At a time only one MOSFET of a leg conducts otherwise it would lead to short circuiting of power supply. The phase through which current enters the motor is connected through the MOSFETs in the upper row and the phase through which current exits is connected through the MOSFETs in the bottom row.

Boost mode

In Fig 1.a, let the phases which are conducting be R and Y. In order to operate the inverter in Boost mode, it follows from switching scheme that MOSFET 1 and 4 conduct. This leads to an increase in motor current with the inductance of the motor acting as inductor present in a boost circuit. Hence the conducting sequence is:

Battery → mosfet 1 → motor → mosfet 4 → Battery

Buck mode

In Fig 1.b, in order to decrease the current, we switch OFF MOSFET 1 and let the current decrease by free-wheeling through the internal diode of MOSFET 2. Hence the conducting sequence is:

Internal diode 2 → motor → mosfet 4 → internal diode 2

Current Sensing

In our approach, we were keen to compute a single value which would be compared to the reference current irrespective of which phase is conducting. For the same reason, we arrived at the following formulation:-

$$I = \frac{|I_R| + |I_Y| + |I_B|}{2} \quad (1)$$

978-1-5386-6624-1/18 $31.00 © 2018 IEEE   841

Let the motor be operating in boost mode as in Fig 1.a. This would mean that phase B current, $I_B = 0$. R phase current, I_R and Y phase current, I_Y are opposite to each other because the current enters through phase R and leaves through phase Y. Hence, sum of the absolute values of I_R and I_Y will give twice the DC link current. When finally divided by 2, we arrive at our intended current value to be sent into the relay. Same line of reasoning applies to buck mode as well.

B. For large machines

Large machines have higher inductances than their smaller counterpart. Hence the current may not decrease fast enough in the free-wheeling mode as desired for hysteresis control. We utilized a strategy where the boost mode remains same as in the case of small machines (Fig 1.a) and buck mode is replaced by a technique where the source voltage polarity is reversed to drive down the current (Fig 1.c). One significant advantage of this implementation is the requirement of only one current sensor, thereby decreasing sensor costs by order of one-third. However, this method would lead to more switching losses [5].

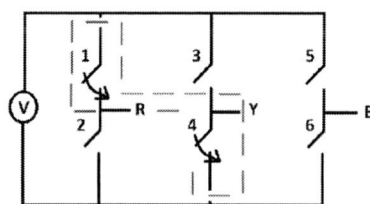

Figure 1.a. Boost mode operation

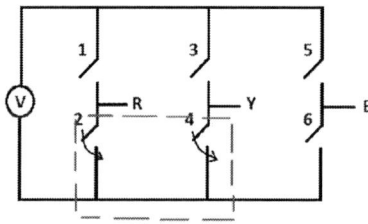

Figure 1.b. Buck mode operation

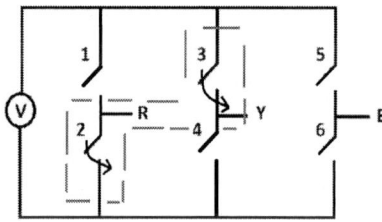

Figure 1.c Voltage polarity reversal for large machines

III. SIMULATION

For the purpose of simulation, we used Simulink software developed by MathWorks. The test setup was based on the block diagram given in Fig 2.a and Fig 2.b. As explained in the next section, hall sensor feedback decides

which leg to switch. A relay is used to compare the phase current (DC link current for Fig 2.b) to a reference current. The relay then actuates the switches in upper and bottom row accordingly to increase or decrease the current.

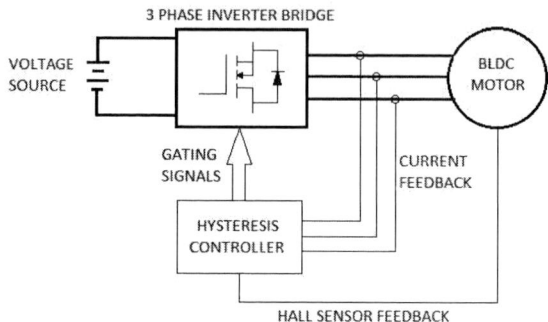

Figure 2.a. Control topology block diagram for small machines

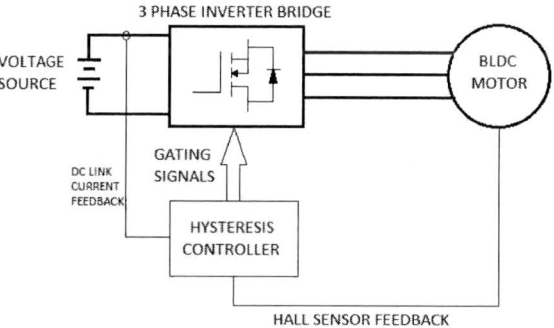

Figure 2.b. Control topology block diagram for large machines

We have used a Permanent Magnet Synchronous Machine model to represent our BLDC motor as it has the same construction and the required trapezoidal back electromotive force, EMF profile. The machine parameters were set corresponding to the specifications of the chosen BLDC motor. The exact value of parameters only influences the hardware components to be used, and not the control strategy thereby rendering our methodology robust. Hall sensor gives output as per the position of the rotor. Table 1 shows how the hall sensor output encodes the maximum positive and maximum negative phase back EMF (Fig 3.a). Current enters through the phase which has maximum positive back EMF and exits through the one with maximum negative back EMF. Using this data, we constructed the Switches Section of Table 1 to generate the necessary gating pulses.

A. Simulink modeling for small machines

It is to be noted that the output labels A, B and C shown in simulation is same as labels R, Y and B in Fig.1. Since the rating of our motor is 3300 revolutions per minute, rpm we simulated different reference speeds ranging from low to high. The reference speed is compared to rotor speed to generate a speed error. This error is then sent into the

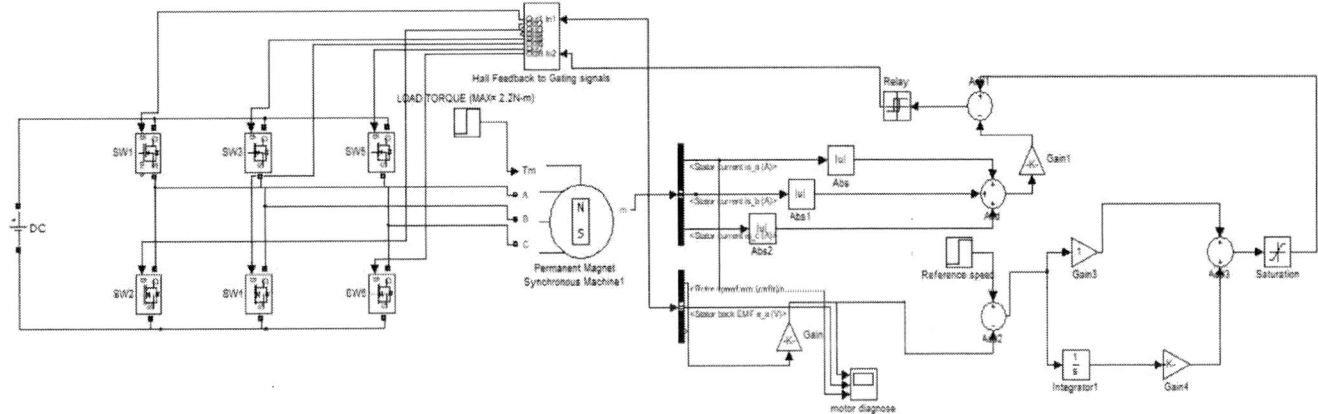

Fig 4.a Simulink model for small machines

proportional integral controller, which produces the current reference. The maximum phase current is limited to 20 Amperes, the rating of our BLDC motor. The reference current is then compared to I (computed from (1)), and their difference is fed to the relay (Fig 3.b). The relay increases phase current by sending gating signals for boost mode operation and decreases the same using buck mode described earlier. Overall, the relay switches such that the phase current is within a band 0.5 Amperes above and below the reference current. The final simulation model of hysteresis current control for small machines has been depicted in Fig 4.a.

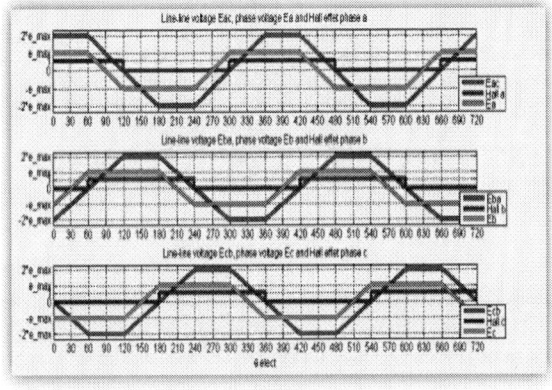

Figure 3.a. back EMF waveforms with Hall sensor outputs. To be read with table 1.a.

HALL OUTPUTS			EMF			SWITCHES					
			E_A	E_B	E_C	1	2	3	4	5	6
0	0	1									
0	1	0	0	-1	+1	0	0	0	1	1	0
0	1	1	-1	+1	0	0	1	1	0	0	0
1	0	0	-1	0	+1	0	1	0	0	1	0
1	0	1	+1	-1	0	1	0	0	1	0	0
1	1	0	0	+1	-1	0	0	1	0	0	1

Table 1.Cell value with +1 indicates the phase with maximum positive back EMF. -1 indicates the phase with maximum negative back EMF. Switches with value are 1 are to be turned on and those with value 0 are to be switched off.

B. Simulink modeling for large machines

In this strategy we measure the DC current and directly compare that to the reference current. From thereon, the relay outputs gating signals on the same basis as described for small machines. The difference comes when current has to be decreased. Voltage polarity to the motor is reversed rather than operating in buck mode. Fig 4.b shows the model.

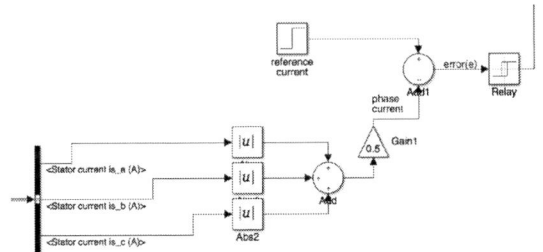

Figure 3.b. current sensing and relay output

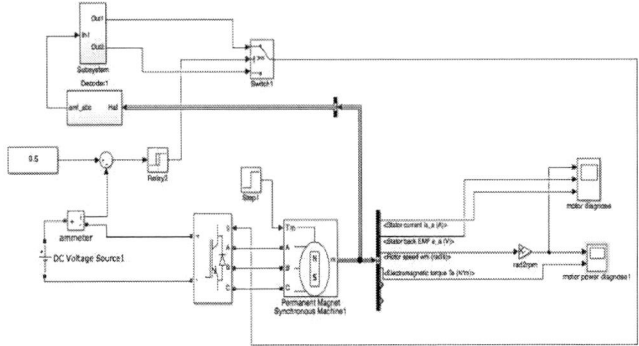

Fig.4.b Simulink model for large machines

978-1-5386-6624-1/18 $31.00 © 2018 IEEE

IV. PERFORMANCE AND RESULTS

A. For small machines

Our model was simulated for two different speed references, i.e., at 2000 rpm (high) and 250rpm (low). In both cases, a load torque of 2.2 Newton-meter was applied. The controller showed good tracking ability as in both instances steady state was reached. The motor each time generated the required electromagnetic torque equal to the load torque, rendering our design as competent. Fig 5 shows the corresponding speed, torque, current and back EMF waveforms for 250 rpm and Fig. 6 is for 2000rpm, with current nicely following a hysteresis profile.

Also, it is essential to focus on the MOSFET switching profile. Fig.7 shows the MOSFET current and Voltage waveforms when the particular switch in a leg is to be actuated (governed by Hall sensor feedback). It depicts the extra power losses that will we have to incur. However, switching frequency of the MOSFET is variable that depends upon the width of the hysteresis band, DC voltage applied and the motor's speed. Also, amount of harmonics on the DC side has increased due to hysteresis current control.

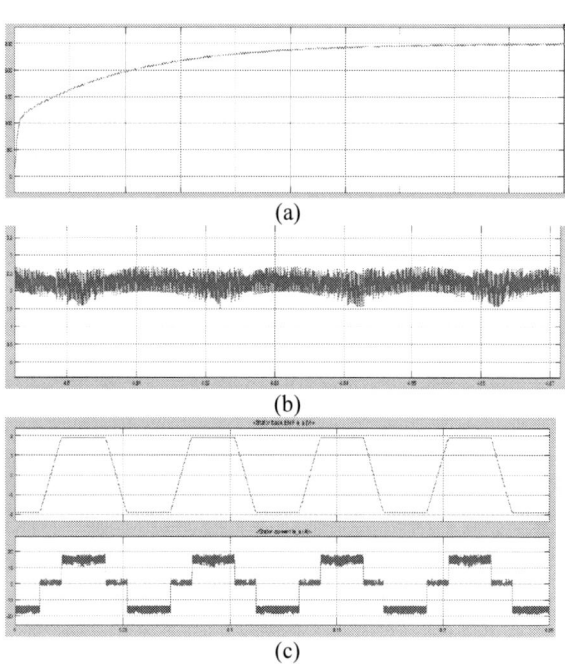

(a)

(b)

(c)

Fig.5
a. Speed profile for 250 rpm reference
b. Steady state electromagnetic Torque
c. Steady state back EMF
d. Steady state phase current

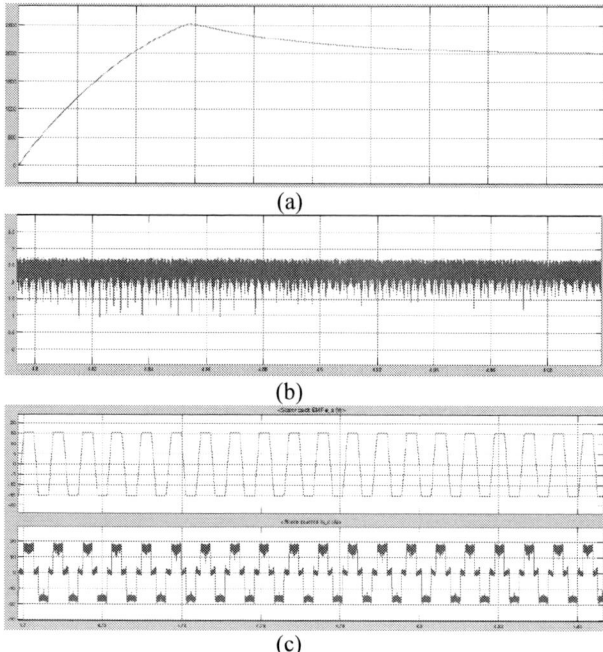

(a)

(b)

(c)

Fig.6
a. Speed profile for 2000 rpm reference
b. Steady state electromagnetic Torque
c. Steady state back EMF
d. Steady state phase current

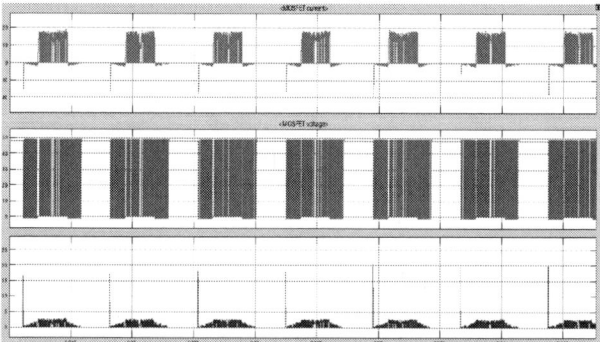

Fig. 7 Mosfet switching characteristics: - the first row scope being current, second being voltage, and third shows power loss.

B. For large machines

For this simulation we applied a load torque of 1 Newton-meter on the motor. Also the tolerance wass set to be 0.25 Amperes above and below the reference current. Satisfactory current waveform and torque control was achieved (Fig.8). Another external loop for speed control can be added as demosntrated in the strategy for small machines.

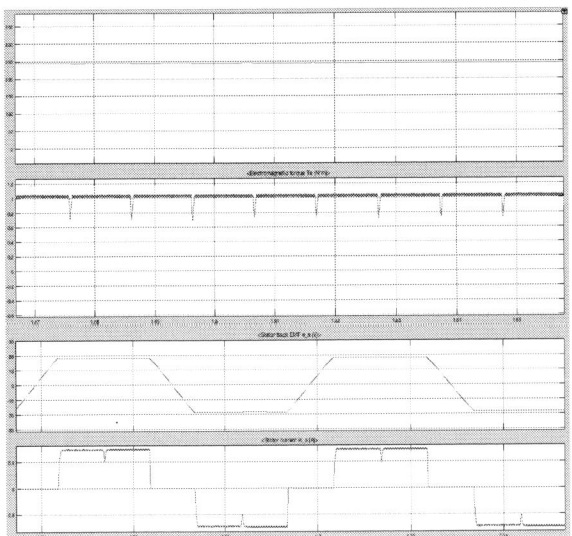

Fig. 8 From top to below :- steady state speed, electromagnetic torque, back EMF and phase current.

V. INVERTER DESIGN

The current sensor and 3-phase Inverter Bridge are essential hardware components for the implementation of our control strategy. Following section provides brief introduction to their design.

A. Current Sensor

Current sensor (suggested: ABB-EL25P1) generates a voltage signal proportional to the current using closed loop hall-effect technology. This voltage is then amplified by an op-amp (OP07) to a desired output voltage level. Fig 9 shows our circuit, and the PCB design for a three phase current sensor.

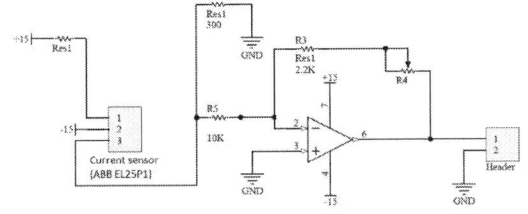

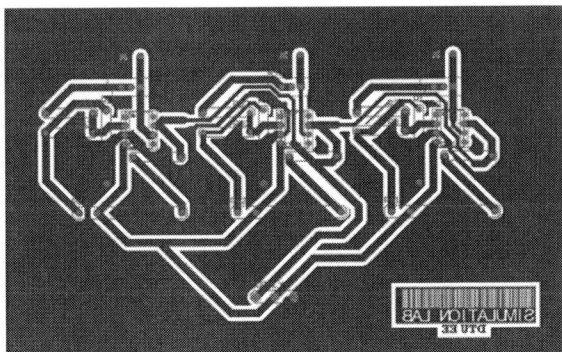

Fig.9 From top to bottom:- Circuit and PCB design for a three phase current sensor.

B. 3-phase Inverter Bridge

For the purpose of this project, a three legged inverter has been designed which uses six MOSFETs. Gating Signals have been provided via the gate Driver ADUM 4223. For MOSFETs, we suggest the use of CSD19531KCS n-channel MOSFETs made by Texas Instruments. The voltage rating of the same is 100V and it can carry a drain current of 100A.

Gate Driver Design

The gate driving circuit is made using ADUM 4223 (in Bootstrapping mode). Bootstrapping allows us to provide sufficient voltage between gate and source for the top row MOSFET, as its source is floating (not connected to ground). The optimum value of the capacitor is calculated using the MOSFET specifications (comes out to be 100 nF). 10K resistors have been used to for current and voltage rate limiting purposes. Fig.10 shows the schematic and PCB design.

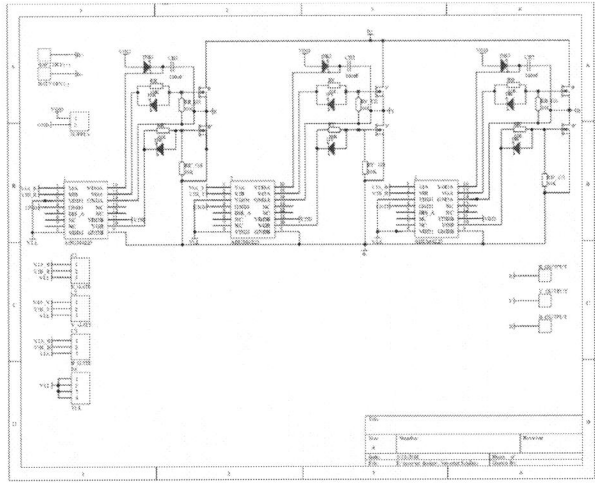

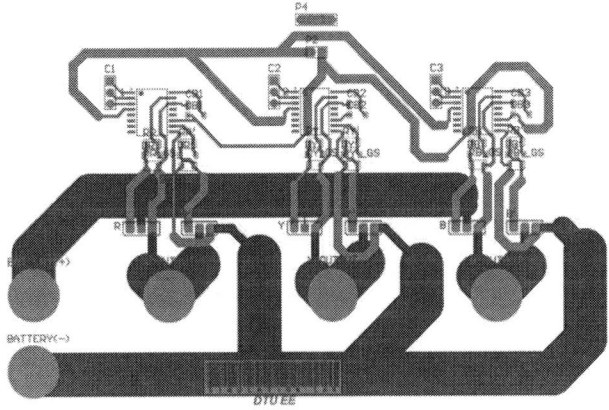

Fig.10 From top to bottom: - Schematic and PCB design

VI. CONCLUSION

The impetus is being given to renewable energy especially for assisting mobility. Naturally electric vehicles have become an active area of research and development. BLDC motors form one of the critical components of the said vehicles and hence designing an appropriate control mechanism needs attention [6]. This relevant because low

978-1-5386-6624-1/18 $31.00 © 2018 IEEE

power vehicles will be used extensively in rural settings and thus any improvement in model design is useful [7]. Keeping this in mind, we have proposed two strategies which make use of hysteresis control mechanism and tested its feasibility using computer-based simulation studies. The control methodology is novel, robust and the cost is comparatively much less than commercially available controllers.

REFERENCES

[1]. S. Faddel, T. Youssef, A. Elsayed and O. Mohammed, "An Automated Charger for Large Scale Adoption of Electric Vehicles," in IEEE Transactions on Transportation Electrification, 2018.

[2]. J.C. Gamazo-Real, E.V. Sanchez, J.G. Gil, "Position and Speed Control of Brushless DC Motors Using Sensor-less Techniques and Application Trends", Sensors 2010.

[3]. Jianwei Zhang, Haitao Yang, Tianshi Wang, Li Li, David G. Dorrell, Dylan Dah-Chuan Lu, "Field-oriented control based on hysteresis band current controller for a permanent magnet synchronous motor driven by a direct matrix converter", IET Power Electronics, 2018, Vol. 11 Iss. 7, pp. 1277-1285.

[4]. K. V. P. Kumar and T. V. Kumar, "Direct torque control of brush less DC motor drive with modified switching algorithm", IEEE International Conference on Power Electronics, Drives and Energy Systems (PEDES), 2016.

[5]. W. Han, T. Y. Lee, Y. J. Kim and S. Y. Jung, "Comparative analysis on efficiency of brushless DC motor considering harmonic component of phase current and iron loss," *2015 18th International Conference on Electrical Machines and Systems* (ICEMS), Pattaya, 2015, pp. 1575-1579.

[6]. A. F. Noor Azam et al., "Current control of BLDC drives for EV application," *2013 IEEE 7th International Power Engineering and Optimization Conference* (PEOCO), Langkawi, 2013, pp. 411-416.

[7]. A. Karthikeyan, K. K. Prabhakaran and C. Nagamani, "Four quadrant operation of direct torque controlled PMSM drive using speed loop PDFF controller," *2018 International Conference on Power, Instrumentation, Control and Computing* (PICC), Thrissur, India, 2018, pp. 1-6.

2nd IEEE International conference on power Electronics, Intelligent Control and Energy systems (ICPEICES-2018)

Multi-Objective Genetic and Adaptive Particle Swarm Optimization Algorithms: A Performance Analysis with Benchmark functions

Sudarshan K.Valluru and Madhusudan Singh

Incubation Center for Control, Dynamical
Systems, Dept. of *Electrical Engineering*
Delhi Technological University
Delhi-110042, India
sudarshan_valluru@dce.ac.in

madhusudan@dce.ac.in

Abstract— The multi-objective genetic algorithm(MOGA) and adaptive particle swarm optimization(APSO) algorithm are new variants of natural population-based search methods with the effective capability to solve extremely nonlinear mixed integer optimized complex engineering problems. This paper attempts to examine the performance and convergence analysis for MOGA and APSO algorithm with benchmarked test functions namely Rosen Brock, Six Hump Camel Back, Goldstein-Price's and Rastrigin. The numerical simulation results indicate the adaptive particle swarm optimization algorithm was able to find the best solutions than a multi-objective genetic algorithm.

Keywords— Multi-Objective Genetic Algorithm; Adaptive PSO Algorithm; Benchmark Functions

I. Introduction

Control Engineers in all robust optimal controller design problems are rapidly accepting bio-inspired optimized algorithm-based optimization; it is common practice to compare several evolutionary search algorithms using a huge test set, particularly when the test comprises function optimization[1]. Though, the efficacy of the search algorithm compared to one more search algorithm cannot be measured by the number of problems that it solves better. In this paper, instead of conventional GA and standard PSO algorithm, new variants of these algorithms such as MOGA and APSO are tested for their efficacy with benchmark functions of Rosen brock function, Six hump Camelback function, Goldstein Price function, and Rastrigin function.

Towards this goal, this paper is organized under the following section headings. Section II presents the design description of the multi objective genetic algorithm. In Section III, design methodologies of adaptive particle swarm optimization are elucidated. Section IV presents the mathematical description about benchmark functions of Rosen brock function, Six hump Camelback function, Goldstein Price function, and Rastrigin function, followed by comparative analysis in Section V. Conclusion is given in section VI.

II. Multi-Objective Genetic Algorithm

Moreover, many variants of GA are used to optimize the complex engineering problems, with different arrangements for chromosoid illustration, ranking, and fitness calculation, selection, crossover, and mutation. For example, self-organization GA highlighted with a cyclic mutation and real coded GA emphasized with a variable crossover and mutation probability. In several control engineering applications, it is common to adopt different control constraints based on a compromise in manifold performance objectives. Fonseca and Fleming have used an approach to develop a population of Pareto-optimal solutions called the Multiobjective genetic algorithm (MOGA)[2]–[4], which is an extension of standard GA. Generally, the MOGA will use for selection, crossover, and mutation which are typical for standard GA as given in pseudo code. The foremost modification between standard GA and a MOGA exist in the assignment of fitness. After fitness has been assigned to individuals, selection can be completed, and genetic operators applied as standard GA. To limit the size of the near Pareto-optimal set of solutions which are non-dominated and which satisfy a set of disparities. An individual j with a set of objective functions $\varphi^j = (\varphi_1^j,...,\varphi_n^j)$ is said to be *non-dominated* if, for a population of N individuals, there are no other individuals $k = 1,..., N, k \neq j$ such that

a) $\varphi_i^k \leq \varphi_i^j$ *forall* $i = 1,..N$ and

b) $\varphi_i^k < \varphi_i^j$ for at least one i

The pseudo steps used in MOGA is as follows.
Step1: Generate a chromosome population of N individuals
Step2: Decrypt chromosomes to attain phenotypes $\mathbf{p}^j$ $\in \mathsf{P}(j = 1,..N)$

Step 3: Estimate index vectors $\varphi^j (j = 1,.., N)$

Step4: Rank individuals and calculate fitness functions $f_j (j = 1,..., N)$

Step 5: Select N individuals based on fitness
Step 6: Perform crossover on selected individuals
Step 7: Perform mutation on some individuals
Step 8: With the new chromosome population, return to **Step2**
The MOGA is ended when M probable points have been initiated.
The MOGA parameters[5], [6] with a size of population is 10, the mutation rate of 0.5 and 1000 iterations are

978-1-5386-6624-1/18 $31.00 © 2018 IEEE

considered. The MOGA were unique category of combinatory optimized search algorithm which needs discretized search space to solve real pronouncement variables of system optimization problems for single objectives as well as multi-objective functions together. The MOGA chooses individuals based on arbitrary selection from the population and comprises of genetic transformations such as mutation, crossover to create a new set of population. Henceforth, at first there is no clue, about the best answer for an optimization problem. Over the continuous generations, the finest among the lot is selected, operators are performed upon on them and appraised till attainment of the global or suboptimal solutions.

III. ADAPTIVE PARTICLE SWARM OPTIMIZATION ALGORITHM

PSO is a population-based metaheuristic method inspired by social behavior and group intelligence of animals and insects such as fish schooling, bird flocking and animal herding in nature. The investigations of the search space starting from some preliminary guess is accomplished by multiple agents known as particles, and the population is called a swarm. Adaptive Particle Swarm Optimization (APSO) algorithm has some comparisons with MOGA, but APSO is much more straightforward since it uses real number of randomness and global communications instead of mutation/crossover. The APSO algorithm[7] is initialized with a group of arbitrary particles in a search space, and then the search for optima begins by a bringing up-to-date generations. Through each iteration, the particles in the search space change their position by using two best values. The first best comes from the most significant fitness it has achieved so far (self-experience) which is stored as *pBest* and the other best is tracked by any particle in the population (social experience) which is called the neighborhood best or *nBest* . When the algorithm finds these two best values, it apprises its velocity and position. A swarm population size contains N particles where each particle $i \in 1, 2, ..., N$ has D-dimensional solution space with velocity and position vectors, such as $V_i = \left[v_i^1, v_i^2, .., v_i^D \right]$ and $X_i = \left[x_i^1, x_i^2, .., x_i^D \right]$ respectively. The velocity and the position of each particle are resetting random vectors with recent updated solution space of dimension d of the particle 'i' respectively. The innovative velocity and position vectors are determined by the equations (1) and (2) respectively.

$$v_i^d = \omega v_i^d + c_1 rand_1^d (pBest_i^d - x_i^d) + c_2 rand_2^d (nBest^d - x_i^d) \quad (1)$$

$$x_i^d = x_i^d + v_i^d \quad (2)$$

Here ω is inertia weighted factor, 'c_1' and 'c_2' be acceleration coefficients and $rand_1^d$ $rand_2^d$ are arbitrarily distributed random numbers generated between 0 and 1. The inertia weight gradually diminishing with iterative generations as per the equation (3)

$$\omega = \omega_{max} - (\omega_{max} - \omega_{min}) \frac{g}{G} \quad (3)$$

Where g and G are generation index and a pre-defined number of maximum generations respectively. In equation (1) $pBest_i$ is the best fitness position for the ith particle and $nBest_i$ is neighborhood best position. For investigating convergence speed and avoiding local optima, the inertia weight has been exchanged with constriction coefficient, and the velocity update in equation(1) can be revised as equations (4) and (5)

$$v_i^d = \chi [v_i^d + c_1 rand_1^d (pBest_i^d - x_i^d) + c_2 rand_2^d (nBest^d - x_i^d)] \quad (4)$$

$$\chi = \frac{2}{|2 - (c_1 + c_2) - \sqrt{(c_1 + c_2)^2 - 4(c_1 + c_2)}|} \quad (5)$$

The parameters chosen in used in the adaptive particle swarm optimizer(APSO)[5], [6] are size of the swarm(N) is 10, dimension(D)is 3 , maximum iterations(t)is 1000, cognitive or individual learning factor(c_1)is 2.0, social learning factor(c_2)is 2.0, and inertia weight factor(ω)is 0.9.

IV. BENCHMARKED TEST FUNCTIONS

For testing, the MOGA and APSO algorithms, the Rosenbrock, Six Hump Camelback, Goldstein-Prices and Rastrigin benchmark functions[8] were chooses to assess the overall computational behavior of algorithms.

The objective function used in these algorithms is $J = (r(n) - y(n))^2 + (max[r] - max[y])^2$ Where, $r(n) = n^{th} input$ and $y(n) = n^{th} output$.

A. Rosenbrock Function

The Rosenbrock function is as well named as banana or valley function and it is common test function for many optimization algorithms. Also, it is unimodal function and its global minimum lying in a narrow parabolic valley. Even though the global minimum is easy to find but finding the convergence is difficult. The function is assumed in equation (6) and shown in Fig.1.

$$f(x) = \sum_{i=1}^{d-1} [100(x_{i+1} - x_i^2)^2 + (x_i - 1)^2] \quad (6)$$

The Rosenbrock function is evaluated on the hypercube $x_i \in [-2.048, 2.048]$, for all i=1…d

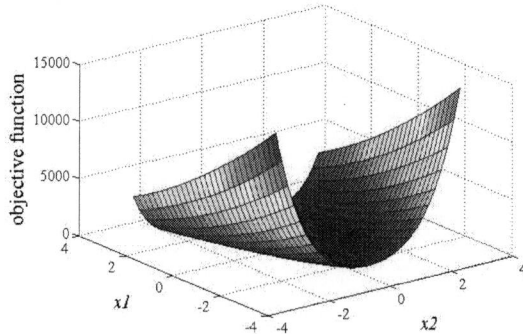

Fig. 1 Rosenbrock Function

978-1-5386-6624-1/18 $31.00 © 2018 IEEE

B. Six Hump Camelback Function

Six Hump camel back function is a two-dimensional multimodal benchmark function which has six minimum points out of which two are the global minima. It is given in equation (7) and shown in Fig.2.

$$f(x_1, x_2) = \left(4 - 2.1x_1^2 + \frac{x_1^4}{3}\right).x_1^2 + x_1.x_2 + (-4 + 4.x_2^2).x_2^2 \tag{7}$$

The Six Hump camel back function is evaluated in the region $-3.0 \leq x_1 \leq 3.0 \quad -2.0 \leq x_2 \leq 2.0$. The global minimum is found at ordinates of $(x_1, x_2) = (0.0898, -0.7126), (-0.0898, 0.7126)$ with the global minima value is -1.03164.

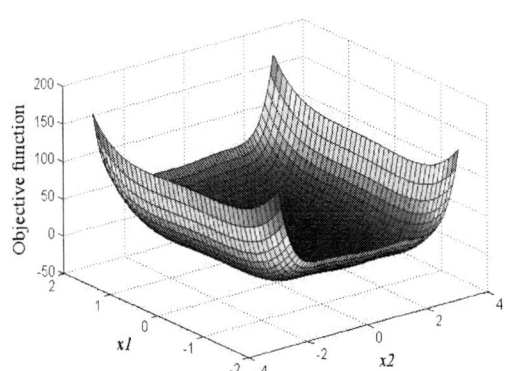

Fig.2 Six Hump Camelback Function

C. Goldstain-Price's Function

The Goldstein price function is a continuous and marginally multi modal function with two variables. The variables are evaluated inside the bounds $-2 \leq x_i \leq 2$. The function is represented in Fig.3 and given in equation (8).

$$f(x_1, x_2) = (1 + (x_1 + x_2 + 1)^2 \times$$
$$(19 - 14x_1 + 3x_1^2 - 14x_2 + 6x_1x_2 + 3x_2^2))$$
$$\times(30 + (2x_1 - 3x_2)^2 \times \tag{8}$$
$$(18 - 32x_1 + 12x_1^2 + 48x_2 - 36x_1x_2 + 27x_2^2))$$

The global minimum is value of $f = 3$ at ordinates of $(x_1, x_2) = (0, -1)$.

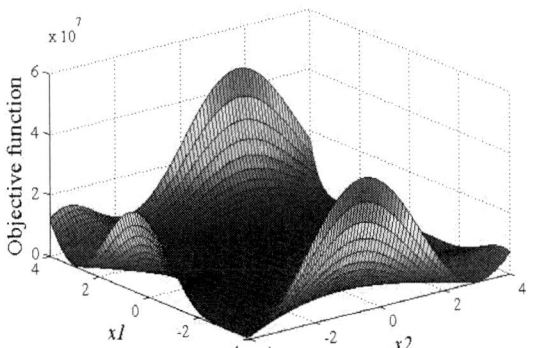

Fig.3 Goldstain-Prices Function

D. Rastrigin Function

The Rastrigin function is extremely a multimodal function with several local minimum points. Though, all the minimum points are scattered evenly. The Rastrigin function is assumed in equation (9) and shown in Fig.4.

$$f(x) = 10d + \sum_{i=1}^{d} [x_i^2 - 10\cos(2\pi x_i)] \tag{9}$$

The function is evaluated in the region $x_i \in [-5.12, 5.12]$ for all i=1,2…d. The global minimum value of objective function $f = 0$ at ordinates $x = (0, 0, \dots .0)$.

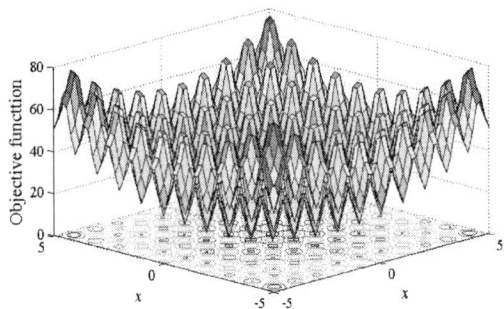

Fig.4.Rastrigin Function

V. ANALYSIS AND RESULTS

The convergence velocity of MOGA with Rosenbrock, Six Hump Camelback, Goldstain-Prices, Rastrigin test functions are shown in Fig.5, Fig.6, Fig.7, and Fig.8 respectively. It is detected that The MOGA is converged at 4th iteration with Rosenbrock, 2nd iteration with Six Hump Camelback, Goldstain-Prices, Rastrigin test functions.

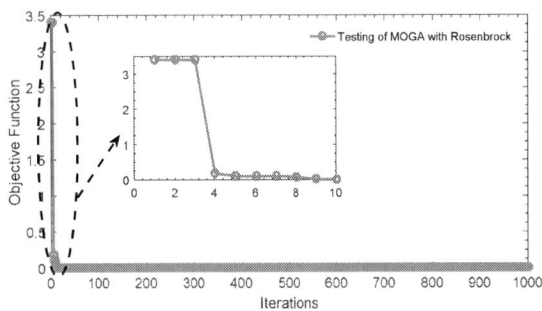

Fig. 5 Convergence Velocity of MOGA with Rosenbrock function

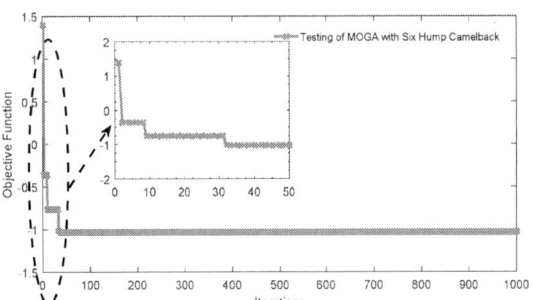

Fig. 6 Convergence Velocity of MOGA with Six Hump Camelback function

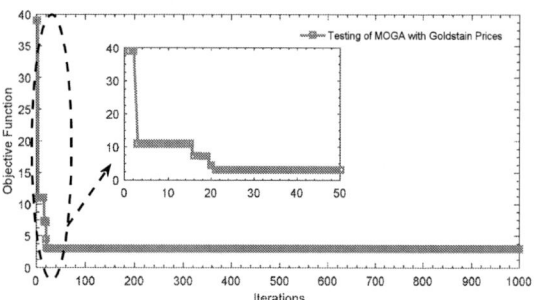

Fig. 7 Convergence Velocity of MOGA with Goldstain function

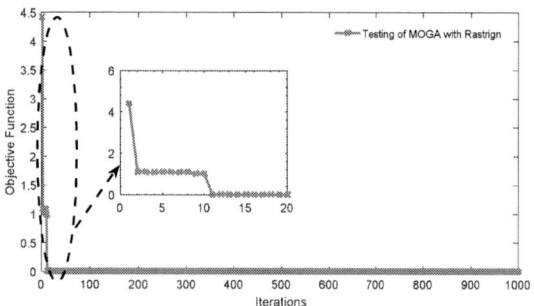

Fig. 8 Convergence Velocity of MOGA with Rastrigin function

The Convergence velocities for APSO with Rosenbrock, Six Hump Camelback, Goldstain-Prices, Rastrigin test functions are shown in Fig.9, Fig.10, Fig.11 and Fig.12 respectively. The APSO converges at 330th iteration for Rosenbrock, 47th Iteration for Six Hump Camelback and 2nd iteration for Goldstain and Rastrigin functions.

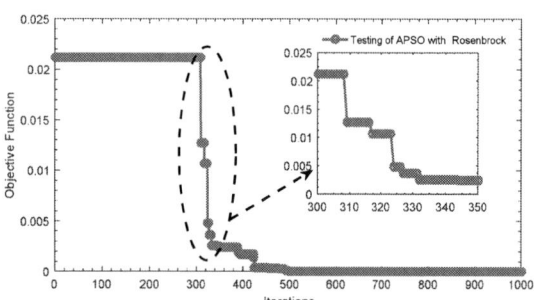

Fig. 9 Convergence Velocity of APSO with Rosenbrock function

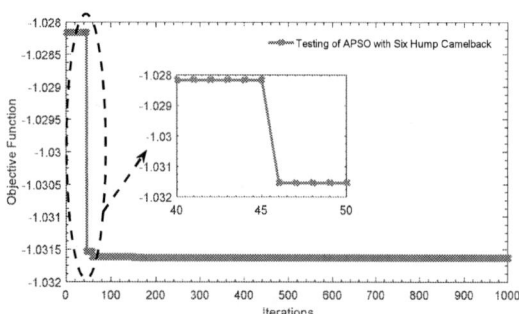

Fig. 10 Convergence Velocity of APSO with Six Hump Camelback function

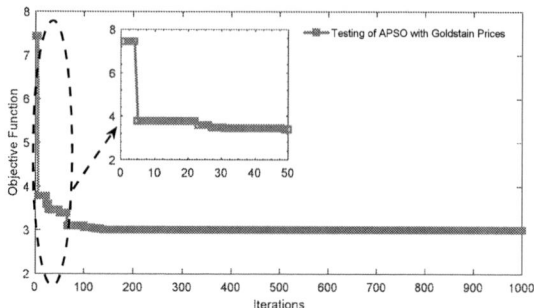

Fig. 11 Convergence Velocity of APSO with Goldstain function

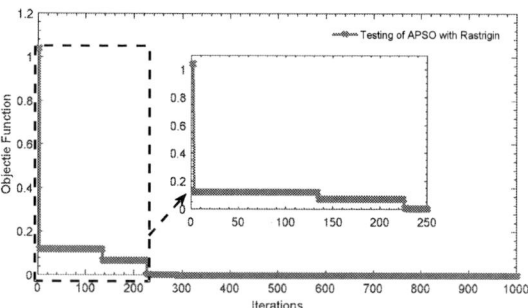

Fig. 12 Convergence Velocity of APSO with Rastrigin function

The solutions for MOGA and APSO are given in Table. I. Even though the APSO converge at more cost of iterations the solution quality of the APSO does not differ than that solution values of bench mark test function values. However, the MOGA has significant difference than bench marked test functions.

TABLE I. SOLUTIONS FOR MOGA AND APSO

Benchmark functions		Rosen brock	Six Hump Camel Back	Goldstein-Price	Rastrigin
Values					
x_1		1	-0.0898	0	0
x_2		1	0.7126	-1	-
Global minimum(f)		0	-1.0316	3	0
MOGA	x_1	1.0410	-0.0895	1.1657e^{-04}	8.388e^{-09}
	x_2	1.0839	0.7128	-1.0001	-
	f	0	-1.0316	3	0
APSO	x_1	1	-0.0898	1.2072e^{-10}	9.960e^{-10}
	x_2	1	0.7126	-1	-
	f	0	-1.0316	3	0

VI. CONCLUSION

A benchmark study was conducted that the performance of MOGA and APSO multi objective optimization algorithms. Even though, the evaluated benchmark functions, the MOGA converges to global minima more co-herently and faster than APSO algorithm but failed to find best solutions as compared to standard benchmarked functions such as Rosenbrock, Six Hump Camelback, Goldstain-Prices, Rastrigin functions.

REFERENCES

[1] G. P. Liu, J.-B. Yang, and J. F. Whidborne, *Multiobjective Optimisation and Control*, First., vol. 4. Hertfordshire, England: Research Studies Press, 2003.

[2] C. M. Fonseca and P. J. Fleming, "Multiobjective genetic algorithms made easy: selection sharing and mating restriction," in *IEE International Conference on Genetic Algorithms in Engineering Systems: Innovations and Applications*, 1995, pp. 45–52.

[3] P. J. Fleming; C. M. Fonseca, "Genetic Algorithms in Control Systems Engineering: A Brief Introduction," in *IEE Colloquium on Genetic Algorithms for Control Systems Engineering*, 1993, pp. 1–5.

[4] C. M. Fonseca and P. J. Fleming, "Multiobjective genetic algorithms," in *IEE Colloquium on Genetic Algorithms for Control Systems Engineering*, 1993, pp. 1–6.

[5] S. K. Valluru and M. Singh, "Metaheuristic Tuning of Linear and Nonlinear PID Controllers to Nonlinear Mass Spring Damper System," *Int. J. Applied Eng. Res.*, vol. 12, no. 10, pp. 2320–2328, 2017.

[6] S. K. Valluru and M. Singh, "Stabilization of Nonlinear Inverted Pendulum System Using MOGA and APSO Tuned Nonlinear PID Controller," *Cogent Eng.*, vol. 4, no. 1, pp. 1–15, 2017.

[7] Z.-H. Zhan, J. Zhang, Y. Li, and H. S.-H. Chung, "Adaptive particle swarm optimization.," *IEEE Trans. Syst. man, Cybern.*, vol. 39, no. 6, pp. 1362–1381, 2009.

[8] Xin-She Yang, *Engineering Optimization An Introduction with Metaheuristic Application.* 2010.

978-1-5386-6624-1/18 $31.00 © 2018 IEEE

2nd IEEE International conference on power Electronics, Intelligent Control and Energy systems (ICPEICES-2018)

Fuzzy Quaternion-based Pixel Wise Information Extraction for Face Recognition

Sudesh Yadav
University School of Information, Communication &
Technology,
Guru Gobind Singh Indraprastha University, New Delhi,
India
yadavsudesh01@gmail.com

Virendra P. Vishwakarma
University School of Information, Communication &
Technology,
Guru Gobind Singh Indraprastha University, New Delhi,
India
virendravishwa@rediffmail.com

Abstract—**In advancement of today's technological era, face recognition is an important task in various security systems, monitoring criminal and fraudulent activities. In this article, we present a new method for information extraction using two new concepts viz. quaternion and fuzzy logics. Quaternion helps in incorporating interrelationship between three color channels of a color face image. After obtaining quaternion vector of a color image we fuzzify it by using membership function (MF) i.e. π-MF, which find out in how much amount a pixel value of a color face image is associated with other face images of different classes. After that principle component analysis and k-nearest neighbor classification is used to obtain error rate in classification accuracy over Georgia Tech and Indian face (females) datasets. Investigational results clearly show the efficacy and superiority of our approach.**

Keywords—Face recognition; fuzzy logic; information extraction; quaternion.

I. INTRODUCTION

As in literature, till now face recognition approaches using fuzzy approaches have succeeded by transforming color face images to gray scale images. To enhance the recognition rate of face recognition approaches, we can use direct color information present in the color face images. So, in today's era of development color face recognition is very hot topic in machine pattern recognition. In this article, we present a new method using two different concepts viz. quaternion and fuzzy logics.

As we know, the color of an image gives us important information about morphological structure of a color image for reconstruction, signal encoding, alignment of images, denoising, image matching and image classification application of pattern recognition. Different color representation of a color image typically shows various discriminatory properties with various benefits and drawbacks, shown in [1]. Color face recognition using fuzzy quaternion in collaboration with discriminant analysis. The distinguishing nature of color face images inspired us to invent a color model to enhance the discriminatory power of face recognition systems. As commonly available color models for face recognition is formed by three main color channels, i.e. RGB (red, green, blue). The latest color model is discriminate color space (DCS) [2].

Gray scale image is a scalar, whereas color image is a vector function. So, there is always a need to define structure of chromatic information as a vector valued function. To deal with vector valued function, a lot of approaches have been available in literature, which process multichannel function in directional as well as magnitude processing [3].

Hypercomplex algebra, a part of mathematics, proves an important means to deal with vector valued functions. Among many number systems available in hypercomplex algebra, quaternion algebra is very first number system which is very closed to familiar systems of real and complex numbers [4][5].

In quaternion color image processing, a color image pixel is denoted by a quaternion matrix of three channel codes (RGB). In past two decades, quaternion concept has been proved to be a significant tool in color image recognition [3]. The main benefit behind using quaternion algebra is that a color image is treated as a vector function irrespective of scalar or binary function. Nowadays, in era of ubiquitous computing, almost all captured images are chromatic in nature irrespective of monochromatic. For the first time, sangwine [6] and pie and change [7] exploits the effect of quaternion algebra in digital color image processing. After that, a lot of classical tools extended to color image processing using various methods viz. fourier transform [6][8][9][10], neural networks [11][12], singular value decomposition (SVD) [13][14], principle component analysis (PCA) [15][16] , independent component analysis(ICA) [17][18], moments [19][20] etc.

In this article, we present a method which incorporates the benefit of interrelationship among RGB color codes of a color image in color face recognition using quaternion concept of hypercomplex algebra of mathematics. Our approach mainly focuses on pixel wise information available in color pixels using fuzzy logic and quaternion concept. Till now, there is no approach available which discovers the effect of class wise belongingness of individual pixel in color face images. The present article discovers the effect of class wise belongingness of individual pixel values of color face images to different classes.

978-1-5386-6624-1/18 $31.00 © 2018 IEEE

II. PROPOSED APPROACH

The present article proposes a new approach of fuzzy quaternion-based pixel wise information extraction (FQPIE) for face recognition. FQPIE incorporates the benefit of interrelationship between RGB color codes of a color image in recognition of color face images. In proposed approach, we consider color face image as a vector function irrespective of scalar function. Therefore, we denote each face image as a vector form in color codes i.e. RGB color code using quaternion number sequence. Next, we do concatenation of each color vector. Thereafter, we do fuzzification process (FPIE) to find degree of relationship between pixels of an image to different subjects. In this, we obtain one quaternion fuzzy vector corresponding to each color face image. For classification to different subjects, we apply k-nearest neighbor classifier. The block diagram showing the complete architecture of our proposed approach is shown in Figure 1.

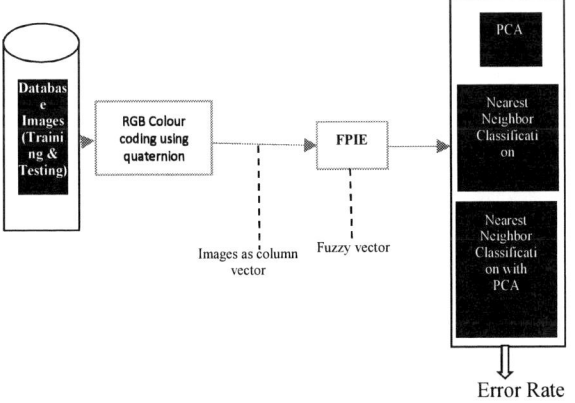

Figure 1 Block diagram of proposed approach

A. Fuzzy Quaternion Pixel Wise Information for colour images:

Let us we represent the vector form of a color image C in each channel as $Y_c \in X^n$, where n is the dimension of the vector, where c= r, g, b denotes the R, G, B channels respectively. So, the final representation of results is a combinatorial of atoms through three different color codes in quaternion matrix whose representation is as follows:

$$Y_C = [X_R X_G X_B]^T \qquad (1)$$

Next using FPIE module, we fuzzifies the pixel values of the color image. FPIE module takes quaternion vector of a color image as an input and using membership function (MF) generates pixel wise association of a color face image to different classes.

The concept of membership functions (MFs) was first given by Prof. L. Zadeh in his first paper in 1965 on Fuzzy Set [21]. MF is basically a curve which represent how the value of genuineness or truth-ness can be mapped to a value i.e.

membership value between 0 and 1 in input space. In simple terms we can say that it is generalization of crisp set. Here a higher value of MF denotes more genuineness of the set. We can represent mapping of fuzzy set F as follows:

$$u_F : x \rightarrow [0,1] \qquad (2)$$

where x is the universal set in input space.

In our proposed approach, for accomplishing the task of fuzzification of quaternion vector of a color face, we use a π - type MF [8][16][17]. This MF consists of a parameter, called controller (c), that helps in adjusting the generalization capability of function as per demand of the system for classification purpose. By changing the value of c, we can control the sharpness of the MF. The MF can be derived as follows:

$$
\begin{aligned}
\pi(x; \rho, \sigma, \tau) = 0, & \qquad x \leq \rho \\
= 2^{(c-1)}[(x-\rho)(\sigma-\rho)]^c & \qquad \rho < x \leq v_1 \\
= 1 - 2^{c-1}[(\sigma-x)(\tau-\rho)]^c & \qquad v_1 < x \leq \sigma \\
= [(x-\sigma)(\tau-\sigma)]^c & \qquad \sigma < x \leq v_2 \\
= 1 - 2^{c-1}[(\tau-x)(\tau-\sigma)]^c & \qquad c_2 < x < \sigma \\
= 0, & \qquad x \geq \sigma
\end{aligned}
$$

$$(3)$$

where v_1, v_2 are two crossover points and ρ, σ and τ represent minimum, maximum and mean value of datapoints on the membership function curve for an intended pixel value respectively. In present scenario, we have chosen value of c equal to 2. The value of ρ, σ and τ can be determined by value of an intended pixel number (p) in the training dataset as follows:

$$
\begin{aligned}
\rho &= minimum(p) \\
\sigma &= maximum(p) \\
\tau &= mean(p)
\end{aligned}
$$

$$(4)$$

This function increases gradually from ρ to τ, and decreases gradually from τ to σ. The value of this function at center is one and at crossover points it will be 0.5 and away from crossover points it gradually decreases. For a color face image represented in quaternion vectorized form given by Eq. 1, the fuzzified membership matrix n obtained after applying FQPIE is shown below:

$$x = [x_1, x_2, \ldots, x_t, \ldots, x_D]^T \qquad (5)$$

where x_t denotes the fuzzified membership value of tth pixel of quaternion face vector x.

B. Principle Component Analysis (PCA):

For dropping the computational price of high dimensional data obtained in above phase, saving memory and increasing the speed of the system, we use PCA. For the first time, it was invented by Karl Pearson [22] in his principle axes theorem used in mechanics. The concept was introduced to FR by Sirvovich and Kirby [23]. As we know each color face image is of three-dimensional nature. So, to include all the benefits in recognizing color face images, we work on these images directly. Therefore,

978-1-5386-6624-1/18 $31.00 © 2018 IEEE

we apply quaternion concept on color face images directly and three channel codes are extracted i.e. RGB color codes each of having two-dimensional nature. Next, we concatenate all the three-color codes. Thereafter, to find a set of principle components we use PCA an orthogonal linear transformation for reducing the computational cost by doing dimension reduction of data matrix obtained above. In which each component is orthogonal to each other and components are obtained where first principle component corresponds to largest possible variance, second to the second largest variance and so on. And a projection metric is derived using process given by Turk et al. [24]. Thereafter recognition accuracy results are obtained using simple Euclidean distance metric.

C. Classification using k-NNC

In this paper, we use *k*-Nearest Neighbor Classifier (k-NNC) for classification of unknown models. The reason behind using this is that it is non-parameterized, unsupervised and approximated locally so applicable to very large-scale datasets. For the first time researcher T.M. Cover and his friends in 1967[25] has given the concept of NNC for the first time. For classification and regression analysis in machine learning tasks *k*-NNC was highly recommended and used. Here *k* is the number of input images of train data. In later section, we give the detailed analysis of experiments performed for testing the efficacy of the approach presented here on two datasets viz. Georgia tech and Indian face databases.

III. RESULTS AND ANALYSIS

For the efficacy of our proposed approach, we have tested it on two datasets viz. Georgia Tech and Indian face dataset for female. The information regarding datasets, the experimental results obtained on Matlab version 15 and comparison to methods viz PCA, FPIE [26], IT2FPIE [27] and Quaternion PCA [28] is given below:

A. Georgia Tech Face dataset

This dataset comprises of 50 different people consisting of 15 example color images per people taken in various condition pertaining to lightening. Example images showing the Georgia tech face dataset are shown in Figure 2. For the efficacy of our proposed approach, we cropped and resized these images to dimension 30 × 40. Investigational results attained with our approach are given in table 1 and shown graphically in Figure 5. The table shows efficient improvement in recognition accuracy obtained using our approach over other methods.

TABLE I.   COMPARISON OF PERCENTAGE RECOGNITION ERROR RATE ON GT FACE DATABASE

Figure 2 Example showing some of the face images Georgia Tech Face database

B. Indian face dataset (female)

This dataset consists of 40 distinct subjects each of having 11 sample images collected at IIT Kanpur, India in 2002. All the images in dataset were taken against different orientation of faces and background of images was taken homogenous always. In our approach: we are doing experiments on female database only, which comprises of 22 subjects with 11 sample images per subject. Some of the captured images were shown below in Figure 3. Investigational results and comparison of our approach to other methods is given in Table 2 and shown graphically in Figure 4.

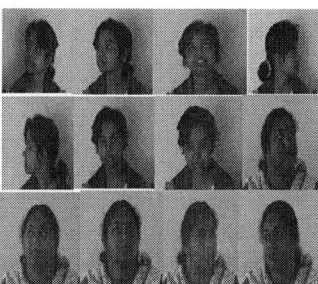

Figure 3 Some of the examples showing images of Indian face dataset.

Methods	Number of images per training sample									
	3	4	5	6	7	8	9	10	11	12
FQPIE	**42.50**	**42.00**	**39.40**	**25.78**	**22.50**	**19.42**	**17.00**	**16.80**	**16.00**	**15.33**
IT2FPIE [27]	51.16	48.36	44.80	33.56	27.50	24.57	21.66	22.00	19.50	-
Quaternion PCA [28]	47.27	45.20	36.67	34.00	30.86	26.67	26.00	22.50	19.33	-
FPIE [26]	48.62	47.00	45.27	42.60	29.55	25.25	24.57	20.33	21.60	20.50
PCA	54.33	50.00	48.20	40.89	36.50	33.43	31.33	29.20	26.50	23.33

TABLE II. Table 1 comparison of percentage recognition Error rate on Indian Face database

Methods	Number of images per training sample						
	1	2	3	4	5	6	7
FQPIE	**11.81**	**8.58**	**7.95**	**7.79**	**7.57**	**7.27**	**3.41**
Quaternion PCA [28]	28.64	23.74	22.73	16.88	15.91	19.09	13.63
IT2FPIE [27]	18.18	13.13	11.36	10.39	9.85	10.00	9.09
FPIE [26]	15.00	12.12	10.79	8.44	11.36	12.72	7.95
PCA	34.55	25.25	23.86	18.18	17.42	19.09	15.91

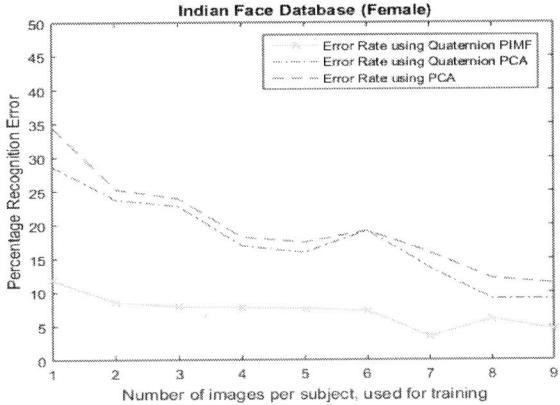

Figure 4 graph showing comparison of %age error using our proposed approach on Indian Face database (female)

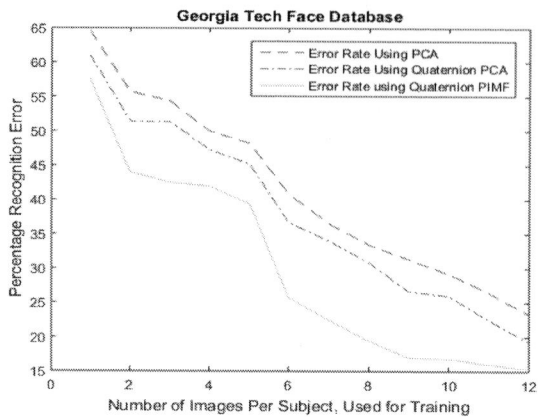

Figure 5 graph showing comparison of %age recognition error rate using our proposed approach on Georgia Tech Face database

IV. CONCLUSION

In this paper, we present a novel and robust approach to information extraction for color face recognition. The key point of our approach is that it includes the complete information of a feature by incorporating the interrelationship between RGB color codes of a color face image by utilizing the benefit of π-MF. It provides an improvement to other fuzzy based information extraction approaches available in literature for color face recognition. Experiments done over Georgia Tech and Indian face databases shows a significant improvement by applying FQPIE.

REFERENCES

[1] J. Yang and C. Liu, "A discriminant color space method for face representation and verification on a large-scale database," in *Pattern Recognition, 2008. ICPR 2008. 19th International Conference on*, 2008, pp. 1–4.

[2] P. Shih and C. Liu, "Comparative assessment of content-based face image retrieval in different color spaces," *Int. J. Pattern Recognit. Artif. Intell.*, vol. 19, no. 7, pp. 873–893, 2005.

[3] R. Lukac, B. Smolka, K. Martin, K. N. Plataniotis, and A. N. Venetsanopoulos, "Vector filtering for color imaging," *IEEE Signal Process. Mag.*, vol. 22, no. 1, pp. 74–86, 2005.

[4] Y. Xu, L. Yu, H. Xu, H. Zhang, and T. Nguyen, "Vector sparse representation of color image using quaternion matrix analysis," *IEEE Trans. image Process.*, vol. 24, no. 4, pp. 1315–1329, 2015.

[5] C. E. Moxey, S. J. Sangwine, and T. A. Ell, "Hypercomplex correlation techniques for vector images," *IEEE Trans. Signal Process.*, vol. 51, no. 7, pp. 1941–1953, 2003.

[6] S. J. Sangwine, "Fourier transforms of colour images using quaternion or hypercomplex, numbers," *Electron. Lett.*, vol. 32, no. 21, pp. 1979–1980, 1996.

[7] S.-C. Pei and C.-M. Cheng, "A novel block truncation coding of color images using a quaternion-moment-preserving principle," *IEEE Trans. Commun.*, vol. 45, no. 5, pp. 583–595, 1997.

[8] T. A. Ell and S. J. Sangwine, "Hypercomplex Fourier transforms of color images," *IEEE Trans. image Process.*, vol. 16, no. 1, pp. 22–35, 2007.

[9] T. A. Ell, "Hypercomplex spectral transformations," 1992.

[10] T. Bülow, "Hypercomplex spectral signal representations for image processing and analysis," *Kiel Univ.*, 1999.

[11] P. Arena, L. Fortuna, L. Occhipinti, and M. G. Xibilia, "Neural networks for quaternion-valued function approximation," in *Circuits and Systems, 1994. ISCAS'94., 1994 IEEE International Symposium on*, 1994, vol. 6, pp. 307–310.

[12] T. Nitta, "A quaternary version of the back-propagation algorithm," in *Neural Networks, 1995. Proceedings., IEEE International Conference on*, 1995, vol. 5, pp. 2753–2756.

[13] N. Le Bihan and S. J. Sangwine, "Color image decomposition using quaternion singular value decomposition," 2003.

[14] J.-H. Chang, J.-J. Ding, and others, "Quaternion matrix singular value decomposition and its applications for color image processing," in *Image Processing, 2003. ICIP 2003. Proceedings. 2003 International Conference on*, 2003, vol. 1, p. I--805.

[15] N. Le Bihan and S. J. Sangwine, "Quaternion principal component analysis of color images," in *Image Processing, 2003. ICIP 2003. Proceedings. 2003 International Conference on*, 2003, vol. 1, p. I--809.

[16] Y. Sun, S. Chen, and B. Yin, "Color face recognition based on quaternion matrix representation," *Pattern Recognit. Lett.*, vol. 32, no. 4, pp. 597–605, 2011.

[17] N. Le Bihan, S. Buchholz, B. ENSIEG, and N. Le-Bihan, "Quaternionic independent component analysis using hypercomplex nonlinearities," *Habilit. À Dir. LES Rech.*, p. 91, 2006.

[18] J. V'\ia, D. P. Palomar, L. Vielva, and I. Santamar'\ia, "Quaternion ICA from second-order statistics," *IEEE Trans. Signal Process.*, vol. 59, no. 4, pp. 1586–1600, 2011.

[19] B. C. B. Chen, H. S. H. Shu, H. Z. H. Zhang, G. C. G. Chen, and L. L. L. Luo, "Color Image Analysis by Quaternion Zernike Moments," *Pattern Recognit. (ICPR), 2010 20th Int. Conf.*, no. 7, pp. 7–10, 2010.

[20] L.-Q. Guo and M. Zhu, "Quaternion Fourier--Mellin moments for

978-1-5386-6624-1/18 $31.00 © 2018 IEEE

color images," *Pattern Recognit.*, vol. 44, no. 2, pp. 187–195, 2011.

[21] L. A. Zadeh, "Fuzzy sets," *Inf. Control*, vol. 8, no. 3, pp. 338–353, 1965.

[22] K. Pearson, "LIII. On lines and planes of closest fit to systems of points in space," *London, Edinburgh, Dublin Philos. Mag. J. Sci.*, vol. 2, no. 11, pp. 559–572, 1901.

[23] L. Sirovich and M. Kirby, "Low-Dimensional Procedure for the Identification of Human Faces," *J. Opt. Soc. Amer. A*, vol. 4, no. 3, pp. 519–524, 1987.

[24] M. Turk and A. Pentland, "Eigenfaces for Recognition," *Journal of Cognitive Neuroscience*, vol. 3, no. 1. pp. 71–86, 1991.

[25] T. Cover and P. Hart, "Nearest neighbor pattern classification," *IEEE Trans. Inf. Theory*, vol. 13, no. 1, pp. 21–27, 1967.

[26] V. P. Vishwakarma, S. Pandey, and M. N. Gupta, "Fuzzy based pixel wise information extraction for face recognition," *Int. J. Eng. Technol.*, vol. 2, no. 1, p. 117, 2010.

[27] S. Yadav and V. P. Vishwakarma, "Interval type-2 fuzzy based pixel wise information extraction: An improved approach to face recognition," in *Computational Techniques in Information and Communication Technologies (ICCTICT), 2016 International Conference on*, 2016, pp. 409–414.

[28] E. S. Jaha and L. Ghouti, "Color face recognition using quaternion PCA," 2011.

2nd IEEE International conference on power Electronics, Intelligent Control and Energy systems (ICPEICES-2018)

Controlling of AVR Voltage and Speed of DC Motor using Modified PI-PD Controller

Vinay Kumar Singh
Department of Electronics and communication PDPM-Indian institute of information technology
Jabalpur, 482005, India
1612211@iiitdmj.ac.in

Sudeep Sharma
Department of Electronics and communication PDPM-Indian institute of information technology
Jabalpur, 482005, India
1612702@iiitdmj.ac.in

Prabin Kumar Padhy
Department of Electronics and communication PDPM-Indian institute of information technology
Jabalpur, 482005, India
Prabin16@iiitdmj.ac.in

Abstract—**This paper explains the tuning method of PI-PD controllers, which is validated using simulation results. The closed-loop control of four-quadrant chopper fed DC motor and Automatic voltage regulator (AVR) are two models taken to show results. The controller parameters are obtained through time domain analysis with the help of plant model in the form of transfer function. The success of the method can be measured in terms of obtained desired lower overshoot, lower IAE, and settling time. The results obtained from the modified relay based PI-PD tuning method are compared with other conventional methods of PID tuning. The proposed PI-PD controller for speed control of DC motor gives better results which is proved by MATLAB simulations and can be useful for controlling or regulating the performances of electric drives. The other application is of automatic voltage regulator (AVR) which is a device used to regulate voltage automatically i.e. to analyze the fluctuations in voltage levels and turn it into a constant voltage.**

Keywords— *AVR, PI-PD controller, PID controller, fractional order systems, DC motor.*

I. INTRODUCTION

Relay based controller design [1], [2], [3], [4], [5], removes the need to have information about the class of plant from which they belong before applying available methods. Concept of decentralized and proportional resonant controllers as given in [6], [7], [8] this gave PID Parameters by approximating values of reference frequencies (ω_r) by making changes in ultimate frequency (ω_u). As it is evident that with increasing requirement for automation in the industries [11], [12], drives having high reliability and accuracy are needed. The speed control of DC motor is very significant especially in applications where precision is of great importance. Due to its ease of controllability the DC motor is used in many industrial applications requiring variable speed and load characteristics. As the precise speed control of a DC motor is very crucial in industries that is why the applications of these drives includes in machine tools, traction and robotics [13], [14]. The automatic voltage regulators [15] are used in power system applications to maintain Constant terminal voltage for reliable supply. DC motors can also be very useful in operations of Hybrid Electric Vehicle (HEV) or EVs. DC motor model is considered as a SISO system having torque-speed characteristics compatible to most of the mechanical loads. Modified relay feedback method for PID tuning and PI-PD tuning can be traced from [16], [17] according to it introduces a fractional order transfer function in series with relay to obtain limit cycle, from it amplitude and

frequency of oscillations can be obtained and by using conventional PID tuning methods and others for example proposed by Alexandre Sanfelice Bazanella, Luís Fernando Alves Pereira, and Adriane Parraga in 2017. The objective of this research study is to minimize both settling time and overshoot in the motor speed response..

Organization of this paper is such that its Section 2 is devoted to four quadrant chopper operation, motor modelling. Controller design methodology along with simulation results are is given in Section 3. Conclusion and references are given in section 4 and 5 respectively.

II. FOUR QUADRANT CHOPPER OPERATION MOTOR MODELLING AND CONTROLLER DESIGN

A chopper is a static device that converts fixed DC input to a variable DC output voltage directly. It's also known as DC-to-DC converter [9]. It's widely used for motor control. It's also used in regenerative braking. Essentially, a chopper is an electronic switch that is used to interrupt one signal under the control of another. The switching operation is explained as below:

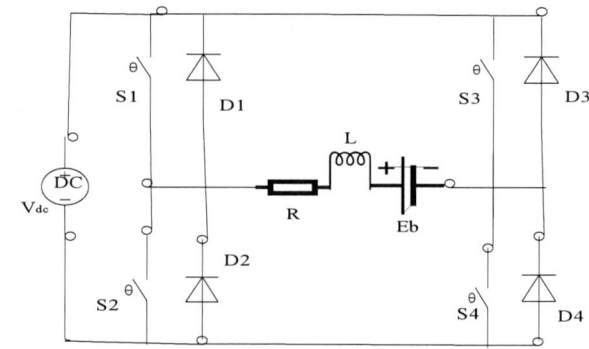

Fig 1: Chopper operational circuit diagram

1. SCR (θ is variable, firing angle of SCR) or IGBT can be taken as switch to explain the four quadrant operation of chopper. The output voltage swings in both directions i.e. from $+V_{dc}$ to $-V_{dc}$ and the PWM method of switching is utilized.

(i) Quadrant I: In first quadrant the power is positive. In this case, the power flows from source to load. In this case initially switch S1 and S4 are in operation making current to flow from R to E_b, after that when S4 is switched off the sored energy of inductor free wheels through diode D2. First quadrant operation of chopper operates motor in forward motoring mode.

978-1-5386-6624-1/18 $31.00 © 2018 IEEE

(ii) Quadrant 2: In this quadrant (second), the voltage is positive but the current is negative. The switch S2 is along with diode D2 operates, when S2 goes into off condition the current flows through D4 via D1, as it is clearly visible that in second quadrant operation current becomes negative while voltage remains positive. Overall power in forward motoring operation remains negative and load feeds back stored energy to source.

(iii) Quadrant 3: In this quadrant (third) the chopper operates such that is rotates dc drive in reverse motoring mode as both the voltage and currents are negative hence the power is positive. In this case, the power flows from source to load. In this case the battery polarity is in reverse direction to previous two cases. In third quadrant switches S2 and S3 are in operating condition and when switch S3 goes off the conduction of stored energy of inductor takes place via D4 and switch S2.

(iv) Quadrant 4: The power is therefore negative and connected device works in reverse braking mode. In the fourth quadrant voltage is negative but current is positive. Here S3 and D2 operates and when S4 is not in operation due to reverse voltage around it, the diodes D2 and D3 conducts and feeds the stored energy of inductor to source(Reverse regenerative braking mode).

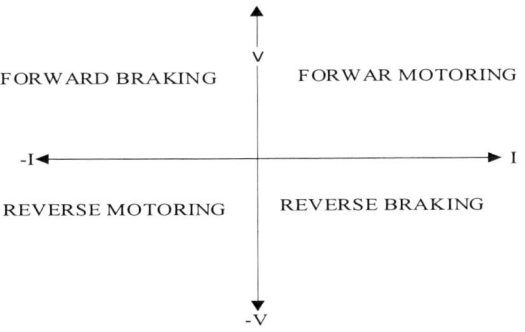

Fig 2: Four quadrant operation of drive

Modeling of DC Motor:

Motor equations:

$$V_s - E_b = L\frac{di}{dt} + I_a R_a \tag{1}$$

$$T_d = J\frac{d\omega}{dt} + B\omega + T_L \tag{2}$$

$$E_B = K_b\omega \tag{3}$$

Laplace Domain representation:

$$V(s)_s - E_b(s) = sLI_a(s) + I_a(s)R_a \tag{2}$$

$$T_d(s) - T_L(s) = sJ\omega(s) + B\omega(s) \tag{3}$$

$$E_b(s) = K_b\,\omega(s) \tag{4}$$

From above equation transfer function of separately excited DC motor [20] can be written as below

$$\frac{\omega(s)}{V_s(s)} = \frac{K_t = K}{(Js+B)(Ls+R)+K^2} \tag{5}$$

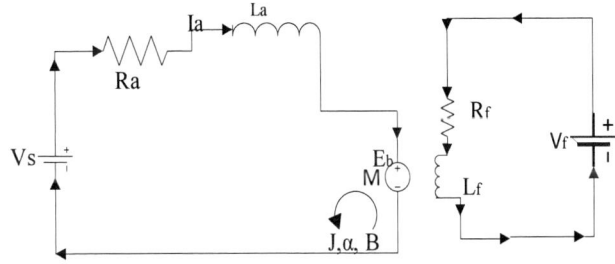

Fig. 3: Modelling circuit of dc motor

Table 1: Parameters for DC drive model

R_a	4.372 ohm
L_a	61.6mH
J	0.0059
K	1.14
B	0.0024

From above table of parameters we can easily extracted to get the required model [21] transfer function as given by equations below:

$$G_p(s) = \frac{1.14}{(0.0059s + 0.0024)(0.0616s + 4.372)} \tag{6}$$

$$G_p(s) = \frac{3137}{(s^2 + 71.167s + 3605)} \tag{7}$$

From above equation in Laplace domain the roots (complex conjugate) of equation are given as below:
(-35.5+73.94j and -35.5-73.94j).

AVR VOLTAGE CONTROLLER MODEL:

The role of an AVR is to hold the terminal voltage magnitude of a synchronous generator at a specified level. A simple AVR system comprises four main components, namely amplifier, exciter, generator, and sensor. For mathematical modeling and transfer function of the four components, these components must be linearized, which takes into account the time constant and ignores the saturation or other nonlinearities. The reasonable transfer function of these components may be represented, respectively. Where V_{ref} is reference voltage and V_t is terminal voltage of device.

Amplifier model

$$\frac{V_r(s)}{V_e(s)} = \frac{K_a}{(1+t_a s)} \tag{8}$$

Normally values K_a is of the range of 10 to 40. The amplifier time constant is very small ranging from (0.02 to 0.01).

Exciter model:

The transfer function of a modern exciter may be represented

978-1-5386-6624-1/18 $31.00 © 2018 IEEE

by a gain and a single time constant as below:

$$\frac{V_f(s)}{V_r(s)} = \frac{K_e}{(1+t_e s)} \qquad (9)$$

Normal values K_a of are in the range of 0.7 t0 10. The time constant is in the range of 0.4 to 1.0 s.

Generator model:
These constants are load dependent, may vary between 0.7 to 1.0, and between 1.0 and 2.0 (t_g) in sec. from full load to no load.

$$\frac{V_t(s)}{V_f(s)} = \frac{K_g}{(1+t_g s)} \qquad (10)$$

Sensor model:
For very small values of t_r ranging from 0.001 to 0.06

$$\frac{V_s(s)}{V_t(s)} = \frac{K_r}{(1+t_r s)} \qquad (11)$$

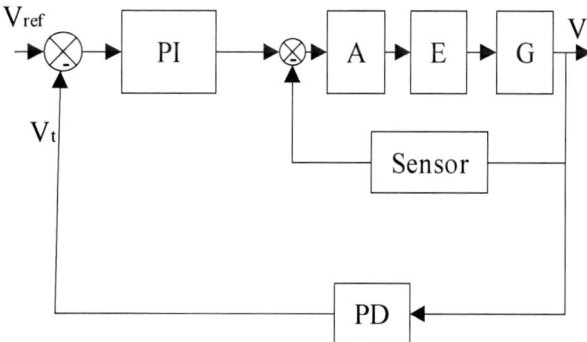

Fig: 4 AVR controller block

III. PROPOSED MODIFIED PI-PD CONTROLLER DESIGN AND SIMULATION RESULTS

The optimized plant as obtained after adding previously given PID parameters in series with original plant, as given in [22].

$$G_{AVR(new)}(s) = \frac{0.0194s^3 + 1.99s^2 + 5.58s + 4.369}{0.0004s^5 + 0.0454s^4 + 0.555s^3 + 3.45s^2 + 6.536s + 4.369}$$

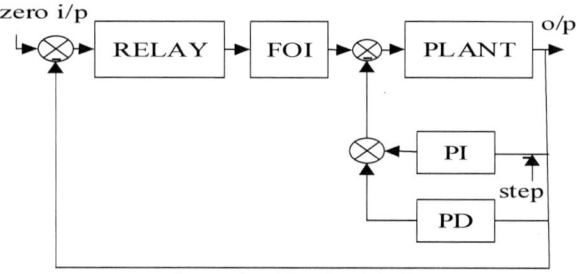

Fig: 5 Identification block

For identification process the inner feedback is provided by PD controller ($G_{cd}(s)$) in parallel to the plant ($G_p(s)$), and PI

controller ($G_{ci}(s)$) which is connected in series with relay and FOI $\left(F(s) = \frac{1}{s^{2/3}} \right)$ [6], [21] as the model shown in fig.5, which produces limit cycle. The PI and PD controller is chosen such as Kc is the proportional gain, T_i, the integral time constant, T_d the derivative time constant and filter constant N is assumed to be 20 [19].

The denominator of $G_{cd}(s)$ is neglected as it is very small for ease of calculations. In this work, initial auto-tuning test is done by setting $T_d = 0$ and $T_i = \infty$, one more advantage of this tuning scheme is that the relay height can be adjusted from zero to some acceptable value when the operator requires to tune the controller as desired.

$$G_{c_i} = PI = K_c \left(1 + \frac{1}{T_i s} \right) \qquad (12)$$

$$G_{c_d} = PD = \frac{\left(K_f (T_d s + 1) \right)}{\left(\left(\frac{T_d}{N} \right) s + 1 \right)} \qquad (13)$$

$$G_c(s) = K_c \left(1 + \frac{1}{T_i s} \right) + K_f (T_d s + 1) \qquad (14)$$

From limit cycle output modified describing function due to FOI can be given as:

$$D_f = \frac{4h}{\pi A} \left(\exp(-\frac{\pi}{2}\alpha j) \right) = \frac{4h}{\pi A}(0.5 - j0.866) \qquad (15)$$

$$G_p(\omega_{cr})\left(D_f + G_c(\omega_{cr}) \right) = -1 \qquad (16)$$

Here, h is relay height (gain), A is magnitude of oscillations where $\alpha = \frac{2}{3}$ and frequency of oscillation is given as ω_{cr}, further from magnitude condition we can get following equations.

From above equations and desired plant (24) we can write\ following equations and solve them using user defined phase and gain margins and it can be observed below:

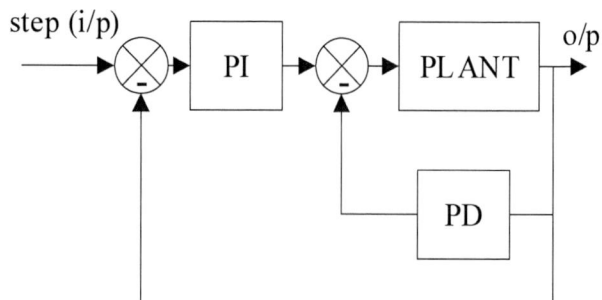

Fig 6: Proposed PI-PD controlling method

978-1-5386-6624-1/18 $31.00 © 2018 IEEE   860

$$D_f G_p(\omega_{cr}) = -1 \qquad (17)$$

$$K_a \left(\left| G_p(\omega_{cr}) (p + jq) \right| \right) = (-1) \qquad (18)$$

Where K_a is (open loop) gain and values of p, q can be given as follows:

$$p = \frac{4h}{\pi A} \left(\cos(\frac{\alpha \pi}{2}) \right) + \left(K_c + K_f \right) \qquad (19)$$

$$q = \left(\omega_{cr} K_f T_d - \frac{K_c}{\omega_{cr} T_i} - \left(\frac{4h}{\pi A} \right) \sin \frac{\alpha \pi}{2} \right) \qquad (20)$$

From above equations time constant T_a can be obtained as:

$$T_a = \frac{\sqrt{\left[K^2{}_a \left(p^2 + q^2 \right) - 1 \right]}}{\omega_{cr}} \qquad (21)$$

$$\theta_a = \pi + \tan^{-1}\left(\frac{q}{p} \right) - \tan^{-1}\left(\omega_{cr} T_a \right) \text{ (for unstable plants)} \quad (22)$$

$$\theta_a = \tan^{-1}\left(\frac{q}{p} \right) + \tan^{-1}\left(\omega_{cr} T_a \right) \text{ (for stable plants)} \quad (23)$$

$$G_p(s) = \frac{K_a e^{-\theta_a s}}{T_a s \pm 1} \qquad (24)$$

The closed-loop transfer function of the inner feedback loop of fig.5 can be written as:

$$F_a(s) = \frac{G_p(s)}{1 + G_{cd}(s) G_p(s)} \qquad (25)$$

Using first order Pade's approximation for the time delay terms, hence taking $T_d = (\theta_a/2)$, $F_c(s)$ becomes:

$$F_a(s) = \frac{K_a e^{-\theta_a s}}{\left[\left(T_a - K_a K_f \theta_a / 2 \right) s + \left(K_a K_f \pm 1 \right) \right]} \qquad (26)$$

It is apparent from the denominator of (14) that $G_{cd}(s)$ is capable of providing robust stability to both stable and unstable processes by relocating the poles.

$$K_f \geq \frac{1}{K_a} \qquad (27)$$

$$\frac{1}{K_a} \leq K_f \leq \frac{2T_c}{K_a \theta_c} \qquad (28)$$

Then, one can choose following as the proportional gain for the PD (feedback) controller:

$$K_f = \frac{1}{K_a} \left(\sqrt{\frac{2T_a}{\theta_a}} \right) \qquad (29)$$

Now, one of the loop transfer function can be given as $L_a(s) = G_{ci}(s) F_a(s)$.

Let the phase crossover and gain crossover frequencies of the outer loop transfer function $L_a(s)$ be as ω_a and ω_b respectively. Similarly, let φ_m and g_m be the phase and gain margins. From the definition of φ_m and g_m, the following set of equations Can be obtained. Using magnitude condition following equation can be written as

$$\left| G_{ci}(j\omega_a) F_a(j\omega_a) = 1 \right| \qquad (30)$$

And using phase condition we can write

$$\arg(G_{ci}(j\omega_b) F_a(j\omega_b)) = -\pi \qquad (31)$$

$$g_m = \frac{1}{\left| G_{ci}(j\omega_a) F_a(j\omega_a) \right|} \qquad (32)$$

Substituting G_i & $\Gamma_a(s)$ and ω_a, in above equations we can further solve using phase margin criteria. With ω_b in eq.(20) we can get following simplification:

$$K_a K_c = \omega_b T_i \sqrt{ \frac{ \omega^2{}_b (T_a - (K_a K_f \theta_a / 2))^2 + (K_a K_f \pm 1)^2 }{ (\omega_b T_i)^2 + 1 } } \qquad (33)$$

$$\left[\varphi_m = \frac{\pi}{2} + \tan^{-1}(\omega_b T_i) - \tan^{-1} \frac{(\omega_b (T_a - (K_a K_f \theta_a / 2))}{K_a K_f \pm 1} - \theta_{\omega_b} \right] \quad (34)$$

On further solving equation (21) with ω_a

$$\left(\frac{\pi}{2} + \tan^{-1}(\omega_a T_i) - \tan^{-1}\left[\frac{(\omega_a (T_a - (K_a K_f \theta_a / 2))}{K_a K_f \pm 1} \right] - \theta_{\omega_a} \right) \qquad (35)$$

$$g_m K_g K_c = \omega_p T_i \sqrt{ \frac{ \omega^2{}_p (T_c - (K_g K_f \theta_c / 2))^2 + (K_g K_f \pm 1)^2 }{ (\omega_p T_i)^2 + 1 } } \qquad (36)$$

To solve these equations numerically one can take approximation for arc (tan) function as:

$$\tan^{-1}(x) = \left(\frac{\pi}{2} - \frac{1}{x} \right) \qquad (37)$$

For $x \geq 1$ or $(1/x) \leq \leq 1$ we can write $\tan^{-1}(x) = (1/x)$ we can simplify above terms as:

$$K_a K_c = \omega_b \left(T_c - \left(K_a K_f \theta_a / 2 \right) \right) \qquad (38)$$

$$\varphi_m = \left(\frac{\pi}{2} - \frac{1}{\omega_b T_i} + \frac{K_a K_f \pm 1}{\omega_b (T_c - (K_a K_f \theta_a / 2))} - \theta_{\omega_g} \right) \qquad (39)$$

$$g_m K_a K_c = \omega_p (T_a - (K_a K_f \theta_a / 2)) \qquad (40)$$

$$K_c = \frac{C_a}{(K_a \theta_c)} (T_a - (K_a K_f \theta_a / 2)) \qquad (41)$$

$$\text{where, } C_a = \frac{2\varphi_m + \pi(g_m - 1)}{2(g^2{}_m - 1)} \qquad (42)$$

$$C_b = \frac{\pi}{2} g_m C_a (1 - (2 g_m C_a / \pi)) \qquad (43)$$

$$\tau_a = \frac{(T_a - (K_a K_f \theta_a / 2))}{(K_a K_f \pm 1)} \qquad (44)$$

978-1-5386-6624-1/18 $31.00 © 2018 IEEE

$$T_i = \frac{\tau_a}{1 + \left(\frac{c_b}{\theta_a}\right)\tau_a} \quad (45)$$

$$T_d = \left(\frac{\theta_a}{2}\right) \quad (46)$$

These formulas can be used for a process (identified values of K_a, T_a, θ_a) and user specified g_m, φ_m to get desired controller parameters can be obtained to tune controller..

Example1: Self excited DC motor model
Having control over the speed of your motors and machinery is vital for you to be able to get the full potential of your tools. Giving yourself full speed control will ensure that you experience little to no tool malfunctioning occurs.

For every machining job, there is a cutting speed, with full speed control it is observed that working at this speed machine can give the optimum result. From the transfer function of motor [14] as given in equation below:

$$G_p(s) = \frac{3136.69}{\left(s^2 + 70.4870s + 3604.6\right)}.$$

In this case for speed control of separately excited DC motor the initial tuning is done with $Kc = K_f = 0.1, T_i = $ inf, $T_d = 0$ from limit cycle obtained after relay test the amplitude (A=0.0294) and critical frequency ω_{cr} (87.9 rad/sec) are obtained.
From these values as input to second iteration the controller parameters with phase margin (1.04 rad/sec) and gain margin (20 rad/sec) are obtained as given in table 4.3 and identified

plant is $G_{pi}(s) = \dfrac{K_a e^{-\theta_a s}}{(T_a s + 1)} = \dfrac{0.6283 e^{-0.2286 s}}{(14.56 s + 1)}$

Motor speed vs Time variation using different methods is shown in fig.7.

Table 2: DC Motor comparison results

Controller	K_p	K_i	K_d	Kc	K_f	T_i	T_d
Conventional PID	3.99	21.04	0.613	0	0	0	0
Modified PID	3.99	21.042	0.6173	0	0	0	0
Modified PI-PD	0	0	0	21.53	94.08	0.00075	0.002

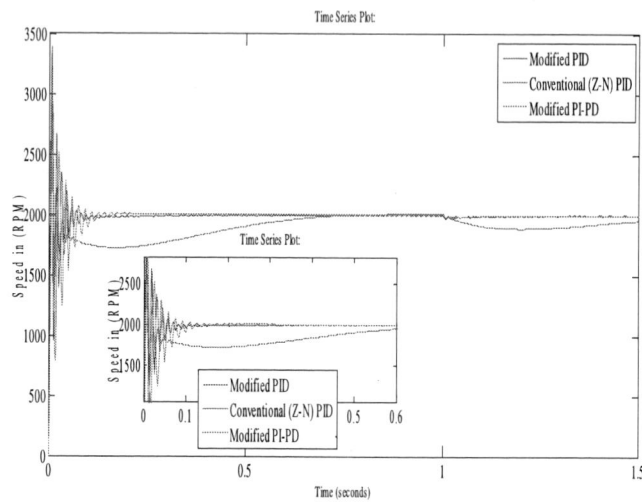

Fig 7: Response diagram of Motor model

Table 3: Comparison results of motor speed response

Controller type	Settling time	overshoot	Error(IAE)
Conventional pid	1.99	25.6%	0.87
Modified pid	2.13	11.26%	1.20
Modified pi-pd	0.49	0.87%	0.32

Example 2. AVR voltage controller model
According to parameters given in [4, 15] the normal plant is given as: (Original plant).

$$G_{AVR}(s) = \frac{0.1s + 10}{0.0004s^4 + 0.0454s^3 + 0.515s^2 + 1.515s + 11}$$

In this example the Optimized plant, after including PID controller as given in $G_{AVR(optimized)}(s)$ has been taken. he initial values for identification (as given in fig 5) and controller parameters are taken where $K_c = 0.01$, $K_f = 0.1$ with $T_i = $ infinite and $T_d = 0$ from these values the amplitude of oscillation (A) and critical frequency (ω_{cr}) from relay test are noted. From these values the controller parameters at phase margin (1.04 rad/sec) and gain margin (5 rad/sec) are obtained as given in table 5 and results are shown in (Figure 8) in second iteration as absolute values. As clearly evident from fig.8 which shows that modified PI-PD controller gives lower settling time lesser overshoot and lower error. AVR model considered is given as:

$$G_{AVR(obtimized)}(s) = \frac{0.0194s^3 + 1.99s^2 + 5.58s + 4.369}{0.0004s^5 + 0.0454s^4 + 0.555s^3 + 3.45s^2 + 6.536s + 4.369}$$

Table 4: Comparison of AVR results

Controller	Settling time	Overshoot %	Error(IAE)
PSO- PID (1&2)	1.2 sec.	2	0.229
Modified PI-PD	0.86 sec	0.4	0.157

Identified plant after 2nd iteration can be given as

$$G_{pi}(s) = \frac{K_a e^{-\theta_a s}}{(T_a s + 1)} = \frac{0.1202 e^{-0.0336 s}}{(0.0521 s + 1)}$$

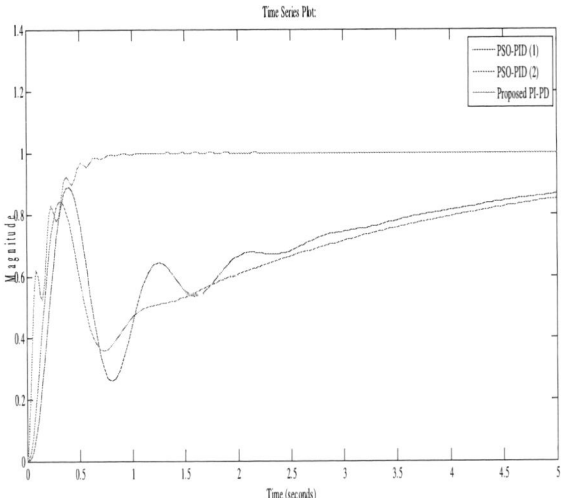

Fig 8: Response of AVR controller

IV. CONCLUSION

Relay based identification has certain limitations in terms of choosing initial values of gains (K_c, K_f) and time constants. Better results can be obtained by performing more than one iterations with new or updated values of parameters. This can be optimally done using optimization techniques like Hybrid (PSO+GA) optimization or normal optimization using MATLAB codes. These optimized values can give updated values of parameters which gives good results, avoiding lengthy mathematical calculations and manual errors.

V. REFERENCES

[1] K. Astrom, "PID controllers: theory, design and tuning," *Instrument Society of America*. p. 343, 1995.

[2] G.K. Dubey, "Fundamentals of Electric Drives", Narosa Publishing House, New Delhi 1995.

[3] Z. Gaing, "A Particle Swarm Optimization Approach for Optimum Design of PID Controller in AVR System," vol. 19, no. 2, pp. 384–391, 2004.

[4] K. H. A. K. H. Ang, G. Chong, and Y. L. Y. Li, "PID control system analysis, design, and technology," *IEEE Trans. Control Syst. Technol.*, vol. 13, no. 4, pp. 559–576, 2005.

[5] P. K. Padhy and S. Majhi, "Relay based PI-PD design for stable and unstable FOPDT processes," *Comput. Chem. Eng.*, vol. 30, no. 5, pp. 790–796, 2006.

[6] L. Campestrini, L. C. Stevanatto Filho, and A. S. Bazanella, "Tuning of multivariable decentralized controllers through the ultimate-point method," *IEEE Trans. Control Syst. Technol.*, vol. 17, no. 6, pp. 1270–1281, 2009.

[7] A. Tepljakov, E. Petlenkov, and J. Belikov, "FOMCON : a MATLAB Toolbox for Fractionalorder System Identification and Control," vol. 17, no. 4, 2011.

[8] M. Beschi, S. Dormido, J. Sanchez, and A. Visioli, "An automatic tuning procedure for an event-based PI controller," *Proc. IEEE Conf. Decis. Control*, pp. 7437–7442, 2013.

[9] Z. Li, C. Yin, Y. Chen, and J. Liu, *Feedback with a Fractional Order*, vol. 47, no. 3. IFAC, 2014.

[10] L. F. A. Pereira and A. S. Bazanella, "Tuning Rules for Proportional Resonant Controllers," *IEEE Trans. Control Syst. Technol.*, vol. 23, no. 5, pp. 2010–2017, 2015.

[11] S. Seeraji, "A Flexible Closed Loop PMDC Motor Speed Control System for Precise Positioning," no. 2, pp. 211–219.

[12] A. Ansaari and H. Mehta, "Four Quadrant Operation of Chopper – Fed Separately Excited DC Motor by Decoupled PWM Control Using Digital Signal Processor," vol. 4, no. 5, pp. 1720–1729, 2015.

[13] M. K. Singh and S. Srivastava, "ANFIS based four quadrant chopper control of separately excited DC motor : A literature review," no. Iccccm, 2016.

[14] B. Kumar, S. K. Swain, and N. Neogi, "Controller Design for Closed Loop Speed Control of BLDC Motor," vol. 9, no. 1, 2017.

[15] Adnan Khalid1, Ahmed Hussnain Shahid2, Kamran Zeb1, Amjad Ali1, Aun Haider1' "Comparative Assessment of Classical and Adaptive controllers for Automatic Voltage Regulator", Proceedings of the 2016 International Conference on Advanced Mechatronic Systems, Melbourne, Australia, November 30 - December 3, 2016.

[16] Z. Li, C. Yin, Y. Chen, and J. Liu, *Feedback with a Fractional Order*, vol. 47, no. 3. IFAC, 2014.

[17] A. S. Bazanella, L. F. A. Pereira, and A. Parraga, "A new method for PID tuning including plants without ultimate frequency," *IEEE Trans. Control Syst. Technol.*, vol. 25, no. 2, pp. 637–644, 2017.

Design of FPI-PD Controller for Brushless DC Motor

Roshan Bharti, Rishika Trivedi, Prabin Kumar Padhy
Department of Electronics and Communication Engineering,
PDPM Indian Institute of Information Technology, Desgn and Manufacturing, Jabalpur, India

[1]bhartiroshan99@gmail.com
[2]rishika.t@iiitdmj.ac.in
[3]prabin16@iiitdmj.ac.in

Abstract— **In this paper, a new fuzzy proportional-integral-proportional-derivative (FPI-PD) controller is designed, in which fuzzy PI controller is in the forward path and conventional PD controller is in inner feedback loop. The gain parameters of the proposed controller are calculated by gradient descent optimization method. The performance evaluation of this controller is performed for speed control of brushless DC motor in MATLAB/Simulink. The obtained responses are compared with conventional PID controller and the existing fuzzy PID controller. It reveals that proposed FPI-PD controller is more efficient than both the above mentioned controllers.**

Keywords— **PID, FLC, FPI-PD, Brushless DC motor, Gradient descent optimization method**

I. INTRODUCTION

Electrical motors are used in almost all electromechanical movement around us. It converts electrical energy into mechanical energy. There are two types of motors: AC and DC type. The conventional DC motor is used in various applications like electric shaver, trimmer, hair dryer, cranes, starting motor in the car and two-wheeler, photocopy or Xerox machine etc., due to its simple design and low cost [1]. In conventional DC motor, brushes are used for the commutation, which is made up of carbon. The brushes tear out after a span of time which leads to sparking and eventually damages the motor. The frequent maintenance of conventional DC motor is required from time to time to avoid wear and tear of the brushes. This is the main reason due to which conventional DC motor is not used for the applications which requires durability and hence, the use of brushless DC motor is increasing day by day. Brushless DC (BLDC) motor is extensively used in numerous applications like heating and ventilation (HVAC), servo robotic positioning actuators, traction, fans, quadcopter, transportation, washing machine, mixer, disc drives, printers, blowers, and drones. The advantages of BLDC motor are low maintenance due to unavailability of brushes, high reliability, high power to weight ratio, high efficiency, high speed and electronic control [2]. BLDC motor is a type of permanent magnet synchronous motor which is driven by DC voltage via three phase inverter but commutation of current is achieved by electronic switches. The instant of commutation is depended on the position of the rotor which is determined by a speed sensor or sensor-less techniques [3]. The back EMF in BLDC motor is of two types either sinusoidal or trapezoidal [4].

Due to a simple structure and low cost, PID controller is widely used in industries even after availability of many advance controllers. There are various kind of methods available for PID controller tuning [5],[6] like Ziegler Nichols, Cohen-Coon, Error-Trial, Chien-Hrones-Reswick, Good gain method, Pole placement method, Kappa-tau method, and Optimization method etc., but among them, Ziegler-Nicholas [7] is most popular and widely used. The problem with conventional PID controller is, it gives a better result only when the parameters are well matched with the system dynamics. However, obtaining the system dynamics of a nonlinear plant is a challenging task. Therefore, intelligent controllers like Artificial Neural Network (ANN) and the fuzzy logic based controllers are widely used for controlling nonlinear plants. The drawback of using ANN is that, it requires input-output data for training purpose [8]. Besides, fuzzy logic does not require exact information about plant model like zeros and poles of the transfer function.

In 1965, Lotfi A Zadeh proposed the fuzzy set theory [9]. In 1974, E.H Mamdani introduced a first fuzzy logic controller for laboratory build steam engine [10]. Fuzzy logic controller essentially contains four parts: - fuzzification, fuzzy inference system, fuzzy rule base and defuzzification. The advantage of the fuzzy logic controller is that it can be easily incorporated with a conventional controller like PI, PD and PID and this combination performance better than a conventional controller. The major issue with Fuzzy logic based controller is the tuning of the gain parameters. Since, the fuzzy logic controller is nonlinear, so, tuning of the gain parameters is crucial as compared to PID controller. If number of gain parameters are less then linear approximation [11] can be used for gain calculation. It is difficult to obtain large number of gain parameters using linear approximation. Consequently, optimization techniques used to calculate the gain parameters of the fuzzy incorporated controller. Various literatures have reported different optimization techniques for the tuning of the fuzzy-based controller such as, Genetic algorithm [12]. Particle swarm optimization [13], Firefly algorithm [14], Bat algorithm [2], Cuckoo search Algorithm [15], Hybrid bacterial-foraging swarm optimization [16] etc.

In this paper, a new FPI-PD controller is proposed for the speed control of BLDC motor where a fuzzy PI controller is in forward path and a conventional PD controller is in inner feedback loop. The gain parameters of the proposed controller are calculated by gradient descent optimization method [17]. The mathematical model of BLDC motor is derived for

978-1-5386-6624-1/18 $31.00 © 2018 IEEE

evaluating the performance of the proposed controller. MATLAB/Simulink environment is used for the simulation. The obtained results are then compared with conventional PID controller and the existing fuzzy PID controller. The results show that proposed controller performs better than both of the above-mentioned controllers in terms of rise time, percentage overshoot and settling time.

II. MATHEMATICAL MODELLING OF BRUSHLESS DC MOTOR

The transfer function of BLDC motor is obtained by:

$$G(s) = \frac{\frac{1}{k_e}}{\tau_m . \tau_e . s^2 + \tau_m . s + 1}, \tag{1}$$

where, τ_m is mechanical time constant and τ_e is electrical time constant

$$\tau_m = \sum \frac{R.J}{k_e.k_t} = \frac{J.\sum R}{k_e.k_t}, \tag{2}$$

and,

$$\tau_e = \sum \frac{L}{R} = \frac{L}{\sum R}. \tag{3}$$

Since, there is a three-phase symmetrical arrangement in brushless DC motor, so mechanical and electrical time constant will become:

$$\tau_m = \frac{J.3R}{k_e.k_t}, \tag{4}$$

and,

$$\tau_e = \frac{L}{3R}, \tag{5}$$

where,

k_e is back EMF constant in v-secs/rad and k_t is torque constant in N-m/A.

The specifications used for modelling of this BLDC motor is obtained from datasheet of EC 45 Ø45 mm, brushless, 30-watt Maxon motor [18].

TABLE I
CHARACTERISTICS OF BLDC MOTOR

Maxon motor data	Value
Terminal resistance (per phase)	1.20 Ω
Terminal inductance (per phase)	0.560 mH
Rotor inertia	92.5 gcm^2
Torque constant	25.5 mNm/A
Mechanical time constant	17.1 ms

From equation (5):

$$\tau_e = \frac{0.560 \times 10^{-3}}{3 \times 1.20} = 155.56 \times 10^{-6}. \tag{6}$$

From equation (4), k_e can be obtained:

$$k_e = \frac{J.3R}{\tau_m . k_t}, \tag{7}$$

where,

$$\tau_m = 17.1 ms = 0.0171 s$$
$$R = 1.20 \Omega,$$
$$J = 92.5 \, gcm^2 = 9.25 \times 10^{-6} \ Kgm^2,$$
$$k_t = 25.5 \frac{mNM}{A} = 25.5 \times 10^{-3} \ Nm / A$$
$$k_e = \frac{9.25 \times 10^{-6} \times 3 \times 1.20}{0.0171 \times 25.5 \times 10^{-3}} = 0.0763 \, v - secs / rad. \tag{8}$$

The transfer function of Maxon motor can be obtained by putting the value of k_e, τ_m and τ_e in equation (1):

$$G(s) = \frac{\frac{1}{0.0763}}{0.0171 \times 155.56 \times 10^{-6}.s^2 + 0.0171.s + 1}, \tag{9}$$

or,

$$G(s) = \frac{13.11}{2.66 \times 10^{-6}.s^2 + 0.0171.s + 1}. \tag{10}$$

III. CONVENTIONAL PID CONTROLLER

PID controller considered is the parallel combination of three control modes, namely, proportional, integral and derivative as shown in Fig. 1. It has combined feature of both PD and PI controller. It improves both transient as well as steady state performance of the plant.

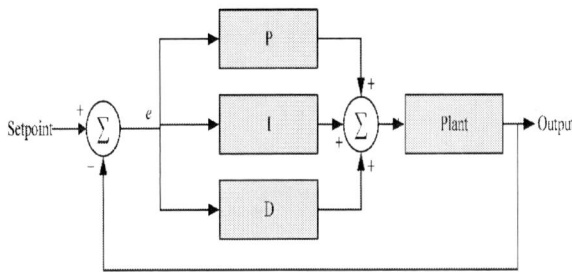

Fig. 1:PID Controller

The output equation of PID controller is:

$$U_{PID}(t) = K_p e(t) + K_i \int e(t) dt + K_d \frac{d}{dt} e(t) \tag{11}$$

or,

$$U_{PID}(s) = K_p e(s) \left(1 + \frac{1}{T_i.s} + sT_d \right), \tag{12}$$

where, e, T_d, T_i, K_p, K_d and K_i represent feedback error, derivative time, integral time, proportional gain, derivative gain and integral gain, respectively.

IV. FUZZY PID CONTROLLER

The architecture shown in Fig. 2 is one of the most popular architecture of fuzzy PID controller (FPID). There are two inputs: error (e) and change in error (ce). There is one output u of FPID controller, which is used for self-tuning the parameters of PID controller. The inputs and the output are divided into seven linguistic variables using five triangular and two gaussian

978-1-5386-6624-1/18 $31.00 © 2018 IEEE

membership functions. The membership functions for both inputs and output is shown in Fig. 3 and Fig. 4, respectively.

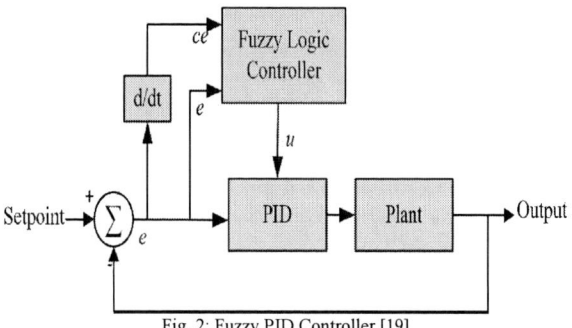

Fig. 2: Fuzzy PID Controller [19]

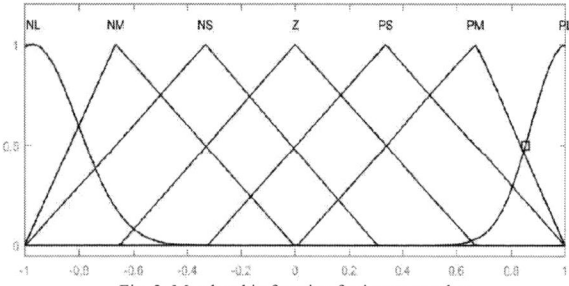

Fig. 3: Membership function for inputs e and ce

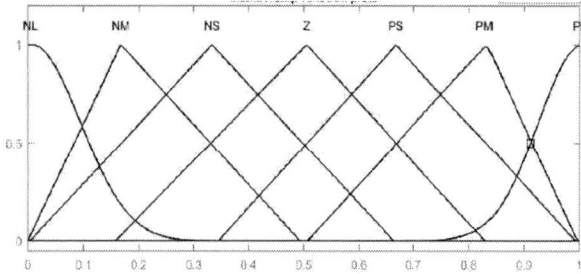

Fig. 4: Membership function for output u.

In this FPID controller, Mamdani model is used for fuzzy inference system and Centre of gravity (COG) method is used for defuzzification. The rule base for FPID controller is

shown in Table II. The linguistic variables are PL, PM, PS, Z, NS, NM and NL, namely, positive large, positive medium, positive small, zero, negative small, negative medium and negative large, respectively.

TABLE II
RULE BASE FOR FPID CONTROLLER [19]

ce /e	NL	NM	NS	Z	PS	PM	PL
NL	PL	PL	PL	PL	PL	PS	Z
NM	PL	PM	PM	PM	PS	Z	NS
NS	PL	PM	PS	PS	Z	NS	NM
Z	PM	PM	PS	Z	NS	NM	NS
PS	PM	PS	Z	NS	NM	NM	NL
PM	PS	Z	NS	NM	NM	NL	NL
PL	Z	NS	NM	NL	NL	NL	NL

V. PROPOSED FPI-PD CONTROLLER

A. Controller Architecture

The proposed structure of FPI-PD controller is shown in Fig. 5. In this structure, fuzzy PI controller is connected in the forward path and conventional PD controller is connected in the inner feedback loop. In this proposed structure, the fuzzy logic controller has two inputs: normalized error (E) and normalized change in error (CE). It has two outputs $u1$ and $u2$ which are used to adjust the proportional and integral gain of PI controller, respectively. The inputs are divided into seven linguistics variable using seven triangular membership functions and the outputs are divided in to five linguistic variables using five triangular membership functions. The rule base for FPI controller is shown in Table III, where S, M, L, VL, PL, PM, PS, Z, NS, NM, and NL are known as small, medium, large, very large, positive large, positive medium, positive small, zero, negative small, negative medium, and negative large respectively. The membership functions for inputs and outputs are shown in Fig. 6 and Fig. 7, respectively. In this controller, Mamdani model is used for fuzzy inference system and Centre of gravity (COG) method is used for defuzzification

There are six scaling factors $\left(k_1 - k_6\right)$ in this proposed controller, where $\left(k_1 - k_2\right)$, $\left(k_3 - k_4\right)$ and $\left(k_5 - k_6\right)$ are the scaling factors of the inputs, PI controller and PD controller,

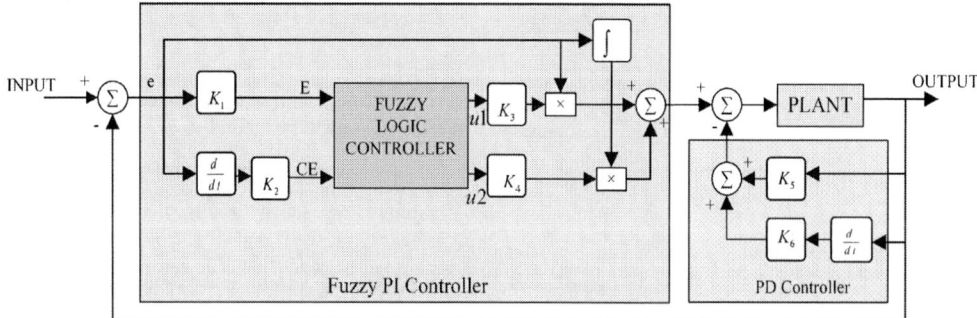

Fig. 5. Proposed FPI-PD controller structure

978-1-5386-6624-1/18 $31.00 © 2018 IEEE

respectively. The calculation of scaling factors $(k_1 - k_6)$ is a difficult task because it is committed with both: fuzzy PI and PD controller. Therefore, gradient descent optimization technique is used to calculate the scaling factors of this proposed controller.

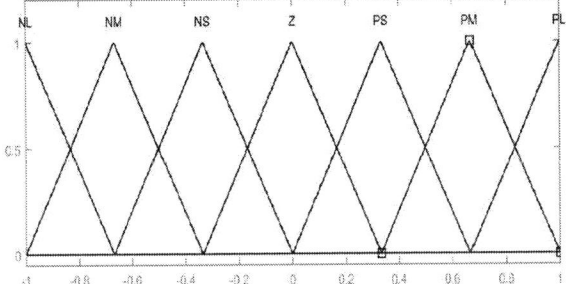

Fig. 5: Membership function for inputs E and CE.

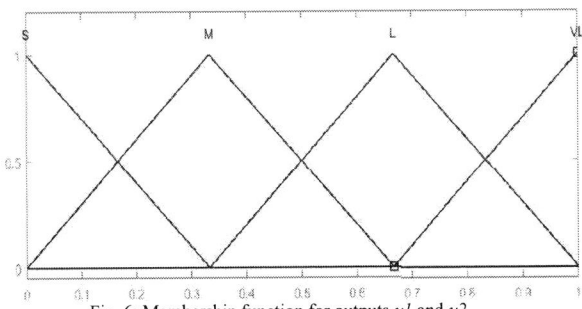

Fig. 6: Membership function for outputs *u1* and *u2*.

TABLE III
RULE BASE FOR FPI CONTROLLER [16]

CE/E	NL	NM	NS	Z	PS	PM	PL
NL	VL	VL	VL	VL	VL	VL	VL
NM	L	VL	VL	VL	VL	VL	L
NS	M	L	VL	VL	VL	M	M
Z	S	M	L	VL	L	L	S
PS	M	L	VL	VL	VL	M	M
PM	L	VL	VL	VL	VL	VL	L
PL	VL	VL	VL	VL	VL	VL	VL

B. Gradient descent optimization method:

In 1847 it was Cauchy who first minimized a function by taking negative of the gradient vector as a search direction. The gradient descent method which is also known as steepest descent method[20] is first order iterative optimization algorithm which is used to determine the local minimum of the function. In gradient descent optimization method we start from an initial point (X_1) and iteratively move towards the direction of negative of gradient vector until the minimum point is obtained. The flow chart of gradient descent optimization method is shown in Fig. 8. Similarly, to find local maximum of the function one needs to consider steps equivalent to positive of the gradient of that function. This method is named as gradient ascent method.

The iteration process has repeated according to the following equation:

$$X_{K+1} = X_K + \lambda_K^* S_K = X_K - \lambda_K^* \nabla f_K \qquad (13)$$

where, S_K is search direction, K is a number of iteration, λ_K^* is optimal step size and ∇f_K is gradient vector of function.

$$S_i = -\nabla f_K = -\nabla f(X_K) \qquad (14)$$

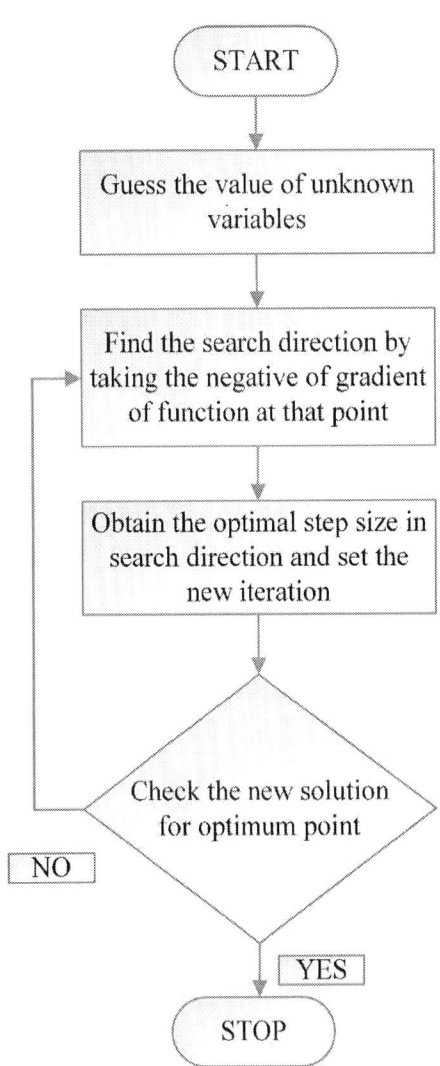

Fig. 7: Flowchart of gradient descent optimization method

VI. SIMULATION RESULTS

The obtained transfer function of BLDC motor is used for simulation. The simulation is run for 0.05 seconds by considering the fixed step size of 0.00001. The gain parameters of the proposed controller are calculated by gradient descent optimization method via Simulink design optimization toolbox in MATLAB/Simulink. There are two set-point operating condition is considered. In the first case, there is a fixed speed in which a load disturbance of 0.3 is given at 0.03

seconds. In the second case, a variable speed is considered in which speed goes down from 1 to 0.5 at 0.025 seconds. The responses of brushless DC motor for both the set-point operating condition is shown in Fig. 9 and Fig.10, respectively.

TABLE IV
GAIN PARAMETERS OF CONTROLLERS FOR BLDC MOTOR

PID Controller (Ziegler-Nicholas Method)	k_P	11.327
	k_I	1381.34
	k_D	0.0232
FPI-PD Controller (Gradient descent optimization method)	k_1	43.7774
	k_2	1.8678
	k_3	338.1774
	k_4	0.2554
	k_5	0.0025
	k_6	0.0062

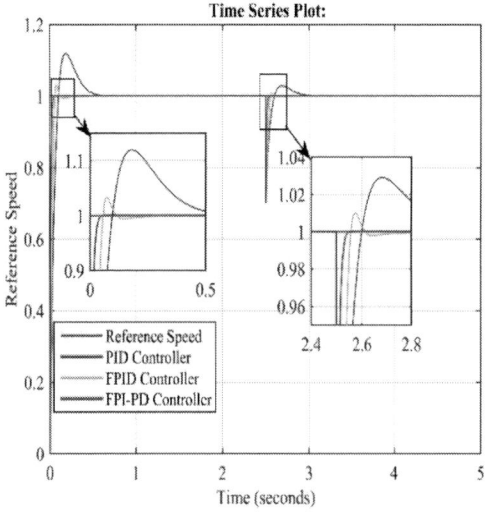

Fig. 8: Response of BLDC motor at fixed speed with load disturbance

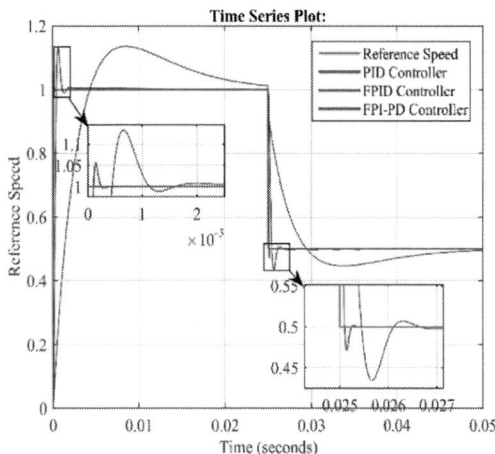

Fig. 9: Response of BLDC motor at variable speed

TABLE V
PERFORMANCE INDICES OF CONTROLLERS FOR BLDC MOTOR AT FIXED SPEED WITH LOAD DISTURBANCE

Controller	Rise time (in ms)	% Overshoot	Settling time (in ms)
PID	2.340	22.619	16.446
FPID	0.296814	13.068	0.7907
FPI-PD	0.059822	6.471	0.148016

TABLE VI
PERFORMANCE INDICES OF CONTROLLERS FOR BLDC MOTOR AT VARIABLE SPEED

Controller	Rise time (in ms)	% Overshoot	Settling time (in ms)
PID	3.744	45.352	22.50
FPID	0.168017	25.556	1.1269
FPI-PD	0.042049	11.224	0.156923

VII. CONCLUSION

The performance of proposed FPI-PD Controller is evaluated by controlling the speed of BLDC motor in MATLAB/Simulink. The mathematical model of BLDC motor is derived by its parameters. Its speed is controlled by conventional PID controller, FPID controller and proposed FPI-PD controller by considering the different set-point operating condition such as the fixed speed with load disturbance and variable speed condition. The performance of all the above mentioned controllers is summarized in Table V and Table VI for both the set-point operating condition in terms of transient performance. The results reveal that the proposed FPI-PD controller outperforms in terms of rise time, percentage overshoot and settling time, etc.

REFERENCES

[1] U. K. Bansal and R. Narvey, "Speed Control of DC Motor Using Fuzzy PID Controller," vol. 3, no. 9, pp. 1209–1220, 2013.

[2] K. Premkumar and B. V. Manikandan, "Bat algorithm optimized fuzzy PD based speed controller for brushless direct current motor," *Eng. Sci. Technol. an Int. J.*, vol. 19, no. 2, pp. 818–840, 2016.

[3] M. A. Shamseldin and A. A. El-samahy, "Speed control of BRUSHLESS DC motor by using PID control and self-tuning fuzzy PID controller," *15th Int. Work. Res. Educ. Mechatronics*, pp. 1–9, 2014.

[4] K. Tabarraee, J. Iyer, S. Chiniforoosh, and J. Jatskevich, "Comparison of brushless DC motors with trapezoidal and sinusoidal back-EMF," in *2011 24th Canadian Conference on Electrical and Computer Engineering(CCECE)*, 2011, pp. 000803–000806.

[5] S. B. Prusty, K. K. Mahapatra, U. C. Pati, and S. Padhee, "Comparative performance analysis of various tuning methods in the design of PID controller," *Michael Faraday IET Int. Summit 2015*, p. 8 (6 .)-8 (6 .), 2015.

[6] P. Cominos and N. Munro, "PID controllers: recent tuning methods and design to specification," *IEE Proc. - Control Theory Appl.*, vol. 149, no. 1, pp. 46–53, Jan. 2002.

[7] B. J. G. Ziegler and N. B. N. Rochester, "Optimum Settings for Automatic Controllers," 1942.

[8] M. A. H. Azman, J. M. Aris, Z. Hussain, A. A. A. Samat, and A. M. Nazelan, "A Comparative Study of Fuzzy Logic Controller and

Artificial Neural Network in Speed Control of Separately Excited DC Motor," no. November, pp. 24–26, 2017.

[9]   L. a. Zadeh, "Fuzzy sets," *Inf. Control*, vol. 8, no. 3, pp. 338–353, 1965.

[10]   E. H. Mamdani, "Application of fuzzy algorithms for control of simple dynamic plant," *Proc. Inst. Electr. Eng.*, vol. 121, no. 12, p. 1585, 1974.

[11]   R. Manikandan, A. Arulprakash, and R. Arulmozhiyal, "Design of equivalent fuzzy PID controller from the conventional PID controller," *2015 Int. Conf. Control Instrum. Commun. Comput. Technol. ICCICCT 2015*, pp. 356–362, 2016.

[12]   Zhihong Xiu and Guang Ren, "Optimization design of TS-PID fuzzy controllers based on genetic algorithms," in *Fifth World Congress on Intelligent Control and Automation (IEEE Cat. No.04EX788)*, vol. 3, pp. 2476–2480.

[13]   Z. Huang and Y. Wang, "Design of a PSO Based Fuzzy Logic Controller for Vessel Mooring Shifting System," in *2008 Fifth International Conference on Fuzzy Systems and Knowledge Discovery*, 2008, pp. 306–310.

[14]   R. Trivedi, P. K. Padhy, S. K. Jain, and R. Base, "Design of Fuzzy PID Controller Using Modified Firefly Algorithm," *Power, Commun. Inf. Technol. Conf. (PCITC), 2015 IEEE*, 2015.

[15]   P. Kumar, S. Nema, and P. K. Padhy, "Design of Fuzzy Logic based PD Controller using cuckoo optimization for inverted pendulum," in *2014 IEEE International Conference on Advanced Communications, Control and Computing Technologies*, 2014, pp. 141–146.

[16]   A. Fereidouni, M. A. S. Masoum, and M. Moghbel, "A new adaptive configuration of PID type fuzzy logic controller," *ISA Trans.*, vol. 56, pp. 222–240, 2015.

[17]   C. C. Soon, R. Ghazali, H. I. Jaafar, and S. Y. S. Hussien, "PID controller tuning optimization using gradient descent technique for an electro-hydraulic servo system," *J. Teknol.*, vol. 77, no. 21, pp. 33–39, 2015.

[18]   O. J. Oguntoyinbo, "PID Control of Brushless DC Motor and Robot Trajectory Planning Simulation with MATLAB/Simulink," *Thesis Bachelor*, p. 90, 2009.

[19]   A. Varshney, D. Gupta, and B. Dwivedi, "Speed response of brushless DC motor using fuzzy PID controller under varying load condition," *J. Electr. Syst. Inf. Technol.*, vol. 4, no. 2, pp. 310–321, 2017.

[20]   S. S. Rao, *Engineering Optimization: Theory and Practice*. 2009.

Condition Monitoring of Photovoltaic Systems Using Machine Learning Techniques

Kurukuru Varaha Satya Bharath
Department of Electrical Engineering,
Jamia Millia Islamia,
New Delhi, India.
kvsb272@gmail.com

Ahteshamul Haque
Department of Electrical Engineering,
Jamia Millia Islamia,
New Delhi, India.
ahaque@jmi.ac.in

Mohammed Ali Khan
Department of Electrical Engineering,
Jamia Millia Islamia,
New Delhi, India.
mak1791@gmail.com

Abstract: Condition monitoring of any system is essential to maintain its healthy operation as it results in getting maximum revenue with minimum maintenance and operation costs. The main objective of this paper is to develop a fault detection algorithm capable of classifying different faults that can be occur in a Photovoltaic (PV) systems. Output characteristics of the PV system are used as valuable information to observe various types of faults and their locations. Wavelet transforms and neural network systems were adapted to filter the non-significant anomalies and make it easier to detect faults that are to be taken care of in a timely manner. The neural network (NN) classification adapts Multilayer perceptron (MLP) to identify the type and location of occurring faults. Wavelet transform (WT) based signal processing technique is utilized in the feature extraction process to provide inputs to the NN. The developed detection algorithm is adapted for 24/7 automated surveillance. The developed algorithm achieved 98.2% accuracy when tested on a predetermined fault data set.

Keywords— Photovoltaic (PV) Systems, Fault Detection, Wavelet Transform, Feature Extraction, Principle component analysis (PCA), Multi-layer perceptron neural network (MLPNN).

I. INTRODUCTION

A standard PV system consists of the PV array, responsible for absorbing sunlight and producing electricity, the inverter, to convert the direct current coming from the PV array to alternating current so that it can be directed to the utility grid, mounting and cabling accessories for assembling a working system. Normally, PV systems also have data acquisition and monitoring systems, but the information they provide is currently not being used as effectively as it could be for maximizing performance and deciding which maintenance practices are the best and when they should be done. Given that this is still a relatively new and very attractive industry, the main focus of most companies is still on installing more systems and to expand, leaving only a minority of them to specialize on maintenance. However, this trend is about to change as the costs of PV modules and installation rapidly decrease and companies look for new ways of increasing revenue [1].

PV monitoring is essential in all kinds of systems, small or big, for reducing costs related to operation and maintenance and increasing both the system's revenue and lifetime. It is also important for the system owners (some of them may have very little knowledge of how the system works) to have a way of seeing how their systems are operating. Since PV systems may keep generating for long periods of time, even if some components are in some way not working properly and some faults are simply temporary due to the intermittency of the resource or due to the automatic corrective actions of the

inverters, it sometimes become difficult for the system operators to decide if they should allocate resources to solve faults as soon as they appear, since current monitoring tools don't offer enough information for such decisions to be easily made. On the other hand, there are cases where some faults that critically affect the system's performance are not promptly resolved.

There is still much to be done in the photovoltaic industry as new services and business opportunities appear. The creation of a tool aimed at supporting the system operator's decision making may be a step in the right direction. Within this context we feel compelled to create a tool that can detect faults by processing the information about the system's parameters and analyzing electrical patterns that may exist in the data provided by the existing data acquisition tools found on the PV systems. Changes in these patterns may represent faults or unwanted behavior within the components of the system. It is our hope that we can minimize maintenance costs and maximize system revenue by helping operators decide exactly which situations require special attention and avoid unnecessary maintenance operations on one side and prevent critical faults from occurring for an extended period of time on the other.

The aim of this paper is to develop methods to detect variations in performance inside the system and ascertain which of them require the most attention from the system operator, more specifically: Identify reference patterns for production throughout days, weeks and months. Identify patterns in the measurements of the inverters. Find deviations in production between the inverters, to help find anomalies on them. Identify reference patterns for global production of the system, to help find variations that may indicate, for example, dirt on the modules or other generic problems. This paper will be structured a brief review of the Literature which Contains a description of typical PV system faults and ways to identify them. Wavelet transform and feature extraction involving discrete wavelet transform. Principal component analysis to minimize the extracted feature data set. And finally, Multi-Layer Perceptron Neural Network to train the extracted features for the purpose of fault detection and classification.

II. LITERATURE SURVEY

Irrespective of the faults at manufacturing and installation stage of photovoltaic modules, other external conditions like Shading [2], delamination [3], and Junction box failure [4] which causes a significant decrease in performance and may damage the arrays, Module soiling, which results in the power loss [5], Failure of DC-DC converters due to MPPT problems [6], and inability of the inverter to connect to the grid due as inverter's DC current decreases to zero, or inverter's DC voltage increases to its maximum value (VOC) or inverter's DC power and AC delivered power decrease to zero in grid-tied systems [7], were the most common faults.

Overall, all the above-mentioned faults can be detected visually or with the help of more advanced methods that rely on computational power. In recent years there has been an increasing interest in developing new, more efficient electrical fault detection methods. This interest is reflected in the literature, which has been visibly expanding on this particular subject. The following paragraphs contain a general overview of the most acknowledged methods in the literature. To begin with, in ref. [8] a method was proposed which relies on satellite-based systems to acquire meteorological data of the desired location. This method allows the detection of constant and variable energy losses caused by degradation of the solar modules, soiling, string failures, mismatch effects, disconnection of the system from the grid, inverter's limitations, MPPT failures or unfavorable temperatures. It can also detect losses due to the presence of snow or due to blackouts. While this method cuts down on computational and simulation costs, it can, on the other hand, lead to high costs for data loggers (electronic devices) and communication systems. As the cost for PV systems decreases rapidly, cheaper monitoring methods are favored. A monitoring circuit built to measure the operating voltage and current, the short-circuit current and the open circuit voltage of PV modules was developed in ref. [9]. The circuit can substitute the junction box and, since it is wireless, does not require any additional cables. This method excels at detecting partial shading on the strings, the number of shaded panels and the efficiency of the MPPT. Some methods do not require climate data, such as the following method based on the Earth Capacitance Measurement (ECM) that was developed in ref. [10] in order to detect the disconnection of a PV module. Additionally, the Time-Domain Reflectometry (TDR) proposed in ref. [11] can not only be used to detect disconnections of PV modules but it can also detect the impedance variation due to degradation. Finally, a statistical method developed in ref. [12] based on the Analysis of Variance (ANOVA) and on the non-parametric Kruskale Wallis (KW) tests shows high levels of accuracy and speed in locating mis operations and forecasting faults. A fault detecting method based on the extended correlation function and the matter element model was proposed in ref. [13] to detect malfunctions in a fast and efficient way and with less memory consumption. It also allows the maintenance teams to confirm the fault types without interrupting the system. A method to detect the partial shadow phenomenon was developed in ref. [14] which relies on the analysis of the current-voltage characteristic (I-V characteristic). There are also methods that use the Maximum Power Point Tracking (MPPT) approach. A fault detection procedure based on the analysis of power losses was developed in ref. [15]. It allows the identification of three groups of faults. The following method [16] is also based on the analysis of power losses but it can detect faults in both the PV array and the inverter. Other methods are based on Artificial Intelligence (AI) techniques. A technique based on Expert Systems was developed in ref. [17] and allows the identification of two types of faults (failures due to shading effects and inverter failures). Observing all the above techniques and their drawbacks a

machine learning technique was developed in further sections for efficient monitoring od PV Systems.

III. FEATURE EXTRACTION USING WAVELET TRANSFORM

PV monitoring is the act of visualizing and constantly checking the information coming from the system about its state and performance in order to maintain a healthy operation. In order to achieve it, the electrical outputs of the PV systems under various faults and operating conditions were simulated using MATLAB/ Simulink and the responses were recorded for development of the algorithm. Once the data regarding all the signal were acquired wavelet transform based feature extraction was performed to observe the different features for all the signals. The graphical representation of the work process id depicted in figure 1.

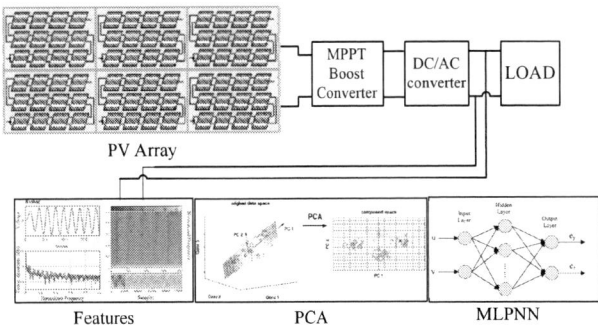

Fig. 1 Graphical Network of PV Monitoring System

A. System Characteristics

The subsequent analysis for different faults were carried out in a 3.5kW PV system at multiple irradiances. The nominal voltage of the dc link is 400V. Irrespective of module faults, various faults in power converters are analyzed, which are observed due to parametric variation in filter components, open circuit and short circuit of different switches in the converters and many more. Under every fault, the components associated with the output of the PV system change their magnitudes with respect to the nominal values as shown in figure 2. These facts can be used for a scheme of fault diagnosis based on the voltage or current analysis.

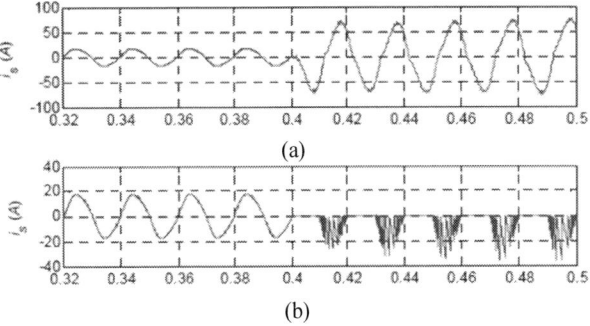

Fig. 2 PV system output currents due to converter faults, (a) Short circuit of Boost Converter Diode, (b) Open circuit of a switch in Inverter.

B. Wavelet Analysis and Feature Extraction

Wavelets are mathematical functions which depicts an oscillation with an amplitude beginning at zero, rises, and then declines back to zero. This results in observing some interesting features which make them advantageous for signal processing applications. In general a wavelet transform is represented as a function of certain orthonormal series [18]. Considering a time-domain signal f(t) which can be expanded by the use of orthonormal wavelet functions [19], such that

$$[W_\psi f](s,\tau) = \int_R f(t)\psi^*\left(\frac{t-\tau}{s}\right)dt \qquad (1)$$

where ψ(t) is the mother wavelet, s = 1, ..., m and τ = 1, ..., n are the scale and translation parameters, respectively; $[W_\psi f](s,\tau)$ is the wavelet transformation of the signal f(t) [20]. Wavelet analysis is an effective tool that extracts two-dimensional information from time-domain signals. The wavelet coefficients form the $m \times n$ matrix from the raw data. The magnitude-square is computed from the wavelet data for mathematical convenience.

Wavelet analysis decomposes signals into a different levels and the sub-band energies which are observed at the last or previous levels are used as features. Based on the dominant frequency components of the signal the decomposition level is selected. The mother wavelet and its decomposition level must be carefully selected. Several mother wavelets can be used [18], however Daubechies (db) is a popular choice [21] due to its near smoothing features that make suitable detecting changes on any given signal and optimal time-frequency localization properties. In this paper wavelet analysis is carried out using Discrete Wavelet Transform. The DWT is less complex and has less frequency resolution than the other transforms resulting in better decomposition and therefore a better resolution by using more filters. By applying DWT to the electrical outputs of the PV system using db5, the signals are decomposed. A multi levl wavelet decomposition is shown in figure: . At each wavelet decomposition level j, the discrete-time signal is decomposed into approximation wavelet coefficients Ca_j (low frequency), and detail wavelet coefficients Cd_j (high frequency). The approximation and the detail wavelet coefficients can be computed using the following:

$$Ca_j(e) = \sum_{l_e} f_{l_o}(l_e - 2e)Ca_{j-1}(l_e) \qquad (2)$$
$$Cd_j(e) = \sum_{l_e} f_{l_1}(l_e - 2e)Ca_{j-1}(l_e) \qquad (3)$$

Where, f_{l_o} is the low pass filter and f_{l_1} is the high pass filter.

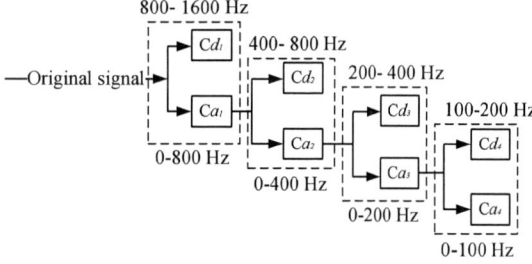

Fig. 3 A multilevel DWT decomposition of signals

Since DWT is a time-frequency domain method, it is suitable to study non-stationary signals like the ones obtained from electrical outputs, Electroencephalography and others. Time domain and frequency domain methods like filters and Fast Fourier Transform approaches are not usually well suited to handle this type of signals. Once the signal is filtered and reconstructed, various features like energy, entropy, mean, skewness, kurtosis, standard deviation, Power spectral density, peak analysis, Total Harmonic Distortion, Signal to noise ratio and many more can be extracted. The mathematical formulations for various features are given in equations below:

$$Mean(\mu) = \frac{1}{N}\sum_{i=0}^{N-1}V_i \qquad (4)$$

Where N= number of samples and V_i is the sampled reconstructed signal
Standard Deviation of the reconstructed signal is calculated by:

$$Standard\ Deviation\ (\sigma) = \frac{1}{N-1}\sum_{i=0}^{N-1}(V_i - \mu)^2 \qquad (5)$$

Skewness and Kurtosis of the reconstructed signal is given by:

$$Skewness\ (S) = \frac{\frac{1}{N}\sum_{i=0}^{N}(V_i-\mu)^3}{\left(\sqrt{\frac{1}{N}\sum_{i=0}^{N}(V_i-\mu)^2}\right)^3} \qquad (6)$$

$$Kurtosis\ (K) = \frac{\frac{1}{N}\sum_{i=0}^{N}(V_i-\mu)^4}{\left(\sqrt{\frac{1}{N}\sum_{i=0}^{N}(V_i-\mu)^2}\right)^4} \qquad (7)$$

Peak to peak value of the reconstructed signal is given by:

$$V_{pp} = 2\sqrt{2}\sigma \qquad (8)$$

Energy of the decomposed signal is formulated using:

$$E = \int_{-\infty}^{\infty}|V(t)^2|dt \qquad (9)$$

Power of the reconstructed signal is given by:

$$P = \lim_{N\to\infty}\frac{1}{2T}\int_{-T}^{T}|V(t)^2|dt \qquad (10)$$

Where V(t) is the reconstructed signal
Entropy, defined as a major tool in information theory. It is also used to estimate the type of wavelet suitable for decomposing and reconstructing a given signal. The entropy of a given signal is found by equation:

$$H(V) = -\sum_{i=1}^{N}p(V_i)\log_{10}p(Vi) \qquad (11)$$

Where $p(V_i)$ is given by probability of sample of voltage signal. To estimate the total harmonic distortion of a sinusoidal signal in time domain we use the equation:

$$y(t) = \alpha_0 + \frac{\alpha_2}{2} + \frac{3\alpha_4}{8} + \left(\alpha_1 + \frac{3\alpha_3}{4} + \frac{10\alpha_5}{16}\right)sin\omega_0 t - \left(\frac{\alpha_2}{2} + \frac{\alpha_4}{2}\right)cos2\omega_0 t - \left(\frac{\alpha_3}{4} + \frac{5\alpha_5}{16}\right)sin3\omega_0 t + \frac{\alpha_4}{8}cos4\omega_0 t + \frac{\alpha_5}{16}sin5\omega_0 t \qquad (12)$$

Where α_i =coefficients of Taylor series.
The signal to noise ratio for a given signal is determined by the ratio of reconstructed signal to original signal:

$$SNR = \frac{Reconstructed\ Signal}{Original\ Signal} \qquad (13)$$

Once the required features of all the faults and operating conditions of PV systems were extracted we apply principal component analysis to minimize the feature set. Some of the extracted features were depicted in **table:1**.

Table 1: Samples for Testing Output Results of the MLP Model

Sample Number	Features extracted					
	Mean	Std. Dev	Skewness	Kurtosis	Entropy	Energy
1	3.7598	30.4455	8.2192	67.5562	0.0899	0.9572
...	...	...	...	...	...	...
189	25.4473	72.3419	3.1593	9.9814	0.3325	0.8005
...	...	...	...	...	...	...
388	15.7644	59.3511	4.0140	16.1122	0.2488	0.8709

...	...	...	...	...	...	...
656	11.8335	52.2702	4.6329	21.4643	0.2061	0.8961

B. Data Reduction Using Principal Component Analysis

The Principal Component Analysis (PCA) is a data reduction method which transforms any particular data set into a lower dimensional feature space [20]. To apply PCA, the winning cells of each class are placed into a matrix X, where $X = [R(\theta), \forall \theta \in \Theta^*]$, such that the rows represent the length of a cell and the columns represent the number of optimal cells $v = |\Theta^*|$. The goal of PCA is to find a reduced feature set Y containing $q < v$ principal components that summarize the most useful information in X. Using the Karhunen-Loeve (KL) algorithm, the uncorrelated features, called Principal Components (PCs) or score vectors, are inferred from the data matrix. PCA is an optimal feature extraction method as the PC's computed are based on the maximization of the data matrix variance. These PC's capture the most useful information in the classes from the original data matrix, as can be seen by the formation of separable clusters in the feature space. The KL algorithm is described as shown in figure 4:

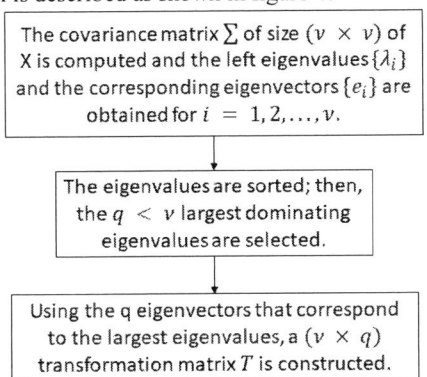

Fig. 4 Karhunen-Loeve Algorithm for Principal Component analysis

Consequently, the original data set X is transformed into the feature set Y, using Equation 14 below.

$$Y = X \times T \tag{14}$$

The Principal component analysis on the features in table 1 resulted in three uncorrelated principal components which are scattered as shown in figure:5.

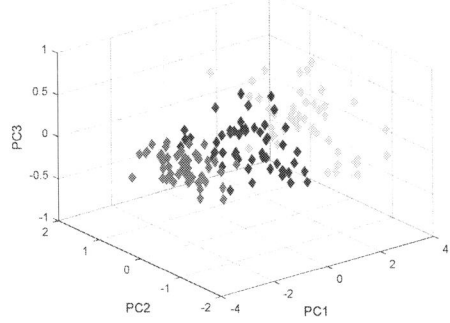

Fig.5 Principle component variable after PCA

IV. FAULT CLASSIFICATION USING MLPNN

In general, fault classification algorithms can learn from the data in the form of parametric methods or non-parametric methods. For parametric designs, a decision boundary is constructed around the different class regions. Non-parametric methods seek to associate posterior probabilities that a feature is assigned to a particular class. Furthermore, most classification algorithms are based on statistical methods. For example, a classifier learns the existence of a fault by observing most frequent patterns that a fault can exhibit. After much consideration of the standard classifiers, the Multi-Layer Perceptron Neural Network (MLPNN) algorithm was used for classification in this paper.

A. Multi-Layer Perceptron Neural Network

Artificial neural network has a remarkable ability to learn and get useful results from complicated and confused data. This characteristic of ANN gives an advantage over the conventional classification methods [23]. There are different types of ANNs, each with its own advantages and limitations depending on the application. Selection and implementation of the network topology should be more desirable with selecting the smallest number of neuron nodes with an appropriate learning algorithm. The purpose of learning algorithm is to minimize NNs structure by clipping unnecessary neurons. The main advantages of that can be reduced runtime cost and physical implementation [24].

In this work, the network of choice for classification is an MLP with a back propagation learning algorithm for supervised classification. An MLP is a feedforward artificial neural network model, characterized by the fact that the information starting from the input is only allowed to pass forward in the network to the output through a specified hidden layer, no feedback is allowed. The MLPs are popular due to their computational simplicity, flexibility, finite parameterization, stability, and smaller size for a problem as compared to other architectures [25]. The architecture of this class of network, besides having the input and the output layers, also has one or more intermediary layer or hidden layer [26].

B. Training of MLPNN

The objective of training an MLP is to produce the desired output when a set of input is applied to the MLP. The proceeding of a neural network starts from the system activated by the input layer where the input data are weighted, and then neurons in the hidden layer perform a user chosen computation method and continue to activate all neurons to the end of this layer. Finally, the output layer determines which characteristics should be read [27].

The MLPs are learned using various algorithms like Back Propagation (BP). Because of the BP network algorithm is simple, a small amount of calculation and parallel advantages. It's currently one of the most used and most mature neural network training algorithms. According to statistics, results, between 80%-90%, researchers are using the BP neural network model [28]. But the limitations of this algorithm are that the traditional BP neural network model in the low learning rate, generalization ability is weak, easy to fall into local minima and the algorithm does not converge. The improved algorithms of traditional back propagation neural network conclude gradient descent Bp algorithms, Quasi-Newton Bp algorithms, and conjugate gradient Bp algorithms [28]. After supervised learning of neurons is over, the trained networks are stored to be used in the algorithm. Whenever an image is taken as input to the algorithm, then simulated by the trained network and from the results; a percentage can be given to which diagnosis should be taken from the data set [27], [29].

For simulation analysis, the input data are arranged, and then entered to the intelligent classification program. The classifier models are trained for functional fault classification. Once trained, the network performance is validated using testing data which is different from the training data. Depending on the behavior of PV modules and various faults affecting them the features are divide in to four classes of Fault Free (FF), Module Failure (MF), Converter Failure (CF), and Inverter Failure (IF) [30].

Furthermore, the whole data set is divided randomly into training data set and testing data set. Among 656 samples, 459 samples are selected as a training data set and the remaining 197 samples as testing data set. The accuracy of classification is defined as the percentage of total samples classified correctly to the total number of samples in the dataset as in equation 6.

$$Classification\ Accuracy =$$
$$\frac{Total\ No\ of\ samples\ classified\ correctly}{Total\ no\ of\ samples\ in\ the\ data\ set} * 100\% \qquad (15)$$

To decide that the chosen MLP network layout is good enough, several layouts are tested with the number of hidden neurons ranging from 4 to 40 in steps of 4. The configuration that produces the minimum performance factors comprises one hidden layer with 12 neurons. The input layer has 6 neurons, which takes the value of the input data while the output layer has 4 neurons corresponding to the fault condition as shown in figure 4.

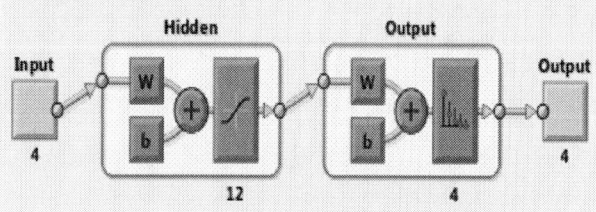

Fig. 6 The proposed neural network architecture layout.

There are several training algorithms, which can be applied to train the MLP. The performance comparison between three main methods of training algorithms is done in [31] and Levenberg-Marquardt backpropagation (Trainlm) is adapted as it converges to the minimum performance factor in the minimum possible time. The training, performance of the algorithm is reduced to 1.0296e-8 after 246 epochs on average. To confirm the developed neural network, a set of input data not used within the training stage, is given to the neural network. Neural network output is then compared with the desired output to confirm the neural network. Data set has characteristics which covers the full operating range for both healthy and faulty operation modes. Output from the neural network is the PV module condition, according to the table 2.

Table 2: Samples for Testing and Output Results of the MLP Model

Input (to MLP)					
PCA Correlated Features					
Feature 1	Feature 2	Feature 3	PV Module	Output (From MLP)	Fault Status
3.509996	2.617815	-0.63745	1	(0100) b	FF
…	…	…	…	…	…
-3.71374	0.248826	0.861656	1	(1000) b	MF
…	…	…	…	…	…
-1.1600	-1.0041	0.1644	2	(0010) b	CF
…	…	…	…	…	…
0.894117	-0.13368	-0.62562	3	(0001) b	IF

If the PV module is working in healthy mode the binary outputs would give a value of (01000) b=FF. Under any fault operation, the corresponding output will classify the fault into four main classes: (1000) b= MF is a fault, (0010) b=CF is a fault, (0001) b = IF. The comparison between the neural network outputs and the desired target outputs shows the correct prediction of the Module condition. The summarized developed MLP network setting parameters in this work can be shown in table 3.

Table. 3 Summarized Settings Parameters for the MLP Network

Parameters	Wavelet-MLPNN Classification Process		Radial Basis Function Kernel Extreme Learning Machine [32]	
	Training Process	Testing Process	Training Process	Testing Process
samples used	459	197	600	400
Network Structures	4-12-4	4-12-4	---	---
Epochs	246	246	300	300
No. of Unclassified samples	4	0	---	---
Performance Criteria	Mean squared error	Mean squared error	---	---
Training algorithm	Levenberg–Marquardt	Levenberg–Marquardt	Simulated Annealing	Simulated Annealing
Average Accuracy of classification	97.82%	98.2%	96.72%	93.55%

Finally, the developed ANN with MLP network present training and testing accuracy about 97.82% and 98.2%, respectively. Therefore, it detects 454 samples out of 459, and 197 samples out of 197 samples after 246 epochs in the training and testing process, respectively.

V. CONCLUSIONS

In this paper, a fault detection system that corresponds to four different operating states in a photovoltaic system is developed. The fault detection system displays information about the system's operating state in a very intuitive way and were designed to be constantly updating in a real scenario, making them ideal for monitoring. The system was developed using two different algorithms that handle the data in different ways - one of them extracts the features of the operating states outputs using discrete wavelet transform and the neural network trains the extracted features to develop the required classification set. Additionally, Principal component analysis was implemented to reduce the feature set for supporting the classification process. In the end it is observed that the system allowed us to detect which faults occurred persistently and which didn't. The final results depicted 97.82% training accuracy and 98.2% testing accuracy for a given unknown feature set.

REFERENCES

[1] S. P. Alliance SuNLaMP O and M. Working Group, 'Best Practices in Photovoltaic System Operations and Maintenance 2 nd Edition NREL/Sandia/Sunspec Alliance SuNLaMP PV O&M Working Group', no. December, 2016.

[2] H. Patel, V. Agarwal, and S. Member, 'MATLAB-based modeling to study the effects of partial shading on PV array characteristics', *Energy Conversion, IEEE Trans.*, vol. 23, no. 1, pp. 302–310, 2008.

[3] J. Tracy, N. Bosco, and R. Dauskardt, 'Encapsulant Adhesion to Surface Metallization on Photovoltaic Cells', *IEEE J. Photovoltaics*, vol. 7, no. 6, pp. 1635–1639, 2017.

[4] J. Kalejs, 'Junction box wiring and connector durability issues in photovoltaic modules', *SPIE Sol. Energy + Technol.*, vol. 9179, p. 91790S, 2014.

[5] R. Hammond, D. Srinivasan, A. Harris, K. Whitfield, and J. Wohlgemuth, 'Effects of soiling on PV module and radiometer performance', *Conf. Rec. Twenty Sixth IEEE Photovolt. Spec. Conf. - 1997*, pp. 1121–1124, 1997.

[6] J. Solórzano and M. A. Egido, 'Automatic fault diagnosis in PV systems with distributed MPPT', *Energy Convers. Manag.*, vol. 76, pp. 925–934, 2013.

[7] A. Transmissions, 'Installation Guide', pp. 1–38, 2007.

[8] A. Drews *et al.*, 'Monitoring and remote failure detection of grid-connected PV systems based on satellite observations', *Sol. Energy*, vol. 81, no. 4, pp. 548–564, 2007.

[9] P. Guerriero, V. D'Alessandro, L. Petrazzuoli, G. Vallone, and S. Daliento, 'Effective real-time performance monitoring and diagnostics of individual panels in PV plants', *4th Int. Conf. Clean Electr. Power Renew. Energy Resour. Impact, ICCEP 2013*, pp. 14–19, 2013.

[10] T. Takashima, J. Yamaguchi, K. Otani, T. Oozeki, K. Kato, and M. Ishida, 'Experimental studies of fault location in PV module strings', *Sol. Energy Mater. Sol. Cells*, vol. 93, no. 6–7, pp. 1079–1082, 2009.

[11] I. N. Tro, 'Fault Detection in a Photovoltaic Plant by Time Domain Refleetometry', vol. 2, no. September 1993, pp. 35–44, 1994.

[12] S. Vergura, G. Acciani, V. Amoruso, G. E. Patrono, and F. Vacca, 'Descriptive and inferential statistics for supervising and monitoring the operation of PV plants', *IEEE Trans. Ind. Electron.*, vol. 56, no. 11, pp. 4456–4464, 2009.

[13] K. H. Chao, S. H. Ho, and M. H. Wang, 'Modeling and fault diagnosis of a photovoltaic system', *Electr. Power Syst. Res.*, vol. 78, no. 1, pp. 97–105, 2008.

[14] M. Miwa, S. Yamanaka, H. Kawamura, H. Ohno, and H. Kawamura, 'DIAGNOSIS OF A POWER OUTPUT LOWERING OF PV ARRAY WITH A (-dI / dV) -V CHARACTERISTIC', pp. 2442–2445, 2006.

[15] A. Chouder and S. Silvestre, 'Automatic supervision and fault detection of PV systems based on power losses analysis', *Energy Convers. Manag.*, vol. 51, no. 10, pp. 1929–1937, 2010.

[16] W. Chine, A. Mellit, A. M. Pavan, and S. A. Kalogirou, 'Fault detection method for grid-connected photovoltaic plants', *Renew. Energy*, vol. 66, pp. 99–110, 2014.

[17] Y. Yagi *et al.*, 'Diagnostic technology and an expert system for photovoltaic systems using the learning method', *Sol. energy Mater. Sol. cells*, vol. 75, no. 3–4, pp. 655–663, 2003.

[18] W. K. Ngui, M. S. Leong, L. M. Hee, and A. M. Abdelrhman, 'Wavelet Analysis: Mother Wavelet Selection Methods', *Appl. Mech. Mater.*, vol. 393, pp. 953–958, 2013.

[19] A. A. Silva, S. Gupta, A. M. Bazzi, and A. Ulatowski, 'Wavelet-based information filtering for fault diagnosis of electric drive systems in electric ships', *ISA Trans.*, 2017.

[20] S. Mallat, *A Wavelet Tour of Signal Processing, Third Edition: The Sparse Way.* 2008.

[21] I. S. Kim, 'On-line fault detection algorithm of a photovoltaic system using wavelet transform', *Sol. Energy*, vol. 126, pp. 137–145, 2016.

[22] 'Oppenheim_Signals_and_Systems.pdf'. .

[23] S. Ben Driss *et al.*, 'A comparison study between MLP and Convolutional Neural Network models for character recognition To cite this version : HAL Id : hal-01525504 A comparison study between MLP and Convolutional Neural Network models for character recognition', 2017.

[24] M. A. Sanz-Bobi, A. Muñoz San Roque, A. De Marcos, and M. Bada, 'Intelligent system for a remote diagnosis of a photovoltaic solar power plant', *J. Phys. Conf. Ser.*, vol. 364, no. 1, 2012.

[25] C. Ozkan and F. S. Erbek, 'The comparison of activation functions for multispectral Landsat TM image classification', *Photogramm. Eng. Remote Sensing*, vol. 69, no. 11, pp. 1225–1234, 2003.

[26] N. S. Shahraki and S. H. Zahiri, 'Inclined planes optimization algorithm in optimal architecture of MLP neural networks', *3rd Int. Conf. Pattern Anal. Image Anal. IPRIA 2017*, no. Ipria, pp. 189–194, 2017.

[27] W. Chen and A. M. Bazzi, 'A generalized approach for intelligent fault detection and recovery in power electronic systems', *2013 IEEE Energy Convers. Congr. Expo.*, pp. 4559–4564, 2013.

[28] S. S. Behera and S. Chattopadhyay, 'A comparative study of back propagation and simulated annealing algorithms for neural net classifier optimization', *Procedia Eng.*, vol. 38, pp. 448–455, 2012.

[29] Haoyang Cui, Yongpeng Xu, Jundong Zeng, and Zhong Tang, 'The methods in infrared thermal imaging diagnosis technology of power equipment', *2013 IEEE 4th Int. Conf. Electron. Inf. Emerg. Commun.*, pp. 246–251, 2013.

[30] P. K. Ray, B. K. Panigrahi, P. K. Rout, A. Mohanty, and H. Dubey, 'Detection of Faults in Power System Using Wavelet Transform and Independent Component Analysis', *First Int. Conf. Adv. Comput. Commun. Electr. Technol.*, no. October, pp. 1–5, 2016.

[31] Ö. Kişi and E. Uncuoğlu, 'Comparison of three back-propagation training algorithms for two case studies', *Indian J. Eng. Mater. Sci.*, vol. 12, no. 5, pp. 434–442, 2005.

[32] Y. Wu, Z. Chen, L. Wu, P. Lin, S. Cheng, and P. Lu, 'An Intelligent Fault Diagnosis Approach for PV Array Based on SA-RBF Kernel Extreme Learning Machine', *Energy Procedia*, vol. 105, pp. 1070–1076, 2017.

A Comparative Study of Fuzzy Systems and Neural Networks for System Modeling and Identification

Karnika Pandey
Electrical and Electronics Engg
Maharaja Surajmal Institute of Tech
New Delhi, India
Karnika0212@gmail.com

Shayari Bhattacharjee
Electrical and Electronics Engg
Maharaja Surajmal Institute of Tech
New Delhi, India
shayarib211196@gmail.com

Shreya Lau
Electrical and Electronics Engg
Maharaja Surajmal Institute of Tech
New Delhi, India
shreyalau@gmail.com

Meena Tushir
Electrical and Electronics Engg
Maharaja Surajmal Institute of Tech
New Delhi, India
meenatushir@yahoo.com

Abstract—In this paper, a comparative study is presented for modelling and identification of non-linear systems. In this paper, Fuzzy modeling is done using Gradient Descent method and Neural Network modeling is done using Back Propagation and their performance index (PI) and identifier output is used as a basis of comparison. In this paper, three case studies are presented to illustrate which method provides better approximation results and robustness. Simulation are done using MATLAB 2015a and simulation results are presented.

Index Terms—Fuzzy modelling, Neural Network, Back Propagation, Gradient Descent

I. INTRODUCTION

System identification is determination of black or grey model of a dynamic system on the basis of its input-output characteristics within a specified class systems [1]. The prerequisites of system identification involves a validation criterion, a set of model structures and an aim which has to be achieved [2]. Various aims of system identification involves improving system performance, analyzing properties, acquiring inner knowledge and forecasting evolution of the system [3].

Zaldeh proposed a fuzzy set based fuzzy modelling in [4]. In this paper, fuzzy relation were build based on the input-out data or the operators knowledge and were expressed in the terms of the linguistic variables. Later on, Mamdani [5][6] applied the concept of Compositional Rule of Inference (CRI) for system modelling. But complex system modelling was difficult with the previous works because of formation of complicated fuzzy relations. Tagachi-sugeno [7] proposed TS-model which brought a revolution in the approach of system modelling of non-linear systems. Their work aimed at partitioning the fuzzy input space into small subspaces which can be represented as a linear equation in the form of IF-Then rules. Thus, a large set of rules can be used to represent a highly non-linear dynamic system. The identification of fuzzy system can be divided into two steps: Firstly, Formation of IF-THEN rules which defines the system behavior (Structure Identification). Secondly, Assigning values to parameters after every rule (Parameter Identification) [8].

For the past three decade, system identification using Neural networks for system identification of non-linear systems [9]. Neural networks can be broadly classified into three forms:- 1. Multilayer perceptron, 2. Recurrent neural networks and 3.Convolutional neural networks. Works regarding system identification using multi-layer neural networks can be found in [10][11]. In this paper, Multi-layer neural network is considered.

This paper is organized as follows. The section II gives the background information regarding TS-model and algorithms such as Gradient Descent and Back Propagation Neural Network (BPNN) which are taken into consideration for the comparative study. In section III, the results and simulation for the comparative study on three non-linear system are presented, which is followed by the conclusion in Section IV.

II. BACKGROUND

A. Fuzzy model: Takagi- Sugeno Model (TS-model)

As an improvement to the existing Mamdani model, Takagi-Sugeno model provided a more systematic approach for rule formation in fuzzy systems. The TS model aims at partitioning the fuzzy input space into small subspaces which can be represented as a linear equation in the form of IF-Then rules. Thus, a large set of rules can be used to represent a highly non-linear dynamic system. Learning algorithms can be used to identify the parameters of the fuzzy system. The identification of fuzzy system can be divided into two steps: 1. Formation of IF-THEN rules which defines the system behavior. 2. Assigning values to parameters after every rule. The weighted sum of every rule is then used to calculate the output of the system.

978-1-5386-6625-8/18/$31.00 ©2018 IEEE

The rules used in TS-model has the following structure:

$$RULE : if \ x^k \ is \ A^k \ THEN \ y^k \ is \ f^k(x^k) \qquad (1)$$

where $f^k(x^k)$ defines the system as

$$f^k(x^k) = b_{K0} + b_{k1} + ... + b_{kn_k}xn_k \qquad (2)$$

The firing strength can be found by considering membership function as follows:

$$\mu^k(x^k) = \mu_1^k(x_1) \lor ... \lor \mu_{kn_k}^k(x_{n_k}) \qquad (3)$$

The normalized output and non-normalized output can be calculated as:

$$y^0 = \sum_{k=1}^{m} \frac{\mu^k(x^k)}{\sum_{j=1}^{m} \mu^j(x^j)} f^k(x^k) \qquad (4)$$

$$y^0 = \sum_{k=1}^{m} \mu^k(x^k) f^k(x^k) \qquad (5)$$

B. Back Propagation Neural Network (BPNN)

Back Propagation Neural Networks (BPNN) are multi-layer feed forward network with activation functions which are continuously differentiable [12]. In this network, the weights are continuously updated till the target is achieved. The training of BPNN can be described in three steps: 1. Input of training pattern 2. Calculation and back propagation of error 3. Layer weight updating.

The algorithm of BPNN can be described as follows:-

1) Phase 1 (Feed Forward):: Here the weights are randomly initialized. The hidden layer output is calculated using the following formula:

$$y_{inj} = b_1 + \sum_{i=1}^{n} x_i w_{ij} \qquad (6)$$

Where b_1 denotes the bias on hidden layer, x_i denotes the input and w_{ij} denotes weight. The output of hidden layer is obtained by applying activation function on y_{inj} as shown

$$y_j = f(y_{inj}) \qquad (7)$$

Similarly, the output layer is calculated using following equations:

$$z_{inj} = b_2 + \sum_{n=1}^{n} y_j w_{jk} \qquad (8)$$

$$z_j = f(z_{ink}) \qquad (9)$$

2) b) Phase 2 (Back Propagation of error): The output layer computes the error correction term (δ) by taking into consideration the desired output (y) and the obtained output (y)

$$\delta = (y - y')f'(y_{ink}) \qquad (10)$$

The correction term for outer layer where α is the learning rate is calculated as

$$\triangle \ w_{jk} = \alpha\delta w_{jk} \qquad (11)$$

Similarly, the correction term for hidden layer can be computed using following equations

$$\delta_{inj} = \sum_{m=1}^{k} \delta_k w_{kj} \qquad (12)$$

$$\delta_j = \delta_{inj} f'(y_{inj}) \qquad (13)$$

$$\triangle \ w_{ij} = \alpha\delta_j x_i \qquad (14)$$

3) Phase 3 (Weight updation): For output layer, following formula is used

$$w_{jk}(new) = w_{jk}(old) + \triangle \ w_{jk} \qquad (15)$$

For hidden layer, the formula is

$$w_{ij}(new) = w_{ij}(old) + \triangle \ w_{ij} \qquad (16)$$

III. RESULTS AND SIMULATIONS

Fuzzy model identification using fuzzy logic and neural networks is applied on three non-linear systems which are discussed in this section. The input data set is generated randomly and output data set is initially calculated using the plant equation. The error function is considered as the Performance Index (PI) in equation (11).

The PI serves as the basis of comparison between the two algorithms used. The PI can be calculated as:

$$PI = \frac{1}{2}(y - y') \qquad (17)$$

The three cases are discussed below:

A. Case Study I

A simple non-linear system is considered whose plant equation is defined as:

$$y(x) = \frac{sinx}{x} \qquad (18)$$

Where $y(x)$ is the system output and x is the system input. The constraints for determining the input data set is

$$-10 < x < 10 \qquad (19)$$

The Plant identifier output using gradient descent is shown in Fig 1. The Plant identifier output using back propagation is shown in Fig 2. The comparison of PI is shown in Fig 3.

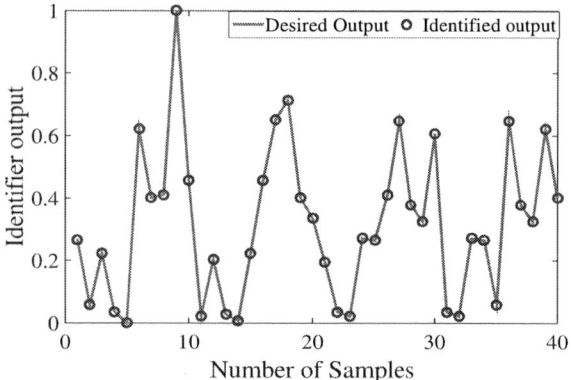

Fig. 1. Identified output of Fuzzy logic using Gradient Descent. (Case I)

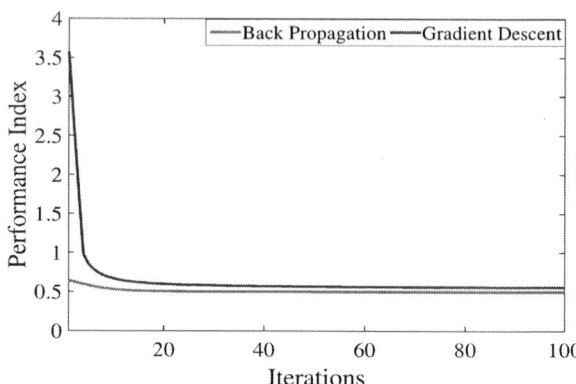

Fig. 3. Performance Index Comparison (Case I)

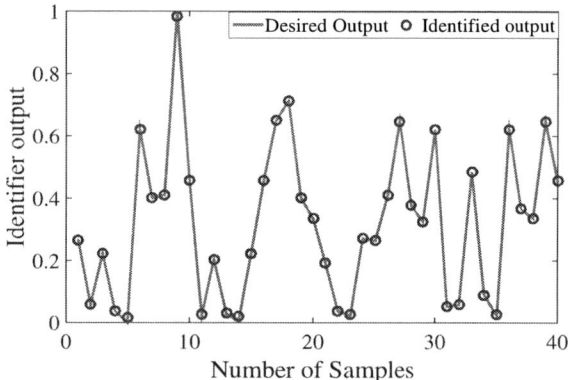

Fig. 2. Identified output of Neural Network using Back Propagation (Case I)

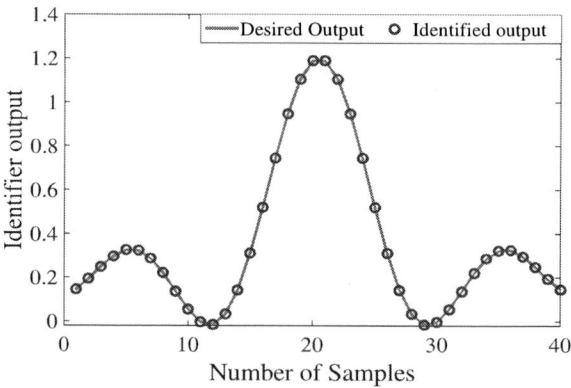

Fig. 4. Identified output of Fuzzy logic using Gradient Descent. (Case II)

B. Case Study II

A non-linear system is considered whose plant equation is defined as:

$$y(P) = 0.6sin(\pi P) + 0.3sin(3\pi P) + 0.1sin(5\pi P) \quad (20)$$

Where $y(x)$ is the system output and P is the system input. The constraints for determining the input data set is

$$-1 < P < 1 \quad (21)$$

The Plant identifier output using gradient descent is shown in Fig 4. The Plant identifier output using back propagation is shown in Fig 5. The comparison of PI is shown in Fig 6.

C. Case Study III

A non-linear system is considered whose plant equation is defined as:

$$y(x) = (1 + x_1^{-0.5} + x_2^{-1} + x_3^{-1.5})^{-2} \quad (22)$$

Where $y(x)$ is the system output and x is the system input.

Table 1 provides a comparison of the Performance Index (PI) for the three cases presented in this paper after 100th iteration using the algorithm of fuzzy logic and neural networks and Table 2 presents the comparison between the times of execution.

IV. CONCLUSION

In this paper a comparative study of the Performance Index (PI) of methods of achieving model Identification of non-linear system is presented. The methods which have been taken for comparison are Gradient Descent and Back Propagation. Both the algorithm are applied on three problems and their results are used for comparison. It has been observed that the convergence rate of Back Propagation is faster and the Performance index value is also relatively smaller in comparison to the Gradient descent method of identification and modelling of non-linear systems.

TABLE I
COMPARISON OF PERFORMANCE INDEX (PI) IN PRESENTED
CASE STUDIES AFTER 100TH ITERATION.

APPROACH USED	CASE STUDIES		
	I	*II*	*III*
Fuzzy Logic	0.23573	0.2046	0.586
Neural Networks	0.21059	0.2006	0.520

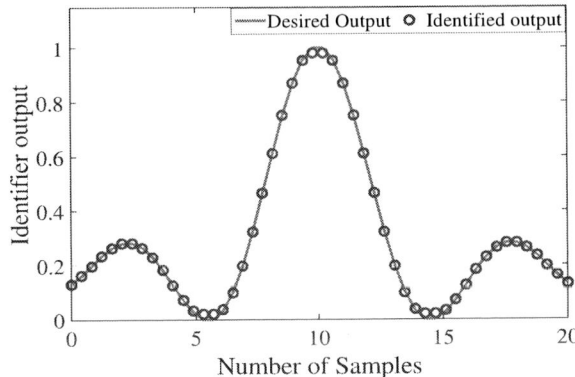

Fig. 5. Identified output of Neural Network using Back Propagation(Case II)

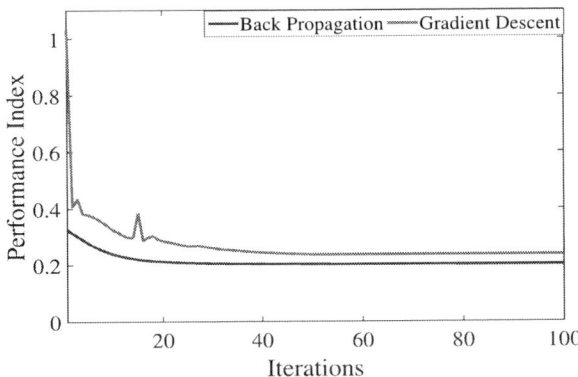

Fig. 6. Performance Index Comparison (Case II)

REFERENCES

[1] Keesman, Karel J. System identification: an introduction. Springer Science and Business Media, 2011.
[2] Eickhoff, P. System identification, Wiley,1974.
[3] Ljung L, Glad T, Modeling of dynamic systems. Prentice Hall, Englewood Cliffs, 1994.
[4] Zadeh, Lotfi A. "Outline of a new approach to the analysis of complex systems and decision processes." IEEE Transactions on systems, Man, and Cybernetics 1 ,1973, pp. 28-44.
[5] [5] Mamdani, Ebrahim H. "Application of fuzzy algorithms for control of simple dynamic plant." Proceedings of the institution of electrical engineers.1974, Vol. 121. No. 12, pp. 1585-1588.
[6] [6] Mamdani EH. Application of fuzzy logic to approximate reasoning using linguistic synthesis. InProceedings of the sixth international symposium on Multiple-valued logic 1976,pp. 196-202.

TABLE II
COMPARISON OF TIME OF EXECUTION IN PRESENTED CASE
STUDIES. (IN SECONDS)

APPROACH USED	CASE STUDIES		
	I	*II*	*III*
Fuzzy Logic	1.498918	7.9732	3.44621
Neural Networks	0.523546	0.5663	0.94885

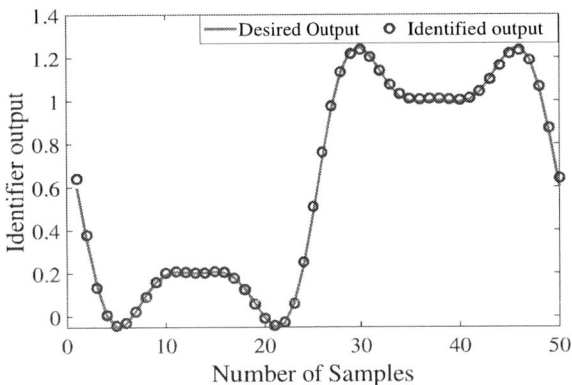

Fig. 7. Identified output of Fuzzy logic using Gradient Descent. (Case III)

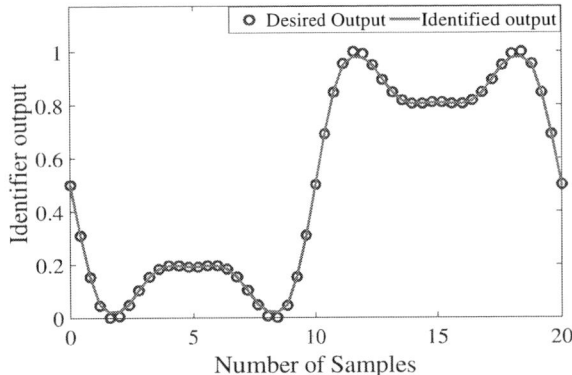

Fig. 8. Identified output of Neural Network using Back Propagation (Case III)

[7] Takagi, T., and Sugeno, M. Fuzzy identification of systems and its applications to modeling and control. In Readings in Fuzzy Sets for Intelligent Systems. 1993, pp. 387-403.
[8] Tushir M. TS Model for Identification and Prediction of Nonlinear System. Journal of Basic and Applied Engineering Research. 2016,Volume 3, Issue 3, pp. 298-301.
[9] Ogunmolu, O., Gu, X., Jiang, S., and Gans, N. Nonlinear sys-

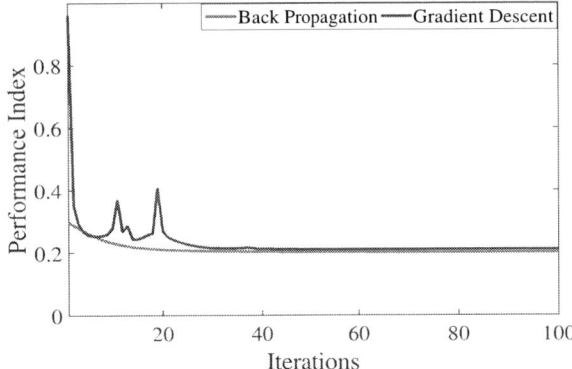

Fig. 9. Performance Index Comparison (Case III)

978-1-5386-6624-1/18 $31.00 © 2018 IEEE

tems identification using deep dynamic neural networks. 2016, arXiv:1610.01439.

[10] K. S. Narendra and K. Parthasarathy. Neural networks and dynamical systems. International Journal of Approximate Reasoning. 1992, vol. 6, no. 2, pp. 109131.

[11] K.S.Narendra and K.Parthasarathy. Identification and Control of Dynamical Systems using Neural Networks. IEEE Transactions on Neural Networks. 1990,Vol. 1, no. 1, pp. 4-27

[12] Sivanandam, S. N., Deepa, S. N. Principles of Soft Computing. John Wiley and Sons, 2007.

Fuzzy C-means Based Model Simplification using Jaya Optimization Algorithm

Chhabindra Nath Singh
EE Deptt.
HBTU Kanpur
Kanpur, India
cnsinghhbti7@gmail.com

Akhilesh Kumar Gupta
EE Deptt.
MNNIT Allahabad
Allahabad, India
akhilesh_ree5213@mnnit.ac.in

Deepak Kumar
EE Deptt.
MNNIT Allahabad
Allahabad, India
deepak_kumar@mnnit.ac.in

Paulson Samuel
EE Deptt.
MNNIT Allahabad
Allahabad, India
paul@mnnit.ac.in

Abstract— **A new mixed approach of model order reduction (MOR) for the simplification of higher order linear time-invariant single-input single-output (SISO) systems is presented in this article by combining the benefits of Fuzzy C-means (FCM) along with Jaya optimization algorithm. The FCM is a pole clustering based method used for the determination of reduced order denominator polynomial while the numerator polynomial is calculated by the Jaya optimization algorithm. Further, two numerical examples are considered and their performance indices are computed to justify the superiority of the proposed approach.**

Keywords: Fuzzy C-Means, Jaya Optimization, Model Order Reduction.

I. INTRODUCTION

Mathematical modeling of the system is very important to know the dynamic performance of the models and designing of controllers for closed loop operations. Nowadays the modern controller like H_∞ and LQG are having same or even higher order than the order of the plant. Therefore, the designing of such controllers for higher order model (HOM) is posing a great challenge. Hence, there is a need to reduce the HOMs into reduced order models (ROM) which provides the identical characteristics as the original system. A significant number of approaches for MOR in time domain [36-38] as well as in frequency domain [1-13] are presented in the literature. Each method has advantages and some disadvantages. Inspite of the availability of several methods for the MOR, no method offers the best result for all the system and it is yet to be investigated. Several conventional methods for MOR are aggregation approach [1], Padè approximation method [2], continued fraction expansion (CFE) techniques [3-6], moment matching method [7], Markov parameters and moment matching approach [8] etc. Although these methods offer computational easiness but do not always promise the stability of the approximant. To overcome this problem, many scholars suggested stability assured method such as Routh approximation approach [9-11], Hurwitz polynomial approximation technique [12] and stability equation approach [13] etc. Later, some mixed methods [14-18] are proposed for the generalization of higher order continuous-linear time invariant models. The recent development of soft computing techniques like particle swarm optimization (PSO) [19-20], genetic algorithm (GA) [21], big bang-big crunch [22], Cuckoo search [23], and Jaya optimization algorithm [24] paves the new direction to the MOR problem. Subsequently, these optimization techniques are applied in conjunction with classical or some stability preserving methods for the simplification of LTI systems [25-28].The approximation of HOM by using these soft computing approaches is based on the minimization of certain performance attributes such as integral square error (ISE) etc. As far as the methods in the time domain are concerned, the Gugercin and Antoulas [34] presented a state-of-art on the balancing related MOR schemes. Later, Ghafoor and Sreeram [35] reviewed the frequency-weighted model reduction schemes whereas authors [36-38] presented several other time domain approaches.

Many methods have been presented in the recent years by merging conventional method with a suitable optimization technique. There is great scope for improvement in approximation error by using this approach. Therefore, in this paper a new reduction approach is developed. The remaining part of this paper is organized as follows: The organization of paper is divided into the five sections with the inclusion of introductory section. A brief introduction of problem formulation is narrated in Section II whereas the proposed method is elaborated in Section III. Furthermore, section IV is related with two case studies to prove the effectiveness of the proposed approach. In the last, the conclusive remarks are made in Section V.

II. PROBLEM FORMULATION

Consider the transfer function (TF) of the nth order SISO system given in the frequency domain as

$$G_n(s) = \frac{N_n(s)}{D_n(s)} = \frac{\sum_{i=0}^{n-1} b_i s^i}{\sum_{i=0}^{n} a_i s^i} \tag{1}$$

where a_i's and b_i's are denominator and numerator polynomials multiplicative factors of the HOM respectively.

The main goal is to find the rth $(r < n)$ order reduced model having characteristics very similar to the original HOM. The TF of the desired ROM is expressed as

$$G_r(s) = \frac{N_r(s)}{D_r(s)} = \frac{\sum_{i=0}^{r-1} l_i s^i}{\sum_{i=0}^{r} k_i s^i} \tag{2}$$

where l_i's and k_i's are coefficients of numerator and denominator polynomial for ROM respectively.

III. PROPOSED METHOD

The method proposed in this paper combines the attributes of the FCM clustering algorithm [33] and Jaya optimization algorithm [24]. The FCM algorithm is applied to find the reduced order denominator polynomial's

978-1-5386-6624-1/18 $31.00 © 2018 IEEE

coefficients whereas the numerator polynomial is computed by Jaya algorithm.

Data clustering is a process of diving data into different classes of comparable data. These classes are called clusters. The popular clustering techniques are K-Means and FCM clustering. To find the cluster center by K-means clustering algorithm, the cost function of dissimilarity is minimized. In this approach, each data belongs only to its specific cluster. For such type of clustering, the membership matrix is called binary membership because it contains either 1 or 0 in its matrix. Therefore, K-Means clustering is also called Hard C-Means (HCM) clustering. Unlike K-Means clustering, in FCM each point belongs to a cluster by a specified degree and it permits same point to belong to more than one cluster. The membership matrix U of FCM contains the element between 0 and 1. The summation of membership matrix's element for a data to all clusters must be unity i.e.

$$\sum_{i=1}^{n} u_{ik} = 1 \quad where \quad k = 1, 2, \dots, m \tag{3}$$

The objective function for FCM is given by (4) as

$$J = \sum_{i=0}^{Q-1} J_i = \sum_{i=1}^{n} \sum_{k=1}^{m} u_{ik}^m d_{ik}^2 \tag{4}$$

where u_{ik} is membership matrix whose value lies between 0 and 1 and d_{ik} is Euclidean distance between the ith cluster center and k^{th} data point and is given by $d_{ik} = \|c_i - x_k\|$. The cluster center is given by

$$c_i = \frac{\sum_{k=1}^{n} u_{ik}^m x_k}{\sum_{k=1}^{n} u_{ik}^m} \tag{5}$$

and

$$u_{ik} = \frac{1}{\sum_{j=1}^{c} \left[\frac{c_i - x_k}{c_j - x_k} \right]^{2/(m-1)}} \tag{6}$$

The FCM works until it find its objective function's value below its termination criterion. The FCM technique is implemented by steps given below:

Step 1: Determine the degree of belongingness of each data points to each cluster by determining the membership matrix U.

Step 2: Compute the fuzzy cluster centers c_i where $i = 1, 2, \dots, c$.

Step 3: Determine the objective function until it has met its stopping criterion.

Step 4: Estimate the new membership matrix U.

A. Determination of ROM Denominator Polynomial

To determine the reduced order denominator polynomial, the following rules must be kept in mind:

i. There should be separate clusters for real and imaginary poles.

ii. For complex conjugate poles such as $\left[(x_1 \pm jy_1), (x_2 \pm jy_2), \dots \right]$ the FCM generates cluster centers

as $Z_{zj} = X_{zj} + jY_{zj}$, where X_{zj} and Y_{zj} are cluster centers for real and imaginary parts.

iii. Poles lie on the imaginary axis and most dominant pole must be presented in the ROM. Therefore, the desired clusters are *(m-r)*.

B. Determination of ROM Numerator Polynomial

The Jaya optimization approach is used to evaluate the numerator coefficients which are described as follows:

Jaya Optimization algorithm

Unlike PSO [19-20] and GA [21] the Jaya optimization algorithm [24] is parameter free algorithm since it does not require the selection operator, mutation probability and tuning of inertia weight for implementation of the algorithm except few tuning control parameters. The Jaya optimization converges the solutions to the optimum point by minimization of the objective function with the tendency to come closer to the better solution and move away from the worst solution.

Let f_x is a fitness function of xth solution which has to be minimized and $p_{x, y, z}$ is a randomly selected solution among K-distributed solution over search space (*i.e. x = 1, 2, ..., K*) with N design variables (*i.e. y = 1, 2, ..., N*) at z^{th} iteration, the Jaya algorithm comprises the following phases:

Step 1: [Initialization] The parameters like number of random solutions (population size), termination criterion or the number of iteration etc. should be initialized in very early stage of the optimization algorithm.

Step 2: [Best and Worst] The worst and best solutions are selected according to the maximum and minimum value of fitness function respectively.

Step 3: [Alteration] The possible existing solutions are modified according to (7)

$$p_{x,y,z}^{new} = p_{x,y,z} + rand_{1,x,z}(p_{x,best,z} - | p_{x,y,z} |) - rand_{2,x,z}(p_{x,worst,z} - | p_{x,y,z} |) \tag{7}$$

where, $p_{x,y,z}^{new}$ is the new value of $p_{x, y, z}$ at z^{th} iteration for x^{th} variables with two random variables $rand_{1, x, z}$ and $rand_{2, x, z}$ which shows the degree of approach toward the best solution and degree to avoid the worst solution respectively, $p_{x, best, z}$ is the best solution for x^{th} variable, $p_{x, worst, z}$ is the value of worst solution for x^{th} variable (both are defined in terms of fitness value).

Step 4: [Substitution] Substitute the present solution with new one former one has higher fitness value in comparison with later one.

Step 5: [Execution] The steps from 2 to 4 are repeated until the maximum number of iteration has completed or some pre-specified stopping criterion has been met.

The flow chart of the Jaya optimization algorithm is shown in Fig. 1.

2nd IEEE International conference on power Electronics, Intelligent Control and Energy systems (ICPEICES-2018)

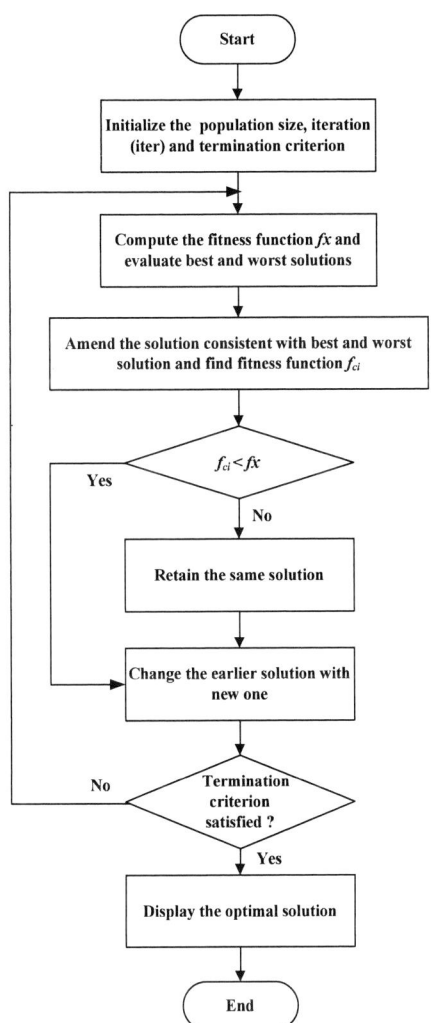

Fig. 1. Flow Chart of Jaya algorithm

IV. CASE STUDIES

Case 1: Let the eighth order continuous-time LTI system [17, 18] as

$$G(s) = \frac{\begin{matrix}18s^7 + 514s^6 + 5982s^5 + 36380s^4 + 122664s^3 \\ + 222088s^2 + 185760s + 40320\end{matrix}}{\begin{matrix}s^8 + 36s^7 + 546s^6 + 4536s^5 + 22449s^4 \\ + 67284s^3 + 118124s^2 + 109584s + 40320\end{matrix}}$$

The poles of the above system are:

$a_1 = -1,\ a_2 = -2,\ a_3 = -3,\ a_4 = -4,$

$a_5 = -5,\ a_6 = -6,\ a_7 = -7,\ a_8 = -8.$

With the retention of dominant pole -1, the other two clusters centers computed by the FCM algorithm are given as -3.1630 and -6.8382.

Hence, the ROM's denominator polynomial is determined as

$$D_3(s) = (s+1)(s+3.1630)(s+6.8383) = s^3 + 11s^2 + 31.63s + 21.63.$$

By proposed method, ROM's numerator polynomial is computed as

$$N_3(s) = 16.8904s^2 + 71.6262s + 21.63.$$

Finally, by proposed method the ROM is obtained as

$$R_3(s) = \frac{16.8904s^2 + 71.6262s + 21.63}{s^3 + 11s^2 + 31.63s + 21.63}.$$

The step and frequency responses of the original 8th-order system and the ROM obtained by the proposed and existing methods [29, 31] are shown in Figs. 2 and 3 respectively. The plots exhibit that, the responses of the proposed ROM are in a close agreement to the responses of original HOM better than [29, 31]. The qualitative comparisons of time response specifications and performance indices are depicted in Tables 1 and 2 respectively. It may be easily concluded from the Table 1 that, the time response parameters for proposed method are very adjacent to the original HOM and it is also comparable to the existing methods [29, 31, 32]. Further, there is a substantial improvement in the errors indices for the proposed approach as compared to other well-known techniques [29, 31, 32] as shown in Table 2.

Case 2: Let us consider the 10th order system [30, 32] as

$$G(s) = \frac{\begin{matrix}9.469\times10^{14} + 2.979\times10^{14}s + 3.478\times10^{13}s^2 \\ + 2.062\times10^{12}s^3 + 7.012\times10^{10}s^4 + 1.445\times10^9s^5 + \\ 1.828\times10^7s^6 + 1.384\times10^5s^7 + 574s^8 + s^9\end{matrix}}{\begin{matrix}9.4\times10^{15} + 5.344\times10^{15}s + 9.713\times10^{14}s^2 + \\ 8.381\times10^{13}s^3 + 4.027\times10^{12}s^4 + 1.165\times10^{11}s^5 + 2.105\times10^9s^6 + \\ 2.381\times10^7s^7 + 1.635\times10^5s^8 + 621.3s^9 + s^{10}\end{matrix}}$$

By proposed approach, the ROM is computed as

$$G_3(s) = \frac{0.6215s^2 + 126.7969s + 1200.7498}{s^3 + 150.5s^2 + 4292s + 11920}$$

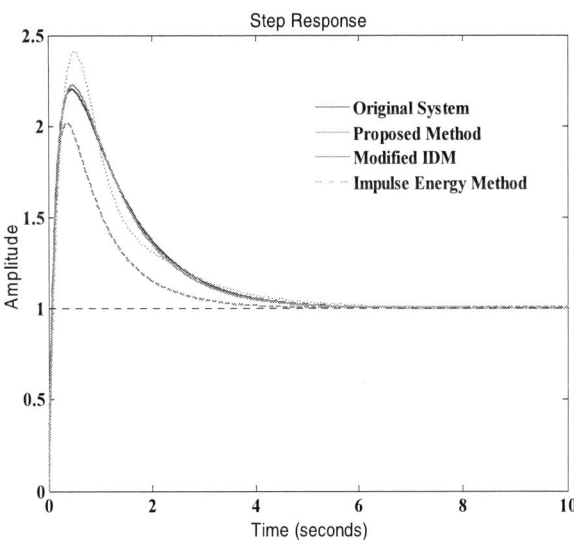

Fig.2.Step responses of the HOM and ROM for Case 1

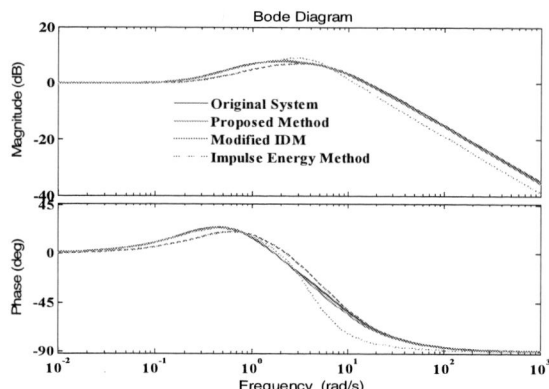

Fig.3.Frequency responses of the HOM and ROM for Case 1

The step and frequency responses of the original 10th-order system and the ROM obtained by the proposed and existing methods [29-30] are shown in Figs. 4 and 5 respectively. The plots exhibit that, the responses of the proposed ROM are in a close agreement to the responses of original HOM better than [29-30].The qualitative comparisons of time response specifications and performance indices are revealed in Tables 3 and 4 respectively. It is concluded from the Table 3 that, the time response parameters for the proposed method are very close to the original HOM and it is also comparable to the existing methods [29-30]. Further, there is a significant upgradation in the errors indices for the proposed scheme as compared to existing techniques [29-30] as shown in Table 4.

TABLE 1: COMPARISON OF TRANSIENT CONSTRAINTS FOR CASE 1

Reduction Techniques	Settling time	Rise Time	Overshoot (%)
Original Model	4.82	0.0569	123
Proposed Method	4.67	0.06	120
Modified IDM [29]	3.75	0.0571	101
Impulse Energy Method [31]	5.24	0.0809	141
FCM and Pade [32]	4.48	0.057	110

TABLE 2: PERFORMANCE INDICES COMPARISON FOR CASE 1

Reduction Techniques	ISE ($\times 10^{-3}$)	ITSE ($\times 10^{-3}$)	IAE ($\times 10^{-2}$)	ITAE ($\times 10^{-1}$)
Proposed Method	1.28	1.92	7.03	1.44
Modified IDM [29]	189.22	258.16	77.74	13.84
Impulse Energy Method [31]	38.87	31.16	31.60	184.85
FCM and Pade [32]	48.7	66.1	39.2	6.88

TABLE 3: COMPARISON OF TRANSIENT CONSTRAINTS FOR CASE 2

Reduction Techniques	Settling time	Rise Time	Overshoot (%)
Original Model	1.16	0.631	0
Proposed Method	1.17	0.641	0
Modified IDM [29]	0.601	0.339	2.22×10^{-14}
Improved clustering Method [30]	1.11	0.653	0

TABLE 4: PERFORMANCE INDICES COMPARISON FOR CASE 2

Reduction Techniques	ISE	ITSE	IAE	ITAE
Proposed Method	7.34×10^{-7}	1.12×10^{-7}	6.24×10^{-4}	2.31×10^{-4}
Modified IDM [29]	1.09×10^{-4}	4.33×10^{-5}	1.02×10^{-2}	6.60×10^{-3}
Improved clustering Method [30]	8.07×10^{-5}	5.55×10^{-6}	4.79×10^{-3}	1.19×10^{-3}

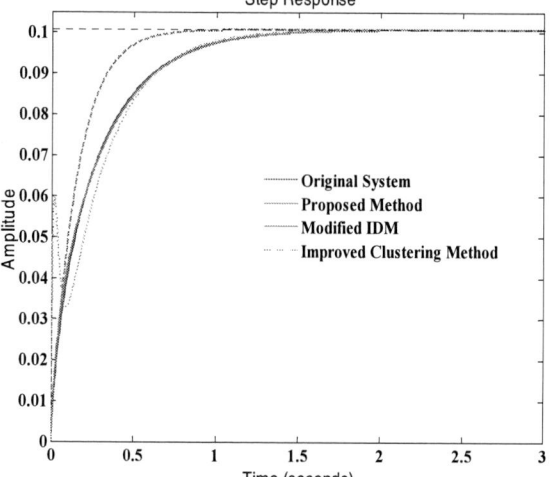

Fig.4.Step responses of the HOM and ROM for Case 2

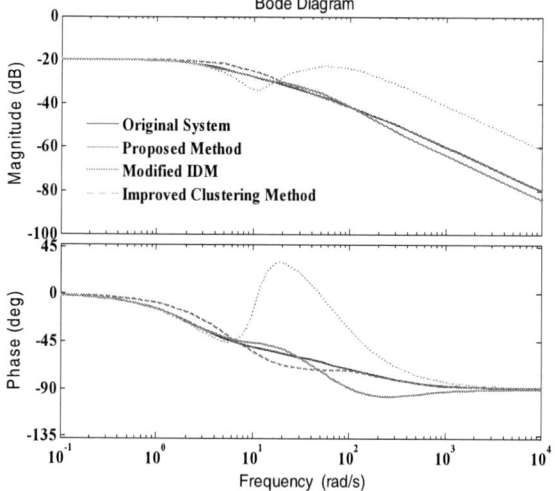

Fig.5.Frequency responses of the HOM and ROM for Case 2

V. CONCLUSION

In this article the Fuzzy C-means clustering and Jaya optimization algorithm are applied to develop a new method to obtain a ROM. The proposed approach is mathematically elegant, simple and very effective. In two cases, the coefficients of denominator of ROM are obtained by the FCM algorithm while the Jaya optimization is used to compute the numerator polynomial of the ROM. The performance indices such as ISE, ITSE, IAE and ITAE are computed to validate the supremacy of the proposed technique over other well-known approaches available in the literature. Apart from this, a significant improvement in the performance indices for the proposed method further validates its efficacy. The proposed methodology can be easily extended for the order simplification of LTI continuous-time MIMO system and discrete-time systems.

REFERENCES

[1] M. Aoki, "Control of large-scale dynamic systems by aggregation," IEEE Trans. Automat. Contr., vol. AC-13, pp. 246-253, 1968.

[2] Y. Shamash, "Stable reduced order models using Padè type approximation," IEEE Trans. Automat. Control, vol. AC-19, pp. 615-616, 1974.

[3] C F. Chen and L. S. Shieh, "A novel approach to Linear model simplification," Int. J. Control, voL 8. pp. 561-570, 1968.

[4] Y.Shamash, "Continued fraction methods for the reduction of linear time-invariant systems," In: Proc. Conf. Computer Aided Control system Design, Cambridge, England, pp. 220-227, 1973.

[5] L. S. Shieh and M. J. Goldman, "A mixed Cauer form for linear system reduction," IEEE Trans. Syst. Man, Cybern. , vol. SMC4, pp. 584-588. 1974.

[6] S. C. Chuang, "Application of continued-fraction method for modelling transfer function to give more accurate transient response," Electron. Lett., vol 6, pp-861-863. 1970.

[7] N. K. Sinha and B. Kuszta, " Modeling and identification of dynamic systems," New York: Van Nostand Reinhold, pp.133-163, 1983.

[8] E. Jonckheere and C. Ma, "Combined sequence of Markov parameter and moments in linear systems," IEEE Trans. Autom. Control, vol. AC-34, pp. 379-382, Mar. 1989.

[9] M.F. Hutton and B. Friedland, "Routh approximation for reducing order of linear time-invariant system", IEEE Trans. Autom. Control, vol. 20, pp. 329-337, 1975.

[10] Y. Shamash, "Model reduction using the Routh stability criterion and the Padè approximation technique", Int. J. Control, vol. 21, pp. 475-484, 1975.

[11] A. S. Rao, S. S. Lamba, and S. V. Rao, "Routh-approximant time domain reduced-order modelling for single-input single-output systems'', IEE Proc. Control Theory Appl., vol. 125, pp. 1059-1063, 1978.

[12] R. K. Appiah,"Padè methods of Hurwitz polynomial approximation with application to linear system reduction", Int. J. Control, , vol. 29, pp. 39-48, 1979.

[13] T .C. Chen, C. Y. Chang, and K. W. Han, "Stable reduced-order Padè approximants using stability equation method", Electron. Lett., vol.16, pp. 345-346, 1980.

[14] V. Singh, D. Chandra, and H. Kar, "Improved Routh-Padè approximants: a computer-aided approach," IEEE Trans. Autom. Control, vol. 49, no. 2, pp. 292-296, 2004.

[15] G. Parmar, S. Mukherjee, and R. Prasad, "System reduction using factor division algorithm and Eigen spectrum analysis," *Appl. Math. Model.*, vol. 31, no. 11, pp. 2542-2552, 2007.

[16] A. Sikander, R. Prasad, "Linear time invariant system reduction using a mixed methods approach," Appl. Math. Model., vol. 39, no. 16, pp. 4848-4858, 2015.

[17] J. Singh, C. B. Vishwakarma, and K. Chatterjee, "Biased reduction method by combining improved modified pole clustering and improved Padè approximants," Appl. Math. Model., vol. 40, pp.1418-1426, 2016.

[18] C. N. Singh, D. Kumar, and P. Samuel, "Improved pole clustering-based LTI system reduction using factor division algorithm," Int. J. Model. Sim., pp:1-13, 2018.

[19] Y. Zhang, A comprehensive survey on particle swarm optimization algorithm and its applications. Mathematical Problems in Engineering: 931256, 2015.

[20] J. Kennedy and R. Eberhart, "Particle swarm optimization," In: IEEE International Conference on Neural Networks, pp.1942-1948, 1995.

[21] D. E. Goldberg, "Genetic algorithms in search, optimization, and machine learning," Addison-Wesley Longman Publishing Co. Inc. Boston, MA, USA, 1989.

[22] O. K. Erol and E. Ibrahim, "A new optimization method: big bang-big crunch," Advances in Engineering Software , vol.37.2, pp. 106-111, 2006.

[23] A. B. Mohamad, A. M. Zain and N. E. N. Bazin, "Cuckoo search algorithm for optimization problems-a literature review and its applications," Applied Artificial Intelligence, vol. 28, pp. 419-448, 2014.

[24] R. V. Rao and G. G. Waghmare, "A new optimization algorithm for solving complex constrained design optimization problems," Eng. Optimization, pp. 1-24, 2016.

[25] A. Narwal and R. Prasad, "A novel order reduction approach for LTI systems using cuckoo search optimization and stability equation," IETE J. Res., vol. 62, no. 2, pp.1-10, 2015.

[26] S. R. Desai and R. Prasad, "A novel order diminution of LTI systems using big bang-big crunch optimization and Routh Approximation," Appl. Math. Model., vol. 37, no. 16-17, pp. 8016-8028, 2013.

[27] A. K. Gupta, D. Kumar, and P. Samuel, "A meta-heuristic cuckoo search and Eigen permutation approach for model order reduction," Sadhana: 1-11, 2018.

[28] G. Parmar, R. Prasad, and S. Mukherjee, "Order reduction of linear dynamic systems using stability equation method and GA," Int. J. Electr. Comput. Energ. Electron. Commun. Eng., vol. 1, no.2, pp. 236-242, 2007.

[29] C. B.Vishwakarma, R. Prasad, "System reduction using modified pole clustering and Pade approximation," In:XXXII National Syst. Conf. (NSC), pp.592-596, 17-19 December 2008.

[30] M. Srinivasan and A. Krishnan, "Transformer linear section model order reduction with an improved pole clustering," European Journal Scient. Research, vol.44, no.4, pp.541-549, 2010.

[31] T. N. Lucas, Scaled impulse energy approximation for model reduction, IEEE Trans. Automat. Control, vol. 33, no.8, pp.791-793, 1988.

[32] A. Narain, D. Chandra, and R. K. Singh, "Model order reduction using Fuzzy C-means clustering," Trans. Inst. Meas. Control, vol. 36, no. 8, pp. 992-998, 2014.

[33] K. Hammouda and K. Fakhareddine, " A comparative study of data clustering techniques," Ph.D. Thesis, University of Waterloo, Ontario, Canada, 2000.

[34] S. Gugercin and A. C. Antoulas, "A survey of model reduction by balanced truncation and some new results," Int. J. Control, vol.77, no. 8, pp.748–766, 2004.

[35] A. Ghafoor and V. Sreeram, "A survey/review of frequency-weighted balanced model reduction techniques," J. Dyn. Syst. Meas. Control, 130:061004-1-16, 2008.

[36] D. Kumar and S. K. Nagar, "Model reduction by extended minimal degree optimal Hankel norm approximation," Appl. Math. Model., vol. 38, pp. 2922–2933, 2014.

[37] D. Kumar, A. Jazlan, V. Sreeram et al., "Partial fraction expansion based frequency weighted model reduction for discrete-time systems," Numer. Algebra Control Optim., vol. 6, no. 3, pp. 329-337, 2016.

[38] D. Kumar, V. Sreeram, X. Du, "Model reduction using parameterized limited frequency interval Gramians for 1-D and 2-D separable denominator discrete-time systems," IEEE Transactions on Circuits and Systems I: Regular Papers, 2018.

A Voltage Controlled Sensorless Speed Control of PMBLDC Motor Drive for an Electric Two Wheeler

Chitra Bhattacharya
Student Member, IEEE
Dept. of Electrical Engineering
Shri G.S. Institute of Technology
&Science   Indore, India
chitrabhattacharya0gmail.com

Shailendra Kumar Sharma
Senior Member, IEEE
Dept. of Electrical Engineering
Shri G.S. Institute of Technology
&Science   Indore, India
ssharma.iitd@gmail.com

H.K. Verma
Member, IEEE
Dept. of Electrical Engineering
Shri G.S. Institute of Technology
&Science   Indore, India
vermaharishgs@gmail.com

Abstract—**The global warming has led to a widespread acceptance for electric vehicles. They are environment friendly in comparison to conventional and hybrid vehicles. This paper focuses on the speed control strategy of a three phase permanent magnet brushless DC (PMBLDC) motor employed in an electric two wheeler. The speed control of the motor is done using output voltage control of bidirectional boost-buck converter. The presented system is modeled, simulated and analysed using MATLAB/SIMULINK for different speed and load condition.**

Keywords—*Bidirectional boost-buck converter, PMBLDC motor drive, Back Electromotive Force (EMF), Voltage Source Inverter (VSI).*

I. Introduction

Electric vehicles are a dignified solution for sustainable transportation systems as they do not emit greenhouse gases. They are the most effective resolution for efforts regarding reduction in usage of fossil fuels. They are becoming popular due to their economy in running cost and efficient performance. The employment of brushed gear and mechanical commutator based motors for such purpose requires frequent maintenance, thereby reducing reliability of the system. Hence, for such applications, there is a need for a motor drive that has simple structure, compact size and thus reduced weight, higher efficiency, higher speed range of operation, high reliability and no maintenance. Owing to these properties, permanent magnet brushless DC motors became a favourite choice among the manufacturers of electric vehicles. The usage of direct drive, in-wheel designs [1-3] reduces the mechanical transmission system required in conventional and hybrid vehicle and improved its performance by manifolds. The efficient speed control mechanisms of such drives are a matter of keen interest among researchers. Hence, position sensorless speed control mechanisms are a superior choice in this regard. Various techniques [4-15] have been dicussed for position sensorless speed control mechanism. Researchers [8-10] discussed about the back EMF sensing method in which back EMF is measured between the silent phase and the implicit neutral point. The zero crossings of the sensed back EMFs are phase shifted by 90° to obtain exact commutation instants. It resulted in high frequency noise due to PWM and has therefore limited the range of speed of the motor. Various schemes [8-9] were developed for properly selecting the PWM and sensing strategy but estimation of a virtual neutral is a difficult task. Gui *et al.* [11] developed a scheme which requires back EMF sensing of only one phase from among the three phases of PMBLDC motor. The microcontroller measures the time instants between two commutation intervals generated by zero crossing detection of back EMF and creates other commutation instants. Although the cost of implementation is reduced, this method cannot be effectively used for variable speed. The zero crossings of the phase back EMFs gives the phase shifted instants. So, to obtain exact commutation instants, the detected zero crossings are delayed by 30° [12]. Exact commutation instants are acquired from the zero crossing instants of the line back EMFs [13]. This method fails to determine the zero crossings at near zero or low speeds as the magnitudes of back EMFs are also quite less during such interval. Thus, a starting mechanism has been proposed for enabling the motor to run from near zero to rated speed.

In this paper, for starting the motor, rotor angle observation is done by filtering the generated back EMFs and deciding commutation signals accordingly. When the speed is more than 60 rpm, the control is shifted to zero crossing detection of line back EMFs which generates virtual hall signals that gives exact commutation instants. The speed of the motor is controlled by controlling the output voltage of bidirectional boost-buck converter. Simulation results are obtained using MATLAB 2015a.

II. System Configuration

Fig. 1 shows the basic arrangement of an electric two wheeler and Fig. 2 shows the system under consideration. It consists of a Li-ion battery which provides the input voltage to bidirectional boost-buck converter. This DC-DC converter is used to boost the input voltage from the level of battery voltage to the level of the rated voltage of motor (48V) during the motoring operation and reduce the voltage to the level of battery voltage (24V) for charging it during the regenerative braking. Interfacing assemblage provides platform for collaboration between sensors and controller. This DC-DC converter is controlled in continuous conduction mode (CCM) to reduce the ripple component of inductor current, reduce the rating of the switches and obtain higher efficiency. Its output is fed to three phase voltage source inverter that drives the motor. The gate pulses for VSI are generated from the rotor angle observation and zero crossing detection of the back EMFs.

978-1-5386-6624-1/18 $31.00 © 2018 IEEE

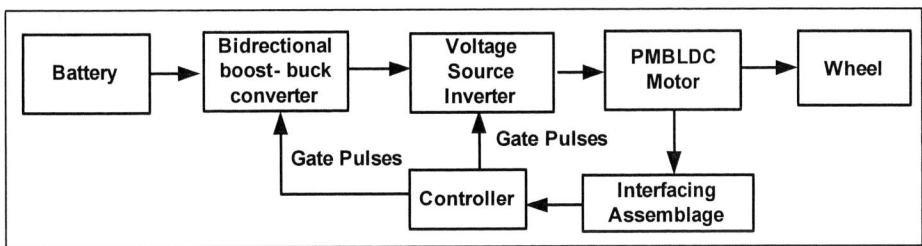

Fig. 1 Schematic of a Two Wheeler Electrical Drive

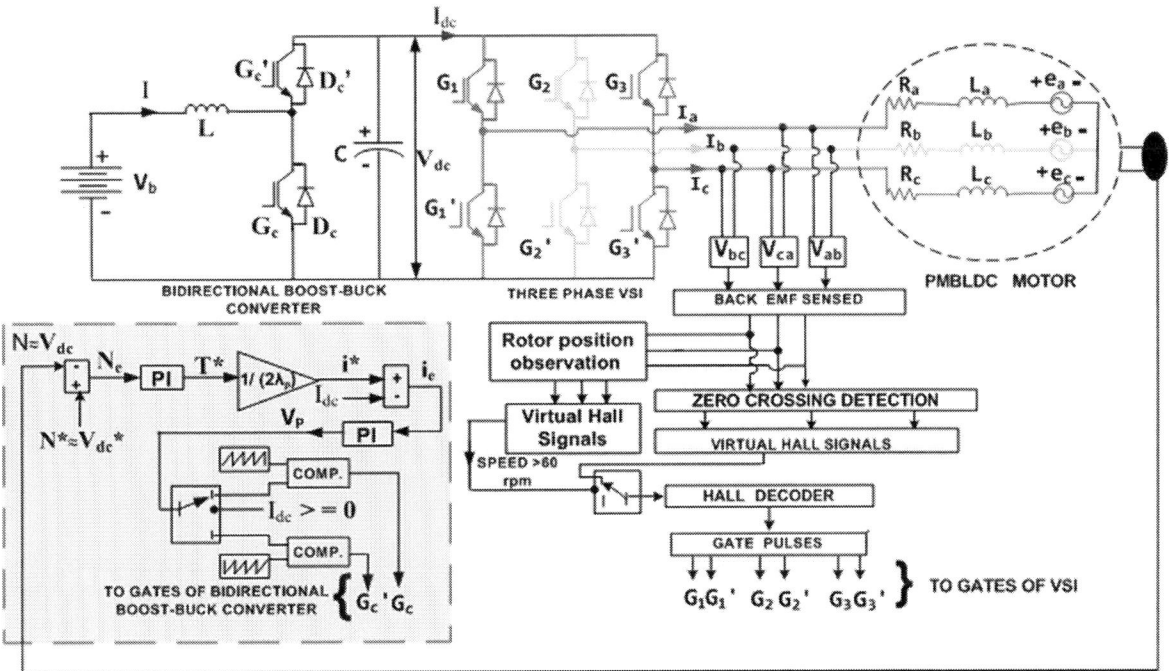

Fig. 2 Position Sensorless Speed Control of PMBLDC motor using voltage control mode

III. DESIGN OF THE SYSTEM

A. Design of Bidirectional Boost-Buck Converter

During motoring operation, diode D_c is reverse biased and switch G_c' is always off. The converter functions as boost converter [16].

Mode 1: Switch G_c is on and diode D_c' is off.

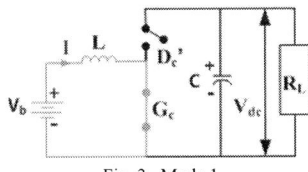

Fig. 3   Mode 1

$$L\frac{di}{dt} = V_b \tag{1}$$

$$L\frac{\Delta i}{t_{on}} = V_b \tag{2}$$

$$C\frac{dV_{dc}}{dt} = -\frac{V_{dc}}{R_L} \tag{3}$$

$$C\frac{\Delta V_{dc}}{t_{on}} = -\frac{V_{dc}}{R_L} \tag{4}$$

Mode 2: Switch G_c is off and diode D_c' is on.

$$L\frac{di}{dt} = V_b - V_{dc} \tag{5}$$

$$L\frac{\Delta i}{t_{off}} = V_b - V_{dc} \tag{6}$$

$$C\frac{dV_{dc}}{dt} = I - \frac{V_{dc}}{R_L} \tag{7}$$

$$C\frac{\Delta V_{dc}}{t_{off}} = I - \frac{V_{dc}}{R_L} \tag{8}$$

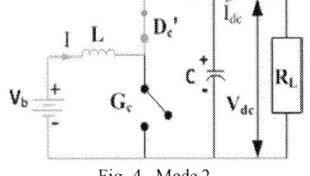

Fig. 4   Mode 2

Averaging above equations over a switching cycle and on solving the equations we get ' D_m ' (duty cycle) as

$$D_m = (1 - \frac{V_b}{V_{dc}}) \tag{9}$$

During braking operation, switch G_c is always off and diode D_c' is reverse biased. The converter functions as buck converter.

Mode 3: Switch G_c' is on and diode D_c is off.

$$L\frac{di}{dt} = V_{dc} - V_b \tag{10}$$

$$L\frac{\Delta i}{t_{on}} = V_{dc} - V_b \tag{11}$$

$$C\frac{dV_{dc}}{dt} = I - \frac{V_{dc}}{R_L} \tag{12}$$

$$C\frac{\Delta V_{dc}}{t_{on}} = I - \frac{V_{dc}}{R_L} \tag{13}$$

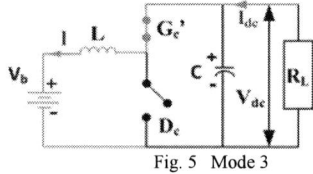

Fig. 5   Mode 3

Mode 4: Switch G_c' is off and diode D_c is on.

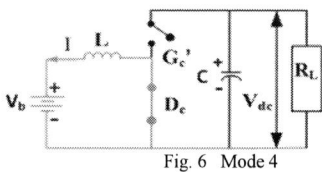

Fig. 6   Mode 4

$$L\frac{di}{dt} = V_b \tag{14}$$

$$L\frac{\Delta i}{t_{off}} = V_b \tag{15}$$

$$C\frac{dV_{dc}}{dt} = -\frac{V_{dc}}{R_L} \tag{16}$$

$$C\frac{\Delta V_{dc}}{t_{off}} = -\frac{V_{dc}}{R_L} \tag{17}$$

Averaging above equations over a switching cycle and on solving the equations we get ' D_{br} ' (duty cycle) as

$$D_{br} = \left(\frac{V_b}{V_{dc}}\right) \tag{18}$$

R_L is the net load resistance acting on the bidirectional boost -buck converter.

We know that, duty cycle = t_{on} f_m, where f_m is the modulating frequency.

Now, if the modulating frequency is quite higher than the fundamental frequency, it can be assumed that Δi and ΔV_{dc} are same for both buck and boost operation.

Hence, considering modulating frequency as 20 kHz, the value of inductor (L) is calculated as 0.3 mH and capacitor (C) as 250 µF.

The motor of 640 W has an efficiency of 91%. Thus, the input power required by the motor is 703.3 kW. Therefore, the bidirectional boost-buck converter is designed for 750 W.

So, the maximum input current from battery can be 31.25 A. Therefore, 400 V, 40 A IGBTs are used for designing the converter.

B. Design of Voltage Source Inverter (VSI)

The IGBTs are selected based on the rated current of the PMBLDC motor. The peak current through IGBT in each phase (I_{peak}) is acquired as [17],

$$I_{peak} = 1.25 \left(I_{p-p'} + \sqrt{2}\, I_{VSI} \right) \tag{19}$$

where $I_{p-p'}$ is 10 % peak-peak ripple current
I_{VSI} is the rated current of PMBLDC which is given as 11 A
$I_{peak} = 20.82$ A
Therefore, 400V, 40A IGBTs are used for the VSI.

C. Mechanical Design for an Electric Two Wheeler

An electric two wheeler requires the power to counteract the following forces [18-19]:

C.1. Rolling Resistance force : The force produced due to frictional movement between the tyres and road is known as rolling resistance and is given as,

$$F_{rr} = C_{rr}\ mg\cos\theta \tag{20}$$

C_{rr} is the coefficient of rolling resistance, 'm' is the net mass of the two wheeler, people and cargo (kg), 'θ' is the slope angle (degree) and 'g' is acceleration due to gravity (m/sec^2).

C.2. Aerodynamic drag force : It is the force produced due to air resistance against the motion of the vehicle and is given as,

$$F_d = 0.5\rho_a\ C_d\ A\ v \tag{21}$$

where 'ρ_a' is the density of air in kg/m^3, 'C$_d$' is the drag force coefficicent, 'A' is the frontal area of the two wheeler in m^2 , 'v' is the velocity of vehicle in m/sec .

C.3. Gravitational force : The force created by gravity and acting on the vehicle when it is moving along the slope. It is given as,

$$F_{grav} = mg\sin\theta \tag{22}$$

The force is positive when ascending the slope and negative when descending the slope.

C.4. Transient or Acceleration force : The force acting on the vehicle due to its acceleration or deceleration is given as,

$$F_a = m.(dv/dt) \tag{23}$$

$$F_a = m.a \tag{24}$$

where 'a' is the acceleration or deceleration of the two wheeler given in m/sec^2 .

Hence, the net force acting on the two wheeler is,

$$F = F_{rr} + F_a + F_{grav} + F_d \tag{25}$$

Hence, power supplied to an electric two wheeler is utilized to overcome force 'F '.

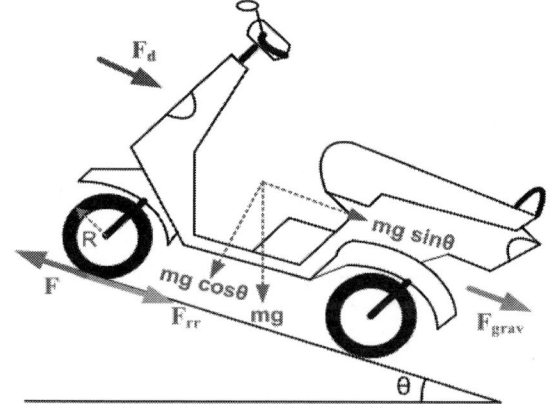

Fig. 7 Free body diagram of an electric two wheeler

The net mechanical torque offered by the vehicle is given as [19],

$$T_l = \alpha\left[I_\omega + \frac{mR_w^2}{3}\right] + \frac{R_w}{3}\left[mg\sin\theta + F\right] \tag{26}$$

'α' is angular acceleration of wheel, 'R_w' is the radius of the wheel, I_ω is the moment of inertia of the wheel.

So, taking the above mentioned details into consideration, a proper mechanical design of an electric two wheeler is done.

IV. CONTROL ALGORITHM

A. Methodology for back EMF zero crossing detection and starting of motor.

The phase to neutral voltage equation for PMBLDC is expressed as :

$$V_{pn} = i_{pn}R + L(di_p/dt) + e_{pn} \qquad (27)$$

p = a, b, c respectively for phase a, phase b and phase c;

V_{pn} = phase terminal voltage, i_{pn} = phase current, e_{pn} = phase back EMF, L = Inductance of windings, R = Resistance of windings.

The relationship between electrical rotor angle (θ_e) and mechanical rotor angle (θ_m) in radians is:

$$\theta_e = \frac{P}{2}\theta_m \ , \ P = \text{No. of poles} \qquad (28)$$

From the equation (26), the line back EMFs can be derived as

$$V_{ab} = (i_a - i_b)R + L\left(\frac{di_a}{dt} - \frac{di_b}{dt}\right) + e_{ab} \qquad (29)$$

Hence, $e_{ab} = V_{ab} - (i_a - i_b)R + L\left(\frac{di_a}{dt} - \frac{di_b}{dt}\right) \qquad (30)$

$$e_{bc} = V_{bc} - (i_b - i_c)R + L\left(\frac{di_b}{dt} - \frac{di_c}{dt}\right) \qquad (31)$$

$$e_{ca} = V_{ca} - (i_c - i_a)R + L\left(\frac{di_c}{dt} - \frac{di_a}{dt}\right) \qquad (32)$$

where V_{ab}, V_{bc}, V_{ca} are line terminal voltages and e_{ab}, e_{bc}, e_{ca} are line back EMFs. The back EMFs derived from the line voltages need to be filtered for removing high frequency components. The zero crossings of e_{ab} give commutation instants of phase back EMF e_b [12]. Similarly, the commutation instants of other phases can also be derived.

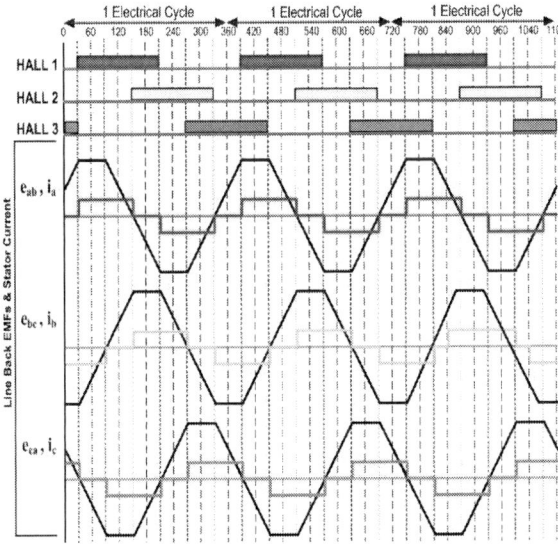

Fig 8. Hall Signals from zero crossings of line Back EMFs

The filtered back EMF is subjected to low pass filter so that the back EMFs are just sinusoidal in nature.

The Clark's transformation is stated as,

$$e_{\alpha\beta} = \frac{2}{3}\begin{bmatrix} 1 & -1/2 & -1/2 \\ 0 & \sqrt{3}/2 & -\sqrt{3}/2 \end{bmatrix}\begin{bmatrix} e_a \\ e_b \\ e_c \end{bmatrix} \qquad (33)$$

So, $e_\alpha = \frac{1}{3}(2e_a - e_b - e_c)$ and $e_\beta = \frac{1}{\sqrt{3}}(e_b - e_c)$ (34)

Hence, $e_\alpha = \frac{1}{3}(e_{ab} - e_{ca})$ and $e_\beta = \frac{1}{\sqrt{3}}(e_{bc})$ (35)

Also, since the EMFs are sinusoidal in nature,

$e_a = \sqrt{2}\,e_{rms}\cos(\omega t)$; $e_b = \sqrt{2}\,e_{rms}\cos(\omega t - (2\pi/3))$

and $e_c = \sqrt{2}\,e_{rms}\cos(\omega t + (2\pi/3))$ (36)

Also, $d\theta_e/dt = \frac{P}{2}\omega'$ (37)

ω' = angular speed in rad/sec,

For solid body rotation around fixed axis

$$\frac{2}{P}\theta_e = \omega't = \omega t \qquad (38)$$

Putting (29) and (31) in equation (27) and solving, we get

$$e_\alpha = \sqrt{2}\,e_{rms}\cos\left(\frac{2}{P}\theta_e\right) \qquad (39)$$

$$e_\beta = \sqrt{2}\,e_{rms}\sin\left(\frac{2}{P}\theta_e\right) \qquad (40)$$

So, $(e_\alpha/e_\beta) = \tan\left(\frac{2}{P}\theta_e\right) = \tan\theta_m$ or

$$\theta_e = \frac{P}{2}\tan^{-1}(e_\alpha/e_\beta) \qquad (41)$$

Thus, during starting, commutation instants can be derived from virtual hall signals generated by estimating the rotor position and after 60 rpm of motor speed; it switches over to line back EMF zero crossing detection method as stated above. This method enables the speed control of PMBLDC drive from near zero to rated speed.

B. Control Algorithm for bidirectional boost-buck converter.

The production of back EMF can be given as,

$$e_{pn} = K\,\omega'\,f(\theta' + \phi) \qquad (42)$$

where ϕ represents the phase difference, K is the back EMF constant, ω' is the angular speed of the rotor (in rps) and θ' is the rotor position angle.

Also, the line to line output voltage (V_{LL}) of a three phase voltage source inverter is given as,

$$V_{LL} = 0.612\,m_i\,V_{dc} \qquad (43)$$

where 'm_i' is the modulation index.

Hence, from equations (26), (41) and (42), we conclude that production of back EMF can be controlled by controlling the DC output voltage (V_{dc}) of bidirectional boost-buck converter. In this way the speed of the motor is controlled. Therefore, the reference speed (N^*) is compared with actual speed (N) and the error speed (N_e) is fed to PI controller.

$$T^*(k) = T^*(k-1) + K_P\{N_e(k) - N_e(k-1)\} + K_i N_e(k) \qquad (44)$$

T^* is multiplied with gain ($1/2\lambda_p$), to get stator current command (i^*), where λ_p is the product of flux and number of conductors in series. The relationship is given as,

$$i^* = T^*/2\lambda_p \qquad (45)$$

The stator current command is compared with DC link current (I_{dc}) and the error current (i_e) is fed to PI controller.

$$V_P(k) = V_P(k-1) + K_P\{i_e(k) - i_e(k-1)\} + K_i i_e(k) \qquad (46)$$

When I_{dc} is positive, switch G_c is controlled by comparing the generated signal (V_P) with a sawtooth carrier wave of 20 kHz and G_c' is off. During negative value of I_{dc}, G_c' is controlled by comparing V_P with an inverted carrier wave of 20 kHz and G_c is off . Thus, during motoring operation, the battery is discharged and the bidirectional boost-buck converter functions as a boost converter. During regenerative braking operation, the converter functions as a buck converter and the battery is charged. The bidirectional boost-buck converter is controlled in CCM mode so as to reduce the current ripples and rating of the switches.

C. Control Algorithm or VSI feeding PMBLDC motor drive

The generated virtual hall signals are decoded and then properly combined to generate gate pulses for the voltage source inverter.

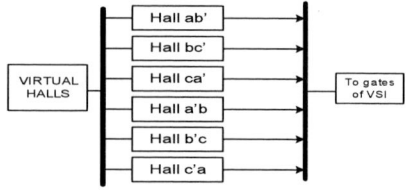

Fig. 9 Generation of gate pulse for VSI

V. SIMULATION RESULTS AND DISCUSSIONS

The simulation model of the proposed system is carried out using MATLAB / SIMULINK .

A. Performance under starting and braking operation at 700 rpm

The performance of the proposed speed control scheme of PMBLDC motor drive is shown under starting (Fig.9) and braking mode (Fig.10). The motor is started at 0.1 sec at rated load. The input current is high. The motor takes very less time to reach the reference speed. The output voltage of the bidirectional converter also varies with the speed. At 0.92 sec, the braking command is released and the power is delivered to battery via bidirectional boost-buck converter. There is a momentary increase in the battery voltage which is obtained due to charging of the battery. At 0.94 sec, the speed and torque of PMBLDC motor becomes zero.

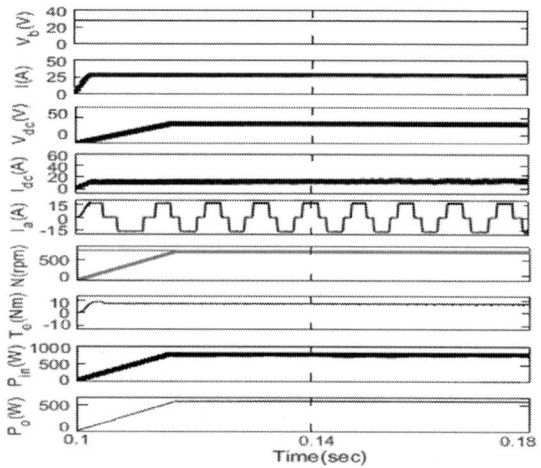

Fig.10 Motor under staring at 700 rpm and 8.73 Nm

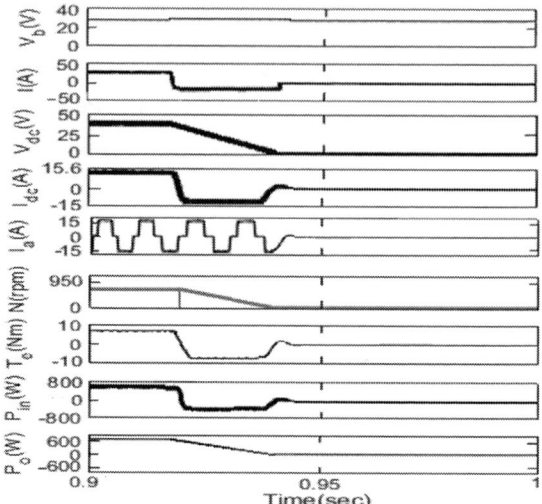

Fig.11   Motor under braking at 700 rpm and 8.73 Nm

B. Performance under change of load torque at 700 rpm

At 0.82 sec, the load torque is changed from rated value to nearly half the rated value (4 Nm). The battery current (I), DC output current of bidirectional boost-buck converter (I_{dc}) and stator current (I_a) reduces to a value corresponding to the changed load torque. The battery voltage (V_b) and output voltage of the DC-DC converter is maintained under this condition.

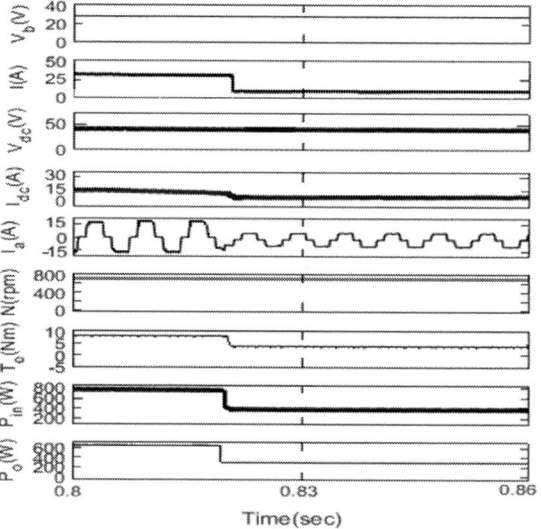

Fig.12   Motor under change of load torque at 700 rpm from 8.73 Nm to 4 Nm

C. Performance under change of speed from 600 rpm to 350 rpm.

At 0.25 sec, the speed is changed from 600 rpm to half the rated speed (350 rpm) at rated load. The battery current (I), DC output current of bidirectional boost - buck converter (I_{dc}) reduces to a value corresponding to the changed load torque for transient period and again settles to value in accordance with the rated torque. The battery voltage (V_b) is maintained. The output voltage of the DC-

DC converter changes corresponding to the change in speed.

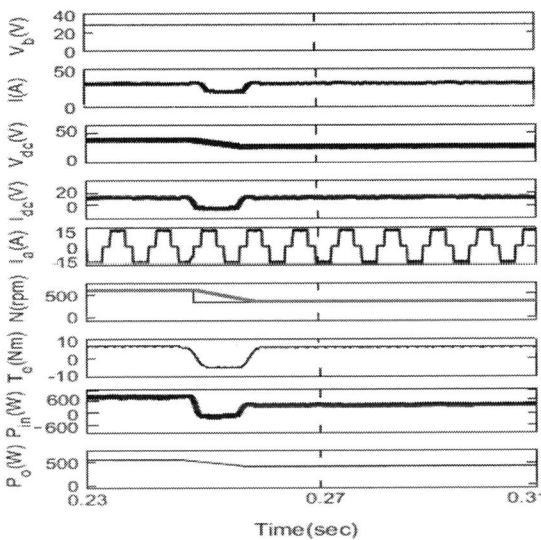

Fig.13   Motor under change of speed from 600 rpm to 350 rpm at rated load

TABLE 1 : Performance of PMBLDC drive under constant rated load and variable speed

$V_{dc}(V)$	Speed (rpm)	Efficiency (%)
6.86	100	85.97
13.71	200	86.04
20.55	300	86.10
27.40	400	86.12
34.32	500	86.14
41.15	600	86.22
48.00	700	86.31

VI. CONCLUSION

A speed control technique of PMBLDC drive is presented for electric two wheeler using voltage control mode. Thus, speed control is proportional to the change in the output voltage of the used DC-DC converter. The use of rate limiter limits the value of input current during starting of the motor. The performance of the system have been analysed for motoring, braking, change of load torque and change of speed of the motor. The proposed PMBLDC drive exhibited an efficient speed control.

APPENDIX

PMBLDC motor rating = 640 W, No. of pole pairs= 4, Rated Voltage = 48 V (DC), Moment of Inertia = 4.5 kg-cm^2, Stator Resistance = 0.4 Ω/phase, Stator Inductance = 0.8mH/ phase, Rated Speed = 700 rpm, Rated Torque = 8.73 Nm, Rated Current = 11 A, Motor efficiency = 91%, Stator winding connection = Star type connection, Battery Specification (Li-ion) = 24 V, 70 Ah.

REFERENCES

[1] B.V. Ravi Kumar and K. Sivakumar, "Design of a new switched stator BLDC drive to improve the energy efficiency of an electric vehicle," in *IEEE ICIT*, pp 532-537, March 2017.

[2] A.Tashakori and M. Ektesabi,"Stability analysis of sensorless BLDC motor drive using digital PWM technique for electric vehicles," *IEEE IECON,* pp 4898 - 4903, Oct 2012

[3] Manuele Bertoluzzo and Giuseppe Buja,"Development of Electric Propulsion System for Light Electric Vehicles, "*IEEE Trans. on Insustrial Informatics*, vol. 7, no.3, pp 428-435, Aug 2011

[4] S. Ogasawara and H. Akagi, "An approach to position sensorless drive for brushless DC motors," in *Proc. IEEE Ind. Applicat. Soc. Annual Meeting*, pp. 443-447, vol.1, Oct 1990.

[5] R. C. Becerra, T. M. Jahns and M. Ehsani, "Four-quadrant sensorless brushless ECM drive," in *Proc. IEEE Applied Power Electronics Conference and Exposition*, pp. 202-209, March 1991.

[6] J. X. Shen, Z. Q. Zhu and D. Howe, "Sensorless flux-weakening control of permanent-magnet brushless machines using third harmonic back EMF ," *IEEE Trans. Indus. Application*, vol. 40, no. 6, pp. 1629-1636, Nov-Dec 2004.

[7] N. Ertugrul and P. Acarnley, "A new algorithm for sensorless operation of permanent magnet motors," *IEEE Transactions on Industry Applications*, vol. 30, no. 1, pp. 126-133, Jan/Feb 1994.

[8] J. Shao, D. Nolan, M. Teissier and D. Swanson, "A Novel Microcontroller-Based Sensorless Brushless DC (BLDC) Motor Drive for Automotive Fuel Pumps," *IEEE Trans.Ind.Applic.*, vol.39, no.6, Nov/Dec 2003.

[9] Jianwen Shao, Dennis Nolan and Thomas Hopkins, "A novel Direct Back EMF Detection for Sensorless Brushless DC (BLDC) motor drives," *Proc. IEEE APEC*, vol. 1,pp 33-37 , March 2002.

[10]Kenichi Iizuka, Hideo Uzuhashi, Minoru Kano, Tsunehiro Endo and KatsuoMohri,"Microcomputer Control for Sensorless Brushless Motor" *IEEE Trans. on Industry Applications*, vol IA-21,no. 6, pp 595-601, Nov 1985.

[11] Gui-Jia Su and J. W. McKeever," Low cost sensorless control of brushless DC motors with improved speed range", *IEEE APEC* ,vol 1, pp 286-292, March 2002.

[12] P. Damodharan and K. Vasudevan, "Sensorless Brushless DC Motor Drive Based on the Zero-Crossing Detection of Back Electromotive Force (EMF) From the Line Voltage Difference," *IEEE Trans. Energy Convers.*, vol. 25, no. 3, pp. 661-668, Sept. 2010.

[13] Gajraj, Yogesh Kr. Chauhan and Bhavnesh Kumar, "Indirect Back EMF Detection based Sensorless Operation of PMBLDC Motor Drive," *IEEE International Conference on Power Electronics, Intelligent Control and Energy Systems*, pp 1-5,July 2016.

[14] Ajmal and M.T. Rajappan Pillai "Back EMF Based Sensorless BLDC Drive Using Filtered Line Voltage Difference," *IEEE International Conference on Magnetics, Machines & Drives*, pp. 1-6, July 2014.

[15] Paul P. Acarnley and John F. Watson "Review of Position Sensorless Operation of Brushless Permanent-Magnet Machines," *IEEE Trans. Industrial Electronics*, vol. 53, no. 2, April 2006.

[16] Abhijeet Sah, Kalpana Chaudhary and V. Venkata Ratnam,"Non-isolated multiphase buck-boost converter design for electric vehicle applications," *IEEE Annual International Conference on Emerging Research Areas: Magnetics, Machines and Drives* , pp 1-6, July 2014.

[17] Bharat Singh Parihar and Shailendra Sharma," Performance Analysis of Improved Power Quality Converter Fed PMBLDC Motor Drive," *IEEE Students' Conference on Electrical, Electronics and Computer Science* , pp 1-6, March 2014.

[18] Farshid Naseri, Ebrahim Farjah and Teymoor Ghanbari,"An Efficient Regenerative Braking System Based on Battery/Supercapacitor for Electric, Hybrid and Plug-In Hybrid Electric Vehicles with BLDC Motor," *IEEE Trans. on Vehicular Technology*, vol. 66, no. 5, pp 3724-3728, May 2017.

[19] Ranjan K. Behera, Rustam Kumar, Srirama Murthy Bellala and P. Raviteja,"Analysis of electric vehicle stability effectiveness on wheel force with BLDC motor drive," *IEEE International Conference on Industrial Electronics for Sustainable Energy Systems* , pp 195-200, Jan-Feb 2018.

Dynamic Performance Analysis of Reactive Power and Improved Rotor Flux Based MRAS for Induction Motor Drives Employing PI and Fuzzy Controller

Mogili Ankarao
Dept.of.EEE, JNTUA
Ananatapuramu, A.P., India
ankaraomogili@gmail.com

M.Vijaya Kumar
Dept.of.EEE, JNTUA
Anantapuramu, A.P., India
mvk_2004@rediffmail.com

Panditi Dinesh
Power and Industial Drives
Dept.of.EEE,JNTUA
Anantapuramu, A.P., India
dineshpanditi07@gmail.com

Abstract— Mainly, the low speed estimation of induction motor (IM) sensorless drive can be done through the MRAS methods effectively due to its simplicity in construction of model design. The most used structures are back emf, rotor flux (RF), reactive power (Q) and torque based. Although reactive power based MRAS operate well at low or near zero speeds, the modification in rotor flux based MRAS method makes more effective than Q-MRAS, it allows adequate at low and near zero speeds for IM drive using improved RF-MRAS based estimator. Conventional PI and fuzzy logic controller (FLC) are used in both Q-MRAS and improved RF-MRAS to estimate speed of the IM drive. This paper shows that the performance of improved RF-MRAS with FLC gives better results over Q-MRAS with fuzzy controller in terms of estimated speed, torque, currents, rotor resistance, and stator resistance. These are evaluated through the MATLAB/Simulink software.

Keywords— *Induction motor Drives, Flux estimation, MRAS, PI controller, Fuzzy logic controller (FLC), Membership function.*

I. INTRODUCTION

Induction motor (IM) drives are rugged, compact and reliable. There are different techniques for estimation of rotor speed for sensorless vector controlled IM drives [1]. Thus speed estimation in sensorless drives requiring terminal voltage and currents values. Several sensorless schemes have been proposed in literature [2].

In observer based method, complete or partial machine model equations. This method affected with instability problems at low speeds. Kalman filter speed estimation methods [3] are robust to noise in the system but computation intensive. Sliding mode methods [4] have the advantage of finite time convergence, eliminates chattering effect but increases complexity in the system. ANN based methods [5], providing stable operation at all speeds if algorithm trained properly. These methods also involve parameter dependencies and large computational overhead.

MRAS is one of the best methods of sensorless schemes for estimation of rotor speed due to its simplicity, performance and ease to implementation [6]. A number of MRAS estimators based on rotor flux [1], back EMF [7], reactive power [8], and electromagnetic torque [9]. The instantaneous and steady-state values of $\vec{V}^* \times \vec{\imath}$ (or X-) are

considered to develop the reference and adaptive models respectively in newly reported X-MRAS where, X is a fictive quantity i.e., neither reactive power nor active power.

The main purpose of this paper to seek the better speed estimator by comparison between reactive power based MRAS and improved RF-MRAS with conventional PI and fuzzy logic controllers. The comparison done with several measuring quantity like speed, toque, currents, rotor resistance, rotor resistance. The performance analysis of these schemes is evaluated by using MATLAB/ Simulink software.

II. MODEL FORMATION

A. Formulation of Q-MRAS

The IM stator voltages in the synchronously rotating reference frame expressed as

$$v_{ds} = R_s i_{ds} + \sigma L_s p\, i_{ds} + \frac{L_m}{L_r} p\psi_{dr} - \sigma L_s \omega_e i_{qs} - \omega_e \frac{L_m}{L_r}\psi_{qr} \tag{1}$$

$$v_{ds} = R_s i_{ds} + \sigma L_s p\, i_{ds} + \frac{L_m}{L_r} p\psi_{dr} - \sigma L_s \omega_e i_{qs} - \omega_e \frac{L_m}{L_r}\psi_{qr} \tag{2}$$

The expression for instantaneous reactive (Q) power is given as

$$Q_1 = Q_{ref} = v_{qs} i_{ds} - v_{ds} i_{qs} \tag{3}$$

The expression of Q_1 is used as reference model. Substituting the values of v_{ds} and v_{qs} eq. (3)

$$Q_2 = \sigma L_s\left(i_{ds}p i_{qs} - i_{qs}p i_{ds}\right) + \sigma L_s \omega_e\left(i_{ds}^2 + i_{qs}^2\right) + \sigma L_s \omega_e\left(i_{ds}^2 + i_{qs}^2\right)\frac{L_m}{L_r}\left(i_{qs}p\psi_{dr} - i_{ds}p\psi_{qr}\right) + \omega_e\frac{L_m}{L_r}(i_{qs}\psi_{qr} - i_{ds}\psi_{dr}) \tag{4}$$

In Q_2, the derivative term in may affects the exactness of the estimated quantity because it introduces noise in the process of computation. However, during steady state, the expression for reactive power becomes

$$Q_3 = \sigma L_s \omega_e \left(i_{ds}^2 + i_{qs}^2\right) + \omega_e \frac{L_m}{L_r}(i_{qs}\psi_{qr} - i_{ds}\psi_{dr}) \tag{5}$$

Furthermore, for vector control operation, and ψ_{qr} =0, the abridged expression of Q independent of any flux terms obtained as

$$Q_4 = \sigma \, L_s \omega_e \left(i_{ds}^2 + i_{qs}^2 \right) + \omega_e \, \frac{L_m}{L_r} i_{ds}^2 \qquad (6)$$

Anyone of Q_2, Q_3 or Q_4 can be used as adaptive model as all of these Q expressions contain rotor speed (ω_r) term.

B. Improved RF- MRAS

Improvement in the classical RF-MRAS estimator, the reference model equation taken from the rotor flux error based which has flux values in terms of voltage and currents. The adaptive model can be taken as state variables X- MRAS. The reference voltages are measured by employing PI controller as adaptive mechanism which is error between adjustable model and reference model. Improved rotor flux based MRAS shows good performance at low speed operation.

Reference model:

$$\rho \, \varphi_{dr}^s = \left(\frac{L_r}{L_m} \right) V_{ds}^s - \left(\frac{L_r}{L_m} \right) \left[R_s + \sigma L_s \rho \right] i_{ds}^s \qquad (7)$$

$$\rho \, \varphi_{qr}^s = \left(\frac{L_r}{L_m} \right) V_{qs}^s - \left(\frac{L_r}{L_m} \right) \left[R_s + \sigma L_s \rho \right] i_{qs}^s \qquad (8)$$

Where total leakage factor $\rho = \left[1 - \dfrac{L_m^2}{L_s L_r} \right]$

Adaptive model:

$$\hat{X} = \widehat{\omega_e} + \omega_{sl} \left(L_s i_{ds}^2 - \sigma L_s i_{ds}^2 \right) + 2 R_s i_{ds} i_{qs} \qquad (9)$$

Where $\omega_{sl} = \dfrac{(R_s i_{qs})}{(L_r i_{ds})}$

III. DESIGN OF CONTROLLER

A. Fuzzy Logic Controller Adaption Mechanism

FLC observes the speed error signal and the corresponding output i_q is updated so that the actual speed w_r matches the reference command speed signal. Two input signals (error in speed and change in error) are fed to the fuzzy logic controller and the output is updated according to the input. For this, rule base is used in FLC. IF-THEN rule base is used. Here, 7×7 rule base was used according to the rules as in Table I i.e., 49 rules are to be developed.

FLC has four parts which are fuzzifier, Fuzzy inference system, knowledge base and defuzzifier. The inputs are crisp data or variables which are converted into fuzzy values

by using fuzzifier. All mathematical and logic operations are performed in FIS.IF-THEN rule base are framed in knowledge base. The fuzzy output is converted into a crisp data by means of defuzzifier.

TABLE I: MATRIX RULES FOR FUZZY LOGIC CONTROL SYSTEM

e \\ ce	NB	NM	NS	Z	PS	PM	PB
NB	NVB	NVB	NB	NB	NM	NS	Z
NM	NVB	NB	NB	NM	NS	Z	PS
NS	NB	NB	NM	NS	Z	PS	PM
Z	NB	NM	NS	Z	PS	PM	PB
PS	NM	NS	Z	PS	PM	PB	PB
PM	NS	Z	PS	PM	PB	PB	PVB
PB	Z	PS	PM	PB	PB	PVB	PVB

IV. SIMULINK DIAGRAMS

A. Reactive Power Based MRAS

From the equations (1) to (6) represents the reactive power based MRAS. Fig4.1 and fig 4.2 represents simulink of Q-MRAS with PI controller and control technique.

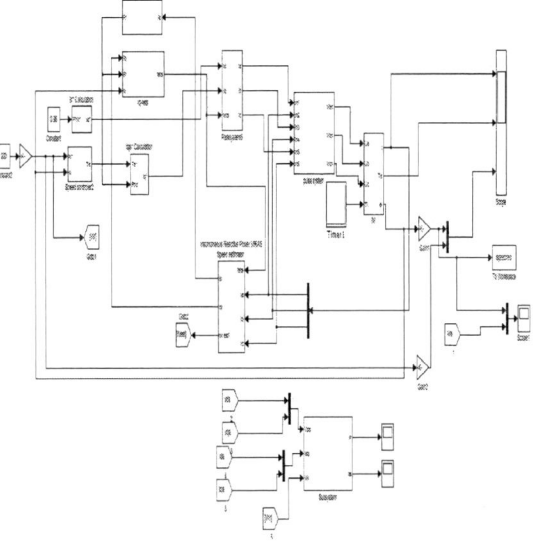

Fig. 4.1. Simulink block diagram of Q- MRAS with PI controller

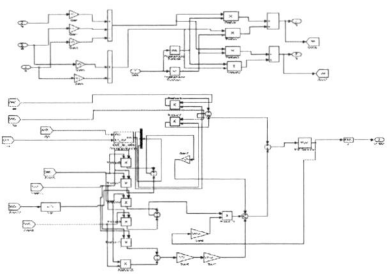

Fig. 4.2. control design Simulink block

978-1-5386-6624-1/18 $31.00 © 2018 IEEE   893

B. Improved RF-MRAS

From equations (7) to (9) represents the improved RF-MRAS. Fig 4.3 and 4.4 represents simulink diagram of the improved RF-MRAS and control design respectively.

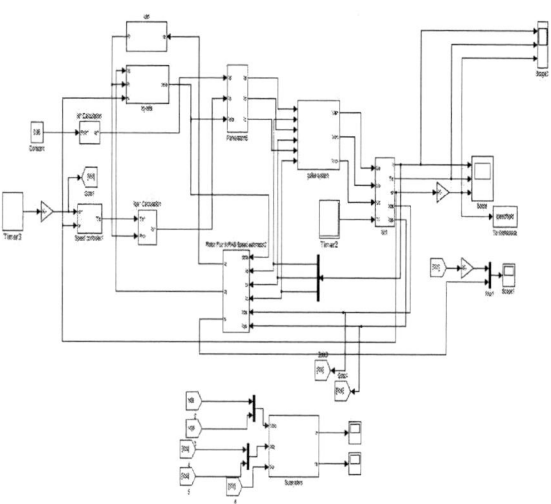

Fig. 4.3. Simulink block diagram of IRF- MRAS with PI controller

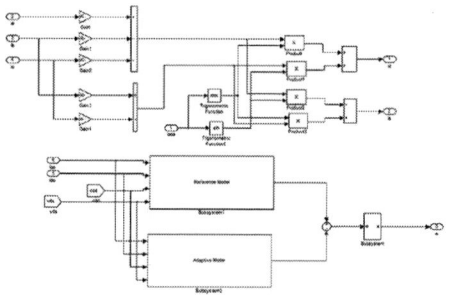

Fig 4.4. Control design of IRF-MRAS

The fuzzy logic controller is replaced with the conventional PI controller the dynamic performance should be carried out for both Q-MRAS and IRF-MRAS, fig.4.4 represents the FLC simulink block.

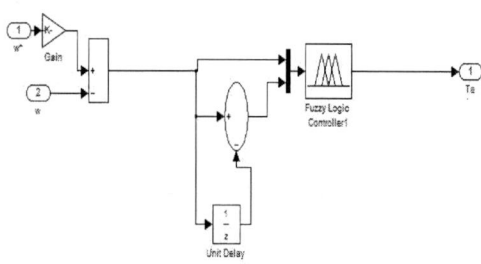

Fig. 4.5. Simulink block of fuzzy controller

V. REASULT AND ANALYSIS

A. Q-MRAS with PI and Fuzzy Controller

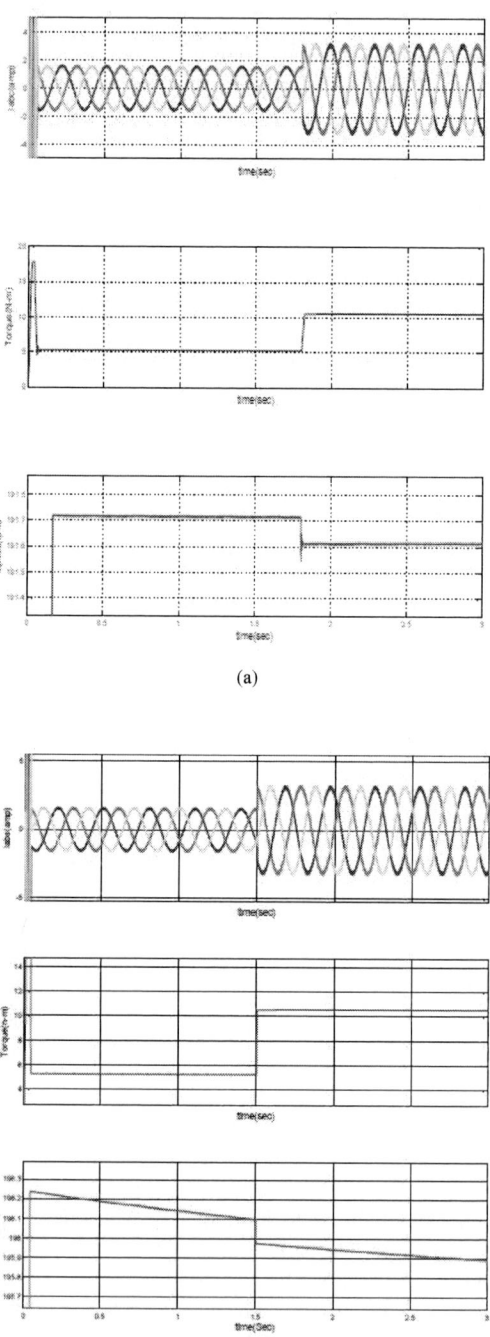

Fig. 5.1. Tracking performance of Q-MRAS at speed n= 200rpm employing (a) PI and (b) FLC

978-1-5386-6624-1/18 $31.00 © 2018 IEEE   894

B. Improved RF-MRAS with PI and FLC

changing the torque value corresponding speed and currents are varying accordingly.

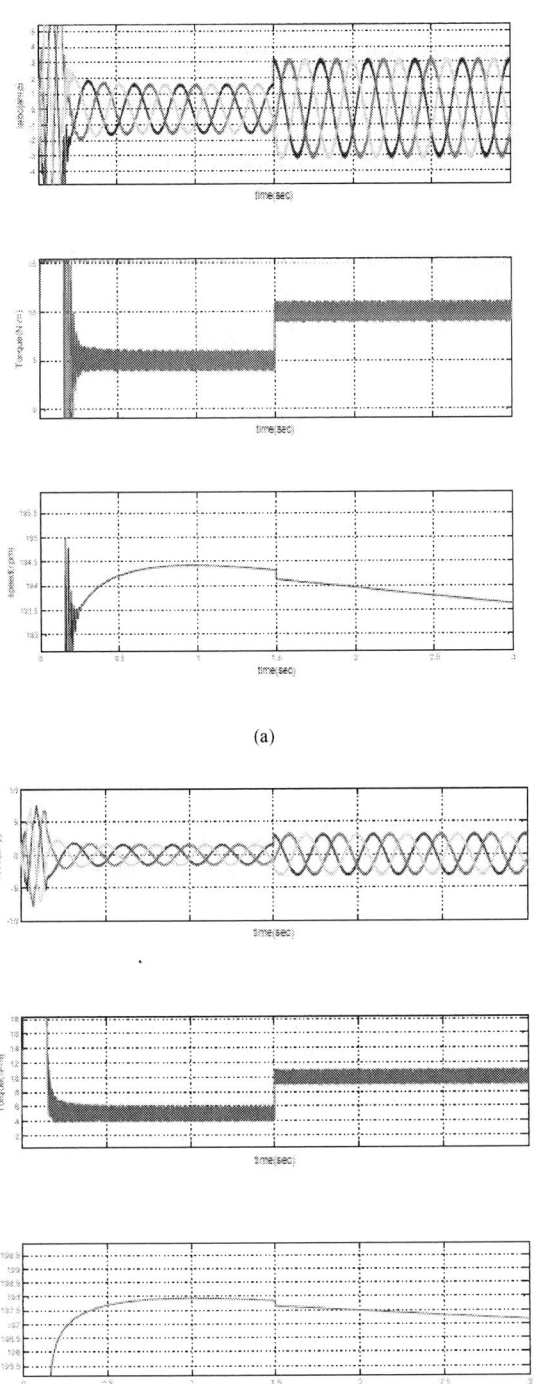

(a)

(b)

Fig. 5.2. Tracking performance of Improved RF-MRAS at speed n= 200rpm employing (a) PI and (b) FLC

The above figure represent the tracking performance of both Q-MRAS and improved RF-MRAS at speed n=200 rpm with application of step torque, here observe that the

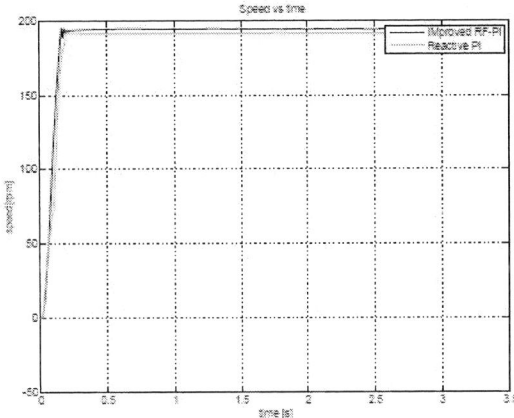

Fig. 5.3. Comparison of estimated rotor speed by Q- power and Improved RF-MRAS with PI controller

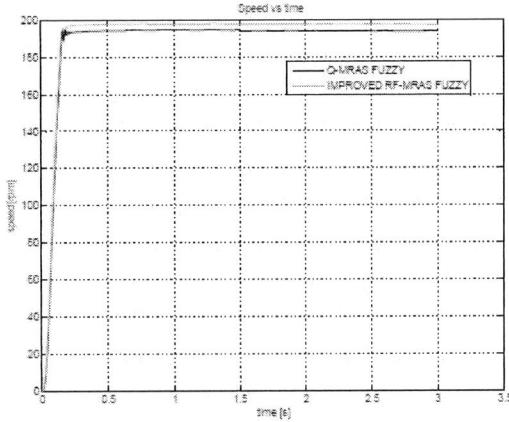

Fig. 5.4. Comparison of estimated rotor speed between Q-MRAS and Improved RF- MRAS with FLC

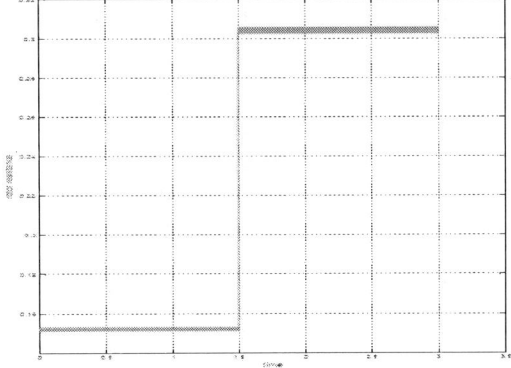

Fig.5.5. Rotor resistance of IM

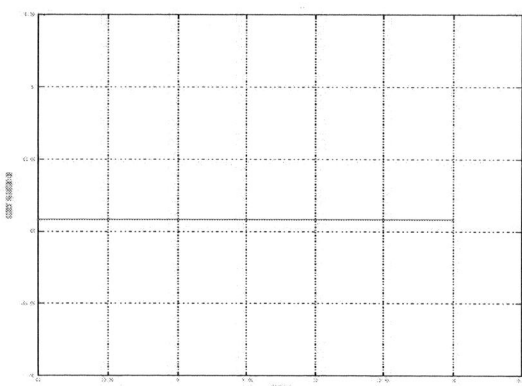

Fig.5.6. Stator resistance of IM

VI. CONCLUSION

This paper presents a comparison between two different MRAS known as Q-MRAS and improved RF-MRAS with PI and Fuzzy controller at low speed with application of step torque. Here also dynamic performances like Torque, speed, currents are evaluated. Also, the comparison of simulated results shown that improved rotor flux MRAS with fuzzy logic controller method is superior among other methods in terms of speed, torque, and currents, rotor resistance and stator resistance. As fig 5.4 shown the rotor speed with improved RF- MRAS with FLC is reached to 198rpm whereas with Q-MRAS with FLC the rotor speed is 196rpm with the reference speed of 200rpm.

REFERENCES

[1] P. Vas, Sensorless vector and direct torque control, New York: Oxford University Press, 1998.

[2] B. Chen, T. Wang, W. Yao, K. Lee and Z. Lu, "Speed convergence ratebased feedback gains design of adaptive full-order observer in sensorless induction motor drives," *IET Elect. Power Appl.,* vol. 8, no. 1, pp. 13-22, Jan. 2014.

[3] F. Alonge, T. Cangemi, F. D'Ippolito, A. Fagiolini and A. Sferlazza, "Convergence analysis of extended Kalman filter for sensorless control of induction motor," *IEEE Trans. Ind. Electron.,* vol. 62, no. 4, pp. 2341-2352, April 2015.

[4] R. P. Vieira, C. C. Gastaldini, R. Z. Azzolin and H. Grundling, "Sensorless sliding-mode rotor speed observer of induction machines based on magnetizing current estimation," *IEEE Trans. Ind. Electron.,* vol. 61, no. 9, pp. 4573-4582, Sept. 2014.

[5] S. Maiti, V. Verma, C. Chakraborty and Y. Hori, "An adaptive speed sensorless induction motor drive with artificial neural network for stability enhancement," *IEEE Trans. Ind. Informat.,* vol. 8, no. 4, pp. 757-766, Nov. 2012.

[6] J. Xiao, B. Li, X. Gong, Y. Sheng and J. Chai, "Improved performance of motor drive using RBFNN-based hybrid reactive power MRAS speed estimator," *in IEEE International Conference on Information and Automation (ICIA),* pp. 588-593, 20-23 June 2010.

[7] M. Rashed and A. F. Stronach, "A stable back-EMF MRAS-based sensorless low-speed induction motor drive insensitive to stator resistance variation," *in IEE Proceedings Electric Power Applications,* vol. 151, no. 6, pp. 685-693, 7 November 2004.

[8] V. Verma, A. V. R. Teja, and C. Chakraborty, "A new formulation of reactive-power-based model reference adaptive system for sensorless induction motor drive," *IEEE Trans. Ind. Electronics.,* vol. 62, no. 11, pp.6797-6808, Nov. 2015.

[9] H. M. Kojabadi, "Active power and MRAS based rotor resistance identification of an IM drive," *Simulation Modeling Practice and Theory,* vol. 17, no. 2, pp. 376-389, 2009.

2nd IEEE International conference on power Electronics, Intelligent Control and Energy systems

Neuro-ELO Based Speed Estimation of Improved Designed Induction Motor Drive for Single Stage Photovoltaic Fed Water Pumping

Khusro Khan, Saurabh Shukla and Bhim Singh fellow,IEEE
Department of Electrical Engineering, *IIT Delhi*, New Delhi-110016, India.
Emails: khusro.abb@gmail.com, saurabh.shukla.ee@gmail.com, bsingh@ee.iitd.ac.in

Abstract— This paper deals with a system with an improved designed induction motor for photovoltaic (PV) array fed water pumping system. The induction motor drive (IMD) is made mechanical sensorless using a proposed adaptive observer based neuro-Luenberger technique to reduce both cost and complexity with the single stage PV array based maximum power point tracking (MPPT) providing cost-effective solution along with simultaneous assurance of optimum power utilization of a PV array. The proposed system is controlled by field-oriented control (FOC). The MPPT as well as DC link voltage, is regulated by three-phase voltage source inverter (VSI). The estimation of motor speed eliminates the use of mechanical sensor and makes the system cheaper and robust. A new robust speed adaptive algorithm is presented, which is less dependent on parameters. A detailed study of various factors affecting the efficiency of the motor, is given to improve the behavior of the IMD for water pumping. The designed motor is tested on the developed prototype in the laboratory and its suitability is judged through various test results under steady state and dynamic conditions of insolation variations.

Index Terms— *PV Array, Speed Adaptation, Field-Oriented Control (FOC), Neuro-Extended Luenberger Observer (ELO), IMD (Induction Motor Drive), Water pump.*

I. INTRODUCTION

Renewable energy resources such as such as sunlight, wind, tidal waves etc. are generally replenished on a human timescale. Renewable energy provides energy for different purposes in which electricity generation is the prime objective.

The results of large scale utilization of renewable energy, and technological assortment of energy sources, would be in the form of energy security and economic benefits [1-2]. The need for the water and energy source management, has become a vital issue during the last decade, and it is becoming more essential in the future. In the meanwhile, PV (photovoltaic) technology is the most popular technology that directly utilizes solar energy and converts it to a direct current [3-4].

DC motors have been used for the PV array fed water pumping because of their virtue of high efficiency and simple structure. However, because of the requirement of its frequent maintenance, BLDC (Brushless DC) motors are used. On the other hand, induction motors are also used in PV water pumping system (PVWPS) as they are reliable and maintenance free alternative [5].

An efficiency improvement of 3-phase induction motors, has been a major concern for these motors used in solar water pumping. Although the energy lost in the pump and control valve pressure drops, are usually much greater than the energy lost in the motor, still it is not insignificant. There is a huge cost saving over life cycle of the pump or the motor by optimizing

the pump motor efficiency. The design of the motor is advised in such a way that both the requirements of optimum efficiency and practicability to manufacture, are met.

A number of sensorless techniques have been reported in the literature. Few of them deal with model reference adaptation schemes (MRAS) [6], while others report extended Kalman filter (EKF) algorithm [7]. One of the disadvantages of MRAS, is its sensitivity to detuning in the stator and rotor inductances. The EKF is a stochastic observer, whereas the extended Luenberger observer (ELO) is a deterministic observer [8]. Few intelligent estimation techniques have also been reported in the literature. Fuzzy logic and neural networks have been a subject of research in recent years [9]. Though these algorithms are quite heavy for basic microprocessors, they are required for effective control with changing model parameters.

In a SCIM (Squirrel Cage Induction Motor), there is no way to introduce a change in rotor resistance once the rotor has been casted. Therefore, to achieve the required resistance and reactance for performance improvement, one has to decide the rotor slot design prior to casting [10]. The performance criterion in the proposed case, is high efficiency. The impedance is increased as the rotor resistance and reactance, are increased causing low starting current and high starting torque. This results in a low efficiency. In SCIM with constant rotor resistance, the rotor design has to be compromised. A low rotor resistance results in high efficiency but on the other hand, it also results in low starting torque and high starting current at a low starting power factor [11-13].

The objective of this work, is to provide direction on the design of an induction motor for PV array fed water pump systems for domestic as well as irrigation purpose. Since the pump motor forms an integral part of the system, the performance improvement of the system by proper speed estimation technique with minimum number of parameters involvement, is required.

To achieve this objective, the following improvements are made in the motor.

- The rotor slot is designed to achieve the required starting torque to fulfill the requirement of high starting torque and high efficiency (like deep bar rotor in starting and shallow bar rotor during full load condition).

- The diameter, length and material of the stampings, are designed for efficiency improvement.

- Its efficiency is obtained through its testing on Ansys RMxprt software. Finite element method and Maxwell stress tensor are used to analyze the magnetic force torque and energy in the three phase squirrel cage induction motor.

978-1-5386-6624-1/18 $31.00 © 2018 IEEE

- Proposed neuro-ELO based speed estimation technique is used for its use in submersible pump.

II. SYSTEM DESIGN

Fig.1 demonstrates the block diagram of proposed scheme with stator voltage and current based speed adaptation mechanism for FOC of IMD (Induction Motor Drive). The system consists of a designed three-phase induction motor of a 2.2 kW, 230 V, 4-pole, powered by a 2.4 kW solar PV array, being used to drive, which is discussed in subsequent sections. The specifications of proposed system, are given in Appendices.

A. DC Link Voltage Selection

The DC link voltage is chosen higher than maximum amplitude of phase voltage of the motor for proper control as [14],

$$V_{dc} > \sqrt{2} \times V_L = \sqrt{2} \times 230 = 325V \qquad (1)$$

Hence, the value of DC link voltage is taken equal to 360 V.

B. DC Link Capacitor Design

The DC link capacitor is designed as per the design equation is given here.

$$\frac{1}{2}C_{dc}\left(V_{dc}^2 - V_{dc1}^2\right) = \sqrt{3} \times a \times V_l \times I \times t \qquad (2)$$

$$C_{dc} = \frac{2\sqrt{3} \times a \times V_l \times I \times t}{\left(V_{dc}^2 - V_{dc1}^2\right)} = \frac{6 \times 1.2 \times 132.7 \times 6.58 \times 0.005}{360^2 - 320^2} \approx 1140\mu F$$

where the lower allowable limit of DC link voltage during transient condition V_{dc1}=340 V, a=1.2, V_l= line voltage=230 V, I=6.32 A, t=0.005s, V_{dc}=360 V. The selected value is 1100 µF.

III. DESIGN OF IMPROVED PERFORMANCE BASED IMD

This section deals with the various design related issues for an induction motor, their causes and methods of improvement. The various factors affecting the performance of an induction motor has been given below.

The starting torque can be optimized by reducing the slot height and increasing the slot width. Stator turns/coil pitch is also reduced and skewing is to be reduced. When the requirement is of high starting torque, deep bar rotors are used. However, they have high reactance values i.e. high magnetic flux leakages causing low efficiency for the same performance parameters. In this solution, there is deterioration in the efficiency of the motor.

The skin effect on rotor side is increased. The starting current is minimized by virtue of proposed control for variable frequency drive. The efficiency of the motor is improved by reducing the stator copper loss. This is achieved by increasing the diameter of copper wire and by decreasing the stator turns/coil pitch and core loss is minimized by reducing the flux density [15]. When there is requirement of high efficiency, shallow bar rotors are used but they have low starting characteristics due to low reactance values i.e. low magnetic flux leakage. Similarly, the rotor copper loss is decreased by increasing the rotor slot area. Therefore, both the requirements

of starting and full load operation, are fulfilled.

The requirement of both high efficiency and high starting torques can be met when the rotor has high reactance in starting and low reactance during operation i.e. it behaves like deep bar rotor in starting and shallow bar rotor during the full load operation and this is achieved by selecting rotor slot geometry.

IV. SPEED ESTIMATION AND SPEED CONTROL OF IMD

This section deals with the speed estimation and speed control of the IMD for PV fed water pumping.

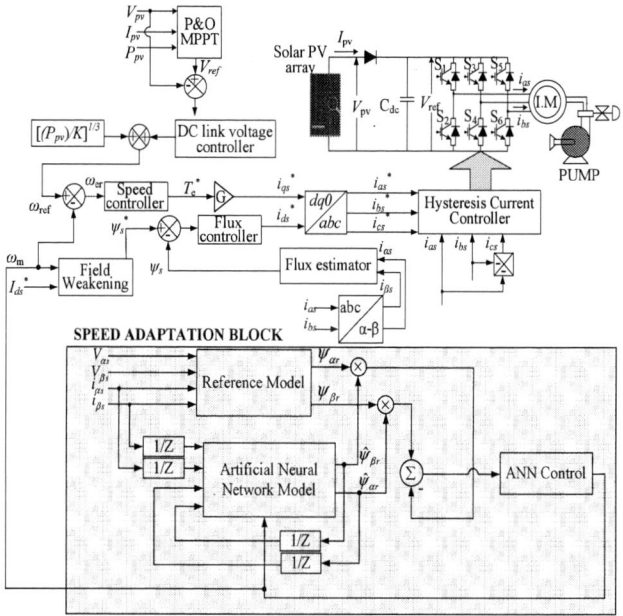

Fig.1. Schematic of proposed system

A. Speed Estimation By Conventional Method

The conventional speed estimation technique is based on stator flux orientation (SFO), which has been explained in [5]. The estimation of slip speed (ω_{sl}) and synchronous speed (ω_e) are given as,

$$\omega_{sl} = \frac{\left(1 + \rho S \tau_r\right) L_s i_{qs}^*}{\tau_r \left(\psi_d - \rho L_s i_{ds}^*\right)} \qquad (3)$$

$$\omega_e = \frac{\left(V_{\beta s} - R_s * i_{\beta s}\right)\psi_{\alpha r} - \left(V_{\alpha s} - R_s * i_{\alpha s}\right)\psi_{\beta r}}{\psi_{\alpha r}^2 + \psi_{\beta r}^2} \qquad (4)$$

Therefore, the motor speed ($\hat{\omega}_m$) can be calculated as,

$$\hat{\omega}_m = \frac{2}{P}\left(\omega_e - \omega_{sl}\right) \qquad (5)$$

The flux angle in this case is calculated as,

$$\hat{\theta} = \int \hat{\omega}_m dt \qquad (6)$$

where i_{ds}^* and i_{qs}^* are current components in synchronously rotating $dq0$ frame, $\rho = 1 - L_m^2 / \left(L_s * L_r\right), \tau_r = L_r / R_r$, L_r=rotor inductance, L_m=magnetizing inductance, R_r=stator referred rotor resistance, R_s=stator resistance.

It is worth mentioning here that the sampled data model has some modeling issues in estimation of ω_e which can be resolved

978-1-5386-6624-1/18 $31.00 © 2018 IEEE

by using LPF. However, this causes delay in transient response. Tis also causes delay in the flux-weakening region and consequently, the flux level increases and current regulation is lost.

B. Speed Estimation of IMD by Proposed Method

The extended Luenberger observer (ELO) method is a deterministic type of observer, which is based on deterministic variable model. ELO is used to find out the rotor flux components. The general model is given as,

$$\dot{X} = AX + BU + \gamma \left(Y - \hat{Y} \right) \qquad (7)$$

$$Y = CX \qquad (8)$$

with [X], [U], [A], [B] and [C] are given as,

$$X = \begin{bmatrix} i_{\alpha s} & i_{\beta s} & \psi_{\alpha r} & \psi_{\beta r} \end{bmatrix}^T \quad U = \begin{bmatrix} V_{\alpha s} & V_{\beta s} & 0 & 0 \end{bmatrix}^T \quad (9)$$

$$A = \begin{bmatrix} -\left(\dfrac{L_m^2}{\rho L_s L_r \tau_r} + \dfrac{R_s}{\rho L_s} \right) & 0 & \dfrac{L_m}{\rho L_s L_r \tau_r} & \dfrac{L_m \omega_m}{\rho L_s L_r} \\[2mm] 0 & -\left(\dfrac{L_m^2}{\rho L_s L_r \tau_r} + \dfrac{R_s}{\rho L_s} \right) & \dfrac{L_m \omega_m}{\rho L_s L_r} & \dfrac{L_m}{\rho L_s L_r \tau_r} \\[2mm] \dfrac{L_m}{\tau_r} & 0 & \dfrac{-1}{\tau_r} & -\omega_m \\[2mm] 0 & \dfrac{L_m}{\tau_r} & \omega_m & \dfrac{1}{\tau_r} \end{bmatrix} \quad (10)$$

$$B = \begin{bmatrix} \dfrac{1}{\rho L_s} & 0 & 0 & 0 \\[2mm] 0 & \dfrac{1}{\rho L_s} & 0 & 0 \end{bmatrix}^T \quad C = \begin{bmatrix} 1 & 0 & 0 & 0 \\ 0 & 1 & 0 & 0 \end{bmatrix} \quad (11)$$

$$\gamma = \begin{bmatrix} k_{11} & k_{12} & k_{21} & k_{22} \\ -k_{12} & k_{11} & -k_{22} & k_{21} \end{bmatrix} \quad (12)$$

where $k_{11} = \left(a_{11} + a_{33} \right) \left(1 - k \right)$, $k_{12} = \omega_m \left(1 - k \right)$

$k_{21} = \left(a_{31} + \dfrac{a_{11}}{a_{14}} \right) \left(1 - k^2 \right) - \dfrac{k_{11}}{a_{14}}$, $k_{22} = \dfrac{k_{12}}{a_{14}}$, L_r and L_s are rotor and stator inductances, respectively. R_r and R_s are rotor and stator resistances, respectively.

The error in estimation of stator current and rotor flux, is given as,

$$\hat{e} = \left(A - LC \right) e + X \Delta A \qquad (13)$$

By considering the Lyapunov's function as,

$$V = e^T e + \dfrac{\left(\Delta \omega_m \right)^2}{\nu} \qquad (14)$$

where $e = X - \hat{X}$ and λ is the positive coefficient.

Therefore, the estimated speed is given as,

$$\omega_r = K_p \left[-\hat{\psi}_{\beta r} e_{i\alpha s} + \hat{\psi}_{\alpha r} e_{i\beta s} \right] + K_I \int \left[-\hat{\psi}_{\beta r} e_{i\alpha s} + \hat{\psi}_{\alpha r} e_{i\beta s} \right] dt \; (15)$$

In this work, the PI controller is replaced by artificial neural network (ANN) based controller, which is explained here.

C. Artificial Neural Network for Current Control

The neural network used in this work for speed tuning, is trained using backpropagation method until mean square error between output and desired pattern is very small [16]. The input function to the neural network in terms of rotor flux error, is given as,

$$v_i = \sum x_i w_i + b = \sum \left(\left(e \times \hat{e} \right) w_i + b \right) \qquad (16)$$

The activation function is sigmoidal, which is given as,

$$y_i(k) = \dfrac{e^{\sum 2 * \left((e \times \hat{e}) w_i + b \right)} - 1}{e^{\sum 2 * \left((e \times \hat{e}) w_i + b \right)} + 1} \qquad (17)$$

This mean square error is calculated using following relation,

$$MSE = E(k) = \dfrac{1}{n} \left[\sum_{i=1}^{n} \left(d_i(k) - y_i(k) \right)^2 \right] \qquad (18)$$

where for the given input x_i, $y_i(k)$ is the actual response by the network, $d_i(k)$ is the desired response, n is the number of input-output training data and k is the number of iteration.

The weights of neurons are updated to minimize the cost function MSE. Equation for weight update is given as,

$$w_{ji}(k+1) = w_{ji}(k) - \xi \dfrac{\partial E(k)}{\partial w_{ji}(k)} \qquad (19)$$

where $w_{ji}(k+1)$ is new weight between i_{th} and j_{th} neurons, $w_{ji}(k)$ is the corresponding old weight, ξ is the learning rate for tuning the speed .

D. Speed Control of IMD

The desired speed with the estimated speed (ω_{ref} and ω_m) of the IMD is passed through a PI (proportional-integral) controller and the output is the reference torque (T_e^*). The equation is given as,

$$T_e^*(n) = T_e^*(n-1) + K_p \left(\omega_{er}(n) - \omega_{er}(n-1) + K_i \omega_{error}(n) \right) \; (20)$$

where ω_{er} is the error between reference speed (ω_{ref}) and estimated speed (ω_m). The torque producing component of current (i_{qs}^*) is derived from T_e^* as,

$$i_{qs}^* = G T_e^* \qquad (21)$$

where G is a gain factor.

The rated flux component is calculated as λ_s^* ,

$$\psi_s^* = \psi_{ds}^* \qquad (for \; \omega_m \leq \omega_{base}) \qquad (22)$$

The field weakening occurs when the motor speed is higher than rated speed. The equation is given as,

$$\psi_s^* = \dfrac{\omega_{base}}{\omega_m} \times \psi_{ds}^* \quad (for \; \omega_{base} > \omega_m) \qquad (23)$$

where $\psi_{ds}^* = \left(\sqrt{2} \times V_l \right) / \left(\sqrt{3} \times \omega_e \right)$ is the rated flux, V_l is the rated line voltage of an induction motor and ω_e is the speed of the synchronously rotating frame in elec-rad/s.

The reference flux producing component of current i_{ds}^* is estimated as the output of flux controller.

$$i_{ds}^*(n) = i_{ds}^*(n-1) + K_p \left(\lambda_{er}(n) - \lambda_{er}(n-1) + K_i \lambda_{er}(n) \right) \; (24)$$

The synchronous speed thus calculated is used to get flux-angle (θ_e) at the kth sampling instant as,

978-1-5386-6624-1/18 $31.00 © 2018 IEEE

$$\theta_{e(k)} = \theta_{e(k-1)} + \omega_e \times T \qquad (25)$$

where T is the sampling period of the signal and ω_e is the synchronous speed.

The value of *q-axis* and *d-axis* current components i_{ds}^* and i_{qs}^* respectively obtained from (21) and (24), are used to obtain reference phase currents i_{as}^*, i_{bs}^*, i_{cs}^*, by following equations,

$$i_{as}^* = i_{ds}^* \sin\theta_e + i_{qs}^* \cos\theta_e \qquad (26)$$

$$i_{bs}^* = i_{ds}^* \sin\left(\theta_e - 120^\circ\right) + i_{qs}^* \cos\left(\theta_e - 120^\circ\right) \qquad (27)$$

$$i_{cs}^* = i_{ds}^* \sin\left(\theta_e + 120^\circ\right) + i_{qs}^* \cos\left(\theta_e + 120^\circ\right) \qquad (28)$$

The sensed stator currents (i_{as}, i_{bs} and i_{cs}), are compared with these reference currents ($i_{as}^*, i_{bs}^*, i_{cs}^*$) and current controller based on hysteresis-band controller is used to control the error signals, which subsequently generates the switching signals for VSI (Voltage Source Inverter) to drive the motor.

E. Magnetic Flux Density Distribution of Stator and Rotor Fluxes

A time-varying magnetic field is produced by time-varying currents flowing in conductors, which is in direction of perpendicular to the conductor. In turn, eddy currents are induced by this magnetic field in the source conductor and in any other conductor parallel to it. The eddy current equation is derived from Maxwell's equations and it is reported in the literature [17].

V. SIMULATION RESULTS

This section deals with performance of the system subjected to various atmospheric condition, realized in terms of changing insolation. The details are given here.

A. Starting and Steady State Performance of PV Array Fed System

Figs.2 (a-b) deal with the starting performance of the system. The system is started at rated insolation. MPPT (Maximum Power Point Tracking) is reached soon after the system is started. The reference speed (ω_{ref}), which is a function of PV array power, achieves its rated value of 150 rad/s as the MPP is reached. The motor speed (ω_m) quickly follows the reference speed (ω_{ref}). By the time, motor phase currents settles to its rated value from its starting value, which is limited to twice its rating.

B. Dynamic Behavior of PV Array Fed System During Insolation Decrement from 1000W/m²-300 W/m²

Figs.3 (a-b) deal with the simulated results for proposed PV array fed system, under dynamic condition of insolation decrement. It can be inferred from the waveforms that motor-pump assembly delivers less volume of water under decreased insolation. This is because of the reduction in reference speed (ω_{ref}), which is the function of PV power (P_{pv}). All other drive parameters follow the decreasing trend in accordance.

C. Dynamic Behavior of PV Array Fed System during Insolation Increment from 300W/m²-1000 W/m²

Figs.4 (a-b) show similar behavior, when the system is operated during dynamic condition of insolation increment. All

PV array indices as well as motor-pump assembly performance indices show satisfactory performance.

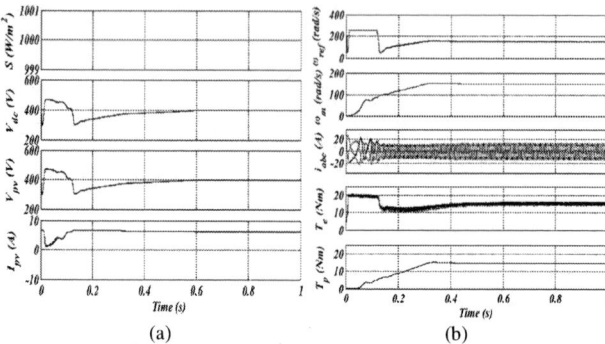

Fig.2. Starting of system (a) PV array indices (b) Induction motor drive

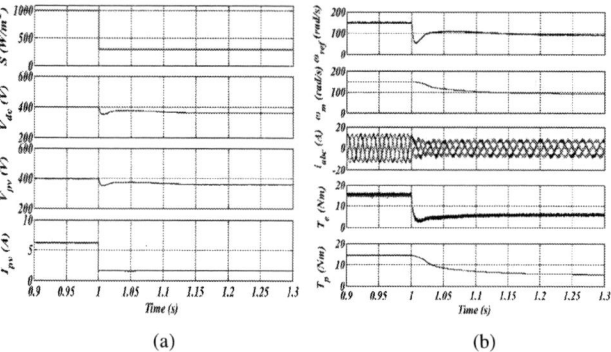

Fig.3. Performance indices of system (a) PV array indices (b) Induction motor drive, during insolation decrement (1000-300) W/m²

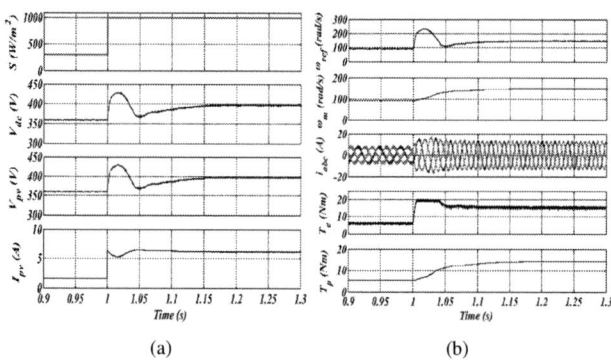

Fig.4. Performance indices of system (a) PV array indices (b) Induction motor drive, during insolation increment (300-1000) W/m²

VI. EXPERIMENTAL RESULTS

The improved designed motor is tested under different conditions. The efficiency improvement of the motor w.r.t. conventional motor of same rating is also tested. The various results for loss reduction and efficiency improvement, are discussed in detail in [18].

A. Efficiency Vs Speed Performance of an Improved Designed Motor

Test results are obtained on Ansys RMxprt software during the design of the motor and is given in Figs.5 (a-b) as efficiency vs. speed curve. It is found that the maximum efficiency of the conventional 3 hp motor is around 81%, while it has reached 87% for the new designed motor.

B. Magnetic Flux Density Distribution

As mentioned earlier, finite element method and Maxwell stress tensor, are used to analyze the magnetic force, torque and energy in the three phase squirrel cage induction motor. The torque is produced by the shear force in the air gap of the motor and by the geometry of the stator and rotor teeth placement. Fig.6 shows the simulation results, which analyzes the magnetic and electrical performance of an improved designed three phase induction motor.

C. Test Results for Tracking Efficiency of MPPT

Figs.7 (a-b) show the tracking efficiency curves at 1000 W/m² and 500 W/m² insolation levels. The desired tracking efficiency of solar PV array of 2.4 kW, is achieved by P&O control technique for a 2.2 kW induction motor feeding a water pump, being realized by separately excited DC generator connected to a resistive load. The P_{pv}-V_{pv} and I_{pv}-V_{pv} curves justify that the tracking efficiency is always more than 99.5% and it reaches almost 100% at lower insolation.

D. Performance of Drive at Starting and Steady-State Condition

Fig.8 (a) presents the smooth starting of the drive at 1000W/m². The various indices V_{pv}, I_{pv}, i_{sa} and the actual motor speed (ω_m) are recorded. This can be observed that the drive is started smoothly and V_{pv} and I_{pv} reach their steady-state values quickly. The drive speed reaches rated value after some time of MPP tracking because of inertia of the motor, which is quite natural. This procedure is repeated for lower insolation and similar trends in the waveforms are recorded for 500W/m² insolation. Figs.8 (b) shows the performance indices, which advocate the suitability of the drive even at lower insolation.

E. Comparative Analysis of Proposed Speed Estimation Technique with Conventional Technique

Figs.9 (a-b) show the comparative results, which justifies that the motor speed (ω_m) estimated by conventional speed estimation technique with stator fluxes, has delayed response during transient case, whereas, it is faster in latter case, when it is achieved by Neuro-ELO technique.

F. Comparative Analysis of Proposed Speed Estimation Technique with Conventional Technique

Figs.10 (a-b) show the comparative results of proposed speed estimation technique with conventional one. It is observed that the torque ripple in the conventional speed estimation technique (Fig.10 (a)) is greatly reduced in the proposed method, as shown in Fig.10 (b).

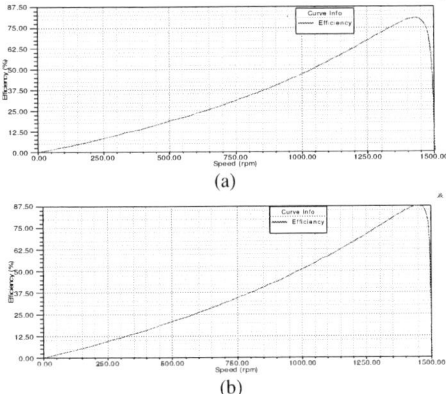

(a)

(b)

Fig.5. Efficiency versus speed for (a) Conventional motor (b) improved designed motor

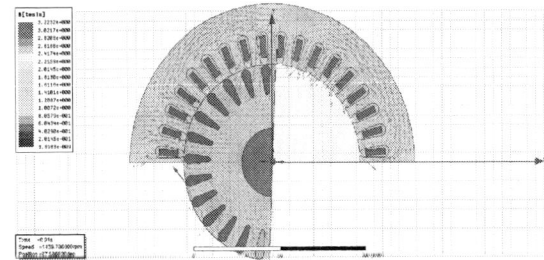

Fig.6. Energy density distributions of the stator and rotor of new model

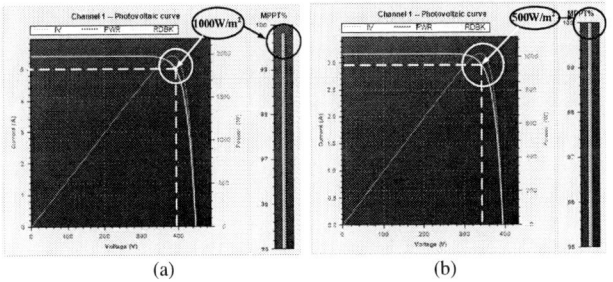

(a)   (b)

Fig.7. Tracking efficiency at different insolation level (1000 W/m²) (b) 500 W/m²

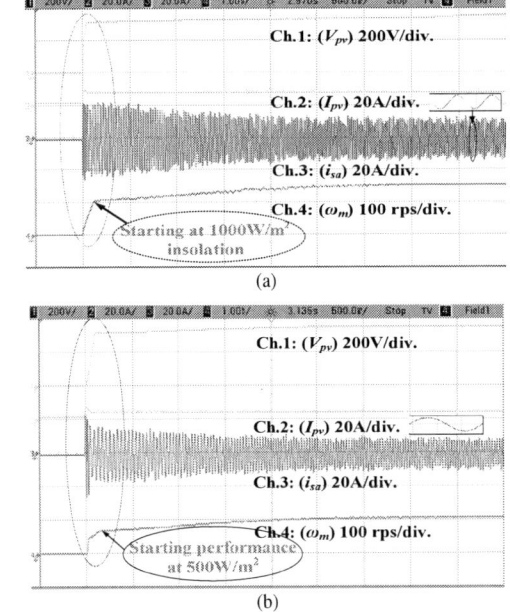

Ch.1: (V_{pv}) 200V/div.

Ch.2: (I_{pv}) 20A/div.

Ch.3: (i_{sa}) 20A/div.

Ch.4: (ω_m) 100 rps/div.

Starting at 1000W/m² insolation

(a)

Ch.1: (V_{pv}) 200V/div.

Ch.2: (I_{pv}) 20A/div.

Ch.3: (i_{sa}) 20A/div.

Ch.4: (ω_m) 100 rps/div.

Starting performance at 500W/m²

(b)

Fig.8. Experimental results: (a) Startng at 1000W/m² (b) Steady-state performance of the drive at 1000W/m²

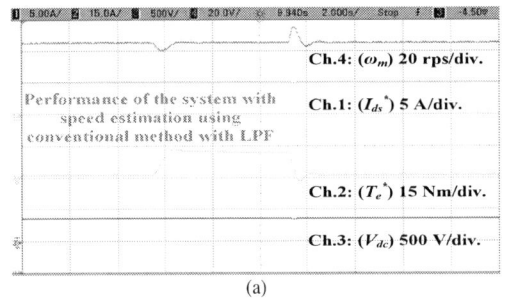

Ch.4: (ω_m) 20 rps/div.

Performance of the system with speed estimation using conventional method with LPF

Ch.1: (I_{ds}^*) 5 A/div.

Ch.2: (T_e^*) 15 Nm/div.

Ch.3: (V_{dc}) 500 V/div.

(a)

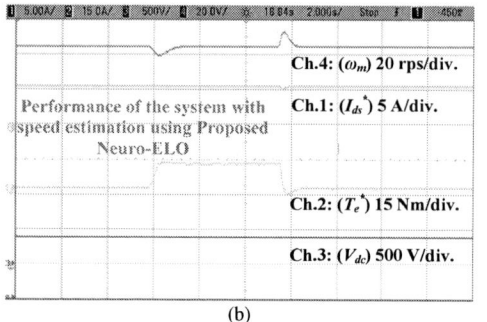

(b)

Fig.9. Comparative Study (a) Conventional speed estimation technique (b) Neuro-ELO based speed estimation technique, during load transition

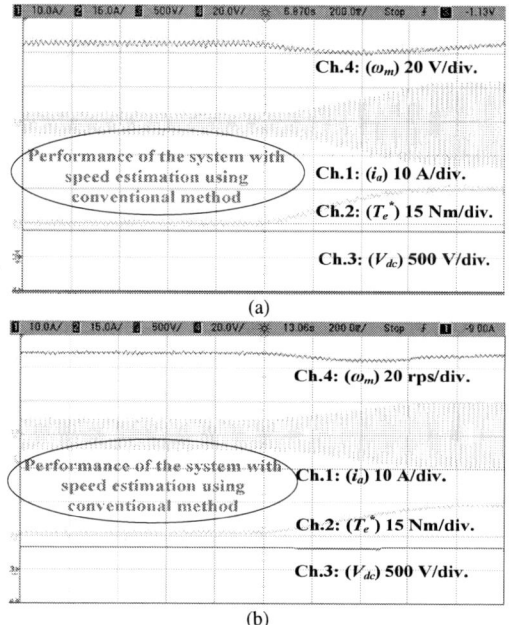

(a)

(b)

Fig.10. Comparative Study (a) Conventional speed estimation technique (b) Neuro-ELO based speed estimation technique, during load transition

VII. CONCLUSIONS

The improved designed motor has been used for PV array fed water pumping. The single stage topology curtailed the cost of the system as well as reduced the complexity. The proposed speed estimation technique eliminated drawbacks of LPF used in conventional speed estimation technique. Moreover, the motor phase current (THD_i) is also greatly reduced. Owing to virtues of simple structure, control, cost-effectiveness, fairly good efficiency and compactness, it can be inferred that the suitability of the system can be judged by deploying it in the field.

APPENDICES

A. Solar PV Array
V_{oc}=400 V, V_{mp}=360 V, I_{sc}=7.1 A, I_{mp}=6.58 A.

B. Improved Designed IM Rating

2.2 kW, 3-phase, 230V, 4 poles, R_s=0.53389 Ω, L_{ls}=0.002862 H, R_r=0.71159 Ω, L_{lr}=0.00411H, L_m=0.1056 H, J=0.011 Kg-m^2.

C. Conventional IM Rating

2.2 kW, 3-phase, 230V, 4 poles, R_s=0.603Ω, L_{ls}=0.00293H, R_r=0.7Ω, L_{lr}=0.00293H, L_m=0.07503H, J=0.011Kg-m^2,

REFERENCES

[1] H. Seok, B. Han, B. H. Kwon and M. Kim, "High Step-Up Resonant DC–DC Converter With Ripple-Free Input Current for Renewable Energy Systems," *IEEE Trans. Ind. Electron.*, vol. 65, no. 11, pp. 8543-8552, Nov. 2018.

[2] G. Ning and W. Chen, "A Hybrid Resonant ZCS PWM Converter for Renewable Energy Sources Connecting to MVDC Collection System," *IEEE Trans. Ind. Electron.*, vol. 65, no. 10, pp. 7911-7920, Oct. 2018.

[3] G. Carannante, C. Fraddanno, M. Pagano and L. Piegari, "Experimental Performance of MPPT Algorithm for Photovoltaic Sources Subject to Inhomogeneous Insolation," *IEEE Trans. Ind. Electron.*, vol. 56, no. 11, pp. 4374-4380, Nov. 2009.

[4] R. F. Coelho, W. M. dos Santos and D. C. Martins, "Influence of power converters on PV maximum power point tracking efficiency," *in Proc. IEEE/IAS International Conference on Ind. Appl.*, Fortaleza, pp. 1-8, 2012.

[5] S. Shukla and B. Singh, "Single-Stage PV Array Fed Speed Sensorless Vector Control of Induction Motor Drive for Water Pumping," *IEEE Trans. Ind. Appl.*, vol. 54, no. 4, pp. 3575-3585, July-Aug. 2018.

[6] M. H. Holakooie, M. Ojaghi and A. Taheri, "Direct Torque Control of Six-Phase Induction Motor With a Novel MRAS-Based Stator Resistance Estimator," *IEEE Trans. Ind. Electron.*, vol. 65, no. 10, pp. 7685-7696, Oct. 2018.

[7] E. Zerdali and M. Barut, "The Comparisons of Optimized Extended Kalman Filters for Speed-Sensorless Control of Induction Motors," *IEEE Trans. Ind. Electron.*, vol. 64, no. 6, pp. 4340-4351, June 2017.

[8] J. Kim, J. Ko, J. Lee and Y. Lee, "Rotor Flux and Rotor Resistance Estimation Using Extended Luenberger-Sliding Mode Observer (ELSMO) for Three Phase Induction Motor Control," *Canadian Journal of Electrical and Computer Engineering*, vol. 40, no. 3, pp. 181-188, 2017.

[9] Yen-Shin Lai and Juo-Chiun Lin, "New hybrid fuzzy controller for direct torque control induction motor drives," *IEEE Trans. Power Electron.*, vol. 18, no. 5, pp. 1211-1219, Sept. 2003.

[10] I. Boldea and S. A. Nasar, The Induction Machines Design Handbook, 2nd Edition, CRC Press, Dec. 2009.

[11] H. J. Lee, S. H. Im, D. Y. Um and G. S. Park, "A Design of Rotor Bar for Improving Starting Torque by Analyzing Rotor Resistance and Reactance in Squirrel Cage Induction Motor," *IEEE Trans. Magnetics*, vol. 54, no. 3, pp. 1-4, March 2018.

[12] S. Toscani, M. Faifer, M. Rossi, L. Cristaldi and M. Lazzaroni, "Effects of the Speed Loop on the Diagnosis of Rotor Faults in Induction Machines," *IEEE Trans. Instrumentation and Measurement*, vol. 61, no. 10, pp. 2713-2722, Oct. 2012.

[13] A.E. Fitzgerald, Charles Kingley and Stephen D Umans, Electric Machinery, 6th edition, Tata McGraw Hill Edition. 2012.

[14] B. Singh and R. Kumar, "Simple brushless DC motor drive for solar photovoltaic array fed water pumping system," *IET Power Electron.*, vol. 9, no. 7, pp. 1487-1495, Aug. 2016.

[15] B. Yan, X. Wang and Y. Yang, "Comparative Parameters Investigation of Composite Solid Rotor Applied to Line-Start Permanent-Magnet Synchronous Motors," *IEEE Trans. Magnetics, Early Access*, 2018.

[16] S. Maiti, V. Verma, C. Chakraborty and Y. Hori, "An Adaptive Speed Sensorless Induction Motor Drive With Artificial Neural Network for Stability Enhancement," *IEEE Trans. on Ind. Inf.*, vol. 8, no. 4, pp. 757-766, Nov. 2012.

[17] J. Wu, J. Wang, C. Gan, Q. Sun and W. Kong, "Efficiency Optimization of PMSM Drives Using Field-Circuit Coupled FEM for EV/HEV Applications," *IEEE Early Access*.

[18] K. Khan, S. Shukla, and B. Singh, "Performance Based Design of IMD for Single Stage PV Fed Water Pumping," Accepted *IEEE, IAS Annual Meeting*, 2018.

978-1-5386-6624-1/18 $31.00 © 2018 IEEE

Transformer Health Monitoring System Using Internet of Things

Divyank Srivastava
Electrical Engineering Department
Delhi Technological University
New Delhi, India
divyanksri.srivastava@gmail.com

M. M. Tripathi
Electrical Engineering Department
Delhi Technological University
New Delhi, India
mmmtripathi@gmail.com

Abstract— Health monitoring of electrical equipment using IOT may help to replace the equipment before failure and continuity of the power will not be disturbed. This paper presents an implementation of this concept which acquires the real-time condition of the distribution transformer remotely with the use of internet implementing IOT. The proposed health monitoring system works in real time and uses temperature-sensor, potential transformer and current transformer for monitoring temperature, voltage and current of the distribution transformer and send these information to a remote server where it can be monitored and necessary action may be taken to avoid the outage of the electricity supply. The proposed system has been implemented in Laboratory and tested. The data collected are validated and it has been found that the proposed system is working satisfactorily.

Keywords—Internet of things, Health monitoring system, Distribution transformer,GSM, Microcontroller, Sensor.

I. INTRODUCTION

A distribution transformer is a step down transformer which delivers power to the users which can be used in houses or industries, etc. Various predicaments can be faced by distribution transformers if their health is not monitored periodically[1]. The operating parameters such as temperature, voltage and current of a distribution transformer are vital measurement criterion which can tell about the health of the transformer. The operating condition depends on several factors such as overloading, loss of supply etc. Life of a transformer is proportional to its health. Better the health of a transformer, longer the life. Issues such as overburden and inadequate heating of a transformer are the main reasons of deterioration of a transformer's health[2]-[3]. Earlier the health monitoring system was not much prevalent in the distribution network and the faults could only be detected after a complete blackout. It caused major penalties for the distribution system as large amount of losses were incurred. Therefore, there was a foremost need for a reliable health monitoring system which would help in creating preventive measures beforehand. The present health monitoring systems pose a few concerns which are listed below: (i) Common systems for transformer measurements merely identify a single transformer parameter. Even if some systems are able to detect multiple parameters, the time consumed in the testing is too lengthy and hence slow speed is attained. (ii) Detection system is erratic. The performance of the detection system is less accurate[4]-[8]. Slower detection of the faults is present and it is unreliable and unstable (iii) The monitoring

system is unable to observe all the user data of transformers to decrease costs.

As per the abovementioned requirements, we need a monitoring system to analyze real time data associated with the distribution transformer to characterize various operating parameters and further provide the information to the monitoring centre in requisite time [9]. This is when Internet of Things(IOT) comes into picture. It helps to invigilate the data online of the key functional factors of the distribution transformers which grants constructive data about the health of distribution-transformers which in turn will facilitate the services to use their transformers in a best possible way and increase the life of a transformer. IOT helps in identifying the problems beforehand i.e. Before the occurrence of any failed mechanism which facilitates cost effective solutions and hence less penalty. IOT serves greater reliability and stability than other conventional systems[10].

II. BLOCK DIAGRAM

The fig.1 represented by a block diagram suggests it is essential that the device being used for invigilating (monitoring) be housed nearby the transformer itself. The blocks represented in the block diagram evaluate different parameters communicating the health of the transformer. Furthermore, the proposed system provides us with a health monitoring system of the distribution transformer which presents us with the data regarding various parameters of the transformer. Additionally, three sensors have been used in the proposed system, namely, voltage sensor, a current sensor, and temperature sensor. For the use of PIC microcontroller, a power supply for supplying power has been used, a Wi-fi modem for internet connectivity has been used and several sensors has been used. Fig.1 shows the different modules used in the proposed model. The data received by the IOT is sensed by the sensors incorporated and displayed on the LCD screen. Furthermore, at the same time, data is also being sent to the user on a given server or on the cellphone via Wi-fi connection. The advantage of having such a system is that if estimation of an unprotected system is made by IOT, system failure can be prevented. The proposed system comprises the following module:

A. Power Supply

As per the standard process we need 5V power supply for our proposed model, in which we have used a bridge rectifier, capacitors and a 7805 voltage regulator. Here we have used a 230/12V transformer to convert the power supply to its desirable value, i.e 12V [11]. A bridge rectifier is connected after a 230/12V transformer which convert the 12V AC to DC, then a 1000µf capacitor is connected in parallel to filter out the ripples from there we give a connection to our GSM module then a 7805 regulator is connected after that two 100µf capacitors are connected which gives a pure 5V dc to give in whole circuit.

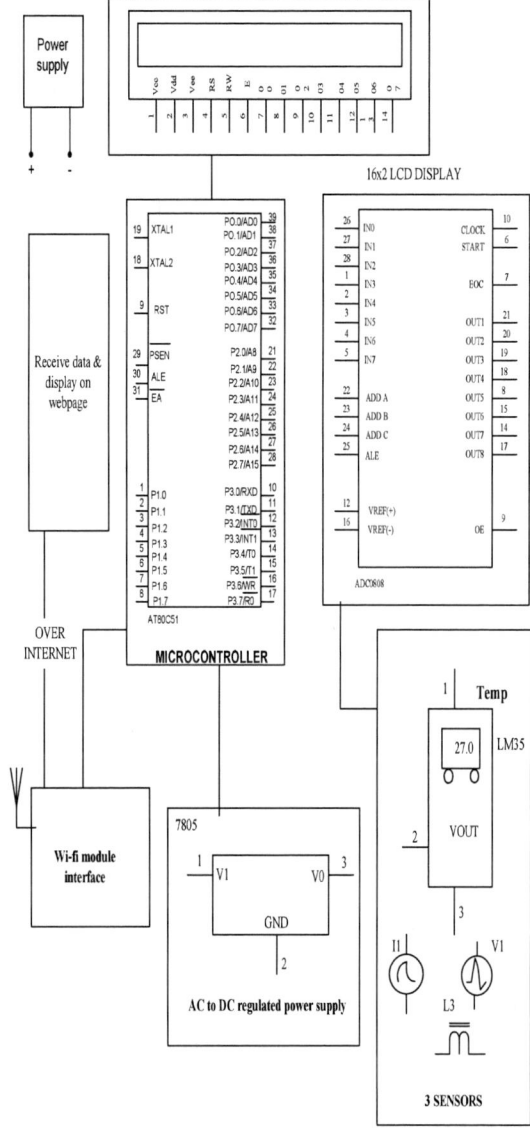

Fig.1: Block Diagram of proposed model

B. LCD Module

One of the commonly used devices connected to a PIC microcontroller is an LCD module. Some common LCDs which can be used with the PIC microcontroller for implementation of any kind of hardware projects are 16x2 type and 20x2 type display LCD, means it can show 16 characters in each line, and 20 characters in each line (both can show output in two line).

C. PIC-Microcontroller 18F4520

PIC-microcontroller was made by Microchip-Technology in 1993.Acronym of PIC is Peripheral Interface Controller. The performance of these microcontrollers are steadfast and they can implement a program very easily compared with other microcontroller of same categories. PIC-microcontrollers are widely used because of their effortless programming, wide obtainability, easily connection with other peripherals, cheapness, large client network and serial-programming ability (reprogramming along with flash-memory), etc. In 2000, Microchip-technology present the PIC18 microcontroller which is distinct from the 17 series, which become very popular and fast selling microcontroller in the market, with a huge number of PIC-microcontroller-17 deviations presently in production. In compare to earlier microcontroller, which mainly used assembly language for programming, Now C has become the major programming language [12].

D. GSM Module

In the village areas, GSM delivers an ultimate communication network by using the web of mobile communication network. With the use of GSM technology we can transmit data with high efficiency, suitability and with low cost. Through GSM the monitoring of distribution-transformer health can be done easily. We used 2.4 GHz GSM module which is a very precise, accurate and high speed data transfer can be achieved .It works with the baud rates 9600, 19200.etc. It is a transceiver part through which information can be simultaneously transmitted and received. It should be connected to any TTL/CMOS logic serial RXD and TXD lines and can support Baud-rate of 9600bps, 19200bps, 38400bps and 57600bps. It can also works with 4 different RF [13].

E. Regulator

The used voltage regulator L7800 series is a three-terminal regulator(+), which is available in D²PAK packages and multiple no of output voltage, because of that we can used it in different and wide number of application. With these regulators local-on card regulation can be achieved, removing the error occur due to single-point regulation.[14] Each regulator type can used in internal current limiting, thermal shut-down and protection of the network, making it fundamentally enduring. If suitable heat-sinking is offered, they can gives over 1A output current. While these devices used as primarily voltage regulator, these can also be used with external hardware circuit for getting constant voltage and desirable current.

F. LM 35 Temperature Sensor

LM35 sensors are used for sensing the temperature and an IC ADC0808 is used to convert the data into digital form. LM35 sensor consist of 3 pin's i.e., VCC, GND and output pin. When LM35 is heated the voltage at output pin increases, it is connected to the analog to digital convertor IC (ADC). This LM35 series sensors are more accurate integrated-circuit temperature sensors, because there output voltage is directly proportional (linear) to Celsius temperature scale. Thus LM35 are more advantageous than other standard linear temperature sensors which are calibrated in kelvin, due to this there is no need to minus a large voltage value from its output to convert in centigrade value. The LM35 don't entail any external regulation to deliver distinctive precision of $\pm 1/4°C$ at room temperature and $\pm 3/4°C$ over a full(-55 to $+150°C$) temperature range.

G. Voltage Sensor

A voltage sensor is nothing but a potential transformer. Voltage sensors sense the voltage of an output terminal and gives a signal directly proportional to it. The sent signal can be in the form of analog current or voltage, which can be converted into a digital output using ADC [15]. The output produced is employed to exhibit the precise voltage by a suitable measuring device or it can be saved for further analytical purpose in the database for controlling of different variables.

H. Current Sensor

A current sensor is equivalent to a current transformer. Current sensors sense the current in an external load and gives a signal directly proportional to it. The produced output could be analog voltage or current which can be converted into digital form using ADC. The produced signal is utilized to display the measured current using a suitable measuring device [15]. Value obtain from current sensors can be saved for further analytical purpose in the database for controlling of different variables. The output of current sensor can be in the form of various quantities such as:

DC input, a tripping output like in a relay, which replicates the value of the sensed current digital output, which tripped when the sensed current surpasses a certain pickup value.

III. CONNECTIONS AND DESIGNING

The connection of LCD , microcontroller , temeperature sensor, current transformer and potential transformer are as follows:

We connect a 16x2 LCD which is having 16 pin , 1st pin is connected to ground , 2nd to 5v supply, 3rd is connected with 2 resistors value 1kΩ and 10kΩ through 5v supply , this pin is used to adjust the contrast, 4th pin is Rs pin, 5th is data read/write, 6th is data enable pin and these 3 goes to 27,28,29th pin of microcontroller. 7th to 14th pins are data pins and connected to 33th to 40th pins of microcontroller, 15th pin is again used for brightness adjustment, connected to 5v supply and the last 16th pin is grounded.

33th to 40th pins are data pins which is connected to 7th to 14th pin of 16x2 LCD , 27th 28th 29th pins are Rs , RD/WR , En pins connected to 4th 5th 6th pin of 16x2 LCD , 11th and 32th pins are power pin connected to 5v supply and 12th and 31th pins are used for grounding. There is a .1µf capacitor is connected between pin 12 and 32. 1st pin is used to reset the microcontroller connected to 5v supply through a 10kµ resistor and a switch. 13th and 14th pins are used to connect the crystal oscillator of 11.0592MHZ frequency, 22pf capacitor are connected to both the pins. 15th 16th 17th 18th pins are used to enter the mobile no on which we have to send the transformer health data, each is connected to 5v supply through a 10kΩ ladder resistor network and a switch. Pins 19th 20th 21th and 22nd are connected to 4 LED to send some data from the remote center to our GSM module and the LED glow according to the message, pins 25th and 26th are transmitter and receiver pins connected to 11th and 12th pins of max 232, microcontroller works on RS232 level, 5V TTL logic.

MAX 232 is used for level conversion because as all the system works on 5V supply but our GSM module works on +10V, +12V, +18V supply so MAX 232 convert +5V to +12V , its 11th and 12th pins are connected to 25th and 26th pins of microcontroller. 16th pin is connected to +5V supply and 15th pin to ground, there is .1µf capacitor connected between these two pins. There is 10µf capacitor connected between 1st -3th pin and 4th – 5th pin, similarly pin 2 is connected to 5v supply through 10µf capacitor and pin 6th to ground through 10µf capacitor.

GSM module (SIM 900) also works on RS232 level its pin 3 is receiver pin which receive the signals from 14th pin of MAX 232 and its 2 pin is transmitter pin which transmit data to pin 13th of MAX 232.

The 12V ac coming from 220/12V transformer is converted to 12V DC through 4 diode (1N4007) bridge converter then a 100µf capacitor is connected in parallel to eliminate the ripple content then 3, 10kΩ resistors are connected in series with a 6.5kΩ resistor and 4.7kΩ variable resistor in parallel. This variable resistance is connected to limit the output to 5V, than this circuit output goes to pin 3rd of ADC.

A phase and neutral wire goes to load from which a C.T is connected to load wire from which a diode (1N4148) is connected , a 10kΩ resistor and a 100µf capacitor are connected in parallel, one 83kΩ resistor is connected in series then a 6.8kΩ resistor along with 4.7kΩ variable resistor is connected in parallel to limit the output current, than this circuit output is goes to pin 4 of ADC.

There is an analog temperature sensor i.e LM35 which can sense 10mV/o , its 1st pin is connected to 5V supply and 3rd pin to ground. 2nd pin is goes to pin 2 of ADC.

For designing a PCB layout we use DIP TRACE software. In this software, for creating a 16, 32 or 40 DIP pin we use pattern template. The toolbox is used to select various components. The selected component is held and dragged into the workspace. The wired network of desired width is used to connect the components in the workspace. We can use pin manager for connection of microcontroller pins with other components. We have Design manager on the left side of our workspace from which properties of selected object or function can be modified. Parameters in this panel change, depending on the selection (component, net, bus etc.). If several objects are chosen, we will see only common parameters for all of them. The schematic diagram of our PCB layout is shown in fig.2.

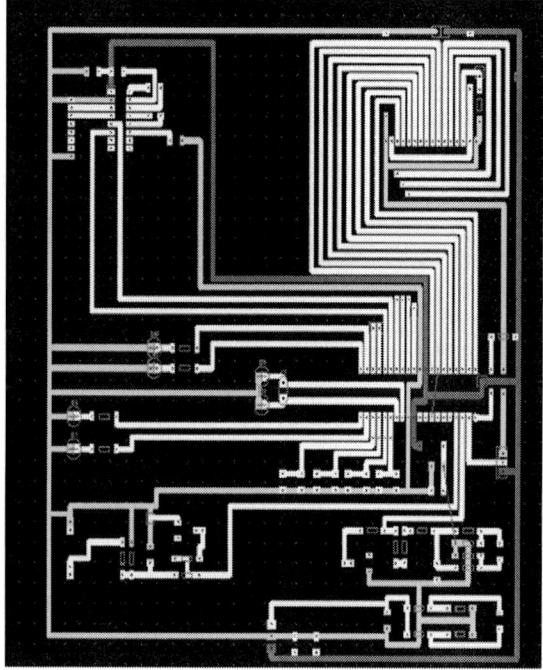

Fig.2: PCB Layout

When we switch on the circuit it takes 7 initialization pulses to display the current, voltage and temperature at the 16x2 LCD and after 9 initialization pulses the data is send to IOT server on which we can see the health data of transformer. There is 199 count counter using which we can see a new response after 199 count, as we increase the load the load current increases in the monitor, and we can vary the transformer voltage with the help of variable resistor. The hardware model is shown in fig.3.

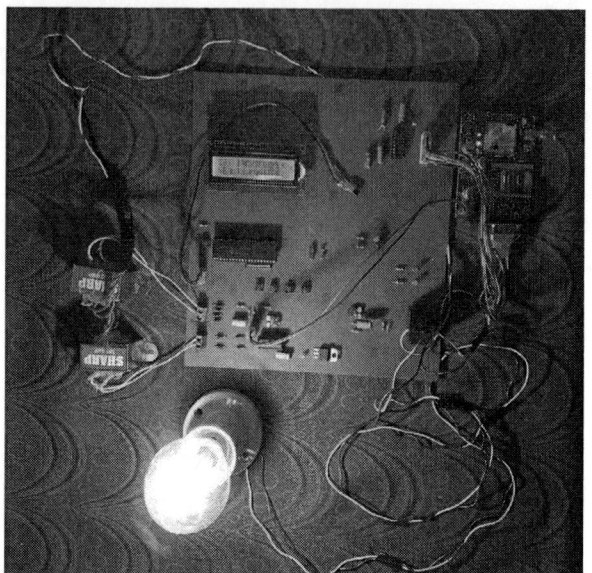

Fig: 3 Actual hardware Model

IV. RESULTS

On the 16x2 LCD, transformer current, voltage and temperature were observed after 7 initialization pulses. After the next 9 pulses the data was sent to the IOT servers on which the health was monitored and necessary actions were taken.

On the IOT servers, three parameters can be viewed i.e. temperature, voltage and current as seen in fig.3, fig.4 and fig.5 respectively. The data entries can also be viewed on the IOT server on JSON file type as shown in fig.6.

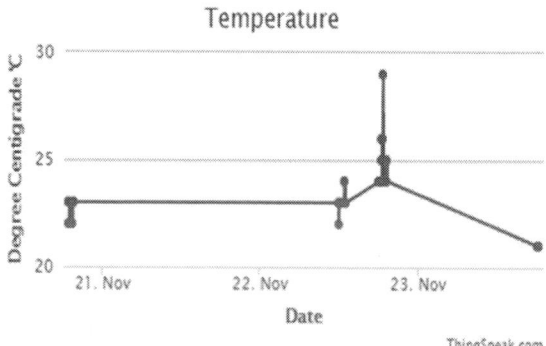

Fig.3: Transformer Temperature

978-1-5386-6624-1/18 $31.00 © 2018 IEEE

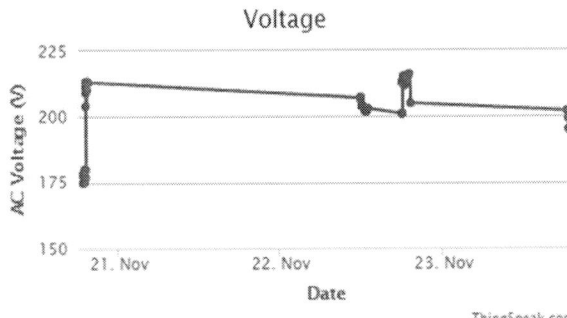

Fig.4: Transformer Voltage

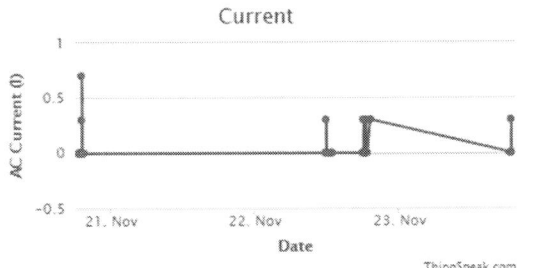

Fig.5: Load current

V. CONCLUSIONS

The IOT wireless open typical technology is being designated in this paper as the energy management and efficiency technology of choice. Employing the system for real time monitoring of power line with an open standard such as IOT helps to keep costs down and condensed power consumption. We can observe from this project that sensors can be employed to conventional as well as smart grid for monitoring of different parameter of the grid. It can be concluded that our model showing results on internet through GSM module as well as on the on-board LCD. Using IOT for monitoring different parameter of distribution transformer, human labour will be minimised and we can also save the data for forecasting as well as for any electricity theft. With the use of IOT our power system would become more accurate and reliable. The advantage of using GSM based monitoring over manual monitoring is that there is no need for an operator to note down the variables after every minute because with the use of IOT the data can be stored with every 90 second After we found out any abnormality in the system, necessary action can be taken instantly and any catastrophic failure can be prevented.

In the same way we can implement wireless sensors on relays of transmission line, RTD of generators and different power system component, and send their data through GSM on internet making our power system network healthy and reliable.

REFERENCES

[1] Chan, W. L, So, A.T.P. and Lai, L., L.; "Interment Based Transmission Substation Monitoring", IEEE Transaction on Power Systems, Vol. 14, No. 1, February 2014, pp. 293-298.

[2] Performance Monitoring of Transformer Parameters in (IJIREEICE) Vol. 3, Issue 8, August 2015.

[3] Gsm based transformer monitoring" in "International Journal of Advance Research in Computer and Communication Engineering", Vol.2, Issue3, JAN 3.

[4] "Distributed Transformer Monitoring System" International Journal of Engineering Trends and Technology (IJETT) - Volume4 issue5- May 2013.

[5] Microcontroller Based Substation Monitoring and Control System with Gsm Modem" IOSR Journal of Electrical and Electronics Engineering (IOSRJEEE) ISSN: 2278-1676 Volume 1, Issue 6 (July-Aug. 2012).

[6] Ravishankar Tularam Zanzad, Prof. Nikita Umare, and Prof Gajanan Patle "ZIGBEE Wireless Transformer Monitoring, Protection and Control System",International Journal of Innovative Research in Computer and Communication Engineering (An ISO 3297: 2007 Certified Organization), Vol. 4, Issue 2, February 2016.

[7] N Maheswara Rao, Narayanan R, B R Vasudevamurthy, and Swaraj Kumar Das, "Performance Requirements of Present-Day Distribution Transformers for Smart Grid", IEEE ISGT Asia 2013 1569815481.

[8] Mohamed Ahmed Eltayeb Ahmed Elmustafa Hayati, and Sherief F. Babiker, "Design and Implementation of Low-Cost SMS Based Monitoring System of Distribution Transformers", 2016 Conference of Basic Sciences and Engineering Studies (SGCAC).

[9] Leibfried, T, "Online monitors keep transformers in service", Computer Applications in Power, IEEE, Volume: 11 Issue: 3, July 1998 Page(s): 36 -42.

[10] http://www.microchip.com/wwwproducts/cn /PIC18F4520.

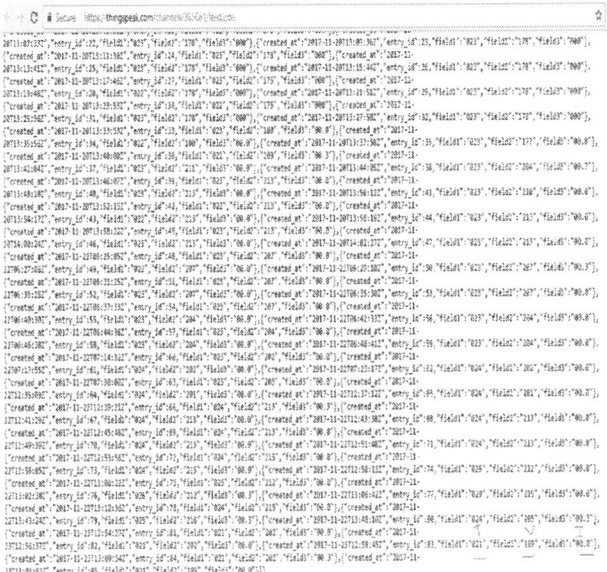

Fig: 6 Real Time Data on IOT Server

[11] Xiao-hui Cheng, and Yang Wang, "The remote monitoring system of transformer fault based on The internet of Things", 2011 International Conference on Computer Science and Network Technology.

[12] Leib fried, T, "Online monitors keep Transformers in service", Computer Applications in Power, IEEE, Volume: 11 Issue: 3, July 1998 Page(s):36-42.

[13] Chan, W. L, So, A.T.P. and Lai, L., L., "Interment Based Transmission Substation Monitoring", IEEE Transaction on Power Systems, Vol. 14,No. 1, February 1999, pp.293 – 298.

[14] Abdul-Rahman AI-Ali, Abdul Khaliq & Muhammad Arshad," GSM-Based Distribution Transformer Monitoring System", IEEE MELECON 2004, May 12-5,2004, Vol 3 Pages-9991002, Croatia.

[15] Muhammad Ali Mazidi , Janice Gillispie Mazidi, Rolin D.Mckinlay, The 8051 Microcontroller And Embedded Systems Using Assembly And C,Second Edition, Pearson Education, 2008, India.

2nd IEEE International Conference on Power Electronics, Intelligent Control and Energy systems(ICPEICES-2018)

Parametric Optimization of Proton Exchange Membrane Fuel Cell using Successive Trust Region Algorithm

Shashwati Ray
Department of Electrical Engineering Bhilai
Institute of Technology, Durg-491001
(C.G.)India
Email:shashwatiray@yahoo.com

Heena Mishra
Department of Electrical Engineering,
C. V. Raman University Bilaspur (C.G.)
India
Email:heena.nitrr@gmail.com

T. V. Dixit
Department of Electrical Engineering
Government Polytechnic College
Kondagaon (C.G.) India
Email:tvdixit@gmail.com

Abstract—In this paper the parameters of a static voltage model of Nexa-1.2 KW proton exchange membrane fuel cell are identified using nonlinear optimization technique. The model is based on semi-empirical formulae defined by parametrical equations characterizing the voltage current relationship of the fuel cell operation. Trust region optimization algorithm is used successively to minimize the objective functions based on two different performance criterion. The results of the optimized model obtained from simulation conform with the experimental data with very small uniform error. The derived model shows significant reduction in the mean square error as compared to the existing optimized PSO model.

Keywords: *Proton Exchange Membrane Fuel Cell, Parameter Identification, Optimization, Trust Region Algorithm.*

I. INTRODUCTION

The global problem faced by the world today is ever increasing population, consequently leading to rise in energy demand. Energy is a prime mover for the wheel of life, and its requirement is increasing with change in life style, comfort level, and faster urbanization and industrialization. The earth is having limited energy in various forms and fossil fuels are largely consumed for generation of power. Along with fuel consumption, there is a proportional rise in the levels of environmental pollution, so some solutions based on renewable energy need to be explored. Therefore, the quest for new sources and new conversion technologies to convert the non conventional energy in usable form has led to fuel cells (FCs) as the new emerging energy source.

Based on the developed technology, electrolyte used in fuel cells are broadly classified as Molten Carbonate Fuel Cells (MCFC), Phosphoric Acid Fuel Cells (PAFC), Alkaline Fuel Cells (AFC), Solid Oxide Fuel Cells (SOFC) and Proton Exchange Membrane Fuel Cell (PEMFC). On the basis of various comparisons - electrolyte material, operating temperature, nature of fuel, efficiency, start up time, advantages and limitations, applications, etc., it has been seen that among all

the available FC technologies, PEMFCs are considered to be most suitable, more so, due to their several attractive features, such as low operating temperatures, relatively low cost and quick start up, simplicity, viability, and high efficiency. However, the performance of a PEMFC system depends on different operating condition such as temperature, pressure and air stoichiometry [1]–[3].

In order to understand completely the fuel cell system behaviour, it is necessary to know the mathematical model for efficient designing of fuel cell based systems through simulation under different conditions of temperature, pressure, stack current, stack voltage etc. [4]. Various fuel cells models have been developed in past few decades and are categorized in three categories: mechanistic, semi empirical and empirical models. Mechanistic models use algebraic equations and rules to explain physical, chemical and electrical phenomena. This approach is very complex, using electrochemical, thermodynamics and fluid mechanics, hence, its application is limited to only online PEMFC model. ANN can simplify nonlinear function with good accuracy, hence ANN technique was implemented by [5] and a comparison of mechanistic and ANN model was carried out.

Different models are proposed in the literature to study the performance of PEMFC depending upon the level of complexity [6], [7]. The Amphlett model is a semi-empirical model of PEMFC that integrates mechanistic and empirical regression modeling to estimate the stack voltage under different stack currents, operating temperatures, partial pressures of hydrogen and oxygen, etc. This model gives high precision for wide operating range of stack current. In [7] two dynamic models were examined using an electrical equivalent and semi-empirical equations, where the authors proposed a parametric analysis method and a simplified transfer function representation for the dynamic response using simulation software.

Several approaches have been reported in the literature to address the fuel cell topologies, ranging from heuristics and

978-1-5386-6624-1/18 $31.00 © 2018 IEEE
909

expert systems to advanced mathematical modeling and optimization techniques, hierarchical-decompositions, combinatorial schemes and model predictive control. Earlier stochastic search techniques were developing efficient methods for solving optimization problem of different mathematical models, whereas, presently various other optimization techniques are used for parameter identification of PEMFC. In recent years many heuristic techniques like genetic algorithm (GA), particle swarm optimization (PSO) [8] and stochastic methods like simulated annealing (SA) [9] have gained attention. Cheng et al., in [10] proposed Genetic algorithm (GA) for improving the accuracy of the fuel cell model parameter identification which is inefficient for complex problems. However, these problems can be solved using PSO algorithm [11] which depends on observation. Li et al., in [12] compared the different forms of PSO and found an Effective Informed Adaptive Particle Swarm Optimization (EIA-PSO) algorithm to improve the drawbacks of basic PSO algorithm. He compared experimental data with simulation data and concluded that EIA-PSO has better precision even in the presence of noise.

The PEMFC system is a nonlinear, multi variable electrochemical system that requires a thorough knowledge of the complex internal phenomena. In [13] a circuit model of PEMFC was developed using mathematical equations and static and dynamic performances were simulated and compared under load step changes. The steady-state models analyse the working of FC at fixed operating point or at slow dynamics, whereas, dynamic models analyse the working of the FC in transient modes. The results obtained using voltage step response is improved by electrochemical impedance spectroscopy (EIS) technique. In [14] a PSPICE based model of PEMFC was used to design and analyze the fuel cell power systems on the basis of static and dynamic characteristics obtained from simulation and later compared with experimental data. In [6] modeling of PEMFC was based on both electrical and thermal model. In electrical model, thermodynamic, activation, ohmic, concentration and double layer phenomena were considered, whereas in thermal, thermal energy balance was considered which includes heat generation, dissipation and FC thermal capacity. The model with empirical parameters associated with activation and concentration phenomena, was implemented in Matlab Simulink and a mean square error less than 0.6 V for the FC voltage prediction was obtained, and a mean square error of less than 1.5°C for the FC operating temperature prediction was obtained. But it did not consider the fluctuations in the power, therefore energy storage system, such as, a bank of battery or super capacitors need to be added to provide quick response to variations in the power.

The PEMFC static model adopted in this paper is based on the semi-empirical equations defined by parametrical equations and a group of parameters in order to characterize and predict the voltage-current characteristics of the fuel cell operation. To provide an improved accurate mechanism model of PEMFC, it is required to identify the parameters of the models using some optimization techniques. The main contribution of this work is the use of Trust region method successively, focused to extract the model parameters so as to minimize the mean square error as well as the maximum error between the simulated and the actual voltage values.

Rest of the paper is organized as follows: In Section II,

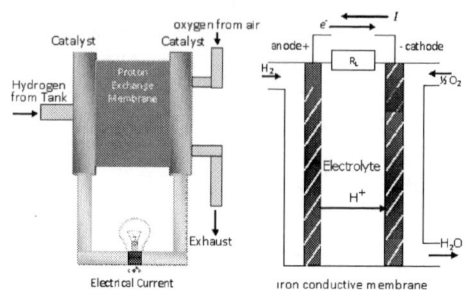

Figure 1.   Operating scheme of a single fuel cell

we present the basic concept and working of the fuel cell, electrical equivalent circuit model and the corresponding mathematical formulations. In Section III we present in detail the proposed optimization method for model parameter extraction along with its algorithm and discuss briefly the experimental setup to get the data used to test our algorithm. Here, we also present the optimization results and validate the same by comparing experimental and simulated results through various graphs. Finally Section IV gives the conclusions of our work.

II.   FUEL CELL MECHANISM AND MODELING

A. Working Principle of Fuel Cell

The basic principle of fuel cell is to convert chemical energy from fuel directly to electrical energy and a number of fuel cells are attached to form a stack to generate higher voltage [15]. The PEMFC consumes hydrogen gas as the primary fuel and oxygen from the air to produce electrical power with high efficiency, low operation noise and emissions in the form of water and heat. The basic operating scheme of Fuel Cell is shown in Figure 1. The PEMFC is supplied with hydrogen and air at the inlet and flow fields, which then diffuse through porous media to the polymer membrane. The membrane in the middle of a cell contains catalyst both with the anode and cathode. The catalyst layer on the anode separates hydrogen molecules into protons and electrons. Thereafter, the membrane allows transfer of protons and enables the electrons to flow through an external circuit before recombining with protons and oxygen at the cathode to form water. This migration of electrons produces electricity. The electrochemical reactions carried out at the anode and cathode is given as [16]:

At Anode: $2H_2 \rightarrow 4H^+ + 4e^-$
At Cathode: $4H^+ + 4e^- + O_2 \rightarrow 2H_2O$
Overall reaction: $2H_2 + O_2 \rightarrow 2H_2O$

Apart from fuel cell stack the PEMFC system have various other components, such as air delivery system, hydrogen delivery system, thermal and water management system to handle temperature and humidity during the operation [17]. PEMFC is a type of fuel cell, which is also applicable for vehicular applications [18].

B. Fuel cell Modeling

To simplify PEMFC modeling, following assumptions are considered [11]:

1) The gases used for chemical reaction (H_2 and O_2) are ideal and uniformly distributed.
2) The temperature inside the cell is uniform.
3) The pressure inside the cell is assumed to be constant.
4) The model is lumped and single dimension.
5) No parasitic reaction is considered.
6) Thermodynamic properties are analyzed at uniform average stack temperature, and specific heat capacity of stack is assumed to be invariable substantially.

The Nernst potential or the thermodynamic potential given by (1) is the voltage across the fuel cell electrodes under thermodynamic balance at no load [19]. It is also known as reversible cell potential or internal potential or open circuit voltage. The Nernst voltage E_{Nernst} depends on cell temperature, partial pressure of hydrogen and oxygen on the surface of catalyst at both cathode and anode.

$$E_{Nernst} = -\frac{\Delta G}{2F} - \kappa_E(T - T_{ref})$$
$$+ \frac{RT}{2F}[\ln P_{H_2} + 0.5\ln P_{O_2}] - E_d \qquad (1)$$

where, ΔG is the change in Gibbs energy, F is Faradays constant, R is the universal gas constant, T_{ref} is the reference temperature, P_{H_2} and P_{O_2} are respective partial pressures of hydrogen and oxygen, T is the cell temperature, κ_E is empirical constant, E_d is a potential that shows the influence of the flow of hydrogen and oxygen delays on output voltage during load transient and its steady-state value is zero. Under standard temperature and pressure (STP) (1) is calculated as:

$$E_{Nernst} = 1.229 - 8.5 \times 10^{-4}(T - 298.15)$$
$$+ 4.308 \times 10^{-5}T(\ln P_{H_2} + 0.5\ln P_{O_2}) - E_d \qquad (2)$$

The losses occuring in the PEMFC model are activation loss, ohmic loss and concentration loss. The ideal standard potential of fuel cell is 1.229 V/cell with water only as byproduct [20].

1) Activation Potential Drop: The activation loss in PEMFC is due to sluggishness in the chemical reactions that takes place on the active surface of the electrode. Based on several physical experimentations and various electrochemical reactions, Tafel has proposed an empirical formula to estimate the activation over-potential V_{act} as [9], [21]

$$V_{act} = -[\xi_1 + T(\xi_2 + 2 \times 10^{-4}\ln A + 4 \times 10^{-5}\ln C_{H_2})$$
$$+ \xi_3 T\ln C_{O_2} + \xi_4 T\ln I] \qquad (3)$$

with

$$C_{O_2} = 1.97 \times 10^{-7}\exp(498/T)P_{O_2} \qquad (4)$$
$$C_{H_2} = 9.17 \times 10^{-7}\exp(-77/T)P_{H_2} \qquad (5)$$

where, I is the stack current and ξ_1, ξ_2, ξ_3 and ξ_4 are the parametric coefficients for each cell. These coefficients are based on electrochemical, thermodynamics, fluid mechanics and dissolved oxygen and hydrogen concentration at the cathode-membrane interface and the effect of these parameters on static $V - I$ characteristics are reported in [7]. The oxygen and hydrogen concentration is estimated using C_{O_2} and C_{H_2}, by considering the pressure drop in the flow channel, the diffusion in the electrode as well as the diffusion through a water film created by presence of water in the electrode.

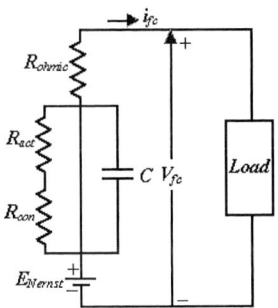

Figure 2. Electrical equivalent circuit of PEMFC

2) Ohmic Potential Drop : The ohmic over-potential V_{ohmic} is caused due to resistance offered in flow of electron through the electrode as well as migration of ions through the electrolyte, i.e., it is generated from membrane resistance to transfer protons and from electrical resistance of the electrodes to transfer electrons. In addition, the ohmic loss also depends on interconnection of bipolar plates. It can be expressed as [12]:

$$V_{ohmic} = \frac{(R_M + R_t)}{I} \qquad (6)$$

with

$$R_M = \rho_M l/A$$

and ρ_M, the resistivity of membrane for electron flow (cm) of Nafion series PEMFC can be estimated using

$$\rho_M = \frac{181.6[1 + .03(\frac{I}{A}) + .062(\frac{T}{303})^2(\frac{I}{A})^{2.5}]}{[\sigma - 0.634 - \frac{3I}{A}]\exp[4.18(\frac{T-303}{T})]} \qquad (7)$$

where, R_M and R_t are the membrane and transfer equivalent resistance of the cell respectively, l is the thickness of membrane and A is active area of membrane. The adjustable fitting parameter σ is affected by membrane manufacturing process. The ohmic losses can be minimized by reducing the internal resistance of the cell. This could be achieved by reducing the thickness of the electrodes and using high conductivity electrodes.

3) Concentration Potential Drop: The main cause of this loss is the reduction in concentration of reactants in the region of electrodes during the fuel consumption. Unfortunately, this loss cannot be calculated analytically. The empirical equation to calculate concentration potential drop V_{con} is expressed as [9], [22]:

$$V_{con} = -\frac{RT}{2F}\ln[1 - (J/J_{max})] \qquad (8)$$

where, J and J_{max} are the actual and maximum current density.

4) Output Voltage: The electrochemical output static voltage of fuel cell V_{fc} depends on Nernst and Tafel equations given by Amphlett and Kim [9] and is the algebraic sum of all the four potentials given as:

$$V_{fc} = E_{Nernst} - V_{act} - V_{ohmic} - V_{con} \qquad (9)$$

The electrical equivalent circuit of the PEMFC is represented in Figure 2.

978-1-5386-6624-1/18 $31.00 © 2018 IEEE

C. Power of PEMFC

The electrical output power P_{fc} supplied by the fuel cell to the load is:

$$P_{fc} = IV_{fc} \tag{10}$$

III. PARAMETER IDENTIFICATION

Since the fuel cell model characteristic is nonlinear, the parameter extraction is done through optimization. The objective function which is obtained through mathematical voltage model of PEMFC is highly nonlinear and we solve to identify the optimal set of parameter values using Trust Region Algorithm which is a nonlinear optimization technique.

A. Trust Region Algorithm

The trust region method is one of the most important numerical optimization method for solving nonlinear unconstrained optimization problems. It is an iterative method which calculates a step to improve the value of the objective function in each iteration [23], subject to the condition that the step is within a proper trust region. Trust region is defined as a spherical area of radius Δ centered at the current iterate, where the approximate model, usually the quadratic model lies [24]. It is a subset of the domain of objective function used in mathematical optimization. On the other hand, the trust region is also adjusted adaptively according to the performance at the current iteration. If the approximate model fits the original problem, then the trust region can be enlarged otherwise it is reduced [25]. The fit is evaluated by comparing the ratio of the actual improvement observed in the objective function with the improvement expected from the model approximation. The convergence can be ensured that the trust region in each iteration depends on the improvement made in previous iterations.

Generally, we require the objective function to be twice differentiable so that in each iteration we can obtain the gradient and the Hessian matrix of the objective function, which are used in the trust region subproblem. The flow chart of the algorithm is shown in Figure 3. The Trust Region Algorithm can be summarized as:

- Set the iteration number $k=0$ and provide the initial starting point x_0 and then the algorithm generates a sequence of iterates $x_1, x_2,...,x_k,...$

- Iteration process is completed when no progress is made or the solution is approximated with sufficient accuracy.

- Assume a model $f(x_k + \delta_k)$ obtained by Taylor series that approximate objective function near x_k. Δ_k is the trust region where the model is a good approximate of $f((x_k + \delta_k))$.

- Choose the step length Δ to move towards approximate minimum of the model in this region and update it at each iteration using heuristics.

- Obtain a quadratic model of $f(x_k + \delta_k)$ as $q(\delta) = f(x_k) + g_k^T \delta_k + \frac{1}{2}\delta_k^T H_k \delta_k$

- The subproblem to be solved to find the step length to be taken during iteration is $\min q(\delta)_{||\delta|| \leq \Delta}$. Also the iteration step is $x_{k+1} = x_k + \delta_k$

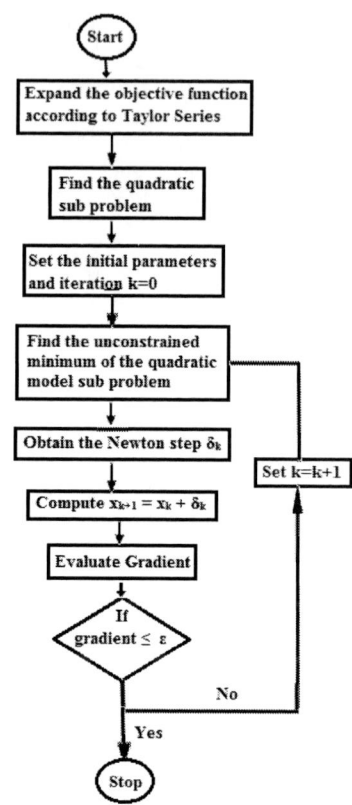

Figure 3. Flow chart of Trust Region algorithm

- The heuristics to update the size of trust region depends on the ratio of expected change in $f(x_k)$ to predicted change, i.e., $p_k = \frac{f(x_k) - f(x_{k+\delta_k})}{q(x_k) - q(x_{k+\delta_k})}$

- If $p_k \approx 1$ then Δ is increased. If p_k is poor or very small then Δ is decreased. If $p_k \approx 10^{-4}$ (smaller than threshold) then the step is rejected, else the new value of x_k is retained.

B. Parameter identification with Trust Region Algorithm

In order to proceed with the optimization to determine the parameter values, we have considered two objective functions which are minimized in succession. The first objective function to be minimized is based on the principle of mean square error as the performance criterion and the second objective function is based on the principle of maximum absolute error. With n as the total number of experimental data which are voltages V_{exp} obtained experimentally from current-voltage characteristic, the objective functions $f_1(\mathbf{x})$ and $f_2(\mathbf{x})$, where $\mathbf{x} = [\xi_1, \xi_2, \xi_3, \xi_4, \sigma]$ are defined as follows:

$$f_1(\mathbf{x}) = \frac{1}{n} \sum_{j=1}^{n} (V_{exp} - V_{fc})^2 \tag{11}$$

and

$$f_2(\mathbf{x}) = \max_n (|V_{exp} - V_{fc}|) \tag{12}$$

subject to the parametric constraints

$$-1 \leq \xi_1 \leq 0$$

$$0 \leq \xi_2 \leq 1$$
$$0 \leq \xi_3 \leq 1$$
$$-0.1 \leq \xi_4 \leq 0$$
$$10 \leq \sigma \leq 23$$

In this paper, with the initial values $\mathbf{x}$ as given in Table I, the objective function (11) is first minimized using the trust region algorithm. Then, taking the obtained values $\mathbf{x}$ as the staring values, the second objective function is minimized to give an uniform error.

C. Implementation and Results

Experimental results were obtained in the laboratory with the Nexa 1.2 KW fuel cell module comprising of 36 cells in series to form a stack that can carry a maximum current I_{max} as 60 A. The length l of the cell is 0.0178 cm and area A is 120 sq cm. A single fuel cell produces 1 V at open circuit or no load and 0.5 V at rated load, thus, this module produces a voltage of 36 V at no load and around 18 V at rated load. The output of fuel cell module depends on different conditions such as hydrogen and oxygen pressures, partial pressure,stack temperature stack current etc. The fuel is 99.99 percent pure hydrogen and oxygen is absorbed from the ambient air. The system is designed for pure gaseous hydrogen application, therefore, it does not need fuel humidification. The air is humidified through a built in humidity exchanger to maintain membrane saturation and enhance the life of the membrane. Other parameters of the fuel cell are given in Table I. The fuel cell system is operated at steady state condition with the stack current I varying from 0.3 V to 57.3 V and the resulting output voltage V_{exp} varies from 33.66 V to 21.0 V.

Table I.   INITIAL PARAMETERS OF THE PEMFC

Parameters	Available/Initial values
R_t	$0.0001\,\Omega$
P_{H_2}	300×10^{-3} bar
P_{O_2}	100×10^{-3} bar
T	316.8 K
T_{ref}	298.15 K
κ_E	8.5×10^{-4} V/K
F	96487 C/mol
R	8.314 J/K mol
ξ_1	-.5
ξ_2	0.5
ξ_3	0.05
ξ_4	-0.05
σ	15

The proposed algorithm is coded and implemented in MATLAB environment with Windows-7, 32 bit operating system. The validation of the proposed method is done by comparing the simulated and the experimental results shown in Table III with the help of Figure 4 and Figure 5. The error curves after minimizing the objective functions using trust region algorithm successively are shown in Figure 6. We compare our results with those obtained by the existing PSO method for the same mechanistic model [26], [27].

Table II.   PARAMETER VALUES OBTAINED BY PSO AND THE PROPOSED ALGORITHM

Parameters	ξ_1	ξ_2	ξ_3	ξ_4	σ
PSO	-0.3676	6.9515e-5	4.5364e-8	-0.0378	15.999
Proposed Algorithm	-0.3759	0.6241	0.0386	-0.00016	22.999

Table III.   COMPARISON OF EXPERIMENTAL AND MODELLED VOLTAGES AT DIFFERENT CURRENTS

	Experimental Data		Estimated Voltage	
Sample	Current	Voltage	PSO Model	Proposed Model
1	0.3	33.66	32.48	33.62
2	0.59	33.34	31.54	32.41
3	0.97	32.74	30.83	31.52
4	1.28	32.15	30.43	31.01
5	1.6	31.49	30.09	30.59
6	2	30.60	29.76	30.19
7	2.13	29.75	29.66	30.07
8	2.47	29.33	29.44	29.79
9	3.72	28.70	28.77	29.00
10	4.43	27.87	28.48	28.66
11	5.2	27.71	28.19	28.34
12	5.74	27.45	28.01	28.14
13	6.22	27.23	27.86	27.97
14	6.53	27.16	27.76	27.87
15	7.35	26.89	27.53	27.62
16	7.99	26.68	27.35	27.44
17	8.58	26.53	27.20	27.29
18	9.31	26.34	27.02	27.11
19	12.28	25.93	26.38	26.46
20	13.16	25.81	26.12	26.29
21	15.94	25.17	25.56	25.81
22	18.31	25.16	25.18	25.44
23	21.92	25.14	24.73	24.93
24	24.52	24.95	24.33	24.59
25	28	24.69	23.82	24.15
26	33.08	23.93	23.10	23.55
27	36.29	23.50	22.66	23.19
28	38.11	23.29	22.40	22.98
29	41.58	23.16	21.91	22.59
30	44.87	22.74	21.44	22.21
31	45.9	22.63	21.29	22.09
32	46.46	22.08	21.25	22.01
33	49.75	21.46	20.74	21.61
34	53.59	21.22	20.08	21.07
35	56.4	21.06	19.50	20.56
36	57.3	21.00	19.26	20.35
Error	-	-	0.98	0.608

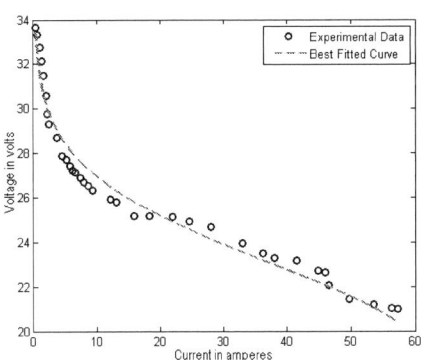

Figure 4.   Voltage curve obtained after optimization with experimental data

The obtained parameter values using successive trust region algorithm and PSO are shown in Table II and Table III gives the current-voltage relationship of the optimized model. From Table III we observe that the proposed method gives a mean square error of 0.608 which is quite less than 0.98 which is obtained using PSO [26].

IV. CONCLUSION

In this paper, a parametric optimization technique is developed for a electro-chemical based PEMFC using trust region algorithm which has been applied successively with different objective functions. The results obtained from trust region algorithm are presented and compared with an existing PSO

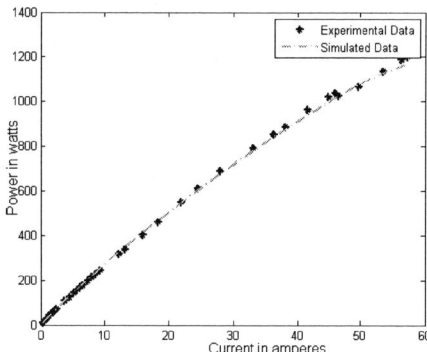

Figure 5. Power curve obtained after optimization with experimental data

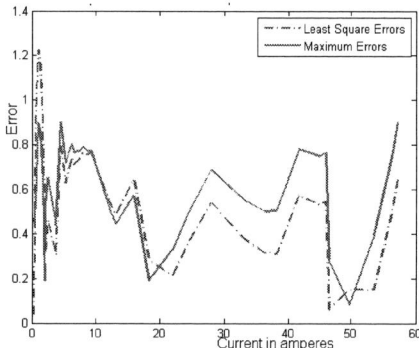

Figure 6. Error curves obtained after successive optimization

optimized similar voltage model where it has been found that the proposed method gives a better optimized model. The obtained results are demonstrated graphically, from where it can be observed that at some values the optimized model and the experimental data do not match. The reason behind this is the reduction in cell voltage due to increase in temperature. Even though certain assumptions are made for simplified analysis, the variations in temperature and pressures can be easily incorporated using interval analysis techniques.

REFERENCES

[1] M. Sarvi and M. Safari, "Estimation the performance of a pem fuel cell system at different operating conditions using neuro fuzzy (anfis)," *World Applied Programming*, vol. 3, 2013.

[2] R. I. Salim, H. Noura, M. Nabag, and A. Fardoun, "Modeling and temperature analysis of the nexa 1.2 kw fuel cell system," *Journal of Fuel Cell Science and Technology*, vol. 12, December 2015.

[3] O. Hegazy, J. V. Mierlo, R. Barrero, N. Omar, and P. Lataire, "Pso algorithm-based optimal power flow control of fuel cell/supercapacitor and fuel cell/battery hybrid electric vehicles," *COMPEL: The International Journal for Computation and Mathematics in Electrical and Electronic Engineering*, vol. 32, no. 1, pp. 86–107, 2013.

[4] N. Benchouia, A. Hadjadj, A. Derghal, L. Khochemane, and B. Mahmah, "Modeling and validation of fuel cell pemfc," *Revue des Energies Renouvelables*, vol. 16, no. 2, p. 365 377, June 2013.

[5] Grondin-Perez, R. B., L. S., B. C., D. M., and C. Kadjo, "Mechanistic model versus artificial neural network model of a single cell pemfc," *Scientific Research Publishing*, vol. 6, 2014.

[6] I. S. Martn, A. Ursa, and P. Sanchis, "Modelling of pem fuel cell performance: Steady-state and dynamic experimental validation," *Energies*, vol. 7, 2014.

[7] P. N. Papadopoulos, M. Kandyla, P. Kourtza, T. A. Papadopoulos, and G. K. Papagiannis, "Parametric analysis of the steady state and dynamic performance of proton exchange membrane fuel cell models," *Renewable Energy*, vol. 71, 2014.

[8] R. I. Salim, H. Noura, and A. Fardoun, "A parameter identification approach of a pem fuel cell stack using particle swarm optimization," in *ASME 2013 11th International Conference on Fuel Cell Science, Engineering and Technology*, 2013.

[9] M. T. Outeiro, R. Chibante, A. S. Carvalho, and A. T. de Almeida, "A parameter optimized model of a proton exchange membrane fuel cell including temperature effects," *Journal of Power Sources*, vol. 185, no. 2, pp. 952–960, December 2008.

[10] C. Bao, M. Ouyang, and B. Yi, "Modeling and optimization of the air system in polymer exchange membrane fuel cell systems," *Journal of Power Sources*, vol. 156, pp. 232–243, June 2006.

[11] P. Hu, G.-Y. Cao, X.-J. Zhu, and J. Li, "Modeling of a proton exchange membrane fuel cell based on the hybrid particle swarm optimization with levenberg-marquardt neural network," *Simulation Modelling Practice and Theory*, vol. 18, pp. 574–588, May 2010.

[12] Q. Li, W. Chen, Y. Wang, S. Liu, and J. Jia, "Parameter identification for pem fuel-cell mechanism model based on effective informed adaptive particle swarm optimization," *IEEE TRANSACTIONS ON INDUSTRIAL ELECTRONICS*, vol. 58, no. 6, pp. 2410–2419, June 2011.

[13] M. Hinaje, S. Ral, P. Noiying, D. A. Nguyen, and B. Davat, "An equivalent electrical circuit model of proton exchange membrane fuel cells based on mathematical modelling," *Energies*, vol. 5, 2012.

[14] D. Yu and S. Yuvarajan, "Electronic circuit model for proton exchange membrane fuel cells," *Journal of Power Sources*, vol. 142, 2005.

[15] P. Thounthong, B. Davat, S. Rael, and P. Sethakul, "An overview of power converters for a clean energy conversion technology," *IEEE Industrial Electronics Magazine*, vol. 3, no. 1, pp. 32–46, 2009.

[16] K. W. Suh, "Modeling, analysis and control of fuel cell hybrid power systems," Master's thesis, The University of Michigan, Ann Arbor, Michigan, Department of Mechanical Engineering, 2006.

[17] A. Emadi, K. Rajashekara, S. S. Williamson, and S. M.Lukic, "Topological overview of hybrid electric and fuel cell vehicular power system architectures and configurations," *IEEE Transcations on Vehicular Technology*, vol. 54, no. 3, pp. 763–770, May 2005.

[18] R. Chamousis, "Hydrogen: Fuel of the future," California State University, Stanislaus, CSU Stanislaus, CA., Tech. Rep., October 2008.

[19] J. Jia, Q. Li, Y. Wang, Y. T. Cham, and M. Han, "Modeling and dynamic characteristic simulation of a proton exchange membrane fuel cell," *IEEE Transactions on Energy Conversion*, vol. 24, pp. 283 – 291, March 2009.

[20] J. M. Correa, F. A. Farret, L. N. Canha, and M. G. Simoes, "An electrochemical-based fuel-cell model suitable for electrical engineering automation approach," *IEEE Transactions on Industrial Electronics*, vol. 51, 2004.

[21] M. H. Wang, M. Huang, W. Jiang, and K. Liou, "Maximum power point tracking control method for proton exchange membrane fuel cell," *IET Renewable Power Genertion*, vol. 10, no. 1, pp. 908–915, 2016.

[22] G. Abraham and P. Pillay, "Implementation of fuel cell emulation on dsp and dspace controllers in the design of power electronic converters," *IEEE Transactions on Industry Applications*, vol. 46, no. 1, pp. 285 – 294, March 2010.

[23] H. Le, "Efficient trust region subproblem algorithms," Master of Mathematics, University of Waterloo, Waterloo, Ontario, Canada, 2011.

[24] N. Börlin, "Nonlinear optimization, trust-region methods," December 2007.

[25] Y. xiang Yuan, "A review of trust region algorithms for optimization," September 1999.

[26] T. V. Dixit, "Real time investigation on performance enhancement of power conditioning units for photovoltaic and fuel cell," Ph.D. dissertation, National Institute of Technology, Raipur, May 2018.

[27] T. V. Dixit, P. T. Bankupalli, A. Yadav, and S. Gupta, "Fuel cell power conditioning unit for standalone application with real time validation," *International Journal of Hydrogen Energy*, vol. 43, pp. 14 629–14 637, 2018.

Dynamic Behavior Control of Induction Motor with STATCOM

Majid Dehghani
Dept. of Electrical Engineering
Amirkabir University of
Technology
Tehran, Iran
majid1369@aut.ac.ir

Peyman Karimyan
Dept. of Electrical Engineering
Amirkabir University of
Technology
Tehran, Iran
peyman.sena @gmail.com

Mehradad Abedi
Dept. of Electrical Engineering
Amirkabir University of
Technology
Tehran, Iran
abedi@aut.ac.ir

Hadis Karimipour
School of Enginering
University of Guelph
Guelph, Canada
hkarimi@uoguelph.ca

Abstract— **STATCOMs is used widely in power systems these days. Traditionally, this converter was controlled using a double-loop control or Direct Output Voltage (DOV) controller. But DOV controller do not function properly during a three-phase fault and has a lot of overshoot. Also, the number of PI controllers used in double-loop control is high, which led to complexities when adjusting the coefficients. Therefore, in this paper, an improved DOV method is proposed which, in addition to a reduced number of PI controllers, has a higher speed, lower overshoots and a higher stability in a wider range. By validating the proposed DOV method for controlling the STATCOMs, it has been attempted to improve the dynamical behaviors of induction motor using Matlab/Simulink, and the results indicate a better performance of the proposed method as compared to the other methods.**

Keywords—STATCOM, DOV Controller, Induction Motors

I. Introduction

Nonlinear loads used in the industry exert a negative impact on the quality of the power delivered to the consumer. Voltage fluctuations are one of those phenomena that have an adverse effect on the system's sensitive equipment and cause problems in the system[2]. In the past, reactive power compensators, such as TCRs or capacitor banks, were used to solve this problem [3]. With the development of electronic devices, a new generation of static compensators (STATCOMs) were introduced that utilize a PWM voltage source converter and connect in parallel to the transmission or distribution system [4]. A STATCOM, in contrast to an SVC, has a better performance in reactive power compensation [5] and is a better voltage stabilizer at PCC point[6]. Accordingly, throughout recent years, the use of STATCOMs has been of interest to researchers to improve the power quality of electrical power systems. A STATCOM, similar to an SVC, can operate at higher speeds and with better dynamic characteristics, and, unlike the SVC, it does not depend on the network voltage. This is important when a fast dynamic response is required or the network voltage is low; in this case, the STATCOM has a better performance than the SVC. The functioning of the STATCOM is not dependent only on topology, but also on the way it is controlled [3]. Researchers focus on topology and control strategy in order to improve the performance of the STATCOM. These strategies should be used for bus voltage regulation, reactive power compensation and power factor correction.

Mehtn and Schauder [7]presented a Conventional Double-Loop Control Strategy. In this method, there are two loops; the external loop of the active and reactive current generates the reference current for maintaining the voltage at the PCC point, and the internal loop is used to control the inverter current[8].

Due to the use of both voltage and current sensors, four PI controllers are used in this method. Adjusting PI controller coefficients is a key factor in the performance of a STATCOM, on which a large number of studies have been conducted [9-14]. In [15], using the genetic algorithm, the neural network and the fuzzy-neural system, the PI controller coefficients for the STATCOM were investigated, and the results demonstrated that the performance of the fuzzy-neural controller was better than in other methods. In [16], a DSTATCOM was first modeled, and then a fuzzy system was employed to determine PI controller coefficients that improved the damping of the power system with a new controller.

However, this control strategy has four PI controllers, which not only complicate the design, but also bring about difficulties when adjusting its control parameters. These parameters are adjusted empirically or through trial and error, which either is time-consuming or cannot improve stability over a wide range of different system conditions [1].

To solve this problem, CHEN and Hsu [17], based on the theory of instantaneous power balance, proposed a Direct Output Voltage Control (DOV) for the STATCOM. This new controller eliminates the active and reactive feedback loop, reduces the number of PI controllers and calculates the STATCOM output voltage using an algebraic algorithm based on the principle of maintaining the power balance [18]. In [19], the DOV controller is used to correct the power factor and to reduce line harmonics. In [20], a direct output voltage control strategy is proposed based on a multi-modulator controller and a neural network. In this method, a neural network is used to adjust the values of the PI controller parameters according to the optimal control rule. Simulation results showed that, in comparison with the traditional PI controller, the PI controller is able to withstand a change in voltage with a higher compensation accuracy.

978-1-5386-6624-1/18 $31.00 © 2018 IEEE

One of the disadvantages of this method is its inappropriate performance during a three-phase fault, which results in a high overshoot and low response speed. In this paper, an improved DOV method is presented to overcome the above problems. This control strategy can not only increase the response speed of the controller, but also reduce unwanted overshoots in the system during a three-phase fault.

In the second part of this paper, STATCOM-based reactive power compensation has been briefly described, and its performance-related characteristics and dynamical equations have been investigated. In the third section, the double-loop controller and the DOV controller have first been explained, and then the controller has been described. In the fourth section, the impact of these controllers on system performance is investigated. Finally, a conclusion has been made in the fifth part.

II. THE STATIC SYNCHRONOUS COMPENSATOR

Static synchronous compensator is one of the instruments of FACTS. As can be seen from Fig. 1, STATCOM is connected to the transmission line in a parallel pattern [14, 21]. Voltage source converter exchanges the reactive power with the line by changing DC voltage to a variable AC voltage in the output. The level and direction of reactive power which is exchanged between STATCOM and the transfer line are identified based on the relative disparity of the level of STATCOM output voltages [22].

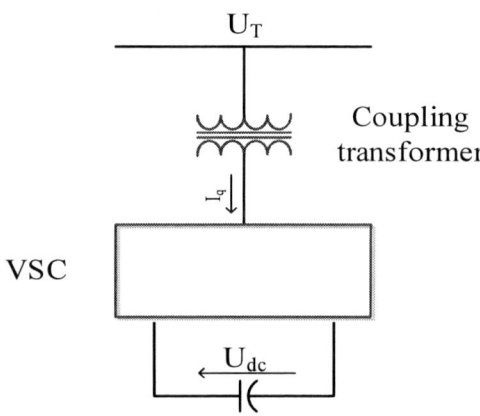

Fig. 1. STATCOM parallel connection to the network

A. Dynamic characteristics of the STATCOM

A STATCOM model is shown in Fig 2. In this circuit, R represents the sum of the ohm losses of the transformer winding and the conductor losses in the converter, and L is the transformer leakage inductance [23]. Sum of the losses of the converter switching and the power loss in the capacitor is modeled with r as resistance, which is parallel to the DC link capacitor.

The voltages e_a, e_b and e_c are the phase voltages at the converter output. Inductors current of STATCOM are

considered as state variable. By writing *KVL* equations at STATCOM output, following equation is obtained [24, 25]:

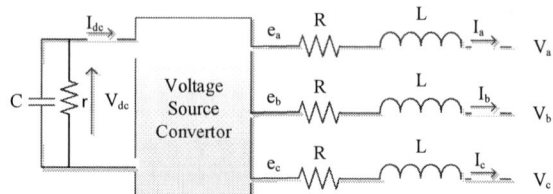

Fig. 2. The STATCOM equivalent model

$$L_S \frac{dI_{as}}{dt} = -R_s I_{as} + V_{as} - V_{al}$$

$$L_S \frac{dI_{bs}}{dt} = -R_s I_{bs} + V_{bs} - V_{bl} \quad (1)$$

$$L_S \frac{dI_{cs}}{dt} = -R_s I_{cs} + V_{cs} - V_{cl}$$

where I_{as}, I_{bs} and I_{cs} are three-phase currents, V_{al}, V_{bl} and V_{cl} are voltage vector of the PCC.

By utilizing Park's transformation for both sides of Eq. (1) and simplifying the relations, state equations for the phase current in the d-q coordinates are obtained as follows [26, 27]:

$$L_s \frac{dI_{ds}}{dt} = -R_s I_{ds} + wL_s I_{qs} + V_{ds} - V_{dl}$$

$$L_s \frac{dI_{qs}}{dt} = -R_s I_{qs} - wL_s I_{ds} + V_{qs} - V_{ql} \quad (2)$$

Where, I_{ds}, I_{qs}, V_{ds}, V_{qs} are d- and q-axis STATCOM output currents and voltages respectively, ω is the synchronously rotating angle speed of the voltage vector of the PCC and V_{dl}, V_{ql} are the load voltages.

III. THE CONVENTIONAL DOUBLE-LOOP CONTROL STRATEGY

In the conventional control strategy, the typical PI controller is used to exchange active and reactive powers between the STATCOM and the network. The block of the conventional controller diagram is shown in Fig. 3.

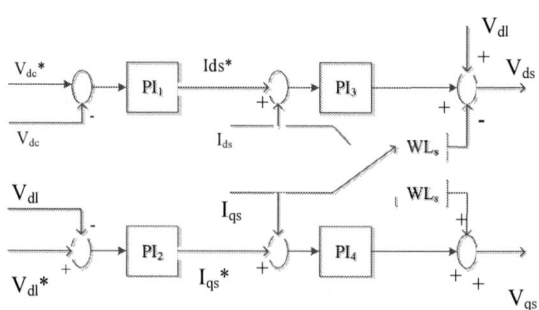

Fig. 3. Schematic configuration of the double-loop control strategy

978-1-5386-6624-1/18 $31.00 © 2018 IEEE

This method has two loops; the external loop of active and reactive currents provides a reference current for maintaining the voltage at the PCC point, and the internal loop is used to control the inverter current [28]. PI controller coefficients have constant values that may not have an acceptable function in systems with time and nonlinear delay. This control strategy also requires four PI controllers for its control system, so it can be described as a tedious and time-consuming task for engineers to carry out trial and error studies to find suitable parameters when the operating conditions of the system have changed considerably [29].

A. The DOV control strategy

Relying on the principle of power balance, the transient output power of STATCOM is equal to the total power consumption of R_s, L_s and load power P_L and Q_L [1].

$$P_S = P_{RL} + P_L$$
$$Q_S = Q_{RL} + Q_L \tag{3}$$

The STATCOM transient output power can also be expressed as follows:

$$P_s = \frac{3}{2}(V_{ds}I_{ds} + V_{qs}I_{qs})$$
$$Q_s = \frac{3}{2}(V_{ds}I_{qs} - V_{qs}I_{ds}) \tag{4}$$

The transient power consumption of R_s and L_s are expressed as follows:

$$P_{rl} = \frac{3}{2}R_s I^2 = \frac{3}{2}R_s(I_{ds}^2 + I_{qs}^2)$$
$$Q_{rl} = \frac{3}{2}wL_s I^2 = \frac{3}{2}wL_s(I_{ds}^2 + I_{qs}^2) \tag{5}$$

And the active and reactive power equations are defined as follows:

$$P_l = \frac{3}{2}(V_{dl}I_{ds} + V_{ql}I_{qs})$$
$$Q_l = \frac{3}{2}(V_{dl}I_{qs} - V_{ql}I_{ds}) \tag{6}$$

These equations are transmitted using the park transformation to the d-q axis, on which the voltage of the PCC point corresponds to that of the d-axis, and the q-axis is perpendicular to the d-axis:

$$V_{dl} = |v|$$
$$V_{ql} = 0 \tag{7}$$

By inserting Eq. (7) in Eq. (6), the following equation is obtained:

$$P_l = \frac{3}{2}(V_{dl}I_{ds})$$
$$Q_l = \frac{3}{2}(V_{dl}I_{qs}) \tag{8}$$

By inserting Eqs. (4), (5) and (8) in Eq. (3), the following equation is obtained:

$$V_{ds} = R_s I_{ds} - wL_s I_{qs} + V_{dl}$$
$$V_{qs} = R_s I_{qs} + wL_s I_{ds} \tag{9}$$

Therefore, the STATCOM reference output voltage (V_{ds}, V_{qs}) can be obtained using the STATCOM reference current control (I_{ds}, I_{qs}), R_s, L_s and V_{dL} as shown in Figure 4.

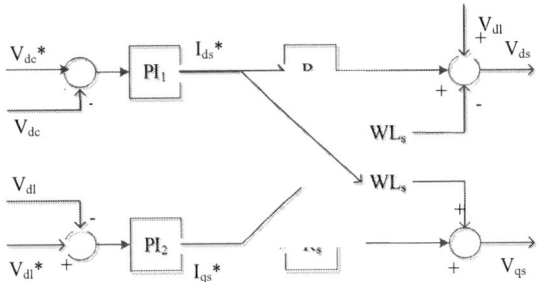

Fig. 4. Schematic configuration of the DOV control strategy[1]

B. The proposed control strategy

In this strategy, the power equations and the STATCOM are first defined and described, and then, using the obtained equations, a new control system is designed.

The active and reactive power equations in a permanent state are defined as follows:

$$P_l = \frac{3}{2}(V_{dl}I_{ds} + V_{ql}I_{qs})$$
$$Q_l = \frac{3}{2}(V_{dl}I_{qs} - V_{ql}I_{ds}) \tag{10}$$

These equations are transmitted using Park's transformation to the d-q axis, on which the voltage of the PCC point corresponds to that of the d-axis, and the q-axis is perpendicular to the d-axis:

$$V_{dl} = |v|$$
$$V_{ql} = 0 \tag{11}$$

And by inserting Eq. (11) in Eq. (10), the following equation is obtained:

$$P_l = \frac{3}{2}(V_{dl}I_{ds})$$
$$Q_l = \frac{3}{2}(V_{dl}I_{qs}) \tag{12}$$

As shown in Eq. (12), in this control method, there is an active power control as well as a reactive power control. The active power can be controlled using the d-axis current component, and the reactive power using the q-axis current component.

By inserting Eq. (11) in Eq. (2), the following equation is obtained:

$$L_s \frac{dI_{bs}}{dt} = -R_s I_{ds} + wL_s I_{qs} + V_{ds} - V_{dl}$$

$$L_s \frac{dI_{qs}}{dt} = -R_s I_{qs} - wL_s I_{ds} + V_{qs} \tag{13}$$

Eq. (13) can be expressed as follows:

$$V_{ds} = R_s I_{ds} - wL_s I_{qs} + L_s \frac{dI_{bs}}{dt} + V_{dl}$$

$$V_{qs} = R_s I_{qs} + wL_s I_{ds} + L_s \frac{dI_{qs}}{dt} \tag{14}$$

Therefore, the STATCOM output voltage (V_{ds}, V_{qs}) can be obtained by controlling the STATCOM current control (I_{ds}, I_{qs}), R_s, L_s and V_{dL} as shown in Figure 5.

This system includes only two PIs that are used to regulate the AC voltage and DC voltage. In the AC voltage regulator, The input is the difference between the terminal voltage and reference voltage and the output is a q-axis current. The DC voltage regulator receives the DC voltage error as an input and generates d-axis current. since the d-axis is always coincident with the voltage vector of the PCC, so it is an effective factor in passing the active power; also the q-axis is in quadrature with the voltage, so it is an effective factor in passing the reactive power through the converter .

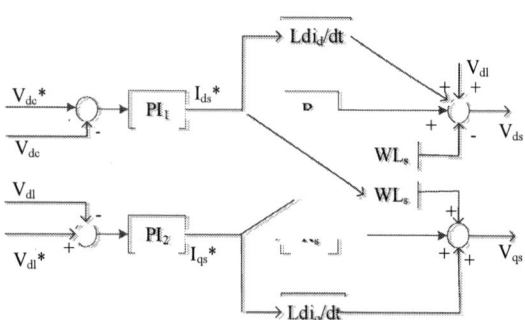

Fig. 5.   The proposed controller of

The active power will change the voltage of the DC bus and the reactive power will have the same effect on the terminal voltage. Then, using I_q and I_d, the reference value of the STATCOM output voltage is obtained through Eq. (14). As can be seen in Fig. 5, this strategy has fewer PIs than the double-loop controller.

IV. SIMULATION RESULTS

In this section, the transient performance of the induction motors is simulated in the presence of a STATCOM in Matlab/Simulink. The studied system consists of 9 induction motors connected through a transformer, and a transmission line to a three-phase network. As shown in Fig. 6, in parallel with the electric load, a three-phase reactive power compensator is used to improve the dynamic behavior of the system.

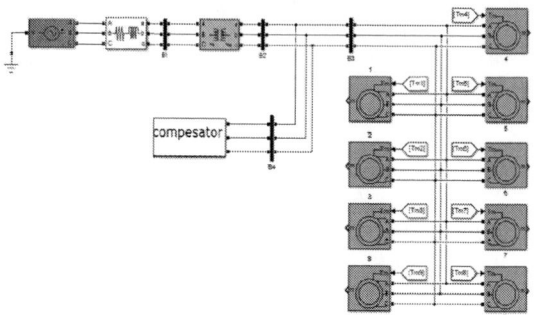

Fig. 6.   The studied circuit system with compensation in Simulink

For accuracy and speed evaluation, the simulation results are compared with DOV and double-loop controllers. A 20-percent voltage swell is applied on load bus for duration of 2 seconds; then, at t = 10s, a 20-percent voltage sag is applied for duration of 2 seconds.

Fig. 7 shows the system terminal voltage variation per unit for different controllers. when using a conventional controller. As can be seen from Fig. 7, proposed controller yields better performance compared to the other controllers. The propose controller has better compensation in comparison with double-loop controller and less overshoot in comparison with DOV method.

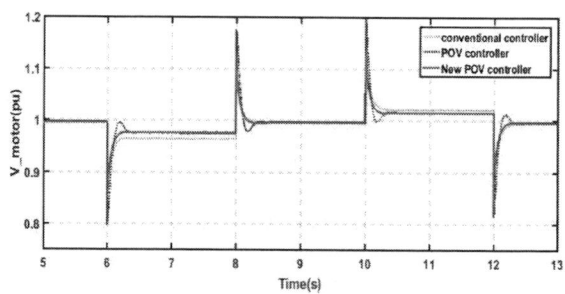

Fig. 7.   Terminal voltage of motor

Fig. 8 shows the changes in the speed of the induction motors. By compensating the voltage drop of the motors, the proposed controller reduces the speed variations in comparison with double-loop controller and less overshoots than with the DOV controller.

Fig. 9 shows the torque of the induction motors. As can be seen from the figure, the proposed controller yields better performance compared to the other controllers.

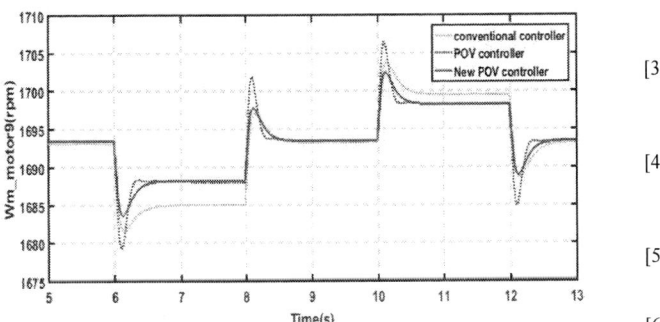

Fig. 8. Rotor speed (motor 3)

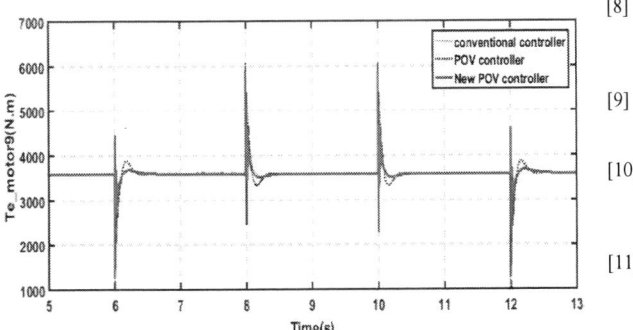

Fig. 9. the torque of the induction motors

V. CONCLUSION

This study shows that there are a number of problems with conventional STATCOM controllers, such as the large number of PI controllers and the difficulties posed when regulating their parameters. DOV controllers, too, present their own set of problems, including frequent overshoots and a slower dynamic response. In this paper, an improved DOV control strategy is presented for use with a STATCOM to overcome these problems. To investigate the proposed control function, the system was tested under a voltage sag and a voltage swell, where the controller was used to reduce overshoots and increase response speed. A comparison of the simulation results of the proposed controller to those of a DOV controller and a conventional controller indicated that this controller performs better under voltage swells and sags and reaches its final value with a higher speed after correcting the fault. This controller has fewer PIs and overshoots than with the conventional controller.

REFERENCES

[1] A. Luo, C. Tang, Z. Shuai, J. Tang, X. Y. Xu, and D. Chen, "Fuzzy-PI-based direct-output-voltage control strategy for the STATCOM used in utility distribution systems," *IEEE Transactions on Industrial Electronics,* vol. 56, pp. 2401-2411, 2009.

[2] P. Puleston, S. Gonzalez, and F. Valenciaga, "A STATCOM based variable structure control for power system oscillations damping," *International Journal of Electrical Power & Energy Systems,* vol. 29, pp. 241-250, 2007.

[3] S. Li, L. Xu, and T. A. Haskew, "Control of VSC-based STATCOM using conventional and direct-current vector control strategies," *International Journal of Electrical Power & Energy Systems,* vol. 45, pp. 175-186, 2013.

[4] A. Sode-Yome and N. Mithulananthan, "Comparison of shunt capacitor, SVC and STATCOM in static voltage stability margin enhancement," *International Journal of Electrical Engineering Education,* vol. 41, pp. 158-171, 2004.

[5] J.-f. ZHOU, Y.-q. GU, and S.-q. WEI, "Comprehensive comparative analysis of SVC and STATCOM," *Electric Power Automation Equipment,* vol. 12, p. 015, 2007.

[6] A. Majed, Z. Salam, and A. M. Amjad, "Harmonics elimination PWM based direct control for 23-level multilevel distribution STATCOM using differential evolution algorithm," *Electric Power Systems Research,* vol. 152, pp. 48-60, 2017.

[7] C. Schauder and H. n. Mehta, "Vector analysis and control of advanced static VAR compensators," in *IEE Proceedings C (Generation, Transmission and Distribution)*, 1993, pp. 299-306.

[8] S. A. Chatterjee and K. Joshi, "A comparison of conventional, direct-output-voltage and Fuzzy-PI control strategies for D-STATCOM," in *Modern Electric Power Systems (MEPS), 2010 Proceedings of the International Symposium*, 2010, pp. 1-6.

[9] Y. Xu and F. Li, "Adaptive PI control of STATCOM for voltage regulation," *IEEE transactions on power delivery,* vol. 29, pp. 1002-1011, 2014.

[10] D. K. Raju, B. S. Umre, A. Junghare, and B. C. Babu, "Improved Control Strategy for Subsynchronous Resonance Mitigation with Fractional-order PI Controller," *International Journal of Emerging Electric Power Systems,* vol. 17, pp. 683-692, 2016.

[11] E. Najafi and A. Yatim, "A novel current mode controller for a static compensator utilizing Goertzel algorithm to mitigate voltage sags," *Energy Conversion and Management,* vol. 52, pp. 1999-2008, 2011.

[12] B. Singh and S. R. Arya, "Adaptive control of four-leg VSC based DSTATCOM in distribution system," *International Journal of Emerging Electric Power Systems,* vol. 15, pp. 93-99, 2014.

[13] M. I. Mosaad, "Model reference adaptive control of STATCOM for grid integration of wind energy systems," *IET Electric Power Applications,* 2018.

[14] H. Rao, S. Xu, Y. Zhao, C. Hong, W. Wei, X. Li, *et al.,* "Research and application of multiple STATCOMs to improve the stability of AC/DC power systems in China Southern Grid," *IET Generation, Transmission & Distribution,* vol. 10, pp. 3111-3118, 2016.

[15] N. K. Saxena and A. Kumar, "Reactive power control in decentralized hybrid power system with STATCOM using GA, ANN and ANFIS methods," *International Journal of Electrical Power & Energy Systems,* vol. 83, pp. 175-187, 2016.

[16] D. Amoozegar, "DSTATCOM modelling for voltage stability with fuzzy logic PI current controller," *International Journal of Electrical Power & Energy Systems,* vol. 76, pp. 129-135, 2016.

[17] W.-L. Chen and Y.-Y. Hsu, "Direct output voltage control of a static synchronous compensator using current sensorless dq vector-based power balancing scheme," in *Transmission and Distribution Conference and Exposition, 2003 IEEE PES*, 2003, pp. 545-549.

[18] S. Wang, L. Li, X. Wang, Y. Zheng, and G. Yao, "Direct output voltage control of a STATCOM using PI controller based on multiple models," in *Industrial Electronics and Applications (ICIEA), 2011 6th IEEE Conference on*, 2011, pp. 2203-2208.

[19] R. Baharom, M. K. M. Salleh, and N. F. A. Rahman, "Active power filter with direct output voltage control of single-phase AC to DC converter," in *Power and Energy (PECon), 2016 IEEE International Conference on*, 2016, pp. 412-416.

[20] C. Zhou, Y. Zheng, X. Wang, L. Li, G. Yao, and N. Xie, "Direct Output Voltage Control Strategy for STATCOM Based on Multi-model and Neural Network PI Controller," in *Unifying Electrical Engineering and Electronics Engineering*, ed: Springer, 2014, pp. 387-394.

[21] Y. Xu, L. Tolbert, J. Chiasson, J. Campbell, and F. Peng, "A generalised instantaneous non-active power theory for STATCOM," *IET Electric Power Applications,* vol. 1, pp. 853-861, 2007.

[22] S. O. Farees, M. Gayatri, and K. Sumanth, "Performance Comparison between SVC and STATCOM for Reactive Power Compensation by Using Fuzzy Logic Controller," *IJITR,* vol. 2, pp. 991-994, 2014.

[23] P. Rao, M. Crow, and Z. Yang, "STATCOM control for power system voltage control applications," *IEEE Transactions on power delivery,* vol. 15, pp. 1311-1317, 2000.

[24] M. R. Tavana, M.-H. Khooban, and T. Niknam, "Adaptive PI controller to voltage regulation in power systems: STATCOM as a case study," *ISA transactions,* vol. 66, pp. 325-334, 2017.

[25] B. Singh and V. S. Kadagala, "A new configuration of two-level 48-pulse VSCs based STATCOM for voltage regulation," *Electric Power Systems Research,* vol. 82, pp. 11-17, 2012.

[26] M. Saeedifard, H. Nikkhajoei, and R. Iravani, "A space vector modulated STATCOM based on a three-level neutral point clamped converter," *IEEE Transactions on Power Delivery,* vol. 22, pp. 1029-1039, 2007.

[27] N. Sahoo, B. Panigrahi, P. Dash, and G. Panda, "Application of a multivariable feedback linearization scheme for STATCOM control," *Electric Power Systems Research,* vol. 62, pp. 81-91, 2002.

[28] M. P. Kazmierkowski and L. Malesani, "Current control techniques for three-phase voltage-source PWM converters: A survey," *IEEE Transactions on industrial electronics,* vol. 45, pp. 691-703, 1998.

[29] K. Padiyar and N. Prabhu, "Design and performance evaluation of subsynchronous damping controller with STATCOM," *IEEE Transactions on Power Delivery,* vol. 21, pp. 1398-1405, 2006.

Impact of Particle Swarm Optimization Parameters on its Convergence

N.K.Jain
Electrical Engineering Dept.
Delhi Technological University
Delhi,INDIA
uma.nangia62@gmail.com

Uma Nangia
Electrical Engineering Dept.
Delhi Technological University
Delhi,INDIA
uma_nangia@rediffmail.com

Jyoti Jain
Electrical and Electronics Engineering Dept.
Maharaja Surajmal Institute of Technology
Delhi, INDIA
jyotijain_in@yahoo.com

Abstract - Particle swarm optimization algorithm is an intelligent optimization technique applied to solve optimization problem in the field of Engineering, optimization design of electrical networks, aircraft, control systems, optimum design of chemical processing equipment, design of civil engineering structure, economic load dispatch, load flow, management system, medical etc. Various parameters of PSO are : population size(P), inertia weight, acceleration constants random number, Maximum number of iteration. In this paper an attempt is made to understand the PSO algorithm by solving Rosenbrock benchmark function manually. The impact of variations of parameters on PSO convergence has been studied. Each parameters is varied systematically keeping all other parameters fixed to some value. Rosenbrock benchmark function has been used to study the impact of parameters variations. It was found that PSO is sensitive to the values of these parameters, combination of various parameters for best accuracy and faster convergence has been determined.

Keywords— convergence, optimization, iteration, Inertia weight

I. INTRODUCTION

Intelligent search methods such as Genetic Algorithm [1-2], Particle Swarm Optimization [3-4] and Ant Colony Optimization [5] etc. have been employed to solve various optimization problems.

Wei-bing Liu and Xian-Jia Wang [6] presented a new particle swarm optimizer based on evolutionary game (EGPSO). Ying-Ping Chen and Wen-Chih Peng [7] suggested to improve the performance of the particle swarm optimizer by incorporate the linkage concept, which is an essential mechanism in Genetic Algorithms and design a new linkage identification technique called dynamic linkage discovery to address the linkage problem in real-parameter optimization problems. Tao Gongand Andrew L. Tuson [8] presented the working mechanism of PSO in a principled manner with forma analysis and investigates the applicability of PSO on the quadratic assignment problem.

Narender Kumar Jain et al. [9-11] developed improved PSO algorithm to reduce the number of function evaluations required to optimize the function. Recently PSO has been applied to solve various power system problems ELD [12], MELD [13] and Load flow [14] problem etc. Review of particle swarm optimization is presented in [15].

In this paper, the working of PSO algorithm has been studied in detail by solving the Rosenbrock function manually for a fixed set of parameters for ten iterations. Also, the effect of parameters of PSO: population size, inertia weight maximum number of iterations on convergence has been studied.

II. PARTICLE SWARM OPTIMIZATION

Particle Swarm Optimization is a self adaptive, population based technique. Initially Population and their velocities are randomly generated. According to mathematical model fitness function is evaluated. Then personal best values and global best value are searched, this process is repeated until all the function values converge at single point. This convergence depends on stopping criteria number of iterations, function value, difference between the function values of consecutive iterations etc. In an n-dimensional search space, position and velocity of particle j is represented by vectors $X_j=(X_{j1},X_{j2},\ldots\ldots X_{jn})$ and $V_j=(V_{j1},V_{ij2},\ldots\ldots V_{jn})$ respectively.

Let X_{pbest} and X_{gbest} be the personal and global best position of particle j. The modified velocity of each particle can be calculated using current velocity and distance from X_{pbest} and X_{gbest} as follows:

$$V_{ij}^{k+1} = W*V_{ij}^{k} + C_p*r_p \ (X_{pbestij}^{k} - X_{ij}^{k}) + C_g*r_g$$
$$(X_{gbestj}^{k} - X_{ij}^{k})$$
$$j=1,2\ldots P \quad i=1,2,3\ldots N \quad \ldots \tag{1}$$

Velocities of particles are updated using (1)

Where
k is iteration count.

N — Number of variables

P — Number of Particles

V_{ij}^{k+1} — is velocity of jth particle of ith variable at k^{th} iteration.

X_{ij}^{k} — is the value of position of jth particle of ith variable at k^{th} iteration.

C_p, C_g — are acceleration coefficients.

$X_{pbestij}^{k}$ — is the value of ith variable of Personal best position of jth particle in k^{th} iteration.

X_{gbesti}^{k} — is the value of ith variable of global best particle until k^{th} iteration.

N — is the total number of variables.

P — is Number of particles in swarm.

r_p, r_g — are two separately generated uniformly distributed random numbers.

W — Inertia weight

ITmax   Maximum number of iteration
Wmax   Maximum value of inertia weight
Wmin   Minimum value of inertia weight
Position of each particle is updated using. (2):

$$X^{k+1} = X^k + V_{ij}{}^{k+1}$$

$$i = 1,2....N; \quad j=1,2.....P \tag{2}$$

In this paper two types of Inertia weight Linearly decreasing using (3) and Simulated annealing inertia weight using (4) have been used to observe the effect of Inertia weight on convergence using Matlab programme.

$$W=Wmax-[(Wmax-Wmin)*k/ITmax \tag{3}$$
$$W=W+(W_{max}-W_{min})*\lambda^{(k-1)} \qquad \lambda=0.95 \tag{4}$$

Flow Chart of Particle Swarm Optimization Technique is shown in Fig.1.

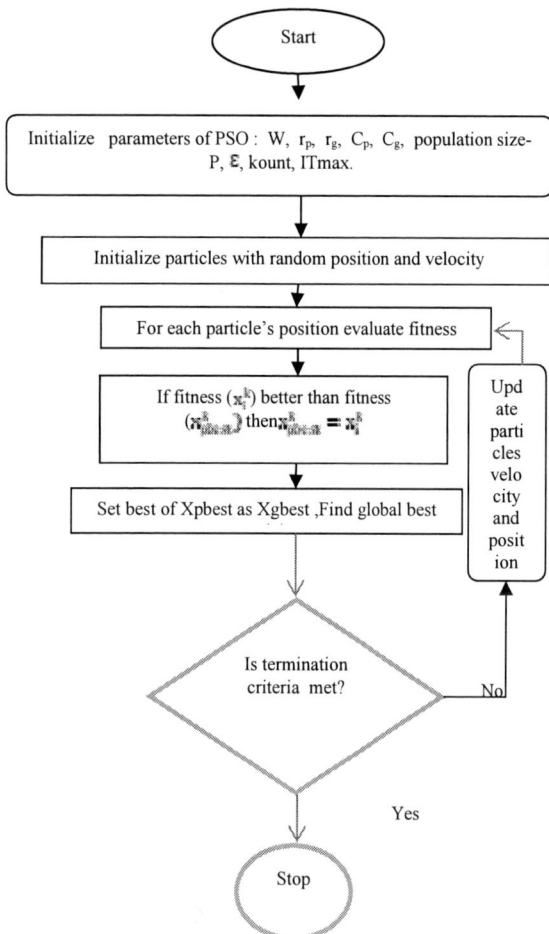

Fig. 1   Flow Chart of Basic PSO

III.   MATHEMATICAL REPRESENTATION OF ROSENBROCK FUNCTION

In mathematical optimization, the Rosenbrock function is a non-convex function used to check the performance of test maintaining the integrity of the specifications problem for optimization algorithm. Mathematically, it is defined as:

$$f=100*(x_1^2 - x_2)^2+(1-x_1)^2$$
$$f(1,1)=0 \qquad \text{Range: } x_1 \geq 0, x_2 \leq 2 \tag{5}$$

It has global minimum at (1,1) where $f(x_1, x_2)=0$.

IV.   COMPUTATIONAL PROCEDURE

The PSO algorithm is applied to optimize the Rosenbrock function. Initially following parameters are considered for PSO algorithm to optimize the function manually as shown in Table 1.

Table 1: Parameters for PSO algorithm

P	r_p	r_g	C_p	C_g	ITmax	Wmax	Wmin
10	0.4	0.5	2	2	10	0.9	0.4

V.   COMPUTATIONAL RESULTS

Initially population size of 10 particles and their velocities are generated randomly and PSO algorithm is applied to Rosenbrock function. The calculations have been done manually to understand the working of PSO.

Results of function value at position of particles X_1 and X_2 for zero[th] iteration is shown in Table 2.

Table 2: Result of Rosenbrock function at Zero[th] Iteration

S. No.	X_1	X_2	V_1	V_2	function value
1	1.2886	0.4155	0.3111	0.5949	155.08329
2	0.7572	0.6025	0.9234	0.2622	0.143913363
3	1.6232	0.9418	0.4302	0.6028	287.0059104
4	1.0657	0.461	0.1848	0.7112	45.52855068
5	0.7015	1.6886	0.9049	0.2217	143.2497888
6	1.878	0.3895	0.9797	0.1174	985.0887203
7	1.7519	0.4518	0.4389	0.2967	685.6193456
8	1.1003	0.3414	0.1111	0.3188	75.5713705
9	1.245	0.4553	0.2581	0.4242	119.9023076
10	1.1741	0.8714	0.4087	0.5079	25.74644817

Results are graphically represented by Fig.2 for zeroth iteration as shown in Table 2. The position of particles at Zeroth Iteration is shown in Fig.2.

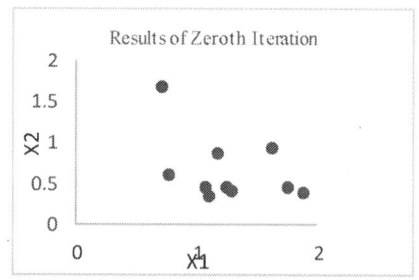

Fig.2   position of particles at Zeroth Iteration

Results of function value for position of particles X_1 and X_2 for First iteration is shown in Table 3.

Table 3: Result of Rosenbrock function at First Iteration

S. No.	X_1	X_2	V_1	V_2	function value
1	1.0216	1.1065	-0.26697	0.692665	0.41555
2	1.5420	0.8253	0.78489	0.22287	241.3727
3	1.1228	1.1148	-0.50033	0.17308	2.14544
4	0.9142	1.2070	-0.15142	0.74602	13.7797
5	1.5263	0.7909	0.824865	-0.89766	237.0814
6	1.5899	0.7022	-0.28806	0.31279	333.6423
7	1.1302	0.8546	-0.62164	0.402895	17.8932
8	0.8516	0.8738	-0.24867	0.53208	2.21821
9	0.9765	0.9630	-0.26842	0.50777	0.00929
10	1.1045	1.0342	-0.0695	0.162815	3.4673

The position of particles X_1 and X_2 for First Iteration is shown in Fig.3.

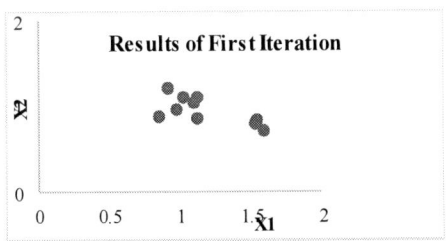

Fig. 3.   position of particles at First Iteration

Results of function value and position of particles X_1 and X_2 for Second Iteration is shown in Table 4.

Table 4: Result of Rosenbrock function at Second Iteration

S. No.	X_1	X_2	V_1	V_2	function value
1	0.7630	1.5172	-0.2586	0.409037	87.4811
2	0.9765	0.9630	-0.5655	0.1377	0.0092
3	0.5763	1.1015	-0.5465	-0.01335	59.3753
4	0.8554	1.5598	-0.05883	0.3528	68.5946
5	0.9765	0.9630	-0.54978	0.172125	0.00929
6	0.7461	1.2133	-0.8438	0.511012	43.17359
7	0.4792	1.2853	-0.65099	0.430691	111.7170
8	0.7776	1.3887	-0.07398	0.515254	61.51340
9	0.7618	1.3692	-0.21473	0.406216	62.28767
10	0.9209	1.0933	-0.18361	0.059107	6.014429

Fig.4 shows results for Second Iteration.

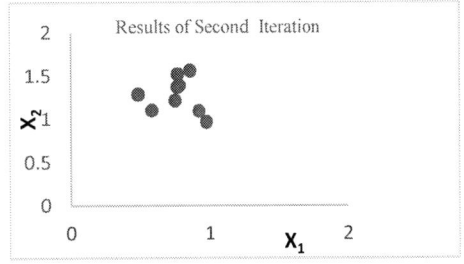

Fig.4. position of particles at Second Iteration

Results for position of particles X_1 and X_2 for Third Iteration is shown in Table 5.

Table 5: Result of Rosenbrock function at Third Iteration

S.No.	X_1	X_2	V_1	V_2	function value
1	0.989516	0.942618	0.226503	-0.57458	0.13351
2	0.552456	1.066345	-0.42413	0.103275	58.1333
3	1.003912	0.963737	0.427591	-0.1378	0.19452
4	0.979527	0.945427	0.124078	-0.61446	0.02015
5	0.56425	1.092164	-0.41234	0.129094	60.0643
6	0.343732	1.346329	-0.40241	0.133027	151.273
7	1.009134	0.941535	0.529857	-0.34385	0.59017
8	0.980284	0.937307	0.202631	-0.45143	0.05632
9	0.987322	0.942759	0.225469	-0.42653	0.013703
10	0.985766	0.960115	0.064785	-0.13321	0.013703

Results of Table 5 are graphically represented in Fig.5. for position of particles for Third Iteration .

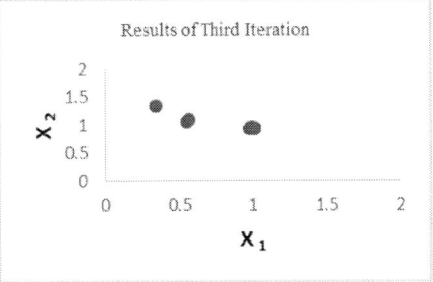

Fig.5.   position of particles at Third Iteration

Results of function value at position of particles X_1 and X_2 for Forth Iteration is shown in Table 6.

Table 6: Result of Rosenbrock function at Fourth Iteration

S. No.	X_1	X_2	V_1	V_2	function value
1	1.01899	0.952743	0.466542	-0.1136	0.73333
2	1.27589	0.866612	0.271987	-0.09712	58.0348
3	1.06343	0.532948	0.083913	-0.41248	35.7589
4	0.68795	1.053436	0.123701	-0.03873	33.7559
5	1.01682	0.949767	0.673094	-0.39656	0.7087
6	1.34748	0.722375	0.338351	-0.21916	119.660
7	1.11842	0.647071	0.138143	-0.29024	36.4723
8	1.13441	0.664501	0.147091	-0.27826	38.755

978-1-5386-6624-1/18 $31.00 © 2018 IEEE

| 9 | 1.02193 | 0.869825 | 0.036169 | -0.09029 | 3.0463 |
| 10 | 1.16441 | 0.654501 | 0.17091 | -0.28826 | 37.7552 |

Results of Table 6 are graphically represented in Fig.6. for position of particles for Forth Iteration .

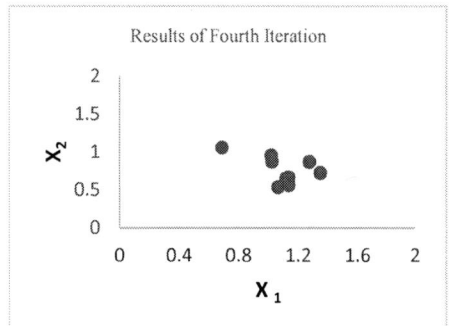

Fig.6.   position of particles at Fourth Iteration

Similarly results for Fifth, Sixth, Seventh, Eighth , Ninth and Tenth Iterations are shown in Fig.7,8,9,10, 11 and 12.

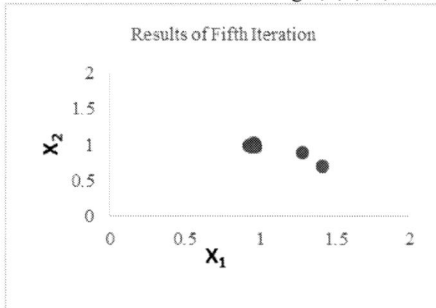

Fig.7.   position of particles at Fifth Iteration

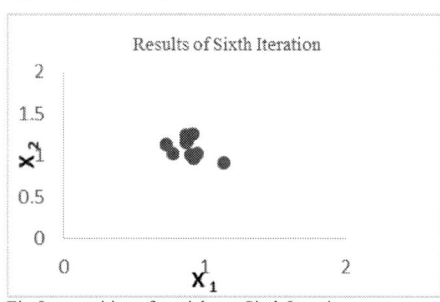

Fig.8.   position of particles at Sixth Iteration

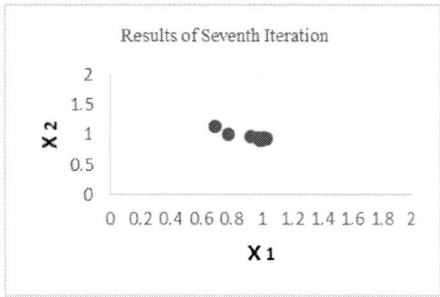

Fig.9.   position of particles at Seventh Iteration

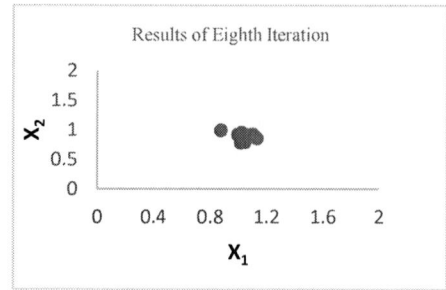

Fig.10.   position of particles at Eighth Iteration

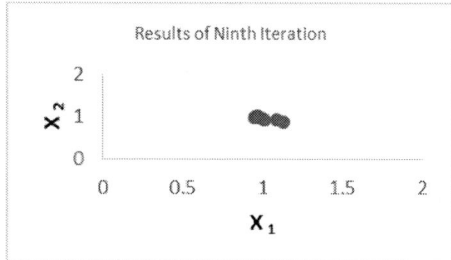

Fig.11.   position of particles at Ninth Iteration

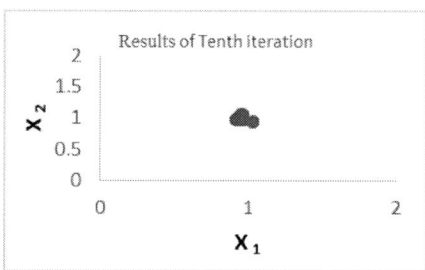

Fig.12.   position of particles at Tenth Iteration

Position of particles in each iteration represented by Figures 2-12 shows that initially particles have large search area, when iteration increases search area reduces and at 10th iteration all particles converged in small area between 0.92923676 and 1.050355942. In this paper results were graphically presented for small populations because calculations are done manually. When population size increases, convergence occurred at exact position (1,1) for Rosenbrock function. Same is true for others functions [10].

VI.   VARIATIONS Of PARAMETERS OF PSO

Population Size (P), maximum iteration (ITmax), Inertia weight (W) on convergence has been studied. One of the parameters is varied while all other are kept fixed. This study has been conducted with the help of Matlab programme in which r_p and r_g are fixed to 0.4 and 0.5 respectively while C_p and C_g are fixed to two(2).

A.   Population Size (P):
In this case, linearly decreasing Inertia weight and ITmax of 1000 has been considered. The population size is varied from 10 to 60.The results are shown in Table 7.

Table 7 : Effect of Population Size

P	k	x1	x2	function value
10	56	1.220362	1.489734	4.85797e-002
20	40	9.53764e-001	9.080800e-001	2.39098e-003
30	62	1	1	3.989808e-020
40	51	1	1	1.828370e-022
50	60	1	1	1.828370e-022
60	58	1	1	8.692590e-020

From Table 7, it is observed that the function does not get optimimized for population size of 10 and 20. Howere,it gets optimized accurately for 30 particle size. It is further observed that for a particle size of 40, maximum accuracy is achieved in minimum number of iteration.

B. Maximum Iteration (ITmax):

In this study, for each fixed value of particle size, Maximum number of iteration (ITmax) is varied from 30 to 200. The particle size has been fixed to 30, 40, 50 and 100. The results are shown in Table 8.

In the table 9, Column (2) and (3) represents particles size and maximum number of iteration for a particle swarm optimization algorithm. Column (4) represents number of iterations required to optimize the function. Column (5) and (6) represents the value of variables X_1 and X_2 at which function is optimized. Columns (7) represents the function value.

Table 8: Effect of ITmax

S.No.	P	ITmax	k	X1	X2	Function value
(1)	(2)	(3)	(4)	(5)	(6)	(7)
1	30	30	30	1	1	1.145576e-011
2	30	40	32	1	1	1.375645e-011
3	30	50	40	1	1	2.5987932e-013
4	30	60	41	1	1	6.940071e-015
5	30	70	39	1	1	1.216929e-013
6	30	80	41	1	1	2.343508e-014
7	**30**	**110**	**55**	**1**	**1**	**4.763705e-020**
8	30	140	50	1	1	1.675177e-019
9	30	200	53	1	1	1.821156e-018
10	40	90	48	1	1	9.198076e-018
11	40	100	45	1	1	1.944461e-017
12	40	120	48	1	1	1.149075e-017
13	40	140	53	1	1	1.216215e-019
14	40	160	52	1	1	2.827675e-019
15	**40**	**180**	**57**	**1**	**1**	**1.093662e-020**

S.No.	P	ITmax	k	X1	X2	Function value
(1)	(2)	(3)	(4)	(5)	(6)	(7)
16	40	200	54	1	1	2.085538e-018
17	50	30	30	1	1	1.415376e-012
18	50	50	36	1	1	3.085142e-014
19	50	60	41	1	1	6.603557e-016
20	50	70	41	1	1	1.532676e-014
21	50	80	47	1	1	1.308451e-015
22	50	100	50	1	1	4.974816e-018
23	50	120	47	1	1	5.543777e-017
24	**50**	**140**	**57**	**1**	**1**	**2.060528e-021**
25	50	160	48	1	1	1.193581e-017
26	50	180	58	1	1	5.891339e-021
27	50	200	56	1	1	2.357654e-019
28	50	220	51	1	1	9.522462e-018
29	100	50	40	1	1	2.65228e-015
30	100	60	45	1	1	3.845000e-019
31	100	70	48	1	1	4.959005e-018
32	100	80	49	1	1	5.829558e-020
33	100	90	49	1	1	9.005843e-019
34	100	100	57	1	1	30268523e-021
35	100	110	43	1	1	2.1000047e-016
36	100	120	57	1	1	3.910962e-020
37	100	140	51	1	1	4.449501e-019
38	100	160	50	1	1	2.142912e-018
39	100	180	53	1	1	1.868118e-019
40	**100**	**200**	**58**	**1**	**1**	**6.037815e-022**
41	100	220	59	1	1	2.402596e-020

From Table 8 it is observed that accuracy is obtained for the following combinations of particle size (P) and ITmax which are shown in Table 9.

Table 9: Best combination of P and ITmax

P	ITmax	k	X_1	X_2	Function value
30	110	55	1	1	4.763705e-020
40	180	57	1	1	1.093662e-020
50	140	57	1	1	2.060528e-021
100	200	58	1	1	6.037815e-022

C. Inertia Weight:

In this paper, a study has been conducted by an PSO algorithm in which, two types of Inertia weight have been considered :Linearely decreasing Inertia weight given by (3) and simulated annealing inertia weight as given by (4).

The other parameters have been fixed to the values shown in Table 10.

Table 10:Parameters of PSO

P	rp	rg	Cp	Cg	ITmax	Wmax	Wmin
10-1000	0.4	0.5	2	2	10-120	0.9	0.4

For both types of Weights, population size is varied from 10 to 1000 and ITmax has been fixed to the corresponding value for which the function got optimized with maximum accuracy. Results for Linearely decreasing weight is shown in the Table 11.

Table 11: Results of Linearly Decreasing Inertia weight

P	k	IT $_{max}$	X$_1$	X$_2$	function value
10	71	200	1	1	3.7e-11
20	74	200	0.99	0.99	3.5e-09
30	80	160	1	1	3.9e-015
40	63	90	1.0	1.0	9.8e-019
50	74	140	1.00	1.00	9.87e-021
100	48	50	1.00	1.00	8.31e-018
200	96	220	1.00	1.00	2.71e-023
1000	54	60	1.00	1.00	4.93e-129

Results of Simulated annealing inertia weight is shown in Table 12.

Table 12: Results of simulated annealing Inertia weight

P	k	IT $_{max}$	X$_1$	X$_2$	function value
10	10	10	0.99	0.99	3.66e-08
20	51	110	1.000	1.0	3.6e-017
30	55	110	0.99	0.99	4.7e-020
40	57	180	1.0	1.0	1.09e-020
50	57	140	1.00	1.00	2.06e-021
100	58	200	1.00	1.00	6.03e-022
200	59	110	1.00	1.00	5.38e-023
1000	60	220	1.00	1.00	1.33e-025

VII. CONCLUSION

The Calculations have been done manually as well as by Matlab Programme to understand the variation of PSO parameters on convergence. Following are the results:

(1) Particles size should not be less than 20 to optimize the function accurately.
(2) For small size of Particles less number of maximum iteration required.
(3) Simulated annealing Inertia weight optimized the function accurately in minimum number of iteration in comparison to Linearly decreasing Inertia weight.
(4) For better accuracy particle size should be less than maximum number of iteration

REFERENCES

[1] N.K.Jain, Uma Nangia, Jyoti Jain,"Effect of Population and Bit Size on optimization of Function by Genetic Algorithm", International Conference on Computing for Sustainable Global Development March, 2016 .

[2] N.K.Jain, Uma Nangia, Jyoti Jain, "GA based Multi-objective Economic Load Dispatch by Maximization Of Minimum Relative Attainments", IEEE 5th India International Conference on Power Electronics (IICPE 2012), December 2012.

[3] J. Kennedy and R. Eberhart, "Particle Swarm Optimization, " in Proc. IEEE Int. Conf. Neural Networks, 1995, pp. 19942-1948.

[4] James Kennedy and Russell Eberhart, "A New Optimizer Using Particle Swarm Theory." Proceeding of sixth International Symposium on Micro Machine and Human Science,1995,pp. 39-43. 0-7803-2676-8/95 IEEE.

[5] Musirin I., Ismail N.H.F., Kalil M. R. et al.: 'Ant Colony Optimization (ACO) Technique in Economic Power Dispatch Problems' Trends in Communication Technologies and Engineering Science. Lecture Notes in Electrical Engineering, vol 33. Springer, Dordrecht pp. 191-203.

[6] Liu Weibing and Wang Xianjia "An evolutionary game based particle swarm optimization algorithm." Journal of computational and Applied Mathematics,2008 (214)4,30-35.

[7] Ying-Ping Chen and Wen-ChihPeng, "Particle Swarm Optimization With Recombination and Dynamic Linkage Discovery."IEEE transaction on systems, man, and cybernetics-Part B:cybernetics, vol. 37, no.6, December 2007.

[8] Tao Gong and Andrew L.Tuson, "Particle Swarm Optimization For Quadratic Assignment Problems –A Forma Analysis Approach" International journal of computational Intelligence Research ,ISSN 09731-1873 Vol.2,No.X(2007).

[9] N.K.Jain, Uma Nangia, Jyoti Jain, "An improved PSO Based On Initial Selection of particles (ISBPSO) for Economic Load Dispatch", IEEE first *International Conference on Power*Electronics, Intelligent Control and Energy (PIECES 2016).

[10] N.K.Jain, Uma Nangia, Jyoti Jain, "An improved PSO Based On Initial Selection of particles (ISBPSO)", IEEE first *International Conference on Power*Electronics, Intelligent Control and Energy (PIECES 2016).

[11] N.K.Jain, Uma Nangia, Jyoti Jain ,"Economic Load Dispatch using Social acceleration Constant based PSO", J.Inst.Eng. India ser.B (2018).

[12] PulkitJain,PranjalBhatia,Rashika Mann, Nidhikumari andJyotijain, "Economic Load Dispatch using PSO", International journal of Technology and science,Vol. 5, issue 2, 2018.

[13] Narender Kumar Jain, Uma Nangia, Aishwary Jain , "PSO for multiobjective Economic Load Dispatch (MELD) for Minimizing Generation Coast and Transmission losses", J.Inst.Eng. India ser.B (2015).

[14] U. Nangia, N.K. Jain, C.L. Wadhwa, "Multiobjective optimal load flow based on ideal distance minimization in 3D space", Electr. Power Energy Syst. **23**, 847–855 (2001).

[15] N.K.Jain, Uma Nangia, JyotiJain ,"Review of particle swarm optimization", J.Inst.Eng. Indiaser.B, August (2018),Vol.99, Issue 4, pp. 407-411.

2nd IEEE International conference on power Electronics, Intelligent Control and Energy systems (ICPEICES-2018)

Audio Watermarking Using Beat Detection and Pitch Estimation

Lavi Tanwar
Department of Electronics & Communication
Engineering Delhi Technological University
Delhi, India
lavi.tanwar02@gmail.com

Radhika Dang
Department of Electronics & Communication
Engineering Delhi Technological University
Delhi, India
radhika_bt2k14@dtu.ac.in

Satvik Maurya
Department of Electronics & Communication
Engineering Delhi Technological University
Delhi, India
satvik_bt2k14@dtu.ac.in

Jeebananda Panda
Department of Electronics & Communication
Engineering Delhi Technological University
Delhi, India
jpanda@dce.ac.in

Abstract- **This paper presents a novel semi-blind Audio Watermarking Algorithm in the Time Domain with the use of Beat detection and Pitch Estimation. Using the estimated values of the beat locations and their respective pitch or fundamental frequencies, a watermark is generated from the binary information that is to be hidden, which is then embedded in the audio file. The robustness of this algorithm has been gauged using standard audio processing attacks such as Resampling, Re-quantisation and Noise attack using AWGN. Apart from these, the watermarked audio file has been subjected to equalisation attacks which are prevalent in the music industry today, using equalisation modes such as Jazz, Rock, Classical and Bass. The performance of the algorithm has been proved to be satisfactory against all such attacks.**

Keywords- Objective Differential Grade, Beat Detection, Pitch Estimation, Audio Equalisation.

I. INTRODUCTION

With the continuous growth of the internet, it is becoming increasingly easier to obtain pirated versions of various types of intellectual property which include Books, Software, Music and even Publications. It has thus become imperative to provide such properties with some form of protection against illegal reproduction.

Digital Audio Watermarking has thus received great interest in recent years because of the lack of means to provide copyright protection of audio content. The watermarking of audio signals is thus characterised by the need to keep the watermark imperceptible i.e., it should not be audible and robust against various malicious and non – malicious attacks. These requirements have resulted in a wide variety of time domain and transform domain watermarking algorithms which can be used effectively for copyright protection of audio content [1], [2]. Other approaches include the use of both time and transform domains and compressed domain watermarking. As mentioned in [1], the use of transform domain algorithms is preferred by researchers even though no conclusive evidence of the superiority of the

transform domain approach over other approaches has been found.

Time domain algorithms for audio watermarking primarily exploit the principle that the Human Auditory System (HAS) cannot detect slight changes in time varying regions of an audio signal [3]. Some noteworthy time domain approaches as mentioned in [4] include the one proposed by Lie et. Al. in [5] in which the amplitudes are varied for embedding the watermark. This method fulfils the imperceptibility criteria for a good watermarking scheme but was shown to exhibit poor robustness as the watermark could easily be removed by using a suitable attenuator. Van Schyndel et. Al. proposed the elementary method of digital watermarking i.e., the Least Significant Bit (LSB) method [6]. In this approach, the watermark is embedded in the LSB of the host signal. A major drawback of this method was the lack of robustness against noise. The use of Empirical Mode Decomposition (EMD) for watermarking has been mentioned in [7] in which the watermark is embedded in the lower order IMFs. However, this scheme was computationally complex. In [13], S. Xiang et. Al. modified the statistical features in time domain and shifted the histogram to embed the watermark. This method is robust to MP3 compression and can embed large number of bits but its imperceptibility is poor than the existing methods. K. Gopalan presented a fragile technique in [14] by modifying the bits of the audio signal. The bit values at selected indices are used to choose the watermark key. This technique is fragile and robust to noise but the careful selection of key is required as wrong key selection may result in loss of information and false detection. D. Bratic et. Al. proposed an algorithm in [15] which generates a support function using spectrogram and that support function is used to create the watermark. The gradient algorithm is used for reconstruction. The perceptuality of this technique is good but its computational complexity is high.

In this paper, a novel watermarking method in the time domain has been presented, which has then been proven imperceptible and robust to a wide range of signal processing attacks. The algorithm proposed here makes use of the estimated values of the beat locations and the fundamental frequencies for the generation of a watermark from a binary image which is then embedded in the audio file. The proposed

978-1-5386-6624-1/18 $31.00 © 2018 IEEE

algorithm is semi-blind in nature and thus the original audio file is not required for the successful recovery of the binary image from the watermarked audio file. To determine the robustness of the watermarking procedure, the watermarked audio file has been subjected to signal processing attacks such as addition of AWGN, Re-quantisation and Resampling, and to several popular Audio Equalisation modes wherein particular frequency components of the audio signal are amplified or attenuated. With the help of readily available audio processing software, Equalisation can be applied easily on an audio file using various pre-determined settings which define different genres such as 'Jazz', 'Rock', 'Classical', 'Pop' and 'Bass'. Thus, to gauge the performance of the algorithm proposed in this paper against such attacks, the watermarked audio was equalised into different modes and the reconstruction of the binary information was performed for each case. The paper thus proceeds to elaborate the process of beat detection and pitch estimation in Section II. Then in Sections III and IV, the watermark embedding and reconstruction algorithms have been presented, followed by the results and performance analysis in Section V. The conclusions have been presented in Section VI.

II. BEAT DETECTION AND PITCH ESTIMATION

A. Beat Detection

A beat, in the simplest of terms, is an instant in the audio file when the listener will tap his foot, and thus it represents the rhythmic nature of an audio file. The domain beat tracking and detection has been researched upon for quite a few years and many efficient algorithms have been formulated including the one proposed by Fillon et. Al. in [8] which used Conditional Random Field (CRF) framework for the efficient detection of beats. Zapata et. Al. proposed a in [9] a beat tracker modelled with multiple input features. Their implementation was shown to be highly accurate and was better than most state-of-the-art beat detectors at the time. In this paper, however, a simple beat detection algorithm proposed by Daniel P.W. Ellis in [10] has been implemented which used dynamic programming for a simple beat detector.

Ellis' system first estimates a global tempo. Then, using dynamic programming, the prospective beat times are evaluated which match the chosen tempo and the best matches are retained. This approach was shown to be simple and computationally efficient and performed satisfactorily. Further elaboration of the algorithm is beyond the scope of this paper.

B. Pitch Estimation

The term 'Pitch Estimation' is most commonly used for the estimation of the Fundamental frequency or the Formant frequency of a periodic waveform. The Fundamental frequency for a periodic waveform can be defined as the lowest frequency component present in that waveform. The Formant frequency on the other hand is defined as the resonant frequency of the waveform and there can be more than one Formant in an audio file. In music, the Fundamental frequency, or Pitch, is the lowest tone present. As this paper deals with music, the terms fundamental frequency and pitch have been used interchangeably.

There are many pitch estimation algorithms that have been devised in recent years for various uses such as pitch scaling, intonation scoring etc. Some of the common methods

for extracting the pitch from an audio file are explained briefly below:

1) *Using Auto-Correlation Function:* The auto-correlation of a signal $x(m)$ with a window function $w(m)$ with a lag parameter 'k' is given by-

$$R_n(k) = \sum_{m=-\infty}^{+\infty} x(m)w(n-m)x(m+k)w(n-(m+k)) \quad (1)$$

This method uses the property that the auto-correlation function is a measure of self-similarity. Thus, when the auto-correlation of an audio file is taken which has a repeating pattern, it will yield maximum values when the entire periodic sequence is overlapped with itself. Hence, by using these local maxima, the period and hence fundamental frequency can be calculated for a sequence of samples.

2) *Using Cepstral Analysis:* The real – cepstrum of a frame of audio $x(m)$ using a window $w(m)$ at a sample n can be defined by-

$$c_s(n) = F^{-1}\big(\log\big(\big|F\big(x(m)\big)\big|\big)\big) \quad (2)$$

Where F is the DTFT operator and F^{-1} is the inverse DTFT operator. In the cepstral domain, the pitch can be calculated by taking the frequency in the cepstral domain which has the highest energy. This frequency is thus the fundamental frequency of the set of samples.

Apart from these methods, there are several other ways to estimate the pitch of an audio file such as the use of Harmonic Product Spectrum, the Average Magnitude Difference Function (AMDF) and Maximum Likelihood estimation. For this paper, the cepstral analysis method for pitch estimation has been used.

III. THE EMBEDDING ALGORITHM

Kirovski and Attias introduced beat detection and block redundant coding for nullifying the effects of desynchronization [11]. Teleşpan et. Al. proposed a watermarking scheme which employed Empirical Mode Decomposition and Beat Detection for the watermark embedding process [12]. Their primary aim was to perform the watermark embedding at locations that are least susceptible to various audio compression techniques, which they successfully executed with the help of beat detection. The proposed approach towards the watermark embedding process is based on the same principle.

It should be noted here that the watermarking procedure has been performed on audio which has a stereo output. Thus, throughout the embedding and reconstruction processes, two distinct sets of sampled data corresponding to two distinct channels will be considered. A flowchart showing the embedding process is shown in Figure 2 followed by a description of all the steps involved.

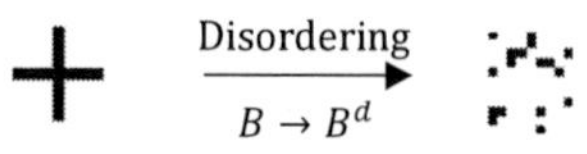

Fig. 1. Disordering of the Binary Image

978-1-5386-6624-1/18 $31.00 © 2018 IEEE

1) Beat Detection: Using the beat detection functions, the beat locations are computed for both channels. Let the number of beats be 'c_1' and 'c_2' for the two channels.

2) Disordering of Binary Information: For additional security in the watermarking procedure, the binary image of size m × n is used for generation of watermark. It is disordered to obtain B^d using Key 1 as shown in Figure 1, which will be used as side information for extraction of watermark.

3) Pitch Estimation: Taking windows of length 'L', where L is the number of samples between the last beat location and the end of the audio file, the fundamental frequencies are calculated for a window starting at every beat location and are stored in vectors f_1 and f_2 of lengths c_1 and c_2 respectively.

$$W_L^M(i,j) = \begin{cases} -1, if\ f_L(i \times j) < f_0\ and\ B^d(i,j) = 0 \\ 1, if\ f_L(i \times j) > f_0\ and\ B^d(i,j) = 1 \\ -2, if\ f_L(i \times j) < f_0\ and\ B^d(i,j) = 1 \\ 2, if\ f_L(i \times j) > f_0\ and\ B^d(i,j) = 0 \end{cases} \quad (3)$$

$$W_R^M(i,j) = \begin{cases} -1, if\ f_R(i \times j) < f_0\ and\ B^d(i,j) = 0 \\ 1, if\ f_R(i \times j) > f_0\ and\ B^d(i,j) = 1 \\ -2, if\ f_R(i \times j) < f_0\ and\ B^d(i,j) = 1 \\ 2, if\ f_R(i \times j) > f_0\ and\ B^d(i,j) = 0 \end{cases} \quad (4)$$

Where, f_L and f_R represent the set of fundamental frequencies corresponding to the two channels (L and R) and f_0 is a pre-defined reference frequency. This Non-Return to Zero (NRZ) scheme was chosen because embedding zero as binary watermark will not cause any change to the sample magnitude of the audio file and this will affect the robustness of the watermarking scheme.

5) Embedding of the Watermarks: W_L^M and W_R^M are embedded in the original audio file to obtain the watermarked audio file. First, a window of length L is taken at every beat location of the two audio channels. In that window, the corresponding Watermark generated by using (3) and (4) is embedded using the relation –

$$x_1(wn:wn + L) = x_1(wn:wn + L)\big(1 + \alpha \times W_L^M(i,j)\big) \quad (5)$$

Where, wn is n^{th} beat location and $(wn:wn + L)$ represents a window starting from wn up to a length L, 'α' represents the strength of watermarking and x_1 represents the set of sample values corresponding to audio channel (L) and x_2 represents the set of sample values corresponding to audio channel (R). The same relation as shown in (5) is applied for the audio channel(R). Thus, the watermarked audio file will contain both watermarked channels, x_1 and x_2.

The value of the reference frequency f_0 is calculated as follows: The fundamental frequency is obtained from L samples of the audio file starting at every beat location. If the mean of all the fundamental frequencies that have been obtained for the two channels are represented as M_L and M_R and the highest fundamental frequencies that have been obtained for both channels are given by P_L and P_R, then the reference frequency f_0 is given by–

$$f_0 = mean(P_L - M_L, P_R - M_R) \quad (6)$$

IV. THE RECONSTRUCTION ALGORITHM

As the watermarking procedure is semi-blind in nature, the original audio file is not required for the reconstruction of the binary image from the watermarked audio file. However, the reference frequency f_0 and the original watermark information derived in (3) and (4) are necessary for the proper reconstruction of the binary image B. The steps are shown in Figure 3 and are explained below:

1) Beat Detection: The beat locations are first calculated for the watermarked audio file (again, the file will contain two data sets for the two channels). Let the sets of beat locations have lengths 'rc_1' and 'rc_2'.

2) Pitch Estimation: Now, for the first channel, at every beat location, a window of length 'L' is taken, where

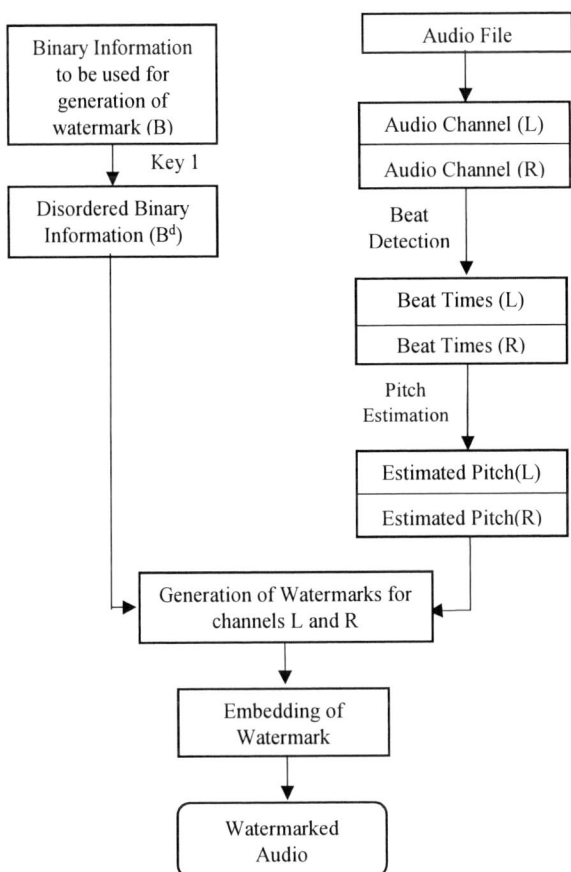

Fig. 2. Flowchart for the embedding algorithm

4) Generation of the Watermarks: To provide redundancy to the watermarking procedure to further improve its security and invisibility, two different watermarks, W_L^M and W_R^M are generated for the two channels, as per the following comparisons:

'L' is the number of samples between the last beat location and the last sample of the audio file, and the fundamental frequency is calculated for that window. This process is repeated for the second channel as well resulting in two sets of data fr_1 and fr_2 of lengths rc₁ and rc₂ respectively containing pitch information.

3) Retrieval of Disordered Binary Image B^d: The modified watermarks W_L^M and W_R^M of the two channels are used for the retrieval of the disordered binary image B^d as per the following comparisons:

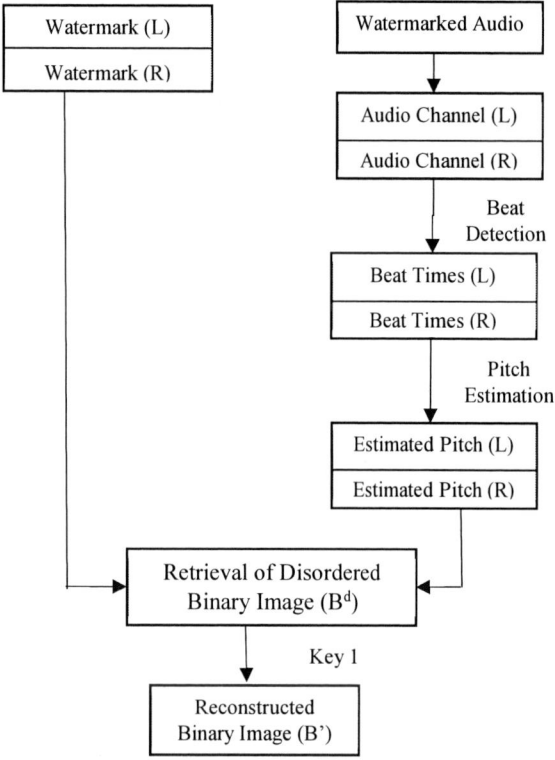

Fig. 3. Flowchart for the reconstruction algorithm

$$B_1^d(i,j) = \begin{cases} 0, if\ fr_L(i \times j) < f_0\ and\ W_L^M(i,j) = -1 \\ 1, if\ fr_L(i \times j) < f_0\ and\ W_L^M(i,j) = -2 \\ 0, if\ fr_L(i \times j) > f_0\ and\ W_L^M(i,j) = 1 \\ 1, if\ fr_L(i \times j) > f_0\ and\ W_L^M(i,j) = 2 \end{cases} \quad (7)$$

$$B_2^d(i,j) = \begin{cases} 0, if\ fr_R(i \times j) < f_0\ and\ W_R^M(i,j) = -1 \\ 1, if\ fr_R(i \times j) < f_0\ and\ W_R^M(i,j) = -2 \\ 0, if\ fr_R(i \times j) > f_0\ and\ W_R^M(i,j) = 1 \\ 1, if\ fr_R(i \times j) > f_0\ and\ W_R^M(i,j) = 2 \end{cases} \quad (8)$$

$$B^d = B_1^d \wedge B_2^d \quad (9)$$

As is apparent from the above relations, the disordered binary image is reconstructed from a logical OR of the disordered binary images obtained from both the channels. As mentioned before, both channels are used to add redundancy to the watermarking procedure.

4) Reordering of the Disordered Binary Image: After obtaining the disordered binary image from (7), (8) and (9), the final binary image is reconstructed by reordering the disordered binary image using Key 1.

V. RESULTS

A binary image of size 11 × 11 pixels, as shown in Figure 4, has been used here for the watermarking procedure described in section III. The implementation of the embedding and reconstruction algorithm was performed in a MATLAB 2015a environment and the equalisation attacks were performed using Simulink models. The algorithm has been implemented on audio files of three distinct genres of music, namely, Rock, Pop and EDM (Electronic Dance Music) of different lengths and tempo sampled at 44.1 kHz with 16-bit quantisation. The reference frequency f_0, as obtained from (8), for the Rock genre was 147.5424 Hz, for the Pop genre, it was 157.8009 Hz and for the EDM genre, it was 164.1590 Hz. Table 1 shows the number of detected beats for the three audio files.

The Rock audio file the window lengths for both channels was 58984 samples and 58278 samples respectively. For the Pop audio file, the window lengths for pitch estimation for the two channels were 326490 samples and 310260 samples while for the EDM audio file, they were 340400 samples and 347980 samples for the two channels.

To measure the robustness of our algorithm, various attacks on the watermarked audio were performed, and the values of the Bit Error Rate (BER) and Normalised Cross-Correlation (NC) as defined in (9) and (10) were analysed for all the cases with attacks [12]. Reconstruction of the binary information without any attacks resulted in a Signal to Noise Ratio (SNR) of infinity, which signifies that the watermark was perfectly retrieved.

$$BER(B, B') = \frac{\sum_{i=0}^{N} B(i) * B'(i)}{N} \quad (10)$$

$$NC(B, B') = \frac{\sum_{i=0}^{N} B(i) * B'(i)}{\sqrt[2]{\sum_{i=0}^{N} B(i)^2} * \sqrt[2]{\sum_{i=0}^{N} B'(i)^2}} \quad (11)$$

Where, B and B' represent the original and reconstructed binary images respectively with length N. To assess the quality and imperceptibility of the watermarked audio, 15 human subjects were asked to use the Objective Differential Grade (ODG) scale (Table 2) to rate a thirty second sample of the watermarked audio file based on the perceptibility of the watermark. Table 3 shows the ODG values obtained for the three genres that have been watermarked.

It can be seen from Table 3 that the watermark has a satisfactory level of imperceptibility. The performance of the watermarking algorithm has been thoroughly analyzed by determining the values of BER and NC of the reconstructed binary image after the watermarked audio is attacked as shown in Table 4.

Fig. 4. Original Binary Image

TABLE 1. NUMBER OF BEATS FOR AUDIO FILES

Channel	Rock	Pop	EDM
L	500	1190	1071
R	501	596	1070

TABLE 2. ODG SCALE

Perception	ODG
Imperceptible	0
Perceptible, but not annoying	-1
Slightly annoying	-2
Annoying	-3
Very Annoying	-4

TABLE 3. ODG VALUES OBTAINED FROM HUMAN SUBJECTS

	Rock	Pop	EDM
ODG	-0.06	-0.13	-0.40

TABLE 4. PERFORMANCE OF THE ALGORITHM

GENRE	ROCK		POP		EDM	
AUDIO TYPE	SNR (dB)					
Original	Inf		Inf		Inf	
ATTACK PERFORMANCE						
ATTACK	BER (%)	NC	BER (%)	NC	BER (%)	NC
Cropping	0.826	0.995	0.826	0.995	0.826	0.995
AWGN	4.959	0.972	0.000	1.000	1.653	0.991
Re-Quantisation (8 bits)	2.479	0.986	0.826	0.995	0.000	1.000
Re-Sampling	0.826	0.995	0.000	1.000	0.000	1.000
MP3 Compression (128 bits)	2.479	0.986	0.000	1.000	0.826	0.995
MP3 Compression (96 bits)	0.826	0.995	0.000	1.000	0.000	1.000
Jazz Equalisation	4.959	0.972	2.479	0.986	0.000	1.000
Bass Equalisation	3.306	0.981	0.000	1.000	0.826	0.995
Treble Equalisation	2.479	0.986	0.000	1.000	0.000	1.000
Rock Equalisation	4.959	0.972	0.000	1.000	0.000	1.000
Classical Equalisation	4.959	0.972	0.000	1.000	0.826	0.995
Pop Equalisation	3.306	0.981	0.000	1.000	3.306	0.981

The binary image 'B' can be reconstructed perfectly when the watermarked audio file is not attacked. In this paper, attacks such as cropping, addition of WGN, Re-quantisation and Re-sampling as well as MP3 compression and equalisation, have been used.

For the cropping attack, the watermarked audio file was cropped to 60% of its original length and then the binary image B was reconstructed. This attack reported similar values of BER and NC for all three genres.

For the AWGN attack, WGN was added to the watermarked audio file till the SNR of the audio file was 30 decibels. This attack had maximum effect on the Rock genre while it didn't have any effect on the Pop genre. Thus, on average, the BER for this attack was about 2.2%.

The watermarked audio file was re-quantised to 8 bits and 24 bits. Re-quantisation to 24 bits resulted in perfect reconstruction of the watermark and hence we refrain from listing those results. Re-quantisation to 8 bits had no effect on the EDM genre and thus the average BER for this attack was about 1%.

For the Re-sampling attack, the watermarked audio file was down-sampled to 22.05 kHz and then re-sampled back to the original sampling frequency of 44.1 kHz. In this case, the genres Pop and EDM were unaffected by the attack and thus, the average BER was about 0.27%.

MP3 compression at 96 kbps resulted in perfect reconstruction of the binary image B for the Pop and EDM genres while Rock was slightly affected, thus giving an average BER of around 0.27%. MP3 compression at 128 kbps did not affect the Pop genre thus yielding an average BER of 1.1%.

The following equalisation attacks have been performed on the watermarked audio file: Jazz, bass, treble, classical, pop and rock equalisation. These modes were chosen because they are generally the most popular and most widely used equalisation modes among average users today. The equalisation attacks had maximum effect on the rock genre while Pop and EDM were relatively unaffected.

VI. CONCLUSION

In this paper, a novel time domain watermarking algorithm has been proposed which makes use of both pitch estimation as well as beat detection. An interesting aspect of this algorithm is that the binary information that is to be hidden is not embedded directly in the audio. Instead, two watermarks are generated using this information which are then embedded in the audio file. These watermarks are then used for the reconstruction of the binary information. The proposed algorithm is semi-blind in nature and is quite imperceptible as indicated by the ODG values. Considerably good results were obtained for the BER and NC, when the watermark was retrieved from a corrupted watermarked audio file, which are also significantly better than the ones reported in [12] which reported an average BER of 7.7% for addition of WGN to the watermarked audio file, and an average BER of 7.5% against re-sampling. Furthermore, it shows better robustness against MP3 compression (both 96 and 128 kbps). The proposed algorithm is computationally efficient as it is a time domain approach without use of any complex mathematical operations.

REFERENCES

[1] G. Hua, J. Huang, Y. Q. Shi, J. Goh and V. L. Thing, "Twenty Years of Digital Audio Watermarking – A Comprehensive Review," Signal Processing, vol. 128, pp. 222-242, 2016.

[2] K. V. Goenka and P. K. Patil, "Overview of Audio Watermarking Techniques," International J. of Emerging Technology and Advanced Engineering , vol. 2, pp. 67-70, 2012.

[3] P. Bassia, I. Pitas and N. Nikolaidis, "Robust audio watermarking in the time domain," IEEE Trans. on Multimedia, vol. 3, pp. 232-241, 2001.

[4] R. Shahriar, S. Cho and U. Chong, "Time-domain audio watermarking using multiple marking spaces," International Conf. on Informatics, Electronics and Vision, 2012.

[5] W. N. Lie and L.-C. Chang, "Robust and high-quality time-domain audio watermarking based on low-frequency amplitude modification," IEEE Trans. on Multimedia, vol. 8, pp. 46-59, 2006.

[6] V. Schyndel, R. G. Tirkel, A. G. and C. F. Osborne, "A Digital Watermark," in IEEE International Conf. on Image Processing, Austin, 1994.

[7] J. Panda, K. R. Gera and A. Bhattacharyya, "Non-Blind Audio Watermarking Scheme Based on Empirical Mode," International J. of Advanced Science, Engineering and Technology, vol. 3, pp. 38-43, 2014.

[8] T. Fillon, C. Joder, S. Durand and S. Essid, "A Conditional Random Field System for Beat Tracking," in IEEE International Conf. on Acoustics, Speech and Signal Processing, 2015.

[9] J. R. Zapata, M. E. P. Davies and E. Gomez, "Multi-feature beat tracking," IEEE/ACM Trans. on Audio, Speech and Signal Processing, vol. 22, pp. 816-825, 2014.

[10] D. P. Ellis, "Beat Tracking by Dynamic Programming," J. of New Music Research, vol. 36, pp. 51-60, 2007.

[11] D. Kirovski and H. Attias, "Audio Watermark Robustness to Desynchronization via Beat Detection," Information Hiding, Springer, pp. 160-176, 2003.

[12] M. Teleşpan and B. M. Schuller, "Audio Watermarking Based on Empirical Mode Decomposition and Beat Detection," IEEE International Conf. on Acoustics, Speech and Signal Processing, Shanghai, 2016.

[13] S. Xiang, L. Yang and Y. Wang, "Robust and Reversible Audio Watermarking by Modifying Statistical Features in Time Domain," Advances in Multimedia, vol. 2017, pp. 1-10, 2017.

[14] K. Gopalan, "An algorithm for fragile audio watermarking by bit modification," IEEE Int. Conf. on Industrial Informatics, France, 2016.

[15] D. Bratic, F. Vesovic and V. Mijanovic, "Audio Watermarking under gradient-based reconstruction attack," IEEE Meditearanean Conf. on Embedded Computing, Montenegro, 2016.

Gap in pagination due to withheld paper.

Pages 933-937

Implementation of V/f Adjustable Speed Drive for Induction Motor Using dSPACE dS1104

Chinmay Tigade[1] and Mini Sreejeth[2]

Dept. of Electrical Engineering, *Delhi Technological University*, Delhi

E-mail: [1]chinmaytigade@gmail.com, [2]minisreejeth@dce.ac.in

Abstract — **In this paper, performance of Open-loop and Closed-loop V/f speed control methods is examined using MATLAB/Simulink at various speed. These methods are implemented on laboratory prototype using dSPACE DS1104 on a 3Hp, 415V, 50Hz Induction Motor fed by VSI. Inverter Voltage and frequency is controlled using SPWM technique. It is observed that the experimental results of these methods almost matches with the simulation results.**

Keywords—**Three-phase Induction Motor; Open-loop and Closed-loop V/f Speed Control Method; dSPACE DS1104;**

I. INTRODUCTION

Induction motors are used in huge volume all over the world due to countless advantages it has. IM are trustworthy, extremely tough, cheap to run and they need less maintenance. Thus, IM are preferred in variety of commercial and domestic applications [1]. It is estimated that more than 50% of total electric energy generated all over the world is utilized by induction motors. Hence it is very important to efficiently control the induction motor drive to reduce environmental pollution and for economical saving [2]. Traditionally for example the flow of fluid coming out of pipe is controlled by manually adjusting the control valve and keeping the pump running at rated speed. The efficiency of such type of control technique used for various applications is very low. Thus to solve this problem various efficient speed control methods were attempted over a period of time such as open-loop V/f, closed-loop V/f, direct vector control, indirect vector control, direct torque control and many more [3]. The open-loop V/f method is low-cost speed control technique which is very easy to implement. It uses parameters which does not depend on each other. But the performance of this methodology is very poor [4]. The problem observed in this method is that the speed of motor cannot be controlled precisely that is because only the frequency of stator supply is controlled i.e. only the synchronous speed is controlled and the rotor speed is always slightly less than synchronous speed. This problem is resolved in closed-loop V/f method in which speed of motor is measured and it is compared with desired speed. The error is processed by PI controller to set the stator supply frequency [6] - [7].

In this paper performance evaluation of open-loop and closed-loop V/f method is performed using MATLAB and real time performance of developed model is tested using laboratory

prototype using dSPACE DS1104. In section II open-loop and closed-loop control strategies are explained in detail, section III includes discussion on the hardware implementation, section IV includes simulation results and discussion and section V contains conclusion.

II. V/F SPEED CONTROL METHODS

In scalar control method also known as V/f method magnitude of control variable is altered to change the speed of motor hence termed as scalar control. The flux of IM can be controlled by adjusting voltage whereas torque can be controlled by changing frequency. If we neglect the voltage drop across stator resistance, the stator flux is given as $F_s = V_s/f_s$ where Vs is stator voltage and fs is stator supply frequency. Thus, if it's voltage is kept constant while reducing frequency to decrease speed of motor then flux will saturate causing excess of stator current which might harm the motor permanently. Hence, to adjust the speed voltage and frequency are altered proportionally in V/f method [5]. Also, at low frequency the voltage drop across the stator resistance might cause flux reduction. The simple method used to compensate this voltage drop is consisting of boosting the stator voltage [6]. Now if the frequency is increased above rated value to increase the speed the flux will decrease as voltage cannot be increased beyond rated value. This region of operation is called as field weakening region. Hence, to operate the motor at rated flux both frequency and voltage need to vary in proportion. There are two types of scalar control techniques viz. open-loop and closed-loop V/f.

A. *Open-loop V/f Method*

Open-loop V/f method is the simplest and economical speed control technique of three-phase induction motor. Fig. 1(a) shows open-loop V/f MATLAB Simulink Model. The main components of this method are DC voltage source and PWM voltage source inverter. The desired speed command ω_m^* is given as input to the system which is then multiplied by V_s/ω_s ratio to obtain corresponding desired voltage and θs is obtained by integrating ω_s. The desired three- phase voltages are then calculated as shown in model which are fed to Sinusoidal Pulse Width Modulator (SPWM) which compares it with triangular carrier wave to generate pulses. These pulses are given to IGBT switches of three-phase inverter. The output of inverter is given to the motor. The IM is connected to the inverter whose output is variable voltage variable frequency supply depending upon input speed command.

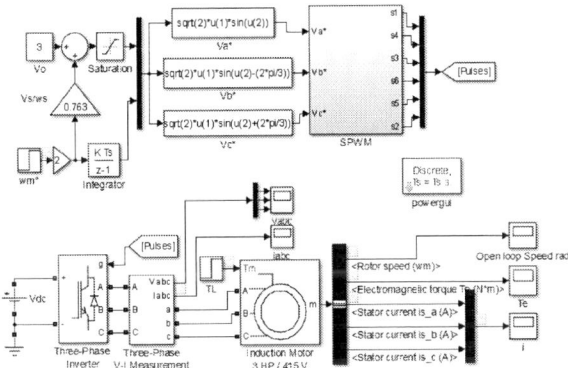

Fig 1.   (a) Open-loop V/f MATLAB Simulink Model

The speed of motor in open-loop method will never be exactly equal to the given speed command as IM can never run at synchronous speed. Also, there is significant drop in motor speed when load torque is applied. The absence of any feedback signal makes this method very simple but at the same time it makes the system prone to instability. If there is any disturbance in the system or speed of motor, system cannot rectify it but still it can be used in the applications where high precision is not required.

B. Closed-loop V/f Method

The issues involved in open-loop method are solved by including speed control loop. Desired speed command ω_m* is converted into electrical rotor speed ω_r* and compared with instantaneous speed ω_r and error is processed by PI controller which generated slip frequency command $\omega_{sl}*$. The addition of slip frequency and instantaneous speed gives desired stator frequency ω_s* as $\omega_{sl} = \omega_s - \omega_r$. This frequency is then fed to V/f controller which then computes the corresponding voltages. The remaining operation is same as the open-loop method. The speed control loop amends ω_{sl} whenever there is drop in speed due to application of load torque or any disturbance and maintains it at desired speed command. Simulink model of closed-loop V/f method is as shown in Fig. 1(b).

III.   HARDWARE IMPLEMENTATION

Open-loop and closed-loop V/f speed control methods are implemented on a laboratory prototype using dSPACE. The laboratory setup consists of a 3 phase induction motor of rating 3HP, 415V, 50Hz fed by a VSI. The block diagram of laboratory setup is as shown in Fig. 2(a), whereas Fig. 2(b) shows the experimental setup used for implementation of these technique. The switching signals for the switches of VSI is obtained using SPWM technique and given through DS1104 dSPACE controller.

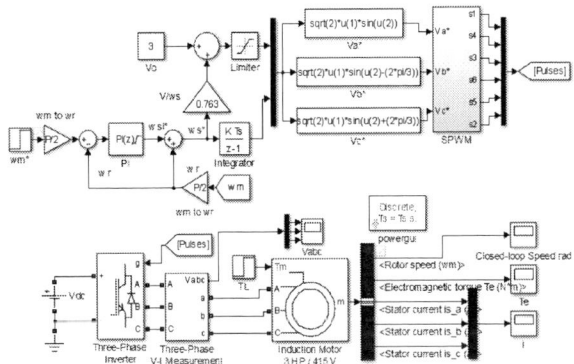

Fig 1.   (b) Closed-loop V/f Simulink Model

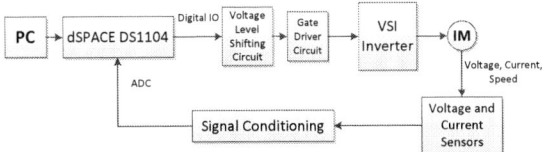

Fig 2.   (a) Block Diagram of Laboratory Setup

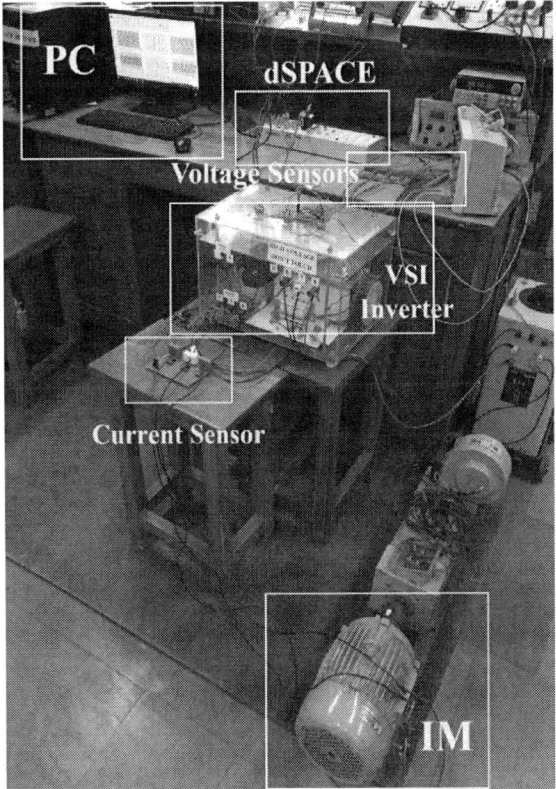

Fig 2.   (b) Experimental Setup

978-1-5386-6624-1/18 $31.00 © 2018 IEEE   939

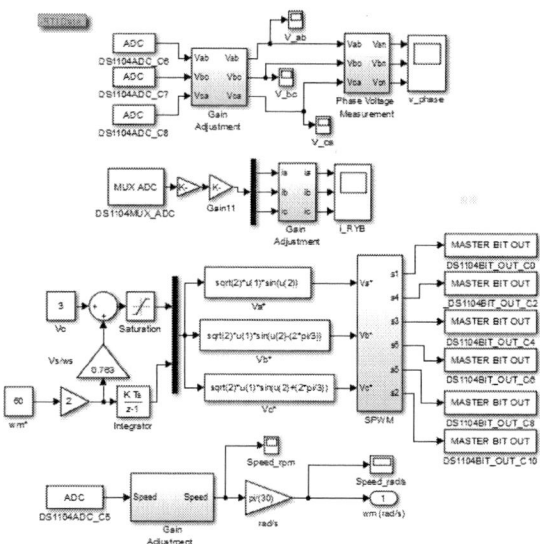

Fig 3. (a) Open-Loop V/f Simulink Model for Hardware Implementation

1. Implementation of Open-loop V/f Method –

For hardware implementation, the SPWM switching signals are obtained using MASTER BIT OUT terminal of dSPACE using channel 0, 2, 4, 6, 8, 10 and given to the IGBT switches of three-phase inverter to get variable voltage variable frequency supply. To observe speed of motor the output of Tachometer generator is given to the voltage sensor which is then fed to ADC channel 5 of dSPACE. Fig. 3(a) shows open-loop V/f model used for hardware implementation and Control Desk layout used for controlling speed of motor is shown in Fig. 5. Time plotter shows mechanical speed of motor in rad/sec. The speed can be changed using slider or by typing value in variable array window.

2. Implementation of Closed-loop V/f Method –

Similar to open-loop V/f method the SPWM signal is obtained using MASTER BIT OUT as shown in Fig. 3(b). Speed signal is converted into rad/sec and fed back to the speed control loop. The control desk for closed-loop V/f is as shown in Fig. 7. Similar to open-loop V/f method we can alter speed using slider or by typing value.

IV. RESULTS AND DISCUSSION

The dynamic performances of open-loop and closed-loop methods are examined for sudden change in speed and load torque. Modelling and simulation of open loop V/f and closed loop V/f IM drive are carried out using MATLAB Simulink. For Simulation parameters mentioned in table 1 are used and for hardware implementation name plate data of IM is used.

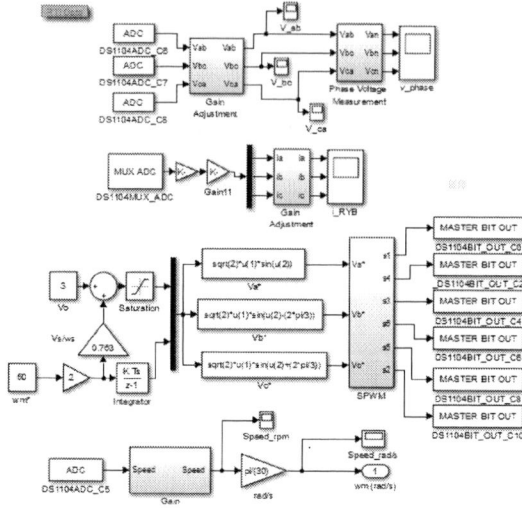

Fig 3. (b) Closed-loop V/f Simulink Model for Hardware Implementation

Rated Power	3 HP
Rated Voltage	415 V
Rated Current	4.8 A
Rated Frequency	50 Hz
Stator Resistance (R_s)	11.66 Ω
Rotor Resistance (R_r)	15.79 Ω
Stator Leakage Reactance (X_{ls})	16.7 Ω
Rotor Leakage Reactance (X_{lr})	16.7 Ω
Mutual Reactance (X_m)	268.27 Ω
Number of Poles (P)	4

A. Simulation Results

Simulation results of open-loop and closed-loop V/f method is as shown in fig. 4(a) and 4(b). Initially under no load, speed command of 50 rad/sec is given to motor, then the speed is increased to 100 rad/sec and 150 rad/sec at time t = 1 sec and t = 2 sec respectively. It is observed that the speed of motor is 47 rad/sec, 94 rad/sec and 140 rad/sec corresponding to given speed command. The desired speed is represented by red line and actual response is represented in blue color. It is found that when speed command is given, speed of motor increases gradually and settle down close to desired speed. Lot of oscillations are observed during starting of motor using open-loop V/f method. When motor is turned on it takes approximately 0.5 secs to reach speed of 50 rad/sec and 0.4 secs to reach speed of 100 rad/sec and 150 rad/sec. In case of closed-loop system the speed of motor increases linearly and more quickly than open-loop system. It takes around 0.2 sec to reach speed of 50, 100 and 150 rad/sec. It is found that in closed-loop method, speed control loop increases the slip frequency to maintain the motor at desired speed.

TABLE I. PARAMETERS OF INDUCTION MOTOR

978-1-5386-6624-1/18 $31.00 © 2018 IEEE

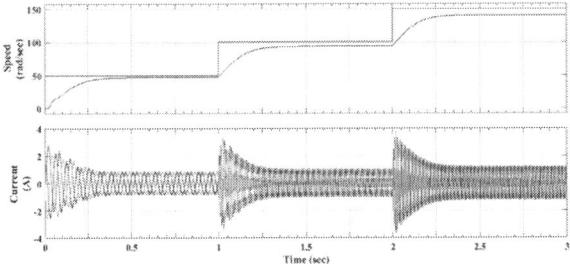

Fig 4.   (a) Open-loop V/f MATLAB Simulation Result

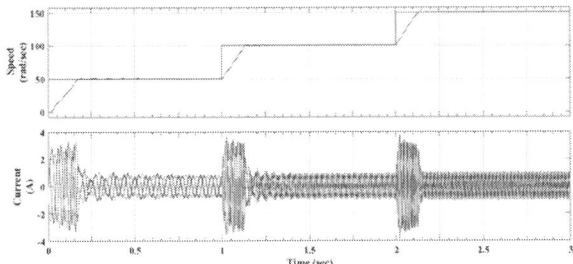

Fig 4.   (b) Closed-loop V/f MATLAB Simulation Result

B. Experimental Results

The experimental results of open-loop and closed-loop methods are obtained using Control Desk. The performance of both methods is observed for step change in speed. Initially speed command of 50 rad/sec is given to motor then it is increase to 100 rad/sec and finally it is raised to 150 rad/sec. The speed of motor is then reduced to 100 rad/sec. Fig. 5

displays control desk view of open-loop method to step change in speed. The scope on left side of screen shows actual speed of motor and right scope shows current waveform. It is observed that speed of motor is 48 rad/sec, 92 rad/sec and 140 rad/sec when speed command is 50 rad/sec, 100 rad/sec and 150 rad/sec respectively. Fig. 6(a) shows current response when speed is increased from 100 rad/sec to 150 rad/sec whereas Fig. 6(b) shows current response when speed is reduced to 100 rad/sec from 150 rad/sec. It is found that when speed change command is given current increases and then reduces gradually as motor attains desired speed. Fig. 6(c) and 6(d) show change in frequency of supply voltage when speed command is given. It is observed that when speed command is changed then frequency of stator supply voltage changes quickly to desired value and there is no fluctuation in frequency. Similarly, in case of closed-loop method speed is increased in step. It is found that, when speed command is given, speed increased linearly and after minute transients it settled to desired speed. Fig. 7 shows control desk view of closed-loop method to speed command of 50 rad/sec, 100 rad/sec and 150 rad/sec. Fig. 8(a), 8(b), 8(c) and 8(d) shows current and voltage response when speed is increased from 100 rad/sec to 150 rad/sec and when speed is reduced from 150 rad/sec to 100 rad/sec. It is observed that when speed is changed current increases and after few transients it settles to steady state value when motor reaches desired speed. The sharp increase in frequency of stator voltage is observed when speed is increased. When speed is reduced it is observed that the frequency of supply reduced significantly and then it increases to desired value which results in fast speed response.

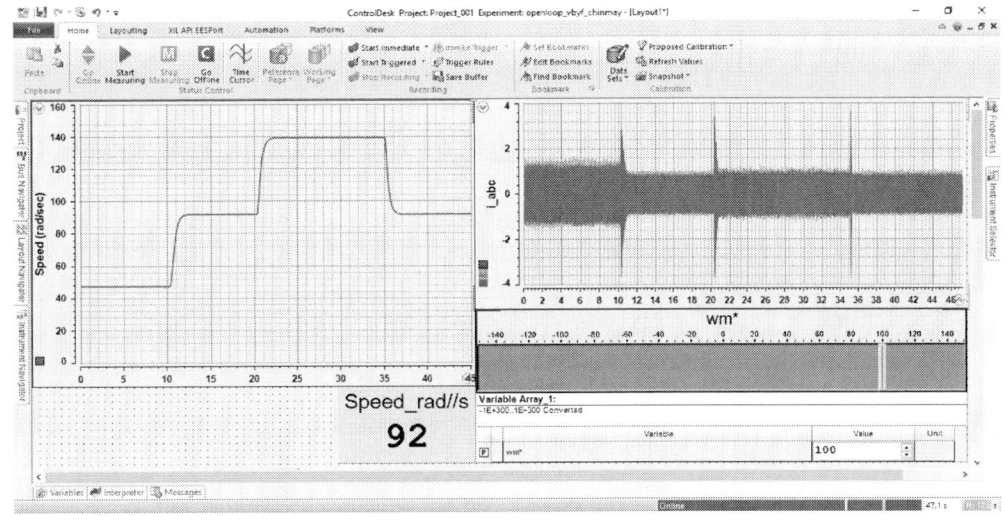

Fig 5.   Control Desk view of Open-loop V/f method for Speed Change Command

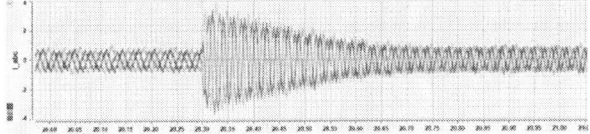

Fig 6.   (a) Current Response of Open-loop V/f When Speed is Increase

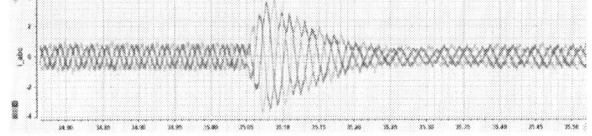

Fig 6.   (b) Current Response of Open-loop V/f When Speed is Reduced

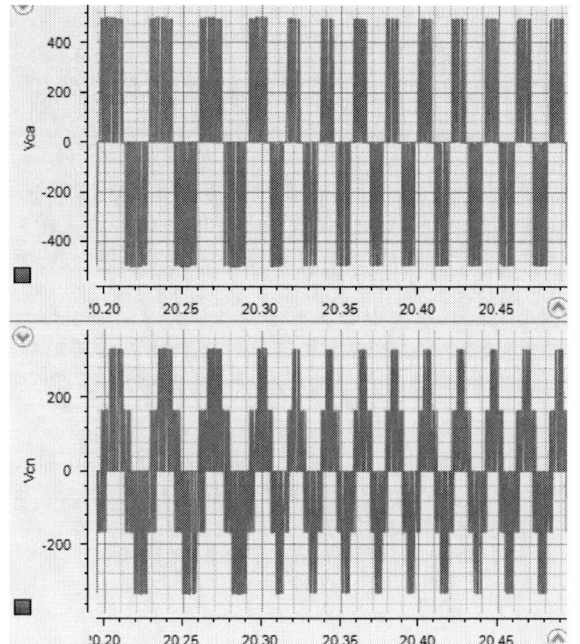

Fig 6.   (c) Voltage Response of Open-loop V/f When Speed is Increased   Fig 6.   (d) Voltage Response of Open-loop V/f When Speed is Reduced

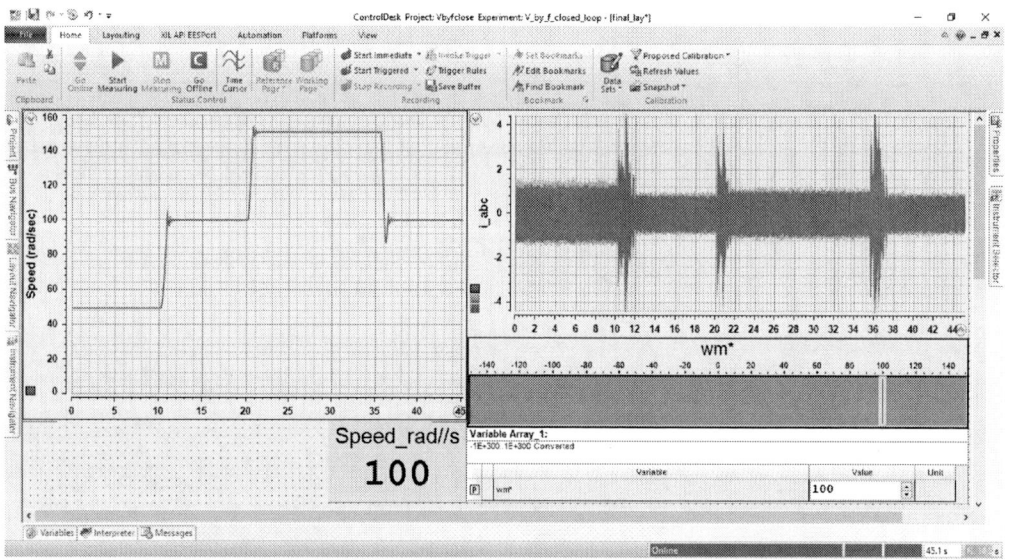

Fig 7.   Control Desk view of Closed-loop V/f Method for Speed Change Command

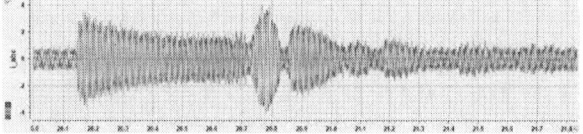

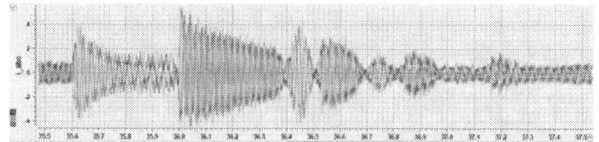

Fig 8.   (a) Current Response of Closed-loop V/f When Speed is Increased   Fig 8.   (b) Current Response of Closed-loop V/f When Speed is Reduced

978-1-5386-6624-1/18 $31.00 © 2018 IEEE   942

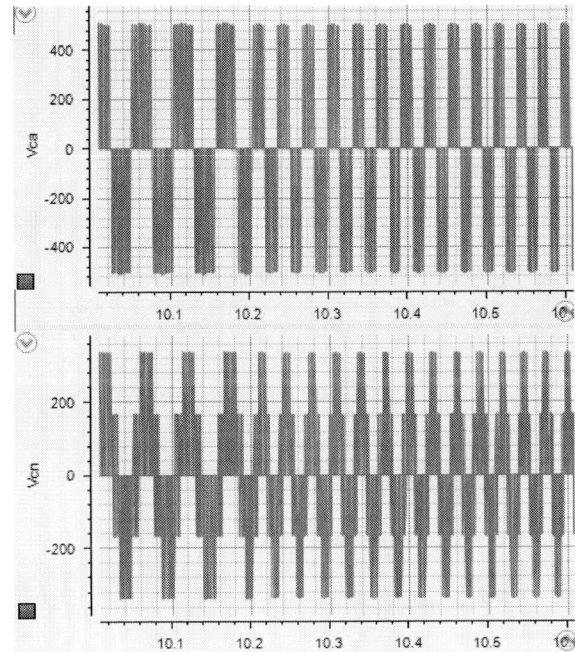

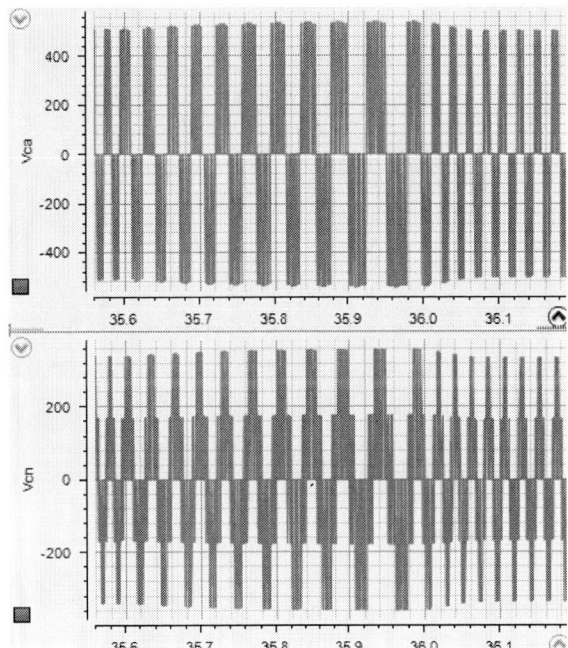

Fig 8.　(c) Voltage Response of Closed-loop V/f When Speed is Increased

Fig 8.　(d) Voltage Response of Closed-loop V/f When Speed is Reduced

V.　CONCLUSION

In this paper, performance analysis of Scalar speed control techniques of IM is performed. It is found that in open-loop and closed-loop method there are lot of oscillations in torque during starting. In open-loop method when load torque is applied there is significant drop in speed whereas in case of closed loop speed is maintained constant at desired speed command. Thus, these methods are suitable where high performance speed control is not required. If simple and less expensive speed control technique is required, then open-loop method is used and if precise speed control is required then closed-loop method is used. The speed of 3Hp, 460V, 50Hz IM is controlled by implementing scalar speed control strategies using dSPACE DS1104. The experimental results prove the effectiveness of both speed control methods.

ACKNOWLEDGEMENT

This work has been carried out in Power Electronics and Drive lab, Electrical department (Delhi Technological University). The authors would like to thank DTU for providing all the required hardware.

REFERENCES

[1]　Martinez-Hernandez M.A., "A Speed Performance Comparative of Field Oriented Control and Scalar Control for Induction Motors," *in 2016 IEEE Conference on Mechatronics, Adaptive and Intelligent Systems (MAIS)*. Hermosillo, Mexico, 20 October 2016. Hermosillo, Mexico: IEEE.

[2]　Hussein Sarhan, "Energy Efficient Control of Three-Phase Induction Motor Drive," Energy and Power Engineering, vol. 3, pp. 107-112, 2011.

[3]　B.K. Bose, *Power Electronics and AC Drives*, Prentice-Hall, New Delhi, 2002.

[4]　J. M. Peña and E. V. Diaz, "Implementation of V/f scalar control for speed regulation of a three-phase induction motor," *in 2016 IEEE ANDESCON*. Arequipa, Peru, 19 October 2016.

[5]　R. K. Bindl and I. Kaur, "Comparative Analysis of Different Controlling Techniques using Direct Torque Control on Induction Motor," *in 2016 2nd International Conference on Next Generation Computing Technologies (NGCT – 2016)*. Dehradun, India, 14 - 16 October 2016.

[6]　H. Akroum, M. Kidouche and A. Aibeche, "Scalar Control of Induction Motor Drives Using dSPACE DS1104," *in International Conference on Systems, Control and Informatics*. Rhodes (Rodos) Island, 16-19 July. 322-327.

[7]　P. K. Behera, M. K. Behera, A. K. Sahoo, "Speed Control of Induction Motor using Scalar Control Technique," *in International Conference on Emerging Trends in Computing and Communication (ETCC – 2014), Journal of Computer Application*, (0975-8887), pp. 37-39.

[8]　The University of Utah (2016) ECE 5671/6671 – Lab 1 dSPACE DS1104 Control Workstation & Simulink Tutorial, Available at: http://www.ece.utah.edu/~bodson/5671/Labs/Lab%201%20-%20Handout.pdf (Accessed: 7th May 2018).

[9]　Cleveland State University (2018) Setting up a real-time digital data acquisition and control interface in dSPACE, Available at: http://academic.csuohio.edu/richter_h/courses/mce484/dspace_guide.pdf (Accessed: 7th May 2018).

[10]　Azad Ghaffari (2012) "dSPACE and Real-Time Interface in Simulink," Availableat: http://flyingv.ucsd.edu/azad/dSPACE_tutorial.pdf (Accessed: 7 May 2018).

Evaluation of Emission and Performance Characteristics of Different Biodiesel Blends with Varying FIP on a Single Cylinder CI Engine

Siddhartha Agarwal
Department of Mechanical Engineering
BML Munjal University
Gurugram, India
13siddharthaagarwal@gmail.com

Gianeshwar Aggarwal
Department of Mechanical Engineering
BML Munjal University
Gurugram, India
www.gianeshwaragg@gmail.com

Archit Gupta
Department of Mechanical Engineering
BML Munjal University
Gurugram, India
archit646@gmail.com

Sumit Roy
Department of Mechanical Engineering
BML Munjal University
Gurugram, India
sumitroy@hotmail.de ;
sumit.roy@bmu.edu.in

Abstract—The present study talks about drawing conclusions by evaluating the trends of emissions and performances characteristics of different biodiesel blends. As an alternative fuel for transport, Biodiesel is used due to easy availability and economic factors. Experiment analysis will be carried out on single cylinder, four stroke diesel engines. For carrying out the experiment palm oil is used as biodiesel and three blends were created one being pure diesel and the rest being proportional mix of diesel, biodiesel and ethanol. Compression ratio of the engine was fixed to 18:1 and the load was fixed to 12kgs with fuel injection pressure (FIP) ranging from 220bars to 260bars. Fuel consumption trends suggested that pure diesel have the least fuel consumption thus having the highest mechanical efficiency.

Keywords— *diesel, ethanol, performance-emission characteristics, dual-fuel mode, emissions.*

I. INTRODUCTION

According to Indian government, by 2030 only we would be able to switch all on-road vehicles to electric mode [1], which still leaves a decade before we finally leave the era of IC engines. Today the global challenge is to reduce the greenhouse gases emissions, and to develop a sustainable environment which could thrive in future. In the United States in 1985, road vehicles were responsible for 70% of CO emissions, 45% of NO_x emissions, and 34% of hydrocarbons [2]. Diversification of transportation fuels, other than conventional non-renewable sources like petrol and diesel, can be achieved by harnessing the power from renewable energy sources. That's why we shifted to biologically generated and renewable energy for transportation called as biofuel.

Biofuel used as an alternative fuel is produced from biomass. For future transportation fuel, biofuels including bioethanol, bio-methanol, biodiesel and biohydrogen are attractive options [3]. The main advantages of biodiesel as a fuel are like they have higher cetane number, higher biodegradability, minimal sulfur, non-toxicity, higher flash point, higher lubricity and aromatic content. Whereas if its disadvantages are considered, it includes their low calorific value, higher viscosity, lower vitality and they are having higher pour point. Also, it is found that their oxidation stability is lower [4]. However, their advantages in reducing the air pollution takes better of when it comes to comparing them with other fuel. Therefore, bio-fuels are being promoted in the transportation sector. Research programs are mainly focusing on alternative fuels, sustainable development due to major concern for the degradation of quality of environment [3].

Bio fuels can be categorized in first generation and second generation. First generation are basically vegetable oils, oil generated from sugar, starch etc. But when used in large amount can cause disruption of stock supply [5]. Vegetable oil such as castor, palm, sunflower oil are all first-generation bio-fuels. However, for this experimental purpose we are using palm oil because carbon dioxide emitted from palm oil in the atmosphere as compared to other biodiesel is way to less and hence prevent global warming [6]. Using palm oil as a major fuel in a long run help in cultivation of palm trees which are one of the major carbon dioxide absorbent trees [6]. Pure biodiesel cannot be used as it is less volumetrically energetic than other fuels and can cause blockage of fuel filters in the engine, as it cleans the entire fuel circuit in the engine. Therefore, we preferred blends of biodiesel for carrying out the experiment. For benchmarking D100 (pure diesel) is also used.

The other possible blend can be diesel + ethanol, which is not possible due poor miscibility of ethanol in diesel. Diesel + ethanol + methanol blends in proportion of 80-5-15 shows a good cetane number. So, following the trends we considered a blend of diesel (70%), biodiesel (20%) and ethanol (10%) [7].

Many studies have been conducted with the different blends of the bio diesel with different oils. The parameters which affects the performance of the engine using different blends has been studied with various emission characteristics and combustion characteristics. In a study the blends with waste cooking oil and diesel is used as a fuel. The performances and variations are checked by varying the compression ratio from 18:1 to 22:1 at 50% load and study is taken out to choose the suitable compression ratio which affects the parameters giving an optimum performance [8].

Using the same, another researcher analyzed that exhaust emissions are decreased by slight change in the fuel consumption which affects the change in NO_x emission [9].

In another study the influence on diesel engine performance is reviewed with the blends of ethanol to check the specific fuel consumption and exhaust emissions (such as CO, CO_2, HC, Smoke and NO_x). Results are compared for CO_2 and NO_x which decreases when compared with injection timings [10].

Here, experiment analysis is carried out on off-road, single-cylinder four-stroke diesel engine, to study the combustion efficiency and emission properties via different blends of biodiesel. The experiment will accompany by varying the injection pressure (220,240,260 bar) keeping the compression ratio constant i.e. 18:1 and load as 12kgf. The trends are plotted between different blends of fuels and injection pressure analyzing the results of different blends.

II. EXPERIMENTAL INVESTIGATION

A. Experimental Setup

For this study, an off-road engine is considered of model Kirloskar Oil Engine, TV1. The tests were conducted on single cylinder four stroke test rig. The complete set up consists of the bed where it is kept for the test, fuel consumption meter and equipments for smoke test and test for exhaust emissions.

The specification for this engine is shown in Table-I.

TABLE I. ENGINE SPECIFICATIONS

Make	Kirloskar Oil Engine, TV1
Compression ratio	18:1
Number of cylinders	Single
Number of Strokes	Four
Combustion principle	Compression ignition
Cooling system	Water cooling
Fuel	Diesel
Speed	1500 rpm
Power	3.50KW
Bore	87.50 mm
Stroke length	110.00 mm
Connecting rod length	234.00mm
Swept volume	661.45 (cc)

To measure the amount of emission constituting opacity and exhaust gas, we use emission measurement system. It consists of the smoke meter to check opacity and exhaust gas analyzer to check emissions.

Table-II shows the specification of AVL 450 di-gas exhaust gas analyzer.

TABLE II. MEASURING RANGES OF EXHAUST GAS ANALYZER

Measured parameter	Specification
Oxygen	0-22% Vol
Carbon Monoxide	0-10%Vol
Carbon dioxide	0-20% Vol
Hydro carbon	0-20000ppm
Nitrogen oxide	0-5000ppm
Engine Speed	400-6000rpm
Lambda	0 to 9.999

Online performance analysis is done using Engine Performance Analysis software package — EnginesoftLV. The software EnginesoftLV serves by entering data and reporting it. The data includes power, efficiencies and fuel consumption. The stored data is converted into tabular and graphical format which can be printed for further analysis

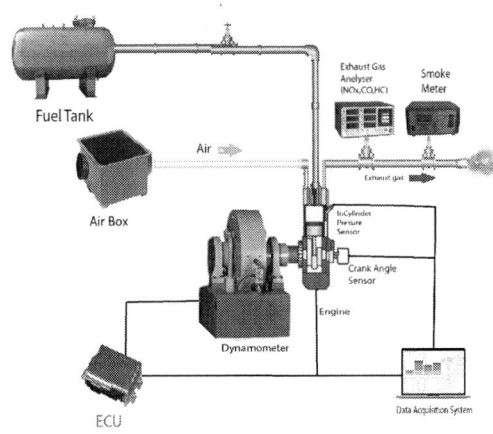

Fig. 1. Schematic Diagram of Experimental Setup

B. Experiment Methodology

Before the starting of the experiment the fuel blends were prepared providing the homogeneous mixture of fuel. The blends prepared are and referred as:

- D100 – 100% diesel

- B20D80 - 20% biodiesel, 80% diesel

- E10B20D70 - 10% Ethanol, 20% biodiesel, 70% diesel

TABLE III. SPECIFICATIONS OF FUELS

FUEL PROPERTIES	DIESEL	BIODIESEL	ETHANOL
Density(kg/m^3)	832	978	785.5
Specific heat (J/ kg. K)	1750	1848	2440
Viscosity (Kg/m. s)	0.00275	0.00410	0.001007
Melting point (k)	304.15	303.35	159
Ignition Temperature (F)	410	600.8	689

978-1-5386-6624-1/18 $31.00 © 2018 IEEE

Enthalpy of vaporization (KJ/Kg)	1651.46	300	918
Saturation Vapor Pressure (kPa)	0.053	0.19 to 2.16	8.773
Calorific Value (kJ/kg)	44,800	23,604.71	29,700

Constituents consisting of Carbon monoxide (CO), Carbon dioxide (CO_2), Oxides of nitrogen (NO_x) and Unburnt Hydrocarbons (HC) are measured by exhaust gas analyzer. Opacity is measured by Smoke meter in terms of Hartrigde Smoke Unit (%) and light absorption coefficient (K expressed).

III. RESULTS AND DISCUSSIONS

This portion deals with the discussion about various trends that these blends show at various loads and fuel injection pressures.

NOx Emissions:

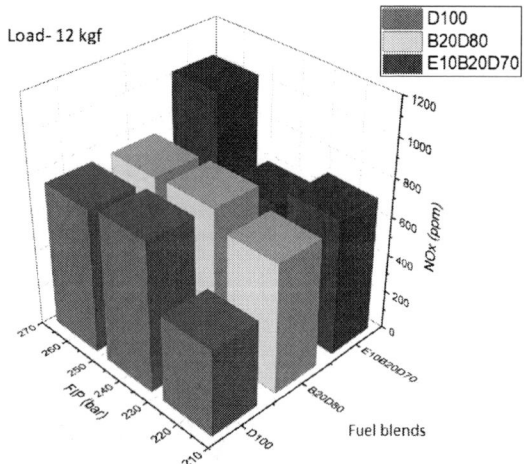

Fig. 2. Effect of FIP and fuel blends on NO_x emissions

For pure diesel (D100) and bio-diesel blend (D80B20) emissions increases first till 240 bar and the gradually decreases and in case of ethanol-biodiesel blend (D70B20E10) it first decreases and then steeply increases.

From the experimentation, it was inferred that at FIP=240bar, ethanol-biodiesel blend shows better performance with 21% of less emissions and hence it is preferred over diesel, but the performance of ethanol-biodiesel blend decreases when we considered 220bar and 260bar FIP.

HC Emissions:

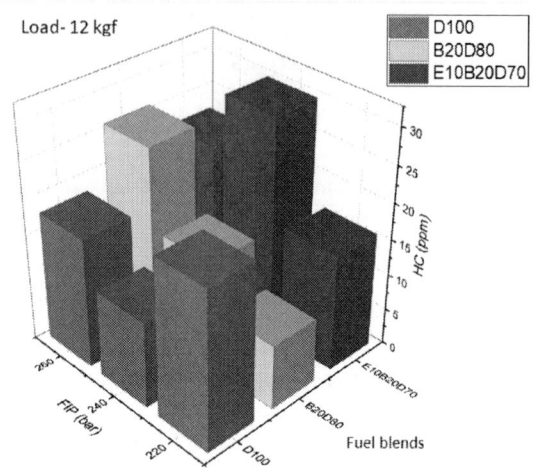

Fig. 3. Effect of FIP and fuel blends on HC emissions

HC emissions for ethanol-biodiesel blend increases till 240bar and gradually decreases at 260bar whereas biodiesel blend steeply increases. Diesel blend decreases at 240bar and then increases till 260bar.

From the experimentation it was found that at 220bar FIP ethanol-biodiesel blend shows the 27% less emissions as compared to diesel fuel whereas at 240bar and 260bar pressure.

CO_2 Emissions:

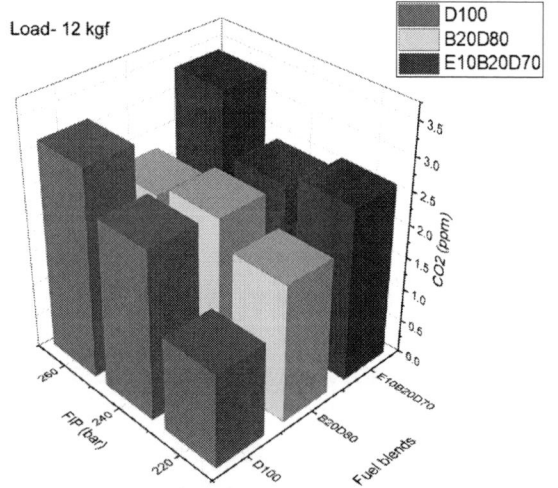

Fig. 4. Effect of FIP and fuel blends on CO_2 emissions

For D100 CO_2 emissions were increasing with increase in FIP. Whereas Biodiesel shows irregular trend with least emission being at 260bar. Ethanol-biodiesel blend also shows irregular trend with least emission at 240bar.

Ethanol-biodiesel blend shows 5% less emissions at 240bar as compared to diesel.

Opacity:

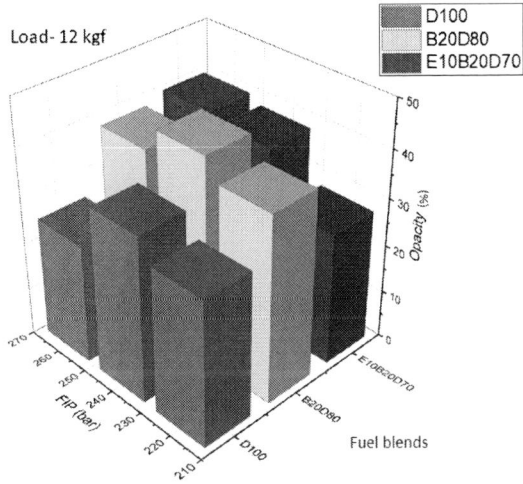

Fig. 5. Effect of FIP and different fuel blends on opacity

Opacity first increases for both pure diesel (D100) and biodiesel blend (D80B20) till 240 bar and then decreases till 260bar. For the case of Ethanol-biodiesel blend opacity was found to increase gradually up to 260bar.

It was observed that when FIP=220bar was taken, ethanol-biodiesel blend shows a better performance than diesel with about 9% less opacity. So, it is more effective than pure diesel at 220 bar FIP.

Brake Specific Fuel Consumption (BSFC):

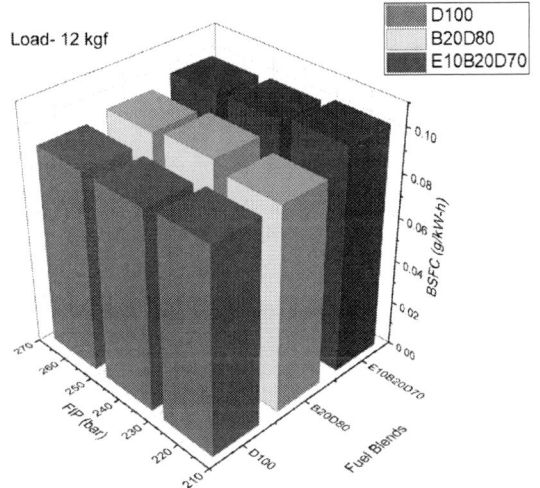

Fig. 6. Effect of FIP and different fuel blends on BSFC

Diesel blend remains almost constant at every FIP. Biodiesel blend first increase till 240 bar and then decreases at 260bar. Well ethanol-biodiesel blend increases slightly by increasing the FIP.

As diesel fuel and ethanol-biodiesel blend shows almost equal results for BSFC. Diesel shows 10% better performance for BSFC, so it is more effective then ethanol-biodiesel blend.

IV. CONCLUSIONS

It was inferred that with increase in injection pressure NO_x emissions decreases to a significant amount for ethanol-biodiesel blend. Similarly, HC emission shows a reducing trend for the same fuel blend. Considering opacity, a decrease was shown by ethanol-biodiesel blend which also shows preference of ethanol-biodiesel fuel over diesel. In case of CO_2 emissions again ethanol-biodiesel blend takes the lead. Overall, ethanol-biodiesel blend (D70B20E10) shows better performance by having lesser emissions and lower fuel consumptions, therefore having better efficiency over both biodiesel blend (D80B20) and pure diesel (D100).

ACKNOWLEDGMENT

We would like to acknowledge our sincere thanks of gratitude to Mr. Gora Banerjee and Ms. Anuradha, who gave us the golden opportunity to carry out analysis of bio-diesel blends on the engine. Secondly, we also acknowledge the constant support of Himanshu Dixit, Tushar Singh Rajput, Jai Mehta, Nitesh Kumar and Shivam Gupta. Lastly, we also acknowledge each other support and mutual understanding throughout the project.

REFERENCES

[1] "Niti Aayog mulls to set electric vehicle target for state public transporters," Economic Times, 8 July 2018.

[2] J. V. Assuncao, "Environmental Effects of toxic Materials from Oil and Gas Combustion", Interactions: Energy/Environment.

[3] M. Balat, H. Balat, "Recent trends in global production and utilization of bio-ethanol fuel", Applied Energy, vol. 86 (11), 2009, Pages 2273-2282.

[4] C. D. Rakopoulos, K. A. Antonopoulos, D. C. Rakopoulos, D. T. Hountalas, E. G. Giakoumis, "Comparative performance and emissions study of a direct injection Diesel engine using blends of Diesel fuel with vegetable oils or bio-diesels of various origins", Energy Conversion and Management, vol. 47, 2006, pp. 3272-3287

[5] R. Banerjee, B. Debbarma, S. Roy, P. Chakraborti, P. K. Bose; "An experimental investigation on the potential of hydrogen−biohol synergy in the performance-emission trade-off paradigm of a diesel engine"; International Journal of Hydrogen Energy; 41 (5), 2016, pp. 3712−3739.

[6] Y. Basiron, "The palm oil advantage in biofuel", New Straits Times, 24 February 2007.

[7] P. Kwanchareona, A. Luengnaruemitchaia, S. Jai-In, "Solubility of a diesel–biodiesel–ethanol blend, its fuel properties, and its emission characteristics from diesel engine", Fuel, vol. 86, 2007, pp. 1053-1061.

[8] K. Muralidharan, D. Vasudevan, "Performance, emission and combustion characteristics of a variable compression ratio engine using methyl esters of waste cooking oil and diesel blends", Applied Energy, vol. 88, 2011, pp. 3959-3968.

[9] S. Kalligeros, F. Zannikos, S. Stournas, E. Lois, G. Anastopoulos, Ch. Teas, F. Sakellaropoulos, "An investigation of using biodiesel/marine diesel blends on the performance of a stationary diesel engine", Biomass and Bioenergy, vol. 24, 2003, pp. 141-149.

[10] C. Sayin, "Engine performance and exhaust gas emissions of methanol and ethanol–diesel blends", Fuel, vol. 89, 2010, pp. 3410-3415.

2nd IEEE International conference on power Electronics, Intelligent Control and Energy systems (ICPEICES-2018)

Design and Simulation of Sensorless Control Algorithms of Brushless DC Motor: A Review

Ashish Dimri
Scientific Officer 'C'
Bhabha Atomic Research Centre, Mysore, India
ashishdimri18@gmail.com

R.D.Kulkarni
Member, IEEE and Scientific Officer 'H'
Bhabha Atomic Research Centre, Mumbai, India
rdk@barc.gov.in

S.R.Gurumurthy Scientific Officer 'H' *Bhabha Atomic Research centre*, Mysore, India srguru@barc.gov.in

J.Nataraj
Scientific Officer 'H'
Bhabha Atomic Research Center, Mysore, India
natarajj@barc.gov.in

Abstract—**The aim of the paper is to review various recently proposed sensor less control algorithms for brushless DC motor. Brushless DC motors have wide applications in industry, consumer appliances and nuclear research due to its, high power density, excellent high speed performance, high efficiency, low maintenance and silent operation. High speed operation of brushless DC motor drive is an utmost critical issue especially by detecting the rotor position without use of any sensor. The brushless DC motors are controlled by using a three-phase inverter which needs a proper commutation sequence to control the switching of inverter. The commutation sequence is obtained, by detecting the rotor position either by sensor or sensor less schematic. In this way rotor position is the salient feature of the brushless DC motor. Based on the rotor position and command signals which may be voltage command, torque command, speed command, etc., the control algorithms generate the gate signal to each semiconductor device in the power electronic converter. Operating principle and working of Back Electro Motive Force (EMF) detection through line voltage difference is demonstrated with the help of Simulink result.**

Keywords—Brushless DC motor; Electromotive force; Digital signal processor; Zero cross detection; Semiconductor devices

I. INTRODUCTION

Brush **L**ess **D**irect **C**urrent (BLDC) motors always been a prime requirement for the applications in field of robotics and automation, electric vehicle, high speed drives and some of the special purposes for long term maintenance free operation [1, 2]. BLDC motor acquire all the operating characteristics of **D**irect **C**urrent (DC) motor by replacing the commutator brush assembly with the electronic commutative switches. These switches gets triggered according to the known position of the rotor. In BLDC motor, electromagnetic armature is stationary whereas **P**ermanent **M**agnet (PM) rotor is rotating. Following the principle of rotating machines, for development of torque, the stator and rotor field should be constant, stationary to each other and there should be an angular displacement between these two fields. Therefore, for rotating the machine, the knowledge of the rotor position is essential because with this knowledge the commutation of the electronic switches can be performed. Position of the rotor can be detected by several ways, however, broadly these are categorized in two methods, (i) Position detection with sensor and (ii) Position detection without sensor or sensor less detection.

Different methods are discussed by the different authors. Position detection through sensors includes Hall sensors, optical encoders, variable reluctance (VR) sensor and accelerometer. The rotor position detection with sensor has certain limitations like low reliability, vulnerable to electromagnetic interference, temperature sensitive, complexity in installation of sensors, high cost of the sensor, etc. [3]. Nowadays sensor less detection scheme has achieved more attention than position detection with sensor because it has overcome the problems emerging by using the sensors. Therefore, the advancement in the sensor less operation of BLDC motor has become a current topic of research in recent years. A comparative study of various algorithms for sensorless control scheme of BLDC motor are discussed in this paper. Basic block diagram showing operation of BLDC motor drive is presented in Fig.1.

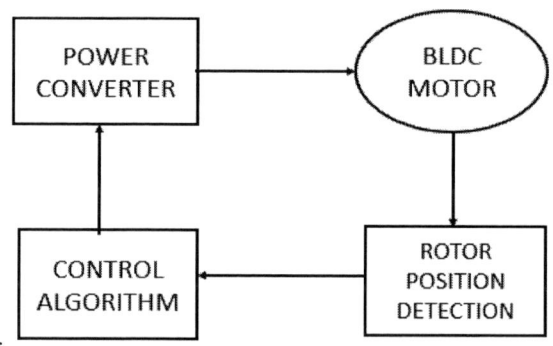

Fig.1 Basic block diagram of BLDC motor drive

The power supply for BLDC motor is fed by three phase inverter. The rotor position is estimated either by sensor or sensor less approach. The detected signal is then interfaced with the microcontroller or **D**igital **S**ignal **P**rocessor (DSP) based controller. By obtaining the knowledge of rotor position, semiconductor switches are commutated sequentially in every 60^0. The BLDC motor is operated in six-step commutation mode and the conduction interval for each phase is 120^0 electrical angle. The phase sequence for commutation is likely to be AB-AC-BC-BA-CA-CB or AB-CB-CA-BA-BC-AC. Each conducting stage is called one step. In this way, at a time only two phases are excited leaving the third phase

978-1-5386-6624-1/18 $31.00 © 2018 IEEE   948

on floating condition. Maximum torque can be achieved in the condition if the motor current is in phase with the back **E**lectro **M**otive **F**orce (EMF) [4, 5]. Thus, the rotor position is the salient features of the BLDC motor. Based on the rotor position and various command signals i.e. torque command, speed command, voltage command and so on, the control algorithms derive the gate triggering signal to each semiconductor devices in the power electronic converter.

II. METHODOLOGY FOR SENSOR LESS POSITION DETECTION

Several papers have been studied and analysed to understand the various methodologies for sensor less rotor position detection of BLDC motor.

A. Third harmonics indirect sensing scheme

The BLDC motor having trapezoidal wave shape of back EMF, contains only fundamental and odd harmonics, due to its odd symmetry. The method to detect rotor flux position using third harmonic is described in [6]. The sum of all three stator phase voltages gives third harmonics and its multiple shows higher order harmonics, however, third harmonic component dominate the sum. High order harmonics can be easily filtered out designing tuned passive filters. This method requires neutral point of stator. Third harmonic rotor flux can be derived by integrating resultant voltage shown in equation (1).

$$\lambda_{\hat{s}} = \int v_{\hat{s}} \cdot dt \tag{1}$$

In order to get maximum torque per ampere, the stator current and rotor flux shall be 90^0 apart. At each 60^0 electrical, the third harmonic component of rotor flux crosses zero. Fig.2 shows the waveforms of (a) Trapezoidal back EMF, (b) Third harmonic voltage, (c) Rotor flux and rotor flux fundamental component and (d) Phase currents of BLDC motor [6].

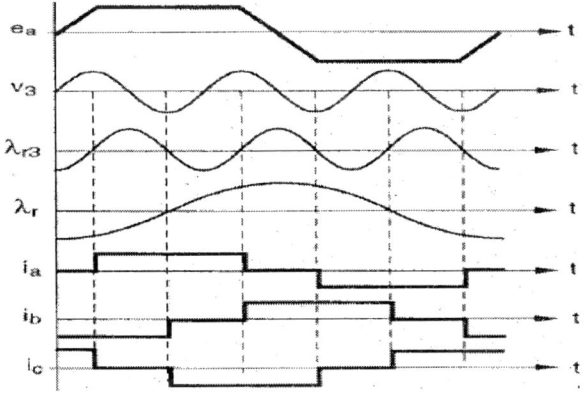

Fig.2 Waveforms of (a) Trapezoidal back EMF, (b) Third harmonic voltage, (c) Rotor flux and rotor flux fundamental component and (d) Phase currents of BLDC motor [6]

The information about the zero crossing of third harmonic signal is taken as reference for detection algorithm where it is synchronized for proper current commutation control. With this control scheme, rotor flux position is estimated and inverter current is controlled such that rotor flux and current shall be in quadrature phase shift. This method is simple to implement, robust and less vulnerable to electromagnetic interference and noise. Although this method has very good control over wide range of speed, however, position sensing using third harmonic signal has some constraints like, it

requires neutral connection of stator phase (three wire method using virtual neutral require complex filtering), stator winding pole pitch shall be greater than 2/3 otherwise third harmonic voltage will not be able to induce in the stator phase winding because there is no linking of third harmonic rotor flux to stator winding. This method also suffers from DC offset error.

B. Position sensing by detecting the conducting interval of freewheeling diode

An indirect method of position sensing is proposed in [7] where the position information is extracted from conducting interval of freewheeling diode connected anti-parallel with the IGBT/MOSFET switches. However, for the maximum torque production, the rectangular shaped current shall be in same phase as the back EMF. Therefore, the inverter gets operated for 120^0 mode of operation i.e. at a time only two devices, IGBT/MOSFET conducts, one from the positive group of first phase and the other from the negative group of second phase, third phase, where no signal is given to any device is known as open phase. Fig.3 depicts a circuit for detection of conduction of freewheeling diode [7].

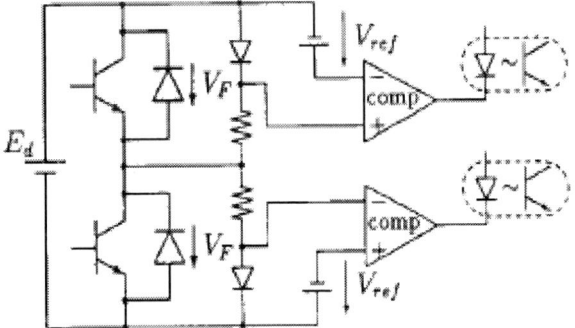

Fig.3 Detection circuit for conduction of freewheeling diode [7]

In the detection circuit for voltage clamping a resistor and diode are connected to the comparator, however, this detection circuit require two isolated power supplies. Consequently, for three phase inverter, it will need six isolated power supplies and six number of comparators. An important point in this scheme is that the detection circuit characteristic does not depends on load because there is no variation of current in open phase. It has been found that the open phase current results from the back EMF. For the better result of algorithm, inverter shall operate in chopping mode.

C. Position sensing using extended kalman filter

A method for rotor position estimation by prediction and correction of state variable and input variable of BLDC motor, which are computed by the **E**xtended **K**alman **F**ilter (EKF) algorithm is described in [8, 9]. This algorithm needs excellent knowledge of motor steady state and dynamic model [10]. Kalman filter is optimal recursive approach to estimate the exact position of rotor by correcting the previous value of the parameter. Basically EKF algorithm has two variants. In first approach, it need constant motor parameters, while in second approach stator resistance is estimated from the motor state variables. The second approach drastically reduces the error estimation in speed. Kalman filter is applied on discrete time based system, therefore all the continuous state space model is deduced to the discrete time based system. This approach is very efficient and accurate because the measurement errors

978-1-5386-6624-1/18 $31.00 © 2018 IEEE   949

and system noise are also included into the error covariance matrix and then calculated from prediction and correction equation. This method cannot be applied at stand still condition of motor, because it estimates infinite error.

D. Position sensing using flux linkage method

An alternate method of rotor position detection from the flux linkage method is described in [11, 12]. Winding flux linkage estimation is possible only if motor terminal voltage and line currents are taken as feedback signal. At each sample from the previous estimated value of position and flux linkage, line current is estimated such that it corrects the new estimated position and new value of flux linkage. Phase voltage equation of balance BLDC motor is given in equation (2) [12].

$$
\begin{bmatrix} v_a \\ v_b \\ v_c \end{bmatrix} = \begin{bmatrix} R & 0 & 0 \\ 0 & R & 0 \\ 0 & 0 & R \end{bmatrix} \begin{bmatrix} i_a \\ i_b \\ i_c \end{bmatrix} + \frac{d}{dt} \begin{bmatrix} \Psi_a \\ \Psi_b \\ \Psi_c \end{bmatrix} \quad (2)
$$

In equation (2), v_a, v_b and v_c are phase voltages, R is stator winding resistance per phase, i_a, i_b and i_c are line currents whereas Ψ_a, Ψ_b and Ψ_c are the flux linkage of the stator winding. Flux linkage of winding is defined in equation (3).

$$
\begin{bmatrix} \Psi_a \\ \Psi_b \\ \Psi_c \end{bmatrix} = \begin{bmatrix} L_{aa} & M_{ab} & M_{ac} \\ M_{ba} & L_{bb} & M_{bc} \\ M_{ca} & M_{cb} & L_{cc} \end{bmatrix} \begin{bmatrix} i_a \\ i_b \\ i_c \end{bmatrix} - \begin{bmatrix} \lambda_m(\theta) \\ \lambda_m(\theta - 2\pi/3) \\ \lambda_m(\theta - 4\pi/3) \end{bmatrix} \quad (3)
$$

Here, λ_m magnetic flux linkage, L_{xx} self-inductance of the winding and M_{xy} mutual inductance of two windings are the function of θ angle. In the star connected motor, the summation of currents is given in equation (4).

$$
i_a + i_b + i_c = 0 \quad (4)
$$

Assuming rotor reluctance is constant and not varying with the angle, then we get as in equation (5) and (6).

$$
L_{aa} = L_{bb} = L_{cc} = L \quad (5)
$$

$$
M_{ab} = M_{ba} = M_{ac} = M_{ca} = M_{bc} = M_{cb} = M \quad (6)
$$

Hence,

$$
\begin{bmatrix} \Psi_a \\ \Psi_b \\ \Psi_c \end{bmatrix} = \begin{bmatrix} L-M & 0 & 0 \\ 0 & L-M & 0 \\ 0 & 0 & L-M \end{bmatrix} \begin{bmatrix} i_a \\ i_b \\ i_c \end{bmatrix} - \begin{bmatrix} \lambda_m(\theta) \\ \lambda_m(\theta - 2\pi/3) \\ \lambda_m(\theta - 4\pi/3) \end{bmatrix} \quad (7)
$$

Substituting the equation (7) in (2) to obtain equation (8) as,

$$
\begin{bmatrix} v_a \\ v_b \\ v_c \end{bmatrix} - \begin{bmatrix} R & 0 & 0 \\ 0 & R & 0 \\ 0 & 0 & R \end{bmatrix} \begin{bmatrix} i_a \\ i_b \\ i_c \end{bmatrix} =
$$
$$
\begin{bmatrix} L-M & 0 & 0 \\ 0 & L-M & 0 \\ 0 & 0 & L-M \end{bmatrix} \frac{d}{dt} \begin{bmatrix} i_a \\ i_b \\ i_c \end{bmatrix} -
$$
$$
\frac{d}{dt} \begin{bmatrix} \lambda_m(\theta) \\ \lambda_m\left(\theta - \frac{2\pi}{3}\right) \\ \lambda_m\left(\theta - \frac{4\pi}{3}\right) \end{bmatrix} \quad (8)
$$

The information about flux linkage can be extracted from the direct measurement of phase voltage and current. From the equations (7) and (8), it can be seen that the change in the time

dependent flux linkage of each phase is obtained in terms of rotor position, line current and motor parameters. It is ability of algorithm to estimates the position continuously, from actual current and voltage, even during the speed transients. Continuous estimation of flux linkage always needed integration process, but this process creates unwanted effects like offset error, which is common problem in implementation of integration. This drift in the integrator limits the accuracy of calculation of both flux position and speed. Low pass filter and some advanced integrator [5] circuit is used to reduce the integrator drift or offset but it limits operating range of speed.

E. Position sensing using fuzzy and neural network

With the knowledge of dynamic model of BLDC motor, an adaptive neural fuzzy algorithm can be implemented [13, 14]. The input parameter, per phase current and phase voltage are passed to an input layer which consists with input membership function. These membership functions are invented on the basis of some control rules, also known as fuzzy rules. Basically these rules are conditional statement. However, there are several algorithms available but among them least mean square is most popular to generate the membership function, and to minimize the error, back propagation algorithm is adopted in backward direction, for the membership function parameter estimation. A basic block diagram of the adaptive neural fuzzy based control system is shown in Fig.4.

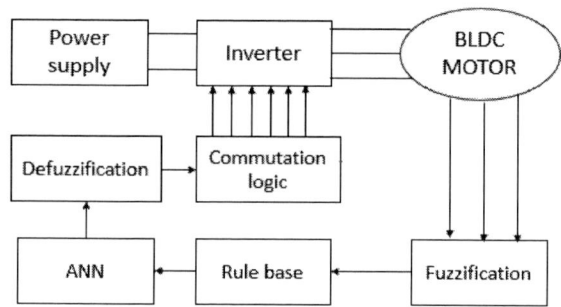

Fig.4 Basic block diagram of adaptive neural fuzzy based control system

In the fuzzification process, measured input or raw data is converted into linguistic variables and they are fed to the rule based, where rules are written on the basis of previous knowledge. The output of this block is given to the **A**rtificial **N**eural **N**etwork (ANN), which is trained from the particular algorithms to select the proper set of rule base. Finally, these variables then defuzzified and fed to commutation logic block.

F. Position sensing using back EMF detection

Position estimation from back EMF is one of the simple and most widely used method amongst all. The zero crossing point of the trapezoidal back EMF wave of floating phase is investigated to get commutation sequence for the switching of the inverter. Back EMF can be sensed either by direct method or by indirect method. In this order to extract the back EMF from the measured motor terminal phase voltage or from the line voltage, been always a challenging task. For the star wound BLDC motor, sensing the back EMF of unexcited phase is simple and most economic method. The zero crossing point of back EMF is detected by comparing this to neutral point voltage and in some cases by comparing to virtual

978-1-5386-6624-1/18 $31.00 © 2018 IEEE

neutral point voltage [15,16,17]. In order to improve motor running performance and to reduce switching noise, different type of **P**ulse **W**idth **M**odulation (PWM) techniques are implemented. These PWM control scheme can be classified on the basis of their method of execution, some of the schemes are unipolar and some are bipolar switching scheme [17-20]. Fig.5 depicts classification of various types of PWM switching scheme.

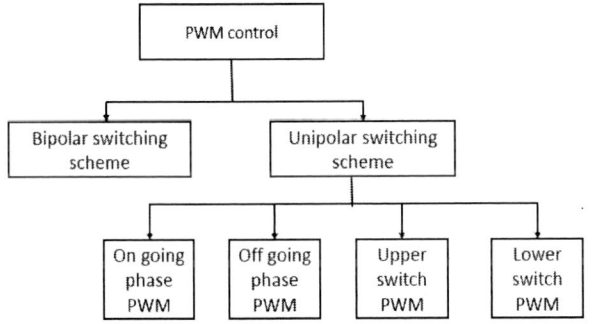

Fig.5 Classification of various types of PWM switching scheme

In [21,22,23] a unique method of zero crossing detection of back EMF from the line voltage difference has been explained. It is well known fact from the motor dynamic model discussed in [3,10] that the back EMF of BLDC motor is proportional to its angular speed. In [24] author explained a precise position detection of rotor from line to line back EMF, which is also extracted from line voltages of the motor. Rotor position detection through difference of line voltages is simulated in MATLAB software. The BLDC motor parameters are shown in table 1 which used for simulation.

TABLE 1 BLDC Motor parameter

Parameters	Values
Input voltage	24 Volts
Per phase resistance	0.7 Ohm
Per phase inductance	1.2 mHenry
Voltage constant	25 V/KRPM
Speed	5000 RPM
Pole	4
Rotor inertia (J)	$4*10^{-6}$ kg-m^2
Damping constant (B)	$7*10^{-5}$ Nm-s
Type of Connection	Star
Power	40 Watt

Ideally back EMF wave is trapezoidal in shape but practically spikes are present in extracted back EMF. These spikes come at each commutation instants and reason of these spike is flow of current through freewheeling diode. In above aforementioned methods, commutation sequence is evaluated from zero crossing point of back EMF. The spikes which present in extracted back EMF give false zero cross detection as in Fig.6.

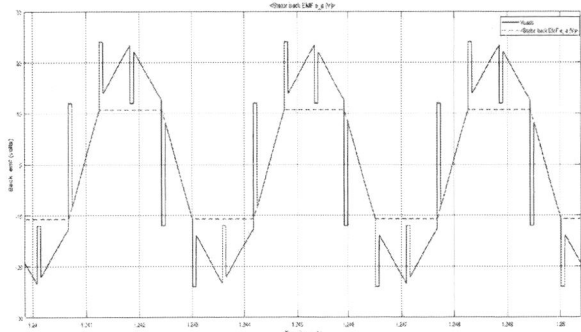

Fig.6 Extracted back EMF and ideal trapezoidal back EMF from Simulink

Many literature and paper presents, methods to eliminate these spikes, in [23] sample and hold circuit is discussed to eliminate false zero crossing. However, [24,25] suggested low pass filter to eliminate these high frequency spikes, but this low pass filter phase shifted the back EMF wave and introduce delay in zero crossing point. The back EMF before passing through **L**ow **P**ass **F**ilter (LPF) and after LPF is seen in Fig.7.

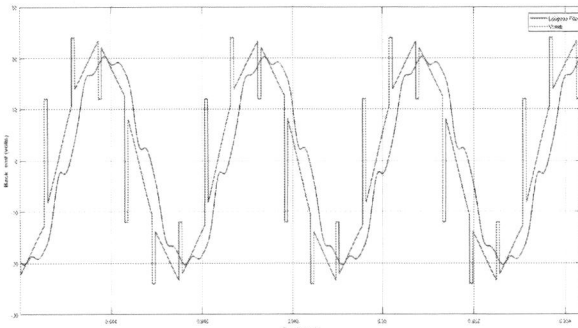

Fig.7 Extracted back EMF waveform before and after passing through LPF

Each device is switched ON for 120^0 and within this interval again it passes through one commutation state where the diode will freewheel the current. Current waveform without mechanical load and with load is shown in Fig.8.

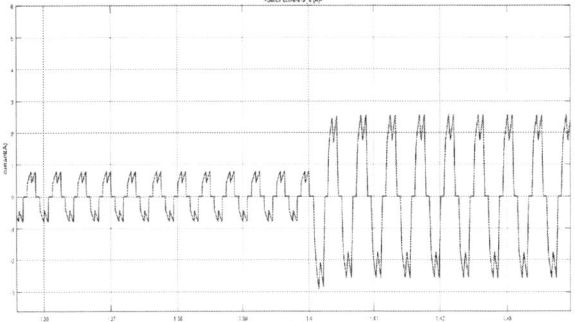

Fig.8 Phase current in BLDC motor during no load to loading from Simulink

Due to the ripples in the current, ripples are also reflected to the speed and torque of BLDC motor. These oscillations in the speed and torque also depend on the rotor inertia and damping constant of the motor. Speed response and torque response during no load to full load condition is shown in the Fig. 9 and Fig. 10 respectively.

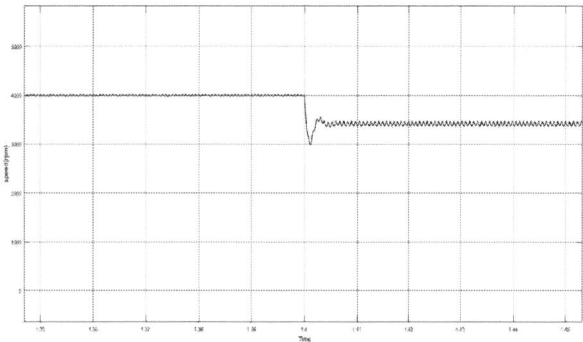

Fig.9 Speed response of BLDC motor during no load to load from Simulink

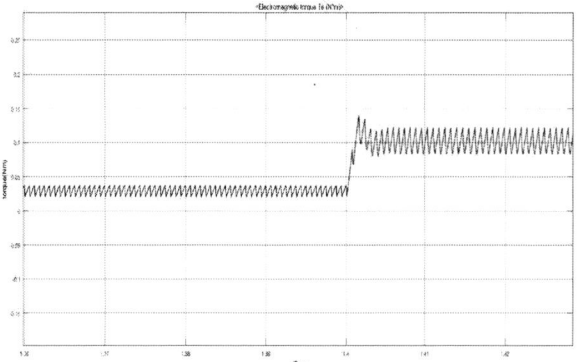

Fig.10 Torque response of BLDC motor from no load to load condition from Simulink

Ripples are introduced from commutation retardation. Now with the implementation of digital filter and phase advancer in DSP, commutation retarding is reduced [26-28]. All extracted data is given to DSP which is very efficient to perform all operation within fraction of second. Speed control of BLDC motor using the PI controller, fuzzy logic controller and hybrid controller i.e. combination of fuzzy and PI controller are discussed in [29]. The results are satisfactory.

III. CONCLUSION

This paper emphasizes the operating principle of BLDC motor, advantages of sensor less control scheme over control through sensors and different methods of sensor less control of BLDC motor. Sensor less control topology is observed to be much superior as compared the circuit with sensor. The Simulink results evaluating back EMF technique is also executed and validated. The detail review of control algorithm is also discussed, compared and presented.

REFERENCES

[1] N. H. Kim, O. Yang and M. H. Kim, "BLDC motor control algorithm for industrial applications using a general purpose processor," J. Power Electronics, Vol. 7, no. 2, pp. 132-139, April 2007.

[2] Adil Usman, Bhakti M. Joshi and Bharat S. Rajpurohit, " A Review of Modelling, Analysis and control methods of Brushless DC motors", IEEE conf. on computation of Power, Energy Information and communication (ICCPEIC), 2016.

[3] R. M. Pindoriya, A. K. Mishra, B. S. Rajpurohit and R. Kumar, "Analysis of Position and Control of Sensorless BLDC Motor using Zero Crossing Back EMF Technique", 1st IEEE International conference on power electronics, Intelligent control and Energy System.

[4] Daune C. Hanselman, "Brushless Permanent Magnet Motor Design," Text Book, ISBN-007026025.

[5] Paul P. Acarnley and John F. Watson, "Review of position sensorless operation of brushless permanent magnet machines," IEEE Transactions on Industrial Electronics, Vol. 53. No.2, April 2006.

[6] Julio C. Moreira, "Indirect sensing for rotor flux position of permanent magnet AC motors operating over a wide speed range," IEEE Trans. on Industry Applications, Vol. 32. No.6, November/December 1996.

[7] Satoshi Ogasawara and Hirofumi Akagi, "An approach to position sensorless drive for brushless dc motors," IEEE Trans. on Industry Applications, Vol. 27. No.5, September/October 1991.

[8] B. Terzic and M. Jadric, "Design and implementation of the Extended Kalman filter for the speed and rotor position estimation of brushless DC motor, " IEEE Trans. on Industrial Electronics, Vol. 48. No.6, Dec 2001.

[9] R. Dhaouadi, N. Mohan and L. Norum, "Design and implementation of an Extended Kalman filter for the state estimation of a permanent magnet synchronous motor," IEEE Trans. on Power Electronics, Vol. 6, pp. 491-497, July 1991.

[10] M. Poovizhi, M. Senthil, P. Ragul, L. Irene Priyadarshini and R. Logambal, "Investigation of mathematical modelling of brushless DC motor (BLDC) drives by using matlab-simulink," IEEE Int. Conf. on Power and Embedded Drive Control (ICPEDC), 2017.

[11] R. Wu and G. R. Slemon, "A permanent magnet motor drive without a shaft sensor," IEEE Trans. on Industrial Application, Vol. 27. No.5, pp. 1005-1011 Sept./Oct. 1991.

[12] Nesimi Ertugrul and Paul Acarnley, "A new algorithm for sensorless operation of permanent magnet motors," IEEE Trans. on Industrial Application, Vol. 30. No.1, January/February 1994.

[13] P. Devendra, Rajetesh G., K. Alice Mary and Ch. Saibabu, "Sensorless control of brushless DC motor using Adaptive Neuro Fuzzy Inference Algorithm, " IEEE Int. conf. on Energy, Automation and signal, 2011.

[14] F. Huang and D. Tian, "A neural network approach to position sensorless control of brushless DC motors," IEEE 22nd Int. Conf. on Industrial Electronics, Control, Instrumentation, Vol. 2 pp. 1167-1171, 1996.

[15] Vinatha U., Swetha Pola and Dr. K. P. Vittal, "Recent Developments in Control Scheme of BLDC Motors," IEEE Int. conf. on Ind. Tech., 2006.

[16] Kenichi Iizuka, Hideo Uzuhashi, Minoru Kano, Tsunehiro Endo and Katsuo Mohri, "Microcomputer control for sensorless brushless motor," IEEE Trans. on Industrial Application, Vol.1A-21, No.4, May/Jun 1985.

[17] Jianwen Shao, Dennis Nolan, Maxime Teisseir and David Swanson, "A Novel Microcontroller based sensorless brushless DC motor drive for automotive fuel pumps," IEEE trans. on industrial application, vol. 39, No.6, pp. Nov/Dec 2003.

[18] Jianwen Shao, Dennis Nolan and Thomas Hopkins, "Improved direct back EMF detection for sensorless BLDC motor drives," IEEE 2003.

[19] Jianwen Shao and Dennis Nolan, "Further improvement of direct back EMF detection for sensorless brushless DC motor drives, " IEEE 2005.

[20] Yen-Shin Lai, Fu-san Shyu and Yung-Hsin Chang, "Novel pulse-width modulation technique with loss reduction for small power brushless dc motor drives," Industry application conf., Vol. 3, pp. 13-18.Oct. 2002.

[21] P. Damodharan and K. Vasudevan, "Indirect back-EMF zero crossing detection for sensorless BLDC motor operation," IEEE PEDS 2005.

[22] P. Damodharan and K. Vasudevan, "Simple position sensorless starting method for BLDC motor," IET Elect. Power Application, pp.49-55,2008.

[23] P. Damodharan and Krishna Vasudevan, "Sensorless brushless DC motor drive based on the zero crossing detection of back electromotive force (EMF) from the line voltage difference," IEEE trans. on Energy Conversion, vol. 25, No. 3, September 2010.

[24] Gang Liu, Chenjun Cui, Kun Wang, Bangcheng Han and Shiqiang Zheng, "Sensorless control for high speed BLDC motor based on the line to line Back emf, " IEEE Trans. on Power Electronics, vol. 31, No. 7, pp. July 2017.

[25] C. H. Chen and M. Y. Cheng, "A new cost effective sensorless commutation method for brushless DC motors without phase shift circuit and neutral voltage, " IEEE Trans. on Power Electronics, vol. 22, No. 2, pp. 647-650 March 2007.

[26] Kuang-Yao Cheng and Ying-Yu Tzou, "Design of a sensorless commutation IC for BLDC motors," IEEE Trans. on Power Electronics, vol. 18, No. 6, Nov. 2003.

[27] Haitao Li, Shiqiang Zheng and Hongliang Ren, "Self correction of commutation point for high speed sensorless bldc motor with low inductance and nonideal back EMF," IEEE Trans. on Power Electronics, vol. 32, No. 1, Jan. 2017.

[28] K.S.Rama Rao, Nagadevan and Soib Taib, "sensorless control of a BLDC motor with back emf detection method using DSPIC," 2nd IEEE Int. conf. on power and energy (PECon 08), Dec. 2008.

[29] Pranoti K. Khanke and Sangeeta D. Jain, "Comparative analysis of speed control of bldc motor using PI, simple FLC and Fuzzy–PI controller," IEEE Int. conf. on Energy System and Application, 2015.

Bifurcation Curves for Electrical DC Motors

Sudarshan K Valluru, Madhusudan Singh, Aditya Verma, Anshul Gupta
Incubation Center for Control, Dynamical Systems and Computation, Dept. of Electrical Engineering Engineering
Delhi Technological University
Delhi-110042,India

Abstract—**This paper studies the numerical bifurcation analysis of three different types of DC Motor namely DC Series Motor, DC Shunt Motor, and permanent magnet DC (PMDC) motor. The bifurcation curves' role is to exemplify the bifurcation behavior in the motors while changing one of the parameters keeping other parameters constant for bifurcation curves. The variable parameter is along the abscissa, while the machine states are along the ordinate of the bifurcation curve. The mathematical models of DC Series Motor, DC Shunt Motor and PMDC Motor are first derived, which are suitable for bifurcation curves. Then, the effect of the input voltage and constant external torque on the motors are studied, and bifurcation curves are plotted with input voltage as a control parameter.**

Keywords—Bifurcation curves, DC series motor, DC shunt motor, Permanent magnet DC motor

I. INTRODUCTION

Bifurcation diagrams illustrate the behavior of the system when the system control parameter is changed. The system states are observed at periodically varying values of the parameter. The bifurcation diagrams are commonly used for the representation of system dynamics and states[1].

DC motor is any device from a variety of rotary devices that converts direct current electrical energy to mechanical energy. DC series motor, DC shunt motor and Permanent Magnet DC motor have been studied in this brief. While studying the dynamic characteristics of different types of DC motors it needs mathematical models of motors which are multivariable, interlinked and non-linear. To begin with, the mathematical models of the motors are derived followed by formulation of steady-state characteristics of the motor while considering the effects of the input voltage and external torque. Determination of the parameters of the motor for different cases is discussed. The bifurcation curves are analyzed through simulation studies. This paper is divided into five sections. Section I contains the introduction; Section II illustrates the mathematical modeling of DC Motors; Section III presents the bifurcation analysis of the DC motors. Results and discussion are given in Section IV, and the paper is concluded in Section V.

II. MATHEMATICAL MODELLING OF DC MOTORS

Mathematical modeling of DC motor systems is vital for analysis and controller design. The mathematical models DC motors are modeled by combining electrical and mechanical elements. It will be useful to predict the future behavior of the system at given conditions. In this section, the dynamical modeling of familiar DC motors are explained through differential equations.

A. DC Series Motor

The differential equations (1) and (2) represent DC Series Motor[2] and its parameters[3] indicated in Table-I.

$$i' = \frac{V}{L} - \frac{Ri}{L} - \frac{M_{af}\omega i}{L} \tag{1}$$

$$\omega' = \frac{M_{af}i^2}{J} - \frac{B\omega}{J} - \frac{\tau_L}{J} \tag{2}$$

TABLE I.   DC SERIES MOTOR NAMEPLATE DETAILS

Rated Power	1.5kW
Maximum Terminal Voltage (V_{max})	200 V
Minimum Terminal Voltage (V)	40V
Rated Armature current	5A
Maximum speed (ω)	5000 rev/min
Rated speed (ω_{max})	600 rev/min
Armature resistance (R)	7.2 Ω
Armature and field inductance(L)	0.0917 H
Mutual Inductance (M_{af})	0.027 H
Moment of Inertia (J)	0.0007046 Kg.m^2
Viscous torque of rotational losses(B)	0.0004 Nm/rpm
Torque constant(A)	0.22 Nm

B. DC Shunt Motor

The differential equations (3),(4) and (5) indicate the dynamical model of a DC shunt motor[4] and its parameters[5] are given in Table-II

$$i'_f = \frac{\left(V_m - R_f i_f\right)}{L_f} \tag{3}$$

$$i'_a = \frac{\left(V_m - R_a i_a - M_{af} i_f \omega\right)}{L_a} \tag{4}$$

$$\omega' = \frac{\left(M_{af} i_f i_a - T_L - B\omega\right)}{J} \tag{5}$$

TABLE II.   DC SHUNT MOTOR NAME PLATE DETAILS

Terminal voltage (V)	120 V
Armature current (I_a)	9.2 A
Shaft speed (n, ω)	1500rpm,157.1rad/sec
Electromagnetic torque (T_m)	6.22 Nm
Armature resistance (R_a)	1.5 Ω
Field resistance (R_f)	100 Ω
Armature self-inductance(L_a)	0.02 H
Field self-inductance(L_f)	0.06 H
Mutual inductance between armature and field(M_{af})	0.518 H
Moment of Inertia(J)	0.02365 Kg.m^2
Viscous torque of rotational losses(B)	0.00025 Nm/rpm
Load torque(T_L)	0.3+0.00039 $\omega^{1.8}$

978-1-5386-6624-1/18 $31.00 © 2018 IEEE

C. Permanent Magnet DC motor

The differential equations (6) and (7) show the dynamical model of a Permanent Magnet DC Motor[6] and its parameters[7] given in Table-III.

$$i'_a = -\frac{R_a}{L_a}i_a - \frac{K_v}{L_a}\omega + \frac{V_a}{L_a} \qquad (6)$$

$$\omega' = \frac{K_v}{J}i_a - \frac{B}{J}\omega - \frac{\tau_L}{J} \qquad (7)$$

TABLE III.   PMDC MOTOR PARAMETERS

Armature Circuit inductance (L_a)	0.1H
Armature Circuit resistance (R_a)	2 Ω
Viscous damping torque constant (B)	0.01 Nm/rad/s
Motor torque constant ($k_v = k_i$)	0.3 Nm/rad/s
Load torque (τ_L)	0.012 Nm
Moment of Inertia(J)	0.1 Kg.m^2
Maximum applied armature voltage(V_a)	110V
Motor Speed(ω)	2800rev/min

III. BIFURCATION ANALYSIS OF DC MOTOR SYSTEMS

Beginning with the determination of the fixed points (equilibrium or critical points) of the differential equations of different DC motors, their stability and characteristics are checked using the linearized models. The determinant, trace, and eigenvalues of the linear matrix define the stability and characteristics of the fixed points. About the bifurcation point the number of fixed points or the stability of the fixed point changes.

A. DC Series Motor

- Fixed Points

The equations (8) and (9) are used to determine fixed points of DC series motor.

At a fixed point $i'_f = 0, i'_a = 0, \omega' = 0$.

$$M_{af}i\omega + Ri - V = 0 \qquad (8)$$

$$M_{af}i^2 - T_l - A - B\omega = 0 \qquad (9)$$

Fixed point is the solution of the equations of (10) and (11)

$$M_{af}^2 i^3 - (T_l M_{af} + AM_{af} - BR)i - BV = 0 \qquad (10)$$

$$\omega = \frac{V - Ri}{M_{af}i} \qquad (11)$$

- Local Stability of Fixed point

The local stability of fixed points is checked using the Linearized model represented in equation (12).

$$\begin{bmatrix} i' \\ \omega' \end{bmatrix} = \begin{bmatrix} \dfrac{-M_{af}\omega - R}{L} & \dfrac{-M_{af}i}{L} \\ \dfrac{2M_{af}i}{J} & \dfrac{-B}{J} \end{bmatrix} \begin{bmatrix} i \\ \omega \end{bmatrix} \qquad (12)$$

The determinant of the equation (12) is given as equation (13)

$$(\Delta) = \frac{(M_{af}\omega + R)B}{LJ} + \frac{2M_{af}^2 i^2}{LJ} \qquad (13)$$

Therefore, for, i.e., $\omega < -0.1067 - 0.054 i_a^2$ the fixed point is a saddle. The stability is determined using the trace given as equation (14)

$$\text{Trace}(\tau) = -\frac{J(M_{af}\omega + R) + BL}{LJ} \qquad (14)$$

The eigen values of the linearized matrix as given by equation (15) are used to further study the characteristic of a fixed point.

$$\lambda_{1,2} = \frac{\tau \pm \sqrt{\tau^2 - 4\Delta}}{2} \qquad (15)$$

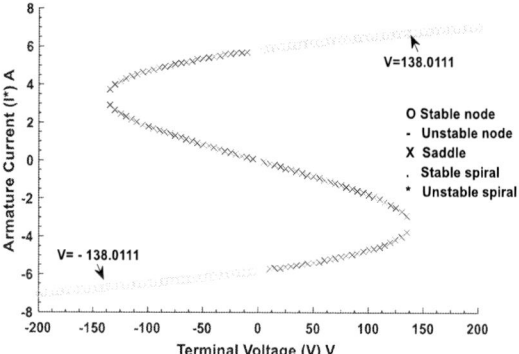

Fig. 1.   Bifurcation curve of DC Series motor

Case I: (-138.0111 < V < 138.0111)
There are three fixed points, one being a stable node ($\Delta > 0$, $\tau < 0$ and corresponding $\lambda_{1,2}$ are real) and the other two being saddle points ($\Delta < 0$) shown in Fig.1.

Case II: (V >= 138.0111 or V <= -138.0111)
There is only one stable node and also shown in Fig.1.

B. DC Shunt Motor

- Fixed Points

The equations (16), (17) and (18) are used to determine fixed points of DC series motor.

At a fixed point $i'_f = 0, i'_a = 0, \omega' = 0$.

$$i_f = \frac{V_m}{R_f} \qquad (16)$$

$$V_m - R_a i_a - M_{af} i_f \omega = 0 \qquad (17)$$

$$M_{af} i_f i_a - T_L - B\omega = 0 \qquad (18)$$

Fixed point is the solution of the equations (19) and (20).

$$M_{af} i_a V_m - (0.3 + 0.00039\omega^{1.8})R_f - BR_f\omega = 0 \qquad (19)$$

$$i_a = \frac{V_m R_f - M_{af} V_m \omega}{R_a R_f} \qquad (20)$$

- Local Stability of Fixed points

The local stability of fixed points is checked using the Linearized model represented in equation (21).

978-1-5386-6624-1/18 $31.00 © 2018 IEEE

$$\begin{bmatrix} i'_f \\ i'_a \\ \omega' \end{bmatrix} = \begin{bmatrix} \dfrac{-R_f}{L_f} & 0 & 0 \\ \dfrac{-M_{af}\omega}{L_a} & \dfrac{-R_a}{L_a} & \dfrac{-M_{af}i_f}{L_a} \\ \dfrac{M_{af}i_a}{J} & \dfrac{M_{af}i_f}{J} & \dfrac{-B}{J} \end{bmatrix} \begin{bmatrix} i_f \\ i_a \\ \omega \end{bmatrix} \qquad (21)$$

Hence $i'_f = \dfrac{-R_f}{L_f}i_f$

This suggests that $i_f(t)$ will have an exponentially decaying behavior.

As i_f is independent of i_a and ω in the linearized model represented by equation (21), it can be ruled out, reducing the matrix to equation (22).

$$\begin{bmatrix} i'_a \\ \omega' \end{bmatrix} = \begin{bmatrix} \dfrac{-R_a}{L_a} & \dfrac{-M_{af}i_f}{L_a} \\ \dfrac{M_{af}i_f}{J} & \dfrac{-B}{J} \end{bmatrix} \begin{bmatrix} i_a \\ \omega \end{bmatrix} \qquad (22)$$

Substituting $i_f^* = \dfrac{V_m}{R_f}$ in equation (22) to check the local stability and behavior of fixed points

$$\begin{bmatrix} i'_a \\ \omega' \end{bmatrix} = \begin{bmatrix} \dfrac{-R_a}{L_a} & \dfrac{-M_{af}V_m}{L_a R_f} \\ \dfrac{M_{af}V_m}{L_a R_f} & \dfrac{-B}{J} \end{bmatrix} \begin{bmatrix} i_a \\ \omega \end{bmatrix} \qquad (23)$$

The determinant of the equation (23) is given as equation (24)

$$\Delta = \frac{R_a B}{L_a J} + \frac{M_{af}^2 V_m^2}{L_a^2 R_f^2} \qquad (24)$$

$\Delta > 0 \, \forall V_m$ eliminating the possibility of saddle nodes. The stability is determined using the trace given as equation (25)

$$\text{Trace}\,(\tau) = -\frac{R_a}{L_a} - \frac{B}{J} \qquad (25)$$

$\tau < 0 \, \forall V_m$ eliminating the possibility of unstable nodes. The eigen values of the linearized matrix as given by equation (26) are used to further study the characteristic of a fixed point.

$$\lambda_{1,2} = \frac{\tau \pm \sqrt{\tau^2 - 4\Delta}}{2} \qquad (26)$$

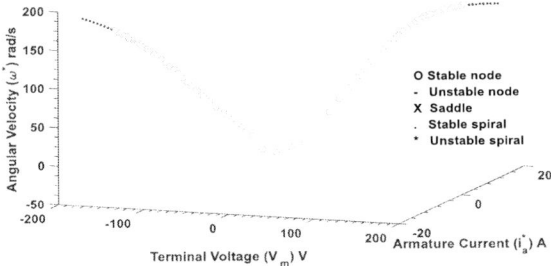

Fig. 2.   Bifurcation curve for DC shunt motor

The fixed point of the equation is locally stable $\forall V_m \in R$, and the point changes from node ($\lambda_{1,2}$ are real) to spiral ($\lambda_{1,2}$ are complex) about $V_m = \pm 157.4239$ as shown in Fig.2.

C. Permanent Magnet DC motor
  • *Fixed Points*

At fixed point $i'_a = 0, \omega' = 0$.

The equations (27) represents the fixed points of the PMDC motor.

Solving for fixed point $P(i_a, \omega)$:

$$P\left(\frac{BV_a + K_v \tau_L}{R_a B + K_v^2}, \frac{K_v V_a - \tau_L R_a}{R_a B + K_v^2}\right) \qquad (27)$$

  • *Local Stability of Fixed points*

The local stability of fixed points is checked using the Linearized model represented in equation (28).

$$\begin{bmatrix} i'_a \\ \omega' \end{bmatrix} = \begin{bmatrix} \dfrac{-R_a}{L_a} & \dfrac{-K_v}{L_a} \\ \dfrac{K_v}{J} & \dfrac{-B}{J} \end{bmatrix} \begin{bmatrix} i_a \\ \omega \end{bmatrix} \qquad (28)$$

The determinant of the equation (28) is given as equation (29)

$$\text{Determinant}\,(\Delta) = \frac{R_a B + K_v^2}{L_a J} \qquad (29)$$

$\Delta > 0 \, \forall V_m$ eliminating the possibility of saddle nodes. The stability is determined using the trace given as equation (30).

$$\text{Trace}\,(\tau) = -\frac{R_a}{L_a} - \frac{B}{J} \qquad (30)$$

$\tau < 0 \, \forall V_m$ eliminating the possibility of unstable nodes. The eigen values of the linearized matrix as given by equation (31) are used to further study the characteristic of a fixed point.

$$\lambda_{1,2} = \frac{\tau \pm \sqrt{\tau^2 - 4\Delta}}{2} \qquad (31)$$

$\tau^2 - 4\Delta > 0 \, \forall V_m$ eliminating the possibility of the spiral.

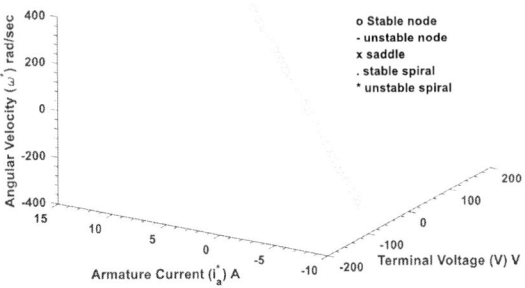

Fig. 3.   Bifurcation curve for PMDC motor

$\lambda_{1,2}$ are real and greater than zero $\forall V_m$ and hence the fixed point is locally stable $\forall V_m$ as shown in Fig.3.

978-1-5386-6624-1/18 $31.00 © 2018 IEEE   955

IV. RESULTS AND DISCUSSION

Numerical simulation is performed in MATLAB environment, and operating characteristics for DC series motor, DC shunt motor and permanent magnet DC motor have been plotted.

A. DC Series Motor

The characteristics of the DC series motor for operating voltage of V=138.0111 have been presented in Fig. 4.

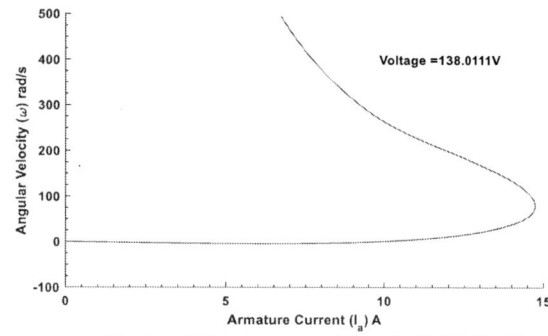

Fig. 4.   DC Series motor trajectory for V=138.0111V

The Fig.4 depicts that the DC series motor is stable when operated at terminal voltage V=138.0111 V.

B. DC Shunt Motor

- Without External Resistance R_e

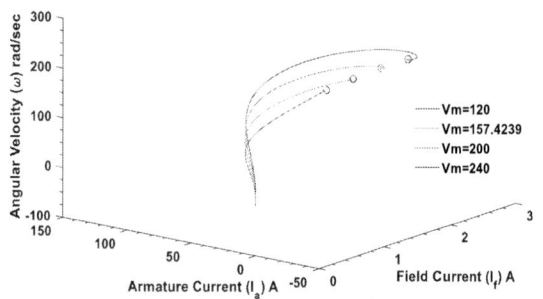

Fig. 5.   DC Shunt motor trajectory for various voltages as it changes from node to spiral about the fixed point

The trajectories for DC motor for different values of V have been plotted in Fig. 5. It is observed from Fig 5 that the armature current has a peak overshoot above the rated current values. Also, a spiral is observed for operating voltages greater than 157.4239.

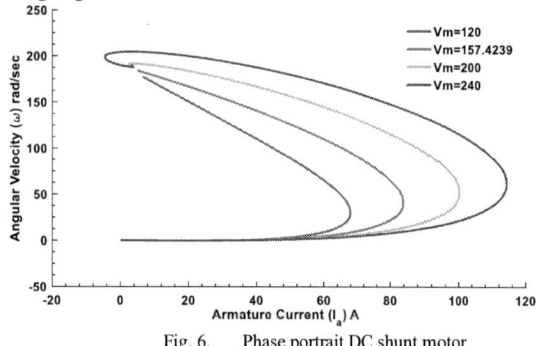

Fig. 6.   Phase portrait DC shunt motor

The phase portrait of a DC shunt motor has been shown in Fig. 6.
As expected the trajectory changes from node to spiral about the fixed point $V_m = 157.4239$ and overshoot are observed in the current in the beginning due to properties of dc shunt motor.

- With external resistance R_e

To reduce overshoot of armature current external resistance R_e is added to the circuit with R_a.

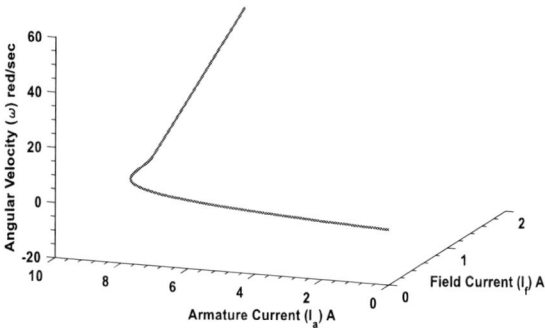

Fig. 7.   The trajectory of DC shunt motor after adding external resistance to reduce initial overshoot in armature current

It can be observed in Fig. 7 that the issue of overshoot in armature current has been resolved by adding the external resistance of appropriate value.

C. Permanent Magnet DC Motor

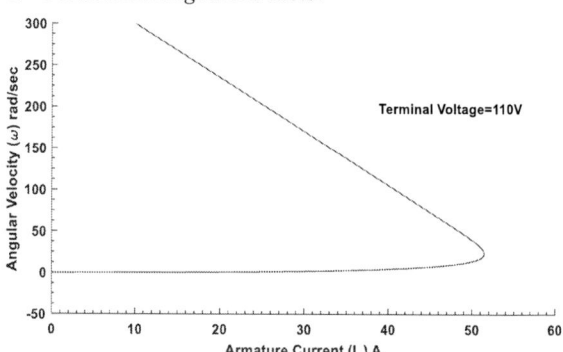

Fig. 8.   Trajectory for PMDC motor at V=110V

The Fig.8 depicts that the DC series motor is stable when operated at terminal voltage V=110 V.

V. CONCLUSIONS

The numerical bifurcation analysis of three different types of DC Motor namely DC Series Motor, DC Shunt Motor, and permanent magnet DC (PMDC) motor has been conducted and the bifurcation curves are plotted with terminal voltage V as a control parameter that represents the type of behavior and local stability of the fixed point which are in order with the corresponding operating characteristics.

978-1-5386-6624-1/18 $31.00 © 2018 IEEE   956

REFERENCES

[1] Rudiger Seydel, *Practical Bifurcation and Stability Analysis*, 3rd ed. NewYork, USA: Springer, 2010.

[2] S. Mehta and J. Chiasson, "Nonlinear Control of a Series DC Motor: Theory and Experiment," *IEEE Trans. Ind. Electron.*, vol. 45, no. 1, pp. 134–141, 1998.

[3] SudarsahnK.Valluru, Madhusudan Singh, and N. Kumar, "Implementation of NARMA-L2 neuro controller for speed regulation of series connected DC motor," in *India International Conference on Power Electronics*, 2012, pp. 1–7.

[4] J. Chiasson and M. Bodson, "Nonlinear Control of a Shunt DC Motor," *IEEE Trans. Automat. Contr.*, vol. 38, no. 11, pp. 1662–1666, 1993.

[5] Sudarshan K. Valluru Madhusudan Singh, "Trajectory Control of DC Shunt Motor by NARMA Level-2 Neuro Controller," in *1st IEEE International Conference on Power Electronics. Intelligent Control and Energy Systems*, 2016, pp. 1–6.

[6] M. A. Obeidat, L. Y. Wang, and F. Lin, "Real-time parameter estimation of PMDC motors using quantized sensors," *IEEE Trans. Veh. Technol.*, vol. 62, no. 7, pp. 2977–2986, 2013.

[7] T. K. Nizami, A. Chakravarty, and C. Mahanta, "Design and implementation of a neuro-adaptive backstepping controller for buck converter fed PMDC-motor," *Control Eng. Pract.*, vol. 58, no. March 2016, pp. 78–87, 2017.

Performance Analysis of FSO Link under Diferent Conditions of Fog in Delhi, India

Aditya Kesarwani
Department of Information Technology
Amity School of Engineering and
Technology, Amity University Uttar
Pradesh
Noida, India
kesarwanikarankumar1998@gmail.com

Anuranjana
Department of Information Technology
Amity School of Engineering and
Technology, Amity University Uttar
Pradesh
Noida, India
aranjana@amity.edu

Sanmukh Kaur
Department of Electronics and
Communication Engineering
Amity School of Engineering and
Technology, Amity University Uttar
Pradesh
Noida, India
skaur2@amity.edu

Manpreet Kaur
Department of Information Technology
Amity School of Engineering and
Technology, Amity University Uttar
Pradesh
Noida, India
preetmoney96@gmail.com

Pawan Singh Vohra
Department of Information Technology
Amity School of Engineering and
Technology, Amity University Uttar
Pradesh
Noida, India
psinghvohra@gmail.com

Abstract— **Free space optical communication (FSO) is one of the most effective technology having advantages like free license spectrum, high data rate, swift ability to expand and easy deploy-ability. In FSO communication atmosphere is used as a channel, therefore atmospheric turbulence leads to the degradation in performance of the communication system. In this paper, one of the atmospheric turbulence condition that is atmospheric fog attenuation has been analyzed to study its effect on the performance of an FSO link. This paper also presents a comprehensive survey on attenuation due to fog conditions in Delhi region. The atmospheric fog attenuation used for simulation in the system is calculated by using Kim's model for two specific months i.e. January and November. Q-factor has been analyzed by varying the transmission power and range of the FSO system. The simulation results show that for error free transmission of data during the fog months, the transmitting power between 25-35dBm is recommended to be used for the optimal transmission range from 6-8Km.**

Keywords— *FSO, Atmospheric Fog Attenuation, Q-factor, Transmission Range.*

I. INTRODUCTION

The foundation of IT Industry is Internet and its services, so need for high-speed internet is increasing day by day, which requires drastic increment in the bandwidth and capacity. In generally used radio frequency (RF) technology, due to increasing data and internet supply, congestion occurs that raises a need to switch from RF technology to free space optical (FSO) technology [1]. In RF, there is restriction in spectrum usage whereas no such restriction exists in case of FSO and hence higher bandwidth is obtained for heavy applications. In contrast with RF technology FSO is efficient in saving the power usage up to 50%. FSO technology is also more secure and encrypted because of its less probability of signal intercepts and anti-jam characteristics [2].

Free Space optical communication is an emerging wireless technology which treats atmosphere as channel for the communication. It uses light propagation to wirelessly transmit the data under certain ranges and most significantly at the places where physical setup is not easier to install and use due to higher cost or some other considerations. FSO

provides Line-Of-Sight communication, which works with both infrared as well as visible spectrum. The fundamental principle of FSO system is almost same as that of optical fiber transmission, in which the data signal transmits through guided channel whereas in FSO, unguided channel is used. Optical transceivers are added at each ends for providing a two-way communication [1, 2]. FSO is the technology which offers many advantages such as high data rate, less power usage, license free spectrum and quick deploy-ability [3].

FSO communication provides many benefits as compared to RF communication. The most important difference between the FSO and RF technology is a larger gap between their wavelengths. In FSO communication, in normal atmospheric condition the wavelength lies between 700 to 1600 nm whereas in case of RF technology it ranges from 30mm to 3m. The wavelengths used in the RF technology are almost thousands of time greater than FSO technology [2].

The features of FSO communication discussed above make it efficient for many applications. Some of the applications are Telecommunication and computer networks, Point-To-Point LOS links, Security applications, Military Application, Storage Area Networks (SANs), etc. [4].

In this work, one of the atmospheric turbulence condition that is atmospheric fog attenuation has been discussed and analyzed to study its effect on the performance of an FSO link. A comprehensive survey on attenuation due to fog conditions has been carried out for Delhi region. The simulation results show that during the fog months, the transmitting power between 25-35dBm is recommended to be used for error free transmission.

II. SYSTEM LAYOUT OF PROPOSED FSO MODEL

The FSO System consists of the following components i.e. transmitter, receiver and channel as shown in Fig. 1. The transmitter component of the FSO system has the work of converting the data into the optical signal, which is then propagated to the receiver. The receiver works for recovering the transmitting data from the received optical signal. The FSO system uses the atmosphere as a channel although the transmission can also take place in vacuum as light waves can travel in vacuum too [5].

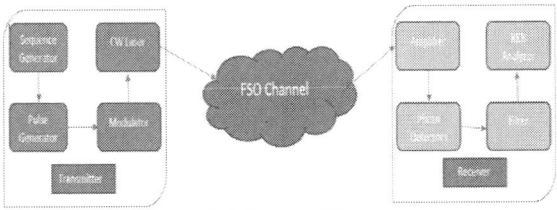

Fig. 1 Block diagram of FSO system

There are two types of parameters on which the performance of the FSO communication depends. They are listed and discussed as: External parameters are those parameters, which are totally depended on the atmosphere such as atmospheric attenuation, which can be due to fog, rain, clouds etc. Internal parameters are those parameters which is all related to the system properties such as power loss, beam divergence etc. [5].

A model for FSO system has been designed by using different system components [Fig. 2]. FSO model is also designed by using a software Optisystem to evaluate the Q-factor by varying the transmission power and range. In the proposed FSO layout, atmosphere has been used as FSO channel and PIN Photodiode has been used to convert optical

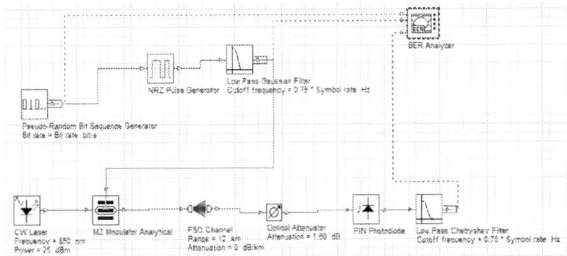

Fig. 2 Proposed layout for FSO system simulation with fog effect

signal into an electrical signal. Optical attenuator has been used to add attenuation in the channel due to fog. Continuous wave (CW) Laser is used to produce a continuous wave optical signal. NRZ pulse generator produces a non-return to zero coded signal. Pseudo-random bit sequence generator produces a binary sequence according to various operation modes. MZ modulator generates an optical non-return to zero coded signal, which is feed to the FSO channel for transmission. BER analyzer is a visualizer used to calculate and display the bit error rate (BER) and Q-factor of the signal at the receiver end [10, 11, 12].

III. DEGRADATION IN PERFORMANCE OF FSO SYSTEM DUE TO FOG IN THE ATMOSPHERE

FSO communication comes with lots of merits but it has some challenges too. As FSO uses atmosphere as a channel, the channel comes with some factors which degrade the quality and performance of the communication.

FSO communication treats atmosphere as a channel for transmitting the data or signals whose properties and characteristics depends on time and area. It makes the FSO technology dependent on the climatic conditions and geographical locations.

The main contributor in the atmospheric attenuation is fog because of its contribution in scattering as well as absorption. When the visibility is less than 50m, the attenuation rises up to more than 350 dB/km. Such situation demands powerful lasers that includes special procedure to enhance communication. Generally, lasers with wavelengths of 1550 nm are preferred for the dense fog conditions due to their high transmitting power.

Beer-Lambert law has been used to calculate the attenuation due to fog according to which the following relation is considered [3]:

$$\alpha_{fog} = \frac{3.912}{V(km)} \left(\frac{\lambda}{550} \right)^{-q} \tag{1}$$

Where V is the visibility in km, λ is the wavelength of signal and q is the size distribution coefficient of scattering. Calculation of the value of q is given by Kim and Kruse model which is given in table 1. For calculating the atmospheric fog attenuation in this paper KIM's model have been used.

Table 1. Kim and Kruse models

Kim model	Kruse model
1.6 for v > 50km	1.6 for V>50km
1.3 for 6km<v<1km	1.3 for 6km<V<1km
0.16v+0.34 for 1km<v<1km	$0.585V^{1/3}$ for V<0.5km
v-0.5 for 0.5km<v<1km	

IV. RESULTS

The FSO system has been designed in such a way to optimize the system parameters to improve overall performance in the regions where atmospheric fog attenuation reduces the effectiveness of the FSO system. The fog data for Delhi region has been referred, which describes visibility in km for each day of every month of a year [6]. From that data, average visibility in the months of January and November for the last six years has been calculated as shown in Table 2. The corresponding attenuation values have also been shown in the table.

Table 2: Visibility and Attenuation due to fog in Delhi region [6].

Years	January		November	
	Visibility (KM)	Attenuation (dB\KM)	Visibility (KM)	Attenuation (dB\KM)
2017	0.96	3.33	1.70	1.30
2016	0.87	3.82	1.06	2.09
2015	1.09	2.03	1.36	1.63
2014	0.90	3.64	1.63	1.36
2013	1.48	1.49	1.60	1.38
2012	1.77	1.25	1.43	1.55

The performance of FSO system is measured by analyzing the Q-factor and received signal power at different ranges. Optisystem 15 has been used for simulating and optimizing the results. The wavelength and bit rate used for analysis have been fixed at 850nm and 10 GB per second respectively.

978-1-5386-6624-1/18 $31.00 © 2018 IEEE

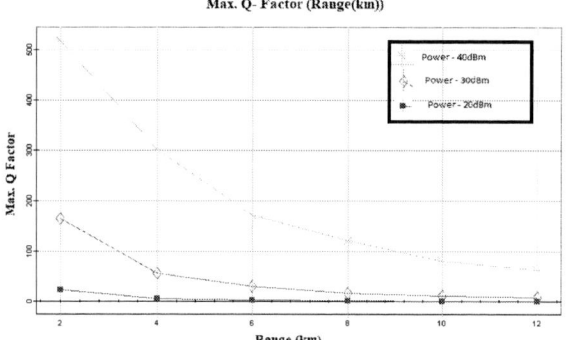

Fig. 3: Q-factor vs transmission range under different transmission power of 20, 30 and 40dBm, January.

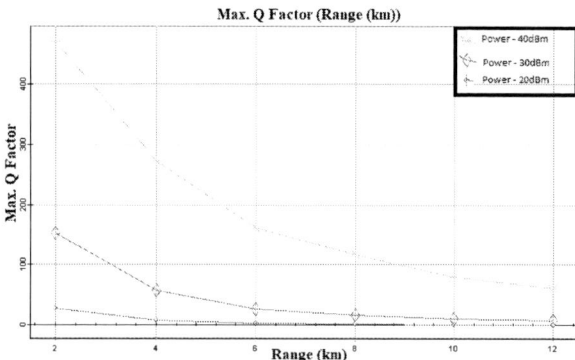

Fig. 4: Q-factor vs transmission range under different transmission power of 20, 30 and 40dBm, November.

For the calculated value of average attenuation [Table 2], Q-factor is analyzed with respect to transmission range for different values of input power for the months of January and November. The transmission range has been varied from zero to 12 km in the graphs shown in Fig. 3 and Fig. 4 respectively. As shown in the graphs, the Q-factor is decreasing with increase in the transmission range. It can be observed that at in the month of January, with the average attenuation of 2.59 dB\km, we should use the power of 30dBm or above to transmit the data in the range of 8 km for the acceptable value of bit error rate at the receiver side. Similarly, in the month of November, with the average attenuation of 1.55 dB\km, we should use the power of 25dBm or above to transmit the data in the range of 8 km for the acceptable value of bit error rate at the receiver side.

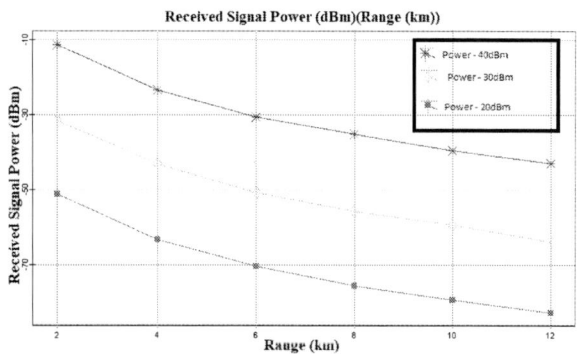

Fig. 5: Received Signal power vs transmission range under different transmission power of 20, 30 and 40dBm, January.

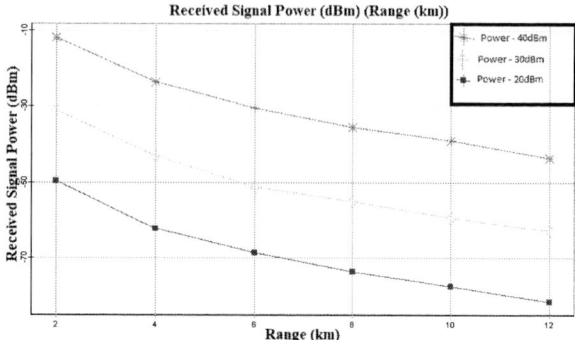

Fig. 6: Received Signal power vs transmission range under different transmission power of 20, 30 and 40dBm, November.

For the calculated value of average attenuation [Table 2], received signal power is measured with respect to transmission range for different values of input power for the months of January and November. The transmission range has been varied from zero to 12 km in the graphs shown in Fig. 5 and Fig. 6 respectively. As shown in the graphs, the received signal power is decreasing with increase in the transmission range.

It can be observed that at in the month of January, with the average attenuation of 2.59 dB\km, we should use the power in the range of 25-35dBm or above to transmit the data up to a distance of 8 km for the acceptable value of the signal power at the receiver. Likewise, for the same range of operation, data should be transmitted with the power above 25dBm for acceptable value of the received signal power in the month of November.

The simulation results show that for error free transmission of data during the fog months, the transmitting power between 25-35dBm is recommended to be used for the optimal transmission range from 6-8Km.

V. CONCLUSION

In this paper, we present the performance analysis of FSO link under different conditions of fog in Delhi region. Kim's model has been used for the calculation of atmospheric fog attenuation for two specific months i.e. January and November.

From the results we conclude that for error free transmission of data during the fog months, the transmitting power between 25-35dBm is recommended to be used for the optimal transmission range from 6-8Km.

REFERENCE

[1] Florence Upendo Rashid, Senzota Kivaria Semakuwa, "Performance Analysis of Free Space Optical Communication Under the Effect of Rain in Arusha Region, Tanzania", Vol .3 Issue 9, September 2014.

[2] Hemani Kaushal, Georges Kaddoum, "Free Space Optical Communication: Challenges and Mitigations Techniques", 16 June 2015.

[3] Sushank Chaudhary, Angela Amphawan, "The Role and Challenges of Free-space Optical Systems" Journal of Optical Communications 35, no. 4, July 24 2014.

[4] Disha Srivastava and Gurjit Kaur, "Optimization of FSO System Parameters Under Varying Meteorological Conditions in Delhi" School of Information and Communication Technology, Gautam Buddha University, Greater Noida 201308, Uttar Pradesh, India, Volume 9 No. 1, 2017.

978-1-5386-6624-1/18 $31.00 © 2018 IEEE

[5] Ghassemlooy Z., Popoola W.O., "Terrestrial Free-Space Optical Communications", Optical Communications Research Group, NCRLab, Northumbria University, Newcastle upon Tyne, United Kingdom, 01 January 2010.

[6] Support of Fog Data: "The Weather Channel Media Solutions Group", https://www.wunderground.com/history/airport/VIDP/2014/1/5/Daily History.html.

[7] Aditi Malik, Preeti Singh, "Free Space Optics: Current Applications and Future Challenges", 2015.

[8] Akhil Gupta, Pankaj Anand, Rohit Khajuria, Sonam Bhagat and Rakesh Kumar Jha, "A Survey of Free Space Optical Communication Network Channel over Optical Fiber Cable Communication", Jammu and Kashmir, India, International Journal of Computer Applications (0975-8887), Volume 105-No. 10, November 2014.

[9] M. Sauer, K. Andrey and G. Jacob, "Radio over fiber for picocellular network architectures", Journal of Lightwave Technology, November 25, 2007: 3301-3320.

[10] S.Bloom, E.Korevaar. J.Schuster, H.Willebrand, "Understanding the performance of free-space optics", Volume 2 June 2003.

[11] A. Polishuk, S.Arnon, "Optimization of a laser satellite communication system with an optical preamplifier", Volume 21 July 2004.

[12] S.Arnon, "Performance of a laser satellite network with an optical preamplifier", Volume 22 April 2005.

[13] Anas Chaaban, Jean-Marie Morvan and Mohamed-Slim Alouini, "Free-Space Optical Communications: Capacity Bounds, Approximations, and a New Sphere-Packing Perspective", Volume 64 March 2016.

Feature Expanded and Weight Selective Model to Classify the Heart Disease Patients

Kapil Juneja
Department of Computer Science and Engineering,
University Institute of Engineering and Technology,
Maharshi Dayanand University,
Rohtak, Haryana,124001, India.
kapil.juneja81@gmail.com,
kapil.juneja.1981@ieee.org

Chhavi Rana
Department of Computer Science and Engineering,
University Institute of Engineering and Technology,
Maharshi Dayanand University,
Rohtak, Haryana,124001, India.
chhavi1jan@yahoo.com

Abstract—In day-to-day life, the stress and work load have increased the chances of heart disease in different age groups. The basic physical and heath associated information for an individual is able to isolate the expected heart patients. The detailed examination and treatment can be recommended for these patients to avoid the heart stroke situations. In this paper, a fuzzy weighted model is presented to identify the heart patients by evaluating the basic heart influencing features. The associated weights are generated on the paired features to expand the feature set. Only the higher weighted and associated features are selected to generate more appropriate and controlled results. The weighted fuzzy features are trained by decision tree and Bayesian networks to identify the significant improvement in existing classifiers. The proposed model is evaluated under 10-fold method and identified a significant gain in accuracy and FMeasure for both decision tree and Bayesian network classifiers.

Keywords—Heart Disease, Classification, Weighted, Fuzzy, Decision Tree, Bayesian Network.

I. INTRODUCTION

In the recent year, lot of advancement is achieved in the health care system to provide the instant and accurate solutions to the people. Various online heath care systems, interfaces and protocols were designed by the researchers to acquire the health information of patients and to answer their health queries. These existing health care systems are generalized as well as specific to the particular health disease. The capabilities of these health care systems can be measured as the query handling capability, robustness, accuracy and processing features. These smart systems reduce the cost and efforts to obtain the health related solutions in real time. These systems are also integrated with artificial intelligence to compile the health care system under some disease specific rules to predict the occurrence of disease in people. The rules were characterized within the system under expert observation and by setting up the relative constraints. The system must be capable to predict the disease without the involvement or the availability of a physician.

The quality of these automated health processing and disease prediction [1] systems depends on the number on the number of rules framed to cover different aspects of diseases. The number of case studies included in rule formulation decides the reliability and integrity of composed system. The data mining methods, artificial intelligence and learning rules were defined by the researchers to process the heath characteristics and disease symptoms. These kind of intelligent data processing systems are capable to perform heavy computation to provide the reliable and accurate results in real time. The association mining methods and feature weights generation methods were used by the researchers to setup the relation between the available features and to improve the relevancy and quality in heart disease prediction. The history based data pattern processing and the composition of symptoms, history features and predictive decisions were used individually and collectively for taking more significant decisions. Various decision driven, probabilistic and predictive classifiers are available to utilize these features. In this research work decision tree and Bayesian network classifiers are used individually on fuzzy weighted datasets to improve the heart disease prediction.

II. RELATED WORK

The heart disease is the most common problem identified even in healthy individuals. The heart disease prediction can be done by performing various tests and information at different levels. The health information based data processing is also used in various online healthcare systems to predict the health status of an individual. Researchers have provided various feature processors and classifiers to predict the existence of heart disease in patients. Sultana et al. [2] has provided the analytical study of various data mining methods including J48, KStar, SMO, Bayesian network and multilayer perceptron for the prediction of heart disease. The observations identified that SMO and Bayesian network provided more effective results. The attribute level evaluation under different classifiers [3] was proposed to observe the symptoms of heart disease. Author identified the attribute dependency as the measure to improve the prediction of heart disease. Another work on associated feature evaluation based prediction rule formation was proposed by Lakshmi et al. [4]. The attribute based rule formation and pruning was defined to strengthen the confidence rules. Later, the dynamic tree was applied to classify the patients of heart disease. A symptom level derivation with class classification was observed by Gandhi et al. [5] to predict the class of heart disease. Author analyzed the features under supervised and unsupervised learning methods to improve the accuracy of disease prediction. A prioritization [6] method was suggested to discover the rules to predict the risk of heart disease. The relative features were processed under decision tree to improve the accuracy of heart disease classification.

Various supervised learning and optimization methods were investigated by the researchers to improve the accuracy of heart disease prediction. Bharti et al. [7] has combined the feature based mapping through different supervised learning

978-1-5386-6624-1/18 $31.00 © 2018 IEEE

methods and optimized them using some evolutionary and swarm based methods. Author also provided the work on feature selection and optimization method to improve the prediction rate. The vector quantization [8] based learning method was applied on data patterns to improve the accuracy of heart disease prediction. The author has provided a controlled method with different configurations to improve the accuracy of disease prediction. The disease symptoms were regulated using self organizing map [9]and applied with multi-layer perceptron to predict the heart disease. Jabbar et al. [10] has used the lazy association between symptoms to generate the rules for heart disease prediction. The global and component selective search method is defined for effective feature selection to improve the structure of controlled classifier. The association rule and probabilistic measures based methods were explored by Sivagowry et al. [11]. Sundar et al. [12] defined a system with multiple classifiers to answer the complex queries for effective recognition of heart disease. The likelihood analysis based performance measure was defined to classify the patients.

Pandy et al. [13] has designed a health care system to extend the database and to generate more effective symptoms based on disorder feature analysis. The rules were defined to build the relationship for effective mapping of symptoms to the disorder exist in the health system. Jabbar et al. [14] has also provided the association mining based rule discovery system to strengthen the health care system for heart disease prediction. An attribute set based processing under association rule mining was provided by Srinivas et al. [15] to improve the prediction rate of heart disease. Krishniah et al. [16] used the SVM (Support vector Machine) as a qualifier to classify the heart disease data.

III. RESEARCH METHODOLOGY

In today's social and professional scenarios and the living and eating habits are affecting the health of an individual. These environmental, behavioral and eating constraints can be reflected directly at the heart problem in a patient. At the earlier stage, the heart problems can be diagnosed by observing the symptoms and reading of cholesterol level, heart rate, etc. In the online health care system, the automated mechanism is established to generate the rules to decide the chances of occurrence of heart disease. The proposed model has trained and analyzed the health features of the expected patients at two levels to generate these rules as shown in figure 1. At the first level, the features are collected and assigned the weight based on the significance of the features. At this level, the info-gain measure is applied to identify the contribution of these features. During this weight generation stage, the feature pruning is also done to identify the most contributing features. The description of the info-gain method is provided in sub-section 3.1. After obtaining the significant individual features with relative weight vectors, the second level of features are formed using the feature expansion method. In this stage, the feature-pairing is done to identify the collective contribution to predict the occurrence of disease. At this level, the association based analysis is performed for two and three features in composite form and generate the relative contribution to generate the decision on disease prediction. The contribution vector is also analyzed under the info-gain measure to validate the significance of the composite associated feature. The most contributing associated-feature pairs are selected from the pool based on this weight vector.

These associated-weighted features are combined with weighted-individual featureset to generate the expanded larger featureset. The weight selected expanded feature set is passed through fuzzy rules in the final stage to generate the decision for the prediction of heart disease. The fuzzy rules transform the numerical quantitative values to the nominal values so that more clear decisions are derived from this expanded featureset. The process of transition of the available featureset to fuzzy-weighted expanded featureset is provided in algorithm 1.

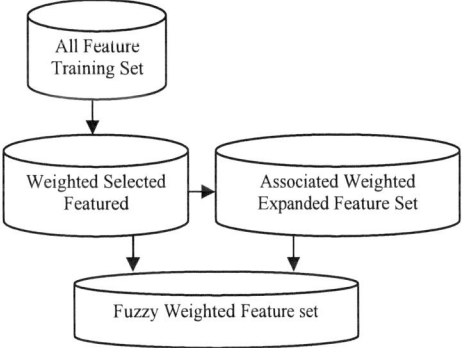

Figure 1 : Proposed Fuzzy Extended Feature Formation Model

Figure 1 has provided the fuzzy-rule transition model applied on the available training set. This rule adaptive featureset is now considered as the trainingset and processed through decision tree and Bayesian network classifiers. These individual classification models are further verified on the testingset using 10-fold method. The comparative results obtained from the each classification model are provided in section 4.

Algorithm 1 : Expanded Fuzzy-Weighted Featurset Generation Model

```
FeatureExpansionModel(Featureset)
/*Featureset is available collected dataset with relative disease class
specification*/
Begin
    1.  Weights=InfoGain(Featureset)
        /*Generate the weights on featurset using InfoGain
        method*/
    2.  Featureset =Pruning(Featureset,Weights, TH)
        /*Identify the most contributing featureset by observing the
        feature significance*/
    3.  A2Featurset=AssociationFeatures(Featureset,2)
        /*Generate the associated pairing of two features from
        pruned featurset*/
    4.  A3Featurset=AssociationFeatures(Featureset,3)
        /*Generate the associated pairing of three features from
        pruned featurset*/
    5.  A2Featurset          A3Featurset]=InfoGain(A2Featurset,
        A3Featurset)
        /*Apply the weights on the expanded features*/
    6.  [A2Featurset          A3Featurset]=Pruning(A2Featurset,
        A3Featurset, TH)
        /*Prune the expanded featurset*/
    7.  ExFeatureset=Featurset U A2Featureset U A3Featurset
        /*Generate the expanded Featureset */
    8.  FExFeaturset=FuzzyRules(ExFeatureset)
        /*Apply the rules on expanded featurset to transform to
        nominal featureset*/
    9.  Return FExFeaturset
End
```

Algorithm 1 has provided the detailed procedure employed in this research to expand the featureset as well as

978-1-5386-6624-1/18 $31.00 © 2018 IEEE

to apply the fuzzy rules on the quantified features. This procedure is applied on the training set to expand the featureset and to generate the rules. Same featureset and rules are also mapped on the testing set to normalize the processing set. The info-gain is using the weight measure to quantify the contribution of each participating feature. Once the weight vectors are generated, the complete classification method model is analyzed for different threshold values (TH) to identify the most effective features. By applying the subsequent observations on different threshold values verified against Bayesian network and decision tree classifiers, .1 is considered as the most significant threshold value. This pruned and selective featureset is expanded under the contribution of paired and group features. This contribution is evaluated under the associatively measure and its mapping to the particular disease class. The group of two and three features is done at this stage and labeled these groups as separate features. These expanded features are also analyzed using the info-gain measure to identify the contributing feature-combinations. Finally, a wide featureset is obtained by combining the weighted features, A2Featureset (2-Feature Associated) and A3 (3-Feature Associated) featureset. The fuzzy rules are applied in the final expansion stage to transform the quantitative values to nominal range values. In section 3.2, the fuzzy rule formulation method used in this model is described. This rule framed feature expansion and transformation method is applied on both training and testing sets. The decision tree and Bayesian network classifiers are applied individually on these transformed featureset to generate the prediction model. The description of the datasets and the classification results are provided in section 4.

A. Info-Gain Weighted Measure

The attribute selection and weight identification process is able to identify the most significant and contributing features from the available pool. In the decision tree and other rule based methods, the weights of the attributes help to decide the order or rule integration in the machine learning model. The information gain is one such measure used in this research work to identify the significant base attributes and associated weight attributes. In this measure, the expected contribution of an attribute to map to the object label can be obtained using equation (1). Let m is the number of object classes and C is the set of c data samples. The information gain measure respective to m class is given as

$$I(c1, c2 \ldots cm) = -\sum_{i=1}^{m} \frac{c_i}{c} log_2 \left(\frac{c_i}{c} \right) \qquad (1)$$

Here, $\frac{c_i}{c}$ represents the probability of a sample data to map to class C_i and m is the number of classes

The Info-gain estimate the entropy value for each attributes relative to the available classes and generate the composite weight to identify the correctness of the mapping. The info-gain in this research is used for selection base attributes as well as for processing attributes after inclusion of compound attributes. The information-gain values of attributes and relative decision cutoff value is shown in figure 2.

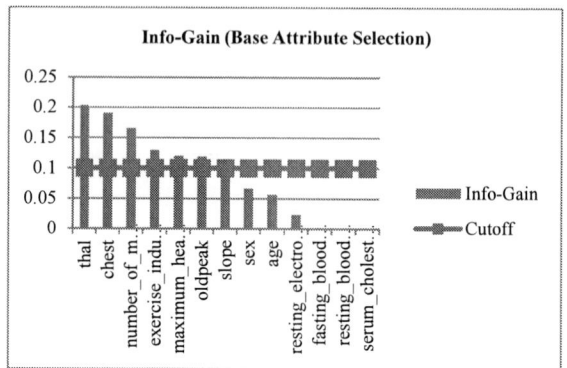

Figure 2 : Info-Gain based Selection of Base Attributes *(Statlog (Heart) Dataset)*

The info-gain is applied in this research at two levels. At first level, the dataset attributes are evaluated to identify the most significant base features. Figure 2 shows the evaluated Info-gain value for heart-statlog dataset. Once the Info-gain value is obtained, the cutoff restriction is applied to discover the base attributes. For this dataset, the cutoff value is .1 as shown in the figure. After obtaining the base attributes, the dataset expansion is done by applying the association rules on composite attributes. At this stage, the grouping of two and three attributes is done to obtain the significance of composite attributes. Finally, this expanded dataset is also measured by Info-Gain measure to identify the most significant processing dataset. This expanded weighted and filtered dataset is finally processed by fuzzy rules to obtain a clearer view of decision rules. The fuzzy integration of the model is described in the next subsection.

B. Fuzzy Frame Rules

Fuzzy logic is capable to tolerate the imprecision and able to translate the data to nominal feature values. The model can use the arbitrary and nonlinear functions to reduce the complexity of the rule formulation. The fuzzy rules are easier to interpret and able to derive more clear decision. The fuzzy reasoning for medical data is proven important as it reflects the data in the way the people really perceive the concept. In this research, the expanded and weighted featureset is processed under fuzzy rules to avoid the effect of imprecision and uncertainty. The attributes are fuzzify to improve the approximation of probability distribution. The numerical values are translated to 'High', 'Medium' and 'Low' with range consideration of 'Less', 'Average' and 'High' factors. The fuzzification process improved the sensing of the attribute values respective to the human thoughts. These rules are interpreted as the in-then-else statements and described as the conditional input. The fuzzification is applied to the final attribute dataset to translate it to the processing data form. The decision tree and Bayesian network classifiers are applied to this rule framed dataset.

C. Decision Tree

Decision tree is a condition based classification method that generates a tree structure based on the sequential and dependent decisions taken to label the data. Each node of this tree is represented by a condition or decision which is derived from one or more parameters. The test condition begins with the identification of the root node and a leading branch node is included with each new condition. The process is derived till all the conditions are not discovered and the labeled leaf node is not derived. Each condition of the decision tree is generated by observing the training set records respectively to one or more attributes. The data of these attributes are evaluated using some computational and suboptimal algorithm to partition the data. Greedy algorithm is one such algorithm which can be applied on attribute specific computation. The element level computation can be applied to generate the contingency matrix and to improve the classification results. In this research work, the fuzzy rule based weighted data is accepted by the decision tree to improve the classification results.

D. Bayesian Network

Bayesian network is the probabilistic evaluation based learning technique used to classify the objects. The main idea of this classifier is based on the Bayes theorem of probabilistic theory. In this method, the computation of the existence of some value x is computed over the available n tuples. The existence and non-existence of that value is measured to generate the decision rule. In case of multiple attributes, the conditional evaluation on the probabilistic values is generated to take decision and to recognize the object label. The probabilistic evaluation of this theorem is represented as $P(a_i,v_j)$. Where a_i is the attribute begin observed and v_j represents the conditional dependency exist for the value of other attribute. This process is repeated on each attribute to generate the composite rules for identification of the object class. In this research work, the fuzzy weighted data is processed under the Bayesian probabilistic model to recognize the heart disease patients accurately.

IV. RESULTS

The proposed fuzzy weighted and associated feature based expanded feature processed model is defined to recognize the heart disease accurately. The proposed model is applied on Statog(Heart) dataset. The multivariate dataset is having 270 instances with 13 attributes. The attributes include the real as well as categorical features. The comparative evaluation of this proposed model is done by applying the 10-fold validation over the dataset. The proposed FWAF(Fuzzy Weighted and Associated Feature) based dataset is analyzed using the Decision tree and Bayesian Network classifiers to predict the heart disease. The comparative evaluations are recorded against the Decision Tree, Naïve Bayes, Random Forest, Random Tree and Decision Table Classifiers. The evaluations are taken in terms of Accuracy Rate, MAE (Mean Absolute Error), and F-Measure parameters. The comparative results are graphically provided in this section.

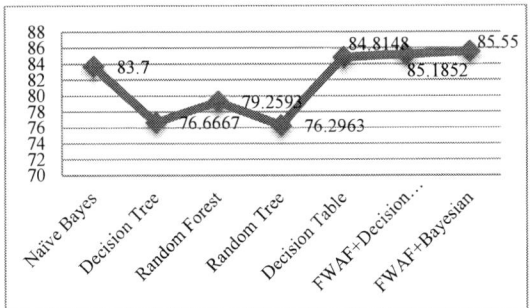

Figure 3 : Accuracy Rate Evaluation

Accuracy is the foremost metric that represents the correctness of the prediction. It is the simplest measure that identifies the number of instances correctly recognized by the proposed model. The results in figure 3 show the comparative evaluation of accuracy rate (%). The results verified that the proposed FWAF+Bayesian and FWAF+Decision Tree models have provided the maximum accuracy rate of 85.5% and 85.2% Whereas the Random itself achieved the least accuracy of 76.3%.

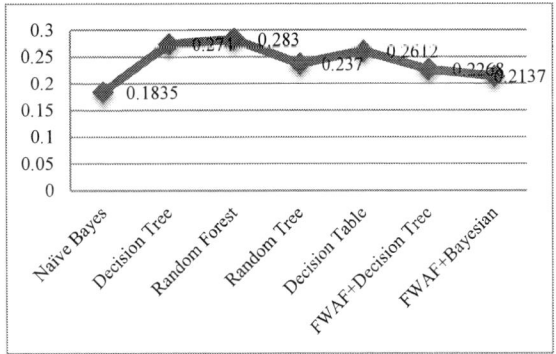

Figure 4 : MSE Rate Evaluation

MSE(Mean Square Error) represents the difference between the expected and predicted results. Higher the MSE value, lesser significant the classification results are considered. It is considered as the complement to accuracy measure. The comparative MSE results shown in line graph of figure 4 verified that the MSE value for proposed FWAF based methods are .2268 and .2137. The Naive Bayes is the only method which provides the lesser MSE to proposed method. The proposed model has outperformed all other listed classifiers under this parameter.

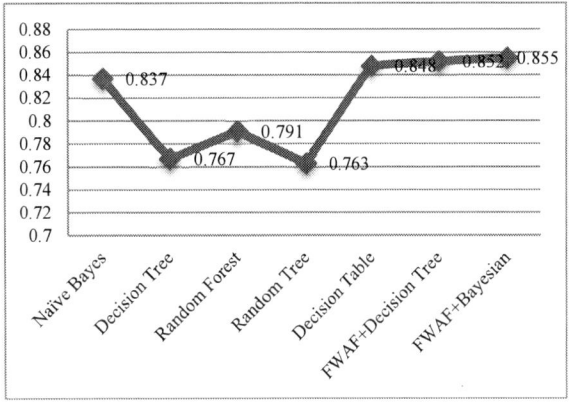

Figure 5 : FMeasure Evaluation

FMeasure is considered as good discriminator over accuracy rate measure. This metric is identified as the harmonic mean between the precision and recall values. Higher the FMeasure value, more significant the classification results are considered. Figure 5 shows that comparative evaluation in terms of FMeasure. The results identified that the proposed FWAF+Decision Tree and FWAF+Bayesian achieved the accuracy rate of .852 and .855. The comparative observations verified that the

978-1-5386-6624-1/18 $31.00 © 2018 IEEE   965

proposed model achieved the higher F-Measure against existing methods. The reliability of the heart disease prediction is improved by the proposed model.

V. CONCLUSION

The paper has provided the expanded featureset based weighted method to predict the chances of heart disease. At the earlier stage, the symptoms and features associated with heart disease are evaluated using Information-gain measure. The adaptive threshold is computed to identify the most significant features from the pool. The selected features are then presented as the associated paired features to identify the associated contribution. Based on these associated two and three paired features, the expanded featureset is composed. The second level information-gain measure is applied to identify the processing featureset. Once the featureset is formed, the fuzzy rules are applied to transform the featureset to decisive form. This weighted and rules framed featureset is finally trained through Bayesian network and decision tree classifiers to predict the heart disease. The evaluation of this proposed model is done under accuracy, MSE and FMeasure parameters. The comparative results identified that the proposed FWAF+Decision Tree and FWAF+Bayesian network model has improved the FMeasure and accuracy significantly and reduced the MSE rate. The overall significance and performance of the proposed model is improved by the proposed model.

REFERENCES

[1] Kapil Juneja and Chhavi Rana, "An improved weighted decision tree approach for breast cancer prediction.," *International Journal of Information Technology (2018)*, pp. 1-8, 2018.

[2] M. Sultana, A. Haider, and M. S. Uddin, "Analysis of data mining techniques for heart disease prediction," in *3rd International Conference on Electrical Engineering and Information Communication Technology (ICEEICT)*, Dhaka, 2016, pp. 1-5.

[3] J. Thomas and R. T. Princy, "Human heart disease prediction system using data mining techniques," in *International Conference on Circuit, Power and Computing Technologies (ICCPCT)*, Nagercoil, 2016, pp. 1-5.

[4] K. P. Lakshmi and C. R. K. Reddy, "Fast rule-based heart disease prediction using associative classification mining," in *International Conference on Computer, Communication and Control (IC4)*, Indore, 2015, pp. 1-5.

[5] M. Gandhi and S. N. Singh, "Predictions in heart disease using techniques of data mining," in *International Conference on Futuristic Trends on Computational Analysis and Knowledge Management (ABLAZE)*, Noida, 2015, pp. 520-525.

[6] Purushottam, K. Saxena, and R. Sharma, "Efficient heart disease prediction system using decision tree," in *International Conference on Computing, Communication & Automation*, Noida, 2015, pp. 72-77.

[7] S. Bharti and S. N. Singh, "Analytical study of heart disease prediction comparing with different algorithms," in *International Conference on Computing, Communication & Automation*, Noida, 2015, pp. 78-82.

[8] J. S. Sonawane and D. R. Patil, "Prediction of heart disease using learning vector quantization algorithm," in *Conference on IT in Business, Industry and Government (CSIBIG)*, Indore, 2014, pp. 1-5.

[9] J. S. Sonawane and D. R. Patil, "Prediction of heart disease using multilayer perceptron neural network," in *International Conference on Information Communication and Embedded Systems (ICICES2014)*, Chennai, 2014, pp. 1-6.

[10] M. A. Jabbar, B. L. Deekshatulu, and P. Chandra, "Heart disease prediction using lazy associative classification," in *International Mutli-Conference on Automation, Computing, Communication, Control and Compressed Sensing (iMac4s)*, Kottayam, 2013, pp. 40-46.

[11] S. Sivagowry, M. Durairaj, and A. Persia, "An empirical study on applying data mining techniques for the analysis and prediction of heart disease," in *International Conference on Information Communication and Embedded Systems (ICICES)*, Chennai, 2013, pp. 265-270.

[12] N. Aditya Sundar, "Performance Analysis of Classification Data Mining Techniques over Heart Disease Data Base," *International Journal of Engineering Science & Advanced Technology*, vol. 2, no. 3, pp. 470-478, 2012.

[13] Mr. Dhiraj Pandey, "Prediction System to Support Medical Information System using Data Mining Approach," *International Journal of Engineering Research and Applications (IJERA)*, vol. 2, no. 3, pp. 1988-1996, 2012.

[14] M.A.JABBAR, "Knowledge Discovery from Mining Association Rules for Heart Disease Predictions," *Journal of Theoretical and Applied Information Technology*, vol. 41, no. 2, pp. 166-174, 2012.

[15] K.Srinivas, "Mining Association Rules from Large Datasets Towards Disease Prediction," in *International Conference on Information and Computer Networks (ICICN 2012) IPCSIT*, 2012, pp. 22-26.

[16] V. V. Jaya Rama krishniah, "Predicting the Heart Attack Symptoms using Biomedical Data Mining Techniques," *The International Journal of Computer Science & Applications (TIJCSA)*, vol. 1, no. 3, pp. 10-18, 2012.

2nd IEEE International Conference on Power Electronics, Intelligent Control and Energy Systems (ICPEICES-2018)

Synchronization and Anti Synchronization of Fractional order system

Lokesh Shankar Singh,
Electrical Engineering Department
National Institute of Technology
Hamirpur (H.P), India
Luckyptpp1990@rediffmail.com

Himesh Handa,
Electrical Engineering Department
National Institute of Technology
Hamirpur (H.P), India
hklhanda@gmail.com

Nitesh gupta
Electrical Engineering Department
National Institute of Technology
Hamirpur (II.P), India
nitesh93.gupta@gmail.com

Abstract— **This paper addresses hybrid projective synchronization and anti-synchronization of fractional order chaotic systems. Master system and slave system is synchronized using a suitable transformation matrix. From past few years chaos synchronization found some intentness in non-linear science and many synchronization techniques are used till now i.e. active control, h-transform, beckstepping approach etc. The stability of the system and potency of the controller has been inveterate by using the Laplace transform. The simulation results of both chaotic system evince the cogency and possibility of projected synchronization techniques.**

Keywords— hybrid projective synchronization, active control, antisynchronization, chaotic systems

I. INTRODUCTION

Fractional calculus has a long history starting from 17th century where podlubny described it. Fractional order has its origin in 1695 by G.F.A. de L'Hôpital (1661-1709) and G.W. Leibnitz (1646-1716) where paradox was created "what does n=1/2 mean in derivative of function" [4,7]. Since then the fractional calculus acquired an interest by many of the scientists and many works have been carried out in this field. It has a wide research field such as neural network, time series analysis of chaotic systems and secure communication etc. Recently application to physics and engineering acquired most interest in fractional calculus [1, 2]. Many systems related to collaborative fields are divinely described using fractional order calculus for specimen viscoelastic systems, complex systems quantum evolution etc. [9, 10, 12].

Master and slave system are used in the synchronization where the mater system's output regulate slave system and output of slave system is tracked by outcome of drive system. All states of drive system should be synchronized to the slave systems states for being completely synchronization [13, 14]. There is endurance of synchronization types like projective synchronization, anti-synchronization, whole synchronization etc. Many investigations were carried out in chaotic field leading to some important results i.e. behaviour of Chua circuit as chaotic attractor when the order is less than 2.7 , Duffing system's chaotic naturein case of order will be less than 2 [15]. Anti-synchronization is a peculiar form of synchronization witnessed in periodic chaotic systems through alliened of chaos synchronization. In anti-synchronization theory, variables of state of synchronized systems taking altered original values have opposite signs identical absolute values. In conclusion, the sums of two signals in anti-synchronization is expected to converge to zero [11, 13, 16].

II. PRELIMINARIES

Simplification of integer type calculus to non-integer type operative aD_t^r where the order q is real no. and a, t are the limits.

$$aD_t^r = \begin{cases} \frac{d^\alpha}{dt^\alpha} & , R(\alpha) > 0 \\ 1 & , R(\alpha) = 0 \\ \int_a^t (d\tau)^{-\alpha} & , R(\alpha) < 0 \end{cases}$$

(1)

Where

α: fractional order(complex number)

$R(\alpha)$: Real part of α

Also $a < t$, a and t are limit of the exertion. Above equation defines the differential and integral operation on the basis of values of R(α). Generally Fractional Calculus has 3 obtrusive definitions:

1) Caputo definition :

$$\mathcal{D}^q f(t) \triangleq \tau^{m-q} \mathcal{D}^m f(t)$$

$$= \frac{1}{\Gamma(m-q)} \int_0^t \frac{f(\tau)^m}{(t-\tau)^{q-m+1}} f(\tau) d\tau$$

(2)

where $\Gamma(n) = (n-1)!$, $t > 0$, $q \epsilon \, \mathfrak{R}^+$.

2) Grunwald-letnikov(GL) definition :

$$aD_t^q f(t) = \lim_{h \to 0} h^{-q} \sum_{r=0}^n (-1)^r \binom{q}{r} f(t-rh)$$

$$= \sum_{k=0}^m (f^{(k)}(a)((t-a)^{-q+k})/\Gamma(-q+k+1) + 1/\Gamma(m-q+1) \int_a^t (t-\tau)^{n-q} f^{(m+1)}(\tau) d\tau$$

(3)

978-1-5386-6624-1/18 $31.00 © 2018 IEEE

Here $m < q < m+1$, and $rh = t - a$.

3) *Riemann-Liouville definition:*

$$aD_t^q z(t) = \frac{d^n}{dt^n} J_t^{n-q} z(t) , q > 0$$

Here n= q and

$$J_T^\vartheta \psi(t) = 1/\Gamma(\vartheta) \int_0^t \psi(v)/(t-v)^{(1-\vartheta)} dv$$

Where $0 < \vartheta \leq -1$ and $\Gamma(.)$ denotes gamma function.

Lemma 1 Let us consider a sovereign system described as
$$D^\alpha z(t) = Bz , z(0) = z_0$$
Where $0 < \alpha < 1$, $z \in R^m$ and $B \in R^{m*m}$

I. For the system to be stable it must satisfy $\left|arg(\lambda_j(B))\right| > \alpha\pi/2$, j=1,2,3,.....m and $arg(\lambda_j(B))$ symbolizes the argument of the eigen values λ_j of matrix B.

II. System is stable only when it is either asymptotically steady or the critical Eigen values that mollifies the condition $\left|arg(\lambda_j(B))\right| > \alpha\pi/2$ must have geometric profusion once [3,8,9].

Lemma 2 If elements d_{ij} $(i,j = (1,2, ... n))$ of matrix D $\in \mathfrak{R}^{n*n}$ satisfies the resulting condition:

1. $d_{ij} = -d_{ji}$ $(i \neq j)$
2. $d_{ii} \leq 0$

$|arg(\lambda_i(D))| > \alpha\pi/2$ (i=1,2...n).

Description 1:
For non-identical fractional order systems description of master and slave systems are:
$$D^\beta w(t) = Bw(t) + g(w(t))$$
$$D^\alpha z(t) = Cz(t) + f(z(t)) + u(t)$$

(4)

Here $0 < \alpha \leq \beta < 1$ is a fractional order with y(t), z(t) $\in \mathfrak{R}^n$ X(t) $\in \mathfrak{R}^m$ is the controller deliberate late B,C $\in \mathfrak{R}^{m*m}$ are relentless matrices of system parameter.
G,f : $\mathfrak{R}^m \to \mathfrak{R}^m$ are two continuous vector functions [5,6,8].

Description 2:
If there exists 2 constant matrices
E,F $\in \mathfrak{R}^{m*m}$ and F$\neq 0$ such that
$$\log_{t \to +\infty} \|Ez(t) - F(w(t))\| = 0$$

(5)

The drive system is said to be synchronized with the response system where $\|.\|$ is the matrix norm. As Result:

a. When E=F=I, where I is identity matrix then system is said to be completely synchronized.
b. When E=I, F=-I then the system is said to be anti-synchronized.

And when the matrix E=I then the synchronization to be projective synchronized.

III. System Description

Volta system is salient chaotic system which was introduced by VOLTA. Volta system is described as:

$$\dot{w_1} = -w_1(t) - lw_2(t) - w_2(t)w_3(t)$$
$$\dot{w_2} = -w_2(t) - mw_1(t) - w_3(t).w_1(t)$$
$$\dot{w_3} = nw_1(t) + w_1(t)w_2(t) + 1$$

(6)

For parameters (l, m, n) = (5, 85, 0.5) and initial condition $(w_1(0)w_2(0)w_3(0)) = = (8, 2, 1)$ the system is exhibits the chaotic behaviour.
The fractional order description of the Volta's system is:

$$D^{q_1}w_1(t) = -w_1(t) - lw_2(t) - w_2(t)w_3(t)$$
$$D^{q_2}w_2(t) = -w_2(t) - mw_1(t) - w_3(t).w_1(t)$$
$$D^{q_3}w_2(t) = nw_1(t) + w_1(t)w_2(t) + 1$$

(7)

Here $q_1 q_2 q_3$ are the order of the chaotic system > 0.95 and l m n are parameters of chaotic system.
When the parameters (l, m, n) = (18, 10, 0.6) and the initial conditions are (7, 4 , 2). The time step taken is h=0.0005 and simulation time is $T_{sim} = 20 \; sec$.

IV. Hybrid Projective Synchronization and Anti Synchronization of Fractional Order Voltas System

Hybrid projective synchronization of fractional order identical drive and response system is described. Here the drive system is described as

$$D^{q_1}w_1(t) = -w_1(t) - lw_2(t) - w_2(t)w_3(t)$$
$$D^{q_2}w_2(t) = -w_2(t) - mw_1(t) - w_3(t).w_1(t)$$
$$D^{q_3}w_3(t) = nw_1(t) + w_1(t)w_2(t) + 1$$

(9)

And response system is defined as

$$D^{q_1}z_1(t) = -z_1(t) - lz_2(t) - z_2(t)z_3(t) + u_1$$
$$D^{q_2}z_2(t) = -z_2(t) - mz_1(t) - z_3(t).z_1(t) + u_2$$
$$D^{q_3}z_3(t) = nz_1(t) + z_1(t)z_2(t) + 1 + u_3$$

(10)

u_1, u_2 and u_3 are the controller notations used in the response system.

From definition 2 equation (5) error of the HPS is defined as
$$e = Ez(t) - Fw(t)$$
Here the E and F are the matrix. E is taken as identity matrix and transformation matrix F is taken as
$$F = \begin{matrix} 1 & 0 & 0 \\ 0 & -1 & 0 \\ 0 & 0 & 1 \end{matrix}$$

Therefore the error dynamics is

$$e_1 = z_1 - w_1$$
$$e_1 = z + w_2$$

978-1-5386-6624-1/18 $31.00 © 2018 IEEE

$$e_3 = z_3 - w_3 \tag{11}$$

e_1, e_2 and e_3 are synchronization errors. These has to be made zero for achieving synchronization.

Now the error dynamics are expressed as

$$D^{q_1}e_1(t) = -e_1(t) - lz_2(t) + w_2(t)w_3(t) + lw_2(t) - z_3(t)z_2(t) + u_1$$
$$D^{q_2}e_2(t) = -e_2(t) - mz_1(t) - w_1(t)w_3(t) - mw_1(t) - z_3(t)z_1(t) + u_2$$
$$D^{q_3}e_3(t) = -ne_3(t) - w_1(t)w_2(t) - z_2(t)z_1(t) + u_2 \tag{12}$$

Controllers (u_1, u_2 and u_3) should be designed in such a way that the synchronization error should tend to zero via active control method. The controllers chosen are

$$u_1 = -B_1 e_1(t) + lz_2(t) - w_2(t)w_3(t) - lw_2(t) + z_3(t)z_2(t)$$
$$u_2 = -B_2 e_2(t) + mz_1(t) + w_1(t)w_3(t) + mw_1(t) + z_3(t)z_1(t)$$
$$u_3 = -B_3 e_3(t) + w_1(t)w_2(t) + z_2(t)z_1(t) \tag{14}$$

Here the controllers demarcated above reduce the error dynamics to

$$D^{q_1}e_1(t) = -(1 + B_1)e_1(t)$$
$$D^{q_2}e_2(t) = -(1 + B_2)\,e_2(t)$$
$$D^{q_3}e_3(t) = (n - B_3)\,e_3(t) \tag{15}$$

Taking the same master and slave system as taken for the synchronization in equation (9) and (10). Choosing the transformation matrix F as

$$F = \begin{matrix} -2 & 0 & 0 \\ 0 & -1 & 0 \\ 0 & 0 & -2 \end{matrix}$$

The error dynamics in equation (11) is reduced to

$$e_1 = z_1 + 2w_1$$
$$e_1 = z_2 + w_2$$
$$e_3 = z_3 + 2w_3$$

Taking derivative of error equations. The error dynamics now become

$$D^{q_1}e_1(t) = -e_1(t) - lz_2(t) - 2w_2(t)w_3(t) - 2lw_2(t) - z_3(t)z_2(t) + u_1$$
$$D^{q_2}e_2(t) = -e_2(t) - mz_1(t) - w_1(t)w_3(t) - mw_1(t) - z_3(t)z_1(t) + u_2$$
$$D^{q_3}e_3(t) = ne_3(t) + 2w_1(t)w_2(t) + z_2(t)z_1(t) + 3 + u_2 \tag{16}$$

Controllers chosen are
$$u_1 = -B_1 e_1(t) + lz_2(t) + 2w_2(t)w_3(t) + 2lw_2(t) + z_3(t)z_2(t)$$
$$u_2 = -B_2 e_2(t) + mz_1(t) + w_1(t)w_3(t) + mw_1(t) + z_3(t)z_1(t)$$

$$u_3 = -B_3 e_3(t) - w_1(t)w_2(t) - z_2(t)z_1(t) - 3 \tag{17}$$

Controllers described above reduce error dynamics to
$$D^{q_1}e_1(t) = -(1 + B_1)e_1(t)$$
$$D^{q_2}e_2(t) = -(1 + B_2)\,e_2(t)$$
$$D^{q_3}e_3(t) = (n - B_3)\,e_3(t) \tag{18}$$

V. SIMULATION RESULTS

Simulation is done by using MATLAB Tool and simulation prameters are listed in table 1.

Table 1 Simulation Parameters and its value

Parameters	Value
(a, b, c)	(18,10,0.6)
$q_1 = q_2 = q_3 = q$	0.95
Master system ($w_1(0)w_2(0)w_3(0)$)	(8, 2, 1)
Slave system ($z_1(0)z_2(0)z_3(0)$)	(7, 4, 2)

The time step taken is h=0.0005 and simulation time is $T_{sim} = 20\ sec.$. The synchronization of the state w_1, z_1 is shown in Fig.1, while synchronization of states w_2, z_2 and w_3, z_3 are illustrated in Fig.2 and Fig.3 respectively. Anti-synchronization of the system states (w_1, z_1), (w_2, z_2) and (w_3, z_3) are shown in the figures 5, 6 and 7 respectively. The synchronization and Anti-synchronization error tends to zero as shown in figures 4 and 5 respectively.

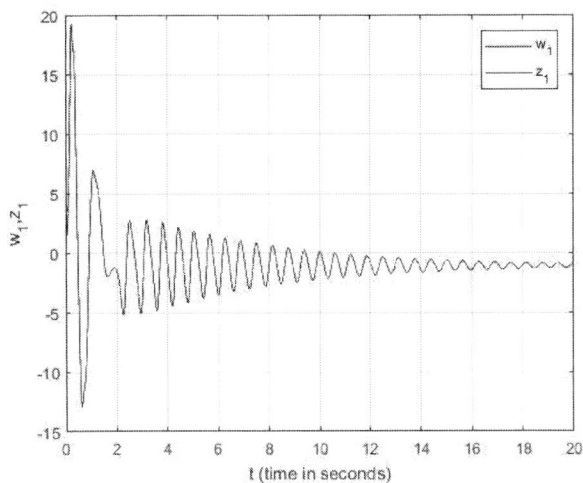

Fig. 1 Hybrid projective synchronization of states w, z_1

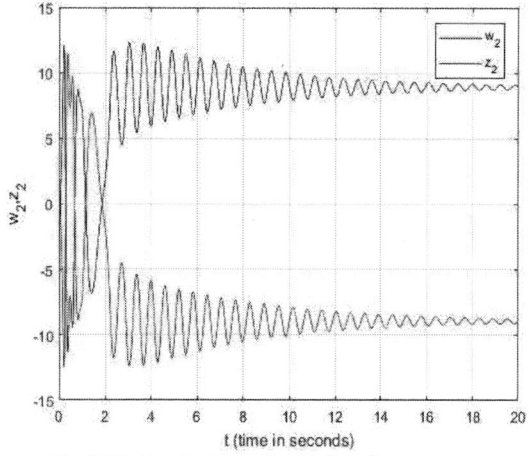

Fig. 2 Hybrid projective synchronization of states w_2, z_2

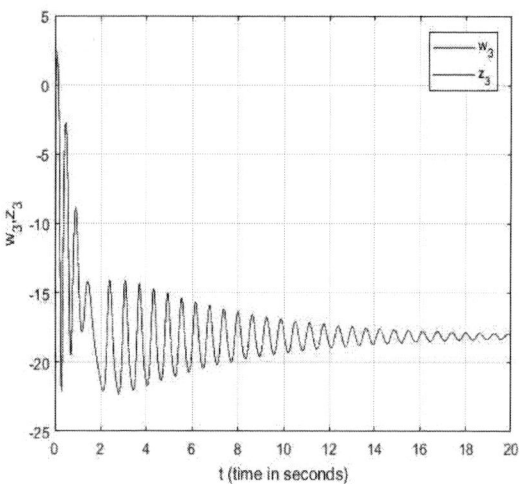

Fig. 3 Hybrid projective synchronization of states w_3, z_3

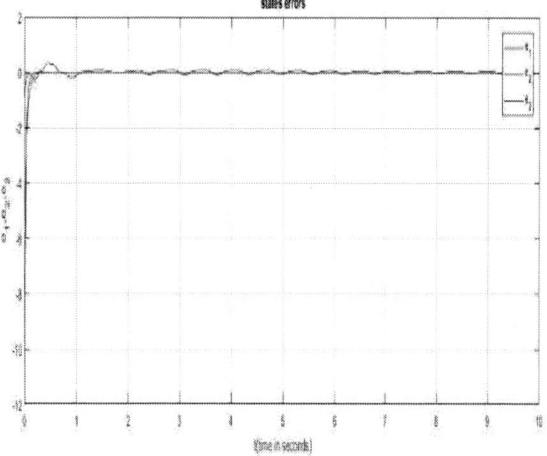

Fig. 4. Confluence of hybrid projective synchronization errors of fractional order Volta's system

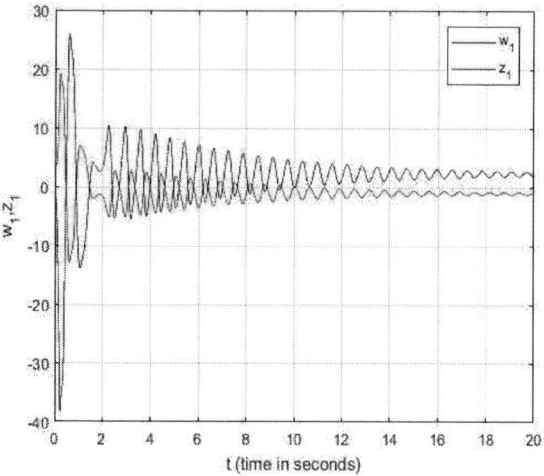

Fig. 5 Anti-synchronization of states w_1, z_1

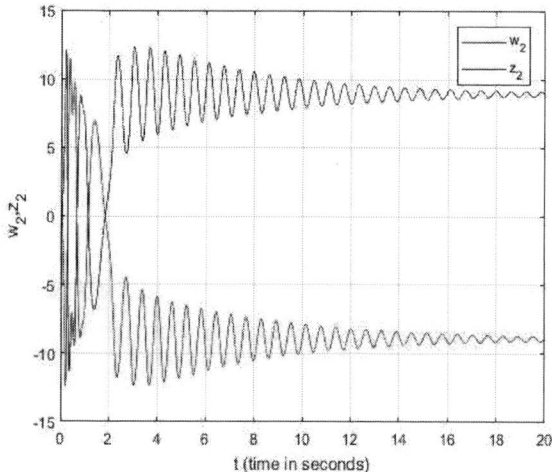

Fig. 6 Anti-synchronization of states w_2, z_2

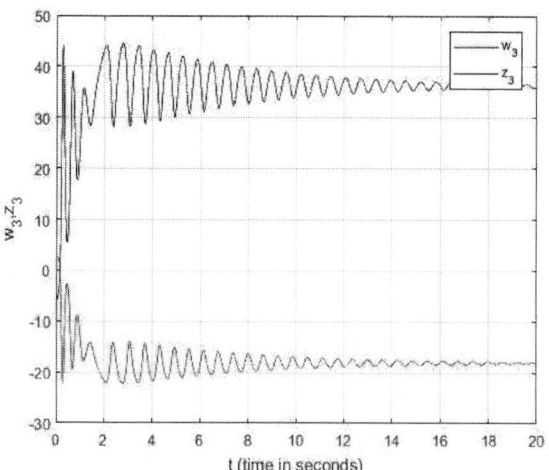

Fig. 7 Anti-synchronization of states w_3, z_3

978-1-5386-6624-1/18 $31.00 © 2018 IEEE

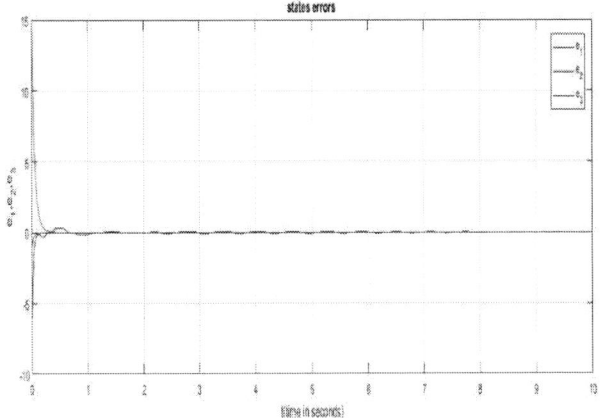

Fig. 8. Confluence of Anti-synchronization errors of fractional order Volta's system

VI. CONCLUSION

This paper illustrates synchronization and anti-synchronization of the Volta's system by using active control mechanism. Simulation results shows synchronization and anti-synchronization as synchronization error is tending to zero.

REFERENCES

[1] P. Zhou, R. Ding, "Modified function projective synchronization between different dimension fractional-order chaotic systems", Abstract and Applied Analysis, Vol.2012, Article ID 862989, 2012.

[2] S. Wang, Y. G. Yu, M. Diao, "Hybrid projective synchronization of chaotic fractional order systems with different dimensions", Physica A, Vol.389, No.21, 4981–4988, 2010.

[3] J. Bai, Y. G. Yu, S. Wang, Y. Song, "Modified projective synchronization of uncertain fractional order hyperchaotic systems", Communications in Nonlinear Science and Numerical Simulation, Vol.17, No.4, 1921-1928, 2012.

[4] X. J. Wu, Y. Lu, "Generalized projective synchronization of the fractional-order Chen hyperchaotic system", Nonlinear Dynamics, Vol.57, No.1-2, 25-35, 2009.

[5] Bai, E.W. and Lonngran, E.E. (1997). "Synchronization of two Lorenz systems using active control". Chaos Solutions and Fractals, 8, 1, 51–58.

[6] Ge, Z.M. and Chen, C.C. (2004). "Phase synchronization of coupled chaotic multiple time scales systems ", Chaos, Solitons and Fractals, 20, 639-647.

[7] Pecora, L.M. and Carroll, T.L. (1990). "Synchronization in chaotic systems". Phys. Rev. Lett., 64, 821-824.

[8] Sudheer, S. K. and Sabir, M. (2009). "Hybrid Synchronization of Hyperchaotic Chen System". National Conference on Nonlinear Systems & Dynamics, Kolkata, India.

[9] Hongyue D., Qingshuang Z., ChanghongW, Mingxiang L., "Function projective synchronization in coupled chaotic systems", Nonlinear Anal. Real World Appl., 2010, 11, 705-712.

[10] Zhenbo L., Xiaoshan Z., "Generalized function projective synchronization of two different hyperchaotic systems with unknown Parameters", Nonlinear Anal. Real World Appl., 2011, 12, 2607-2615.

[11] Al-Sawalha M. M., Noorani M. S. M., "On anti-synchronization of chaotic systems via nonlinear control", Chaos Soliton & Fract., 2009, 42, 170-179.

[12] Ho, M.C. & Hung, Y.C. (2002) "Synchronization of two different chaotic systems by usinggeneralized active control", Physics Letters A, Vol. 301, pp 424-428.

[13] Sundarapandian, V. & Karthikeyan, R. (2011) "Anti-synchronization of the hyperchaotic Liu and hyperchaotic Qi systems by active control," International Journal on Computer Science and Engineering, Vol. 3, No. 6, pp 2438-2449.

[14] J. Yu, J. jhang, Li jhang, "Improved full state hybrid projective synchronization of chaotic systems with the different order" ,9th international conference for young computer scientists, 2009

[15] J. Liu, S. Liu and C. Yuan, "Modified generalized projective synchronization of fractional-order lu systems", advances in difference equations, 2013, pp 578-587.

[16] C.-M. Jiang, S.-T. Liu, C. Luo, "A new fractional-order chaotic complex system and its antisynchronization", Abstr. Appl. Anal. 2014.

Energy Harvesting using d_{33} Mode by Insole Embedded Low Cost Piezo-Sensors

Sumit Balguvhar
Department of Civil Engineering
Indian Institute of Technology
Delhi New Delhi, India
ersumitest@gmail.com

Sidhartha Singhal
Department of Civil Engineering
Indian Institute of Technology
Delhi New Delhi, India
sidharthasinghal25@gmail.com

Suresh Bhalla
Department of Civil Engineering
Indian Institute of Technology
Delhi New Delhi, India
sbhalla@civi.iitd.ac.in

Abstract—**Harvesting mechanical energy using piezo-transducers from human motion is an attractive approach for acquiring clean and sustainable electric energy to power wearable gadgets or recharge batteries, which can be used for body area network such as health monitoring, activity recognition and so on. This paper investigate the feasibility of harvesting energy using d_{33} mode from low cost ceramic PZT sensors of different sizes and shapes, locally available in the market. It was found that 25 mm sensor produced maximum power. The maximum average power of magnitude 8.21 µW at the toe and 3.62 µW at the heel was generated by 95 kg subject (BMI -30) at a speed of 9 km/hr when 25mm EJSs were used. The insole EJSs gave better results than the surface bonded sensors.**

Keywords—*piezoelectricity, energy, human motion, low cost sensors*

I. INTRODUCTION

Energy harvesting is a developing technology and is a fast growing tool to produce portable and wireless microelectronics. In addition, one can feel the need in today's realm where we can see depletion of our natural resources at an alarming rate, thus developing renewable source of energy becomes more significant and substantial. Energy harvesting can be done, from solar energy, wind energy, wave power, geothermal energy etc. Energy harvesting from human movement is one the most attracting approach to power wearable devices and thereby replacing batteries. Starner [1] characterized the power available from various sources such as arms, biceps curls, respiration etc. He concluded that the energy available from bicep curls is 24 W, compared to 60 W available from arm lifts, 0.83 W from respiration, 67 W from walking. Paradiso et al.[2] designed a system to harness foot strike energy by flattening curved, pre-stressed spring metal strips laminated with a semi-flexible form of piezoelectric lead zirconate Titanate (PZT) under the heel. Numerous research has been done in harnessing human energy for powering wearable gadgets [3], though almost all commercial wearable devices are still powered by a coin cell battery. Piezo ceramic sensors such as PZT patches operates in two mode i.e. d_{31} (length) mode and d_{33} mode (thickness), limited research has been done utilizing d_{33} [4]. Das [5] studied the foot pressure at fifteen plantar pressure points for different conditions and found out that due to the escalation in walking speed, the pressure increases at heel and the ball of foot region whereas it decreases at mid foot and metatarsals regions.

The result of this study is used to decide the exact location for placing of the PZT patch. Although a number of studies can be found dovetailed to piezoelectric energy harvesting from walking, there is lack of a dedicated and packages shoe sole power generator that can be widely used on large scale for any type of shoe [6-7]. This paper is based on the work carried out in Master's thesis [8] which investigates the feasibility of harvesting energy using d_{33} mode from low cost ceramic PZT sensors of different sizes and shapes, locally available in the market.

II. FABRICATION OF PIEZO-HARVESTER

The fabrication of harvester was carried out, by using low cost ceramic piezo sensors locally available in the market to check their feasibility for harvesting energy. Three types of sensors were procured from CEL, Sahibabad. (see Table 1).

Table 1

S No.	Type of Sensors	Diameter	Thickness	Cost
1.	Disc	25 mm	1.85mm	Rs 325
2	Disc	20 mm	1.85mm	Rs 325
3.	Disc	10 mm	1.00mm	Rs 300

The sensors were sandwiched by a layer of epoxy adhesive hence known as Epoxy Jacket Sensors (EJS) shown in Fig. 1. The uneven surface of the sensor was flattened and smoothened using a grinder, so as to make the impact area free of any undulation. The grinding needs to be done very carefully without damaging the soldering of wires and maintaining the requisite thickness of epoxy required for protection.

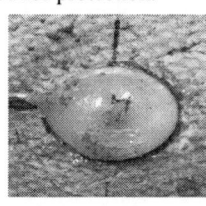 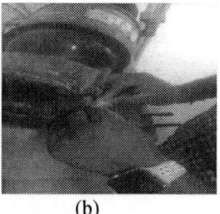

(a) (b)

Fig 1. Fabrication of Sensor using Epoxy
(a) Epoxy Jacketed Sensor (EJS)
(b) Grinding of Excess Epoxy

III. EXPERIMENTATION OF HARVESTER

A. Study of Insole embedded sensors by walking subjects of different weights.

The fabricated EJSs were embedded into the sole of the shoe at two positions i.e. at the ball and the toe as shown in Fig. 2. Two subjects of different weights were selected and made to wear the EJS fitted shoes. The experiment was conducted in the laboratory for 20 mm and then 25 mm size sensors by both the subjects and the reading was acquired across 1 MΩ resistance using the Power Measuring Device (PMD). The peak and average power was recorded for both the subjects and were compared in Table 2.

Table 2

Type of Sensors	Peak Power Across 1MΩ		Peak Power Across 1MΩ	
	Subject weight 65 kg		Subject weight 85 kg	
	Heel (µW)	Toe (µW)	Heel (µW)	Toe (µW)
20 mm	1.55	0.72	1.16	6.48
25 mm	1.56	6.21	7.38	10.76

Table 3

Type of Sensors	Avg. Power Across 1MΩ		Avg. Power Across 1MΩ	
	Subject weight 65 kg		Subject weight 85 kg	
	Heel (µW)	Toe (µW)	Heel (µW)	Toe (µW)
20 mm dia	0.68	0.61	0.81	3.49
25 mm dia	0.82	2.22	4.49	6.33

The peak open circuit voltage of 2.2 V was observed in 25 mm PZT in comparison with 20 mm PZT which was half of it at 1.1 V. The maximum average power of magnitude 4.49 µW at heel and 6.33 µW at toe was generated by the 25 mm PZT across 1 MΩ resistance. It was observed the average power was higher in the toe region in comparison to the heel as the same was observed for peak open circuit voltage. From the Table 2 and 3 it is also inferred that with increase in weight of the subjects, there is increase in the peak and average voltage respectively. The EJSs when embedded in the sole gave good result as the pressure applied on the sensors were uniform and due to the compressive nature of the rubber sole pressure was applied from both top and bottom of sensors.

B. Study of Insole embedded sensors worn by subjects of varying weight / BMI running on treadmill.

The next experiment was carried out by using only 25 mm EJSs to study the variation in energy harvesting by insole embedded sensors, worn by subjects of different weight, walking and running at different speed (see Fig 2).

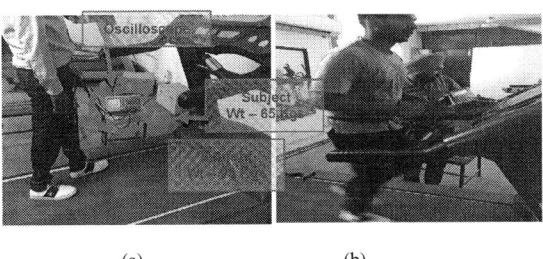

(a)   (b)

Fig 2 : Experimental Setup – Sujbects Running at varying Speeds on a Treadmill
(a) 65 Kg Subject – Walking on the Treadmill
(b) 95 Kg Subject – Running on the Treadmill

Three different subjects were selected, having three different weights and BMI (see Table 4). Each one of the subjects were made to wear the fabricated shoe with insole embedded EJSs at the foot and ball of the shoe. Each subjects were made to run on the same treadmill wearing the shoe and the readings were recorded using the oscilloscope across the PMD using 1 MΩ resistances. While conducting the experiment few limitations were inherent, i.e. different gait of each subject, different running style, non-linearity in increase of weight of subjects, use of same prototype for each subject. With these limitations in mind, the experiments were conducted. The readings were recorded (see Table 5) processed and plotted (see Fig 3 and 4 and).

Table 4

Subject	Weight	BMI
1	65 Kg	25
2	85 Kg	28
3	95 Kg	30

Table 5

Walking / Running Speed (Km/Hr)	Average Power Toe			Average Power Heel		
	65 Kg (BMI 25)	85 Kg (BMI 28)	95 Kg (BMI 30)	65 Kg (BMI 25)	85 Kg (BMI 28)	95 Kg (BMI 30)
3	0.69	0.68	0.82	1.33	2.1	2.38
6	1.06	1.68	2.272	1.85	2.8	3.45
7.5	1.31	1.80	2.27875	2.425	3.14	4.34
9	1.58	3.43	3.68	3.02	4.0	8.21

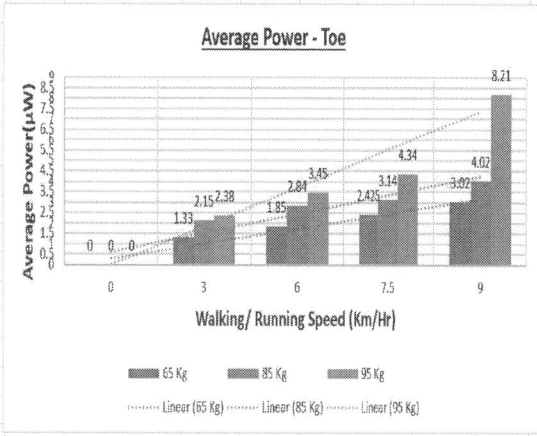

Fig 3: Average Power generated at the Toe

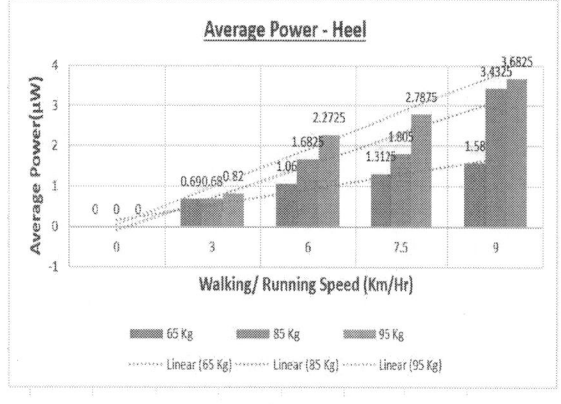

Fig 4: Average Power Generated at the Heel

The peak power 3.98 μW at heel and 5.81 μW at toe was generated by subject weighing 95 kg at a running speed of 9 km/hr whereas the minimum power 1.3 μW at heel and 1 μW at toe was generated by subject weighing 65 Kg at a walking speed of 3 km/hr. The maximum average power of magnitude 3.6825 μW at heel and 8.21 μW at toe was generated by subject weighing 95 kg at a running speed of 9 km/hr, which was measured using the PMD. It was observed that the average power was higher in the toe region in comparison to the heel due to the running style of subjects where the toe was landing first thus resulting in more pressure. From the above data it was also highlighted that with increase in weight and BMI of the subjects, there is increase in the peak and average voltage respectively where linear increase pattern is observed. The plantar pressure is maximum under the heel and the ball of foot and this pressure can be used to produce substantial energy to power devices of having low power requirement. The maximum average power of magnitude 8.21 μW at the toe and 3.62 μW at the heel was generated by 95 kg subject (BMI -30) at a speed of 9 km/hr when 25mm EJSs were used.

IV. CONCLUSIONS AND RECOMMENDATIONS

After detailed analysis from the above experiments it was found out that the 25 mm dia disc ceramic sensor was the best suited for energy harvesting by human motion by d_{33} effect as it generates more power than the 10mm and 20 mm dia sensors. The 25mm dia sensor is 1.85 mm thick and has larger contact area, resulting in more power generation. The sensor was surface bonded to the sole and further embedded within the sole of a shoe. The insole EJSs gave better results than the surface bonded sensors. The embedding of the EJS also enhance the life of the harvesters as the sole provided additional cover and thus protects it from the harsh rugged conditions. The EJS, can easily be installed without harming the comfort and design of the shoe, thus allowing the user to wear the harvester with ease.

The maximum average power of magnitude 3.6825 μW at heel and 8.21 μW at toe was generated, by subject weighing 95 kg at a running speed of 9 km/hr and it can be observed that with increase in weight and BMI there is linear increasing trend of the peak and average voltage. Also with increase in walking and running speed the peak and average voltage increased. The power generated can be stored subsequently to power sensors and gadgets which have low voltage requirement such as temperature sensors, LEDs, etc. capability of a soldier in the battlefield and reducing its requirement of batteries required to power his gadgets and equipment.

REFERENCES

[1] T. Starner, "Human-powered wearable computing," in *IBM Systems Journal*, vol. 35, no. 3.4, pp. 618-629, 1996.

[2] N. S. Shenck and J. A. Paradiso, "Energy scavenging with shoe-mounted piezoelectrics," in *IEEE Micro*, vol. 21, no. 3, pp. 30-42, May-June2001.

[3] T. Xue, X. Ma, C. Rahn, and S. Roundy, "Analysis of upper bound power output for a wrist-worn rotational energy harvester from real-world measured inputs," *J. Phys. Conf. Ser.*, vol. 557, no. 1, 2014.

[4] N. Kaur and S. Bhalla, "Feasibility of energy harvesting from thin piezo patches via axial strain (d31) actuation mode," *J. Civ. Struct. Heal. Monit.*, vol. 4, no. 1, pp. 1–15, 2014.

[5] A. k. Das "Bio-mechanical Application of Piezo Sensors for Plantar Pressure Measurement". Master's Thesis, Department of Civil Engineering, Indian Institute of Technology Delhi, New Delhi, 2015.

[6] S. R. Anton and H. A. Sodano, "A review of power harvesting using piezoelectric materials (2003-2006)," *Smart Mater. Struct.*, vol. 16, no. 3, 2007.

[7] A. H. Abdul Razak, A. Zayegh, R. K. Begg, and Y. Wahab, "Foot plantar pressure measurement system: A review," *Sensors (Switzerland)*, vol. 12, no. 7, pp. 9884–9912, 2012.

[8] Maj Sidhartha Singhal "Performance of embedded piezo-sensors under low and high strain loading". Master's Thesis, Department of Civil Engineering, Indian Institute of Technology Delhi, New Delhi, 2018.

Gain Enhancement of Circularly Polarised Microstrip Patch Antenna using Metasurface

Shubham Kumar Mangal
Electronics and Communication Engineering
Galgotias College of Engineering and Technology, Greater Noida, India
Email: smngl1995@gmail.com

Sujay Kumar
Electronics and Communication Engineering
Galgotias College of Engineering and Technology, Greater Noida, India

Email : websujay@gmail.com

Tanuj Agarwal
Electronics and Communication
Engineering
Galgotias College of Engineering and Technology, Greater Noida, India
Email : tanujagarwalk97@gmail.com

Ujjawal Sarkar
Electronics and Communication Engineering
Galgotias College of Engineering, and Technology, Greater Noida-201308 Email : ujjwalsarkar10@gmail.com

Ankit Sharma
Assistant Professor, Electronics and Communication Engineering
Galgotias College of Engineering and Technology, Greater Noida, India
Email : ankit.sharmaece@galgotiacollege.edu

Lakshamanan.M
Professor, Electronics and Communication Engineering
Galgotias College of Engineering and Technology, Greater Noida, India
Email : lakshmanan.m@galgotiacollege.edu

Abstract— In this paper, a circularly polarized microstrip patch antenna is designed using metasurface as superstrate to achieve enhanced gain, so as to make it desirable for practical communication purposes. A parametric study of the effect of variation in the air gap between two substrates and variation in the truncation length of metasurface patches with antenna characteristics is conducted. The results from the parametric study are used to determine optimized values of the different parameters of antenna to get the optimum results and performance. The design and simulation of the antenna is carried out using CST Microwave Studio 2016 simulation software. The return loss is below -10dB from 6.91GHz to 7.94GHz with an impedance bandwidth of 8%. The gain obtained for antenna without superstrate is 4.33dB. A maximum gain of 6.08dB is achieved at 7.24GHz for proposed antenna with superstrate. A significant increase of 1.7dB in gain is hence achieved by the proposed stacked layer. The axial ratio is less than 3dB within the operating frequency range of 6.91 GHz to 7.57GHz which comes under the impedance bandwidth of antenna. The VSWR is less than 2 within the frequency range of 6.91GHz to 7.57GHz. The proposed antenna prototype has been fabricated and tested using VNA and anechoic chamber facility in order to verify the simulated results. Within fabrication tolerances, the agreement with the simulations has been satisfactory.

Keywords— Circular Polarization (CP), Metasur*face, Truncated Square Patch, Parasitic Patches*

I. INTRODUCTION

Microstrip patch antennas found great applications in modern wireless communication systems since they account for large number of advantages like they are low in cost, small in size, light in weight, ease of fabrication, frequency agility, feedline flexibility, beam scanning omnidirectional patterning, but they also suffer from problems of narrow

bandwidth and low gain. Several techniques by researchers and scientists have been employed to overcome this problem such as using dual feed network based structures[1],stacked patches[2]-[3],antennas with sequential feeding structure[4]

and multi-layered structures[5].Lately, researchers are also investigating the use of metamaterial to enhance the

bandwidth and gain within the desired operating frequency range. A metamaterial is a two dimensional artificial sheet that is being used to alter the properties of electromagnetic waves. For reduction in size, researchers have used RIS (Reactive impedance substrates) and AMC (Artificial magnetic conductors) for designing purposes [6]-[7].However such designs were limited to generate linearly polarized waves. For enhancing the axial ratio, a C-type single feed stacked microstrip antenna has been proposed having 3-dB AR bandwidth of 13.5% and gain of 7.5 dBi [8]. To achieving the CP (circular polarization), a square patch is excited through a microstrip feed line of negligible thickness. Having 3dB AR and good impedance matching has been the requirement of an antenna which is fulfilled by this method [9]-[10]. Recently, a metasurface consisting of an array of annular ring unit cells has been proposed for the enhancement of bandwidth of a single feed circularly polarized antenna [11]. Similarly, gain of the circularly polarized slot antenna is increased using metasurface [12]. Furthermore, metasurface in the stacked form are used by the researchers to increase the gain of antenna, axial ratio bandwidth of antenna and impedance bandwidth of the antenna [13-17].So, it is clear that metasurface can be used to improve the overall radiation characteristics of the antenna. In this paper, a metasurface consisting of an array 8x8 corner truncated rectangular ring unit cells have been used as parasitic patches.

II. ANTENNA GEOMETRY AND DESIGN

Fig.1 shows the cross-sectional view of proposed CPMA. The antenna is oriented such that the centre of the square patch is at the origin of the coordinate. It consists of a ground plane of dimensions 40 X 40 mm. An FR4 (ϵ = 4.3, tanδ = 0.025) substrate layer has been placed over the ground plane with height h_1= 0.8mm.

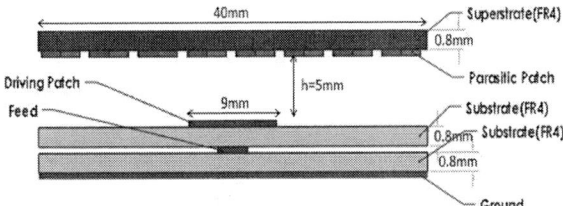

Fig 1: Cross sectional view of proposed antenna

Another layer of FR4 substrate with height h_2= 0.8mm is placed on top of it to create proximity coupled feeding. The feedline 17 X 3.1 mm² is sandwiched in between the two substrates.

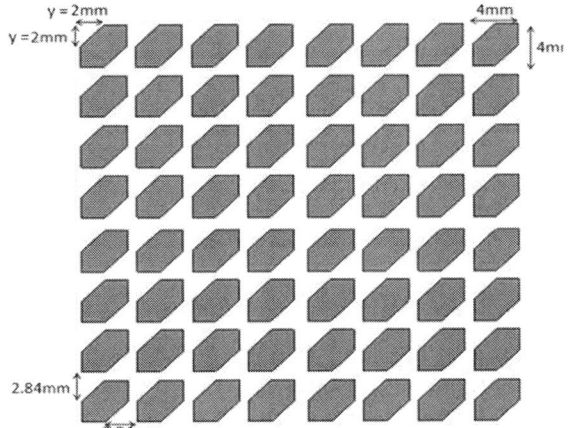

Fig 2: An array of 8 X 8 parasitic patch unit cells

The feedline reaches upto y = 3 mm. A square, truncated patch radiator of dimension 9 X 9 mm² is placed on top of upper FR4 substrate layer. The corners of the driving patch are truncated in the form of isosceles triangle of side 1.6mm each. When truncation is done along the two opposite ends of the square patch, it helps to get resonance frequency along the diagonal to be higher as compared to that of unchopped diagonal [Parik]. The truncation helps to achieve circular polarization. An array of parasitic patches (8 X 8) is placed beneath the superstrate at a height of h = 5mm above the driving patch as shown in Fig.2. The unit cell array is placed symmetrically about the coordinate axes. The dimension of each parasitic patch unit cell is 4 X 4 mm² .The corner ends of the parasitic patches have been truncated in the form of isosceles triangle with side 2mm each. The horizontal and vertical distance between the adjacent parasitic patches is 0.84mm. The superstrate is of FR4 material with height h_3= 0.8mm.

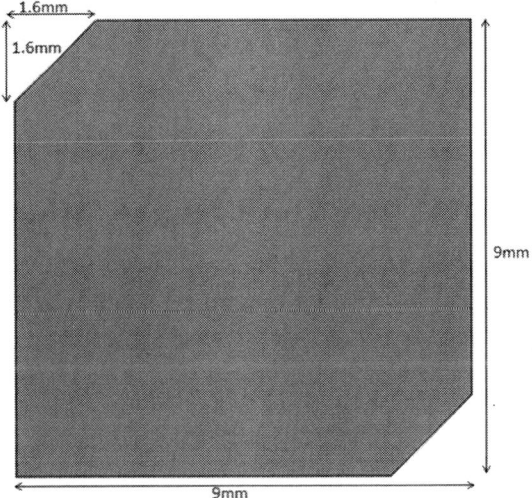

Fig 3: Square driving patch antenna with truncation

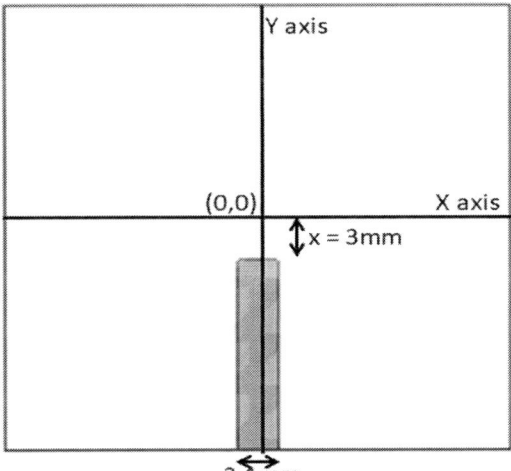

Fig 4: Feed position with respect to coordinate axes

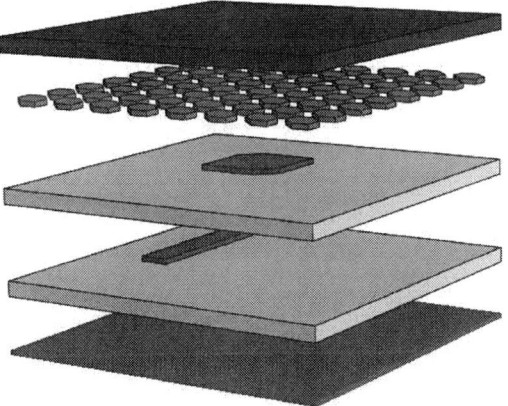

Fig 5: Exploded view of proposed antenna

III. PARAMETRIC STUDIES

Parametric Studies were conducted to understand the effect of the antenna's geometrical parameters on its return loss, gain, CP radiations performance and to provide reader with detail information about the proposed antenna design and optimization. The studies are carried out by simulation such that only one parameter is changed at a time, while others are kept constant. The design dimensions are the same as listed in previous section.

A. Variation of Air Gap (h)

The air gap height (h) between the driving patch and superstrate layer is varied to study its effect on all the antenna parameters. Fig.6 shows the variation of return loss (S_{11}) magnitude with different values of h = 1, 3, 5, 7 mm. Fig.7 Shows the variation of Gain (dB) with different values of h = 1, 3, 5, 7 mm. And Fig.8 shows the variation of axial ratio (dB) with different values of h=1, 3, 5, 7 mm.

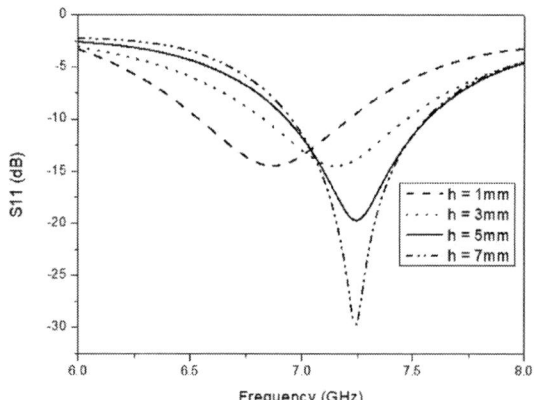

Fig.6: Return loss (S_{11}) variation with h

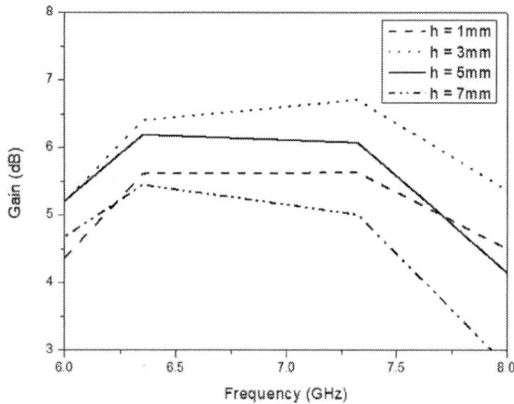

Fig.7: Gain (dB) on variation of h

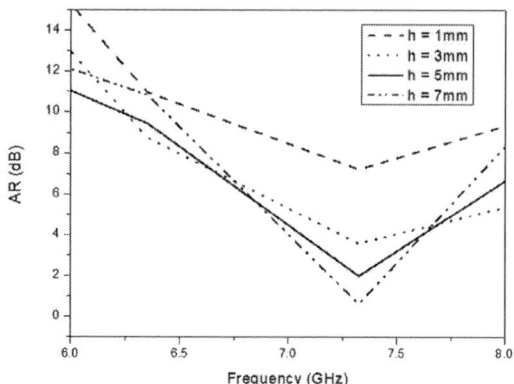

Fig.8: Axial Ratio (dB) on variation of h

From the above figures we can observe that the return loss magnitude (S_{11}) improves with increase in the value of air gap height (h). The best value of S_{11} is for h = 7mm. But the magnitude of gain improves with decrease in the value of air gap height (h). The best value of gain is for h = 3mm. The axial ratio is below 3dB for h = 5, 7mm. we can say that the optimum value of air gap (h) for which we obtained favorable return loss (S_{11}) magnitude, gain and circular polarization is h = 5mm. This value of h is fixed for further designing in the letter.

B. Variation of Truncation Length of Parasitic Patch (y)

The truncation length of the parasitic patches attached on lower surface of superstrate is varied to obtain circular polarization. Fig.12, Fig.13 & Fig.14 shows the effect of variation of truncation length of the parasitic patches y = 1, 1.5, 2, 2.25 mm on S_{11} parameter, gain and axial ratio respectively.

The effect of truncation length variation on gain (dB) is negligible. The return loss magnitude (S_{11}) and axial ratio (dB) improves with increase in the value of truncation length of the parasitic patches (y). The optimum value of y = 2mm is selected after observing the graphs. This value of h is fixed for further designing in the letter.

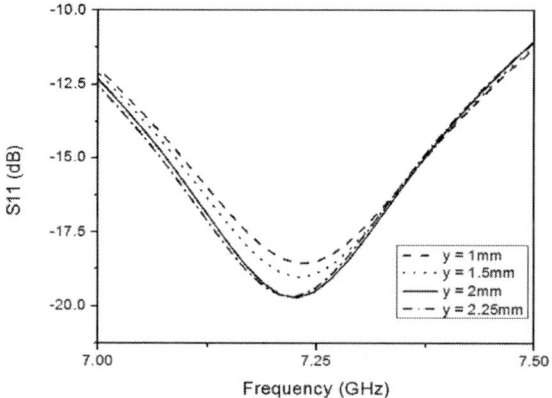

Fig.12: Return loss (S_{11}) variation with y

978-1-5386-6624-1/18 $31.00 © 2018 IEEE

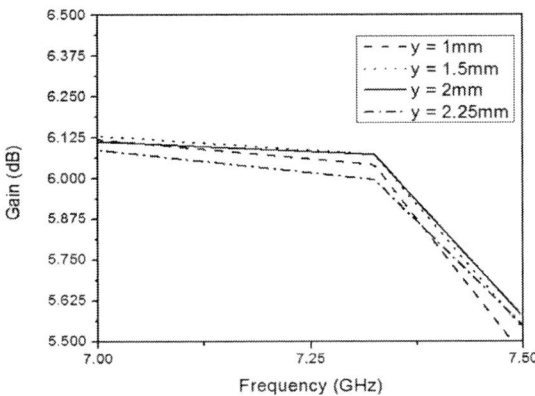

Fig.13: Gain (dB) on variation of y

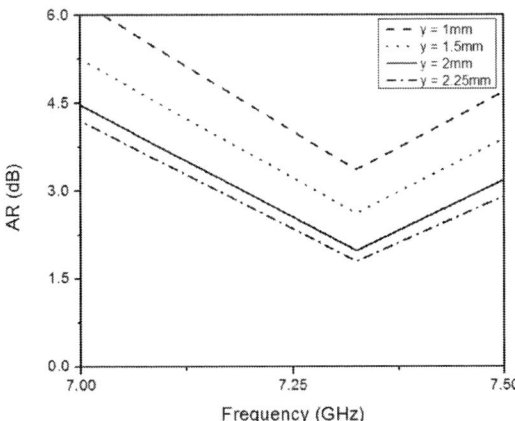

Fig.14: Axial Ratio (dB) on variation of y

IV. RESULTS

The proximity feed microstrip antenna is designed and simulated in CST microwave studio. The effect of variation of air gap between driving patch and superstrate layer, variation of feed position and variation of truncation on antenna parameters is studied in detail in previous section. The initial antenna designed without the superstrate operates at 7.55GHz frequency, offers a bandwidth of 600MHz (8%), a gain of 4dB and is not circularly polarized. A significant improvement in gain of the antenna is observed after adding the superstrate surface with parasitic patch array. The improved antenna with superstrate operates at 7.23 GHz frequency, offers a bandwidth of 632MHz (8.7%), a gain of 6.08dB and is circularly polarized.

A. Comparison of Antenna With and Without Superstrate

A graphical comparison of S_{11} parameter, gain and axial ratio between antenna without superstrate and antenna with superstrate is shown below in Fig.15, Fig.16 and Fig.17 respectively. We can easily observe the significant improvement in circular polarization and in gain of almost 2dB.

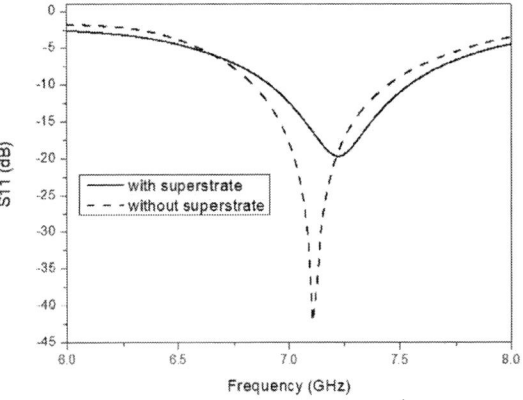

Fig.15: Return loss (S11) comparison

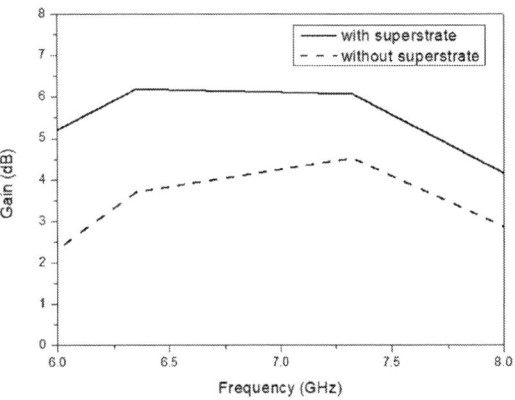

Fig.16: Gain (dB) comparison

Fig.18 and Fig.19 show the radiation patterns of an antenna with and without the superstrates. When the superstrate is introduced, the magnitude of back lobe is increased as compared to that without the superstrate in which the magnitude of back lobe is negligible. The magnitude of the main lobe is thus, enhanced with the introduction of the superstrate. Thereby, the overall gain of the antenna is increased.

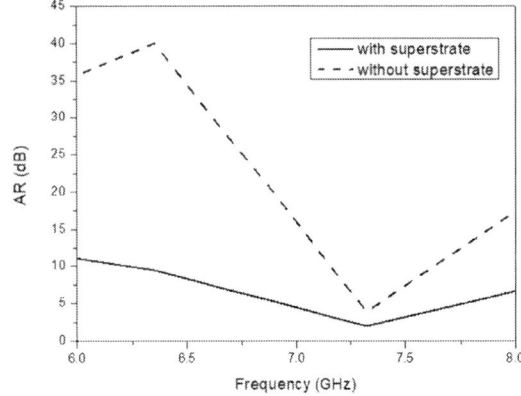

Fig.17: Axial Ratio (dB) comparison

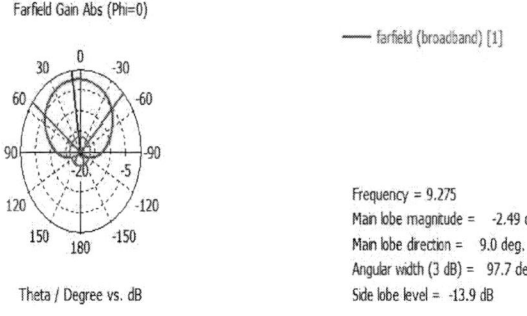

Farfield Gain Abs (Phi=0)

—— farfield (broadband) [1]

Frequency = 9.275
Main lobe magnitude = -2.49 dB
Main lobe direction = 9.0 deg.
Angular width (3 dB) = 97.7 deg
Side lobe level = -13.9 dB

Theta / Degree vs. dB

Fig.18: Simulated Radiation pattern without superstrate

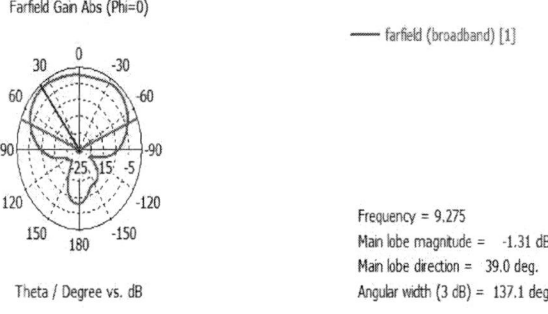

Farfield Gain Abs (Phi=0)

—— farfield (broadband) [1]

Frequency = 9.275
Main lobe magnitude = -1.31 dB
Main lobe direction = 39.0 deg.
Angular width (3 dB) = 137.1 deg

Theta / Degree vs. dB

Fig.19: Simulated Radiation pattern with superstrate

B. Comparison of Simulation and Measured Results

The proposed antenna is fabricated and the antenna parameters are measured in the chamber to compare the simulation results with actual measured results. This gives us a necessary insight of how the antenna will perform in real life practical scenarios. The measured values of return loss (S_{11}) magnitude, gain (dB) and axial ratio (dB) is compared with the simulated values. Fig.4.10, Fig.4.11 and Fig.4.12 shows the comparison of return loss, gain and axial ratio respectively.

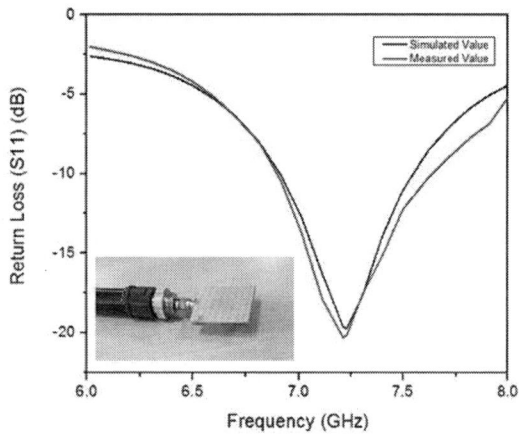

Fig.20: Return Loss (S_{11}) Vs Frequency (inset: fabricated antenna inside anechoic chamber)

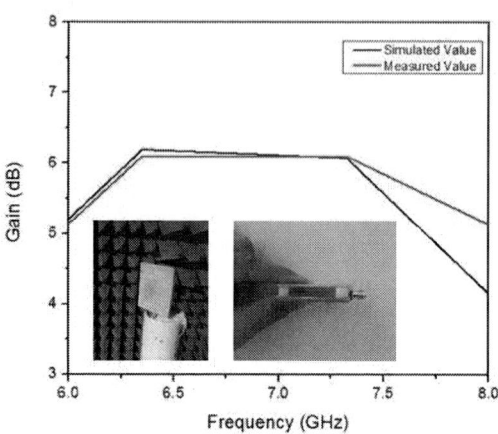

Fig.21: Gain (dB) Vs Frequency (inset: fabricated antenna at laboratory)

The measured values of return loss and gain are replicative of the simulated values obtained in CST Studio Suite. The measured AR value is slightly deviant due to practical losses. The overall comparison shows that the proposed antenna is capable of replicating the simulated values and it can be used for practical applications.

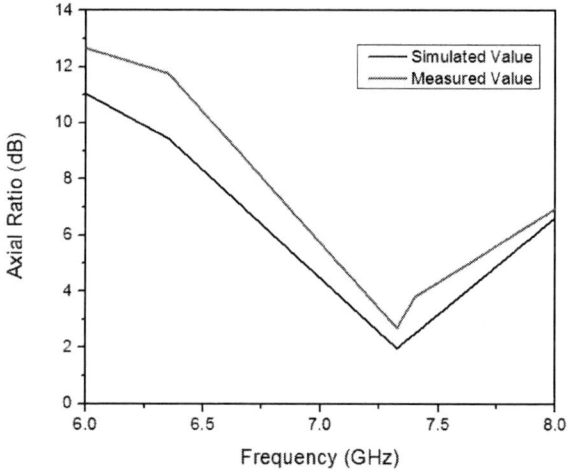

Fig.22 Axial Ratio (dB) Vs Frequency

V. CONCLUSION

In this communication, the gain of a circularly polarized microstrip patch antenna is enhanced using metasurface. The metasurface consists of an array of corner truncated square patches which is placed as superstrate at an optimized distance over the antenna. The proposed antenna achieves peak gain of 6.08 dB at 7.24 GHz whereas the peak gain of the reference antenna is around 4 dB at the same frequency. The minimum gain enhancement of around 1.7 dB is obtained over the entire working band of the proposed antenna when compared with reference antenna. The proposed antenna also shows improvement in impedance and 3-dB ARBW as compared to reference antenna. So, it can be concluded that the overall radiation performance of the proposed antenna is improved by the employment of metasurface. With the advantages of low profile, and circular polarization, the proposed antenna can be implemented for satellite communication purposes, cordless telephones and weather radar systems.

REFERENCES

[1] S. D. Targonski and D. M. Pozar, "Design of wideband circularly polarized aperture coupled microstrip antennas," IEEE Trans. Antennas Propagat., vol. 041, no. 2, pp. 214–220, Feb. 1993.

[?] N. Herscovici, Z. Sipus, and D. Bonefacic, "Circularly polarized single-fed wide-band microstrip patch," IEEE Trans. Antennas Propagat., vol. 51, no. 6, pp. 1277–1280, June 2003.

[3] N. Nasimuddin, K. P. Esselle, and A. K. Verma, "Wideband circularly polarized stacked microstrip antennas," IEEE Antennas Wireless Propagat. Lett., vol. 6, pp. 21–24, 2007.

[4] H. Oraizi and R. Pazoki, "Wideband circularly polarized aperture-fed rotated stacked patch antenna," IEEE Trans. Antennas Propagat., vol. 61, no. 3, pp. 1048–1054, Mar. 2013.

[5] S. Raut and A. Petosa, "GPS wideband circularly polarized microstrip antenna array," in Third European Conf. Antennas and Propagat., Berlin, 2009, pp. 2990–2993.

[6] Haneishi, M., Yoshida, S.: A design method of circularly polarized rectangular microstrip antenna by one-point feed. Electron. Commun. Jpn. (Part I: Communications) 64(4), 4654 (1981)

[7] Sharma, P., Gupta, K.:Analysis and optimized design of single feed circularly polarized microstrip antennas. IEEE Trans. Antennas Propag. 31(6), 949–955 (1983)

[8] Nasimuddin, Esselle, K.P., Verma, A.K.: Wideband circularly polarized stacked microstrip antennas. IEEE Antennas Wirel. Propag. Lett. 6, 21–24 (2007)

[9] Gao, S., Qin,Y., Sambell,A.: Low-cost broadband circularly polarized printed antennas and array. IEEE Antennas Propag. Mag. 49(4), 57–64 (2007)

[10] Wang, M.-Z., Zhang, F.S.:Acircularly polarized elliptical-ring slot antenna using an L-shaped coupling strip. Prog. Electromagn. Res. Lett. 35, 29–35 (2012)

[11] Nasimuddin, N., Chen, Z.N., Qing, X.: Bandwidth enhancement of a single-feed circularly polarized antenna using a metasurface: metamaterial-based wideband CP rectangular microstrip antenna. IEEE Antennas Propag. Mag. 58(2), 39–46 (2016)

[12] Ankit Sharma, Deepak Gangwar, Binod Kumar Kanaujia, Santanu Dwari," Gain enhancement and broadband RCS reduction of a circularly polarized aperture-coupled annular-slot antenna using metasurface," Journal of Computational Electronics, May 2018,

[13] N. Nasimuddin, Zhi Ning Chen, and Xianming Qing," Bandwidth Enhancement of a Single-Feed Circularly Polarized Antenna Using a Metasurface Metamaterial-based wideband CP rectangular microstrip antenna," IEEE Antennas & Propagation Magazine, Vol. 58, pp. 39-46, April 2016.

[14] Son Xuat Ta and Ikmo Park," Compact Wideband Circularly Polarized Patch Antenna Array Using Metasurface," IEEE Antennas and Wireless Propagation Let., Vol. 16 ,pp. 1932 – 1936, March 2017.

[15] Yuandan Dong, Hiroshi Toyao, and Tatsuo Itoh," Compact Circularly-Polarized Patch Antenna Loaded With Metamaterial Structures," IEEE TRANSACTIONS ON ANTENNAS AND PROPAGATION, VOL. 59, NO. 11, NOV. 2011.

[16] Kush Agarwal, Nasimuddin and Arokiaswami Alphones," Wideband Circularly Polarized AMC Reflector Backed Aperture Antenna," IEEE Transactions on Antennas and Propagation , Vol. 61, no. 3, pp. 1456 – 1461, March 2013.

[17] Parikh, H., Pandey, S., & Modh, K. (2012, December). Wideband and high gain stacked microstrip antenna for Ku band application. In Engineering (NUiCONE), 2012 Nirma University International Conference on (pp. 1-5). IEEE.

978-1-5386-6624-1/18 $31.00 © 2018 IEEE

2nd IEEE International Conference on Power Electronics, Intelligent Control and Energy Systems (ICPEICES-2018)

Efficient Feature Extraction using DWT-DCT for Robust Face Recognition under varying Illuminations

Virendra P. Vishwakarma[1], Sahil Dalal[2] and Varsha Sisaudia[3]
University School of Information, Communication & Technology
Guru Gobind Singh Indraprastha University, Sector 16-C, Dwarka, New Delhi, India
[1]virendravishwa@rediffmail.com ,vpv@ipu.ac.in, [2]dalalsahil22@yahoo.co.in , [3]sisaudia.varsha@gmail.com

Abstract—**Face recognition is an important aspect of computer vision since past many decades under uncontrolled variations such as illumination, pose and expression. In this paper, an algorithm is proposed for efficient face recognition under varying illumination by extracting robust features from the illumination normalized face images. This is performed by integrating discrete wavelet transform (DWT) with discrete Cosine transform (DCT) as the feature extraction technique. Combination of DWT and DCT is exploited so that redundancy which is not extracted by DCT alone, is firstly extracted using DWT and subsequently, the local correlation is utilized by DCT. DWT also helps in extracting the global features of the face image. The algorithm is implemented and tested over Yale, Yale B and CMU PIE face databases. As it can be seen here, promising results have been achieved by proposed approach compared to the results of the existing papers.**

Keywords—*Face Recognition, Discrete wavelet transform (DWT), Discrete Cosine transform (DCT), Principal component analysis (PCA), k nearest neighbour (kNN) Classifier.*

I. INTRODUCTION

Every human being has a unique face which helps in differentiating one person from another. Classification of face images is a point of interest since past many decades [1][2]. It is used in security based systems, for the surveillance and monitoring purposes. Classification can be done using support vector machine (SVM) [3], extreme learning machine (ELM) [4], artificial neural network (ANN) [5], nearest neighbour (NN) [6] etc. For this, feature extraction is also an important task that needs to be performed before the classification. As in [3], kernel principal component analysis was used for feature extraction.

Illumination variations on the face of the person make the recognition difficult. It is a very challenging factor in face recognition. The performance of face recognition algorithm varies with the change in illumination level on the face images. Hence, illumination should be normalized for achieving better results in face recognition. Various algorithms have already been proposed for minimizing the effect of illumination on the face images. The state of art illumination normalization methods were given as fuzzy filter with adaptive histogram equalization along with logarithmic transform (AHE+Log) [7], estimation sets and multi-level matching metric [8], discard low frequency DCT (LFDCT) coefficients [9], Gradientfaces [10], 9PL [11], small and large scale (S&L) (discard-LFDCT Coef) [12], S&L (NPL-QI) [12] etc. They have used various methods to nullify the effect of varying illumination on the face images. Various feature extraction techniques were also proposed such as PCA

[13], kernel PCA [3] etc. In the existing techniques, only single tool is utilized for feature extraction. Hybridization of two techniques has not been used for feature extraction in face recognition till now [7], [9], [14], [15]. Infact, combination of DWT and DCT is utilized by Yu et.al, but it was not used for illumination normalization as ORL database does not contain any illumination effects [16]. So, here, this novel method, which utilizes DWT and DCT hybridization for efficient feature extraction under varying illuminations, is proposed for robust face recognition.

The remaining sections are structured as follows: Section II tells about the preliminaries of the concepts used in the proposed approach with a brief description about the datasets used. Section III gives the details about the proposed robust feature extraction method for classification of face images. Section IV provides the experimental results followed by conclusions in Section V.

II. PRELIMINARIES

A. Discrete Wavelet Transform

DWT [17] helps in transforming the gray-scale images to spatial and frequency domain at the same time. In this, successive filtering (using low and high pass filters) is performed so that decomposition of the images can be achieved. This decomposition [17] can be formulated in (1) and (2) as:

$$s_{hi}[k] = \sum x[n].q_l[2k-n] \quad (1)$$

$$t_{lo}[k] = \sum x[n].q_h[2k-n] \quad (2)$$

where x is any signal whose wavelet decomposition to be performed. q_l and q_h are low pass and high pass filters with half the cut-off frequency from the previous one in dyadic DWT respectively.

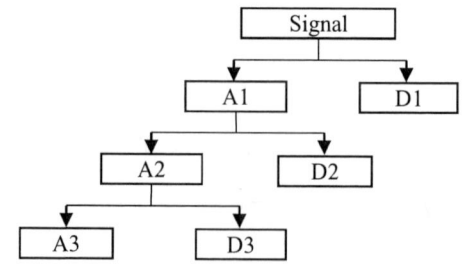

Fig. 1. Wavelet decomposition into approximation and detailed coefficients

978-1-5386-6624-1/18 $31.00 © 2018 IEEE   982

In the wavelet transform, scaling function is represented using (3) and (4) so that low and high pass filters can be mathematically represented.

$$\phi(n) = \sum_{j=0}^{N-1} c_j \phi(2n - j) \tag{3}$$

$$\phi_{j,k}(t) = \sqrt{2^j} . \phi(2^j t - k) \tag{4}$$

Wavelet function [17] for the wavelet transform can be formulated as:

$$\psi_{j,k}(t) = \sqrt{2^j} . \psi(2^j t - k) \tag{5}$$

where $t, k \in Z$. Fig. 1 shows the decomposition of the input signal on the basis of wavelet transform. It is dyadic decomposition by using successive low pass and high pass filters to decompose the input signal into low frequency (A) and high frequency (D) coefficients respectively. It is shown in Fig. 1. *Haar* wavelet is the basic wavelet transform which is used in the present investigation. Scaling and wavelet functions for the *haar* is given by:

$$\varphi(t) = \begin{cases} 1 & if\ 0 \le t < 1 \\ 0 & otherwise \end{cases} \tag{6}$$

$$\psi(t) = \begin{cases} 1 & 0 \le t < 1/2 \\ -1 & 1/2 \le t < 1 \\ 0 & otherwise \end{cases} \tag{7}$$

Here, *haar* wavelet transform is used to extract the low frequency (or approximation) and high frequency (or detailed) coefficients on which further processing is performed using DCT. DWT splits a 2-D image into four sub-images as approximation (LL), horizontal (LH), vertical (HL) and diagonal (HH).

B. Discrete Cosine Transform

DCT is similar to the real part of the discrete Fourier transform. For an $R \times S$ size face image, the 2-D DCT [17] can be defined as:

$$D(p,q) = \beta(p)\beta(q)\sum_{r=0}^{R-1}\sum_{s=0}^{S-1} g(r,s) \times$$

$$\cos\left\lfloor \frac{\pi(2r+1)p}{2R} \right\rfloor \times \cos\left\lfloor \frac{\pi(2s+1)q}{2S} \right\rfloor \tag{8}$$

for, $p = 0, 1, 2, 3, ..., R-1$ and $q = 0, 1, 2, 3, ..., S-1$.

The inverse discrete Cosine transform (IDCT), now, can be defined [17] as:

$$g(r,s) = \sum_{p=0}^{R-1}\sum_{q=0}^{S-1} \beta(p)\beta(q)D(p,q) \times$$

$$\cos\left\lfloor \frac{\pi(2r+1)p}{2R} \right\rfloor \times \cos\left\lfloor \frac{\pi(2s+1)q}{2S} \right\rfloor \tag{9}$$

for, $r = 0, 1, 2, 3, ..., R-1$ and $s = 0, 1, 2, 3, ..., S-1$.

where,

$$\beta(p) = \begin{cases} \dfrac{1}{\sqrt{R}} & ; p = 0 \\ \sqrt{\dfrac{2}{R}} & ; p = 1,2,3,...,R-1 \end{cases} \tag{10}$$

$$\beta(q) = \begin{cases} \dfrac{1}{\sqrt{S}} & ; q = 0 \\ \sqrt{\dfrac{2}{S}} & ; q = 1,2,3,...,S-1 \end{cases} \tag{11}$$

r and s represent the rows and columns of the face image matrix respectively and, p and q denote the rows and columns of the DCT matrix. It is used to transform the data from spatial/time domain to frequency domain. It allows the data to be represented in such a way that operations can be performed on the reduced data and further processing can be done.

C. Fuzzy Set and Logic

The fuzzy set and logic utilized [18] to relate the real world problems with the approximate mathematical reasoning, is a useful tool for solving many real world problems. Here, fuzzy membership functions (MFs) are utilized to nullify the effect of illuminations. It is represented in (12) by the universal set (z), for the real world F [19]:

$$\mu_F : z \to [0,1] \tag{12}$$

In this paper, this universal set is the low frequency DCT coefficients [7].

D. Principal Component Analysis

Principal component analysis (PCA) is a tool which helps in transforming the data linearly into a new frame of reference [15], [20], [21]. PCA attains this by taking values of the variances of the dataset and project the variances such that the largest one will be treated as first principal component as it lies on the first coordinate. Similarly, the second coordinate gives the second principal component, and so on. Hence, this orthogonal tool reduces the number of parameters of the dataset and maps the dataset to a new space of lower dimensionality.

E. kNN Classifier

Nearest neighbour classifier is utilized here to give classification results on the feature vectors. This classifier is the simplest classifier among various kinds of linear and non-linear classifiers. It exploits Euclidean distance (minimum value of distance) for the classification [22]. This parameter can be computed using (13):

$$d(x,y) = \sqrt{\sum_{i=1}^{n} (x_i - y_i)^2} \tag{13}$$

F. Databases Used

a. YALE face database

YALE face database [23] is one of the three databases used for checking the consistency of the proposed approach. This face database has 15 subjects with 11 face images of each subject (i.e. 165 in total) with various face expressions including normal, happy, surprised, wink, sad, occlusion (with or without glasses) and with varying direction of illumination [23]. All of the images in this database are in gif format with 320×243 size of each image. The images are cropped to the resolution of 220×175

978-1-5386-6624-1/18 $31.00 © 2018 IEEE

pixels which is again resized for the proposed algorithm to 138 × 110 of resolution. The number of images for training dataset is varied from 1 to 9 per subject and the remaining images are used for testing purpose.

b. Yale B face database

Yale B face database consists of 10 subjects each having 5 subsets [24]. All the images are divided into these subsets on the basis of the illumination variations. With 0°-12° illumination variation images lies in the Subset 1. Similarly, with 12°-25°, 26°-50°, 51°-77° and above 77° illumination variations, face images lies in Subset 2, 3, 4 and 5 respectively. Before cropping, the original size of the database images is 640 × 480 pixels which is cropped to 192 × 168 and further sub-samples to 120 × 105. Subset 1 with frontal illumination is considered for training dataset and rest subsets are used for testing purpose.

c. CMU PIE database

CMU PIE database consists of 1386 images which are affected by illumination from different directions [25]. This database contains images with pose and expression variations as well. For now, only focus is put on illumination face images. 21 face images with illumination variations are there in each of the 66 subjects. In these 21 images, only one image is uniformly (frontal) lightened and is taken for training purpose. The rest (20) face images are under varying lighting from different directions and are considered as test images for analysis on this database. Consequently, the number of images for training is 66 and 1320 images for testing. These images are cropped to 237 × 197 pixels to include only face with some hair and background in the image which is further resized to 119 × 98 pixels for this investigation.

III. PROPOSED APPROACH

The proposed technique gives a robust method to efficiently extract the features for the recognition of the face images which are illuminated with lights from varying directions. In this paper,

a fixed number of low frequency DCT coefficients which are highly affected by the illumination are modified using the fuzzy filter and then, the resultant illumination normalized face image is used to extract the features with the help of DWT-DCT hybridization. The Yale, Yale B and CMU PIE face databases are utilized here to evaluate the performance of proposed approach. The explanation about the databases is already given in the previous section. As it is already said that the databases contain face images with varying illumination from different directions, they all must be illumination normalized. These images are of low contrast which is pre-processed by histogram equalization (HE). The contrast stretching can also be performed by the adaptive histogram equalization along with logarithmic transform (AHE+LT). Although AHE+LT gives better contrast stretching compared to HE, the computational complexity of AHE+LT is higher than that of HE [17]. DCT of the contrast stretched image is obtained and after arranging the DCT coefficients in zigzag manner, low frequency DCT coefficients are selected. Variation of illumination on the face images is given by (14) and fuzzy filter i.e. utilized to nullify that effect of illumination is formulated in (15) [7].

$$I(z) = 1 - \frac{z^{\beta}}{Z^{\beta}} \tag{14}$$

where, Z denotes the total number of low frequency DCT coefficients that are considered for modifying using fuzzy filtering, $I(z)$ as the illumination component in frequency domain and β as any positive constant.

$$\mu_F = \frac{z^{\beta}}{Z^{\beta}} \tag{15}$$

After the illumination normalization is performed, the modified DCT coefficients can be obtained by using (16):

$$\sum_{z=1}^{RS-1} FD(z) = \sum_{z=1}^{N} D(z) \times M + \sum_{z=N+1}^{RS-1} D(z) \tag{16}$$

Algorithm of the Proposed Method:
Variables: Training Data (*TrnData*), Testing Data (*TstData*), Face Image (*Img*), Illumination Normalized Image (*IllNormImg*)
Result: Percentage error rate as classification outcome
for *TrnData&&TstData* do
    for *Img* do
        Apply HE or AHE+LT on *Img*
        Compute DCT (say *DCT_matrix* is obtained)
        Convert *DCT_matrix* to vector form by scanning *DCT_matrix* in zigzag manner
        Select Z low frequency DCT coefficient from that matrix
        Compute membership grades of fuzzy filter for these Z coefficients using (15)
        Modify the low frequency DCT coefficients using fuzzy membership grades
        FD is obtained using (16)
        Convert fuzzy filtered vector *FD* into 2-D matrix form (un-zigzag)
        Now, compute IDCT to obtain *IllNormImg*
        Apply 1-level DWT on *IllNormImg* (for obtaining LL, LH, HL, HH sub-images)
        LL sub-image is subdivided into blocks of $u \times u$ size and compute DCT over these blocks
        LH, HL and HH sub-images are subdivided into blocks of $v \times v$ size and compute DCT over these blocks
    end
end
Feature vector is obtained
PCA is applied over the feature vector
Classification using *k*NN classifier using (13)

where *D(z)* represents the DCT coefficient obtained by zig-zag scanning of DCT coefficients (5). *M* represents the membership grades of the low frequency DCT coefficients. After taking inverse DCT (6), the output represents illumination normalized face image.

Selection of a fixed number of low frequency DCT coefficients is done optimally so that efficient face recognition can be achieved. With the help of fuzzy filter [7], [15], given by (15) illumination normalization is performed and the illumination normalized face image is used for further processing.

After the face images are illumination normalized, feature extraction is performed. Before feature extraction, all the illumination normalized face images are sub-sampled by 1.6 for Yale and Yale B face databases; and by 2 for CMU PIE face database. DWT (1-level) is performed on these rescaled illumination normalized rescaled face images. It will split the image into four sub-images as horizontal, vertical, approximation and diagonal. These four sub-images coefficients are exploited for feature extraction using DCT. For reducing the redundancy within the sub-images, $u \times u$ and $v \times v$ non-overlapping sub-blocks are made from the approximation sub-image (LL band) and remaining sub-images (horizontal, vertical and diagonal) respectively. DCT is applied on these sub-blocks which helps in obtaining the maximum redundancy from the respective sub-blocks. Only first DCT coefficient is used as a feature value (representative) of the block. This process generates a feature vector of very low dimension corresponding to one face image. Hence the proposed approach of feature extraction is providing significant dimension reduction. These feature vectors from each face image are further used for classification.

The feature vector generated on the outcome of DWT_DCT hybridization, PCA is employed. Further the classification based on these feature vectors is performed using *k*NN classifier with the help of Euclidian distance.

IV. EXPERIMENTAL RESULTS

In this section, a detailed analysis of the experimental results is shown based on the proposed method when applied on the face image databases (Yale, Yale B and CMU PIE). Firstly, Yale face database is experimented.

A. Experiments on Yale face database:

This database has illumination variations along with expression variations, misalignment etc. Illumination factor affects the low frequency DCT coefficients and the variations in this are small, therefore, few of the low frequency coefficients are sufficient to modify by the fuzzy filter so that illumination normalization can be done. Here, 6 low frequency DCT coefficients are modified using the fuzzy filter. This is performed using equations (15)-(16). This efficiently nullifies the illumination effect from the face images.

Illumination normalized face images are used to extract the features from them. Before feature extraction, all the illumination normalized face images are sub-sampled by 1.6. DWT (1-level) is applied on the rescaled illumination normalized face image to decompose it into four sub-images. Only LL sub-image is sufficient enough for finding optimal results on this database. 4×4 ($u = 4$) non-overlapping sub-blocks

are made from LL band. On these sub-blocks DCT is applied. This generates a feature vector of size 252 by taking only first DCT coefficients from each block.

These feature vectors are mapped to eigen-space of 70 principal components using PCA so that dimensionality of the feature vector can be further reduced. *k*NN classifier is used for classification of the feature vectors obtained after PCA. The classification has been performed on different testing dataset by varying the number of images per subject used for training of the YALE face database. As already explained in the Section II, YALE face database contains 165 face images of 15 subjects having 11 images each. For training dataset, images per subject are varied from 1 to 9 and correspondingly for testing dataset, images varied from 10 to 2, hence the testing dataset have 150 to 30 images on which percentage error rate has been measured using *k*NN classifier.

These are represented in the Table I showing variation between percentage error rate and number of images per subject used for training. This table also shows the comparison of the proposed method results with the existing techniques on this database. The proposed approach of feature extraction has been applied on this database with two variants of pre-processing, one with HE and other with AHE+LT. It is clearly evident from this table that the proposed approach when used with AHE+LT significantly outperforms the existing approaches for all the variations of the training datasets.

B. Experiments on Yale B face database:

The proposed approach has been further tested on Yale B database. The results on this database are shown in Table II. Here also, one-level DWT has been applied on the sub-sampled face images. DCT has been applied on three sub-images which are LH, HL and HH on $v \times v$ non-overlapping sub-blocks. Various values of *v* are experimented. Most promising results have been obtained for $v = 2$. All approximation band coefficients have been used in the feature vector, thus the value of *u* is 1. Subset 1 images are used as the training dataset and results are computed by considering Subset 3, 4 and 5 as the testing datasets. Result on Subset 2 is almost same as Subset 3 which is containing comparatively less variation of illumination than Subset 3. Hence the result on Subset 2 has not been shown. Two pre-processing techniques which are HE and AHE+LT have been used. The % error rates of the proposed approach of feature extraction with HE (without illumination normalization) and with AHE+LT (without illumination normalization) are 0, 15, 22.11 and 0, 5, 16.32 respectively on Subset 3, 4, 5. When comparing these error rates of the proposed approach with that of the approach given in [7], it is found that the proposed approach significantly outperforms. It can also be seen from the results that the proposed approach of feature extraction when used with fuzzy filter, gives zero error rate on this database for all the test subsets. The number of principal components for the experiments on Yale B database is 50.

C. Experiments on CMU PIE face database:

Similarly, when experimentation is performed on CMU PIE face database, 100% accurate results are achieved with this database as well. In this database also, image with frontal illumination (one image per subject) is considered for training dataset and remaining all images (20 images per subject) are used for testing dataset. Results on CMU PIE database is shown in Table III.

978-1-5386-6624-1/18 $31.00 © 2018 IEEE

TABLE I. COMPARISON OF THE PROPOSED APPROACH ON YALE DATABASE

Number of images used for training	Error Rate (%)				
	Without illumination normalization using kNNC (PCA) [7]	With fuzzy filter based illumination normalization using kNNC (PCA) [7]	Without illumination normalization using DLM [26]	Proposed approach (with HE)	Proposed approach (with AHE+LT)
1	51.33	43.33	44	22.0	22.67
2	22.22	17.04	12.59	11.11	8.89
3	17.50	10.0	11.67	8.33	2.50
4	16.19	4.76	9.52	7.62	1.90
5	15.56	5.56	8.89	6.67	1.11
6	13.33	4.0	5.33	5.33	1.33
7	15.0	1.67	5.00	5.00	0
8	17.78	2.22	-	6.67	0
9	3.33	0.0	-	3.33	0

TABLE II. COMPARISON OF THE PROPOSED APPROACH ON YALE B DATABASE

Subset	Error Rate (%)					
	Without illumination normalization, only HE using kNNC (PCA) [7]	With fuzzy filter based illumination normalization using kNNC (PCA) [7]	Proposed approach of feature extraction (Without illumination normalization, only HE)	Proposed approach of feature extraction (Without illumination normalization, only AHE+LT)	Proposed approach of feature extraction (With illumination normalization, HE & fuzzy filter)	Proposed approach of feature extraction (Without illumination normalization, only AHE+LT & fuzzy filter)
Subset 3	7.5	0	0	0	0	0
Subset 4	60	0	15.0	5.0	0	0
Subset 5	48.95	0	22.11	16.32	0	0

TABLE III. COMPARISON OF THE PROPOSED APPROACH ON CMU PIE DATABASE

Training and test images per subject	Error Rate (%)					
	Without illumination normalization, only HE using kNNC (PCA) [7]	With fuzzy filter based illumination normalization using kNNC (PCA) [7]	Proposed approach of feature extraction (Without illumination normalization, only HE)	Proposed approach of feature extraction (Without illumination normalization, only AHE+LT)	Proposed approach of feature extraction (With illumination normalization, HE & fuzzy filter)	Proposed approach of feature extraction (Without illumination normalization, only AHE+LT & fuzzy filter)
One training and remaining (20) images for testing	25.98	0	4.39	0.53	0	0

In this database also, DCT has been applied on LH, HL and HH bands on 2 × 2 non-overlapping sub-blocks. Here also the % error rate of the proposed approach of feature extraction with HE only (without illumination normalization) and with AHE+LT only (without illumination normalization) are 4.39 and 0.53 respectively which is very less in comparison to this value (25.98) for the approach given in [7]. The number of principal components in this case is 40.

The number of principal components is taken as 70, 50 and 40 for Yale, Yale B and CMU PIE face databases respectively. These values have been taken to make a fare comparison of the proposed approach with existing state of art approach, as these values were used in the implementation [7] on these databases.

Summary of the the parameters taken in the present investigation are:

i) 1-level DWT has been applied on all the sub-sampled images of the databases. The wavelet function for DWT implementation is 'haar'.

ii) The linear membership function has been used in fuzzy filter represented by (15). The value of β is 1.

iii) The number of low frequency DCT coefficients processed by fuzzy filter is 6, 903 and 10 for Yale, Yale B and CMU PIE face databases respectively.

iv) The number of principal components used in PCA is 70, 50 and 40 for Yale, Yale B and CMU PIE face databases respectively.

V. CONCLUSIONS

It can be seen from the results that, with the proposed approach, a very good classification results have been obtained in classifying Yale, Yale B and CMU PIE face databases. Feature extraction using DWT-DCT hybridization makes the recognition more accurate than that of the existing techniques, as it is clearly evident from the comparison table. The proposed approach is much efficient in reducing the percentage error rate to 0 (for the cases when the number of images used for training are 7 or more) for Yale face database. For Yale B and CMU PIE face databases, the percentage error rate has been reduced to 0 for all the testing datasets. The % error rate for HE and AHE+LT only (without illumination normalization) has also been significantly reduced by the proposed approach of feature extraction as compared to the existing techniques. In future, the proposed approach can be explored with non-linear classification techniques on other databases for achieving more efficient recognition rate.

REFERENCES

[1] T. Goel, V. Nehra, and V. P. Vishwakarma, "An adaptive non-symmetric fuzzy activation function-based extreme learning machines for face recognition," *Arab. J. Sci. Eng.*, vol. 42, no. 2, pp. 805–816, 2017.

[2] W. Zhao, R. Chellappa, P. J. Phillips, and A. Rosenfeld, "Face recognition: A literature survey," *ACM Comput. Surv.*, vol. 35, no. 4, pp. 399–458, 2003.

[3] K. I. Kim, K. Jung, and H. J. Kim, "Face recognition using kernel principal component analysis," *IEEE Signal Process. Lett.*, vol. 9, no. 2, pp. 40–42, 2002.

[4] R. Mehta, N. Rajpal, and V. P. Vishwakarma, "Adaptive Image Watermarking Scheme Using Fuzzy Entropy and GA-ELM Hybridization in DCT Domain for Copyright Protection," *J. Signal Process. Syst.*, 2015.

[5] P. Latha, L. Ganesan, and S. Annadurai, "Face recognition using neural networks," *Signal Process. An Int. J.*, vol. 3, no. 5, pp. 153–160, 2009.

[6] R. Jensen and C. Cornelis, "Fuzzy-rough nearest neighbour classification and prediction," *Theor. Comput. Sci.*, vol. 412, no. 42, pp. 5871–5884, 2011.

[7] V. P. Vishwakarma, "Illumination normalization using fuzzy filter in DCT domain for face recognition," *Int. J. Mach. Learn. Cybern.*, vol. 6, no. 1, pp. 17–34, 2015.

[8] Y. Cheng, L. Jiao, Y. Tong, Z. Li, Y. Hu, and X. Cao, "Directional Illumination Estimation Sets and Multi-Level Matching Metric for Illumination-Robust Face Recognition," *IEEE Access*, 2017.

[9] W. Chen, M. J. Er, and S. Wu, "Illumination compensation and normalization for robust face recognition using discrete cosine transform in logarithm domain," *IEEE Trans. Syst. Man, Cybern. Part B*, vol. 36, no. 2, pp. 458–466, 2006.

[10] T. Zhang, Y. Y. Tang, B. Fang, Z. Shang, and X. Liu, "Face recognition under varying illumination using gradientfaces," *IEEE Trans. Image Process.*, vol. 18, no. 11, pp. 2599–2606, 2009.

[11] K.-C. Lee, J. Ho, and D. J. Kriegman, "Acquiring linear subspaces for face recognition under variable lighting," *IEEE Trans. Pattern Anal. Mach. Intell.*, vol. 27, no. 5, pp. 684–698, 2005.

[12] X. Xie, W.-S. Zheng, J. Lai, P. C. Yuen, and C. Y. Suen, "Normalization of face illumination based on large-and small-scale features," *IEEE Trans. Image Process.*, vol. 20, no. 7, pp. 1807–1821, 2011.

[13] J. Yang, D. Zhang, A. F. Frangi, and J. Yang, "Two-dimensional PCA: a new approach to appearance-based face representation and recognition," *IEEE Trans. Pattern Anal. Mach. Intell.*, vol. 26, no. 1, pp. 131–137, 2004.

[14] Z. M. Hafed and M. D. Levine, "Face recognition using the discrete cosine transform," *Int. J. Comput. Vis.*, vol. 43, no. 3, pp. 167–188, 2001.

[15] V. P Vishwakarma, S. Pandey, and M. N. Gupta, "An illumination invariant accurate face recognition with down scaling of DCT coefficients," *J. Comput. Inf. Technol.*, vol. 18, no. 1, pp. 53–67, 2010.

[16] M. Yu, G. Yan, and Q. Zhu, "New face recognition method based on dwt/dct combined feature selection," in *Machine Learning and Cybernetics, 2006 International Conference on*, 2006, pp. 3233–3236.

[17] R. Gonzalez and R. Woods, *Digital image processing*. India: Pearson Education, 2006.

[18] L. A. Zadeh, "The concept of a linguistic variable and its application to approximate reasoning—I," *Inf. Sci. (Ny).*, vol. 8, no. 3, pp. 199–249, 1975.

[19] L. A. Zadeh, "Fuzzy sets," *Inf. Control*, vol. 8, no. 3, pp. 338–353, 1965.

[20] M. Turk and A. Pentland, "Eigenfaces for Recognition," *Journal of Cognitive Neuroscience*, vol. 3, no. 1. pp. 71–86, 1991.

[21] P. N. Belhumeur, J. P. Hespanha, and D. J. Kriegman, "Eigenfaces vs. fisherfaces: Recognition using class specific linear projection," *IEEE Trans. Pattern Anal. Mach. Intell.*, vol. 19, no. 7, pp. 711–720, 1997.

[22] H. Samet, "K-nearest neighbor finding using MaxNearestDist," *IEEE Trans. Pattern Anal. Mach. Intell.*, vol. 30, no. 2, pp. 243–252, 2008.

[23] "YALE Face Database." [Online]. Available: http://cvc.yale.edu/projects/yalefaces/yalefaces.%0Ahtml.

[24] "YALE B Face Database." [Online]. Available: http://cvc.yale.edu/projects/yalefacesB/%0AyalefacesB.html.

[25] T. Sim, S. Baker, and M. Bsat, "The CMU pose, illumination, and expression (PIE) database," in *Automatic Face and Gesture Recognition, 2002. Proceedings. Fifth IEEE International Conference on*, 2002, pp. 53–58.

[26] V. P. Vishwakarma, "Deterministic learning machine for face recognition with multi-model feature extraction," in *Contemporary Computing (IC3), 2016 Ninth International Conference on*, 2016, pp. 1–6.

978-1-5386-6624-1/18 $31.00 © 2018 IEEE

2nd IEEE International Conference on Power Electronics, Intelligent Control and Energy Systems (ICPEICES-2018)

ECG Classification using Kernel Extreme Learning Machine

Sahil Dalal[1], Virendra P. Vishwakarma[2] and Varsha Sisaudia[3]

University School of Information, Communication & Technology
Guru Gobind Singh Indraprastha University, New Delhi, India

[1]dalalsahil22@yahoo.co.in,[2]virendravishwa@rediffmail.com, vpv@ipu.ac.in,[3]sisaudia.varsha@gmail.com

Abstract—**Almost all humans have different ECG waveforms though the frequency range is quite narrow which is confined within few hundred hertz only. The doctors, especially cardiologist use Electro-Cardio-Graphy (ECG) to acquire first-hand knowledge about the well-being of human heart. Generally, ECG provides indicative information for most of the cardiac ailments. ECG can also be used effectively for other non-medical applications such as biometrics, person identification etc., but in such cases large amount of ECG data is required to be processed and classified to obtain desired results. This process is quite complicated, tedious and time consuming. The proposed method classifies UCI repository dataset of ECG signals based upon the parameters that include P, Q, R, S, T peaks, their envelopes width etc. This method is a comprehensive measure of non-linearity and gives better results in terms of time complexity. It provides a quality detection technique in comparison to the other methods, used earlier in this field.**

Keywords—**Electro-Cardio-Graphy (ECG), Kernel Extreme Learning Machine (KELM),Arrhythmia.**

I. INTRODUCTION

The improvement of Electro-Cardio-Gram (ECG) analysis which is part of bio-signal processing to obtain the ECG classification has been studied by many researchers in the past[1]. ECG waveform with its characteristic points [2]is shown in the Fig. 1. These studies have been carried out through experimental and quantitative works. In the ECG analysis, the main idea is to enhance the degree of accuracy in classifying the ECG. By utilizing the suitable analysis methods, ECG classification can be achieved accurately at a faster rate through the analysis process. Therefore, the past studies of ECG algorithms are an important aspect, hence needs in-depth review. However, starting with the earliest diagnosis and treatment based upon ECG has been used to observe these signals. The ECG analysis techniques require the feature extraction and classifier stages. For the pattern recognition of ECG waveform, there are different solutions presented in the literature which have been proposed during the last decade and are under constant evaluation using the methods like support vector machine (SVM) [3], principal component analysis (PCA)+SVM [4], neural networks (NN) [5], [6], *k*-nearest neighbor (*k*NN) [7], Fuzzy+SVM [8], *k*NN+SVM [9], random forest algorithm [10], naïve bayes [11] and other methods with each approach exhibiting its own advantages and disadvantages. SVM, used in [3], [4], [8] and [9] gave better results only when utilized with some other tool. However, beside these methods, kernel extreme

learning machine (KELM) [12] can also be used for the classification purposes. Earlier this method had been used by Wan-Yu Deng et al. for cross-person activity recognition [12], Onguz Findik et al. in color image watermarking [13] etc.

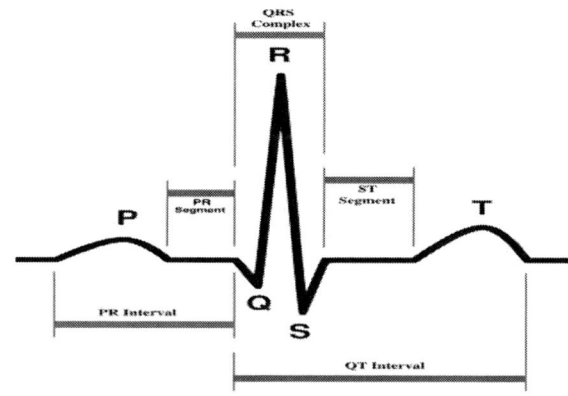

Fig. 1. ECG waveform with its characteristic points

KELM had not been utilized in the ECG waveform analysis and for ECG classification till now. From the literature, it is found that the ECG analysis systems developed by using hybrid algorithms are too complex. Such techniques need a very precise algorithm in order to achieve good efficiency and accuracy. In this paper, a novel approach of ECG classification using KELM is presented. Proposed approach overcomes the complexity of hybrid systems and it is sufficient to provide good classification accuracy of ECG alone, even without fusion with any other tool. ECG is a non-linear signal and requires a technique which can be operated on the non-linear signals. Hence, for fulfilling the same purpose, KELM [14] is utilized.Therefore, the present work in this paper is an attempt to simplify and maximize the classification accuracy of non-linear signals (ECG).

In the proposed method, ECG database with various parameters depicting various conditions of heart (452 in total) are utilized by the KELM to classify the ECG database. Results are also compared with the other existing methods. The paper is organised as follows. Section II provides a brief overview of the preliminaries used. Section III provides details about the proposed method. Section IV provides the classification results followed by conclusion in Section V.

978-1-5386-6624-1/18 $31.00 © 2018 IEEE

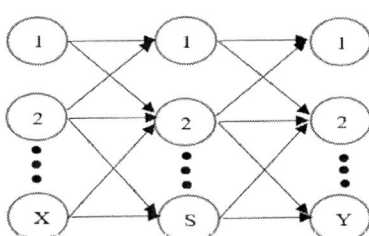

Fig. 2. ARCHITECTURE OF SINGLE LAYER FFNN

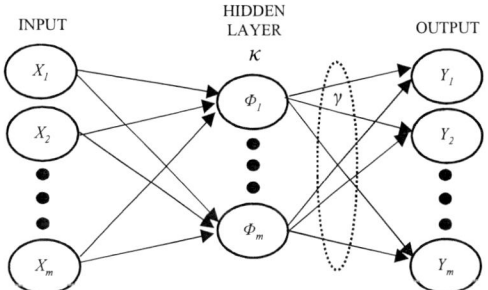

Fig. 3. ARCHITECTURE OF KELM IN WHICH HIDDEN LAYER BECOMES THE KERNEL MATRIX [16]

II. PRELIMINARIES

A. Extreme Learning Machine

Extreme learning machine (ELM) is an algorithm that is simply a single hidden layer feed forward neural network (FFNN) [15]. Unlike NN, ELM randomly assigns its hidden layer weights and need not to be learned iteratively. This algorithm requires less human supervision and can be considered to be a learning process formed as a non-linear mapping process of input space (R^D) to a high dimensional feature space (R^H).Here, D is the dimensions of input patterns and H is dimensions of the high dimensional feature space [17]. This is followed by an optimization scheme which is employed for the linear projection determination on the high (hidden layer) to low dimensional data feature space (output layer).Then, classification is performed at this low dimensional feature space (R^L) by any linear classifier [18], [19].L represents the number of classes in the problem for which ELM has been applied. The architecture for the single layer FFNN is shown in Fig. 2. It is having a hidden layer having S number of neurons. The input weights and bias are selected on random initialization of the parameters of the hidden layer. Compute the output φ for the hidden layer matrix[19] as follows:

$$Y = \sum_{m=1}^{M} \gamma_m \varphi(a_m, b_m, x) = \gamma.H(x) \qquad (1)$$

where $H(x) = [\varphi(a_1,b_1,x)..., \varphi(a_M,b_M,x)]$ is the output matrix obtained from the hidden layer with respect to the input x. a_M and b_M are the input weight and bias respectively which are randomly generated. The output weight γ which connects output and hidden nodes is obtained analytically and can be obtained by:

$$\gamma = H^{-1}Y = H^T\left(\frac{I}{C} + HH^T\right)^{-1}T \qquad (2)$$

where T is the target class and C is user defined parameter for regularization. Therefore, ELM model can be formulated as:

$$Y_{ELM}(x) = H(x)H^T\left(\frac{I}{C} + HH^T\right)^{-1}T \qquad (3)$$

ELM is to minimize the training errors and the output weights norm i.e. $||\gamma||^2$. It also allows to reduce the computational time that is required in optimizing the parameters. Gradient-descent

methods or global search method stake a much longer time as compared to ELM. With these advantages, ELM also has some disadvantages. It has local minima issue and has easy over-fitting [17].

B. Kernel Extreme Learning Machine

To overcome the limitations of ELM, another algorithm is proposed in terms of kernel matrix. The kernel matrix [16]can be obtained using (4):

$$\kappa = \sum_{c=1}^{M} H(a_c, b_c, x_m).H(a_c, b_c, x_m) \qquad (4)$$

where, $x_m \in R^N$ denotes the training data and M represents the number of training data used. This kernel matrix κ that can be obtained using Mercer's condition

$$\kappa = HH^T \qquad (5)$$

where,

$$\kappa = K(x_j, x_k) \qquad j, k = 1, 2, 3, ..., n$$

Here, K is the kernel function used to obtain the kernel matrix. Therefore, utilizing the equations (3) and (5), formulation for KELM can be done as:

$$Y_{KELM}(x) = H(x)H^T\left(\frac{I}{C} + \kappa\right)^{-1}T \qquad (6)$$

This approach is named as kernel extreme learning machine (KELM), can be used as a binary as well as for multi-class classification. This is basically an arrangement of KELM kernel matrix and some standard optimization method that is utilized to find the solution. This optimization method is expressed as[16]:

$$\min F = \frac{1}{2}\sum_{p=1}^{M}\sum_{q=1}^{M}\tau_p\tau_q\kappa(x_p, x_q)\delta_p\delta_q - \sum_{p=1}^{M}\delta_p \qquad (7)$$

subject to $0 \leq \delta_p \leq C, p = 1,2,3,...., M$

The architecture for KELM is shown in Fig. 3. In this, hidden layer becomes the kernel matrix [16]. Here, X are the inputs, Y are the outputs and κ is the hidden layer kernel i.e. computed using (4).

978-1-5386-6624-1/18 $31.00 © 2018 IEEE

C. Database Used

UCI repository database [20] is used for checking the consistency of the proposed approach. This arrhythmia database contains 452 number of subjects and 279 attributes in terms of different parameter values and 280th attribute tells about the respective class to which these all attributes belong. These 452 subjects represent the arrhythmia type to which these classes relate. Table I shows different arrhythmia classes with number of instances belonging to each of the classes in the dataset.

TABLE I.
COMPARISONARRHYTHMIA CLASSES WITH CORRESPONDING NUMBER OF INSTANCES IN THEDATASET[20]

Class	Class Name	No. of instances
1	Normal	245
2	Ischemic changes (Coronary Artery Disease)	44
3	Old Anterior Myocardial Infarction	15
4	Old Inferior Myocardial Infarction	15
5	Sinus tachycardy	13
6	Sinus bradycardy	25
7	Ventricular Premature Contraction (PVC)	3
8	Supraventricular Premature Contraction	2
9	Left bundle branch block	9
10	Right bundle branch block	50
11	1. degree AtrioVentricular block	0
12	2. degree AV block	0
13	3. degree AV block	0
14	Left ventricule hypertrophy	4
15	Atrial Fibrillation or Flutter	5
16	Others	22

Here, Class 1 refers to the normal ECG and classes 2 to 15 represents the different classes of arrhythmia. Class 16 is the remaining which are unclassified ones. From all the attributes, 206 attributes are linear valued and the rest are nominal. In this database, there are many missing values in the 11th to 15th attributes of each class[4].

III. PROPOSED APPROACH

The proposed approach gives a novel method of classifying ECG arrhythmia by exploiting KELM as a classifier. In this paper, UCI repository arrhythmia database is used to check the efficiency of the proposed method. As it is already explained in the above sub-section, this database has 452 different classes of 16 different arrhythmic situations with each situation represented by 279 attributes which includes age, sex, height, weight, duration of QRS complex, duration between onset P and Q waves, between Q and offset T waves, duration between two consecutive P waves etc. These all duration values are in milliseconds and represent average values. First four values in these attributes represent the general description of the subject

and the rest attributes gives the information about the ECG taken from a standard 12 lead ECG recording[6], [9].

As it is stated above in UCI repository database sub-section, attributes 11th to 15th contains some missing values[4], therefore, there is a strong need to pre-process this data to maintain the reliability and relevance of the arrhythmia database. In the previous researches, it is done by removing those columns which contains all missing values or all zeroes. They have also removed those rows which contained missing values in the database and also removed those classes which have insignificant number of classes (attributes belonging to 16th class). Here, pre-processing is done by replacing all the missing values with the standard deviation value of the remaining attributes of that class in the dataset. The dataset was a multiclass problem and highly biased towards the normal arrhythmia (as shown in Table I) and hence, to remove that problem all the 2 to 16 classes are considered as one single class and named as abnormal arrhythmia. Now the problem becomes the binary classification with 245 normal arrhythmia and 207 abnormal arrhythmia classes[6]. It is represented in Table II.

TABLE II.   ARRHYTHMIA (NORMAL AND ABNORMAL) CLASSES WITH CORRESPONDING NUMBER OF INSTANCES

Class	Class Name	No. of instances
1	Normal	245
2	Abnormal	207

After the pre-processing of the arrhythmia database, classification of this database is performed using KELM. It is performed by taking various percentage of database for the training purpose. It is depicted in the Table III. Here, UCI repository arrhythmia database is distributed in five different situations. This includes distribution of training and testing datasets. It is done by varying the percentage of training dataset from 50% to 90%. And accordingly testing dataset is varied. Say, when training dataset is 50% means 226 UCI repository arrhythmia subjects are used for training, remaining 226 subjects are considered for testing. Similarly, for 60%, 70%, 80%, 85% and 90% of total database is used for training and remaining subjects (40%, 30%, 20%, 15% and 10%) are used for testing respectively. These datasets are applied over KELM for classification. KELM contains mainly three variables on which the reliability of this classifier depends i.e. regularization coefficient, type of the kernel used and kernel parameters. Values of regularization coefficient and kernel parameters are decided as the optimal value which can give best classification results. Types of kernel can be RBF, linear, polynomial and wavelet kernel. Here, RBF kernel is used for the proposed approach.

IV. EXPERIMENTAL RESULTS

In this section, KELM evaluation over the UCI arrhythmia database[6], [7], [20] is presented. UCI arrhythmia database is, firstly, converted to a binary classification problem, as this database is highly biased towards normal arrhythmia with 245 classes among 452 in total. Therefore, rest 207 different arrhythmias are grouped together in single class and named as abnormal arrhythmia. It is represented in Table II.

Using this modified database, KELM is utilized to classify the normal and abnormal arrhythmias in their respective classes. Before classification, missing values, present in 11th to 15th attributes of the database, are replaced by the standard deviation of the remaining values of the remaining attributes (other than 11th to 15th attributes) belong to that class. Hence, all the 279 attributes have some values in them and no value is remained missing as it was in the original database before pre-processing is performed.

For the classification process, there are some parameters in KELM which need to be selected and there is no appropriate method existing in the literature for finding them. Hence, the values of the variables used in the KELM are optimally selected based upon experiments which give best classification accuracy. These variables includes: regularization coefficient, kernel parameter and kernel type. Regularization coefficient is utilized in KELM to remove the problems of over-fitting [17] or to solve the ill-posed problems. It is selected optimally as 5 for the proposed method. Kernel parameter of KELM is related to the non-linear mapping from lower dimensional to higher dimensional feature space and its value is selected as 1 optimally. The kernel type, which includes linear, polynomial, wavelet and RBF, RBF kernel is selected for the present investigation. Results for the proposed method are shown in the Table III.

TABLE III. PROPOSED APPROACH ACCURACY ON UCI ARRHYTHMIA DATABASE USING KELM

%age of Data used for training	Training Data Used	Testing Data Used	Accuracy (%)
50	226	226	63.27
60	271	181	61.33
70	316	136	63.97
80	361	91	73.63
85	384	68	77.94
90	406	46	78.26

TABLE IV.

COMPARISON OF THE PROPOSED METHOD WITH THE EXISTING TECHNIQUES

Techniques Used	Accuracy (%)
Proposed Approach (KELM) (90% training data)	78.26
Proposed Approach (KELM) (85% training data)	77.94
Proposed Approach (KELM) (80% training data)	73.63
Wrapper method [11]	74.47
Kernel Difference weighted KNN [7]	70.66
SVM+Logistic regression [3]	70-77
Modular Neural Network [6]	75-80

In Table III, five different situations are considered in which the efficiency of the KELM is tested over this arrhythmia database. Dataset selected for the training of KELM is varied from 50% to 90% and accuracy is measured for the same. Distribution of the database, in training and testing sets, is shown in Table III. When the training data is 50% of the database, 63.27% of the tested data is correctly classified using the proposed method. This accuracy starts increasing when the training data is increased. From Table III, when the training dataset used for KELM training are 60%, 70%, 80%, 85% and 90%, correctly classified data over the remaining test dataset are 61.33%, 63.97%, 73.63%, 77.94% and 78.26% respectively.

The results achieved are also compared with some existing techniques. It is shown in the Table IV. Proposed approach has an advantage over the existing techniques that whole of the database in the experimental analysis has been used in the proposed approach while other techniques do not use the whole arrhythmia database for classification. They utilized less number of attributes than 279 (in total). Proposed approach is applied and tested over all attributes and gives an accuracy of 78.26% (maximum) in classifying the UCI arrhythmia database.

V. CONCLUSION

The proposed approach classified the arrhythmias of UCI repository database with promising results. Comparison with previous studies clearly revealed that the proposed method is more efficient and faster. Replacing the missing values with standard deviation of the remaining attributes provides a much accurate way of pre-processing. The classification using KELM on this processed database performed efficiently with 78.26% of accurate results. Even this method gave promising results with all the attributes (279 attributes), there is still an improvement needed in terms of accuracy. Proposed method requires a large amount of training data for achieving good results. It will be further explored with other efficient feature extraction techniques to improve the performance of the proposed method. It can also be explored with some other pre-processing approaches on the missing data, so that better results on the UCI arrhythmia database can be achieved. This approach can also be applied on the EEG signals to identify the type of disorder so that proper diagnosis of brain can be possible.

REFERENCES

[1] E. J. da S. Luz, W. R. Schwartz, G. Cámara-Chávez, and D. Menotti, "ECG-based heartbeat classification for arrhythmia detection: A survey," Comput. Methods Programs Biomed., vol. 127, pp. 144–164, 2016.

[2] S. Dalal and R. Birok, "Analysis of ECG Signals using Hybrid Classifier," Int. Adv. Res. J. Sci. Eng. Technol., vol. 3, no. 7, pp. 89–95, 2016.

[3] A. Uyar and F. Gurgen, "Arrhythmia classification using serial fusion of support vector machines and logistic regression," in Intelligent Data Acquisition and Advanced Computing Systems: Technology and Applications, 2007. IDAACS 2007. 4th IEEE Workshop on, 2007, pp. 560–565.

[4] K. Polat and S. Güne\cs, "Detection of ECG Arrhythmia using a differential expert system approach based on principal component analysis and least square support vector machine," Appl. Math. Comput., vol. 186, no. 1, pp. 898–906, 2007.

[5] R. D. Raut and S. V Dudul, "Arrhythmias classification with MLP neural network and statistical analysis," in Emerging Trends in Engineering and Technology, 2008. ICETET'08. First International

Conference on, 2008, pp. 553–558.

[6] S. M. Jadhav, S. L. Nalbalwar, and A. A. Ghatol, "Modular neural network based arrhythmia classification system using ECG signal data," *Int. J. Inf. Technol. Knowl. Manag.*, vol. 4, no. 1, pp. 205–209, 2011.

[7] W. M. Zuo, W. G. Lu, K. Q. Wang, and H. Zhang, "Diagnosis of cardiac arrhythmia using kernel difference weighted KNN classifier," in *Computers in Cardiology, 2008*, 2008, pp. 253–256.

[8] N. O. Ozcan and F. Gurgen, "Fuzzy support vector machines for ECG arrhythmia detection," in *Pattern Recognition (ICPR), 2010 20th International Conference on*, 2010, pp. 2973–2976.

[9] K. A. K. Niazi, S. A. Khan, A. Shaukat, and M. Akhtar, "Identifying best feature subset for cardiac arrhythmia classification," in *Science and Information Conference (SAI), 2015*, 2015, pp. 494–499.

[10] P. Shimpi, S. Shah, M. Shroff, and A. Godbole, "A Machine Learning Approach for the Classification of Cardiac Arrhythmia," in *Computing Methodologies and Communication (ICCMC), 2017 International Conference on*, 2017, pp. 603–607.

[11] A. Mustaqeem, S. M. Anwar, M. Majid, and A. R. Khan, "Wrapper method for feature selection to classify cardiac arrhythmia," in *Engineering in Medicine and Biology Society (EMBC), 2017 39th Annual International Conference of the IEEE*, 2017, pp. 3656–3659.

[12] W.-Y. Deng, Q.-H. Zheng, and Z.-M. Wang, "Cross-person activity recognition using reduced kernel extreme learning machine," *Neural Networks*, vol. 53, pp. 1–7, 2014.

[13] O. Findik, I. Babaoğlu, and E. Ülker, "A color image watermarking scheme based on hybrid classification method: Particle swarm optimization and k-nearest neighbor algorithm," *Opt. Commun.*, vol. 283, no. 24, pp. 4916–4922, 2010.

[14] F. Cao, Z. Yang, J. Ren, M. Jiang, and W.-K. Ling, "Linear vs. Nonlinear Extreme Learning Machine for Spectral-Spatial Classification of Hyperspectral Images," *Sensors*, vol. 17, no. 11, p. 2603, 2017.

[15] G.-B. Huang, Q.-Y. Zhu, and C.-K. Siew, "Extreme learning machine: a new learning scheme of feedforward neural networks," in *Neural Networks, 2004. Proceedings. 2004 IEEE International Joint Conference on*, 2004, vol. 2, pp. 985–990.

[16] C. M. Wong, C. M. Vong, P. K. Wong, and J. Cao, "Kernel-based multilayer extreme learning machines for representation learning," *IEEE Trans. neural networks Learn. Syst.*, 2016.

[17] V. P. Vishwakarma, "Deterministic learning machine for face recognition with multi-model feature extraction," in *Contemporary Computing (IC3), 2016 Ninth International Conference on*, 2016, pp. 1–6.

[18] G.-B. Huang, Q.-Y. Zhu, and C.-K. Siew, "Extreme learning machine: theory and applications," *Neurocomputing*, vol. 70, no. 1–3, pp. 489–501, 2006.

[19] G.-B. Huang, H. Zhou, X. Ding, and R. Zhang, "Extreme learning machine for regression and multiclass classification," *IEEE Trans. Syst. Man, Cybern. Part B*, vol. 42, no. 2, pp. 513–529, 2012.

[20] M. Lichman, "UCI Repository Arrhythmia Database," *{UCI} Machine Learning Repository*, 2013. [Online]. Available: https://archive.ics.uci.edu/ml/datasets/Arrhythmia

978-1-5386-6624-1/18 $31.00 © 2018 IEEE

2nd IEEE International Conference on Power Electronics, Intelligent Control and Energy Systems (ICPEICES-2018)

Design and Experimental analysis of Integer Order PID Controller for Ceramic IR heating system

Vineet Shekher
Department of Electrical Engineering,
Noida Institute of Engineering and Technology,
Greater Noida, Utter Pradesh,201306, India
vshekher2407@gmail.com

Ajendra Singh
Department of Electrical Engineering
Delhi Technological University
Delhi-110042, India
ajendrasingh25@gmail.com

J. N. Rai
Department of Electrical Engineering
Delhi Technological University
Delhi-110042, India
jnrai@dce.ac.in

Abstract -This paper shows the tuning methods for determination of parameters of proportional-integral- derivative (PID) controllers with reference to fractional order model with unity power function. The study focused to the step response of real time hardware in loop of ceramic temperature control to the response of a MATLAB Simulink of the reference model for the precise and smoother control experience. This is much benefited when worked with model and uncertain parameters in real time control application. Real time Experimental study shows that the reference model driven by auto-tuning of the HIL temperature loop.

Keywords -Integer Order PID Controller, Ceramic Infrared heater, First Order plus Time Delay System, Data Acquisition

I. INTRODUCTION

Due to having simple tuning rules and design, standard PID controller has great success. In literature, several schemes and design methods were presented. Zeigler-Nichols, Cohen-Coon and Åström-Hugglund tuning method for PID being proposed in [1, 4, 5]. The main module available in MATLAB, for the usage information for the users in [1, 4, 8]. Fractional calculus introduced for differentiation and integration of non-integer orders. Podlubny introduced the basics of fractional controller having generalized PI Dμ controller, in which and μ are the order of integrator and differentiator. Using integer order PID controllers, which have five tuning parameters, in which differentiation and integration parameters are taken as unity whereas fractional order controller provides flexibility in the design process, being based on five tuning parameters (Kp, Ki, Kd, ,μ), but the calculation and setting of these parameter are more complex Over the last few decades, theoretical research on fractional order integration and derivative for many application proposed based on fractional calculus based controller in various fields such as transmission lines, modelling of thermal system, signal processing, system identification and control system[6,7].

In this section of the paper a short review of fractional calculus based IOPID controllers, as presented in the literature is presented. In chapter 2, the concepts related to the fractional calculus and IOPID controllers are presented. In chapter 3, simulation and real time implementation of various PID controlled systems, tuned by Zeigler-Nichols method, Cohen

Coon, Åström-Hugglund and IOPID method are tested and final result are considered and include in conclusion section.

II. FRACTIONAL CALCULUS

The fractional order calculus (FOC) deals with the branch of mathematics, in which differentiation and Integration work under real or even complex number, not only the integer one. The FOC has great consequences in modern research and real-world applications.

The fractional order differential equation for the continuous time or time invariant system shown by the equation (1)

$$a_n D^{\alpha_n} y(t) + a_{n-1} D^{\alpha_{n-1}} y(t) + ---- + a_0 D^{\alpha_0} y(t)$$

$$= b_m D^{\beta_m} u(t) + b_{m-1} D^{\beta_{m-1}} u(t) + ---- + \beta_0 D^{\beta_0} u(t) \quad (1)$$

Where input signal is $u(t)$, output signal is y (t) , fractional derivative represented as $D^\gamma \equiv {}_0 D_t^\gamma$, constants are denoted as a^K where (K=0,----n) and b^K where (K=0,-----,m) and α_K with (k=0,-----,n) and β_K with (K=0,-----,m) are arbitrary real numbers.

Stability is very fundamental and critical requirement for control system design. The stability of an integer order continuous time linear time invariant system is stable if and only if all roots of its characteristics equation have negative real parts.

The differ-integral operator, denoted by ${}_a D_t^q$, commonly used in fractional calculus and expressed in single notation by (2)

$$_a D_t^q = \begin{cases} \dfrac{d^q}{dt^q} & q > 0 \\ 1 & q = 0 \\ \displaystyle\int_a^t (de)^{-q} & q < 0 \end{cases}$$

$$(2)$$

Where q is the fractional order and a and t are the limits of operation. The Grunwald- Letnikov (G-L) definition [18] is given as:

$$_a D_t^q f(t) = \frac{d^q f(t)}{d(t-a)^q} = \lim_{N \to \infty} \left[\frac{t-a}{N} \right]^{-q} \sum_{j=0}^{n-1} (-1)^j \binom{q}{j} f\left(t - j \left[\frac{t-a}{N} \right] \right)$$

978-1-5386-6624-1/18 $31.00 © 2018 IEEE

The Riemann- Lowville (R-L) definition is the simplest and easiest definition to use. This definition is given as (3):

$$_aD_t^q f(t) = \frac{d^q f(t)}{d(t-a)^q} = \frac{1}{\lceil(n-q)} \frac{d^n}{dt^n} \int_0^t (t-z)^{n-q-1} f(z)dz \tag{3}$$

Where, 'n' is the integer which is not less than q i.e. $n-1 \leq q < n$ and $\lceil$ is the gamma function.

$$\overline{\lceil(z)} = \int_0^\infty t^{z-1} e^{-t} dt$$

For function $f(t)$ having n continuous derivative for $t \geq 0$ where $n-1 < q < n$, the G-L and R-L definitions are equivalent. The Laplace transform of the Riemann- Lowville fractional integral and derivative are given as follow (4):

$$L\left\{_0D_t^q f(t)\right\} = s^q F(s) - \sum_{k=0}^{n-1} s^k \, _0D_t^{q-k-1} f(0) \quad n-1 < q \leq n \in N \tag{4}$$

It is observed that transform technique is not suitable for R-L fractional derivative because it should have the knowledge of the non-integer derivatives of the functions at t=0..

The equations with R-L operators are equivalent to Caputo operator with homogeneous initial conditions assumptions and Laplace transform of the Caputo fractional derivative is given as (5)

$$L\left\{_0D_t^q f(t)\right\} = s^q F(s) - \sum_{k=0}^{n-1} s^{q-k-1} f^{(k)}(o) \quad n-1 < q \leq n \in N \tag{5}$$

The Caputo fractional derivative in the form of Laplace transform for zero initial condition, above equation takes the form as below:

$$L\left[_0D_t^q f(t)\right] = s^q F(s)$$

The Fractional Order differentiation has the following properties

a. $_0D_t^\alpha f(t) = f(t)$

b. $_0D_t^\alpha[af(t)+bg(t)] = a\,_0D_t^\alpha f(t) + b\,_0D_t^\alpha g(t)$

c. $_0D_t^\alpha[_0D_t^\beta f(t)] = _0D_t^\beta[_0D_t^\alpha f(t)] = _0D_t^{\alpha+\beta} f(t)$

d. $L[_0D_t^\alpha f(t)] = s^\alpha L[f(t)] - \sum_{k=1}^{n-1} s^k [_0D_t^{\alpha-k-1} f(t)]_{t=0}$

In particular, if the derivatives of the function $f(t)$ are all equal to 0 at t = 0, so laplace transform of the fractional derivative is given as follow (6)

$$L[_0D_t^\alpha f(t)] = s^\alpha L[f(t)] \tag{6}$$

III. INTEGER ORDER PID CONTROLLER

The differential equation of Fractional PID is represented by

$$u(t) = K_p e(t) + K_i(_aD_t^\lambda e(t)) + K_d(_aD_t^\mu e(t)) \tag{7}$$

The continuous transfer function of fractional PID controller is obtained through the properties used in equation (7), which is given by

$$\frac{U(s)}{E(s)} = C_{FOPID}(s) = K_p + \frac{K_i}{s^\lambda} + K_d s^\mu, (\lambda, \mu > 0) \tag{8}$$

Figure 1(a) shows the block diagram of FOPID having =1 and μ=1, and takes an form of an integer order Proportional-I and Derivative (PID) Controller as shown in figure 1(b).

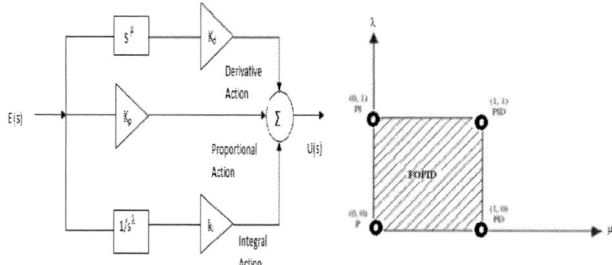

Fig: 1(a) FOPID Block Diagram
Fig. 1(b) Expansion of FOPID in Plane

From the figure1 (b), the fractional order PID controller generalizes the integer order PID controller and expands it from point to plane which shows the flexibility in the Integer order PID control design.

Design Specification:

1. Phase Margin & Gain Crossover frequency

$$\left|G(j\omega_{cp})\right| = \left|C(j\omega_{cp})P(j\omega_{cp})\right|_{dB} = 0 \; dB$$

$$Arg[G(j\omega_{cp})] = Arg[C(j\omega_{cp})P(j\omega_{cp})] = -\pi + \phi_{pm} \tag{9}$$

Where cp is gain crossover frequency, •pm is phase margin, $C(j\omega_{cp})$ is the PID Controller in frequency domain with crossover frequency and $P(j\omega_{cp})$ is the IR plant (FOPID) in frequency domain with crossover frequency.

2. Robustness to gain variation in the gain of the plant

$$\left(\frac{d(Arg(G(j\omega)))}{d\omega}\right)_{\omega=\omega_c} = 0 \tag{10}$$

The variation in gain of the plant shows that the phase order with respect to the frequency is zero i.e. the bode plot of phase is flat at a certain gain crossover frequency.

IV. SIMULATION EXPERIMENT WITH CONTROLLER

In recent era an increasing interest by researcher for the integration of controllers in industrial application are been observed .For closed loop system, four tuning methods are studied and compare with real time implementation, as follows;

- FOPTD plant with ZN Controller;
- FOPTD plant with Åström - Hägglund Controller;

2nd IEEE International conference on power Electronics, Intelligent Control and Energy systems (ICPEICES2018)

- FOPTD plant with Coon-Cohen Controller;
- FOPTD plant with IOPID Controller.

$$r_b e^{j(\pi+\phi_b)} = r_a . r_c e^{j(\pi+\phi_c+\phi_a)} \qquad (12)$$

Therefore, $r_c = \dfrac{r_b}{r_a}$ and $\phi_c = \phi_b - \phi_a$

Using the Zeigler- Nichols process reaction method, the plant (infrared heater, 400watt) was found to have an approximately transfer function by open loop experiment data (reaction curve method) expressed by equation (11)

$$G_{IR}(s) = \frac{[2.6770]e^{-46s}}{262s+1} \qquad (11)$$

Case 1. Zeigler Nichols Closed loop Controller

A ZN-PID tuning rule is the classical tuning rules for closed loop, in which the gain (Kcr) that will cause the closed loop output to oscillate at constant amplitude. Corresponding to the Ultimate gain (Kcr), oscillation ultimate period (Pcr) can be determined. Zeigler – Nichols recommended the following ZN-PID tuning rule presented in the table (I)

Table I. Zeigler Nichols closed loop tuning rule

Controller	Gain (Kp)	Integral Gain(Ki)	Derivative Gain(Kd)
PID	0.6 Kcr	2/Pcr	Pcr/8

Case 2. Åström- Hägglund tuning

A little improvement of the Zeigler- Nichols closed loop method is provided by Åström- Hägglund method having relay feedback system. Åström- Hägglund recognized that the Zeigler- Nichols continuous cycling method actually identifies the point $\left(-\dfrac{1}{K_u}, 0\right)$ on the nyquist curve, and moves it to a predefined point.

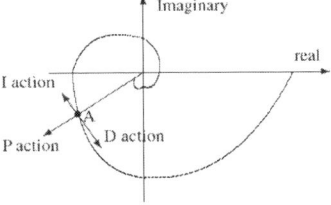

Fig.2 Effect of Changing the PID parameter on Nyquist plot

To move an arbitrary point A to a predefined position B, a convenient choice for point A is the ultimate point. Point B can be determined by a desired gain margin and phase margin, and is written in polar co-ordinates as $B = r_b e^{j(\pi+\phi_b)}$. Assuming that the PID controller at frequency ω_0 is $G_c(s) = r_c e^{j(\phi_c)}$. Then,

PID Controller	Gain(Kp)	Integral Time(Ti)	Derivative Time(Td)
PID	$k_u r_b \cos(\phi_b)$	$\dfrac{-Tu}{\pi}\left(\dfrac{1+\sin(\phi_b)}{\cos(\phi_b)}\right)$	$\dfrac{-Tu}{4\pi}\left(\dfrac{1+\sin(\phi_b)}{\cos(\phi_b)}\right)$

Table II. Åström-Hägglund Tuning values

This method identifies only one point on the Nyquist curve and moves to a desired position.

Case 3. COHEN- COON METHOD

The Cohen- Coon method used for the First Order plus Delay time (FOPDT) model to develop the tuning parameters and it is similar to Zeigler-Nichols reaction method. The parameters involved have more mathematical operations.

Table III. Cohen-Coon Tuning values

PID Controller	Gain(Kp)	Integral Time(Ti)	Derivative Time(Td)
PID	$\dfrac{1}{K}*\left(\dfrac{T}{L}\right)\left(\dfrac{4}{3}+\dfrac{L}{4T}\right)$	$L\left(\dfrac{32+6(L/T)}{13+8(L/T)}\right)$	$L\left(\dfrac{4}{11+((2*L)/T)}\right)$

Case 4. IOPID Method

The phase margin and gain margin of the process plant as per Bode as shown in figure 3and the value of ω_{cg} and ϕ_{pm} as 0.008 rad/sec and 80^0 respectively.

2nd IEEE International conference on power Electronics, Intelligent Control and Energy systems (ICPEICES2018)

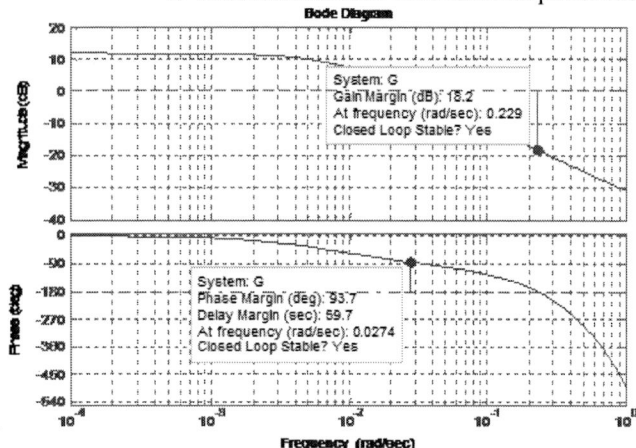

Fig.3 Bode plot of FOPTD transfer function of Ceramic IR heater

Åström-Hägglund	0.5184	0.0038	17.7915
Cohen-Coon	2.8593	0.0271	46.3487
IOPID	0.9768	0.0063	25.6246

V. RESULTS AND SIMULATION

4.1 Simulation studies

The simulation runs are carried out with ZNC-PID, AH-PID, CC-PID and IOPID controllers in Matlab Simulink. The performance indices calculated in terms of error (ISE and IAE) and performance parameters (Tr, Tp, %Mp and Ts) and are reported in table (V). From the result table, IOPID yield a fair response with minimum error values, acceptable overshoot and settling time.

Table IV. IOPID Parameters

PID Controller parameter	IOPID Parameters
Kp	$K_p = \dfrac{1}{K}\sqrt{\dfrac{B_1}{1+A_1^2}}$
Ki	$K_i = \dfrac{1}{2K}\sqrt{\dfrac{1+A_1^2}{B_1}}\left(T\omega_{cp}^2 + LB_1\omega_{cp}^2\right) - A_1\omega_{cp}\sqrt{\dfrac{B_1}{1+A_1^2}}$
Kd	$K_d = \dfrac{1}{2K}\left[\sqrt{\dfrac{1+A_1^2}{B_1}}\left(T+LB_1\right) + A_1\omega_{cp}^{-1}\sqrt{\dfrac{B_1}{1+A_1^2}}\right]$

Where $\dfrac{K_d\omega_{cp}^2 - K_i}{K_p\omega_{cp}} = A_1$ $B_1 = 1 + \omega_{cp}^2 T^2$ and

The PID controller setting (Kp,Ki,Kd) for the above said controllers and worked out based on FOPTD system is provided in table (IV).

Table IV. Calculated PID Controlling parameter for FOPTD

Controller Tuning	Kp	Ki	Kd
ZN-Closed loop	2.15	0.027389	42.193

Table V. Performance parameter using controller tuning

Performance	Rise Time (T_r)	Peak Overshoot(M_p)	Peak Time (T_p)	Settling Time(T_s)	ISE	IAE
ZNC-PID	25.39	74.70	116.93	389.36	80.55	116.4
CC-PID	21.10	103.57	105.07	1270	143.8	269.8
AH-PID	207.91	11.82	504.27	850.55	96.58	183.2
IOPID	112.41	9.8	341.91	613.86	66.92	114.8

4.2 Real time experiment studies

In this section, experiments made it possible and the performance of ZNC-PID, AH-PID, CC-PID and IOPID controller is tested in hardware in loop for ceramic IR heater. A ceramic infrared heater plate temperature shown in figure (4) is controlled using analog power controller made by Libratherm LTC-16 shown in figure (4), having variable control signal range of 0-5v,4-20mA or 10K potentiometer. The output voltage varies proportionally to the input variable signal. LTC-16 has back-back connected SCR to control load of 2kW, 10A, 230V. The firing phase angle controls the gradual and smooth voltage control for the load.

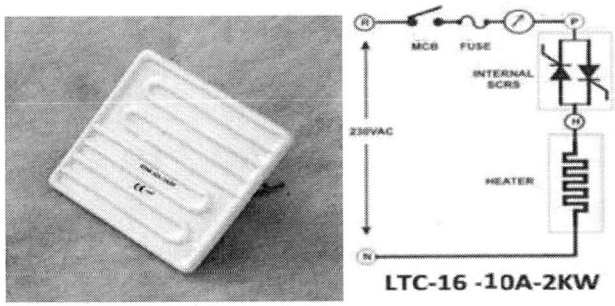

Fig.4   Ceramic IR Heater Plate and LTC 16 wiring diagram

The laboratory setup and wiring diagram as shown in Figure (5) consists of Ceramic IR heater, DAQ card, Analog power controller and temperature sensor (K-type thermocouple). The power delivered to the IR heater is controlled by an analog signal using analog power controller. DAQ card is used for the analog signal generation, where output is 0-5 V DC and it provides to analog power controller which get in firm of fixing angle to back to back SCR of analog power controller to maintain the desired analog power or voltage to IR Heater.

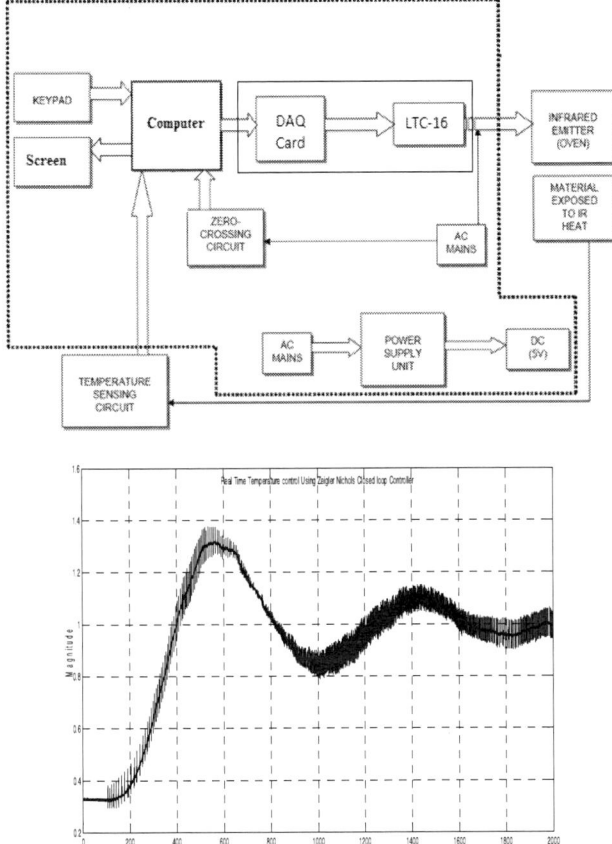

Fig.5 Block Diagram of the temperature control of ceramic IR heating System

Block diagram shown in figure (5) consists of Ceramic IR heater, DAQ card, Analog power controller and temperature sensor (K-type thermocouple). The power delivered to the IR heater is controlled using an analog signal using analog power controller. For the analog signal generation, we use DAQ Card, where output is 0-5 V DC and it provides to analog power controller which get in firm of fixing angle to back to back SCR of analog power controller to maintain the desired analog power or voltage to IR Heater.

Simulation Conducted on Infrared heater to show warm up temperature shown in figure (6) with respect to time was conducted. The type of heater used in the study was 400 watt and has a maximum operating temperature of 1200^0C.

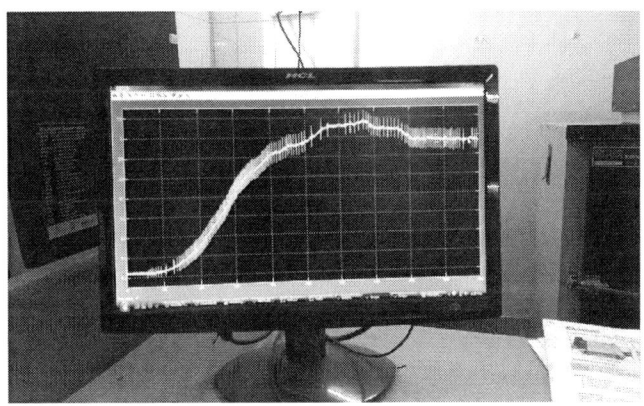

Fig. 6 Experimental output step response

Real time experimental performance of Zeigler Nichols closed loop, Åström-Hägglund loop, Coon-Cohen and IOPID method and their Controller output as shown in figure (7).

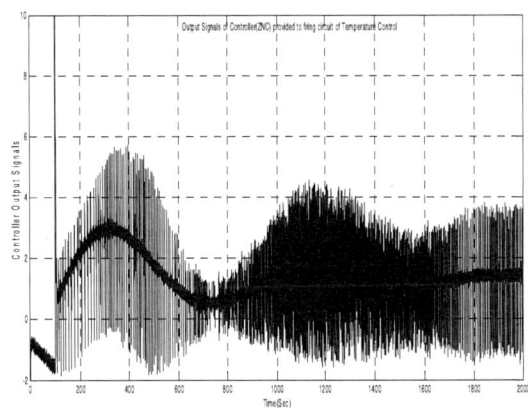

978-1-5386-6624-1/18 $31.00 © 2018 IEEE

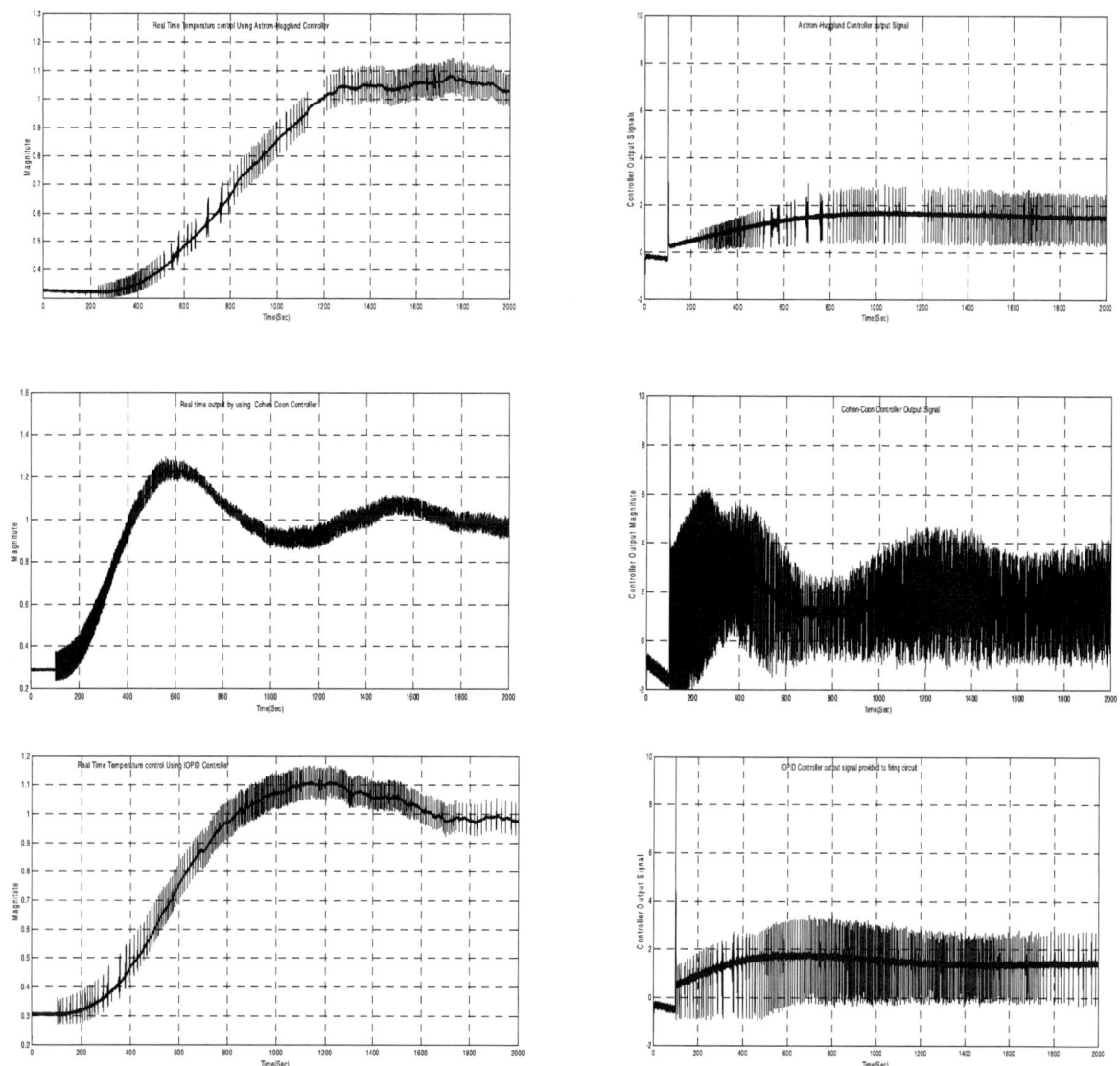

Fig. 7 Real Time Temperature control and Output Signal generate by using ZN, AH, CC and IOPID Controller

The comparative analysis of the temperature control of ceramic Infrared HFE system

Table VI. Performance parameter of different Controller

Performance Parameter	ZNO	ZNC	AH-PID	CC-PID	IOPID
Rise Time	241.3	226.16	632.06	259.2	484.47
Settling Time	1997	1999	1997	1993	1989

Peak Time	538.7	536.3	1749	559	1129
Peak Overshoot	34.72	38.15	11.25	34.6	20.07

VI. CONCLUSION

By observing the result in Table, it is found that the IOPID method is generally better than other conventional tuning method controllers. In the experimental set up the process control includes the time required by the heater to be settled on the initial set up temperature. The real time controller includes the signal transfer from the temperature sensor for the real time action. The temperature warm up is slow due the resistance heating element used in ceramic infrared heater. So the settling time is very high in comparison with the simulation result.

REFERENCES

[1] C. A. Monje, Y. Q. Chen, B. M. Vinagre, D. Xue and V. Feliu, Fractional-order systems and controls. fundamentals and applications,Springer,2010.

[2] H. S. Ahn, V. Bhambhani and Y. Q. Chen, Fractional-order integral and derivative controller design for temperature profile control,Control and Decision Conference, 2008, CCDC 2008, Chinese, IEEE, 4766-4771 (2008).

[3] A. Tepljakov, E. Petlenkov, J. Belikov and S. Astapov, Tuning and digital implementation of a fractional-order PD controller for a position servo, Int. J. Micr. Comput. Sci. 4(3), 116-123 (2014).

[4] C. A. Monje, B. M. Vinagre, Y. Q. Chen, V. Feliu, P. Lanusse and J. Sabatier, Proposals for fractional PIl Dm tuning, Proceedings of The First IFAC Symposium on Fractional Differentiation and its Applications (FDA04), 115-120 (2004).

[5] S. Das, S. Saha, S. Das and A. Gupta, On the selection of tuning methodology of FOPID controllers for the control of higher order processes, ISA Transact. 50(3), 376-388 (2011).

[6] V. Bhambhani, Y. Han, S. Mukhopadhyay, Y. Luo and Y. Q. Chen, Hardware-in-the-loop experimental study on a fractional order networked control system testbed, Commun. Nonlin. Sci. Numer. Simul. 15(9), 2486-2496 (2010).

[7] I. Podlubny, Fractional differential equations: an introduction to fractional derivatives, fractional differential equations, to methods of their solution and some of their applications, vol. 198, Academic Press, 1998.

[8] K. H. Ang, G. Chong and Y. Li, PID control system analysis, design, and technology, Control Systems Technology, IEEE Transactions 13(4), 559-576 (2005).

[9] C. A Monje, B. M. Vinagre, V. Feliu and Y. Q. Chen, Tuning and auto-tuning of fractional order controllers for industry applications, Contr. Engin. Pract. 16(7), 798-812 (2008).

[10] D. Valerio and J. Sa da Costa, A review of tuning methods for fractional PIDs, 4th IFAC Workshop on Fractional Differentiation and its Applications, FDA, vol. 10, 2010.

[11] I. Petras, The fractional-order controllers: Methods for their synthesis and application, arXiv preprint math/0004064 (2000).

[12] Y. Q. Chen, I. Petras and D. Xue, Fractional order control-a tutorial, American Control Conference, 2009 ACC09, IEEE, 1397-1411 (2009).

[13] I. Podlubny, Fractional-order systems and fractional-order controllers, Institute of Experimental Physics, Slovak Academy of Sciences, Kosice 12 (1994).

[14] I. Podlubny, L. Dorcak and I. Kostial, On fractional derivatives, fractional-order dynamic systems and PIl Dmcontrollers, Proceedings of the 36th Conference on Decision & Control, 5, 4985-4990 (1997).

[15] K. J. Astrom, PID controllers: theory, design and tuning, Instrument Society of America (1995).

[16] Z. Li, L. Lu, S. Dehg, Y. Q. Chen and X. Ding, A review and evaluation of numerical tools for fractional calculus and fractional order controls, Int. J. Contr.1-17(2016).

[17] H. Panagopoulos, K. J. Astrom and T. Hagglund, Design of PID controllers based on constrained optimisation, IEE Proceedings- Control Theory and Applications 149(1), 32-40 (2002).

[18] D. Valerio and J. S da Costa, Time-domain implementation of fractional order controllers, IEE Proc. Contr. Theor. Appl.152, 539-552 (2005).

[19] B. M. Vinagre, I. Podlubny, A. Hernandez and V. Feliu, Some approximations of fractional order operators used in control theory and applications, Fract. Calc. Appl. Anal. 3 3, 231-248 (2000).

[20] A. Oustaloup, F. Levron, B. Mathieu and F. M. Nanot, Frequency-band complex noninteger differentiator: characterization and synthesis, Circuits and Systems I: Fundamental Theory and Applications, IEEE Transactions 47(1), 25-39 (2000).

[21] A. Tepljakov, E. Petlenkov and J. Belikov, FOMCON: Fractional-order modeling and control toolbox for MATLAB, Proc. 18th Int Mixed Design of Integrated Circuits and Systems (MIXDES) Conference, 684-689 (2011).

[22] A. Tepljakov, E. Petlenkov, J. Belikov and M. Halas, Design and implementation of fractional-order PID controllers for a fluid tank system, American Control Conference (ACC), 2013, IEEE, 1777-1782 (2013).

[23] A. Tepljakov, E. Petlenkov and J. Belikov, A flexible MATLAB tool for optimal fractional-order PID controller design subject to specifications, Control Conference (CCC), 2012, 31st Chinese, IEEE, 4698-4703 (2012).

978-1-5386-6624-1/18 $31.00 © 2018 IEEE

2nd IEEE International Conference on Power Electronics, Intelligent Control and Energy Systems (ICPEICES-2018)

Design and Mathematical Analysis of Microstrip Bandpass Power Divider for Ku Band Communications

Karteek Viswanadha[1]

Delhi Technological University
New Delhi, India.
karteekviswanath@gmail.com

N.S.Raghava[2]

Delhi Technological University
New Delhi, India.
nsraghava@dce.ac.in

Abstract— **Many power dividers such as Wilkinson, Geysel, etc. are proposed to achieve wide bandwidth and the ideal value of insertion loss for N-way arrangement. These devices also perform filtering functions. Most of the power dividers are integrated with bandpass filters in order to suppress the harmonics. This paper presents the design of a novel power divider consisting of series and shunt microstrip stubs of quarter wavelength embedded on the output branches of the power divider thereby forming bandpass filters of the third order. The bandpass filters along with the power dividers not only perform the filtering function but also power division function. Embedding the higher-order bandpass filters in the branches of power divider heavily suppresses the harmonics. Overall dimensions of the proposed power divider are 19.1 x 13.25 x 1.524mm³ including ground plane. The proposed power divider further possesses wide bandwidth of 3.5GHz along with the insertion loss of-3.32dB at 14GHz. The proposed power divider and its equivalent lumped element model are analyzed mathematically. It is found that the mathematical analysis of the proposed power divider and its lumped equivalent model have good agreement with their simulation results.**

Keywords—power dividers, stubs, bandwidth, insertion loss, lumped elements, microstrip, bandpass, filter, quarter wavelength, third order.

I. INTRODUCTION

Power Dividers are important front-end devices in many wireless devices like antenna array systems, transponder etc. Power dividers are designed using microstrip line, strip line and waveguides. Most of the Power dividers are designed using microstrip lines as microstrip lines enjoy the advantage of planar nature, easy fabrication. The design of power dividers depends on the frequency of operation, bandwidth of operation [1]. The shift of applications towards the extremely higher frequencies will miniaturize the power dividers. Power dividers are designed to achieve set of specifications such as wide bandwidth, good insertion loss, harmonic suppression, low transmission losses while minimizing the size and fabrication cost. Different types of power dividers are proposed based on the designer's specifications especially bandwidth enlargement with low insertion loss. Conventional way of increasing the Bandwidth of operation is to cascade the

power dividers. But, the cascading of power dividers not only increases the bandwidth of operation but also the insertion loss. Hence, there should be tradeoff between the bandwidth and insertion loss. Based on the power division, power dividers are classified as equal and unequal power dividers [14]. Equal and unequal power divisions are achieved with different types of power dividers like T-junctions, Wilkinson and Geysel. T-junctions are basic transmission line power dividers of appropriate width and electrical length. But, these suffered from very poor insertion loss between the input and output ports. Apart from T-junctions, Wilkinson type power dividers are widely used in many wireless applications as they provide good insertion loss, low insertion and reflection

losses. Power divider of this type consists of an internal insertion loss resistor connected between the two output ports which absorbs the reflected waves from the output ports and thus providing good insertion loss [2].Practically, Wilkinson power divider suffer from insertion and reflective losses and as a result different types of power dividers such as 'H', 'G', 'E' shaped power divider are proposed. These types of power dividers provide good insertion loss with bandpass response. The required low VSWR, high insertion loss can also be achieved by varying the electrical length of output branches of a power divider. The suppression of harmonics can be achieved by embedding the higher-order bandpass filters in the output branches of the power divider. In [3-4], a novel wideband power divider integrated with bandpass filter is designed and fabricated which is different from conventional half-wavelength slotted line structure and lumped resistor and capacitor are bridged between the two E-shape units, which makes the designed power divider show good performance on impedance matching at all ports, insertion loss and filtering response over a wide frequency range. Good insertion loss can be achieved by using insertion loss resistances which are connected between two output ports. To optimize the design of a power divider, insertion loss resistors are embedded inside the power divider itself. Power dividers with multiple internal insertion loss resistors improve in-band insertion loss, bandwidth and impedance matching [5].Power dividers designed using short-circuited transmission lines, stepped impedance transmission lines and coupled sections not only

978-1-5386-6624-1/18 $31.00 © 2018 IEEE

provide good out-of-band rejection but also a good insertion loss by introducing unique loading point for enhancing the insertion loss [9]. Power dividers are also designed using waveguides especially Substrate Integrated Waveguide (SIW) [6] as they show very high performance with low insertion and return losses. Reflection and insertion loss losses are improved by out-of-phase interference. These devices can also be realized using microwave resonators line stepped impedance resonators, ring resonators etc. Resonators have the capability to suppress the harmonics as they perform filtering operations and hence no significant power loss occurs when the power dividers are designed using microwave resonators. Many resonator power dividers are proposed like Stepped impedance resonator power dividers (SIR), Ring resonator power dividers to obtain the ideal value of insertion loss for an N-way power dividers practically. Stepped impedance resonator (SIR) power dividers [7] have the capability to suppress the harmonics due to their different geometries. In [10], SIR power dividers are realized using short-circuited SIR sections in order to obtain good input-output port insertion loss and harmonic suppression. Ring resonators are used in to design power dividers with open circuit stubs and series of coupled lines connected around the ring resonators. Ring resonator power dividers introduce transmissions zeroes in the frequency bands and thereby enhancing the stop band range [8]. Power dividers are fabricated on Defective Ground Structures (DGS) as these structures reduce the surface and thereby increasing the transmission factor. In [11], defective ground structures are used at the feed lines in order to suppress the harmonics without affecting the passband characteristics. To control the transmission losses, programmable power dividers are proposed using Radio Frequency Micro Electro Mechanical Switches (RF-MEMS) for antenna systems [15]. Power dividers are also proposed using lumped elements like spiral inductors, metal-insulator-metal capacitors (MIMC) which will ease the fabrication process [18-20]. Power dividers are often used as the backend devices in wireless communication systems. They are used as power feeding devices for antenna arrays [12]. There are applications where power dividers are used as impedance matching devices [21].

This paper presents the design and mathematical analysis of a wideband microstrip power divider. The proposed power divider possesses wide bandwidth of 3.5GHz and the insertion losses of -3.32dB at output ports.

II. Design of Power Divider

The proposed power divider is designed using RT-Rogers 5880 substrate ($\varepsilon_r = 2.2$) with the thickness of 60mil(1.524mm).The third order power divider with the centre frequency of 14GHz consists of third-order bandpass filters at the output branches. The bandpass filters are designed using open circuit and short circuit stubs of a quarter wavelength. Open circuit stubs are chosen to have the characteristic impedance of 50 ohms. The characteristic impedance of short circuit stub is chosen according to the formula:

$$Z_{on} = \frac{\pi \Delta Z_o}{4g_n}, n = 1,2 \text{ and } 3.$$

The calculations are carried out by considering 0.5db ripple in the passband.

Figure.1 shows the detailed design of the proposed third order bandpass power divider. The main line is designed to have an electrical length of 180 degrees. The input impedance and the load of the main line are automatically matched irrespective of the characteristic impedance. The power divider possesses the dimensions of 19.1 x 13.25 x 1.524mm^3.

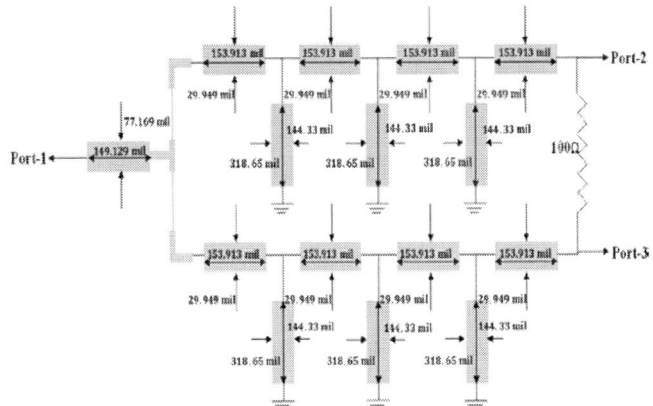

Fig.1.Proposed power divider

III. Results and Mathematical Analysis

The proposed power divider is designed and simulated in Agilent ADS 2009. Wilkinson power divider with Bandpass filter response and the reflection coefficient are plotted from 0-20GHz as shown in Fig.2. The red plot indicates the reflection coefficient. The insertion loss is constant between 12.30 GHz to 15.80 GHz. Therefore, the bandwidth of the power divider is 3.5GHz which is quite high.

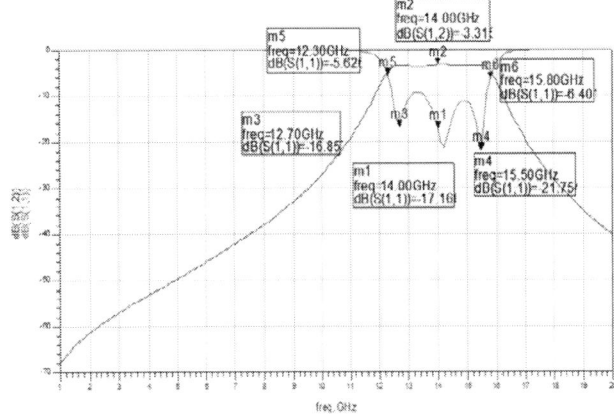

Fig.2.The consolidated plot of Insertion loss (marked in blue colour) and Reflection coefficient (marked in red colour).

-10dB is considered as the reference level for reflection coefficient. The power divider functions fruitfully at the frequencies which are having reflection coefficient below -10dB. The reflection coefficient graph has three dips at 12.70 GHz, 14.2 GHz and 15.50 GHz respectively. This indicates that the power divider is having three poles.

Even mode ABCD Parameters $\begin{bmatrix} A_e & B_e \\ C_e & D_e \end{bmatrix}$ are given by:

$$\begin{bmatrix} j\frac{Z_{oin}Z_{o2}Z_{o3}}{Z_{o1}Z_{stub2}Z_{o4}} & j\frac{Z_{oin}Z_{o2}Z_{o4}}{Z_{o1}Z_{o3}} + j\frac{Z_{oin}Z_{o2}Z_{o3}Z_{o4}}{Z_{o1}Z_{stub2}Z_{stub3}} \\ j\frac{Z_{o3}}{Z_{o2}Z_{o4}} + \frac{Z_{o1}Z_{o2}Z_{o3}}{Z_{oin}Z_{stub1}Z_{stub2}Z_{o4}} & j\frac{Z_{o1}Z_{o2}Z_{o4}}{Z_{o2}Z_{stub1}Z_{o3}} + \frac{Z_{o3}Z_{o4}}{Z_{o2}Z_{stub3}} + \frac{Z_{o1}Z_{o2}Z_{o3}Z_{o4}}{Z_{oin}Z_{stub1}Z_{stub2}Z_{stub3}} \end{bmatrix}$$

Finding even T-Parameters in terms of even ABCD parameters using the below equation,

$$T_e = \frac{2}{(A_e + D_e) + (B_e Y_o + C_e Z_o)}$$

Finding even mode reflection co-efficient using below equation,

$$\Gamma_e = \frac{(A_e - D_e) + (B_e Y_o - C_e Z_o)}{(A_e + D_e) + (B_e Y_o + C_e Z_o)}$$

Substituting all the values of impedances in both the ABCD matrices,

$$\begin{bmatrix} A_e & B_e \\ C_e & D_e \end{bmatrix} = \begin{bmatrix} 0 & j50 \\ \frac{j}{50} & 0 \end{bmatrix}$$

$T_e = -j$ and $\Gamma_e = 0$

S $-$ Parameters are given by:

$S_{11} = \frac{1}{2}(\Gamma_e) = 0$

$S_{12} = \frac{1}{2}(T_e) = -\frac{j}{2} = S_{21}$

$S_{13} = \frac{1}{2}(T_e) = -\frac{j}{2} = S_{31}$

Return Loss of the device is given by: $20\log \log_{10}|S_{11}| = -\infty$ dB and the mathematical result of -17.16dB is obtained at 14GHz.

The insertion loss of the device is given by $10\log \log_{10}|S_{21}| = -3$ dB and the mathematical result of -3.32 dB is obtained at 14GHz. These values are calculated by not considering conductor and dielectric losses.

Final S-parameters of the device are given by:

$$\begin{bmatrix} 0 & -\frac{j}{2} & -\frac{j}{2} \\ -\frac{j}{2} & 0 & 0 \\ -\frac{j}{2} & 0 & 0 \end{bmatrix}$$

The lumped element equivalent model is derived by considering the open circuit stubs as series L-C circuits and short circuit stubs as parallel L-C circuits. The length of all the stubs is $\frac{\lambda}{4}$.

The input impedance of a transmission line is given:

$$Z_{in} = Z_o \frac{Z_L + jZ_o Tan\beta l}{Z_o + jZ_L Tan\beta l}$$

Where, Z_o is the characteristic impedance of the transmission line, Z_L is the load of the transmission line and βl is the electrical length of the transmission line.

Figure.3 represents the lumped element equivalent mode for the bandpass power divider shown in Figure.1

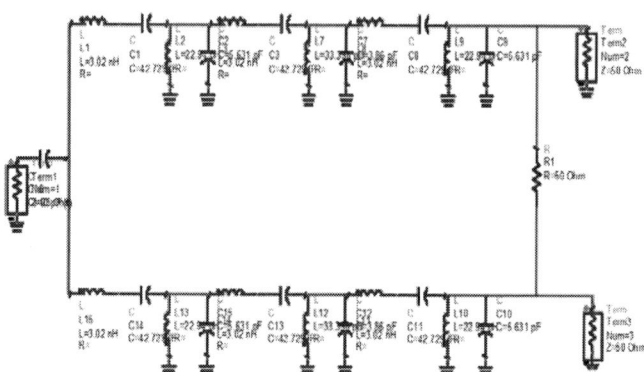

Fig.3.Lumped elements equivalent circuit of the bandpass power divider.

In Figure.4, red graph indicates insertion loss and blue graph indicates the reflection coefficient. An insertion loss of -3.32 dB is observed and the reflection coefficient of -44.227 dB is observed. The reflection coefficient graph has three dips at 13.4 GHz, 14 GHz and 14.6 GHz respectively.

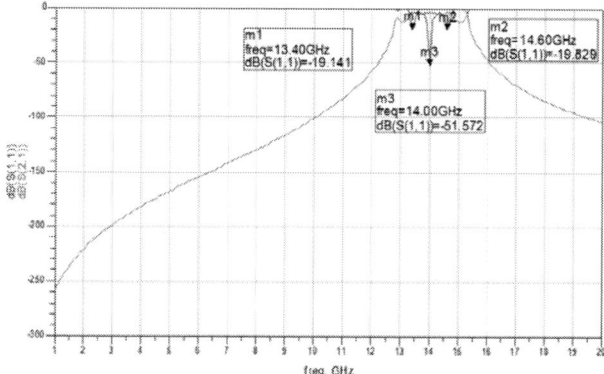

Fig.4. The consolidated plot of insertion loss and reflection coefficient for lumped element equivalent.

Even and odd mode analysis is carried out on the lumped element equivalent circuit.

Even mode ABCD $\begin{bmatrix} A_e & B_e \\ C_e & D_e \end{bmatrix}$ Parameters are given by:

$$\begin{bmatrix} 1 & \frac{-j}{C_{in}\omega}\frac{-j}{C_1\omega}\frac{-j}{C_3\omega}\frac{-j}{C_5\omega}\frac{-j}{C_7\omega} \\ 0 & 1 \end{bmatrix}$$

Substituting all the values of inductors and capacitors, we get

$$\begin{bmatrix} A_e & B_e \\ C_e & D_e \end{bmatrix} = \begin{bmatrix} 1 & -j3.9 \\ 0 & 1 \end{bmatrix}$$

The three dips are obtained at 13.4 GHz, 14 GHz and 14.6 GHz respectively. The insertion loss is observed to be -3.15dB at 14GHz and reflection coefficient is observed to 44.2dB at 14GHz.

Finding even T-Parameters in terms of even ABCD parameters using the below equation:

$$T_e = \frac{2}{(A_e + D_e) + (B_e Y_o + C_e Z_o)}$$

Find even mode reflection co-efficient using below equation:

$$\Gamma_e = \frac{(A_e - D_e) + (B_e Y_o - C_e Z_o)}{(A_e + D_e) + (B_e Y_o + C_e Z_o)}$$

Substituting all the values of impedances in both the ABCD matrices,

$$\begin{bmatrix} A_e & B_e \\ C_e & D_e \end{bmatrix} = \begin{bmatrix} 1 & -j3.9 \\ 0 & 1 \end{bmatrix}$$

$T_e = 0.999 + j0.03$ and $\Gamma_e = -j0.039$

S − Parameters are given by:

$$S_{11} = \frac{1}{2}(\Gamma_e) = -j\,0.0195$$

$$S_{12} = \frac{1}{2}(T_e + T_o) = 0.495 + j0.015 = S_{21}$$

$$S_{13} = \frac{1}{2}(T_e - T_o) = 0.495 + j0.015 = S_{31}$$

Return Loss of the device is given by: $20\log \log_{10} |S_{11}|$ = -34 dB and the simulated result of   -44dB is obtained at 14GHz.

The insertion loss of the device is given by $20\log \log_{10} |S_{21}|$ = -3dB and the simulated result of -3.32 dB is obtained at 14GHz.

These values are calculated by not considering conductor and dielectric losses

Final S-parameters of lumped element equivalent power divider are:

$$\begin{bmatrix} -j\,0.0195 & 0.495 + j0.015 & 0.495 + j0.015 \\ 0.495 + j0.015 & 0 & 0 \\ 0.495 + j0.015 & 0 & 0 \end{bmatrix}$$

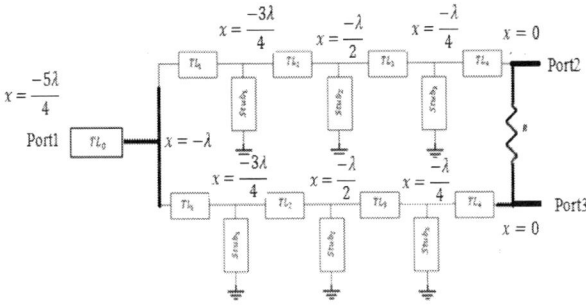

Fig.5.Bandpass power divider

The electric fields and magnetic fields are obtained as follows:

$$E_x = \frac{2}{3}V_o \sum_{n=odd}^{\infty} \frac{n\pi}{w}\,\text{Sin}\,\frac{n\pi}{w}x * \text{Sinh}(\frac{n\pi}{l}y)$$

$$E_y = \frac{2}{3}V_o \sum_{n=odd}^{\infty} \frac{n\pi}{l}\,\text{Cos}\,\frac{n\pi}{w}x * \text{Cosh}(\frac{n\pi}{l}y)$$

$$H_y = -\frac{2}{3Z_o}V_o \sum_{n=odd}^{\infty} \frac{n\pi}{w}\,\text{Sin}\,\frac{n\pi}{w}x * \text{Sinh}(\frac{n\pi}{l}y)$$

$$H_x = \frac{2}{3Z_o}V_o \sum_{n=odd}^{\infty} \frac{n\pi}{l}\,\text{Cos}\,\frac{n\pi}{w}x * \text{Cosh}\left(\frac{n\pi}{l}y\right)$$

Table.1 shows the comparison of the proposed power divider with the existing power dividers.

TABLE. I. COMPARISON OF THE PROPOSED POWER WITH THE DIFFERENT POWER DIVIDERS

Reference	Dimensions (mm³)	Bands (GHz)	Insertion Loss (dB)	Bandwidth (GHz)
[22]	20 x30x0.508	2.1~11	-3.3	8.9
[23]	66.3 x 100 x 1.57	3~8	-7.5~-10	5
[24]	11 x 10.5 x 0.5	0.1~4	-3.5	3.9
[25]	33 x 46 x1.524	5~7	-1.2~-3.2	2
[26]	90 x 90 x 1.6	0.7~1.4	-4.2	0.7
Proposed work	19.1 x 13.25x 1.524	12.3 ~15.8	-3.32	3.5

IV. CONCLUSION

Design and detailed mathematical analysis of a wideband power divider are presented in this paper. It has been observed that by introducing higher order bandpass filters in the upper and lower branches of the power divider, the harmonics are suppressed and the bandwidth is highly improved. By embedding the power dividers with the bandpass filters, the design of the power divider can be optimized as well as the gain bandwidth product can be increased. The wide bandwidth of the power divider makes it suitable for Ku-band satellite communications.

REFERENCES

[1]. Pozar, D.M, *Microwave engineering*. 4th edn. Oxford, United Kingdom: Wiley, John & Sons, 2011.

[2]. A.Nassiri, "Power dividers and couplers", *Massachusetts Institute of Technology*, U.S.Particle Accelerator School, pp.1-66, 2010

[3]. Xiao, L., Peng, H. and Yang, T. ,"A novel power divider integrated with one bandpass filter", *Progress In Electromagnetics Research C*, Vol.52, pp. 115–124. doi: 10.2528/pierc14052606, 2014.

[4]. Gao, L. and Yin Zhang, X., "Novel 2: 1 Wilkinson power divider integrated with bandpass filter", *Microwave and Optical Technology Letters*, Vol.55, issue.3, pp. 646–648, doi: 10.1002/mop.27395, 2013.

[5]. Lu Y.L. and Dai, "Novel filtering power divider using multiple internal resistors", *Progress In Electromagnetics Research Letters*, Vol.45, pp. 75–80. doi: 10.2528/pierl114022703, 2014.

[6]. Kim,Kyeongmin, Jindo Byun and Hai-Young Lee, "Substrate integrated waveguide Wilkinson power divider with improved isolation performance", *Progress in Electromagnetics Research Letters*, Vol.19, pp. 41-48, 2010.

[7]. Avrillon, "Dual-Band Power Divider Based On Semiloop Stepped-Impedance Resonators", *IEEE Transactions on Microwave Theory and Techniques*, pp.1269-1273, 2003.

[8]. Gao, S.S., Sun, S. and Xiao, S, "A novel Wideband Bandpass power divider with harmonic-suppressed ring Resonator", *IEEE Microwave and Wireless Components Letters*, 23(3), pp. 119–121, 2013.

[9]. Zhang, B., Yu, C. and Liu, Y,"Compact power divider with bandpass response and improved out-of-band rejection", *Journal of Electromagnetic Waves and Applications*, Vol.30, issue.9, pp. 1124–1132, 2016.

[10]. Deng, Yijing, Jia-lin Li, and Jianpeng Wang, "Design Of Compact Wideband Filtering Power Divider With Extended Isolation And Rejection Bandwidth", *Electronics Letters*, Vol.52, pp. 1387-1389, 2016.

[11]. Chaudhary, Girdhari, Yongchae Jeong, and Jongsik Lim, "Harmonic Suppressed Dual-Band Bandpass Filters With Tunable Passbands", *IEEE Transactions on Microwave Theory and Techniques*,Vol.60, pp.2115-2123,2012.

[12]. Natani, P., Kapoor, S., Saha, C. and Kumar.S, "Design of Slotted Waveguide Array Antenna Fed by H-Plane Power Divider" *Indian Institute of Space Science & Technology*, pp. 1–2, 2014.

[13]. Yang. Y., Wang, Y. and Fathy A.E, "Design of compact vivaldi antenna arrays for UWB see through wall applications", *Progress In Electromagnetics Research*, Vol.82, pp. 401–418,2008.

[14]. Li, B., Wu, X., Yang, N. and Wu, W, "Dual band Equal/Unequal Wilkinson power dividers based on coupled line section with short circuited stubs", *Progress In Electromagnetics Research*, Vol.111, pp. 163–178,2011.

[15]. Ocera, Gatti, Mezzanotte, Farinelli and Sorrentino, "A MEMS programmable power divider/combiner for reconfigurable antenna systems", *IEEE 2005 European Microwave Conference*, pp. 1–2, 2006.

[16]. N.S. Raghava, Asok De, Pushkar Arora, Sagar Malhotra, Rishik Bazaz, Sahil Kapur, and Rahul Manocha, "A Novel Patch Antenna for Ultra Wideband Applications", 2011 IEEE, International Conference on Communication and Signal Processing, pp. 276-279, 2011.

[17]. P.Dawar., N.S.Raghava,. and Asok De, "A novel metamaterial for miniaturization and multi-resonance in antenna", *Cogent Physics*, Vol.2, issue.1, pp.1-4, 2015.

[18]. Liang-Hung Lu, P. Bhattacharya, L. P. B Katehi and G. E Ponchak, "X-band and K-band lumped Wilkinson power dividers with a micromachined technology", *IEEE Microwave Symposium Digest, IEEE MTT-S International*, pp. 1–3, 2002.

[19]. Bahl, I. and Wysocki, R.K, *Lumped elements for RF and microwave circuits*. Boston, MA: Artech House Publishers, 2003.

[20]. Ahn, H.-R. *Asymmetric passive components in microwave integrated circuits*. New York, NY, United States: Wiley, John & Sons

[21]. Kaymaksut, E., Gürbüz, Y. and Tekin.I, "Impedance matching Wilkinson power dividers in 0.35 μm SiGe BiCMOS technology", *Microwave and Optical Technology Letters*, Vol.51, issue.3, pp. 681–685, 2009.

[22]. Lihua Wu, Shanqing Wang, Luetao Li, and Chengpei Tang, "Compact Microstrip UWB Power Divider with Dual Notched Bands Using Dual-Mode Resonator", *Progress In Electromagnetics Research Letters*, pp.39-45, 2018.

[23]. Faraz Ahmed Shaikh, Sheroz Khan, AHM Zahirul Alam, Mohamed Hadi Habaebi, Othman Omran Khalifa and Talha Ahmed Khan, "Design and Analysis of 1-to-4 Wilkinson Power Divider for Antenna Array Feeding Network", *IEEE International Conference on Innovative Research and Development (ICIRD)*, pp.1-4, 2018.

[24]. Tae-Hyeon Lee and Ki-Cheol Yoon, "Compact size of the Wilkinson power divider with transmission-line using π-model transformation method", *Microwave and Optical Technology Letters*, Vol.60, issue.7,pp.1632-1638, 2018.

[25]. Duolong Wu, Adriana Serban, Magnus Karlsson, and Shaofang Gong, "Highly Unequal Three-Port Power Divider: Theory and Implementation", *International Journal of Antennas and Propagation*, Vol.2018, pp.1-8, 2018.

[26]. Jin Guan, Min Gong, and Bo Gao, "A Novel Three-Way Gysel Power Divider/Combiner on Plane Structure", *Progress In Electromagnetics Research Letters*, Vol.75, pp.113-117, 2018.

978-1-5386-6624-1/18 $31.00 © 2018 IEEE

2nd IEEE International Conference on Power Electronics, Intelligent Control and Energy Systems (ICPEICES-2018)

Contrast Enhancement and Pseudo Coloring Techniques for Infrared Thermal Images

Mandeep Kaur
Department of Computer Science and Engineering
(Student, M.Tech)
SLIET Longowal
Sangrur, India
gagankaur005@gmail.com

Manminder Singh
Department of Computer Science and Engineering
(Assistant Professor)
SLIET Longowal
Sangrur, India
manminderfzr@yahoo.com

Abstract— **Infrared thermography is a non-invasive and non-contact type radiometric approach that is widely used in medical diagnosis and many other fields. The thermal camera provides the temperature variation information in the form of 2D false color image (or thermogram). These thermograms are either grey scale image or pseudo color image. Grey scale images have poor contrast and difficult for human interpretation. So, there is need of some algorithms to process the image to improve the geometric information of the Region Of Interest (ROI). Various pseudo coloring and contrast enhancement algorithms are developed to improve the visual appearance of the thermogram. In this paper, the overview of some contrast enhancement and pseudo-coloring algorithms with future directions are discussed. In addition, the basic introduction of infrared radiations, thermal imaging, contrast enhancement and pseudo coloring are included.**

Keywords— *Infrared thermography, contrast enhancement, pseudo coloring, image segmentation, grayscale images*

I. INTRODUCTION

A. Infrared Radiations

The invisible Electromagnetic (EM) radiations were discovered by William Herschel, a British astronomer [1]. All living and non-living things that are placed above -273.15°C temperature, emit Infrared (IR) radiations from its surface due to molecular motion. These radiations lie in the wavelength of 0.75μm to 1000μm of the EM spectrum.

In Figure 1 the presence of IR radiations is detected by a thermometer, but not visible to the naked eyes, as the temperature increases the number of radiations are detected by the thermometer. The Planck gives us the law for wavelength distribution and energy that any object emits [2]. The total amount of IR energy emitted per unit area of an object is given

by Stefan Boltzmann law, the law describes the amount of energy emitted in terms of emissivity of an object and temperature.

$$E = \sigma \varepsilon T^4 Jm^{-2}s^{-1} \qquad (1)$$

Where, E=total energy radiated

σ(Stefan-Boltzmann constant)=5.6704*10^{-8} Wm^{-2}K^{-4}

ε = emissivity

T = temperature in Kelvin

Clean and un-oxidized bare metal surface have a quite low emissivity as compared to non-metals. The human body is a homoeothermic, i.e. able to keep the body temperature stable even though the temperature of the surrounding is changed. Humans generally, increase their comfort and used clothes to protect the body from different conditions. The core of the body is most stable in temperature, and regulatory process is formed by the shell of the body [3]. Usually, human skin behaves as an almost blackbody with an emissivity of 0.96-0.98 at wavelength ranges from $2 - 20 \ \mu m$[4].

B. Thermal Imaging

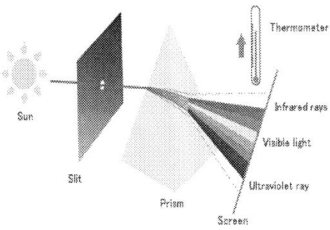

Figure 1: EM Spectrum

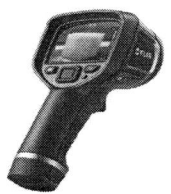

Figure 2(a): Thermal Camera

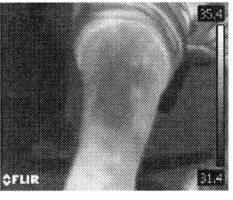

Figure 2(b): Thermograms

978-1-5386-6624-1/18 $31.00 © 2018 IEEE

In Figure 2(a) thermal camera is shown that is used for capturing infrared thermal images. An IR thermal imaging is a non-invasive technique for capturing the emitted infrared radiations from the surface, which are not visible to naked eyes.IR radiations are detected by the thermal camera which converts it into electrical impulses and maps the surface temperature distribution called the thermogram as shown in Figure 2(b). Thermograms are basically grey-level images or false color image. IR Thermography (IRT) is non-ionizing and non-contact type technique. IRT is used to determine areas of the body that have abnormal temperature values.

IR thermography is a physiological test used to measure specific physiological constructs in an individual. The physiologic changes occur with aging in all organ systems or due to burns, fractures [5]. With the advancement, utility of thermal imaging is increasing in every field like defense, medical diagnosis, security and surveillance. It is highly preferable in the clinical applications as compared to X-rays and mammography because of its non-ionizing nature. We make the use of IRT so that problems associated with fever screening [9, 10], kidney transplantation [8] and gynecology [6, 7] can be easily detected.

C. Pseudo Coloring

False color image given by thermal camera is in generalized form and not appropriate for human interpretation because abnormal areas is not distinguished from normal areas. For a better interpretation of the image pseudo coloring algorithm are used. Various pseudo coloring algorithms are available such as winter, lava, rainbow, iron bow and many more. To diagnose the abnormal areas of the thermal images easily; pseudo coloring of thermograms is an important factor and coloring should be such that the target (abnormalities) is easily detectable from non-target elements.

Colors mapped in pseudo coloring algorithm do not increase any image information, it only provides us better visualization in terms of temperature variations occur in thermograms and single color is not used for one value of temperature in all pseudo coloring algorithms [11].

D. Color Enhancement

Enhancement of the image describes to improve the quality of the image by removing noise, edge enhancement and color enhancement so that visual effects and clarity of the image is improved and an image become more conducive for the computer to process [12]. The visual appearance of the image is enhanced by improving the supremacy of some features or by degrading the inexactness between regions of the image [13].

Contrast enhancement is an important factor for image enhancement. Contrast enhancement refers to changing the pixels of each possible bin and helps in better analysis of the image. Many contrast enhancement techniques are present that can be categorized into global and local enhancement techniques. Although global techniques are fast and simple like global histogram equalization but have some limitations that it cannot include the local features of an image because it only contains the global information over whole image and limits the contrast ratio of the image. On the other hand, local enhancement techniques can improve the overall contrast more accurately; taking each input pixel sequentially. Limitation of the Local enhancement technique is that it increases the computational constant and enhances the noise effect in an image as well.

II. RELATED WORK

The image of same person appears to be different due to varying lightening and different viewing conditions [21], but infrared images are robust against varying illumination and pose conditions. The authors have detected the likeness of an image using the morphological operations, as these operations are robust against noise and illumination changes [22]. Vijay et al. [13] from Maharashtra described different local and global contrast enhancement techniques like histogram equalization (HE),Dualistic Sub Image Histogram Equalization (DSIHE), Brightness Bi-Histogram Equalization (BBHE),Minimum Mean Brightness Error Bi-Histogram Equalization (MMBEBHE), Brightness Preserving Dynamic Histogram Equalization (BPDHE), Local Transformation Histogram Equalization (LHE), Global Transformation Histogram Equalization (GHE), Local Transformation Histogram Equalization (LHE). The authors claimed that most of the contrast enhancement techniques results in losses the brightness of the original image while enhancing the image. At last, the authors analyzed that for preserving the brightness of the image and to improve the visual effects the mixture of global and local contrast enhancement techniques is superior to other techniques.

Jishan et al. [14] designed algorithm to enhance the images that are compressed by the JPEG standard. The algorithm was used to find the dissimilarities of the image defined within the discrete cosine transform (DCT) domain. The algorithm designed to be applied to any compressed image based on the DCT standard like H.261, MPEG and JPEG. The main advantage of the algorithm was that it does not affect the compressibility of the original image because it enhances the image on the decompression stage. The results of the proposed algorithms are based on visual appearance and shown in Figure 3(a) and Figure 3(b).

Figure 3(a): Decompressed JPEG image method image

Figure 3(b): Proposed contrast-measure-based

Zahedi et al. [11] from Isfahan University of Technology, Iran proposed different types of pseudo-coloring algorithms for IR thermograms. For pseudo coloring of IR images, the authors implemented nonlinear function transform based on the relevant properties of the human eye and compared the results of various proposed algorithms on visual experience.

Singh et al. [17] proposed the non-training contrast enhancement algorithms. The authors have developed pseudo coloring algorithm for converting the gray level image into an RGB image. The authors have proposed the iterative method for optimal Thresholding to eliminate the background from the image as shown in figure 4.

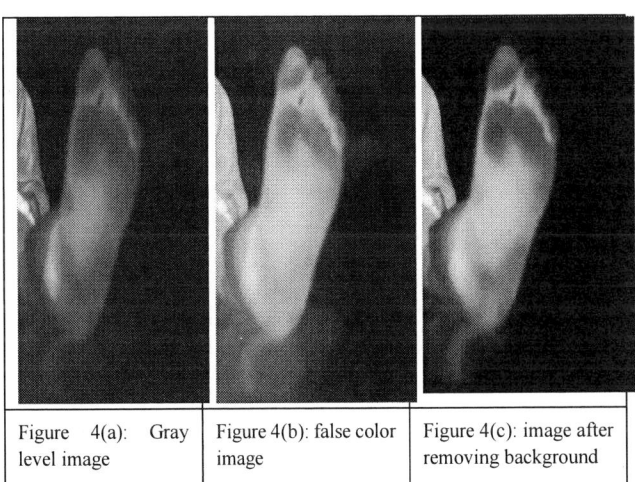

| Figure 4(a): Gray level image | Figure 4(b): false color image | Figure 4(c): image after removing background |

Figure 4: Depicting the implementation of proposed pseudo-coloring algorithm on plantar view of foot thermogram. (a) Normalized gray level image (b) false color image of gray level image (c) false color image of plantar view of foot after removing the background of the image

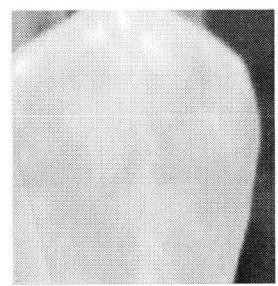 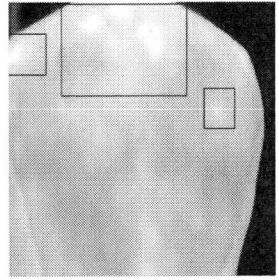

Figure 5(a): Thermal image

Figure 5(b): Thermal enhanced image

Figure 5(a): Representing thermogram after pseudo-coloring and Figure 5(b) represents image enhancement on previous thermogram.

Color enhancement has been done by the authors using decor relation contrast stretching (DCS) as shown in Figure 5. The authors evaluate the result of proposed methodology based on the entropy values; more the entropy difference between the two stages, better the color enhancement is achieved [17].

Choi et al. [20] have proposed low resolution thermal image enhancement algorithms based on the convolution neural network (CNN) called in Thermal Enhance Network (TEN) with the concept of RGB guidance as shown in Figure 6. In this study, a single low-resolution image learns a direct end to end mapping to the desired high-resolution image. The authors demonstrated that the method has powerful capabilities to enhance the visibility of the image for pedestrian detection and visual odometry.

Figure 6(a): Without TEN

Figure 6(b): with TEN

Gillespie et al. [18] have proposed two techniques for improvement of highly correlated images. The first technique is the de correlation stretch i.e. transformation of data into basic components based on the Principal Component Analysis (PCA) to overemphasize the deviations from the grey or the false color image. The second technique is a stretch of Hue Saturation Intensity (HSI), in which data is represented in spherical coordinates and then by stretching the color saturation is adjusted directly. As the result, the authors claim that the most effective techniques are based on the transformation of the

978-1-5386-6624-1/18 $31.00 © 2018 IEEE

image data into the new coordinates, retransformation to the original image and enhancement of the transformed image.

Apart from enhancing the image, Duarte et al. [16] developed the segmentation algorithms for identifying the ROI with specific geometrical shapes. Most of the times, lucrative software utilizes regular prismatic shapes like rectangle, circle etc., which either include some irrelevant data or exclude essential information of the thermal images. The authors developed segmentation algorithm for choosing any ROI and then optimizing for further use. As a result, the authors have compared the untreated ROI and optimized ROI.

Villasenor et al. [19] discussed different preprocessing method to improve the contrast of the image. The authors applied different innocuous substances like Vaseline, baby oil, ultrasound imaging gel, sunscreen, ethylic alcohol and penicillin cream in the region of interest for better visualization of the image after applying the substances the authors claimed that initial image contrast is enhanced by 400%. The original is shown in Figure 7(a) and the image after applying the substance is shown in Figure 7(b).

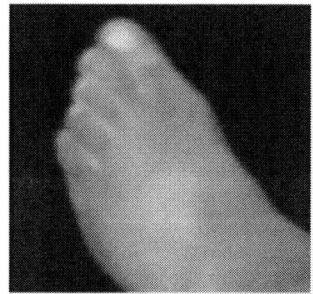 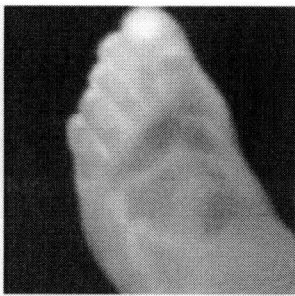

Figure 7(a): Before applying substance   Figure 7(b): After applying substance

Hariharan et al. proposed the approach for image fusion and enhancement by using Empirical Mode Decomposition (EMD). In this study, the authors decomposed images into Intrinsic Mode Functions (IMFs) and then the fusion was performed at the decomposition level. Device weighting schemas emphasize features and increased the visual content of the fused image by minimizing the mutual information between the IMFs. Finally, the author evaluates the result against pixel by pixel averaging.

III. CONCLUSIONS AND FUTURE DIRECTIONS

Contrast enhancement is the major component of the thermal image processing. The original brightness of the image gets diminished with some contrast enhancement algorithms like Histogram Equalization (HE). In future contrast enhancement algorithms must be developed that could be maintained the brightness of the original image. On the other hand, more attention should be given for the contrast enhancement algorithms to improve the visualization of the thermal image rather than applying any substance to identify the region of interest from the image. Further research is required for preserving the various features (visibility) of the original image. Knowledge of body temperature profile helps in developing the better algorithm to distinguish the ROI from other irrelevant body parts. In future, the thermal image can be enhanced by using some better contrast stretching algorithms. Contrast stretching algorithms should be applied only on the ROI instead to the whole thermal image by r Planck emoving the irrelevant background of the image. The outcomes of method should be analyzed statistically instead of visual appearance.

REFERENCES

[1] S. C. Liew, "Electromagnetic waves Centre for Remote Imaging, Sensing and Processing", pp. 10–27, 2006.

[2] M. Planck, "On the theory of the energy distribution law of the normal spectrum", Ann. Phys. 4(553), 237–245, 1901.

[3] E F. Ring, J Ringand K Ammer, "Infrared thermal imaging in medicine",Physiol. Meas. 33R33–R46, 2012

[4] J. Steketee, "Spectral emissivity of skin and pericardium". Phys. Med. Bio, 18, 686–694, 1973.

[5] Qi. Hairong, N. A. Diakides, "Infrared Imaging in Medicine", pp.1-10,2007

[6] G.H. Cohen, Brueschke, "Obstetric and gynecologicThermography, Obstetrics and Gynecology", pp.842–847, 1965.

[7] C. Loriaux, "Role of thermography in gynecology", Journal de Radiologie dElectrologie de Medecine Nucleaire 56 (Suppl.), pp.57–58, 1975.

[8] W. Oosterlinck, W. S. De, "Avascular nephrotomy by means of thermography", European Urology 7, pp.25–26, 1981.

[9] F. J. Ring, "Thermal Imaging for fever screening", ISO Focus, pp.33–35, February 2007.

[10] E. Y. K. Ng, G. Kaw, W. M. Chang, "Analysis of IR thermal imager for mass blindFever screening", Microvascular Research 68, pp.104–109, 2004.

[11] Z. Zahedi, M. Soltani, S. Sadri, A. Moosavi, "Breast thermography and pseudo-coloring presentation for improving gray level images", In: Photonics Global Conference (PGC), pp. 1–5. IEEE, 2012.

[12] He. Renjie, S. Luo, Z. Jing and Y. Fan, "Adjustable Weighting Image Contrast Enhancement Algorithm and Its Implementation", 6th IEEE Conference on Industrial Electronics and Applications, pp.1750-1754, 2011.

[13] A. Vijay, S. S. Gharde, "Review of various image enhancement Techniques", Vol. 2, Issue 7, July 2013.

[14] J. Tang, E. Peli, S. Acton, Senior Member, IEEE, "Image Enhancement Using a Contrast Measure in the Compressed Domain", VOL. 10, NO. 10, OCTOBER 2003.

[15] H. Hariharan, A. Gribok, M.A. Abidi, A. Koschan, "Image fusion and enhancement via empirical mode decomposition", Journal of Pattern Recognition Research, pp 16-32, 2006.

[16] L. Duarte, M. Espanha, T. Viana, D. Freitas, P. Bártolo, P. Faria,H.A. Almeida, "Segmentation algorithms for thermal images", Procedia Technology 16, 2014

[17] J. Singh, and A.S. Arora, "Contrast Enhancement Algorithm for IR Thermograms Using Optimal Temperature Thresholding and Contrast Stretching", 2017

[18] R. A. Gillespie, A. B. Kahle and R. E. Walker, "color enhancement of highly correlated images I. Décor relation and HSI Contrast Stretches", pp. 209-235, 1986.

[19] C. V. Mora, J. Francisco, S. Marin, E. G. Sevilla, "Contrast enhancement of mid and far infrared images of subcutaneous veins", pp 221–228, 2008.

[20] Y. Choi, N. Kim, S. Hwang and In. S. kweon, "Thermal Image Enhancement using Convolutional Neural Network", pp. 1-08, October 2016.

[21] M. Singh and A.S. Arora, Varying illumination and pose conditions in face recognition, Procedia Computer Science, 2016 – Elsevier.

[22] M. Singh and A.S. Arora, A robust anti-spoofing technique for face liveness detection with morphological operations, OPTIK, 2017 – Elsevier.

[23] B.F. Jones and P. Plassmann: "Digital infrared thermal imaging of human skin", IEEE engineering in medicine and biology Magazine vol. 21, Issue 6, pp 41-48, 2002

[24] A.S. Arora and J. Singh,: "Para-nasal sinusitis detection using thermal imaging" IEEE in science and information conference (SAI), pp.184-188, 2015

Laguerre Function Based Model Predictive Control for Van-de-Vusse Reactor

Akansha Jain
Department of Electronics Engineering
Rajasthan Technical University
Kota, Rajasthan 324010
Email: akanshajain328@gmail.com

Rajashree Taparia
Department of Electronics Engineering
Rajasthan Technical University
Kota, Rajasthan 324010
Email: rtaparia@rtu.ac.in

Abstract—**This paper presents the discrete time Laguerre functions based Model Predictive Control (MPC) for Van de Vusse reaction in a continuous stirred tank reactor (CSTR). The Model Predictive Control effectively handles the constraints on the state and the input. The set of Laguerre network is introduced to describe the future control input trajectory which has fewer parameters to adjust. The orthonormal functions for representing the control trajectory is used. Model Predictive Control does the efficient parametrization of the difference of the control signal. Simulations are done to validate the theory and results are discussed.**

Keywords—***Model Predictive Control (MPC), Laguerre functions, constraints, continuous stirred tank reactor, Van de Vusse reaction.***

I. INTRODUCTION

The idea of Model Predictive Control was first advocated in the late 1970s. Model Predictive Control has wide applications in process control industries because it has the ability to handle challenging problems as time delays, right-half-plane zeros and variable constraints [1]. The continuous stirred tank reactor (CSTR) with Van de Vusse reaction is a highly nonlinear reactor system and exhibits interesting property such as the inverse response [2] and complex reaction networks [3]. It is considered as a benchmark example of non-minimum phase process to demonstrate various control methodologies. For nonlinear Process control systems two approaches exact feedback linearization and gain scheduling are designed in [4]. In the isothermal case of continuous stirred tank reactor, the temperature in the reactor is assumed to be constant. The substance concentration is the only variable to be con-trolled with input feed as the manipulated variable [5]. Linear controllers usually give unsatisfactory performance for such plants and nonlinear approaches become more attractive. For nonlinear systems if optimization problem is solved without any redeeming features, implies that global optimum can not be assured [6]. In Van de Vusse reactor, if different operating regimes are used then it shows steady-state multiplicity and dynamic change to handle these nonlinear problems develop a multiple Model Predictive Control strategy [7]. But when the same control performance is needed to be optimized, optimal

control is the common approach found in the literature. There are constraints on the input and states for such a plant. The traditional Linear Quadratic Regulator (LQR) control poses these constraints after deriving the control law whereas MPC considers the constraints while deriving the control law, thus improving the controller performance [8]. MPC uses the model of the plant to predict the behavior of the system during generation of the control signal. The suitable cost function is optimized over the prediction horizon considering the constraints. The first input of generated control sequence is applied to the system and the optimization is again done over a shifted horizon and the procedure is repeated for the entire simulation cycle [9] [10]. An improved version of Model Predictive Control is based on containing it with Laguerre functions. The key idea used with the design of discrete MPC is deriving the control signal or the difference of the control signal by forward shift operators [11]. If the process dynamics is complicated using the conventional technique then large number of forward shift operators are required. If a discrete orthonormal function is used to represent the control trajectory, the number of parameters used for description is reduced [12] [13]. For the large dimensional problems the conventional quadratic programming is unsuitable, to overcome Laguerre function in conjunction with quadratic programming is used in [14] to reduce the coding complexity or computational effort. In [15] dynamic positioning control of vessel is studied by using Laguerre function based MPC. It's application in the field of electromechanical system include the electric servo drive control of RT-70 antenna using Laguerre function based MPC [16]. In this paper we present the design methodology of Laguerre function in discrete time frame for the linearized model of Van de Vusse reactor. By using Laguerre function the number of terms used in optimization is reduced to a fraction of that required in normal MPC.

This paper has been organized as follows. In Section II the problem formulation for Van de Vusse reactor is formulated. In Section III Model Predictive Controller using Laguerre functions is designed. In Section IV simulation results with proposed controllers are given. The concluding remarks are presented in Section V.

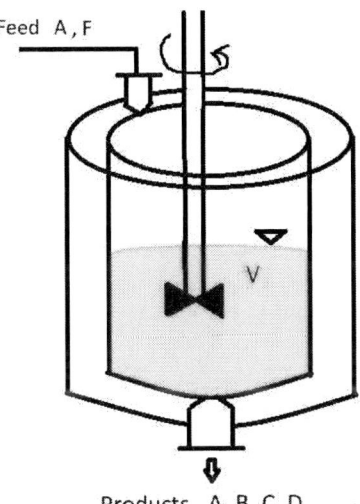

Feed A, F

V

Products A, B, C, D

Fig. 1. Schematic representation of the Continous Stirred Tank Reactor

II. VAN DE VUSSE REACTOR

We consider the nonlinear isothermal continuous stirred tank reactor with Van de Vusse series of reactions. The equations that govern the system are [2].

$$
\begin{aligned}
\dot{C_a} &= -k_1 C_a - k_3 C_a^2 + \frac{F}{V}(C_{af} - C_a) \\
\dot{C_b} &= k_1 C_a - k_2 C_b - \frac{F}{V}C_b
\end{aligned} \tag{1}
$$

where C_a and C_{af} are the concentrations in the reactor and in the feed respectively. k_1, k_2 and k_3 are the reaction rate constants. The manipulated variable is the dilution rate F. The reactor volume V is constant. C_b is the concentration of substance B. Choosing the state variables as,

$$
x_1 = C_a; \ x_2 = C_b; \ x_{af} = C_{af}; \ u = F/V.
$$

Therefore the dynamics for the considered system is,

$$
\begin{aligned}
\dot{x_1} &= -k_1 x_1 - k_3 x_1^2 + (x_{af} - x_1)u \\
\dot{x_2} &= k_1 x_1 - k_2 x_2 - x_2 u
\end{aligned} \tag{2}
$$

The Van de Vusse series of reactions are described by the following reaction scheme [17].

$$
\begin{array}{ccccc}
A & \xrightarrow{k_1} & B & \xrightarrow{k_2} & C \\
2A & & \xrightarrow{k_3} & & D
\end{array} \tag{3}
$$

with x_1 and x_2 are the concentrations of A and B respectively and $u = F/V$ is the dilution (feed) rate. The control objective is to design the improved Model Predictive Controller using the Laguerre functions to maintain the concentration of substance B at a set point of one.

III. LAGUERRE FUNCTION BASED MPC DESIGN

In this section, the Laguerre function based MPC design is discussed. To implement Laguerre function based MPC, the augmented model of the considered system is formulated as discussed below.

A. Model Predictive Control

Consider the discrete time state-space model,

$$
\begin{aligned}
x(k+1) &= A_n x(k) + B_n u(k) \\
y(k) &= C_n x(k)
\end{aligned} \tag{4}
$$

where A_n, B_n and C_n are the state, input and output matrices respectively. x is the state vector; u is the control signal and y is the system output. By taking the difference of the state variable vector,

$$
\Delta x(k+1) = A_n \Delta x(k) + B_n \Delta u(k) \tag{5}
$$

where

$$
\begin{aligned}
\Delta x(k) &= x(k) - x(k-1) \\
\Delta u(k) &= u(k) - u(k-1)
\end{aligned}
$$

We modify the state vector as,

$$
x(k) = \begin{bmatrix} \Delta x(k) & y(k) \end{bmatrix}^T.
$$

the augmented model becomes,

$$
\begin{aligned}
x(k+1) &= A x(k) + B \Delta u(k) \\
y(k) &= C x(k)
\end{aligned}
$$

where

$$
\begin{aligned}
\begin{bmatrix} \Delta x(k+1) \\ y(k+1) \end{bmatrix} &= \begin{bmatrix} A_n & 0_n \\ C_n A_n & 1 \end{bmatrix} \begin{bmatrix} \Delta x(k) \\ y(k) \end{bmatrix} \\
&\quad + \begin{bmatrix} B_n \\ C_n B_n \end{bmatrix} \Delta u(k) \\
y(k) &= \begin{bmatrix} 0_n & 1 \end{bmatrix} \begin{bmatrix} \Delta x(k) \\ y(k) \end{bmatrix} \tag{6}
\end{aligned}
$$

To design MPC based controller we need the future plant output, which is calculated from future state variables. If the sampling instant is k_r, the future control signals will take the form as,

$$
\Delta u(k_r), \Delta u(k_r + 1), \cdots, \Delta u(k_r + n_c - 1),
$$

where n_p and n_c are the prediction and control horizon respectively. The state vector $x(k_r)$ denotes the present state of the plant, from which the future state variables are predicted as below,

$$
x(k_r + 1|k_r), x(k_r + 2|k_r), \cdots, x(k_r + n_p|k_r)
$$

Based on the augmented model (6), the future state variables are calculated sequentially as in (7) and from the predicted state variables the predicted output variables are calculated as in (8). So the predicted output variables at the instant k_r is obtained as,

$$
Y = \Psi x(k_r) + \Upsilon \Delta U \tag{9}
$$

978-1-5386-6624-1/18 $31.00 © 2018 IEEE   1011

$$x(k_r + 1|k_r) = Ax(k_r) + B\Delta u(k_r)$$
$$x(k_r + 2|k_r) = Ax(k_r + 1|k_r) + B\Delta u(k_r + 1)$$
$$= A^2 x(k_r) + AB\Delta u(k_r) + B\Delta u(k_r + 1)$$
$$\vdots$$
$$x(k_r + n_p|k_r) = A^{n_p} x(k_r) + A^{n_p-1} B\Delta u(k_r) + A^{n_p-2} B\Delta u(k_r + 1|k_r) + \cdots +$$
$$+ A^{n_p-n_c} B\Delta u(k_r + n_c - 1). \tag{7}$$
$$y(k_r + 1|k_r) = CAx(k_r) + CB\Delta u(k_r)$$
$$y(k_r + 2|k_r) = CAx(k_r + 1|k_r) + CB\Delta u(k_r + 1)$$
$$= CA^2 x(k_r) + CAB\Delta u(k_r) + CB\Delta u(k_r + 1)$$
$$\vdots$$
$$y(k_r + n_p|k_r) = CA^{n_p} x(k_r) + CA^{n_p-1} B\Delta u(k_r) + CA^{n_p-2} B\Delta u(k_r + 1|k_r) + \cdots + \cdot$$
$$+ CA^{n_p-n_c} B\Delta u(k_r + n_c - 1). \tag{8}$$

where

$$Y = \begin{bmatrix} y(k_r + 1|k_r) \\ y(k_r + 2|k_r) \\ \cdots \\ y(k_r + n_p|k_r) \end{bmatrix}$$

$$\Delta U = \begin{bmatrix} u(k_r) \\ u(k_r + 1) \\ \cdots \\ u(k_r + n_c - 1) \end{bmatrix}$$

$$\Psi = \begin{bmatrix} CA \\ CA^2 \\ \vdots \\ CA^{n_p} \end{bmatrix}$$

$$\Upsilon = \begin{bmatrix} CB & 0 & \cdots & 0 \\ CAB & CB & \cdots & 0 \\ \vdots & \vdots & \ddots & \\ CA^{n_p-1}B & CA^{n_p-2}B & \cdots & CA^{n_p-n_c}B \end{bmatrix}$$

B. Discrete Laguerre Function

The core idea of discrete-time MPC is based on optimizing the future control trajectory. The difference of the control input signal is computed for the control horizon as $\Delta U(k)$ for $k = 0, 1, \cdots, n_c - 1$ and the rest of terms from $k = n_c, \cdots, n_p$ are taken as zeros. Due to rapid sampling, complicated process dynamics and high demand on closed-loop performance, the ΔU may require a very large number of parameters. These lead to poor numerical condition and heavy computational load when implemented on-line [10] [18]. As a consequence, an improved MPC using Laguerre function is used. Furthermore, a long control horizon can be realized without using a large number of parameters.

The discrete-time z-transform of Laguerre networks represented by the following set of equations.

$$\tau_1(z) = \frac{\sqrt{(1-p^2)}}{1 - pz^{-1}}$$
$$\vdots$$
$$\tau_N(z) = \frac{\sqrt{(1-p^2)}}{1 - pz^{-1}} \left(\frac{z^{-1} - p}{1 - pz^{-1}}\right)^{N-1} \tag{10}$$

where p is the pole of the discrete-time Laguerre network with $0 \le p < 1$. N is the number of terms used in computing the control signal. From the inverse z-transform of $\tau_1(z), \tau_2(z), \cdots, \tau_N(z)$, the set of discrete-time Laguerre function is expressed as,

$$L(k) = \begin{bmatrix} l_1(k) & l_2(k) & \cdots & l_N(k) \end{bmatrix}^T \tag{11}$$

Besides the Laguerre function satisfies the following orthonormal properties:

$$\frac{1}{2\Pi} \int_{-\pi}^{\pi} \tau_i(e^{j\omega}) \tau_i(e^{j\omega})^T d\omega = 1$$
$$\frac{1}{2\Pi} \int_{-\pi}^{\pi} \tau_i(e^{j\omega}) \tau_j(e^{j\omega})^T d\omega = 0 \quad i \neq j \tag{12}$$

The dynamics of Laguerre function is defined as,

$$L(k+1) = A_l L(k) \tag{13}$$

where the matrix A_l is expressed as below, which is a function of parameters p and $\delta = (1 - p^2)$.

$$A_l = \begin{bmatrix} p & 0 & \cdots & 0 \\ \delta & p & \cdots & 0 \\ (-p)\delta & \delta & \cdots & 0 \\ \vdots & \vdots & \ddots & \\ (-p)^{N-2}\delta & (-p)^{N-3}\delta & \cdots & p \end{bmatrix} \tag{14}$$

978-1-5386-6624-1/18 $31.00 © 2018 IEEE

with the initial condition vector as,

$$L(0) = \sqrt{\delta} \begin{bmatrix} 1 & -p & \cdots & (-p)^{N-1}) \end{bmatrix}^T \quad (15)$$

By using Laguerre network based MPC, the control horizon n_c is replaced with the number of terms N. The parameters N and p are used to describe the complexity of the set point tracking. In the optimization procedure, for the long control horizon we use large value of p with smaller number of terms N. When $p = 0$, $N = n_c$, the Laguerre based MPC becomes the normal MPC. The future control increment is obtained by a set of Laguerre functions upto the future sample instant k_r as,

$$\Delta u(k + k_r) = L(k_r)^T \sigma \quad (16)$$

where $\sigma = \begin{bmatrix} e_1 & e_2 & \cdots & e_N \end{bmatrix}^T$ is a vector of Laguerre coefficients. The prediction of state vector at sampling instant k_r can be written as,

$$
\begin{aligned}
x(k + k_r | k) &= A^{k_r} x(k) + \sum_{r=0}^{k_r - 1} A^{k_r - r - 1} B L(r)^T \sigma \\
&= A^{k_r} x(k) + \Theta(k_r)^T \sigma \quad (17)
\end{aligned}
$$

and the output variable can be written as,

$$y(k + k_r | k) = C A^{k_r} x(k) + C \Theta(k_r)^T \sigma \quad (18)$$

where

$$\Theta(k_r)^T = \sum_{r=0}^{k_r - 1} A^{k_r - r - 1} B L(r)^T$$

The main objective of the predictive control system is the predicted output should be as close as possible to the set-point. So the system is designed such that an error function between the reference and the predicted output is minimized. The cost function J is formulated as follows,

$$J = \sum_{r=1}^{n_p} x(k + k_r | k)^T Q x(k + k_r | k) + \sigma^T R \sigma \quad (19)$$

where Q is positive semi-definite and R is positive definite matrices. The optimal solution for the vector σ is obtained by optimizing the cost function and setting $\frac{\delta J}{\delta \sigma} = 0$.

$$\sigma = - \sum_{k_r=1}^{n_p} (\Theta(k_r) Q \Theta(k_r)^T + R)^{-1} \sum_{k_r=1}^{n_p} (\Theta(k_r) Q A^{k_r}) x(k) \quad (20)$$

Let

$$
\begin{aligned}
\Omega &= \sum_{k_r=1}^{n_p} (\Theta(k_r) Q \Theta(k_r)^T + R)^{-1} \\
\varphi &= \sum_{k_r=1}^{n_p} (\Theta(k_r) Q A^{k_r})
\end{aligned}
$$

Therefore,

$$\sigma = -\Omega^{-1} \varphi x(k) \quad (21)$$

The control $\Delta u(k)$ is designed using a linear state feedback control law strategy.

$$\Delta u(k) = -K_{mpc} x(k) \quad (22)$$

where the state feedback gain matrix is obtained as,

$$K_{mpc} = L(0)^T \Omega^{-1} \varphi$$

Model Predictive Control handles the constraints in an effective way. The control signals are parameterized using Laguerre functions. This makes us possible to choose the constraints at any point of time in the system dynamics. Thus the flexibility of introducing future constraints is increased. The constraints can be on the states, the control input and the difference of the control input. The constraint on the difference of the control input is considered as a hard constraint, as it limits the size of the control signal movement. If Δu_{max} and Δu_{min} are the upper and lower bounds on the difference of the control input signal respectively, then the constraint can be specified as,

$$\Delta u_{min} \leqslant \Delta u(k) \leqslant \Delta u_{max} \quad (23)$$

IV. NUMERICAL SIMULATION

The MPC based on Laguerre network design is simulated for the nonlinear isothermal continuous stirred tank reactor with Van de Vusse series of reactions (3). Equation (2) is linearized about the desired steady state and on discretizing the result with a sampling time of $\Delta T_s = 0.002$ [19] we get,

$$A_n = \begin{bmatrix} 0.95123 & 0 \\ 0.08833 & 0.81873 \end{bmatrix}; B_n = \begin{bmatrix} -0.0048771 \\ -0.0020429 \end{bmatrix}$$

$$C_n = \begin{bmatrix} 0 & 1 \end{bmatrix} \quad (24)$$

The control objective is to keep the concentration of substance B at the set-point of one. The reaction rate constants are chosen as $k_1 = 50$, $k_2 = 100$ and $k_3 = 10$. The weighing matrices are taken as $Q = C^T C$ and $R = 1$. The prediction horizon $n_p = 90$ and the control horizon is considered as $n_c = 50$. n_c is always taken less than n_p. The Laguerre parameters are chosen as $p = 0.9$ and $N = 3$ and the initial condition as $x(0) = \begin{bmatrix} 0.5 & 0.1 \end{bmatrix}^T$.

The control input u is the feed rate and the output y is the concentration of substance B which is also the state x_2. The constraint is considered on the difference of control input variable as,

$$-0.5 \leq \Delta u(k) \leq 0.1 \quad (25)$$

Fig. 2 and Fig. 3 show the results of Van de Vusse reactor with normal MPC and Laguerre based MPC. From the plots it is observed that the set point one of the concentration of B is achieved. The corresponding control input U for both the normal MPC and Laguerre based MPC are shown. The plots of the difference of the control input show that the constraints are satisfied for both normal MPC and Laguerre MPC. The number of iterations required in optimization for Laguerre based MPC is reduced considerably. In normal MPC the control horizon $n_c = 50$ whereas for Laguerre based MPC the number of terms $N = 3$. The same response is achieved

978-1-5386-6624-1/18 $31.00 © 2018 IEEE   1013

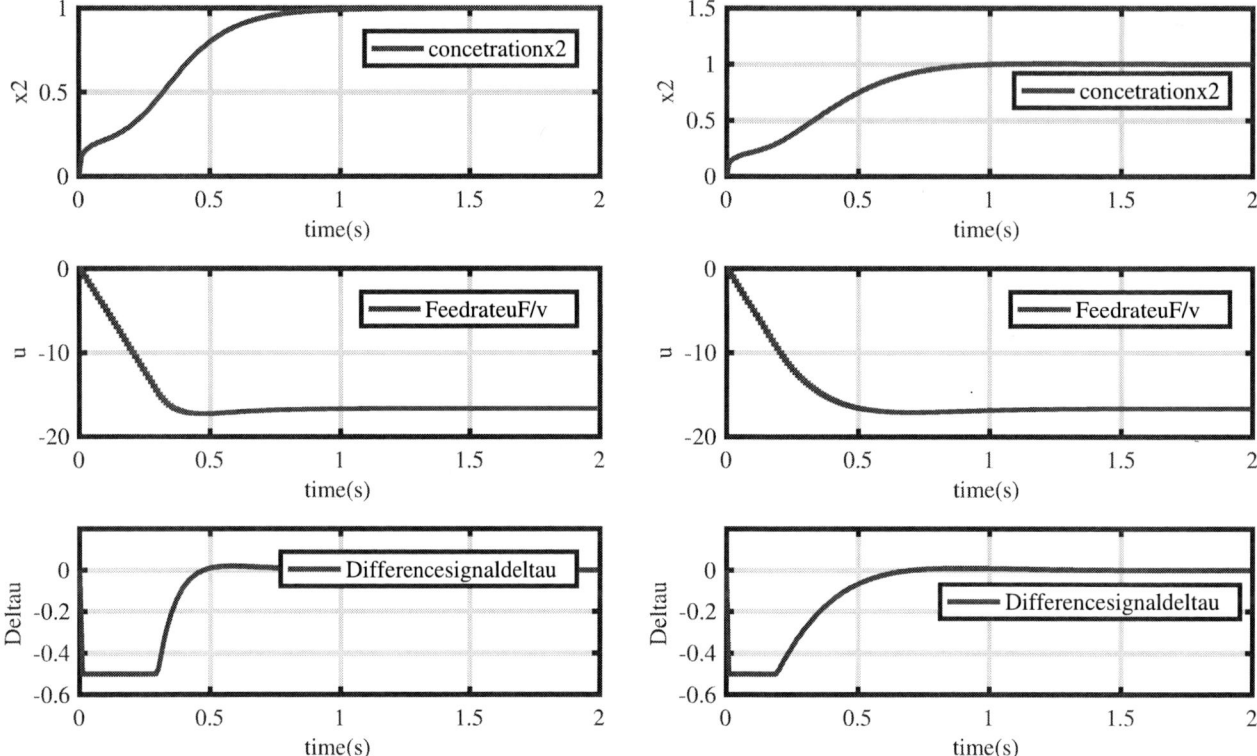

Fig. 2. The concentration, the feed rate and difference of control signal with normal MPC

Fig. 3. The concentration, the feed rate and difference of control signal with Laguerre functions based MPC

in both the cases with substantial reduction in optimization iterations.

V. CONCLUSIONS

In this paper, the Model Predictive Control for nonlinear isothermal continuous stirred tank reactor with Van de Vusse series of reactions is discussed. The Laguerre based MPC design for the CSTR model with Van de Vusse reactions is discussed. The orthonormal function for representing the control trajectory is used. The set point of one for the concentration of substance B is achieved while maintaining the constraints on the difference of the control input signal. From the simulations it is observed that the optimization iterations required for Laguerre based MPC are considerably reduced while keeping the same controller performance for both the cases.

REFERENCES

[1] P. B. Sistu and B. W. Bequette, "Model predictive control of processes with input multiplicities," *Chemical engineering science*, vol. 50, no. 6, pp. 921–936, 1995.

[2] S. Kuntanapreeda and P. M. Marusak, "Nonlinear extended output feedback control for cstrs with van de vusse reaction," *Computers & Chemical Engineering*, vol. 41, pp. 10–23, 2012.

[3] S. J. Parulekar, R. S. Waghmare, and H. C. Lim, "Yield optimization for multiple reactions," *Chemical engineering science*, vol. 43, no. 11, pp. 3077–3091, 1988.

[4] K.-U. Klatt and S. Engell, "Gain-scheduling trajectory control of a continuous stirred tank reactor," *Computers & Chemical Engineering*, vol. 22, no. 4-5, pp. 491–502, 1998.

[5] H. Chen, A. Kremling, and F. Allgöwer, "Nonlinear predictive control of a benchmark cstr," *Proceedings of 3rd European control conference*, pp. 3247–3252, 1995.

[6] M. Morari and J. H. Lee, "Model predictive control: past, present and future," *Computers & Chemical Engineering*, vol. 23, no. 4-5, pp. 667–682, 1999.

[7] B. Aufderheide and B. W. Bequette, "A variably tuned multiple model predictive controller based on minimal process knowledge," *American Control Conference, 2001*, vol. 5, pp. 3490–3495, 2001.

[8] D. Q. Mayne, J. B. Rawlings, C. V. Rao, and P. O. Scokaert, "Constrained model predictive control: Stability and optimality," *Automatica*, vol. 36, no. 6, pp. 789–814, 2000.

[9] E. F. Camacho and C. B. Alba, *Model predictive control.* Springer Science & Business Media, 2013.

[10] L. Wang, *Model predictive control system design and implementation using MATLAB®.* Springer Science & Business Media, 2009.

[11] A. Dubravić and Z. Šehić, "Using orthonormal functions in model predictive control," *Tehnički vjesnik*, vol. 19, no. 3, pp. 513–520, 2012.

[12] P. Sanila and J. Jacob, "Simultaneous tracking and vibration control of flexible joint manipulator using laguerre network based composite fast-slow mpc," *Signal Processing, Informatics, Communication and Energy Systems (SPICES), 2015 IEEE International Conference on*, pp. 1–6, 2015.

[13] M. Spasic, D. Miti, M. Hovd, and D. Antic, "Tube model predictive

978-1-5386-6624-1/18 $31.00 © 2018 IEEE   1014

control based on laguerre functions with an auxiliary sliding mode controller," *Intelligent Systems and Informatics (SISY), 2017 IEEE 15th International Symposium on*, pp. 000 243–000 248, 2017.

[14] G. Valencia-Palomo and J. Rossiter, "Using laguerre functions to improve efficiency of multi-parametric predictive control," in *American Control Conference (ACC), 2010*. IEEE, 2010, pp. 4731–4736.

[15] X. Qian, Y. Yin, X. Zhang, X. Sun, and H. Shen, "Model predictive controller using laguerre functions for dynamic positioning system," in *35th Chinese Control Conference (CCC), 2016*. IEEE, 2016, pp. 4436–4441.

[16] T. H. Phuong, M. P. Belov, and T. D. Khoa, "Model predictive controller based on laguerre functions for large radio telescope servo control system," in *Young Researchers in Electrical and Electronic Engineering (EIConRus), 2018 IEEE Conference of Russian*. IEEE, 2018, pp. 1003–1007.

[17] J. Van de Vusse, "Plug-flow type reactor versus tank reactor," *Chemical Engineering Science*, vol. 19, no. 12, pp. 994–996, 1964.

[18] S. H. HosseinNia and M. Lundh, "A general robust mpc design for the state-space model: Application to paper machine process," *Asian Journal of Control*, vol. 18, no. 5, pp. 1891–1907, 2016.

[19] P. O. Scokaert and J. B. Rawlings, "Constrained linear quadratic regulation," *IEEE Transactions on automatic control*, vol. 43, no. 8, pp. 1163–1169, 1998.

2nd IEEE International Conference on Power Electronics, Intelligent Control and Energy Systems (ICPEICES-2018)

Highly linear Current Follower Transconductance Amplifier (CFTA) design and its filter application

Shweta Kumari
Division of Electronics and Communication Engineering
Netaji Subhas Institute of Technology
New Delhi, India
Shwetagupta20009@gmail.com

Maneesha Gupta
Division of Electronics and Communication Engineering
Netaji Subhas Institute of Technology
New Delhi, India
Maneeshapub@gmail.com

Abstract— **In this paper, High linearity low voltage Current Follower Transconductance Amplifier (CFTA) design has been proposed. The proposed circuit operates at ±0.6V symmetric supply voltage. The linearity of proposed CFTA circuit is enhanced by using parallel connected PMOS and NMOS differential stage for current follower (CF) and source degeneration technique is used in transconductance amplifier (TA) stage. Proposed CFTA circuit can provide ±350μA input current linear range, 1mS transconductance and 198MHz bandwidth. TSMC 180nm process technology parameters are used in design of proposed CFTA and its application.**

Keywords— *CFTA, linearity, Source Degeneration Technique, Current mode biquad filter*

I. INTRODUCTION

In the area of integrated circuits, current mode circuit have become more demanded due to its better performance in terms of architectural simplicity, slew rate, wide bandwidth, high linearity, low power consumption as compare to voltage mode circuits. Many current mode circuits such as CCII (Second Generation Current Conveyor), CDBA (Current Differencing Buffered Amplifier), OTA (Operational Transconductance Amplifier) have proposed in literature [1-3]. CCII and OTA circuits have some problems like terminals of these devices have serious parasitic effects, impedance matching is required at the time of cascading and do not have low impedance features. CDTA is one of the novel current mode active element reported by Biolek [4].

CDTA (Current Differencing Transconductance Amplifier) is frequently used in the field of current mode analog signal processing applications as its input and output signals are currents. CDTA consists of Current Differencing Unit (CDU) as first stage and Transconductance Amplifier (TA) as second stage. CDTA offers electronic tuning of transconductance (G_m). Due to tuning capability of CDTA, it is extensively useful in synthesis of current mode analog signal processing circuit. Various CDTA based applications such as filters, oscillators, rectifiers and multiplier/square rooter circuit are also reported in literature [5-10]. On the other hand, these CDTA based employment do not use both input ports (n or p), that can cause noise injection into monolithic circuit. This problem can be overcome by using current follower as first stage in place of CDU of CDTA block. A new active building block called as Current Follower Transconductance Amplifier

(CFTA) is reported in literature [11] after combining CF as first part and TA as second part. CFTA architectures which operate on different supply voltages are available in literature [12-18]. CFTA based employments such as oscillators, rectifier, modulator, filter and Frequency-dependent negative resistance (FDNR) etc are also reported in literature [19-24]. Major focus of this work is to design Highly linear low voltage CFTA circuit. CF stage of proposed CFTA is based on parallel connected PMOS and NMOS differential stage which provide high linearity [25] and TA stage is based on Current mirror OTA with source generation technique in order to improve the linearity of proposed circuit [26]. In order to justify the performance of proposed CFTA circuit, current mode biquad filter has taken as application.

The paper is arranged as follows. Section II includes basics of CFTA, proposed CFTA and its small signal analysis. Section III describes the filter application based on proposed CFTA. Section IV includes proposed CFTA simulation results and biquad filter results. Section V includes conclusion of paper.

II. PROPOSED CFTA AND ITS CHARACTERISTIC

CFTA has two stages, first is current follower which send input current (i_F) to Z port and second is transconductance amplifier. Transconductance amplifier stage changes voltage at Z port (V_Z) into currents i_{X+} and i_{X-} at ports X+ and X- respectively. Fig. 1(a) and (b) show the circuit symbol and behavioral model of CFTA, where F is current input port, Z is an intermediate port and X± are current output ports. The hybrid matrix given below presents the terminal relation of CFTA:

$$\begin{bmatrix} v_F \\ i_Z \\ i_{X+} \\ i_{X-} \end{bmatrix} = \begin{bmatrix} 0 & 0 & 0 & 0 \\ 1 & 0 & 0 & 0 \\ 0 & +G_m & 0 & 0 \\ 0 & -G_m & 0 & 0 \end{bmatrix} \begin{bmatrix} i_F \\ v_Z \\ v_{X+} \\ v_{X-} \end{bmatrix} \quad (1)$$

where G_m is transconductance which can be electronically controllable by bias current I_B of CFTA. Fig. 2 shows the circuit diagram of proposed CFTA circuit. It has two parts, input stage as a CF and output stage as a TA.

978-1-5386-6624-1/18 $31.00 © 2018 IEEE

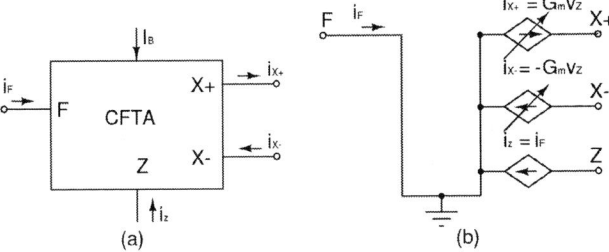

(a)

(b)

Fig. 1. (a) Circuit symbol (b) behavioral model of CFTA.

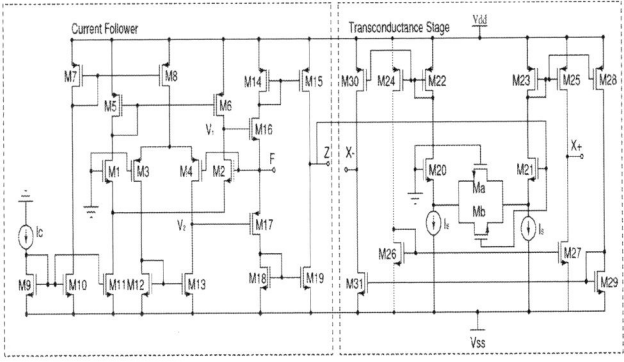

Fig. 2. Proposed CFTA circuit diagram.

CF stage is designed using parallel connected PMOS and NMOS differential stage which may provide rail to rail swing. The PMOS differential input pairs allow input signal to work near negative supply rail whereas NMOS differential input pairs allow input signal to work near positive rail. Two source followers M16 and M17 are used, in order to provide low resistance at terminal F. Small signal equivalent circuits of proposed CFTA are shown in Fig. 3 (a) and Fig. 3 (b) and g_{mi}, r_{0i}, gi, di, and si present transconductance, output resistance, gate, drain and source terminals of corresponding MOSFETs Mi respectively (where i=1-31) which shown in Fig. 2. The current i_Z and i_F of proposed CFTA (see Fig. 3 (a)) are given by Eq (2) and Eq. (3) respectively (neglect channel length modulation):

$$i_z = g_{m15}V_{gs15} + g_{m19}V_{gs19} \qquad (2)$$

$$i_F = -g_{m16}V_{gs16} - g_{m17}V_{gs17} \qquad (3)$$

The value of V_{g15} and V_{g19} have been evaluated as:

$$Vg_{15} = -\frac{g_{m16}V_{gs16}}{g_{m14}}, \quad Vg_{19} = -\frac{g_{m17}V_{gs17}}{g_{m18}} \qquad (4)$$

By using Eq. (2) - (4) and assuming $g_{m14} = g_{m15}$, $g_{m18} = g_{m19}$, current transfer ratio (i_Z/i_F) is given as:

$$\alpha = \frac{i_Z}{i_F} = \frac{g_{m16}V_{gs16} + g_{m17}V_{gs17}}{g_{m16}V_{gs16} + g_{m17}V_{gs17}} \cong 1 \qquad (5)$$

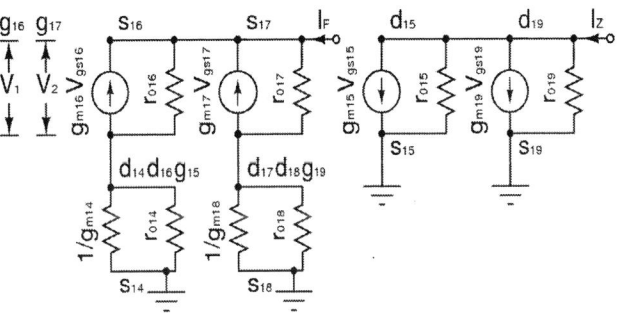

Fig. 3(a). Small signal equivalent circuit for CF stage.

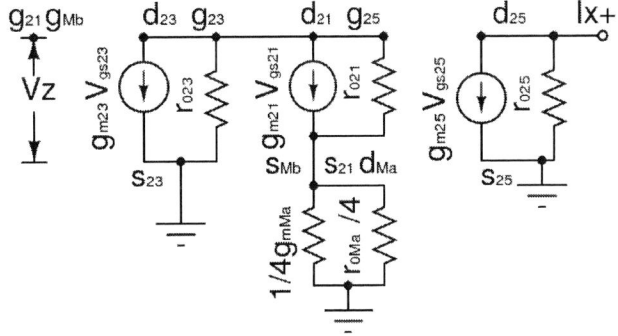

Fig. 3 (b). Small signal equivalent circuit of half circuit for TA stage.

For evaluation of terminal resistance F, a test current I_F is joined at terminal F. Current I_F can be calculated (see Fig. 3(a)) as:

$$I_F = \left\{ -g_{m16}(V_1 - V_F) - g_{m17}(V_2 - V_F) \right\} + \frac{r_{016} + r_{017}}{r_{016}r_{017}}V_F \qquad (6)$$

where expression for V_1 and V_2 can be directly calculated from Fig. 2 (assume $r_{0N} = r_{02} \parallel r_{06}$, $r_{0P} = r_{04} \parallel r_{013}$) as:

$$V_1 \cong -\frac{r_{0N}}{2}(g_{m2}V_F), \quad V_2 \cong -\frac{r_{0P}}{2}(g_{m4}V_F) \qquad (7)$$

By using Eq. (6) and Eq. (7), the terminal resistance R_F is given as:

$$R_F = \frac{V_F}{I_F} = \frac{1}{g_{m16}\left(1 + \frac{r_{0N}g_{m2}}{2}\right) + g_{m17}\left(1 + \frac{r_{0p}g_{m4}}{2}\right) + \frac{r_{016} + r_{017}}{r_{016}r_{017}}} \qquad (8)$$

978-1-5386-6624-1/18 $31.00 © 2018 IEEE   1017

Terminal Z resistance is high (see Fig. 3 (a)) and is obtained as:

$$R_Z = \frac{r_{O15}r_{O19}}{r_{O15}+r_{O19}} \tag{9}$$

Fig. 3(b) presents small signal equivalent circuit of half circuit for TA stage of proposed circuit. Straightforward circuit analysis provides the transconductance (assume g_{mMa} = g_{mMb}) of proposed circuit can be written as:

$$G_m = \frac{g_{m21}}{1+g_{m21}/4g_{mMa}} \tag{10}$$

The linearity of proposed circuit is improved by source degeneration technique at the expense of decrease in transconductance gain that can be concluding from Eq. (10).

IV APPLICATION BASED ON PROPOSED CFTA

To validate the performance of proposed circuit, a biquad filter has been designed and simulated. The simulation results of biquad filter are included in subsequent section 5.

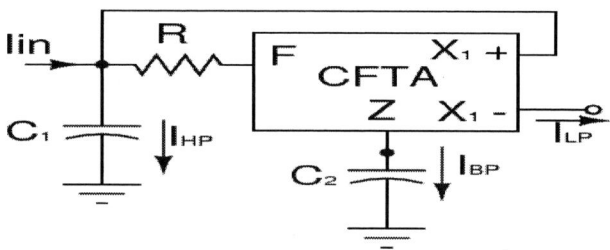

Fig. 4. Circuit diagram of biquad filter [27].

Biquad filter presented in [27] is designed in order to justify the performance of proposed CFTA. Current mode biquad filter based on proposed CFTA is shown in Fig. 4. Circuit analysis of biquad filter gives current transfer function as follows:

$$\frac{I_{LP}(s)}{Iin(s)} = \frac{G_m / RC_1C_2}{D(s)}$$

$$\frac{I_{BP}(s)}{Iin} = \frac{s / RC_1}{D(s)}$$

$$\frac{I_{HP}(s)}{Iin} = \frac{s^2}{D(s)} \tag{11}$$

$$D(s) = s^2 + \frac{s}{RC_1} + \frac{G_m}{RC_1C_2} \tag{12}$$

The centre frequency ω_0 can be calculated from Eq. (12) as follows.

$$\omega_0 = \sqrt{\frac{G_m}{RC_1C_2}} \tag{13}$$

V SIMULATION RESULTS

TSMC 0.18μm CMOS technology has been used for designing of circuits to verify their characteristics. The circuit operates with ±0.6V supply voltage. Table 1 shows the aspect ratio of proposed CFTA. Resistor R_Z = 1KΩ is connected to terminal Z for computation of simulation results of proposed CFTA.

TABLE I. ASPECT RATIO OF PROPOSED CFTA

Structure	Proposed CFTA
Transistor	W(μm)/L(μm)
M1, M2,	9/0.36
M3, M4	27/0.36
M5, M7, M16, M18, M19, M20, M21	50/0.36
M6, M8, M14, M15, M17	100/0.36
M9, M10, M11, M12, M13	18/0.36
M22, M23, M24, M25, M28, M30	40/0.36
M26, M27, M29, M31	20/0.36
Ma, Mb	6.5/0.36

DC response for CF stage of proposed CFTA at bias current Ic = 100μA is presented in Fig. 5. Current I_Z is linear in range of +420μA to -436μA with change of I_F from 500μA to -500μA. Fig. 6 presents AC response of CF stage. Current transfer ratio (α) and bandwidth of CF stage are 0.973 and 268MHz respectively.

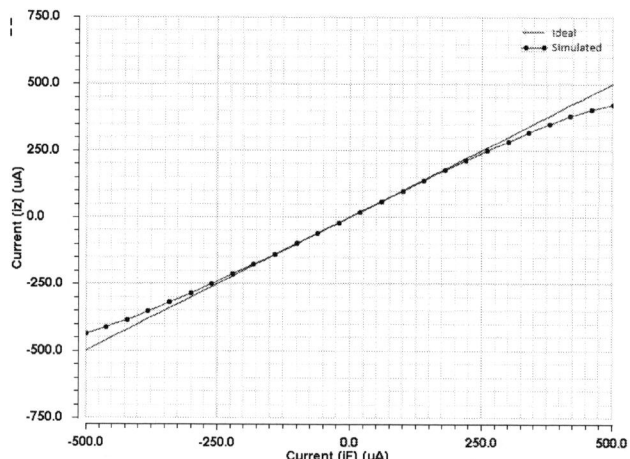

Fig. 5. DC response of CF stage.

Transconductance range of 66μS -1.03mS can be achieved by proposed CFTA for variation in I_B of 10 μA - 250μA which is shown in Fig. 7. AC and DC current transfer response from terminal F to terminal X+ are shown in Fig. 8 and Fig. 9 respectively. Current I_{X+} is linear with variation of input current I_F from +350μA to -350μA at bias current I_B = 250μA and bandwidth of proposed CFTA is 198MHz .

978-1-5386-6624-1/18 $31.00 © 2018 IEEE   1018

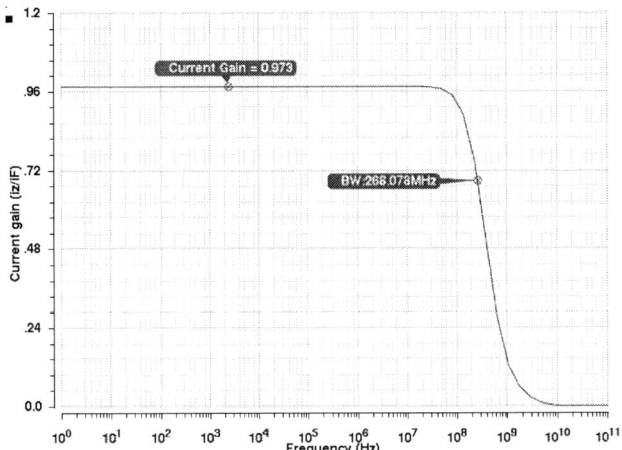

Fig. 6. AC response of CF stage.

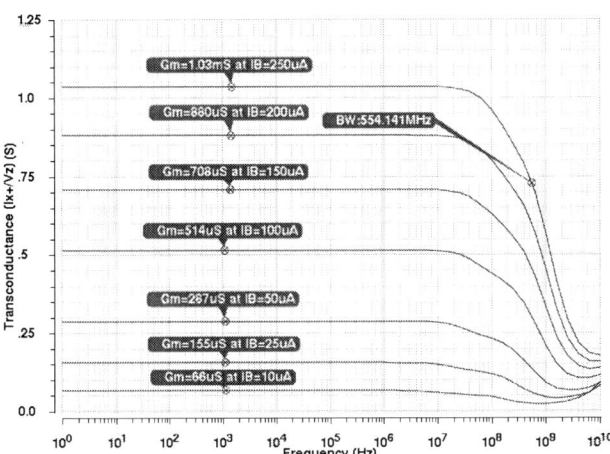

Fig. 7. AC transfer characteristic of transconductance stage at different bias currents (I_B).

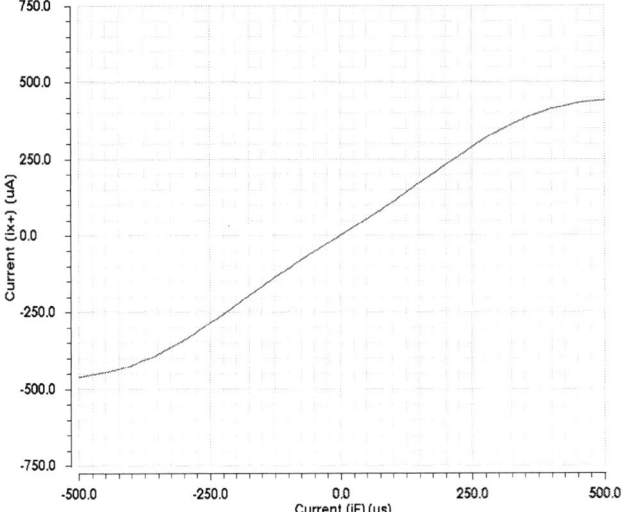

Fig. 8. DC response for current I_F vs current I_{X+}.

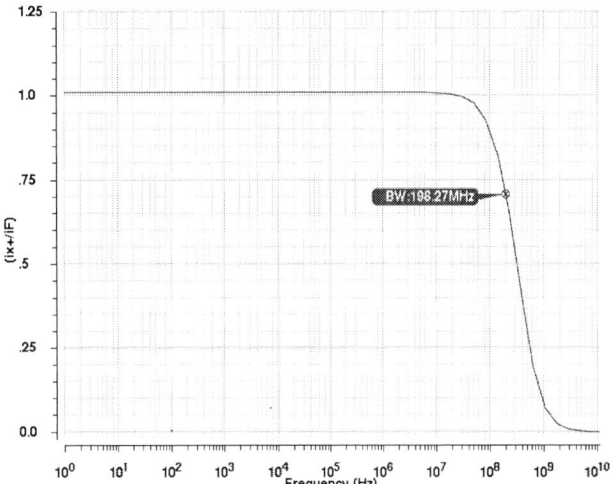

Fig. 9. Frequency response of i_{x+}/i_F .

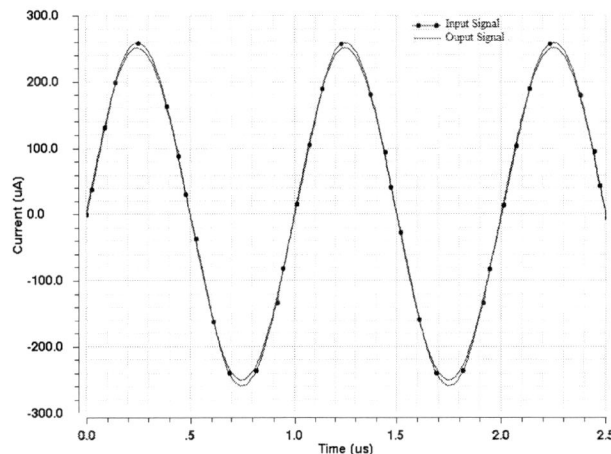

Fig. 10. Transient characteristics for proposed CFTA at 1MHz.

TABLE. II. SUMMARY OF SIMULATION RESULT OF PROPOSED CFTA

Parameters	Proposed CFTA
Technology	180nm
Supply Voltage	±0.6V
Bias current (I_B)	250µA
Transconductance (Gm)	1mS
Current gain of CF	0.973
Linearity of CF stage	+420µA to -436µA
Bandwidth of proposed CFTA	198MHz
Linearity of proposed CFTA	-350µA to 350µA
DC Power Dissipation	2.3mW

For transient analysis, a sinusoidal current signal of 250µA amplitude at 1MHz is applied to proposed CFTA (G_m is 1.03mS). Fig. 10 presents input and output sinusoidal current signals and THD (Total harmonic distortion) of output signal is 0.68%. The power dissipation of proposed circuit is 2.3mW. Performance comparison of proposed CFTA with

978-1-5386-6624-1/18 $31.00 © 2018 IEEE   1019

other CFTA circuits are presented in Table. 3. The proposed CFTA provides high linearity as contrast to other CFTA circuits. It also operates at low supply as compared to ref. [14, 15, 24] which can be seen from table 3.

TABLE III. COMPARISON OF PROPOSED CFTA WITH PREVIOUSLY REPORTED CFTA CIRCUITS

Parameter/Ref.	This Work	[17]	[18]	[14]	[24]	[15]
Technology (nm)	180	180	180	180	180	180
Supply Voltage (V)	±0.6	±0.6	±0.6	±1.5	±1.85	±1
Bias Currents (µA)	250	300	300	300	300	300
Transconductance (mS)	1 .03	8.5	11.3	0.459	0.156	1.8
Linearity (µA)	±350	±50	±20	±65	±90	±100
Current gain (α)	0.973	0.969	0.980	1.0	1.0	0.997
Power dissipation(mW)	2.3	1.7	1.8	3	4	2.2

The biquad filter which is shown in Fig. 4 is designed using proposed CFTA for centre frequency $f_0 = 15.9$MHz. Passive and active components value used during biquad filter design are $G_m = 1.03$mS, $R = 1$KΩ and $C_1 = C_2 = 10$pF. Fig. 11 presents the AC responses of current mode biquad filter. The centre frequency f_0 of current node biquad filter is 15.4MHz which is close to theoretical value. During transient analysis, a sinusoidal input current signal of 250µA peak value at 100MHz is applied to the biquad filter. Transient response for biquad filter is shown in Fig. 12. THD of high pass output signal is 0.88% and power consumption of biquad filter is 2.5mW.

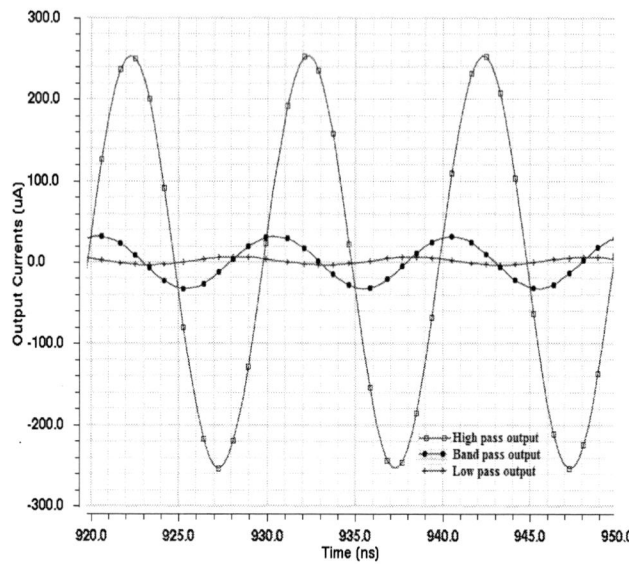

Fig.12. Transient response of biquad filter for 250µA sinusoidal current input signal at 100MHz.

V. CONCLUSION

Highly linear CFTA architecture has been presented in this paper. To enhance the linearity performance, a parallel connected PMOS and NMOS differential stage in CF along with source degeneration technique in TA stage are used in proposed CFTA. Proposed CFTA circuit provides ±350µA input dynamic range and operates at ±0.6V supply voltage along with 198MHz bandwidth. To justify the performance of proposed CFTA, current mode biquad filter is selected and simulated. For low voltage high linearity analog signal processing applications, the proposed CFTA circuit can be useful.

REFERENCES

[1] G. Ferri, N.C. Guerrini, Low-voltage low-power CMOS current conveyors, Springer Science & Business Media, 2003.

[2] E. Sanchez-Sinencio, J. Silva-Martinez, "CMOS transconductance amplifiers, architectures and active filters: a tutorial, IEEE proceedings-circuits, devices and systems," vol.147, pp.3-12, 2000.

[3] A. U. Keskin, E. Hancioglu, "CDBA-Based Synthetic Floating Inductance Circuits with Electronic Tuning Properties," ETRI journal, vol.27, pp.239-242, 2005.

[4] D. Biolek, "CDTA-building block for current-mode analog signal processing, In Proceedings of the ECCTD," vol.3, pp.397-400, 2003.

[5] A. U. Keskin, D. Biolek, E. Hancioglu, V. Biolková, "Current-mode KHN filter employing current differencing transconductance amplifiers," AEU-International Journal of Electronics and Communications, vol. 60, pp. 443-446, 2006.

[6] L. I. Yong-an, "Forth Order Current Mode Band Pass Filter with Coupled Tuned by Current Using CCCDTAs," Journal of Electron Devices, vol.7, pp.210-213, 2010.

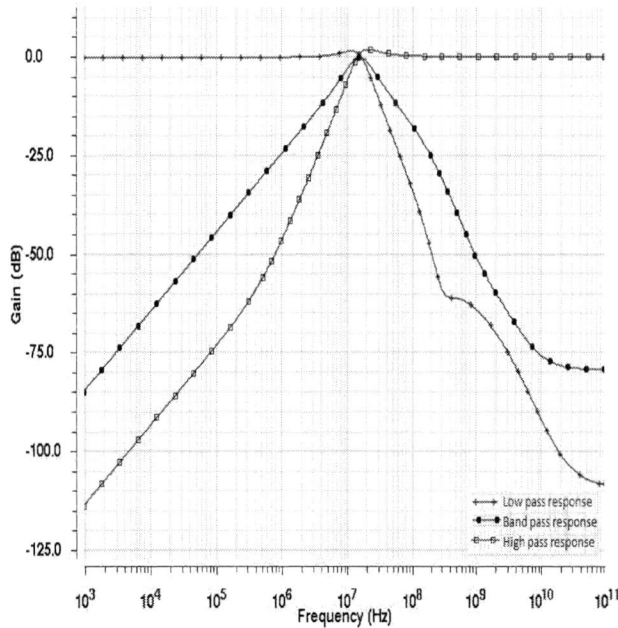

Fig. 11. AC transfer characteristic of current mode biquad filter.

[7]   Y. Li, "A new single MCCCDTA based Wien-bridge oscillator with AGC," AEU-International Journal of Electronics and Communications, vol. 66, pp. 153-156, 2012.

[8]   Y. Li, "Electronically tunable current-mode quadrature oscillator using single MCDTA," Radioengineering, vol.19, pp.667-671, 2010.

[9]   F. Kacar, M. E. BAŞAK, "A new mixed mode full-wave rectifier realization with current differencing transconductance amplifier," Journal of Circuits, Systems, and Computers, vol. 23, p.1450101, 2014.

[10]   W. Tangsrirat, T. Pukkalanun, P. Mongkolwai, W. Surakampontorn, "Simple current-mode analog multiplier, divider, square-rooter and squarer based on CDTAs," AEU-International Journal of Electronics and Communications, vol. 65, pp. 198-203, 2011.

[11]   N. Herencsár, J. Koton, K. Vrba, I. Lattenberg, J. Misurec, "Generalized design method for voltage-controlled current-mode multifunction filters," Proceedings of the 16th Telecommunications Forum TELFOR, pp. 400-403, 2008.

[12]   N. Herencsar, J. Koton, K. Vrba, J. Misurec, "A novel current-mode SIMO type universal filter using CFTAs," Contemporary Engineering Sciences, vol. 2, pp. 59-66, 2009.

[13]   R. Sotner, J. Petrzela, J. Slezak, "Current mode tunable KHN filter based on controlled MO-CFTAs," Proceedings of the 3rd International Conference on Signals, Circuits and Systems (SCS), pp. 1-4, 2009.

[14]   N. Herencsar, J. Koton, K. Vrba, A. Lahiri, "Novel mixed-mode KHN-equivalent filter using Z-copy CFTAs and grounded capacitor," Proceedings of the 4th international conference on Circuits, systems and signals, World Scientific and Engineering Academy and Society (WSEAS), pp.87-90, 2010.

[15]   R. S. Tomar, S.V. Singh, D. S. Chauhan, "Cascadable low voltage operated current-mode universal biquad filter," WSEAS Transactions on Signal Processing, vol.10, pp.345-353, 2014.

[16]   W. Tangsrirat, "Voltage-mode analog PID controller using a single z-copy current follower transconductance amplifier (ZC-CFTA)," Informacije MIDEM-Journal of Microelectronics Electronic Components and Materials, vol.45, pp.175-179, 2015.

[17]   S. Kumari, M. Gupta, "Design and analysis of high Transconductance Current Follower Transconductance Amplifier (CFTA) and its applications," Analog Integrated Circuits and Signal Processing, vol.93, pp.489-506, 2017.

[18]   S. Kumari, M. Gupta, "A new CMOS design of high transconductance current follower transconductance amplifier and its applications," Analog Integrated Circuits and Signal Processing, vol.95, pp.325-349, 2018.

[19]   W. Kongnun, P. Silapan, "A single MO-CFTA based electronically/temperature insensitive current-mode half-wave and full-wave rectifiers," Advances in Electrical and Electronic Engineering, vol.1, pp .275-283, 2013.

[20]   N. Herencsár, K. Vrba, J. Koton, A. Lahiri, "Realisations of single-resistance-controlled quadrature oscillators using a generalised current follower transconductance amplifier and a unity-gain voltage-follower," International Journal of Electronics, vol.97, pp. 897-906, 2010.

[21]   J. Satansup, W. Tangsrirat, "Single-input five-output electronically tunable current-mode biquad consisting of only ZC-CFTAs and grounded capacitors," Radioengineering, vol.20, pp. 650-655, 2011.

[22]   W. Kongnun, A. Aurasopon, "A novel electronically controllable of current-mode level shifted multicarrier PWM based on MO-CFTA," Radioengineering, vol. 22, pp. 907-915, 2013

[23]   Y. A. Li, "A series of new circuits based on CFTAs," AEU-International Journal of Electronics and Communications, vol. 66, pp. 587-592, 2012.

[24]   N. Herencsar, J. Koton, K.Vrba, A. Lahiri, O. Cicekoglu, "Current-controlled CFTA-based current-mode SITO universal filter and quadrature oscillator," Proceedings of International Conference on Applied Electronics (AE), pp.1-4, 2010.

[25]   E. Arslan, A. Morgül, "Wideband current conveyor with rail to rail input stage," Proceedings of the 5th International Conference on Electrical and Electronics Engineering, ELECO 2007, pp. 66-70, 2007.

[26]   K.C. Kuo, A. Leuciuc, "A linear MOS transconductor using source degeneration and adaptive biasing," IEEE Transactions on Circuits and Systems II: Analog and Digital Signal Processing, vol.48, pp.937-943, 2001

[27]   W. Tangsrirat, "Novel current-mode and voltage-mode universal biquad filters using single CFTA," Indian Journal of Engineering and Materials sciences, Vol.17, pp. 99-104, 2010.

High Gain Transimpedance Amplifier Using Self Cascode Structure

Preeti Singh
Department of Electronics and Communication Engineering
Netaji Subhas Institute of Technology
Delhi, India
preeti.vlsi@gmail.com

Maneesha Gupta
Department of Electronics and Communication Engineering
Netaji Subhas Institute of Technology
Delhi, India
maneeshapub@gmail.com

Abstract: **High gain transimpedance amplifier (TIA) with low power dissipation is presented in which Self cascode structure (also known as split length transistor structure) is used to achieve high output impedance which leads to gain enhancement. Simulations are done in TSMC 0.18μm CMOS technology with 50fF photodiode capacitance. In the proposed TIA, the transimpedance gain is 56.8dBΩ which is increased by 11.2% and the bandwidth is 2.7 GHz with 0.42mW power dissipation.**

Keywords: Transimpedance amplifier, self cascode, split length, gain.

I. INTRODUCTION

The continuous increasing demand of high data rates is fulfilled due to optical communication systems. Transimpedance amplifier (TIA) is used in optical communication systems at the receiver end. TIA converts small photodiode current into appropriate voltage level which motivates researchers to design high gain transimpedance amplifier. Several techniques are reported to enhance the gain such as positive feedback technique [1], cascading of gain stages [2][3] and self cascode structure [4][5]. Positive feedback network generates the negative conductance which reduces the overall output conductance and thereby increases the gain but it has stability issues. Cascading of gain stages also enhances the gain but it needs frequency compensation. To attain high gain without degrading high frequency performance, self cascode (SC) structure is used in this paper.

Self cascode structure is reviewed in section II. The small signal analysis of simple TIA and proposed TIA are discussed in section III. In section IV, simulation results are discussed and finally Section V concludes the paper.

II. SELF CASCODE STRUCTURE

Self cascode structure comprises of transistors M_1 and M_2 (shown in Fig.1) which can be worked as single merged transistor. It has longer effective channel length. Since both the transistors use the same gate biasing, it is called self cascode structure. In Fig.1, SC transistor is used to design common source (CS) amplifier in which M_2 transistor drives in saturation region and M_1 transistor typically drives in triode region. Therefore, transistor M_1 acts as a resistor which is dependent on the input. For optimal working, the aspect ratio of M_2 transistor must be larger than the aspect ratio of M_1 transistor. The main

advantage of SC structure is its high output impedance without additional voltage requirement [6].

The output impedance of Fig.1 is given as in [5]:

$$Z_{out} = \frac{1+(g_{m2}+g_{m1})r_{o2}+sC_{gs2}r_{o2}}{g_{m1}+sC_{gs2}} \tag{1}$$

where g_{m1} and g_{m2} are the transconductance of transistors M_1 and M_2 respectively. r_{o2} is the output resistance and c_{gs2} is the gate to source capacitance of transistor M_2.

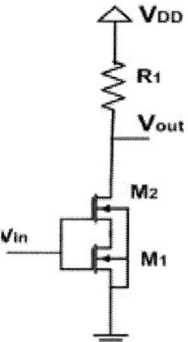

Fig. 1. SC based CS amplifier.

After putting s=0 in Eq. (1), the output resistance is given as in [5]:

$$R_{out} = \frac{1+(g_{m2}+g_{m1})r_{o2}}{g_{m1}}. \tag{2}$$

From Eq. (2), it is clearly seen that SC based CS amplifier has higher output impedance which in result increases the gain.

III. THE PROPOSED TIA

The schematic of simple TIA [7] and proposed TIA are shown in Fig. 2 and Fig. 3 respectively in which C_{pd} is the photodiode capacitance and R_1 is feedback resistance to provide dc current path. Transistors M_2 and M_3 comprise current mirror. In the proposed TIA, SC structure is used to get the advantage of high output impedance. The small signal analysis for output impedance calculation of simple TIA and the proposed TIA are discussed in section III-A and III-B respectively.

978-1-5386-6624-1/18 $31.00 © 2018 IEEE

A. Output impedance of simple TIA

Fig. 4 shows the small signal equivalent circuit of simple TIA to calculate output impedance. In the circuit, g_{m1} is the transconductance and v_{gs1} is the gate to source voltage of transistor M_1. r_{o1} and r_{o2} are output resistances of transistor M_1 and the current mirror respectively. C_1 is the total capacitance at the input which consists of gate to source capacitance and photodiode capacitance. Kirchhoff's current law (KCL) is applied at input node (V_1) and output node (V_{out}) to calculate output impedance.

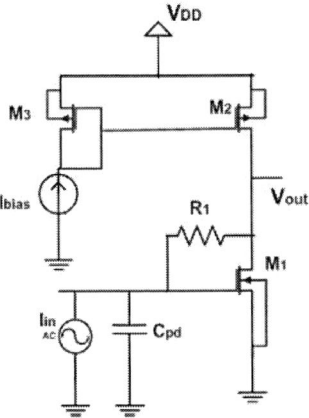

Fig. 2. Circuit diagram of simple TIA.

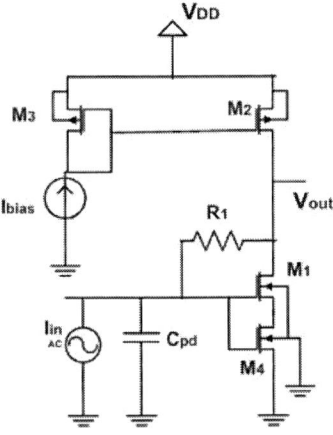

Fig. 3. Circuit diagram of the proposed TIA.

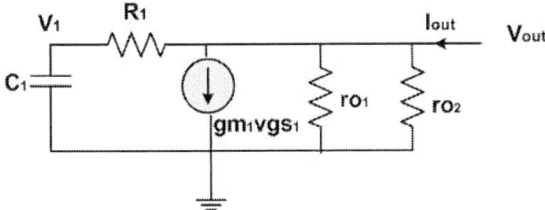

Fig. 4. Small signal equivalent circuit of simple TIA.

At node V_1 :

$$\frac{V_1 - V_{out}}{R_1} + V_1 s\, C_1 = 0. \qquad (3)$$

At node V_{out} :

$$I_{out} = g_{m1}V_1 + \frac{V_{out}}{R} + \frac{V_{out}-V_1}{R_1} \qquad (4)$$

where $R = r_{o1}||r_{o2}$.

After solving Eqs. (3) and (4), Output impedance of TIA is calculated as:

$$Z_{out} = \frac{RR_1(1+sR_1C_1)}{(1-g_{m1}R)R_1+s(R+R_1)R_1C_1}. \qquad (5)$$

After putting s=0, output resistance is given as:

$$R_{out} = \frac{R}{(1-g_{m1}R)}. \qquad (6)$$

B. Output impedance of SC based TIA

The small signal circuit is shown in Fig. 5. All the transistors are operating in saturation region except the transistor M_4 which is operating in linear region. Therefore, the small signal equivalent of the transistor M_4 is equal to the resistance whose value is $1/g_{m4}$ where g_{m4} is the transconductance of transistor M_4. KCL is applied at node V_1, V_{out} and V_X to calculate output impedance.

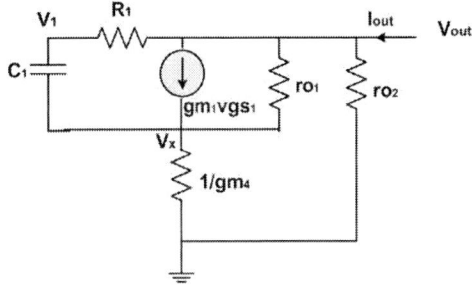

Fig. 5. Small signal equivalent circuit of the proposed TIA.

At node V_1:

$$(V_1 - V_X)sC_1 + \frac{V_1-V_{out}}{R_1} = 0. \qquad (7)$$

At node V_{out}:

$$\frac{V_{out}-V_1}{R_1} + g_{m1}(V_1 - V_X) + \frac{V_{out}-V_X}{r_{o1}} + \frac{V_{out}}{r_{o2}} = I_{out}. \qquad (8)$$

At node V_X:

$$(V_1 - V_X)sC_1 + g_{1m1}(V_1 - V_X) + \frac{V_{out}-V_X}{r_{o1}} = V_X g_{m4}. \qquad (9)$$

After solving Eqs. (7) - (9), output impedance of SC based TIA is calculated by:

$$Z_{out} = \frac{R_1 r_{o1}(1+sR_1C_1)(sC_1+sC_1R_1g_{m4}+g_{m1}+g_{m4}+sR_1C_1g_{m1})}{(g_{m1}+sC_1R_1g_{m4})(r_{o1}||r_{o2})(1+sR_1C_1)+R_1(sC_1+g_{m1}+g_{m4}+sr_{o1}C_1g_{m1})}. \qquad (10)$$

After putting s=0 in Eq. (10), output resistance of SC based TIA is obtained as:

$$R_{out} = \frac{R_1 r_{o1}(g_{m1}+g_{m4})}{g_{m1}R+R_1(g_{m1}+g_{m4})}. \quad (11)$$

On comparing Eqs. (6) and Eq. (11), it can be concluded that output impedance of the proposed TIA is increased by using SC structure which eventually increases the gain.

IV. SIMULATION RESULTS

The simulations are carried out in CMOS technology using 1.8V power supply. The transistors M_1-M_3 are working in saturation region while the transistor M_4 is operating in triode region. All circuit components, aspect ratio of transistors and bias current are mentioned in Table 1.

TABLE I. CIRCUIT PARAMETER OF THE PROPOSED TIA

Parameter	Values
Aspect ratio	$M_2, M_3, M_4 = 10/0.18$
	$M_1 = 20/0.18$
Bias current	$I_{bias} = 10\mu A$
Resistance	$R_1 = 1K\Omega$

Fig. 6 shows the comparison between the frequency responses of simple TIA and SC based TIA. It is shown that the transimpedance gain of proposed TIA is increased from 51.1dBΩ to 56.8dBΩ without affecting the bandwidth which is 2.7GHz with 0.42mW power dissipation.

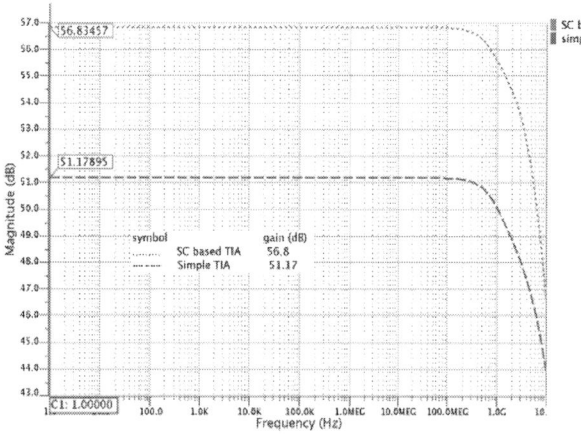

Fig. 6. Frequency response of simple TIA and the Proposed TIA.

Fig. 7 and Fig. 8 show the effect of variations in photodiode capacitance and temperature respectively. It can be observe that the transimpedance gain is increasing by increasing the temperature. The input current noise spectral density is shown in Fig. 9 which is about 11.2pA/√Hz. Fig. 10 and Fig. 11 present the dc and transient response of the proposed TIA respectively.

The Monte Carlo simulation is also carried out by selecting a set of 100 samples on applying mismatch in the width of the transistors and presented in Fig.12. Fig. 13

shows the frequency response of proposed TIA at different process corners and it is seen that, for the sequence of process corners FF, FS, SF and SS, gain is increasing while bandwidth is reducing.

Table 2 illustrated the performance comparision between exsisting TIAs and the proposed TIA.

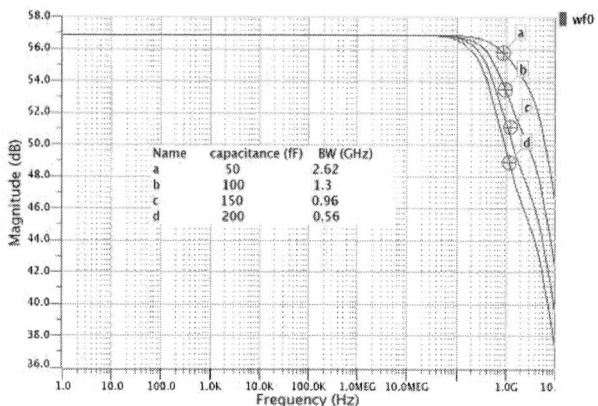

Fig. 7. Frequency response of the proposed TIA at different values of capacitance.

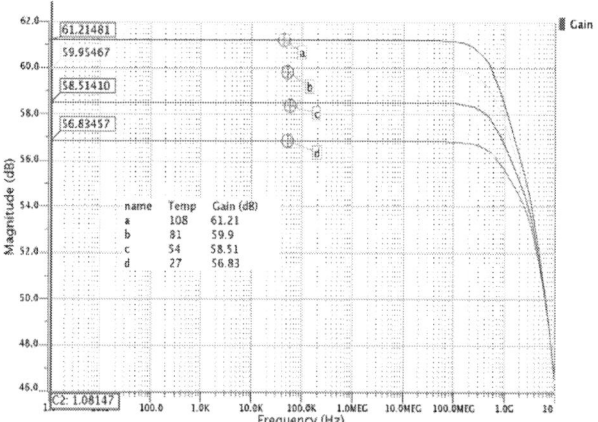

Fig. 8. Frequency response of the proposed TIA at different temperatures.

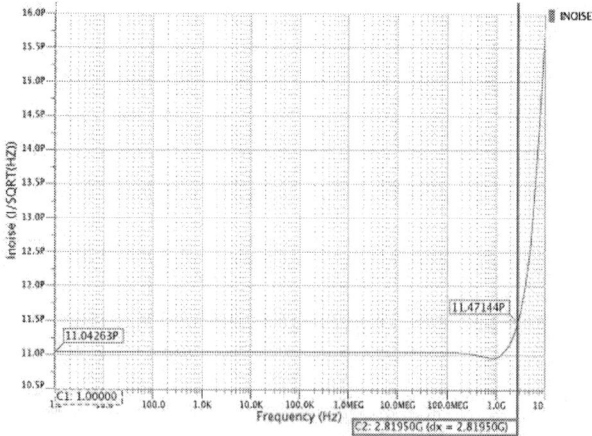

Fig. 9. Input referred noise of the proposed TIA.

TABLE 2. PERFORMANCE COMPARISION SUMMARY

Ref.	Technology (µm)	Gain (dBΩ)	BW (GHz)	Power (mW)
This work	0.18	56.8	2.7	0.42
[9]	0.18	56.0	1.5	27
[10]	0.18	48.8	5.2	31.5
[11]	0.18	54.3	5.3	3.48
[12]	0.1 3	50	7.5	4.1

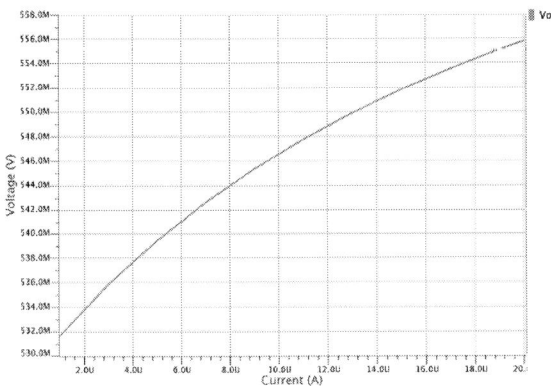

Fig. 10. DC response of the proposed TIA.

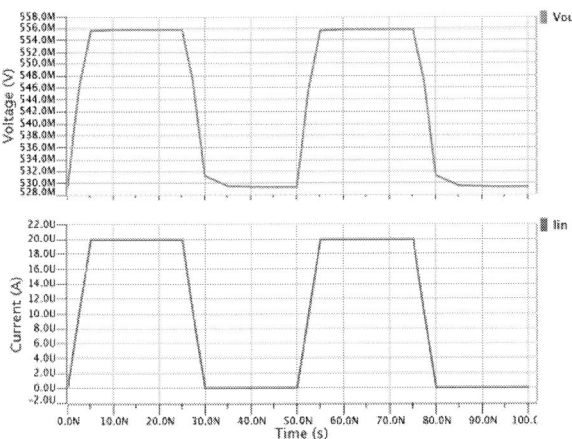

Fig. 11. Transient analysis of the proposed TIA.

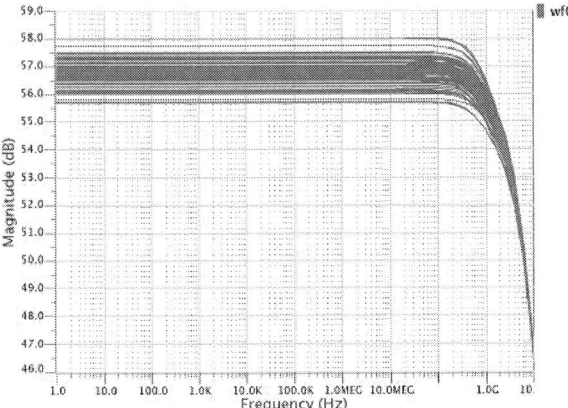

Fig. 12. The frequency response of Monte Carlo simulation (100 runs) .

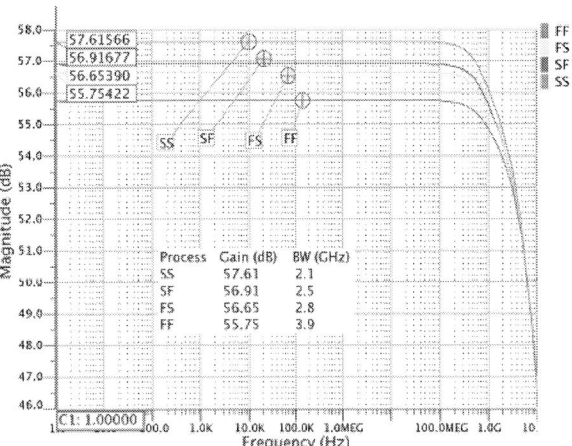

Fig. 13. Frequency response of the proposed TIA at different process corners.

V. CONCLUSION

The proposed TIA is designed which has improved performance as compared to simple TIA. In this paper, SC structure is used to design TIA to get the benefit of high output impedance which leads to gain enhancement. Power dissipation and noise of the proposed TIA are also very less. Hence, it can be concluded that the proposed TIA can be used in many optical receiver circuits where high gain and low noise are required.

REFERENCES

[1] M. M. Amourah, and R. L. Geiger, "A high gain strategy with positive-feedback gain enhancement technique," IEEE International Symposium on Circuits and Systems, vol. 1, pp. 631 - 634, May 2001.

[2] C. Laber, and P. Gray, "A positive-feedback transconductance amplifier with applications to high-frequency, high-Q CMOS switched-capacitor filters", IEEE J. Solid-state Circuits, vol. 23, pp. 1370-1378, Dec. 1988.

[3] K. Bult, and G. Geelen, "A fast-settling CMOS Op Amp for SC circuits with 90-dB DC gain", IEEE J. Solid-state Circuits, vol. 25, pp. l379-1384, Dec. 1990.

[4] S. Yan, and E. Sanchez-Sinencio, "Low voltage analog circuit design techniques: A tutorial", IEICE Transactions on Analog Inegrated Circuits and Systems, vol. 2, pp. 1–17, Feb. 2000.

[5] U. Singh, M. Gupta, and R. Srivastava, "A new wideband regulated cascode amplifier with improved performance and its application", Microelectronics J., vol. 46, pp. 758–776, June 2015.

[6] S. S. Rajput, S. S. Jamuar, "Low voltage analog circuit design techniques", Circuits and Systems Magazine, vol. 2, pp. 24–42, Aug. 2002.

[7] B. Razavi, Design of Integrated Circuits for Optical Communications, New York:McGraw-Hill, 2003.

[8] David M. Binkley, Tradeoffs and optimization in analog CMOS design, England:John Wiley & Sons Publication, 2008

[9] G. Royo, C. Sánchez-Azqueta, C. Aldea, S. Celma, and C. Gimeno, "CMOS transimpedance amplifier with controllable gain for RF overlay", Proc. 12th Ph. D. Research in Microelectronics and Electronics (PRIME), 2016..

[10] D. Chen, K. S. Yeo, X. Shi, M. A. Do, C. C. Boon, and W. M. lim, "Cross-Coupled Current Conveyor Based CMOS Transimpedance Amplifier for Broadband Data Transmission", IEEE Transactions on very large scale of integration (VLSI) systems, Vol. 21, pp. 1516-1525, 2013.

[11] R. Raut, and O. Ghasemi, "A power efficient wide band transimpedance amplifier in sub-micron CMOS integrated circuit technology", proc. 6th Int. IEEE Northeast Workshop on Circuits and Systems, 2008.

[12] T. H. Ngo, T.W.L., and H. H. park, "4.1 mW 50 dBΩ 10Gbps transimpedance amplifier for optical receiver in 0.13μm CMOS", Microwave and optical technology Letters, Vol. 53, pp. 448-451, 2011.

Application of Linear Quadratic Methods to Stabilize Cart Inverted Pendulum Systems

Sudarshan K Valluru
*Incubation Center for
Control,Dynamical Systems
Electrical Engineering,Delhi
Technological University*
Delhi 110042, India
sudarshan_valluru@dce.ac.in

Madhusudan Singh
*Incubation Center for
Control,Dynamical Systems
Electrical Engineering,Delhi
Technological University*
Delhi-110042, India
madhusudan@dce.ac.in

Mayank Singh
*Incubation Center for
Control,Dynamical Systems
Electrical Engineering,Delhi
Technological University*
Delhi-110042, India
mayank202244@gmail.com

Abstract—**This paper presents the application of linear-quadratic methods of optimal control for a class of reference tracking problem. Specifically, linear quadratic regulator and linear-quadratic Gaussian methods are outlined to synthesize a desired closed loop system. Selection criteria of weights for both state weighting matrix, Q, and control weighting matrix, R, is addressed for the LQR design to satisfy the robustness measures for a closed loop system. The design is extended to employ linear quadratic estimator for the LQG design. An unstable linearized model of the cart inverted pendulum is considered to validate the proposed methodology.**

Keywords— *LQR controller, LQG controller, Cart Inverted Pendulum System(CIPS).*

I. INTRODUCTION

This paper utilizes the theory of deterministic linear regulator to design a desired closed loop system satisfying a stringent set of conditions. The cart inverted pendulum system often considered a benchmark system[1], [2] for controller design is considered to test the proposed design strategies. A simple tracking problem for the system is encountered where the pendulum angle with the vertical tracks the cost position to balance the pendulum in the upright position. The inherent instability of the system due to its under-actuated nature presents conceptual as well as practical difficulties for the control system design and real-time implementation. This paper presents linear quadratic regulator and linear quadratic Gaussian design methods to stabilize this unstable SIMO system.

The LQR controller obtains a closed loop system with optimal performance[3]. This is done by minimizing a cost function comprising the state vector and the control input which establishes a direct relationship with the desired performance indices. This process results in the optimal state feedback gain, K. The value of this gain, K, is closely related to the values/weights selected for the state weighting matrix, Q, and control weighting matrix, R. Weights of Q and R, are selected so as to design the closed loop system to ensure the robustness properties for both LQR and LQG to small parametric changes, i.e., reasonable changes in the parameters of open-loop transfer functions which can be accommodated in the closed loop design with ensured stability.

Towards this goal, this paper is organized under the following section headings. Section II presents the mathematical description of the system for both open loop and closed loop for real-time implementation. Controllability and observability measures are also discussed in this section. In Section III, design methodologies are elucidated with the determination of parameters for experimental implementation. Section IV presents the obtained results for both design methods, followed by conclusion in Section V.

II. MATHEMATICAL DESCRIPTION OF CART INVERTED PENDULUM SYSTEM

A. Open Loop Analysis

The cart inverted pendulum system is sown in Fig.1, which is considered for closed-loop design. The system parameters are specified in Table- I.

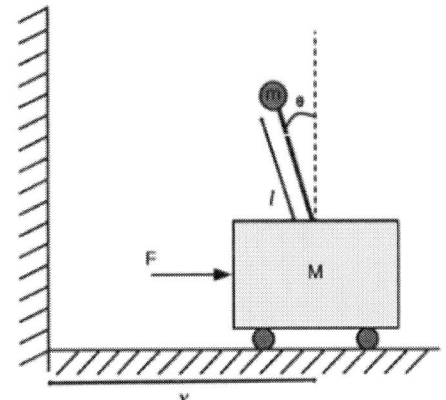

Fig.1 Cart Inverted Pendulum System

TABLE I. C I P S PARAMETERS[4]

Parameters	Value
F - Force applied to cart ±24N	±24N
x - Cart position from the reference ±0.5 m	±0.5m
θ - Pendulum angle w.r.t vertical axis 0.1 rad	0.1 rad
g - Gravity 9.81m/s2	$9.81m/s^2$
l - Length of the pole 0.38m	0.38m
M - Cart mass 2.4 Kg	2.4kg
m - Pole mass 0.23 Kg	0.23kg
I - Moment of inertia of pole 0.099Kg/m2	0.099kg/m2
b - Cart friction coefficient 0.05Ns/m	0.05Ns/m
d - Pendulum damping coefficient 0.05 s /rad	0.05s/rad

The standard non-linear differential equations can be designed using Newtonian mechanics or Lagrange's equations. For state vector as $x = [x(t) \quad \dot{x}(t) \quad \theta(t) \quad \dot{\theta}(t)]^T$, and output vector as $y = [x \quad \theta]^T$, which is given as a non-linear differential equation in (1).

$$
\begin{bmatrix} \dot{x}_1 \\ \dot{x}_2 \\ \dot{x}_3 \\ \dot{x}_4 \end{bmatrix} = \begin{bmatrix} x_2 \\ \frac{(mlcosx_3)(dx_4 - mglsinx_3) + (I + ml^2)(mglsinx_3)(x_4)^2 + bx_2}{\Delta(\theta)} \\ x_4 \\ \frac{(M + m)(dx_4 - mglsinx_3) + (mlcosx_3)(mglsinx_3(x_4)^2) + bx_2}{\Delta(\theta)} \end{bmatrix} + \begin{bmatrix} 0 \\ \frac{I + ml^2}{\Delta(\theta)} \\ 0 \\ -\frac{mlcosx_3}{\Delta(\theta)} \end{bmatrix} F
$$

(1)

where $\Delta(\theta) = (ml)^2 - (I + ml^2)(M + m)cosx_3$. Also, for the experimental setup force F(s) is related to control voltage U(s) which is given in equation (2).

$$\frac{F(S)}{U(S)} = 18 \tag{2}$$

The system can be linearized at θ=0, cos θ≈1, sin θ≈0, and $(\dot{\theta})^2 = 0$, the system can be rewritten as equations (3) and (4)

$$\dot{x} = Ax + Bu \tag{3}$$
$$y = Cx + Du \tag{4}$$

where $A \in \mathbb{R}^{n \times n}$, $B \in \mathbb{R}^{n \times m}$, $C \in \mathbb{R}^{p \times n}$, $D \in \mathbb{R}^{p \times m}$ are dynamic matrix, input matrix, output matrix, feed through the matrix, respectively, which are obtained as

$$A = \begin{bmatrix} 0 & 1 & 0 & 0 \\ 0 & -0.019 & 0.22 & -0.001 \\ 0 & 0 & 0 & 1 \\ 0 & -0.013 & 6.6 & -0.04 \end{bmatrix} \tag{5}$$

$$B = \begin{bmatrix} 0 \\ 6.99 \\ 0 \\ 4.625 \end{bmatrix} \tag{6}$$

$$C = \begin{bmatrix} 1 & 0 & 0 & 0 \\ 0 & 0 & 1 & 0 \end{bmatrix} \tag{7}$$

$$D = \begin{bmatrix} 0 \\ 0 \end{bmatrix} \tag{8}$$

Cart Position transfer function $T_x(s)$, and pendulum angle transfer function $T_\theta(s)$ are obtained[5] from the state-space representation is given in (9) and (10)

$$\frac{x(s)}{U(s)} = T_x(s) = \frac{6.99s^2 + 0.246s - 45.38}{s^4 + 0.05811s^3 - 6.63s^2 - 0.1261s} \tag{9}$$

$$\frac{\theta(s)}{U(s)} = T_\theta(s) = \frac{4.626s^2}{s^4 + 0.05311s^3 - 6.63s^2 - 0.1261s} \tag{10}$$

B. Controlability and Observability

A constant coefficient linear system is completely controllable if the matrix M_c is

$$M_c = [B \quad AB \quad A^2B \quad \ldots \ldots A^{n-1}B] \tag{11}$$

The M_c has rank n, where $x(t) \in \mathbb{R}^n$. A constant coefficient is linearly observable if M_o has rank n.

$$M_o = [C' \quad A'C' \quad (A^2)'C' \quad \ldots \ldots (A^{n-1})'C'] \tag{12}$$

For the cart inverted pendulum system, the controllability and observability matrices are given in equations (13) and (14).

$$M_c = \begin{bmatrix} 0 & 6.99 & -0.14 & 1.02 \\ 6.99 & -0.14 & 1.02 & -0.11 \\ 0 & 4.62 & -0.268 & 30.68 \\ 4.62 & -0.268 & 30.68 & -2.98 \end{bmatrix} \tag{13}$$

$$M_o = \begin{bmatrix} 1 & 0 & 0 & 0 & 0 & 0 & 0 & 0 \\ 0 & 0 & 1 & 0 & -0.019 & -0.013 & 0 & 0 \\ 0 & 1 & 0 & 0 & 0.22 & 6.6 & -0.013 & -0.26 \\ 0 & 0 & 0 & 1 & 0 & -0.03 & 0.22 & 6.6 \end{bmatrix}^T \tag{14}$$

Here the rank of $(M_c) = (M_o) = 4 = n$, which proves the system is fully observable and controllable.

C. Closed Loop Analysis of CIPS

The Fig. 2 describes the closed-loop architecture comprising the two PID controller where gain values are computed from proposed methodologies.

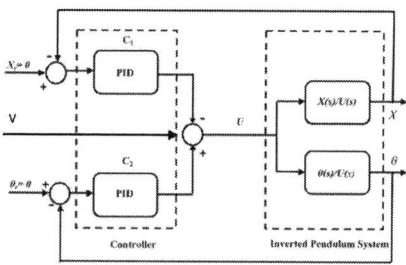

Fig. 2 Closed loop architecture for CIPS with Controllers

The closed-loop system can be summarised as follows in (15), (16), and (17).

$$C_1(s) = K_{p1} + \frac{K_{i1}}{s} + K_{d1}s \tag{15}$$

$$C_2(s) = K_{p2} + \frac{K_{i2}}{s} + K_{d2}s \tag{16}$$

$$\begin{bmatrix} T_{x\_CL} \\ T_{\theta\_CL} \end{bmatrix} = \frac{1}{1 + T_x C_1 + T_\theta C_2} \begin{bmatrix} T_x \\ T_\theta \end{bmatrix} \tag{17}$$

Where $T_{x\text{-cl}}$ and $T_{\theta\text{-cl}}$ are the closed loop cart position and pendulum angle transfer function respectively, and C_1 and C_2 are the controller parameters. The net characteristic equation from (17) is obtained as equation (18)

$$
s^5 + (6.99K_{d1} + 4.626K_{d2} + 0.053)s^4 +
$$
$$
(6.99K_{p1} + 0.246K_{d1} + 4.626K_{p2} - 6.63)s^3
$$
$$
+(0.246K_{p_1} + 6.99K_{i1} - 45.38K_{d1} + 4.626K_{i2} - 0.1261)s^2
$$
$$
+(-45.38K_{p1} + 0.246K_{i1})s + (-45.38)K_{i1} = 0
$$

(18)

The characteristic equation (18) be similar to a fifth order characteristic equation (19). Which is compactly rewritten as equation (20).

$$s^5 + p_1 s^4 + p_2 s^3 + p_3 s^2 + p_4 s + p_5 = 0 \quad (19)$$

$$\begin{bmatrix} 0 & 0 & 6.99 & 0 & 0 & 4.626 \\ 6.99 & 0 & 0.246 & 4.626 & 0 & 0 \\ 0.246 & 6.99 & -45.38 & 0 & 4.626 & 0 \\ -45.38 & 0.246 & 0 & 0 & 0 & 0 \\ 0 & -45.38 & 0 & 0 & 0 & 0 \end{bmatrix} \begin{bmatrix} K_{p1} \\ K_{i1} \\ K_{d1} \\ K_{p2} \\ K_{i2} \\ K_{d2} \end{bmatrix} = \begin{bmatrix} p_1 - 0.058 \\ p_2 + 6.63 \\ p_3 + 0.1261 \\ p_4 \\ p_5 \end{bmatrix}$$

$$(20)$$

The equation (20) is compared with equation (21)

$$SR = p \quad (21)$$

The least norm solution for equation (21) can be calculated as equation (22)

$$R = (S^T S)^{-1}(S^T P) \quad (22)$$

The desired characteristic equation (19) can be obtained from the proposed control strategies.

III. Design of Controllers

A. Design of LQRController

The linear quadratic regulation (LQR) method assumes full state feedback for linear controller design[6]. For a continuous time, the linear dynamical system as represented by equations (3)and (4), the optimization around certain constraints are posed on the minimization of a cost index defined in equation (23).

$$J = \int_0^\infty [x^T Q x + u^T R u] \, dt \quad (23)$$

Where Q is the stable weighting matrix, and R is the control weighting matrix. Both Q and R, are symmetric, where the former is positive semi-definite, and the latter is positive definite. The steady state solution of the regulator problem in the linear time-invariant, state feedback control law is given in equation (24).

$$u(t) = -Kx(t) \quad (24)$$

where $K \in \mathbb{R}^{m \times n}$ is the control gain matrix, which is calculated by equation (25)

$$K = R^{-1}B^T P \quad (25)$$

Moreover, P is symmetric positive semi-definite, which is calculated the solution by using algebraic Riccati equation (26).

$$PA + A^T P + Q - PBR^{-1}B^T P = 0 \quad (26)$$

The selection of weights for Q and R matrices and then their evolution of cost function represents the subjective part of the problem. These weights are selected to ensure a good degree of robustness, chacharacterized LQ optimal design by a gain margin, GM > 2, and a phase margin, PM > 60 degrees. The following state control weighting matrices in (27) and (28) were selected.

$$Q = \begin{bmatrix} 50 & 0 & 0 & 0 \\ 0 & 1 & 0 & 0 \\ 0 & 0 & 100 & 0 \\ 0 & 0 & 0 & 1 \end{bmatrix} \quad (27)$$

$$R = [1] \quad (28)$$

Based on these matrices and using the solution of algebraic Riccati equation, the control gain matrix, K is calculated by equation (29)

$$K = [-3.16 \quad -3.73 \quad 22.91 \quad 8.48] \quad (29)$$

To obtain a fifth order characteristics equation, a fifth pole was added at 4-10 times the distance of the farthest pole. The closed loop characteristics equation is obtained from LQR design in equation (30).

$$s^5 + 40.4s^4 + 576s^3 + 3733s^2 + 7755s + 6177 = 0 \quad (30)$$

B. Design of LQG Controller

When the LQR controller is combined with linear quadratic estimator (Kalman filter), it results into the linear quadratic Gaussian controller. To overcome the assumption of full state feedback, a full state observer is designed to estimate all states from available state measurements. A stochastic model given in equations (31) and (32) are considered to account for system uncertainties, i. e., process noise, and sensor noise.

$$\dot{x} = Ax + Bu + Fd \quad (31)$$

$$y = Cx + n \quad (32)$$

Where d is the process disturbance vector and x is the sensor noise vector. Both noises are assumed to be zero mean Gaussian white noises with $E[d(t)] = E[n(t)] = 0$. The disturbance and noise covariances are given by V_d and V_n respectively, and given in equations (33) and (34).

$$E[d(t)d(\tau)^T] = V_d \delta(t - \tau) \quad (33)$$

$$E[n(t)n(\tau)^T] = V_n \delta(t - \tau) \quad (34)$$

The state equation for the Kalman filter is given by equation (35).

$$\dot{x}_0 = Ax_0 + Bu + L(y - Cx_0) \quad (35)$$

Where x_0 is the estimated state vector and L is the optimal filter gain. The objective of Kalman filter is to minimize the covariance of the estimation error, given by (36).

$$R_e(t) = E[e(t)e(t)^T] \quad (36)$$

Where e(t) is the estimation error given by x(t)-x_0(t) and R_e(t) is the covariance of the estimation error which has to

978-1-5386-6624-1/18 $31.00 © 2018 IEEE

be minimized. The error dynamics equation is obtained in (37)

$$\dot{e}_o = (A - LC)e_o + Fd - Ln \tag{37}$$

The optimal Kalman gain filter is given by (38)

$$L = YC^T V_n \tag{38}$$

Where Y is the solution to the Riccati equation (39)

$$YA^T + AY - YC^T V_n^{-1} CY + V_d = 0 \tag{39}$$

The state space representation of the system using the LQG compensation is given by equations (40) and (41)

$$\dot{x}_o = (A - BK - LC)x_o + Ly \tag{40}$$

$$u = -Kx_o \tag{41}$$

Where K is the LQR gain. To find the Kalman filter gain matrices (42) and (43) are used for V_d and V_n respectively.

$$V_d = \begin{bmatrix} 25 & 0 & 0 & 0 \\ 0 & 25 & 0 & 0 \\ 0 & 0 & 25 & 0 \\ 0 & 0 & 0 & 25 \end{bmatrix} \tag{42}$$

$$V_n = \begin{bmatrix} 0.1 \\ 0.1 \end{bmatrix} \tag{43}$$

The Kalman filter gain matrix is obtained in equation (44)

$$L = \begin{bmatrix} 4.8328 & 0.0288 \\ 3.865 & 0.2647 \\ 0.0288 & 6.6196 \\ 0.0651 & 14.0976 \end{bmatrix} \tag{44}$$

The closed loop poles were obtained for the matrix $[A - BK - LC]$ and given by G in equation (45)

$$G = \begin{bmatrix} -10.36 + i9.36 \\ -10.36 - i9.36 \\ -3.46 \\ -0.41 \end{bmatrix} \tag{45}$$

From the LQG design, the closed loop characteristic equation is obtained in equation (46)

$$s^5 + 33s^4 + 344s^3 + 1700s^2 + 3600s + 3000 = 0 \tag{46}$$

IV. EXPERIMENTAL RESULTS AND COMPARATIVE ANALYSIS

A. Experimental time response for CIPS with LQR

The experimental responses position, angle and control voltage of CIPS with LQR controller are shown in Fig.3, Fig.4 and Fig.5 respectively. The cart and pendulum responses are noisy even though stabilized.

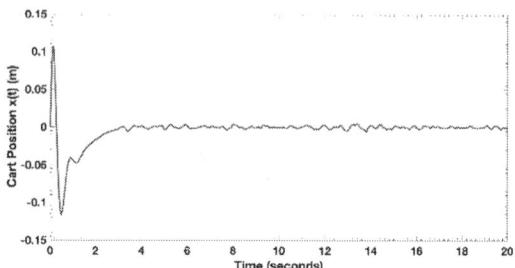

Fig. 3 Experimental time response for cart position x(t)

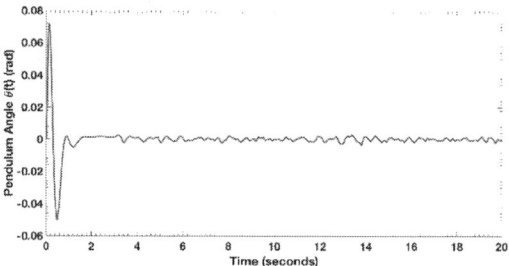

Fig. 4 Experimental time response for pendulum angle θ(t)

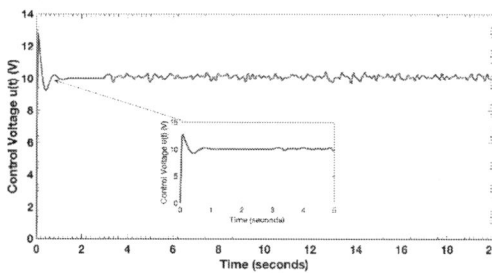

Fig. 5 Experimental time response for control voltage u(t)

B. Experimental time response for CIPS with LQG

The experimental responses position, angle and control voltage of CIPS with LQR controller are shown in Fig.6, Fig.7 and Fig.8 respectively. It is observed that from implementing LQG controller with cart inverted pendulum system responses are lesser noisy as compared to LQR controller.

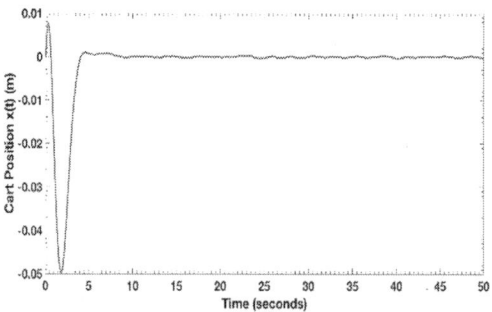

Fig. 6 Experimental time response for cart position x(t)

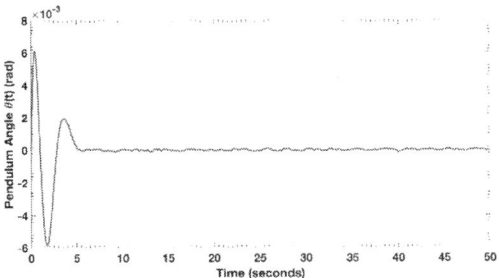

Fig. 7 Experimental time response for pendulum angle θ(t)

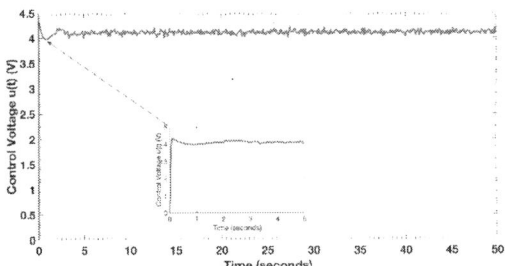

Fig. 8 Experimental time response for control voltage u(t)

C. Comparison of PID Gain Values

The characteristic equation obtained from LQR and LQG design in (30) and (46) are utilized to determine the PID gain values using equations (21) and (22). Table-II enlists the obtained PID gain values.

TABLE II. PID GAIN VALUES FOR
LQR AND LQG CONTROLLER

Controller Parameters	LQR	LQG
K_{p1}	-172	-80
K_{i1}	-136	-66
K_{d1}	-416	-728
K_{p2}	407	158
K_{i2}	606	611
K_{d2}	637	503

D. Comparison by performance indices

Table-III presents the comparison between proposed methodologies based on standard performance indices.

TABLE III. COMPARISON OF PERFORMANCE INDICES

Performance Indices	LQR		LQG	
	x(t)	θ(t)	x(t)	θ(t)
Settling Time (sec)	3.12	2.81	5.23	5.18
Gain Margin (dB)	50.4	61.2	78	82
Phase Margin (deg)	60.1	62.3	76.4	79.8

From the obtained time responses, the closed loop system designed by the LQG method has a higher settling time w.r.t. the closed loop system designed by the LQR method. However, the LQG method leads to a system with higher tolerance to disturbances as greater PM and GM values are obtained for the respective closed loop

system. Also, the experimental implementation by LQG method has lower PID gain values than the LQR counterpart.

V. CONCLUSION

In this paper, LQR and LQG methods are used to design a robust closed loop system validated a cart inverted pendulum system. Proposed methodologies were used to predetermine the PID gain values for experimental implementation. Experimental time responses were presented which indicates a more robust closed loop system designed via the LQG method.

ACKNOWLEDGMENT

The authors would like to thank Govt. of India and Govt. NCT Delhi for providing funding through Delhi Technological University under TEQIP-II to procure Digital Pendulum System Experimental Setup.

REFERENCES

[1] E.J.Davison, "Benchmark Problems for the Control System Design: Report of the IFAC Theory Committe," Schlossplatz 12,A-2361 Laxenburg, Austria, 1990.

[2] Jonathan P. How, "Benchmarks," *IEEE Control Systems Magazine*, pp. 6–7, 2015.

[3] A. Ghosh, T. R. Krishnan, and B. Subudhi, "Robust Proportional-Integral-Derivative Compensation of an Inverted Cart-Pendulum System: An Experimental Study," *IET Control Theory Appl. Theory Appl.*, vol. 6, no. 8, pp. 1145–1152, 2012.

[4] FeedbackInstruments, "Digital Pendulum Control Experiments Manual," 2016.

[5] S. K. Valluru and M. Singh, "Stabilization of Nonlinear Inverted Pendulum System Using MOGA and APSO Tuned Nonlinear PID Controller," *Cogent Eng.*, vol. 4, no. 1, pp. 1–15, 2017.

[6] Sudarshan K. Valluru Madhusudan Singh S.Singh, "Prototype Design and Analysis of Controllers for One Dimensional Ball and Beam System," in *1st IEEE International Conference on Power Electronics. Intelligent Control and Energy Systems*, 2016, pp. 1–6.

2nd IEEE International Conference on Power Electronics, Intelligent Control and Energy Systems (ICPEICES-2018)

Implementation of High Performance Clock-gated Flip-flops

Prakash Kumar
Division of Electronics and Communication Engineering,
Netaji Subhas Institute of Technology
New Delhi, India
kmrprakash01@gmail.com

Kunwar Singh
Division of Electronics and Communication Engineering,
Netaji Subhas Institute of Technology
New Delhi, India
kunwar.singh@nsit.ac.in

Abstract— **In this paper, optimum transistor sizing of clock-gated master slave flip-flops viz. Gated Master Slave Latch (GMSL) and clock-gated Transmission Gate Flip-flop (CG-TGFF) has been implemented for high performance based on Logical Effort (LE) theory. In contrast to the previous work where clock-gating was merely utilized only for reduction of total dynamic power dissipation, we have also optimized the delay caused due to addition of extra transistors for clock-gating circuit. A comparative study of power-delay product (PDP) and power-delay-area product (PDAP) for the aforementioned circuits is performed. Simulation results using 180nm/1.8V TSMC CMOS technology have indicated that GMSL offers upto 45.8% and 49.6% improvement in PDP and PDAP respectively when compared to CG-TGFF at 0% switching activity (all ones). Results have been verified for various switching activities of the data input with respect to clock signal frequency.**

Keywords— *Master–Slave TGFF, CMOS, Clock-gating, Logical Effort Theory, Power-Delay-Area Product, GMSL*

I. INTRODUCTION

The concern of digital integrated circuit (IC) designers in the past has been focused majorly on reducing size, cost and increased performance on chip. As a result, device size has been scaled down to the nanometer range, thereby fitting more and more number of components in a single IC package. With this increased number of transistors on single IC package and also the faster speed of operation (increased clock speed) the problem of enhanced dynamic power dissipation has become the main concern. Modern day wearable electronic devices running on batteries also need to be operated for increased performance and lower power consumption per cycle of operation. Flip-flops (FFs) represent the fundamental building block of memory/sequential elements in electronic circuits. FFs are extensively used for design of multi-core processors and lead to 30-70% of the total system power dissipation when coupled with the clocking network [1], [2].

Clock signal is used for synchronization of various operations throughout the chip. The operational frequency of systems has increased to several Gigahertz thereby accounting for more than 40% of total power dissipation in the present scenario [1]. The power consumption critically depends on device sizes and frequency of operation at internal nodes of the FF. Therefore, optimizing the transistor sizes provides an extra dimension for reduction of total dynamic power consumption.

There are several techniques for reduction of power dissipation in VLSI domain. Clock gating in particular helps in reduction of dynamic power dissipation by disabling the signal whenever there is no change in the data present at the input and output of a FF. Different types of gating circuits have already been proposed in the literature [3]. Traditionally, the main focus of circuit designers remains only in reduction of total power consumption on chip whereas addition of extra transistors due to insertion of gating circuit causes increased delay and area which leads to performance degradation. Conventionally, the sizing of devices on critical path of a FF is performed using the method of LE but transistor sizing of clock-gating circuit has remained unnoticed which causes undue losses in the overall performance. To overcome this problem, the clock gating circuit has been sized using LE theory [4] in this work.

This paper is structured as follows. Section II presents a brief review of different clock gating circuits. Section III presents simulation results and discussion. The LE theory based optimization of TGFF (non-clock gated) and clock-gated configurations GMSL and CG-TGFF is explained in this section. Conclusion is presented in section IV of the paper.

II. CLOCK GATING CIRCUITS

The work done in this paper is restricted to master-slave FF circuits having transmission gates in their critical path. The circuit comprises of two latches, master and slave, connected in cascade. The first latch is transparent at negative level of the clock while second latch functions on the positive level of clock signal. Hence, the final transition of data from input port to output port is achieved only when the clock transits from the negative level to the positive level i.e., the positive edge of the clock signal. Different clock-gating circuits have been used by researchers in the past to restrict power dissipation in FFs. Some of the FF circuits with clock-gating topologies are now discussed briefly.

978-1-5386-6624-1/18 $31.00 © 2018 IEEE

A. Gated Master Slave latch (GMSL)

Fig. 1 shows positive edge triggered gated master–slave latch (GMSL) circuit [5]. The circuit comprises of two parts, the TGFF and the clock-gating circuit. The clock-gating circuit consists of eleven transistors. This gating circuit is responsible for generating the clock enable signal and the circuit operation can be understood as explained below.

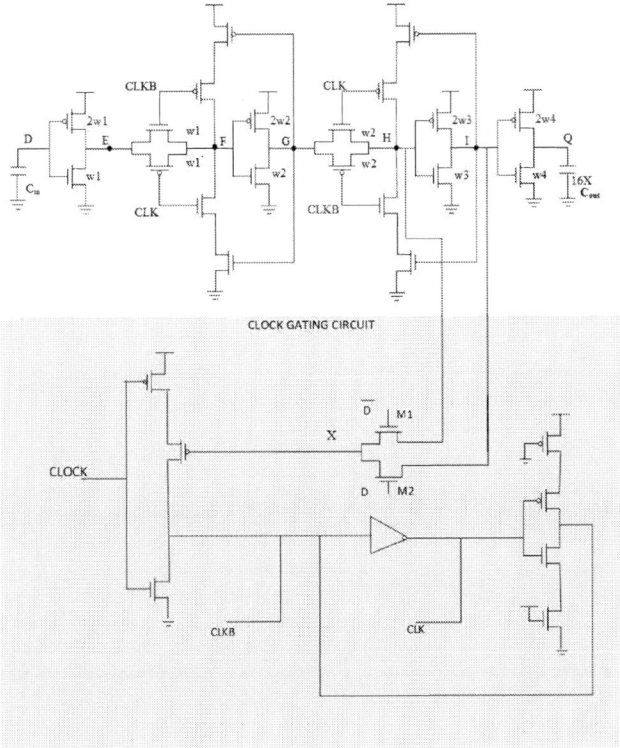

Fig. 1.   Gated-Master Slave Latch (GMSL)

Comparison of input and output data is performed by pass transistors M1 and M2, and as a result signal 'X' is generated which helps in enabling or disabling the clock signal accordingly. The last set of inverters are used to realize a static loop.

B. Clock-gated Transmission Gate Flip-flop (CG-TGFF)

The circuit in Fig. 2 represents the block diagram of another clock-gated FF circuit CG-TGFF. Traditionally, an EXOR gate is used as the comparator while AND gate is used for enabling/disabling the clock signal depending upon the output of the comparator. However, this circuit arrangement is not synchronized with the triggering edge of the clock signal. To eliminate this problem a positive pulse generating circuit has been introduced in this work which helps in synchronization with the clock edge. The detailed clock-gating circuit of configuration is shown in Fig. 3.

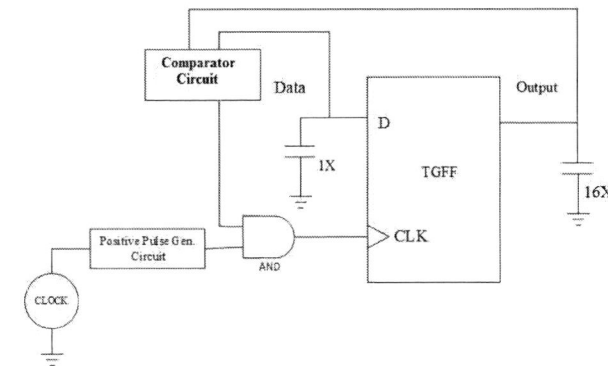

Fig. 2.   Block level description of CG-TGFF

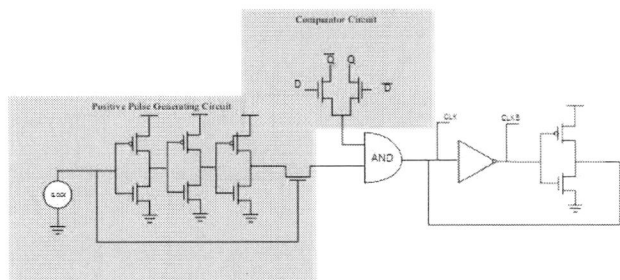

Fig. 3. Detailed clock-gating circuit for CG-TGFF

The pulse is generated with the help of delay produced by a series of inverters and a pass transistor. Comparator circuit uses only two pass transistors. The output of comparator goes high only when there is difference in signal level at the input and the output ports of the FF. Both the circuit configurations viz. GMSL and CG-TGFF provide exactly the same output waveforms for similar simulation conditions. The only difference lies in the clock-gating logic and the number of transistors in the clock-gating circuitry.

III. SIMULATION RESULTS AND DISCUSSION

In order to compare the performance of both GMSL and CG-TGFF, a series of steps have been followed as indicated in Fig. 4. Firstly, the FF has been optimized for minimum delay by sizing the transistors on the critical path using LE method. Since the basic FF employed in both the cases is TGFF, therefore, the optimization process for delay minimization of TGFF using LE theory is explained next.

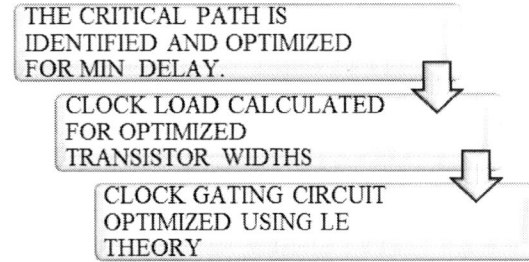

Fig. 4. Design flow for optimization

978-1-5386-6624-1/18 $31.00 © 2018 IEEE   1033

A. Delay optimization of TGFF

The optimization of TGFF starts with fixing the output load which has been restricted to 16X. The value of capacitance per unit length for TSMC/180nm is determined by following the procedure as mentioned in [6] and the relationship is obtained as 1nm = 2.31×10^{-18} Farads. The value of input capacitance of minimum sized inverter ($W_p = 720nm$ and $W_n = 360nm$) at 0.18μm technology node is hence determined as 2.49fF which represents 1X capacitance (C_{min}) and therefore the absolute value of 16X capacitive load becomes 39.84fF.

In order to optimize the TGFF for minimum delay, the critical path is divided into four stages and the transistor sizing ratios are evaluated using LE method for each stage as mentioned in [7]. The absolute transistor sizes for a particular load in the network are determined using the capacitance transformation formula as indicated below [4].

$$C_{in-x} = (C_{out-x} \times g_x) / f \qquad (1)$$

Where, C_{in-x} is the input capacitance, C_{out-x} is the output capacitance and g_x is the load capacitance of the x_{th} stage in the critical path while f represents the optimum stage effort for minimum delay.

The input capacitance is progressively increased in fixed increments of C_{min} till the delay saturates and the difference between two consecutive delay readings is less than 5%. For the purpose of simulation TSMC 180nm CMOS technology parameters have been utilized at a power supply voltage of 1.8V. The various parameters used for simulation are listed in Table 1.

TABLE I. SIMULATION PARAMETERS

Signal slope	Clock Frequency	Power Supply	Capacitive load (16X)	W_{min}
50ps	0.5GHz	1.8V	39.84fF	360nm

The simulation results for TGFF are demonstrated in Table 2. It can be concluded from the table that increasing input capacitance (widths of transistors in the network) reduces the delay but also increases the power and area consumed. Data-to-output delay (T_{DQ}) has been used for characterization which is the sum of set-up time (T_{setup}) and clock-to-output delay (T_{CQ}).

$$T_{DQ} = T_{setup} + T_{CQ} \qquad (2)$$

TABLE II. VARIATION IN POWER AND DELAY CHARACTERISTICS WITH C_{IN} FOR TGFF

C_{in} (fF)	T_{DQ} (ps)	Power (μW)	PDP (fJ)
2.49	250.43	92.68	23.2
4.98	230.76	108.14	24.9
7.48	**219.03**	**120.62**	**26.4**
9.97	210.85	141.33	29.8
12.46	206.31	143.32	29.5

It is to be observed that there is no significant improvement (the difference between two consecutive delay readings is less than 5%) in speed of the FF beyond Cin = 7.48fF. All the device sizes in the network are evaluated using LE method and capacitance transformation formula as mentioned earlier and the clock load corresponding to minimum delay is evaluated. In order to determine the clock load, the total width of transistor connected to the clock signal is evaluated and converted to the equivalent capacitance value.

Furthermore, for these transistor sizes simulations are performed for varying switching activities and the results are presented in Table 3. It can be observed that dynamic power consumption is directly proportional to the input data switching activity.

TABLE III. VARIATION IN POWER DISSIPATION AND PDP AT DIFFERENT SWITCHING ACTIVITIES

Input Data Switching Activity	Power (μW)	PDP (fJ)
50%	120.62	26.4
25%	78.76	17.2
10%	46.79	10.2
5%	38.76	8.48
0% (All Ones)	31.41	6.87
0% (All Zeroes)	27.82	6.09

B. Performance optimization of GMSL and CG-TGFF

The GMSL circuit shown in Fig.1 displays enhanced delay due to the addition of clock-gating circuitry. The total clock-to-output delay can be minimized by optimizing the clock path delay to minimum. Hence, the clock load value is firstly determined for the optimized critical path. The value is 6.8fF.

Table 4 demonstrates the variation in power consumption, delay and area at different switching activities for TGFF and the clock-gated FFs viz. GMSL and CG-TGFF. To maintain a fair ground for comparison TGFF is optimized for minimum delay and after evaluating the clock load, clock-gating circuits in both the cases are optimized for minimum delay using LE theory.

Simulation results clearly indicate that GMSL consistently consumes less power at all switching activities under consideration. For e.g., the power dissipation of GMSL is 27.3% (all zeroes) and 34.6% (all ones) lower than that of CG-TGFF respectively at zero percent switching activity. The improvement in data-to-output delay is about 17.5% in case of GMSL when compared to CG-TGFF. It is noteworthy that GMSL occupies 7% lesser area which is marginally beneficial.

TABLE IV. SIMULATION RESULTS FOR DIFFERENT FF CONFIGURATIONS @ 16X LOAD

Circuit	Power at Different Data Switching Activity (μW)						Delay (ps)	Area (μm²)
	50%	25%	10%	5%	All Zeros	All Ones		
TGFF (NON-CLOCK GATED)	120.62	78.76	46.79	38.76	27.82	31.41	219	4.61
GMSL	171.84	100.75	40.61	23.34	4.88	5.32	296	6.36
CG-TGFF	206.31	107.48	47.23	25.99	6.71	8.13	359	6.83

TABLE V. PDP and PDAP values for FFs at varying switching activities

Input Data Switching Activity	TGFF		GMSL		CG-TGFF	
	PDP (fJ)	PDAP (fJ.μm²)	PDP (fJ)	PDAP (fJ.μm²)	PDP (fJ)	PDAP (fJ.μm²)
50%	26.4	121.7	50.8	323.1	74.1	506.1
25%	17.2	79.3	29.8	189.5	38.5	399.5
10%	10.2	47.02	12.02	76.4	16.9	115.4
5%	8.5	39.18	6.91	43.9	9.3	63.5
All Zeroes	6.1	28.12	1.44	9.15	2.4	16.39
All Ones	6.9	31.81	1.57	9.98	2.9	19.81

The values of PDP and PDAP are shown in Table 5. GMSL exhibits considerable improvement in PDP and PDAP by 40% and 44.1% (all zeroes). Conversely, the improvements obtained in PDP and PDAP for all ones are 45.8% and 49.6% respectively. Fig. 5 indicates the percentage improvements exhibited by GMSL over CG-TGFF by means of a histogram at all switching activities considered in this work.

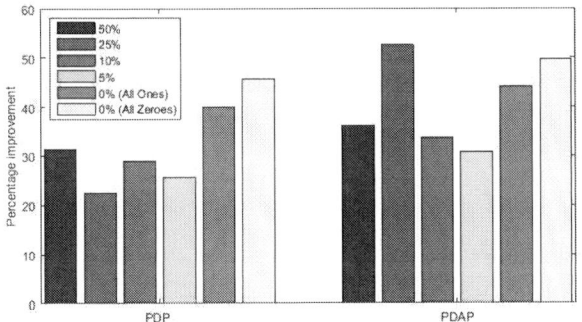

Fig. 5. Percentage improvement in GMSL over CG-TGFF

IV. CONCLUSION

This work presents a technique for performance optimization of clock-gated FFs. The methodology has been tested for design optimization of GMSL and CG-TGFF. A comparative study of GMSL and CG-TGFF based on the proposed optimization process shows that GMSL offers better performance than CG-TGFF and is more suitable for low power applications at low switching activities. GMSL shows significant improvement in terms of PDP and PDAP which have been chosen as the figures of merit.

REFERENCES

[1] G. Yeap, "Practical Low Power Digital VLSI Design", pp. 140, Springer Science+ Business media, LLC, 1998.

[2] H. Kawaguchi., T. Sakurai "A reduced clock-swing flip-flop (RCSFF) for 63 percent power reduction". IEEE Journal of Solid-State Circuits, Vol. 33, No. 5, May 1998

[3] A. G. M. Strollo, E. Napoli and D. De Caro, "New clock-gating techniques for low-power flip-flops," Low Power Electronics and Design, 2000. ISLPED '00. Proceedings of the 2000 International Symposium on, Rapallo, Italy, 2000, pp. 114-119.

[4] I. E. Sutherland, B.F. Sproull, D.L.harris, "Logical Effort: Designing Fast CMOS Circuits", Morgan Kaufmann Publishers, Inc, May 19. 1988.

[5] D. Markovic, B. Nikolic, and R. Brodersen, "Analysis and design of low-energy flip-flops," in Proc. Int. Symp. Low Power Electron. Des. Aug. 2001, pp. 52–55.

[6] N.H.E. Weste, D.M Harris "CMOS VLSI Design, A Circuits and System perspective", 4th Edition Addison-Wesley.

[7] M. Alioto, E.Consoli, G. Palumbo, "Analysis and Comparison in the Energy-Delay-Area Domain of Nanometer CMOS Flip-Flops: Part I—Methodology and Design Strategies" IEEE transactions on very large scale integration (vlsi) systems, vol. 19, No. 5, pp 725-736, May 2011.

An Efficient Unscented Kalman Filter for Joint Angles Estimation and Control of Omni bundle with Noise

Rohit Rana, Prerna Gaur, Vijyant Agarwal and Harish Parthasarathy
ICE Division, MPAE Division, ECE Division, NSIT (University of Delhi), New Delhi, India
E-mail: [1]rohitrana982007@gmail.com, [2]prernagaur@yahoo.com,
[3]vijyant.agarwal@gmail.com, [4]harisignal@yahoo.com

Abstract—**This paper presents the efficient Unscented Kalman filter (UKF) based estimation and control of a 3-DOF Omni bundle robot. The angular trajectory tracking becomes more difficult under highly non-linear measurement noise and process noise. A rapidly converging Unscented Kalman Filter (UKF) is presented in the paper. The extended Kalman filter (EKF) provides less accurate results in the highly non-linear system and sometime may become unstable because of convergence related issues, whereas in UKF the results in the non-linear system are better than EKF. UKF is more accurate in state and noise estimation. The computed torque technique is taken as a model to apply UKF. After, estimation the noise is removed. The experiment is conducted on Omni bundle with process and measurement noise. The results prove the superiority of presented UKF over conventional UKF and EKF.**

Index Terms—**EKF, UKF, Omni Bundle, Torque computation**

I. INTRODUCTION

The filtering processes for stochastic systems has been applied in many domains like Image processing, navigation, and tracking etc. The working space coordinates X, Y, Z are highly important while performing various tasks. Any deviation from in these coordinates may lead to an uncontrollable situation in the robot workspace. These coordinates are controlled by robot angular positions q_1, q_2, q_3 Further, the deviation is caused due to the noises present in the system. In [1] a location-based algorithm using UKF was proposed to estimate the position of the beacon. The results were better than EKF in the localization of beacon. A similar method was proposed in [8]. In [2] the comparison between various localization and mapping is shown and UKF based FastSLAM is implemented with RMS robot position and orientation error. The tuning of UKF filter in stochastic filtering is shown in [3] using the unscented transform. [5] provided a robust method in which system model error is constructed by the model prediction, then it is used to rectify the UKF process to obtain the estimate of the real system state. The similar method for GPS is provided by [4]. The wavelet-based filtering is shown in [9], where a tremor signal is estimated using the wavelet technique. In [10] it is shown the use of disturbance observer to reduce the system noise including the disturbance. The EKF method was used to estimate disturbance as a state.

In this paper, An efficient UKF method is proposed and implemented on an Omni Bundle Robot. The UKF estimates the joint angular positions in the highly nonlinear noisy system. The highlights of this paper are: 1) Computationally more efficient UKF is proposed. 2) Real-time implementation on Omni bundle. Sect. II provides the mathematical formulation, the UKF modeling is presented in II-B. Application of UKF on Omni Bundle robot is given in Sect. III. The results are discussed in Sect. IV. Final conclusions are made in sect.V.

II. PROBLEM FORMULATION

A. Dynamics of Omni Bundle

The dynamics model of Omni Bundle [6]- [7] robot is given by

$$M(q, \dot{q})\ddot{q} + N(q, \dot{q}) + G(q) = \tau \qquad (1)$$

where $N(q, \dot{q})_{(3\times3)}$ is the Coriolis and Centrifugal forces, $M(q, \dot{q})_{(3\times3)}$ is the Inertia matrix, $G(q)_{(3\times1)}$ is the gravity vector and τ is the torque. The q is the joint angle matrix having three angles q_1, q_2, q_3. $\dot{q}$ is the velocity component and $\ddot{q}$ shows the acceleration. The inverse dynamics or computed torque acceleration controller equation with PD control is given by

$$f_c = M(q)(\ddot{q} + K_d(\dot{q}_d - \dot{q}) + K_p(q_d - q)) + N(q, \dot{q}) + G(q) \quad (2)$$

Here K_p & K_d are the gain matrix.

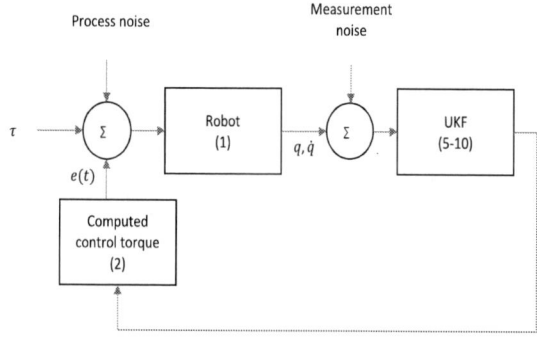

Fig. 1. Block Diagram with equation number.

2nd IEEE International conference on power Electronics, Intelligent Control and Energy systems (ICPEICES-2018)

B. The UKF

In this section, the state equation of the UKF is derived as follows: The variables q and $\dot{q}$ need to be estimated and is defined by a state vector X

$$X = [q, \dot{q}]^T \qquad (3)$$

If $\omega = \dot{q}$, then $\dot{\omega} = F(q, \omega, t)$. $F(q, \omega, t)$ is written as

$$F(q, \omega, t) = M^{-1}(q)(\tau - N(q, \dot{q}) - G(q)) \qquad (4)$$

Let $\xi_1, \xi_2, \ldots$ be iid $N(0, I)$ random vectors. If P is a positive definite matrix, then $\sqrt{P}\xi_k, k = 1, 2, \ldots$ are iid $N(0, P)$ random vectors, Where P is the error covariance matrix. Now the state equations are

$$X(t+1) = F(t, X(t)) + W(t+1) \qquad (5)$$

and the measurement equation is

$$Z(t) = h(t, X(t)) + V(t) \qquad (6)$$

where W, V are discrete time iid normal random vectors. Let

$$Y(t) = \{Z(s) : s \le t\} \qquad (7)$$

i.e the aggregate of all measurements collected upto time t. In the EKF, we write $\hat{X}(t+1|t) = \mathbb{E}(X(t+1)|Y(t)) = \mathbb{E}(F(t, X(t))|Y(t)) \approx F(t, \mathbb{E}(X(t)|Y(t))) = F(t, \hat{X}(t|t))$ which is not a good approximation when F is highly nonlinear. We cannot take the expectation operator inside a highly nonlinear function without making a big error. In the UKF, we replace this by

$$\hat{X}(t+1|t) = K^{-1} \sum_{a=1}^{K} F(t, \hat{X}(t|t) + \sqrt{P(t|t)}\xi_a) \qquad (8)$$

This is justified because $e(t|t) = X(t) - \hat{X}(t|t)$ is zero conditional mean with conditional covariance $P(t|t)$ and so has $\sqrt{P(t|t)}\xi_a$. The UKF computation of $P(t+1|t)$ likewise proceeds as

$$P(t+1|t) = cov(X(t+1) - \hat{X}(t+1|t)|Y(t))$$
$$= cov(F(t, X(t)) + W(t+1) - \hat{X}(t+1|t)|Y(t))$$
$$= cov(F(t, X(t)) - \hat{X}(t+1|t))$$
$$= cov(F(t, \hat{X}(t|t) + e(t|t)) - \hat{X}(t+1|t)) \qquad (9)$$

Therefore, $P(t+1|t)$ is

$$= K^{-1} \sum_{a=1}^{K} (F(t, \hat{X}(t|t) + \sqrt{P(t|t)}\xi_a)$$
$$- \hat{X}(t+1|t))(F(t, \hat{X}(t|t) + \sqrt{P(t|t)}\xi_a)$$
$$- \hat{X}(t+1|t))^T \qquad (10)$$

A more convenient way to write this is

$$P(t+1|t) = K^{-1} \cdot \sum_{a=1}^{K} e^a(t+1|t) \cdot (e^a(t+1|t))^T \qquad (11)$$

In the UKF to calculate $\hat{X}(t+1|t+1)$ and $P(t+1|t+1)$ approximately, we use the well known formula given in equation (12) that if U, V are jointly Gaussian random vectors, then

$$\mathbb{E}(U|V) = \mathbb{E}(U) + cov(U, V) \cdot cov(V)^{-1}(V - \mathbb{E}(V)) \qquad (12)$$

In our context, we shall use it in the following form:

$$\hat{X}(t+1|t+1) = \mathbb{E}(X(t+1)|Y(t), Z(t+1)) \approx$$

$$\hat{X}(t+1|t) - cov(e(t+1|t), e_Z(t+1|t))$$
$$\cdot cov(Z(t+1)|Y(t))^{-1}(Z(t+1) - \hat{Z}(t+1|t)) \qquad (13)$$

where we define

$$e_Z(t+1|t) = h(t+1, X(t+1)) + V(t+1)$$
$$- K^{-1} \sum_{a=1}^{K} h(t+1, \hat{X}(t+1|t) + e^a(t+1|t)) \qquad (14)$$

$$= h(t+1, \hat{X}(t+1|t) + V(t+1) + e(t+1|t))$$
$$- K^{-1} \sum_{a=1}^{K} h(t+1, \hat{X}(t+1|t) + e^a(t+1|t)) \qquad (15)$$

and hence

$$cov(e(t+1|t), e_Z(t+1|t)) \approx K^{-1} \sum_{b=1}^{K} e^b(t+1|t) \cdot (e_Z^b(t+1|t))^T \qquad (16)$$

where

$$e_Z^b(t+1|t) = h(t+1, \hat{X}(t+1|t) + e^b(t+1|t)) -$$
$$K^{-1} \sum_{a=1}^{K} h(t+1, \hat{X}(t+1|t) + e^a(t+1|t)) \qquad (17)$$

Also, in the UKF approximation,

$$cov(Z(t+1)|Y(t)) = cov(e_Z(t+1)|Y(t))$$
$$= P_V + K^{-1} \sum_{b=1}^{K} e_Z^b(t+1|t) \cdot (e_Z^b(t+1|t))^T \qquad (18)$$

and

$$\hat{Z}(t+1|t) = \mathbb{E}(h(t+1, X(t+1))|Y(t))$$
$$= \mathbb{E}(h(t+1, \hat{X}(t+1|t) + e(t+1|t))|Y(t)) \qquad (19)$$
$$= K^{-1} \cdot \sum_{a=1}^{K} h(t+1, \hat{X}(t+1|t) + e^a(t+1|t)) \qquad (20)$$

Finally, in the UKF approximation,

$$P(t+1|t+1) = cov(X(t+1) - \hat{X}(t+1|t+1)|Y(t), Z(t+1))$$
$$= cov(X(t+1) - \hat{X}(t+1|t)|Y(t), Z(t+1))$$
$$= cov(e(t+1|t)|Y(t), Z(t+1))$$
$$= cov(e(t+1|t)|Y(t)) - cov(e(t+1|t), Z(t+1)|Y(t))$$
$$\cdot cov(Z(t+1)|Y(t))^{-1} \cdot cov(e(t+1|t), Z(t+1)|Y(t))^T \qquad (21)$$

978-1-5386-6624-1/18 $31.00 © 2018 IEEE

2nd IEEE International conference on power Electronics, Intelligent Control and Energy systems (ICPEICES-2018)

where

$$cov(e(t+1)|Y(t)) = P(t+1|t), cov(e(t+1|t), Z(t+1)|Y(t))$$
$$= cov(e(t+1|t), e_Z(t+1|t)|Y(t))$$
$$= K^{-1}. \sum_{a=1}^{K} e^a(t+1|t).(e_Z^a(t+1|t))^T, cov(Z(t+1)|Y(t))$$
$$= cov(e_Z(t+1)|Y(t))$$
$$= K^{-1}. \sum_{a=1}^{K} e_Z^a(t+1).(e_Z^a(t+1))^T + P_V \quad (22)$$

This completes the description of the UKF

III. APPLICATION OF UKF ON OMNI BUNDLE ROBOT

The proposed algorithm is implemented on 3-DOF Omni Bundle and is illustrated below The workspace tool coordinates X, Y, Z are given by

$$X = (0.5 \times a_2 cos(q_1 + q_2) - 0.5 \times q_3 sin(q_1 - q_2 - q_3)$$
$$+ 0.5 \times a_2 cos(q_1 - q_2) + 0.5 \times a_3 sin(q_1 + q_2 + q_3))$$
$$Y = 0.5 \times a_2 cos(q_1 + q_2) - 0.5 \times q_3 sin(q_1 - q_2 - q_3)$$
$$+ 0.5 \times a_2 cos(q_1 - q_2) + 0.5 \times a_3 sin(q_1 + q_2 + q_3))$$
$$Z = (a_3 cos(q_2 + q_3) - a_2 sin(q_2) \quad (23)$$

Where $a_1 \& a_2$ are the link lengths and both are $132\ mm$. The link angular positions for measurement model is given as:

$$q_1 = tan^{-1}(\frac{y}{z}) \quad (24)$$

$$q_3 = \frac{3\pi}{2} - cos^{-1}\left(\frac{L_1^2 - k^2 + L_2^2}{2L_1 L_2}\right) \quad (25)$$

$$q_2 = \phi - \gamma \quad (26)$$

where $k = \sqrt{x^2 + y^2 + z^2}$, $\gamma = sin^{-1}\left(\frac{L_2 sin(\frac{3\pi}{2}) + q_3}{k}\right)$,

$$\phi = \begin{cases} cos^{-1}\frac{d}{k} & , z \leq 0 \\ -cos^{-1}\frac{d}{k} & , z > 0 \end{cases} \quad \text{and } d = \sqrt{x^2 + y^2}$$

However, the measurement matrix provided in equation 6 can be taken in either X, Y, Z form or in q_1, q_2, q_3 form. We have considered the q_1, q_2, q_3.

The state transition matrix calculation is given as follows: The inertia matrix is given as

$$MI = \begin{bmatrix} MI11 & MI12 & MI13 \\ MI21 & MI22 & MI23 \\ MI31 & MI32 & MI33 \end{bmatrix}$$

where
$MI11 = k_11 + k_22 \times cos(2\theta_2) + k_3 \times cos(2\theta_3) + k_4 \times cos(\theta_2) \times sin(\theta_3), MI12 = k5 \times sin\theta_2, MI13 = 0, MI21 = k5 \times sin\theta_2, MI22 = k6, MI23 = -0.5 \times k4 \times sin\theta_2 - \theta_3, MI31 = 0, MI32 = -0.5 \times k4 \times sin\theta_2 - \theta_3, MI33 = k7.$

The $N(q, \dot{q})$ matrix and parameters $k_1 ..... k_7$ value is taken as per [6]. Note the state vector is defined as $X =$

$[q_1, q_2, q_3, \omega_1, \omega_2, \omega_3]^T$ thus the system model state equation can be written as : $\dot{X} = [\omega_1, \omega_2, \omega_3, \dot{\omega}_1, \dot{\omega}_2, \dot{\omega}_3] + [\xi_{q1}, \xi_{q2}, \xi_{q3}, \xi_{\omega_1}, \xi_{\omega_2}, \xi_{\omega_3}]^T$ which is equal to

$$\dot{X} = f(X, \tau) + P(\xi) = \left[M^{-1}\left(-N \begin{bmatrix} \omega_1 \\ \omega_2 \\ \omega_3 \end{bmatrix} + \begin{bmatrix} \tau_1 \\ \tau_2 \\ \tau_3 \end{bmatrix} \right) \right] + P \begin{bmatrix} \xi_{q1} \\ \xi_{q2} \\ \xi_{q3} \\ \xi_{\omega_1} \\ \xi_{\omega_2} \\ \xi_{\omega_3} \end{bmatrix} \quad (27)$$

This equation (27) is implemented in equation (8). The process noise in each link angle is taken as zero mean with a finite variance and the measurement noise is taken as zero mean with high variance. The sampling time is taken as T_s.

IV. RESULTS

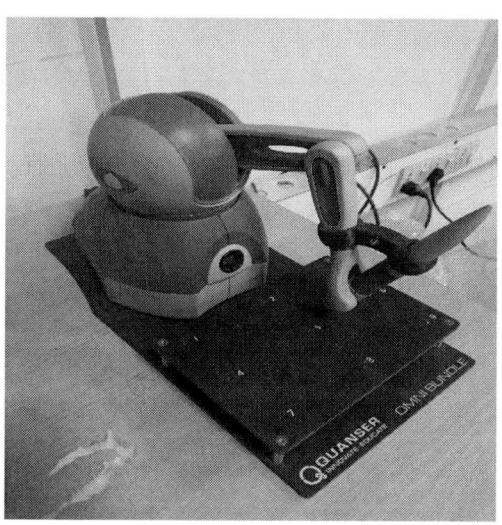

Fig. 2. Picture of Omni bundle

The experimental setup is connected as shown in Fig. 1. The Omni bundle robot is connected to a Matlab installed computer through USB port with Quanser software installed on the computer. The Fig. 3 shows Omni bundle. The Omni bundle robot is set up in 3-DOF configuration. The experiment is conducted as per the section II of this paper. The signal provided to q_1 is 0, to q_2 is $\pi/4$ and q_3 is a sine wave. Fig. 3 shows the input angular trajectory corrupted with process noise. Fig. 4 Shows the output angular positions of q_1, q_2, q_3 from the robot corrupted with measurement noise and process noise. Also, it may be noted here that from time $0 - 2$ sec the robot is trying to reach the final position from home position. Afterward, it is following the trajectory provided in Fig. 3 The connection of both the noises is the same as shown in Fig. 1. The UKF estimated angular positions are shown in Fig. 5 for all three links. The error is calculated as the difference between the positions from the robot Fig. 4 and estimated positions from UKF Fig. 5 and is shown in Fig. 6. The RMS values of all the signals are shown in Table.I.

978-1-5386-6624-1/18 $31.00 © 2018 IEEE

2nd IEEE International Conference on Power Electronics, Intelligent Control and Energy Systems (ICPEICES-2018)

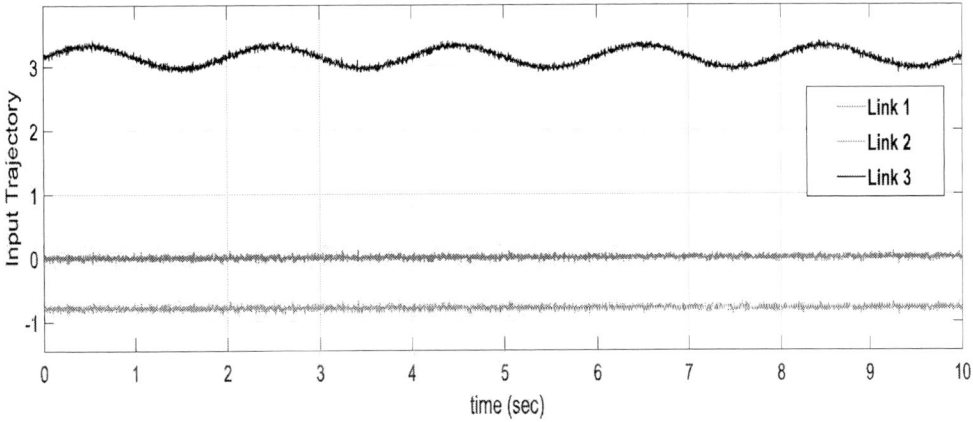

Fig. 3. Input angular trajectory corrupted with process noise

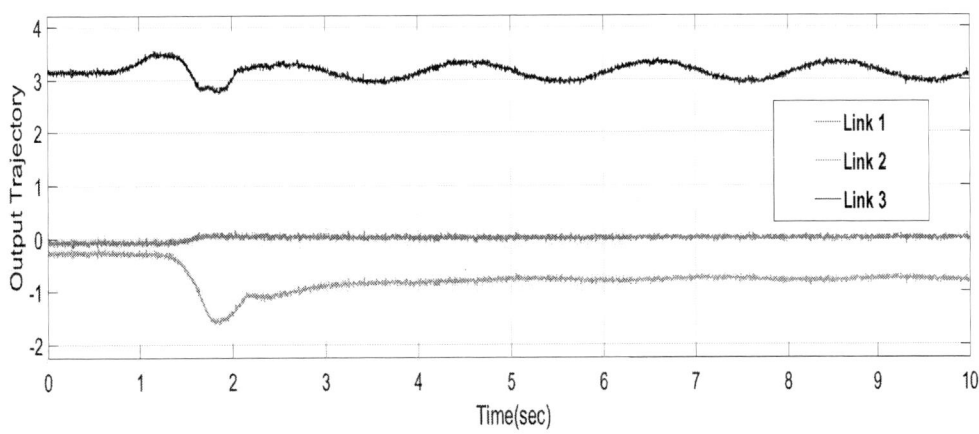

Fig. 4. Output angular trajectory corrupted with measurement noise

TABLE I
RMS VALUES[a]

Link Name	Link Angle	RMS values of angular positions at different stages				
		Input trajectory without noise	Input trajectory with noise	Output trajectory	UKF output	Error
Link1	q1	0.00e+00	3.144e−02	4.587e−02	3.148e+00	3.362e−03
Link2	q2	7.854e−01	7.865e−01	8.131e−01	8.130e−01	3.379e−03
Link3	q3	3.144e+00	3.144e+00	3.148e+00	3.148e+00	5.299e−03

[a] RMS values measured through Matlab simulink scope.

The entire simulation is done on Matlab Simulink with the process noise in each link angle is taken as zero mean with variance 10^{-3} and the measurement noise is taken as zero mean with variance 10^{-3}. The sampling time is taken as $T_s = 0.001s$.

V. CONCLUSION

The advantage of UKF compared to EKF is that we still retain the Gaussian assumption after it passes through non-linear systems and in some sense, it is a Gaussian approximation for non-Gaussian approximation. The result is, therefore better than the EKF and classical UKF. Also, we can add neural state filter in the future scope of work to investigate it further.

REFERENCES

[1] Yoon BD., Yoon HN., Choi SH., Lee JM. (2013) UKF Applied for Position Estimation of Underwater-Beacon Precision. In: Lee S., Cho H., Yoon KJ., Lee J. (eds) Intelligent Autonomous Systems 12. Advances in Intelligent Systems and Computing, vol 193. Springer, Berlin, Heidelberg.

[2] Z. Kurt-Yavuz and S. Yavuz, "A comparison of EKF, UKF, Fast-SLAM2.0, and UKF-based FastSLAM algorithms," 2012 IEEE 16th International Conference on Intelligent Engineering Systems (INES), Lisbon, 2012, pp. 37-43. doi: 10.1109/INES.2012.6249866

[3] Scardua, L.A. & da Cruz, J.J. J Control Autom Electr Syst (2016) 27: 10. https://doi.org/10.1007/s40313-015-0223-1

978-1-5386-6624-1/18 $31.00 © 2018 IEEE

2nd IEEE International conference on power Electronics, Intelligent Control and Energy systems (ICPEICES-2018)

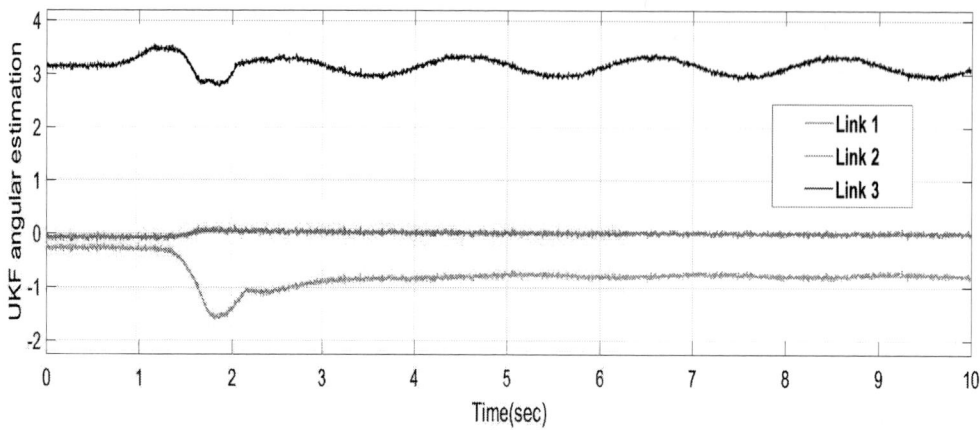

Fig. 5. UKF estimated angular position

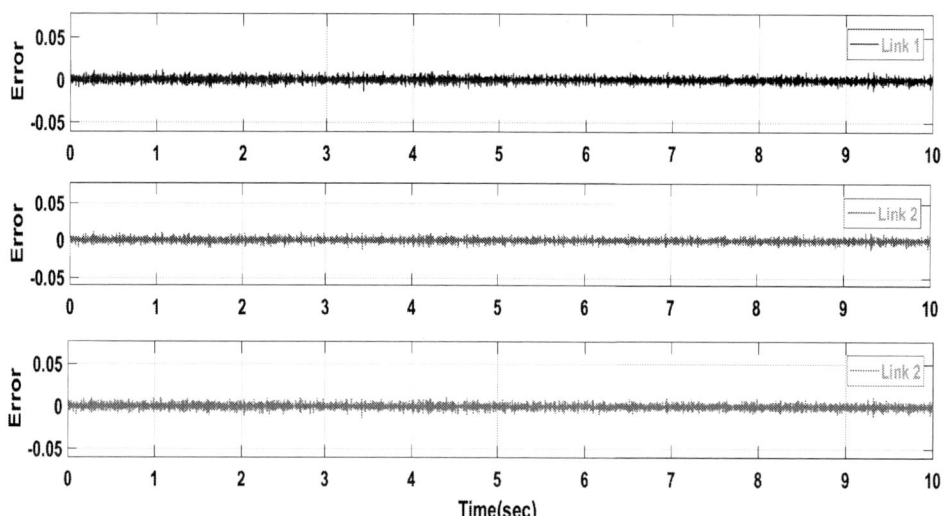

Fig. 6. Error between UKF angular estimate and the output angular trajectory with process and measurement noise

[4] Jwo, DJ. & Lai, CN. GPS Solut (2008) 12: 249. https://doi.org/10.1007/s10291-007-0081-9

[5] Zhao, Y., Gao, S., Zhang, J. et al. Int. J. Control Autom. Syst. (2014) 12: 996. https://doi.org/10.1007/s12555-013-0048-2

[6] V. Agarwal, "An extended Kalman filter for real-time estimation and control of Omni robot with stochastic noise," 2016 3rd International Conference on Computing for Sustainable Global Development (INDI-ACom), New Delhi, 2016, pp. 2456-2460.

[7] N. Nagpal, V. Agarwal and B. Bhushan, "A Real-Time State-Observer-Based Controller for a Stochastic Robotic Manipulator," in IEEE Transactions on Industry Applications, vol. 54, no. 2, pp. 1806-1822, March-April 2018. doi: 10.1109/TIA.2017.2785339.

[8] Luigi DAlfonso, Walter Lucia, Pietro Muraca, Paolo Pugliese, Mobile robot localization via EKF and UKF: A comparison based on real data, Robotics and Autonomous Systems, Volume 74, Part A, 2015, Pages 122-127, ISSN 0921-8890, https://doi.org/10.1016/j.robot.2015.07.007.

[9] R. Rana, V. Agarwal and H. Parthasarthy, "Wavelet transformation based tremor removal," 2015 International Conference on Computer, Communication and Control (IC4), Indore, 2015, pp. 1-3. doi: 10.1109/IC4.2015.7375544

[10] Agarwal, V. & Parthasarathy, H. Nonlinear Dyn (2016) 85: 2809.

https://doi.org/10.1007/s11071-016-2864-4

2nd IEEE International Conference on Power Electronics, Intelligent Control and Energy Systems (ICPEICES-2018)

Application of Wireless Technology to Enhance Safety and Productivity in Steel Plant

Indranil Banerjee, M Shrujan, S K Das, S Singh
Research and Development Centre for Iron & Steel,
Steel Authority of India Limited Ranchi, India
indranil@sail-rdcis.com

Abstract—Indigenously developed innovative scheme for transmission of signals, to remotely control mudgun and drill machine at Blast furnace in Rourkela Steel Plant (RSP), has been designed and commissioned. Control cable and switchgear based conventional automation scheme resulted in breakdowns, amounting to loss of production. In hazardous steel plant locations, cable laying or on-line fault rectification is extremely difficult at times. It also sometimes creates life threatening situations for the operators due to proximity of hot liquid iron coming out of tap holes of a Blast Furnace. Industrial Wireless Controls have rapidly replaced wired automation systems. Industrial Wireless Controls send signals through radio waves of varying frequencies. These electromagnetic waves are received by radio receivers at other ends. This hassle-free means of data transfer has made Industrial Wireless Controls popular across in different walks of life. This developmental activity was taken up with objectives firstly to improve safety parameters for operator by providing flexibility to operate from safe distance secondly to reduce production loss due control cable and associated switchgear related breakdowns and consequently to improve availability of cast house equipment like mudgun and drill machine . The system comprises of radio remote control (RRC), valve stand, Programmable Logic Controllers (PLC) and special panels suiting the harsh ambiance. PLC software has been developed with in –house expertise. The system is in continuous operation at RSP, Blast Furnace since 2015.

Keywords—*mud gun,dril machine,BlastFurnace, radio remote control*

I. INTRODUCTION

In any steel plant, Blast furnace is the most critical unit responsible for liquid iron production. Liquid iron from Blast Furnace is transferred to Steel Melting Shop for further processing. So the availability of Blast Furnace (BF) is the key factor for sustained productivity. The BF # 4of RSP has two tap holes from where liquid iron and slag come out, once the process of iron making is complete. Blast furnace #4 cast house is shown in house "Fig.1".

Mud gun and drill machine are considered as most important equipment of Blast furnace cast house. Mud gun is used to pack the tap hole with mud. Then burden is charged from the top and the process of iron making starts. Once the furnace is ready with the iron making, drill machine is used to drill through the tap hole and the liquid iron comes out. Since there are two tap holes in BF#4, there are correspondingly two mud guns and one drill machine. The mud gun and drill

Fig. 1. Blast Furnace #4 cast house

machine are operated from two control desks, one for tap hole#1, and the other for tap hole # 2. Close proximity of the control desks from the tap hole caused dangerous situation to the operating personnel. Also, splinters from flowing metal caused damage to the associated cables of the control desk, which affected the machine availability, leading to reduced blast in Blast furnace subsequently leading to loss of productivity. Blast Furnace # 4 of RSP is one of the highest production furnaces with around 2400 Ton /day average production. With every tapping around 250 Ton metal comes out of the tap hole.

In the earlier scheme"Fig.2" two control desks were used to control operation of the mud guns and drill machine. Distance of the control desks were approximately 5 meters from the tap holes. A bunch of control cable approximately 200 meters, ran between the desk and the relay panel in the hydraulic room. Sequencing and interlocking for the mud gun and drill machine was realized by relay logic in the relay panel. Hydraulic and pneumatic valves were used for various operations of mud gun and drill machine. Valve stand for mud gun # 2 was mechanico-hydraulic type having control handles on the desk. Valve stand for control of mud gun # 1 and drill machine was electrical proportional type. Earlier scheme was not PLC based, so any modification or trouble shooting was extremely difficult and time consuming.

Keeping above difficulties in view, an innovative scheme was designed and implemented to operate mud gun and drill machine of the Blast Furnace # 4 from remote and safe

978-1-5386-6624-1/18 $31.00 © 2018 IEEE   1041

location. A RRC based automation system was designed and implemented, successfully at RSP.

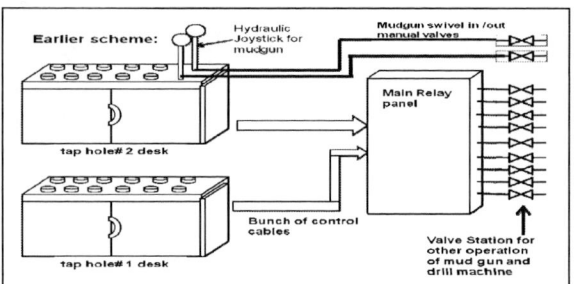

Fig. 2. Earlier system of signal transmission through wired cables

II. STRATEGY FOR RADIO SELECTION

Various technical papers / journals were referred to know more about industrial wireless applications, protocols, frequency hopping techniques, and time division multiple accesses etc [1], [2], [3].A new system capable of transmitting all the digital and analog signals while retaining the functionality of control desk and switchgear was developed jointly by Research & Development Centre for Iron and Steel (RDCIS) and RSP.

The prime objective was to achieve trouble free operation of cast house equipment using, by increased availability. Apart from equipment breakdown, operator safety was also a major issue. The existing system was thoroughly examined. Various key issues were considered before the design was finalized, they are mentioned below. Since the radio transmission system was going to be installed in a critical application at an extremely hazardous location, inside plant, selection criteria had to be stringent.

Mentioned below are some criteria based on which RRC was selected,

a) Wireless planning coordination (WPC) guidelines for frequency / Effective Radiated Power (ERP) selection

According to guidelines provided by WPC, some frequency bands are licensed, whereas some are license free depending on frequency and ERP. Penetration is better and attenuation is less in lower frequencies whereas range shall be better if higher frequencies are used in a free line of sight situation. Keeping in view, the actual site condition, it was decided to go for lower frequency. Lower frequency bands are mostly licensed, but

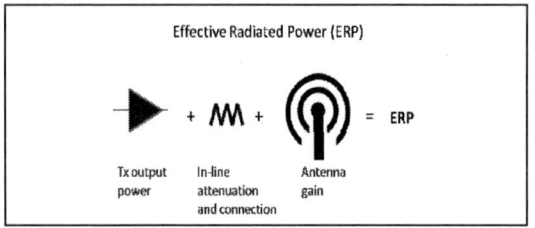

above a certain ERP. According to WPC notification frequency band 433-434 MHz with a maximum ERP of 10 mW and channel bandwidth within 10 KHz with built-in antenna comes under license free zone.

b) International standards for safety of operation

Safety features like Listen before Talk (LBT) and Stop Function should conform to international standards for mission critical applications. It was decided to go for Category 3 performance level d (PL_d) as mentioned in ISO 13849-1:2008. A particular performance level is achieved depending upon architecture of Safety-Related Part of a Control System (SRP/CS), reliability of components and effectiveness of error detection system. In PL_d conforming systems, average probability of dangerous failure per hour falls in the range of 10^{-7} and 10^{-6} per hour.

b) Ingress protection

Since the system was being designed for dust laden industrial location Ingress Protection (IP) played a crucial parameter. It was decided to have minimum IP 65 for all radio equipment. IP ratings are defined in international standard EN 60529 (British BS EN 60529:1992, European IEC 60509:1989). IP 65 provides total protection from dust and also low pressure water jet from any direction.

d) Signal attenuation and reliability

Radio signal attenuation [4] at Blast Furnace cast house was considered for possible obstructions. Suitable place for antenna was identified so as to achieve minimum obstructed Line Of Sight (LOS). Objective was to design a scheme which is reliable enough and operates in un-interruptible manner [5], [6], [7]. Minimum receiver sensitivity has to be such that it is able to listen to the transmitter. Attenuation of radio waves "Fig.3"can be calculated as given below:

Free space loss = 32.4 + 20xLog F(MHz) + 20xLog R(Km) F is the RF frequency expressed in MHz. R is the distance between the transmitting and receiving antennas. At 2.4 Ghz, this formula is: 100+20xLog R (Km)

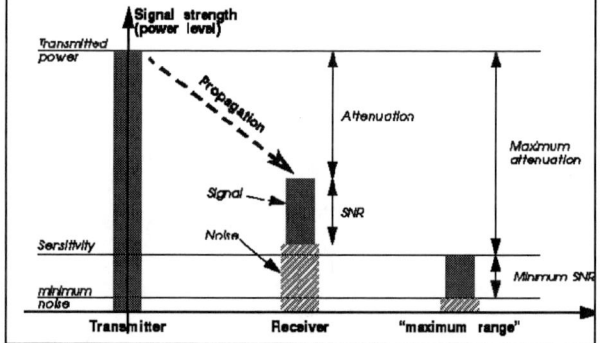

Fig.3 Signal attenuation

d) On-board I/O capability

RRC was going to replace existing control desk for mudgun and drill machine of BF#4. Selected transmitter must be able to accommodate push buttons/joysticks/toggle switches such that complete functionality was achieved.

e) Interference

Unprecedented growth of wireless application in industry is likely to cause potential interference problems. Sources of interference could include anything that radiates electromagnetic (EM) energy such as machinery and undesirable radio devices. Frequency Hopping Spread Spectrum (FHSS) has been used to obtain interference immunity. FHSS is a process by which the carrier frequency is changed during or between transmissions accordingly to a pre-defined synchronized method. With FHSS, transmitter and receiver must tune to the same frequency at precisely the same time. FHSS adds a layer of complexity to the system but also makes interception or disruption of the wireless system more difficult, thereby making the system more reliable.

f) Site-survey

Site-survey was done to determine appropriate antenna size and location such that antenna could be mounted clear of any metal obstruction. Also the cable length between antenna and receiver was kept at a minimum of 3 mtrs.

After deliberating on all above factors, Radio remote control (RRC) was selected with following features as mentioned in "Table I"

TABLE I.   **SPECIFICATION OF RRC**

Sl No	Parameter	Specifications
1.	Stop Function	Cat. 3 - PLd
2.	Ingress protection	IP65/NEMA4
3.	Frequency Band	433.050-434.775 MHz
4.	Maximum ERP	1 mW
5.	Working Temperature Range	-20C +70C (-4F +158F)
6.	Display	128x64 Graphic LCD
7.	Range Limiter	Yes
8.	Interference mitigation	FHSS
9.	Topology	Peer-to-peer single hop

III.   IMPLEMENTED SCHEME

Challenge was to replicate complete functionality of the two fixed control desks in one portable operator station. Also, we had to integrate new signal transmission from RRC with existing relay system. Mud gun and drill machine are huge equipment operated hydraulically and pneumatically. One new hydraulic valve stand had to be designed to operate mud gun because the old valve stand was purely mechanical type and having no provision for electrical signals. It was yet another challenge to integrate the new valve stand and the old mechanical equipment i.e mud gun # 2. One new PLC was used in the project. Algorithm was developed in the PLC to realize sequencing interlocking of mud gun and drill machines. Joysticks, toggle switches, selector switches were mounted on the specially designed transmitter. Interlocking scheme was created in the PLC, taking care of all possible modes of

operation like remote / local. PLC program has been developed with in-house expertise keeping in mind, ease of trouble shooting and future expansions.

Digital signals are transmitted from the transmitter / portable operator console. After receiving the signals at the receiver end, they were integrated with the PLC. Algorithm of sequencing and interlocking is executed in the PLC and finally command is generated for corresponding valve actuation. In the PLC panel valve amplifier cards are mounted and wired with the PLC output. Output of the amplifier card is taken to the valve station. Total automation system has been designed in such a way that old system of operation has been retained for the time being. Since, for the first time a remote control system was being implemented, it needed some time to get accepted and tuned. As the new system became stable and reliable, old system has been phased out. The new system has been immensely beneficial and giving very good result consistently since commissioned. It has resulted in improved productivity and enhanced safety for the operating personnel. Now the operator can operate mud gun and drill machine standing at a safe distance anywhere in the cast house.

"Fig. 4" represents the schematic layout for transmission of wireless signals.

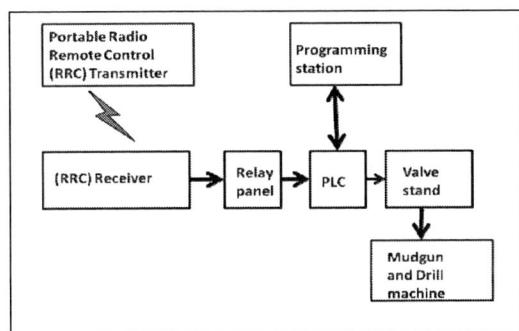

Fig.4 Schematic of designed system for wireless communication

RRC based operation system consists a portable console with full functionality of the two control desks used earlier. PLC is used to create necessary interlocking scheme so that when mud gun is operated drill machine does not operate. When operator is operating one tap hole, the other is disabled automatically.   One new valve stand has been designed specifically for mud gun # 2 operations in wireless mode. New and old valve stand operates in standby mode with leak proof arrangement between them.

IV.   MODELLING OF PROPOSED SYSTEM

To design appropriate error control schemes, effective channel models to predict the estimate bit / packet loss rates and channel behavior were necessary. Modeling methods considered were stochastic channel models, channel impulse responses and deterministic channel models.

978-1-5386-6624-1/18 $31.00 © 2018 IEEE

To calculate the, the received power for the chosen frequency at a certain distance, d, model used was:

$$Pr(d) = P0 - 10\gamma \log 10\ d/d0 + X\sigma + Y$$

where P0 is the power at the reference distance d0, γ is the path loss exponent, Xσ and Y are the large- and small-scale fading contributions respectively.

The noise in industrial wireless can be modeled as the superposition of Additive White Gaussian Noise (AWGN) and impulse noise created due to motors, heavy machineries, ignition systems, inverters, voltage regulators, electric switch contacts, welding equipments, etc. and expressed as

$$n(t) = w(t) + b(t) \cdot k(t),\ \text{for}\ t \in \{1, 2, \ldots, T\}$$

where w(t) and k(t) are zero-mean Gaussian distributed processes, b(t) is a {0, 1}-random variable which describes the state of the channel and t is the time index.

Models considered to mathematically calculate fading were Rician, Rayleigh, Nakagami and lognormal.

Following simplified model for path loss as a function of distance has been used for system design:

$$Pr = Pt\ K\ [d0\ /d]\ \gamma$$

The dB attenuation is thus

$$Pr(dBm) = Pt(dBm) + K(dB) - 10\gamma \log 10\ [d\ /d0]$$

In this approximation, K is a unit less constant which depends on the antenna characteristics and the average channel attenuation, d0 is a reference distance for the antenna far-field, and γ is the path loss exponent.

V. EXPERIMENTAL VALIDATION

The Radio Remote Control based automation system has been commissioned and is working satisfactorily. Response of the system was found to be much faster and reliable with respect to earlier system. With the implementation of wireless signal transmission system, accuracy and reliability of automation system has increased. Dependence on bunch of control cable and junction boxes has been substantially reduced. This has helped in reduced breakdown time leading to increase in production. Operator safety factor has been improved. Availability of cast house equipments i.e mud gun and drill machine has been enhanced. Various benefits reaped "Fig. 5" by introduction of the new innovation is discussed below:

A. Technological

Introduction of the new system resulted in

a) Migration from wired to wireless signal transmission aided availability of blast furnace cast house equipment, because of reduced breakdowns.

b) Reliabile digital signal transmission enhanced response of cast house equipment.

c) It was a common phenomena to encounter, that mudgun or drill machine getting stuck to the tap hole and operator unable to control the equipment because of some switchgear/cable burnout failure. This resulted in either the entire equipment getting damaged or increased consumption of spares like drill rod,nozzle

and poking rod etc. Due to enhancement in reliability of operation, those kind of situations could be avoided and a reduction in spare consumption has been noticed.

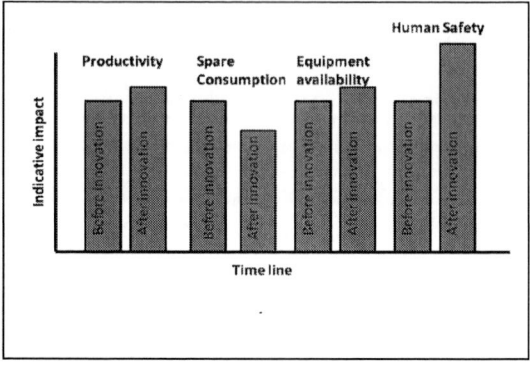

Fig.5 Validation of results

B.Financial

a)Financial benefit achieved due to increased level of hot metal production, reduction in spare consumption and saving in hydraulic oil has been approx Rs 3.0 crores per annum, Table II

b)Benefit to cost ratio is approximately 4:1

TABLE II.   **IMPROVEMENT IN BLAST FURNACE PARAMETER AFTER INNOVATION**

Sl No	Parameters	Improvement
1.	Hot metal production	Increased by 0.25%due to better availability of cast house equipment
2.	Reduction in Consumable like drill rod, mud gun nozzle, poking rod etc	Due to better control of mudgun and drill machine savings to the tune of Rs75000/Ton could be achieved
3.	Reduction in hydraulic oil consumption	Design of new valve stand with leak proof arrangement, savings to the tune of 300litre/month could be achieved

C. Expertise development

A huge amount of expertise [8], [9], [10] has been developed by Research & Development Centre for Iron and Steel (RDCIS) Steel Authority of India limited (SAIL). With this developed expertise SAIL has embarked on many ambitious projects involving wireless signal transmission in hazardous locations. Technology is horizontally transferable to all existing Blast Furnaces.

D. Safety aspect

Manpower is the real asset of any company. We can not in any case ignore safety of our operating personnel. Whatever

978-1-5386-6624-1/18 $31.00 © 2018 IEEE

new technology we adopt must have some imprint towards improvement of life quality for our operators. With implementation of this scheme, safety aspect for the operating personnel has immensely improved. Earlier system demanded operator to stand in a small cabin close to the tap hole and operate the machines, sometimes causing life threatening situations. Sometimes the operator either got injured or consumed huge amount of poisonous gas on a regular basis. Now the operator operates the machine from a safe distance.

VI. CONCLUSION

Wireless technology is being aggressively adopted in steel plant having harsh and dangerous operating environment that includes close proximity to hot metal, dusty ambiance, mechanical vibration, electromagnetic interferences, and extremely high temperatures. The system developed has huge possibilities of horizontal technology transfer in areas like Raw Materials Handling Plant, cranes, Coke ovens, Sinter Plants and other Blast furnaces etc.

Radio Remote Control based automation system has been successfully implemented in one of the most hazardous area of iron making namely cast house equipment of any Blast Furnace. The system comprises of RRC based communication system, newly designed proportional hydraulic valve stand, PLC based automation system and special dust free panels. The sequencing and interlocking software has been developed with in- house expertise of PLCs. Implemented system has been successful in transmitting electrical signals without depending on the fixed control desk and bunch of control cable, routed through fire prone areas. Wireless based, automation system thus implemented has resulted in significant reduction in breakdown subsequently resulted in enhanced availability of cast house equipment, hence increase in hot metal production at Blast Furnace of Rourkela Steel Plant. Most important aspect is the enhanced operator safety parameter and creating positive impact on life of operating personnel.

During the process of implementing this technology substantial amount of expertise in the field of Industrial application of wireless transmission has been developed. With the knowhow created from this project, SAIL has embarked on many areas of Steel making where wireless signal transmission is being implemented.

ACKNOWLEDGMENT

The authors express their heartfelt gratitude to the collective of Electrical, Mechanical and Operation department of Blast Furnace, and Research &Control Lab of RSP. The authors are sincerely thankful to the management of RDCIS SAIL for their encouragement, financial support, technical guidance, and the necessary infrastructural and logistic support for smooth commissioning of this project.

REFERENCES

[1] Bill Conley "Solving Industrial Monitoring Challenges through Wireless I/O" White paper Industrial Wireless, B&B Electronics

[2] Åkerberg et al; licensee Springer. 2011"Efficient integration of secure and safety critical industrial wireless sensor networks" EURASIP Journal on Wireless Communications and Networking 20112011:100

[3] William L. Mostia, PE, Fellow, SIS-TECH Solutions"When to Use Wireless in Safety Application. It Comes Down to Reliability, and Simple Differences Add Up to Significant Limitations ", Jul 21, 2014

[4] Wei Liang, Shuai Liu, Yutuo Yang, Shiming Li"Research of Adaptive Frequency Hopping Technology in WIA-PA Industrial Wireless Network" Volume 334 of the series Communications in Computer and Information Science pp 248-262,Springer

[5] J.S.Prasath "Security in industrial wireless networks", Int.J.Computer Technology & Applications,Vol 5 (3),1302-1308 IJCTA | May-June 2014

[6] Li Peng "How to Choose the Right Industrial Firewall: The Top 7 Considerations" White paper, MOXA

[7] Amit shah "Wireless Device Networking in Process Industry - Challenges and Applications" A White Paper E-Senza Technologies GmbH

[8] Vehbi C. Gungor, Gerhard P. Hancke "Industrial Wireless Sensor Networks: Challenges, Design Principles, and Technical Approaches", Vol 56 issue 10, IEEE Transactions on Industrial Electronics

[9] Kamran Khakpour, M. H Shenassa "Industrial control using Wireless Sensor Networks" IEEE conference ICTTA 2008

[10] Abdulrahman Yarali "Wireless Mesh Networking technology for commercial and industrial customers "Electrical and Computer Engineering, 2008. CCECE 2008. DOI:10.1109/CCECE.2008.4564493

Observer Based Controller Design for Inverted Pendulum System

Shahida Khatoon
Electrical Engineering
Jamai Millia Islamia
New Delhi(India)
Skhatoon@jmi.ac.in

Devendra kumar Chaturvedi
Electrical Engineering
Dayalbagh Educational
Institute Agra(India)
Dk.foe@gmail.com

Naimul Hasan
Electrical Engineering
Jamai Millia Islamia
New Delhi(India)
nhasan@jmi.ac.in

Md Istiyaque
Electrical Engineering
Jamai Millia Islamia
New Delhi(India)
istijmi@yahoo.com

Abstract—**An inverted pendulum system is an inherently unstable and possess highly non- linear behavior, it is tested and used for validation for several control algorithm, it is having one control input and two degrees of freedom. In this paper, detailed mathematical modeling of such system is discussed. State space model is then analyzed for stability parameter such as peak overshoot, damping ratio, settling time etc. Pole placement method is applied for controller design using Ackermann's formula as all system states, is not measurable. On comparing the real state with their estimated values, the controller is validated for various dynamic response characteristics.**

Keywords—*Inverted Pendulum, Parameter Estimation Identification and Ackermann formula*

I. INTRODUCTION

Inverted pendulum itself is an unstable system [2] and when we talk about an unstable system on another unstable system like on cart will be an unstable system [3]. It has open loop and close loop instability and it also shows high level of non-linearity. It is likely to the non-linearity presented by a rocket launcher and PSLV. Balancing inverted pendulum system in horizontal position is not a simple task but we can control the inverted pendulum by applying force on the cart in the area of automatic control system [4,5]. A control force can be used to keep this system unscathed, for the controlling purpose it has two degrees of freedom and one input [8]. So that the system associated to the field of mechanical system we call it underactuated mechanical system. That will do the controlling task more challenging.

This system which is used in this paper must have two output one for straight movement of cart and the other one for movement of the inverted pendulum arm with some angle or angular rotation [8]. The purpose of designing this paper is to maintain the position of the cart and stability of the pendulum on their track as quick as possible because it is so difficult for the pendulum to maintain their position and stabilize themselves in inverted position during such movement. The aim is to design the mathematical modeling for the highly nonlinear and unstable system so that it will be upright after producing the disturbance and design an optimal controller for the system so that the position of the cart can be controlled the system quickly.

II. MODELLING OF PHYSICAL SYSTEM

For detailed study of dynamic behavior of any system or subsystem or related controller and there parameter and design it is really necessary to analyze and identify system first, in real life it is very difficult to design a simple and suitable algorithm that can provide robust and stable controller to get ideal system output and optimal performance. We used PID controller and LQR in this paper to steady the inverted pendulum. Using Lagrange equation of second kind of mathematical model of system is derived. The PID controllers are used in real life application just because its simplicity. Full state observer is designed using Acker function, comparison of real states and there estimation is shown by graphically well. The result of the proposed controller is verified in MATLAB Simulink environment.

Let us take these parameters for the system

TABLE 1 PARAMETERS OF THE SYSTEM

PARAMETER	VALUE	UNIT
Mass of the inverted pendulum cart (M)	0.5	Kg
Mass of the inverted pendulum (m)	0.2	Kg
Length of the inverted pendulum (l)	0.3	M
Coefficient of friction of cart (b1)	0.2	N/m/s
Coefficient of friction of cart (b2)	0.002	N/Rad/sec
Mass moment of inertia of the pendulum (I)	0.006	Kg/m^2
Gravitational Force (g)	9.8	m/s^2
Force applied to cart (F)	---	N
Position of cart (X)	---	m
Pendulum angle (φ)	---	Rad

Kinetic energy of the cart is

$$E_{KV} = \frac{1}{2} M \dot{x}^2$$

Coordinates x_k and y_k show the position of the inverted pendulum

$$x_k = x + l \sin \psi$$

$$y_k = l \cos \psi$$

The square of velocity of the inverted pendulum is

$$V_k^2\big| = V_{kx}^2 + V_{ky}^2$$

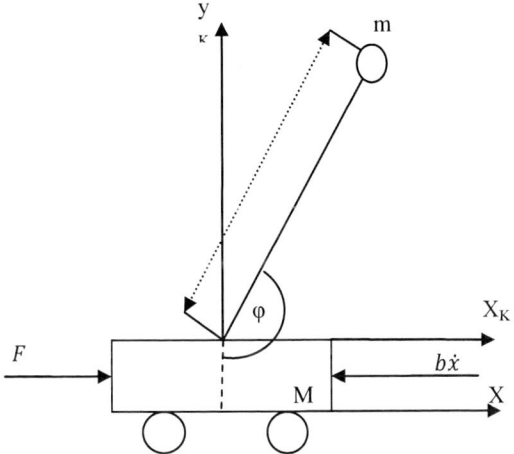

Fig. 1 Inverted pendulum system

$$V_K^2\big| = \dot{x}^2 + 2l\dot{x}\dot{\psi}\cos\psi + l^2\dot{\psi}^2\cos^2\psi + l^2\dot{\psi}^2\sin^2\psi$$

After simplifying

$$V_k^2\big| = \dot{x}^2 + 2l\dot{x}\dot{\psi}\cos\varphi + l^2\dot{\psi}^2$$

Finally kinetic energy of the system is

$$E_{kk} = \frac{1}{2}m\dot{x}^2 + ml\dot{x}\dot{\psi} + \frac{1}{2}ml^2\dot{\psi}^2$$

And system total kinetic energy is

$$E_k = L = E_{kv} + E_{kk}$$

$$E_k = \frac{1}{2}(M+m)\dot{x}^2 + \frac{1}{2}(l+ml^2)\dot{\psi}^2 + ml\dot{x}\dot{\psi}\cos\psi$$

Separate derivations of Lagrange equation for position $\dot{x}$ of the cart are

$$\frac{\partial E_k}{\partial x} = (M+m)\dot{x} + ml\dot{\psi}\cos\psi$$

$$\frac{d}{dt}\frac{\partial E_k}{\partial \dot{x}} = (M+m)\ddot{x} + ml\ddot{\psi}\cos\psi - ml\dot{\psi}^2\sin\psi$$

$$\frac{\partial E_k}{\partial x} = 0$$

$$Q_x = F - b_1\dot{x}$$

And also respective derivations of Lagrange equation for the angle ψ are

$$\frac{\partial E_k}{\partial \dot{\varphi}} = (I + ml^2)\dot{\psi} + ml\dot{x}\cos\psi$$

$$\frac{d}{dt}\frac{\partial E_k}{\partial \dot{\psi}} = (I + ml^2)\ddot{\psi} + ml\ddot{x}\cos\psi - ml\dot{\varphi}\sin\psi$$

$$\frac{\partial E_k}{\partial \psi} = -ml\dot{x}\dot{\psi}\sin\psi$$

$$Q_\psi = -mgl\sin\psi - b_2\dot{\psi}$$

The general Lagrange's equation for the cart velocity x and pendulum angle φ are

$$\frac{d}{dt}\frac{\partial E_k}{\partial \dot{x}} - \frac{\partial E_k}{\partial x} = Q_x$$

$$\frac{d}{dt}\frac{\partial E_k}{\partial \dot{\psi}} - \frac{\partial E_k}{\partial \varphi} = Q_\psi$$

Using these two above equations and putting the individual derivative, we get after simplifying the resultant equations of motion.

$$\ddot{x} = -\frac{ml}{(M+m)}\ddot{\psi}\cos\psi - \frac{ml}{(M+m)}\dot{\psi}^2\sin\psi - \frac{b_1}{(M+m)} + \frac{F}{(M+m)}$$

$$\ddot{\psi} = -\left(\frac{ml}{I+ml^2}\right)\ddot{x}\cos\psi - \frac{b_2}{(I+ml^2)}\dot{\psi} - \frac{mgl}{(I+ml^2)}\sin\psi$$

The above equations show the system dynamics

III. Linear Differential Equations

For future work, we have to linearize the differential equations of the system. This is a linearization around the symmetric point, in this case $\theta = \pi$ about vertical position. The essential in this system remains in the velocity of this position. Here in this case we assume φ a discrepancy from Symmetric point, we obtain $\theta = \pi + \varphi$ when pendulum is in vertical position (upright).

$$\cos\theta = \sin(\pi + \emptyset) = -1$$

$$\sin\theta = \sin(\pi + \emptyset) = -\emptyset$$

$$\dot{\theta}^2 = \dot{\emptyset}^2$$

Using these equations we get the linearized equations of the systems

$$\ddot{x} = \frac{ml}{M+m}\ddot{\emptyset} - \frac{b_1}{M+m}\dot{x} + \frac{F}{M+m}$$

$$\ddot{\psi} = \frac{ml}{I + ml^2}\ddot{x} - \frac{b_2}{I + ml^2}\dot{\varnothing} + \frac{mgl}{I + ml^2}\varnothing$$

IV. STATE SPACE

By putting the value of linearized equation $\ddot{\varphi}$ into $\ddot{x}$ and rearranging them we get

$$\ddot{x} = \frac{F(I + ml^2) - (I + ml^2)\dot{x} - mlb_2\dot{\psi} + m^2l^2g\psi}{q}$$

And for $\ddot{x}$ into $\ddot{\psi}$

$$\ddot{\psi} = \frac{mlF - mlb_1\dot{x} - b_2(M + m)\dot{\psi} + mlg(M + m)\psi}{q}$$

Where

$$q = [(I + ml^2)(M + m) - (ml^2)]$$

Now state variables, x_1 can be easily define through x_4

$$x_1 = x$$

$$x_2 = \dot{x}_1 = \dot{x}$$

$$x_3 = \psi$$

$$x_4 = \dot{\psi}$$

$$y_1 = x_1$$

$$y_2 = x_3$$

State space model is finally represented as

$$x(t) = Ax(t) + Bu(t)$$
$$Y(t) = C^T x(t) + Du(t)$$

$$A = \begin{bmatrix} 0 & 1 & 0 & 0 \\ 0 & -\frac{b_1(I+ml^2)}{q} & \frac{m^2l^2g}{q} & -\frac{mlb_2}{q} \\ 0 & 0 & 0 & 1 \\ 0 & -\frac{mlb_1}{q} & \frac{mlg(M+m)}{q} & -\frac{b_2(M+m)}{q} \end{bmatrix}$$

$$B = \begin{bmatrix} 0 \\ \frac{I + ml^2}{q} \\ 0 \\ \frac{ml}{q} \end{bmatrix}$$

$$C^T = \begin{bmatrix} 1 & 0 & 0 & 0 \\ 0 & 0 & 1 & 0 \end{bmatrix}, \qquad D = \begin{bmatrix} 0 \\ 0 \end{bmatrix}$$

V. TRANSFER FUNCTION

Initially we have to rearrange the linearized equation of the system

$$(I + ml^2)\ddot{\psi} + b_2\ddot{\psi} - mgl\psi = ml\ddot{x}$$

$$(M + m)\ddot{x} + b_1\dot{x} - ml\ddot{\psi} = u$$

We must take the Laplace transform of transfer function that we obtain assuming zero initial condition.

$$I + ml^2)\varnothing(s)s^2 + b_2\varnothing(s)s - mgl\varnothing(s) = mlX(s)s^2$$

$$M + m)X(s)s^2 + b_1X(s)s - ml\varnothing(s)s^2 = U(s$$

For finding the transfer function $\varnothing(s)$ which is output and $U(s)$ which is input we need to seperate $X(s)$

$$X(s) = \left[\frac{I + ml^2}{ml} + \frac{b_2}{mls} - \frac{g}{s^2}\right]\varnothing(s)$$

By substituting the value of X(s) in second equation we get

$$\frac{\varnothing(s)}{U(s)} = \frac{\frac{ml}{q}s}{s^3 + \frac{b_1(I+ml^2)+b_2(M+m)}{q}s^2 - \frac{b_1b_2+mlg(M+m)}{q}s - \frac{b_1mlg}{q}}$$

The transfer function and the output of the cart position $X(s)$ can be stated as

$$\frac{X(s)}{U(s)} = \frac{\frac{(I+ml^2)}{q}s^2 + \frac{b_2}{q}s - \frac{mlg}{q}}{s^4 + \frac{b_1(I+ml^2)+b_2(M+m)}{q}s^3 - \frac{b_1b_2+mlg(M+m)}{q}s^2 - \frac{b_1mlg}{q}s}$$

$$P_K(s) = \frac{\varnothing(s)}{U(s)}\frac{rad}{N}\Bigg], P_V(s) = \frac{X(s)}{U(s)}\frac{m}{N}$$

VI. BEHAVIOR ANALYSIS

$$rank(Q_{co}) = rank[b \quad Ab \quad A^2b \quad A^3b] = n$$

$$rank(Q_{OB}) = rank \begin{bmatrix} C^T \\ C^T A \\ C^T A^2 \\ C^T A^3 \end{bmatrix}$$

Eig = [0.0000 + 0.0000i, 0.1338 + 0.0000i, 0.0430 + 3.8196i, 0.0430 - 3.8196i]

The behaviors that show in MATLAB by step command for the output are unstable in open loop, From *p-z map* it is confirmed [10,11].
By checking the function *ctrb* in MATLAB we find the rank of matrix Q_{CO} is 4 it means the system can be controllable.

Accordingly, the system can be completely observable as the observability matrix must have rank n,
And we find here the system is observable, using the function *obsv* as the rank of the matrix Q_{OB} is 4.

$$rank(Q_{OB}) = rank \begin{matrix} C^T \\ C^T A \\ C^T A^2 \\ C^T A^3 \end{matrix} = n$$

VII. CONTROLLER DESIGN

The inverted Pendulum system can be easily controlled by the full state feedback system shown below is determined by calculating feedback vector *K*. That could be calculated by the method called pole placement method [9,10]

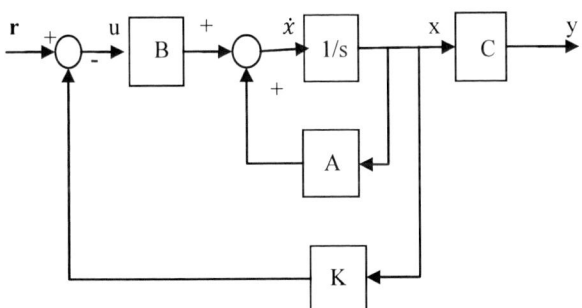

Fig.2 Full state feedback

The dynamics of the controller will be

$$u = r - KX$$
$$x = Ax + B(r - Kx) = (A - BK)x + Br$$

$$y = Cx$$

This method is completely based on some criteria like peak overshoot settling time that is used in this design. The criteria of design for the system are settling time for both the outputs Ts < 5s and angle over shoot %*OS* < 20.

Complex domain specification is use to locate dominant pole roots of $s^2 + 2\xi\omega s + w^2 = 0$. This is done by using the formula

$$\xi = \frac{-\ln(\%os/100)}{\sqrt{\pi^2 + \ln^2(\%os/100)}},$$

$$T_s < \frac{4}{\xi w_n}, \quad w_d = w_n\sqrt{1-\xi^2}$$

$$\sigma = \xi w_n, \quad \beta = \cos^{-1}\xi$$

The resulting values

$$\xi = 0,45594, w_n = 1,75 \, rad, \beta = 62,87, w_d = 1.558$$

The complex conjugates dominant poles are

$$s_{1,2} = -0,798 \pm 1,558i$$

The LQR response of the following gain vector "*K*" was calculated for poles at different sets

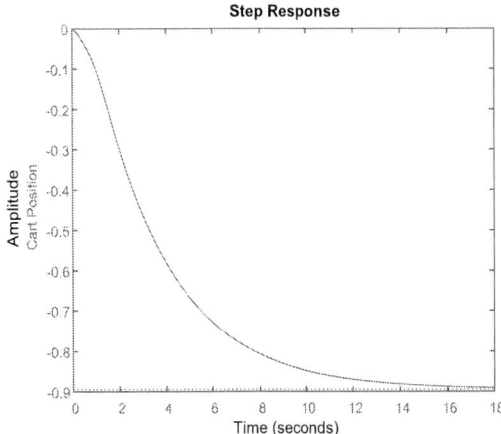

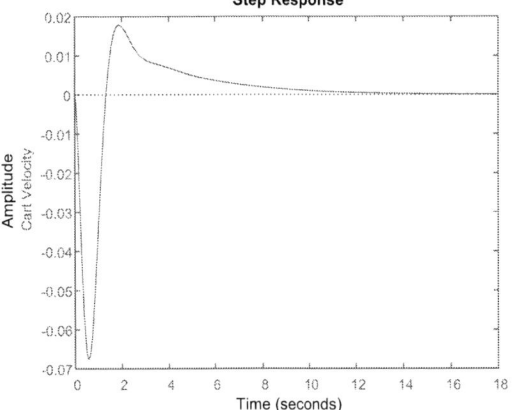

Fig. 3 Step Response of LQR

$$K = [-1.1180 \quad -4.4649 \quad -4.7290 \quad -1.7541];$$

$Eig(Ac) = [-3.0478 + 0.0000i, \ -0.3194 + 0.0000i, \ -1.9725 + 2.7056i, \ -1.9725 - 2.7056i];$

Using *acker* function following gain vectors calculated for different sets of P

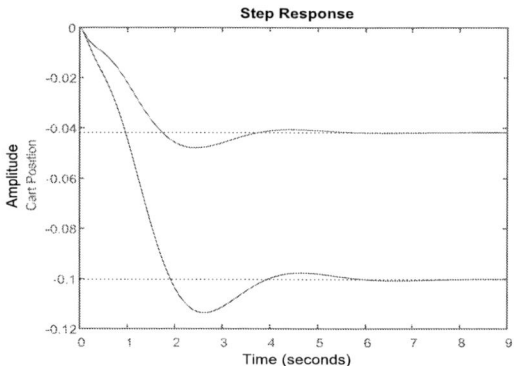

Fig. 4 Step Response of pole placement/Acker Function

$$P2 = [-8.0000+0.0000i, -4.0000+0.0000i, -0.8000+1.5500i, -0.8000 - 1.5500i];$$

$$K2 = [-9.9753 \quad -9.2205 \quad -14.7846 \quad -2.8071];$$

VIII. FULL ORDER STATE OBSERVER DESIGN

An observer is used to estimate the state of system, sometimes the state variables of dynamic system cannot be measured easily. Once a full state observer is designed as it is done here in this paper successfully it means it is not possible to measure any state variable.

For this system the poles of the observer is $p1= -7$, $p2= -8$, $p3=-10$ and $p4= -12$. The function *place* was used for the observer matrix.

Characteristics polynomial can be easily calculated using the formula

$$A_L = A - lC^T$$

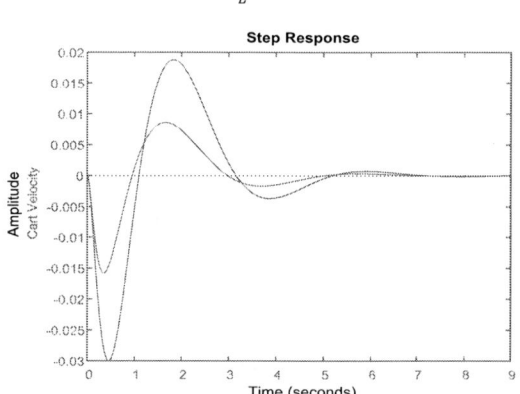

Fig.5 Step Response of full state observer/Acker Function

$$P3 = [-9.6000 + 0.0000i \; -8.0000 + 0.0000i \; -0.8000 + 1.5500i \; -0.8000 - 1.5500i];$$

$$K3 = [-23.9408 \quad -18.3579 \quad -34.6082 \quad -1.7842];$$

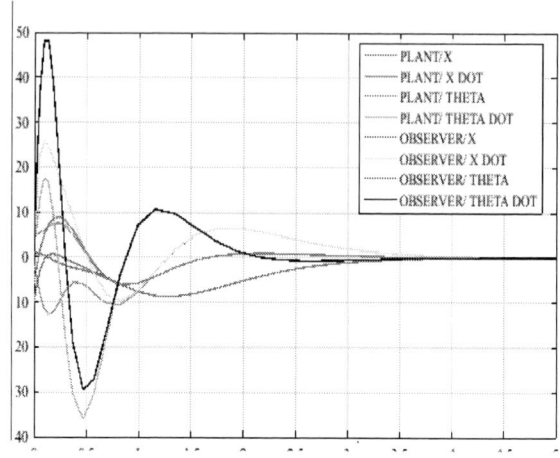

Fig.6 Comparison of real states and their estimations
Angle[rad]/estimate[rad]; cart position[m]/estimate[m]; angular velocity[rad's]/estimate[m/s]

IX. CONCLUSION

In this paper, the mathematical modeling of inverted pendulum system is done using the Euler-Lagrange method. The present system is a nonlinear system; therefore it is linearized close to the vertical position. The modeling is done with the help of state space and the transfer function. The system is checked for controllability and observability. It is revealed that the system is unstable but controllable and observable. The controller design is done using pole placement technique with full state vector feedback. An optimal controller is designed using a linear quadratic regulator. The performance index is set with the help of Ricatti equation. Full state observer is designed perfectly and show the comparison of real state and there estimation. The dynamic characteristics of the present controllers show satisfactory performance.

REFERENCES

[1] Michael Muehlebach, Raffaello D Andrea "Nonlinear Analysis and Control of a Reaction-Wheel-Based 3-D Inverted Pendulum" in IEEE Transactions on Control Systems Technology, pp. 235-246, Volume 25, issue 1, 2017

[2] Shahida Khatoon, D. K. Chaturvedi, Naimul Hasan and Md Istiyaque "Optimal control of a double inverted pendulum by linearization technique" IEEE conference on Multimedia, Signal Processing And Communication Technologies IMPACT pp 123-127, 2017.

[3] Mohammad Shahbazi, Robert Babuska, and Gabriel A. D. Lopez "Unified Modelling and Control of Walking and Running on the Spring Loaded Inverted Pendulum" IEEE Transactions on Robotics, pp1178-1195, Volume 32, Issue: 5, 2016.

[4] Premysl Strakos, Jiri Tuma "Mathematical Modelling and Controller Design of Inverted Pendulum" in 18th international Carpathian control conference(ICCC), IEEE, pp 388-393, 2017

[5] Zhao jie and Ren Sijing "Sliding Mode Control of Inverted Pendulum Based on State Observer" IEEE sixth international Conference on Information Science and Technology(ICIST) pp322-326, 2016,

[6] Rupali Khairnar, Chandrakant Kadu "Design of Controller for Inverted Pendulum System" international journalof scientific and engineering

research, volume 6 , pp.1544- 1549, issue 8, ISSn 2229-5518, August-2015.

[7] Aman Jacknoon, M. A. Abido "Ant Colony Based LQR and PID tuned Parameters for Controlling Inverted Pendulum" international conference on communication, control computing and electronics engineering (ICCCCEE), pp 1-8, 2017.

[8] M. Gopal, Digital Control and State Variable Methods: Tata McGraw-Hill, 2008

[9] Jian Huang, Feng Ding, Toshio Fukuda and Takayuki Matsuno "Modeling and Velocity Control for a Novel Narrow Vehicle Based on Mobile Wheel Inverted Pendulum" IEEE Transactions on Control System Technology. , pp 1607-1617, vol 21 issue 5, 2013

[10] MODRLAK, O. Non-liner Systems. Studing Material. Libere. Technical University on Liberec, Faculty of Mechatronics, 2004. 38 s.

[11] KURECKOVA, E. Selected Methods of Non-Linear Control. Zlin. Tomas Bat'a University in Zlin, Faculty of Applied Informatics, 2013. 71 s. Diploma thesis, Thesis head: Dostal, P

[12] KUPKA, L.Matlab & Simulink, studijni study materials for Fundamentals of Cybernatics. Liberec: Technical University of Liberec. 2008.224 s.

[13] VITECKOVA, M. VITECEK, A.Fundamentals of Automatic Control. Ostrava : VSB – Technical University of Ostrava. 2008. 244 s. ISBN 978-08-248-1924-2,2. Edition.

[14] NOSKIEVIC, P . Modelling and Syatem Identification. 1999. Ostrava : MONTANEX, 1999.278 s. ISBN 80-7225-030-2,1.reprint.

[15] MAKOVY, V. Design and Relization on Inverted Pendulum: master thesis. Ostrava: VSB – Technical University of Odtrava, faculty of Mechanical Engineering, The Department of Control System and Instrumentation, 2014, 48 p. Thesis head: Wagnerova, R.

Design of Fractional Order Butterworth Filter using Genetic Algorithm

Ashu Soni
Advanced Electronics Lab
Division of Electronics and Communication Engineering
Netaji Subhas Institute of Technology
New Delhi, India
soniashu.14@gmail.com

Maneesha Gupta
Advanced Electronics Lab
Division of Electronics and Communication Engineering
Netaji Subhas Institute of Technology
New Delhi, India
maneeshapub@gmail.com

Abstract—**This paper proposes the designing of fractional order low pass Butterworth filter using Genetic algorithm. The simulated magnitude response obtained using MATLAB is analyzed and verified at circuit level. The optimized results for the order 1.1, 1.5 and 1.9 are designed and verified using Tow Thomas biquad with SPICE simulations. Fractional order capacitor is used which is approximated using continued fraction expansion method. MATLAB and SPICE simulation results are compared for maximum attenuation in pass band. Good matching between MATLAB and SPICE results are achieved that shows the reliability of proposed filter.**

Keywords—fractional order, Genetic Algorithm, biquad, SPICE

I. INTRODUCTION

Filters are very important block of analog and digital signal processing for various applications. Filter is a network that changes the magnitude and phase response of a signal with frequency. The most common way of classification of filters are based on frequency selectivity. Filters are classified into four categories low pass, high pass, band pass and band reject where each name indicates how a band of frequencies is affected. An ideal filter allows a specific band of frequency of interest while attenuates the specific unwanted frequency band. Various techniques are available to design low pass filter of particular specifications and the high pass or bandpass filter is possible to implement using transformation techniques.

Nowadays, Fractional calculus is used in various fields like electromagnetics, agriculture, chaotic systems, electrochemical studies, noise analysis, control theory, image processing, signal processing etc. Definition of fractional order derivatives are given by many researchers. The Riemann-Liouville definition of fractional derivative is as follows

$$_{a}^{RL}D_{t}^{\alpha}f(t)=\frac{d^{m}}{dt^{m}}\left[\frac{1}{\Gamma(m-\alpha)}\int_{a}^{t}(t-\tau)^{m-\alpha-1}f(\tau)d\tau\right],$$
$$m-1<\alpha\le m, m\in N. \tag{1}$$

The caputo definition is as follows-

$$D_{0}^{\alpha}f(t)=\frac{1}{\Gamma(m-\alpha)}\int_{0}^{t}f^{(m)}(u)(t-u)^{m-\alpha-1}du \tag{2}$$

The Grunwald- Letnikov definition is as follows

$$_{a}^{GL}D_{t}^{\alpha}f(t)=\lim_{h\to 0}\frac{\sum_{k=0}^{\left[\frac{t-a}{h}\right]}(-1)^{k}\binom{\alpha}{k}f(t-kh)}{h^{\alpha}}, \alpha \in R, t-a=nh \tag{3}$$

where Γ (.) is gamma function, m is an integer and α is fractional order [17], [18].

Designing of analog filters in fractional domain gives more design flexibility and degree of freedom. Accurate requirement of frequency response, roll off rate and attenuation can be achieved using fractional order filters as compared to integer order filters [12]. Various methods have been used to approximate Butterworth, Chebyshev, Inverse Chebyshev, Cauer and Bessel filters. The Butterworth filter has maximally flat response in passband. It has a slower roll-off as compared to Chebyshev or an Elliptic filter, therefore it will need a higher order to implement a particular stopband specification. These filters have a more linear phase response in the pass-band than Chebyshev and Elliptic filters. The Butterworth polynomial has an all-pole transfer function with no finite zeros present. It is the approximation of choice when moderate selectivity and low phase distortion are essential. Significant amount of work has been done to approximate the behavior of fractional order low pass filter. Different optimization techniques have been used to estimate the behavior of low pass filter. In paper [1] nonlinear least square technique has been used to investigate the response of fractional order low pass Butterworth filter. In paper [2] optimization in fractional order RLC filter has been investigated. In paper [3] optimization technique has been used to study fractional order low pass filter. In paper [4] fractional order inverse Chebyshev filter has been analyzed using nonlinear least square technique. In paper [5] fractional order Butterworth filter has been examined using active and passive realization. In paper [6] fractional order filter has been studied using two fractional capacitors of different order. In paper [7] CCII has been used to investigate different order fractional order filters. In paper [8] fractional order Chebyshev filter has been investigated using optimization technique. In paper [9] various research areas of fractional order system have been discussed. In paper [10] basic theory of fractional order circuits and systems, its design aspects and applications have been discussed. In paper [11] first order filters have been investigated in fractional order domain.

This work focuses on the design of fractional low pass Butterworth filter of order $(1+\alpha)$ with fractional step α. Here α lies in $0 < \alpha < 1$. Design of these filters uses an integer-order approximation of the fractional order laplacian operator s^{α}. Here, genetic algorithm has been used to analyze and design fractional order low pass Butterworth filter. Genetic algorithm has already been used to analyze different approximations of low pass filter but not for the

978-1-5386-6624-1/18 $31.00 © 2018 IEEE

fractional order filters. Here, fractional order low pass Butterworth filter is optimized using genetic algorithm (GA) and verified using Tow Thomas biquad topology.

The paper is structured as follows. Section II deals with the description of Genetic algorithm (GA). Section III focuses on the use of GA for finding the filter coefficients of fractional order low pass Butterworth filter. Section IV presents the design of fractional order low pass Butterworth filter through Tow Thomas biquad using fractional capacitor to validate the proposed filter. Section V shows the comparison of maximum attenuation of proposed fractional order Butterworth filter using MATLAB and SPICE. Finally Section VI concludes the work.

II. GENETIC ALGORITHM

GA generates the solutions of problems by using methods motivated by inheritance, mutation, selection and crossover. It is a robust search which based on natural selection and natural genetics. It is not certain in nature, sometimes initial parent population gives best result that would not be achieved after many generation.

GA works on iterations and starts from the initial solution. Each iteration step is carefully chosen an operation and used for the current solution. If the new solution is appropriate e.g. if it is better than the current solution then it becomes the current solution for the next iteration, otherwise, it is rejected. This process stops when the requested solution or the number of iterations is achieved [19]. The flow chart of Genetic algorithm is shown in Fig.1.

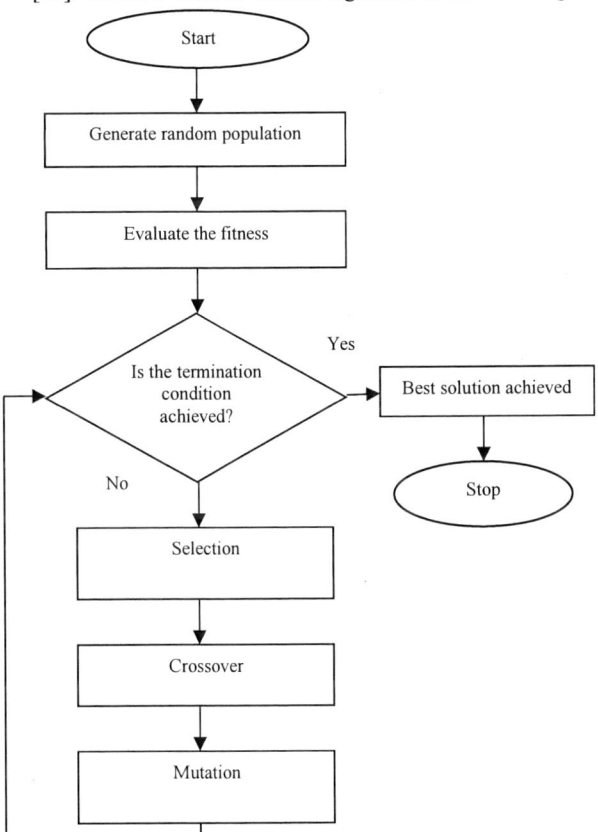

Fig. 1. Flow chart of Genetic algorithm

III. OPTIMIZED FRACTIONAL ORDER BUTTERWORTH FILTER

In proposed work, GA has been chosen to approximate the characteristics of fractional order low pass Butterworth filter. Further, GA optimized filter coefficients of proposed filter are utilized to design fractional order low pass Butterworth filter using Tow Thomas biquad.

The $(1+\alpha)$ order low pass transfer function is known by following equation

$$T(s) = \frac{c}{as^{1+\alpha} + bs^{\alpha} + 1} \tag{4}$$

where c, a and b are filter coefficients.

Butterworth second order transfer function with normalized frequency value of 1 rad/sec is presented by the following equation [20]

$$B_2(s) = \frac{0.5012}{s^2 + 1.414s + 1} \tag{5}$$

In MATLAB, GA is used within the frequency range from $\omega = 10^{-5}$ rad/s to 1 rad/s to get the coefficients of c, a and b of (4). Following equation is used to reduce the square of the error among the low pass $(1+\alpha)$ order transfer function and Butterworth second order transfer function.

$$\min_a \sum_{i=1}^{k} (abs(T(x,\omega_i)) - abs(B_2(\omega_i)))^2 \tag{6}$$

Equation (4) and (5) are used in (6) to get the GA optimized filter coefficient values. Optimized filter coefficients of $(1+\alpha)$ order low pass Butterworth filter are shown in Table I for α ranging from 0.1-0.9 in the steps of 0.1. The GA simulated magnitude responses are plotted using filter coefficients c, a and b from Table I for α equals to 0.1 to 0.9 is shown in Fig.2.

TABLE I. Simulated filter coefficients using GA

α	c	a	b
0.1	0.9107	2.3404	1.8097
0.2	0.4324	5.3838	0.6436
0.3	0.9376	68.4905	9.0643
0.4	0.5312	1.6539	0.3513
0.5	0.5169	2.1057	0.3782
0.6	0.5312	34.9333	4.2446
0.7	0.5581	1.4824	1.3613
0.8	0.4852	80.1854	8.4364
0.9	0.5312	0.4487	2.9385

978-1-5386-6624-1/18 $31.00 © 2018 IEEE

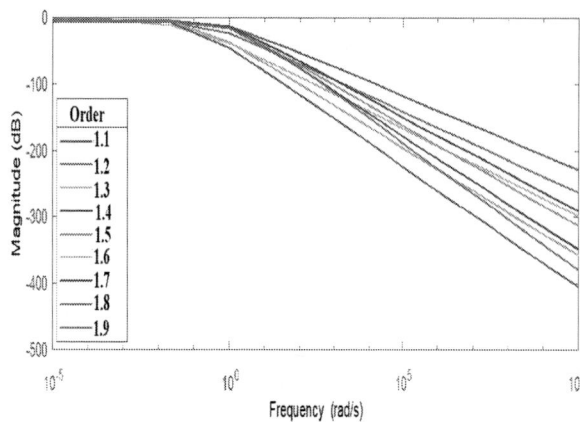

Fig. 2. GA simulated magnitude response of fractional order low pass Butterworth filter

IV. SPICE SIMULATION OF FRACTIONAL ORDER LOW PASS FILTER

GA optimized filter coefficients are used in (4) to analyze and design $(1+\alpha)$ order low pass Butterworth filter. Equation (4) with the GA optimized values of c, a and b for α equals to 0.1 to 0.9 is validated using Tow Thomas biquad. Fig.3 shows the Tow Thomas biquad using two capacitors C_1 and C_2, where C_2 is replaced by fractional capacitor [15], [16]. Fractional order capacitors can be realized using various methods like continued fraction expansion (CFE), rational approximation etc. C_2 is fractional order capacitor using fourth order integer approximation with continued fraction expansion method is revealed in Fig.4. Tow Thomas biquad using C_2 as a fractional capacitor is shown in Fig.5.

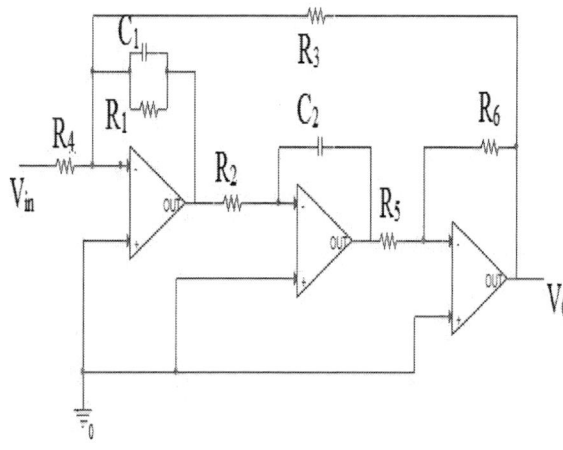

Fig. 3. Tow Thomas biquad topology

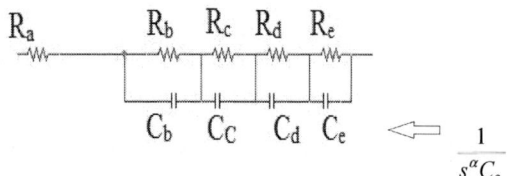

Fig. 4. Fractional order capacitor using fourth order integer approximation

The transfer function of Tow Thomas biquad is given by (7). In Fig.5, by keeping $R_4=R_5=R_6=1\,K\Omega$, $C_1=C_2=1\,F$ and remaining component values can be attained by comparison of (4) with (7). The frequency of proposed $(1+\alpha)$ order low pass Butterworth filter is shifted to 1 KHz and magnitude is scaled by a factor of 1000.

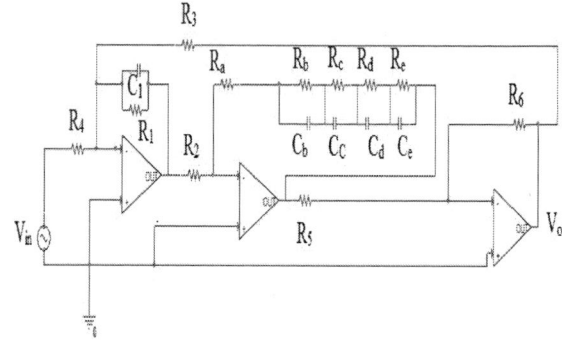

Fig. 5. Fractional order capacitor using fourth order integer approximation

$$\frac{V_0(s)}{V_{in}(s)}=\frac{R_3R_5/R_4R_6}{S^{1+\alpha}R_2R_3C_1C_2+S^{\alpha}(R_2R_3C_2/R_1)+1} \quad (7)$$

Comparing the coefficients of (4) and (7) to get the component values of R_1, R_2 and R_3 of biquad. The design equations are as follows-

$$R_1 = a/b \quad (8)$$

$$R_2 = a/c \quad (9)$$

$$R_3 = c \quad (10)$$

Equations (8)-(10) are used to find the values of R_1, R_2 and R_3. Table II shows the component values for realization of 1.1, 1.5 and 1.9 order low pass Butterworth filter. Fractional capacitor element values are given in Table III for α value equals to 0.1, 0.5 and 0.9. Table II and Table III component values are used to verify the design of $(1+\alpha)$ order low pass Butterworth filter [13], [14].

TABLE II. Component values for $(1+\alpha)$ order low pass Butterworth filter

Component	Order		
	1.1	1.5	1.9
$C_1 (\mu F)$	0.159	0.159	0.159
$C_2 (\mu F)$	415	12.6	0.381
$R_1 (\Omega)$	1293	559	152
$R_2 (\Omega)$	2569.8	4073.7	844.6
$R_3 (\Omega)$	910.7	516.9	531.2
$R_4,R_5,R_6(K\Omega)$	1	1	1

978-1-5386-6624-1/18 $31.00 © 2018 IEEE

TABLE III. Component values for fractional capacitor

Component	Order		
	1.1	1.5	1.9
$R_a(\Omega)$	659	111	6.8
$R_b(\Omega)$	195.8	252	43
$R_c(\Omega)$	135	378.8	130.8
$R_d(\Omega)$	158.5	889	670
$R_e(\Omega)$	370	7.4k	146k
$C_b(nF)$	69	84	704
$C_c(\mu F)$	0.627	0.29	1.1
$C_d(\mu F)$	2.1	0.54	1
$C_e(\mu F)$	6.6	0.69	0.20

Fig.6 depicts the SPICE simulated magnitude response of fractional order low pass Butterworth filter for α equals to 0.1, 0.5 and 0.9.

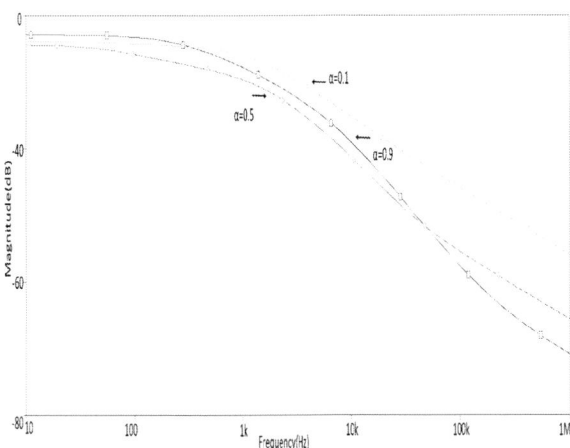

Fig. 6. SPICE simulated magnitude response of $(1+\alpha)$ order low pass Butterworth filter

V. RESULT DISCUSSION

Table IV shows maximum attenuation (dB) of 1.1, 1.5 and 1.9 order proposed filter by simulating it in both MATLAB and SPICE. The absolute error obtained by MATLAB and SPICE attenuation results are very low. It shows very good matching between results.

TABLE IV. Comparison of pass band gain (dB) using MATLAB and SPICE

α	MATLAB Attenuation(dB)	SPICE Attenuation(dB)	Absolute Error (dB)
0.1	2.05	7.65	5.6
0.5	16.1	8.82	7.28
0.9	5.45	5.67	0.22

VI. CONCLUSION

Genetic algorithm is used for designing 1.1, 1.5 and 1.9 order low pass Butterworth filter. Optimized fractional order

proposed filter is verified using Tow Thomas biquad topology using fractional capacitor. Good matching between MATLAB and SPICE results are achieved. This work can be further extended for other approximations.

REFERENCES

[1] T. J. Freeborn, "Comparison of $(1+\alpha)$ fractional-order transfer functions to approximate low pass Butterworth magnitude responses," Circuits Syst. Signal Process, vol. 35, pp. 1983-2002, 2016.

[2] A. G. Radwan and M. E. Fouda, "Optimization of fractional-order RLC filters," Circuits Syst Signal Process, vol. 32, pp. 2097-2118, 2013.

[3] L. A. Said, S. M. Ismail, A. G. Radwan, et al., "On the optimization of fractional order low pass filters," Circuits Syst. Signal Process, vol. 35, pp. 2017-2039, 2016.

[4] T. J. Freeborn, A. S. Elwakil and B. Maundy, "Approximated fractional order Inverse Chebyshev low pass filters," Circuits Syst. Signal Process, vol. 35, pp. 1973-1982, 2015.

[5] A. S. Ali, A. G. Radwan and A. M. Soliman, "Fractional order Butterworth filter: active and passive realizations," IEEE J. Emerging Sel. Top. Circuits System, vol. 3, pp. 346-354, 2013.

[6] M. C. Tripathy, K. Biswas and S. Sen, "A design example of a fractional –order Kerwin-Huelsman-Newcomb biquad filter with two fractional capacitors of different order," Circuits Syst. Signal Process, vol. 32, pp. 1523-1536, 2013. doi: 10.1007/s00034-012-9539-2

[7] A. Soltan, A. G. Radwan and A. M. Soliman, "CCII based fractional filters of different orders," J. Adv. Res., vol. 5, pp. 157-164, 2014.

[8] T. J. Freeborn, B. Maundy and A. S. Elwakil, "Approximated fractional order Chebyshev low pass filters," Hindawi Publishing Corporation, pp. 1-7, 2015.

[9] A. S. Elwakil, "Fractional-order circuits and systems an emerging interdisciplinary research area," IEEE Circuits and Systems Magazine, vol. 10, pp. 40-50, 2010.

[10] C. Psychalinos, A. S. Elwakil, A. G. Radwan, et al., "Guest editorial: fractional –order circuits and systems: theory, design, and applications," Circuits Syst. Signal Process, vol. 35, pp. 1807-1813, 2016.

[11] A. G. Radwan, A. M. Soliman and A. S. Elwakil, "First-order filters generalized to the fractional domain," Journal of Circuits, Systems, and Computers, vol. 17, pp. 55-66, 2008.

[12] T. Helie, "Simulation of fractional order low pass filters," IEEE/ACM IEEE/ACM Transactions on Audio, Speech, and Language Processing, vol. 22, pp.1636-1647, 2014.

[13] T. J. Freeborn, B. Maundy and A. Elwakil, "Fractional resonance based RLβCα filters," Hindawi Publishing Corporation, 2013. doi: 10.1155/2013/726721.

[14] M. Li, "Approximating ideal filters by systems of fractional order," Hindawi Publishing Corporation: Computational and Mathematical Methods in Medicine, 2012. doi:10.1155/2012/365054.

[15] B. T. Krishna and K. V. V. S. Reddy, "Active and passive realization of fractance device of order ½," Hindawi Publishing Corporation, 1-5, 2008.

[16] A. Soltan, A. G. Radwan and A. M. Soliman, "Fractional order Sallen key and KHN Filters: Stability and Poles Allocation," Circuits Syst. Signal Process, vol. 34, pp. 1461-1480, 2015.

[17] A. Acharya, S. Das and I. Pan, "Extending the concept of analog Butterworth filter for fractional order systems," Signal Processing, vol. 94, pp. 409-420, 2014.

[18] K. S. Miller and B. Ross, "An Introduction to the fractional calculus and fractional differential equations," Wiley, New York, 1993.

[19] I. Elkhetali Said and Y. Aldabiski Ibrahim, "Analog filter design by Genetic algorithm," in ACMOS'09 Proceedings of the 11th WSEAS international conference on Automatic control, modelling and simulation, Istanbul, Turkey, 2009. ISBN: 978-960-474-133-5.

[20] T. Khanna and D. K. Upadhyay, "Design and realization of fractional order Butterworth low pass filters," in International Conference on Signal Processing, Computing and Control (ISPCC), Waknaghat, India, 2015. doi.10.1109/ISPCC.2015.7375055.

Analysis of Aqueous Supercapacitor for Variation in Separator Wetness.

Ravi Giri
Dept. of Electrical Engr.
College of Engineering,
Pune-5, India.
ravigiri041@gmail.com

Saurabh Chaudhari
Dept. of Mechanical Engr
AISSMS College of Engineering,
Pune-01, India.
csaurabh108@gmail.com

Utkarsh Mishra
Dept. of Electronics &
Telecomm.,
Army Institute of Technology,
Pune-15, India,
utkarshmishra2210@gmail.com

A.P.Deshpande
Dept. of Electrical Engr.,
College of Engineering,
Pune-5, India.
apd.elec@coep.ac.in

P. B. Karandikar
Dept. of Electronics & Telecomm.,
Army Institute of Technology,
Pune-15,India,
pkarandikar@aitpune.edu.in

Sahil Sharma
Dept. of Electronics and Telecomm.,
Army Institute of Technology,
Pune-15,India,
sahilsharmajarvis@gmail.com

Abstract— **The supercapacitor is an emerging technology that promises to play an important role in meeting the demands of electronic devices and systems, both now and in the future. Supercapacitor has properties which can be integrated with properties of other energy storage technologies in the hybrid electric energy system. Due to behavioral characteristics of supercapacitor like fast charging and discharging ability as well as high power density, a supercapacitor and battery can operate efficiently to avoid charging and discharging of the battery. This helps to achieve longer battery life and enable higher system peak power performance thus improves the efficiency of the system. In the work carried out, efforts are put to fabricate a cost-effective model/prototype of supercapacitor using steel wire mesh, polyethylene, isopropyl alcohol, potassium sulfate, and manganese dioxide. The results are obtained in the laboratory to find out capacitance of the supercapacitor. To get the best result in terms of capacitance and internal resistance best carbon is selected. Various trials on the concentration of the electrolyte, effect of wetness of the separator on capacitance and internal resistance have been presented in the paper.**

Keywords—Supercapacitor, electrode, carbon, separator, electrolyte.

I. INTRODUCTION

Now a days, no doubt vehicles have made our life easier. The very first step to start a vehicle is starting its engine first. Initially to start engine human powered techniques were used, but in present starter motor is used. Batteries are used to start these starter motor [1] and generally lead acid batteries are used. But batteries have certain limitations, such as large size which have to be increased further as demand increases, poor response to sudden power demand. They also needs maintenance and life is also very less. Due to the introduction of more electrical equipment, there is a significant rise in power demand in vehicles to increase comfort and making vehicle operation easier, such as power steering, power window, air conditioning, heater, de-ice, etc. These all increase the current demand from the battery.

During starting of an engine sudden large current is drawn by starter motor from the battery. Due to which heating of battery takes place and heating reduces life as well as capacity of the battery. Current demand for starting of an Internal Combustion Engine increases with the size of an engine and decrease in temperature and vary according to the type of engine. Generally, voltage per cell in a battery is 2.35 volts which are practically not achieved [2], 6 cells i.e.12 V battery is used for starting the engine, if the voltage drops below 10 volts a functional failure occurs and below 7.2 Volt total failure occurs. In the market, there are some power batteries available such as Li-ion, Ni-Cd, Nickel metal hydride battery, but the cost makes them out of option [4].

Application of supercapacitor in automobile industry provides a maintenance free and eco-friendly power source. Other benefits are increased life of existing energy sources, large environmental working range, efficient regenerative braking, increase the scope of stop and start technology, power buffer in electrical drive train, reduced weight and volume of energy storing devices [1]. Supercapacitor has emerged as new technology in energy storage. It has a same fundamental equation as that of conventional capacitor. They are also known as Ultra capacitors and electrochemical dual layer capacitor.
The fundamental equation for a capacitor is,

$$C = \varepsilon_0 \varepsilon_r A / D \tag{1}$$

Where, A is the surface area of an electrode and D is the distance between the two electrodes.

$$C = \frac{Q}{V} \tag{2}$$

Where, Q is the charge and V is the voltage applied.

978-1-5386-6624-1/18 $31.00 © 2018 IEEE

In equation (1), value of area cannot be stated as one has to take accessible surface area of electrode, due to various pores which are not possible to calculate or predict. Same in the case of distance between the two electrodes i.e. equation (1) is applicable but not useful for the work carried out.

From the above equation (1) capacitance is directly proportional to area of electrodes and inversely proportional to distance between electrodes. Supercapacitors have low energy density and high power density as compared to batteries, so supercapacitors can be used where there is a sudden power demand. Supercapacitor can be constructed from two carbon based electrodes, separator between electrodes and electrolyte.

In available literature, it is found that many researchers have worked on design, development, fabrication, testing, modeling and application of supercapacitors.[1]-[3]. Material aspects have been investigated thoroughly in various ways [8]. Many have done the work in stacked and rolled type aqueous supercapacitor. Nobody ever investigated about the effects of wetness of separator used in the supercapacitor. Wetness of separator in the supercapacitor affects the capacitance and internal resistance is explained in this paper. Concentration of an electrolyte determines capacitance and internal resistance of supercapacitor [10].

Therefore, problem statement converged to "Analysis of aqueous supercapacitor for variation in separator wetness", considering a prototype/model fabrication, testing through statistical method and available measuring instruments to develop the relationship with reference to physical and electrical parameters of supercapacitor.

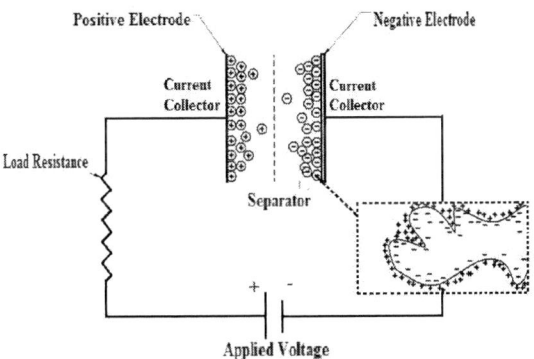

Fig.1.Schematic representation of supercapacitor [2]

This paper is organized as follows; Section II consists of supercapacitor module development, Section III has selection carbon in electrode, Section IV gives description of experimentation with electrolytes, Section V presents the concluding remarks.

II. SUPERCAPACITOR MODULE DEVELOPMENT

After going through the literature and in-depth study on various aspects of supercapacitor technology, it was necessary to choose an appropriate methodology for various electrode and electrolyte materials for the fabrication of supercapacitor prototype. In this section, the work is carried out in developing a supercapacitor prototype has been presented.

Supercapacitors are of various types, based on electrolyte used, they can be classified as aqueous or non-aqueous. Commercially available supercapacitors are of non-aqueous type. From construction point of view, they are stacked type or rolled type. For higher capacitance value, rolled type construction is more suitable. However, laboratory research cannot be done with this type of construction as it requires capacitor manufacturing. Generally, stacked type of supercapacitor prototype can be prepared in lab and then same materials can be used in rolled type of supercapacitor fabrication at industry. In this paper, construction of stacked type aqueous supercapacitor and its evaluation is presented.

Various basic materials/components used in assembly of supercapacitor module are as follows:
a) Steel wire mesh – 3 square cm.
b) Number of separators – 3 Polyethylene of suitable size.
c) Composition of carbon (Various carbon materials) and manganese dioxide in weight ratio of 1:1.
d) Electrolyte - Different concentration of potassium sulfate in distilled water.

For developing the supercapacitor, electrode material required can be available easily. To develop the electrode, the paste of mixture of active carbon Vulcan X-72 and manganese dioxide of equal proportion by weight using IPA (Isopropyl Alcohol) as solvent is made and coated at rate of at $20mg/cm^2$ on stainless steel wire mesh which acts as a current collector [7] using painting brush. The electrode prototype is shown in diagram below

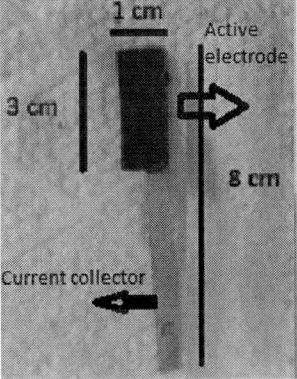

Fig.2.Electrode prototype

To prevent the occurrence of electrical contact between two electrodes, layer of polyethylene is used as a separator. But it is ion-permeable, allowing ionic charge transfer to take place. The polymer or paper separators can be used with organic electrolytes and ceramic or glass fiber separators are often used with aqueous electrolytes.

For best performance, the separator should have high electrical resistance, a high ionic conductance, and low thickness. Once the electrodes are developed, pasted by using glue on polyethylene, three such polyethylene are used and the developed prototype after putting separator in between is shown in fig. 3

978-1-5386-6624-1/18 $31.00 © 2018 IEEE

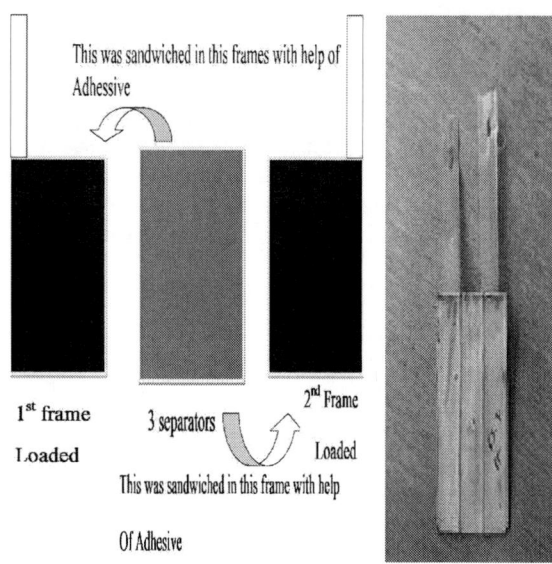

Fig.3.Development of supercapacitor module.

Developed module is put in potassium sulphate of 0.65mol concentration to operate it as electro chemical capacitor. Test set up used for measurement of capacitor and internal resistance is shown in Fig4.

III. SELECTION OF CARBON IN ELECTRODES.

The performance of supercapacitor must be able to be evaluated quantitatively in order to make comparison between different device and technologies, and also to be able to ascertain the suitability of particular device. Various tests will be performed on the capacitor to check its internal resistance, capacitance. Various grades of activated carbons available in market were tested to get best carbon for prismatic type of lab grade supercapacitor. Five types of carbons as described in Table 1 are tested for values of capacitance and internal resistance.

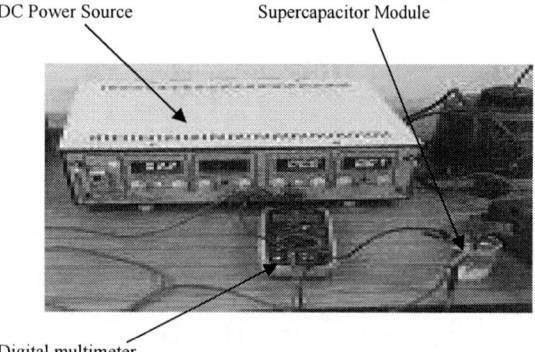

Fig.4.Test bench of supercapacitor

Testing of supercapacitor is done by supplying it with constant 2 V DC supply for some specific time I.e., charging time. Thus, it is allowed to charge completely to its maximum voltage of 2 V. Then it is discharged through short circuit to measure peak current using milli-ammeter.

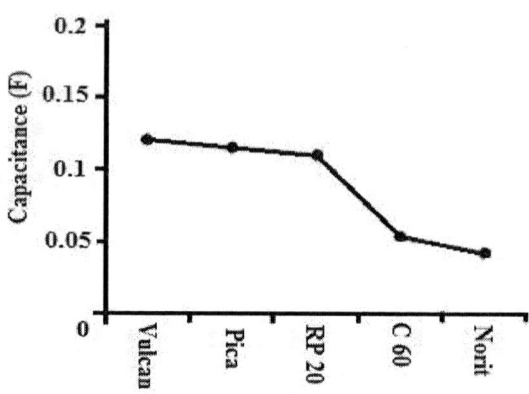

Fig.5.Variation in capacitance with various carbons.

From Fig.5, Vulcan and pica give capacitance more than other different carbon materials.

TABLE I
VARIOUS CARBON USED IN TESTING

Name of carbon	APPROXIMATE SSA IN SQ CM PER GRAM	Make of carbon [a]
Vulcan XC 72 (Vulcan)	250	Cabot corporation, US
PICA	300	Vierzon, France
RP 20	1300	Kuraray, Japan
C60	200	Sisco, India
Norite	150	Cabot Norite AC, US

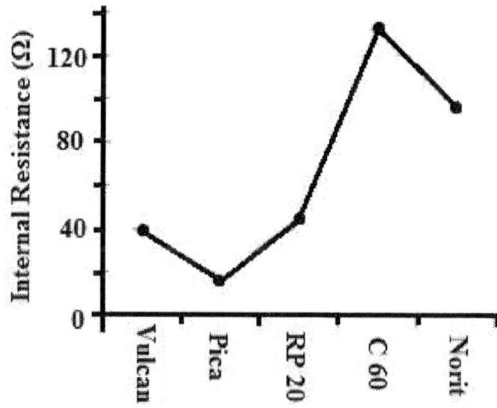

Fig.6.Variation in internal resistance with various carbons.

From Fig.6, Vulcan and pica give internal resistance less than other different carbon materials. Vulcan and pica give best result among the other carbon materials. Vulcan and pica are selected as best carbons for preparing aqueous supercapacitor.

IV. EXPERIMENTATION WITH ELECTROLYTES.

Prepared supercapacitor is dipped in potassium sulfate electrolyte. Then charged and discharged it for about 15 to 20 times by supplying it with constant 2V DC supply. The dipped supercapacitor is then taken out, put outside an electrolyte and water is allowed to evaporate. Then the weight of the supercapacitor is measured using weighing machine in specific time interval. Different readings of wetness of separator in terms of weight of supercapacitor are taken. Fig.7 shows capacitance as the function of wetness of the separator of the supercapacitor. It can be clearly observed that as the separator wetness reduces the capacitance also gets reduced.

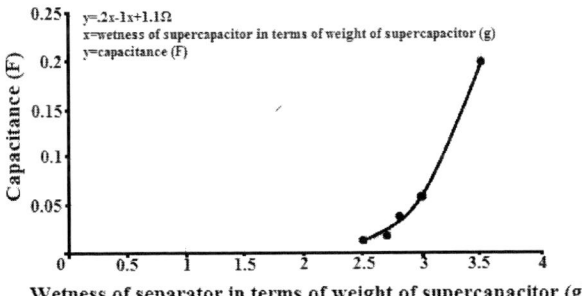

Fig.7 Variation in capacitance with wetness of separator.

As the water content gets evaporated the capacitance value will gradually decrease and nature of the characterstics is not linear. Hence, 2nd order equation as model of this characteristics is presented in graph. At one stage the supercapacitor gets dry and stop showing any value of capacitance. This shows the importance of adding at regular interval in aqueous supercapacitor.

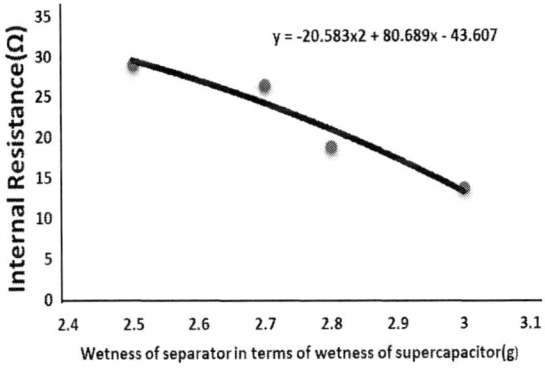

Fig 8 Internal resistance variation with wetness of separator.

Internal resistance as the function of wetness of separator the supercapacitor is shown in Fig.8. Here also the nature of characteristics is not linear and hence 2nd order equation is presented as model of internal resistance.

It was decided to check effect of concentration of electrolyte on capacitance and internal resistance of

supercapacitor. The molar concentration of potassium sulphate is increased step by step, the effect on capacitance and internal resistance is calculated. After one stage, as an electrolyte gets saturated. Even if the molar concentration of potassium sulphate is increased, there will be no effect on capacitance and internal resistance. Fig.9 shows capacitance as the function of concentration of the potassium sulfate in an electrolyte.

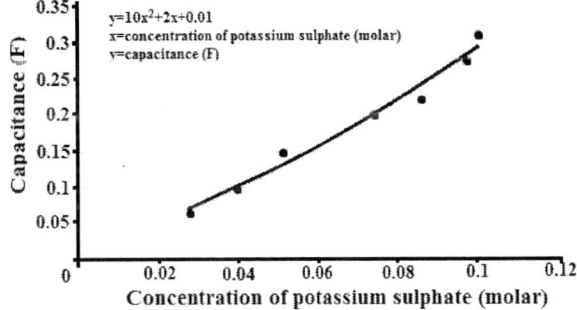

Fig.9.Variation in capacitance with concentration of potassium sulfate.

As concentration of potassium sulfate increases the internal resistance decreases. As the saturation limit of an electrolyte reached by adding potassium sulfate the internal resistance will not get affected. It is shown in Fig.10. In this graph also, non-linearity exists, and hence 2nd order model is proposed.

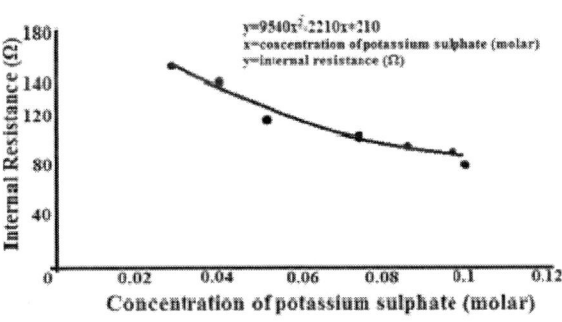

Fig.10.Variation in internal resistance with concentration of the potassium sulphate.

From Fig.10 internal resistance as the function of concentration of the potassium sulfate in an electrolyte, as the concentration of potassium sulfate increases internal resistance gets reduced, which is represented by 2nd order model.

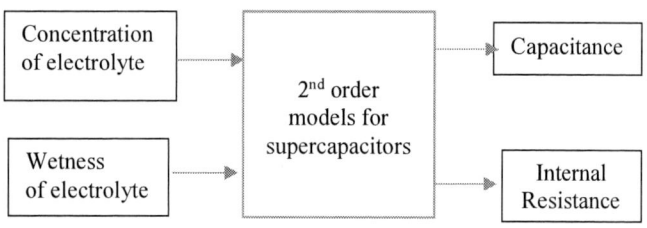

Fig11 Multi input multi output system of supercapacitor

Table 2 Proposed second order wetness-based models of supercapacitor

Sr. No.	Parameter	Proposed second order Equation
1	X=wetness of separator in terms of weight of supercapacitor (g) Y=capacitance (F)	$Y=0.2x^2-1x+1.1$
2	X= wetness of separator in terms of weight of supercapacitor (g) Y=internal resistance (Ω)	$Y=-20.583x^2+80.689x-43.607$
3	X=concentration of potassium sulfate (molar) Y=capacitance (F)	$Y=10x^2+2x+0.01$
4	X=concentration of potassium sulfate (molar) Y=internal resistance (Ω)	$Y=9540x^2-2210x+210$

Fig11 shows the multi input multi output system presented in this paper. From the Fig.7, 8, 9 and 10 second order equations are found as best curve fit to take into account the non-linearity in the characteristics. Table 2 shows the proposed model of supercapacitor for concentration of electrolyte and wetness of separators. Thus, at different concentration of potassium sulfate and wetness of separator, capacitance and internal resistance can be predicted in the accuracy range of 5% to 10%.

V. CONCLUSIONS

The supercapacitors using various carbon materials and manganese dioxide using potassium sulfate as an electrolyte has been studied. It is found that Vulcan X-72 has better characteristics than other activated carbon materials for supercapacitor. Maximum value of capacitance of supercapacitor obtained is about 0.183 F for proposal supercapacitor module at 2V. Low cost materials are used in developing supercapacitor module maximum capacitance and minimum internal resistance has been targeted throughout the experimentation. All trials were conducted on aqueous supercapacitor.

Aqueous supercapacitor can also be operated like lead-acid battery by adding distilled water frequently in an electrolyte through inlet vent which can be provided. Gel based potassium sulfate may be a solution for problem of frequent adding of distilled water. Moisturized electrolyte gives better performance as seen from the trails. Molarity selection is based on acceptable maximum voltage that electrolyte can sustain.

ACKNOWLEDGMENT

Authors would like to thank Dr. D.B.Talange, HoD-Electrical Engineering Department, College of Engineering Pune for useful discussions. Authors also thank the management of Army Institute of Technology for providing research facilities.

REFERENCES

[1] T. Brousse, M. Toupin, D. Belanger, "A hybrid activates carbon-manganese dioxide capacitor using a mild aqueous electrolyte", *Journal of the Electrochemical Society*, vol. 151, Issue no. 4, 2004, pp 614-622.

[2] Burke, Andrew. "Ultracapacitors: why, how, and where is the technology." Journal of power sources Vol 91.Issue no 1, 2000, pp 37-50.

[3] P.B.Karandikar, D.B.Talange, Uday Mhaskar, Ramesh Bansal, Development, characterization and modeling of aqueous metal oxide based supercapacitor, Energy, Elsevier, April 2012, Vol 40, Issue no 1, pp 131-138.

[4] P.B.Karandikar, D.B.Talange, Uday Mhaskar, Ramesh Bansal, Validation of Capacitance and ESR Model of Manganese Oxide Based Aqueous Supercapacitor, Electric Power Components and Systems, Taylor and Francis, July 2012, Vol 40, Issue no 01, pp 1105-1118.

[5] T. Brousse, M. Toupin, D. Belanger, "A hybrid activates carbon-manganese dioxide capacitor using a mild aqueous electrolyte", *Journal of the Electrochemical Society*, Vol. 151, Issue no. 4, 2004, pp A614-A622.

[6] Wei T, Qi X, Zhiping Qi. "An improved ultracapacitor equivalent circuit model for the design of energy storage power systems, " Proceedings of International Conference on Electrical Machines and Systems 2007, October 8-11, Seoul,Korea, pp 550-559.

[7] P B Karandikar, Akanksh Dixit, Amal Paul Shailendra, "Optimization of electrode parameters of stacked structured ultracapacitor", *Fourth International Conference on Advances in Energy Research at IITB*,Vol 01,Issue no 01, 2013, pp 135-136.

B. E. Conway, "Transition from 'Supercapacitor' to 'Battery' behavior in electrochemical energy storage," Proceedings of the 34th International, Power Sources Symposium, 2009, pp 319-327.

[8] Uzunogla M, Alam MS. Dynamic modeling design and simulation of a combined PEM Fuel cell and ultracapacitor system for standalone residential applications. IEEE Transactions on Energy Conversion September 2006, Vol 21, Issue no 01, pp767-775.

[9] Chan, C. C., A. Bouscayrol, and K. Chen, Electric, hybrid, and fuel-cell vehicles: Architectures and modeling, IEEE Transactions on Vehicular Technology, Vol. 59, No. 2, 2010, 589-598.

978-1-5386-6624-1/18 $31.00 © 2018 IEEE

[10] R. Kötz, M. Carlen, "Principles and applications of electrochemical Capacitors", *Electrochimica Acta*, Vol 45, Issue no 15, 2001, pp. 2483-2498.

[11] Seema Mathew, P B Karandikar, G Shekhar, D S Chavan, "A novel Neem based supercapacitor and its modeling using artificial neural network," International conference on Power and Advanced Control Engineering, Bengaluru, India, Vol 1, Issue no 1, August 2015, pp 55-59.

[12] Karandikar P B, Rathod D, Talange DB. "Photovoltaic chargeable electrochemical double layer capacitor, " Proceedings of 2nd International Conference on Computer and Electrical Engineering, vol. 2, 28-30 December 2009, Dubai, pp. 217-220.

[13] Niraj Mahulkar, P.B. Karandikar, "Comparison of different types of carbon materials as electrodes of supercapacitor", *International journal of innovative research in engineering and technology*, vol. 01, no. 1, 2017, pp. 12-18.

[14] Song R Y, Park J H, Shivakumar S R, Kim S H, Ko J M, Park DY, et al., "Supercapacitor properties of polyaniline/ Nafion/hydrous ruthenium oxide composite electrodes," Journal of Power Sources, 2007 Vol no 166, Issue 01, pp 297-301.

2nd IEEE International Conference on Power Electronics, Intelligent Control and Energy Systems (ICPEICES-2018)

Analysing Stability of Time Delayed Synchronous Generator and Designing Optimal Stabilizer Fractional Order PID Controller using Partical Swarm Optimization Technique

Mohammad Ali Daftari
Faculty of Electrical Engineering
K N Toosi University of Technology
Tehran, Iran
m_daftari@email.kntu.ac.ir

Mohammad Ali Nekoui
Faculty of Electrical Engineering
K N Toosi University of Technology
Tehran, Iran
manekui@eetd.kntu.ac.ir

Abstract: **This paper investigates designing of optimal fractional order PID controller for a time delayed generator excitation control system. Time delays may result to weaken performance and instability in system. In this research, we aim to design fractional order PID (FOPID) controller for time delayed synchronous generator which preserve performance and make a suitable delay margin for the system. The controller will be designed using Particle Swarm Optimization (PSO) algorithm by defining a suitable cost function. At the end, a comparison between results of FOPID controller and PID controller proves advantages of fractional order controller.**

Keywords: synchronous generator; delay margin; fractional controller; particle swarm optimization

I. INTRODUCTION

In electrical power systems, load frequency control (LFC) and automatic voltage regulator (AVR) equipments are installed for each generator to control the system frequency and generator output voltage magnitude. Figure 1 shows the schematic block diagram of a typical excitation control system for a large synchronous generator. Generator excitation control system consists of an exciter, a phasor measurement unit (PMU), a rectifier, a stabilizing transformer (rate feedback stabilizer), and a regulator. Regulator consists of a controller and an amplifier [1]. The regulator processes and amplifies input control signals to a level and controls the exciter. In this paper The FOPID controller is used as a regulator to improve the dynamic response as well as eliminating the steady-state error. The PMUs are units that measure dynamic data of power systems, such as voltage, current, angle, and frequency using the discrete Fourier transform (DFT).

This paper also investigates the effect of the time delay on the stability of the generator excitation control system. The time delays are due to the use of measurement devices and communication links for data transfer. The delay have a destabilizing impact on the inevitable time delays in power

system dynamics and lead to unacceptable performance. Therefore, they could not be ignored. In the design of a controller, time delays must be taken into account and analytical tools should be developed to study the complicated dynamic behavior of delayed power systems.

The measurement and communication delays involved between the instant of measurement and that of signal being available to the controller are the major problem in the power system control. This delay can typically be in the range of 0.5–1.0 second [2].

We will present a method for optimum tuning of practical fractional order PID controllers for automatic voltage regulator system using particle swarm optimization (PSO) algorithm considering delays in system.

This paper is organized as follows. Sect. 2, discusses a brief review to fractional calculus especially fractional-order PID controller. In Sect. 3, AVR system model with time delay and stability is given. Sect. 4, is allocated to design of $PI^\lambda D^\mu$ controller using PSO and brief introduction to particle swarm optimization algorithm. In Sect. 5, numerical and simulation results are presented. Section 6 is the conclusion.

II. REVIEW ON FRACTIONAL CALCULUS

A. Fractional-Order PID Controllers (FOPID)

The fractional PID controller is a generalization of the PID controller. The transfer function of this controller is given by the following function:

$$C(s) = k_p + \frac{k_i}{s^\lambda} + k_d s^\mu \qquad (1)$$

Where $s = j\omega$ is the complex frequency, k_p is the proportional constant, k_i is the integration constant, k_d is differentiation constant, and λ and μ are positive real

978-1-5386-6624-1/18 $31.00 © 2018 IEEE

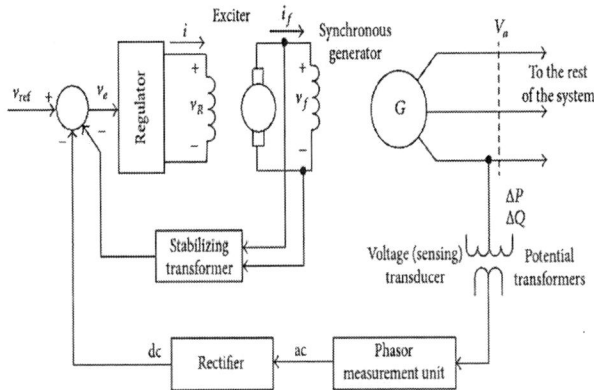

Fig. 1. The schematic block diagram of the generator excitation control system

numbers. The Grünwald-Letnikov expressed the fractional-order derivative by the following equation:

$$^{GR}_aD^r_t\, f(t) = \lim_{h \to 0} h^{-r} \sum_{h=0}^{\left[\frac{t-a}{h}\right]} \frac{(-1)^r \binom{r}{j} f(t - rh)}{h^\alpha} \quad (2)$$

Where [.] means the integer part and $r \in \mathbb{R}^+$ [3]. Reimann–Liouville (RL) expression for fractional-order derivative is given by:

$$^{RL}_aI^r_t\, f(t) = \frac{1}{\Gamma(n-r)} \frac{d^n}{dt^n} \int_a^t \frac{f(\tau)}{(t-\tau)^{r-n+1}} d\tau \quad (3)$$

For $(n - 1 < r < n)$ and $\Gamma(.)$ is the *Gamma* function and $r \in \mathbb{R}^+$ [3].

B. Oustaloup Approximation

Oustaloup's approximation method uses a band-pass filter to approximate the fractional-order operator s^λ based on frequency-domain response [4]. The approximate transfer function of a continuous fractional order operator s^λ with Oustaloup algorithm is as follows:

$$H_N(s) = k \prod_{n=1}^N \frac{1 + s/\omega_{z,n}}{1 + s/\omega_{p,n}} \quad (4)$$

Where the zeros, poles and the gain can be evaluated, respectively, as:

$$\omega_{p,n} = \mu\omega_{z,n}, \qquad n = 1, \dots, N$$

$$\omega_{z,n+1} = \eta\omega_{p,n} \qquad n = 1, \dots, N-1$$

$$\omega_{z,1} = \sqrt{\eta}\omega_l \;\; ; \;\; \eta = \left(\frac{\omega_h}{\omega_l}\right)^{(1-\lambda)/N} \;\; ; \;\; \mu = \left(\frac{\omega_h}{\omega_l}\right)^{\lambda/N}$$

$$k = \frac{|s^\lambda|}{\left|\prod_{n=1}^N \dfrac{1 + s/\omega_{z,n}}{1 + s/\omega_{p,n}}\right|}; \quad s = \sqrt{\omega_l\omega_h}\, j \quad (5)$$

$\omega \in [\omega_l, \omega_h]$ and ω_l is lower bound and ω_h is upper bound of frequencies. N is number of zeros and poles determining accuracy of approximation. For negative power of s^λ, inverse of positive power Oustaloup approximation will used.

III. AVR SYSTEM MODEL WITH TIME DELAY AND STABILITY

A. AVR System Model with Time Delay

For load frequency control and excitation control systems, linear or linearized models are commonly used to analyze the system dynamics and to design a controller. Figure 2 shows the block diagram of a generator excitation control system with delays. Note that each component of the system, namely amplifier, exciter, generator, sensor, and rectifier is modeled by a first-order transfer function [5&6]. The transfer function of each component is given in the following:

$$G_G = \frac{K_G}{1 + T_G s}; \qquad G_E = \frac{K_E}{1 + T_E s}$$

$$G_A = \frac{K_A}{1 + T_A s}; \qquad G_R = \frac{K_R}{1 + T_R s} \quad (6)$$

Where K_A, K_E, K_G, and K_R are the gains of amplifier, exciter, generator, and sensor, respectively, and T_A, T_E, T_G, and T_R are the corresponding time constants. As illustrated in Fig. 2, using exponential terms, the total of measurement and communication delays (τ) is placed in the feedback part. The characteristic equation of the excitation control system can be easily obtained from:

$$\Delta(s,\tau) = 1 + e^{-\tau s}G_R(s)G_C(s)G_A(s)G_E(s)G_F(s) \quad (7)$$

As

$$\Delta(s,\tau) = P(s) + Q(s)e^{-\tau s} = 0 \quad (8)$$

P(s), Q(s) are polynomials in s with real coefficients given in the following:

$$P(s) = s^\lambda(1 + T_A s)(1 + T_E s)(1 + T_G s)(1 + T_R s) \quad (9)$$

$$Q(s) = K_A K_E K_F K_R e^{-\tau s}(k_p s^\lambda + k_i + k_d s^{\mu+\lambda}) \quad (10)$$

A necessary and sufficient condition for the system to be asymptotically stable is that all the roots of the characteristic equation of (8) lie in the left half of the complex plane. The main objective of this section is to present a new method for the evaluation of stability and determination of the unstable roots of a fractional order delay system. In this approach, we are trying to calculate the amount of where there exists a crossing of poles through the imaginary axis.

B. Delay Margin Computation

In order to find the points on the imaginary axis, where the crossing takes place, the auxiliary variable described in the previous section and the bilinear substitution are used and as a result, the transcendental characteristic equation is converted into an algebraic equation. System (8) will be asymptotically stable if, and only if, all its roots are on the open left-half complex plane. Characteristic equation (8) is transcendental. To simplify this characteristic equation, the transcendental term is substituted by substitution [7], which has been defined as follows:

$$e^{-\tau s} = \frac{e^{-\tau s/2}}{e^{\tau s/2}} = \frac{1 - jT}{1 + jT} \qquad (11)$$

It is important to note that, this substitution is an exact expression of $e^{-\tau s}$ for purely imaginary roots $s = j\omega$. By examining the amplitude and phase of equation (11), the relationship between T and τ^* can be obtained as follow:

$$\tau^* = \frac{2(\tan^{-1}(T + k\pi))}{\omega}, \qquad k = 0,1,2,\dots \qquad (12)$$

By substituting (11) into (8), we get:

$$\Delta(s,\tau) = (1 + jT)P(j\omega) + Q(j\omega)(1 - jT) = 0 \qquad (13)$$

This equation is an algebraic expression in s of which the coefficients are polynomial functions of T. The real and imaginary parts of the equation (14) should be separated as follows:

$$Re(\Delta(s)) = Re((1 + jT)P(j\omega) + Q(j\omega)(1 - jT)) = 0$$

$$Im(\Delta(s)) = Im((1 + jT)P(j\omega) + Q(j\omega)(1 - jT)) = 0 \qquad (14$$

Now, the set of all ω and T which makes both equations (14) zero can be obtained. Then for every ω and T, delay margin is determined correspondingly (11).

IV. DESIGN OF $PI^\lambda D^\mu$ PSO-CONTROLLER

A. Particle Swarm Optimization (PSO)

PSO was developed in 1995 by James Kennedy (social psychologist) and Russell Eberhart (electrical engineer) [8]. It uses a number of agents (particles) that constitute a swarm moving around in the search space looking for the best solution. Jth particle (x_j) is treated as a point in a N-dimensional space $(x_j = x_{j,1}, x_{j,2}, \dots, x_{j,N})$ which adjusts its "flying" according to its own flying experience as well as the flying experience of other particles. Each particle keeps track of its coordinates in the solution space which are associated with the best solution (fitness) that has achieved so far by that particle. This value is called personal best, $pbest_j = (pbest_{j,1}, pbest_{j,2}, \dots, pbest_{j,N})$ is previous position of the jth particle in a N-dimension space. Another best value that is tracked by the PSO is the best value obtained so far by whole particles. This value is called global best (gbest). The basic concept of PSO lies in accelerating each particle toward its pbest and the gbest locations, with a random weighted acceleration at each time step. The modification of the particle's position can be mathematically modeled according the following equations:

$$v_{j,N}^{(t+1)} = \omega v_{j,N}^{(t)} + c_1 r_1 (pbest_{j,N} - x_{j,N}^{(t)}) \qquad (15)$$
$$+ c_2 r_2 (gbest - x_{j,N}^{(t)})$$

$$x_{j,N}^{(t+1)} = x_{j,N}^{(t+1)} + v_{j,N}^{(t+1)} \qquad (16)$$

$$v_{j,N}^{(t+1)} = \begin{cases} V_N^{max} & v_{j,N}^{(t+1)} > V_N^{max} \\ V_N^{min} & v_{j,N}^{(t+1)} < V_N^{max} \end{cases} \qquad (17)$$

$$x_{j,N}^{(t+1)} = \begin{cases} x_N^{max} & x_{j,N}^{(t+1)} > x_N^{max} \\ x_N^{min} & x_{j,N}^{(t+1)} < x_N^{max} \end{cases} \qquad (18)$$

Where $j = 1,2,\dots,n$ and $N = 1,2,\dots,m$ and n number of particles in the population (population size); m dimension of problem (number of members in a particle) that there is five $(k_p, k_i, k_d, \lambda, \mu)$; t pointer of iterations (generations); $v_{j,N}^{(t)}$ velocity of particle j at iteration t, $V_N^{min} < v_N^{(t)} < V_N^{max}$; where V_N^{min} and V_N^{max} represent the lower and upper bounds of member N of the particle j respectively; c1, c2 acceleration factors; r_1, r_2 uniformly distributed random numbers between 0 and 1; $x_{j,N}^{(t)}$ Current position of particle j at iteration t; $x_N^{min} < x_{j,N}^{(t)} < x_N^{max}$ where x_N^{min} and x_N^{max} represent the lower and upper bounds, respectively, of member N of the particle j; $pbest_{j,N}$ personal best position of particle j ; gbest global best position of swarm; ω is weighting function computed from following equation [8]:

$$\omega = \omega_{max} - \frac{\omega_{max} - \omega_{min}}{iter_{max}} \times iter \qquad (4)$$

ω_{max} and ω_{min} are lower and upper band of weighting function respectively. $iter_{max}$ is maximum iterations of PSO and iter is number of accomplished iteration until right now. This decreasing helps the convergence of algorithm.

B. Purposed Fitness Function

We define a fitness function as follow:

$$J(k_p, k_i, k_d, \mu, \lambda, \tau^*) = e^{-\beta}(T_s + T_r)$$
$$+ (1 - e^{-\beta})(ITSE + M_p) + 10e^{-\frac{2.3025(\tau^* - \tau)}{\tau}} \qquad (20)$$

Where T_s settling time, T_r rise time, M_p overshoot, β is weighting factor chosen 1.5 in practical usages. ITSE is integral of time multiplied square error criterion given by:

$$ITSE = \int_0^{t_{sim}} te^2(t)\, dt \qquad (21)$$

Where t_{sim} total simulation is time and $e(t) = v_{ref} - v_s$ is the tracking error. τ is amount of delay in system and τ^* is delay margin of the system. Our goal is obtaining the best step response. We minimize $J(k_p, k_i, k_d, \mu, \lambda, \tau^*)$ using PSO algorithm and obtain optimum parameters of $PI^\lambda D^\mu$ PSO-controller. To ensure the stability of closed loop, the final cost function is defined with a penalty function $P(k)$ given as:

978-1-5386-6624-1/18 $31.00 © 2018 IEEE

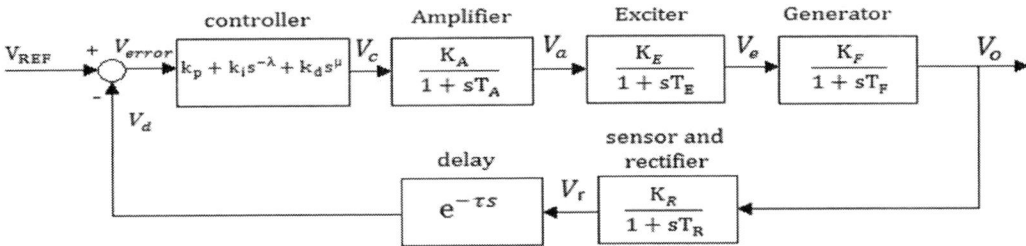

$$P(k) \qquad (22)$$
$$= \begin{cases} 10^5 \times ITSE & \text{if } k \text{ is unstable} \\ J(k_p, k_i, k_d, \mu, \lambda, \tau^*) & \text{else} \end{cases}$$

This penalty function helps to Converge answers of algorithm when response of the system is unstable.

C. Controller design steps

The design steps are expressed as follows:

(1) Randomly initialize the particles of the population including searching points and velocities.

(2) For each initial particle k of the population, calculate the values of the performance criterion in (22).

(3) Compare each particle's evaluation value with its personal best $pbest_{j,N}$. The best evaluation value among $pbest_j$ is denoted as gbest.

(4) Modify the member velocity of each particle according to equations (15) and (17).

(5) Modify the member position of each particle j according to equations (16) and (18) .

(6) If the number of iterations reaches the maximum, then go to Step 7, otherwise go to Step 2.

(7) The latest gbest is the optimal controller parameter.

After defining transfer function of the system, PSO algorithm designs a random FOPID controller for AVR system and approximates s^λ and s^μ by Oustaloup approximation. Then computes penalty function for step response of the synchronous generator system. Afterwards, second controller is designed by PSO and penalty function for step response of second controller is computed. If penalty function of new controller was less from previous controller, memorize new controller as optimal controller and finds better controllers. This Process repeats until PSO algorithm fetch up. Finally best controller is given as optimal controller.

V. NUMERICAL AND SOLUTION RESULTS

A. Study system parameters

A practical high-order AVR is used to verify the efficiency of the proposed FOPID controller. The system parameters are: $K_A = 10$, $K_E = 1$, $K_G = 1$, $K_R = 1$, $T_A = 0.1$, $T_E = 0.4$, $T_G = 1$, $T_R = 0.01$. inspired from

practical requirements, the lower bounds of the five controller parameters are zero and their upper bounds are set to two. The following Parameters are used for carrying out the FOPID design using for carrying out the FOPID design using PSO:

- The members of each particle are $k_p, k_i, k_d, \mu, \lambda$ (m=5).

- Population size = 40.

- Inertia weight factor ω is set   where $\omega_{max} = 0.09$ and $\omega_{max} = 0.4$

- The limit of change in velocity is set as:

$$V_{max} = \alpha(x_{max} - x_{min})$$
$$V_{min} = -V_{max} \qquad (23)$$

α is the factor of velocity step and is chosen 0.1 in practical requirements.

- Acceleration constants $c_1 = c_2 = 2$.

- Maximum iteration is set to 200.

- ω_l And ω_h in (4)–(5) are set to 10^{-3} and $10^3 \, rad/s$, respectively.

- Order of approximation in (4) is set to $N = 8$.

B. Designed FOPID controllers Representation

We have supposed 0.7s delay for generator system and designed FOPID controller. Coefficients of the FOPID controller are represented in table 1. In order to emphasize advantages of the proposed FOPID controller, a PID Controller is also designed using the same method. Characteristics of step response for FOPID and PID controllers are represented in table 2. A Comparison between Step response and penalty function of FOPID and PID controller shows than FOPID controller has better performance and optimal response than PID controller. Also delay margin of the system is more than three times of delay in system.

C. Simulation results

In this section, we examine the performance of PSO-FOPID. Terminal voltage step response of the AVR system controlled by optimum controllers and without controller is shown in Fig3 when delay of system (τ) is equal to 0.7s. It can be seen from Figure that the step response of the AVR system controlled by PSO-FOPID controller has very good

TABLE I. PERFORMANCE INDICES OF THE AVR SYSTEM WITHOUT CONTROLLER AND WITH PSO-FOPID AND PSO-PID

performance.

controller	$\tau(delay\ of\ system)$	k_p	k_i	k_d	λ	μ
FOPID	0.7s	0.1512	0.789	0.061	1	1.2
PID	0.7s	0.1282	0.06741	0.06742	1	1

TABLE II. THE OPTIMUM PSO-FOPID AND PSO-PID CONTROLLER PARAMETERS

controller	$\tau(delay\ of\ system)$	T_r	T_s	$M_p(\%)$	$\tau^*(delay\ margin)$	$penalty\ function$
without	0	0.27	6.99	50.6	0.064	41.707
FOPID	0.7s	0.83	1.21	0.133	1.95	0.81
PID	0.7s	0.83	3.77	0.002	2.36	1.203

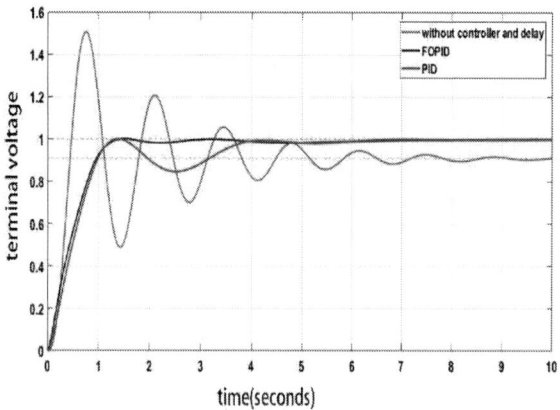

Fig. 3. Terminal voltage step response of an AVR system without controller and delay, with FOPID and PID controller for $\tau = 0.7s$

VI. CONCLUSION

This paper presents a design method for determining the FOPID controller parameters using PSO algorithm. The proposed method involves a new frequency-domain performance criterion. Application of the method to a practical AVR shows that the proposed algorithm can perform an efficient search for the optimal FOPID controller parameters. Furthermore, it can be concluded from the above simulations that the proposed FOPID controller has better performance characteristics than the PID controller applied to the AVR. Finally the AVR system has acceptable delay margin witch increases system robustness versus increasing time delay.

REFERENCES

[1] S. Ayasun, U. EminoLlu, and F. Sönmez, "Computation of Stability Delay Margin of Time-Delayed Generator Excitation Control System with a Stabilizing Transformer" Nigde University, Turkey, 13 May 2014.

[2] S. Ayasun, A, Gelen, "Stability analysis of a generator excitation control system with time delays", Nigde University, Turkey, 18 December 2009.

[3] Y. Chen, I. Petr´a˘s and D. Xue "Fractional Order Control - A Tutorial" American Control Conference- June 10-12, 2009.

[4] A. Oustaloup, X. Moreau and M. Nouillant, "The CRONE suspension" .Control Engineering Practice, 4(8), 1101–1108.

[5] Saadat H "Power system analysis". McGrawHill Inc,New York 1999

[6] P. Kundur. "Power system stability and control" .McGraw-Hill Inc, New York ,1994.

[7] M. Ali. Nekoui , M. Ali. pakzad "Stability Analysis Of Time Delayed Fractional Order Generator Excitation Control System" K.N.T.U Univercity,Tehran October 2015.

[8] J. Kennedy, R. C. Eberhart, " Particle swarm optimization. In Proceedings of IEEE international conference on neural networks".Perth, Australia, pp. 1942–1948 ,1995.

Investigation of Effect of Charcoal Particle Size on Earth's Resistance

Meghna R Yashwante
Department of Engineering Sciences
Marathwada MitraMandal's Institute of Technology,Lohgaon, Pune, India
ymeghna@gmail.com

P. B.Karandikar
Department of Electronics And Telecommunication Engineering
Army Institute of Technology, Pune, India
pbkarandikar@gmail.com

N.R.Kulkarni
Department of Electrical Engineering
P.E.S's MCOE
Pune, India
nrkmcoe@gmail.com

Sushil B. Dhembare
Department of Electronics And Telecommunication Engineering
Army Institute of Technology, Pune, India
sushildhembare95@gmail.com

Abhijit B. Bhosle
Department of Electronics And Telecommunication Engineering
Army Institute of Technology, Pune, India
abhijit.bhosle11@gmail.com

Abstract— **Earthing is vital to achieve equipment as well as personnel protection. Moreover, the stability of any electrical or electronic system is greatly affected by the condition of earthing. The paper includes investigation of effect of charcoal size on earth resistance. Earth resistance value depends on many factors such as electrode, back fill and soil in which earthing is done. Each of these parameters has sub-parameters on which earthing depends. Due to some unsuitable soil conditions at the site of earthing, meeting the requirement of the earth resistance according to the application becomes difficult. To overcome this problem, backfill materials such as charcoal salt mixture is used to acquire the required value of the earth resistance. The value of earth resistance depends on the size of charcoal. This paper presents mathematical modeling involving variation of the earth resistance with the change in the size of charcoal used as a backfilling material in the process of earthing.**

Keywords—Earthing, Charcoal, Particles and Soil

I. INTRODUCTION

Safety is major concern when dealing with electricity. Earthing is used to achieve safety for people and instruments. Good earthing system must have low earth resistance. Factors which affect earth resistance most are earthing electrode and soil conditions around earth electrode [1-2]. Electrode resistance is lowered by selection of suitable material, increasing depth, structure change or by parallel connection of more electrodes. Contact resistance is lowered by compact soil and electrode structure. Deep driven electrodes or soil treatments are done to reduce earth resistance [3-9]. In case of hard, rocky soil deep driven electrodes are not viable option. In such cases soil improvements like backfilling are done [9-11]. Researchers have found several different methods to lower down the earth resistance in an unfavorable condition few methods among them are lengthening the earth electrode in the earth, suggested use of multiple rods and by the treatment of the soil with chemicals. There are some drawbacks of these methods. Lengthening the earth rod is possible to some extent; it becomes difficult in presence of the rock beneath the surface of soil. Use of multiple rods reduces the earth resistance to some extent. Minimum separation between the rods for reduction in the earth resistance should be twice the length of the rod underground for proper discharge of rod. So, large area would be required to meet the required earth resistance. Use of multiple rod technique is not suitable if very less area is available for earthing. The next method that is treatment of soil with chemicals has advantages that it avoids the variation of the earth resistance due to periodical wetting and the drying of the soil around the earthing. But chemical treatment has got its own drawbacks. Chemicals gradually wash away due to adequate rainfall and the porosity of the soil, chemical is to be replaced after a particular period so the period of the replacement of the chemical varies depending on these parameters. Chemical treatment is expensive, sometimes cause corrosion of electrodes. Chemicals may leach and give different results in dry condition. To lower soil resistance most common methods are by adding alternate layers of salt, charcoal. Various other additives like bentonite powder, dead - sea water, various salts magnesium sulfate or copper sulfate are mixed with soil [13-18]. Variation in the size of the backfill material used in earthing would have significant change in the value of earth resistance. Charcoal is one of the important components of backfill material used in earthing. The purpose of this study is to predict the behavior of the ground resistance of a single rod by varying charcoal size.

In this paper the first section comprises of detailed information about the backfill material which is followed by the second section consisting some information about the backfill material and its importance. The third section explains about different types of charcoal, their properties and its effect on earthing. The fourth section consists of experimental details regarding change in the ground resistance with the variation in the size of charcoal. The same section consists of a diagram representing the different layers used for earthing, a part of this section also explains about the method employed to measure the earth's resistance and finally has details about the results obtained for the experiments being conducted. The fifth section consists of the conclusions followed by acknowledgement and references.

II. BACKFILL MATERIALS

Backfill materials have an extensive importance in the field of construction as well as electrical earthing. Backfill materials are used to fill the voids after any excavation, where this material is used to decrease the resistivity of the system where earthing system are to be installed. Backfill material provides stability to the soil and low resistance path between earthing electrode and surrounding earth masses [19].

Charcoal is among one of the major elements used as a backfill in the earthing. Charcoal contains carbon element. Plants, animals and other organisms have abundant carbonic compounds. These carbonic compounds are ultimately converted to other carbonic compounds when they die. When water and other volatile substances are removed from the carbonic compounds, charcoal is produced. Majorly pyrolysis is used to produce charcoal where organic materials are decomposed at high temperatures in the absence of oxygen. So, the physical phase of the matter and chemical compositions will change rapidly.

Material charred and the charring temperature decides the properties of the charcoal produced. Charcoal properties depend on varying amounts of hydrogen and oxygen as well as ash and other impurities, the structure.

The approximate composition of charcoal is sometimes empirically described as C_7H_4O. Carbon atoms are arranged in layers made up of in layers made up of hexagonal rings with interstitial spaces in form of pores. This porous nature of charcoal increases the cation exchange capacity of the soil thus increasing conductivity.

Apart from the fact that carbon has pores that retain water in the soil to increase conductivity, when mixed with sodium chloride in the soil, various reactions take place to produce the very hygroscopic $NaNO_3$. These reactions are shown below -

$$C\ (s) + O_2\ (g)\ \rightarrow\ CO_2\ (g) \qquad (1)$$

Oxidation of sodium to form sodium oxide
$$Na\ (s) + O_2\ (g)\ \rightarrow\ Na_2O\ (s) \qquad (2)$$
Oxidation of sodium to form sodium oxide
$$Na_2O\ (s) + H_2O\ (l)\ \rightarrow\ 2NaOH\ (s) \qquad (3)$$

Reaction of sodium hydroxide and carbon dioxide to form sodium carbonate

$$2NaOH\ (s) + CO_2\ (g)\ \rightarrow\ Na_2CO_2\ (s) + H_2O\ (l) \qquad (4)$$

$CaNO_3$ salt is present in soil which reacts with sodium carbonate to produce
$$Na_2CO_3\ (s) + CaNO_3\ (aq)\ \rightarrow\ CaCO_3\ (aq) + 2NaNO_2\ (s)\ (5)$$

Double decomposition reaction between sodium carbonate and calcium nitrate to produce calcium carbonate and sodium nitrate is as follows -
$$2Cl^-\ (aq)\ \rightarrow Cl_2\ (g) + 2e^- \qquad (6)$$

The cathode will accept hydrogen ions from water
$$2H^+\ (aq) + 2e^- \rightarrow H_2\ (g) \qquad (7)$$

Copper gives up 2 electrons into solution which will result in turning electrolyte into blue/green, the released electrons help in the conductivity.
$$Cu\ (s)\ \rightarrow 2e^- + Cu_2 +\ (aq) \qquad (8)$$

Therefore, the salt water will move the charged ion to the rod of the salt water and will move its charged ion to the grounding rod since the copper grounding rod has the ability to attract charged particles. Thus, charcoal is used to retain the moisture for long duration because it is an adsorbent and salt is added to increase the conductivity both these products help to bring soil resistance down [28]. Backfill materials are also called as the earth enhanced compounds provide several different features. They are as follows:

- They are not soluble and are part of the earth pit
- They are hydroscopic in nature
- They enhance the performance of electrode - Electrolytic Gel solution
- They maintain moisture and enhances conductivity around the electrode
- They leach into the normal soil
- They reduce soil resistivity around the electrode [29]

III. TYPES OF CHARCOAL AND THEIR PROPERTIES

This section consists of information about the various types of charcoal and their properties. The properties of various charcoals were compared and the best that can be used for earthing was identified. For the properties mentioned for various charcoals in this section refer Table 1 of this section.

A. Lump Charcoal

It is the original charcoal, made by burning trees or logs in a kiln, sealed cave, or underground. Unlike briquettes, lump charcoal is pure wood free of binders or petroleum-based accelerants [20]. The ash content in this type of charcoal is less as compared to briquettes charcoal. It has ash content approximately equal to 1.5%, a moisture content of about 7.5% and fixed carbon content of about 75% of its total composition. Due to such high moisture content as compared with other types, lump charcoal provides less resistance to electricity flow.

B. Charcoal Briquettes

Traditional briquettes contain wood scraps, sawdust, coal dust, borax, and petroleum binders. It has ash content of about 1.5%, moisture content approximately as less as 4% and a fixed carbon content of about 68% of the total composition. It has less moisture content as compared with lump charcoal, so the resistance provided by it is higher as compared to lump charcoal.

C. Binchotan

Japan's super premium lump charcoal binchotan is used in top yakitori parlors throughout Japan and in the United States. *Binchotan* is traditionally made from oak in mud-sealed caves in southwest Japan. It has ash content of about 2%, moisture content approximately equal to 3.5%, and a fixed carbon content of about 81% of its total composition. It has the moisture content less than lump and briquettes charcoals, so the resistance provided by it is higher as compared with both of them. That is why this charcoal is less preferred in the earthing. While on the other hand it has many other applications in industrial and household work.

D. Coconut Shell Charcoal

The drum method and pit method are the most used for manufacturing this type of charcoal. It is made by burning coconut shells in a limited supply of oxygen as oxygen could destroy shells if the air will not be limited. Clean, fully dried and mature shells should be used in order to get high quality charcoal. It has a moisture content of about 4%, ash content approximately equal to 1.5% and a fixed carbon content equal to 83% of its total composition. Because of its less moisture content as compared with lump charcoal so, it provides higher resistance compared to it.

TABLE I.   CHARCOAL PROPERTIES

Charcoal Type	Properties				Price Rs/kg
	Moisture %	Ash %	Fixed Carbon %	Resistance	
Lump Charcoal	7.5	1.5	75	Highest	12 - 55
Charcoal Briquettes	4	1.5	68	Medium	9 - 19
Binchotan	3.5	2	81	Low	18 - 25
Coconut Shell Charcoal	4	1.5	83	Medium	20 - 45

From the above discussion and Table it is clear that among all these charcoals for earthing purpose Lump charcoal has the highest moisture content and hence the least resistance. Hence, lump charcoal was used in the experiments performed for earthing.

IV. EXPERIMENTAL DETAILS AND RESULTS

For investigation the outdoor experiments were carried out. Different size of charcoal was used for experimentation. Charcoal was crushed and separated. Size was categorized as small, medium and large. GI rod was used as earth electrode. Pit was filled with alternate layers of charcoal and salt. Various pits had different composition of backfill materials. Water was poured additionally and allowed to settle for 24 hours. And then the resistance of every pit was measured using the three-point method. Resistance of each pit was measured on various days and their mean was taken as the final value.

The pit size used for experiment was about 1feet in length, width and height. But for actual earthing the size of pit used is about 5 feet in length, width and height. Therefore, resistance offered while experimentation was higher as compared to original earthing system. The resistance from experiments can be compared with the resistance obtained in original earthing by use of the scaling factor. The scaling factor for the experiment was 1/8. Therefore, to compare the readings with original earthing this scaling factor should be multiplied with the actual obtained resistance in this experiment.

Every earthing pit had five different levels of Backfill material surrounding a GI electrode. The first layer was a mixture of charcoal and soil. The second layer was salt layer. The third level had charcoal. Again, the fourth level had salt. The top layer had fine crushed charcoal mixed with soil. The layer diagram of pit is shown below in the Fig 1.

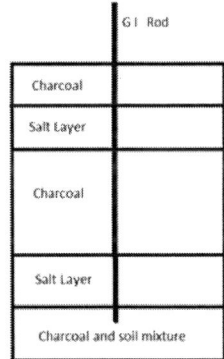

Fig1. Various layers in earthing pit

Three point method to measure earth's resistance –

The three-point or fall of potential method is the technique carried out to determine the ground resistance. Refer the Fig 3 given in this section to understand the method from following lines of explanation. This method uses three poles where the potential difference is measured on points P1 and P2 while current is applied to points P1 and P3. From Ohm's Law, E=RI the ground stake resistance can be obtained [21-23].

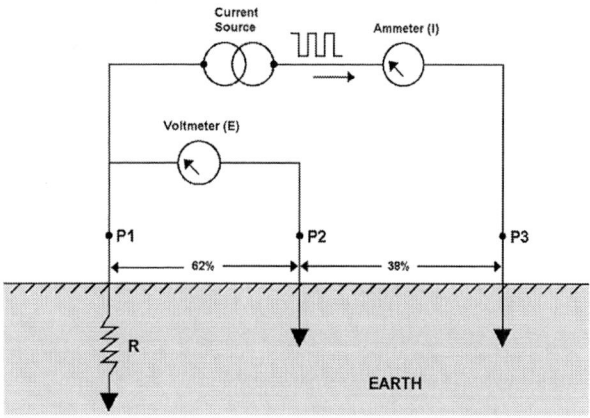

Fig 2. Three-point method circuit diagram to measure earth's resistance

The proper spacing of the electrodes is the key in precise ground resistance measurement. The auxiliary electrode P3 should be far enough from the ground pole P1 under test. This is to ensure that the potential probe P2 is outside the effective areas of resistance of both the auxiliary current rod P2 and ground rod P1. To determine if the potential rod P2 is outside an effective resistance area, move between P1 and P3 and take a reading at each point. The values should not significantly vary and should be close to each other [30].

The earth's resistance was measured using three-point method throughout the trials and various experiments. The earth resistance was measured on different days. The different pits had different composition and size of charcoal used. The resistance was measured on different days for every pit and the mean value was taken as the resistance offered by the pits. The graphs in Fig 3 and Fig 4 shows the resistance measured for different pits containing different sized charcoal.

The size of charcoal used for the experimental purpose varied from 2mm to 50mm. This was classified into various categories as small, medium and large charcoal. The charcoal particles from the range of 2mm to 8mm were considered as small sized charcoal. The charcoal particles in the range of 9mm to 28mm were considered medium sized charcoal and the particles in the range of 29mm to 50mm were considered large sized charcoal. Accordingly, the charcoals were separated and were used in the experiment.

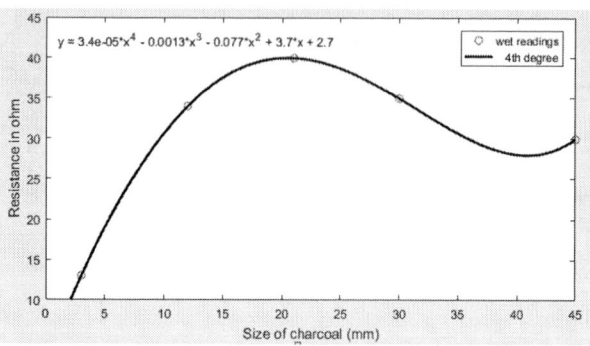

Fig 3. Variation of earth resistance with charcoal particle size for wet pits

The Fig 3 graph shows the resistance measured of the pits with different charcoal sized backfill. These readings were taken after 24 hours of watering the pits. It was observed that the equation representing the curve was of fourth order. The obtained equation is –

$$y = -0.5x^4 - 0.0013x^3 + 3.7x + 2.7 \tag{9}$$

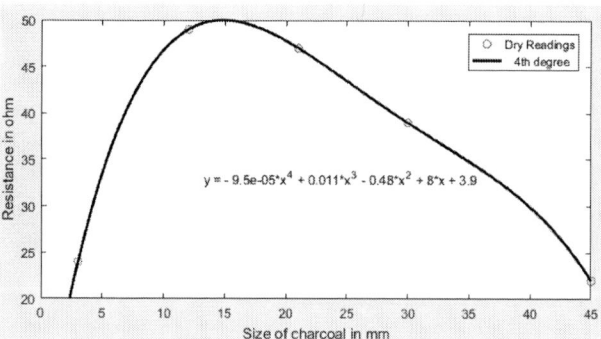

Fig 4. Charcoal Size Vs Resistance Graph for dry pits

These readings were taken without watering the pits. It was observed that the equation representing the curve was of fourth order. The obtained equation is –

$$y = -0.5x^4 + 0.011x^3 - 0.48x^2 + 8x + 3.9 \tag{10}$$

It was observed that the reason for less value of earth resistance for small charcoal particle size is due to formation of region which will act like several resistors connected in parallel. Therefore, the equivalent resistance will be a small value. Accordingly, the resistance offered by the large particle size was less due to the fact that it provides larger area for the current to flow and discharge into soil. Hence the earth resistance is less for both small as well as large particle size of the charcoal. Fig 5 shows the earthing for large charcoal particle size.

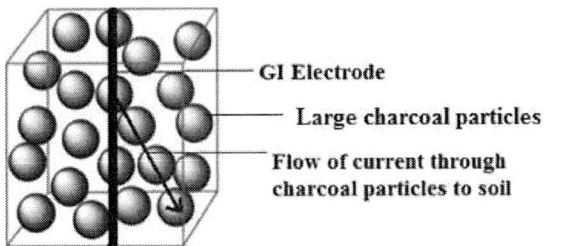

GI Electrode

Large charcoal particles

Flow of current through charcoal particles to soil

Fig 5. Flow of current through large charcoal particles in pit to earth

V. CONCLUSIONS

Earthing resistance and soil resistivity are the main parameters characterizing electrical properties of the earthing system. Earthing is a multidimensional design as electrode geometry, soil conditions, and climatic conditions are important parameters to be considered, with the environmental issues.

The experiment on effect of size of coal particle on the earth's resistance showed that resistance offered by different sized coal particle is different. Various sized coal particles can be used as per the requirement to reduce the resistance offered by the soil of a particular area.

Observations were out carried on a pit of smaller size as compared to the standard size of the pit required for the earthing. Required standard value of the earth resistance can be obtained by multiplying the resistance by scaling factor. It has been observed that the relation between the resistance and the size of charcoal is non-linear. Resistance offered by earthing in case of dry readings are higher as compared to wet readings. As in soil under dry condition the moisture content is low. Resistance offered by earthing in case of wet readings are low as the moisture contained of the soil is high.

Plot of resistance with respect to size of charcoal showed a nonlinear relationship, the curve exhibits equation of order four. Relation of charcoal particle size and earth resistance was found to be non-linear. The relationship was expected to be linear theoretically but in practical case it was obtained to be non-linear. It was observed that wet pits offered less resistance as compared to dry pits. Also, it was observed that the small as well as extra-large size charcoal are more suitable for earthing as the resistance offered by both is less as compared to medium sized charcoal.

ACKNOWLEDGEMENT

Authors thanks management of Army Institute of Technology for providing research facilities. Authors would also thank Savitribai Phule Pune University for providing funding support for this research work.

REFERENCES

[1] Standard IS 3043," Code of practice for earthing", 1987.

[2] Standard IEEE 80-2000," IEEE Guide for Safety in AC Substation Grounding",1986.

[3] Sundaravaradan N. A. and M. Jaya Bharata Reddy. "How Is Earthing Done?" IEEE Potentials 37(2):42-46March 2018,pp.42-46.

[4] Ahmeda, M., N. Ullah, N. Harid, H. Griffiths and A. Haddad, "Current and voltage distribution in a horizontal earth electrode under impulse conditions, "44th International Universities Power Engineering Conference (UPEC) IEEE", 2009,pp.1-4.

[5] Elmghairbi, A., M.Ahmeda, N. Harid, H. Griffiths and A. Haddad. "A technique to increase the effective length of horizontal earth electrodes and its application to a practical earth electrode system,"7th Asia-Pacific International Conference on Lightning IEEE", 2011, pp.690-693.

[6] Yamane, Hiroshi, Tsuyoshi Ideguchi, MasamitsuTokuda and Hiroaki Koga. "Long-term stability of reducing ground resistance with water-absorbent polymers," IEEE International Symposium on Electromagnetic Compatibility", 1990, pp.678-682.

[7] Ufer, H. G. "Investigation and testing of footing - type grounding electrodes for electrical installations." IEEE Transactions on Power Apparatus and Systems 83, no. 10, 1964, pp.1042-1048.

[8] Hallmark Clayton L., "Horizontal strip electrodes for lowering Impedance to ground", 19th International Telecommunications Energy Conference (INTELEC) IEEE, 1997, pp.368-375.

[9] Morgan, P. D. and H. G. Taylor, "The resistance of earth electrodes", Journal of the Institution of Electrical Engineers72, no. 438, 1933, pp.515-518.

[10] Meng, Qingbo, Jinliang He, F. P. Dawalibi and J. Ma. "A new method to decrease ground resistances of substation grounding systems in high resistivity regions." IEEE Transactions on Power Delivery 14, no. 3, 1999, pp.911-916.

[11] HeJinliang, Gang Yu, Jingping Yuan, RongZeng, Bo Zhang, Jun Zou, and Zhicheng Guan. "Decreasing grounding resistance of substation by deep-ground-well method", IEEE transactions on power delivery 20, no. 2, 2005, pp.738-744.

[12] El-Tous, Yousif, and Salim A. Alkhawaldeh. "An Efficient Method for Earth Resistance Reduction Using the Dead Sea Water", Energy and Power Engineering 6, 2014, pp.47.

[13] Manikandan, P. "Characterization and comparison studies of Bentonite and Flyash for electrical grounding", Electrical, Computer and Communication Technologies (ICECCT) IEEE International Conference, 2015, pp.1-4.

[14] Rowland Phillip W. "Industrial system grounding for power, static, lightning and instrumentation, practical applications", Textile, Fiber and Film Industry Technical Conference IEEE, 1995, pp.1-6.

[15] Kostic M. B., Z. R. Radakovic, N. S. Radovanovic and M. R. Tomasevic-Canovic, "Improvement of electrical properties of grounding loops by using bentonite and waste drilling mud", IEEE Proceedings-Generation, Transmission and Distribution146, no. 1, 1999,pp.1-6.

[16] MaJinxi and Farid P. Dawalibi. "Effect of Backfill on the Performance of Substation Grounding Systems", Asia-Pacific Power and Energy Engineering Conference (APPEEC) IEEE, 2012, pp.1-4.

[17] Chen S-D., "Granulated blast furnace slag used to reduce grounding resistance." IEE Proceedings-Generation, Transmission and Distribution 151, no. 3, 2004,pp.361-366.

[18] Eduful George, Joseph Ekow Cole and F. M. Tetteh, "Palm kernel oil cake as an alternative to earth resistance-reducing agent", Power Systems Conference and Expositions IEEE, 2009, pp.1-4.

[19] CoelhoVilsonLuiz, AlexandrePiantini, Hugo AD Almaguer, Rafael A. Coelho, Wallace do C. Boaventura and José Osvaldo S. Paulino, "The influence of seasonal soil moisture on the behavior of soil resistivity and power distribution grounding systems", Electric Power Systems Research 118, 2015, pp.76-82.

[20] Taulbee Darrell Neal, "Method for producing fuel briquettes from high moisture fine coal or blends of high moisture fine coal and biomass", U.S. Patent Application 12/704,895, filed August 18, 2011.

[21] IEEE Guide for Measuring Earth Resistivity, Ground Impedance and Earth Surface Potentials of a Grounding System, IEEE Std. 81, 2012.

978-1-5386-6624-1/18 $31.00 © 2018 IEEE

[22] Blattner C. J., "Study of driven ground rods and four point soil resistivity tests", IEEE Transactions on Power Apparatus and Systems 8, 1982,pp.2837-2850.

[23] "Getting Down to Earth" (A practical guide to Earth Resistance Testing), Megger Publication.

Design and Analysis of High Performance Line Started Permanent Magnet Synchronous Motor

Arvind Kumar
Dept.of Electrical Engineering, *G.B.P.U.A.&T,*
Pantnagar, US Nagar, Uttarakhand, India
arvindee69@gmail.com

Ajay Srivastava
Dept. of Electrical Engineering
G.B.P.U.A.&T, Pantnagar, US Nagar, Uttarakhand, India
drajay16@gmail.com

Abstract– **This paper considers the design and performance investigation of a novel line started permanent magnet synchronous motors (LSPMSM) including transient as well as steady state performance with the help of advanced finite element analysis (FEA) using ANSYS Maxwell 2D for the design. Here, various steady state as well as transient characteristics such as speed, starting torque and load torque characteristics are used to decide the performance of the machine.**

Keywords- finite element analysis, speed, starting torque and load torque characteristics, ansoft maxwell 2-D.

I. INTRODUCTION

Energy is a prerequisite to economic development of any country. Energy consumption of a country is the indicator of its development. Electricity is one of the greatest technological innovations of mankind. Due to simple and robust construction, easy to manufacture and line start capability, the three phase induction motors are widely utilized in various applications, such as pumps, fans and compressors. On the other hand, some disadvantages of small rated power induction motors are small efficiency and power factor. Contrarily, permanent magnet synchronous motors (PMSM) are distinguished by higher torque per unit current, efficiency and power factor. The main disadvantage of PMSM is their inability to line-start from zero speed under load. Permanent magnet synchronous motors are, therefore, equipped with different position sensors and fed from current controlled voltage converters.

In 1955, a first design of line LSPMSM was suggested by F. W. Merrill. But poor magnet properties at that time prevented the immediate commercial use of Merrill's promising idea. After this initial work, several research groups, leading motor manufacturers followed the path stated by Merrill. Those days, ferrite and alnico alloys were the main constituents of permanent magnets. These materials suffered from various problems and during operation, they get easily demagnetized due to strong non-linear demagnetization. With each major advance in PM material technology, interest in these motors has increased. LSPMSM is a hybrid or combination of an IM and a PMSM and may be called as "induction-start" and "synchronous-run" motor. It may be called as a PMSM with added squirrel cage bars in the rotor for the induction starting. In other words, it is an induction motor with added permanent magnets in the rotor for synchronous running. By combining the two different rotor designs into a single rotor, the negative aspects of both motors are eliminated and a more efficient motor may be obtained which may be used in place of the induction motors for applications like, fan, pumps and compressors.

The role of squirrel cage bars is to provide line started property and dynamic oscillations damping to the motor. This type of motor is capable of line-starting under load; it exhibits higher efficiency as well as higher power factor values when comparing to an induction motor of equal dimensions and rated data; also the position sensor is no longer needed. The NdFeB magnets do not bring about the limitation of large brake torque during synchronization and run-up of PM motors.

The purpose of this research is to develop a high performance LSPMSM by which the induction motors can be replaced for applications, such as pumps, fans and compressor. In this work, main emphasis is on the design of LSPMSM of 5 kW rating which is widely used in many applications. For economic purpose, the stator and windings of the LSPMSM are identical to that of same power rating of induction motor. So, in LSPMSM only rotor configuration is altered as compared to the induction motor.

II. DESIGN PROCEDURE

It is difficult to arrive at the synchronous speed for permanent magnet synchronous motor. The coupled field-circuits technique is used to calculate the transient electromagnetic field of a LSPMSM. The permanent magnet dimensions greatly affect the transient performance of the LSPMSM. A small variation in dimension create problem to achieve synchronous speed by the motor. The structure of stator and rotor laminations and windings are designed. The current, torque and speed curve are obtained. The starting performances are essential for LSPSM.

During starting process, starting torque should be high and starting current should be low without any damage to the electrical equipment. The LSPMSM suggested in this paper meet these specifications.

III. ROTOR CONFIGURATION

There may be various configurations to place the permanent magnet in the rotor. It may be either interior type or surface type motors are available in literature. In this paper, interior type permanent magnet rotor is considered. A squirrel cage winding is placed on the rotor to start the motor as induction motor. Transient performance of LSPMSM mainly depends on characteristics of squirrel cage winding. Line start permanent magnet synchronous motors mainly utilize ferrite or rare-earth magnets to provide synchronous torque. Permanent magnet materials are solely responsible for steady state performance of LSPMSMS. During the selection of a particular permanent magnet material, there should be a balance between transient as well as steady state performance. The stator core and stator winding are identical to that of conventional induction machine or synchronous machine.

Fig. 1 demonstrates the basic structure of the proposed LSPMSM whose specifications, that are decided on the basis of engineering data, are given in Table I.

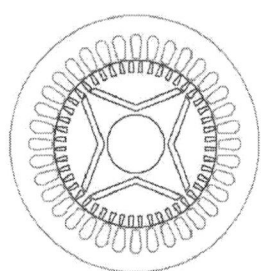

Fig. 1. Schematic diagram of LSPMSM

Table. 1		
Design data of the proposed LSPMSM		
	Output power	5 kW
	Number of poles	4
	Reference speed	1500
	Voltage	415 V
Stator	Outer Diameter	260 mm
	Inner Diameter	170 mm
	Stack Length	90 mm
	Steel type	M19_24G
	Number of Slots	36
Rotor	Outer diameter	169.35 mm
	Inner diameter	60 mm
	Number of slots	48
	Magnet Material	NdFeB35
	Magnet thickness	3 mm
	Magnet width	18 mm

To obtain the rated performance of the motor, a magnetic design of the machine was needed. A theoretical technique may be utilized to predict the flux per pole. As a result, the thickness magnet can be computed.

Since, the permanent magnet can be provided on many places in the rotor, the new magnet type will be different from that of actual theoretical approach. In that situation, the normal theoretical approach might not be so precise. A much valid estimate can be achieved using finite element analysis. Finite element analysis is utilized for simulation of PM rotor of LSPMSMS for different magnet positions.

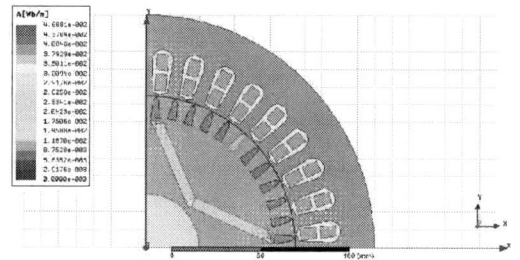

Fig. 2. Schematic of 2-D design model of LSPMSM

Various magnets with varying grade and height are employed. Other variable parameters are the air gap length and magnet pitch in the design. A suitable magnet position was decided on the basis of completion of several iterations. At the end a trade-off was made between both mechanical and performance considerations [17].

IV. MODELING OF LSPMSM USING FEM

Due to the complex rotor configuration of LSPMSM, it is not easy to obtain more accurate designed performance from conventional analysis. Advanced electrical and mechanical finite element analysis (FEA) has to be applied for more accurate result evaluations during the design stage. Normally, more considerations are concentrated on the starting-up performance, cogging torque eliminating, rotor mechanical stress, thermal flow and temperature rise, no-load voltage and harmonics, d- and q-axis reactance, etc.

The magnetic circuit or permeance method is a very useful for calculating approximate magnetic field for devices with simple geometry. But for more precise calculation the computer programs using finite element method are necessary. ANSYS Maxwell software uses the finite element method for design analysis. There is no requirement of any assumption of flux paths or related empirical factors in above mentioned software.

ANSYS Maxwell

ANSYS Maxwell software is one of the commercial electromagnetic field simulation software that uses finite element analysis (FEA) to solve electromagnetic problems. For the solution of electromagnetic field problems Ansys Maxwell software uses Maxwell equations in a finite region with boundary conditions and user-specified initial conditions to obtain a solution with high accuracy. The designer may use RMxprt option by using user defined primitives (UDP) or may create his own design by using 2-D/3-D option for drawing a model. The RMxprt provide user defined primitive's option only for conventional machines. Other than conventional and any design can be created by using 2-D/3-D design options and at the same time the design variables can be declared, Both Maxwell 2-D and Maxwell 3-D use Finite Element Approach for the analysis of design.

A three-phase LS-PMSM 5 kW, 415 V, four-poles is designed using ANSYS Maxwell® software. The transient solver with time integration method using backward Euler is used to compute the quantities of LS-PMSM. Several simulations are carried out for selecting the magnet shape, cage shape, and the dimensions. The cage bar and magnets give the best choices for compromise between starting and synchronous performance.

V. DYNAMIC STARTING PERFORMANCE

For a good design of LSPMSM should satisfy the three main requirements. Firstly, it must provide enough starting torque for full speed range during starting. The transient performance of LSPMSM is very essential to rotate the motor from standstill to synchronous speed. The rotor cage is responsible to achieve asynchronous positive start torque same as in induction motor. On the other hand, the permanent magnet in the rotor produce a negative torque called generator torque, which resist the rotor to speed-up. Also, some other components of torques such as reluctant torque, harmonic torque, etc. may be useful or harmful for the motor to start-up. Here, 2-D FEA models are utilized to investigate the dynamic starting performance. A balanced three-phase voltage at 415 V and 50 Hz is provided to the stator.

For LSPMSM, the equation of motion may be written as,

$$J \frac{d^2\delta}{dt^2} = T_s(\delta) - T_d \frac{d\delta}{dt} - T_L$$

For LSPMSM, the sum of magnet alignment torque and reluctance torque is the synchronous torque. The equation of the synchronous torque is as follow:

$$T_s(\delta) = \frac{3pE_0V}{2\omega_s X_{sd}} sin\delta + \frac{3pV^2(X_{sd} - X_{sq})}{4\omega_s X_{sq} X_{sd}} sin2\delta$$

Fig.-(3) represents the speed versus time curve of LSPMSM. The motor achieve the synchronous speed in

very short time, 40 ms. The stable average speed is 1500 rpm.

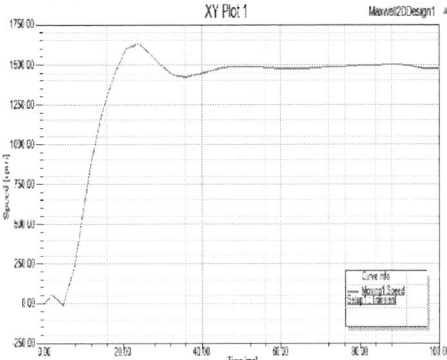

Fig.3- Speed-time characteristics of LSPMSM

Fig.- (4) demonstrates the predicted dynamic responses when the motor is started directly from the supply mains. The simulated speed-time characteristic shows that the motor is pulled into synchronism at around 0.25 s after the start.

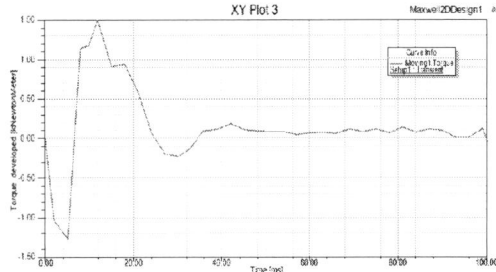

Fig.4- Starting torque-time characteristics of LSPMSM

When the motor is synchronized, it starts running under a steady synchronous torque. At this point, the cage torque becomes zero with the cage acting as a set of damper bars. The torque versus time characteristic is shown in fig, (5).

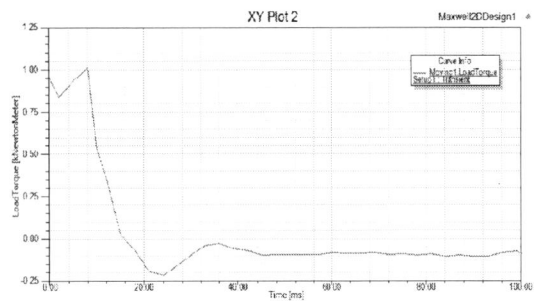

Fig.5- Load torque-time characteristics of LSPMSM

VI. BACK EMF

In LSPMSM, the back EMF is induced due to the presence of permanent magnets. It is a key parameter in meeting the simultaneous requirements. The mechanical transients are derived using the phase terminal voltage. The profile of the phase back EMF is shown in fig. (6).

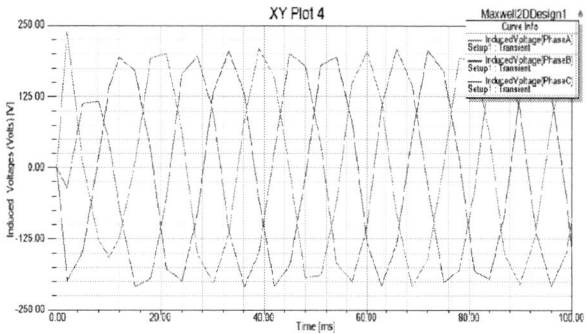

Fig. 6- Back EMF induced by the LSPMSM

Here, 2D finite element model of LSPMSM is utilized for the verification of the obtained results. A quarter geometry of the machine is subdivided into triangular elements.

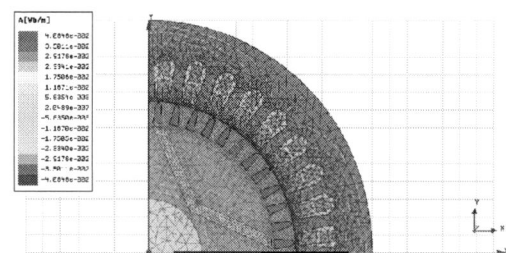

Fig.7- Finite element mesh of LSPMSM

Fig. (8) demonstrates the magnetic flux density distribution of the machine, based on which the flux paths and the corresponding magnetic equivalent circuit are extracted.

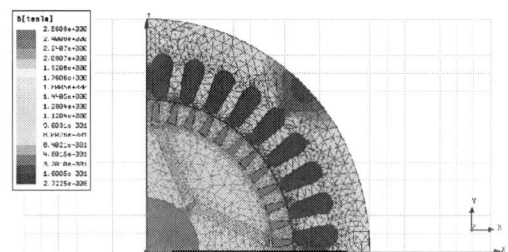

Fig.8- Finite element mesh of LSPMSM

Figure (9) shows the magnetic flux lines in LSPMSM.

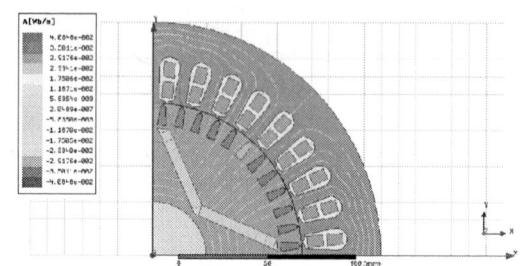

Fig. 9- Flux lines in LSPMSM

VII. AIR GAP FLUX DENSITY ANALYSIS

For LSPMSM, the open circuit air gap flux density waveform is shown in Fig.10. The slots near the magnets experience higher field due to contribution by both the magnets. The air gap flux density waveform has peaks and deeps in it; those are due to the stator and rotor slot and teeth surface non uniformity.

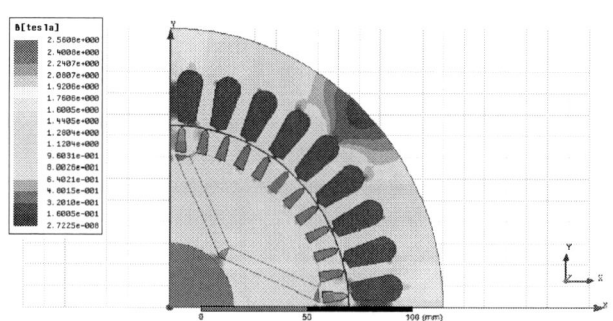

Fig. 10- Flux density distribution of LSPMSM

VIII. CONCLUSIONS

FEM plays a vital role in designing and studying of LSPMSM. A very rigorous investigation has been done for magnet shapes and dimensions. By doing so, we have designed the various structures of stator and rotor laminations and windings. By using this technique, LSPMSM has been successfully developed and simulated under given conditions. To study the steady state and transient performance of the motor, the 2-D time stepping FEA model is used. The results came out after simulation confirms the good hold on the validity of the designed model. The magneto static analysis is performed to obtain various steady state performance parameters. The transient field is utilized to study the starting process of LSPMSM.

978-1-5386-6624-1/18 $31.00 © 2018 IEEE

REFERENCES

[1] F. W. Merrill, "Permanent magnet excited synchronous motor," Transaction of AIEE, vol. 74, part III, 1955, pp.1754-1760.

[2] Tongshan Diao, Xiuhe Wang ,Yufang Wang and Ying Pei, "Finite Element Analysis of Transient Behavior of Permanent Magnet Synchronous Motor", 2011 international conference on machines and systems, 2011.

[3] K. H. Ha and J. P. Hong, "Dynamic rotor eccentricity analysis by coupling electromagnetic and structural time stepping FEM," IEEE Trans. Magn., vol. 37, no. 5, pp. 3452–3455, Sep. 2001.

[4] Y. Bao, L. Liu, Y. Zhang and X Feng, "Performance Investigation and Comparison of Line Start-up Permanent Magnet Synchronous Motor with Super Premium Efficiency", 2011 international conference on machines and systems, 2011.

[5] A.H Isfahani, 2011. Effects of Magnetizing Inductance on Start-Up and Synchronization of Line-Start Permanent-Magnet Synchronous Motors," IEEE Transaction on magnetics, vol. 47, no. 4, pp. 823-829.

[6] M. J. Melfi, S. D. Rogers, S. Evon, and B. Martin, "Permanent-magnet motors for energy savings in industrial applications,"IEEE Trans. Ind.Appl., vol. 44, no. 5, pp. 1360–1366, Sep./Oct. 2008.

[7] V. B. Honsinger, "Performance of polyphase permanent magnet machines," IEEE Transaction on Power Apparatus and Systems, Vol. PAS-99, No.4, pp.1510- 1518.

[8] T. J. E. Miller, "Synchronization of line start permanent magnet AC motors," IEEE Trans., vol. PAS-103, No.7, July 1984, pp. 1822-1828.

[9] A. I. de Almeida, F. Ferreira, J. F. Busch, and P. Angers, "Comparative analysis of IEEE 112-B and IEC 34-2 efficiency testing standards using stray load losses in low-voltage three-phase, cage induction motors," Industry Applications, IEEE Transactions on, vol. 38, no. 2, pp. 608–614, 2002.

[10] W.H. Kim, K.C. Kim, S.J. Kim, D.W. Kang, S.C. Go, H.W. Lee, Y.D. Chun and J. Lee, "A study on the optimal rotor design of LSPM considering the starting torque and efficiency,"IEEE Transactions on Magnetics, 45(3): 1808-1811, March 2009.

[11] Lefevre L, Soulard J. Finite element transient start of a line-start permanent magnet synchronous motor. In: Proceedings of International Conference on Electrical Machines, ICEM Aug 2000. pp. 1564–8.

[12] Boldea, and S. Scridon,"Electric propulsion systems on HEVs: review and perspective," EVER 2010, Monaco, 25-28 March, pp. 1-8, 2010.

[13] Adkins, B. and Harley, R.G. "The general theory of alternating current machines", Chapman & Hall, London, 1978.

[14] Popescu, M.; Miller, T.J.E.; McGilp, Malcolm I.; Strappazzon, G.; Trivillin, N. and Santarossa, R. 2003. Line-Start Permanent-Magnet Motor: Single-Starting Performance Analysis, *IEEE Transactions on Industry Appl.*, Vol. 39, No.4, pp. 1021-1030.

[15] Kurihara, K. and Rahman, M. Azizur, 2004. High efficiency line-start interior permanent magnet synchronous motors, *IEEE Transactions on Industry Applications*, Vol. 40, Issue: 3, pp. 789- 796

[16] Ding T, Takorabet N, Sargos FM. Design and analysis of different line-start PM synchronous motors for oil-pump applications. In: Proceedings on Electromagnetic Field Computation, 2008 13th Biennial IEEE Conference, Athen, Greek, 11–15 May, 2008. p. 356.

[17] R. T. Ugale, B. N. Chaudhari and A. Pramanik, "Overview of research evolution in the field of line start permanent magnet synchronous motors," IET Electr. Power Appl. vol. 8, no. 4, pp. 141-154, Apr. 2014.

Fractional Order Tchebichef Moment and its Invariants

Vishal Kumar Pandey, Jyotsna Singh and Harish Parthasarathy
Multimedia Research Lab, Division of ECE
N.S.I.T, New Delhi, India-110078
vishalkp@nsitonline.in, jsingh.nsit@gmail.com, harisignal@yahoo.com

Abstract—Discrete orthogonal moments are computationally efficient as compared to its continuous counterpart. The Tchebichef moment shows less computation cost among moments defined in discrete coordinate system. In this paper, a generalized order of Tchebichef polynomial is presented. Moment and its translation, scale and rotation invariants are proposed and derived respectively. Various experiments are performed and the results demonstrate the robustness of proposed invariants.

Index Terms—Fractional order moment, Image reconstruction, Moment invariants, Tchebichef moment

I. INTRODUCTION

Various types of moments and moment invariants have been used in image and pattern analysis applications. The projection of image is taken onto polynomial basis function to calculate image moments. Hu [1] was the first researcher to introduce geometric moments and their translation, scale and rotation (TSR) invariants. The description of image with very less redundancy can be achieved from orthogonal moments [2]. Teague [3] suggested representation of image in terms of continuous orthogonal moments. Legendre and Zernike are examples of such kind of orthogonal moments that provide compressed form of image description having low redundancy of information. Other major issues that need to be investigated [4] are image reconstruction from their corresponding moment functions, computational complexity and numerical stability. Later it was found that discrete orthogonal moments have better image representation capability as compared to the continuous moments. Mukundan et al. [5] proposed Tchebichef moments (TM) in this category and performed the comparative analysis with the already existing continuous moments. The analytical properties, such as invariance and orthogonality, of the moment functions may get affected by discrete approximation of integrals. Discrete moment has some advantages over continuous moment. Discrete orthogonal moments remove the discrete approximation error which generally comes while implementing the continuous integrals and image coordinate system transformation [6]. All these moments can be derived either in cartesian or polar coordinate system [7]. The moments can be made invariant to its geometric transformations by applying mathematical concepts on the polynomial basis functions [8]–[10].

In recent years, a great amount of progress can be seen in fractional order signal analysis. Some of these applications in the field of image analysis can be given as fractional fourier transform [11]–[16], fractional hartley transform and fractional DCT [17], fractional DWT [18], fractional transforms in watermarking application [19]. The orthogonal moments were restricted to integer order because the order of orthogonal polynomial were taken as integer. Zhang et. al. [20] proposed pattern recognition using orthogonal fourier mellin moment of fractional order. Then, a general framework for real order of orthogonal and non orthogonal moments were proposed by Xiao et. al. [21]. In the survey, the superiority of Tchebichef moment has been found in various applications but the order of moment was restricted to integer.

In this paper, the generalized order of discrete orthogonal Tchebichef moment and its invariants are proposed for image analysis. The required moment is derived from fractional order Tchebichef polynomials using recursive scheme in order to reduce the computational complexity. Also, numerical stability of fractional order moment is investigated for different values of fraction coefficient. The translation and scale moment invariants of fractional order Tchebichef moment (FOTM) are derived in cartesian coordinate system using geometric moments. To prove rotation invariance, polynomial is defined using polar coordinates. The varying deviation coefficients are used to measure the variation of moment invariants from the mean value.

The organization of this paper is as follows. Fractional order Tchebichef moment and its invariants are proved in Section 2. Section 3 present the experimental results evaluated from the derived relation of invariants, image reconstruction using matrix method and plots of fractional order Tchebichef polynomial for different order. Last section describes the concluding remarks.

II. FRACTIONAL ORDER TCHEBICHEF MOMENT AND ITS INVARIANTS

In this section, the generalized order Tchebichef moment and its invariants are provided. Tchebichef moments can be represented in terms of geometric moment. These geometric moments are invariant to different transformations, like translation, scale, rotation e.t.c., so indirectly Tchebichef moments can be proved invariant to these transformations. Fractional order moments and its invariants can be defined using the same technique.

978-1-5386-6624-1/18 $31.00 © 2018 IEEE

A. Tchebichef moment (TM)

Tchebichef moments are defined using discrete Tchebichef orthogonal polynomial. The Tchebichef polynomial, $t_p(x)$ can be given as :

$$t_p(x) = (1-N)_p \sum_{k=0}^{p} \frac{(-p)_k (p+1)_k}{(1-N)_k (k!)^2} (-x)_k. \quad (1)$$

$(b)_j$ is Pochhammer symbol, defined as $(b)_j = b(b+1)\cdots(b+j-1)$, for $b \geq 1$. The definition of Pochhammer symbol is mathematically same as rising factorial.

Rising factorial $(b)_k$ and falling factorial $\langle b \rangle_k$ are related as given in [22], can be written as

$$\langle b \rangle_k = (-1)^k (-b)_k = b(b-1)(b-2)...(b-k+1), k \geq 1$$
$$= 1 \qquad\qquad\qquad\qquad\qquad k = 0$$

Now (1) can be rewritten as :

$$t_p(x) = \sum_{k=0}^{p} a_{pk} \langle x \rangle_k. \quad (2)$$

here $a_{pk} = \frac{(1-N)_p \langle p \rangle_k (p+1)_k}{(1-N)_k (k!)^2}$

Based on the given polynomial in (1), the Tchebichef moment [5] for an image, $f(x,y)$ of size $N \times N$ is defined as :

$$T_{pq} = \frac{1}{\beta(p,N)\beta(q,N)} \sum_{x=0}^{N-1} \sum_{y=0}^{N-1} t_p(x) t_q(y) f(x,y) \quad (3)$$

where $\beta(p,N)$ is defined as : $\beta(p,N) = \sqrt{(2p)! \binom{N+p}{2p+1}}$ for $p = 0, 1, ..., N-1$.

The falling factorial, $\langle x \rangle$, in (2) is also related with the signed stirling numbers, having relation

$$\langle x \rangle_k = \sum_{i=0}^{k} s(k,i) x^i \quad (4)$$

recurrence relation [23] for Stirling number is defined as :

$$s(k,i) = s(k-1,i-1) - (k-1)s(k-1,i), \quad (5)$$

for $k \geq 1$, $i \geq 1$. The initial values

$$s(k,0) = s(0,i) = 0, \quad (6)$$

for $k \geq 1$, $i \geq 1$ and s(0,0) = 1.

Now, substituting the value of $\langle x \rangle_k$ from (4) to (2) and the expression can be represented in terms of stirling number as :

$$t_p(x) = \sum_{k=0}^{p} B_{pk} \langle x \rangle_k$$
$$= \sum_{k=0}^{p} \sum_{i=0}^{k} B_{pk} s(k,i) x^i$$
$$= \sum_{i=0}^{p} \sum_{k=0}^{p-i} B_{pk} s(p-k,i) x^i$$
$$= \sum_{i=0}^{p} C(p,i) x^i \quad (7)$$

where

$$C(p,i) = \sum_{k=0}^{p-i} B_{pk} s(p-k,i)$$
$$= \sum_{k=0}^{p-i} C_k(p,i) \quad (8)$$

and

$$C_k(p,i) = B_{pk} s(p-k,i) \quad (9)$$

Similar to (7), Tchebichef polynomial can also be defined in y axis direction.

$$t_q(y) = \sum_{j=0}^{q} C(q,j) y^j \quad (10)$$

Let discrete form of geometric moment, G_{pq}, for an image $f(x,y)$ of size N $\times$N of order $p+q$ is :

$$G_{pq} = \sum_{x=0}^{N-1} \sum_{y=0}^{N-1} x^p y^q f(x,y) \quad (11)$$

where $p, x = 0, 1, 2, \cdots, N-1$ and $q, y = 0, 1, 2, \cdots, N-1$. Using (7) and (10), Tchebichef moment defined in (3), can be represented in terms of geometric moment [24] :

$$T_{pq} = \frac{1}{\beta(p,N)\beta(q,N)} \sum_{x=0}^{N-1} \sum_{y=0}^{N-1} \sum_{i=0}^{p} C(p,i) x^i \sum_{j=0}^{q} C(q,j) y^j f(x,y)$$
$$= \frac{1}{\beta(p,N)\beta(q,N)} \sum_{i=0}^{p} \sum_{j=0}^{q} C(p,i) C(q,j) \sum_{x=0}^{N-1} \sum_{y=0}^{N-1} x^i y^j f(x,y)$$
$$= \frac{1}{\beta(p,N)\beta(q,N)} \sum_{i=0}^{p} \sum_{j=0}^{q} C(p,i) C(q,j) G_{ij} \quad (12)$$

B. Fractional order Tchebichef moment (FOTM)

FOTM can be constructed using fractional order Tchebichef polynomial (FOTP). FOTP is also an orthogonal polynomial for $0 < \gamma < 1$. It is defined as [21] :

$$t_p(\gamma, x) = \sqrt{\gamma} x^{\frac{\gamma-1}{2}} t_p(x^\gamma)$$
$$= \sqrt{\gamma} \sum_{i=0}^{p} C(p,i) x^{i\gamma + \frac{\gamma-1}{2}} \quad (13)$$

In similar to (13), FOTP in y axis direction, for the same fractional coefficient γ, can be defined as :

$$t_q(\gamma, y) = \sqrt{\gamma} y^{\frac{\gamma-1}{2}} t_q(y^\gamma)$$
$$= \sqrt{\gamma} \sum_{j=0}^{q} C(q,j) y^{j\gamma + \frac{\gamma-1}{2}} \quad (14)$$

Now, FOTM can be easily defined using (13) and (14)

$$T_{pq}^\gamma = \frac{1}{\beta(p,N)\beta(q,N)} \sum_{x=0}^{N-1} \sum_{y=0}^{N-1} t_p(\gamma, x) t_q(\gamma, y) f(x,y)$$
$$= \frac{\gamma}{\beta(p,N)\beta(q,N)} \sum_{i=0}^{p} \sum_{j=0}^{q} C(p,i) C(q,j) G_{ij}^\gamma \quad (15)$$

978-1-5386-6624-1/18 $31.00 © 2018 IEEE

where, G_{ij}^γ is geometric moment of order $(i\gamma+\frac{\gamma-1}{2}, j\gamma+\frac{\gamma-1}{2})$

$$G_{ij}^\gamma = \sum_{x=0}^{N-1} \sum_{y=0}^{N-1} x^{i\gamma+\frac{\gamma-1}{2}} y^{j\gamma+\frac{\gamma-1}{2}} f(x,y)$$

Reconstruction of the image can be done using

$$f(x,y) = \sum_{x=0}^{N-1} \sum_{y=0}^{N-1} t_p(\gamma,x) T_{pq}^\gamma t_q(\gamma,y)$$

a) Recursive scheme for FOTP: Recurrence relation is proposed to remove the computational complexity. The following equation can be used to calculate FOTP :

$$t_{p+1}(\gamma,x) = t_p(\gamma,x)\frac{(2p-1)}{p}\frac{(2x^\gamma+1-N)}{N}$$
$$- t_{p-1}(\gamma,x)\frac{(p-1)}{p}(1-\frac{(p-1)^2}{N}) \qquad (16)$$

with initial values

$$t_0(\gamma,x) = \sqrt{\gamma} x^{\frac{\gamma-1}{2}}$$
$$t_1(\gamma,x) = \sqrt{\gamma} x^{\frac{\gamma-1}{2}} \frac{(2x^\gamma+1-N)}{N}$$

C. Fractional order Tchebichef moment invariants (FOTMI)

In this section, Translation, scale and rotation invariance of FOTM are proved. Using (15), one can easily prove translation and scale invariance with the help of geometric moments. Translation invariance can be achieved by subtracting image centroid from the corresponding coordinate values i.e. shifting the origin of coordinate system to image's centroid. Central moment is used for translation invariance. Central moment, μ_{pq}, is defined as :

$$\mu_{pq} = \sum_{x=0}^{N-1} \sum_{y=0}^{N-1} (x^\gamma - x_o)^p (y^\gamma - y_o)^q f(x,y) \qquad (17)$$

here, x_o and y_o are centroid location of the image.
For scale invariance, a rational function [24] is defined to cancel out the effect of scaled image. The rational function is given as :

$$\nu_{pq} = \frac{G_{pq}^\gamma (G_{00}^\gamma)^{\zeta+1}}{G_{(p+\zeta)0}^\gamma G_{0(q+\zeta)}^\gamma} \qquad \zeta \in R \qquad (18)$$

Now, translation and scale invariance are proved by replacing G_{pq}^γ with μ_{pq} and ν_{pq} respectively.

To prove rotation invariance, polynomial should be defined in polar coordinate system. So, according to [21], FOTP may be defined in polar coordinate system.

$$t_p(\gamma,r) = \sqrt{\gamma} r^{\frac{\gamma-2}{2}} t_p(r^\gamma)$$
$$= \sqrt{\gamma} \sum_{i=0}^{p} C(p,i) r^{i\gamma+\frac{\gamma-2}{2}} \qquad (19)$$

The definition of fractional order radial Tchebichef moment is given as :

$$T_{pq}^{r\gamma} = \frac{1}{N} \sum_{r=0}^{N-1} \sum_{\theta=0}^{N-1} t_p(\gamma,r) exp(-i2\pi q \frac{\theta}{N}) f(r,\theta) \qquad (20)$$

where, N is size of the image.
Let $f_{r_o}(r,\theta)$ is the rotated form of the image function $f(r,\theta)$ by an angle of ϕ.

$$f_{r_o}(r,\theta) = f(r,\theta-\phi)$$

The fractional order radial Tchebichef moment of the rotated image $f_{r_o}(r,\theta)$ is

$$T_{pq}^{r_o\gamma} = \frac{1}{N} \sum_{r=0}^{N-1} \sum_{\theta=0}^{N-1} t_p(\gamma,r) exp(-i2\pi q \frac{\theta}{N}) f_{r_o}(r,\theta)$$
$$= \frac{1}{N} \sum_{r=0}^{N-1} \sum_{\theta=0}^{N-1} t_p(\gamma,r) exp(-i2\pi q \frac{\theta+\phi}{N}) f(r,\theta)$$
$$= exp(-i2\pi q \frac{\phi}{N}) T_{pq}^{r\gamma} \qquad (21)$$

Eq. (21) proves that the magnitude of moment for rotated and unrotated images are same.

III. Experimental Results

In this section, various experiments are performed on an image of chinese character *shu.bmp* as shown in Fig. 2, to validate the proposed theory. All simulation work have been performed on MatLab R2015b with PC intel core I5, 3.33 GHz, 2GB RAM. Image reconstruction is performed in the first section. Subsequent part is based on Translation, Scale and Rotation (TSR) invariant moments, described in section II-C are verified. Fig. 1 shows the plot of polynomial for different order value (n). Fig. 1 (a)-(d) are shown for different fractional coefficients 'γ'. Fig. 1 (a)-(c) are plotted for $\gamma \leq 1$ and Fig. 1 (d) for $\gamma > 1$. It can be seen from the plots that polynomial is of converging nature for $\gamma \leq 1$ i.e. plot lies between (-1, 1) and diverging nature for $\gamma > 1$. Thus, stable moments can be formed using $\gamma \leq 1$. Thus, the value of fractional coefficient (γ) is randomly selected as 0.5 for all the experiments.

a) Image reconstruction: Matrix method is computationally fast to reconstruct the image. Thus, the expression for matrix method of image reconstruction using FOTM is given as :

$$f^{(N\times N)} = t_p^{(N\times 1)} T_{pq}^{\gamma(1\times 1)} t_q^{(1\times N)} \qquad (22)$$

here $f^{(N\times N)}$ is the image of size $N \times N$, $t_p^{(N\times 1)}$ and $t_q^{(1\times N)}$ are p^{th} and q^{th} order vector of size $N \times 1$ and $1 \times N$ respectively and $T_{pq}^{\gamma(1\times 1)}$ is $(p + q)^{th}$ order FOTM of size 1×1. The image given in Fig. 2 (a) is perfectly reconstructed, as shown in Fig. 2 (b).

978-1-5386-6624-1/18 $31.00 © 2018 IEEE

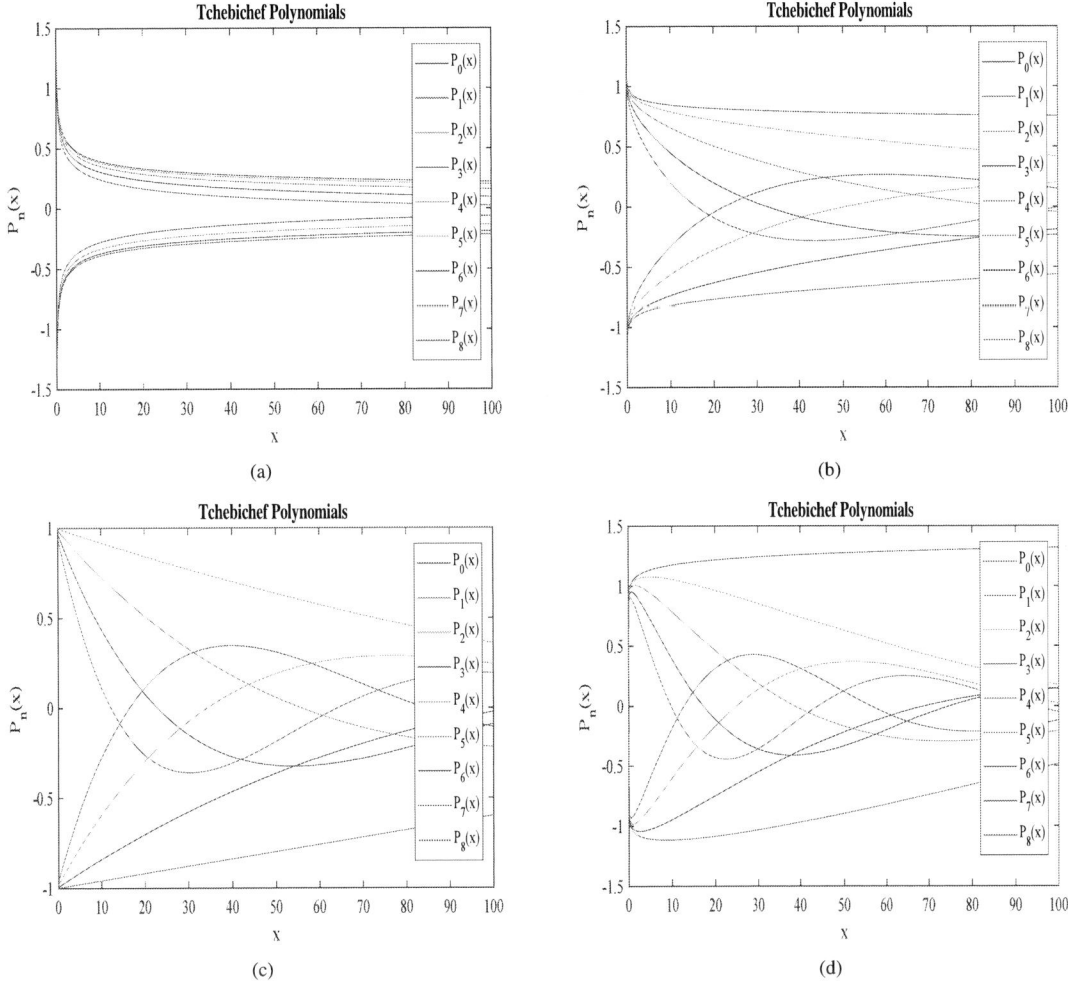

Fig. 1. Polynomial plot for given γ values (a) 0.5 order, (b) 0.9 order, (c) 1 order, (d) 1.1 order.

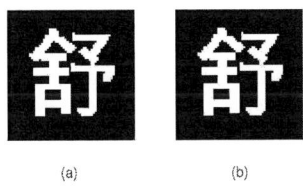

Fig. 2. Reconstruction of Shu image (a) Original image (b) Reconstructed image.

b) Test of TSR moment invariance: Translation, scale and rotation moment invariants of image are simulated and results are shown, in this section. The image *shu.bmp* is displaced at four distinct locations i.e. (+1, +1), (-1, +1), (-1, -1) and (+1, -1) and the calculation of different order of FOTM invariants are performed using (15) and (17). The results are tabulated in Table I. Deviation coefficient $\frac{\sigma}{\mu}$ is calculated to check the deviation from the mean value of moment invariants. Here, σ and μ are standard deviation and mean respectively in $\frac{\sigma}{\mu}$.

In Table I, the value of $\frac{\sigma}{\mu}$ is zero for a distinct value of order. Thus, it shows perfect invariance with respect to image translation.

Order $\rightarrow$ $(\Delta i, \Delta j)\downarrow$	$(1,3)$ $\times 10^{-5}$	$(2,2)$ $\times 10^{-4}$	$(2,3)$ $\times 10^{-6}$	$(4,2)$ $\times 10^{-9}$
(+1, +1)	8.2532	1.2699	1.1768	8.4219
(+1, -1)	8.2532	1.2699	1.1768	8.4219
(-1, -1)	8.2532	1.2699	1.1768	8.4219
(-1, +1)	8.2532	1.2699	1.1768	8.4219
$\frac{\sigma}{\mu}$	0	0	0	0

TABLE I

TRANSLATION $(\Delta i, \Delta j)$ INVARIANT OF FRACTIONAL ORDER TCHEBICHEF MOMENT, FOR IMAGE *Shu.bmp* OF SIZE 64×64

To evaluate scale invariant FOTM, *shu.bmp* image is equally scaled by 0.25, 0.5, 1 and 2 in both axis directions. Then, scale invariant FOTM is calculated using (15) and (18) for different values of order (i.e. (1, 3), (2, 2), (2, 3) and (4,

978-1-5386-6624-1/18 $31.00 © 2018 IEEE

2)). As it can be seen from Table II that the value of $\frac{\sigma}{\mu}$ is very small. The small value of $\frac{\sigma}{\mu}$ shows scale invariance.

Order→ (a,b) ↓	$(1,3)$	$(2,2)$	$(2,3)$ $\times 10^{-4}$	$(4,2)$ $\times 10^{-7}$
$(0.25, 0.25)$	0.0257	0.0589	4.7341	9.5823
$(0.5, 0.5)$	0.0384	0.0714	4.0483	1.4296
$(1, 1)$	0.0440	0.0782	2.4477	5.1968
$(2, 2)$	0.0464	0.0811	1.3205	1.4892
$\frac{\sigma}{\mu}$	0.2073	0.1181	0.4262	0.7562

TABLE II

SCALE (a, b) INVARIANT OF FRACTIONAL ORDER TCHEBICHEF MOMENT, FOR IMAGE $Shu.bmp$ OF SIZE 64×64

To calculate rotation invariant FOTM, (15) and (21) are used. The rotation invariant FOTM is calculated for four different rotation angles (i.e. 40°, 60°, 75° and 100°). The calculated values of rotation invariant FOTM are tabulated in Table III. The values of $\frac{\sigma}{\mu}$ are very small which proves FOTM is rotationally invariant for any angle of rotation.

Order→ Angle ↓	$(1,3)$	$(2,2)$ $\times 10^{-4}$	$(2,3)$ $\times 10^{-4}$	$(4,2)$ $\times 10^{-6}$
40°	0.0081	9.6093	8.2795	3.4360
60°	0.0085	8.0066	7.6998	2.8219
75°	0.0079	9.2471	7.2748	1.9354
100°	0.0068	9.1645	4.8005	1.5908
$\frac{\sigma}{\mu}$	0.0008	0.0667	0.1891	0.2160

TABLE III

ROTATION INVARIANT OF FRACTIONAL ORDER TCHEBICHEF MOMENT, FOR IMAGE $Shu.bmp$ OF SIZE 64×64

CONCLUDING REMARKS

In this paper, Tchebichef moment and its invariants are derived for generalized order of orthogonal Tchebichef polynomial using fractional coefficient (γ). It is found from the plots of FOTP that stable moment invariants can be formed only for $\gamma \leq 1$. Effectiveness of RST invariance are investigated for the proposed discrete orthogonal moments using deviation coefficient. It can be inferred from the results that the proposed moment and its invariants possess better descriptive capability.

REFERENCES

[1] M. K. Hu, "Visual pattern recognition by moment invariants", IRE Transaction on Information Theory, vol. 8, pp. 179-187, 1962.

[2] Y. S. Abu-Mostafa and D. Psaltis, "Recognitive Aspects of Moment Invariants", in IEEE Transactions on Pattern Analysis and Machine Intelligence, vol. PAMI-6, no. 6, pp. 698-706, Nov. 1984.

[3] M. R. Teague, "Image analysis via the general theory of moments", Journal of the Optical Society of America, 70, pp. 920-930, 1980.

[4] C. H. Teh and R. T. Chin, "On image analysis by the methods of moments", in IEEE Transactions on Pattern Analysis and Machine Intelligence, vol. 10, no. 4, pp. 496-513, Jul 1988.

[5] R. Mukundan, S. Ong, P. Lee, "Image analysis by Tchebichef moments", IEEE Transactions on Image Process. vol. 10, no. 9, pp. 1357-1364, 2001.

[6] R. Mukundan, "Some computational aspects of discrete orthonormal moments", IEEE Transactions on Image Processing 13 (8), pp. 1055-1059, 2004.

[7] J. Flusser, B. Zitova, T. Suk, "Moments and moment invariants in pattern recognition", Wiley, Chichester, 2009.

[8] H. Q. Zhu, H. Z. Shu, T. Xia, L. M. Luo, J. L. Coatrieux, "Translation and scale invariants of Tchebichef moments", Pattern Recognition 40, 2530-2542, 2007.

[9] V. K. Pandey, J. Singh, H. Parthasarathy, "Translation and scale invariance of 2D and 3D Hahn moments", In 3rd IEEE International Conference on Signal Processing and Integrated Networks (SPIN). Noida, India, 2016.

[10] V.K. Pandey, J. Singh, H. Parthasarathy, "Algebraic technique for computationally efficient Hahn moment invariants", Multidimensional System and Signal Processing, 2017.

[11] Lang, J., Zhang, Z., "Blind digital watermarking method in the fractional Fourier transform domain", Opt. Lasers Eng., 53, pp. 112-121, 2014.

[12] Tang, L.L., Huang, C.T., Pan, J.S., et al, "Dual watermarking algorithm based on the fractional Fourier transform", Multimedia Tools Appl., 74, (12), pp. 4397-4413, 2015.

[13] H. M. Ozaktas, Z. Zalevsky, and M. A. Kutay, The Fractional Fourier Transform with Applications in Optics and Signal Processing, John Wiley, Chichester, New York, 2001.

[14] V. Namias, The fractional order Fourier transform and its application to quantum mechanics, J. Inst. Math. Appl. 25, pp. 241-265, 1980.

[15] R. Tao, B. Deng, and Y. Wang, Fractional Fourier Transform and Its Applications, Tsinghua University Press, Beijing, China, 2009.

[16] B. L. Almeida, The fractional Fourier transform and time-frequency representations, IEEE Transactions on Signal Processing, 42 (11), pp. 3084–3091, 1994.

[17] S. C. Pei, J.J. Ding, Fractional cosine, sine, and Hartley transforms, IEEE Transactions on Signal Processing, vol. 50 (7), pp. 1661–1680, 2002.

[18] G. Bhatnagar, Q.M. Jonathan Wu, B. Raman, Discrete fractional wavelet transform and its application to multiple encryption, Information Sciences 223, pp. 297–316, 2013.

[19] Qi, M., Li, B.Z., Sun, H.F., "Image watermarking via fractional polar harmonic transforms", J. Electron. Imaging, 24, (1), pp. 1-12, 2015.

[20] H. Zhang, Z. Li, and Y. Liu, "Fractional orthogonal fourier-mellin moments for pattern recognition", in Communications in Computer and Information Science, vol. 662, pp. 766-778, 2016.

[21] Bin Xiao, Linping Li, Yu Li, Weisheng Li, Guoyin Wang, "Image analysis by fractional-order orthogonal moments", Information Sciences, vol. 382–383, pp. 135–149 ,2017.

[22] Spanier, J. and Oldham, K. B., "The Pochhammer Polynomials $(x)_n$." Ch. 18 in An Atlas of Functions. Washington, DC: Hemisphere, pp. 149–165, 1987.

[23] Comtet, L., "Advanced Combinatorics: The Art of Finite and Infinite Expansions", rev. enl. ed. Dordrecht, Netherlands: Reidel, p. 113, 1974.

[24] Hongqing Zhu, Huazhong Shu, Ting Xia, Limin Luo, Jean Louis Coatrieux, "Translation and scale invariants of Tchebichef moments," Pattern Recognition, vol. 40, Issue 9, pp. 2530–2542, 2007.

Microcontroller Based Load Prioritization Technique in Residential Sector

MohiniYadav
Department of Electrical Engineering
Jamia Millia Islamia
New delhi , India
mohiniyadav565@gmail.com

Majid Jamil
Department of Electrical Engineering
Jamia Millia Islamia
New delhi , India
majidjamil@hotmail.com

M.Rizwan
Department of Electrical Engineering
Delhi Technical University
New delhi , India
rizwan@dce.ac.in

Abstract—**With the continuous increase in load demand, it will be problematic at generation side to keep on varying the power supply to match the pace of increasing customer's demand. Hence, it causes large power cuts and power shortages in residential areas. As a solution of above problem, we have proposed the automation based load prioritization technique in the residential sector. For the above study, we have classified the loads into two categories as Flexible and non-Flexible loads. Flexible loads are one which can be treated as lowest priority loads and its operating time can be shifted easily to reduce the energy consumption. Whereas, Non-flexible loads are one which can be treated as highest priority loads and they are not shifted from their scheduled time. As a result, this approach shifts the load according to user's priority index to manage the consumption effectively. Study has been demonstrated with the use of ATmega 2560 microcontroller powered by relay circuitry. This proposed automation technique will reduce the man work by automatically shedding of load based on priority mechanism.**

Keywords—ATmega 2560 Microcontroller, Relay, Load shedding, IR transmitter, IR receiver

I. INTRODUCTION

Owing to the increase in global population, the energy demand keep on rising day by day. Compared to fossil fuels, Renewable energy sources (RES) has become the significant part due to technical and economical problems associated with establishment of grid and its expansion in different remote areas as well as in rural fields [1,2]. Consequently, RES has been a significant source to overcome all kinds of energy crisis and to fulfill the demand of consumer in rural areas as well [3]. Amongst all, solar and wind energy has been playing a significant role for electricity generation in different sectors across the globe [4,5]. In residential sectors, the solar system gains a paramount importance due to less noise ambiguity, higher reliability, lower maintenance cost [6,7]. Despite from the above mentioned benefits of solar energy to fulfill the energy demand in residential sectors, this has faced some vagaries in some aspects like sudden fluctuations due to unpredictable weather conditions. So, with the implementation of PV systems, it is required to explore new techniques in order to improvise its efficient use across the globe. Different researchers have introduced techniques to overcome fluctuations in solar irradiance to procure maximum energy benefits from these systems. They have implemented probability based approaches [8,9].

Artificial intelligence techniques [10] with new methodologies for deciding the optimal size of PV system [11-13] based on deterministic models [14-16] has been developed. Some authors have presented the combined analysis of PV systems with benefits of battery system [17,18]. Some researchers have proposed the demand side management techniques to limit the cost of electricity by improving the generation from renewable sources of energy. Different demand response management techniques [4] have been developed to increase the battery life while reducing the $CO2$ radiations for hybrid solar systems. This will improve the load profile by implementation of load scheduling approaches [19]. Load scheduling with dual control has been proposed for efficient energy management while maintaining the users comfort at low cost. Authors have designed a multi-agent system to provide safeguard design of integrated PV with micro-grid [20]. Researchers have used old optimization techniques like penalty functions to classify the loads on the basis of priority. They have also introduced Time of Use techniques for energy management and effective manage of residential loads to gain the reduced consumption cost. A fuzzy based controller [21] has been implemented to manage the residential loads effectively. This power based controlling for hybrid standalone solar system has been developed for maintaining the load demand by balancing demand supply scenario. Few authors have shown the load prioritization in a distribution system using fuzzy theories [22]. The fuzzy control systems have been developed to enhance the battery life while maintaining state of charge (SOC) of battery.

From the above studies, it is apparent that the whole work of load prioritization has presented the approaches based on hybrid PV systems [23]. These aforementioned techniques have intermittency issues with solar and besides this it has higher installation cost due to implementation of batteries and inverters. The PV panels may get damaged easily with time. All these limitations get overcome by the use of cost efficient automation technique. In this concept, we prioritize the loads according to usage and user comfort level with the use microcontroller with relay. The significance of our proposed work lie in the fact that loads are prioritized based on automation technique. Our main objective is to make the power available at each consumer end without any undesirable fluctuations in energy supply.

The study is arranged in following sections as; Section II deals with Loads classification. Section III discusses the implementation of automation based Load prioritization design. Section IV shows the result obtained with study. Section V concludes the whole work. Section VI presents the references taken under consideration for study.

II. CLASSIFICATION OF LOAD

Loads are categorized in two forms as flexible and non-flexible loads. Flexible loads are termed as the lowest priority loads, so their scheduling can be done at any point of time. Whereas, the non-flexible loads are highest priority loads that require continuous supply of power without any delay.

At point of peak demand, the highest priority is given to the non-flexible loads and the least priority loads get switched off. Table I shows the Categorization of load based on its priority level and Table II shows the Mean Energy Consumption of different loads with its usage frequency. Table III shows loads with different power consumption.

TABLE I. CATEGORIZATION OF LOAD

Type of load	Priority level	Different Loads
Flexible	1	Washing Machine
	2	Dryer
	3	Heater
	4	Iron Rod
	5	Dishwasher
Non-Flexible	1	Refrigerator, Freezer
	2	Lights
	3	Entertaining equipments (TV,DVD)
	4	Laptops /Desktops
	5	Kitchen appliances (Microwave, Toast Machine, Kettle, etc)

TABLE II. MEAN ENERGY CONSUMPTION WITH USAGE FREQUENCY FOR DIFFERENT LOADS

Loads	Mean Energy Consumption on monthly basis (W)	Usage frequency
Washing Machine	3006.4	One time in a week
Refrigerator	24,567	Whole day
Oven	15,675	One or two per week
Dish Washer	2100	Five per month
Laptop	13,564	Whole day
Printer	2456.9	15 per month
TV	9567	Weekday – 21 per month Weekend – 9 per month
Iron Rod	1567	Two per week

TABLE III. DIFFERENT LOADS WITH DAILY POWER CONSUMPTION

Loads	Wattage consumed	Hours of operation	Daily Power Consumption
Washing Machine	800	1	800
Refrigerator	110	24	2640
Oven	800	1	800
Dish Washer	50	1	50
Laptop	65	3	195
Printer	25	2	50
TV	80	6	480
Iron Rod	1100	1	1100
Light	16	12	192

III. IMPLEMENTATION OF AUTOMATION BASED DESIGN

In our proposed work, microcontroller is the heart of design. Its aim is to control the relay operation and provide serial communication among the devices. Wi-Fi is the communication protocol used to establish a communication between microcontroller and relay. The RF remote receives the commands on/off signals from PC and transfers it to air. Every appliance has its own priority which is pre-defined depending on user comfort level. Microcontroller receives the load data from meter regarding its operation scenario and verifies the base value or threshold condition. As load exceeds base value, microcontroller passes that signal to relay. And then relay will switch on/off the loads depending on the priorities decided in different modes of its operation. RF receiver receives the signal from antenna and pass to the computers.

A. ATmega 2560

It is a microcontroller having 54 digital input/output pins, 4 hardware serial ports (UARTs), 16 analog inputs with USB connection. It has 10 bits of resolution.

It supports four hardware serial ports with power jack, ICSP header and a reset button. To store the code, it has 256KB of flash memory in which 8KB is for boot loader. It has different facilities to establishing a communication with computers and other microcontrollers.

When data is transmitted from microcontroller to the PC, receiver (Rx) and transmitter (Tx) LED's get lighten up. Arduino software is used to program microcontroller.

Each pin receives maximum current of 40mA with 20-50 k Ohm as internal pull up resistor. This can be powered by external supply of 6-20 V. The appropriate range of supply is 7–12 V. Fig. 1 shows the Pin diagram of ATmega 2560.

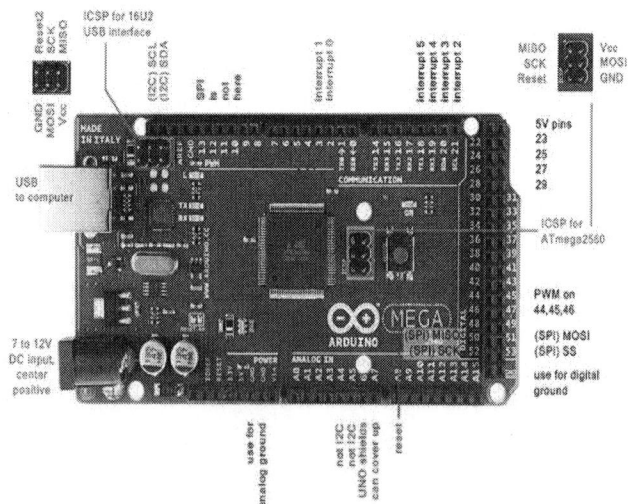

Fig.1.Pin diagram of ATmega 2560.

B. 4-Channel Relay Module

The JQC-3FF-S-Z four channel relay module is used. Each channel requires 15-20 mA current for driving the circuit and also provides the switching current of 10A at

250Vac and 15A at 125 Vac. This four channel module is powered by DC.

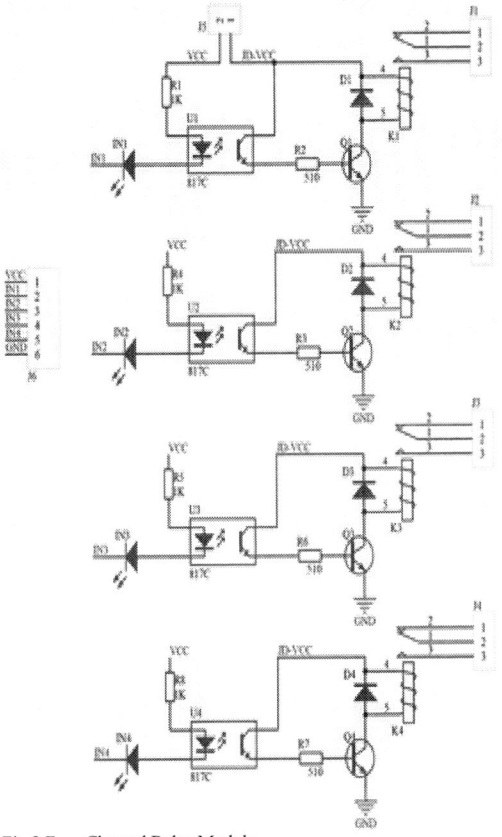

Fig.2 Four-Channel Relay Module

C. IR Transmitter and Receiver

IR Transmitter as shown in Fig. 3 is powered with 5 V supply, which is directly connected to microcontroller having three pins of Vcc, GND and output. It provides 2400 bits per second data transfer by providing immunity to electrical disturbances. It carrier frequency is 38 KHz design with 555 timer. The capacitor is used to reduce ripples in the circuitry. In order to determine the time period of oscillation, consider capacitor C2 and resistors R1 and R2.

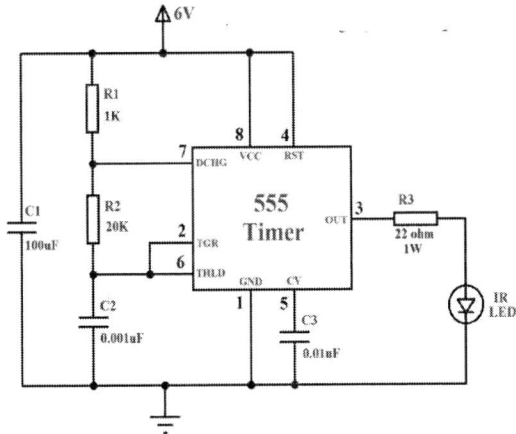

Fig.3. IR Transmitter Circuit Diagram

The modulated signal is received by RF receiver which then is then demodulated. Fig.4 shows the circuit diagram of transmitter block and receiver block in a complete system.

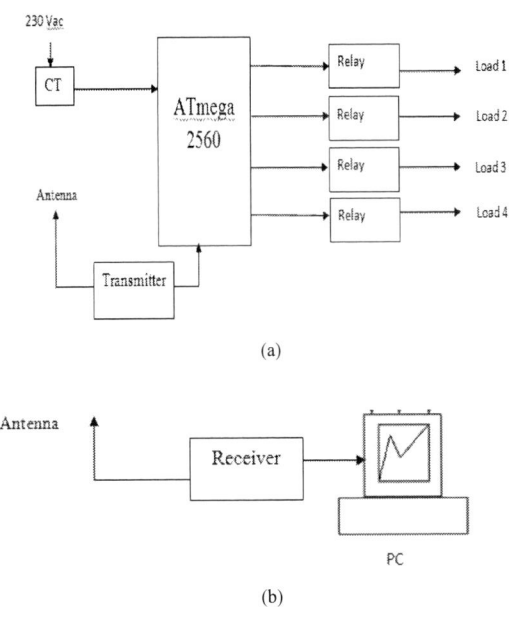

(a)

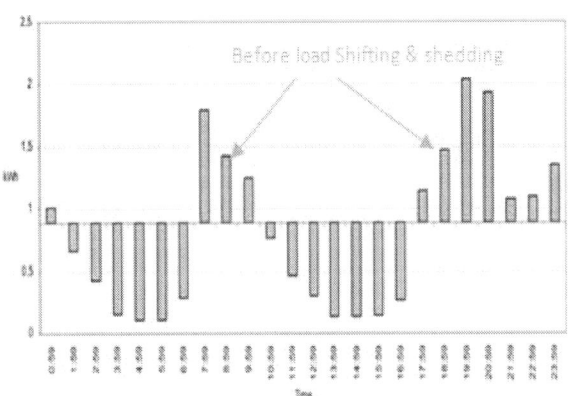

(b)

Fig.4 (a) Transmitter diagram (b) Receiver diagram

IV. RESULT

Fig.5 shows the variation of average power consumption with time before load shifting and shedding for a day. This graph shows the sinusoidal variations of load with time according to customer requirements.

Fig.5. Load profile before load shedding and shifting

Fig. 6 shows the comparative study before and after shifting the load according to user's priority index for a day. It is depicted from graph that before load shifting the power consumption at 1:00 am is 2.6 kWh but after effect of load shedding, shows its drop to 1kWh. This reduction is based on user's priority level design chart as given in Table I. Similarly in afternoon at 1:00 pm, the load demand before

978-1-5386-6624-1/18 $31.00 © 2018 IEEE   1085

load shifting is 0.15kWh and after load shifting it is increased to 0.3kWh.

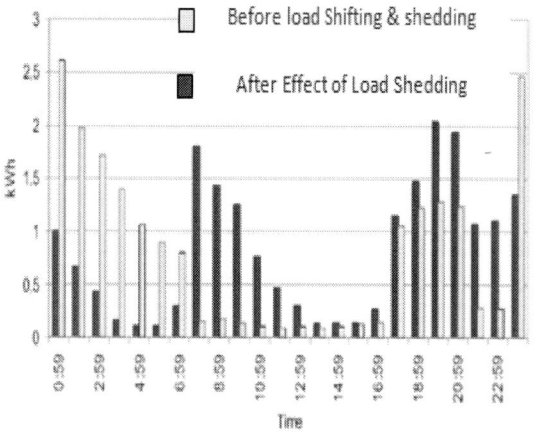

Fig.6. Comparison between before load shedding and after load shedding

V. CONCLUSION

This proposed study has shown the automation based load prioritization technique with the use of ATmega 2560 microcontroller. The Wi-Fi based communication link has been set up between microcontroller and relay. Previously, all loads are controlled manually but with this proposed technique the loads can be controlled automatically while sending the signal between RF transmitter and receiver. As shown graphically in Fig.6, the shifting of load according to user's priority index has effectively balanced the consumption scenario and achieves the target of effective energy management design.

VI. REFERENCES

[1]   Ayodele TR. Sustainable electricity generation in rural communities using hybrid energy system: The case study of Ojataye Village. Journal of Renewable Energy and Smart Grid Technology. 2016;11(1):43-56.

[2]   Ogunjuyigbe AS, Ayodele TR. Techno-economic analysis of stand-alone hybrid energy system for Nigerian telecom industry. International Journal of Renewable Energy Technology. 2016;7(2):148-62.

[3]   Malatji A, Popoola O, Binini G. Load prioritization for maximation of solar energy source in rural households. InIndustrial and Commercial Use of Energy (ICUE), 2017 International Conference on the 2017 Aug 15 (pp. 1-7). IEEE.

[4]   Kallel R, Boukettaya G, Krichen L. Demand side management of household appliances in stand-alone hybrid photovoltaic system. Renewable Energy. 2015 Sep 1;81: 123-35.

[5]   Ogunjuyigbe AS, Ayodele TR, Akinola OA. Optimal allocation and sizing of PV/Wind/Split-diesel/Battery hybrid energy system for minimizing life cycle cost, carbon emission and dump energy of remote residential building. Applied Energy. 2016 Jun 1;171: 153-71.

[6]   Tsai HL, Tu CS, Su YJ. Development of generalized photovoltaic model using MATLAB/ SIMULINK.

InProceedings of the world congress on Engineering and computer science 2008 Oct 22 (Vol. 2008, pp. 1-6).

[7]   Ogunjuyigbe AS, Ayodele TR, Akpeji KO. Optimum selection of photovoltaic modules using probabilistic approach based on capacity factor estimation. International Journal of Ambient Energy. 2018 Jan 2;39(1):11-6.

[8]   Khallat MA, Rahman S. A probabilistic approach to photovoltaic generator performance prediction. IEEE Transactions on Energy Conversion. 1986 Sep(3):34-40.

[9]   Salameh ZM, Borowy BS, Amin AR. Photovoltaic module-site matching based on the capacity factors. IEEE transactions on Energy conversion. 1995 Jun;10(2):326-32.

[10]   Koutroulis E, Kolokotsa D, Potirakis A, Kalaitzakis K. Methodology for optimal sizing of stand-alone photovoltaic/wind-generator systems using genetic algorithms. Solar energy. 2006 Sep 1;80(9):1072-88.

[11]   Balouktsis A, Karapantsios TD, Antoniadis A, Paschaloudis D, Bezergiannidou A, Bilalis N. Sizing stand-alone photovoltaic systems. International Journal of Photoenergy. 2006; 2006.

[12]   BELHADJ J. Optimal sizing of a stand-alone photovoltaic system using statistical approach. International Journal of Renewable Energy Research (IJRER). 2014 Jun 20; 4(2):329-37.

[13]   Sidrach-de-Cardona M, Lopez LM. A simple model for sizing stand alone photovoltaic systems. Solar Energy Materials and Solar Cells. 1998 Aug 24; 55(3):199-214.

[14]   Rekioua D, Matagne E. Optimization of photovoltaic power systems: modelization, simulation and control. Springer Science & Business Media; 2012 Jan 5.

[15]   Ishaque K, Salam Z, Taheri H. Modeling and simulation of photovoltaic (PV) system during partial shading based on a two-diode model. Simulation Modelling Practice and Theory. 2011 Aug 1; 19(7):1613-26.

[16]   Tsai HL, Tu CS, Su YJ. Development of generalized photovoltaic model using MATLAB/SIMULINK. In Proceedings of the world congress on Engineering and computer science 2008 Oct 22 (Vol. 2008, pp. 1-6).

[17]   Ani VA, Ani EO. Simulation of Solar-Photovoltaic Hybrid Power Generation System with Energy Storage and Supervisory Control for Base Transceiver Station (BTS) Site Located in Rural Nigeria. International Journal Of Renewable Energy Research. 2014;4(1):23-30.

[18]   Lim JH. Optimal combination and sizing of a new and renewable hybrid generation system. International Journal of Future Generation Communication and Networking. 2012 Jun;5(2):43-59.

[19]   Wu Z, Tazvinga H, Xia X. Demand side management of photovoltaic-battery hybrid system. Applied Energy. 2015 Jun 15;148:294-304.

[20]   Wu Z, Xia X. Optimal switching renewable energy system for demand side management. Solar Energy. 2015 Apr 1;114: 278-88.

[21]   Reddy GH, Chakrapani P, Goswami AK, Choudhury NB. Prioritization of load points in distribution system considering multiple load types using fuzzy theory. In Fuzzy Systems (FUZZ-IEEE), 2017 IEEE International Conference on 2017 Jul 9 (pp. 1-6). IEEE.

[22]   Saravanan S, Thangavel S. Fuzzy logic controller based power management for a standalone solar/wind/fuel cell fed hybrid system. Journal of Renewable and Sustainable Energy. 2013 Sep;5(5):053147.

[23]   Ayodele TR, Ogunjuyigbe AS, Akpeji KO, Akinola OO. Prioritized rule based load management technique for residential building powered by PV/battery system. Engineering science and technology, an international journal. 2017 Jun 1; 20(3):859-73.

2nd IEEE International Conference on Power Electronics, Intelligent Control and Energy Systems (ICPEICES-2018)

Visual Representation of Change in Vegetation Area of Dehradun, Uttarakhand, India using Normalized Difference Vegetation Index (NDVI)

Amit Kumar Shakya
Department of Electronics and
Communication Engineering
Graphic Era (Deemed to be University)
Dehradun, Uttarakhand, India
{xlamitshakya.gate2014@ieee.org}

Ayushman Ramola
Department of Electronics and
Communication Engineering
Graphic Era (Deemed to be University)
Dehradun, Uttarakhand, India
{ayushi4ramola@gmail.com}

Kunal Sawant
Department of Civil Engineering
*Graphic Era
(Deemed to be University)*
Dehradun, Uttarakhand, India
{kunalsawant94@gmail.com}

Shalini Tiwari
Department of Electronics and
Communication Engineering
Graphic Era (Deemed to beUniversity)
Dehradun, Uttarakhand, India
{stiwarib01@gmail.com}

Shamshul Aarfin
Department of Civil Engineering
Graphic Era (Deemed to be University)
Dehradun, Uttarakhand, India
{shamshula0@gmail.com}

Prag Mittal
Department of Electronics and
Communication Engineering Graphic
Era (Deemed to be University)
Dehradun, Uttarakhand, India
{pragmittal2@gmail.com}

Abstract: The backbone of the Indian economy is the agricultural sector. Today agricultural fields are getting converted in concrete societies and barren lands. Due to heavy pollution in the environment, crops and fields are losing their productivity. So it's become very essential to have a continuous check on the crops and the agricultural region of the country. In this research work we have used the Normalized Difference Vegetation Index (NDVI) technique to check the vegetation area in the Dehradun region of the Uttarakhand, India. We have calculated the NDVI values for the three regions of the Dehradun using multi-spectral Landsat 8 data. We have found the values of NDVI in three different areas and concluded that the region of Dehradun is fairly good for crop productivity and agricultural purpose till date.

Keywords: Multispectral; Pollution; NDVI; Crop productivity; Landsat 8.

I INTRODUCTION

In the World of today agriculture sector is facing a tremendous amount of problems and challenges, especially in the developing countries like India, where more efforts are done making the country a global superpower instead of the work that need to be done in the conservation of the existing resources. Today, the agriculture sector is facing severe

problems, some of them are natural and some of them are man-made. Problems related to storage facilities, quality seeds, fertilizers, soil erosion, drought, floods, lack of mechanism are some major problems that are affecting the quality and quantity of the of the agricultural production. In the developing country like India, it is the agriculture sector, which absorbs 23% of the total damage and losses [1]. In this research work we have used NDVI technique for the assessment of the quantity of vegetation area of the Dehradun region of Uttarakhand, India. NDVI stands for Normalized Difference Vegetation Index. It works on the principle of estimation of the greenness in the land cover on the basis of the near infrared light reflected from the plants and trees [2]. In NDVI when sunlight collides with the objects some wavelength of the light spectrum are absorbed and while some wavelength get reflected back, in the plant there is a green pigment called chlorophyll which strongly absorbs the visible spectrum of the light i.e. from 0.4 to 0.7 μm and re-reflects the light of the other spectrum [2]. NDVI technique is used by various researchers for the estimation of the greenness of the agricultural canopy. Here we can observe some of the notable work done in the field of agriculture through NDVI. S. Estel [3] developed a method of estimating the active and fallow land which can be reused for the agriculture purpose. The MODIS (Moderate Resolution Imaging SpectroRadiometer) satellite data was used in this research work. Y. Shao [4] used asymmetric

978-1-5386-6624-1/18 $31.00 © 2018 IEEE

Gaussian, double-logistic, whittaker smoother and discrete Fourier transformation with MODIS data and used NDVI technique to estimate greenness around the area across Great lake basins (GLB). B Zheng [5] used stratified random approach and intelligent selection approach fused with support vector machines (SVM) to identify the various crop types in a complex cropping system. The data used in their research was multi-temporal Landsat data blended with NDVI technique. S. Eckert [6] used NDVI technique fused with the MODIS data for detecting the land degradation in the Mongolia area. Y.Ke [7] used and compared Landsat 8 Operational Land Imager (OLI) and Landsat 7 Enhanced Thematic Mapper Plus (ETM+) data with continuous '8' day observation to calculate Top of Atmosphere (TOA) reflectance, surface reflectance and NDVI. Z.Pan [8] used HJ-1A/B satellite data to map the crop canopy of Guanzhong Plain, China. The time series data acquired from the satellite is used against NDVI technique to calculate the greenness in the Guanzhong Plain. X Tong [9] used NDVI technique to monitor the change developed in the agricultural regions of Sahel dry lands of Western Niger from 2000 to 2004. X. Zhu [10] used the Landsat time series data to estimate the increase in biomass and forest land cover in the Ohio, United States. M.Gason [11] used NDVI technique to mark the greenness in the surrounding region followed by the development of a relationship between building environment characteristics and NDVI. T.Chen [12] used satellite data to measure the content of moisture in the soil and later used NDVI technique to monitor the areas where moisture is present (as in the areas with more moisture content, the amount of greenness present in that area is also more as compared to the less moisture exposed area).

II. LANDSAT 8 SPECTRUM

National Aeronautics and Space Administration (N.A.S.A) and United States Geological Survey (U.S.G.S) together, came forward to develop one of the most successful satellite program 'Landsat'. This program was started in the year 1972. Since then NASA has successfully launched '7' Landsat missions and one failed Landsat 6 satellite missions. Today, Landsat 8 is operational it was launched on February 11, 2013 [13]. It is carrying two sensors operational land imager (OLI) and a thermal infrared sensor (TIRS) for the purpose of earth monitoring and imaging. OLI sensor collect information from the nine spectral bands. These bands are coastal / aerosol, blue, green, red, near infrared, short wave infrared 1, shortwave infrared 2, panchromatic, and cirrus. Landsat 8 works on two mission objectives command and control of satellite and transmission of the received data back to Earth data centre [14].

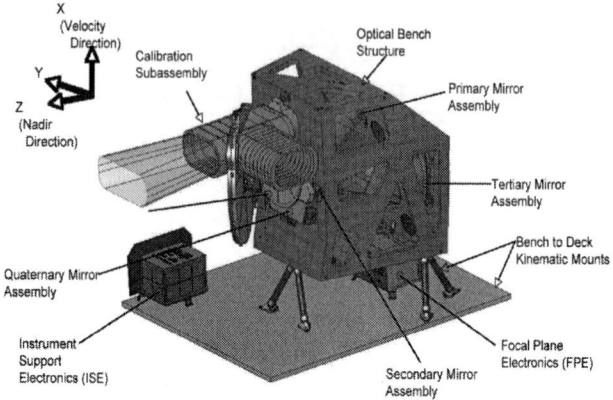

Fig. 1 Operational Land Imager (OLI) sensor setup (Image Courtesy: USGS)

III. MATHEMATICAL MODEL OF NDVI

The Normalized Difference Vegetation Index (NDVI) is a universal measure of the plant 'greenness' or photosynthetic activity. It is considered as one of the most reliable methods to estimate the amount of greenness of the plant. Vegetation indices are normally based on the observations that different surfaces reflect for the different types of lights. Photosynthetically active vegetation usually absorbs most of the red light that collides with the plant surface and reflects much of the near infrared light. Vegetation that is dead reflects more red lights and less near infrared light. Likewise, non-vegetated surfaces have a much evener reflectance across the light spectrum. As a result of this the area under more vegetation seems to appear redder in colour, as the plants and trees absorb water the amount of red light appearing will be denser in colour. Now the amount of greenness present in the crop is estimated by taking the ratio of the red and near-infrared bands. The Normalized Difference Vegetation Index (NDVI) is probably the most common of these ratio indices for vegetation. The NDVI is calculated on a per-pixel basis as the normalized difference between the red and near-infrared bands.

The mathematical expression of the calculation of the NDVI is

$$NDVI = \frac{NIR-Red}{NIR+Red} \qquad (1)$$

Where 'NIR' is the near-infrared band and 'Red' is the red band. Since we are using Landsat 8 image in our investigation so band no: 4 and band no: 5 are used for RED and Near Infrared respectively. NDVI techniques constitute all of the above special features with it, but at the same time it has some limitations also, the limitation of the NDVI techniques are listed below.

Temporal Resolution: In NDVI calculation, it is very essential to maintain the right balance between the temporal resolution of the data and the time scale of the quantity to be measured.

Landcover Types: During the calculation of the NDVI from the satellite data it is very essential to have a proper data under investigation as improper land cover may provide wrong interpretation of the NDVI values.

Sparse vegetation and Soil types: The radiations reflected back from the soil surface can produce a significant effect on the NDVI values. These variations can go up to 20% so it becomes essentially important to consider these effects while estimating the final NDVI values.

Sensor Degradation: In this relationship between the sensor values and pixel values is related to each other. Since the satellite radiometers degrade with the increasing time. So it is important to correct the data before evaluating the NDVI values.

Off Nadir effects: Radiometers are designed to scan the Earth, during the scanning process, there is a single point only located at the centre of the scanning device, that is located directly below the scanner. The distance between this point and radiometers sometimes increases and sometimes decreases. This results in the increase in the atmospheric interference in the system, resulting in the decrease in NDVI values.

Atmospheric interferences: Scattering increases the quantity of red radiation received by satellite as we are aware that red is most readily scattered in the atmosphere than near infrared. This effect results in reducing the NDVI values.

IV STUDY AREA

In this study, we have taken the satellite data of Dehradun, Uttarakhand which is the 27th State of the Union of India. The state comprises the central Himalaya, which has spread over 53, 483 square km. The state of Uttarakhand is having a population of 8.48 million populations (Census, 2001). The state is known for its scenic beauty and is also known as "Devbhoomi" due to its shrines, temples & places of worship and meditation. In this research work, we have taken Landsat 8 multispectral images of Dehradun, Uttarakhand, India, which is also the capital of Uttarakhand State. The state of Uttrakhand consists of two different regions Garhwal region and the Kumaon region. Dehradun falls under the Garhwal region. The latitude and longitude of the Dehradun region is 30.3165° N, 78.0322° E. It has an elevation of 653 meters, which is equivalent to 2142 feet above the sea level. In the fig. 2 we can observe the location of Dehradun region in the state of Uttarakhand.

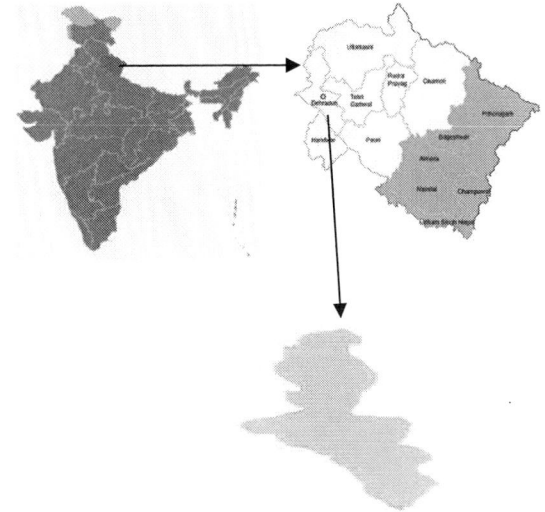

Fig. 2 Geographical location of India, Uttarakhand, Dehradun

The major crops grown in the state of Uttarakhand include basmati rice, wheat, soybeans, groundnuts, coarse cereals, pulses, and oilseeds. Thus the agriculture sector of the Uttarakhand needs to be carefully monitored for any sought crop related problem.

V EXPERIMENT

In the NDVI calculation of the area of Dehradun, we have obtained images of the Dehradun region in an individual band format from United States Geological Survey (U.S.G.S) all the bands are stacked in a single multispectral image with the assistance of ENVI 4.8 software. The stacked image is shown in the fig. 3.

Fig. 3 Original Landsat 8 stacked image of the Dehradun 2018

978-1-5386-6624-1/18 $31.00 © 2018 IEEE

Now creating the false colour composite (FCC) image of the original image, in FCC we replace the NIR band with Red band, Red band with Green band and Green band with Blue band. The FCC image also provides us information regarding denser, less dense and no agricultural area.

Fig. 4 False colour composite image of the original image of Dehradun

Now the false colour composite image presents the deep red, light red, and snow cover over the mountains in white. The red in the false colour composite represents the heavy absorbed moisture content in the vegetation region. As the region is mostly mountainous we can observe the spread of red colour in most of the region. Now applying band math to create a single band image of the region, which is the ratio of the (NIR-Red) and (NIR+Red).

This ratio is the NDVI values obtained after the differencing of the NIR and Red band.

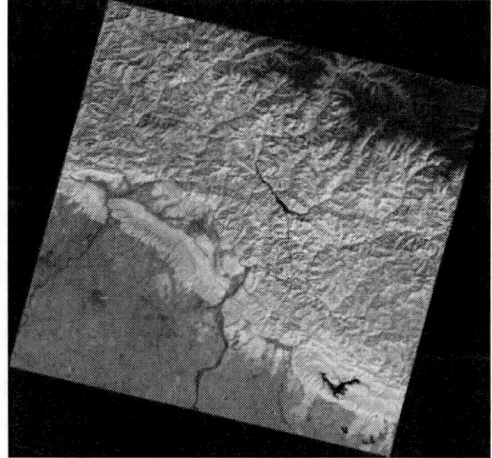

Fig. 5 NDVI single band image

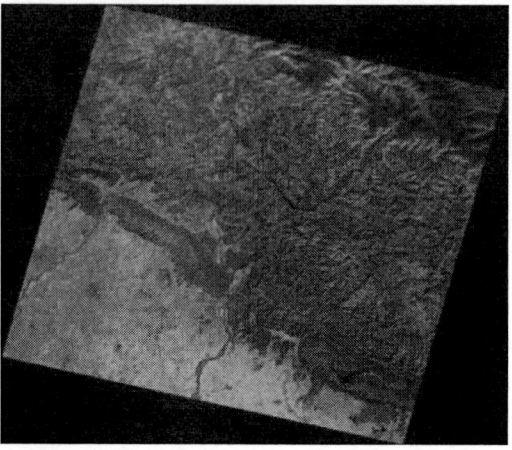

Fig. 6 NDVI colour band image

Now in the fig.6, we have highlighted the three different locations. The green area represents the less dense area which is actually the region of agricultural purpose. The blue area represents the densest forest cover area. The highlighted brown area represents the mountainous region which is usually covered with snow so there is no agriculture activity.

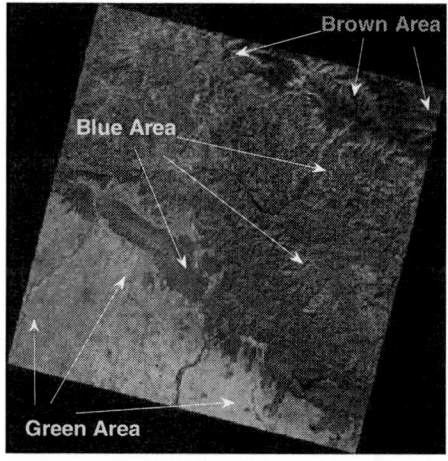

Fig. 7 NDVI colour band image with Area 1, 2 & 3

Now the NDVI values obtained in the three different areas are shown in the table. 1 the colour representation of the area under investigation can be observed in the fig. 7. From here we have observed that the Green Area 1 has moderate NDVI values, Blue Area 2 have highest NDVI values and Brown Area 3 have NDVI values in the negative which symbolizes that there is no vegetation or greenery in that area.

978-1-5386-6624-1/18 $31.00 © 2018 IEEE

TABLE.1 NDVI VALUES FOR THE THREE DIFFERENT LOCATIONS

S.No	Area under Investigation	Area 1 GREEN	Area 2 BLUE	Area 3 BROWN
1	Location 1	0.1046	0.2530	-0.0497
2	Location 2	0.1703	0.3013	-0.0558
3	Location 3	0.1662	0.2789	-0.0526
4	Location 4	0.1294	0.2348	-0.0703
5	Location 5	0.1591	0.2413	-0.0405
6	Location 6	0.1448	0.2175	-0.0487
7	Location 7	0.1611	0.2512	-0.0146
8	Location 8	0.1245	0.2617	-0.0112
9	Location 9	0.1211	0.2258	-0.0069
10	Location 10	0.1647	0.2322	-0.0023

Now we have computed the variation in the values of NDVI for the different locations in Area 1, 2 and 3.

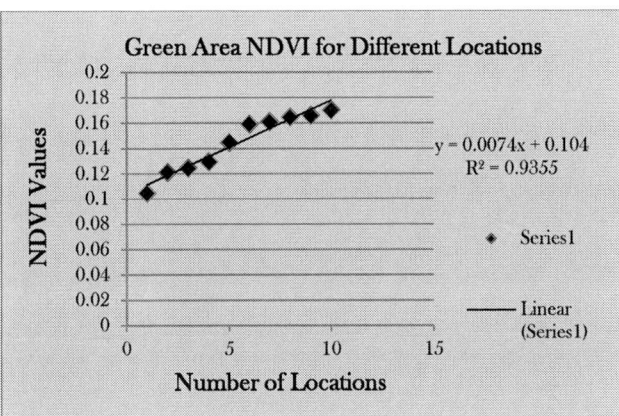

Fig. 8 Relationship plot between obtained NDVI values and different location of Area 1

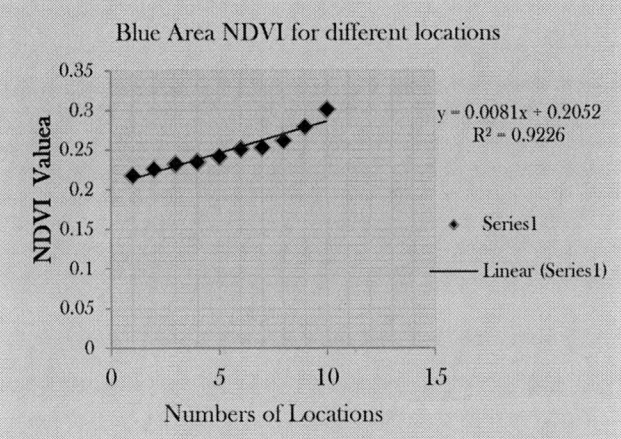

Fig. 9 Relationship plot between obtained NDVI values and different location of Area 2

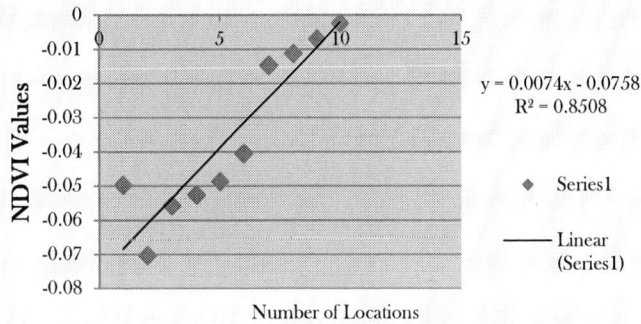

Fig. 10 Relationship plot between the obtained NDVI values and different location for Area 3

Now from the plots shown in fig. 8, 9and 10 we can observe that on shorting the obtained NDVI values a linear relationship is obtained have R^2= 0.9355, 0.9226 and 0.8508 respectively.

V CONCLUSION

The result of this research work has proved that NDVI is an efficient technique for the estimation of the greenness in the crops and trees. One of the important observations concluded from this research work is NDVI values are directly proportional to the greenness present in the crops. As the amount of greenness increases the NDVI values will approach towards '1' and for no agriculture region, NDVI values will approach a negative value.

ACKNOWLEDGEMENT

Authors wish to express their deep gratitude of thanks to United States Geological Survey (U.S.G.S) and U.S.G.S Earth Explorer for providing multispectral Landsat images of Dehradun, Uttarakhand, India used in this research work.

REFERENCES

[1] United Nations, "The impact of disasters on agriculture: Addressing the information gap," Food and Agriculture Organization of the United Nations, Rome, 2017.

[2] NASA. (2018, June) NASA Earth Observatory. [Online]. https://earthobservatory.nasa.gov/Features/MeasuringVegetation/measuring_vegetation_2.php

[3] Stephan Estel, Tobias Kuemmerle, Camilo Alcántara, and Christian Levers, "Mapping farmland abandonment and recultivation across Europe using," *Remote Sensing of Environment*, vol. 163, no. 2, pp. 312–325, March 2015.

[4] Yang Shao, Ross S.Lunetta, Brandon Wheeler, John S.Iiames, and James B.Campbell, "An evaluation of time-series smoothing algorithms for land-cover classifications using MODIS-NDVI multi-temporal data," *Remote Sensing of Environment*, vol. 174, no. 1, pp. 258-265, March 2016.

[5] Baojuan Zhenga, Soe W.Myint, Prasad S.Thenkabail, and Rimjhim M.Aggarwal, "A support vector machine to identify irrigated crop types using time-series Landsat NDVI data," *International Journal of Applied Earth Observation and Geoinformation*, vol. 34, no. 1, pp. 103-112, Feburary 2015.

[6] Sandra Eckert, Fabia Hüsler, Hanspeter Liniger, and Elias Hodel, "Trend analysis of MODIS NDVI time series for detecting land degradation and regeneration in Mongolia," *Journal of Arid Environments*, vol. 113, no. 1, pp. 16-28, Feburary 2015.

[7] Yinghai Kea, Jungho Im, Junghee Leeb, Huili Gong, and Youngryel Ryu, "Characteristics of Landsat 8 OLI-derived NDVI by comparison with multiple satellite sensors and in-situ observations," *Remote Sensing of Environment*, vol. 164, no. 1, pp. 298-313, July 2015.

[8] Zhuokun Panab, Jingfeng Huang, Qingbo Zhou, Limin Wang, and Yongxiang Cheng, "Mapping crop phenology using NDVI time-series derived from HJ-1 A/B data," *International Journal of Applied Earth Observation and Geoinformation*, vol. 34, no. 1, pp. 188-197, Feburary 2015.

[9] Xiaoye Tong et al., "Revisiting the coupling between NDVI trends and cropland changes in the Sahel drylands: A case study in western Niger," *Remote Sensing of Environment*, vol. 191, no. 1, pp. 286-296, March 2017.

[10] Xiaolin Zhu and Desheng Liu, "Improving forest aboveground biomass estimation using seasonal Landsat NDVI time-series," *ISPRS Journal of Photogrammetry and Remote Sensing*, vol. 102, no. 1, pp. 222-231, April 2015.

[11] Mireia Gascon, Marta Cirach, David Martínez, and Payam Dadvand, "Normalized difference vegetation index (NDVI) as a marker of surrounding greenness in epidemiological studies: The case of Barcelona city," *Urban Forestry & Urban Greening*, vol. 19, no. 1, pp. 88-94, September 2016.

[12] T.Chen, R.A.M.de Jeu, Y.Y.Liu, G.R.van der Werf, and A.J.Dolmana, "Using satellite-based soil moisture to quantify the water driven variability in NDVI: A case study over mainland Australia," *Remote Sensing of Environment*, vol. 140, no. 1, pp. 330-338, January 2014.

[13] Wikipedia. (2018, March) Landsat 8. [Online]. https://en.wikipedia.org/wiki/Landsat_8

[14] U.S. Department of the Interior and U.S. Geological Survey. (25/04/2018, April) USGS Science of Changing the World. [Online]. https://landsat.usgs.gov/landsat-8

978-1-5386-6624-1/18 $31.00 © 2018 IEEE

Electronically Tunable Current Mode Universal Filter using a Single MX CCCII

Deepak Agrawal, Sudhanshu Maheshwari
Department of Electronics Engineering,
Z. H. College of Engineering and Technology, AMU,
Aligarh 202002, India.
deepaka.38@gmail.com, maheshwarispm@gmail.com

Abstract—**In this paper an electronically tunable current-mode universal filter is proposed, which employing only a single active element namely Multi-X current controlled conveyor (MX-CCCII), and two grounded capacitors. The proposed circuit using the minimum component, so it is easily integrable on IC. Also, the circuit provide the low input impedance for the input current and high output impedance for output currents so, it is easily cascadable at the input and output node. The circuit provides the filter functions: LP, HP, BP, and BR simultaneously without need of any matching condition. The angular frequency (ω_o) is electronically tunable and do not depend on the quality factor (Q). The effects of non-idealities and parasitics of active element on the circuit performances are discussed. The results are verified through PSPICE simulation using a 0.25 μm TSMC model with the supply voltages of ±1.25 V.**

Keywords—*current conveyors, universal filter, MX-CCCII*

I. INTRODUCTION

Universal biquadratic filters are the most important analog function which is widely used in communication and instrumentation systems [1-5]. In the literature, second generation current conveyor (CCII) based filters require RC networks [1-3]. Later, the active-C filters have been realized by using current controlled conveyor (CCCII) [4-6]. Recently, an extra X current controlled conveyor (EX-CCCII) have been used to minimize or reducing the complexity of the existing biquadratic filters [7]. The current mode (CM) filter that is current signal at input and current signal at output node this require low impedance at the input node and high impedance at the output node so, the circuit can be cascadable [6-15]. Owing to the popularity of EX-CCCII, this paper presents a current conveyor with three X terminals namely MX-CCCII, which is then effectively applied to realize a CM universal filter. The paper is organized as follows: in Section 2 the CMOS implementation of MX-CCCII is described; and in the same section CM universal filter is realized with the study of parasitic effects. In Section 3, simulation results are given. In section 4, comparative study is done with the available circuits in the literature [8-14]. Conclusion of this study is presented in Section 5.

II. CIRCUIT DESCRIPTION

The CMOS implementation of MX-CCCII is shown in Fig. 1. The transistors M1-M8 realize mixed translinear loop, transistors M9-M11 and M18, M19 provide a DC biased to translinear loop, whereas M12-M17, and M20-M25 realize

simple current mirror which transfer the current from Xi (i=1, 2, 3) terminals to Zi (i=1, 2, 3) terminals. The port relationships of MX-CCCII are given as:

$$
\begin{bmatrix} I_Y \\ V_{X1} \\ V_{X2} \\ V_{X3} \\ I_{Z1} \\ I_{Z2} \\ I_{Z3} \end{bmatrix} = \begin{bmatrix} 0 & 0 & 0 & 0 & 0 & 0 & 0 \\ \beta_1 & R_{X1} & 0 & 0 & 0 & 0 & 0 \\ \beta_2 & 0 & R_{X2} & 0 & 0 & 0 & 0 \\ \beta_3 & 0 & 0 & R_{X3} & 0 & 0 & 0 \\ 0 & \alpha_1 & 0 & 0 & 0 & 0 & 0 \\ 0 & 0 & \alpha_2 & 0 & 0 & 0 & 0 \\ 0 & 0 & 0 & \alpha_3 & 0 & 0 & 0 \end{bmatrix} \begin{bmatrix} V_Y \\ I_{X1} \\ I_{X2} \\ I_{X3} \\ V_{Z1} \\ V_{Z2} \\ V_{Z3} \end{bmatrix} \quad (1)
$$

Where, β_i is the voltage transfer gain from Y terminal to Xi terminals and α_i is the current transfer gain from Xi terminals to Zi terminals (indices i denote the 1, 2, and 3). The negative Zi stages are implemented through cross-inverted current mirrors. The current gain of negative Zi terminals is denoted as $-\alpha_{i1}$ which is not shown in equation (1). Next, R_{X1}, R_{X2} and R_{X3} are the intrinsic resistances at X1, X2 and X3 terminals, respectively with their value as:

$$
R_{X1} = R_{X2} = R_{X3} = \frac{1}{\sqrt{8\mu C_{OX}(\frac{W}{L})I_o}} \quad (2)
$$

Where, μ is the mobility of electron, C_{OX} is the oxide capacitance per unit area of MOS transistor, I_o is the biasing current of MX-CCCII and (W/L) refers to ratio of transistors forming translinear loops.

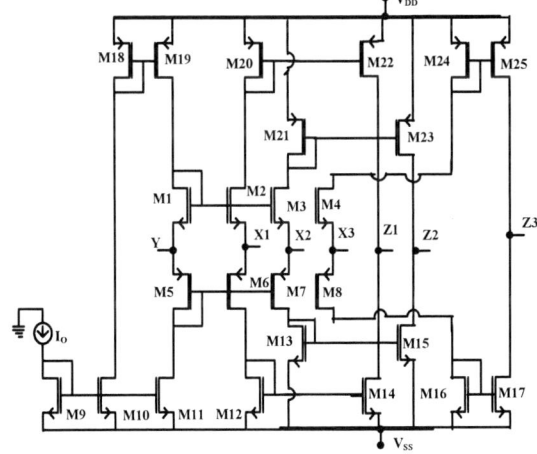

Fig. 1 CMOS implementation of MX-CCCII

978-1-5386-6624-1/18 $31.00 © 2018 IEEE

The proposed electronically tunable CM filter is shown in Fig. 2. The proposed circuit employs only a single MX-CCCII, and two grounded capacitors so, it is appropriate for IC implementation. The circuit provides the cascadibility and tunability features as well as all filter functions are obtained simultaneously. Now, the circuit is analyzed by using the equation (1) and yields as:

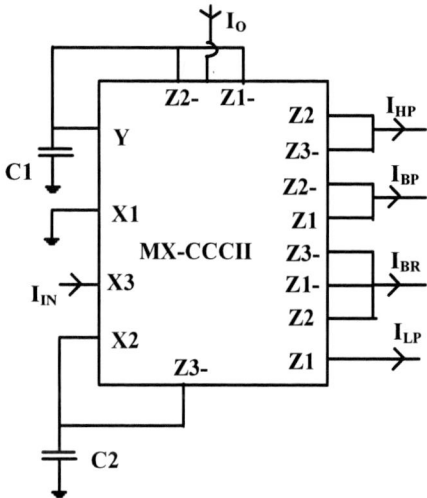

Fig. 2 Proposed electronically tunable CM universal filter

$$I_{LP} = \frac{-\alpha_1 \alpha_2 \beta_1}{D(s)} \tag{3}$$

$$I_{HP} = \frac{A(s)}{D(s)} \tag{4}$$

$$I_{BP} = \frac{-B(s)}{D(s)} \tag{5}$$

$$I_{BR} = \frac{C(s)}{D(s)} \tag{6}$$

Where;

$$A(s) = \alpha_{31} s^2 C_1 C_2 R_X^2 + sR_X \{ C_2 (\alpha_{31} - \alpha_2) - C_1 \alpha_{31} (\alpha_2 \beta_2 - \alpha_{11} \beta_1) \}$$
$$+ (\alpha_2 + \alpha_{31})(\alpha_2 \beta_2 - \alpha_{11} \beta_1)$$

$$B(s) = \alpha_1 \alpha_2 \beta_1 + \alpha_{21} \alpha_{11} \beta_1 - 2\alpha_2 \alpha_{21} \beta_2 - \alpha_{21} s C_2 R_X$$

$$C(s) = \alpha_{31} s^2 C_1 C_2 R_X^2 + s C_2 R_X (\alpha_{31} - \alpha_2) - \alpha_{31} s C_1 R_X (\alpha_2 \beta_2 - \alpha_{11} \beta_1)$$
$$+ (\alpha_2 + \alpha_{31})(\alpha_2 \beta_2 - \alpha_{11} \beta_1) + \alpha_{11} \beta_1$$

$$D(s) = s^2 C_1 C_2 R_X^2 + sR_X \{ C_2 - C_1 (\alpha_2 \beta_2 - \alpha_{11} \beta_1) \} + (\alpha_2 \beta_2 - \alpha_{11} \beta_1) + \frac{\alpha_2 \beta_2}{\alpha_3}$$
$$\tag{7}$$

From, the above equations (3)-(7) the obtained filter parameters as:

$$\omega_o = \sqrt{\frac{\alpha_2 \beta_2 - \alpha_{11} \beta_1 + \frac{\alpha_2 \beta_2}{\alpha_3}}{C_1 C_2 R_X^2}} \tag{8}$$

$$Q_o = \frac{\sqrt{\left(\alpha_2 \beta_2 - \alpha_{11} \beta_1 + \frac{\alpha_2 \beta_2}{\alpha_3} \right) C_1 C_2}}{C_2 - C_1 (\alpha_2 \beta_2 - \alpha_{11} \beta_1)} \tag{9}$$

Now, consider the unity value of the non-idealities in the above equations thus the equations (9-10) becomes:

$$\omega_o = \frac{1}{R_X} \sqrt{\frac{1}{C_1 C_2}} \tag{11}$$

$$Q_o = \sqrt{\frac{C_1}{C_2}} \tag{12}$$

Thus, the ω_o can be tuned by varying a bias current without affecting the value of Q_o.

Parasitic effects

The parasitics present at the terminals of MX-CCCII is described as follows: (i) the resistance R_Y is in parallel with capacitor C_Y at the Y terminal; (ii) the resistance R_{Xi} is at X_i terminals; (iii) the resistance R_{Zi} is in parallel with capacitor C_{Zi} at the Z_i terminals. Considering these parasitics the circuit (Fig. 2) is reanalyzed which yields as:

$$I_{LP} = \frac{-1}{D(S)} \tag{13}$$

$$I_{HP} = \frac{s^2 C_1' C_2' R_X^2 + \dfrac{sR_X^2 (C_2' R_Y + C_1' R_Z)}{R_Y R_Z} - \dfrac{R_X^2}{R_Y R_Z}}{D(s)} \tag{14}$$

$$I_{BP} = \frac{s C_2' R_X + \dfrac{R_X}{R_Y}}{D(s)} \tag{15}$$

$$I_{BR} = \frac{s^2 C_1' C_2' R_X^2 + 1 + \dfrac{sR_X^2 (C_2' R_Y + C_1' R_Z)}{R_Y R_Z} + \dfrac{R_X^2}{R_Y R_Z}}{D(s)} \tag{16}$$

Where,

$$D(s) = s^2 C_1' C_2' R_X^2 + s C_2' R_X + \frac{sR_X^2 (C_2' R_Y + C_1' R_Z)}{R_Y R_Z} + 1 + \frac{R_X (R_X + R_Z)}{R_Y R_Z} \tag{17}$$

$C_1' = C_1 + C_{Z3-}$; $C_2' = C_2 + C_Y + C_{Z1-} + C_{Z2}$; $R_Y = R_Y // R_{Z1-} // R_{Z2}$; $R_Z = R_{Z3-}$.

From, the above equations the obtained filter parameters as:

$$\omega_o = \sqrt{\frac{1}{C_1' C_2' R_X^2} + \frac{R_X^2 + R_X R_Z}{C_1' C_2' R_X^2 R_Y R_Z}} \tag{18}$$

$$Q_o = \frac{\sqrt{C_1' C_2' R_Y R_Z (R_X^2 + R_Y R_Z + R_X R_Z)}}{C_2' R_Y R_Z + C_2' R_Y R_X + C_1' R_X R_Z} \tag{19}$$

From the above equations it is concluded that ω_o and Q_o are slightly affected by the parasitics.

III. SIMULATION RESULTS

The port relationship of proposed MX-CCCII is verified through PSPICE simulation using the transistor parameters of TSMC MOSIS 0.25 μm CMOS technology with supply voltages of $V_{DD} = -V_{SS} = 1.25$ V. Fig. 3 shows the voltage transfer gain (βi) from Y terminal to X_i terminals with their value of 0.9977. Next, Fig. 4 shows the current transfer gain (αi) from X_i terminal to Z_i terminals with their value of 1.0057. Next, the proposed electronically tunable CM filter is verified. The circuit in Fig. 2 is designed for frequency

978-1-5386-6624-1/18 $31.00 © 2018 IEEE

f_o= 3.09 MHz, for this C_1=C_2= 50 pF and bias current I_o=50 μA to provide the intrinsic resistances R_{X1}= R_{X2}= R_{X3} = 1.03 kΩ were used. Fig. 3 shows the gain response for LP, HP, BP and BR filters. The obtained centre frequency for band pass filter is 3.35 MHz which is slightly higher than the theoretical value due to the parasitics as shown in equation (18). Next, the Fig. 6 shows the tunabilty of the centre frequency of BP filter which is varying with the bias currents of 10 μA, 50 μA, and 100 μA. The centre frequency are obtained as 1.58 MHz, 3.35, and 4.35 MHz correspond to the above biasing currents. Next, the quality factor of BP filter can be tuned by changing the values of capacitors C1 and C2 such that the ratios are changed but their product remains constant. Fig. 7 shows the tunability of Q with the value of 1, 2, and 5. Next, the 10% tolerance is considered in the capacitances value to studied the effect on centre frequency of BP filter. For this, the Monte Carlo analysis is done for 100 runs. Fig. 8 shows the obtained mean centre frequency is 3.35 MHz. The value of the output impedances are obtained as 0.33 MΩ, 0.17 MΩ, 0.17 MΩ, 0.10 MΩ at the LP, HP, BP and BR respectively through the simulations.

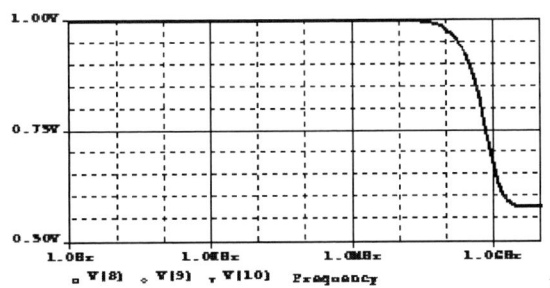

g. 3 Magnitude response of voltage transfer functions

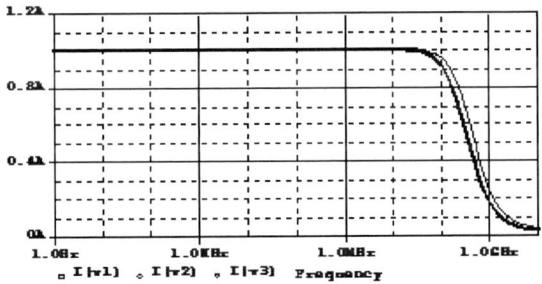

Fig. 4 Magnitude response of current transfer functions

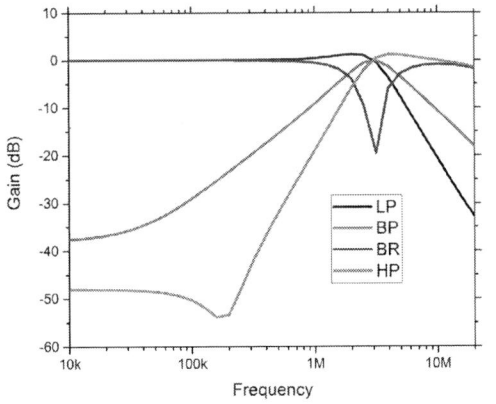

Fig. 5 Gain response for CM filters

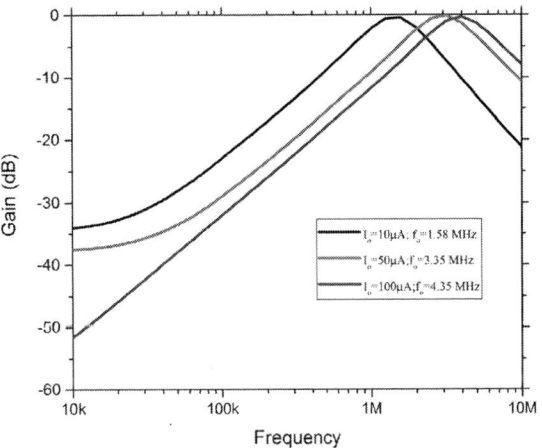

Fig. 6 Tunability of centre frequency (BP) for different bias current I_o.

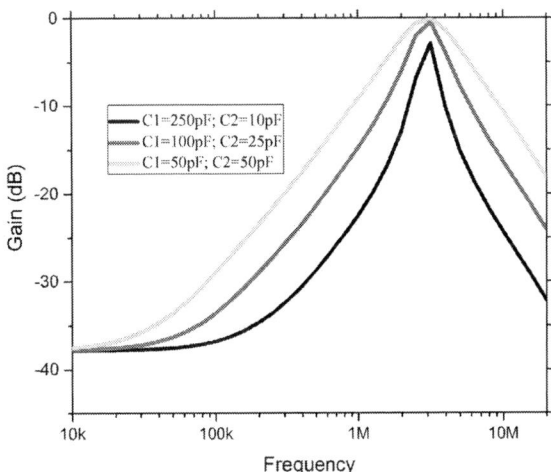

Fig. 7 Tunability of Quality factor

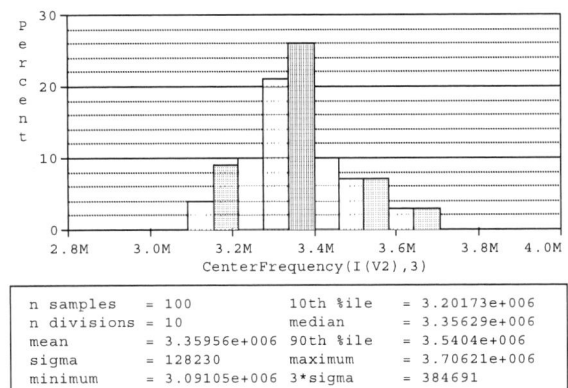

n samples = 100	10th %ile	= 3.20173e+006
n divisions = 10	median	= 3.35629e+006
mean = 3.35956e+006	90th %ile	= 3.5404e+006
sigma = 128230	maximum	= 3.70621e+006
minimum = 3.09105e+006	3*sigma	= 384691

Fig. 8 Monte Carlo analysis for the centre frequency of BP filter

IV. COMPARATIVE STUDY

Now, the comparative study is done with the available circuits [8-14] working in CM operation and employing only a single active element. The circuits presented in [8-

11, 13, 14] employs three or more than three passive elements. The circuits presented in [8, 9, 11] do not provide the tunability feature whereas; the circuits [8-10, 12, 14] do not provide the cascadibility feature. Moreover, the circuits [9, 11, 14] requires the passive matching condition to realizing the filter functions. The CMOS implementation of the circuit presented in [14] require more number of current sources. Next, the circuits [8, 10, 12, 13] provide some filter functions current output signals through the capacitor which is not feasible.

TABLE I.   COMPARATIVE STUDY BASED ON SINGLE ACTIVE ELEMENT

Ref.	Active element	Passive element	Tunab ility	Casca dibility	Matching condition require
[8]	CDTA	2C(G), 2R(F)	No	No	No
[9]	FDCCII	2C(G), 3R(G)	No	No	Yes
[10]	VDTA	2C(G), 1R(G)	Yes	No	No
[11]	MO-DXCCII	2C(G), 2R(G)	No	Yes	Yes
[12] fig.2	EX-CCCII	2C(G)	Yes	No	No
[13]	DX-CCTA	2C(G), 1R(G)	Yes	Yes	No
[14]	CFTA	2C(G), 1R(F)	Yes	No	Yes
Propo sed	**MX-CCCII**	**2C(G)**	**Yes**	**Yes**	**No**

V. CONCLUSION

In this work, an electronically tunable CM universal filter is proposed which employs a single MX-CCII, and two grounded capacitors. The circuit provides the input current signal at low input impedance of 1.03 kΩ and output current signals at high output impedances as 0.33 MΩ, 0.17 MΩ, 0.17 MΩ, 0.10 MΩ at the LP, HP, BP and BR respectively; thereby there is no need of buffer at input and output stages. The circuit provides the filter functions namely low pass, high pass, band pass and band reject simultaneously. Also, the angular frequency is electronically tunable and independent of the quality factor. Also, the proposed work is compared with the available circuits in the technical literature. PSPICE verifications using 0.25 μm CMOS parameters and supply voltage of ±1.25 V support the validity of the proposed circuit.

ACKNOWLEDGMENT

The research is supported by "Visvesvaraya Ph.D scheme" under the Ministry of Electronics & Information Technology (MeitY) Government of India.

REFERENCES

[1] S.Özoğuz, A. Toker, and O. Çiçekoğlu, "New current-mode universal filters using only four (CCII+)s", *Microelectronics Journal, 30*(3), 1999, pp.255-258.

[2] D.R. Bhaskar, V.K. Sharma, , M. Monis, and S.M.I Rizvi, "New current-mode universal biquad filter", *Microelectronics Journal, 30*(9), 1999, pp.837-839.

[3] F. Yucel, and E. Yuce, "Grounded capacitor based fully cascadable electronically tunable current-mode universal filter", *AEU-International Journal of Electronics and Communications, 79*, 2017, pp.116-123.

[4] I.A. Khan, and M.H. Zaidi, "Multifunctional translinear-C current-mode filter", *International Journal of Electronics, 87*(9), 2000, pp.1047-1051.

[5] W. Tangsrirat, "Current-tunable current-mode multifunction filter based on dual-output current-controlled conveyors", *AEU-International Journal of Electronics and Communications, 61*(8), 2007, pp.528-533.

[6] N. Pandey, S.K. Paul, and S.B. Jain, "A new electronically tunable current mode universal filter using MO-CCCII", Analog Integr Circ Sig Process, 58, 2009; pp.171–178.

[7] R.S. Tomar, S.V. Singh, and D.S. Chauhan, "Cascadable low voltage operated current-mode universal filter", *WSEAS Transactions on Signal Processing, 10*, 2014, pp.345-353.

[8] D. Prasad, D.R. Bhaskar and A.K. Singh, "Universal current-mode biquad filter using dual output current differencing transconductance amplifier", *AEU-International Journal of Electronics and Communications, 63*(6), 2009, pp.497-501.

[9] R. Senani, K.K. Abdalla, and D.R. Bhaskar, "A State Variable Method for the Realization of Universal Current-Mode Biquads", *Circuits and Systems, 2*(4), 2011, pp.286-292.

[10] D. Prasad, D.R. Bhaskar, and M. Srivastava, Universal current-mode biquad filter using a VDTA. *Circuits and Systems, 4*(01), 2013, pp. 29-33.

[11] J. Mohan and S. Maheshwari, "Generalized current-mode configuration with low input and high output impedance", *IU-Journal of Electrical & Electronics Engineering, 16*(1), 2016, pp.1971-1979.

[12] D. Agrawal, and S. Maheshwari, "Current mode filters with reduced complexity using a single EX-CCCII", *AEU-International Journal of Electronics and Communications, 80*, 2017, pp.86-93.

[13] A. Kumar, and B. Chaturvedi, "Novel CMOS dual-X current conveyor transconductance amplifier realization with current-mode multifunction filter and quadrature oscillator", *Circuits, Systems, and Signal Processing, 37*(6),2017, pp. 2250-2277.

[14] P. Mongkolwai, T. Pukkalanun, and W. Tangsrirat, "Three-Input Single-Output Current-Mode Biquadratic Filter with High-Output Impedance Using a Single Current Follower Transconductance Amplifier", *IAENG International Journal of Computer Science, 44*(3), 2017,pp. 1-5.

[15] D. Agrawal, and S. Maheshwari, "Cascadable Current mode instrumentation amplifier", *AEU-International Journal of Electronics and Communications, 94*, 2018, pp. 91-101.

2nd IEEE International Conference on Power Electronics, Intelligent Control and Energy Systems (ICPEICES-2018)

Design and Implementation of Cache Coherence Protocol for High-Speed Multiprocessor System

Daman Preet Kaur
*Academic and Consultancy Services
Division
Centre for Development of
Advanced Computing*
Mohali, India
dpkaur.165@gmail.com

V. Sulochana
*Academic and Consultancy Services
Division
Centre for Development of
Advanced Computing*
Mohali, India
vemus@cdac.in

Abstract—To maintain data consistency between the cache memories in centralized and distributed shared-memory multiprocessor system, particular protocols are used known as cache coherence protocols. The performance of a multi-core computer system is strongly influenced by the type of cache coherence protocol used. In this paper, the snoopy bus cache coherence protocols using 3-state, 4-state and 5-state are designed and implemented using the write-invalidate approach in a shared memory dual processor system. The simulation results show that the MOESI protocol reduces the load misses by 48%, memory latency by 35%, power consumption by 18%, increases the gate count by 34% and improves the execution time by 22%. Thus, the overall performance of the MOESI is better than the MESI and MSI cache coherence protocols in a shared memory dual processor system.

Keywords—Cache coherence protocols, MSI, MESI, MOESI, Shared memory multiprocessors

I. INTRODUCTION

In today's computer architectures, shared memory multi-core processors are becoming dominant. But, the processor takes too many cycles to access the main memory in multi-core architectures because of the speed difference between the main memory and processor [1]. So, for tuning up the performance of multi-core architectures large multilevel caches are used in the system that helps to reduce the memory access time of the processor. These caches also help to reduce demand of memory bandwidth of a processor [2]. Due to the reduction in the main memory bandwidth demand of a single processor, multiple processors are able to share the same memory. Such symmetric shared memory multiprocessors are extremely cost-effective that can perform caching of both shared and private data as shown in Fig. 1. The private data is cached only by the single processor so the program behavior is similar to that in the uniprocessor while shared data is replicated in the multiple caches [3].

Along with the reduction in memory bandwidth and access latency, the reduction in the contention is also provided by the replication of data. The contention takes place when the multiple processors read the shared data simultaneously [4]. However, in spite of all these advantages, the hindrance called cache coherence problem was raised due to caching of shared data.

The cache coherence problem may occur due to data inconsistency between the caches and the shared memory

during sharing of modified data, process migration and I/O operation [5]. When the cache memories of multiple processors maintain local copies of the same data object taken from the main memory and even if any one of the caches modifies the value of same data object then that will result in globally inconsistent view of the shared data between caches and shared memory [6]. This problem is called cache coherency. This problem may occur when multiple processors operate independently and asynchronously. To eliminate this problem in the memory system defined protocols are used [7]. These protocols are known as cache coherence protocols. Different kinds of cache coherence protocols can be implemented according to the different multiprocessors based systems [8].

In this paper, the classification of cache coherence protocols and snoopy bus protocols are described in section II. The design and implementation of MSI, MESI, and MOESI protocols using dual-core architecture are shown in section III and finally in section IV conclusion and future scope is presented.

II. CACHE COHERENCE PROTOCOLS

The specific set of rules executed for maintaining coherency in multi-core processors are known as cache coherence protocols. The numbers of cache coherence protocols are available for maintaining coherency in the multiprocessor system. Coherency helps in managing the read and write operations in the shared cache to have the consistent view among all the processors. The fundamental action of a cache coherence protocol is to find the state of shared data blocks. Cache coherence protocols are classified into two categories that are directory-based protocols and snoopy bus protocols [8] [9].

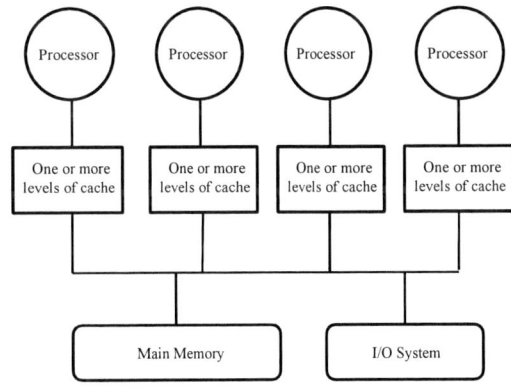

Fig. 1. The basic structure of a shared memory multiprocessor system

978-1-5386-6624-1/18 $31.00 © 2018 IEEE

Directory-based protocols are used in the multistage network in which hundreds of processors are connected by network [10]. These protocols have one location called the central directory which records the state and presence information for each cache block. They support distributed shared-memory architectures.

Snoopy bus protocols are used by the bus-based memory system. Unlike directory protocols, there is not any specified region in snoopy bus protocols that stores the information of data items that are being shared but instead of that these protocols make use of the global bus [10] [11]. The bus is a very convenient source for maintaining cache coherency by allowing all multiple processors to observe ongoing memory transactions [12]. In the bus-based memory system, each cache has its own cache controller that continuously monitors or snoops on the global bus to detect the requested data block copy on the bus [13]. These protocols support centralized shared-memory architectures.

The two main approaches used to implement snoopy bus protocols are write-update and write-invalidate. In write invalidate approach if any one local cache updates its value then the data copies of all other caches are invalidated. In write update approach, if the data item is written in any one cache then it will get updated in all caches. There is more bandwidth consumption in write-invalidate approach because of the broadcast of the write operation in all caches [14] [15]. That is why the use of the write invalidate approach is always preferred in many bus protocols.

In this work, MSI, MESI, and MOESI snooping protocols explained in detail below are designed and implemented in a dual-core system consist of two caches and the main memory. The main cache events and actions that are triggered by the main memory and processor in protocols are explained in Table I. According to all these cache actions, the transitions between the different states of the cache take place [16] [17].

A. MSI (Modified-Shared-Invalid) Protocol

MSI protocol is called classic invalidate-based protocol. It is composed of three states. The state diagram for the MSI protocol is shown in Fig.2. The cache is in a modified state when the data block of the cache is modified and is different from the value of the data block that is in the main memory

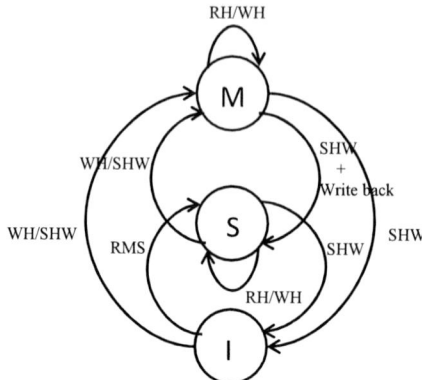

Fig. 2. State transition diagram of MSI Protocol

and other caches. This is the only cache whose data copy is valid [18]. The cache is in a shared state when the data block is shared among all the caches taken from the main memory. The cache is in the invalid state when the data block of the cache is of no use.

B. MESI (Modified-Exclusive-Shared-Invalid) Protocol

The most commonly used protocol that supports write-back cache is the MESI protocol. This is an extension of the MSI protocol in which the fourth state is added known as the *Exclusive* state that helps to reduce the number of bus transitions from an invalid state to a modified state [19]. The state transition for the MESI protocol is shown in Fig.3. When only one cache has a valid data block then that cache is known to be in the exclusive state and that valid data block is also updated in the main memory but not in the other caches. The use of the exclusive state is that without any need for snooping from other caches, local cache can modify itself independently.

C. MOESI (Modified-Owned-Exclusive-Shared-Invalid) Protocol

MOESI protocol is an improved version of the MESI

TABLE I. CACHE EVENTS AND THEIR ACTIONS

Cache Events	Actions	
	Source	*Function*
Write hit (WH)	Processor	Write data in the cache
Read hit (RH)	Processor	Read data from the cache
Snoop hit on write (SHW)	Bus	Hit from another cache during write
Snoop hit on read (SHR)	Bus	Hit from another cache during read
Read Miss Exclusive (RME)	Bus	Request data from cache or memory
Read Miss Shared (RMS)	Bus	Request data from the other cache

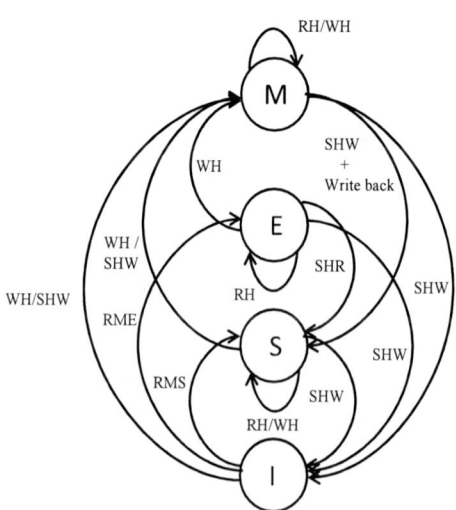

Fig. 3. State transition diagram of MESI Protocol

978-1-5386-6624-1/18 $31.00 © 2018 IEEE

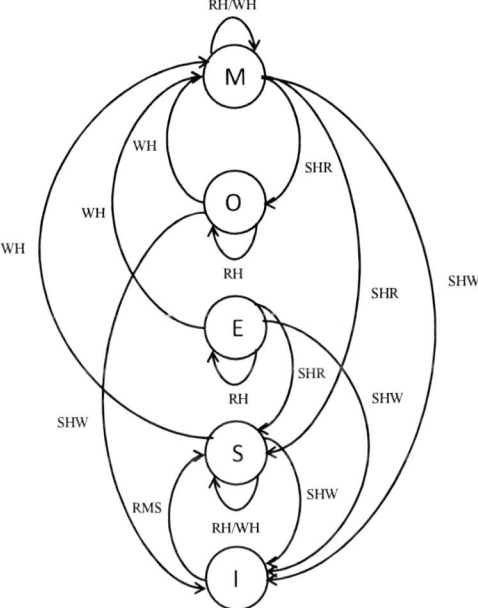

Fig. 4. State transition diagram of MOESI Protocol

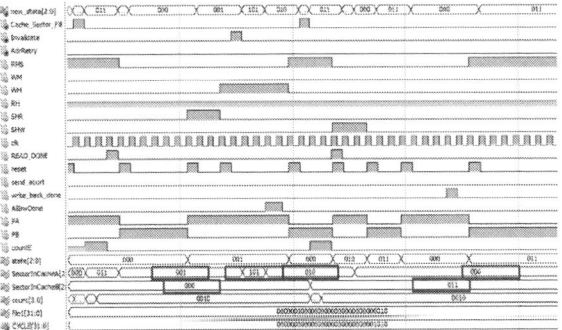

Fig. 5. State transitions of MSI Protocol

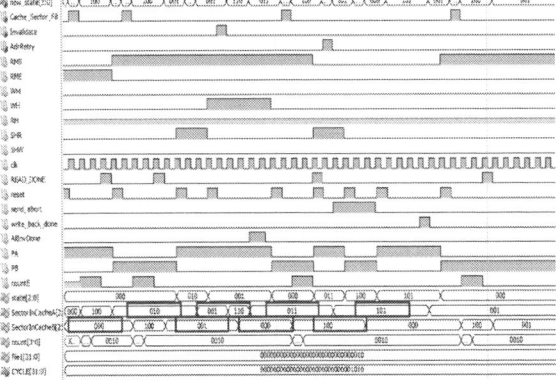

Fig. 6. State transitions of MESI Protocol

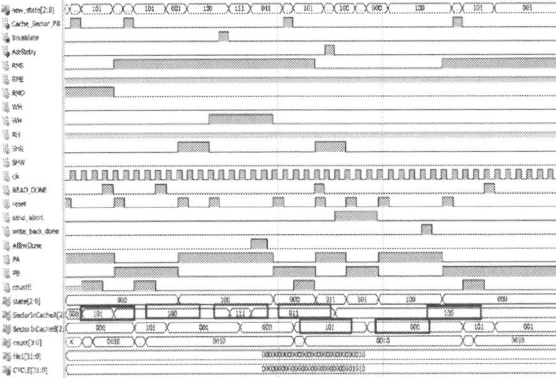

Fig. 7. State transitions of MOESI Protocol

in which the fifth state is added called the owned state that helps to reduce the number of memory accesses [20]. The state diagram of the MOESI protocol is shown in Fig.4. In the multiprocessor system, the owned state cache containing the valid data block is only one and that cache can copy the valid data block to other caches directly in spite of sharing with the main memory to minimize access to the main memory. Thus, the owned state represents both the modified state and shared state. The dirty values can also be shared with other caches because the data in the main memory is not updated [21]. It means that owned state cache has a valid data block and other caches and the main memory may have a valid data block too at the same time. This is another advantage of this protocol.

III. RESULTS

A. Design of Protocols

The state machines of write-invalidate snooping cache coherence protocols viz. MSI, MESI, and MOESI have been first designed using write-back cache block in Verilog HDL and simulated using ISim simulator. The different input signals like write hit (WH), snoop hit on write (SHW), read hit (RH) and, snoop hit on read (SHR) are applied in the cache to run the different protocols state wise. Fig.5, Fig.6, and Fig.8 show the successful state wise transition of MSI, MESI, and MOESI protocols in the simulation waveform. The different states of the two caches are represented by the three-bit signals named as SectorInCacheA and SectorInCacheB respectively.

The designed MSI, MESI and MOESI protocols are implemented in a dual processor system. The general design of shared memory dual processor system used to verify the functionality of snooping protocols consists of processor unit (PU), memory mapping unit (MMU), cache, memory bus controller (MBC) and memory as shown in Fig.8. All

modules of the dual-core system are simulated in ModelSim PE Student Edition. There are two caches – Cache A and Cache B in the design. The caches are two-way associative.

Each cache has 8 entries of 11 bits that includes data, tag, update, dirty and valid bits. The two processor unit Processor A and Processor B are used to perform write actions and read actions in the caches. The shared memory used in the design has 32 entries and each entry is of 10 bits as shown in Fig.9.

Memory mapping unit is used to convert virtual address to physical address by cutting the two most significant bits of the virtual address. Memory bus controller is used to make the synchronization between different modules that access the shared bus at the same time. The simulations of cache

978-1-5386-6624-1/18 $31.00 © 2018 IEEE

and memory are carried out with the help of the memory mapping unit and memory bus controller. The simulation results of read and write operations in the cache are shown in Fig.10 and the memory read and writes operations are shown in Fig.11.

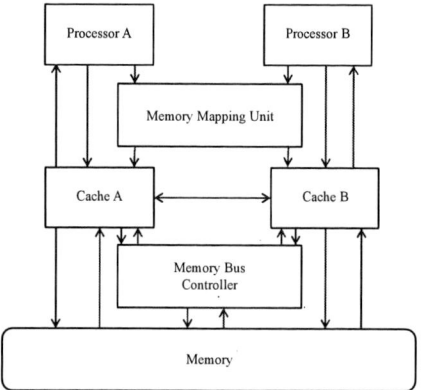

Fig. 8. General design of shared memory dual processor system

```
sim:/Memory/memArray @ 0 ps
31 : 1111000000 1111000001 1110100010 1110000011
27 : 1101100100 1101000101 1100100010 1100000111
23 : 1011101000 1011001001 1010101010 1010001011
19 : 1001101100 1001001101 1000101110 1000001111
15 : 0111110000 0111010001 0110110010 0110010011
11 : 0101110100 0101010101 0100110110 0100010111
 7 : 0011111000 0011011001 0010111010 0010011011
 3 : 0001111100 0001011101 0000111110 0000011111
```

Fig. 9. Memory consists of 32 entries at reset condition

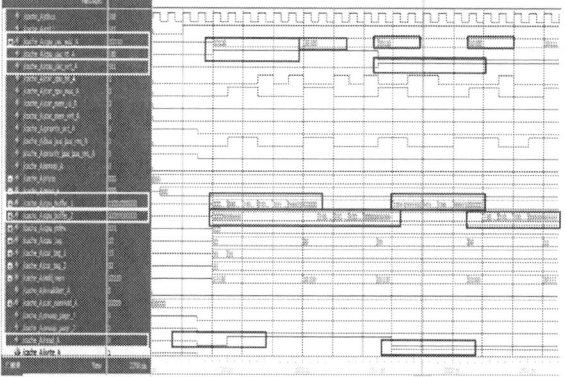

Fig. 10. Simulation showing read and write operations of cache A

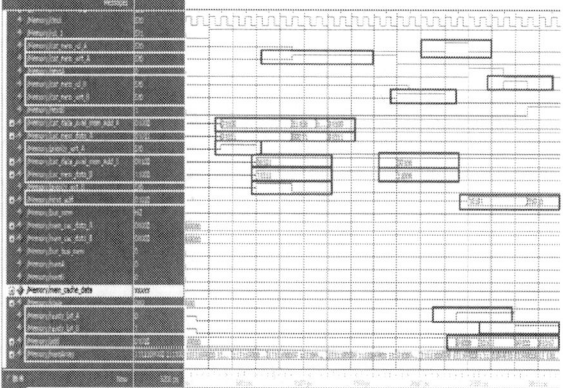

Fig. 11. Simulation showing read and write operations of the memory

B. Implementation of Protocols

All the designed modules of shared memory dual processor system are synthesized and implemented in Xilinx ISE Design Suite after successful simulation of each part of shared memory dual processor system. The register transfer level schematic of each part of the shared memory dual processor system is created that shows the various input and the output ports of each module. Internally it contains the connections among the different logics and the registers used to design the modules. The RTL schematics of cache A, cache B, memory, memory mapping unit and memory bus controller are shown in Fig.12.

The three different test benches are designed to verify the functionality of the MSI, MESI and MOESI snoopy bus protocols in the designed environment of shared memory dual core system. The designed environment is created as the top module by calling the instance of each module of the dual processor system. The required different input commands and cache requests are given in the test bench at the proper time intervals to run the transitions between the different states of the cache.

The write invalidate approach of a dual-core shared memory system is verified in test benches by changing the data of cache A that automatically invalidates the data of cache B according to the logic set up in the caches and shared memory during designing. The write-back approach of each cache block is also verified in the test bench by changing the data of each cache at the regular intervals but the same data is not updated in the main memory. Thus, both the write-invalidate approach and write back cache blocks helps out for improving the execution time of different protocols in the designed environment.

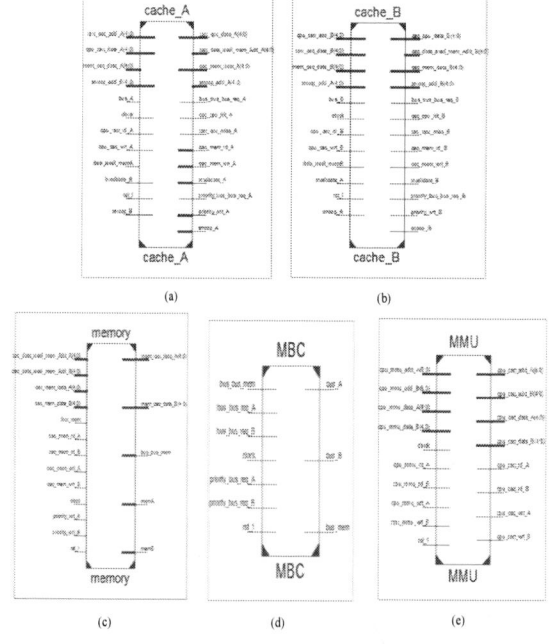

Fig. 12. RTL schematics of (a) cache A, (b) cache B, (c) memory, (d) memory bus controller (MBC), (e) memory mapping unit (MMU)

The different cache requests that are run within the system like read hit, write hit, read miss and write miss and

978-1-5386-6624-1/18 $31.00 © 2018 IEEE   1100

the different state transitions like modified, shared, invalid, exclusive and owned that takes place according to the cache action of MSI, MESI and MOESI protocols are displayed in Fig. 13. as the transcript to show the implementation of protocols in shared memory dual processor system.

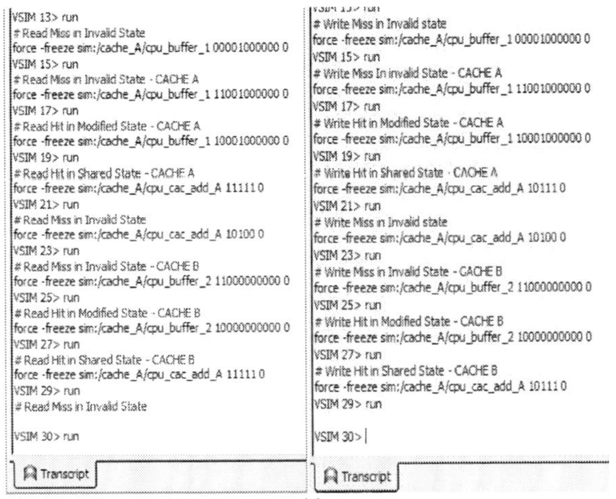

(a)

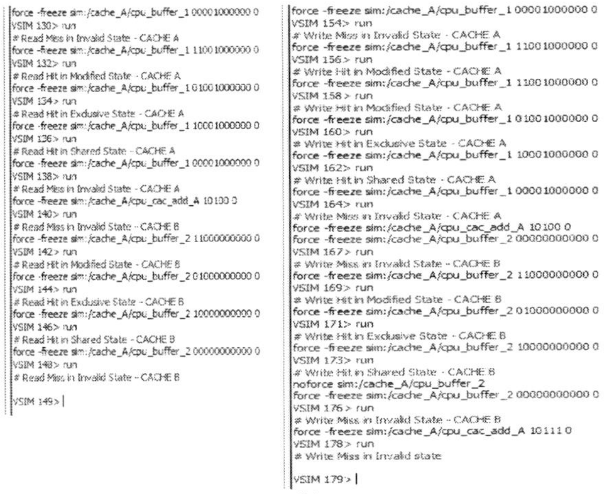

(b)

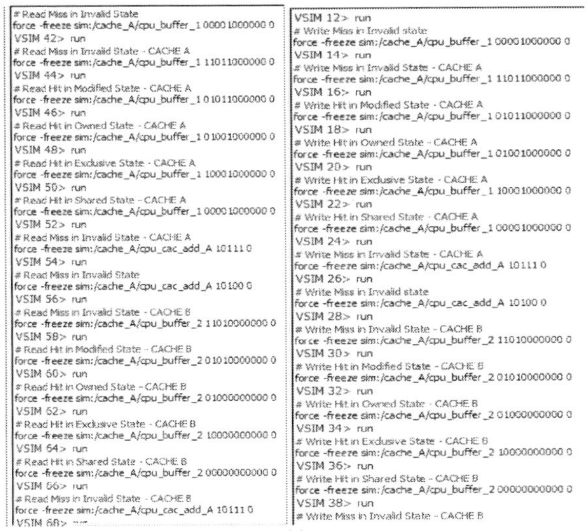

(c)

Fig. 13. Transcripts showing different read and write actions and different

state transitions of cache A and cache B blocks running in shared memory dual processor system of (a) MSI protocol, (b) MESI protocol and (c) MOESI protocol

C. Comparison of Protocols

The comparison section examines the performance of MSI, MESI and MOESI protocol in terms of load misses, memory latency, power consumption, gate count and execution time on the basis of the observation of the simulation and implementation results of MSI, MESI, and MOESI snooping protocols in shared memory dual processor system.

The comparison results as in Table II show that the MOESI protocol reduces the load misses by 48% and memory latency by 35% as the total number of main memory accesses are reduced by the owned state of MOESI protocol. The power consumption is also reduced by 18% and overall execution time is improved by 22% because of the use of both the write-invalidate approach and write back cache blocks in the MOESI protocol. However, the gate count is increased by 34% because the numbers of states are more in the MOESI protocol. Therefore, results demonstrate the superiority of the MOESI protocol in comparison with the MSI and MESI protocol because it has low power consumption and less delay.

TABLE II. COMPARISON OF MSI, MESI AND MOESI PROTOCOLS

Protocol	Comparison Parameters				
	Load Misses	Gate Count	Memory Latency	Power consumption	Execution Time
MSI	128	6681	120 ns	61 %	372 ns
MESI	105	8264	93 ns	52 %	290 ns
MOESI	61	10192	42 ns	43 %	82 ns

IV. CONCLUSION

The performance of multi-core computer systems is strongly influenced by the type of cache coherence protocol used. For the different kinds of shared memory multiprocessor architectures, the suitable cache coherence protocol should be used because it has a direct impact on memory latency, power consumption and execution time of the system.

The results provided in this paper are according to shared memory dual-core system. The performance of dual processor system is observed by using write-invalidate MSI, MESI and MOESI snoopy bus protocols using write- back cache blocks. Based on the simulation and implementation results it is observed that MOESI protocol which is an add-on of MESI and MSI protocol minimizes the memory latency, power consumption and execution time. However, the area utilization of MOESI protocol is more as compare to MESI and MSI protocol. Therefore it is concluded that the overall performance of MOESI protocol is better than MESI and MSI protocols in a shared memory dual processor system.

Further, this work can be expanded by implementing

MSI, MESI, and MOESI snooping protocols on the real-time system like FPGA boards. The future scope of this work is to assess the performance of the shared memory system by increasing the number of processors or by implementing other snoopy bus protocols like dragon and firefly write update based protocols.

REFERENCES

[1] Hennessy, J.L., Patterson, and D.A, "Multiprocessor and Thread-Level Parallelism," *in Computer Architecture: A Quantitative Approach,* 4th ed., San Francisco,California, USA: M K Publishers, 2007, pp. 195-286

[2] Hesham Altwaijry and Diyab S.Alzahrani, "Improved-MOESI Cache Coherence Protocol," Arab J Sci Eng. Springer-Verlag, Berlin Heidelberg New York, vol.39., pp. 2739-2748, 2014.

[3] Kyueun Yi, Won W. Ro, and Jean-Luc Gaudiot., "Importance of Coherence Protocols with Network Applications on Multicore Processors," IEEE Transactions on Computers, IEEE Press, New York vol.62., pp. 6-15, 2013.

[4] Amit D. Joshi, S.V.; B.Sham, and N.Rama, "Performance Analysis of Cache Coherence Protocols for Multi-core Architectures" AICTC, ACM, New York, 2016.

[5] Somdip Dey, and Mamatha S. Nair "Design and Implementation of a Simple Cache Simulator in Java to Investigate MESI and MOESI Coherency Protocols," International Journal of Computer Applications, IEEE Press, New York, vol.87, pp. 0975 – 8887, 2014.

[6] Alberto Ros, Manuel E.Acacio, and Jose M.Garcıa, "A Direct Coherence Protocol for Many-Core Chip Multiprocessors" IEEE Transactions On Parallel And Distributed Systems, IEEE Press, New York, vol.21, pp. 1779-1792, 2010.

[7] Daniel J Sorin, Mark D Hill, and David A Wood, "A primer on memory consistency and cache coherence," in *Synthesis Lectures on Computer Architecture:* Morgan & Claypool Publishers, 2011,pp. 1 - 214

[8] Kai Hwang , "Multiprocessors and Multicomputers," *in Advanced Computer Architecture,* 2nd ed., New Delhi, India: Tata McGraw-Hill Edu. Publishers, 2011, pp. 281-340

[9] Suh, S., "Integration and Evaluation of Cache Coherence Protocols for Multiprocessor SOCS.Ph.D," Thesis Georgia Tech, USA, 2006.

[10] Lluc Al, Lluís V., Marc G., Xav. M., and Eduard Ayg, "Hardware–Software Coherence Protocol for the Coexistence of Caches and Local Memories," IEEE Transactions on Computers, IEEE Press, New York, vol.64, pp. 152-165, 2015.

[11] Hackenberg, D.; Molka, D., Nagel, and W.E "Comparing cache architectures and coherency protocols on x86-64 multicore SMP systems In," Micro-42 Proceedings of the 42nd Annual IEEE/ACM International Symposium on Microarchitecture, New York 2009.

[12] Felix Garcia-C., Jesus Carr, Alejandro Cal., Jose M. Perez, and Jose D. G "An Adaptive Cache Coherence Protocol Specification for Parallel Input/output Systems" IEEE Transactions On Parallel And Distributed Systems, IEEE Press, New York, vol.15, pp. 533-545, 2004

[13] Diana Keen, Mark Oskin, Justin Hensley, and Frederic T. Chong, "Cache Coherence in Intelligent Memory Systems," IEEE Transactions on Computers, IEEE Press, New York, vol.52, pp. 960-966, 2003.

[14] Sabela Ramos, and Torsten Hoefler, "Cache Line Aware Algorithm Design for Cache-Coherent Architectures," IEEE Transactions On Parallel And Distributed Systems, vol.27, pp.2824-2837, October 2016.

[15] Daehoon Kim, Chang Hyun Park, Hwanju Kim, and Jaehyuk Huh, " Virtual Snooping Coherence for Multi-Core Virtualized Systems," IEEE Transactions On Parallel And Distributed Systems, vol.27, pp.2155-2167, July 2016.

[16] Meng Zhang, Jesse D.Bingham, John Erickson, and Daniel J. Sorin, " PVCoherence: Designing Flat Coherence Protocols for Scalable Verification," IEEE Micro, vol.35, pp.84-91, 2015.

[17] Abdullah Kayi, Olivier Serres, and Tarek El-Ghazawi, "Adaptive Cache Coherence Mechanisms with Producer–Consumer Sharing Optimization for Chip Multiprocessors," IEEE Transactions on Computers, vol.64, pp.316-328, February 2015.

[18] Hesham Altwaijry and Diyab S.Alzahrani, " Improved-MOESI Cache Coherence Protocol," Arabian Journal for Science and Engineering, Springer, vol.39, pp.2739-2748, April 2014.

[19] Christian Fensch, Nick Barrow-Williams, Robert D. Mullins, and Simon Moore, "Designing a Physical Locality Aware Coherence Protocol for Chip-Multiprocessors IEEE Transactions on Computers, vol.62, pp.914-928, 2013.

[20] Taeweon Suh, Hsien-Hsin S. Lee, and Douglas M. Blough, "Integrating Cache Coherence Protocols for heterogeneous Multiprocessor Systems, Part 1," IEEE Micro, vol.24, no.4, pp.33-41, 2004.

[21] Daniel J. Sorin, Manoj Plakal, Anne E. Condon, Mark D. Hill, Milo M.K. Martin, and David A. Wood, "Specifying and Verifying a Broadcast and a Multicast Snooping Cache Coherence Protocol," IEEE Transactions On Parallel And Distributed Systems, vol.13, pp.556-578, June 2002.

2nd IEEE International Conference on Power Electronics, Intelligent Control and Energy Systems (ICPEICES-2018)

Model Reduction of Continuous-Time Interval Systems using Eigen Spectrum analysis

Chhabindra Nath Singh
EE Deptt.
HBTU Kanpur
Kanpur, India
cnsinghhbti7@gmail.com

Deepak Kumar
EE Deptt.
MNNIT Allahabad
Allahabad, India
deepak_kumar@mnnit.ac.in

Paulson Samuel
EE Deptt.
MNNIT Allahabad
Allahabad, India
paul@mnnit.ac.in

Ankit Sachan
EE Deptt.
IIT (B.H.U)Varanasi
Varanasi, India
ankitmnnit91@gmail.com

Abstract— **This paper presents mixed methods based on Eigen spectrum approach for order reduction of higher-order linear time-invariant continuous-time interval systems. The proposed methods are developed as a combination of Eigen spectrum approach along with the moment matching and differentiation methods by employing Kharitonov's theorem. The system stiffness and pole centroid of higher-order interval systems are preserved in the reduced-order interval models. Therefore, the stability of resulting reduced-order interval model is ensured for a stable higher-order interval system. A comparative analysis of the proposed approaches with well-known existing approaches is included to demonstrate the efficacy of the proposed methods.**

Keywords— *Interval systems, model reduction, Eigen spectrum.*

I. INTRODUCTION

The approximation of a higher-order system (HOS) by a reduced model is an active area of research for reducing complexity and cost in many control engineering applications. In recent years, several methods are proposed for order reduction of the fixed coefficient and interval systems. Some conventional frequency domain methods are the continued-fraction expansion method [1-2], Padè approximation [3], time moment and Markov parameters matching [4-5] and so on. The problems such as instability, non-minimum phase behaviour and low accuracy in the mid and high-frequency ranges of reduced model limit the applicability of [1-3]. Further, the main drawback of [3] is that the resulting approximant may be unstable even if the HOS is stable. To overcome the instability problem, several stability preservation methods [6-14] are developed. Recently, some mixed methods [15-17] are also proposed by combining optimization algorithms with conventional methods [1-3] and stability based methods [6-14] to obtain reduced-order approximants. The uncertainty due to actuator constraints, unmodelled dynamics, sensor noise and parameter variations paves the way to a higher-order interval system (HOIS). These perturbations may be bounded by the upper and lower limits of system parameters. Hence, it is important to develop model reduction methods for HOISs to reduce complexity. Therefore, many methods [18-20] are presented using interval arithmetic to derive the approximants. Subsequently, Hwang and Yang [21] pointed out the instability problem associated with [18-19]. Dolgin and Zeheb [22] modified [18] to ensure the stability of the reduced model. Later, Dolgin [23] formulated two additional conditions to overcome the limitation of [22] pertaining to the inconsistency of the newly calculated row in the Routh table. Further, Bandyopadhyay et al. [24] proposed another method using Kharitonov's polynomials [25] for interval

systems such that the resulting approximant is robustly stable. Sastry and Rao [26] proposed a method which does not formulate $\gamma-\delta$ tables and compute time moments beforehand unlike [18-19]. Later, Yang [27] showed that [22] cannot ascertain the stability of reduced models. Hote et al. [28] showed that the formulation and testing of all Kharitonov's polynomials are not mandatory for determining the stability of HOISs. Later, Hote et al. [29] presented another method without using interval arithmetic. Jaiswal et al. [30] presented a mixed method for continuous-time HOISs by combining Eigen spectrum and factor division algorithm. Further, Kumar et al. [31] proposed a method based on Kharitonov's polynomials and Routh approximation which preserves the stability and impulse response energy. Recently, Choudhary and Nagar [32] carried out a critical survey of the various Routh approximation based methods.

Since the present article is related to the order reduction of continuous-time interval systems (CTIS), therefore, a survey of discrete-time interval systems is not included. Several methods have been presented till date for model reduction of CTIS but still, there is a lot of scope for improvement in approximation error. Further, no method has been presented using Eigen spectrum along with moment matching and differentiation methods as per the best of authors' knowledge. In this paper, we propose two mixed methods for order reduction of linear time-invariant (LTI) CTIS by extracting the advantages of existing methods. The Eigen spectrum technique [33] is employed along with moment matching [1] and differentiation method [13]. The proposed methods are computationally simple, elegant and computer-oriented. Further, the proposed methods preserve the input-output behaviour and always produce stable reduced models for a stable HOIS. The proposed techniques are compared with [19], [34], [35] in terms of performance indices, step and frequency response plots to validate the effectiveness.

II. PROBMEN STATEMENT

Consider the transfer function (TF) of a HOIS as

$$H(s) = \frac{N(s)}{M(s)} = \frac{\sum_{i=0}^{n-1}[N_i^-, N_i^+]s^i}{\sum_{i=0}^{n}[M_i^-, M_i^+]s^i}, \qquad (1)$$

The objective is to find a rth order reduced interval model with

978-1-5386-6624-1/18 $31.00 © 2018 IEEE   1103

$$h(s) = \frac{n(s)}{m(s)} = \frac{\sum_{i=0}^{r-1}[n_i^-, \, n_i^+]s^i}{\sum_{i=0}^{r}[m_i^-, \, m_i^+]s^i}, \tag{2}$$

where N_i^-, M_i^- and N_i^+, M_i^+ are the lower and upper bounds for the coefficients of $H(s)$.

III. PROPOSED METHOD

In this section, the proposed methods for order reduction of LTI-CTIS are discussed. Both methods follow the Eigen spectrum analysis for computation of reduced-order denominator polynomial whereas differentiation and time moment matching approaches are employed for deriving the numerator polynomials in the proposed mixed methods 1 and 2 respectively.

A. Computation of reduced-order denominator polynomial

The steps to obtain the reduced-order denominator polynomial are as below:

Step 1: Construct the fixed coefficient transfer functions from the numerator and denominator polynomials of the given HOIS using [25] as

$$H^I(s) = \frac{N^I(s)}{M^I(s)}, \quad I = 1, \, 2, \, 3, \, 4 \tag{3}$$

Step 2: If $-\lambda_i^I \, (i = 1, 2, 3, \cdots, n)$ are the poles of $H^I(s)$ located at $-\left(\operatorname{Re}\lambda_i^I \pm \operatorname{Im}\lambda_i^I\right); (i = 1, \cdots, v)$, then, the zone covered by lines crossing the nearest and farthest real poles vertically and farthest imaginary pole pairs horizontally is Eigen spectrum zone (ESZ) as shown in Fig.1. Similarly, the ESZ for reduced model $h^I(s) = n^I(s)/m^I(s)$, is shown in Fig. 2 for the poles $-\delta_i^I \, (i = 1, 2, \cdots, r)$ located at $-\left(\operatorname{Re}\delta_i^I \pm \operatorname{Im}\delta_i^I\right); (i = 1, 2, \cdots, k)$.

Step 3: The pole centroid λ_{pc}^I and system stiffness λ_s^I for $H^I(s)$ are calculated as

$$\lambda_{pc}^I = \frac{\sum_{i=1}^{v}\operatorname{Re}\left(\lambda_i^I\right)}{v}, \tag{4}$$

$$\lambda_s^I = \frac{\operatorname{Re}\left(\lambda_1^I\right)}{\operatorname{Re}\left(\lambda_v^I\right)}. \tag{5}$$

Step 4: Similarly, the pole centroid δ_{pc}^I and stiffness δ_s^I for the reduced-order case are computed as

$$\delta_{pc}^I = \frac{\sum_{i=1}^{k}\operatorname{Re}\left(\delta_i^I\right)}{k}, \tag{6}$$

$$\text{and } \delta_s^I = \frac{\operatorname{Re}\left(\delta_1^I\right)}{\operatorname{Re}\left(\delta_k^I\right)}. \tag{7}$$

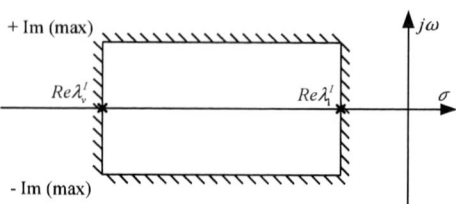

Fig. 1. ESZ for HOIS

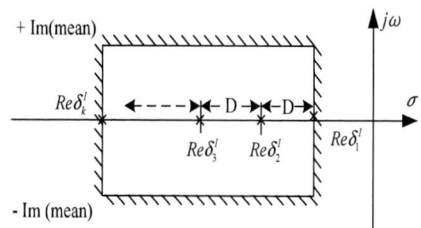

Fig.2. ESZ for reduced order model

If the pole centroid and stiffness of the higher-order interval system are preserved in the reduced-order interval model then, we can write

$$\delta_{pc}^I = \frac{\sum_{i=1}^{k}\operatorname{Re}\left(\delta_i^I\right)}{k} = \lambda_{pc}^I \quad \text{and} \quad \delta_s^I = \frac{\operatorname{Re}\left(\delta_1^I\right)}{\operatorname{Re}\left(\delta_k^I\right)} = \lambda_s^I.$$

Step 5: The Eigen spectral points are evaluated by using the following equations

$$\operatorname{Re}\left(\delta_k^I\right) = D^I(k-1) + \operatorname{Re}\left(\delta_1^I\right), \tag{8}$$

$$S^I = \operatorname{Re}\left(\delta_1^I\right)(k-1) + \operatorname{Re}\left(\delta_k^I\right) + QD^I, \tag{9}$$

where $S^I = \lambda_{pc}^I k, QD^I = D^I + 2D^I + \cdots + (k-2)D^I$. If $\operatorname{Re}\left(\delta_1^I\right) = \lambda_s^I \operatorname{Re}\left(\delta_k^I\right)$, then (8) and (9) may be rewritten as

$$\operatorname{Re}\left(\delta_k^I\right)(1-\lambda_s^I) + D^I(1-k) = 0, \tag{10}$$

$$\operatorname{Re}\left(\delta_k^I\right)\{\lambda_s^I(k-1)+1\} + D^I Q = S^I. \tag{11}$$

Step 6: Now, $\operatorname{Re}\left(\delta_k^I\right)$ and D^I are used to locate the Eigen spectral points and thereby the reduced-order denominator polynomials of fixed coefficient $m^I(s)$, $I = 1, 2, 3, 4$ are computed as

$$m^I(s) = (s - \operatorname{Re}(\delta_1^I))\cdots(s - \operatorname{Re}(\delta_r^I)) \, ; \, I = 1, 2, 3, 4 \tag{12}$$

where $m^1(s) = m_0^- + m_1^- s + m_2^+ s^2 + \cdots$, $m^2(s) = m_0^+ + m_1^+ s + m_2^- \times s^2 + \cdots$, $m^3(s) = m_0^+ + m_1^- s + m_2^- s^2 + \cdots$, and $m^4(s) = m_0^- + m_1^+ \times s + m_2^+ s^2 + \cdots$.

Step 7: Finally, the reduced-order interval denominator polynomial is determined from the fixed coefficient polynomials computed using (12) and Kharitonov's theorem as

$$m(s) = [m_0^-, \, m_0^+] + [m_1^-, \, m_1^+]s + \ldots + [m_r^-, \, m_r^+]s^r. \tag{13}$$

B. Reduced-order numerator for mixed method 1

The reduced-order numerator polynomial of the proposed mixed method 1 is obtained as given below:

Step 1: Obtain the reciprocal transformation of higher-order interval numerator polynomial $N(s)$ as

$$\hat{N}(s) = \frac{1}{s} N\left(\frac{1}{s}\right). \tag{14}$$

Step 2: Differentiate the transformed higher-order interval numerator polynomial $\hat{N}(s)$ as

$$\hat{n}(s) = \frac{d^{n-r}}{ds}\left(\hat{N}(s)\right). \tag{15}$$

Step 3: The reciprocal transformation of $\hat{n}(s)$ is performed to obtain the reduced-order interval numerator polynomial as

$$n(s) = \frac{1}{s} \hat{n}\left(\frac{1}{s}\right). \tag{16}$$

Step 4: By steady-state matching, the reduced-order interval model using mixed method 1 is obtained as

$$h_d(s) = x \frac{n(s)}{m(s)}, \tag{17}$$

where $x = H(s)\big|_{s=0} / h(s)\big|_{s=0}$, and $m(s)$ is obtained by (13).

C. Reduced-order numerator for mixed method 2

For a given HOIS, the reduced-order numerator polynomial by proposed mixed method 2 is obtained as follows:

Step1: Obtain $n^I(s)$, $I = 1, 2, \cdots 4$ by cross-multiplication of original and reduced-order models to match the time moments as follows:

$$n^I(s) = \frac{N^I(s)}{M^I(s)} m^I(s) \; ; \;\; I = 1, 2, \cdots 4, \tag{18}$$

where $N^I(s)$ and $M^I(s)$ are the numerator and denominator polynomials of HOIS and $m^I(s)$ is the reduced-order denominator polynomial that is obtained by using the method discussed in Section III. Therefore, the fixed coefficient reduced-order numerator polynomials are obtained as

$$n^1(s) = n_0^- + n_1^- s + \cdots, \; n^2(s) = n_0^+ + n_1^+ s + \cdots,$$
$$n^3(s) = n_0^+ + n_1^- s + \cdots, \text{ and } n^4(s) = n_0^- + n_1^+ s + \cdots.$$

Step 2: Now, obtain the reduced-order interval numerator polynomial from $n^I(s)$, $I = 1, 2, 3, 4$ by applying inverse-Kharitonov's theorem as

$$n(s) = [n_0^-,\, n_0^+] + [n_1^-,\, n_1^+]s + \ldots + [n_{r-1}^-,\, n_{r-1}^+]s^{r-1}. \tag{19}$$

IV. NUMERICAL EXAMPLE

Consider a third-order interval system [34] as

$$H(s) = \frac{[1200, 2000]s^2 + [6800, 8000]s + [3500, 4000]}{[1000, 1500]s^3 + [4800, 5800]s^2 + [6000, 6600]s + [1800, 2200]}.$$

By applying Kharitonov's theorem, the fixed coefficient transfer functions for the given 3rd order interval system are obtained as

$$H^1(s) = \frac{2000s^2 + 6800s + 3500}{1500s^3 + 5800s^2 + 6000s + 1800},$$

$$H^2(s) = \frac{2000s^2 + 8000s + 3500}{1000s^3 + 5800s^2 + 6600s + 1800},$$

$$H^3(s) = \frac{1200s^2 + 6800s + 4000}{1500s^3 + 4800s^2 + 6000s + 2200},$$

$$H^4(s) = \frac{1200s^2 + 8000s + 4000}{1000s^3 + 4800s^2 + 6600s + 2200},$$

where the Eigen values of above transfer functions are:

$\lambda^1 = -0.555, \; -0.8942, \; -2.4173; \; \lambda^2 = -0.41, -1, \; -4.39; \; \lambda^3 = -0.602, \; -1.3 \pm 0.8651i, \; \lambda^4 = -0.49, \; -1.7593, \; -2.55.$

The pole centroid and system stiffness for $H^I(s)$, $I = 1, 2, 3, 4$ are calculated as

$\lambda_{pc}^i = -1.289, \; -1.933, \; -1.067, \; -1.69, \; \lambda_s^i = 0.2296, \; 0.0934, \; 0.4631, \; 0.1922$.

Now, the 2nd order reduced denominator polynomials of fixed coefficient $m^I(s)$, $I = 1, 2, 3, 4$ are obtained through the Eigen spectral points computed using Step 5 of Section III-A as

$$m^1(s) = (s + 0.4814)(s + 2.0965) = s^2 + 2.5779s + 1.009,$$

$$m^2(s) = (s + 0.3302)(s + 3.5358) = s^2 + 3.866s + 1.675,$$

$$m^3(s) = (s + 0.6758)(s + 1.4576) = s^2 + 2.1333s + 0.985,$$

$$m^4(s) = (s + 0.5159)(s + 2.6841) = s^2 + 3.2s + 1.3847.$$

Finally, the reduced-order interval denominator polynomial from the above polynomials is obtained by inverse Kharitonov's theorem as

$$m(s) = s^2 + [2.1333, 3.866]s + [1.009, 1.3847].$$

For obtaining the reduced interval numerator polynomial by mixed method 1, the original interval numerator polynomial $N(s)$ is transformed as

$$\hat{N}(s) = [3500, 4000]s^2 + [6800, 8000]s + [1200, 2000].$$

By differentiating the transformed original interval numerator polynomial, we have
$$\hat{n}(s) = [7000, 8000]s + [6800, 8000].$$

Now, the reciprocal transformation of $\hat{n}(s)$ is performed to obtain the reduced-order interval numerator polynomial $n(s)$ as

$$n(s) = [6800, 8000]s + [7000, 8000].$$

By steady-state matching with HOS, the reduced interval model by proposed mixed method 1 is obtained as

$$h_{mm1}(s) = \frac{[2.0346, 2.3937]s + [2.0945, 2.3937]}{s^2 + [2.133, 3.866]s + [1.009, 1.3847]}.$$

2nd IEEE International conference on power Electronics, Intelligent Control and Energy systems (ICPEICES-2018)

The reduced-order fixed coefficient numerator polynomials by the second method are

$n^1(s) = 2.2847s + 1.9619$, $n^2(s) = 4.382s + 2.2702$, $n^3(s) = 1.6997s$

$+ 1.9153$ and $n^4(s) = 3.3s + 2.6925$ for $H^I(s)$, $I = 1, 2, 3, 4$.

Thereby, the reduced-order interval numerator polynomial is constructed as

$n(s) = [1.6997, 4.382]s + [1.9619, 2.6925]$.

Finally, the reduced-order interval model by proposed mixed method 2 is obtained as

$$h_{mm2}(s) = \frac{[1.6997, 4.382]s + [1.9619, 2.6925]}{s^2 + [2.1333, 3.866]s + [1.009, 1.3847]}.$$

A comparison of performance indices between the reduced models by proposed and existing methods [19], [34], [35] is given in Tables 1-4 respectively. It is clear from the Tables that the errors obtained by the proposed methods are well comparable to [19], [34], [35]. The step responses (lower and upper bound) for the proposed and existing methods [19], [34], [35] are plotted in Figs. 3 and 4 respectively. It is observed from the responses that, the proposed reduced models preserve the important characteristics of the HOIS. Further, the frequency responses (lower and upper bound) for the proposed methods and [19], [34], [35] are plotted in Figs.5 and 6. It is observed from the responses that, the proposed approximants preserve the important characteristics of the HOIS.

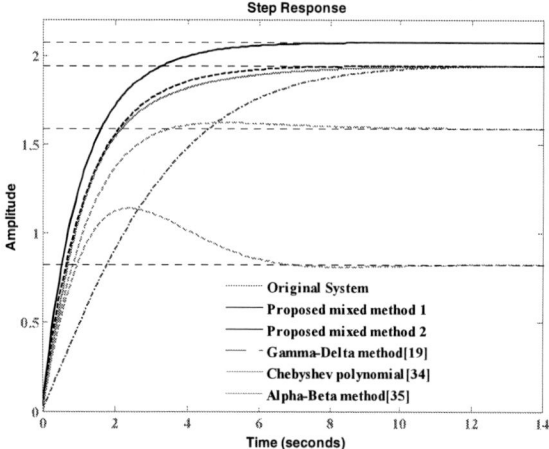

Fig. 3. Step responses of reduced models (lower bound)

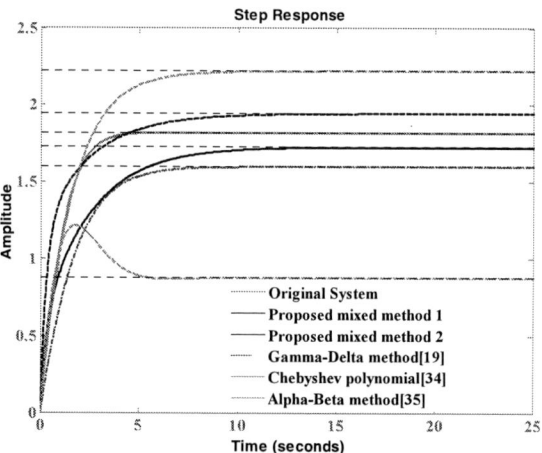

Fig.4. . Step responses of reduced models (upper bound)

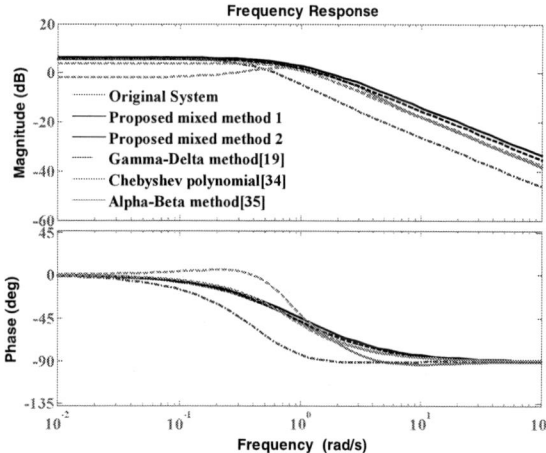

Fig. 5. Frequency responses of reduced models (lower bound)

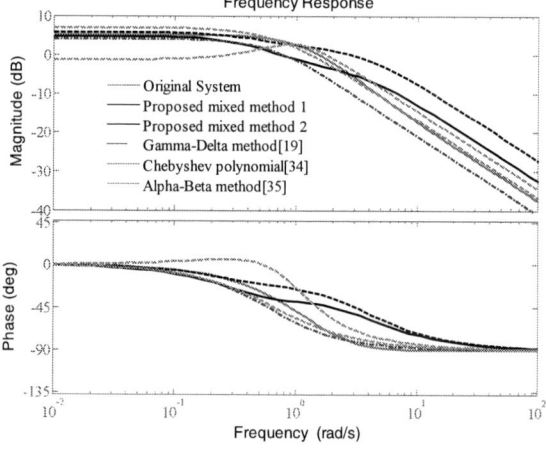

Fig. 6. Frequency responses of reduced models (upper bound)

V. CONCLUSION

In this paper, two new mixed methods are proposed for order reduction of LTI continuous-time interval systems using Kharitonov's theorem. The Eigen spectrum analysis is followed to compute the reduced-order denominator while the reduced-order numerator polynomials are calculated by using differentiation and time moment matching techniques

in the proposed mixed methods 1 and 2 respectively. The resulting reduced models are stable for a stable higher-order interval system as the system stiffness and pole centroid of higher-order interval system are preserved in the reduced-order models. The proposed methods are compared with the existing techniques and relatively lower values of ISE, ITSE, IAE and ITAE are obtained in most of the cases. The proposed methods may easily be extended to discrete-time interval systems as well.

978-1-5386-6624-1/18 $31.00 © 2018 IEEE

2nd IEEE International conference on power Electronics, Intelligent Control and Energy systems (ICPEICES-2018)

TABLE 1: COMPARISON OF INTEGRAL SQUARE ERRORS

Reduction Method	Reduced Models	ISE (Lower)	ISE (Upper)
Proposed mixed method 1	$G_2(s) = \dfrac{[2.0346, 2.3937]s + [2.0945, 2.3937]}{s^2 + [2.133, 3.866]s + [1.009, 1.3847]}$	0.2061	0.5402
Proposed mixed method 2	$G_2(s) = \dfrac{[1.6997, 4.382]s + [1.9619, 2.6925]}{s^2 + [2.1333, 3.866]s + [1.009, 1.3847]}$	0.0078	0.1828
Bandyopadhyay et al. [19]	$G_2(s) = \dfrac{[0.4829, 0.8941]s + [0.5233, 1.1388]}{s^2 + [0.9868, 1.7082]s + [0.2691, 0.7118]}$	1.1683	0.8012
Ismail [34]	$G_2(s) = \dfrac{[1.5909, 2.2222] + [2.8528, 4.6774]s}{[1,1] + [2.7272, 3.667]s + [2.1818, 3.2222]s^2}$	0.3018	0.3748
Sharma et al. [35]	$G_2(s) = \dfrac{[1.2303, 1.8823]s + [0.4749, 0.9134]}{s^2 + [1.0855, 1.5529]s + [0.5756, 1.0404]}$	3.7957	3.5430

TABLE 2: COMPARISON OF INTEGRAL TIME SQUARE ERRORS

Reduction Method	Reduced Models	ITSE (Lower)	ITSE (Upper)
Proposed mixed method 1	$G_2(s) = \dfrac{[2.0346, 2.3937]s + [2.0945, 2.3937]}{s^2 + [2.133, 3.866]s + [1.009, 1.3847]}$	0.7195	1.7827
Proposed mixed method 2	$G_2(s) = \dfrac{[1.6997, 4.382]s + [1.9619, 2.6925]}{s^2 + [2.1333, 3.866]s + [1.009, 1.3847]}$	0.0239	0.1519
Bandyopadhyay et al.[19]	$G_2(s) = \dfrac{[0.4829, 0.8941]s + [0.5233, 1.1388]}{s^2 + [0.9868, 1.7082]s + [0.2691, 0.7118]}$	2.6730	2.4373
Ismail [34]	$G_2(s) = \dfrac{[1.5909, 2.2222] + [2.8528, 4.6774]s}{[1,1] + [2.7272, 3.667]s + [2.1818, 3.2222]s^2}$	1.3967	2.0131
Sharma et al. [35]	$G_2(s) = \dfrac{[1.2303, 1.8823]s + [0.4749, 0.9134]}{s^2 + [1.0855, 1.5529]s + [0.5756, 1.0404]}$	19.0353	17.3313

TABLE 3: COMPARISON OF INTEGRAL ABSOLUTE ERRORS

Reduction Method	Reduced Models	IAE (Lower)	IAE (Upper)
Proposed mixed method 1	$G_2(s) = \dfrac{[2.0346, 2.3937]s + [2.0945, 2.3937]}{s^2 + [2.133, 3.866]s + [1.009, 1.3847]}$	1.1911	1.8019
Proposed mixed method 2	$G_2(s) = \dfrac{[1.6997, 4.382]s + [1.9619, 2.6925]}{s^2 + [2.1333, 3.866]s + [1.009, 1.3847]}$	0.2188	0.6716
Bandyopadhyay et al. [19]	$G_2(s) = \dfrac{[0.4829, 0.8941]s + [0.5233, 1.1388]}{s^2 + [0.9868, 1.7082]s + [0.2691, 0.7118]}$	2.5077	2.2685
Ismail [34]	$G_2(s) = \dfrac{[1.5909, 2.2222] + [2.8528, 4.6774]s}{[1,1] + [2.7272, 3.667]s + [2.1818, 3.2222]s^2}$	1.3658	1.3085
Sharma et al. [35]	$G_2(s) = \dfrac{[1.2303, 1.8823]s + [0.4749, 0.9134]}{s^2 + [1.0855, 1.5529]s + [0.5756, 1.0404]}$	4.5862	4.3568

TABLE 4: COMPARISON OF INTEGRAL TIME ABSOLUTE ERRORS

Reduction Method	Reduced Models	ITAE (Lower)	ITAE (Upper)
Proposed mixed method 1	$G_2(s) = \dfrac{[2.0346, 2.3937]s + [2.0945, 2.3937]}{s^2 + [2.133, 3.866]s + [1.009, 1.3847]}$	4.1917	6.3180
Proposed mixed method 2	$G_2(s) = \dfrac{[1.6997, 4.382]s + [1.9619, 2.6925]}{s^2 + [2.1333, 3.866]s + [1.009, 1.3847]}$	0.7428	1.2436
Bandyopadhyay et al. [19]	$G_2(s) = \dfrac{[0.4829, 0.8941]s + [0.5233, 1.1388]}{s^2 + [0.9868, 1.7082]s + [0.2691, 0.7118]}$	6.8425	7.5240
Ismail [34]	$G_2(s) = \dfrac{[1.5909, 2.2222] + [2.8528, 4.6774]s}{[1,1] + [2.7272, 3.667]s + [2.1818, 3.2222]s^2}$	5.7260	6.4610
Sharma et al. [35]	$G_2(s) = \dfrac{[1.2303, 1.8823]s + [0.4749, 0.9134]}{s^2 + [1.0855, 1.5529]s + [0.5756, 1.0404]}$	20.7503	19.8350

978-1-5386-6624-1/18 $31.00 © 2018 IEEE

REFERENCES

[1] C. F. Chen and L. S. Shieh, "A novel approach to linear model simplification," Int. J. Control, vol. 8, pp. 561-570, 1968.

[2] Y. Shamash, "Continued fraction methods for reduction of constant linear multivariable systems," Int. J. Syst. Sci., vol. 7, pp. 743-758, 1976.

[3] Y. Shamash, "Stable reduced-order models using Padè type approximation," IEEE Trans. Automat. Contr., vol. AC-19, pp.615-616, 1974.

[4] A. Narain, D. Chandra, and R. K. Singh, "Model order reduction using fuzzy C-means clustering," Trans. Inst. Meas. Control, pp.1-7, 2014.

[5] J. Singh, C. B. Vishwakarma, and K. Chatterjee, "Biased reduction method by combining improved modified pole clustering and improved Padè approximations," Appl. Math. Model., vol.40, pp.1418-1426, 2016.

[6] M. F. Hutton and B. Friedland, "Routh approximations for reducing order of linear time-invariant systems," IEEE Trans. Automat. Contr., vol.25, no.3, pp.329-337, 1975.

[7] Y. Shamash, "Linear system reduction using Padè approximation to allow retention of dominant modes," Int. J. Control, vol. 21, pp. 257-272, 1975.

[8] R. K. Appiah, "Linear model reduction using Hurwitz polynomial approximation," Int. J. Control, vol. 28, pp.477-488, 1978.

[9] J. Pal, "Stable reduced-order Padè approximants using Routh Hurwitz array," Electron. Lett., vol. 15, no.8, pp. 225-226, 1979.

[10] T. C. Chen, C. Y. Chang, and K. W. Han, "Reduction of transfer functions by stability equation method," J. Frank. Inst., vol. 308, pp. 389-404, 1979.

[11] T. N. Lucas, "A tabular approach to the stability equation method," J. Frank. Inst., vol. 129, pp.171-180, 1992.

[12] B. W. Wan, "Linear model reduction using Mihailov criterion and Padè approximation technique," Int. J. Control, vol. 33, pp. 1073-1089, 1981.

[13] P. O. Gutman, C. F. Mannerfelt, and P. Molander, "Contributions to model reduction problem," IEEE Trans. Automat. Contr., vol. AC-27, no. 2, pp. 454-455, 1982.

[14] T. N. Lucas, "Factor division: a useful algorithm in model reduction," IEE Proc., vol. 130, no.6, pp.362-364, 1983.

[15] S. R. Desai and R. Prasad, "A novel order diminution of LTI systems using big bang-big crunch optimization and Routh approximation,". Appl. Math. Model., vol. 37, pp. 8016-8028, 2013.

[16] S. R. Desai and R. Prasad, "A new approach to order reduction using stability equation and big bang-big crunch optimization," Syst. Sci. Control Eng., vol. 1, no. 1, pp. 20-27, 2013.

[17] A. Sikander and R. Prasad, "A novel order reduction method using cuckoo search algorithm," IETE J. Res., vol. 61, no. 2, pp.1-8, 2015.

[18] B. Bandyopadhyay, O. Ismail, and R. Gorez, "Routh-Padè approximation for interval systems," IEEE Trans. Automat. Contr., vol. 39, pp. 2454-2456, 1994.

[19] B. Bandyopadhyay, A. Upadhye, and O. Ismail, "γ-δ Routh approximations for interval systems," IEEE Trans. Automat.Contr., vol. 42, pp. 1127-1130, 1997.

[20] G. V. K. Sastry, G. R. R. Rao, and P. M. Rao "Large scale interval system modelling using Routh approximants," Electron. Lett., vol. 36, no.8, pp.768-769, 2000.

[21] C. Hwang and S. F. Yang, "Comments on the computation of interval Routh approximants," IEEE Trans. Automat. Contr., vol. 44, no.9, pp.1782-1787, 1999.

[22] Y. Dolgin and E. Zeheb, "On Routh-Padè model reduction of interval systems," IEEE Trans. Automat. Contr., vol. 48, no.9, pp.1610-1612, 2003.

[23] Y. Dolgin, "Author's reply [to comments on 'On Routh-Pade model reduction of interval systems']," IEEE Trans. Automat. Contr., vol. 50, no.2, pp. 274-275, 2005.

[24] B. Bandyopadhyay, V. Sreeram, and P. Shingare, "Stable γ-δ Routh approximation of interval systems using Kharitonov polynomials," Int. J. Info. Syst. Sci., vol. 4, pp. 348-361, 2007.

[25] V. L. Kharitonov, "Asymptotic stability of an equilibrium position of a family of systems of linear differential equations," Differentsial'nye Uravneniya, (14), pp. 2086-2088, 1978.

[26] G. V. K. R. Sastry and P. M. Rao, "A new method for modeling of large-scale interval systems," IETE J. Res.., vol. 49, no. 6, pp. 423-430, 2003.

[27] S. F. Yang, "Comments on Routh-Padè model reduction of interval systems," IEEE Trans. Automat. Contr., vol. 50, no. 2, pp. 273-274, 2005.

[28] Y. V. Hote, D. R. Choudhury, and J. R. P. Gupta, "A robust test of uncertain linear systems," J. Control Theory Appl., vol. 7, no. 3, pp.277-280, 2009.

[29] Y. V. Hote, A. N. Jha, and J. R. Gupta, "Reduced order modelling for some class of interval systems," Int. J. Model. Sim., vol. 34, no. 2, pp. 63-69, 2014.

[30] A. Jaiswal, P. K. Singh, S. Gangwar, S. Manmatharajan, and D. Kumar, "Order reduction of interval systems using eigen spectrum and factor division algorithm," 3rd Int. Conf. Advances in Control Optimi. Dynamic. Syst., pp.363-36, 2014.

[31] M. S. Kumar, N. V. Anand, and R. S. Rao, "Impulse energy approximation of higher-order interval systems using Kharitonov's polynomials," Trans. Inst. Meas. Control, pp.1-11, 2015.

[32] A. K. Choudhary and S. K. Nagar, "Order reduction techniques via Routh approximation: a critical survey," IETE J. Res., pp.1-15, 2018.

[33] G. Parmar, S. Mukherjee, and R. Prasad, "System reduction using factor division algorithm and eigen spectrum analysis," Appl. Math. Model., vol. 31, pp. 2542-2552, 2007.

[34] O. Ismail, "Model reduction of linear structured uncertain systems using Chebyshev polynomlal techniques," *EUROCON*, Serbia and Montenegro, Belgrade, 2005.

[35] M. Sharma, A. Sachan, and D. Kumar, "Order reduction of higher-order interval systems by stability preservation approach," Int. Conf. Power, Control Embedded Syst., pp.1-6, 2014.

Comparative study and Investigation of the Broadband Powerline Channel Model

Zainul Abdin Jaffery
Electrical Engg. Dept.
Jamia Millia Islamia,
N. Delhi, INDIA
zjaffery@jmi.ac.in

Ibraheem
Electrical Engg. Dept.
Jamia Millia Islamia,
N. Delhi, INDIA
ibraheem_2k@yahoo.com

Mukesh Kumar Varma
Electrical Engg. Dept.
Jamia Millia Islamia
N. Delhi, INDIA
mkvsuni@gmail.com

Abstract—**This paper presents the characterization and modeling of several correlative powerline channel models, namely, the Phillips model and Zimmermann and Dostert model. In this work we describe some important features and obtain the analytical expressions of the power line channel, explore the validity of the channel transfer function, reports the important applications to contribute to the future consensus of the broadband powerline communication. The proposed modeling approach is based on the radial structure of the low voltage electrical networks and this is an extension of the Zimmermann model.**

Keywords- Broadband powerline, Transmission line, Channel modeling, Transfer Function

I. INTRODUCTION

In Today's digital world, the world wide uses of broadband communication are growing very fast. The common technologies earlier being used comprising of phone wires, fiber optic, Ethernet cabling, wireless and satellite technologies. Each method has its limitations to spread the maximum consumers. Broadband power line communication is an emergent technology for, broadband communication over the existing power line infrastructure and smart grid. The communication technology used is a combination of Electrical communication and telecommunication. The power line is fundamentally constructed for power delivery at low frequencies (50 or 60 Hertz) so that the electrical energy could be delivered conveniently to the consumers, however at the same time, it can be used as a communication medium at high frequencies (1-30MHz). However, the powerline communication channel exhibits unfavorable transmission properties for high-frequency communication signals as this electrical wiring is not shielded, radiofrequency signals passing along it are in part, and similar to all other technologies, this technology also faces its own set of complications and technical challenges. The parameters line impedance, signal attenuation and noise are changeable with location, time and frequency and hence still there is no reference channel model has been designed and recognized yet. However, in recent time, Lots of research activity are in progress, for the

broadband power line communication channel modeling. There are some channel modelling methods were proposed over the years for channel modelling are discussed in this paper. Zimmermann et.al. [2], gives an analytical time–domain approach, describing complex transfer function of typical electrical network based on signal propagation effects in mains network. H. meng et.al.[3], provides a modelling approach in which the intrinsic line parameters are derived first and then an echo-model is developed. While in H. meng et.al.[3], modelling is based on scattering Metrix is derived. H. Philips [1] has provided a statically analysis of some measurement obtained foe echo powerline model is similar to descriptions of mobile radio channels. another approach in which equivalent two-port network is obtained and is represented by means of ABCD matrices for Indoor powerline channel, presented by S. Galli [4,5] provides a channel modelling based on transmission line theory of multi-conductor's line.

Two channel modelling methodologies (bottom-up and top-down models) were found in literature, the first one, The Bottom-Up methodology starts from the theoretic calculation of model parameters. This methodology defines the behavior of a network by a great quantity of distributed constituents using matrices (scattering parameter matrices or four pole impedance and admittance matrices), means fundamentally by Developing a deterministic model for defining the transfer function hypothetically without taking authenticated dimensions of the transmission line. The bottom-up modelling approach derives all the intrinsic parameters (cable parameters, load impedances, etc.) of transfer function from a theoretical basis. For a given channel, it necessitates detailed knowledge of all components and their respective characteristics [7]. In contrast, The Top-Down, or parametric methodology, the model constraints are obtained from measurements [9]. Some additional top-down random channel generation procedures were existing in the literature [1,2,6]. The proposed model is based on indoor powerline network from the particular cable model, simple expression derived for the characteristic impedance and the propagation constant of the powerline network. The effort of this method is on the house indoor Powerline network placed inside the conduit in the frequency range 1-30 MHz .

In section II, we introduce the transmission line characteristics. Subsequently, Section III contains the theoretical of existing channel models and an analysis for a proposed channel model. All the performance assessment, the transfer function response compared and respective results are discussed in sections IV. Section V, describes the application and advantages of broadband powerline communication. And Finally the paper work is concluded in Section VI.

II. TRANSMISSION LINE CHARACTERISTICS

Broadband over Power Line is a novel access technology that raises to the transmission (sending and receiving) of high frequency broadband data over the standing power cables and electricity distribution structures. The Broadband over Power Lines system does not utilize the whole network of power grid, but it bypasses the substations and medium, high-voltage network and emphasis on the low voltage transmission lines (carrying voltages in the range from 230 volts and 440 volts) and the transformers that convert the electrical low voltage in the range from 220 to 440 volts used to supply the consumer's premises. The transmission line having a pair of wires-cable. An identical transmission line has two conductors that has uniform cross-section and are parallel to each other. Figure 1 shows the intuitive view of the per-unit length line parameters for a two conductor line.

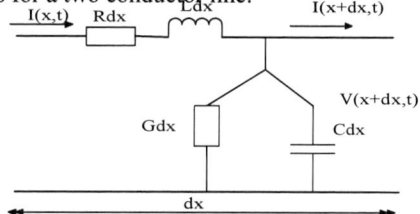

Fig.1- The per unit length equivalent circuit for a two conductor line

In this diagrams, x symbolizes the longitudinal direction of the line and R, L, G and C are denoted the p.u. Resistance (Ω/m), p.u. inductance (H/m), conductance p.u.(S/m) and capacitance p.u. (F/m) correspondingly. The electric elements are reliant on the formal parameters. The p.u. length equivalent circuit for a section of line of length dx is displayed in figure 1, now applying the Kirchhoff's voltage law (KVL) about the loop, we get [3]

$$V(x + dx, t) - V(x, t) = -RdxI(x, t) - Ldx\frac{\partial I(x,t)}{\partial t} \quad -(1)$$

Dividing both sides by dx and taking the limit as dx$\rightarrow$ 0 yield the first transmission line equation

$$\frac{\partial V(x,t)}{\partial x} = -R . I(x, t) - L . \frac{\partial I(x,t)}{\partial t} \quad -(2)$$

Writing Kirchhoff's current law (KCL) at the upper node gives

$$I. (x + dx, t) - I. (x, t) = Gdx. V. (x + dx, t) - Cdx\frac{\partial V(x+dx,t)}{\partial t} \quad -(3)$$

Dividing both the sides by dx and taking the limit as dx$\rightarrow$ 0 yield the next transmission line equation

$$\frac{\partial I(x,t)}{\partial x} = -G. V(x, t) - C. \frac{\partial V(x,t)}{\partial t} \quad -(4)$$

Hence, the time domain equation (2) and (4) of the transmission line can be expressed as

$$-\frac{\partial V(x,t)}{\partial x} = R. I(x, t) + L. \frac{\partial I(x,t)}{\partial t} \quad -(5)$$

And $$-\frac{\partial I(x,t)}{\partial x} = G. V(x, t) + C. \frac{\partial V(x,t)}{\partial t} \quad -(6)$$

These per unit parameters will, in general, be frequency dependent and will depend to varying degrees on the frequency of excitation of the line, $\omega=2\pi f$, which are denoted as R(ω),L(ω),C(ω) and G(ω). So to include these frequency dependent per unit parameters, equation (5) and (6) are rewritten in phasor form in the frequency domain.

Transforming the time-domain transmission line equation with the Fourier transform yields the frequency domain transmission line equations. These may be obtained by simply replacing $\frac{\partial}{\partial t} \rightarrow j\omega$ (where $\omega=2\pi f=$ angular frequency in radian and f= frequency in Hz) to give

$$\frac{\partial V(x,\omega)}{\partial x} = -[R (\omega) + j \omega L]. I(x, \omega) \quad -(7)$$

And $$\frac{\partial I(x,\omega)}{\partial x} = -[G (\omega) + j \omega C]. V(x, \omega) \quad -(8)$$

On the basis of the lumped element constraint, shown in figure 1,the two inherent line parameters for the transmission line, i.e. the characteristic impedance Z_0 and propagation constant γ,can be written as

$$Z_0 = \sqrt{\frac{(R+j\omega L)}{(G+j\omega C)}} \quad -(9)$$

And $\gamma = \sqrt{(R + j\omega L)(G + j\omega C)} = \alpha + j\beta \quad -(10)$

Where, the real part α of the propagation constant is the attenuation constant (in N_p /m) and imaginary part β is the phase constant (in rad. /m). Both the characteristic impedance Z_0 and propagation constant γ are the characteristic properties of a transmission line even if the line is extremely long. By modeling the power line as a transmission line, its characteristic impedance Z_0 and propagation constant γ will direct the wave behavior laterally, the line, the parameter to model the transfer function of the channel [3] ,The p.u. length Resistance R, p. u. inductance L, p.u. length Capacitance C and P.u. length conductance G can be expressed as

$$R = \frac{1}{\pi a} \sqrt{\frac{\pi \mu f}{\sigma_c}} \qquad \text{-(11)}$$

$$\text{where } \delta = \text{skin depth} = \delta = \sqrt{\frac{1}{\pi f \mu \sigma_c}} \qquad \text{-(12)}$$

$$L = \frac{\mu_0 \mu_r}{\pi} \cosh^{-1} \frac{D}{2a} \qquad \text{-(13)}$$

$$C = \frac{\pi \varepsilon}{\cosh^{-1} \frac{D}{2a}} \qquad \text{-(14)}$$

$$\text{And } G = 2\pi f C \tan \delta \qquad \text{-(15)}$$

Where a is the radius of the conductors, f – supply frequency, D- the distance between the wires, σ_c- wire conductivity, μ_r- permeability of the dielectric and ε is the relative permittivity of dielectric ε_r =4 for pvc and 2.3 for Xlpe.

The structure of a typical single-phase low voltage power cable as shown in Figure- 2, under study in this paper, which is commonly found in India. The cables are made up of two copper conductors with PVC insulation, are usually laid inside PVC wall. Normally, the live and neutral cables are used as the power line communication transmission channel.

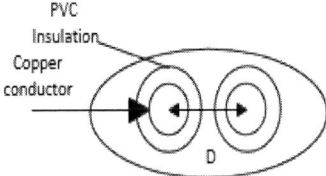

PVC
Insulation
Copper
conductor
D

Fig.2- structure of a typical single-phase low voltage power cable [11]

Distance between two conductors D= 2 t + 2 t +2 a

Where t = insulation thickness = 1.2 mm

  a = conductor radius =1.5 mm,

  Therefore, D= 7.80 mm,

  Conductivity of copper σc = 5.8 x10^7 s/m,

  Relative permittivity of dielectric material ε_r =4 for pvc and 2.3 for Xlpe,

  Conductivity of dielectric σ d = 1x10^-5 s/m,

  Length of the transmission line to be l =200m.,

  Dissipation of pvc tan δ =0.025,

  Relative permeability of copper μ r =1.0,

  Relative permeability of free space μ0 =1.256x 10^-6 H/m.

III.POWER-LINE CHANNEL MODELS

In this section we describe different power line channels models available in the literature and compare their results. In all the simulation works it follows the frequency domain solution first then Fourier transform is made. The block diagram of the channel is given in figure 3.

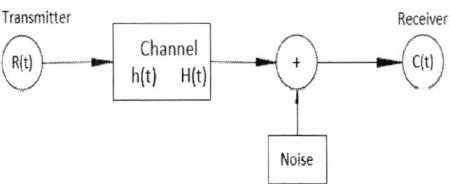

Transmitter
R(t)
Channel
h(t)   H(t)
+
Noise
Receiver
C(t)

Fig.3- Block diagram of Powerline communication channel

According to the approach of Phillips [1,10], the transfer function of the power line channel model is specified by equation (16). In (16), n is the count of probable paths of signals flow, delayed in each path by time τ_i is multiplied by a complex factor ρ_i. The parameter ρ_i is the product of the transmission factor and the reflection factors. In (16) the parameters f, c0, di and ε_r are frequency, velocity of light, path length and relative permittivity correspondingly.

$$\text{Transfer function } H(f) = \sum_{i=1}^{N} \rho_i \, e^{-2\pi f \tau_i} \qquad \text{-(16)}$$

Where $\rho_i = |\rho_i| \, e^{j\emptyset_i}$ and

$$\emptyset_i = \arctan \frac{Im \, (\rho_v)}{Re \, (\rho_v)} \qquad \text{-(17)}$$

$$\tau_i = \frac{d_i}{v_p} \qquad \text{-(18)}$$

$$\text{And } v_p = \frac{c_0}{\sqrt{\varepsilon_r}} \qquad \text{-(19)}$$

The second technique of Zimmerman and Doestert [2,10], has been dedicated for the power channel model, to determine the transfer function, considering the Signal does not flow, through direct path from a transmitter to a receiver in a power-line network. It means multipath scenario with frequency as, similar to as in the radio channel. the power line can be considered a multipath channel that is caused by reflections generated at the cable branches because of impedance discontinuities resulting from various electrical loads. Depending on the primary cable parameters R,L,C and G,and voltage V(x) at line length d_i,(as in fig.1-2). The transfer function can be stated as follows

$$H(f) = \frac{V(x=d_i)}{V(x=0)} = e^{-\gamma d_i} = e^{-\propto(f)d_i} e^{-j\beta(f)d_i} \qquad \text{-(20)}$$

an approaching procedure for the attenuation factor α is initiate in the form

$$\alpha(f) = a_{0+}a_1f^K \qquad -(21)$$

With (20) and (21), the attenuation of a powerline cable can be considered by

$$e^{-\alpha(f)d_i} = e^{-(a_0+a_1f^K)d_i} \qquad -(22)$$

The parameter of attenuation a_0 (offset of attenuation), a_1 (increase of attenuation), and K (exponent of attenuation) can be attained from the magnitude of the frequency response.

Thus the transfer function is given by the equation [2]

$$H(f) = \sum_{i=1}^{N} g_i \times e^{\left[-(a_0+a_1f^K)d_i\right]} \times e^{\left[-2\pi f\left(\frac{d_i}{v_p}\right)\right]} \qquad -(23)$$

Where N is the number of multi paths, the g_i is weighting factor, d_i is path length of I'th path, a_0 and a_1 is attenuation parameters, and v_p phase velocity, f-frequency and k- exponent of attenuation factor (0.2-1.0). Part of equation $e^{\left[-(a_0+a_1f^K)d_i\right]}$ is an attenuation portion and the part $e^{\left[-2\pi f\left(\frac{d_i}{v_p}\right)\right]}$ is known as delay portion.

IV. PERFORMANCE ASSESSMENT AND DISCUSSION OF RESULTS

The system architecture of a proposed Channel model network overlapping the electricity distribution network as described in Figure 4.

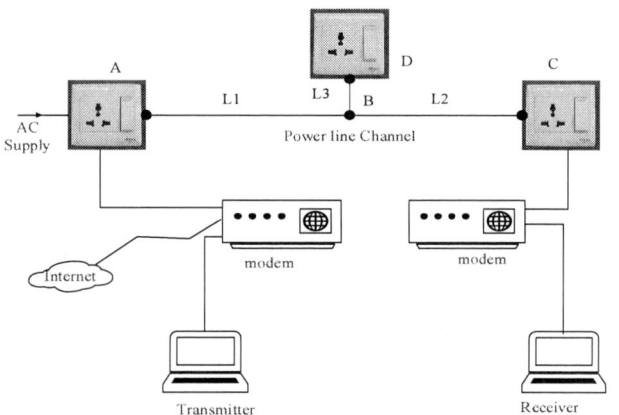

Fig.4. Proposed indoor powerline channel network

The connection has four nodes A, B, C and D, Having only one branch and with the lengths L_1, L_2 and L_3. Now Consider the power line network as shown in Fig. 4, Z_S, Vs, ZL_1 and ZL_2 are the impedance, voltage of the source, node C load impedance and node D-load impedance correspondingly. The line segment length AB, BC and BD was deliberated as 30m, 170m and 11.20 m separately. P.u. length inductances and capacitances was taken as 0.44388 micro Hanery/m and 0.61734 pico Farad /m correspondingly for each line segments. Fig. 5 shows the simulations results for the Phillips model [1,10], by using equation (16) of transfer function.

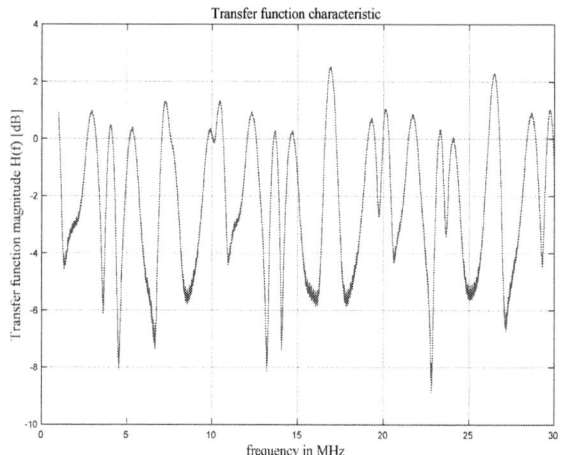

Fig. 5- Channel Transfer function response N=5

Another performance evaluation test results for the broadband powerline channel model introduced by Zimmerman and dostert [5], considering fig.4, is presented here. For the proposed network, the signal transmission routes for such type grid can be inscribed as: (i.e., A→B→C, A→B→D→B→C, A→B→D→B→D→B→ C, and so on). for 4-path power line channel, the estimated Parameters result are shown in Table I, for 4-paths .Attenuation parameters K=1, a_0=0, a_1=7.8x10^{-10}s/m.

TABLE I-PATH PARAMETER FOR N=4

Path number i	weighing factor g_i	Path length d_i (m)
1	0.64	200
2	0.38	222.4
3	-0.15	411
4	0.05	490

We now consider the proposed structure of the indoor powerline channel network, as given in figure 4; The direct length of the network, from transmitter to receiver was kept constant at 30m plus170m. The branched length was varied as (L_3 = 30 m, 40 m, and 50 m branches). Next we calculated the transfer

978-1-5386-6624-1/18 $31.00 © 2018 IEEE

characteristics, Figure-6, show the corresponding transfer function responses for various branch line lengths using equation (23).

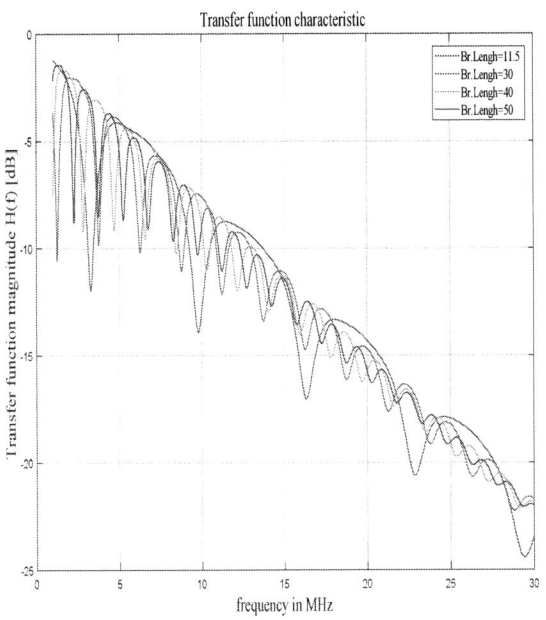

Fig.6- Transfer function response for different branch length, N=4

V. BROADBAND POWERLINE COMMUNICATION APPLICATIONS

The Broadband powerline communication has been known to be very promising communication technology, it provides the internet access, wherever the powerline is available. The remote monitoring and management of user equipment and machines installed along the powerline network. The surveillance of smart grid, data management and recording, VoIP etc. The smart grid communications that is used for the automatic meter reading (AMR), energy management service (EMS) and distribution automation (DA). As it also used in home area network (HAN), intelligent building system (IBS), for different electrical appliances operation and control as like IOT technology. Another very important application of broadband powerline communication is in supervisory control and data acquisition (SCADA). Fig. 7 shows simple application for indoor powerline communication network.

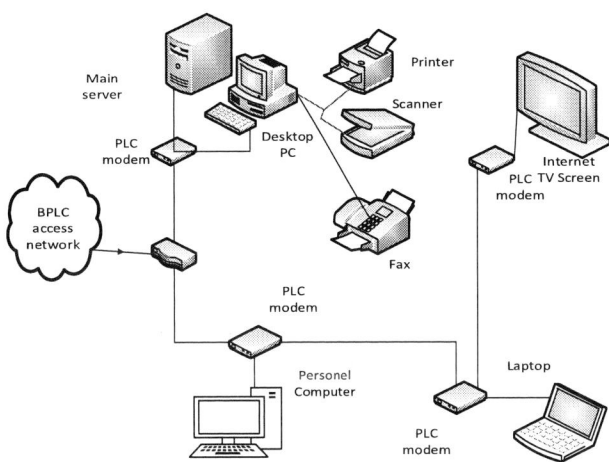

Fig. 7. Application of the Broadband powerline communication scheme

CONCLUSIONS

The broadband powerline communication has grown as a substitute technology for broadband communication. we have described the primary Philips channel model, derived the expression of transfer function. Also studied and investigated the channel transfer function of the another channel model of Zimmerman and Dostert multipath propagation model. The transfer function response of each model was compared. The performance improvement was also demonstrated with the simulation results of channel transfer function plot, and we have found that the position of the deep notches and peaks in the magnitude responses are largely affected with change in the branch length and the line impedance. Finally, we indicate that the model is anticipated for indoor powerline communication.

REFERENCES

[1] H. Phillips, "Modelling of powerline communication channels," in *Procedia. IEEE International Symposium of Power Line dCommunication* Application, Mar. 1999, pp.14–21.

[2] M. Zimmermann and K. Dostert, "A multipath model for the powerline channel," *IEEE Trans. Communication*, vol. 50, no. 4, pp. 553–559,April 2002.

[3] H. Meng, Y. L. Guan, C. L. Law, P. L. So, E. Gunawan, and T.Lie,"Modeling of transfer characteristics for the broadband power line communication channel," *IEEE Trans. Power Delivery.*, vol. 19, no. 3, pp. 1057–1064, Jul. 2004

[4] S. Galli and T. C. Banwell, "A deterministic frequency-domain model for the indoor power line transfer function," *IEEE Journal of Selected Areas of Communication*, vol. 24, no. 7, pp. 1304–1316, Jul. 2006.

[5] S. Galli, "A novel approach to the statistical modeling of wireline channels" *IEEE Trans. Communication*, vol. 59, no. 5, May 2011, pp.1332–1345.

[6] A. M. Tonello, "Wideband impulse modulation and receiver algorithms for multiuser power line communications," *EURASIP Journal Of Advance Signal Processing*, vol. 2007, pp. 1–14.

978-1-5386-6624-1/18 $31.00 © 2018 IEEE

[7] A.M.Tonello and Fabio Versolatto "Bottom-Up statistical PLC channel modeling-part II: Inferring the statistics" IEEE Trans. On Power Delivery, vol.25, N0.4, October 2010, pp 2356-2363.

[8] Apostolos N.M., Konstantinos N.S., Georgios T.A. and Dimitrios P.L. "Modeling of medium-voltage power-line communication systems Noise levels" *IEEE Transactions on power delivery,* vol.28, n0.4, October 2013, pp 2004-2013

[9] A.M. Tonello, F. Versolatto and B. Bejar "A top-down random generator for the in-home PLC channel".in Procedia IEEE global. communication conference, Dec.2011, pp 1-5.

[10] Justinian Anatory and Nelson Theethayi "Comparison of different Channel Modeling Techniques used in the BPLC Systems " World Academy of Science, Engineering and Technology, Vol:56, August 2011,pp 848-854.

[11] Mukesh Kumar Varma, Zainul Abdin Jaffery and Ibraheem "Advances of Broadband power line communication and its application" 12'th IEEE International Conference INDICON, December17-20 2015.

[12] Mukesh Kumar Varma, Zainul Abdin Jaffery and Ibraheem "Broadband power line communication: Solution for the next generation access network" proceeding *IEEE sponsored 41'st national system Conference, Dayalbagh Educational Institute, Agra, India*, December 1-3, 2017

[13] Mukesh Kumar Varma, Zainul Abdin Jaffery and Ibraheem "A comprehensive study on channel modeling of broadband communication over low voltage power line" proceeding IEEE sponsored 5'th International Conference on Signal Processing and Integrated Networks, SPIN 2018, Amity University, Noida, India, February 22-23, 2018

[14] Clayton R.Paul" Analysis of multiconductor transmission lines" second edition, John wiley and sons,Inc. publication.2007.

2nd IEEE International Conference on Power Electronics, Intelligent Control and Energy Systems (ICPEICES-2018)

Cognitive Radio Technology in 5G Wireless Communications

Shruti Bhandari
Dept. of Electronics and Communication Engineering,
Maharana Pratap University of Agriculture and Technology Udaipur, India
shrutibhandari@live.com

Sunil Joshi
Dept. of Electronics and Communication Engineering
Maharana Pratap University of Agriculture and Technology Udaipur, India
suniljoshi7@rediffmail.com

Abstract—The fifth Generation (5G) of wireless communication standards and Cognitive radio (CR) are believed to be the solution for present day data intensive applications. The 5G wireless networks are expected to provide higher data transfer rates, ubiquitous connectivity, low end-to-end latency, much higher system capacity and improved energy efficiency. Cognitive radio network (CRN) offers dynamic spectrum sharing to achieve higher spectrum efficiencies as required in 5G architecture. Present report encompasses concise information about mobile cellular system, 5G requirements, 5G enabling technologies and cognitive radio paradigm for 5G heterogeneous multi-tier networks.

Keywords— *5G, 5G enabling technologies, cognitive radio, heterogeneous multi-tier networks*

I. INTRODUCTION

Wireless communication has experienced exponential traffic growth in the past decade owing to the recent popularity of internet based data intensive applications and their increased demand in various sectors. This has initiated intensive R&D activities towards the next generation of the mobile cellular system, catering to high capacity and capable of adapting to the variation in user demands. Since the initiation of mobile wireless communication in late 1970s from analog voice calls to present day high quality mobile broadband services, a new cellular system brings in a new generation roughly every ten years. First generation (1G) communication system are dated 1982, while the 2G ones were commercially deployed in 1992, followed by 3G systems in early 2000s and 4G systems are being fully exploited since 2010. Considering the chronological rule of thumb for development of a new cellular system every decade, research communities and industries hypothesize that the introduction of the next generation i.e. 5G standards might take place in early 2020s. For successful deployment of 5G, innovative technologies are investigated to provide increased capacity, higher speed and improved spectral and energy efficiency. A number of academicians, researchers and telecommunication companies have highlighted the key requirements, enabling technologies, communication standards, research challenges and resource management for upcoming 5G cellular networks [1-4]. Considering the huge amount of users, devices and systems, the 5G network is expected to have capabilities to overcome the limitations of previous system generations, particularly in view of the system capacity, data rate, latency, network reliability, network availability and energy efficiency in the present era of IoT (internet of things). With this vision, intensive research is going on for incorporating advancements in

existing 3G/4G architectures and systems to enable interoperability and compatibility.

Conventional cellular networks built on expensive licensed bands have introduced a paradoxical situation, where on one side, there is scarcity of spectrum due to increasing demand and on other side there is spectrum underutilisation. In 5G paradigms, the Cognitive Radio (CR) technology is being proposed to facilitate dynamic spectrum sharing between 5G network entities. In context of CR for enabling 5G vision, Badoi et al. [5] have highlighted a novel term 'WISDOM' (Wireless Innovative System for Dynamic Operating Mega-communications), for an integration of 5G wireless access networks through CR technology. The 5G architecture will include a large number of different network entities mostly sharing a common spectrum resource, formed by licensed and unlicensed frequency bands, via Dynamic Spectrum Access (DSA) approach, instead of traditional static band assignment. This solution is expected to significantly increase the overall spectrum efficiency; however it may also introduce the challenge of optimizing the co-existence between different devices, systems and technologies [6, 7].

Present paper encompasses concise information about mobile cellular system, 5G requirements, 5G enabling technologies, and cognitive radio paradigm for 5G network with special reference to heterogeneous multi-tier networks.

II. KEY FEATURES OF 5G

The features of 5G communication systems shall enable a wireless infrastructure for human-centric communications and upcoming industrial requirements at data rates as high as 1 Tbps (Terabits per second). The important 5G expectations catering to these requirements are summarized here [8, 9].

- *Huge network capacity* is expected to cater the exponential growth of traffic volume. More than 100-fold increase of area capacity from 0.1 Mbits/s/m^2 to 10–100 Mbits/s/m^2 is being targeted as compared to the existing LTE-A architectures of 4G.

- *Ultrahigh reliability and availability* with extremely low loss rate is expected from 5G network. Network availability of 99.999% is expected particularly in critical control services, traffic safety, automated industrial applications and medical services.

- *Ultralow latency* is envisioned for 5G networks for real-time applications, such as internet, traffic control, industrial processes and other vital infrastructures. 5G is expected to provide latency of

978-1-5386-6624-1/18 $31.00 © 2018 IEEE

1ms or less and nearly "zero latency" in case of cloud-based applications.

- *Ubiquitous connectivity* is required to cater expected ten to hundred fold increment in the number of connected users resulting in connection density to rise from 10^5 devices/km^2 to $10^6 - 10^7$ devices/km^2.

- *High data rate* of 1 Gbps (Gigabits per second) throughput everywhere, including sparsely populated rural areas is expected. The peak data rate more than 10 Gbps is expected for indoor and dense outdoor environments.

- *Energy efficiency* is more compelling for future communication technologies with millions additional base stations (BS) and billions of connected user equipments (UE). A 100-fold increase in bits/Joule along with 10-times prolonged battery life of devices is being targeted for 5G network.

The requirements of future 5G networks are not going be easy to achieve. There are multifold challenges in context of 5G, since it is not just handling the huge communication traffic volume, the increasing randomness and diversity in mobile data is also to be taken care off. These challenges are expected to be addressed by modifying existing technologies and introducing newer enabling technologies. Table I. includes features of 5G vision vis-a-vis possible technological solutions for the 5G requirements.

TABLE I   5G FEATURES AND ENABLING 5G TECHNOLOGIES

Key Features of 5G	Enabling Technologies
High capacity & connectivity	• Heterogeneous multi-tier network • Massive MIMO technology • Millimeter-Wave communications • Cloud-Based Radio Access Network
High data rate everywhere	• Heterogeneous multi-tier network • Massive MIMO technology • Millimeter-Wave communications • Full-Duplex communications
Low latency	• Full-Duplex • Cloud-Based Radio Access Network • Device-To-Device (D2D) links
Improved energy efficiency	• CognitiveRadio technology • Spectrum Sharing • Energy Harvesting • Device-To-Device (D2D) links
High reliability and availability	• Cloud-based Radio Access network • Heterogeneous multi-tier network • Wireless Network Virtualization
Network management & IoT	• Machine-To-Machine (M2M) communication • Cognitive Network

III.   5G ENABLING TECHNOLGIES

The defining features of 5G vision are global wireless communication, convergence, connectivity and co-operation. This section highlights the limitations and advancements in existing system as well as newer 5G enabled technologies. Fig. 1 depicts 5G enabling technologies.

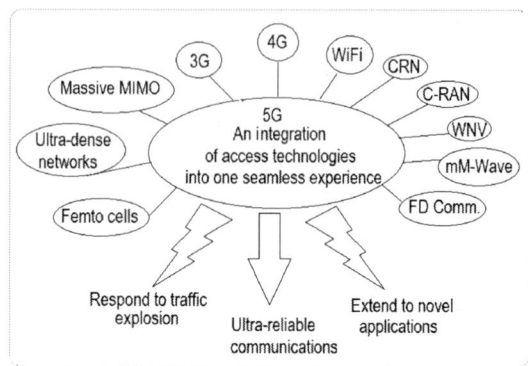

Fig. 1   Enabling technologies for 5G vision

A.   Advancements in 3G/4G technologies

Researchers and telecom companies are on a continuous lookout for new modulation and channel multiplexing techniques that fulfills the need of 5G. The characteristics of 4G networks like high mobility and data rates; number of different applications and services; cooperative access with unified architecture; detecting and picking of the best network; technology and topology independence; seamless handover and continuity in service, makes the 4G technology as the right precursor of next generation, 5G technology. Accordingly, the advancement in the existing technologies is being pointed out in this section to make them suitable for 5G implementation.

1.   Multiplexing technology

Most preferred multiplexing approach in 4G mobile transmission is orthogonal frequency division multiplexing (OFDM), where subcarriers are orthogonal to each other. The flat response of the channel improves the system efficiency in terms of increased data-rate and high capacity without much Inter Symbol Interference (ISI). In OFDM the symbols are close to each other and to reduce the ISI, cyclic prefix (CP) is inserted between the symbols. However addition of CP leads to loss of bandwidth which in turn will reduce the spectral efficiency. Additional data in subcarriers reduces the peak average power ratio (PAPR) and thus system performance is reduced to a great extent. Hence, OFDM in its conventional form is not suitable for 5G. An advanced version of OFDM is Orthogonal Frequency Band Multi Carrier (OFBMC) which does not use cyclic prefix. OFBMC integrates spatial division multiplexing (SDM) with space time block codes (STBC) and space time trellis codes (STPC) to enhance the performance of modulation techniques resulting in improved utilization of available bandwidth and adding robustness to the system [10]. Another advancement of OFDM is Filter Band Multi Carrier (FBMC), where a set of parallel data is transmitted through an array of filters. Universal Filter Multi Carrier (UFMC) is a versatile technique for 5G communication. UFMC is an integration of OFDM and FBMC, where side lobe is efficiently reduced to improve the performance of the system. This technique utilizes very short length of filter and therefore useful for short burst communication [11]. A non-orthogonal modulation of OFMD has emerged as Generalized Frequency Domain Multiplexing (GFDM) technique, which can be applied for non-orthogonal waveform in 5G domain.

978-1-5386-6624-1/18 $31.00 © 2018 IEEE

2. Multiple access technology

An advanced multiple access technology termed as Beam Division Multiple Access (BDMA) is proposed to replace the prevalent technologies like Frequency Division Multiple Access (FDMA) and Time Division Multiple Access (TDMA) that are unable to cater the frequency and time division with increased number of users [12]. As a radio interface, BDMA is independent of frequency and time resources. An orthogonal beam is allocated to each mobile station and beam of antenna is divided to provide multiple access to different locations. The mobile BS and the user are in line-of-sight (LOS) allowing simultaneous data transmission.

3. Massive MIMO technology

The prevailing MIMO (Multiple-Input Multiple-Output) communication technology is upgraded as Massive MIMO to cater 5G requirements. The current MIMO system utilizes 2 to 4 antennas, whereas massive MIMO exploits larger arrays of antennas to achieve the goal of higher capacity gains. The arrays of antenna containing very large number (10s or even 100s) of antennas which serve many of the user terminals simultaneously, in single frequency slot. Arrays of a number of receiving and transmitting antennas can be placed on a single device or distributed among many devices. A large number of simultaneously operating antennas improves the system performance significantly with respect to data rate, link reliability and energy efficiency due to multiplexing and array gains. The computational processing of phase coherent signals from all the antennas of Massive MIMO at the BS is possible. By using simple linear precoding and detection algorithms, the effect of noise, fast fading and intracell interference in transmission may be avoided. Massive MIMO systems obtain a high multiplexing gain while eliminating the problems of unfavorable propagation environments. A number of survey papers have outlined the characteristics and advantages of Massive MIMO for 5G vision [13-15]. Towards energy efficiency and power optimization, Björnson et al. [16] have shown that on increasing the number of antennas and small cell access (SCA) point at the BS, the total power per subcarrier decreases tenfold in comparison to the case of SCA point without antenna. Human intervention in the networking can be minimized by introducing the concept of SCA point in self organizing network technology [17]. Thus massive MIMO can be considered as evolving technology of Next generation networks, with robust, secure and spectrum efficient communication.

4. Modulation Techniques

The spectrum efficiency, speed and performance of communication system depend upon the reliable modulation and filtration technique. Quadrature Amplitude Modulation (QAM) technique is extensively used in radio communications. Different types of QAM are identified by an integer associated with it. The integer shows the number of points on the QAM constellation diagram. QAM-256 is a high order modulation technique, wherein a carrier signal such as an LTE waveform, transmits data and information. In QAM-256, 8 bits per symbol are transmitted, generating a mobility speed of 40 mbps and 33% increased efficiency vis-a-vis QAM-64. However, with increase in the order of modulation schemes, resilience to noise and interference decreases. Spatial modulation (SM) is another recently developed promising transmission technique that uses a third dimension, known as index of antennas which transmits the multiples data simultaneously to improve the system performance [18]. Spatial modulation technique coupled with MIMO gives the best performance.

B. Newer Technologies for 5G Vision

Apart from above mentioned advancements in existing communication technologies, newer technological solutions are also imperative to achieve the 5G vision. The objective of developing these technologies is to cater exponential increased capacity with efficient utilization of all possible resources in 5G network domain. Based on the well-known Shannon theory, the total system capacity, C_{sum} is equivalent to the sum capacity of all sub channels and heterogeneous networks as expressed by eq1.

$$C_{sum} \approx \sum_{Hetnets} \sum_{Channels} B_i \log_2 \left[1 + (P_i / N_p) \right] \qquad (1)$$

Where, B_i refers to the bandwidth of the ith channel, P_i is the signal power of the ith channel, and N_p denotes the noise power. Accordingly, the network coverage can be increased by right combination of network densification, increased number of sub channels through massive MIMO, SM, interference management, dynamic spectrum management through cognitive radio networks, millimeter-wave communications and power efficiency through energy-harvesting. The most investigated enabling technologies for the 5G vision are outlined here.

1. Full-duplex communication system

Full-duplex (FD) communication system is able to transmit and receive the signals at the same frequency & time, mitigating the self-interference (SI) in radio communications [19]. FD systems can provide high data rate than conventional half duplex communication system. Moreover, the RF interference cancellation techniques can facilitate in-band FD radios. FD communication is envisioned to be applied to 5G, to increase the spectrum efficiency at physical layer by maintaining the same frequency band or time slot for uplink and downlink transmission.

2. Energy Harvesting

Recently, energy harvesting is viewed as a prospective solution to achieve energy efficiency and prolong the battery life of equipments. User equipment (UE) can harvest energy from a number of renewable natural energy sources viz. solar and wind. However, the available energy levels from such environmental sources may vary significantly over climatic and geographic conditions, affecting the reliability and quality-of-service (QoS). The radio frequency -powered energy harvesting network (i.e. RF-EHN) using ambient radio signals are also being analyzed as possible energy harvesting solutions for processing and transmitting the information [20]. The characteristic feature of RF-EHN i.e. low power and long distance transfer makes it a promising approach for energy efficiency in 5G wireless networks.

978-1-5386-6624-1/18 $31.00 © 2018 IEEE

3. Millimeter-wave communications

A large amount of spectrum in the high frequency band, known as millimeter wave (30-300GHz) is lying idle and is being exploited gainfully in 5G era. Rappaport group [21] has envisioned that mm-wave mobile communication will work well in 5G era. Preliminary studies have shown that 28-38 GHz frequency bands can be used with steerable antennas. In mm-wave frequency bands, the use of high gain, steerable antennas and the CMOS technology can operate well. The use of mm-wave as a carrier frequency is estimated to allow high data transfer rates, increased capacity and low latency. The small frequencies of mm-wave will utilize the polarization and different spatial processing techniques. The challenges of using the mm-wave spectrum are high penetration loss, high phase noise, penetration through obstacles, poor diffraction and expensive equipments.

4. Cloud-based Radio Access Network and Wireless Network Virtualization

Two potential methodologies to be applied in 5G communications are cloud-based radio access network (C-RAN) and wireless network virtualization (WNV). The characteristics of a radio access network (RAN) are quite akin to those of a cloud computing platform as it connects individual devices to other parts of a network through radio connections. Hence, a useful integration of cloud computing with RAN as Cloud-based Radio Access Network (C-RAN) for 5G vision is being proposed [22]. Cloud computing platform can provide on-demand, scalable access to resources on shared basis for a mobile base station (MBS) without installing expensive network devices [23]. C-RAN enables a distributed architecture for a single base station (BS) whereas Wireless network virtualization (WNV) simplifies the resource sharing among many users with easier migration to newer products and technologies. WNV enables sharing of infrastructure and spectrum resources, making the system cost effective [24]. Use of cognitive radio is expected to gives more options for better virtualization.

5. Network densification: Heterogeneous multi-tier networks

The 5G architecture is going to be ultra dense multi-tier network with the co-existence of many different classes of BSs including small cells, macrocells, and low power nodes like relays and remote radio heads; along with the provision for device-to-device (D2D) or machine-to-machine (M2M) communication [25]. This kind of heterogeneous network with provides flexible coverage, improved spectral efficiency, capacity and power consumption. However, robust mechanisms for the inter-tier and intra-tier interferences avoidance must be provided. The interferences related challenges associated with network densification can be addressed by cognitive radio, as discussed in next section.

IV. COGNITIVE RADIO AND HETEROGENEOUS NETWORKS (HETNETS)

Cognitive Radio (CR) proposed by J. Mitola [26] is a path breaking innovation in radio technology that has opened up gates for immense possibilities for improving the utilization of the congested radio spectrum. CR, built upon software defined radio (SDR), has been introduced as "context-aware intelligent radio, capable of autonomous reconfiguration by applying cognitive intelligence in learning from and adapting to the communication environment". The incorporation of CR in 5G networks is expected to solve the problem of underutilization of a large chunk of radio spectrum [27, 28]. The key features of CR are:

1. Cognitive capability refers to the ability of the radio technology to sense the information from its environment and identify the portions of the spectrum that is unused or idle at a specific time or location (white spaces).

2. Re-configurability, which enables the radio to be dynamically programmed for transmission and reception of a variety of frequencies using different transmission access technologies supported by its hardware design.

Thus, CR integrates and adds cognitive intelligence to wireless communication system. Collecting information on the spectrum resource is usually referred to as spectrum sensing and results of spectrum sensing are then used for optimizing the sharing of spectrum, among network entities. Despite advanced spectrum sensing techniques such as energy detection, matched filter detection and cyclo-stationary detection, some challenges are there in terms of detection of ideal spectrum and detection under dynamic transmission-reception at a minimum interval of time.

Cognitive-radio-inspired spectrum sharing improves spectrum efficiency and utilization. By definition, spectrum sharing is "usage of same spectrum by more than one user with respect to frequency, time and location" [29]. Spectrum sharing is one of the fundamental aspects of the CR networks to provide access channels and sharing the resources without changing the existing spectrum allocation policies. Dynamic Spectrum Access (DSA) or Dynamic Spectrum Management (DSM) integrates various approaches of spectrum reforms to address the issue of underutilization of spectrum. In hierarchical access approach of DSA, a licensed band is shared between licensed primary users (PUs) and unlicensed secondary users (SUs) without much interference to the PUs. Basically three different approaches can be identified within hierarchical access [30]. In *overlay approach*, SUs access the network opportunistically i.e. SU's are allowed to borrow spectrum resources only when PU's do not use them. Thus, spectrum sharing is on interference-free basis. In *underlay approach*, SU's can share the spectrum resource with the PU's at the same time, while keeping the interference below a threshold limit i.e. *Interference Temperature*. Interference-tolerant CR networks can achieve enhanced spectrum utilization and better energy efficiency by opportunistically sharing the radio spectrum resources with licensed users. *Hybrid* hierarchical access approach combines overlay and underlay access approaches.

Recently, dense deployment of small cells over rich portions of low radio frequency is being investigated as a potential approach for spectrum sharing and to address the energy efficiency in 5G architecture [31, 32]. The network densification results in partial channel sharing and co-channel deployment, with macro cells as primary (licensed users and small cell tiers as cognitive or secondary users.

Deployment of such a varied environment with co-existence of several classes of cells, power nodes and user equipments creates a heterogeneous network (HetNet) scenario, which has opened up enormous possibilities in addressing 5G requirements

TABLE II   SMALL CELLS AND THEIR CHARACTERISTICS

| Small Cells | Power Range | | Range | Deployment Environment | No. of concurrent users |
	Indoor	Outdoor			
Femto	10-100 mW	0.2-01 W	tens of meters	Residential and Commercial	04-32
Pico	100-250 mW	01-05 W	tens of meters	Public place	64-128
Micro	—	05-10 W	few Kilometers	Urban area to fill the macro coverage	128-2568
Macro	—	10-20 W	Kilometers	Urban area to increase the capacity	> 250
WiFi	20-100 mW	0.2-01 W	few tens of meters	Residential and Commercial	< 50

Small cells are available in a variety of size ranging from femto to macro cells. They are low power, wireless access points operating in licensed spectrum. The coverage profiles and power characteristics of small cells are as shown in Table II. Dense deployments of IEEE 802.11 based WLANs (Wireless Local Area Networks) also fall in broad definition of small cells; however they operate over unlicensed bands. Small cells can provide improved cellular coverage, capacity and applications for residential, commercial and public urban places and rural spaces. Fig. 2 depicts a typical heterogeneous multi-tier network with concurrent operation of macrocell, picocell, and femtocell base stations to facilitate a varied coverage area and improved spectral usage. In such a heterogeneous architecture, spectral efficiency can be achieved by higher frequency re-use and reducing the transmit power.

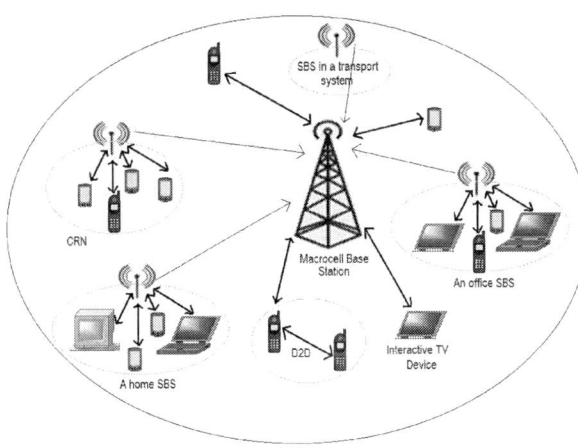

Fig. 2   Heterogeneous multi-tier networks

The major challenges of applicability of 5G networks in HetNets are to use different types of networking nodes.

Nodes may vary from stand-alone BS to systems with varied degrees of centralized processing as per available backhaul technology. Challenge on integration of Wi-Fi, LTE-A (long term evolution-advanced) and HSPA (high speed packet access) comprising a HetNet, in 5G networks are also highlighted [33]. According to Caso et al. [34] the application of CR capabilities depends on mode of heterogeneous networks (HetNets) usage scenarios. In non-cooperative HetNets, the cognitive (or secondary) small cells provide support to the macrocell tiers while working on the same spectrum resource. In case of cooperative HetNets, the cognitive (or secondary) relays and repeaters select the optimum relying and amplifying strategy, for increased cellular coverage. In HetNet scenario, centralized control for spectrum access and network interference (NI) mitigation is hard to implement due to scalability of control message exchange among intra-tier cells and inter-tier cells. Thus, Cognitive Radio inspired spectrum resource management (CR-SRM) mechanisms in are in high demand. Further, the hierarchical access model of dynamic spectrum access is most suited for HetNets scenarios

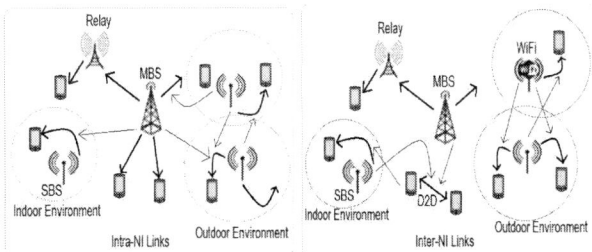

Fig.3   Interference among the heterogeneous 5G network entities

The interference arising due to the coexistence of macrocells and small cell tiers is referred to as cross-tier interference (Cr-TI). While, the interference arising due to the dense and random deployment of small cell base stations (SBSs) is referred to as co-tier interference (Co-TI). The cross-tier or co-tier interference are cases of intra-network interference (Intra-NI) [34]. Inter-network interference (inter-NI) generally affects the transmission/reception data link, which causes interference to participating transmitter and receiver. Fig. 3 shows the intra-NI scenario and inter-NI scenario for downlink transmissions in cellular networks and Wi-Fi in the presence of device to device communication links. Such HetNet scenarios call for an efficient spectrum resource management to limit the network interferences. An integrated platform, C-MIANS (Cognitive Media Independent Access Network Selection), has been highlighted [35] to address the massive deployment, heterogeneity, coexistence and energy efficiency as required for 5G architecture. Platform, C-MIANS may prove to be useful within the IEEE 1900.6 spectrum sensing protocol towards the integration of context awareness into dynamic spectrum management, network selection and resource allocation.

V.   CONCLUSION

A summary on 5G envisioned requirements, enabling technologies, challenges and the application of the CR paradigm towards 5G has been described in the light of recent literature. 5G systems will be supported by newer 5G standards as well as numerous updated releases of 3G/4G

978-1-5386-6624-1/18 $31.00 © 2018 IEEE

systems integrated across many spectrum resources. Within this vision, the use of unlicensed bands through cognitive radio opens new opportunities and challenges. These scenarios will require efficient resource management in order to limit the network interferences. The cognitive radio and cognitive networking are being proposed to enable dynamic spectrum sharing in heterogeneous environment to achieve high spectrum efficiencies using small cell of different sizes in 5G network architecture. Cognitive radio adapts and works with different the wireless technologies whereas the 5G communication networks integrates and interconnects various the wireless technologies. Thus cognitive radio represents the technological "tool" to implement 5G convergence concept.

REFERENCES

[1] M-G Di Benedetto, A. F. Cattoni, J. Fiorina, F. Bader and L. De Nardis (Editors), Cognitive Radio and Networking for Heterogeneous Wireless Networks: Recent Advances and Visions for the Future, Springer, 2016.

[2] 5G: Challenges, Research Priorities, and Recommendations, NetWorld2020 ETP European Technology Platform for Communications Networks and Services, August 2014.

[3] C-I Badoi, N. Prasad, V. Croitoru and R. Prasad, "5G Based on cognitive radio," Wireless Pers. Commun., vol. 57, pp. 441–464, 2011.

[4] N. Panwar, S. Sharma and A. K. Singh, "A survey on 5G: The next generation of mobile communication," Phys. Comm., vol 18, (2), pp. 64-84, March 2016.

[5] E. Hossain and M. Hasan, "5G Cellular: Key enabling technologies and research challenges," IEEE Instrumentation & Measurement Magazine, vol. 18 (3), pp.11-21, 2015.

[6] S. Chen and J. Zhao, "The requirements, challenges, and technologies for 5G of terrestrial mobile telecommunication," IEEE Commun. Mag., vol. 52 (5), pp. 36–43, 2014.

[7] A. Kumar and M. Gupta, "A review on activities of fifth generation mobile communication system," Alexandria Engg. J., in press. Available online 20 February 2017 DOI: 10.1016/j.aej.2017.01.043.

[8] Ericsson (2016) 5G radio access – capabilities and technologies. White paper Uen 284:23–3204. Rev C.

[9] Qualcomm Technologies Inc (2016) Leading the world to 5G. White paper.

[10] D. Roque, C. Siclet, "Performances of weighted cyclic prefix OFDM with low-complexity equalization," IEEE Commun. Lett., vol.17 (3), pp. 439–442, 2013.

[11] V. Vakilian, T. Wild, F. Schaich, S. Brink, J-F. Frigon, "Universal filter multi carrier technique for wireless system beyond LTE," IEEE Global Comm. Workshop, p. 223–228, 2013.

[12] C.-X. Wang et al., "Cellular architecture and key technologies for 5G wireless communication networks," IEEE Commun. Mag., vol. 52 (2), pp. 122-130, Feb. 2014.

[13] E.G. Larsson, O. Edfors, F. Tufvesson, T.L. Marzetta, "Massive MIMO for next generation wireless systems," IEEE Commun Mag., vol. 52(2), 186–195, 2014.

[14] L. Lu, G. Y. Li, A. L. Swindlehurst, A. Ashikhmin, and R. Zhang, An overview of massive MIMO: Benefits and challenges, IEEE J. Sel. Topics Signal Process., vol. 8, no. 5, pp. 742-758, Oct. 2014.

[15] C.-X. Wang, S. Wu, L Bai, X. You, J. Wang and C-L I, "Recent advances and future challenges for massive MIMO Channel

measurements and models," Sci. China Inf. Sci., vol.59 (2) pp.1-16, 2016.

[16] E. Björnson, M. Kountouris, and M. Debbah, "Massive MIMO and small cells: Improving energy efficiency by optimal soft-cell coordination," in Proc. 20th Int. Conf. Telecommun. (ICT), pp. 1-5, May 2013.

[17] M. Peng, D. Liang, Y. Wei, J. Li, and H.-H. Chen, "Self-configuration and self-optimization in LTE-advanced heterogeneous networks," IEEE Commun. Mag., vol. 51, no. 5, pp. 36-45, May 2013.

[18] E. Basar, "Index modulation techniques for 5G wireless networks," IEEE Comm. Mag. vol. 54 (7) pp 1-9, 2016.

[19] S. Hong, J. Brand, J. Choi, M. Jain, J. Mehlman, S. Katti and P. Levis, "Applications of self-interference cancellation in 5G and beyond," IEEE Commun. Mag. vol.52(2), pp. 114–121, 2014.

[20] L. Xiao, P. Wang, D. Niyato, D. Kim and Z. Han "Wireless networks with RF energy harvesting: a contemporary survey," IEEE Commun. Surv. Tutorials, vol. 17(2), pp.757–789, 2014.

[21] T. S. Rappaport et al. "Millimiter wave mobile communications for 5G cellular: It will work!," IEEE Access, vol 1, pp335–349, 2013.

[22] P. Rost et al., "Cloud technologies for flexible 5G radio access networks," IEEE Commun. Mag., vol. 52(5), pp. 68-76, May 2014.

[23] A. Checko et al. "Cloud RAN for mobile networks – a technology overview," IEEE Commun. Surv. Tutorials, vol 17 (1), pp. 405–426, 2015.

[24] C. Liang and F. Yu, "Wireless network virtualization: a survey, some research issues and challenges," IEEE Commun Surv Tutorials, vol. 17(1), pp.358–380, 2015.

[25] N. Bhushan et al. "Network densification: The dominant theme for wireless evolution into 5G," IEEE Commun. Mag. vol. 52(2), pp.82–89, 2014.

[26] J. Mitola and G.Q. Maguire, "Cognitive radio: making software radios more personal," IEEE Pers. Commun., vol. 6(4), pp.13–18, 1999.

[27] S. Haykin, "Cognitive radio: Brain-empowered wireless communications," IEEE J. Sel. Areas Commun., vol. 23, pp. 201–220, 2005.

[28] I.F. Akyildiz, W-Y Lee, M. C. Vuran and S. Mohanty, "Next generation/dynamic spectrum access/cognitive radio wireless networks: a survey," Elsevier Comput Netw., vol. 50, pp.2127–2159, 2006.

[29] ECC, "Licensed Shared Access (LCA)", Repor 205, Feb 2014.

[30] R. Etkin, A. Parekh and D. Tse, "Spectrum sharing for unlicensed bands," IEEE J. Sel. Areas Commun., vol. 25(3), pp.517–528, 2007.

[31] X. Hong et al., "Capacity Analysis of Hybrid Cognitive Radio Networks with Distributed VAAs," IEEE Trans.Vehic. Tech., vol. 59(7),, pp. 3510–23, Sept 2010.

[32] 5G: Challenges, Research Priorities, and Recommendations, NetWorld2020 ETP European Technology Platform for Communications Networks and Services, August 2014.

[33] H. Zhang, X. Chu, W. Guo and S Wang, "Coexistence of Wi-Fi and heterogeneous small cell networks sharing unlicensed spectrum," IEEE Commun. Mag., vol. 53(3), pp.158–164, 2015.

[34] G. Caso, Mai T. Phuong Le, Luca De Nardis, and M-G Di Benedetto, "Non-Cooperative and Cooperative Spectrum Sensing in 5G Cognitive Networks," in Handbook of Cognitive Radio, W. Zhang (ed.), Springer Nature, Singapore , 2017.

[35] G. Caso, L.De Nardis and M-G. Di Benedetto, "Toward context-aware dynamic spectrum management for 5G," IEEE Wireless Communications, 24, (5) pp 38-43, October 2017.

2nd IEEE International Conference on Power Electronics, Intelligent Control and Energy Systems (ICPEICES-2018)

Design and Development of Digital Signal Controller based Motorized Zoom Controller for 16X Zoom Thermal Imager

Himanshu Singh[1], Millie Pant[2], Sudhir Khare[1], Ranabir Mandal[1], Kanchan Chandra[1], Hirdesh Gangoli[1]

[1]Instruments Research and Development Establishment, Dehradun, India
[2]DASE, IIT Roorkee, India
himanshu@irde.drdo.in

Abstract—**A parfocal indigenous zoom lens system with motorized control has been designed and developed and an advanced 3rd Generation Thermal Imager has been realized using this 16x continuous Infrared Zoom. This paper covers in detail the design philosophy, control algorithms, application software and electronics hardware of the motorized zoom controller.**

Keywords—Zoom; Thermal Imager; Digital Signal controller (DSC); Integrated Development Environment (IDE)

I. INTRODUCTION

Any research and development activity should not only pitch for strong theoretical foundation but also culminate it with practical implementation. Trend in commercial and defence industry is towards Thermal Imagers with continuous zoom optics. Zoom lenses provide hassle-free continuous zooming of infrared images of interest. They are being used in still, motion picture cameras, projectors, binoculars, telescopes, telescopic rifle sights and other optical instruments. Under the ongoing research activities, a 16x Infrared zoom lens based Thermal Imaging sight[1] has been designed and developed which provide incessant variation in the magnification of image. Zoom lenses[2] maintain the focus even when its focal length varies. The designed zoom lens has focal length variation catering for 16x magnification mounted in front of an advanced 640 x 512 InSb material Infrared detector based Thermal Imager. Zoom lens design comprises of several key/ critical technologies such as optical design, opto-mechanical design, fabrication and control electronics. Digital Signal Controller (DSC) based motorized control electronics plays a major role in its proper focusing and effortless zooming. One-time calibration based approach for auto focusing which does not consume computational resources relentlessly is followed here. This paper discusses about the control algorithm used, robust software implementation using CodeWarrior Integrated Development Environment (IDE) for Freescale DSC MC56F8255, hardware electronics including flash, LCD, power supply, I2C based serial link, encoder, DC micromotor and Graphical User interface (GUI). Finally, the results and conclusion are given.

II. CONTROL ALGORITHM AND EQUATIONS

In discussing the mathematical formulation of zoom controller design[3], it is worth mentioning that several logic and algorithms have been implemented to ensure best performance. Fig. 1 shows the general block diagram of digital control system. The DSC has in-built A/D and D/A converter for proper handling of digital/ analog signals. The plant is an electromechanical system involving a motor-gear-load mechanism. A mathematical model was developed for this and is shown in fig. 2 and equation 1 as given under [4]:-

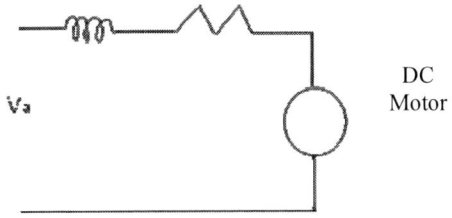

Fig. 2: Model of dc motor

$$V_a = \frac{LI}{K_t}\ddot{\omega} + \frac{RI}{K_t}\dot{\omega} + K_e\omega$$

(1)

where:

V_a = applied voltage
L = motor inductance

R = motor resistance
I = current
K_t = Torque constant
K_e = Back Emf constant
ω = angular speed of motor

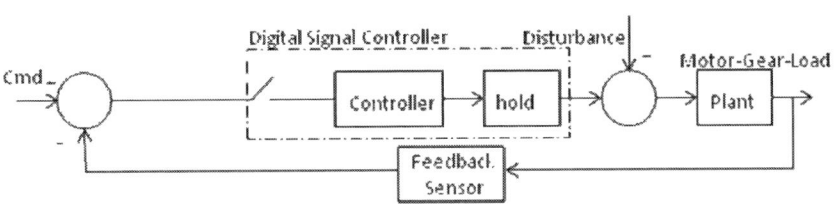

Fig. 1: General block diagram of digital control system

978-1-5386-6624-1/18 $31.00 © 2018 IEEE

Based on this mathematical model, simulation was done in MATLAB using simulink block set and the model is shown in fig. 3. Varying values of different parameters were fed in this simulink model[5]. A "Proportional Controller" is implemented in this controller[6]. It is simple and easy to tune and serves the purpose. The command is provided by a step input. Error is found as difference of command & feedback from encoder. This error is scaled by the single control law gain constant K_p to be given to the H-bridge amplifier. The proportional gain constant is tuned to 500. Sample time is 100 µs. The fig. 4 shows the response to the step input.

III. HARDWARE ELECTRONICS

A. Digital Signal Controller

The selection of right processing element for Zoom Controller application was an important issue. The main candidates of choice were Digital Signal Processor (DSP), Microcontroller (µC) and Field Programmable Gate Array (FPGA). Various attributes like performance, input/ output interfacing, throughput & latency, development ease and last but not the least power consumption were considered. The single µC, DSP or FPGA cannot meet the demands of all kinds of products or applications alone. So, a hybrid solution of DSP, µC or FPGA is the ideal solution and it is the primary trend in the market. In our control application, a hybrid DSP and µC i.e. a Digital Signal Controller (DSC) was selected and used because it fulfills the present requirement of low level control, real world I/O interfacing and data processing but at lower cost and low power consumption. The main processor used is a 32 bit, 100 MIPS Freescale MC56F8255 which is a Digital Signal Controller (DSC) and it combines, on a single chip, the processing power of a DSP and functionality of microcontroller with a flexible set of peripherals to create an extremely cost-effective solution with vibrant architecture. [7].

B. Keypad Interface module

A small menu driven keypad (3 buttons) is interfaced to the card externally to operate the zoom. A key Debouncing time of 20 ms is provided through software means.

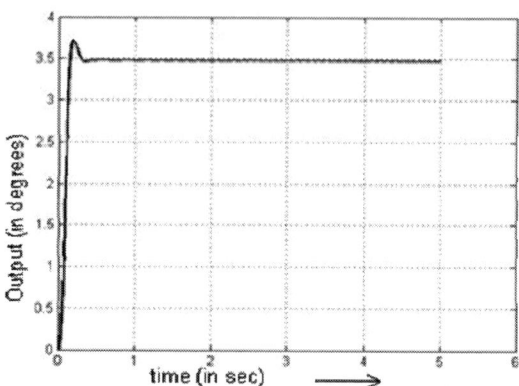

Fig.4: Step response output

In Manual mode, a scheme is devised to calibrate the zoom based on Intelligent homing and span calculation method for incremental quadrature encoder. While in Auto mode, the zoom moves to the precalibrated positions and its retentivity was tested.

C. Zoom & Focus motors and associated drive circuit

The motors used for focusing and zoom are DC micro motors with 2 channel in-bulit linear encoder and reduction gears[9]. The power supply design includes H-bridge configuration to drive sufficient power to motors with over current and overvoltage protection. Advantage of using H bridge in an integrated circuit form is low component count and safety from accidental short circuit. Based on the design inputs, torque and speed requirements of zoom and focus motors were determined and a motor with torque (without gear head) of 3.8 mNm and planetary gear head ratio of 66:1 was selected which will provide a torque at shaft output of 3.8 mNm x 66 = 250.8 mNm that fulfills the requirement of 250 mNm torque.

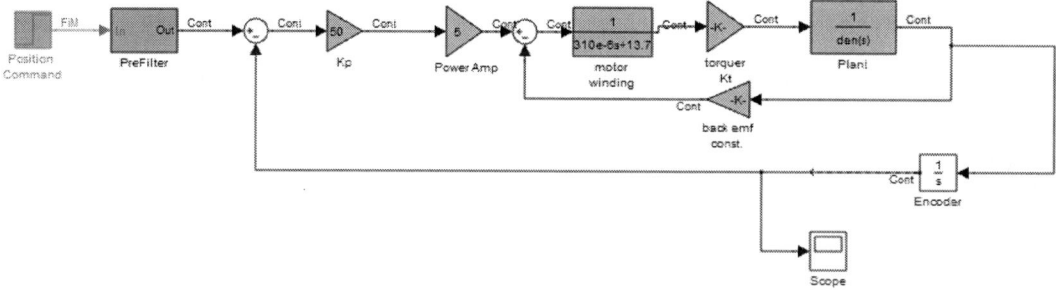

Fig. 3: Modeling using Simulink in MATLAB

978-1-5386-6624-1/18 $31.00 © 2018 IEEE

1) Angular acceleration and torque calculation for focusing assembly

First, the minimum distance or depth of focus is defined over which the image appears to be sharp and unchanged. At the image plane, depth of focus is the small axial distance over which the focus appears to be sharp and unchanged.

$$D = 4 \times \lambda \times (f/\#)^2 = 514.25 \ \mu m \qquad (2)$$
$$\text{where } \lambda = 4.25 \ \mu m,$$
$$f/\# = 5.5$$

Hence, 250 µm may be taken as the minimum step size for the focusing motor. Assume some of the data as:-
i) total linear travel for focusing motor is 10 mm in 5 s
ii) groove angle= 82°
iii) turn ratio between focusing CAM and motor= 4.85 :1
10 mm linear travel corresponds to (82°x4.85) rotary motion, 0.25 mm linear travel corresponds to 9.9425° or 0.1735 rad rotary motion.
Angular Acceleration,

$$\alpha = \frac{\theta}{T_r^2} \qquad (3)$$

where
θ = rotary motion in rad= 0.1735 rad
T_r = Rise time in s
ξ = damping factor =0.5 (given)
T_s = Settling time (assumed to be 0.2 s
ω_n = natural frequency of oscillation=40 rad/s $\qquad (4)$

ω_d = damped frequency of oscillation

$$= \omega_n \sqrt{1 - \xi^2}$$

$$\beta = \tan^{-1}\left(\frac{\omega_d}{\xi \omega_n}\right) \qquad (5)$$

$$T_s = \frac{(\pi - \beta)}{\omega_d} \qquad (6)$$

$$\alpha = \frac{\theta}{T_r^2} \qquad (7)$$

Moment of Inertia of focusing cylinder
$$J = 7424.51 \ gm\text{-}mm^2$$

Torque required at focusing motor
$$J*\alpha = 7424.51 \times 47.588 x 10^{-9} = 0.353 \ mNm \qquad (8)$$

2) Liquid crystal display module

The LCD is used to display the output visual information. Three control signals are used to control its operation, they are Enable (E), read/write(R/W), Register select (RS). The display is used to show all the control commands and mode selection for zoom and focussing motors.

3) Encoder Interface Module

An encoder is an electromechanical device used to convert the mechanical position of shaft or axle to an analogue or digital signal. In the present work, an incremental quadrature encoder has been used. Incremental encoders are position feedback devices that provide incremental counts. Thus, an incremental encoder provides relative position, where the feedback signal is always referenced to a start or home position. Incremental Encoder (IE) generates a stream of binary pulses proportional to the rotation of the shaft or distance travelled. 2 channel IE provides two streams of output pulses which are 90° out of phase with each other. These dual channel designs are called Quadrature encoder[6] due to four rise/ fall points of signal output. There are 2 speed limitations in rotary IE- one is mechanical speed and the other is electrical speed. Mechanical speed limit is that speed above which encoder can break. As per data sheet, the selected motor has max mech. speed upto 12000 rpm while the max electrical speed is calculated as here under:-
Max electrical speed (in rpm) = Max freq response x 60 $\quad (9)$
Pulses per revolution = 24000
So, the mechanical speed limits the encoder and motor motion and cannot be operated above 12000 rpm.
Number of counts of encoder required for 1 step movement of motor is given by:
= (Lines per revolution) x (Counts per line) x (motor rotation due to CAM) x (Depth of focus) $\quad (10)$

4) Athermalization Circuit

An athermalization circuit is implemented (as shown in fig.5) which will make the assembly insensitive to temperature changes. The lenses used in the assembly are temperature sensitive as their refractive index changes with temperature.

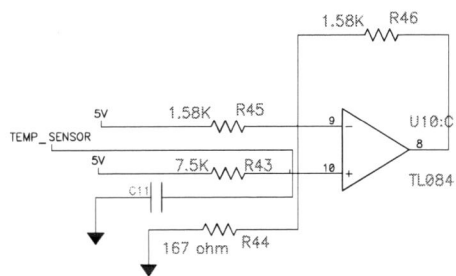

Fig.5: Athermalization Circuit

Change in temperature causes defocusing of image. Focusing can be achieved by moving the focusing lens according to athermalization calibration table. For this, a temperature sensor PT1000 is mounted on the lens assembly to measure temperature and accordingly reposition the moving lens assembly for athermalization. PT1000 is a temperature sensor with linear characteristics and high precision. Its resistance changes linearly with temperature (3.85Ω/°C). As the temperature varies, the resistance of PT1000 sensor changes, which is input to a wheatstone bridge network with three constant resistances and PT1000 as the fourth resistance of the bridge. The output of this circuitry is given to CPU. A relation between the temperature value and focusing parameter (look-up table) is implemented in the software. Depending upon this relation, movement of lens will be controlled. Fig. 6 shows the hardware schematic diagram of the motorized zoom controller card.

978-1-5386-6624-1/18 $31.00 © 2018 IEEE

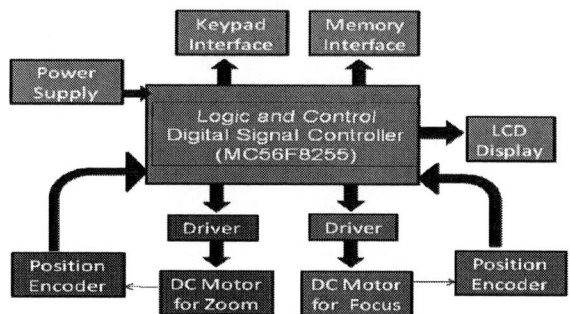

Fig. 6: Hardware schematic diagram of the motorized zoom

controller card

IV. APPLICATION SOFTWARE

The Zoom controller application software is written in C language using the Freescale CodeWarrior IDE[8]. The processor expert feature of this IDE helps in rapid coding and debugging. The basic design philosophy of the software is described in terms of a flowchart as shown in fig. 7. In the software, various interface codes were written to implement the intended zoom control application. Quadrature incremental encoder which outputs pulses in order to determine rotation is interfaced to determine rotation of zoom and focus motors, LCD is interfaced to display various supervisory commands, a 64k flash is interfaced to save the different positions of zoom and focusing motors, a user friendly Graphical User Interface (GUI) useful in calibration of zoom & focus position has also been programmed.

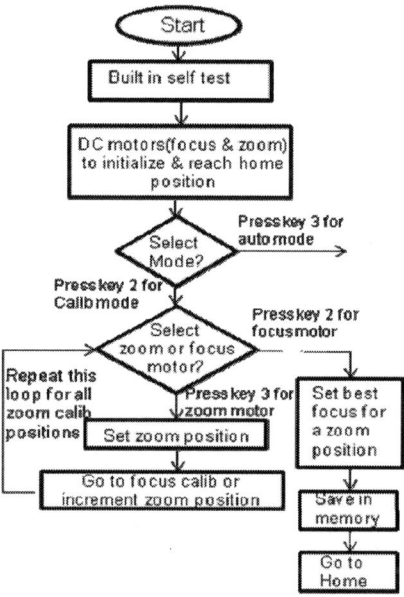

Fig. 7: Flowchart of the zoom controller software

V. RESULTS AND CONCLUSION

The indigenously built 16X Zoom Controller has been successfully realized. The following images in fig. 8-10 show the results of zoom controller at 3 positions with and without focus control.

Fig. 8: Wide Field of View image with and without Focus control

Fig. 9: Mid Field of View image with and without Focus control

Fig. 10. Narrow Field of View image with and without Focus control

The calibration of this zoom optics was initially done at 13 points and later at 100 points by using Integrated Computerized Test Setup (ICTS) and both zoom and focus positions were stored in EEPROM memory. Employing this zoom, an advanced 640 x 512 Infrared detector based Thermal Imager has been realized as shown in Fig. 11.

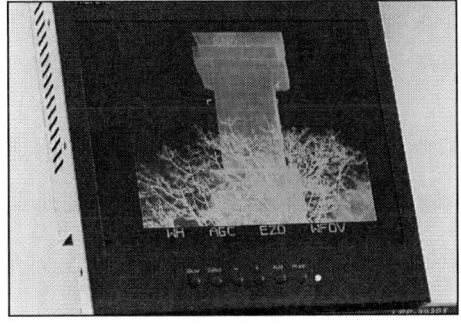

Fig. 11: Photograph showing Thermal Imager with 16x Zoom and image

VI. Acknowledgements

The authors would like to thank Shri Benjamin Lionel, Director I.R.D.E. for the continuous guidance and support for this work and for permission to publish the paper.

VII. References

[1] R.G.Driggers, M.H.Friedman, J.M. Nichols, 2012. Introduction to Infrared and Electro-optical Systems, Second Ed. Artech House, London

[2] A.Mann, 2009. Infrared Optics and Zoom Lenses, Second Revised Ed. SPIE Press

[3] George Ellis, 2004, Control System Design Guide, 3rd Ed. Elsevier Academic Press, USA

[4] D. Ibrahim, 2006. Microcontroller based Applied Digital Control, John Wiley & Sons Ltd., UK

[5] MATLAB & Simulink Reference manual

[6] Tim Wescott, "*PID without a PhD,*" Flir Systems (USA), EETimes India, 741-755 (Oct. 2000)

[7] Freescale Technical Reference Manual, Product Data Sheets and Reference Manuals

[8] Freescale Codewarrior Develeopment Studio IDE 5.6 User's Guide

[9] Faulhaber product catalogue, application notes and user manuals

22nd IEEE International Conference on Power Electronics, Intelligent Control and Energy Systems (ICPEICES-2018)

Millimeter Wave Reconfigurable Vivaldi Antenna using Power Divider for 5G Applications

Akhilesh Verma
Electronics & Communication
Delhi Technological University
Delhi, India
akhilesh.verma6388@gmail.com

N. S. Raghava
Electronics & Communication
Delhi Technological University
Delhi, India
nsraghava@dce.ac.in

Abstract—**In this research paper, a reconfigurable Vivaldi antenna array is designed using a power divider for 5G applications. The proposed antenna consists of an array of the Vivaldi antenna which is fed through a power divider. The power divider splitting the power equally into two branches of the Vivaldi antenna. In this work, two antennas are designed, one with pin diode to shift the resonant frequency and another one without using a pin diode. By implanting pin diode, desired frequency bands are achieved. The proposed antenna1 without pin diode resonate at 25.4 GHz which covers from 24.91 -25.89 GHz bands with the bandwidth of .98 GHz. The gain and the radiation efficiency of the antenna are 5.2203 dB, 71.42 %. The VSWR of the antenna is 1.1421 which is suitable for the mobile phone application. The proposed antenna 2 with pin diode obtained multiband which are given as 19 GHz (17.06-20.81 GHz), 38.7 GHz (33.04-43.06 GHz). The gain of the antenna at these frequency bands are 4.725 dB, 7.369 dB, 7.9878 dB. All parameters of the antennas have been simulated in HFSS (High-frequency structure simulator).**

Keywords—*Vivaldi, Power Divider, Pin Diode, Millimeter Waves,*

I. INTRODUCTION

In wireless communication technology, everyone is working on the capacity to reduce the traffic of the channel. Today's world the need for high bandwidth, low latency, high speed is increasing very rapidly. There were many limitations in previous wireless technologies in respect of high traffic, narrow bandwidth, more latency, less speed [1], [2]. Mostly telecom companies are working on 5G wireless technology which will be as a boon for wireless communication to overcome all limitations of previous technologies. The attractive features of 5G communication are having 1-10 Gbps data rate, 1 ms round-trip latency energy reduction by 90 % and 100 % coverage at any time and any place, that means the network will be connecting every time [3], [4]. There are some spectrums which are looking for 5G communication. The expected spectrum for 5G is 6 GHz, above 6 GHz, 24 GHz, 27-29 GHz, 37-40 GHz and 60 GHz bands which are under investigation as shown in Fig. 1 [5]-[6]. And while the more traditional sub-6 GHz spectrum of 5G will see some improvements as well, it's a much smaller gain tempered by limited spectrum and decreasing spectral efficiency gains. At mm-wave, the whole spectrum is having the very wide bandwidth. At these waves network traffic is very less and the unused spectrum may be used to implement new coming wireless technology [7]. The millimeter wave and massive mimo have drawn the attention to implementing 5G communication. Millimeter wave is a critical component of the 5G communication, as

it's going to be these high frequencies that are going to give 5G enough spectrum to hit its multi-gigabit target speeds. Massive mimo is a very attractive technology in which a large number of arrays are placed in the small area at mm-wave fulfill the requirements of 5G technology. Massive mimo increase the spectral efficiency which is one of the requirements of 5G communication [8]-[11]. The reconfigurable antenna is also a very interesting antenna to achieve 5G challenges. In this type of antenna, the frequency of operation can be shifted to an interested band of operation in which pin diode, varactor diode are used to get desired outputs.

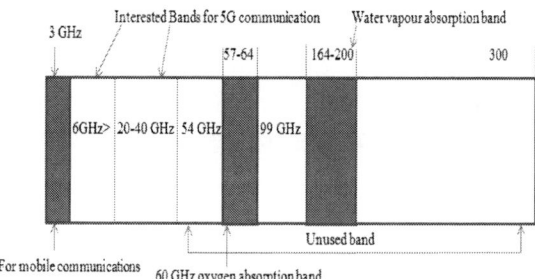

Fig. 1. Millimeter wave 5G communication spectrum [6].

The Vivaldi antenna is also called tapered slot antenna can be printed on a printed circuit board for any frequency. The shape of the antenna is in the form of two exponential flares just mirror image of one another [12]. In [13], the better impedance matching has been achieved with the help of antipodal Vivaldi antenna in the low-frequency band. Furthermore, to improve the gain performance in the higher band the proposed design is surrounded by a horn antenna. In [14], to achieve better coupling a dual-stub is used to coplanar Vivaldi antenna. The gain of the proposed antenna is achieved with the help of slits in the outer edges and the radiation characteristics of the antenna, beamwidth is controlled by the optimization of the shape of the parasitic element. For the matching purposes, the two-stage coupled filter has been used to two open shunt stubs. In [15], four cross-shaped Vivaldi antennas are designed which reduced the grating lobes at a high frequency of 5.4 GHz. In [16], for spectrum monitoring and cognitive radio applications, a frequency agile Vivaldi antenna has been designed in which frequency of operation can be switched between two selected bands with the help of RF switches. In [17], switched based reconfigurable antennas, non-switched based reconfigurable antennas, neural networks based reconfigurable antennas have been discussed.

978-1-5386-6624-1/18 $31.00 © 2018 IEEE

In this research work, an array of two Vivaldi antennas fed with power divider have been designed to achieve 5G requirements. One of the design is simply the array of Vivaldi antenna using a power divider which achieved moderate gain and bandwidth. To improve the performance of the antenna, a pin diode is used in the second design. Because of the pin diode introduced between the arms of the power divider, multi-band has been achieved with high gain and wide bandwidth.

II. DESIGN OF TWO ELEMENT ARRAY VIVALDI ANTENNA USING POWER DIVIDER WITHOUT PINDIODE

Fig. 2. Shows the 3D view of the proposed design. The dielectric substrate used to design this antenna is RT Duroid 5880 has dielectric constant $\varepsilon_r = 2.2$ and loss tangent value is $\delta = .0009$. The length and width of the substrate are L = 15 mm, W = 10 mm. The thickness of the substrate is given as h = 1.524 mm.

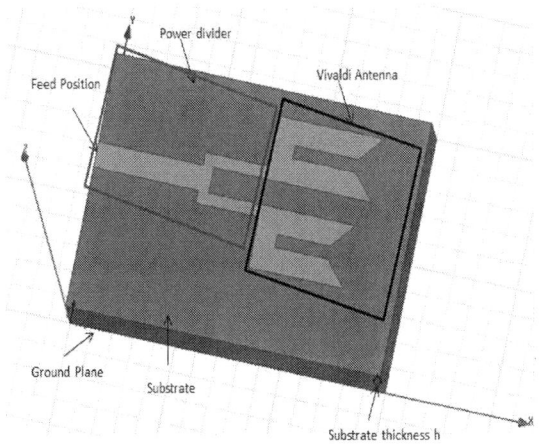

Fig. 2. 3D view of the proposed antenna.

The arrays of the two Vivaldi antenna are fed by a power divider which distributes the power equally to each element. The detailed dimensions of the proposed antenna are shown in Fig. 3.

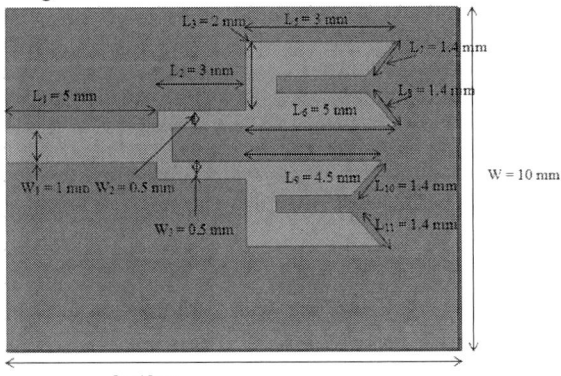

Fig.3. Detailed dimensions of the proposed antenna.

The dimensions of the power divider to the arrays of Vivaldi antenna as shown in Fig. 3. are $L_1 = 5$ mm, L2 = 3 mm, W_1 = 1mm , $W_2 = 0.5$ mm, W3 = 0.5 mm respectively. Fig. 4. shows the 3dB T-branch power divider in which the incident

and reflected waves are marked as i, I_1, I_2 and o, O_1 and O_2. The relationship between the incident and reflected waves can be established with the help of S-matrix to calculate overall return loss of the network.

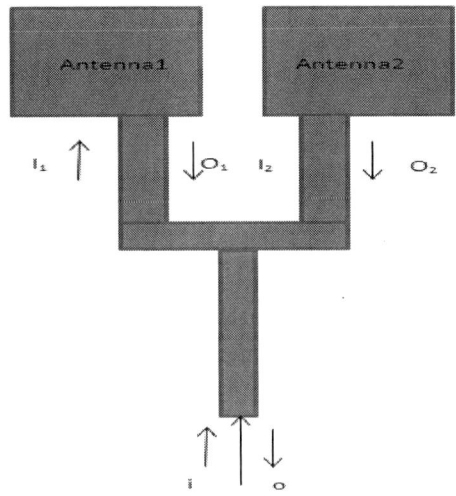

Fig. 4. T-branch power divider fed to two Vivaldi antennas.

We know that for lossless power divider [17].

$$
\begin{bmatrix} O \\ I_1 \\ I_2 \end{bmatrix} = \begin{bmatrix} 0 & \frac{1}{\sqrt{2}} & \frac{1}{\sqrt{2}} \\ \frac{1}{\sqrt{2}} & \frac{-1}{2} & \frac{1}{2} \\ \frac{1}{\sqrt{2}} & \frac{1}{2} & \frac{-1}{2} \end{bmatrix} \begin{bmatrix} I \\ O_1 \\ O_2 \end{bmatrix} \tag{1}
$$

If the elements of the array are identical, then we have

$$
\begin{bmatrix} O_1 \\ O_2 \end{bmatrix} = \begin{bmatrix} S_{11} & S_{12} \\ S_{12} & S_{11} \end{bmatrix} \begin{bmatrix} I_1 \\ I_2 \end{bmatrix} \tag{2}
$$

Because of array symmetry and power divider, we can calculate the relationship between s- parameters, incident and reflected power respectively.

$$
I_1 = I_2 = \frac{1}{\sqrt{2}} I, \tag{3}
$$

$$
O_1 = O_2 = \frac{1}{\sqrt{2}} I(S_{11} + S_{12}) \tag{4}
$$

Overall the reflection coefficient of the network is

$$
\Gamma = S_{11} + S_{12} \tag{5}
$$

The proposed work consists of arrays of the Vivaldi antenna which is also called a tapered slot antenna. This kind of antenna is used for ultra-wideband applications. Vivaldi antenna considered as 2-D exponential horn antenna. The size of the Vivaldi antenna is very small at mm-wave scale. The length and width of the antenna are 5 mm and 2.5 mm and the thickness of the copper sheet on a dielectric substrate is .025 mm. Mostly Vivaldi antenna is linearly polarized, but due to the arrays which are fed by power divider, it has been converted into elliptically polarized. The dimensions of the Vivaldi antenna1 are $L_2 = 3$ mm, $L_3 = 2$

978-1-5386-6624-1/18 $31.00 © 2018 IEEE

mm, $L_5 = 3$ mm, $L_6 = 5$ mm, $L_7 = L_8 = 1.4$ mm respectively. The dimensions of Vivaldi antenna2 are $L_9 = 4.5$ mm, $L_{10} = L_{11} = 1.4$ mm also shown in table 1. All dimensions of the whole network is less than equal to $\dfrac{\lambda}{4}$.

TABLE I. DETAILED DIMENSIONS OF THE PROPOSED ANTENNA

PARAMETERS	Value (mm)
L_1	5
W_1	1
L_2	3
W_2	0.5
W_3	0.5
L_3	2
L_5	3
L_6	5
L_7	1.4
L_8	1.4
L_9	4.5
L_{10}	1.4
L_{11}	1.4
L	15
W	10
h	1.6

III. DESIGN OF TWO ELEMENT ARRAY VIVALDI ANTENNA USING POWER DIVIDER WITH PIN DIODE

As shown in Fig. 5. pin diode introduced in between the two arms of the power divider through which antenna can be made reconfigurable. This is one of the methods to achieve reconfiguration of the frequency accordingly for what application antenna is designing. The pin diode or varactor diode is used as ON/OFF switches through which wideband and desired frequency band can be achieved. In this work, a pin diode is used to get multiband with wide bandwidth and high gain respectively. The proposed antenna using a pin diode generates 3 bands, high gain, and wide bandwidth to serve 5G applications. The achieved bands can be shifted by

varying the value of pin diode capacitance. The polarization reconfigurability is also tuned with the help of a pin diode.

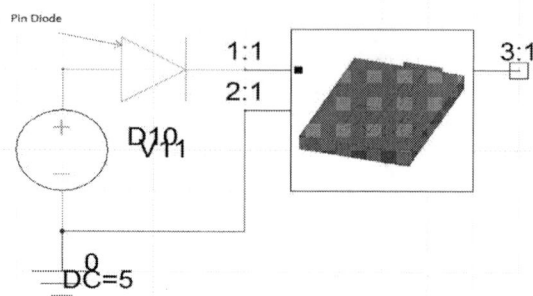

Fig. 5. Proposed antenna using a PIN diode.

IV. RESULTS AND DISCUSSION

The parameters of the proposed antenna are simulated with help of high-frequency structure simulator (HFSS). The comparison of the reflection coefficient of both antennas using pin diode and without using pin diode as shown in Fig. 6.

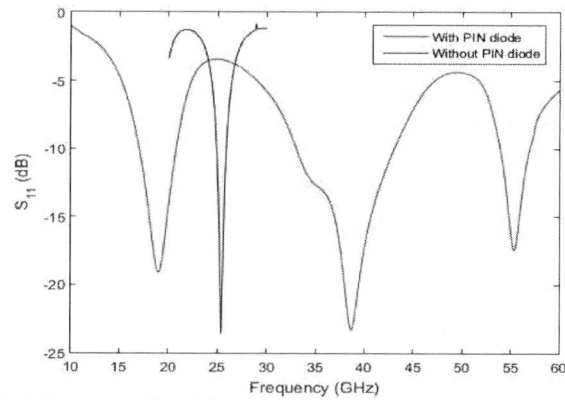

Fig. 6. S-parameters (S_{11}) of the proposed antenna.

The proposed antenna without using pin diode is resonate at 25.4 GHz (24.91 -25.89 GHz) with return loss -23.5674 dB < - 10 dB. The proposed antenna using pin diode generates three bands 19 GHz (17.06-20.81 GHz), 38.7 GHz (33.04-43.06 GHz) with return loss -19.0933 dB, -23.0164 dB, -17.4261 dB respectively. The radiation pattern of the proposed antenna is shown in Fig. 7.

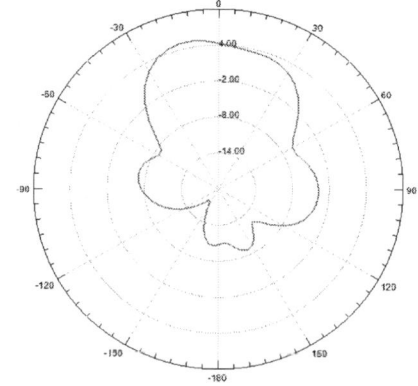

Fig. 7. Radiation pattern of the proposed antenna without a PIN diode.

978-1-5386-6624-1/18 $31.00 © 2018 IEEE

The VSWR of the antenna without using PIN diode at 25.4 GHz is 1.14 and using pin diode VSWR at all three bands are 1.2497, 1.1521, 1.3108 which is suitable for the mobile phone applications.

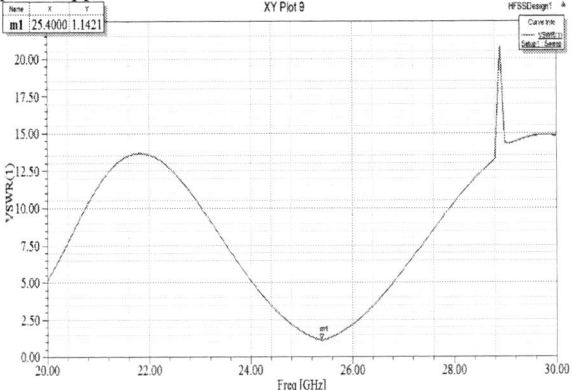

Fig. 8. VSWR of the proposed antenna without PIN diode.

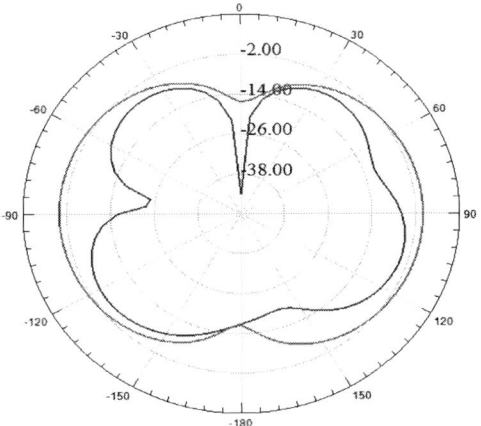

Fig. 9. Co and Cross Polarisation of the proposed antenna without PIN diode.

Fig. 10. (a), (b), (c) Co and Cross Polarisation of the proposed antenna using a PIN diode at 19 GHz, 38.5 GHz, 55.4 GHz

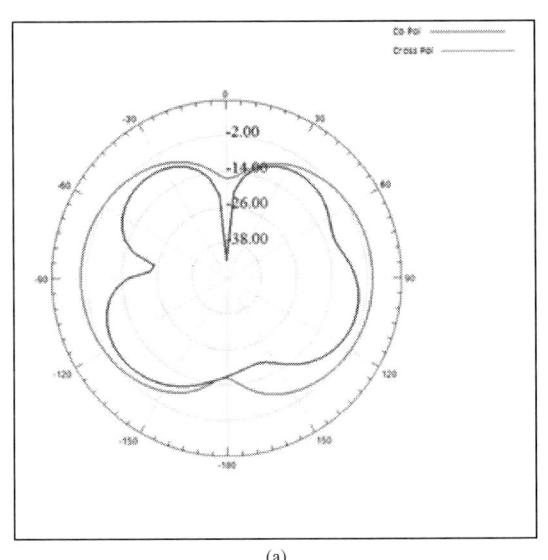

(a)

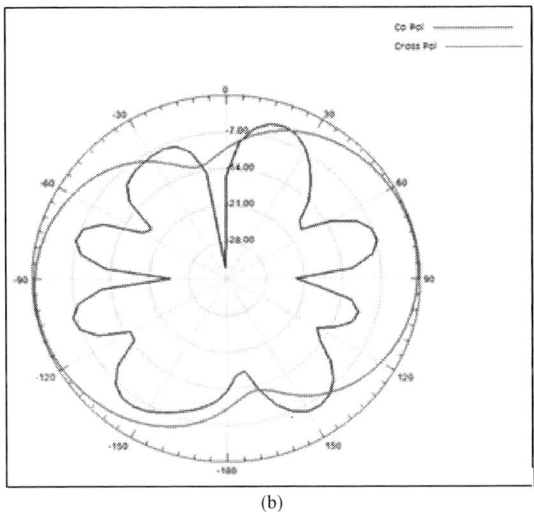

(b)

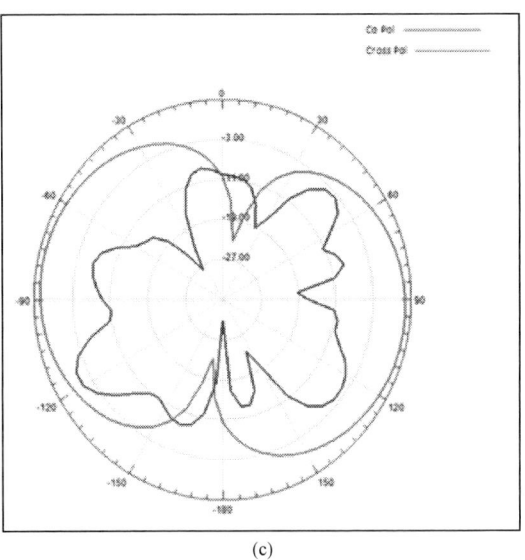

(c)

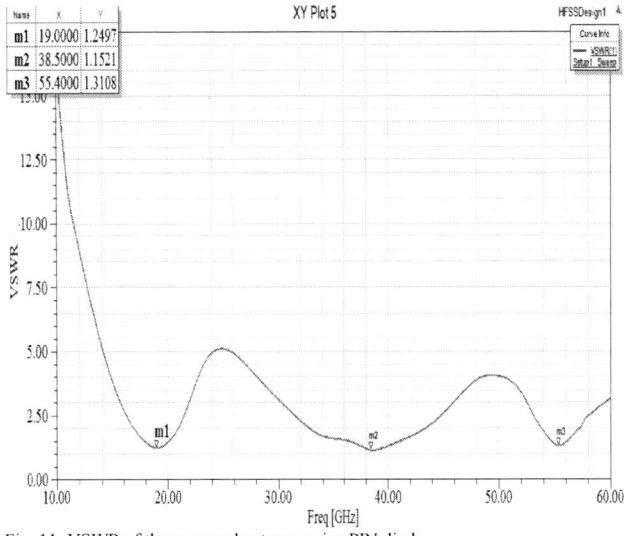

Fig. 11. VSWR of the proposed antenna using PIN diode.

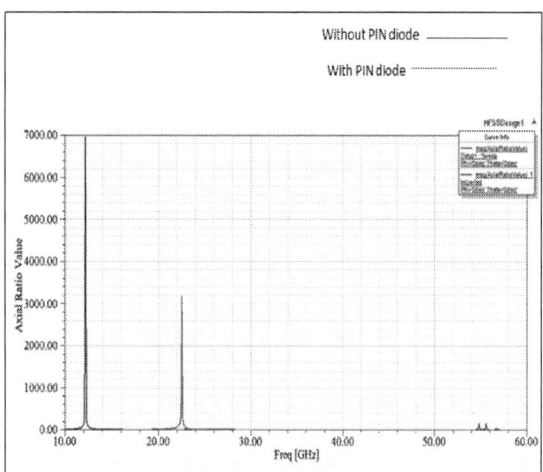

Fig. 12. Axial ratio of the proposed antenna with pin diode and without pin diode.

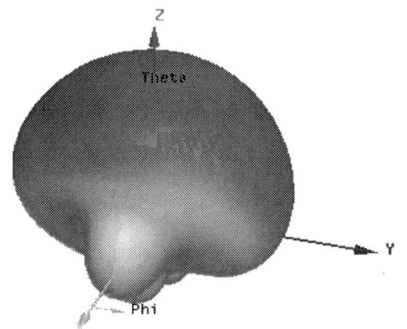

Fig. 13. 3 D radiation pattern of the proposed antenna without using pin diode at 25.4 GHz.

The radiation pattern of the antenna is omnidirectional as shown in Fig. 13. The omnidirectional antennas are very important to radiate and receive energy equally from all directions. The maximum gain achieved of this antenna is 5.2202 dB.

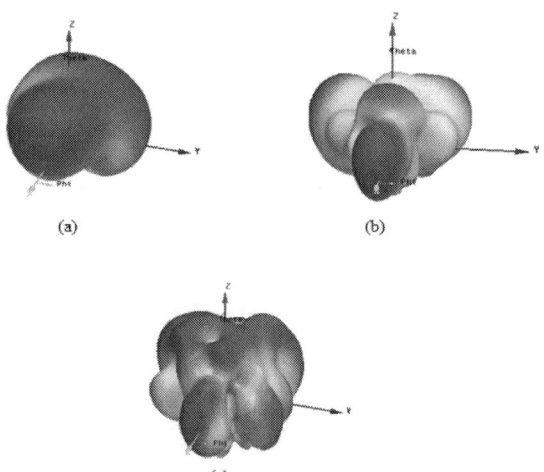

Fig. 14. 3 D radiation pattern of the proposed antenna using pin diode at (a) 19 GHz, (b) 38.5 GHz, (c) 55.4 GHz.

The gain of the proposed antenna using pin diode are 4.7253 dB, 7.4151 dB, 6.8419 dB.

V. CONCLUSION

The Vivaldi antenna fed by T-branch power divider with and without pin diode for 5G applications is presented. The proposed antenna without pin diode is resonating at 25. 4 GHz (24.91 -25.89 GHz) achieving very less bandwidth of 0.98 GHz and gain is 5.2202 dB. To further improve the performance of the antenna, pin diode has been used in between the two arms of the power divider. The frequency of the antenna is basically reconfigured through pin diode. It is also observed that after using pin diode, three frequency bands are generated. The proposed antenna with pin diode resonating at 19 GHz (17.06-20.81 GHz), 38.7 GHz (33.04-43.06 GHz) with return loss -19.0933 dB, -23.0164 dB, -17.4261 dB respectively. The gain of the antenna using a pin diode is 4.7253 dB, 7.4151 dB, 6.8419 dB. It seems that the pin diode generates multiband with wide bandwidth and high gain which make it suitable for 5G applications.

REFERENCES

[1] T. S. Rappaport, W. Roh, and K. Cheun, "Wireless engineers long considered high frequencies worthless for cellular systems. They couldn't be more wrong," IEEE Spectr., vol. 9, pp. 34-58, Sep. 2014.

[2] T. S. Rappaport, Yunchou Xing, George R. MacCartney, Jr., Andreas F. Molisch, Evangelos Mellios, Jianhua Zhang, "Overview of Millimeter Wave Communication for Fifth-Generation (5G) Wireless Networks-with a focus on Propagation Models," IEEE Transactions on Antennas and Propagation, vol. 65, no. 12, Feb 2017.

[3] Mamta Agival, Abhishek Roy, and Navrati Saxena, "Next Generation 5G Wireless Networks: A Comprehensive Survey," IEEE COMMUNICATIONS SURVEYS and TUTORIALS, vol.18, no. 3, pp. 1617-1655, Feb 2016.

[4] GSMA Intelligence, "Understanding 5G: Perspectives on future technological advancemnets in mobile," White paper, 2014.

[5] Pekka Pirinen, "A brief overview of 5G research activities," 1st International Conference on 5G for Ubiquitous Connectivity, pp. 17-22, Nov 2014.

[6] Zhouyue Pi, Farooq khan, "An Introduction to Millimeter-Wave Mobile Broadband Systems," IEEE Communication Magzine, vol. 49, no. 6, pp. 101-107, June 2011.

[7] Zhouyue Pi, Farooq Khan, "An introduction to millimeter-wave mobile broadband systems," IEEE Communication Magazine, VOL. 49, Issue. 6, pp. 101-107, June 2011.

[8] Theodore S. Rappaport, Felix Gutierrez, Eshar Ben-Dor, James N. Murdock, Yijun Qiao, Jonathan I.Tamir, "Broadband Millimeter-Wave Propagation Measurements and Models Using Adaptive –Beam Antennas for Outdoor Urban Cellular Communications," IEEE Transactions on Antennas and Propagation, vol. 61, no. 4, April 2013.

[9] Jie Wu, Yu Jian Cheng, Yong Fan, "Millimeter-Wave Wideband High –Efficiency Circularly Polarized Planar Array Antenna," IEEE Transactions on Antennas and Propagation, vol. 64, no. 2, Feb 2016.

[10] P. Adhikari, "Understanding millimeter wave wireless communication," Loea Corp., White paper, 2008.

[11] A. Nordrum. (2016, May). "5G researchers set new world record for spectrum efficiency," IEEE Spectr., [Online]. Available: http://spectrum.ieee.org/techtalk/telecom/wireless/5gresear chers- achieve-new-spectrum-efficiency-record

[12] Qian Chen, Hongtao Zhang, Xiolin Zhang, Moupin Jin, Wei Wang, "Wideband RCS Reduction of Vivaldi Antenna Using Electromagnetic Band Gap Absorbing Structure," International Symposium on Antennas and Propagation (ISAP), Nov 2017.

[13] Tae Heung Lim, Jong-Eon Park, and Hosung Choo, "Design of a Vivaldi-Fed Hybrid Horn Antenna for Low-Frequency Gain Enhancement," IEEE Transactions on Antennas and Propagation, vol. 66, no. 1, Jan 2018.

978-1-5386-6624-1/18 $31.00 © 2018 IEEE

[14] Kansheng Yang, Manh-Ha Hoang, Xiulong Bao, Patrick, McEvoy, Max J. Ammann, "Dual-stub Ka-band Vivaldi antenna with integrated bandpass filter," IET Microwaves, Antennas & Propagation, vol. 12, no. 5, pp. 668-671, April 2018.

[15] Reid, E.W., Ortiz-Balbuena, L., Ghadiri, A, "A 324-element Vivaldi antenna array for radio astronomy instrumentation," IEEE tans. Instrum. Meas., vol. 61. No. 1, pp. 241-250, Jan 2012.

[16] Cristina Borda-Fortuny, Kin-Fai Tong, Kevin Chetty, "Low-cost mechanism to reconfigure the operating frequency band of a Vivaldi antenna for cognitive radio and spectrum monitoring applications," IET Microwaves, Antennas & Propagation, vol. 12, no. 5, pp. 779-782, 2018.

[17] Ming Wang, Wen Wu, and Zhongxiang Shen, "Bandwidth of Antenna Arrays Utilizing Mutual Coupling between Antenna Elements," International Journal of Antennas and Propagation, vol. 2010, pp. 1-10, March 2010.

2nd IEEE International Conference on Power Electronics, Intelligent Control and Energy Systems (ICPEICES-2018)

Generating Electricity on Roadside Using INVELOX

Abdullah Abu Sayed
Dept. of EEE
*American International
University-Bangladesh*
Dhaka, Bangladesh
sayeddip@gmail.com

Md. Zyed Ibn Sadiq
Dept. of EEE
*American International
University-Bangladesh*
Dhaka, Bangladesh
zyed_sadiq@outlook.com

Quazi Nasrul Rudaba
Dept. of EEE
*American International
University-Bangladesh*
Dhaka, Bangladesh
rudaba_sneha@outlook.com

Shihab Khondokar
Dept. of EEE
*American International
University-Bangladesh*
Dhaka, Bangladesh
222shihab333@gmail.com

Abu Hena Md. Shatil
*Dept. of EEE American
International University-
Bangladesh*
Dhaka, Bangladesh
abu.shatil@aiub.edu

Abstract— the future generation society all over the world is promised to get a sustainable green environment to live in. In this modern world, researchers are finding some alternative power source instead of conventional way of power generation as demand of electricity is increasing. Only renewable technology can secure the energy demand. Using the renewable technology new path has been created by "Generating Electricity on Roadside Using INVELOX". It is mainly funnel tube wind turbine. The main purpose of this project is to make proper use of the roadside for producing green energy. It captures the air when vehicles are in moving condition and air pass through a narrow space path in the INVELOX. In this path generator is placed. It converts mechanical energy into electrical energy. Actually, these types of wind turbine do not consume much more space. On the roadside it is quite difficult to generate electricity because of scattered air. But INVELOX can work properly and generate electricity enormously. We have implemented it in different roads, collected data and tried to improve more for the project system configuration.

Keywords— *Renewable energy, Wind turbine, PMDC generator, Venturi effect, Betz limit.*

I. INTRODUCTION

Limitation of fossil fuel and challenging of carbon emitting, renewable energy is one of the most popular energy now a day. Among all renewable energy wind turbine is the 2nd most popular. INVELOX is one kind of wind turbine which captures the wind from all directions, uses funnels to channel the wind to the ground based turbine and finally convert it into electrical energy. Capturing the air flow which is created by moving vehicles and this air will rotate the turbine which will convert into electrical energy from mechanical energy. The street lamps can be lightened by the produced electrical energy. Major advantage of this project is, it can operate at low wind speed and maintenance cost is low. This project is based on Venturi effect. In this project permanent magnet DC generator has been used. This can be used in vehicles by changing the design for charging the battery of vehicles. It will save the fuel. Solar energy can also be integrated to make it hybrid system. In these types of design multiple turbines can be installed in series.

II. BASIC BLOCK DIAGRAM

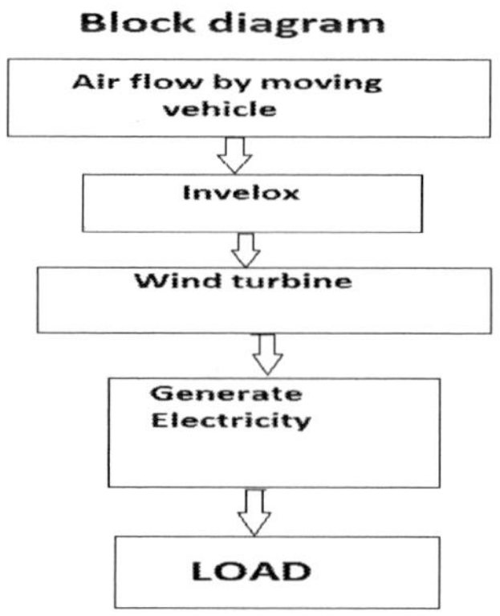

Block diagram

Air flow by moving vehicle
↓
Invelox
↓
Wind turbine
↓
Generate Electricity
↓
LOAD

Fig. 1.   Block diagram

When vehicles will move, air flow will be created which is captured by INVELOX. In the INVELOX there is a turbine which will convert mechanical energy into electrical energy. This electricity will supply to the street lamp on the road side.

III. THEORITICAL STUDIES

This project is the combination of physics law and electrical mechanism. Our full project consists of following these things. A turbine blade is connected with a permanent magnet DC generator. This generator is high rpm. That's why no need to use any gear. This word INVELOX means increase velocity. INVELOX is one kind of funnel tube wind turbine. This word "INVELOX" means increase velocity. It captures wind then it accelerates and converts mechanical power into electrical power. INVELOX is mainly based on the principle of Venturi effect. It is comparatively cost effective and highest performing of wind power generation. Here in INVELOX, turbine is near to the ground level. That's why maintenance cost is low. An Italian scientist name Giovanni B Venturi (1746-1822) proved that when

978-1-5386-6624-1/18 $31.00 © 2018 IEEE   1132

pressure is low like constricted area of pipe velocity of the fluid will increase. This phenomenon can occur because of pressure difference. Venturi effect can be proved by Bernoulli's famous equation. [1]

The Bernoulli's equation,

$$P_1 - P_2 = \frac{\rho}{2}(V_2^2 - V_1^2)$$

(1)

Here

P_1 = Low pressure in large area of pipe

P_2 = High pressure in constricted area of pipe

V_1 = Low velocity in the large area of pipe.

V_2 = High velocity in constricted area of pipe

ρ = density of the fluid

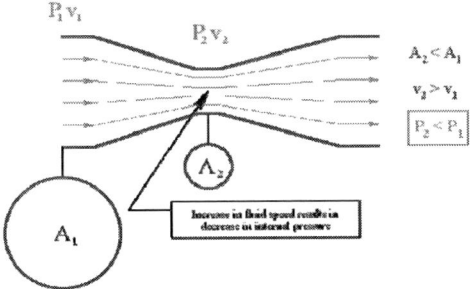

Fig. 2.   Venturi principle [2]

In wind turbine, mechanical energy of the air flow is converted into electrical energy. But there is a limitation. German physicist Albert Betz proved, the maximum theoretical efficiency that wind turbine can achieve is 59.26% conversion of kinetic energy to mechanical energy. [3].

IV. WORKING PRINCIPLE

This INVELOX is placed on the divider of the roads. When vehicles are moving air flow will create. This flow of air will be captured by this INVELOX in both directions. Then air passes through a narrow path. Where air pressure will decrease and velocity will increase. In this narrow path a turbine is placed. Increased air flow will rotate the turbine blade and from turbine electricity will be generated.

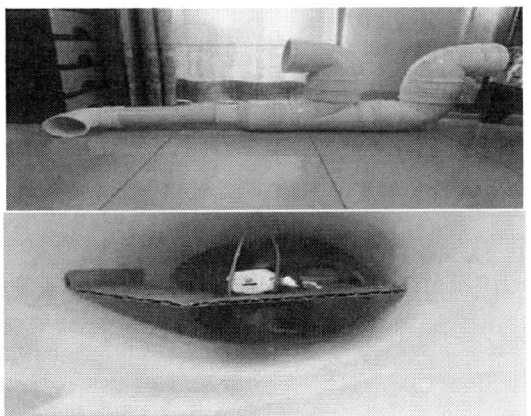

Fig. 3.   Practical structure of INVELOX

V. SIMULATED RESULT AND PRACTICAL DATA

We have made our design and simulated it by using the SOLIDWORKS software. In our design there have two inlets path and one outlet path. In these two inlets paths INVELOX can capture the flow of air in both directions and this air will pass through one narrow outlet path, where turbine has been placed.

A. Simulation

In this simulation we have selected the boundary condition. We have selected inlet velocity as 4.5 m/s and atmospheric pressure is 1 atm or 101325pa. Using this boundary condition and based on Bernoulli's equation (1) this simulation has been completed. Here we have simulated the velocity of the air as well as pressure of the air.

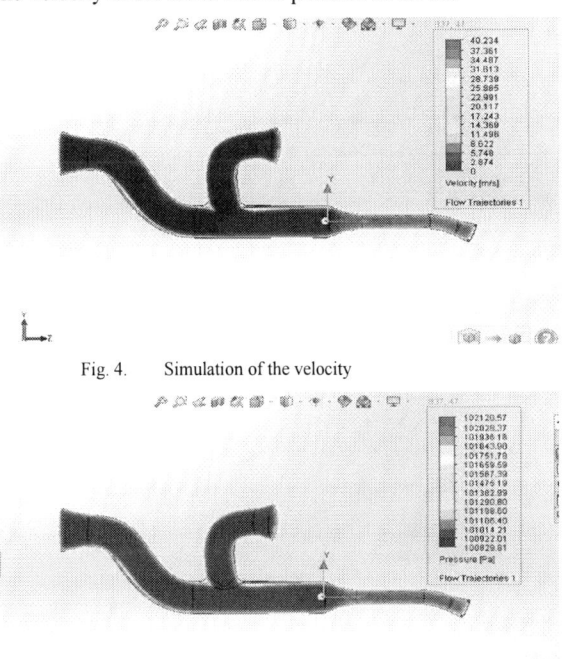

Fig. 4.   Simulation of the velocity

Fig. 5.   Simulation of the pressure

B. Practical data

We are able to collect data in three different road of Dhaka city. First we have collected data from Purbachal Highway. Then Airport road and last we have collected from Dhaka Aricha Highway in Saver. Velocity of the air was measured by Anemometer.

TABLE. I.   DATA OF PURBACHAL HIGHWAY ROAD

Reading number.	velocity(Km/h)	Voltage(mV)	Current(mA)
1	18	67	0.39
2	17	62	0.38
3	15	60	0.35
4	13	58	0.28
5	12	56	0.24

TABLE. II.   DATA OF AIRPORT ROAD

Reading number.	velocity(Km/h)	Voltage(mV)	Current(mA)
1	28	97	0.6
2	25	93	0.58
3	23	85	0.47
4	22	82	0.45
5	19	75	0.41

TABLE. III.   DATA OF DHAKA ARICHA HIGHWAY

Reading no.	velocity(Km/h)	Voltage(mV)	Current(mA)
1	28	96	0.62
2	24	84	0.46
3	22	82	0.45
4	20	75	0.42
5	16	69	0.39

Fig. 6.   Practical data collection in Purbachal Highway

VI. COMPARISON

This project is compared with the traditional wind turbine. In this project, why INVELOX is better than traditional wind turbine have been proved. For this comparison, data was collected using fan in the room. First generator has been placed in the INVELOX and measured data. Then same generator has been placed outside the INVELOX and measured data. There have enormous different between with INVELOX and without INVELOX.

In INVELOX flow of air is almost increased more than 1.4 times average.

TABLE. IV.   VOLTAGE AND CURRENT WITH INVELOX

Reading	Velocity(m/s)	Voltage(mV)	Current(mA)
1	4.5	79	0.2
2	6.4	120	0.4
3	9	190	0.9

TABLE. V.   VOLTAGE AND CURRENT WITHOUT INVELOX

Reading	Velocity(m/s)	Voltage(mV)	Current(mA)
1	3.2	31	0.1
2	4.2	84	0.3
3	7.3	140	0.4

By plotting these data in the graph, it will be more clarify that why INVELOX is better than the traditional wind turbine.

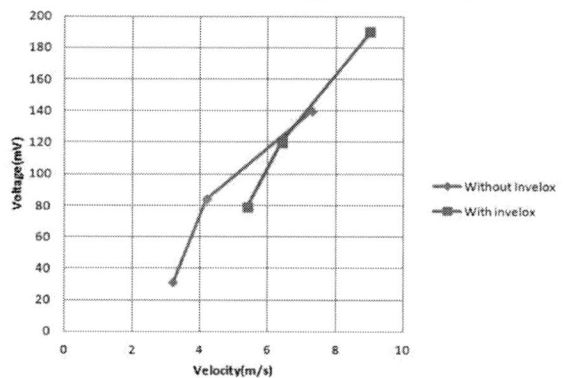

Fig. 7.   Voltage comparison with respect to velocity

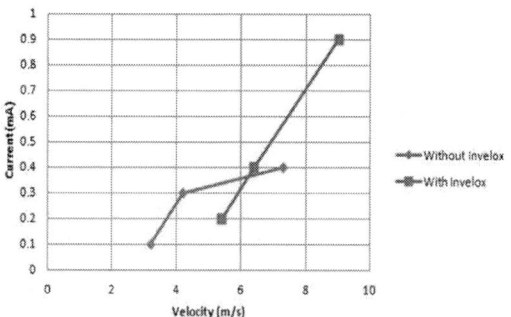

Fig. 8.   Current comparison with respect to velocity.

Figure 6 and Figure 7 will give a crystalline conception that why INVELOX is better. Both voltage and current is much better than normal wind turbine. Not only increased the voltage and current but also increased the generator efficiency. Power extraction coefficient or Cp of the is

19.82%, when generator has been placed in the INVELOX and Power extraction coefficient or Cp of the is 12.97%, when generator has been placed outside of the INVELOX.

VII. FUTURE SCOPE

This project can be used in different purpose by changing its shape. Multiple turbines can be used in a series for getting better output. There is a scope for making it hybrid by installing Solar cells. It can also used in vehicles, where continuous air flow is available. In summer season electricity can be saved by INVELOX. Because an artificial air flow will be produced by the fan may flow through INVELOX and rotate turbines. It can also be placed in between high stores buildings where continuous air flow is available. By using more suitable generator this INVELOX can be upgraded.

VIII. CONCLUSION

This project is created a new path for producing electricity in renewable energy. In previous divider of our road is only used for solar energy for producing electricity. But this project is shown that by using air flow, electricity can be produced. The main advantage of this project is cost efficient and high efficiency. Another advantage is that in low air flow it can operate smoothly. But for traditional wind turbines a minimum wind flow is needed. Only these types of renewable energy project can meet the demand of energy in this modern world. As we know fossil fuel is sufficient. As a developing country renewable energy can play a great role for meeting up with energy demand.

REFERENCES

[1]"Venturi principle and how do Venturi work". Available at: http://www.hendersons.co.uk/wms/venturi_principle.html
[Retrieved: 11-02-18]

[2]"VenturiFlowmeter - FlowMaxx Engineering". Available at: http://www.flowmaxx.com/venturi.htm
[Retrieved: 14-02-18]

[3]Himani Kala, K.S. Sandhu."Effect of change in power coefficient on the performance of wind turbines with different dimension " **INSPEC Accession Number:** 16177660. Publisher: IEEE. Date added to IEEE Xplore: July, 2016. Available at: http://ieeexplore.ieee.org/document/7522487/
[Retrieved: 14-02-18]

Compensating a Third Order Process having Inverse Response

Gaurav Kataria
Department of Chemical Engineering
Malaviya National Institute of Technology
Jaipur, India
gkataria64@gmail.com

Kailash Singh
Department of Chemical Engineering
Malaviya National Institute of Technology
Jaipur, India
ksingh.mnit@gmail.com

Abstract—**Inverse response in a process output is encountered because of the two opposing processes working simultaneously causing positive zeroes in the process transfer function. In this paper, a technique which is based on Smith predictor is used to solve the inverse response problem for the third order processes having a second order and a first order opposing processes working at the same time. A case study of inverse response in packed reactive distillation column has been taken for the implementation of the compensator. When the closed loop response of a process with and without compensator are compared, it is observed that there is a decrease in the overshoot and integral errors while using the compensator in the loop.**

Index Terms—**Inverse Response, Smith Predictor, Third Order Process, Control**

I. INTRODUCTION

During a step change in an open loop when the response of the output starts in a direction opposite to that of the final steady state direction, the response is known as an inverse response in a system. The inverse response has a characteristic of having atleast one zero which is a positive real number or imaginay pole with positive real part in the process transfer function. These processes are difficult to control as initial corrections made by the controller are in the wrong direction, leading to sluggish response of the closed loop [1]. The general technique of removing inverse response in the process is similar to the Smith predictor technique which is used to compensate for the processes having dead time [2].

The inverse response can be observed in chemical units like distillation columns [3], [7], chemical reactors [4], [5], and municipal incinerators [6] because of two opposing processes working simultaneously. To eliminate the effect of inverse response, two methods are mainly used either compensator is added for the removal of positive zeroes or controller is tuned differently keeping inverse response into the focus. An extension of a Smith predictor was used by Zhang and Sun [8] to remove the effect of inverse response in the process. They designed smith predictor using H_∞ control theory making the predictor to be more robust than basic one. Some researchers [9], [11] used internal model control (IMC) for curing the inverse response. Skogestad [11] and Chen et al. [10] converted the positive zeroes into dead time and than used the controller accordingly. Majority of work is done on PID tuning to remove the inverse response in the process. Luyben

[4] removed inverse response from a CSTR reactor using PI controller. He tuned the controller parameters by making them as a function of dead time and positive zeroes. While Sree and Chidambaram [5] used PI, which was derived by matching the s coefficients in the numerator and denominator of the closed loop transfer function of a CSTR. Chien et al. [3] proposed PID controller with tuning derived from direct synthesis controller design method to remove effect of inverse response along with dead time in the process. Alfaro and Vilanova [12] and Martinez et al., [13] used model reference technique for tunning the 2 degree of freedom (DoF) PID and a PI controller. Jeng and Lin [14] combined smith predictor with advanced tuning technique for removal of positive zeroes in the process. They tuned their PID using Maclaurin series approach along with the compensator.

All the authors given above has considered either inverse response in two first order opposing processes or they have converted CSTR and distillation problems into two first order opposing processes. It has been mentioned in Iinoya and Alpeter [1] and Stephanopoulos [2] that inverse response can also take place in a third order process when a first order and second order opposing processes are working simultaneously. In this paper a compensator is proposed for a third order process having inverse response due to the first order and second order opposing processes working simultaneously. For the study of proposed technique, a case study of the inverse response in a reactive distillation column has been considered from Peng et al. (2003) [15].

II. INVERSE RESPONSE

Assume that the following third order process without time delay is showing the inverse response.

$$G(s) = \frac{K(\tau s + 1)}{(as^3 + bs^2 + cs + 1)} \qquad (1)$$

The process has atleast one zero which is positive or is a imaginary number with positive real part which is a cause of inverse response in the process output. This third order process is fractioned into two opposing processes having a first order process and a second order process.

$$G_1(s) = \frac{K_1(\tau s + 1)}{(\tau_1 s^2 + 2\tau_1 \zeta s + 1)} \qquad (2)$$

978-1-5386-6624-1/18 $31.00 © 2018 IEEE

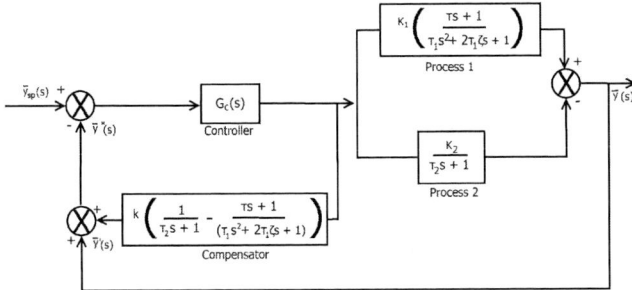

Fig. 1. Closed clop with compensator

$$G_2(s) = \frac{K_2}{(\tau_2 s + 1)} \qquad (3)$$

Where, K_1 and K_2 are process gains, τ_1 and τ_2 are process time constants and ζ is a damping factor for second order process.

The overall open loop response of the process is

$$y(s) = G_c(s)\big(G_1(s) - G_2(s)\big)\bar{y}_{sp}(s) \qquad (4)$$

Where, $G_c(s)$ is a transfer function of a controller and $\bar{y}_{sp}(s)$ is set point value of the required output.

$$y(s) = G_c(s)\frac{(K_1\tau\tau_2 - \tau_1^2 K_2)s^2 + (K_1(\tau + \tau_2) - 2K_2\tau_1\zeta)s + (K_1 - K_2)}{(\tau_1^2 s^2 + 2\tau_1\zeta s + 1)(\tau_2 s + 1)}\bar{y}_{sp}(s)$$
$$(5)$$

The roots of the numerator of the above equation may be positive causing the process to have a inverse response. To eliminate the positive zeroes of the process, a compensator is added as shown in fig: 1.

$$y'(s) = G_c(s)k\left(\frac{1}{(\tau_2 s + 1)} - \frac{\tau s + 1}{(\tau_1 s^2 + 2\tau_1\zeta s + 1)}\right)\bar{y}_{sp}(s)$$
$$(6)$$

where k is a compensator gain. After adding compensator, the overall open loop response changes to

$$y^*(s) = y(s) + y'(s) \qquad (7)$$

$$y^*(s) = G_c(s)\Bigg(\frac{(K_1\tau\tau_2 - K_2\tau_1^2 + k(\tau_1^2 - \tau\tau_2))s^2}{(\tau_1^2 s^2 + 2\tau_1\zeta s + 1)(\tau_2 s + 1)}$$
$$+ \frac{(K_1(\tau + \tau_2) - 2K_2\tau_1\zeta + k(2\tau_1\zeta - (\tau + \tau_2))s + (K_1 - K_2)}{(\tau_1^2 s^2 + 2\tau_1\zeta s + 1)(\tau_2 s + 1)}\Bigg)\bar{y}_{sp}(s) \quad (8)$$

To eliminate the inverse response of the process, the k value has to be optimized in such a way that zeroes of the transfer function, $y^*(s)$, lies on the negative left plane.
If α and β are the roots of the numerator (zeroes) of $y^*(s)$ then for the negative zeroes

$$\alpha\beta > 0 \quad and \quad \alpha + \beta < 0 \qquad (9)$$

Where,

$$\alpha\beta = \frac{(K_1 - K_2)}{(K_1\tau\tau_2 - K_2\tau_1^2 + k(\tau_1^2 - \tau\tau_2))} \qquad (10)$$

$$\alpha + \beta = -\frac{(K_1(\tau + \tau_2) - 2K_2\tau_1\zeta + k(2\tau_1\zeta - (\tau + \tau_2)))}{(K_1\tau\tau_2 - K_2\tau_1^2 + k(\tau_1^2 - \tau\tau_2))}$$
$$(11)$$

This leads to the two cases, $K_1 > K_2$ and $K_1 < K_2$

1) Considering the first case, i.e. when second process is dominating in the starting and finally first process prevails. If $K_1 > K_2$ then according to Eq: 9 denominator of Eq: 10 has to be greater than zero and numerator of Eq: 11 has to be less than zero.

$$K_1\tau\tau_2 - K_2\tau_1^2 + k(\tau_1^2 - \tau\tau_2) > 0 \qquad (12)$$

and

$$-(K_1(\tau + \tau_2) - 2K_2\tau_1\zeta + k(2\tau_1\zeta - (\tau + \tau_2))) < 0 \quad (13)$$

As sign shift occurs because of coefficients of k in both the equations, further cases are considered as
if $\tau_1^2 > \tau\tau_2$

$$k > \frac{K_2\tau_1^2 - K_1\tau\tau_2}{\tau_1^2 - \tau\tau_2} \qquad (14)$$

else

$$k < \frac{K_2\tau_1^2 - K_1\tau\tau_2}{\tau_1^2 - \tau\tau_2} \qquad (15)$$

and, if $\tau + \tau_2 > 2\tau_1\zeta$

$$k < \frac{K_1(\tau + \tau_2) - 2K_2\tau_1\zeta}{\tau + \tau_2 - 2\tau_1\zeta} \qquad (16)$$

else

$$k > \frac{K_1(\tau + \tau_2) - 2K_2\tau_1\zeta}{\tau + \tau_2 - 2\tau_1\zeta} \qquad (17)$$

2) In second case first process is dominating in the initial stage and second process in the end i.e. $K_1 < K_2$. Then according to Eq: 9 denominator of Eq: 10 has to be less than zero and numerator of Eq: 11 has to be greater than zero.

$$K_1\tau\tau_2 - K_2\tau_1^2 + k(\tau_1^2 - \tau\tau_2) < 0 \qquad (18)$$

and

$$-(K_1(\tau + \tau_2) - 2K_2\tau_1\zeta + k(2\tau_1\zeta - (\tau + \tau_2))) > 0 \quad (19)$$

Again due to coefficients of k, sign shift can occur,so further cases can be considered as
if $\tau_1^2 > \tau\tau_2$

$$k < \frac{K_2\tau_1^2 - K_1\tau\tau_2}{\tau_1^2 - \tau\tau_2} \qquad (20)$$

else

$$k > \frac{K_2\tau_1^2 - K_1\tau\tau_2}{\tau_1^2 - \tau\tau_2} \qquad (21)$$

and, if $\tau + \tau_2 > 2\tau_1\zeta$

$$k > \frac{K_1(\tau + \tau_2) - 2K_2\tau_1\zeta}{\tau + \tau_2 - 2\tau_1\zeta} \qquad (22)$$

else

$$k < \frac{K_1(\tau + \tau_2) - 2K_2\tau_1\zeta}{\tau + \tau_2 - 2\tau_1\zeta} \qquad (23)$$

Using the optimized value of compensator gain, k, inverse response is eliminated from the process.

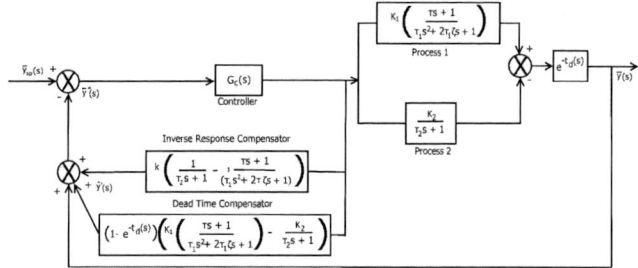

Fig. 2. Process with Inverse Response and Dead Time Compensator

This technique can also be extended to third order processes with inverse response and a dead time, t_d.

$$G(s) = \frac{K(\tau s + 1)}{(as^3 + bs^2 + cs + 1)} e^{-t_d(s)} \quad (24)$$

The compensator as shown in Fig. 2 is modified by removing dead time as well as inverse response factor of the process.

$$y'(s) = G_c(s)\left((1 - e^{-t_d(s)})\left(\frac{K_1(\tau s + 1)}{(\tau_1 s^2 + 2\tau_1 \zeta s + 1)} - \frac{K_2}{(\tau_2 s + 1)}\right)\right. $$
$$\left. + k\left(\frac{1}{(\tau_2 s + 1)} - \frac{\tau s + 1}{(\tau_1 s^2 + 2\tau_1 \zeta s + 1)}\right)\right)\bar{y}_{sp}(s) \quad (25)$$

While value of k can be optimized using same method given above.

III. CASE STUDY

The case study for studying the Smith predictor has been taken from Peng et al., (2003). They observed an inverse response in the product conversion in a reactive distillation column during -10% step change in a distillate flow rate. The data has been extracted from the figure and an approximate transfer function has been estimated using MATLAB software.

$$G_p(s) = \frac{0.1642s^2 - 51.23s + 48.24}{(s^3 + 26.39s^2 + 152.4s + 115.7)} \quad (26)$$

The transfer function of second order characteristic equation gave a percent fit of 97.85% and mean square error (MSE) of 1.692×10^{-07}. But a transfer function with third order characteristic equation gave a percent fit of 99.47% and MSE of 1.012×10^{-08}. The third order fit between deviated output of transfer function and time is shown in fig: 3. The figure also shows the inverse response in the output of the transfer function and has the positive zeroes 0.945 and 311.05.

The third order process was converted to a combination of a first order and second order process.

$$y(s) = G_c(s)\left(\frac{-0.5621(9.74 \times 10^{-03}s + 1)}{(7.7 \times 10^{-03}s^2 + 0.197s + 1)} + \frac{0.979}{(1.1205s + 1)}\right)\bar{y}_{sp}(s) \quad (27)$$

Where, K_1 and K_2 are -0.5621 and -0.979, τ, τ_1 and τ_2 are 9.74×10^{-03}, 0.088 and 1.1205 and ζ is 1.12. The transfer function of the compensator for the process will be

$$y'(s) = G_c(s)k\left(\frac{1}{(1.12057s + 1)} - \frac{(9.74 \times 10^{-03}s + 1)}{(7.7 \times 10^{-03}s^2 + 0.197s + 1)}\right)\bar{y}_{sp}(s) \quad (28)$$

The value of k can be estimated using above mentioned method. In this case, as $K_1 > K_2$, $\tau_1^2 < \tau\tau_2$ and $\tau + \tau_2 > 2\zeta\tau$, so

$$k < \frac{K_2\tau_1^2 - K_1\tau\tau_2}{\tau_1^2 - \tau\tau_2} < 0.442 \quad (29)$$

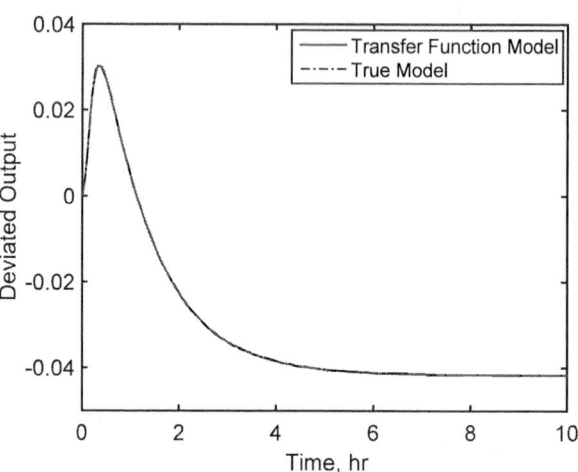

Fig. 3. Model fitting

$$k < \frac{K_1(\tau + \tau_2) - 2K_2\tau_1\zeta}{\tau + \tau_2 - 2\tau_1\zeta} < -0.474 \quad (30)$$

The common solution for k from Eq: 29 and 30 was found to be $k < -0.474$.

The case study has also been extended to the process with inverse response and a dead time. Let us assume that process is having a dead time of 0.1 hr.

IV. RESULTS AND DISCUSSION

The feedback loop for controlling the product conversion was developed using PI controller. The controller was tuned using Ziegler-Nichols method and the controller parameters are P = 1.268 and I = 1.725. To eliminate the inverse response the compensator gain was taken as -0.475. The overall transfer function of compensator is

$$y'(s) = \left(\frac{1.525^{-03}s^2 + 0.44s}{8.64 \times 10^{-03}s^3 + 0.228s^2 + 1.317s + 1}\right) \quad (31)$$

while for both the inverse response and dead time compensation the overall transfer function of compensator changes to

$$y'(s) = (1 - e^{-0.1s})\left(\frac{0.1642s^2 - 51.23s + 48.24}{s^3 + 26.39s^2 + 152.4s + 115.7}\right)$$
$$+ \left(\frac{1.525^{-03}s^2 + 0.44s}{8.64 \times 10^{-03}s^3 + 0.228s^2 + 1.317s + 1}\right) \quad (32)$$

A. Closed loop response

The servo response was studied by considering two cases i.e. process with and without compensator for a step change in set point of $\pm 10\%$. It can be seen in fig: 4 that process with compensator gives better results with less overshoot and less oscillations for both the cases. The integral errors for the closed response is shown in Table: I from which it is observed that errors get reduced on using the compensator.

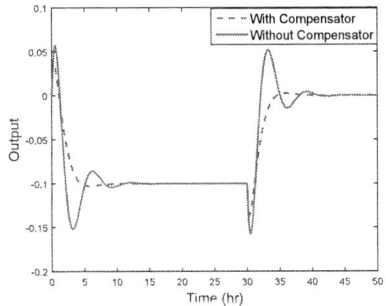 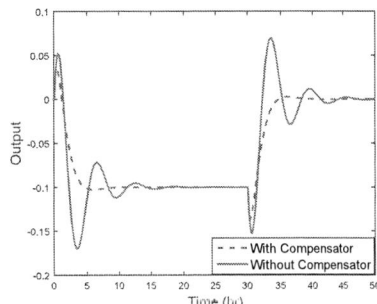

Fig. 4. Closed loop response for set point change by $\pm 10\%$ (a) Only inverse response (b) Both inverse response and dead time

TABLE I
INTEGRAL ERRORS FOR $\pm 10\%$ STEP CHANGE

.	Inverse Response		Inverse Response & Dead Time	
	With Compensator	Without Compensator	With Compensator	Without Compensator
ISE	0.046	0.060	0.049	0.073
IAE	0.507	0.675	0.537	0.904
ITAE	8.130	11.470	8.784	16.457

B. Change in compensator gain

The compensator gain which was decided to be as -0.475 is given $\pm 20\%$ change for checking robustness of both the compensators. Fig: 5 shows the response when gain is changed from -0.475 to -0.38 and -0.57. In both the cases, when k is -0.38 the process still shows inverse response as per k value range, which is the reason of showing higher overshoot than other responses. But every time the process shows better response than that without compensator making the compensator to be robust for changes in compensator gain. Integral errors for the change in compensator gain are given in Table: II.

TABLE II
INTEGRAL ERRORS FOR $\pm 20\%$ CHANGE IN COMPENSATOR GAIN

	Inverse Response			Inverse Response & Dead Time		
	k = -0.475	k = -0.380	k = -0.570	k = -0.475	k = -0.380	k = -0.570
ISE	0.024	0.024	0.024	0.025	0.025	0.025
IAE	0.260	0.253	0.272	0.270	0.263	0.282
ITAE	0.354	0.340	0.392	0.381	0.366	0.419

C. Process model mismatch

The robustness of the compensator was studied by changing the process gain, K, and time constants of both the processes (1st and 2nd order processes).

1) Uncertainty in Process Transfer Function Gain: The model mismatch was studied by giving change in both the process gains by $\pm 25\%$. The mismatch was studied by giving negative step change in setpoint for both the loops having just inverse response compensator and loop with inverse response plus dead time compensator. It can be seen in fig. 6 the process is getting controlled in both the positive and negative changes with some variations from the normal gain value. In the case of K_2, as shown in the fig. 7, the negative change is bringing high overshoot and sluggishness in the response. However every

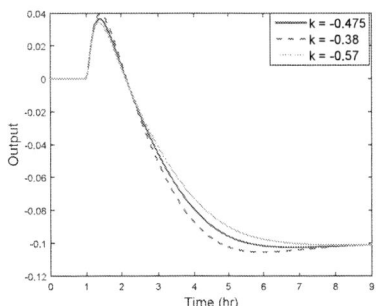

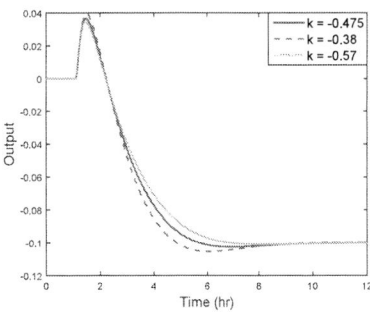

Fig. 5. Change in compensator gain by $\pm 20\%$ (a) Only inverse response (b) Both inverse response and dead time

TABLE III
INTEGRAL ERRORS FOR UNCERTAINTY IN PROCESS GAINS OF THE PROCESSES

	Inverse Response			Inverse Response and Dead Time		
	$K_1 = -0.562$	$K_1 = -0.422$	$K_1 = -0.703$	$K_1 = -0.562$	$K_1 = -0.422$	$K_1 = -0.703$
ISE	0.023	0.014	0.045	0.025	0.015	0.048
ITE	0.254	0.185	0.395	0.268	0.191	0.421
ITAE	0.338	0.215	0.649	0.384	0.227	0.779
	$K_2 = -0.979$	$K_2 = -0.734$	$K_2 = -1.224$	$K_2 = -0.979$	$K_2 = -0.734$	$K_2 = -1.224$
ISE	0.023	0.058	0.014	0.025	0.059	0.015
ITE	0.254	0.584	0.167	0.268	0.588	0.179
ITAE	0.338	1.583	0.168	0.384	1.573	0.201

time process is getting controlled making the process robust to the changes in process gain. Integral errors for the process model mismatch are given in Table: III.

978-1-5386-6624-1/18 $31.00 © 2018 IEEE

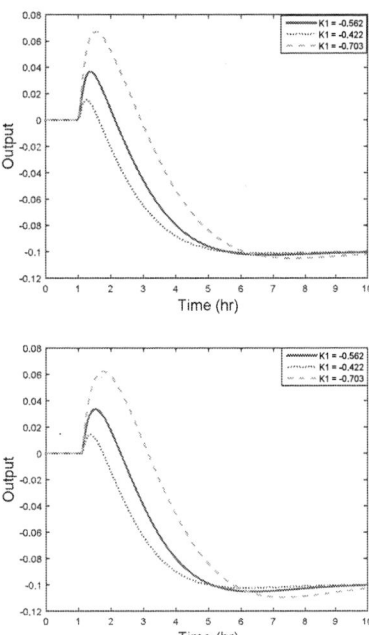

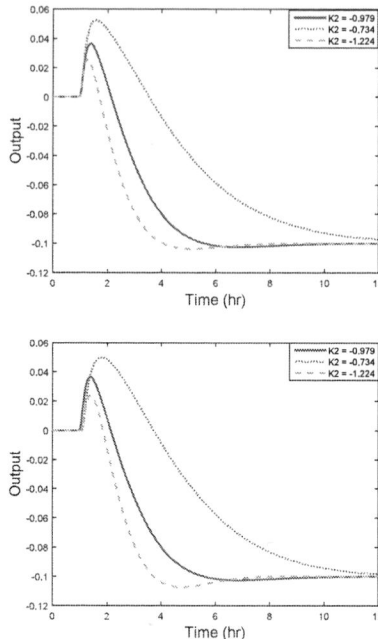

Fig. 6. Uncertainty in Process Gain K_1 by $\pm 25\%$ (a) Only inverse response (b) Both inverse response and dead time

Fig. 7. Uncertainty in Process Gain K_2 by $\pm 25\%$ (a) Only inverse response (b) Both inverse response and dead time

TABLE IV
INTEGRAL ERRORS FOR UNCERTAINTY IN TIME CONSTANTS OF THE PROCESSES

	Inverse Response			Inverse Response and Dead Time		
	$\tau_1 = 0.088$	$\tau_1 = 0.11$	$\tau_1 = 0.066$	$\tau_1 = 0.088$	$\tau_1 = 0.11$	$\tau_1 = 0.066$
ISE	0.023	0.022	0.024	0.025	0.023	0.026
ITE	0.260	0.259	0.260	0.270	0.269	0.270
ITAE	0.354	0.371	0.341	0.381	0.395	0.368
	$\tau_2 = 1.120$	$\tau_2 = 1.4$	$\tau_2 = .84$	$\tau_2 = 1.120$	$\tau_2 = 1.4$	$\tau_2 = .84$
ISE	0.023	0.033	0.018	0.025	0.034	0.019
ITE	0.260	0.350	0.245	0.270	0.364	0.255
ITAE	0.354	0.711	0.417	0.381	0.771	0.447

2) Uncertainty in Process Transfer Function Time Constants: As model is carrying two processes simultaneously, the model robustness is verified by changing time constants of both the processes by $\pm 25\%$. The response during change in process time constants for both the cases are shown in Fig. 8 and 9. In both the cases, response is showing controlled results with variable overshoot and settling time. The integral errors for the responses are shown in Table. IV.

V. CONCLUSIONS

An, inverse response for a third order process with the combination of a first and second order processes has been considered for designing a compensator. It was designed to convert the positive zeroes into negative zeroes. Another compensator was also designed for the compensation of third order processes with inverse response and a dead time. The designed compensator was applied on a case study of inverse response in reactive distillation column and dead time was also considered for the same. The results obtained showed better control results with less integral errors. The compensator

also proved to be robust to the changes in compensator gain, process gain and time constant values of first and second order processes.

REFERENCES

[1] K. Iinoya, and R. J. Altpeter,"Inverse Response in Process Control,"Industrial & Engineering Chemistry Fundamentals, vol. 54(7), pp.39–43, 1962.

[2] G. Stephanopoulos , Chemical process control : an introduction to theory and practice, Prentice-Hall, 1984.

[3] I. L. Chien, Y. C. Chung, B. S. Chen, and C. Y. Chuang, "Simple PID Controller Tuning Method for Processes with Inverse Response Plus Dead Time or Large Overshoot Response Plus Dead Time," Industrial & Engineering Chemistry Research, vol. 42(20), pp. 4461–4477, Oct 2003.

[4] W. L. Luyben, "Tuning Proportional Integral Controllers for Processes with Both Inverse Response and Deadtime," Industrial & Engineering Chemistry Research, vol. 39(4), pp. 973–976, 2000.

[5] R. P. Sree, and M. Chidambaram, "Simple method of tuning PI controllers for stable inverse response systems," Journal of the Indian Institute of Science, vol. 83, pp. 73–85, 2003.

[6] M. Rovaglio, D. Manca, G. Pazzaglia, and G. Serafini, "Inverse response compensation for the optimal control of municipal incineration plants: Model synthesis and experimental validation," Computers & Chemical Engineering, vol. 20(96),pp. S1461–S1467, 1996 .

[7] S. R. V. Raghavan, T. K. Radhakrishnan, and K. Srinivasan, "Soft sensor based composition estimation and controller design for an ideal reactive distillation column," ISA Transactions, vol. 50(1), pp. 61–70, 2011.

[8] W. Zhang, X. Xu, and Y. Sun, "Quantitative Performance Design for Inverse-Response Processes," Industrial & Engineering Chemistry Research, vol. 39(6), pp. 2056–2061, 2000.

[9] C. Scali, and A. Rachid, "Analytical Design of Proportional-Integral Derivative Controllers for Inverse Response Processes," Industrial & Engineering Chemistry Research, vol. 37(4), pp. 1372–1379, 1998.

978-1-5386-6624-1/18 $31.00 © 2018 IEEE

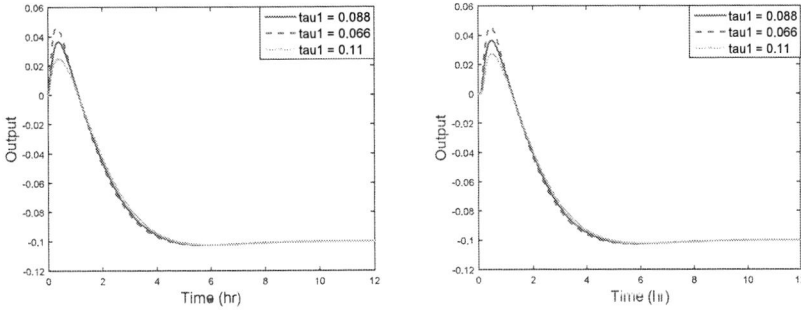

Fig. 8. Uncertainty in τ_1 for process with (a) Inverse response (b) Inverse response and dead time

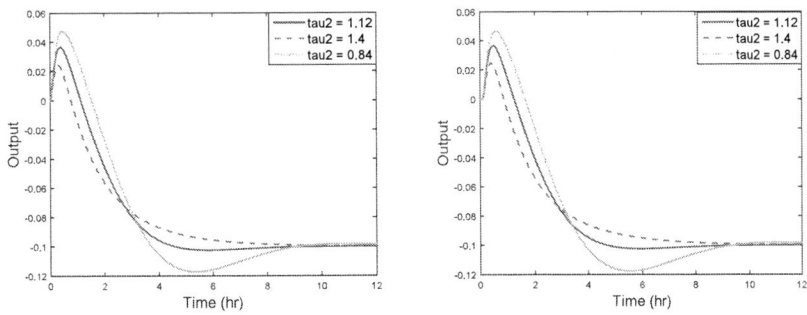

Fig. 9. Uncertainty in τ_2 for process with (a) Inverse response (b) Inverse response and dead time

[10] P. Y. Chen, Y. C. Tang, Q. Z. Zhang, and W. D. Zhang, "A new design method of PID controller for inverse response processes with dead time," Proceedings of the IEEE International Conference on Industrial Technology, pp. 036–1039, 2005.

[11] S. Skogestad, "Simple analytic rules for model reduction and PID controller tuning," Journal of Process Control, vol. 13(4), pp. 291-309, 2003.

[12] V. M. Alfaro, and R. Vilanova, "Robust tuning of 2DoF five-parameter PID controllers for inverse response controlled processes," Journal of Process Control, vol. 23(4), pp. 453–462, 2013.

[13] J. A. Martinez, O. Arrieta, R. Vilanova, J. D. Rojas, L. Marin, and M. Barbu, "Model reference PI controller tuning for Second Order Inverse Response and Dead Time Processes," Proceeding in IEEE International Conference on Emerging Technologies and Factory Automation, ETFA, Nov 2016.

[14] J. C. Jeng, and S. W. Lin, "PID controller tuning based on Smith-type compensator for second-order processes with inverse response and time delay," Proceeding in 2011 8th Asian Control Conference (ASCC), 2011.

[15] J. Peng, T. F. Edgar, and R. B. Eldridge, "Dynamic rate-based and equilibrium models for a packed reactive distillation column," Chemical Engineering Science, vol. 58(12), pp. 2671–2680, 2013.

2nd IEEE International Conference on Power Electronics, Intelligent Control and Energy Systems (ICPEICES-2018)

Design of PD-PID Controller with Double Derivative Filter for Frequency Regulation

Priyambada Satapathy
Department of Electrical Engineering
Siksha 'O' Anusandhan University
Bhubaneswar, Odisha, India
lirasatapathy@gmail.com

Manoj Kumar Debnath
Department of Electrical Engineering
Siksha 'O' Anusandhan University
Bhubaneswar, Odisha, India
mkd.odisha@gmail.com

Pradeep Kumar Mohanty
Department of Electrical Engineering
Siksha 'O' Anusandhan University
Bhubaneswar, Odisha, India
pkmohanty68@rediffmail.com

Abstract— In this research proposal a newly invented Proportional derivative-Proportional integral derivative controller along with double derivative filter (PDPID+DDF) has been introduced over a reheat system. To tune the constraints of the proposed controller sine-cosine algorithm is involved with integral time absolute error (ITAE) as evaluative function in this research work. To analyze system reaction a sudden load disorder of 0.01 p.u. is applied in thermal area-1. The system usefulness is demonstrated by presenting a time delay in each control areas of the system. The vigorous analysis of the system is placed to do evaluation with standard PID controller and the efficiency is proven toughly.

Keywords— Load Frequency Control, Reheat Based Thermal System, Sine-Cosine Algorithm, Controller Design, PDPID plus double derivative filter controller.

I.   INTRODUCTION

Basically in power generating station heat energy and force of water towards downward are transformed into electrical energy. Similarly wind turbines helps to extract wind energy from the air. The consistency supply in voltage and frequency reveals the good quality of power supply during alteration in load demand. But maintaining the constraints of power system in their prescribed value is a very challenging and complex task [1-2].

A extensive eminence of work has been done of numerous research job in this field. The research exploration exposes that in the past era the artificial neural network has been introduced on a hybrid system [3]. To obtain the finest gains craziness based particle swarm optimization [4] is implemented over thermal power system. A super conducting magnetic energy storage (SMES) scheme is organized for PID to retain steadiness over hydro-thermal power system [5]. Fuzzy logic is added to the PI controller [6], generation rate constraints (GRC)[7], reinforced learning neural network controller [8] are considered to verify the steadiness of the hydro-thermal system [9]. The past analysis also conveys that optimization methods and novel secondary controllers show an important role in AGC. In 2017 Jagatheesan introduced novel particle swarm optimization to normalize the power system restraints within bounds [10], fractional order PID to construct TCSC-based damping controller [11], PID controller in addition to

the fuzzy logic [12], two degree of freedom fractional order controller [13] employed over multi area reheat based thermal system. A proportional derivative controller is cascaded with PID controller tuned with the help of Bat algorithm [14] for LFC. A non-reheat and reheat based thermal system has employed chaotic optimization technique [15] and artificial bee colony algorithm [16] to tune the gains of PID controller. Further to attain finest optimized value of fuzzy logic based PID controller researchers implemented a hybridized optimization technique over a multi-source interconnected system [17].   The existence of time delay can vitiate the dynamic enactment of the system which leads to the instability. The communication delays [18] are situated on the measured frequency and power tie-line flow can easily transferred to area controller error through real terminal unit (RTU).

A profundity investigation has also resolved that PID controller is employed in utmost of the cases due to simplicity but PID controllers fail to eliminate noise from the reference signal. So on to achieve more benefits a filter is associated with derivative part called as PID controller with double derivative filter (PIDF) to solve the noise problem related to high frequency .This modelled system is used sine-cosine algorithm to optimize the fourteen gains of this novel profound controller and also a communication time delay is introduced to verify system instabilities and uncertainties.

II.   SYSTEM INSPECTED

PDPID along with double derivative filter are established over two equal area reheat thermal is depicted Fig.1.Reheat thermal system with proposed two distinctive PDPID plus double derivative controllers are being intricate for each generating unit to normalize the disturbances. The minimal gains for reheat based thermal system are enumerated in [16]. In case of any unsteadiness of the designed model the controller uses the area controller error (ACEs) as the actuating signals show a character to minimize the system disruption and accomplishing the steady state value. The controller factors should be appropriately elected in order to confirm the accurate performance of the designed model. A communication delay having exponential function $e^{-s\tau}$ is

978-1-5386-6624-1/18 $31.00 © 2018 IEEE

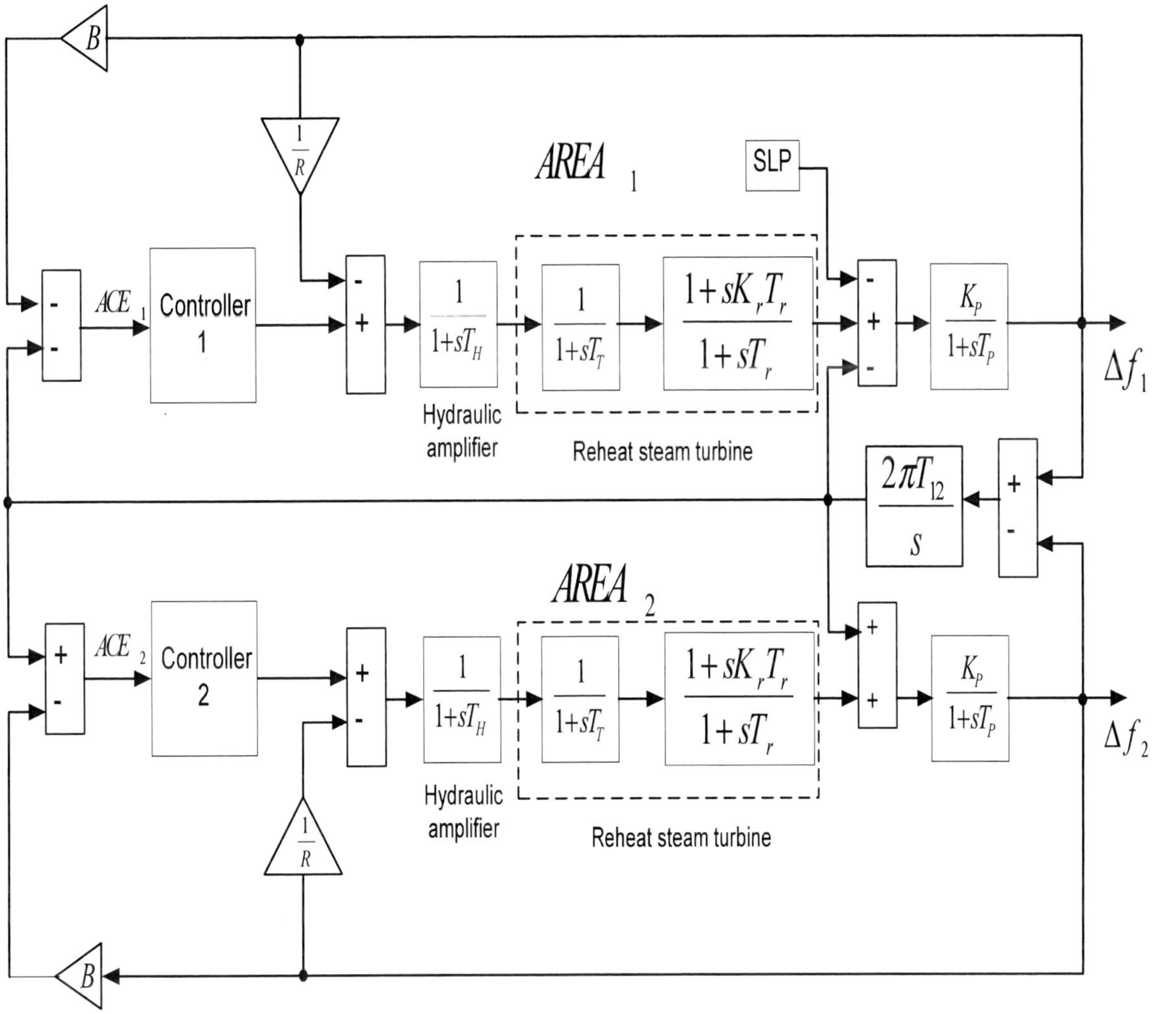

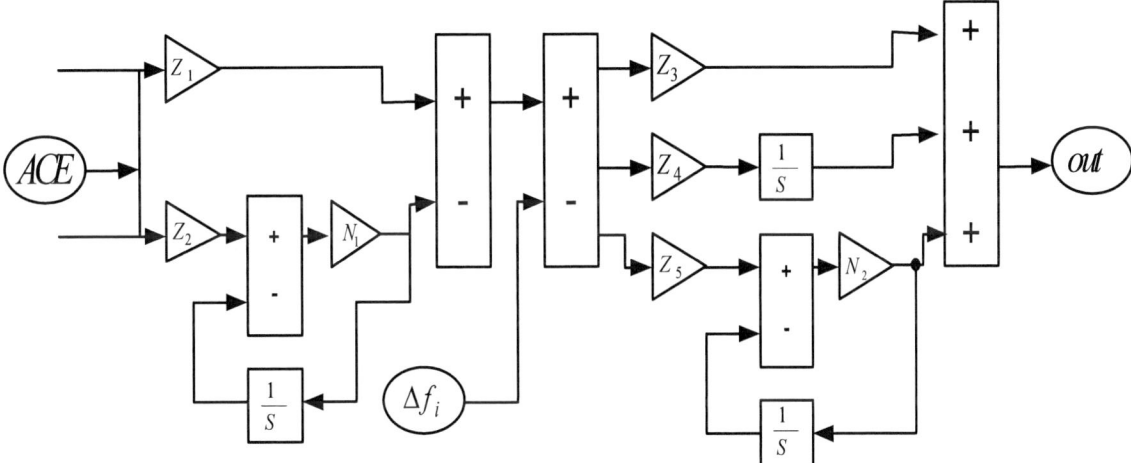

Fig.1. Two area reheat based thermal type system with PDPID plus double derivative filter.

Fig.2. Architecture of PDPID plus double derivative filter controller.

978-1-5386-6624-1/18 $31.00 © 2018 IEEE 1143

introduced in this designed model to verify the system instabilities [18]. The frequency and the interline power exhibits an alteration in transient responses and a suitable control signal is made by feedback control mechanism to make the signal flow to the load.

III. SUGGESTED PROCEDURE

A. Controller Arrangement

PID controller is very attractive due to its feasibility and easy to be implemented nature. Basically in PID controller has to be balanced all three gains of PID controller which may influence the transient responses of the system. Without the help of proper value of system parameters, the uncertainties and disturbances could not resist of a system. The PID controller have low robust ability in various operating circumstances like temperature, weather, etc. Appropriate filtering is very much crucial to protect the actuator from destruction. Basically the adding of filter with the derivative term does not alter the location of zeros in PID controller. To overcome the problems in complex power system with PID controller, an artificial intelligent based PDPID controller with double derivative filter is deliberated for AGC.Fig.2 explains the arrangement of designed PDPID+DDF controller.

B. Sine-Cosine Algorithm

Dated back Mirjalili developed a population based optimization algorithm which include Initialisation, Evaluation and updation process. The below expression is introduced to update the locations of fitness function.

$$X_{i,new} = \begin{cases} X_i + r_1 \times \sin(r_2) \times |r_3 P_i - X_i| \, if \, r_4 < 0.5 \\ X_i + r_1 \times \sin(r_2) \times |r_3 P_i - X_i| \, if \, r_4 \geq 0.5 \end{cases}$$

Here the current position is symbolised by X_i, P_i imitates the finest solution of the present location and is the objective point. r_1, r_2, r_3 and r_4 are acknowledged as major factors, where the direction of movement is ruled by r_1. The following updated location is ruled by $r1$, which falls in the area in the middle of the objective and solution. Here, r_2 is within 0 to 2π which symbolises up to what variety the movement should be just before or away the objective and r_3 assigns an arbitrary performer for the objective so as to highlight ($r_3 > 1$) or de-emphasise ($r_3 < 1$) the consequence of objective to delineate the distance. Between the cosine and sine modules, r_4 consistently shifts. So as to obtain the global resolution, appropriate convergence and appropriate balance within exploitation and exploration, the border of sine and cosine in beyond equation is speckled consistently as arranged by next equation.

$$r_1 = a - t\frac{a}{T}$$

Where the existing iteration is signified by 't', T symbolises the entire numbers of iteration and 'a' embodies fixed significance.

Though the sine and cosine functions are in the restrictions [1, 2] and [−2, −1], exploration is attained. On the other hand, when the assortments have the intermission [−1, 1], exploitation is attained. The optimisation method in SCA initiates with an initialization of a random resolution and then the finest solution is kept as objective and the other entities update their location established on this objective value. During the iterative method the bound of cosine and sine functions are restructured in order to discover the resolution in the search province. The optimisation method is concluded when profound number of iterations is achieved.

IV. SIMULATION AND RESULT

Sine-cosine optimization algorithm has been effectively instigated in the suggested effort to acquire the optimum standards of PDPID plus double derivative controller. The fourteen controller factors have been optimized by SCA procedure to rise the steadiness in terms of settling time, peak overshoots and undershoots. The objective function (J) is known as integral time absolute error (ITAE). This ITAE objective function is expressed as follows

$$ITAE = \int_0^t \left(|\Delta f_1| + |\Delta f_2| + |\Delta P_{tie}| \right) t . dt$$

Simulation and tuning of two equal areas reheat based thermal system has been modelled in Matlab/Simulink environments for the proposed system by allowing the above objective function. The SCA method deliberated a population of 100 and extreme iterations of 50 for the optimization method. The upper and lower bounds of the all the twelve controller parameters were set at 3.0 and 0.1 respectively and [10-200] for co-efficient of N. A relative study has been completed by introducing typical PID controller and novel healthy PDPID plus double derivative filter along with time delay over designed reheat power system with a disruption of 1% in area 1. The simulation consequences of fluctuations in frequency of respective control areas and fluctuations in interline power with traditional type PID [16] and recommended PDPID+DDF controller of reheat system are shown in Figs. (3-5).

Table I specifies the optimum SCA regulated tuned gains of reheat based two area system modelled with PDPID plus double derivative filter and traditional type PID controller [16]. Table II exhibits the dynamic outcomes of suggested PDPID+DDF and traditional type PID controller [16] of examined reheat type system.

Further a communication delay of 0. 02 sec is subjected to the recommended system to verify the system nonlinearities and transient responses which are depicted in fig. (6-8). The time delay accounts for various delays introduced by different elements of the system. The delay is applied before the control signal to account for various delays in transportation. The response graph (Figs.6-8) shows that the implemented control technique successfully maintains the system stability even if we consider the delay in the system.

978-1-5386-6624-1/18 $31.00 © 2018 IEEE

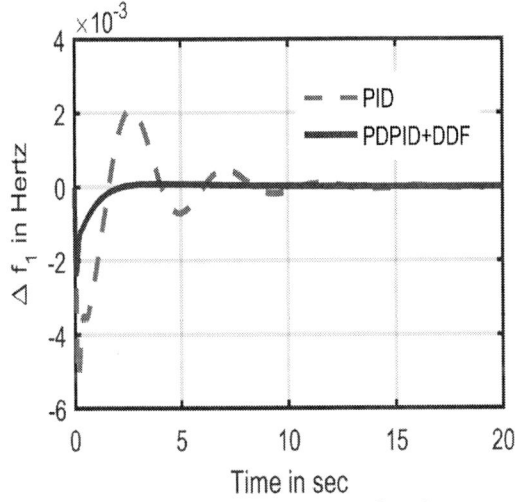

Fig.3. Fluctuation in Frequency of area 1.

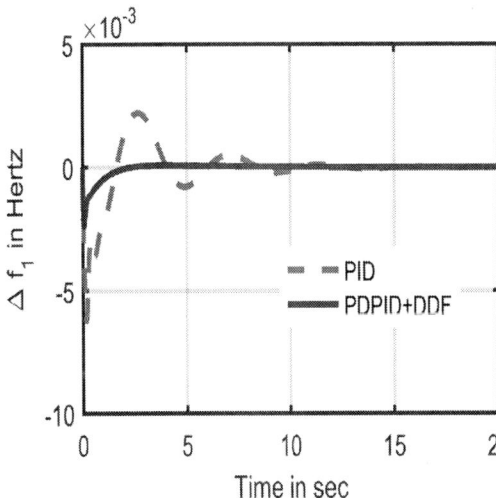

Fig.6. With delay fluctuation in Frequency of area 1.

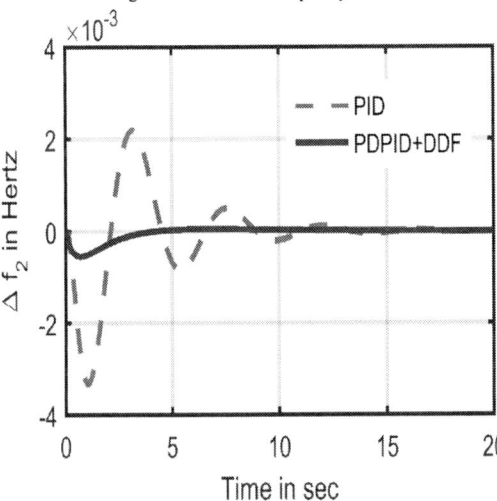

Fig.4. Fluctuation in Frequency in area 2.

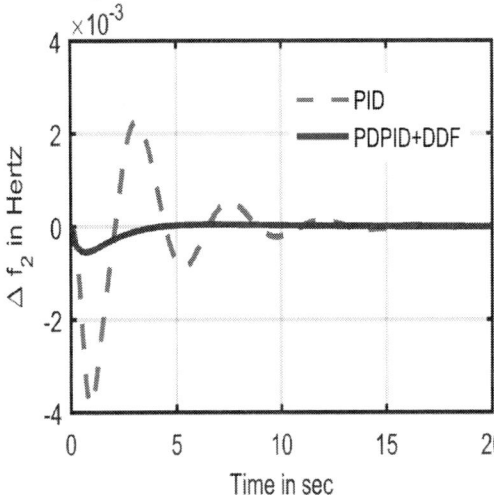

Fig.7. With delay fluctuation in Frequency of area 2.

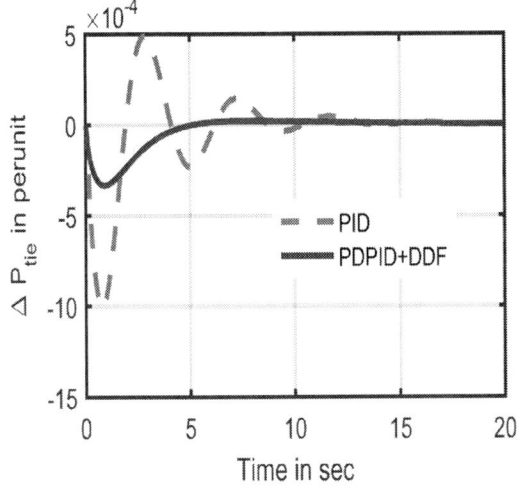

Fig.5. Fluctuation in tie-line power.

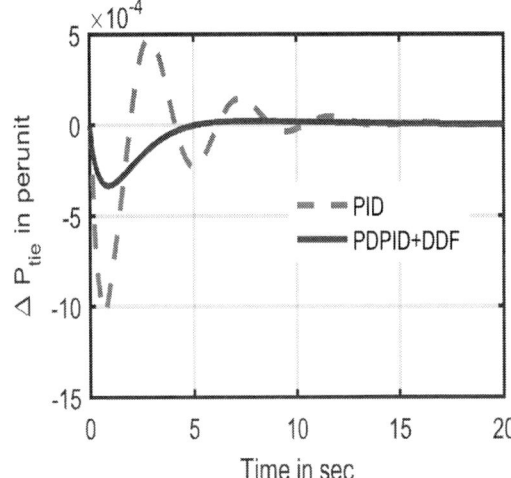

Fig.8. With delay fluctuation in tie-line power.

978-1-5386-6624-1/18 $31.00 © 2018 IEEE

TABLE I. THE FINEST GAIN PARAMETERS OF PDPID PLUS DOUBLE DERIVATIVE FILTER /PID CONTROLLER.

PDPID+DDF Controller													
Area 1							Area 2						
Z_1	Z_2	Z_3	Z_4	Z_5	N_1	N_2	Z_1	Z_2	Z_3	Z_4	Z_5	N_1	N_2
3.0000	3.0000	2.7604	3.0000	0.1000	200.0000	193.8029	2.1068	1.9485	2.6603	0.1000	1.0589	10.0000	128.9502

PID controller [16]					
Control Area 1			Control Area 2		
Z_P	Z_I	Z_D	Z_P	Z_I	Z_D
1.966	9.5902	3.9320	0.710	0.6827	0.7419

TABLE II. EVALUATIVE FACTORS OF THE RESPONSES

Abnormalities	Evaluative Factors	PDPID+DDF Controller	PID Controller[16]
Δf_1	U_{sh} in Hz	-0.0023	-0.0052
	T_s in sec(0.05 band)	7.3900	12.3700
	$O_{sh} \times 10^{-3}$ in Hz	0.08975	2.1
Δf_2	U_{sh} in Hz	-0.0005	-0.0033
	T_s in sec(0.05 band)	3.4600	12.9300
	$O_{sh} \times 10^{-3}$ in Hz	0.04846	2.2
ΔP_{tie}	U_{sh} in Hz	-0.0003	-0.0010
	T_s in sec(0.05 band)	3.8400	8.3400
	$O_{sh} \times 10^{-3}$ in Hz	0.02301	5.0

V. CONCLUSION

This investigated work introduced an advance newly PDPID series controller along with Double derivative filter to regulate the frequency over an equal two area reheat interconnected power system with the presence of time delay. Fourteen controller factors were tuned successfully with the help of sine-cosine algorithm. An exploration has been considered to verify the governance of suggested controller over PID controller under a disruption of 0.01p.u. The vigorous analysis reveals that the recommended recently developed PDPID plus derivative controller with the presence of time delay accomplishes better outcomes with a lesser amount of instability of system in addition to less overshoot and undershoot as compare to PID controller.

VI. REFERENCES

[1] Kundur, Prabha, Neal J. Balu, and Mark G. Lauby. Power system stability and control. Vol. 7. New York: McGraw-hill, 1994.

[2] Bevrani, Hassan, and Takashi Hiyama. Intelligent automatic generation control. CRC press, 2016.

[3] Zeynelgil, H. L., A. Demiroren, and N. S. Sengor. "The application of ANN technique to automatic generation control for multi-area power system." International journal of electrical power & energy systems 24.5 (2002): 345-354.

[4] Gozde, Haluk, and M. Cengiz Taplamacioglu. "Automatic generation control application with craziness based particle swarm optimization in a thermal power system." International Journal of Electrical Power & Energy Systems 33.1 (2011): 8-16.

[5] Abraham, Rajesh Joseph, D. Das, and Amit Patra. "Automatic generation control of an interconnected hydrothermal power system considering superconducting magnetic energy storage." International Journal of Electrical Power & Energy Systems 29.8 (2007): 571-579.

[6] Chang, C. S., and Weihui Fu. "Area load frequency control using fuzzy gain scheduling of PI controllers." Electric Power Systems Research 42.2 (1997): 145-152.

[7] Das, D., et al. "Automatic generation control of a hydrothermal system with new area control error considering generation rate constraint." Electric Machines & Power Systems 18.6 (1990): 461-471.

[8] Saikia, Lalit Chandra, et al. "Automatic generation control of a multi area hydrothermal system using reinforced learning neural network controller." International Journal of Electrical Power & Energy Systems 33.4 (2011): 1101-1108.

[9] Parmar, KP Singh, S. Majhi, and D. P. Kothari. "Automatic generation control of an interconnected hydrothermal power system." India Conference (INDICON), 2010 Annual IEEE. IEEE, 2010.

[10] Jagatheesan, K., et al. "Particle swarm optimisation-based parameters optimisation of PID controller for load frequency control of multi-area reheat thermal power systems." International Journal of Advanced Intelligence Paradigms 9.5-6 (2017): 464-489.

[11] Morsali, Javad, Kazem Zare, and Mehrdad Tarafdar Hagh. "Applying fractional order PID to design TCSC-based damping controller in coordination with automatic generation control of interconnected multi-source power system." Engineering Science and Technology, an International Journal 20.1 (2017): 1-17.

[12] Sahu, Binod Kumar, et al. "Teaching–learning based optimization algorithm based fuzzy-PID controller for automatic generation control of multi-area power system." Applied Soft Computing 27 (2015): 240-249.

[13] Debbarma, Sanjoy, Lalit Chandra Saikia, and Nidul Sinha. "Automatic generation control using two degree of freedom fractional order PID controller." International Journal of Electrical Power & Energy Systems 58 (2014): 120-129.

[14] Dash, Puja, Lalit Chandra Saikia, and Nidul Sinha. "Automatic generation control of multi area thermal system using Bat algorithm optimized PD–PID cascade controller." International Journal of Electrical Power & Energy Systems 68 (2015): 364-372.

[15] Farahani, Mohsen, Soheil Ganjefar, and Mojtaba Alizadeh. "PID controller adjustment using chaotic optimisation algorithm for multi-area load frequency control." IET Control Theory & Applications 6.13 (2012): 1984-1992.

[16] Gozde, Haluk, M. Cengiz Taplamacioglu, and Ilhan Kocaarslan. "Comparative performance analysis of Artificial Bee Colony algorithm in automatic generation control for interconnected reheat thermal power system." International Journal of Electrical Power & Energy Systems 42.1 (2012): 167-178.

[17] Debnath, Manoj Kumar, Ranjan Kumar Mallick, and Binod Kumar Sahu. "Application of Hybrid Differential Evolution–Grey Wolf Optimization Algorithm for Automatic Generation Control of a Multi-Source Interconnected Power System Using Optimal Fuzzy–PID Controller." Electric Power Components and Systems 45.19 (2017): 2104-2117.

[18] Bevrani, Hassan, and Takashi Hiyama. "Robust decentralised PI based LFC design for time delay power systems." Energy Conversion and Management 49.2 (2008): 193-204.

978-1-5386-6624-1/18 $31.00 © 2018 IEEE

Design and Simulation of Low-power Conditional-Discharging Flip Flop

Nidhi Gupta
Dept. of Electronics and Communication Engineering
University School of Information & Communication technology, GGSIP, University
Delhi, India nidhi08gupta@gmail.com

Krishna Singh
Dept.of Electronics and Communication
Engineering *G. B. Pant Engineering College, GGSIP, university, GGSIP University*
Delhi, India nidhi08gupta@gmail.com

Abstract— In this seminal, a low-power conditional discharging-flip flop (Modified CD-FF) has been presented. A clock-pulse by single feedback through scheme is proposed for use in the Modified CD-FF. Two inverters in clock-pulse generator circuit have been removed to optimize delay in clock path. In this low-power circuit design, the clock pulse is designed by a single inverter and pass-transistor logic (PTL). The proposed circuit (MCD-FF) reduces the additional switching activity of certain internal nodes as well as generates less glitch in the output with a lesser transistor count. As a consequence, the width of transistor is increased to manage the delay for clock pulse. Overall performances of circuit design provides better optimized area, delay and power consumption (average power) than earlier flip flop designs compared to 90nm technology.

Keywords—flip flop, Delay, Average power, peak power.

I. INTRODUCTION

The explosive growth of the microelectronics industry is characterized by features like higher performance, higher reliability, higher package density, dwindle high voltage and low power dissipation. [1]. All these parameters also define characteristics feature of the CMOS technology. In microelectronics industry, there is a trade-off between speed of operation, cost and size. The clock-pulse system used in FF is one of the best options to save power in VLSI circuit design when used in combination with clock-frequency inter-connection and timing elements. The overall power-consumption in any system is varied from 20%-70% [1]. Additionally, in order to maintain optimized performance, throughput and timing issues have been employed not only in data path sections, but also in comprehensive pipe-lining. Flip flop, being a crucial and repetitive building block in almost every digital system, has a great impact on overall power dissipation and performance of digital systems. Even small reduction achieved in delay and power dissipation of flip flop actually results in significant improvements in power and delay at system level. Therefore, circuit design of flip flop has a profound effect, both in terms of power consumption and higher performance of systems. All above reasons are the main ideology for flip flop circuit design and analysis. A vast selection of various types of flip flop is found in existing papers [2]–[17]. Many contemporary microprocessors use a pulse triggered and master slave flip flops [2]. There are two edge properties

in the flip flop. First is hard edge and the other is soft edge. The master slave flip flop is characterized by hard edge property. Two stages of the master slave flip flop reduce into one stage in pulse triggered flip flop which is characterized by the soft edge property. Inside the pulse triggered flip flop, the circuit complexity and number of stages are reduced, leading to small D-to-Q delay. Timing issues in the pulse triggered flip flop provide more desirable performance than the master slave flip flop. Master Slave flip flop has drawbacks of poor latency and less power efficiency. Pulse triggered flip flop overcomes the issues of master slave flip flop in terms of latency and power efficiency, however it has less race tolerance than master slave flip flop. The pulse triggered flip flop architecture is better for complex logic design, like process designs in the range of GHz. On the basis of pulse triggering, pulse trigger flip flop can be implicit type (ip-FFs) and explicit type (ep-FFs) [14]. In implicit pulse triggered flip flop, triggering clock pulse is generated internally. Examples of implicit pulse triggered flip flops are semi-dynamic flip flop (SDFF) [18], hybrid-latch flip flop (HLFF) [19] , implicit data close to output flip flops (ip-DCO) [20] etc. In explicit pulse triggered-flip-flops, pulse is generated externally. Examples of explicit pulse triggered-flip-flops are explicit-data close to output flip flop (ep-DCO) [20], Hybrid latch flip flop (HLFF) and semi dynamic flip flop(SDFF) [21,22]. As compared to ip-FFs, ep-FFs consume more power due to its explicit nature. If a pulse generator is shared among group of flip flops in ep-FFs, then consumed power can be decreased.

Although the clock frequency is laid down by system specifications, the use of DETFFs (dual edge-triggered flip-flops) can help to reduce the clock frequency to half that of the SETFFs (single edge-triggered flip-flops), while keeping certain data throughput [5].The most efficient way of reducing the over-all energy is by selecting the proper design architecture that has inherent low probability switching event in the local node with reduced number of clock cycles. Mostly static circuits consume less energy corresponding to the dynamic circuits due to their pre-charging and discharging operations of local switching nodes in every clock triggering [6].

The explicit pulse data close to output flip flops (ep-DCOFF) [4] are semi-dynamic flip flops and are good for high speed operations. However, power may be consumed in the

978-1-5386-6624-1/18 $31.00 © 2018 IEEE

operation of pre-charge and evaluation phase even when the input is not switching at dynamic stage in the circuits and the explicit pulse static flip flops (ep- SFF) [23] are used to overcome these issues.

In this article: Section II presents a Literature review of the previously published flip flops and classifies them into three groups. In Section III, a modified version of the CD-FF (MCD-FF) is proposed that has a less number of internal-nodes, minimum transistors, lesser power dissipation and lesser delay as compared to the existing flip flop with clock-pulse generator. Section IV presents the simulation results graphically.

Average and Peak Power: The power p(t) consumed by a circuit element is a product of current and voltage through the circuit.

$$P(t) = V(t)*I(t) \qquad (1)$$

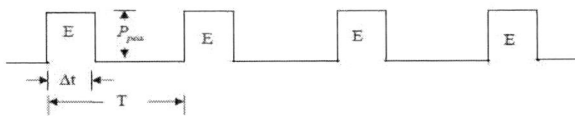

Fig.1. Clock Pulse

Energy consumed over a cycle is integration of the instantaneous power, where T is time period of cycle

$$E = \int_0^T P(t)dt \qquad (2)$$

The average power is time rate of change energy flow:

$$Pavg = \frac{E}{T} = E*f = \frac{1}{T}\int_0^T P(t)\,dt \qquad (3)$$

Where $f = \frac{1}{T}$, the energy implied in each clock pulse is constant. The energy content of the clock pulse is equal to the peak power level of the pulse multiplied by the clock pulse width. P_{avg} of clock pulse is:

$$Pavg = \frac{E}{T} \qquad (4)$$

Equating the energy (E)

$$Ppeak *\Delta t = Pavg *T \qquad (5)$$

Rearrangement of all the variables allows to define a new quantity called Duty cycle.

$$Duty\ Cycle = \frac{\Delta t}{T} = \frac{Pavg}{Ppeak} \qquad (6)$$

The peak power of a pulse can be easily calculated:

$$Ppeak = \frac{Pavg}{Duty\ Cycle} \qquad (7)$$

II. LITERATURE REVIEW

A. *Reducing the Switching Activity and Single Edge-Triggered Clock Pulse Generator*

Most of the flip flop circuits are dynamic in nature and reducing some internal-nodes. When an input is stable, without any useful activity at the output, each cycle is pre-charged and evaluated. Reducing the unnecessary switching

activity is found to be effective in reducing the power dissipation. Other flip flop techniques have been presented in these references [20-25]. In this paper, a brief literature survey of new techniques, clock pulse generator is conducted and also reduces the internal-node switching activity in modified conditional discharge flip flops. Reducing the clock pulse or clock frequency has more effect on the dynamic power dissipation.

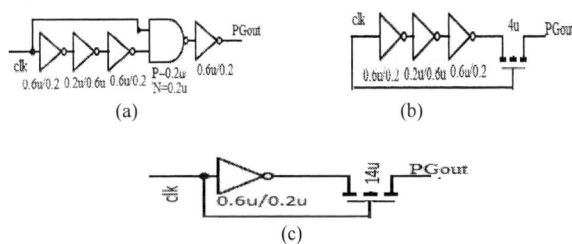

Fig. 2: (a) & (b) Clock Pulse (c) Prop. Clock pulse generator.

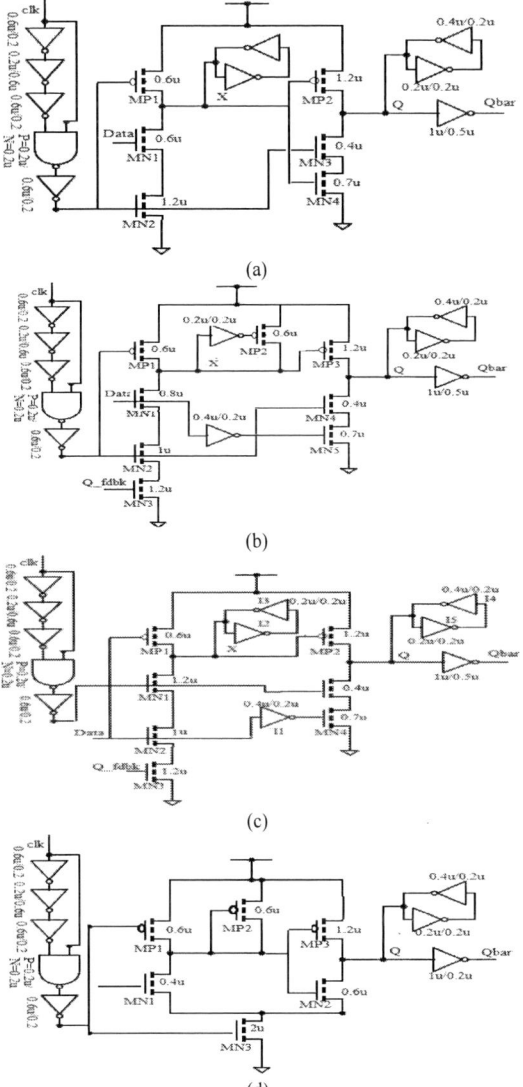

Fig. 3: Previous P-FF designs. (a) ep-DCO [15]. (b) CD-FF [9]. (c) Static-CD-FF [24]. (d) MHL-FF [19].

B. The single edge-triggered Pulse Flip Flops

Single edge-triggered Pulse Flip Flops (SET-FFs) consists of a pulse generation circuit and a latch that capture the pulse. In this paper, an evolution of a pulsed triggered flip flop circuit is constructed by using recent circuits like as ep-DCO-FF [15], ep-S-FF [12] and CD-FF [9]. A well established proposed circuit of single edge pulse trigger generator is shown in fig.2(c). It is suitable for single edge sampling and has been applied to all the existing flip flops [9], [13], [14], and proposed MCD-FF design. All the flip flop designs have been simulated and have been analyzed to obtain proper evaluation performances.

In fig.3(a) [15], author says that, the first phase for the circuit shown is dynamic in nature and its second phase is static in nature, which is regarded as the fastest flip flop due to its semi dynamic-nature. The clock pulse generator circuit drives the pre-charge pseudo nMOS transistor MP1 and two evaluation nMOS transistors MN2 & MN3. MN1 transistor works in capturing the input data. The flip flop becomes transparent during single edge sampling period and input data is propagated to the output. After this period, the pull-down (nMOS logic) network path in both phases is turning off via nMOS transistors MN2, & MN3. Therefore, no changes in the data input can be presented at the output. When data input does not change so that the node x is high and low in every clock pulse. The power is being consumed by charging and discharging path. When the output is 1(high), then the glitches are caused by the repeated or discharging path of the internal node X in each clock pulse to appear at the output. These glitches are propagating, not only driven by the gates, but also increase the switching activity, and create the noise problems.

CD-FF structure is similar to the ep-DCO-FF circuit. In this structure, the additional MN3 transistor controlled by the output Q_fdbk signal is working so that no discharging path occurs, if Data is 1. The node X is high because pull up MP1 is ON. Charging and discharging of the node is not performed on each clock pulse, so that there are no glitches visible at node Q. When the input stays high, less switching noise is generated. In Most of the cases node X stays 1 (high) and remains pre-charged, which will help to reduce the output capacitive load.

III. MODIFIED CONDITIONAL CHARGING FLIP FLOP WITH PROPOSED CLOCK GENERATION

This seminal review has five existing flip flop circuits. All of these face the same worst-case timing occurring during data transition from 0 to 1. The modified design shown in Fig. 4(b) attains a signal to improve power saving and delay of a modified clock pulse generator. The modified design avoids the unnecessary additional switching activity at the internal nodes. There are general differences to make the modified flip flop design better than the existing flip flop designs. Firstly, pull-up MP1 gate is connected to the GND and is always ON. It is known as nMOS pseudo-logic for the node X to be kept at intact high. Second, MN2 PTL (pass transistor logic) is controlled by the clock-pulse generator and it contains input data that can be driven at node Q directly through the latch circuit. Pull-up MP2 in the second phase of an inverter latch

circuit facilitates accessory signal driven from the input to the node Q. The main role of the transistor MN2 is two folded, it provides an additional driving signal at node Q during its charging, data transition 1 from 0 and 1 to 0 data transition during node Q discharging. The modified design circuit is employed as follows: the redundant switching power is reduced by discharging, control by the transistor MN3 in the discharging path of the first stage. If Q = 0 and node X = 1, and if nMOS transistor MN3 is on, then consequently enabling the discharging path. If Data input makes 0 to 1 transition and clk is 1, transistor MN1, MN2 & MN3 are switched ON, the node X is discharged to 0.Single path is used in the flip flop, to make sure that the Data input low-to-high transition is sampled. The output rise time transition is lead to be a slow by employing a single path, the capacity of node X decrease, and thus it can reduce the delay transition 0 to 1.

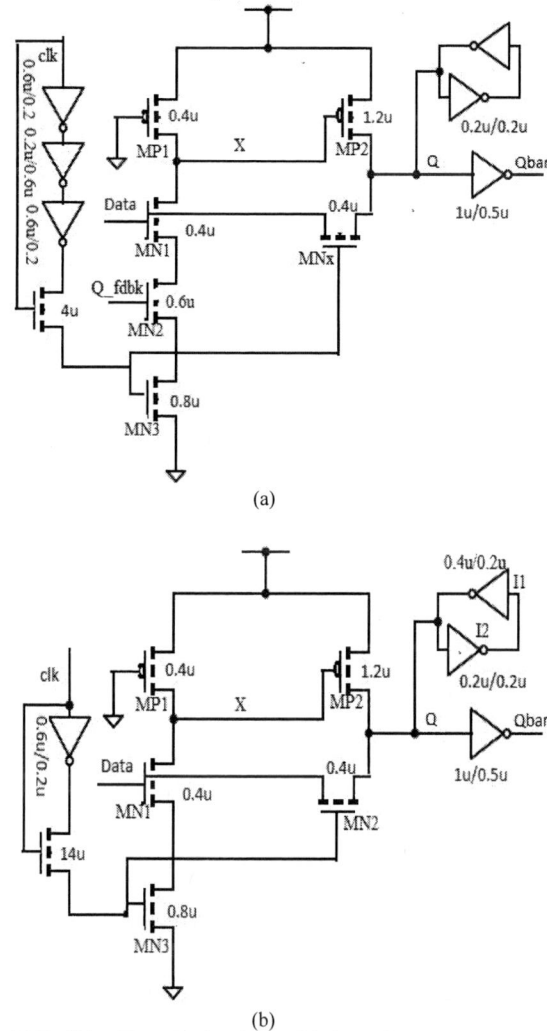

Fig. 4: (a) Traditional P-FF Design (b) Modified Conditional Discharging Flip flop (proposed)(MCD-FF).

III. SIMULATION RESULT

Performance of the modified conditional discharge flip flop (MCD-FF) circuit has been assessed against the existing flip

978-1-5386-6624-1/18 $31.00 © 2018 IEEE   1150

flops. Comprehensive simulation is done by using 90nm CMOS technology. The input voltage employed in the simulation results is 1.8V and clock-frequency fclk of 100MHz is used. To achieve better results, we have simulated all the flip flop circuits in HSPICE environment and the flip flop inputs (Data, clock) are driven by a fixed input. The transistor width is optimized to minimize the power consumption of the flip flop. They are simulated with various test patterns. Table I, II, III, & IV summarize the circuit features and the leakage power of all flip flop designs under various combinations of clock fclk and input signal Data at varied voltage level. All circuit features in the modified circuit design are obtained by using minimum number of transistors, and hence small silicon layout chip area is obtained.

The modified flip flop circuit manipulates the transistor width on the discharging path and clock-pulse generator. The power saving varies in different combinations of the test pattern. Modified design is the best in terms of power dissipation while four patterns have been tested. In the first stage of the modified flip flop design, the pseudo-nMOS logic MP1 is always-on and it will get the full swing output voltage at node X. But, it does not depend on the result in any dc power dissipation problem. Two inverters provide a charge keeper and thus a full swing is available at node Q. In the modified flip flop at node Q, the voltage-level remains at intact value of voltage Vdd. In Table I, it can be seen that the modified design results in terms of peak power, average power and delay value that are close to those for other flip flop designs. But one of those MHL-FF designs suffers from large dc average power consumption, which cause a non-full swing output [13]. The MCD-FF requires a relevant signal, which is driven from input data to the node Q and calculates the power consumption by the transistor MN2. The power saving is varied in various combination of test pattern and flip flop design. The power saving against the different flip flops, CDFF, SCD-FF, ep-DCO, and P-FF are 30.2%, 38%, 35.2% and 16.3% respectively at 1.8V. The S-CDFF design consumes maximum power because of an additional discharging path at the internal node. The layout design of modified condition discharging flip flop is shown in fig.7. The layout is designed by 90nm technology.

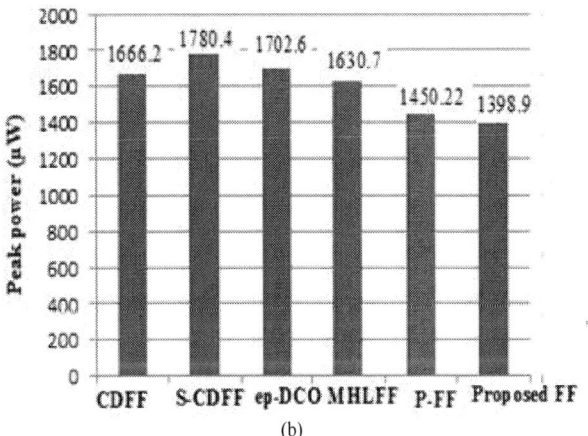

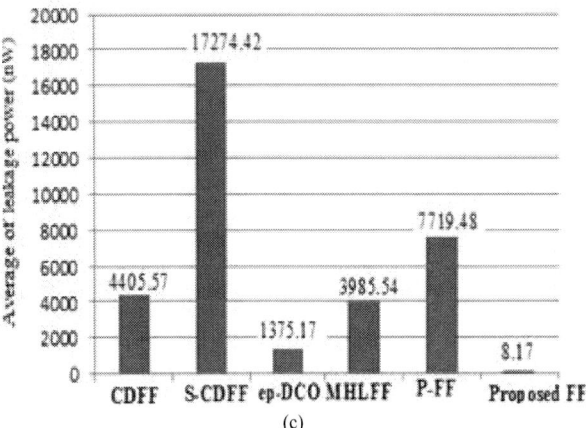

Fig. 5 (a) Results of Average Power (Power consumption) (b) Results of Peak Power (c) Results of leakage power (nW)

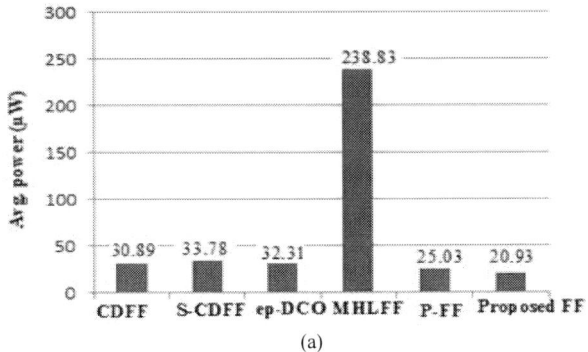

IV. CONCLUSION

In this article, we presented a low power MCD-FF using PTL and mixed design style. The conditional discharging flip flop is driven by the modified pulse generation using a single inverter and PTL. This design successfully solves the unnecessary problem of discharging path to achieve better performance. The design avoids an unnecessary Q_fdbk transistor of Fig.4a. The comparison of results of few parameters like delay, average power (power consumption), and peak power with the conventional flip flops are summarized. In the MCD-FF design, the internal node switching activities are reduced and less glitch is achieved. It also provides additional driving facility to reduce transition time, increase speed, reduce power and improve performance. Hence, power dissipation is very low. All the simulations have been conducted to substantiate the claims of the modified design in various performance aspects.

TABLE I: Results varied Flip Flop Designs at 1.8V

FF Design	CDFF	S-CDFF	ep-DCO	MHLFF	P-FF	Prop.FF
No. of Transistor	30	31	28	19	22	14
Rising Delay(ns)	25.01	45.03	25.06	20.08	25.08	25.07
Falling Delay(ns)	34.06	34.04	34.06	14.88	34.06	34.87
Avg. power(uW)	30.89	33.78	32.31	238.83	25.03	20.93
Peak power(uW)	1666.2	1780.4	1702.6	1630.7	1450.22	1398.9

TABLE II: Results varied Flip Flop Designs at 2.5V

FF Designs	CDFF	S-CDFF	ep-DCO	MHLFF	P-FF	Prop.FF
No. of Transistor	30	31	28	19	22	14
Rising Delay (ns)	19.50	19.50	20.00	18.69	18.19	19.46
Falling Delay (ns)	14.02	14.00	14.01	12.14	13.14	12.88
Avg. power (uW)	108.14	116.26	109.24	78.83	81.38	72.04
Peak power (uW)	1953.0	4048.50	3962.4	2567.3	1716.9	1702.60

TABLE III: Results varied Flip Flop Designs at 3.5V

FF Designs	CDFF	S-CDFF	ep-DCO	MHLFF	P-FF	Prop. FF
No. of Transistor	30	31	28	19	22	14
Rising Delay (ns)	19.55	10.00	19.96	18.69	19.11	19.14
Falling Delay (ns)	13.95	13.95	13.96	12.14	12.54	12.40
Avg. power (uW)	289.89	316.61	305.06	688.35	291.78	216.23
Peak power (uW)	8307	10357	9570.1	3608.8	5367.5	2567.3

TABLE IV: Leakage-Power results in Standby Mode (nW)

FF Design	CDFF	S-CDFF	ep-DCO	MHLFF	P-FF	Prop.FF
(CLK, DATA)=(0,0)	50.09	56.03	41.24	42.78	35.14	27.99
(CLK, DATA)=(0,1)	4174.3	451119	41.67	60831.1	3767.2	0.0795
(CLK, DATA)=(1,0)	4628.9	4625.6	41.67	2524.7	21991	1.41
(CLK, DATA)=(1,1)	8769	13297	5376.1	7291.6	5084.6	1.23
Average	4405.57	17274.42	1375.17	3985.54	7719.48	8.17

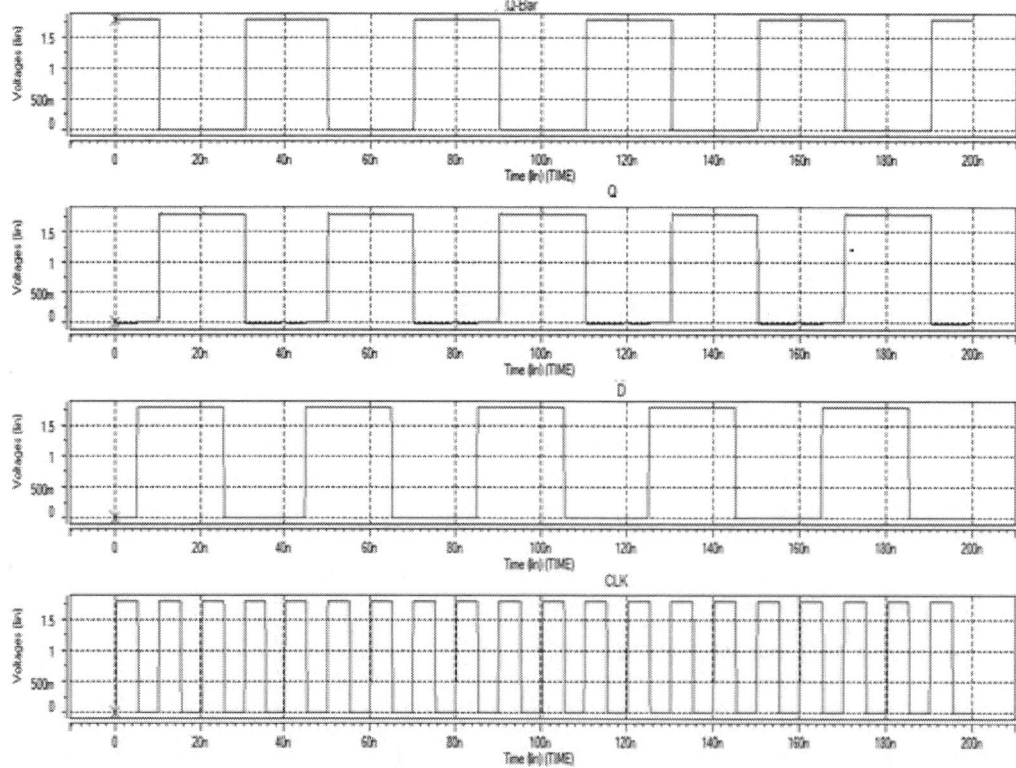

Fig. 6: Waveform of modified CD-FF

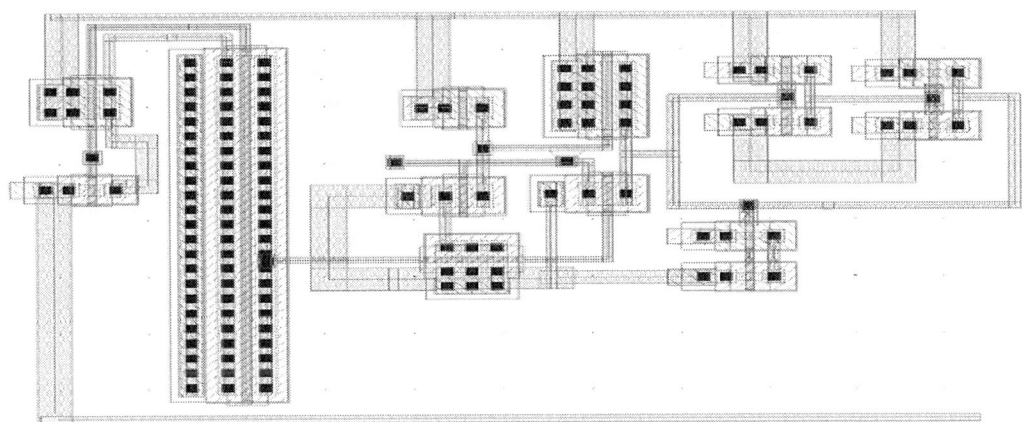

Fig. 7: Layout of modified CD-FF

REFERENCES

[1] A. P. Chandrakasan, W. J. Bowhill, and F. Fox, Design of highperformancemicroprocessor circuits. Wiley-IEEE press, 2000.

[2] A. P. Chandrakasan, W. J. Bowhill, and F. Fox, Design of high performance microprocessor circuits. Wiley-IEEE press, 2000.

[3] N. Nedovic, M. Aleksic, and V. G. Oklobdzija, "Conditional techniques for low power consumption flip flops," vol. 2, pp. 803–806, 2001.

[4] M. Alioto, E. Consoli, and G. Palumbo, "General strategies to design nanometer flip-flops in the energy-delay space," IEEE Transactions on Circuits and Systems I: Regular Papers, vol. 57, no. 7, pp. 1583–1596, 2010.

[5] P. Zhao, J. McNeely, W. Kuang, N. Wang, and Z. Wang, "Design of sequential elements for low power clocking system," IEEE Transactions on very large scale integration (VLSI) systems, vol. 19, no. 5, pp. 914–918, 2011.

[6] W. Chung, T. Lo, and M. Sachdev, "A comparative analysis of low power low-voltage dual-edge-triggered flip-flops, " IEEE transactions on very large scale integration (VLSI) systems, vol. 10, no. 6, pp. 913–918, 2002.

[7] V. G. Oklobdzija, V. M. Stojanovic, D. M. Markovic, and N. M. Nedovic, Digital system clocking: high-performance and low-power aspects. John Wiley & Sons, 2005.

[8] M. Alioto, E. Consoli, and G. Palumbo, "Analysis and comparison in the energy-delay-area domain of nanometer cmos flip-flops: Part-I methodology and design strategies," IEEE Transactions on Very Large Scale Integration (VLSI) Systems, vol. 19, no. 5, pp. 725–736, 2011.

[9] Y.-T. Hwang, J.-F. Lin, and M.-H. Sheu, "Low-power pulse-triggered flip-flop design with conditional pulse-enhancement scheme," vol. 20, no. 2. IEEE, 2012, pp. 361–366.

[10] M. Alioto, E. Consoli, and G. Palumbo, "Analysis and comparison in the energy-delay-area domain of nanometer cmos flip-flops: Part iiresults and figures of merit," IEEE Transactions on Very Large Scale Integration (VLSI) Systems, vol. 19, no. 5, pp. 737–750, 2011

[11] P. Zhao, T. K. Darwish, and M. A. Bayoumi, "High-performance and low-power conditional discharge flip-flop," IEEE transactions on very large scale integration (VLSI) systems, vol. 12, no. 5, pp. 477–484, 2004.

[12] U. Ko and P. T. Balsara, "High-performance energy-efficient d-flip-flop circuits," IEEE Transactions on Very Large Scale Integration (VLSI) Systems, vol. 8, no. 1, pp. 94–98, 2000.

[13] M. W. Phyu, W. L. Goh, and K. S. Yeo, "A low-power static dual edge triggered flip-flop using an output-controlled discharge configuration," pp. 2429–2432, 2005.

[14] H. Kawaguchi and T. Sakurai, "A reduced clock-swing flip-flop (rcsff) for 63% power reduction," vol. 33, no. 5. IEEE, 1998, pp. 807–811

[15] H. Partovi, R. Burd, U. Salim, F. Weber, L. DiGregorio, and D. Draper, "Flow-through latch and edge-triggered flip-flop hybrid elements," in Solid-State Circuits Conference, 1996. Digest of

Technical Papers. 42nd ISSCC., 1996 IEEE International. IEEE, 1996, pp. 138–139.

[16] F. Klass, "Semi-dynamic and dynamic flip-flops with embedded logic," in VLSI Circuits, 1998. Digest of Technical Papers. 1998 Symposium on. IEEE, 1998, pp. 108–109.

[17] J. Tschanz, S. Narendra, Z. Chen, S. Borkar, M. Sachdev, and V. De, "Comparative delay and energy of single edge-triggered & dual edgetriggered pulsed flip-flops for high-performance microprocessors," in Proceedings of the 2001 international symposium on Low power electronics and design. ACM, 2001, pp. 147–152.

[18] Y. Zhang, H. Yang, and H. Wang, "Low clock-swing conditional precharge flip-flop for more than 30% power reduction," Electronics Letters, vol. 36, no. 9, pp. 785–786, 2000.

[19] F. Klass, "Semi-dynamic and dynamic flip-flops with embedded logic," in VLSI Circuits, 1998. Digest of Technical Papers. 1998 Symposium on. IEEE, 1998, pp. 108–109.

[20] P. Zhao, T. Darwish, and M. Bayoumi, "Low power and high speed explicit-pulsed flip-flops," in Circuits and Systems, 2002. MWSCAS-2002. The 2002 45th Midwest Symposium on, vol. 2. IEEE, 2002, pp. II–II.

[21] H. Partovi, R. Burd, U. Salim, F. Weber, L. DiGregorio, and D. Draper, "Flow-through latch and edge-triggered flip-flop hybrid elements," in Solid-State Circuits Conference, 1996. Digest of Technical Papers. 42nd ISSCC., 1996 IEEE International. IEEE, 1996, pp. 138–139.

[22] J. Tschanz, S. Narendra, Z. Chen, S. Borkar, M. Sachdev, and V. De, "Comparative delay and energy of single edge-triggered & dual edge triggered pulsed flip-flops for high-performance microprocessors," in Proceedings of the 2001 international symposium on Low power lectronics and design. ACM, 2001, pp. 147–152.

[23] B.-S. Kong, S.-S. Kim, and Y.-H. Jun, "Conditional-capture flip-flop for statistical power reduction," IEEE Journal of Solid-State Circuits, vol. 36, no. 8, pp. 1263–1271, 2001.

[24] N. Nedovic and V. G. Oklobdzija, "Hybrid latch flip-flop with improved power efficiency," in Integrated Circuits and Systems Design, 2000. Proceedings. 13th Symposium on. IEEE, 2000, pp. 211–215.

[25] N. Nedovic, M. Aleksic, and V. G. Oklobdzija, "Conditional precharge techniques for power-efficient dual-edge clocking," in Low Power Electronics and Design, 2002. ISLPED'02. Proceedings of the 2002 International Symposium on. IEEE, 2002, pp. 56–59.

[26] J. Yuan and C. Svensson, High-speed cmos circuit technique, IEE Journal of Solid-State Circuits, vol. 24, no. 1, pp. 6270, 1989.

2nd IEEE International Conference on Power Electronics, Intelligent Control and Energy Systems (ICPEICES-2018)

Composite Nonlinear Feedback Control for Inverted Pendulum with Input Saturation

Bhavna Agarwal
Department of Electronics Engineering
Rajasthan Technical University
Kota, India
bhavna.agarwal1994@gmail.com

Manisha Bhandari
Department of Electronics Engineering
Rajasthan Technical University
Kota, India
manisha.rtu@gmail.com

Abstract—In this paper, we develop and implement Composite Nonlinear Feedback (CNF) control law for an inverted pendulum subject to actuator saturation. The inverted pendulum mounted over a cart is inherently unstable and nonlinear system. Our aim is to balance pendulum near to its vertically upright position while tracking the cart position for a given setpoint. CNF control law is proposed for asymptotical tracking of cart position and to make the process faster and smoother with fewer oscillations. The key to CNF control technique design is the combination of linear and nonlinear control where linear control gives quick rising time due to low damping ratio and nonlinear control provides high damping ratio to remove the overshoot resulting from linear part during tracking the reference. Simulations are performed to validate the theory and results are discussed.

Index Terms—Composite Nonlinear Feedback, control applications, inverted pendulum, actuator saturation.

I. INTRODUCTION

The Inverted Pendulum is an inherently unstable and nonlinear system, it is one of the most difficult control problem. The inverted pendulum consists of a cart which can move horizontally, a hinged free moving pendulum mounted over the cart in an inverted position which has to be balanced actively in order to remain in vertically upright position and resistant to disturbance. A dc servo motor is used to provide the required control force to the cart so as to balance the inverted pendulum. Inverted pendulum finds its application in Segway vehicles, rockets, robots [1]. Due to its importance in control engineering, it is a topic of research for many years during which many control techniques have been developed, in [2], pole placement approach to developing the control law is discussed. Proportional integral derivative (PID) controller for an inverted pendulum is discussed in [3]. PID is a traditional technique for control systems. It is the most simple, applicable and easy to develop a controller. Linear quadratic regulator (LQR) technique for the inverted pendulum developed in [4], it is an optimal control technique. A modification of this technique is proposed in [5], by combining PID control and LQR. However, there is still a possibility for the improvement in closed-loop system performance.

The Control law is developed under the assumption of ideal system i.e neglecting nonlinearities, however, in the real world, systems do have nonlinearities that cannot be neglected also

only a few can be overcome. One of such nonlinearities is actuator saturation (or input saturation) [6]- [7], if ignored can cause serious performance deterioration and instability. Saturation occurs in motion control systems when any of its components reach their maximum capacity. Inverted Pendulum system has nonlinearities such as friction, air disturbance, and actuator saturation. Here we work to overcome the actuator saturation(or input saturation).

In [8], Low gain feedback (LGF) was first introduced for linear systems with actuator saturation to establish semi-global stabilizability it is based on the concept of asymptoticity. As proposed in [8], LGF belongs to a certain family of feedback laws where gain $(K(\varepsilon))$ is obtained from a parameterized gain matrix, which approaches zero as parameter (ε) tends to zero. Over the past decades, many contributions have been made in this area. In [9], eigenvalue assignment approach is studied. In [10] algebraic riccati equation approach is discussed. In [11] parametric Lyapunov equation based design is studied. We propose low and high gain feedback law [12] for an inverted pendulum. It improves tracking response of the cart, also it provides input within saturation limits.

The Inverted pendulum has two outputs one gives cart position and other is for an angle of the pendulum, as much we are interested in balancing the pendulum in an inverted position we also required the cart to track the reference position in a quick and smooth way. In this paper, the inverted pendulum with improvement in performance is proposed by using Composite Nonlinear Feedback control technique. CNF control law can be seen as a modified type of low and high gain feedback law with nominal linear feedback as low gain feedback and nonlinear feedback law as scheduled high gain feedback law by tuning the high gain parameter as nonlinear function. The CNF control law was first introduced in [13] since then a lot of work has been done towards developing the alternate design methods. Over the years, the study shows that CNF is an effective technique for asymptotical tracking and minimum overshoot. In [14], the results of [13] extended for higher order systems for SISO system. CNF for MIMO system was discussed in [15] and successfully applied to HDD servo system for tracking reference signal. In [16], CNF was developed for a full order and reduced order system. CNF for DC motor speed control described in [17]. CNF consists of two

978-1-5386-6624-1/18 $31.00 © 2018 IEEE

parts linear and nonlinear feedback control, construction of the controller is sequential and without any switching element, the nonlinear feedback controller is designed based on the result from linear feedback control. Linear feedback control results quick response and low damping ratio followed by nonlinear feedback law which gives fast settling time and high damping ratio to overcome overshoot resulting from the linear part. Control design is verified in simulation results which shows the improvement in the response indices of inverted pendulum system with CNF and is compare with [5] and [12].

This paper is divided into five sections next one is about problem formation for inverted pendulum system. In section 3 Low and High gain feedback theory is being discussed. CNF theory is discussed in section 4. In section 5 results from the designed controllers for the system are compared and plotted followed by conclusive remarks.

II. PROBLEM STATEMENT

A linear system with input saturation is designed by the controller. Consider the system as:

$$
\begin{aligned}
\dot{x} &= Ax + Bsat(u) \\
x(0) &= x_0 \\
y &= Cx
\end{aligned} \tag{1}
$$

where state, control input and output of plant are represented as $x \epsilon R^n$, $u \epsilon R^p$, $y \epsilon R$ respectively. $A, B,$ and C are appropriate dimension constant matrices, and actuator saturation represents as $sat : R \to R$,

$$
sat(u) = sgn(u)min(u_{max}, |u|) \tag{2}
$$

with u_{max} being the maximum saturation level of the input. The aim is to design a CNF control, to track a given set point r quickly with minimum overshoot. The following assumptions are taken into consideration while solving the closed-loop system:
1) (A, B) is stabilizable;
2) (A, B, C) is invertible and has no zeros at s=0.
An inverted pendulum is a classic control system which is nonlinear and unstable. During cart tracking the give set point, pendulum mounted on the cart moves from its position. The idea here is to balance the system vertically upright while tracking down the position. With the position as command input, we are interested in the angle of pendulum and trajectory as the control output. While implementing the control law we consider actuator saturation as it is inevitable in the feedback control system, if ignored, can cause instability and result in degraded performance. To overcome this problem we design CNF control and Low-High gain feedback law to take care of input saturation.

III. DESIGN OF THE LOW AND HIGH GAIN FEEDBACK LAW

The main idea of LGF is peak magnitude of control signal approaches zero as ε approaches zero for any given initial condition, if the linear system is asymptotically null controllable by bounded controls (ANCBC).The Low and high

gain feedback law design technique is based on [12]. LGF is calculated based on ARE approach and used to obtain HGF. The design of Low-High gain feedback is sequential i.e we first calculate LGF following HGF which is based on design of LGF for closed loop system. Adding LGF and HGF gives Low-High gain feedback law. Consider a linear system.

$$
\dot{x} = Ax + B(u) \tag{3}
$$

Step 1.1: Design of low gain feedback law,

$$
u_l = K_l(\varepsilon)x \tag{4}
$$

Now, let $T(\varepsilon) > 0$ such that,

$$
(A + BK_l(\varepsilon))^T T + T(A + BK_l(\varepsilon)) \leq -Q(\varepsilon) \tag{5}
$$

for $Q(\varepsilon) > 0$
we design LGF based on ARE approach, then $T(\varepsilon)$ is the solution of ARE .
Step 1.2: Design of high gain feedback law,

$$
u_h = (-\rho)B^T T(\varepsilon)x \tag{6}
$$

where ρ is any non-negative parameter termed as a high gain parameter.
Step 1.3:Design of Low-High gain feedback,
Addition of LGF and HGF gives the family Low and High gain feedback law

$$
u_f = u_l + u_h \tag{7}
$$

this design method is termed as additive low and high gain design [18].

$$
u_f = -(1 + \rho)B^T T\tilde{x} \tag{8}
$$

where, $\tilde{x} = x - x_e$ is co-ordinate transformation.

$$
u_L = -B^T T(x - x_e) \tag{9}
$$

is low gain feedback law.

$$
u_H = (-\rho)B^T T(x - x_e) \tag{10}
$$

is high gain feedback law.

IV. DESIGN OF THE CNF CONTROL LAW

The CNF control technique is designed based on [13]. The design procedure is done in 3 steps.
Step 2.1: Design of linear feedback law,

$$
u_L = Kx + Gr \tag{11}
$$

where r is a command input and K is the feedback gain, calculated such that $A + BK$ has stable eigenvalues, whereas $C(sI - A - BK)^{-1}B$ has low damping ratio which results in a quick response of the system.
K can be obtained by pole placement and LQR approach, and G is a scalar quantity given as,

$$
G = -[C(A + BK)^{-1}B]^{-1}. \tag{12}
$$

978-1-5386-6624-1/18 $31.00 © 2018 IEEE

Note since (A, B, C) is invertible and at $s = 0$ has no invariant zeros. G is well defined because $A + BK$ is stable.
Let,

$$
\begin{aligned}
H &:= [I - K(A + BK)^{-1}B]G \\
and\ x_e &:= G_e r \\
&:= -(A + BK)^{-1}BGr
\end{aligned}
\tag{13}
$$

Also let, $P > 0$ be the solution of the Lyapunov equation with $Q \epsilon R^{(n*n)}$ be the positive definite matrix of the equation.

$$
(A + BK)^T P + P(A + BK) = -Q \tag{14}
$$

such a P exists since $(A + BK)$ has stable eigenvalue values.
Step 2.2: Design a Nonlinear feedback law

$$
u_n = \rho(r, y)B^T P(x - x_e) \tag{15}
$$

where $\rho(r, y)$ is any non-positive function locally Lipschitz in y, it is used to increase the damping ratio and to reduce the overshoot in the control design of the closed-loop system as the output approaches the command input.
Step 2.3: Finally, combining the above feedback laws to form CNF controller.

$$
u = u_L + u_N = Kx + Gr + \rho(r, y)B^T P(x - x_e) \tag{16}
$$

Next, we find the magnitude of r which can be tracked by control output without exceeding control input limits [19], which shows that the closed-loop system with (1) and (16) is asymptotically stable.
Consider $\tilde{x} = x - x_e$ as co-ordinate transformation. Using equation of H the linear control law can be defined as:

$$
\begin{aligned}
u_L(t) &= K\tilde{x}(t) + [1 - K(A + BK)^{-1}B]Gr \\
&= K\tilde{x}(t) + Hr
\end{aligned}
\tag{17}
$$

where $|Kx| \leq u_{max}(1 - \delta) \quad \forall\ x \epsilon X_\delta := x : x^T Px \leq c_\delta$ such that $\delta \epsilon(0, 1)$ and $c_\delta > 0$ be the largest positive scalar satisfying above situation.

$$
x_0 := (x_0 - x_e)\epsilon X_\delta \qquad |Hr| \leq \delta u_{max} \tag{18}
$$

Hence, the closed-loop system is linear for all $\tilde{x} \epsilon X_\delta$ and, given $|Hr| \leq \delta u_{max}$, $|Kx + Hr| \leq u_{max}$ and is termed as,

$$
\dot{\tilde{x}} = (A + BK)\tilde{x} + Ax_e + BHr \tag{19}
$$

Further solving, we get.

$$
Ax_e + BHr = 0 \tag{20}
$$

Solving the closed loop system equation (19).

$$
\dot{\tilde{x}} = (A + BK)\tilde{x} \tag{21}
$$

Similarly, the closed-loop dynamics comprising with (1) and (16) given as:

$$
\dot{\tilde{x}} = (A + BK)\tilde{x} + B\omega \tag{22}
$$

where

$$
\omega = sat(K\tilde{x} + Hr + u_N) - K\tilde{x} - Hr. \tag{23}
$$

Note if $\rho = 0$, (16) is reduced to (11). Next to prove the stability of closed-loop system (22) inspired by the stability proof in [15], we define a Lyapunov function $V = \tilde{x}^T P\tilde{x}$ and examine the derivative of V.

$$
\begin{aligned}
\tilde{V} &= \dot{\tilde{x}}P\tilde{x} + \tilde{x}^T P\dot{\tilde{x}} \\
&= \tilde{x}^T(A + BK)^T P\tilde{x} + \tilde{x}^T P(A + BK)\tilde{x} + 2\tilde{x}^T PB \\
&= \tilde{x}^T W\tilde{x} + 2\tilde{x}^T PB\omega
\end{aligned}
\tag{24}
$$

note that for all,

$$
\tilde{x}\epsilon X_\delta = \tilde{x} : \tilde{x}^T P\tilde{x} \leq c_\delta \Rightarrow |F\tilde{x}| \leq u_{max}(1 - \delta) \tag{25}
$$

Next, there are three cases for $\dot{V}$, based on different values of saturation function for stability analysis given as:
Case 1) If $|K\tilde{x} + Hr + u_N| \leq u_{max}$, then $\omega = u_N = \rho B^T P\tilde{x}$ and thus

$$
\dot{V} = \tilde{x}^T W\tilde{x} + 2\rho\tilde{x}^T PBB^T P \tag{26}
$$

Case 2) If $K\tilde{x} + Hr + u_N \leq u_{max}$, and by construction $|K\tilde{x} + Hr| \leq u_{max}$, we have

$$
0 < \omega = u_{max} - K\tilde{x} - Hr \leq \rho B^T P\tilde{x} \tag{27}
$$

which gives $\tilde{x}^T PB < 0$ and hence $\dot{V} = -\tilde{x}^T W\tilde{x} + 2\tilde{x}^T PB\omega \leq -\tilde{x}W\tilde{x}$
Case 3) Finally, if $K\tilde{x} + Hr + u_N < -u_{max}$, we have

$$
\rho B^T P\tilde{x} = u_N < \omega = -u_{max} - K\tilde{x} - Hr < 0 \tag{28}
$$

So, from above cases it is clearly seen that

$$
\dot{V} \leq -\tilde{x}^T W\tilde{x} \qquad \tilde{x} \epsilon X_\delta \tag{29}
$$

which implies that X_δ set of closed-loop system (22) does not vary along the trajectories which start from inside X_δ will approach origin as $Q > 0$. This indicates that for all initial states x_0 and step command input of amplitude r that satisfy (18).

$$
\lim_{x \to \infty} x(t) = x_e \tag{30}
$$

therefore

$$
lim_{x \to \infty} y(t) = Cx_e = -C(A + BK)^{-1}BGr = r \tag{31}
$$

From the above proof, it can be seen that by adding nonlinear control law to the linear control law ability to track the reference input does not get affected. To the contrast, this additional nonlinear control u_N can be used to acquire the improvement in system's performance by reducing the overshoot and giving high damping ratio. This is the key property of the CNF control technique, also for the case $x_0 = 0$, command output can asymptotically track reference of amplitude r provided,

$$
|r| \leq [c_\delta(G_e^T PG_e)^{-1}]^{1/2} \qquad |Hr| \leq \delta u_{max} \tag{32}
$$

978-1-5386-6624-1/18 $31.00 © 2018 IEEE

A. Tuning of nonlinear gain $\rho(r, y)$

The function $\rho(r, y)$ is chosen such that the control law are tuned in a way to improve the closed-loop system performance as the tracking error approaches zero. $\rho(r, y)$ can be written as,

$$\rho(r, y) = -1.58198(\alpha)(\exp^{-(\beta)|1-(y/r)|} - \exp^{-1}) \quad (33)$$

where $\alpha \geq 0$, $\beta \geq 0$ and $y_o = y(0)$ are tuning parameters and choice of $\rho(r, y)$ is non-unique. Selection of $\rho(r, y)$ depends on the state of output y, if it's far away from a set point then $|1 - (y/r)|$ approaches 1 which makes $\rho(r, y)$ small and limited effects of nonlinear part on closed loop system. On the hand when $|1 - (y/r)|$ approaches 0, $\rho(r, y) \simeq -\alpha$ also nonlinear part becomes effective providing high damping ratio and fast settling time.

V. APPLICATION

Here in this section, we consider a numerical example of Inverted pendulum and simulations are performed by applying the theory of CNF control, such that while tracking the horizontal position of cart Inverted pendulum does not deviate from its vertical position. An Inverted pendulum model based on [20]. With the state-space model, represented as,

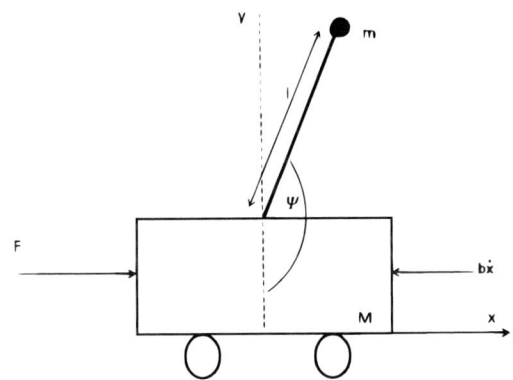

Fig. 1. Inverted Pendulum

$$\begin{aligned} x(t) &= Ax(t) + Bu(t) \\ y(t) &= Cx(t) + Du(t) \end{aligned} \quad (34)$$

where

$$A = \begin{bmatrix} 0 & 1 & 0 & 0 \\ 0 & -b_1(I + ml^2)/q & m^2l^2g/q & -mlb_2/q \\ 0 & 0 & 0 & 1 \\ 0 & -mlb_1/q & mlg(M_1)/q & -b2(M_1)/q \end{bmatrix}$$

$$B = \begin{bmatrix} 0 \\ (I + ml^2)/q \\ 0 \\ ml/q \end{bmatrix}$$

$$C = \begin{bmatrix} 1 & 0 & 0 & 0 \\ 0 & 0 & 1 & 0 \end{bmatrix}$$

Parameter	Constant
Cart mass(M)	0.5kg
Pendulum mass(m)	0.2kg
Pendulum length(l)	0.3m
Coefficient of friction of cart(b_1)	$0.2Nm^{-1}s^{-1}$
Coefficient of friction of pendulum(b_2)	$0.02Nrad^{-1}s^{-1}$
Mass moment of inertia of pendulum(I)	$0.006kg/m^2$
Gravitation force(g)	$9.81m/s^2$
Cart position	x(m)
Angle of pendulum	φ(rad)

TABLE I
PARAMETER VALUES FOR INVERTED PENDULUM

$$D = \begin{bmatrix} 0 \\ 0 \end{bmatrix} \quad (35)$$

Now, for the system (33) with $u_{max} = 1$ and initial condition as $x_0 = 0$, where reference set point is $r = 0.7m$. Firstly, Low and high gain feedback law is developed based on ARE based low gain feedback design with design parameters as $\varepsilon = 0.5$ which gives $K_l = [0.7071\ 0.5686\ 0.4456\ 0.0533]$, and high gain parameter as $\rho = 0.5$. Secondly, we design CNF controller with linear feedback control based on LQR technique with deign parameters as

$$Q1 = \begin{bmatrix} 1.5 & 0 & 0 & 0 \\ 0 & 0 & 0 & 0 \\ 0 & 0 & 75 & 0 \\ 0 & 0 & 0 & 0 \end{bmatrix} \quad (36)$$

and $R1 = 1$ which gives $K = [1.2247\ 0.7800\ 6.1903\ 0.3075]$ and $G = -1.2247$, and for the nonlinear feedback law, where P is obtained by solving Lyapunov equation taking $Q = 1.25 * I_4$ and nonlinear function ρ chosen as $-1.587 * (\alpha)(exp^{-(\beta)|1-(y/r)|} - 0.3678)$ where tuning factors are $\alpha = 0.15$ and $\beta = 0.6$. For comparing the result with [5], the design parameters for LQR as

$$Q = \begin{bmatrix} 2 & 0 & 0 & 0 \\ 0 & 0 & 0 & 0 \\ 0 & 0 & 100 & 0 \\ 0 & 0 & 0 & 0 \end{bmatrix} \quad (37)$$

and R=1, and for PID equation

$$K_p e(t) + K_i \int e(t) + K_d \frac{de(t)}{dt} \quad (38)$$

as $K_p = -0.53$, $K_i = -1.86$ and $K_d = -0.11$ by trial and error method.

	Settling time	Rising time
CNF	2.367sec	808.044msec
Low and high gain	3.454sec	1.094sec
PID+LQR	3.799sec	1.207sec

TABLE II
CART TRACKING

Step response of cart with different controllers is plotted in fig.3 and corresponding disturbance in pendulum is plotted in fig. 3. Also, performance indices are tabulated, from Table 2 it

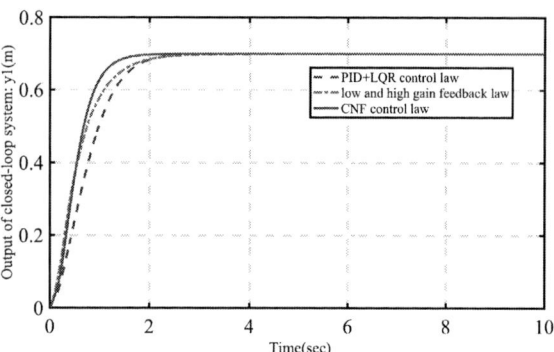

Fig. 2. Cart tracking

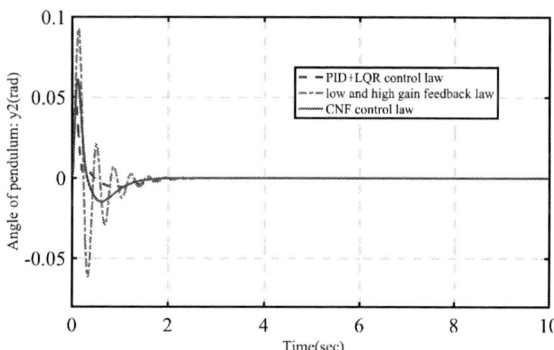

Fig. 3. Angle of pendulum

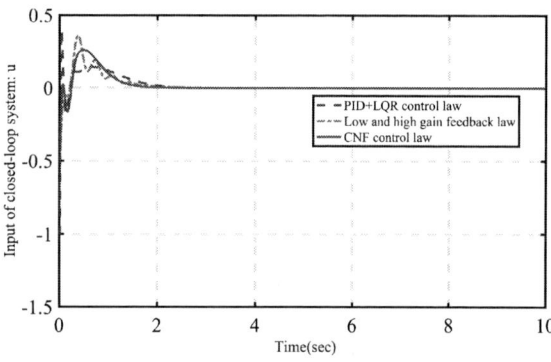

Fig. 4. Input of system

	Overshoot	Setting time
CNF	0.625%	2.25sec
Low and high gain	16.716%	2.61sec
PID+LQR	42.063%	2.79sec

TABLE III
PENDULUM RESPONSES

can be clearly seen that the settling time (i.e the time that takes the control output to enter the target region ±1% of y) for cart tracking is faster in CNF control than PID+LQR control and Low and high gain feedback control. Also, from Table 3 we can see the pendulum stabilizes faster in CNF control with

minimum oscillations.

VI. CONCLUSION

In this paper, we designed the CNF control technique for an inverted pendulum system. CNF achieves better performance to justify, the comparison analysis has been performed between the CNF control law, PID+LQR control law and Low and high gain feedback law for the system. The linear gain for CNF control law is calculated by LQR method and tuning of PID is done by trial and error method. Simulation results validate the comparative advantage of CNF method. Cart tracks the given trajectory with quick rising time and less settling time and the pendulum stabilizes in an upright position with minimum oscillations.

REFERENCES

[1] Y. Ha and S. Yuta. Trajectory tracking control for navigation of self-contained mobile inverse pendulum. In *Intelligent Robots and Systems '94. 'Advanced Robotic Systems and the Real World', IROS '94. Proceedings of the IEEE/RSJ/GI International Conference on*, volume 3, pages 1875–1882 vol.3, Sep 1994.

[2] Andrej Rybovic, Martin Priecinsky, and Marek Paskala. Control of the inverted pendulum using state feedback control. In *ELEKTRO, 2012*, pages 145–148. IEEE, 2012.

[3] Jia-Jun Wang. Simulation studies of inverted pendulum based on pid controllers. *Simulation Modelling Practice and Theory*, 19(1):440 – 449, 2011. Modeling and Performance Analysis of Networking and Collaborative Systems.

[4] Wende Li, Hui Ding, and Kai Cheng. An investigation on the design and performance assessment of double-pid and lqr controllers for the inverted pendulum. In *Control (CONTROL), 2012 UKACC International Conference on*, pages 190–196. IEEE, 2012.

[5] Lal Bahadur Prasad, Barjeev Tyagi, and Hari Om Gupta. Optimal control of nonlinear inverted pendulum dynamical system with disturbance input using pid controller & lqr. In *Control System, Computing and Engineering (ICCSCE), 2011 IEEE International Conference on*, pages 540–545. IEEE, 2011.

[6] Zongli Lin and Ali Saberi. A low-and-high gain approach to semi-global stabilization and/or semi-global practical stabilization of a class of linear systems subject to input saturation via linear state and output feedback. In *Decision and Control, 1993., Proceedings of the 32nd IEEE Conference on*, pages 1820–1821. IEEE, 1993.

[7] Ali Saberi, Zongli Lin, and Andrew R Teel. Control of linear systems with saturating actuators. *IEEE Transactions on Automatic Control*, 41(3):368–378, 1996.

[8] Zongli Lin and Ali Saberi. Semi-global exponential stabilization of linear systems subject to input saturation via linear feedbacks. *Systems & Control Letters*, 21(3):225–239, 1993.

[9] Zongli Lin and Ali Saberi. Low-and-high gain design technique for linear systems subject to input saturation— a direct eigenstructure assignment approach. *International Journal of Robust and Nonlinear Control*, 5:381–398, 1994.

[10] Zongli Lin, Anton A Stoorvogel, and Ali Saberi. Output regulation for linear systems subject to input saturation. *Automatica*, 32(1):29–47, 1996.

[11] Bin Zhou, Guangren Duan, and Zongli Lin. A parametric lyapunov equation approach to the design of low gain feedback. *IEEE Transactions on Automatic Control*, 53(6):1548–1554, 2008.

[12] Zongli Lin. An overview of the development of low gain feedback and low-and-high gain feedback. *Journal of Systems Science and Complexity*, 22(4):697, 2009.

[13] Zongli Lin, Meir Pachter, and Siva Banda. Toward improvement of tracking performance nonlinear feedback for linear systems. *International Journal of Control*, 70(1):1–11, 1998.

[14] M. C. Turner, I. Postlethwaite, and D. J. Walker. Non-linear tracking control for multivariable constrained input linear systems. *International Journal of Control*, 73(12):1160–1172, 2000.

[15] B. M. Chen, T. H. Lee, Kemao Peng, and V. Venkataramanan. Composite nonlinear feedback control for linear systems with input saturation: theory and an application. *IEEE Transactions on Automatic Control*, 48(3):427–439, Mar 2003.

[16] T.H. Lee B.M. Chen and V. Venkataramanan. *Hard Disk Drive Servo Systems*. Springer, 2002.

[17] W. Lan and Q. Zhou. Speed control of dc motor using composite nonlinear feedback control. In *2009 IEEE International Conference on Control and Automation*, pages 2160–2164, Dec 2009.

[18] Zongli Lin. Low gain feedback: Properties, design methods and applications. In *Control Conference (CCC), 2013 32nd Chinese*, pages 26–32. IEEE, 2013.

[19] Yingjie He, Ben M. Chen, and Chao Wu. Composite nonlinear control with state and measurement feedback for general multivariable systems with input saturation. *Systems & Control Letters*, 54(5):455 – 469, 2005.

[20] P. Strako and J. T?ma. Mathematical modelling and controller design of inverted pendulum. In *2017 18th International Carpathian Control Conference (ICCC)*, pages 388–393, May 2017.

2nd IEEE International Conference on Power Electronics, Intelligent Control and Energy Systems (ICPEICES-2018)

Integrating Wavelet Coefficients and CNN for Recognizing Handwritten Characters

Madhuri Yadav
USIC&T
GGSIP, University
New Delhi, India
madhuri.usict.900100@ipu.ac.in

Ravindra Kr. Purwar
USIC&T
GGSIP, University
New Delhi, India
ravindra@ipu.ac.in

Abstract—Convolutional Neural Network (CNN) based image recognition has shown significant progress in recent years. It has achieved state-of-art results in field of pattern recognition. The proposed work uses convolutional network for Hindi handwritten character recognition. In traditional CNN, the raw images are fed as input to the network, which along with relevant information also contain redundant data. This redundant data makes feature extraction complex and increases the training time of the network. To improve the feature learning capacity of CNN, this work integrates wavelets and convolutional network for recognizing Hindi characters. The wavelet coefficients in three directions i.e. horizontal ,vertical and diagonal are extracted from raw images and fed as input to the network. These coefficients are trained independently on different networks and the extracted features are merged at dense layer. The proposed framework outperforms the traditional CNN as demonstrated by experiments. It achieves significant results in much lesser time as compared to traditional CNN.

Keywords—handwritten characters, CNN, wavelet coefficients, Hindi character recognition, hindi database

I. INTRODUCTION

Handwritten character recognition involves recognition of scanned character images. It is an interesting and popular area of research because of its wide range of applications in post offices, banks, mail sorting, address recognition and various others. The major challenges in offline handwritten character recognition are similar shape characters and different writing styles. Similar shape characters make inter-class classification task very complex and different writing styles reduces the cohesiveness between intra class characters. The literature works of different languages have tried to overcome the above challenges by using variety of feature extraction and classification techniques. Traditionally, the literature works of character recognition followed a basic procedure of pre-processing, feature extraction and classification. Most of the works of hindi character recognition are based on traditional methods of recognition. Hanmandlu et.al.[11] used reinforcement learning on fuzzy sets. S.Behle [12] proposed online hindi word recognition system by segmenting words into vowels, matras, syllables etc using Hmm models and giving recognized word probabilities by symbol trees. In [13], shape and texture features are extracted from isolated hindi characters using gradient masks and LBG vector quantization, respectively. Structural and statistical features such as end points, intersection points, branch points, and

quadratic polynomial coefficients are used as features in [14]. The proposed work in [15] used curvelet transform for feature extraction and k-nearest neighbor (k-nn) as classifier. One of the most recent work [16], used Hu-geometric moments and histogram of oriented gradients for hindi handwritten characters. The authors exploit the geometric invariant property of moments and used image gradients for spatial correlation. The performance of these features was evaluated on SVM and MLP. Shitala [17] proposed multi-lingual character recognition using wavelet, curvelet and ridgelet multi-resolution transforms. SVM and k-nn classifiers were used for classification purposes. The work in [18] used projection profile histogram features and compared performance of different classifiers such as MLP, SVM, Bagged trees for Hindi isolated characters.

Thus, the widely used features for hindi characters are curvelet transforms, statistical features such as end points, branch points, distance vectors, histogram of oriented gradients, hu-moments, complex moments, quadratic coefficients, and regular expressions. The classifiers used were support vector machines(SVM), k-nearest neighbor, multi-layer perceptron(MLP), multi-quadractic discriminant function (MQDF). Although these works are of great importance and have significant impact on hindi character recognition, but end-to-end learning methods have shown significant progress as compared to traditional methods.

End-to-end learning methods are based on deep learning concept. They automate the procedure of feature extraction and reduces the overhead of manual feature extraction. In recent years, Convolutional neural networks (CNN) which are based on deep learning have provided solution to many complex pattern recognition problems. It has shown its potential benefits in scene recognition, character recognition [20], image recognition [19], object recognition. In [21], a multi-column multi-scale convolutional neural network is proposed for various indic scripts. It achieves an accuracy of 95.18% on devnagari basic characters. Adarsh et.al. [22] proposed hybrid convolutional network by using genetic algorithm for weight optimization. In [23], the directional feature maps are integrated with 9-layer CNN for in-air character recognition. It is evident from the success of convolutional networks that they have outperformed traditional techniques. The deep learning systems have numerous benefits, but they come along with few disadvantages. In recent years, CNN architectures have increased recognition accuracies but they incur high computational cost. Even the most famous architectures such as Alexnet [24], Zfnet [25], VGG net [26], GoogleNet [27],

978-1-5386-6624-1/18 $31.00 © 2018 IEEE

and Microsoft ResNet [28] take 5 to 6 days, 12 days, 2 weeks, 2 to 3 weeks to train, respectively. The proposed work try to overcome the problem of higher training time and achieve results comparable to state-of-arts. This work uses an integration of wavelet coefficients and convolutional network for achieving significant recognition rates.

II. PROPOSED WORK

A. Database

The proposed work used an offline database containing 4428 characters collected using 108 persons. The procedure for data acquisition is explained in [16].

B. Feature Extraction and Classification

B.1 Wavelet coefficients

The wavelet transform is a tool that segments data into different frequency components and then studies each component with multi-resolution on different scales [7]. It is a mathematical function which divides data into small scale components. It has been used for feature extraction in heart rate monitoring [8], text image localization [1], bank note classification [2], sign language recognition [3], face recognition [4],character recognition [5], [6] and many more. Continuous wavelet transform are used for signals whereas discrete wavelet transform(DWT) are used for images. Wavelet functions are spatially localized, dilated, translated and scaled versions of mother wavelets such as haar and daubechies. The DWT of a signal or image (x) is calculated by using it through a set of filters. The image is passed through low pass and high-pass filters resulting in convolution of images and filter. It is important to note that the approximation coefficients are given by low pass filters and output of high-pass filters give detail coefficients i.e. horizontal, vertical and diagonal coefficients. These filters reomve half of the frequencies and thus, images are downsampled by 2, firstly row-wise and then column-wise. Fig. 1 illustrates the concept of using low-pass and high-pass filters on a signal.

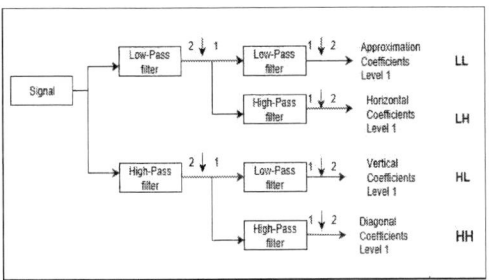

Fig.1. LL,LH,HL and HH components of image after wavelet decomposition.

This figure represents LL, LH, HL and HH components. Here, L stands for low-pass filtering, and H stands for high-pass filtering. The LL band corresponds to down-sampled version of the original image. The LH band tends to preserve localized horizontal features, while the HL band tends to preserve localized vertical features in the original image. Finally, the HH band tends to isolate localized high-frequency point features or diagonal features in the image.

LL component can be further decomposed into higher decomposition levels and obtain higher level detail coefficients. Fig. 2 represent the vertical, horizontal, and diagonal coefficients of original character image.

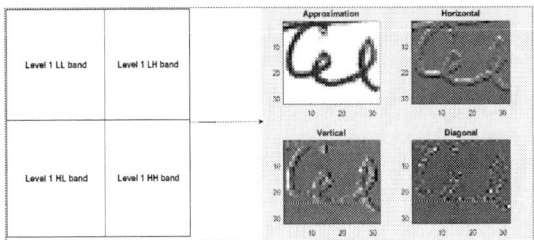

Fig.2. Level 1 detail coefficients of original character image.

B.2 Convolutional Neural Network

Convolutional networks [9] are based on deep learning architecture. They are specialized type of neural networks for processing data with grid like topology e.g. image can be considered as 2-D grid. As the name suggests, a mathematical operation called "convolution" is applied in this neural network. Fig. 3 demonstrates how convolution is applied on an image. The figure shows the convolution operation applied only on 3x3 part of the image. The kernel is slided by more or more pixels over the image and the output is computed. The commonly known layers of this network are convolutional layer, activation layer, max pooling layer and dense layer. This article only discusses the layers used in this work, detailed information about CNN can be found in [10].

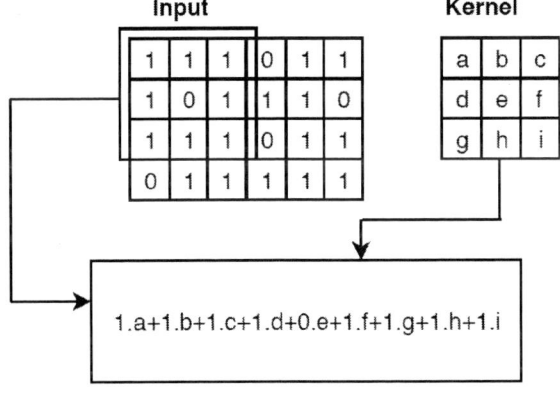

Fig.3. Convolution operation on segment of input image using 3x3 kernel.

The input to the convolutional layer are the raw images on which filter or kernel is convolved to produce feature maps. These feature maps acts as input to next level layers. The convolutional network leverage three basic concepts: sparse interactions, parameter sharing and translation invariance. Sparse interactions or weights allow to extract small meaningful features such as edges, points. It is accomplished by the help of small-sized kernels. Thus, only the relevant features are fed as input to next layer. Parameter sharing allows for local connectivity. In convolutional networks, each member of the kernel is used at every location of the image, thus the connected pixels learn from same set of parameters. This type of learning helps in understanding of complex features in deeper layers. Parameter sharing also

978-1-5386-6624-1/18 $31.00 © 2018 IEEE   1161

makes the network equi-variance to translation. It means if the input changes, the output also changes in the same way. Pooling layer replaces the intensities of the image at the certain location with some neighbouring statistics operation applied at the same image. The max statistic operation give rise to max pooling layer, other operations which can be applied are min, average etc. Pooling layer makes the network invariant to small translations. This property is important when we want to detect a feature irrelevant to its position in the image. The last layer of the network is dense layer or fully connected layer. A network can have two or more dense layers. The fully connected layer looks at the activation maps generated by previous layers and helps in determining character classes. At the end, the fully connected layer contain neurons equivalent to the number of classes in the database.

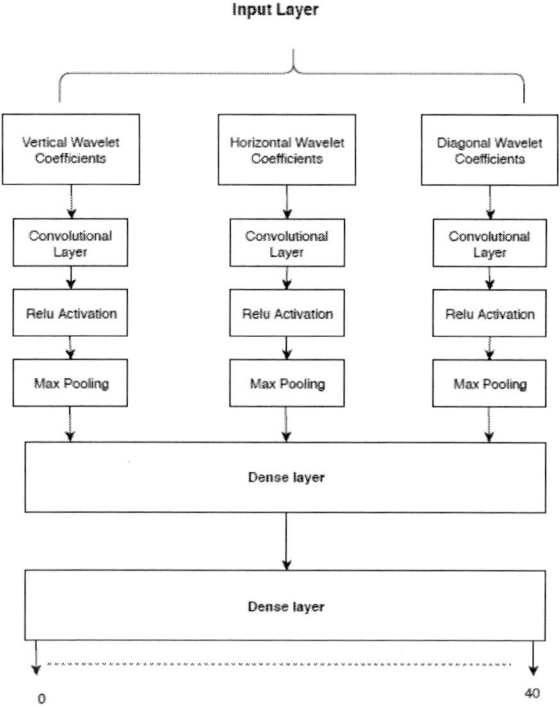

Fig.4. Proposed CNN architecture for Hindi character recognition.

B.3 Combination of Wavelet Coefficients and CNN

The proposed work uses combination of wavelet coefficients and convolutional network for feature extraction. The proposed framework is shown in Fig. 4. This figure shows three different inputs which are fed as input to three different convolutional layers. The first input is vertical wavelet coefficients, second is horizontal wavelet coefficients and the third is diagonal coefficients. The vertical wavelet coefficients contain the information relevant to vertical components of an image. The convolutional layer adjusts the weight of its filters according to vertical features of an image. The convolutional layer corresponding to horizontal and diagonal coefficients work on horizontal and diagonal components of an image, respectively. The basic advantage of using these coefficients as input is that the network receives only the relevant information as features. Thus, the filters of the convolutional layers can be more precise and focused towards extracting

stable features. In contrary, basic CNN receives raw images as input which along with relevant information also contain redundant data or pixels. Due to presence of these extra pixels, filters take much more time to extract relevant features which could uniquely identify a character. Thus, using wavelet coefficients with CNN reduces training time of the network. They also eliminate the need of deeper networks as the first layer itself has different type of filters working on three different types of images. To conclude, the combination of wavelet coefficients and CNN achieves significant recognition results as illustrated in Section III. They also reduce training time of the network, which results in reducing computational cost.

III. EXPERIMENTAL RESULTS

The experiments of the proposed work are performed with Intel dual core i5 processors 7th gen, 8 GB RAM, and a NVIDIA GeForce 750 Ti graphics card with 1TB internal memory, having 640 CUDA cores. Python has been used as programming environment. The main objectives of this work are as follows:

- To propose a deep learning based convolutional network that improves the recognition rate for isolated handwritten Hindi characters, and
- To experimentally demonstrate that the wavelet coefficients in combination with convolutional network improve the recognition results as compared to basic CNN.

Fig.5 illustrates the dimensions of the different layers used in the network. Each convolution layer has 64 filters with "Relu" as activation function. The activation maps obtained as output form convolutional layer are fed as input to max pooling layer. As shown in Fig.4 the output from each max pooling layer is fed as input to dense layer which contains 256 neurons. The output of this layer is finally fed to fully connected layer which contains 41 neurons which are equivalent to number of character classes. The learning parameters are only found in convolutional layer and dense layers of the network as shown in Fig.5. The total learning parameters are 6,975,913.

Layer (type)	Output Shape	Param #	Connected to
input_1 (InputLayer)	(None, 32, 32, 1)	0	
input_2 (InputLayer)	(None, 32, 32, 1)	0	
input_3 (InputLayer)	(None, 32, 32, 1)	0	
conv2d_1 (Conv2D)	(None, 30, 30, 64)	640	input_1[0][0]
conv2d_2 (Conv2D)	(None, 30, 30, 64)	640	input_2[0][0]
conv2d_3 (Conv2D)	(None, 30, 30, 64)	640	input_3[0][0]
max_pooling2d_1 (MaxPooling2D)	(None, 15, 15, 64)	0	conv2d_1[0][0]
max_pooling2d_2 (MaxPooling2D)	(None, 10, 10, 64)	0	conv2d_2[0][0]
max_pooling2d_3 (MaxPooling2D)	(None, 10, 10, 64)	0	conv2d_3[0][0]
flatten_1 (Flatten)	(None, 14400)	0	max_pooling2d_1[0][0]
flatten_2 (Flatten)	(None, 6400)	0	max_pooling2d_2[0][0]
flatten_3 (Flatten)	(None, 6400)	0	max_pooling2d_3[0][0]
merge_1 (Merge)	(None, 27200)	0	flatten_1[0][0] flatten_2[0][0] flatten_3[0][0]
dense_1 (Dense)	(None, 256)	6963456	merge_1[0][0]
dense_2 (Dense)	(None, 41)	10537	dense_1[0][0]

Total params: 6,975,913
Trainable params: 6,975,913
Non-trainable params: 0

Fig.5. Dimensions and framework for CNN architecture.

Fig.6 represents the recognition accuracy achieved for the proposed method and Fig.7 represents the recognition accuracy for the basic CNN with raw images as input.

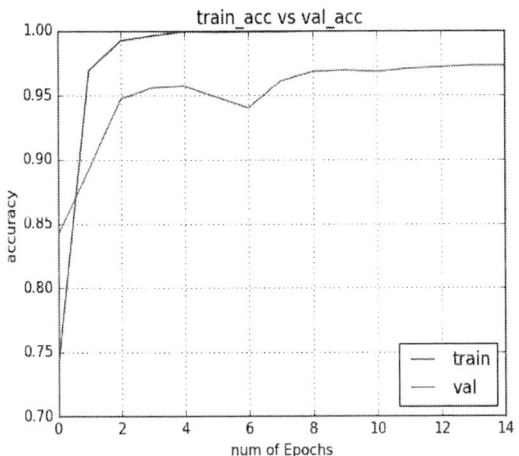

Fig. 6. Recognition accuracy achieved by proposed method (wavelet+CNN)

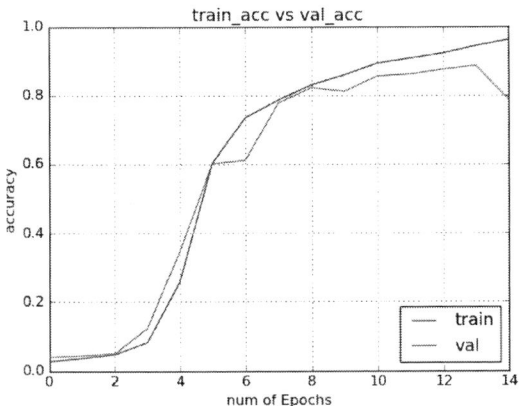

Fig.7. Recognition accuracy achieved by basic CNN.

Fig.8 shows the comparison between the proposed approach and the basic CNN architecture. It shows that the training time of the proposed method is much less than the training time of basic CNN. It also highlights that proposed CNN architecture reaches stable state much faster than basic CNN and achieves high recognition rate as compared to basic CNN. Table I summarizes the recognition rates achieved by most recent literature works and the proposed method.

TABLE I.   SUMMARY OF RECOGNITION RATES ACHIEVED BY LITERATURE WORKS

Methods	Recognition Rates(in %)
Deepti et. al.[21]	93.4
Ritesh et. al. [14]	95.18
Proposed Method	96.58

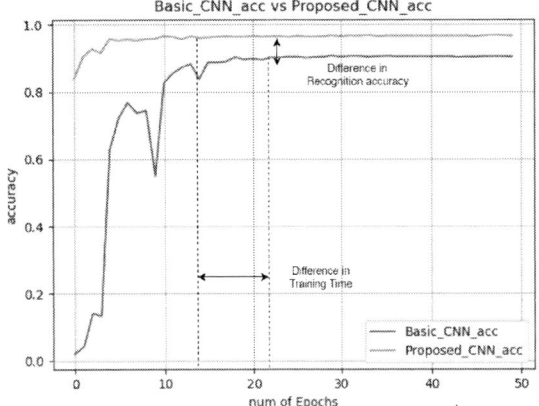

Fig.8. Comparison of recognition accuracies achieved by basic CNN and proposed CNN architectures.

REFERENCES

[1] Suhad A. Ali, Ashwaq T. Hashim, "Wavelet transform based technique for text image localization", Karbala International Journal of Modern Science, vol. 2, Issue 2, pp. 138-144, 2016.

[2] Shan Gai, Guowei Yang, Minghua Wan, "Employing quaternion wavelet transform for banknote classification", Neurocomputing, vol. 118, pp. 171-178, 2013.

[3] Ala addin I. Sidig, Hamzah Luqman, Sabri A. Mahmoud, "Transform based Arabic sign language recognition", Procedia Computer Science, vol. 117, pp. 2-9, 2017.

[4] Chandan Singh, Ali Mohammed Sahan, "Face recognition using complex wavelet moments", Optics & Laser Technology, vol. 47, pp. 256-267,2013.

[5] Parshuram M. Kamble, Ravinda S. Hegadi, "Handwritten Marathi Character Recognition Using R-HOG Feature", Procedia Computer Science, vol. 45, pp. 266-274, 2015.

[6] P. Malik and R. Dixit, "Handwritten Character Recognition Using Wavelet Transform and Hopfield Network," 2013 International Conference on Machine Intelligence and Research Advancement, Katra, 2013, pp. 125-129.

[7] Daubechies, I. (1992), Ten lectures on wavelets, CBMS-NSF conference series in applied mathematics. SIAM Ed.

[8] Ashish Kumar, Rama Komaragiri, Manjeet Kumar, "Heart rate monitoring and therapeutic devices: A wavelet transform based approach for the modeling and classification of congestive heart failure", ISA Transactions, vol. 79,pp. 239–250, 2018.

[9] Y. LeCun, B. Boser, J. S. Denker, D. Henderson, R. E. Howard, W. Hubbard and L. D. Jackel, "Backpropagation Applied to Handwritten Zip Code Recognition", Neural Computation, vol. 4, pp.541–551, Winter 1989.

[10] Ian Goodfellow, Yoshua Bengio , Aaron Courville, Deep Learning, 1st ed., vol. 1. MIT press, England, 2016, pp.321–359.

[11] Hanmandlu, M., Grover, J., Madasu, V. K., and Vasikarla, S., "Input Fuzzy Modeling for the Recognition of Handwritten Hindi Numerals", Fourth International Conference on Information Technology, IEEE, LasVegas, NV, USA , pp. 208-213, 2007.

[12] Belhe, S., Paulzagade, C., Deshmukh, A., Jetley, S., and Mehrotra, K.," Hindi Handwritten Word Recognition Using HMM and Symbol Tree", In Proceeding of the Workshop on Document Analysis and Recognition, ACM, 2012, pp.9-14.

[13] Kekre, H. B., Thepade, S. D., Sanas, S. P., and Shinde, S., "Devnagari Handwritten Character Recognition using LBG vector quantization with gradient masks", International Conference on Advances in Technology and Engineering (ICATE), 2013, pp. 1-4.

[14] Deepti Khanduja, S. P., Neeta Nain, "Hybrid Feature Extraction Algorithm for Devanagari Script", ACM Transactions on Asian and Low-Resource Language Information Processing (TALLIP), pp. 15:2:1-2:10, 2015.

[15] Gyanendra K.Verma, P. K., Shitala Prasad, "Handwritten Hindi Character Recognition Using Curvelet Transform", In proceedings of Information Systems for Indian Languages, Springer, 2011, pp.224–227.

[16] Madhuri Yadav, Ravindra K. Purwar, "Hindi handwritten character recognition using oriented gradients and Hu-geometric moments", Journal of Electronic Imaging, vol.27, pp. 051216.1–051216.11, 2018.

[17] Prasad, S., Verma, G., Singh, B., and Kumar, P., 2012. "Basic handwritten character recognition from multi-lingual image dataset using multiresolution and multi-directional transform". International Journal of Wavelets, Multiresolution and Information Processing, vol. 10,5, pp.1-28, 2012.

[18] Yadav, M. and Purwar, R., "Hindi handwritten character recognition using multiple classifiers", 7th International Conference on Cloud Computing, Data Science Engineering - Confluence, 2017, pp.149–154.

[19] Jiajia Zhang, Kun Shao, Xing Luo, "Small sample image recognition using improved Convolutional Neural Network", Journal of Visual Communication and Image Representation, vol. 55, pp. 640–647, 2018.

[20] Mohamed Elleuch, Rania Maalej, Monji Kherallah, "A New Design Based-SVM of the CNN Classifier Architecture with Dropout for Offline Arabic Handwritten Recognition", Procedia Computer Science, vol. 80, pp. 1712–1723, 2016.

[21] Ritesh Sarkhel, Nibaran Das, Aritra Das, Mahantapas Kundu, Mita Nasipuri, "A multi-scale deep quad tree based feature extraction method for the recognition of isolated handwritten characters of popular indic scripts", Pattern Recognition, vol. 71, pp. 78–93, 2017.

[22] Adarsh Trivedi, Siddhant Srivastava, Apoorva Mishra, Anupam Shukla, Ritu Tiwari, "Hybrid evolutionary approach for Devanagari handwritten numeral recognition using Convolutional Neural Network", Procedia Computer Science,vol. 125, pp. 525–532, 2018.

[23] Xiwen Qu, Weiqiang Wang, Ke Lu, Jianshe Zhou, "Data augmentation and directional feature maps extraction for in-air handwritten Chinese character recognition based on convolutional neural network", Pattern Recognition Letters, vol. 111, pp. 9-15, 2018.

[24] A. Krizhevsky, I. Sutskever, and G. E. Hinton, ImageNet classification with deep convolutional neural networks, in Advances in Neural Information Processing Systems vol. 25, pp. 1097-1105, 2012.

[25] M. D. Zeiler and R. Fergus, Visualizing and understanding convolutional networks, CoRR abs/1311.2901 , 2013.

[26] K. Simonyan and A. Zisserman, Very deep convolutional networks for large-scale image recognition, CoRR abs/1409.1556 , 2014.

[27] C. Szegedy et al., "Going deeper with convolutions", In Computer Vision and Pattern Recognition (CVPR 15) , 2015 .

[28] K. He et al., Deep residual learning for image recognition, arXiv:1512.03385 , 2015.

Review of Different Transforms used in Digital Image Watermarking

Lavi Tanwar
Department of Electronics & Communication
Engineering Delhi Technological University
Delhi, India
lavi.tanwar02@gmail.com

Jeebananda Panda
Department of Electronics & Communication
Engineering Delhi Technological University
Delhi, India
jpanda@dce.ac.in

Abstract- **Digital watermarking is the process of inserting the secret information, related to the identity of the owner and the source, into the original digital data such as image, text, audio or video. The secret information is known as a watermark. The insertion can be done in two domains which are spatial domain and frequency domain. In the spatial domain, direct pixels of the digital data are considered for the watermark embedding while in the frequency domain the transform of the original data is taken first and then the watermark is embedded into its transform coefficients. The frequency domain digital watermarking is more robust than the spatial domain digital watermarking due to which today researchers are showing great interest in frequency domain digital watermarking. This paper summarizes some of the most commonly used frequency domain techniques with their characteristics, advantages and disadvantages.**

Keywords- Digital Watermarking, Wavelet Transform, Contourlet Transform, Shearlet Transform

I. INTRODUCTION

Today everyone is making the extensive use of internet for information sharing. One can share the documents, images, videos and audio data by making the use of different methods via internet. But with the increase of information sharing, the threat of data forgery, authentication, security of data and copyright protection is also increasing. For example, a person X is having an original video and he shared it with some other person Y. That person Y made its multiple copies and sold it. Then there must be some technique that can prove the ownership of person X. Consider another example of sonar waveform transmission [1]. Sonar transmissions are generally not provided with source origin and broadcast platform information. So, these sonar waveforms may be a threat for underwater acoustic channels. To prove the authentication multiple waveforms are sent with different frequency and different bandwidth. There is a need for an intelligent technique that can take care of the authentication of the waveform. Similarly, there are numerous examples where we need efficient and intelligent techniques to prevent forgery, to authenticate the source and to protect the copyright.

There already exist many techniques to prevent data forgery and for copyright protection. But there is a need for more efficient methods and techniques that can help one to share data securely without any loss of information. One such method is Digital Watermarking. Digital Watermarking is the process of inserting some secret information in the host data and that secret information is known as watermark. Now if one tries to forge or manipulate the watermarked data then the complete data will be destroyed as he doesn't have any knowledge of watermark. But the care should be taken while inserting the watermark that it doesn't interfere with the original host data. If the watermark interferes with the host data then the original information will be lost and the whole process will be a complete waste of time and resources. Also, the watermark must be inserted in such a way that it can be retrieved effectively later on the receiver side. To incorporate these points in the digital watermarking technique, there are basically two domains where the digital watermark can be inserted. These two domains are: Spatial domain and Frequency domain. In the spatial domain, the watermark is embedded in the pixels of the host image signal while in the frequency domain the watermark information is embedded into the spectral components of the host signal. Some commonly used techniques are Least Significant Bit (LSB), Additive Watermarking, SSM Modulation based technique, Texture Mapping coding technique, Patchwork algorithm and Correlation based technique [2], [4], [6] and Commonly used techniques are Discrete Fourier Transform, Discrete Cosine Transform, Discrete Wavelet Transform, Singular Value Decomposition, Contourlet Transform, and Discrete Shearlet Transform [2], [5], [6], [7] which has been discussed in detail in the next section Fig. 1 shows the classification of digital watermarking on different basis [2],[3].

II. FREQUENCY DOMAIN TECHNIQUES OF DIGITAL WATERMARKING

As discussed above, digital watermarking can be done in two domains, namely spatial domain and frequency domain. Both the domains have their own advantages and disadvantages which are given below [2]:

- The computational cost of spatial domain techniques is lower than that of frequency domain techniques.
- The frequency domain techniques are more robust than the spatial domain techniques.
- The computational complexity of spatial domain techniques is lower than that of frequency domain techniques.
- The computational time of spatial domain techniques is less than that of frequency domain techniques.
- The capacity of spatial domain techniques is higher than that of frequency domain techniques.

As frequency domain watermarking techniques are more robust therefore today researchers are implementing watermarking algorithms based on these techniques. These techniques are discussed in detail below.

978-1-5386-6624-1/18 $31.00 © 2018 IEEE

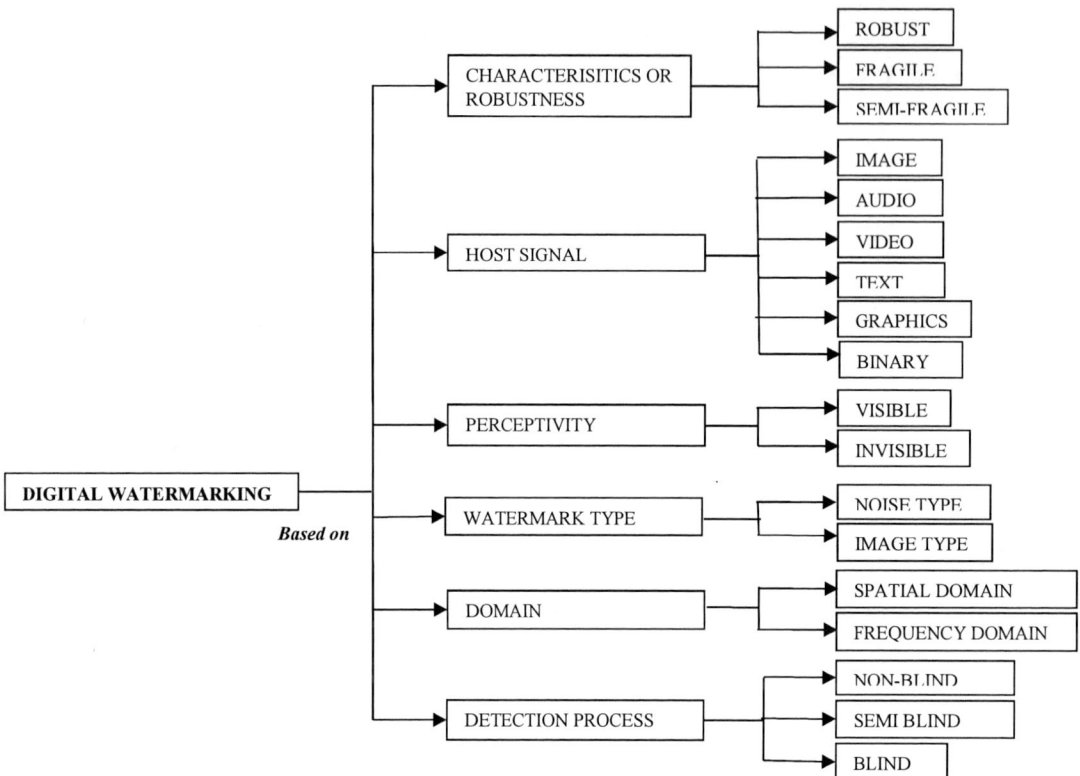

Fig. 1. Classification of Digital Watermarking

A. Discrete Fourier Transform

The DFT of N-point signal x[n], n=0,1,2,...,N-1 can be defined as [8] [9]:

$$X(k) = \sum x[n]e^{-j2\pi kn/N}, \ k = 0,1,2,\dots,N-1 \quad (1)$$

where, k is frequency domain ordinal, n is time domain ordinal and N is length of the sequence to be transformed.

The inverse DFT (IDFT) can be defined as:

$$x[n] = \sum X(k)e^{\frac{j2\pi kn}{N}}, \ k = 0,1,2,\dots,N-1; \ n = 0,1,2,\dots,N-1 \quad (2)$$

DFT doesn't get affected by translation and rotation and hence it is robust to the geometrical attacks [10], [11], [12]. This is due to the fact that these attacks do not alter the phase content of the Fourier coefficients and magnitude and phase are two components of any Fourier coefficient [13]. In [14] it is shown that the robustness against geometrical attacks highly depends on the orientation of that vector in which the data bit is to be inserted. Since it exhibits energy distribution property and robust to geometrical attacks, it can be used for the Print-Scan (PS) process [10]. The different variants of DFT like quaternion DFT (QDFT) can be used for color images. One of such examples is given in [15].

But Fast Fourier Transform is generally affected by round-off errors. Due to this the quality gets degraded and hence the watermark can't be extracted faithfully [10]. Also, DFT is very sensitive to cropping [10].

B. Discrete Cosine Transform

DCT is the sum of cosine functions that oscillates at different frequencies to represent the finite sequence of data points. The 2-D DCT of and image f can be defined as [16], [17]:

$$DCT(u,v) = c(u).c(v).\sum_{i=1}^{S}\sum_{j=1}^{S} f(i,j).\cos\left(\frac{i+\left(\frac{1}{2}\right)}{S}u\pi\right).\cos\left(\frac{j+\left(\frac{1}{2}\right)}{S}v\pi\right) \quad (3)$$

where, i is sample value in the space domain with size S, u ∈ [1,S] is sample frequency in frequency domain and c is static coefficient function and it is defined in equation (4) as follows:

$$c(u) = \begin{cases} \sqrt{\frac{1}{S}}, & u = 0 \\ \sqrt{\frac{2}{S}}, & u \neq 0 \end{cases} \quad (4)$$

The 2-D inverse DCT (IDCT) can be defined as [16][17]:

$$f(i,j) = \sum_{u=1}^{S}\sum_{v=1}^{S} c(u).c(v).DCT(u,v).\cos\left(\frac{i+(1/2)}{S}u\pi\right).\cos\left(\frac{j+(1/2)}{S}v\pi\right) \quad (5)$$

DCT exhibits the property of fine energy compaction as given in [18]. JPEG compression technique makes the use of DCT to compress the image and hence due to this fact DCT shows a great robustness to JPEG compression but it doesn't provide robustness against geometrical attacks like rotation, scaling and translation [10], [19]. As shown in [16], [20] its compression ratio is high, its computational complexity is high and error rate is low. DCT shows a good performance in case of low cost hardware design [21].

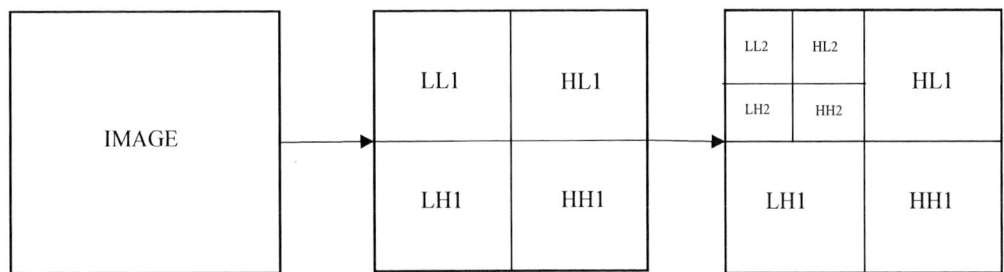

Fig. 2. 2-level Discrete Wavelet Decomposition

C. Discrete Tchebichef Transform

DTT is a derivative of orthonormal Tchebichef polynomials [22]. The forward DTT can be defined as [22]:

$$T_{ij} = \sum_{x=0}^{7} \sum_{y=0}^{7} r_u(x) s_v(y) f(x,y) \qquad (6)$$

The inverse DTT can be defined as [51]:

$$f(x,y) = \sum_{u=0}^{7} \sum_{v=0}^{7} T_{ij} r_u(x) s_v(y) \qquad (7)$$

where T_{ij} is 8 X 8 block basis matrix and is defined as [22]:

$$T_{ij} = \begin{bmatrix} r_u(0)s_v(0) & r_u(0)s_v(1) & ... & r_u(0)s_v(7) \\ r_u(1)s_v(0) & r_u(1)s_v(1) & ... & r_u(1)s_v(7) \\ ... & ... & ... & ... \\ r_u(7)s_v(0) & r_u(7)s_v(1) & ... & r_u(7)s_v(7) \end{bmatrix} \qquad (8)$$

DTT exhibits the similar properties that DCT have like orthogonality, symmetry and separability. But as compared to DCT it has high energy compaction ability. Since the polynomials of DTT closely matches the HVS characteristics, therefore the perceptual quality of DTT based watermarking schemes is better than DCT based watermarking scheme as shown in [23]. It has been shown in [24] [25] that the computational complexity and memory used for running an image processing program is much less than that is required by DCT. Due to this the time required for the computation also reduced significantly in case of DTT. The better reconstruction accuracy for gray level images and higher gain can be achieved using DTT as compared to DCT [26]. It can also be used for the implementation in hardware similar to DCT. Although DTT exhibits so many advantages over DCT but it shows poor performance at low bit rates.

D. Discrete Wavelet Transform

DWT decomposes the image, in a hierarchical structure, into sub bands with different frequencies and limited duration. The decomposed components of a signal are known as wavelets. The wavelet transform coefficients give the information of position. The implementation of DWT is done by low pass filtering and high filtering of the image which gives coarse approximation image pixels and detailed image pixels, respectively and hence it shows robustness against the attacks like low pass filtering and median filtering [10], [27]. The image can be decomposed into any number of levels. When the DWT is applied to the image for the first time, it decomposes the image into four multi resolution sub bands, i.e. LL1, HL1, LH1 and HH1 which will acquire the next coarser wavelet coefficients. When again DWT is applied then it will decompose the LL sub band into further four sub bands which are LL2, HL2, LH2 and HH2. So, at each level, LL sub band of previous level is considered for the decomposition. The low frequency sub band is always decomposed further because of the fact that significant energy of any signal is always present in its low frequency components. Fig. 2 shows the process of 2-level DWT.

DWT provides good multiscale and directional representation of an image [28]. It exhibits very good spatio-frequency localization property. Due to this it is easy to determine those areas of image where one can hide the watermark. As shown in [29] there is no requirement of original image for the watermark detection purpose using this technique and hence it can be used for audio and video compression and noise removal. JPEG-2000 compression technique makes the use of DWT to compress the image and hence it shows great robustness to compression attacks [19]. It exhibits good multiresolution property that basically resembles Human Visual System (HVS) characteristics. One can embed the watermark in both high frequency and low frequency coefficients. High frequency coefficients provide the edge information of the image. If watermark is embedded in high frequency coefficients then it won't be visible to human eye as these areas are not much sensitive to human vision. But during attacks like compression or scaling, the high frequency components are removed. Hence the chances of loss of watermark are high. If watermark is inserted in low frequency coefficients then the watermark would be more robust under high capacity as shown in [30] but its invisibility would be poor as these areas are sensitive to human vision. It has the demerit of not identifying the smoothness of the contour [31] and also it is not much robust to the geometrical attacks. It shows lack of shift invariance, i.e. small shift in input leads to a large change in the filter coefficients and it shows poor directional selectivity as its separable filters are not able to discriminate the edge features of the opposing diagonals [32] [33] [34].

E. Dual-Tree Complex Wavelet Transform

The disadvantages of DWT like lack of shift invariance and poor directional selectivity has been resolved by Complex Wavelet Transform (CWT). In CWT, limited redundancy has been introduced in it. The coefficients of the filters used in it are complex and hence complex output samples are generated by it. But this makes the perfect reconstruction difficult beyond the decomposition level 1 [33]. To overcome this limitation, Dual-Tree Complex Wavelet Transform (DT-CWT) has been introduced by N. Kingsbury in [35]. In this technique, two DWTs acting in parallel are used which works on the same data. One DWT works on the even samples of the data and other on the odd part and then two trees of the DT-CWT are produced using the sum and difference of the decompositions of these two DWTs as shown in [36]. Since DT-CWT uses the fundamental difference between the two filters to apply in the two trees, its directional selectivity has been improved as shown in [34]. The implementation of 1-D DT-CWT has been shown in fig. 4 and it is given by [33] [34]:

$$f(t) = \sum_{l \in Z} s_{j_0,l} \emptyset_{j_0,l}(t) + \sum_{j \geq j_0} \sum_{l \in Z} c_{j,l} \Psi_{j,l}(t) \qquad (9)$$

978-1-5386-6624-1/18 $31.00 © 2018 IEEE   1167

where, $\Psi(t)$ is dilated mother wavelet, $\emptyset(t)$ is scaling function, $s_{j0,l}$ is scaling coefficient and $c_{j,l}$ is complex wavelet coefficient

To implement 2D DT-CWT, separable wavelet transforms are required which can perform separate filtering along rows and then columns. The process of 2D DT-CWT can be described as [32]: First the input image is decomposed up to desired level by separable DWT which will generate six high-pass sub bands at each level which are oriented at $\pm 15°$, $\pm 45°$ and $\pm 75°$. Next, the corresponding sub bands which have the same pass bands are combined linearly using averaging or differencing. DT-CWT is more robust to geometrical attacks as compared to DWT.

F. Walsh Hadamard Transform

WHT is non-sinusoidal and orthogonal transform which can be described as a matrix of plus ones and minus ones. The rows and columns of the matrix are orthogonal to each other. The first row and first column of the matrix give DC component where no sign is changed whereas other elements where the large number of sign change occurs give AC values and correspond to high frequency [37]. According to [37] 2D-Hadamard Transform can be defined as:

$$[V] = \frac{H_n[U]H_n}{N} \tag{10}$$

where, $[U]$ is original image, $[V]$ is transformed image and H_n is N X N Hadamard Matrix, $N=2^n$, $n=1,2,3,....$ with element values either +1 or -1

According to [37] inverse 2D-Hadamard Transform can be defined as:

$$[U] = H_n^{-1}[V]H_n^* = \frac{H_n[V]H_n}{N} \tag{11}$$

The Hadamard Transform H_m for $m>0$ can also be defined as:

$$H_m = \frac{1}{\sqrt{2}} \begin{pmatrix} H_{m-1} & H_{m-1} \\ H_{m-1} & -H_{m-1} \end{pmatrix} \tag{12}$$

WHT exhibit the advantage of less computational complexity than the other orthogonal transforms like DWT and DCT as shown in [38]. Since the rows and columns of the matrix are independent to each other, therefore, WHT shows more robustness to image modification than other transforms at low quality factor [39]. As shown in [40] it is better suited for real time implementation of watermarking in hardware. As it provides a wide range of middle frequency bands with low processing noise, hence it shows robustness to compression attack [41]. But according to [42] WHT shows less compression ratio than FFT.

G. Singular Value Decomposition

SVD is a technique that is used to transform the set of correlated variables into a set of uncorrelated variables and this can be used to explore the new dimensions of the relationships among the original data [43]. The algebraic features of an image can be extracted using SVD. The luminance of the image can be determined using the singular values and using the corresponding pair of singular vectors, the geometry of the image can be specified. The SVD of an image A of size m x m can be described as [44]:

$$A = USV^T \tag{13}$$

$$A = [u_1, u_2, ..., u_m] \text{ x} \begin{pmatrix} \lambda1 & 0 & \cdots & 0 \\ 0 & \lambda2 & \cdots & 0 \\ \vdots & 0 & \ddots & 0 \\ 0 & 0 & \cdots & \lambda m \end{pmatrix} \text{x } [v_1, v_2, ..., v_m]$$

$$A = \sum_{i=1}^{r} \lambda_i u_i v_i^T \tag{14}$$

where, U and V are orthogonal matrix. Columns of U are left singular vectors of image A while column of V are right singular vectors of image A. S is diagonal matrix of singular values λ_i, I = 1,2, ..., m arranged in decreasing order and r is rank of A.

SVD packs the maximum energy of the signal into fewer number of coefficients which can be used for the purpose of image compression [45]. Its information embedding quality is also good [46]. While implementing the SVD there is a requirement of extensive computation and hence it is preferred to use SVD along with some other techniques as shown in [47]. However, many researchers implemented some hybrid techniques using SVD which are computationally fast. As shown in [48] SVD can offer many advantages such as variable size of matrices obtained from SVD transformation and not much variation in singular values if information is added to an image.

H. Contourlet Transform

The Contourlet Transform (CT) is used to represent 2D signals with smooth contours and better sparseness. Its implementation process is divided into two stages: first is sub band decomposition which makes the use of Laplacian Pyramid (LP) filters and second stage is directional transform which makes the use of Directional Filter Banks (DFB) [49]. LP filter is used to capture the point singularities contained in the residual signal. Then DFB decomposes that residual signal in different directions to obtain its directional information. This process is repeated at multiple scales [31]. In this way it provides higher degree of directionality. The wavelet transform can also provide directionality but it works efficiently only for 1D signals [50]. Curvelet transform can give the same information in continuous domain but it fails for the discrete domain [51]. Fig. 3 represents the process of analysis part of the contourlet transform.

CT exhibits the spreading property which means that if watermark bits are embedded into any low pass or high pass sub band then these bits will be spread out in all the sub bands during the reconstruction of the watermarked image. This makes it more robust against the attacks [52]. CT is more computationally efficient than other transform as it makes the use of iterated filter banks and also it has a 2D frequency partitioning on concentric rectangles [52]. Since CT can capture geometric structures in images and their smooth contours, it can be used in wide variety of image processing applications like feature extraction, image watermarking, image denoising and text retrieval [53], [54].

The characteristics of CT can be modeled using non-Gaussian distributions like Cauchy PDFs, Generalized Gaussian and alpha stable. But Normal Inverse Gaussian PDF can model its characteristics in a better way as shown in [31] because this distribution resembles the CT distribution. In contourlet distribution there is a large peak around zero and its tails are heavier than Gaussian PDF [31].

978-1-5386-6624-1/18 $31.00 © 2018 IEEE

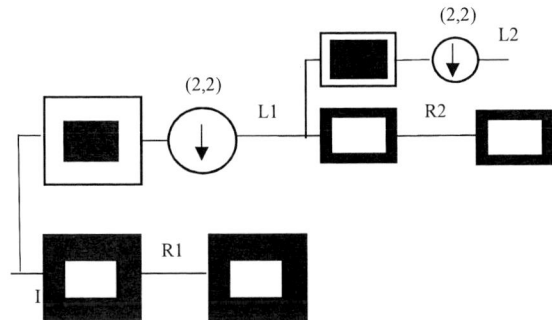

Fig. 3. Analysis Part of Contourlet Transform

I. Discrete Shearlet Transform

Shearlet Transform, used to capture directional and anisotropic features of the image, contains a single mother Shearlet function whose parameters are scaling, shear and translation. Shear parameters captures the direction of singularities. It is implemented using Laplacian Pyramid scheme and directional filtering [55]. By sampling the Continuous Shearlet Transform on the discrete subset of Shearlet group, Discrete Shearlet Transform (DST) can be computed [55]. It doesn't restrict the number of directions for the shearing which makes it better than other transforms like wavelet transform and contourlet transform. It can efficiently capture the geometric information of the image.

TABLE 1. ADVANTAGES AND DISADVANTAGES OF FREQUENCY DOMAIN DIGITAL IMAGE WATERMARKING TECHNIQUES

Technique	Advantages	Disadvantages
Discrete Fourier Transform	• Robust against geometrical attacks	• Highly sensitive to cropping • Affected by round off errors • Watermark extraction not much faithful • Rotate signal in frequency domain only by $\pi/2$
Discrete Cosine Transform	• Fine energy compaction property • Robust to JPEG compression attacks • Low error rate • Good performance in case of low cost hardware design	• Not very robust to geometrical attacks • High Computational Complexity
Discrete Tchebichef Transform	• high energy compactness • less computational complexity • Faster than DCT • better reconstruction accuracy for gray level images	• Poor performance at low bit rates
Discrete Wavelet Transform	• Good multiscale and directional image representation • Exhibit Good spatio-frequency localization property and multiresolution property • Robust against attacks like low pass filtering, median filtering and compression	• Doesn't identify the smoothness of contours and sparseness of image • Not very robust against geometrical attacks • Lack of shift invariance • Poor directional selectivity
Dual-Tree Complex Wavelet Transform	• Good directional selectivity • No lack of shift invariance • No redundancy • Perfect reconstruction • More robust to geometrical attacks than DWT	• Computational Complexity increases
Walsh Hadamard Transform	• Less computational complexity • Good for real time implementation of watermarking in hardware	• Difficult to analyze • Compression ratio less than FFT
Singular Value Decomposition	• Packs maximum energy into a smaller number of coefficients • Good Information embedding quality • Not much variation in singular values while inserting information in the image	• Require extensive computation
Contourlet Transform	• Provide smooth contours and better sparseness • Provides higher degree of directionality • Multiscale and directional image representation • Computationally efficient	• Restricts number of directions • Restricts the support size
Discrete Shearlet Transform	• Captures directional and anisotropic features of the image • Doesn't restrict the number of directions • Doesn't restrict the support size	• Complexity high

III. CONCLUSION

This paper summarizes the various aspects of the Digital Image Watermarking. It can be seen that the two domains of digital watermarking have their own advantages and disadvantages. Spatial domain techniques are simple, fast and cheap whereas on the other hand, frequency domain techniques are more robust, secure and reliable. So, depending upon the requirements of the user watermarking can be done in any one of these domains. Also, the user can make use of techniques from both the domains to implement more secure and robust hybrid algorithms. There is a lot of research work is going on around the world in this field to develop more reliable algorithms for watermark embedding and its faithful detection.

REFERENCES

[1] B.G. Mobasseri and R.S. Lynch, "Information Embedding in Sonar by Modifications of Time-Frequency Properties," IEEE J. Oceanic Engg., vol. 41, pp. 139-154, January 2016.

[2] P. Singh and R.S. Chadha, "A Survey of Digital Watermarking Techniques,Applications and Attacks," Int. J. Engg & Innovative Tech. (IJEIT),vol. 2, pp. 165-175, March 2013.

[3] L.K. Saini and V. Shrivastava, "A Survey of Digital Watermarking Techniques and its Applications," Int. J. Computer Science Trends & Tech. (IJCST), vol. 2, pp. 70-73, May-Jun 2014.

[4] I.J. Cox, J. Kilian, F.T. Leighton and T. Shamoon, "Secure Spread Spectrum Watermarking for Multimedia," IEEE Trans. Image Process., vol. 6, pp. 1673-1687, December 1997.

[5] S. Tyagi, H.V. Singh, R. Agarwal and S.K. Gangwar, "Digital Watermarking Techniques for Security Applications," IEEE Int. Conf. Emerging Trends in Electrical, Electronics and Sustainable Energy Systems, 2016.

[6] A. Rashid, "Digital Watermarking Applications and Techniques:A Brief Review," Int. J. Computer Applications Technology and Research, vol. 5, pp. 147-150, 2016.

[7] R. Vasudev, "A Review on Digital Image Watermarking and Its Techniques," J. Image & Graphics, vol. 4, pp. 150-153, 2016.

[8] Z. He, M. Bystrom and S.H. Nawab, "Bidirectional Conversion Between DCT Coefficients of Blocks and Their Subblocks," IEEE Trans. Sig. Process., vol. 53, pp. 2835-2841, 2005.

[9] N. Senthilkumaran and S. Abinaya, "Digital Image Watermarking Using DFT Algorithm," Adv. Computing Int. J., vol. 7, pp. 9-17, 2016.

[10] A. Poljicak, L. Mandic and D. Agic, "Discrete Fourier Transform-based watermarking method with an optimal implementation radius," J. Electronic Imaging, vol. 20, pp. 1-8, SPIE and IS&T, 2011.

[11] M. Urvoy, D. Goudia and F. Autrusseau, "Perceptual DFT Watermarking with improved detection and robustness to geometrical distortions," IEEE Trans. Info. Forensics & Security, vol. 9, pp. 1108-1119, 2014.

[12] V. Solachidis and I. Pitas, "Circularly Symmetric Watermark Embedding in 2-D DFT Domain," IEEE Trans. Image Proc., vol. 10, pp. 1741-1753, 2001.

[13] T.K. Tsui, X.P. Zhang and D. Androutsos, "Color Image Watermarking Using Multidimensional Fourier Transforms," IEEE Trans. on Info. Forensics and Security, vol. 3, pp. 16-28, 2008.

[14] A. Abbas, S. Aslam and M.K. Samee, "Robust and Reversible Watermarking Scheme in DFT Domain," IEEE Int. Symp. Sig. Proc. and Info. Tech., 2014.

[15] B. Chen, G. Coatrieux, G. Chen, X. Sun, J.L. Coatrieux and H. Shu, "Full 4-D quaternion discrete Fourier transform based watermarking for color images," Dig. Sig. Proc., ELSEVIER, vol. 28, pp. 106-119, 2014

[16] S. Liu, Z. Pan and H. Song, "Digital image watermarking method based on DCT and fractal encoding," IET Image Proc., vol. 11, pp. 815-821, 2017.

[17] M. Paunwala and S. Patnaik, "Biometric template protection with DCT-based watermarking," Machine Vision and Applications, Springer, vol. 25, pp. 263-275, 2014.

[18] L. Dong, Q. Yan, Y. Lv and S. Deng, "Full band watermarking in DCT domain with Weibull model," Multimed. Tools Appl.,Springer, vol. 76, pp. 1983-2000, 2017.

[19] A. Phadikar, S.P. Maity and B. Verma, "Region based QIM digital watermarking scheme for image database in DCT domain," Comp. and Electrical Eng., ELSEVIER, vol. 37, pp. 339-355, 2011.

[20] S.D. Lin, S.C. Shie and J.Y. Guo, "Improving the robustness of DCT-based image watermarking against JPEG compression," Comp. Standards & Interfacing, ELSEVIERE, vol. 32, pp. 54-60, 2010.

[21] S.A. Parah, J.A. Sheikh, N.A. Loan and G.M. Bhat, "Robust and blind watermarking technique in DCT domain using inter-block coefficient differencing," Dig. Sig. Proc., ELSEVIER, vol. 53, pp. 11-24, 2016.

[22] D.R.I. M. Setiadi, T. Sutojo, E. H. Rachmawanto and C. A. Sari, "Fast and Efficient Image Watermarking Algorithm using Discrete Tchebichef Transform," IEEE International Conference on Cyber and IT Service Management (CITSM), August 2017.

[23] J. L. D. Shivani and R. K. Senapati, "A CRT Based Robust Image Watermarking Using Discrete Tchebichef Transform," ARPN Journal of Engineering and Applied Sciences, vol. 12, pp. 3600-3606, June 2017.

[24] R.K. Senapati, U.C. Pati and K.K. Mahapatra, "Reduced memory, low complexity embedded image compression algorithm using hierarchical listless discrete Tchebichef transform," IET Image Processing, vol. 8, pp. 213-238, 2014.

[25] P.A.M. Oliveira, R.J. Cintra, F.M. Bayer, S. Kulasekera and A. Madanayake, "Low-complexity Image and Video Coding Based on an Approximate Discrete Tchebichef Transform," IEEE Trans. on Circuits & Systems for Video Technology, vol. 27, pp. 1066-1076, January 2016.

[26] P. Jayanthi and C.T. Kavitha, "FPGA Implementation of 1-D Discrete Tchebichef Transforms," World Engineering & Applied Sciences J., vol. 8, pp. 7-11, 2017.

[27] M. Malonia and S.K. Agarwal, "Digital Image Watermarking using Discrete Wavelet Transform and Arithmetic Progression Technique," IEEE Conf. Electrical, Electronics & Comp. Science, 2016.

[28] M. Narang and S. Vashisth, "Digital Watermarking using Discrete Wavelet Transform," Int. J. Comp. Appl., vol.74, pp. 34-38, 2013.

[29] H. Lala, "Digital Image Watermarking using Discrete Wavelet Transform," Int. Research J. Eng. & Tech, vol. 4., pp. 1682-1685, 2017.

[30] S.T. Chen, C.Y. Hsu and H.N. Huang, "Wavelet-domain audio watermarking using optimal modification on low-frequency amplitude," IET Sig. Proc., vol. 9, pp. 166-176, 2015.

[31] H. Sadreazami, M.O. Ahmad and M.N.S. Swamy, "Multiplicative Watermark Decoder in Contourlet Domain Using the Normal Inverse Gaussian Distribution," IEEE Trans. Multimed., vol. 18, pp. 196-207, 2016.

[32] M.S. Sudha and T.C. Thanuja, "A Robust Image Watermarking Technique using DTCWT and PCA," International Journal of Applied Engineering Research, vol. 12, pp. 8252-8256, 2017.

[33] J. Adabala1 and K.N. Prakash, "Dual Tree Complex Wavelet Transform For Digital Watermarking," International Journal of Advances in Engineering & Technology (IJAET), vol. 4, pp. 482-492, 2012.

[34] J. Panda, S. Maurya, R. Dang, B.L. Narayanapuram, "Analysis of Robustness of an Image Watermarking Algorithm Using the Dual Tree Complex Wavelet Transform and Just Noticeable Difference," IEEE International Conference on Signal Processing and Communication (ICSC), 2016.

[35] N. Kingsbury, "Shift Invariant Properties Of The Dual-Tree Complex Wavelet Transform," IEEE International Conference on Acoustics, Speech, and Signal Processing. Proceedings (ICASSP99), 1999.

[36] A.I. Thompson, A. Bouridane, F. Kurugollu and C. Tanougast, "Watermarking for Multimedia Security Using Complex Wavelets," J. Multimedia, vol. 5, pp. 443-457, 2010.

[37] A.T.S. Ho, J. Shen, S.H. Tan and A.C. Kot, "Digital Image-in-Image Watermarking For Copyright Protection of Satellite Images Using the Fast Hadamard Transform," IEEE Int. Geoscience and Remote Sensing Symposium, June 2002.

[38] R.C.Gonzalez and R.E.Woods, "Digital Image Processing," 3rd Edition, pp. 461-510.

[39] M. Ramkumar and A.N. Akansu, "Capacity Estimates for Data Hiding in Compressed Images," IEEE Trans. Image Processing, vol. 10, pp. 1252-1263, August 2001.

[40] K. Meenakshi, C.S. Rao and K.S. Prasad, "A Hybridized Robust Watermarking Scheme based on Fast Walsh-Hadamard Transform and Singular Value Decomposition using Genetic Algorithm," Int. J. Computer Applications, vol. 108, pp. 1-8, December 2014.

[41] E. Moeinaddini and R. Ghasemkhani, "A novel image watermarking scheme using blocks coefficient in DHT domain," IEEE Int. Symposium on Artificial Intelligence and Signal Processing (AISP), 2015.

[42] A.B. Roy, D. Dey, D. Banerjee and B. Mohanty, "Comparison of FFT, DCT, DWT, WHT Compression Techniques on Electrocardiogram & Photoplethysmography Signals," Int. J. Computer Applications Special Issue on International Conference on Computing, Communication and Sensor Network (CCSN), 2013.

[43] M. Ali M and C.W. Ahn, "An optimized watermarking technique based on self-adaptive DE in DWT-SVD transform domain," Sig. Proc., ELSEVIER, vol. 94, pp. 545-556, 2014.

[44] C.C. Lai, "An improved SVD-based watermarking scheme using human visual characteristics," Optics Communications, vol. 284, pp. 938-944, 2011.

[45] R.A. Sadek, "SVD Based Image Processing Applications: State of the Art, Contributions and Research Challenges," Int. J. Advanced Comp. Science & Appl., vol. 3, pp. 26-34, 2012.

[46] C.C. Chang, Y.S. Hu and C.C. Lin, "A Digital Watermarking Scheme Based on Singular Value Decomposition," In: Chen B., Paterson M., Zhang G. (eds) Combinatorics, Algorithms, Probabilistic and Experimental Methodologies, Lecture Notes in Computer Science, Springer, vol. 4614, pp. 82-93, 2007.

[47] N.M. Makbol, B.E. Khoo, T.H. Rassem, "Block-based discrete wavelet transform-singular value decomposition image watermarking scheme using human visual system characteristics," IET Image Proc., vol. 10, pp. 34-52, 2016.

[48] C.C. Lai, "A digital watermarking scheme based on singular value decomposition and tiny genetic algorithm," Dig. Sig. Proc., ELSEVIER, vol. 21, pp. 522-527, 2011.

[49] M.A. Akhaee, S.M.E. Sahraeian and F. Marvasti, "Contourlet-Based Image Watermarking Using Optimum Detector in a Noisy Environment," IEEE Trans. Image Proc., vol. 19, pp. 967-980, 2010.

[50] S. Ghannam and F.E.Z.A. Chadi, "Contourlet Versus Wavelet Transform: A Performance Study for a Robust Image Watermarking," IEEE Int. Conf. Appl. of Dig. Info. & Web Tech., 2009.

[51] Y. Zhou and J. Wang, "Image denoising based on the symmetrical normal inverse Gaussian model and non-subsampled contourlet transform," IET Image Proc., vol. 6, pp. 1136-1147, 2012.

[52] H. Sadreazami, M.O. Ahmad and M.N.S. Swamy, "A Study of Multiplicative Watermark Detection in the Contourlet Domain Using Alpha-Stable Distributions," IEEE Trans. Image Proc., vol. 23, pp. 4348-4360, 2014.

[53] H. Sadreazami, M.O. Ahmad and M.N.S. Swamy, "Contourlet domain image modeling by using the alpha-stable family of distributions," IEEE Int. Symp. Circuits & Systems, 2014.

[54] H. Sadreazami, M.O. Ahmad and M.N.S. Swamy, "Contourlet Domain Image Denoising using Normal Inverse Gaussian Distribution," IEEE Canadian Conf. Electrical & Comp. Engg.,2014.

[55] B. Ahmaderaghi, F. Kurugollu, J.M.D. Rincon and A. Bouridane, "Blind Image Watermark Detection Algorithm based on Discrete Shearlet Transform Using Statistical Decision Theory," IEEE Trans. Comput. Imag., vol. 4, pp. 46-59, 2018.

2nd IEEE International Conference on Power Electronics, Intelligent Control and Energy Systems (ICPEICES-2018)

State Estimation of Single-Phase Rectifier Based Load Circuit using Unscented Kalman Filter

Amit Kumar Gautam
ECE Department, *DTU*
New Delhi, 110042, India
amitgautam.cicdu@gmail.com

Sudipta Majumdar
ECE Department, *DTU*
New Delhi, 110042, India
korsudipta@rediffmail.com

Abstract— **The state estimation of single-phase full wave rectifier circuit using unscented Kalman filter (UKF) is presented in this paper. The state space model has been obtained using Kirchhoff's laws (KCL and KVL). UKF has been used to estimate the capacitor voltage and diode current through the circuit. The performance of UKF has been compared with extended Kalman filter (EKF). The simulation results show larger signal to noise ratio (SNR) using UKF as compared to EKF method. As the UKF uses unscented transformation (UT), whereas EKF uses Taylor series expansion for linearization of nonlinear model, UKF presents smaller linearization error as compared to EKF. Also, UKF avoids the computation of Jacobian matrix as required in EKF method.**

Keywords— *Unscented Kalman filter, Kirchhoff's voltage and current law, rectifier circuit.*

INTRODUCTION

Various techniques have been presented in literature for parameter estimation such as least square (LS) method [1], maximum likelihood (ML) method [2], wavelet transform (WT) based method [3]-[4], genetic algorithm (GA) [5] and Kalman filter (KF) [6]-[9] etc. for linear and nonlinear systems. The LS method has convergence problem. The ML method is an optimum estimator but, it results in highly nonlinear equations. WT based method cannot be used for real time estimation. GA requires large computational time for optimum solution. KF has been widely used for different types of application, but the limitation is that it can be used for linear systems only. The KF evaluates the minimum mean square error estimate of the random vector that represents the system states. The KF dynamics are derived in the frame work of Gaussian probability density function (pdf). These dynamics result from iterative use of prediction and filtering. The KF can be implemented for systems having linear state dynamics and observation dynamics. But, nonlinearity in system model or observation model results in non Gaussian pdfs. For nonlinear systems EKF has been used [10]-[14] which is an extension of KF. EKF approximates the nonlinear system before implementing the Kalman filter equations on the linearized system. The new states are estimated using the nonlinear model and the new measurements. This linearization of nonlinear functions has the disadvantage of complex Jacobian calculation and filter instability. The linearization of nonlinear systems introduces large errors in state estimation. To overcome these disadvantages of EKF another version called UKF was developed [15]-[20]. which uses the fact that pdf approximation is easier than nonlinear function approximation.

Single phase rectifier based load has been used in various applications including voltage clamper for ac-dc power conversion [21], as pulse width modulation rectifier [22], railway electrical traction system [23] etc.

This paper is organized as follows. Section II gives brief theory of EKF and UKF. Section III presents state space modelling of single phase rectifier circuit. Simulation and conclusions are presented in section IV and section V respectively.

NONLINEAR FILTERS ALGORITHM

In general, a discrete time nonlinear system is represented using state and measurement equations as:

$$x_k = f(x_{k-1}, u_{k-1}) + w_{k-1} \tag{1}$$

$$y_k = h(x_k, u_k) + v_k \tag{2}$$

where u_k is a known input. f and h denote the nonlinear functions. w_k and v_k are the system and measurement noise respectively. These are uncorrelated Gaussian noise having following statistics:

$$E[w_k] = 0, \quad E[w_k, w_k^T] = Q_k, \quad E[w_k, w_j^T] = 0 \forall k \neq j,$$

$$E[w_k, x_0^T] = 0 \forall K, \quad E[v_k] = 0, \quad E[v_k, v_k^T] = R_k,$$

$$E[v_k, v_j^T] = 0 \forall k \neq j, \quad E[v_k, x_0^T] = 0 \forall k, \quad E[w_k, v_j^T] = 0 \forall k \, \& \, j.$$

EKF [10] includes time update and measurement update at each time step k after initialization. EKF uses Taylor's series expansion to linearize the nonlinear functions f and h. F_k and H_k denotes the Jacobian matrices and calculated as:

$$F_k = \frac{\partial f_i(x, u_k)}{\partial x_j}\Big|_{x = \hat{x}_{k|k}} \tag{3}$$

$$H_k = \frac{\partial h_i(x, u_k)}{\partial x_j}\Big|_{x = \hat{x}_{k|k}} \tag{4}$$

Table I summarizes the EKF steps.

The Unscented Kalman Filter (UKF) [10] removes the limitations of EKF by using unscented transform (UT) for nonlinear system representation via sigma points. UKF uses sigma points for representing the mean and variance of the state discretization function of dynamic state equations. The steps for UKF algorithm are:

978-1-5386-6624-1/18 $31.00 © 2018 IEEE

TABLE I.   ALGORITHM OF EKF

Algorithm 1: Extended Kalman Filter
1. *Initialization step:* $\hat{x}(0), P(0), Q$ and R at $k=0$
2. *State prediction:* Compute matrices f as: $$F_k = \frac{\partial f_i(x,u_k)}{\partial x_j}\Big|_{x=\hat{x}_{k|k}}$$ Compute state $\hat{x}_{k|k-1}$ as: $\hat{x}_{k|k-1} = f_{k-1}(\hat{x}_{k-1|k-1},u_{k-1})$ Predicted covariance error computation: $$P_{k|k-1} = F_k P_{k-1|k-1} F_k^T + Q_{k-1}$$
3   *Measurement update:* Compute matrices H_k as: $$H_k = \frac{\partial h_i(x,u_k)}{\partial x_j}\Big|_{x=\hat{x}_{k|k}}$$ Compute the Kalman gain as: $K_k = P_{k|k-1} H_k^T S^{-1}$ where $S = (H_k P_{k|k-1} H_k^T + R_k)$. Update the state estimation: $$\hat{x}_{k|k} = \hat{x}_{k|k-1} + K_k(y_k - h_k(\hat{x}_{k|k-1}))$$ Compute covariance error matrix: $P_{k|k} = (I - K_k H_k) P_{k|k-1}$

1. *Prediction Step:*
    a) *Initialize the state and covariance matrices as:*

$$\hat{x}_0 = E[x_0] \qquad (5)$$

$$P_0 = E[(x_0 - \hat{x}_0)(x_0 - \hat{x}_0)^T] \qquad (6)$$

    b) *Sigma points calculation:* Sigma points are

$$\mathcal{X}_{k-1}^{(i)} = \begin{cases} \hat{x}_{k-1}, & i=0 \\ \hat{x}_{k-1} + (\sqrt{(n_x+\lambda)}P_{k-1})_i, i=1,...,n_x \\ \hat{x}_{k-1} - (\sqrt{(n_x+\lambda)}P_{k-1})_i, i=n_x+1,...,2n_x \end{cases} \qquad (7)$$

n_x denotes state dimension and scaling factor λ is given as:

$$\lambda = \alpha^2(n_x+q) - n_x \qquad (8)$$

Here α and q represent positive scaling factor respectively. W_m is the mean weight given by:

$$W_m^{(i)} = \begin{cases} \dfrac{\lambda}{\lambda+n_x}, i=0 \\ \dfrac{\lambda}{2(\lambda+n_x)}, i=1,...,2n_x \end{cases} \qquad (9)$$

    c) *Sigma points propagation:* Sigma points are calculated as:

$$\mathcal{X}_{k|k-1}^{(i)} = f(\mathcal{X}_{k-1}^{(i)}, u_{k-1}) \qquad (10)$$

    d) *Priori state and error covariance calculation:* A priori state can be estimated as

$$\hat{x}_{k|k-1} = \sum_{i=0}^{2n_x} W_m^{(i)} \mathcal{X}_{k|k-1}^{(i)} \qquad (11)$$

W_c is expressed as:

$$W_C^{(i)} = \begin{cases} \dfrac{\lambda}{\lambda+n_x} + (1-\alpha^2+\beta), & i=0 \\ \dfrac{\lambda}{2(\lambda+n_x)}, i=1,...,2n_x \end{cases} \qquad (12)$$

where β denotes a parameter in state estimation. A priori covariance error is:

$$P_{k|k-1} = \sum_{i=0}^{2n_x} W_c^{(i)}(\mathcal{X}_{k|k-1}^{(i)} - \hat{x}_{k|k-1})$$
$$\times(\mathcal{X}_{k|k-1}^{(i)} - \hat{x}_{k|k-1})^T + Q_k \qquad (13)$$

2. *Process update:*
    a) *Sigma points calculation:* The updated sigma points are:

$$\mathcal{X}_{k|k-1}^{(i)} = \begin{cases} \hat{x}_{k|k-1}, & i=0 \\ \hat{x}_{k|k-1} + (\sqrt{(n_x+\lambda)}P_{k|k-1})_i, i=1,...,n_x \\ \hat{x}_{k|k-1} - (\sqrt{(n_x+\lambda)}P_{k|k-1})_i, i=n_x+1,...,2n_x \end{cases} \qquad (14)$$

    b) *Prediction of system output:* New sigma points are computed as:

$$Y_{k|k-1}^{(i)} = h(\mathcal{X}_{k|k-1}^{(i)}, u_k) \qquad (15)$$

Output is written as

$$\hat{Y}_{k|k-1} = \sum_{i=0}^{2n_x} W_m^{(i)}(Y_{k|k-1}^i) \qquad (16)$$

    c) *Calculate Kalman gain K_k:*

$$K_k = P_{y_k y_k} P_{x_k y_k}^{-1} \qquad (17)$$

$$\hat{P}_{y_k y_k} = \sum_{i=0}^{2n_x} W_c^{(i)}(Y_{k|k-1}^{(i)} - \hat{Y}_{k|k-1})$$
$$\times(Y_{k|k-1}^{(i)} - \hat{Y}_{k|k-1})^T + R_k \qquad (18)$$

$$P_{x_k y_k} = \sum_{i=0}^{2n_x} W_c^{(i)}(\mathcal{X}_{k|k-1}^{(i)} - \hat{x}_{k|k-1})$$
$$\times(Y_{k|k-1}^{(i)} - \hat{Y}_{k|k-1})^T + R_k \qquad (19)$$

    d) *Computing a posteriori state and error covariance:* Estimate of posteriori state and error covariance use following expressions

$$\hat{x}_k = F_k \hat{x}_{k|k-1} + K_k(y_k - \hat{Y}_{k|k-1}) \qquad (20)$$

$$P_k = P_{k|k-1} - K_k P_{y_k y_k} K_k^T \qquad (21)$$

where y_k is the measurement at step k. Table II summarizes the UKF algorithm.

TABLE II.   ALGORITHM OF UKF

Algorithm 2: Unscented Extended Kalman Filter
1. Initialize λ, $\hat{x}(0)$ and $P(0)$.
2. Compute sigma point $a\sqrt{P_{k-1}}$, for $i=1,...,n_x$.
3. Prediction step: At **i=0**
i. Set central point: $\mathcal{X}_{k-1}=\hat{x}_{k-1|k-1}$
ii. Set central weight: $W_m=\dfrac{\lambda}{\lambda+n_x}$
At each iteration i
iii. Calculate $\mathcal{X}_{k-1}$
iv. Weights are assigned as: W_{k-1}
End
v. Calculate $\hat{x}_{k|k-1}$
vi. Calculate $P_{k|k-1}$
4. Update step: At each iteration i
i. Compute $(\mathcal{X}_{k|k-1})_i=f_k(\mathcal{X}_{k-1})_i$
ii. Weights are assigned as: $(W_{k|k-1})_i=(W_{k-1})_i$
End
iii. Measurement is estimated as: $\hat{y}_{k|k-1}$
iv. Compute covariance of the prediction observation P_{yy}
v. Cross-covariance computation of the prediction P_{xy}
vi. Calculate Kalman gain K_k
vii. Update state $\hat{x}_{k|k}$
viii. Calculate covariance of updated estimate $P_{k|k}$

STATE SPACE MODEL FOR SINGLE PHASE RECTIFIER CIRCUIT

Single-phase full wave rectifier circuit [24] is shown in Fig. 1. $v_i(t)$ is the input voltage. The circuit consists of inductor L_s and resistor R_s. The capacitor C is used at the output, which is in parallel with the load resistance R_L. We assumed that D_1 to D_4 are identical diodes with voltage drop equal to v_D. $i_D(t)$ and $v_c(t)$ are the diode current and capacitor voltage drop respectively.

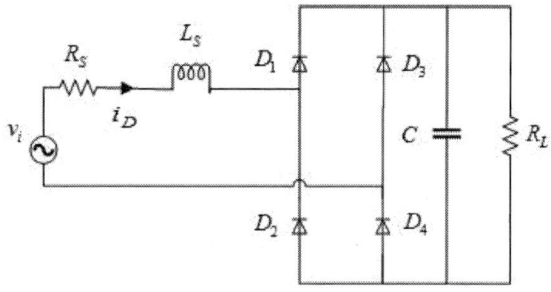

Fig. 1.   Circuit diagram of single-phase rectifier based load.

The circuit equations of rectifier have been obtained using KVL and KCL. They are:

$$C\frac{d}{dt}v_c(t)+\frac{v_c(t)}{R_L}=i_D(t) \tag{22}$$

$$v_i(t)=R_S i_D(t)+L_S\frac{d}{dt}i_D(t)+2v_D+v_c(t) \tag{23}$$

Where $i_D(t)=I_0(e^{v_D/v_T}-1)$. Representing equations (24) and (25) in terms of state equations, we have

$$\frac{d}{dt}v_c(t)=-\frac{1}{R_L C}v_c(t)+\frac{1}{C}i_D(t) \tag{24}$$

$$\frac{d}{dt}i_D(t)=-\frac{1}{L_S}v_c(t)-\frac{(R_S+2V_T/I_0)}{L_S}i_D(t)$$
$$+\frac{V_T}{L_S I_0^2}i_D^2(t)-\frac{1}{L_S}v_i(t) \tag{25}$$

Here, V_T and I_0 are the thermal voltage and reverse saturation current of diode respectively. Where, $x_1=v_c(t)$ and $x_2=i_D(t)$. The nonlinear model is:

$$\frac{d}{dt}x(t)=F(x(t))+Bv_i(t)+w(t) \tag{26}$$

$$y(t)=H(x(t))+v(t) \tag{27}$$

where

$$F(x(t))=\begin{bmatrix}-\dfrac{1}{R_L C} & \dfrac{1}{C}\\ -\dfrac{1}{L_S} & -\dfrac{(R_S+2V_T/I_0)}{L_S}\end{bmatrix}+\begin{bmatrix}0 & 0\\ 0 & \dfrac{2V_T^2}{I_0}\end{bmatrix} \tag{28}$$

$$B=\begin{bmatrix}0 & \dfrac{1}{L_S}\end{bmatrix}^T \tag{29}$$

In (28), both the matrices on the right side represent the linear and nonlinear part respectively. The measurement model is:

$$y(t)=Hx(t) \tag{30}$$

Where $H=\begin{bmatrix}1 & 0\end{bmatrix}$. The discrete time equations are:

$$x_{k+1}=F_k x_k+B_k v_{ik}+w_k \tag{31}$$

$$y_k=H_k x_k+v_k \tag{32}$$

The matrices F_k, B_k and H_k are

$$F_k=\begin{bmatrix}1-\dfrac{T_s}{R_L C} & \dfrac{T_s}{C}\\ -\dfrac{T_s}{L_S} & 1-T_s(\dfrac{(R_S+2V_T/I_0)}{L_S}+\dfrac{2V_T^2}{I_0})\end{bmatrix},$$

$$B_k=\begin{bmatrix}0\\ \dfrac{T_s}{L_S}\end{bmatrix},\qquad H_k=\begin{bmatrix}1\\ 0\end{bmatrix}^T.$$

SIMULATION RESULTS

The capacitor voltage and diode current of single-phase rectifier have been estimated for sinusoidal input voltage. Fig. 2 shows the applied sinusoidal input with maximum amplitude of 10 V and frequency 50 KHz. The white Gaussian noise of zero mean and different variances have been used for estimation purpose. The system noise and measurement noise are white Gaussian noise of zero mean with variance 0.5 and 0.01 respectively. The parameters used for simulations are $R_L = 750\Omega$, $R_S = 17.5\Omega$, $L_S = 91.9mH$ and $C = 100\mu F$. D1N4002 diode model of PSPICE has been used for simulations. The PSPICE simulated values have been considered as the actual value.

The capacitor voltage v_c of single-phase rectifier circuit has been estimated using UKF and EKF. Fig. 3 - Fig. 5 show the capacitor voltage estimation for different noise values. Also, the estimation of diode current i_D using UKF and EKF are shown in Fig. 6 - Fig. 8. Table III and Table IV show the comparison of signal to noise ratio (SNR in dB) value and root mean square error value (RMSE) for UKF and EKF methods.

$$RMSE = \sqrt{\frac{\sum_{i=1}^{n}(\hat{y}_i - y_i)^2}{n}} \tag{33}$$

$$SNR = 10log_{10}\left[\frac{\sum_{i=1}^{n}(\hat{y}_i)^2}{\sum_{i=1}^{n}(\hat{y}_i - y_i)^2}\right] \tag{34}$$

where, $\hat{y}$ and y denote the estimated value and actual value. n is the number of samples.

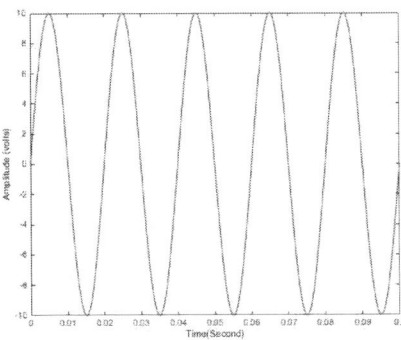

Fig. 2. Input sinusoidal voltage.

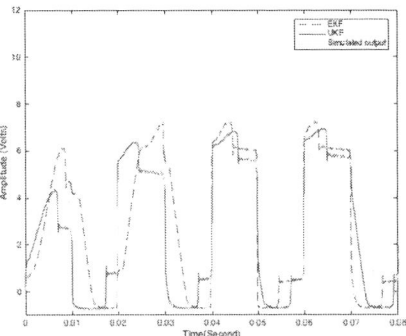

Fig. 3. Comparison of capacitor voltage estimation using noisy input with UKF and EKF methods. Noise variance= 0.1.

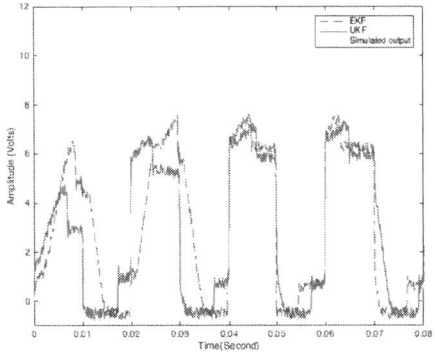

Fig. 4. Comparison of capacitor voltage estimation using noisy input with UKF and EKF methods. Noise variance= 0.5.

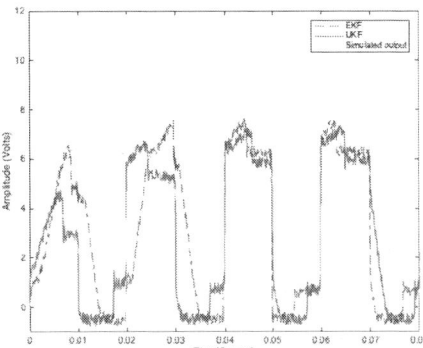

Fig. 5. Comparison of capacitor voltage estimation using noisy input with UKF and EKF methods. Noise variance= 1.0.

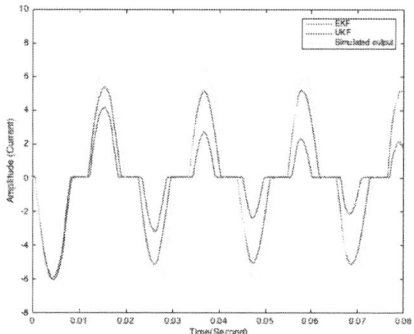

Fig. 6. Comparison of capacitor current estimation using noisy input with UKF and EKF methods. Noise variance= 0.1.

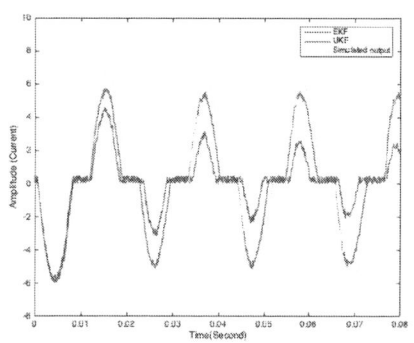

Fig. 7. Comparison of diode current estimation using noisy input with UKF and EKF methods. Noise variance= 0.5.

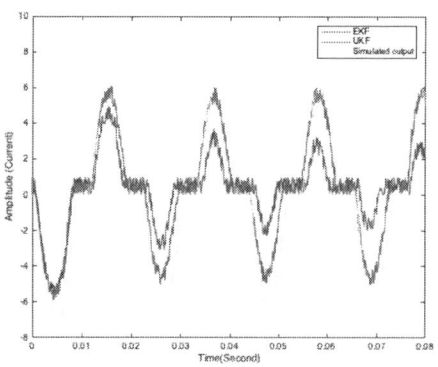

Comparison of diode current estimation using noisy input with UKF and EKF methods. Noise variance= 1.0.

TABLE III. COMPARISION OF CAPACITOR VOLTAGE (v_C) ESTIMATION USING DIFFERENT METHODS

Noisy input signal	Parameter		UKF method	EKF method
Noiseless input		*SNR (dB)*	2.42	1.07
		RMSE	0.4	0.86
Noisy signal with variance =	0.1	*SNR(dB)*	2.40	1.01
		RMSE	0.50	0.96
	0.5	*SNR(dB)*	1.35	1.00
		RMSE	0.70	1.01
	1.0	*SNR(dB)*	1.26	0.90
		RMSE	0.75	1.02

TABLE IV. COMPARISION OF CAPACITOR VOLTAGE (i_D) ESTIMATION USING DIFFERENT METHODS

Noisy input signal	Parameter		UKF method	EKF method
Noiseless input		*SNR (dB)*	2.5	1.290
		RMSE	0.10	0.336
Noisy signal with variance =	0.1	*SNR(dB)*	2.20	1.05
		RMSE	0.40	0.66
	0.5	*SNR(dB)*	2.16	1.01
		RMSE	0.45	1.56
	1.0	*SNR(dB)*	2.10	1.0
		RMSE	0.55	1.82

CONCLUSIONS

The estimation of parameters of a single phase rectifier using UKF is presented in this paper and compared with EKF method. Simulation results show the better closeness of estimated values using UKF with PSPICE simulated values as compared to the EKF method. The SNR value of UKF is better than EKF. Also, RMSE values using UKF are smaller than EKF method due to small linearization error of UKF. Simulation results demonstrate the superiority of the UKF method.

ACKNOWLEDGMENT

First author thanks to UGC.

REFERENCES

[1] K. Wang, J. Chiasson, M. Bodson and L. M. Tolbert, "A nonlinear leastsquares approach for identification of the induction motor parameters," IEEE Transactions on automatic Control, vol. 50, no. 10, pp. 1622-1628, October 2005.

[2] A. Blomqvist and B. Wahlberg, "On the relation between weighted frequency-domain maximum-likelihood power spectral estimation and the prefiltered covariance extension approach," IEEE transactions on signal processing, vol. 55, no. 1, pp. 384-389, January 2007.

[3] S. Majumdar and H. Parthasarathy, "Wavelet based transistor parameter estimation using second order Volterra model," Circuits, Systems, and Signal Processing, vol. 30, no. 6, pp. 1289-1311, December 2011.

[4] D. Yong and Z. He, "Parameter estimation of analog circuits based on the fractional wavelet method," Journal of Semiconductors, vol. 36, no. 3, pp. 035006(1-8), Mar 2015.

978-1-5386-6624-1/18 $31.00 © 2018 IEEE

[5] N. H. Eklund, "Using genetic algorithms to estimate confidence intervals for missing spatial data," IEEE Transactions on Systems, Man, and Cybernetics, Part C (Applications and Reviews), vol. 36, no. 4, pp. 519-523, June 2006.

[6] M. Ahmeid, M. Armstrong, S. Gadoue, M. Al-Greer and P. Missailidis, "Real-Time Parameter Estimation of DCDC Converters Using a SelfTuned Kalman Filter," IEEE Transactions on Power Electronics, vol. 32, no. 7, pp. 5666-5674, July 2017.

[7] H. Fang, N. Tian, Y. Wang, M. Zhou and M. A. Haile, "Nonlinear Bayesian estimation: from Kalman filtering to a broader horizon," IEEE/CAA Journal of Automatica Sinica, vol. 5, no. 2, pp. 401-417, March 2018.

[8] R. Cisneros-Magaa, A. Medina, V. Dinavahi and A. Ramos-Paz, "Time-Domain Power Quality State Estimation Based on Kalman Filter Using Parallel Computing on Graphics Processing Units," IEEE Access, vol. 6, pp. 21152-21163, 2018.

[9] D. Hodgson, B. C. Mecrow, S. M. Gadoue, H. J. Slater, P. G. Barrass, D. Giaouris, "Effect of vehicle mass changes on the accuracy of Kalman filter estimation of electric vehicle speed," IET Electrical Systems in Transportation, vol. 3, no. 3, pp. 67-78, September 2013.

[10] P. Stano, Z. Lendek and J. Braaksma, "Parametric Bayesian filters for nonlinear stochastic dynamical systems: A survey," IEEE transactions on cybernetics, vol. 43, no. 6, pp. 1607, 2013.

[11] N. Hoffmann and F. W. Fuchs, "Minimal invasive equivalent grid impedance estimation in inductiveresistive power networks using extended Kalman filter," IEEE Transactions on Power Electronics, vol. 29, no. 2, pp.631-641, Febuary 2014.

[12] S. Nadarajan, S. K. Panda, B. Bhangu and A. K. Gupta, "Online modelbased condition monitoring for brushless wound-field synchronous generator to detect and diagnose stator windings turn-to turn shorts using extended Kalman filter," IEEE Transactions on Industrial Electronics, vol. 63, no. 5, pp. 3228-3241, May 2016.

[13] M. Yazdanian, A. Mehrizi-Sani and M. Mojiri, "Estimation of electromechanical oscillation parameters using an extended Kalman filter. IEEE Transactions on Power Systems" vol. 30, no. 6, pp. 2994-3002, November 2015.

[14] K. Bogdanski and M. C. Best, "Kalman and particle filtering methods for full vehicle and tyre identification," Vehicle System Dynamics, vol. 56, no. 5, pp. 769-790, May 2018.

[15] W. He, N. Williard, C. Chen and M. Pecht. "State of charge estimation for electric vehicle batteries using unscented Kalman filtering," Microelectronics Reliability, vol. 53, no. 6, pp. 840-847, June 2013.

[16] E. Ghahremani and I. Kamwa, "Online state estimation of a synchronous generator using unscented Kalman filter from phasor measurements units," IEEE Transactions on Energy Conversion, vol. 26, no. 4, pp. 1099-1108, December 2011.

[17] A. Lalami, R. Wamkeue, I. Kamwa, M. Saad and J.J. Beaudoin, "Unscented Kalman filter for non-linear estimation of induction machine parameters," IET electric power applications, vol. 6, no. 9, pp. 611-620, November 2012.

[18] M. Huang, Z. Wei, G. Sun, Y. Sun, H. Zang and K. W. Cheung, "A historical data-driven unscented Kalman filter for distribution system state estimation," InPower and Energy Society General Meeting, IEEE, pp. 1-5, July 2017.

[19] P. Pichlik and J. Zdenke, "Locomotive Wheel Slip Control Method Based on an Unscented Kalman Filter," IEEE Transactions on Vehicular Technology, Febuary 2018.

[20] F. Deng, J. Chen and C. Chen, "Adaptive unscented Kalman filter for parameter and state estimation of nonlinear high-speed objects," Journal of Systems Engineering and Electronics, vol. 24, no. 4, pp. 655-665, August 2013.

[21] W. Zhu, K. Zhou, M. Cheng and F. Peng, "A high-frequency-link singlephase PWM rectifier," IEEE Transactions on Industrial Electronics, vol. 62, no. 1, pp. 289-298, January 2015.

[22] M. S. Ortmann, T. B. Soeiro and M. L. Heldwein, "High switches utilization single-phase PWM boost-type PFC rectifier topologies multiplying the switching frequency," IEEE Transactions on Power Electronics, vol. 29, no. 11, pp. 5749-5760, November 2014.

[23] B. Gou, X. Ge, S. Wang, X. Feng, J. B. Kuo and T. G. Habetler, "An open-switch fault diagnosis method for single-phase PWM rectifier using a model-based approach in high-speed railway electrical traction drive system," IEEE Transactions on Power Electronics, vol. 31, no. 5, pp. 3816-3826, May 2016.

[24] A. Tokic, A. Jukan and J. Smajic, "Parameter Estimation of SinglePhase Rectifier-Based Loads: Analytical Approach," IEEE Transactions on Power Delivery, vol. 31, no. 2, pp. 532-540, April 2016.

A Novel Method for Predicting Attenuation Caused by Clouds for Higher Frequency Bands

Hitesh Singh
Dept. of RCVT
Technical University of
Sofia Sofia, Bulgaria
hitesh.singh.85@gmail.com

Boncho Bonev
Dept. of RCVT
Technical *University of*
Sofia Sofia, Bulgaria
bbonev@tu-sofia.bg

Peter Petkov
Dept. of RCVT
Technical University of
Sofia Sofia, Bulgaria
pjpetkov@tu-sofia.bg

Ravinder Kumar
Dept. of CSE, *HMRITM*
Affiliated with GGSIPU
Delhi, INDIA
ravinder_y@yahoo.com

Abstract— **Higher frequencies especially from 10 GHz to 100 GHz rages opens new avenues to meet the high bandwidth demands of current exponentially growing telecom sectors. These higher frequencies also come with challenges like attenuation caused by atmospheric impairments eg. Rain, cloud, dust, gas, etc. this paper discusses the attenuation of radio awes caused due to clouds. A novel method is proposed to predict attenuation caused by clouds. In this method a novel techniques is introduced to calculate a parameters dielectric constant of water at different temperature and frequencies. This method is then compared with the ITU- Model.**

Keywords— *Satellite Communication, Millimeter Wave, Cloud Attenuation, Dielectric constants of water, Temperature.*

I. INTRODUCTION

The exponential growth of mobile communications forces industries and researchers for higher frequencies bands. Frequencies above 10 GHz provide new paradigms to meet the high bandwidth demands of telecom sector. These high frequencies provide higher bandwidth greater than 1 GHz which is useful for higher data rates transmissions.

The overall radio wave propagation is divided into two main parts namely terrestrial and satellite communications. Each part has different impact of different meteorological conditions. The conditions which has impact on terrestrial communications are urban environments, rural environments and other environmental conditions like fog, dust, rain, gas and vegetation. In case of satellite communications the different meteorological conditions are cloud, rain, dust, scintillation, solar effects and wet antenna. Out of these effects clouds has very complex structures [1]. The presence of clouds in atmosphere has significant impact on radio links. Various researchers has done research on attenuation caused by cloud especially on higher frequencies. Some has proposed models for predicting attenuation caused by clouds.

Gunn [2] has proposed theoretical results in the form of tables, graphs and equations. His work is compared with other published work. The important parameters used in this model are liquid water content in g/m3, wavelength in cm and imaginary parts of absorption coefficient k of water.

Another model was proposed by Staelin [3]. Integrated information about liquid water content of clouds along with water vapor distribution was discussed in his works. The common observation was made that accuracy of results for water vapor distribution is more than distribution of liquid water content.

In Slobin model [4] brief review of types of clouds, liquid water density, cloud noise temperature and attenuation caused by clouds are discussed. They have done experiments at 15 different sites of America for calculation of attenuation and noise temperature. Data are captured annually for different sites. Elevation angles measurements are need to be performed.

Another model was proposed after extensive experimentation from the ranges of 15 GHz to 35 GHz by Altshuler et. al. [5]. They have used 29 different elevation angles profile fr The aim of this special session on Intelligent Information processing and Security is to present a unified platform for advanced and multi-disciplinary research towards design of intelligent information processing systems. The theme focuses on various innovation paradigms in biometrics system design, intelligent computing, security and privacy for interconnected society that may be applied to provide realistic solution to variegated problems in society, environment and industries. From 1 degree to 20 degree each. It has been observed that attenuation is proportional to elevation angle. They have also proposed the method for calculating radius and height of cloud cover. They also shows the relationship between attenuation and humidity. The parameters used by them were elevation angle, frequency and humidity.

Libe [6] has done experimentation for the prediction of attenuation caused by clouds, fog, rain, water vapor, dry air at frequencies up to 1000 Ghz. Based on the results obtained from experiments an empirical model was derived. Parameters used by the are pressure, temperature, relative humidity, water droplets and rate of rainfall.

A model for calculating attenuation caused due to clouds at satellite communications was presented by Salonon and Uppala [7]. They validate the results obtained from model with the experimental results performed with radiometer at different locations of Europe. there model is applicable for mid altitude with an elevation angle of 15 to 40 degrees.

A new model has developed for calculating attenuation caused by Dissanayake et. al. [8]. They had done an empirical study of experimental results obtained at elevation angle below 10 degree. The data for slant path are obtained from experiments performed by radiometers and beacons at different sites. There model observed very less attenuation of 1.5 dB for clouds.

A semi empirical model has been developed by Dintelmann [9] for cloud attenuation. A statistical data has been collected for humidity at ground and surface temperature. Although they have modified slobin [4] model but some parameters have been derived from experimental results conducted at 20 GHz and 30 Ghz.

International Telecommunication Union has recommended their own model ITU-R P840 [10], based on various experimental data obtained from all over the world. It is the mst widely used and trusted model. The maps, curves and equations presented in the models are used to predict attenuation caused by clouds and fog.

II. PROPOSED MODEL

In the previous section various models proposed for predicting attenuation caused due to cloud are discussed. Models used for predicting the cloud attenuation are divided into three main categories. First is the one in which attenuation is computed using Rayleigh Approximations and Mie Theory. Gunn and Libe Models come under this category. In the second category attenuation is directly related to the surface absolute humidity as shown by Altshuler Model. In the third category studies are done using meteorological data and computations are made by using these data to calculate cloud liquid water content in order to calculate attenuation due to clouds. Dintelmann and Slobin models belong to this category.

Their mathematical form and prediction vary over a large range. A parameter common to all models is the liquid water content of clouds. But this parameter is also most difficult to predict and measure. While predicting the attenuation due to clouds liquid water content is not only the deciding factor for propagation studies. In clouds drop size of water and temperature are also the parameters which affect the attenuation considerably. Cloud drop size also plays an important role as the size varies with height and other conditions like place of occurrence of clouds and weather conditions. Temperature of clouds also plays an important role. We have also proposed a new model for calculating attenuation due to clouds whose detailed derivation is published in [11]. In this model the parameters used were real and imaginary parts of dielectric constants. In [11] the values are taken from the formulas present in ITU-R model [9].

This work proposed a new method for calculating real and imaginary parts of water and by using them attenuation is calculated. From the model presented in [11-15], specific attenuation is calculated as:

$$\Delta_{total} = \frac{rf}{u}(l + mr^2 f^2 + nr^3 f^3) \tag{1}$$

Where r is the radius of water droplets in meters, f is the frequency in GHz and u, l, m, n, are constants given as.

$$u = 1.002(-0.071C^2W + 2.213 * C * W + 141.56W)^{0.6473} \tag{2}$$

$$l = \frac{1886\varepsilon''}{(\varepsilon'+2)+\varepsilon''^2} \tag{3}$$

$$m = 137 \times 103\varepsilon'' \left\{ \frac{6}{5} \left(\frac{7\varepsilon'^2 + 7\varepsilon''^2 + 4\varepsilon' - 20}{[(\varepsilon'+2)^2 + \varepsilon''^2]^2} \right) + \frac{1}{15} + \frac{5}{3[(2\varepsilon'+3)^2 + 4\varepsilon''^2]} \right\} \tag{4}$$

$$n = 379 + 104 \left\{ \frac{(\varepsilon'-1)^2(\varepsilon'+2) + [2(\varepsilon'-1)(\varepsilon'+2)-9] + \varepsilon''^4}{[(\varepsilon'+2)^2 + \varepsilon''^2]^2} \right\} \tag{5}$$

The complex dielectric permittivity of water is given by:

$$\varepsilon' = af^b \tag{6}$$

$$\varepsilon'' = cf^d \tag{7}$$

where ε' and ε'' are real and imaginary parts of dielectric constants and f is the frequency I GHz. a, b, c and d are constants which depends on Temperature in Kelvin are given as:

$$a = pX^{4.7} + q \tag{8}$$

where P=4.2224*10-6; q=18.75;

$$b1 = rT + s \text{ [From -10o C to 20o C]} \tag{9}$$

$$b2 = uT^2 + vT + w \text{ [From -30o C to -10o C]} \tag{10}$$

where r=-0.01929; s=4.776; u=-0.0002485; v=0.12314; w=-15.5068;

$$c = gX^3 + hX^2 + iX + j \tag{11}$$

where g=-0.0125269; h=0.89; i=-4.286; j=105.8;

$$d = kX^3 + lX^2 + mX + n \tag{12}$$

Where k=3.796*10-6; l=-1.151*10-4; m=-0.006538; n= -0.6667;

$$X = T - 243.15 \tag{13}$$

This new model presented in equation 6 and 7 are obtained by approximation of values obtained from ITU cloud model [9]. The method for calculating real and imaginary parts of dielectric constants of water ε' and ε'' are given in ITU model [9]. The equations of ITU cloud model are very complex in nature. Therefore we have designed the simpler model to calculate the dielectric constants. As this dielectric constant are dependent on both temperature and frequency. So first we find the values of dielectric constants for different frequencies from 10 GHz to 100 GHz at different temperature from 243.15 K to 293.15K. We choose this temperature ranges because researchers shows that the average temperature of clouds are lies at these ranges. From these calculated values we draw the graph of frequency in GHz vs ε' and ε'' at different temperatures ranges. Then we take an approximations using MATLAB curve fitting tool box. From there we got the approximations curve by using power function in the form of x=aYb. From this approximation function when applied to different temperatures ranges we got different values of a and b constants. Then again applied the approximations on the values of a and b differently we get the approximation equations 6 and 7. As we can see that these equations are much simpler then present in ITU cloud model for calculation of dielectric constants. The ε' and ε'' are obtained using new model and compared with the ITU cloud model.

III. RESULTS AND DISCUSSIONS

The implementation results of ITU model and New model are shown from Fig.2.1 to Fig.2.11. The models are implemented for different frequencies and for different temperatures. It has been observed that the model is very closely matches the ITU-R cloud model.

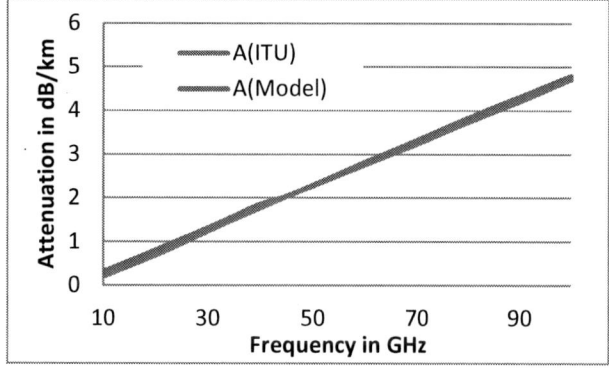

Fig. 1. ITU vs Proposed model for 243.15 K temperature

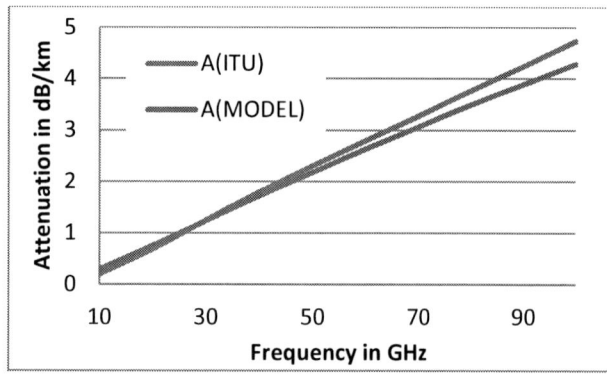

Fig. 2. ITU vs Proposed model for 248.15 K temperature

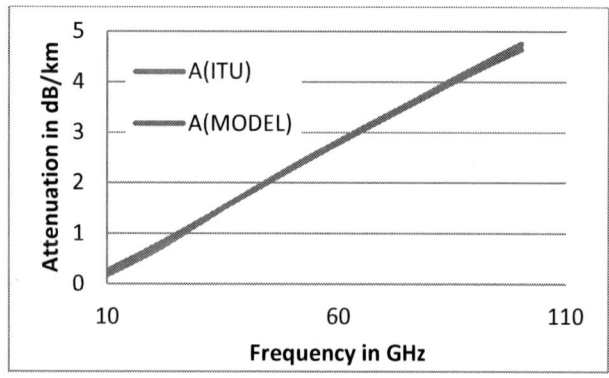

Fig. 3. ITU vs. Proposed model for 253.15 K temperature

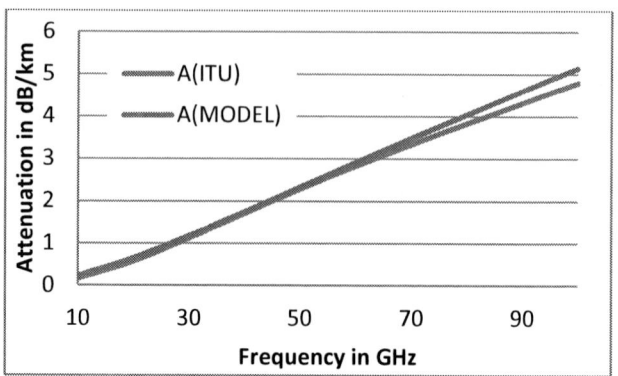

Fig. 4. ITU vs Proposed model for 258.15 K temperature

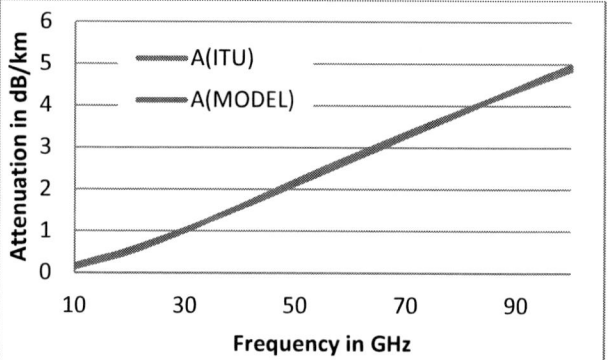

Fig. 5. ITU vs Proposed model for 263.15 K temperature

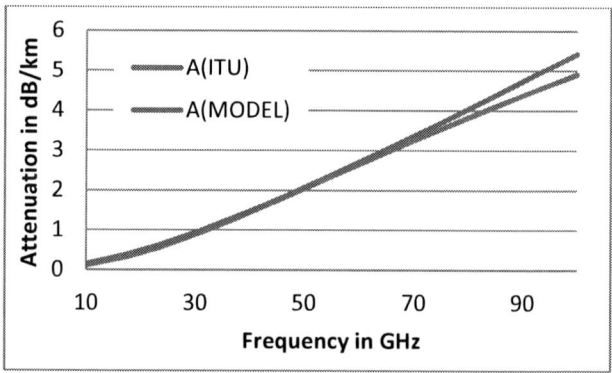

Fig. 6. ITU vs Proposed model for 268.15 K temperature

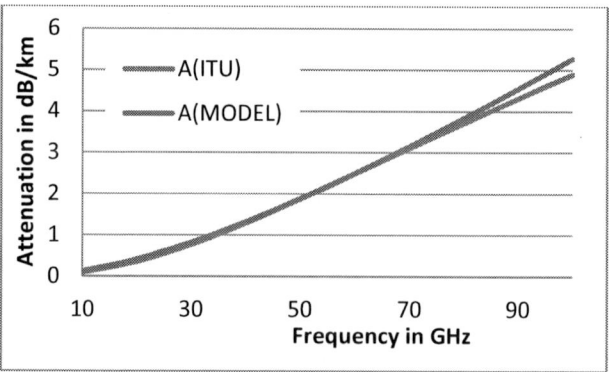

Fig. 7. ITU vs Proposed model for 273.15 K temperature

978-1-5386-6624-1/18 $31.00 © 2018 IEEE

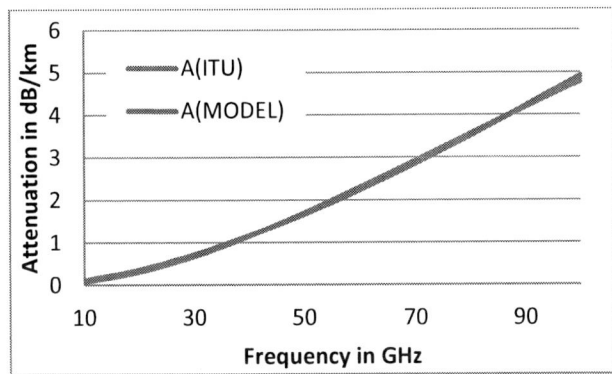

Fig. 8.   ITU vs Proposed model for 278.15 K temperature

For different temperature ranges it has shown that attenuation caused by clouds is around 2.5 dB for 60 GHz. For higher frequencies attenuation is higher. The results shows that the proposed model closely match the ITU model. The difference between proposed model and ITU model can be further studied with error analysis. It can be done by the following formula:

$$error = \sum_{i=1}^{n} \frac{|x_{i\,apr} - x_{i\,ITU}|}{x_{i\,ITU}} \cdot 100\% \qquad (14)$$

Where i is current value number (for given frequency) and n is the number of calculation points. Table 1 gives the results of the error analysis. From the results it has been observed that for the frequency 10 GHz the error is higher at 243.15 K around 49%. And it decreases with the increase in temperature and become 22% at 293.15 K. it is also observed that for frequencies above 20 GHz the error rate is within 10% and less.

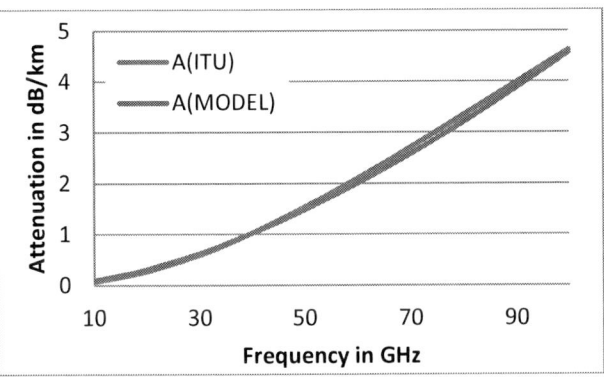

Fig. 9.   ITU vs Proposed model for 283.15 K temperature

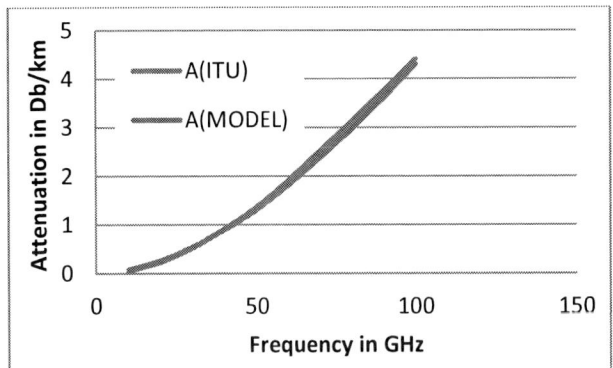

Fig. 10. ITU vs Proposed model for 288.15 K temperature

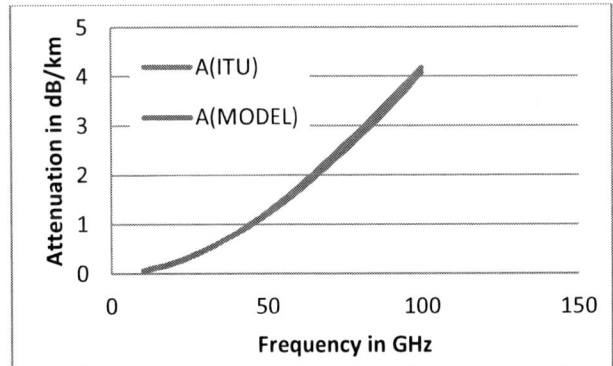

Fig. 11. ITU vs Proposed model for 293.15 K temperature

TABLE I.   ERROR ANALYSIS OF RESULTS

	Average % Error at Different Temperature in Kelvin										
f(GHz)	243.15	248.15	253.15	258.15	263.15	268.15	273.15	278.15	283.15	288.15	293.15
10	49.73	44.79	44.03	38.54	32.99	32.47	32.53	32.11	30.18	26.61	22.03
20	16.06	9.69	13.12	13.96	11.42	14.16	15.06	14.62	12.99	10.82	9.14
30	5.51	1.12	2.62	4.88	2.05	5.86	6.54	5.79	4.35	3.02	2.58
35	4.37	3.32	0.40	2.93	0.43	3.61	4.00	2.99	1.56	0.49	0.42
40	4.11	4.52	0.78	1.97	1.97	2.20	2.24	0.92	0.56	1.44	1.24
45	-0.02	5.16	1.32	1.69	2.83	1.45	1.09	0.57	2.14	2.89	2.51
50	0.58	5.50	1.48	1.85	3.19	1.21	0.45	1.58	3.28	3.96	3.46
55	1.18	5.72	1.44	2.30	3.18	1.37	0.22	2.19	4.05	4.70	4.13
60	1.70	5.92	1.32	2.90	2.91	1.84	0.32	2.47	4.51	5.17	4.57
65	2.10	6.15	1.20	3.59	2.46	2.56	0.71	2.47	4.71	5.40	4.80
70	2.36	6.44	1.11	4.28	1.90	3.46	1.32	2.23	4.67	5.41	4.85
75	2.48	6.81	1.10	4.95	1.28	4.50	2.12	1.78	4.43	5.25	4.73
80	2.47	7.25	1.16	5.56	0.62	5.64	3.07	-1.16	4.02	4.92	4.46
85	2.34	7.76	1.31	6.10	0.03	6.84	4.14	-0.38	3.45	4.44	4.06
90	2.11	8.33	1.53	6.55	0.66	8.08	5.30	0.52	2.75	3.82	3.54
95	1.79	8.94	1.84	6.91	1.26	9.34	6.55	1.53	1.92	3.09	2.91
100	1.41	9.59	2.21	7.17	1.80	10.59	7.84	2.62	0.99	2.25	2.17

IV.   CONCLUSION

From the proposed model for cloud attenuation at different frequencies above 10 GHz at various temperatures we concluded that the attenuate is significant at higher frequencies above 60 GHz. In this model simpler equations are used to calculate real and imaginary parts of dielectric constants as compared with the ITU model. This simpler model will help researchers and engineers to predict the attenuation caused due to clouds in less amount of time.

ACKNOWLEDGEMENT

This work was supported in part by the Grant DN07/19/15.12.2016 "Methods for Estimation and

Optimization of Electromagnetic Radiation in Urban Areas" of the Bulgarian Science Fund.

REFERENCES

[1] World Meteorological Organization, ed(1975). Cirrus, International Cloud Atlus.

[2] Gunn, Kenrich Lewis Stuart, and Thomas William Russell East. "The microwave properties of precipitation particles." Quarterly Journal of the Royal Meteorological Society 80.346 (1954): 522-545.

[3] Staelin, David H. "Measurements and interpretation of the microwave spectrum of the terrestrial atmosphere near 1□centimeter wavelength." Journal of Geophysical Research 71.12 (1966): 2875-2881.

[4] Slobin, Stephen D. "Microwave noise temperature and attenuation of clouds: Statistics of these effects at various sites in the United States, Alaska, and Hawaii." Radio Science 17.6 (1982): 1443-1454.

[5] Altshuler, Edward E., and Richard A. Marr. "Cloud attenuation at millimeter wavelengths." IEEE Transactions on antennas and propagation 37.11 (1989): 1473-1479.

[6] Liebe, Hans J. "MPM—An atmospheric millimeter-wave propagation model." International Journal of Infrared and millimeter waves 10.6 (1989): 631-650.

[7] Salonen, E., and S. Uppala. "New prediction method of cloud attenuation." Electronics Letters 27.12 (1991): 1106-1108.

[8] Dissanayake, Asoka, Jeremy Allnutt, and Fatim Haidara. "A prediction model that combines rain attenuation and other propagation impairments along earth-satellite paths." IEEE Transactions on Antennas and Propagation 45.10 (1997): 1546-1558.

[9] Dintelmann, F., and G. Ortgies. "Semiempirical model for cloud attenuation prediction." Electronics Letters 25.22 (1989): 1487-1488.

[10] Attenuation due to cloud and fog, Recommendation ITU-R P.840-5,P Series Radio wave propagation .

[11] Hitesh Singh, Prof.Peter Z. Petkov, Prof.Boncho G. Bonev and Sarang M. Patil," Cloud Attenuation Model at Millimeter wave frequency Bands ,"IEEE International Conference on Infocom Technologies and Unmanned Systems (ICTUS'2017)",Amity Directorate of Engineering & Technology (ADET),Dubai during 18th-20th December 2017

[12] H. Singh, R. Kumar, B. Bonev, P. Petkov, "Cloud Attenuation issues in Satellite Communications at millimetre Frequency Bands- State of Art", International Journal of Scientific & Engineering Research Volume 8, Issue 7,pp. 858-862, July-2017.

[13] H. Singh, R. Kumar, B. Bonev, P. Petkov, "The Studies of Millimeter Waves at different Frequencies in different Environmental conditions for 5G Applications – A State of Art", International Journal of Scientific & Engineering Research Volume 8, Issue 7,pp. 851-857, July-2017.

[14] Singh, Hitesh, Ramjee Prasad, and Boncho Bonev. "The Studies of Millimeter Waves at 60 GHz in Outdoor Environments for IMT Applications: A State of Art." Wireless Personal Communications 100, no. 2 (2018): 463-474.

[15] Singh, Hitesh, Boncho Bonev, and Ashoak Chandra. "Effects of Atmospheric Impairments of Satellite Link Operating in Ka Band." Wireless Personal Communications 101, no. 1 (2018): 425-437.

2nd IEEE International Conference on Power Electronics, Intelligent Control and Energy Systems (ICPEICES-2018)

An Efficient Weighted Trust Method for Malicious Node Detection in Clustered Wireless Sensor Networks

Bhavnesh Jaint
Department of Electrical Engineering
Delhi Technological University
Delhi, India
bhavneshmk@gmail.com

S.Indu
Department of Electronics and
Communication Engineering
Delhi Technological University
Delhi, India
s.indu@dce.ac.in

Vishwamitra Singh
Department of Electronics and
Communication Engineering
Delhi Technological University
Delhi, India
vishwamitra_bt2k15@dtu.ac.in

Neeta Pandey
Department of Electronics and
Communication Engineering
Delhi Technological University
Delhi, India
neetapandey@dce.ac.in

Lalit Kumar Tanwar
Department of Electronics and
Communication Engineering
Delhi Technological University
Delhi, India
lalittanwar55@yahoo.com

Abstract—**In this paper, we consider a wireless sensor network (WSN) that consists of sensor nodes (SN), cluster head (CH), forward node (FN) and a base station (BS). The information acquired by the sensor node is sent to the CH, all the CH's send the information to a FN which forwards it to the BS. Fast detection of malicious nodes is imperative to the performance of a WSN and therefore we study the weighted trust method for malicious node detection. We have considered two scenarios one with single cluster head without grid and other with multiple cluster head with non-overlapping grid. The results indicates that the scenario with multiple cluster heads with non-overlapping grid requires less time for malicious node detection with better accuracy as compared to the scenario with single cluster head without grid.**

Keywords—**weighted trust, sensor nodes, malicious nodes, cluster cased WSN,**

I. INTRODUCTION

Wireless Sensor Network (WSN) have a huge domain in applications varying from emergency response system, energy management, medical monitoring, logistics management, inventory management, and battlefield management [11]. WSN consists of a group of small devices which are called sensor nodes. These nodes are portable devices equipped with sensors, processing unit, memory unit, transceiver and power supply as shown in Fig.1. The sensor nodes are application specific and are intended to monitor specific parameters like temperature, pollutant levels at various locations, particle concentration in chemicals, pressure, intensity of sound/light, etc. These sensors are deployed randomly in a geographical region of interest. The sensor nodes collects information and send it to base station (BS) with the help of cluster head (CH) node and forward node (FN). The sensor nodes have limited power supply, usually a battery and therefore possess limited capabilities. The forwarding nodes are assumed to have high power and it collects and processes the data from lower level sensor nodes (SNs) and Base Stations (BSs) that act as media between Wireless Sensor Network (WSN) and wired network. The clusters are formed by equipartitioning the area under operation in four clusters. Each cluster head processes the data from all sensor points under it and pass the result to forwarding node. This scheme is based on

assumptions that the forwarding node and base stations are not malicious and thus are always trusted.

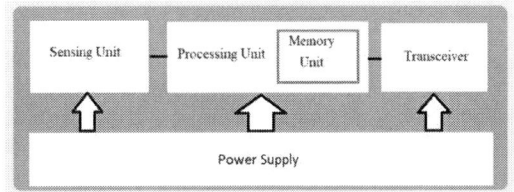

Figure 1.   WSN Sensor Node Components

Due to unattended nature of the sensor networks, an attacker could launch various attacks and compromise sensor nodes. The network should be robust against these attacks, and if an attack succeeds, its impact should be minimized. Compromising of one or few sensor node should not crash the entire network. Therefore the critical issues of security and performance have to be studied for wireless sensor networks.

There have been many techniques presented in literature for detection of malicious node in wireless sensor networks.

There have been many techniques presented in literature for detection of malicious node in the wireless sensor networks. Wireless Sensor Networks can be compromised due to many reasons such as finite battery life, finite memory space, and finite computing capabilities [1], [2], [3]. It is very essential to detect the malicious node and isolate it to prevent it from generating wrong results. Ad hoc networks without a definite structure are rarely good against any types of attacks, which can lead to a node being compromised easily [4]. They presented a method that if a node is treated as trusty by its neighboring, then the node will be declared fault free, and is not a malicious node. However, for it to work it must have a minimum number of nodes nearby, which is not guaranteed in sensor networks.

So the better idea as proposed by W. Du, L. Fang and P. Ning [6] to get the node to compare the data from the other nodes nearby around it with the data generated by the node itself. If the difference between the data is marginal, then the nodes are said to be trusty. This method gave a better result for the localized detection of malicious node.

978-1-5386-6624-1/18 $31.00 © 2018 IEEE

Weighted Trust Evaluation, used primarily for the detection of the malicious node is quite reliable. The entire sensor nodes are assigned a weight or amount of trust that is evaluated frequently. The trust decreasing every time the node provides wrong information. Once the trust decreases below the fixed threshold, the node is declared malicious and isolated [5] [7] [8] demonstrated the method by using a three-level hierarchical network with components as: Sensor Nodes (SN) whose function is to sense the parameter. SNs sends its data to Forwarding Node. Forwarding nodes are high powered that collects data from SNs and processes it and transmits to base station.

This scheme is based on one critical assumption, the forwarding nodes and Base station are never faulty as once a faulty forwarding node is entered it can launch attack in network [9] [10] [7]. [15] Presents extensive simulation in MATLAB for Extensive weighted trust evaluation emphasizing on response time and detection ratio [13] [14]. Focused on the security issues in WSN and described the various kinds of attacks [17] described various clustering algorithms in the network, these clustering algorithms describes a power efficient way to localize sensor nodes and increase reliability.

In this paper we study the propagation time required to detect the malicious sensor nodes using weighted trust evaluation scheme. We compare the time taken by a traditional weighted trust evaluation scheme [5] (one with three level hierarchical network consisting of Sensor Nodes, Forwarding Node and Base Stations) with time taken by one proposed cluster based weighted trust evaluation (one with hybrid topology sensing nodes, cluster head, forward node and base station).

The rest of the paper is organized as follows. Section II describes the security in WSN and related work. Network model and architecture to be used throughout the paper is presented in Section III. In Section IV Cluster based weighted trust evaluation scheme is presented. Simulation results are shown in Section V. This paper concludes in Section VI.

II. SECURITY IN WSN

Security in WSN is paramount importance and the following are the challenges faced while designing WSN:

Unfavorable harsh environment: Sensor nodes faces extremely harsh environmental conditions and are susceptible to damage through it or capture by attackers as the sensor nodes are exposed in an open area. The attackers can capture a sensor node and access to data or transmit false data through them.

Unattended operations: The sensor nodes are deployed in a hostile area which expose them to physical tempering and attacks.

Limited resources: The security of sensor nodes requires resources like energy, memory and storage capacity as the data are constantly monitored and/or recorded the shortage of resources of sensors brings out challenges to resource-intensive security mechanisms. These resources are very limited in a small sensor node.

Reliability in Wireless Communication: In WSN the sensor data may be distorted due to channel errors which may lead to conflicts, and at highly busy node the data may also be and thus Denial-of Service (DoS) attack can be easily launched. Due to the greater congestion at a single node the overload results in increasing the latency in the sensor network thus causing synchronization errors and lag in the system, including sensor nodes. Attackers can launch an attack in sensor network through malicious nodes. These attacks are broadly classified as Passive attack and Active attack. The passive attack is easier to analyze and is difficult to detect. In passive attack, attacker does not modify or exchange information, this type of attack is basically for finding the knowledge of some classified information [12].

In an active attack the attacker tries to alter the data transferred by the sensor node or overload the sensor by inserting the traffic to start denial of service attack.

III. NETWORK MODEL AND ARCHITECTURE

There are basically two components of network model i.e. WSN components and WSN networking topology. The components are explained as follows:

A. Components of WSN

Wireless Senor Network for environment monitoring for air pollution system consists of the sensor points or sensor nodes that are spread over the field. These sensor nodes consist of sensors that detects and monitors the concentration of pollutants in the environment. These sensor nodes are connected together through a wireless link to the cluster head that processes the collected readings from every individual node and forwards these data to the forwarding node that again processes the collected readings from individual cluster heads and propagates it to a base station shown in Fig.2.

B. Network Topology

In Wireless Sensor Network we present a clustered WSN architecture that uses a hybrid network topology which combines star, mesh and ring topologies together. These three topologies are used to increase the versatility of the system while maintaining reliability taking into account the available resources as following [16]:

- A group of the sensing nodes are connected to a central node i.e. cluster head that has other abilities like connectivity and data exchange from other cluster heads this includes the star topology thus resulting is a set of clusters. Star topology has advantages over other topologies that includes its scalability and power usage reduction.

- Cluster heads are connected together as a mesh to provide a reliable communication, clusters heads had additional resources that facilitate the use of such topology. All clusters heads are connected directly to the forwarding node which acts as a path from wireless to wired network i.e. to the outside world. Fig.2 depicts the overall architecture of WSN using hybrid network topology.

978-1-5386-6624-1/18 $31.00 © 2018 IEEE

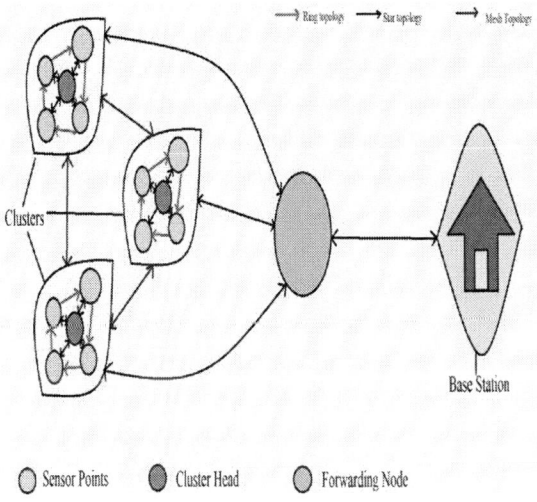

Figure 2. System Architecture

C. Error Detection and Weight Reduction Formula

In detecting malicious nodes, we employ trust values of sensor node to reflect their track record in decision making process. Each cluster head maintains trust values of its associated sensor nodes. The trust value (V_w) lies in the range (0,1) and is initialized to 1 for each sensor node. The weight represents the sensor node dependability and the sensor node with a higher weight is more trustworthy. Updating the trust values is important to maintain the fidelity of the readings obtained from the sensors. To determine the non-erroneous node the weighted threshold value should be greater than the minimum threshold value.

$$V_{th} \leq V_w$$

$$V_w = \frac{M - |D|}{M}$$

Where, V_{th} is the minimum threshold weight and V_w is the weighted threshold. M is the Average or mean value of the sensor at a point. (M is the user defined value). D is the deviation from the M. D is dependent on sensor input value and can be calculated as ($M - I$). Where I is the Input value from the sensor.

For example in a tropical rainforest thee temperature varies from 21 to 30 degrees then the value of mean (M) would be 25.5 degrees. If a particular sensor detects the temperature at that time the detected value would be I. Let the detected value be 34 degrees. Now the calculated threshold value would be

$$V_w = \frac{25.5 - |25.5 - 34|}{25.5}$$

Thus, V_w is calculated to be 0.67. Assuming the minimum threshold to be set to 0.7, the equation $V_{th} \leq V_w$ is not satisfied. Thus the corresponding sensor node is an erroneous node.

Weight Modification

If the sensor node is found as erroneous then the weight is reduced according to following formula.

$$W' = W + F \times W$$

Where W' Is the modified weight, W is the current weight and F is weight penalty factor.

IV. CLUSTER BASED WEIGHTED TRUST EVALUATION

The goal was to come up with a prototype of the enhanced weighted trust evaluation scheme that detect malicious Sensor Nodes and Cluster Heads. The performance requirements that the scheme should meet are short response time, high detection ratio and low misdetection ratio. Response time refers to the average number of cycles required by the scheme to correctly detect malicious nodes, detection ratio refers to the ratio of malicious nodes correctly detected by the scheme to the total number of malicious sensor nodes present in the WSN. Several sensor nodes "n" are deployed randomly in the field, a subset of them are elected as the forwarding nodes whereas the rest become the ordinary sensor node (SN). The sensor nodes organize themselves to form a clustered operational network.

V. SIMULATION RESULTS

The WTE based detection algorithm was installed in the cluster node for monitoring of all the member sensor nodes.

TABLE I. Simulation Parameters

Minimum threshold weight	0.7
Number of clusters	4
Number of repetitions of samples	5
Network field dimension	100x100
Sensor Nodes in field	100
Weight Penalty Factor	20%
Malicious nodes at deployment	20%

Though the field parameters are user inputs but heterogeneous network of hundred sensor nodes are deployed randomly in the area with a dimension of [100x100] the simulation is as shown in Fig.4

The Sensors are grouped in clusters and the weighted trust evaluation algorithm performs over it. As shown in figure 5 the malicious nodes are displayed in respective clusters. This method is performed for 5 times and the propagation time is compared in figure 6. It is shown that the propagation time decreases with clustering and increases without clustering as the number of samples increases.

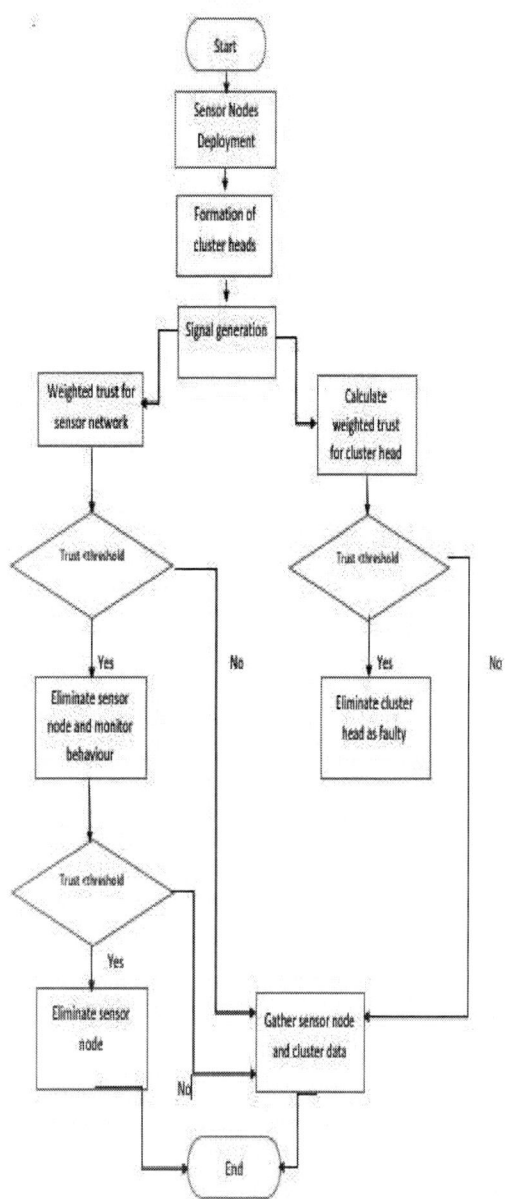

Figure 3.   Algorithm for Weighted Trust Evaluation

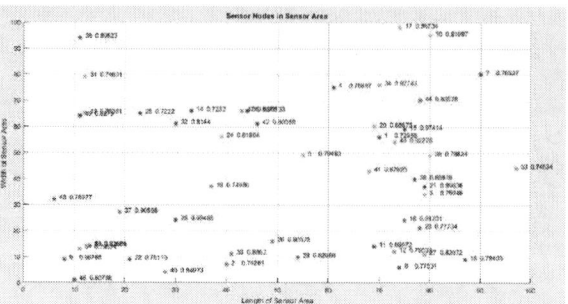

Figure 4.   Plot of all the Sensor Nodes in an area

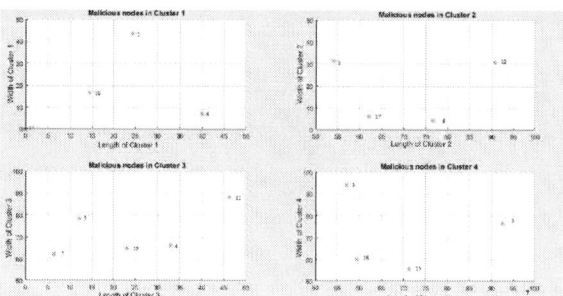

Figure 5.   Plot of all the Malicious Nodes in respective forwarding node

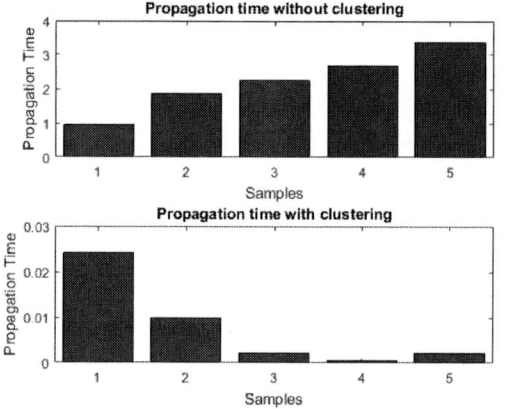

Figure 6.   Time taken to find malicious nodes in

    i.   Whole area at once and then find propagation time.

    ii.   Considering individual clusters and then find propagation time including all forwarding nodes.

Though the field parameters are user inputs but heterogeneous network of hundred sensor nodes are deployed randomly in the area with a dimension of [100x100] the simulation is as shown in Fig.4

The Sensors are grouped in clusters and the weighted trust evaluation algorithm performs over it. As shown in figure 5 the malicious nodes are displayed in respective clusters. This method is performed for 5 times and the propagation time is compared in figure 6. It is shown that the propagation time decreases with clustering and increases without clustering as the number of samples increases.

Detection Ratio:

Detection Ratio (DR) is the ratio between the number of malicious nodes detected and the total number of malicious nodes in the network (set at the time of deployment). In one of the simulation runs, the percentage of malicious nodes, is set to 20% ($m = 0.2$). This means that:

Malicious nodes = 20/100 * n

Where n = number of deployed sensor nodes i.e. =100.

Malicious nodes =20

There is one more set of malicious nodes that is cluster head

Malicious cluster head = m *(p*n)

=20/100 *(4/100 * 100) = 0.8=1

Where p= percentage of cluster head in the network =4/100.

978-1-5386-6624-1/18 $31.00 © 2018 IEEE

TABLE II. Parameter Vs Nodes

Total Malicious Nodes (At Deployment)	20	40	60	80
Detected Malicious Nodes (Simulation)	18	33	38	29
Detection Ratio (DR)	0.9	0.83	0.63	0.36

Malicious ordinary sensor nodes = m *(n- (p*n))
=20/100 * (100 – (4/100* 100)) = 19.2= 19

The number of detected malicious ordinary sensor nodes is 17 out of the 19 (Through Simulation) that had been set as malicious whereas all the malicious cluster heads are detected by the method.

Detection Ratio = No. of correctly detected malicious nodes/Total no. of malicious nodes in the network.

DR = (17 + 1) / 20 = 0.90

In figure 7, we plot the number of malicious nodes vs detection ratio. It is observed that the detection ratio decreases with the increase in malicious nodes.

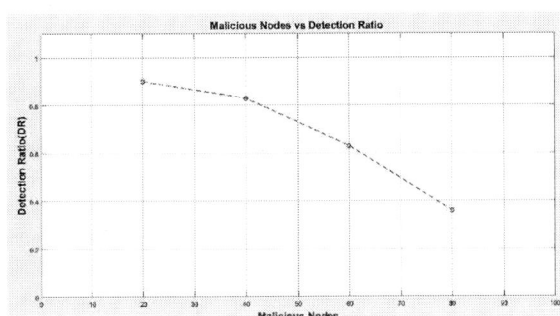

Figure 7.   Malicious Nodes vs Detection Ratio

VI. CONCLUSION

The simulation clearly shows that the propagation time in second case (when individual clusters are considered independently instead of whole area at once) is much smaller (nearly1times less).Also 90% malicious nodes are detected when 20 malicious nodes at deployment are considered. If the number of clusters are increased then the Detection Ratio would increase.

In this paper simulation results are being reported and they shows that algorithm can be applied to a flexible number of sensor nodes that operate under a cluster head, it thus has achieved better scalability with a considerable detection rate and less propagation time.

REFERENCES

[1] E. Ayday, F. Delgosha, and F. Fekri, "Location-Aware Security Services for Wireless Sensor Networks using Network Coding," *Infocom*, May 2007.

[2] S. Zhu, S. Setia, and S. Jajodia, "LEAP: Efficient Security Mechanisms for Large-Scale Distributed Sensor Networks," *CCS'03*, October 2003.

[3] C. Karlof, N. Sastry, and D. Wagner, "TinySec: A Link Layer Security Architecture for Wireless Sensor Networks," *ACM Sensys*, November 2004.

[4] Haiyun Luo, Petros Zerfos, Jiejun Kong, Songwu Lu, Lixia Zhang, "Self-securing Ad Hoc Wireless Networks," IEEE ISCC (IEEE Symposium on Computers and Communications) 2002, Italy.

[5] Idris M. Atakli, Hongbing Hu, Yu Chen, Wei-Shinn Ku and Zhou Su, "Malicious Node Detection in Wireless Sensor Networks using Weighted Trust Evaluation", SpringSem, pp. 836-843, 2008.

[6] W. Du, L. Fang, and P. Ning, "LAD: Localization Anomaly Detection for Wireless Sensor Networks," the 19th International Parallel and Distributed Prioccessing Symposium (IPDPS'05), April 3 – 8, 2005, Denver, Colorado, USA.

[7] I. M. Atakli, H. Hu, Y. Chen, W. S. Ku and Z. Su, Malicious Node Detection in Wireless Sensor Networks, The Symposium on Simulation of Systems Security (SSSS"08), Ottawa, Canada, p. 838, 2008.

[8] S. Zhao, K. Tepe, I. Seskar and D. Raychaudhuri, Routing Protocols for Self-Organizing Hierarchical Ad-Hoc Wireless Networks, Proceedings of the IEEE Sarnoff Symposium, Trenton, NJ,, March 2013.

[9] K. Sumathi and D. M. Venkatesan, A Survey on Detecting Compromised Nodes in Wireless Sensor Networks, (IJCSIT) International Journal of Computer Science and Information Technologies, vol. 5, pp. 7720-7722, 2014.H. Hu, Y. Chen, W.-S. Ku, Z. Su and C.-H. J. Chen, Weighted trust evaluation-based malicious node detection for wireless sensor networks, Int. J. Information and Computer Security, vol. 3, no. 2, p. 148, 2009.

[10] R. Sharma and N. Tripathi, Comprehensive Review on Wireless Sensor Networks, Oriental Journal of Computer Science & Technology, Vol. 8, No. 1, pp. 59-64, April 2015.

[11] D. G. Padmavathi and M. D. Shanmugapriya, A Survey of Attacks, Security Mechanisms and Challenges in Wireless Sensor Networks, International Journal of Computer Science and Information Security,, vol. 4, 2009.

[12] Rajkumar, Vani B. A., G. Rajaraman, Dr. H. G. Chandrakanth, "Security Attacks and its Countermeasures in Wireless Sensor Networks", Int. Journal of Engineering Research and Applications, Vol. 4, Issue 10 (Part-1), pp. 04-15, October 2014.

[13] Mohamed-Lamine Messai" Classification of Attacks in Wireless Sensor Networks" International Congress on Telecommunication and Application'14University of A.MIRA Bejaia, Algeria, 23-24 APRIL 2014.

[14] Koriata P. Tuyaa and W.Okelo-Odongo, "Enhanced weighted Trust Scheme for detection of malicious nodes in Wireless Sensor Networks", International journal of computer applications (0975-8777), Vol 155-No.4, December 2016.

[15] T. Sathyamoorthi, D. Vijayachakaravarthy, R. Divya And M. Nandhini, A Simple and Effective Scheme To Find Malicious Node In Wireless Sensor Network, International Journal of Research In Engineering And Technology, Vol. 03, No. 02, 2014.

[16] X. liu and J. Shi, "Clustering Routing Algorithms in Wireless Sensor networks: an overview",KSII Transactions on Internet and Information Systems, Vol 6, No.7, 2012 :pp.1735-1755.

978-1-5386-6624-1/18 $31.00 © 2018 IEEE

Performance Analysis of Non-coherent Modulations over FTR Fading Model

Veenu Kansal
Department of ECE
Punjabi University
Patiala, Patiala, India.
veenukansal@outlook.com

Harpreet Kaur
Department of ECE
Punjabi University
Patiala, Patiala, India.
hksaini55@gmail.com

Simranjit Singh
Department of ECE
Punjabi University
Patiala, Patiala, India.
simranjit@live.com

Abstract—In this research paper, an exact error performance analysis for non-coherent modulation schemes over fluctuating two ray fading model (FTR) has been analyzed. Non-coherent modulation schemes can be the preferable mean to analyze the error probability in a practical scenario. Here the focus is done also on the comparative analysis of various non-coherent modulation techniques for FTR model. This model has been considered as a generalization of two wave diffused power (TWDP) fading model in which two specular components are considered to be fluctuating because of variation in amplitude and random phases along with other diffuse components instead of constant amplitudes. This model effectively deals with the problem of spectrum shortage because of its accurate measurements at millimetre (mm) scale of the wave spectrum. In this paper, system performance is studied by calculating average bit error rate (ABER) over FTR fading model by observing variation in various parameters such as K, m and Δ. The derived results are compared with the results of special cases calculated in the literature.

Keywords—FTR fading model; non-coherent modulations;average bit error rate.

I. INTRODUCTION

With the fastest emerging trends in the field of wireless communication system, a shortage in the wireless spectrum and degradation in transmission environment is quite natural. Along with these complexities, channel phase estimation and tracking a coherent demodulation reference signal seems to be a bit difficult. In a practical scenario, all the information regarding phase may be completely or partially lost because of fading, usage of unreliable or cheaper local oscillators at receiver structures and other obstructions. Due to these difficulties of exact phase estimation and to avoid the circuit complexity analysis for non-coherent detection is comparatively easier. Hence the non-coherent detection may be considered as preferable mean to analyze BER in a practical scenario [1].

Non-coherent detection is the case in which the receiver has neither a cognition of phase and nor a user's channel envelope. In this paper, the expressions have been derived for the ABER of FTR model for various non-coherent modulation techniques such as differential phase shift keying (DPSK), non-coherent frequency shift keying (NCFSK) and non-coherent M-ary frequency shift keying (NCMFSK).

FTR model is considered as an accurate model for the measurements at mm wave spectrum level. With the emergence of remarkable inventions in the field of wireless communication available spectrum seems to be shorter somewhere to fulfil the requirements of users. So, to deal with such types of problems a new fading model i.e. FTR fading model has been recently proposed. Millimetre wave spectrum was earlier used for satellite communication and for military applications. But to meet up the needs of today's era this spectrum is now used for local communication. A problem of high attenuation of millimetre waves (mmW) and effect of attenuation due to rain can be tackled with high gain antennas [2-3].

At this range of wireless spectrum Rayleigh and Rician fading model do not seem to fit to measure the random fluctuations occur in amplitudes. Also, these fading models do not capture the bimodality of mmW. At this stage, FTR is presumed to be well suited to capture the bimodality and unimodality.

To analyze the system performance in most basic kind of fading environment the error probability is calculated for the Rayleigh fading model for the M-ary phase shift keying and DPSK without using any diversity [4]. Similarly to detect the system performance for M-ary DPSK affected by Nakagami fading, average symbol error rate (ASER) expressions have been derived. Here, fading is considered to be slow and non-selective in order to avoid the amplitude variations in received signal for each symbol duration [5].

In the similar way, to examine the system performance in the line of sight environment, closed-form BER expressions for Rice fading channel have been studied for different modulation schemes by using moment generating function (MGF) approach in [6]. Further, to scrutinize the effect of diversity in general fading environment BER performance for optimum combining diversity over Rayleigh fading channel for NCFSK has been examined in [7]. Also to examine the TWDP fading scenario with NCFSK and DPSK, expressions for ABER have been derived by using post detection equal gain combining (EGC) with the MGF approach in [8].

To analyze the system performance of TWDP fading for M-ary PSK modulation the expressions for ABER have been derived in [9]. Similarly to check the system performance in fluctuating dominant power components environment, the analysis for ABER of coherent modulations is presented in the newly developed FTR fading model in [10]. But no work is done on the performance of FTR fading channel for non-coherent modulation schemes. So, this paper accentuates on the performance analysis under FTR fading scenario with non-coherent modulations.

II. SYSTEM MODEL

FTR fading model is considered as an effective model for analyzing the small scale fluctuations occurred in the amplitudes of transmitting signals. The fluctuations occurred in the amplitude of a received signal sent over a communication channel can be examined through the superposition of cluster of specular power components along with other diffused components.

The probability density function (PDF) of this FTR model can be evaluated by [10-12]:

$$f_\gamma(x) = \frac{1}{2^{m-1}} \times \frac{1+K}{\bar{\gamma}} \times \left(\frac{m}{\sqrt{(m+K)^2 - K^2\Delta^2}} \right)^m \times$$

$$\sum_{q=0}^{\left\lfloor \frac{m-1}{2} \right\rfloor} (-1)^q C_q^{m-1} \left(\frac{m+K}{\sqrt{(m+K)^2 - K^2\Delta^2}} \right)^{m-1-2q} \times \quad (1)$$

$$\times \Phi_2^{(4)} \begin{pmatrix} 1+2q-m, m-q-\frac{1}{2}, m-q-\frac{1}{2}, 1-m; 1; -\frac{m(1+K)x}{(m+K)\bar{\gamma}}, \\ -\frac{m(1+K)x}{(m+K(1+\Delta))\bar{\gamma}}, -\frac{m(1+K)x}{(m+K(1-\Delta))\bar{\gamma}}, -\frac{m(1+K)x}{\bar{\gamma}} \end{pmatrix}$$

where K indicates the ratio of average power of two line of sight components to remaining power of other diffused components ,parameter $\Phi_2^{(4)}$ represents the confluent series of 4 variables, Δ indicates relative strength of two line of sight components and $\lfloor . \rfloor$ represents the floor function.

In similar way, the cumulative distribution function (CDF) of FTR model is evaluated by [10]:

$$F_\gamma(x) = \frac{x}{2^{m-1}} \times \frac{1+K}{\bar{\gamma}} \times \left(\frac{m}{\sqrt{(m+K)^2 - K^2\Delta^2}} \right)^m \times$$

$$\sum_{q=0}^{\left\lfloor \frac{m-1}{2} \right\rfloor} (-1)^q C_q^{m-1} \left(\frac{m+K}{\sqrt{(m+K)^2 - K^2\Delta^2}} \right)^{m-1-2q} \times \quad (2)$$

$$\Phi_2^{(4)} \begin{pmatrix} 1+2q-m, m-q-\frac{1}{2}, m-q-\frac{1}{2}, 1-m; 2; -\frac{m(1+K)}{(m+K)\bar{\gamma}}x, \\ -\frac{m(1+K)}{(m+K(1+\Delta))\bar{\gamma}}x, -\frac{m(1+K)}{(m+K(1-\Delta))\bar{\gamma}}x, -\frac{m(1+K)}{\bar{\gamma}}x \end{pmatrix}$$

A coefficient in both (1) and (2) can be represented in given form as below:

$$C_q^{m-1} = \frac{\left| (2m-2-2q) \right|}{\left| q \right| \left| (m-1-q) \right| \left| (m-1-2q) \right|} \quad (3)$$

III. ABER PERFORMANCE ANALYSIS

In this section, a generalized closed-form expression has been derived for various non-coherent modulation schemes such as DPSK, NCFSK and NCMFSK over FTR fading model. An ABER can be calculated by averaging conditional BER for a given modulation scheme over the PDF of given fading channel. Mathematically it can be represented as:

$$P_e = \int_0^\infty P_e(x) f_\gamma(x) dx \quad (4)$$

where $P_E(x)$ represents the conditional BER of a modulation scheme and $f_\gamma(x)$ represents the PDF of fading channel.

For simplicity, (1) and (2) can also be rewrite by using below identity:

$$P_n(z) = \frac{1}{2^n} \times \sum_{q=0}^{n/2} C_q^n z^{n-2q} \quad . \quad (5)$$

where $P_n(z)$ is is Legendre polynomial of degree n.

Alternatively, ABER can also be evaluated by doing integration by parts of (4) in terms of CDF as:

$$P_e = -\int_0^\infty P_e'(x) F_\gamma(x) dx \quad (6)$$

where $P_e'(x)$ represents the first order derivative of conditional BER and $F_\gamma(x)$ represents the CDF of given fading channel.

The conditional BER for a non-coherent modulation is expressed by [13]:

$$P_e(x) = a \exp(-bx) \quad (7)$$

where a and b are dependent on modulation schemes.

TABLE I.   VALUES OF 'a' AND 'b' BFOR DIFFERENT NON-COHERENT MODULATION SCHEMES

Modulation Type	Parameters	
Non-coherent	a	b
NCFSK	0.5	0.5
DPSK	0.5	1
NCMFSK	$\frac{1}{2(M-1)}\sum_{h=2}^{M}(-1)^h C_h^m$	$\frac{h-1}{h} \times \log_2 M$

978-1-5386-6624-1/18 $31.00 © 2018 IEEE

$$\overline{P_e} = \frac{a.b}{2^{m-1}} \times \frac{1+K}{\overline{\gamma}} \left(\frac{m}{\sqrt{(m+K)^2 - K^2\Delta^2}} \right)^m \sum_{q=0}^{\left\lfloor \frac{m-1}{2} \right\rfloor} (-1)^q C_q^{m-1} \left(\frac{m+K}{\sqrt{\left(m+K\right)^2 - K^2\Delta^2}} \right)^{(m-1-2q)} \times$$

$$\int_0^\infty x \times \exp\left(-bx\right) \times \Phi_2^{(4)} \left(1+2q-m, m-q-\frac{1}{2}, m-q-\frac{1}{2}, 1-m; 2; \frac{-m(1+K)}{(m+K)\overline{\gamma}}x, \frac{-m(1+K)}{(m+K(1+\Delta))\overline{\gamma}}x, \frac{-m(1+K)}{(m+K(1-\Delta))\overline{\gamma}}x, \frac{-m(1+K)}{\overline{\gamma}}x \right) dx \quad , \quad (9)$$

$$\overline{P_e} = \frac{a}{2^{m-1}b} \times \frac{1+K}{\overline{\gamma}} \times \left(\frac{m}{\sqrt{(m+K)^2 - K^2\Delta^2}} \right)^m \times \sum_{q=0}^{\left\lfloor \frac{m-1}{2} \right\rfloor} (-1)^q C_q^{m-1} \left(\frac{m+K}{\sqrt{(m+K)^2 - K^2\Delta^2}} \right)^{(m-1-2q)} \times$$

$$\left(1 + \frac{m(1+K)}{b(m+K)\overline{\gamma}} \right)^{-(1+2q-m)} \times \left(1 + \frac{m(1+K)}{b\left(m+K\left(1+\Delta\right)\right)\overline{\gamma}} \right)^{-\left(m-q-\frac{1}{2}\right)} \times \quad . \quad (10)$$

$$\left(1 + \frac{m(1+K)}{b\left(m+K\left(1-\Delta\right)\right)\overline{\gamma}} \right)^{-\left(m-q-\frac{1}{2}\right)} \times \left(1 + \frac{1+K}{b\overline{\gamma}} \right)^{m-1}$$

The first order derivative of (7) can be calculated as:

$$P_e^{'}(x) = -a.b\exp(-bx) \quad (8)$$

By introducing (2) and (8) in (6), ABER can be obtained as in (9). Using [14, 4.24.5] in (9), the ABER of generalized non-coherent modulation for FTR fading model is calculated as in (10). The expression calculated in (10) is the final generalized ABER expression over FTR fading model where '*a*' and '*b*' are dependent on given modulation schemes as given in Table I.

IV. NUMERICAL RESULTS

The analytical derived closed-form expressions of ABER in Section 3 are drawn for several values of *K*, *m* in this section. In figs.1-5, the ABER plots against average SNR are plotted for various non-coherent modulation schemes over FTR model. Fig.1 shows the ABER plot for NCFSK modulation for diverse values of *m* and for fixed values of *K*=5 and for Δ= 0.2. From the plotted results it is noticed that as we increase the value of *m*, the ABER decreases more rapidly. So, the higher value of *m* leads to better system performance. Similarly to observe the significant variation in the ABER, similar results have been drawn in Fig.2 for varying values of at *K*=*m*=5 and in Fig.3 by varying *K* for *m*=5 and Δ = 0.2. In Fig.2 it has been noticed as the power of dominant power components increases, the fading severity decreases.

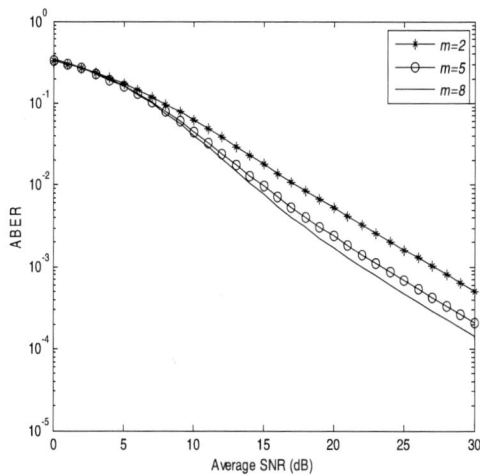

Fig. 1. ABER of NCFSK versus average SNR

To observe the effect of given parameters for DPSK modulation, ABER is plotted for varying values of *m* and for fixed value of *K*=5 in Fig.4. The plotted results demonstrate that *m* and *K* play a vital role to improve the system performance as their value increases. Also, it has been observed that smaller values of yield better performance.

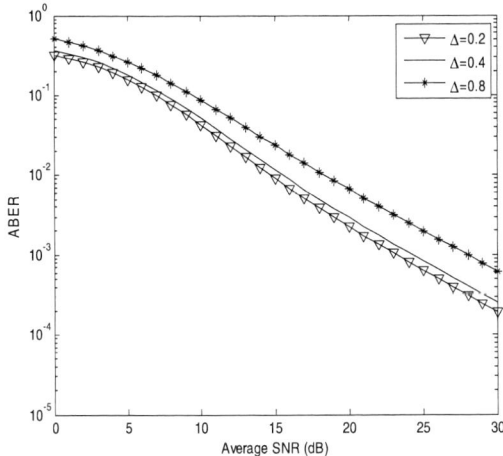

Fig. 2. ABER versus average SNR for distinct values of Δ and for K=5, m=5 for NCFSK modulation.

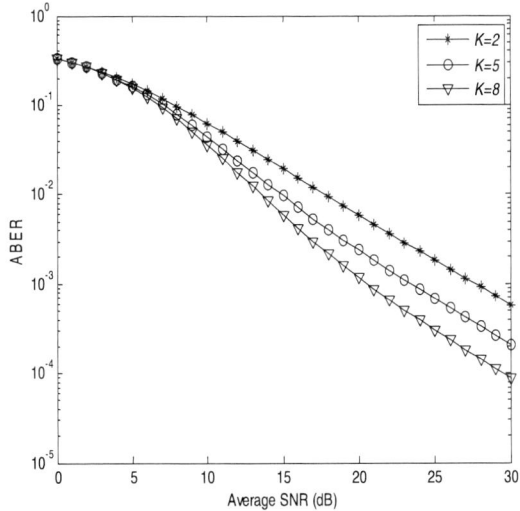

Fig. 3. ABER of NCFSK modulation against average SNR.

Fig.5 depicts the ABER plot for NCMFSK for several values of M and m and for constant K=5 and Δ= 0.2. Similar response of ABER variation as in DPSK and NCFSK has been observed for variation of m. Fig.6 shows the comparison plot for above discussed modulations. From the plot it is proclaimed that DPSK gives better results than NCFSK and NCMFSK for a given values of m and K and Δ. These observations agree with the special cases of FTR discussed in [10].

V. CONCLUSION

In this paper, ABER expressions have been evaluated for various non-coherent modulation schemes as DPSK, NCFSK and NCMFSK. From the divulged results it can be proclaimed

that ABER decreases significantly as the value of average SNR increases. Also augmentation of m and K leads to better system performance. Apart from these two parameters, Δ also plays a pivot role to analyze the system performance. ABER decreases much rapidly as the value of Δ decreases. From the plotted results it has been concluded that the dissimilar dominant power components experience lighter fluctuations than similar specular components and thus they yield better performance. The comparison plot shows that DPSK lead to better system performance in comparison with NCFSK and NCMFSK for given parameters.

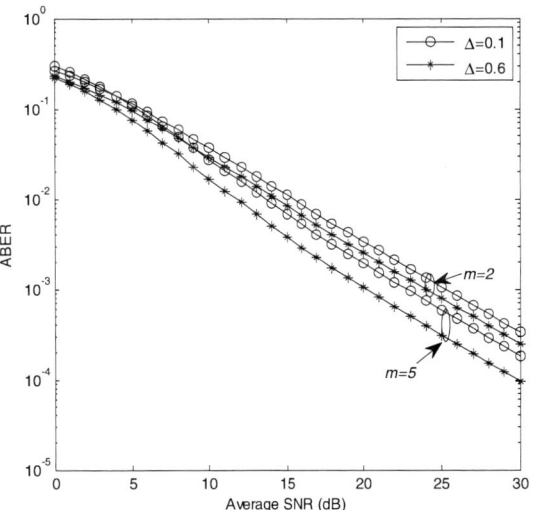

Fig. 4. ABER of DPSK modulation against average SNR.

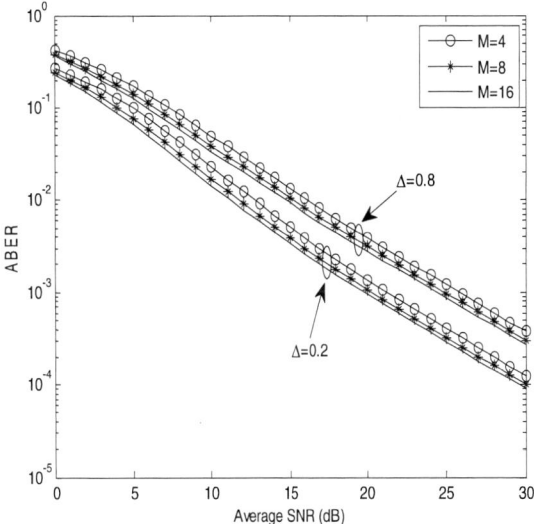

Fig. 5. ABER of NCMFSK modulation against average SNR for segregate values of M and Δ with K=5, m=5.

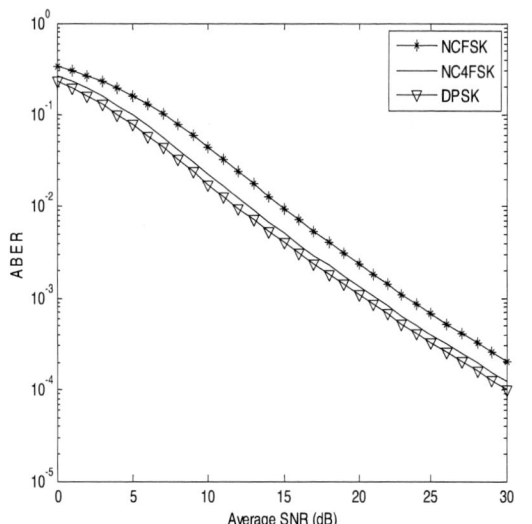

Fig. 6. Comparison plot for the given non-coherent modulations for K=5, m=5 and Δ = 0.2.

REFERENCES

[1] A. A. Nasir, H. Mehrpouyan, D. Matolak, S. Durrani, "Non-coherent FSK: An attractive modulation set for millimetre-wave communications," in Proc. IEEE Wireless Commun. and Networking Conf., Sept. 2016, pp. 1-7.

[2] M. R. Akdeniz, et al., "Millimeter Wave Channel Modeling and Cellular Capacity Evaluation," IEEE J. Sel. Areas in Commun., vol. 32, no. 6, pp. 1164–1179, Jun. 2014.

[3] T. S. Rappaport, G. R. Mac, M. K. Samimi, S. Sun, "Wideband Millimeter-Wave Propagation Measurements and Channel Models for Future Wireless Communication System Design," IEEE Trans. on Commun., vol. 63, no. 9, pp. 3029-3056, Sep. 2015.

[4] C. K. Paw, D. L. Schilling, "Probability of error for M-ary PSK and DPSK on a Rayleigh fading channel," IEEE Trans. on Commun., vol. 36, no. 6, pp. 755-756, Jun. 1988.

[5] G. Fedele, "Error probability for detection of M-DPSK signals in slow non selective Nakagami fading," IEEE Elect. Lett., vol. 30, no. 8, pp. 620-622, Jan. 1994.

[6] Y. Ma, T. J. Lim, "Bit error probability for MDPSK and NCFSK over arbitrary Rician fading channels," IEEE J. on Sel. Areas in Commun., vol. 18, no. 11, pp. 2179-2189, Nov. 2000.

[7] M. K. Simon, M. S. Alouini, "Average bit-error probability performance of optimum combining diversity of non-coherent FSK over Rayleigh fading channels," IEEE Trans. on Commun. vol. 51, pp. 566-69, Apr. 2003.

[8] W. S. Lee, "Performance of post detection EGC NCFSK and DPSK systems over Two-wave with diffuse power fading channels," in Proc. IEEE Int. Symp. on Commun. and Inf. Tech., pp. 911-915, Dec. 2007.

[9] S. Singh, V. Kansal, "Performance of M-ary PSK over TWDP fading channels," Int. J. Elect. Letters, vol. 4, no. 4, pp. 433-37, Aug. 2015.

[10] J. M. Romero, F. J. Martinez, J. Paris, A. J. Goldsmith, "The Fluctuating Two-Ray Fading Model: Statistical Characterization and Performance Analysis," IEEE Trans on Wireless Commun., vol. 99, pp. 1-11, May 2017.

[11] D. Dixit, P. R. Sahu, "Performance of Multihop Detect-And-Forward Relayimg System over Fluctuating Two-Ray Fading Channels," Trans. on Telecommun. Emerging Technol., pp. 1-14. 2018.

[12] W. Zeng, J. Zhang, S. Chen, K. Peppas, "Physical Layer Security over Fluctuating Two-Ray Fading Channels," IEEE Trans. on Veh. Technol., pp. 1-5, 2018.

[13] M. K. Simon, M. S. Alouini, Digital Communication over Fading Channels (1st ed.). New York: Wiley, 2000.

[14] M. Abramowitz, I. A. Stegun, Handbook of Mathematical Functions with Formulas, Graphs, and Mathematical Tables, 1972.

2nd IEEE International Conference on Power Electronics, Intelligent Control and Energy Systems (ICPEICES-2018)

A Novel Approach for Predicting Attenuation of Radio Waves caused by Rain

Hitesh Singh
Dept. of RCVT
Technical University of Sofia
Sofia, Bulgaria
hitesh.singh.85@gmail.com

Boncho Bonev
Dept. of RCVT
Technical *University of Sofia* Sofia, Bulgaria
bbonev@tu-sofia.bg

Peter Petkov
Dept. of RCVT
Technical University of Sofia Sofia, Bulgaria
pjpetkov@tu-sofia.bg

Ravinder Kumar
Dept. of CSE,
HMRITM Affiliated with GGSIPU Delhi, INDIΛ
ravinder_y@yahoo.com

Abstract— **The radio wave propagation studies at higher frequencies above 10 GHz plays an important role for outdoor environmental conditions. Attenuation caused due to the rain plays a significant role while designing the radio links. This paper discusses the ITU-R model of rain and its limitations and proposes the new model. As in ITU model in order to calculate attenuation values of regression coefficients k and α should be known, which are dependent on frequency and polarization. In proposed model these coefficients are not required in order to calculate attenuation. The results of proposed model also match with the ITU model.**

Keywords— **Rain Attenuation, satellite communication, Millimeter waves, precipitation.**

I. INTRODUCTION

The higher frequency especially from 10 GHz to 100 GHz is more sensitive to environmental conditions. Studies of radio wave propagation in the conditions like precipitation are very important. Various researchers have done the propagation studies for precipitation out of which rain rain is the important factor.

Rain is the type of precipitation which is a product of condensation of atmospheric water vapor that fall under gravity. There are two types of rainfall namely stratiform and convective. Convective rain usually falls from cumulus and cumulonimbus clouds while stratiform rain precipitate from nimbostratus clouds [1].

In order to study the attenuation caused due to rain, researchers has also know about the area confined to rain. It is called rain cell. The term "cell" is preferred when there is a reference of dynamic nature of rain is used [2].

Another important factor for calculating attenuation due to rain in the falling velocity and rain drop shape. Experimental study was conducted by Gunn and Kinzer [3] in order to calculated the falling velocity of rain drop. Raindrop different shapes and size at various height levels are discussed in [4]. Besides raindrop velocity and size, raindrop canting angle is also important parameter which was discussed in [5].

Raindrop size distribution (N(D)) is another very important parameter in order to calculate attenuation due to rain. It is the number and size of drop within a unit volume described as

[number/m^3/mm]. Various researchers has described the methods for calculating DSD [6-12].

Rest of the paper is described in such a way that section II describe about ITU-R model and followed by proposed model in section III describes the results obtained while implementation of ITU model and Proposed model.

II. PROPOSED MODEL

A. ITU-R Model

The recommendations ITU-R Rec. P. 618-10 rain attenuation model is the widely used method for the predictions of rain effects on satellite communications system. The attenuation is calculated for 99.99% fade depth by:

$$Att = kR\alpha dr \ dB \tag{1}$$

Where

R= 99.99% rain rate for rain region in mm/h

kR^α = specific attenuation in dB/km

d = link distance in km

$$r = 1/(1 + d/d0) \tag{2}$$

where

$$d0 = 35e(-0.015R) \ km \tag{3}$$

where d0 is the effective path length and r is called the distance factor. k and α are the regression coefficient for frequencies and polarization [14-16].

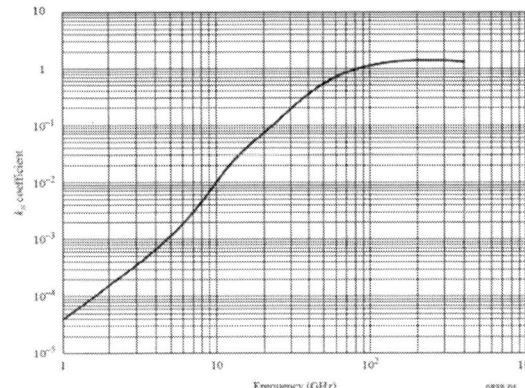

Fig. 1. k coefficient for horizontal polarization as a function of frequency

978-1-5386-6624-1/18 $31.00 © 2018 IEEE

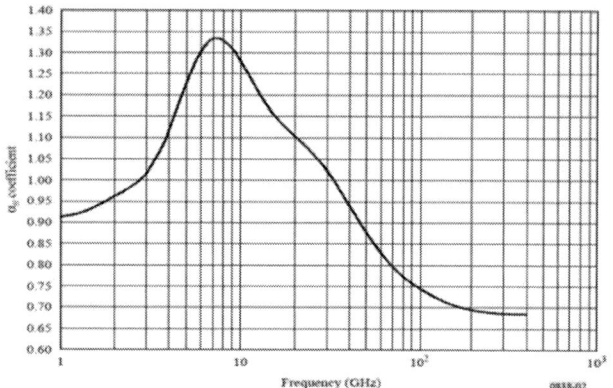

Fig. 2. α coefficient for horizontal polarization as a function of frequency

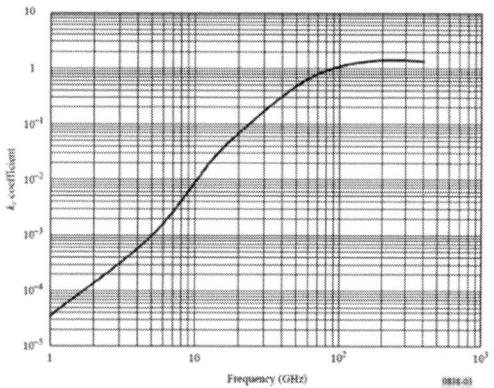

Fig. 3. k coefficient for vertical polarization as a function of frequency

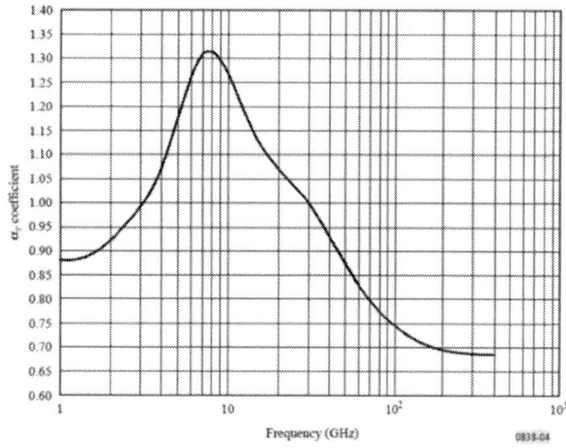

Fig. 4. α coefficient for vertical polarization as a function of frequency

The fig 1 to 4 shows the value of k and α over frequency in GHz for horizontal and vertical polarization. From the following figures it is clearly observed that the value of k and α very uneven for frequencies 1 to 10 GHz. For finding the values of coefficients for a given frequency is very difficult task. Some methodology is needed to be developed in order to find these values. In this work Least Square method is used which is implemented in MATLAB.

B. Proposed Approximation Model

From the previous section we have seen that the ITU-R rain model is very complex one. As this model is purely depends on the coefficients a and b. from the graphs shown in Fig 1 to Fig 4 it has been clearly shown that the curves are complicated in nature. It is very difficult for the researchers and engineers to calculate the attenuation caused by rain for a particular rain rate.

In this section a new model for calculating specific attenuation due to rain for both horizontal and vertical polarization is introduced. In this model there is no need to know the values of regression coefficients. In order to design the model step by step procedures are explained:

Step 1: From the ITU-R rain model the specific attenuation for different frequencies from 1 GHz to 100 GHz at different rain rate is calculated.

Step 2: Now graphs are drown for attenuation Vs frequency in GHz at different rain rate in mm/h. from these graphs approximation curves are drown using polynomial approximation method. From that approximation curves equations are obtained. Each equation from each rain rate is obtained. The example of one curve with approximation equation for horizontal polarization is given in Fig 5 for 1mm/h rain rate. Rest is calculated likewise.

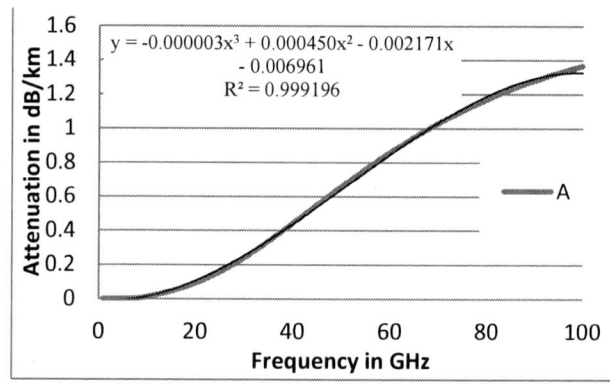

Fig. 5. Approximation curve for Frequency Vs Specific attenuation for 1mm/h

Step 3: from the Step 2 cubic equation is obtained of form:

$$A(dB/km) = af3 + bf2 + cf + d \qquad (4)$$

Step 4: From the above equation obtained for different rain rate the values of different rain rates, the values of constant a, b, c and d for horizontal polarization are stored in Table 1.

TABLE I.   DIFFERENT VALUES OF CONSTANTS AT DIFFERENT RAIN RATE

Rain Rate (mm/h)	a	b	c	D
1	-0.000002952	0.0004502	-0.00217	-0.006961
10	-0.00001	0.00105	0.061375	-0.3523
30	-0.00001044	0.000002	0.2494	-1.1721
50	-0.00000477	-0.001951	0.4523	-1.9601
100	0.00002029	-0.008368	0.9798	-3.782
150	0.00005331	-0.01587	1.5182	-5.426
200	0.0000909	-0.02397	2.0605	-6.9242
300	0.0001745	-0.04125	3.1474	-9.5547

Step 5: From the values obtained in step 4 of different constants, draw the curve between rain rate in mm/h and constants a, b, c and d. so that we got the four curves as shown in Fig 6 to Fig 9.

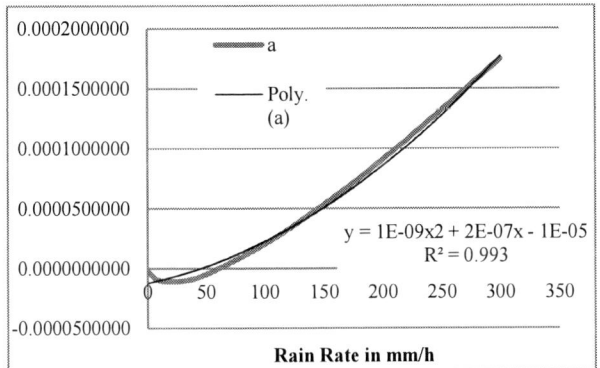

Fig. 6.   Approximation curve for Rain Rate Vs Constant a

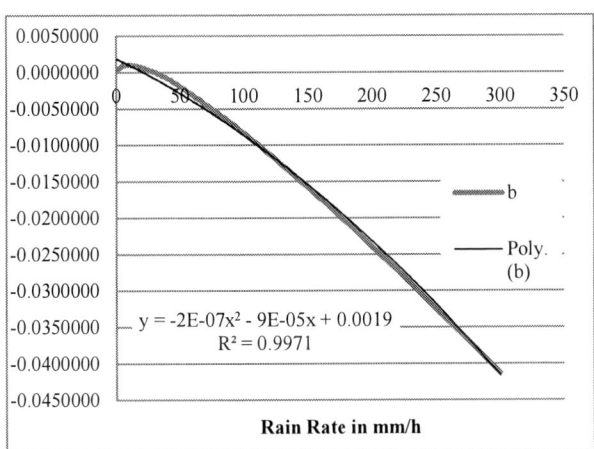

Fig. 7.   Approximation curve for Rain Rate Vs Constant b

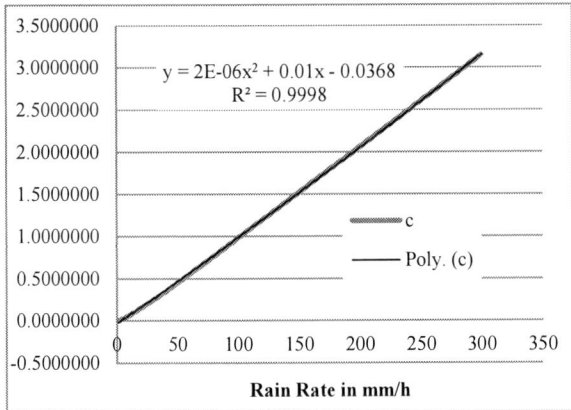

Fig. 8.   Approximation curve for Rain Rate Vs Constant c

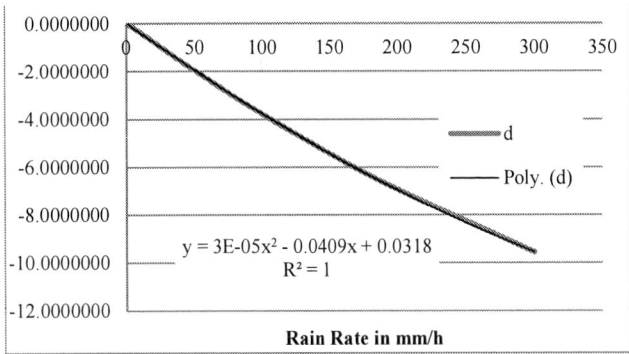

Fig. 9.   Approximation curve for Rain Rate Vs Constant d

Step 6: From each graph obtained in step 5 draw an approximation curves. From the polynomial approximation method we got the square equation for constants for horizontal polarization and cubic equations for vertical polarization (by following the step 1 to step 5 for vertical polarization) like:

$$ah = 1.422 \times 10 - 9x2 + 2.03 \times 10 - 7x - 1. \tag{5}$$

$$bh = 1.963 \times 10 - 7x2 + 8.618 \times 10 - 7x + 0.001 \tag{6}$$

$$ch = 2.114 \times 10 - 6x2 + 0.01x - 0.036 \tag{7}$$

$$dh = 3 \times 10 - 5x2 - 0.040x - 0.031 \tag{8}$$

$$av = -5.520 \times 10 - 12x3 + 3.26 \times 10 - 9x2 - 1.21 \times 10 - 7x - 6 \times 10 - 6 \tag{9}$$

$$bv = 8 \times 10 - 10x3 - 4.552 \times 10 - 7x2 - 3.03 \times 10 - 5x + 0.001 \tag{10}$$

$$cv = -5.71 \times 10 - 9x3 + 6 \times 10 - 7x2 + 8.707 \times 10 - 3x - 0.018 \tag{11}$$

$$dv = -1.073 \times 10 - 7x3 + 1.068 \times 10 - 4x2 - 0.0598x + 0.0442 \tag{12}$$

where x Rain Rate in mm/h

Step 7: From the equations obtained in step 6 we will calculate the new constant ah, bh, ch and dh for horizontal polarization and av, bv, cv, and dv for vertical polarization at different rain rate from above equations. The values are stored in Table II and Table III.

TABLE II.   DIFFERENT VALUES OF NEW CONSTANTS AT DIFFERENT RAIN RATE FOR HORIZONTAL POLARIZATION

Rain Rate in mm/h	ah	bh	ch	dh
1	-0.00001210	0.001895	-0.03681	0.031714
10	-0.00001210	0.001895	-0.03682	0.031746
30	-0.00001210	0.001895	-0.03684	0.031823
50	-0.00001210	0.001896	-0.03686	0.031907
100	-0.00001210	0.001896	-0.03692	0.032143
150	-0.00001211	0.001897	-0.03699	0.03242
200	-0.00001211	0.001897	-0.03706	0.032737
300	-0.00001211	0.001899	-0.03725	0.033492

TABLE III. DIFFERENT VALUES OF NEW CONSTANTS AT DIFFERENT RAIN RATE FOR VERTICAL POLARIZATION

Rain Rate in mm/h	av	bv	cv	dv
1	-0.0000060953	0.0009431	-0.009726	-0.01545
10	-0.0000068680	0.0006254	0.06860	-0.543
30	-0.0000068253	-0.0003256	0.2428	-1.6569
50	-0.0000045719	-0.001583	0.4172	-2.6927
100	0.0000089947	-0.005816	0.8515	-4.9759
150	0.0000305822	-0.01112	1.2803	-6.8859
200	0.0000560506	-0.01691	1.6992	-8.5031
300	0.0001020703	-0.0275	2.4903	-11.1811

Step 8: From different values of new constants from Table II and Table III specific attenuation in dB/km for different polarization is obtained for different rain rates by substituting these coefficients in equation:

$$A(dB/km) = a1f3 + b1f2 + c1f + d1 \qquad (13)$$

This equation is our final rain model.

III. IMPLIMENTATION RESULTS AND DISCUSSION

The results obtained from new model are compared with the ITU-R rain model and results are described in Fig 10 to Fig 16 for Horizontal Polarization and from Fig 17 to Fig 22 for Vertical Polarization.

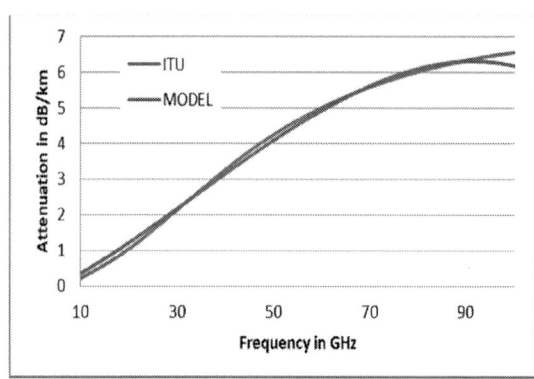

Fig. 10. New Rain Model Vs ITU Model at 10mm/h rain rate

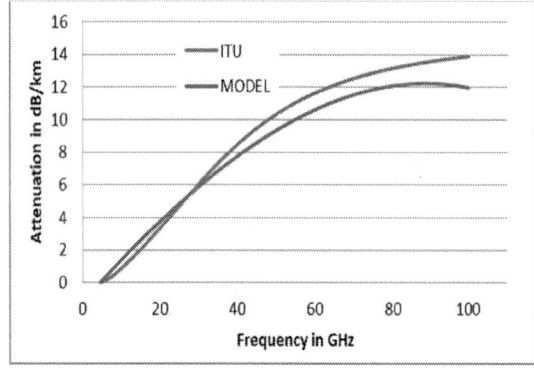

Fig. 11. New Rain Model Vs ITU Model at 30mm/h rain rate

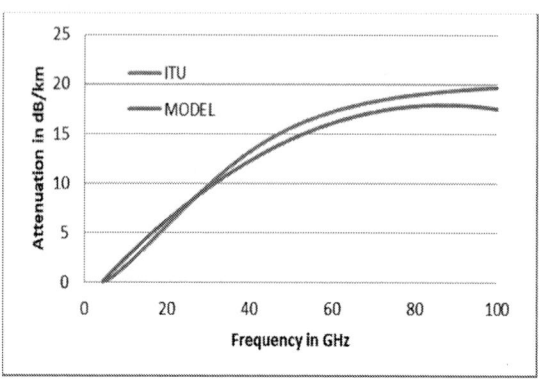

Fig. 12. New Rain Model Vs ITU Model at 50mm/h rain rate

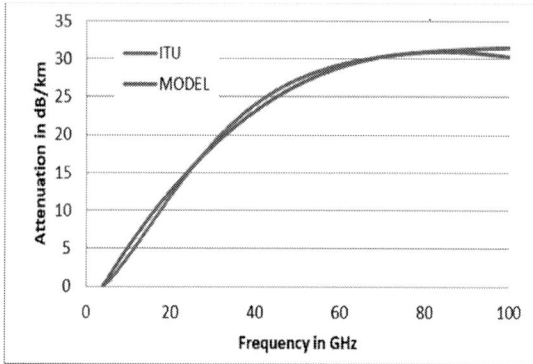

Fig. 13. New Rain Model Vs ITU Model at 100mm/h rain rate

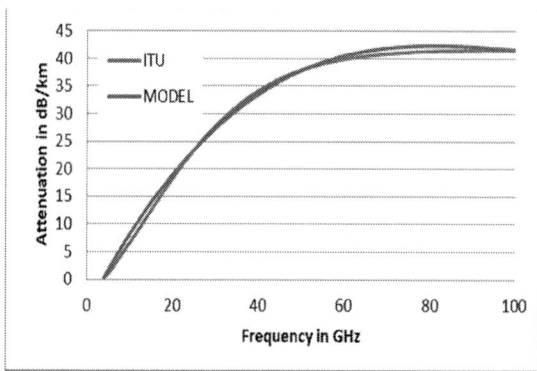

Fig. 14. New Rain Model Vs ITU Model at 150mm/h rain rate

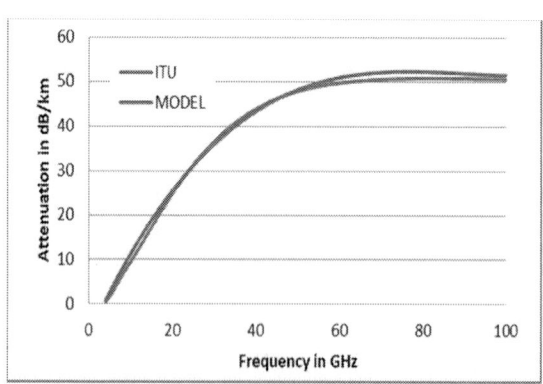

Fig. 15. New Rain Model Vs ITU Model at 200mm/h rain rate

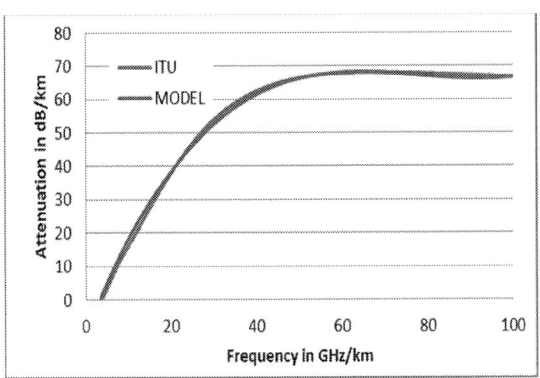

Fig. 16. New Rain Model Vs ITU Model at 300mm/h rain rate

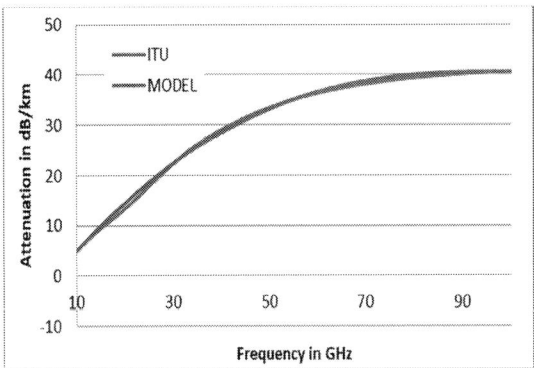

Fig. 20. New Rain Model Vs ITU Model at 150mm/h rain rate

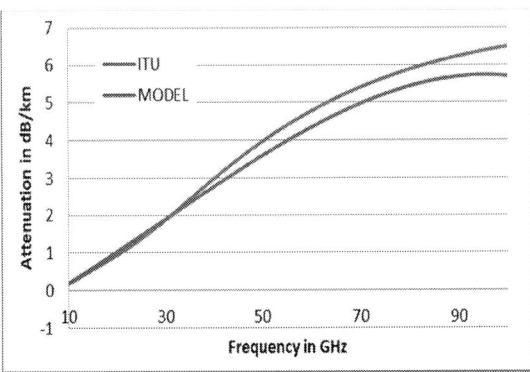

Fig. 17. New Rain Model Vs ITU Model at 10mm/h rain rate

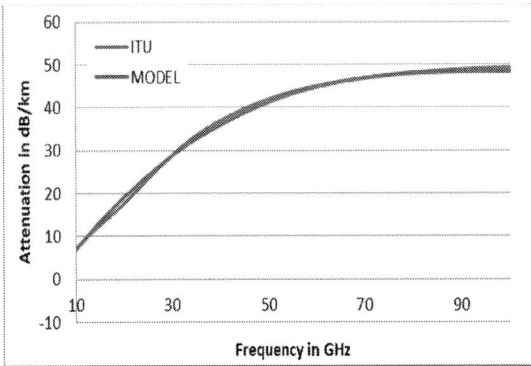

Fig. 21. New Rain Model Vs ITU Model at 200mm/h rain rate

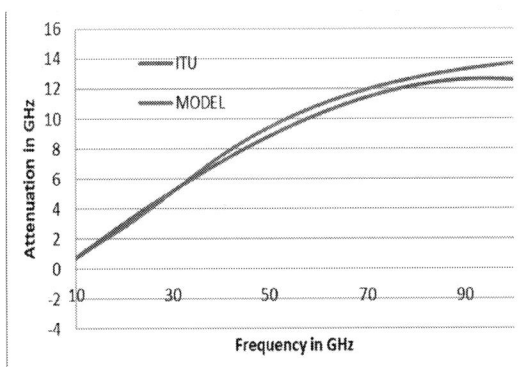

Fig. 18. New Rain Model Vs ITU Model at 30mm/h rain rate

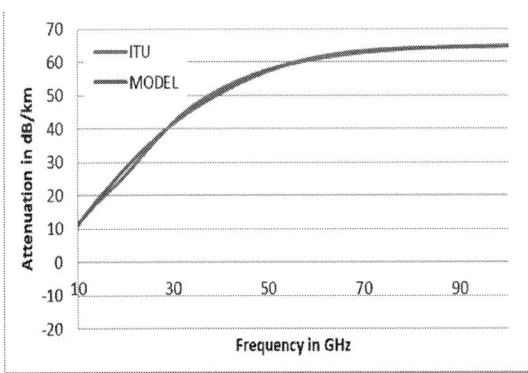

Fig. 22. New Rain Model Vs ITU Model at 300mm/h rain rate

The results obtained from above figures clearly shows that our new approximation model is closely matching with the ITU-R rain model. It has also been observed that our model is simpler to calculate for different rain rates by using proposed model as compared to complex ITU model.

IV. CONCLUSION

Various researchers have proposed different models for rain attenuations but ITU rain model is widely acceptable model. In the ITU model different values of regression coefficients k and α are given for different frequencies. In this work approximation model is designed for calculating specific attenuation caused by rain at different rain rates. From the simulation work it has been observed that approximation

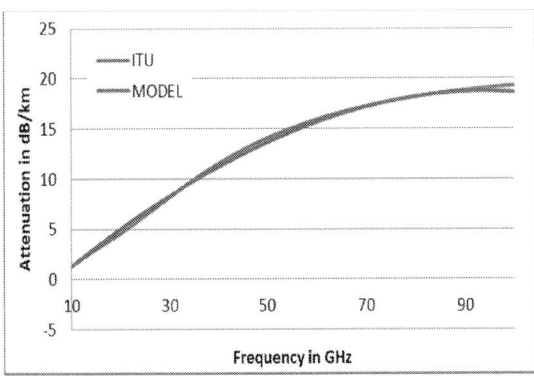

Fig. 19. New Rain Model Vs ITU Model at 50mm/h rain rate

978-1-5386-6624-1/18 $31.00 © 2018 IEEE

model is closely matched with the original ITU rain model. The equations present in this model are very simple as compared to ITU model. It helps researchers and engineers to identify coefficients for a given frequencies.

ACKNOWLEDGEMENT

This work was supported in part by the Grant DN07/19/15.12.2016 "Methods for Estimation and Optimization of Electromagnetic Radiation in Urban Areas" of the Bulgarian Science Fund.

REFERENCES

[1] Houze Jr RA. Cloud dynamics, 573 pp. San Diego: Academic. Google Scholar. 1993.

[2] Henri Sauvageot, L Castanet, J Lemorton, "HYCELL – A new hybrid model of the rain horizontal distribution for propagation studies: Modelling of the rain cell", Radio Science, Journal, Wiley, Vol. 38, Issue 3, June 2003.

[3] Gunn, Ross, and Gilbert D. Kinzer. "The terminal velocity of fall for water droplets in stagnant air." Journal of Meteorology 6.4 (1949): 243-248.

[4] Spilhaus, Athelstan F. "Raindrop size, shape and falling speed." Journal of Meteorology 5.3 (1948): 108-110.

[5] Brussaard, G. "Rain-induced crosspolarisation and raindrop canting." Electronics Letters 10.20 (1974): 411-412.

[6] Kathiravelu, Gopinath, Terry Lucke, and Peter Nichols. "Rain drop measurement techniques: a review." Water 8.1 (2016): 29.

[7] Williams, Christopher R., and K. S. Gage. "Raindrop size distribution variability estimated using ensemble statistics." Annales geophysicae: atmospheres, hydrospheres and space sciences. Vol. 27. No. 2. 2009.

[8] Marshall, J. S. and Palmer, W. M. K.: The distribution of raindrops with size, J. Meteor., 5, 165–166, 1948.

[9] Illingworth, A. J. and Blackman, T. M.: The need to represent raindrop size spectra as normalized gamma distributions for the interpretation of polarization radar observations, J. Appl. Meteorol., 41, 286–297, 2002.

[10] Zhang, G., Vivekanandan, J., Brandes, E., Meneghini, R., and Kozu, T.: The shape-slope relation in observed gamma raindrop size distributions: Statistical error or useful information?, J. Atmos. Oceanic Technol., 20, 1106–1119, 2003.

[11] Feingold, G. and Levin, Z.: The lognormal fit to raindrop spectra from frontal convective clouds in Israel, J. Appl. Meteorol., 25, 1346–1364, 1986.

[12] Baltas, E. A., and M. A. Mimikou. "The use of the Joss-type disdrometer for the derivation of ZR relationships." Proceedings of ERAD. Vol. 291. No. 294. 2002.

[13] "Propagation data and prediction methods required for the design of Earth-space telecommunication Systems", ITU-R P.618-10, 2009

[14] "Specific attenuation model for rain for use in prediction methods",ITU-RP.838-3,2003.

[15] H. Singh, R. Kumar, B. Bonev, P. Petkov, "Cloud Attenuation issues in Satellite Communications at millimetre Frequency Bands- State of Art", International Journal of Scientific & Engineering Research Volume 8, Issue 7,pp. 858-862, July-2017.

[16] H. Singh, R. Kumar, B. Bonev, P. Petkov, "The Studies of Millimeter Waves at different Frequencies in different Environmental conditions for 5G Applications – A State of Art", International Journal of Scientific & Engineering Research Volume 8, Issue 7,pp. 851-857, July-2017.

2nd IEEE International Conference on Power Electronics, Intelligent Control and Energy Systems (ICPEICES-2018)

Real time analysis of MAC based and Level based Routing Protocol for Wireless Sensor Network

Aditi Gaur
Department of Electronics and Communication Engineering
Maharana Pratap University of Agriculture & Technology
Udaipur, India
aditigaur075@gmail.com

Sunil Joshi
Department of Electronics and Communication Engineering
Maharana Pratap University of Agriculture & Technology
Udaipur, India
suniljoshi7@rediffmail.com

Shweta Jaroli
Department of Electronics and Communication Engineering
Maharana Pratap University of Agriculture & Technology
Udaipur, India
shweta.jaroli21@gmail.com

Shruti Bhandari
Department of Electronics and Communication Engineering
Maharana Pratap University of Agriculture & Technology
Udaipur, India
shrutibhandari@live.com

Abstract— **Wireless Sensor network is the combination of wireless communication, Micro Electro Mechanical Systems (MEMS) and digital electronics which together adds up to the feasible development and design of multi-feature and multi-functional sensor nodes. Wireless Sensor networks follow the particular process of sending information from source to destination which is known as routing protocol. The performance of wireless sensor networks highly depends on the technique to provide route to the data based on IEEE 802.15.4 standard. Routing protocols have been categorized into much of the energy efficient and quality of service based design to yield better communication performances. This paper provides a performance analysis of two of the routing protocols: Mac based routing and Level based routing. Mac based routing protocol has been adopted as it exhibits the feature of collision avoidance and Level based routing protocol exhibits the feature of two way communications with introduction of Mac support approach. The analysis is done on the basis of certain performance metrics: delay, total energy consumption, average jitter, throughput, packet delivery ratio and packet loss. The received results depict that level based routing has proved to be better than compared to Mac based routing.**

Keywords— Average jitter, delay, energy consumption, IEEE 802.15.4, packet delivery ratio, packet loss, throughput and Wireless sensor network.

I. INTRODUCTION

Wireless Sensor Network (WSN) is a set up in which communication takes place between components of the network. These components are placed in an area either inside physical phenomenon or its proximity and are directly linked to each other. Basically, these small sized wireless sensor nodes perform important functionalities such as data processing, sensing and communication. Based on these functionalities Wireless Sensor (WS) nodes are broadly classified into (a) sensor nodes and (b) sink nodes as shown in Fig. 1.

The sensor nodes are small and provide unhindered short distance communication. The tiny sensor nodes exhibit characteristics like sensing, data processing, communicating and thus, enable the realization of WSN based on

collaborative effort of plenty of sensor nodes. These sensor nodes exhibit two functionalities that are both data originators and data routers [1].

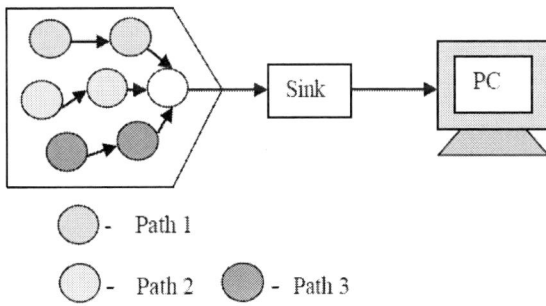

Fig.1. Wireless Sensor Network

Hence, communication is performed for two reasons: (i) Source function – Event informed source nodes perform communication functionalities in order to transmit their packets to sink. (ii) Router function – Participation of sensor nodes in forwarding the packet received from other nodes to the next destination in the multi-hop path to the sink. In this manner, data is transmitted by intermediate nodes to establish routes with the multiple links or hops to the sink.

WSN uses IEEE 802.15.4 standard [2], it defines the standard for Wireless personal area network (WPAN) for low area devices which specifies the standard for physical and Mac layer. IEEE 802.15.4 works at a data rate of less than 250 kbps. The Mac sub layer controls the activities on the radio link. The tasks included are acknowledgment, data re-transmission, flow control and network synchronization. Mac controls the access to the radio channel and employs the services of Carrier Sense Multiple Access – Collision avoidance (CSMA-CA) to avoid packet collision on the RF link. The two types of nodes are used in the network:

- **PAN coordinator-** There must be only one PAN coordinator in the network. It acts as a receiver. It finds the radio frequency channel for network operation [2].

- **Coordinator-** There may be one or more coordinators in the network. It acts as transmitter and relays message from one node to another. Handles network joining requests.

978-1-5386-6624-1/18 $31.00 © 2018 IEEE   1199

In this paper, we have worked on protocols based on Mac layer and physical layer of WSN. We have focused on real time analysis of Mac based and Level based routing protocols. On the basis of these, comparative analysis is presented in terms of performance parameters. The rest of the paper is organized as follows: Section II presents Routing Protocols in WSN. Section III presents System Model. Section IV presents Simulation Results & Discussion. We have concluded paper in section V.

II. ROUTING PROTOCOLS IN WSN

A. Routing Protocols

The role of WSN is governed by routing protocols. The routing task is responsible for determining the path followed in the establishment of communication between source and sink [3]. The routing protocol in charge maintains the route of the network. Thus, it specifies the router function in a manner in which they communicate and distribute information. This information enables them to select routes between the nodes. Each router contains knowledge of the network attached to it. This information is shared among the intermediate nodes and then throughout the network by the help of routing protocol. They address important aspects like energy efficiency, quality of service, delay transmissions, network security etc.

The topology determines the interconnected patterns of network elements which present an efficient performance measure of the network. The topology information is also received by the routers. Hence, Routing is one of the key features of the Wireless Sensor Network and plays a significant role in designing a sensor network. Each routing protocol has its own functionality and these are categorized on the basis of data centric mechanism, hierarchical distribution and location information. Fig. 2 below mentions the categorization of routing protocols:

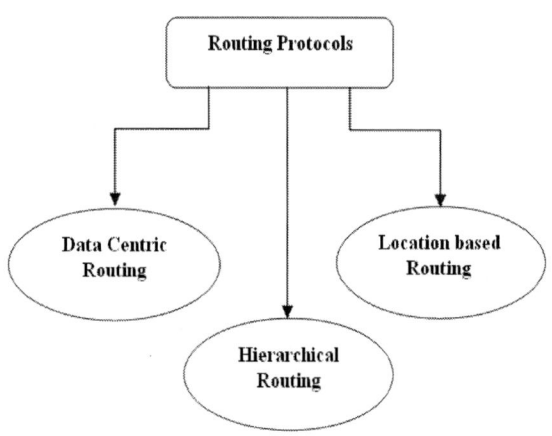

Fig. 2. Categories of routing protocols

B. Types of Routing Protocols

The Data centric routing protocols and Hierarchical routing protocols are undertaken for the research work. In Data centric routing protocol, the data is sent from source to sink through intermediate nodes and this process helps in

energy saving. The Mac based routing comes under the Data centric category. In Hierarchical routing, the nodes are distributed in multi level and these levels broadcast information to lower level, carrying the network topology. Both these routing protocols are explained further in the paper along with their real time analysis on a WSN platform.

C. Mac Based Routing

MAC Based routing (MBR) is the routing technique in Wireless Sensor networks which is based on IEEE 802.15.4 protocol specifying the PHY and media access control for low-rate wireless personal area networks (LR-WPANs) [4].

Network setup messages - MLME (MAC layer management entity) and data packets - MCPS (MAC common port layer) are generated by MAC and higher layers respectively [5]. During network setup, each node gets associated with a coordinator and every node sends the packet to the associated coordinator. This process repeats until the packet is received by the destination node. The source node transmits data to its own coordinator until the line breaks and finds a new coordinator in its range at the time of line breakage. MAC based routing works only with the data packet and no control packets are used in this. It sends data packets in the hop by hop manner. So, this protocol can be used as Multi-hop protocol. The flow diagram for the working of MAC based routing is shown below in Fig. 3. We can understand the working by following steps:

- In Mac Based routing, Routing is done on IP Level.

- The router will receive packets for its own MAC address but for the different IP address.

- The router will then check if it can directly reach the target IP address.

- It then sends the packet to target. Otherwise, the router itself has an upstream router configured and will send the packet to the target.

- The packet is sent to target (Pan-coordinator).

- The destination is always Pan –coordinator in MAC based routing.

Merits:
- MAC based routing router receives data of its own MAC address, so it avoids collision with other networks.
- It manages the communication traffic on a shared medium

Demerits:
- Energy consumption, a delay is more in MAC based Routing.
- MAC based routing require more power to transmit data at larger distances.

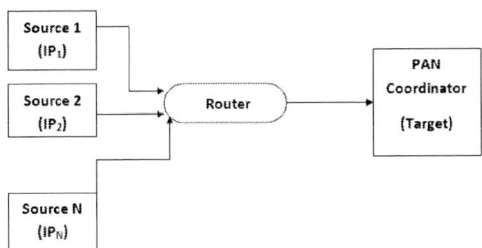

Fig.3. Block diagram of Mac based routing technique

D. Level Based Routing

Level Based Routing (LBR) is used in the multi-hop scenario of the wireless sensor networks. The technique followed up in the level based routing is the round robin technique [6]. In round-robin technique when there are two or more tasks having the same priority, it then lets the kernel to run a particular task for a predetermined amount of time, in this type of routing PAN Coordinator can be used as a source.

During the setting up of the network, the sensor nodes as the elements of the network are divided into the number of levels or layers of different levels which is based on the number of hops to sink. Each level node saves information of its co-nodes and thus using this level information, a sensor node can send data messages collected from the environment to a sink node in a more efficient way as shown in Fig. 4.

i) MAC support approach in LBR

In this protocol we have introduced the Mac support approach which in case of route availability failure provides a backup path to the protocol. LBR can work with both data packet and control packet. We can understand the working as shown below:

- This protocol works in levels when it initiates it sets up MAC layer throughout the network.

- The Pan-Coordinator of the network occupies LEVEL 0.

- Now packet is sent to the nearby nodes of Pan-coordinator.

- These receiving nodes then set their level as LEVEL 1.

- In case of route failure, Mac support comes into action and provides a backup path for the packets.

- Each node acting either as FFD or RFD saves information of maximum 4 previous nodes.

- The nodes of level 1 rebroadcast a packet to level 0.

- Also in the next hop, the nodes receive the packet from level 1's multiple nodes and set its level as LEVEL 2.

- Level 2 will rebroadcast only first packet of level 1.

Merits:
- The energy consumption is minimal and thus it offers better network lifetime.
- When modified any node can be made sink.

Demerits:
- Packet dropping is a major issue in this protocol.
- This protocol deals with a little issue in managing stored information of node if traffic increases.

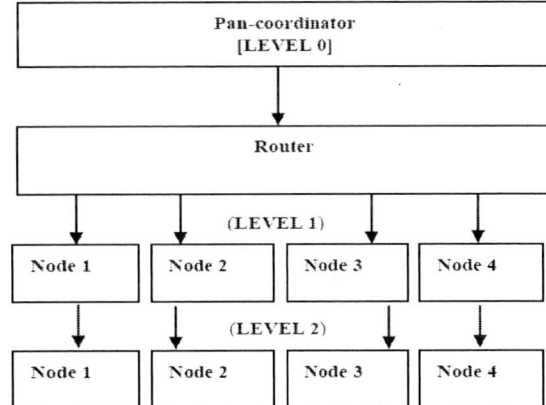

Fig.4. Level based routing technique

III. SYSTEM MODEL

The system model is created to analyze the routing protocols and implemented on Sensenuts – a WSN Platform. This system model uses IEEE 802.15.4 which can be deployed based on parameters like low bandwidth requirement, low range, on-off behavior and low battery consumption. This model consists of a Pan-coordinator or sink node and coordinators with sensor nodes deployed in a laboratory for monitoring its room temperature and light intensity.

Area of the lab monitored is around 56 m². The Fig. 5 depicts the system model with one Pan - coordinator and N sensor nodes (where N = 10). The data gathered from sensor nodes are transmitted to the pan-coordinator through coordinator and thus displayed on PC (user interface).

A. Hardware

The elements of hardware section are nodes, radio modules, gateway modules and sensor modules. A radio module has JN5168 microcontroller and a transceiver performing tasks of processing, sending and receiving the information. The gateway module [7] plays the role in programming the microcontroller and interfaces the network and PC. Sensor module senses the light and temperature data from the physical environment. It consists of two types of nodes (i) coordinators (for transmitting) and (ii) Pan – Coordinator (for receiving) respectively. The coordinator is made up of Sensor module, Gateway module and Radio module. The Pan-Coordinator is made up of Gateway module and Radio module.

978-1-5386-6624-1/18 $31.00 © 2018 IEEE

B. Software

The software used in the system model is Sensenuts which is integrated with Eclipse IDE. The application code for the protocol is written in C language. The real time analysis of data is done on the Graphical User Interface which is a java and SQL based application.

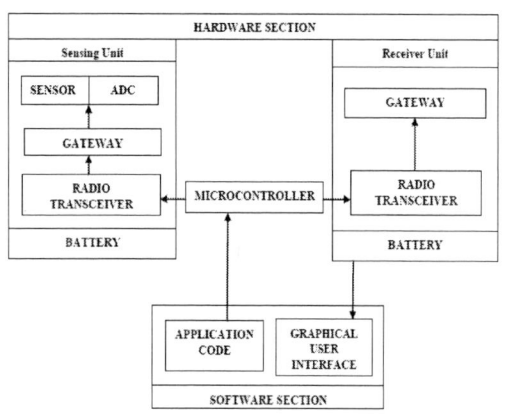

Fig.5. System model

C. Performance Characteristics in Routing Protocols

The analysis of various performance characteristics in this research work is presented. These parameters and also their influence on the performance of the system according to the network, application and users are mentioned below.

1. Throughput - It is the amount of data effectively handed from source node to destination node within certain duration over a communication channel. It is measured in bits per second, data packets per second or data packets per time slot.

Throughput is resembled to channel efficiency and channel bandwidth utilization i.e. it increases throughput value by using lots of amounts of packets. By using IEEE 802.11 and IEEE 802.15.4 MAC protocol we achieve high throughput.

$$Th = \text{Received data*no. of bits/Data } T^x \text{ period} \quad (1)$$

2. Packet delivery ratio (PDR) - This network performance metric is defined as the ratio between numbers of data packets successfully delivered to the destination and the number of packets transmitted by the source.

$$PDR= \text{(Received packets/Generated packets)*100} \quad (2)$$

3. Delay – The processing delay across the network based on the transmission [8]. Thus, the time the router takes for transmission of a packet from an interface into an output interface is called a delay.

It depends upon factors like speed and utilization of the CPU, IP switching mode, architecture of the router. It is a configured feature on both input and output interface of the router.

$$Delay = \text{Receiving time} - \text{Transmitting time} \quad (3)$$

$$Total\ delay = \text{Delay/Number of packets received} \quad (4)$$

4. Average jitter – It is calculated as the mean variance of the delay obtained for each node [9]. It is calculated by following steps –

$$S^2 = \sum (x_i - x')^2/(n-1) \quad (5)$$

where, S^2 is the variance
x' is the mean &
n is number of nodes

5. Energy – Energy is termed as the flow of transmission. In Wireless Sensor Networks, the energy consumption rate for sensors shows immense variation and this depends on the protocols used for communications. The energy consumption formula [10] for transmitting and receiving data between two sensors is as follows:

$$Etx = E_t * k * d^2 \quad (6)$$

$$\&\quad Erx = E_r * k \quad (7)$$

Where,
Etx = Transmitter energy in J/bit
E_t = Energy required to transmit per bit in J/bit
k = Packet size
d = Distance (in meters)
E_r = Energy required to receive per bit in J/bit
Erx = Receiver electronics energy

Energy Analysis- E_t is calculated by multiplying current, voltage and time values of microcontroller JN5168 [10] during transmission i.e.

$$E_t = \text{Tx current*Tx voltage*Tx time} \quad (8)$$

Similarly, E_r is calculated similarly by taking reception values

$$i.e.\ E_r = \text{Rx current*Rx voltage*Rx time} \quad (9)$$

$$\text{TOTAL ENERGY CONSUMED (TEC)} = \sum Etx + \sum Erx \quad (10)$$

6. Packet Loss – It determines the number of packets dropped during transmission and reception of data in a network. Packet loss can be due to many reasons like delay, network traffic, sensor node failure etc. It is given by:

$$Packet\ loss = \text{Total number of packet sent} - \text{Number of packets received} \quad (11)$$

978-1-5386-6624-1/18 $31.00 © 2018 IEEE

D. Flowchart

The Fig.6 shows the generalized working of network set-up of MBR and LBR routing protocol on software.

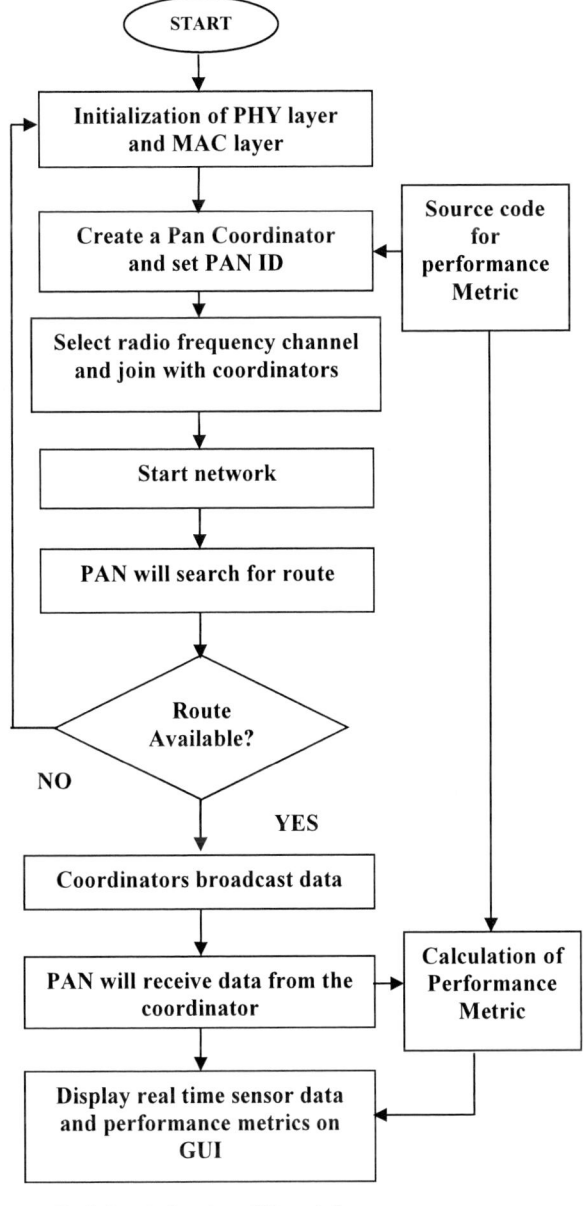

Fig.6. Generic flowchart of Network Set-up

IV. SIMULATION RESULTS & DISCUSSIONS

The following tabular data and graphs depict the results of light and temperature values along with various performance parameters namely - Throughput, PDR, Delay, Energy, average jitter and packet loss respectively.

(a) The real- time analysis of light and temperature data: The data received from temperature and light sensors deployed in the lab at distance of 1m. The light intensity is

calculated in terms of lux and temperature in degree Celsius as shown in Fig. 7.

TABLE I SIMULATION PARAMETERS

S.NO.	PARAMETERS	VALUES
1.	No. of nodes	10
3.	MAC	IEEE 802.15.4
4.	Transmission Current	15.3×10^{-3} A
5.	Transmission Voltage	2.8 V
6.	Transmission Time	2.528×10^{-3} s
7.	Reception Current	17×10^{-3}A
8.	Reception Voltage	3V
9.	Reception Time	2.528×10^{-3} s
10.	Packet size	6 Bytes
11.	Packet Interval	1 second
12.	Total number of packets	100

a. Specifications used in Simulations

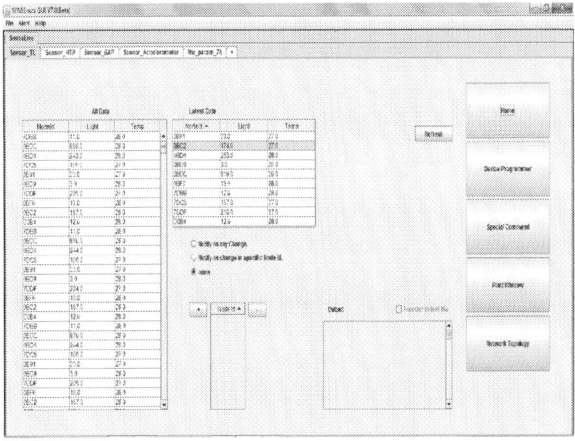

Fig.7. Real time temperature & light data

• **The graphical representation of Light intensity and temperature data:** The given graph in Fig.8 depicts the values of temperature and light at each node. The node with ID 0BC2 is displaying the highest light intensity and the room temperature is constant.

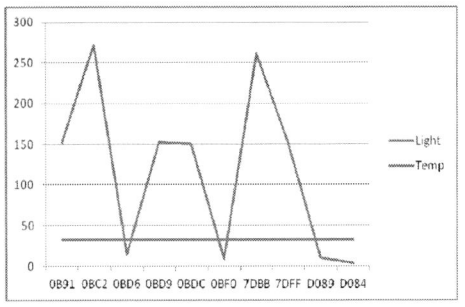

Fig.8. Graphical representation of Light and temperature

(b) Performance parameters in Mac Based routing protocol: In the Fig.9 below, the data with 10 nodes recorded as a room monitoring is shown. The routing protocol followed is Mac Based Routing. The overall throughput (th) is 46.53 bits/sec, PDR (pd_r) is 0.954, delay (de_l) is 1.04 seconds, Total energy (etx & erx) is 5.38 J/bit-m^2 and packet loss (p_lo) is 4.6. The negative values in p_lo given below in Fig. 9 represent re-transmission of packets.

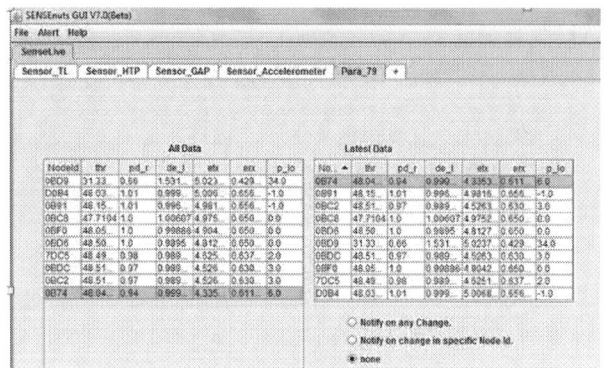

Fig.9. MBR's data representation with 10 nodes

(c) Performance parameters in Level Based Routing:
In the Fig.10 below, the data with 10 nodes recorded as a room monitoring is shown. The routing protocol followed is Level Based Routing. The overall throughput is 49.10 bits/sec, PDR is 0.817, delay is 0.977 seconds, Total energy is 4.04 J/bit-m^2 and packet loss is 18.3.

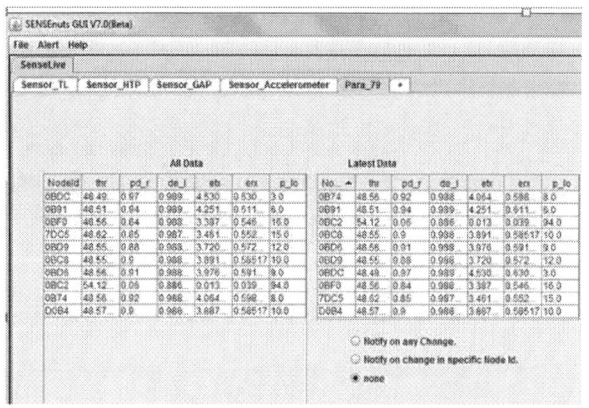

Fig.10. LBR's data representation with 10 nodes

(d) The comparative analysis of performance parameters of Level Based Routing Vs Mac Based Routing: The following Table II gives an analysis of overall data. In Fig.11graphical representation of the data obtained by MBR and LBR on the real-time basis is shown.

TABLE II OVERALL COMPARISONS OF PARAMETERS

Routing Protocol	Parameters					
	Through put (Bits/ Second)	Energy (Joules /bit-m²)	Packet Delivery Ratio (%)	Packet Loss	Delay (seconds)	Average Jitter (seconds)
MBR	46.53	5.38	95.4	4.6	1.04	0.028
LBR	49.10	4.04	81.7	18.3	0.977	0.001

b. Average data of 10 nodes for MBR & LBR

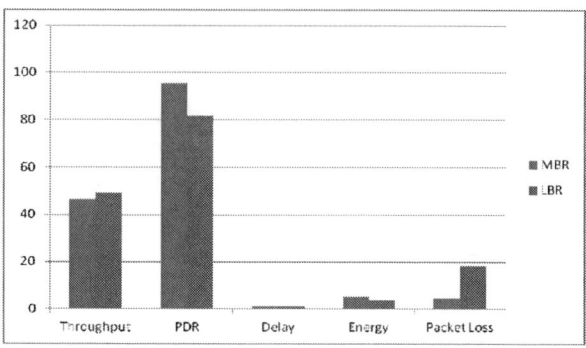

Fig. 11. Overall comparison of MBR and LBR

V. CONCLUSION

On the basis of overall comparison, we conclude that Level based routing is showing better results in terms of throughput, total energy consumption, delay and average jitter than Mac Based routing. The packet loss and packet delivery ratio in Level based routing technique is compromised. Hence, the Mac support approach provided in case of Level based routing has proved to be beneficial for the better execution and performance.

REFERENCES

[1] I.F. Akyildiz and M.C. Varun, Wireless Sensor Network. John Wiley and Sons Ltd. Publication, pp. 1-10, 2010.

[2] IEEE, IEEE Standard for Information Technology-Telecommunications and Information Exchange Between Systems-Local and Metropolitan Area Networks-Specific Requirements Part 15.4: Wireless Medium Access Control (MAC) and Physical Layer (PHY) Specifications for Low-Rate Wireless Personal Area Networks (WPANs), IEEE Std 802.15.4-2006 (Revision of IEEE Std 802.15.4-2003).

[3] A. Sarkar, and T. Murugan, "Routing protocol for wireless sensor network: What the literature says?," Elsevier Alexandria Engineering Journal, vol. 55, no. 4, pp.1-11, 2016.

[4] S. Agrawal, A. Payal, and B.V.R Reddy, "Design and implementation of mediumaccess control based routing on real wireless sensor networks testbed," International Journal of Computer, Electrical, Automation, Control and Information Engineering, vol. 11, no. 6, pp.665-673, 2017.

[5] NXP Laboratories, Data-sheet JN516x, JN-3024 IEEE 802.15.4 stack user guide Semiconductors Integrated Phripherals API user guide, and manuals, (NXP Laboratories, UK, 2012)

[6] D. Das, I. Roy, and S. Bhakta, "Introduction to SENSEnuts programming," Technical Report, Indian Statisitical Institute, pp.1-42, 2015.

[7] Sensenuts reference modules and guides.

[8] V. Kumar and A. K. Gupta, "Measuring parameters of quality of service in Wireless Sensor Network," International Journal of Advanced Research in Computer Engineering & Technology (IJARCET), vol. 3, no. 11, pp.4040-4045, 2014.

[9] J. Birla, and B. Sah, "Performance metrics in Ad-hoc Network," International Journal of Latest Trends in Engineering and Technology (IJLTET)," vol. 1, no. 1, pp.46-49, 2012.

[10] N. Kamyabpour, and D.B. Hoang, "Modelling overall energy consumption in Wireless Sensor Networks," IEEE International Conference on Parallel and Distributed Computing, Applications and Technologies (PDCAT), pp.1-8, 2007.

978-1-5386-6624-1/18 $31.00 © 2018 IEEE

2nd IEEE International Conference on Power Electronics, Intelligent Control and Energy Systems (ICPEICES-2018)

Evaluation of Emission Characteristics of Green Diesel in A Single Cylinder CI Engine

Manu J Nair
Department of Mechanical
Engineering
BML Munjal University
Gurgaon, India.
manujnair007@gmail.com

Vaibhav Pahuja
Department of Mechanical
Engineering
BML Munjal University
Gurgaon, India.
vaibhav.pahuja123@gmail.com

P. Suvesh
Department of Mechanical
Engineering
BML Munjal University
Gurgaon, India.
suvesh98.p@gmail.com

Sumit Roy
Department of Mechanical
Engineering
BML Munjal University
Gurgaon, India.
sumitroy@hotmail.de

Abstract— This experimentation: Emission characteristics of Green Diesel is a comparative analysis between the emission characteristics of Baseline Diesel and Green Diesel – which is a blend of sunflower biodiesel, diesel, water and additives. The aim is to verify whether or not this blend of alternative fuel can give a better result in terms of the emissions of NOx, HCs, CO2 and in terms of Opacity. To carry out the experiment Sunflower Biodiesel is used as the primary biodiesel and blended with Diesel and Water to create 'Green Diesel'. Compression ratio of the
engine was fixed to 18:1 and the load was varied to check the emission characteristics. The emission characteristics suggested that the Green Diesel performs better in terms of NOx emissions and CO2 emissions while the emission of HCs and the Opacity were on the higher side when compared to Baseline Diesel.

Keywords— Green diesel, Emissions, Dual-fuel, Baseline diesel

I. INTRODUCTION

The interest in vitality is expanding at an exponential rate because of the exponential development of total populace. Propelled vitality effectiveness innovations lessen the vitality expected to give vitality administrations, consequently decreasing natural and national security expenses of utilizing vitality and possibly expanding its unwavering quality. By and large, a maintainable vitality framework incorporates vitality proficiency, vitality dependability, vitality adaptability, vitality improvement and progression, consolidated warmth and power (CHP) or cogeneration, fuel neediness, and ecological effects. The natural effects of vitality utilize are not new. For a considerable length of time, wood consuming has added to the deforestation of numerous regions [1]. Then again, the average qualities of a maintainable vitality framework can be gotten from political definitions. A reasonable vitality framework can be characterized likewise by looking at the execution of various vitality frameworks regarding maintainability pointers. Since, by definition, manageable vitality frameworks must help both human and biological community wellbeing over the long haul, objectives on decent discharges should look

well into what's to come. They ought to likewise think about the general population's propensity to request more. A vitality framework is comprised of a vitality supply area and vitality end-utilize advances. The protest of the vitality framework is to convey to shoppers the advantages that vitality offers. Vitality assumes a crucial part in our regular day to day existences. The vitality framework regularly comprises of vitality assets and generation, security, transformation, utilize, dispersion, and utilization. The vitality sources have been part into three classifications: petroleum derivatives, sustainable sources, and atomic sources. Vitality assets will assume an essential part on the planet's future. Sustainable power source assets, for example, biomass, sunlight based, geothermal, and wind vitality can possibly supply a significant part of vitality necessities in the coming years. Sustainable power source alludes to fuel sources more reliably accessible than their fossilized partners. Vitality change for the most part comprises of thermochemical and bio-concoction transformation advancements, control plant, fuel refinery, and vitality stockpiling. Vitality security implies the accessibility of vitality constantly in different structures, in adequate amounts, and at moderate costs. Vitality utilize is firmly connected to a scope of social issues, including neediness easing, populace development, and urbanization. Existing innovation for creating diesel fuel from plant oils, for example, rapeseed, soybean and palm has focused on transesterification of oils with methanol to deliver unsaturated fat methyl esters or biodiesel. The general accord of these examinations demonstrates that substituting biodiesel for oil diesel brings about a critical diminishment in both petroleum product utilization and in addition ozone harming substance outflows [2]. While it is protected to state that biodiesel, as a substitute for oil diesel, decreases GHG outflows; there are quality issues related with its across the board utilize [3]. Although it has been reported that the methyl esters have numerous attractive fuel characteristics, for example, great cetane number and lubricity, the measure of biodiesel added to an oil diesel mix is regularly constrained to 5% or less because of poor stockpiling dependability, minor frosty stream properties, extreme

978-1-5386-6624-1/18 $31.00 © 2018 IEEE　　　　1205

dissolvability and motor similarity issues [4]. Notwithstanding these quality issues, a general vitality adjust that incorporates the life cycle of the crude materials utilized as a part of biodiesel generation demonstrates that oil determined diesel is still more vitality productive. Future far reaching utilization of biofuels relies upon growing new process innovations to deliver amazing transportation energizes from organically determined feedstocks. These new biofuels should be perfect with the current fuel and transportation foundation. The requirement for various handling courses to change over vegetable oils into a top-notch diesel fuel or diesel mix stock that is completely perfect with oil inferred diesel fuel [5]. The two organizations began a communitarian explore exertion in 2005 to grow such a procedure in view of traditional hydro-processing innovation that is as of now broadly sent in refineries and uses the current refinery foundation and powers dispersion framework. The consequence of this exertion is the UOP/Eni Ecofining TM process to deliver green diesel. This innovation uses broadly accessible vegetable oil feedstock to create an isoparaffin-rich diesel substitute [6]. This item, alluded to as green diesel, is a fragrant and sulfur free diesel fuel which has a high cetane mixing esteem. The cool stream properties of the fuel can be balanced in the process to meet atmosphere cloud point determinations in either the perfect or mixed fuel.

II. EXPERIMENTAL INVESTIGATION

A. Experimental Setup

For this study, an off-road engine is considered of model Kirloskar Oil Engine, TV1. The tests were conducted on single cylinder four stroke test rig. The complete set up consists of the bed where it is kept for the test, fuel consumption meter and equipment for smoke test and test for exhaust emissions.

The specifications of the engine used is as shown in the table below:

TABLE I.   ENGINE SPECIFICATIONS

Make	Kirloskar Oil Engine, TV1
Compression ratio	18:1
Number of cylinders	Single
Number of Strokes	Four
Combustion principle	Compression ignition
Cooling system	Water cooling
Fuel	Diesel
Speed	1500 rpm
Power	3.50KW
Bore	87.50 mm
Stroke length	110.00 mm
Connecting rod length	234.00mm
Swept volume	661.45 (cc)

To measure the amount of emission constituting opacity and exhaust gas, we use emission measurement system. It consists of the smoke meter to check opacity and exhaust gas analyzer to check emissions.Table II shows the specification of AVL 450 di-gas exhaust gas analyzer.

TABLE II.   MEASURING RANGES OF EXHAUST GAS ANALYZER

Measured parameter	Specification
Oxygen	0-22% Vol
Carbon Monoxide	0-10%Vol
Carbon dioxide	0-20% Vol
Hydro carbon	0-20000ppm
Nitrogen oxide	0-5000ppm
Engine Speed	400-6000rpm
Lambda	0 to 9.999

Online performance analysis is done using Engine Performance Analysis software package — Enginesoft LV. The software Enginesoft LV serves by entering data and reporting it. The data includes power, efficiencies and fuel consumption. The stored data is converted into tabular and graphical format which can be printed for further analysis.

The following image shows the schematic presentation of the setup:

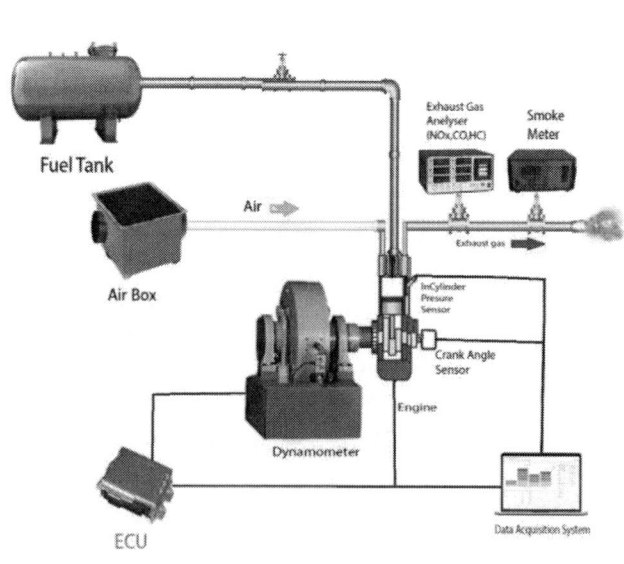

Fig. 1.   Schematic Diagram of Experimental Setup

978-1-5386-6624-1/18 $31.00 © 2018 IEEE   1206

B. Experiment Methodology

Before the starting of the experiment the fuel blend was prepared providing the homogeneous mixture that forms the green diesel. The blends prepared are and referred as:

- D100 – 100% diesel *(Baseline Diesel)*

- D80BD20W20 - 20%Biodiesel (Sunflower), 80%diesel, 20% Water, Additives *(Green Diesel)*

TABLE III. SPECIFICATIONS OF FUELS

FUEL PROPERTIES	DIESEL	FUEL PROPERTIES	GREEN DIESEL
Density(kg/m^3)	832	Density(kg/m^3)	910
Specific heat (J/ kg. K)	1750	Kinematic Viscosity(m^2/s)	72
Viscosity (Kg/m. s)	0.00275	Gross CV (J/g)	38548
Melting point (k)	304.15	Cetane Number	54
Ignition Temperature (F)	410	C% (w/w) H%(w/w) N%(w/w)	70.12 10.2 0.2
Enthalpy of vaporization (KJ/Kg)	1651.46	S(ppm)	25
Saturation Vapor Pressure (kPa)	0.053	Flashpoint	65°C
Calorific Value (kJ/kg)	44,800		

Constituents consisting of Carbon monoxide (CO), Carbon dioxide (CO_2), Oxides of nitrogen (NO_x) and Unburnt Hydrocarbons (HC) are measured by exhaust gas analyzer. Opacity is measured by Smoke meter in terms of Hartrigde Smoke Unit (%) and light absorption coefficient (K expressed).

III. RESULTS AND DISCUSSIONS

This portion deals with the discussion about trends in various emissions that this particular blend shows at various loads.

NOx
Emissions

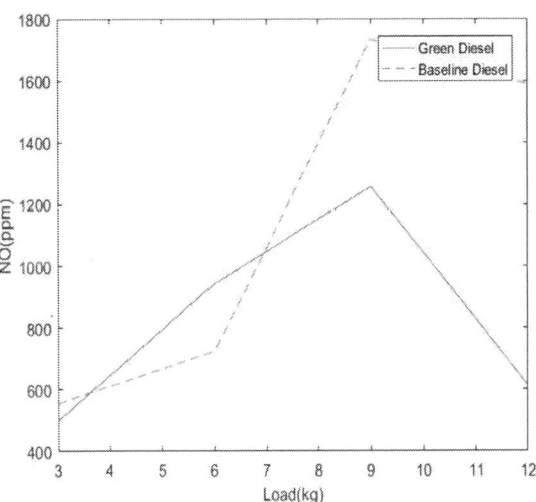

Fig. 2. Effect of variation of loads on NO_x emissions with both Baseline Diesel and Greendiesel

For pure diesel (D100) and green diesel (D80BD20W20) emissions increases first till a load of 9 kg and gradually decreases after that in the case of pure diesel (D100) but steeply decreases in the case of green diesel (D80BD20W20).

From the experimentation, it was inferred that pure diesel shows better performance with less NOx emissions till a load of 7 kg but has a decline in performance with higher NOx emissions than that of green diesel (D80BD20W20), especially at 12 kg load. Hence green diesel (D80BD20W20) is better than pure diesel (D100) in terms of NOx emissions at loads higher than 7 kg.

HC Emissions

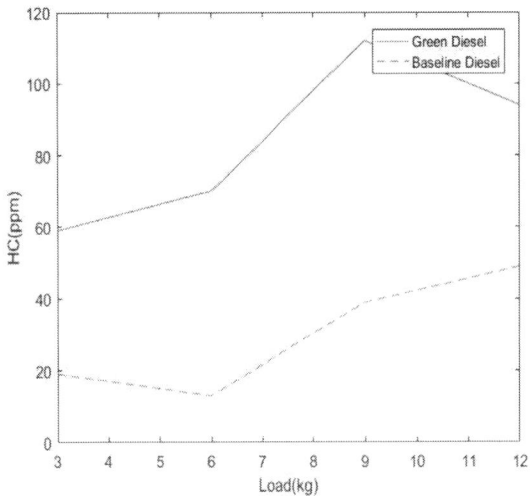

Fig. 3. Effect of variation of loads on HC emissions for Green Diesel.

HC emissions for pure diesel (D100) is observed to be considerably lower than that of green diesel (D80BD20W20) throughout the variation of load.

Hence from the experiment it was found that pure diesel (D100) is better than green diesel (D80BD20W20) in terms of HC emissions, given the load varies from 3 kg to 12 kg.

CO_2 Emissions

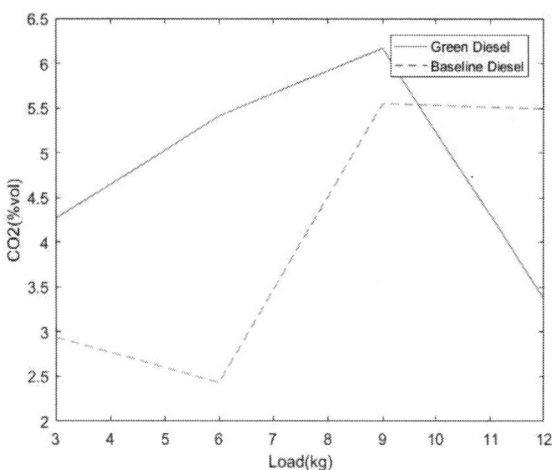

Fig. 4.   Effect of variation of loads on CO_2 emissions for Green Diesel.

For pure diesel (D100) CO_2 emissions were decreasing till a load of 6 kg and then steeply increased till a load of 9 kg. It then gradually declined till a load of 12 kg. Whereas, in the case of green diesel (D80BD20W20) the CO_2 emissions were higher and increased till a load of 9 kg and then steeply decreased till a load of 12 kg, eventually resulting emissions of CO_2. Hence it is found from the experiment that green diesel (D80BD20W20) is better in terms CO_2 emissions at loads higher than 9 kg.

Opacity

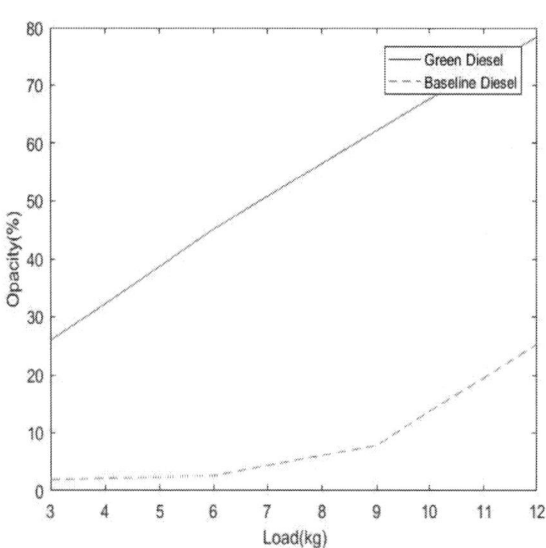

Fig. 5.   Effect of variation of load on the opacity of Green Diesel and Baseline Diesel.

Opacity gradually increases for both pure diesel (D100) and green diesel (D80BD20W20) throughout the variation of load from 3 kg to 12 kg. It was observed that the opacity of green diesel (D80BD20W20) was considerably higher than that of pure diesel (D100) throughout the experiment.

IV. CONCLULSION

It was inferred from this experiment that with the variation of load, the NOx emissions from Green Diesel (D80BD20W20) is significantly lower than the Baseline Diesel (D100), and similar traits of lesser emissions can be noted in the case of CO2 as well; The CO2 emissions of Green Diesel (D80BD20W20) is lesser than that of Diesel (D100). These lower values of emissions in CO2 and NOx can be noted in loads greater than 9 Kilograms which is the usual case in an engine. Hence, this seemed to be a good trait in this unique blend. However, with the variation of loads, when the opacity and emissions of HCs is observed, we see that the Baseline Diesel (D100) performs better than Green Diesel (D80BD20W20). With variation in blends these characteristics vary and hence we conclude that a better blend of this alternative fuel could result in a lesser emission of HCs as well as a lower opacity level. Overall, the emission levels were low and comparable except in the case of HCs.

ACKNOWLEDGMENT

We would like to extend our gratitude to Mr. Gora Banerjee and Ms. Anuradha, who guided and helped us throughout project. We would also like to acknowledge the support of PK Raju, Akhilesh Devarkonda, Priyanka Gangadhar and Tarun Kumar. For the last we would like to thank each other for equal efforts and understanding through the course of the project.

REFERENCES

[1]   D. Evan Mercer and John Soussan, Fuelwood Problems and Solutions, 1986, pp. 177-213

[2]   Hill J et al, Environmental Economical and Energetic costs and benefits of biodiesel and ethanol biofuels, Proc Natl Acad Sci U S A., vol. 103, July 2016

[3]   Ayhan Dermibas, Biofuels sources, biofuels policy, biofuel economy and global biofuel projection, Energy conversion and Management, pp. 2106-2116, Volume 49 Issue 8, August 2008

[4]   Hill J et al, Environmental Economical and Energetic costs and benefits of biodiesel and ethanol biofuels, Proc Natl Acad Sci U S A., vol. 103, July 2016

[5]   Ayhan Dermibas, Biofuels sources, biofuels policy, biofuel economy and global biofuel projection, Energy conversion and Management, pp. 2106-2116, Volume 49 Issue 8, August 2008

[6]   DYC Leung, X Wu, MKH Leung, A review on Biodiesel production using catalysed transesterification, Applied Energy- Elsevier, pp. 1083-1095, Volume 87 Issue 4, April 2010

Gap in pagination due to formatting issues.

Pages 1209-1212

2nd IEEE International Conference on Power Electronics, Intelligent Control and Energy Systems (ICPEICES-2018)

Comparative Research for Managing Delay in Signal Processing via Multipliers

Aniket Kumar
Dept. of Electronics & Electrical Engg.
Shobhit Deemed University, Modipuram,
Meerut, India
aniket.kumar@shobhituniversity.ac.in

Ekta Gupta
Department of Physics, SBAS,
Shobhit Deemed University,
Modipuram, Meerut, India

R.P. Agarwal *Shobhit Deemed Univerity, Modipuram,* Meerut
India

R.K. Jain
Department of Physics, SBAS,
Shobhit Deemed University,
Modipuram, Meerut, India

Abstract— **In the present era , need of compact devices with low power consumption & portably has created the need of Integrated circuits and when these devices are used for signal processing, multipliers plays a vital role & has become dominant functional blocks in processors and microcontrollers. Thus it has become crucial to explore a algorithm for fast multiplier that enhance structural design to increase speed & minimize area.**

This paper introduces the Array, Vedic, Wallace & Dadda Algorithms & a performance comparison between them. In this manuscript simulation has been done for the mentioned algorithms for different bit lengths i.e. two, four, eight & sixteen bit on Model-Sim using VHDL language and then their implementation them on Xilinx for comparative analysis.

Keywords—**Array; Vedic; Wallace; Dadda; FPGA; Urdhav triyakbhyam sutra; Delay.**

I. OVERVIEW

Multipliers are the essential circuit block of any microprocessor, micro-controller & DSP that are used to process signal has appealed layout designers. Therefore selection of a multiplier algorithm cannot be underestimated. Algorithm differs in terms of partial product terms & the addition procedure to calculated product. A basic multiplier contains of a array of AND gates to generate product terms & adders to add them to obtain product term [1]. For high order multiplication, an enormous no. of AND & adder circuits are needed that becomes quite significant to decide the performance of a system.

The delay constraint is mainly connected with the multiplier algorithm which determines architecture leads to delay due to adders & logical gates involved. To meet the requirement of high speed processor, a fast multiplier is must. Such proficient unit is logically designed in terms of minimum delay, power consumption & area and high speed. The traditional mathematical algorithms can be streamlined and optimized by the usage of different algorithms[2].

There are generally three stages to do multiplication: The first stage generates partial products that are generated through an array of AND gates; Second stage reduces the partial products by the use of partial product reduction schemes; and lastly the product is obtained by adding the partial products.

A Array Multiplier

Add & shift algorithm is the basis of Array multiplier. The partial product terms are generated by the multiplication of the multiplicand with one multiplier bit [5]. Considering two binary numbers A and B, of n bits each, being multiplicand and multiplier respectively which requires n x n multiplier, n (n-2) full adders, n half-adders and n^2 AND gates and delay approx. (2n+1) td.

In order to simplify the concept, of 2X2 bit multiplication, assuming A = a (1) a (0) and B= b (1) b (0), the various bits of the final product term P can be written as:-

$$P(0) = a(0)b(0) \qquad (1)$$
$$P(1) = a(1)b(0) + b(1)a(0) \qquad (2)$$
$$P(2) = a(1)b(1) + C1; \qquad (3)$$

Where, $C1$ is the carry generated in eq. (2), $P(3) = C2$; where $C2$ is the carry generated in eq. (3),

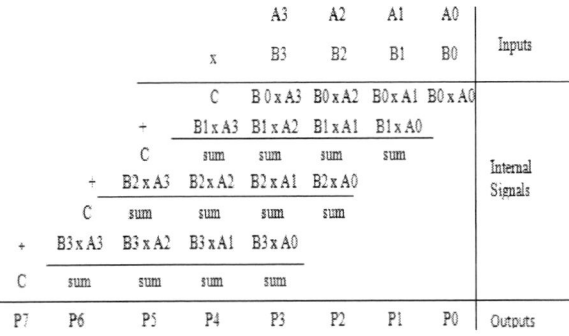

Fig.1 Four bit Array Multiplier

In fig.1, 4x4 multiplication process is described using array multiplication method, say $A = A(3) \, A(2) \, A(1) \, A(0)$ and $B = B(3) \, B(2) \, B(1) \, B(0)$. The output line for this multiplication is $P = P(7) \, P(6) \, P(5) \, P(4) \, P(3) \, P(2) \, P(1) \, P(0)$.

B. VEDIC MULTIPLIER

An old system of mathematics is the Vedic mathematics having a unique method for fast calculations grounded on 16 sutras. Vedic mathematics is a part of four Vedas[8], Urdhva

978-1-5386-6624-1/18 $31.00 © 2018 IEEE   1213

Tiryakbhyam an algorithm of ancient Indian Vedic Mathematics. "Urdhva" and "Tiryagbhyam" words are derived

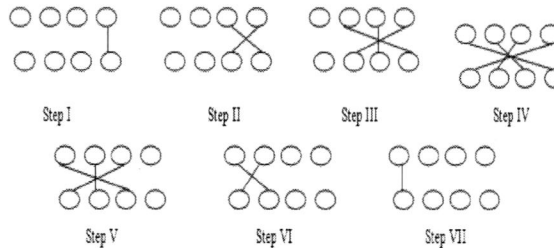

from Sanskrit literature. Urdhva means "Vertically" and Tiryagbhyam "crosswise", this multiplier is based on

Fig 2. Four Bit Vedic Multiplier

vertical & crosswise multiplication. Fig.2 shows the logical diagram to multiply of two 4-bit numbers.

The 16 bit Vedic Multiplier has been designed using four 8- bit Vedic Multipliers blocks and three Ripple Carry Adders blocks as well as Adders blocks of 16 bit, 24 bit, 24 bit. Using Cascade adders & look ahead adder (LDA) algorithm has been analysed.

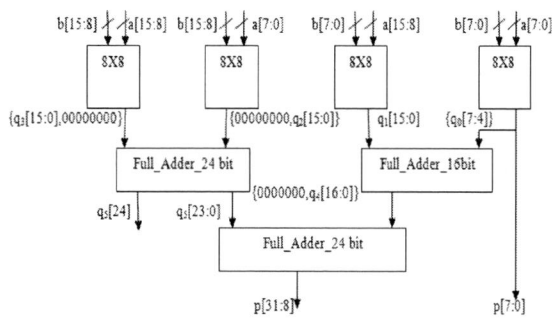

Fig.3   16x16 bit Vedic basic element

C.   Wallace Tree Multiplier

This type of multiplier uses tree structure to reduce number of additions in critical path to O(logn) rather than O(n).

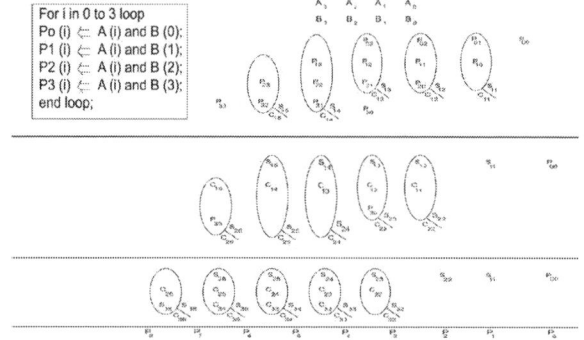

Fig.4 4x4 bit *Wallace Tree* Multiplier

It has difficult structure to layout and integrate with partial product crossbar [9].The logical illustration for multiplication of two 4-bit numbers is as shown in figure 5 using Wallace Tree Multiplier.

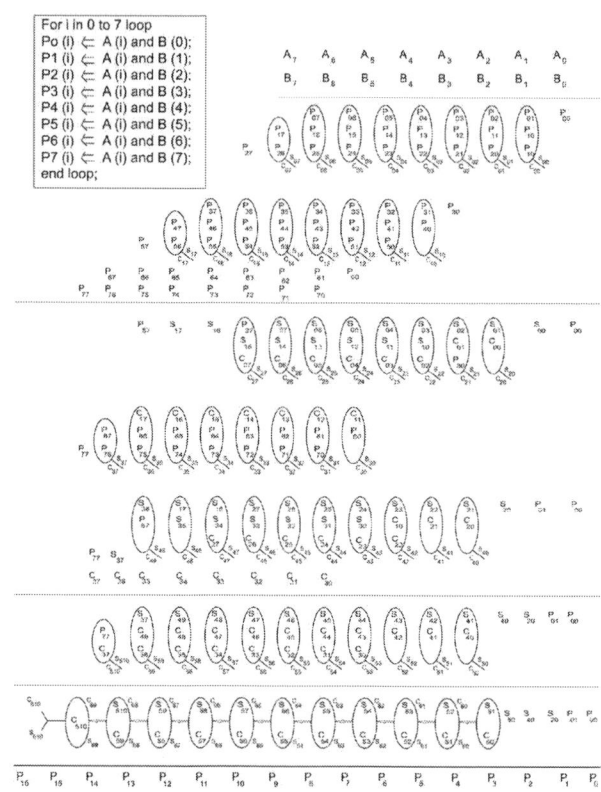

Fig.5   8x8 bit *Wallace Tree* Multiplier

For large operands, faster performance is deled by a Wallace tree multiplier. Contrasting an array multiplier the partial product matrix for a Wallace tree multiplier is reorganized in a tree-like format, reducing both the critical path and the number of adder cells needed [15].

D.   Dadda Multiplier

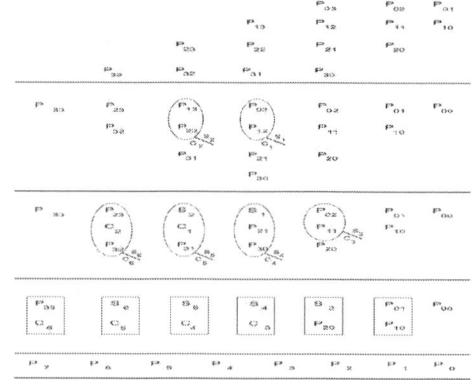

Fig.6 4x4 bit *Dadda* Multiplier

978-1-5386-6624-1/18 $31.00 © 2018 IEEE   1214

Dadda Multiplier is a multiplier designed similar to Wallace multiplier. As Wallace perform reductions as much as possible on each layer, Dadda multipliers do as few reductions as possible [17].

The logical illustration for multiplication of two 4-bit numbers is revealed in figure 6 using Dadda Algorithm.

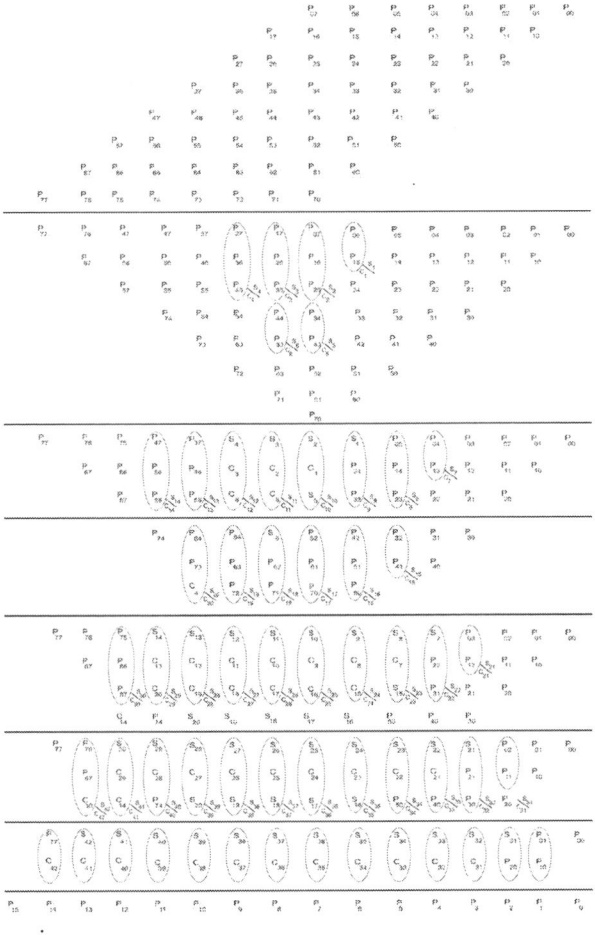

Fig.7 8x8 bit *Dadda* Multiplier

The logical illustration for multiplication of two 8-bit numbers is revealed in fig. 7 using Dadda Algorithm.

II. SYTHESIS & SIMULATION

Model-Sim Altera Edition 6.6d has been used for simulation purpose & Xilinx 14.4 with family Spartan6, device as xc6slx45, package csg324 with speed grade of -3 has been used for Implementation.

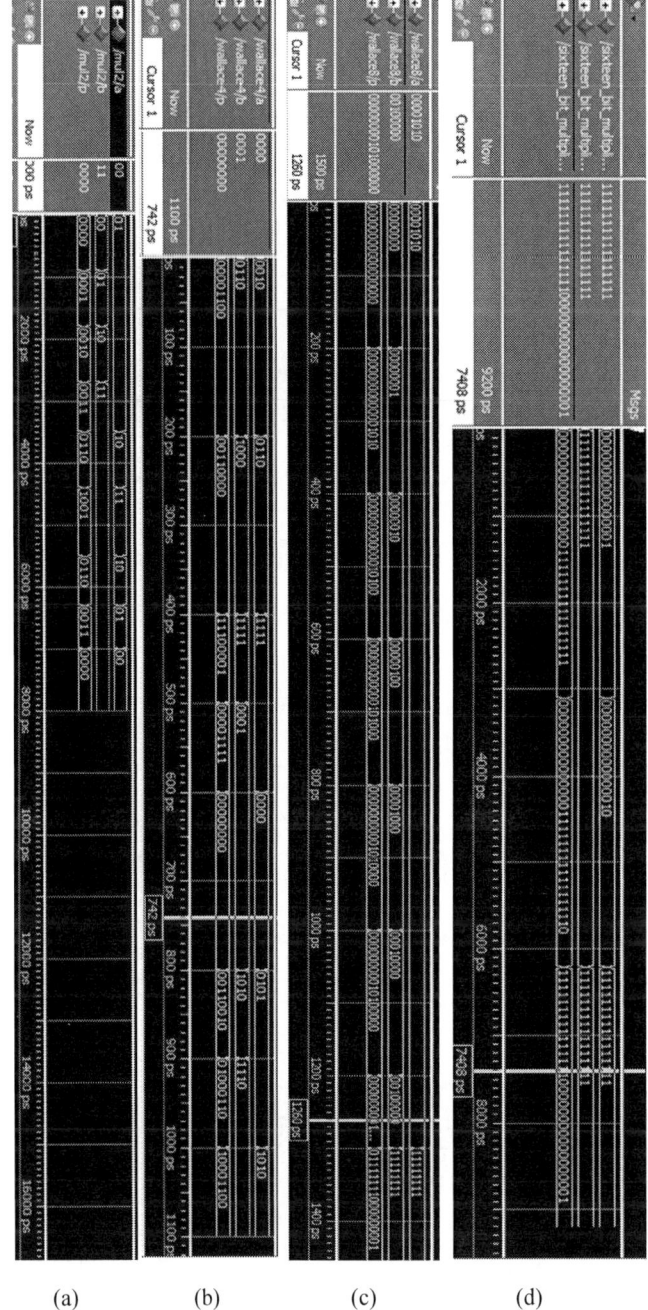

Fig.8 Simulation results of 2-bit (a), 4-bit (b), 8-bit (c) and 16-bit (d) of Wallace Tree Algorithm

Similarly hdl codes of Array, Vedic, & Dadda Algorithms are simulated for $2^1, 2^2, 2^3$ and 2^4 bit respectively.

III   SYTHESIS & SIMULATION

Xilinx 14.4 with family Spartan6, device as xc6slx45, package csg324 with speed grade of -3 has been used for Implementation.

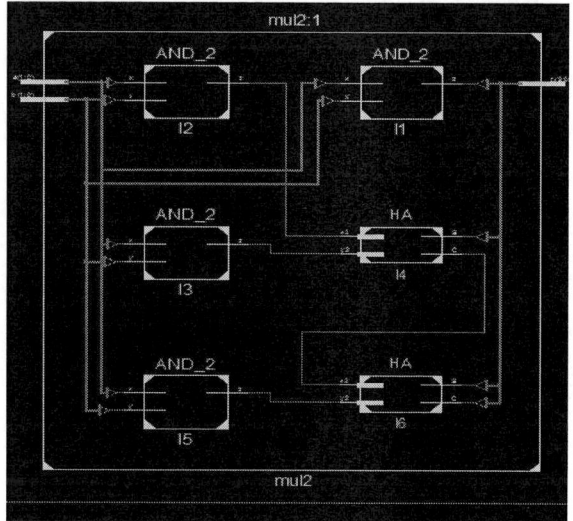

Fig. 9 2-bit Dadda multiplier, RTL Schematic

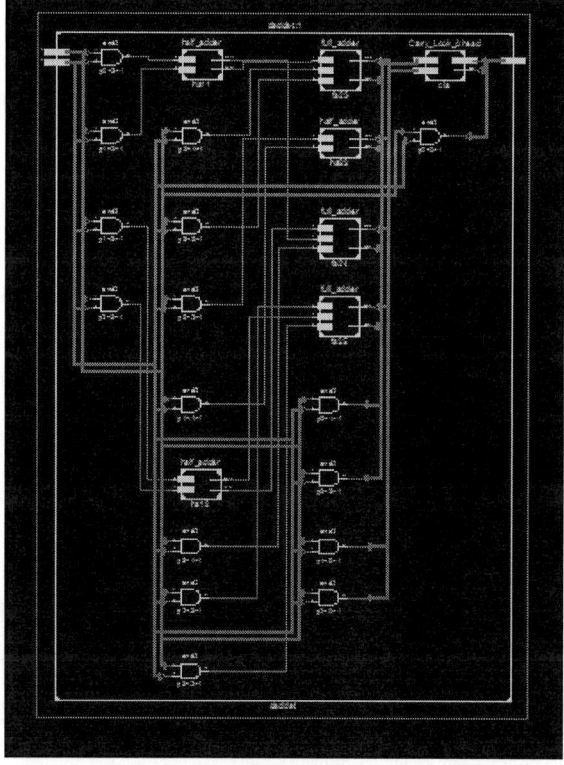

Fig. 10 4-bit Dadda multiplier, RTL Schematic

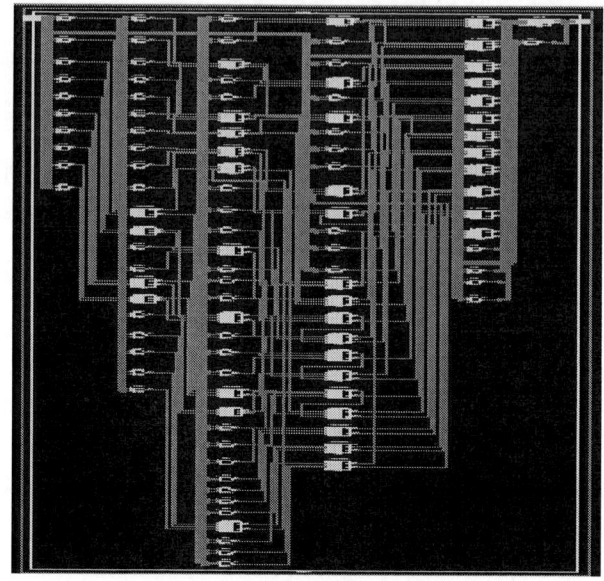

Fig. 10 8-bit Dadda multiplier, RTL Schematic

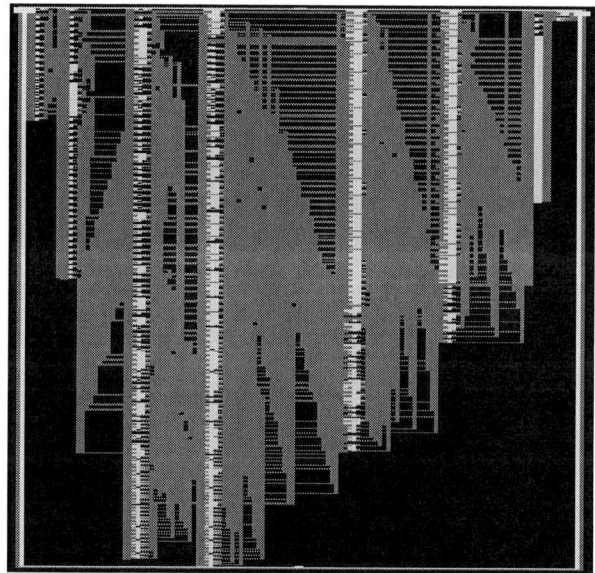

Fig. 11 16-bit Dadda multiplier, RTL Schematic

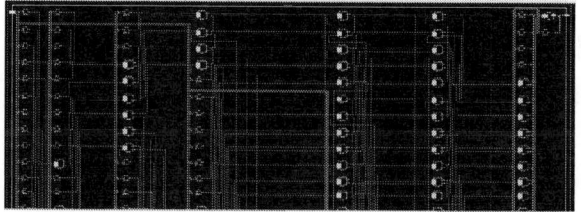

Fig. 12 Technological View (i.e. In terms of LUTs) of 16-bit Dadda multiplier

978-1-5386-6624-1/18 $31.00 © 2018 IEEE

Similarly HDL codes of Array, Vedic, & Wallace Tree Algorithms are simulated & then Implemented for 21,22, 23 and 24 bit respectively.

IV. COMPARATIVE ANALYSIS

Simulation & then implementation of a particular algorithms for different bit length generates synthesis report with significant parameters like delay, memory usage , levels of logic, no. of look up tables (LUTs) and many more that may be used to deduce the ultimate algorithm for particular requirement

Table No. I Comparison Analysis

Bit Length	Type of Algorithm		Delay (ns)	Levels of Logic	No. of slice LUTs	Memory (KB)
2	Vedic	Cascade	6.494	4	6	258052
		LDA	5.505	3	4	258764
	Array		6.494	4	6	258052
	Wallace		6.494	4	6	258052
	Dadda		6.494	4	6	258052
4	Vedic	Cascade	7.942	5	7	255024
		LDA	10.950	8	22	258380
	Array		55.591	40	38	257284
	Wallace		17.00	7	23	257036
	Dadda		14.02	7	19	257292
8	Vedic	Cascade	18.270	15	121	257988
		LDA	14.195	10	106	258892
	Array		120.485	175	82	259588
	Wallace		19.829	15	128	257612
	Dadda		13.907	11	114	258508
16	Vedic	Cascade	27.278	23	656	259076
		LDA	23.353	19	418	260108
	Array		263.072	734	178	260868
	Wallace		36.731	29	499	295564
	Dadda		22.124	19	453	261004

V. CONCLUSION

From From implementation, simulation & comparative results, it is concluded that Dadda is best in terms of speed as its delay for 16 bit is 22.124ns whereas delay for Vedic is 23.353ns when using LDA structure for same bit length. If memory usage is prime concern then Vedic leads all other multipliers as it uses 259076 KB of memory when using cascade structure whereas Wallace uses 295564KB for same that is 16 bit length. If packing is concerned, then Array uses lesser LUT's then Vedic for 16 bit length, Hence this comparison can help us to select a suitable multiplier for a particular task or application.

REFERENCES

[1] Akhilesh G. Naik and Dipankar Pal, "Delay estimation, chip-power analyses and comparision of single-level and multi-level recursive vedic algorithm with conventional algorithms for digital multiplier" Proceeding IEEE International Conference on IC Design & Technology, Otanto, Italy, pp 29-32, June, 4th -6th , 2018.

[2] W. Liu, et al., "Design of Approximate Radix-4 Booth Multipliers for Error-Tolerant Computing", *IEEE Trans. Computers*, vol. 66, no. 8, pp.1435-1441, 2017.

[3] H. Jiang, et al., "Approximate Radix-8 Booth Multipliers for Low-Power and High-Performance Operation", *IEEE Trans. Computers*, vol. 65, no. 8, pp. 2638-2644, Aug. 2016

[4] Jasbir Kaur, Sumit K, "Analysis and Comparison of Different Multiplier,"International Journal on Recent and Innovation Trends in Computing andCommunication ISSN: 2321-816975-78. Volume: 4 Issue: 6. June 2016.

[5] Ekta Gupta, Aniket Kumar and R.K. Jain, "Simulation &Comparative Analysis of Booth Multiplier", *International Journal of Electronics Engineering (ISSN: 0973-7383)*, UGC J. No. 2946, Vol. 10, No.2, June-Dec, 2018, pp. 404-412.

[6] Aniket Kumar and Vishikha, "Comparative Analysis of Vedic & Array Multiplier", *International Journal of Electronics and Communication Engineering and Technology*, ISSN: 0976-6464(P),0976-6472, Vol. 8, Issue. 3, pp.17-27, May - June 2017.

[7] Kavitha Priya N, and Karthikeyan KV, "Analysis of Mac Unit Using VedicMultiplier and Sklansky Adder", Research Journal of Pharmaceutical, Biologicaland Chemical Sciences, ISSN: 0975-8585, May – June 2016.

[8] Vishal Galphat1, Nitin Lonbale, "The High Speed Multiplier by using PrefixAdder with MUX and Vedic Multiplication," International Journal of Science andResearch (IJSR), ISSN (Online): 2319-7064, Volume 5 Issue 1, January 2016.

[9] Mahendra Tiwari & Aniket Kumar, Implementation of High Speed and Low Power Novel Radix 2 BoothMultiplier using 2248 BECConverter , International Journal of Engineering Science and Computing, Volume 7,Issue No. 3, pp. 4861-4863, March 2017.

[10] Vishika Sharma and Aniket Kumar, "Design, Implementation & Performance of Vedic Multiplier for Different Bit Lengths", *International Journal of Innovative Research in Computer and Communication Engineering*, ISSN : 2320-9801(O) , 2320-9798(P), Vol. 5, Issue. 4, pp.7912-7919, April 2017.

[11] Srini Devadas, "Introduction to Algorithms," Lecture 11: Integer Arithmetic, Karatsuba Multiplication, MIT Open Course Ware, Massachusetts Institute of Technology, 6.006, Fall 2011. Available at: https://www.youtube.com/watch?v=eCaXIAaN2uE.

[12] A. Mehta, C. B. Bidhul, S. Joseph and Jayakrishnan P., "Implementation of Single Precision Floating Point Multiplier using Karatsuba Algorithm", *International Conference on Green Computing, Communication and Conservation of Energy (ICGCE)*, pp. 254-256, 2013.

[13] K. Shruthilaya and M. Vinoth, "Power Estimation of Modified Booth Recoder for Efficient Add-Multiply Operator", *IEEE Conference on Computing for Sustainable Global Development (INDIACom)*, pp. 1684-1689,2015.

[14] R. Pratibha, P. Sandhya, and R. Varun, "Design of High Performance and Low Power Multiplier using Modified Booth Encoder", *IEEE International Conference on Electrical, Electronics and Optimization Techniques (ICEEOT)*, pp. 794-798, 2016.

[15] B. Ramkumar, H.M. Kittur, and P. M. Kannan, "ASIC implementation of modified faster carry save adder," *Eur.J. Sci. Res.*vol. 42, no. 1, pp. 53–58, 2010.

[16] K.S.Rao, M.Reddy, A.V.Babu and P.Srinivasulu, "Energy and area efficient carry select adder," *IJERA, vol. 34* issue 2, pp.436-440, Mar-Apr 2012.

978-1-5386-6624-1/18 $31.00 © 2018 IEEE

[17] Y.Kim and L.-S. Kim, "64-bit carry-select adder with reduced area," *Electron. Lett., vol. 37, no. 10, pp. 614–615,*May 2001.

[18] Z.Huang, "High-Level Optimization Techniques for Low-Power Multiplier Design," *PhD dissertation, Univ. of*California Los Angeles, June 2003

[19] M. Morris mano: Computer system architechture Prentice Hall Date: 1992-10-29,pp 344.

[20] Jagadguru Swami Sri Bharti Krishna Tirthaji Maharaja "Vedic Mathematics or Sixteen Simple Mathematicle Formulae from the Veda, Delhi(1965)", Motilal Banarsidas Varanasi, India.

[21] Ch. Harish Kumar, Implementation and Analysis of Power, Area and Delay of Array, UrdhvaandNikhilam Vedic Multipliers, International Journal of Scientific and Research Publications, 2013, 3 (1), 1-5.

[22] PushpalataVerma and KK Mehta, Implementation of an Efficient Multiplier based on Vedic Mathematics Using EDA Tool, International Journal of Engineering and Advanced Technology, 2012,1(5), 75-79

[23] G Kumar, Design of High Speed Vedic Multiplier using Vedic Mathematics Techniques, International Journal of Scientific and Research Publications, 2012, 2 (3), 1-5.

[24] PrabirSaha, Arindham Banerjee, ParthaBattacharyya andAnupDhandapat, High Speed Design of Complex Multiplier using Vedic Mathematics, Proceedings of IEEE Students Technology Symposium, IIT Kharagpur, India, 2011, 237-241.

978-1-5386-6624-1/18 $31.00 © 2018 IEEE   1218

2nd IEEE International Conference on Power Electronics, Intelligent Control and Energy Systems (ICPEICES-2018)

Perturbation Based Nonlinear Analysis of MOSFET Circuit

Rahul Bansal[1] and Sudipta Majumdar[2]

[1,2]Electronics & Communication Engineering Department

Delhi Technological University, Shahbad Daulatpur, Delhi, 110042, India

Email: [1]rahulbansal1591@gmail.com, [2]korsudipta@rediffmail.com

Abstract—This paper presents the implementation of per-turbation technique for analysis of Enz-Krummenacher-Vittoz (EKV) modeled metal oxide field effect transistor (MOSFET). Use of perturbation technique presents more accurate closed form nonlinear relation between input and output of MOSFET. The linear and nonlinear output voltage expressions have been simulated in MATLAB software. The distortion error due to use of linear term only shows the importance of nonlinear expression. The proposed model can be used when large input amplitude is applied. Also, it can be used for parameter estimation purpose.

Keywords—MOSFET, perturbation method, Taylor series expansion.

I. INTRODUCTION

Various models have been presented in literature for MOS-FET. In [1], Hasani proposed a three port model for MOSFET used in millmeter wave band which captures the longitudinal distributed effect. Chunha *et al.* [2] presented physics based model of MOSFET that can be used for design and analysis of integrated circuits. Allam *et al.* [3] proposed high frequency model for MOSFET that can be used for saturation region. The model used the transmission line behavior of the gate region. In [4], Andersson *et al.* proposed model for switching characteristics of MOS devices. Wang *et al.* [5] proposed a model for high voltage MOS transistors, which is useful for high voltage IC simulations. In [6], Abbasian *et al.* presented a method using genetic algorithm to find important elements of MOSFET and used it to present a model for I-V characteristics of MOSFET. Kushaa *et al.* [7] proposed charge based model for the intrinsic capacitances in MOS transistor which is based on quasi static approximation. Shen *et al.* [8] proposed mod-elling of tunnel FET, which is based on Poisson's equation.

The paper is organized as follows:- Section II presents brief theory of perturbation method. Deterministic modeling of MOSFET circuit using perturbation method is presented in Section III. Section IV presents simulation results. Section V concludes the work.

II. PERTURBATION THEORY

The perturbation technique is applied by adding a small term to exactly solvable problem as

$$A = A_0 + \epsilon A_1 + \epsilon^2 A_2 + ... \qquad (1)$$

where A_o is the known solution to the exactly solvable problem. $A_1, A_2, ...$ are higher order nonlinear terms [9]-[11].

First author thanks CSIR, India for financial support.

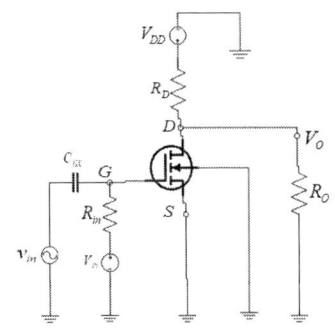

Figure 1. MOSFET circuit diagram.

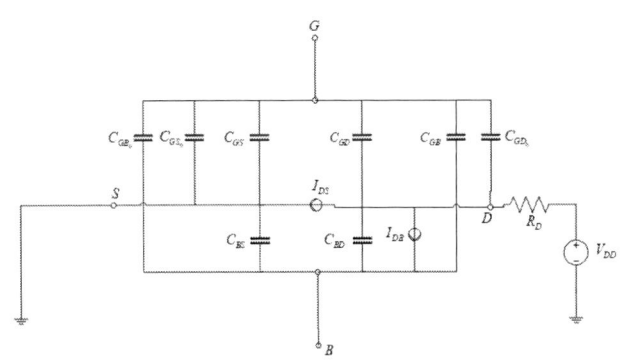

Figure 2. EKV equivalent circuit for MOSFET.

III. DETERMINISTIC MODELING OF MOSFET CIRCUIT USING PERTURBATION METHOD

MOSFET circuit is shown in Fig. 1. MOSFET equivalent EKV model is shown in Fig. 2. Applying Kirchhoff's voltage law (KVL) and Kirchhoff's current law (KCL) and replacing the MOS transistor by EKV mode, we have:

978-1-5386-6624-1/18 $31.00 © 2018 IEEE   1219

$$
(C_{GS} + C_{GS_0}) \left(\frac{dv_S(t)}{dt} - \frac{dv_G(t)}{dt} \right) + (C_{GD} + C_{GD_0})
$$
$$
\times \left(\frac{dv_D(t)}{dt} - \frac{dv_G(t)}{dt} \right) + (C_{GB} + C_{GB_0}) \times \left(\frac{dv_B(t)}{dt} \right)
$$
$$
- \frac{dv_G(t)}{dt} = 0 \tag{2}
$$

$$
(C_{GD} + C_{GD_0}) \left(\frac{dv_G(t)}{dt} - \frac{dv_D(t)}{dt} \right) + C_{BD} \left(\frac{dv_B(t)}{dt} \right)
$$
$$
- \frac{dv_D(t)}{dt} = I_{DS} + I_{DB} \tag{3}
$$

$$
(C_{GS} + C_{GS_0}) \left(\frac{dv_G(t)}{dt} - \frac{dv_S(t)}{dt} \right) + C_{BS} \left(\frac{dv_B(t)}{dt} \right)
$$
$$
- \frac{dv_S(t)}{dt} = -I_{DS} \tag{4}
$$

$$
C_{GX} \left(\frac{dv_{in}}{dt} - \frac{dv_G(t)}{dt} \right) + \frac{V_{IN} - v_G(t)}{R_{in}} = 0 \tag{5}
$$

where $v_G(t)$, $v_S(t)$, $v_D(t)$ and $v_B(t)$ are the state variables. C_{GD}, C_{GS} and C_{GB} are the drain to channel capacitance, source to channel capacitance and base to channel capacitance respectively. C_{OX} is oxide capacitance. Drain current I_D is

$$
I_D = I_{DS} + I_{DB} \tag{6}
$$

As $I_{DB} \cong 0$, therefore $I_D \cong I_{DS}$. From the EKV model of weak inversion, we have

$$
I_D = I_{DS} = I_0 \frac{W}{L} e^{\frac{v_{GB}(t) - V_{T_0}}{\eta U_T}} \left(e^{-\frac{v_{SB}(t)}{U_T}} - e^{-\frac{v_{DB}(t)}{U_T}} \right) \tag{7}
$$

where $\frac{W}{L}$ is the aspect ratio, V_{T_0} and U_T are the the equilibrium threshold voltage and thermal voltage respectively. I_0 is unary specific current. η is subthreshold slope factor.

Expanding (7) using Taylor series and retaining up to quadratic terms, we have

$$
I_D = I_0 \frac{W}{L} \left\{ \left(\frac{v_D - v_S}{U_T} \right) \left(1 - \frac{V_{T_0}}{\eta U_T} + \frac{V_{T_0}^2}{2\eta^2 U_T^2} - \frac{1}{6} \frac{V_{T_0}^3}{\eta^3 U_T^3} \right) \right\}
$$
$$
+ I_0 \frac{W}{L} \left\{ (v_S^2 - v_D^2 - 2v_S v_B + 2v_D v_B - v_D v_S + v_S v_D) \right.
$$
$$
\left. \times \left(1 - \frac{V_{T_0}}{\eta U_T} + \frac{V_{T_0}^2}{2\eta^2 U_T^2} - \frac{1}{6} \frac{V_{T_0}^3}{\eta^3 U_T^3} \right) \right\} + \frac{I_0 W}{U_T L} \left\{ (v_G v_D \right.
$$
$$
\left. - v_B v_D - v_G v_S + v_B v_S) \left(\frac{1}{\eta U_T} - \frac{2V_{T_0}}{\eta^2 U_T^2} + \frac{3V_{T_0}^2}{\eta^3 U_T^3} \right) \right\} \tag{8}
$$

Using $C_{GS} + C_{GS_0} = C_S$, $C_{GD} + C_{GD_0} = C_D$ and $C_{GB} + C_{GB_0} = C_B$ and separating linear and first order nonlinear terms of state variables as:

$$
v_G(t) = v_G^{(0)}(t) + \epsilon v_G^{(1)}(t) \tag{9}
$$
$$
v_D(t) = v_D^{(0)}(t) + \epsilon v_D^{(1)}(t) \tag{10}
$$
$$
v_S(t) = v_S^{(0)}(t) + \epsilon v_S^{(1)}(t) \tag{11}
$$
$$
v_B(t) = v_B^{(0)}(t) + \epsilon v_B^{(1)}(t) \tag{12}
$$

where linear state variables are $v_G^{(0)}(t)$, $v_D^{(0)}(t)$, $v_S^{(0)}(t)$, $v_B^{(0)}(t)$ and first order nonlinear terms of state variables are $v_G^{(1)}(t)$, $v_D^{(1)}(t)$, $v_S^{(1)}(t)$, $v_B^{(1)}(t)$. Thus from (2)-(5) and (7)-(12), we have

$$
\{v_G^{(0)}(t) + \epsilon v_G^{(1)}(t)\} \left(\frac{d}{dt} + \frac{1}{R_{in} C_{GX}} \right) = \frac{V_{IN}}{R_{in} C_{GX}} + v_{in}' \tag{13}
$$

$$
C_D \frac{d\{v_G^{(0)}(t) + \epsilon v_G^{(1)}(t)\}}{dt} + \{v_D^{(0)}(t) + \epsilon v_D^{(1)}(t)\} \left(-C_D \frac{d}{dt} \right.
$$
$$
\left. - C_{BD} \frac{d}{dt} - k_1 \right) k_1 \{v_S^{(0)}(t) + \epsilon v_S^{(1)}(t)\} + C_{BD} \frac{d\{v_B^{(0)}(t)\}}{dt}
$$
$$
+ C_{BD} \frac{d\{\epsilon v_B^{(1)}(t)\}}{dt} = k_1 v_T \{v_S^{2(0)}(t) - v_D^{2(0)}(t) - 2v_S^{(0)} v_B^{(0)}
$$
$$
+ 2v_D^{(0)}(t) v_B^{(0)}(t)\} + k_2 \{v_G^{(0)}(t) v_D^{(0)}(t) - v_B^{(0)}(t) v_D^{(0)}(t)
$$
$$
- v_G^{(0)}(t) v_S^{(0)}(t) + v_B(t) v_S(t)\} \tag{14}
$$

$$
C_S \frac{d\{v_G^{(0)}(t) + \epsilon v_G^{(1)}(t)\}}{dt} + \{v_S^{(0)}(t) + \epsilon v_S^{(1)}(t)\} \left(-C_S \frac{d}{dt} \right.
$$
$$
\left. - C_{BS} \frac{d}{dt} - k_1 \right) + k_1 \{v_D^{(0)}(t) + \epsilon v_D^{(1)}(t)\} + C_{BS} \frac{d\{v_B^{(0)}(t)\}}{dt}
$$
$$
+ C_{BS} \frac{d\{\epsilon v_B^{(1)}(t)\}}{dt} = -k_1 v_T \{v_S^{2(0)}(t) - v_D^{2(0)}(t) - 2v_S v_B
$$
$$
+ 2v_D(t) v_B(t)\} - k_2 \{v_G^{(0)}(t) v_D^{(0)}(t) - v_B^{(0)}(t) v_D^{(0)}(t)
$$
$$
- v_G^{(0)}(t) v_S^{(0)}(t) + v_B^{(0)}(t) v_S^{(0)}(t)\} \tag{15}
$$

$$
\{v_G^{(0)}(t) + \epsilon v_G^{(1)}(t)\} (-C_G - C_D - C_B) + C_D \{v_D^{(0)} + \epsilon v_D^{(1)}\}
$$
$$
+ C_S \{v_S^{(0)}(t) + \epsilon v_S^{(1)}(t)\} + C_B \{v_B^{(0)}(t) + \epsilon v_B^{(1)}(t)\} = 0 \tag{16}
$$

where

$$
k_1 = \frac{I_0}{U_T} \frac{W}{L} \left(1 - \frac{V_{T_0}}{\eta U_T} + \frac{V_{T_0}^2}{2\eta^2 U_T^2} - \frac{1}{6} \frac{V_{T_0}^3}{\eta^3 U_T^3} \right)
$$

$$
k_2 = \frac{I_0}{U_T} \frac{W}{L} \left(\frac{1}{\eta U_T} - \frac{V_{T_0}}{\eta^2 U_T^2} + \frac{V_{T_0}^2}{2\eta^3 U_T^3} \right)
$$

A. Zeroth Order Approximation

Comparing $\epsilon^{(0)}$ terms in (13)-(16) to obtain zeroth order terms

$$
v_G^{(0)}(t) \left(\frac{d}{dt} + \frac{1}{R_{in} C_{GX}} \right) = \frac{V_{IN}}{R_{in} C_{GX}} + v_{in}' \tag{17}
$$

$$
C_D \frac{dv_G^{(0)}(t)}{dt} + v_D^{(0)}(t) \left(-C_D \frac{d}{dt} - C_{BD} \frac{d}{dt} - k_1 \right) + k_1 v_S^{(0)}(t)
$$
$$
+ C_{BD} \frac{dv_B^{(0)}(t)}{dt} = 0 \tag{18}
$$

$$C_S \frac{dv_G^{(0)}(t)}{dt} + v_S^{(0)}(t)\left(-C_S \frac{d}{dt} - C_{BS}\frac{d}{dt} - k_1\right) + k_1 v_D^{(0)}(t)$$
$$+ C_{BS}\frac{dv_B^{(0)}(t)}{dt} = 0 \tag{19}$$

$$v_G^{(0)}(t)(-C_G - C_D - C_B) + C_D v_D^{(0)}(t) + C_S v_S^{(0)}(t)$$
$$+ C_B v_B^{(0)}(t) = 0 \tag{20}$$

Laplace transformed equation is

$$\mathbf{A}_1(s)\mathbf{x}^{(0)}(s) = \mathbf{B}_1(s)u_1(s) \tag{21}$$

where s is complex variable and $u_1(s) = \left(\frac{V_{IN}}{R_{in}C_{GX}} + v'_{in}\right)$.
State vector $\mathbf{x}^{(0)}(s)$ is

$$\mathbf{x}^{(0)}(s) = \left[\begin{array}{cccc} v_G^{(0)}(s) & v_D^{(0)}(s) & v_S^{(0)}(s) & v_B^{(0)}(s) \end{array}\right]^T,$$

Where $\mathbf{A}_1(s)$ is

$$\mathbf{A}_1(s) = \left[\begin{array}{cccc} A_{1_{11}} & 0 & 0 & 0 \\ sC_D & A_{1_{22}} & k_1 & sC_{BD} \\ sC_S & k_1 & A_{1_{33}} & sC_{BS} \\ A_{1_{41}} & C_D & C_S & C_B \end{array}\right], \tag{22}$$

where
$A_{1_{11}} = \left(s + \frac{1}{R_{in}C_{GX}}\right)$, $A_{1,22} = (-sC_D - sC_{BD} - k_1)$,
$A_{1_{33}} = (-sC_S - sC_{BS} - k_1)$, $A_{1_{41}} = -C_G - C_D - C_B$
Where $\mathbf{B}_1(s)$ is

$$\mathbf{B}_1(s) = \left[\begin{array}{cccc} 1 & 0 & 0 & 0 \end{array}\right]^T$$

Solution of (21) is given by

$$\mathbf{x}^{(0)}(s) = \mathbf{A}_1^{-1}(s)\mathbf{B}_1(s)u_1(s) \tag{23}$$

Impulse responses of linear model (small signal equivalent) are

$$v_G^{(0)}(t) = \frac{A_{11}}{|\mathbf{A}_1|} * u_1(t) \tag{24}$$

$$v_D^{(0)}(t) = \frac{A_{12}}{|\mathbf{A}_1|} * u_1(t) \tag{25}$$

$$v_S^{(0)}(t) = \frac{A_{13}}{|\mathbf{A}_1|} * u_1(t) \tag{26}$$

$$v_B^{(0)}(t) = \frac{A_{14}}{|\mathbf{A}_1|} * u_1(t) \tag{27}$$

where $*$ is convolution operator. A_{11}, A_{12}... are the cofactors of matrix $\mathbf{A}_1$ and $|\mathbf{A}_1|$ is the determinant of matrix $\mathbf{A}_1$.

$$v_G^{(0)}(t) = e^{-\frac{1}{R_{in}C_{GX}}t}u(t) * u_1(t) \tag{28}$$

$$v_D^{(0)}(t) = \left[\frac{k_5}{k_3}m_1 e^{-\frac{1}{R_{in}C_{GX}}t}u(t) + \frac{k_5}{k_3}m_2 e^{-\frac{k_1 k_4}{k_3}t}u(t)\right]$$
$$* u_1(t) \tag{29}$$

$$v_S^{(0)}(t) = \left[\frac{k_7}{k_3}m_3 e^{-\frac{1}{R_{in}C_{GX}}t}u(t) + \frac{k_7}{k_3}m_4 e^{-\frac{k_1 k_4}{k_3}t}u(t)\right]$$
$$* u_1(t) \tag{30}$$

$$v_B^{(0)}(t) = \left[\frac{k_9}{k_3}m_5 e^{-\frac{1}{R_{in}C_{GX}}t}u(t) + \frac{k_9}{k_3}m_6 e^{-\frac{k_1 k_4}{k_3}t}u(t)\right]$$
$$* u_1(t) \tag{31}$$

where
$k_3 = (C_D C_B C_S + C_D C_B C_{BS} + C_S C_B C_{BS} + C_D C_{BD} C_{BS}$
$\quad + C_D C_S C_{BS} + C_S C_{BD} C_{BS})$
$k_4 = \{C_B C_D + C_B C_{BD} + C_B C_S + C_B C_{BS} + C_S C_{BS}$
$\quad + C_D C_{BS} + C_S C_{BD} + C_D C_{BD}\}$
$k_5 = \{C_D C_B C_S + C_D C_B C_{BS} + C_D C_S C_{BS} + C_{BD}$
$\quad (C_S + C_{BS})(C_S + C_D + C_{DD}) - C_S^2 C_{DD}\}$
$k_6 = \{C_D C_B + C_S C_B + C_S C_{BS} + C_B C_{BS} + C_D C_{BS}$
$\quad + C_{BD}(C_S + C_D + C_{BD})\}$
$k_7 = \{-C_D^2 B_S + (C_D + C_{BD})(C_S C_B + C_S C_{BS} + C_B C_{BS}$
$\quad + C_D C_{BS}) + C_S C_D C_{BD}\}$
$k_8 = (C_B C_D + C_B C_S + C_S C_{BS} + C_B C_{BS} + C_D C_{BS}$
$\quad + C_S C_{BD} + C_B C_{BD} + C_D C_{BD})$
$k_9 = (C_S C_D C_{BS} + C_S C_{BD} C_{BS} + C_S C_D C_{BD}$
$\quad + C_D C_{BD} C_{BS} + C_S C_B C_D + C_B C_D C_{BS}$
$\quad + C_S C_{BD} C_B + C_B C_{BD} C_{BS})$
$k_{10} = \{C_S C_{BD} + C_S C_{BS} + C_D C_{BD} + C_D C_{BS} + C_D C_B$
$\quad + C_B C_{BD} + C_S C_B + C_B C_{BS}\}$

$$m_1 = \frac{\left(-\frac{1}{R_{in}C_{GX}} + \frac{k_1 k_6}{k_5}\right)}{\left(-\frac{1}{R_{in}C_{GX}} + \frac{k_1 k_4}{k_3}\right)}, \qquad m_2 = \frac{\left(-\frac{k_1 k_4}{k_3} + \frac{k_1 k_6}{k_5}\right)}{\left(\frac{1}{R_{in}C_{GX}} - \frac{k_1 k_4}{k_3}\right)},$$

$$m_3 = \frac{\left(-\frac{1}{R_{in}C_{GX}} + \frac{k_1 k_8}{k_7}\right)}{\left(-\frac{1}{R_{in}C_{GX}} + \frac{k_1 k_4}{k_3}\right)}, \qquad m_4 = \frac{\left(-\frac{k_1 k_4}{k_3} + \frac{k_1 k_8}{k_7}\right)}{\left(\frac{1}{R_{in}C_{GX}} - \frac{k_1 k_4}{k_3}\right)},$$

$$m_5 = \frac{\left(-\frac{1}{R_{in}C_{GX}} + \frac{k_1 k_{10}}{k_9}\right)}{\left(-\frac{1}{R_{in}C_{GX}} + \frac{k_1 k_4}{k_3}\right)}, \qquad m_6 = \frac{\left(-\frac{k_1 k_4}{k_3} + \frac{k_1 k_{10}}{k_9}\right)}{\left(\frac{1}{R_{in}C_{GX}} - \frac{k_1 k_4}{k_3}\right)}.$$

B. First Order Approximation

Comparing $\epsilon^{(1)}$ terms in (13)-(16) to obtain first order nonlinear terms, we have

$$v_G^{(1)}(t)\left(\frac{d}{dt} + \frac{1}{R_{in}C_{GX}}\right) = 0 \tag{32}$$

$$C_D \frac{dv_G^{(1)}(t)}{dt} + v_D^{(1)}(t)\left(-C_D \frac{d}{dt} - C_{BD}\frac{d}{dt} - k_1\right) + k_1 v_S^{(1)}$$
$$+ C_{BD}\frac{dv_B^{(1)}(t)}{dt} = k_1 v_T \{v_S^{2(0)}(t) - v_D^{2(0)}(t) - 2v_S^{(0)}v_B^{(0)}$$
$$+ 2v_D^{(0)}(t)v_B^{(0)}(t)\} + k_2\{v_G^{(0)}(t)v_D^{(0)}(t) - v_B^{(0)}(t)v_D^{(0)}(t)$$
$$- v_G^{(0)}(t)v_S^{(0)}(t) + v_B^{(0)}(t)v_S^{(0)}(t)\} \tag{33}$$

$$C_S \frac{dv_G^{(1)}(t)}{dt} + v_S^{(1)}(t)\left(-C_S \frac{d}{dt} - C_{BS}\frac{d}{dt} - k_1\right) + k_1 v_D^{(1)}$$
$$= + C_{BS}\frac{dv_B^{(1)}(t)}{dt} - k_1 v_T\{v_S^{2(0)}(t) - v_D^{2(0)}(t) - 2v_S^{(0)}v_B^{(0)}$$
$$+ 2v_D^{(0)}(t)v_B^{(0)}(t)\} - k_2\{v_G^{(0)}(t)v_D^{(0)}(t) - v_B^{(0)}(t)v_D^{(0)}(t)$$
$$- v_G^{(0)}(t)v_S^{(0)}(t) + v_B^{(0)}(t)v_S^{(0)}(t)\} \tag{34}$$

$$v_G^{(1)}(t)(-C_G - C_D - C_B) + C_D v_D^{(1)}(t) + C_S v_S^{(1)}(t)$$
$$+ C_B v_B^{(1)}(t) = 0 \tag{35}$$

Laplace transformed equation is

$$\mathbf{A}_2(s)\mathbf{x}^{(1)}(s) = \mathbf{B}_2(s)u_2(s) \qquad (36)$$

State vector $\mathbf{x}^{(1)}(s)$ is

$$\mathbf{x}^{(1)}(s) = \begin{bmatrix} v_G^{(1)}(s) & v_D^{(1)}(s) & v_S^{(1)}(s) & v_B^{(1)}(s) \end{bmatrix}^T$$

Where $\mathbf{A}_2(s)$ is

$$\mathbf{A}_2(s) = \begin{bmatrix} A_{2_{11}} & 0 & 0 & 0 \\ sC_D & A_{2_{22}} & k_1 & sC_{BD} \\ sC_S & k_1 & A_{2_{33}} & sC_{BS} \\ A_{2_{41}} & C_D & C_S & C_B \end{bmatrix}, \qquad (37)$$

where
$A_{2_{11}} = \left(s + \frac{1}{R_{in}C_{GX}}\right)$, $\quad A_{2_{22}} = (-sC_D - sC_{BD} - k_1)$,
$A_{2_{33}} = (-sC_S - sC_{BS} - k_1)$, $A_{2_{41}} = -C_G - C_D - C_B$
and $\mathbf{B}_2(s)$ is

$$\mathbf{B}_2(s) = \begin{bmatrix} 0 & 1 & -1 & 0 \end{bmatrix}^T$$

Solution of (36) is

$$\mathbf{x}^{(1)}(s) = \mathbf{A}_2^{-1}(s)\mathbf{B}_2(s)u_2(s) \qquad (38)$$

Impulse response of the first order nonlinear output is

$$v_D{}^{(1)}(t) = \frac{A_{22}}{|\mathbf{A}_2|} * u_2(t) - \frac{A_{32}}{|\mathbf{A}_2|} * u_2(t) \qquad (39)$$

$$= -\frac{k_{11}}{k_3}e^{-\frac{k_1 k_4}{k_3}t} * u_2(t) \qquad (40)$$

where
$$u_2(t) = \Big[k_1 v_T\{v_S^{2(0)}(t) - v_D^{2(0)}(t) - 2v_S^{(0)}(t)v_B^{(0)}(t) + 2v_D^{(0)}\}$$
$$\times v_B^{(0)}(t) - k_2\{v_G^{(0)}(t)v_D^{(0)}(t) - v_B^{(0)}(t)v_D^{(0)}(t)$$
$$-v_G^{(0)}(t)v_S^{(0)}(t) + v_B^{(0)}(t)v_S^{(0)}(t)\}\Big]$$
$k_{11} = C_B C_S + C_B C_{BS} + C_S C_{BS} + C_B C_{BD}$.

IV. SIMULATION RESULTS

The linear and nonlinear expressions derived have been simulated in MATLAB for different input amplitude values and different frequencies. Parameters used for simulations are: $U_T = 0.0256V$, $V_{T_0} = 0.5V$, $R_{in} = 5k\Omega$, $C_{GX} = 6.0 \times 10^{-10}$, $C_S = 0.5 \times 10^{-9}F$, $C_D = 1.0 \times 10^{-9}F$, $C_B = 1.5 \times 10^{-10}F$, $C_{BS} = 1 \times 10^{-11}F$, $C_{BD} = 1.15 \times 10^{-9}F$, $I_0 = 1.0 \times 10^{-9}A$ and $\eta = 1$. Fig. 3 shows linear and nonlinear output voltage for $1V$ peak to peak input. Fig. 4 shows output voltage for $0.5V$ peak to peak input voltage amplitude. Table I shows the percentage distortion error due to use of linear term only. Percentage distortion is calculated using the following expression:

$$\text{Distortion error} = \frac{v_o - v_o^{(0)}}{v_o}\% \qquad (41)$$

where v_o represents the sum of linear and nonlinear output voltage. $v_o^{(0)}$ denotes only linear output voltage.

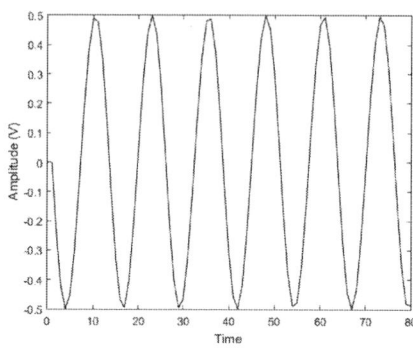

(a) Input to MOSFET circuit.

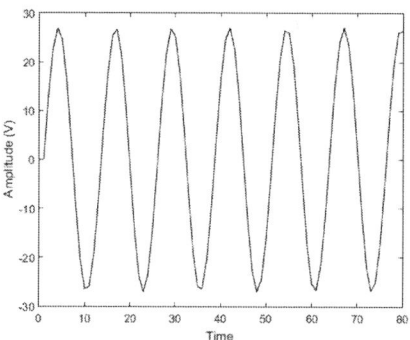

(b) Zeroth order output voltage.

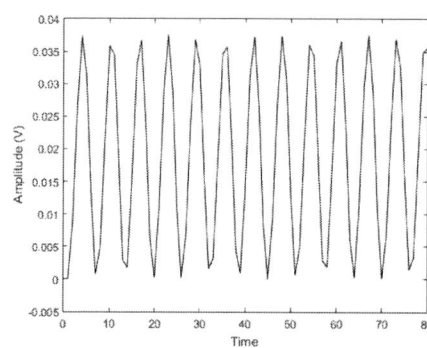

(c) First order output voltage

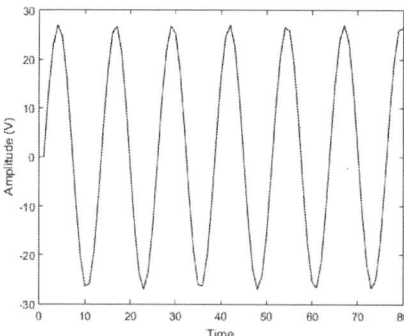

(d) Zeroth and first order output voltage

Figure 3. MOSFET output voltage for sinusoidal input with peak to peak value 1V and frequency 100Hz.

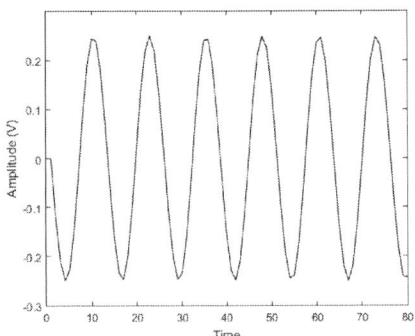

(a) Input to MOSFET circuit.

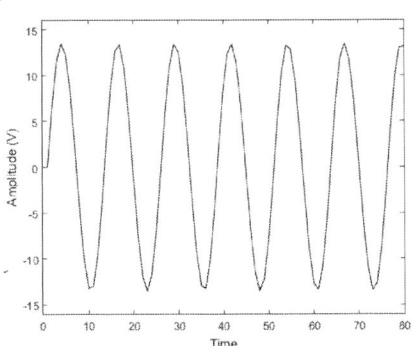

(b) Zeroth order output voltage.

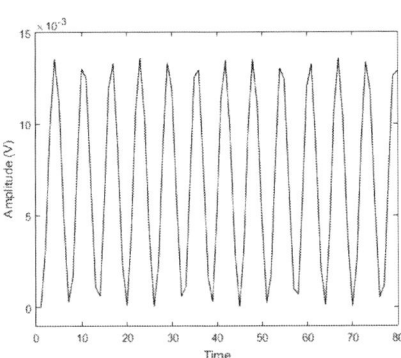

(c) First order output voltage

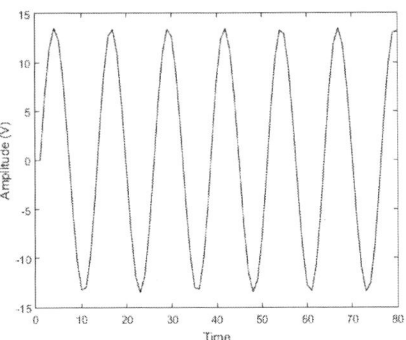

(d) Zeroth and first order output voltage

Figure 4. MOSFET output voltage for sinusoidal input with peak to peak value 0.5V and frequency 100Hz.

Table I
PERCENTAGE DISTORTION WHEN DIFFERENT VOLTAGE AMPLITUDE AND FREQUENCY IS GIVEN AT INPUT OF MOSFET CIRCUIT.

S. No.	Input voltage (V)	Input frequency	Percentage distortion
1.	0.25	100	0.062%
2.	0.25	1000	0.063%
3.	0.25	10000	0.066%
4.	0.50	100	0.067%
5.	0.50	1000	0.068%
6.	0.50	10000	0.071%

V. CONCLUSIONS

This paper presents closed from nonlinear expression for MOSFET using perturbation method and EKV model of MOSFET. Although, for small signal analysis, linear model can be used but, for large signal analysis, higher order nonlinear terms can not be neglected. The expressions obtained can be used for parameter estimation of MOSFET circuit.

Acknowledgement First author thanks to CSIR India.

REFERENCES

[1] J. Y. Hasani, "Three-port model of a modern MOS transistor in millimeter wave band, considering distributed effects", *IEEE Trans. on CAD of Integrated Circuits and Systems*, vol. 35, no. 9, pp. 1509-1518, 2016.

[2] A. I. A. Cunha, M. C. Schneider and C. Galup-Montoro, An MOS transistor model for analog circuit design. *IEEE Journal of solid-state circuits*, vol. 33, no. 10, pp. 1510-1519, 1998.

[3] E. Abou-Allam and T. Manku, "An improved transmission-line model for MOS transistors." *IEEE Trans. on Circuits and Systems II: Analog and Digital Signal Processing*, vol. 46, no. 11, pp. 1380-1387, 1999.

[4] M. Andersson, P. Kuivalainen and H. Pohjonen, "Circuit simulation models for MOS-gated power devices: Application to the simulation of an electronic lamp ballast circuit", IEEE *Applied Power Electronics Conference and Exposition, APEC-1993. 1993*, pp. 498-503.

[5] W. Wang, B. Tudor, X. Xi, W. Liu and F. J. Lee, "An accurate and robust compact model for high-voltage MOS IC simulation", *IEEE Trans. on Electron Devices*, vol. 60, no. 2, pp. 662-669, 2013.

[6] A. Abbasian, M. Taherzadeh-Sani, B. Amelifard and A. Afzali-Kusha, "Modeling of MOS transistors based on genetic algorithm and simulated annealing", *ISCAS-2005, IEEE International Symposium on Circuits and Systems, 2005*, pp. 6218-6221.

[7] A. A. Kushaa and M. El Nokali, "A CAD model for MOS transistors valid in all regions of operation," *Proceedings of the 34th Midwest Symposium on Circuits and Systems*, 1991, pp. 364-367.

[8] C. Shen, S. L. Ong, C. H. Heng, G. Samudra and Y. C. Yeo, "A variational approach to the two-dimensional nonlinear Poisson's equation for the modeling of tunneling transistors," *IEEE Electron Device Letters*, vol. 29, no. 11, pp. 1252-1255, 2008.

[9] S. Majumdar and H. Parthasarathy, "Perturbation approach to EbersMoll equations for transistor circuit analysis," *Circuits, Systems and Signal Processing*, vol. 29, no.3, pp. 431-448, 2010.

[10] A. Buonomo and A. L. Schiavo, "Perturbation analysis of nonlinear distortion in analog integrated circuits," *IEEE Trans. on Circuits and Systems I: Regular Papers*, vol. 52, no. 8, pp. 1620-1631, 2005.

[11] A. Biswas, Y. Yildirim, E. Yasar, Q. Zhou, S. P. Moshokoa M. and Belic, "Optical soliton perturbation with resonant nonlinear Schrodinger's equation having full nonlinearity by modified simple equation method," *Optik*, 160, pp. 33-43, 2018.

2nd IEEE International Conference on Power Electronics, Intelligent Control and Energy Systems (ICPEICES-2018)

Resistorless Electronically Controllable Quadrature Sinusoidal Oscillator Employing VDIBA

Kanhaiya Lal Pushkar
Deaprtment of Electronics &
Communication Engineering,
Maharaja Agrasen Institute of
Technology
Rohini, New Delhi-110086, India
klpushkar17@gmail.com

Komal Rohilla
Department of Electronics &
Communication Engineering,
Guru Jambheswar University
Hisar, Haryana, India
komalrohilla243@gmail.com

Sushil Kumar
School of Engineering & Technology,
Noida International University
Gautam Budh Nagar, U. P. - 203 201,
India
sushilkumar0108@gmail.com

Abstract—**A novel resistorless electronically controlled quadrature sinusoidal oscillator (ECQSO) employing two voltage differencing inverting buffered amplifiers (VDIBAs) and two capacitors have been proposed. The proposed quadrature oscillator offers electronic control of frequency of oscillation (FO), low active and passive sensitivities. Effects of non idealities of the VDIBA on the proposed oscillator have also been investigated in detail. The validity of the proposed structure has also been confirmed by SPICE simulation with TSMC 0 .18 μm process parameters.**

Keywords—electronically controlled quadrature sinusoidal oscillator, voltage differencing inverting buffered amplifier, Voltage-mode, and quadrature sinusoidal oscillator

I. INTRODUCTION

The quadrature sinusoidal oscillators (QSOs) are crucial building blocks when it comes to the synthesis of modern day transceiver systems. A QSO generates two sinusoids with a 90^0 phase difference. The QSOs find various applications in telecommunications for quadrature mixers and single sideband generators [1], in direct-conversion receivers which are used for measurement purposes in vector generators and selective voltmeters [2]. Due to these applications, numerous QSOs have been realized by employing different active building blocks in the open literature [3]-[11]. The VDIBA is one of the active building blocks among the various active building blocks introduced in reference [12] which is emerging as a very flexible and versatile building block for analog signal processing/signal generation and also has been used earlier for realizing a number of functions. The VDIBA has been used in single resistance/element controlled oscillators, simulation of inductors, and realization of active filters [13]-[18]. Recently, VDIBA has also been used in the realization of QSO where the circuit has used four active devices (two VDIBAs and two n-MOS) [19]. Therefore, the purpose of this communication is to propose a new QSO having electronic control of FO after fixing CO by either C_1 or C_2. The proposed configuration also offers low active and passive sensitivities. The feasibility of the proposed

sinusoidal oscillators has been verified by SPICE simulation with TSMC 0.18μm process parameters.

II. PROPOSED NEW CONFIGURATION

In [12] number of possible variants of the various active building blocks were introduced for the first time, VDIBA [13] is a modified form of VDBA which was introduced in [12]. Fig.1 demonstrates the symbolic notation and equivalent circuit of VDIBA [13]. The terminal equations of VDIBA, using the standard notations, can be given by the following set of equations.

$$\begin{pmatrix} I_+ \\ I_- \\ I_z \\ V_w \end{pmatrix} = \begin{pmatrix} 0 & 0 & 0 & 0 \\ 0 & 0 & 0 & 0 \\ g_m & -g_m & 0 & 0 \\ 0 & 0 & -\beta & 0 \end{pmatrix} \begin{pmatrix} V_+ \\ V_- \\ V_z \\ I_{w-} \end{pmatrix} \quad (1)$$

Where β is a non-ideal voltage gain of the VDIBA. The value of β in an ideal VDIBA is unity and g_m is the transconductance of the VDIBA.

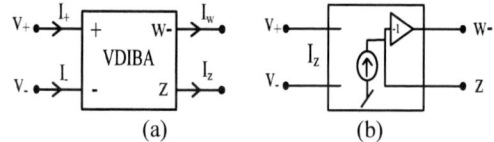

Fig. 1. (a) Symbolic notation (b) equivalent model of the VDIBA

The proposed new QSO is shown in Fig. 2.

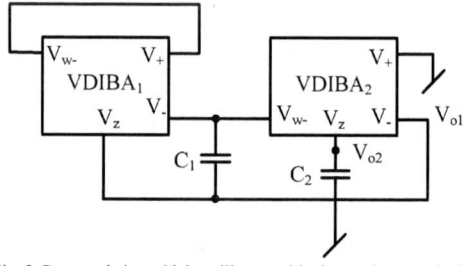

Fig. 2. Proposed sinusoidal oscillators with electronic control of FO

978-1-5386-6624-1/18 $31.00 © 2018 IEEE

A routine circuit analysis of the circuit of Fig. 2 yield the following characteristic equation (CE), for the proposed quadrature sinusoidal oscillator:

$$CE : s^2 + s\left(\frac{g_{m_1}}{C_1} - \frac{g_{m_2}}{C_2}\right) + \frac{g_{m_1}g_{m_2}}{C_1 C_2} = 0 \tag{2}$$

For $g_{m_1} = g_{m_2} = g_m$ (equal bias current i.e. $I_{b1} = I_{b2}$), CE becomes

$$s^2 + s g_m\left(\frac{1}{C_1} - \frac{1}{C_2}\right) + \frac{g_m^2}{C_1 C_2} = 0$$

CO and FO given as

$$CO: \ (C_2 - C_1) \le 0 \tag{3}$$

$$FO: \ \omega_0 = \frac{g_m}{\sqrt{C_1 C_2}} \tag{4}$$

From Eq. (3) and Eq. (4), it is seen that FO is electronically controllable by the transconductance g_m, after fixing CO either by C_1 or C_2. Thus, FO is electronically controllable by transconductance ($g_{m1} = g_{m2} = g_m$) of VDIBAs.

III.   NON-IDEAL ANLAYSIS   AND SENSITIVITY PERFORMANCE

Let R_z and C_z denote the parasitic resistance and parasitic capacitance of the Z-terminal and R_w denotes the parasitic resistance of the W-terminal of VDIBA. Taking the non-idealities into account namely the voltage of W-terminal $V_{W-} = (-\beta^+ V_z + I_w R_w)$ where $\beta^+ = 1-\varepsilon_p$ ($\varepsilon_p \ll 1$) denotes the voltage tracking error of Z-terminal of VDIBA. The expressions (neglecting the effect of R_w) for CE, CO and FO respectively become:

$$CE: \ s^2(C_1+C_z)(C_2+C_z) + s\left\{ \begin{matrix} (C_2+C_z)\left(\dfrac{1}{R_z}+\beta^+ g_{m_1}\right) + \\ \dfrac{(C_2+C_z)}{R_z} - \beta^+ C_1 g_{m_2} \end{matrix} \right\} \tag{5}$$

$$+\frac{1}{R_z}\left(\frac{1}{R_z} + \beta^+ g_{m_1}\right) + \beta^+ g_{m_1} g_{m_2} = 0$$

$$CO: \left\{ \begin{matrix} (C_2+C_z)\left(\dfrac{1}{R_z}+\beta^+ g_{m_1}\right) \\ +\dfrac{(C_2+C_z)}{R_z} - \beta^+ C_1 g_{m_2} \end{matrix} \right\} \le 0 \tag{6}$$

$$FO: \ \omega_0 = \sqrt{\frac{1+R_z\beta^+ g_{m_1} + R_z^2\beta^+ g_{m_1} g_{m_2}}{R_z^2\left(C_1+C_z\right)\left(C_2+C_z\right)}} \tag{7}$$

The various active and passive sensitivities of FO are given

by:

$$S_{C_1}^\omega = -\frac{1}{2}\frac{C_1}{C_1+C_z} \tag{8a}$$

$$S_{C_z}^\omega = -\frac{1}{2}\left(\frac{1}{C_1+C_z} + \frac{1}{C_2+C_z}\right) C_z \tag{8b}$$

$$S_{R_z}^\omega = -\frac{1}{2}\frac{2+R_z\beta^+ g_{m_1}}{1+R_z\beta^+ g_{m_1} + R_z^2\beta^+ g_{m_1} g_{m_2}} \tag{8c}$$

$$S_{C_2}^\omega = -\frac{1}{2}\frac{C_2}{C_2+C_z} \tag{8d}$$

$$S_{g_{m_1}}^\omega = S_{g_{m_1}}^\omega = \frac{1}{2}\left(1 - \frac{1}{1+R_z\beta^+ g_{m_1} + R_z^2\beta^+ g_{m_2}}\right) \tag{8e}$$

$$S_{\beta^+}^\omega = \frac{1}{2}\left(1 - \frac{1+R_z\beta^+ g_{m_1}}{1+R_z\beta^+ g_{m_1} + R_z^2\beta^+ g_{m_2}}\right) \tag{8f}$$

In the ideal case, the various sensitivities of FO with respect to C_1, C_2, C_Z, R_Z, g_{m1}, g_{m2}, and β^+ are found to be:

$$S_{C_1}^{\omega_0} = S_{C_2}^{\omega_0} = -\frac{1}{2}, \tag{9}$$

$$S_{C_z}^{\omega_0} = S_{R_z}^{\omega_0} = 0, \ \ S_{g_{m_1}}^{\omega_0} = S_{g_{m_2}}^{\omega_0} = S_{\beta^+}^{\omega_0} = \frac{1}{2}$$

Considering the typical values of various parasitic as given in [13] e.g. C_z= 0.367 pF, R_z= 131.93 kΩ, R_w = 42.36 Ω, $\beta^+ = 1$, $g_{m1} = 440$ μS and $g_{m2} = 440$ μS along with $C_1 = 0.912$ nF, $C_2 = 0.9$nF the various sensitivities are found to

$$S_{C_1}^{\omega_0} = -0.499, \ S_{C_2}^{\omega_0} = -0.499, \ S_{C_z}^{\omega_0} = 0.004, \ S_{R_z}^{\omega_0} = -0.0086,$$

$$S_{g_{m_1}}^{\omega_0} = S_{g_{m_2}}^{\omega_0} = 0.499, S_{\beta^+}^{\omega_0} = 0.491$$

which are all low.

IV.   SPICE SIMULATION RESULTS

The proposed QSO was simulated using CMOS VDIBA (as shown in Fig. 3) to confirm the theoretical analysis. The CMOS VDIBA is implemented using 0.18μm TSMC real transistor models which are listed in Table 1. Table 2 shows the aspect ratios of transistors used in Fig. 3. The passive elements were selected as C_1=0.912 nF and $C_2 = 0.9$ nF. The transconductance of VDIBAs was controlled by the bias currents. SPICE generated output waveforms indicating transient and steady state responses of circuit of Fig. 2 are shown in Fig. 4 and Fig. 5 respectively. These results, thus, confirm the validity of the proposed configuration. Fig. 6 shows the output spectrum of circuit shown in Fig. 2; whereas the total harmonic distortion (THD) for both the outputs V_{o1} and V_{o2} are found to be 2.5 % and 0.37 % respectively. Fig. 7 shows the lissajous pattern for the circuit of Fig. 2.

978-1-5386-6624-1/18 $31.00 © 2018 IEEE   1225

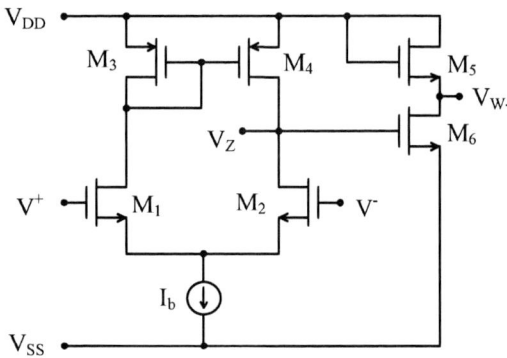

Fig. 3. A CMOS implementation of VDIBA [13], $V_{DD} = V_{SS} = 0.9$ V

Table 1: 0.18μm TSMC CMOS model parameters

DEL N NMOS (LEVEL = 7
VERSION=3.1 TNOM=27 TOX=4.1E-9 XJ=1E-7 NCH=2.3549E17
VTH0=0.3725327 K1=0.5933684 K2=2.050755E-3 K3=1E-3
K3B=4.5116437 W0=1E-7 NLX=1.870758E-7 DVT0W=0 DVT1W=0
DVT2W=0 DVT0=1.3621338 DVT1=0.3845146 DVT2=0.0577255
U0=259.5304169 UA=−1.413292E-9 UB=2.229959E-18 UC=4.525942E-
11 VSAT=9.411671E4 A0=1.7572867 AGS=0.3740333 B0=−7.087476E-
9 B1=−1E-7 KETA=−4.331915E-3 A1=0 A2=1 RDSW=111.886044
PRWG=0.5 PRWB=−0.2 WR=1 WINT=0 LINT=1.701524E-8 XL=0
XW=−1E-8 DWG=−1.365589E-8 DWB=1.045599E-8 VOFF=−0.0927546
NFACTOR=2.4494296 CIT=0 CDSC=2.4E-4 CDSCD=0 CDSCB=0
ETA0=3.175457E-3 ETAB=3.494694E-5 DSUB=0.0175288
PCLM=0.7273497 PDIBLC1=0.1886574 PDIBLC2=2.617136E-3
PDIBLCB=−0.1 DROUT=0.7779462 PSCBE1=3.488238E10
PSCBE2=6.841553E-10 PVAG=0.0162206 DELTA=0.01 RSH=6.5
MOBMOD=1 PRT=0 UTE=−1.5 KT1=−0.11 KT1L=0 KT2=0.022
UA1=4.31E-9 UB1=−7.61E-18 UC1=−5.6E-11 AT=3.3E4 WL=0 WLN=1
WW=0 WWN=1 WWL=0 LL=0 LLN=1 LW=0 LWN=1 LWL=0
CAPMOD=2 XPART=0.5 CGDO=8.53E-10 CGSO=8.53E-10 CGBO=1E-
12 CJ=9.513993E-4 PB=0.8 MJ=0.3773625 CJSW=2.600853E-10
PBSW=0.8157101 MJSW=0.1004233 CJSWG=3.3E-10
PBSWG=0.8157101 MJSWG=0.1004233 CF=0 PVTH0=−8.863347E-4
PRDSW=−3.6877287 PK2=3.730349E-4 WKETA=6.284186E-3
LKETA=−0.0106193 PU0=16.6114107 PUA=6.572846E-11 PUB=0
PVSAT=1.112243E3 PETA0=1.002968E-4 PKETA=−2.906037E-3)
.MODEL P PMOS (LEVEL=7
VERSION=3.1 TNOM=27 TOX=4.1E-9 XJ=1E-7 NCH=4.1589E17
VTH0=−0.3948389 K1=0.5763529 K2=0.0289236 K3=0 K3B=13.8420955
W0=1E-6 NLX=1.337719E-7 DVT0W=0 DVT1W=0 DVT2W=0
DVT0=0.5281977 DVT1=0.2185978 DVT2=0.1 U0=109.9762536
UA=1.325075E-9 UB=1.577494E-21 UC=−1E-10 VSAT=1.910164E5
A0=1.7233027 AGS=0.3631032 B0=2.336565E-7 B1=5.517259E-7
KETA=0.0217218 A1=0.3935816 A2=0.401311 RDSW=252.7123939
PRWG=0.5 PRWB=0.0158894 WR=1 WINT=0 LINT=2.718137E-8 XL=0
XW=−1E-8 DWG=−4.363993E-8 DWB=8.876273E-10
VOFF=−0.0942201 NFACTOR=2 CIT=0 CDSC=2.4E-4 CDSCD=0
CDSCB=0 ETA0=0.2091053 ETAB=−0.1097233 DSUB=1.2513945
PCLM=2.1999615 PDIBLC1=1.238047E-3 PDIBLC2=0.0402861
PDIBLCB=−1E-3 DROUT=0 PSCBE1=1.034924E10
PSCBE2=2.991339E-9 PVAG=15 DELTA=0.01 RSH=7.5 MOBMOD=1
PRT=0 UTE=−1.5 KT1=−0.11 KT1L=0 KT2=0.022 UA1=4.31E-9
UB1=−7.61E-18 UC1=−5.6E-11 AT=3.3E4 WL=0 WLN=1 WW=0
WWN=1 WWL=0 LL=0 LLN=1 LW=0 LWN=1 LWL=0 CAPMOD=2
XPART=0.5 CGDO=6.28E-10 CGSO=6.28E-10 CGBO=1E-12
CJ=1.160855E-3 PB=0.8484374 MJ=0.4079216 CJSW=2.306564E-10
PBSW=0.842712 MJSW=0.3673317 CJSWG=4.22E-10
PBSWG=0.842712 MJSWG=0.3673317 CF=0 PVTH0=2.619929E-3
PRDSW=1.0634509 PK2=1.940657E-3 WKETA=0.0355444
LKETA=−3.037019E-3 PU0=−1.0227548 PUA=−4.36707E-11 PUB=1E-
21 PVSAT=−50 PETA0=1E-4 PKETA=−5.167295E-3)

Table. 2: The aspect ratios of MOSFETs

Transistor	W(μm)	L (μm)
M1- M4	18	1.08
M5, M6	54	0.18

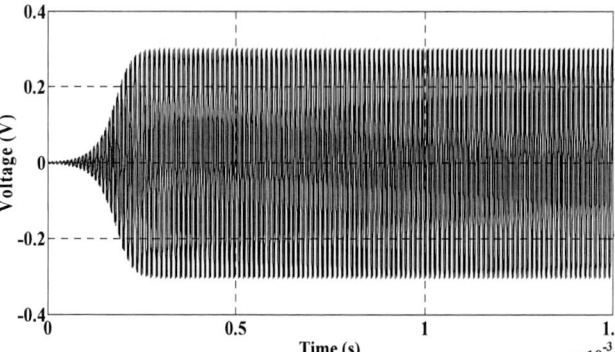

Fig. 4. Transient response of proposed quadrature sinusoidal oscillator

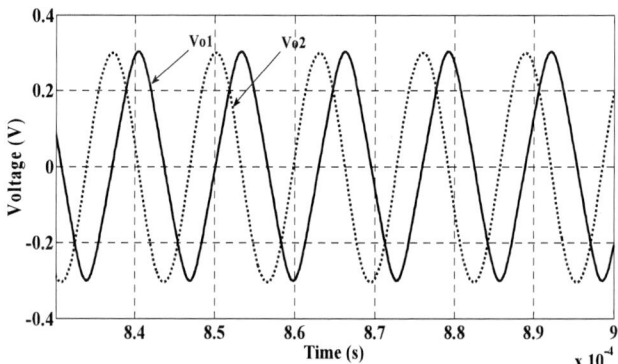

Fig. 5. Steady state response of proposed quadrature sinusoidal oscillator

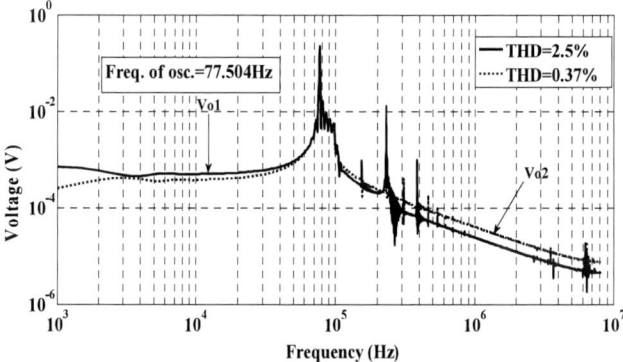

Fig. 6. Spectrum of output waveform of proposed quadrature sinusoidal oscillator

978-1-5386-6624-1/18 $31.00 © 2018 IEEE

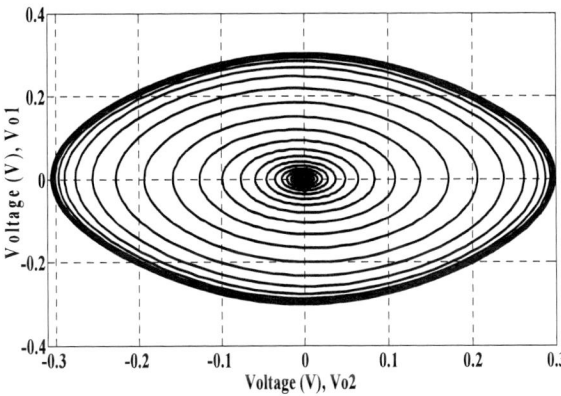

Fig.7. Lissajous pattern of proposed quadrature sinusoidal oscillator

V.   CONCLUSION

This communication demonstrates a new circuit configuration employing two VDIBAs along with a minimum number of passive elements (i.e. two capacitors). The proposed quadrature sinusoidal oscillator offers electronic control (for $g_{m1} = g_{m2}$) of FO after fixing CO and low active and passive sensitivities. The feasibility and workability of the quadrature sinusoidal oscillator has also clearly been established by SPICE simulation with 0.18µm TSMC process technology. Moreover, the experimental results are in accordance with the expected results.

REFERENCES

[1]   J. W. Horng, C. L. Hou, C. M. Chang, W. Y. Chung, H. W. Tang, and Y. H. Wen, "Quadrature oscillators using CCIIs," *International Journal of Electronics*, vol. 92, pp. 21–31, 2005.

[2]   J. D. Gibson, "The communication handbook," *CRC Press, Boca Raton, Fla*, USA, 1997.

[3]   W. Tangsrirat, W. Surakampontorn, "Single-resistance controlled quadrature oscillator and universal biquad filter using CFOAs," *AEU-International Journal of Electronics and Communications*, vol. 63, no. 12, pp.1080–1086, 2009.

[4]   J. W. Horng, "Current differencing buffered amplifiers based single resistance controlled quadrature oscillator employing grounded capacitors", *IEICE Transactions on Fundamental of Electronics, Communications and Computer Sciences*, vol. E85A, pp. 1416–1419, 2002.

[5]   S. Ozcan, A. Toker, C. Acar, H. Kuntman, O. Cicekoglu, "Single resistance-controlled sinusoidal oscillators employing current differencing buffered amplifier," *Microelectronics Journal*, vol. 31, pp. 169-174, 2000.

[6]   P. Prommee and K. Dejhan, "An integrable electronic controlled quadrature sinusoidal oscillator using CMOS operational transconductance amplifier," *International Journal of Electronics*, vol. 89, pp. 365–379, 2002.

[7]   A. Rodriguez-Vazquez, B. Linares-Barranco, J. L. Huertas, and E. Sanchez-Sinencio, "On the design of voltage controlled sinusoidal oscillators using OTA's," *IEEE Transactions on Circuits and Systems*, vol. 37, pp. 198–211, 1990.

[8]   R. Holzel, "A simple wide-band sine wave quadrature oscillator," *IEEE Transactions on Instrumentation and Measurement*, vol. 42, pp. 758-760, 1993.

[9]   K. L. Pushkar, "Electronically controllable quadrature sinusoidal oscillator using VD-DIBAs," *Circuits and systems*, vol. 9, no. 3, pp. 41-48, 2018.

[10]   D. R. Bhaskar, D. Prasad and K. L. Pushkar, "Fully uncoupled electronically controllable sinusoidal oscillator employing VD-DIBAs," *Circuits and systems*, vol. 4, no. 3, pp. 264-268, 2013.

[11]   K. L. Pushkar, "Voltage-mode third order quadrature sinusoidal oscillator using VDBAs," *Circuits and systems*, vol. 8, no.12, pp. 285-292, 2017.

[12]   D. Biolek, R. Senani, V. Biolkova and Z. Kolka, "Active elements for analog signal processing: classification, review, and new proposals," *Radioengineering*, vol. 17, no. 4, pp. 15-32, 2008.

[13]   N. Herencsar, S. Minaei, J. Koton, E. Yuce And K. Vrba, "New resistorless and electronically tunable realization of dual-output VM all-pass filter using VDIBA," *Analog Integrated Circuits and Signal Processing*, vol. 74, pp. 141-154,2013.

[14]   K. L. Pushkar, Ghanshyam Singh and R. K. Goel, "CMOS VDIBAs-based single-resistance-controlled voltage-mode sinusoidal oscillator," *Circuits and Systems*, vol. 8, no. 1, pp. 14-22, 2017.

[15]   K. L. Pushkar and D. R. Bhaskar, "New single-element-controlled sinusoidal oscillator using single VDIBA," *Journal of Engineering Technology*, vol. 6, no. 1, pp. 595-604, Jan. 2018.

[16]   K. L. Pushkar, "Electronically controllable sinusoidal oscillator employing VDIBAs," *Advances in Electrical and Electronics Engineering*, vol. 15, no.5, pp. 799-805, Dec. 2017.

[17]   K. L. Pushkar, D. R. Bhaskar and D. Prasad, "Voltage-mode new universal biquad filter configuration using a single VDIBA," *Circuits, Systems, and Signal Processing*, vol. 33, no. 1, pp. 275-285,2013.

[18]   W. Tangsrirat, "Synthetic grounded lossy inductance simulators using single VDIBA," *IETE Journal of research*, vol. 63, no. 1, pp. 134-141, 2017.

[19]   O. Channumsin and W. Tangsrirat, "VDIBA-based sinusoidal quadrature oscillator," *Przeglad Elektrotechniczny*, vol. 93, no. 3, pp. 248-251, 2017.

2nd IEEE International Conference on Power Electronics, Intelligent Control and Energy Systems (ICPEICES-2018)

Design of Multi-Loop L-PID and NL-PID Controllers: An Experimental Validation

Sudarshan K. Valluru, Madhusudan Singh, Arnav Goel, Manpreet Kaur, Daksh Dobhal, Kumar Kartikeya, Aditya Verma and Anshul Gupta

Incubation Center for Control, Dynamical Systems and Computation
Department of Electrical Engineering, Delhi Technological University
Delhi-110042, India

Abstract—The three-term PID controllers are significantly used in most of the benchmarked systems as controllers. However, the benchmarked Single Input Multiple Output (SIMO) systems may require more than one PID controllers in the control loops, the way to design PID controllers is not an easy task. Although, the linear PID (L-PID) controllers are not yielded a satisfactory control results for a benchmarked system due to the presence of inherent nonlinearities. In order to improve performance, the linear PID controllers are modified by nonlinear characteristics. In this paper, the linear and nonlinear PID (NL-PID) controllers are implemented for the stabilization and control of Gantry Crane System as a test problem. The experimental results prove that the multi-loop NL-PID controllers exhibit more robustness to outer large with fast external disturbances. The L-PID and NL-PID controllers are compared to find out the best controller, which gives optimum responses and can also handle external disturbances.

Keywords— Multiloop L-PID; Multiloop NL-PID; Gantry Crane Control

I. INTRODUCTION

The Linear PID is one of the conventional controllers used in Control System industry and is widely accepted for providing generic and efficient solutions to real-world control problems[1], [2]. Few reasons for its popularity include simplicity in simulation, being inexpensive and efficacious in performance characteristics. Zeigler and Nichols pioneered significant work on the tuning of PID controller. Ziegler-Nichols type rules to much extent[3] are now widely used in PIDs. The design of an L-PID controller for a particular problem is a tedious task. This is because when the system under consideration has nonlinearity or the system parameters vary over a wide range, the performance of linear PID deteriorates considerably.

When used practically, L-PID controllers[4] quite often exhibit unexpected behavior including undesired oscillations and haphazard cart movement. This happens because the reference inputs are usually not continuous or smooth as they are subjected to the disturbance or noise signal, while the output of the system is required to be continuous and smooth. Since the smooth continuous output is taken as the direct objective, the inertia influence of the system is neglected which causes abrupt oscillations when used practically. Also, the reference signals are usually non-differentiable signals which makes it problematic to achieve the differential signal of the error. Therefore, If not properly implemented it can cause the system to collapse causing damage to the apparatus. Hence, care needs to be taken while designing the controller.

In recent years, there has been an upsurge in finding alternatives for Linear PID. The need for substitutes has arisen because the transient performance and parameters of Linear PID are usually degraded. Non-Linear PID has emerged to be an effective alternative to L-PID. In this context, the NL-PID designed uses a nonlinear sigmoid function. The overall response is more fast and robust than that of L-PID.

The paper is divided into five sections. Section I contains the introduction; Section II illustrates the design of Multi PID controllers. The benchmarked gantry crane system is explained in Section III. Experimentation validation is presented in Section IV, and the paper is concluded with the discussion presented in Section V.

II. DESIGN OF MULTI-LOOP PID CONTROLLERS

For effective minimization of swing angle and cart tracking of the input signal, two PID controllers are required, one each for the angle and position error. First, the PID controller for angle error was tuned, then keeping the earlier obtained values fixed; the position controller was tuned. Experimentation was carried out using the following two approaches:

A. Multi-Loop L-PID Controller

The L-PID takes in error, e(t) which is the difference between the desired output and real-time output, and processes it according to the equation (1) and its schematic structure is shown in Fig.1.

$$u(t) = K_p e(t) + K_i \int_{t_1}^{t_2} e(t)dt + K_d \frac{de(t)}{dt} \qquad (1)$$

Where K_p, K_i, and K_d are the proportional, integral and derivative gains respectively. This gives the control signal, $u(t)$. Each term in the above equation has its own significance. The proportional term leads to an increase in the speed of the system response, but it can be increased only up to a certain limit varying from system to system. A further increase will lead to oscillation of the variable being processed. The major steady error reducing term in the equation is the Integral term. The Derivative term results in a decrease of the output variable if the error is increasing and minimization of `oscillations.

978-1-5386-6624-1/18 $31.00 © 2018 IEEE   1228

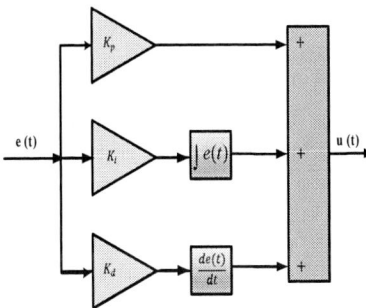

Fig. 2 Schematic structure of L-PID Controller[5]

B. Multi-Loop NL-PID Controller

In order to alleviate the problems associated with the L-PID, a novel method involving nonlinearities is proposed. The P, I & D error terms are processed through a non-linear function[6] to get the control signal, $u(t)$ is given in equation (2) and its schematic diagram is shown in Fig.2.

$$u(t) = e_p(t) + e_i(t) + e_d(t) \qquad (2)$$

where $e_p = f(K_p, e), \; e_i = f(K_i, e), \; e_d = f(K_d, e)$

Here, f is a non-linear function, which is defined as equation (3)

$$f\left(K_\alpha, e\right) = \frac{1}{1 + \exp(-K_\alpha e)} \qquad (3)$$

This function maps high values to a value of 1 and a low value to 0. This behavior leads control error to be constrained in a bounded region and so doesn't permits voltage input to rise to high value. Further, in both the controllers a velocity filter is added in series with a derivative block in order to decrease oscillations and smooth out movement of the pendulum. The filter follows the equation (4) in the frequency domain.

$$d(s) = \frac{\exp(4)}{s^2 + 70.7s + \exp(4)} \qquad (4)$$

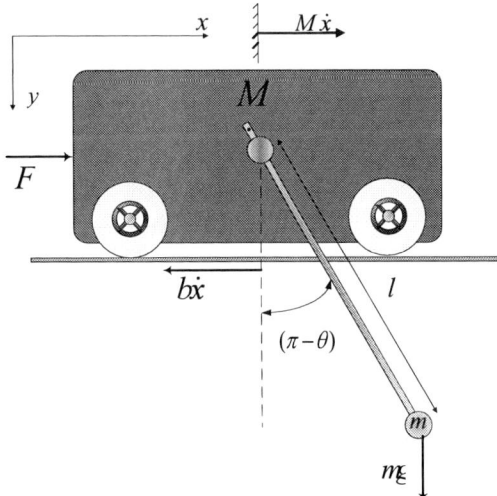

Fig. 2 Schematic structure of NL-PID Controller

III. BENCHMARKED GANTRY CRANE SYSTEM

Despite its simple structure, the crane system[7], just like the inverted pendulum model has many complexities owing to its non-linear nature. Due to these problems, it is regarded as one of the ten benchmarks of control systems and its Phenomenological model is shown in Fig.3. Assuming the payload to be a lumped-mass distribution rather than a distributed-mass system, we can model the gantry crane system as a simple pendulum. The Fig.2. Illustrates the approximated structure of a gantry crane system with the assumptions of (1) payload has a point mass distribution (2)

cable is massless (3) cable is not flexible, and its length unvarying and(4) damping coefficient of the pendulum is neglected.

Let M, m, l, b and I be the mass of the trolley, mass of the pendulum bob, length of the cable, the coefficient of friction of cart and the MOI of the bob respectively. Also, x and θ be the position of trolley along the belt and angular position of the pendulum.

Fig.3 Phenomenological model of Gantry Crane System

The crane system is an under-actuated system as the degrees of freedom for this system is more than the actuators, i.e., we have trolley position, and vertical pendulum angle controlled by changing the voltage (V) applied to the DC motor. The force F (Tension force) on the trolley mass M is a linear function of the voltage applied to the servo DC motor coupled to the socket wheel.

The crane system is an under-actuated system as the degrees of freedom for this system is more than the actuators, i.e., we have trolley position, and vertical pendulum angle controlled by changing the voltage (V) applied to the DC motor. The force F (Tension force) on the trolley mass M is a linear function of the voltage applied to the servo DC motor coupled to the socket wheel.

The two equations obtained for the system are:

$$F = (M + m).\ddot{x} + b\dot{x} - ml\sin\theta.\dot{\theta}^2 + ml\cos\theta.\ddot{\theta} \qquad (5)$$

$$0 = (I + ml^2).\ddot{\theta} - mgl\sin\theta + ml\cos\theta.\ddot{x} \qquad (6)$$

The system parameters as specified in equation (5) and (6) are given in Table –I.

TABLE I. GANTRY CRANE SYSTEM PARAMETERS

S No.	System Parameters	Symbol	Value
1.	Mass of the Cart	M	2.4 kg
2.	Mass of the Pole	m	0.23 kg
3.	Friction Coefficient of the Cart	b	0.05 Ns/m
4.	Length of the Pole	l	0.32 m
5.	Moment of Inertia of the pole	I	0.099 kgm^2
6.	Gravity constant	g	9.81 ms^{-2}
7.	Range of operating voltage of D.C. Servo Motor	V	$\pm$ 24 Volts

978-1-5386-6624-1/18 $31.00 © 2018 IEEE

For $\theta \to 0$, $\sin\theta \approx -\theta$, $\cos\theta \approx -1$, $\dot{\theta}^2 \approx 0$

For the lucid understanding of carrying out the stability analysis of the dynamics of the Gantry Crane System, the system equations (5) and (6) linearized and given as equation (7) and (8).

$$F = (M + m).\ddot{x} + b\dot{x} - ml.\ddot{\theta} \qquad (7)$$

$$0 = (I + ml^2).\ddot{\theta} + mgl\theta - ml.\ddot{x} \qquad (8)$$

Alternatively, the above two equations can also be derived using Lagrange equation.

$$\frac{d}{dt}\left(\frac{d\mathcal{L}}{d\dot{q}}\right) - \frac{d\mathcal{L}}{dq} = 0 \qquad (9)$$

Where q is the degrees of freedom of the system (x, θ) and $\mathcal{L}$ is a difference between total and potential energy.

IV. EXPERIMENTAL VALIDATION AND COMPARATIVE ANALYSIS

The controller maintained a controlled voltage within the rated limits of $\pm$ 24V for DC servo motor. Real-time experimentation was done on the Feedback Instruments made Digital Control Pendulum model. The experimental diagram of the gantry crane system is shown in Fig.4

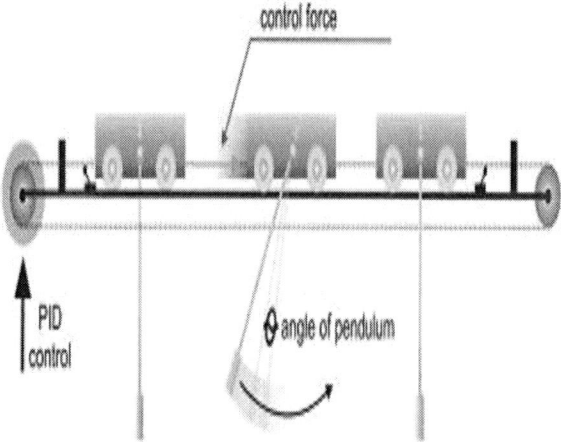

Fig. 4 Block Diagram for Gantry Crane System[8]

The gain values for angle and cart control were tuned manually by the trial-and-error method. The runtime for the experiments was taken as 100secs. The best gain values obtained are given in Table-II.

TABLE II.   CONTROLLERS GAIN PARAMETERS

S. No.	Gain	L-PID		NL-PID	
		Cart	Pendulum	Cart	Pendulum
1.	K_p	4.30	8.00	5.00	6.00
2.	K_i	0.50	0.15	0.15	0.10
3.	K_d	6.00	0.30	5.00	0.30

A. Multi-Loop L- PID Controlled Gantry Crane System

The responses of the L-PID controller for the stabilization of the cart and pendulum of the gantry crane system is shown in Fig.5. The gantry crane system is operated for 100 sec and for every 20 sec interval some external disturbances are applied manually to the pendulum.

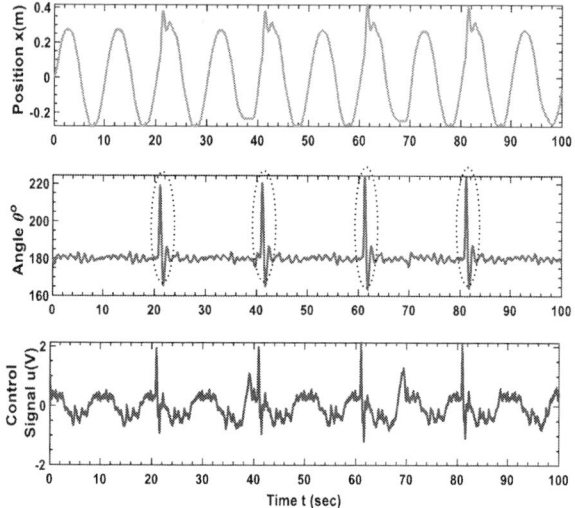

Fig.5 Responses of multi-loop L-PID Controller

B. Multi-Loop NL- PID Controlled Gantry Crane System

The responses of NL-PID controller for the stabilization of the cart and pendulum of the gantry crane system is shown in Fig.5. The gantry crane system is operated for 100 sec and for every 20 sec interval some external disturbances are applied manually to the pendulum.

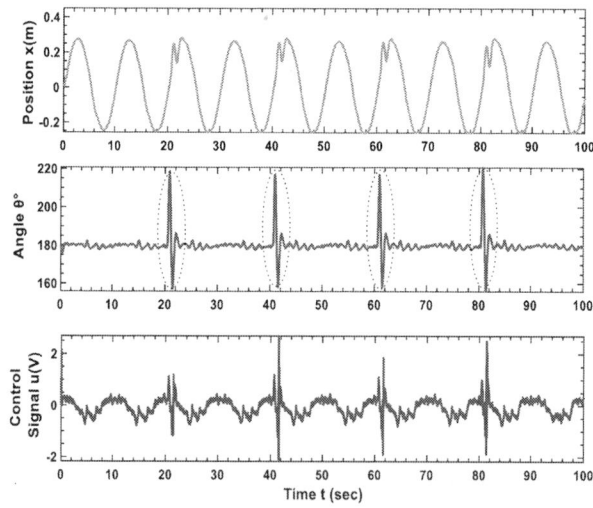

Fig.6 Responses of multi-loop NL-PID Controller

C. Comparative Analysis of L-PID and NL-PID controllers

The L-PID and NL-PID responses to stabilization of the pendulum angle against the external noise/ disturbance is given in Table-III and Table-IV respectively. From the tables, it is observed that the NL-PID stabilizes (θ about π) in the shorter period than the L-PID controller.

TABLE III. PENDULUM ANGLE STABILIZATION RESPONSE FOR THE L-PID CONTROLLER

Sl.	Disturbance initiating time (sec) For L-PID (a)	Instant of Stabilization (sec) For L-PID (b)	Time taken for stabilization(sec) by L-PID (c =b-a)
1	20.04	23.13	3.09
2	39.99	43.74	3.75
3	60.4	63.24	2.84
4	79.89	83.17	3.28

TABLE IV. PENDULUM ANGLE STABILIZATION RESPONSE FOR THE NL-PID CONTROLLER

Sl.	Disturbance initiating time (sec) For NL-PID (a)	Instant of Stabilization (sec) For NL-PID (b)	Time taken for stabilization(sec) by NL-PID (c =b-a)
1	20.5	22.69	2.19
2	40.5	42.76	2.26
3	60.51	62.74	2.23
4	80.37	82.63	2.26

V. RESULTS AND DISCUSSION

In this paper, multi-loop L-PID and multi-loop NL-PID schemes are designed and validated on a benchmarked gantry crane system. It is observed that the multi-loop NL-PID controller quickly brings the pendulum to the desired position under the external disturbances as compared to the multi-loop L-PID controller.

ACKNOWLEDGMENT

The authors would like to thank Govt. of India and Govt. NCT Delhi for providing funding through Delhi Technological University under TEQIP-II to procure Digital Pendulum System Experimental Setup.

REFERENCES

[1] K. J. Astrom and P. R. Kumar, "Control: A perspective," *Automatica*, vol. 50, no. 1, pp. 3–43, 2014.

[2] Sudarshan K. Valluru Madhusudan Singh Bharat Bhushan, "Comparative Analysis of PID, NARMA L-2 and PSO Tuned PID Controllers for Nonlinear Dynamical System," *J. Autom. Syst. Eng.*, vol. 9, no. 2, pp. 94–108, 2015.

[3] Astrom.K.J. and Hagglund.T, "The Future of PID Control," *Control Eng. Pract.*, vol. 9, pp. 1163–1175, 2001.

[4] S. K. Valluru and M. Singh, "Stabilization of Nonlinear Inverted Pendulum System Using MOGA and APSO Tuned Nonlinear PID Controller," *Cogent Eng.*, vol. 4, no. 1, pp. 1–15, 2017.

[5] S. K. Valluru and M. Singh, "Investigation of NARMA L-2 and Artificial Bee Colony Tuned PID Controllers for Bench Scaled Nonlinear Dynamical System," Int. J. Control Theory Appl., vol. 10, no. 6, pp. 363–374, 2017.

[6] G. Zaidner, S. Korotkin, E. Shteimberg, A. Ellenbogen, M. Arad, Y. Cohen, and B. Sheva, "Non Linear PID and its application in Process Control," in *IEEE 26-th Convention of Electrical and Electronics Engineers in Israel*, 2010, pp. 574–577.

[7] L. Ramli, Z. Mohamed, A. M. Abdullahi, H. I. Jaafar, and I. M. Lazim, "Control strategies for crane systems: A comprehensive review," *Mech. Syst. Signal Process.*, vol. 95, pp. 1–23, 2017.

[8] FeedbackInstruments, "Digital Pendulum Control Experiments Manual," 2016.

2nd IEEE International conference on power Electronics, Intelligent Control and Energy systems (ICPEICES-2018)

Radio Frequency Based (RF) Control & Operation of Electrical/Electronic Appliances in Home/Offices

T. Ramachandran, Sanjiv Kumar, Ajay Kumar, Ravi Agarwal
Department of Electrical & Electronics Engineering
Subharti Institute of Technology and Engineering
Meerut (UP)-250005, India
Email: ramspowerthangamugam@gmail.com, activesanjiv007@rediffmail.com

Abstract— In this technological era, automation is one of the very important sectors where dramatic progress and development is takes place. There is no field without automation let it be industry, domestics, agriculture etc,. The automation is reducing the physical work of human being and also it reduces the error in executing the work/operation. In this paper home/office automation with RF based control technology has been used. This technology is the one of the simple and easy technology for implementing the automation of the electrical / Electronic appliance in home/office. This technology can be used to control/operate any home/office appliance like TV, tube light, fan, motor, refrigerator, charging device etc,. However, in this paper this RF based control technology is used to control the light, motor, fan and charging point.

Keywords— *RF Technology, Proposed Methodology, Microcontroller programming, Power Relay, Electrical appliance*

I. INTRODUCTION

Nowadays considering any industry it may be manufacturing/ service providing, it goes for fully and or partially automation in their field of service. Based on the nature of the industry it may go for process automation and or product manufacturing automation. The automation implemented in the industry also reflects in home/office appliance automation as many appliances also used in home/office. At this present technological era, 80% of the homes have more than four/five electrical/electronic appliance. This appliance are control by manually at present on need base. But now there are many technologies which can be used to control or operate the Electrical/Electronic appliances in home/ office for effective and efficient utilization [1]. The main purpose of the automation is to use the appliance in an efficient and effective manner. The automation will save the energy and money as it will work according to the program set up.

The automation will also reduce the human physical work. The product which will be made with this technology can help the people who are in old age and may not be able to move here and there for ON/OFF of the Electrical /Electronic appliance used in the home/office. This paper present the control of appliance used in home/office will the help of RF technology. There are four appliance are taken in this paper to control/ ON/OFF as and when required. This RF module which includes RF transmitter, RF receiver,

and microprocessor are the main module/ devices which are used in this RF technology based home /office automation.

II. WIRELESS BASED DEVICE CONTROL

A. *Wireless RF Based Device Control*

Wireless RF Based Device Control is an elementary RF monitoring circuit that can be controlled remotely using primarily the RF mode. The RF remote control has the advantage of adequate range (up to 200 meters with proper antennae) besides being directional. On the other hand, an IR remote would function over a limited range of about 5 meters and the remote transmitter has to be oriented towards the receiver module quite precisely.

B. *Operation of Holtek HT12E and HT12D*

RF based remote control use the Holtek encoder-decoder pair of HT12E and HT12D employing for RF as well as IR principles. Both of these are 18pin DIP ICs. HT12E and HT12D are CMOS ICs with working voltage ranging from 2.4V to 12V. Encoder HT12E has eight address and another four address/data lines. The data set on these twelve lines (address and address/data lines) is serially transmitted when the transmit enable pin TE is taken low. The data output appears serially on the D_{out} pin. The data is transmitted four times in succession. It consists of differing length of positive-going pulses for '1' and '0' the pulse-width for '0' being twice the pulse-width for '1'. The frequency of these pulses may lie between 1.5 and 7 kHz depending on the resistor value between OSC1 and OSC2 pins.

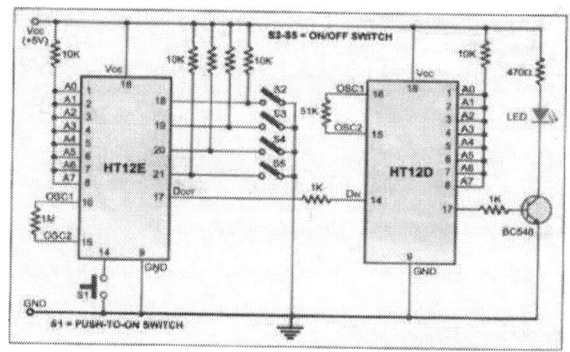

Fig. 1 Test circuit remote controller

978-1-5386-6624-1/18 $31.00 © 2018 IEEE

The internal oscillator frequency of decoder HT12D is 50 times the oscillator frequency of encoder HT12E. The resistor values used in the circuits here are chosen for approximately 3 kHz frequency for the encoder (HT12E) and 150 kHz for decoder HT12D at Vdd of 5V. The HT12D receives the data from the HT12E on its DIN pin serially. If the address part of the data received matches the levels on A0 through A7 pins four times in succession, the valid transmission (VT) pin is taken high. The data on pins AD8 through AD11 of the HT12E appears on pins D8 through D11 of the HT12D. Thus the device acts a receiver of 4-bit data (16 possible codes) with 8-bit addressing.

The test circuit given above in Fig. 1 will help in checking the functional serviceability and synchronization of the frequency of operation. Once the frequency of the pair is aligned, on pressing of push switch S1 on the encoder, LED on the decoder should glow. You can also check the transfer of data on pins AD8 through AD11 (the data pins of the encoder can be set as high or low using switches S2 through S5), which is latches on pins D8 through D11 of the decoder once TE pins is taken low momentarily using push switch S1. This completes the testing of encoder decoder pair of HT12E and HT12D.

C. RF transmitter and receiver

The RF transmitter and receiver modules have been employed for RF remote control. The RF transmitter TX-433 is an AM/ASK transmitter. Its features include that 5V-12V single supply operation, On-off-keying (OOK) / amplitude shift keying (ASK) data format, Up to 9.6kbps data rate, +9dBm output power (about 200m range), SAW-based architecture, For antenna, a 45cm wire is adequate

Table 1. Technical Specifications of Tx-433

VCC	O/P	CURRENT
5V DC	-0dBm	1.0mA
12V DC	+9dBm	3mA

Table 2 Technical Specifications of Rx-433

Parameter	Value
Bandwidth	12MHz
Sensitivity	-103 dBm
Data rate	48000 bps
Max data rate	9600 bps
Standby current	1.2Ma
Antenna	Whip, strip line or helical
Voltage	4.5V-5.5V DC

D. RF encoder and Decoder

The RF transmitter –Encoder is shown in Fig. 2. When any switch is open the pin connected to that switch is at logic 1, and when it is closed the respective pin is at logic '0'. The data pins are pulled high via resistors R2 through R5. In this condition, if TE pin is taken low (by depressing STOP switch), the binary data transmitted via pins AD8 through AD11 will be '1111' (decimal 15). When any other data pin

marked area-1, area-2, area-3 and area-4 alone is pressed, a '0' will be sent at that data position, while other data pins will represent logic '1' state. The logic circuitry at the receiver-decoder end will decode the data appropriately for indication through LED

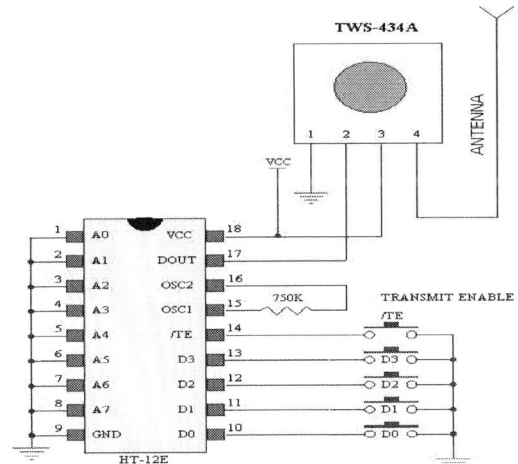

Fig. 2 RF Transmitter-Encoder

The complete RF receiver-decoder circuit employing HT12D is shown in Fig.3 Assuming that identical address is selected on the encoder and the decoder, when any of the switches on the transmitter (marked as area-1, area-2, area-3, area-4) is depressed, the corresponding data pin of the demodulator will go low. The data outputs of HT12D are fed to 8-bit priority encoder CD4532 via inverters to generate appropriate logic outputs in conformity.

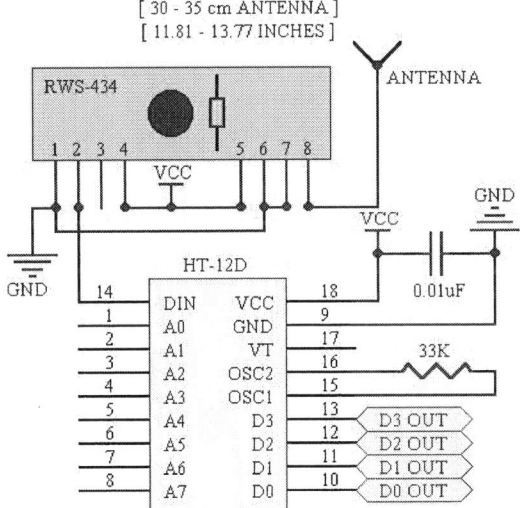

Fig. 3 RF Receiver -Decoder

2nd IEEE International conference on power Electronics, Intelligent Control and Energy systems (ICPEICES-2018)

III. PROPOSED RF BASED AUTOMATION SYSTEM

The aim of this paper is to design an RF based Home Automation system using 8051 microcontroller, in which different home appliances are remotely controlled using RF technology [2]. For the proper working of this proposed system, the RF transmitter and receiver modules are switched ON before turning on the microcontroller. This is to avoid the junk data, received by the receiver module, when it is pairing with the transmitter. Once the modules are successfully paired, the LED attached to VT pin of the receiver module will glow. From now onwards, the data out pins of the decoder will continuously give logic high as there is no button pressed in the transmitter. Once we turn on the microcontroller, all the loads connected to it are switched OFF as it receives logic high continuously from the receiver. The status of the loads is displayed on the LCD. If any of the buttons is pressed in the transmitter, the corresponding pin in receiver will become low. This transition will help the microcontroller to understand that a key is pressed and will turn ON the corresponding load as mentioned in the program. If the same button is pressed once again, the microcontroller will turn OFF the load.

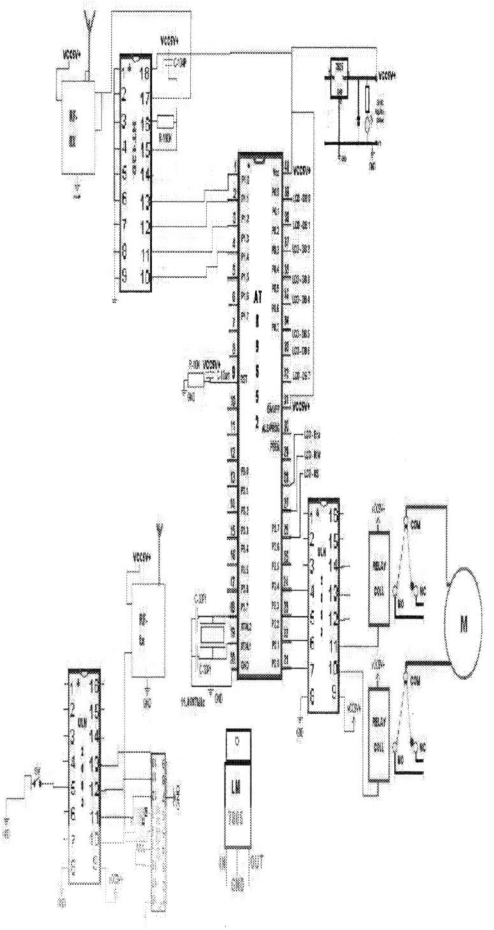

Fig. 4 Schematic diagram of the proposed method

A. *Microcontroller AT89C51*

The figure shows in Fig.5 is the basic architecture of 8051 family of microcontroller [3]. It has the features with Compatible with MCS-51™ Products, 4K Bytes of In-System Reprogrammable Flash Memory, Endurance: 1,000 Write/Erase Cycles, Fully Static Operation: 0 Hz to 24 MHz, Three-Level Program Memory Lock, 128 x 8-Bit Internal RAM, 32 Programmable I/O Lines, Two 16-Bit Timer/Counters

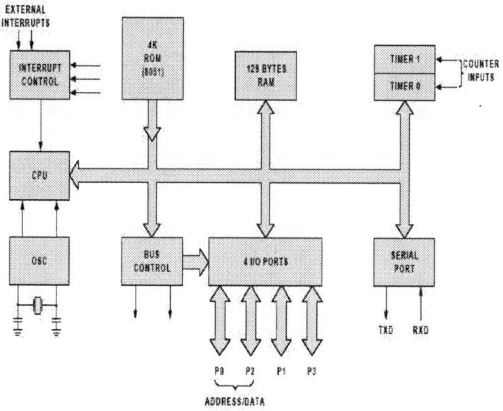

Fig. 5 Microcontroller AT89C51

The AT89C51 is a low-power, high-performance CMOS 8-bit microcomputer with 4K bytes of Flash Programmable and Erasable Read Only Memory (PEROM). The device is manufactured using Atmel's high density non-volatile memory technology and is compatible with the industry standard MCS-51™ instruction set and pinout. The on-chip Flash allows the program memory to be reprogrammed in-system or by a conventional nonvolatile memory programmer. By combining a versatile 8-bit CPU with Flash on a monolithic chip, the Atmel AT89C51 is a powerful microcomputer which provides a highly flexible and cost effective solution to many embedded control applications. The AT89C51 provides the following standard features: 4K bytes of Flash, 128 bytes of RAM, 32 I/O lines, two 16-bit timer/counters, five vector two-level interrupt architecture, a full duplex serial port, and on-chip oscillator and clock circuitry.

In addition, the AT89C51 is designed with static logic for operation down to zero frequency and supports two software selectable power saving modes [4]. The Idle Mode stops the CPU while allowing the RAM, timer/counters, serial port and interrupt system to continue functioning. The Power down Mode saves the RAM contents but freezes the oscillator disabling all other chip functions until the next hardware reset.

B. Programming Algorithm

Before programming the AT89C51, the address, data and control signals should be set up according to the Flash

978-1-5386-6624-1/18 $31.00 © 2018 IEEE   1234

programming mode table and Figures 3 and 4. To program the AT89C51, take the following steps [5].

1. Input the desired memory location on the address lines.
2. Input the appropriate data byte on the data lines.
3. Activate the correct combination of control signals.
4. Raise EA/VPP to 12V for the high-voltage programming mode.

5. Pulse ALE/PROG once to program a byte in the Flash array or the lock bits. The byte-write cycle is self-timed and typically takes no more than 1.5 ms. Repeat steps 1 through 5, changing the address and data for the entire array or until the end of the object file is reached.

(a) Data Polling and Ready/Busy: The AT89C51 features Data Polling to indicate the end of a write cycle. During a write cycle, an attempted read of the last byte written will result in the complement of the written datum on PO.7. Once the write cycle has been completed, true data are valid on all outputs, and the next cycle may begin. Data Polling may begin any time after a write cycle has been initiated.

The progress of byte programming can also be monitored by the RDY/BSY output signal. P3.4 is pulled low after ALE goes high during programming to indicate BUSY. P3.4 is pulled high again when programming is done to indicate READY.

(b) Program Verify and Chip Erase: If lock bits LB1 and LB2 have not been programmed, the programmed code data can be read back via the address and data lines for verification [7]. The lock bits cannot be verified directly. Verification of the lock bits is achieved by observing that their features are enabled.

The entire Flash array is erased electrically by using the proper combination of control signals and by holding ALE/PROG low for 10ms. The code array is written with all "1"s. The chip erase operation must be executed before the code memory can be re-programmed [6].

(c) Reading the Signature Bytes and programming Interface: The signature bytes are read by the same procedure as a normal verification of locations 030H, 031H, and 032H, except that P3.6 and P3.7 must be pulled to logic low. The values returned are as follows:

(030H) = 1EH indicates manufactured by Atmel
(031H) = 51H indicates 89C51
(032H) = FFH indicates 12V programming
(032H) = 05H indicates 5V programming

(d) Every code byte in the Flash array can be written and the entire array can be erased by using the appropriate combination of control signals. The write operation cycle is self timed and once initiated, will automatically time itself to completion [8]. All major programming vendors offer worldwide support for the Atmel microcontroller series. Please contact your local programming vendor for the appropriate software revision.

IV. HARDWARE IMAGE OF THE PROPOSED METHOD

The hardware of the proposed system have the two modules namely on is the main module and the other one is the remote module of the proposed system

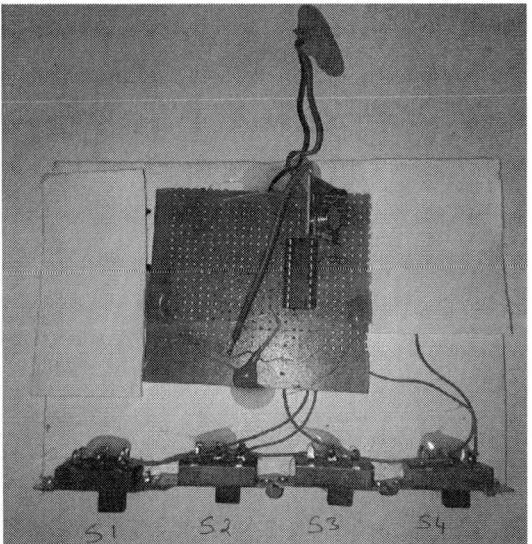

Fig. 6 Hardware image of the proposed method-main module

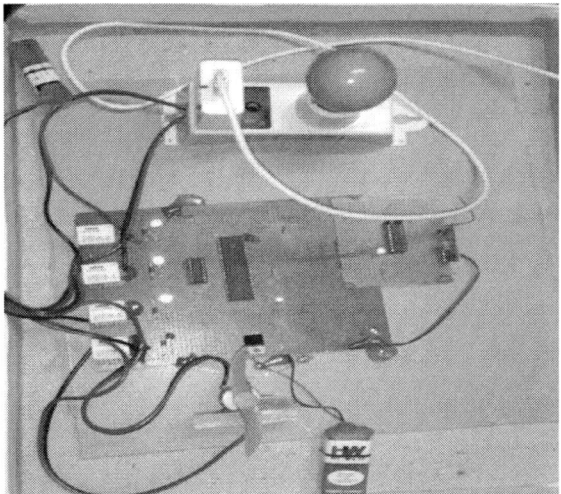

Fig. 7 Hardware image of the proposed method -RF remote control module

V. CONCLUSION

RF based device control technology has been used in this paper to control the home appliance. There are four home appliances (motor, mobile charging, bulb, blank supply) has been controlled using the proposed RF based device control technology module. Microcontroller AT89C51 is used as part of programming and controlling device. Wireless RF Based Device Control is an elementary RF monitoring circuit that can be controlled remotely using primarily the RF mode. The RF remote control has the advantage of adequate range (up to 200 meters with proper antennae) besides being directional. On the other hand, an IR remote would function over a limited range of about 5 meters and the remote transmitter has to be oriented towards the receiver module quite precisely [9]-[11]. This proposed technology can also extended to control more electrical appliance as part of the future work of this paper.

REFERENCES

[1] J. Chandramohan, R. Nagarajan, K.Satheesh kumar, N. Ajithkumar, P.A. Gopinath, S.Ranjithkumar, "Intelligent Smart Home Automation and Security System Using Arduino and Wi-fi" International Journal of Engineering and Computer Science, Vol. 6, No. 3, 2017.

[2] Abdullah, R., Rizman, Z. I., Dzulkefli, N. N. S. N., Ismail, S. I., Shafie, R., & Jusoh, M. H. (2016). Design an automatic temperature control system for smart TudungSaji using arduino microcontroller. ARPN Journal of Engineering and Applied Sciences. Vol. 11, No. 16 , 2016.

[3] Nawi B, Sulaini B, Mohd Z A R, Shamsul A Z, Zairi I R.PID voltage control for DC motor using MATLAB Simulink and Arduino microcontroller. Journal of Applied Environmental and Biological Sciences, Vol. 5, No.9, pp.166-173, 2015

[4] Jebelli A, Yagoub M C. Development of sensors and microcontrollers for small temperature controller systems. Journal of Automation and Control Engineering, Vol .3, No.4, pp.322-328, 2015

[5] Mohamed Abd El-Latif Mowad, Ahmed Fathy, Ahmed Hafez "Smart Home Automated Control System Using Android Application and Microcontroller" International Journal of Scientific & Engineering Research, Vol. 5, No. 5, May-2014.

[6] Jeetender Singh Chauhan, Sunil Semwal " Microcontroller Based Speed Control of DC Geared Motor Through RS-232 Interface With PC" . International Journal of Engineering Research and Applications, Vol. 3, No1, pp.778-783, 2013

[7] Hsien-Tang Lin" Implementing Smart Homes with Open Source Solutions" International Journal of Smart Home, Vol. 7, No. 4, pp 289-295.

[8] Gowthami, Dr. Adiline macriga "Smart Home Monitoring and Con trolling System Using Android Phone" International Journal of Emerg-ing Technology and Advanced Engineering, Vol. 3, No. 11, November 2013.

[9] Hsien-Tang Lin" Implementing Smart Homes with Open Source Solutions" International Journal of Smart Home Vol. 7, No. 4, pp 289-295, July, 2013

[10] Sanjiv Kumar, Narendra Kumar, "Alleviation SSR and low frequency power oscillations in series compensated transmission line using SVC supplementary controllers," Springer Journal of The Institution of Engineers (India), vol. 98, No. 3, pp. 255-266, 2017.

[11] Narendra Kumar, Sanjiv Kumar, "Alleviation of transient torsional torque stresses of turbine generator shaft segments using CBVLC supplementary controller," Int. J. of Power and Energy Conversion, vol. 7, No. 1, pp. 42-56, 2016.

2nd IEEE International Conference on Power Electronics, Intelligent Control and Energy Systems (ICPEICES-2018)

Multiple-Input Single-Output Universal Biquad Filter Using Single Output OTAs

Ajishek Raj
Department of Electrical Engineering
Delhi Technological University
New Delhi, India
ajishekaerotek@gmail.com

D. R. Bhaskar
Department of Electronics and
Communication Engineering
Delhi Technological University
New Delhi, India
drbhaskar@dtu.ac.in

Pragati Kumar
Department of Electrical Engineering
Delhi Technological University
New Delhi, India
pragati_1964@yahoo.co.in

Abstract— **A new voltage-mode multi-input single-output(MISO) universal biquad filter configuration employing five single output operational transconductance amplifiers (OTAs), two grounded capacitors (ideal for integrated circuit implementation) with three input voltage signals has been proposed. The circuit has versatility to realize all five second-order filter responses namely, low pass (LPF), high pass (HPF), band pass (BPF), band reject (BRF) and all pass (APF) using appropriate selection(s) of three inputs. The biquad enjoys low active and passive sensitivities. The workability of this multifunctional biquad filter topology has been established through SPICE simulation results.**

Keywords—Operational Transconductance Amplifier (OTA), Analog Circuit Design, Voltage Mode Filters.

I. INTRODUCTION

Operational transconductance amplifiers (OTAs) based analog filters (used for continuous-time signal processing) have exhibited some advantages due to their structural simplicity, high frequency capability, electronic controllability and monolithic integrability both in Bipolar and CMOS technologies. MISO-type multifunction biquad filters are especially versatile due to the fact that the same topology may be used to describe various filter responses by the appropriate choice of input signal(s). In reference [1] advantages and applications of such filters have been reported. Number of voltage-mode (VM) MISO-type OTA-C universal biquad filter topologies have been presented in [2-7] using multiple-output OTAs. VM MISO-type filter structures presented in [8-16] use only single output OTAs which are commercially available as off-the self ICs.

Recently, Psychalinos, Kasimis and Khateb [16] presented an OTA-C universal biquad filter employing six single output OTAs, two grounded capacitors with four input voltage signals for use in discrete applications since commercially available OTAs are only single output type. This filter configuration is capable of realizing all the standard second-order filter responses by proper selection of four input voltage signals therein. In this communication we propose a new circuit structure which by contrast, employs only five OTAs and two grounded capacitors, to realize all the five filter functions with appropriate choice of only three input voltage signals. Workability of this biquad filter has been verified through SPICE simulations and some sample results have been presented.

II. THE PROPOSED VOLTAGE-MODE FILTER CONFIGURATION

The proposed new MISO-type VM universal biquad filter structure for realizing low pass (LPF), high pass (HPF), band pass (BPF), band reject (BRF) and all pass (APF) using appropriate selection(s) of three inputs is shown in Fig. 1. A routine analysis of the circuit of Fig. 1 with ideal OTAs yields the expression for the output voltage in terms of the three input voltages as:

$$V_0(s) = \frac{AV_{in1}(s) + BV_{in2}(s) + s^2 V_{in3}(s)}{s^2 + s\dfrac{g_{m2}g_{m3}}{C_2 g_{m4}} + \dfrac{g_{m1}g_{m2}g_{m3}}{C_1 C_2 g_{m4}}} \quad (1)$$

Where

$$A = \left(s^2 \frac{g_{m3}}{g_{m4}} - s \frac{g_{m3}g_{m5}}{C_2 g_{m4}} + \frac{g_{m1}g_{m2}g_{m3}}{C_1 C_2 g_{m4}} \right)$$

$$B = s\left(\frac{g_{m3}g_{m5}}{C_2 g_{m4}} \right)$$

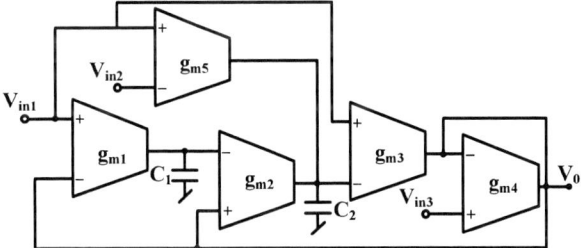

Fig. 1. Proposed Biquad Filter Configuration

Thus, the various second-order filter functions can be derived from equation (1) by proper selection of input voltages as follows:

(i) LPF: if $V_{in1} = V_{in2} = -V_{in3} = V_{in}$ (input voltage signal), and $g_{m3} = g_{m4}$,

$$\frac{V_0(s)}{V_{in}(s)} = \frac{\left(\dfrac{g_{m1}g_{m2}}{C_1 C_2} \right)}{D(s)} \quad (2)$$

(ii) BPF: if $V_{in1} = V_{in3} = 0$ and $V_{in2} = V_{in}$,

978-1-5386-6624-1/18 $31.00 © 2018 IEEE   1237

$$\frac{V_0(s)}{V_{in}(s)} = \frac{s\left(\dfrac{g_{m3}g_{m5}}{C_2 g_{m4}}\right)}{D(s)} ; g_{m2} = g_{m5} \qquad (3)$$

(iii)   HPF: if $V_{in1} = V_{in2} = 0$ and $V_{in3} = V_{in}$,

$$\frac{V_0(s)}{V_{in}(s)} = \frac{s^2}{D(s)} \qquad (4)$$

(iv)   BRF: if $V_{in3} = 0$ and $V_{in1} = V_{in} = V_{in2}$,

$$\frac{V_0(s)}{V_{in}(s)} = \frac{\left(s^2 + \dfrac{g_{m1}g_{m2}}{C_1 C_2}\right)}{D(s)} ; g_{m3} = g_{m4} \qquad (5)$$

(v)   APF: if $V_{in1} = V_{in}, V_{in3} = V_{in2} = 0$,

$$\frac{V_0(s)}{V_{in}(s)} = \frac{s^2 - s\left(\dfrac{g_{m5}}{C_2}\right) + \left(\dfrac{g_{m1}g_{m2}}{C_1 C_2}\right)}{D(s)} ; g_{m3} = g_{m4} \qquad (6)$$

where

$$D(s) = s^2 + s\left(\frac{g_{m2}g_{m3}}{C_2 g_{m4}}\right) + \frac{g_{m1}g_{m2}g_{m3}}{C_1 C_2 g_{m4}} \qquad (7)$$

The various filter parameters namely, resonance frequency (ω_o) bandwidth (BW) and quality factor (Q_o) of the proposed filter configuration can be obtained as:

$$\omega_o = \sqrt{\frac{g_{m1}g_{m2}g_{m3}}{g_{m4}C_1 C_2}} \qquad (8)$$

$$Q_o = \sqrt{\frac{C_2 g_{m1}g_{m4}}{C_1 g_{m2}g_{m3}}} \qquad (9)$$

$$BW = \frac{g_{m2}g_{m3}}{C_2 g_{m4}} (for\ BPF\ and\ BRF) \qquad (10)$$

From equations (8) and (10), it is evident that ω_o and BW are orthogonally programmable through the transconductance g_{m1} (i.e., after fixing the value of BW, the ω_o can be controlled with g_{m1}).

The sensitivities of ω_o, and Q_o can be determined as:

$$S_{gm1}^{\omega_o} = S_{gm2}^{\omega_o} = S_{gm3}^{\omega_o} = S_{gm1}^{Q_o} = S_{gm4}^{Q_o} = S_{C_2}^{Q_o} = 0.5$$

$$S_{gm4}^{\omega_o} = S_{C_1}^{\omega_o} = S_{C_2}^{\omega_o} = S_{gm2}^{Q_o} = S_{gm3}^{Q_o} = S_{C_1}^{Q_o} = -0.5$$

Thus, all active and passive sensitivities are small.

III. SIMULATION RESULTS

The workability of the proposed configuration has been verified through SPICE simulations using the macro model of the OTAs in the IC LM13700 [17]. The DC power supplies of $\pm15V$ were used to bias the OTAs and equal bias currents $I_{bias1} = I_{bias2} = I_{bias3} = I_{bias4} = I_{bias5} = 548\mu A$ ($g_{mi} = 10.5mS$) ($i = 1-5$) were used to simulate the circuit of Fig. 1. The capacitors used were of values $C_1 = 47$ nF and $C_2 = 23.45$ nF

for the cut-off frequency $f_o = 50$ kHz and quality factor $Q_o = 0.707$. Fig. 2 shows the simulated magnitude responses of all the filters. Transient response and phase response of APF at cut-off frequency 50 kHz are shown in Fig. 3 and Fig. 4 respectively.

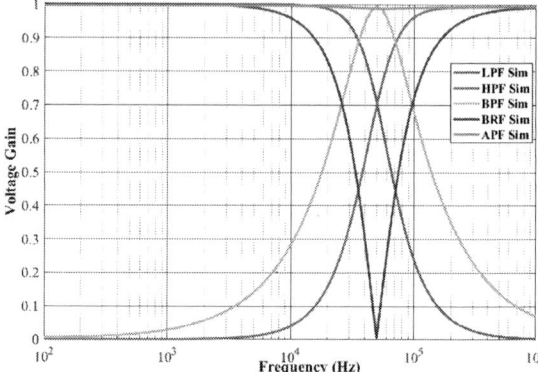

Fig. 2 Frequency responses of the proposed configuration

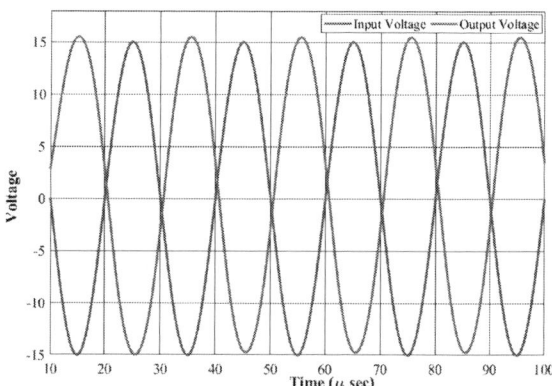

Fig. 3 Simulated transient response of APF

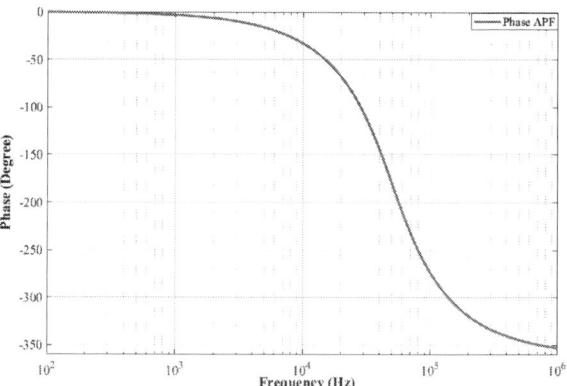

Fig. 4 Simulated phase response of APF

IV. ORTHOGONAL TUNING OF RESONANT FREQUENCY AND BANDWIDTH

It is observed from equations (8) and (10) that the BW and the resonant frequency are orthogonally tunable, i.c., the resonant frequency can be tuned without changing the BW by varying g_{m1}. The biquad filter with the nominal value of the cut-off frequency f_o = 50 kHz, BW = 71.49 kHz, quality factor = 0.707 and gain = 1 was designed by taking g_{mi} (i=1-5) = 10.5mS, C_1 = 47 nF and C_2 = 23.45 nF. I_{bias} of 548 μA (for all the OTAs) results in fixing this value of g_m for all the OTAs.

To confirm the controllability of f_o without changing the bandwidth, we have used I_{bias1} = 276 μA, 368 μA, 552 μA and 1.104mA to get the variations in cut-off frequencies 36.76 kHz, 41.12 kHz, 50.429 kHz and 69.589 kHz respectively for the BPF and BRF. These results have been shown in Fig. 5 and Fig. 6 for BPF and BRF respectively. It may be noted that the BW remained constant at 71.3 kHz for both BPF and BRF. A comparison with previously published MISO type OTA-C biquads has also been presented in Table 1.

TABLE I. COMPARISION RESULTS OF PROPOSED FILTER

Features	Circuit Reference					
	[4]	[5]	[8]	[9]	[16]	Fig. 1
Numbers of OTAs	7	6	6	7	6	5
Number of resistors and capacitors	2C	2C	2C+ 2R	2C	2C	2C
Grounded passive components	Y	Y	Y	Y	Y	Y
High impedance at the inputs	Y	Y	Y	Y	Y	Y
Availability of orthogonal tunability of ω_o and BW	N	N	Y	Y	Y	Y

From the Table 1, it is observed that the proposed circuit uses minimum number of single output OTAs while providing tunability of ω_o and BW for BPF and BRF.

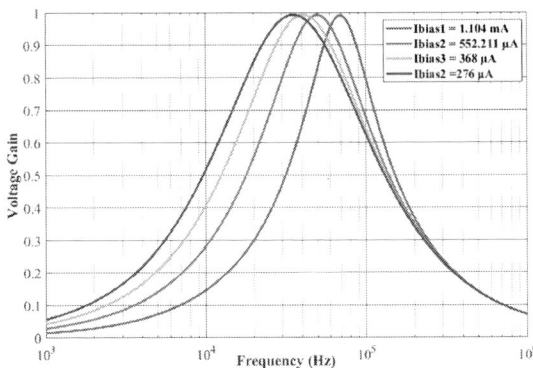

Fig. 5 Electronic tunability of BPF

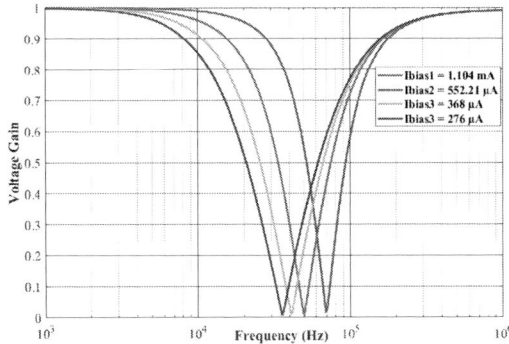

Fig. 6 Electronic tunability of BRF

V. CONCLUSION

In this paper, a new voltage-mode electronically controllable three input, one output universal biquadratic filter configuration using five single output OTAs and two grounded capacitors has been presented. The LPF, BPF, HPF, BRF and APF can be realised by appropriate choice(s) of the input signals from the same topology. The input terminals of the OTAs offer high input impedance level which enables easy cascadability of the voltage-mode operation. The cut-off frequency (f_o) and bandwidth are orthogonally tunable through bias current of the OTA1 for BPF and BRF.

REFERENCES

[1] Horng JW. Voltage-mode universal biquadratic filter with one input and five outputs using OTAs. International Journal of Electronics. vol. 89, no. 9, 2002, pp. 729-37.

[2] Horng, JW. "Voltage-mode universal biquadratic filter using two OTAs." *Active and Passive Electronic Components* vol. 27, no. 2, 2004, pp. 85-89.

[3] Chang, Chun-Ming. "New multifunction OTA-C biquads." *IEEE Transactions on Circuits and Systems II: Analog and Digital Signal Processing* vol. 46, no. 6, 1999, pp. 820-824.

[4] Lee, Chen-Nong. "Multiple-mode OTA-C universal biquad filters." *Circuits, Systems and Signal Processing* vol. 29, no. 2, 2010, pp. 263-274.

[5] Kumngern, Montree, Boonying Knobnob, and Kobchai Dejhan. "Electronically tunable high-input impedance voltage-mode universal biquadratic filter based on simple CMOS OTAs." *AEU-International Journal of Electronics and Communications*, vol. 64, no. 10, 2010, pp. 934-939.

[6] Wu, Jie, and Chang-Yan Xie. "New multifunction active filter using OTAs." *International Journal of Electronics* vol. 74, no. 2, 1993, pp. 235-239.

[7] Chen, Hua-Pin, Yi-Zhen Liao, and Wen-Ta Lee. "Tunable mixed-mode OTA-C universal filter." *Analog Integrated Circuits and Signal Processing* vol. 58, no. 2, 2009, pp. 135-141.

[8] Kumngern, Montree, Pichai Suksaibul, and Boonying Knobnob. "High-input impedance four-input one-output voltage-mode universal filter using OTAs." In *Information and Communication Technology, Electronic and Electrical Engineering (JICTEE), 2014 4th Joint International Conference on*, pp. 1-4. IEEE, 2014

[9] Abuelma'atti*, Muhammad Taher, and Abdulwahab Bentrcia. "A novel mixed-mode OTA-C universal filter." *International Journal of Electronics* vol. 92, no. 7, 2005, pp. 375-383.

[10] Lee, Chen-Nong, and Chun-Ming Chang. "High-order mixed-mode OTA-C universal filter." *AEU-International Journal of Electronics and Communications* vol. 63, no. 6, 2009, pp. 517-521.

[11] JIE, WU EZZ I. "Universal voltage-and current-mode OTAs based biquads." *International Journal of Electronics* vol. 85, no. 5, 1998, pp. 553-560.

[12] Sanchez-Sinencio, Edgar, Randall L. Geiger, and H. Nevarez-Lozano. "Generation of continuous-time two integrator loop OTA filter structures." *IEEE Transactions on Circuits and Systems* vol. 35, no. 8, 1988, pp. 936-946.

[13] Sun, Yichuang. "Second-order OTA-C filters derived from Nawrocki-Klein biquad." *Electronics letters* vol. 34, no. 15, 1998, pp. 1449-1450.

[14] Wattikornsirikul, Natchayathorn, and Montree Kumngern. "Three-input one-output voltage-mode universal filter using simple OTAs." In *ICT and Knowledge Engineering (ICT and Knowledge Engineering), 2014 12th International Conference on*, pp. 28-31. IEEE, 2014.

[15] Safari, Leila, Shahram Minaei, and Bilgin Metin. "A low power current controllable single-input three-output current-mode filter using MOS transistors only." *AEU-International Journal of Electronics and Communications* vol. 68, no. 12, 2014, pp. 1205-1213.

[16] Psychalinos, Costas, Chrysostomos Kasimis, and Fabian Khateb. "Multiple-input single-output universal biquad filter using single output operational transconductance amplifiers." *AEU-International Journal of Electronics and Communications* vol. 93, 2018, pp. 360-367.

[17] Texas Instruments, LM13700: Dual Operational Transconductance Amplifier with Linearizing Diodes and Buffers, URL http://www.ti.com/product/LM13700; 2015.

AUTHOR INDEX

Aarfin, Shamshul 1087
Abedi, Mehradad 915
Acharya, Swastik 516
Agarwal, Bhavna 1154
Agarwal, Pramod 809
Agarwal, Ravi 1232
Agarwal, Siddhartha 944
Agarwal, Tanuj 976
Agarwal, Vijyant 1036
Agarwal, Vivek 599
Aggarwal, Archna 403, 408
Aggarwal, Gianeshwar 944
Agrawal, Deepak 1093
Agrawal, Seema 374, 483
Ahmad, Aijaz 149, 498
Ahmad, Samiuddin 143
Ahuja, Rajesh Kr. 414
Akhtar, Iram 455
Allahloh, Ali S. 815
Anan, Ashique 361
Anand, P. 287
Ankarao, Mogili 892
Anup, Sunitha 160
Anuranjana 958
Anzar, Masood 570
Aree, Pichai 837
Arora, Abhishek 249
Arora, Ankita 558
Arora, Parul 200
Arora, Sudha 605
Arora, Swati 170
Arya, Ravi Vardhan 492
Arya, Yogendra 325
Azad, Abhishek 398
Azam, Farooque 102, 443
Azeem, Abdul 636
Bagai, Komal 78
Bagha, Gurudutt 267
Balguvhar, Sumit 972
Baliyan, Arjun 477
Ballal, Makarand S. 754
Banerjee, Indranil 1041
Banerjee, Subrata 760
Bansal, Praveen 343, 355
Bansal, Rahul 1219
Basak, Prasenjit 449
Bath, S. K. 287
Bazaz, Mohammad Abid 108
Bhadouria, Vivek Pratap Singh 23
Bhalla, Suresh 972
Bhandari, Manisha 1154
Bhandari, Shruti 1115, 1199
Bharath, Kurukuru Varaha Satya 552, 870
Bharatiraja, C. 636
Bhargav, Allu 558

Bhargava, Vishal 734
Bharti, Roshan 864
Bhaskar, D. R. 1237
Bhaskar, Mukesh Kumar 132
Bhattacharjee, Ankur 546
Bhattacharjee, Shayari 876
Bhattacharya, Chitra 886
Bhattacharyya, Sudip 659
Bhatti, T. S. 160, 783
Bhosle, Abhijit B. 1067
Bhowmick, S. 154
Bindu, S 206
Biswal, Gyan Ranjan 449
Blaabjerg, Frede 689
Bonev, Boncho 1178, 1193
Celeita, David 33
Chaithra, A 39
Chakraborty, Tapan Kumar 361
Chandak, Sheetal 431
Chandra, Kanchan 1121
Chandran, Vineet P. 582
Chankaya, Mukul 498
Chari, Shrivatsan K 841
Chaturvedi, Devendra Kumar 1046
Chaudhari, Saurabh 1056
Chaudhary, Priyanka 305
Choubey, V. K. 370
Choudekar, Pallavi 44
Choudhury, Anwesha 743
Choudhury, Shreeram 114
Dabas, Shivam 78
Daftari, Mohammad Ali 1062
Dahiya, Pankaj 488
Dalal, Sahil 982, 988
Dang, Radhika 927
Das, Debabrata 546
Das, P. Vipin 689
Das, S K 1041
Debnath, Manoj Kumar 114, 1142
Dehghani, Majid 915
Deshmukh, Rohit R. 754
Deshpande, A. P. 1056
Dey, Anamika 743
Dhembare, Sushil B. 1067
Dhiman, Rishav 841
Dimri, Ashish 948
Dixit, T. V. 909
Dmesh, Panditi 892
Dobhal, Daksh 1228
Dogra, Priyanka 743
Dua, Vaibhav 73
Dube, Anirudh 576
Dubey, Garima 809
Dwivedi, Nikhil 85
Fatima, Mehtab 44

AUTHOR INDEX

Fernandez, E. 331
Gangoli, Hirdesh 1121
Ganguly, Sourish..........................546
Gao, S. 437
Garai, Rabindranath 743
Garg, Rachana 96, 671, 717
Gaur, Aditi.....................................1199
Gaur, Prerna 1036
Gautam, Amit Kumar 1172
Ghose, Udayan 827
Ghosh, Ahana 546
Ghosh, Swapnendu Narayan.............234
Ghosh, T. 370
Giri, Ravi 1056
Goel, Arnav 1228
Goel, Nidhi......................................317
Goel, Nitin.......................................386
Goel, P. K.......................................504
Gupta, Abhishek Kumar 92
Gupta, Akhilesh Kumar 881
Gupta, Anshul 953, 1228
Gupta, Archit 944
Gupta, Atul......................................349
Gupta, C. P. 331
Gupta, Ekta 1213
Gupta, Himanshu...............................29
Gupta, Maneesha 1016, 1022, 1052
Gupta, Nidhi 273, 325, 1148
Gupta, Nitesh..................................967
Gupta, Rajeev..................................471
Gupta, Ritesh K...............................729
Gupta, Ritesh Kant 743
Gupta, S. K..50
Gupta, Saket 280
Gupta, Sapna...................................471
Gupta, Shilpa....................................311
Gupta, Shubham Kumar...............29, 343
Gupta, Srishti 8
Gupta, Tripurari Nath.......................630
Gupta, Vijay Kumar...............374, 483
Gurumurthy, S. R..............................948
Handa, Himesh.................................967
Haque, Ahteshamul............461, 552, 870
Hasan, Naimul.................61, 143, 1046
Hussain, Ikhlaq................................683
Hussain, Shahzad..............................471
Hussan, Reyaz..................................293
Ibraheem............................61, 1109
Indu, S..1183
Istiyaque, Md....................................1046
Iyer, Parameswar K..........................729
Iyer, Parameswar Krishnan...............743
Jaffery, Zainul Abdin 229, 461, 1109
Jain, Akansha.................................1010
Jain, Jyoti.......................................921

Jain, N. K. 921
Jain, Pranav 85
Jain, R. K. 1213
Jain, Vandana..................................528
Jain, Vertika 96
Jaint, Bhavnesh 1183
Jalindar, Dubal Amol........................337
Jamil, Majid..............455, 477, 576, 1083
Jaroli, Shweta.................................1199
Jena, Anuj 516
Joshi, D. 154
Joshi, Dheeraj 386, 809
Joshi, Sunil 1115, 1199
Juneja, Kapil 962
Kala, Peeyush 605
Kamal, Md. Mostofa 361
Kamra, Rakhi 8
Kanaujia, Anoop Kumar 831
Kandpal, Tara C 777
Kansal, Veenu 1188
Kansara, H R 739
Kant, Piyush 771
Karandikar, P. B. 1056, 1067
Karimipour, Hadis 915
Karimyan, Peyman............................915
Karthikeyan, V. 689
Kartikeya, Kumar 1228
Kashif, Mohd. 765
Kataria, Gaurav 1136
Kaur, Daman Preet 1097
Kaur, Harpreet 1188
Kaur, Mandeep................................1005
Kaur, Manpreet..................958, 1228
Kaur, Sandeep 170
Kaur, Sanmukh 958
Kaur, Simar Preet 792
Kesari, J. P 102, 443
Kesarwani, Aditya 958
Khan, Imran 546
Khan, Kuhsro 897
Khan, Mohammed Ali............552, 870
Khanna, Rintu 170
Khare, Sudhir 1121
Khatoon, Shahida 1046
Khondokar, Shihab 1132
Khunte, Kiran Vijay 539
Kirmani, Sheeraz................127, 455
Kiruthika, M. 206
Kothari, D. P. 477
Kulkarni, N. R. 1067
Kulkarni, R. D. 948
Kumar, Ajay 1232
Kumar, Akshiv 56
Kumar, Amit 267
Kumar, Amritesh 492

AUTHOR INDEX

Kumar, Aniket 1213
Kumar, Arvind 1073
Kumar, Astitva 212
Kumar, Bhavnesh 56
Kumar, Deepak 881, 1103
Kumar, Gangavarapu Guru 188
Kumar, Himanshu 200
Kumar, Jagdish 398
Kumar, Jitender 65
Kumar, Kelam Sudheer 534
Kumar, M. Vijaya 892
Kumar, Manoj 120
Kumar, Mukesh 50
Kumar, Narendra 65, 257, 273, 280, 287, 311, 325, 570, 623
Kumar, Pragati 1237
Kumar, Prakash 1032
Kumar, Punit 617
Kumar, Ravinder 1178, 1193
Kumar, Sanjiv 92, 831, 1232
Kumar, Shailendra 522, 659, 665, 677, 706, 777, 783
Kumar, Sujay 976
Kumar, Sushil 788, 1224
Kumaravel, S. 188
Kumari, Shweta 1016
Kundu, Sourabh 760
Kurm, Shashank 599
Kushwaha, Satendra Kr Singh 420
Lakshamanan, M 976
Lau, Shreya 876
Lodi, Kaif Ahmed 636
Lone, Ashiq Hussain 108
Madichetty, Sreedhar 539
Mahabubunnabi, Md. 361
Mahajan, Priya 671, 717
Maheshwari, Sudhanshu 1093
Majumdar, Sudipta 1172, 1219
Mallick, Pradeep Kumar 218
Mandal, Ranabir 1121
Mangal, Punit 223
Mangal, Shubham Kumar 976
Manohar, T. Gowri 253
Maurya, Satvik 927
Meena, Duli Chand 821
Meena, Manish Kumar 257
Meena, Rakesh K. 729
Mehta, Jaimin 349
Mishra, A 809
Mishra, Ashish 648
Mishra, Heena 909
Mishra, Manohar 431
Mishra, Sukumar 261, 522, 539, 611, 642
Mishra, Utkarsh 1056
Mitra, A. 370
Mittal, Ayush 748

Mittal, Prag 1087
Modi, Gaurav 665
Modi, Sangeeta 39
Mohammad, Sarfraz 815
Mohanty, Pradeep Kumar 1142
Mohanty, S. R. 420
Mukhija, Pankaj 488
Murshid, Shadab 582, 630, 765
Nagarajan, S. T. 96
Nagpal, Divya 408
Naik, K. A. 331
Nair, Manu J 1205
Nangia, Uma 212, 921
Naqvi, Syed Bilal Qaiser 706
Narayana, O. V. L. 798
Narayanan, Vivek 722
Narula, Aditya 492
Nasiruddin, Ibraheem 143, 182
Nataraj, J. 948
Nath, Shabari 729
Nekoui, Mohammad Ali 1062
Nirmal, A V 739
Padhy, Prabin Kumar 165, 534, 858, 864
Padmanaban, Sanjeevikumar 102, 443
Padmini, N 44
Pahuja, Vaibhav 1205
Pal, Kanwar 777
Pal, Nidhi Singh 132
Pal, Subhrasish 546
Palwalia, D. K. 374
Panchal, Tejas H. 804
Panda, Jeebananda 927, 1165
Pandey, Anand K 127
Pandey, Karnika 876
Pandey, Neeta 1183
Pandey, Vishal Kumar 1078
Pandey, Yudhishthir 61
Panigrahi, B. K. 200
Panigrahi, Saswat 588
Pant, B. K. 13
Pant, Millie 1121
Pant, Peeyush 700
Parthasarathy, Harish 1036, 1078
Patel, Amit N. 804
Patel, Nimai Charan 114
Patel, Rajesh M. 804
Pathak, Om 176
Pattanaik, Shreeva 588
Perveen, Gulnar 317
Petkov, Peter 1178, 1193
Pillai, Swathy 349
Prakash, Prem 176, 239
Prakash, Surya 108
Pranith, Sai 783
Prasad, Dinanath 623

AUTHOR INDEX

Prasad, Jeetendra................712
Priya, B. K. 798
Priyadarshi, Neeraj...............102, 443
Prodhan, Md. Imran................361
Puchalapalli, Sambasivaiah.........504
Purwar, Ravindra Kr. 1160
Pushkar, Kanhaiya Lal...........788, 1224
Quadri, A. H. 287
Quadri, Imran Ahmad..............154, 229
Raghava, N. S. 734, 1000, 1126
Rahman, Obaidur 717
Rahul, K. 729
Rai, J. N. 85, 993
Rai, Jitendra Nath.................73
Raj, Ajishek 1237
Rajamanickam, R. 349
Rakib, Sakhawat Hossen 361
Ram, Atma....................414
Ramachandaramurthy, Vigna K. 102, 443
Ramachandran, T. 1232
Ramaramy, Sudha 510
Ramesh, T. K. 798
Ramola, Ayushman.............1087
Ramos, Gustavo 33
Rana, Chhavi.................962
Rana, Kailash 821
Rana, Rohit.................1036
Rao, A. V. Koteswara 182
Rathore, Kuldeep.............29, 355
Ray, Shashwati.................909
Reddy, A. V. Sudhakra...........244
Reddy, B. Bhargava Reddy 244
Reddy, Damodhar..............510
Reddy, K. Jyotheeswara 465
Reddy, Madhav.................798
Reddy, Y. Praveen Kumar 244
Reddy, Y. V. Krishna.............244
Rizwan, M............ 1, 287, 305, 317, 477, 570, 576, 1083
Rizwan, Mohammad.............212
Rohilla, Komal...............1224
Rout, Pravat Kumar 431
Roy, Amit Kumar 449
Roy, Sumit.................944, 1205
Rudaba, Quazi Nasrul 1132
Sabarinath., G 253
Sachan, Ankit...............1103
Sadiq, Md. Zyed Ibn 1132
Sahoo, Harish Kumar...........516, 588
Sahoo, Sandeep Kumar 677
Saini, R. P 120
Sampath, N. 798
Samuel, Paulson.............420, 881, 1103
Sarathi, Arun 188
Sarkar, Ujjawal 976
Sarwar, Adil.................293

Sarwar, Md 229
Satapathy, Priyambada...........1142
Savita 249
Sawant, Kunal 1087
Saxena, A. K. 13
Saxena, Anmol Ratna...........488
Saxena, Ravi.................92
Saxena, Rohit.................841
Sayed, Abdullah Abu 1132
Seema 722
Sehwag, Vinay 73
Selvi, M. Vetri 261
Shaik, Abdul Gafoor 17
Shaik, Mahmood 17
Shakya, Amit Kumar 1087
Sharma, Abhishek 8
Sharma, Akhilesh 437
Sharma, Amarjeet Kumar 102, 443
Sharma, Ankit.................976
Sharma, Deepika 483
Sharma, Himanshu 461
Sharma, Paras Ram 386, 414
Sharma, Parsh Ram 365, 380
Sharma, Rakhi.................623
Sharma, Richa.................299
Sharma, S V 739
Sharma, Sahil.................748, 1056
Sharma, Shailendra Kumar 886
Sharma, Shivangni.............392, 426
Sharma, Sudeep...............858
Shatil, Abu Hena Md. 1132
Shekher, Vineet 73, 85, 993
Shivangi 13
Shobhit, R. P. Agarwal...........1213
Shrivastava, Sumit 249
Shrivastava, Tanmay 29
Shrujan, M 1041
Shubhra 564
Shukl, Pavitra.................593
Shukla, Saurabh...............897
Siddiqui, Anwar Shahzad 229
Siddiqui, M. Shadab 13
Simlai, J. 370
Singh, Ajendra.............73, 85, 993
Singh, Alka.................558, 792
Singh, Amresh Kumar 683
Singh, Ashish.................743
Singh, Bhim...............504, 522, 528, 564, 582, 593, 611, 630, 653, 659, 665, 677, 683, 706, 722, 765, 771, 777, 783, 897
Singh, Bipin 165
Singh, Chhabindra Nath...........881, 1103
Singh, Deepak.................437
Singh, Ghanshyam 788
Singh, Himanshu 1121

AUTHOR INDEX

Singh, Hitesh 1178, 1193
Singh, Indu Prabha 648
Singh, Jaspreet 671
Singh, Jyotsna 1078
Singh, Kailash 1136
Singh, Kamaljeet 739
Singh, Krishna 1148
Singh, Kunwar 1032
Singh, Lokesh Shankar 967
Singh, Madhusudan 847, 953, 1027, 1228
Singh, Manminder 1005
Singh, Mayank 1027
Singh, Nisha 273, 311, 325
Singh, Preeti 1022
Singh, Pukhrambam Devachandra 437
Singh, Rajveer 92
Singh, S 1041
Singh, Simranjit 1188
Singh, Upma 1
Singh, Vinay Kumar 858
Singh, Vishwamitra 1183
Singh, Yashi 611
Singhal, Sidhartha 972
Singla, Deepshikha 365, 380
Sinha, S. K. 182
Sisaudia, Varsha 982, 988
Sivaprasad, A. 188
Somani, R. K. 374, 483
Soni, Ashu 1052
Soni, K. M. 182
Sreejeth, Mini 938
Sriavstava, S P 809
Srikakulapu, Ramu 138
Srinivas, Vedantham Lakshmi 522
Srivastava, Ajay 1073
Srivastava, Divyank 194, 903
Srivastava, Kriti 403
Srivastava, Laxmi 23, 280, 299
Srivastava, Neha 827
Subudhi, Umamani 516, 588
Sudhakar, N. 465
Sulochana, V. 1097
Sumedh, N. 798
Surabhi, Jagriti 305
Suryawanshi, H. M. 754
Suthar, Bhavik N. 804
Suvesh, P. 1205
Swaroopa, Manuhar 8
Talapur, Girish G. 754
Tanwar, Lalit Kumar 1183
Tanwar, Lavi 927, 1165
Taparia, Rajashree 1010
Tariq, Mohd 636
Tej, M. Arun 729
Teron, Biki 743

Tharani, Kusum 78
Tigade, Chinmay 938
Tiwari, S. K. 504
Tiwari, Shalini 1087
Tiwari, Sheela 337
Tomar, Anuradha................................... 642, 748
Torres, Esperanza S. 33
Tripathi, M. M. 194, 903
Tripathi, Ramesh Kumar 712
Trivedi, Rishika 864
Tushir, Meena 876
Tyagi, Arjun.. 218
Valluru, Sudarshan K. 847, 953, 1027, 1228
Varma, Mukesh Kumar................................. 1109
Varshney, Vikas 249
Veerachary, M. 694
Veerachary, Mummadi 617
Venkatesh, V. 739
Verma, Aditya 953, 1228
Verma, Akhilesh 1126
Verma, Anjeet 653
Verma, Ashu 160, 218
Verma, Bharat 165, 534
Verma, H. K. 886
Verma, Vimlesh 392, 426
Verma, Vishal 700
Vinatha, U. 138
Vishwakarma, Virendra P.............827, 852, 982, 988
Viswanadha, Karteek 1000
Vohra, Pawan Singh 958
Yadav, Madhuri 1160
Yadav, Mohini 1083
Yadav, N. K. 50
Yadav, Sandeep Kumar 17
Yadav, Sudesh 852
Yadav, Vinod Kumar 132
Yashwante, Meghna R. 1067
Yousuf, Viqar 108, 149

IEEE
445 Hoes Lane
Piscataway, NJ 08854-4141

ISBN 978-1-5386-6624-1